D1741692

1 MONTH OF
FREE
READING

at
www.ForgottenBooks.com

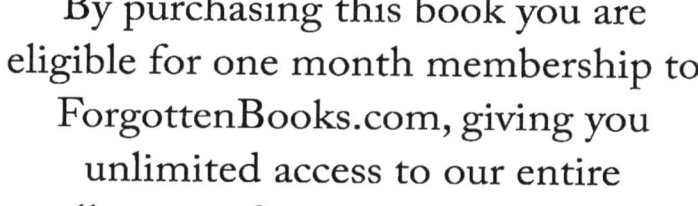

By purchasing this book you are eligible for one month membership to ForgottenBooks.com, giving you unlimited access to our entire collection of over 700,000 titles via our web site and mobile apps.

To claim your free month visit:
www.forgottenbooks.com/free720500

* Offer is valid for 45 days from date of purchase. Terms and conditions apply.

ISBN 978-0-483-52899-4
PIBN 10720500

This book is a reproduction of an important historical work. Forgotten Books uses
state-of-the-art technology to digitally reconstruct the work, preserving the original format
whilst repairing imperfections present in the aged copy. In rare cases, an imperfection in
the original, such as a blemish or missing page, may be replicated in our edition. We do,
however, repair the vast majority of imperfections successfully; any imperfections that
remain are intentionally left to preserve the state of such historical works.

Forgotten Books is a registered trademark of FB &c Ltd.
Copyright © 2017 FB &c Ltd.
FB &c Ltd, Dalton House, 60 Windsor Avenue, London, SW19 2RR.
Company number 08720141. Registered in England and Wales.

For support please visit www.forgottenbooks.com

Nature,
December 13, 1900]

Nature

A · WEEKLY

ILLUSTRATED JOURNAL OF SCIENCE

VOLUME LXII

MAY to OCTOBER 1900

" To the solid ground
Of Nature trusts the mind which builds for aye."— WORDSWORTH

𝔏𝔬𝔫𝔡𝔬𝔫

MACMILLAN AND CO., LIMITED

NEW YORK: THE MACMILLAN COMPANY

[*Nature,*
December 13, 1900

Q
1
N2
v.62
cop.2

RICHARD CLAY AND SONS, LIMITED,
LONDON AND BUNGAY.

INDEX

B

RICHARD CLAY AND SONS, LIMITED, LONDON AND BUNGAY.

A WEEKLY ILLUSTRATED JOURNAL OF SCIENCE.

" To the solid ground
Of Nature trusts the mind which builds for aye."—WORDSWORTH.

THURSDAY, MAY 3, 1900.

MOUNT ST. ELIAS.

La spedizione di sua Altezza Reale il Principe Luigi Amedeo di Savoia, Duca degli Abruzzi al Monte Sant' Elia (Alaska) 1897. Da Dottore Filippo de Filippi; illustrata da Vittorio Sella. Pp. xvii + 273; with 34 plates, 4 panoramic views, 2 maps and 115 figures in text. A beneficio delle Guide Alpine Italiane. (Milano: U. Hoepli, 1900.)

MOUNT ST. ELIAS, with an altitude, as now ascertained, of 18,092 feet, stands—a majestic cornerpost—exactly at the angle where the Alaskan boundaryline ceases to run parallel to the coast and strikes northward along the 141st meridian, and its summit is now generally acknowledged to lie on the Canadian side of the frontier. Whether it maintain its supposed preeminence among the mountains of the North American continent, or whether it eventually prove to be overtopped by Mount Logan, its great neighbour on the north, or, as the most recent explorations seem to indicate, by Mount McKinley, one of the yet unvisited peaks to the westward, it must, from its commanding position on the verge of the open ocean, always impress the imagination as the grandest of the Alaskan Chain. In recalling the fact that the mountain was for long erroneously supposed to be a volcano, Mr. Douglas Freshfield has told us, on the authority of the poet himself, that Tennyson had Mount St. Elias in mind when he described the landscape of a volcano among snow as one of the pictures on the walls of "The Palace of Art" (see *The Alpine Journal,* vol. xix. p. 174).

So far as our present knowledge goes, nowhere else on the face of the globe is there so great a vertical range of snow and ice as among these Alaskan mountains. On Mount St. Elias the permanent snow-line comes down to within about 3000 feet of the sea, while the enormous glaciers nourished by the excessive humidity of the climate not only descend to sea-level, but unite and spread out in a vast plain of ice covering an estimated area of 1500 square miles between the foot of the mountain and the ocean.

Thus entrenched in ice sheets, so that even its base is defended, it is not surprising that the mountain withstood several attacks before it was conquered. The first attempt was made, in 1886, by Messrs. Libbey, Schwatka and Seton-Karr; the next, in 1888, by Messrs. Topham, Broke and Williams; the third and fourth, by Prof. I. C. Russell, in 1890 and 1891; and the fifth, by Prof. H. G. Bryant, in 1897, almost simultaneously with the successful Italian expedition. Of these explorers, Prof. Russell achieved in every way the most important results, bad weather alone preventing his complete success in 1891, after an altitude of 14,500 feet had been attained and the practicability of the ascent had been demonstrated. Prof. Russell correctly determined the height of the mountain, and carried out investigations upon its physical characteristics which proved of high scientific importance. Lieut. Seton-Karr had previously called attention to the singular condition of the glaciers at the foot of the mountain, where immense piles of morainic debris, in places overgrown with dense vegetation, hide the marginal surface of the ice; but it was not until Prof. Russell published his more adequate descriptions that geologists fully recognised the value of the phenomena of the "piedmont" ice in elucidating the conditions of ice-covered lowlands in general during the Glacial Period, and especially during its closing stages. So closely has Prof. Russell's name become associated with the mountain, that one cannot stifle a regret that the satisfaction of being first upon the summit did not fall to his lot. In the volume before us, however, we are glad to find a graceful acknowledgment of the work of previous explorers, in a chapter having for its motto this quotation, from Mr. D. Freshfield :—

"Those who went first and opened the way are not less entitled to credit than those who came afterwards and reaped the fruit of their predecessors' labours."

The leader of the Italian expedition, H.R.H. Prince Louis of Savoy, in planning an ascent higher than the Alps could offer, had at first contemplated an attack upon one of the great peaks of the Himalayas. Forced by unfavourable circumstances to abandon this idea, he turned for consolation to Mount St. Elias. He could not have selected a more princely amusement, or a better exercise in skilful organisation and patient endurance. That he achieved his object without mishap, and that he and his little band of fellow-countrymen had the patriotic satisfaction of planting the Italian flag on a summit assuredly never before trodden by the foot of man, was due to the careful forethought with which all the preliminary arrangements of the expedition were planned.

A graphic account of the ascent was communicated by one of the party, Dr. Filippo de Filippi, to the English Alpine Club a few months afterwards, and was published in the *Alpine Journal* for May 1898. As chronicler of the expedition, Dr. Filippi has now expanded his story in the handsome and portly volume before us, in which he deals at full length with the conditions of the climb, and describes in glowing language the wild grandeur of the scenery. The beautiful photogravure plates and the illustrations in the letterpress, with which the book is so bountifully provided, are reproduced from photographs taken, for the most part, by Vittorio Sella, who was also of the party, and these are especially valuable as a faithful and permanent record of the physical characters of this seldom-visited region, and particularly of the untraversed wilderness of snow and mountain peaks to the northward of St. Elias. Among many that are excellent, there is one plate (P. 136), showing the snowy eastern spurs of St. Elias delicately fluted by innumerable avalanches, which seems to us peculiarly impressive.

Of the ten chapters of the book, only five deal with the actual ascent ; the first three, and also the final chapter, are devoted to the outward and homeward journeys ; the fourth to the previous history of the mountain ; and appendices, covering seventy-four pages, to the equipment and scientific results of the expedition.

As Russell had foretold, the mountaineering difficulties encountered during the climb were slight, and the adventure resolved itself into a long, arduous struggle upward for thirty days, usually in wretched weather, over interminable snow-fields and glaciers. The character of the climb was pithily given by one of the guides, in answer to inquiries after his return :—" C'est comme le Breithorn, seulement beaucoup plus haut." The expedition, consisting of the Prince with four compatriots, five Italian guides, and ten Americans who acted as porters, landed safely near Point Manby, at the foot of the moraine of the Malaspina ice-field, on the evening of June 23, and on July 1 started forward across the ice, all subsequent encampments being upon snow. Traversing the Malaspina in three days, partly in dense fog, the party struck upward along the eastern flank of its great tributary the Seward Glacier, which was afterwards crossed, and the Agassiz Glacier gained, at an altitude of only 3480 feet, by Russell's previous route through the Dome Pass, on July 13. Thence the explorers forced their way slowly up the Newton Glacier, through labyrinths of crevasses and ice-falls, for thirteen days, of which only three were fine. Fortunately the fog and snow which fell to their lot were unaccompanied by either wind or electrical disturbance; nor did the party suffer from cold, the temperature ranging steadily between 25° and 35° F. Their progress along this glacier averaged only about 1 mile 500 yards daily. At this stage they received news that the expedition led by Prof. Bryant, which had set out a few days ahead of them, had been compelled to

return to the coast owing to the illness of one of its members, after having reached the foot of the Newton Glacier.

The Italians were greatly impressed with the vivid colouring of the névé and ice even in the thickest weather, the tints ranging from brilliant turquoise and azure to the deepest blue, without the greenish tinge familiar to them in the Alps. This and the weird atmospheric effects in these mountain solitudes are eloquently described by Dr. Filippi.

Having left the American porters behind, and established their advance camp on the col at the head of the Newton valley, at an altitude of 12,287 feet, the success of the mountaineers depended solely upon the weather. Fortunately this proved more favourable than at the lower elevations, and they were able, without delay, to attack the summit. It was absolutely calm and clear on July 31, when after a heavy climb of 5800 feet from their last bivouac, during which the majority of the party were more or less affected by mountain-sickness, the Prince and his comrades reached the crest just before noon. The thermometer registered a temperature of 10·5° F., and the barometer stood at 15 inches 15 lines. The height of the mountain as determined by the barometer was 18,092 feet, which is in remarkably close agreement with Russell's figures, 18,100 feet, obtained by triangulation.

From the summit they saw the majestic mass of Mount Logan to the north-eastward, sinking north-westward into a very intricate lower chain, while to the westward was a chaos of low ridges, névés and glaciers, overtopped at a distance of some hundred miles or so by three great snowy giants as yet unexplored, which proffer substantial work for the future.

Then came the descent and the return to the coast, which was safely reached in ten days. Some of the lower ridges overlooking the Malaspina Glacier, where they had previously found snow, were now knee-deep in blossoming plants.

The appendices to the volume are, from a scientific standpoint, not particularly important. The first describes the equipment of the expedition in detail, and should be of service to explorers of similar regions. The excellent plan was adopted of packing the supplies in tin boxes, each containing sufficient material of every kind for twenty-four hours. The second appendix consists of meteorological tables, giving the simultaneous observations made daily between June 25 and August 3 by the expedition, and by the Rev. C. J. Hendricksen, of the Swedish Mission at Yakutat, at the foot of the mountain. The third deals with the health of the party. The absence of colds, rheumatism, or other ill results from the trying conditions of the journey is made the subject of comment ; and the symptoms which effected most of the explorers during the final stages of the ascent are fully discussed, but it is thought that these might be in part attributed to excitement and want of sleep. The only case of real illness was that of one of the American porters, who, after having passed a night, during the return, on ground covered by vegetation, on the Hitchcock Hills, was seized with an attack of malaria while crossing the Malaspina Glacier. The terrible plague of mosquitoes on the coastal strip of forest is especially mentioned. Another appendix, on the zoological material, is principally devoted to the description of a new arachnid and of a new oligochæte annelid collected on the snow. An appendix on the rocks and minerals is for the most part a discussion of Russell's previous work, but contains the information that the outcrops near the summit of the mountain consist of typical diorite passing locally into hornblendite.

The ascent of Mount St. Elias was an achievement worthy of a prince, and this handsome volume is worthy of the achievement. Beautifully printed, magnificently illustrated and tastefully bound, it reflects credit upon all concerned in its production. But (alas! the inevitable but!) it has no index.   G. W. L.

A HYDRODYNAMICAL THEORY OF ACTION AT A DISTANCE.

Vorlesungen über hydrodynamische Fernkräfte nach C. A. Bjerknes' Theorie. Von V. Bjerknes. Band i. Pp. 338 ; with 40 figures. (Leipzig : Johann Ambrosius Barth, 1900.)

THEORIES of matter—or should we not rather call them theories of *force*, since, in "explaining" the properties of matter, we are mainly concerned with those manifestations which we say are due to "force"—naturally fall into two distinct classes. The first class includes those hypotheses which regard continuous matter as being built up of discrete particles, and the direct action of finite portions of matter as being due to action at a distance of these particles. The second class includes those hypotheses which regard these particles as singularities in a continuous medium, and which attribute their action at a distance to the direct agency of the medium. In a certain sense, these two theories are reciprocal. In both, certain attributes are localised at points, and it is necessary to bridge over the distance between these points. According to the first hypothesis, a field of force pervades the intervening gaps ; according to the second, they are filled with a distribution of mass. The belief that both hypotheses are possible, enables us to imagine that there may be no limit to the smallness of the scale on which Nature conducts her operations, the phenomena occurring in any region being made to depend in their turn on others occurring in the far more minute regions which are regarded as constituting its ultimate elements, and these elements being in turn capable of further subdivision, and so on indefinitely.

In 1852, Lejeune-Dirichlet, being unacquainted with the works of Green and Stokes on this subject, published a paper containing the solution of the problem of the motion of a sphere in an incompressible fluid. In a course of lectures given at Göttingen in 1855-56, Dirichlet gave the corresponding solution for a sphere fixed in a steady current, and invited his pupils to attempt the solution for an ellipsoid. Among these pupils were Schering, who solved the problem, and C. A. Bjerknes, who gave a generalisation for space of *n* dimensions. At this time the doctrine of action at a distance may have been said to be at its zenith, and Göttingen had given birth only a few years previously to the last brilliant product of that doctrine, Weber's Law. As a foreigner,

Bjerknes was, however, less influenced by the views then prevailing in the Göttingen school, and a volume of Euler's correspondence falling into his hands caused him to oppose the doctrine of action at a distance. A fresh light was thrown on the hypothesis of a continuous all-pervading medium by Dirichlet's discovery that a sphere moving in an incompressible perfect fluid experiences no retardation from the fluid, and an impetus was given to Bjerknes to develop Dirichlet's investigations in a direction widely differing from anything then contemplated by his professor.

From the effects of purely translational motions of two spheres, Bjerknes was led on to consider the mutual actions of two pulsating spheres, and discovered that such spheres attract or repel one another according as their phases are the same or opposite, the law of force being that of the inverse square. Bjerknes found, moreover, that the expressions for the forces acting on a sphere moving in liquid consisted of two terms, which he distinguished as "inductional forces" and "energy forces," a result which he arrived at by considering the expressions for the pressures on the spheres, but which might have been found more readily had Thomson and Tait's application of Hamilton's principle been then known to him. About 1875, Bjerknes published a paper in which he established the hydrodynamical law of action and reaction, and the analogy with electric and magnetic action at a distance ; and in the following year he gave an independent investigation based on the Hamiltonian principle.

From 1875 onwards, Bjerknes appears to have occupied himself chiefly with the terms of lowest order in the expressions for the forces ; and in 1878 he discovered the law of rotation for oscillating spheres. Since then he seems to have devoted his attention mainly to electric and magnetic analogies, and in the middle of his eightieth year he completed the discussion of the "inductional forces," and by this means pushed the analogy between hydrodynamic action at a distance and electromagnetic phenomena as far as it could be pushed without departing from the fundamental hypotheses.

A complete account of these investigations was never published, and it remained for his son, Prof. V. Bjerknes, to embody them in the present volume. For three years Prof. V. Bjerknes has delivered courses of lectures on the subject at the University of Stockholm, and the book is practically based on these lectures. It is divided into four parts : the first, an introductory part, dealing with the general principles of vector fields and hydrodynamical equations ; the second, dealing with the motion of the liquid surrounding a system of moving spheres treated from a kinematical standpoint ; the third, dealing with the influence of the pressures on the motion of the spheres themselves ; and the fourth, with the theory of apparent actions at a distance, of hydrodynamical origin. In the second part, the diagrams of the stream lines due to a moving, oscillating or pulsating sphere in various fields of force are noticeable for their elegance.

It is to be wished that the courses of lectures which Prof. V. Bjerknes delivered on the work of his father could be taken as models of what university lectures should be, for the development of a theory such as the present affords an excellent and not difficult insight into the methods of mathematical analysis. So long as our English university colleges are, to a great extent, in the hands of oligarchies, who attach more importance to such trifles as the handwriting and spelling of matriculation or medical preliminary students than to higher scientific study, such courses of training will only be accessible to those who seek them in countries more enlightened in the matter of scientific education than Great Britain. We can readily imagine that Bjerknes' theories may find their way into many transatlantic universities among the "classics of science." They have, indeed, no small claim to be regarded as classical. It is true, as Prof. V. Bjerknes points out, that his father's and Kirchhoff's work in several cases somewhat overlapped, but it would appear that in developing the theory of motion of spheres in liquids as a basis for explaining the properties of matter, Bjerknes stood entirely on ground of his own making. Other theories involving the conception of a continuous medium have sprung up ; we have had the vortex-atom theory before us, and we now find it necessary to postulate the existence of an ether, whose attributes resemble those of an elastic solid rather than a fluid. At the present time few will regard the hypothesis of pulsating spheres as of more than classical interest. As having been first developed in the face of a prevalent belief in the doctrine of action at a distance, and as ingenious methods of replacing this action at a distance by the action of an intervening medium, the application of the term "classical" to these investigations of C. A. Bjerknes may not be altogether without justification.

G. H. BRYAN.

PHOTO-MICROGRAPHY.

Photo-micrography. By Dr. Edmund J. Spitta. Pp. xi + 163. (London : The Scientific Press, Ltd., 1899.)

A QUARTER of a century has now elapsed since the renaissance of the art and science of photomicrography. Up to that time much of the best work in this direction was accomplished in America by Lieut.-Colonel Woodward, of Washington, whose successful photographs of diatoms excited the admiration of all microscopists who saw in his productions the faithful delineations of those "markings" on them, on which many hours of microscopical manipulation had been spent in bringing their delicate tracery to a correct definition. From that time to the present the fascination of transferring the minutest details of histological and biological science to the photographic plate has found many ardent votaries, with the result of improved apparatus and lenses corrected to such a degree of accuracy for this work that sharp and well-defined images can now be obtained in a manner that would have been a boon and a revelation to workers twenty-five years ago.

Amongst the latest exponents of this branch of microscopical science we must name that of the author of the book under consideration.

Dr. Spitta in this work on photo-micrography has dealt with the subject very fully and from a scientific standpoint, so that the student who takes up this branch of the photographic art is thoroughly furnished with all

the information necessary to the accomplishment of perfect work, leaving, however, only that amount of *personal experience* to be obtained and which will be demanded of every one who first embarks on this art, and without which he is liable to be landed in many difficulties.

In Chapter i. the author deals with illuminants, a by no means unimportant point for consideration ; for although we have several good illuminants for low power work, it is when we come to work with the highest power objectives that either the lime-light or that of the electric arc lamp must be employed to produce the best possible results. These lights are not always readily accessible ; but as the aspiring student most probably will try his 'prentice hand on low power work, the single wick lamp burning the best paraffin oil will furnish him with a light sufficiently rich in actinic rays that, provided the proper length of exposure be given, will result in a very successful negative. Dr. Spitta in Chapter ii. proceeds to give directions for obtaining photo-micrographs by low power objectives, dealing with this in such a lucid manner that the student who closely follows his clear description cannot fail in being rewarded by satisfactory results, being assisted in his work by algebraical formulæ and illustrations of simple but effective apparatus.

Chapter iii. deals with medium power photo-micrography, and contains some very necessary cautions relative to the avoidance of vibrations in the apparatus, for, as the author observes, "when photographing at 1000 diameters, 1/1000 of an inch shake in the specimen makes a shift of one inch in the photographic plate," or he might have said in the photographic *image ;* therefore the absolute necessity of the most perfect stability, not only in the apparatus but even in the studio, can be readily understood and provided for—even a heavy tread on the floor of an adjoining room being sufficient to disturb the steadiness of the optical arrangement. Dr. Spitta describes different methods whereby this difficulty may be overcome. Allowance must also be made for the expansion of the metal of the microscope from the heat of the illuminant, for even in low power work, say of 250 diameters, the heat from the oil lamp must not be considered a negligible quantity, and must be considered so far that no photographic exposure should be attempted till the metal has had time to become fully expanded.

Chapter iv. is overloaded with woodcuts of different makes of microscopes valuable as affording the student a choice of various instruments, but by no means necessary to his work, as any one of these is sufficient for attaining good medium power work. This chapter also deals with the subject of lenses and eyepieces and the accessory fittings of the microscope generally ; but there is one point that must have the greatest attention, and that is the fine adjustment, and Dr. Spitta does well in laying great stress upon its importance ; nothing is more embarrassing to the operator, when perhaps everything else in the apparatus is working well, to find that the fine adjustment by which he hopes to obtain that sharp definition without which his work is valueless, is altogether useless from faulty construction, and Dr. Spitta describes the various forms of this all-important addition to the photo-micrographic installation.

The remaining three chapters of this work treat of such subjects as substage fittings, coloured screens, and the various subsidiary apparatus useful in high power or "critical" photo-micrography. These particulars do not bear the condensation that is necessitated by the space allotted to this report, but are full of information for the guidance of the photo-micrographic student and will materially assist him in his work. A valuable feature is included in the appendices, and is headed "25 common faults in photo-micrography; their causes and means of cure"; by a reference to p. 152 every error that may present itself in the beginner's work is described, the reason for it given, and the remedy indicated. Added at the end of the book are five plates of representative work in photo-micrography, the work of the author, while a copious index brings the work to a conclusion.

GEORGE KINGSLEY'S LIFE AND WRITINGS.

Notes on Sport and Travel. By George Henry Kingsley. With a memoir by his daughter, Mary H. Kingsley. Pp. viii + 544. (London: Macmillan and Co., Ltd., 1900.)

THIS is a book, we venture to think, that most readers will lay down with deep regret—regret that a very talented writer, an acute observer, and an ardent sportsman (in the best sense of the word) should have bequeathed so little of his experiences to the world. For George Kingsley, a member of a clever family (or, as his biographer will have it, a member of a clever generation of an ancient family), was evidently a man far above the ordinary intellectual level, and enjoyed unrivalled opportunities of adding to our store of knowledge by travel in distant lands at a time when they were still, to a great extent, populated by their native denizens and unspoiled by the march of civilisation. Unfortunately, however, he seems to have been devoid of those regular and methodical habits of work by which alone the results of a life of exploration and travel can be properly recorded, and we have consequently to be content with mere scraps and fragments of a vast store of information.

From such scraps and fragments as the editor, who is to a great extent also the author, of the present volume has been able to save from oblivion, we glean how keen an observer and how true a lover of nature was Dr. Kingsley. Whether among the coral-girt isles of the South Pacific, when they were yet in great part free from the "beachcomber," or on the prairies of the "wild west," at a time when the bison were still to be numbered by hundreds, if not by thousands, his descriptions of scenery and animals are life-like pictures.

The greater part of the account of the author's travels is given in the memoir by his daughter, which occupies more than a third of the whole volume, and is, in great measure, in the form of letters or of extracts from the same. And here we take the opportunity of expressing our sense of the excellent manner in which Miss Kingsley —herself a traveller and writer of world-wide repute— has discharged what must evidently have been a task of no ordinary difficulty.

Kingsley (in company with the late Lord Pembroke)

visited the South Seas in the late "sixties"—a time when yachting in those latitudes had not come into vogue; and such descriptions as he has left of the natives and natural products only make us regret that they were not fuller. Fish seem especially to have attracted his attention; but when he states that he disbelieves the story of a *Chaetodon* [1] shooting water at a fly, the editor should have added that the only fish which performs this feat is a species of *Toxotes*, whose southern range only extends to North Australia, so that it could not have come under the ken of the author.

The travels in Canada and the United States were undertaken in company with Lord Dunraven, between 1870 and 1875; parts of them being described by the latter in "The Great Divide."

Of the various collected papers of Dr. Kingsley, perhaps the most interesting to the naturalist is the one entitled "Among the Sharks and Whales." Here the author graphically describes, as an eye-witness, certain encounters between the larger Cetaceans and smaller members of the same order, together, perhaps, with other denizens of the deep. We are told, for instance, how some of these creatures, of thirty feet or so in length, were seen to leap clean out of the water, and then to fall with a sounding "smack" that could be heard half a mile off. But whether the creatures in question were attacking a whale, or leaping for mere fun, the author was unable to determine. Neither could he say definitely whether or no they were "killers." And he seems, indeed, to be somewhat confused between "killers" and "threshers"; although, as to the sharks commonly called by the latter name, he denies that they ever attack whales, adding that he has never even known a shark of any kind throw itself out of the water.   R. L.

OUR BOOK SHELF.

Irrigation and Drainage, Principles and Practice of, their Cultural Phases. By F. H. King, Professor of Agricultural Physics in the University of Wisconsin, author of "The Soil." The Rural Science Series. Pp. xxi + 502. (New York: The Macmillan Company London: Macmillan and Co., Ltd., 1899.)

THE object of this book, as stated in its preface, is "to present, in a broad yet specific way, the fundamental principles which underlie the methods of culture by irrigation and drainage," and we may say that we consider the author successfully does this.

The introductory chapter treats of the importance of water in cultivation, and in it a number of interesting experiments on the amount of water absorbed by cereals and other plants, and the weight of dry matter produced are described, from which it appears that with cereals the amount of water used varies from about 300 to 500 lbs. per pound of dry matter produced. The general result of these experiments is considered to show "that well-drained lands in Wisconsin, and in other countries having similar climatic conditions, are not supplied naturally with as much water during the growing season as most crops are capable of utilising, and hence that all methods of tillage which are wasteful of soil moisture detract by so much from the yield per acre."

1 The editor avows a difficulty in deciphering some of the MS. which came into her hands, and therefore suggests the possibility of a certain amount of mis-spelling. Some naturalist friend would, however, doubtless have corrected the following errors, viz.:—P. 61, *Chetadons* for *Chaetodons*; p. 222, *Haroldus* for *Harelda*; p. 414, *Megaptera australis* for *Balaena australis*; p. 421, *Ovulit* and *Mutras* for *Olivies* and *Mitras*; and p. 424, *Orcus* for *Orca*.

Similar experiments have been made with other crops, as, for instance, potatoes, and the importance of such experiments is, as stated further on in the book, " because only such knowledge as this can show how economical or how wasteful our methods of tillage may be, and how nearly we are realising the largest profits which are possible to the business."

The conditions of rainfall under which irrigation is practised in different parts of the world are discussed, and the means of " conserving the moisture of the subsoil " by proper tillage pointed out. An excellent account is given of the depth of root penetration in the soil, which is illustrated, as is the rest of the book, by some very good and instructive engravings. A short account is given of sewage irrigation ; and the idea that the milk of cows fed on sewage produce is in any way detrimental is disposed of by quotations from Sir Henry Littlejohn, and from Mr. Spier, the Scottish Dairy Commissioner. Methods of diverting streams for irrigation are carefully described and fully illustrated, as also are the methods of applying the water to the ground. In Part ii. (a small portion at the end of the book) the necessity for soil drainage is insisted on, and the methods of carrying it out are described.

The book altogether is very readable, although the spelling of some of the words seems curious to an English reader. It is also well printed, and the only misprint noticed is on p. 403, where the word "denitrification" is used instead of "nitrification." W. H. C.

The Refraction of the Eye, including a Complete Treatise on Ophthalmometry. A Clinical Text-book for Students and Practitioners. By A. Edward Davis, A.M., M.D. Pp. 431. (New York : The Macmillan Co., 1900.)

THIS volume should prove a valuable addition to the library of the ophthalmic surgeon, for though several books on retinoscopy have been published, this is the only work on ophthalmometry yet written in English.

It comprises a description of Javal and Schiötz's modification of Helmholtz's ophthalmometer, together with full instructions in the use of the instrument ; the necessity of forming a clear mental picture of the state of the eye from the results of an experiment being rightly insisted upon.

One hundred and fifty illustrative cases are included in the text, and a comprehensive index has been appended, so that the student can readily find a parallel to any case which may give him trouble. One hundred and nineteen diagrams, including a clear and well-drawn woodcut of the ophthalmometer of Javal and Schiötz, are distributed throughout the text.

Although the advantages which may be gained by the use of the ophthalmometer are insisted upon, the author has taken great pains to indicate the limitations of its usefulness. By its aid we may determine with accuracy the radii of curvature of the cornea in various meridians ; but the author endorses the generally accepted opinion that there is no definite relation between the curvature of the cornea and the refractive condition of the eye, as far as hypermetropia or myopia are concerned. Myopia usually depends upon an elongation, and hypermetropia upon a shortening of the axis of vision. Strangely enough, in cases of extreme myopia, a somewhat flattened cornea is generally met with. Nevertheless, in cases of simple hypermetropia and myopia, the ophthalmometer eliminates the question of corneal astigmatism. The routine of examination followed by the author is (1) use the ophthalmometer ; (2) use trial lenses and test cards ; (3) use the ophthalmoscope ; (4) if after two tests on different days the result is still unsatisfactory, employ a mydriatic and use the retinoscope in addition to the other tests. It is stated that (1) to (3) suffice for 99 per cent. of uncomplicated cases.

In the use of test glasses, it is recommended that a series of positive lenses, gradually increasing in power, should first be employed. By this means spasmodic accommodation is avoided. The fact that the use of atropine can so often be dispensed with is of great importance, since many men might hesitate to have their eyes examined if this necessitated a temporary cessation of their business duties.

A number of instructive cases are included, showing the serious results which may follow on the prescription of unsuitable glasses for a patient. Not only severe pain and inability to use the eyes for any length of time, but even personal disfigurement may be produced. Thus a case is recorded (p. 307) of a patient whose eyes were being forced into a divergent squint by the use of prismatic glasses. After a careful examination, the prisms were discarded and suitable lenses were ordered, with the result that, after two weeks, complete comfort and the possibility of working with satisfaction were enjoyed for the first time for many years.

Altogether this book gives us a good idea of the vast advantages to the human race which have resulted from the optical researches of Helmholtz, culminating in the invention of the ophthalmometer and the ophthalmoscope. E. E.

A Key to the Birds of Australia and Tasmania, with their Geographical Distribution in Australia. By R. Hall. Pp. xii + 116 ; plate and map. (Melbourne : Mullen and Slade ; London : Dulau and Co., 1899.)

WERE it nothing more than a synopsis of Australian birds, with just sufficient in the way of description to enable the different species to be easily recognised, this well-printed little " Key " would be to a great extent of merely local interest. But since the author has very wisely made geographical distribution its leading feature, the work appeals to a much wider circle of students than would otherwise have been the case.

In his Report on the Zoology of the Horn Expedition, Prof. Baldwin Spencer recently divided Australia into three zoological sub-regions ; namely, (1) the Torresian, embracing the northern and eastern districts as far as South Queensland ; (2) the Barsian, comprising eastern New South Wales, Victoria and Tasmania ; and (3) the Eyrean, including the remainder of the mainland. These sub-regions are further split up into " areas," and the fact that bird-distribution accords with such a parcelling-out of the continent from other lines of evidence affords important testimony in support of Prof. Spencer's views. It is noteworthy that the South Queensland area forms the headquarters of the Australian Passeres, a fact for which there must surely be some adequate physical reason, if only it could be discovered. The total number of species recorded is 767, among which the black emu is believed to be extinct ; and, so far as we have been able to verify them, the diagnoses of the various groups and species seem well adapted to their purpose. The work appears singularly free from errors and misprints, and ought to be in the hands of every Australian bird-lover. R. L.

Pages Choisies des Savants Modernes. By A. Rebière. Pp. viii + 620. (Paris : Nony et Cie, 1900.)

THIS is a series of extracts (translated into French when not written in that language) from the works of eminent men of science. It appeals mainly to the general reader, and the best that can be hoped of it is that it may induce some members of this class to study the works of one or other man of science seriously. A scientific writer does not appear to the best advantage in " tit-bits " selected from his works ; and, except as a possible stimulus, the value of such a miscellany as this cannot be reckoned very high. The portraits, of which there is a considerable number, will probably be found, by scientific readers, the most interesting feature of M. Rebière's compilation.

Les Vieux Arbres de La Normandie. By Henri Gadeau de Kerville. Fasc. iv. Pp. 219 + 352. (Paris : J. B. Baillière et fils, 1899.)

THIS instalment of M. de Kerville's careful monograph contains twenty views of trees from photographs by the author, accompanied by detailed descriptions and historical notes. The work is well and conscientiously done, whilst the illustrations are well selected and admirably reproduced in collotype. The trees here shown include ten oaks, six yews, two beeches, a lime and a poplar. As the photographs of the deciduous trees have been taken in very early spring, before the opening of the buds, their ramification and general architecture are shown to the greatest advantage. With this volume, *à propos* of a notable oak-tree growing at Isigny-le-Buat, the author includes an interesting account of recorded cases of mistletoe upon oaks in Normandy. He is able to produce evidence in support of some twenty-seven recorded instances. The book will appeal to all tree-lovers ; may it stimulate some to similar studies. We remember to have seen something of the kind for Northumberland nearly thirty years ago in the *Transactions* of the Tyneside Naturalists' Field Club.

LETTERS TO THE EDITOR.

[The Editor does not hold himself responsible for opinions expressed by his correspondents. Neither can he undertake to return, or to correspond with the writers of, rejected manuscripts intended for this or any other part of NATURE. No notice is taken of anonymous communications.]

The Nature of the Solar Corona.

I SEE in the recently-published number of *Science Abstracts*, No. 802, that there is every reason to think that the corona line is not represented by any dark line in the solar spectrum. I write to call attention to the way this confirms the suggestion that the corona is an aurora round the sun. In the March number of the *Annalen der Physik* for this year, p. 462, Herr Cantor describes experiments from which he concludes that there is no absorption corresponding to the emission of light by a gas which is caused to radiate by an electric discharge. He makes certain deductions as to the temperature of the gas which emphasise the difficulty of defining "temperature" in the case of a non-steady state ; but, whatever is to be deduced from his observation, it certainly lends weight to the suggestion that the corona is due to an emission of a similar character to that of a gas transmitting an electric discharge.

April 30.   GEO. FRAS. FITZGERALD.

Rock-structures in the Isle of Man and in South Tyrol.

MR. LAMPLUGH's recent paper referred to in his letter in NATURE of April 26 (p. 612) is devoted to an elucidation of the "relations of the Carboniferous limestone to the Carboniferous volcanic rocks" in the Isle of Man (*Q.J.G.S.* 1900, p. 71). From Mr. Lamplugh's description, these relations are very similar to the relations which I described as subsisting between the Mid-Triassic dolomitic limestone ("Mendola Dolomite") and the tufaceous "Wengen" beds of Enneberg. The "Buchenstein Agglomerate" of Enneberg, which I mentioned in my letter (NATURE, March 22), had been described in geological literature as a "Middle Triassic agglomerate" of local occurrence above "Mendola Dolomite," in the neighbourhood of eruptive outbursts of that age. My map and sections showed that the agglomerate had a limited occurrence in fault-zones and overthrust-planes where differential movement had taken place between the harder, more resisting "Mendola Dolomite" and the yielding, mixed "Wengen" series "comprising dust-tuffs and lavas, as well as fossiliferous shales and shaly limestones." I therefore explained the so-called "Triassic agglomerate as a subsequent structure, of the nature of a shear-breccia, produced by the earth-movements of the later Alpine upheaval (*Q.J.G.S.* 1899, pp. 567, 584. Figs. 1, 4, 9, 10).

Mr. Lamplugh describes in the Carboniferous series of the Isle of Man rock-structures of brecciated limestone, tuffs with contained strips of limestone, and coarse agglomerate which had previously been referred to the effects of Carboniferous eruptive

action. Mr. Lamplugh's explanation is that the various complexities in the structure of these rocks "have not been caused by the volcanic outburst, but have been brought about at a later date by the differential movement of segments of the eruptive rocks upon their original floor of limestone" (*Q.J.G.S.* pp. 15, 19, Figs. 3, 4). The parallelism between the two cases is self-evident. In 1894, I had explained on precisely the same principle of subsequent differential movement, the occurrence of certain anomalous phenomena at the *upper* limit of the Wengen-Cassian series in Enneberg, *i.e.* the limit of this plastic and compressible series against the higher horizon of Triassic calcareo-dolomitic rock, termed "Schlern Dolomite" ("Coral in the Dolomites," *Geol. Mag.* 1894, p. 55).

The parallelism in the general sequence of events in the Isle of Man and in South Tyrol is as follows :—

Isle of Man.	*Enneberg.*
Pre-Carboniferous Movement.	Pre-Triassic Movement.
Lower Carboniferous Deposition.	Triassic Deposition.
Subsequent Movement.	Subsequent Movement.

The crust-movement immediately antecedent to Triassic deposition in South Tyrol was that which accomplished the upheaval of the Permian Alps, post-Triassic crust-movement culminated in the upheaval of the present Alps (*ant. Q.J.G.S.* 1899, p. 628, and NATURE, Sept. 7, 1899, pp. 445-6).

The farther issues of my paper in showing how differential movements twist the rocks by taking place in cross-directions were not touched in my letter of March 22, for the reason that Mr. Lamplugh did not in his paper enter into the torsional results of differential movements. But, as I have elsewhere expressed, rock-torsion or "warping" goes on all the time in crust-folding, and clearly, where from any cause whatsoever there is the greatest complexity in the differential movements, there will be the greatest complexity in the torsional phenomena.

MARIA M. OGILVIE GORDON.

POMPEII AND ITS REMAINS.[1]

THE city of Pompeii is one which will ever maintain a hold upon the imagination of cultured man, as much for what it represented in the history of civilisation, as for being the victim of one of the most awful visitations of the powers of nature which have ever befallen the abiding place of a great society of men. It is not the place here to descant upon the wealth and luxury of its

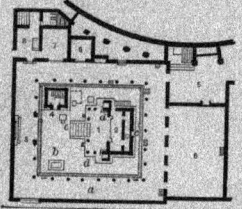

FIG. 1.—Plan of the Temple of Isis.
a, Portico ; *b*, cella ; *c*, shrine of Harpocrates ; *d*, purgatorium ; *e*, hall of initiation ; *h*, hall of mysteries ; *z*, *k*, *o*, abodes of priests ; *n*, colonnade ; *s*, refuse pit ; *v*, niche for statue of Bacchus ; *d d*, niches ; *x*, large altar.

inhabitants, on the bright and reckless lives which they led, on the splendour of its buildings, or even the fancied security wherein men and women lulled themselves, notwithstanding the violent shock of earthquake which shook the city to its very foundations on February 5, A.D. 63, for all these things are the commonplaces of history ; but we are concerned with the remains left by the awful catastrophe which took place on August 24,

[1] "Pompeii, its Life and Art." By August Mau. Translated by F. W. Kelsey. Pp. xxii + 509. (New York : The Macmillan Co., 1899.)

A.D. 79, and buried the cities of Herculaneum and Pompeii in a layer of mingled mud, lava, pumice stone, dust and wet ashes. In less than thirty-six hours Vesuvius had completely blotted out these towns and had covered the ground around for miles with pumice stones, barely as large as walnuts, to the depth of ten

![Fig. 2.—View of the Temple of Isis.]

Fig. 2.—View of the Temple of Isis.

feet. Of the twenty thousand people who are estimated to have been in Pompeii when destruction came upon the doomed country, about two thousand perished, the rest saved themselves by flight ; but fortunately for the people of our own time they were compelled to leave behind them most of the things which describe to the student and antiquary the manner of their lives, and reveal the high standard in luxury and artistic civilisation to which they had attained. The blow fell so suddenly, and the overwhelming of the city was so swiftly and effectively performed, that men and animals had no time to die in the usual manner, and the ashes which caked round them have preserved forms and scenes which, though belonging to the dead and dying, are replete with unerring suggestions of life.

Soon after the city of Pompeii was buried, the survivors came back and began to dig out the objects of value belonging either to themselves or their friends which they knew to be in the houses. As the upper parts of many of the houses still stood above the pumice stone and ashes, they were able to locate them in many instances with convenient accuracy, and as a result there remained in Pompeii, when the searchers had finished work, but few houses which had not been partly or wholly explored. Anything like a systematic search, however, was never made, and the excavators worked most in the places which seemed to promise the best results. Among others, the builders' labourers

made themselves very busy, for the costly stones and marble used in the construction of porticos, vestibules and baths, not to mention the pillars, were eagerly sought after for the building of new villas and houses. When such human vultures had battened on the remains of the town, they left what they could not, or would not, carry away to decay and desolation. For fifteen hundred years, Pompeii and its dead slept in peace, and certain pious folk comforted themselves with the view that its inhabitants, like those of the Cities of the Plain, richly deserved their punishment. About A.D. 1600, D. Fontana, who was occupied in bringing water from the Sarno to Torre Annunziata, cut a conduit through a part of the site of Pompeii, and two inscriptions were found in the course of the work. In 1719, Count Elbeuf's workmen sank a shaft on the site of Herculaneum, and reached a level corresponding with the stage of the theatre. In 1754, a number of tombs at Pompeii were discovered by the road-makers who were working to the south of the city, but no systematic attempt to leave what had been excavated uncovered and visible to all was made until 1763, when the discovery of the inscription of Suedius Clemens definitely proved that the site was that of Pompeii. A year later, the theatres, the Street of Tombs, and the villa of Diomedes were uncovered, and general interest in the work was at last awakened. Between 1806 and 1815, under Joseph Napoleon and Murat, the Herculaneum Gate and Forum were excavated ; and between 1825 and 1848, a large number of beautiful houses were cleared out and made accessible to

Fig. 3.—The Temple of Isis restored.

the curious and the learned. Up to this period, the work of excavation, though carried on with skill and zeal, was exceedingly unscientific ; indeed, judged by the canons of the excavator of to-day, it would be pronounced to possess no system at all. In 1860, however, explorations and excavations on the site of Pompeii were entrusted to the hands of G. Fiorelli, and most of the excellent results which have attended the excavations made during the last forty years are due to the plan inaugurated by him. At the present time, about one-half of the site of Pompeii has been excavated, and, according to the calculations which he made as far back as 1872, the work of clearing the undisturbed parts in the western half of the ancient city, and the whole of the eastern half, will not be completed much before the year 2000. The above facts will enable the reader to grasp the magnitude of the undertaking, and to appreciate the help which is forthcoming from Prof. Mau's exhaustive work, of which we must now speak briefly.

It is well known that Prof. Mau has for more than a score of years devoted all his winters to the study of the antiquities of Pompeii, and there is little doubt that he is *facile princeps* among the experts in this special branch of Roman archæology. His articles and papers in the scientific periodicals have secured for him a high position among savants, even in his own country, and his "Mittheilungen" are at once the product of good scholarship and enthusiasm. The volume before us is not a mere translation of one previously issued, but is to all intents and purposes a new work, now published for the first time in English. Mr. Kelsey, who is responsible for the English work bearing Prof. Mau's name, is more than the translator, for he has abridged the German manuscript which he had to work from in many places, and a number of additions to the text are due to him. He has done his part of the work faithfully, and the English visitor to Pompeii has now available in his own tongue a volume in which lucidity of treatment goes hand in hand with erudition and scholarship. The English text is accompanied by twelve plates, six plans, and two hundred and sixty-three cuts, which are inserted as near as possible to the subject-matter illustrated by them. We have only one fault to find with the book—it is a little heavy to carry about. Thus having said our worst, we proceed to describe very briefly its contents. The six first chapters really form the introduction, which they are actually called, and they treat of the early history and general situation of Pompeii, the overwhelming of the city, and the excavations undertaken during the last hundred and fifty years. The last chapter of the section on building materials and architectural periods is particularly instructive, and will be read by more than the tourist. Part i. contains twenty-five chapters, which deal exhaustively with the public buildings and places of Pompeii, including the Forum, the Basilica, the Comitium, the theatres, the temples of Jupiter, Apollo, Zeus Milichius, and, strangest of all, the temple of the Egyptian goddess Isis. It will be remembered that the Ptolemies, by the help of Manetho, an Egyptian priest, and of Timotheus, a man who had peculiarly perfect knowledge of the Eleusinian Mysteries, associated certain Egyptian religious ceremonies with those of the Mysteries, in the hope of binding his Greek and Egyptian slaves together in the bonds of a common form of worship. The new cult, though it was abominated by the philosophers, was very

popular, and it spread from Alexandria by way of the Delta into Syria, and from the same centre to Rome. As a result, we find that a college of priests of Isis, or Pastophori, was founded at Rome in the time of Sulla, about B.C. 80. The Romans objected to the introduction of the Egyptian gods, and three times in the space of eleven years was their temple destroyed. Oddly enough, a temple in honour of Osiris and Isis was built in Rome about B.C. 44, and before the end of the century their festival was recognised by the public calendar. But other cities of Italy were more tolerant than Rome, for a temple in honour of Serapis was standing at Puteoli B.C. 105, and not long after this date the temple to Isis was built at Pompeii. In the earthquake which took place A.D. 63 this temple suffered greatly, but it was rebuilt by Numerius Popidius Celsinus at his own expense "from the foundation." From the view given by Prof. Mau on p. 166, we see enough to show us

Fig. 4.—The adoration of the holy water of the Nile during the worship of Isis.

that although the building bore slight resemblance to an Egyptian temple, there was, notwithstanding, a wish on the part of the architect to produce an unwonted effect on the mind of the beholder. The deities Osiris, Isis, Anubis and Harpocrates were represented by statues, and as they have never been found, it is probable they were carried off by the faithful on that awful day in August, A.D. 79. We know little of the ceremonies connected with the initiation into the Mysteries, but two skulls, a marble hand, two small boxes, a gold cup, a small glass vessel, and a statuette of the god nearly one inch in height seem to have played a prominent part in them. We have not space to follow Prof. Mau through his description of all the various parts of this interesting temple, but we may note that the existence of the hieroglyphic sepulchral inscription, set up for the scribe Hat on a pillar to the right of the altar, indicates the adoption in Pompeii of a widespread Egyptian custom. The worship of Isis attracted large numbers to her temple, and the principal services took place before daybreak. The curtains were drawn aside and the statue of the goddess was presented to her worshippers, who straightway prayed to her; an hour after sunrise a hymn was sung to the rising sun, typified by Harpocrates, and the service was over. The second service of the day was held two hours after noon, and it seems to have consisted in the adoration of water in a vessel which was supposed to have been taken from the Nile. Whatever the details may have been, the services certainly had reference to scenes connected with the finding of the dead body of Osiris by his wife Isis, and they were intended to urge the beholder to renounce the present life and to prepare for a second birth into a purified and beatified state of existence in a new world. The temple of Isis at Pompeii is a remarkable relic of the adoption of a remarkable religion by the Romans, and we hope that Prof. Mau will add any new facts which he may glean from subsequent researches to the future editions of his work. The second part of Prof. Mau's volume deals with the houses of Pompeii, and it seems to us to be the best in the book, for it recalls the scenes and occurrences in the daily household life of the Pompeians in a most realistic fashion. The mind's eye has so many facts supplied to it with such lucid explanations that a street of houses appears before it without fatigue, and as the result of but little effort. Parts iii.-vi. deal with trades and occupations, the tombs, Pompeian art and inscriptions; the chapters of these sections are written in the same easy style, but at the same time the reader feels that he is being led along an interesting path by the hand of a master of his craft.

THE UNVEILING OF THE HUXLEY MEMORIAL STATUE.

THE statue, by Mr. Onslow Ford, R.A., of the late Right Hon. Thomas Henry Huxley, now placed in the first right-hand recess of the Great Hall of the Natural History Museum, was unveiled by H.R.H. the Prince of Wales on Saturday last, April 28, the ceremony being performed, by his Royal Highness's desire, immediately after the meeting of the Trustees appointed for that day.

Seating accommodation had been provided for the Huxley family, the Trustees of the British Museum, the members of the Memorial Committee, and other distinguished guests and chief subscribers to the Memorial Fund, in front of the statue; and a still greater number of persons, most of whom were subscribers also, assembled in the corridors overlooking the Great Hall, and on the staircases.

There were from 700 to 800 persons present, adequately representative of all branches of science, art, law, music, and politics, and of several foreign nations. The following is a classified list of the persons more directly concerned in the ceremony :—

Trustees of the British Museum.

H.R.H. the Prince of Wales.
Earl of Elgin, K.G.
Earl of Hopetoun.
Viscount Cross.
The Bishop of Winchester.
The Lord Walsingham.
The Right Hon. Sir George Trevelyan, Bart.
The Right Hon. John Morley, M.P.
Sir Nathaniel Lindley, Master of the Rolls.

Dr. W. S. Church, President of the Royal College of Physicians.
The Rev. F. H. Annesley.
Mr. Cavendish-Bentinck.
The Duke of Devonshire, K.G.
Lord Russell of Killowen.
Lord Avebury.
Viscount Peel.
Viscount Dillon.
Sir John Evans, K.C.B.
Sir Richard Webster.

Executive Committee of the Memorial Fund and others.

Lord Shand (Chairman).
Sir Joseph Fayrer, Bart., K.C.S.I., F.R.S.
Sir Henry Thompson, Bart.
Sir Joseph Hooker, G.C.S.I., C.B., F.R.S.
Sir John Donnelly, K.C.B.
Sir Norman Lockyer, K.C.B., F.R.S.
Sir Michael Foster, K.C.B., M.P., F.R.S.
Sir Spencer Walpole, K.C.B.
Sir A. Geikie, F.R.S.
Mr. Briton Riviere, R.A.

Dr. P. L. Sclater, F.R.S.
Prof. G. B. Howes, F.R.S. (Hon. Secretary).
Mrs. Huxley and members of the Huxley family, to the number of thirty-two.
Sir E. Maunde Thompson and Officers of the British Museum, Bloomsbury.
Prof. E. Ray Lankester, the Director, and the Officers of the British Museum (Natural History).

Among other persons who were seated in the central enclosure were the following :—

Sir F. Abel, Bart., F.R.S.
Prof. T. Clifford Allbutt, M.D., F.R.S.
Sir L. Alma-Tadema, R.A.
Sir Edwin Arnold, K.C.I.E., C.S.I.
The Attorney-General.
Mr. Alfred Austin.
Sir Squire Bancroft.
Hon. Edmund Barton, Q.C.
Prof. Bastian, F.R.S.
Sir Lowthian Bell, Bart., F.R.S.
Mr. Horace Brown, F.R.S.
Sir T. Lauder Brunton, M.D., F.R.S.
Rt. Hon. L. Courtney, M.P.
Sir Wm. Crookes, K.C.B., F.R.S.
Mr. Francis Darwin, F.R.S.
The Earl of Ducie, F.R.S.
Sir W. Thiselton - Dyer, K.C.M.G., F.R.S.
Mr. R. Etheridge, F.R.S.
Prof. J. B. Farmer, M.A.
Lady Flower.
Prof. Le Neve Foster, F.R.S.
Dr. R. Garnett, C.B.
Dr. J. H. Gladstone, F.R.S.
Lieut.-Col. Godwin-Austen, F.R.S.
Dr. A. Günther, F.R.S.
Mr. G. Henschel.

Lord Hobhouse, K.C.S.I., C.I.E.
Prof. Victor Horsley, F.R.S.
Prof. J. W. Judd, C.B., F.R.S.
Right Hon. W. E. H. Lecky, M.P.
Sir Hugh Low, G.C.M.G.
Dr. P. Manson.
Dr. Ludwig Mond, F.R.S.
Prof. R. Meldola, F.R.S.
Sir Francis Mowatt, K.C.B.
Sir Andrew Noble, K.C.B., F.R.S.
Admiral Sir Erasmus Ommanney, Bart., C.B., F.R.S.
Prof. J. Perry, F.R.S.
Sir W. C. Roberts-Austen, K.C.B., F.R.S.
Sir Henry Roscoe, F.R.S.
Prof. A. W. Rücker, F.R.S.
Sir J. S. Burdon-Sanderson, Bart., F.R.S.
Dr. D. H. Scott, F.R.S.
Sir G. G. Stokes, Bart., F.R.S.
Prof. G. Johnstone Stoney, F.R.S.
Mr. J. J. H. Teall, F.R.S.
Prof. T. E. Thorpe, F.R.S.
Prof. W. A. Tilden, F.R.S.
Rev. Canon Tristram, F.R.S.
Sir William Turner, F.R.S.
Prof. W. F. R. Weldon, F.R.S.

Foreign nationalities were represented by :—

Dr. F. P. Moreno (of the Argentine Republic).
Major Dr. von Wissmann (Germany).
Mons. L. Geoffray (France).
Mons. F. Fuchs (Congo Free State).
Prof. Batalha Reis (Portugal).

Prof. G. Paladino (of Naples).
Prof. G. Gilson (of Louvain).
Señor Don Pedro Jovar y Tovar (Spain).
Count Bottaro Costa (Italy).
Plenipotentiaries at the International Conference for the preservation of wild animals in Africa.

Punctually at the time appointed (1.15 p.m.), his Royal Highness took up a position to the spectators' left of the statue, supported by the Standing Committee of the Trustees of the Museum, with Sir Maunde Thompson and Prof. Ray Lankester; while Sir Joseph Hooker, similarly supported by the members of the Executive of the Memorial Committee, stood on the right; the sculptor, Mr. Onslow Ford, being in proximity to the statue.

The proceedings were opened by Prof. Ray Lankester, with the following introductory statement :—

YOUR ROYAL HIGHNESS, MY LORDS, LADIES AND GENTLEMEN,—The duty of briefly explaining the nature of the present proceedings has devolved upon me. I feel it to be a great privilege to discharge this duty on the occasion designed to do honour to my venerated master, Prof. Huxley. This celebration would have been no less dear to Huxley's fellow-worker and friend, the late director of this museum, Sir William Flower, who unhappily is no longer with us to witness the completion of the memorial statue which he, especially, desired to see placed in this hall.

A few months after Prof. Huxley's death in 1895, a committee was formed for the purpose of establishing a memorial of the great naturalist and teacher. At a meeting of that committee, held on November 27, 1895, at which 250 members were present, and at which his Grace the Duke of Devonshire presided, the following resolution was carried :—

"That the memorial do take the form of a statue, to be placed in the Museum of Natural History, and a medal in connection with the Royal College of Science ; and that the surplus be devoted to the furtherance of biological science in some manner to be hereafter determined by the committee, dependent upon the amount collected."

From all parts of the world, besides our own country, from every State of Europe, from India and the remotest Colonies, and from the United States of America, subscriptions have been received for the Huxley memorial, amounting in all to more than 3380*l.*

Three years ago the committee commissioned and obtained the execution of a medal bearing the portrait of Huxley, and has established its presentation as a distinguished reward in the Royal College of Science. The re-publication of the complete series of Huxley's scientific memoirs, which was proposed as one of the memorials to be carried out by the committee, has been undertaken by Messrs. Macmillan, without assistance from the committee. I am glad to be able to state that two large volumes of these richly illustrated contributions to science have been already published.

Whilst these other memorials were in progress under the auspices of the executive committee, they secured the services of Mr. Onslow Ford, R.A., to execute the statue which it had been decided by the general committee to regard as the chief object of the subscriptions entrusted to them. On the completion of the statue, the trustees of the British Museum agreed to receive it and to place it in the great hall where we are now assembled.

On behalf of the vast body of subscribers to the memorial, Sir Joseph Hooker, Huxley's oldest and closest friend, himself the survivor of that distinguished group of naturalists, including Charles Lyell, Richard Owen and Charles Darwin, who shed so much lustre on English science in the Victorian age, will hand over the statue of Huxley to the trustees of the British Museum. Your Royal Highness has been graciously pleased, as one of the trustees, to represent them on the present occasion, and to receive the statue on their behalf. The memorial statue of Huxley is the expression of the admiration, not only of the English people, but of the whole civilised world, for one who as discoverer, teacher, writer and man, must be reckoned among the greatest figures in the records of our age.

Sir Joseph Hooker then stepped forward from among the committee, and presented the statue in the following words :—

May it Please Your Royal Highness,—I have the honour of being deputed, by the subscribers to the statue of my friend the late Prof. Huxley, to offer it to your Royal Highness, on behalf of the trustees of the British Museum, with the intent that it should be retained in this noble hall as a companion to the statues of Prof. Huxley's distinguished predecessors, Sir Joseph Banks, Mr. Darwin and Sir Richard Owen. It would be a work of supererogation, even were I competent to do so, to dwell upon Prof. Huxley's claims to so great an honour, whether as a profound scientific investigator of the first rank, as a teacher, or as a public servant ; but I may be allowed to indicate a parallelism between his career and those of two of the eminent naturalists to whom I have alluded, which appears to me to afford an additional argument in favour of retaining his statue in proximity to theirs. Sir Joseph Banks, Mr. Darwin and Prof. Huxley all entered upon their effective scientific careers by embarking on voyages of circumnavigation for the purpose of discovery and research under the flag of the Royal Navy. Sir Joseph Banks and Prof. Huxley were both Presidents of the Royal Society, were both trustees of the British Museum ; and, what is more notable by far, they were their scientific services estimated by the Crown and their country, that they both attained to the rare honour of being called to seats in the Privy Councils of their respective Sovereigns.

With these few words I would ask your Royal Highness graciously to accede to the prayer of the subscribers to this statue, and receive it on behalf of the trustees of the British Museum.

He was followed by Sir Michael Foster, who pronounced the following *éloge* on Huxley's work and influence :—

May it Please Your Royal Highness,—Before you unveil this statue it is my duty and privilege to add a few words to those which have just been spoken by the beloved Nestor of biological science. Sir Joseph D. Hooker, born before Huxley was born, a sworn comrade of his in the battle of science, standing by him and helping him like a brother all through his strenuous life, may perhaps be allowed to shrink from saying what he thinks of the great work which Huxley did.

We of the younger generations, Huxley's children in science, who know full well that anything we may have been able to do springs from what he did for us, cannot on this great occasion be wholly silent.

Some of us have at times thought that Huxley gave up for mankind much which was meant for the narrower sphere of science ; but if science may seem to have been thereby the loser, mankind was certainly the gainer ; and indeed it was a gain to science itself to be taught that her interests were not hers alone, and that not by one tie or by two, but by many was her welfare bound up with the common good of all.

To many perhaps the great man whose memory we are here met to honour was known chiefly as the brilliant expositor of the far-reaching views of that other great man who through his statue is now looking down upon us. Your Royal Highness is doubtless at this moment thinking of that interesting occasion, fifteen years ago, when you unveiled that statue of Darwin, and you are calling to mind the weighty words then spoken by him, whose own statue brings us here to-day.

Huxley it is true fought for Darwin, and indeed "he was ever a fighter." But he fought not that Darwin might prevail ; he fought for this alone—that the views which Darwin had brought forward might be examined solely by the clear light of truth, untroubled by the passion of party or by the prejudice of preconceived opinion. As he never claimed for those views the infallibility of a new gospel, so he always demanded that they should not be peremptorily set aside as already proved to be wrong.

Huxley worked for his fellow men in many ways other than the way of quiet scientific research. Had we not known this we should have thought that his whole life had been given up to original scientific investigation, so much has the progress of biologic science, since he put his hand to it, been due to his labours. On the sands of many a track of biologic inquiry he has left his footprint, and his footprint has ever been to those coming after him a token to press on with courage and with hope. The truths with which he enriched science are made known in his written works ; but that is a part only of what he did for science. No younger man, coming to him for help and guidance, ever went empty away ; and we all—anatomists, zoologists, geologists, physiologists, botanists, and anthropologists—came to him. The biologists of to-day, all of us, not of this country alone, but of the whole world of science, forming, as it were, a scattered fleeting monument of this great man, are proud at the unveiling of this visible lasting statue here.

In conclusion, Sir M. Foster, facing the Prince, added the words :—May I crave your Royal Highness's permission to seize this opportunity to assure you incidentally, but none the less from the bottom of our hearts, on the part of men of science that we, in common with all Her Majesty's subjects, are rejoicing that you escaped the dreadful peril to which a few days back you were exposed, and to express to you our continued esteem and respect ?

On Sir M. Foster's return to his seat among the committee, the Duke of Devonshire, speaking from in front of the veiled statue, said he had the honour nearly five years ago of presiding over the committee formed for the purpose of establishing a memorial to Prof. Huxley. He had now to report to his Royal Highness that the labours of that committee had terminated, and to say that the committee desired to present the statue to his Royal Highness on behalf of the trustees of the British Museum. They felt, however, that the real memorial to the deceased man of science was to be found in the writings which had already been referred to, and still more in the scientific work he accomplished or helped to promote, and in the influence he exercised and was still exercising upon the minds of younger men, many of whom they trusted might at some future time emulate his distinguished example. On behalf the committee he begged to tender his Royal Highness their thanks for having come to give a final sanction to their proceedings, and for having undertaken the duty of unveiling the statue that day.

The Prince of Wales then withdrew the covering from the statue, and brought the proceedings to a close with the following words :—

MY LORDS, LADIES AND GENTLEMEN,—I consider it a very high compliment to have been asked by the Huxley Memorial Committee to unveil and receive this statue, and to do so in the name of the trustees of the British Museum, of whom I have the honour to be one. I have not forgotten that fifteen years ago I performed a similar duty in connection with the fine statue of the celebrated Charles Darwin, which is at the top of the stairs, when it was similarly handed over to the British Museum. We have heard to-day most eloquent and interesting speeches with reference to that illustrious man of science and the great thinker, the late Prof. Huxley. It would, therefore, be both superfluous of me, I may even say unbecoming in me, to sound his praises here in the presence of so many men of science, who know more about all his work than I do. I can only, on my own behalf, endorse everything that has fallen from the lips of those gentlemen who have spoken, and I beg to repeat the expression of the great pleasure it has given me for the second time to have performed the interesting ceremony of taking over the statue of another great and illustrious man of science.

The statue is a colossal seated one of white marble, the figure being represented in a doctor's gown, with the right hand clasping one arm of the chair, and the left lying across the other with the fist clenched. The pedestal is of Verona marble on a black base, and bears upon its face the name and dates of birth and death in simple bronze letters.

The statue is a thoroughly successful work of art, and stands out in bold relief to the dim mystery of the recess in which it is placed. Though the expression of the face is perhaps a little severe, the features are true to nature ; and when it is considered that the artist was never privileged with a sitting in life, and that the only material available to him were the death mask and an assemblage of none too favourable photographs, it must be admitted he has done well. Great praise must be given to the modelling of the hands, in which those who knew the great philosopher intimately will recognise a faithful portrayal of well-defined characteristics.

The first and main object of the Memorial has thus been successfully achieved. As for those which remain, the award at the Royal College of Science is to be known as the "Huxley Gold Medal," for the "promotion of science in the directions in which Huxley was distinguished," and especially for research to be carried on in the laboratory which bears his name. It has been further arranged that the use of the obverse die shall be granted to the Anthropological Institute (of which Huxley was practically the founder), in connection with the establishment by that body of a Huxley Lecture-ship, and a medal, for which they will furnish the reverse. Huxley's labours as an anthropologist are among the most important of his scientific career, and it may be questioned whether his "Man's Place in Nature," published against the advice of some of his friends, who feared his "ruin" did it appear, does not now rank among the best and most enduring of his works. His influence as an anthropologist was great, and devotees to that branch of science will hail with satisfaction this decision to perpetuate his memory.

PRELIMINARY NOTES ON THE RESULTS OF THE MOUNT KENYA EXPEDITION, 1899.

THE Mount Kenya Expedition left Nairobi, the then head of the Uganda Railway, on July 26, 1899, and returned to Naivasha, a station on the Uganda Road, on September 29. Considerable difficulties were experienced in the matter of commissariat, on account of the drought

and famine prevalent throughout East Africa. For this reason a longer sojourn on the mountain would have been impracticable, even if other circumstances had permitted of it.

Previous accurate knowledge of Mount Kenya rested chiefly on the work of Captain G. E. Smith, R.E., who had fixed the position of the peak, by triangulation along the Uganda Road, and of Dr. J. W. Gregory, who, in 1893, ascended the south-western slope to a height which appears to have been nearly 16,000 feet. An account of the 1899 journey is given in the May number of the *Geographical Journal*.

Mount Kenya is a vast flattened dome, seered with radiating valleys. It rises from a plateau, the level of which is 5000 to 7000 feet above the sea. Upon the crown of the dome is a precipitous pyramid, the cleft peak of which has an altitude of 17,200 feet. The entire *massif* measures about fifty miles from east to west and forty miles from north to south. Its northern slopes are crossed by the equator.

We made a plane table survey of the central portion of the mountain, and connected it by route surveys with Nairobi and Naivasha. The altitude of the central peak was determined by boiling point and theodolite, combined in four different ways, with an average result practically the same as that obtained by Captain Smith at a distance of ninety miles.

The central pyramid is the core of the denuded and dissected volcano, a fact first suggested by the late Joseph Thomson, who saw the mountain from the Laikipian plateau. Although not yet examined in section, the holocrystalline rock on the summit may probably be identified with the nepheline syenite obtained by Gregory at a lower level. The core must, therefore, have risen considerably above the present peak, and if allowance be made for still loftier crater-walls, the original height of Kenya may have equalled that of the still complete Kibo summit of Kilimanjaro.

The most significant point in the structure of the mountain is the fact that, while the major axis of the peak strikes west-north-westward and throws the glaciers down northern and southern slopes, the chief water-parting runs in a direction at right angles to this, past the eastern foot of the central peak, with the effect that the valleys are thrown off eastward and westward, and that all the existing glaciers belong to the westward drainage. From a series of rock specimens obtained at widely separated spots on the summit of the craggy ridge constituting the divide, it appears that the lie of the water-parting has been determined by a system of great dykes, which must almost have split the mountain in two.

There are fifteen existing glaciers, of which two are a mile in length, and the remainder are small. Their lower ends descend to about 14,800 feet. Everywhere and at all hours at the time of our visit the surfaces were dry and crisp. Comparatively little water flowed from them, and the stream banks below gave small indication of floods. The ice was intensely hard, and fed by fine hail rather than snow. These facts may be explained by the meteorological conditions. Although the air-temperatures were not very low at night, there was then great radiation into the cloudless sky. In the after-noon, on the other hand, a cloud cap regularly warded off the sunshine. The air was usually dry, the relative humidity falling on more than one occasion to 54 per cent.

Evidence of past glaciation was frequent down to 12,000 feet both in the eastern and western valleys, and there were occasional traces down to about 9000 feet. The whole of the central part of the mountain, with the exception of the peak and the dividing ridge, must have been buried under a sheet of glacier, more than comparable to that of Kilimanjaro, at a time later than the erosion of the existing valleys.

Snow was absent from the summit, and several species of brilliantly coloured lichen were collected there. Everlasting flowers grew in the rock chinks up to 16,500 feet. In the upper Alpine zone were two distinct species of giant groundsel and two of giant lobelia, seeds of which have been brought home. The greater part of our dried plants was lost, but the mosses and lichens were saved. A series of photographs of the Alpine vegetation in various stages of growth was taken by my colleague, Mr. C. B. Hausburg.

Mr. Oldfield Thomas has described, before the Zoological Society, the skulls and skins of the mammals collected by us. The most interesting is a new species of Rock Dassy (*Procavia Mackinderi*), whose nearest relative has recently been sent home from the Eldoma Ravine by Mr. F. J. Jackson (*P. Jacksoni*). Apart from these two species, no Rock Dassies have been found in any part of East Africa, nor are they known further south. *P. Mackinderi* appears to be isolated above the forest-zone (7000-10,000 feet) on Mount Kenya. A new Forest Dassy was obtained from a lower level.

This mountain block and the Rift Valley may be the necessary complements of one another.

Only a small collection of insects was obtained, chiefly in Kikuyu, but Prof. Poulton informs me that it includes new species of Coleoptera, Forficulidæ and Hymenoptera.   H. J. MACKINDER.

THE DUKE OF ARGYLL.

AMONG the losses which science is from time to time called upon to deplore, not the least serious arise from the death of men of prominent public position who have taken an active personal interest in the advance of natural knowledge, and have done their best to promote it. The late Duke of Argyll was an eminent example of this type of man. Heir of a long line of illustrious ancestors, who for many generations have played a leading part in the stormy annals of their native country, called early in life to the legislature where he mingled conspicuously in the political conflicts of his time, full of

Kenya Peak, from the south-west.

The collection of birds has been described by Dr. Bowdler Sharpe. It includes a new eagle owl, as large as the European species, which feeds on the rats of the Alpine zone of Kenya, and there are three other new species. Generally the birds are similar to those of Mount Elgon, and in a lesser degree to those of Kilimanjaro. This is strikingly indicated by the fact that if Mr. Jackson had not explored Mount Elgon in 1890, nearly every bird we obtained would have been new.

The few human inhabitants of Kenya are Wandorobo, elephant hunters, who live in the forest up to its higher limit. On one occasion a party of them was seen at over 12,000 feet.

To west of Mount Kenya is the so-called Aberdare Range, traversed for the first time by the members of our expedition. It consists of two much denuded volcanic stumps, Nandarua and Sattima, rising to 12,900 and 13,200 feet respectively, and of a raised block, 9000 feet high, defined by parallel fault scarps, which strike in the same direction as the scarps of the Great Rift Valley.

wide and generous sympathies which prompted him to speak or to write on most of the great questions that agitated the public mind during his long and brilliant career, the Duke yet found time to read much and widely in science, and to keep himself acquainted with the progress of scientific discussion and achievement. He was happily gifted with a marvellous versatility, so that he could turn rapidly from one sphere of thought and activity to another far removed. Hence, amid the cares of State and of the administration of a great domain, as well as in the sorrow of domestic bereavement, he was often to be found immersed in the perusal of some recent treatise, or carrying on a research of his own in those parts of the scientific field which more specially interested him. Whether as an acute critic of the labours of others, or as an observer of nature himself, his devotion to these pursuits remained a characteristic feature of his life from the beginning to the end. It is difficult at present to define with precision the extent and value of the services of such a man in the progress of the science of his time. His own original contributions may be little in amount or importance, but his example and his enthusiasm, together with his political activity and his social rank, combine to make him a force in the land, which powerfully aids any good cause which he espouses. The death of the Duke of Argyll is thus an event which must be chronicled with sincere regret in the pages of a scientific journal.

It was through geology that the Duke first came practically in touch with science, and it was in geological pursuits and criticisms that he found the most congenial employment of his leisure moments. It is just half a century since, on a visit to his property in the Island of Mull, he found that one of his tenants had gathered a number of fossil leaves and plants from the rocks of the neighbourhood. At once appreciating the geological significance of these remains, he investigated their mode of occurrence, and recognised their association with sheets of lava and volcanic ashes. The plants were pronounced by Edward Forbes to be probably of Miocene age, and thus was securely laid the first stone of the edifice that has since been reared in illustration of the volcanic history of the Inner Hebrides. It is matter for regret that the Duke never followed up this important discovery.

Other geological fields attracted him, where he found ampler material for the exercise of that critical acuteness and the display of that forensic style of argument which made his writings so lively and so pungent. He had imbibed his earliest ideas of geological causation in the school of the cataclysmists, and to these ideas he adhered to the last. When the earlier views of Hutton and Playfair with regard to the denudation and sculpture of the land were revived and began to spread among the younger men, the Duke raised his protest against them, and poured on them the contempt and ridicule which they seemed to him to deserve. As they grew in acceptance, both in this and other countries, and as their advocates increased in number and in confidence, his vehemence of declamation seemed to augment in proportion.

Nor was this the only line along which the modern tendency in geological speculation seemed to the Duke to be running in an entirely wrong direction. When he began to interest himself in these questions, Agassiz' doctrine, that not only Britain but a large part of Europe was once buried under land-ice, had not been generally accepted. The geologists of this country preferred to account for the phenomena by supposing that the land had been submerged in a sea across which floating ice drifted. The Duke of Argyll was never able to accept the modern doctrine, except in a limited degree. He admitted the former existence of local valley-glaciers, but could not recognise the force of the evidence adduced to show that not only the valleys, but the surrounding hills had once been over-ridden by a vast sheet of ice.

The rise of the modern school of evolution afforded the Duke full scope for the exercise of his acute reasoning power and keen critical faculty. In article after article, address after address, and volume after volume, he subjected the doctrines of that school to the closest scrutiny. It may be freely admitted that he detected here and there a fallacy, and pointed out a conclusion different from, but not less probable than, that which his opponents had drawn. But perhaps his most valuable service lay in that border-land of philosophy and science in which he specially loved to exercise his thoughts and his pen. Even when men of science differed widely from his his conclusions, they could not but admit that in his "Reign of Law" and his "Unity of Nature," he showed the wide range of his reading, the clearness and vigour of his reasoning powers, the force and eloquence of

his style, the grasp he had of some of the more difficult scientific problems of his day, the strong bent of his nature towards metaphysics, and, above all, the lofty tone of his sentiments in regard to the moral nature and destiny of man.

The Duke of Argyll was essentially a man of action, to whom the stir of conflict and the stimulus of controversy were not uncongenial. Even in his scientific discussions he could not always quite forego the style in which he vilipended the opposite party in the House of Lords or in the public prints. He seemed sometimes hardly to realise the full extent and meaning of the evidence which he was criticising. In conversation, indeed, he might appear for a time to be impressed by the force of this evidence, and be willing to admit that the truth might, perhaps, lie somewhere between his own views and those to which he was opposed. But the force of early conviction or prepossession would, in the end, be too strong for him, and possibly the next morning his opposition would be found to be as complete and confident as ever. Unflinching and resourceful as an antagonist, enforcing with almost passionate enthusiasm what he held to be the truth, independent and self-reliant alike in his opinions and his actions, dignified and courteous after the manner of an older time, he formed altogether a striking and picturesque personality.

But the energy of the doughty debater was combined with much personal kindliness even towards those from whom he most seriously differed. Above all the other features of his character there shone out an intense love of nature and an eager desire to know more of her processes and laws. Year after year the Duke would spend weeks at a time in his yacht among the Western Isles, which he loved with all the enthusiastic devotion of one who was born and spent his youth among them. He was familiar with that western coast from one end to the other, under every change of sunlight and shadow. He had sketched every peak and crag and island, and he delighted to recall from his sketch-books the charm with which these scenes had fascinated him. To all their obvious attractions for the ordinary visitor his geological knowledge enabled him to join the fresh interest which is given to them by an acquaintance with the history of their remote past. In this way he kept himself in touch with some of the aspects of nature that most vividly appealed to his imagination. His poetic temperament found refreshment in these frequently renewed sojourns amid the varied scenery of the West of Scotland. As shown by his published writings, his wide acquaintance with modern English poetry furnished him with many an apt quotation and allusion. Tennyson's poetry seemed to be particularly familiar to him, insomuch that a casual citation of a line or expression from that poet by one of the company would sometimes lead the Duke to quote from memory the whole passage.

As the head of a great historic clan, the Duke of Argyll was a true Scot, who had studied his country's history both geological and political, and had made himself personally acquainted with a large part of its surface. The geological problems that more particularly engaged his attention were largely those which his own Highland hills and glens had suggested to his mind. Now and then, in the midst of an eager conversation, a Scottish word or expression would come most readily to his lips as conveying the meaning he wished to express. Of his general services to the country at large this is not the place to speak. But we may confidently anticipate that when some future historian shall review the various forces which have furthered the advance of science in this country during the Victorian age, a well-marked place will be assigned to the services rendered by the Duke of Argyll.   A. G.

PROF. A. MILNE-EDWARDS.

IT is with sincere regret that we have to record the death, at the age of sixty-four, of Prof. Alphonse Milne-Edwards, the Director of the Paris Museum of Natural History, which took place at Paris on Saturday, April 21, after a brief illness. The late professor was of English descent, being the grandson of Mr. Bryan Edwards, M.P., a West Indian planter who settled at Bruges ; and, with this ancestry, it is curious to note how extremely imperfect was his colloquial knowledge of the English language. His father, Prof. Henri Milne-Edwards, was the well-known eminent zoologist of Paris, who died in 1885 ; and father and son were for many years associated in zoological work.

Born in Paris in 1835, Alphonse Milne-Edwards took his medical degree in 1859, and was nominated Professor at the School of Pharmacy in 1865. In 1876 he acted as deputy for his father as Professor of Zoology at the Jardin des Plantes ; in the following year he succeeded the late Prof. P. Gervais as a member of the Institute of the Paris Academy of Sciences ; and in 1885 he entered the Academy of Medicine. In 1891, being already Professor of Zoology, he was appointed Director of the Paris Museum of Natural History and of the Menagerie in the Jardin des Plantes ; his official title as regards the latter post being *Administrateur chargé de la Direction de la Ménagerie au Musée d'Histoire naturelle.*

Having published, in 1864, an important memoir on the anatomy and affinities of the Chevrotains, and a second, in 1866, on the osteology of the Dodo, in 1867 Milne-Edwards issued the first fasciculus of his magnificent work, entitled "Recherches Anatomiques et Paléontologiques pour servir à l'Histoire des Oiseaux Fossiles de la France," which was completed in four volumes (two of text and two of plates) in 1872. As mentioned by Prof. A. Newton, this monumental work marked an epoch in ornithology, for it showed the possibility of forming a classification of birds by means of their "long bones." Much interest was excited by the identification in this work of remains of peculiar existing African and Malagasy genera of birds in the French Tertiaries. While this work was in progress, Alphonse Milne-Edwards was associated with his father in bringing out the "Recherches pour servir à l'Histoire naturelle des Mammifères," which was commenced in 1868 and completed in 1874. A large proportion of the latter was devoted to the description of new types of mammals from Central Asia, among them being the many strange forms, like *Aeluropus*, then recently obtained by Père David in the Moupin district of Eastern Tibet. The period from 1866 to 1874 also saw the issue of "Recherches sur la Faune ornithologique éteinte des Îles Mascareignes et de Madagascar." And the late professor's interest in the Malagasy fauna was likewise shown in a paper on the embryology of the Lemurs, published in 1871, and in his contributions to Grandidier's "History of Madagascar," still in course of publication.

But it would be a mistake to suppose that the researches of Prof. Milne-Edwards were by any means restricted to mammals and birds. From an early period in his career his attention had been directed to the study of zoophytes and crustaceans ; and later on he had attentively studied the animals adherent to submarine cables, which had been raised after a sojourn at the bottom of the sea. With this latter subject the study of the ocean floor was intimately connected. And in 1880 he brought before his Government the advisability of fitting out an expedition for submarine surveying, with the result that in the following year a party of savants, under his own direction, embarked on the *Travailleur* to survey the Gulf of Gascony. The results obtained were so important that the same vessel was again put at the disposal

of the professor, who completed the survey of the Gulf of Gascony, and explored the sea-bottom of the Strait of Gibraltar and of a considerable portion of the Mediterranean. In 1882 the *Travailleur* undertook a surveying voyage of the Atlantic as far as the Canaries. The year following the *Talisman* took the place of the *Travailleur*, and carried Prof. Milne-Edwards and his associates to the coasts of Portugal, Morocco, and the Canary and Cape Verde Islands, and then on to the Sargasso Sea, whence it returned by way of the Azores. The results of these dredging expeditions were published under the title of "Expéditions scientifiques du *Travailleur* et du *Talisman* pendant les années 1881, 1882 et 1883."

For these deep-sea explorations, Milne-Edwards was awarded the gold medal of the Royal Geographical Society. In 1876 he was elected a Foreign Member of the Zoological Society of London, and in 1882 a Foreign Correspondent of the Geological Society. He paid several visits to England, the last on the occasion of the Zoological Congress at Cambridge in 1898. 　　R. L.

NOTES.

THE funeral of the Duke of Argyll will take place at the family burial ground, Kilmun, on the Holy Loch, on Tuesday next, May 8.

THE annual conversazione of the Institution of Electrical Engineers will be held at the Natural History Museum, South Kensington, on Tuesday, June 26.

THE Duke of Cambridge, president of the Sanitary Institute, will occupy the chair at the Institute dinner on Friday, May 11.

THE University of Göttingen has awarded the Volbrecht prize for scientific research to Dr. Gegenbaur, professor of anatomy at Heidelberg. The prize is of the value of 12,000 marks (600*l.*)

To commemorate the foundation of the k. k. geologischen Reichsanstalt of Vienna, in 1849, a jubilee meeting will be held in the great hall of the Institute on June 9, and representatives of science or of scientific institutions are invited to be present.

THE *Botanical Gazette* records the death by drowning, in September last, of Prof. Kyokichi Yatabe, the founder of the Botanical Society of Japan.

THE annual meeting of the American Association for the Advancement of Science will be held at Columbia Universtity, New York, from June 25 to June 30.

WE learn, from the *American Naturalist*, that the herbarium and the principal part of the botanical library of Columbia University have been transferred to the New York Botanic Garden, and that, in future, the advanced work in botany of the University will be carried on in the laboratory of the Garden.

THE *British Medical Journal* states that the tenth award of the Riberi prize of 20,000 lire (800*l.*) will be made by the Royal Academy of Medicine of Turin on December 31, 1901, for the best printed or manuscript work, or the most important discovery, during the quinquennium 1897-1901, in the domain of experimental pathology, hygiene, or forensic medicine.

THE Franklin Institute has awarded John Scott medals and premiums to Mr. A. V. Groupe for his improved braiding machine, to Messrs. C. A. Bell and S. Tainter for their invention of the graphophone, and to Mr. A. M. Hopkins for his pneumatic system for preventing the bursting of water-pipes by freezing. Elliott Cresson medals have been awarded to Mr.

L. E. Levy for his acid-blast method of etching metal plates ; and to Prof. W. O. Atwater and Mr. E. B. Rosa for their respiration calorimeter.

THE *Daily News* states that Lieut. R. E. Peary has forwarded some interesting relics to the Royal Naval College, Greenwich. These consist of the sextant left behind in Repulse Harbour by Lieut. Beaumont in 1876, and subsequently recovered by Lieut. Peary, and the original record deposited in a cache by Sir George Nares on Norman Lockyer's Island in 1875. The great meteorite which Lieut. Peary brought back from his last Arctic expedition still remains on the Cob Dock of the Brooklyn Navy Yard. The meteorite weighs 200,000 pounds, and Lieut. Peary wishes to obtain 15,000l. for it.

THE Trinity House steam vessel *Irene*, with the deputy master, Captain G. R. Vyvyan, on board, accompanied by a committee of the Elder Brethren and their scientific adviser, Lord Rayleigh, has proceeded to the Bristol and English Channels in order that special surveys in connection with new lighthouse works, and observations on both English and French lights from seaward, may be made.

THE death is announced of Mr. G. V. Ellis, who succeeded Prof. Quain as professor of anatomy in University College, London, in 1850, an appointment which he held for twenty-seven years, resigning in 1877, when he was appointed Emeritus Professor. Mr. Ellis was co-editor with the late Dr. William Sharpey of the sixth edition of "Quain's Elements of Anatomy," published in 1856, and the author of several works for students of anatomy.

AMONG the items included in the Prussian Budget is a sum of 7,300,000 marks, for the purchase of lands in Berlin, on which is to be erected a building for the Academy of Sciences and the Royal Library. The value of the land is estimated at more than 11,000,000 marks, but about 3,000,000 marks is obtained by the exchange of other property, and 1,000,000 marks is to be voted next year.

A SUMMER meeting of the Anatomical Society of Great Britain and Ireland will be held at the Owens College, Manchester, on Thursday and Friday, June 21 and 22. Opportunities will be afforded to members of seeing things of local interest during their visit to Manchester. An excursion to the Lake District will be arranged, and members who desire to join the party are requested to inform the local secretary, Dr. Peter Thompson, the Owens College, Manchester.

A COMMITTEE composed of many eminent men of science in France has been formed for the purpose of obtaining funds for the erection of a modest monument at Langres in honour of Auguste Laurent, the renowned chemist. Laurent was born at La Folie, near Langres, in 1808, and in 1831 became assistant to Dumas, under whom he acquired a special knowledge of organic chemistry, and carried on his original researches on naphthalene and carbolic acid, together with their derivatives. After filling various posts, the last of which was a chemical professorship at Bordeaux, Laurent became Warden of the Mint at Paris, where he remained in intimate connection with Gerhardt until his death in 1853. Subscriptions for the proposed monument should be sent to the treasurer of the Committee, M. Caublot, 45 rue de Belleville, Paris.

MR. JAMES MANSERGH has been elected president of the Institution of Civil Engineers, in succession to Sir Douglas Fox. Sir William White, K.C.B., F.R.S., Mr Charles Hawksley, Mr. J. C. Hawkshaw, and Mr. F. W. Webb have been elected vice-presidents. The following awards have been made for papers read and discussed before the Institution during the past session :—A George Stephenson medal and a Telford premium

to Sir Lowthian Bell, Bart., F.R.S. ; Telford medals and premiums to Messrs. H. H. Dalrymple-Hay, B. M. Jenkin, F. W. Bidder and F. D. Fox ; a Watt medal and a Telford premium to Mr. J. Dewrance ; a Crampton Prize to Sir Charles Hartley ; and Telford premiums to Messrs. C. N. Russell and R. A. Tatton. The presentation of these awards, together with those for papers which have not been subject to discussion, and will be announced later, will take place at the inaugural meeting of next session.

AMERICAN ethnology has been deprived of a prominent worker by the death of Mr. Frank H. Cushing. Mr. Cushing, says the *Scientific American*, was born in 1857, at Northeast, Pa., and when he was only eighteen years of age his work was brought to the attention of the late Mr. Spencer F. Baird, who was then Secretary of the Smithsonian Institution, and in 1875 he went to Washington as an assistant in that institution. He had charge of the ethnological exhibit at the Centennial Exposition of 1876, and in 1879 he accompanied an expedition from the Smithsonian Institution to investigate the Pueblos of New Mexico, and at his request was left at the Pueblo of Zuni, where he lived almost continuously for six years. He returned to Washington in 1884 and began to work up his voluminous notes. Two years later he was made Director of the Hemenway South-western Archæological Expedition. Extensive excavations were made in South Arizona and New Mexico, and the large collection of objects of prehistoric art which he gathered is in the Peabody Museum at Cambridge, Mass. This work took up two and one-half years of his time, and then Mr. Cushing returned to the United States Bureau of Ethnology to supervise a memoir on the Zuni myths printed by the Bureau. Three years later he became director of the expedition fitted out by Mrs. Phœbe A. Hearst and the late Dr. William Pepper, conducted under the auspices of the National Museum, the Bureau of Ethnology and the University of Pennsylvania.

THE motion for the second reading of the Sea Fisheries Bill in the House of Commons, on Monday, resulted in a lively discussion. The Bill prohibits the sale of flatfish below a specified size, and its rejection was moved on the grounds that it would not have the effect of preventing the destruction of immature fish, or of increasing the supply of fish. In the course of the discussion, an honourable member said that the whole of the trouble arose from the institution of a number of committees composed of farmers, lawyers, and captains of the horse, foot, and artillery, who knew little of fishing, and who ventilated strange theories and supported them with portentous and irrelevant statistics. This remark was used as an argument against the Bill, but it may also be taken to mean that if fishery matters were controlled by scientific men familiar with the natural history of the sea, and questions concerning fisheries were referred to marine biologists, recommendations might be based. Board of Trade statistics prove that there is a large destruction of immature fish, and that the quantity of fish landed has decreased during recent years. The Government, wishing to preserve a great national industry, have put forward the present Bill, which is really the Undersized Fish Bill of last year, and has appeared under various other titles in previous years. The discussion upon the Bill was not completed when the House adjourned on Monday.

THE report presented at the anniversary meeting of the Zoological Society, held on Monday, stated that the number of Fellows of the Society at the end of last year was 3246. The total income of the Society during the past year was 28,880l. The average annual receipts of the Society for the previous ten years have been 26,370l., so that the receipts for 1899 exceeded that average by 2509l. The number of visitors to the Society's

Gardens in 1899 was 696,707. The number of animals now living in the Gardens is 2753, of which 821 are mammals, 1471 birds, and 461 reptiles and batrachians. Amongst the additions made during the past year, thirteen were specially commented upon as being of remarkable interest, and in most cases new to the Society's collection. Of these, by far the most noticeable objects exhibited for the first time were the pair of Grévy's zebras placed under the care of the Society by Her Majesty the Queen. These animals, which had been presented to Her Majesty by the Emperor Menelik of Abyssinia, were brought down to Zeila, on the coast of Somaliland, under the care of Captain J. L. Harrington, the British Political Agent. At Zeila they were handed over to the Society's assistant superintendant, Mr. Arthur Thomson, who had been sent there by the Council at the request of the Foreign Office on purpose to receive them, and by him they were landed safely in London on August 14 last year. The Council also called special attention to the young male giraffe, acquired in April 1899, by purchase, for the sum of 800*l*. It is believed that this animal, together with the female purchased in 1895, form the only pair of young giraffes now to be found in any of the zoological gardens in Europe. The works in connection with the new bore at the well and the new machinery for pumping were completed last year. The new water supply has been further improved by the construction of a second and larger reservoir, so that an excellent supply of water will henceforth always be available in every part of the Gardens.

MR. F. W. HASELGROVE sends us a photograph of a robin's nest in a water-can, with the bird sitting upon its eggs, now to be seen at Finchley Cemetery. Robins are well known to build

their nests frequently in curious places, one of the most remarkable instances on record being that of a nest in a battered beer-can between the rails over which trucks were continually passing at Worthing railway station. Flower-pots and water-cans appear to be favourite nesting-places of the birds.

THERE are reasons for believing that the Scandinavians discovered America and settled in Massachusetts in pre-Columbian days. The evidence consists in the occurrence of certain ruins which correspond closely with ruins of the Saga-time in Iceland, but which differ from native dwellings and early European

ruins on the coast; and also in the correspondence in the physical features of the Massachusetts coast with the description of the country called Vinland in Icelandic literature. This evidence has recently been brought together in an illustrated article by Miss Cornelia Horsford in *Appleton's Popular Science Monthly*.

THE recent Norwegian earthquakes are studied by Mr. J. Rekstad in a paper published in the *Bergens Museums Aarbog* (1899, No. iv.). During the four years 1895–1898, the number of recorded earthquakes is 77, the corresponding number for Great Britain being 24, and for Greece, 1652.

A SLIGHT earth-shake occurred near Manchester at about 1.17 a.m. on April 7. It was felt at Pendleton, Pendlebury, Seedley, Salford and other places in the immediate neighbourhood of the Irwell Valley fault. The small disturbed area and the rather marked intensity of the shock point to a local origin, probably connected either directly or indirectly with the extensive coal-workings of the district. On February 27, 1899, a similar earth-shake was felt at the same places (see NATURE, vol. lxi. p. 546).

WE have received a reprint of a paper, published in the *Bulletin* of the Geographical Society of Philadelphia, "On the Nicaragua Canal in its Geographical and Geological Relations," by Prof. Hugelo Heilprin. The paper, which is illustrated by maps and photographs, discusses (1) the volcanic phenomena of the region of the proposed canal; (2) an assumed inconstancy in the level of Lake Nicaragua; and (3) the deformation of the Nicaragua coast-line. After pointing out the marked deficiency of trustworthy information concerning the region, especially with regard to lake and river topography and hydrography and dynamical geology, the author concludes that "the facts that are known render doubtful, or at least open to question, the advisability of constructing, or even the practicability, of a canal such as is contemplated." . . . "It may, perhaps, be properly questioned whether, if the canal had been constructed a hundred years ago, along the site that is now being contemplated, it would be in existence to-day."

THE current number of *La Geographie* contains a suggestive paper on the variation of the limits of the Mediterranean region, by M. Gaston Bonnier. It is pointed out that attempts to define the boundary from geological considerations have proved unsatisfactory, and that the region is more clearly distinguished by its climate. This may be traced in the flora, the Mediterranean region being roughly taken in France as the region of the olive. M. Bonnier contends that it can be more closely followed, especially in certain regions, by reference to other plants, and discusses a number of interesting observations with regard to exposure and elevation.

AT the close of an address recently delivered as president of the Anthropological Society of Washington, Mr. W. J. McGee enunciated the cardinal principles of science as follows:—"The indestructibility of matter, the contribution chiefly of chemistry; the persistence of motion, the gift mainly of physics; the development of species, the offering of the biotic sciences; the uniformity of nature, the guerdon of geology and the older sciences; and the responsivity of mind, the joint gift of several sciences, though put in final form by anthropology." These principles are comprehensive enough, but they will not satisfy all students of epistemology, so much depends upon the point of view occupied.

THE manufacture of silk cord from spiders' web seems likely to attain commercial importance, for we learn through the *Board of Trade Journal* that one of the most novel exhibits in the Paris Exposition will be a complete set of bed-hangings manufactured in Madagascar from the silk obtained from the balahe, an enormous spider that is found in great numbers in certain districts of the island. The matter has been taken up by M. Nogue, the head of the Antananarivo Technical School. The results he has already achieved show that the production of spider silk should quickly become a highly important industry. Each spider yields from three to four hundred yards of silk. After the thread has been taken from the spiders they are set free, and ten days afterwards they are again ready to undergo the operation. The silk of these spiders, which is of the most extraordinary brilliant golden colour, is finer than that of the silk-worm, but its tenacity is remarkable, and it can be woven without the least difficulty.

WE have received from the Agricultural Department of the Economical Society of Youriev (Dorpat) a report upon the results of rainfall and temperature observations made in the Baltic Provinces of Livonia and Esthonia during 1898. This is the thirteenth year of publication ; the report contains a large amount of very useful statistics, including monthly and yearly means and the number of rainy days at no less than 203 stations. The same information is also shown very clearly in a graphical manner, together with a comparison of the year's results with a ten years' average. We note that the publication of the results for the year 1899 may be expected very shortly.

THE Weather Bureau of the United States has published a valuable discussion of the climate of San Francisco (*Bulletin* No. 28), by Messrs. A. G. McAdie and G. H. Willson. The work is based upon observations collected during the last thirty years, and the results are given in considerable detail on account of the important position of the town and the peculiarity of its climate. The authors state that if a native of San Francisco were asked which was the coldest month of the year, he might be unable to answer, and if asked which was the warmest, he might say November. This arises from the comparative small range of temperature ; the mean annual temperature is about 56·2°. May and November have practically the same temperature ; the warmest month is September, 60·9°, and the coldest January, 50·1°. The highest temperature recorded was 100°, in June, 1891, and the lowest 29°, in January, 1888. The mean of the three consecutive warmest days has never exceeded 76·3°, and the mean of the three coldest days was 40·7°. The annual rainfall is 23 inches. July and August are practically without rain, while December and January together have nearly 10 inches.

A SERIES of Lower Silurian fossils from Baffin Land, in the region between Hudson Bay and Davis Strait, has been described and figured by Mr. Charles Schuchert (*Proc.* U.S. Nat. Museum, vol. xxii. 1900). The fossils belong to the Trenton group, and the strata rest unconformably on old crystalline rocks. The author notes the early introduction in the Baffin fauna of Upper Silurian genera of corals, such as *Halysites*. He also remarks that the corals, brachiopods, gasteropods and trilobites have a wide distribution, and are less sensitive to differing habitats than the cephalopods or lamellibranchs.

IN an article on the Dwyka Coal-measures (*Trans.* S. African Phil. Soc. vol. xi.), Mr. E. J. Dunn points out that the Dwyka conglomerate, which occurs at the base of the coal-bearing series, is a most valuable horizon, and that its length of outcrop exceeds 2000 miles. This is shown on an accompanying map. Within this outcrop coal *may be present* at varying depths over an immense area, extending from the southern part of the Transvaal to Kimberley and near East London. Borings alone can decide if profitable seams occur, and if so, at what depths.

IT is well known that the blood of animals that have been poisoned with carbonic oxide loses its power of absorbing oxygen. Dr. Adolfo Moutuori, writing in the *Rendiconto* of

the Naples Academy, describes experiments tending to explain the fact that dogs are capable of surviving the injection into their veins of a quantity of carbonic oxide far greater than would poison them if inhaled. It is found that the poisoned blood reacquires its power of absorbing oxygen when it is brought into contact with the pulmonary tissues, but not otherwise.

THE statics and dynamics of pseudospherical space in three dimensions form the subject of a memoir by Prof. D. de Francesco in the *Rendiconto* of the Naples Academy. Defining the co-ordinates of a point as the hyperbolic sines of the perpendiculars on three principal orthogonal planes of reference, the author introduces the conception of the moment of a force with regard to a point, analogous to the moment in ordinary statics, and, in addition, the new notion of the co-moment, of which an analytic expression is given. By representing forces by the hyperbolic sines of segments measured on their lines of action the equation of virtual work is established, and by applying this equation to a rigid system the author determines the six characteristics, the central axis, and the invariants. Starting from the conception of the co-moment, the problem of dynamics is treated by the method of Poinsot. Two invariants are found, and the conditions for their vanishing lead to remarkable geometrical properties, which do not exist in ordinary mechanics. The same author discusses in the *Atti dei Lincei* the kindred problem of integration of the differential equations of free motion of a rigid body in space of constant curvature.

MR. J. E. GRIFFITH, of Bangor, author of the "Flora of Carnarvonshire and Anglesey," proposes to publish a series of photographic reproductions of the cromlechs of these two counties of Wales. The series will contain forty-three photographs of thirty-six different cromlechs.

A RUMOURED project of reclaiming Wicken Fen in Cambridgeshire, forming the subject of a recent leader in the *Standard*, once more raises the question as to the desirability of acquiring by public subscription this last remaining habitat of the old fauna and flora of the Fen district, and thus saving them from extinction. Such a project was suggested some time ago by Mr. Carrington, the editor of *Science Gossip*, and it is much to be hoped that a movement may be set on foot for the purpose before it is too late.

WE have received No. 3 (vol. i.), for April, of *Climate*, "a Quarterly Journal of Health and Travel," edited by Dr. C. F. Harford-Battersby. The periodical is the organ of the Travellers' Health Bureau, the object of which is to supply to inquirers information of every kind connected with the health and comfort of travellers and of residents in unhealthy climates. Among the original articles in the number before us is a very interesting one on "Gardening in West Africa," by Miss Kingsley, and a *résumé* of the facts at present ascertained connecting malaria fever with the pa asite of the mosquito. A short paper on "European Children in Tropical Climates," by Dr. G. D. McReddie, will be read with interest by many.

A "FLORA OF BOURNEMOUTH" is announced for early publication by the Rev. E. F. Linton, of Bournemouth (subscription price, 7*s*. 6*d*.). The area taken is a radius of twelve miles, and includes portions of the counties of Hants and Dorset, with the Isle of Wight. The total number of flowering plants and Pteridophytes is stated at 1137.

IN the *Naturwissenschaftliche Wochenschrift* for April 15, Prof. M. Möbius gives an interesting account of pigments in the vegetable kingdom. Commencing with the colouring matters of fungi and lichens, he proceeds to those in the various groups of Algæ, and then to the pigments of Muscineæ, Pteridophytes, and Phanerogams, contained in the stem, root, leaves, flowers, and fruits. He regards chlorophyll and hæmoglobin as antagonistic substances, the one characteristic of the vegetable, the other of the animal kingdom.

To the *Sitzungsberichte* of the Berlin Academy for March 15, Dr. K. von Möhius, the director of the Zoological Museum, communicates a suggestive paper on our perception of the æsthetic proportions of various mammals.

THE April number of the *Journal of Anatomy and Physiology* contains the full text of the paper read by Dr. Albert Gray at the last meeting of the British Association on Helmholtz's theory of hearing. The author proposes a modification of the theory of the German investigator, according to which a remarkable analogy between the senses of hearing and touch is shown to exist.

IN the last issue of the *Transactions* of the South African Philosophical Society, Dr. R. Marloth gives the results of his investigations as to the mode of growth of the barnacle infesting the Southern Bight Whale. Were it not for some special provision, the growth of the epidermis beneath, coupled with the wearing away of the outer layer, would soon cause the parasite to be shed, and, as a matter of fact, this actually takes place with the dead shells. The living barnacle cannot, however, be discarded in this manner, since it dissolves the part of the epidermis with which its skin is in contact at the same time at which fresh epidermal tissue is formed below. Consequently the layer of epidermis between the barnacle and the true skin never varies in thickness, and the parasite accordingly retains its position, the shell disintegrating at the apex at the rate at which it grows at the base.

MM. GAUTHIER-VILLARS, Paris, have published the third revised edition of the "Traité élémentaire d'Electricité avec les principales Applications," by M. R. Colson.

MR. FELIX L. DAMES, Berlin, has issued a catalogue of books and papers on astronomy, geodesy, meteorology and related sciences, which he has acquired from the library of the late Dr. H. Romberg, and offers for sale.

THE seventh edition of the late Prof. Milnes Marshall's well-known and practical manual on "The Frog : an Introduction to Anatomy, Histology, and Embryology," edited by Dr. G. Herbert Fowler, has been published by Mr. David Nutt. The chief addition consists of a new series of woodcuts in illustration of the development and metamorphosis of the frog.

THE "Handbook of Jamaica," compiled by Mr. T. L. Roxburgh and Mr. J. C. Ford, and published by Mr. Edward Stanford, is filled with historical, statistical and general information concerning the island. We notice that the magnetic declination, which was 6° 30' E. at the end of last century, and has been steadily decreasing since then, is now only 1° 24' E., and in 1910 its value will be zero.

IN the course of a few weeks, Mr. Gustav Fischer, Jena, will commence the publication of "Aus den Tiefen des Weltmeeres," an elaborate work in which Prof. Carl Chun will describe and illustrate the German deep-sea expedition to Antarctic waters. The work will be published in twelve parts, the first of which will appear during this month and the last in November.

A SIXTH edition, revised and enlarged, of "A Text-book of Assaying," by C. and J. J. Beringer, has just been published by Messrs. Charles Griffin and Co. Mr. J. J. Beringer is responsible for the revision of this handy book for assayers ; and he remarks in the preface : "The principal changes in this edition are additions to the articles on gold, cyanides and nickel, and a much enlarged index. The additional matter covers more than forty pages."

SCIENTIFIC students and investigators in Melbourne should be grateful to Mr. T. S. Hall for the "Catalogue of the Scientific and Technical Periodical Literature in the Libraries in Melbourne," which he has prepared. Besides periodicals, the list includes reports of scientific societies, as well as Government reports and Parliamentary papers of scientific import. The catalogue will be a very useful guide to scientific literature accessible in Melbourne and its suburbs.

THE sixteenth part of Mr. Oswin A. J. Lee's fine work, "Among British Birds in their Nesting Haunts, illustrated by the Camera," has just been published by Mr. David Douglas, Edinburgh. The birds illustrated and described are the black-cap, bullfinch, short-eared owl, yellow wagtail, stock dove, pintail, wryneck, and lesser whitethroat. The present part completes the fourth volume, and it is hoped that the whole work will be finished in the course of a few months.

AT the meeting of the Chemical Society on June 1, 1899, Prof. Sydney Young, F.R.S., described a series of tests made by him to determine the relative efficiency of various forms of still-heads for fractional distillation. The design of several new still-heads, superior in many respects to those in common use, was an outcome of the investigation ; and chemists will be glad to know that Messrs. J. J. Griffin and Sons have now placed these improved forms upon the market.

THE additions to the Zoological Society's Gardens during the past week include a Mozambique Monkey (*Cercopithecus pygerythrus*, ♀) from Uganda, presented by Lady Ashburnham ; two Leopards (*Felis pardus*, ♂ ♀) from India, presented by Mrs. C. Simpson ; a Tawny Owl (*Syrnium aluco*) from Scotland, presented by Mrs. C. M. Blackwood ; six Common Vipers (*Vipera berus*) from Dorsetshire, presented by Mr. A. Old ; nine Natterjack Toads (*Bufo calamita*) from Norfolk, presented by Mr. J. B. Thornhill ; a Sykes's Monkey (*Cercopithecus albigularis*, ♀), a Flap-necked Chameleon (*Chamaeleon dilepis*) from East Africa, a Cactus Conure (*Conurus cactorum*) from Bahia, deposited ; two Gold Pheasants (*Thaumalea picta*, 2 ♀), two Silver Pheasants (*Euplocamus nycthemerus*, 2 ♀), two Cabot's Horned Tragopans (*Ceriornis caboti*, ♂ ♀) from China, two Germain's Peacock Pheasants (*Polyplectron germaini*, ♂ ♀) from Cochin China, two Japanese Pheasants (*Phasianus versicolor*, ♂ ♀), two Sœmmerring's Pheasants (*Phasianus soemmerringi*, ♂ ♀) from Japan, three White-backed Trumpeters (*Psophia leucoptera*) from the Upper Amazons, four Wonga-Wonga Pigeons (*Leucosarcia picata*) from New South Wales, a Musky Lorikeet (*Glossopsittacus concinnus*) from Australia, three Blue-crowned Hanging Parrakeets (*Loriculus galgulus*) from Malacca, an Ural Owl (*Syrnium uralense*), North-east European ; a Great Wallaroo (*Macropus robustus*, ♂) from South Australia, a Barbary Wild Sheep (*Ovis tragelaphus*, ♂) from North Africa, purchased ; a Yak (*Poephagus grunniens*, ♂), born in the Gardens.

OUR ASTRONOMICAL COLUMN.

NEW VARIABLE IN TAURUS.—In the *Astronomische Nachrichten* (Bd. 152, No. 3635) M. W. Ceraski, of Moscow, announces the discovery of another new variable by Madame Ceraski during her examination of the plates taken by M. S. Blajko. The star's position is :—

R.A.		Decl.	
h. m. s.			
5 33 17·33	...	+26 18 58·3	(1900)
5 30 29·56	...	+26 17 7·9	(1855)

The star is not found in the B.D. At its maximum it is of 9·0–9·5 mag. ; at minimum, about 12 mag. or less. On 1900 March 29, it was at the limit of visibility in a telescope of 4·5 inches aperture.

SEARCH EPHEMERIS FOR EROS.—In view of preparing for observations of this minor planet during the coming opposition, the following ephemeris has been prepared by J. B. Westhaver from the elements computed by H. N. Russell (*Astronomical Journal*, No. 479, vol. xx. p. 185).

Ephemeris for 12h. Greenwich Mean Time.

1900.		R.A.		Decl.		Mag.
		h. m. s.		° ' "		
May 3	...	23 2 0·1	...	−4 0 25	...	13·4
5	...	5 46·7	...	3 28 2	...	
7	...	9 32·3	...	2 55 29	...	13·3
9	...	13 16·9	...	2 22 45	...	
11	...	17 0·5	...	1 49 52	...	13·3
13	...	20 43·1	...	1 16 48	...	
15	...	24 24·9	...	0 43 35	...	13·2
17	...	28 5·8	...	−0 10 11	...	
19	...	31 45·8	...	+0 23 22	...	13·2
21	...	35 25·0	...	0 57 4	...	
23	...	39 3·3	...	1 30 55	...	13·1
25	...	42 40·8	...	2 4 55	...	
27	...	46 17·5	...	2 39 4	...	13·1
29	...	49 53·3	...	3 13 22	...	
31	...	23 53 28·4	...	+3 47 48	...	13·0

RELATION BETWEEN SOLAR ACTIVITY AND EARTH'S MOTION.—In the *Astronomische Nachrichten* (Bd. 152, No. 3635), Mr. W. G. Thackeray criticises the recent paper by Dr. J. Halm (*Astr. Nach.* Bd. 151, No. 3619, NATURE, March 8, p. 445), deducing certain relations between the sun-spot cycle, the changes in the obliquity of the ecliptic and the variations of the terrestrial latitude. Mr. Thackeray states first, that continuous observations of sun spots have only been made since 1825, so that the sixty years period lacks sufficient evidence; secondly, that Dr. Halm has ignored some of the systematic errors of observation, particularly those depending on the corrections for temperature in the transit circle reductions, although in some cases their amount affects the value of the obliquity by as great a quantity as the whole amplitude of Chandler's long period inequality of latitude variation. The paper includes a table showing the annual corrections to Leverrier's obliquity from 1836–1896, with corresponding yearly means of Wolf's spot numbers. These differ from the values adopted by Dr. Halm, and the resulting plotted curves show little or no resemblance.

DETERMINATION OF SOLAR PARALLAX FROM OPPOSITION OF EROS.—In the *Astronomical Journal* (No. 480, vol. xx. pp. 189–191), Prof. S. Newcomb directs attention to the favourable opportunity for determining the Solar Parallax which will be afforded by the coming opposition of the minor planet Eros, in December 1900, the conditions being conducive to more accurate direct measurements than have ever before been presented. As another such favourable opportunity will not occur for more than thirty years, several suggestions are made for determining the best combination of observations.

The period during which determinations may be made is remarkably long, as during the five months from 1900 October 15 to 1901 March 15, the distance of the planet will be less than 0·50 astronomical unit.

The high degree of precision attainable in late years by photography indicates this as the best method, an additional point in favour of this plan being that photographic telescopes are already in use at various stations, and need only devoting to this work. In arranging the programme of observations three objects should be kept in view :—

First, the station and hours of exposure should be so chosen as to secure the maximum of parallactic angles.

Secondly, endeavour should be made to secure simultaneous exposures at different stations, in order to lessen the uncertainties arising from differences of scale, changes in relative position of planet among stars, and in the reduction of the position of the planet from hour to hour. Series of independent determinations should also be made, each within an interval of twenty-four hours.

Thirdly, the relative displacement should, as nearly as possible, be in a direction at right angles to the motion of the planet among the stars.

Prof. Newcomb then describes four charts included in the paper, showing projections of the earth as seen from Eros at the Epochs (1) middle of October to end of November ; (2) about December 16 ; (3) about January 10 ; (4) about February 1. On these are marked the sunset and sunrise lines, and parallels

of latitude corresponding to the principal observatories : Helsingfors, Pulkowa, 60° lat. ; Greenwich, Paris, Potsdam, &c., 50° lat. ; Jamaica, Madras, 15° lat. ; Arequipa, −15° lat. ; Cape of Good Hope, −35° lat. On these projections the direction of the planet's motion for different dates is indicated, so that observers may find by inspection the relative importance of observations at various stations and at various times of night.

Respecting the degree of precision it may be possible to attain in the final result, it is noticeable that the course of the planet throughout the entire period will lie along the borders of the Milky Way, ensuring more and nearer comparison-stars than would otherwise be available. An element of uncertainty is the probable error of measurement from the plates. From a consideration of Kapteyn's investigation of the Helsingfors parallax plates, and those at Potsdam, it is likely that the probable error of the solar parallax from a pair of simultaneous plates at Arequipa and Helsingfors would be ±0"·02, and even this might be reduced were it not for the uncertainty arising from the motion of the planet.

WORKING SILICA IN THE OXY-GAS BLOWPIPE FLAME.

THE plastic state of silica, and the elasticity of fine threads of vitreous silica, were first observed by M. Gaudin (*Comptes rendus*, viii. 678, 711) in 1839 ; but his observations seem to have attracted but little attention, and the valuable qualities of " quartz threads " remained unutilised till they were independently rediscovered and applied by Prof. C. V. Boys in 1887.

Similarly, M. A. Gautier succeeded, in 1869, in making very narrow tubes of silica, and showed such tubes in Paris in the year 1878, but he failed to make further progress, even with the aid of M. Moissan's electric furnace (*Comptes rendus*, cxxx. 816, March 26), and his early work was so completely forgotten, both in France and England, that the latest French worker on the subject, M. A. Dufour, was evidently unaware of its existence a few weeks ago (*Comptes rendus*, cxxx. 775, March 19).

But though it thus appears that Prof. Boys was not, as has been supposed, actually the first physicist to draw silica into threads, or work it into fine tubes, there can be no doubt but that his observations, methods of working and experiments have formed the basis of all that has been done since the publication of his first paper in 1887.

In June 1899, one of the authors of this article exhibited (in conjunction with W. T. Evans), at the Royal Society's soirée, a tube of vitreous silica, about 12 cm. in length and 1 cm. in diameter, and at the same time showed the process by which it had been made. Since that date we, the present writers, have made a good deal of further progress. We have succeeded in making longer tubes of various thicknesses, and in joining such tubes both end to end and at right angles. On February 22, we filled and sealed an ungraduated mercury thermometer made entirely of vitreous silica[1] ; and what is equally important, we have entirely overcome the difficulty caused by the great tendency of quartz to splinter when suddenly thrust into the oxy-gas flame. We therefore now publish a short account of our methods in the hope that they may enable others to take advantage of the new material without undertaking a tedious preliminary investigation into its properties and the methods of working it. We may perhaps be permitted to add that we have already commenced experiments intended to test the suitability of silica for use in mercury and air thermometers, especially in regard to the fixity, or otherwise of their zero points, that M. A. Dufour is engaged on similar work, especially in relation to high temperature thermometers, and that we are also studying the fitness of silica apparatus for researches on the properties of pure gases.[2]

To prepare Non-splintering Silica.—The best form of silica for use before the blowpipe is rock crystal. This may be obtained in the form of chippings, or in masses which have proved unsuitable for optical work. We have experimented with the lighter particles of Kieselguhr, after well washing them with strong hydrochloric acid, and also with well-washed precipitated silica ; but, though these can be worked before the blowpipe without much difficulty, they have not proved satisfactory in our hands, as they yield an opaque product which is only suitable for a few purposes.

[1] NATURE, April 5, p. 540.
[2] This will obviously involve a careful investigation into its power of condensing gases and vapours.

In order to prepare non-splintering silica from native masses of rock crystal, the latter must be heated in a Bunsen flame, unless they are already perfectly clean, until the outer impure layers can be removed easily by a blow from an iron pestle or hammer. The clean masses of silica must then be heated in a vessel containing boiling water for some time, and dropped whilst hot into clean cold water. This treatment will cause the masses to crack to such an extent that they may easily be broken into fragments of convenient dimensions by sharp blows from a clean hammer. When the material has thus been broken up, the fragments must be examined one by one, and all those which contain foreign matter must be rejected. Finally, the selected fragments must be heated to a yellow-red heat in a platinum dish, and then quickly thrown into deep cylinders containing cold distilled water. After the quartz has been treated in this manner twice, it will be found to be semi-opaque and very much like a white enamel in appearance. It may now be brought safely into the oxy-gas flame, or be pressed suddenly against masses of white-hot plastic silica without any preliminary heating, such as is necessary in the case of the natural quartz. These processes do not occupy much time, and the use of the prepared material saves a great deal of time and trouble at the subsequent stages. We have tried unprepared opal and natural cloudy quartz, but both these splinter badly.

The Blowpipe.—We have worked silica both in the flame of an ordinary " blow through " jet, and in the flame of a good " mixed gas" burner. We find the latter gives by far the more satisfactory results. The large " blow through " burners, such as may be used for welding and melting iron, or for melting platinum, do not give satisfactory results, from an economical point of view, with silica.

Some necessary precautions.—In working silica it is necessary to use very dark glasses to protect the eyes. The darkest glasses usually supplied by spectacle makers are not, in our experience, satisfactory. We use spectacles made specially from glass so strongly darkened, that it is difficult at first to work with them at all. We lay some stress on this matter, as we are satisfied that want of care in selecting the spectacles would be likely to result in injury to the sight of any one who should work silica before the blowpipe frequently, and for long spells.

Relative difficulty of working Glass and Silica.—The fashioning of apparatus from silica before the blowpipe is expensive, for the consumption of oxygen is large, and it demands some patience to build up large pieces of apparatus from shapeless masses of quartz. But owing to the remarkable fact that properly prepared silica, and also silica rendered vitreous by fusion, may be plunged directly into the hottest part of the oxy-gas flame, and afterwards be suddenly cooled, and reheated and recooled, apparently as frequently as one pleases, without any risk of its cracking, it is really very much easier to manipulate silica than any variety of glass. The most careless and most inexperienced worker runs no risk of breaking his apparatus through want of skill in managing the flame, or through the exigencies of his affairs compelling him to put aside half-finished work. It is important, however, to apply the flame to the opaque prepared silica, in the first instance, in such a way as to avoid the forming of air bubbles. Our practice is to heat first the lowest surface of each fresh mass of silica, and to take care that fusion proceeds regularly from below upwards. If this be done, a perfectly clear glass-like product is obtained. Silica is very liable to exhibit a phenomenon resembling devitrification, especially at the earlier stages before the traces of sodium and lithium, which seem to be present in most quartz, have been expelled. In order to avoid permanent injury to the finished work from this cause, care must be taken to employ a quiet flame. If this be done, any devitrification that may appear will be removed easily by reheating the disfigured surfaces.

To make Silica Tubes.—Before one commences to construct apparatus of silica it is well to prepare a stock of the vitreous material in the form of rods about 1 mm. in diameter. These are made by holding a small lump of non-splintering silica in the flame, by means of forceps with platinum tips, so as to melt one corner of the mass, pressing a second fragment of the material against the heated spot till the two adhere, heating the second portion from below upwards until it assumes a clear vitreous appearance, then adding a third fragment of silica to the second, a fourth to the third, and so on, until an irregular rod has been formed. Finally, this irregular rod must be reheated in small sections at a time, and drawn out to the desired extent.

These rods are easily made by any one ; a capable laboratory boy will produce about a score of rods 20 cm. long in an hour, after a few days' practice at the work ; but his consumption of oxygen must be watched closely. The platinum tongs do not suffer much if one works in the manner described, for after the first start off they are only used to press *cold* fragments of silica against the fused ends of the growing rods. Our forceps have been used by four beginners, and are quite unharmed after several years.

When a supply of the rods of vitreous silica has been prepared, bind a few of them, at their ends, with fine platinum wire round a rod of platinum 1 to 1·5 mm. in diameter ; heat the silica cautiously till the rods adhere to one another, and then withdraw the platinum core. If the tube is not perfect, add bits of silica at the defective places and reheat them. Close one end of the rough tube thus produced and blow a small bulb upon the closed end, proceeding in the manner employed for producing glass bulbs. Heat the bottom of the bulb, attach a rod of silica to it, reheat the whole bulb and then draw it out into a tube. Blow a fresh bulb at one end of the fine tube thus made, and draw this out in its turn until the tube is six or seven cm. in length. By the time this is accomplished the worker will have discovered that the hottest spot in his oxy-gas flame is just inside the tip of the inner cone, but not too near the orifice of the jet ; and after this, if he can perform these simpler operations of glass working, he will, with a few weeks' practice, find it easy to make larger apparatus by following the simple instructions given below.

The chief difficulty met with when one wishes to make large bulbs, tubes, &c., is due to the fact that the only thoroughly satisfactory burners give comparatively small flames, and that it is only the hottest parts of these flames that give the desired results. There is no doubt, however, that suitable combinations of small burners could be contrived if they should be demanded, for the production of apparatus of really considerable dimensions.

In order to convert a small bulb of silica into a large tube, proceed as follows :—Heat one end of a fine rod of vitreous silica, and when it is in the plastic state apply it to the bulb at the point C. Then soften the adjacent parts of the rod and allow them to fall upon the bulb so as to form a ring C B, attached to the bulb. Heat the end of the bulb and C B till the silica softens, then blow out the end in the usual manner. If this process

is repeated the bulb will first become ovate and then form a short tube which can be lengthened, practically speaking, indefinitely. Tubes of 1·5 cm. diameter and of considerable length are easily made in this way by a patient person. It does not answer to add lumps of silica at E and then to blow them out ; we had no success in working silica till we abandoned that method. The sides of a tube formed in that way are too thin, and blow-holes constantly form in them. The tubes are easily thickened, when necessary, by adding rings of silica, reheating these, and blowing them to spread the material as one would do when working glass. It is best to blow through a chamber containing potash. If this is connected to the end of the silica tube by india-rubber " valve " tube, one is able to move the silica tube with sufficient freedom. If a large tube is being made, it is best to blow out the softened material whilst it is still in the hottest part of the flame, but smaller objects may be transferred to the less hot parts of the flame with advantage at the moment of blowing. When a comparatively large object must be uniformly heated, it is convenient to place a sheet of silica in front of the flame a little beyond the object to be heated, in order that it may throw back the flames upon those parts of the material which are turned away from the chief source of heat. A suitable plate of silica is easily made by sticking together small, rounded masses of vitrified quartz.

We find that it is not difficult to produce tubes of various thicknesses and various internal diameters by heating and collapsing thin tubes made as described above, and that fine capillaries, " thick millimetre tubes," and tubes of two or three millimetres bore, of moderate thickness, can be produced in this way. Thermometer stems are best made by adding rings of silica to small bulbs, thickening them in the flame till their cavities are very small, and then quickly drawing them out whilst soft. Finally, we may add that tubes of silica can as readily be sealed to one another as tubes of glass, and that T-pieces and side tubes generally may be formed by fixing rings of silica in the positions to be occupied by the side tubes and extending them by blowing as already described, or by attaching tubes of suitable dimensions, previously prepared, to short side tubes blown as just described. It is therefore possible to construct such apparatus as Geissler tubes, small distilling tubes, and thermometers with stems of the German type, &c. We feel sure that small flasks could easily be made also by means of suitable combinations of several oxy-gas burners, though doubtless they would be rather expensive.

Finally, solid rods of silica five or six millimetres in diameter can be made by putting together small masses of prepared silica, or better by pressing together in the flame the softened ends of the fine rods already described.

Notes on some Properties of Vitreous Silica.[1]—A good many of the properties of silica have already been described by Prof. Boys, but a knowledge of the following, some of which are, we think, now described for the first time, will be found useful :—

(1) Vitreous silica is a very poor conductor of heat ; hence it is possible to hold a thick rod of silica very close to a strongly ignited zone.

(2) Our colleague, the Rev. H. Pentecost, finds that vitreous silica is less hard than chalcedony, but harder than felspar. Its surface appears to be about equally hard after it has been heated as strongly as possible and cooled suddenly, and after it has been heated and cooled in the air. Tubes of silica may be readily cut by means of a cutting diamond, and also with a good file of hardened steel.

(3) It has already been stated that cold vitreous silica can be plunged safely into the hottest part of an oxy-gas flame, and that the heating and cooling process can be repeated with impunity. Hot vitreous silica bears sudden cooling equally well. We have repeatedly plunged thick rods and large tubes of silica, heated till plastic, into cold water and even into fusible metal below 100°, without any injury to the material, for when afterwards cut with a diamond it did not fly.[2]

On the other hand, threads of silica become rotten when heated to the highest temperature of an ordinary blowpipe.[3] Large objects seem to be affected to a much less degree ; and we suspect that this phenomenon may be due to surface devitrification. When silica is in this friable state it can be re-annealed by again softening it in the oxy-gas flame. According to Gaudin, wires of silica heated to a suitable temperature (" rouge-blanc ") acquire great cohesion and become very elastic.

We have not yet succeeded in fixing platinum electrodes securely into silica tubes. But we have reason to hope that this may be found to be practicable by the use of kaolin, or some other natural silicate. Meanwhile, it seems possible that they might be soldered into the silica if necessary (see " Laboratory Arts," by R. Threlfall).

We may add that, according to M. Gaudin, emerald gives threads which are even more tenacious than those of silica.

W. A. SHENSTONE.
Clifton College.　　　　　　　　　　　　H. G. LACELL.

---

[1] See also Gaudin, *loc. cit.*
[2] Gaudin obtained similar results with drops of *liquid* silica.
[3] Gaudin observed a similar phenomenon in the case of fine threads, and so also, we believe, did Boys.

UNIVERSITY AND EDUCATIONAL INTELLIGENCE.

CAMBRIDGE.—The following is the Speech delivered, on April 26, by the Public Orator (Dr. Sandys) in presenting Mr. CHARLES HOSE for the degree of Doctor in Science, *honoris causa.*

Insulam Borneonem orbis terrarum inter insulas omnes prope maximam esse constat. Insulae autem illius insulis nostris fere duplo maioris in parte septentrionali patet regio quae unum e Britannis regem suum esse gloriatur. In eadem vero regione provincia quaedam, fluviorum ingentium infra confluentes, abhinc annos decem alumno nostro tradita est, qui barbarorum animos bellicosos ad foedera vocavit, et armorum certamina saeva certaminis nautici in ludum mutavit. Idem non modo in foedere inter barbaros sanciendo victimarum caesarum haruspex sollertissimus, sed etiam avium in silvis volantium augur et

auspex admirabilis exstitit. Ergo alumni nostri auspiciis et Helvetiae et Bataviae et Germaniae et Galliae et Britanniae musea avium et animalium exemplis eximiis aucta et suppleta sunt, et insulae ipsius zoologia, anthropologia, geographia, novo lumine illustratae. Talia propter merita alumnus noster non modo inter nosmet ipsos a regia geographiae societate praemio singulari donatus est, sed etiam inter Europae gentes tum aliis honoribus ornatus est, tum praesertim inter Germanos falconis albi eques iure optimo nominatus est. Nostra denique zoologiae, anatomiae, archaeologiae musea iam plus quam decimum per annum alumni nostri liberalitatem loquuntur. Ergo nos quoque insulae tantae non modo avium et animalium venatorem assiduum, sed etiam montium et fluminum exploratorem intrepidum, ob scientiarum fines etiam imperii Britannici prope terminos feliciter propagatos, laurea nostra hodie libenter coronamus.

Duco ad vos museorum nostrorum patronum liberalissimum, exploratorum nostrorum hospitem benignissimum, CAROLUM HOSE.

The General Board propose the establishment of a lectureship in ethnology, to which Dr. Haddon may be appointed ; and a lectureship in bacteriology and preventive medicine for Dr. Nuttall. Both have unofficially given valuable instruction in their respective subjects, and the recognition now suggested will probably be readily accorded by the University. New lectureships in experimental physics and in agricultural chemistry are also proposed.

The Board of Agricultural Studies, at the close of their first financial year, make a highly satisfactory report. Their income is sufficient for the provision of a complete course of instruction, which has now been organised under the direction of Prof. Somerville. They now ask the University to establish a special examination in agricultural science (botany, chemistry, physics and geology) for the ordinary B.A. degree.

THE history of the University of London, from the time of Sir Thomas Gresham's bequest, in 1575, of his house and garden in Bishopsgate, for the purposes of education, down to the completion of the work of the commissioners appointed under the University of London Act, 1898, is traced in an interesting article in the current number of the *Quarterly Review.* The large part the University has taken in the renascence of natural science, which will hereafter be regarded as the main characteristic of intellectual progress in the nineteenth century, is pointed out, as well as the fact that London degrees in science were the first conferred by British universities.

WE learn from *Science* that the University of Chicago has secured the 2,000,000 dollars needed to meet the requirements of Mr. Rockefeller's gift of an equal amount. At the recent convocation of the University, President Harper gave some details in regard to the gifts received since January 1st. They have come from more than 200 different persons, and 90 per cent. of them were unsolicited. The largest items appear to be the Gurley palæontological collection, 30,000 dollars from Mrs. Delia Gallup, and, given anonymously, 60,000 dollars for a commons, 50,000 dollars and 25,000 dollars for a students' club-house, 20,000 dollars towards a women's hall, and 30,000 dollars with specific use to be designated later. President Harper stated that the total assets of the University are now not far from 11,000,000 dollars.

THE Technical Education Board of the London County Council will proceed shortly to award five senior county scholarships, each of the value of 60*l.* a year for three years, with free tuition fees up to 30*l.* a year. These scholarships are intended to assist young men and women to pursue a course at some University or at a technical college of University rank. Some of the scholars who have been elected in previous years are holding their scholarships at Oxford and Cambridge, others are studying at technical colleges in different parts of England, while others are pursuing courses of study on the Continent. The scholarships are open only to candidates who are under twenty-two years of age, and whose parents are in receipt of not more than 400*l.* a year. In addition to the senior scholarships, the board has in past years made a certain number of grants of smaller value to assist students in pursuing advanced education, and the board has at its disposal a certain number of free places at University College, London, King's College, London, and Bedford College, London. The scholarships and grants are awarded, not on the result of a set examination, but on the consideration of the past achievements and promise of the candidates. Application forms may be obtained from the secretary of the Technical Education Board, 116, St. Martin's Place, W.C., to whom they should be returned not later than May 14. The board is also offering scholarships for the encouragement of horticulture and gardening. Two of these, tenable at the Swanley Horticultural College, Kent, give free board and tuition for two years, and may be reckoned as of the value of 60*l*. a year. They are open to candidates between the ages of sixteen and twenty, and one will be awarded to a young man and the other to a young woman as the result of a competitive examination. No candidate is eligible whose parents are in receipt of more than 400*l*. a year.

SOCIETIES AND ACADEMIES.

London.

Physical Society.—Ordinary meeting held by the invitation of Sir Norman Lockyer, F.R.S., in the Solar Physics Observatory, South Kensington, on April 27.—Mr. T. H. Blakesley, Vice-President, in the chair.—Sir Norman Lockyer gave a short account of the physical problems now being investigated at the Solar Physics Observatory, and their astronomical applications. The chief work carried on at the observatory is the comparison of stellar spectra with spectra obtained from lights emitted by laboratory sources. The light from a star (or the sun) and from an arc (or a spark) are focussed alternately upon the slit of a spectroscope, and the two spectra are photographed side by side upon the same plate. The number of lines in the arc spectrum depends upon which part of the arc is focussed on the slit. The image of the centre is rich in lines, the image of the edge gives a few single lines. Changes in spectra are also dealt with. The thickening and thinning of lines depends upon several things. In the first place, it depends upon the density of the substance, and thus the hydrogen lines in the spectrum of Sirius are much broader than those in a Cygni, the hydrogen being denser in the former star. Changes may also be produced by variations in quantity. A reduction in the quantity of a substance generally simplifies its spectrum, the longest line disappearing last. The motion of a luminous body to or from the spectroscope alters the wave-length of the light emitted and produces a shift in the lines of the spectrum. The amount of deviation is a measure of the velocity in the line of sight. In the case of Nova Aurigæ, we have dark and bright lines of the same substance side by side. This shows that there are two bodies involved, moving with different velocities, the one giving a radiation and the other an absorption spectrum. Another change in the lines depends upon temperature. In general an increase in temperature produces a greater number of lines, a notable exception being sodium, which gives its full number of lines at the temperature of an ordinary Bunsen flame. The spectra of metals obtained from the arc, and by sparking, are often quite different. Those lines which make their appearance, or are intensified in passing from the arc to the higher temperature of the spark, are known as enhanced lines. The comparison of stellar spectra with laboratory spectra is often easy. For instance, the presence of iron in the sun and hydrogen in Sirius is easily seen. Several lines in the spectrum of Bellatrix have been shown to be due to helium, the position of the lines being exactly the same as those due to the gases from clevite. In many cases it is possible to build up the spectrum of a star from the spectra of its constituents taken at the proper temperatures. For instance, the spectrum of γ Orionis can be closely imitated by means of oxygen, nitrogen, and carbon together with the well-marked lines of hydrogen and helium. We can roughly estimate by the character of the spectra of stars, the temperatures of those stars, and thus arrive at a stellar thermometry. Starting with a hot star like Bellatrix, and passing through β Persei, γ Lyræ, Sirius, Castor, Procyon to Arcturus, a cold star, we have a gradual change in the character of the lines which appear in the spectrum of any constituent. The widening of the lines in the case of spectra of sun spots enables us to trace changes in temperature of the sun, and we can compare these temperature changes with a variety of terrestrial phenomena, such as variation in latitude. The extraordinary number of lines exhibited by many metals suggests that what we are accustomed to call chemical elements are really complex bodies which are made up of simpler ones. Attempts have been made to build up the spectra of metals by superimposing

simple sets of lines upon one another. In many cases a great number of series would be required to represent things completely. In the case of hydrogen it would be necessary to have at least twenty-seven series to give the structure spectrum only. Taking the atomic weight of hydrogen as unity, the atomic weight of the little masses which might give rise to any one of these series would be about ·0019. This is of the order of magnitude of the small bodies, of which the existence has been suggested by Prof. J. J. Thomson from his work on ions.

Paris.

Academy of Sciences.—M. Maurice Lévy in the chair.—The President announced to the Academy the death of M. Alphonse Milne-Edwards, and gave an account of his work.—On linear partial differential equations of the second order, and on the generalisation of the problem of Dirichlet, by M. Émile Picard.—On the heats of combustion and formation of some iodine compounds, by M. Berthelot. A redetermination of the heats of combustion of fourteen typical iodine derivatives. In spite of preconceived notions to the contrary derived from the incomplete combustion of such compounds as iodoform in air, no difficulty was experienced in completely burning any of the substances in the calorimetric bomb.—On rifling in cannon, by M. Vallier. A discussion on the best form of curve for the rifling of cannon, and an extension of the work of M. Zaboudski upon the same subject.—On the upright trunks, stems and roots of Sigillaria, by M. Grand'Eury. A study of the Sigillaria existing in a quarry in the neighbourhood of St. Étienne. From the fact that the stems (*Syringodendron*) found in a vertical position are not distributed at random, but are usually found in groups near each other forming well marked colonies, and from other characters of their growth, the author concludes that the hypothesis of R. P. Schmitt that they have been transported by water and deposited in the position found, is untenable. The view of Dawson that they have grown upon unsubmerged soil is also held to be untenable, all the facts noted by the author pointing to the Sigillaria having grown in the place in which they are found in marshy soil; under water varying from 1 metre to 7 or 8 metres in exceptional cases.—Reply to a reclamation of priority of M. Curie, by M. Gustave le Bon.—Reply by M. Th. Tommasina to a reclamation of priority, by MM. Ducretet and Popof.—Note by M. L. M. Bulfier replying to M. Geelmuyden on a question of priority.—On the complementary terms in the criterium of Tisserand, by M. Gruey.—On differential equations of any order whatever with fixed critical points, by M. Paul Painlevé.—On the generalisation of analytical prolongation, by M. Émile Borel.—The theoretical cycle of gas engines, by M. A. Witz. A discussion of the remarks and criticism of M. Marchis.—On the dielectric constant and the dispersion of ice for electromagnetic radiations, by M. C. Gutton. The value of the refractive index for electromagnetic waves was found to vary with the wave-length from 1·76 for a wave-length of 14 cm. to 1·50 for 2088 cm., ice thus presenting normal dispersion for electromagnetic waves.—Two applications of Govi's camera lucida, by M. A. Lafay.—On the maximum sensitiveness practically employed in coherers for wireless telegraphy, by MM. A. Blondel and G. Dobkévitch. The increase of sensibility observed by M. Tissot to occur when the coherer is placed in a magnetic field, is ascribed by the authors to purely mechanical causes, the increase of contact between the powder and the electrodes produced by their mutual attraction.—On the radiations of radium, by M. E. Dorn. The author draws attention to the fact that he published a note on the deviation of the rays emitted by radio-active barium bromide in an electric field on March 11, independently of M. Becquerel.—On a new thermo-calorimeter, by M. G. Massol. Two improvements on Regnault's thermo-calorimeter are suggested, the replacement of alcohol by sulphuric acid, giving a large increase in the range of the instrument, and the use of a reservoir at the upper end of the instrument as in Walferdin's maximum thermometer by which the sensitiveness of the thermo-calorimeter is increased without undue lengthening of the stem. The instrument thus modified has been of especial service in the study of superfused liquids.—A new indicator in acidimetry, and its application to the estimation of boric acid, by M. Jules Wolff. The indicator proposed is a solution of ferric salicylate in sodium salicylate, which passes from violet to orange when the solutions become alkaline. Data are given showing the results obtainable with borates.—On the selenides and chloroselenides of lead, by M. Fonzes-Diacon.—Crystallised lead selenide, PbSe, is obtained by reduction of a selenate by hydrogen or by carbon, by the action of hydrogen selenide upon the vapours of lead chloride, and by the direct fusion in the electric furnace of precipitated lead selenide.—On the alkaline selenio-antimonites, by M. Pouget. Selenio-antimonites can be obtained of analogous composition to the sulpho-antimonites already known ; mixed sulphur and selenium compounds, thioantimonites in which the sulphur is only partially replaced by selenium have also been prepared.— Micro-chemical researches on yttrium, erbium and didymium, by MM. M. E. Pozzi-Escot and H. C. Couquet.—Mechanism of the senility and death of nerve cells, by M. G. Marinesco. As the result of a study of nerve cells from the brain and spinal column of individuals of ages ranging from 60 to 110, it was found that the modifications constituting the old age of the nerve cell do not only consist of the diminution, more or less marked, of this body, but include other more interesting changes, some of which, tangible to the microscope, are described.—Hetero-plastism, by M. Nicolas-Alberto Barbieri.—A determination of the conditions under which tissue from one mammal can be grafted on to another, to replace similar tissue. The results of experiments are given on the grafting of muscular, vascular, and nervous tissue.

DIARY OF SOCIETIES.

CONTENTS.

THURSDAY, MAY 10, 1900.

MECHANISM, IDEA, OR—NATURE?

Naturalism and Agnosticism. By James Ward, Sc.D.
2 vols. Pp. xviii+302 and xiii+291. (London: A.
and C. Black, 1899.)

THE distinguished writer of the well-known article on
psychology in the "Encyclopædia Britannica"
could not but be sure of a welcome for any contribution
towards the establishment of a world-formula that he
found it in him to offer. Prof. James Ward displays
analytical power of quite first-rate quality, even when he
uses it perversely. He has an insight more than common
into the bearings of scientific methods and naturalistic
speculations, even when he is disputing their competency
or restricting their range. If his lucidity is that of the
successful teacher, his earnestness and his often eloquence
mark the great one. Finally, in meeting the apostles of
naturalism within the jurisdiction of their own categories,
and without the mystification of an alien esotericism, he
has set an example of hopeful augury. "Naturalism and
Agnosticism" is for these reasons one of the books that
must count.

On the other hand, while future attempts at construction cannot neglect the reasonings of this very considerable work, Prof. Ward's is not a mind "to nestle in."
His attack upon Naturalism with Agnosticism must, we
venture to think, be held to have failed, his conclusion to
Spiritualistic Monism to be illicit.

Prof. Ward's book embodies his Gifford lectures, in
defence of theism, delivered at Aberdeen in the years
1896 to 1898. As they "take it for granted that till an
idealistic (*i.e.* spiritualistic) view of the world can be
sustained, any exposition of theism is but wasted labour,"
they are in effect a critique of Naturalism and Agnosticism
singly and together, followed by a brief development of
a Monism of Spirit, in whose interests they are assailed.
Their "demurrer" to theistic inquiries is ruled out,
because they themselves, it is claimed, have failed.

We may pass the question whether a Naturalism that
dares to say that it sees no way of access to knowledge
of a certain kind, "demurs" to theism in any manner in
which "spiritualistic monism," with its implicit pantheism,
does not, and consider rather the development of Dr.
Ward's attack on Naturalism. He tries a fall with it in
three fields—Mechanism, Evolution taken as the working
of Mechanism, and Psychophysical Parallelism as the
device by which the mechanical view disposes of the
importunate facts of consciousness. It seems to follow
that unless Naturalism must be identified with Mechanism
our author's thesis fails.

As regards Mechanism, Dr. Ward disclaims any pretensions to specialism in physics, but he shows such
intellectual communion with the studies that are the
great glory of his university, that he fully sustains his
right to be heard. His fundamental point is the abstract-
ness and hypothetical character of modern physics. He
shows how they pass from the perceptual and actual into
what has been happily called "conceptual shorthand."
He finds mathematical physics "idealistic" in their
procedure—epistemologically, we presume, not ontologi-

cally—and he claims that they do not set before us
"what verily is and happens." Matter, mass, energy—
what are these? We seem driven to modify our ideas of
them again and again, till we end either in Nihilism, with,
for instance, Kirchhoff, or in some highly artificial
formula, such as, *e.g.* the "hydrokinetic ideal" of Lord
Kelvin. From the point of view of logic, the inverse
methods of abstract physics are such that our ultimate
principle will not necessarily be a *vera causa* in the sense
of one who can say *hypotheses non fingo*. If, then, we
accept it as ultimate reality, we are simply Neopytha-
goreans. Can we construct from it a cosmos of qualitative
variety? much less an organic world.

Yet, starting from mechanism, such an attempt at
edification has been made by Mr. Herbert Spencer. To
him the sciences in their evolutionary gradation appear
to offer a closed system, a polity, a synthesis which is
philosophy. The absence of the two volumes essential
for the bridging of the gap from inorganic to organic,
and especially of that famous chapter in which, "were
it written," the transition is actually made, puts Mr.
Spencer's high claims out of court. But further, by playing
off dissipation of energy against conservation, the doctrine
of "First Principles" can be shown to be inadequate.
And Mr. Spencer's demand for instability of the homogeneous at the start, instability of the heterogeneous at
the finish, shows his construction to be arbitrary. To
get evolution to the point of a working process, we need,
says Dr. Ward, a teleology—"evolution with guidance,"
or plan, or purpose. And this to our author implies
something incompatible with mechanism, and in some
transcendental sense, god. There is perhaps a lacuna
in the inference, but so comes the god into the mechanism

But if mind is thus to be, at the very least, the predominant partner in the world-system—it is to be much
more—Dr. Ward must get rid of Psychophysical Parallelism. Psychosis cannot be epiphenomenon, nor, to
use Huxley's unfortunately loose phrase, a "collateral
product." Nor can man—though this is not the same
thing—be a conscious automaton. We cannot have any
implication of "the impotence of mind to influence
matter." We must admit "interaction," because "invariable concomitance and absolute causal independence
are incompatible positions."

But if there is not only room for god, as Brahma so to
speak, at the beginning, but also both room for and need
of, as it were, Vishnu, throughout the working of the
mechanism, to save it from nihilism, to supply "guidance" to the evolutionary process, to infuse new energy,
or, as "the sorting demon of Maxwell," restore wasted
energy, to account for life, to work as immanent sustaining
force throughout, we need only to refute dualism in favour
of a "duality of subject and object," and the way is clear
for idealism.

But is this so? Is naturalism really refuted? Is neutral
agnosticism illicit, or, in the alternative, so unstable, as
to be necessarily materialist or mechanical in bias? Or
has Dr. Ward haply shown that certain physicists, like
certain idealists, have no right to their creed? Those,
namely, who fail to take their symbols as formulæ, abstractions, averages, or to see that where explanatory, the
range of their power of explanation is limited. Or has
he perhaps overthrown much in the hasty constructions

C

of Huxley and Mr. Spencer, made in the first flush of the reconnaissance in force of militant science, but left Naturalism the while untouched?

Dr. Ward's polemic against Mechanism is, we take it, justified with some qualifications, as against those who hold that the synthesis of naturalism is complete, and that the law of its continuity implies the resolution of all phenomenal realities into terms of the modern substitutes for matter in motion, conceived of as having no qualitative but only quantitative determinations. Again, the unnecessarily contemptuous criticism of Evolution as the working of mechanism is valid against Mr. Spencer's "First Principles." Mr. Spencer's mastership happily does not rest upon the soundness of the too early stereotyped foundation, nor on the claim that the edifice is complete to its coping-stone. Further, if "science" is at the standpoint of the materialism of Laplace, or even if it has taken the Huxley of the early sixties, with his undoubted materialist bias, as guide in all things, it will have to retrace the steps it has taken in its advance towards a creed. If it abstracts the known from the knower, and maintains that the act of abstraction makes no difference, it can be convicted of positing a phenomenal world *per se*. If, in the faith of continuity, it says that the inorganic, as it is conceived by mathematical physics, not only conditions but also constitutes the organic, in the sense that we must not, in order to explain the organic, look for anything in the inorganic other than those mathematical determinations of which alone abstract physics take account, then it is against that long patience which is the chief of discoverers, and is attempting an "anticipation" of experience. If it treats regulative as constitutive principles, and attributes agency to formulæ, it is guilty of what we had thought was specifically the idealist's fallacy. But must Naturalism do these things?

We might instance Dr. Hodgson's experientialism and Prof. Münsterberg's transformism among types of naturalism able to "let the galled jade wince." Surely, too, the specialist, finding in his own department, recognised as partial and abstract, the immanence of law, and learning that his colleagues in other departments find law there too, and so throughout, is justified in believing that out of nature—human nature, and specifically the nature of human thought included—the solution must come. Unable to find a mediating term between his "non-matter in motion" or what not, and psychic process, he accepts the "parallelism," with hypothetical connection as co-aspects or, since Prof. Ward, despite of Kant and Mr. Bradley, prefers the causal relation, co-effects of a unitary system. And if to the knowledge of this he sees no road from the human standpoint, wherein lies the illicitness of the union, always stigmatised by Dr. Ward as an evil *liaison*, between his positive treatment of his facts of science and his agnostic neutral attitude, without bias either materialist or spiritualist, towards the ultimate real?

Dr. Ward thinks, in terms of the quotation on his titlepage, that law implies teleology, and that teleology implies spiritualistic monism. We do not see the steps by which he establishes this latter point. And he thinks neutral agnosticism unstable in the direction of one bias or other. We do not see why.

Surely in taking Naturalism "to designate the doctrine

that separates Nature from God, subordinates Spirit to Matter, and sets up unchangeable law as supreme," Dr. Ward has imposed upon it three characters—the first an ambiguity, the second a mechanical bias which is not essential to it, the third its pride, or what it would repudiate, according to the meanings attached to the words "law" and "supreme." It is he who has conjured up what, by a curious slip, he calls "a novel Frankenstein."

We cannot accept Dr. Ward's criticism of psycho-physical parallelism. Mr. Stout, who also "carried Cambridge to Aberdeen," is to the point here. He treats it as the best mode of formulating the facts, but needing for explanation something beyond itself. That he finds this something in an idealist metaphysic makes his witness the more impartial. Prof. Ward hankers after "interaction," or at least "activity of mind." The first, in the form in which he demands it, involves him, to our thinking, in a dualism, which is not a duality of subject and object, and for which his own "refutation of dualism" is enough. The second is spiritualism, which, if monistic, precisely inverts material monism and makes man a conscious automaton from the other point of view.

We may note in this connection a sceptical argument of Dr. Ward's. In what seems to be a misapplication of the formula of "introjection," which he applies elsewhere with signal success, he insists that my *psychoses* are experience only for me, my *neuroses* experience only for the physiologist. The inference surely must be to solipsism or to nothing. Does Dr. Ward mean to deny the accompaniment of my psychical phantasmagoria with brain change?

The quality of Dr. Ward's idealism is perhaps to be doubted. Where does he get the "voluntary movement" which is essential to our perception of space? We are not quite sure that his "intellective synthesis" gives him a right to a world of "intersubjective intercourse" at all. It is, to use an illustration of his own, a case of genii each hermetically sealed in his bottle, but collectively at large. Or it is natural realism. Again, his mental "activity" is in collision with the teaching of Mr. Bradley.

Dr. Ward must have creative agency for thought if "nature is spirit" (though if this be so in a plain, straightforward sense, then why naturalism is wrong from the point of view of spirit is hard to see). But all thought that we know is accompanied with body, and does not create. Huxley "quite rightly refuses to convert invariable concomitance into necessary conjunction." If that is so, what becomes of Dr. Ward's formula as to parallelism and causal independence, apart from his fallacious use of it to establish interaction, when the "community" need not imply more than that they are aspects or, if Dr. Ward will have it so, co-effects of the same real?

Prof. Ward declines to allow analysis to be adequate unless you can find your way back to complete synthesis. Judged by this test, what becomes of Spiritualistic Monism? Indeed, the double edge of Dr. Ward's arguments is one of the marked characteristics of his book. What is good for "non-matter in motion" is good for Green's "relations without *relata*." What is good for Lord Kelvin's Plenum is good for Mr. Bradley's Reality. A dialectical process, which must take place in a time considered to be riddled with self-contradictions and *aufgehoben*, is analogous to the form of evolution that

Dr. Ward eviscerates. In truth, mechanism inverted is spiritualistic monism. The naturalism not yet fully formulated, which has allied itself provisionally and in no way illegitimately with neutral agnosticism, is happily neither materialism nor idealism.   H. W. B.

THE EVOLUTION OF EUROPEAN PEOPLES.

The Races of Europe: a Sociological Study. By William Z. Ripley, Ph.D. Pp. 624; and bibliography, pp. 160. 222 portrait types; 86 maps and diagrams, and other illustrations. (London: Kegan Paul, Trench, Trübner and Co., Ltd., 1900.)

IT has been reserved for an American anthropologist to give us the first comprehensive work on the races of Europe, a subject which is as fascinating as it is important.

The first two chapters of this comprehensive work deal with general questions, among others the problem of environment *versus* race in determining ethnic characters is touched upon, and the error of confusing community of language with identity of race is pointed out; nationality may often follow linguistic boundaries, but race bears no necessary relation to them.

As the main arguments in the book are derived from a consideration of three main sets of comparative data—the head-form, hair- and eye-colour, and stature—it was necessary to discuss their value, and in doing so the author has passed in brief review various races of man in all parts of the world. As the shape of the head, that is the length-breadth, or cephalic index, is not liable to be affected by environment as pigmentation appears to be, and stature certainly is, it takes the first rank as a criterion of race, the colour of the hair and eyes comes second, while stature is relegated to the third rank.

Dr. Ripley states as a proposition that is "fairly susceptible of proof":

"The European races, as a whole, show signs of a secondary or derived origin; certain characteristics, especially the texture of the hair, lead us to class them as intermediate between the extreme primary types of the Asiatic and the negro races respectively."

Surely the wavy-haired group of mankind has as much a claim to be considered primitive as are the frizzly- or the straight-haired groups. That certain characters are intermediate does not imply that a mixture has taken place. In some respects each of these three main groups of mankind is nearer to, and in others further from, the higher apes than the other two groups; the wavy character of the hair of the Europeans, for example, is probably an ancestral feature that has been retained by them and the other Cymotrichi.

The earliest and lowest strata of population in Europe were extremely long-headed, and the author regards the living Mediterranean race as most nearly representative of them. He considers it highly probable that the Teutonic race of Northern Europe is merely a variety of the primitive long-headed type of the Stone Age; both its distinctive blondness and its remarkable stature having been acquired in the relative isolation of Scandinavia, through the modifying influence of environment and artificial selection. It is certain that, after the partial occupation of Western Europe by a dolichocephalic type in

the Stone Age, an invasion by a broad-headed race of decidedly Asiatic affinities took place. This intrusive element is represented to-day by the Alpine type of Central Europe.

It is the play of these three groups, Teutonic or Nordic, Alpine and Mediterranean, upon one another, together with the effect of environment, the potency of which varies locally, occasional isolation and sexual selection, which has resulted in the complexity of the ethnology of modern Europe.

Dr. Ripley deals with the various countries of Europe, and endeavours to unravel the anthropological history of each. It is a humiliating fact how often political or religious bias has crept into ethnological arguments; but our author approaches the subject with an unprejudiced mind, and looks at the problem from a broad point of view.

The most remarkable trait of the population of the British Isles is the uniformity of its head-form; the prevailing type is that of the long and narrow cranium, accompanied by an oval rather than broad or round face. The length-breadth indices all lie between 77 and 79, with the possible exception of the middle and western parts of Scotland, where they fall to 76. This index alone proves little in the present instance, and recourse must be made to other characters, such as hair-colour and stature.

These distinctly prove a dual element in the population, one of which is the persistent Neolithic stock, a branch of the Mediterranean race; the other is the northern race, composed of Saxon, Danish and Norwegian elements. Immigrants belonging to the Alpine race, not pure, but as a mixed people, overran all England and part of Scotland, bringing with them bronze implements, the art of pottery-making, and other cultural advantages; but their physical influence was transitory, for at the opening of the historic period the earlier types had considerably absorbed the new-comers, and the Teutonic invasion completed their submergence. Dr. Ripley, however, is scarcely correct in stating that the Alpine immigrant type never reached Ireland, as traces of them have been recorded (*cf. Proc. Roy. Irish Acad.* (3), iv. 1898, p. 570). The distribution of stature bears out a distinction between the Goidels and the Brythons; but the high stature found in South-west Scotland is anomalous, and requires further study.

It is impossible to deal with all the controversial problems in the book, but an author can generally be gauged by his treatment of critical cases, and of these it is no exaggeration to say that Dr. Ripley always takes a sane position. The origin of the Etruscans is a case in point. The different views of various authors are briefly stated, but the author inclines to Sergi's theory that the Etruscans were really compounded of two ethnic elements, one from the north bringing the Hallstatt civilisation of the Danube Valley; the other Mediterranean, both by race and culture. The sudden outburst of a notable civilisation being the result of the meeting of these two streams of human life, the author appears to have overlooked the probability of a similar history for early Greece.

A whole chapter is given to a discussion of the Basques, and Collignon's deductions are adopted. The French

Basques of to-day are more pure than the Spanish ; but they originally came from Spain. Although the Basque face is extraordinarily narrow, the head is broad ; but this is not due to a mixture with the Alpine race, as the Basque head is essentially dolichocephalic, the breadth occurring pretty far forwards near the temples. We have here, in fact, an example of a local modification (a sub-species of the Mediterranean stock) evolved by long-continued and complete isolation, in-and-in breeding primarily engendered by peculiarity of language, and perhaps intensified by artificial selection.

After having analysed the various European groups, Dr. Ripley devotes a couple of chapters to European origins and others to social problems, such as environment *versus* race, acclimatisation, and urban selection ; in the latter he discusses the tendency to long-headedness, shortness of stature and brunetness that characterises most large towns.

Dr. Ripley has presented us with a very valuable and most interesting study of the origins and physical characteristics of various European peoples, which is as indispensable to students of history and sociology as it is to anthropologists. The clearness with which he states and illustrates his facts leaves nothing to be desired; and we offer him our congratulations on having coped so successfully with an intricate problem, and on having brought his laborious researches to such a satisfactory conclusion.

The book is handsomely "got up," and is sumptuously illustrated. There are 222 carefully-selected portrait types, and 86 maps and diagrams. The selection of the portraits could have been no easy task, and the construction of the distributional maps must have entailed an infinitude of labour. The volume concludes with a bibliography on the anthropology and ethnology of Europe, which is as appalling as it is invaluable.

A. C. HADDON.

A REVISION OF CERTAIN CELL PROBLEMS.

Histologische Beiträge, Heft VI.: Ueber Reduktionstheilung, Spindelbildung, Centrosomen und Cilienbildner im Pflanzenreich. Von E. Strasburger, o.ö. Professor an der Universität Bonn. Pp. xx + 224. Mit. vier litho. Tafeln. (Jena : Gustav Fischer, 1900.)

IT is with no small degree of pleasure that we have perused this, the latest, addition to the five series of "Histologische Beiträge," by Prof. Strasburger. The new volume, like some of its predecessors, deals almost exclusively with cell problems, and anything which its author may have to say on such matters must always command special respect. Breadth of treatment and open-mindedness, no less than thoroughness, have always characterised the work of this great investigator, and perhaps few who are not familiarly acquainted with the cell literature up to the early seventies can realise the extent to which our modern knowledge of cytological phenomena is indebted to the pioneer researches of the author of "Zellbildung und Zelltheilung."

In the volume before us, amongst other topics, the whole subject of what are now familiarly known as "Reduction-divisions" is treated afresh, and emphasis is

laid on the need for a wider basis of comparison before we can return a satisfactory answer to the question as to whether the reduction is only *quantitative*, or whether as Weismann and his followers have supposed, it is *qualitative* also.

The majority of English and German botanical cytologists have decided in favour of the former view, and the researches of Flemming, Brauer, Meves and others on the animal side have shown that the opposite view is, at least, not always tenable. The case of the Salamander especially appears to be impossible of interpretation from the standpoint of the " Qualitative " hypothesis, and now Strasburger shows that the vitally important feature in the Salamander mitosis, viz. the *longitudinal* fission of the retreating chromosomes during the diaster stage of the first reduction division, is closely paralleled by the behaviour of the nucleus in the pollen mother cells of *Tradescantia*. Such a discovery is of the highest value as supporting the evidence already accumulated in favour of the merely quantitative character of these mitoses. The explanation which Strasburger gives of the structure of the ordinary V-shaped chromosome in the first reduction diaster will not, perhaps, gain general acceptance till it has been tested afresh. He believes that the original rod-shaped chromosome, divided longitudinally in two planes cutting each other at right angles, first splits completely into two daughter-chromosomes upon the spindle, and then that each of them opens out along the second plane of cleavage, only cohering at one end thus giving rise to the V-shaped chromosomes of the diaster. During the second division, the later finish their longitudinal fission by complete separation of the limbs at the apex of the V, and thus what would appear to be a transverse fission proves to be merely the finish of a longitudinal splitting incepted at a much earlier period.

The author also discusses the nature of the causes which have brought about the difference of sex, and dismisses the " hunger " and autophagy hypothesis of Dangeard, which is, perhaps, a rather crude form of the less tangible but familiar theories of rejuvenescence. The view is supported that one important factor lies in the comparative absence of kinoplasm from the female, and of trophoplasm from the male, gamete. But it may, perhaps, be questioned whether the study of the evolution of sex in such forms as the green algae does not favour the conclusion that such a difference is a result rather than a cause of sex-difference.

Incidentally, the view recently advocated by Němec, that the reproductive mitoses, in their early multipolar character, contrast with the universally bipolar vegetative divisions, is shown to be without foundation. Multipolar spindles occur both in pollen mother cells and in those of the root apex in *Vicia*, and the present writer has also observed them in the apical meristem of *Equisetum*.

The frequent connection of spindle fibres with extra-nuclear nucleoli is admitted, and is utilised to support the contention that these enigmatical bodies stand in a close relation to the kinoplasm which they are regarded as "activating." Many of the bodies which have, by different writers, been described as centrosomes are certainly nothing else than these escaped nucleoli, and in the example of *Nymphaea*, in which the presence of centrosomes has recently been insisted on, Strasburger shows that not only can the occasional granules not be identified as centrosomes, but that the spindle often reaches to and ends on the peripheral layer of the cytoplasm in a multipolar fashion.

Naturally the bodies known as blepharoplasts are also brought under discussion. These structures have by some been identified with centrosomes, but they seem really to be but remotely related to them. The fact that, as was shown by Webber, the true spindle often becomes multipolar, notwithstanding the presence of blepharoplasts, tells strongly against their centrosomic character, whilst the fact that in the earliest stages radiations start from them proves absolutely nothing at all. Fischer has shown how heterogeneous bodies may serve as starting-points for radiations in fixed specimens of albumin ; and Guignard has described and figured, in the case of the lily, similar radiations having the entire nucleus as their common centre. Much more definite is the relationship existing between the blepharoplasts and cilia. Strasburger, who regards them as essentially consisting of kinoplasm, adduces a series of observations in support of the view that they, or bodies like them, are constantly associated with cilia. Certainly it is a fact of no small significance that whilst, in ferns and cycads, they should be absent from all the other nuclear divisions, they are constantly present in those which directly lead to the formation of the motile antherozoids. Moreover, R. Hertwig has found an analogous relation to hold good for Actinosphaerium, stating that "centrosomes" only occur in connection with the polar (*Richtungs*) mitoses, whilst they are quite absent from the somatic divisions.

It is not possible to touch, even briefly, on all the points raised and illustrated in Prof. Strasburger's book ; it is hoped, however, that enough has been said to indicate its importance as embodying, not only a considerable number of new facts, but also many new and suggestive points of view.

And throughout the volume one is struck, not only by the full recognition accorded to the work of other investigators in the same field, but by the invariable courtesy which characterises the author's criticism of their results even when these do not accord with those obtained by himself.　　　　J. B. FARMER.

MODERN POWER LOOMS.

Mechanism of Weaving. By T. W. Fox. Pp. xxii + 514. (London : Macmillan and Co., Ltd., 1900.)

THE second edition of this excellent book, on the construction and working of the power loom, has been carefully revised by the author. It has justly been recognised as a standard text-book on the subject of loom mechanism. The work treats of tappet, dobbie, and Jacquard or harness looms. In the first place, a full exposition is given of the tappet shedding motion, reference being made to the Yorkshire tappet loom, Woodcroft and segment tappets, and also to the different under motions for the depression of the heald shafts. Proceeding, Mr. Fox deals with some of the principal types of dobbies, such as the Blackburn, Keighley, Burn-

ley and American. By means of sectional drawings, the somewhat intricate mechanism of these dobbies is clearly described. The work would have been enhanced to the manufacturer of heavy fabrics, such as linen, woollen and worsted textures, if fuller descriptions had been given of the dobbies employed in the weaving of these fabrics. Still, to the student of cotton weaving and the manufacturer of light fabrics, the information supplied will be found invaluable, and even the makers of heavier cloths might consult the pages on dobbies with profit. It is open to dispute whether the best method of treatment has been adopted, from a student's standpoint, in dealing fully with shedding motions, including the Jacquard, and card stamping, and the methods of tieing up the harness, before reference is made to other essential motions of the loom ; but the plan of the author is evident on only a casual examination of the book, namely, to treat of each distinct motion in all its various forms in succession, excluding the possibility of affording the reader at the outset even a general notion of the combination of movements in power-loom weaving. This explains why some 280 pages, or more than half the book, should be devoted to the principles of shedding, card stamping and harness mounting, prior to any descriptive reference being made to the picking, the warp let-off, fabric take-up, shuttle, and other motions.

In dealing with the Jacquard loom, the single-lift machine—the basis on which all Jacquards are constructed—is first treated of ; then follow descriptions of the double-lift, centre-shed, open-shed, twilling, Bessbrooke and cross-border machines. The doup and gauze harness are very clearly explained. Other systems of tie-up, more elaborately illustrated, might have been advantageously incorporated into this section of the work ; but sufficient data is afforded to enable the student to grasp the principles on which the complex mountings are effected, necessary in the weaving of tapestry and decorative silk fabrics.

Lappet weaving receives adequate attention, especially as worked by means of lappet wheels and the Scotch method ; but only brief details are given on other forms of this motion, in which lags are used and pegs of different lengths, and also in which the frames for carrying the lappet threads operate on the upper side of the fabric.

In regard to picking, Mr. Fox gives some interesting information on the magnitude of the force expended in propelling the shuttle from side to side of the loom. Perhaps there is no motion in weaving in which improvement is so desirable as in picking. This is more obvious in heavy looms, where large shuttles have to be used, travelling at high speed. Under the head of "Warp Protectors," fast and loose reeds are considered, as well as shuttle guards. Many attempts have been made at automatic warp-stop motions, such as those applied to the Northrop and Poyser looms, but probably the author has not mentioned these on account of their not having come into general use in this country ; still, there are principles in both interesting to the student of "Mechanism of Weaving."

The chapter on "Multiple Box Motions" is one very typical of the author's skill in the exposition of difficult mechanical problems. Revolving, as well as drop-box motions, with suitable illustrations, are fully explained. On "beating-up," the author has some instructive information respecting the movement of the crank for carrying the batten or going part against the fell of the fabric. He supplies a table showing the motion of the crank, and treats of the length of the crank-arm and the eccentricity of movement. The concluding portions of the book are devoted to weft-stop motions, mechanism for governing the warp and taking-up of the fabric, the construction of temples and selvage motions. There is also a chapter on the arrangement of weaving-rooms or sheds, with a plate illustrative of the positions of the looms and other machinery. The book should be in the possession of all those interested in the construction of power looms.

OUR BOOK SHELF.

Leçons d'Optique géométrique à l'Usage des Élèves de Mathématiques spéciales. Par E. Wallon, Professeur au Lycée Janson-de-Sailly. Pp. 343. (Paris : Gauthier-Villars, 1900.)

THIS book has been written at the desire of Prof. Wallon's students, to whom a graceful tribute is paid, in the preface, for the assistance which their questions, doubts and objections have rendered in developing the author's methods of teaching. To look on one's students as collaborators, that is certainly the secret of successful teaching ; and, as here presented, Prof. Wallon's lectures are certainly successful in giving a systematic and clearly defined outline to the science of geometrical optics. The diagrams are well drawn and numerous, and the mathematical proofs are simple and yet sufficient. There is, however, little that is novel to be found in the course of these lectures ; indeed, in a few cases it might be objected that there was a tendency to lag behind the times. Thus, in discussing refraction equivalents, Newton's law, that $\frac{n^2 - 1}{d}$ = constant, and Gladstone and Dale's law, that $\frac{n - 1}{d}$ = constant, are alone mentioned (n being the refractive index, and d the density of the substance). Lorenz's law, that $\frac{n^2 - 1}{(n^2 + 2)d}$ = constant, is now most generally accepted. For gases, in which n is nearly equal to unity, all three laws hold with about equal accuracy. But Lorenz's law appears to hold in passing from the gaseous to the liquid state, and must therefore be accepted as the most general.

An interesting chapter is devoted to the subject of the human eye, in which the most well known optical properties of that organ are discussed. In the ensuing chapter, on optical instruments, a particularly good account is given of the optical systems comprised in telescopes and microscopes of various patterns. It is surprising, however, that the ophthalmoscope and ophthalmometer are not mentioned, and are in fact so seldom found described in works on geometrical optics. Both instruments involve interesting optical arrangements, and their practical usefulness would render a description of their details still more interesting. E. E.

Therapeutic Electricity and Practical Muscle Testing. By W. S. Hedley, M.D., M.R.C.S. England. Pp. ix + 278 ; 3 plates ; 99 illustrations. (London : J. and A. Churchill, 1899.)

THE increased use of electro-therapeutic methods renders the appearance of Dr. Hedley's book welcome. The profession have for some time looked somewhat askance at this departure in therapeutics, and are, in many branches of this practice, rather inclined to regard the good effect of the treatment as moral and not actual. The work before us considers the whole subject from a scientific standpoint, and any one interested in it will gain considerable profit from its perusal.

The reader must be warned at once that the book contains no mention of radiography or the application of the Röntgen rays to the healing art, either from a diagnostic or therapeutic standpoint. The author, in his preface, admits that the work is a therapeutical one, and to some extent apologises for the description of such instruments as the cystoscope, &c. No doubt he thinks the profession is in possession of sufficient literature upon the subject of radiography, which may or may not be true ; the sphere of usefulness of the book would, however, certainly have been increased by the inclusion of this subject.

The work is divided into three parts. The reason for this classification is not quite evident ; a part as a classification unit seems, in the author's hands, to differ to no material extent from a chapter. Further, each part is chaptered separately, which, without some very special object is to be gained, is a bad plan ; from this it follows that the book contains three Chapters i., &c.

The first part is mostly concerned with those general physical considerations which have a special bearing upon what the author in the first chapter of Part ii. calls the electro-therapeutic outfit. A good account is given in Chapter vii. (p. 65) of currents of great frequency and high potential, which, as has been frequently shown, are of great therapeutic value. Much technical detail is given, both of a purely electrical and electro-physiological character.

One of the most useful chapters from the standpoint of the general physician is Chapter v. Part ii., upon the action of muscles and the consequences of their paralysis. In Chapter x. Part iii., an interesting account of cataphoresis is given. Very frequent mention is made of authors' names and no reference added, nor is there an index of authors at the end, or anything in the shape of a bibliography. Mere chance or whim has apparently guided the author in giving or omitting the full reference of a work cited ; in some cases the full reference of important monographs is withheld, in others that of trivial ones given. This method cannot be too severely deprecated.

To sum up our remarks, it is with the manner and not the matter of the book we find fault. It is full of useful and, indeed, essential information to those working in this field ; the author has spared no pains to collect fact bearing upon and elucidating his subject.

Lessons in Botany. By Prof. George F. Atkinson, Ph.B. Pp. xv + 365. (New York : Henry Holt and Co., 1900.)

THE present volume is, in a sense, an abridged edition of an excellent text-book by the same author, which appeared a year or two ago. The subject-matter is carefully arranged to suit the convenience of teacher and pupil, and altogether the book is one which should prove useful in this country as well as in America. Naturally, from the British point of view, the difficulty of obtaining the needful specimens occasionally may turn up, though this would not recur very often. We can confidently recommend Prof. Atkinson's book to the notice of teachers.

Outlines of Plant Life, with special reference to Form and Function. By Prof. Charles Reid Barnes. Pp. vi + 308. (New York : Henry Holt and Co., 1900.)

THIS is a work intended for school use. It has some points of merit, especially the special part on ecology, in which the examples are well chosen and fully illustrated. The illustrations, though almost all are (with due acknowledgment) borrowed from other works, are distinctly good. We think the book a useful one, and the exercises which are interspersed through the volume add to its value.

LETTERS TO THE EDITOR.

[The Editor does not hold himself responsible for opinions expressed by his correspondents. Neither can he undertake to return, or to correspond with the writers of, rejected manuscripts intended for this or any other part of NATURE. *No notice is taken of anonymous communications.]*

Note on some Red and Blue Pigments.

THE following data are placed on record because interesting in themselves, and in the hope that they may be useful to others who have the opportunity to make further investigations.

(1) A little boraginaceous plant called *Eremocarya micrantha* (Torrey) is common in sandy places at Mesilla Park, New Mexico, flowering in April. A few days ago, Prof. E. O. Wooton called my attention to the fact that its roots are deep red, and stain herbarium paper. Curious to learn more about this peculiar coloration, I made some tests, with the following results :—The pigment is not soluble in water, but it readily dissolves in cold alcohol, forming a beautiful red solution. The roots, after being treated with alcohol, become white, showing that the pigment is entirely superficial, and is apparently an excretion from the root. The red colour is that of the normal or acid state of the pigment, but on adding enough liquor potassæ to make the solution alkaline, the colour immediately becomes a beautiful blue. An excess of strong caustic potash does not destroy the pigment until after a considerable time. Prof. A. Goss tested the delicacy of the colour-reaction in the presence of acids and alkalis, and found that a very small excess of one or the other would give the characteristic colour. The pigment is, of course, an anthocyan, very similar, at least, to litmus ; and it may be that it can be utilised for the same purposes.

(2) It has been remarked more than once that whereas the hind wings of Acridiidæ (grasshoppers) are sometimes blue, sometimes red, and sometimes yellow, species living in the same locality, though of very diverse genera, will often have similarly coloured wings. In the Mesilla Valley we have common species with red and with yellow wings ; but in the Organ Mountains, not far away, I found two species very abundant, both having blue wings, and otherwise coloured much alike, though of totally different genera. These were *Leprus wheeleri* and a *Trimerotropis* which I took for *T. cyaneipennis*, but which Mr. S. H. Scudder tells me is distinct and apparently undescribed. As the blue of the wings appeared to be certainly a pigmentary colour, and much resembled the vegetable anthocyans, I detached one of the wings of *Leprus wheeleri*, and boiled it in dilute hydrochloric acid. As I had hoped, but hardly ventured to expect, the blue at once became red. Heating the thus reddened wing in liquor potassæ did not change it back to blue, but caused it to turn yellow. I infer that the blue pigment has a red (acid) phase, but that strong alkali will destroy it altogether, leaving a yellow coloration which is of a different character. It is difficult to avoid the conclusion that the redness or blueness of the wings in these grasshoppers may result from the action of some environmental factor (*e.g.* the juices of plants eaten) upon the pigment, and that this accounts for the colour-similarity of diverse species living at the same place. Of course, this is not supposed to account for the similarity of the colours of the tegmina and thorax, of which the various shades of grey, red and brown resemble those of the rocks and ground.

T. D. A. COCKERELL.
Mesilla Park, New Mexico, U.S.A., April 17.

Valve Motions of Engines.

IN your number of December 14, 1899, Prof. John Perry mentions a diagram by Mr. Harrison. This diagram is the same as " Das bizentrische polare Exzenterschieberdiagramm " of F. A. Brix in the *Zeitschrift des Vereins Deutscher Ingenieure,* April 10, 1897.

There is only a small difference, as Mr. Harrison finds the distance OC by means of a circle with radius = length of connecting-rod, and Mr. Brix finds that distance by calculating it out of $\frac{R^2}{2L}$, (R = length of crank, L = length of connecting-rod). Now OC has not exactly that value, but the fault made therewith is much smaller than the fault made by describing the circle. Therefore the method of Mr. Brix is preferable to that of Mr. Harrison.   F. J. VAES.
Rotterdam, April 14.

MR. BRIX seems to have solved only the simple case of a valve worked by an ordinary eccentric. There are other good graphical solutions—for example, by Coste and Maniquet in a modified form of the Reauleaux diagram, which gave accurate results. Mr. Harrison's diagram is more general and is applicable to link and radial valve-gears and to all motions which are composed of a simple harmonic vibration with a small octave superposed. It may be used for velocities and accelerations as well as mere displacements. As to calculating the distance OC by the formula $\frac{R^2}{2L}$, instead of using the construction of the circular arc, this is a matter of no importance because there is no appreciable difference in the answers.
April 28.   JOHN PERRY.

Drunkenness and the Weather.

I NOTICE in your issue of March 15 a communication from Mr. R. C. T. Evans, of Gray's Inn-road, W.C., calling attention to a probable error in my deductions in the paper which appeared in your issue of February 15, under the title "Drunkenness and the Weather." He says, "When a man is intoxicated and commits an assault, the result is entered in the police reports as 'assault,' the more serious offence overshadowing the less ; so that in all probability many of the cases of assault referred to in the statement were also cases of drunkenness, but were not tabulated as such. Studying Prof. Dexter's curves in this light, we may reasonably conclude that the number of those arrested for drunkenness or its results, varies but little throughout the year."

Although his supposition seems a reasonable one, a fuller statement of the conditions of the study will show that the fluctuations of the "drunkenness" curve cannot be so easily accounted for.

First, the monthly occurrence of arrests for drunkenness for New York City is more than twice that for assault, even in the summer, when the former are at the minimum and the latter at the maximum for the year, and if we suppose that every person arrested for assault in the summer was also intoxicated and would have come into the hands of the law for that crime if he had not for the other, even this would not bring the drunkenness curve up to its normal for the winter months.

Second, the method of recording crime by the New York City Department of Police makes this practically impossible. Misdemeanours are there classified and recorded under 183 different headings. The two which I have compared are "assault and battery" and "intoxication." There are, however, four other classes of assault besides, one for "intoxication and disorderly conduct," equalling that of "assault and battery" in the annual number of arrests, besides one for "fighting." A letter just received from the Clerk of Police says, "The crime of intoxication and fighting—a drunken brawl—would be classified in the statistics as 'intoxication and disorderly conduct.'" A careful analysis of all the conditions would make it seem that only occasionally would arrests for "assault and battery" encroach upon the data of drunkenness. I believe they might sometimes do so, but not sufficiently often to materially influence the curve.   EDWIN G. DEXTER.
Greeley, Colo., April 17.

SOME SPECULATIONS AS TO THE PART PLAYED BY CORPUSCLES IN PHYSICAL PHENOMENA.

IN some experiments described in the *Phil. Mag.* October 1897, I showed that in the kathode rays there were present bodies whose mass was exceedingly small compared with the masses of ordinary atoms ; these masses, which carry a charge of negative electricity, I called "corpuscles." Ever since then I have indulged in speculations as to the possibility of these corpuscles existing in a free state in ordinary matter not under the influence of the very intense electric field which are associated with the kathode rays. As recent work has produced some evidence of the free existence of these corpuscles, I have thought that these speculations might be of some interest to a wider circle than that to which they have hitherto been addressed. In the *Phil. Mag.*

(February 1900), I showed that these corpuscles existed in the neighbourhood of a hot wire and of a metal plate illuminated by ultra-violet light, and recently the discovery by Giesel, Curie and Becquerel of the magnetic deflection and electric charge carried by part of the radium radiation may be interpreted as indicating the existence of corpuscles in this substance.

I suppose, then, that there is a certain amount of what may be called corpuscular dissociation taking place in bodies ; that some of the molecules of the substance are continually breaking up by the detachment of a corpuscle, and are being reformed by the arrival of another corpuscle ; the result of this is that at each instant there are a certain number of free corpuscles with negative charges distributed throughout the body, while the corresponding positive charges are on the molecules of the body, the corpuscles are much more mobile than the molecules ; indeed, in solids and liquids, the latter may be regarded as almost fixed in comparison with the former. We thus get the conception of a body permeated with corpuscles which are able under forces to move from one part of the body to another. We must remember that, as the particles are charged, any movement will be accompanied by electrical effects and, in general, a volume density of electrification.

The actual number of corpuscles free at any instant is the result of an equilibrium between the number of corpuscles produced by dissociation and the number which recombine. Thus if q is the number of corpuscles produced by dissociation in unit volume in one second, τ the time during which a corpuscle is free (i.e. the time which elapses between its departure from one molecule and its entry into another), n the number of free corpuscles in unit volume, then when there is equilibrium $q = n/\tau$ or $n = \tau q = \lambda q/u$, if λ is the mean free path of the corpuscle and u its velocity of translation. In non-conductors we suppose that there are very few corpuscles, but that they are abundant in metallic conductors. Let us now trace some of the consequences of the existence of these corpuscles in a solid, and suppose for the moment that the positively charged molecules are fixed ; if the corpuscles are acted upon by gravity (of which point we have no evidence), then in a vertical bar of metal the number of corpuscles in unit volume will be greater at the bottom of the bar than at the top, for just the same reason as the density of the air gets less as we go higher ; thus in this case gravity would produce a displacement of electricity, the bottom of the bar being negatively and the top positively electrified. Again, in a rotating mass of metal the centrifugal force would tend to drive the corpuscles towards the surface ; there would thus from this effect be an excess of the corpuscles near the surface and a deficit near the axis. Thus the outer parts of the metal would be negatively and the inner parts positively electrified, the rotation of the negatively electrified corpuscles being no longer completely balanced by that of the positively electrified molecules would give rise to a magnetic field ; thus a large mass of rotating metal would act as a magnet. Again, suppose we place a piece of metal in a magnetic field, the action of the magnet on the moving corpuscles will make them describe curved paths, and we can easily see that the magnetic effect due to the particles moving in this way is in the opposite direction to that of the external magnetic field. Thus a metal containing these corpuscles would tend to act like a diamagnetic substance. Again, suppose the metal is exposed to an electric force X, the corpuscles will acquire an average velocity along x equal to $X\epsilon e/2m$, where m is the mass of a corpuscle and e its charge. Let us call this velocity vX, then the electric current across unit area is nevX ; thus nev or $qe^2\lambda^2/2mu^2$ is the specific conductivity of the substance. If we suppose that u, the mean velocity of translation of the corpuscles, varies with the temperature in the same way as the velocity of translation of the molecules of a gas, mu^2 would be pro-

portional to the absolute temperature, and the specific resistance would, considered as a function of the absolute temperature θ, vary as θ/q ; if q, the amount of ionisation increases as the temperature increases, the resistance will vary more slowly than the absolute temperature ; if q diminishes as the temperature increases, the resistance would vary more rapidly than the temperature. These corpuscles moving from place to place would carry not merely electric charges, but energy from one part to another ; and since the coefficient of diffusion of these corpuscles is proportional to v, the thermal and electric conductivities would be proportional to each other. Again, when we have conduction of heat we have unequal streams of these corpuscles in opposite directions ; thus the unequal deflection of their paths produced by a magnet would give rise to an electric displacement, and we should have an electromotive force at right angles to the magnetic force and to the temperature gradient, an effect discovered by v. Ettinghausen and Nernst. From the conductivity of the gas we can deduce the value of nev. We know the value of e, and hence another equation would enable us to determine n and v ; for this purpose we turn to the Hall effect, but here the results are disappointing, for we can easily prove that when E^1 and E are the transversal and longitudinal electric forces and H the magnetic force, $E^1/EH = \dfrac{v_1 k_2 - v_2 k_1}{k_1 + k_1}$, where v_1 and v_2 are respectively the velocities of the negative corpuscles and positive molecules under unit electric force, and k_1 and k_2 the values of k for these ions where k = pressure $\div$ number of systems in unit volume. If both the negative corpuscles and the positive molecules behave like perfect gases, $k_1 = k_2$ and $E^1/EH = \frac{1}{2}v_1$, since v_2 is very small ; thus, on this supposition, the Hall effect would give us the value of v ; but there seems no reason to suppose that the positively electrified molecules in the solid would produce the same pressure as an equal number of molecules in the gaseous state, and thus though v_2 is small compared with v_1, k_2 may be so small compared with k_1 that k_1v_2 cannot be neglected in comparison with k_2v_1, and in this case the Hall effect would not be sufficient to determine v. The fact that the Hall effect is of different signs for different substances shows that we have to take into account both terms in the expression for E^1/EH.

Again, if different parts of a metal bar were at different temperatures, the "pressure" as it were of these corpuscles would be different at different parts of the bar, so that the corpuscles would tend to flow from one part of the bar to the other, and cause an electric displacement ; thus difference of temperature would cause an electric displacement. This is the Thomson effect, measured by the "specific heat of electricity." The value of the "specific heat of electricity" will on this theory depend not only on the variation with temperature of the kinetic energy of a single corpuscle, but also on the way the dissociation constant q varies with the temperature. There are many other phenomena which can be interpreted in terms of these corpuscles, but these I must leave for another occasion.   J. J. THOMSON.

Cavendish Laboratory, Cambridge, April 30.

SCIENCE IN RELATION TO ART AND INDUSTRY.

AT the annual banquet of the Royal Academy on Saturday evening, Sir Norman Lockyer, in replying on behalf of science, made the following remarks upon the intimate relation between intellectual progress and the study of nature, and also upon the necessity for a more liberal provision for scientific work if England wishes to compete successfully with other nations struggling for industrial supremacy. Though the public mind may be disturbed by the statement of the principle that the provision made for scientific and technical study and research should be as great as that given by any two other nations, the comparison will serve a useful purpose in directing attention to a view of the claims of science worthy of consideration.

It is a very great honour for a student of science to be called upon in such an august assembly as this to say a few words ; but if I am to be accepted as the representative of science I do not wish to be fettered by your suggestion, Sir, that I should refer to the dependence of art on science. I am sure that I may frankly say for every man of science that we acknowledge freely the firm brotherhood between art and science—a brotherhood founded upon a common object, the study of Nature, "the mistress of all the masters," and carried on by a common method, the proper co-ordination of brain, hand and eye. In every case with which a man of science or a man of art has to deal, imagination is required, and so science and art meet upon terms of mutual helpfulness. I think I may also say that this feeling is thoroughly reciprocated by men of art, for many of them honour me with their friendship, and therefore I know their sentiments. I am the more anxious to say this because some twenty years ago, when I was privileged to attend this anniversary dinner, I heard a distinguished representative of literature express a totally different sentiment. He told us that "before their sister, Science, now so full of promise and pride, was born, there were Art and Literature like twins together," and it was suggested that the sooner art and literature formed an alliance offensive and defensive against the interloper, the better it would be for them. I do not believe in this. For me science is as old as art. They have both advanced together. Let us take the position of things 6000 years ago, to begin at the beginning of things, if we can. Then the priest-mummifiers of Memphis had to be profound anatomists. If you go to the Gizeh Museum you find magnificent specimens in those statues of Chepren in diorite, other statues in wood, and the plaques, veritable Memlings in stone, which clearly show that this knowledge was also possessed by their sculptors. If you come down to a comparatively modern period, something like 600 B.C., and compare those wonderful metopes of Solinunto with the marbles of the Parthenon, which are of a later date, you will find an enormous advance in the latter. You will find that Hippocrates had lived in the interval, and, indeed, that he and Phidias were contemporaries and fellow-townsmen. Carrying the matter down to the introduction of Universities into Northern Italy in the thirteenth century, we find that the difference between the art of Cimabue and Giotto depends on the fact that anatomy had been introduced in the meantime. Science, then, is no new interloper, seeking to detract from the importance of art and literature. What was new twenty years ago was that the work of the late Prince Consort, whose name will always be revered by those who know the benefits he conferred on our country, was then beginning to tell. He showed us that in order to secure industrial progress we must have, above all things, instruction and practice in science and art. In war, being well assured of the valour and endurance of our sailors and soldiers, the chief thing we have to do is to see that they are properly supplied with the engines and munitions of war, and, more than these, the scientific spirit. In peace, for the beauty of a nation's life and a perfect record of it, we must look chiefly to the sweetening and ennobling influences of art and the enduring works of its masters ; but for a nation's continued welfare and progress both science and art are necessary. We are in face of industrial struggles, and we must utilise both science and art to supply the wants of our own and other countries, and to provide commodities made in England, besides handling

> "Things of beauty, things of use,
> That one fair planet can produce,
> Brought from under every star."

We are in face of a struggle for existence in which we know full well that only the fittest will survive. How are we going to carry on the struggle? What are our weapons? Our first line of defence in this direction can only consist of our Universities and our teaching centres. Have we enough of them? We know already that we have not enough of them, because we have already lost several important engagements in these industrial battles. Are there no means by which we can judge of their sufficiency ? In relation to non-peaceful international struggles in which also defeat has to be guarded against, a clear and

universally approved policy has been enunciated ; this is, that the future of our empire, an empire the real unity and strength of which are developing under our eyes at this moment, can be secured if we see to it that our first line of defence, our fleet, shall be equal in strength to the fleets of two other possibly contending powers. The second answer then, I think, is that this principle should be applied to our first line of defence in those industrial conflicts the results of which are much more enduring. Do our teaching and research centres at present outnumber in the same proportion, as do our ships, those of any two nations which are actually contending with us in peaceful enterprise ? And, also, are they equally efficient in every respect ? I believe, and I know that this view is held by many representative men of science, that until our Universities, our science schools, our art schools, and our technical institutions bear the same relation both in number and efficiency to those of other nations as do our battleships, cruisers, and small craft, we shall not be justified in regarding the future of the empire with that freedom from care which is the attribute of a strong man armed.

NOTES.

PROF. E. SUESS, professor of geology in the University of Vienna, has been elected a Foreign Associate of the Paris Academy of Sciences, in succession to the late Sir Edward Frankland. Sir John Burdon-Sanderson, Bart., has been elected a Correspondant of the Academy, in succession to the late Sir James Paget.

DR. S. L. TÖRNQUIST, of Lund (Sweden), has been elected a Foreign Member of the Geological Society, and Prof. F. Sacco, of Turin, has been elected a Foreign Correspondent.

WE much regret to see the announcement of the death of Lieut.-General Pitt-Rivers, F.R.S., the distinguished anthropologist, on Friday last.

THE annual conversazione of the Society of Arts will be held at the Natural History Museum, South Kensington, on Wednesday evening, June 20.

THE adjourned debate on the Sea Fisheries Bill was resumed in the House of Commons on Monday. After a long discussion, a division was taken, and a majority was obtained in favour of the second reading. The Bill was then referred to a Select Committee.

IT is reported that Vesuvius has shown signs of increased activity during the past few days. Explosions have taken place in the crater of the volcano, and masses of rock and lava have been ejected. The huts of the guides and the topmost station of the funicular railway are threatened. Reuter reported that four Englishmen who ascended Vesuvius on Tuesday went beyond the limit indicated as dangerous by the guides and gendarmes, and were seriously injured by a mass of ejected material striking them. This however has since been denied by Reuter's Naples Correspondent.

THE U.S. National Academy of Sciences has decided to award the Barnard medal to Prof. Röntgen for his discovery of the X-rays. This medal is awarded at the close of every quinquennial period for a discovery in physical or astronomical science, or novel application of science to purposes beneficial to the human race. The first presentation of the medal was to Lord Rayleigh and Prof. Ramsay for their joint discovery of argon.

REUTER'S AGENCY learns that Dr. Louis Sambon and Dr. G. C. Low, who has been awarded the Craggs research scholarship of 300*l.* per annum, are about to experiment with a view to proving that malaria is spread by mosquito bites, and expect to begin work seriously on June 1, by which time they will have all their arrangements completed. A suitable spot has been chosen for the erection of their mosquito-proof house in the Campagna, on the line of the railway running from Rome to Tivoli.

A MEETING of the International Association for the Advancement of Science, Arts and Education will be held at the Society of Arts to-morrow (May 11), at 4 p.m. Sir Archibald Geikie, F.R.S., vice-president of the British Committee, will preside. The secretary, Prof. Patrick Geddes, will deliver an address on the nature and aims of the Association and its forthcoming assembly at the Paris Exhibition.

IN connection with the International Congress of Physics to be held in Paris from August 6 to 12, a preliminary programme of papers has been issued. Over sixty reports have already been promised, and among the names of contributors we notice those of Amagat, Arrhenius, d'Arsonval, Battelli, Becquerel, Blondlot, Bouty, Boys, Branly, Brillouin, Broca, Cornu, Curie, Exner, Griffiths, Hurmuzescu, Lippmann, Lorenz, Poincaré, Potier, Poynting, Pringsheim, Righi, Spring, J. J. Thomson, Villard, Warburg and Wien.

THE next meeting of the Comité International des Poids et Mesures is fixed for September 10, 1900. Owing to the death of M. Joseph Bertrand, and the resignation of Prof. Thalen, two of the original members of the Comité, the number of members is now limited to eleven. Great Britain will be represented at the forthcoming meeting by Mr. H. J. Chaney, a member of the Comité.

THE death of M. Edouard Grimaux, at the age of sixty-five, occurred during the past week. M. Grimaux succeeded Cabours as professor in the École Polytechnique at Paris, and also held a chair at the Agronomic Institute. He made numerous and valuable contributions to organic chemistry, and was the author of several chemical treatises. He will be gratefully remembered by chemists for an admirable biography of Lavoisier, which he published in 1884. M. Grimaux lately became prominent in connection with the Dreyfus case. At the Zola trial he expressed his belief in the innocence of Dreyfus. For this he was deprived of his professorship by General Billot, notwithstanding the fact that he had rendered devoted service to the army in 1870. In 1894 M. Grimaux was elected to the Academy in the place of Frémy.

REPLYING to a question in the House of Commons on Monday, Mr. Akers-Douglas stated that the new National Physical Laboratory is not to be erected, as has been reported, in the Queen's Cottage grounds, or in any other grounds attached to Kew Gardens. It will stand quite outside those Gardens on Crown land. The only part of the scheme which might possibly be supposed to affect the amenities of the Gardens is a small building which will not, at the outside, cover a quarter of an acre. This building will be so placed as not to interfere with the views from the Gardens over the Old Deer Park, and it will not be opposite to that part of the Gardens round the Queen's Cottage which is reserved in a wild state. The building will only be used for delicate scientific work which will not disturb the seclusion of the neighbourhood of the Queen's Cottage, and which, in fact, itself requires as much quiet and privacy as can be obtained.

THE Paris correspondent of the *Chemist and Druggist* states that science is represented at the Salon by several portraits of average merit. The best is that of Dr. Vaillard, head army surgeon and professor at the Val de Grace Military Hospital, where he is known to two or three generations of army pharmacists who have followed his lectures. Dr. Vaillard is of middle age, and is shown standing, in regimental dress, with the Cross of the Legion of Honour on his tunic. His left hand is leaning on a laboratory-bench, on which are a microscope and a variety

of analytical appliances. To his right is a lecture-blackboard, and one can dimly see his written demonstration. The artist is M. Paul Bourdier. The portrait of M. Hautefeuille, chemist, and member of the French Institute, is the work of a lady artist. She shows him in everyday attire in a corner of his laboratory, sitting at a table, with a collection of scientific apparatus near at hand ; in the background is a furnace, at which an assistant in a white blouse is working. M. Tisserand, of the French Institute, is another portrait of fair merit. One would like to see more of this class of picture, but must suppose artists find no market for them.

THE death of Dr. Edmund Atkinson on the 4th inst., after a very short illness, will be a matter of deep regret to his large circle of friends. He was born at Lancaster in 1831, and was a student of Owens College, Manchester, in the early days of that institution. There he became assistant to the late Sir Edward Frankland, the first professor of chemistry in the College, and was associated with him in organising the laboratory which has since become so well known. About 1854 he went abroad for some years and continued his scientific studies at the Universities of Marburg, Göttingen and Heidelburg, and at the École de Médecine in Paris under Wurtz. On his return to England he became private assistant to Sir Benjamin Brodie at Oxford, then science master at Cheltenham College, and afterwards professor of experimental science at the Royal Military College, Sandhurst, and at the Staff College. He was several times elected upon the council of the Chemical Society, and was one of the founders of the Physical Society, of which Society he was treasurer from the beginning until the last anniversary meeting, with the exception of a short interval a few years ago. Dr. Atkinson rendered great service to science by his numerous translations into English of foreign scientific works ; among these the best known are Ganot's "Elements of Physics," von Helmholtz's "Popular Scientific Lectures" and Mascart's "Treatise on Electricity and Magnetism." He was a man of excellent judgment in practical affairs, and of late years he gave much time as a magistrate to the local affairs of his neighbourhood. He was always ready to undertake onerous duties for those in need of help, and was a most generous and steadfast friend.

THE council of the Royal Geographical Society have awarded the two Royal medals for this year to Captain H. H. P. Deasy and Mr. James McCarthy. The Founders' medal has been awarded to Captain Deasy for the exploring and survey work accomplished by him in Central Asia. Mr. McCarthy is the Government surveyor of Siam, and the Patron's medal has been awarded to him for his great services to geographical science in exploring all parts of the kingdom of Siam, for his laborious work during twelve years in collecting materials for a map, to form the basis of a survey system, and for his admirable map of Siam just completed. The other awards have been made as follows :—The Murchison award to M. Henryk Arctowski for the valuable oceanographical and meteorological work which he performed on the Belgian Antarctic expedition ; the Gill memorial to Mr. Vaughan Cornish for his researches, extending over several years, on sea-beaches, sand-dunes, and on wave-forms in water ; the Back grant to Mr. Robert Codrington for his journeys in the region between Lakes Nyassa and Tanganyika, during which he removed, on behalf of the Society, the section containing the inscription from the tree under which Livingstone's heart was buried ; and the Cuthbert Peek grant to Mr. T. J. Alldridge for his journeys during the past ten years in the interior of Sierra Leone, during which he has done valuable geographical work.

THE following opportunities for the study of botany during the ensuing summer season in the United States are mentioned in the *Journal* of the New York Botanic Garden for April :—Columbia University, New York, has instituted a summer session, beginning July 2 and ending August 10. The department of botany will be under the charge of Prof. Lloyd, who will offer courses in ecology, general botany, and research work in select subjects. Students in these courses will have access to the museum and collections of the Botanic Garden. The Woods Holl Laboratories will be open from July 5 to August 16, and the botanical staff includes Dr. B. M. Davis, Mr. G. T. Moore, Dr. R. H. True, Miss Rhoda A. Esten, and Miss Lillian G. MacRae. Courses in cryptogamic botany, plant physiology, and plant cytology will be offered. The biological laboratory at Cold Spring Harbour will be open from July 2 to August 25, the botanical staff including Dr. D. S. Johnson, Dr. H. C. Cowles, and Mr. W. C. Coker. Courses of lectures will be offered in cryptogamic botany, ecology, and bacteriology.

THE Annual Summary of the *U.S. Monthly Weather Review* for 1899 contains a very interesting account of the climate of St. Christopher, by Mr. W. B. Alexander. The island lies in latitude 17° 20′ N. and longitude 65° 45′ W. ; its length is 23 miles, and the breadth of the main body is about 5 miles. The central part is occupied by a range of mountains, the highest of which, Mount Misery, rises to a height of about 4100 feet. Tables and diagrams are given showing the barometric pressure for 35 years, and the rainfall for 44 years at Basseterre, which is situated in a spacious and fertile valley. The climate, generally speaking, is dry and healthy, being tempered and purified by frequent thunderstorms. The mornings and evenings of the hottest days, which occur in August, are agreeably cool ; the coldest months are January and February. The mean annual temperature is about 81°, of August, 83°, and February, 78°. The mean annual rainfall is about 51·6 inches ; 37 per cent. of the amount occurs during the first half of the year, and 63 per cent. during the last half. The rainfall is more frequent than heavy ; it has only reached or exceeded 5 inches in 24 hours eleven times in 44 years.

IN the *Proceedings* of the South African Philosophical Society, vol. xi., Mr. J. R. Sutton publishes an important discussion of the winds of Kimberley. The results are obtained from three years' hourly observations with Osler and Robinson anemometers. The period is admittedly short ; but the excellence of the position and the scarcity of hourly observations in South Africa are quoted as reasons for not delaying the appearance of the paper. The observatory is situated at Kenilworth, about three miles N.N.E. of Kimberley, at an altitude of nearly 4000 feet. It has been supposed that there was an overwhelming excess of northerly winds, and theories have been propounded why this is the case ; but the conclusion to be drawn from the paper is that while sometimes one and sometimes another direction may preponderate from year to year, a definite prevailing wind does not exist. Of the 25,898 hours of wind analysed throughout the three years, the final resultant contains the small components of only 50 hours to the north and 100 hours to the west. The diurnal curve of wind velocity contains two maxima (2h. p.m. and 10h. 45m. p.m.) and two minima (5h. a.m. and 7h. 30m. p.m.). The mean hourly velocity is 6·6 miles per hour.

AN interesting illustration of Doppler's principle is noted by Prof. F. Richarz, of Greifswald. The writer was standing by the Brenner Pass near a curve where a railway train was approaching him, the line being backed by a wall of mountain. On the engine giving a short whistle, an echo was heard, the pitch of which was at least half a tone lower than the original sound.

THE American Museum of Natural History, New York, as we learn from a note recently published by Mr. J. A. Allen, has recently obtained a specimen of the head of the wood-bison

(*Bison americanus athabascae*), which is still in existence in the forests near Great Slave Lake. Compared with the bison of the plains (now extinct in a wild state) the woodland bison is stated to be rather larger than the former, and to have the bases of the horn-cores relatively thicker. In 1894 the herd of wood-bisons in the Great Slave Lake district was estimated to be some hundreds in number, but in 1899 it was reduced to about fifty. A very few years more will probably witness the complete extinction of this animal.

AT a recent meeting of the Geographical Society of France, the well-known naturalist, M. Grandidier, the author of the great work upon the natural history of Madagascar, gave an account of his last expedition to that Island, in 1898–99. M. Grandidier landed at Tuléar, on the south-western coast of the Island, and thence made an adventurous journey through the interior to Fianarantsoa, in the Betsileo country; in the south-eastern district. M. Grandidier on his way visited the well-known deposits of Ambolisatra, about 35 kilometres north of Tuléar, where numberless fragments of *Aepyornis*, and almost entire skeletons of the small Madagascar hippopotamus, besides remains of many lemurs of gigantic size and other extinct animals were obtained. From Fianarantsoa, M. Grandidier proceeded north through a well-known country to Antananarivo, the capital of the Island.

As in the case of other larger mammals, the process of dividing the giraffe (*Giraffa camelopardalis*) into "sub-species" is now proceeding apace. Mr. de Winton (*P.Z.S.* 1897, p. 273) first showed, on good grounds, that the giraffe of South Africa was, in certain points of structure, different from the giraffe of the Sahara and Nubia, and proposed to call the former *Giraffa capensis*, leaving the old name *Giraffa camelopardalis* for the northern form. Since then, Mr. O. Thomas (*P.Z.S.* 1898, p. 40) has separated the giraffe of Upper Nigeria from the northern form under the title *Giraffa camelopardalis peralta*. Still more recently, Herr Matschie, of Berlin (*Sitzb. ges. Nat. Fr.* Berlin, 1898, p. 75), has added two new names to the list of giraffes, and called them after their discoverers, *G. tippelskirchi* and *G. schillingsi*, the former being from German East Africa, and the latter from British East Africa. It is curious that these two closely adjoining districts should not agree even in having the same form of giraffe !

THE *Quart. Journ. Micr. Science* for April contains an account by Monsieur P. Bouvier of the results of his examination of the specimens of the primitive Arthropods, commonly known as *Peripatus*, in the collection of the British Museum. The author, who adopts the generic divisions proposed by Mr. Pocock, names one new Andean form after the Director of the Museum, and shows that, with the exception of one from the Congo and a second from Sumatra, all the representatives of the typical genus *Peripatus* are American. To the same journal Mr. E. Warren communicates a paper on the individual differences exhibited by one of the water-fleas (*Daphnia magna*) in its power of withstanding the introduction of salt into the water in which it lives. The physiological condition of the individual is found to have a great effect on its salt-resisting powers.

IN the last issue of the *Zeitschr. Wiss. Zool.*, Dr. R. Gast relates the life-history of a rotifer of the genus *Apsilus*, specimens of which were recently found in an aquarium at Leipzig. This paper is followed by one on the development of a sponge of the group Sycones by Dr. O. Maas, which is worthy of special notice on account of the beauty of the illustrations.

MENTION has already been made in these columns of the description in the *Notes* from the Leyden Museum of the crustaceans collected during the Dutch Expedition to Central Borneo. In the March issue of the same serial this is followed by an account of the birds, which have been worked out by Dr. Büttikofer; Although expectations were entertained that many new forms would be obtained, out of 269 species collected all were previously known, and only two were new to Borneo.

THE April number of the *Journal of the Quekett Microscopical Club* contains the description, by Mr. J. G. Waller, of a new marine British sponge, obtained some twenty years ago at Torbay, for which the name *Raphiodesma affinis* is suggested. Another addition to the British fauna is a new species of Hymenoptera (*Prosopis palustris*), from Wicken Fen, Cambridgeshire, described by Mr. R. C. L. Perkins, in the *Entomologist's Monthly Magazine* for March. This discovery should strengthen naturalists in their opposition to the proposed draining of the fen in question.

MANY strange objects are worn by savage peoples, and for various reasons, also, as with us, rarity usually enhances value. In the Pelew Islands the rubbed-down first vertebra (atlas) of the dugong is worn as a bracelet by the more important men, for it is not often that the vertebra in question is large enough to be so worn. The "klilt," as it is called, has recently been fully described and figured by Dr. O. Finsch (*Globus*, lxxvii. 1900, p. 153). In the Timor Group a wooden imitation is employed; but in Timorlaut the second vertebra (axis) of the dugong is employed; but, although the dugong is greatly hunted in Torres Straits and in South-eastern New Guinea, no ornaments are made from its bones or tusks.

DR. HERMANN MEYER gives an account of a second journey to explore the head waters of the Xingu, in the *Verhandlungen* of the Berlin Gesellschaft für Erdkunde. The route taken was from Cuyaba, reached by ascending the Parana-Paraguay from Buenos Ayres, over the watershed and down the Ronuro to its junction with the main stream, and back to Cuyaba up the course of the Kulischu; practically the same as the former journey of 1896-97, except that the Ronuro was followed throughout its length instead of the Jatoba, a tributary joining it in its lower course. Dr. Meyer concluded that later expeditions will avoid the Ronuro; the Kulischu gives the best access to the region, an exploration of which as far as the Paranayuba would give valuable scientific results.

THE new number of the *Mittheilungen von Forschungsreisenden und Gelehrten aus den deutschen Schützgebieten* contains some interesting papers from the German East African region. Captain Kannenberg gives the first part of an account of a journey through the Marénga Makáli region, with a map. The pendulum expedition under Dr. Fulleborn and Lieut. Glanning reports progress. A summary of the results of the geological expedition in the region north of Lake Nyassa under Dr. Danz is given, and Lieut. Baumstark contributes a paper on the Warangi.

DR. H. NAGAOKA has contributed a valuable paper on the elastic constants of rocks and the velocity of seismic waves to the *Publications* of the Japanese Earthquake Investigation Committee (No. 4 in Foreign Languages). His experiments were made on about eighty specimens of different rocks, cut into prisms 15 cm. long and nearly 1 cm. square in section. They showed at once that Hooke's law does not hold even for very small flexure and torsion, the deviation being prominent in certain specimens of sandstone, and more marked in torsion than in flexure experiments. On releasing the rocks from stress, the return to the original state is extremely small. The elastic constants of archæan and palæozoic rocks (whether of igneous origin or otherwise) are far higher than those of cainozoic rocks, though the velocity of elastic waves in them is not higher in the same proportion. So far as the experiments go, the elastic constants increase more rapidly than the density, so that the velocity

must be greater in the interior than at the surface of the earth's crust.

WE have received, the Twenty-fourth Annual Report of the Geological and Natural History Survey of Minnesota for the years 1895-98; a report which is stated by Mr. N. H. Winchell, the State Geologist, to be his final one. As he remarks: "It ought not to be supposed that by the closing of active work by the present survey, and the publication of its final report, the geology of the State is a finished thing. Geology is a progressive science, and requires continual work." Other States have had surveys which have been hurried to "completion," and have naturally had to enter upon re-surveys, more careful and elaborate. Our own Geological Survey has experienced this as much as any of those abroad; where impatience to see the work "completed" and smallness of revenue have hampered and retarded real progress. The report before us contains a synopsis of the field-work done in Minnesota since 1894, and a useful alphabetical index to the entire series of annual reports of the Survey. Mr. Winchell also notes some of the more important economic and scientific researches which should be carried on in a future survey of the State.

IN the "Palæontologia Indica" for 1899, there is a description of the Cambrian fauna of the Eastern Salt-range, by Dr. K. Redlich, who has supplemented the work of Waagen with more detailed information. A new genus, *Hoeferia*, is now established for the specimens previously referred to *Olenellus*. Among other fossils described are *Hyolithes*, *Lingulella* and *Pseudotheca*. The name *Cylindrites* is applied to "long cylinders, which are often arranged in a fan-shaped aggregate," and appear to be worm-tracks; but it may be pointed out that the name was long ago applied to a genus of Gasteropods. None of the Cambrian fossils from the Salt-range can, in the author's opinion, be referred to a later horizon than the *Paradoxides*-zone. Dr. F. Noetling contributes notes on the morphology of the Pelecypoda, dealing with the hinge of some Miocene and recent bivalves. He endeavours to show that the shape and the delicate and minute variations in the shells can to some degree be expressed better by figures than by words. Dr. C. Diener describes the Anthracolithic fossils of Kashmir and Spiti. In studying the collections made by the Geological Survey of India, he came to the conclusion that fossils both of Permian and Carboniferous ages were included in the series; and he uses the term Anthracolithic as a convenient one for a Permo-Carboniferous group, which appears to be intimately connected stratigraphically and palæontologically. Among the specimens described, the presence of many European types of Carboniferous Brachiopoda is noted, and there are also affinities with the Australian Carboniferous fauna.

IN the form of "Appendix No. 2" for 1900 to the *Kew Bulletin*, we have the usual list of new species of plants brought into cultivation for the first time during last year, or re-introduced after having been lost from cultivation.

THREE of the photographs in natural colours, taken by Mr. H. J. Mackinder in his journey to the summit of Mount Kenya, are reproduced by a three-colour process in the May number of the *Geographical Journal*. Colour photography has thus been brought into the service of geographical exploration, and we may expect to see further developments of its use.

THE May number of the *Journal* of the Chemical Society, which now appears with a regularity worthy of emulation by the publications of other scientific societies, contains Sir Henry Roscoe's memorial lecture on Bunsen, accompanied by a photogravure of the lamented chemist, and Prof. Thorpe's presidential address on some characteristics of the study and progress of chemistry in Great Britain during the present century.

THE current number of *The Builder* (May 5) contains reproductions of Mr. Aston Webb's drawings of the proposed buildings to be erected in the Imperial Institute Road, South Kensington, to accommodate the physics and chemistry departments of the Royal College of Science. The original drawings are on view at the Royal Academy.

THE "Statesman's Year-book," edited by Dr. J. Scott Keltie, with the assistance of Mr. I. P. A. Renwick (Macmillan), has been accepted as a trustworthy authority upon all matters of political geography for so many years, that people familiar with its pages, and therefore conscious of the extent and accuracy of the information contained in them, regard it as one of the few essential annuals. The volume for 1900, which has just been published, is larger than any previous edition, and the numerous rearrangements of territories which were made last year have necessitated many changes in the text, several of the sections having been almost rewritten. Four specially prepared coloured maps are included, dealing with (1) the partition of North-east Africa ; (2) the reorganisation of British Nigeria and the French West African territories ; (3) the political partition of the Pacific ; (4) the final arrangement of the boundary between British Guiana and Venezula. The "Year-book" is thus an epitome of recent geographical events as well as a manual of statistical and historical information concerning the states of the world. So long as the volume is kept so completely up to date as it is at present, it is not likely to be superseded.

IN a short note in the current number of the *Berichte*, Dr. Marckwald discusses some peculiarities shown by picric acid and its solutions, in the light of the ionic hypothesis. Picric acid, as usually obtained, has an intense yellow colour, but on recrystallising from strong hydrochloric acid it becomes nearly colourless. If this white crystalline mass is sucked nearly dry at the filter pump and washed with a little water to remove the adhering hydrochloric acid, the yellow colour at once returns. The mother liquor, which at first has only a pale yellow colour, also becomes more intensely coloured as water is added. Dr. Marckwald shows that if it be assumed that picric acid is itself colourless, but that the ions, $C_6H_2(NO_2)_3O$, are coloured, all these somewhat perplexing phenomena find an immediate explanation in terms of the theory of electrolytic dissociation.

THE confirmation of the relations deducible by thermodynamics as existing between the freezing-point and vapour pressures of a very dilute solution, although of considerable importance for the electrolytic theory of solution, presents great experimental difficulties, especially as regards the vapour pressure determinations. An ingenious method attacking this problem is described in the current *Zeitschrift für physikalische Chemie*, by Dr. R. Gahl. A measured volume of air is drawn through the solution, such as hydrochloric acid, and this is passed through pure water, the change of electrical conductivity of which is measured. The number of cases in which such a method can be applied is obviously restricted, but the accuracy attainable appears to be of the order of ·001 mm. of mercury.

THE additions to the Zoological Society's Gardens during the past week include a Grys-bok (*Raphiceros melanotis*) from South Africa, a Yellow-whiskered Lemur (*Lemur xanthomystax*) from Madagascar, presented by Mr. J. E. Matcham ; a Violet-necked Lory (*Eos riciniata*) from Molluccas, presented by Mr. H. R. Filliner ; two Australian Rails (*Rallus pectoralis*) from New Holland, presented by Mr. C. J. Fox ; a Common Boa (*Boa constrictor*) from South America, an Egyptian Eryx (*Eryx jaculus*) from Egypt, presented by Mr. C. W. Lilley ; two Eyed Lizards (*Lacerta ocellata*), European, presented respectively by Miss Robinson and Miss Ash ; two Edible Frogs (*Rana esculenta*) from Biskra, presented by the Hon. Mrs. A. Cadogan ; a

Crowned Lemur (*Lemur coronatus*), a Black Lemur (*Lemur macaco*), two Blackish Sternotheres (*Sternothoerus nigricans*), a Radiated Tortoise (*Testudo radiata*) from Madagascar, a Slender Loris (*Loris gracilis*) from Ceylon, two Amherst's Pheasants (*Thaumalea picta*, ♂ ♀), ten Reeve's Terrapins (*Damonia reevesi*), a Three-banded Terrapin (*Cyclemmys trifasciata*) from China, a Grooved Tortoise (*Testudo calcarata*) from Khartoum, too Roofed Terrapins (*Kachuga tectum*), a Hamilton's Terrapin (*Damonia hamiltoni*) from India, a Derbian Sternothere (*Sternothoerus derbianus*), two Black Sternotheres (*Sternotherus niger*) from West Africa, three Chequered Elaps (*Elaps leminiscatus*) from South America, a Glass Snake (*Ophiosaurus apus*), European ; six Kentucky Blind Fish (*Amblyopsis speloea*) from Kentucky, deposited ; a Brazilian Tapir (*Tapirus americanus*, ♂) from South America, a Cape Hunting Dog (*Lycaon pictus*, ♀) from South Africa, two Siamese Pheasants (*Euplocamus proelatus*, ♂ ♀) from Siam, two Rufous-tailed Pheasants (*Euplocamus erythrophthalmus*, ♂ ♀) from Malacca, purchased ; a Crowned Lemur (*Lemur coronatus*), an English Wild Cow (*Bos taurus*), born in the Gardens.

OUR ASTRONOMICAL COLUMN.

COMET GIACOBINI (1900 a).—This comet has been in an unfavourable position for observation during the past few weeks, but is now rapidly leaving the sun, and may be searched for in the early morning. The following ephemeris is an abridgment from one given by Herr A. Berberich, of Berlin, in the *Astronomische Nachrichten* (Bd. 152, No. 3636) :—

Ephemeris for 12h. Berlin Mean Time.

1900.		R.A.		Decl.
		h. m. s.		° '
May 21	...	1 17 22	...	+ 24 21·8
22	...	16 27	...	24 41·4
23	...	15 30	...	25 1·2
24	...	14 31	...	25 21·4
25	...	13 29	...	25 41·8
26	...	12 26	...	26 2·6
27	...	11 21	...	26 23·6
28	...	10 13	...	26 44·9
29	...	9 3	...	27 6·5
30	...	7 50	...	27 28·4
31	1	6 34	...	+ 27 50·6

At present the comet is moving slowly in a north-westerly direction through the constellation Pisces, almost in a line between β Arietes and α Andromedæ.

COLOUR SCREENS FOR REFRACTING TELESCOPES.—The *Astronomische Nachrichten* (Bd. 152, No. 3636) contains a description of some experiments undertaken by Messrs. T. J. J. See and G. H. Peters, at the United States Naval Observatory, to determine the utility of viewing celestial objects through variously coloured screens. It was thought that if a suitable screen was chosen which would cut off the violet light of the secondary spectrum shown by the lens, that a considerable improvement of the definition might be expected, and after trial of several types of light filter, several were found which did materially improve the seeing. The screen specially recommended consists of a solution of picric acid and chloride of copper in alcohol. This is applied in a small cell made to fit as a cap outside the eyepiece of the telescope. It is thought that the method may improve meridian work by furnishing better defined star-discs, and also planetary micrometer measurements on account of the diminution of irradiation.

PHOTOMETRIC REVISION OF HARVARD PHOTOMETRY.—The Harvard Photometry, showing the brightnesses of stars north of declination −30°, and of the sixth magnitude or brighter, was compiled from observations made during the period 1879-82. In 1891, on the return of the photometer to Cambridge from Peru, it was decided to redetermine the magnitudes of these stars, and by the end of 1894 the work was almost completed. Nearly all the observations were made by Prof. E. C. Pickering, the Director of the Observatory of Harvard College, and the results of the revision now form Part i. of the last issue of the *Annals of Harvard College Observatory*, vol. xliv.

FITZ GERALD'S "HIGHEST ANDES."[1]

IN the book entitled "The Highest Andes," Mr. E. A. Fitz Gerald relates the experiences of himself and his party upon the journey which he made in 1896-97 in the neighbourhood of Aconcagua, the highest mountain at present known in South America, which it was his aim to map and to ascend. He describes in considerable detail the various operations of the expedition, and recounts with rare frankness the sensations of himself and of his assistants at low atmospheric pressures. Various other matters of considerable public interest are introduced incidentally in his volume, such as the Trans-Andine Railway and the Boundary dispute between Chili and the Argentine Republic; but the attention of the reader will be mainly engrossed by his history of the attacks upon the two great mountains Aconcagua and Tupungato, neither of which was conquered easily.

Although it is visible from Valparaiso, Aconcagua can scarcely be said to have been known at the beginning of the nineteenth century. Humboldt was certainly unacquainted

Engineer, got up and said of Aconcagua that "he believed it to be little less than 15,000 feet high. Admiral Fitzroy had described it as being higher than any of the Himalayan peaks; but he must have been mistaken in his calculations, no doubt in consequence of the difficulty in getting a suitable base for a trigonometrical measurement. He (Mr. Miers) had often seen it void of snow, and as the snow-line in that latitude is, about 15,000 feet, *it is manifest that the mountain cannot exceed that height.*" Though Sir Clements Markham (the present President of the Royal Geographical Society) was at the meeting, it does not appear that either he or any one else entered a protest against this startling statement (see *Proc.* R. Geog. Soc., December 9, 1872, pp. 66-7). Subsequently, Aconcagua rose to a height exceeding 24,000 feet in the pages of the *Daily Chronicle* (January 18, 1897), and it has now, according to Mr. Fitz Gerald, dropped to 23,080 feet, or to almost exactly the height assigned to it by Admiral Fitzroy. This appears to be the greatest elevation that any one has hitherto reached upon a mountain.

Mr. Fitz Gerald's Expedition sailed from Southampton on

FIG. 1.—Looking down Horcones Valley from glacier.

with its name when he was travelling in Peru. He said many years afterwards that, at that time, Chimborazo was everywhere accounted to be the loftiest mountain in the world. But in his "Aspects of Nature," published in England in 1849, he knew differently, and referred to the Great Andes of Peru and Bolivia which were brought to light by Mr. Pentland; and to Aconcagua, which had been found by the officers of the *Adventure* and *Beagle* on Fitzroy's expedition to be between 23,000 and 24,000 feet in elevation. Since then the mountain has had its ups and downs; or, to employ the language of the geologist, it has had its periods of elevation and subsidence. It got to its lowest level about twenty-seven years ago at a meeting of the Royal Geographical Society. After the reading of a paper by Mr. R. Crawford, C.E., upon a projected railway route across the Andes, Mr. J. W. Miers, another Civil

October 15, 1896; left Buenos Aires November 29; and on December 7 arrived at Punta de las Vacas (7858 feet), the terminal station in Argentina of the Trans-Andine Railway.[1] This terminus is only a little more than twenty miles to the south-east of the summit of Aconcagua. No other mountain in the world of anything like its magnitude is approached so closely by railway.[2] An abortive attempt to get to it was first of all made *via* the Vacas Valley, which runs a little west of north from the Terminus and leads to the eastern side of the mountain; and it was subsequently found that the true way towards the summit was by the Horcones Valley, the upper part of which lies to the *west* of the main peak. After some pre-

[1] This line is intended to connect Buenos Aires and Valparaiso. Its construction has been suspended for several years, but it has been quite recently stated that progress will shortly be resumed. About 44 miles remain to be made.

[2] The railway which is being constructed towards Chamonix terminates at present at the village le Fayet, which is less than *ten* miles distant from the summit of Mont Blanc. The summit of Mont Blanc is 13,875 feet above le Fayet, and that of Aconcagua is 16,222 feet above Punta de las Vacas—

[1] "The Highest Andes; a Record of the First Ascent of Aconcagua and Tupungato in Argentina, and the Exploration of the Surrounding Valleys." By E. A. Fitz Gerald. 8vo, pp. 390. (London: Methuen and Co., 1899.)

liminary exploration, a camp was established at the head of this valley, at a height of about 14,000 feet. "The lack of pasturage," it is said, "made it impossible to take the mules any farther," and thenceforward all transport had to effected by men.

Besides Mr. Fitz Gerald, the Expedition at this time consisted of Messrs. Vines, de Trafford and Gosse ; Zurbriggen, the brothers Joseph and Louis Pollinger, Lanti and Weibel. Matthias Zurbriggen, who was born in Switzerland and lives in Italy, is termed guide, and the rest of the men are called porters ; although the two Pollingers and Lochmatter are actually guides, and amongst the best of the younger ones of the Zermatt district. Lanti, who is also called a porter, appears to have been a miner. Mr. Lightbody, an engineer of the Trans-Andine Railway, joined the party at a later date.

On the first day (December 23), Fitz Gerald and Zurbriggen, with four porters and twelve horses or mules, started from the mouth of the Horcones Valley (8948 feet), and went to the spot at its head that has been already mentioned, which was about 14,000 feet above the sea ; and, leaving the animals

difficulty I had in breathing, and partly on account of the dreadful snoring of the men. They would begin breathing heavily, and continue on in an ascending scale till they almost choked. This would usually wake them up, and they were quiet for ten minutes or so, till gradually the whole performance recommenced" (pp. 55-6).

On the following day (December 25) they continued the ascent ; and, although the distance that they mounted was small, the effects became more marked. Mr. Fitz Gerald says of himself and also of Zurbriggen : "We were feeling distinctly weak about the knees, and were obliged to pause every dozen steps or so to catch our breath, and frequently we sat down for about ten minutes to recover" (p. 56). On the next night they encamped on the desired spot, which is said to have been 18,700 feet above the sea. During the day, Zurbriggen advanced (according to his estimate, 2000 feet above the camp), and returned "late in the evening, completely exhausted." On the 27th Mr. Fitz Gerald and the rest retreated to 12,000 feet in the Horcones Valley, in doing which, it seems to me, they made a

FIG. 2.—Saddle on which the 18,700 ft. camp was situated.

there, some of the party pushed on, with the view of arriving at a depression upon the ridge which leads from the summit towards the north-west ; but when an altitude of 16,000 feet is supposed to have been reached, a halt was called on account of the lateness of the hour. "Being much fatigued, we decided not to pitch our tent, but simply to crawl into our sleeping-bags. No one had the energy even to make for himself a smooth place. . . . During the night, one of my Swiss porters, a tall, powerfully-built man, Lochmatter by name, fell ill. He suffered terribly from nausea and faintness." Next day they progressed upwards, but still did not reach the spot for which they were aiming, and passed the night at some elevation that is not mentioned. It was now Mr. Fitz Gerald's turn to feel the effects of diminution in atmospheric pressure.

"I had suffered acutely," he says, "during the afternoon from nausea, and from inability to catch my breath, my throat having become dry from continual breathing through my mouth. . . . I was unable to sleep at all, partly because of the

mistake, and sacrificed some of the advantages which had been gained by considerable labour.

On December 30 they re-started, reached the 18,700 feet camp at the end of the day, and left at 5.45 the next morning with the view of reaching the summit. "At that time we little knew what lay before us ; the summit looked so very near that we even talked of five or six hours as a possible time in which to reach it. We set out towards our peak over the loose, crumbling rocks that covered the north-west face ; the steepness was too great for a direct line of march, and we were obliged to twist and zigzag."

"I noticed Zurbriggen was going very fast ; I was obliged to call to him several times, and ask him to wait for me, as I did not wish to exhaust myself by pressing the pace so early. I was surprised at his hurrying in this way, as it is generally Zurbriggen who urges me to go slowly at first. However, I soon discovered the reason for this ; he was suffering bitterly from cold. Seeing that his face was very white, I asked him if he felt quite well. He answered that he felt perfectly well, but that he was so cold he had no sensation whatever left in his feet; for a few moments he tried dancing about, and kicking his feet against the rocks, to get back his circulation. I began to get alarmed, for frozen feet are one of the greatest dangers one has to contend against in Alpine climbing. The porters who had been lagging behind now came up to us; I at once told Zurbriggen to take his boots off, and we all set to work to rub his feet. To my horror I discovered that the circulation had practically stopped. We continued working hard upon him, but he said that he felt nothing. We took off his stockings, and tried rubbing first with snow and then with brandy; we were getting more and more alarmed, and were even beginning to fear that the case might be hopeless, and might even necessitate amputation. At last we observed that his face was becoming pallid, and slowly and gradually he began to feel a little pain. We hailed this sign with joy, for it meant, of course, that vitality was returning to the injured parts, and we renewed our efforts; the pain now came on more and more severely; he writhed and shrieked and begged us to stop, as he was well-nigh maddened with suffering. Knowing, however, that this treatment was the one hope for him, we continued to rub, in spite of his cries, literally holding him down, for the pain was getting so great that he could no longer control himself, and tried to fight us off. The sun now rose over the brow of the mountain, and the air became slightly warm; I gave him a strong dose of brandy, and after a great deal of trouble induced him to stand up. We slipped on his boots without lacing them, and supporting him between two of us, we began slowly to get him down the mountain side. At intervals we stopped to repeat the rubbing operation, he expostulating with us vainly the while. After about an hour and a half, we succeeded in getting him back to our tent, where he threw himself down, and begged to be allowed to go to sleep. We would not permit this, however, and taking off his boots again we continued the rubbing operations, during which he shouted in agony, cursing us volubly in some seven different languages. We then prepared some very hot soup, and made him drink it, wrapping him up warmly in all the blankets we could find and letting him sleep in the sun. In the afternoon he seemed quite right again, and was able to walk about a little " (pp. 61–2).

This episode brought that day's attempt to an end, but the next morning (January 1) they started again at 8 a.m., with temperature at 26° F., passed the place where they had turned back on December 31, and then encountered great and steep slopes of loose, rolling stones, which, so far as the mountain itself was concerned, seem to have formed the greatest difficulty on the ascent. "The first few steps we took caused us to pause and look at one another with dismay. Every step we made, we slipped back, sometimes the whole way, sometimes more. . . . We continued plodding on for some time, our breath getting shorter and shorter as we struggled and fought with the rolling stones in our desperate attempts not to lose the steps we gained. . . . There was nothing to fix our attention upon except the terrible, loose, round stones, that kept rolling, rolling as if to engulf us." Now another one became ill. "Louis Pollinger" (who is an unusually sturdy and powerful young fellow) "was turning a sickly, greenish hue. All the colour had left his lips, and he began to complain of sickness and dizziness." They went on until 2.15 p.m., and then turned back. "Zurbriggen, I think, could have gone a little farther, but even he admitted that he did not think he would be capable of reaching the summit. . . . The temperature had now dropped to 17° F., and the sun gave us no warmth to speak of. Coming down was almost worse than going up. Fatigued as we were, and chilled and numb to the bone, we constantly fell down, and it was four o'clock before we reached our encampment. . . . We were all of us suffering from splitting headaches."

Although Mr. Fitz Gerald speaks frequently of heat and cold, he does not often quote actual temperatures; but at this point he remarks that the temperature fell to 5° F. during the night, that the maximum in the sun had only been 47° F. during the previous three days, and that it had barely reached 29° F. in the shade. Though the cold which was experienced was not at all lower than might have been expected, they found it trying. "The cold at this altitude seems absolutely unendurable after sunset. I have seen the men actually sit down and cry like children, so discouraged were they by this intense

cold " (p. 57); and he says, truly, at another place, that "with the barometer standing at fifteen inches, the rarefied atmosphere lowers all the vital organs to such an extent that 20° of frost feels more like 60° " below freezing-point (p. 63). There were four of them in their miserable little tent, packed so close that each time one turned over he was obliged to wake the rest." "A terrible and stunning depression had taken hold upon us all, and none of us cared even to speak. At times I felt almost as if I should go out of my mind. . . . All ambition to accomplish anything had left us, and our one desire was to get down to our lower camp, and breathe once more like human beings" (p. 67); and so down they went, this time to Puente del Inca, 8948 feet, at the mouth of the Horcones Valley, and waited there a week.

On January 9 they started again, passed that night half-way up the Horcones Valley, and on the next day went up to the 18,700 feet camp, ascending from 14,000 feet at the rate of 854 feet per hour! "We all seemed so well that I thought it better not to make an attempt on the mountain next day, but to see what a few days of rest and good food would do for us. My hope was that the system would accustom itself to the rarefied air." The minimum of that night was 1° F., which is the lowest temperature recorded in the volume. At 9 a.m. on January 12, Mr. Fitz Gerald set out once more for the summit, accompanied by Zurbriggen and Joseph Pollinger. "For my own part I knew, after the first quarter of an hour, that the attempt would be fruitless. However, I pushed along, hoping against hope that by some chance I might feel better as we went on. I had barely reached 20,000 feet, when I was obliged to throw myself on the ground, overcome by acute pains and nausea," and he returned to the tent, while Zurbriggen pushed on alone. He did not, however, reach the summit; and, when he was returning, was watched through a field-glass.

"He was apparently quite exhausted; he could only take a few steps at a time, and then seemed to stumble forward helplessly. We watched him thus slowly descend for about an hour and a half; first he sat down for four or five minutes, then he slowly plodded onward again. At last he reached a large patch of snow, where, by sliding, he was able to make better time. He did not reach the tent till after sunset, and then he was speechless with thirst and fatigue" (p. 78).

On January 13, another attempt gave a similar result; but at night preparations were made for a renewed assault on the morrow; and on the 14th, Zurbriggen, Joseph Pollinger, Lanti and Mr. Fitz Gerald started at 7 a.m., "all in excellent spirits —so far as it is possible to be cheerful at 19,000 feet." Things went well until 12.30, when they had reached an elevation which was estimated to be about 22,000 feet, and then Mr. Fitz Gerald collapsed. It is to the credit of the head of the Expedition that he writes so frankly, and one cannot but regret that his perseverance did not meet with success. This is his own description :

"I got up, and tried once more to go on, but I was only able to advance from two to three steps at a time, and then I had to stop, panting for breath, my struggles alternating with violent fits of nausea. At times I would fall down, and each time had greater difficulty in rising; black specks swam across my sight; I was like one walking in a dream, so dizzy, and sick that the whole mountain seemed whirling round with me. The time went on; it was growing late, and I had now got into such a helpless condition that I was no longer able to raise myself, but had to call on Lanti to help me. . . . Lanti was in good condition, and could, I feel sure, have reached the summit. He was one of the strongest men we had with us. For a long time past he had been begging me to turn back, assuring me that our progress was so slow, that even should I keep it up I could not reach the top before sunset. I was right under the great wall of the peak, and not more than a few hundred yards from the great couloir that leads up between the two summits. I do not know the exact height of this spot, but I judge it to be about a thousand feet below the top. Here I gave up the fight and started to go down.

"I shall never forget the descent that followed. I was so weak that my legs seemed to fold up under me at every step, and I kept falling forward and cutting myself on the shattered stones that covered the sides of the mountain. I do not know how long I crawled in this miserable plight, steering for a big patch of snow that lay in a sheltered spot, but I should imagine that it,was about an hour and a half. On reaching the snow I lay down, and finally rolled down a great portion of the mountain side. As I got lower my strength revived, and the nausea that I had been suffering from so acutely disappeared, leaving me with a splitting headache. Soon after five o'clock I reached our tent. My headache was now so bad that it was with great difficulty I could see at all.

"Zurbriggen arrived at the tent about an hour and a half later. He had succeeded in gaining the summit, and had planted an ice-axe there; but he was so weak and tired that he could scarcely talk, and lay almost stupefied by fatigue. Though naturally and justifiably elated by his triumph, at that moment he did not seem to care what happened to him. At night, in fact, all hope and ambition seemed to depart, after four days spent at this height, and that night we got little sleep, every one making extraordinary noises during his short snatches of unconsciousness—struggling, panting, and choking for breath, until at last obliged to wake up and moisten his throat with a drop of water " (pp, 82-3).

affected by the diminution in the atmospheric pressure which they experienced, and they were sometimes rendered almost incapable. Upon the map, Tupungato is credited with a height of 21,550 feet, but I have not been able to find in the volume the data from which this elevation has been derived. If it has no better foundation than readings of an aneroid barometer, it is probable that the height has been considerably over-estimated. The elevation assigned to Aconcagua is obtained from the railway-levels as far as the terminus at Punta de las Vacas (7858 feet), carried on by levelling and triangulation up the Horcones Valley, and may be considered authoritative. Notwithstanding its great height, the mountain bears little snow in the middle of the summer; and in this respect the observations by Mr. Miers which are quoted at the beginning of this article are supported. Mr. Fitz Gerald, indeed, says that "when Zurbriggen made the ascent of Aconcagua, he went to the summit of the mountain without placing his foot upon snow; the side of the mountain was bare to the top on the north-west slopes" (P 34). The apex of Tupungato was also bare rock. From

FIG. 3.—Seracs of the Horcones Glacier.

Thus, Zurbriggen alone reached the highest point in the world which has hitherto been ascended; and it is not the least curious fact in this interesting journey that he should have done so, for he was not the most nimble of the party, and in appearance and gait is not the one who might have been expected to be the most successful. That he did succeed was proved on the following 13th of February, when Mr. Vines and Lanti again ascended the mountain, and found an ice-axe on the summit and a substantial pyramid of stones which he had built. The cairn might have been erected by any one, but the axe could have been put there only by himself.

The position assigned to Aconcagua on the map which accompanies Mr. Fitz Gerald's volume is long. 69° 59' west of Greenwich, 32° 39' south latitude, and Tupungato is placed about 57 miles to its south. This latter mountain was ascended by Mr. Vines and Zurbriggen on April 12, 1897, but only after three attempts which ended in failure. Upon it, as on Aconcagua, all those who got to considerable elevations were strongly

the absence of great snow-fields and large glaciers in this elevated region, it would appear that the annual snow-fall there is inconsiderable.

Mr. Fitz Gerald's book will give abundant food for reflection to those who think that the loftiest mountains in the world can be scaled, and scaled easily. He confirms the observations of others, that the greatest heights are reached painfully and laboriously, and that there is a pretty constant diminution in pace the higher one ascends. The illustrations in the volume are reproductions of photographs, and out of the forty-five views of scenery which are given, thirty-three are by Mr. Lightbody. The appendix contains notices of the rocks, by Prof. T. G. Bonney; of the reptiles, scorpions and spiders, by Messrs. Boulenger and Pocock; and of the plants, by Mr. Burkill. The collections seem meagre, and nothing except a few rock specimens appears to have been brought from the greatest heights.

EDWARD WHYMPER.

POTTERY AND PLUMBISM.

DR. T. E. THORPE, F.R.S., gave a lecture on Friday evening, May 4, at the Royal Institution, on the results of an experimental inquiry which he has made, at the instance of the Home Office, on the hygienic questions involved in the use of lead compounds in the manufacture of pottery.

After explaining how lead poisoning occurs in connection with pottery manufacture, he described the conditions which a perfect glaze must fulfil, and named the various forms in which lead compounds enter into the composition of the glazing material as ordinarily employed. He pointed out that experience amply demonstrated, both in this country and on the Continent, that "raw" lead is more generally mischievous in its action than "fritted" lead, that is, lead in the form of a complex silicate associated with alumina, lime, &c. This depends on the more ready solubility of the various modifications of raw lead in the animal secretions, and more particularly in the gastric juice. This fact, indeed, is now generally recognised, and in the inquiry which was instituted by the Home Office in 1893, manufacturers whose names deservedly carry authority in the pottery districts strongly urged the substitution of fritted lead for raw lead in all glazes. Unfortunately, however, this recommendation was not enforced. This may have been due, partly at least, to the circumstance that cases of plumbism occurred from time to time in works where fritted lead was exclusively used. The fact is there is fritted lead and fritted lead.

Dr. Thorpe then proceeded to explain the results of a recent inquiry into the conditions which determine the ease with which lead may be dissolved out from a fritt by dilute acids such as are present in gastric juice. In the first place, it was found that, speaking generally, English fritts yielded a far larger amount of lead to solvents than those made in Holland, Belgium, Germany or Sweden. Indeed, some English specimens of fritted lead were found to be hardly less soluble than raw lead, as shown by the following numbers :—

	Lead oxide dissolved, expressed as percentage of total lead oxide present.
Lead silicate, Specimen I.	99·6
„ „ „ II.	99·6
Glaze A, made with lead silicate	99·2
„ B, „ „	99·2
Various forms of { Litharge	100·0
"raw" lead { Red lead	100·0
{ White lead	100·0

Next, the inquiry showed that there was no necessary relation between the amount of lead oxide in a fritt and the extent to which it would yield lead to solvents comparable, as regards their action, with animal secretions. Some of the compounds richest in lead were, in fact, among those least attacked by solvents. This is illustrated by the following series of numbers :—

I. *Solubilities practically the same, amounts of lead oxide in the fritt very different.*

	Percentage of lead oxide in fritt.	Solubility per cent. on fritt.
Dutch fritt	18·0	traces
English fritt, A	40·4	0·2
Belgian fritt	22·4	0·7
English fritt, B	41·3	0·7
„ „ C	52·3	0·4

II. *Solubilities very different, amounts of lead oxide in fritt practically the same.*

English fritt, D	37·9	28·0
„ „ E	36·2	1·4
„ „ F	45·8	10·8
Swedish fritt	44·1	2·1

Further inquiry elicited the fact that the extent to which the fritt gave up lead to the solvent depended upon two conditions :—

(1) The existence of a definite numerical relation between the basic and acidic oxides in the fritt, and
(2) Complete chemical union.

The definite numerical relation thus alluded to may be stated in the following terms :—If the sum of the equivalent percentages of basic oxides, expressed as lead oxide, is not more than double

the sum of the equivalent percentages of acidic oxides, expressed as silica, the solubility of the fritt, as regards lead, is rarely more than 2 per cent. Any increase in this ratio is attended by an increase in the amount of lead dissolved, and the amount of soluble lead increases very rapidly with even a slight increase in the ratio. The following figures serve to illustrate this fact :—

	Percentage of lead oxide.		Solubility per cent. on fritt.		Ratio.
Dutch fritt, No. 1	18·0	...	traces	...	1·34
Belgian fritt, No. 1	21·8	...	„	...	1·44
Dutch fritt, No. 2	19·0	...	1·2	...	1·50
Belgian fritt, No. 2	22·4	...	0·7	...	1·52
Swedish fritt	44·1	...	2·1	...	1·56
English fritt	24·0	...	0·2	...	1·57
„ „	40·4	...	0·2	...	1·68
„ „	24·5	...	0·6	...	1·70
„ „	36·2	...	1·4	...	1·79
„ „	36·4	...	2·3	...	1·87
„ „	45·8	...	10·8	...	2·61
„ „	37·9	...	28·0	...	2·92
„ „	70·4	...	67·3	...	3·26

It was further found that, provided the ratio of acids to bases is below 2, the nature of the basic oxides has little or no effect upon the amount of the lead oxide dissolved. This may be illustrated by the following numbers :—

	Lead oxide.	Alumina.	Lime.	Alkalis.	Solubility per cent. on fritt.
Dutch fritt	19·0	8·1	9·0	4·9	1·2
English fritt	16·2	10·3	8·5	9·2	1·7
Swedish fritt	44·1	5·5	0·9	3·4	2·1

Further evidence of the fact that the insolubility of a complex silicate is determined by the ratio of acids to bases, and is independent of the specific nature of the bases, is afforded by the case of flint glass, which consists essentially of a silicate of alkali united with a silicate of lead. Separately, these silicates are readily attacked by dilute acids. When united, as in flint glass, the compound is only very sparingly soluble. Merely to flux together the ingredients of a fritt, with no regard to its composition as a definite chemical compound, and with no regard to the time or temperature needed to complete the chemical changes, is not the proper way to make a fritt.

In the course of the inquiry it was found that the Continental fritts, which conformed to the above ratio, and were distinguished by their comparative solubility, were very difficult to break up by the action of acids, and yielded only minute portions of soluble matter (much of which, however, consisted of lead) to solvents, whereas the English fritts were, for the most part, very easily decomposed by the same treatment, and gave up the greater part of their lead to solution. This led to the surmise that the Continental fritts consisted, in the main, of comparatively stable chemical compounds, the minute quantity of lead dissolved being due to some lead compound—oxide or silicate—in a state of incomplete chemical union. Experiment showed that this surmise was correct. By treating a fritt, compounded so as to be within the limiting ratio, with dilute acid, by far the greater portion of the soluble or incompletely fixed lead may be removed, and a highly insoluble complex lead silicate is obtained. A fritt, for example, containing upwards of 53 per cent. of oxide of lead, and of which the limiting ratio of acids and bases was about 2, had this ratio lowered to 1·8 and the solubility diminished from 2 per cent. to four-tenths of a per cent., the amount of lead oxide in the product so treated being upwards of 52 per cent.

A number of manufacturers and professional fritt makers, acting in conformity with the suggestions which have been put forward, and in response to the invitation of the Home Secretary to have their glazes tested in the Government Laboratory, are now producing lead fritts having a solubility which is even below the standard provisionally suggested in the Home Office Circular of December last.

Although measures based upon the above facts will no doubt largely minimise the evil of lead poisoning, Dr. Thorpe stated that he was not sufficiently sanguine to suppose that they would altogether stamp out plumbism in the Potteries. It must be clearly understood that complete immunity from lead poisoning can never be obtained so long as lead compounds continue to be used. The true solution is to be found in the more general adoption of leadless glazes. That leadless glazes of a high brilliancy, covering power and durability, and adapted to all kinds of table, domestic and sanitary ware, to china furniture, to tiles, insulators and electric fittings of the most varied kind, are perfectly practicable, was illustrated by reference to the numerous examples of leadless glazed ware which, thanks to the liberality of a number of the manufacturers, were exhibited to the audience. Among them were specimens from Mintons, from the Worcester Royal Porcelain Company, Burgess and Leigh, Barker and Read, Bernard Moore, the Crystal Glaze Company, Hawley Brothers, Defries, and others. Telegraph insulators of Doulton's and Buller's make were exhibited by the Post Office.

Dr. Thorpe stated that leadless glazed ware was now being supplied to a number of the Government Departments and to certain of the London Clubs. He further stated that the London School Board had resolved to insert a clause in all specifications for new works strictly prohibiting the use of any pottery goods involving lead glaze wherever practicable. The fact that the application of leadless glazes has passed beyond the experimental stage is so obvious that the Secretary of State now proposes to relax the Special Rules, issued by the Factory Department, in regard to the pottery industry in the case of factories or processes in which no compounds of lead are used.

Dr. Thorpe concluded by remarking that every intelligent potter must concede that there is an ample field for investigation by modern methods of attack into problems connected even with the first principles of his art. The craft of the potter largely depends upon the intelligent application of scientific principles. Whether, however, modern science enters into it to the extent that might be desired is perhaps open to question.

There is probably no industry in the world, certainly none in England, so conservative in its operations as that of the potter. The best of English earthenware still enjoys, no doubt, the pre-eminence which the skill and aptitude of Wedgwood and his immediate followers imparted to it. The great potter was fully abreast of the physical science of his day, and was quick to test or take advantage of any discovery which seemed to promise to be of service to his art. But perhaps it may be doubted whether the spirit of Wedgwood actuates his successors to the extent that might be desired. It is at least certain that the exercise of this spirit, that is, the intelligent application of simple chemical principles, would years ago have obviated, to a large extent at least, this evil of plumbism among the pottery workers.

APPLICATIONS OF ELECTRICAL SCIENCE.[1]

I FEEL very much honoured by having been placed in the position I now occupy, and by having to deliver this opening address to the Dublin Branch of the Institution of Electrical Engineers. I believe that we are one of the first branches that has developed into the meeting stage of our existence, and may congratulate ourselves on having passed through our larval transformations safely and rapidly, and on our having been the first to emerge into an imago.

The action of the parent Institution in founding these local branches is worthy of our grateful commendation. We are left perfectly free to develop our own life untrammelled by any rules except such as we would ourselves have necessarily chosen to govern our actions. We have the great advantage of being a branch of a most distinguished Institution of world-wide reputation, and that without paying any extra subscription. I hope that we will add to the life and work of that Institution, and thereby promote both our own interests and the welfare of mankind. Papers and discussions here will be taken as delivered to the Institution of Electrical Engineers, and, if of sufficient merit, will be published in its *Proceedings*, thus securing to us a world-wide publication, while at the same time ensuring that Ireland is credited with the work done.

The history of electricity in the nineteenth century is far too large a subject for an occasion like the present one, but certain aspects of this history convey valuable lessons for the future and may well engage our attention in this last year of the century, and may help us to lay the foundations for further advance in the next. The aspect of the history of electricity during the nine-

[1] Inaugural address to the Dublin Section of the Institution of Electrical Engineers, delivered by Prof. G. F. Fitzgerald, F.R.S. Abridged from the *Journal* of the Institution, April.

teenth century to which I desire to direct your attention is an object-lesson of how to apply science to further the well-being of mankind. The history of any applied science might be considered in this aspect, but the history of applied electricity is particularly appropriate for being thus considered, for several reasons. The history is condensed within a few years ; the discoveries of science have followed one another with extraordinary rapidity, and within a few years after the discoveries were made they have been applied to the use of man. It is just a hundred years since Volta discovered how to make continuous electric currents. Within a few years of that discovery their chemical actions were discovered and electric lights produced, both arc and incandescent. Twenty years afterwards the magnetic effect of an electric current was discovered by Œrsted, its mathematical theory evolved by Ampère, and the law of its intensity worked out by Ohm. Some fifteen years afterwards, Faraday discovered how to produce electric currents by magnetism. Immediately after the discovery of the principle of the conservation of energy it was applied to electro-magnetism, and the foundation of our whole system of electro-magnetic measurement was laid. Faraday's belief in the correlation of electricity and light, following lines suggested by Lord Kelvin, was forged into a consistent theory by Clerk Maxwell, and this theory confirmed experimentally by Hertz. Such, in brief, is the scientific history of electro-magnetism during the expiring century, and on this science practically all the applications of electricity depend.

I may pause for an instant to consider where this theory now lands us. The all-pervading ether has been realised as the means of transmitting light, electricity and magnetism, and we are looking forward to its properties explaining chemical actions and gravitation. We are still looking for a theory of its structure which will give a dynamical explanation of its properties. We know how to express these properties by quantities we call electric and magnetic force, whose laws we know, but whose laws we are, as yet, unable to explain by any structure working on dynamical principles. So far as we know, the properties of electric and magnetic force are explicable upon dynamical principles ; so far there is no known necessity for seeking for a dynamical properties in the ether ; so far we may hope to explain electro-magnetism upon the dynamical principles of Newton's laws without invoking any other principles than those of force and inertia, as expounded in these laws. Until, however, a satisfactory theory of the nature of the ether has been actually invented, there will remain some doubt as to the adequacy of these fundamental dynamical laws to explain all its properties. The direction in which it is most probable that an explanation will be found is in the hypothesis that the ether is of the nature of a perfect liquid full of the most energetic motion. We know that a gas consists of separate molecules in intensely energetic irregular motion. I expect that the ether is a perfect liquid in intensely energetic irregular motion : much more rapid than that of any gas : with a rapidity of internal motion comparable with the speed of light : maybe with enough energy in each cubic centimetre to keep hundreds of horse-power going for a year, if only we could get at it. So far as this hypothesis has been worked at there seems nothing impossible about it, but, on the contrary, much possibility in it, and, to my mind, its inherent simplicity confers on it a great probability.

Be that as it may, we now know that in the electric lighting of our cities, in electric tramways and railways, in electric furnaces and electrolytic vats, and in the other innumerable applications of electricity, we are harnessing the all-pervading ether to the chariot of human progress, and using the thunderbolt of Jove to advance the material welfare of mankind.

Having thus shortly considered the progress of electrical science, the history of the *applications* of electricity may be thus summarised. Shortly after Œrsted discovered the magnetic effect of an electric current this discovery was applied to telegraphy, and Faraday's discovery of how to generate electric currents by magnetism was almost immediately applied to the same use. Telegraphy developed rapidly, and many subsequent discoveries were due to the observations made in the practical application of electricity to telegraphy. This has been developing ever since, accumulating knowledge and applying the accumulations to produce more knowledge and more applications, till all this has resulted in the perfection of the multiplex telegraph and the wonders of the telephone and wireless telegraphy. No other department of applied electricity has had such a continuous development, hardly any interval elapsing between discovery and application in its case, while in almost every other case years have elapsed between discoveries and their application. It is especially the object of this address to call attention to the cause of this and to the lessons to be learnt from it.

Within the first decade of the century, electrolysis and the electric light were discovered ; but, except on a small scale in electro-plating, it was reserved for the last quarter of the century to see their application to the general use of mankind. Before her Majesty began to reign, Faraday had discovered how to generate electric currents by magnetic actions ; but, except to generate currents to light a couple of lighthouses, no applications of Faraday's discovery to generate electric currents on a large scale was made till Wilde, Gramme and Siemens worked at it, more than thirty years after its discovery. The application of electric currents to transmit power on a small scale was made in the electric telegraph years before any applications were made on a large scale. Except for a few experiments by Jacobi and others, the transmission of power by electric currents on a large scale is the work of the last twenty—one might almost say of the last ten—years.

. Consider now what are the characteristics of the applications which developed continuously, and what were those of the applications which lay dormant for years. Maybe we can learn from this consideration how to arrange that, in the future, our discoveries may not lie for years dormant.

The most noticeable difference between the applications of electricity that developed and those that lay dormant is that those that developed were useful on a small scale, while those that lay dormant were not useful until developed on a large scale. Electro-plating and telegraphy were useful on quite a small scale. Experiments as to their efficiency could be conducted on the laboratory scale with quite cheap apparatus, and thus they were actually developed.

A recognised authority, who is fond of poking paradoxical fun at professors, has recently stated that "the progress of telegraphy and telephony owes nothing to the abstract scientific man." I do not know exactly what he means by the abstract scientific man; but I do know that telegraphy owes a great deal to Euclid and other pure geometers, to the Greek and Arabian mathematicians who invented our scale of numeration and algebra, to Galileo and Newton who founded dynamics, to Newton and Leibnitz who invented the calculus, to Volta who discovered the galvanic cell, to Œrsted who discovered the magnetic action of currents, to Ampère who found out the laws of their action, to Ohm who discovered the law of the resistance of wires, to Wheatstone, to Faraday, to Lord Kelvin, to Clerk Maxwell, to Hertz. Without the discoveries, inventions and theories of these abstract scientific men, telegraphy as it now is would be impossible.

We have seen that electro-plating and telegraphy were capable of development on a small scale, and were consequently largely developed by laboratory research. The development of dynamos from Faraday's discovery required expensive experiments, and to test their efficiency on a large scale required very expensive experiments indeed. It was not possible to conduct experiments that would be of much practical use on the small scale on which laboratory experiments have to be conducted, on account of the miserable pittance that is at the command of scientific laboratories. The only opportunity of conducting experiments on a large scale is when an inventor can control capital, as, for example, if he himself is in the position of any engineer to some wealthy body whose money he can employ on experiments. Jacobi and others spent a good deal of money, no doubt, on experiments in power distribution by electro-magnetic engines, but their expenditure, though quite considerable as compared with the usual run of laboratory experiments, was as nothing compared with the enormous sums spent by the pioneers of modern electro-magnetic machinery on *their* experiments.

What we have found, then, is that development depended on whether or no people experimented energetically upon how to render each discovery of practical utility ; where experimenting was energetic, development was rapid ; where experimenting was not energetic, development was slow. We have further found that the energy of experimenting depended on the money available ; where little money was required, development was rapid ; but it was slow where large sums of money were required in order to perform valuable experiments.

We may further inquire how it happened that money and time became available for costly experiments. Money is available for laboratory experiments by the beneficence of private and public endowment, and time is available by the devotion of scientific men to the advancement of natural knowledge. These have been available because some few men have had faith in the desirability of knowledge both for its own sake and for the material and moral advantage of mankind. Money has been available in England on a large scale in the past because of the enthusiastic faith of some very few men in the possibilities of scientific discoveries. One of the most remarkable instances of this faith was in the case of the great experiment of laying the Atlantic cable. A few men with strong faith impressed their belief on a few capitalists, and after years of most expensive experimental work they at last brought their great undertaking to a successful issue ; the general body of capitalists meanwhile looking on with amused incredulity. The development of the dynamo depended similarly upon the strong faith of individuals, who spent immense sums of money and much time and energy on the subject because they had faith in its possibilities. It is remarkable how many of the developments of scientific discoveries of the latter years of the century have been due to foreigners or firms with foreign leaders, such as Siemens Brothers. This has been largely due to the fact that foreigners are far in advance of us over here in their faith in the possibility of using scientific discoveries. The rapid advance of the applications of science in the last quarter of this century has been very largely due to the growth of this faith. It has grown to a strong conviction in the ordinary public of America and the Continent, and is growing daily stronger over here, but is still far weaker here than in other parts of the civilised world. The result of this has been that while the germs of many of the greatest inventions have been made within the British Isles, we have not been pioneers in any great advance in the applications of electricity since the development of submarine telegraphy. Possibly another cause has been our obstinate retention of our abominable series—one cannot call it system—of weights and measures. It is with great hopefulness that I see public opinion gradually growing in favour of the metric system.

How does it happen that one of the foremost countries in advancing science has been one of the last to appreciate the possibilities of applied science ? This has been due, partly, no doubt, to our great success as manufacturers and as mere mechanical inventors. No doubt Watt was a truly scientific inventor, and even mere mechanical inventors are appliers of scientific knowledge that was discovered, in the most part, by scientific men centuries ago ; but most of our success as manufacturers has been due to mechanical inventions and to our well-trained and expert artisans, and not to the useful application of recent scientific discoveries. This great success, and the absence of scientific training in our schools, and the want of contact between manufacturing and scientific society, have all contributed to prevent a due appreciation of the value of scientific discovery and experiment as a means of advancing the material wealth of society.

When can we expect the country or generous benefactors to learn that science on a large scale is at the basis of the material prosperity of the country, and that science on a large scale is very expensive? Of what use is 200*l*. a year in making experiments on a commercial scale ? Ten thousand pounds a year would be more like the figure required ; and 10,000*l*. a year could be most profitably spent on experimental work here in Ireland, on the one subject of utilising our bogs. It is most probable that the energy of their combustion could be transmitted to our towns to provide them with light and power ; but the preliminary experiments are far beyond the capabilities of a scientific laboratory.

Then there are the questions of three-wire tramways, leaky telegraph lines, submarine relays, sun engines, of flying machines which Lord Rayleigh considers can be constructed if money enough were forthcoming, and of vacuum tubes as a means of illumination, and of numberless other matters already ripe for application, to say nothing of the innumerable discoveries that have not yet been even suggested as having practical applications.

Besides these industrial laboratories, all our Government departments, such as the army and navy, should have large experimental organisations where any invention that promised success would be developed on the money available. The decision of what to try should not be left to mere officials, however distinguished, but should be referred to independent scientific advisers—persons who were not trammelled by official traditions, but were in touch with scientific advance and enthusiastic believers in it. If the country spent a couple of millions per annum on experimental work of this kind it would bear much fruit, and we should not find ourselves out-shot by semi-barbarous farmers.

Hope is the great incentive to exertion. Without it a nation is dead. Without it we lose all belief in the possibility of improvement, and improvement at once becomes impossible. The history of electrical engineering, the utilisation of the all-pervading ether for the service of man, should strengthen our hope and our belief in the possibility of improvement. For has it not revolutionised society and enabled high and low, rich and poor, to lead better lives, by making life less hard and grimy, and thus improved the well-being of man both materially and, what is far more important, morally as well?

UNIVERSITY AND EDUCATIONAL INTELLIGENCE.

OXFORD.—The following are the principal lectures announced for this term :—Prof. Clifton, practical physics ; Mr. Baynes, elementary electricity and magnetism ; Mr. Jervis-Smith, dynamo and motor machinery, with electrical testing ; Prof. Odling, silicon compounds ; Dr. Fisher, metals and organic chemistry ; Mr. Watts, organic chemistry ; Mr. Marsh, practice of organic chemistry ; Mr. Hartridge, aromatic compounds ; Mr. Vernon Harcourt, subjects of the preliminary examination ; Mr. Elford, the elements treated in the periodic order ; Mendeleef's periodic system, Groups vii. and viii. ; great chemists and their work ; Mr. Walden, synthetical methods in organic chemistry ; Mr. Wilderman, equilibrium and velocity of physical and chemical reactions in heterogeneous systems ; Prof. Miers, the new theories of crystal structure ; Mr. Bowman, the crystallography of optically active substances ; Prof. Sollas, history of the earth ; Mr. Mackinder, the natural regions of the Old World ; Mr. Dickson, the climatic regions of the globe ; Mr. Herbertson, mountain types ; Prof. Weldon, general course of morphology ; variation, inheritance, and natural selection ; Mr. Goodrich, annelids ; Mr. Jenkinson, vertebrate embryology ; Mr. Günther, arthropoda ; Mr. Barclay Thompson, mammalian morphology ; mammalian palæontology ; Prof. Gotch, the central nervous system ; Prof. Gotch and Mr. Ramsden, advanced course of physiology ; Mr. Mann, advanced histology of nervous system ; Mr. Burch, physiological physics ; Mr. Mann, practical histology ; Prof. Vines, elementary course of botany ; Prof. Tylor, early stages of civilisation (arts of subsistence and protection) ; Sir J. Burdon Sanderson, general pathology ; Dr. Ritchie, pathological bacteriology ; Dr. Collier, medical diagnosis ; Mr. Symonds, fractures and dislocations ; Prof. Thomson, vascular and respiratory systems ; Mr. Smith Jerome, medical pharmacology and materia medica ; Prof. Esson, the synthetic geometry of conics ; Prof. Love, hydrostatics and hydronamics ; Prof. Elliot, the theory of functions.

Mr. William Hatchett Jackson, science tutor of Keble College, who has been elected to the post of Radcliffe's librarian, vacant by the resignation of Sir Henry Acland, has entered on his duties. The new Radcliffe Library, erected for the University by the Drapers' Company, is meanwhile approaching completion.

Scholarships in natural science are announced by the following colleges :—Merton and New, July 3 ; Balliol, Christ Church and Trinity, December 4 ; Magdalen, December 11.

It has been decided that diplomas in geography shall be granted by the University ; the details of the scheme have yet to come before Congregation and Convocation.

CAMBRIDGE.—Honorary degrees are to be conferred on the Hon. Edmund Barton, delegate from New South Wales in connection with the Australian Commonwealth Bill, and on H.M. the King of Sweden and Norway.

There are vacancies at the University Tables in the Zoological Stations of Naples and Plymouth. Applicants should write to Prof. Newton before May 24.

It is proposed to affiliate the University of Tasmania. Bachelors of Arts and Bachelors of Science of that University will thereby be entitled to proceed to Cambridge degrees after two years' residence.

The Financial Board estimate that, owing to the loss of fees, &c., consequent on the absence of many members of the University in South Africa, the income of the Chest will next year fall short of the necessary expenditure by 650*l.*

Seventeen additional freshmen were matriculated on May 5.

Mr. Thomas Andrews, F.R.S., has presented to the Chemical Laboratory a valuable echelon spectroscope, for which the special thanks of the University have been ordered.

DR. TUNNICLIFFE has been appointed to the chair of materia medica and pharmacology in King's College, London.

DR. JOHN WYLLIE has been elected to succeed the late Sir Thomas Grainger Stewart in the chair of medicine in the University of Edinburgh.

IN order to enable Essex dairy-farmers, and ladies engaged in dairy-work, to gain an insight into the organisation and practice of the agricultural industries of Denmark, the Essex Technical Institution Committee have made arrangements for a party to visit that country. Visits will be made to a number of schools and other institutions, farms, and manufactories concerned with dairying, and a valuable insight will be obtained into Danish methods. Full particulars of the programme can be obtained from Mr. T. S. Dymond, County Technical Laboratories, Chelmsford.

THE growth of municipal technical schools in England during the ten years which followed the passing of the Technical Institution Act, 1889, formed the subject of an inquiry made by the National Association for the Promotion of Technical and Secondary Education a short time ago. The results showed that a capital sum of 2,340,651*l.* had been spent on technical schools, and that there were 239 such schools (including agricultural and dairy schools and domestic science schools) in existence or in course of establishment. Since the conclusion of the inquiry, technical schools had been erected, or it had been decided to erect them, in several other towns, and the latest report shows that the total amount incurred for 272 schools under municipal and public bodies is now at least 2,643,172*l.*

THE progress of science and education in the United States is largely due to the interest taken in the work of colleges and universities by private benefactors. Scarcely a week passes without affording instances of generous gifts to institutions of this kind, by persons who desire to promote the development of national character and industries. As an example of this public spirit, we have the case of Dr. D. K. Pearson, of Chicago, who, on attaining his eightieth birthday recently, decided to add 525,000 dollars to the 2,000,000 dollars he had previously given to colleges. Then we have the announcement in *Science* that Mr. Andrew Carnegie has promised the trustees of the Carnegie Institute, Pittsburg, Pa., to become responsible for 3,000,000 dollars, the amount estimated as necessary for the proposed extension and enlargement of the building at the entrance of Schenley Park. The new building will be nearly six times as large as the present one. We should be glad to be able to record many similar gifts to institutions devoted to science and education in this country.

ONE of the good effects of the technical education movement during the past ten years is that many secondary schools, such as grammar and endowed schools, which formerly excluded science from their curricula, have had to adapt themselves to modern requirements as a condition of receiving assistance from technical education authorities. The annual report of the National Association for the promotion of Technical and Secondary Education refers to an inquiry undertaken to determine the extent of the changes which have been brought about in this way, both by the establishment of new secondary schools and by the adaptation of existing secondary schools for the purposes of technical education. The facts revealed by the inquiry go to show that in England alone, since 1889, 81 new public secondary schools have been established, while 215 existing schools have been extended mainly for the purposes of science teaching. As regards the schools in the latter category, the extensions to 195 of them have resulted in the addition of 251 physical and chemical laboratories, 77 workshops for manual training, 76 lecture-rooms, and 50 class-rooms. The total sum of money involved by these developments is 764,449*l.* By their capital grants to secondary schools, County Councils have exerted a direct influence in the reorganisation, and have secured a voice in the management and control of the schools. By the Councils' annual maintenance grants, the work of reorganisation has been gradually consolidated, and the permanence of proper management and control has become assured. It is not surprising, therefore, that the latter, as a continuous source of income to secondary schools, have been increasing in number and in value during recent years.

SOCIETIES AND ACADEMIES.

LONDON.

Royal Society, February 1.—"Researches on Modern Explosives: Second Communication." By W. Macnab, F.I.C., and E. Ristori, Assoc. M.Inst.C.E., F.R.A.S. Communicated by Prof. Ramsay, F.R.S.

The object of the experiments was to endeavour to find a means of determining more accurately than has hitherto been done the temperature reached when an explosive is fired in a closed vessel.

A modification of the method developed by Sir W. C. Roberts-Austen was employed. A thin platinum wire was melted by the heat of the explosion, but a thick wire was unaltered. This showed that the temperature reached was above the melting point of platinum, and also that the duration of the maximum temperature was very short. From this it was argued that if rhodium-platinum couples of different diameters, sufficiently thick not to be melted during explosion, were used in a bomb, the deflections of the galvanometer indicated would' vary inversely with the sizes of the wires forming the couples; that in this way data might be obtained from which might be calculated the deflection of an infinitely thin couple, which could be capable of taking up the heat in an infinitely short time, and that this deflection expressed in degrees would represent the actual temperature reached.

Couples formed of wires of pure platinum and platinum alloyed with 10 per cent. of rhodium, varying in diameter from 0·01 to 0·044 of an inch, were employed. Each couple was successively fixed inside the bomb, and on firing the explosive the deflection of a spot of light reflected from the mirror galvanometer was photographically recorded.

These records show the uniformity of the results, and also the time occupied in heating each couple to its maximum, and that the deflections are in inverse order to the thickness of the couple used.

Two series of experiments made with two different explosives —ballistite (composed of 30 per cent. nitroglycerine and 70 per cent. gun-cotton) and gelatinised gun-cotton—were carried out with a number of different couples, and the results expressed as curves show the gradual rise of the deflections as the thickness of the couple diminishes; but all through the gun-cotton curve is below the ballistite curve, thus indicating that the temperature reached during explosion of the gun-cotton is lower than that of the ballistite.

Experiments made with the following explosives showed that the relative temperature can be easily ascertained. Gun-cotton gave the lowest temperature, and in order came cordite, ballistite (composed of 70 per cent. soluble nitro-cotton and 30 per cent. nitroglycerine) and ballistite (composed of 50 per cent. soluble nitro-cotton and 50 per cent. nitroglycerine).

Another series of experiments is in progress for determining the other necessary elements which will be required before the value of these deflections of the galvanometer can be accurately expressed in degrees of temperature.

April 5.—"Über Reihen auf der Convergenzgrenze." Von Emanuel Lasker, Dr. philos. Communicated by Major MacMahon, F.R.S.

Linnean Society, April 19.—Dr. A. Günther, F R.S., President, in the chair.—On behalf of the Hon. Charles Ellis, the President exhibited photographs of a large tree, *Taxodium distichum*, growing at Oaxaca in Mexico, and of another gigantic tree, a native of Cambodia. The circumference of the former, at a height of 3 feet from the ground, was stated to be 143 feet, while the height was estimated to be not more than 100 feet. The native name for this tree is *Sabino*. Mr. Daydon Jackson read an account of it, quoting from Loudon's *Mag. Nat. Hist.* vol. iv. (1831), p. 30, and Humboldt's "Views of Nature," p. 274. The second gigantic tree, which could not be satisfactorily determined from the photograph, had been observed growing on the Makong River, near the celebrated ruins of the great city of Angkorwat in Cambodia.—Messrs. W. B. Hemsley and H. H. W. Pearson read a paper on some collections of high-level plants from Tibet and the Andes. Mr. Hemsley first gave a brief history of the botanical exploration of Tibet, followed by an account of the unpublished collections presented to Kew by Captain Wellby and Lieut. Malcolm, by Captain Deasy and Mr. Arnold Pike, and by Dr. Sven Hedin. These collections were all made at great altitudes in Central and Northern Tibet; few of them below 15,000 feet,

and some of them at 19,000 feet and upwards. The highest point at which flowering plants had been found was 19,200 feet above the level of the sea. The plants recorded by Deasy and Pike at altitudes of 19,000 feet and upwards are :—*Corydalis Hendersoni, Arenaria Stracheyi, Saxifraga parva, Sedum Stracheyi, Saussurea bracteata, Gentiana tenella, G. aquatica,* an unnamed species of *Astragalus,* and an unnamed species of *Oxytropis.* These are the greatest altitudes on record for flowering plants. Deep-rooting perennial herbs having a rosette of leaves close to the ground, with the flowers closely nestled in the centre, are characteristic of these altitudes. The predominating natural orders are :—Compositæ, Leguminosæ, Cruciferæ, Ranunculaceæ and Gramineæ. The Compositæ largely predominate, and the genus *Saussurea* is represented by numerous species. Specimens of about a dozen species were shown to illustrate the great diversity exhibited by this genus in foliage and inflorescence. Liliaceæ and the allied orders were very sparingly represented. Two or three species of onion occur; one of them, *Allium Semenovii,* in great abundance up to 17,000 feet. None of the collections contained any species of orchid.—Mr. H. H. W. Pearson described the Andine flora, with special reference to Sir Martin Conway's small collection of plants brought from Illimani in the Bolivian Andes in 1898. In consequence of the labours of d'Orbigny, Pentland, Meyen, Weddell, Mandon and other botanists, the high-level flora of the mountains of Bolivia is better known than that of any other equally elevated region of the Andes. Weddell's collections form the nucleus of the materials from which the "Chloris Andina "—the classic work on the flora of the High Andes—was prepared. Many collectors have obtained plants in various parts of the Andes at elevations stated to be greater than 17,000 feet. Colonel Hall states that he saw four plants on Chimborazo in 1831 at "nearly 18,000 feet." These were two species of *Draba,* one of which was *D. aretoides,* H. B. K., and two Composites, one being a *Culcitium.* Mr. Whymper and others have thrown some doubt upon the determination of this elevation, and it is probable that it was over-estimated. Out of forty-six species of flowering plants obtained by Sir Martin Conway, seven are from 18,000 feet or above it, two being as high as 18,700 feet. These, the highest Andine plants on record, are *Malvastrum flabellatum,* Wedd., and *Deyeuxia glacialis,* Wedd. Thirty-nine species in this collection were found above 14,000 feet; these belong to thirty-four genera and twenty-one natural orders; fifteen (*i.e.* about three-eighths of the collection) are Compositæ. Of the thirty-four genera, one only —*Blumenbachia*—is endemic to South America. The species, with one exception, are confined to the Andes, eight or nine of them not being found outside Bolivia. In the collection made by Mr. Fitzgerald's expedition in the Aconcagua valleys between 8000 and 14,000 feet, ten genera (*i.e.* one quarter of the whole) are endemic in South America. The contrast between this and the small endemic element in the Conway collection from above 14,000 feet gives additional support to the generalisation that the flora of high levels is more cosmopolitan than that of low levels.—A paper was read by Mr. E. S. Salmon on some mosses from China and Japan.

MANCHESTER.

Literary and Philosophical Society, April 24.—Prof. Horace Lamb, F.R.S., President, in the chair.—The following gentlemen were elected honorary members of the Society :— Prof. James Dewar, F.R.S., London; Prof. J. A. Ewing, F.R.S., Cambridge; Prof. A. R. Forsyth, F.R.S., Cambridge; Prof. James Geikie, F.R.S., Edinburgh; Prof. Ernst H. P. A. Haeckel, Jena; Prof. H. A. Lorentz, Leyden; Mr. Robert Ridgeway, Washington, U.S.A.; and Mr. Beauchamp Tower, London. The following were elected officers of the Society for the session 1900-1 :—President, Prof. Horace Lamb, F.R.S.; vice-presidents, Prof. O. Reynolds, F.R.S.; Mr. Charles Bailey; Prof. W. Boyd Dawkins, F.R.S., and Mr. J. Cosmo Melvill; hon. secretaries, Mr. Francis Jones and Prof. A. W. Flux; treasurer, Mr. J. J. Ashworth; hon. librarian, Mr. W. E. Hoyle.

PARIS.

Academy of Sciences, April 30.—M. Maurice Lévy in the chair.—On the telescopic planets, by M. C. de Freycinet. The ideas of Laplace upon the distribution of the telescopic planets in concentric spherical layers round the sun are developed analytically and confirmed. If the asteroids are divided into three groups according to their inclination, the mean distance of the planets of these groups from the sun is sensibly constant. —On the transparency of aluminium for the radium radiation, by M. Henri Becquerel. A study of the penetration of thin aluminium sheet by the radium rays, the latter being placed in a strong magnetic field and the effects of the deviable and non-deviable rays being studied separately.—Study of manganous fluoride, by MM. Henri Moissan and Venturi. Pure anhydrous manganous fluoride, MnF_2, was obtained in four ways : by the action of a solution of hydrofluoric acid upon metallic manganese, by the interaction of gaseous hydrogen fluoride and the metal, by heating manganese fluosilicate in a current of HF at $1000°$, and by dissolving manganese carbonate in the acid. The crystallised MnF_2 could not be prepared from aqueous solution, on account of the sparing solubility of the salt in water, but is readily obtainable by fusing a mixture of the salt with manganese chloride.—Agricultural maps of the Canton of Redon. The composition of the soil from the point of view of lime, magnesia, potash and nitrogen, by M. G. Lechartier. An account of the work carried out at the agricultural station of Rennes.— On the vertical trunks, stems and roots of *Cordaites*, by M. Grand'Eury. The view is put forward that *Cordaites*, like *Sigillaria* and other fossil plants dealt with in previous papers, actually grew in the place where they are found, many ligneous trees commonly regarded as growing only on dry land flourishing well with their lower portions constantly submerged in water.—Prof. Suess was nominated a Foreign Associate in the place of the late Sir Edward Frankland.—On a relation between the theory of continuous groups and the differential equations with fixed critical points, by M. Paul Painlevé.—On the function S introduced by M. Appell into the equations of dynamics, by M. A. de Saint Germain.—An improved and simplified solar microscope, by M. A. Deschamps.— The telemicroscope, by M. A. Deschamps.—On an experiment of M. Jaumann, by M. P. Villard. In an experiment described by M. Jaumann, a charged glass rod was brought near a tube immersed in oil, in which kathode rays were being developed, the bundle being repulsed. As these results were not in agreement with the usual hypotheses concerning the kathode, an attempt was made to repeat the experiments, but no deviation of the rays in the opposite direction to that predicted by the theory could be obtained.—On the radium radiation, by M. P. Villard. The rays not deviable in a magnetic field have much greater penetrative power than the deviable rays. The ordinary X-rays from a Crookes' tube behave similarly.—Luminescence of rarefied gases round a metallic wire communicating with one of the poles of an induction coil, by M. J. Borgman.—On the hysteresis and viscosity of dielectrics, by M. F. Beaulard. From the results of the experiments given, the author concludes that dielectrics do not present the phenomenon of hysteresis, but are only endowed with viscosity.—On samarium, by M. Eug. Demarçay. The properties of the samarium isolated by the method of double magnesium nitrates previously described are so well defined that it would appear to be a simple substance analogous to other elements and not a mixture. The pale yellow colour of the oxide is apparently not due to any impurity. The atomic weight, as determined by the sulphate method, is about $147·5$.—On the combination of sulphur dioxide with metallic iodides, by M. E. Péchard. Potassium iodide, either in solution or in the solid state, rapidly absorbs sulphur dioxide, the compound $KI.SO_2$ being formed. This compound is easily dissociated into its constituents, its dissociation pressure at $0°$ being 60 cm. of mercury, at $30°$, 238 cm. Other iodides form similar compounds.—On the gases emitted by the Mont Dore springs, by MM. F. Parmentier and A. Hurion. The gas is carbon dioxide containing $0·49$ per cent. of nitrogen and $·01$ per cent. of argon. —Bromination with aluminium bromide, by M. Ch. Pouret. Organic chlorinated compounds, heated to their boiling points for some time with aluminium bromide, give good yields of the corresponding bromine derivatives. The preparation of bromoform, methylene bromide, methyl bromide, ethyl bromide, pentabromethane, ethylene, ethylidene and acetylene bromides is described in detail.—The action of monochloracetic esters upon the sodium derivative of acetylacetone, by M. F. March. The compounds $(CH_3.CO)_2.CH.CH_2.CO.OC_2H_5$ and $(CH_3.CO)_2.CH.CH_2.CO.OCH_3$ are described ; and also the products of the reaction between these bodies and phenylhydrazine.—Action of ethylidene chloride upon phenols in presence of potash, by MM. R. Fosse and J. Ettlinger.—On the

presence of tyrosine in the water of contaminated wells, by M. H. Causse. The water from contaminated wells at Lyons gave an orange coloration with the chloromercurate of sodium paradiazobenzenesulphonate which proved not to be due to cystine. Tyrosine was then extracted and identified by analysis.—On some changes which occur in plants grown in the dark, by M. G. André. A set of comparative analyses of maize and lupin plants grown in sunlight and in the dark.—Studies in development of *Petromyzon Planeri*, by M. E. Bataillon.—Modifications in structure observed in cells undergoing a true fermentation, by MM. L. Matruchot and M. Molliard. The fermentation of the fruit of *Cucurbita maxima* was carried out under conditions excluding the possibility of intervention of any foreign organisms. Every cell in a state of true fermentation shows a very clear nucleus, a small amount of chromatine arranged on the periphery of the nucleus, a protoplasm full of vacuoles, and numerous minute drops of essential oil formed in the protoplasm.—Botanical zones in French Western Africa, by M. A. Chevalier.—On the granites and syenites of Madagascar, by M. A. Lacroix.—On the Gothlandian of the Peninsula of Crozon (Finisterre), by M. F. Kerforne.—Influence of temperature on the fatigue of the motor nerves of the frog, by M. J. Carvallo. Temperature has a considerable influence upon the activity of motor nerves, the excitability increasing up to $20°$ C.—The functions of the crystalline tube of the Acephala, by M. Henri Coupin. The function of this tube appears to be digestive, a storehouse of diastases.— Topography of the mouth as regards sensitiveness of taste, by MM. Ed. Toulouse and N. Vaschide.

Royal Academy of Sciences, March 31.—Prof. H. G. van de Sande Bakhuyzen in the chair.—On orthogonal comitants, by Prof. Jan de Vries.—On indigo fermentation, by Prof. Beyerinck. Indigo fermentation is the decomposition of the glucoside indican into indoxyl and glucose by the action of a cell. This is effected in two ways : first, by katabolism, *i.e.* by the direct action of the living protoplasm on the indican ; secondly, by specific enzymes. All the indican splitting bacteria examined act by katabolism, and are quite inactive when dead. The indican plants and some kinds of yeast contain indigo enzymes, and so are still active when dead. The indigo enzymes of *Indigo leptostachya*, *Polygonum tinctorium*, *Phajus grandiflorus*, *Saccharomyces sphaericus*, and the emulsion of sweet almonds, which also acts feebly on indican, proved to be quite different enzymes with optima of activity at $61°$, $42°$, $53°$, $44°$ and $55°$ C. respectively. The action of all of them is increased by acid to the amount of $0·5$ c.c. normal per 100 c.c. of indican solution ; more acid as well as alkali decrease their activity. In indigofera there is no katabolism, whilst in *Polygonum* there is a slight katabolism at low, in *Phajus* a very strong katabolism at high, temperatures. Hence the last two decompose indican in both ways at once, while indigofera does so by enzyme action only. In the leaves of *Phajus* indican is localised in the protoplasm both of the cells of the epidermis and of the mesophyll ; the indigo enzyme occurs in the chlorophyll granules.—Prof. Hoogewerff presented on behalf of Mr. J. Hazewinkel, manager of the "experimenting station" for indigo at Klaten (Java), a paper, entitled " Indican, its splitting up, and the enzyme which brings this about." This paper contains the results of inquiries, made in 1898, which for technical reasons were not intended for publication. Beyerinck's publication makes further withholding useless. Mr. Hazewinkel observed that when all enzyme actions are excluded, an aqueous solution might be obtained from leaves of *Indigofera leptostachya*, which solution by the action of enzymes and subsequent oxidation yielded indigo. The glucoside-indican found in this solution appeared to be a fairly stable substance (also at boiling heat and when acted upon by alkalis), provided it was not exposed to the action of enzymes (indimulsin, emulsin) and of acids. Mr. Hazewinkel proved in various ways, among others by the formation of indirubine (with isatin), that the indigo-forming splitting-product of indican is indoxyl, and inquired into various circumstances influencing the detection of indoxyl in those solutions and the formation of indigo from indoxyl, and also observed that during the so-called fermentation of indigo leaves, no indican, but indoxyl is present in the fermentation fluid.— Prof. Hoogewerff also made a communication on behalf of Mr. H. ter Meulen and himself, entitled " A Contribution to the Knowledge of Indican." Basing their inquiries upon the above-mentioned inquiries by Mr. Hazewinkel and those made by Prof. Beyerinck, Prof. Hoogewerff and Mr. ter Meulen prepared pure indican from leaves of *Polygonum tinctorium*, cultivated by Prof. Beyerinck, and from indican solutions received from Mr. Hazewinkel. Indican crystallises out of an aqueous solution with 3 mol. H_2O, probably in rhombic crystals, melting at 51° and decomposing, when heated, to a higher temperature with the formation of violet vapours ; it tastes bitter and is optically active, exerting a left-handed rotation. Over sulphuric acid *in vacuo* it loses its water of crystallisation ; its melting point is then 100°-102°. It dissolves pretty readily in water, methyl alcohol, ethyl alcohol and acetone, and very slowly in benzole, carbon disulphide, ether or chloroform. It is represented by the formula $C_{14}H_{17}NO_6$, corresponding to the formula proposed by Marchlewski. The result obtained was 56·7 per cent. C, 5·8 per cent. H, 4·7 per cent. N ; the molecular weight was determined cryoscopically. On decomposition with HCl and oxidation with air, indican yielded indirubinous indigotine. No difference was observed between indican out of *Indigofera* leaves and that obtained from *Polygonum* leaves. Further investigations were promised.—The following papers were also presented for publication in the *Proceedings* : On a special case of Monge's differential equation, by Prof. W. Kapteyn.— On the locus of the centres of hyperspherical curvature for the normal curves of *n*-dimensional hyperspace, by Prof. Schoute.— On the power of resistance of the red·blood corpuscles, by Mr. Hamburger.—(1) On behalf of Mr. J. D. van der Waals, junr., a paper on equations, containing functions for different values of the independent constant ; (2) on behalf of Dr. J. Verschaffelt, a paper on the critical isotherm and the densities of saturated vapour and liquid in the case of isopentane and carbonic acid, by Prof. van der Waals.—On the 14-monthly period of the motion of the earth's pole, with determinations of the azimuth of the meridional signs of the Leyden Observatory in the years 1882-1896, by Prof. H. G. van de Sande Bakhuyzen, on behalf of Mr. J. Weeder.—Prof. Hoffman presented for publication in the *Transactions* a paper, entitled " Zur Entwicklungsgeschichte der Sympathicus."

DIARY OF SOCIETIES.

THURSDAY, May 10.

Royal Society, at 4.30.—On the Diffusion of Gold in Solid Lead at the Ordinary Temperature : Sir W. Roberts-Austen, F.R.S.—On Certain Properties of the Alloys of Gold and Copper : Sir W. Roberts-Austen, F.R.S., and Dr. T. K. Rose.—Experiments on the Value of Organic Sensation as Contributory to Emotion : Prof. Sherrington, F.R.S.—On the Brightness of the Corona of April 16, 1893. Preliminary Note : Prof. Turner, F.R.S.—The Radio-Activity of Uranium : Sir W. Crookes, F.R.S.

Royal Institution, at 3.—A Century of Chemistry in the Royal Institution : Prof. J. Dewar, F.R.S.

Mathematical Society, at 5.30.—Special Meeting.—The Differential Equation whose solution is the Ratio of Two Solutions of a Linear Differential Equation : M. W. J. Fry.—A Congruence Theorem relating to Eulerian Numbers and other Coefficients : Dr. Glaisher, F.R.S.—Linear Substitutions Commutative with a given Substitution : Dr. L. E. Dickson.

Institution of Electrical Engineers, at 8.—A Frictionless Motor Meter : S. Evershed.

Iron and Steel Institute, at 10.30.—Ingots for Gun Tubes and Propeller Shafts : F. J. R. Carrulla.—The Manufacture and Application of Water-Gas : Carl Dellwik.—The Equalisation of the Temperature of Hot Blast : Lawrence Gjers and Joseph H. Harrison.—The Manganese Ores of Brazil : H. Kilburn Scott.—The Utilisation of Blast-furnace Slag : Ritter Cecil von Schwarz (Liége).

FRIDAY, May 11.

Royal Astronomical Society, at 8.—On the Alleged Rotation of the Spiral Nebula M 51 Canum Venat : H. H. Turner.—Observations of Minor Planets at Windsor, New South Wales : John Tebbutt.—The Duration of the Greater Sun-spot Disturbances for the Years 1881 to 1899 : Rev. A. L. Cortie.—Note on Measures by Prof. Barnard of Two Standard Points on the Moon's Surface : S. A. Saunder.—Micrometrical Measures of Double Stars : W. Coleman.—Diagrams for Planning Photographic Observations of Eros : A. R. Hinks.

Physical Society, at 5.—Discussion of Prof Lodge's Paper on the Controversy concerning Volta's Contact Force.—The Heat of Formation of Alloys : Mr. J. B. Tayler.—On the Want of Uniformity in the Action of Copper-Zinc Alloys on Nitric Acid ; Dr. Gladstone, F.R.S.—An Electromagnetic Experiment, and Experiments illustrating the Aberration called Coma : Prof. S. P. Thompson, F.R.S.

Malacological Society, at 8.—On a New Species of *Desporus*, Newton (*Proserpina*, Gray) ; with Notes on some Allied Forms : E. R. Sykes.— On some New Mollusca from the Philippines : G. B. Sowerby.—On some Lamellibranch Remains occurring in a Sandstone from the Malay Peninsula : R. Bullen Newton.

SATURDAY, May 12.

Royal Institution, at 3.—South Africa : Past and Future : Dr. Alfred P. Hillier.

MONDAY, May 14.

Society of Arts, at 8.—The Incandescent Gas Mantle and its Use : Prof. Vivian B. Lewes.

Royal Geographical Society, at 8.30.—Nature and Man in British New Guinea : Prof. A. Haddon, F.R.S.

TUESDAY, May 15.

Royal Institution, at 3.—Brain Tissue considered as the Apparatus of Thought : Dr. Alex Hill.

Anthropological Institute, at 8.30.

Royal Statistical Society, at 5.—Municipal Finance and Municipal Enterprise : Sir H. H. Fowler.

WEDNESDAY, May 16.

Society of Arts, at 8.—A National Repository for Science and Art Prof. Flinders Petrie.

Royal Meteorological Society, at 4.30.—The Wiltshire Whirlwind of October 1, 1899 : the late G. J. Symons, F.R.S.—The Variations of the Climate of the Geological and Historical Past and their Causes : Dr Nils Ekholm.

Royal Microscopical Society, at 7.30.—Exhibition of Microscopic Pond Life.— At 8.—On the Lag in Microscopic Vision : E. M. Nelson.

THURSDAY, May 17.

Royal Society, at 4.30.

Royal Institution, at 3.—A Century of Chemistry at the Royal Institution : Prof. J. Dewar, F.R.S.

Zoological Society, at 4.30.—The Freshwater Fishes of Africa : G. A. Boulenger, F.R.S.

Society of Arts (Indian Section), at 4.30.—The Industrial Development of India : J. A. Baines.

Institution of Electrical Engineers, at 8.—Alternating Current Induction Motors : A. C. Eborall.

Chemical Society, at 8.—Chlorine Derivatives of Pyridine. VI. The Orientation of some Aminochloropyridines : W. J. Sell and F. W. Dootson.

FRIDAY, May 18.

Royal Institution, at 9.—The Structure of Metals : Prof. J. A. Ewing, F.R.S.

Epidemiological Society, at 8.30.

SATURDAY, May 19.

Royal Institution, at 3.—South Africa : Past and Future : Dr. Alfred Hillier.

CONTENTS.

THURSDAY, MAY 17, 1900.

BIOLOGY AS AN "EXACT" SCIENCE.

The Grammar of Science. By Karl Pearson, M.A.,
F.R.S., Professor of Applied Mathematics and
Mechanics, University College, London. Second Edi-
tion, revised and enlarged, with 33 figures in the text.
Pp. xviii + 548. (London : Adam and Charles Black,
1900.)

THE sciences of life are marked off for practical purposes from those concerned with inorganic matter
by obvious differences in the nature of the material with
which they respectively deal. But in addition to distinctions of this kind, it has been customary to look upon
biology as having a lower claim to the title of an
"exact" science than that enjoyed, for example, by
chemistry and physics. This view has been emphasised
by the practice of calling biology a merely "descriptive"
science, with a kind of implication that other sciences
are that and something more. The distinction, however,
is at best an artificial one, resting mainly on the fact that
the conditions of life are often so complex, and the data
so difficult of access, that the use of those quantitative
methods of induction which in other sciences have been
fruitful of important results, so far as biology is concerned
has to a great extent remained in abeyance.

It could not be expected that this state of things
should be allowed to continue. "Every science," said
Stanley Jevons, "and every question in science, is first a
matter of fact only, then a matter of quantity, and by
degrees becomes more and more precisely quantitative."
In those parts of biology which come into relation with
chemistry and physics, the quantitative methods have
long since gained a footing. Physiology tends increasingly to become a science of exact measurement, and
there is abundant scope for the exercise of mathematical
power in the investigation of its present data. With regard,
however, to many problems of what is known as "general
biology," especially those which gather round the central
doctrine of evolution, it is no doubt true that until recently
measurements have either not been applied at all, or
have been used only in the simplest and crudest form.
That general biology has now ceased to deserve the
reproach of neglecting quantitative methods is largely
due to the labours of Mr. Francis Galton, Prof. Weldon,
and Prof. Karl Pearson, the way towards a greater precision of method having also been in some degree prepared by other workers, such as Milne Edwards, J. A.
Allen and A. R. Wallace.

In the second edition of his well-known "Grammar of
Science," Prof. Pearson has included two new chapters,
which contain a semi-popular account of his recent work
on the mathematical aspects of evolutionary theory. The
ground covered is extensive, comprising quantitative
investigations of variation, correlation, selection in its
various forms, heredity and reversion. Those readers
who may be deterred by the length and elaboration of
Prof. Pearson's papers in the *Proceedings* and *Philosophical
Transactions* of the Royal Society will here find a clear
account of the various problems concerned, together with a

tolerably easy explanation of the mathematical processes involved in their attempted solution, and a useful summary
of the results so far arrived at. The author states his
main position as follows :—

"What we need in the theory of evolution is quantitative measurement following upon precise definition of our
fundamental conceptions. Biologists, even as physicists
have done, must throw aside merely verbal descriptions,
and seek in future quantitative precision for their ideas."

In the same spirit, Prof. Weldon remarked in his Presidential Address to Section D at the Bristol meeting of
the British Association : "Numerical knowledge of this
kind is the only ultimate test of the theory of natural
selection, or of any other theory of any natural process
whatever." That these dicta are substantially true will
hardly be questioned, though it may be objected to Prof.
Pearson that he somewhat overstates his case. All concrete science is in its essence descriptive, and it is not
improbable that parts at least of biological study will
have to remain indefinitely in the condition of "merely
verbal description." It would appear, too, that in his
eagerness to denounce the putting forward of inadequate
hypotheses, the author allows himself to undervalue those
rough preliminary generalisations which have frequently
formed so useful a step in the completion of a great induction. It is possible to attach too much importance
to Faraday's famous saying. If every "suggestive
thought" which has eventually turned out to be imperfect, or even erroneous, had been "crushed in silence"
instead of being given to the world, the cause of scientific
progress would have suffered. We must often, for practical purposes, be content to proceed by the method of
successive approximation. The work of Darwin himself
was only to a limited extent quantitative.

Evolutionists of what may perhaps without offence
be called the "orthodox" type, will find Prof. Pearson's
attitude towards most controverted points sufficiently
correct. Thus, without denying the possibility of a
bathmic element in evolution, he does not countenance
the "inherent growth-forces" that find favour with Neo-
Lamarckians. Demonstration of the inheritance of
acquired characters, he holds to be still wanting ;
tradition, on the other hand, is probably an important
factor in what are called the "instincts" of the lower
animals. He finds no quantitative evidence for tele-
gony, the occurrence of which alleged phenomenon
"seems both mechanically and physiologically inconceivable." The reality of natural selection as a factor
in evolution is quantitatively demonstrable, and sexual
selection is rehabilitated.

It would be impossible within the limits of a notice
like the present to do justice to the lucidity of Prof.
Pearson's explanations, the ingenuity of his mathematical
devices, and the care with which he has avoided possible
sources of error in his calculations. Examples may be
found in his exposition of the technical terms "modal
value" and "standard deviation" ; in his determination
of the coefficient of regression ; and in his discussion of
the relative value of selective and non-selective death-
rates for organs of different sizes. Among the most
valuable of his suggestions are those on the importance
of correlation ; on selective mating in its various forms

(including autogamy, endogamy, homogamy, preferential mating or "sexual selection" in Darwin's sense, and heterogamy); and on "genetic selection" or the inheritance of fertility. The last-named principle promises to be of special weight as a factor in evolution, though the proof of definite correlation of other physical characters with that of fertility must still be considered incomplete. The analysis of natural selection into autogeneric, heterogeneric and inorganic selection ("intraselection" being ignored) is useful, and might have been carried still further.

A contribution to the theory of evolution so original and stimulating as Prof. Pearson's must necessarily run the gauntlet of much adverse criticism. This will probably take the form rather of objection to certain points of detail than to the general drift of his method. Certainly some passages and expressions seem capable of amendment. It is, for instance, scarcely allowable to speak of the approach of the coefficient of correlation to unity as "the transition of correlation into causation." As the author himself elsewhere points out, correlation does not imply causation, though the converse is no doubt true enough. The principle of recognition-marks in their widest sense seems again to deserve more consideration than it receives at his hands. They are requisite to ensure the actual effectiveness of the impulse towards preferential mating. It is worth notice in this connection that the author's view as to the species-forming tendency of differential fertility (which is distinct from "physiological selection," as understood by Romanes) is well exemplified by Dr. Jordan's work on "mechanical selection." In speaking of hybridisation with reference to atavism, the "Grammar" does less than justice to observed facts. The evidence afforded by crosses, such as those so carefully investigated by Standfuss at Zürich and by Prof. Cossar Ewart at Penicuik, has a bearing on heredity and atavism which cannot safely be ignored. Prof. Pearson contents himself with saying that in such cases, "from physiological and mechanical reasons, the gametes produce a zygote which does not give an individual blending the ancestry. Here any singularity almost may be expected." This statement, to say the least, seems wanting in precision. Again, a severe critic might allege that the author is apt to assume theoretical values (as in the case of the resemblance of first cousins) which have not stood the test of rigid proof.

We have not yet learned to like the new term "apolegamy," nor such a phrase as "a comparative few zygotes" (P. 453). The remarkable form of a sentence on p. 461 is probably due to a printer's error, as also the substitution of DAG for FAG at the bottom of p. 447. These, however, are small matters, and do not detract from the value of the book.

We must not be led into a discussion of the earlier chapters, a notice of which appeared in these columns at the time of their original publication. There is, however, one point on which we cannot refrain from noticing. Prof. Pearson takes biologists to task for the loose way in which they often use such terms as "matter," "force" and "motion," as if no important questions lay behind them. Now it is certain that, in their employment of these expressions, biologists have no desire whatever to

prejudice any philosophical problems. When metaphysicians and physicists are agreed about the definition of these terms, the biologist will doubtless be quite ready to follow suit. Meanwhile he must be allowed the use of ordinary language. But Prof. Pearson maintains that if these words are used in their everyday, or, as he calls it, their "figurative" sense, they ought to be defined. Why so? No definition is required for the particular end in view. Supposing an opponent were to say that the "matter" of the argument was not "attractive," and that there was no "force" in this or that contention, would the Professor waste time in making him define his terms? Can we not "beat about the bush" without entering into explanations that would satisfy the schoolmaster and the botanist? It would seem that here the Professor once more overshoots his mark.

It will be convenient to give, in conclusion, a summary of the main contention of these new chapters in the author's own words, as follows :—

"It is not absence of explanations, but rather of the quantitative testing of explanations, which hinders the development of the Darwinian theory." "The problem of the near future is not whether Darwinism is a reality, but what is quantitatively the rate at which it is working and has worked."

It is noteworthy to find him adding :—

"If that problem should be answered in a way that is not in accordance with the age of the earth, as fixed by certain physicists, it by no means follows that it is biology which will have to retrace its steps. When the rate is determined, it will be as exact in its nature as physical appreciations; and it will be a question of superior logic, and not of the superiority of the 'exact' over the 'descriptive' sciences which will have to settle any disagreement of biology and physics." . . . "It is a question of the *rate* of effective change, and when the biologists are in a position to make a definite draft on the bank of time, their credit will be just as substantial as that of the so-called exact sciences."

These last sentences, as coming from a mathematician, are highly significant; and we cannot but admire the courage that has given them expression.   F. A. D.

HERTZ'S MECHANICS.

The Principles of Mechanics presented in a New Form. By Heinrich Hertz. Authorised English Translation, by D. E. Jones and J. T. Walley. Pp. xxviii + 276. (London : Macmillan and Co., Ltd., 1899.)

GREAT expectations were aroused by the publication, in 1894, of a book by Heinrich Hertz, with the title, "Die Principien der Mechanik in neuem Zusammenhange dargestellt." Perhaps it would set out the received theory of dynamics in strictly logical sequence; perhaps it would present a complete theory of energy independent of the notion of force; perhaps it would bridge the gap between the molecular and mechanical standpoints. Whether it would do any of these things or not, what Hertz might have to say would certainly be worthy of attention. Hertz died before the work was printed, and the task of seeing it through the Press was entrusted to Dr. P. Lenard. He tells us that the author had devoted the last three years of his life to the book, the last two being spent in perfecting its form; and, although there are indications that he was not even then completely satisfied, the work may fairly be regarded as the mature expression of his deliberate thought on the subject.

The book opens with a preface by Helmholtz, followed by the author's preface ; then there is an introduction, and the author's theory is formulated in two books :— Book i. : Geometry and kinematics of material systems ; Book ii. : Mechanics of material systems. Helmholtz's preface contains an account, which might be called an appreciation, of the scientific work of Hertz, and is further remarkable for the statement that, while Kelvin, Maxwell and Hertz appear to have derived fuller satisfaction from explanations of physical facts founded on some simple general conception, such as Hertz's " straightest path," he, for his part, has felt safer in adhering to the representation of physical facts and laws by systems of differential equations. In his own preface the author tells us that his object was " to fill up the existing gaps, and to give a complete and definite presentation of the laws of mechanics which shall be consistent with the state of our present knowledge, being neither too restricted nor too extensive in relation to the scope of this knowledge " ; and that what he hoped was new in his work was " the arrangement and collocation of the whole —the logical or philosophical aspect of the matter."

In the introduction the author criticises the received theory of dynamics and the more modern doctrine of energetics, and proceeds to explain the character of the new theory which he proposes. The novelty consists in this : whereas the other two theories started from four fundamental concepts—space, time, mass and force, or energy—he requires only three—space, time and mass— and the hypothesis of concealed masses. In Book i. relations concerning spaces and times are considered, and we have a generalisation of ordinary kinematics, including definitions of the path and velocity of a material system, and its shortest and straightest paths. By a material system is meant what in the ordinary presentation of dynamics would be called a system of particles with invariable connections. Some of the definitions referred to contain arbitrary elements, but they are, at any rate, simple. The definition of *mass* might have been omitted with advantage. In Book ii. the author enunciates his " fundamental law "—that every free system moves in a straightest path. This law may be looked upon as an interpretation of the principle of least action for systems of which all the energy is kinetic, or as an extension of Gauss's principle of least constraint. He proceeds to show how the motions of systems which are not free can be brought under the fundamental law by means of the hypothesis of concealed masses—the visible system is regarded as linked on to another system by invariable connections—and it is proved that the equations of motion of the system contain terms which correspond to the " forces " of ordinary dynamics. It is, perhaps, not remarkable that the dynamics of distant gravitating bodies, which was the immediate object of the received theory, should offer special difficulties from the present point of view (§ 469) ; on the other hand, it is claimed that the new minimum principle is applicable to invariable connections of the type of pure rolling, in which the velocities are connected by non-integrable equations, and that it thus includes more phenomena than the principle

of least action. A considerable portion of Book ii. is taken up with the consideration of cyclical systems. Hertz has here developed important conceptions due to Helmholtz. Throughout both books the " older synthetic method," that of a chain of propositions, has been adopted in order that the logical purity of the theory might be beyond dispute.

Whatever may be the influence exerted on the progress of mechanics by Hertz's kinematical generalisations and fundamental law, there cannot be any doubt of the value of his criticisms of existing dynamical theories. He has explained, in the clearest manner, the object of physical theories, and stated the conditions which such theories must satisfy. He has tested the received theory of dynamics—that which is associated with the names of Galilei, Newton, d'Alembert and Lagrange—in respect of logical permissibility, and in respect of appropriateness as an expression of facts. Concerning this representation of physical experience, he asks : " Is it perfectly distinct ? Does it contain all the characteristics which our present knowledge enables us to distinguish in natural motions ? " And his answer is " a decided—No." He has put his finger on the weakest part of the theory—the relation of the notion of internal stress to that of equal and opposite distance-actions. He makes the supposition that the theory can, even here, be rendered rigorous, and prefers to base his attack on the complexity of the various actions which the theory needs to assume. In a somewhat similar spirit he discusses the representation of physical facts by means of the theory of energy, although it is rather the logical permissibility than the appropriateness of this representation that is called in question.

The translators have done their work well on the whole. Here and there they have been too literal, or not literal enough ; they have left some obvious misprints in the German text, and some in the translation, uncorrected ; but these are slight blemishes, and we must be grateful to them for a rendering which admirably conveys the spirit of the original. Their translation should serve to make more widely known a book which certainly ought to be read by all who wish to have clear ideas concerning the most fundamental of the physical subjects.　　　　　　　　　　　　　　　A. E. H. L.

ASSYRIAN AND BABYLONIAN ASTROLOGY.

The Reports of the Magicians and Astrologers of Nineveh and Babylon. By R. C. Thompson. Vol. i. Pp. xviii + 85 plates of cuneiform text. Vol. ii. Pp. xci + 148. (London : Luzac and Co., 1900.)

IT is now about thirty-five years ago since the late Edward Hincks, whose name will be honourably coupled with the history of cuneiform decipherment, astonished many folk by declaring that he had discovered in the British Museum tablets which related to the pseudo science of astrological astronomy. And it is not surprising that such a declaration evoked general interest, because reasonable grounds existed for hoping that when the texts on the tablets had been deciphered, some trustworthy information about Chaldean astronomy might be forthcoming. The labours of Hincks were followed by those of Lenormant and Oppert, but they had little result, because neither of these scholars was able to devote sufficient time to the study of original texts in the British Museum. Great impetus was given to the study when the late Sir Henry Rawlinson published the third part of the "Cuneiform Inscriptions," and Prof. Sayce found therein material for his paper on the "Astronomy and Astrology of the Babylonians," which appeared in 1873. During the last twenty-five years the astronomy of the Babylonians has been discussed by Strassmaier, Jensen and others, but little has been done for the older, sister subject of astrology. In the two volumes before us Mr. Campbell Thompson gives us the cuneiform text of what is, practically, the complete series of the Astrological Reports of the Royal Library at Nineveh—that is to say, copies of about two hundred and eighty tablets, and transliterations of about two hundred and twenty duplicates, without reckoning the transliterations of the texts of the original series. In addition, we find a translation of the tablets in English, and a vocabulary, with references, and a subject index. The work in each of these sections has been carefully done, and we welome Mr. Thompson in the little band of English Assyriologists, because his pages, somehow, suggest that he intends to try to justify his position as assistant in the British Museum. The study of Biblical parallels and the making of Biblical comparisons are interesting and useful enough in their way, but it is useless to dogmatise about any branch of Assyriology as long as the literature relating to it remains unpublished. Mr. Thompson's book is a good proof of this contention. Many have talked glibly and written vaguely about Chaldean astrologers, but now that we have before us the actual texts of the documents which they drew up, we shall find that most of what has been written on the subject before is incorrect.

The study of astrological astronomy in Western Asia is very ancient, and an old tradition, referred to by Pliny, states that the Babylonians possessed records of calculations which covered a period of 490,000 years ; there is no doubt that we now possess texts of this class which are as old as the reign of Sargon of Agadhe (about B.C. 3800) ; but nothing older than this date has yet been unearthed. The principal astronomical schools in Assyria in the seventh century B.C. were at Ashur, Nineveh and Arbela, and at a later period Sippar, Borsippa and Orchoe, in Babylonia, were famous for their schools. The chief duty of the astrologer in Assyria was to calculate times and seasons, which he did either by observation or by the help of an instrument called *abkallu shikla—i.e.* "master of measure" (*or* reckoning). This instrument may be the clepsydra, which Sextus Empiricus says was known to the Chaldeans. The time measure was called *kasbu*, and contained two hours ; the month was one of thirty days, and the year contained twelve months. The Assyrians employed one intercalary month (second Adar), and the Babylonians two (Elul and Adar). Both nations had a year of lunar months, and much of the time of the Chaldean star-gazer was spent in observing the sun and moon, with the view of determining when the months began and ended. The seven planets were called Sin (moon), Shamash (sun), Umunpauddu (Jupiter), Dilbat (Venus), Kaimânu (Saturn), Gudud (Mercury), and Mushtabarrû-mûtânu (Mars). From these, and the Signs of the Zodiac, and indeed most heavenly bodies,

omens were deduced, and from the horns of the moon many portents were derived. Another source of omens were the halos, two of which were known ; the one was of 22°, and the other of 46°. Dark halos always portended rain, and were well known, and Mr. Thompson suspects that the astrologers were acquainted with mock suns also. That they were good weather prophets is tolerably clear from many indications ; indeed, it would be surprising if they were not. The omens derived from eclipses are very interesting, but the train of reasoning which guided the composition of birth portents cannot always be followed. Thus, in text No. 277, it is related that a certain butcher, called Uddanu, reported to an astrologer that when his sow littered, one of the young pigs had eight legs and two tails, and that he had preserved the animal in brine ; from this birth the astrologer deduced the omen that the Crown Prince of the day would "grasp power." But why? Many of the reports sent to the king are interesting, chiefly because of the variety of their contents. When the astrologer had reported the astrological fact asked for, he added any little detail concerning mundane affairs which he might have room for on the tablet, or which he thought it would amuse the king to have knowledge of. Sometimes there is nothing of special astrological importance in the report at all—*e.g.* No. 22, whereon the writer wishes the king power and riches, and says that as the gods Ashur, Shamash, Nebo and Merodach have delivered Kush and Egypt into his hands, even so will they deliver the Cimmerians and the Mannai. Again, in No. 124, more than one-third of the report is occupied with the discussion of some private affair, in which the writer says, "Now the king knows I hold no land in Assyria." From the literal translations which Mr. Thompson gives in the second volume of his book, it is clear that the writers of these reports wilfully obscured their meaning by using obscure and difficult words, and that they intended to make it necessary for their recipients, royal or otherwise, to call in the professional astrologer. If the Assyrians found it difficult to get out a meaning from such documents, there is small wonder that we, in these days, have a difficulty in understanding them also, and as many of the allusions must necessarily be unknown to us, we may have to wait for new texts which will help us to clear them up. Meanwhile, Mr. Thompson has dealt carefully with his texts, and has erred rather on the side of being too literal than too paraphrastic in his translations. It is to be hoped that he will find time to continue his investigations, and to give us accurate editions of original documents, which may serve as the foundation of a superstructure of facts rather than theories.

THE SCIENCE OF NUMBER.

Éléments de la Théorie des Nombres. Par E. Cahen. Pp. viii + 404. (Paris : Gauthier-Villars et Fils, 1900.)

TO the contemplative mind the science of arithmetic offers irresistible, if tantalising, attractions. The abstract notion of number underlies all scientific knowledge and theory whatever ; and it is in terms of it alone that we are compelled to seek for the ultimate statement of the facts of the sensible world. It is most unfortunate that arithmetic should be so often confounded with the vulgar art of logistic—the necessary, but ignoble, reckonings of the exchange and the market-place. Even those who are aware of the distinction often fall into another error, which is almost equally pernicious. To most of them scientific arithmetic means the "Theory of Numbers," a term which they vaguely associate with an unknown, mysterious branch of mathematics with which only a few eccentric specialists have any concern.

The facts of the case are very different. It is true, of course, that the exact and logical foundation of the very rudiments of arithmetic has required the efforts of a series of the greatest intellects ; that in order to follow its numerous ramifications, and appreciate its relation to other parts of analysis, demands a large amount of ability and perseverance ; and that many of its truths have, as yet, only been proved by elaborate, one may even say artificial, methods ; while other theorems, almost certainly true, still baffle all attempts at demonstration. But, in spite of all this, it may be asserted that arithmetic requires less apparatus and less preliminary training than any other branch of mathematics ; and that, whether as a recreation or as a field for research, it amply rewards a very moderate degree of application.

It is not without reason, therefore, that Prof. Cahen addresses himself deliberately to amateur mathematicians ; and, in fact, any one gifted with common sense, unspoiled by a vicious course of school instruction, ought to profit by his lucid and entertaining pages. In six chapters he deals in sufficient detail, and with appropriate numerical illustration (a most important point), with the elementary definitions and laws of operation, with linear and quadratic congruences, and with the elementary theory of binary quadratic forms. After this come a series of notes, ranging from scales of notation to an outline of the properties of Gauss's complex integers and their nearest allies ; and, finally, a very useful set of tables, which afford the reader material for those applications to particular cases, without which the general theory cannot possibly be mastered.

The appearance of this work, as well as of others with a similar object in view (for instance, M. J. Tannery's excellent "Leçons d'Arithmétique"), encourages the hope that some improvement may be effected in the teaching of arithmetic in schools, and that a sound knowledge of its first principles may cease to be the monopoly of a very small minority of University graduates. It is, unfortunately, true that a very large proportion of class-books, both in arithmetic and in algebra, contain half-informed, misleading attempts at expounding theory which are really worse than the old-fashioned bundles of "Rules" ; and unless these are replaced by something better, the efforts of reformers will have the lamentable result of producing a state of things worse than the old routine : a mere jargon of pseudo science, a barbarous patchwork of sham "Principles."

M. Cahen's work will be found of interest, not only by the amateur in search of recreation, but by intelligent teachers and arithmeticians of every degree of proficiency ; while the professed devotees of the science will look with pleased anticipation for the more extended work on the same subject which the author appears to be preparing.

G. B. M.

OUR BOOK SHELF.

Atlas of Urinary Sediments, with special reference to their Clinical Significance. By Dr. Hermann Riedel. Translated by F. C. Moore, M.Sc., M.D. Victoria. Edited and Annotated by Sheridan Delépine, M.B., C.M. Edinburgh, B.Sc. Pp. viii + 111, and 36 plates. (London : C. Griffin and Co., Ltd., 1899.)

THE work before us, as is evident from its title, is an atlas, and will be of interest rather on account of its plates, which are very beautiful, than of its letterpress ; this latter, however, which is situated at the end of the book, covers more than a hundred pages, and is provided with a bibliography and an index of authors and subjects. The text is sub-divided into an introduction and two parts. The introduction deals with methods of collection and examination, &c. Part i. is devoted to unorganised, Part ii. to organised sediments. The editor has added considerably to the original text, his remarks being indicated by parentheses : he occasionally differs with Dr. Riedel concerning fact. The large additions to the text made by the editor have rather altered the character of the work, and have probably increased the sphere of its usefulness.

Under organised sediments bacteria are considered. A useful chapter is to be found at the end concerning the making of permanent specimens of urinary sediments.

The book should be of value to urinologists, and the plates certainly to physicians in general. The thanks of the profession are due to the translator and the editor for making the work available to English readers, and amplifying its contents.

Dante. By Edmund G. Gardner, M.A. "The Temple Primers." Pp. vi + 159. (Dent, 1900.)

A VERY admirable book, by the author of Dante's "Ten Heavens." Dante was a master of the science of his time, and Mr. Gardner has shown that he has not only carefully studied the "Divina Commedia" from the point of view of literature, but has taken pains to carefully annotate all the references to the then *systema mundi* on which so much of the action of the poem depends. Diagrams and explanations are given at the end of the book, which will be found most useful by the student.

The Farmstead. By Prof. J. P. Roberts, Director of the College of Agriculture, Cornell University. Pp. vi + 350. (New York : The Macmillan Company. London : Macmillan and Co., Ltd., 1900.)

THIS is a very readable compendium of suggestions in regard to providing a beautiful, economical, and healthy rural home. Although written for American farmers, it contains much that is of interest to all who are concerned with a country life, and few will peruse the book without gleaning some useful hints. There are special chapters on house-furnishing, decoration, and sanitation by Prof. Mary Roberts Smith, who writes pleasantly on the lighter sides of a farmer's life. A strong case is made out for the educational opportunities of the farm, which are shown to be ample enough to satisfy the most exacting advocate of Nature Study.　　W. S.

Object Lessons in Botany from Forest, Field, Wayside and Garden. Book ii., for Standards iii., iv. and v. By Edward Snelgrove, B.A. Pp. xviii + 297. (London : Jarrold and Sons.)

THIS is a meritorious little book, and ought to well serve its purpose of inculcating habits of accurate and precise observation in the young pupils for whom it is designed. Although we notice a few slips here and there, they are not serious ones, and are quite eclipsed by the excellent character of the book as a whole. The author is convinced, as he says in the preface, of the value of elementary botany in the education of children, and we think his book justifies his contention.

LETTERS TO THE EDITOR.

[*The Editor does not hold himself responsible for opinions expressed by his correspondents. Neither can he undertake
to return, or to correspond with the writers of, rejected
manuscripts intended for this or any other part of* NATURE.
No notice is taken of anonymous communications.]

Percussion Caps for Shooting in Schools.

THE extraordinary explosive power of fulminate of mercury
is known to all chemists, but it is not generally known that the
explosion of a percussion cap on a gun will cause a current of
air sufficient to extinguish a candle at a distance of ten or fifteen
feet. The distance, of course, varies with the length and bore of
the gun, and with the nature and the size of the candle. The
gun must be pointed at the lower part of the wick, and in order
to blow out the candle the aim at this distance requires to be
nearly as accurate as would be required to make a centre with a
rifle at a hundred yards. In a speech to the Primrose League
on May 9, Lord Salisbury mentioned the expediency of every
man having the chance to learn to handle a rifle within reach of
his own cottage. By beginning with percussion caps children
might be taught to handle a gun at such an early age, that, in
case of invasion of this country, boys of fourteen might be able
to act as soldiers, as they are said to be doing amongst the Boers
at the present time. The objections to training children to
handle a rifle are, first of all, the danger of the child shooting either itself or some one else ; and secondly, the expense.
But the inclination of children to play soldiers might readily be
utilised by teaching them to handle first of all a toy gun, and then
to practice shooting at a candle with caps. For those who
shoot best with caps, the practice with a saloon rifle might be
held out as a reward. One single-barrelled old muzzle-loading
gun would suffice for many children, and as 240 caps cost a
shilling, the expense of providing a gun and material for practice
would be very small.   LAUDER BRUNTON.

Escape of Gases from Planetary Atmospheres.

IN NATURE of March 29 (P. 515), Dr. Stoney, in referring to
a paper by the writer in the January number of the *Astrophysical Journal*, raises the question as to the correctness of
the use of Maxwell's distribution of velocities in computing the
escape of gases from the earth's atmosphere. He maintains
that this distribution does not hold at its attenuated limits. In my
paper I have not taken conditions which may exist there, but
boundary conditions, which are much more favourable to the
escape of the molecules of a gas, and certainly compatible with
the kinetic theory, if we are to accept such a theory at all.
    Of the four conditions discussed in my paper, I will only refer
to the third, the data for which are based on direct observation,
namely, – 66° C. at a height of 20 kilometres (the mean of several
ascensions really giving – 65° C. to – 70° C. for a height of only
16 kilometres). The pressure is calculated from the usual exponential formula, which agrees closely with observations to
this height. At these temperatures and pressures there can be
no question as to the validity of the kinetic theory.
    Let us assume now that the atmosphere abruptly terminates
at this height, and at this temperature the loss would certainly
be greater (in fact, very much greater) than under the actual
conditions, where the temperature and pressure are much lower.
It should also be noticed that in my tables I have assumed the
atmosphere to be entirely made up of one gas—for example,
helium or hydrogen. Even then only 26·73 x 10⁻²³ c.c. of helium
would escape in 10⁷ years. Hence the assumption that helium
is now escaping from our atmosphere is without foundation. In
the case of a hydrogen atmosphere only 0·54 c.c. will escape in
one year. ·If the total amount of air in the atmosphere be taken
approximately at 10²⁴ c.c., and if the actual density and temperature at the outer limits of the atmosphere be also considered,
it will be evident how baseless the supposition is that either
helium or hydrogen is escaping. It should be further noted that
Maxwell's distribution of velocities from zero to infinity is the
only one giving a sufficient velocity for any escape at all, Clausius'
theory not being adequate.
    It was the assumption that helium is escaping from the at-
mosphere—since it had not been detected—that first led me to
verify it on the kinetic theory of Maxwell. The discovery, by
Ramsay, of helium as a constituent of our atmosphere only tends
to confirm the results of my calculations of the impossibility
of its escape.   S. R. COOK.
    Physical Laboratory, University of Nebraska, April 26.

Racket Feathers.

YOUR able reviewer of Meyer and Wiglesworth's " Birds of
Celebes" (NATURE, April 26), criticising the arguments used to
account for the formation of the racket tail feathers of the parrot,
Prioniturus (as an inherited effect of mechanical attrition on
objects against which the tail is liable to be brushed—boughs,
walls of nesting-hole, &c.), asks the pertinent question, why so
few exposed feathers, such " as the external rectrices and remiges
of all birds, and specially the lengthened feathers of wedge-
shaped tails (*Dicrurus*) are neither bare nor racket-shaped nor
incipiently so." The insignificant length of the outer rectrices
of *Dicrurus* perhaps safeguards them ; when these feathers are
longer, as in the closely-allied *Bhringa* and *Dissemurus*, they
are racket-shaped. As to the remiges and rectrices of birds
generally, one feather overlies and to a great extent protects the
next ; but still, the outer webs *are* always very much narrowed
in the outermost and most exposed feathers, less narrowed in the
next, and so on till in the middle of the wing and tail (where
they are well protected on both sides) they are not narrowed at
all. But, while normal wing and tail feathers are exposed to
attrition on one web only, long feathers standing well out from
the rest are liable to have the web frayed on both sides of the
shaft as far as they project beyond the other feathers, and to
some extent where they rest upon the other feathers through.
friction against the latter. It is assumed that at some period
earlier in the history of the race these elongated feathers were
of the usual simple shape, but they are now known to issue from
the follicles displaying peculiarities which are often much the
same as those obtained by scraping an ordinary feather with a
knife—namely, if the shaft is stiff and not very long, a small
terminal spatule is formed (as in *Prioniturus, Parotia*) ; if the
shaft is long and weak, a large spatule (as in *Tanysiptera,
Loddigesia*). A difficulty, perhaps, to the acceptance of the
theory is its apparent consequence—that epidermal (in a sense,
dead) structures, like feathers, possess the power of transmitting
mutilations to posterity. For my own part, I think that the
modification of shape of the feathers is communicated to the
sensitive tissues (much in the same way as the shape of a stick
placed in the hand of a blind man is comprehended by him after
touching other things with it), and that a corresponding physiological adjustment is made and gradually inherited. The result
is probably not an exact recapitulation of the mutilation, but it
sometimes appears to be very nearly so.
                                            L. W. WIGLESWORTH.
    Castlethorpe, Stony Stratford, April 30.

MR. WIGLESWORTH in the above note hardly does more than
recapitulate the (?) arguments advanced in the "Birds of
Celebes." He does not offer any explanation of the crucial
difficulties indicated in the review ; why "mechanical attrition
on objects," or by the wind, is effective only in so few cases
throughout the class *Aves* when so many species are subject to
the necessary conditions ; why, for instance, the species of *Pala-
ornis* (belonging to the same sub-family as *Prioniturus*), or those
of the genus *Trissor*, do not conform to the "law" ; and why
one sex of a species may have "sabre wings," or spatulate ornaments in various situations, and the other sex not.
    The question may also be asked *apropos* of Mr. Wigleworth's
statement above, why in *Paradisea rubra* the *long* and *weak-
shafted* tail feathers have the *small* spatule (which eventually
vanishes) instead of a *large* one, if the knife-scraping analogy
holds good ?
    The reasons for the *exceptions* to the author's rule is what
chiefly demands an explanation, in the opinion of
                                                THE REVIEWER.

THE APPROACHING TOTAL ECLIPSE OF THE SUN.

THE approaching total solar eclipse, on the 28th of
the present month, promises to contribute some
valuable additions to our scientific knowledge of the
centre of our system, inasmuch that the track of the moon's
shadow on the earth's surface passes, in an unusual extent, through regions which are easily accessible. Entering the North American continent near New Orleans,
in Louisiana, the central line of eclipse traverses the
States of Mississippi, Alabama, Georgia and Carolina,

passing on to the Atlantic from the shore of Virginia, near Norfolk. The track is thus crossed by many of the numerous railway systems of the Southern States, exceptional facilities being thereby offered to observers with large instruments. Information supplied by the U.S. Weather Bureau indicates that stations in Alabama and Georgia are most likely to be favoured with an un-clouded sky; hence the expeditions from the chief American observatories will go there. Congress has voted 5000 dollars to the Naval Observatory, and 4000 dollars to the Smithsonian Institution, for the necessary equipment. The Naval Observatory staff will organise two expeditions, one going to North Carolina, the other to Georgia, so that the stations will be some 200 miles apart, and will furnish valuable evidence as to the changes to which the solar surroundings are subject.

The Smithsonian Institution will be represented by Prof. S. P. Langley, and the Princeton Observatory by Prof. Young, who will make a redetermination of the wave-length of the green corona line. Prof. Stone will conduct a party from the University of Pennsylvania, and although details are as yet unknown here, it is expected that expeditions from the Yerkes (Profs. Hale, Barnard and Frost) and Lick (Prof. Campbell) Observatories will endeavour to obtain complete spectroscopic records of the various stages of the eclipse. The latter will again use the 40-foot coronograph, giving a 4-inch disc on plates 14 × 17 inches. Prof. Pickering, of the Harvard College Observatory, proposes to make a systematic search for an intra-Mercurial planet, and will probably occupy a station in Alabama.

By the kindness of Prof. Young, the Rev. J. M. Bacon has been enabled to organise an expedition to the States, and his observations will be made in the neighbourhood of Wadesborough, near the boundary between North and South Carolina. The party will consist of the Rev. J. M. Bacon, Miss Bacon, and Mr. and Mrs. Maskelyne. The two latter observers will expose a telescopic kinematograph on the corona during totality, and also an ordinary kinematograph on the landscape during and after totality, in the hope of recording the sweep of the moon's shadow. The Rev. J. M. Bacon, using a telescopic camera, will photograph the corona at definite times with respect to mid-totality, for determining the positions of sun and moon, and will expose a long film, continuously driven, to the zenith before, during and after totality, for recording the relative brightness of the sky during and without eclipse. By means of a kite, he will also compare the temperature of the air at an altitude of several hundred feet and at ground-level. Miss Bacon will attempt to photograph the outer corona and extensions, and also a series of landscape photographs showing the gradual diminution of illumination. Special attention will also be devoted to the "shadow bands," and to making standard photographic comparisons of the light of the corona with that of the full moon.

Prof. Burckhalter, of the Cbarbo Observatory, will photograph the corona by means of a camera provided with revolving screens, so adjusted as to give varying exposures for the different regions.

As the eclipse will occur at the American stations at times from 1h. 30m. to 1h. 50m., we in England will be able to hear of the results obtained there before the observers in Spain have commenced operations.

After leaving the American coast, the moon's shadow crosses the Atlantic in a westerly direction, and reaches the coast of Portugal, near Ovar, about 4.0 p.m. Thence it rapidly passes across the peninsula, leaving the mainland some little distance south of Alicante, and crossing the Mediterranean to Algiers. Most of the European expeditions will have stations along this line, chiefly at Ovar, Santa Pola and Algiers. Taking the stations in the order suggested by the progress of the eclipse, the

distribution of the various parties and their plan of operations will be as follows :—

Ovar.—At this place, some twenty miles south of Oporto, and five miles from the coast, will be stationed one of the three official expeditions sent out by the British Government, the observers being the Astronomer Royal and Mr. Dyson, his chief assistant. The former has arranged to take large scale photographs of the corona with the 9-inch object-glass of the Thomson photoheliograph at Greenwich, the primary image being enlarged by a *concave* secondary magnifier to a scale of about 4 inches to the sun's diameter, on plates 15 × 15 inches ; and also photographs with the double camera used in previous eclipses, having a 4-inch rapid rectilinear lens of 33 inches focus, and another of 13 inches focus, for recording the extensions of the coronal streamers.

Mr. Dyson's programme is purely spectroscopic. He will have two slit spectroscopes belonging to Captain Hills, and will endeavour to obtain photographs of the spectrum of the "flash" and of the corona.

Prof. Müller, of Potsdam, will from this station determine the albedo of Mercury from direct photometric comparisons with Venus, which will then be near its greatest brilliancy.

Santa Pola.—The second British official expedition will be stationed here, some distance south of the town of Alicante, on the east coast of Spain. The party will be under the direction of Sir Norman Lockyer, who will be primarily assisted by Mr. A. Fowler, Dr. W. J. S. Lockyer and Mr. H. Payn. On their arrival at Gibraltar, they will be taken on board H.M.S. *Theseus*, of the Mediterranean squadron, which will then convey them to their destination. As at Viziadrug in 1898, and Norway in 1896, volunteers will be selected from the ship's company, and parties detailed out for every character of observation it is possible to make during a total solar eclipse ; and in the interval between their landing and the final day, besides the erecting and adjusting of the instruments, the principal observers will have their time fully occupied in giving lectures, practical demonstrations, and rehearsals to the host of volunteers who will undoubtedly offer themselves.

Sir Norman Lockyer will make visual observations with a 4-inch Cooke photo-visual telescope equatorially mounted, and will give the signals for the whole of the remaining human and mechanical machinery to be set in motion. The following are the chief sections of the observers :—

20-foot Prismatic Camera.—This will be manipulated by Mr. Fowler, and consists of a Cooke photo-visual triplet lens, of 6 inches aperture and 20 feet 3 inches focal length. Outside this will be placed the objective prism, of 9 inches aperture and 45° angle, which was used at Viziadrug in 1898. The instrument will be fixed horizontally, and fed with light from a 12-inch siderostat. It is proposed to obtain instantaneous photographs of the chromospheric spectrum at both internal contacts, and long-exposure photographs of corona spectrum during totality. It is hoped that the greatly increased dispersion given by this instrument will increase the contrast between the *line* and *continuous* spectra of the corona, and so render more accurate measurements of wavelength possible. The plates used will be 15 × 2½ inches.

6-inch Prismatic Camera.—This is the same instrument which was used with success by Mr. Fowler in 1898, and will be under the charge of Dr. Lockyer. It consists of a 6-inch object-glass by Henry, of 7 ft. 6 in. focus, outside which are adjusted two objective prisms, each of 6 inches aperture and 45° angle. The programme with this instrument is similar to that of the 20-foot.

Coronographs.—Several coronographs of varying power are being taken, the largest being under the charge of Mr. Howard Payn, a gentleman who has generously volunteered his services for the expedition. This instrument has a Cooke photo-visual lens of 4 inches aperture and 16 feet focal length, the primary image being used on plates 12 × 12 inches.

In addition, the De la Rue coronograph (4⅜ inches aperture and 72 inches focal length), Graham coronograph (3 inches aperture, 21 inches focal length), and Dallmeyer coronograph (6-inch aperture rapid rectilinear, 48 inches focal length) will be used. Parties of the volunteers will be engaged in one or other of the following observations :—

Disc drawings of corona	...	about 19 volunteers.
Observations of ring spectra	...	,, 5 ,,
Observations with pocket slit spectroscopes...	... ...	,, 4 ,,
Observations of shadow bands...		,, 6 ,,
Observations of stars and other celestial objects visible during totality	... ...	,, 20 ,,
Shadow phenomena, both atmospheric and terrestrial ...		,, 6 ,,
Colours of landscape, &c.	...	,, 12 ,,
Meteorology, temperature, pressure, &c. ...	... ...	,, 15 ,,
Photographs of landscape	...	,, 5 ,,
Natural history effects on men and animals ...	...	,, 3 ,,

In addition to these instruments, several of the observers will obtain photographs of the eclipse spectra by means of diffraction gratings and prisms fixed in front of their own small cameras. Those with gratings are likely to be specially useful, as the dispersion is sufficiently great to render it possible for the bright line spectra to show up from the continuous spectrum, and there is the further advantage of the large field given by an ordinary rectilinear, so that the spectrum of the streamers may also be obtained.

Prof. Copeland, Astronomer Royal of Scotland, will also occupy a station at Santa Pola, using a telescope of 40 feet focus.

The British Astronomical Association and the French Astronomical Society will each send parties to both Alicante and Algiers. As, however, the former place is so well occupied by Sir Norman Lockyer's party, the third official party from the British Government will occupy a station at Algiers, and will consist of Prof. Turner, Mr. Newall, Mr. Evershed and Mr. Wesley.

Prof. Turner will photograph the corona with one of the double cameras used in previous eclipses, one of which is arranged to polarise the coronal light before it reaches the photographic plate, and thereby determine the extent to which this light is initially polarised. In addition, he also hopes to repeat his work of 1893 and 1898 for determining photometrically the relative brightness of the corona at varying distances from the limb.

Mr. Newall will have three instruments under his charge, viz. :—(1) A four-prism slit spectrograph for obtaining the spectrum of the "flash," and of the corona. In the latter he hopes to obtain material for showing the difference, if any, between the spectrum of the coronal rays and the other portions. (2) An objective grating camera for photographing the spectrum of the corona in monochromatic light. (3) A polariscopic camera for photographing the corona, special attention being devoted to the study of any differences between the darker and brighter rifts.

Mr. Wesley, the assistant secretary of the Royal Astronomical Society, has for many years critically studied the minute structure of the corona, he being the draughtsman who has engraved the reproductions of many of the corona photographs of past eclipses for publication, but has not hitherto had an opportunity of studying it from nature. By the kindness of M. Trépied, the Director of the French Government Observatory at Algiers, Mr. Wesley will be enabled to examine the corona with the powerful "equatorial coudé" (about 8 inches aperture).

Mr. Evershed will not be stationed at Algiers itself, but intends to observe from a place near the limiting line of totality, about twenty miles south of Algiers, so that he may photograph the "flash" spectrum with somewhat longer exposure than near the central line.

Mr. and Mrs. Maunder will repeat at Algiers their programme so successfully carried out at Buxar, India, in 1898, but with larger apparatus. This will include short exposure photographs of the inner corona, and others with long exposure for extensions and streamers.

Mr. and Mrs. Crommelin will go to Algiers, and take photographs of the corona and of the shadow as projected on the atmosphere.

It is also stated that Mr. Percival Lowell, of Arizona, and Prof. Todd, of Amherst College Observatory, U.S.A., will occupy stations near Tripoli, in North Africa. It is to be hoped that favourable weather will enable the latter astronomer to successfully use his electrical control, by means of which he has arranged that a great number of photographic cameras shall be automatically exposed for varying times, all of which are operated from one revolving drum with delicately fitted electrical contacts.

The eclipse occurs at the European stations about 4.0 p.m. Greenwich time, so that it may be possible to communicate the results of the various expeditions to the evening papers of the same day.

Mention should be made of the generous arrangements which have been made by the authorities of all the Governments concerned, whereby the usual customs tariff and examination will be dispensed with, provided the observer is furnished with a certificate showing that his baggage is really for eclipse observation. The railway companies in Spain have also consented to convey passengers at half the usual fares.

CHARLES P. BUTLER.

THE ROYAL SOCIETY SELECTED CANDIDATES.

FIFTEEN candidates were selected by the Council of the Royal Society on Thursday last for election into the Society. The following are the names and qualifications of the new Fellows :—

GEORGE JAMES BURCH,

M.A. (Oxon). Lecturer at the University Extension College, Reading. Author of the following papers :—(1) "Experiments on Flame" (NATURE, 1885-86); (2) "A Perspective Microscope" (*Proc. Roy. Soc.*, vol. xiii.); (3) "Researches on the Capillary Electrometer" (*Proc. Roy. Soc.*, vol. xlviii., *ibid.*, vol. lix., *Phil. Trans.*, vol. clxxxiii(A)., *The Electrician*, July, 1896). "On a Method of drawing Hyperbolas" (*Phil. Mag.*, Jan., 1896). Also joint author of the following papers :—(1) "Dissociation of Amine Vapours" (with Mr. J. E. Marsh) (*Journ. Chem. Soc.*, 1889); (2) "E.M.F. of certain cells containing Nitric Acid" (with Mr. V. H. Veley) (*Phil. Trans.*, vol. clxxxii(A).; (3) "Effect of Injury in Muscle" (with Prof. Burdon-Sanderson) (*Proc. Physiol. Soc.*, 1893); (4) "Action of Concentrated Acids on Metals in contact" (with Mr. S. W. Dodgson) (*Proc. Chem. Soc.*, 1894); (5) "D'Arsonval Physical Theory" (with Mr. L. E. Hill) (*Journ. Physiol.*, 1894); (6) "The Electromotive Properties of *Malapterurus electricus*" (with Prof. Gotch) (*Phil. Trans.*, 1896).

Supplementary Certificate.

Author of the following scientific papers in addition to those stated in the first certificate :—"On Prof. Hermann's Theory of the Capillary Electrometer" (*Proc. Roy. Soc.*, vol. lx., p. 328); "The Tangent Lens-gauge" (*Phil. Mag.*, 1897, p. 256); "An Inductor-Alternator for Physiological Experiments" (*Journ. of Physiology*, vol. xxi., 1897; "An Account of Certain Phenomena of Colour Vision with Intermittent Light" (*ibid.*); "Artificial Colour Blindness, with an Examination of the Colour-Sensations of 109 Persons" (*Phil. Trans.*, vol. clxli., 1899); joint author with Prof. Gotch, F.R.S., of the following scientific papers :—"The Electrical Response of

Nerve to a Single Stimulus as investigated by the Capillary Electrometer" (*Proc. Roy. Soc.*, vol. lxiii., 1898, p. 300); "The Electrical Response of Nerve to Two Stimuli" (*Journ. of Physiol.*, vol. xxvi., 1899); "The Electromotive Force of the Organ Shock, &c., in *Malapterurus electricus*" (*Proc. Roy. Soc.*, vol. lxv., p. 434, 1900).

T. W. EDGEWORTH DAVID,

B.A. (Oxon.), F.G.S. Professor of Geology in the University of Sydney, N.S.W. Formerly Senior Geologist to the Geological Survey of New South Wales, and author of many reports and maps issued by the Survey. Has published many papers dealing with Glacial action in recent, as well as ancient, geological periods; among others:—"Evidences of Glacial Action in S. Brecknock and E. Glamorgan" (*Quart. Journ. Geol. Soc.*, vol. xxxix., pp. 39–58, 1882); "On Evidences of Glacial Action in the Carboniferous and Hawkesbury Series, N.S.W." (*ibid.*, vol. xliii., pp. 190–197, 1887); "On Glacial Action in Australia in Permo-Carboniferous Times" (*ibid.*, vol. lii., pp. 289–302, 1896); also many papers and addresses dealing with Petrology, Vulcanology, and Stratigraphical Geology in the Southern Hemisphere, published in the *Journals* of the Royal, Linnean, and the Societies of New South Wales. Has superintended and conducted to a successful issue the work of boring the Coral Atoll of Funafuti, undertaken by the Royal Society and the Geographical Society of New South Wales, with the assistance of the Admiralty.

JOHN BRETLAND FARMER,

M.A. (Oxon.), F.L.S. Professor of Botany, Royal College of Science, London. Formerly Fellow of Magdalen College, Oxford. Distinguished for his Botanical and Biological researches. Author of the following papers:—"On the Development of the Endocarp in *Samlucus nigra*" (*Ann. of Bot.*, vol. ii.); Contribution to the "Morphology and Physiology of Pulpy Fruits" (*ibid.*, vol. iii.); "The Stomata in the Fern *Iris Pseudacorus*" (*ibid.*, vol. iv.); "On *Isoetes lacustris*" (*ibid.*, vol. v.); "On Abnormal Flowers in *Oncidium splendidum*" (*ibid.*, vol. vi.); "On the Occurrence of two Prothallia in an Ovule of *Pinus silvestris*" (*ibid.*); "On the Embryogeny of *Angiopteris evecta*" (*ibid.*); "On Nuclear Division in the Pollen-mother-cells of *Lilium martagon*" (*ibid.*, vii.); "On the Relations of the Nucleus to Spore-formation in certain Liverworts" (*Proc. Roy. Soc.*, vol. liv.); "Studies in Hepaticæ" (*Ann. of Bot.*, vol. viii.); "On Spore-formation and Nuclear Division in the Hepaticæ" (*ibid.*, vol. ix.); "Further Investigations on Spore-formation in *Fegatella conica*" (*ibid.*); "Respiration and Assimilation in Cells containing Chlorophyll" (*ibid.*, vol. x.); "Ueber Kerntheilung in Lilium" ("Flora," 1895); "On the Structure of a Hybrid Fern" (*Ann. of Bot.*, vol. xi.). Joint Author of:—with J. Reeves, "On the Occurrence of Centrospheres in *Pellia epiphylla*" (*ibid.*, vol. viii.); with J. H. Williams, "On Fertilisation and the Segmentation of the Spore of Fucus" (*Proc. Roy. Soc.*, vol. lx.); with T. Waller, "Observations on the Action of Anæsthetics on Vegetable and Animal Protoplasm" (*ibid.*, vol. lxiii.); with J. Brentland, "Contributions to our Knowledge of the Fucaceæ, their Life-History and Cytology" (*Phil. Trans.*, vol. cxc.).

LEONARD HILL,

M.B. Lecturer on Physiology, London Hospital Medical College. Distinguished as a Physiologist. Author of the following works:—"On Poisoning by Phosphorus" (*Lancet*, 1890); "On Intra-Cranial Pressure" (*Roy. Soc. Proc.*, vol. lv.); "On Effects of Compression of the Common Carotid Artery" (with Moore) (*Brit. Med. Journ.*, 1894); "On Formation of Heat in the Salivary Glands" (with Bayliss) (*Journ. of Phys.*, vol. xvi.); "On D'Arsonval's Physical Theory of the Negative Variation" (with Birch) (*ibid.*); "On a Simple Form of Gas Pump" (*ibid.*, xvii.); "Exchange of Blood-Gases" (with Nabarro) (*ibid.*); "On Exchange of Blood-Gases in Brain and Muscle" (*ibid.*, xviii.); "On the Influence of Gravity on the Circulation" (*ibid.*); "On Intra-Cranial Pressure and the Circulation" (with Bayliss) (*ibid.*); "The Physiology and Pathology of the Cerebral Circulation," Hunterian Lectures, Churchill, 1896; "On Nervous Pressure and the Pulse" (with Barnard and Sequeira) (*Journ. Physiol.*, xxi.); "Influence of Gravity on the Circulation" (with Barnard) (*ibid.*); "The Causation of Chloroform Syncope" (*Brit. Med. Journ.*, 1897); "A Simple

Form of Sphygmometer" (*ibid.*); "On Arterial Pressure in Man" (*Journ. Phys.*, xxii.); "On Rest, Sleep and Work on Arterial Pressure" (*Lancet*, 1898); "On Syncope and the Influence of Posture on Rabbits" (*Journ. Phys.*, xxii.); "On the Effects of Cerebral Anæmia produced by Ligation of the Cerebral Arteries" (with Mott) (*ibid.*, 1898); "On Human Cerebro-Spinal Fluid" (*Proc. Roy. Soc.*, 1898). In the press:—"Mechanism of the Circulation" (Schäfer, "Text-Book of Phys."); "Cerebral Circulation" (Allbutt's "System of Medicine").

JOHN HORNE,

F.G.S., F.R.S.E. One of the Senior Geologists on the Staff of the Geological Survey of Scotland. Has been engaged for more than thirty years in the Geological Survey. From 1868 to 1876 he personally studied and mapped large areas of the Silurian uplands of Scotland. From 1876 to 1883 he surveyed extensive tracts in the counties of Nairn, Inverness, Banff and Aberdeen. From 1884 till the present time he has taken an important share in the investigation and mapping of the complicated geology of the North-West Highlands. In addition to these official researches he has devoted his intervals of holiday to original exploration, and has made important contributions to our knowledge of the glacial and volcanic geology of the Orkney and Shetland Isles. Among his papers are the following:—"A Sketch of the Geology of the Isle of Man," and the "Post-Pliocene Formation of the Isle of Man" (*Edin. Geol. Soc. Trans.*, ii., 1174, pp. 323, 329); "The Geology of the Island of Unst" (*Edin. Phys. Soc. Proc.*, iv., 1878, p. 274); "The Volcanic History of the Old Red Sandstone Period North of the Grampians" (*Glas. Geol. Soc. Trans.*, vii., 1881, p. 77). Most of his investigations have been worked out in conjunction with Mr. B. N. Peach, F.R.S., but the results have been arranged and described by Mr. Horne. Some of this conjoint work has been of the highest value, both in regard to British geology and to the theoretical treatment of the science. Special reference may be made to the "Report on the Recent Work of the Geological Survey in the North-West Highlands of Scotland" (*Quart. Journ. Geol. Soc.*, xliv., 1888, p. 378), in which the detailed structure of one of the most intricate geological regions in Europe was worked out and illustrated; to a paper on "The Olenellus-Zone in the North-West Highlands" (*ibid.*, xlviii., 1892, p. 227), which demonstrated the existence and stratigraphical relations of Lower Cambrian Rocks in Scotland; and to the large volume recently published by the Geographical Survey, on "The Silurian Rocks of Scotland" (p. 749), which gives the detailed results of a prolonged and laborious investigation by Messrs. Peach and Horne of the whole Silurian region of southern Scotland. In 1888 was awarded the Wollaston Fund by the Geological Society, and in 1899 received from the same Society, in association with his friend and colleague, Mr. Peach, a duplicate Murchison medal. Received, in 1893, the Neill medal from the Royal Society of Edinburgh, in recognition of the value of his contributions to Geology.

JOSEPH JACKSON LISTER,

M.A., F.Z.S. Demonstrator of Comparative Anatomy in the University of Cambridge. Distinguished as a Zoologist. Was Naturalist on board H.M.S. *Egeria* in two cruises, one to Christmas Island (Indian Ocean), the fauna of which he was the first to investigate, and another in the Pacific among the Tonga, Union and Phœnix Islands, during which he made himself well acquainted with the fauna of those islands, and of the Seychelles. His researches on the Foraminifera have thrown important light on the life-history and reproduction of that group. Author of the following papers:—"On the Natural History of Christmas Island in the Indian Ocean" (*Proc. Zool. Soc.*, 1888, p. 512); "On some Points in the Natural History of Fungia" (*Quart. Journ. Micros. Soc.*, vol. xxix., p. 359); "A Visit to the Newly-Emerged Falcon Island, Tonga Group, S. Pacific" (*Proc. Roy. Geograph. Soc.*, March 1890); "Notes on the Birds of the Phœnix Islands, Pacific Ocean" (*Proc. Zool. Soc.*, 1891, p. 289); "Notes on the Natives of Fakaofu (Bowditch Island), Union Group" (*Journ. Anthrop. Inst.*, 1891, p. 43); "Notes on the Geology of the Tonga Island" (*Quart. Journ. Geol. Soc.*, vol. xlvii., p. 590); "Contributions to the Life-History of the Foraminifera" (Abstract, *Proc. Roy. Soc.*, vol. lvi., p. 155. Full paper, *Phil. Trans.*, vol. clxxxvi., 1895b, p. 401); "A Possible Explanation of the Quinqueloculine Arrangement of the Chambers in the Young of the

Microspheric Forms of Triloculina and Biloculina" (*Proc. Camb. Phil. Soc.*, vol. ix., pt. v.) ; with J. J. Fletcher, "On the Condition of the Median Portion of the Vaginal Apparatus in the *Macropodidae*" (*Proc. Zool. Soc.*, vol. lxiii., 1881, p. 976).

Supplementary Certificate.

Author of "*Astroclera Willeyana*, the representative of a New Family of recent Sponges," in the Zoological Results of Dr. Willey's Expedition, 1899.

JAMES GORDON MACGREGOR,

D.Sc. (Lond.), 1876. M.A. (Dalh.) Professor of Physics, Dalhousie College, Halifax, N.S. Well known for his long-continued Researches on Electrolytic Conductivity, on Solutions, on Resistance of Metals, and on Thermo-electricity. Author of numerous Memoirs contributed to the Royal Society of Edinburgh, the Royal Society of Canada, the Physical Society, and the British Association, including the following:—"Note on the Electrical Conductivity of Saline Solutions" (*Proc. Roy. Soc.*, Edin., 1875) ; "On the Electrical Conductivity of Stretched Silver Wires" (*ibid.*, 1878) ; "On the Variation with Temperature of the Electrical Resistance of certain Alloys" (*Trans. Roy. Soc.*, Edin., 1880) ; "On the Measurement of the Resistance of Electrolytes by means of Wheatstone's Bridge" (*Trans. Roy. Soc.*, Canada, 1882) ; "On some Experiments showing that the Electromotive Force of Polarisation is independent of the difference of Potential of the Electrodes" (*ibid.*, 1883) ; "On a Test of Ewing and MacGregor's Method of Measuring the Electrical Resistances of Electrolytes" (*ibid.*, 1890, with Prof. Ewing) ; "Note on the Volumes of Solutions" (*Brit. Assoc. Report*, 1877, with Dr. Knott) ; "On the Thermo-electric Properties of Charcoal and certain Alloys, with a Supplementary Thermoelectric Diagram" (*Edin. Trans.*, 1879, with Dr. Knott and Prof. Michie Smith) ; "The Thermo-electric Properties of Cobalt" (1876, *Proc. Roy. Soc.*, Edin., 1878) ; "On the Absorption of Low Radiant Heat by Gaseous Bodies" (*ibid.*, 1882-83) ; "On the Resistance to the Passage of the Electric Current between Amalgamated Zinc Electrodes and Solutions of Zinc Sulphate" (*Trans. Nov. Scot. Inst. Nat. Sci.*, 1883) ; "On the Density and the Thermal Expansion of Solutions of Sulphate of Copper" (*Trans. Roy. Soc.*, Canada, 1884) ; "On the Relative Bulk of certain Aqueous Solutions and their Constituent Water" (*Trans. Nov. Scot. Inst. Nat. Sci.*, 1886) ; "A Table of Cubicle Expansions" (*Trans. Roy. Soc.*, Canada, 1888) ; "On the Variation of the Density with the Concentration of Weak Aqueous Solutions of certain Salts" (*ibid.*, 1889, vol. ix., 1891) ; "On the Calculation of the Conductivity of Electrolytes" (*ibid.*, 1896) ; "On the Relation of the Physical Properties of Aqueous Solutions to their State of Ionisation" (*Phil. Mag.*, 1897) ; "On the Hypothesis of Abstract Dynamics and the question of the number of Elastic Constants" (*ibid.*, 1896, with Mr. E. A. Archibald) ; "On Calculation of Conductivities of Electrolytes" (*ibid.*, February 1898). Author of "An Elementary Treatise on Kinematics and Dynamics" (1887).

PATRICK MANSON,

C.M.G., M.D. (Aberd.). F.R.C.P. (Lond.). LL.D. (Aberd.). Physician and Medical Adviser to the Colonial Office. Lecturer on Tropical Medicine, St. George's Hospital, Charing Cross Hospital and London School of Tropical Medicine. Distinguished as a Physician and Parasitologist. Discoverer of Filarial Periodicity of the *rôle* of the Mosquito in Filarial Metamorphosis ; of Filarial Ecdysis ; and of many other points in connection with the life-history of the Filaria nocturna. Discoverer of three other blood-worms of man, viz. *Filaria diurna*, *Filaria perstans* and *Filaria Demarquaii*. Discoverer of the disease known as Endemic hæmoptysis and of its Parasitic cause. Discoverer of *Bothriocephalus Mansoni* and of many points in connection with human and animal helminthology. Was the first to describe accurately and to name *Finea imbricata*, and to prove experimentally its dependence on a vegetable parasite. Was the first to point out the significance of the flagellated body as the initial stage of the extra-corporeal cycle of the malaria parasite, and to enunciate the hypothesis that the mosquito was the host of the parasite at this stage, and therefore an active agent in diffusing malaria, an hypothesis since proved by Major Ross to be correct. Author of a work on *Filaria sanguinis hominis* and some Parasitic Diseases of Warm Climates, 1883 ; of "Tropical Diseases," 1898 ; of the Goulstonian Lectures on

NO. 1594, VOL. 62]

the Life-History of the Malaria Parasite Outside the Human Body, 1896 ; of Papers on the Metamorphosis of *Filaria sanguinis hominis* in the Mosquito (*Trans. Linn. Soc.*, 1883) ; "On the Nature and Significance of the Flagellated Body in Malarial Blood" (*Brit. Med. Journ.*, 1894) ; and of many other papers on the subjects mentioned above and allied matters.

THOMAS MUIR,

LL.D., M.A., F.R.S.E. Superintendent-General of Education in the Cape Colony. Distinguished as a Mathematician and Educationist. Author of "A History of Determinants," and fifty-eight original mathematical papers, including "Continuants : a New Special Class of Determinants" (*Proc. Roy. Soc. Edin.*, 1875) ; "On the Transformation of Gauss' Hypergeometric Series into a Continued Fraction" (*Lond. Math. Soc.*, 1876) ; "New General Formulæ for the Transformation of Infinite Series into Continued Fractions" (*Trans. Roy. Soc. Edin.*, 1876) ; "On Eisenstein's Continued Fractions" (*ibid.*, 1879) ; "On a Systematic Determinant connected with Lagrange's Interpolation Problem" (*Lond. Math. Soc. Proc.*, 1881-2) ; "On New and Recently Discovered Properties of certain Symmetric Determinants" (*Quart. Journ. Math.*, 1882) ; "On the Phenomena of 'Greatest Middle' in the Cycle of a Class of Periodic Continued Fractions" (*Proc. Roy. Soc. Edin.*, 1884) ; "The Theory of Determinants in the Historical Order of its Development" (*ibid.*, vol. xiii.–xvi.) ; "On Some Hitherto Unproved Theorems in Determinants" (*ibid.*, 1891) ; "A Problem of Sylvester's in Elimination" (*ibid.*, vol. xx.) ; "New Relations between Bipartite Functions and Determinants, with a Proof of Cayley's Theorem in Matrices" (*Lond. Math. Soc. Proc.*, vol. xvi.). Has rendered services of the highest importance to education in the Cape Colony, and in his capacities of trustee of the South African Museum and member of the Geological Commission has greatly promoted original scientific research in South Africa.

ARTHUR ALCOCK RAMBAUT,

M.A., Sc.D. (Dublin). Radcliffe Observer. Late Royal Astronomer of Ireland. Late Andrews Professor of Astronomy in the University of Dublin. Author of the following researches in Astronomy and Physics :—Catalogue of the Mean Places of 1012 Southern Stars" (" Astronomical Observations and Researches of Dunsink," part vi.) ; "Catalogue of the Mean Places of 717 Stars" (part vii., *ibid.*) ; "On the Determination of Double Star Orbits from Spectroscopic Observations of the Velocity in the Line of Sight" (*Monthly Notices, Roy. Astron. Soc.*, vol. li., No. 5) ; "To Adjust the Polar Axis of an Equatorial Telescope for Photographic Purposes" (*ibid.*, liv., No. 2) ; "On the Inequality in the Apparent Diurnal Movement of Stars due to Refraction, and a Method of Allowing for it in Astronomical Photography" (*ibid.*, vol. lvii., No. 2) ; "On a Geometrical Method of Finding the most Probable Apparent Orbit of a Double Star" (*Proc. Royl. Dubl. Soc.*, vol. vii., part 2) ; "On the Distortion of Photographic Star Images due to Refraction" (*ibid.*, vol. viii., part 2) ; "On the Rotation Period of the 'Garnet' Spot on Jupiter" (*ibid.*, vol. viii., part 5) ; "On the Relative Positions of 223 Stars in the Cluster χ Persei, as Determined Photographically" (*Trans. Roy. Irish Acad.*, vol. xxx., part 4) ; "On the Possibility of Determining the Distance of a Double Star by means of the Relative Velocity of the Components in the Line of Sight" (*ibid.*, 2nd series, vol. iv., No. 6) ; "The Absorption of Heat in the Solar Atmosphere" (in conjunction with W. E. Wilson) (*ibid.*, 3rd series, vol. iii., No. 4).

WILLIAM JAMES SELL,

M.A. Senior Demonstrator of Chemistry, University of Cambridge. Author of the following papers :—"Volumetric Determination of Chromium" (*Trans. Chem. Soc.*, 1879) ; "On a Series of Salts of a Base containing Chromium and Urea," Nos. 1 and 2 (*Proc. Roy. Soc.*, 1882 and 1889) ; "Anhydro-Derivatives of Citric and Aconitic Acids" (*Trans. Chem. Soc.*, 1892) ; "Salts of a new Platinum Sulphurea Base" (*Brit. Assoc. Rept.*, 1893) ; "Studies on Citrazinic Acid," Pts. I.–V. (*Trans. Chem. Soc.*, 1893–1897) ; "Note on the Action of Chlorine on Pyridine" (*Trans. Chem. Soc.*, 1898) ; "The Chlorine Derivatives of Pyridine," Pts. I.–II. (*ibid.*, 1898) ; "Interaction of Ammonia and Pentachlorpyridine" (*ibid.*) ; "Constitution of Glutazine" (*ibid.*).

W. BALDWIN SPENCER,

B.A. (Oxon.), M.A. (Melbourne). Professor of Biology in the Melbourne University ; formerly Fellow of Lincoln College, Oxford ; Hon. Sec. of the Royal Society of Victoria ; Corr. Member Zool. Soc., Lond. Distinguished as an original investigator in Zoology and Comparative Anatomy ; and as a teacher and organiser. Graduated at Oxford twelve years ago. Has published more than thirty memoirs, among which are :— "On a New Family of Hydroidea Ceratellidæ" (*Trans. Roy. Soc. Vict.*, 1890) ; "The Anatomy of *Megascolides Australis*," and other papers on Australian Earthworms and Planarians (*ibid.*) ; "On New Crustacea and New Mammals," in Report of the Horn Expedition to Central Australia (which he organised) ; "On the Pineal Eye in Lacertilia" (*Quart. Journ. Micro. Sci.*, 1887) ; "On the Habits, Blood-vessels and Lungs of *Ceratodus Fosteri*" ; "On a New Genus of Marsupials from Central Australia " (*Proc. Roy. Soc. Vict.*, vol. ix.) ; "On the Cranial Nerves of Scyllium" (*Quart. Journ. Micros. Sci.*, 1881) ; "On the Early Development of *Rana temporaria* (*ibid.*, 1885) ; "The Fauna and Zoological Relationships of Tasmania," (Presidential Address to Sect. D., Austr. Assoc. Adv. Sci., 1892).

JAMES WALKER,

D.Sc. (Edin.), Ph.D., Leipzig. Professor of Chemistry, University College, Dundee. An active and successful worker in chemistry, especially physical and organic. Author of numerous papers, of which the following are among the most important :— "Zur Affinitätsbestimmung Organischer Basen" (*Zeit. Physikal. Chem.*, iv., p. 319, 1889) ; "Ueber Löslisskeit und Schmelzwarme" (*ibid.*, v., 193, 1890) ; "The Dissociation Constants of Organic Acids" (*Journ. Chem. Soc.*, lxi., p. 696, 1892) ; "The Methyl Salts of Camphoric Acid" (*ibid.*, lxi., p. 1088, 1892) ; "The Electrolysis of Sodium Ethyl Camphorate (*ibid.*, lxiii., p. 495, 1893) ; "The Boiling Points of Homologous Compounds" (*ibid.*, part i., lxv., p. 193, 1894 ; part ii., lxv., p. 725, 1894) ; "Hydrolysis in some Aqueous Solutions" (*Proc. Roy. Soc. Edin.*, vol. xx., p. 255, 1894). Along with Prof. Crum Brown, "Electrolytic Synthesis of Dibasic Acids" (parts i. and ii. *Trans. Roy. Soc. Edin.*, vol. xxxvi., p. 211, 1891, and vol. xxxvii., p. 361, 1893). Along with J. Henderson, "Electrolysis of Potassium Allo-Camphorate" (parts i. and ii. *Journ. Chem. Soc.*, vol. lxviii., p. 337, 1895 ; vol. lxix., p. 748, 1896). Along with F. I. Hambly, "Transformation of Cyanate into Urea" (*Journ. Chem. Soc.*, vol. lxvii., p. 746, 1895). Along with J. R. Appleyard, "Transformation of Methylammonium Cyanates into the Corresponding Ureas" (*Journ. Chem. Soc.*, vol. lxix., p. 193, 1896).

PHILIP WATTS,

Naval Architect and Director of the War-Shipbuilding Department of Sir W. G. Armstrong, Whitworth and Co. Distinguished for his knowledge of the science and practice of Naval Architecture. Responsible designer of a considerable number of the swiftest and most powerful war-ships. Has done much original scientific and experimental work in connection with investigations of the stability of ships and floating bodies ; the oscillations of ships in still water and amongst waves ; the propulsion and manoeuvring powers of ships. Was appointed by the Admiralty and acted for some years as assistant to the late Mr. W. Froude, F.R.S., on the analytical and experimental work carried on by that investigator. In that capacity he took part in the device and application of the process of "graphic integration" by which the oscillations of ships can be approximately determined under assumed conditions of wave motion, including the effect of fluid resistance. Has independently proposed a method of reducing the rolling of ships at sea, by the introduction of free water into a suitably formed chamber. This plan was adopted by the Admiralty for several important ships, after mathematical and experimental demonstration of its efficiency. Was entrusted with the experimental investigation of the turning powers of H.M.S. *Thunderer* made in connection with the work of the *Inflexible* Committee. Devised and applied methods for determining exactly the path traversed by the C.G. of the ship, the rate of acquisition of angular velocity, the angle of heel and other phenomena incidental to turning under the action of the rudder. This investigation led to subsequent modifications in the under-water form of ships, tending to increase their handiness. Is author of the following papers printed in the *Trans-*

actions of the Institution of Naval Architects :— "On a Method of Reducing the Rolling of Ships at Sea" (1883) ; "The Use of Water Chambers for Reducing the Rolling of Ships at Sea " (1885) ; "The Italian Cruiser *Piemonte*" (1889) ; "The Steering Qualities of the *Yashima*" (1898) ; "Elswick Cruisers Built during the last Ten Years" (1899).

CHARLES THOMSON REES WILSON,

M.A. (Cantab.), B.Sc. (Vict.). At present engaged in Investigations on Atmospheric Electricity on behalf of the Meteorological Council. Author of the following papers :— "On the Formation of Cloud in the absence of Dust" (*Cam. Phil. Soc. Proc.*, vol. viii., p. 306) ; "The effect of Röntgen's Rays on Cloudy Condensation" (*Roy. Soc. Proc.*, vol. lix., p. 338) ; "Condensation of Water Vapours in the Presence of Dust-free Air and other Gases" (*Phil. Trans.*, A., (1897), pp. 265-307) ; "On the Action of Uranium Rays on the Condensation of Water Vapour" (*Camb. Phil. Soc. Proc.*, vol. ix., pp. 333-338) ; "On the Production of a Cloud by the Action of Ultra-Violet Light on Moist Air" (*ibid.*, vol. ix., p. 392) ; "Condensation Nuclei produced in Gases by the Action of Röntgen Rays, Uranium Rays, Ultra-Violet Light and other Agents" (*Phil. Trans.*, A., 192, pp. 403-453) ; "Comparative Efficiency as Condensation Nuclei of positively and negatively charged Ions" (*ibid.*, A., 193, pp. 289-308).

LIEUT.-GENERAL PITT-RIVERS, F.R.S.

BY the death of Lieut.-General Augustus Henry Lane-Fox Pitt-Rivers, F.R.S., on May 4, anthropology has lost one of her most prominent and enthusiastic students, and one whose place it will be impossible to fill.

Augustus Henry Lane-Fox was born in 1827. He served with distinction in the Crimea, at Alma and Sevastopol, being during that campaign an officer in the Grenadier Guards, and on the staff. As Lieut.-Colonel Lane-Fox he was the earliest and principal associate of Colonel, afterwards Lieut.-General, Hay, the first Commandant and Inspector-General of Musketry, and about 1855 he wrote and delivered the series of lectures which then, and since, formed a principal part of the Hythe curriculum. He had thus the honour and distinction of being prominently associated with the inauguration of one of the most important reforms in our military system. He had the unusual reputation in those days of military dandies of being an able, studious and scientific officer ; but his career at Hythe was not a long one. While he was there he had the practical training and instruction of those who came to qualify as musketry instructors ; and he added to, if he did not originate, the interesting collection of ancient arms and weapons and projectiles in that establishment. General Pitt-Rivers never lost his interest in military matters, and as late as 1893 he was appointed Colonel of the South Lancashire Regiment.

Few men have had the collecting instinct so strongly developed as had General Pitt-Rivers, but in his case not only were his interests extremely wide, but he had always some method in his collecting ; there was invariably some principle or theory that the objects were designed to illustrate. Consequently he bought with judgment, and what in most collections are "curios" or trophies, under his arrangement became links in a chain of scientific argument, or clever suggestions of stages in the evolution of human thought or handicraft.

The spoils of over twenty years of intelligent collecting were exhibited, in 1874, in the Bethnal Green Museum, and the catalogue of this collection was published by the Science and Art Department. It is no exaggeration to say that this collection was a revelation to many people, and it and the catalogue initiated a new departure in the study of handicrafts. It was, in fact, the first practical application of the theory of evolution to objects made by man. As Colonel Lane-Fox he was, fo example, the first to demonstrate the evolutionary history of patterns, or of certain decorative features from realistic originals. He placed together, side by side, analogous objects from all parts of the world, and often he was enabled to demonstrate the origin and modifications of modern weapons, utensils, and the like. This system has its dangers; analogy may often be mistaken for homogeny, and it must be admitted that mistakes were occasionally made or wrong inferences suggested; but with care these may be greatly reduced, and this system of studying human productions appeals alike to the general public and to scientific men. We believe that the collections exhibited in 1874 were first offered to the University of Cambridge, but now they find a final resting-place in the museum at Oxford, where they have since been greatly added to and further elaborated.

Owing to the death of the sixth Baron Rivers in 1880, Mr. Lane-Fox succeeded to large estates in Wiltshire and Dorsetshire, and he assumed the name of Pitt-Rivers. This gave him his chance; many years previously his keen eye had noted the numerous earthworks and tumuli on Cranborne Chase, but he little thought that fortune would hand them over to his keeping.

In 1881 the General commenced excavating, and in 1887 he published the first of his four quarto memoirs on the results of his digging. Many burrows had been rifled before by antiquaries, but never had excavations been so systematically and thoroughly studied in this country. These memoirs are monuments to the princely liberality, technical skill, and conscientious attention to details that characterised General Pitt-Rivers.

In order to display the finds obtained in his excavations, Pitt-Rivers built a new museum at Farnham in Dorsetshire, and once more he gave rein to his passion for collecting, and soon an extensive and valuable ethnographical museum sprang up in this remote country village. Here, systematically arranged and described, may be seen models of the sites and excavations, and every specimen and fragment thence obtained. In order to illustrate the pottery which is found in various diggings, a comparative collection of pottery and ceramics was started which now forms a very valuable epitome of this industry in all ages and climes. In the same manner, a large comparative collection of agricultural implements has been collected. Here also is the collection of locks, upon which he based the memoir he published in 1883. The collections of general ethnography are surprisingly rich, and his well selected specimens of Benin metal work constitute perhaps the most representative series extant. Words fail to express one's surprise at finding this wonderful museum buried in the depth of the country.

At Tollard Royal, near Farnham, the General very carefully restored a thirteenth century house, which is known as King John's House—this he converted into a museum mainly designed to illustrate the rise of the art of painting; and with characteristic thoroughness he began with paintings of the twentieth and twenty-sixth dynasties.

Not far off are the Larmer Grounds, a park which has been beautifully laid out and provided with numerous picturesque large summer-houses for the use of excursionists. During the warm weather a band plays on Sunday afternoons, and large numbers of people avail themselves of the General's hospitality. In this effort to provide free and innocent enjoyment to the multitude, General Pitt-Rivers received much opposition from well-meaning but misguided sabbatarians; but in this as in so many other matters, he pursued what he considered to be his duty without being influenced by the opinions or opposition of others. He was very fond of joining the happy throngs, and he was never more pleased than when many thousands assembled on

great occasions, such as the annual races. It is gratifying to know that his liberality was never abused by unseemly conduct.

General Pitt-Rivers' written contributions to anthropological literature were very numerous, and in his time he took an active part in the work of various societies. General Pitt-Rivers was a Fellow of the Royal Society; on more than one occasion he was President of the Anthropological Institute; and he was a Vice-President of the Society of Antiquaries. His last public appearance was when he read an address as Vice-President of the Royal Archæological Institute at Dorchester in 1893. He was Inspector under the Ancient Monuments Protection Act of 1882, and in this capacity he visited the scheduled monuments; but even his energy was powerless to counteract the restricted powers and scope of the Act.

It would be difficult to detail the wide range of subjects that interested General Pitt Rivers, and the remarkable knowledge he had on so many subjects. He was by no means a man whose sympathies narrowed with age. His strong physique, indomitable energy and imperious will enabled him to accomplish an immense amount of work, and his trained mind, combined with wide knowledge and sympathy, rendered that work of especial merit. Possessed of an abundance of means, he spent lavishly on his beloved science. His strenuous life was devoted to the advancement of knowledge and to the instruction and recreation of the populace.

A. C. H.

NOTES.

THE council of the Society of Arts attended at Marlborough House, on May 8, when his Royal Highness the Prince of Wales, K.G., President of the Society, presented the Albert medal of the Society to Sir William Crookes, F.R.S., "for his extensive and laborious researches in chemistry and in physics; researches which have, in many instances, developed into useful practical applications in the arts and manufactures."

THE Paris correspondent of the *Times* states that the committee of the Paris Academy of Sciences has selected as candidates for election as permanent secretary, in place of the late M. Joseph Bertrand, M. Cornu, professor at the École Polytechnique, and M. Darboux Jean, of the Faculty of Sciences in Paris.

BY the will of the late Mr. G. J. Symons, F.R.S., a valuable bequest is made to the Royal Meteorological Society. Mr. Symons was a great lover of old books, and had succeeded in getting together an extensive meteorological library. He bequeathed to the Society all his books, pamphlets, maps and photographs a copy of which is not already in its library. He also bequeathed his Cross of the Legion of Honour, his Albert medal, and other decorations, as well as the testimonial album presented to him by the Fellows of the Society in 1879. In addition to the above he also bequeathed the sum of 200l.

MR. GOSCHEN made an important announcement at the annual dinner of the Iron and Steel Institute last week. He said that, with a view to developing the power of English guns by means of improving the propellant agent, a committee has been appointed, with Lord Rayleigh as chairman, to investigate the whole subject. The reference to the committee is to carry out trials to ascertain what are the best smokeless propellants for use in existing guns of all natures and in existing small arms, and to report as to whether any modification in the existing designs of guns is desirable with a view to developing to the full the powers of any propellant which may be proposed.

. THE "King of the Belgians has created Mr. E. Windsor Richards, past-president of the Iron and Steel Institute of Great Britain, a Knight Commander of the Order of Leopold.

THE Royal Commissioners who were recently appointed to inquire into and report upon the condition of the salmon fisheries of England, Wales and Scotland, commenced their inquiry on Tuesday.

FORTY-SIX of the sixty-five automobile vehicles which left London on April 23 for a 1000-mile trial, returned on Saturday last. From a report in the *Times*, we gather that the mechanical results of the trial have been very much what they were expected to be. That is to say, the established type of machine has proved itself entirely trustworthy, and between the Daimler, Napier and Panhard motors there has been, in the matter of "staying power," practically nothing to choose. Of the cars which entered, only four were driven by any other motive power than petroleum spirit, and of these one steamer only survived.

LETTERS received from Mr. Moore and Mr. Fergusson announce that the Tanganyika Expedition arrived at Lake Kivu on December 7, 1899, having left two of their party (Messrs. Berridge and Mathews) at the head of Lake Tanganyika. They had ascended the active volcano of Karunga, north of Lake Kivu (11,350 feet), but found only steam without lava issuing from the orifice. They arrived on the shores of Lake Albert-Edward on January 21, and on February 12 were at Fort Gerry, the English post in Zoru, near Mount Ruwenzori, which they were proposing to ascend.

SOME living specimens of the very curious blind fish of the caves of Kentucky, U.S.A. (*Amblyopsis spelaea*), may now be seen in the Zoological Society's fish-house, where they have been deposited by the Hon. Walter Rothschild. They are of a nearly uniform pale flesh colour. When exposed to the light these creatures hide themselves among the stones in the tank in which they are placed, though when shaded they seem to swim about pretty freely like other fishes, but usually keep near the surface.

THE Ancient Monuments Protection Bill was read a second time in the House of Lords on Tuesday. The measure extends the provisions of the Ancient Monument Act of 1882, and proposes that local authorities should be empowered to take over the charge of national monuments and to receive voluntary contributions towards the cost of maintaining and preserving them. Some of the London open spaces had been preserved in this manner, and there seemed no reason why the same principle should not be applied to monuments of archæological interest.

THE U.S. Congress has under consideration a Bill for the conversion of the present Office of Standard Weights and Measures into a National Standardising Bureau similar to the Reichsanstalt at Charlottenburg, and the National Physical Laboratory. The clause dealing with the work of the bureau reads as follows :—The functions of the bureau shall consist in the custody of the standards ; the comparison of the standards used in scientific investigations, engineering, manufacturing, commerce, and educational institutions with the standards adopted or recognised by the Government ; the construction when necessary of standards, their multiples and subdivisions ; the testing and calibration of standard measuring apparatus ; the solution of problems which arise in connection with standards ; the determination of physical constants, and the properties of

materials when such data are of great importance to scientific or manufacturing interests, and are not to be obtained of sufficient accuracy elsewhere.

THE gold medal of the Linnean Society of London, which is annually presented alternately to a zoologist and to a botanist, has this year been awarded to Prof. Alfred Newton, F.R.S., in recognition of his important contributions to zoological science. To the general public Prof. Newton's name will be best known in connection with the latest addition of Yarrell's " British Birds " (vols. i. and ii. of which were revised and edited by him), and the " Dictionary of Birds," an admirable compendium of ornithology. As editor of the *Ibis* (1865-70), and of the " Zoological Record " (1870-72), to which for some years previously he had supplied the annual record of the literature relating to Aves, he has placed ornithologists of all nations under great obligations to him, as he has done, also, by his publications on the avifauna of Iceland, Greenland, the West Indies, the Mascarene and Sandwich Islands, and by his articles in the " Encyclopædia Britannica " and the " Dictionary of National Biography." As chairman for many years of a committee of the British Association he has been instrumental in securing the publication of valuable reports on the migration of birds and in obtaining legislative protection for the more useful species by the appointment of a close time. The medal will be presented at the Anniversary Meeting of the Linnean Society.

THE Académie Royale de Belgique announces the following prize subjects for 1900 :—*Mathematical and Physical Sciences :* A description of researches on critical phenomena in physics, together with new researches upon this subject ; new researches on the viscosity of liquids ; study of the derived carbonates of an element of which the compounds are still little known ; the variation of latitude, together with a discussion of the reasons which have been put forward to account for it ; a contribution to the algebraic and geometric study of *n* linear forms, *n* being greater than 3 ; new researches on the thermal conductivities of liquids and solutions. *Natural Sciences :* The determination of the limits of the Comblain-au-Pont formation, and the place it should occupy in geological classification. Is it Devonian or Carboniferous ? ; researches on the modifications produced in minerals by pressure ; researches on the organisation and development of a flat-worm with the object of determining whether there exist any phylogenic relationships between Platyhelminthes and Enterocœla ; does a nucleus exist in Schizophytes (Schizomycetes) ? if so its structure and mode of division ? ; researches on Devonian plants of Belgium, from the point of view of description, stratigraphical position, and, if possible, anatomical characters. The value of the gold medal to be awarded for each subject is six hundred francs. Memoirs may be written in French or in Flemish, and must be sent to the secretary of the Academy by August 1.

WE learn from the *Electrician* that an instrument called the telephonograph, which is a modification of the phonograph, was recently inspected and tested by the German Postmaster-General and several engineers. Its inventor, Herr Paulsen, a Dane, has replaced the wax cylinder of the Edison phonograph by a steel band, and the style by a magnet energised by a telephone. Currents transmitted by the telephone pass through the electromagnet and create consequent poles on the steel band, and more or less the converse operation is employed for reproducing the sound. A long line can, of course, intervene between the transmitting telephone and the phonograph itself, and it is suggested that a telephone subscriber on leaving his office can set such a telephonograph to receive telephoned messages during his absence.

EXPERIMENTS on the exposure and development of photographic plates in ordinary light have recently been made by Prof. F. E. Nipher, and are described in *Science.* It appears that if a photographic plate in a camera is greatly over-exposed it may be developed in the light. A plate which should for ordinary work have an exposure of a second and a half for street or outdoor photography, may be exposed for two hours. When developed with weak hydrokinone by the light of a lamp, it gives a good positive. If the plate is held too near the lamp the light will dissolve a picture already appearing. If held too far away the plate begins to fog. By moving toward or from the lamp the proper illumination may be soon secured. It is remarkable that a street scene taken in this way shows not a moving thing on the streets. In Prof. Nipher's pictures, tram-cars passing every two minutes, waggons, horses, pedestrians, left no trace upon the plate. But the fixed objects are shown perfectly, with their proper shadows and high lights. Prof. Nipher points out that lantern slides and transparencies may be made directly by this method without re-photographing from a negative. Röntgen ray pictures can also be obtained upon plates which have been exposed to the light of an ordinary room for a few days, by developing in the manner described. Good radiographs have been thus produced upon plates which were uncovered during exposure to the rays.

THE usual proof that the arithmetic mean of any number of positive quantities is greater than their geometric mean consists in showing that if any two of the quantities be replaced by their semi-sum, the new series has the same arithmetic mean and a greater geometric mean. This proof, however, involves the assumption that if this process of substitution be repeated indefinitely, the ultimate result will be a series of quantities each equal to the arithmetic mean of the original series. We have never seen this property proved, and it is certainly by no means an obvious truth in the general case, for the result of the repeated operations is always a fraction whose denominator is a power of 2, while the arithmetic mean of n quantities has n for its denominator. We are glad to see that Mr. G. E. Crawford, writing in the *Proceedings* of the Edinburgh Mathematical Society, recommends an alternative proof in which the number of steps is finite, and the above assumption is not made. Two such proofs are possible, both of which run on somewhat parallel lines, and Mr. Crawford refers to a text-book which appeared a few years ago for the alternative to the proof now given.

WE have received from Prof. A. Klossovsky, the energetic director of the meteorological system of South-west Russia, a very valuable contribution to climatology. The work consists of two volumes, text and charts, and embraces the large area running from about the latitude of London to the northern shores of the Black Sea and the Sea of Azov, and bounded on the east by the River Dnieper. The observations used in the discussion include those made at the stations belonging both to the Central Meteorological Service of St. Petersburg and to the South-west Russian system, and embrace a period of twenty-five years (1871–1895). The tables exhibit monthly, yearly and five-yearly values of all the principal elements, and the distribution of thunderstorms and hail. The charts are coloured, and show very clearly the mean annual distribution of rainfall, the number of days of thunderstorms and hail, mean and extreme temperatures, and the distribution of cloud and humidity. The tables are arranged in various ways, and furnish most useful statistics for agriculturists and for men of science generally.

IN describing some Neocene corals of the United States (*Proc. U.S. National Museum*, vol. xxii. 1900), Dr. H. S. Gane remarks that a majority of the corals in these Eocene, Miocene

and Pliocene formations belong to extinct species. They do not, however, present any close kinship with the corals of a like age in the West Indies, but are more nearly related to tho e now living in the Caribbean Sea and Atlantic Ocean.

MR. CECIL B. CRAMPTON, who has for some time been assistant to Prof. Boyd Dawkins in the museum at Owen's College, Manchester, has been appointed an assistant geologist on the Geological Survey of Scotland.

MR. LESTER F. WARD gives an account of the wonderful " Petrified Forest " or " Chalcedony Park " of Arizona (Report to Department of the Interior, U.S. Geol. Survey, 1900). Countless logs of silicified wood occur over a wide area in Arizona, but they are especially abundant in a particular tract known as the " Petrified Forest," east of Holbrook, between the Little Colorado and Rio Puerco. Here the logs lie in the greatest profusion, " while the ground seems to be everywhere studded with gems, consisting of broken fragments of all shapes and sizes, and exhibiting all the colours of the rainbow." These silicified blocks are not *in situ*, but have been derived from a bed of conglomeratic sandstone of Triassic age, which is exposed on the margin of a high plateau. Mr. Ward refers also to a well-known " Natural Bridge," which consists of a petrified trunk lying across a canyon, and forming a footbridge, and he observes that the trunk here is *in situ*. He advocates that means be taken to preserve these natural phenomena.

A REPORT on the proposed railway from the Commune des Houches, Bonneville, in Haute-Savoie, to the summit of Mont Blanc, has been published by M. Joseph Vallot, Director of the Mont Blanc Observatory, and M. Henri Vallot, engineer. This great undertaking was projected by M. Saturnin Fabre, but various routes have been suggested. These are fully discussed by the authors, who give reasons for recommending a route which starts from the valley of the Arve at an elevation of about 3000 feet, and proceeds by the Aiguille du Goûter and the Dôme du Goûter to a terminus at the Petits Rochers Rouges, where the elevation is about 15,000 feet. The total length of the railway would be about seven miles, and from an elevation of about 4000 feet to its upward termination, the line would for the most part be subterranean. There would be several openings, and also stations giving access to the mountain, at points of special interest and beauty. M. Joseph Vallot contributes chapters on the geology, including the glacial phenomena, and these are illustrated by a section showing the nature of the solid rocks through which the railway would be carried, and the thickness of the glacier-ice above. For a short distance the tunnelling would be made through Liassic slates and Trias with gypsum, and then wholly through various crystalline schists.

IN his memoir recently published in the *Philosophical Transactions* (see NATURE, April 19, p. 595), Mr. Oldham has shown that, in recording the movements due to distant earthquakes, the heavy vertical pendulums employed in Italy answer most readily to the early tremors, while the light horizontal pendulums of Rebeur-Paschwitz and others are most affected by the later-arriving surface-undulations. Dr. G. Agamennone has discussed the same subject independently in a note read on February 18 before the R. Accademia dei Lincei of Rome. At the Rocca di Papa Observatory, of which he is director, are two horizontal pendulums provided with mechanical registration. It is found that these instruments fail to indicate small local shocks, while in recording distant earthquakes they lag behind the vertical pendulums with stationary masses. But, by increasing the weight which the horizontal pendulums carry from 25 to 60 kg., this defect almost disappears. Dr. Agamennone therefore proposes to erect an additional pair of horizontal pendulums at Rocca di Papa in which the masses shall be at least 500 kg., the period of oscillation 10 to 15 seconds, and the magnifying ratio of the writing stiles 50, and possibly 100. He also suggests a double system of registration ; the stiles at one end are to write in ink on white paper moving with a velocity of about 50 cm. per hour, and at the other end on an endless strip of smoked paper, which, on the occurrence of a shock, will be made to travel still more rapidly. The former record will enable the initial and final epochs to be determined, and the latter the period of the individual oscillations.

WE have received the *Transactions* and *Report* of the Manchester Microscopical Society for 1899, which contains a good account of the late Zoological Congress, and also some well illustrated papers on various biological subjects of current interest.

DR. C. S. MINOT has favoured us with a copy of a paper from vol. xxix. of the *Proc. Boston Soc. Nat. Hist.*, entitled "On a hitherto unrecognised form of blood circulation without capillaries in the organs of Vertebrata." "Sinusoids" is the name proposed for the newly-discovered vessels, which are said to differ in man y respects from true capillaries.

THE "Descriptive Guide" to the collection of corals now on view at the South London Art Gallery, Peckham Road, Camberwell, may be described as a wonderful "pennyworth." Not only does it contain two excellent photographic plates of corals, but the text is an excellent popular introduction to the study of these beautiful structures. The collection in question is the property of Mr. J. Morgan, of St. Leonards, who has kindly loaned them for public exhibition.

FROM its last *Report* we are glad to notice that the Felsted School Scientific Society appears in a flourishing condition. Physics and chemistry receive a larger share of attention than is usually the case in school societies, owing, doubtless, to the fact that their president, Mr. A. E. Munby, teaches these subjects in the school. Dr. Charles Hose has kindly offered to present a selection from his Bornean treasures if proper accommodation can be obtained for their display.

"CODIUM" is the title of the fourth issue of the Liverpool Marine Biological Committee's memoirs. The remarkable branching alga originally described as *Fucus tomentosus*, but now designated *Codium tomentosum*, is one of three British representatives of the group *Codiaceae*, but the only one found within the area treated of in this series of memoirs. Although widely distributed, it occurs within the district only in shallow rock-pools at the south end of the Isle of Man. The plants are perennial, and fruit in winter ; the season of fructification apparently extending from November till February. The authors of the memoir have worked out the life-history of the organism so far as their materials admitted of this being done ; but there are certain problems connected with the reproduction which require further investigation.

DR. R. W. SHUFELDT contributes a paper on the psychology of fishes to the April number of the *American Naturalist.* In general it may be said that fishes possess excellent visual power, even to the exact discrimination between objects ; and there is also reason to believe that they are extremely sensitive to any disturbance of their native element, when such disturbance is within the range of appreciation of their nervous organs. Whether, however, any fish has the extreme sensibility of the leech *Clepsine*, which, when the experiment is conducted with

proper precautions, will be conscious of the touch of a needle-point on the surface of the water of the dish in which it is placed, is more than doubtful. The peculiar sensitiveness to teasing exhibited by the fish known as the snowy grouper (*Epinephelus niveatus*) is instanced by the author as a phenomenon requiring special explanation. When much disturbed, this fish displays a spasm, or fit, much resembling the contortions of death, eventually floating belly-upwards, and at the same time changing colour. The author suggests that these movements are for the purpose of warning off predatory fish, which prefer to take their prey in a healthy condition after an exciting chase.

A PRELIMINARY REPORT on the Klondike Goldfields, Yukon District, Canada, has been published by Mr. R. G. McConnell. (Printed in advance from the Summary Report, Geol. Surv. Canada for 1899, 1900). The rocks consist of stratified and foliated rocks, mostly Palæozoic, and of granites and other eruptive rocks of Tertiary age. Of the older rocks, the Klondike series constitutes the country rock along the productive portions of all the richer creeks. It mainly comprises micaceous schists, greatly crushed and altered, which pass in places into a granitoid rock. Quartz veins are very abundant, and these occasionally contain free gold. The placer deposits have derived their gold from the quartz veins and silicified schists of the district ; and it is considered probable that productive veins, or zones of country rock, will eventually be discovered. A fairly full account is given of the placer deposits, so far as they are known, but the author remarks that the work of the prospector will not be completed for many years. The valleys known to be productive in gold are shown on a map.

"NOTES on the Fossil Flora of South Gippsland," by Mr. James Stirling, Government geologist, have been published by the Department of Mines, Victoria (1900). The plants, which include ferns, cycads and conifers, were obtained from Jurassic strata. They are illustrated in six plates, which accompany the notes, and which were prepared under the direction of the late Sir Frederick McCoy.

MESSRS. ARCHIBALD CONSTABLE AND CO. will publish in a few weeks Mr. W. Worby Beaumont's new and comprehensive work, "Motor Vehicles and Motors : Their Design, Construction and Working by Steam, Oil and Electricity."

PROF. J. A. EWING'S standard work on "Magnetic Induction in Iron and Other Metals" (The *Electrician* Printing and Publishing Co.) has reached a third edition. A chapter has been added on practical magnetic testing, and important advances made since the book was originally published have been taken into consideration.

MESSRS. NEWMAN AND GUARDIA, LIMITED, have just placed a new quarter-plate pocket camera—the "Nydia"—upon the market. The camera is fitted with a special 5¼-inch lens, either Zeiss, Wray or Ross make ; it carries twelve plates or eight plates ; when folded it measures only 7¼ × 4¼ × 1¾ inches, and it only weighs, when loaded, 1¾ lbs. Photographers requiring a portable and efficient hand camera at a moderate price should see the "Nydia."

THE fourth edition of "Psycho-therapeutics," by Dr. C. Lloyd Tuckey, has been issued by Messrs. Ballière, Tindall and Cox. The third edition was published nearly ten years ago and since that time hypnotism and suggestion have become recognised forms of medical treatment. Dr. Tuckey's work is a useful statement of the development of the system of psycho-therapeutics, both from the theoretical and practical sides, and it will show practitioners what can be accomplished by hypnotic suggestion.

ALTHOUGH a large amount of work has been published upon the physical properties of dilute solutions of single electrolytes, the experimental study of solutions of mixed electrolytes, notwithstanding its great interest from the point of view of the electrolytic theory of dissociation, has not been worked at so extensively. The theoretical discussion of such mixtures leads to a set of equations somewhat difficult to solve ; but since Prof. MacGregor, of the Dalhousie College, Halifax, Nova Scotia, showed how to solve these equations by a simple graphical method, systematic researches have been carried on at this college on the properties of such mixed solutions of electrolytes. A recent paper, by Mr. J. Barnes, in the *Transactions* of the Nova Scotian Institute of Science, deals with the depression of the freezing point in salts containing a common ion ; and the results show that in the case of mixtures of potassium chloride and sodium chloride, and of sodium chloride and hydrochloric acid, and of all three, it is possible, with the ionisation coefficients obtained by Prof. MacGregor's method, and on the assumption that the molecular depression of an electrolyte in a mixture is the same as it would be in a simple solution of the same total concentration, to predict the depression of the freezing point within the limits of the error involved in observation and calculation.

THE additions to the Zoological Society's Gardens during the past week include a Bonnet Monkey (*Macacus sinicus*, ♂) from India, presented by Lady Malcolm, of Poltalloch ; a White-crested Tiger Bittern (*Tigrisoma leucolophum*) from West Africa, presented by Mr. W. F. Marshal ; four Chaplin Crows (*Corvus capellanus*) from Southern Persia, presented by Mr. B. T. Ffinch ; a Cinereous Vulture (*Vultur monachus*), European, presented by Mr. W. E. Found ; a Common Boa (*Boa constrictor*) from South America, presented by Mr. F. H. Preston ; two Egyptian Foxes (*Canis niloticus*) from North Africa, two Prevost's Squirrels (*Sciurus prevosti*) from Malacca, a Ring-tailed Coati (*Nasua rufa*) from South America, two Porto Rico Pigeons (*Columba squamosa*) from the West Indies, a Sclater's Cassowary (*Casuarius sclateri*), two Red-sided Eclectus (*Eclectus pectoralis*) from New Guinea, four Logger-head Turtles (*Thassochelys caretta*) from Tropical seas, twelve Elegant Terrapins (*Chrysemys scripta elegans*), seventeen Lesueur's Terrapins (*Malacoclemmys lesueuri*) from North America, twelve Adorned Terrapins (*Chrysemys ornata*) from Central America, seven Reeves's Terrapins (*Damonia reevesi*) from China, deposited ; five Hairy Armadillos (*Dasypus villosus*) from La Plata, four Common Indian Starlings (*Sturnus menzbieri*), a Bengal Fox (*Canis bengalensis*) from India, two Meyer's Parrots (*Poeocephalus meyeri*) from South-east Africa, four Australian Sheldrakes (*Tadorna tadornoides*), five —— Wood Swallows (*Artamus sordidus*) from Australia, six Sulphury Tyrants (*Pitangus sulphuratus*), a Black-pointed Tegue_{x}in (*Tupinambis nigropunctatus*) from South America, purchased ; a Crowned Lemur (*Lemur coronatus*), six Common Wolves (*Canis lupus*), a Llama (*Lama peruana*) born in the Gardens.

OUR ASTRONOMICAL COLUMN

UNPUBLISHED OBSERVATIONS AT RADCLIFFE OBSERVATORY, 1774-1838.—In a pamphlet containing a reprint of an article in *Monthly Notices*, vol. lx. pp. 265-293, Dr. A. A. Rambaut, Radcliffe observer at Oxford, calls attention to a very valuable collection of astronomical observations which are pre-

served at the Radcliffe Observatory, but have not been reduced or published. Two of the Oxford astronomers, Profs. Hornsby and Robertson, spent a large amount of labour in reducing Bradley's observations made at Greenwich from 1750-1762, and further continued his work by themselves maintaining a systematic and regular series of observations for sixty-five years, from 1774-1838. These were all made with the instruments supplied by Bird to the Radcliffe Observatory at its installation, consisting of two quadrants each of 8-feet radius, a transit instrument of 8-feet focal length, and a zenith sector of 12-feet focus. The observations have all been methodically copied in a similar form to their printed edition of Bradley's observations, and contain altogether about 130,000 transits and 60,000 zenith distances. Dr. Rambaut states that his staff at present could not undertake the reductions ; but, in order to show the extreme importance of the data available, he has made a selection of them, giving the probable errors compared with other observers.

The planets and sun have received considerable attention, there being about 8000 observations of the sun alone, a number little less than that on which Leverrier's tables were founded, and, moreover, covering the period when the corrections to the mean longitude of the sun, as deduced at Greenwich, Paris and Königsberg, are most discordant.

The working list of stars includes about 4870 of those observed by Flamsteed and Bradley, so that direct comparisons could be made in the reductions. Their great value would be specially apparent in the question of proper motions, filling up as they do the long gaps between Bradley and Piazzi, or Bradley and Pond. Specimens of Dr. Rambaut's reductions are given in the paper to show the high degree of accuracy attained by the observations.

MAXIMUM DURATION FOR A TOTAL SOLAR ECLIPSE.— Mr. C. T. Whitmell, president of the Leeds Astronomical Society, recently read a paper showing the results of calculations he had made in the endeavour to ascertain what is the maximum duration possible for a total solar eclipse (*Monthly Notices*, R.A.S., vol. lx. pp. 435-441). After considering the several effects of the varying distances of sun and moon from the earth in determining size of umbra and velocity of shadow, he cites the following five conditions as required for maximum duration of totality :—

(1) The new moon, at or very near a node, must also be at the most favourable perigee possible ; (2) the sun must be at apogee ; (3) during totality, which should be observed at local noon, the moon's shadow should run along a parallel of latitude, in order that the diurnal movement of the observer may be for the time parallel to the motion of the moon, thereby producing its full effect in detaining him within the umbra ; (4) the sun and moon should be in the zenith, so that the umbra may be as large as possible ; (5) the observer should be on the equator, so that his linear velocity may be as great as possible.

Of these, owing to the sun and moon *not* moving in the plane of the celestial equator, it is impossible that (4) and (5) can be simultaneously fulfilled ; (5) is more favourable than (4).

Taking the moon's horizontal parallax as 61' 22",

,,   ,,   earth's radius as 3963 miles,
,,   ,,   moon's   ,,   1081·5 miles,

and using the present accepted *eclipse* values of the diameters of the sun and moon, the maximum totality will occur near the middle of July, at noon, in geocentric north latitude about 4° 52', and will last about 7m. 40s., the sun being at apogee with a parallax of 8"·70. This is on the assumption that the declinations of the two bodies are considered practically constant during totality. The author gives the following list of long duration eclipses, calculated by Mr. Crommelin from Oppolzer's data :—

	Duration at noon	Position of noon line		
Date		Longitude	Latitude
	h. m.		
1901 May 18	... 6 41·6	... 97 E.	... 2 S.
1919 May 29	... 7 5·9	... 18 W.	... 4 N.
1937 June 8	... 7 19·9	... 131 W.	... 10 N.
1955 June 20	... 7 24·5	... 117 E.	... 15 N.
1973 June 30	... 7 19·6	... 6 E.	... 19 N.
1991 July 11	... 7 10·7	... 105 W.	... 22 N.

THE FRESH-WATER LOCHS OF SCOTLAND.[1]

THE introduction to this paper, published in the *Geographical Journal*, includes the correspondence that passed between the Royal Societies of London and Edinburgh and Her Majesty's Treasury in 1883 and 1884, relative to the survey by a Government Department of some of the inland lakes of Scotland.

The weighty arguments brought to bear upon the Government by these learned societies failed in their object, and the Government declined to undertake the proposed surveys. In these circumstances the authors determined a few years ago to make a systematic survey of all the fresh-water lochs of Scotland, and the present paper is the first instalment in the publication of their results, dealing with a compact series of lakes directly or indirectly connected with the water-supply to the city of Glasgow, viz. Lochs Katrine, Arklet, Achray, Vennachar, Drunkie, Voil, Doine and Lubnaig, which form part of one united drainage system having its outflow by the River Teith.

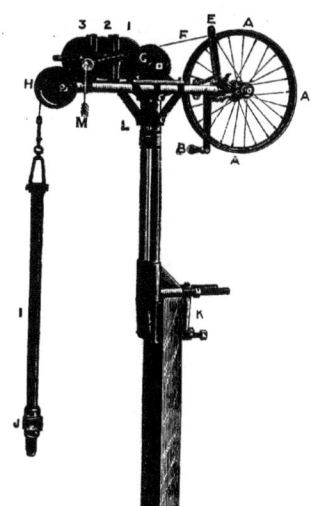

FIG. 1.—F. P. Pullar's sounding machine.

In order to overtake the large amount of work contemplated, involving an immense number of soundings, within a reasonable time, it was necessary to have a portable wire sounding machine adapted for rapid work in small rowing boats. Mr. Pullar designed, and had made, a sounding machine (see Fig. 1), which is described and figured ; this apparatus is admirably adapted for the purpose in view, and with it all the soundings in the different lakes were taken.

The total number of soundings recorded in the paper, taken in the eight lochs mentioned, was 2422, the number varying from 775 in the largest (Loch Katrine) to 90 in the smallest (Loch Doine). These soundings were laid down in position on the large scale (six-inch) Ordnance Survey maps, and contour-lines of depth drawn at certain intervals, from which with the aid of the planimeter the cubic mass of water in each loch was

[1] "A Bathymetrical Survey of the Freshwater Lochs of Scotland." By Sir John Murray, K.C.B., D.Sc., F.R.S., and Fred. P. Pullar, Esq., F.R.G.S. Part I. The Lochs of the Trossachs and Callander District.

calculated. The soundings show that Lochs Katrine, Arklet, Achray, Voil and Doine form each a single basin, while in Lochs Lubnaig, Drunkie and Vennachar the irregularities of the bottom cut up the deep parts of the lochs into separate basins.

The most important of the lakes under consideration is the well-known Loch Katrine, which is eight miles in length, one mile in maximum width, with an area of 4¾ square miles. The greatest depth, 495 feet (82½ fathoms), was found much nearer the eastern than the western end, so that a section drawn down the centre of the loch from west to east (see Fig. 2) shows a gradual increase of depth for nearly four-fifths of the total length, and then a more rapid rise of the bottom towards the eastern end. A section across the loch from north to south (see Fig. 3) shows the deeper part at the point chosen for the section nearer the southern than the northern shore. The mean depth of the loch, *i.e.* the cubic mass of water divided by the area, is 199 feet. The surface of the loch lies at an elevation of 364 feet above the sea, hence a considerable portion of the bottom (over one square mile) falls below the level of the sea ; in this respect Loch Katrine differs from all the other lochs treated of. In connection with the water-supply to Glasgow, Loch Katrine was raised four feet above its previous level, and it is now in process of being raised an additional five feet.

Loch Arklet is a small Highland loch situated between Lochs Katrine and Lomond, at an elevation of 455 feet above the sea. It is over a mile in length, nearly half a mile in maximum width, and covers an area of about one-third of a square mile. The greatest depth, 67 feet, was found nearer the western than the eastern end ; the mean depth is 24 feet. Loch Arklet at present belongs to the watershed of Loch Lomond, but the Corporation of Glasgow have power to divert its waters into the Loch Katrine watershed, with the view of increasing the supply of water to that city.

Loch Achray is situated between Lochs Katrine and Vennachar, at an elevation of 276 feet above the sea. It receives the outflow from Loch Katrine and flows into Loch Vennachar, the level of which is six feet lower. Loch Achray is about 1¼ miles in length, and one-third of a mile in maximum width, covering an area of about one-third of a square mile. The greatest depth, 97 feet, was recorded in two places approximately in the centre of the loch ; the mean depth is 36 feet.

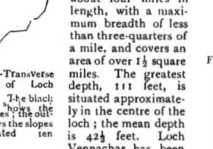

FIG. 3.—Transverse section of Loch Katrine. The black portion shows the true slopes ; the outline shows the slopes exaggerated ten times.

Loch Vennachar is about four miles in length, with a maximum breadth of less than three-quarters of a mile, and covers an area of over 1½ square miles. The greatest depth, 111 feet, is situated approximately in the centre of the loch ; the mean depth is 42½ feet. Loch Vennachar has been raised five feet nine

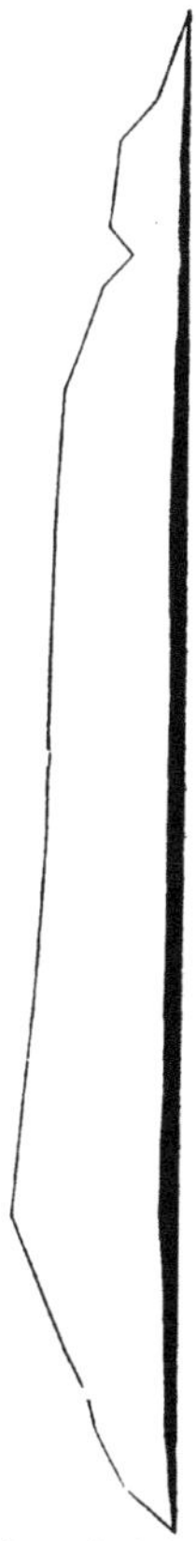

FIG. 2.—Longitudinal section of Loch Katrine along the axis of maximum depth. The black portion shows the true slopes ; the outline shows the slopes exaggerated ten times.

inches in connection with he Glasgow water-supply, for the purpose of providing compensation water to the River Teith.

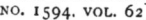

Loch Drunkie is a peculiar irregular little Highland lake, shut in on all sides by high hills, situated, at an elevation of 416 feet above sea-level, a quarter of a mile to the south of Loch Vennachar, into which it flows. It is remarkable in shape, a quadrangular body throwing out three arms in different directions; the maximum length is over a mile, the maximum width of the body over a quarter of a mile, and the area nearly a quarter of a square mile. The greatest depth, 97 feet, was found near the base of the north-eastern arm; the mean depth is 36 feet. Loch Drunkie was raised twenty-five feet in connection with the Glasgow water-supply, for the purpose of supplying compensation water to the River Teith.

Lochs Voil and Doine formed at no distant date a continuous loch, which has been divided into two portions by the material deposited by the rivers. The level of these lochs being fifty feet higher than that of Loch Katrine, it has been suggested that an additional supply of water to Glasgow can be obtained by means of a tunnel from Loch Doine to Loch Katrine through the intervening hills. Loch Voil is over 3½ miles in length, about one-third of a mile in maximum width, and covers an area of nearly nine-tenths of a square mile. The greatest depth, 98

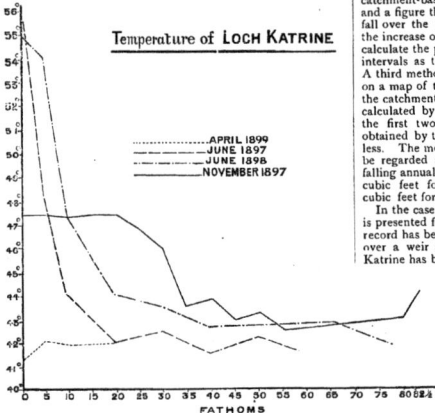

Temperature of LOCH KATRINE

.... APRIL 1899
————— JUNE 1897
———————— JUNE 1898
——————— NOVEMBER 1897

FIG. 4.—Temperature of the water in Loch Katrine.

feet, was found towards the western end; the mean depth is 41 feet. Loch Doine is nearly a mile in length, and over a quarter of a mile in maximum breadth, with an area of nearly a quarter of a square mile. The greatest depth, 65 feet, was found towards the eastern end; the mean depth is 33 feet.

Loch Lubnaig is one of the most interesting of the lochs under consideration, the configuration of the bottom being much more irregular than in any of the others. It receives the outflow from Lochs Doine and Voil, its level being nine feet lower (405 feet above sea-level). The greatest depth, 146 feet, was found approximately near the centre of the loch; the mean depth is 42¾ feet. The contour-lines of depth do not follow the contour of the loch, hollows and ridges alternate with each other, and in some places comparatively deep water is found close to the shore, while in other places shallow water extends a considerable distance from shore.

The deposits forming on the floor of these lochs are described in detail, and the numerous temperature observations taken at the surface and at various depths down to the bottom (some of which were taken as recently as March 1900) are fully discussed. The serial temperatures taken in Loch Katrine are shown

graphically in Fig. 4. The relation between the variation of temperature and the size and depth of the lochs is pointed out, and shows how much more suitable a large deep lake is, as a source of water-supply, than a shallow basin, ensuring a relatively low temperature in summer and a relatively high temperature in winter. Interesting observations were made on the pelagic and other organisms in the various lochs, and on their variation with the season, certain species being obtained in abundance at certain seasons and absent or rare at other seasons.

The amount of rain falling annually on the drainage areas of these lochs was estimated by three methods, the lochs being grouped into two series, viz. Lochs Katrine, Achray, Drunkie and Vennachar, which have their outlet at the eastern end of Loch Vennachar, forming one series; and Lochs Doine, Voil and Lubnaig, which have their outlet at the southern end of Loch Lubnaig, forming the other series. The readings of the rain-gauges at all the observing stations within and near the catchment-basins of these lochs were grouped together into two series corresponding with the two series of lochs indicated, and the average annual rainfall at the average height of the gauges calculated for each series. By the first method, 2½ per cent. of annual rainfall was added for each 100 feet of mean height of the catchment-basin above the average height of the rain-gauges, and a figure thus obtained representing the average annual rainfall over the entire catchment-basin. By the second method, the increase of 2½ per cent. per 100 feet of height was used to calculate the probable annual rainfall at the same heights and intervals as the contour-lines on the Ordnance Survey maps. A third method was afforded by drawing lines of equal rainfall on a map of the district. The total amount of rain falling on the catchment-basins of the two series of lochs indicated was calculated by these three methods, and the results obtained by the first two methods agree very closely, while the results obtained by the third method were in each case considerably less. The mean obtained by the three methods may probably be regarded as a close approximation to the amount of rain falling annually on these catchment-basins, viz. 14,100 million cubic feet for the Vennachar catchment, and 14,700 million cubic feet for the Lubnaig catchment.

In the case of the Vennachar catchment a unique opportunity is presented for comparing the outflow with the rainfall, for a record has been taken twice a day of the depth of water flowing over a weir at the east end of Loch Vennachar since Loch Katrine has been made use of by the Glasgow Corporation as the source of the water-supply to that city. The readings of the depth of the outflowing water during the year 1869 have been calculated out, and the outflow for that year has been estimated at 9572 million cubic feet. The quantity of water supplied to Glasgow during the year 1869 is estimated at 1660 million cubic feet. There is thus an excess of rainfall, according to the mean of the three methods, over the overflow of 2860 million cubic feet. This difference must be accounted for by absorption, evaporation and loss of water by underground channels. The readings of the outflow for a series of twenty-five years or more would be a far more satisfactory basis for calculation than a single year's readings (and the year 1869 was not an average, but a very dry year), and it would be very interesting to have the average annual outflow calculated over the whole period during which the record has been kept, in the same way as the mean annual rainfall is calculated for a particular station. Records of the rainfall at high elevations in different parts of the catchment-basins would also be desirable in comparing the average annual rainfall with the average annual outflow.

Appended to the paper is an account of the geology of the district by Messrs. B. N. Peach, F.R.S., and J. Horne, F.R.S., based on materials collected during the Geological Survey of that region, and published by permission of Sir Archibald Geikie, the Director-General. A brief sketch is given of the geological structure of the area embracing the various lochs, which has an important bearing on the question of the evolution of the valley-system. It is shown that, along the Highland border, there is a great development of conglomerates, grits and greywackes, belonging partly to the crystalline schists and partly to the Old Red Sandstone. These strata, being vertical or nearly so, would be much less easily eroded than the gently inclined schistose rocks lying to the north-west. Such an arrangement would naturally lead to the formation of narrow and comparatively flat-bottomed valleys behind rocky gorges, the latter being cut through the vertical beds of hard grit and conglomerate along the Highland border. Evidence is adduced to show how this remarkable structure likewise contributed to the erosion of rock basins during the glacial period. The glacial phenomena of the region are reviewed, which indicate at least two periods of glaciation ; one, when the ice-shed lay to the north of the area under consideration, when the ice-movement was independent of the existing valley-system, and when even the highest mountains were overridden by the ice. This great development was followed by a period of local glaciation, when the glaciers were confined mainly to the existing valleys. Lastly, the soundings of the various lochs are viewed in relation to the geological history of the area, and with reference to the question of the origin of the various lakes. It is shown that some of the lochs are typical examples of rock basins, that in some instances the deepest soundings occur in front of the rocky barriers at or near their outlets. Reference is made to all the important faults traversing the region, which have led to the more rapid disintegration of the materials, but though they have in certain cases produced modifications of the floors of the lakes, they cannot account for the excavation of the rock basins. The soundings of Loch Lubnaig reveal the striking fact that one of the deep basins in that lake lies on the upthrow side of the most powerful fault traversing the crystalline schists of that region. Messrs. Peach and Horne believe that the soundings of the various lakes in the basin of the Teith above Callander furnish strong evidence in support of Ramsay's theory of their excavation by ice-action.

The paper is illustrated by seven coloured maps, the first three showing on a small scale the orography and drainage areas, the surface geology and the rainfall of the district, the other four showing on a larger scale the bathymetry of the various lochs and the relief of the surrounding country. There are also numerous woodcuts, some of which are reproduced in this review.

IRON AND STEEL INSTITUTE.

THE annual meeting of the Iron and Steel Institute was held on Wednesday and Thursday, May 9 and 10, in the hall of the Institution of Civil Engineers, under the presidency of Sir William Roberts-Austen, K.C.B., F.R.S. The attendance of members was larger than at any previous gathering. The report of the council, which was read by the secretary, Mr. Bennett H. Brough, showed that the Institute is in a flourishing condition. The receipts last year were greater than in any previous year, 110 new members were added to the roll, the supply of original papers was well maintained, and a Royal Charter of Incorporation had been granted. After the usual formal business, the president presented the Bessemer gold medal for 1900 to Mr. Henri de Wendel, the eminent French metallurgist, in recognition of his great services to metallurgy in developing the iron-ore resources of French and German Lorraine. Mr. de Wendel having expressed his appreciation of the honour conferred upon him, Mr. Stead announced that he had decided to postpone the reading of his paper until the autumn meeting in Paris.

Mr. J. Riley then described the various attempts that have been made to use fluid metal in the open-hearth furnace. The results he obtained at Wishaw, in 1898, were encouraging, and experience over a considerable period show that great advantages are derived from the adoption of this method. The best future open-hearth practice, he considers, will include the use of fluid metal direct from the blast furnaces.

The next paper read was one of most conspicuous novelty, by Mr. B. Talbot, on the open-hearth continuous steel process. This process was introduced at the Pencoyd steelworks in Pennsylvania. The furnace used is a basic-lined tilting furnace of seventy-five tons capacity. Many thousands of tons of steel have been made by this method with very satisfactory results, all grades of steel having been produced. The cost and delay in charging cold material is avoided. There is a saving in fuel in charging molten pig iron. The demand for a large supply of good scrap is dispensed with. A regular supply of steel in any desired quantity and at frequent intervals is insured. There is an increased output, an increased yield, and a saving in repairs and in labour charges. At the same time it is possible to use

very large furnaces, with consequent reduction in cost of production, without the necessity for very large cranes and ladles. A long discussion followed the reading of the paper, the opinion being general that the process is an important advance in open-hearth steel practice.

Mr. A. Greiner gave an account of the results obtained at the Cockerill works, Belgium, with the first blowing-engine worked by blast-furnace gas ever employed in any ironworks. This 600 horse-power engine has been running since November 20 last with unpurified gas taken from the Seraing blast-furnaces.

Baron H. von Jüntner submitted a further instalment of his researches on the theory of solution of iron and steel. He discussed the application of the laws of chemical mechanics in the case of iron carbon alloys, and showed what an important bearing thermo-chemistry possesses for a knowledge of the constitution of the alloys of iron and their alterations of state.

The meeting then adjourned until May 10, when Mr. C. Dellwik dealt with the manufacture and application of watergas, describing the production of the gas by means of a simple apparatus with a degree of economy surpassing that of other less valuable gas. Whilst in the old processes the gas leaving the generator during the blow contains principally carbon monoxide and nitrogen, in the author's process it consists chiefly of carbon dioxide and nitrogen.

The subject of utilising blast-furnace slag is a fruitful source of inquiry, and a recent important development was dealt with by Mr. C. von Schwarz. This is a successful method of manufacturing cement from blast-furnace slag, recently employed in Germany and Belgium. The cement thus made obtains a higher price in the market than ordinary Portland cement.

Mr. L. F. Gjers and Mr. J. H. Harrison described an apparatus for equalising the varying temperatures of hot blast. Hitherto the hot blast has been allowed to enter the furnace as it left the stove, and in order to obviate the interference with the steady working of the furnace, the authors have devised an apparatus consisting practically of another small stove with a central division wall. It is filled with chequer work ; and the hot blast, entering at one side of varying temperature, is delivered out at the other side at an even mean temperature.

The form of ingot that would seem to be the most natural for the manufacture of a gun-tube or a propeller shaft is one with a circular section. Mr. F. J. R. Carulla, however, pointed out the drawbacks of this form, and showed that a polygonal ingot with concave sides answers the required conditions.

Mr. H. K. Scott contributed a paper on manganese ore deposits and mining in Brazil, giving a detailed account of the geological structure of the deposits, and of the economic development of the industry.

After the usual votes of thanks to the Institution of Civil Engineers, proposed by Sir John Alleyne, Bart., and to the president for his conduct in the chair, proposed by Mr. Carnegie, the proceedings terminated. Incidentally, Mr. Carnegie announced his intention of founding a scholarship in connection with the Iron and Steel Institute for the advancement of research in connection with iron and steel.

THE ROYAL SOCIETY CONVERSAZIONE.

THE general opinion of the scientific company at the Royal Society on Wednesday, May 9, on the occasion of the first of the two soirées held annually, was that novel and striking exhibits were not so numerous as in some exhibitions of previous years. The following were among the most noteworthy exhibits :—

Mr. Richard Kerr showed a clock controlled at a distance by wireless telegraphy of the Hertzian wave system. Mr. J. Wimshurst, F.R.S., exhibited an influence machine, constructed with twelve plates of vulcanite. Prof. Silvanus P. Thompson, F.R.S., showed some pretty electromagnetic experiments, one being the converse of De La Rive's experiment, using floating magnet instead of floating battery, and others showing new varieties of the De La Rive experiment (see p. 71). Prof. Minchin, F.R.S., showed that luminous flashes could be induced in a helium tube by Hertz waves.

An electric micrometer was shown by Mr. P. E. Shaw. The instrument was designed primarily to measure the small movements of a telephone diaphragm. A screw abuts on a system of three levers, set up on a strong wooden frame. By turning the screw, the far end of the levers moves to and fro through distances which can be controlled and measured. This end of the levers carries a rod, and the diaphragm a small plate, both of iridio-platinum; if these two surfaces touch one another, a flow of a small amount of electricity occurs, producing a sound in a telephone held by the observer; at the same time he reads by a telescope a graduated circular scale fixed on the screw. Since the screw and levers can be moved at will by the observer, he can, by this contact method, find the position of the diaphragm, and follow its movements. Precautions against vibrations are taken by having indiarubber suspensions, and against temperature changes by covering the working parts with boxes wrapped in felt. Movements as small as $\frac{1}{2000}$th of a wave-length of sodium light have been measured by this apparatus.

Mr. Killingworth Hedges exhibited jointing boxes and aigrettes used in the rearrangement of the lightning conductors of St. Paul's Cathedral. The original system for the protection of the Cathedral from lightning was installed under the advice of the Royal Society in about 1756. This was replaced in 1872 by what was then considered the most improved method, when the unsoldered joints were found to be very defective; in some cases they were quite loose; also the earths, originally made by laying the cable in a drain which had become disused, were in some cases insulated from the ground. New earths have been substituted. The method adopted to protect the structure unites the old system and the new cables to a horizontal conductor run on the top of the parapet, entirely round the building; to this copper aigrettes as shown are teed at intervals.

Other electrical exhibits were models illustrating leakage from electric tramways, shown by Mr. A. P. Trotter, and improved forms of standard resistance coils made by the Cambridge Scientific Instrument Co., Ltd.

Dr. Isaac Roberts, F.R.S., exhibited his magnificent volume of photographs of stars, star-clusters and nebulæ, recently reviewed in NATURE (vol. lxi. p. 533). The volume contains seventy-two photographs, which have been enlarged by mechanical processes from the original negatives, and they furnish evidence of the evolution of stellar systems from nebulous matter as seen in the convolutions of spiral nebulæ. They also furnish a foundation for the inference that the system of the *Milky Way* is not unlimited in extent, and that the numerous aggregations of stars, seen in lines and curves in the stellar regions, indicate their development from spiral nebulæ.

Mr. Thomas Thorp exhibited some of his grating films and their application to diffraction colour photography, on Prof. Wood's principle. Dr. Downing, F.R.S., exhibited maps illustrating the track of the total eclipse of the sun of May 28.

Mr. W. A. Shenstone, F.R.S., and Mr. H. G. Lacell showed a quantity of non-splintering silica, suitable for use in the oxy-gas flame. The method of converting this into tubes and other forms of apparatus, as recently described in NATURE (May 3), was demonstrated practically, together with experiments to illustrate the behaviour of vitreous silica under sudden and great changes of temperature. The following apparatus, constructed of silica, was also exhibited. A long tube for use with a platinum thermometer; a mercury thermometer; bulbs and stems for thermometers; a Giessler tube; a small distilling tube; and rods and tubes of various sizes for various purposes. Some examples of leadless glazed ware were shown by Dr. T. E. Thorpe, F.R.S.

Mr. H. B. Hartley and Mr. H. L. Bowman gave a demonstration of the properties of crystals yielding doubly-refracting liquids on fusion. Certain crystalline organic compounds, viz. *p*-Azoxyanisol, *p*-Azoxyphenetol, and Cholesteryl benzoate, have been found by Prof. Lehmann, of Carlsruhe, to give on melting (at temperatures of 116°, 134° and 145° respectively) liquids possessing the properties of double-refraction and dichroism, even under conditions in which a state of strain is impossible. When these anisotropic liquids are further heated, they change at definite temperatures of transition (134°, 165° and 178° respectively) into ordinary isotropic liquids. The intermediate bodies have been called "liquid crystals," for, although the evidence of their elasticity, viscosity, and dielectric capacity shows them to be undoubtedly liquids, yet nevertheless they possess, like crystals, both double refraction and dichroism.

Specimens from the reefs of Funafuti were exhibited by Prof. J. W. Judd, C.B., F.R.S., on behalf of the Coral-Reef Committee of the Royal Society. The exhibits included :—
(1) Specimens illustrating the rate of growth of corals and

calcareous algæ from the reefs of Funafuti. Experiments made by Mr. A. E. Finckh, of Sydney, who in 1898 carried the boring made by Prof. T. E. David in the previous year from the depth of 698 ft. to 1114 ft., have thrown much new light upon this important question. Specimens illustrating these experiments are exhibited. (2) New and interesting forms of Foraminifera, which have been described by Mr. F. Chapman. These include :—(a) *Cycloclypens*, a genus previously regarded as being very rare, but now shown to exist abundantly at Funafuti. The two species formerly described are now shown to be dimorphic forms of the same organism. (b) A curious form of *Polytrema*, which occurs encrusting various objects in alternate layers with the marine alga *Lithothamnion*, thus forming loose nodules. (c) The newly described *Haddonia*, first obtained from Torres Straits, &c.

Prof. H. G. Seeley, F.R.S., showed drawings of restorations of Dimorphodon. The drawings, of the natural size, are based upon fossil remains from the Lias, in the British Museum. They represent the skeleton as in the quadruped and biped positions; and show the contours of the body at rest, walking, and preparing for flight, to illustrate proportions of the skeleton.

Dr. C. I. Forsyth-Major exhibited remains of extinct gigantic and lesser lemurs from Madagascar, and living forms for comparison. Some beautiful examples of chalk fossils were exhibited by Dr. Arthur W. Rowe.

Dr. Manson exhibited longitudinal sections of filariated mosquitoes (*Culex ciliaris*), showing that *Filaria nocturna*, like the malaria parasite, leaves its mosquito host *viâ* the proboscis.

A collection of living marine worms (Annelids) from the neighbourhood of Plymouth, designed to illustrate, as far as possible, the prominent features in the habits of life of the different types of this class of animals, and such modifications of form as are related thereto, formed the exhibit of the Marine Biological Association.

Prof. E. Ray Lankester, F.R.S., on behalf of the Archæological Survey of the Egypt Exploration Fund, showed reproductions of paintings and sculptures in tombs of Ancient Egypt, representing domestic and wild animals and birds. The tombs of Ancient Egypt contain abundant representations of animal life. In spite of the artists' ignorance of perspective and occasional faulty colouring, the outlines are rendered with remarkable fidelity to nature, often enabling the species to be identified. Among domestic animals, the dogs are perhaps the most interesting, as showing that extreme development of various breeds had already taken place. The monuments from which the drawings exhibited were copied are of two periods :—
(1) Tombs at Beni Hasan, of the XIIth Dynasty (*circa* 2000 B.C.); (2) the Tomb of Ptah-hetep at Saqqareh, of the Vth Dynasty (*circa* 3000-2500 B.C.).

Prof. A. C. Haddon, F.R.S., showed specimens illustrating the decorative art of the Sea Dayaks of Sarawak. The carved and painted designs of the Dayak men are entirely different from the woven and embroidered patterns made by the women. The former are chiefly plant derivatives, while the latter are mainly greatly modified animal forms. The significance of the distinction and the real meaning of the patterns themselves are not yet elucidated. The method by which the women make the patterns in their woven fabrics was also illustrated. The warp is stretched on a frame, and numerous strands are tied tightly with strips of leaves; the whole is removed and then submerged in a dye. The lashing is then undone, and the tied-up portions are found to be undyed. The whole process is repeated if a three-colour pattern is required.

Ethnographical objects from Malay Peninsula (Malay and Sakai) were shown by Mr. W. W. Skeat. The phonographic records of songs of the Pangan tribe, a wild aboriginal tribe of Negrito stock, received much attention.

A collection of anthropometric instruments was shown by Dr. J. G. Garson.

The Royal Geographical Society exhibited a section cut from the tree on Lake Bangweulu, Central Africa, under which Livingstone's heart was buried, and containing the inscription carved by his native followers.

In the course of the evening, short discourses and demonstrations were given by Sir Andrew Noble, K.C.B., F.R.S., on modern explosives; Dr. Arthur W. Rowe, on the photomicrography of chalk fossils by reflected light; and Mr. F. Enock, on photographs from living insects, showing the metamorphoses of one of the Odonata.

UNIVERSITY AND EDUCATIONAL INTELLIGENCE.

, CAMBRIDGE.—The King of Sweden and Norway was on Monday admitted by the Chancellor to the honorary degree of LL.D. The ceremony was witnessed by a vast assembly, and the King gave much pleasure by his gracious bearing and evident interest in the proceedings.

The General Board are about to appoint a lecturer in experimental physics in succession to Mr. W. N. Shaw. Applications are to be sent to the Vice-Chancellor by May 19.

The Master of Downing and Dr. Barclay-Smith announce a course of instruction in practical histology, to be given during the long vacation, beginning on July 7.

The trustees of the late Miss R. F. Squire have offered the University a sum of about 13,500l. for the erection of a law library in connection with the new law school, and adjoining the Sedgwick Memorial Museum. This timely benefaction will probably facilitate the speedy erection of the the Botanical and Medical Schools, plans for which are now under the consideration of the Senate.

The proposal to establish a special examination in the sciences bearing on agriculture, as a qualification for the ordinary B.A. degree, was favourably received in the Senate on May 10, and a grace for its adoption has been sanctioned by the council.

An examination for minor scholarships in Natural Science will be held at Downing College in March 1901. Application for particulars should be made to the tutor.

A VERY satisfactory side of technical education is the work carried on in the lecture-rooms and laboratories of University Colleges, in connection with Technical Instruction Committees. Two short courses of evening lectures to teachers, just arranged by the Technical Instruction Committee of Liverpool with Prof. Oliver Lodge and Prof. Herdman, are instances in point. The lectures deal with some recent developments of physical and natural science, Prof. Lodge taking for his subject "Electric Vibrations," and Prof. Herdman "Oceanography." The lectures are free to teachers who can give evidence that they are able to profit by them.

VISITS to museums, and outdoor lessons, are counted as school attendances by the Board of Education, with the result that they are now given a definite place in the scheme of instruction of many schools. In a similar way, the National Zoological Park at Washington is used to place great object-lessons before the hundreds of thousands of visitors to the national capital from all parts of the United States. The pupils in the public schools of Washington benefit greatly by these opportunities. It has become a part of their routine to visit, under the care of a teacher, the Smithsonian Institution and National Museum buildings, as well as the park ; while those outside the city benefit indirectly through the numerous excursions of teachers, and the stimulus and suggestion they may thus receive.

' Two new buildings in connection with the Yorkshire College, Leeds, to be devoted to the development of clothworkers' research and dyeing, &c., were formally opened, on Friday last, by the Master of the Clothworkers' Company, Mr. A. C. Cronin. Principal Bodington, of the Yorkshire College, the professors and students, and mayors of various boroughs, also attended on the occasion. It was explained that it was intended to raise the tone of dyeing, and that the outlay on the extensions is likely to yield a tenfold return. Mr. Cronin, in declaring the new buildings open, expressed a hope that increased knowledge in the industries would be the result of these extensions. At a luncheon which followed, in responding to the toast of "The Clothworkers' Company," he said it was the intention of the Company that the Yorkshire College should become the first and most complete example of a textile and dyeing school not only in Europe, but in the world. There is now hardly any manufacturing town of any size in Yorkshire which has not its technical school or institute, and with which the Clothworkers' Company has not been or is not still connected.

THE medical school of the future was the subject of an address delivered before the fifth triennial Congress of American Physicians and Surgeons, on May 2, by the president, Prof. H. P. Bowditch, of Harvard. According to Prof. Bowditch, we may expect that a medical school of the first rank will, in the immediate future, be organised and administered somewhat as follows :—(1) It will be connected with a university, but will be

so far independent of university control that the faculty will practically decide all questions relating to methods of instruction and the personnel of the teaching body. (2) It will offer advanced instruction in every department of medicine, and will therefore necessarily adopt an elective system of some sort, since the amount of instruction provided will be far more than any one student can follow. (3) The laboratory method of instruction will be greatly extended, and students will be trained to get their knowledge, as far as possible, by the direct study of nature, but the didactic lecture, though reduced in importance, will not be displaced from its position as an educational agency. (4) The work of the students will probably be so arranged that their attention will be concentrated upon one principal subject at a time, and these subjects will follow each other in a natural order. (5) Examinations will be so conducted as to afford a test of both the faithfulness with which a student performs his daily work and of his permanent acquisition of medical knowledge fitting him to practise his profession.

THE first official ceremony of the University of London in the new home at South Kensington was the presentation of degrees by the Prince of Wales on Wednesday, May 9. The University has thus entered upon a new phase of its career. As the Chancellor of the University remarked in his address, nothing has been more striking within the last few years than the progress of new universities in different parts of the country. The University of Wales, of which the Prince of Wales is Chancellor, has been founded, and, although very young, it is already making notable progress and will ultimately be a great success. Besides this, there is the Victoria University, of which Lord Spencer is Chancellor, and which has made remarkable progress ; and also the completely new University of Birmingham. What does all this mean ? It means that the country is stirred up on the subject of education ; and among all classes and places there is a greater sense of the importance of it than ever there has been before. As to the University of London, the Chancellor quoted figures to show the great progress which has taken place, and made special reference to the great stimulus to the improvement of the education of women throughout the country arising out of the action of the University in obtaining a supplementary charter to enable women to be admitted to the examinations. Up to the present time the University has been only an examining body. It has by its examinations done a good work for the education of the people, and it has set an example which has had a very important effect upon all the schools throughout the country. But it is now a teaching University, and with its large list of faculties its work will be very widespread. The Prince of Wales then made a few remarks, in the course of which he said : "No one wishes more sincerely than I do happiness and prosperity to this University ; and from all that we have heard from the Chancellor I think the University is in a fair way of becoming one of the greatest importance, and one that will hold its own, no doubt, with many of the others which are of more ancient origin. I am glad to think that, as the result of somewhat difficult, and I may say somewhat delicate, negotiations, the London University has now found a home in this large building, better known as the Imperial Institute, in which, as you all know, I take a deep interest. We are very grateful to Her Majesty's Government for all they have done, and for having facilitated the arrangements which I hope are now complete. It only rests with me to express the fervent wish that the London University will not regret having come to a more distant part of London, and that they will find that they have ample room for all their requirements in this University."— Sir Michael Foster, M.P., then addressed those who had received awards at the hands of the Chancellor. He reminded them that the value of the degree was not in the degree itself, but in the labour which had led up to it. The degree might be the guinea stamp, but it was the work and the mental discipline which was the real gold.

SCIENTIFIC SERIALS.

Bulletin of the American Mathematical Society, April.— Prof. F. N. Cole summarises the *Proceedings* of the February meeting of the Society, and abstracts a few of the papers communicated. The bye-laws were revised. By this amendment it is provided that the ex-presidents shall be life-members of the council, and that the presidential term of office shall be extended to two years.—Prof. J. Pierpont gives an interesting account of the summer meeting of the Deutsche Mathematiker-Vereinigung held at Munich in September of last year.—Some theorems concerning linear differential equations of the second order is an abstract by Prof. M. Bôcher of certain results which he communicated at the February meeting (see *supra*).—A paper by Dr. M. B. Porter, read at the same meeting, is entitled " A note on the enumeration of the roots of the hypergeometric series between zero and one." It is a continuation of a note published in the May (1897) number of the *Bulletin.*—Dr. J. Sommer reviews Hilbert's " Grundlagen der Geometrie," and Prof. E. O. Lovett does the same for Kœnig's " Leçons de Cinématique."—The longer papers read before the Society will, we presume, be printed in the new *Transactions.*—The notes are very full, and there is a fair list of publications.

Bulletin de l'Académie des Sciences de St. Pétersbourg, vol. vii. No. 3.—On the rotation of Jupiter and his spots, by Th. Bredikhin. An analysis of the observations made by the author himself at Moscow, and of some later observations at Pulkova. A comparison of the times of rotation of spots situated in the same latitudes shows that some of them are formed in the lower, and some in the higher strata of Jupiter's atmosphere. Prof. Joukovsky's formulæ hold good as a rule; but a more careful discussion shows that the law of friction must be altered; the latter is proportional to the square or even to a higher degree of velocity. But it would be extremely difficult to make a theoretical discussion if the law be altered in this sense.—The scientific results of the Black Sea expedition, by A. Ostroumov: iii. Fishes of the Sea of Azov.—Materials for the hydrology of the White Sea and the Murman Sea (Arctic Ocean along the Norman coast), by N. Knipovitch: i. Lists of the Observations. Vol. vii. No. 4 —The series of Jean Bernoulli, by N. Sonin.— New researches into the spectrum of β Lyræ and η Aquilæ, by A. Belopolsky. These new researches were made with the aid of the 30-in. refractor of Pulkova. The spectroscopic velocities of η Aquilæ showed a periodicity very near to the periodicity of the variations of magnitude, *i.e.* 7 days 4 hours, and it was possible to calculate its orbit. Similarly, as for δ Cephei, it was proved that the changes of brilliancy in η Aquilæ cannot be explained by eclipses of the star. As regards β Lyræ, the former suppositions of the author are now fully confirmed. This star represents a system of two bodies, having at any instant opposite spectroscopic velocities, and one of the two bodies eclipses the other during their revolutions.—Preliminary communication on applications of Rykatschew's method for studying the relations between rainfall and height of water in rivers, by Dr. Harry Gravelius.—The third international balloon ascents of May 1, 1897, by Ed. Stelling.—Observations of the satellites of Mars with the 30 in. refractor at Pulkova, by F. Renz; and on the photographs of Mars, by S. Kostinsky.

Vol. vii. No. 5.—On the changes of pressure under the piston of the air-pump, by Prince Galitzin. Theoretical discussion is compared with direct observation.—Some remarks on the sensibility of the eye, by the same author.—Abstract from the yearly report for 1896 of the Central Physical Observatory, by M. Rykatschew.—On the excretory organs of *Ascaris megalocephala,* by S. Metalnikoff.—On the routes of the cyclones over Russia in 1890-92, preliminary communication, by P. Rybkin.

SOCIETIES AND ACADEMIES.

LONDON.

Physical Society, May 11.—Prof. O. J. Lodge, F.R.S., President, in the chair.—A discussion of Prof. Lodge's paper on the controversy concerning Volta's contact force was commenced by Prof. Armstrong. Prof. Armstrong expressed his indebtedness to the president for putting forth clearly what we are trying to understand, and said that it was hardly time for chemists to enter the discussion when physicists themselves differed. There has apparently been a change in front since the time when the effect was supposed to be due either to (1) chemical action between the metals, or (2) oxidation. Prof. Lodge's view is intermediate, but approximates to the second. Prof. Armstrong said that from a practical point the existence of the effect was unknown, because sufficient precautions had never been taken to prevent chemical action. He urged the continuance of experiments similar to those carried out by Mr. Spiers, and stated that modern ideas of chemistry were favourable to the view which

Prof. Lodge had taken up with regard to the Volta effect. —Mr. Glazebrook made some remarks upon the meaning of the term E which occurs in the expression for the Peltier effect at the junction of two metals. If we confine our attention to an infinitesimal cycle at the junction of two metals at slightly different temperatures, we get the equation for the Peltier effect in which E is the potential difference at the point considered. If then, assuming reversibility, we sum up all the infinitesimal cycles round a circuit and get a finite cycle, the E.M.F. of the circuit is a function of the two temperatures between which it is working. Differentiating with respect to temperature the total E.M.F. of the circuit, we get an equation which applies to the circuit as a whole, and in which E is the total E.M.F. round the circuit. Mr. Price asked if any critical experiment could be suggested to settle the question.—Dr. Lehfeldt called attention to some experiments which had been performed to measure the potential difference between an electrolyte and a gas. The electrolytes considered were chiefly aqueous solutions, and the potential differences observed varied largely. The surface tensions of the liquids were measured, and it was shown that the variations in the potential difference were very similar to those in surface tension. This suggests, in the case of electrolytes, true physical surface effects, and not chemical action.—The chairman remarked that Dr. Lehfeldt evidently looked upon the metal-ether boundary as being the effective one. The experimental evidence is not sufficient to say exactly which is the effective contact, but it seems to show that the metal-ether effect is of the same order of magnitude as the oxygen layer effect. According to Helmholtz they ought to be related, and they apparently are. —The chairman then read a paper, by Mr. J. B. Tayler, on the heat of formation of alloys. Experiments have been made upon alloys of lead with tin, bismuth and zinc, and of zinc with tin and mercury. The method employed consisted in dissolving (1) the alloy, and (2) the corresponding mixture of metals in mercury, and measuring the heat of solution in each case. On the assumption that the solutions obtained are identical, the difference between the heat of solution of the mixture and that of the alloy is the heat of formation of the latter. The calorimeter was a thin glass tube silvered on the outside and supported by a stouter tube silvered on the inside. Suitable arrangements were adopted for the introduction of the metals or alloys, which were used in the form of filings. Solution was often complete in less than a minute, and rarely took more than two minutes and a half. The alloys first experimented upon contained their constituents in equivalent proportions, and the heats of formation were found to be small in comparison with those found for brass by Galt and Baker. It was thought that only a small percentage of the atoms present had entered into definite chemical combination, and that more reliable results would be obtained by dissolving a small quantity of one metal in an excess of the other, and calculating from the experimental results the heat of formation of the gramme-molecular weight of compound upon the supposition that the whole of the small quantity of metal had entered into chemical combination by the exercise of its normal valency. Using the numbers so obtained to find, by Kelvin's theory, the potential difference which should exist between the metals concerned when put in contact, results were arrived at which agreed neither with the Volta effect nor the Peltier effect, but which were considerably nearer the former than the latter. A paper on the want of uniformity in the action of copper-zinc alloys on nitric acid was read by Dr. J. H. Gladstone. Experiments have been made by dissolving copper-zinc alloys in nitric acid, following the method of Dr. Galt, and adopting the precautions mentioned by him. The reaction between nitric acid and these metals or alloys is very complicated, and there is a difference between the products in the case of an alloy and in the case of the equivalent mixed metals. The gases evolved being small in the experiments performed, attention was directed to the determination of the substances remaining in solution, *i.e.,* the nitrous acid and ammonia. The alloys gave much more nitrous acid and less ammonia—in fact, two of the alloys employed produced no ammonia. Discrepancies in results may be due to the fact that the zinc and copper in contact form a zinc-copper couple which in the presence of acid sets up a vigorous action and produces a different evolution of heat. Difficulties arise in the investigation because the alloys used may not be definite chemical compounds, but mixtures of two or more alloys with uncombined zinc and copper. The alloy with 38·38 per cent. of copper appears to be fairly uniform. Different observers disagree as to the amount of heat produced by any reaction, but the excess of calories in a zinc reaction over those in a copper reaction appears to be fairly constant. Starting with 640 calories, the value, according to Galt, when copper is dissolved in nitric acid of sp. gr. 1·360, we should have 1357 calories when zinc is dissolved, provided the chemical action is the same in each case. All the calorimetrical results from the different specimens of alloys would theoretically lie upon the straight line drawn between 604 and 1357. This is practically so from pure copper to the copper 70 per cent. alloy, but beyond that there is less heat produced than that indicated by the straight line law, the maximum deviation lying at about copper 37 per cent. The specimen containing 38·38 per cent. copper, which is not far from the alloy $CuZn_2$, shows a loss of 32 calories. The only way in which this deficit can be accounted for is by supposing that the action of this alloy on nitric acid produces a larger quantity of nitric oxide than in the case of pure copper. But, allowing full force to this argument, it cannot account for as much as 10 calories of the deficit. There is, therefore, a residual deficit as yet unaccounted for on chemical grounds. The author states that it is desirable that experiments should be conducted on the zinc-copper alloys with solvents which give a simpler chemical action than that produced by nitric acid. The chairman pointed out that the results obtained by Galt for an alloy which appeared to be a chemical compound, were in close agreement with what would be expected from the existence of agreement with what would be expected from the existence of the Volta contact force. Prof. Armstrong said that the action of nitric acid on brass or zinc and copper was a function of the quantity of acid present, its strength, the temperature and the pressure ; and that, therefore, it was unsatisfactory to conduct experiments using nitric acid as a solvent. He suggested the use of a solution of bromine in which finely-powdered zinc, copper and brass are easily soluble with a simple chemical reaction. Mr. Tomlinson pointed out that it was impossible to use the ordinary formula for the calculation of the Volta effect from the heat of formation of alloys, unless we know exactly the chemical composition of the alloy which is produced. Mr. W. R. Cooper, referring to Mr. Tayler's paper, said it was difficult to see that anything could be proved by the application of the Kelvin theory to a metallic contact, unless there is ground for believing that some particular alloy of fixed composition is always formed. There is also a further difficulty in converting heat of formation into E.M.F. in cases where the metals have different valencies, for there is no reason why one valency should be selected rather than the other. Referring to Dr. Gladstone's paper, Mr. Cooper said that it was possible that the difference in the reducing powers of mixtures and alloys might be due to local action, which would be more pronounced in the case of alloys. More hydrogen would be evolved, and the reduction would be more complete.—Prof. S. P. Thompson then showed an electromagnetic experiment. A circular coil capable of carrying a strong current was placed with its axis horizontal in a tank of water. Into the tank were also placed some small magnets in sealed glass tubes so adjusted as to have a density approximately equal to that of water. The magnets just floated or just sank. On running a current through the coil it was possible to "fish" for the magnets, which, acted upon by the magnetic field, immediately made their way to the coil. When the current was carefully reversed upon the approach of a magnet, repulsion instead of attraction took place, and the magnet retreated. In general, however, reversal of the current produced reversed polarity in the magnet, and attraction still persisted.—The Society then adjourned until May 25.

Chemical Society, May 3.—Prof. T. E. Thorpe, President, in the chair.—The following papers were read :—The substituted nitrogen chlorides and nitrogen bromides derived from ortho- and para-acet-toluide, by F. D. Chattaway and K. J. P. Orton. Hypochlorous and hypobromous acids convert ortho- and para-acet-toluide into substituted nitrogen chlorides and bromides, which readily undergo transformation into the isomeric substituted toluides.—The estimation of hypo-iodites and iodates ; and the reaction of iodine monochloride with alkalis, by K. J. P. Orton and W. L. Blackman. The authors' method of estimating hypoiodites is based on the oxidation of sodium arsenite by hypoiodites, but not by iodates. The initial reaction of iodine monochloride and alkalis is represented by the equation $ICl + 2MOH = MIO + MCl + H_2O$; conversion of the metallic hypoiodite into iodide and iodate becomes complete after twenty-four hours.—Products of the action of sulphur dioxide on ammonia, by E. Divers. Amongst the products of spontaneous decomposition of ammonium amidosulphite is

found a substance of acid properties to which the author assigns the constitution

$$NH\begin{cases}SO_2.O \\ S-NH_2\end{cases}$$

—On brazilin (iv.), by A. W. Gilbody, W. H. Perkin, jun., and J. Yates. From a study of the reactions of brazilin and tri-methylbrazilin, the authors conclude that brazilin probably has the constitution

$$HO.C.CH : C.CO—— ·CH.CH_2·C : CH.C.OH$$
$$HC.CH : C.CH(OH)—CH——C : CH.C.OH$$

—On hæmatoxylin (v.), by W. H. Perkin, jun., and J. Yates. The study of the oxidation products of tetramethylhæmatoxylin leads to the view that hæmatoxylin has the constitution

$$HO.C.C(OH) : C.CO ———CH.CH_2·C : CH.C.OH$$
$$HC.CH══C.CH.CH(OH).CH——C : CH.C.OH$$

—Note on the function of the characteristic meta-orientating groups, by A. Lapworth.

Anthropological Institute, April 24.—Mr. C. H. Read, President, in the chair.—Dr. W. H. R. Rivers described a genealogical method of collecting social and vital statistics which he had used with success when in Torres Straits with the Cambridge Anthropological Expedition. Genealogies of the inhabitants of Murray Island and Mabuiag were compiled which went back for three to five generations, and included nearly all the families at present on those islands. In working out these genealogies, the only terms of relationship used were father, mother, child, husband and wife, and care was taken to limit those terms to their English sense. The chief difficulties were the prevalence of adoption in Murray Island and the custom of exchanging names in Mabuiag. The trustworthiness of the genealogies was guaranteed by the fact that nearly every detail was derived independently from several informants. These genealogies afford material for the exact study of numerous sociological problems ; thus the system of kinship can be worked out very thoroughly by ascertaining the native terms which any individual applies to other members of his family, i.e. the subject can be investigated entirely by means of concrete examples, and abstract terms of relationship derived from European sources avoided. The genealogies also afford material for the study of totemism, marriage customs, naming customs, &c. By this method also vital statistics may be collected, both of the present and the past. The genealogies compiled in Torres Straits supply data for the study of the size of families, the proportion of the sexes, the fertility of mixed marriages, &c. The method has the further advantage of bringing out incidentally many facts in the recent history of the community, to which it gives increased definiteness and concreteness. The paper was discussed at some length by the President, Mr. Gomme and Dr. Jayne.—Dr. A. C. Haddon, F.R.S., exhibited a large number of lantern slides illustrating various native industries in British New Guinea ; the photographs were taken during the recent Cambridge Anthropological Expedition. The most complete series was one showing all the stages in the manufacture of pottery by Port Moresby women ; other slides illustrated the manufacture of canoes at Keapara with stone implements. Photographs were shown of the process of pile-driving and the erection of buildings, as well as of fire-making, and various women's industries, such as tattooing, making string, &c.—Mr. Gowland pointed out a number of parallels from Korea to the mode of pottery-making described by Dr. Haddon.—The secretary laid before the meeting a brief account of the proceedings of the Cretan Exploration Fund, and of the discovery by Mr. A. J. Evans, at Gnossus, of a collection of clay tablets inscribed with pictographic signs.

PARIS.

Academy of Sciences.—M. Maurice Lévy in the chair.—The President announced to the Academy the death of M. E. Grimaux, member of the Section of Chemistry.—Preparation of the β-alkyloxy-α-cyanocrotonic esters, isomers of the acetylalkylcyanacetic esters, by M. A. Haller. The true acetylalkylcyanacetic esters, $CH_3.CO.CR(CN).CO.OC_2H_5$, have been prepared by Held by acting with cyanogen chloride upon the sodium derivative $CH_3.CO.CR.Na.CO.OC_2H_5$; the isomeric ester of the enol form, $CH_3.C(OH)=C(CN).CO.OC_2H_5$, are obtained by first converting the sodium into the corresponding silver derivative, and then treating this with the alkyl iodide. The reactions of the ester so obtained are clearly those of the enolic ester, the alkyl group not being directly united to carbon. —The arable earths of the Canton of Redon from the point of view of phosphoric acid, by M. G. Lechartier. The analyses given show how it is that certain lands in the Canton have been successfully cultivated from time immemorial, without the use of phosphatic manures.—Geographical positions and magnetic observations on the eastern coast of Madagascar, by M. P. Colin. The latitude and longitude of Vatomandry and Mahanoro have been redetermined, and also the values of the magnetic elements at those places. The results show that the existing maps require correction in some respects.—Prof. Burdon-Sanderson was elected a Correspondant for the Section of Medicine and Surgery in the place of the late Sir James Paget.—Positions of fundamental polar stars determined at the Observatory of Lyons, by M. F. Gonnessiat.—Shooting stars observed at Athens during the year 1899, by M. D. Eginitis.—On the method of Neumann and the problem of Dirichlet. by M. A. Korn.—On an application of the method of successive approximations, by M. A. Davidoglou.—On the distribution of prime numbers, by M. Helge von Koch.—On gas engines, by M. L. Marchis. A reply to the criticisms of M. Witz.—An electrically driven pendulum, by M. Ch. Féry. The mechanism described is arranged so as to leave the pendulum as far as possible unconstrained.—The heat of neutralisation of hydrogen peroxide by lime, by M. de Forcrand.—Solubility of a mixture of salts having a common ion, by M. Charles Touren. The curve showing the relation between the solubility of potassium bromide in solutions of potassium bromide of different concentrations is not coincident with the corresponding curve for potassium nitrate and chloride. Hence the law proposed by Nernst, that equivalent solutions of nitrate and bromide should lower the solubility of the chloride to the same extent, is not verified. The author notes as an interesting application of the phase rule that the study of the solubility of a mixture of salts may show that they are isomorphous, when direct proof may be difficult.—The action of phenyl isocyanate and of aniline upon some γ-ketonic acids, by M. T. Klobb.—Some new compounds of antipyrine with mercury halides, by MM. J. Ville and Ch. Astre.—On acetyl-phenylacetylene and benzoyl-phenylacetylene, by MM. Ch. Moureu and R. Delange. Acetyl-phenylacetylene is quantitatively decomposed by alcoholic potash into phenylacetylene and potassium acetate ; benzoyl-phenylacetylene reacts differently, acetophenone being produced.—On the stability of saccharose solutions, by M. Œchsner de Conninck.—Study of the hydrolysis of fibrous tissue, by M. A. Etard. The fibrous tissue of beef, hydrolysed with sulphuric acid, gives a polysaccharide, but practically no.—On some fresh-water *Palæmonidæ* of Madagascar, by M. H. Coutière.—On a new edible tuber from the Soudan, the Ousoulify, by M. Maxime Cornu. The Ousoulify is a tuber resembling the potato in taste, which is cultivated and sold in the Soudan. It is a labiate, and is provisionally named *Plectranthus Coppini.* It has the advantage over the potato that it can be grown in a truly tropical climate.—On the mineralogical composition of the teschenites, by M. A. Lacroix. The hornblende teschenites of Madagascar are analogous, both in structure and mineralogical composition, to the teschenites from Portugal and the Pyrenees, but they contain the nepheline intact. The teschenites from both regions were probably originally identical from the mineralogical point of view.—On the excitement of the electrical nerve of the gymnotus by its own current, by M. Mendelssohn. The electric nerve of the torpedo fish may be excited by its own current.—On the southern aurora observed during the wintering of the Belgian Antarctic expedition.—Barometric deviations produced on the parallel on successive days of the synodic revolution, by M. A. Poincaré.

DIARY OF SOCIETIES.

THURSDAY, MAY 17.

ROYAL SOCIETY, at 4.30.—The Circulation of the Surface Waters of the North Atlantic Ocean : H. N. Dickson.—(1) On Cerebral Anæmia and the Effects which follow Ligation of the Cerebral Arteries ; (2) The Influence of Increased Atmospheric Pressure on the Circulation of the Blood. Preliminary Note : Dr. Leonard Hill.—Contributions to the Comparative Anatomy of the Mammalian Eye, chiefly based on Ophthalmoscopic Examination : Dr. Lindsay Johnson.
ROYAL INSTITUTION, at 3.—A Century of Chemistry at the Royal Institution : Prof. J. Dewar, F.R.S.

ZOOLOGICAL SOCIETY, at 4.30.—The Freshwater Fishes of Africa : G. A. Boulenger, F.R.S.
SOCIETY OF ARTS(Indian Section), at 4.30.—The Industrial Development of India : J. A. Baines.
INSTITUTION OF ELECTRICAL ENGINEERS, at 8.—Alternating Current Induction Motors : A. C. Eborall.
CHEMICAL SOCIETY, at 8.—Chlorine Derivatives of Pyridine. VI. The Orientation of some Aminochloropyridines : W. J. Sell and F. W. Dootson.

FRIDAY, MAY 18.

ROYAL INSTITUTION, at 9 —The Structure of Metals: Prof. J. A. Ewing, F.R.S.
EPIDEMIOLOGICAL SOCIETY, at 8.30.

SATURDAY, MAY 19.

ROYAL INSTITUTION, at 3.—South Africa : Past and Future : Dr. Alfred Hillier.

MONDAY, MAY 21.

SOCIETY OF ARTS, at 8.—The Incandescent Gas Mantle and its Use : Prof. Vivian B. Lewes.
ROYAL GEOGRAPHICAL SOCIETY, at 3.—Anniversary Meeting.
VICTORIA INSTITUTE, at 4.30.—Ethics : Rev. Dr. Wace.

TUESDAY, MAY 22.

ROYAL INSTITUTION, at 3.—Brain Tissue and Thought : Dr. A. Hill.
ZOOLOGICAL SOCIETY, at 8.30.—On the Development of the Skeleton of the Tuatara, *Sphenodon (Hatteria) punctatus* : Prof. G. B. Howes, F.R.S., and H. H. Swinnerton.—On Crustaceans from the Falkland Islands collected by Mr. Rupert Vallentin : Rev. T. R. R. Stebbing, F.R.S.—The Significance of the Hair-slope in certain Mammals : Dr. Walter Kidd.
ROYAL PHOTOGRAPHIC SOCIETY, at 8.—Hydroquinone and Colour Impressions : Alfred Watkins.

WEDNESDAY, MAY 23.

SOCIETY OF ARTS, at 8.—Salmon Legislation : J. Willis-Bund.
GEOLOGICAL SOCIETY, at 8.—The Igneous Rocks of the Coast of County Waterford : F. R. C Reed.—On a New Type of Rock from Kentallen and elsewhere, and its Relations to other Igneous Rocks in Argyllshire : J. B. Hill and H. Kynaston.

THURSDAY, MAY 24.

LINNEAN SOCIETY at 3.—Anniversary Meeting.
INSTITUTION OF ELECTRICAL ENGINEERS, at 8.—Annual General Meeting.

FRIDAY, MAY 25.

ROYAL INSTITUTION, at 9.—The Great Alpine Tunnels : Francis Fox.
PHYSICAL SOCIETY, at 5.—Experiments illustrating the Aberration called Coma : Prof. S. P. Thompson, F.R.S.—Notes on the Measurement of some Standard Resistances : R. T. Glazebrook, F.R.S.—On the Strength of Ductile Materials under Combined Stresses : J. J. Guest.

CONTENTS.

THURSDAY, MAY 24, 1900.

THE SCIENCE OF BACTERIOLOGY.

The Principles of Bacteriology. By Dr. Ferdinand Hueppe. Authorised translation from the German by Dr. E. O. Jordan. Pp. x + 467. (Chicago : The Open Court Publishing Company. London : Kegan Paul, Trench, Trübner and Co., Ltd., 1899.)

IN order to fully appreciate the aim and object of the talented author of this work, it is necessary to quote a few passages from his preface. Prof. Hueppe points out that the natural history side of bacteriology has in the past been kept too much in the foreground, while the scientific side has been relegated almost exclusively to the sections dealing with protective inoculations.

"This mode of treatment." continues the author, "no longer suffices to meet a growing and legitimate demand. In this book I wish to present an attempt at a critical and comprehensive exposition of bacteriology, basing it clearly and solidly upon scientific conceptions. I make this essay in order that our knowledge of the causes of putrefaction, fermentation and disease, together with the methods of the prevention and cure of infection, may develop in a way free from all ontology. It is sometimes of use to restate things which are axiomatic. The 'entities' or 'essences,' which, even in the age which has discovered the law of the conservation of energy and the evolution of living things by means of the struggle for existence, still haunt the mind of the physician who remains sunk in the ontological contemplation of diseased cells and disease-producing bacteria, are a mere remnant of priest medicine, and can have no place in any scientific conception of biology, pathology or hygiene."

The first chapter (pp. 1–49) in the book deals with "The structure of bacteria." No greater authority on this subject than the author could be named ; yet, in view of the highly important questions discussed in Chapters iv.–vii., one is led to doubt whether this portion of the book is not a little out of keeping with the scope of the work as a whole.

The "Vital phenomena of bacteria" are discussed in Chapter ii. (pp. 50–138). Although the subject is most ably dealt with, most of the information given may be found in nearly every text-book of bacteriology. Considering the important character of the rest of the book, this chapter seems unduly long.

In Chapter iii. (pp. 146–219) a brief description of the most important pathogenic bacteria is given. Here the author paves the way for the discussion of the important questions which crop up later in the book. It is curious to note that Prof. Hueppe, although considering that the evidence is most in favour of *B. typhosus* and *B. coli communis* being two distinct species, is by no means dogmatic on the point. Thus he says, on p. 193 :—

"There are, in fact, at present two opposing views. The one, which to me seems to be the better founded, is that the bacteria of typhoid fever and *B. coli communis* are two distinct species. The other view is that the common intestinal saprophyte, *B. coli communis*, is an æco-parasite which, under special conditions, may become able to invade the body and penetrate into the living organism, where it undergoes transformation into the typhoid bacterium."

At the end of this chapter Mr. Jordan contributes a brief *résumé* of Sanarelli's recent papers upon yellow fever. The summary is concisely and well written, and enables one to comprehend without difficulty the extent and value of Sanarelli's researches. The remaining chapters are full of originality, and invite most careful reading and serious attention.

In Chapter iv. (pp. 221–274) the "Cause of infectious disease" is discussed with conspicuous ability. The author endeavours to show what is false and what is scientifically tenable in the different conceptions of the true and sufficient cause of epidemic disease upheld by such authorities as Koch, Virchow and Pettenkoffer.

"Virchow finds an internal cause in the diseased cells ; his opponents see an external cause in the germs that bring about the disease ; and Pettenkoffer sees a cause in those external conditions which play no particular *rôle* either in the eyes of Virchow or in those of Virchow's chief opponents."

If the writer does not altogether succeed in his object, he at all events widens our horizon of thought to an extent which is quite remarkable. It will not be out of place to quote a single paragraph—

"If the facts are considered in a scientific spirit rigorously and without prepossession, it is seen that the sum of the qualities of a disease germ is only apparently the 'essence' of an infectious disease, that, in reality, here as elsewhere, a true internal cause is to be found, inherent in the internal organisation of man. Just as in all natural processes, without exception, so here, the disease germs act as liberating impulses, and are able to set free only what in the form of a predisposition toward disease is in some way prefigured both in nature and amount in the human body."

In Chapter v. (pp. 275–294) the author asks the question—"Can disease be cured by combating the cause ?"

In speaking of Hahnemann's doctrine of the value of small doses, the author passes the following criticism on homœopathy :—

"Even the childish extravagance which found vent in homœopathy could not impair the sound kernel of truth which the doctrine contained."

Although Prof. Hueppe's whole book ought to be read by all those physicians who are modest enough (happily, the great majority) to believe that there is something still to be learnt in the theory and practice of medicine, this chapter is especially full of suggestions and original observations, which the thoughtful practitioner would do well to study.

Chapter vi. (pp. 295–397) treats of "Immunity, protective inoculation, and curative inoculation." It is, perhaps, the most important chapter in the book, and it is impossible in the limits of this notice to do the author full justice. It may, however, be said that it deals with a most difficult and complex subject in a way that is to be highly commended. That it is "stiff" reading cannot be denied, but that is not the fault of the writer, but of the subject. A careful perusal of this portion of the book will well repay the physician as well as the bacteriologist.

The "Prevention of infectious diseases by combating the cause of the disease" is the text of Chapter vii. (pp. 398–439). Here we are not altogether in sympathy with the writer, although his views are clearly and forcibly expressed, and are in the main in touch with the teachings of modern sanitarians.

It is to be regretted that in this chapter the author allows his personal antagonism to Koch's doctrine of disinfection to weaken his arguments and conclusions. That the followers of Koch sometimes carried disinfection too far does not detract from the value of Koch's original observations.

Prof. Hueppe lays peculiar stress on the importance of making infectious disease impossible by removing the predisposition to disease, but he scoffs at the idea of combating disease by warring directly with the germs of disease. Although there is a great deal to be learnt from this chapter, it seems a pity that so able a writer should have marred his own work by a captious criticism of Koch's able investigations.

The last chapter (pp. 440–455) deals with the "History of Bacteriology." Ably written though it is, it, like the first chapter, appears to be foreign to the general scope of the book.

In summary of the book as a whole, it may be said that it affords more ground for serious thought and reflection than perhaps any of the works on bacteriology hitherto published. The original and able manner in which the author attacks biological problems of great difficulty and complexity deserves all praise, and we can cordially recommend the book, not only to bacteriologists pure and simple, but also to those physicians who recognise the limitations of medical science.

Much praise is due to the translator. Mr. Jordan's worth as a bacteriologist is well known and fully appreciated. By giving us this translation of Hueppe's work he has added to his reputation. A. C. HOUSTON.

SUNSHINE AND WINE-GROWING.

Vinification dans les Pays chauds—Algérie et Tunisie. Par J. Dugast. Pp. 281 ; 58 figures. (Paris : Carré et C. Naud, 1900.)

ACCORDING to the preface, valuable scientific and technical works on the production of wine in temperate climates have been published both in France and elsewhere ; but so far the special problems which are encountered by wine-growers in the warm climates of such countries as Algeria and Tunis have remained unnoticed. The present work is intended by the author to fill this blank. But although it has been written specially with a view to describe the difficulties peculiar to wine-making in a warm climate and the means of overcoming them, the author has done more than this, for he has found it advisable, in order to make his purpose quite clear, to embody his special subject in a general scientific and technical description of wine-making. As he has had very considerable practical and scientific experience in his subject, the result is a work well worth the attention of all interested in the making of wine.

The most common difficulty of the Algerian wine-grower, and one which is very rare in the more temperate climate of France, is due to the must, or grape juice, very frequently containing too little acid and too much sugar as a result of very active plant assimilation induced by excessive solar radiation. Deficiency of acid is apt

not only to affect injuriously the flavour of the resulting wine, but also to induce unsoundness ; the latter effect being caused by the low acidity of the wine favouring the growth of injurious bacteria, which the higher acidity of a normal wine tends to inhibit, owing to the well-known fact that an acid medium is unfavourable to the development of most ferment bacteria.

The means employed to remove the difficulty of want of acidity, which are described by the author, let us into secrets of wine-making which some may perhaps be inclined to think border on sophistication. Plastering is one which is undoubtedly objectionable. It consists in adding calcium sulphate to the crushed grapes, which results in the formation, from the cream of tartar present in the must, of sulphate of potash. But this method, though evidently made use of by many wine-growers, is condemned by the author, and also discouraged by the French law, which limits the amount of sulphate of potash to two grammes per litre.

Other methods for increasing the acidity of the must are : crushing a certain quantity of unripe sour grapes with the ripe ones ; the addition of tartaric acid to the must previous to fermentation ; and sprinkling the grapes in the wine-press with, what the author styles, di-calcic phosphate. The latter treatment is said to result in the formation of acid phosphate of potash, a salt considered by the author to be less objectionable than sulphate of potash.

Excess sugar in the must acts detrimentally by throwing too much work on the yeast, which is itself apt to be crippled in the hot climate of Algeria by an exceedingly high fermentation temperature. Mention is made of the fermentation temperature at times rising to upwards of 115° F.—which in itself is sufficient to arrest the fermentation functions of most yeasts.

About 20 per cent. of sugar is considered the most favourable amount for a wine must to contain, and if the saccharometer shows that it exceeds this amount, the best remedy appears to be the simple and inexpensive use of the pump.

An interesting point, about which much has been said of late years, is raised by the author when he deals with the question of the use of pure selected yeasts in the fermentation of wine. It has been advanced by certain upholders of this system that the characteristic flavour or bouquet of most well-known wines is produced in the main by the variety or species of yeast natural to the grapes of the district, and that, if pure cultures of such yeasts are made use of in the fermentation of foreign musts, the flavour of the resulting wines assume the character of the wines of the district from which the yeasts were obtained.

The idea is evidently one of the greatest importance to the wine industry, as it holds out hopes of improving the wine of poor districts into something like, let us say, first quality clarets or Burgundies. The author of this book states that selected yeasts have been much used by the wine-growers of Algeria, and he claims to have had ample opportunities for studying the results. The conclusion he arrives at is that the yeast from a noted growth of wine, when added to an ordinary must, is quite powerless to confer on it the special qualities of the wine from which it comes ; and he further concludes that yeast has little, if any, influence on the bouquet of wine. The true character of a wine, he maintains, is due to numerous factors, among which the variety of grape and the character of the soil and climate preponderate ; if the yeast does produce any flavour, it is indistinguishable among these.

If, however, the author passes adverse judgment on selected yeasts regarding their power of conferring flavour, he does not do so with regard to their use for setting up a rapid and healthy fermentation in wine must. For this purpose he advocates their use warmly, but insists on the employment of a selected indigenous yeast as more calculated to be in harmony with the environment than if it was derived from a foreign source.

The valuable results which have accrued from Emil C. Hansen's remarkable studies on yeast have already led to so many successful results in technical practice, that we still feel inclined to suspend judgment regarding the non-efficiency of wine yeast in the matter of flavour until M. Dugast's interesting observations are confirmed in other quarters.

In conclusion, we call special attention to this book as likely to be useful to our Colonial wine-growers of Australia and the Cape ; the climate of these countries is somewhat similar to that of Algeria, and no doubt some of the special difficulties discussed in this book are also met with in these countries. A. J. B.

THE FAUNA OF THE SHETLANDS.

A Vertebrate Fauna of the Shetland Islands. By A. H. Evans and T. E. Buckley. Pp. xxix + 248. (Edinburgh : D. Douglas, 1899.)

ALTHOUGH it would be too high a meed of praise to say that the authors have done for the Shetlands what Gilbert White did for Selborne (the systematic treatment of the fauna not being favourable to colloquial writing), there is no doubt that they have succeeded in producing a very interesting volume, and one which should be indispensable to every visitor to the most northern group of the British Islands, whether or no he be specially interested in birds. For in place of restricting themselves to a detailed account of the various members of their vertebrate fauna, Messrs. Evans and Buckley have furnished a very interesting description of the more striking physical features of these islands, together with numerous notes on the people and their mode of life. But perhaps the most generally attractive feature of the work will be the exquisite views of Shetland scenery with which it is adorned ; these illustrations reflecting the highest credit alike on the photographer and on the artist responsible for their reproduction in the present form. In introducing these scenic pictures, in place of figures of the birds recorded as members of the fauna, the authors have undoubtedly exercised a wise discretion. In only one instance have they made a natural history object the chief feature of an illustration ; the one exception being the beautiful plate of the nest and young of the great skua—a bird of all-absorbing interest to the naturalist in the Shetlands.

And here it is proper to mention that the volume before us forms a part of the vertebrate fauna of Scotland, of which several volumes by Messrs. Harvie-Brown

and Buckley have already appeared. It seems that Mr. Evans, who has an extensive personal acquaintance with the Shetlands, had an idea of writing an independent work on its animals. The securing his services as a contributor to the larger undertaking will commend itself to all.

After devoting fifty-four pages to a well-written description of the physical features of the country, the authors proceed to their proper subject—the detailed account of the vertebrates, which includes both the terrestrial and the marine forms. In the classification of the birds they follow in the main the scheme of Mr. H. Saunders, and though they suggest that some amendments might perhaps have been made had it not been for the sake of uniformity with the "Fauna of Orkney," yet we are glad to know from his volume in the Cambridge "Natural History" that Mr. Evans, at least, is no friend to the plan of unnecessarily multiplying the genera of British birds, nor to the "*Scomber scomber*" principle.

In the classification of mammals, especially when we note the statement that Mr. Eagle Clarke has *carefully* revised the proofs, it is somewhat surprising to find the narwhal included among the *Physeteridae*. Neither do we see the necessity of regarding the rorquals as the representatives of a family by themselves. But, altogether apart from such trivial details, we must take exception to the practice of including introduced species among mammalian faunas. In the present instance the authors note five species of rodents as belonging to the Shetland fauna, whereas only one of these—*Mus sylvaticus*—is really indigenous. The trouble such methods cause to those who have occasion to write on the geographical distribution of animals is best known to themselves. If introduced forms are mentioned at all, their foreign origin ought to be indicated in such a manner that it will catch the eye of the reader at the first glance. In the case of birds, such as the ruff, which but rarely visit the islands, some conspicuous notification of the fact would likewise be advantageous, although we are ready to acknowledge that the line between regular visitors and accidental stragglers is very hard to draw.

The above mention of *Mus sylvaticus*—the long-tailed field-mouse—reminds us that one of the most important objects of histories of island faunas is to point out whether the indigenous animals are in any way distinguishable from those inhabiting the nearest mainland. In the case of birds of strong flight such differences are not likely to occur, but they should be looked for in birds that never leave their island home, and in the indigenous mammals. On the special characters of the Shetland field-mouse the authors are silent, which in view of Mr. Barrett-Hamilton's recent recognition of a peculiar representative of this type in St. Kilda is distinctly to be regretted. In the case of the common wren, which has likewise a peculiar local race in St. Kilda, the authors state that the Shetland form differs to a certain extent from the one found on the Scottish mainland, although not, in their opinion, sufficiently so as to be entitled to be regarded as representing a distinct race. If this be so, and the field-mouse be indistinguishable from the mainland form, it suggests that the Shetlands have been separated from the mainland at a later date than have the Hebrides ;—but this is just one of the cases where we should have liked a well-considered opinion from the authors!

In an area like the Shetlands the great interest, from a faunistic point of view, centres on the birds; and among these the great skua holds the foremost place, since its only British breeding-stations are on these islands. So much has been of late years written on this subject, both in newspapers and in ornithological journals, that it is one with which the public are tolerably well acquainted. Nevertheless, the account given by the authors of the almost complete extermination of this fine species, and its subsequent rehabilitation by the efforts of various members of the Edmonston family and Mr. Scott, of Melby, will be read with interest, and forms a concise summary of the whole affair. We should, however, like to know more with regard to the meaning of the statement that "protection for the skuas implies some measure of protection also for the gulls; but unless the latter greatly increase, the former cannot be expected to do so."

Some interest also attaches to the specimen of the collared pratincole killed by Bullock in 1812, as being the only example of the species hitherto shot in North Britain. In the fourth edition of "Yarrell" the skin is stated to be in the British Museum, but the investigations of the authors fail to confirm this statement.

Greater attention is, however, merited by the account of the nesting of the storm petrel, which sometimes lays its eggs among large stones on the shore, and in other cases selects deserted rabbit-burrows for its home. The crofters, knowing the value set on the eggs of this bird by collectors, and being likewise extremely partial to young petrels as a *bonne bouche*, are extremely reluctant to indicate the rabbit-holes in which the birds nest to strangers.

To many it will come as a surprise to learn that ravens are still common in the islands; so numerous, indeed, as in certain districts to prove very destructive to the poultry and stock, on which account war is waged against them by the crofters. In contrast to the abundance of these birds is the scarcity of rooks, which are, indeed, little more than casual visitors to the islands.

The weakest point about the book is undoubtedly, as the authors themselves are fain to confess, the section on fishes, the classification followed being altogether obsolete and discredited.   R. L.

PHYSICAL CHEMISTRY.

Introduction to Physical Chemistry. By James Walker, D.Sc., Ph.D. Pp. x + 332. (London: Macmillan and Co., Ltd.)

IT is now nearly ten years since Prof. Walker placed English students under obligation by his admirable translation of Ostwald's "Outlines of General Chemistry." Since that time "little Ostwald" has been the source from which most students have taken their first draught of information about physical chemistry in its modern form. The phrases and paraphrases of the book, the diagrams, the perpetual motions "which are impossible" have become almost painfully familiar to the examiner. The present writer is one of those who believe that

Ostwald's book has been of the highest service to chemistry. At the same time, it must be admitted that it is one to be used with care. There is an illusory appearance of simplicity about it, and if care be not taken the use of the book is eminently calculated to lead to a learned smattering. It is, in fact, a book which forms the summary of a course of instruction, and for beginners it must be supplemented by an extended commentary by an experienced teacher.

These observations arise inevitably in connection with Prof. Walker's new book, which, in size, appearance and typography, as well as in its topics, bears so striking a resemblance to Ostwald's "Outlines." The first question that the reader will ask is—Where lies the difference between the two books? This question is soon answered as one reads; Prof. Walker's book is more limited in range and incomparably simpler. To quote the author's words, it "makes no pretension to give a complete or even systematic survey of physical chemistry"; the aim is to give a full discussion of some of the chief principles of modern physical chemistry, and to show their application to ordinary laboratory chemistry.

Dr. Walker has achieved his purpose in a most satisfactory manner, and has produced a book which will be a real boon to students of physical chemistry. He writes with the knowledge of a specialist and the experience of a teacher, and is very striking to any one who knows the difficulties of students to see how perfectly Dr. Walker appreciates them. Not less striking are the expository power and resourcefulness with which the difficulties are handled. Whilst the whole book is clear, readable, and abreast of the times, some chapters deserve special attention. The one on chemical equations is amongst these. It gives a rational account of the art of constructing chemical equations by dissection and summation, a subject which has been strangely neglected by text-book writers. The chapter on fusion and solidification is made very clear by a thorough discussion of the mutual relations of salt, ice and water. The wide generalisation, or group of generalisations known as the Phase Rule, is expounded within reasonable limits. Hitherto there has been nothing concise on this subject in the English language. The chapters relating to the modern theories of solution are, it need scarcely be said, written with fulness of knowledge and in the spirit of a true believer in the doctrine of electrolytic dissociation. Chemical dynamics is treated succinctly, and admirably illustrated by examples. There is a distinct gain here in departing from the strict historical development of the subject, which is apt to confuse beginners by the series of fresh starts which it involves. The concluding chapter on thermodynamical proofs is made as clear as it well could be. At the end of each chapter references are given to original articles which have appeared in English journals and to English books. The list of these is quite gratifying, but the wisdom of confining the references to English publications seems questionable. The extraordinary backwardness of students in acquiring a reading knowledge of German is condoned by such a restriction; and, besides this, it would have been a service to many students who have some knowledge of the language if Dr. Walker had helped them to select the really important pioneering papers from the vast periodical literature that has arisen in Germany during the past ten years.

In concluding this notice, one is naturally led to reflect upon the attitude which appears to be still maintained by a number of English chemists in regard to the modern theories of solution. There can be no doubt that a student reading Dr. Walker's book will become imbued with these theories, and will acquire convictions that will be difficult to eradicate. If these theories are wrong, if they are even strongly suspect, the responsibility of the teacher becomes serious It is true Dr. Walker gives here and there some indications of the objections which have been urged against them, but there is no explicit statement of the opposition case. The question arises whether an opposition case can be explicitly stated. The theory of ionic dissociation has been applied to explain and co-ordinate a very large number of chemical facts, and has thrown light on matters that were previously dark. The contention of the objectors appears to be mainly that this light is illusory. The present writer is far from claiming judicial functions in the matter; but he ventures to think that the opposition to the dissociation theory would be more respected, both here and on the Continent, if it were of a more positive character, and if a more tangible alternative theory could be presented which should prove itself not less comprehensive and practically productive than the one which is assailed. The history of science shows plainly enough that a comprehensive theory with some weak points will hold its ground until a not less comprehensive theory with fewer weak points makes its appearance. It is probably on this ground that Prof. Walker takes his stand in freely imparting the doctrine of electrolytic dissociation to elementary students of physical chemistry. ARTHUR SMITHELLS.

OUR BOOK SHELF.

Catalogue of the Lepidoptera Phalaenae in the British Museum. Vol. ii. Catalogue of the Arctiadæ (Nolinæ, Lithosianæ) in the collection of the British Museum. By Sir George F. Hampson, Bart. Pp. xx + 589, and plates xviii–xxxv. (London: Printed by order of the Trustees, 1900.)

THE first volume of this series, containing the Syntomidæ, was published in 1898, and we have now to welcome the appearance of the second, comprising two groups, which the author treats as sub-families of the Arctiadæ; the typical Arctianæ being reserved for the third volume. 1193 species are described in the second volume, all of which, except 162, belong to the Lithosianæ, the Nolinæ being a comparatively small sub-family.

The enormous extent of the insect-world is but little realised, even by naturalists, unless they are entomologists; but, considering the progress already made, we are probably well within the mark in saying that it may well take fifty volumes, and the whole of the new century, to complete the Catalogue before us; and yet the moths are only a portion of one of the seven principal orders of insects, and one which is probably far surpassed in number of species by at least three other orders.

The descriptions of the species are necessarily brief, but are arranged on a uniform plan which admits of easy comparison; and their determination is further facili-

tated by comprehensive tables of genera and species, and by the large proportion which have been figured, either in the crowded coloured plates, or in text-illustrations. We are glad to see that space has been found for notices of larvæ, when known. Space has also been devoted to phylogeny; but it is, perhaps, an open question whether it is worth while to deal with this subject in a descriptive work at all. At best, it can only express the momentary and necessarily fluctuating opinions of an individual author on the affinities of genera and species from the very imperfect materials at present available; for until the earlier stages of a considerable number of forms have been carefully studied and tabulated for comparison, it is impossible for us to judge of them completely or accurately. We would therefore prefer to treat this branch of the subject tentatively, in ephemeral publications, rather than to introduce a necessarily fluctuating factor, of merely temporary value at best, into a standard work of reference, of such great and permanent value to all lepidopterists as the present. We must also object to the author's tendency to dogmatise on the subject, especially as our knowledge of fossil insects is at present practically nil, and of the early stages of the great majority no better. Such a phrase as [the Arctiadæ form] "a family of moths derived from the Noctuidæ," seems to us quite out of place in a scientific book at the present state of our knowledge; though a formula which we find a little further on is less objectionable; "the *Nolinae* probably arose from a very early Arctian form which had affinities in the *Noctuidae* to *Hypenae* and *Sarrothripae*."

But these are details of individual taste or judgment; while there cannot be two opinions respecting the value and importance of the work. W. F. K.

Giordano Bruno, zur erinnerung an den 17 Februar, 1600. Von Alois Riehl. Zweite neu bearbeitete Auflage. Pp. iv + 56. (Leipzig: Engelmann, 1900.)

EARLY in 1600 Giordano Bruno went to the stake in the cause of free speech and thought. The ashes of martyrdom have ere now kept evergreen even reputations and names that were otherwise of little worth. But Bruno's life and work are alike memorable. Few, however, of those to whom the romantic wander-years and heroic death appeal, have leisure and training to grapple with the technical Latin and hard Italian of the versatile and stormy Nolan. The tercentenary, therefore, of Bruno's tragedy can have no memorial more fitting than Prof. Alois Riehl's "Giordano Bruno." Would that it were in English! Dating originally 1889, Prof. Riehl's brochure has undergone revision thorough and throughout. It puts Bruno in his right setting of time and place. It resumes, with brevity and lucidity quite noteworthy, the principles for which Bruno gave his life. Bruno originated neither Copernican physics nor pantheist metaphysics. His debt to one close forerunner at least is not small. Yet in taking the new astronomy as a scientific basis, and only therefrom passing to such metaphysical conceptions as infinity and unity, while reaching out ultimately to a monistic principle, it is Bruno and not his precursors, physicist and revived neoplatonist, that may claim to father modern naturalism. Prof. Riehl characterises the system as "theocentric," since nature is, for Bruno, *deus in rebus.* Bruno is said to have met the process which resulted in his condemnation by equivocating between what he accepted *secundum fidem* and what he affirmed *secundum rationem.* At any rate, whatever human weakness he may have shown, he lost no opportunity of reaffirming his principles. He recanted nothing. He could have saved himself would he but have prostituted his pen to apologetics on behalf of the reigning orthodoxy. He chose not *propter vitam vivendi perdere causas.* And he died a knight-errant of the free spirit. H. W. B.

LETTERS TO THE EDITOR.

[The Editor does not hold himself responsible for opinions expressed by his correspondents. Neither can he undertake to return, or to correspond with the writers of, rejected manuscripts intended for this or any other part of NATURE. *No notice is taken of anonymous communications.]*

Escape of Gases from Atmospheres.

I ASK for space to reply to Mr. Cook's letter in last week's NATURE.

There are two ways in which the rate at which gases escape from atmospheres may conceivably be investigated, viz. the *a priori* method, which seeks to determine from the kinetic theory of gas what proportion of molecules attain the requisite speed; and the *a posteriori* method, which seeks to ascertain from the observed effects of escape where and on what scale it has actually taken place.

I tried the *a priori* method more than thirty years ago, but had to abandon it, having satisfied myself that *in the present state of our knowledge it cannot be made to furnish a valid investigation.* I came to this conclusion upon grounds which are fully stated in a paper of which the first part will appear in the May number of the *Astrophysical Journal*, and the second and more important part probably in the June number. I then turned to the *a posteriori* method, and endeavoured to develop it in the memoir which Mr. Cook has criticised (see *Scientific Transactions* of the Royal Dublin Society, vol. vi. (1897), p. 305, or *Astrophysical Journal* for January 1898, p. 25).

Both methods, if correctly applied, should lead to the same results: but the *a priori* method, as handled by Mr. Cook and Prof. Bryan, furnishes a different rate of escape from the *a posteriori* method. In such cases there must be a mistake or mistakes somewhere, and in the above-mentioned paper sent to the *Astrophysical Journal* I have endeavoured to trace out where the mistakes are.

The principal errors seem to be three.

The number of molecular speeds which lie between v and $v + dv$

$$= N(\pi + \delta)dv$$

where N is the number of molecular speeds whose distribution is under consideration, π is the probability function (in this case Maxwell's law), and δ may be called the deviation function, as it furnishes the difference which exists between the actual number and that computed by Maxwell's law. In all cases of probability laws the deviation fuction δ is large when N is small; but when the events whose distribution is sought are independent of one another and have causes all of one kind, then δ becomes inconspicuous when N is sufficiently large, and the distribution law may *in such cases* be reduced to $N\pi dv$ without sensible error. This reduction, however, is not always legitimate when, as in gases, the events are so associated with one another and with other agencies that cumulative effects can arise. Then δ may become larger than π in reference to those values of v which make π small. The first omission seems to be the omission to take this into account.

Another omission is the omission to take the size of the element of volume $dx\,dy\,dz$ into account. As experiment shows, may at the bottom of the earth's atmosphere be as small as the cube of one-tenth of a millimetre. But in the regions from which the escape of gas is possible, it has a volume of many cubic miles. This circumstance, which largely increases the opportunity which molecules have of escaping from that situation, has not been taken into account.

But perhaps the most serious error is overlooking the fact that Maxwell's law holds good only of a portion of isotropic gas surrounded by similar gas. That the gas shall be isotropic is one of the data employed by Maxwell in his proof of the law. Another law (which may be, and in fact is, very different from Maxwell's) is the law of distribution of the molecular speeds in a portion of gas as anisotropic as that of the regions from which the actual escape takes effect. The deductions from Maxwell's law may be correctly derived, but the premiss being wrong the conclusion has no significance.

It would be very satisfactory if we had two ways—the *a priori* method as well as the *a posteriori*—of investigating the problem; but with our present limited knowledge of molecular physics, this does not seem to be within our reach.

Mr. Cook at the end of his letter supposes that "the discovery by Ramsay of helium as a constituent of our atmosphere

only tends to confirm the results of my (Mr. Cook's) calculations of the impossibility of its escape." This is so far from being the case that the quantitative determinations made in Prof. Ramsay's laboratory are now sufficiently advanced to lead with much increased emphasis to the opposite conclusion. This appears from the following data, which have been generously placed at my disposal by my friend, Prof. Ramsay:—

(1) The proportion by volume of argon in dry atmospheric air is about 1 per cent. of the whole, the volume of neon (to which the present note will not further refer) may be taken as about a thousandth part of the volume of argon, and the volume of helium as about 1/10 to 1/20 of the volume of neon. Accordingly, the volume of helium in dry air is something like from 1/10,000 to 1/20,000 of the volume of argon, or from 1/1,000,000 to 1/2,000,000 of the whole volume of the air.

(2) Both argon and helium are supplied to the atmosphere by hot springs; argon generally by all hot springs which contain atmospheric gases, and helium by some of them.

(3) The argon in such springs, like the oxygen and nitrogen, may be simply gas which had previously been removed from the atmosphere by water. A litre of water under ordinary conditions will absorb about 45 c.c. of the oxygen of the air in contact with it; about 15 c.c. of its nitrogen; about 40 c.c. of its argon; and about 14 c.c. of its helium.[1]

Hence in rain we should expect to find the following proportions preserved:—

$$\frac{20 \cdot 9}{100} \times 4 \cdot 5 \text{ of } O_2 ;$$
$$\frac{78 \cdot 1}{100} \times 1 \cdot 5 \text{ of } N_2 ;$$
$$\frac{1}{100} \times 4 \text{ of } A ; \text{ and}$$
$$\left. \begin{array}{l} \text{from } \dfrac{1}{1,000,000} \times 1 \cdot 4 \\ \text{to } \dfrac{1}{2,000,000} \times 1 \cdot 4 \end{array} \right\} \text{ of He,}$$

So far as oxygen, nitrogen and argon are concerned, these proportions are sufficiently nearly those in which the gases are present in the springs referred to. But in those springs in which helium also has been detected, it seems to be present in quantities about 1/10 of the argon—that is, in a quantity which is nearly from 3000 to 6000 more than we can attribute to its having been derived from the atmosphere.

(4) This great excess of helium in some springs has doubtless a mineral origin, some minerals, chiefly uranium compounds, containing much helium which they give up when heated. On the other hand, there does not appear to be any comparable mineral source of argon.

(5) Hence, on the whole, the argon which is being supplied to the atmosphere by hot springs seems to be argon which had previously been withdrawn from the atmosphere and which is being restored to it. Whereas, in contrast to this, there seems to be a continuous transfer of additional helium from the solid earth to the atmosphere always going on.

Thus the facts seem to warrant our inferring :—

(a) That the excessively small quantity of helium in the atmosphere is helium on its way outwards.

(b) That it would have become a much larger constituent of the atmosphere, by reason of the influx from below, if there had been no simultaneous outflow from above.

(c) That the rate of this outflow is presumably equal to the rate of supply; and therefore such as would suffice in a few thousand, or at least in a few million, years to drain away the small stock of helium in the earth's atmosphere, if the source of supply from below could be cut off.[2]

[1] See the determinations made by Herr Estreicher in Prof. Ramsay's laboratory, as recorded in the *Zeitschrift für physikalische Chemie*, vol. xxxi. (1899), p. 184.

[2] If the proportion of helium in the atmosphere is assumed to be something between 1/1,000,000 and 1/2,000,000 of the whole atmosphere (which rather tends to be an over-estimate, since it does not take into account the increased diminution of the density of the helium as it ascends, which is a consequence of its escape from the top of the atmosphere), then the helium in the whole of the earth's atmosphere would, if reduced to standard temperature and pressure, occupy a volume somewhere between a cube of ten miles, and half that space. Now, so far as can be judged from the imperfect observations as yet made on the rate at which helium is being filtered into the atmosphere, it would appear that the present rate of supply is such as would yield this quantity of helium in something like one or two thousand years, and perhaps in a less time.

It thus appears that the recent more exact determinations have raised what was probable when I wrote my memoir into being now almost certain, by showing with greatly increased clearness—

(1) That argon is unable to escape from the earth.

(2) That helium is slowly escaping, and presumably was in a position to escape more freely in the distant past.

It is interesting to observe that another moot question in astronomy seems to be resolved by Prof. Ramsay's work. It is known that the dynamical relation of the vapour of water to Mars is nearly the same as that of helium to the earth. We are accordingly now justified in presuming with greater confidence that water cannot remain upon Mars, that accordingly the polar snows of that planet are probably carbon dioxide, and that some of the other appearances which have been observed are due to the shifting of low-lying fogs of this vapour as they travel alternately towards the two poles.

G. JOHNSTONE STONEY.

8, Upper Hornsey Rise, N., May 20.

"Plotosus canius" and the "Snake-stone."

POSSIBLY the following facts may possess interest for some of your readers :—

A good many years ago, when sea-bathing in the Old Straits of Singapore (*i.e.* those separating the island from the Malay Peninsula), I put my foot in a slight muddy hollow in the sandy sea-bed; the moment I did so, I received an agonising stab near the ankle (from some red-hot poisoned blade, it seemed) which drove me in hot haste ashore, where a Malay constable, on hearing what had happened, and on examining the wound, pronounced my assailant to be the "ikan sĕmbilang" (sĕmbilang fish), *Plotosus canius*, one of the siluroids, I am informed by Mr. Boulenger of the British Museum. The fish is armed with three powerful spines on the head, one projecting perpendicularly from the top, and one projecting horizontally from each side.

The Malay lost no time in running to the barracks near by, whence he shortly returned with a little round charcoal-like stone about the size of a small marble. This he pressed on to the wound, to which it adhered, and remained there by itself, without any continuation of pressure, for a minute or more. Then it fell off, and black blood began to flow, which, after a little, was succeeded by blood of normal colour. The pain, which had been excessively acute, began to diminish soon after this, and in an hour had practically disappeared. The wound gave me no further trouble, but a fortnight afterwards I noticed a hole about the size of a pea where the wound had been.

Another gentleman, who, curiously enough, had suffered in the same way in another part of Singapore the same day, was not so fortunate in his cure, being completely laid up for six weeks.

The black stone applied by the Malay to the wound came, he alleged, from the head of a snake, and claimed, therefore, to be a bezoar stone. It was, no doubt, a snake-stone, probably made of charred bone, and therefore porous in character, which would account for the adhesive and absorptive powers it displayed in my case.

In his "Thanatophidia of India," Sir J. Fayrer (quoted by Yule in "Hobson-Jobson") expresses entire disbelief in the efficacy of these stones as remedies "in the case of the *real bite* of a *deadly* snake," owing to the extreme rapidity with which, in such a case, the venom pervades the system.

However this may be, the late Prof. Faraday, after examination of one of these stones, supplied by Sir Emerson Tennent (quoted by Yule), credits it with certain absorbent powers, and it would seem a pity that the undoubted value of such stones, at all events in minor cases, where they may save a great deal of suffering, should be discredited.

Another remedy, considered of some value by Malays for the stab of *Plotosus canius* is the sap of *Henslowia Lobbiana,* which grows freely on the coasts of the Malay Peninsula.

Among other marine offenders of this class dreaded by Malays are several varieties of the skate or sting-ray, " pari " as they are generically called, and some of the "lĕpu," of which the only dangerous one, I have Mr. Boulenger's authority for saying, is the "lĕpu" proper, viz. *Synancia horrida.* When the skate reaches a large size, he will drag a fisherman's canoe a long way.

Among the Medusæ, one much dreaded is known as "ampai,"

from its long fringes. The effects, unless a remedy can speedily be found, are painful and trying to a degree, seeming to penetrate the whole frame, as it were, electrically, at once specially affecting the seat of any ailment, and even the teeth and the hair. I have never suffered from it myself, but am enabled to speak to these points from two cases which came under my personal observation. A valuable remedy for this sting, if applied soon, is the juice of the young fruit of the papaw (*Carica papaya*).

A further illustration of the value of some native remedies is supplied by a case which occurred some years ago at Malacca, during my residence there, though I cannot state what the remedies employed were.

A young gentlemen in the office of the Telegraph Company went out to bathe in the sea one night from the end of the pier (in any case rather a rash proceeding, if only for the occasional presence of crocodiles !), when he found himself in the embrace of some creature with long tentacles, from which, after desperate struggles, he eventually succeeded in freeing his legs and his arms, and in regaining the pier. The Colonial surgeon could do nothing for him, and he was in such tortures that for a time he seemed to have lost his mental balance, but nine or ten days after the occurrence a native practitioner, being called in, cured him completely.

The Elms, Aldeburgh, May.   D. HERVEY.

Microphotography, Isophotography, Megaphotography.

I HAVE read with much interest your article on microphotography (P. 4) at its best. Possibly some of your photographer readers may be glad to know that microphotography of sorts is within the reach of all who possess a microscope with suitable substage-condenser and a camera. The results may not compete with the best, but they are very useful. I find that any transparent object which can be conveniently seen in the microscope can be reproduced in the camera. If the fine adjustment is good enough for ordinary work, it is good enough for photographic work.

One of my earliest attempts was to photograph fluid inclusions in quartzes with ordinary sunlight, and rock-sections polarised. The only difficulty was that the sun would not keep still, and without a heliostat the work was most troublesome, not to say aggravating. In one case, a mere movement of the condenser-diaphragm made the bubble in the inclusion fly backwards and forwards. A negative was taken in each position, and a lantern slide taken of each negative. With a little device in the double lantern the motion of bubbles in inclusions can be shown on a nine-foot screen. These negatives were taken with a 1/16th immersion, the camera being extended with a brown paper tube, and the extra apparatus did not cost one shilling.

Up to a ½-inch objective, ordinary gas, with isochromatic plates, does very useful work. The only difficulty to surmount is to handle the focusing apparatus, and see the focusing screen at the same time. A hand mirror solves the problem. But a fine adjustment is really scarcely necessary, as it is easy to focus with the camera as in ordinary photography.

It is often desirable to photograph objects their exact size. Before the Kent's Cavern Collection was divided, I photographed the choicest examples for the Torquay Natural History Society. The implements were fixed with beeswax on a piece of plate-glass, which could be placed in any position and backed by any desired background. I sent a couple of prints to the International Amateur Photographic Exhibition at Vienna, and the jury, much to my surprise, awarded them a diploma. The extra apparatus certainly did not cost 10s., and the negatives were taken in the lecture-room of the Natural History Society under some disadvantages.

Of megaphotography I have but a single experience. While observing the transit of Venus, I thought I would try a photograph. I drilled a hole in the telescope cap for diaphragm ; took off the eye-piece and stuffed the telescope into a common camera, with a red cloth to make it light-tight ; exposed six negatives with hand exposure on instantaneous plates. Result : four passable negatives and one good one. This quite unlooked-for success was due to some back volumes of NATURE which propped up the camera. The success was really a downright "fluke" ; for, knowing the exposure must be hundreds of times too much, I added a quantity of bromide of potassium to the developer, and the amount chanced to be correct. All photography is done with objective and camera. In photographing the sun, the·object is some ninety millions of miles off; in photographing a fluid inclusion in quartz, it is the 1/16th of an inch off—a mere question of detail. Most of these scientific photographs are·far easier than the simplest everyday landscape.

A. R. HUNT.

Comets and Corpuscular Matter.

REFERRING to Prof. J. J. Thomson's article on "corpuscles" in your issue of May 10, it occurs to me that the behaviour of corpuscular matter described therein may have some bearing on cometary phenomena.· May not the structure of comets to some extent be explained by assuming that their tails are composed of aggregations of negatively charged particles of extremely minute size, answering to the free corpuscular matter as defined by Prof. Thomson, and which to a large degree may be formed by a sort of "corpuscular dissociation," or detachment, taking place in the comet's nucleus when its temperature is elevated upon nearing the sun? Since Prof. Thomson's experiments indicate the presence of negatively charged matter in kathode rays having a much smaller mass than ordinary atoms, there is reason to believe that matter in this state has properties quite apart from matter in a much coarser state of atomic division. Postulating an electrostatic field as existing in interplanetary space, with the sun as a negative centre or source of electrostatic radiation, and assuming that a comet's tail is composed of these corpuscles, the gravitational force it may suffer, when in proximity to the sun, would perhaps be very small in comparison with the electrostatic force existing throughout the vast congregation of these extremely minute particles, and thereby account for the repulsion of the tails of comets when they approach the sun.

The nuclei of comets may be composed of matter in a much coarser state of subdivision, which, though endowed with positive or opposite electricity, is subject to gravitational influences which determine their course in the neighbourhood of the sun.

While the above is a partial re-statement of existing hypotheses, it may, I venture to suggest, be of interest in connection with Prof. Thomson's remarkable experiments on matter smaller than atoms.　　　　F. H. LORING.

1 Champion Grove, Denmark Hill, S.E., May 18.

A NEW INSTRUMENT TO MEASURE AND RECORD SOUNDS.[1]

A DIRECT, absolute measurement of the intensity of sound at any point in the air must determine in ordinary units, such as kilogram-metres, the energy involved in the condensations and rarefactions of which the propagation of sound consists. But these pulsations follow each other so rapidly, and the amount of energy involved in even the loudest sound is so infinitesimal, that such measurement is attended with considerable difficulty; so much, indeed, that probably not a half-dozen laboratories in the world have any instrument whatever purporting to make direct, absolute measurements of the energy of sound.

We owe to Helmholtz ("Wissenschaftliche Abhandlungen," vol. i. p. 378) a mathematical theory by which we can determine the ratio between the energy of the pulsations of a tone just without, and that within a spherical Helmholtz resonator; to Lord Rayleigh we owe an expression for the energy of sound in terms of the condensation ("Theory of Sound," vol. ii. Sec. 245). Upon these two results this instrument (like Wien's, *Wied. Ann.* 1898, p. 834) is founded.

A pure tone is received into a spherical Helmholtz resonator, a portion of the walls of which is replaced by a small, circular, extremely thin glass plate, situated just opposite the mouth of the resonator. The pulsations within force this plate to vibrate with the tone's

1 This instrument is described somewhat more fully than it is here in the *Monthly Weather Review*, July 20, 1899, published by the U.S. Department of Agriculture. We are indebted to the courtesy of its editor, Prof. Cleveland Abbe, for the accompanying illustrations.

frequency; and if the natural pitch of the plate is made to approximate that of the resonator and tone, the amplitude of the·plate's vibrations are rapidly multiplied. To make this amplitude a definitely measurable quantity, the sensitive plate carries at its centre a tiny mirror, which forms one of a system of mirrors constituting Michelson's refractometer (*Phil. Mag.* 1882, xiii. p. 236). A displacement of the little mirror from its position at rest amounting to a half wave-length of light will cause a corresponding shifting to one side of the interference bands, so that each dark band will take the position before occupied by the next dark band. The width of the bands may be so ·adjusted that a telescope with micrometer eyepiece can easily subdivide each band into a hundred parts. Hence the displacement of the sensitive plate, while a tone is sounding, could be observed with great precision, if the eye could act with sufficient rapidity to mark the oscillation of any one band.

That, of course, is out of the question.· But it is easy to compound this motion of the bands with another motion perpendicular to it (also in the focal plane), and thus to make the displacements visible. To do this, the interference bands are made to stand vertically in the field, and a screen with a narrow, horizontal slit is interposed in the line of sight; consequently the bands during silence appear in the telescope as a ·narrow, horizontal strip, composed of the bands reduced to

FIG. 1.—The refractometer. The resonator has· been unscrewed from the supporting bracket, leaving the sensitive plate and tiny mirror in place.

square spots of dark and light. ·Now a small lens, forming ·the object-glass of· the· telescope, is ·mounted upon the end of one tine of·a tuning fork, electrically driven, and having the pitch of the tune to·be measured. During silence, the vertical vibration of the object-glass·· stretches out the strip of ·spots into a·rectangle ·of long, vertical bands. But when the tone sounds, these bands arrange themselves diagonally across the same rectangle, the slope of the bands increasing with ·the intensity of the tone.

The micrometer eyepiece can be rotated on its optical axis, and it is provided with a tangent screw ·for close adjustment. As it is rotated a vernier moves over a graduated arc, so that the angle of the slope (a) may be measured, as well as the height (o) of the strip, and the width· of five double bands. Putting $B = Q - o$, and $P =$ the displacement of a band, we have $P = B \tan a$. The intensity of the tone is proportional to P^2, which is thus determined in mean wave-lengths of white light.

Thus far it has been tacitly assumed that the source of tone is at just the right distance from the receiving resonator for the vibrations of the sensitive plate to be in phase with those of the fork carrying the object-·glass. But in ordinary work this agreement in phase rarely occurs, so a further modification is important. However, by simply loading the lens fork very slightly, we make the phases of the one oscillation overtake those of the other as slowly as we please. During agreement in phase the appearance of the bands will be that already described, with the slope (let us say) downward to the right. Two or three minutes after, when the two phases are opposite, the slope of the bands will be downward to the left. Between these two appearances confusion will reign, for the rectangle is then occupied by overlapping ellipses of changing eccentricity. But whenever the two oscillations are composed into a straight line there is abundance of time to measure the slope of the interference bands.

We have now attained only a relative measurement of intensity. But if we knew what maximum pressure within the resonator produced the observed amplitude of the sensitive plate, Rayleigh's expression together with Helmholtz's ratio would yield us the absolute intensity of the tone just outside the mouth of the resonator, which we seek. This pressure we do not know ; we can, however, make a pretty close approximation to it. Let us be content, provisionally, with an error of about four parts in a thousand. Accordingly we will remove the sensitive plate from the resonator, in order to substitute for it a thicker plate, of natural pitch four octaves higher. Then we will cork the resonator, and produce a series of pressures within it by means of an air-pump. These pressures, measured statically with a water manometer, together with the corresponding displacements, furnish a table of the degree of approximation sought ; so that by interpolation, when necessary, we may assign the pressure that has caused the amplitude, P, in any particular case, and thence obtain the energy of the tone in absolute units.

Of course, much pains must be taken to exclude all disturbing vibrations from the sensitive plate, whether

For experimentation we require a source of sound that will produce a tone of great constancy and purity, but one whose intensity may be varied at will between wide limits. Moreover, the tone should issue from a small and definitely located area. It will be convenient, also, to have this instrument easily portable, so that it may be moved freely even while sounding. Such a source is

FIG. 3.—The source of tone, with its box removed.

obtained by causing a tuning fork to transmit its vibrations to a thin iron plate, which forms a portion of the walls of another spherical resonator ; for the middle of one tine is rigidly connected with the centre of the plate. This combination is carefully tuned to give the tone required, and it is boxed so that only the mouth of the resonator protrudes. The fork is driven electrically, but

FIG. 2.—The refractometer boxed and ready for use. The resonator is covered with felt.

FIG. 4.—The open camera. The motor is shrouded to prevent its sparking from fogging the film. Adjustment of speed is accomplished by the aid of stroboscopic observation of the disc of black and white sectors, inspected through the square of ruby glass opposite. The electromagnet operates the arm which carries the shutter.

transmitted through the air or through the floor and supports. Moreover, even the waves of the tone to be measured must be allowed to beat only upon the side of the plate which is within the resonator. Accordingly, heavy, padded boxes and piers of soft rubber are employed for the refractometer, for the tuning fork which carries the object-glass ; and also for the instrument which produces the tone, as well as for the camera, both of which remain to be described. With these precautions, however, the result desired is very well attained, as is shown by careful tests. Moreover, the constancy and sensitiveness of this instrument promise to be highly gratifying.

its current is interrupted by the vibrations of a second fork, the two being in relay. The intensity of the tone depends, of course, upon the strength of the current which drives the source-fork, and this we may vary at will. Moreover, the intensity at the mouth of the source-resonator may be defined in terms of the current effective in producing it. These intensities are determined by means of the damping factors of the arrangement. The theory of this source as an independent, absolute measure of intensity is an extension (Sharpe, *Science*, 1899, p. 810) of that given by Lord Rayleigh for the tuning fork (*Phil. Mag.* 1894, vol. xxxviii. p. 365). This instrument makes a very pure and effective source of tone, simple in construction, and useful for a variety of purposes. A feeble current of a few hundredths of an ampere produces a tone that can be distinctly heard in every part of a building, 204 × 114 feet, four stories high, and containing ninety rooms. It may also be used under water.

To photograph and thus record for analysis a sound of any kind whatever, the resonator is removed by simply screwing it off, without disturbing the sensitive plate ; and a camera is substituted for the telescope and eye. The window of the camera now forms the narrow slit, and a lens, placed between the window and the refracto-meter, focuses a narrow, horizontal strip of interference bands upon the photographic film. This film is wound about a cylinder (*cf.* Raps, *Wied. Ann.* 1893, p. 194) kept in rapid rotation by a small electric motor within the camera. The speed of this motor is kept constant by Lebedew's method (*Wied. Ann.* Band 59, p. 118). Con-

NOTES.

As we go to press, a message from Sir Norman Lockyer at Santa Pola informs us that 130 volunteer observers have been obtained from H.M.S. *Theseus*, the instruments have been adjusted, and the Spanish authorities are assisting splendidly. The weather prospects are good.

MR. J. S. BUDGETT left Liverpool on Saturday last on his second expedition to the Gambia, where he is going in order to complete his studies of the fish-fauna of that colony, and especially to investigate the life-history and development of the abnormal fishes *Polypterus* and *Protopterus*. On reaching Bathurst, Mr. Budgett will proceed up the River Gambia to his former quarters on M'Carthy's Island, in the neighbourhood of which he has already ascertained that these fishes are found breeding during the rainy season. A memoir on some points in

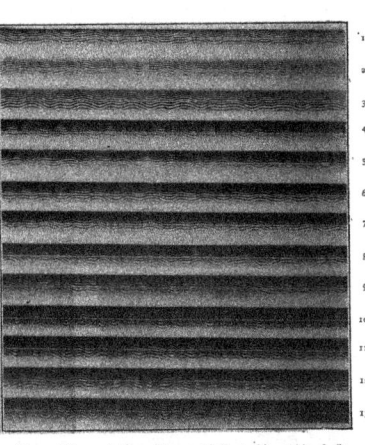

1. Quiet.   2. Fanning I.   3. Fanning II.   4. Noise.   5. Flageolet.   6. Fork C_{128}.   7. Fork c_{256}.   8. Fork c'_{512}.   9. Forks C+c.   10. Forks C+c+c'.   11. Forks g+a.   12. Forks c+e+g+c'.   13. Tone source.

1. (a)h.   2. (o)h.   3. p(oo)l.   4. (a)te.   5. m(ee)t.   6. s(e)t.   7. (a)t.   8. (i)t.   9. (au)ght.   10. (e)re.   11. (u)se.   12. (u)rn.   13. Fork c_{256}.

FIG. 5.—Analyses of Fork Tones and Vowel Sounds.

sequently the lateral vibration of the bands caused by the sound, combined with the steady, vertical motion of the exposed portion of the film, is recorded in parallel, wavy lines. The shutter is opened for the time required for a single rotation of the cylinder by an electrical device. After each exposure the cylinder is moved in the direction of its axis by turning a screw from without. Thus a fresh portion of the film is brought under the shutter, without stopping the motion or opening the camera. In this way were taken the photographs of fork tones and vowels here given (Fig. 5). The photograph of a single tone from the source, whose intensity at the sensi-tive plate has been determined by the first method, affords a standard (viz. its amplitude) for determining the absolute intensity of every other sound photographed ; while comparison with the wave-length appearing in the photograph of the tone of a standard fork gives the pitch of other sounds.   BENJAMIN F. SHARPE.

the anatomy of *Polypterus*, based on specimens obtained by Mr. Budgett during his first expedition, was read before the Zoo-logical Society on May 8, and will shortly be published in the Society's *Transactions*.

AT a recent meeting of the British Ornithologists' Union and Club, under the presidency of Mr. F. D. Godman, F.R.S., the following resolution was unanimously adopted :—" That any member of the union directly or indirectly responsible for the destruction of nests, eggs, young or parent birds of any species mentioned below should be visited with the severest censure of the union and club." The birds referred to are the chough, golden oriole, hoopoe, osprey, kite, white-tailed eagle, honey buzzard, common buzzard, bittern and ruff.

THE committee of the Liverpool School of Tropical Diseases have decided to despatch, at an early date, an expedition to the Amazon to investigate yellow fever. The expedition will probably in the first instance proceed to Baltimore to confer with the yellow fever experts at the Johns Hopkins University, afterwards going to Para and other places on the South American coast.

AN expedition, under the auspices of the Royal Dublin Society and the Royal Irish Academy, conjointly, has left Dublin for Spain, to observe the solar eclipse on May 28. The party consists of Prof. C. J. Joly, Sir Howard Grubb, F.R.S., Dr. A. Rambaut, F.R.S., Mr. W. E. Wilson, F.R.S., Prof. W. Bergin, Mr. S. Geoghegan and Mr. Rudolph Grubb. The observers have selected as their station the hill of Berro-calillo at Placencia, near Madrid, and have already had valuable assistance afforded them by Prof. Iniquez, director of the Observatory at Madrid, and his staff, who will them-selves observe the eclipse at the same station.

PALÆONTOLOGISTS will be glad to know that the King of the Belgians has just made M. L. Dollo, Conservator of the Brussels Museum, a Chevalier of the Order of Leopold.

THE annual meeting of the Italian Botanical Society will be held at Venice on September 9-15, under the presidency of Sig. Sommier.

THE committee of the International Botanical Congress, to be held in Paris from October 1 to 10, has issued a fresh invita-tion to foreign botanists to enrol themselves as members. The subscription fee of members has been fixed at 20 fr., which will include the cost of the publications of the Congress. The following have already been fixed on as subjects for discussion at the Congress :—Monographic studies ; species and hybrids ; unification of micrometric measures ; influence of the nature of the soil, and of the plants growing in it, on the development of fungi ; and other suggestions are invited. The president of the Congress will be M. E. Prillieux ; the general secretary, M. E. Perrot ; and the treasurer, to whom subscriptions should be sent, M. H. Hua, rue de Villersexel 2, Paris.

A DEPARTMENTAL committee has been appointed to inquire into the conditions under which agricultural seeds are at present sold, and to report whether any further measures can with advantage be taken to secure the maintenance of adequate standards of purity and germinating power. The committee consists of the following members, viz. :—The Earl of Onslow, G.C.M.G., chairman ; Sir W. T. Thiselton-Dyer, K.C.M.G., C.I.E. ; Sir Jacob Wilson ; Mr. R. A. Anderson, secretary of the Irish Agricultural Organisation Society ; Mr. R. Stratton ; Mr. Martin J. Sutton ; Mr. James Watt and Mr. David Wilson. Mr. A. E. Brooke-Hunt, of the Board of Agriculture, will act as secretary to the committee.

AN excursion to Malvern and district has been arranged by the Geologists' Association for Whitsuntide. The director will be Prof. T. T. Groom, and during the stay at Malvern, from Saturday, June 2, to Tuesday, June 5, a number of inter-esting geological sections and structures will be examined.

THE tenth International Congress of Hygiene and Demo-graphy will be held in Paris this year, on August 10-17, under the presidency of Dr. Brouardel, Dean of the Faculty of Medicine of Paris. Programmes and forms of application for membership can be obtained from the secretary of the British Committee, Dr. Paul F. Moline, 42, Walton Street, Chelsea, S.W.

A MEETING of the Institution of Mining Engineers will be held in London on June 14-16. The members have been invited to attend the International Congress of Mining and Metallurgy which will be held in Paris on June 18-23, with the object of collecting together engineers and others, who

in various parts of the world are engaged in forwarding the progress of mining and metallurgy. The Congress, like that of 1889, is under the direct patronage of the French Government.

SWEDISH metallurgy has suffered a severe loss by the death, on May 12, of Mr. G. F. Göransson, at the age of eighty-one. Without his help, the Bessemer process might perhaps never have been perfected. In 1858, at Edsken, he increased the area of the tuyeres, and succeeded in shortening the process so as to produce sufficient heat in the converter to allow of the proper separation of the slag from the metal, and thus to con-vert pig-iron into good steel, which having been exported to England encouraged the capitalists who were supporting Sir Henry Bessemer. At the Swedish meeting of the Iron and Steel Institute in 1898, Mr. Göransson, although very infirm, welcomed the members, in an English speech, to the Sandvik works, of which he was chairman and founder.

AT the anniversary meeting of the Royal Geographical Society, on Monday, the medals and other awards already an-nounced (P. 34) were presented. The president, Sir Clements Markham, in the course of his anniversary address, said that a committee has been formed to obtain funds for the erection of a suitable memorial to Dr. Livingstone, on the spot where the tree stood under which the heart of the great explorer was buried. The materials will be conveyed, free of expense, from the mouth of the Zambesi to Lake Bangweolo, by the kindness of the African Lakes Corporation and the British South Africa Company. The prospects of the Antarctic expedition, from a financial point of view, have been somewhat clouded by the war. At least 30,000l. more than has already been raised is required. Apart from the finances, the affairs of the expedition are in a flourishing state, and everything seems hopeful. The keel of the exploring ship is now laid at Dundee. She will be the best polar exploring vessel that has ever left these shores, and the first that has ever been built in this country specially for scientific work in polar regions.

WE regret to record the death, at the age of seventy-seven, of Mr. James Thomson, F.G.S., of Shawlands, Glasgow. Among the many enthusiastic workers at Scottish geology, none had plied his hammer with more zeal. He had been an active member of the Geological Society of Glasgow for up-wards of forty years, and was a frequent attendant at the meet-ings of the British Association. Although he had written on the geology of Islay, and on parts of Arran and the Outer Hebrides, his special researches were on the Scottish Carboni-ferous corals ; and his contributions on this subject, carried on partly in conjunction with the late Prof. H. A. Nicholson, were numerous. He had formed an exceedingly fine collection of fossil corals, which he presented to his native town, Kilmarnock.

A POSSIBLE method of prevention of horse-sickness, which is endemic in the Orange Free State, Transvaal, Natal, Rhodesia. and Bechuanaland, and also occasionally occurs in Cape Colony, is described in the *Cape Times* (April 24) by Dr. G. C. Purvis. Fortified serum, derived from immune horses, almost invariably produces fatal hæmoglobinuria when injected into horses suffer-ing from horse-sickness. Dr. Purvis finds, however, that if the animal is gradually accustomed to the toxin, until it can receive an injection of 100 c.c. or 200 c.c. of serum, virulent blood can be injected without danger. It appears that fortified serum is a useful agent if used in a proper way, and that it is capable of preventing the onset of horse-sickness. Moreover, if, in spite of precautions, an animal acquires the disease, judicious treatment with the serum will assist in bringing about a cure.

A BACTERIOLOGICAL method of exterminating rats, proposed by M. J. Danysz of the Pasteur Institute of Paris, is described in the *British Medical Journal.* M. Danysz has found a microbe which, if introduced into a population of rats, may be trusted to breed a pestilence among them that will wipe them out, or at least make them a negligible quantity. From field-mice suffering from a spontaneous epidemic disease he isolated a cocco-bacillus presenting the general characters of *B. coli,* and thus resembling Loeffler's *B. typhi murium.* By an elaborate process of repeated cultures of this micro-organism passed through series of mice and afterwards through rats, he succeeded in intensifying its virulence so as to make it, when eaten, certainly pathogenic for the latter rodents. Having satisfied himself of the fatal effect of the cultures in the laboratory, he had them tried in a large number of farms, warehouses, and other places infested by rats. From the reports of these experiments, amounting to several hundreds, it appears that in 50 per cent. of cases the method resulted in a complete disappearance of the rats, while in 30 per cent. their number notably diminished ; in 20 per cent. the method failed.

SOME interesting figures showing the high estimation in which technical knowledge is held in certain branches of industry by German manufacturers, have recently been published in the *Zeitschrift für angewandte Chemie,* from a lecture on "Technical Education and the Importance of Scientific Training," delivered before the German Emperor by Prof. J. Bredt. The following statistics, corrected to the end of last year, refer to three of the most important factories in Germany where aniline dyes are made, viz. the Badische Anilin-und Sodafabrik, of Ludwigshafen ; the Farbewerke vorm. Meister Lucius und Brüning, of Höchst am Main ; and the Farben-fabriken vorm. Fr. Bayer and Co., of Elberfeld.

				Ludwigshafen.	Höchst.	Elberfeld.
Workmen	...	...	...	6207 ...	3670 ...	3900
Staff	...	...	...	— ..	128 ...	886
Chemists	...	...	...	146 ...	130 ...	130
Engineers	...	...	...	75 ...	37 ...	29

Of course, the *Engineer* remarks, conditions are somewhat different in Germany from those which obtain in this country, because these dye works own the patents for various highly lucrative proprietary articles, and manufacture numerous pharmaceutical preparations ; but we should be interested to learn how many "chemical" factories in Great Britain employ over 100 skilled chemists.

AN enterprise, similar to the Edison works at Paderno, where energy of some 13,000 horse-power is derived from the River Adda, and employed for producing electricity, which is carried by overhead cable to Manzo and Milan, but on a larger scale, is, states the *Board of Trade Journal,* now on the eve of completion in Northern Italy. A report of H.M. Consul at Milan (*Foreign Office, Annual Series,* 2413) states that the Società Lombardia per distribuzione di energia Elettrica, obtained a concession from the Government on the River Ticino, at Vizzola, some miles below its issue from Lake Maggiore, and immediately set about constructing works for the development of hydraulic power of no less than 20,000 horse-power (theoretical), which will give 10,000 effective horse-power of electric energy for industrial purposes, after making full allowance for loss in transmission. Since the works were begun, the sanction of the Government has been obtained to a project for the construction of a movable dam across the river some distance higher up, which would enable the company to increase its volume of water, and allow of the same being constantly maintained during all seasons of the year. The theoretical hydraulic power would then be 24,000 and the effective electric energy 12,000 horse-power. This dam has

not yet been commenced, but the works have been constructed on the basis of the larger supply of water. Seven turbines and seven dynamos, giving three-phase alternating currents, have been put up. The dynamos and all the other electrical plant have been supplied by Germany. It was originally intended to bring all this electric energy into Milan, a distance of twenty-five miles, but the whole of it has now been disposed of in and about the manufacturing towns of Gallarate, Busto, Arsizio, Legnano and Sarsuno, which lie between Vizzola and Milan, a district which already, for the cotton industry alone, uses steam to the extent of 10,000 horse-power. This enterprise is said to be the most important of its kind in Europe. The plans are due to the initiative of Italian engineers and were made as far back as 1887, but their execution must be attributed in a large measure to the assistance of a German firm which has subscribed a considerable part of the capital of the company.

IN a recent number of NATURE (March 1, p. 421) reference was made to a paper by Dr. Lüdeling, in which diurnal variations of terrestrial magnetism were shown graphically " with the aid of von Bezold's vector diagrams." Though von Bezold appears to have been the first to use the convenient term " vector diagram " to designate the curves referred to, Dr. Chree pointed out in NATURE of March 22 (p. 490) that the curves were employed by Airy in 1863, and since then by several people in this country, including Lloyd and himself, and were not used by von Bezold until 1897. Dr. Lüdeling now sends us a letter in which he states that both von Bezold and himself were well aware of the previous use of the curves, and that acknowledgment of earlier work was made in the paper briefly mentioned in NATURE.

PROF. J. JOLY has discussed " The Theory of the Order of Formation of Silicates in Igneous Rocks " (*Proceedings,* Roy. Dublin Soc. ix. [N.S.] 1900). He has lately found that the softening point of quartz is far below what is currently thought. Observations indicate that silica is a body possessing an extraordinary range of viscosity. It is a thick liquid at about 1500° C. At a temperature of about 800° C. it is plastic, and yields with considerable rapidity to distorting forces. Perhaps it never crystallises very vigorously. The author's experiments show that a silicate containing a small quantity of silica crystallises out at a higher temperature than a silicate with a larger percentage of silica ; and this, according to his theory, is because the crystallising power of the one is less affected by the silica than that of the other.

IN a short article on " The Formation of Minerals in Granite" (*Memoirs,* Manchester Lit. and Phil. Soc. xliv. 1900), Mr. C. E. Stromeyer brings forward some facts and suggestions which lead him to conclude that there is no necessity to limit the temperature of granite formation, as propounded by Dr. Sorby, nor to assume that the earth's interior is solid. Not only temperature and rate of cooling, but also pressure have combined to influence the mineral composition of granites. Where the solid rock resting on the molten material is of a low specific gravity and a bad conductor of heat, the depth at which granite rock would commence to solidify would not be great, and most probably the quartz would crystallise first, forming, say, quartz-porphyry. Where the rock resting on the molten mass is heavy, containing perhaps much iron-oxide, and acting as a good conductor of heat, the depth at which the granite would commence to solidify would be much greater than in the previously-mentioned case, the pressure would be much greater, and most probably the quartz would remain fluid long after the felspars had crystallised, forming, say, felspar-porphyry. In the author's opinion, every intermediate condition is conceivable.

THE second volume of the *Annales* of the National Observatory of Athens contains a catalogue of the earthquakes felt in Greece during the years 1893-1898. Its value will be evident from the facts that it occupies more than 150 quarto pages and contains entries of 3187 shocks. Taking area into account, it therefore appears that earthquakes are about twice as frequent in Greece as they are in Japan. M. Eginitis, the director of the observatory, adds an interesting discussion of the catalogue. For the six years of the records, earthquakes were most numerous during the months of April and May; there is the usual apparent maximum during the early hours of the morning; and the usual doubt as to the existence of any connection between the frequency of earthquakes and the positions of the earth and moon in their orbits. There seems to be no part of the country entirely free from earthquakes, but their distribution is most irregular, 2018 shocks having been recorded in Zante alone. The volume also contains the meteorological tables for 1896, and essays by M. Eginitis on ancient observations of meteor showers, the increase of the discs of the sun and moon at the horizon, and the solar eclipse of August 8, 1896.

TWO observers in the May number of the *Zoologist* note the effect of the unusually cold and late spring on the bird-life of the country. Mr. W. W. Fowler states that after a careful search, on April 10, in the neighbourhood of Chipping Norton, he was unable to discover a single specimen of the summer migrants which ought by that time to be numerous. Mr. W. Wilson, on the other hand, comments on the late pairing of lapwings and partridges in Scotland.

IN the April number of the *Victorian Naturalist*, Mr. D. Le Souef gives an interesting account of the plants and animals met with during a visit to Western Australia. In several passages he comments on the diminution in the number of wild mammals. The rabbit-bandicoot, for example, has disappeared from districts where it was formerly numerous, owing to "ringing" the timber and cultivation; while the common phalanger, or "opossum," has been practically exterminated from the settled districts.

To the *Revue générale des Sciences* of May 15, Monsieur P. Glangeaud contributes a notice of the biological laboratory recently established among the extinct volcanoes of the Auvergne. The principal object seems to be the investigation of the fauna and flora of the numerous lakes, several of which are of great depth and cover a large area. Already important observations have been made with regard to the "plankton" of the lakes. On the salt plains the existence of a marine fauna has long been known, and the discovery is now announced of the survival there of a marine fauna.

WE have received the fourth number of the *News Bulletin* of the Zoological Society of New York, which contains a popular illustrated account of some of the new buildings in the menagerie, as well as of several of the most notable animals. Some of the photographs, especially those of polar bears, of a group of wapiti (elk), and of a bull bison, are exquisite productions. We are, however, sorry to note that there is a deficiency of funds for the support of the zoological park; and an earnest appeal is made by the Board of Managers to induce more of the residents of New York to become members of the Society.

MR. J. K. BARTON has sent us a copy of a paper on the anatomy of the digestive tract of the salmon, published in the April number of the *Journal of Anatomy and Physiology*. The object of the investigation was to determine the truth of the statement that when salmon enter our estuaries they are suffering from a degenerative catarrh of the mucous membrane of the intestines, which subsequently spreads upwards to the stomach. The examination of a considerable number of specimens is

stated to refute this assertion, and that previous observers have been misled by the effects of the methods employed in their investigations.

PART III. of "A Manual of Surgical Treatment," by Dr. W. Watson Cheyne, F.R.S., and Dr. F. F. Burghard, has been published by Messrs Longmans, Green and Co. The subject is the treatment of the surgical affections of the bones, and amputations. We propose to review the work when the six parts of which it will be composed have been published.

THE fifth, revised and enlarged edition of Dr. Richard Hertwig's "Lehrbuch der Zoologie" has just been published by the firm of Gustav Fischer, Jena. As with other zoological text-books, many alterations and additions have had to be made in order to bring it in touch with the present state of knowledge.

THE material collected by Dr. Arthur Willey from New Britain, New Guinea, Loyalty Islands and elsewhere, when in search of the eggs of the Pearly Nautilus, during the years 1895-97, has proved exceptionally rich in subjects of study. Part iv. of the "Zoological Results". (Cambridge: University Press) contains ten papers upon various forms of life, and Part v. is in the press. The original intention was to complete the work in five or six parts.

A NEW sugar has been discovered by M. Gabriel Bertrand, by the action of the sorbose bacterium upon erythrite, and is described by him under the name of erythrulose in the *Comptes rendus* for May 14. By its reactions it appears to be a ketone of the composition $CH_2(OH).CO.CH(OH).CH_2OH$, thus being a lower homologue of levulose. Erythrulose is not fermentable by yeast, but forms a well crystallised osazone; it resists oxidation by bromine water, and hence is probably a ketone.

A NEW general method of preparing secondary and tertiary alcohols, which, on account of the excellent yields obtainable, promises to be of considerable service, is described by M. V. Grignard in the current number of the *Comptes rendus*. Magnesium turnings react but slowly with methyl iodide at ordinary temperatures, but in presence of ether a violent reaction takes place, resulting in a clear solution probably containing $CH_3.MgI$. If to this solution an aldehyde or ketone is added, and the product treated with dilute acid, about 70 per cent. of the theoretical amount of the corresponding secondary or tertiary alcohol can be isolated. Thus methyl iodide and acetaldehyde give isopropyl alcohol; benzaldehyde and isobutyl bromide give phenylisobutyl-carbinol; methyl iodide and acetophenone, dimethyl-phenyl-carbinol.

THE additions to the Zoological Society's Gardens during the past week include a Squirrel Monkey (*Chrysothrix sciurea*) from Guiana, presented by Mr. Percy L. Isaac; an Ocelot (*Felis pardalis*) from South America, presented by Mr. M. A. French; an Allen's Porphyrio (*Hydrornia alleni*) captured at sea, presented by Captain J. C. Robinson; a Snowy Owl (*Nyctea scandiaca*, ♀) from Bylott Island, Lancaster Sound, presented by Mr. A. Barclay Walker; two Long-eared Owls (*Asio otus*), European, presented by Mr. D. F. Campbell; six Long-nosed Crocodiles (*Crocodilus cataphractus*) from West Africa, presented by Mr. J. S. Budgett; four Blood-rumped Parrakeets (*Psephotus haematonotus*), two Rose Hill Parrakeets (*Platycercus eximius*), two Crested Pigeons (*Ocyphaps lophotes*), two Plumed Ground Doves (*Geophaps plumifera*), two Black and White Geese (*Anseranas semipalmata*) from Australia, two African Tantaluses (*Pseudotantalus ibis*), two Senegal Touracous (*Turacus persa*) from West Africa, purchased; two King Snakes (*Coronella getula*), a Coralline Snake (*Coronella gentilis*), two American Black Snakes (*Zamenis constrictor*), ten Pennsylvanian Mud Terrapins (*Cinosternum pennsylvanicum*), four Adorned Terrapins (*Chrysemys orna'a*), thirteen Elegant Terrapins (*Chrysemys scripta elegans*), six Lesueur's Terrapins (*Malacoclemmys lesueuri*), six Red Newts (*Sperlepes ruber*) from North America, a Garnett's Galago (*Galago garnetti*) from East Africa, a Serval (*Felis serval*) from Africa, a Common Teguexin (*Tupinambis teguexin*), three Annulated Terrapins (*Nicoria annulata*) from South America, four Blue Wall Lizards (*Lacerta muralis*, var. *cœrulea*) from Faraglione, five Schlagintweit's Frogs (*Rana cyanophlyctis*) from Southern Asia, deposited ; a Barbary Wild Sheep (*Ovis tragelaphus*, ♂), born in the Gardens.

OUR ASTRONOMICAL COLUMN.

THE DARK FRINGES OBSERVED DURING TOTAL SOLAR ECLIPSES.—We have received a communication from Señor V. Ventosa, astronomer at the Madrid Observatory, concerning the appearance and probable cause of the dark fringes—or "shadow bands" as they are generally called—which are always observed some few seconds before and after totality during the progress of a total eclipse of the sun. The chief points of his communication are here summarised.

These alternating dark and bright fringes are parallel to each other, all moving in the same direction, but the velocity varies greatly from time to time. Several reasons have been advanced to account for their appearance, chief of which are those regarding them as (1) diffraction fringes bordering the actual shadow of the moon on the earth's surface ; (2) shadow phenomena produced in the body of our own atmosphere, and affected by the direction of the wind. The examination of the observed facts appears to support to some extent those holding the latter view, as while the bands may be well seen in one place, they may be invisible in a neighbouring locality ; their form, generally rectilinear or slightly undulating, is also variable, while their breadth has been variously estimated from 1 cm. to 50 cm., although this will, of course, partly depend on the inclination of the surface on which they are observed. Sometimes they move with about the speed of a man walking, at others with the speed of an express train, the velocity always being less, however, than that of the shadow itself. (During the coming eclipse the shadow will move through 800 kilom. in 12 minutes.)

Señor Ventosa has been occupied for some time in studying the currents in the higher regions of our atmosphere by observing the undulations round the sun and stars with a telescope, and thinks that these upper atmospheric currents may possibly have some bearing on the question of the eclipse shadow bands ; the movement of these higher portions showing through the quieter lower strata and being rendered visible on account of different refractive powers. He thinks it would be useful to determine the velocity of these currents by anemometers at various altitudes, and also to observe the undulations round the limb of the sun at the time of eclipse, comparing them with the shadow bands in direction and velocity of movement. To ascertain if any experimental illustration of this hypothesis could be presented, he states that bands may be produced by passing diffuse light reflected from a sheet of corrugated glass through a circular aperture representing the sun, over which an opaque disc, representing the moon, is made to slide. When the segment left uncovered is about 5 mm. in width, alternate bright and dark bands can be observed on a white screen held near, if the length of the segmental opening is approximately parallel to the undulations of the glass, but if at right angles they entirely disappear. He trusts, however, that his putting forward this hypothesis for establishing a connection between eclipse shadow bands and atmospheric undulations will show the advisability of recording the direction and velocity of the wind during eclipses, so that more definite data may be available for discussion.

PHOTOMETRY OF CORONA, APRIL 16, 1893.—In a communication recently made to the Royal Society, Prof. H. H. Turner, F.R.S., gives the details of procedure and results obtained in photometric observations of the corona during the total eclipse of the sun in April 1893. The visual brightness of

the corona was determined by Prof. T. E. Thorpe at the eclipses of 1886 August 29, and 1893 April 16, by a method arranged by Sir W. Abney (*Phil. Trans.* A, 1889, p. 363, and 1896, p. 433), but soon after the first of these, Sir W. Abney devised a method of comparing the coronal light with that of a fixed standard by photographic means. This method was first put into practice at the eclipse of 1889, and has been repeated systematically since. Part of the photographic plate, before being taken for eclipse use, is exposed to a graduated series of exposures from a standard source of light in the laboratory, and then without development is afterwards used to receive the impression of the corona, the part carrying the previous standard exposures being protected from further light action. On subsequent development there results a picture of the corona, and a series of squares of graduated densities on the same plate, so that the brightness of any part of the coronal structure may be directly compared with the brightness of the standard light of the laboratory.

The 1889 photographs have not yet been measured, but Prof. Turner has reduced several of the plates taken in 1893 by Sergeant Kearney at Fundium, Africa. These were obtained with the "double tube" apparatus, giving pictures of two sizes, the moon's disc being 0·6 inch and 1·5 inches in diameter. Examples of each scale image were examined, one of the large scale photographs, taken with an exposure of 50 seconds, being specially carefully measured along four radii extending due N., S., E., W., from the limb respectively, and the resulting table of comparison measures is included in the present paper. This table shows :—

(1) That the accuracy of the method is such that the intensity of the light is determinable within a very small error.

(2) The intensity of the coronal light falls off in nearly the same manner in all four directions (1893 was near a sun-spot *maximum*, with corona of symmetrical form). There is a marked difference, however, between the intensities along the north and south radii.

(3) The falling off in intensity is at first exceedingly rapid, becoming very gradual at distances more than 45 minutes from the limb.

(4) The absolute brightness of the corona in terms of the "moon" by using a conversion factor.

Prof. Turner then compares the brightness thus determined photographically with that obtained visually by Abney and Thorpe, and presents two curves showing the combined observations, which show a marked agreement between the results arrived at in such different ways. No measures of brightness, however, were made *visually* within 0·6 of a radius from the limb, and it would be useful if this were done at the coming eclipse.

MAXIMUM DURATION OF TOTALITY FOR SOLAR ECLIPSE.— Mr. C. T. Whitmell sends us the following corrections to the data given in the abstract of his paper last week (p. 64) :—

Earth's radius to be taken as 3963·296 miles.
Moon's ,,   ,, ,, ,,   ,,   ,, 1080·000 ,,

The eclipse for which the totality will be a maximum will take place at noon about the beginning, not the middle, of July.

SOME MODERN EXPLOSIVES.[1]

NEARLY thirty years ago, in the Royal Institution, I had the honour of describing the great advances which had then recently been made both in our knowledge of the phenomena which attend the decomposition of gunpowder, and in its practical application to the purposes of artillery.

I described the uncertainty which up to that date had existed as to the tension developed by its explosion, the estimates varying enormously from the 101,000 atmospheres (about 662 tons on the square inch) of Count Rumford to the 1000 atmospheres (6·6 tons per square inch) of Robins, or, taking more modern estimates, from the 24,000 atmospheres (158 tons per square inch) of Piobert and Cavalli to the 4300 atmospheres (about 29 tons per square inch) of Bunsen and Schischkoff.

These uncertainties which I may say, set to rest by certain experiments carried out both in guns and close vessels at Elswick, by the labours of the Explosive Committee appointed

---

[1] A Discourse delivered at the Royal Institution on Friday, March 23, by Sir Andrew Noble, K.C.B., F.R.S.

by the War Office, and by researches conducted by Sir F. Abel and myself. These researches were conducted on a large scale with the view of reproducing as nearly as possible in experiment the conditions that exist in the bore of a gun. You may judge of the magnitude of the experiments when I tell you that I have fired and completely retained in one of my cylinders a charge of no less than 28 lbs. of ordinary powder.

The result of the discussion of the whole series of experiments led to the following conclusions:—

(1) That the tension of the products of combustion at the moment of explosion when the powder practically filled the space in which it is fired—that is, when the density is about unity—is a little over 40 tons on the square inch, or about 6400 atmospheres.

(2) Although changes in the chemical composition of powder, and even changes in the mode of ignition, cause a very considerable change in the metamorphosis experienced in explosion, as evidenced by the proportions of the products, the quantity of heat generated, and the quantity of permanent gases produced, being materially altered, it is somewhat remarkable that the tension of the products in relation to the gravimetric density is not nearly so much affected as might be expected from the considerable alteration in the above factors.

(3) The work that gunpowder is capable of performing in expanding in the bore of a gun was determined both by actual measurement and by calculation, and the results were found to accord very closely.

(4) The total potential energy of exploded gunpowder supposed to be fired at the density of unity was found to be about 332,000 gramme units per gramme, or 486 foot tons per lb. of powder.

I must confess that when I gave the lecture I have referred to, seeing the many centuries during which gunpowder has held its own as practically the sole propelling agent for artillery purposes, seeing also that gunpowder differs in certain important points from the explosives to which I shall presently call your attention, I had serious doubts as to whether it would be possible so far to modify these latter as to permit of their being used in large charges and under the varied conditions required in the naval and military services.

Gunpowder is not like gun-cotton, cordite, nitro-glycerine, lyddite, and other similar explosives, a definite chemical combination in a state of unstable equilibrium, but is merely an intimate mixture of nitre, sulphur and charcoal, in proportions which can be varied to a very considerable extent without striking differences, in results. These constituents do not, during the manufacture of the powder, suffer any chemical change, and being a mixture it cannot be said under any condition truly to detonate. It deflagrates or burns with great rapidity, varying very largely with the pressure and other circumstances under which the explosion is taking place, a train like that to which I set fire taking as you see an appreciable time to burn ; while, in the bore of the gun, a similar length of charge would be consumed in less than the hundredth part of a second.

You will further have observed the heavy cloud of smoke which has attended the deflagration you have seen. Nearly sixtenths of the weight of the powder, after explosion, remains as a finely divided solid, giving rise to the so-called smoke familiar to many of you, and of which a good illustration is shown in this instantaneous photograph. By way of comparison I burn similar lengths of gun-cotton in the form (1) of cotton, (2) of strand, (3) of rope, and you will observe the different rates at which these varied forms of the same material are consumed, the rate depending in this case upon the greater aggregation and higher density, consequently higher pressure, of the successive samples.

Although the names of cordite and ballistite are probably familiar to all of you, the appearance may not be so familiar, and I have here on the table samples of the somewhat Protean forms which these explosives, or explosives of the same nature, are made to assume.

Here, for instance, are forms of cordite, the explosive of the service, for which we are indebted to the labours of Sir F. Abel and Prof. Dewar. This, which is in the form of fine threads, is used in small arms, and here are successive sizes, adapted to successive larger calibres, until we reach this size which is that employed for the charge of the 12-inch, 50-ton guns.

A couple of the smaller cords I burn, both for purposes of comparison and to draw your attention to the entire absence of smoke.

The smoke of the gunpowder you see still floating near the ceiling, but little or no trace of smoke can be seen from such explosives as gun-cotton, cordite or ballistite, their products of combustion being entirely gaseous.

You will have observed that in the combustion which you have just seen there is no smoke, but I must explain, and I shall shortly show you, that this combustion is not quite the same as that which takes place, for instance, in the chamber of a gun. Here the carbonic oxide and hydrogen, which are products of explosion, burn in the air, giving rise, with the aid of a little free carbon, to the bright flame you see, and somewhat increasing the rate of combustion. In a gun, however, owing chiefly to pressure, the cordite is consumed in a very small portion of a second.

In order to illustrate the effect of pressure upon the rate of combustion, I venture to show you a very beautiful experiment devised by Sir F. Abel. It has been shown in this room before, but it will bear repetition.

In this globe there is a length of cordite. I pass a current through the platinum wire on which it is resting and you see the cordite burns. I now exhaust the air and repeat the experiment. The wire is red-hot, but the cordite will not burn. That the failure to burn is not due to the absence of oxygen is shown by plunging lighted cordite into a jar of carbonic acid, where, although a match is instantly put out, the cordite continues to burn—but observe the difference. There is no longer any bright flame, although the cordite is being consumed at about the same rate as when burned in air ; and when a sufficient quantity of the CO_2 is displaced, I can make the inflammable gases ignite and burn at the mouth of the jar.

Another illustration is also instructive. I have here a stick of cordite wrapped round with filter paper ; I dip it in water and light the end ; you may note that at first you see the bright flame. But as the combustion retreats under the wet filter paper, there appears a space between the flame and the cordite, the flame finally disappears, hot gases with sparks of carbon alone showing.

One other pretty experiment I show. I have here a stick of cordite which I light—when fairly lighted I plunge it in this beaker of water. The experiment does not always succeed at the first attempt, but you now see the cordite burning under the water much as it did in the jar of carbonic acid. The red fumes you observe are due to the formation of nitric peroxide caused by the decomposition of the water by the heat.

I have on the table samples of certain other smokeless explosives of the same class. Here is a ballistite used in Italy. Here is some Norwegian ballistite. Here again is ballistite in the tubular form, and in these bottles it is seen in the form of cubes. Here is some gelatinised gun-cotton in the tubular form, and here are some interesting specimens with which I have experimented, and which up to a certain pressure gave good results, but which exhibited some tendency to violence when that pressure was exceeded. Here also are some samples of the French B.N. powder, consisting of nitro-cellulose partially gelatinised and mixed with tannin, and with barium and potassium nitrates. Lastly, I show you here a sample of picric acid, a substance which has been used for many years as a colouring material, but which will be of interest to you because it is used as the explosive of lyddite shell, concerning which I shall presently have more to say ; it differs from all the other explosives in being, in the crystalline form, exceedingly difficult to light. I fuse, however, in this porcelain crucible, a small quantity. I pour a little on a slab, and on dropping a fragment into a red-hot test-tube you see with how much violence the fragment explodes. I also burn a small quantity, and you will observe that, unlike gun-cotton, cordite and ballistite, it is not free from smoke, the smoke in this case being simply carbonaceous matter. You will observe also how much more slowly it burns.

The composition of these various explosives (although in the case of both cordite and ballistite I have experimented with samples differing widely in the proportion of their ingredients) may be thus stated.

The gun-cotton I employed was of Waltham Abbey manufacture, and, when dried, consisted of 4·4 per cent. of soluble cotton and 95·6 per cent. of insoluble ; as used, it contained 2·25 per cent. of moisture.

The service cordite consists of 37 per cent. trinitro-cellulose, with a small proportion of soluble gun-cotton, 58 per cent. of nitro-glycerine and 5 per cent. of the hydrocarbon vaseline.

The ballistite I principally used was composed of 50 per cent. dinitro-cellulose (collodion cotton) and 50 per cent. of nitroglycerine. The whole of the cellulose was soluble in ether alcohol, and the ballistite was coated with graphite.

The French B.N. powder consisted of nitro-cellulose partly gelatinised, and mixed with tannin, and with barium and potassium nitrates. The transformation experienced by some of these explosives is given in Table I., while the pressures in relation to the gravimetric densities of some of the more important are shown in Fig. 1.

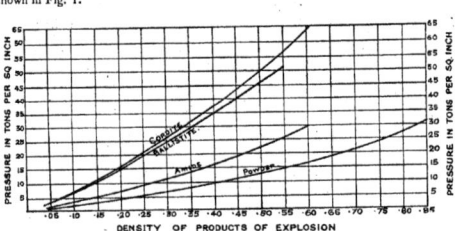

Fig. 1.—Pressures observed in closed vessels with various explosives.

TABLE I.

Constituents				Cordite	Ballistite	B.N.	Lyddite
				vols.	vols.	vols.	vols.
CO_2	...	...	...	20·5	29·1	21·1	12·8
CO	...	...	...	23·3	21·4	24·2	49·7
H	...	...	...	16·5	15·0	16·4	13·8
N	...	...	...	14·6	10·1	12·6	19·6
H_2O	...	...	...	23·6	24·4	25·0	3·8
CH_4	...	...	...	1·5	trace	0·6	0·3
Quantity of gas in c.c. per gramme	...			890·5	807	822	960·4
Units of heat				1272	1365	1003	856·3

The decomposition experienced by these high explosives on being fired is of much greater simplicity than that experienced by the old powders, and is, moreover, not subject to the considerable fluctuations in the ultimate products exhibited by them.

The products of explosion of gun-cotton, cordite, ballistite, &c., are at the temperature of explosion entirely gaseous, consisting of carbonic anhydride, carbonic oxide, hydrogen, nitrogen and aqueous vapour, with generally a small quantity of marsh gas.

The water collected, after the explosion vessel was opened, always smelt, occasionally very strongly, of ammonia, and an appreciable amount was determined in the water.

In examining the gaseous products of the explosion of various samples of gunpowder, it was noted that as the pressure under which the explosion took place increased, the quantity of carbonic anhydride also increased, while that of carbonic oxide decreased. The same peculiarity is exhibited by all the explosives with which I have experimented. I show in Table II. the result of a very complete series of a sample of gun-cotton fired under varying pressures, and it will be noted that the volumes of carbonic oxide and carbonic anhydride are, between the highest and lowest pressures, nearly exactly reversed.

TABLE II.

Constituents	Under pressure of explosion, tons per square inch						
	2 tons	8 tons	12 tons	18 tons	20 tons	45 tons	50 tons
	vols.						
CO_2	21·44	25·06	26·27	27·21	26·75	28·13	29·27
CO	29·66	26·31	25·08	25·24	24·53	23·19	22·31
H	15·92	15·33	16·03	14·56	14·77	14·14	13·56
N	13·63	13·80	13·22	13·13	13·43	12·99	13·07
H_2O	19·09	19·09	19·09	19·09	19·09	19·09	19·09
CH_4	·26	·41	·31	·77	1·47	2·46	2·70

There are slight changes as regards the other products, but they do not compare in importance with that to which I have referred.

But before drawing your attention to other points of interest, it is desirable to give you an idea of the advances in ballistics which have been made both by improvements in the manufacture of the old powders and by the introduction of the new.

On Fig. 2 is placed the results as regards velocity of nine explosives, commencing with the R.L.G. powder, which was in use in the latter part of the fifties, and terminating with the cordite of the present day.

The experiments I am now referring to were made in a gun of 100 calibres in length, and were so arranged that in a single round the velocities could be measured at 16-points of the bore. The chronoscope with which these velocities were taken has been already described, and I will now only say that it is capable of registering time to the millionth of a second with a probable error of between two and three millionths. One curious fact connected with the mode of registration I may mention. In the early experiments with the old powders, where the velocities did not exceed 1500 or 1600 feet-seconds, the arrangement for causing the projectile to record the time of its passing any particular point was effected by the shot knocking down a small steel knife or trigger which projected slightly into the bore ; but when the much higher velocities, with which I subsequently experimented, were employed, this plan was found to be unsatisfactory, the steel trigger, instead of being immediately knocked down by the shot, frequently preferred instead to cut a groove in the shot, sometimes nearly its whole length, before it acted. Hence another arrangement for cutting the primary wires had to be adopted.

The diagram I am now showing you is, however, both interesting and instructive. The intention, among other points, was to ascertain for various calibres in length in a 6-inch gun the velocities and energies that could be obtained, the maximum pressures, whether mean or wave, not exceeding about 20 tons on the square inch. The horizontal line or axis of abscissae represents the travel of the shot in feet, the ordinates or perpendiculars from this line to the curve represents the velocity at that point.

The lowest curve on the diagram gives, under the conditions I have mentioned, the velocities attainable with the powder which was used when rifled guns were first introduced into the service, and you will note that with this powder the velocity attained with 100 calibres was only 1705 foot-seconds, while with 40 calibres it was 1533 foot-seconds. Next on the diagram comes pebble powder with a velocity of 2190 foot-seconds ; next comes brown prismatic with a velocity of 2529 foot-seconds.

The next powder is one of considerable interest, and one which might have arisen to importance had it not been superseded by explosives of a very different nature. It is called Amide powder, and in it ammonium nitrate is substituted for a large portion (about half) of the potassium nitrate, and there is also an absence of sulphur. You will observe the velocity in the 100 calibre gun is very good, 2566 foot-seconds. The pressure also was low and free from wave action. It is naturally not smokeless, but the smoke is much less dense and disperses much more rapidly than does the smoke of ordinary powder. Its great advantage, however, was that it eroded steel very much less than any other powder with which I experimented, while its great disadvantage was due to the deliquescent properties of ammonium nitrate necessitating the keeping of the cartridges in air-tight cases.

Next on the diagram comes B.N. or Blanche Nouvelle powder, an explosive which, while free from wave action, is remarkable, as you will note if you follow the curve, in developing a much higher velocity than the other powders in the first few feet of motion, and less in the later stages of expansion.

Thus, if you compare this curve with the highest curve on the diagram, that of the four-tenths cordite, you will note that the B.N. curve for the first eight feet of motion is the higher, and that at about eight feet the curves cross, the B.N. giving a final velocity of 2786 foot-seconds, or 500 feet below the cordite curve.

Then follows ballistite, which, with much lower initial pressure, gives a velocity of 2806 foot-seconds, or somewhat higher than that of B.N. Then follow three different sizes of cordite, the highest of which gives a muzzle velocity of 3284 foot-seconds, or a velocity nearly double that of the early R.L.G.

Table III.—*6-inch Gun*, 100 *Calibres long. Velocities and Energies realised with High Explosives. Weight of Projectile,* 100 *lbs.*

Nature and Weight of Explosive.	Length of Bore, 40 Calibres.		Length of Bore, 50 Calibres.		Length of Bore, 75 Calibres.		Length of Bore, 100 Calibres.	
	Velocity.	Energy.	Velocity.	Energy.	Velocity.	Energy.	Velocity.	Energy.
	f. s.	ft.tons	f. s.	ft tons	f. s.	ft.ton	f. s	ft.ton
Cordite, '4 in. (27.5 lbs.)	2794	5413	2940	5994	3166	6950	3284	7478
Cordite, 0.35 in. (22 lbs.)	2444	4142	2583	4626	2798	5430	2915	5892
Cordite, 0.3 in. (20 lbs.)	2495	4316	2632	4804	2821	5518	2914	5888
Ballistite, 0.3 in. cubs. (20 lbs.)	2416	4047	2537	4463	2713	5104	2806	5460
French B.N. (25 lbs.)	2422	4068	2530	4438	2700	5055	2736	5182
Amide prism (32 lbs.)	2225	3433	2331	3768	2486	4285	2566	4566
Brown prism (30 lbs.)	2145	3190	2257	3532	2435	4111	2509	4483
Pebble powder (36 lbs.)	1885	2464	1980	2718	2110	3087	2190	3326
R.L.G₂ (23 lbs.)	1533	1630	1592	1757	1668	1929	1705	2016

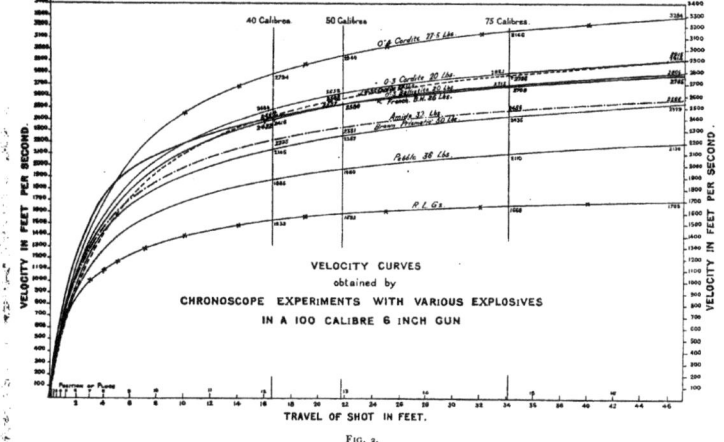

VELOCITY CURVES
obtained by
CHRONOSCOPE EXPERIMENTS WITH VARIOUS EXPLOSIVES
IN A 100 CALIBRE 6 INCH GUN

TRAVEL OF SHOT IN FEET.

Fig. 2.

In the somewhat formidable-looking Table III. I have placed on the wall are exhibited the velocities and energies realised in a 6-inch gun with the various explosives I have named, and the Table, in addition, shows the velocities and energies in guns of the same calibre but of 40, 50 and 75 calibres in length, as well as in that of 100 calibres.

If you compare the results shown in the highest and lowest lines of this table, that is, the results given by the highest and lowest curves on the diagram, you will see that the velocity of the former is nearly twice as great as that of the latter, while its energy and capacity for penetration is nearly four times as great.

I need hardly remind most of you that in artillery matters it is the energy developed, not the velocity alone, that is of vital importance. I venture to insist upon this point, because so many of those who desire to instruct the authorities, write as if velocity were the only point to be considered. In a given gun with a given charge, if the weight of the shot, within reasonable

limits, be made to vary, the ballistic advantage is greatly on the side of the heavier shot, and for three principal reasons.

(1) More energy is obtained from the explosive.
(2) Owing to the lower velocity the resistance of the air is greatly reduced.
(3) The heavier shot has greater capacity for overcoming the reduced resistance.

You will observe that on this velocity diagram, upon which I have kept you so long a time, is shown, not only the travel of the shot in feet, but the position of the plugs which gave the velocities. Further, on the higher and lower curves, the observed velocities are shown where it is possible to do so. Near the origin of motion the points are so close that is is not possible to insert them without confusing the diagram.

At the risk of fatiguing you, I show, in Fig. 3, curves showing the pressure existing in the bore at all points, these pressures being deduced from the curves of velocity.

You will note the point to which I drew your attention, with regard to the powder called B.N. You will remember that in the early stages of motion it gave velocity to the, shot much more rapidly than did the other powders. You see the effect in the pressure curves, the maximum being considerably higher than any of the other pressures, while the pressure towards the muzzle is, on the other hand, considerably below the average.

I fear you may think I have kept you unnecessarily long with these somewhat dry details, but I have had reasons for so doing.

In the first place I desire to demonstrate to you the enormous advances which have been made in artillery by the introduction of the new explosives, and which we in a great measure owe to the distinguished chemists and physicists who have occupied themselves with these important questions.

Secondly, I desire to show you that the explosive which has been adopted by this country, and which we chiefly owe to the labours of Sir F. Abel and Prof. Dewar, is in ballistic effect inferior to none of its competitors. I might go further and say that it is decidedly superior.

add that in the present war it appears to have been handled in a way worthy of the reputation of the corps.

I fear the causes of some of our military failures at the commencement of the war must be looked for in other directions, and the present unfortunate war will turn out to be a blessing in disguise, if it should awaken the Empire to the necessity of correcting serious defects in our organisation, possibly the natural result of our constitution ; and, in that case, the invaluable lives that have been lost will not have been sacrificed in vain. ,

(*To be continued.*)

THE USE OF STEEL IN SHIPBUILDING.[1]

MANY changes and developments in the construction of ships for the mercantile marine have taken place during the last forty years. At the commencement of this period wood was still the principal material employed for shipbuilding, and although iron had been introduced for general shipbuilding

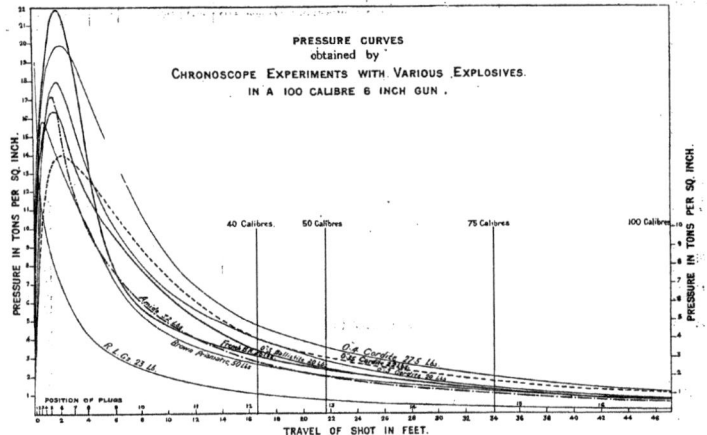

PRESSURE CURVES
obtained by
CHRONOSCOPE EXPERIMENTS WITH VARIOUS EXPLOSIVES.
IN A 100 CALIBRE 6 INCH GUN .

FIG. 3.

Lastly, at a time when the efficiency of all our arms, and especially our artillery, is a question which has been deeply agitating the country, I may do some good by pointing out that the authorities are well aware that any practicable velocity or energy they may desire for their guns is at their disposal.

They have such guns, I mean guns with high velocity and high energy—whether they have enough of them, and whether they are always in the right place, is another matter, for which perhaps the military authorities are not altogether responsible. But velocity and energy is not the only thing that is required under all circumstances in war, and I ask you to believe that if the War Office authorities have, for their field guns, fixed on a velocity very much below what is possible, they have had sound and sufficient reasons for so doing.

My firm and I, individually, have had much to do with the introduction of the larger high-velocity and quick-firing guns into our own and other services ; but as an old artillery officer, in no way responsible for our field guns, I may perhaps be allowed to say that, whether as regards materiel or personnel, our field artillery is inferior to none anywhere ; and I venture to

purposes some twenty years earlier, the record of new tonnage added to the British Register in 1860 shows only about 30 per cent. to have been built of iron.

The general adoption of iron for shipbuilding on the Wear dates from about the year 1863, and by 1880 it had, in that district, entirely taken the place of wood. On the Clyde, Mersey and Tyne, iron shipbuilding was adopted at an even earlier date. So far back as 1855, iron had largely taken the place of wood for shipbuilding on the Clyde.

The difficulty of preventing the fouling of the bottoms of iron ships due to corrosion or marine growths, and the consequent loss of speed, led to various attempts being made to sheath the bottoms of iron ships and cover the wood sheathing with copper, yellow metal, or zinc sheets. The result was the introduction of the system of construction known as "Composite," in which the framing was of iron, with wood planking wrought on the iron frames, and sheathed with copper or yellow metal.

1 Abstract of a paper read before the Institution of Naval Architects by Mr. B. Martell.

The early composite ships were classed as experimental, and subject to biennial survey, in order that the condition of the fastenings might be examined, and the effects of the galvanic action set up by the iron framing and yellow metal sheathing ascertained from time to time.

So far back as 1862, applications were made for vessels to be classed which were to be built with puddled steel, but in the absence of experience regarding the durability of steel, the Committee of Lloyd's Register felt it was not in their power to sanction the proposal.

In 1864, however, a steam yacht of 2400 tons was built for the Viceroy of Egypt under the survey of Lloyd's Register Surveyors, and constructed partly of steel. A reduction of about one-fourth was allowed in the steel scantlings from those required for an iron ship of the same size.

In April, 1876, Mr. James Riley, then manager of the Siemens Steel Works at Landore, read a paper before the Institution of Naval Architects on the production of mild steel, setting forth the results of experiments that had been made with steel manufactured by the Siemens-Martin or open hearth process, and showing the qualities of this material as to ductility and tensile strength.

These results were placed before the Committee of Lloyd's Register, and in 1877 plans from Messrs. J. Elder & Co. were approved for the construction of two paddle steamers to be

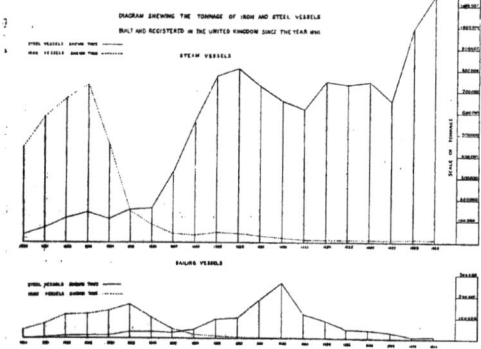

DIAGRAM SHEWING THE TONNAGE OF IRON AND STEEL VESSELS BUILT AND REGISTERED IN THE UNITED KINGDOM SINCE THE YEAR 1880.

built of this material for the English Channel service, with a reduction of about 20 per cent. in the scantlings which had been adopted for iron vessels.

In the same year, in consequence of a report which may be found in the volume of *Transactions* of this Institution for 1877, it was decided to admit steel with scantlings 20 per cent. lighter than prescribed for iron, in vessels building for classification, subject to the material having a tensile strength of from 26 to 30 tons per square inch, and an elongation of 20 per cent. on a length of eight inches. These limits of tensile strength have since been raised to 28–32 tons.

The progress in the use of mild steel for shipbuilding purposes may be judged from the fact that while in 1878 seven steel vessels, of 4470 tons, were classed in Lloyd's Register, and 435 iron vessels, of 517,692 tons, the record for the year 1885 showed 118 steel vessels, of 165,437 tons, as compared with 260 iron vessels, of 290,429 tons. As wood was superseded by iron as a material for shipbuilding, so in its turn iron has given place to steel. Of the total output of the United Kingdom during the past year, 98.8 per cent. of the tonnage was built of steel, and 1.1 per cent. of iron. The iron tonnage was principally made up of trawlers, and comprised no vessel of more than 303 tons.

The accompanying diagram shows the relative changes in the tonnage of steamers and sailing vessels of iron and steel built and registered in the United Kingdom since 1880.

Soon after the introduction of mild steel for shipbuilding purposes, attention was given to the making of heavy steel castings to take the place of iron forgings for stern frames, rudders, propeller brackets, stems, quadrant tillers, &c. These castings are required to be subjected to certain tests, and at the present time are often adopted in place of iron forgings.

It may be here remarked that, notwithstanding the early doubts as to the durability of steel, experience has shown that where proper care is taken to thoroughly clean and paint the surfaces, the deterioration is not appreciably greater than that of iron. In some parts, however, such as thin deck plating, and plating of inner bottom and floors under boilers, steel appears to be more liable to deteriorate, and in consequence of this, iron is often used for these parts in vessels otherwise constructed of steel.

UNIVERSITY AND EDUCATIONAL INTELLIGENCE.

OXFORD.—An appointment will ere long be made to the new Wykeham professorship of physics, which will be endowed in accordance with statute by New College. It is understood that a portion of the space to be vacated in the University Museum by the removal of the Radcliffe Library will be utilised, at least temporarily, as a laboratory for the teaching of electricity.

Merton College proposes to contribute, out of its University Purposes Fund, the sum of 700*l*. towards the cost of fitting up, and 500*l*. towards that of maintaining for two years, the new electrical laboratory, provided that no further liability be hereby undertaken by the College. This proviso is intended to guard against the College University Purposes Fund being regarded as a permanent source of income. Messrs. W. Peterson, principal of McGill University, and John Fletcher, professor of Latin in the University of Toronto, have been appointed as representatives of the University at the centenary of the University of New Brunswick, and Mr. W. R. Morfill, reader in Russian, has been appointed representative at the five-hundredth anniversary of the University of Cracow.

The statute making the degrees of B.C.L. and D.C.L. accessible to persons who have obtained a degree in arts in other Universities, and study law in Oxford although they have not been admitted to the degree of B.A., has been approved by Congregation and Convocation; and also the decree instituting the new research degrees of Doctor of Letters and Doctor of Science.

It is proposed that the necessary qualification for intending candidates for the diploma in Geography shall be that candidates give satisfactory evidence that they have received a good general education, and not, as at first suggested, that they should have passed the examination for the B.A. degree.

On May 22 the honorary D.C.L. degree was conferred upon the following colonial representatives:—The Hon. Alfred Deakin, the Hon. James R. Dickson, C.M.G., and the Hon. Sir Philip O. Fysh, K.C.M.G.

The 212th meeting of the Oxford University Junior Scientific Club was held on Friday, May 11. Papers were read by Mr. S. A. Ionides, Balliol, on "Microphotography," and by Mr. P. Elford, St. John's, on "Chemists of the Nineteenth Century." The following papers will be read during the course of the present term:—"Musical Tetanus," Prof. Sir John Burdon Sanderson, F.R.S.; "The Labile Hydrogen Atom," Mr. A. F. Walden, New College; "A Method for Measuring the Diameter of the Earth," Rev. T. C. Porter.

CAMBRIDGE.—Dr. J. W. L. Glaisher, F.R.S., has been appointed by the council of the Senate a governor of St. Paul's School.

Sydney University, New South Wales, has been placed on the list of recognised schools of medicine.

The Rev. T. Wiltshire has founded a prize to be awarded annually for proficiency in geology and mineralogy. The prize is open to members of the University who have passed Part i. of the Natural Sciences Tripos, and are not of more than ten terms' standing.

Prof. J. Ward and Prof. R. Adamson, of Glasgow, are appointed electors to the Gerstenberg studentship in philosophy, open to students of natural science.

Prof. Leon Guignard has been appointed director of the Paris School of Pharmacy.

Prof. Ludwig Boltzmann, of Vienna, has accepted the invitation to the chair of physics in the University of Leipzig.

The *Chemist and Druggist* announces that Prof. Moissan has been elected a member of the Paris Superior Council of Public Instruction, in succession to the late M. Planchon, deceased. He has also accepted the important post of professor of chemistry at the Paris Sorbonne, in place of M. Troost, who retires on account of advancing age.

SCIENTIFIC SERIALS.

American Journal of Science, May.—Notes on [the geology of the Bermudas, by A. E. Verrill. The present Bermuda Islands are the remnant of a very much larger island, covering an area of about 300 to 400 square miles. A subsidence of at least 80 to 100 feet took place at a comparatively recent period. The Greater Bermuda, as well as the present Bermudas, are composed of shell sand drifted from the sandy flats by the winds in former times into hills, and afterwards consolidated by infiltration and exposure into what is known as Aeolian limestone. The shell sand is constantly increasing in amount, chiefly by the annual growth and death of small shells, as in former periods, so that the total mass of the islands is probably still increasing beneath the sea. The "red soil" of Bermuda is mainly the residue left after the destruction and solution of the limestones. The islands rest on the hidden summit of an ancient volcano. —Some boiling point curves, by C. L. Speyers. The author shows that the equation

$$\frac{n}{N+n} = \frac{p - p'}{p}$$

accounts for the boiling point curves of every mixture for which the partial pressures of the constituents are known at some temperature not very far from the boiling point of the mixture under consideration.—Action of ammonium chloride upon natrolite, scolecite, prehnite and pectolite, by F. W. Clarke and G. Steiger. The authors show how the ammonium chloride reaction can be used for studying the chemical structure of these minerals, and that the orthosilicate formulæ for natrolite and scolecite must be discarded.—Siliceous calcites from the Bad Lands, Washington County, South Dakota, by S. L. Penfield and W. E. Ford. The calcites obtained from the new locality have a peculiar crystallisation, being steep hexagonal pyramids instead of rhombohedra.—Studies in the Cyperaceæ, by T. Holm. This paper deals with the segregates of *Carex filifolia*, Nutt.—Mineralogical notes, by A. F. Rogers. Describes various peculiar forms of gypsum and calcite. Twinned gypsum crystals from Lebo, Kansas, possess hemimorphic orthorhombic symmetry rather than monoclinic.—The Hayder Creek, Idaho, meteoric iron, by W. E. Hidden. This meteorite, weighing 870 grammes, was found at the bottom of a twelve-foot shaft. No companions have been found.—Explorations of the *Albatross* in the Pacific, by Alexander Agassiz. This is the author's fourth and last letter to the U.S. Fisheries Commissioner on the cruise of the *Albatross*. It describes the work in the Ellice, Gilbert and Marshall Islands, as well as the Carolines and Ladrones. The Truk Archipelago was perhaps the most interesting of the island groups of the Carolines, and it is the only group of the volcanic islands surrounded by an encircling reef which the author has seen in the Pacific, which at first glance lends any support to the theory of the formation of such island groups as Truk by subsidence. But a closer examination shows that this group is not an exception to the general rule thus far obtaining in all the island groups of the Pacific visited during this trip, that we must look to submarine erosion and to a multitude of local mechanical causes for our explanation of the formation of atolls and of

barrier and encircling reefs, and that, on the contrary, subsidence has played no part in bringing about existing conditions of the atolls of the South and Central Pacific.

American Journal of Mathematics, vol. xxii. 2.—Remarks concerning the expansions of the hyperelliptic sigma-functions, by Oskar Bolza, are supplementary to two papers, by the same writer, in vol. xxi. pp. 107-125 and pp. 175-190.—On a certain class of groups of transformation in space of three dimensions, by H. F. Blichfeldt, is the carrying on of an investigation (by S. Lie) of groups of transformations in 3 variables, defined by the properties : two points have one, and only one, invariant ; $s > 2$ points have no invariants independent of such two-point invariants. This class belongs to a wider class in n variables defined by the properties : not less than $m > 1$ points may possess invariants, while s points, $s > m$, may have no invariants independent of the m-point invariants. The wider class includes the group of Euclidean motions in space of 2 or 3 dimensions, the group of translations in space of n dimensions, the group of Euclidean motions and similar transformations in space of 3 dimensions, &c. Certain groups are discussed and their general properties stated.—Dr. L. E. Dickson, in a paper on the canonical form of a linear homogeneous substitution in a Galois field, gives a short proof by induction of a result which M. Jordan had previously obtained by a rather lengthy analysis.— Dr. E. O. Lovett writes on families of transformations of straight lines into spheres. If a plane σ containing two points E and E_1 moves upon a coincident plane σ_1 containing two straight lines g and g_1, so that E remains upon g and E_1 upon g_1, the two planes form a mechanism possessing the following well-known properties : Every point of σ traces an ellipse upon σ_1, and every point of σ_1 traces a limaçon upon σ (*cf.* Chasles, Aperçu, p. 49), a circle c of radius a in σ rolls upon the inner side of a circle c_1 of radius $2a$ in σ_1. Every point of c describes a straight line passing through the centre of c_1. Any two of these lines, with the points which generate them, can be taken for g, g_1 and E, E_1 in defining the movement. Mr. E. M. Blake's object, in his article on the Ellipsograph of Proclus, is to study (1) the curves generated by the points of σ and σ_1 ; (2) the ruled surfaces generated by any straight line carried by σ or σ_1 and not parallel to them ; (3) the curves enveloped by any straight line of σ or σ_1 ; (4) the developables enveloped by carried planes (*cf.* Cayley, on the kinematics of a plane, *Q.J.* xvi. 1878 ; Schell, "Theorie der Bewegung und Kräfte," i. pp. 227-230, and articles by Burmester).—Mr. N. J. Hatzidakis, in displacements depending on one, two, . . . k parameters in a space of n dimensions, extends to the general case results obtained for 4 dimensions by Prof. Craig (vol xx. 2) and M. Darboux.—The main object of Dr. G. A. Miller on the product of two substitutions is to prove the following theorem :—If l, m, n are any three integers greater than unity, of which we call the greatest k, it is always possible to find three substitutions (L, M, N) of $k + 2$ or some smaller number of elements, and of orders l, m, n respectively, such that $LM = N$.

Annalen der Physik, No. 4.—Temperature and potential gradient in rarefied gases, by G. C. Schmidt. When a vacuum tube is heated, the positive light becomes stratified. The stratifications increase in breadth as the temperature increases. Eventually, the positive light retires towards the anode, so that the discharge becomes dark. At the kathode, on the other hand, an increase of the temperature produces an extension of the glow light, such as is produced by an increase of the current strength. When the dark discharge has set in, the potential gradient is greatest at the anode, and is proportional to the distance from the kathode.—Mechanical motions under the influence of kathode rays and Röntgen rays, by L. Graetz. Rotations similar to those produced by Quincke in liquids may be produced in air ionised by X-rays, by mounting light dielectric bodies provided with agate caps on needle points in the space between two condenser plates exposed to the rays. The sense of the rotation depends upon the initial tendency, except when the rotating body contains a metallic substance, in which case the direction of rotation depends upon the direction of the rays and the electric field. The rotations are explained by the electrostatic forces between the wall of the tube and the parts of the body charged by the ions. The author believes that these rotations furnish an explanation for the rotations under the influence of kathode rays first observed by Crookes.— Atomic and molecular magnetism, by S. Meyer. Special investigations of the magnetic susceptibilities of copper compounds have shown that there is no essential difference between cupric and cuprous compounds. Wherever the formation of a molecule out of its constituent atoms leads to a considerable contraction of volume, the molecular magnetism is increased, so that the result may even be a paramagnetic compound. Where, on the other hand, there is expansion, the diamagnetism increases.— Energy of kathode rays, by W. Cady. The author discusses the various methods of determining the energy of kathode rays. The thermopile and the bolometer have undoubted advantages as compared with the calorimeter, but it is necessary to know how much of the energy incident upon them is reflected, and how much energy is lost in the process of reflection. The author bases his calculations upon the supposition that 40 per cent. of the kathode energy is reflected, and that the amount of energy lost during reflection is 30 per cent.—Electric arc between metallic electrodes in nitrogen and hydrogen, by L. Arons. The electromotive forces necessary to produce an arc between metallic electrodes depends upon the nature of the surrounding gas. In air, silver electrodes give a fine arc, but no arc can be produced with them in nitrogen. Iron electrodes, which require a high voltage in air, require only a very low voltage in nitrogen.—Electrolytic records of electric currents, by P. Gruetzner. The author gives details of the method of recording alternating currents of high frequency with the aid of iodine paper, and shows that for low voltages it offers decided advantages over the dust-figure method.—Change of volume of rubidium during fusion, by M. Eckardt. The fusing point of rubidium is 37·80°. During melting, 1 gramme of rubidium expands by 0·01657 c.cm.

Symons's Monthly Meteorological Magazine, May.—Meteorological extremes : series of articles ; the first two referred to pressure and temperature. The difficulties are far greater than in the other cases, as in determining wind-force observations no homogeneity exists either as regards the instruments employed, or the units of the various scales in which the results, either instrumental or estimated, are expressed. The instrument most generally used is Dr. Robinson's cup-anemometer, the few others being chiefly Osler's or Dine's pressure anemometers. In the velocity instruments the factor for obtaining the true velocity of the wind depends upon the length of the arms and the size of the cups. Until recently the factor used has been 3, but some recent experiments have shown that the speed at the cups should be multiplied by the factor 2·2, so that some very high velocities formerly recorded should be reduced by nearly one-third. Among the highest velocities recorded in this country (reduced by the new factor 2·2), we may mention a severe gale in the Irish Sea in January 1899, in which a rate of 90 miles per hour was recorded in one gust ; the maximum mean force for an hour at Fleetwood was 75 miles. The highest recorded velocity in a gust was recorded by Dines's anemometer at Rousdon, in South Devon, in March 1897; viz. at a rate of 101 miles per hour. At Greenwich a pressure of 51½ lbs. on the square foot was recorded on January 18, 1881, which is equivalent to a velocity of about 130 miles per hour, but there is good reason for believing that in strong winds the records of these pressure plate anemometers are occasionally much too high. It is still a moot question, what is the strongest force that the wind attains, and whether the force in some of the gales which visit our exposed shores from the Atlantic is much exceeded in tropical cyclones.

Bollettino della Società Sismologica Italiana, vol. v. 1899-1900, No. 7.—List of earthquakes observed in the East, and specially in the Ottoman Empire, during the year 1896, by G. Agamennone. An extract from a paper noticed in NATURE, vol. lxi. p. 400.—The Etnean earthquake of May 14, 1898, by A. Riccò. The epicentre was at S. Maria di Licodia on the south-west slope of Etna, and the focus must have been shallow, the shock was strong enough to damage many buildings near the centre of a small disturbed area.—Notices of earthquakes recorded in Italy (October 11-November 19, 1898), by A. Cancani ; the most important being earthquakes in Sicily on November 1, 2 and 3, Dalmatia on November 8, the Ionian Sea November 9, and distant earthquakes on October 12 and November 17.
No. 8.—The Modena-Bologna earthquake of the night of February 1-2, 1900, by G. Agamennone. A slight shock, with a disturbed area of about 60,000 sq. km., but recorded by a seismograph at Lubiana (330 km. from the epicentre).—On an electrothermic phenomenon in electrical contacts with slight pressure, by A. Cancani.—Latian earthquake of July 19, 1899,

by A. Cancani, a paper noticed in NATURE, vol. lxi. p. 573.— Notices of earthquakes recorded in Italy (November 21-December 31, 1898), by A. Cancani, the most important being distant earthquakes on December 1 and 3.—On a new form of multiplication applicable to seismic movements and on a new seismoscope founded on the same, by G. Pericle.

Bulletin de la Société des Naturalistes de Moscou, 1899, No. 1. —Meteorological observations at Moscow in 1898, by E. Leyst.— On the development of green algues under conditions excluding the assimilation of carbon-dioxide, by Dr. A. Artari.—On the *Hedysarum* species (15) found in European Russia, Crimea and Caucasia, by B. Fedtschenko.—On the Hydrachnids of the neighbourhood of Moscow, by A. Croneberg (plate). Forty-nine species, several of which are new, are described.—On the iron-ores (*turjit*) of the South Urals, by J. Samoiloff. All these articles, with the exception of the last one, are in German, or contain German *résumés*.—Notes on Coleoptera of European Russia and Caucasia, by A. Semenoff.

Memoirs of the Mathematical Section o, the Novorossian (Odessa) Society of Naturalists, vol. xix.—Foundations of a theory of analytical functions, by J. Timtchenko, continued from vols. xii. and xvi. This part contains the history of certain special questions, the discussion of which has mainly contributed to the development of the theory of these functions.

Memoirs of the Kazan Society of Naturalists, vols. xxxii. and xxxiii.—Materials relative to the flora of the northern boundaries of the black-earth region, by S. Griegorieff.—The corals of the Devonian deposits in the Urals, by N. Bojartsen (one plate). Fifty-six species are enumerated, several of these, as also one genus (*Nicholsonia*), being new.—On the saliva glands of *Periplaneta orientalis*, by A. Lebedeff (plate).—The Ranunculaceae of Russian Turkestan, by Olga and Boris Fedtschenko. One hundred and fifty-eight species are enumerated, forty-three species being endemic, and thirty-eight species belonging to the Alpine region. A suggestion for the determination of the Turkestan species is given. All articles are summed up either in French or in German.

SOCIETIES AND ACADEMIES.

LONDON.

Royal Society, May 10.—"On Certain Properties of the Alloys of Gold and Copper." By Prof. Sir W. C. Roberts-Austen, K.C.B., F.R.S., and T. Kirke Rose, D.Sc.

The alloys of gold and copper, which are of great industrial importance owing to their use in coinage, have not been subjected hitherto to systematic examination. It has been assumed that they differ widely from the silver-copper series, which has been studied from different points of view, but there is very little evidence on which this view can be based.

Examination with the aid of a thermo-couple and autographic recorder shows that the freezing-point curve of the gold-copper series consists of two branches setting out from the points of solidification of the pure metals and meeting at a point, which is the freezing point of the eutectic. The eutectic contains about 82 per cent. of gold and 18 per cent. of copper, or about 60 atoms of gold to 40 of copper, and solidifies at 905°. The general shape of the curve therefore resembles that of the silver-copper series when the abscissæ give the relative number of atoms.

Under the microscope, alloys containing more than 82 per cent. of gold show a minutely granular structure in which it is not certain that two constituents can be distinguished. The section of standard gold containing 91·6 per cent. of gold bears a close resemblance to that of standard silver prepared in the same way. The alloy with 80 per cent. of gold shows the characteristically-banded eutectic structure almost exclusively, and the alloys with less gold consist of crystals of copper set in a matrix of the eutectic.

Another point of similarity between the gold-copper and silver-copper series is that both the eutectics are brittle and show scarcely any extensibility ; they differ in these respect from most other eutectics. Analysis of various portions of ingots of standard gold reveals the fact that liquation takes place as, definitely as in standard silver, the difference in composition, between the centre and the outside of similar ingots being, however, three or four times greater in standard silver than in standard gold. In the latter case, the centre contains from 0·3 to 1·0 part per 1000 less gold than the outside.

It follows from these results that the gold-copper series of alloys presents many points of similarity with the silver-copper series, and that the main difference is only one of degree, copper being apparently more soluble in gold in the solid state than in silver.

Geological Society, April 25.—J. J. H. Teall, F.R.S., President, in the chair.—The President read the following resolution which had been passed unanimously by the Council : " That this Council desire to place on record their deep sense of the loss which both science and literature have sustained in the death of the Duke of Argyll, who was the oldest surviving past-President of the Geological Society " ; and stated that on behalf of the Council he proposed to communicate a copy of the resolution to the Duchess of Argyll, coupled with an expression of respectful sympathy.—On a complete skeleton of an Anomodont reptile from the Bunter Sandstone of Reichen, near Basel, giving new evidence of the relation of the Anomodontia to the Monotremata, by Prof. H. G. Seeley, F.R.S. The author discusses various views which have been expressed with regard to the position of the Labyrinthodonts. He has already separated these animals from the Amphibia and combined them with the Ichthyosauria in a group of reptiles named Cordyiomorpha, and he enumerates a series of characters which constitute so close a link between the two types "that it is not possible, in the absence of evidence, to conceive of their being referred to different classes of animals." In conclusion, the author argues that the points of structure are so few in which Monotreme mammals make a closer approximation to the higher mammals than is seen in the fossil described and other Anomodontia, that the Monotreme resemblances to fossil reptiles become increased in importance. He believes that a group Theropsida might be made to include Monotremata and Anomodontia, the principal differences (other than those of the skull) being that Monotremes preserve the marsupial bones and the atlas vertebra. *Ornithorhynchus* shows pre-frontal and post-frontal bones, and has the malar arch formed as in Anomodonts and some other reptiles.—On Longmyndian Inliers at Old Radnor and Huntley (Gloucestershire), by Dr. Charles Callaway. The grits, with some associated slaty bands, forming a ridge near Old Radnor were considered by Sir Roderick Murchison to be May Hill Sandstone. The author has discovered that one of the beds of Woolhope Limestone, dipping westward, is crowded with rounded and angular fragments of grit bearing a general resemblance to the arenaceous parts of the Old Radnor Group.

May 9.—J. J. H. Teall, F.R.S., President, in the chair. —The Pliocene deposits of the East of England. Part ii. : The Crag of Essex (Waltonian), and its relation to that of Suffolk and Norfolk, by F. W. Harmer, with a report on the inorganic constituents of the Crag by Joseph Lomas. Three divisions of the Red Crag are proposed, namely, Waltonian, Newbournian and Butleyan, which are distinguished alike by the difference of their faunas, and by the position which they occupy. The first, with its southern shells, is confined to the county of Essex ; the second, containing a smaller proportion of southern and extinct, and a larger proportion of northern and recent species, occupies the district between the Orwell and Deben, and a narrow belt of land to the east of the latter river ; the third, in which Arctic forms such as *Cardium groenlandicum* are common, is found only farther north and east. All these beds are believed to have originated in shallow and land-locked bays, successively occupied by the Red Crag sea as it retreated northward, which were silted up, one after the other, with shelly sand. The conditions under which the Red Crag beds originated seem to exist at the present day in Holland, where sandy material brought down by rivers, with dead shells in great abundance from the adjacent sea, is being thrown against and upon the coast, principally by means of the westerly winds now prevalent. From meteorological considerations, it seems probable that strong gales from the east may have prevailed over the Crag area during the latter part of the Pliocene epoch. —A description of the Salt-Lake of Larnaca in the Island of Cyprus, by C. V. Bellamy. After a brief description of the general geology and geography of the island, the author proceeds to deal with the topography of the lake, which covers an area basin shut off from the sea, its deepest part being about to feet below sea-level. The barrier between the salt-lake and the sea is made of stiff calcareous clay associated with masses of conglomerate resting on plastic clay, that on watery mud, and that again on stiff calcareous clay. The sea-water appears to percolate through the highest deposits, meeting with checks in the conglomerates, and thus reaches the basin somewhat slowly, where it is evaporated to dryness by the summer heat and deposits its salt. Artificial channels have been made, to carry the flood-water from the land direct to the sea, so that it does not dilute the brine of the lake. The rainfall in the catchment-area round the lake is at the most only enough to supply 223 million gallons, and as the lake contains 480 million gallons when full, the balance of 257 million gallons must be derived from the sea. The lake is probably situated on what was an extensive arm of the sea at the close of the Kainozoic era. The salt-harvest begins in August, at the zenith of summer heat, and it is reported that a single heavy shower at that time of year suffices to ruin it. Observations are given on the density of the water, the plants and animals in the water, and the lake-shore deposits.

Zoological Society, May 8.—Dr. W. T. Blanford, F.R.S., Vice-President, in the chair.—Mr. Sclater exhibited a mounted specimen of a male reedbuck, which had been obtained by Mr. Ewart S. Grogan on the Songwé River, north of Lake Nyasa. The specimen was of about the same size as the common reedbuck (*Cervicapra arundinum*), but differed from that species in several important points. Mr. Sclater considered it referable to a new species, and proposed to name it *Cervicapra thomasinae.*—Mr. C. Davies Sherborn made some remarks on the progress of his " Index Generum et Specierum Animalium," of which he expected the first portion (1751-1800), containing about 60,000 entries, to be ready for publication at the end of this year.—Mr. G. A. Boulenger, F.R.S., read a paper on the batrachians and reptiles collected by Mr. G. L. Bates in the Gaboon (French Congo), among which were specimens of ten new species and five new genera of the former, and of one new species of the latter, which were described. These descriptions were incorporated with a list of the previously known species from the Gaboon, by which it was shown that the batrachians known from this country reached thirty-nine in number and the reptiles eighty.—Mr. W. R. Ogilvie Grant read a paper on the birds of the Hainan, based on a collection sent home by the late Mr. John Whitehead from the Five-Finger Mountains in the interior of the island. Examples of many interesting species had been procured, which were either new to science or to the fauna of the island. Among the former, which numbered eleven, were mentioned a splendid silver pheasant, a remarkable night-heron, and a peculiar brown-and-white jay of the genus *Urocissa.* The paper contained a complete account of the avifauna of Hainan as known at the present time.—Mr. Philip Crowley read a paper on the Rhopalocera collected by the late Mr. John Whitehead on the Five-Finger Mountains in the interior of Hainan. Specimens of 108 species were contained in the collection, of which eight were described as new, and many others were recorded from that island for the first time.—Mr. J. S. Budgett read a paper, entitled "Some points in the anatomy of *Polypterus,*" as deduced from an examination of specimens lately procured by the author in the River Gambia.—Mr. G. A. Boulenger gave a list of the fishes collected by Mr. J. S. Budgett during his recent expedition to the Gambia. Among these were examples of two new species, which were proposed to be named *Clarias budgetti* and *Synodontis ocellifer.* Altogether specimens of forty-two species of fishes were obtained by Mr. Budgett from the river.

Mathematical Society, May 10.—Prof. Elliott, F.R.S., Vice-President, in the chair.—The chairman having read the by-laws bearing upon the subject of the special meeting, announced that it was proposed " that by-law iv. 1, be amended by substituting the words 'half-past five o'clock in the afternoon' for ' eight o'clock in the evening.' " The motion having been seconded by Dr. J. Larmor, F.R.S., was carried unanimously. —At the ordinary meeting, Dr. Glaisher, F.R.S., communicated a congruence theorem relating to Eulerian numbers and other coefficients.—Prof. Lamb, F.R.S., spoke briefly on a property of the wave-system due to the free vibrations of a nucleus in an extended medium.—Prof. Love, F.R.S., gave a description of some diagrams illustrating a paper, by Mr. J. H. Michell, which treats of distributions of stress in two dimensions.—The following papers were communicated by their titles :—The differential equation whose solution is the ratio of two solutions of a linear differential equation, by Mr. M. W. J. Fry ; Note on a quinquisectional equation, by Prof. L. J. Rogers ; On the differentiation of single theta functions, by the Rev. M. M. U.

Wilkinson ; and linear substitutions commutative with a given substitution, by Dr. L. E. Dickson.—Lieut.-Colonel Cunningham, R.E., V.P., showed that numbers which are expressible in the *two* forms $N = \frac{\mu x^2 + \nu y^2}{a} = \frac{\mu' x'^2 + \nu' y'^2}{a'}$ are always *composite*, when $\mu\nu = \mu'\nu'$; and showed how to reduce them to the forms $N = X^2 + \mu\nu Y^2 = X'^2 + \mu\nu Y'^2$, the factorisation of which is known from Euler's researches.

Royal Meteorological Society, May 17.—Dr. C. Theodore Williams, President, in the chair.—A paper was read on the Wiltshire whirlwind of October 1, 1899, which had been prepared by the late Mr. G. J. Symons, F.R.S., a few days before he was stricken down with paralysis. This whirlwind occurred between 2 p.m. and 3 p.m., commencing near Middle Winterslow and travelling in a north-north-easterly direction. The length of the damage was nearly twenty miles, but the average breadth was only about 100 yards ; in this narrow track, however, buildings were blown down, trees were uprooted, and objects were lifted and carried by the wind a considerable distance before they were deposited on the ground. Fortunately the greater part of the district over which the whirlwind passed was open Down, otherwise the damage and perhaps loss of life would have been considerable. At Old Lodge, Salisbury, the lifting power of the whirlwind was strikingly shown by several wooden buildings being lifted up and dropped down several feet north-west of their original position. At a place eighteen miles from its origin the whirlwind came upon a rick of oats, a considerable portion of which it carried right over the village of Ham and deposited in a field more than a mile and a half away. —A paper by Dr. Nils Ekholm, of Stockholm, was also read on the variations of the climate of the geological and historical past and their causes. In this the author attempts to apply the results of physical, astronomical and meteorological research in order to explain the secular changes of climate revealed by geology and history.

DUBLIN.

Royal Dublin Society, February 21.—Prof. G. F. Fitzgerald, F.R.S., in the chair.—Prof. W. N. Hartley, F.R.S., communicated his papers on the action of heat on the absorption spectra and chemical constitution of saline solutions, and on the occurrence of cyanogen compounds in coal-gas, and of the spectrum of cyanogen in that of the oxy-coal-gas flame.—Prof. E. J. McWeeney gave an account of the recently demonstrated connection between mosquitoes and malaria, with a lantern demonstration of the life-history of the former.—Prof. T. Johnson communicated a note on Sclerotium disease of Jerusalem artichoke grown at Greystones, county of Wicklow. March 21.—Dr. F. T. Trouton, F.R.S., in the chair.—Dr. G. H. Pethebridge read a paper, entitled "Contributions to the knowledge of the action of inorganic salts on the structure and development of plants."—Mr. R. J. Moss (in the absence of Dr. W. E. Adeney) communicated a paper, by Prof. E. A. Letts and Messrs. Blake, Caldwell and Hawthorne, on the nature and speed of the chemical changes which occur in mixtures of sewage and sea-water.—Prof. J. Joly read a paper on the theory of the order of formation of silicates in igneous rocks, which was illustrated by a diagram (see p. 84). April 25.—Prof. J. Emerson Reynolds, F.R.S., in the chair. —Prof. J. Emerson Reynolds, F.R.S., read a paper on recent analyses of the Dublin gas supply, and observations thereon.— Prof. G. A. J. Cole communicated a paper by himself and Mr. J. A. Cunningham on certain rocks styled "felstones," occurring as dykes in the county of Donegal.

EDINBURGH.

Royal Society, May 7.—Sir Arthur Mitchell, K.C.B., in the chair.—Mr. John Aitken, F.R.S., read a paper on the dynamics of cyclones and anticyclones, Part ii., which was illustrated by an ingenious experiment showing the production of vortex columns in the air. Over the upper metal surface of a flat box through which steam was blown was spread a sheet of brown paper thoroughly soaked with hot water. A steady gentle blast of air was driven across this steaming surface by means of a rotating fan ; and when a barrier was intercepted so as to cut off half of the surface from the direct effect of the blast, a succession of whirls was started at the boundary between the sheltered and unsheltered parts of the surface. These whirls were plainly visible in the columns of rotating cloud, and showed on a small

scale some of the characteristics of cyclones. According to Mr. Aitken's mode of looking at the phenomenon, the blast of air produced by the fan is analogous to the anticyclonic marginal wind which is regarded as driving the cyclone. The relations between the upper and lower currents in a cyclonic movement were also illustrated in the experiment.—Mr. R. C. Punnett communicated a paper on certain Nemerteans from Singapore, in which several facts of morphological interest were brought to light, notably the presence, in one species, of ducts placing the anterior portion of the alimentary canal in communication with the excretory system, and so with the exterior ; and the different features shown in the termination of the lateral nerve-cords in a single genus where there might be a commissure either above or below the rectum, or else no commissure at all.—Mr. R. T. Omond, in a paper on the reduction to sea-level of the Ben Nevis barometer, pointed out that, using the ordinary reduction formula, we get an appreciable difference between the observed sea-level pressure at Fort William and the reduced Ben Nevis reading. Leaving out of account all cases in which strong winds were blowing, Mr. Omond had worked out in detail the hourly readings for a period of time extending over six years, and gave reasons for his belief that the discrepancy noted above was due to a false estimate of the average temperature of the air between Fort William and Ben Nevis. This average was not the mean of the bottom and top temperatures.

Mathematical Society, May 11.—Mr. Muirhead, President, in the chair.—A theorem in continued fractions (Prof. Steggall), on certain elementary inequality theorems (Prof. Gibson), note sur un problème de géométrie (Mons. Ed. Collignon) ; communicated by Dr. Mackay.

PARIS.

Academy of Sciences, May 14.—M. Maurice Lévy in the chair.—On a zenitho-nadiral apparatus designed to measure the zenithal distances of stars near the zenith, by M. A. Cornu. In front of a horizontal telescope carrying a wire micrometer is placed a special arrangement of two mirrors making an angle with each other of 90°. Four images can be seen simultaneously, that of the star near the zenith, the cross wires, the reflection of these wires in the mercury bath, and the image of the wires from the special reflector. When the image of the movable wire coincides with its reflected image from the mercury bath, the nadiro-zenithal image of this wire passes through the zenith, whatever may be the deviation from a right angle of the angle between the mirrors. The arrangement possesses important advantages over the methods at present in use.—Remarks on a meteor which fell in Bolivia on November 20, 1899, by the French Chargé d'Affaires at La Paz.—On divergent series, by M. Le Roy.—On the representation of non-uniform functions, by M. L. Desaint.—On a modification which metallic surfaces undergo when submitted to light, by M. H. Buisson. Under the influence of light, the metallic surface changes its state as measured by the rate at which it loses a charge of electricity, this change not being permanent but gradually disappearing when the radiant energy is cut off.—On the thermoelectric properties of some alloys, by M. Émile Steinmann. The alloys studied were ten nickel steels, four samples of platino-iridium, three of aluminium bronze, five telegraphic bronzes, five brasses, and four of German silver, at temperatures ranging from 0° to 260° C. In the binary alloys the observed electromotive forces are arranged in the order of magnitude of one of the components, but no simple relation could be deduced between the electromotive force and chemical composition in the case of nickel steel or of ternary alloys.—Duplex and diplex transmission by electric waves, by M. Albert Turpain.—Experiments in wireless telegraphy from a free balloon, by MM. J. Vallot, J. Lecarme and L. Lecarme. It was found to be possible to transmit messages to the balloon without an earth wire, even up to a distance of six kilometres and a vertical height of 800 metres.—An arrangement designed to prevent the interception of despatches in wireless telegraphy, by M. D. Tommasi. —On the hydrated calcium peroxides, by M. de Forcrand. A thermochemical paper.—On the allotropic transformations of the alloys of iron and nickel, by M. L. Dumas.—Preparation of some aluminium compounds and of the corresponding hydrogen derivatives, by M. Fonzes-Diacon. Details are given of the preparation of aluminium sulphide, selenide, phosphide, arsenide and antimonide. By the decomposition of these substances the hydrides H_2S, H_2Se, PH_3 and AsH_3 are obtained in a very pure state. SbH_3 can also be prepared in considerable quantity.—The estimation of thallium, by M. V. Thomas. The oxidation of thallous to thallic salts is carried out with bromoauric acid, the precipitated gold being weighed. Provided that the quantity of thallium present is not too small, the results are very exact.—Action of anhydrous aluminium chloride upon acetylene, by M. E. Baud. The aluminium chloride absorbs nearly four times its weight of acetylene, hydrogen, marsh-gas and ethylenic hydrocarbons being evolved. Complicated condensation products are formed, which are being further examined.—Some new organometallic combinations of magnesium and their application to syntheses of alcohols and hydrocarbons, by M. V. Grignard (see p. 85).—Santalenes and santalols, by M. M. Guerbet. A description of the isolation of two alcohols and two hydrocarbons from essence of sandalwood, together with the products resulting from the action of acetic acid, hydrochloric acid, and nitrosyl chloride upon the hydrocarbons and acetic anhydride, and phosphorus pentoxide upon the alcohols.—On tyrosinase, by M. C. Gessard. Tyrosinase is a ferment isolated from fungi, it possessing oxidising powers, giving a red oxidation product with tyrosin.—On the oxidation of erythrite by the sorbose bacterium and production of a new sugar, erythrulose, by M. Gabriel Bertrand (see p. 85).—On the amount of iron in hæmoglobin from the horse, by MM. L. Lapicque and H. Gilardoni.—On a method allowing of the extraction of the sugar from molasses by means of the ordinary boiling apparatus, by M. Paul Lecomte.—Chlorophyll assimilation in plants confined in rooms, by M. Ed. Griffon.— A new self-registering apparatus for continuous currents, by MM. Auguste and Louis Lumière.

GÖTTINGEN.

Royal Society of Sciences.—The *Nachrichten* (mathematico-physical section), Part iii., for 1899, contains the following memoirs communicated to the Society :—

October 28, 1899.—E. Riecke : Lichtenberg figures in the interior of Röntgen tubes.—W. Voigt : on a problem of Kohlrausch's in thermodynamics.—P. Gordan : new proof of Hilbert's theorem on homogeneous functions.

November 25.—W. Kaufmann : outlines of an electrodynamical theory of gaseous discharges (Part i)..

December 14.—W. Kaufmann : the same (Part ii.).

December 9.—G. Bohlmann : a problem concerning the "smoothing-out" of statistical curves.

January 13, 1900.—S. Kantor : a theorem in determinants. —A. Schoenflies : a proposition in the analysis of position.— E. Neumann : on Robin's method for determining electrostatic potential.

February 8.—W. Voigt : remarks on the theory of so-called thermomagnetic effects.

February 3.—E. Zermelo : on the motion of a system of points in relation to inequations of condition.

February 17.—A. von Koenen : on the age of North German Wealden formation (*Wälderthon*).

AMSTERDAM.

Royal Academy of Sciences, April 21.—Prof. H. G. Van de Sande Bakhuyzen in the chair.—The following papers were read :—Prof. Kluyver on approximation formulæ concerning the prime numbers, not exceeding a given limit. The author shows that it is possible to express the approximate value of the sum of the $(-r)$th powers of these prime numbers, if only their total number be given. A similar formula gives an approximation to the value of the logarithm of the least common multiple of all integers below a given number.—Prof. Winkler, on behalf of Mr. M. A. van Melle, on some reflexes in respiration in connection with Laborde's method of re-establishing respiration stopped by narcosis by rhythmically pulling the tongue.—Prof. Franchimont, on behalf of Dr. Greshoff, on Echinopsine, a new crystalline vegetable base. This communication was accompanied by remarks by Prof. Kobert, of Rostock, and Prof. Verschaffelt, of Amsterdam.—Prof. van Bemmelen, on behalf of Dr. F. A. H. Schreinemakers, on the composition of the vapour phase in the system of water and phenol with one and with two liquid phases.—Prof. Bakhuis Roozeboom, (a) on behalf of Dr. A. Smits, on decreases in vapour tension and rises of the boiling point in the case of diluted solutions ; (b) on behalf of

Dr. Ernst Cohen, on thermodynamics of Clark's normal element.—All the above papers will be inserted in the *Proceedings*.—The following papers were presented for publication in the *Proceedings* :—(a) One by Prof. Schoute, entitled "Joachimsthal Theorem for Normal Curves" ; (b) one by Prof. Bakhuis Roozeboom, on behalf of Dr. Ernst Cohen, entitled "Studies on Inversion (I.)."

DIARY OF SOCIETIES.

CONTENTS.

THURSDAY, MAY 31, 1900.

A LIFE OF SCHÖNBEIN.

Christian Friedrich Schönbein, 1799–1868. Ein Blatt zur Geschichte des 19. *Jahrhunderts.* Von Georg W. A. Kahlbaum und Ed. Schaer. I. Theil. Pp. xix+230. (Leipzig : Johann Ambrosius Barth, 1900.)

THIS work forms the fourth part of the series of monographs on the history of chemistry being published under the editorship of Prof. Kahlbaum, of Bâle, whose qualifications for the task have already been made known to English men of science in the notices of two of his earlier volumes published recently in these columns.[1] The present instalment covers the period from the time of Schönbein's birth to the year 1849, and is divided into four sections, which comprise respectively the intervals 1799–1820, the "Wanderjahre" 1820–1828, the residence at Bâle from 1828 till the discovery of the passivity of iron towards the end of 1835, and the prosecution of the researches on the latter subject and on cognate electrical subjects from 1836 till 1849. There is a supplementary section dealing with Schönbein as a teacher and friend, which is by no means the least interesting part of the present volume. A perusal of the work will not only convince its readers that Schönbein was altogether a remarkable man as a thinker and experimenter, but that his personality and work could not have fallen for delineation and estimation into any better or more appreciative hands than those of Dr. Kahlbaum and his colleague.

The subject of the present biography was born at Metzingen, in Schwabia, on October 18, 1799. Passing over his boyhood, it appears that in his fourteenth year he made his first start in life as a pupil in the chemical and pharmaceutical factory ʾof Metzger and Kaiser at Böblingen, so as to become a practical chemist. He suffered much at first from home-sickness, which, the authors tell us, is a purely German ailment :—

" Das bittere Leid des Heimwehs, dieser ächt deutschen Krankheit, die Engländer und Franzosen haben kaum ein eigenes Wort dafür, &c."

After seven years in this factory, he went, in 1820, into Dr. J. G. Dingler's factory for chemical products at Augsburg, on which occasion it is noteworthy that he underwent his only examination, and obtained his only certificate from Dr. Kielmeyer, of Stuttgart. The original document, which has been obtained by Dr. Kahlbaum, testifies that at that time Schönbein was possessed of a good scientific and practical knowledge of chemistry. Dr. Dingler's letter, setting forth the qualifications which he expected on the part of the young man whom he was thinking of engaging, is dated March 20, 1820, and as a revelation of the state of affairs in a German establishment during the early part of the nineteenth century, it will repay careful perusal. The chemist required by him was to have scientific rather than ordinary routine chemical knowledge ; he was to have at the same time something more than a superficial acquaintance with chemistry, in order that he might be able to carry out the analytical work required of him ; he was to have a knowledge of

[1] NATURE, February 8, p. 337; and March 29, p. 513.

languages, so as to be able to translate, at least from French ; he was to be possessed of moral rectitude, and to be entirely worthy of confidence. He was to come on probation for fourteen days, and if not found suitable he was to be sent back " carriage paid " (" bei Vergütung der Reisekosten "). If found suitable, he was to be boarded and lodged, and to receive from 200 to 300 florins per annum, with an increase to follow.

As the editor points out, Schönbein must have made good use of his time at Böblingen, since he seems to have come up to Dingler's requirements, and was appointed to the Augsburg factory ; but before entering upon his duties he drew a fatal conscription number, and had to undergo a short term of military service. It appears, however, that he was soon discharged from this duty, through the intervention of the King, and in May 1820 he was " militärfrei."

The eight years from 1820–28 must have been years of great activity in Schönbein's early life. He remained only a few months at Augsburg, and then travelled from one University to another. His name is associated during this period with the Universities of Tübingen and Erlangen. While studying at the latter place, where he had Liebig for a contemporary, he was also holding the appointment of director in Adam's factory at Hemhofen, but finding that his factory work interfered with the prosecution of a regular course of study at the University, Mr. Adam relieved him of this work, and assisted him pecuniarily by appointing him tutor in his family.

Among many other interesting episodes in Schönbein's career at this period is his sojourn in England, which appears to have been the outcome of a taste for pedagogy inspired by his friend Christian Friedrich Wurm, who subsequently became professor of history in the Hamburg gymnasium—a man of many parts, a master of the English language and an ardent disciple of Pestalozzi, whose works he had translated into English. It was in 1826 that the young Schönbein entered the service of Dr. Mayo, who kept a school at Epsom, where Wurm was already engaged, for the purpose of imparting instruction in mathematics and natural philosophy on Pestalozzian principles in return for " 50l. sterling per annum ; with board, lodging and washing." The description of Dr. Mayo's establishment given by Wurm, and the criticisms which he makes upon the English educational methods of that time, are preserved in a letter to Schönbein written from Epsom in 1825, and published by Dr. Kahlbaum in the present volume. The editor comes to the conclusion that the Epsom academy was as far removed from the ideal Fröbel institute as the classical establishment of Dr. Blimbers at Brighton, in which young Paul Dombey was "forced" to death. Schönbein appears, however, to have made the best of his opportunities while in England, and to have paid visits to London and to Scotland, making friends and acquaintances, and gleaning knowledge wherever he went. In 1827 he left for Paris, and a long extract from his diary of travel, reprinted in the present work, is full of most interesting comparisons of English with French modes of travelling, and of the personal characteristics of the two nations. While Anglophobia, judging from some of the correspondence received by Schönbein at that time,

F

appears even then to have existed in Germany, it is satisfactory to learn that he was never influenced by it :—

"Dann war gerade ihm der afflammende Strohfeuer-enthusiasmus der Franzosen nicht sympathisch, sein deftiges, bedachtes Wesen war sehr wohl, wir haben das ja gesehen, begeisterungsfähig und hingebend, aber, wie sein Humor nichts von dem sprühenden Feuerwerk französischen Esprits hatte, so wenig trat sein Enthusiasmus als schnell verrauschende Schwärmerei auf. Die langsame niedersächsische Art der Engländer war ihm, dem Schwaben, darum viel herzwärmender als das griechische Feuer der Franzosen."

It is, in fact, quite remarkable to find throughout this biography how warmly Schönbein felt himself in sympathy with England and English people. Faraday, Grove and Graham were his intimate and life-long friends. He appears to have gone to Paris under the same conditions and for the same purpose that he came here—to acquire a more intimate knowledge of the language, and to gain some insight into French pedagogy. The school in Paris, kept by a M. Rivail, in which he temporarily became a teacher, was unsatisfactory from every point of view, and on the whole the young German seems to have had anything but a pleasant time in the French capital about that period. But there, as elsewhere, he made the best of his opportunities by attending lectures at the Sorbonne, where he came under the influence of Gay-Lussac and Thénard, Biot, Dumas, Pouillet, Brongniart, &c., and by the time he returned to England to stay with his friend and Epsom colleague, Barron, at Stanmore, his appreciation of France and the French had considerably increased. Schönbein's views on the nature and constitution of Polytechnics, and his letters to Wurm written from Paris, and giving his experience of the Sorbonne and its professors, are full of interest.

In 1827, Merian, the professor of physics and chemistry at Bâle, was taken ill, and a substitute had to be found to carry on his duties. The post was first offered to Schönbein's friend, Engelhart, then also in Paris, who was unable to accept it, and afterwards to Schönbein, who was in England, and who finally undertook the duties, thus severing himself from this country, apparently to his regret, and becoming attached, in 1828, to that University, on which he ultimately shed such lustre. The first years of his connection with Bâle were unsettled by the provisional character of his appointment, and were further troubled by political disturbances, during which Schönbein himself bore arms, and it was not till February 1832 that he made his first communication to the scientific society of that town. This paper dealt with the classification of the elements into metals and non-metals, the former being defined as those elements which form basic oxides, and the latter those which form acid oxides. A few other papers followed during the years 1833–1835 ; one on the Pepys gas-holder, one on polarised light, one on an *ignis fatuus* observed at Bärenthal in the Black Forest, and one on the isomerism of chemical compounds. With the clearing of the political atmosphere and the cessation of hostilities, the University of Bâle underwent reorganisation, and Schönbein was appointed ordinary professor of physics and chemistry in 1834. His marriage took place the following year, towards the end of

which (December 23, 1835) he made known to the " Naturforschenden Gesellschaft" his memorable work on the behaviour of tin and iron towards nitric acid, later communications on the passive state of iron and other metals having been made on January 21 and March 3, 1836.

The observation which formed the starting-point of Schönbein's researches appears to have been made by many previous investigators, among whom our own countryman, James Keir, F.R.S. (*Phil. Trans.* 1794) is given the priority. The period covered by the next section of the present work, viz. from 1836 to 1849, was full of activity and productiveness on the part of Schönbein, whose development of ideas, from his first experiments on the " passive" state of metals through all their ramifications into the various fields of electro-chemistry, is followed out and set forth by Dr. Kahlbaum with a masterly hand. As we are at this period well within what might be called the public aspect of Schönbein's work, when his results were being continuously published and discussed throughout the scientific world, it is unnecessary to dwell at any greater length upon the contents of the present instalment of his biography. It will interest English readers particularly to find how skilfully the authors trace the influence of Schönbein's correspondents, and particularly Faraday, upon his work. This work centred round the subjects of the origin of the electric current and the polarisation of the electrodes. The great controversy between the " chemical" and the " contact" theories of electromotive force was then raging, and it is now a matter of history how ably and staunchly Schönbein advocated the former. Most clearly are his views expressed in the extracts from his correspondence with Faraday, Poggendorff, Grove, De la Rive and others which the authors have brought together in this biography. Now and again passages occur which are really prophetic, such, for example, as his statement concerning the possible utility of the " Voltaic cell" in organic chemical investigation,[1] and his remarks[2] on the desirability of there being a more frequent blending of physics and chemistry in the same individual, as exemplified by Berzelius, Gay-Lussac, De la Rive, Becquerel, Daniel and Grove. Dr. Kahlbaum points to the modern school of physical chemistry as the embodiment of this wish.

In the concluding section, Schönbein's position in the world of science, as deduced from his own statements, is most instructively summed up. He was something more than a " physicist" or " chemist" :—

" Also Schönbein war nichts weniger als ein kritikloser Anhänger der Naturphilosophie im gewöhnlichen Sinne des Wortes, als der er im allgemeinen verschrieen ist, aber er war eine durch und durch philosophisch angelegte Natur mit gefülltem philosophischem Schulsack und gut geschultem Denken, die eben immer aus theoretischen Ansichten heraus ihre Arbeiten unternahm."

This judgment is borne out by an extract relating to his work on ozone contained in a letter written to Liebig in 1866, in which he states that, although the detection of a peculiar smell in electrolytic oxygen was accidental, all that has since proceeded from this observation cannot be ascribed to accident.

[1] *Pogg. Ann.* 1839, xlvii. 583.
[2] *Beiträge zur physikalischen Chemie*, 1844.

Among the many interesting aspects of Schönbein's life and work dealt with in this section is his dislike for organic chemistry already referred to in his correspondence with Faraday. Dr. Kahlbaum, we may add, endorses this opinion with some very strong remarks of his own (pp. 204-205), which will, no doubt, be forgiven by the "Herren Organiker" in view of the very important service to the history of nineteenth century science which he is rendering by these biographical contributions. Then, again, one cannot but be struck by the versatility of Schönbein's genius as revealed by the narration of his connection with journalism. That the illustrious Bâle professor was possessed of great literary power is made clear by his biographer. It is worthy of record that Schönbein attended the Birmingham meeting of the British Association in 1839, and the Cambridge and Southampton meetings in 1845 and 1846 ; of the first of these he gave an account in his "Reisetagebuch eines deutschen Naturforschers," of which extracts in English were published in the *Athenæum*. As an excellent example of his literary style may be mentioned his charming description of Easter festivities in Germany, written in English to Faraday in 1856. With respect to the literary style and method of publishing his scientific writings, there is a long and interesting critical letter from De la Rive in 1839, in which he reproaches Schönbein for being too diffuse, for writing too much and at too great a length, for introducing too often unverified suppositions, and, in fact, as we should say at the present time, for transferring the contents of his laboratory notebooks to the pages of his published memoirs :—

"C'est une voie tentative, à la tête de laquelle est Faraday dans ce moment, qui publie, publie le journal de ses expériences, aussi voyez le peu d'effet que font ses travaux sur le continent."

This criticism, by the way, is endorsed by Dr. Kahlbaum, who regrets that the Germans, "on account of its foreign origin," should have imitated a style which he characterises as incivility (Unhöflichkeit) to the readers.

Enough has been gleaned from this volume to show our readers that as a contribution to the history of the science of the nineteenth century, it is in no way inferior to its predecessors.   R. MELDOLA.

PROFESSOR TAIT'S SCIENTIFIC PAPERS.

Scientific Papers. By Peter Guthrie Tait, M.A., Sec. R.S.E. Vol. ii. Pp. xiv + 500. (Cambridge : At the University Press, 1900.)

PROF. TAIT is to be congratulated on the energy with which this reprint is being pushed forward. The first volume, noticed in NATURE, vol. lx. p. 98, is already followed by a second, so that the completion of the work at an early date may be anticipated.

The present instalment contains two considerable experimental investigations ; one of these, on the compressibility of water at very high pressures, was suggested by a previous research on the *Challenger* thermometers ; for the second, on impact, we are indebted to the author's well-known interest in golf. There is also a very interesting discussion of the cause of the "soaring" flight of a golf ball.

The most important theoretical research consists of a revision of the kinetic theory of gases, from the old standpoint of elastic spheres. All students of this intricate subject will be glad to have Prof. Tait's acute examination of it in the present compact form. It is interesting to note, by the way, the author's frank confession : "I have . . . . abstained from reading the details of any investigation (be its author who he may) which seemed to me to be unnecessarily complex. Such a course has, inevitably, certain disadvantages, but its manifest advantages far outweigh them !" Let us hope that no indolent reader will be tempted to turn against Prof. Tait himself a *dictum* which conveys a very salutary warning to authors !

One of the most useful features of this reprint is the number of short papers which to many readers will now become known for the first time. There are also included a few biographical notices, as well as articles from the "Encyclopædia Britannica." In a note to the article on "quaternions" we are told that the sketch of the subject recently given by Prof. Klein in the "Theorie des Kreisels" rests on a misapprehension. This is one disappointment the more for those students who have vainly striven time after time to get a clear notion of what a quaternion really is, and who hoped that they had found at last something like a clear and compact and intelligible account of the matter. If, in spite of the fact that "the grandest characteristic of quaternions is their transparent intelligibility," men like Cayley and Klein are declared to have gone astray, one may be excused for asking whether there may not be something wanting after all in the official presentations of the subject ?

The paper on the laws of motion hardly addresses itself to points on which a modern reader would seek enlightenment. Instead, we have verbal questions as to the meaning of "force" and the proper translation of certain phrases of Newton. Are not such questions disposed of once for all by the simple statement that since the time of Newton scientific people have specialised their usage of the word "force"? Although this has not been an unmixed advantage, it is probably now irrevocable. Still, one may reasonably urge that it is hardly fair to take a popular term, used in a great variety of senses, to attribute it for special purposes one and only one of these, and then to denounce as ignorant any one who continues to use it in its former latitude. The scorn, for example, which has been called forth by the term "centrifugal force" has often been most unjust, the physical notions of the users being clear enough, although they were not expressed in the conventional phraseology. The endless discussions which have been inflicted on us as to the meaning of the word "weight," furnish another instance of the trouble which may be wrought by specialists attempting to usurp functions which do not properly belong to them.

The last paper in the volume, on the teaching of natural philosophy, contains matter which probably hardly any one would question. Yet it well deserves reprinting, if only for the passage near the end which speaks of "the fatal objections to the school-teaching of physical science," based on the intrinsic difficulties of the subject, and the maturity of mind required to overcome them. Any one who is aware of the futility and the pedantry of a good deal that goes on in schools under the name of science-teaching will thank Prof. Tait for this courageous utterance. The mischief is that school-teaching is dominated by examinations, and that the kind of science-teaching which it is possible, and highly desirable, to have in schools does not readily lend itself to examination-tests of the ordinary kind.

The volume is marked by the same beauty and accuracy of printing as the former one. It is intimated that a third volume will complete the work.

HORACE LAMB.

WYATT'S BRITISH BIRDS.

British Birds; with some Notes in reference to their Plumage. By C. W. Wyatt. Coloured Illustrations. (London: William Wesley and Son, 1899.)

WHETHER the beautifully illustrated work on the same subject by the late Lord Lilford leaves room for the present volume and its predecessor, is a question for the publisher rather than for the reviewer to answer ; but, if the stream of books on the subject be any criterion, the appetite of the British public for natural histories of the avifauna of their own country seems insatiable. Apart from all this, the present work, of which the first volume was issued in 1897, has high claims on the consideration of the public, the large size (4to.) of the paper on which they are printed permitting the plates to be on a scale of greater magnitude than in the work above-mentioned, while their excellence from an artistic point of view, as well as their apparent fidelity to nature, leaves little or nothing to be desired from the point of view of the connoisseur in animal painting. In too many instances we have either an inartistic but truthful portrait of the creature depicted, or an artistic picture in which details of coloration are sacrificed to the general effect ; but in the present case, the happy mean appears to have been attained in these respects. The plates are signed with the initials "C. W. W.," but we are told in the preface that the colouring has been done by the daughters of Dr. Bowdler Sharpe, whose training is a sufficient guarantee for its accuracy.

It must, indeed, be understood that the book stands or falls by the plates, as the letterpress is restricted in the main to details concerning the plumage of the specimens figured, or to generalities relating to seasonal changes of colour, nothing in the way of description being given.

When the scientific names applied to the different species are those of almost universal acceptation, no references to other works are added ; but in the case of those where uniformity is by no means general, a reference is made to the synonyms used in standard manuals, such as the fourth edition of "Yarrell." It may be added that the reference to the latter work in the case of the Hen-Harrier appears to have been introduced by mistake, as the nomenclature employed is the same. As regards generic nomenclature, the author adopts a middle course, avoiding the inordinate "splitting" followed by some ornithologists, as he does the excessive "lumping" favoured by others.

The first volume was devoted to the resident Passeres of the British Islands, and as the present commences with

the migratory members of the same order, it will be evident that the author does not confine himself to a strictly systematic arrangement. In excluding the casual visitors, which, in our own opinion, have no right whatever to the title of British Birds, the author differs from the plan followed by some of his brother ornithologists, whose object seems to be to draw up as long a list as possible, without any regard to the facts of geographical distribution. The other groups included in this volume include the Picarians, Owls, Hawks, and Pigeons, so that the Game Birds, Waders, and Water-Birds alone remain for its successor.

As a handsome, and at the same time an accurate, series of volumes for the drawing-table, the work may be heartily commended to all bird-lovers with whom "money is no object."

R. L.

OUR BOOK SHELF.

Our Native [American] Birds, how to protect them, and attract them to our homes. By D. Lange. Pp. x + 162. (New York: The Macmillan Co. ; London: Macmillan and Co., Ltd., 1899.)

LEST our readers should be misled into thinking that the present little volume is but another item in the already large literature of British ornithology, we have ventured to indicate its birth-place by a bracketed interpolation in the title.

The author, to whom the love of birds is evidently second nature, starts with the assertion that, with the exception of a few counties, the number of song-birds has of late years been steadily decreasing in the United States, and then proceeds to consider in detail—firstly, how this unfortunate state of things has been brought about, and, secondly, how it may best be remedied. Nor are song-birds alone considered, a certain amount of space being devoted to game-birds (inclusive of the *Anatidae*), many of which have likewise suffered severely.

The fact of the decrease in the former group seems to rest on conclusive evidence ; the main causes assigned being lack of suitable nesting-places, want of water and food, the abundance of cats (domestic and feral), the ravages committed by boys, collectors, and plume-hunters, the aggressive habits of the English sparrow, and the use of poison in gardens and farms.

As regards legislative protection, the author wisely leaves this to the various "Audubon Societies," which have been established in the States, and other suitable agencies ; devoting his attention mainly how to supply to his feathered friends such objects as are essential to their well-being, and how to guard them from the attacks of their chief foes. As our readers are aware, many towns and villages in the States are located on the open prairie, where the absence of cover renders the birds especially liable to destruction ; while even in districts more favoured by nature there seems to be a great tendency to make the gardens of residents as open and bare of shrubbery as possible. Old hollow trees, too, which form the nesting-places of so many species, have likewise been ruthlessly felled, so that the unhappy birds have literally no retreats wherein to hide.

Accordingly, the planting of trees, vines and shrubs (especially kinds which afford good cover and edible berries) is strongly urged, while beds of suitable kinds of flowers, such as gladioli, should be planted to attract humming-birds. For species building in hollow trees, nesting-boxes should be provided in suitable sites ; while drinking and bathing vessels should be furnished in the dry season, and abundance of suitable food at all times. The noxious sparrow is to be hustled out of the usurped nesting-places, while coils of barbed wire, or suitable wire fences, must be used to balk prowling cats. As to the best means of dealing with the human foes of birds, these, as already said, are mainly left to the powers that be ; but the formation of "bird-leagues," by members of the female sex who are willing to forego the ornamentation of their head-gear by the plumes of songsters, is strongly urged, as is the repression of the ordinary collector. Education, and the establishment of an annual "bird-day," are also regarded as important factors in the scheme.

The author has performed his task in a manner calculated to interest his readers, and his work should be acceptable to those on both sides of the Atlantic who love to hear bird-music around their homes.　　　R. L.

Der Ursprung der Kultur. Von L. Frobenius. Bd. i. Der Ursprung der Afrikanischen Kulturen. Mit 26 Karten, 9 Tafeln, sowie ca. 240 Text illustrationen. Pp. xxxi + 368. (Berlin : Gebrüder Borntraeger, 1898.)

THIS is the first volume of an ambitious work. The author proposes to seek out the Origin of Civilisation on what he considers to be a new plan. But in reality Mr. Frobenius can only work on the old lines ; he can only compare one custom with another, and use the same old weak argument from analogy to prove connection between tribes who have similar customs : " er lehrt alte Weisheit als neue." (" Programm " : p. xii.). He proclaims the virtues of his " new plan," however, in the very manner of the Teutonic *Gelehrte* : he considers himself to be laying the foundations of a new science (p. xiv.) : " Was bedeuten alle Entbehrung und Entsagung, wenn sie auch noch so herb sein mögen, gegenüber dem grossen Glücke, schaffend und Schöpferisch bei der Gründung einer Wissenschaft teilnehmen zu können. Ich habe die bitteren Stunden und herben Übel nie so stark empfunden, wie die Freude über die Erfolge, *das stolze Gefühl des selbständigen Schöpfers.* Und ich habe den herzlichen Wunsch, dass etwas von jener Spannkraft, die Müdigkeit und alle sonst vielleicht verzeihlichen und berechtigten Wünsche vergessen lässt, aus diesen Blättern dem Leser bemerkbar werden und in ihn übergehen möge." The italics are our own : we greatly fear that Mr. Frobenius, like so many of his " Fachgenossen," has no sense of humour. He does not forget to castigate his predecessors in ethnological study, some of whom are apparently prone to set fool's caps on their heads and give them out to be academical costume (p. ix.). The whole " Programm " which precedes the book is a typical product of what the author himself calls the " überhitzten Gelehrtenkopf " (p. ix.).

Apart from the rather ridiculous pretensions of its introduction, the book as a whole is useful enough as a series of essays on various phases of African ethnology, which are often very interesting, *e.g.* the chapter on building-styles (p. 194 ff.). They cannot, however, be said to prove much with regard to the origin of African civilisation, which is presumably what they are intended to do. The author's arguments in favour of his theory of the " Malayonigritish " origin of West African culture are interestingly put forward.

Absolutely nothing whatever is said about *Ancient Africa* : not a word with regard to the Zimbabye ruins, which we had expected to find exhaustively discussed here : not a word about the wonderful civilisation of Egypt, with the earliest beginnings of which we have now, thanks to the energy of Prof. Petrie and Messrs. Quibell, De Morgan and Amélineau, been brought into close contact, and which appears more and more African in character the further we go back. Not a single comparison of the Zulu and Egyptian head-rests even, to take the instance which first comes to mind ; but a curious misapprehension on p. 97, where Fig. 60 is described as a " Sceptermesser der Pharaonen," whatever that may be : the object in question is merely the well-

known and commonly-used Egyptian sword called *Khepesh* (on account of its resemblance to the shape of an animal's *thigh*, e.g. *khepesh*), which had nothing in particular to do with either Pharaohs or sceptres.

Of the illustrations, while the majority are good, some are certainly very bad, *e.g.* Plate iv. and Figs. 137, 139.

The Amateur's Practical Garden Book. (" The Garden Craft Series.") By C. E. Hunn and L. H. Bailey. Pp. vi + 250. (New York : The Macmillan Co. ; London : Macmillan and Co., Ltd., 1900.)

THE sub-title of this book very aptly indicates the nature of its contents, " The simplest directions for the growing of the commonest things about the house and garden."

The subjects dealt with are arranged alphabetically, beginning with Abobra and ending with Zinnia. It must not, however, be concluded that the book is merely a dictionary of plant names. It is much more.

Thus, under the heading " Annuals," we have an explanation of the term, the cultural details necessary for their proper growth, together with lists classified according to the colour of the flower, or the purpose the flowers have to serve.

The book is written for the climate of New York, but with the requisite modifications it is suitable for gardeners in this country also. It is severely practical, and principles, though perceptibly diffused, are not so much as mentioned.

Man and his Ancestor: a Study in Evolution. By Charles Morris. Pp. 238. (New York : The Macmillan Company. London : Macmillan and Co., Ltd., 1900.)

THE author has written this little book for the purpose of providing the intelligent person with a good and sufficient reason for the evolutionary faith that is in him. It is true that there is no book of a non-technical nature that quite covers the ground taken by the author, and it is only fair to him to state that he has filled this gap in a most creditable manner. It is obvious that many stages in the evolutionary history of man can only be guessed at by us, and that there is much room for discussion in these hypotheses as well as in the interpretation of accepted facts ; but Mr. Morris is not aggressively dogmatic, nor has he striven to be sensational. There are, however, several statements to which exception can be taken in the chapter on the " Vestiges of Man's Ancestry." If Mr. Morris thinks the function of the thyroid is a " minor and obscure one," let him have his own excised and then he will know. Club foot is not generally regarded as a reversion to the anthropoid foot. Taking it all round, the book may be safely recommended to that class of readers for whom it was intended, and it may lead such to consult the recognised works on the various topics on which he touches. Owing to no references being given, inquirers will have to seek elsewhere for an introduction to the literature of human evolution. The author has not considered his little book worthy of an index.

A First Geometry Book. By J. G. Hamilton and F. Kettle. Pp. ii + 91. (London : Edward Arnold, 1900.)

THIS little book contains a series of elementary exercises in geometry based on the method of allowing the pupil to deduce as many principles as possible after, and from the results of, experiments or exercises dependent on them. The deductions are drawn from the pupil's own measurements of his drawings to scale of the usual geometrical figures. From this it will be understood that the book really consists of a series of graduated exercises which appear to be well chosen and arranged, and likely to prove suggestive to teachers and useful to students beginning their first studies of the subject.

LETTERS TO THE EDITOR.

[The Editor does not hold himself responsible for opinions expressed by his correspondents. Neither can he undertake to return, or to correspond with the writers of, rejected manuscripts intended for this or any other part of NATURE. *No notice is taken of anonymous communications.]*

A Third Specimen of the Extinct "Dromaius ater," Vieillot; found in the R..Zoological Museum, Florence.

IN January 1803, a French scientific expedition, under Baudin, visited the coast of South Australia and explored Kangaroo Island, called by them "Isle Décrès." One of the naturalists attached to the expedition was the well-known F. Péron, who wrote an interesting narrative thereof. He noticed that Decrès Island was uninhabited by man, but, although poor in water, was rich in kangaroos and emus (*Casoars* he calls the latter), which in troops came down to the shore at sunset to drink sea-water. *Three* of these emus were caught alive, and safely reached Paris; we learn from the "Archives du Muséum" that one was placed in the Jardin des Plantes, and two were sent to "La Malmaison," then the residence of the Empress Josephine. We learn later that two of these birds lived to 1822, when one was mounted entire and placed in the ornithological galleries of the "Muséum," the other was prepared as a skeleton and placed in the comparative anatomy collections. No mention is made of the ultimate fate of the third specimen.

Péron was unaware that the emu he had found on the Kangaroo Island was peculiar and specifically quite distinct from the New Holland bird ; this was found out much later, and *too late* ;'for after Péron and his colleagues no naturalist evermore set eyes on the pigmy emu of Kangaroo Island in its wild condition ! It appears that when South Australia was first colonised, a settler squatted on Kangaroo Island and systematically exterminated the small emu and the kangaroos. When the interesting fact was ascertained that Péron's emu was a very distinct species quite peculiar to Kangaroo Island and found nowhere else, *Dromaius ater* had ceased to exist ; and the only known specimens preserved in *any* museum were the *two* mentioned above, in Paris.

For some years past my attention had been drawn to a small skeleton of a Ratitæ in the old didactic collection of the R. Zoological Museum under my direction ; it was labelled "Casoario," but was in many ways different from a cassowary ; but other work kept me from the proposed closer investigation, and it was only quite recently, during a visit of the Hon. Walter Rothschild, on his telling me that he was working out the cassowaries, that I remembered the enigmatical skeleton. A better inspection showed us that it is, without the least doubt, a specimen of the lost *Dromaius ater*. I afterwards ascertained that it had been first catalogued in this museum in 1833 ; that most of the bones bore written on them in a bold round hand, very characteristic of the first quarter of the nineteenth century, the words "Casoar mâle ;" and lastly, that during the latter part of Cuvier's life, about 1825-30, an exchange of specimens had taken place between the Paris and the Florence Museums. I have thus very little doubt that our specimen is the missing *third* emu brought alive to Paris by Péron in 1804-5.

This highly interesting ornithological relic is now on loan at the Tring Museum, and can be seen there by any ornithologist in England who may wish to examine it. I intend shortly to give a fuller notice of this valuable specimen.

HENRY H. GIGLIOLI.

R. Zoological Museum, Florence, May 15.

Chlorophyll a Sensitiser.

IT was with a feeling of great satisfaction that I read the concluding lines of Dr. H. Brown's highly interesting presidential address (NATURE, September 14, 1899), as I was glad to see that this distinguished chemist, to whom the physiology of plants is so much indebted, adopts certain views on the chlorophyll function, which I have been defending for more than a quarter of a century against the leading authorities of the German Physiological School (Sachs and Pfeffer). But since some slight errors seem to have crept into Dr. Brown's statements of my opinions on the subject, I may, perhaps, be allowed to bring forward the following corrections.

Dr. Brown seems to believe that the analogy between the action of chlorophyll and that of.a chromatic sensitiser was "first pointed out by Captain Abney" and "more fully elaborated " by me ; and secondly, that I give "a far too simple explanation of the facts" by admitting a *"mere physical* transference of vibrations of the right period from the absorbing chlorophyll to the reacting carbon dioxide and water."

To begin with the less important question of priority, I must confess that up to this date I am not aware of Captain Abney's claims. Had I known them, I should have been the first to acknowledge my debt to that accomplished investigator, whose brilliant achievements in this line of research I have never omitted to admire. The fact that the dissociation of the carbon dioxide in the green leaf is affected by the rays of light absorbed by chlorophyll was for the first time established by my researches in 1873, and an account of these experiments presented to the International Congress of Botany in Florence (May 1874).[1] At the same date (1873) Prof. H. Vogel made his important discovery of the chromatic sensitisers, and in November 1875, E. Becquerel applied it to the chlorophyll-collodion plates. . In May 1875 appeared my Russian work on the chlorophyll function, of which the French article[2] in the *Annales de Chimie et de Physique* of 1875, as expressly stated, is but an extract. In this French translation the idea that chlorophyll may be considered as a sensitiser is fully discussed. Consequently any claim of priority may be fairly advanced, only in favour of a paper having appeared in the short interval of a year—from May 1874, when I *announced the fact*, to May 1875, when I *interpreted it* in the light of H. Vogel's recent discovery. On consulting the *R. S. Catalogue of Scientific Papers*, I could not find any paper of Captain Abney's for this period 1874-1875.[3].

So far concerning the priority question. Passing to the second point, I am sorry to say Dr. Brown is decidedly in the wrong, for in my French paper just cited, and which probably escaped his notice, after discussing the quite recent discoveries of H. Vogel and Edmond Becquerel, I conclude : "Ou ne saurait pour le moment décider la question de savoir si cet effet serait dû uniquement à un phénomène physique, ou bien si la matière colorante prendrait part à la transformation chimique. Cette dernière manière de voir ferait rentrer l'action de cette matière (chlorophylle) dans la règle générale de l'action accélératrice des matières organiques dans les réactions photochimiques, car c'est généralement en absorbant les produits de la dissociation, effectué par la lumière, que les substances organiques détruisent cet equilibre qui tend à s'établir entre le corps décomposé et les produits de décomposition et c'est ainsi qu'une dissociation partielle aboutit à une décomposition complaite."[4] At a later date, in a report presented to the International Congress of Botany in St. Petersburg (1884), taking in account the subsequent photographical work on the sensitisers, I brought forward experimental proof that chlorophyll may be considered a sensitiser in Captain Abney's sense of the word : "La chlorophylle est un sensibilisateur régénéré à mesure qu'il se décompose et qui provoque en éprouvant une décomposition partielle la décomposition de l'acide carbonique."[5]

From all these quotations it may be inferred that I always kept in view the chemical aspect of the chlorophyll function, now advocated with such stress by Dr. Brown.[6]

But I did not content myself with such purely theoretical considerations, and ever since have been in search of what Dr.

[1] *Atti del Congresso Botanico tenuto in Firenze*, 1875, p. 108. At a still earlier date (*Botanische Zeitung*, 1869, No. 14), I found out the source of T. W. Draper's error, and proved that the process is chiefly due to the red rays of light.
[2] "Recherches sur la décomposition de l'acide carbonique dans le spectre solaire par les parties vertes des végétaux" (Extrait d'un ouvrage "Sur l'assimilation de la lumière par les végétaux," St. Petersbourg, 1875, publié en langue Russe) *Annales de Chimie et de Physique*, 5 serie, t. xii. 1877.
[3] Prof. Pfeffer, in his account of the whole subject ("Pflanzenphysiologie." Zweite Auflage, pp. 325-341), goes so far as to attribute this sensitiser theory of the chlorophyll function to Prof. Reinke, whose paper appeared ten years later.
[4] *L.c.* p. 40. In a footnote I add that certain physiological facts seem to agree with this point of view.
[5] "État actuel de nos connaissances sur la fonction chlorophyllienne" (*Annales des Sciences Naturelles Botanique*, 1885, p. 119).
[6] At a still earlier date (in a Russian work on the "Spectrum Analysis of Chlorophyll," St. Petersburg, 1871) I even expressed Dr. Brown's present point of view in the form of an equation :

$$XO + CO_2 = XCO + O_2$$
$$+ H_2O$$
$$= \overline{XO + CH_2O} + O_2$$

X being Dr. Brown's hypothetical "reduced constituent of chlorophyll."

Brown so appropriately terms the "reduced constituent of chlorophyll." My persistent endeavours resulted in the discovery of *protophylline*, a substance obtainable through the action of nascent hydrogen on chlorophyll solutions.[1] Some years later I discovered this substance in the living plant.[2]

The existence of a *reduced constituent of chlorophyll* may be consequently considered as a perfectly established fact, and will be probably brought to account by the chemical theory of the chlorophyll function. I conclude my French paper with the following words :—" L'étude de ces substances ne manquera pas à jetter une vive lumière sur le *co't chimique* de la fonction chlorophillienne *qui à été étudié dans ce dernier temps presque exclusivement au point de vue physique.*"

To sum up : though it may be clearly seen that for nearly thirty years I have been considering chlorophyll as a *chemical* sensitiser (or, strictly speaking, an *absorbent* of the products of dissociation of CO_2 and H_2O), still even now I must confess that this theory lacks direct experimental proof and may be considered only as a matter for further research, whereas the physical aspect of the question (*i.e.* that CO_2 and H_2O are decomposed through the agency of those rays of the spectrum, which are absorbed and somehow transformed by chlorophyll) is but the expression of a fact, put beyond any doubt by my researches, both on the decomposition of CO_2 and on the production of starch in the living plant.[3] But I do not abandon the hope that the discovery of the *protophylline* may turn out some day to be a step in the direction of a *chemical theory* of the chlorophyll function, somewhat similar to that of the colouring matter of the blood—an analogy which has been present to my mind ever since I became acquainted with the classical researches of Sir G. G. Stokes in that direction.

University, Moscow.	CLEMENT TIMIRIAZEFF.

I REGRET that M. Timiriazeff should regard the concluding lines of my presidential address as doing him some injustice.

No one can be more impressed than I have been with the extreme beauty and importance of M. Timiriazeff's work, which cleared away many illusions, and for the first time prominently brought out the fact that the rays corresponding to the principal absorption band of the chlorophyll spectrum are those which are mainly active in the assimilatory process.

I have always regarded M. Timiriazeff's paper of 1885 (*Ann. des Sciences Nat.* [*Bot.*], vol. ii. p. 99) as being one of the most convincing and eloquent expositions in scientific literature, and the final proof of the proposition there laid down was given by the author in 1890 (*Compt. rend.* 110, 1346), when he succeeded in showing that the reappearance of starch in a depleted leaf exposed to a pure spectrum only takes place in the region of the red corresponding exactly to the principal absorption band of chlorophyll.

With regard to the first point raised in M. Timiriazeff's letter, I may say that when preparing my address I experienced a difficulty in ascertaining who it was that first drew attention to the existing analogy between chlorophyll and a chromatic sensitiser.

There is no complete list of Sir William Abney's papers, and knowing that he has sent many communications on this and cognate subjects to photographic journals in various parts of the world, I applied to Sir William Abney before writing what I did. There can be no doubt that chromatic sensitisers were very much "in the air" immediately after Vogel's discoveries of 1873, and it is probable that the application of these new ideas to chlorophyll occurred independently to Abney, Timiriazeff and Becquerel.

M. Timiriazeff's second objection is that I have not sufficiently taken into account his views of the function of chlorophyll as a *chemical sensitiser*. On this point I may say that I had in view his paper of 1885 : " État actuel de nos connaissances sur la fonction chlorophyllienne," which it was fair to imagine fully embodied the author's view up to that date. It is certainly the *physical rôle* of chlorophyll which is there insisted upon, as the following quotation indicates : " Le rôle de la chlorophylle dans le phénomène de la décomposition de l'acide carbonique peut donc être résumé ainsi : elle absorbe les radiations qui possèdent

[1] The first description of this curious substance was given in two short notes communicated to these columns : "Colourless Chlorophyll" (NATURE, 1885, p. 342) and "Chlorophyll" (NATURE, 1886, p. 52). For more ample details, see *Comptes rendus*, 1889.
[2] "La protophylline dans la plante vivante" (*Comptes rendus*, 1889).
[3] "Enregistrement photographique de la fonction chlorophyllienne par la plante vivante" (*Comptes rendus*, 1884).

la plus grande énergie et transmet cette énergie aux molécules de l'acide carbonique qui, à elles seules, n'éprouveraient pas de décomposition, étant transparentes pour ces radiations énergiques."

That the physical conception was certainly uppermost in M. Timiriazeff's mind at that time is further shown by the diagram and remarks immediately following, in which he regards the molecules of carbon dioxide as suffering "shipwreck" in the luminous undulations corresponding to maximum amplitude.

It is, however, quite clear from M. Timiriazeff's references to his paper of 1877, and especially to his Russian paper of 1871, neither of which I have seen, that he has expressed views which are practically identical with those contained in the concluding remarks of my address. It is to be regretted that these ideas were not again clearly brought forward in the 1885 paper, which purported to give the author's latest views on the whole question, and that the physical idea of the immediate transference of the energy of radiation was there made the dominant one.

52, Nevern Square, Kensington.	HORACE T. BROWN.

A Simple Experiment on Thermal Radiation.

THE following experiment, which has been successfully performed by our students for several years, may be of interest to teachers of physics.

Three chemical thermometers are chosen of equal size and shape. The bulb of one is silvered, of the other covered with dead black paint by dipping it into a mixture of lamp-black and alcohol, whilst the third is left unchanged. For silvering, any of the well-known solutions and processes will be applicable. The thermometers indicate the same temperature if there is no source of radiation near them.

But if a gas flame, for example, an Argand burner, be placed at a distance of 20 centimetres from them, so that the thermometers, hanging from a stand, are at equal distances from the flame, the temperature rises at a different rate, and to a different, though in each thermometer constant, height. The silvered

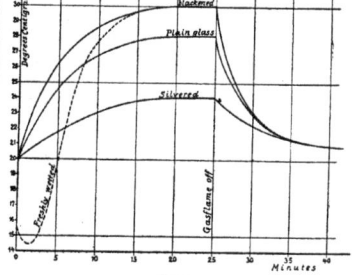

FIG. 1.

thermometer gives the lowest reading, and the blackened the highest, whilst the thread of the uncovered one stops at some point between these readings nearer to that of the blackened than the silvered ; for the different surfaces of the thermometers absorb the radiation of heat generated in the flame in different proportion. The blackened thermometer bulb almost completely absorbs the rays falling on it ; the silvered and polished bulb reflects the radiation reaching it ; the plain glass bulb partly reflects and partly absorbs the rays. Thus, none but the silvered bulb thermometer indicates the temperature of the air communicating heat to it by conduction. As the other thermometers rise in temperature, they emit radiation ; and when the amount of heat emitted from them equals the amount received through radiation from the gas flame, they are in the final stationary state, which is, of course, reached by the thermometers at different temperatures.

If the gas flame is put out, the temperatures of the three

thermometers fall at different rates. Observations made simultaneously on them every minute, and plotted on squared paper, illustrate fairly well Kirchhoff's law enunciating that a body emits those rays best which it absorbs best. When the gas flame is replaced by a freezing mixture, the greatest fall of temperature is experienced by the blackened thermometer, and the least by the silvered one, for the same reasons.

The same arrangement of thermometers may be used to show the cold produced by evaporation. For this purpose, the bulb to be blackened has to be wetted immediately before starting the radiation experiment. First the temperature of this thermometer falls, even though the gas flame be lighted, but after a few minutes its temperature rises very quickly to reach the same state of equilibrium as when taken with dry black paint. In Fig. 1 the dotted curve represents the behaviour of the freshly wetted blackened thermometer.

For silvering the thermometer bulbs, we use most successfully the process described first by A. Martin in *Poggendorff's Annalen* (cxx. 1863, p. 335), and reprinted in many of the books on practical physics.   K. T. FISCHER.

München (Bavaria), Kgl. Technische Hochschule.

THE TOTAL ECLIPSE OF THE SUN.

THE last total solar eclipse of this century appears to have been successfully observed all along the line of totality. The weather conditions were favourable at all the observing stations, and numerous photographic and visual observations have been made of the phenomena revealed during a total eclipse. Elaborate arrangements were made to study the eclipse in all its aspects, and it has fortunately been possible to carry them out in a most satisfactory manner.

A code telegram received at the Solar Physics Observatory, South Kensington, from Sir Norman Lockyer, states :—"At the time of the eclipse the weather was excellent, and all the instruments were satisfactorily employed. There was a fall in temperature during the eclipse of from 4° to 6° C. The eclipse was not a dark one, and very few stars were seen. The corona exhibited large equatorial extensions and distinct polar tracery as expected. Observations of the shadow band were fully made in two planes. In the fixing up of the instruments, and the making of the observations, assistance was given by about 150 of the officers and crew of H.M.S. *Theseus*, which conveyed the eclipse party from Gibraltar to Santa Pola, a few miles south-west of Alicante."

The corona was similar to those observed during the eclipses of 1878 and 1889—both epochs of sunspot minimum—and thus supplies additional support to the probability of a real connection between the coronal structure and the state of solar activity. There were two long equatorial streamers, the western one being bifurcated and extending about two solar diameters. Several observers note that the inner corona was visible for at least *five* seconds after totality.

The eclipse was a short, and therefore a bright, one, which accounts for the general report that no shadow was seen either on land or in the atmosphere, and that very few stars were visible. Mercury and Venus were, however, observed. All the reports agree in estimating the duration of totality as shorter than was expected, so that the lunar tables will need slight revision for future computations.

Important observations were made of the shadow bands, which are stated to be very different in many respects from those previously observed. From one of the American stations it is reported that the bands were about one inch in breadth, their general direction being south 56½° E. ; before totality their motion was at right angles to this—that is, almost north-east ; and in the opposite direction after totality. Superposed on the linear bands, however, were certain dark patches previously unnoticed, having a motion at right angles to that of the bands. Baily's beads were well seen at the instants of second and third contact.

Prof. Todd, at Tripoli, is reported to have successfully employed twenty photographic cameras, one of which was furnished with a lens of 24 inches aperture.

The party at Pinehurst, from the U.S. Naval Observatory, under Prof. Skinner, obtained a good series of spectrum photographs, including five with plane and concave gratings and four with an objective prism ; also five large scale photographs of the corona with a lens of forty feet focus.

Prof. Pickering obtained a good series of photographs with the new large instrument he had specially made for searching for an intra-Mercurial planet.

As we go to press, the following description of the observing parties at Santa Pola has been received from Sir Norman Lockyer.

PREPARATIONS AT SANTA POLA.

Santa Pola, Friday, May 25.

The party from the Solar Physics Observatory arrived here on May 17, and now, thanks to the assistance so freely rendered by the Spanish authorities of all grades, and the strong working parties furnished by H.M.S. *Theseus*, the instruments are all in order and we are ready for the eclipse.

At Gibraltar the Captain of the *Theseus* sent off Mr. Daniels, torpedo gunner, to meet the Expedition, and the sixty-nine cases of instruments were carefully transferred to a lighter, and so soon as they were landed here those belonging to each instrument were at once brought alongside the piers which had already been erected for them on a site as near the landing stage as possible, thanks to the diligence of Mr. Howard Payn, a volunteer assistant who had preceded the party by rail and had secured the necessary bricks at Alicante.

The prismatic cameras, and those of the ordinary kind, fed by cœlostats and siderostats, with all prisms and mirrors, were in adjustment by the 21st, and drills were begun on the 22nd.

The parties of observers are as follows ; and careful notes of the arrangements made are being kept, as some improvements have been made on those adopted in 1898.

Parties on Shore.—Prismatic cameras. (1) One prism (20 ft.) ; (2) two prisms (7 ft. 6 in.). Coronagraphs. (3) Graham (f. 6·5) ; (4) Dallmeyer (f. 8·0 about) ; (5) De la Rue (f. 17·5) ; (6) long focus (f. 48). (7) Discs. (8) Shadow bands. (9) Meteorology. (10) Stars. (11) Landscape colours.

Parties on Board.—(1) Stars. (2) Shadow. (3) Meteorology. (4) Landscape.

The whole party is in robust health, thanks to the glorious climate and any amount of work in the open air. We live in a little inn, which since the Queen's birthday has blossomed into the "Victoria Hotel," kept by one Frasquito Dols, a Spanish sailor and sea-cook, a regular "handy man," who has put up mosquito curtains, and rigged up a lift to carry our well-cooked food and excellent local wine to the first floor where we reside ; in rooms which, though furnished with unparalleled simplicity, are absolutely clean. It seems a pity that more do not know of this delightful climate so near home, in which the winter months may be so pleasantly spent in the shadow of date palms.

The ship is much further away from the shore than in 1898—some 2000 yards—and the winds rise very suddenly in the open roadstead. The administration of the camp, therefore, devolves upon Lieut. Doughty, R.N., who, with Lieuts. Andrews and Patrick, remain constantly on shore in a pile-dwelling—a bathing establishment which has seen better days, and has been rechristened "Theseus Villa."

The "Scotch Commission," as it is called here—that is,

Dr. Copeland's party—has chosen a site up the hill behind the town some distance from the jetty.

Elaborate arrangements have been made for the observation of shadow bands, two walls, E. and W. and N. and S., composed of first-class volley targets 16 × 6 ft., having been erected on a level space which has been whitewashed.

Six discs have been set up on spars, and most careful drills have taken place. I have been quite astonished at the exact reproduction of all the features of a dummy corona set up on each occasion.

It appears that the east wind is the best for us, and it is blowing now ; a cloudy morning generally is followed by a cloudless sky in the afternoon. The weather chances are good, but they are not perfect.

NORMAN LOCKYER.

FIFTY YEARS OF GEOLOGICAL SURVEY IN INDIA.

THOUGH the Honourable East India Company had showed their interest in the advancement of geological science by the appointment, so long ago as 1818, of a geologist to the Great Trigonometrical Survey, it is but fifty years since the first "Report of the Geological Survey of India for 1848-49," by Dr. John McClelland, was published. In 1851 Dr. McClelland was relieved by Dr. Thos. Oldham, who, on his arrival in Calcutta, found the Geological Survey represented in the capital of India by a room, a box and a messenger. One assistant, Mr. W. Theobald, was already in the employment of the Company, and during the following five years seven assistants were appointed, of whom but Mr. H. B. Medlicott and Dr. W. T. Blanford, names cut deep in the record of Indian geology, survive.

It was not, however, till 1856 that the Geological Survey was established as a regularly organised service, with a sanctioned establishment of superintendent (now styled director), fifteen graded assistants and a palæontologist. In spite of the increased area over which British rule extends, the establishment sanctioned in 1856 remained the same, with some minor, temporary changes, and an alteration of nomenclature, till 1892, when, instead of an increase, the permanent staff was reduced by three, and to compensate for this reduction arrangements were made for the employment of two "specialists" for terms of years, who were expected to devote their services more especially to economic geology. From one cause and another, this scheme has not received a full trial yet, and it is only during the present year that the full sanctioned staff is at work. The experiments so far made in the temporary employment of assistance to the Geological Survey, for the special purpose of economic work, cannot be regarded as successful, and the result of the present trial will be watched with interest, as it is likely to have great influence in the shaping of the future course and policy of the Survey.

The concrete results of less than half a century's work with this inadequate staff are a geological map of nearly the whole of India proper, which is accurate as regards its main features for this large area, and as regards details for a large proportion of it ; and a considerable acquaintance, largely accompanied by maps, with the mountainous country to the north-west and north, and of the countries to the east, which are included in the Indian Empire. The published results are contained in thirty volumes of the "Records," twenty-nine of the "Memoirs," and twenty volumes, not counting those only partly published, of the "Palæontologia Indica."

Besides this collection of separate memoirs there was prepared, with the approval and sanction of the Government of India, a "Manual of the Geology of India," in two volumes, by Messrs. H. B. Medlicott and W. T.

Blanford, published in 1879, to which were subsequently added a volume on the "Economic Geology," by the late Dr. Valentine Ball, and one on the "Mineralogy," by Mr. F. R. Mallet. These volumes contained not only much information collected by the Survey, which it had not been possible to publish previously, but for the first time, by collecting scattered information into one general review, made the geology of India generally accessible and intelligible. The need for, and value of, these volumes is shown by the fact that they soon went out of print, and in 1894 a revised version of the first two volumes was issued. The progress of the Survey in the period intervening between these two issues had been so great that a totally different scheme could be adopted, and instead of the series of separate descriptions of isolated areas, which was to a large extent inevitable in 1879, it was possible to treat the geology of India as an harmonious whole in 1894. A re-issue of the third volume of the original edition, the volume on "Economic Geology," has also been commenced, but though nominally a re-issue, it is, even more than in the case of the stratigraphical and structural geology, a new book, being different in scope and in aims, and containing no part of the original work.

The results of the Geological Survey, apart from its publications, are to be looked for both in India and out of it. In India, in the economic development of the Empire ; and out of India, in the influence they have had on the advancement of geological science. The former of these is naturally that to which the Administration attaches the greater importance, and in this connection the existence of the Survey is amply justified in the fact that two of the coal-fields, which yield an important part of the coal-supply of India, were discovered and explored by the Geological Survey. Singarenni, surveyed by Dr. W. King, and Umaria, by Mr. T. W. Hughes, have, from their geographical position, a much greater importance than would appear merely from a numerical statement of the number of tons of coal raised in them, for they serve to supply a large area with cheap fuel which would otherwise be deprived of that advantage. These two fields in themselves would justify the existence of the Survey, from an economic point of view, apart from other benefits ; but besides this the existence of a band of trained advisers, and of the observations accumulated by them, has frequently been instrumental in preventing the useless expenditure of large sums of money, and in this way alone the Survey has rendered an ample return for its maintenance.

Though the Administration is naturally most interested in the economic aspects of the work of the Geological Survey, there has never been any attempt to convert it into a mere prospecting or mining department. The Government of India has always recognised purely scientific work as an important duty of the Survey, and regarded the advancement of science not only as a thing to be desired and encouraged on its own account, but as furthering and rendering more valuable the economic results of the Survey, by improving the instrument with which it works. It is this portion of the work of the Survey which is of the greater interest outside India, and more especially to the readers of NATURE.

First among the results which have influenced the course of geological science may be placed the recognition of the importance of deposits formed on land, in which Indian survey took an early and important part. It was shown that the Gangetic alluvium, formerly looked upon as a marine deposit, was, as regards its upper layers at least, a land deposit ; it was shown that the great series of sandstones and conglomerates of which the foot-hills of the Himalayas are composed were formed, not in the sea, but on land, by rivers which were the ancestors of those now draining the Himalayas ; the great Gondwana system was shown to be exclusively a dry land deposit, and later the same origin was attributed to the great Vindhyan system.

In the next place we have the recognition of the Permian glacial epoch. The first description of these beds was published in the *Memoirs* of the Geological Survey in 1856, and their glacial origin proclaimed in 1875 by the late Mr. H. F. Blanford. Though the idea of glaciation in Permian times and in what are now low latitudes has met with great opposition, it has gradually made progress, and it is now generally recognised that the Permian boulder-beds of India, though extending into regions that are now within the tropics, are relics of a bygone glacial epoch. In Africa, the glacial origin of similar beds has been accepted by more than one observer; and in Australia—where the traces of glacial action in the marine Permian or Permo-Carboniferous beds, below the principal coal-measures, was first recognised by a member of the Indian Geological Survey who had been deputed by the Indian Government to study the Australian coal-measures—the existence of glacial action on a large scale has been fully confirmed by workers in that country. In South America, too, it seems that there are similar beds, of apparently the same age, and the evidence of this widespread glacial epoch, more remarkable in many ways even than the post-Tertiary extension of glaciation, must be reckoned with in any speculations attempting to account for the great climatic changes of which the past sediments bear witness.

The labours of the Indian Geological Survey have had important results in geological science in other minor points, too numerous to detail in the limited space of an article, but a mention of the great earthquake of 1897 cannot be omitted. This earthquake was the greatest of which there is historic record, exceeding the great Lisbon earthquake of 1755; but even before this was known the Indian Government had ordered the Survey to make a complete scientific investigation of it. Being the greatest earthquake of which there is historic record, the visible effects were on an unprecedented scale, and its investigation has consequently yielded results which must be taken into account in all future seismological research. Nor must mention be omitted of one of the most recent suggestions, which appears likely to be fruitful of results, made in 1898 by Mr. T. H. Holland, that much of the decomposition, and more especially hydration, of the minerals composing igneous rocks was submarine, and that the undecomposed state of similar rocks, even of perishable minerals like olivine and nepheline, in certain regions, is due to these being ancient land-areas which have not been submerged beneath the sea since a remote geological period.

Such, briefly stated, is the record of the Geological Survey of India, a record which reflects credit on all who have been concerned in the making of it. Yet it must not be forgotten that credit is due also to the Civil Administration of India, which has not only maintained the staff by whom the record has been made, but has given the further pecuniary assistance, modest in amount but steadily continued, which has enabled the Survey to form a museum fully illustrating the geology of India in all its branches, to establish a well-equipped laboratory, and to collect a library which, as a geological working library, is probably unsurpassed by any and equalled by few.

NOTES.

M. DARBOUX, Dean of the Faculty of Sciences of Paris, has been elected permanent secretary of the Paris Academy of Sciences, in succession to the late M. Joseph Bertrand. Prof. J. Willard Gibbs, professor of mathematical physics in Yale University, has been elected a correspondant of the Academy in the section of mechanics. Prof. J. Chatin, assistant professor

of histology at the Sorbonne, has been elected a member of the section of anatomy of the Academy, in succession to the late M. Blanchard.

THE recommendations of the international conference which recently met in London to determine the steps which might usefully be taken for the preservation of wild animals, birds and fish in South Africa, have now been published as a Parliamentary Paper. The zone within which it is proposed to apply the provisions of the Convention is bounded on the north by the 20th parallel of north latitude, on the west by the Atlantic Ocean, on the east by the Red Sea and by the Indian Ocean, on the south by a line following the northern boundary of the German possessions in South-Western Africa, from its western extremity to its junction with the River Zambesi, and thence running along the right bank of that river as far as the Indian Ocean. To preserve the various forms of animal life existing in a wild state within this zone, it is proposed to prohibit the hunting and destruction of certain animals, especially females when accompanied by their young or capable of being otherwise recognised, of which the protection, whether owing to their usefulness or to their rarity and threatened extermination, may be considered necessary by each local government. The establishment, as far as it is possible, of reserves within which it shall be unlawful to hunt, capture or kill any bird or other wild animal except those specially exempted from protection by the local authorities, is recommended, and also of close seasons with a view to facilitate the rearing of young. It is proposed to put export duties on the hides and skins of giraffes, antelopes, zebras, rhinoceroses and hippopotami, on rhinoceroses and antelope horns, and on hippopotamus tusks, and to prohibit the hunting or killing of young elephants. Measures are to be taken for ensuring the protection of the eggs of ostriches, and for the destruction of the eggs of crocodiles, of those of poisonous snakes, and of those of pythons. It is, however, understood that some of the principles laid down may be relaxed, either in order to permit the collection of specimens for museums or zoological gardens, or for any other scientific purpose.

PROF. J. PERRY, F.R.S., has been elected president of the Institution of Electrical Engineers for the session 1900-1901.

MR. BORCHGREVINK, who recently returned from his explorations in the Antarctic, will, it is expected, give a lecture before the Royal Geographical Society on June 18.

THE American Academy of Arts and Sciences has decided to award the Rumford medal to Prof. Carl Barus, of Brown University, for his researches in heat.

WE learn from *Science* that the Committee of Coinage, Weights and Measures of the U.S. House of Representatives has unanimously agreed to report as an amendment to the Sundry Civil Bill the measure establishing a United States Standardising Bureau, referred to in NATURE of May 17 (p. 61).

WE regret that a part of the edition of last week's NATURE appeared without the announcement that the names of Dr. D. Gill, F.R.S., and Dr. T. E. Thorpe, F.R.S., were included in the list of Birthday Honours. The former has been promoted to the rank of K.C.B., and the latter has been created a C.B.

THE third Liverpool expedition for the study of tropical diseases, referred to last week, will start in the first week in July. The members of the expedition are Drs. Durham and Walter Myers. The object of the expedition is to study yellow fever, malaria and dysentery.

AN excursion to Malvern and district has been arranged by the Geologists' Association for Whitsuntide. The director will be Prof. T. T. Groom, and during the stay at Malvern, from Saturday, June 2, to Tuesday, June 5, a number of interesting geological sections and structures will be examined.

A MEETING of the Yorkshire Naturalists' Union will be held at York on Whit-Monday for the investigation of the natural history of Askham Bog, and for the geological investigation of the morainic ridges of Askham and Bilbrough. Askham Bog is one of the very few undrained spots left in the Vale of York ; hence the naturalist values it much as the palæontologist values one bone of an extinct animal, for from it he can draw such a true and interesting picture of a stage in the development of the district.

A FISHERIES Exhibition will be held at Salzburg, Austria, on September 2, and the eight following days. The exhibits are divided into nine classes, and include sections for artificial breeding apparatus, preserving methods, tackle, and the literature and statistics of fishing.

THE *Times* announces that the appointment of the commanding officer of the National Antarctic Expedition has been made by the joint committee of the Royal and Royal Geographical Societies. The officer selected is Lieut. Robert F. Scott, now torpedo-lieutenant of the *Majestic*. He has been fifteen years in the Navy, has a record of service of the highest class, and will shortly be promoted to commander. The head of the scientific staff will be Dr. J. W. Gregory, recently appointed professor of geology in the University of Melbourne. Though he has only just entered upon his duties at Melbourne, the authorities have granted him leave of absence to serve with the Antarctic Expedition. He will come to England in October to prepare for his new work.

AN exhibition of photographs, by Dr. P. H. Emerson, will be open at the Royal Photographic Society, 66 Russell Square, W.C., from May 30 until June 30.

BY the will of the late Prof. Piazzi Smyth, the executors are instructed to repay to the Government Grant Committee of the Royal Society all of the advances, estimated at 300*l.*, made by the Society to Prof. Smyth for the purchase of scientific instruments after he went to Ripon. The will bequeaths to the Royal Society of Edinburgh the portrait of Prof. Smyth, by Faed, R.S.A., and all his books of original drawings and journals, and his boxes of glass photographs. The residuary estate is to be in trust for certain legatees for life, and subject to their life interest for the Royal Society of Edinburgh if agreeable to receive the same as a trust, whereof the income is to be employed by that Society, first, in printing for a limited free distribution and a small sale to the public, at a cost of about 600*l.*, the spectroscopic MSS. offered by Prof. Smyth to the Government in October 1857, and then to assist or promote every ten or twenty years an exceptional expedition for the study of some particular branch of astronomical spectroscopy in the purer air of some mountain elevation of not less than 6000 feet above the sea-level, as tried and found feasible by him in the first experiment on the Peak of Teneriffe in 1856. If the residuary estate is not accepted by the Royal Society of Edinburgh, it is to be distributed amongst the pecuniary legatees.

LAST week the Royal Horticultural Society held its thirteenth "Temple Show." In every respect, apart from the uncertainty of the weather, the great annual exhibition more than fulfilled the expectation of lovers of flowers and of horticulturists generally. On the other hand, the botanist was greeted by no species that was not already known. The student of evolution might, nevertheless, have made the acquaintance of many new artificial races, and hours might have been spent in examining fresh garden "varieties," produced by hybridisation and cross-breeding. Even when some striking variation has been chanced upon, and "fixed" by careful selection, judicious crossing may be resorted to, in order that further "improvements" may be brought about. To take a case,

Messrs. Laing and Sons showed some begonias, in which the development of a "crest" or tuft of small outgrowths from the petals was very much marked. This appeared sporadically and slightly at first in a plant with flowers of the same colour as those of its parents, but since the establishment of the crested race it has been crossed with others, and now crested petals may be had of many tints. The cactus-flowered zonal pelargonium may be mentioned on account of its vivid colouring and numerous narrow petals. Its rearer, Mr. E. S. Towell, obtained it from the seed of a "semi-double" *Pelargonium*, which he crossed with pollen from many different flowers. Among these was that of *Lychnis chalcedonica*; and Mr. Towell, though not absolutely certain of the fact, considers that the last-named species is the father of his "Fire Dragon." The particular tint of scarlet shown by the petals, the time these persist, and their divided appearance favour this view.

THE Sugar-Beet Committee of the Central Chamber of Agriculture have completed arrangements for a limited number of experiments in the growth of sugar-beet during the forthcoming season, each experimental plot being at least one acre in extent. In all, there will be about thirty-three different experiments, of which twenty-five are situated in England, four in Scotland, and four in Ireland. The English counties in which one or more experiments will be made are Wilts, Hants, Berks, Oxon, Beds, Kent, Suffolk, Hereford, Worcester, Warwick and Lancaster. As previous experiments have, in certain cases, demonstrated the value of sugar-beet for the feeding of live stock (independently of its value for the manufacture of sugar), it has been decided to keep this point specially in view in connection with the experiments of the present year.

AN interesting feature of the Paris Exposition is the elevated moving pavement. The line, which is described in the *Scientific American*, forms a complete circuit, running along the side of the Champ de Mars, the Quai d'Orsay, the Esplanade des Invalides and the Avenue de la Motte-Picquet, the total length of its course being 3500 metres. The platform is supported on an elevated structure, to which access is given from a number of stations situated within the Exposition grounds. The substructure supports three platforms, one fixed and two movable, these having a speed of eight and four kilometres per hour. To enable the platform to pass around the curves, the different sections are dovetailed into each other by large circular portions, forming a kind of horizontal hinge. Each of the platforms carries an I-beam running along under the centre ; these rest upon a series of rollers placed at intervals, operated by electric motors. Upon the shaft of the motor is mounted a large roller for the high-speed platform and a roller of one-half the diameter for the slow speed. The friction of the platform is sufficient to cause its adhesion to the rollers. The platform was put into operation on April 14, and has proved a great success, as by its means an easy passage through the grounds is afforded, as well as a series of interesting views. The tour is made in twenty-six or fifty-two minutes.

PARTICULARS of the short electric line—about 5000 feet in length—between Earl's Court and High Street, Kensington, which has just been opened on the Metropolitan District Railway, are given in the current number of the *Electrician*. The engineers, Sir John Wolfe Barry and Sir W. H. Preece, were required to equip this line electrically without any interference with the permanent way, without any interference with the running of the ordinary train service, and without allowing any electric current to pass through the permanent way or the subsoil, lest such should interfere with the signalling arrangements of the line. In accordance with these stringent regulations, it became necessary to adopt an insulated system throughout, and to do the whole of the construction work in the few midnight hours when the trains were not running. The system may be termed a four-rail system. It includes the two ordinary track rails, which are not used for any electrical purpose whatever, and two electrical rail conductors placed on either side outside the track. A special type of train has been designed for the line, the design being such as to adapt it specially to the experimental conditions. There is no separate locomotive, the train being worked in block, and a motor carriage being placed at either end. Only one motor carriage, however, is used at a time—viz. that one in the front in the direction in which the train is moving. This arrangement, while duplicating the amount of electric motor plant, is convenient, as it obviates shunting the motor carriage. It is intended to carry out a series of careful experiments on electric traction upon this line, and for this purpose a dynamometer car will sometimes be attached to the train. Already certain experiments have been made. In his evidence before the Select Committee of the House of Commons considering the Manchester-Liverpool Express Railway, Sir William Preece recently stated that the train, fully loaded, had started on the very difficult gradient of 1 in 43—a feat which an ordinary steam locomotive was unable to perform when hauling a similar load. Moreover, in a tug-of-war between the electric train and a steam locomotive, the electric train readily overcame the steam engine.

IT has been said that every person is mentally a little unbalanced, and that education from this point of view is simply the attempt to secure and maintain mental equilibrium, which, however, is never actually attained. Lapses of thought, inadvertencies in expression, and other slips in speaking or writing (*lapsus linguae* and *lapsus calami*) are thus of interest to the psychologist as useful guides to the understanding of mental processes. Every one has experienced unaccountable lapses of this kind, and the lapse often comes as a surprise to the speaker or writer himself. During a lecture, a professor inadvertently referred to the "tropic of Cancercorn," intending to say "the tropics of Capricorn and of Cancer." Many similar instances might be cited, for example, the man who was going for a walk to "get a breash of freth air," the person who inquired for the "portar and mestle," and another who said "the pastor cut the shermon sort." A physicist is recorded to have said that he feared he should "get the instrument out of needle," when he intended to say he feared he would "get the instrument out of level and deflect the needle." This is curious, but it is not so amusing as the order of "beggs and acon" for breakfast, or the remark of a nervous churchman to a stranger in his seat, "Excuse me, but you are occupying my pie." Mr. H. Heath Bawden has made a detailed study of similar mental lapses, both oral and graphic, and his results are described in a monograph of the *Psychological Review*. It is suggested that the aberrations dealt with are due to incipient aphasia or agraphia, and the similarity between them is held to show that our ordinary experience borders at every point on what is called the abnormal or pathological condition.

FOR some time past peat has been largely used in this country as litter for stables in the place of straw. This material is now likely to have a much more extended use, and the peat bogs of this and other countries made to assume a value never before realised. For the past twelve years Herr Zschörner, of Vienna, has been investigating the properties of peat, and has shown its possibilities. In the Vienna Exhibition of last year was a building in which everything, from the carpets on the floor to the curtains on the windows, and the paper on the walls, had all been made from peat. Herr Zschörner's investigations have shown that, although the fibres of the remains of the reeds and grasses of which peat is composed have become

altered in their physical and chemical character, yet they have not suffered any anatomical change; and while nothing capable of fermentation or decay is left, the fibrous structure remains intact; that they are very durable, elastic, good non-conductors of heat and non-combustible. Fabrics woven from them are found to have the toughness of linen with the warmth of wool. There is no textile fabric that cannot be woven from these fibres. Blankets and other coverings used for horses and cattle have been found in use to excel in warmth and cleanliness. The unspun fibre is found to be a good substitute for absorbent cottons possessing strong antiseptic properties. Paper of several qualities has been made, and the uses to which peat fibre has already been applied indicate possibilities that may render the peat bogs of Ireland a valuable addition to the resources of that country, and give full occupation to the inhabitants of the "congested" districts.

THE *Rendiconto* of the Naples Academy for March and April contains a complete list of the mathematical works of the late Prof. Beltrami.

IN the *Bulletin de la Classe des Sciences* of the Belgian Academy, M. Vandenberghe continues his researches on the dissociation of substances in solution. The author, by new experiments conducted with the use of solvents belonging to the same homologous series, establishes the conclusion that the influence exerted on the decomposition of molecular associations by the solvent does not materially influence the effects due to elevation of temperature.

A PRELIMINARY note on the magnetic observations made during the *Belgica* Antarctic expedition is given by M. G. Lecointe in the *Bulletin de la Classe des Sciences* (Brussels). For the measurements of declination Neumayer's apparatus was used, the declination being the difference between the magnetic azimuth of a star and the true azimuth calculated from the local time. The Neumayer apparatus was also found suitable than the theodolite for measuring the horizontal component, the instability of the theodolite as its feet began to sink into the ice rendering observations made with it of little value. In determining the inclination the great sensitiveness of Gambey's compass could not be utilised regularly on account of the ice-movements, and here again Neumayer's apparatus proved the most serviceable. The paper consists chiefly of a table of the recorded observations.

IN his Wilde Lecture, published in the *Manchester Memoirs*, 1899, No. 5, Lord Rayleigh discusses the mechanical principles and possibilities of flight, both natural and artificial. The problem of the sailing bird is treated from the three alternative points of view, which attribute its source of energy to upward currents, variation of wind-velocity with the altitude and pulsating gusts of wind. Lord Rayleigh then considers the law of dependance of the aerial resistance of a plane surface on its obliquity, and describes experimental methods whereby the resistances at different obliquities may be compared by an "astatic" arrangement, in which pairs of vanes are so adjusted that the moments of the oppositely turned vanes balance each other. In connection with the expenditure of power required to support a given weight, Lord Rayleigh has calculated that, in order for a man to support himself by a *vertical* screw by working at the power an average man can maintain for eight hours a day, he would require a screw ninety metres in diameter, and in this estimate no account has been taken of the weight of the mechanism or of frictional losses. In conclusion, the effects of flapping wings are briefly discussed.

A FURTHER addition to Mr. H. C. Russell's interesting current papers (No. 4), containing the tracks of 124 bottles received during a year ending with September last, has been published. The comparatively large number of bottles received appears to be owing to the prevalence of southerly winds ; the north-west winds being found to alter the direction of the drifting bottles, so that they pass to the south of Australia. The suggestion made in the previous paper that bottles thrown over on the east coast drifted first to the east in Tasman Sea, and then northwards until they reached the great current from the east, which passes south of New Caledonia, is supported in a remarkable way by the drift of the *Perthshire* after she was disabled in the Tasman Sea ; her general direction for 640 miles was N.E. by N., at an average daily rate of 13·6 miles. To-wards the end of the drift she travelled rapidly to the west. Two bottles floated near Cape Horn came over to Australia at the daily rates of 12·2 and 9·5 miles respectively. There are also some very interesting bottle tracks in the North Atlantic Ocean. One of these, floated in the Gulf of Mexico, made a run of 6300 miles in a south-easterly direction—the longest hitherto recorded in that ocean by Mr. Russell. The propor-tion of bottles received to those thrown overboard appears to be very disappointing ; out of 48 bottles thrown from ss. *Gulf of Bothnia*, to take an extreme case, only one was received.

THE resolutions passed at the International Congress for Marine Research held at Stockholm last summer are published *in extenso* in the April number of the *Scottish Geographical Magazine*. An important feature in these resolutions is the recognition that the primary object of the investigations recom-mended to be undertaken is the improvement and promotion of fisheries by means of international agreements.

IN *Appleton's Popular Science Monthly* for May, Prof. E. S. Morse gives a full account of the observations made by himself many years ago as to the manner in which the larval insect known as the "cuckoo-spit" forms the mass of froth in which it is concealed. If the insect be cleared from the mass of froth and allowed to settle upon some succulent plant-stem, it will soon thrust its piercing organs through the outer layers and commence sucking the juices. After a short time a clear fluid exudes from the abdomen, and after flowing over the body eventually fills up the spaces between the latter, the legs and the stem, so that the entire creature is soon totally enveloped. For about half an hour the insect will remain quiescent in this condition, when it suddenly begins to "blow bubbles" by turning its tail out of the fluid, opening the terminal segment, which appears like claspers, and then bending down the tail into the fluid with an attached air-bubble, which is instantly allowed to escape. These movements are repeated at the rate of 70 or 80 a minute till the entire envelope of fluid is converted into the mass of froth with which we are all familiar.

Bulletin No. 23 of the Division of Entomology of the U.S. Department of Agriculture is devoted to a series of articles, by Mr. F. H. Chittenden, dealing with some of the insects in-jurious to garden crops. Sixteen different species of such pests are described, with the devastation they cause. Out of these, the most generally interesting is the invasion of the "fall army-worm" in 1899. This caterpillar (*Laphygma frugiperda*) derives its name from the circumstance that, unlike the true "army-worm," it is seldom observed, except perhaps in the most Southern States, to travel in large hosts until the autumn, or, at least, before August. During 1899 these caterpillars appeared in vast swarms over a large area of the States, where they in-flicted much damage on crops of various kinds. Properly speaking, the "fall army-worm" is a grass-feeder, but when it makes its appearance in such numbers as to consume all acces-sible pasture in the neighbourhood, as was the case last season, it turns its attention to gardens, orchards and greenhouses. The crops affected last year, in addition to grass and clover, in-cluded rice, maize, wheat, oats, cabbage, beet, peas, turnips and

even tobacco. Unfortunately, the "fal army-worm" differs from the true "army-worm" in that its hosts may reappear the year after a visitation ; and destructive measures, such as poisoning by kerosene or arsenic, are accordingly essential.

WE have received the *Proceedings* of the South London Entomological and Natural History Society for 1899, which includes the President's address and several original communi-cations on entomological subjects.

THE latest issue of the *Natural History Transactions* of Northumberland, Durham, &c., contains a catalogue of the unique and unrivalled collection of British birds presented in 1883 to the trustees of the Natural History Society of those counties by the late John Hancock. The catalogue has been drawn up by Mr. J. Howse.

THERE are already several excellent editions of Gilbert White's "Selborne," but a welcome will be extended to the splendid volumes, the first of which has just been published by Mr. S. T. Freemantle. In this edition we shall have in two volumes a superb "Natural History and Antiquities of Selborne, and a Garden Kalendar," edited by Dr. R. Bowdler Sharpe, with an introduction to the Garden Calendar by Dean Hole, and numerous plates and other illustrations.

"LA SPÉLÉOLOGIE" is the title of a little handbook by M. E. A. Martel on the science of caverns. It belongs to the "Scientia" series, published by MM. Carré et Naud (Paris, 1900 ; pp. 126). The author gives an account of the origin of fissures and caverns, of the action of subterranean waters and all matters connected with them. He deals also with the phe-nomena of ice-caves (*glacières*), and again with the relations between rock cavities and metalliferous deposits. The various prehistoric and historic remains found in caverns are somewhat briefly dealt with ; and finally the author discourses on the plants and animals found living in subterranean regions.

MR. W. ENGELMANN, of Leipzig, has just commenced the publication of an elaborate work, by Prof. W. Wundt, en-titled "Völkerpsychologie : Eine Untersuchung der Entwick-lungsgesetze von Sprache, Mythus, und Sitte." The work will be completed in three volumes—the first dealing with language as the expression of the emotions by signs and speech, the second with myths and religions, and the third with ceremonies and customs. Each volume will be complete in itself, and will be separately indexed. The second (and concluding) part of the first volume will be published in the autumn of this year, and will then be reviewed with the part which has just appeared.

DR. ROBERT MUNRO's "Rambles and Studies in Bosnia-Herzegovina and Dalmatia" (Blackwood) is not only an excellent book of travel, but a very valuable contribution to archæological literature. An appreciative notice of the work appeared in these columns four years ago (vol. liv. p. 78), and we have now to announce the publication of a second, revised and enlarged edition. An account is given of the proceed-ings of the Congress of Archæologists and Anthropologists held at Sarajevo in August 1894, and as the Government of Bosnia-Herzegovina have departed from their original intention to publish a report of the congress, Dr. Munro's volume has the distinction of being the only record, in book form, of the important problems which were considered. A number of additions have been made to the original volume, and a much-wanted index has been supplied.

THE second and third parts of the second volume of the unique "Encyklopädie der mathematischen Wissenschaften" in course of publication by the firm of B. G. Teubner, Leipzig, have just been issued. The scope of this great undertaking is so extensive that several years must elapse before the work is completed. There will be seven volumes in all, having the following subjects and editors :—Arithmetic and algebra, Prof. W. F. Meyer ; analysis, Prof. H. Burkhardt ; geometry, Prof. Meyer ; mechanics, Prof. F. Klein ; physics, Prof. A. Sommerfeld ; geodesy and geophysics, Prof. E. Wiechert ; astronomy (under arrangement) ; history, philosophy, and didactic questions, Prof. Meyer. The work is published under the auspices of the Munich and Vienna Academies of Science, and the Göttingen Society of Sciences, and no mathematical library will be complete without it.

SIR JOHN LUBBOCK'S book on "The Scenery of Switzerland, and the causes to which it is due" has been translated into Italian by Dr. L. Scotti, and is published by Signor U. Hoepli, of Milan, as "Le Bellezze della Svizzera, Descrizione del Paesaggio e sue Cause geologiche." The first English edition was noticed in NATURE of September 10, 1896 (vol. liv. p. 439) ; the translation is from the third edition, published in 1898.

THE use of acetylene for lighting rooms upon a commercial scale renders its purification from sulphuretted and phosphuretted hydrogen imperative, on account of the injurious effects of the products of combustion of these impurities in a confined space. Numerous substances have been put forward by different inventors as effecting the desired purification, among which may be mentioned ferric chloride, chromium sulphate, petroleum, benzene, chromic acid, bleaching powder, and cuprous chloride. The ideal purifier should remove the impurities as completely as possible, should not absorb acetylene itself, and should not communicate any objectionable properties to the purified gas. The current number of the *Moniteur Scientifique* contains abstracts of numerous papers upon this subject. From these it would appear that solutions of metallic salts do not wholly remove the impurities, chromic acid and chloride of lime solutions being the only substances that effect a complete purification, and of these the former is preferable, as with the latter explosions have occurred, probably owing to the formation of chloro-acetylene.

THE additions to the Zoological Society's Gardens during the past week include a Diana Monkey (*Cercopithecus diana*) from West Africa, a Common Squirrel (*Sciurus vulgaris*), British, presented by Mrs. Morris ; a Common Paradoxure (*Paradoxurus niger*) from Java, presented by Mr. E. E. Hewens ; a Boddaert's Snake (*Drymobius boddaerti*), a Chequered Elaps (*Elaps lemniscatus*), a Rat-tailed Opossum (*Didelphys nudicaudata*) from Trinidad, presented by Mr. Leon Bernstein ; a Summer Snake (*Contia oestiva*), a Mexican Snake (*Coluber melanoleucus*), six Menobranches (*Necturus maculatus*), five American Green Frogs (*Rana halecina*) from North America, deposited.

OUR ASTRONOMICAL COLUMN.

ASTRONOMICAL OCCURRENCES IN JUNE.
June 2.   8h. 33m. to 9h. 35m.   Moon occults κ Cancri (mag. 5·0).
    4.   9h. 49m. to 10h. 45m.   Transit of Jupiter's Satellite III. (Ganymede).
    7.   9h. 58m. to 10h. 55m.   Moon occults the star D.M. – 10°, 3570 (mag. 6·0).
   11.   8h.   Jupiter in conjunction with moon.   Jupiter 1° 29' North.
   11.' 11h. 23m. to 12h. 43m.   Transit of Jupiter's Satellite III. (Ganymede).
   12.   Partial eclipse of the moon.
         13h. 16·2m.   First contact with penumbra.
         15h. 24·2m.   First contact with shadow.
         15h. 27·6m.   Middle of the eclipse.
         15h. 31·0m.   Last contact with the shadow.

June 12.   17h. 39·0m.   Last contact with the penumbra.
         It will be a very small eclipse, the proportion of the moon's surface covered by the earth's shadow being equal to only one-thousandth part. The fainter outlying shadow will, however, cover a large region, but will be only faintly discernible.
   13.   7h.   Mercury in conjunction with ε Geminorum. Mercury, 0° 3' South.
   13.   9h. 40m. to 10h. 52m.   Moon occults the planet Saturn.
   15.   Venus.   Illuminated portion of disc, 0·144.   Mars, ·· 0·962.
   16.   8h. 48m.   Jupiter's Satellite IV. (Callisto) in conjunction south of planet.
   19.   Saturn.   Polar semi-diameter, 17"·0.   Outer minor axis of outer ring, 18"·87.
   23.   5h.   Saturn in opposition to sun.

SEARCH-EPHEMERIS FOR EROS.—The following is continued from the ephemeris by J. B. Westhaver (*Astronomical Journal*, No. 479, vol. xx. p. 185).

Ephemeris for 12h. Greenwich Mean Time.

1900.		R.A.			Decl.			Mag.
		h. m. s.			° ′ ″			
June 2	...	23 57 2·6	...	+4 22 22		...		
4	...	0 36·1	...	4 57 6		...		12·9
6	...	4 8·7	...	5 31 57		...		
8	...	7 40·7	...	6 6 57		...		12·9
10	...	11 11·9	...	6 42 6		...		
12	...	14 42·5	...	7 17 24		...		12·8
14	...	18 12·4	...	7 52 51		...		
16	...	21 41·7	...	8 28 27		...		12·7
18	...	25 10·3	...	9 4 12		...		
20	...	28 38·1	...	9 40 5		...		12·7
22	...	32 5·3	...	10 16 8		...		
24	...	35 31·8	...	10 52 19		...		12·6
26	...	38 57·5	...	11 28 39		...		
28	...	42 22·4	...	12 5 8		...		12·5
30	...	45 46·6	...	12 41 47		...		
July 2	...	49 10·0	...	13 18 34		...		12·5

Prof. Howe is reported to have discovered the planet in the constellation Aries.

OXFORD UNIVERSITY OBSERVATORY.—In the twenty-fifth annual report of the Savilian professor at Oxford, Prof. H. H. Turner briefly reviews the history of the institution. The late Prof. Pritchard, in 1873, successfully appealed to the University for facilities to institute the means of carrying on astronomical research, but the plans originally projected being modified by the presentation of Dr. De la Rue's instruments, the building was not finished until 1875. However, notwithstanding his advanced age, Prof. Pritchard carried out before his death two important researches, the *Uranometria Nova Oxoniensis*, and the determination of stellar parallaxes ; and initiated a third, the share of the Observatory in the International Astrographic Chart. During the six years of Prof. Turner's directorship the energies of the Observatory have been chiefly directed to carrying out, as expeditiously and economically as is consistent with the necessary accuracy, this great work of fundamental astronomy. One or two more years will be required to complete it, but the work is at present as well advanced as at any of the other eighteen observatories which are collaborating. In addition, the Observatory has been utilised as an educational institution for the benefit of the students of the University. For the Astrographic Catalogue, 736 plates are now measured, and 705 completely reduced, out of the 1180 falling to the share of the Observatory. Measurements have been made on a plate supplied by Prof. E. C. Pickering to determine the optical distortion of a photographic doublet. A preliminary discussion of these measures indicates a distortion varying as the cube of the distance from the centre of the plate ; this somewhat surprising result, if confirmed, will enable the reduction of photographs of star fields of wide angle to be made with great accuracy.

ROUSDON OBSERVATORY, DEVON.—Sir C. E. Peek sends us another of his pamphlets (No. 6), containing the detailed particulars of the observations of variable stars during the past decade. The observations of T Cassiopeiæ extend over the ten years 1889–1898, and those of R Cassiopeiæ from 1887–1898. At the end of the observation the light curves of the two stars are shown.

SOME MODERN EXPLOSIVES.[1]

II.

I NOW pass to points which have to be considered when weighing the comparative merits of explosives for their intended ends.

You will easily understand that between explosives which are intended to be used for propelling purposes, and those which are intended to be used, say for bursting shell, a wide difference may exist.

In the former case, facility of detonation would be an insuperable objection ; in the latter, the more perfect the detonation the better, certain special cases, to which I have not time to refer, excepted.

There exists, I think, considerable diversity of opinion as to what does, and what does not, constitute true detonation. I find many persons speak of a detonation, when I should merely consider that a very high pressure had been reached. This gun-cotton slab on the table affords me, I think, a fair opportunity of explaining my meaning. Were I to set fire to it, except for the large volume of flame and the great amount of heat generated, we in this room would not suffer ; we should probably experience more inconvenience did I fire a similar slab of gunpowder, as detached burning portions would probably be projected to some distance.

But if I fired this same slab with two or three grammes of fulminate of mercury, a detonation of extreme violence would follow. The detonation would be capable of blowing a hole in a tolerably thick iron plate, and would probably put an end to a considerable portion of the managers in the front row.

I mentioned to you some time ago the time in which a charge would be consumed in the chamber of a gun—if a charge of 500 lbs. of these slabs were effectively detonated, this charge would be converted into gas in less than the 20,000th part of a second.

No such result would follow were I to try a similar experiment with a slab of compressed gunpowder of the same dimensions. I do not say the experience would be pleasant, but there would be nothing of the instantaneous violent action which marks the decomposition of the gun-cotton.

To give you an idea of the extraordinary violence which accompanies detonation, I have fired, for the purpose of this lecture, with fulminate of mercury, a charge of lyddite in a cast-iron shell, and those who are sufficiently near can see for themselves the result. By far the greater part of the cast-iron shell, weighing about 10 lbs., is reduced to dust, some of which is so fine that I assumed it to be deposited carbon until I had tested it with a magnet. I may add that the indentation of the steel vessel by pieces of the iron which were not reduced to powder would appear to indicate velocities of not less than 1200 feet-seconds, and this velocity must have been communicated to the fragments in a space of less than two inches.

For the sake of comparison, I place beside it a cast-iron shell burst by gunpowder. You will observe the extraordinary difference. I also have on the table two small steel shells exploded, one by a perfectly detonated, the other by a partially detonated charge. I may remark that in the accounts of correspondents from the seat of war, frequent mention is made of the green smoke of lyddite. This appearance is due probably to imperfect detonation—to a mixture, in fact, of the yellow picric with the black smoke. I do not say, however, that imperfect detonation is necessarily an evil.

To another experiment I draw your attention.

For certain purposes I caused to be detonated, in the chamber of a 12-pounder, a steel shell charged with lyddite. The detonation was not perfect, but the base of the shell was projected with great violence against the breech screw. You may judge of how great that violence was when I tell you that the base of the shell took a complete impression of the recess for the primer, developing great heat in so doing ; but, what was still more remarkable, the central portion of the base also sheared, passing into the central hole through which the striker passes. This piece of shell is upon the table, and open to your inspection.

One other instance to illustrate the difference between combustion and detonation I trouble you with. Desiring to ascertain the difference, if any, in the products of explosion between combustion and detonation, I fired a charge of lyddite in such a manner that detonation did not follow. The lyddite merely

[1] A Discourse delivered at the Royal Institution on Friday, March 23, by Sir Andrew Noble, K.C.B., F.R.S. Continued from p. 90.

deflagrated. But a similar charge differently fired shortly afterwards detonated with such extreme violence as to destroy the vessel in which it was exploded. The manner in which the vessel failed I now show you (Fig. 4), and I have on the table the internal crusher gauge which was used, and which was also totally destroyed.

The condition of this gauge is very remarkable, and the action on the copper cylinder employed to measure the pressure was one to which I have no parallel in the many thousand experiments I have made with these gauges. The gauge itself is fractured in the most extraordinary way, even in some places to which the gas had no access, and the copper cylinder, which when compressed usually assumes a barrel-like form (that is, with the central diameter larger than that at the ends as shown in Fig. 5) ; but in this experiment, and in this only, the cylinder was bulged closed to the piston, as you see. It would appear as if the blow was so suddenly given that the laminæ of the metal next the piston endeavoured to escape in the direction of

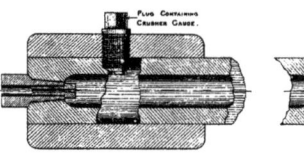

FIG. 4.—Explosion vessel.

least resistance, that being easier than to overcome the inertia of the laminæ below.

The erosive effect of the new explosives is another point of first-rate importance in an artillery point of view. The cordite of the service is not, if the effect be estimated in relation to the energy impressed on the projectiles, more erosive than, for example, brown prismatic, which was itself a very erosive powder ; but as we are able to obtain, as you have seen, very much higher energies with cordite than with brown prismatic, the erosion of the former is, for a given number of rounds, materially higher.

There is, however, one striking difference. By the kindness of Colonel Bainbridge, the Chief Superintendent of Ordnance Factories, I am enabled to show you a section of the barrel of a large gun eroded by 137 rounds of gunpowder, and a barrel of a 4·7-inch quick-firing gun eroded by 1087 rounds of gunpowder, and another eroded by 1292 rounds of cordite. You

1.   **2.**   **3.**

FIG. 5.—Copper cylinders.

will observe the difference. In the former case the erosion much resembles a ploughed field. In the latter the appearance is more, as if the surface was washed away by the flow of the highly heated gases.

But take it in what way you please, the heavy erosion of the guns of the service, if fired with the maximum charges, is a very serious matter, as with the large guns, accuracy, and in a smaller degree energy, are rapidly lost after a comparatively small number of rounds have been fired.

Cordite was first produced for use in small arms only, where, owing to the small charges employed, the question of erosion is not of the same importance as with large guns ; but its employment, from the great results obtained with it, was rapidly extended to artillery, and the attention of my friends, Sir F. Abel and Prof. Dewar, has for some time been devoted in conjunction with myself to investigating whether it is not possible materially to reduce this most objectionable erosion.

With this object I made the following series of experiments.

I had cordite of the same dimensions prepared with varying proportions of nitro-glycerine and gun-cotton. The nitro-glycerine being successively in the proportions of 60, 50, 40, 30, 20 and 10 per cent., and with each of these cordites I determined the following points :—

(1) The quantity of permanent gases generated.
(2) The amount of aqueous vapour formed.
(3) The heat generated by the explosion.

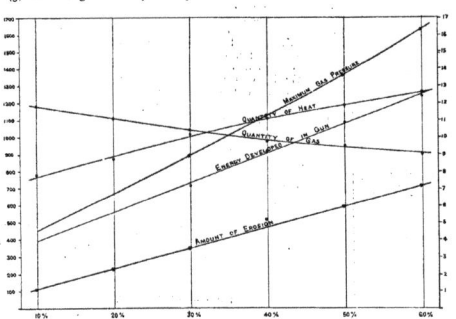

FIG. 6.—Energy in foot tons ; heat in units ; gas in c.c. ; erosion in inches ; pressure in tons.

(4) The erosive effect of the gases.
(5) The ballistic energy developed in a gun, and the corresponding maximum pressure.
(6) The capacity of the cordite to resist detonation when fired with a strong charge of fulminate of mercury.

The results of these experiments were both interesting and instructive.

To avoid wearying you with a crowd of figures, I have placed on Fig. 6 the results of the first five series of experiments.

On the axis of abscissæ are placed the percentages of nitro-glycerine, while the ordinates show the quantities of the gases generated, the amount of heat developed, the erosive effect of the explosive, the ballistic energy exhibited in a gun, and the maximum gaseous pressure.

You will note that with the smallest proportion of nitro-glycerine the volume of permanent gases is a maximum, and that the volume steadily decreases with the increase of nitro-glycerine. On the other hand, the heat generated as steadily increases with the nitro-glycerine, and if we take the product of the quantity of heat and the quantity of gas as an approximate measure of the potential energy of the explosive, the higher proportion of nitro-glycerine has an undoubted advantage ; but in this case, as in the case of every other explosive with which I have experimented, the potential energies differ less than might be expected from the changes in transformation, as the effect of a large quantity of gas is to a great extent compensated by a great reduction in the quantity of heat generated.

This effect is, of course, easily explained; and was very strikingly exhibited in the much more complicated transformation experienced by gunpowders of different compositions, a long series of which were very fully investigated by Sir F. Abel and myself.

Looking at this diagram you will have observed that the energy developed in the gun is very much smaller with the

smaller proportions of nitro-glycerine, but if you will look at the corresponding maximum pressure-curve you will note that the pressures have decreased nearly in like proportion. Hence it is probable that the lower effect is mainly due to a slower combustion of the cordite, and it follows that this effect may be, to a great extent, remedied by increasing the rate of combustion by reducing the diameter of the cordite to correspond with the reduction in the quantity of nitro-glycerine.

To test this point I caused to be manufactured a second series of cordites of the same composition, but with the diameters successively reduced by ·03. as you see with the samples I hold, and this diagram (Fig. 7) shows at a glance the result. The energies you see are, roughly, practically the same, but if you look at the pressure-curve you will observe that I have obtained a curve in which, on the whole, the pressures vary in the contrary direction, that is to say, in this case the pressures increase as the nitro-glycerine diminishes.

Taking the two series into account, they show that by a proper arrangement of amount of charge and diameter of cord it would be possible to obtain the same ballistics and approximately the same pressure from any of the samples I have exhibited to you.

But I have to draw your attention to another point. From the curve showing the quantities of heat you will note that in passing from 10 per cent. nitro-glycerine to 60 per cent., the heat generated has increased by about 60 per cent. But here is the curve indicating the corresponding amount of erosion, and you will see that while the quantity of heat is only greater by about 60 per cent., the erosion is greater by nearly 500 per cent.

These experiments entirely confirm the conclusion at which I have previously arrived, viz. that heat is the principal factor in determining the amount of erosion.

In experimenting with a number of alloys of steel, the greatest resistance was shown by an alloy of steel with a small proportion of tungsten, but the difference between the whole of these amounted only to about 16 per cent.

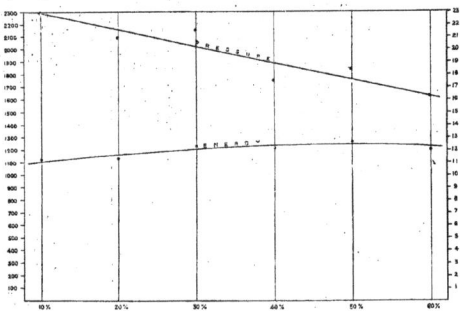

FIG. 7.—Energy in foot tons ; pressure in tons.

The whole of these cordites were, as I have mentioned, subjected to detonation tests. None of them, so far as my experiments went, exhibited any special tendency in this direction.

I will now endeavour to describe to you a most interesting and important series of experiments which, I regret to say, is a long way from completion.

The objects of these experiments were (1) to ascertain the

time required for the combustion of charges of cordite in which the cordite was of different thicknesses, varying from 0·05 inch to 0·60 of an inch ; (2) the rapidity with which the explosives part with their heat to the vessel in which the charge is confined ; and (3) to ascertain, if possible, by direct measurement, the temperature of explosion, and to determine the relation between the pressure and temperature at pressures approximating to those which exist in the bore of a gun, and which are, of course, greatly above any which have yet been determined.

As regards the first two objects I have named, I have had no serious difficulties to contend with, but as regards the third, I have so far had no satisfactory results, having been unable to use Sir W. Roberts Austen's beautiful instrument owing to the temperature at the moment of explosion being greatly too high, high enough indeed to melt and volatilise the wires.

I am, however, endeavouring to make an arrangement by which I hope to be able to determine these points when the temperature is so far reduced that the wires will no longer be fused.

If the piston be left free to move the instant of the commencement of pressure, the outside limit of the time of complete explosion will be indicated ; but, on account of the inertia of the moving parts, the pressure indicated will be in excess of the true pressure, and the excess will be, more or less, inversely as the time occupied by the explosion.

If we desire to know the true pressure, it is necessary to compress the gauge beforehand to a point closely approximating to the expected pressure, so that the inertia of the moving parts may be as small as possible—the arrangement by which this is effected is not shown in the photograph, but the gauge is retained at the desired pressure by a wedge-shaped stop, held in its place by the pressure of the spring, and to the stop a heavy weight is attached—when the pressure is relieved by the explosion, the weight falls and leaves the spring free to act.

I have made a large number of experiments with this instrument, both with a variety of explosives and with explosives fired under different conditions. Time will not permit me to do more than to show you on the screen three pairs of experiments to

Fig. 8.

The apparatus I have used for these experiments is placed on the table. The cylinder in which the explosives were made is too heavy to transport here, but this photograph (Fig. 8) will sufficiently explain the arrangement. The charge I used is a little more than a kilogramme, and it is fired in this cylinder in the usual manner.

The tension of the gas acting on the piston compresses the spring, and indicates the pressure on the scale here shown. But to obtain a permanent record, the apparatus I have mentioned is employed.

There is, you see, a drum made to rotate by means of a small motor. Its rate of rotation is given by a chronometer acting on a relay, and marking seconds on the drum, while the magnitude of the pressure is registered by this pencil actuated by the pressure-gauge I have just described.

To obtain with sufficient accuracy the maximum pressure, and also the time taken to gasify the explosive, two observations— that is, two explosions—are necessary.

illustrate the effect of exploding cordite of different dimensions, but of precisely the same composition.

I shall commence with rifle cordite. In this diagram (Fig. 9) the axis of abscissæ has the time in seconds marked upon it, while the ordinates denote the pressures, and I draw your attention to the great difference, in the initial stage, between the red and the blue curves. You will notice that the red curves show a maximum pressure some 4½ tons higher than that shown by the blue curve ; but this pressure is not real. It is due to the inertia of the moving parts. The red and blue curves in a very small fraction of a second come together, and remain practically together for the rest of their course. The whole of the charge is consumed in something less than fifteen thousandths ('015) of a second.

In the case of the blue curve the maximum pressure indicated is obtained in the way I have described, and is approximately correct—about nine tons per square inch. The rapidity with which this considerable charge parts with its heat by communication to the explosion vessel is very striking. In four seconds after the explosion the pressure is reduced to about one-half, and in twelve seconds to about one-quarter.

I now show you (Fig. 10) similar curves for cordite 0·35 inch

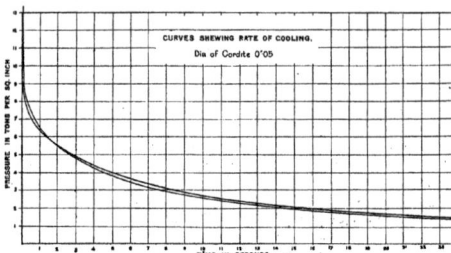

FIG. 9.

in diameter, or about fifty times the section. Here you see that the time taken to consume the charge is longer. The effect of inertia is still very marked, although much reduced. The true maximum pressure is little over 8 tons, but after the first third of a second the two curves run so close together that they are indistinguishable.

Again, you see the pressure is reduced by one-half in four seconds, and in a little more than twelve seconds again halved.

The last pair of curves I shall show (Fig. 11) you was obtained with cordite 0·6 inch in diameter, or nearly 150 times the section of the rifle cordite. With this cordite the combustion has been so slow that the effect of inertia almost disappears; it is reduced to about half a ton per square inch. The maximum being nearly the same as in the last set of experiments. The time of combustion indicated I have called slow, but it is about ·06 of a second, and the whole of the experiments show a most remarkable regularity in their rate of cooling, the pressures at the same distance of time from the explosion being in all cases approximately the same—as, indeed, they ought to be. The density being the same and the explosive the same, the only difference being the time in which the decomposition is completed.

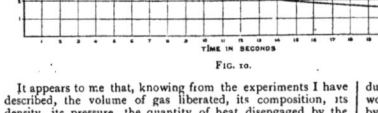

FIG. 10.

It appears to me that, knowing from the experiments I have described, the volume of gas liberated, its composition, its density, its pressure, the quantity of heat disengaged by the

explosion; and knowing all these points with very considerable accuracy, we should be able, from the study of the curves to which I have drawn your attention, and which can be obtained from different densities of gas, to throw considerable light upon the kinetic theory of real, not ideal gases, at temperatures and pressures far removed from those which have been the subject of such careful and accurate research by many distinguished physicists.

The question, as I have said, involves some very considerable difficulties; nevertheless, I am not without hope that the experiments I have been describing may, in some small degree, add to our knowledge of the kinetic theory of gas.

That wonderful theory faintly shadowed forth almost from the commencement of philosophic thought, was first distinctly put forward by Daniel Bernoulli early in the last century. In the latter half of the century now drawing to a close the labours of Joule, Clausius, Clerk Maxwell, Lord Kelvin and others have placed the theory in a position analogous and equal to that held by the undulatory theory of light.

The kinetic theory has, however, for us artillerists a special charm, because it indicates that the velocity communicated to a projectile in the bore of a gun is due to the bombardment of that projectile by myriads of small projectiles moving at enormous speeds, and

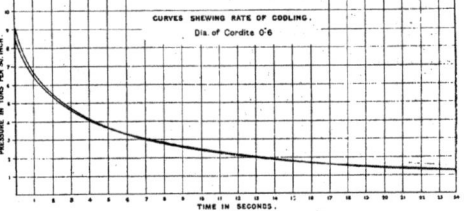

FIG. 11.

parting with the energy they possess by impact to the projectile.

There are few minds which are not more or less affected by the infinitely great and the infinitely little.

It was said that the telescope which revealed to us infinite space was balanced by the microscope which showed us the infinitely small; but the labours of the men to whom I have referred have introduced us to magnitudes and weights infinitesimally smaller than anything that the microscope can show us, and to numbers which are infinite to our finite comprehension.

Let me draw your attention again to this figure (Fig. 2) showing the velocity impressed upon the projectile, and let me endeavour to describe the nature of the forces which acted upon it to give it its motion. I hold in my hand a cubic centimetre, a cube so small that I daresay it is hardly visible to those at a distance. Well, if this cube were filled with the gases produced by the explosion at 0° C. and atmospheric pressure, there would be something over seven trillions, that is, seven followed by eighteen cyphers, of molecules. Large as these numbers are, they occupy but a very small fraction of the contents of the cubic centimetre, but yet their number is so great that they would, if placed in line touching one another, go round many times the circumference of the earth, a pretty fair illustration of Euclid's definition of a line.

These molecules, however, are not at rest, but are moving, even at the low temperature I have named, with great velocity, the molecules of the different gases moving with different velocities dependent upon their molecular weight. Thus, the hydrogen molecules which have the highest velocity move with about 5500 feet-seconds mean velocity, while the slowest, the carbonic anhydride molecules, have only 1150 feet-seconds mean velocity, or about the speed of sound.

But in the particular gun under discussion, when the charge was exploded there were no less than 20,500 cubic centimetres of gas, and each centimetre at the density of explosion contained 580 times the quantity of gas—that is, 580 times the number of molecules that I mentioned. Hence the total number of molecules in the exploded charge is 8¼ quadrillions, or let us say approximately for the total number eight followed by twenty-four cyphers.

It is difficult for the mind to appreciate what this immense number means, but it may convey a good idea if I tell you that if a man were to count continuously at the rate of three a second, it would take him 265 billions of years to perform the task of counting them.

So much for the numbers; now let me tell you of the velocities with which, at the moment of explosion, the molecules were moving. Taking first the high-velocity gas, the hydrogen, the molecules of the gas would strike the projectile with a mean velocity of about 12,500 feet-seconds. You will observe I say mean velocity, and you must note that the molecules move with very variable velocities. Clerk Maxwell was the first to calculate the probable distribution of the velocities. A little more than one-half will have the mean velocity or less, and about 98 per cent. will have 25,000 feet-seconds or less. A very few, about one in 100 millions, might reach the velocity of 50,000 feet-seconds.

The mean energy of the molecules of different gases at the same temperature being equal, it is easy from the data I have given to calculate the mean velocity of the molecules of the slowest moving gas, carbonic anhydride, which would be about 2600 foot-seconds.

I have detained you, I fear, rather long over these figures, but I have done so because I think they throw some light upon the extraordinary violence that some explosives exhibit when detonated. Take, for instance, the lyddite shell exploded by detonation I showed you earlier in the evening. I calculate that that charge was converted into gas in less than the 1/60,000th part of a second, and it is not difficult to conceive the effect that these gases of very high density suddenly generated, the molecules of which are moving with the velocities I have indicated, would have upon the shell.

The difference between the explosion of gunpowder fired in a close vessel, and that of gun-cotton or lyddite when detonated, is very striking. The former explosion is noiseless, or nearly so. The latter, even when placed in a bag, gives rise to an exceedingly sharp metallic ring, as if the vessel were struck a sharp blow with a steel hammer.

But I must conclude. I began my lecture by recalling some of the investigations I described in this place a great many years ago. I fear I must conclude in much the same way as I then did, by thanking you for the attention with which you have listened to a somewhat dry subject, and by regretting that the heavy calls made on my time during the last few months have prevented my making the lecture more worthy of my subject and of my audience.

EXTENSIONS OF THE DYEING DEPARTMENT OF YORKSHIRE COLLEGE.

THE opening of the extensions in the Clothworkers' Departments of Yorkshire College, Leeds, has already been referred to (P. 69). The new buildings, which are shown in the accompanying illustration, comprise practical and pattern dyehouses and a research laboratory; and, as with several other parts of Yorkshire College, they owe their erection to the generous interest taken in technical education by the Clothworkers' Company of London.

The Clothworkers' Departments of the Yorkshire College consist of textile industries, dyeing and art. The buildings occupied by these departments have been erected by the Clothworkers' Company at a cost of about 60,000l.; they are spread over an area of about one-and-a-half acres, and have been specially arranged and equipped for the teaching of all the subjects connected with the designing and manufacturing of woven fabrics.

The Dyeing Department of the Yorkshire College was established in 1880, and the head of the department is Prof. J. J. Hummel. Although the accommodation at first provided was extremely limited, it nevertheless sufficed to show that a demand for instruction in dyeing really existed, and that a continuous supply of students for this subject was available. In due time

New Buildings of the Dyeing Department of Yorkshire College.

it was found desirable to increase the facilities for experimental work, and in 1885 the Clothworkers' Company of London erected and equipped, at an expense of about 12,000l., the front portion of the handsome and commodious building at present occupied.

It was felt some years ago that the work of the different departments might be connected. It was considered desirable, for example, that the coloured yarns employed in the weaving department should be dyed by the students in the dyeing department, so that, if at the same time these yarns could also be manufactured on the premises by the establishment of a spinning department, it would become possible to teach the whole routine of clothworking, from the wool in the raw state to the finished cloth. Acting upon this idea, the Clothworkers' Company decided to make the necessary provision for carrying out the scheme suggested, and to extend both the weaving and dyeing departments, at a cost of about 25,000l. In connection with the dyeing department, it was arranged to build a three-storied building, to provide two additional dyehouses in which practical dyeing could be carried on, and also a research laboratory for the prosecution of scientific investigations connected with dyestuffs and dyeing.

In July 1896, the foundation stone of the new Clothworkers' Research Laboratory and the other extensions was laid by the

Master of the Clothworkers' Company, and the completed buildings were opened on May 11.

At the present time, therefore, the dyeing department of the Yorkshire College is represented by a building of considerable dimensions, and so comprehensive in character and equipment as practically to meet every requirement for the purpose of giving a complete theoretical and practical instruction in the art of dyeing in all its branches.

Some idea of the magnitude of the work done in the dyeing department of Yorkshire College may be gained from the fact that each session over 200,000 dyed patterns are distributed. Each student, according to the time spent in the dyehouse, receives during his course of instruction from 2000 to 20,000 patterns, each of which conveys a definite piece of information on some point connected with the application of this or that colouring matter. Not only is the behaviour in the dyebath of each colouring matter investigated, but notes of the results obtained are made by the students during the progress of the work. Further, each student enters in his own book all the patterns received, together with notes of the materials employed, and the results of each experiment. Hence the students not only learn how to experiment and discover the capabilities of each colouring matter for themselves, but they also acquire the useful habit of observing and of making notes, while their pattern books contain a fund of information which is invaluable to them in their after career. The systematic training which they receive also prepares them to deal with the variable conditions of work in actual practice, such as the character of the water, the nature of the textile material employed and its ultimate uses, and many other points which must always be taken into account in dyeing.

In the practical and pattern dyehouses the students are shown how they are expected to apply in practice the principles they have learned in the course of their experimental work. Moreover, the solution of difficulties which naturally arise under the slightly altered conditions from those obtaining in the experimental dyehouse, the greater confidence inspired by dealing with the larger quantities of material, and the knowledge that the products of their labour are really to be employed in the manufacture of cloth, are all factors of inestimable value in the training of the students before they enter into actual practice, to which they are as it were brought indeed one step nearer by the character of the work pursued. Altogether, the students are able, in the College dyehouses, to gain at least some insight as to the meaning and value of practical experience, and an influence is exerted which reacts by giving life and vigour to the work of the whole department.

The art of dyeing owes much to science, and in a University College like the Yorkshire College, it is not unreasonable to expect that students of the art should, in return, contribute something to science, more particularly to that branch of it which pertains to dyeing. If in the experimental and practical dyehouses the students are taught the *art* of dyeing, in the Clothworkers' Research Laboratory they are also urged to study the *science* of dyeing. The aim here is to assist in the work of gaining a fuller and truer knowledge of the fundamental laws and principles connected with dyestuffs and dyeing, and so help to raise, as far as possible, the whole tone and level of the dyeing trade, by infusing into it the traits of an exact science. The carrying on of original research by advanced students has already become, indeed, a marked feature of the department, and the Clothworkers' Company have, in a special way, recognised the value of such work by establishing a lectureship, the holder of which devotes his whole time to co-operating with the professor in introducing students to this higher form of study.

This research work, too, has an intimate connection with Prof. Hummel's lectures, in the course of which are described the methods employed in preparing the coal-tar colours, in isolating the pure colouring principles of dyewoods, and in studying the chemistry of mordanting, dyeing, &c. By allowing the students to carry out similar experiments themselves, the College enables them to understand, in a clearer manner than is otherwise possible, how our knowledge concerning dyestuffs and dyeing has been acquired, and it is hoped that by reason of the practical experience thus gained in the art of research, some students may, in due time, become independent investigators.

The Clothworkers' Research Laboratory is an addition which gives completeness to the means of instruction in dyeing already furnished. The advanced students are thereby provided with the facilities for extending the boundaries of science connected

with dyeing, and it is hoped that many young men will take advantage of the opportunity thus given. If in the pursuit of this object the authorities at Yorkshire College can succeed in attracting and training a band of earnest workers; if a well-recognised and successful School of Research in Dyeing is established, side by side with the School of Practical Dyeing, it cannot but be of inestimable value from an educational as well as from a practical point of view, for, if the students, before they leave the College, are taught to contribute to the general sum of knowledge it it surely education in the truest and best sense of the term.

MR. NIKOLA TESLA'S RECENT ELECTRICAL EXPERIMENTS.

A REMARKABLE paper, by Mr. Nikola Tesla, appears in the June number of the *Century Magazine*. The subject is "The Problem of Increasing Human Energy," with Special Reference to the Harnessing of the Sun's Energy"; and though metaphysical and sociological questions receive a large share of attention, the article contains an account of some very interesting electrical experiments, now described for the first time, illustrated by several very striking photographs, two of which are here reproduced. Mr. Tesla has bee years in further investigating the properties of alternate currents of high potential and frequency, with which he astonished audiences at the Royal Institution in 1892 (see NATURE, vol. xlv. p. 345). The following abstract of a part of his paper shows that his work has led to results of scientific interest and significance.

Electrical discharges capable of making atmospheric nitrogen combine with oxygen have recently been produced. Experiments made since 1891 showed that the chemical activity

FIG. 1.—Combustion of atmospheric nitrogen by the discharge of an electrical oscillator giving twelve million volts and alternating 100,000 times per second. The flame-like discharge shown in the photograph measured 65 feet across.

of the electrical discharge was very considerably increased by using currents of extremely high frequency or rate of vibration. This was an important improvement, but practical considerations soon set a definite limit to the progress in this direction. Next, the effects of the electrical pressure of the current impulses, of their wave-form and other characteristic features, were investigated. Then the influence of the atmospheric pressure and temperature and of the presence of water and other bodies was studied, and thus the best conditions for causing the most intense chemical action of the discharge and securing the highest efficiency of the process were gradually ascertained. The flame grew larger and larger, and its oxidising action more and more intense. From an insignificant brush discharge a few inches long it developed into a marvellous electrical phenomenon, a roaring blaze, devouring the nitrogen of the atmosphere and measuring sixty or seventy feet across (Fig. 1). The flame-like discharge visible is produced by the intense electrical oscillations which pass through the coil shown, and violently agitate the electrified molecules of the air. By this means a strong affinity is created between the two normally indifferent constituents of the atmosphere, and they combine readily, even if no further provision is made for intensifying the chemical action of the discharge.

Under certain conditions the atmosphere, which is normally a high insulator, assumes conducting properties, and so becomes capable of conveying any amount of electrical energy. The discovery of the conducting properties of the air, though unexpected, was only a natural result of experiments in a special field carried on for some years previously. It was during 1889 that certain possibilities, offered by extremely rapid electrical oscillations, led to the design of a number of special machines adapted for their investigation. One of the earliest observations made with these new machines was that electrical oscillations of an extremely high rate act in an extraordinary manner upon the human organism. Thus, for instance, powerful electrical discharges of several hundred thousand volts, which at that time were considered absolutely deadly, could be passed through the body without inconvenience or hurtful consequences. Another observation was that by means of such oscillations light could be produced in a novel and more economical manner, which promised to lead to an ideal system of electric illumination by vacuum-tubes, dispensing with the necessity of renewal of lamps or incandescent filaments, and possibly also with the use of wires in the interior of buildings.

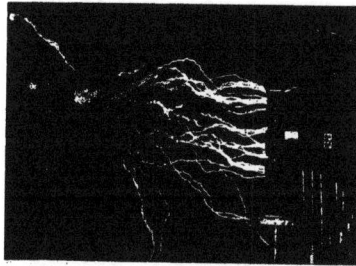

FIG. 2.—The coil, partly shown in the photograph, creates an alternating current of electricity at the rate of 100,000 alternations per second. The discharge escapes with a deafening noise, striking an unconnected coil 22 feet away, and creating such an electrical disturbance that sparks an inch long can be drawn from a water-main at a distance of 300 feet from the laboratory.

The investigations led to other valuable observations and results, one of the more important of which was the demonstration of the practicability of supplying electrical energy through one wire without return. To what a degree the appliances have been perfected since the demonstrations in 1892, when the apparatus was barely capable of lighting one lamp, will appear from the fact that as many as four or five hundred lamps have been lighted in this manner.

The success of this method of transmission suggested that the earth could be used as a conductor, thus dispensing with wires. The earth was regarded as an immense reservoir of electricity, which could be disturbed effectively by a properly designed electrical machine. Accordingly efforts were directed toward perfecting a special apparatus which would be highly effective in creating a disturbance of electricity in the earth, and a novel kind of transformer or induction-coil, particularly suited for this special purpose, was designed. By means of this apparatus, it is practicable, not only to transmit minute amounts of electrical energy for operating delicate electrical devices, but also electrical energy in appreciable quantities.

However extraordinary the results exemplified by Fig. 2 may appear, they are but trifling compared with those which are attainable by apparatus designed on these same principles.

Electrical discharges have been produced, the actual path of which, from end to end, was probably more than one hundred feet long; but it would not be difficult to reach lengths one hundred times as great. Electrical movements occurring at the rate of approximately one hundred thousand horse-power have been obtained, but rates of one, five, or ten million horse-power are easily practicable.

The most valuable observation made in the course of these investigations was the extraordinary behaviour of the atmosphere toward electric impulses of excessive electromotive force. The experiments showed that the air at the ordinary pressure became distinctly conducting, and this opened up the wonderful prospect of transmitting large amounts of electrical energy for industrial purposes to great distances without wires, a possibility which, up to that time, was thought of only as a scientific dream. Further investigation revealed the important fact that the conductivity imparted to the air by these electrical impulses of many millions of volts increased very rapidly with the degree of rarefaction, so that air strata at very moderate altitudes, which are easily accessible, offer, to all experimental evidence, a perfect conducting path, better than a copper wire, for currents of this character.

The experiments have indicated that, with two terminals maintained at an elevation of not more than thirty thousand to thirty-five thousand feet above sea-level, and with an electrical pressure of fifteen to twenty million volts, the energy of thousands of horse-power can be transmitted over distances which may be hundreds and, if necessary, thousands of miles. Investigations are now being carried on with the object of reducing considerably the height of the terminals now required.

SOME SCIENTIFIC ASPECTS OF TRADE.

A REPORT on the trade and commerce of Leghorn, for the year 1899, by Mr. Vice-Consul Carmichael, has just been received at the Foreign Office and published as No. 2714 of the Annual Series. The following extracts from the report are of interest as showing the various points at which scientific work and knowledge touch industry.

The proportion of sulphate of copper imported from Great Britain in 1898 was 96 per cent.; it had in 1899 fallen to 76 per cent. The explanation of this unwelcome fact appears to be due to keen United States competition. Italian manufacture is likely to become an even more formidable danger in the near future. Manufacturers appear as a rule to have gone to England for the greater part of the raw material, and that of itself was a handicap. Now, however, the flourishing and influential Società Metallurgica of Leghorn is busily erecting the necessary plant for the manufacture of sulphate of copper on a large scale. Italy produces some 26,000 tons of copper annually, and it is said that the company can depend upon securing its material at home. Should this be the case it will at once be seen how formidable a competitor is entering the field. In any case the more satisfactory days of the English trade in sulphate seem to be over.

As this series of reports is yearly obtaining a larger circulation it may perhaps be necessary to state that the wood from which briar pipes are made is not the root of the briar rose, but the root of the large heath known in botany as the *Erica arborea*. Our "briar" is but a corruption of the French "bruyère." The briar-root industry has had a somewhat curious history. First begun in the Pyrenees some 50 years ago, it travelled along the French Riviera and the Ligurian coast (taking Corsica by the way) to the Tuscan Maremma, and has now reached Calabria in the south, which is at present its most flourishing centre. By the very nature of the business, when a certain district has been exhausted of all its roots, the industry must come to an end there, and I have heard the opinion expressed that the Italian branch of it cannot last much more than another ten years. Leghorn has always been the centre of the export of Tuscan briar-root since the Maremma industry came into existence, but as the South Italian briar is of admittedly superior quality, a large quantity of the Calabrian root is also imported into Leghorn for selection and subsequent export.

The olive oil crop throughout Tuscany, small as it promised to be, has, I regret to say, been more than half destroyed by the ravages of the olive fly. Hence the quantity of olive oil obtained this season in Tuscany has been insignificant, while the quality of most of it is distinctly inferior. A full crop of olive oil may be reckoned at a money value of some 10,000,000*l.*

The olive maggot—which subsequently develops into the olive fly—destroys the pulp of the fruit, and so potent are the ravages of this pest that it is capable of diminishing the yield of oil by one-half, and seriously injuring the quality of the remainder. It will therefore be seen that the fly may actually cause damage in one year amounting to · 5,000,000*l.* Notwithstanding the urgency of the matter, no means of destroying the insect appear so far to have been discovered, nor has the State suggested any practical remedy. The subject is recommended to the notice of English men of science, as any discovery which should exterminate the plague ought certainly to be profitable. What seems to be wanted is that entomologists of experience should carefully study the habits of the fly with a view to finding out the hitherto undiscovered winter habitat. Then alone could proper steps be taken for its destruction. It has been hazarded with some likelihood that the winter habitat of the fly must be in the bark of the olive trees. If that were the case, all that would be needed would be to paint the trees during the winter with a simple solution of lime, which, though it might spoil the beauty of the Italian landscape, would rid the country of a very formidable enemy to its agricultural prosperity.

UNIVERSITY AND EDUCATIONAL INTELLIGENCE.

CAMBRIDGE.—Two University lectureships in experimental physics are now vacant. The appointment is for five years, and the stipend 50*l.* a year. Applications should reach the Vice-Chancellor by June 2.

The researches of Mr. L. N. G. Filon, advanced student, of King's College, in relation to certain problems in applied mechanics, have been approved as a qualification for the B.A. degree.

Sixty-three men and nineteen women have acquitted themselves so as to deserve honours in the Mathematical Tripos, Part I.

Honorary degrees will, on June 12, be conferred on the Earl of Rosse, F.R.S., Sir Benjamin Baker, F.R.S., Sir W. L. Buller, F.R.S., Prof. S. P. Langley, Prof. W. M. Flinders Petrie and Prof. H. Poincaré.

The graces for the establishment of a new special examination in agricultural science for the B.A. degree was opposed on May 24, but it was carried by a large majority. The first examination will take place at the end of the year.

ONE of the chief difficulties which has to be overcome by Technical Education Committees is the defective character of elementary education, respecting which lament is very general. Several instances of this difficulty are given in the current number of the *Record of Technical and Secondary Education.* The Durham committee have been compelled for some years to give financial assistance to preparatory classes now formed in all but twenty-one districts of the county. The committee have by such means paved the way for their new regulation of 1899 that there must be "the production of evidence of preparatory training on the part of all new applicants on whom attendance grants would be claimed." This action already appears to be having a satisfactory effect. The Cambridgeshire, Nottinghamshire and Staffordshire committees also deal at some length with the question of defective elementary education. The Cambridgeshire committee go so far as to say :—"The very backward state of elementary education has made it extremely difficult, if not impossible, to establish a system of technical education in the proper sense of the term." The Staffordshire committee speak of it in its relation to "the early age at which pupils leave the elementary schools," and this has thrown upon them "much elementary and preparatory work which otherwise would have been unnecessary." The importance of promoting the efficiency of the work of evening continuation schools cannot be too strongly urged, as they largely constitute the foundation of the work of Technical Education Committees and thus lead on to higher and specialised instruction.

SOCIETIES AND ACADEMIES.

LONDON.

Physical Society, May 25.—Prof. J. D. Everett, F.R.S., Vice-President, in the chair.—Prof. S. P. Thompson showed some experiments illustrating the aberration called Coma. If a converging lens is placed obliquely in a parallel beam of light, instead of giving a point image, it produces unilateral distortion, and the bright central spot is accompanied by a pear-shaped tail, which is known as a coma. The direction in which this tail points depends upon the side of the lens which is presented to the light. With a concavo-convex lens the convex surface gives an inward pointing coma, and the concave surface an outward pointing coma. The existence of this phenomenon is due to unequal magnification from different zones of the lens, a fact which was shown by covering the lens with a zone-plate of three or four rings and viewing on a screen the distorted images of the several zones. The form of a coma varies greatly with the distance of the screen from the lens. A parallel beam of light which has passed obliquely through a convex lens is capable of producing some curious shadows. The shadow of a rod can be obtained as a circular spot, and that of a grating, made by stretching threads between two rods, as concentric circular rings. Prof. Thompson also showed a stringed model illustrating the paths of light-rays in the formation of a coma. —Mr. R. T. Glazebrook then read some notes on the measurement of some standard resistances. Three methods have been employed by the author for building up multiples of a standard resistance, such as a one-ohm coil. The first method consists in making as accurately as possible three three-ohm coils. These in parallel can be compared directly with the standard by Carey Foster's method. Their resistance in series is very approximately nine times that in parallel, and hence an accurate determination of a resistance about nine ohms can be obtained. If, then, this resistance is put in series with the standard, an accurately-known ten-ohm resistance is obtained. By a similar process, a hundred- or a thousand-ohm coil can be built up. The second method consists in calibrating a resistance-box. The one-ohm coils of the box are compared directly with the standard, and the other resistances determined accurately by a building-up process, using a subsidiary resistance-box. In comparing the high resistances, the difference between the two boxes may be so great as to send the balance off the bridge wire. In these cases the third method is employed. The equal arms of the bridge are accurately known, and one of them is shunted with a resistance, which need not be accurately known, until the reading is brought back on to the wire. The coils chiefly used throughout the experiments are made. of platinum-silver. —Mr. J. J. Guest read a paper on the strength of ductile materials under combined stresses. The author throughout his experiments has used the "yield point" of a material as the true criterion of its strength, and has rejected the elastic limit as being modified by local yielding. At present, two theories are used in the calculation of strengths of materials. The first is that the material yields when one of the principal stresses reaches a certain amount. This theory, which was adopted by Rankine and is used by engineers in England and America, is not in accord with recent experiments. The second theory is that the material yields when the greatest strain reaches a certain amount. This was advocated by St. Venant, and is used by engineers on the Continent. Besides these there is a third theory of elastic strength, in which the condition of yielding is the existence of a shearing stress of a specific amount. In the case of a solid bar subjected to torsion, there is a variation in the strain from the axis outwards, and consequently the materials have been used in the form of thin tubes. This allows the application of an internal fluid pressure. The specimens were of steel, copper and brass, the state of set caused by drawing having been removed by annealing. The tubes were subjected to (1) torque, (2) torque and tension, (3) tension only, (4) tension and internal pressure, (5) torsion and internal pressure, and (6) internal pressure only. The axial elongation, the twist, and occasionally the circumferential strain were measured. Towards the end of the experiments observations were made on bending. The results disprove the maximum stress theory, and are at variance with the maximum strain theory. The maximum shearing stress developed, and the corresponding shearing strain were comparatively constant throughout the experiments, and no other simple relation between the stresses or strains was even approximately constant. The results of the experiments have been plotted synoptically on a curve, and the several lines have been drawn upon which these points should lie, according to the various theories. It is readily seen that the points cluster round the line which represents the existence of a specific shearing stress. The author, therefore, favours the existence of this stress for any material. The chairman read a communication upon the subject from Dr. Chree. Mr. Guest, in his paper, has regarded the shearing stress theory as a little known one. As the shearing stress is half the difference between the greatest and least principal stresses, this theory is the same as Prof. G. H. Darwin's maximum stress-difference theory. All the theories suppose that the stress-strain law is linear, and that strains are so small that their squares and products can be neglected. Mr. Guest concludes that in ordinary materials the law is linear to the elastic limit, which answers to a stress lower than that which answers to the yield point, and that yield point phenomena arise between these. Nevertheless, he focusses attention on the yield point as the criterion of strength, and assumes that Hooke's law holds up to it.

Entomological Society, May 2.—Mr. W. L. Distant, Vice-President, in the chair.—Mr. W. L. Distant exhibited the cocoon, measuring nearly three and a half inches each way, of a Coprid beetle—probably belonging to the genus *Heliocopris*—found at Pretoria in the Transvaal.—The Rev. Theodore Wood exhibited a specimen of *Carabus auratus*, L., taken in either June or September 1898 by Mr. Ferrand, of Littlefield House, Exmouth, on the Haldon Hills in the neighbourhood of that town.—Mr. McLachlan exhibited an example of *Rhinocyphea fulgidipennis*, Guérin, a brilliant little dragon-fly of the subfamily Calopteryginae, a native of Cochin China, which, so far as he knew, had not been captured since prior to 1830. It had been in M. Guérin's hands, and Mr. McLachlan had received it from M. René Oberthür.—Mr. T. A. Chapman exhibited various specimens illustrating *Acanthopsyche opacella*.—Mr. Barrett exhibited specimens of Heterocera destructive to the fruit crops of South Africa. Among them *Sphingomorpha monteironis*, Butl., known as the Fruit Moth in Cape Colony—a bold and powerful insect, with a sucking tongue strong enough to pierce the sound skin of a peach or fig. The presence of a light does not appear to disturb it, so that examination of its methods can be readily made, when it can be seen that it does not take advantage of the natural opening into a fig, or of a crack or other injury to a peach, but deliberately pierces a hole which afterwards shows as a small round spot, from which decay invariably results. It seems a matter of indifference to the moth whether the fruit has fallen, or is on the tree, ripe or unripe. With regard to *Achaea lienardi* and *Serrodes inara*, the two species are restless and timid, and therefore more difficult to observe. In the present season, however, both have been extremely abundant, and have been seen at apparently uninjured fruit, so that it seems they are capable of equal destruction. Several others, feeding mainly on damaged fruit, were also taken with the aforesaid species, among them some new to science, and recently described by Sir George Hampson. Mr. Jacoby exhibited *Callomorpha wahlbergi* from Africa, and *Spilopyra sumptuosa* from Australia.—A paper was communicated on "New Palæarctic Pyralidæ," by Sir George F. Hampson, Bart.

Anthropological Institute, May 15.—Mr. C. H. Read, President, in the chair.—The president alluded to the severe loss which the Institute and anthropology in general had sustained in the loss of its former president, the late General Pitt-Rivers.—Mr. F. Haverfield contributed a note on certain stone objects discovered on a Roman site at Clanville, in Hants, and a discussion ensued from which it appeared improbable that they were of human workmanship.—Mr. J. Allen Brown described a collection of stone implements brought from Pitcairn Island by Lieut. Pike, R.N. The implements are of two types, both formed of the volcanic rocks of the island. The first series consists of stone axes of analogous forms to those of other islands of the Pacific. The other is peculiar, being large, and with incurved sides and broad cutting edge, more or less ground as well as chipped. A third form is that of a cylindrical chisel. The author mentioned also the discovery of rock carvings of sun, moon, birds, &c., tombs with pottery and human skulls, and of carved stone figures like those of Easter Island. The fact that the implements were found below the surface of the ground, and that from the time of its discovery by Carteret until its occupation by the mutineers of the *Bounty*, makes it probable that the remains in question are of considerable age.—Mr. H. Stopes

exhibited a number of unclassified stone objects which he had collected from the river gravels of the Thames valley, and discussed the purposes for which he believed them to have been shaped. He also produced specimens of *Neritina flaviatilis* found in the same gravels, which he regarded as an indication of their age.

Zoological Society, May 22.—Dr. Albert Günther, F.R.S., Vice-President, in the chair.—A communication was read from Prof. G. B. Howes, F.R.S., and Mr. H. H. Swinnerton, on the development of the skeleton of the Tuatera, *Sphenodon* (*Hatteria*) *punctatus*, which was stated to be the outcome of eighteen months' work on materials supplied to the authors by Prof. Dendy, of Christchurch, N.Z. An account was given of the egg, the hatching, and the habits of the hatched young, which the authors reared till four months old.—The Malacostracan Crustacea collected by Mr. Rupert Vallentin at the Falkland Islands, from December 1898 to February 1899, formed the subject of a paper by the Rev. T. R. R. Stebbing, F.R.S. Many of the species had long been known, as several scientific expeditions had been made to these islands during this century. This carefully made collection, however, had afforded a much needed opportunity for discussing and clearing up obscure points in some of the earlier descriptions of the Crustacean fauna.—Mr. L. A. Borradaile read the fourth instalment of his memoir on Crustaceans from the South Pacific. This part contained an account of the crabs, of which 77 species were enumerated. Seven new species were described, and a scheme of classification of the swimming crabs (*Portunidae*) was put forward.—A communication was read from Dr. R. Bowdler Sharpe, which contained an enumeration of the birds—56 species in all—collected during the Mackinder Expedition to Mount Kenya, accompanied by field-notes of the collectors.—Mr. F. E. Beddard, F.R.S., read a paper, entitled "A Revision of the Earthworm Genus *Amyntas*." According to the author, this genus comprised 102 species, which were enumerated and commented upon.—Mr. Beddard also read a paper on the structure of a new species of earthworm, which he proposed to name *Benhamia budgetti*, after its discoverer, Mr. J. S. Budgett, who had obtained two specimens of it at M'Carthy's Island during his recent visit to the Gambia.

PARIS.

Academy of Sciences, May 21.—M. Maurice Lévy in the chair.—Researches on the formation of nitric acid during combustions, by M. Berthelot. The compressed oxygen used in combustions in the calorimetric bomb always contains a small quantity of nitrogen, up to 8 per cent. A portion of this is oxidised during the combustion, and the amount of nitric acid so formed has been regularly measured in order to correct the calorimetric data. The author now attempts to trace the relation between the nature of the organic substance under combustion and the quantity of nitric acid formed, details being given in the present paper of experiments on amorphous carbon, graphite and diamond.—Limits of combustibility by red-hot copper oxide of hydrogen and methane diluted with large volumes of air, by M. Armand Gautier. When combustible gases, such as hydrogen or marsh gas, are mixed with large quantities of air and are passed over columns of red-hot copper oxide, the difficulty of completely burning the gas increases with the dilution. Thus with a dilution of 20 parts in 100,000, hydrogen is not completely burnt on passing over a column of 35 centimetres of red-hot copper oxide, but combustion is complete when this length is doubled. Methane is more difficult to burn; thus at a dilution of 7 in 100,000 nearly half the carbon escaped unburnt after passing over a column of oxide 70 cm. long.—Publications of the Observatory of Besançon from 1886 to 1896, by M. Loewy.—Action of hydrogen bromide upon dextro-rotatory benzylidene camphor, by MM. A. Haller and J. Minguin. Benzylidene camphor combines with hydrobromic acid to form mono-bromo-benzyl-camphor. If the combination is carried out at 100°, two other products are obtained, benzylidene-campholic acid,

$$(COOH).C_8H_{14}.CH:CH.C_6H_5,$$

and phenyloxy-homocampholic acid,

$$(COOH).C_8H_{14}.CH_2.CH(OH)C_6H_5,$$

derivatives of which are described.—On fossil forests and the vegetative soils of the coal-measures, by M. Grand'Eury. Further arguments in favour of the author's view that the vegetable fossils have really grown in the places where they now occur, and have not been deposited there by water.—

Report on works presented by M. Marx.—Remarks on an eruption of the volcano Mayon in the Isle of Lucon, by the French Consul for the Philippines.—On the convergence of the coefficients in the development of the perturbation function, by M. A. Férand.—Remarks on a memoir of M. Massau on the graphical integration of some partial differential equations, by M. J. Coulon.—On a remarkable point in relation wi h the Joule-Thomson effect, by M. Daniel Berthelot. The point at which the inversion of the Joule-Thomson effect occurs is deduced by a graphical construction from the data of Amagat, and the result compared with some recent calculations of M. Van der Waals.—On the distribution of currents and potentials in the periodic state set up in the length of a symmetrical polyphase line presenting capacity, by M. Ch. Eug. Gaye.—On resonance in wireless telegraphy, by M. A. Blondel. —Communication by wireless telegraphy with the aid of radio-conductors with polarised electrodes, by M. C. Tissot.—On anhydrous calcium dioxide and the constitution of its hydrates, by M. de Forcand. A thermochemical paper.—On some properties of aluminium, and on the preparation of gaseous hydrogen phosphide, by M. Camille Matignon. Details are given showing how to burn aluminium in steam, carbon monoxide and dioxide, oxides of nitrogen, formic acid, sulphur dioxide and other gases and vapours. The preparation of crystallised aluminium phosphide is described, from which pure PH_3 can be readily obtained.—Combinations of lithium bromide with ammonia, by M. J. Bonnefoi.—The compounds $LiBr.NH_3$, $LiBr.2NH_3$, $LiBr.3NH_3$, $LiBr.4NH_3$ are indicated by a study of the dissociation pressures. The application of the Clapyron formula to these data gives values for the heats of dissociation of these compounds in good agreement with the direct thermochemical measurements.—On two polysulphides of lead and copper, by M. F. Bodroux. The compounds PbS_2 and Cu_2S_2 are described.—On a mercury chlorosulphide, by M. F. Bodroux. A chlorosulphide, $Hg_2S_3.HgCl_2$, can be prepared, which is stable at ordinary temperatures.—Action of water upon mercurous sulphate, by M. Gouy.—Partial synthesis of lævo-rotatory erythritol, by M. L. Maquenne. Wohl's method is applied to xylose, the steps being xylosoxime, acetylxylonic nitrile, erythose-acetamide, and erythrite.—Preparation of the dialkylamido-dichloran-thraquinones, by M. E. C. Severin.—On a monoiodohydrin of glycol, by MM. E. Charon and Paix-Séailles.—On γ-chloro-crotonic acid, by M. R. Lespieau. A description of the properties of $CH_2Cl.CH:CH.CO_2H$, its nitrile and ethyl ester.— On the composition of the albumen of the St. Ignatius bean and of the *nux vomica* bean, by MM. Em.Bourquelot and J. Laurent. The albumen from these beans yields the same carbohydrates, a mixture of a mannane and a galactane, as the albumen of leguminous beans previously studied. In *nux vomica* the proportion of galatose found on hydrolysis is very high. These beans, in fact, serve as an advantageous source of crystallised galatose. —Experimental researches upon the evolution of the lamprey, by M. E. Bataillon.—Remarks upon certain points in the life-history of the lower organisms, by M. J. Kunstler.—On some new Synclavellæ in the compound Ascidians, by M. Maurice Caullery.—Analysis of marine deposits collected off Brest, by M. J. Thoulet.—The mineral matters in the human fœtus during the last five months of pregnancy, by M. L. Hugonenq. —Identity of a bacillus from milk with the pneumobacillus of Friedlænder, by MM. L. Grimbert and G. Legros. The complete identity of these bacilli was shown by a comparative study of their general biology and morphology, and of their action upon carbohydrates.

DIARY OF SOCIETIES.

THURSDAY, MAY 31.

ROYAL SOCIETY, at 4.30.—Palæolithic Man in Africa : Sir John Evans, F.R.S.—On the Estimation of the Luminosity of Coloured Surfaces used for Colour Discs : Sir W. de W. Abney, F.R.S.—The Sensitiveness of Silver and of some other Metals to Light : Major-General Waterhouse.— The Crystalline Structure of Metals (Second Paper): Prof. Ewing, F.R.S., and W. Rosenhain.—Vapour-density of Bromine at High Temperatures (Supplementary Note): Dr. E. P. Perman and G. A. S. Atkinson.

FRIDAY, JUNE 1.

ROYAL INSTITUTION, at 9.—Bunsen : Sir Henry Roscoe, F.R.S.
GEOLOGISTS'ASSOCIATION,at 8.—Our Older Sea Margins : Sir Archibald Geikie, F.R.S.

TUESDAY, JUNE 5.

ANTHROPOLOGICAL INSTITUTE, at 8.30.—The Metric System of Identification used in Great Britain : Dr. J. G. Garson.

WEDNESDAY, JUNE 6.

GEOLOGICAL SOCIETY, at 8.—Mechanically-formed Limestones from Junagadh and other Localities : Dr. J. W. Evans.—Note on the Consolidated Æolian Sands of Kathiawad : Frederick Chapman.—On Ceylon Rocks and Graphite : A. K. Coomara Swamy.
SOCIETY OF PUBLIC ANALYSTS, at 8.—The Determination of Oxygen in Copper by Ignition in Hydrogen : Leonard Archbutt.—Uniformity in the Conduct of Soil Analysis : A. D. Hall.—The Adulteration of Wheaten Flour with Maize : G. Embrey.—A New Colour Reaction for distinguishing between certain Isomeric Allyl and Propenyl Phenols : Alfred C. Chapman.
ENTOMOLOGICAL SOCIETY,at 8.

THURSDAY, JUNE 7.

LINNEAN SOCIETY, at 8.—On a Viviparous Syllid Worm (E. S. Goodrich. —On the Genera Phæoneuron, Gilg., and Dicellandra, Hook f. : Dr. A. Itapf.—On the Structure and Affinities of *Echinrus unciuctus* : Miss Embleton.
CHEMICAL SOCIETY, at 8.—Diphenyl- and Dialphyl-ethylenediamines, their Nitro-derivatives, Nitrates, and Mercurochlorides : W. S. Mills. —Condensation of Ethyl Acetylenedicarboxylate with Bases and β-ketonic Esters : Dr. S. Ruhemann and H. E. Stapleton.—The Constitution of Pilocarpine : Dr. H. A. D. Jowett.—The Nitrogen Chlorides derivable from ω-Chloroacetanilide and their Transformations : Dr. F. D. Chattaway, Dr. K. J. P. Orton, and W. H. Hurtley.—Derivatives of Cyanocamphor and Homocamphronic Acid : Dr. A. Lapworth.
RÖNTGEN SOCIETY (St. Bartholomew's Hospital), at 8.—Dr. Lewis Jones will show an Influence Machine of American design.—Mr. James Wimshurst, F.R.S., will give a short statement of his work in the design and the perfecting of the several forms of his Influence Machine.—Dr. Rémy, of Paris, will show a new Localising Apparatus.

FRIDAY, JUNE 8.

ROYAL INSTITUTION, at 9.—The Effect of Physical Agents on Bacteri Life : Dr. Allan Macfadyen.
PHYSICAL SOCIETY, at 5.—On the Magnetic Properties of Iron and Aluminium Alloys, Part II. : Dr. S. W. Richardson.—Note on Crystallisation produced in Solid Metal by Pressure : W. Campbell.— On the Viscosity of Mixtures of Liquids and of Solutions : Dr. C. H. Lees.
ROYAL ASTRONOMICAL SOCIETY, at 8.

CONTENTS.   PAGE

THURSDAY, JUNE 7, 1900.

MODERN PHYSICAL CHEMISTRY.

The Theory of Electrolytic Dissociation. By H. C. Jones. Pp. xii + 289. (New York : The Macmillan Company. London : Macmillan and Co., Ltd., 1900.)

THE theory of electrolytic dissociation is only some fifteen years old, but in that short time its growth has been very great, and its suggestiveness most marked. We gladly welcome a volume on the subject by one who has himself done much to promote its advance, and to render more secure some of the positions it has taken.

The author's preface explains his object ; he has been asked from time to time where an account of the newer developments of physical chemistry is to be found. Original memoirs are not always accessible to a student, and in many cases explanation is wanted, and further development of an argument or line of thought may prove helpful ; and so Mr. Jones has given us a book based in the main directly on the work of van 't Hoff, Ostwald, Arrhenius and the others who have made the theory, but in which the numerous developments of Arrhenius' original idea are skilfully brought together, and the bearing of the theory on phenomena, apparently widely diverse, is clearly shown. The plan of the book is in the main a good one. The first chapter is devoted to the earlier physical chemistry with the object of showing its relation to that which was to follow ; accounts are given of Kopp's work on the boiling points of liquids and on molecular volumes, of the researches of Lorentz, Gladstone and Dale, Le Bel and van 't Hoff and Perkin on optical properties. The investigations of Favre and Silbermann, Berthelot and Julius Thomsen into thermal chemistry, the electrolytic work of Faraday and Clausius, Hittorf and Kohlrausch, are described ; and an interesting and important discussion of the development of chemical dynamics and chemical statics concludes the chapter.

In the explanation of Guldberg and Waages' Law of Mass Action, there is a vague and somewhat unsatisfactory use of the word *force ;* we are told that

"if we represent the active masses of two substances by m and n, and the coefficient depending on the nature of the substance, &c., by c, the force of the chemical reaction is expressed by mnc."

"Force" has no meaning used in this connection, the mass of compounds, or the number of molecules of compounds produced, can clearly be put equal to mnc, and the condition of equilibrium will be reached when this mass is equal to the mass of matter combining to form the original substance. There is an obvious misprint on page 62 ; k'/k is clearly written for c'/c of page 61.

In Chapter ii. we are introduced to the main subject of the book. An account is given of van 't Hoff's original paper "The Rôle of Osmotic Pressure in the Analogy between Solutions and Gases" (*Zeitschrift für physikalische Chemie,* i. p. 481), and the grounds for believing that in certain solutions the osmotic pressure conforms exactly to the three gaseous laws of Boyle, Gay Lussac and Avogadro are stated. Attention is here drawn to the large class of compounds, all the acids, all the bases, and all the salts, which form exceptions to the above statement. For these the law is no longer $PV = RT$, but

$PV = iRT$, where i is a coefficient always greater than unity, to which a meaning is given when we consider the work of Arrhenius.

According to this the molecules in an electrolyte, or some of them, are dissociated into ions. The electrolytic effects depend on the dissociated or active molecules ; let there be n in number, and suppose each is divided into k parts ; suppose also that there are m molecules remaining inactive or undissociated, then the total number of molecules is $m + n$, the number of inactive molecules and ions is $m + kn$, and the value of van 't Hoff's coefficient i is shown to be $(m + kn)/(m + n)$.

Arrhenius' theory of electrolysis is an extension of that of Clausius. Clausius had shown that in an electrolyte it was necessary to suppose some molecules of the dissolved salt were broken up into ions ; Arrhenius explained how to determine from observations on osmotic pressure the number of such molecules in a given solution.

From this we are led on to two interesting chapters— "Evidence for the Theory" and "Applications of the Theory" ; the evidence which the author has accumulated is most valuable, while the fertility and resourcefulness of the theory are strikingly shown. The book will be very useful ; at the same time, in one respect, it is open to criticism of some importance. A student not unnaturally asks, What is osmotic pressure due to ? Why, under certain circumstances, does liquid run into a closed vessel apparently against the pressure ? What is the mechanism by which such a process is managed ? It may be answered, We do not know ! The author may fairly wish to use language independent of any molecular theory, and not bind himself down in any way ; it is enough for many purposes, it may be said, to know that there is a definite pressure in such a solution without inquiring how that pressure is caused. At the same time it is impossible to avoid alluding to molecular impacts and the like ; there is no evidence that Mr. Jones does wish to avoid it, and in places, *e.g.* p. 95 and elsewhere, he refers to the modern kinetic theory of gases, and we think—this is the criticism—that it would have made the book clearer if he had based his explanation throughout on the extension of that theory to liquids. When an ordinary experiment for measuring osmotic pressure is started with a solution inside a vessel closed with a semipermeable membrane, the number of water molecules which strike a unit of area of the interior surface in any given time is less than the number striking the same area on the outside ; thus more molecules of water enter the space than leave it ; the molecules of the salt cannot pass the membrane, hence the pressure inside increases. The tendency is both for the water and the dissolved substance to distribute themselves uniformly ; the pressure inside is due to the impacts (1) of the water, (2) of the dissolved substance that outside arises from the impacts of the water only ; ultimately the pressures due to the water balance, and the excess of pressure inside measures the effect due to the impacts of the dissolved substance.

The whole is merely an example of the first proposition of Mr. Jones' third chapter. The physical properties of completely dissociated substances should be additive. This is all implied in the book ; it might with advantage be stated more precisely.

G

Another matter with regard to which a greater definiteness seems desirable, even at the expense of some generality, is the theory of the action of the voltaic cell.

Nernst's theory of the electrolytic "solution-tension" of a solid—solution-pressure is a preferable term—is stated in his own words, but they are vague :

"We must ascribe," it is said, after a reference to osmotic pressure, "to a dissolving substance in contact with a solvent, similarly, a power of expansion, for here also the molecules are driven into a space in which they exist under a certain pressure. It is evident that every substance will pass into solution until the osmotic partial pressure of the molecules in the solution is equal to the 'solution-tension' of the substance" (pp. 231–232).

We may put the whole theory slightly differently, thus :—

In the case when a substance is being dissolved in such a way that *molecules* pass from the solid into the liquid, the pressure rises in consequence of the impacts of these molecules on the walls of the containing vessel ; now when molecules also pass from the liquid into the solid, the "evident" fact is that the steady state is reached when the numbers entering and leaving the liquid are the same. In such a case the osmotic pressure measures the solution-pressure, and no electrical action is involved.

But now let us suppose that a metal is passing, not in the form of *molecules*, but in that of *ions* into water. Each of these ions carries with it a positive charge ; the water therefore tends to become positive, the metal negative, and an electrical double layer is formed over the surface of separation. The charged ions are not free to move throughout the water, but few escape from the surface ; hence the additional pressure due to the impacts of the metallic ions—the solution-pressure, as it is called—is small.

Again, let us take the case of a metal, such as copper, in a solution of one of its own salts, say copper sulphate ; here, also, if there were no electrical effects, we might suppose that copper *molecules* would be deposited out of the sulphate on to the metal, while other molecules would leave the metal ; the steady state would be reached when these two sets of molecules became equal in number, and the osmotic pressure would become—in reality, unless the solution were very weak, would *fall* to—the solution pressure. But according to the theory, the copper passes as *ions* which carry with them out of the solution their positive charge ; this they give up to the metal on becoming molecules. And since we suppose that, unless the solution be very weak, the number of copper ions leaving it is, to start with, greater than those entering, the metal becomes positive, the negative ions of the solution are attracted to it, the positive ions driven off, a double layer is again formed ; a difference of electrical potential is established between the metal and the solution—the metal being positive, the solution negative.

If, however, we consider a metal, such as zinc, which has a high solution pressure when immersed in, say, zinc sulphate, we must suppose that at the start more metallic ions leave the metal than enter it, the solution thus becomes positive, the metal negative, and the double layer formed is one which tends to prevent the positive metallic ions from leaving the zinc, and is thus opposite to that formed on the copper.

NO. 1597, VOL. 62]

In both these cases we must suppose, when the steady state is reached, that the ions leaving the metal leave it under the solution pressure of the metal in the liquid. This may be seen as follows : If there were no electrical force called into action, the pressure would go on changing in the liquid up to the solution pressure, when the number of metallic ions leaving the surface would balance those entering.

Thus the solution pressure measures the whole amount of momentum which the ions of the metal tend to transfer per second across unit area of the surface. Now according to the theory this momentum depends on the metal only, and the tendency to transfer momentum remains the same, however the transfer be stopped ; in reality, the electrical forces acting across the double layer stop it, not the opposing momentum of the liquid ions, and the pressure exerted by these electrical forces must be therefore equal to the solution pressure of the solid, *i.e.* when a current is flowing the positive ions start from the metal at the solution pressure of the metal, and become, when in the solution, ions at the osmotic pressure of the liquid.

Now, however, let us suppose that a piece of copper is connected to the zinc, the two being dipped into zinc sulphate ; and suppose further, for simplicity, that there is no action at the interfaces zinc-copper or copper-liquid, then negative electricity from the zinc passes over to the copper through the zinc-copper junction, attracting to itself the positive ions in the solution and destroying the double layer at the zinc-liquid junction ; thus a current of positive electricity passes through the solution from zinc to copper. The source of the E.M.F. is at the zinc-liquid junction, arising from the fact that more zinc ions pass from the zinc into the solution than from the solution into the zinc ; or, as Nernst would put it, that the solution pressure of the zinc is greater than the osmotic pressure of the liquid. In reality, of course, there may be actions at both the other junctions similar in character to that which we have supposed to go on at the junction of the zinc and the liquid, and the resultant E.M.F. depends on all of these.[1]

In this simple case the energy of the cell is obtained from the passing of the zinc ions from the saturation pressure of the zinc to the osmotic pressure of the liquid, and we obtain at once Nernst's expression for the electromotive force, varying as $RT \log., P/p$, where P is the saturation pressure, p the osmotic pressure.

But an article which started as a notice of Mr. Jones' most useful book is in danger of becoming a dissertation on the seat of the electromotive force of a voltaic cell, a result to be avoided.   R. T. G.

MESOZOA AND ENANTIOZOA.

Traité de Zoologie Concrète. T. ii. 1ᵉ partie. *Mésozo-aires—Spongiaires.* By Yves Delage and Edgard Hérouard. Pp. ix + 244. (Paris : C. Reinwald, 1899.)

AS might have been anticipated, this part of the massive "Traité de Zoologie," which is now in course of publication, contains matter of exceptional interest. One-fifth of the present issue is devoted to the Mesozoa,

[1] A reference should be made to Prof. Lodge's article in the May number of the *Philosophical Magazine*, which has appeared since the above was written.

and the remainder to the Sponges. The Mesozoa are classified provisionally under four divisions :— (1) Mesocœlia for *Salinella* ; (2) Mesenchymia for *Trichoplax* and *Treptoplax* ; (3) Mesogonia for *Dicyemidæ* (parasitic in the renal sacs of dibranchiate Cephalopods) and *Orthonectidæ* (parasites of Nemertines, Ophiurids and Polychaets) ; (4) Mesogastria for *Pemmatodiscus*.

Salinella has been regarded as the incarnation of an ideal promorph, the true *Mesozoon*, or link between unicellular and multicellular animals. The minute creature which has been saddled with so grave a responsibility was found, in 1892, by the late Dr. Frenzel in a jar of 2 per cent. salt solution containing mud taken from the salt works of Cordoba, in the Argentine Republic. The jar had been exposed for a long time, and some iodine washings had been thrown into it by mistake. The authors of the " Traité " give a full account of *Salinella*, and admit that, if it really exists, "c'est le vrai Mésozoaire."

When the complex character of the structure and lifehistory of the higher Protozoa is considered, the imputed simplicity of *Salinella* becomes almost grotesque, and it seems impossible to assign a cosmic importance to it, even should its autonomy become, in future years, an established fact.

"On ne le dit pas, mais il règne une certaine méfiance vis-à-vis de cet être venu si à propos, recueilli dans des conditions si étranges, observé si loin de nous et une seule fois. Ce vase contenant un liquide artificiel, exposé à l'air et aux poussières, qui a reçu les rinçures de la verrerie d'une table d'histologiste, ce pays lointain, tout cela ne prouve rien d'une manière positive contre la Salinelle."

Trichoplax and *Treptoplax* are likewise aquariumproducts, the former having been found at Trieste, in 1883, by Prof. F. E. Schulze, and the latter at Naples, in 1892, by Prof. F. S. Monticelli. These forms, which superficially resemble an acœlous Turbellarian, are riddles of the aquarium, like *Salinella* in this respect, and it seems premature to draw far-reaching conclusions from them until they are themselves solved.

The authors of the "Traité" introduce new matter into their account of the Mesogonia derived from a work written in Russian by N. A. Keppen, in which the spermatozoa of *Dicyema* are described and figured for the first time. Attention is drawn to the mystery surrounding the dissemination of the Dicyemid parasites from one host to another, since it is only the infusoriform males which can endure immersion in sea-water, this being quickly fatal to the vermiform females.

Pemmatodiscus is a gastruliform organism found by Monticelli (1895), living in closed sacs in the jelly of a Medusa, *Rhizostoma pulmo*. It would no doubt have excited enthusiasm twenty years ago. Its right to be regarded as an independent type is founded upon three considerations, namely, its parasitic habit, its inability to endure immersion in sea-water, and its power of multiplying by division. The first and last of these reasons are by no means conclusive, since parasitic larvæ, as well as embryos contained in brood-pouches, are known among Medusæ, as is also the phenomenon of embryonic fission.

An account of Haeckel's Gastræadæ is given on pp. 38

and 39, by way of appendix. One might almost have expected to find that the apocryphal Physemaria would have been allowed to go the way of *Bathybius* and *Eozoon*.

A second appendix (pp. 40-45) is devoted to the ciliated urns found in the body-cavity of Sipunculids. These are regarded by M. M. Kunstler and Gruvel, whose original drawings are here published for the first time, as being certainly parasites, and not forming part of the organisation of the Sipunculid. Two genera are described, *Kunstleria* n.g. from *Phymosoma* ; and *Pompholyxia*, Fabre-Domergue, from *Sipunculus*.

In their treatment of the Sponges, the authors tread on firmer ground, and the result of their labours is a most satisfactory performance. As promorph (type morphologique) of the entire group, they select for preliminary description the *Olynthus* of Haeckel. *Olynthus* is a generalised abstraction which has its embodiment in concrete zoology. Admitting that a treatise on Sponges at present could hardly be introduced in any other way, it may be pointed out that there are reasons for doubting whether the phyletic value of the *Olynthus* is as great as its undoubted morphological and didactic importance.

In the section devoted to the calcareous sponges (pp. 66-82), the authors quote freely from the researches of our compatriots, Prof. E. A. Minchin and Mr. G. P. Bidder. The classification recently suggested by Bidder is given *in extenso* on p. 67, although not adopted in the body of the work.

The sextets of actinoblasts which secrete the triradiate spicules of Ascons, as discovered and described by Minchin, are duly recorded, but the figure reproduced on p. 67 gives no idea of the excellence of the illustrations contained in Minchin's monograph.

The complete inversion of the layers, which takes place at the metamorphosis (pp. 60, 69, 106, 159), marks one of the most interesting phases of sponge-life. The primitive endoderm of the larva gives rise to the permanent epidermis of the adult, while the primitive flagellated ectoderm sinks in to form the flagellated chambers of the adult. This fact of inversion has induced Delage to separate the Sponges, under the designation Enantiozoa, from all other Metazoa.

The metamorphosis of the parenchymula-larva is accompanied by phenomena which have an interest extending beyond the limits of sponge-lore. The account given on pp. 110-111 shows the following succession of events which occurs in some cases during the conversion of the flagellated ectoderm of the larva into the choanocytes (collar-cells) of the adult :—

I.	II.	III.	IV.
Flagellated Ectoderm.	Histolysed Ectoderm.	Syncytial Ectoderm.	Choanocytes.

The reconstructions on the coloured plates, which elucidate the increasing complexity of the inhalent and exhalent canal systems throughout the group, are well executed, and produce a satisfying impression of solidity and reality. If there is a complaint to be made, it is that, in not a few cases, the authors have omitted to add in brackets the name of the generic type to which the diagrams and text-figures may be taken to refer.

Textual errors and inconsistencies are rare, and obvious

when they occur. A few examples will suffice. On pp. 2 and 36, the terms "cœlomique" and "cœlome" refer to a blastocœlic space; on p. 60, "gemmules" is given as an alternative expression to "bourgeons," which arise as outgrowths involving all the layers of the body (*e.g. Lophocalyx*), whereas on p. 177 the endogenous "gemmules" of *Spongilla* are rightly described as special formations, quite distinct from ordinary lateral or exogenous buds, although the buds of *Tethya* (p. 167) seem to be intermediate between the exogenous and endogenous varieties. On p. 91 (footnote), Sollas's term *collenchyme* is branded, with other related terms, as "bien inutile," but on p. 152 the superficial cortex of *Geodia* is characterised as "collenchymateuse."

In a footnote on page 203, we are reminded that H. J. Carter instituted a comparison between the flagellated chambers of sponges and the branchial sac of Ascidians. The authors add that this comparison "nous semble bien singulière aujourd'hui où ces êtres sont mieux connus." On the contrary, the comparison is appropriate, the analogy between the flagellated chambers of a sponge (in respect of their respiratory and nutritive functions and of their relations to the inhalent and exhalent canals) and of the Ascidiozooids in a compound Tunicary (*cf.* especially the Didemnidæ) being an extraordinarily close one; but of course Carter was innocent of the distinction between homology and homoplasy. What is very singular indeed is the fact that, in these latter days, the same fatal confusion between actual physiological conditions and abstract genetic relationships is constantly being repeated.   A. W.

THE DURATION OF THE BRITISH COAL-FIELDS.

Les Charbons Britanniques et leur épuisement. By E. Lozé. Pp. ix + 559, and vii + 562 to 1229. (Paris : C. Béranger, 1900.)

IN France, as in the rest of Europe, consumers have during the past winter been complaining of the difficulty of obtaining an adequate supply of coal, the chief cause of the increased demand having been the activity in the iron and steel trades. At the same time, prolonged strikes in Austria and elsewhere, and the temporary cessation of the production of the collieries of Natal and Cape Colony, have lessened the supplies usually available. The prevailing scarcity of coal is a matter of serious moment to France, where, owing to the increasing depth of the collieries and the costly nature of mining operations, the quantity of coal that has to be imported from other countries grows larger every year. At the present time about two-thirds of the coal consumed in France is raised in the country; and last year the imports amounted to 10,500,000 tons, of which quantity 6,000,000 tons were obtained from Great Britain. France being so largely dependent on Great Britain, it will readily be seen that the duration of the British coal-fields is a subject of no little importance to French economists. M. Lozé has, therefore, been induced to devote two bulky volumes, covering together 1229 pages, to a critical consideration of the investigations of Prof. Stanley Jevons, the Right Hon. Leonard H. Courtney, Mr. R. Price-

Williams, Mr. T. Forster Brown, Prof. E. Hull and othe English writers.

The results of his studies are grouped in four sections. The first contains an account of the geography of the British Isles, with historic, geological and economic details. The second section contains a detailed description of each of the British coal-fields, with a chapter on the coal resources of the Colonies. The third section deals with commercial geography, water and railway transport, and the principal industrial centres. The fourth and last section contains an estimate of the coal supplies of the United Kingdom, with a summary of the views expressed as to their probable duration. The work concludes with a lengthy appendix dealing with cognate matters, the production and consumption of mineral fuel in various parts of the world, the constitution of the British Colonial empire, the navy and the army.

In discussing the views of the various authorities, the author prefers to accept the pessimistic forecast of Mr. T. Forster Brown rather than the optimistic estimate of Prof. Hull. Mr. Forster Brown calculates that the amount of coal of good quality remaining in the United Kingdom at a depth not exceeding 2000 feet, the depth that he regards as the limit of economical mining, is 15,000 million tons. Such is the supply on which Great Britain must base its hopes in the inevitable economic conflict with the United States. In spite of the care and accuracy with which the divergent views on the subject are set forth, it may be doubted whether the author has made out a clear case for rejecting Prof. Hull's estimates, which show that the amount of coal remaining within a depth of 4000 feet is 81,683 million tons. The criticism of Prof. Hull's views is not convincing, inasmuch as M. Lozé, who does not appear to possess a practical knowledge of geology and mining, has not followed the recent investigations as to the limits at which mining may be carried on with profit. At the present time the greatest depth at which in Great Britain mining operations may be carried on has been reached at the Pendleton colliery, near Manchester, where the deepest workings are nearly 3500 feet below the surface. This enormous depth has, moreover, been exceeded in other countries, notably in the Lake Superior district, where a shaft of the Calumet and Hecla copper mine has now attained the record depth of 4900 feet, and in Belgium, where a colliery at Mons is 3937 feet deep. Depths such as these show that the limit of depth of 4000 feet assumed by Prof. Hull is well within the bounds of possibility. In view of the marvellous efficiency of modern winding-engines, no considerations of a mechanical nature need limit the prospective depth of shafts. By far the most important obstacle to very deep mining is the increase of temperature in proportion to the depth. Here, again, the author is apparently not familiar with recent observations. Since 1848 and 1854, the dates of observations cited by him, methods of determining earth temperatures have been greatly improved, and the results recently obtained at the Paruschowitz borehole in Silesia, put down by the Prussian Government to a depth of 6573 feet, show an increase of temperature of 1° F. for every 62·1 feet. This rate of increase would not present an insuperable obstacle to mining at a depth of 4000 feet.

The author gives in tabular form an estimate of the population, coal output, export and consumption for the years 1899 to 1950, by which date the 15,000 million tons assumed to be now remaining will be exhausted. The prosperity assured by the coal of the country to navigation, manufactures and commerce will then gradually disappear, and the historian of a powerful empire will conclude, the author prophesies, his account of a remarkable period by the words : *Finis Britanniae !* Happily, however, the array of statistics, the copious particulars of the coal-seams, and the faithfully translated estimates of eminent experts do not altogether justify the author's Cassandra-like attitude.

The work has been compiled with great care, and the author deserves high praise for the accuracy with which the names of English places and persons have been presented. On p. 564 there is a curious slip. Speaking of the introduction of railways in 1844, the author says : "Mine, aubergiste *of the George*, pleurait la fin des diligences." The archaic expression "Mine host" has proved too severe a test for the author's undoubtedly extensive knowledge of the English language.

BENNETT H. BROUGH.

OUR BOOK SHELF.

Ueber den Bau und die Entwicklung der Linse. By Dr. Carl Rabl. Pp. 324 ; plates 14. (Leipzig, 1900.)

IN the "Notes" column brief mention has recently been made of the concluding portion of Dr. Rabl's important investigations on the structure and development of the crystalline lens of the eye, which appeared in the *Zeitscrhift fur wiss. Zoologie.* The author has now reproduced the entire monograph as a separate work, with the original coloured plates ; and since it is a most elaborate treatise on a very difficult subject, its appearance in this form should be welcomed by all students of this branch of anatomy.

There are, perhaps, few phenomena in the developmental history of animals more astounding to the ordinary mind than the fact that a structure seated so comparatively deep as the crystalline lens of the human eye should arise from the outer, or epiblastic, layer of the embryo, and attain its permanent position, first by invagination, and then by separation from its parent layer. Nevertheless, it is a fact about which there can be no possibility of dispute ; and the more superficial position occupied by the spherical lens of fishes serves, in a measure, to indicate the manner in which the conditions obtaining in the mammalian eye have been gradually evolved.

By means of the beautiful series of plates illustrating Dr. Rabl's work the student is enabled to comprehend at a glance, firstly, the mode of development of the lens respectively characteristic of fishes, amphibians, reptiles, birds and mammals ; and, secondly, the different histological peculiarities presented by the lens itself in the same groups. Within the limits of a notice in this column, it is out of the question to discuss any details of the work before us ; but it may be mentioned that in the concluding section the author enters into the abstruse speculation as to what may have been the degree of development of the eye in *Archaeopteryx* and other extinct animals, and also as to the gradations which may have formerly existed between the present differentiated types of lens-structure. Very interesting, too, are his observations with regard to the lens in the aborted eye of the mole. Here the rudimentary condition of the lens does not commence in the course of development, or in the fully adult animal ; but it is distinctly ob-

servable in the earliest stages, when it is relatively smaller and contains fewer cells than in other mammals. Hence we have evidence of the extreme antiquity of the mole's adaptation to its present state of existence—evidence fully supported by palæontological facts.

The work may be characterised as a masterpiece of patient and careful investigation in an abstruse and difficult line of research.   R. L.

Building Construction for Beginners. By J. W. Riley. Pp. vi + 255. (London : Macmillan and Co., Ltd., 1899.)

THIS is an addition to the increasing number of works on Elementary Building Construction, which all have for their ultimate goal the preparation of students for the May examinations of the Department of Science and Art.

Commencing with the inevitable introductory remarks on drawing instruments and scales, the student is taken through all the various building trades, and at the end of each are added questions in the form of examination papers which should test the student's knowledge as he advances.

As the author observes, isometric projection is a very valuable means of showing the beginner exactly what is intended, as it gives in one view the plan, elevation and section of the object portrayed. We are glad to see that an extensive use is made of such a form of illustration.

We may also congratulate our author on abstaining from confusing his illustrations by figuring with too many dimensions. Some authors refer with pride to their use of such a system, but as Mr. Riley observes, it is very confusing, and tends by its complication to hinder the very object for which it is introduced.

In a new edition several small slips can be attended to, such, for instance, as the wall-plate surroundings in Fig. 384. The brickwork in this case should be taken up to the underside of the tiles. The "summary" at the end of each trade is an excellent innovation, and the book can be confidently recommended as the best of its class.

Catalogue of the Fossil Bryozoa in the Department of Geology, British Museum (Natural History). The Cretaceous Bryozoa. Vol. i. By Dr. J. W. Gregory. Pp. xiv + 457, and plates. (London : Printed by the order of the Trustees, 1899.)

WE may congratulate Dr. J. W. Gregory in having completed this volume before he left this country to take up the geological professorship at Melbourne. The value of this, and similar works, is inestimable to palæontologists in all parts of the world. The book itself is naturally a list of hard names ; but it is something nowadays to know which is the correct name to apply to any particular fossil, and Dr. Gregory gives as far as possible the synonymy, diagnosis, dimensions and geological distribution of each species. A number of woodcuts in the text and seventeen excellent plates illustrate a great many of the species. We should have been glad of a table of the Cretaceous strata, to inform or remind us of the approximate British equivalents of such divisions as Rhodanian, Campanian, Hauterivian, &c., and also to indicate the sense in which the terms Neocomian and Cenomanian are used.

The volume deals with the various families which are included under the sub-orders Tubulata, Cancellata and Dactylethrata. All these are ranged under the order Cyclostomata, the sub-class Gymnolæmata, the class Ectoprocta and the group Bryozoa. It will be remembered that in the catalogue of recent marine forms in the British Museum, by Busk, that author employed the term Polyzoa instead of Bryozoa. The effort to secure a fixity in zoological nomenclature is one of the trials which beset the path of the worker. Dr. Gregory's carefully prepared catalogue will, we hope, have a permanent value in this respect.

LETTERS TO THE EDITOR.

[*The Editor does not hold himself responsible for opinions expressed by his correspondents. Neither can he undertake to return, or to correspond with the writers of, rejected manuscripts intended for this or any other part of* NATURE. *No notice is taken of anonymous communications.*]

The Kinetic Theory of Planetary Atmospheres.

IN the paper which I communicated to the Royal Society on April 5, I examined the logical conclusions obtained on the hypothesis that the atmosphere of a planet is distributed according to the generalised form of the Boltzmann-Maxwell distribution applicable to a gas in a field of external force, with the further generalisation required to take account of the effects of axial rotation. As regards the effects of the planet's attraction on the distribution of density, the expressions assumed to represent these were of the form now generally accepted by writers on the kinetic theory (*e.g.* Watson and Burbury), and the modifications required in taking account of centrifugal force were investigated by me in 1894, and are in harmony with the conclusions to which Maxwell's investigations led. In the aforementioned paper I showed how to calculate a superior limit to the rate at which a planet is losing its atmosphere. and obtained the results that helium would be permanently retained at all ordinary temperatures by terrestrial gravitation and vapour of water by the gravitation on Mars ; conclusions with which those deduced by Mr. Cook would appear to be identical, so far as I judge from his letter.

The objections which naturally suggest themselves to the mode of treatment in this paper are that the distribution in question is that which would be brought about exclusively as the result of molecular encounters, and of the free paths of the molecules between these encounters ; and that it therefore represents the distribution in an atmosphere of uniform temperature. In an actual atmosphere the equilibrium of the lower strata is largely modified by convection currents, so that the adiabatic law, rather than the isothermal law, is applicable. This point I hope to discuss at full length in the second part of the paper ; in the meanwhile, it is hardly likely that any one will suggest that helium escapes from our atmosphere because the upper strata are at a *low* temperature, but that it would cease to escape if the upper strata were heated up to the same temperature as the lower ones. The point at issue between Dr. Johnstone Stoney and Mr. Cook and myself appears to be how far the Boltzmann-Maxwell distribution represents what happens in the *upper* strata of the atmosphere. To assert " that in the present state of our knowledge it " (the *a priori* method as Dr. Stoney calls it) " cannot be made to furnish a valid investigation," seems to me tantamount to striking at the very foundations of our kinetic theories of matter. It may be that these theories will not resist such an attack, but the consequences of the onslaught cannot be properly traced, except by making mathematical determinations in the way that I have done. It appears to me to be just in this very problem of planetary atmospheres that the fundamental assumptions of the kinetic theory are least open to objections. Experiments on the relation of diffusion to temperature led Maxwell to abandon the notion that the molecules of a gas behave as elastic spheres and to consider the effects of finite intermolecular forces. So far as I am aware, (1) every attempt at a kinetic explanation of the thermodynamical properties of gases on the latter view involves some assumption which restricts its validity to the limiting case of attenuated gases, where the number of molecules within each other's sphere of influence is a negligible proportion of the whole number, and the duration of an encounter is negligible in comparison with the time of free motion between encounters. On the other hand, (2) it is amply proved by Watson and Burbury that the Boltzmann-Maxwell distribution, if it hold at any instant, will hold at all future instants in the *absence* of molecular encounters. (3) Boltzmann's minimum theorem tells us that if encounters take place at random, the molecules tend towards the distribution in question. (4) We are told on good authority that we must regard the Boltzmann-Maxwell law as a theorem in probability. Now the divergence between actual conditions and the assumptions required under heading (1) gets less and less as we ascend in the atmosphere ; (3) gives us reason for believing that the Boltzmann-Maxwell distribution holds at the highest altitudes where encounters not unfrequently take place ; (2) shows that the molecules which are projected from these strata and ascend to still greater altitudes

without encountering other molecules remain distributed according to the same law ; and (4) removes the necessity of taking the size of the element of volume $dx\,dy\,dz$ into account by telling us that the law represents not merely the number of molecules having given limits of velocity occurring in the element, but also the *probability* of a molecule coming within these limits, and this probability may be as small as we please.

If helium really does escape from our atmosphere, either there must be a fallacy in the assumptions underlying (1), (2), (3), or (4), and this fallacy must affect numerous previous writings on the kinetic theory, or else our preconceived notions as to the relation between temperature and kinetic energy are at fault. With regard to (4), it may be objected that the error-law fails to apply to events of exceptional occurrence, and therefore that we cannot apply it to calculate the probability of a molecule escaping from the atmosphere when the velocity required would represent an abnormal divergence from the mean. This point was carefully considered by me. It appears, however, to be the accepted view that abnormal divergences are excluded because in practice they never occur, not because their occurrence is far more frequent than the error-law would lead us to suppose. If the methods of the kinetic theory should prove to be inapplicable to rarefied gases as well as to dense assemblages of molecules, and they do not altogether agree with experiment for distributions of intermediate density, the position is indeed a serious one. In face of such a possibility, instead of abandoning our mathematical calculations we ought to push them to their ultimate consequences, in order to arrive at a better understanding of the true state of the case. The escape of gases from the atmospheres of planets is a phenomenon probably more directly dependent on the translational kinetic energy of the molecules than any other property of gases. The prevailing doctrine that not only is the mean value of this translational kinetic energy proportional to the absolute temperature, but the conceptions of temperature and kinetic energy are physically identical, has always seemed to me to require closer investigation than it has as yet received, and it may well be that the kinetic theory of planetary atmospheres furnishes one means of putting this doctrine to a test.

Plâs Gwyn, Bangor, May 26.   G. H. BRYAN.

The Severn Bore.

No one who suffers from scientific curiosity should miss seeing a tidal bore at least once in his life. The locality and conditions under which the Severn Bore can be seen make it an ideal object for a pleasurable excursion. The time to be selected is about twenty-four hours after new or full moon ; the largest spring tides should be chosen, if possible, and an occasion when the light permits both evening and morning bore to be seen. They occur at about 7.30 to 9 o'clock, a.m. and p.m. The visits should therefore be either when the days are long or at full moon. During a recent excursion, I stayed at Newnham-on-Severn, below Gloucester. This is about 3 hours 20 minutes from Paddington station, and it is possible to leave this station at 3.15 p.m. and be in time for the evening bore, see the morning bore next day, and be back at Paddington by 2.20 p.m.

On April 29, twelve hours after full moon, I awaited the bore at the south-east corner of Newnham Churchyard. The churchyard is the summit of a cliff situated on the outer bank, and near the center of the base of a U-shaped bend of the Severn, the limbs of the U being four miles long, and the width between the limbs two miles. The prospect is one of the most pleasing in the South of England ; the broad, winding river, emerald pastures abandoned by the wandering channel, miles of rich champagne country, with apple and plum orchards, and the distant range of the Cotswolds. At 6.45 p.m. the bore was sighted as a line of white foam between Aure and Fretherne, rather more than three miles down the river. For a quarter of an hour I watched its march up stream, first wheeling by the left, then advancing up the straight reach, and finally wheeling by the right round the last bend. The wheeling movement is most fascinating to watch. I now hurried down to the ferry, and shoved off the boat into deep water to meet the bore, which was now roaring like a railway train. The water channel was about 200 yards wide ; at high water it is double that width. On the sands of the opposite convex, shallow shore the bore discharged itself obliquely as a curling breaker. Against our rocky shore it was a bursting surge. A rise of level was perceptible about ten yards in front of this. In the deep channel we rode easily over a smooth wave. Against the rocky promontory which protects the landing-stage the water surged up violently, then subsided 3 or 4 feet, and surged up again more than once. We now put in behind the shelter of this promontory. At 7.15 p.m. the bore was 300 yards past the ferry, having travelled 3¼ miles, or a little more, in thirty minutes. It was due at the ferry, according to the tide-table, at 7.17 p.m. At 7.21 a steady torrent of water was pouring past the promontory. At 7.29 the torrent was roaring, and the standing waves appeared to be 3 feet high. At 7.44 the waves were

FIG. 1.—The Bore approaching.

smoothed out; the current appeared to be quite as swift, but the greatly increased depth diminishes the surface effect of the rough bottom. The boatmen tell me that the current was "logged" when a bridge was in contemplation, a velocity of 11 knots being registered. Owing to dark clouds and a lurid sunset, I took no photographs. After the passing of the bore there was half an hour's gossip at the ferry, with reminiscences of many bores.

Next morning, April 30, I got into the dog-cart at 7.30 a.m., and drove 6¼ miles, much of the way through plum orchards in

FIG. 2.—Wave Surface at Back of Bore.

full blossom, to Denny, 9¼ miles by river above Newnham Ferry. Owing to the difference of distance by road and river, it is possible to see the same bore at both places by cycling or driving; but I required spare time to arrange for photography. The clouds were heavy and a little fine rain fell at times, hence the necessarily instantaneous photographs are not as bright as they should be for successful reproduction. The spot for observation is a cottage garden by the Denny Brook. The river here is little more than 50 yards wide, flowing between steep banks, slime-covered between tide marks, with no sandy shoals. The bore appeared at 8.47 a.m., and disappeared round a bend

at 8.53 a.m. Taking its time at Newnham Ferry from the tide-table, this gives 70 minutes for time of traversing 9¼ miles, or close on 8 miles an hour. The speed below Newnham was one mile per hour less than this. I guessed the height of the bore at 3 feet in the deep water, 4 feet where bursting on the outer bank. The broken water flew higher than this. These, however, are not trustworthy estimates, as I was busy photographing, obtaining seven exposures in all. Fig. 1 was taken as the bore approached; in Fig. 2 it is seen in passing, showing the wave-surface at back of the bore itself. When in the boat at Newnham I recorded the same thing as a rising and falling of the water against the bank. High water was reached about 9.40 a.m. The current continued to flow up stream; at 10 a.m. it was slack on the concave (western) shore, but still flowing in mid-stream; at 10.18 a.m. it was distinctly running down even in midstream, the water-level having already fallen nearly 3 feet. From the arrival of the bore to the complete turn of the current may be taken to be 1½ hours.

Ordinary photographs show the form of a bore, but its character does not lie so much in its form as in its motion, which combines the mysterious, ghost-like movement of a wave with the rushing steadiness of a railway train. I hope the phenomenon may soon be cinematographed.

1, Savile Row, W.　　　　　VAUGHAN CORNISH.

Bamboo Manna.

THE recent occurrence of a sweet secretion on the stems of bamboos growing in the Central Provinces is a most interesting fact to students of antiquarian medicine. Bamboo manna derives its name from the Sanskrit words—Tvak-kshira, "bark milk"; Vansa-sarkara, "bamboo sugar"; and Vansa-karpura, "bamboo camphor." Vansa-lochana is the name by which it is known by Indian physicians at the present day. These terms would signify a manna-like substance exuding from the stem of the tree, but what is known and used as Vansa-lochana all over India is quite a different article.

That bamboo manna is not a sugar, but a white, gritty body, now called Tabáshir by Europeans, is gathered from the account of Dioscorides, and from the fact that no kind of sugar prepared from the sugar cane answering to this description was known in India in his time. Dioscorides writes: "What is called σάκχαρον is a kind of concrete honey, found in reeds in India and Arabia Felix, in consistence like salt, and brittle between the teeth like salt." Tabáshir, or bamboo manna, was known to the early Arab travellers in the East, and the port of Thana, on the western coast of India, was famous for this product in the twelfth century. Tabáshir is employed as a medicine for its cooling, tonic, aphrodisiac and pectoral properties. In its crude state, when taken from the inside of the bamboo stems, it is mixed with insect remains, and has a blackish appearance; but on gently calcining it becomes quite white, with a pearly lustre. It consists of about 80 per cent. of pure silica, with variable proportions of alkalis, water and organic matter. The history and properties of tabáshir have been very fully discussed by Sir David Brewster (*Philos. Trans.* 1819; *Edin. Journ. Science*, vol. viii. p. 286); Sir George Birdwood (*Bombay Products*, pp. 95–96); Dr. F. A. Flückiger (*Zeit. des Allg. Österr. Apoth. Ver.* 1887, No. 14), and by Sir D. Brandis (*Indian Forester*, March 1887).

The only modern work which alludes to a sugar in the bamboo is the "System of Botany," by La Maout and Decaisne. The authors remark:—"The young shoots of these two trees (*Bambusa arundinacea* and *B. verticellata*) contain a sugary pith which the Indians seek eagerly; when they have acquired more solidity, a liquid flows spontaneously from their nodes, and is converted by the action of the sun into drops of true sugar. The internodes of the stem often contain silicious concretions, of an opaline nature, named tabáshir." Here a distinction is made between the manna forming on the outside of the stem and the tabáshir found inside, but no reference is made to any record where the first named exudation was observed or examined. Dr Watt, when writing the article on Bambusa for his "Dictionary of Economic Products of India," sums up the general experience with regard to this point, and says: "nor has the spontaneous excretion of sugar on the outside of the stem ever been recorded by Indian travellers."

The strange appearance of manna on the stems of the bamboo was reported last March by the Divisional Forest Officer, Chanda, Central Provinces, and notices of this phenomenon have been published in the local papers. The bamboo forests of Chanda consist of *Dendrocalamus strictus*, the male bamboo, a bushy plant from 20 to 30 feet in height, and affecting the cooler northerly and westerly slopes of Central and Southern India. This is said to be the first time in the history of these forests that a sweet and gummy substance has been known to exude from the trees. The gum has been exuding in some abundance, and it has been found very palatable to the natives in the neighbourhood, who have been consuming it as a food. The occurrence of the manna at this season is all the more remarkable, since the greatest famine India has known is this year visiting the country, and the districts where the scarcity is most keenly felt are in the Central Provinces.

An authentic specimen of this bamboo manna was sent to Dr. Watt, Reporter on Economic Products, Calcutta, and was subsequently handed to me for examination. It occurred in short stalactiform rods about an inch long, white or light brown in colour, more or less cylindrical in shape, but flattened or grooved on one side where the tear had adhered to the stem. It was pleasantly sweet, without the peculiar mawkish taste of Sicilian manna (*Fraxinus rotundifolia*). It was soluble in less than its own weight of water, and the solution when allowed to repose deposited white, transparent crystals of sugar. The manna contained 2·66 per cent. of moisture, 0·96 per cent. of ash, 0·75 per cent. of a substance reducing Fehling solution, and a small quantity of nitrogenous matter. The remainder consisted of a sugar which became inverted in twenty minutes when boiled with dilute hydrochloric acid (1 per cent.), and from its solubility, melting-point and crystalline nature, appeared to be a saccharose, related to, if not identical with, cane sugar. It contained no mannite, the saccharine principle peculiar to true manna.

The bamboos and sugar canes belong to the same natural order of grasses, and perhaps it is not unnatural to expect them to yield a similar sweet substance which can be used as a food; but it is a coincidence that the culms of the bamboo, hitherto regarded as dry and barren, should in a time of great scarcity afford sustenance for a famine-stricken people.

Indian Museum, Calcutta, May 3.   DAVID HOOPER.

Solution Theory Applied to Molten Iron and Steel.

I AM pleased to notice that the theory of solution of iron and steel has recently received attention, and that valuable work has been placed before us for consideration by Baron von Juptner (see recent proceedings of the Iron and Steel Institute).

Will you, however, permit me to state that many years ago, in a contribution to the Institute (*Iron and Steel Inst.* 1881), I advocated the theory of solution in the following words :—

"The solution theory is directly applicable to fluid iron and steel, as it is to water. Carbon, phosphorus, &c., are more or less soluble in the fluid metal, just as salts are soluble in water; in both cases the same forces are at work; water, however, at the normal temperature of 60° Fahr., fluid iron about 2500°-3500° Fahr."

"Further, the physical or gaseous theory of solution best explains the facts; the so-called chemical theory of solution is not so applicable. It is difficult to give satisfactory reasons for the union of stable bodies such as carbon and iron, but the gaseous theory of solution apparently does so.

"The difficulty of its complete or further application becomes one of degree only, for no definite distinction can be drawn between gases, liquids and solids, more especially when the latter are heated.

"The quantity of matter dissolved in a given time is simply a function of temperature, and at low osmotic pressure is comparable with that of a liquid evaporating under the pressure of its own vapour" (NATURE, 1892).

"Moreover, it is remarked that ordinary soft steels for sheets, rails, &c., should be so manipulated as to produce a colloid, or, as near as possible, a non-crystalline material, avoiding always the formation of large crystals" (*Iron and Steel Inst.* 1881).

In my practice I have always adhered to the solution theory, finding that it gave the key to the solution of many discrepancies observed in the manufacture of steel, which ordinary analysis, and the usual theoretical deductions therefrom, sometimes failed to explain.

It appears to me, however, that the solution theory requires extension. We have, I think, up to the present only touched upon the surface of the matter, and more extended and deeper

research will amply repay those who have already done work in this direction.

In connection with this subject, although perhaps not exactly bearing upon it, there is what may be termed the theory of the crystallisation of steel and iron. A sheet of ice, as is well known, shows, when heated, beautiful structural, or more correctly crystalline, changes. Why should not a steel plate exhibit changes of this kind if similarly treated?

It is evident, as has been remarked of others, that if the sheet of either ice or iron be suddenly cooled at a given temperature, the structure or grain at that temperature will be approximately retained, and that steel of a given chemical composition may give a material of varying physical properties practically governed by the applied temperature, but not, strictly speaking, in accordance with its chemical composition, as usually assumed.

I have lately found that this happens, and have produced steel of four degrees of hardness by mere temperature manipulation, with metal containing only one-tenth per cent. of carbon together with low per cents. of sulphur and phosphorus. I believe also that this has been done to a certain extent by others, but the facts have not, so far as I know, met with the attention of the practical manufacturers of steel.

Newport, Mon., May 16.   JOHN PARRY.

THE BACTERIAL TREATMENT OF SEWAGE.

THE discovery made by Schwann, in 1839, that a putrefying liquid swarmed with microscopic living organisms, gave occasion to a long series of remarkable investigations as to the general nature and the life-history of these organisms, and the chemical changes which they produced.

Prominent amongst the names of those who prosecuted these investigations stands that of Pasteur, who, in 1857, drew attention to the nature and causes of fermentative changes produced upon sugar solution, of the putrefactive changes in liquids containing animal substances, and of disease changes in the blood of the living animal, which were produced in the presence of various minute living organisms. He showed that, if these liquids were sterilised by heat, and were then duly protected against receiving solid particles from the air, or from other sources, these changes did not occur; and that contact with air which had passed through a red-hot tube, or had been filtered through a cotton-wool plug, was incompetent to introduce the organisms and to start the above changes.

These researches drew attention to the important part played by the air as a vehicle of the organisms or of their spores, and was supplemented by the researches of Tyndall (1876), who proved that air which had been allowed to remain at rest until its motes had subsided was incompetent to produce putrefaction. Tyndall also proved that boiled sterilised broth, when opened in Alpine air, did not usually putrefy, and that the air near the earth's surface in different localities, and even in the same locality at different times, possessed infective power varying from nil to something considerable. The inference is that the distribution of these organisms and of their spores varies very considerably in any horizontal plane near the earth's surface.

Percy Frankland (1886) determined the number of these living organisms which could be developed from equal volumes of air collected at varying heights from the earth's surface. He made use of hills and cathedral towers for the purpose of collecting his samples, and noted a regular decrease in the number of the organisms which were in the air at greater and greater distances from the earth's surface.

These typical researches render it evident that the organisms and their spores, which are produced at or near the earth's surface, are wafted by natural atmospheric movements to some height, but are constantly tending to subside, and to sow the organisms broadcast as they descend.

It has been shown by more recent bacteriological investigations that many of these minute organisms are normally present in the living organism, and make their appearance in large numbers in the dejecta. It is therefore not remarkable that sewage, which contains the dejecta of men and animals, as well as the washings of considerable road and other surfaces, should contain micro-organisms and their spores in large number.

The fact that animal dejecta and sewage are inoffensively and gradually resolved into simple chemical compounds by contact with different kinds of soil has long been known, but this resolution has, until recently, been attributed to the purifying action of the earth itself, or of the organisms which it may contain. It is now abundantly proved that the resolving or purifying agents are, in the main, the micro-organisms which were originally present in the dejecta themselves, although undoubtedly organisms derived from the air, and those already present in the soil, contribute to the change when they are present.

The experimental purification of sewage by letting it stand in tanks filled with flints, gravel, coal, coke or other mineral substances, proves that there is no special virtue in soil. These experiments, originally commenced by the Massachusetts Board of Health, in 1887, have been repeated by many public sanitary authorities, and their results have been abundantly verified; and in various localities broken stone, broken slate, broken clay vessels, " ballast " or burnt clay have been successfully employed in the tanks in place of the materials which were originally used.

For the successful and inoffensive treatment of sewage by this means, a preliminary " priming " of the material is necessary. This is effected by allowing it to remain immersed in sewage for several hours daily for a few weeks. Sewage, which is then introduced and allowed to remain for a few hours in the tank containing the " primed " coke or other material, has the amount of its putrescible dissolved matters considerably and rapidly reduced, while its solid, finely-divided fæcal matter is brought into solution, and caused to undergo, in large measure, inoffensive resolution into simple compounds.

In order that these changes may be completed inoffensively, it is necessary that the " primed " coke surfaces shall be frequently placed in contact with air, and the process is therefore an intermittent one. The coke-bed is first filled with sewage, which is then allowed to flow out from the bottom and to draw air into the interstices of the coke. After the coke surfaces have been for several hours in contact with the air, the cycle of processes is then repeated. The treatment of fresh quantities of sewage in the same coke-bed may apparently be continued indefinitely.

The effluent from one coke-bed undergoes a considerable further purification if it is made to undergo similar treatment in a second coke-bed; and if this second contact with the coke surfaces is followed by ordinary sand filtration, such as is usually applied to river-water which is to be used for drinking purposes, an effluent of extraordinary purity is obtained.

The original methods introduced by the Massachusetts experiments, and known as the intermittent aërobic treatment, is sometimes preceded by a preliminary anaërobic treatment. This consists in allowing the sewage to remain quiescent in, or to flow very slowly through, a large tank or channel. A thick, tough scum soon forms upon its surface, and protects the liquid from the air. Under these conditions many of the solid suspended particles of an organic nature pass into solution, and are thus rendered rapidly resolvable by subsequent aërobic intermittent treatment.

The above general description of the bacterial treatment

of sewage has been subjected to modification as to details to suit the conditions of particular localities. Thus the sewage is in some places subdivided by suitable mechanical arrangements into drops, and allowed to fall continuously like rain upon the surface of the coke-bed. The bed never becomes full of liquid, since when the sewage has trickled through the coke, and has been exposed to the coke surfaces and to the interstitial air, it is at once allowed to flow away from the bottom of the bed.

That these methods of purifying sewage are correctly described as bacterial has been placed beyond doubt. Any conditions which are unfavourable to bacterial life at once retard the purification, while any treatment of the sewage which sterilises it arrests the purification entirely.

The bacteria in the sewage are considered to be the active agents, producing the changes either directly or indirectly through their products or enzymes. Bacteria and their spores are found to be present in great numbers in sewage. London sewage has been shown by Dr. Houston and others to contain very large numbers of bacteria, varying from about three to six million per

Fig. 1.—*Proteus vulgaris.* Impression preparation from "swarming islands" on gelatine; 20 hours' growth at 20° C. × 3000. (Houston.)

cubic centimetre. It seems probable that many of these bacteria form films, or "swarming islands," on the coke surfaces, similar to those which are produced by their growth upon the surface of a gelatine film (Fig. 1); the period of formation of these films may be assumed to be the period of "priming" already referred to. Probably the coke-bed aids bacterial action largely by furnishing surfaces of attachment to the bacteria, upon which they may alternately be exposed to air and to the sewage. The useful effect of solid surfaces in promoting bacterial action in the case of other similar changes is well-known, and it may be connected with the effect which the surfaces exert in preventing the settling of the bacteria to the bottom of the liquid.

Sewage contains many different species of bacteria, some of which have been described and figured by Dr. Houston.[1] As is seen in Figs. 2, 3, 4, some of these

1 The illustrative figures in this article have been selected from Reports on "The Bacteriology of London Crude Sewage" and on "The Bacterial Treatment of Crude Sewage," by Dr. Clowes and Dr. Houston, issued by the London County Council (F. S. King and Son); they were originally produced from micro-photographs taken by Dr. Norman from Dr. Houston's cultivations.

bacteria possess motile tail-like flagella, and by the movement of these the minute organisms maintain a rapid progress through the liquid. Bacteria which are devoid of flagella, and cannot traverse free paths in the liquid, are shown in Figs. 5, 6, 7 and 8. In Fig. 9, the spores of these minute vegetable organisms are seen interspersed amongst the organisms themselves. The organisms have

metabiotic, or possibly of both kinds. The organisms seem to establish and maintain a condition of equilibrium amongst themselves in the coke-bed, since attempts to artificially increase the number of certain species have thus far failed.

It appears that in the above processes there is no separation of the bacterial action which takes place in

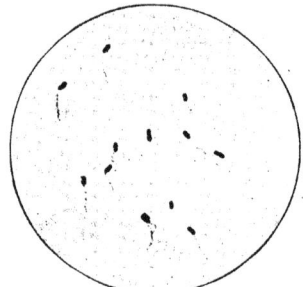

Fig. 2.—"Sewage proteus." Microscopic preparation stained by V. Ermengem's method, showing one flagellum at the end of each rod ; from a 24 hours' growth agar culture at 20° C. × 1000.

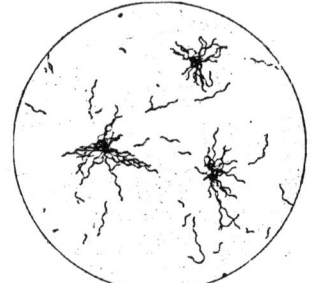

Fig. 4.—*B. mesentericus.* Sewage variety I. Microscopic preparation stained by V. Ermengem's method, showing numerous flagella ; from a 20 hours' agar culture at 20° C. × 1000.

two methods of multiplying, by fission and by producing spores : the spores have great power of retaining vitality. It is found that none of these bacteria are selectively retained by a coarse coke-bed during the treatment, but that all the species make their appearance in but slightly diminished numbers in the purified effluent from the coke-bed. The average reduction in number of bacteria

the presence of air from that which occurs only in the absence of air, and both processes probably proceed side by side in the open coke-bed. The anaërobic, or so-called "septic," treatment, during which cellulose is slowly resolved with separation of hydrogen and methane, is, however, sometimes made to precede the more truly aërobic treatment.

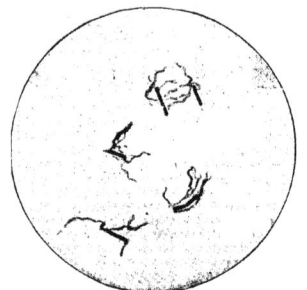

Fig. 3.—*B. mesentericus.* Sewage variety E. Microscopic preparation stained by V. Ermengem's method, showing numerous flagella, from a 20 hours' agar culture at 20° C. × 1000.

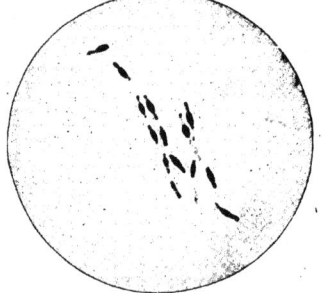

Fig. 5.—*Bacillus subtilis.* × 1500.

suffered by the sewage by one treatment in a coarse coke-bed amounted to only 27·7 per cent. It would therefore appear that the different species of bacteria assist one another in the purifying action, and by producing either contemporaneous or consecutive effects upon the sewage secure its purification : in bacteriological language, their action is either symbiotic or

One result of the anaërobic treatment is the liberation of large volumes of combustible gas, and this gas has been employed at some works for illuminating purposes on the incandescent principle.

The general products from both processes of bacterial action are carbon dioxide, water, ammonia, nitrogen, hydrogen and methane ; and in the aërobic changes the ammonia is subsequently oxidised into nitrite and nitrate.

The experience obtained from several years' experimental bacterial treatment of sewage at several of our largest cities has recently been published.

In 1893 the London County Council constructed an acre coke-bed about three feet in depth at the Barking Outfall of the North London Sewage. This bed has been

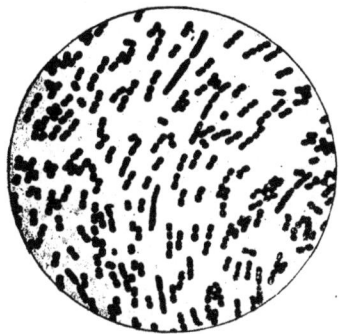

FIG. 6.—*B. subtilissimus.* Impression preparation from a gelatine plate culture. × 1000.

receiving screened and sedimented sewage up to the present time, the process of sedimentation having been assisted by the addition of a small proportion of solutions of lime and of ferrous sulphate. Two years ago the bed was deepened to about six feet. Its purifying action, as measured by the amount of oxidisable matter present in

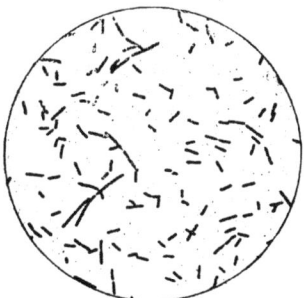

FIG. 7.—*B. mesentericus.* Sewage Variety E. Microscopic preparation from a 20 hours' agar culture at 20° C. × 1000.

the raw sewage and in the clear effluent, amounts to 92 per cent., and if the purification is calculated from the clear sewage and effluent, it amounts to 84 per cent. More recent experiments have proved that the treatment of raw, roughly-screened sewage in such coke-beds is satisfactory, but that the capacity of the bed becomes

continuously reduced by the deposition upon the coke of mineral matter from road detritus, of particles of straw, chaff and woody matter from the horse-traffic and from the wood pavements. It was, therefore, evident that these matters must be deposited by sedimentation before the sewage was brought into the coke-beds. A comparatively

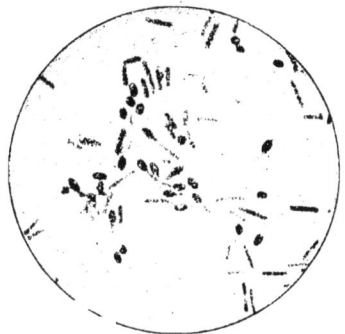

FIG. 8.—*B. enteritidis sporogenes* (Klein). Microscopic double-stained preparation, from a serum culture, showing spores. × 2000.

rapid process of sedimentation suffices to remove these matters, since even the cellulose matters arrive in the sewage in a heavy and waterlogged condition.

It was found advantageous to use coke in comparatively large fragments, about the size of walnuts, since this facilitated the rapid draining of the liquid from the coke,

FIG. 9.—*Proteus vulgaris.* × 1000.

and at the same time increased the sewage capacity of the bed and promoted its efficient aëration. The depth of the beds has been augmented from 4 to 13 feet, and the increase of depth seems to be attended with increase of efficiency. The 13-foot bed has for long periods given a purification from dissolved oxidisable matter of over

60 per cent. It has maintained a most satisfactory state of aëration, since the air drawn from the bottom has contained, on an average, 17 per cent. of oxygen.

About 60 per cent. of the matter which settles from the sewage under ordinary conditions is combustible, and could, therefore, very well be dealt with by a destructor.

The tendency of the coke bacteria beds is undoubtedly to improve in their purifying power with age, provided they are not overworked. A bed which had given for some time a 50 per cent. purification, gradually increased in efficiency until its purifying effect reached nearly 70 per cent. The effluent from this bed underwent an additional purification of 20 per cent. by treatment in a second similar bed.

The effluent from a single coke-bed worked on the intermittent principle was clear and odourless, and remained in this condition when it was kept in open or closed bottles in a warm laboratory. It maintained the life of gold-fish, roach, dace and pike indefinitely : it was therefore not only well aërated, but was able to maintain its aërated condition. This proves that it was free from any rapidly oxidisable matter. It was undoubtedly, however, undergoing gradually further purification by the action of the bacteria which it contained, and with the assistance of dissolved oxygen. Such an effluent would be quite suitable for introduction into the tidal part of the river, where the water is too salt and muddy to be used for drinking purposes.

Bacteria are present in large numbers in the river-water itself, and undoubtedly exert a most useful purifying effect upon the water during its flow. The relation between the number present in the sewage and in the water of the River Thames, below and above locks, is shown by the following estimations made by Dr. Houston. The number of liquefying bacteria included in the total number of bacteria present in one cubic centimetre, and the number of spores of bacteria, are also stated :—

	Bacteria.	Liquefying bacteria.	Spores.
Raw sewage from North London, Feb. to April 1898	3,899,259	430,750	332
Raw sewage from South London, Feb. to April 1898	3,526,667	400,000	365
May to Aug. 1898	6,140,000	860,000	407
Effluent from coke-bed, South London, May to Aug. 1898 ...	4,437,500	762,500	252
[Percentage reduction by passing through coke-bed...	[27·7]	[11·4]	[38]
Lower Thames water, Greenhithe, half ebb-tide, Oct. 1898...	10,000	—	63
Lower Thames water, Barking, low tide, Nov. 1898	34,400	—	89
Upper Thames water, between Sunbury and Hampton, Nov. 1898	5,100	—	56
Upper Thames water, Twickenham, Nov. 1898	3,000	—	18

The results obtained by the experimental bacterial treatment of sewage at Manchester during the last two or three years bears out generally those which have been obtained in London. The treatment has differed in some details from that adopted in London. The particles of coke constituting the coke-beds have been smaller. The coke-beds have been subjected to a larger number of intermittent fillings per day ; and the preliminary treatment in an open anaërobic tank has been carried out with advantageous results. The scientific experts who have suggested and watched the experiments state their conviction that bacterial treatment is the treatment which is most suitable for Manchester sewage, but that in order to secure the most effective purification, the coke-beds

must have sufficiently frequent and prolonged periods of rest, and must be fed with sewage as free as possible from suspended matter, and as uniform in quality as may be. Preliminary anaërobic treatment is referred to as the best means of securing uniformity in quality of the sewage, and of adapting it to rapid subsequent aërobic purification. Four fillings in 24 hours have been found suitable, if one day's rest in seven is given to each coke-bed ; the number of fillings, however, may exceed this without detriment to the bed or to the character of the effluent.

Town sewage is found to arrive at the outfalls at an almost constant temperature throughout the year. It rarely falls below 13° C. And this temperature not only prevents the possibility of the coke-beds being stopped by the freezing of the sewage, but also secures to the bacteria one condition favourable to their action. When a bed is too freely aërated by the passage of frosty air constantly through the interstices of the coke, this favourable condition is, however, seriously interfered with, and the bed may even become stopped by the freezing of the sewage.

In the more recent experiments carried out in America by the State Board of Health, Massachusetts, the tendency has been to use fine coke, and to allow the effluent from the coke to pass through sand. The passage of the liquid has either been allowed to take place with the outflow widely opened, so that the bed never fills ; or the sewage has been allowed to fill the bed and to remain quiescent in contact with the coke for a time, as in the English experiments. The conclusions arrived at seem to be that the degree of purification obtained by the use of fine coke and sand is very satisfactory, but that the volume of sewage dealt with in a given time is smaller than when larger coke fragments are used, and the tendency seems to be to adopt the larger coke in order to expedite the more rapid drainage away of the effluent.

It will be seen from what has already been said that it is well not to speak of this system of treatment as one of filtration. Filtration ordinarily implies a process of mechanical separation of material suspended in a liquid. The fact that the coke-beds only commence their purifying action after they have been "primed" by repeated contact with sewage, and that this purifying action keeps increasing as the bed "matures," is sufficient to show that the action is by no means of a mechanical nature. It would be well, therefore, to speak of it as a process of bacterial treatment, and thus to indicate that the purifying agents are bacteria, which are acting under control, and are placed under conditions favourable to the development of their full activity.

It would be rash to say that the methods of bacterial treatment have as yet reached their most effective state ; but it is significant that these methods have secured converts wherever they have received careful and air trial, and that those who are their warmest advocates who have had the widest experience of their working. It is even probable that further improvements will be made in the means of treating sewage bacterially, but it is quite certain that the processes at present in use are able to secure the economical and satisfactory purification of ordinary town sewage.   FRANK CLOWES.

THE TOTAL ECLIPSE OF THE SUN.

SINCE the first series of telegrams was received announcing successful observations of the total eclipse of the sun on May 28, all the more detailed reports to hand confirm the universal satisfaction of the various parties at the results. As, however, most of the parties having a definite programme arranged to obtain photographic records, complete details cannot be known until the development of the whole of the plates, and in some cases this will not be until the observers return home.

Rough prints from several of the negatives obtained with the prismatic cameras used by Sir Norman Lockyer's party show as great amount of detail as those taken in

FIG. 1.—Two of the cameras fed by a cœlostat.

1898. From a cursory examination of the negatives few differences appear in the chromospheric spectrum ; the "1474" corona ring seems, however, slightly more feeble than before.

FIG. 2.—The 20-foot prismatic camera and siderostat.

In a letter received from Mr. Fowler he states that the negatives obtained by Mr. Payn with the 16-foot Cooke coronagraph are excellent, especially one showing the inner corona.

The accompanying illustrations, received too late for reproduction with Sir Norman Lockyer's letter last week, show some of the arrangements made for observing the eclipse at Santa Pola. Particulars concerning the various instruments will be found in NATURE of May 17.

Prof. H. H. Turner, at Bonsarea, near Algiers, successfully carried out his programme of photographing the corona, obtaining seven ordinary pictures and seven with interposed polarising apparatus. The polarisation indicated was decidedly radial.

Mr. H. F. Newall obtained the "first flash" and "corona" spectra with both slit spectroscope and objective grating, those taken with the latter, however, being weak. With Mrs. Newall he also made polariscopic observations.

Mr. W. H. Wesley made an excellent drawing from his observations with the eight-inch Coudé equatorial placed at his disposal by M. Trépied, director of the French observatory at Algiers. He reports that very little structural detail was discernible in the inner corona.

The need for correction of the lunar tables is indicated by the universal experience that totality was some seconds *less* than that previously computed. The American observers estimate the difference as three seconds, while at Ovar, in Portugal, Mr. W. H. M. Christie, the Astronomer Royal, gives the time of totality as 85 seconds, whereas the calculated value was 93 seconds. Several observations indicate that the discrepancy is to be looked for in the moon's diameter being taken too large.

The most unfortunate victim of this error appears to have been Mr. Evershed, who journeyed to an outlying station, near Mazafran, so close to the limiting line of totality as was considered safe. He did this with the object of photographing the "flash" spectrum with as long duration as possible ; this will be understood when it is considered that exactly on the central line the duration of the flash will be merely momentary, but as the observer recedes from the central line the line of sight to the moon's limb becomes more oblique, until on the limiting line of totality the so-called "flash" is visible for the whole time of totality at that point. Owing to this ambiguity of the data, the station chosen was evidently somewhat further from the central line than was anticipated, and consequently Mr. Evershed had the unpleasant experience of less than one second totality. His preparations must have been exceedingly perfect, however, for he reports having obtained a good photograph at the proper instant, though it will fall short of expectation for the reason stated.

Prof. Howe, of Denver, has already determined the position of the planet Eros, which he was fortunate enough to discover on his photographic plates during the eclipse, and has circulated his result. The co-ordinates of the planet will be found in the "Astronomical Column."   C. P. B.

NOTES.

SIR ARCHIBALD GEIKIE, F.R.S., has been elected a Foreign Honorary Member of the American Academy of Arts and Sciences in the section of Geology, Mineralogy and Physics of the Globe, in succession to the late Carl Friedrich Rammelsberg.

FIG. 3.—Discs on spars, for naked eye observations of the corona.

PROF. FOUQUÉ has been elected vice-president of the Paris Academy of Sciences for the year 1900, in succession to the late Prof. Milne-Edwards. Prof. Boltzmann has been elected to succeed the late Prof. Beltrami, in the mechanics section of the Academy.

PROF. PAUL GROTH, of Munich, has been elected a Foreign Member of the Geological Society, and Prof. A. Issel, of Genoa, a Foreign Correspondent.

THE annual conversazione of the Institution of Electrical Engineers will be held on Tuesday, June 26, at the Natural History Museum, South Kensington. The guests will be received by the president (Prof. Perry), and Mrs. Perry.

THE Croonian Lectures for 1900 will be delivered by Dr. F. W. Mott, F.R.S., before the Royal College of Physicians of London, on June 19, 21, 26 and 28. The subject is "The Degeneration of the Neurone."

IT is stated that Captain W. Bade di Wismar has organised an expedition to the east coast of Spitsbergen and Franz Josef Land to seek for traces of Andrée, and also to obtain intelligence of the Duke of the Abruzzi. No apprehension is felt about the Duke of the Abruzzi, as a long interruption in his communications with the rest of the world was foreseen.

A MEETING was held at the Meteorological Society on Thursday last to consider the question of a memorial of the late Mr. G. J. Symons, F.R.S. It was resolved that the memorial should take the form of a gold medal, to be awarded from time to time by the council of the Royal Meteorological Society for distinguished work in connection with meteorological science. An executive committee was appointed to take the necessary steps to raise a fund for this purpose. Contributions will be received by the assistant secretary, Mr. W. Marriott.

LORD LISTER will open the new clinical laboratories at the Westminster Hospital on Tuesday, June 12, at 4 p.m. He will be received by Sir J. Wolfe-Barry, chairman of the committee, and supported by Lord Kelvin, Sir Michael Foster, M.P., Dr. Church, president of the Royal College of Physicians, Sir William MacCormac, president of the Royal College of Surgeons, and the Dean of Westminster.

THE completion of the twenty-fifth year of teaching by Prof. Luciani, Rector of the University of Rome, was celebrated on May 3 in the physiological laboratory of the University. The *British Medical Journal* states that the theatre was crowded with admirers of the well-known physiologist, conspicuous among whom was Prof. Baccelli. An address was delivered by Prof. Todaro, to which Prof. Luciani, who was much moved, replied. Prof. Baccelli also spoke, and ended by embracing Prof. Luciani, who was the object of enthusiastic congratulations from the assembly.

THE decision of the Trinity House authorities to remove the wireless telegraphy installation between the South Goodwin lightship and the South Foreland was discussed by the Dover Chamber of Commerce on Friday. It was decided to memorialise the Trinity Board and to request the Chambers of Commerce of the ports of the United Kingdom, as well as Lloyd's and other shipping bodies, to support the memorial, with a view to the establishment of a connection between lightships and the shore on dangerous sands.

THE president of the Board of Education has approved of a Committee, which is now sitting, "to inquire into the organisation and staff of the Geological Survey and Museum of Practical Geology; to report on the progress of the Survey since 1881; to suggest the changes in staff and arrangements necessary for bringing the Survey in its more general features to a speedy and satisfactory termination, having regard especially to its economic importance; and, further, to report on the desirability, or otherwise, of transferring the Survey to another public department." The members of the Committee are:—The Right Hon. J. L. Wharton, M.P. (chairman), Mr. Stephen E. Spring

Rice, C.B., Mr. T. H. Elliott, C.B., General Festing, C.B., Dr. H. F. Parsons, Mr. W. T. Blanford, F.R.S., and Prof. C. Lapworth, F.R.S., with Mr. A. E. Cooper as secretary.

THE announcement of the death of Miss Mary H. Kingsley, at Simonstown, on Friday, will be received with deep regret by geographers, ethnologists, and many others who are familiar with her works. Miss Kingsley was the elder of the two children of the late Dr. G. H. Kingsley, and quite recently (May 3) her memoir of her father, published with his "Notes on Sport and Travel," was noticed in these columns. Miss Kingsley will chiefly be remembered for her explorations in West Africa, and her works upon them. The first volume in which she recorded her experiences was "Travels in West Africa," published in 1897. Last year, a further volume of "West African Studies" appeared, and a few weeks ago her "Story of West Africa" was published in the Empire Series. Miss Kingsley's books are marked by a sincerity and humour which make them of deep interest even to readers who may not always agree with her forcibly-expressed convictions. Her interest in West Africa, as an obituary notice in the *Times* points out, was partly scientific, partly sociological, partly political. She made numerous contributions to our knowledge of the fishes of some of the West African rivers, and of the reptiles in that part of the continent. In both her books on West Africa she made valuable additions to our knowledge of the native mind and character, and her studies in fetish bring out in a remarkable manner the sympathetic insight which enabled her to project herself into the mind of the negro races. In "West African Studies" Miss Kingsley set forth, with much array of facts and arguments, a strong indictment of the system of government by Crown colonies in West Africa. Personally, Miss Kingsley was of a modest and retiring disposition; but the frequent journeys that she made up African rivers and through the bush with none but native attendants afforded undoubted testimony to her pluck, powers of endurance and fertility of resource.

AT the last meeting of the General Medical Council, the report of the Pharmacopœia Committee, referring to the subject of a proposed international Pharmacopœia limited to drugs of a drastic nature, was adopted. If an international conference on the subject in question is arranged, the Council will appoint representatives to participate in it, and one or more members will be appointed to act as delegates. Communications have been opened with the United States authorities with a view to bringing about greater uniformity in the official preparations contained in the British Pharmacopœia and the United States Pharmacopœia respectively; and it is hoped that, by mutual concessions, important approximations and assimilations in the contents of the two works may be ultimately secured. Further communications have been received with reference to the Indian and Colonial "Addendum," and important suggestions from Canada have been considered by the committee in detail. It is hoped that the addendum will be authorised for issue by the end of the year. By the efforts of Dr. Leech, a valuable collection of British and foreign works bearing on the history and development of the Pharmacopœia has been collected and deposited in the office of the Council.

THE widespread invasion and persistent devastations of locusts in so many parts of Africa give interest to all trials and experiments, as well as the ordinary remedies, employed for the alleviation of this ruinous plague of the farmer. The following notes from Mr. W. C. Robbins, Stock Inspector of the Lower Tugela and Mapumulo Districts, are published in the Cape official *Agricultural Journal*:—"For the past three days I have been over the ground where my men have been infecting locusts with Government fungus, and the result was that I found dead locusts everywhere. I send you a sample ; you will notice they are full of worms, and we know from experience than when locusts are found in this state whole swarms die off. Some, you will see, are half eaten ; these were eaten by their fellows. I have seen many clusters of locusts eating dead ones." The feeding upon bodies of dead locusts suggests that diseased locusts may be utilised as a substitute for locust fungus. Tests are being made to determine whether a preparation from diseased dead locusts will infect a swarm in the same way as locust fungus made in the Government laboratory.

IN a paper on "The Standardisation of Electrical Engineering Plant," published in the *Journal* of the Institution of Electrical Engineers, Mr. R. Percy Sellon arrives at the following general conclusions :—(a) Standardisation to a greater degree than at present exists is in the interest of the manufacturer, as a means of facilitating repetition and production, and of meeting the competition of standardising foreign manufacturers. (b) Standardisation of "ends" or "performance" as distinct from "means" or "constructional details" is equally in the interest of the user, by securing for him low purchase cost, prompt delivery, freedom from the risks of experimental designs, and full manufacturers' guarantees. (c) The relative absence of standardisation in Great Britain, in contrast with other countries, is mainly traceable to the prevailing system wherein the user's engineer specifies "means" or "constructional" details instead of confining himself to "ends" or "performance." (d) The determination of standards by organised effort rather than by the slow and costly process of "trial and error" is desirable, and should be undertaken under the auspices of the Institution of Electrical Engineers, as representing the interests of both producer and user.

IF standardisation is important for the electrical engineer, it is none the less urgently needed in connection with scientific literature. Although the pages of a large number of journals and transactions, both in this country and on the Continent, are of uniform sizes, both quarto and octavo, this is by no means the universal rule ; and proceedings, especially of local societies in remote districts, as well as the more popular class of scientific journals, show almost every possible variation in the dimensions of their pages. We have before us a pile of such publications, arranged in order of size, and increasing gradually from $7 \times 4\frac{1}{2}$ inches at the top to 12×10 inches at the bottom. They include many papers which it is desirable to bind up with other literature on the same subjects, but which have had to be relegated to "the pile" on account of their inconvenient sizes. This is the more unfortunate because journals of this particular character often contain reports on current research, the inclusion of which in bound volumes of reprints, easy of reference, might often save those repetitions of investigations which involve much loss of time, and only lead to disappointment, accompanied by unpleasant—not to say undignified—controversies as to priority.

THE U.S. Department of Agriculture has issued a Bulletin, No. 74, containing "Organisation Lists of the Agricultural Colleges and Experiment Stations in the United States, with a list of Agricultural Experiment Stations in Foreign Countries." Thirty-six pages are occupied by notes on the courses of study and the names of the boards of instruction at fifty-nine colleges exclusively devoted to agricultural teaching, or with agricultural departments ; while twenty-one pages give the names of the governing board and staff at fifty-six experiment stations. Then follows a list of foreign experiment stations, with the names of the directors, to which is added a most useful statement of the more important publications issued in 1899 by the various stations of the United States. Some notes on the relationship of the colleges and stations to the United States

Treasury complete this exhaustive record. Probably, the information, so far as it concerns the United States, is trustworthy, but the same cannot be said in regard to the British stations, for this section of the work is defective alike as regards accuracy and completeness. It would be well to have the British section thoroughly revised in any future issue.

WE have received from Dr. W. van Bemmelen a memoir on the deviation of the magnetic needle from the end of the fifteenth century to the year 1750, with isogonic charts for the epochs 1500, and subsequent half centuries down to 1700. The work is published as a supplement to vol. xxi. of the "Batavia Meteorological and Magnetical Observations," and is the outcome of researches made during several years in various libraries and archives in the Netherlands and other European countries prior to the author's appointment to the Batavia Observatory. The work is a laborious compilation of all the most trustworthy observations, commencing with the voyage of Columbus in 1492, and is a most valuable contribution to terrestrial magnetism, containing between five and six thousand observations in all parts of the world, with references to the positions and the sources whence the information has been obtained. The value of the work is much enhanced by numerous critical remarks and by explanatory text ; the language used is German.

As attention has recently been much directed to the enormous drafts that are being made on the coal supply of the world for power purposes, the following description of one of the most recent attempts to obtain power by utilising the hitherto wasted resources of nature may be of interest. A company called the Saint Lawrence Power Company, composed of English and American shareholders, some time since obtained a tract of 2000 acres of land at Massena, adjacent to the Saint Lawrence and Grasse rivers. On this land an electrical installation of considerable magnitude is in course of construction. The works, which it is expected will be completed next autumn, are intended to develop ultimately 110,000 horse-power. The plant is situated on the River Grasse, a tributary of the Saint Lawrence, from which the water for driving the machinery will be diverted through a canal three miles long, 200 feet wide at the bottom, and 25 feet deep. The bottom of this canal at the river end will be 60 feet above the ordinary water-level in the River Grasse, which will form the tail-race for the turbines. The preliminary mechanical equipment will be eight units of 5000 horse-power, each obtained by three twin turbines and dynamos. The land adjacent to the works which is to be utilised for manufacturing and allied purposes will be accessible by branches of the New York Central Railway and by the canal to the Saint Lawrence, which will be large enough to take vessels of considerable draught.

THE means of overcoming the difference of level of the country through which canals pass is in most cases overcome by locks placed either singly or in flights, depending on the height to be overcome. About twenty-five years ago, the locks between the Trent and Mersey Canal and the River Weaver, where there is a difference of 50 feet, were superseded by the hydraulic lift at Anderton. The boats here are floated into iron troughs which are raised or lowered by hydraulic power, one boat ascending and another descending at the same time. This system was subsequently adopted on other canals in France and Belgium, and, with some modifications, in Germany. What is claimed as an improvement on this system is now being carried out on the Erie Canal in America, at Lockport, by what is termed a "Pneumatic Balance Canal Lock." A description of this lift was given in a paper contributed to the Franklin Institute by Mr. Chauncey N. Dutton. The existing stone locks were erected in 1836, and overcame a lift of $62\frac{1}{2}$ feet by means of five flights. The lock which is being erected to supersede these consists of two steel chambers, one for ascending and the other for descending boats. These chambers are divided into two parts, the upper one containing water to receive the boats, and provided with gates, as in the case of the Anderton lift ; and beneath this a second chamber containing compressed air on which the lock-chamber floats. The air-chambers are so proportioned that they automatically differentiate the air-pressure. The water in the lock-chamber which contains the boat at the upper level is so adjusted that its weight, with the boat it contains, is 200 tons greater than that of the lower one. Each of these locks weighs 1500 tons and contains 4500 tons of water, the weight in motion, when the boats are ascending and descending, exceeding 12,000 tons. The advantages claimed by the use of compressed air are a saving in cost, safety in working, and great economy in water. The power for compressing the air is furnished by a 36-inch turbine working a four-cylinder pump. This also drives the dynamos which operate the gates and light the lift.

A RECENT report by Prof. Le Neve Foster upon the number of persons employed, and the number of fatal accidents, in mines and quarries in the United Kingdom, shows that in 1899 the death-rate of the workers at mines under the Coal Mines Act, taking underground and surface workers as a whole, was 1·26, whilst that of 1898 was 1·28. At the mines under the Metalliferous Mines Act, the death-rate of the underground and surface workers as a whole was 1·59, a figure decidedly higher than that of 1898, which was only ·96. The inside workers in quarries had a slightly smaller death-rate from accidents in 1899 than they had in the previous year.

A RECENT consular report (No. 2418) on the trade of Corsica states that of the few industries at present carried on, that of extracting tannic acid from chestnut wood is now perhaps the most flourishing in the island. This industry is carried on in Bastia, which is the commercial centre of Corsica, by two large factories which export together about 4000 tons of extract per annum, in concentrated liquid form. To prepare this quantity requires nearly 20,000 tons of wood of the sweet chestnut tree yearly. The immense forests are equal to supplying the demand for many years ; but this tree not being under the control of the Administration of Woods and Forests its wholesale destruction without compulsory replanting will, it is feared, in time not only influence adversely the climate of large districts, but cause much misery in those districts where chestnut flour forms the staple food of the peasants. It is prepared from the dried fruit of the sweet chestnut.

ACCORDING to the Acting British Consul at Samoa, rubber has been introduced there, and is being grown by several of the planters. It appears to thrive, and as far as can be seen the soil is admirably adapted for the growth of this most valuable product.

WE have received the official edition of the Fourth Annual Report of the New York Zoological Society, the substance of which is given in more popular form in a publication alluded to a short time ago in our "Notes" column.

THE Marlborough College Natural History Society, in its *Report* for 1899, sets an admirable example to institutions of this nature in publishing a list of the Lepidoptera of the district, the elaboration of which has been a work of years. It is by the thorough working out of local faunas that provincial natural history societies can alone properly justify their existence.

REFERRING to a remark in the review of the "Vertebrate Fauna of the Shetland Islands " in NATURE of May 24 (p. 75),

Mr. Eagle Clarke writes to say that though he revised some of the proofs, the revision of the Cetaceans was undertaken by Mr. James Simpson, and that he did not revise either the MS. or the proof relating to that order. "Mr. Simpson, who had a special knowledge of the group, has passed from among us, but I have little doubt that his inclusion of the Narwhal in the *Physeteridæ* was the result of a mere slip."

THE *Entomologist* for June contains the first instalment of the translation of an article by Prof. Max Standfuss on experiments in hybridisation, and on the influence of temperature on the development of the Lepidoptera. As an instance of the line of investigation followed, we may quote the case of the map-butterfly (*Vanessa levana*), in which the difference between insects bred from the summer and winter pupæ is so great as to have formerly led to the belief that they belonged to different species. By placing the summer pupæ in an ice-house the winter imago was produced ; but, on the other hand, it was found much more difficult to change by warmth the winter pupæ into the summer imago. This led to the inference that the winter form was the original one ; and this is confirmed by the circumstance that the only near relatives of this insect are four species from northern Asia.

THE second edition, revised and largely rewritten, of Dr. Julius Wiesner's work, "Die Rohstoffe des Pflanzenreiches," is in course of publication by the firm of W. Engelmann, Leipzig ; and the second and third parts have just appeared. The work will be completed in two volumes, and will probably be completed towards the end of this year.

A NEW edition of Thompson's "Gardener's Assistant," which has for many years been accepted as a trustworthy repository of information on the science and art of gardening in all its branches, is in course of publication by the Gresham Publishing Company. The work has been completely revised and entirely remodelled under the direction and general editorship of Mr. William Watson, of the Royal Gardens, Kew, and contains contributions by many eminent horticulturists. The first volume has just been published.

THE first volume of a "Cyclopedia of American Horticulture"—a work described as "comprising suggestions for cultivation of horticultural crops, and descriptions of the trade species of fruits, vegetables, flowers, and ornamental plants, together with geographical data and biographical sketches," has just been published by Messrs. Macmillan and Co., Ltd. It is edited by Prof. L. H. Bailey, whose fertility in the production of excellent botanical books is really astonishing, assisted by Mr. W. Miller. The present volume extends from A to D, and contains 509 pages and 743 illustrations. The work will be completed in four volumes.

A COMPLETE and convenient cabinet of glass-blowing apparatus and materials, arranged especially for students or others using Mr. T. Bolas's book on "Glass Blowing," has been put on the market by the Camera Construction Company. Exercises in the manipulation of glass cultivate delicacy of touch and perception, and are therefore excellent as manual training for young people. In scientific work, and more especially in physical and chemical sciences, the ability to work glass is a very valuable accomplishment, and a cabinet which provides a ready means of obtaining practice in this art is a desirable possession for laboratories as well as private students.

THE question as to whether strontium and barium can replace calcium in plants has been made the subject of inquiry by more than one experimenter. The February number of the *Bulletin* of the College of Agriculture of Tōkyō contains an interesting contribution to this question by Dr. U. Suzuki. Experiments were carried out with several species of plants and in soils containing varying amounts of calcium. The results show that strontium and barium can never replace calcium in phanerogams, as they are strongly poisonous, although the poisonous action may be lessened to a certain extent by the addition of lime salts. The *Bulletin* also contains papers by the same author on arginin, and its formation in coniferous plants; and by K. Asō, on the chemical composition of the spores of *Aspergillus Oryzae*.

AMONGST the products of the action of fluorine upon sulphur recently investigated by M. H. Moissan (see NATURE, April 19, vol. lxi. p. 597), thionyl fluoride, SOF₂, the existence of which was first indicated by M. Meslans, was noticed. MM. Moissan and Lebeau have now made this fluoride the subject of a more detailed study, and have succeeded in obtaining it in a pure state by two different methods—by the action of fluorine upon thionyl chloride, and by the interaction of fluoride of arsenic upon thionyl chloride. Thionyl chloride is a colourless gas, fuming slightly in moist air, and possessing an unpleasant odour resembling carbonyl chloride. It is easily condensed by a mixture of solid carbon dioxide and acetone, giving a liquid boiling at $-32°$. In the absence of moisture, glass is not attacked by the gas at temperatures below 400° C.; above this temperature silicon tetrafluoride and sulphur dioxide are produced. Water decomposes thionyl fluoride slowly at ordinary temperatures, giving hydrofluoric and sulphurous acid. Indications were obtained of another oxyfluoride of sulphur, not absorbed by water and possessing a much lower boiling point.

THE additions to the Zoological Society's Gardens during the past week include two Wild Swine (*Sus scrofa*, ♀ ♀), European, presented by the Lord Carnegie; three Chaplain Crows (*Corvus capellanus*) from Southern Persia, presented by Mr. B. T. Ffinch; a Herring Gull (*Larus argentatus*), European, presented by Mr. J. W. Berry; two Red Howlers (*Mycetes seniculus*, ♂ ♀) from Colombia, a Great Kangaroo (*Macropus giganteus*, ♂) from Australia, an American Flying Squirrel (*Sciuropterus volucella*), three American Box Tortoises (*Cistudo carolina*), a North American Trionyx (*Trionyx ferox*), three Changeable Tree Frogs (*Hyla versicolor*) from North America, a Black Sternothere (*Sternothaerus niger*) from West Africa, two Greek Tortoises (*Testudo graeca*), South European; six Argentine Tortoises (*Testudo argentina*) from the Argentine Republic, a Red and Yellow Macaw (*Ara chloroptera*) from South America, two Black-headed Caiques (*Caica melanocephala*) from Demerara, a Chough (*Pyrrhocorax graculus*), British, deposited; two Brown Mynahs (*Acridotheres fuscus*) from India, a Brown Mock Thrush (*Harporhynchus rufus*) from North America, an Occipital Blue Pie (*Urocissa occipitalis*) from the Western Himalayas, purchased; two Thars (*Capra jemlaicus*), five Swinhoe's Pheasants (*Euplocamus swinhoii*), bred in the Gardens.

OUR ASTRONOMICAL COLUMN.

PHOTOGRAPHIC OBSERVATION OF EROS.—A circular from the Centralstelle at Kiel furnishes particulars of the photograph of the planet Eros obtained by Prof. Howe, of Denver Observatory, U.S.A., during the recent total eclipse. The position determined was :—

R.A.   23h. 47m. 3·9s. } 1900 May 27·9129.
Decl. + 2° 46′ 33″ }   Greenwich Mean Time.

OCCULTATION OF SATURN.—There will be an occultation of Saturn by the moon on Wednesday evening, June 13, the particulars of which are as follows :—

	Sidereal Time.	Mean Time.	Angle from North Point.	Vertex.
	h. m.	h. m.	°	°
Disappearance	... 15 7	... 9 40	... 89	... 116
Appearance	... 16 19	... 10 52	... 265	... 283

The planet rises about 8.55 p.m., so that the conditions for observation will not be very favourable.

HARVARD COLLEGE OBSERVATORY.—In *Circular* No. 50 issued from the Harvard College Observatory, Prof. E. C. Pickering reviews the methods adopted in the measurement of photographic light intensities. Since 1887 all the photographs obtained at the Observatory have had the image of a standard light impressed upon them for comparison. The methods now adopted have been developed by Mr. E. S. King, under whose direction the photographs are taken at Cambridge, and his description of the plan followed occupies the greater part of the circular. All sources of light, that of the sun, moon, sky, Milky Way, aurora and stars are to be referred to one standard, given by the meridian photometer, with which Polaris has a magnitude of 2·15. The artificial standard for practical convenience is that given by an Argand burner behind a small aperture; but this is compared with Polaris every month, when a series of tests are made on a 8 × 10 in. plate, the various parts of which are then cut and stored for future inspection. These monthly comparisons in addition furnish a valuable check on the constancy of the plate and the developer used, and will, moreover, as the several parts of the divided plate are developed at different periods, furnish data concerning any change in the image dependent on the interval between exposure and development. Spectroscopic photometry is also adopted to record the photographic intensity in terms of light of a particular wavelength.

Prof. W. H. Pickering has evolved a method of reducing the standard of comparison to the actual radiation received from a certain star shining directly on the plate. This unit, however, being so small, secondary and tertiary standards have been made from it by using lenses of known aperture and focal length. Thus, with a simple plano-convex lens of 8·2 cm. aperture, the image of a Ursæ Minoris was received on a piece of ground glass placed 3 cm. from the photographic plate. The "sensitive unit" was produced after twenty minutes' exposure, and the intensity of the light was calculated to be thirty times greater than the direct radiation from the star. For lights of great intensity this secondary standard is still too small, and then recourse is had to the Argand burner constant.

LIVERPOOL OBSERVATORY.—We have received the report of Mr. W. E. Plummer, the director of the Liverpool Observatory at Bidston, Birkenhead, on the work done in the year 1899. Although the seismograph has not been in use all the year, it is intended to commence keeping a continuous record of earth movements by means of the present instrument and one to be supplied by the Earthquake Committee of the British Association. The two will be placed so as to record movements in planes at right angles to each other.

The report contains detailed results of all meteorological observations during the year, including temperature, barometric pressure, rainfall, sunshine and cloud, wind velocity, humidity, &c.; and an appendix is added containing a summary of the mean values of many of these quantities during the past thirty years.

TEMPERATURE CONTROL OF SPECTROGRAPH.—In the *Astro-Physical Journal* (vol. xi. pp. 259-261, 1900), Prof.W. W. Campbell describes the arrangement he has finally adopted for securing as complete uniformity as possible of the temperature of the various parts of the spectrograph used at the Lick Observatory for determining stellar velocities in the line of sight. The whole instrument is first enclosed in two thicknesses of thick grey blanket, the prism case having an additional two thicknesses over it. Outside the whole is then fitted a case of cedar, lined with felt, in which is embedded a length of German silver wire. This latter is heated by an electric current, the strength of which is so regulated that a thermometer placed in the prism box shows as constant reading as possible. The efficiency of the device is clearly shown by a table giving the actual variations observed during a night's work. From 8.28 p.m. to 4 a.m. the temperature of the air in the dome varied from 17°·2 C. to 19°·0 C., but the extreme readings of the thermometer in the prism box were only 18°·70 C. and 18·84° C., so that the maximum variation was less than one-fifth of a degree.

ADVANCEMENT OF ELECTRICAL CHEMISTRY.

IN a previous article (March 1, p. 428) upon the advancement of electrical chemistry, various developments of electro-metallurgy or electrical deposition of metals were described. Electrolytic processes for obtaining the non-metallic elements and for the preparation of inorganic and organic compounds were left for consideration in a separate article, and are now dealt with.

In the year 1800, Nicholson and Carlisle showed that water could be decomposed into oxygen and hydrogen by means of a "volta pile"; since that time the electrolytic decomposition of water has been employed as a lecture experiment to show the composition of water. It is, however, only quite recently that oxygen and hydrogen have been produced on a manufacturing scale by the electrolysis of dilute solutions of caustic soda or sulphuric acid. The hydrogen so obtained is usually almost absolutely pure, but the oxygen is generally mixed with about 3 per cent. of hydrogen, which, however, can be removed by passing it through red-hot tubes.

The powerful oxidising action of ozone has through the ad-vancement of electrical science been pressed into the service of the manufacturer. The methods employed for its production are all more or less based upon the well-known Siemen's tube. Generally speaking, air and not oxygen is ozonised, the air to be ozonised being freed from dust and from excess of moisture, the last of which causes formation of hydrogen peroxide. The temperature should be kept as low as possible, because at low temperatures oxides of nitrogen are less liable to be formed and the quantity of atmospheric oxygen converted into ozone is increased; indeed, some manufacturers cool the air down to 4° C. before subjecting it to the electric discharge. For con-venience of use the ozone is generally compressed into iron cylinders under a pressure of from four to five atmospheres. It is used for refining and bleaching linseed and palm oils, and for the manufacture of oxidised oil for linoleum. Brewers are often troubled with fouling of the beer barrels; this seems to be due to the growth of a fungus which often penetrates the wood to a considerable depth, so that ordinary methods of cleansing fail to remove it. The oxidising action of ozone has been successfully employed to remove this growth, the method being to alternately steam and ozonise the casks. It has also been utilised to remove fusel oil from alcohol, in the purifying of water, the refining of sugar in place of animal charcoal, and in a great variety of other manufacturing processes.

It is well known that synthetical diamonds have been obtained by means of the electric furnace; charcoal obtained from sugar is rammed into a wrought iron cylinder, which is then closed with a plug. The cylinder so filled is placed in a bath of molten iron kept at a high temperature in an electric furnace, after which the crucible which contains the iron is rapidly cooled by immersion in melted lead. On dissolving the iron in acid minute diamonds are obtained. It was a question whether here we had a case of simple crystallisation of the carbon from the molten metal on cooling, or whether the enormous pressure which was exerted upon the interior of the mass by the rapid cooling of the outside acting upon the carbon at a high tempera-ture caused the formation of crystals of diamond. An ex-ceedingly ingenious experiment which has been carried out by Majorana shows that at any rate the influence of high pressure and high temperature combined is sufficient to convert amorphous carbon into the crystalline variety. Majorana's experiment is as follows:—

A cylindrical chamber, A (Fig. 1), is hermetically closed at the top by a solid block of iron, E, the bottom by a solid piston, s. The sides of the chamber are made of tempered steel, and to further strengthen it the chamber is surrounded by fifteen iron rings 1 cm. thick, which are bolted together. The whole system is placed within an hexagonal frame, K, also made from iron plates. The piston, s, has a small solid iron cylinder about 1 cm. in diameter attached to it, at the end of which is fastened a small piece of carbon, C, about 2 grms. in weight. Directly below the piston a thick block of iron, ρ, is fixed, into which a hole exactly the size of the small end of the cylinder has been drilled. In carrying out the experiment the carbon is heated by means of the two carbon poles, D, D', with a current of 25 amperes and 100 volts. When the carbon has become white-hot, 70 grms. of gunpowder contained in the chamber A is exploded, the piston being driven down, carrying the heated

carbon before it and compressing it with enormous force. On taking the system to pieces the carbon is found to have been partially converted into microscopic diamonds, which when freed from unchanged amorphous carbon are found to possess all the characteristics of natural diamonds.

Reference has already been made to the importance of the manufacture of calcium carbide; another carbide, that of silicon, is now being manufactured in considerable quantities, and, owing to its extreme hardness, is being employed in place of emery for polishing steel and making grindstones. This carbide, which goes under the name of "carborundum," is manufactured by means of the electric furnace. An American company at Niagara Falls employs furnaces capable of dealing with ten tons of material, consisting of coke, sand, common salt and sawdust, which yield two tons of carborundum in twenty-four hours. In the first half of the year 1897 it is stated that in America alone 750,000 lbs. of carborundum were manu-factured. Since the introduction of electricity to chemistry the carbides of almost all the metals have been obtained, the majority naturally being more of theoretical than of commercial interest.

From the days of Leblanc, the founder of the soda industry, perhaps no branch of inorganic chemistry has been more worked at, or has better shown the results of patient toil and inventive genius, than the alkali and bleaching industry. Only after many attempts and many failures has the seemingly simple task of electrolysing sodium and potassium chloride yielded results

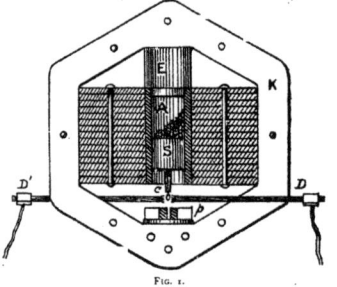

FIG. 1.

which have enabled electricity to enter into competition with the former methods of manufacture.

According to the manner in which the electrolysis is con-ducted, a solution of potassium chloride may be converted into chlorine and caustic potash, potassium hypochlorite, or into potassium chlorate. If the electrolysis takes place at low temperatures, a solution of hypochlorite is obtained, which may without further treatment be used for bleaching purposes. The difference in price between this solution and a solution of bleach-ing powder is not very great, but the greater cleanliness and purer bleaching action of potassium (sodium) hypochlorite make it, when electric power can be easily obtained, at least a power-ful competitor with bleaching powder. When the electrolysis is conducted at temperatures from 60° and upwards, the bath being kept slightly alkaline by addition of potassium bicarbonate or lime, potassium chlorate is produced, which, owing to its slight solubility in water, crystallises out, and by washing readily freed from adhering chloride.

If caustic potash and chlorine are required, some arrangement must be made to prevent the liberated chlorine from again reacting with the caustic potash formed at the same time. Formerly, and still to a small extent, this was arranged by means of a diaphragm which separated the anode from the kathode. Owing, however, to the difficulty of obtaining a *pervious impervious* diaphragm, *i.e.* one which allows the current to pass, but is impervious to the solution, it is now more general to electrolyse without a diaphragm. The method em-ployed is one which was originally employed by Castner and

Kellner, the kathode being a layer of mercury at the bottom of the bath. On the current being passed, the potassium liberated at the kathode dissolves in it, forming an amalgam, which as it is formed is drawn off and run into pure water, the water being decomposed, assisted by an auxiliary current, with evolution of hydrogen and formation of caustic potash, which is obtained in the pure condition by evaporation. Carbonate of potash may be prepared by passing a stream of carbonic acid gas into the caustic liquors before evaporation. In commerce, naturally, it is more general to electrolyse the cheaper sodium chloride, at any rate, in this country. Strontium and barium chlorate are also manufactured by electrolysis of their chlorides.

It has been found possible to prepare painters' colours by electrolysis, *e.g.* "white lead" is obtained in a very pure condition by electrolysing a dilute solution of sodium chlorate and carbonate, the electrodes being of lead. If the sodium carbonate is replaced by sodium chromate, a neutral lead chromate is produced, an acid chromate being formed by cautious addition of a solution of chromic acid during the electrolysis.

To attempt to mention, much less to describe, the enormous number of inorganic compounds and elements which have been prepared or isolated by the aid of electricity, would be, in an article such as this, impossible ; but sufficient examples have been given to show the importance of electrical processes in this branch of chemistry.

Turning now to organic chemistry, we notice that, although a vast amount of work has been done, it is more of theoretical interest than of technical value. But now that the initial difficulties have been to a large extent overcome, and the manner in which the reactions take place is better understood, it is probable that shortly this branch of manufacturing chemistry will also undergo a revolution in the hand of the electro-chemist. As a means of synthesis and of proving the formula of substances, electrolysis has been, and is, of great value to the organic chemist. Thus on electrolysing a solution of an alkaline acetate, ethane is produced ; whereas by employment of a succinate, ethylene is formed, a solution of fumaric acid yielding acetylene. These are, of course, simple cases ; but even that seemingly unsolvable problem, the constitution of camphoric acid, has been attacked by Walker, and by means of electrolysis of it and its derivatives he has obtained results which must be of great value in ultimately deciding what is the correct formula for this substance.

Iodoform can now be produced in a state of great purity by electrolysing a solution of potassium iodide and sodium carbonate to which alcohol has been added. On electrolysis, employing electrodes of platinum, iodine is continually set free at the anode, and coming in contact with the alcohol at the moment of its liberation produces iodoform. As the reaction proceeds some of the iodine becomes converted into hydriodic acid, and this combines with the alkali liberated at the kathode, or which has been added to the bath, potassium iodide being regenerated, which by the further passage of the current is again split up. The process is a continuous one, the iodoform being from time to time removed and a further quantity of alcohol, potassium iodide and sodium carbonate added. It is interesting to note that the alcohol cannot be replaced by acetone, as in this case only a very small quantity of iodoform is produced. Chloroform and bromoform have not been successfully prepared by this method. Chloral can, however, be produced by electrolysis of a solution of potassium chloride at 100°, to which alcohol is from time to time added.

By the electrolysis of nitrobenzene in a strongly acid alcoholic solution, aniline is produced. If the nitrobenzene is suspended in concentrated hydrochloric acid, ortho- and para-chloraniline are formed. By electrolysis under other conditions, azobenzene, hydrazobenzene or azoxybenzene are obtained.

By the electrolytic oxidation of aniline, dye products can be obtained the nature of which depends upon the solution employed, the strength of the current and the material of the electrode, *e.g.* if an aqueous solution of aniline hydrochloride, which may be either acidified with sulphuric acid or be practically neutral, is electrolysed, platinum electrodes being employed, a green precipitate is produced at the anode, which becomes violet, bluish-violet and finally almost black, practically the theoretical quantity of aniline black having been formed. If the aniline contain toluidine, then mauveaniline, rosaniline, &c., are produced.

Attempts have been made to obtain alizarine by electrolysis of anthraquinone in strong alkaline solution ; indeed, small quantities are said to have been obtained.

It has even been found possible to utilise electricity in the dyeing of cloth. When, *e.g.* a piece of cloth is soaked in a solution of aniline sulphate and placed between two metal plates, which are connected with opposite ends of a dynamo, and the current passed, the aniline sulphate is converted into aniline black ; indeed, by altering the strength of the solution and the density of the current, shades varying from green to deepest black can be produced.

In the case of indigo the cloth is thoroughly impregnated with a thin paste of indigo-blue and caustic alkali ; it is then placed between two metal electrodes. On the current being passed, the insoluble indigo-blue is converted by reduction into the soluble indigo-white, which on exposing the cloth to the action of air becomes again oxidised and the material dyed blue. Patterns may be printed on the cloth by cutting or stamping the plates in relief, or by connecting one pole to a metal plate and the other to a metallic pencil, when patterns, &c., can be readily sketched upon the material (Goppelsroeder).

Such processes as have been described in these articles appear, when seen in print, as extremely simple. Theoretically they may be so ; but in practice, the carrying out of these seemingly simple operations is often attended by great difficulties. For example, the temperature must not be allowed to rise too high or fall too low. The quantity of current and its potential require often to be kept within extremely narrow limits, as the following simple example illustrates.

Silver and copper can be separated by means of electrolysis, the silver alone being deposited if a very low current density ('10 ampere) is employed, whereas with a higher density ('50-1'0 ampere) the copper is deposited. Naturally, then, if at the commencement of the operation the higher current density is employed, both metals will be deposited together. Many of the difficulties to be overcome are to a large extent engineering. And it is to a considerable extent due to collaboration of chemists and engineers that the science of electro-chemistry has become what it is.

Electro-chemistry is quite in its childhood, but it is making marvellous and rapid progress. Works dealing with technical chemistry but a few years old require to be revised, owing to the alterations which this branch of chemical industry has brought about.

It is humiliating to realise that in this country there is hardly a book on the subject to be obtained, and in most cases even these are only translations from Continental works. And it is to be feared that unless this branch of chemistry becomes more studied than it has been up to the present, we shall find in the near future that electro-chemistry, both theoretical and practical, is the property of America and the Continent.

F. MOLLWO PERKIN.

ROAD LOCOMOTION.[1]

THE author commenced by saying that the subject of mechanical propulsion upon common roads had now reached a point when it deserves the very careful consideration of mechanical engineers.

For many years the uses and importance of the traction engine have become more and more recognised, but its work covers only a portion of the field for mechanical propulsion on roads, and he went on to consider what has led to a general revival of a movement for lighter road-locomotives which about seventy years ago, in the days of Hancock and Gurney, reached a point that for a time appeared to be leading to permanent results of the most important kind, but which ended in complete failure. In one sense this revival is undoubtedly due to the passing of the Locomotives on Highways Act in 1896, previous to which, for more than twenty years, a law had existed which made it impossible for any self-propelled vehicle to proceed at a rate of more than four miles an hour. The immediate cause of the passing of this Act was the attention aroused in this country by the successful introduction of the motor vehicle for purposes of pleasure in France. The real causes of the present movement were probably to be traced to the gradual feeling amongst all classes of the community that modes of transport, both for purposes of pleasure and business, on the roads had not kept pace, or indeed had made little progress at all, compared with the great changes which had been effected

[1] Abstract of a paper read before the Institution of Mechanical Engineers by Prof. Hele-Shaw, F.R.S., April 26.

in speed, comfort and convenience, in the direction of locomotion by rail.

The author went on to consider at some length the question of the conveyance of goods, and by means of a graphical diagram showed that up to forty miles motor vehicles, owing to the terminal charges, might compete with railways. The hygienic question and growing difficulties of traffic in large cities was next touched upon. Next the difficulties of the problem were considered, and it was shown that these difficulties were very great; and so far from the success of the railway system being an argument in favour of the immediate development of locomotion by road, the conditions of the problem were such as to involve improvement exactly in the opposite direction to that in which the railway locomotive has been successfully developed.

No doubt the progress of invention would enable a greater amount of power from a given weight of motor to be obtained ; but the surface to be moved over, which is the real difficulty of the road locomotive, would remain the chief factor of the problem.

The first section of the paper was therefore devoted to the mechanical problem of the behaviour of the wheel upon the road, and the progress which has been made in this direction.

Under this head the whole question of resistance upon the road was treated, and the author made a strong point of the fact that there was very little data available for determining resistance upon roads at the high speeds now permissible, and with different kinds of tyres now in use. He gave diagrams of horse-power curves of resistance adapted to English measures from the "Tableaux Numerique et Graphique" of MM. Boramé and Julien, and a series of graphical results taken with the Viagraph of Mr. Brown, showing the nature of the shocks to a vehicle by roads of macadam, stone, asphalte and wood.

The subject of pneumatic tyres was then discussed, and also the question of wheels suitable for heavy traffic, and illustrations of a number of improved types of wheels for this purpose, which had been invented during the last year or so, were given. With any existing system in which four wheels are used, it was shown that the problem of road locomotion was difficult because of the concentration of the load upon such a limited area of support. Even for heavy traffic the pneumatic tyre might come into use in the future as it extends the area of resistence by yielding, so that the surface in contact is much greater than in the case of an iron-rimmed wheel, especially when running over stone sets or hard ground.

Beyond this, it is quite conceivable that, just as in railways the number of wheels has been largely increased until a modern bogie carriage has commonly twelve wheels supporting it, it may be found economical to support a motor vehicle also upon a much greater number than at present.

The steering of motor vehicles, which was the subject of the next section of the paper, is evidently a very important part in their design, and it may at once be said that with one or two exceptions the great majority of motor vehicles are steered upon the principle which was invented by Ackermann as long ago as 1818. The essential principle of the Ackermann system consists in replacing the pivoted fore carriage of an ordinary vehicle which has one axle for the two wheels, by two short pivoted axles each carrying one of the steering wheels. The conditions of correct running of the wheels are that when their plane is turned, their normals intersect on a point on the line of the axles of the driving wheels. The paper then went on to describe the various modifications of the Ackermann system.

The next section of the paper was devoted to a consideration of motive power and its transmission to the wheels, and under this head the particular uses and advantages of oil, steam and electricity were considered ; although these various agencies have at the present moment fairly well recognised spheres of operation this must be by no means regarded as the final condition of things, or as giving a limitation to the employment of each of these types of motive power. Thus, although at present oil-engines are used for light motor vehicles and steam for heavy traffic, there are very ingenious steam motor-cars both in this country and abroad, while light oil-engines have been applied in France and also in this country in connection with heavy traffic.

Oil-engines, or internal-combustion engines, have by a process of the survival of the fittest been found so far best suited for light motors and pleasure vehicles. The cycle of the gas-engine is really complex, so that these motors have been brought to a high state of perfection, so that upon being started they are found to work for long distances without any attention. If really well designed and constructed, and used with a moderate

amount of care, they need little repairs or adjustment, while the objection of smell, vibration, and danger from the use of light petroleum spirit with a low flash point, have all been much reduced, while each year sees an increasing number of places in town and country where petroleum spirit can be obtained. Still the objections above-mentioned must be admitted to exist, and this, together with the great expense of pleasure vehicles, has to a certain extent hitherto prevented their introduction becoming general. Again, an oil-engine, which has little elasticity in regard to an increased demand for power when ascending a hill, requires elaborate gearing for change of speed, which may be after a time, if not at first when the car is new, a very noisy and objectionable feature. Heavy-oil engines for internal combustion have been tried for motor vehicles, but the difficulties of starting and smell have not yet been satisfactorily overcome.

Steam, or external-combustion motors, require not only a generator or boiler, but also a condenser, in addition to the steam-engine itself. The latter is not used with all motors, but in winter the cloud of steam which must be visible in damp cold weather at a little distance from the exhaust, even if the steam is superheated, really contravenes the Act, which states, "No smoke or visible vapour must be emitted, except from any temporary or accidental cause." Steam introduces a more complicated array of pipes and fittings, and requires more attention and skill in working, but it is highly probable that such improvements will be made in connection with steam motors, that no skilled attendant will be necessary. There is already at least one steam system which is entirely automatic, whilst others are to a great extent so. It is not too early to speak as to the practical and commercial success of any of the systems using steam, but if a condensing steam-engine, automatic in action, with a boiler which is perfectly safe from any fear of explosion, can be produced, it may safely be predicted that there is a great future before it, both for light and heavy traffic, as it would have the advantages of great power and elasticity, freedom from smell, and if using heavy oil, or even coal or coke, would be free from the danger and trouble incidental to the employment of light oil, especially abroad. Moreover, the ease with which a steam motor can be started and stopped, and more particularly reversed, cannot be over-estimated.

Fuels, other than coal, coke or oil, have been the matter of careful consideration by motor-car designers. The most promising of these is acetylene, which, as derived from calcium carbide, enables a much greater quantity of energy to be obtained from a given weight of fuel ; but although it only requires one-fourth of the weight of calcium carbide to produce a given amount of work as compared with coke, the expense at present makes its use commercially impossible.

Electrical motors are clean, extremely convenient and simple, free from all vibration and danger and altogether an ideal type of motor. The limitations in the use of electricity are, however, very serious, and are discussed later in the paper.

The details of internal combustion motors are discussed under the six headings upon which their success more or less depends, viz. : (1) carburisation, (2) ignition, (3) starting, (4) governing, (5) balancing, and (6) cooling.

The subject of steam is then treated at considerable length, and types of the more important steam heavy motor vehicles, such as those of Thorneycroft, Simpson and Bodman, Musker, Coulthard, Bayley and Clarkson and Capel are given, together with various examples of water-tube and flash boilers, which may be said to be the two types of boilers specially suitable for motor vehicles on account of their high steaming capacity in proportion to their small weight.

In considering the actual results which have been obtained by motor vehicles, a distinction is made between pleasure vehicles and those for the conveyance of goods. For the former, the actual cost of working is not by any means the first consideration : in a large number of cases, in fact, the cost is comparatively of small importance. Questions of comfort, durability and safety, as well as freedom from liability to break down, are the chief points to be considered. These matters can scarcely be summed up except as the result of lengthy experience, and now undoubtedly that experience is gradually being acquired.

When we come to the question of goods traffic, the matter is of course entirely one of cost, including not merely the outlay, working and upkeep, but deterioration, which in road vehicles is exceptionally heavy. Extended trials of actual working are necessary for any final opinion of the relative merits of different types of heavy motor vehicles, and the author has fortunately been able to secure much valuable testimony of this sort on the subject.

A great deal, however, can be ascertained by careful trials, such as those which have been undertaken on two occasions at Liverpool (1898 and 1899), since measurements and data can be obtained with a staff of observers for a limited period, which could scarcely be secured in continuous working. The results of these trials are given in tables and also statements by the Chief Mechanical Engineer of the Lancashire and Yorkshire Railway, on the working of a Thornycroft motor wagon; the Engineer-in-Chief, Mersey Docks and Harbour Board; and the City Engineer of Liverpool, on the working of Leyland motor wagons; and by Mr. Bryan Donkin, on the tests of motor carriages at Richmond and Birmingham.

Looking at the whole question, it may be safely said that the motor vehicle has come to stay, and that its uses, both in peace and war, will rapidly and enormously develop. The public interest which is now seen partly by the immense number of patents taken out in connection with the industry, partly by the great growth of literature on the subject, and by the formation of automotor clubs, is not a mere transient thing, and although the motor vehicle is at present still somewhat of a *rara avis* upon our roads, it may not be going too far to think that the coming century will see a development of locomotion upon roads comparable with the development of locomotion of the railway in the century which, according to our individual views of chronology, is either past or so very nearly past.

THE UNIVERSITY OF BIRMINGHAM.

THE present position of the scheme for the establishment of a Midland University was explained by Mr. Chamberlain, Chancellor of the new University, at the first meeting of the Court of Governors, held on Thursday last. In the course of his remarks, Mr. Chamberlain is reported by the *Times* to have said that it was desired to create a great teaching University, in which all who came to them for it should find efficient and complete instruction in every branch of knowledge. Again, they desired that their University should be a school of research. They were firmly convinced that that was necessary if it was to maintain its dignity and great position. They believed that those were the best teachers who were themselves constantly learning, and that without adding continuously to the common stock of knowledge they would not be fulfilling their duties. In order to secure those objects they ventured to ask for a further endowment of a quarter of a million sterling. To-day they were able to announce that they had already received promises of 330,000*l.*, the amount having been largely increased by the munificent donations of Mr. Carnegie, of an anonymous benefactor, of Mr. Charles Solcroft, and of Mr. George Kenrick.

They intended that their University should be a distinctive University. In what he had hitherto indicated there was nothing original, nothing in which they were likely to specially differentiate themselves from the other great Universities, especially from the modern Universities in this country and the older Universities of Scotland; but they hoped that their University would take some colour from its environment, that not only would it be a school of general culture, but that it would also practically assist the prosperity and welfare of the district in which it was situated by the exceptional attention which it would give to the teaching of science in connection with its application to local industries and manufactures; and this portion of their task had turned out to be much greater, much more responsible, than they anticipated. They were encouraged in undertaking it by the gift of Mr. Carnegie, which was specially to be devoted to the creation of a college of science, following somewhat the example which had been set by the great colleges in the United States of America; and Mr. Carnegie followed this up by a proposal that a deputation from the intended University should visit the chief seats of learning across the water.

Those who had read the report of the committee that had visited Canada and the United States would begin to understand how it was that we were behindhand in the preparation for that great struggle which must come, that commercial competition between nations in which the weakest would inevitably go to the wall. For what did they find established both in the United States and in our own colony of Canada? They found great institutions connected with a general University, with colleges of science occupying large spaces, in which the area

was counted by many acres, fully equipped with proper buildings, with the most modern and complete machinery, with the latest scientific appliances, with laboratories for every conceivable scientific purpose; and in those great colleges a training was given such as they desired to see imitated in this country—a training based, as all education ought to be, upon a foundation of general culture, but specialised in its course, highly specialised according to particular and separate work which each student intended to undertake in life. As a result of this they began to see how it was that in America the great commercial and industrial undertakings, the manufacturers and inventors, found no difficulty whatever in obtaining the services of as many young men as they might require to manage and complete and develop their undertakings, all of them ready when they left college, not merely to deal with the ordinary routine and management of a business, but to bring to it the latest discoveries and to apply the highest science to its development. That was what they wanted in Birmingham, and they would not have the University which they all had in their minds until they had accomplished it.

All that was wanted was money. The committee had pointed out that to carry out this scheme with any completeness a further sum, partly for endowment, partly for buildings and machinery and appliances, of 155,000*l.* was required. He was quite convinced, even from an incomplete examination of the project, that they had under-estimated the cost. He thought himself that another quarter of a million was the smallest sum which they would require in order to put this portion of their undertaking upon a thoroughly satisfactory basis. Well, they must get it, and he anticipated that they would obtain it. He anticipated that they would obtain it from two sources. Nothing he thought was more striking to any one who had studied educational progress in America and in our great colonies than the readiness, the eagerness, with which men who had acquired great wealth had been willing to devote a considerable portion of it in sums to which we here, he was sorry to say, were almost unaccustomed, to the promotion of the higher education. It was the case in Canada, in the great Universities of Montreal and Toronto; it was the case in America, in Cornell, in the Stamford University, in the Chicago University, in the Columbia University; and it was also visible in the great donations which had been made to the older Universities of Harvard and of Yale. He could not doubt that the feeling that no better application than this could be found for wealth would grow among them about Birmingham, and that although they lived in a district which had hitherto not been remarkable for exceptional fortunes, yet which did contain many men of great wealth. They also would find a tendency, from which the University would derive advantage in the future, to make their contributions towards such purposes as he had described. He hoped that this might be the case, and he thought he might say that he had confidence that it would be the case, and they might expect before long that their funds would be largely increased from some such source.

UNIVERSITY AND EDUCATIONAL INTELLIGENCE.

CAMBRIDGE.—Mr. Chawner, Master of Emmanuel College, has been re-elected Vice-Chancellor.

Mr. Frederick Harrison will deliver the Rede Lecture in the Senate House on June 12, at noon. The honorary degrees referred to last week will be conferred on the same day, at 3 p.m.

The Knightbridge Professorship is vacant by the resignation of Dr. Sidgwick, who has been seriously ill.

Mr. L. R. Wilberforce, of Trinity College, has been elected a University Lecturer in Experimental Physics in the place of Mr. W. N. Shaw.

A grant of 50*l.* from the Balfour Fund has been made to Mr. J. S. Budgett in aid of his researches on the development of Polypterus.

Dr. Allbutt and Dr. Collingridge have been appointed delegates to represent the University at the International Congress of Hygiene and Demography to be held at Paris next August.

THE 500th anniversary of the foundation of the University of Cracow will be celebrated to-day, June 7. Representatives will be present from most of the European universities.

MR. W. T. A. EMTAGE, principal of the Wandsworth Technical Institute, has been appointed Director of Public Instruction in Mauritius. The post has been newly created, and Mr. Emtage will have the oversight of all the educational work under Government in the Colony. His first task will be the organisation of a system of technical education.

AT University College, London, Andrews Entrance Scholarships of 30l. each have been awarded to Mr. L. Graham, of Mason College, Birmingham, and to Mr. C. E. K. Mees, of St. Dunstan's College, Catford. The Atchison Scholarship of 55l. per annum for two years has been awarded to Mr. R. E. Lloyd for the greatest proficiency as a student of the medical faculty and the hospital during the past two years. The Bruce medal has also been awarded to Mr. R. E. Lloyd for proficiency in pathology and surgery.

THE Senate of the University of London has resolved that one sum of 100l. be offered as the Rogers Prize open for competition to all the members of the medical profession in Great Britain and Ireland, for an essay upon the production of immunity in specific infective diseases generally; and with particular reference to any one disease on which the writer of the essay has made original investigations. The essay is to be sent to the Registrar, University of London, South Kensington, S.W., on or before June 1, 1901.

THE Yorkshire College, Leeds, is now one of the university centres that grant a degree to students of agriculture. At a meeting of the Court of Victoria University (on May 3) a report of the Council recommending the inclusion of agriculture as a subject for the B.Sc. degree was adopted. Among other requirements, the scheme provides that students before taking their degree must conduct at an experimental farm controlled by a College of the University an experiment on some agricultural subject, and submit a report of the same. Only those students who, before entering the University, have attended an agricultural school for two years will be exempted from this rule.

AN illustrated prospectus of the courses of chemistry and chemical engineering at the Massachusetts Institute of Technology has recently been received. The prospectus includes descriptions of the various chemical laboratories, and the accompanying illustration of the main laboratory of industrial chemistry is of interest as indicating the provision made, in one of the foremost technical institutions in the United States, for work by students taking a general course in chemical industries. The ordinary course in chemistry in the Institute extends over a period of four years, and embraces almost all branches of chemical science. The aim throughout the whole course of instruction is not only to impart the necessary professional knowledge, but also to teach the student self-reliance, to accustom him to habits of accurate thought and work, and to instruct him in the methods of investigation of new problems. The course is designed primarily to prepare students for actual work in connection with manufactures based on chemical principles, but it provides also for those who expect to become teachers of chemistry, and for those who intend to devote themselves to scientific research. The object of the instruction in industrial chemistry is to set before the students as fully as possible the present status of the chemical industries. The laboratory instruction includes the preparation of pure chemicals, and the refinement or purification of technical products, by industrial processes. Among the processes carried out in the laboratory are the manufacture of dyers' mordants, soaps, phosphates from bone ash, and soda crystals; and also the preparation of salts of ammonium, barium, calcium, iron, copper, tin, chromium, &c., from minerals or other crude material. In addition, about eighty lectures are given on the most important industrial processes, and excursions are frequently made to manufacturing establishments.

FIG. 1.—Laboratory of Industrial Chemistry of the Massachusetts Institute of Technology.

SOCIETIES AND ACADEMIES.

LONDON.

Chemical Society, May 17.—Prof. Thorpe, President, in the chair.—The following papers were read :—The chlorine derivatives of pyridine. VI. The constitution of some amino-chloropyridines, by W. J. Sell and F. W. Dootson.—Ortho-substituted nitrogen chlorides and bromides, and the entrance of halogen into the ortho-position in the transformation of nitrogen chlorides, by F. D. Chattaway and K. J. P. Orton. When phenylacetyl nitrogen chloride undergoes transformation, a mixture of 95 to 96 per cent. of para- with 4 to 5 per cent. of ortho-chloroacetanilide is produced.—Ammonium imidosulphite, by E. Divers and M. Ogawa. A crystalline ammonium imidosulphite, $NH(SO_2NH_4)_2$ is obtained on allowing ammonium amidosulphite to decompose below 35° in a current of hydrogen or nitrogen.—The constitution of ethyl sodio-cyanacetate and of ethyl methylsodiocyanacetate, by J. F. Thorpe. The reactions of ethyl sodiocyanacetate and of ethyl methylsodiocyanacetate are best represented by the formulæ $CN.CH:C(ONa)OEt$ and $CN.CMe:C(ONa)OEt$ respectively. — The $aa_1\beta\beta$-tetramethylglutaric acids, by J. F. Thorpe and W. J. Young. Ethyl sodiocyanacetate reacts with ethereal iodine solution, yielding ethyl iodocyanacetate, and under certain conditions gives an unstable diiodide which reacts with the excess of ethyl sodiocyanacetate giving ethylic

$$\text{dicyanosuccinate,} \quad \begin{matrix} CH(CN).CO_2Et \\ | \\ CH(CN).CO_2Et \end{matrix} \quad \text{Ethyl methylsodiocyan-}$$

acetate reacts with iodine, forming ethyl methyliodocyanacetate which condenses with ethyl dimethylacrylate, giving the salt $CO_2H.CH(CN).CMe_2.CHMe.CO_2Et$; the latter on distillation gives ethyl a-methyl-$\beta\beta$-dimethyl-γ-cyanobutyrate,

$$\text{.} \quad \text{.} \quad \text{.} \quad \text{.} \quad CH_2(CN):CMe_2.CHMe.CO_2Et,$$

and this when hydrolysed yields $a\beta\beta$-trimethylglutaric acid. The preparation of cis- and trans-$a\beta\beta a_1$-tetramethylglutaric acid is also described.—β-Isopropylglutaric acid and the cis- and trans-methylisopropylglutaric acids, by F. H. Howles, J. F. Thorpe and W. Udall.—Methyl iodide acts on the sodio-derivative of the product resulting from the condensation of ethyl sodiocyanacetate with ethyl β-isopropylacrylate, yielding ethyl a-cyano-a-methyl-β-isopropylglutarate. The latter on hydrolysis yields ultimately trans-a-methyl-β-isopropylglutaric acid and its imide ; the imide is converted into the cis-acid by heating with sulphuric acid.—The racemisation of optically active tin compounds. Dextromethylethylpropyl tin dextrobromocamphorsulphonate, by W. J. Pope and S. J. Peachey. Optically inactive methylethyl-propyl tin iodide is wholly converted into dextromethylethyl-propyl tin dextrobromocamphorsulphonate by treatment with the silver salt of the acid and evaporating the filtered solution ; the new salt has the molecular rotatory power $[M]_D = +318°$ in dilute aqueous solution, but after heating and cooling the solution the value $[M]_D$ falls to $+273°$, which is the value $[M]_D$ of the acid in aqueous solution. After evaporating the solution to dryness and making up to the original volume by dissolving the residue in cold water, the value $[M]_D = +315°$ was obtained. It is thus proved that the asymmetric tin radicle

$$(CH_3)(C_3H_5)(C_2H_5)Sn -$$

can be easily racemised and easily converted into one optically active component.—Racemic and optically active forms of isoamarine, by H. L. Snape. The author has resolved optically inactive isoamarine into its optically active components by crystallising its tartrate ; the dextro-base has the specific rotatory power $[a]_D = +62·02°$. The crystals of the optically active bases are orthorhombic and sphenoidally hemihedral.

Linnean Society, May 3.—Mr. C. B. Clarke, F.R.S., Vice-President, in the chair.—Mr. H. E. Smedley exhibited a number of botanical wax models prepared on an enlarged scale to show the morphological structure and also the process of reproduction in various types of plants.—Mr. J. E. Harting exhibited and made remarks on some skins of willow grouse collected by Prince Demidoff on the N.W. border of Mongolia between Alta Mountains and the Kobdo River.—On behalf of Miss E. S. Barton, the Botanical Secretary read a paper on a new species of *Halimeda* from Funafuti ; and on behalf of Miss A. L. Smith, a paper on some West Indian fungi, with descriptions of a new genus and species.

May 24. Anniversary Meeting.—Dr. A. Günther, F.R.S., President, in the chair.—The following were elected into the

Council :—Mr. Clement Reid, Dr. D. H. Scott, Rev. T. R. R. Stebbing, Prof. S. H. Vines, and Mr. A. Smith Woodward ; and as President, Prof. Sydney Howard Vines, F.R.S. ; Treasurer, Mr. Frank Crisp ; Secretaries, Mr. B. Daydon Jackson and Prof. G. B. Howes, F.R.S.—The retiring President then delivered his annual address, choosing for his subject, " The unpublished correspondence of William Swainson with contemporary naturalists (1806–1840)," lately acquired by the Society.—The Gold Medal of the Society was then presented to Prof. Alfred Newton, F.R.S., in recognition of his important contributions to zoological science.

Royal Microscopical Society, May 16.—Mr. Carruthers, F.R.S., President, in the chair.—Mr. Chas. Baker exhibited two microscopes ; one made specially for critical work was fitted with eye-pieces of the Society's new Standard gauge, No. 3, of 1·27 in. The other instrument, named the " Plantation " microscope, was designed for use in the tropics for the purpose of discovering the ova of internal parasites. Dr. Hebb said a paper had been received from Mr. Millett, being Part viii. of his report on the Foraminifera of the Malay Archipelago. This, as on former occasions, would be taken as read.— E. M. Nelson read a paper on the lag in microscopic vision, which he illustrated by diagrams and a series of tables showing the proportionate values of the performance of various objectives under eye-pieces of different powers. In the case of an apochromatic objective of fine quality, the degree of merit was shown to range from 14·7 with a low eye-piece, to 7·7 with a deep one, but the difference was more marked with ordinary dry achromatic lenses. Mr. Nelson's experiments had shown that in respect to the lag, microscopes with short tubes had some advantage over those with long tubes. Mr. Nelson also read a paper, for Mr. E. B. Stringer, on a new form of fine adjustment, a microscope by Messrs. Watson and Son, fitted with the arrangement, being exhibited. Mr. Nelson said that its working seemed exceedingly good. As the fine adjustment was placed just behind the body, the limb could be made of any length without putting additional strain upon the screw, a matter which would be of great advantage in microscopes made for examining large sections.—In announcing the adjournment of the meeting until Wednesday, June 20, the president said he hoped then to be able to submit and explain a series of lantern slides representing minute structure of some Palæozoic plants.

PARIS.

Academy of Sciences, May 28.—M. Maurice Lévy in the chair.—Formation of nitric acid in combustions, by M. Berthelot. When sulphur is burnt in the calorimetric bomb in compressed oxygen under a pressure of twenty-five atmospheres, some nitrogen also being present, nitric acid is formed in quantities amounting to about ·001 of the sulphur present. At atmospheric pressure the amount of nitric acid formed is much reduced. With metals such as iron and zinc no nitric acid is formed.—Preparation, properties and analysis of thionyl fluoride, by MM. H. Moissan and P. Lebeau (see p. 137). —On the laws of specific heats of fluids, by M. E. H. Amagat. The formula $\dfrac{dC}{dp} = -AT\dfrac{d^2v}{dt^2}$ is applied to find the relation between the specific heat and pressure of carbon dioxide. The values o $\dfrac{dv}{dt}$ and $\dfrac{d^2v}{dt^2}$ were found graphically from the experimental data, and the results are given in the form of curves.—On some remarkable sub-groups of a group of substitutions or transformations of Lie, by M. Edmond Maillet. —On partial differentiation of the third order which admit of an intermediate integral, by M. A. Guldberg.— Formulæ giving the volumes of saturated vapour and the maximum pressure, by M. H. Moulin. The formulæ deduced from theoretical considerations by the author are compared with the experimental data of Young, Tate and Amagat for benzene, fluorbenzene, carbon tetrachloride, ether, acetic acid, methyl alcohol, water and carbon dioxide with satisfactory results.— The energy absorbed by condensers submitted to a sinusoidal difference of potential, by MM. H. Pellat and F. Beau-lard.—The transparency of some liquids for electrostatic oscillations by M. A. de Heen.—On some photochemical effects produced by the wire radiating Hertzian waves, by M. Thomas Tommasina.—On a lithium peroxide, by M. de Forcrand. Since the combustion of lithium in oxygen gives only traces of a peroxide, attempts were made to prepare lithium peroxide in the wet way, by the action of hydrogen peroxide upon solutions of lithium salts. A thermochemical study of the products showed that some Li_2O_2 is formed in this way.—On the unknown earths contained in crude samaria, by M. Eug. Demarçay. The oxide isolated contains neither samarium nor gadolinium, and is of an atomic weight between these two elements. The chief lines of the spark and absorption spectrum are described.—The reduction of erythrulose and the preparation of a new erythrite, d-erythrite, by M. Gabriel Bertrand. Ordinary erythrite is easily oxidised by the sorbose bacterium to the ketone erythrulose,

$$CH_2(OH).CO.CH(OH).CH_2OH,$$

and this on treatment with sodium amalgam gives a mixture of two erythrites, one identical with the original inactive erythrite, the other, separated by means of its acetal, is active, possessing a rotatory power $[a]^D = -4°·76$.—Action of cyanogen chloride upon acetone-dicarboxylic ethyl ester, by M. Juvénal Derôme. The cyano-derivative produced,

$$(CO_2C_2H_5).CH.CN.CH_2(CO_2C_2H_5),$$

readily forms metallic salts, the hydrogen adjacent to the cyanogen group being replaced.—On the metallic combinations of diphenylcarbazone, by M. P. Cazeneuve.—Osmotic pressure of the egg and experimental polyembryony, by M. E. Bataillon.—On the sub-fossil Lemuridæ of Madagascar, by M. Guillaume Grandidier.—On the discovery of a cave containing animal remains at Bains-Romains, near Algiers, by MM. E. Ficheur and A. Brives. The remains found include the bones of the species *Bubalus, Bos, Cervus, Antilope, Hippopotamus, Rhinoceros* and *Equus*. The presence of man was indicated by a molar, flint heads, and the presence of calcined bones.—Mode of action of antileucocytic serums upon the coagulation of the blood, by M. C. Delezenne. The mode of action appears to be identical with that of a peptone, the intravenous injection of a leucolytic agent being the same in all cases, the destruction of the white corpuscles circulating in the blood.—On the restoration to life obtained by the rhythmical compression of the heart, by MM. Tuffier and Hallion. A claim for priority against M. Batelli.

DIARY OF SOCIETIES.

CONTENTS.

THURSDAY, JUNE 14, 1900.

MALAY MAGIC.

Malay Magic; being an Introduction to the Folklore and Popular Religion of the Malay Peninsula. By W. W. Skeat. Pp. xiv + 685, and numerous plates and illustrations. (Macmillan and Co., Ltd., 1900.)

THE object of this interesting and important work is set forth on the title-page with such clearness that the reviewer and reader are spared some trouble in defining it, and it is pleasing to be able to say that the author exhibits the same clearness throughout the hundreds of pages which he has devoted to the discussion of his subject. Speaking broadly, Mr. Skeat's volume is divided into six sections or chapters, which indicate by their length the relative importance of the matters of which they treat, and the well-chosen illustrations do much to enlighten the reader of the work on many points which do not fall naturally under the heading of facts of folklore. Mr. Skeat's book differs greatly from the works on folklore which appear from time to time, for it contains, not only what seems to us to be an exhaustive statement of facts which he has collected and arranged with care and discretion, but a series of deductions made after due consideration of the general principles which have, consciously or unconsciously, guided man in all ages and in all countries in working out theories as to the relations which exist between the animate and inanimate in nature. Many travellers and sojourners in foreign lands and remote islands have written books on the folklore of their inhabitants, but the greater number of them have been characterised by haste, and by a lack of knowledge of the fundamental facts of primitive anthropology. Moreover, it has frequently happened that, although their writers have given their facts correctly, they have not given all that might have been given had their own knowledge of them at first hand been sufficiently good to draw forth from the natives all that might have been extracted from them. Mr. Skeat has given abundant time to his subject, and as he has relied for guidance in difficult matters upon such works as Prof. E. B. Tylor's "Primitive Culture," the non-expert will feel that he is in safe hands. Mr. Skeat's years of residence in the Malay States gave him unwonted facilities for collecting information, and his official position and knowledge of the native dialects enabled him to make the fullest use of his opportunities. Another fact must be remembered. The influence of the West upon the East grows stronger every year, and the systems of the white man and government according to modern Western ideas, which, sooner or later, he invariably succeeds in imposing upon the coloured man, are not favourable to the preservation of native superstitions and beliefs. Little by little they are set aside, and eventually they disappear; thus frequently it happens that information which the student of comparative folklore would consider priceless for his studies is lost for ever. Mr. Skeat has done well in collecting such information in the Malay Peninsula whilst it is still to be obtained, and we can only hope that other officials who have the time and opportunity for collecting an-

thropological facts may emulate his devotion and industry.

According to Malay views in general, the earth and the sea were formed, each in seven stages, after the light, which was an emanation from the Deity, had become the "world-ocean." The earth was surrounded by a ring of mountains which kept it in its place, and served as the abode for legions of spirits. This mountain is, of course, the old Arab mountain of Kâf, from which, as Yâkût says, "all other mountains are derived." Certain sages, however, hold other views, and describe how the Kâbah, or home of the famous Black Stone at Mekka, the navel of the earth, was made immediately after God made himself manifest by his tokens the sun and moon. Next, the angel Gabriel killed the great serpent Sakâtimuna, and the description of the subsequent disposal of her body forcibly recalls the Babylonian account of the fight between Merodach and Tiâmat. In fact, it seems pretty clear that Semitic cosmogonies have been drawn upon by the Malay theologians for several of the above theories. In shape, the earth is oval, and it revolves upon its own axis once every three months. Day and night are caused by the sun, which is a circular body moving round the earth. The sky is made of stone or "bed-rock," and the stars are merely holes which let light through from the place of light above. An earthquake is caused by the buffalo which supports the earth on its horns, throwing it from the tip of one horn on to the tip of the other; this buffalo stands on an island in the midst of the nether ocean. The tides are caused by a huge crab moving in and out of his cavern, which is situated at the root of the Pauh Janggi tree. Eclipses are the result of a monster dragon trying to swallow the sun and moon; and indeed any untoward movement in nature is attributed to the movements of beasts of enormous size or dragons.

The appearance of man upon the earth is accounted for in various ways, but it appears that all Malay explanations of his origin are based upon Arabic legends of the creation of man by Allah, who is said to have fashioned him out of earth, air, fire and water. The version of one legend, printed by Mr. Skeat on pp. 19-20, with its mention of Michael, Gabriel and Izrafel, proclaims the source from which it was derived. The body is composed of earth, air, fire and water, and with these elements are connected four essences—the soul or spirit with air, love with fire, concupiscence with earth, and wisdom with water. But the works of Arabic writers on such matters were not the only authorities consulted by the early Malay philosophers, for Greek authors of treatises on the composition of man are often quoted. passing over the consideration of the sanctity of the body for want of space, we come to the mention of the soul, which is described as a thin, unsubstantial human image or mannikin, which is temporarily absent from the body in sleep, trance and disease; after the death of the body, the soul departs from it for ever. It is usually invisible, but it is supposed to be as big as the thumb, and to resemble the body in shape, proportion and complexion; it is of an impalpable, filmy, shadowy substance, and causes no displacement in the body into which it enters. It possesses all the attributes of the body to

H

which it gives life, and it suffers from all its disabilities ; sickness is supposed to be caused by its absence from the body, and the soul may be abducted from it by unlawful means. The human soul is seven-fold, and it seems, at times, as if each was independent, for in certain ceremonies an abode is provided for each. The idea that a man possesses several souls is very old, and in Egyptian religious texts it may be traced back to the period of the earliest dynasties, about six thousand years ago. The number seven is, of course, and always has been, a magical number, and in ceremonies which are intended to do good, as well as those in which the object is to do evil, it plays a prominent part. In Babylonian and Assyrian magical texts we find the seven evil spirits of the deep, and the Mesopotamian underworld possessed seven gates ; it must not be forgotten, too, the famous temple of Nebo at Borsippa, which tradition identifies with the Tower of Babel, was built in seven stages.

When we come to discuss Malay gods, we find the subject to be one of some difficulty. In the old religion, which the Malay professed to throw off when he adopted Muhammadanism, his ideas had formulated the existence of a large number of nature powers which closely resemble the Hindu gods found in Brahmanism ; and before he adopted these as the objects of his worship, he seems to have peopled heaven and earth with myriads of spirits. To this day, when in trouble, he cries out, not to the Allah preached by Muhammad, nor to the deities which the Brahman religion made known to him, but to the evil spirits which his ancestors worshipped and feared untold centuries ago. It has been the same in all ages and in all countries, and the nations which become "converted" to a new religion in reality only drop the observances connected with their old faiths ; and although they may tear down the shrines of old gods and build others to new ones, they do not succeed in uprooting from their minds the beliefs and ideas of which the overthrown shrines were the outward and visible signs. In spite of the teaching of Muhammad and the Brahmans, the Malay still believes that every department of nature is presided over by a "god" who must be propitiated by man, and to be specially honoured and revered are such gods as Batara Guru, Batara Kala, Batara Indra, and Batara Bismu ; the greatest of this group is the first. It is interesting to note that native influence has succeeded in introducing into the Malay pantheon a number of gods of the sea, which from certain aspects are identified with older terrestrial gods. Many of the Jinn, or evil spirits of the Arabs, have been identified with old Hindu spirits, and the view held by the Malay on the importance of such beings may be gathered from the fact that it was believed to be possible to buy them from the Shêkh of the Jinh at Mekka, at prices varying from ninety to a hundred dollars each !

More than three-quarters of Mr. Skeat's volume are occupied with a description of the magic rites which the Malay connects with the various departments of nature, and with the life of man. This is not to be wondered at, for it is clear at a glance that there is no event in his life, however trivial and apparently unimportant, which, unless properly protected by magic rites and ceremonies, may give hostile devils and fiends an opportunity for

doing undreamed-of mischief to the wretched mortal whom accident or design has left unguarded. We regret that we cannot follow Mr. Skeat through his description of birth-spirits and birth-ceremonies, and through the whole period of a man's life from the cradle to the grave, as sketched by him, for our space is exhausted, and the reader can study for himself the curious Malay customs which concern betrothals, marriages and deaths. Many of them have their counterparts in other countries, but not a few are peculiar to the Malay. As we read of them we cannot help wondering how, if the pious Malay fulfils all his religious obligations, he ever finds time to do anything else. It is improbable in these days that many men are found who are able to carry out all the religious performances enumerated by Mr. Skeat, and it is much to be hoped that the influence of the English will drive many of them out of existence. Meanwhile a good and careful record of Malay sorcery, witchcraft and demonology, which is invaluable for the study of comparative religion and folklore, has been given us by Mr. Skeat, and there is no doubt that he has laid anthropologists and ethnographers and Oriental archæologists under a heavy debt of gratitude.

THE NANSEN NORTH POLAR EXPEDITION.

The Norwegian North Polar Expedition, 1893–96 : Scientific Results. Edited by Fridtjof Nansen. Vol. i. *The Jurassic Fauna of Cape Flora, Franz Josef Land.* By J. F. Pompeckj. With a geological sketch of Cape Flora and its neighbourhood by Fridtjof Nansen. Pp. 147 ; with 3 plates. *Fossil Plants from Franz Josef Land.* By A. G. Nathorst. Pp. 26 ; with 2 plates. *An Account of the Birds.* By R. Collett and F. Nansen. (London : Longmans, Green and Co., 1900.)

THE second chapter of the first volume of the "Scientific Results" of the Nansen North Polar Expedition opens with a geological sketch of Cape Flora and its neighbourhood by the leader of the expedition. It was a wise determination, on the part of those responsible for the publication of the results, to issue the several articles in English. The policy, too, frequently followed, of writing important scientific papers in the language of the country where they are published, tends to place serious obstacles in the way of those who endeavour to follow the researches of Continental investigators. It is narrowness of view, rather than true patriotism, that compels authors to publish their results in languages which cannot be read by the great majority of scientific workers.

The geological investigation of Cape Flora, Franz Josef Land, was undertaken by Dr. Reginald Koettlitz, the geologist of the Jackson-Harmsworth Expedition, during the years 1894–97. Dr. Nansen's residence at "Elmwood," as the guest of Mr. Jackson, during a period of rather less than two months, afforded him an opportunity of visiting the most important localities in company with Dr. Koettlitz ; the information he collected bears testimony to the good use which was made of this short visit. Nansen has given us a clear account, accompanied by diagrammatic sketches and photographs, of the geology of Cape Flora. This portion of Franz Josef

Land has the character of a plateau with a basaltic cap, 150 metres thick, composed of sheets of lava arranged in regular and almost horizontal terraces, which present a striking resemblance to the familiar basalt sheets in the cliffs of the Western Isles of Scotland.

From the face of the basalt a talus-slope extends to near the shore-line, where it passes into almost horizontal raised beaches, which occur at approximately the same level on both sides of the Cape, and point to a uniform and recent elevation. The volcanic rocks rest on Jurassic sedimentary strata, consisting for the most part of soft shale or clay containing numerous nodules of hard stone. From a "nunatak" protruding through the glacier, about 600 or 700 feet above sea-level, several fossil plants were found in fragments of shale spread over the surface of the rock within two small areas. The important question as to whether the shale was actually *in situ* and represented the remnant of an interbasaltic bed, or whether it had been broken off from a lower stratum and carried up by the intrusion of igneous material, has not been definitely settled. Nansen is of opinion that the plant-bed was *in situ*, and may be looked upon, therefore, as throwing important light on the age of the basaltic sheets ; if this view is correct, the basalt must be assigned, on palæobotanical evidence, to an Upper Jurassic or Lower Cretaceous age. Very little is known as to the Jurassic deposits of Northbrook Island beyond Cape Flora ; the beds examined at Cape Gertrude have yielded no fossils beyond fragments of wood and lignite. Nansen inclines to the view expressed by Messrs. Newton and Teall,[1] that the beds at Cape Gertrude were deposited under varying conditions and during oscillations of level ; while the argillaceous sediments of Cape Flora, which are more uniform in composition, appear to have been laid down in a shallow sea during a period of comparative tranquillity.

The marine Jurassic fossils collected by Nansen from the rocks of Northbrook Island in the Franz Josef Archipelago are described by Dr. J. F. Pompeckj, whose work bears the stamp of thoroughness and accuracy. An account is given of previous literature relating to the Jurassic rocks of Franz Josef Land, special prominence being naturally given to the description by Mr. Newton of the fossils brought to England by the Jackson-Harmsworth Expedition. Some portions of the Cape Flora strata are fairly rich in fossils, but the fragmentary nature of the material renders accurate determination a matter of considerable difficulty, and in many cases the fragments are indeterminable. Dr. Pompeckj has performed his task with ability, and his conclusions have been arrived at as the result of careful sifting of the meagre evidence at his disposal. A glance at the comparative table of the Cape Flora fossils collected by the Jackson-Harmsworth Expedition and by Nansen shows that in several instances Pompeckj's determinations do not agree with those of Newton ; considering the fragmentary nature of many of the specimens, it would be strange indeed if there were no discrepancies in the lists of the two palæontologists.

The fauna, as described by Pompeckj, is represented

[1] *Quart. Journ. Geol. Soc.* vol. liii. (1897), p. 477 ; *ibid.* vol. liv. (1898), p. 646.

by the following genera :—*Pentacrinus, Serpula, Lingula, Discina, Pseudomonotis, Pecten, Lima, Leda, Macrodon, Amberleya, Macrocephalites, Cadoceros, Quenstedtoceras,* and *Belemnites.* The Ammonites appear to be abundant as compared with other groups, the genus *Cadoceros* being specially prominent as regards both species and the number of specimens.

Of the twenty-six species collected by Nansen, seventeen are new to the region, and five are considered to be new species. As the author points out, his results "differ in no slight degree from those which Newton arrived at from his examination of the Jackson-Harmsworth material." The sedimentary strata of Cape Flora are classed by Pompeckj as Lower Bajocian, Lower, Middle and Upper Callovian.

In the concluding palæo-geographical remarks, attention is drawn to the importance of the Cape Flora fossils as coming from the most northerly development of Jurassic rocks so far investigated. The occurrence of marine Bajocian species demonstrates "the existence of a Bajocian Sea in the north of the Eurasian-Jura continent." The extent of this northern sea cannot be determined, but the Jurassic sediments of Cape Flora afford evidence of deposition in shallow water near the shore-line of an Arctic continent. Neumayr's fascinating theory of climatic zones in the Jurassic period does not receive support from the palæontological results of Pompeckj and Newton ; the scanty evidence at present available points to the existence of a decided central European facies in the fauna of Cape Flora, a fact opposed to the conclusions of Neumayr.

The patches of sedimentary rock from which Nansen obtained several fragmentary remains of plants have already been referred to as either portions of strata preserved *in situ*, or conceivably derived from lower strata and carried to a higher level by igneous forces. It is unfortunate that the history of the vegetation which flourished on the site of Franz Josef Land during the Mesozoic period is not represented by more legible records, but we may congratulate Prof. Nathorst on having exercised caution and care in the interpretation of the imperfect documents at his disposal.

Among the genera recognised by Nathorst are the following :—*Cladophlebis* and *Sphenopteris* fragments represent the ferns, small specimens referred to *Podozamites* and *Pterophyllum* may be accepted as evidence of the existence of Cycadean plants : *Ginkgo, Czekanowskia, Phoenicopsis, Feildenia, Taxites, Abietites, Pityanthus* and *Pityostrobus* demonstrated the occurrence of Ginkgoales and Coniferæ.

The fairly numerous examples of small *Ginkgo* leaves are the most interesting fossils dealt with by Nathorst ; they enable us to extend the range of the Mesozoic species of this isolated genus, which is to-day represented by the maiden-hair tree of China and Japan. The leaves named by Nathorst *Ginkgo polaris* bear a close resemblance to *Ginkgo digitata,* a species which played a prominent part in the Jurassic vegetation of several regions ; the Franz Josef Land specimens are characterised by the small size of the leaves, and may possibly be regarded as a northern variety of the larger-leaved *Ginkgo digitata* of the Inferior Oolitic rocks of East Yorkshire. As regards the question of geological age, we agree with Nathorst's verdict that the plant-bearing beds must be assigned either to an Upper Jurassic or to a Lower Cretaceous horizon. Several of the plants suggest a comparison with Inferior Oolite species from the rich plant-beds of the Yorkshire coast, and it is not improbable that in the fragmentary fossils from Cape Flora we have the remains of a flora but slightly younger than that which has left abundant traces in the Lower Oolite strata of more southern latitudes. While admitting the danger of attempting to assign an exact geological date to the fragmentary and imperfect specimens, there can be no doubt that they must be referred to a period anterior to the Tertiary, and in all probability they are remnants of an Upper Jurassic flora.

While regretting that the fossils from Franz Josef Land are not more numerous and less fragmentary, we may offer a hearty welcome to the two able palæontological memoirs by Dr. Pompeckj and Prof. Nathorst ; these authors, in carrying out their difficult tasks with thoroughness and good judgment, have set a standard of efficiency which promises well for the succeeding volumes of the " Scientific Results " of the Nansen Expedition.   A. C. S.

As might have been expected, no birds new to science were obtained during the voyage of the *Fram* ; nevertheless, some interesting observations were made on the range and distribution of bird-life in the high north, while naturalists have, apparently for the first time, been made fully acquainted with the early plumage of the roseate gull. In the course of the expedition birds were observed in the highest latitudes in which they are definitely known to be able to exist. During the summer of 1895, when the vessel was between 84° and 85° 5′ north lat., in the neighbourhood of Franz Josef Land, ten species were from time to time observed, although none occurred in any numbers. The one found farthest north was the Fulmar petrel, which was seen in lat. 85° 5′ ; in the last edition of "Yarrell" the extreme range of this bird is given as 82° 30′.

During the summer of 1896, when the *Fram* was north of Spitzbergen, the first herald of returning bird-life was a snow-bunting, which made its appearance on April 25. From the observations made during the same season, it is now evident that to the north of Spitzbergen, between lat. 81° and 83°, the Arctic Ocean is the resort of large numbers of birds, belonging, however, to comparatively few species. Apparently these consist for the most part of immature individuals, in the first plumage, which spend the summer among the open channels in the ice. The little auk and the ivory gull were among those most numerously represented ; Sabine's gull having only been seen on a single occasion. Although swimming birds were by far the most numerous in these high latitudes, shore-haunting species were represented by the ringed plover and the grey phalarope, which were seen running about on the ice by the side of the open water.

The fasciculus is illustrated by an artistic plate of the roseate gull in its first plumage, which is mainly brown on the upper-parts, and therefore quite unlike that of the adult.   R. L.

THE CYANIDE PROCESS.

The Cyanide Process of Gold Extraction. By James Park. Pp. viii + 127. (London : Griffin and Co., Ltd., 1900.)

THE great success which has attended the introduction of potassium cyanide for the extraction of gold has created a widespread interest in this chemical process, and given rise to several books and papers on the subject from various authors. When we consider that at one large works 500 tons of gold are treated in twenty-four hours, we understand on what a colossal scale the cyanide method is worked. The process, like many others, has grown up from small beginnings, and it is largely owing to Messrs. MacArthur and Forrest that cyanide of potassium is now successfully applied to the treatment of gold ores in different parts of the world.

It is a most significant sign of the times that men who have been practically engaged in an enterprise are willing to communicate the results of their experience to the public at large, and from the manner in which the literature of the subject is growing, every detail requisite for economic working will soon be widely known and utilised. Therefore, one is not inclined to analyse the text too minutely, with the object of finding small flaws, provided the information is broadly reliable and accurate. It was inevitable that electricity should be brought into play in connection with such an important process, and we find Messrs. Siemens and Halske early in the field, with a method of depositing the gold on lead by means of electrolysis. There are two sides to this subject, as to most others, viz. the economic, or practical as it is termed, and the scientific. Now the former seems to be fairly well treated, but what is wanted is much greater attention to be paid to the latter, as it is possible that, with fuller and more intimate knowledge, potassium cyanide may be equally useful in the treatment of other metals besides gold, especially as it is now so largely used in the electro-deposition of gold, silver, copper, brass, &c. The work under notice has passed through three editions in New Zealand, and this is the first English edition. It is intended for the use of students, metallurgists and cyanide operators. Several new illustrations and tables are added, and the information relating to the treatment of slimes and the analysis of solutions has been greatly extended. It is gratifying to learn that wet crushing and cyanide treatment have been followed with as much success in New Zealand as in South Africa, although the ores are of a complex character.

The arrangement of the contents of the book is admirable. After a brief introduction and a general statement as to the limitations of the subject, the chemistry of the subject is wisely introduced, so that the student is at once brought face to face with the various reactions that occur, and led to see the reason for loss of cyanide, which is sometimes so excessive. Valuable information is given on pp. 10–13 on the action of potassium cyanide on metallic sulphides. A very useful chapter on laboratory experiments will be appreciated by teachers and students of metallurgy, as well as by the chemist and works manager ; indeed, a commodious and well-equipped laboratory is one of the most important and necessary parts of a cyanide plant. The control, testing, and analysis of solutions is treated in a fuller manner than is usual with books of this class, and of the three methods given we prefer the silver nitrate test. The tables for the assay of cyanide solutions are a useful addition to this chapter. The appliances for cyanide extraction are briefly described, and although accompanied by several good scale drawings, certain details are omitted which might have been profitably included.

The synopsis of the process for the actual extraction by potassium cyanide is well written, and the conditions for successful treatment, such as strength of cyanide solution, &c., are stated as clearly as one could wish. Chapter vii. deals with the applications of the processes at different works. Leaching and precipitation are succinctly dealt with in Chapters viii. and ix. These are followed by a short description of the Siemens-Halske electrical process, which not only deposits the gold, but gives rise to the production of a number of valuable commercial bye-products, such as lead, copper, litharge and paint. For all those who wish to obtain a sound knowledge of the cyanide process, as conducted at the present time, we heartily commend Park's handbook.

OUR BOOK SHELF.

The Cause and Prevention of Decay in Teeth. By J. Sim Wallace, M.D., B.Sc., L.D.S. Pp. 101. (London: J. and A. Churchill, 1900.)

THIS is a reproduction in book form of a series of articles published in the *Journal* of the British Dental Association.

The subject has been dealt with in the light of the now universally accepted chemico-parasitic theory of dental caries, but the author treats less of exciting or immediate causes than of those remote and predisposing. He attributes the great and increasing prevalence of dental caries among civilised nations to the elimination of the coarser and more fibrous parts of foodstuffs from the diet, and points out that this may act in two ways. Firstly, owing to the absence of mechanically detergent constituents of food, more of the fermentable, acid-producing and germ-sustaining parts of the latter remain in contact with the teeth for some time after meals. Secondly, that the tongue, being less actively employed during the act of chewing and swallowing, fails to attain its full size and exercise its normal important function in modelling the dental arches, so that irregularities arising from crowding and malposition of the teeth serve to intensify their predisposition to caries.

The subject is, on the whole, efficiently dealt with, and the book may be recommended to the medical practitioner or intelligent layman.

It is a pity, however, that the author lays such persistent stress upon what he considers the daring heterodoxy of his opinions, as these are at most modifications of those currently accepted. It is somewhat irritating, too, to see set forth for the instruction of the dentist, and with an air of great originality (as on p. 94), certain points in the operative treatment of caries which are among the very first impressed upon all students in schools of dental surgery.

Surely, too, the accusation of ignorance of the causes of the diseases he attempts to combat, and empiricism in practice, are undeserved by the educated dental surgeon of to-day.　　　　　HAROLD AUSTEN.

LETTERS TO THE EDITOR.

[*The Editor does not hold himself responsible for opinions expressed by his correspondents. Neither can he undertake to return, or to correspond with the writers of, rejected manuscripts intended for this or any other part of* NATURE. *No notice is taken of anonymous communications.*]

Atmospheric Electricity.

IN a letter on this subject in NATURE of March 29, Mr. Aitken criticises the theory which attributes the prevalence of positive electrification in the atmosphere to the superiority in efficiency as nuclei for the condensation of water vapour, of the negative ions over the positive.

That any difference in the degree of supersaturation necessary to make water condense on positively and on negatively charged ions would result under suitable conditions in the production of an electric field was pointed out by Prof. J. J. Thomson (*Phil. Mag.* vol. xlvi. p. 533), and it was suggested by him that this might be a source of atmospheric electricity. Experiments made by the present writer proved that there is such a difference, and that water vapour condenses much more readily on negative than on positive ions; while Elster and Geitel (and independently, Lenard) have recently brought forward evidence based on their own experiments and those of Liuss, tending to show the existence of free ions in the atmosphere.

There remains the question whether the necessary degree of supersaturation can ever occur in the atmosphere. Mr. Aitken contends that there is no such thing as dust-free air in the atmosphere, and that therefore any considerable degree of supersaturation is impossible.

Air practically dust-free does, however, seem to have been met with on Ben Nevis, accompanied by something very like supersaturation (Rankin, *Journ. Scot. Met. Soc.* vol. ix. p. 131). In Mr. Aitken's own papers, too, records of small numbers of dust particles (sometimes considerably less than 100 per c.c.) are not rare ; and the lowest values are met with just under the conditions where their occurrence is of most significance. For "most of the low numbers in the tables were observed during rainy weather, and the very low ones in misty rain, when the clouds were at or near the surface of the earth" (Aitken, *Edin. Trans.* xxxvii. p. 664). Again, the purest air met with by Mr. Aitken was that blowing from off the Atlantic Ocean, the mean number of dust particles in a series of 258 observations extending over nearly five years amounting to 338 per c.c. ; on one occasion the number was as low as 16 per c.c. (*Edin. Trans.* xxxvii. p. 666). Air coming from such a region can hardly be considered as abnormal. Moreover, such observations are necessarily made in air within a few feet of the ground ; at a greater height it is likely to be less contaminated.

Consider a mass of air occupying 1 c.c. and saturated with water-vapour at 10° C., and let it expand till, say, 3×10^{-6} gram. (less than one-third of the total water) has condensed to form 100 drops. Let us suppose the drops to be equal in size and let us calculate the volume and thence the radius of each drop, and from this obtain the rate at which they will fall relatively to the air (assuming the velocity $= \frac{2}{9} g \frac{r^2}{\mu}$, the viscosity μ being taken as 1.8×10^{-4}). We obtain for the radius of each drop the value 1.9×10^{-3} centim., and for the rate of fall through the air, $v = 4.4$ cms. per second.

In a rising current of moisture-laden air containing 100 dust particles per c.c. there is thus no difficulty in seeing how the drops as they ascend may grow large enough to lag behind the air at the rate of 4.4 cms. per second (= 160 metres per hour) ; while the greater part of the moisture in the surrounding air is still retained as vapour. If then the upper surface of the cloud is carried to such a height that the drops reach the size $r = 1.9 \times 10^{-3}$ cm., it will there be lagging behind the rising air at the rate named, and a dust-free layer must exist immediately above it, increasing in vertical thickness at the rate of something like 180 metres per hour. Even if 1000 drops were formed in each c.c. of the cloud, the rate of growth of the dust-free layer would, as a similar calculation shows, when the same quantity of water had separated, amount to 34 metres per hour.

A difficulty raised by Mr. Aitken in connection with the removal of dust particles by condensation of water upon them is this : "When a cloud forms in ordinary impure air, only a small proportion of the dust particles become active centres of condensation, whilst many receive no charge of vapour." Instead of being an addition to our difficulties, does not this rather suggest a method by which, even if the air entering the base of a cloud be very impure, it may become freed from its dust? For it follows that even in such air a comparatively small number of drops will be formed in each c.c. when the saturation level is reached. What becomes of the nuclei which do not there form active centres of condensation? If the presence of a few slightly more efficient nuclei has prevented them from coming into play, the same number of actual drops will be at least equally effective in this respect. Will the dust particles then remain free until they are carried up beyond the reach of the drops, and there become active centres of condensation as Mr. Aitken suggests? It seems to me that, after a considerable vertical thickness of cloud has accumulated, this is highly improbable; such a cloud is likely to act as a very efficient air filter. For if even very impure air be kept in a small vessel with wetted walls the dust particles are removed in a comparatively short time—the shorter the smaller the vessel—by coming in contact with the walls. Dust particles in air travelling through a cloud must be very favourably situated for removal by contact with the drops. They are thus not likely to survive as free nuclei long enough to be able to come into play at the upper surface of the cloud, unless the time taken to traverse the cloud has been comparatively short. A cloud, due to an ascending air current containing near its lower surface as many dust particles (7700 per c.c.) as that encountered by Mr. Aitken on one occasion on the Rigi Kulm, even if it receive a continuous supply of equally or more impure air from below, may thus have no dust particles left in its upper portions beyond what are contained in the drops; while the number of drops per c.c. may amount to only a small fraction of the number of dust particles originally present, the size of each being correspondingly greater.

Mr. Aitken refers to the possible re-evaporation of drops due to the tendency of the larger ones to grow at the expense of the smaller. But all drops which have survived the great tendency to evaporate which accompanies the initial stages of their growth will surely continue to grow so long as the rate of expansion remains the same, or even if it be much reduced. The effect of the size of the drops on the vapour pressure necessary to cause water to condense on them is in fact relatively unimportant except in the case of very small drops; if we apply Lord Kelvin's formula to the case of drops even as small as 10^{-4} cm. in radius we find that the vapour pressure exceeds by only about one part in a thousand that over a flat surface of water; the evaporation from the drop of one part in 30,000 of its mass would cool it sufficiently to counterbalance this difference.

With respect to the power of sunshine to manufacture nuclei in air containing various gaseous impurities specified by Mr. Aitken, it may be observed that there is no evidence of such an effect of sunlight in normal atmospheric air, and that all the substances mentioned by Mr. Aitken (ammonia, nitric acid, &c.) being very soluble in water would be dissolved out of the air in passing through a cloud of water drops. It is true that sunshine does appear to produce in pure air nuclei (which however require a fourfold supersaturation to make water condense on them), and that strong ultra-violet light produces large nuclei like dust particles (*Phil. Trans.* 192, p. 403); but these effects have not, so far as I can see, any immediate bearing on the subject of the possibility of supersaturation in the atmosphere.

I do not know of any evidence to show whether the small drops in clouds tend to coalesce to form larger ones or not. Such coalescence would tend to hasten the process of separation of dust-free air from the cloud, by increasing the downward velocity of the drops relatively to the air; but it is unnecessary to assume its occurrence.

We have now seen reason for believing that the drops in the upper portion of a cloud produced in ascending air are likely, before the air around them has lost any very large proportion of its vapour, to have grown large enough to lag behind the ascending air at quite an appreciable rate; and that the air between them is likely to be dust-free. Under these conditions a dust-free layer will be formed above the cloud, and will continually increase in vertical thickness. This layer will be saturated with moisture at its lower edge, above this it will be supersaturated; the amount of supersaturation being greatest near its upper limit, and depending on the vertical distance through which the air has risen since escaping from the cloud. Now to produce in air initially saturated the supersaturation (approximately fourfold) necessary to cause water to condense on negative ions, it is

sufficient to let the volume of the air increase adiabatically to 1·25 times its initial value (*Phil. Trans.* A, vol. cxciii. p. 289); an expansion which will result from an ascent of the air through a vertical distance of 2500 metres, if we suppose the air on escaping from the cloud to be at a temperature of 10° C. (at lower temperatures a smaller elevation would suffice). Thus, when the air in the uppermost layers of the supersaturated stratum has reached a height of about 2500 metres above the level at which it escaped from the cloud, a sudden change will result; condensation will there take place on the negative ions. The thickness of the supersaturated stratum (*i.e.* the vertical distance which the upper surface of the cloud has lagged behind the air), when the condensation on the negative ions begins, may vary greatly; it may be very small if the drops are small and the ascent of the air rapid; it may amount to nearly the whole 2500 metres in the case where the drops grow large enough to acquire a velocity relative to the air as great as the upward velocity of the air, so that the upper surface of the cloud has ceased to ascend. Above any cloud in an ascending air current, however numerous and small the drops, we should expect to find a supersaturated layer (possibly of very small vertical thickness), provided its upper surface has risen high enough for all dust particles to have either come into play as condensation nuclei, or to have been removed by coming in contact with drops already formed; provided also that the heating effect of sunshine on the drops at the upper surface of the cloud is not sufficient to counterbalance the cooling effect of the expansion and cause them to evaporate. And if the ascending current continues till a level about 2500 metres higher is reached, we get condensation taking place in the dust-free layer. It is difficult to avoid connecting this process with the sudden appearance of "false cirrus" at the top of a cumulo-nimbus cloud at the commencement of a shower.

We must now consider what will happen to the drops condensing from the supersaturated layer. Mr. Aitken takes the view that if condensation ever did take place on the ions, the drops formed would fall at once as rain, and that a cloud would never result. He remarks that the supersaturated air will be, as it were, in an "explosive" condition, which will cause the extremely rapid growth of any drop that may begin to form, thus preventing condensation on neighbouring ions. There is, however, no obvious reason for supposing the rate of increase of size of a drop in supersaturated air to be of a different order from that of the diminution in size of a similar drop in an unsaturated atmosphere. In neither case is there anything of the nature of an explosion. In the one case evaporation causes the lowering of the temperature of the drop below that of the surrounding air (to the wet-bulb temperature), the evaporation being thereby retarded; in the other case, the condensation on the drop at once raises its temperature above that of the surrounding supersaturated air, the rate of growth being mainly determined by the rate at which the drop can give out to the surrounding air the heat developed in it by the condensation. I do not think we have the data for determining whether the drops will fall at once as rain or remain in suspension till they have travelled into regions where the ascending current is insufficient to support them. In either case, if the drops fall through a supersaturated layer of some thickness, they are likely to reach the ground as negatively charged rain. I see, however, no reason to conclude that negatively charged clouds may not also be produced by condensation on the negative ions.

The foregoing considerations contain a theory of the origin of rain such as I had in view when the paper, criticised by Mr. Aitken, on the difference between the positive and negative ions as condensation nuclei was written (*Phil. Trans.* A, vol. cxciii. p. 289). That rain may sometimes at least have its origin in supersaturated portions of the atmosphere has indeed been held by v. Bezold, Cleveland Abbe, and other meteorologists.

I do not propose to consider what is likely to happen after the rain has begun to fall. It may be pointed out, however, that we are likely then to have a reduction in the supply of dust particles, especially if the rain extends over a considerable area; for the inflowing air is likely to have a considerable proportion of its dust particles carried down by the rain before it has penetrated any great distance into the rain-washed area. In Mr. Aitken's papers may be found references to the apparent dust-removing power of rain.

Mr. Aitken considers that the positive ions would not remain in the atmosphere, because a slightly greater supersaturation than was necessary to cause condensation in the negative ions would bring them down also. It is conceivable that they may sometimes be removed in this way; but if we consider that a greatly increased supersaturation (six-fold instead of four-fold) is necessary, and that the production of ions is continually going on, so that negative ions as well as positive are always present, we can hardly consider it a likely occurrence. What then is the subsequent history of the positive ions after being carried up out of reach of the drops formed on the negative ions? They will, under the action of the electric field produced by this separation, tend to travel downwards relatively to the air with a velocity of the order of one centimetre per second for a field of too volts per metre, as the measurements of Rutherford and others have shown. After being carried beyond the region of ascending air-currents, they will travel downwards towards the earth's surface; but long before reaching it they will become attached to cloud particles or to the dust particles of the lower layers of the atmosphere, where the positive charge will accumulate.

It is not claimed that the process described above is the only source of rain or the only source of atmospheric electricity. It should be pointed out, for example, that another way in which rain may possibly acquire a negative charge is by falling through ionised air. For according to Zeleny (*Phil. Mag.* vol. xlvi. p. 135) a body suspended in a current of ionised air becomes negatively charged in virtue of the slightly greater velocity of the negative than of the positive ion under a given force. Elster and Geitel make use of this difference between the positive and negative ions to account for the normal positive electrification of the atmosphere, by the passage of air through the vegetation on the earth's surface. Whether, however, the charged particles, the presence of which near the surface of the earth their experiments seem to prove, are really free ions whose velocity under a given force is that of the ions produced by Röntgen and other rays and not comparatively slow-moving masses (the nuclei called dust particles by Mr. Aitken) to which ions have attached themselves remains as yet undecided. In air charged with dust even to the extent to which clear air near the surface of the ground is shown by Mr. Aitken's observations to be, it is likely, since the rate of ionisation in the atmosphere is certainly slow, that an ion would be in dust-free air, where it is determined merely by the rate of recombination of the ions.

In conclusion, it must be confessed that if the rate at which the electric field of the earth is being destroyed by leakage through the air is anything like so great as is given by Elster and Geitel's interpretation of their experiments (*i.e.* of the order of 1 per cent. per minute), no theory which attributes the normal fine weather electricity to the effect of precipitation at a distance is sufficient to explain the facts.　　C. T. R. WILSON.

Cambridge Laboratory, Cambridge, May 16.

Specimens of "Dromæus ater."

IN reference to Prof. Giglioli's note (*suprà*, page 102), I may perhaps be allowed to remark that Bullock's Museum appears to have contained a specimen of the extinct *Dromaeus ater*. The twelfth edition of the "Companion" to that Museum, published in 1812, has the following entries (page 80) :—

　"Great Emeu, or New Holland Cassowary . . .

　"Lesser Emeu, not half the size of the above, and a distinct species."

At the dispersal of his collection the sale Catalogue includes both specimens as lots 97 and 98 on the eleventh day of the sale (May 18, 1819), the latter as

　"Lesser Emew, a distinct species from the last,"

and my annotated copy of the Catalogue shows that both were bought by the Linnean Society—for 10*l*. 10*s*. and 7*l*. 10*s*. respectively. I have traced the latter specimen, but in vain. It may still exist unrecognised.　ALFRED NEWTON.

Magdalene College, Cambridge, June 4.

Effect of Iron upon the Growth of Grass.

SOME years ago NATURE published a short letter of mine from India, noticing the way in which laying out iron (famine) tools on the ground brought on grass upon very dry surfaces. Any one who looks now under the rows of iron chairs, and round the railings, of the band-stand on the east side of the Green Park, will see the same stimulating effect produced.　A. T. F.

London, June 4.

SOURCES AND PROPERTIES OF BECQUEREL RAYS.

IN the following article a general account is given of a few of the more striking phenomena connected with Becquerel rays, including some of the recent developments of the subject at the hands of Becquerel, M. and Mme. Curie and others.

Among a large number of papers which have lately been published, dealing with properties of these rays, two are worthy of especial notice, as giving a comprehensive view of the phenomena. For those who propose to study the subject more fully, no better guide can be found than Prof. Elster's report in Eder's *Jahrbuch für Photographie und Reproductionstechnik* for 1900. The footnote references to original papers form a complete bibliography of the literature of the subject existing at the time when the article appeared, and it is surprising that Prof. Elster should have succeeded in summarising so large an amount of matter in eleven very small pages. Dr. B. Walter's article in the *Fortschritte auf dem Gebiete der Röntgenstrahlen* is somewhat less condensed and more popular; the chief phenomena, especially the photographic and fluorescent properties, are dealt with at greater length, and the article is illustrated by a plate of radiographs showing the difference between the actions of Becquerel and Röntgen rays. Already Walter's paper and, to a less degree, Elster's report, have become out of date on the subject of magnetic deviation, and for this and other later developments no better guide could be found than the well-condensed summaries contained in the current monthly parts of *Science Abstracts*.

The discovery of these rays in 1896 was a natural sequence of the discovery of the Röntgen rays, and was led up to, on the one hand, by the attempts of M. Henry to intensify the action of Röntgen rays by the use of phosphorescent substances; and, on the other hand, by the theory, since abandoned, that the Röntgen rays were themselves the result of phosphorescence of the vacuum tube. Becquerel and other physicists made numerous experiments to test whether phosphorescent substances emitted rays capable of acting on a photographic plate that was enveloped in opaque paper, and it was found that rays which produce actinic action were emitted by the phosphorescent salts of uranium, not only when these salts had been exposed to the action of sunlight or of Röntgen rays, but even after they had been kept in the dark for months, the "radio-activity" showing no perceptible falling off.

The next step was the discovery, by Mme. Curie, that Bohemian pitch-blende—a black, shiny ore of uranium—possessed a higher degree of radio-activity than uranium itself, and this result naturally suggested the view that the ore contained, besides uranium, some other substance to whose presence the increased action was due. By separating the pitch-blende into its constituents, M. and Mme. Curie were led to discover the existence of two sources of radio-activity, one associated with the compounds of bismuth, and the other with those of barium occurring in the ore. Seeing that barium and bismuth obtained from other sources do not emit Becquerel rays, these radiations were attributed to the existence of two new substances, that associated with bismuth being named polonium, a name derived from the Polish nationality of Mme. Curie, while the other substance associated with barium chloride was called radium. The separation of these two substances has led to the production of rays of sufficient intensity to excite fluorescent screens, discharge electrified conductors, and, indeed, to reproduce, with differences, most of the properties of Röntgen rays. A third radio-active substance, produced from the residues of pitch-blende, is recorded by Debierne, who names it actinium. It is precipitated by the principal agents for titanium, and it emits rays which reproduce the same phenomena as the rays emitted by radium and polonium, and are 100,000 times the intensity of ordinary uranium rays. Certain thorium compounds are also radio-active, a property first established in these by G. C. Schmidt and Mme. Curie, and subsequently investigated by R. B. Owens and Rutherford.

Since this article was in the printer's hands a paper by Sir W. Crookes on the radio-activity of uranium, read before the Royal Society on May 10, has been received. The author records an entire absence of radio-active effects in all the barium minerals in his cabinet from which uranium was absent, while pitch-blende and other minerals containing uranium and thorium excited a photographic plate. Arrangements were then made for working up half a ton of pitch-blende, and the radio-activity of the uranium salts was definitely traced to the presence of a foreign body, which Sir W. Crookes has christened for the time UrX (*i.e.* the unknown quantity in uranium), following a fashion initiated by Röntgen, and which has previously led to the introduction into our vocabulary of such terms as "Xd air" (Italian "aria Xata" or *ixata*). We would suggest the name "Crookesium" as a substitute. Whether uranium-X is or is not identical with radium seems not fully decided, but it appears to be distinct from polonium. It is now proposed to try to separate the radio-active component of thorium.

Le Bon, who claims to have anticipated the Becquerel rays in his "lumière noire," has expressed the opinion that the properties attributed to radium and polonium do not prove the existence of new elements, and may be accounted for by supposing the radio-active substances to be mere allotropic modifications of bismuth and barium. On this view there is no more fundamental difference between the properties of radio-active and ordinary barium than between phosphorescent and ordinary sulphuret of lime. Giesel, of Brunswick, also has adopted the terms "radio-active barium" and "radio-active bismuth" in preference to "radium" and "polonium." In support of the opposite view, Demarçay has proved that radium possesses a characteristic spectrum, and M. and Mme. Curie find that the atomic weight of radio-active barium chloride is greater than that of ordinary chloride, amounting in one specimen to as much as 146 as against 137.

The pitch-blende used in the preparation of these substances is obtained from Joachimsthal, in Bohemia. Under the direction of Giesel, working in co-operation with Profs. Elster and Geitel, the firm of E. de Haën, of List, near Hanover, have undertaken the preparation in small quantities of radio-active barium emitting rays that are unequalled in intensity, and have also placed on the market cheaper by-products which also emit rays of sufficient intensity to visibly excite a fluorescent screen. The solid radio-active compounds of barium increase in activity from the time of solidification, but do not reach their maximum for more than a month. The barium preparations are all luminescent, the chloride and bromide especially so when dry. According to Giesel, the bismuth or polonium preparations lose their radio-activity in a few weeks, and this property is also cited by Elster.

The radio-activity of barium bromide is found by Elster not to be destroyed by continuous heating for twenty-four hours *in vacuo.* After cooling, the strength is much reduced, but is restored after the lapse of a few days to nearly the original intensity.

Becquerel rays resemble Röntgen rays in their power of "ionising" air, a property they possess to such a degree as to discharge all conductors within a considerable distance of the radio-active substance. Their action on electric sparks has been studied by Elster and Geitel. A spark gap 1 cm. wide, consisting of a positive knob and a negative disc, was exposed to the radiations from a barium

preparation. The sparks or brushes were converted into a violet glow-discharge, but the former discharge was re-established on interposing a plate of lead. With discs made of semi-conducting card the radium affected the discharge at the distance of over 1 metre. According to Elster, heating a small trace of a radio-active substance in air in a Bunsen flame increases the electric dispersion of the air of the room.

Becquerel finds many bodies acquire the temporary power of discharging conductors under the influence of the rays, thus affording proof that these rays involve a continuous emission of energy. The bodies do not, however, act on a photographic plate, and their activity is lost on heating. This property is not assumed by the double sulphate of uranium and potassium.

There appears at present no prospect of utilising Becquerel rays as a substitute for Röntgen rays in surgery. The difference of behaviour of the two kinds of rays is well shown by two radiographs of the human hand accompanying Dr. Walter's paper. In the one taken with Röntgen rays the outlines of the bones are remarkably clear and sharp; in the other, taken with the rays emitted by Giesel's most powerfully radio-active preparations, a dark, ill-defined shadow of the outline of the hand is seen, but not a trace of the bones is visible. This latter radiograph, which was taken with the relatively short exposure of an hour, shows clearly the shadows of a needle and of a coin that were placed under the middle of the hand, proving that a certain proportion of the rays had actually passed through the hand, but without differentiating the bones from the rest. Experiments undertaken by Walter to account for the hazy outline of the Becquerel rays point to the conclusion that the Becquerel rays, when passing through substances of small atomic weight, experience a far greater diffuse scattering than Röntgen rays. Further, the secondary radiations emitted by both light and heavy substances under the influence of the Becquerel rays differ far less from the incident rays in intensity and penetrability than in the case of the secondary rays investigated by Sagnac in connection with Röntgen rays. A further difference lies in the far greater absorption of Becquerel rays by specifically light substances, such as those forming the flesh of the human hand. With the use of a platino-cyanide of barium screen, Walter observed the same absence of all traces of bones as with photographic methods, although the shadow of the hand was clearly seen on the screen.

The composite nature of Becquerel rays is suggested by experiments on phosphorescence and selective absorption, as well as on magnetic deviation. Mme. Curie has found that Becquerel rays are more easily absorbed when they have already penetrated an absorbing layer than when they have not. One aluminium disc absorbed a certain proportion of the rays; a second aluminium disc absorbed an even greater proportion of the remainder. According to the note on Mme. Curie's paper in *Science Abstracts*, "this is due to the fact that the less penetrative rays are absorbed in the first absorptive layers," but such a view would more naturally lead one to expect that the proportion of absorbed rays would be less at the second screen than the first, instead of greater; the phenomena can, however, be accounted for by the hypothesis that the first screen transforms the rays into secondary rays of lower penetrating power. The existence of such secondary rays has been supported by Villard, Meyer and Schweideler, Dorn and others. Becquerel has, however, shown that in the case of polonium rays from the Curies' preparations, no secondary rays are emitted by aluminium. The phenomenon of selective absorption has been studied by Becquerel, who exposed various substances to the action of radio-active barium chloride, including hexagonal blende, platinocyanide of barium, diamond, and double sulphate of uranium and potassium. The phosphorescence varied in different cases. When different screens were interposed—namely, aluminium, mica, black paper, glass, ebonite and copper—the absorptions of the radiations which excite phosphorescence in different substances by the same screen were found to be unequal. R. B. Owens has shown that thorium radiations resemble those associated with the derivatives of uranium ore, but possess greater variety. There are indications that they are not confined to so few distinct types, if, indeed, the number of types is limited. Becquerel shows that the absorption of "radium" rays by screens is variable according to the distance of the screens from the source, and that the intensity of the radiation decreases with the distance more rapidly than it would do according to the law of the inverse square ; both of these are results of absorption by the air. The view advanced by Le Bon two years ago, that Becquerel rays could not be polarised, has been confirmed by Rutherford.

The magnetic deviation of Becquerel rays has absorbed a large amount of attention during the last few months, and conclusions from recent experiments have in several instances been in contradiction with the inferences from earlier investigations. Thus a survey of the literature of the subject shows that amongst others the following views have been advanced : (1) that Becquerel rays are not deviated ; (2) that they are deviated in air but not *in vacuo ;* (3) that the deflection gives rise to phenomena which are more marked with polonium than with radium ; (4) that both radium and polonium rays are deviated *in vacuo ;* (5) that radium rays show marked deviation, but polonium rays show no deviation whatever. The first negative result was obtained by Elster and Geitel ; Giesel proved the magnetic deflection of the rays in air, and attributed the previously observed absence of deflections to the experiments having been performed *in vacuo.* Elster, by repeating the experiments with a different arrangement of apparatus, using the same radio-active bismuth and barium as in Giesel's experiments, has discovered the cause of his previous failure, and has established the magnetic deflection of the rays *in vacuo.* Giesel used a strongly radio-active bismuth preparation, and got more marked effects than with his barium compound ; Elster, using a similar bismuth preparation and a relatively feeble one of barium, was led to infer that the barium radiations were the most deflected. In these experiments the rays are received on a photographic plate or fluorescent screen ; P. Curie, on the other hand, has described an apparatus for comparing the magnetic deviation by means of the electro-dispersion produced by the rays. When not deviated the rays pass out normally between two lead blocks, and traverse the space between the plates of a condenser, causing a current to flow ; when deflected the rays are absorbed by the lead blocks, and the current ceases.

Both Curie and Becquerel find that the magnetic deflection varies with different substances. According to Becquerel's paper of December 26, polonium showed no deflection, while radium showed a strong deflection. The absence of deflection in polonium rays has been observed by Mme. Curie, who states that they travel in a straight line. ·In comparing these results with the different conclusions obtained by Elster, reference must be made to Dorn's hypothesis, according to which it is suggested that the primary rays are not deflected, but are transformed into deviable secondary rays. But in a recent paper Becquerel finds that the Curies' polonium rays are neither deflected by a magnetic field of 10,000 C.G.S. units, nor are they transformed into deviable secondary rays. He has also made experiments to test whether the curvature of radium rays is affected by interposing a screen, as would occur if the transmitted rays were secondary rays moving with lower velocity. No such effect has been as yet observed. The most probable inference at present is that there are two kinds of rays, one deviable and the

other not. The Curies find both forms coexist in radium rays ; and from Giesel's experiments the deviable rays certainly exist in some preparations of polonium, but were doubtless not present to an appreciable extent in the samples experimented on by the Curies and Becquerel. According to Curie, the rays from radioactive barium carbonate are deflected to a very different extent. Those rays which have the greatest penetrative power are the most easily deflected, and those rays which are not deflected only penetrate air to a distance of 6 or 7 mm. Becquerel finds that magnetically deviable rays are absorbed by different screens up to a certain inferior limit of distance, while they penetrate a screen that is placed sufficiently near the source.

When the magnetic field is uniform and the direction of the rays is perpendicular to the lines of force, they describe circles and return to the starting point ; when the rays start in a direction oblique to the lines of force, the paths are helices. These results have been recently verified by Becquerel, and from them it is possible to form a general prediction of the corresponding effects produced in a non-uniform field, such as that produced by a horseshoe magnet, which effects we now proceed to describe.

In Giesel's experiments, the sensitive plate was laid on the poles of the magnet, film downwards, the polonium being placed below and in contact with the film. Between the black patch produced above the substance and the dark zone produced by the deflected rays, a number of dark traces were observed, resembling wavy hair or like the ramifications in Lichtenberg's figures. Becquerel has shown that when the radio-active barium is placed on one pole of an electromagnet and a fluorescent screen on the other, the effect of exciting the magnet is to concentrate and contract the luminous area, a result unaltered by reversing the poles. When the rays pass across the lines of force, they, after proceeding upwards, are bent round and impinge on the plate along a curve, which extends from one pole to the other, bending out of the way of the radiant substance in the centre. ·When a piece of radium preparation is placed on a plate in a uniform field near a plane normal to the lines of force, the result is an intense impression limited by a spiral whose sense is that of the current which produces the field. This spiral is the trace, deformed by the field, of the line of intersection of the vertical plate and the plate on which the radium rests.

In the *Journal de Physique* for April, Becquerel shows that different radio-active compounds of barium emit rays that are equally deviated, and he establishes the fact that the deviation conforms to laws similar to those which apply to kathodic rays. The phenomenon of dispersion is established, and by interposing strips of paper, aluminium and platinum against the gelatine plate, on which the deflected rays are received, a kind of absorption spectrum is obtained, showing that the most deviable rays are the most readily absorbed under the conditions of the experiment. By calculating an inferior limit to $H\rho$ (the product of the magnetic force and the radius of curvature of the path) for the rays transmitted by various substances, the absorption by different substances is compared, and the results are of the same order of magnitude as for the kathodic rays. These and other facts suggest that part of the radiation is of similar nature to the kathodic rays, where small negatively-charged masses are transported with great velocity, and the Curies' experiments prove the existence of such charges, which, however, are exceedingly feeble. According to this view, the magnetic deviation is given by the formula $vm/e = H\rho$, and in an electrostatic field of intensity F the rays ought to undergo a deviation, $\theta = Fl \div (v^2 m/e)$, l being the length of the path. It appeared, at first, that the electrostatic force required to make any such deviation visible would exceed the limit for which disruptive discharge would take place in air, and could only be obtained *in vacuo*. In a footnote, however, Becquerel tells us that he has since observed the electric deviation in air with a field of about 10^{12} C.G.S. units, and has found for certain rays which pass through black paper the values $m/e = 10^7$ and $v = 1\cdot6 \times 10^{10}$.

The chemical effects of Becquerel rays have been examined by M. and Mme. Curie and Becquerel; they may be briefly summarised here. The rays from active salts of barium transform oxygen into ozone, a process involving a continuous expenditure of energy. Potassium iodide is coloured blue. Glass in contact with the salts is coloured violet, ultimately becoming nearly black, and the colour penetrates the glass; this phenomenon is analogous to the coloration of fluorspar by kathodic rays. Platinocyanide of barium screens gradually turn yellow, then brown, and finally lose their fluorescence, which, however, is restored by exposure to sunlight. Fluorine continues to phosphoresce for twenty-four hours after being excited, and calcined fluorspar which has lost its phosphorescence regains its luminosity in the presence of radium. Chemical activity is confined to those radio-active preparations which are luminous, but is not always proportional to the luminosity.

According to the Curies' experiments, powerfully radio-active compounds of radium and polonium, when they act on inactive substances, are able to communicate radio-activity to them. This induced radio-activity increases with the time of exposure up to a certain limit. If the inducing substance is 5000 to 50,000 times the activity of uranium, the induced activity may amount to fifty times that of uranium. It is reduced to one-tenth of its amount in an hour after removal, but it may persist for many days, finally disappearing. The emanation of radio-active particles from thorium compounds, investigated by Rutherford, is remarkable. This emanation ionises the gas in its neighbourhood, and it will pass through thin layers of metal, through thicknesses of paper, or through a plug of cotton wool. It is also unaffected by bubbling through hot or cold water, weak or strong sulphuric acid. The emanation retains its radio-active power for some minutes, gradually losing it. The positive ion produced in the gas by the emanation was found to possess the power of inducing radio-activity in all substances on which it fell, this power of giving radiation lasting several days. Whether the emanation be a vapour of thorium is doubtful.

The question as to the amount of energy emitted by the Becquerel rays has already been referred to in NATURE, and need not therefore occupy our space further now. The problem of discovering the seat of this energy would seem of late to have taken another form. At first it was supposed that a difficulty would exist in reconciling the continuous emission of these rays with the principle of conservation of energy; now, however, that the amount of the emitted energy has been estimated, the difficulty is seen to lie in the experimental observation of changes of such inappreciable magnitude as would suffice to generate this energy.

Before 1896 physicists were just beginning to grasp Maxwell's theories, and to realise more clearly the simplification introduced into notions electric and optical by the conception of the ether. The discovery of rays capable of discharging electrified bodies in air has not only shown the fallacy of our preconceived dogmatic notions as to the division of substances into conductors and dielectrics, but has taught us that the properties of the ether are not so simple as we had anticipated. We can only wonder whether Maxwell would have been able to develop his electromagnetic and electro-optic theories had the complications arising from Becquerel and other rays been before him, and the want now makes itself felt of a second Maxwell, who shall co-ordinate the newly-accumulated mass of experimental facts into the form of a connected mathematical theory.   G. H. BRYAN.

MODERN MICROSCOPES.[1]

IN spite of the attention which has of late years been paid to the improvements of every detail of microscope construction, it is remarkable how Powell's No. 1 stand has now existed, practically unchanged, for some fifty years. It may therefore be considered a permanent type, and it is one to which the best modern instruments conform more and more. Its most obvious peculiarity, however, a tripod base, has not yet become general. The heavy horseshoe foot is still in all but universal favour on the Continent, although Powell's base is occasionally imitated. Thus the Leitz firm in 1893, and the Hartnack firm in 1898, brought out large model microscopes on a tripod base; Greenough's low-power stereoscopic binocular microscope (1898) is similarly equipped. This last instrument, which is the most recent binocular novelty, is highly esteemed. It is made by Zeiss, is fitted with porro prisms, and, among other advantages, affords views of the *under* as well as of the *upper* side of an object.

English makers have lately paid much attention to the perfecting of cheaper stands with some excellent results. In their new model and educational microscope, Messrs. Ross have reintroduced the principle of a reversing and locking foot, which was first invented by Cuff (*circa* 1765). By this means the instrument acquires great stability when used horizontally. The same firm, in their bacteriological microscope, use a tripod stand, of which the hind toe is made to fold forward between the two fixed front toes when not in use, thereby economising space in packing. The stage of this, as well as of Baker's microscopes, is fitted with the Nelson horseshoe perforation. The advantage of this device is that in high-power work, when the objective necessarily works very close to the cover glass, the slide can be tilted with the finger, and the focus gradually attained with far less risk to the object than if the slide rested immovably on the stage.

Messrs. R. and J. Beck's student's microscope and Messrs. W. Watson's " Fram " microscope are other examples of really good, small, cheap microscopes. Economy is obtained, not by sacrificing quality of work, but by simplifying the design. Every step in the direction of reducing the cost of a good instrument is too obviously desirable to require demonstration. Some designs strive after cheapness by using a fine adjustment, and trusting to a push-tube motion for the coarse. But if a microscope is to have only one adjustment, most microscopists will prefer a good coarse to an indifferent fine adjustment. This is the principle of Messrs. Watson's school microscope, which has a coarse adjustment only (diagonal rack and pinion), so good that a ⅛-inch objective can be accurately focussed with ease. The cost, with eye-piece and objectives, is only three pounds.

The practical difficulty is, of course, that the great amount of wear upon the coarse adjustment affects in time the evenness of the racking, and produces loose action. But an important piece of progress towards obviating this trouble has been made by Mr. E. M. Nelson, who has applied the principle of stepped rackwork (Fig. 1). The two similar racks are placed so that their teeth are slightly out of step, the amount of divergence being regulated by the upper right-hand screw. The two screws in the centre of the pinion regulate the pressure by which the pinion is forced into the rack. The advantage of the arrangement is not only compensation for wear and tear, but rapidity and smoothness of action, for the tube obeys the slightest movement of the milled heads. If experience confirms the favourable opinion with which this novelty has been received, the necessity of a fine adjustment in cheaper stands will disappear.

1 Fuller accounts of all the instruments referred to will be found in the *Journal* of the Royal Microscopical Society for 1897, 1898, 1899 and 1900.

Another coarse adjustment improvement has been made by Herr Reichert. Its principle is a rack and pinion of specially hardened gun-metal, and a very important feature is the springing with adjusting screws for tightening up. Of these there are three : one being used for regulating the grip on the tube-mount, the others for tightening the pinion ; thus the unavoidable wear and tear can be compensated for.

Fine adjustments, owing to the rigorous requirements of high-power photomicrography, have received very great attention.[1] Modern advances may, however, be reduced to some five main types : (1) the direct-acting

FIG. 1.

screw, (2) the same with lever interposed, (3) the differential screw, (4) Reichert's lever fine acting, (5) Berger's endless screw. The first of these is seen in the Zeiss' microscope (1886), where a left-handed micrometer screw with a hardened steel point presses on a hardened steel plate. In this, one revolution of the milled head causes a movement of 1/101 inch. The second is met with in Messrs. Watson's microscope, where a lever of the first kind with unequal arms is interposed. This arrangement not only greatly reduces the weight bearing on the fine adjustment, but slows the speed down to 1/350 inch for one revolution. This speed is not at all too slow, as any lens possessing a fair optical index is excessively

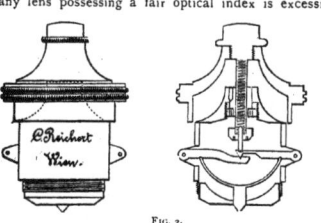

FIG. 2.

sensitive to focal adjustment when a ¾ illuminating cone is used. The differential screw fine adjustment occurs in the microscopes of Messrs. Baker and Messrs. Swift. Its advantage is that fine movements are obtained by the use of coarsely threaded screws, thus reducing the liability to wear and tear. Any degree of slowness may be obtained, but 1/200 inch is adopted by the makers. Herr Reichert's fine adjustment is especially suited to the Continental model, and consists of an ingenious adaptation of a double lever of the second order to the usual direct-acting screw (Fig. 2). It is arranged for a movement of 1/200 inch.

[1] For fuller information see " On the Evolution of the Fine Adjustment," by Mr. E. M. Nelson (*Journal* R.M.S. 1899, pp. 366-375).

Herr Berger's (Zeiss) is another great improvement in fine adjustments, and is on an entirely novel plan. The extremity of the micrometer screw comprises a horizontal toothed-wheel, which is actuated by an endless screw, terminating in the usual micrometer knobs. This arrangement permits of any degree of sensitiveness, and is ingeniously packed away in a hollow part of the limb, so that the fine adjustment is both dustproof and out of risk from any accidental injury. A valuable feature in the contrivance is a device for preventing strain on the fine adjustment from over-winding. This particular microscope is intended for the highest class of work, and possesses another good feature, in that the arm can be made to extend to any length over the stage without increasing the pressure on the fine adjustment. Thus an error in the original design of the Continental model, viz. the shortness of the distance between the limb and the optic axis, is corrected.

In Pillischer's international microscope a serious attempt has been made to reduce the inconveniences inseparable from stage-clips. He secures these to a bar, which, by means of a flange piece, raises or lowers them *simultaneously*, so that their points press on the object slide with any desired degree of pressure.

The use of large slides has led to a corresponding increase in the size of the stage, as well as in improved stage-finders and mechanical movements. There is a marked tendency to make the stages completely rotatory.

Among special stages, that of Herr Kraus (Reichert and Co.) is perhaps the most notable. It is heated by an electric current passed through a coil of platinum immersed in the liquid paraffin with which the stage box is filled. Regulation is accomplished by an ingenious contact thermometer, a device which resembles an ordinary thermometer, except that it is open at the top end. A platinum thread can be set in the tube at any desired temperature, and, when the mercury has risen and has met this thread, an automatic arrangement switches off the supply current. If the platinum and the mercury break contact, the current re-enters. The apparatus is said to be capable of rapidly producing and sustaining a temperature constant to 0·1° C.

In the department of lenses, Messrs. Watson have brought out their holoscopic eye-pieces, in which one lens-mount slides within the other, telescope fashion, thus forming a very convenient arrangement, easily adjustable as an over-corrected or under-corrected ocular as desired. The appropriate graduations are engraved on the tube.

Messrs. Zeiss have issued, at Dr. Hartwig's suggestion, a very useful series of low-power objectives, called "planktonsearchers," which are to be immersed in a trough of water, and so used for exploring it. They are made of Jena glass, and present an image completely plane and free from astigmatism quite close up to the periphery. The microplanar series of objectives by the same firm are used for projecting micro-slides on a screen, and give a wonderfully good ten-feet picture remarkably plane, sharp and well-defined, even to the limits of the field.

Another important set of objectives is Messrs. Leitz' series of achromatics. These are made of Jena glass free from fluorite. They are so well designed and corrected that they give results but very little inferior to the best apochromatics. Their freedom from fluorite renders the risk of atmospheric deterioration insignificant. Moreover, they are sold at so low a price that Messrs. Leitz must be admitted to have outstripped all other opticians in this particular detail.

Mr. H. J. Grayson, of Melbourne, has by an unknown process produced some very fine examples of ruled test-plates, the rulings being executed in various gradations of fineness, as far as 60,000 lines per inch, and 2000 lines per millimetre. The accuracy of the work is remarkable and the mounting in realgar ($\mu = 2\cdot5$) makes the lines stand out with a distinctness and brilliancy hitherto unknown.

The Royal Microscopical Society have, after careful consideration and after full consultation with microscope makers, drawn up a code of standard sizes for eye-pieces and sub-stage fittings. It is to be hoped that this important and useful step towards universality will be generally adopted.   ALFRED N. DISNEY.

THE FORTHCOMING MEETING OF THE BRITISH ASSOCIATION.

THE meeting of the British Association, which is to be held in Bradford this year, promises to be an unusually large and important one. Bradford being mid-way between London and Edinburgh, serves as a common meeting-ground for scientific men from the south of England and from Scotland and Ireland, and it is within easy reach of the Midland and Northern University Colleges. Bradford and Leeds are so close together that for such a purpose as this they are almost one city, and the Bradford Committee, therefore, have the advantage of the Yorkshire College being practically on the spot. The last meeting of the British Association in Bradford was held in 1873, but since that time the city (which, by the way, was then only a town) has practically been re-built, and has grown and developed in a manner resembling the progress of an American rather than that of an English town.

It is, therefore, much better provided now with hotel accommodation and with public buildings suitable for reception and sectional rooms. It is probable that the number of visitors will be far above the average; already some sixty or seventy Fellows of the Royal Society have announced their intention of being present, and professors and eminent lecturers from nearly every University in England, Scotland and Ireland have promised to attend. The Church will be represented by the Bishop of Ripon, the legal profession by the Master of the Rolls and Lord McLaren, and the names of over a score of members of both Houses of Parliament have been sent in.

The meeting will commence on Wednesday, September 5, when the new President, Prof. Sir Wm. Turner, of Edinburgh, will deliver his address in St. George's Hall. On the following evening the Mayor of Bradford will give a conversazione in St. George's Hall, at which it is hoped there will be exhibits illustrating the most recent scientific work. On Friday evening the lecture will be delivered in St. George's Hall by Prof. Gotch, F.R.S., on "Animal Electricity." The lecture to artisans on the Saturday will be given by Prof. Silvanus Thompson, F.R.S., and it is expected that there will be an audience in St. George's Hall of 4000 to 5000 working men. On Monday afternoon the Mayor and Corporation will give a garden-party in Lister Park, and in the evening an address will be given by Prof. W. Stroud on "Range-Finders." The Mayor and Corporation will give another large conversazione on Tuesday evening, and on the Wednesday evening a concert will take place in St. George's Hall with the Permanent Orchestra and the Festival Choral Society, under the conductorship of Mr. Fredk. Cowan. There will also be one or two eminent vocalists.

During the week there will be a textile exhibition at the Technical College, which will illustrate the various processes of the local industries, and the machinery employed can be seen in motion. There will be a reception at the College on Thursday afternoon, September 6, and the smoking concert in honour of the President will also be given at the Technical College, after Prof. Gotch's lecture on Friday.

Excursions to places of interest in the neighbourhood will be made on Saturday, the 8th, and on Thursday, the 13th; among the places selected are Bolton Priory, Ripon and Fountains Abbey, Malham, Clapham and Ingleton, the Nidd Valley, Farnley Hall, Haworth, Ilkley, Knaresboro' and Harrogate.

It has become an established custom to publish locally in the towns visited by the British Association a handbook containing a review of the objects of scientific interest and of the social and industrial conditions which prevail in the district. Many of these handbooks have been excellent in their character, and have covered ground altogether unexplored by the ordinary guide-book.

There is every prospect that the handbook published in connection with the Bradford meeting of the Association will be marked by the same width of view and thoroughness of execution, and may prove of permanent value. The work has been taken up with energy by the Publications Committee, of which Mr. Harry Behrens is chairman, and Mr. Mortimer Wheeler is honorary secretary and editor.

The book will be divided mainly into three sections. The earlier will deal with the history of Bradford and the development of the Bradford trade, under the following sub-heads:—(1) Prehistoric archæology, for which Mr. Butler Wood, chief librarian of the Bradford Free Libraries, is responsible : (2) the historical record of Bradford through mediæval times to the middle of the 18th century, which has been undertaken by Mr. Harry Speight; and (3) the social life and development of Bradford from the middle of the 18th century to the present, which is in the hands of Mr. Halliwell Sutcliffe, the novelist. To this section there will be addenda on the local dialect, local folklore, and the local place-names, including street-names, which will be written by Mrs. Wm. Wright (of Oxford) and Mr. Butler Wood.

The second section of the volume, of which Mr. John Bacchus is sectional editor, will deal comprehensively with the Bradford industries and institutions. The following are sub-heads:—Introductory notice and description of the staples employed in the Bradford trade; description of the processes in combing, spinning, weaving, dyeing and finishing; merchanting in the textile trade; the iron trade and machinery-making; the stone trade and minor industries; the Chamber of Commerce and the Exchange.

The third part of the book, which is under the sectional editorship of Mr. J. E. Wilson, will deal with the scientific material of the locality, the flora, fauna, geology, meteorology, climate and public health. There will be in addition a number of pages devoted to the topography of the district, for which Mr. J. H. Hastings is responsible.

In addition to the handbook, the Publications Committee have in preparation a series of small guides to the various places to which excursions are to be made. These will be issued in the form of a small portfolio, and each guide is being prepared on a scheme similar to that of the handbook, local specialists being called upon to describe the archæology, geology, botany and zoology of the various districts.

In regard to the accommodation of visitors, it is not anticipated that, in spite of the large influx of strangers, there will be any difficulty in finding comfortable quarters for everybody. Bradford is well provided with hotels, the two largest of which, the Midland and the Great Northern, can put up a great number of guests. All the available accommodation at the Royal Hotel has been secured by the local committee in order to provide for the secretaries of the different sections, who, of recent years, have been in the habit of lodging together. A large amount of private hospitality will be provided by the inhabitants of Bradford, and the Hospitality Committee is also drawing up a list of furnished apartments, which can be had on application. It is important, however, that all persons proposing to attend the meeting should give a long notice of their intention, in order to facilitate the arrangements which the Committee wish to make for their comfort. RAMSDEN BACCHUS.

NOTES.

A CONFERENCE of delegates for the International Catalogue of Scientific Papers was held at the Royal Society on Tuesday and Wednesday.

THE second of the two soirées held annually at the Royal Society will take place on Wednesday next, June 20. This is the soirée to which ladies as well as gentlemen are invited.

MR. C. E. BORCHGREVINK will give an account of his Antarctic work at the meeting of the Royal Geographical Society on Monday, June 25, instead of June 18, as previously announced.

THE annual visitation of the Royal Observatory, Greenwich, will take place on Tuesday, June 26. The visitation has previously been held on the first Saturday in June, and the change of the customary date is due to the absence of the Astronomer Royal, and other astronomers, for the purpose of observing the solar eclipse. This does not, however, explain the change of day.

ON the occasion of the retirement of Sir Frederick Bramwell from the office of honorary secretary of the Royal Institution, the managers of the Institution unanimously resolved " to place on permanent record an expression of their high appreciation of the admirable way in which he has performed the duties of that office and of his signal services to the Institution generally."

THE death is announced of Dr. Julius Althaus, the distinguished physician and neurologist. He was the author of works on "Diseases of the Nervous System," "Failure of Brain Power," "Diseases of the Spinal Cord," "Medical Electricity," "Influenza" and "The Spas of Europe," and was an authority upon the use of electricity in medical practice.

THE next lecture of the Zoological Society of London will be delivered at the Society's Meeting Room, on Thursday, June 21, at 4.30 p.m., by Prof. E. Ray Lankester, F.R.S. The subject will be the gigantic sloths of Patagonia.

AT the last meeting of the Royal Society of Edinburgh, the following were elected as British Honorary Fellows :—Dr. Edward Caird, Master of Balliol College, Oxford ; Dr. David Ferrier, professor of neuro-pathology, King's College, London ; Dr. G. F. Fitzgerald, professor of natural and experimental philosophy, Trinity College, Dublin ; Dr. Andrew Russell Forsyth, Sadlerian professor of pure mathematics in the University of Cambridge ; Dr. Archibald Liversidge, professor of chemistry in the University of Sydney ; Dr. T. E. Thorpe, principal of the Government Laboratories, London ; and, as Foreign Honorary Fellows :—Dr. Arthur Auwers, secretary, Royal Prussian Academy of Sciences ; Prof. Wilhelm His, Leipzig ; and Prof. Adolf Ritter von Baeyer, Munich.

THE celebration of the centenary of the Royal College of Surgeons of England will commence on July 25 with a conversazione at the College. On Thursday, July 26, a centenary meeting will be held at the University of London, when an address will be delivered by the president, Sir William Mac-Cormac, and honorary fellowships will be conferred. On Friday, July 27, there will be a conversazione at the Mansion

House. The Committee have issued invitations to foreign and colonial surgeons, and propose to issue invitations to certain persons of distinction residing in Great Britain and Ireland. A short history of the College, with eight illustrations, has been prepared, and will be presented to guests invited to the centenary celebrations.

THE new clinical laboratories of Westminster Hospital were opened by Lord Lister on Tuesday, in the presence of a distinguished company. The laboratories have been added to the hospital to provide for a more scientific and systematic examination of disease than can be carried out satisfactorily in the wards. A few particulars concerning the work of the hospital were given by Sir J. Wolfe-Barry, K.C.B., and are reported in the Times. Westminster Hospital was, he said, one of the oldest hospitals in London, having been founded in 1719. 30,000l. had been spent in bringing the hospital up to modern requirements, and in 1899 it was decided to add clinical laboratories and improve the isolation wards and nursing accommodation. In time they hoped to institute an electrical laboratory fitted with the apparatus for the Röntgen rays and micro-photography. To meet these expenses 10,000l. was wanted.—Dr. Lazarus-Barlow, in giving a detailed account of the laboratories, said the hospital tried to keep in front of all research and modern improvements, scientific and clinical. The laboratories contained all the most recent apparatus for the clinical work of the hospital.— Lord Lister said it was no less a pleasure than an honour to him to take part in that day's ceremony. The beautiful clinical laboratories they had inspected would give the physicians of the hospital an opportunity of bringing to bear on their cases the most advanced knowledge and the most refined methods of investigation. Many a diagnosis which would otherwise be obscure would be rendered clear in those rooms. He need hardly say that the correct diagnosis was directly proportioned to successful treatment. In respect of what Sir J. Wolfe-Barry had said, he himself felt convinced that those who had worked in the laboratories would not only benefit patients in the hospital, but would also, unfailingly, be able to extend the boundaries of knowledge and promote the now rapid advance of pathological and therapeutic knowledge. The laboratories would also be of service as a powerful means of affording sound practical knowledge to the student.

DURING the early part of the present week a wave of unusual heat has passed over parts of England, accompanied by brilliant sunshine. In the neighbourhood of London, the shade thermometer rose to 89 on Monday, the 11th inst., and thunderstorms occurred over several parts with heavy rainfall, amounting to an inch in the Midland Counties. So high a temperature has not occurred at so early a period of the summer in the neighbourhood of London for more than fifty years. A sharp thunderstorm also visited London about 10 a.m. on Tuesday, and another occurred in the evening ; there was also a renewal of severe thunderstorms over a large part of England. The temperature on Wednesday was considerably lower than on the preceding days.

FROM St. Petersburg to Vladivostok by way of the Arctic Ocean is (says the National Geographic Magazine) the plan of itinerary of an exploring party that early in June leaves the former city on the steamer Aurora. Six men of science and twelve sailors, all experienced in Arctic travel and led by Baron Toll, make up the party. Their special object is the careful exploration of the Arctic regions north of Siberia. After a brief stop at Tromsö, Norway, and at the new Russian port of Catherine Harbour, on the Lapland coast, they will proceed to the Taimur Peninsula, west of the Yenisei River, and there establish their winter headquarters. The neighbouring territory is to be explored during the winter of 1900-1901. On the breaking up of the ice, about August 1901, it is proposed to push on to Sannikoff Land, discovered by Baron Toll in 1886 and as yet unexplored, and later farther northward to Bennett and De Long Islands, following the routes of the *Jeannette* in 1881 and of the *Fram*. The winter of 1901-1902 will be devoted to determining whether this group of islands extends to the Pole. When the water route reopens in 1902 they will resume their voyage to Bering Strait and reach Vladivostok in the autumn of the same year.

The application of science to the great problem of mechanical traction is revealing the fact that at no distant date the motor car, or automobile, will be regarded a decided success in every respect. Electricity, steam and oil are still fighting for the paramount position of best agent for propulsion, and, on this account, trials and experiments always prove of interest. The *Engineer* (May 25) describes a series of trials for touring vehicles at the Paris Exhibition, and out of thirty-seven cars competing we find that were driven by steam, and all the remainder propelled by petroleum motors. Among the more recent improvements on the heavier classes are smaller driving wheels; the motor still develops about six horse-power for cars carrying four or six passengers, their lower centre of gravity, owing to their smaller wheels, also proving an advantage when rounding corners at high speed. After exhaustive trials embracing distance, manipulation, grades, &c., medals were awarded to the makers of the following vehicles: the "Peugot" car, the "Delahaye" car, the "De Dietrich" car, and the "Panhard et Levasser" wagonette. All these are driven by petrol motors; for this kind of work, therefore, petrol stands in good stead, and it will be of interest to see whether this agent or steam is adopted for freight vehicles of a much heavier description.

Though the articles upon scientific subjects in popular magazines can often only be called scientific by courtesy, yet we like to think that their presence in increasing numbers indicates a growth of public interest in the progress of science. *Pearson's Magazine* usually contains contributions which are instructive as well as interesting; and the reader who is no longer thrilled by episodes in the lives of freebooters, mysterious knights and similar personages over whom the glamour of the past may be thrown, must find relief by turning to the articles in which imagination is tempered with truth. In the June number of the magazine, we find an account of the destruction of the jack rabbit of the United States, by driving them into a corral, as described and illustrated in Nature several years ago. M. Flammarion's experiments on the growth of plants under different coloured glasses, also described in these columns, form the subject of another interesting article. Mr. George Griffith concludes his story of imaginary visits to other worlds by means of a machine moved by a force with peculiar properties. If we may venture a criticism of this series of contributions, it is that Mr. Griffith's ideas are too anthropomorphic, and too limited by the present state of knowledge of the objects visited by his interplanetary travellers. Some of the work of the U.S. Fish Commission in pisciculture forms the subject of a short illustrated article; and an interview with Prof. Milne, illustrated by several seismograms, contains much interesting popular information upon earthquake waves. Finally, a number of reproductions of photographs of faces of athletes at the moment of victory are reproduced. The photographs are interesting to students of facial expressions, and a curious point revealed by them is that only in one case of the hundreds of photographs from which the selection was made is a pleasant expression upon the face of the winner.

A NOVEL way of making building land is being carried out not far from New York. The rapidly growing population of this city has made ground scarce on which to build villas and houses for the summer resort of the inhabitants; but the enterprise of the American builder is equal to the emergency, and land is now being literally pumped up from the sea, on which it is intended to erect houses, and to create a new suburb. The site chosen for this venture is the Nassau Beach, on the shore of Jamaica Bay, in Long Island, not far from Brooklyn. The salt marshes bordering on this coast, which for centuries have been overflowed by the tides, and which, of course, while in this condition were utterly unfit for building purposes, are being raised from four to six feet above high water by pumping up the sand, shells and gravel which form the floor of the bay, and delivering this on to the land to be reclaimed. The process adopted to attain this end is as follows:—A powerful suction dredger raises the material from the bed of the bay at the rate of 18,000 cubic yards a day, and with this five times the volume of water, which is sufficient to carry the sand and gravel along the twelve-inch pipes which deliver it on the low land. The water flows off by ditches along a more or less circuitous route back to the bay, the dredged material settling and quickly drying, and forming solid land. The thickness of the material when first deposited averages about eight feet, but there is shrinkage as it dries and consolidates. Ten acres have thus been raised since the pumping began a few months ago. A raised road and promenade two miles long and seventy feet wide, and an electric railway, will connect this new suburb with the railway to Brooklyn and New York.

We have received the first numbers of the *Boletim Mensal* of the Rio de Janeiro Observatory. The work contains much useful information, chiefly contributed by Sr. L. Cruls, the able director of the Observatory, and it will form a welcome addition to current meteorological literature. As long ago as 1887, Sr. Cruls issued a large number of circulars to all meteorological organisations with the object of collecting data for a climatological dictionary. For want of adequate resources, this valuable compilation has not been published; but we are glad to see that he intends to utilise the bulletin for the publication of some of the principal results. The number for February contains the annual means and extremes for twenty-nine stations in Japan. Another paper worthy of special note is that by Sr. Pereira da Costa on the earliest observations made in Brazil.

READERS of Mr. Fitzpatrick's "Transvaal from Within" will recollect his reference in an appendix to a discussion which took place in President Krüger's Volksraad on the wickedness of firing guns in order to bring down rain. We learn from the *Corriere della Sera* that the practice of firing cannon as a preventive of hail has been adopted lately in Italy with successful results. On May 25, at about 17 o'clock or 5 p.m., three successive storms collected in the neighbourhood of Rogeno (Como), the clouds of which were evidently charged with hail. At a given signal fourteen cannon opened fire on the clouds, with the result that nothing fell except a little sleet, here and there. On the same day a vast amount of damage was done in the vicinity of Alessandria by hailstorms passing over the districts of Rocchetta, Tanaro, Masio, Felizzano and Quattordio about 16 o'clock (4 p.m), the hailstones in some places massing to a depth of 50 cm. In the districts where cannon were adopted for dispersing the hail, the results exceeded all expectations; while in many parts of the district where this precaution was not taken the vineyards were completely destroyed

ACTUARIAL experience is of distinct value in connection with the application of the statistical method to biological problems; therefore it is of interest to learn from the annual report of the Institute of Actuaries that the mortality investigation, which is being conducted jointly by the Institute and the Faculty of Actuaries, has made material progress under the honorary supervision of Mr. T. G. Ackland. The volume containing the unadjusted data of the annuity experience has already been published. It has been decided to include in one volume the unadjusted data relating to endowment assurances and minor classes of assurance (male and female); and the council report that this volume, forming a second of these series, is now completed, and on the eve of publication. The extensive tables comprising the unadjusted data for whole-life assurances (male lives), are in the printer's hands, and will, when ready, form a third volume of the series. These will be followed, so far as the unadjusted data are concerned, by a fourth and final volume, which will contain the experience of whole-life assurances (female lives). The tables to be included in this final volume are finished, and are in course of being copied for the printer.

WE learn from *Science* that the University of Illinois has fallen heir to the Bolter Collection of Insects. The collection numbers approximately fifteen thousand species, represented by about seventy thousand specimens, besides thirty thousand duplicates not in the systematic collection. This collection, accumulated during the last fifty years by the late Andreas Bolter, is remarkable for the excellence of the material and for the exquisite care with which it has been prepared and arranged. It represents all orders of insects and North America in general, and contains also a considerable amount of exotic material. The gift was made by the executors of Mr. Bolter, in accordance with the terms of his will, conditional upon its maintenance as a unit, under the name of the "Bolter Collection of Insects," in a fire-proof building.

IN the *Irish Naturalist* for June, Prof. R. J. Anderson endeavours to account for the circumstance that in certain breeds of domesticated fowls the keel of the breastbone is crooked. It is somewhat curious to find that these crooked keels occur only in pure bred birds, the ordinary barndoor fowls having the keel straight. After consulting with a number of poultry breeders, the author comes to the conclusion that in-and-in breeding, the nature of the perches, the season, early hatching, defective food and cold may all contribute to the production of the abnormal condition.

IN the Christiania *Nyt Magasin for Naturvidenskaberne*, vol. xxxviii. Pt. 1, Dr. G. Guldberg publishes some observations on the body-temperature of the Cetacea, in which he shows how extremely imperfect is our knowledge of this subject. As he remarks, it is a matter of extreme difficulty to obtain the temperature of living Cetaceans, although this has been taken in the case of a white whale and a dolphin, which some years ago were kept in confinement in a pond in the United States. With the larger whales such a mode of procedure is, however, obviously quite impracticable, and we have accordingly to rely on *post-mortem* observations. The layer of blubber by which all Cetaceans are protected from cold renders the *post-mortem* refrigeration of the blood a much slower process than in most mammals, so that such observations have a much higher value than might at first be supposed to be the case. Indeed, the blood-temperature of a specimen of Sibbald's rorqual three days after death still stood at 34° C. The various observations that have been taken have afforded the following results in *individual cases*:—Sperm-whale, 40° C.; Greenland right whale, 38·8° C.; porpoise, 35·6° C., liver of a second individual, 37·8° C.; common rorqual, 35·4° C.; dolphin, 35·6° C. The average blood-temperature of man is 37° C., and that of other mammals 39° C.; while that of birds is 42° C. The record of 40° C. in the case of the sperm-whale seems to indicate that at least some Cetaceans have a relatively high temperature.

PROF. F. E. NIPHER, in a communication to the Academy of Science of St. Louis, has recently described some experiments that he has made in photographic "reversal," one of his aims apparently being to find a useful method of manipulating photographic plates without the need for the exclusion of light from them. The advantage of such a method is obvious when the experiments incur the possibility of light being accidentally produced, as in electrical work. He exposes the plates to light for a few days before use—"to the light of an ordinary room." The other descriptions of exposures are also vague, the time being given, for example, of camera exposures, but with no record of the lens aperture or indication of the character of the light. It is, therefore, not possible to follow the experiments described in other than a qualitative way. Prof. Nipher has taken street views, Röntgen ray photographs, and "electrographs," on plates that have received preliminary exposures, and developed them by the light, for example, of a sixteen candle-power incandescent lamp, at a distance of from about 1 to 5 feet. The exposure on the object was, in one case, about forty times the exposure that would have been required for making an ordinary negative; the over-exposed and pre-exposed plate giving, of course, a positive image. A good deal of work has already been done in this direction, but the uncertainty of the reversal, and the great difficulty of getting rid of mixed results of reversal and non-reversal, have so far prevented any practical use being made of the possibilities of these methods.

SUCH experiments as Prof. Nipher describes are interesting qualitatively, but before any process of the kind can be safely recommended for general use in cases where it might be advantageous, it will be necessary to determine the range of conditions that can be relied on to give simple, that is, unmixed results, and this can never be done by working with such objects and processes as are described in this communication. We would suggest the use of a series of graduated exposures, with a measurement of the opacities produced, and then the ordinary negative image and the reversed image could both be traced. The character of the reversed image could be judged of, and the range of exposures through which its production could be relied upon could be determined. Until some definite information of this kind is available, experiments in reversal will remain more curious than useful. It appears from Prof. Nipher's communication that he is still pursuing the subject, and we hope that he will succeed in placing the method on a firm foundation.

IN a paper in the *Berichte* of the German Chemical Society, Herr G. Kramer and Herr A. Spilker make the suggestion that an important source of petroleum beds may be the oil which is always diffused through the protoplasm of diatoms.

THE Report of the Botanical Exchange Club of the British Isles for 1898 has just been issued under the editorship of Mr. James Groves. The occurrence of *Stachys alpina* in Gloucestershire is regarded as an undoubted addition to the native flora of Great Britain.

WE have received a prospectus of the "Scientific Roll and Magazine of Systematized Notes," to consist of three volumes of about 500 pages each, which will be supplied to subscribers for 10s. per volume, at the rate of one volume a year, commencing in September 1900. The first part will be devoted to the literature of the Schizomycetes. The compiler is Mr. A. Ramsay, of 4, Cowper Road, Acton.

THE Annual Report of the Royal Botanic Gardens, Trinidad, for the year 1899, by the superintendent, M. J. H. Hart, gives evidence of work done in the Gardens in connection with the acclimatisation of foreign economical plants, and the study of diseases of fruits and other crops, with the assistance of the Kew establishment. The *Bulletin of Miscellaneous Information*, from the same Gardens for April, contains, in addition to some natural history notes, a continuation of the descriptive list of West India and Guiana ferns.

A'-SHORT obituary notice of the late Franz Ritter von Hauer appeared in NATURE for April 13, 1899. A full account of the life and labours of this distinguished geologist has now been published by Dr. E. Tietze (*Jahrb. k.k. geol. Reichs.* Wien, Bd. 49). It is accompanied by a portrait, and by a list of geological papers and books dating from 1846 to 1897.

MESSRS. WILLIAMS AND NORGATE have just published a sixth revised edition of "Prehistoric Times as illustrated by Ancient Remains and the Manners and Customs of Modern Savages," by Lord Avebury (Sir John Lubbock). The first edition was published more than thirty-five years ago.

PROF. PRANTL's "Lehrbuch der Botanik," upon which Prof. Sydney Vines' "Students' Text-book of Botany" is based, has reached an eleventh edition. The new edition has been revised and enlarged by Dr. Ferdinand Pax, professor of botany, and director of the Botanical Gardens, at Breslau. Mr. W. Engelmann, Leipzig, is the publisher.

PROF. VIVIAN B. LEWES has in the press an exhaustive work on acetylene gas—a handbook for the student and manufacturer. The book will contain over 250 illustrations, and comprises a history of acetylene, its preparation, properties and chemical reactions, together with a complete list of legal enactments in full concerning its manufacture, patents, and other important data. Messrs. Archibald Constable and Co. are the publishers.

Two publications of interest to botanists will be issued from the Clarendon Press before long—the first part of the authorised English edition by Prof. J. B. Balfour of Dr. K. Goebel's "Organography of Plants," and Dr. A. Coppen Jones's translation of Prof. Alfred Fischer's "Structure and Functions of Bacteria." The former brings within reach of English students the only book of recent years upon its special subject ; the latter is the only work on bacteriology of similar scope and mode of treatment that has appeared in England since Dr. A. de Bary's "Lectures on Bacteria," a second edition of which appeared in 1887. This translation of Prof. Fischer's "Vorlesungen über Bakterien" should be welcome in pathological laboratories.

THE Orient Company announce that the cruise to Norway, Spitsbergen and Iceland will be repeated this summer. Their steamship *Cuzco*, 3912 tons register, is appointed to leave London on July 3, and to arrive back on August 4. After visiting some of the most interesting Norwegian fiords, the *Cuzco* will proceed to Spitsbergen, thus affording an opportunity of viewing the midnight sun, as for five days and nights after the ship leaves the North Cape the sun will be continuously above the horizon. Thereafter the *Cuzco* will proceed to Iceland, and her contemplated stay there of three days will enable passengers to see some of the most interesting sights in this remote island. The Faroe Islands will also be visited on the way back to London, *viâ* Leith.

THE purification of mercury is frequently necessary in physical and physico-chemical work, the process generally relied upon being distillation *in vacuo*. The apparatus in general use for this purpose, although convenient, has the disadvantage of being somewhat fragile, and requires large quantities of mercury. Some doubts, moreover, have been thrown on the efficacy of distillation as a purifying process, as Victor Meyer, in 1887, showed that traces of foreign metals passed over even after repeated redistillations. According to G. A. Hulett (*Zeitschrift für physikalische Chemie*, xxxiii. p. 611), these traces of foreign metals are carried over mechanically during the bumping of the boiling mercury ; and if measures are taken to prevent this bumping, perfectly pure mercury can be obtained in one distillation. Instead of the complicated apparatus of Weinhold, or its various modifications, a slight modification of the arrangement of two distilling flasks, with a capillary tube for admitting air, as commonly employed in organic work, was found to work perfectly.

PROF. RICHARDS, of Harvard, continuing his valuable re-determination of atomic weights, has lately published, in conjunction with Mr. G. P. Baxter, a preliminary paper on the atomic weight of iron. He points out that the value Fe = 56, which is now used, is practically based on work of fifty years since—being Wackenroder's corrected value of Berzelius' result, which was based upon the conversion of metallic iron into ferric oxide. In their preliminary determinations, Messrs. Richards and Baxter have reduced ferric oxide to the metal. The ferric oxide was prepared in the first case from ferric hydrate, which itself was prepared with elaborate precautions from very pure iron ribbon. The mean of two closely agreeing determinations gave Fe = 55·900. In the second case, ferric oxide was prepared with equal care from ferric nitrate. The mean value of five determinations gave Fe = 55·883. Further determinations are promised, but meanwhile the higher value of the older number (Fe = 56) is explained as probably due to one or more of the following causes :—The possible presence of magnetic oxide in the ferric oxide ; the possibility of incomplete reduction during the analysis of the substance ; the possible presence of alkaline, siliceous or other non-reducible material. At the present stage of the work 55·88 may be taken as the most probable value.

THE additions to the Zoological Society's Gardens during the past week include a Sykes's Monkey (*Cercopithecus albigularis*) from British Central Africa, presented by Mr. C. H. Ambruster ; a Barbary Ape (*Macacus inuus*) from Algeria, presented by Mr. R. S. Allen ; a Large Red Flying Squirrel (*Pteromys inornatus*) from Northern India, presented by Mr. A. Dudley Yorke ; three Goshawks (*Astur palumbarius*), European, presented by Mr. John Simonds ; a Little Egret (*Ardea garzetta*) from North-west Africa, presented by Mr. J. H. Yates ; an Allen's Porphyrio (*Hydrornia alleni*), captured at sea, presented by Miss Wallace ; a West African Python (*Python sebae*) from West Africa, presented by Francis E. Colenso ; a Green Lizard (*Lacerta viridis*), European, presented by Miss Mabel A. Heaton ; a Common Snake (*Tropidonotus natrix*), British ; two Mocassin Snakes (*Tropidonotus fasciatus*) from North America, presented by Mr. W. H. St. Quintin ; a Lion (*Felis leo*, ♂) from Kattiwar ; a Nylghaie (*Boselaphus tragocamelus*, ♂), two Four-horned Antelopes (*Tetraceros quadricornis*), three Indian Gazelles (*Gazella bennetti*) from India ; four Bearded Lizards (*Amphibolurus barbatus*), two Stump-tailed Skinks (*Trachysaurus rugosus*) from Australia, five American Box Tortoises (*Cistudo carolina*), six Stink-pot Mud Terrapins (*Cinosternum odoratum*) from North America, deposited ; a Rocky Mountain Goat (*Haplocerus montana*, ♂) from British Colombia, two Cunning Bassaris (*Bassaris astuta*) from Mexico, five Gentoo Penguins (*Pygoscelis taeniatus*) from the Falkland Islands, a Three-toed Sloth (*Bradypus tridactylus*) from British Guiana, purchased ; two Japanese Deer (*Cervus sika*), a Burchell's Zebra (*Equus burchelli*, ♂), born in the Gardens.

OUR ASTRONOMICAL COLUMN.

ROTATION PERIOD OF VENUS.—In the *Astronomische Nachrichten* (Bd. 152, No. 3641), Prof. A. Belopolsky gives the detailed measurements of the photographs of the spectrum of Venus taken during the recent favourable disposition, from which he has been enabled to confirm the short rotation period of the planet.

The spectrograms have been made with the 30-inch refractor at the Observatory of Pulkowa, using two different spectrographs, one of which was provided with two simple prisms, the other having three compound prisms. The spectra being obtained, the inclination of the spectral lines and the difference of wave-length of the light coming from the two opposite equatorial limbs of the planet is measured, and after corrections being applied for the inclination of the planet's equator to the line of sight, the resulting displacement indicates the equatorial velocity. As the light from the planet is reflected sunlight, the value measured is, of course, double the actual velocity.

The complete measures were obtained from fourteen plates taken with the two-prism spectrograph, and from five obtained with the instrument furnished with three compound prisms, are given. The values adopted are the means of measurements of from six to sixteen spectrum lines on each plate.

The photographs were obtained on the evenings of March 25, 30; April 4, 6, 7, 8, 10, 11, 20, 28; May 4, 5, 13, with exposures varying from 7m. to 60m.   The angular diameter of the planet varied from 8".6 to 11".0. With the 30-inch refractor, of about 40 feet focal length, the linear diameter at the principal focus was 1.2 mm., and this was further reduced by the relative foci of collimator and camera objectives to 0.8 mm. on the photographic plate.

From the difficulty of the determination it is to be expected that the several means should vary for the different plates; but the extreme values given still prove the short rotation period. Taking the diameter of Venus to be 12,700 km., the values of the equatorial velocity (*v*) are as follows, the corresponding time of rotation (T) being placed under each :—

$$v = 0.7 \quad 0.5 \quad 0.462 \quad 0.45 \quad 0.3 \text{ km. per sec.}$$
$$T = 15.9 \quad 22.1 \quad 24.0 \quad 24.6 \quad 37.0 \text{ hours.}$$

The author expresses the hope that the astronomers having the control of the large telescopes at the Potsdam, Lick and Yerkes Observatories will repeat his observations for confirmation or revision.

NEW VARIABLE IN AURIGA.—Dr. T. D. Anderson, of Edinburgh, announces in the *Astronomische Nachrichten* (Bd. 152 No. 3642), the detection of a new variable star in Auriga. It is not charted in the B.D., and has the following position :—
$$\left. \begin{array}{l} \text{R.A.} = \quad 6h \ 0.9m. \\ \text{Decl.} = +50° \ 14' \end{array} \right\} (1855.0)$$
The changes in brightness during April and May 1900 were from 8.3 to 8.8 magnitude.

PHOTOGRAPHIC OBSERVATIONS OF SATELLITE OF NEPTUNE.—In the *Astronomische Nachrichten* (Bd. 152 No. 3642), M. S. Kostinsky gives the particulars relating to a series of determinations of the satellite of Neptune, obtained from measures of photographs taken with a telescope of 13 inches aperture at Pulkowa. Many of the difficulties encountered in the photographic delineation of two neighbouring objects of very different brightness have been previously discussed by the author (*Bull. de l'Acad. Impér. des Sc. St. Pétersbourg*, vol. vii. November 1897). In the present case of Neptune the problem is rendered slightly less difficult by the feeble brightness of the planet and the slow movement of the satellite.

The photographs described were obtained during the period 1899 February 4–March 25. the plates having exposures varying from 20m. to 60m. A table giving the corresponding calculated and observed values shows the method to be very accurate.

SOME NOTES ON THE LATE PROF. PIAZZI SMYTH'S WORK IN SPECTROSCOPY.

LAMENTING, as we must do, that time has stolen from us a mighty Ajax in the field of science, a sturdy, patient Atlas who through more than half of this fast waning century robustly upheld on his strong shoulders the growing spires and architraves of science's ever-increasing edifice, it is with keenest sorrow that the writer of these notes turns over the ample pages, rich to profusion in details and superb in colour, of the monumental works of spectroscopy left to us by the late Prof. C. Piazzi Smyth, with the nearly hopeless intention of endeavouring to give a short account of some of his most conspicuously important contributions to that branch of science. The late Prof. Smyth was, indeed, no *dilettante* in the intricate and difficult but fertile and alluring byways of science to which his leisure moments were devoted ; and he was far from conceitedly or affectedly pedantic in the grasp of science which he brought to bear upon his philosophical investigations. Although these

embraced a range of astronomical and meteorological subjects which would singly engage all the energies of most men, and their whole lifetimes to study with success, yet his mastery of the state of science in the questions which he set himself to solve or to explore, was acquired with so much inventive skill, unsparing pains and ardour, as always to make the character of the work which he accomplished in them permanent and thorough. Well accustomed as he was from his youth, and trained from boyhood,[1] to delicate telescopic, angular and micrometrical measurements by eye and hand, he further possessed a gift of great artistic skill in committing to paper, canvas, and even to frescoes, beautiful drawings, photographs and coloured paintings of the scenes of travel which he witnessed, and of sights which clouds, the heavens, or his laboratory experiments disclosed to him. This accomplishment, well illustrated, long ago, by his publication in the Edinburgh *Philosophical Transactions* (vol. xx. pt. iii.) of a scene of darkness on the coast of Norway, near Bergen, during the Total Solar Eclipse of July 1851, contributed again in colours from his original, carefully kept paintings of the scene, together with a similar view of the Zodiacal Light as seen at Palermo in April 1872, to a new illustrated work on astronomy published by Messrs. Cassell and Co. in 1894, led him to leave to others the study of the actinic spectrum-regions with the aid of photography, and to restrict his spectrum-measurements entirely to all that could be seen and measured by the eye alone, of the solar spectrum, or of the characteristic features of gaseous bright-line spectra, in the whole visible portion of the spectrum only.

In his keen perception of all the grand sublimities of law and order by which Nature's works are everywhere controlled and guided and sustained, and in the constant intentness of his mind to seek out these nature's workings, and to promulgate lucidly and clearly his own perceptions and interpretations of them, Kirchhoff's great discovery, in 1859, that the chemical constitution of the sun could be read in its light's prismatic spectrum, constrained him like a spell, as it quickly did many other physicists, to devote much of his leisure time and abilities to spectrochemical researches. New striking truths were taught in 1860–61 by Sir William Huggins' not less surprising discovery from observations of their spectra, of the gaseous conditions of certain nebulæ, and by Sir David Brewster's and Dr. J. H. Gladstone's majestically mapped separation from the really solar dark lines in the sun's spectrum, of its low-sun, or terrestrial atmospheric lines, soon afterwards distinguished by Secchi, Ångström and the first detector of the "rain-band" near solar D, in America, Dr. J. P. Cooke. and especially by Dr. Janssen's observations among the high Alps of Switzerland and experiments with a long steam-tube in Paris, in 1866–7, into "aqueous-vapour" and "dry-air" telluric lines. Kirchhoff's and Hofmann's first chemical investigation of the solar spectrum was rapidly extended in the years from 1859 to 1868, with tables of metallic and other elementary line-spectra by Huggins and Miller, Mascart, Plücker, Ditscheiner, Van der Willigen, Thalen, Lockyer and others, into a wonderfully novel panorama field of spectrum-analysis, chiefly applicable at first to celestial chemistry and physics, but in such skilled hands as those of Bunsen, Crookes, Reich and Richter, and later of Lecoq de Boisbaudran and other able chemists, to the discovery also of new terrestrial elements. The appearance at Upsala, in 1868, of Ångström and Thalen's classically accurate and chemically expounded "Normal Solar Spectrum" map, with its line-places in a natural diffraction-spectrum order of wave-length progression reckoned in "tenth-metres," or (10)[th] parts of a metre as scale-units of wave-length,[2] and the detection with spectroscopes in the total solar eclipse of the same year in India, of the hydrogen-flame nature of the sun's red prominences, seen in full sunshine there by Dr. Janssen and almost simultaneously also by Sir J. N. Lockyer in England, afforded to the new

[1] Under Sir Thomas Maclear's care, in 1836, at the age of seventeen ; at the famous Observatory at the Cape of Good Hope, where, during the last three years, the presence of oxygen was discovered by its line-spectrum in certain southern stars by the indefatigable English amateur astronomer, Mr. F. Maclean ; and where both that discovery and another by Sir J. N. Lockyer of the presence of silicium in the same stars, have been confirmed, and made independently by its energetic Director, Sir David Gill, with a noble spectrophotographic 24-inch refracting telescope presented to the Observatory under his own directions and liberal care for its completeness by the same munificent explorer of stellar spectra in the northern and the southern heavens, Mr. Frank Maclean.

[2] It has now become a common usage in spectroscopy, microscopy and molecular physics, to reckon such small quantities as light wave-lengths in a tenfold larger unit than the Ångström one, denoting it by "$\mu\mu$," the thousandth part of "μ," the thousandth part of a millimetre, "mm."

study of spectroscopy at once a sound philosophic basis for spectrum-definitions, and a new territory of interesting astronomical investigations in the sun's glowing atmosphere, upon which it was not remiss or slow to grow up in strength and stature, expanding itself largely in new observational, practical and theoretical directions, in the next following interval of ten or fifteen years.

From the graphic mementos which he kept, as we have seen, of the total solar eclipse of 1851, and from his successful attempts, described in 1858 in his well-known and most attractively illustrated volume on Teneriffe, to prove by visits (in that year, and again in 1868) to the Island and Peak of Teneriffe, the practical benefit to be obtained in astronomical observations, of avoiding in great part the atmosphere's absorbing action on the light of stars and planets by establishing observatories on mountain heights, it cannot be doubted that these discoveries with spectroscopes concerning the bright, ruddy light-flakes seen round the sun, or round the moon's disc when the sun is totally eclipsed, and regarding the particular rays of the sun's light which undergo absorption in the earth's atmosphere, would, as important contributions to our extremely circumscribed knowledge of the materials and physical conditions of planetary and stellar atmospheres, greatly impress and interest him.

In the preface of his "Spectre Normale du Soleil," Ångström pointed out that the spectrum of the aurora, as he had frequently observed it in the winter months of 1867-8, consisted almost entirely, as Dr. O. Struve at Pulkova, on hearing from Prof. Ångström of this discovery also confirmed it in May 1868, and as Prof. Ångström had previously found to be the case with the spectrum of very bright appearances of the zodiacal light at Upsala in March 1867, of a nearly solitary bright yellow line. Many exact corroborations of this line's conspicuousness, and detections of several less constant bright and faint auroral lines were thereupon made by observers of a series of fine red auroras which at the time of the marked maximum of sunspot frequency in 1870 appeared during the years from 1869 to 1871. A paper recommending Prof. Swan's well-known blue gas-flame, or blowpipe flame spectrum with its five well-determined line or bandedge positions as a most suitable one for reference in mapping auroral spectra, was sent in 1870 by the late Prof. Piazzi Smyth to the Royal Astronomical Society in London ; but owing to his describing the flame-spectrum as Prof. Swan had done in 1856, and as did also Ångström in the introduction to his "Solar Spectrum Atlas" in 1868, as the spectrum of hydrocarbons, or of acetylene, it remained unpublished on account of the doubtfully correct chemical apellation given in the paper to this important set of spectrum-bands. Yet the same chemical origin, describing it as probably that of acetylene, was attributed to this spectrum both by Profs. Liveing and Dewar at Cambridge, and by Ångström and Thalen in their "Spectres des Metalloïdes" at Upsala, in 1875 ; and Prof. Smyth never felt induced to resign the view which he held in such good company, by the contrary opinion steadfastly maintained by many not less skilled and experienced and at least as chemically well versed spectroscopists, that the blue candle-flame's spectrum of delicately fluted bands was not really due to any chemical compound of carbon with other elements, but to carbon itself in one of the modes of molecular aggregation into which, like the materials of some other metalloids at least, the substance of carbon in its volatilised state is liable, by temperature or by some sufficient chemical or electrical powers of dissociation to be broken up.

Another very similar band-spectrum to the Bunsen-flame one, agreeing in the positions of its two brightest (citron and green) band-edges pretty closely with two corresponding bright band-edges of the latter spectrum, but differing from it throughout in its more numerous remaining bands' positions, and with all its bands evenly shaded, instead of (as in the other spectrum) both fluted and shaded off towards the blue direction, only too often seen mingling with the latter spectrum to some extent in nearly all electrically excited vacuum-tubes, can be very readily produced in its natural purity with ordinary induction-sparks in carbon oxide or di-oxide vacuum-tubes ; and it was described on that account, in their "Spectres des Metalloïdes" in 1875, by Ångström and Thalen, and after some hesitation about its possible chemical nature in his first paper on "Gaseous Spectra in Vacuum-tubes under Small Dispersion" in 1880,[1] it was after-

wards regarded also by the late Prof. Piazzi Smyth as belonging to carbonic oxide. Appearing as these two spectra do almost ubiquitously as impurities in ordinary gas-vacuum-tubes, their precise discrimination from each other, and the resolution of their many hazy bands into as many ranks of scores upon scores of accurately measured linelets, was a work of exact spectrometry in which the great light-intensities of his vacuum-tubes and the powerful train of prisms finally used by Prof. Smyth for the maps of gaseous spectra which he constructed in 1884,[1] accomplished some of the most wonderfully perfect and beautiful achievements. The much debated experimental evidence as to the chemical origins of these two spectra, moreover, prepared the way for some most embracing views of the modes of production of stellar and celestial spectra, which, besides providing astronomers with the means of classifying stars and the lesser lights of nebulæ and comets methodically, also afforded chemists an imposing outline of problems for consideration, of apparently successive stages of subdivision of the elementary forms of matter from dense into light-atomed elements.

In Sir J. N. Lockyer's hands the condensed spark of a Leyden-jar introduced into the vacuum-tube circuit (which Prof. Smyth never used, having decided to confine himself to weak-spark or low temperature excitations only in his spectrum-measures), supplied a method of transition from the oxy-carbon spectrum in carbon-oxide and dioxide vacuum-tubes, directly to the hydrocarbon or blue gas-flame one,[2] showing that only a rise of temperature was needed, from that of the nearly continuous induction-spark or simple brush-discharge in rarefied gas-tubes, to the vastly hotter disruptive spark (instantly volatilising gold-leaf or thin metallic wires), of a Leyden-jar and air-gap in the outer circuit, to furnish a new spectrum, not, we must conclude, by any chemical change of substance, but by disgregation, it seemed evident, of cool and dense into hot light molecules of pure carbon, which could thus be made at pleasure to give either of these two spectra in succession. The flame, and tube spectra, or the hydrocarbon and carbonic oxide ones are therefore now usually referred to, by Sir J. N. Lockyer, as the "hot carbon" and the "cool carbon" spectrum, respectively.[3] But all the best means that can be used to obtain, on the one hand, an evenly ascending scale of temperature and uniform intensities of action of discharges of the electric arc and spark (the only sufficient known means which can be used to reach the high temperatures demanded), and on the other hand the requisite chemical purity of the substances submitted to spectroscopic examination, are so very liable to unsuspected failures from the many lurking sources of deception which most insidiously waylay and falsify the observations and conclusions, that although, on both sides, these sources of error have been unremittingly sought out and often most startlingly disclosed and very skilfully eliminated, it is difficult to say yet whether the distinctive attributes in which the substances which give the different banded carbon spectra really differ fundamentally from each other, are either, as was at first supposed, simply chemical, or else, according to a subsequent suggestion, attributing to pure carbon spectroscopic properties which are at least not at-variance with those of oxygen, hydrogen, sulphur, selenium and phosphorus, of an entirely structural kind ; that is to say, gaseously allotropic, or molecularly disgregational under the action of increasing temperature. New discoveries and fresh discussions of these bands must doubtless be awaited before we can be definitely sure to what extent the views expressed by different observers as to the chemically compound or elementary dissociated natures of the material sources of special series of shaded or fluted bands seen in banded carbon spectra can be fully trusted.

Besides the two chief ranks already mentioned, there is another

[1] Edinburgh *Philosophical Transactions*, vol. xxx. (1883), pp. 99, 104. In a letter to NATURE (vol. xx. p. 75), in May 1879, on "End-on Vacuum-tubes brought to bear on the Carbon and Carbohydrogen question," the late Prof. Piazzi Smyth also adopted at first, without reserve the view of this spectrum that it is produced by carbon simply.

[1] "Micrometrical Measures of Gaseous Spectra under high Dispersion," Edinburgh *Philosophical Transactions*, vol. xxxii. Pt. 3 (1886). The end-on Vacuum-tubes used in these measures and in those of the earlier paper, were devised by the late Prof. Smyth himself, as described in a paper, "End-on Vacuum-tubes in Private Spectroscopy," read before the Royal Scottish Society of Arts in 1879. The eminent spectroscopist of Ghent, in Belgium, Dr. van Monckhoven, had, however, invented and used such tubes a few years earlier. An ingenious arrangement of electrodes which he applied to them in 1882 (one electrode at the foot, and one at the summit of each upright leg), for passing two discharges of different strengths, either simultaneously or alternately through the connecting capillary tube, in a research on the effects of temperature and pressure in widening gas-spectrum lines, was described in an interesting paper by Dr. van Monckhoven, in *Comptes rendus*, vol. xcv. (*ème semestre*, 1882).
[2] "Carbon and Carbon-compounds," by Prof. A. S. Herschel, NATURE, vol. xxii. p. 320, August, 1880.
[3] "Researches on the Spectra of Meteorites." *Proceedings* of the Royal Society, vol. xliii. pp. 118, 133, Map 3, November, 1887.

group of two carbon-bands, usually accompanied by a preceding one close-following the gas-flame spectrum's blue beam, near the Fraunhofer rays *h*, H, at the violet confines of the visible spectrum. All three are seen brightly in the spectrum of the electric arc between carbon poles, where the furthest member of this blue, violet, and ultra-violet array produces a just ocularly visible *pharos*-like mass of grey-looking light slightly beyond the spectrum-place of the furthest visible pair of dark lines H, K, of the solar spectrum. This strong outlying pair and its near preceding blue band were referred by Profs. Living and Dewar,[1] and also, when he found the two chief bands bright by themselves in his spectrum photographs of the Comet 1881, III., by Sir William Huggins,[2] to cyanogen. But the same strong pair's violet, or first colonnade, between G and *h*, when he first traced it beautifully distinct and bright with the ordinary coil-spark in an end-on marsh-gas vacuum-tube,[3] was coupled on by Prof. Piazzi Smyth, in his "Measures of Gaseous Spectra with High Dispersion," 1884, as it was also grouped by Dr. W. M. Watts in his "Index of Spectra," 1872 (and where Sir J. N. Lockyer classed all these three flutings provisionally together,[4] from a careful survey made in 1880[5] of their deportments, in an unstable-systemed extension, "B," of the "hot carbon's" ordinary set of flutings, "A"), as a sixth or extra fluted-band transcending in refrangibility all the five commonly seen ones of the "hydrocarbon," or blue gas-flame series.

Six of the seven main lines of the blue band of this set were marked as measured lines distinctly, by Prof. Piazzi Smyth, in a hazy glow of light immediately following the fourth, or blue band of his "High Dispersion Spectrum" Paper's full-length map of the "hydrocarbon" spectrum, but as considerably weaker lines than those of the violet or "marsh-gas" band. As the blue band, however, is in fact the weakest one of a group of bands which only the exceedingly hot flame of cyanogen, or the intense heat of the electric arc, or jar-spark usually produces, its exact indication there, precisely in its natural inferiority of strength to the violet array, and with only one line missing of its seven, at the beginning of the hazy glow, is a speaking testimony to the faithful accuracy of the late Prof. Piazzi Smyth's spectrum records, as well as to the watchful care with which all the spectra which he mapped were guarded against contaminating admixtures of interloping gas-spectra ; since with the modest 2-3 inch sparks which he was content to use, of a simple induction coil, nothing but the lamp-like brightness of the Salleron and Casella end-on tubes examined, and the chemical purity of the contents of those used in taking final spectrum measures, could have been expected to show the weakest of the three "cyanogen" bands so equally free from other spectrum-glares, and almost as sharply well-defined in position, as its bright violet companion tier of "marsh-gas" lines was seen and measured.

The fifth (faint violet) band—or the latter part of it—seen under high dispersion to contain only hazy linelets, with no strong lines or sharp-edged flutings, is the only visible light-beam in the Bunsen-flame spectrum which Sir J. N. Lockyer seems willing to admit,[6] can be described correctly as a "hydrocarbon" band ; but in his splendid series of discriminations of celestial spectra, that brightest portion of the Bunsen flame's violet band forms the whole system of spectroscopic bands which in those analyses of celestial spectra is usually indicated as characteristically denoting hydrocarbon radiation. Two small bands, or fluted line-groups, however, sometimes occur also in this unstable violet, or "Carbon-B" region, of which one is classed by Sir J. N. Lockyer together with the *pharos*-like band beyond H, K, as an invariable accompaniment (much more refrangible than the four "Carbon-A" bands) of the "hot-carbon" spectrum. This small three-lined band falls exactly in place and width on the Bunsen-flame spectrum's fifth, or violet band's preceding zone of weak hazy light, as the late Prof. Piazzi

Smyth, in his full-length map and micrometric measures of that spectrum pictured it, surrounding the place of the violet line Hγ almost as closely as its bright following "hydrocarbon" light-zone surrounds the dark solar line G's position, with a curiously prominent solitary bright line in the dark partition space between them. A fairly satisfactory explanation of this fifth band's construction might thus, with no material need of any reconciling adjustment to the "Carbon-B" band's line-places, be extracted from the Edinburgh spectrum record, by supposing the first and second portions of its divided light-field to belong really to different radiant sources, and to be due, independently of each other, respectively to "hot carbon" and to "hydrocarbon gas's" incandescence. But the near agreement in position between the flame-band's feeble front-domain of shapeless light-haze, and the "hot carbon's" small three-ribbed fluting lacks far too much from being well affirmed by exact coordinations to be any certain evidence of a real spectroscopic or physical connection ; and the weak preceding portion of the violet flame-band has thus been very appropriately consorted by Sir J. N. Lockyer with the following bright portion of this violet haze-band, as belonging both together to the hydrocarbon spectrum.[1] Another small violet-region band was traced by Sir W. Huggins in the spectrum of the Comet 1881, III., where it lay between the violet and the ultra-violet "cyanogen" bands, a little beyond *h* towards the line H of the solar spectrum.

Among this "Carbon-B" *suite* of bands, suspected by Sir J. N. Lockyer at an early stage of his spectroscopic observations of the sun, in 1874, to have counterparts in the dark lines of the solar spectrum, the strong *pharos*-like ultra-violet fluting's delicate train of bright lines and linelets was at length photographically proved by Sir J. N. Lockyer, in 1878,[2] to coincide precisely with a close-packed orderly array of faint, exceedingly fine dark lines at the same place in the solar spectrum ; and the same coincidence of about thirty serrations of this band in ten Ångström's wave-length units, with as many exactly corresponding ripplings of light and darkness at the ultra-violet confines of the sun's visible spectrum was again very abundantly well proved by Profs. Trowbridge and Hutchins at Hartford, U.S., in 1887.[3] It was also pointed out by Profs. Living and Dewar at the close of the second of their above-quoted papers, in 1880, on the "Spectra of Compounds of Carbon with Hydrogen and Nitrogen," that a fluted ultra-violet band in the spectrum of the cyanogen-flame, of which they photographed many in an ultra-violet region extending far beyond this grey one's position, exactly coincided in spectrum-place with the remarkably fluted ultra-violet dark band P, in the solar spectrum. After Sir W. Huggins and Padre Secchi had independently detected the "hydrocarbon's" or low gas-flame's bands in the spectrum of Winnecke's Comet, in 1868, and some ten or twelve comets in as many following years were found to show the same bands in their spectra,[4] together with occasional traces of the oxy-carbon or "cool-carbon" spectrum, a far wider range of the "hot" and "cool carbon" bands was presently discovered for them by Sir J. N. Lockyer among the spectra of celestial bodies, and in his "Researches on the Spectra of Meteorites," in 1887,[5] and in the Bakerian Lecture to the Royal Society on a "Suggested Classification of the various Species of Heavenly Bodies,"[6] the low gas-flame's or hot carbon spectrum's bands were clearly shown to exhibit themselves, with rarer excrudescences of cool carbon bands, not only in comets, but alike in sun-like and fluted, and bright-line and temporary stars, and even in nebulæ, the aurora, and sometimes in lightning-flashes, as a sort of torch-light glow of colliding meteorites, condensed meteoric swarms, and electrically gasified and illumined meteoritic dust, throughout the universe.

It surely needed then only the recent discovery by Prof. G. E. Hale and his coadjutors, Mr. W. S. Adams and Prof. Frost, in

[1] "On the Spectra of the Compounds of Carbon with Hydrogen and Nitrogen." Two Papers, Nos. i., ii. *Proceedings* of the Royal Society, vol. xxx. p. 252 and 404, February–June, 1880.

[2] *Proceedings* of the Royal Society, vol. xxxiii. p. 1, November 1881.

[3] It was also seen by Dr. W. M. Watts ("On the Spectrum of Carbon," **Nature**, vol. xxiii. p. 197, December, 1880,) "very bright" in a pure Marsh-gas vacuum-tube ; and in a methyl vacuum-tube by Dr. Plücker.

[4] The Bakerian Lecture, "Suggestions on the Classification of the various Species of Heavenly Bodies,"—"Radiation Flutings" : *Proceedings* of the Royal Society, vol. xliv. p. 53, April, 1888 ; and "Appendix to the Bakerian Lecture," Section vi. "General Statement with regard to Carbon," *Proceedings* of the Royal Society, vol. xlv. p. 186, November, 1888.

[5] "Further Note on the Spectrum of Carbon," *Proceedings* of the Royal Society, vol. xxx. p. 461, May, 1880.

[6] "Researches on the Spectra of Meteorites," *Proceedings* of the Royal Society, vol. xliii. p. 118 ; November, 1887.

[1] The Bakerian Lecture :—"Radiation Flutings," *Proceedings* of the Royal Society, vol. xliv. p. 53, April 1888.

[2] "Note on the existence of Carbon in the Coronal Atmosphere of the Sun," *Proceedings* of the Royal Society, vol. xxvii. p. 308, April, 1878.

[3] "On the Spectrum of Carbon compared with that of the Sun," *Proceedings* of the American Academy of Arts and Sciences, vol. xxiii. p. 10, 1887–8 ; and *American Journal of Science*, Series 3, vol. xxxiv. p. 345, 1888 ; **Nature**, vol. xxxvii. p. 114, December, 1887.

[4] "The Meteoritic Hypothesis," by Sir J. N. Lockyer (Macmillan and Co., 1890), p. 176 :—Table of Carbon-Spectrum Comets.

[5] *Proceedings* of the Royal Society, vol. xliii. pp. 117–156, November, 1887.

[6] *Proceedings* of the Royal Society, vol. xliv. pp. 1–93, April, 1888 ; and ("Appendix to the Bakerian Lecture") vol. xlv. pp. 157–262, November, 1888.

America,[1] with the giant telescope of the Yerkes Observatory's enormous power, that the green and citron hydrocarbon's chief band-lines can be observed dark on the photosphere at the sun's edge, and close by, bright in the chromosphere to a height from the sun's edge which they estimated not to exceed 1″ of arc, or about 500 miles, to completely ratify the foregoing views that those low gas-flame's fluted bands are produced by carbon vapour at an exceedingly high temperature ; and fully to justify the first observers in England and America of the presence of carbon in the sun, in the opinion which they independently expressed, that the temperature of the glowing region of the sun's atmosphere where this carbon vapour is produced and made to incandesce, must certainly approach nearly to, and at the same time not much exceed, that of the electric arc.

Carbon substance furnishes yet another known form of gaseous spectrum, which consists only of a few sharp bright lines, quite free from bands of shaded light, or flutings ; and this linear form of its spectrum may be pretty certainly ascribed to carbon vapour in its simplest molecular, perhaps even monatomic, state of aggregation, since it is only obtained by heating carbon in a condensed electric spark to the highest possible artificial temperatures. No indications, however, of carbon's occurrence at this exceedingly high temperature in any celestial spectra, appear as yet to have been met with. Although no gaseous spectra produced at such high temperatures as those of the condensed electric spark were spectroscopically measured by the late Prof. Piazzi Smyth, yet a depiction of this carbon spectrum as it was first seen in the Leyden-jar spark between carbon poles by Ångström in 1863, and as it was represented by Ångström and Thalen in their "Spectres des Metalloides" in 1875, is given with the line-spectra of common air, hydrogen, nitrogen and oxygen and of vapour of mercury, by different authors, in the Plate of full-length spectra of his "High Dispersion Spectrum" Paper of 1884, by the late Prof. Smyth, to compare with his own measurements of low temperature spectra of the same elementary gases or their compounds. The map of the carbon line-spectrum given by Ångström and Thalen shows a spectrum-field extremely large of lines, but terminated, at its two ends, by two very bright ones, a red, closely double line almost coincident with Fraunhofer's C, or Hα, and a violet one close-following G and the violet hydrogen line Hγ, and like the hydrogen-lines appearing to be easily widened into a diffuse, broad line by taking the spark in gases at increasing pressures.[2]

A faint single line near E, and two groups of three and four moderately bright, pretty close-packed lines near the beginnings of the two brightest (green and citron) flutings of the Bunsen-flame, or "hot-carbon" band-series, are all the remaining visible portions of this spectrum figured, as they had excellently observed and studied it, by Profs. Ångström and Thalen. But the latter two isolated line-groups appear to fit on remarkably well to the view already apparently borne out and substantiated by what precedes, that with rising temperatures and increasing disgregation of carbon-vapour molecules, the interval between the beginnings of the green and citron flutings becomes wider in passing from the "cool" to the "hot" carbon-band series. For while those band-beginnings are respectively at λ = 519·70 and 560·75 (distance = 41·05μμ) in the cool, or oxy-carbon set, and at λ = 516·40 and 563·34 (distance = 46·94μμ) in the hot, or hydrocarbon set of bands, the front-lines of the "Excelsior" Carbon-spectrum's (as the late Prof. Piazzi Smyth casually termed it), or still hotter and more broken-up carbon-vapour molecules' two small solitary line-groups, are at λ = 515·05 and 569·41, in Ångström and Thalen's Table of these carbon lines ; both shifted again slightly in position in the same left and right directions as before, and with the interval between them again increased a little, now, from its last measures, 41·05 and 46·94, to 54·36μμ.

But a most industrious explorer, and a describer and recorder unsurpassed in the skill of his depictions of the surprising beau-

[1] *Bulletins* of the Yerkes Observatory, No. 12, 1899.
[2] This widening of the carbon violet line to a "broad band" at λ=4272 (Ångström, 4366·0) is very distinctly recorded in Dr. W. M. Watts' "Index of Spectra," 1872, "Carbon-Spectrum, No. IV."; where the groups and single lines, α, β (*plus* the two next lines), γ, and ι, compose together Ångström and Thalen's line-spectrum of pure carbon. With four or five exceptions, all the many lines contained in the several other line-groups besides these, in the same Carbon-Spectrum Table, can, however, be readily identified with lines of the oxygen line-spectrum mapped by Ångström and Thalen on the same Spectrum-Plate ("Spectres des Metalloides," Upsala *Nova Acta*, vol. ix. 1875) with their line-spectrum of carbon, and also with lines in Dr. Schuster's map (*Philosophical Transactions* of the Royal Society, 1879) of the line-spectrum of oxygen.

ties of all this rich domain of matter's spectroscopic radiations, we must again here grieve to note, has passed away. Besides his already-mentioned extremely perfect measurements of "gaseous spectra," the late Prof. Smyth's published spectrum-maps and spectroscopic writings comprised long descriptions too of not less than five full series of measurements with high dispersion, in southern skies, and with great magnifying powers, of the dark lines of the solar spectrum.[1] These graphic solar-spectrum maps and those of the "gas-spectra," and separate papers treating also of the oxygen-gas spectrum singly, and of the dark line group "*b*" in the solar spectrum by itself,[2] together form a lasting store of precious materials for spectroscopic study too variously instructive and often suggestive of interesting theoretical deductions from their well-recorded details, to be here dealt with shortly and concisely. It is with a sense of doing only very partial justice to the exceedingly high merit and scientific value of those other important spectrum records and researches, that as much space as could be accorded to these short notes has been devoted here to pourtraying only the increasing cosmical significance and the widely-spreading applications in spectroscopic astronomy, of his valuable investigations of the ordinary forms of carbon spectra. In his effectual unravelling of the mazy linelet systems of those familiar spectra's bands, a plain and simple law of sequence in the linelets' spectrum-places was disclosed, which some years later also proved the proper clue to elicit order from the complex-looking linelet structures of the dark absorption-bands, "A" and "B" (both due to oxygen in our terrestrial atmosphere), at the red end of the solar spectrum. Although those shaded groups' constructions were only perfectly made known at last in 1893, by Mr. G. Higgs, of Liverpool,[3] from the beautiful figures of them given in his then published "Photographic Studies of the Solar Spectrum," yet the drawings of those bands in Prof. Smyth's Madeira and Winchester Solar Spectrum Plates in 1881 and 1884, only second to Mr. Higgs' photographs in their clear discriminations and accurate positions of the bands' details, would have certainly afforded ripe enough materials to establish at least the major portions of their simple featured laws of linelet sequence by themselves, if they had been searchingly examined, and carefully enough discussed and studied for the purpose.

Further examples of the same simple law of linelet intervals in such "shaded" bands (where each distinguishable *suite* or tier of linelets exhibits simply a fixed and uniform excess or growth of interval—of each *suite's* own amount or proper measure—in every pair of adjoining lines, over that of its immediately preceding line-pair, as the rank of lines advances from the brighter to the dimmer region of the shading) occur, moreover, not only in the brightest, green, but also in the citron and the blue band-figures, very plainly, of Prof. Smyth's full-plate "high dispersion" maps of those three most notable light-ridges in the "carbon oxide" (or "cool carbon") spectrum. Another interesting indication of line-systems also can be traced in his full-length mapped array of the four then known low temperature lines of oxygen, three of which he discovered to be finely triple, and to which he contributed three more just similarly triple lines. Two Balmer's series of three lines each can be pretty certainly distinguished in this strikingly peculiar group of six mapped triplets, converging approximately to a nearly common progression-head, or series-limit, at about λ = 430 μμ. Possibly these two line-sequences which his much extended range and finely multiplied line-features of the ordinary tube-spectrum of oxygen appear to show, may have been already recognised and fittingly comprised by Messrs. Kaiser and Runge among the many such line-series which they have found indications of in the spectrum-field of oxygen. But these and many more such philosophical results may be looked for to be richly gleaned and brought to light by coming years' discussions of the minute and copious information which with Mr. T. Heath's skilful assistance in their draughtsmanship and computations, is lucidly unfolded in Prof. Smyth's noble works of well resolved and accurately measured ranks of lines both in the solar and in gaseous spectra. In those several sound and stalwart *opera*

[1] At Lisbon, in 1877–8, with glass prisms (the whole visible solar spectrum), Edinburgh *Philosophical Transactions*, vol. xxix. 1880 ; in Madeira, in 1881, with a Rutherford's diffraction grating (21 special "subjects," or small regions of the solar spectrum) ; "Madeira Spectroscopic," Edinburgh, 1882 ; and at Winchester, in 1884, with a Rowland's diffraction grating (the whole visible spectrum, mapped *thrice*), Edinburgh *Philosophical Transactions*, vols. xxxii. 1886.
[2] Edinburgh *Philosophical Transactions*, vols. xxx. Part 1, and xxxii. Part 1.
[3] *Proceedings* of the Royal Society, vol. liv. p. 200, October 1893.

magna of spectroscopic explorations, we may surely feel convinced that after-times will neither fail to be gratified with results of scientific consequence, nor find it easily possible to overlook the great accessions made by the late Prof. Piazzi Smyth to spectroscopic science, by his boldly-planned recourses to, and ingeniously contrived employments of, great optical power and very high dispersions.　　　　　A. S. HERSCHEL.

UNIVERSITY AND EDUCATIONAL INTELLIGENCE.

OXFORD.—At a meeting of the Junior Scientific Club, held on Friday, May 18, Dr. Mann gave a *résumé* of the history of the nerve cell from Malpighi's and Leeuwenhoek's time up to the present. He showed the advance due to the introduction of new methods, viz. chromic acid by Hannover, Golgi's and Ehrlich's methods, and the Picro-corrosive sublimate method of the author. Ehrenburg in 1833 discovered the nerve-cell, Beale in 1863 the nerve-fibril, Flamming and Nissl the basophil substance of the cell, which, as Mann was the first to show, becomes used up during functional activity. Hohngren's observations on material fixed by Mann's methods had demonstrated the universal occurrence of intracellular lymph channels in nerve-cells. Finally it was suggested that the basophil substance (Nissl's bodies) should rather be regarded as the homologues of Zymogen granules than as reserve material in the strict sense of the word.—Mr. A. D. Darbishire (Balliol) showed a number of living crustaceæ by microscopic projection on to a transparent screen.

CAMBRIDGE.—The honorary degreès in law, science, and letters, were conferred in presence of a large and brilliant assemblage on June 12. Prof. Langley was unable to arrive 'from America in time to be present, but the American Ambassador, Lord Rosse, Sir Benjamin Baker, Sir Walter Buller and Prof. Poincaré attended, and received a cordial welcome. The following are the speeches delivered by the Public Orator, Dr. Sandys, in presenting to the Vice-Chancellor the under-mentioned recipients of honorary degrees for distinction in science :—

THE RIGHT HON. THE EARL OF ROSSE, LL.D.

Assurgit proximus Universitatis Dubliniensis Cancellarius, cuius pater munere eodem ornatus atque etiam Regiae Societatis praepositus, Hiberniâ in mediâ instrumentum ingens stellis observandis olim construxit ; cuius de fratre autem, navis cellerimae inventore sollertissimo, omnibus notum est "quo turbine torqueat" undas. Ipse famam inter peritos adeptus est, non modo de lunae calore subtilius inquirendo, sed etiam stellarum nebulis remotissimis (ut Aristophanis a Nubibus aliquantulum mutuemur) δμματι τηλεσκόπῳ observandis. Habetis exemplar domus praeclarae scientiarum amore conspicuae, cuius caput dignitatis Academicae heres dignissimus exstitit. Qui abhinc annos octo Universitatis suae inter ferias saeculares tot honores in alios contulit, hodie fortasse nobis ignoscet, quod bonos ipsi olim debitus praecepto illi Horatiano nimium paruisse visus est :—"nonum prematur in annum."

SIR BENJAMIN BAKER, SC.D.

Quantum miratus esset historiarum scriptor, Gaius Cornelius Tacitus, si providere potuisset fore aliquando, ut Caledoniae fretum, Bodotriae nomine nihi notum, duobus deinceps pontibus immensis iungeretur ! Quantum miratus esset historiae pater ipse, Herodotus, si audivisset fore aliquando, ut vir quidam, ab insulis Britannicis sibi prorsus ignotis oriundus, fluminis Nili aquas redundantes duplici mole et aggere magno coerceret et Aegypti regioni immensae irrigandae conservaret ! Operis utriusque magni conditorem magnum hodie praesentem contemplamur, qui non pacis tantum triumphis contentus, velut alter Archimedes, etiam Martis tormentorum inventor et machinator admirabilis exstitit. Atqui ne Martis quidem inter opera pacis causam revera deseruit ; etenim scriptoris antiqui de re militari monitum non ignotum est :—"qui desiderat pacem, praeparet bellum."

SIR WALTER LAWRY BULLER, SC.D.

Coloniae nostrae remotissimae, Novae Zealandiae, inter decora conspicua numeratus, adest hodie vir regionis illius indigenarum linguae imprimis peritus, cui, propter merita eius egregia, gratiae saepenumero publice sunt actae. Adest vir qui etiam regionis illius avibus summâ curâ describendis atque arte eximiâ depingendis opus magnum dedicavit. Quantum autem liberalitati eius etiam nostra Academia debeat, Musei

nostri parietes, avium et animalium aut prorsus aut prope extinctorum exemplis ornati, satis clare loquuntur. Ergo quem ipsa Regina, quem et Gallia et Germania et Italia honoribus cumulaverunt, eundem etiam nosmet ipsi, tot munerum non immemores, laudis nostrae diademate decoramus.

M. HENRI POINCARÉ, SC.D.

Sequitur scientiarum Academiae Gallicae socius illustris, scientiae mechanicae caelestis inter Parisienses professor insignis, societatis Regiae Londinensis inter socios exteros olim numeratus, astronomorum denique a societate Regia numismate aureo nuper donatus. Quantam laudem meritae sunt investigationes illae subtilissimae, de aestuum maritimorum natura universa, de molium liquidarum sese rotantium aequilibrio, de planetarum denique et satellitum cursu vario, ad exitum felicem ab hoc viro perductae ! Studiorum mathematicorum in utrâque provinciâ, et analyticâ et physicâ, propter scientiae suae prope infinitam varietatem inter principes numeratus, quam egregie nuper ostendit quantum provinciae illae vicinae invicem inter sese deberent ! Quam pulchre studiorum suorum voluptatem cum artis musicae et artis pingendi voluptate comparavit ! Quam ingenue mathematicam physicam confitetur novam quandam linguam desiderare ; linguam cotidianam nimis exilem, nimis ambiguam esse, quam ut aliquid tam delicatum, tam subtile, tam varium, possit exprimere. Et nos idem hodie libenter confitemur : viro tali pro meritis eius tam variis laudando lingua nostra vix sufficit. "Conamur tenues grandia."

THE Knightsbridge Professorship, vacant by the resignation of Dr. Sidgwick, will be filled up on Saturday, June 30. Candidates are required to send their names to the Vice-Chancellor by June 25.

Mr. G. F. C. Searle has been appointed a university lecturer in experimental physics ; and Dr. G. H. F. Nuttall, university lecturer in bacteriology and preventive medicine.

The first award of the Raymond Horton-Smith Prize has been made to Mr. A. B. Green, for his M.D. thesis on amyloid disease.

Mr. W. A. Macfarlane-Grieve, M.A., of Oxford and Cambridge, of Impington Park, has offered to the University a farm of about 145 acres near Cambridge, free of rent till Michaelmas 1909, for the purposes of the Department of Agriculture. This handsome offer has been gratefully accepted on behalf of the Board of Agricultural Studies, to whom the management of the experimental farm is assigned.

The *University Reporter* for June 12 contains an interesting report on Mr. W. W. Skeat's exploring expedition to the Malay provinces of Lower Siam.

The Right Hon. A. J. Balfour, M.P., will deliver the inaugural address to the students of the Vacation Courses, arranged by the University Extension Syndicate, at 12 noon on Thursday, August 2.

The Arnold Gerstenberg Studentship in moral philosophy for graduates in natural science has been awarded to Mr. T. J. Jehu, of St. John's College, who holds a Heriot Fellowship in geology from the University of Edinburgh.

The Mathematical Tripos list is unusually short this year. The sixteen wranglers are headed by Mr. J. E. Wright of Trinity, Mr. A. C. W. Aldis of Trinity Hall being second wrangler. An Indian student, Mr. Balak Ram of St. John's is fourth ; and Miss W. M. Hudson of Newnham College, sister of the senior wrangler of 1898, is bracketed eighth wrangler. Miss E. Greene, also of Newnham, is equal to tenth. St. John's claims five of the wranglers, Trinity four, Clare two.

In Part II., the bracketed senior wranglers of last year, Mr. Birtwistle of Pembroke, and Mr. Paranjpye of St. John's, are placed with two others in the first division of the first class.

MR. CHAMBERLAIN, Chancellor of Birmingham University, has received a letter from Mr. W. H. Foster, Apley Park, Bridgnorth, offering a donation of 2000*l*. towards the endowment fund.

MR. J. G. CLARK, whose death is announced from Worcester, Massachusetts, was a generous promoter of the interests of education in the United States. By his efforts the Clark University at Worcester, Mass., was founded "to increase human knowledge and transmit the perfect culture of one generation to the ablest youth of the next ; to afford the highest education and opportunity for research." He gave a close study to the subject of the higher education, and was anxious to include in his proposed University the best features to be found in institutions in America and elsewhere. At the foundation of the University in 1889, Mr. Clark gave it an endowment of one million dollars, to which he added a like amount later on. By his death, the institution receives his magnificent library of rare and costly books. Clark University is perhaps unique among the educational institutions of the United States. It is devoted entirely to post-graduate studies, and recently celebrated its tenth anniversary.

As announced last week, a series of festivities began at Cracow, on June 7, in commemoration of the 500th anniversary of the foundation of Cracow University. A Reuter correspondent states that a large number of men of science, including representatives of most of the European universities and colleges, attended the celebration. The Austrian and foreign investigators went in procession on Thursday morning to the Church of St. Mary, where a Papal Brief in reference to the celebration was read. The graves of the founders of the University were visited and wreaths deposited upon them. At the special commemorative meeting subsequently held, speeches in Latin were delivered by Prof. Tarnowski, the rector, and Dr. von Hartel. An illuminated address was presented by a deputation from Oxford University. The proceedings terminated with the distribution of the diplomas of honour to those upon whom the honorary degree of doctor has been conferred.

THE new Directory of the Board of Education, South Kensington, containing regulations for establishing and conducting science and art schools and classes, has just been published. Many of the regulations have been modified, more particularly those referring to administrative matters and practical work. The syllabus of practical mathematics has been revised, but the subjects remain much the same as were prescribed in last year's syllabus. A syllabus of an advanced stage of practical mathematics has been added. The syllabus of mineralogy has been slightly modified and recast. The laboratories in a School of Science are to be available for preparation work by students of the school beyond the school hours of the time-table. Courses of work for Schools of Science in rural districts have been added. The obligatory subjects of the elementary course for men are :— (1) mathematics ; (2) chemistry (with practical work) ; (3) physiography (Section I.) or elementary physics (with practical work) ; (4) biology (Section I.) or elementary botany (practical work may be in the field or garden) ; (5) drawing, practical geometry, or practical mathematics. Manual instruction in its application to workshop and garden must also form part of the course, which is intended to cover two years. The elementary course for women in Schools of Science differs slightly from the foregoing. Physics and chemistry are optional for the second year, and hygiene may be taken instead of botany. Practical mathematics is not included. Separate advanced courses of work are prescribed for men and women who have passed through the elementary courses. Managers of schools are now allowed the option of having the grant for practical chemistry in the advanced course assessed by the inspector or on the results of examination in the advanced stage. The announcement made last year that examinations in the elementary stages of science and art subjects will only be held upon special application, in which case a fee for each paper asked for will be charged, is ratified. This probably means the abolition of examination in the elementary stages; for apparently nothing can be gained by arranging for the examination of candidates.

SOCIETIES AND ACADEMIES.
LONDON.

Physical Society, June 8.—Dr. J. H. Gladstone, F.R.S., Vice-President, in the chair.—A paper on the magnetic properties of alloys of iron and aluminium (Part ii.), by S. W. Richardson and L. Lownds, was read by Dr. Richardson. Experiments have been made to ascertain in what way the hysteresis loss between given limits of the field strength is connected with the temperature for an alloy containing 3·64 per cent. of aluminium. The experiments show that the hysteresis loss attains a maximum value at a temperature considerably higher than the temperature of maximum induction. The changes produced in the magnetic properties of the alloy by heating and subsequent cooling have also been investigated.

The properties depend largely on the previous history of the specimen, but there does not appear to be any essential difference between the behaviour of the alloy during heating and cooling (except near the temperature of minimum permeability). Experiments have also been conducted on the abrupt change in the permeability that takes place at a temperature of about 650° C. The conclusions arrived at are as follows :—(1) The hysteresis loss at first diminishes as the temperature rises. It then increases and reaches a maximum at about 550° C. On further heating it falls off rapidly, and is negligible at 700° C. (2) The magnetic properties of the specimen depend largely on its previous history. (3) There is no essential difference between the behaviour during heating and cooling except near the temperature of minimum permeability. (4) An abrupt increase in the permeability takes place at about 650° C. during heating, followed by an equally abrupt diminution on further heating. (5) This abrupt change is more marked with falling than with rising temperatures. (6) Continued heating and cooling diminish the permeability. (7) The curve connecting temperature of minimum permeability and percentage of aluminium is a straight line. (8) The microscopic examination of the specimens shows the presence of crystals. Prof. S. P. Thompson asked if the specimens had been kept for any length of time at a high temperature, because crystals changed and grew in metals at temperatures even far below their melting points. Prof. Reinold asked if any specimens had been examined where the crystalline structure had not been observed. Mr. Blakesley asked if any explanation of the orientation of the crystals could be given. The chairman said it was difficult to know exactly what substances were being dealt with. They might be pure alloys or mixtures of two or three alloys with iron or aluminium. Dr. Richardson in reply said the crystals might be dissolved in nitric acid and analysed, but at present he did not know their composition.—Mr. W. Campbell then read a note on crystallisation produced in solid metal by pressure. In the preparation of sections of tin, particles cling to the file and, if allowed to remain, tend to tear the surface of the metal. The effect is not immediately noticeable, but on etching the polished surface there appears, besides the usual structure of the tin, lines of much smaller crystals with irregular boundaries but possessing different orientation. The effect is only superficial because it can be removed by polishing. The same behaviour is noticed in some alloys, and it would thus appear that the pressure of a file is sufficient to cause a metal or an alloy to re-arrange itself. Prof. S. P. Thompson suggested that the effect might be due to local heating caused by tearing rather than to pressure. Mr. Campbell said that the effect was not due to the heating of the file, because if the file were kept perfectly clean no crystals formed. Prof. S. P. Thompson asked if scratching the surface with a diamond produced crystallisation. The author said he had tried with a sharp knife without success, but cutting with a blunt chisel produced crystallisation along the chisel mark.—A paper on the viscosities of mixtures of liquids and solutions was read by Dr. C. H. Lees. Three formulæ have been suggested for expressing the viscosity of a mixture in terms of the percentages and viscosities of its constituent parts. The first of these represents the viscosity as being the sum of a number of terms, each one of which is the product of the percentage of any constituent and its viscosity. The second formula represents the logarithm of the viscosity in a similar manner, and the third one the reciprocal. None of these formulæ represent the viscosity of a mixture with closeness. The author suggests a formula in which the mth power of the reciprocal of the viscosity of a mixture is equal to the sum of a number of terms, each one of which is the product of the percentage of any constituent, and the mth power of the reciprocal of the viscosity of that constituent. This formula gives satisfactory agreement, and, moreover, leads to Slotte's formula for the variation of viscosity with temperature.—The secretary read a note from Prof. Wood on an application of the method of striæ to the illumination of objects under the microscope. The object chosen was powdered glass immersed in cedar oil of the same refractive index. The glass particles were almost invisible under ordinary conditions of illumination. The illuminating system here was arranged as follows :—A screen, bounded by a straight edge, was placed in front of an incandescent gas lamp, so as to cut off half of the mantle, and give a source of light bounded by one perfectly straight edge. A small lens of very short focus was placed below the stage as close as possible to the object. The lamp was at a distance of six feet, and the light reflected from the mirror was brought to a focus by this lens, passing through the object on its way. An image of the lamp was formed in space and viewed by the microscope. A little strip of thin brass, with a carefully cut straight edge, was fastened to the stand carrying the bull's eye condenser, and moved into position between the objective and object so as to cut off the flame-image with the exception of a narrow thread of light along the straight edge. The brass screen must be in the plane of the flame-image, with its edge parallel to the straight edge of the flame. The brass was then advanced over the flame until nearly all the light was cut off. Upon lowering the microscope until the object was in focus, and carefully advancing the brass strip until practically all the flame-image was cut off, it was found that the glass particles suddenly appeared with great sharpness, showing as distinctly as if in air. Two photographs of glass in oil were shown, one taken with ordinary illumination, and the other by the Schlieren-Methode.

Geological Society, May 23.—J. J. H. Teall, F.R.S., President, in the chair. — The igneous rocks of the coast of County Waterford, by F. R. Cowper Reed. The first part of this paper is devoted to a discussion of the field-evidence, as shown by the coast-sections from Newtown Head to Stradbally. The igneous rocks there exposed are divided into the following five categories :—(*a*) The felsitic rocks ; (*b*) necks of non-volcanic materials ; (*c*) the basic sills and vents ; (*d*) intrusions of dolerite ; (*e*) intrusions of trachyte, andesite, &c. ; (*f*) intrusions of other types. In regard to the age of the rocks, there appear to have been two main periods of volcanic activity : the first, in Ordovician times, was marked solely by outpourings of a felsitic nature ; the second, post-Ordovician but pre-Upper-Old-Red-Sandstone, was characterised by a succession of several distinct types of igneous rocks. The relative age of some of the peculiar types of intrusive rocks is indicated in the paper in those cases in which it can be determined. That those rocks which are later in date than the folding of the Ordovician are older than the Upper Old Red Sandstone is shown (1) by the unconformity of the Upper Old Red Sandstone ; (2) by the fact that the latter rock does not contain any interbedded igneous rocks ; and (3) by the absence of felsitic or other intrusive rocks from the Old Red Sandstone of the district. The second part of the paper is devoted to petrological notes on the different rock-types.—On a new type of rock from Kentallen and elsewhere, and its relation to other igneous rocks in Argyllshire, by J. B. Hill, R.N., and H. Kynaston. [Communicated by permission of the Director-General of H.M. Geological Survey.] A rock originally described by Mr. Teall from Kentallen is used by the authors as a type round which they group a peculiar series of basic rocks discovered in several localities. The rocks consist essentially of olivine and augite with smaller amounts of orthoclase, plagioclase and biotite, while apatite and magnetite are accessory. The peculiar feature of the rocks is the association of alkali-felspar with olivine and augite, and the group is related to the shonkinite of Montana and the olivine-monzonite of Scandinavia.

Anthropological Institute, May 29.—The President in the chair.—Prof. Oscar Montelius, of Stockholm, made a communication on the earliest communications between Italy and Scandinavia. Beginning with the evidence derived from the discovery, in Denmark and Central Europe, of bronze bowls and other objects of Roman date, coming from the workshops in Italy, as similar to examples found at Pompeii and elsewhere, he traced the active and copious intercourse thus demonstrated, step by step backwards in time through the period of early Greek commerce at the beginning of the Iron Age in the Mediterranean, into the later and earlier Bronze Age ; and illustrated his conclusions by a variety of classes of objects, which though originally of Italian origin and manufacture are found widely distributed in Central Europe, in Denmark and in Sweden ; and can be shown by numerous examples to have been imitated by the bronze working industry of these northern areas. Among these objects, he regarded the attenuate sword-hilts and the bucket-like *situlæ* as demonstrating this intercourse for the early centuries of the first millennium B.C. ; the transversely-grooved sword-hilts, and the simple bow-fibula as proving the same for the later half of the second millennium, corresponding with the Mycenæan Age in the Mediterranean ; the triangular dagger-shaped blades, and the imitations of open spiral torques and bracelets, as representing the earlier half of the same ;

and the rude hour-glass shaped types of cups and vessels as carrying the same argument back beyond the date 2000 B.C.— The paper was followed by a discussion.

PARIS.

Academy of Sciences, June 5.—M. Maurice Lévy in the chair.—The eclipse of the sun of May 28, 1900, at Paris, by M. Lœwy. The observations were interfered with by the state of the sky.—Total eclipse of May 28, by M. J. Janssen. An account of the work done in Spain by observers from the Observatory of Paris.—On the calorific equilibrium of a closed surface radiating outwardly, by M. Émile Picard.—Observation of the solar eclipse of May 28 at Marseilles and Algiers, by M. Stéphan. As the atmospheric conditions were extremely favourable at these two stations, good observations of the times of contact, and of the corona and solar protuberances, were taken.—Observations of the partial eclipse of the sun of May 28 at the Observatory of Bordeaux, by M. G. Rayet.—Observations of the planet (FG) (Wolf-Schwassmann, May 22), made with the large equatorial of the Observatory of Bordeaux, by MM. G. Rayet and Féraud.—On the curve of rifling in fire-arms, by M. Vallier.—The formation of the coal-measures, by M. Grand'Eury. In contrasting the two theories current for explaining the formation of coal deposits, the drift theory and the peat bog theory, the author cites instances in which, from his own observations, both influences must have been at work simultaneously.—The total eclipse of the sun of May 28 observed at Hellin, in Spain, by M. Hamy. The observations, which were entirely successful, include seven photographs of the corona. The characteristic green line of the corona, although falling within the sensitive region of the orthochromatic plates employed, gave no trace of impression in the spectrum photographs.—The total eclipse of the sun of May 28. Observations made at the Observatory of Algiers, by M. Ch. Trépied. The plan of operations included observations of the four contacts, visual study of the corona, photography of the partial eclipse, photography of the corona and of the spectrum of bright lines in the chromosphere, the photography of the spectrum of the corona, and thermo-actinometric observations. Under the very favourable atmospheric conditions, all the results were good, the only failure being in the attempt to photograph the spectrum of the corona.—On the solar eclipse of May 28, by MM. Meslin, Bourget and Lebeuf. Results of observations made at Elche. The photograph of the spectrum of the corona, obtained with a Rowland concave grating, shows circles corresponding to the lines H, K and G.—Observations of the eclipse of the sun of May 28, by M. de La Baume Pluvinel. Nine photographs of the corona were taken, but the instrument set apart for the special study of the coronal line gave no result.—On the proportion of polarised light in the solar corona, by M. J. J. Landerer. The proportion found was 0'52.—The eclipse of the sun of May 28 observed at Besançon, by M. Gruey.—The partial eclipse of the sun of May 28, at the Observatory of Lyons, by M. Ch. André. The scheme of operations included the comparison of the time of direct observation of contact with that made by projection upon a white screen, and the examination of the dark line noted in the eclipse of 1882.— The solar eclipse of May 28, observed from a balloon, by Mdlle. D. Klumpke.—On the theory of the moon, by M. H. Andoyer. —On the congruences of circles and spheres which are multicyclic, by M. C. Guichard.—On divergent series, a correction of an earlier note, by M. Le Roy.—On the decomposition of continued finite groups, by M. Edmond Maillet.—On the integration of the equation $\Delta u = fu$, by M. J. W. Lindeberg.—On the electrical state of a Hertz resonator in activity, by M. Albert Turpain.—Researches on the existence of a magnetic field produced by the movement of an electrified body, by M. V. Crémieu. According to Maxwell, a charged electrified body in motion should produce magnetic effects, the magnitude of which can be calculated from the charge and velocity of the moving body, and experiments by Rowland in 1876, and Rowland and Hutchinson in 1889, gave results in agreement with Maxwell's views. Lippmann, applying the principle of the conservation of energy to Rowland's experiment, shows that magnetic variations ought to produce a movement in electrified bodies situated within the field, but experiments by the author would appear to show that such an effect is not produced. A repetition of Rowland's original experiments, under conditions more favourable to accuracy, also leads the author to conclude that the motion of an electrified body produces no magnetic effect.—Measurement of the quantity of electricity and of electrical energy distributed by continuous currents, by MM. A. and V. Guillet.—On a mode of decomposition of some metallic chlorides, by M. Œchsner de Coninck. Gold can be completely removed from solutions of auric chloride by filtering through animal charcoal; solutions of the perchlorides of platinum and iron are also decomposed on filtration through animal charcoal. No such decomposition could be observed with the chlorides of nickel, cobalt, manganese, zinc, copper and magnesium.—On the conditions of stability of rotatory power, by M. J. A. Le Bel. It is found that at temperatures of 100° or thereabouts, many optically active bodies tend to lose their rotatory power by race-misation; on the other hand, if the asymmetric radicals grouped round a central atom are increased in volume, the stability is increased.—On the dihydroxylates, by M. de Forcrand.—Addition of hydrogen to acetylene in presence of copper, by MM. Paul Sabatier and J. B. Senderens. A mixture of hydrogen and acetylene passed over reduced copper at a temperature of 130°–200°, reacts readily, forming ethane, ethylene and other hydrocarbons, no acetylene remaining unchanged if the hydrogen is in excess.—On the copper and mercury organo-metallic compounds of diphenylcarbazone.—On acidimetry, by M. A. Astruc. A study of the behaviour of isethionic, sulphanilic, meconic and mellic acids with indicators.—On a new species of subterranean Isopod, *Caecosphaeroma Faucheri*, by MM. Adrien Dollfus and Armand Viré.—Gregarinæ and intestinal epithelium, by MM. L. Léger and O. Duboscq.—On the animal fossils collected by M. Villiaume in the carboniferous strata near Nossi-Bé, by M. H. Douville. The whole of the carboniferous strata in the region of Nossi-Bé belongs to the Upper Lias, and is to be classified with the carboniferous strata of the same age in the north of Persia.—On the vegetable fossils collected by M. Villiaume in the carboniferous beds in the north-west of Madagascar, by M. R. Zeiller. The conclusions drawn are in harmony with those drawn by M. Douville in the previous paper from a study of the animal fossils.—The volcano of Gravenoire and the mineral springs of Royat, by M. P. Glangeaud.

DIARY OF SOCIETIES.

THURSDAY, June 14.

Royal Society, at 4.—Election of Fellows.—At 4.30.—Some New Observations on the Static Diffusion of Gases and Liquids, and their Significance in certain Natural Processes occurring in Plants: H. T. Brown, F.R.S., and F. Escombe.—The Electrical Effects of Light upon Green Leaves (Preliminary Communication): Dr. A. D. Waller, F.R.S.—The Nature and Origin of the Poison of Egyptian Lotus (*Lotus Arabicus*): W. R. Dunstan, F.R.S., and T. A. Henry.—The Exact Histological Localisation of the Visual Area of the Human Cerebral Cortex: Dr. J. S. Bolton.—Data for the Problem of Evolution in Man. V. On the Correlation between Duration of Life and the Number of Offspring: Miss M. Beeton, G. U. Yule, and Prof. K. Pearson, F.R.S.—The Diffusion of Ions produced in Air by the Action of a Radio-active Substance, Ultra-violet Light and Point Discharges: J. S. Townsend.—On an Artificial Retina and on a Theory of Vision, Part I. : Prof. J. C. Bose.

Mathematical Society, at 5.30.—Some Multiform Solutions of the Partial Differential Equations of Physical Mathematics and their Applications, Part II.: H. S. Carslaw.—Some Quadrature Formulæ : W. F. Sheppard.—Notes on Concomitants of Binary Quantics : Prof. Elliott, F.R.S.—Extensions of the Riemann-Roch Theorem in Plane Geometry : Dr. Macaulay.—On the Invariants of a certain Differential Expression connected with the Theory of Geodesics : J. E. Campbell.—On the Constants which occur in the Differentiation of Theta Functions : Rev. M. M. U. Wilkinson.—On the Transitive Groups of Degree *n* and Class *n*−1 : Prof. W. Burnside, F.R.S.—The Invariant Syzygies of Lowest Order for any Number of Quartics : A. Young.—Further Notes on Bilinear Forms : T. J. I'A. Bromwich.

MONDAY, June 18.

Royal Geographical Society, at 8.30.—The Country between Lake Rudolf and the Nile Valley : Captain M. S. Wellby.

TUESDAY, June 19.

Zoological Society, at 8.30.—On the Significance of the Hair-slope in certain Mammals : Dr. Walter Kidd.—On the Anatomy of *Bassaricyon alleni*: F. E. Beddard, F.R.S.—Observations on the Habits and Natural Surroundings of Insects and other Animals, made during the "Skeat" Expedition to the Siamese Malay States : Nelson Annandale.

Royal Statistical Society, at 5.—The Defence Expenditure of the Empire : The Right Hon. Sir Charles W. Dilke, Bart.

Mineralogical Society, at 8.—On Conchite, a New Variety of Calcium Carbonate : Miss Agnes Kelly.—On the General Determination of the Three Principal Indices of Refraction from Observations made in any Arbitrary Zone : G. F. Herbert Smith.—On Monazite from Tintagel : H. L. Bowman.—On the Oxidation of Pyrites by Underground Water : Dr. J. W. Evans.—Petrological Notes : G. T. Prior.—A Quantitative Determination of the Action of Hydrochloric Acid and Soda Solution on the Enstatite and Felspar of the Mount Zomba Meteorite : L. Fletcher, F.R.S.

WEDNESDAY, June 20.

Geological Society, at 8.—On the Skeleton of a Theriodont Reptile from the Baviaans River (Cape Colony): Prof. H. G. Seeley, F.R.S.—On

Radiolaria from the Upper Chalk at Coulsdon (Surrey): W. Murton Holmes.—Fossils in the Oxford University Museum. IV. Notes on some Undescribed Trilobites : H. H. Thomas.

Royal Meteorological Society, at 4.30.—Rainfall in the West and East of England in Relation to Altitude above Sea-level : William Marriott.—Description of Halliwell's Self-recording Rain Gauge : Joseph Baxendell.

Royal Microscopical Society, at 8.—Demonstration on the Structure of some Palæozoic Plants, with Sections of the Plants shown by the Lantern : W. Carruthers, F.R.S.

THURSDAY, June 21.

Royal Society, at 4.30.

Linnean Society, at 8.—On some Scandinavian Crustacea : Dr. A. G. Ohlin.—The Subterranean Amphipoda of the British Islands : Chas. Chilton.—On certain Glands of Australian Earthworms : Miss Sweet.—Notes on Najas : Dr. A. B. Rendle.

Zoological Society, at 4.30.—The Gigantic Sloths of Patagonia : Prof. E. Ray Lankester, F.R.S.

Anatomical Society (Owens College, Manchester), at 10.30.—Lantern Demonstration on the Comparative Anatomy and Histology of the True Cæcal Apex—the Appendix Vermiformis : Dr. R. J. Berry.—Lantern Demonstration of some Surface Markings of the Calvaria, and their Significance : Prof. Dixon.—Lantern Demonstration of Microphotographs of the Maturation Stages in the Ovum of Echinus : Dr. T. H. Bryce.—Some Points in the Anatomy of the Digestive System : Prof. Birmingham.—(*a*) Two Cases of Absent Vermiform Appendix ; (*b*) A Specimen showing Direct Continuity between the Long External Lateral Ligament of the Knee-joint and the Peroneus Longus Muscle ; (*c*) A Supernumerary Bone in the Carpus connected with the Trapezium : Prof. Fawcett.—A Note on the Genital Apparatus of the Jerboa : Dr. Armour.

Chemical Society, at 8.—Ballot for the Election of Fellows.—Notes on the Chemistry of Chlorophyll : Dr. L. Marchlewski and C. A. Schunck.—Researches on Morphine, I.: Dr. S. B. Schryver and F. H. Lees.—A New Series of Pentamethylene Derivatives, I. : Prof. W. H. Perkin, jun., F.R.S., Dr. J. F. Thorpe, and C. W. Walker.—Experiments on the Synthesis of Camphoric Acid. III. The Action of Sodium and Methyl Iodide on Ethyl-dimethyl-butanetricarboxylate ; Prof. W. H. Perkin, jun., F.R.S., and Dr. J. F. Thorpe.—On the Oxime of Mesoxamide and some Allied Compounds : Miss M. A. Whiteley.—The Oxyphenoxyn and Phenylenoxy-acetic Acids : W. Carter and Dr. W. T. Lawrence.—(1) The Condensation of Ethyl α-Bromo-isobutyrate with Ethyl Malonates and Ethyl Cyanacetates : α-Methyl-α-isobutylglutaric Acid ; (2) Methylisoamylsuccinic Acid, II.: Dr. W. T. Lawrence.

FRIDAY, June 22.

Physical Society, at 5.—Notes on Gas Thermometry : Dr. P. Chappuis.—A Comparison of Impure Platinum Thermometers : H. M. Tory.—On the Law of Cailletet and Mathias and the Critical Density : Prof. J. Young, F.R.S.

Anatomical Society (Owens College, Manchester), at 10.30.—Note on the Configuration of the Heart in a Man and some other Mammalian Groups : Dr. C. J. Patten.—On the Arrangement of the Pelvic Fascia and their Relationship to the Levator Ani : Dr. Peter Thompson.—(*a*) A Preliminary Note on the Development of the Sternum ; (*b*) Specimens of Diaphragmatic Hernia and of a Left Inferior Vena Cava : Prof. Paterson.—Preparations and Lantern Slides illustrating : (*a*) The Anatomy of the Subclavian and Axillary Arteries ; (*b*) The Position and Relations of the Eustachian Tubes : (*c*) Stereoscopic Views of Anatomical Preparations : Dr. Arthur Robinson.—A Series of Microscopical Preparations illustrating the Development of the Posterior End of the Aorta : Prof. Young and Dr. Arthur Robinson.—Demonstration of a Series of Preparations of the Posterior End of the Adult Aorta : Prof. Young.

CONTENTS.

THURSDAY, JUNE 21, 1900.

THE REMINISCENCES OF A VETERAN OF SCIENCE.

Erinnerungen aus meinem Leben. By A. Kölliker. Pp. x + 399 ; with 7 plates, 10 text figures, and portrait of the author. (Leipzig : W. Engelmann, 1899.)

THE memoirs of the venerable Professor of Anatomy at Würzburg will interest a wide circle of readers in this country, whether amongst the older generation of scientific men, whose privilege it has been to know the author as a genial friend and colleague, or amongst the juniors in rank and years, to whom the name of Kölliker has been one which from their youth upwards they have learnt to respect as that of a great leader in scientific thought and discovery. Many of the latter class may, perhaps, learn from this book, for the first time, how much modern zoology owes to its author. So rapid has been the advance of biological science in the latter half of the nineteenth century, and so great is the interval, judged not by time, but by the progress of knowledge, which separates the science of to-day from that of fifty years ago, that there is always considerable danger of the merits of those who have grown grey in the ranks of science being overlooked or insufficiently realised by the younger generation. Students are taught at an early stage in their career facts or principles which seem so well established or even self-evident, in the light of current knowledge, that it is quite an awakening to find that the man who first enunciated them is still living in our midst. To give one instance, a student of zoology is taught, probably in the very first lecture he attends, the distinction between Protozoa and Metazoa based upon the essentially unicellular nature of the individual in the former sub-kingdom. If he reflects at all on the matter, a truth so obvious and so easily demonstrated will seem to him one which has been recognised by mankind perhaps from a remote antiquity. Yet it was Kölliker who first, in 1845, pointed out the existence of unicellular animals, and brought forward the Gregarines as instances, and who later, in conjunction with von Siebold, expressed the opinion that all the Infusoria, with the exception of such forms as the Rotifers, consisted of single cells. In a further work upon *Actinophrys* this conclusion was extended to the Rhizopods, and so a great generalisation was established, the truth of which is now never called in dispute.

Quite apart, however, from the great interest which these memoirs possess from the scientific point of view, their appearance at the present time is welcome for other reasons. At a period of strained political relations, when our country appears isolated in aims and sympathies from the rest of Europe, when international antipathies and prejudices seem in a fair way to spread from the official to the personal sphere, it is a refreshing change to read the narrative of one who was a frequent and a welcome visitor in our midst. To judge, at least, from the prevailing tone of this book, its author is no "Britenfresser." He refers constantly with warmth, we might say with affection, to the hospitality of his many friends in this country and to the pleasant times he spent

in their homes, feelings which, we can be sure, were as warmly reciprocated by those about whom he writes.

The book is divided into two parts, the one personal, the other scientific. Part i. contains a general account of his life, with details of his many scientific and other journeys, and a brief account of his relations to various learned societies. Part ii. may be described as a *catalogue raisonnée* of his works, and is a marvellous record of many-sided scientific activity. His publications, amounting to nearly 250 memoirs, are arranged under the headings of histology, anatomy, physiology, embryology, evolution, comparative anatomy and zoology, and other miscellaneous items. Under each subdivision is given a historical account of his work, its main results, the ideas which guided him, and the conclusions which he upheld. Here much will be found of great value to the student—using the word in its widest sense—which cannot be dealt with adequately within the limits of a brief review. We turn, therefore, with greater interest to the personal narrative set forth in Part i.

Rudolf Albert Kölliker was born at Zürich on July 6, 1817. His boyhood and schooldays were passed in his native town, and he was intended at first for a business career, but, fortunately for science, this idea was given up, and he entered Zürich University, in 1836, as a medical student. At the University his attention was first given to botany, a subject in which he had as fellow-student his intimate friend, Carl Nägeli, and his first publication (1839) was a list of the phanerogams of Zürich. Besides other medical and scientific courses he attended the stimulating lectures of Oken on zoology and nature-philosophy. In 1839 he spent a semester at Bonn, and attended lectures on surgery and kindred subjects which were still delivered in Latin. He next went to Berlin for three semesters, from 1839 to 1841, a period which he describes as a turning-point in his life, since here he came under the influence of two great masters, whose courses he attended—namely, Johannes Müller and Jakob Henle. Of the former, he writes : "the comprehensive outlook by which he connected forms widely separated, and showed what they had in common, was especially stimulating and, for me, new." From Henle on the other hand, he received his first introduction to the cell-theory of Schwann, and his attention was directed to the structure of the animal body in a number of lectures and demonstrations which he describes with enthusiasm :—

"Now when the youngest medical student is acquainted with all this and much more from pictures of all kinds, and the facts concerning the minutest structure of the body are in every one's mouth even at school, it is not easy to realise the impression made upon the student at that time by the first sight of a drop of blood, a ciliated lining, a section of bone or a striped muscle fibre, and the impress of these experiences remains permanently in the memory."

Besides Müller and Henle, he attended many other eminent teachers at Berlin, including Ehrenberg and Remak. From the latter he received his first demonstrations of the embryology of the chick. In spite, however, of his ardent medical and scientific pursuits, he found time to attend lectures on ethics and Hegelian philosophy. It was a result doubtless of Henle's influence that his first

T

anatomical memoir was an investigation upon spermatozoa, published in 1841, with which he took his degree of philosophy at Zürich in 1841, and of medicine at Heidelberg in 1842. In the former year he passed his State examination, of which he records the following contretemps :—

" I, who had at my fingers' ends the finest ramifications of the cranial nerves, the structure of the auditory labyrinth, of the eye, the brain, and so forth, was unable to answer a question on the portal vein."

This is an experience which will assuredly come home to many, and while hence eliciting our sympathies, will at the same time afford no slight consolation to those who reflect on the subsequent achievements of the unfortunate examinee.

In 1841 Kölliker was appointed assistant to Henle, who had received the chair of anatomy at Zürich. In the following year he took a trip to Naples, where he made the acquaintance of Delle Chiaje, Costa and Krohn, and occupied himself with, amongst other things, his well-known studies on the development of Cephalopods. In 1843 he became docent at Zürich, and was prosector to Henle from 1842 until the latter's promotion to Heidelberg in 1844. Henle's chair was then divided into one of anatomy and one of physiology, and Kölliker received the latter; but in 1847 he accepted a call to Würzburg. His departure from Zürich, which was much regretted there, was largely caused by political intrigues in the faculty of the University.

At Würzburg he occupied, at first, the chair of comparative anatomy, but in 1849 he received that of anatomy, which he has now filled for more than fifty years, in a way that needs no praise. The names of many of the most eminent professors of anatomy in Germany, past or present, are to be found in the lists of his pupils or assistants, of whom it is only necessary to mention C. Gegenbaur, Fr. Leydig, R. Wiedersheim, H. Grenacher and Th. Eimer. In 1848 he was associated with von Siebold in founding the *Zeitschrift für wissenschaftliche Zoologie,* of which famous journal he is still one of the editors.

The accounts of his many journeys are compiled, for the most part, from letters written by him at the time to his relations or friends. There is much of interest to be found in them, especially in his visits to England. His first acquaintance with this country was made in 1845, and renewed on many subsequent occasions. In his letters he gives his impressions of England and English life. He quickly made for himself a large circle of intimate scientific friends, amongst whom he mentions, particularly in his earlier letters, the names of Todd, Bowman, Grant, Sharpey, Edward Forbes and Wharton Jones. His time in London seems to have been very well filled up, as he writes in one letter that in the last twelve days he had gone through nine dinners and two breakfasts, some of which do not seem to have been very entertaining. " I took part yesterday in a fearfully wearisome dinner, enough to kill one (etwas ganz totmach-endes)," he writes ; and further on he complains that " these everlasting dinners, lasting from 6 to 11 o'clock, have taken me *en grippe,* as the French say ; but what can one do?" But in other cases he seems to have been happier. In London he is presented at Court, and

finds that "the Queen is really pretty, and Prince Albert is also a handsome man." On the eve of his departure, he expresses himself almost as much at home in London, in spite of its size, as in Zürich, and considers it "very interesting, often pleasant, but for the most part fatiguing." He visited this country again in 1850 and 1857, on both of which occasions he spent some, or most, of the time in Scotland, where he became intimately acquainted with John Goodsir and Allen Thomson, and in London with Queckett. His letters from Scotland to C. Th. von Siebold contain some interesting remarks about English science and scientific men.

" The English doctors and physicians are, above all, practical men, and all that pertains to the theoretical side takes with them the second place. This is partly owing to the fact that the English are a people occupied chiefly with commerce, but only partly so ; the chief cause of the phenomenon in question is the fact that science does not hold the place it deserves in popular estimation, nor is it supported by the Government in such a way that a man who devotes himself to it can be free from care."

This is the reason, he thinks, why so many men full of enthusiasm for science remain in practice, and finally lose themselves in it ; while others regard theoretical studies merely as an advertisement to gain them more clients, since practice in England is golden, and procures for the practitioner a position which contrasts vividly with that of a professor.

" I know only three anatomists and physiologists in England," he adds, " who do not practise—namely, Owen, Sharpey and Grant, of whom Owen alone has a position at all equal to his merits."

In 1850 he also paid a short visit to Oxford, where he met Acland, Strickland and J. V. Carus, but found little that attracted him, and he returned, he tells us, to noisy but infinitely more stimulating London, well satisfied that he was not obliged to spend all his days in "this most peculiar of all university towns."

Space does not permit of reference to the many interesting personal reminiscences or amusing incidents which recur so frequently in this book, especially that detailed in two letters on p. 162, of which we lose nothing by its being to a large extent veiled in the obscurity of the English tongue. It can only be said that the book affords delightful reading, and gives pleasing glimpses of a warm-hearted and charming personality as well as of a great man of science.   E. A. M.

DIFFERENTIAL EQUATIONS.

Theory of Differential Equations. By A. R. Forsyth, Sc.D., F.R.S. Part i. (1890). Pp. xiv + 340. Part ii. (1900). Pp. xii + 344, and x + 392. (Cambridge : At the University Press.)

ALTHOUGH these volumes contain more than a thousand pages, it would be premature to express an opinion upon the plan and proportions of Prof. Forsyth's work as a whole ; so much of his vast subject still remains unrepresented. Thus the reader will find nothing, except incidentally, of the theory of partial differential equations ; and, what is more remarkable, the subject of ordinary linear equations has been reserved for a future volume. However, the two parts which have now been published are so far complete in themselves that it is possible to give some account of their contents, and to appreciate, to some extent, the author's method and point of view.

Part i. treats of exact equations and the problem of Pfaff. Of the two chapters on exact equations it is enough to say that they contain an excellent summary, with well-chosen examples, of the various methods which have been suggested ; the most interesting part is that which deals with Mayer's very remarkable extension of Natani's procedure.

The rest of vol. i. is devoted to Pfaff's problem. A chapter on the history of the problem is followed by ten others, which give, in the order of their discovery for the most part, the principal results of Pfaff, Jacobi, Natani, Clebsch, Grassmann, Lie and Frobenius. This plan has its advantages, especially for those who wish to become familiar with the literature of the subject ; and mathematical experts will duly appreciate the service which Prof. Forsyth has done them. But if we look at the result as a text-book for mathematical students, it is a question whether the course taken is the best one. A chapter which is an excellent guide to a reader who has before him the original book or memoir upon which it is based, may be simply puzzling to a student unfamiliar with the subject, and unable to refer to the primary sources. It is doubtful, for instance, if any one who has not mastered the *Ausdehnungslehre* will be able to appreciate the chapter on Grassmann's method ; and in the same way, the chapters on tangential transformations and Lie's method will not, we fear, do much, in themselves, to arouse an interest in Lie's magnificent discoveries. It is unfortunate that Prof. Forsyth's exclusively analytical attitude has prevented him from utilising Lie's geometrical or quasi-geometrical conceptions. It is quite true that intuitional methods require to be controlled by strict analysis ; but they often vivify a mathematical theory in a very instructive and fruitful way. Take, for instance, the question of the "integral equivalent" of the differential relation $Pdx + Qdy + Rdz = 0$, where P, Q, R are functions of x, y, z. If we take x, y, z as ordinary Cartesian co-ordinates, this relation associates with any point $A(x, y, z)$ a flat pencil of elementary line-elements, concurrent at A, and lying in a definite plane $P(\xi - x) + Q(\eta - y) + R(\zeta - z) = 0$. Thus we may take the "content" of the differential relation to be either a manifold of ∞^1 line-elements, or of ∞^3 plane-elements. If the given relation is an "exact equation" $d\phi = 0$, the integral $\phi = c$ gives us a family of ∞^1 surfaces, each of which contains ∞^3 line-elements of the content and ∞^2 plane-elements of it. Moreover, every continuous curve made up of line-elements lies (in general) on one of the integral surfaces $\phi = c$, and the line and plane elements of the surfaces exhaust the corresponding elements of the content. These considerations justify us in saying that $\phi = c$, with c an arbitrary constant, is a complete integral equivalent of the differential relation. But in a case like $xdx + zdy - ydz = 0$, we cannot construct an integral equivalent of this kind ; and the question arises, what integral equivalent, if any, exists, and what will be the nature of its equivalence ? To Prof. Forsyth, this is a purely analytical question ; he simply inquires what functional

relations connecting x, y, z are consistent with the given relation. Of the degree and nature of the equivalence to be expected he says very little ; and the gist of what he does say is relegated to a note on p. 250. The geometrical theory at once suggests the possibility of constructing "integral curves" by linking line-elements of the content ; a complete integral equivalent may be conceivably constructed by a system of ∞^3 integral curves together exhausting all the line-elements of the content, or again by ∞^2 integral curves, each with ∞^1 associated plane-elements of the content. As an example of the latter kind of integral equivalent, the system of lines

$$x = a, \quad y = bt, \quad z = t$$

where a, b are arbitrary constants, and t is a variable parameter, are integral curves derived from the content of $xdx + zdy - ydz = 0$; and if with each point (a, bt, t) we associate the elementary flat-pencil which lies in the plane $a(x - a) + t(y - bz) = 0$, we have a complete integral equivalent, all the elements of the content being taken into account. If we take the two analytical relations $x = a, y = bz$, involving arbitrary constants only, we get, it is true, a kind of integral equivalent ; but this is not complete, in any sense analogous to the complete integral of an exact equation.

Part ii. deals with ordinary equations, not linear ; and the point of view is almost entirely that of function-theory. The coefficients in the equation are analytical functions, in Weierstrass's sense ; and the main problem is that of discussing the functional nature of the dependent variable or variables. The discussion is necessarily based upon the work of Cauchy, Briot and Bouquet, Weierstrass and Fuchs ; the analysis is simple enough in essence, but the details, unfortunately, are unavoidably lengthy, and tend to be monotonous, owing to the necessity of considering different cases and establishing a set of typical forms. The results are so important that the student is bound to make himself familiar with them ; but the judicious reader will do well to use his privilege of skipping. The fact is, that the demonstrations fall naturally into a very few types ; and it is as profitless to study every one of them minutely as to attempt a detailed examination of every kind of singularity of an algebraic curve. There are, of course, many points in the analysis which cannot fail to arouse interest and admiration ; for instance, the use of a dominant function in proving the existence-theorem, and the employment of a sort of extended Puiseux diagram in the applications.

Then, again, there are those surprisingly general and definite results which have been deduced, almost as corollaries, from this somewhat unattractive analytical theory. It must suffice to refer to Painlevé's theorem (ii. p. 211), that the points of indeterminateness of every integral of a single equation of a certain very general type are *fixed points* determined by the differential equation itself ; and to the result established by Bruns (iii. p. 311 and following), that every algebraic integral of the differential equation of the problem of three (or more) bodies can be constructed algebraically from the long-known classical integrals. But the reader will find other results of almost equal interest due to Poincaré, Fuchs, Picard and others. The reaction of the Weierstrassian function-theory upon other branches of analysis, and in particular upon the problems of celestial mechanics, is truly remarkable.

It is to be hoped that the publication of Prof. Forsyth's work will make English mathematicians better acquainted with current research on the subjects with which he deals. The value of his treatise for really competent readers is evident, and needs no commendation. But we may, perhaps, regret that he has not more definitely considered the interests of the rising generation. It is most important that new ideas and recent methods should be introduced to young men of ability while their minds are keen and susceptible ; and their interest is seldom aroused in the first instance by a treatise which aims at being exhaustive. To take an example in point ; few readers, we imagine, to whom the subject was new, would persevere in the study of Lie's great work on transformation-groups ; yet what mathematical student could fail to be delighted with his lectures on differential equations with known infinitesimal transformations, as edited by Dr. Scheffers ?

No doubt the task of writing an introductory, and thoroughly didactic, treatise on the modern aspects of this theory is very difficult ; more so, very likely, than the one to which Prof. Forsyth has applied himself. The selection, combination and assimilation required would demand a great deal of care and judgment ; a certain lightness of touch would also be desirable, and this is not easy to maintain after a course of reading in the extremely ponderous memoirs which are so often found in the literature of the subject. But a work of this kind might do more than the most conscientious handbook to encourage a living interest in the theory of differential equations. There is some appearance of a tendency to over-elaboration in English treatises presumably written for students ; to authors as well as to lecturers may be commended the maxim " Above all, do not be dull."   G. B. M.

OUR BOOK SHELF.

Origin and Character of the British People. By Nottidge Charles Macnamara. Pp. 242 ; 33 figures. (London : Smith, Elder and Co., 1900.)

MR. MACNAMARA seeks, in a small compass, to indicate the origin of the component parts of the British people, and to account for the differences of local moral character by proportionate inheritance from the original races, all of which are assumed to have their mental and moral peculiarities as fixed as their physical characters. He believes that the Iberians, as he prefers to call the Mediterranean or Afro-European race, formed the primary stock from which the existing inhabitants of Great Britain and the West of Europe are derived ; and that they are the modified descendants of Palæolithic man. The tall fair Aryans originated in Western Asia. The pioneer migration of the Aryans into Europe formed the Cro-Magnon race ; then came the dolmen-builders, the South Mediterranean branch extending from the Amorites to the " fair Libyans " ; the migrants into Central Europe mixed with the brachycephals and constituted the " Celts." A distinct northern migration formed the Teutonic Aryans.

The author also believes that dolmens and long barrows are everywhere the work of the Aryan race. The pre-

historic tall brachycephals of Northern Europe were a branch of the Northern Mongolian or Turanian race. The short dark brachycephals of Central Europe brought the art of working in bronze from Asia, presumably from Burmah. The Formorians of Ireland were Iberians ; in North-west Ireland are still to be found descendants of the Northern Mongoloid race ; the Firbolgs were Celtic Aryans or dolmen-builders. The Southern Mongoloids arrived in the bronze age ; these are the Tuatha de Danann. A second invasion of Aryan Celts, or Milesians, arrived in Ireland also during the bronze age. This abstract gives a fair idea of the scope and views of the author.

The Geography of the Region about Devil's Lake and the Dalles of the Wisconsin. By Prof. R. D. Salisbury and Mr. W. W. Atwood. Pp. x+151. (Madison, Wisconsin : Geological and Natural History Survey, 1900.)

THIS is the first number of an " Educational Series " to be published by the Wisconsin Geological and Natural History Survey. The region to which attention is now particularly called is in the south-central part of Wisconsin, and it is of interest because it well illustrates many points in the geographical evolution of land-surfaces. It comprises an undulating plain chiefly of Potsdam Sandstone, with some areas of magnesian limestone, and with a northern and southern range of bold quartzite hills. The southern range rises from 500 to 800 feet above the surrounding land, or up to 1600 feet above sea-level, and in the bottom of a deep gap, which divides this range, lies Devil's Lake. This is a lake which, in glacial times, occupied an enclosure between the ice on the one hand and the quartzite ridge on the other : a gorge which originally was the work of a pre-Cambrian stream. The melting of the ice supplied abundant water, and the lake rose perhaps 90 feet above its present level. In this and in many other cases the irregular deposition of glacial drift gave rise to many depressions without outlets, in which surface-waters collected after the ice had disappeared. Few of these lakes now remain in the region, but Devil's Lake, which is more than a mile in length and half a mile wide, occupies an unfilled portion of an old river valley, isolated by great morainic dams from its surface-continuations on either hand. Streams originate beyond these dams. The " Dalles " are sandstone cliffs which form a gorge along the Wisconsin River for a length of about seven miles, and a depth of 50 to 100 feet. The effects of weathering by atmospheric agents, and of erosion by the river, are well exhibited, and the views remind us of the rock-scenery along the Eden near Corby Castle.

The volume, which, with its index, extends to 151 pages, is in reality an essay on the origin of scenery treated from a geological point of view. The authors deal with the pre-Cambrian history of the quartzite, from its origin in loose sand to its uplift and deformation ; and they deal similarly with the other strata. They contribute also a fairly full account of the phenomena of the Glacial period, and of the work of rain and rivers. Numerous excellent photographic representations of the scenery are given, including views of various natural arches, tors, and needles.

Monistische Gottes- und Weltanschauung. Von J. Sack. Pp. viii + 278. (Leipzig : Engelmann, 1899.)

IN Herr Sack's view all particular existences are modes of one spirit-substance—God. He calls this doctrine monism, and not pantheism, because he thinks the latter not incompatible with polytheism. Be this as it may, the distinguishing mark of his thesis is that it works to an Hegelian doctrine of being along the lines of a naturalistic theory of becoming that might satisfy Mr.

Spencer. The result is a form of vitalism. The movement which is to be found in the inorganic world is not merely continuous with, but synonymous with life and consciousness. Matter is not only the revelation of spirit, but body and spirit are one and the same. His method, which consists simply in the assumption that human spirit is an *analogon* of the world-principle, will not bear this conclusion. And his superstructure is rather in the air.

In his view of evolution there is nothing novel. It is, of course, teleological. Its real dynamic, as opposed to its formal occasions, is the all-inclusive being as principle of organisation. The working of this is elucidated quite after the manner of Mr. Spencer, by what Herr Sack oddly calls "antinomies"—viz. the antithesis of individuality and community, and the like.

It is when he comes to deal with art, morals and religion that Herr Sack is most at home. These are man's adumbrations of the contents of the intellectual intuition of the universal spirit : Art, like ethics, is a social product. Ethics are treated in a manner on the whole definitely Spencerian, even to the condemnation of the social-democratic movement. In his discussion of religion, Herr Sack is opposed to Mr. Spencer, and, while owing a good deal to Prof. Max Müller, is original. Not in dreams with their presentment of the dead, not in natural phenomena like sunrise and sunset, not in anything so symbolic as totemism, does the matter of religion arise. They might confirm its sublimity ; they are most of them too habitual and ordinary phenomena to create it. It is rather what suggests the invisible, the beyond, the infinite, that originates religious feeling—the horizon, the movement of the wind, the breath of life. Infinite space and infinite movement, and the *anima mundi*, are the elements of the religion of monism, and primitive religion was monistic. Cult degrades it into polytheism, and an interested priestcraft corrupts it ; but monism has never been without a witness.

A world of spirits, in the spiritualist's sense, is of course incompatible with such a view. As is individual immortality. In truth, personality other than relative can belong only to the *Allwesen*, "in whom we live, and move and have our being."

In description, Herr Sack often shows a good deal of power. His views in the field of *Religionsforschung* doubtless express something of the truth, though not to the exclusion of other explanations. Indeed, the horizon, and the wind, and breathing are habitual too ! Herr Sack's monistic formula, if true, must be established on other lines than his. Its only value here is that of any unverified vaticination that has brought peace to some of our fellow-men.　　　　　　　　　　　H. W. B.

First Stage Hygiene. By Robert A. Lyster, B.Sc.Lond. Pp. viii + 199. (London : W. B. Clive, 1900.)

IN general character this book resembles those already available for students of elementary hygiene and public health. It is intended more particularly for students receiving lessons upon the lines of the syllabus of the Department of Science and Art, now the Board of Education, but it may also be used by other students. The order of treatment differs from that usually adopted, but it may be doubted whether in some cases the change is an improvement. A noteworthy point, however, is that, so far as possible, the physiological facts required to intelligently consider hygienic principles are dealt with as they are required, instead of being described in a separate section devoted to physiology. Another characteristic of the book is that simple experiments illustrating the points described are given at the ends of some of the chapters. There is still room for a book containing not only lecture experiments, but a good course of laboratory work to be done by individual students of hygiene.

LETTERS TO THE EDITOR.

[The Editor does not hold himself responsible for opinions expressed by his correspondents. Neither can he undertake to return, or to correspond with the writers of, rejected manuscripts intended for this or any other part of NATURE. *No notice is taken of anonymous communications.]*

Measurements in Schools. Collateral Heredity.

I AM at present engaged on an investigation into the strength of collateral heredity, *i.e.* the degree of resemblance for a variety of mental and physical characters of pairs of brothers, pairs of sisters, and pairs of brothers and sisters. In this matter I cannot seek the aid of parents, for they are scarcely unbiased observers, but I have to appeal for aid to those who teach in schools, and have thus an independent and often extensive knowledge of their pupils' characters. This is very frequently combined with the scientific training and caution which renders the teacher's aid of special value. As it is necessary to obtain measurements and observations of both sexes, I have appealed to both men and women teachers, and as it is also needful to combine the sexes (in the brother-sister measurements) to those working in elementary schools, as well as in boys' public schools and in girls' high schools. The result of my appeal has been to bring me a great deal of most valuable aid. Several high schools have been dealt with, four of our chief public schools have been, or are being measured, and a considerable variety of private, elementary and other schools. But a single public school (even of 500 to 700 boys) will often have only ten to twenty pairs of brethren, not, perhaps, as many as in a village national school, and I am most desirous of getting further help. The determination of the strength of collateral heredity is a problem of great scientific importance, and it can only be achieved by co-operative action. I have found so many teachers in all classes of schools willing to give disinterested aid in the cause of science that I venture to make a further appeal through NATURE for more assistance. Besides observations of physical and mental characters, which can be recorded without measurement, my data papers ask for certain head-measurements, which can, following the printed instructions, be taken quite easily. I shall be most glad to send sample papers to any one willing to assist, and if, after considering these, they find themselves able to assist, say by filling in data papers for ten or more pairs of brothers or sisters, I will at once despatch a head-spanner, of which I have several at the present time, free. The head-spanner should not be retained (unless under special circumstances) for more than a few weeks. Where the school is a small one, one master has, as a rule, filled in the papers entirely ; in larger schools, one of the science masters, or even the medical officer, has done the head-measurements, and the other data have been provided by house, form or consulting masters. In the ultimate publication of the statistics all aid will be duly acknowledged, but I make the appeal for help simply on the ground that the investigation of heredity is to-day one of the most important scientific problems, and that its exact quantitative determination is well within the reach of co-operative observation.　　　　　　　　KARL PEARSON.
University College, London.

The Perseid Meteoric Shower.

IN the years from 1893 to 1899 inclusive, about 120 determinations of the Perseid radiant were made. With the exception of three or four positions, the dates of the observations ranged from August 1 to 16, while the majority were for August 10 and 11 only.

It seems of little use to continue accumulating observations of the radiant point on and near the date of its maximum. What we essentially require are observations of the earlier stages of the shower during the last half of July, and as the present year offers a good prospect for obtaining them, I trust observers will make a special effort in this direction. The moon will reach her last quarter on July 19, and will prove a very slight hindrance to observation during the ensuing fortnight. When the sky is clear it should be watched all night, the paths of such meteors as are visible carefully recorded, and the results for each date kept separate, so that the place of the Perseid radiant may be traced in its diurnal motion of about 1° to the E.N.E. Some really good determinations of the radiant in July would be valuable, for very few have ever been made owing to the comparative feebleness of the shower in this month. An observer, however, who extends his watch over a long period, if not over the whole of the night, will find little difficulty in mapping a sufficient number of Perseids to indicate a good radiant.

Bishopston, Bristol, June 10.　　　　　　　　W. F. DENNING.

Variations in Plants of the Herb Paris.

THE enclosed table, showing the variations in 200 plants of Herb Paris, picked this month in the woods near Wells, may be of interest to some of your readers, especially if looked at in connection with the memorandum written by Sir Edward Fry, which he is kind enough to allow me to send with it.

L. ELEANOR JEX-BLAKE.

HERB PARIS.

Plants	Leaves	Sepals	Petals	Stamens	Cells of Ovary	Styles
96 ⎫ normal flowers	4 ⎧	4	4	8	4	4
44 ⎬	5 ⎨	4	4	8	4	4
2 ⎭	6 ⎩	4	4	8	4	4
13	5	4	4	9	4	4
8	5	4	4	9	4	4
5	4	4	4	9	4	4
2	4	4	3	8	4	4
2	4	4	4	8	4	5
2	5	5	5	10	5	5
2	5	5	4	10	4	4
2	5	4	3	7	3	3
2	6	4	4	10	4	4
2	7	4	4	9	4	4
1	3	4	4	9	4	4
1	3	4	3	8	5	5, and one rudimentary
1	4	4	3	8	4	5
1	4	5	3	9	4	4
1	4	5	4	9	4	4
1	4	4	4	10, one double	4	4
1	4	4	4	8	4	3
1	5	5	3	8	4	4
1	5	4	3	8	4	5
1	5	4	4	8, one double	4	4
1	5	4	4	8	3	2, and one rudimentary
1	6	4	4	8	4	4
1	6	6	4	8	4	4
1	6	5	3	8	3	3
1	6	5	4	9	5	5
1	6	5	4	9	4	4
1	6	4	3	9	4	4
1	5½ two halves grew together	4½	3	8	4	4

[Miss Jex-Blake's table seems to me to show many points of interest.

The Herb Paris has long been known to be very variable in the number of its parts; this table quantifies (I use the word, though it used to make a friend of mine very angry) the variability of the plant. It shows that, taking the 96 plants as exhibiting the normal form, more than one-half, *i.e.* 104 out of 200, vary from the standard; and that looking at the flowers alone, 142 plants out of 200 are normal, 58 only abnormal; that the 58 thus varying plants fall into no less than 28 groups; that not only do the plants vary as wholes, but that parts usually the same in number, or multiples of the same number, do not maintain this relation, *e.g.* that in 13 plants you get 5 sepals, 4 petals and 9 stamens, and so on.

The plant being thus given over to variability and belonging to the great group of monocotyledons, in which 3 and multi-

ples of 3 are the dominant number for the parts of the flower, a systematist might expect that the variations of the Herb Paris would oscillate round 3, or a multiple of 3, as the standard form; but, in fact, they oscillate round 4 as the dominant number, the 96 normal plants having that number, or a multiple of that number, everywhere, and 44 plants having that number and multiple everywhere in the leaves. Nature, therefore, disappoints our reasonable expectation.

It has, I believe, been suggested that the flower of Herb Paris is ideally of 6 and 12 parts, and that it has been reduced to 4 and 8 parts by atrophy and suppression of 2 and 4 parts respectively. If this were a true theory, you would expect to find here and there a reversion to the ancestral form; but the table shows that the number 6 occurs in the floral parts once, and once only, viz. in the sepals, and the number 12 never occurs in the stamens or elsewhere, so that the suggestion of suppressed parts becomes highly improbable.

The Herb Paris wanders from the ordinary type of monocotyledons, not only in the number of the floral parts, but in having ramifying veins of the leaves in the place of parallel veins; there are other monocotyledons which have this variation in the leaf from the standard. Do they, too, show any tendency to vary in the number of the floral parts? or to put it in other words, is there any correlation of the two variations? I have not looked into the subject, but it might prove worth consideration.—E. F.]

May 25.

Quaternion Methods applied to Dynamics.

I SHALL be obliged if any of your readers can give me the titles of any works on statics, or dynamics, or any physical science which are based on Quarternion methods and use nothing but Quaternion symbols.

The end chapters of P. G. Tait's "Quaternions" give examples; Kelland and Tait work out the theory of strains using Quaternion methods, but neither of these suffice for the purpose I have in view, namely, to put into the hands of a student a text-book on dynamics, &c., written in Quaternion language.

Jubbelpore, June 1.　　　　　　　　W. G. BARNETT.

PLANT HYBRIDS.

HORTICULTURISTS have recognised that as time goes on they must look more and more to hybridisation for "new plants." Biologists are already pointing out that, if anything can, breeding experiments will add to our knowledge of "the species." For both of these reasons the current volume [1] of the Royal Horticultural Society's *Journal* is of very particular interest, seeing that it is in fact the detailed report of the Conference on Hybridisation and Cross-breeding held last summer. The present writer has already summarised in these pages [2] the chief facts of importance brought out in the two days' proceedings; but several of the papers have been elaborated and illustrated, while many further contributions have been sent in and are now published. The latter in particular call for further comment.

Speaking generally of the report, it may be said that it is of very great value as a record of parentage, as a storehouse of many facts, and as putting forth several interesting theories. Furthermore, among the contributors are amateur and professional horticulturists, scientific workers pure and simple, as well as men who combine the interests of both, and this is a decided step in the right direction. It is not to be expected that the collection of papers forms a complete treatise to guide the practical or theoretical student; useful points are only to be found among cases at present not to be reconciled together and along with striking differences of opinion.

The very discrepancies are, however, to be welcomed, for from them can be learned the work to which attention should be most ungrudgingly given in the future; and by the publication of the "Hybrid Conference Report" the Royal Horticultural Society will earn the gratitude of a larger circle than ever. In the present account it will be

[1] *Journ. R.H.S.* vol. xxiv. (April 1900), pp. 1-348; 123 Figs.
[2] NATURE, vol. lx. (No. 1552, July 27, 1899), pp. 305-307.

best to touch upon individual papers rather than to attempt to discuss them together under special headings, as was done before.

One must not pass over without mention the list of some hundreds of hybrid plants exhibited at Chiswick on the first day of the Conference. In this are given the names of the parent species and of the raisers, as well as notes as to characteristics and habits, and points in which the hybrids most resemble their father or their mother. The generic headings are arranged alphabetically, while several plants and the pitchers of some *Nepenthes* are figured. A page is further devoted to the interesting series of mixed grafts which were also shown. In these the branches of both scion and stock retain their foliage, and in all cases the component plants belong to different genera. The title of "Hybrid Grafts," given to them in the report, does not seem to be a satisfactory one, being open as it is to misinterpretation or to confusion with "graft hybrids."

Dr. Masters' introductory address has already appeared in NATURE (vol. lx. p. 286). Among Mr. Bateson's contentions as to the origin of species, which have not been previously alluded to, is his statement that most professed botanists and zoologists are agreed that no natural species, whether animal or plant, has arisen by direct hybridisation. This may be mentioned, as another contributor to the report expresses the opposite opinion. Furthermore, Mr. Bateson's remarks as to the benefits that many horticulturists might confer upon the student of evolution, by recording even rough statistics, are very much to the point.

The genus Anthurium.—M. de la Devansaye says : (1) that in this genus, pollen to be of value must come from plants springing from a different batch of seeds from that giving rise to the ovule-bearing individuals ; (2) that pollen from allied genera has a beneficial effect ; (3) that variations may not be seen in the first or second generation of hybrids, and yet may appear in the third or fourth. Hence experiments should not be abandoned too soon.

Monstrosities.—Prof. Hugo de Vries' paper, read under the title of "Hybridisation as a means of Pangenetic Injection," now appears as "Hybridisation of Monstrosities." There is plenty in it, however, which does not refer to monstrous plants. Variation among hybrids of the first generation, as regards the colour of the flower, in a case considered by the professor, is put down by him as justifying the supposition that they simply inherited their variability from their mother. He lays down as a rule of horticultural practice the choosing of forms to hybridise, of which at least one is known to be very variable. The well-known multiformity of hybrids is stated to arise from this, but the fact—abundantly proved by the Conference—is also noted, that many hybrids can hardly be distinguished exteriorly from one or other of their parents, and therefore may be often mistaken for true species.

Hybridisation and its Failures.—Physiological affinity, says Prof. Henslow, it would seem, must be neglected

altogether in purely systematic work. He gives many cases where plants that botanically are placed in separate genera or families, on the strength of a single character, will not breed together, and he contends that genera that can be crossed should not be united for this reason alone, for if interbreeding is to be the test, polymorphic forms of the same species would logically have to be separated. The many "failures" recorded by Prof. Henslow must not be all put down as definitely proved to be such, as in many cases adverse conditions, of which the experimenters were ignorant, may have prevailed.

One would be interested to know whether the professor gained much information from the answers to the question set by him at the Royal Horticultural Society's examination last year, which ran, "Give any instances of failures, and state your opinion as to their causes, in crossing distinct species."

FIG. 1.—True and false hybrids of *Citrus*.

Official Work of the United States.—In the previous notice were mentioned the difficulties met with by Mr. Webber and Mr. Swingle, owing to the ovule of *Citrus* producing more than one embryo. In the accompanying illustration (Fig. 1), reproduced from the report by the courtesy of the Royal Horticultural Society, pots 1 and 4 each contain two seedlings of *Citrus trifoliata* type, arising from a single seed, and which show no effects of any cross. In the second pot are three young plants, again rising from a single seed, and which germinated. The seed was the result of a cross between *C. trifoliata*, ♂, and the Tangerine orange, ♀. One of the seedlings has trifoliate leaves of larger size than the typical *C. trifoliata*, and this is the true hybrid from the egg-cell proper, while the other plantlets with unifoliate leaves, and resembling the Tangerine, are from adventive embryos.

In No. 3, where the parents of the seed were the sweet orange, ♂, and *C. trifoliata*, ♀, two seedlings have grown both with trifoliate leaves, and that having these larger and more abundant may be put down as the hybrid. The other, which is like the mother in every respect, is looked upon as the product of the nucellus. Mr. Webber's other remarks and illustrations apply to the hybridisation of cotton and maize.

The Structure of Certain Hybrids.—Dr. Wilson contents himself chiefly with the external structure of hybrid *Passiflora, Albuca, Ribes* and *Begonia.* His figures bring out very forcibly the intermediate nature of many hybrids. The grades between ideal "tuberous" and "non-tuberous" conditions in hybrid *Begonia* are remarkable, joints of the stem falling away in several instances. We reproduce his illustration (Fig. 2), a

Fig. 2.—Flowering shoot of *Ribes nigrum,* ♀, × *R. grossularia,* ♂, (nat. size).

flowering shoot of a hybrid between the gooseberry, ♂, and black currant, ♀, and his sections of the ovary walls of the young plant and its two parents (Figs. 3, 4, 5). Several experimenters have obtained the cross and fruit from it, but no seeds. It is interesting that no odour of the black currant is possessed by the leaves, and that the caterpillars of the gooseberry saw-fly attack them without hesitation.

Self-sterility.—It is well that the importance of determining whether a plant may not be self-sterile has been

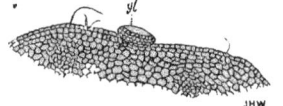

Fig. 3.—Transverse section of part of the ovarian wall of the black currant.

brought out by Dr. F. Ludwig. All the individuals of a species which is propagated vegetatively may, in a particular neighbourhood, be practically the same plant, and incapable of fertilising one another (compare the case of *Crocus sativus* on p. 276). Hence the importance in bringing pollen from physiologically independent individuals at a distance, mentioned in the discussion by the Rev. G. H. Engleheart with regard to daffodils, but not explained by him. Among a series of his opinions summarised conveniently by Dr. Ludwig is one with regard

to the springing up of races within the same species, which may be self-sterile and self-fertile. Another is of a very practical nature, and deals with the advisability, when introducing a new species of plant into a garden, to obtain at least two examples of it as of different origin as possible, or to procure the seed of such.

Work at the Paris Natural History Museum, 1887–99. —M. L. Henry contributes a list of plants supposed to be hybrids, which he suggests might have their origin

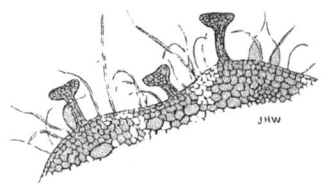

Fig. 4.—Transverse section of part of the ovarian wall of the hybrid.

proved by hybridisation experiments. He adds a record of his work during recent years, giving most details with regard to lilacs.

Graft Hybrids.—This account of the Bronvaux medlar, by M. E. Jouin, appeared originally in *Le Jardin* (January 20, 1899) ; and M. Daniel (" La Variation dans la Greffe," *An. Sci. Nat. Bot.* Series 8, vol. viii. (1898), pp. 1–226 ; pls. i.–x.) has figured and given some details in

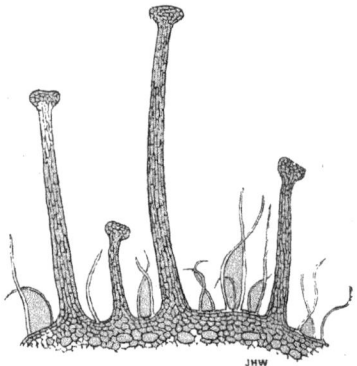

Fig. 5.—Transverse section of part of the ovarian wall of the gooseberry.

his recent paper of the remarkable branches pushed out by the whitethorn stock below the graft.

Branch No. 1 is intermediate between the whitethorn and medlar. It is, however, thorny, and bears corymbs of as many as twelve flowers, instead of solitary blossoms. The fruits are medlars, but small and much flattened.

Branch No. 2.—The young leaves resemble those of the whitethorn ; the older, those of the medlar, being hardly, or not at all, lobed. The flowers are like those of the former plant, and arranged in corymbs, but a trifle larger. The fruits are not medlars.

Branch No. 3.—The base is simply whitethorn, but the extremity is practically like Branch No. 2.

Similar cases are instanced by M. Jouin, who puts down the now celebrated *Cytisus adami* as having arisen in the way that the branches above described have done.

Drosera Hybrids.—This paper, by Prof. Macfarlane, has already appeared in the publications of the University of Pennsylvania ; it deals with the structure of a batch of natural hybrids. It may be noted that several instances of what the author has called bi-sexual hybridity occur in the plants considered (p. 248) ; for instance, instead of finding structures intermediate between the elongated glandular hairs of *Drosera filiformis* and the sessile two-celled glands of *D. intermedia*, both appear on the calyx of the hybrid between them. This fact calls forth some interesting speculations of a cytological nature, which the Professor hopes to see verified. As showing the growing opinion in favour of graft hybrids being realities, it might be mentioned that *Cytisus adami* is referred to as such in the paper.

The Influence of each Parent.—From experiments with cereals and Bromeliaceæ, Dr. Wittmack concludes that "the mother has the more influence upon the habit ; the father the more upon the inflorescence ; at least, upon its colour." The contrary opinion of M. Duval is given, who also holds that to reduce the volume of the plant the larger must be fertilised by the small parent species. According to Mr. Tropp, the same holds good usually, but not always, with orchids.

Principles.—The laws given by Herr Max Leichtlin may be quoted in full :—

(1) The female parent gives to the offspring the form and shape of the flowers ; also certain qualities.

(2) The male parent gives more or less of the colouring of the flowers, and if it be richer and blooms more freely than the female, this property is transferred to the offspring.

(3) Artificially produced offspring give larger flowers than either of their parents.

(4) The more distant the habitats of the species intended to hybridise, the more difficult is it for them to be fertilised with each other's pollen.

(5) The offspring becomes infertile and delicate if the form of the flowers of their parents is widely different in shape and outline.

Breeding Staple Food Plants.—In alluding to the cost in labour and money of developing hybrids when the immense number of plants that should be dealt with are used, Prof. Willett Hays points to the importance of selecting carefully the parental individuals. The best flower, he says, too, should be chosen from the best part of the plant.

An Improved Variety of Crocus Sativus.—It was not till after many experiments with examples from many places that the saffron could be got to produce seed, except very meagrely (compare the remarks already made on self-sterilising above). After a wild plant of *Crocus graecus* was obtained from the island of Syra, as much seed as was wanted was obtained. In the variety produced by M. Chappellier there is a proliferation of stigmas, sometimes thirty, and even bracts and sheaths have been converted into them.

Experiments with Dioscorea.—In an attempt to obtain a tuber which was short enough for one to dig up easily, a plant was obtained by M. Chappellier bearing both male and female flowers. This worker also contributes a note on *Mirabilis*.

Hybrid Lilacs.—M. E. Lemoine sends an account of how he proved the Varin lilac to be a hybrid between *Syringa persica laciniata* and *S. vulgaris*, a piece of work which M. Henry would also have succeeded in if his plants had not died before flowering.

Hybrid Clematis are dealt with by M. Morel and Mr. Jackman. M. Duval treats of *Anthurium scherzerianum*, of Bromeliads and of Gloxinias. This hybridist points out how important it is to know the pedigree of plants experimented with, and says that the male parent should be most carefully selected, as being the one whose influence greatly preponderates. Mr. Meehan and Mr. Smythe have written a few general notes. Mr. Weekes has a little to say about Chrysanthemums, while Mr. James Lye, when discussing the cross-fertilisation of the *Fuchsia*, states that he uses the tip of a squirrel's tail to transfer the pollen, and prefers muslin bags to those made of paper for enclosing the chosen blossoms.

Mr. Wilks, the secretary of the Royal Horticultural Society, must be complimented upon the successful production of the report.   WILFRED MARK WEBB.

OUR NORTHERN BIRDS.[1]

MR. DIXON is a prolific writer, and confines himself almost entirely to one subject. Nevertheless he always succeeds in interesting his readers, and contrives to say something fresh even upon such a trite and thread-

FIG. 1.—Rough-legged Buzzard (From Dixon's "Among the Birds in Northern Shires.")

bare theme as British birds. In a former volume Mr. Dixon took as his subject "Bird-life in a Southern County"; and in the present work he dwells on the great difference between the bird-fauna of the more northern counties of England and Scotland from that of the south

[1] "Among the Birds in Northern Shires." By Charles Dixon. Pp. x + 303. (London and Glasgow: Blackie and Son, 1900.)

of England. Not only are many of the birds of the northern districts normally strangers to the south, either at one season of the year or at all times, but notable differences in their habits are observable. Species, for instance, that sing during winter in the south are silent at that season further north ; while others that are permanent residents in the former area are migratory in the latter. And it is certain that from an ornithological point of view the northern counties are more interesting than the southern—and more especially the south-western counties.

In the treatment of his subject, Mr. Dixon has acted wisely in abandoning a systematic classification in favour of a grouping by means of "station," so that we have chapters on the birds of the upland streams, of the moors, the mountains, the heaths and marshes, the forests and copses, the farm and the garden, the river and pool, the sea and the beach, and the crag and sea-cliff. By this arrangement a much more discursive and "chatty" style of writing is permissible than would otherwise have been the case. The reader is accordingly spared a repetition of the descriptions of the various species of which we

FIG. 2.—The Dotterell. (From Dixon's "Among the Birds in Northern Shires.")

have already more than enough ; and the author has also seen fit to abandon the use of technical names, and to content himself with the English titles of the birds mentioned.

Much of the attraction of the book is due to the exquisite portraits of birds in their haunts from the accomplished pencil of Mr. C. Whymper. Where all are good it is difficult to select examples for special commendation, but the black-cock crowing is one that strikes our own fancy, and the two cuts that we are enabled, by the courtesy of the publishers, to reproduce, will serve as good examples of the general style of illustration. If we mistake not, the figure of the raven is very like one that has appeared elsewhere.

In the chapter on the birds of the upland streams an excellent account is given of the habits of the water-ouzel or dipper ; and here the author appears, for once, to have caught Prof. Newton "tripping." In his "Dictionary of Birds" the origin of the name "dipper" is attributed by the Cambridge ornithologist to the writer of the letter-press of Bewick's "British Birds," but Mr. Dixon points

out that it occurs in a work published as far back as 1771, and a later issue of which was actually edited by the learned professor himself !

An especial feature of all the author's works is his own practical experience of birds in their native haunts ; and all who have had bird-nesting adventures in their own early days will read with delight the description, on p. 136, of his ascent of a lofty oak to secure a clutch of buzzard's eggs, which were safely carried down. In making friends, during his youthful days, with both poachers and game-keepers, Mr. Dixon seems indeed to have had an almost unique experience, and one whereby his practical acquaintance with the ways of birds was largely augmented. He has many stories to tell of the wanton manner in which he has known keepers fire into the nests of brooding birds and otherwise inflict destruction on what they are pleased to denominate "vermin." In regard to these latter, he urges (p. 151) that our indigenous native game-birds would thrive all the better if hawks, crows, magpies, owls and the like were left unmolested. The pheasant, he thinks, however, might not fare so well ; but, he adds in effect, perish the pheasant ! This, however, we venture to suggest, is scarcely a practical way of looking at things. If pheasant-preserving were to be given up, our coverts would not be guarded at all, and many of the other birds would be ten times worse off than they are under the present régime.

Not the least interesting chapter in a very interesting book is the final one on bird migration in the northern counties, where, instead of a learned discussion on the theory of migration, we have an actual account of the manner in which the swarms of autumn and spring migrants reach and leave our coasts. Here the author remarks that the short-eared owl and the woodcock frequently reach the Wash together, making the passage from the Continent during the same night, although the one just skims the water while the other flies high in the air. And many other passages attests the author's close observation of the movements of birds. It is to be wished, however, that he would employ the familiar name hedge-sparrow in place of the pedantic hedge-accentor. The term sparrow, as Prof. Newton remarks, was probably originally applied to all our smaller birds ; and it is surely an unwarrantable assumption on the part of ornithologists to traverse popular usage and attempt its restriction to the members of the restricted genus *Passer*.   R. L.

NOTES.

THE delegates to the third biennial conference in connection with the International Catalogue of Scientific Literature were entertained at dinner by the Royal Society as we went to press last week. In the course of the evening several interesting speeches were made in proposing and responding to toasts. Prof. Forsyth, in proposing "International Science," referred to the great empire of science, the possessions and achievements of which are intended for the welfare of all men. Prof. Darboux responded for France, Prof. Klein for Germany, and Prof. Weiss for Austria. The toast of "The Delegates to the Conference" was proposed by Sir John Gorst and responded to by Prof. Ciamician (Italy), Dr. Graf (Switzerland), and Dr. Brunchorst (Norway). Sir Michael Foster proposed the toast of "The Guests," which was responded to by Lord Strathcona ; and the health of "The President" was proposed by Dr. Schwalbe and acknowledged by Lord Lister. We have not yet received from the Royal Society the *procès verbal* of the meetings of the delegates, but it is understood that the support promised will warrant a commencement of the Catalogue at the time fixed.

THE third biennial Huxley lecture at the Charing Cross Hospital Medical School will be delivered by Lord Lister on Tuesday, October 2. The two former lecturers, in 1896 and 1898, were Sir Michael Foster and Prof. Virchow.

THE Society of Arts has awarded its Albert medal for the present year to Mr. Henry Wilde, F.R.S., "for the discovery and practical demonstration of the indefinite increase of the magnetic and electric forces from quantities indefinitely small." This principle is the one on which the invention of the dynamo machine is based, and is utilised in the construction of all modern dynamos.

AT the annual general meeting of the Royal Statistical Society, held on Tuesday, Lord Avebury was elected president for the ensuing session. It was announced that the subject of the essays for the Howard medal, which will be awarded in 1901, with 20*l*. as heretofore, is "The History and Statistics of Tropical Diseases, with especial reference to the Bubonic Plague." The essays should be sent in on or before June 19, 1901.

THE annual meeting of the Marine Biological Association will be held at the Royal Society on Wednesday next, June 27.

WE learn with regret that Mr. W. Percy Sladen, for years an honorary secretary of the Linnean Society, died at Florence on June 11.

THE death is announced of M. Boutan, one of the founders of the French Physical Society, and the author of an excellent "Traité de Physique" as well as other works.

PROF. EDMUND PERRIER has been appointed to succeed the late Prof. Milne-Edwards as director of the Paris Natural History Museum.

THE autumn meeting of the Iron and Steel Institute will be held at Paris under the auspices of the Société d'Encouragement pour l'Industrie Nationale, on Tuesday and Wednesday, September 18 and 19.

THE summer meeting of the Institution of Mechanical Engineers will be held in London on June 27 and 28. The programme includes an adjourned discussion on road locomotion (a short supplementary paper dealing with the recent trials will be submitted by Prof. H. S. Hele-Shaw, F.R.S.); recent locomotive practice in France, by M. Edouard Sauvage; polyphase electric traction, by Prof. C. A. Carus-Wilson; observations on an improved glass revealer, for studying condensation in steam-engine cylinders and rendering the effects visible, by Mr. Bryan Donkin.

AN interesting exhibition of objects illustrating the population, monuments, customs, and native industries of the Chawi and Kabyle tribes of Algeria will be on view in the rooms of the Anthropological Institute, 3 Hanover Square, W., until June 23, from 11 a.m. to 5 p.m. The objects were collected in the course of a recent journey in Algeria by Mr. D. MacIver, student of Egyptology at Worcester College, Oxford, and Mr. Anthony Wilkin, of King's College, Cambridge.

THE Advisory Committee appointed by the Board of Trade in connection with the business of the Intelligence Branch of the Commercial, Labour, and Statistical Departments of that office met on Thursday last, Sir Courtenay Boyle being in the chair. There were present, among others, Lord Avebury, Sir Frederick Abel, F.R.S., Prof. Wyndham R. Dunstan, F.R.S., and Mr. C. A. Harris. C.M.G.

THE grant of 1000*l*. in aid of the work of the Marine Biological Association ; the site of the National Physical Laboratory at Kew ; and the grant to the British School at Athens, were brought before the House of Commons on Friday last, upon the vote to complete the sum of 50,724*l*. for scientific investigation. It was urged by Mr. Gibson Bowles that the grant to the Marine Biological Association should be largely increased ; and by Lord Balcarres that the vote of 7000*l*. for building and equipping the National Physical Laboratory should not bind the Treasury to adhere to the site which has been proposed. Mr. Hanbury said it should be borne in mind that the grant of 1000*l*. to the Marine Biological Association was not the only grant made in connection with the fisheries of the United Kingdom. A grant was given to the Fishery Board of Scotland for the purpose of scientific investigation, and similar assistance was given to the Irish fisheries. Under present conditions there did not seem to be any urgent necessity to increase the grant. The Treasury had very little voice in the matter of the Physical Laboratory ; it had acted on the recommendation of a committee of the Royal Society. It was absolutely necessary to find a site near Kew Observatory, and after looking at every possible site the committee strongly reported that no other site would answer the purpose so well as that which adjoined Kew Gardens. He agreed that nothing ought to be done which would interfere with the amenities of Kew Gardens, and this point had been considered in the selection of the site. The two buildings, one for the machinery and the other for carrying on the more delicate scientific operations, were to be placed in positions which would not mar the views from the gardens or injure their amenities. The voting of the 7000*l*. would in no way prejudice the consideration of the case against the proposed site.

REFERRING more particularly to the British School at Athens, Mr. Balfour said that the only ground for the alarm expressed was that the original grant was for five years, and that this term was drawing to a close. The question of Governmental subvention of scientific investigation was a very important subject, and there was no doubt that this country had, from a traditional policy, lagged greatly behind other nations in this respect. It never occurred to us to do what the Germans, the French, or the Americans did in making certain grants for investigations ; and whether we were right or wrong he did not undertake to say. His own personal inclination was rather in the direction of Governmental aid in cases where they could not expect private aid to come forward ; but at the same time he confessed that he often thought how strange it was in a very rich country like ours there were not found some people who, in a difficulty to find other and more profitable investments, did not attempt to earn glory for themselves by carrying on those investigations with the money that was required. He could only say that certainly the grant would not be discontinued without a generous consideration of the facts and interests involved.

A MEETING of the Röntgen Society was held on Thursday, June 7, at St. Bartholomew's Hospital, by the invitation of Dr. Lewis Jones. A large American Holtz machine has recently been presented to the hospital, and it was chiefly to allow the members to have an opportunity of seeing this machine at work in connection with X-ray tubes that Dr. Jones invited the Society to meet at the hospital. Large Holtz machines, though used considerably in America, are rarely seen in this country, where the Wimshurst pattern is more commonly employed. A dark room has been fitted up in the electrical department specially for X-ray work. The wires for bringing the current to the tubes are passed through a partition to the machine, which is on the other side. The observer and the patient are thus in no way disturbed by the movements of the machine, or of the motor or other appliances connected with the working of the same. The light given by the tubes was perfectly steady, which is one of the advantages usually claimed for the influence machine over the coil, although this has been somewhat diminished since the advent of the Wehnelt and Caldwell electrolytic breaks.

MR. J. WIMSHURST, F.R.S., also read a short paper on his work in connection with the design and perfecting of the several forms of his influence machine, describing, among others, the large machine made for and presented to the Science and Art Department, with plates 84 inches in diameter; and another, with twenty-four plates 36 inches in diameter, shown at the Earl's Court Exhibition, and which is now, we believe, in the possession of Dr. J. Macintyre, of Glasgow.—Dr. Rémy, of Paris, showed a new localising apparatus. This consists of a vertical support moving in a socket fixed to the table. The tube is supported by a cross piece at the lower end, under the table, while the fluorescent screen is attached to the upper end, together with two pointers representing the paths of the rays, and held in supports moving in slots or grooves. Two observations are made and the pointers adjusted, after which the apparatus is turned round away from the table, and the pointers lowered until their points meet, thus indicating the depth of the hidden object below the surface. A bullet hidden in a loaf of bread was found in a minute or so by Dr. Rémy.

THE *Scientific American* contains an account of Count von Zeppelin's projected navigable balloon now under construction on Lake Constance. The balloon is to be 416 feet in length and 38 feet in diameter, divided into seventeen compartments, and supported on an aluminium framework. It is to carry two cars and motors, and to be propelled by screws placed in pairs at the side of the balloon and geared to the driving-shaft by two diagonal shafts. In the preliminary experiments for testing the efficiency of the motors, a launch was driven on the lake at from 6·8 to 9·2 miles an hour by aërial propellers. The fuel is benzene, and it is calculated that the balloon will carry sufficient fuel to perform a journey of over 179 miles.

THE cutting of the *sudd* on the Upper Nile and the consequent release of large volumes of stagnant water has, we learn from the *Times*, had an unanticipated influence on the condition of the river at Assuan. From reports received by Sir Benjamin Baker from the engineering staff it would appear that the absence of free oxygen in the water has caused wholesale destruction of the fish. Within a hundred yards of the resident engineer's office at least a million dead fish, ranging in size from minnows to six feet in length, are to be found. This result is consistent with London experience when it was usual to pour crude sewage into the stream. The filtered water, though clear and odourless, was drunk with impunity, but, having no free oxygen, eels plunged into it would struggle violently and finally die of suffocation, as has apparently been the case with the fish in the Nile in the special circumstances resulting from the long-deferred cutting of the *sudd* this year.

SOME notes on New Zealand volcanoes are contributed to the latest volume (1899) of *Transactions and Proceedings* of the New Zealand Institute, by Dr. B. Friedländer. A description of an eruption of Te Mari witnessed by him is of interest. The eruption began with an explosion, and masses of ash-bearing steam were ejected. There were at least four different light-phenomena:—(1) the reflection of incandescent matter upon the dark clouds; (2) a large number of red-hot boulders, which were shot high up and fell down in parabolic curves; (3) light-

ning, due to electricity produced by friction. The lightning appeared in masses of ash-bearing steam; and the ashes were coarse, the single grains being about the size of a pin's head. (4) Blue flames, and probably reddish flames. Some red flames were apparently distinct from the light due to illumination of steam, and the blue flames must have been real. Dr. Friedländer suggests, to account for the flames, that during the explosion there escaped combustible gases which at a certain height above the crater met the oxygen necessary for taking fire. He considers that vaporised sulphur would explain his observations better than hydrogen, the flames of which are less brilliant and less distinctly blue. Another paper of interest to students of vulcanology, in the volume referred to, is a detailed description of the volcanoes of the Pacific, by Mr. Coleman Phillips.

AN official report by Captain R. H. Elliot, upon his researches into the nature and action of snake-venom, is referred to in the *Madras Mail*. Captain Elliott confirms the fact that the mongoose is not immune in the fullest sense of the expression, seeing that it may succumb to a snake-bite, if sufficiently severe, like any other animal. His researches go to show, however, that the mongoose does enjoy a partial and comparative immunity from snake-poison—that is to say, a mongoose takes from ten to twenty-five times as much cobra venom to kill it as a rabbit does, and five to twelve times as much as a dog. M. Calmette gives a somewhat lower estimate than this; but he made only a few experiments, and it is noteworthy that the mongooses that he experimented with were obtained from Guadeloupe, where venomous snakes are unknown. The mongoose was introduced into Guadeloupe (and Barbadoes) some twenty-five years ago with a view to the destruction of rats. Captain Elliott thereupon remarks:—"We are thus led to the interesting conclusion that the introduction of the mongoose into a country in which venomous snakes are unknown has resulted, in so short a period as a quarter of a century, in a very appreciable reduction of the animal's resistance to snake-venom. This fact points strongly to the farther conclusion that the immunity is an acquired one, and inasmuch as the acquired characteristic has been so rapidly and easily diminished, it would appear likely that it must be maintained from generation to generation. Be it remembered that a quarter of a century probably means about fifty generations."

THE growing necessity of obtaining greater speed on railways has of late been freely discussed, and different designs have been put forward favouring very much the idea of a single-rail system resting on supports and its train suspended below. Hitherto all railways of this nature have been propelled by steam or electric motors, but *Fielden's Magazine* (June) describes and illustrates a still later application of this suspended car system, patented by Mr. H. S. Halford under the name of the Halford gradient railway. The remarkable feature about his system is the fact that no locomotive or electric motors are carried, as the train derives its motion by gravitation imparted by raising, as long as this is required, to a slight incline, the section of line upon which the train is running. The track of this railway, which is supported, is divided into sections, the extremities of which can be raised or lowered by hydraulic (or other) power, the operation being performed either automatically by the carriage in its transit or by the driver at will. It is stated that the cars ride smoothly going from one section to another, and also the change of incline is so small and gradual that the lifting of the track is almost unperceived. The Halford gradient railway has yet to see a more practical test, but the following advantages claimed make the devise feasible in many ways:—(1) In all other systems, the greater the load the less the speed; in this, the greater the load the greater the speed. (2) There is no need to stop for coal or water. (3) Its natural tendency must be to increase in speed. A photograph of a working model and diagrams illustrate the article.

CONSIDERING the advances that have been made in the rate of travelling during the present century, it would be unsafe to say that a speed of 120 miles an hour is not attainable The evidence that was produced before a Committee of the House of Commons, in a Bill recently brought forward, was not, however, sufficient to satisfy the Committee that such a rate of speed can be attained with safety to the passengers. The scheme of the inventor, Mr. Behr, was for an electric railway to run between Liverpool and Manchester, and to perform the journey of 35 miles, including stoppages, in 17½ minutes, which would mean a speed of more than double that now attained by the best express trains. The carriages were to be suspended on a single rail resting on A-shaped iron tressels, with two side rails to keep the carriages in place. The idea of carriages being suspended from rails is not new, an electric railway on this principle, 8 miles long, having for some time been in use between Barmen and Elberfeld. The trolley rails there are double, and the speed attained is only 18½ miles an hour. In the Lartigue system, which has also been in use for some years, the carriages are suspended from a single rail, but no high speed is attempted. An experimental railway was constructed at Brussels on the Behr system, when a speed of 80 miles an hour was said to be attained. The sensation produced by the sudden pulling up of an ordinary express train is sufficiently uncomfortable, to say the least, to create considerable doubt as to the safety of stopping a train within any reasonable distance travelling at double this rate. Anyway, the Committee, in the interest of the public, declined to give their sanction to the scheme as presented to them ; and while admitting that the mono-rail system when properly matured might. make an important development in railway traffic, yet as regards the method of applying the brake-power to trains running at such high speeds, they were not satisfied that the safety of the public was sufficiently provided for.

THE action which the vestry of the Parish of Hammersmith has taken to make known the nature of consumption, and the measures which should be adopted to prevent its spread, is altogether praiseworthy, and other public authorities should emulate it. At the request of the vestry, the medical officer of health, Mr. N. C. Collier, has prepared a report upon the causes and prevention of consumption, and it has been distributed in the form of a leaflet. It is pointed out that " there is now no doubt that consumption is caused by a minute living organism, the bacillus of tubercle, and that the presence in the body of the tubercle bacillus is most rarely inherited, but becomes introduced from without. What is inherited is the non-resistile condition of the vitality of certain cells in the body which are unable to destroy the tubercle bacillus, when it has become accidentally introduced into the system. To prevent consumption it is necessary, firstly, to avoid all those means by which the tubercle bacillus may be introduced into the body ; and, secondly, to avoid all those causes which enfeeble the vitality of the cells of the body, and so render them unable to destroy the tubercle bacillus should it become introduced." The hygienic principles to be borne in mind in order to prevent the spread of the disease are briefly summarised, and the information given cannot be too widely known to the public.

THE recent report of progress of the observatory at Colába (Bombay) shows the large amount of work accomplished during the year ending March 1900. In addition to the usual magnetic and meteorological instruments, one of Prof. Milne's horizontal pendulums has been in action for nearly two years. During the year covered by the report, twenty-seven earthquakes were

registered, besides 1398 small and local movements. A second horizontal pendulum, designed and made locally, was erected last March. It is similar in principle to the other, but much more sensitive. The record is made mechanically by means of an ordinary crow quill and glycerine ink, writing on paper driven at the rate of five inches an hour, excellent open diagrams being thus obtained.

· THE Deutsche Seewarte has recently issued its twenty-second volume of *Aus dem Archiv* for the year 1899, containing valuable discussions relating to the motions of air and sea. Among the most popular subjects we may mention a paper, by Dr. van Bebber, on a scientific basis of weather prediction for several days in advance. The same subject has been treated of by the author, in a preliminary way, in periodical publications, and has already been noticed in our columns. The question is one of great importance, and we therefore refer our readers to the pre sent more elaborate discussion. An examination of the weather conditions of twenty years, as shown by the daily weather charts of the Deutsche Seewarte, has led the author to distinguish five principal types, under one of which the actual conditions may be classed, with a fair degree of probability that the behaviour of the weather (on the Continent) will conform in its general features to that of the type in question. The types all refer to the more persistent areas of high barometric pressure, in contradistinction to the more mobile areas of lower barometric pressure. The paper is accompanied by sixteen charts and diagrams printed in the text, and will repay careful study by those interested in weather prediction.

WE have received a double number of the *Journal* of the Scottish Meteorological Society, completing vol. xi. (3rd Series) of this useful publication. It contains the usual valuable meteorological returns from the Scottish lighthouses, and from a large number of stations belonging to the Society. These observations (which refer to the years 1897 and 1898) have been carefully examined ; and monthly means have been calculated and utilised in the preparation of the annual reports on the meteorology of Scot. land. In addition to this routine work, the number contains several special discussions, e.g. the " Annual Rainfall of Scotland from 1800 to 1898." This is a comprehensive and laborious compilation, by Dr. Buchan, and will be of the greatest utility in any inquiry bearing upon the rainfall of this part of the United Kingdom. The tables are divided into two parts, showing (1) the annual amounts, and (2) the average rainfall for the whole period, the heaviest and least yearly amount, the height of the station, and other particulars. Among the other papers may be mentioned " Barometric and Thermometric Gradients, 1704-1898," showing the differences in the mean monthly and annual values of these elements at London and Edinburgh, by Mr. R. C. Mossman. We are glad to see that the important work of the Ben Nevis Observatories will be completed in the way desired by the directors, thanks to the magnificent donations of two of the members of the Society.

THE Wisconsin Geological Survey sends us the third number of an " Economic Series," a " Preliminary Report on the Copper-bearing Rocks of Douglas County, Wisconsin," by Dr. Ulysses S. Grant. The copper occurs mainly as the native metal, and most commonly in the upper amygdaloidal parts of the old lavas belonging to the Lower Keweenawan (pre-Cambrian) formation. It occurs also in small particles scattered through both igneous and stratified rocks, in minute seams and in veins. It was deposited in its present position by circulating waters. At times, at the surface the native copper is not discernible, and its presence may then be detected by the green and blue alteration products or stains, malachite or azurite. Areas where the rock is highly charged with epidote are of a yellow or yellowish-green colour, and it is recommended that these be searched for particularly, as in them copper is likely to occur. The rocks of Douglas county are the same in nature, in origin and in age, as the copper-bearing rocks of Keweenaw Point, Lake Superior ; but at present it has not been determined in Douglas county that any deposit of copper of sufficient richness is extensive enough to be of economic value. Some of the recent explorations are, however, very encouraging.

THE *Quarterly Journal* of the Geological Society for May is a bulky number, which contains the address of the ex-president, Mr. Whitaker, and fourteen papers dealing with a variety of subjects. Perhaps the most important of these are contributed by women, who, by the by, are not at present eligible to become Fellows of the Society. There can be no doubt, however, that the essay by Miss Gertrude Elles on the zonal classification of the Wenlock shales of the Welsh borderland, and that by Miss Ethel Wood on the Lower Ludlow formation and its graptolite-fauna, make very great advances on our previous knowledge. The papers bear evidence of long-continued and critical research on the Silurian strata and on the difficult subject of the zoological characters of the graptolites which characterise successive stages in the rocks.

THE volume of "Geological Literature," which since 1895 has been separately published by the Geological Society of London, reflects great credit on the compiler, Mr. W. Rupert Jones, and on the editor, Mr. Belinfante. In this work the titles are given of the books and of all the geological papers contained in periodicals which have been added to the Society's library during the year 1899. This list occupies over a hundred pages, while the subject-index brings the total to 176 pages. As a work of reference it is indispensable to all geologists.

THERE is now in the press, and will shortly be published by Messrs. Young, Liverpool, and Messrs. Porter, London, the report of the conjoint expedition to Sokotra and Abd el-Kuri, conducted in 1898-99 by the British Museum (represented by Mr. Ogilvie-Grant, of the Zoological Department) and the Liverpool Museums (represented by the director of museums to the Corporation, Dr. H. O. Forbes). The expense of its publication is borne by the Museums Committee of the Liverpool City Council, and the volume is edited by Dr. Forbes. It will be illustrated by between twenty-five and thirty plates, chiefly coloured, depicting the zoological and botanical discoveries of the expedition, the ethnography of the islands, &c. The introductory chapters by the editor give an account, fully illustrated by blocks, of the journey, of the geography of the islands and of their inhabitants. The scientific chapters are contributed by Lord Walsingham, F.R.S., Prof. I. B. Balfour, F.R.S., Mr. Boulenger, F.R.S., Dr. Forbes, Mr. Ogilvie-Grant, Mr. A. E. Smith, Colonel Godwin-Austen, F.R.S., Mr. De Winton, and other well-known naturalists.

THE last number of the *Zeitschrift für wissenschaftliche Zoologie* contains an elaborate paper, by Dr. E. Zander, on the male reproductive organs of the Hymenoptera. It is illustrated by a remarkably well-executed coloured plate.

A CORRESPONDENT of *Nature Notes* asks for some good reason why a lover of animals should not wear the stuffed head of a bird or other creature as an ornament. The query appears to us a pertinent one.

WE have received the Report of the South African Museum for 1899, in which the Trustees express themselves generally well satisfied with the progress of the institution. They record an addition to the edifice of a large block of new building to receive the art collections.

IN No. 3 of *Marine Investigations in S. Africa*, Mr. G. A. Boulenger describes an example of the rare unicorn-fish (*Lophotes cepedianus*) from the Cape of Good Hope, where it has not hitherto been definitely known to exist. The specimen was considerably over a yard in length. The unicorn-fish, which is an ally of the ribbon-fishes, takes its name from the peculiar filamentous process arising from the front of the elevated head.

THE Annual Report of the Field Columbian Museum, containing an excellent portrait of its founder, Marshall Field, is likewise to hand. The museum appears to be making extremely rapid progress, its ethnological series having been very largely increased by the acquisition of the rich collection acquired by the Stanley McCormick Expedition among the Hopi Indians. An especial feature of the report is the introduction of a number of photographs of recent acquisitions. Among these, we may call attention to the portrait of a Hopi bride, and also to a plate of a group of the extraordinary gigantic spiral fossils known as "devil's corkscrew," or *Daemonelix*, which have lately excited so much interest.

THE Imperial Department of Agriculture for the West Indies has just issued a small pamphlet on the best means of destroying that troublesome insect, the "moth-borer" (*Diatraea saccharalis*), which inflicts so much damage, while in the caterpillar stage, on sugar-cane. It appears that a considerable number of the eggs of the moth-borer (which are laid in patches on the leaves of the sugar-cane) are attacked by parasites which prevent the development of the caterpillars, and in due course come forth as flies. These parasite-infested eggs are readily distinguished from healthy eggs by being black, instead of yellow or orange. · It is recommended to destroy all the yellow and orange eggs that can be collected, but to leave the black ones, in order that they may breed flies to destroy other clutches of eggs. If this remedy were adopted as soon as the young cane commences to show, and continued as long as it is sufficiently small, the loss of the best shoots would be avoided. If, however, the caterpillars are allowed to hatch out and bury themselves in the cane, there is nothing for it but to cut out the "dead hearts," and this to a considerable depth. When cut out, they must forthwith be destroyed, or the caterpillar will either complete its development in them, or crawl out to other canes.

Bulletin No. 2 of the West of Scotland Agricultural College is a report by Prof. R. Patrick Wright on experiments on the manuring of rye-grass and clover-hay in 1899.

THE following are the most recent official botanical publications which have reached us from the United States :—The germination of seeds as affected by certain chemical fertilisers, by Mr. G. H. Hicks (U.S. Department of Agriculture, Division of Botany, *Bulletin* No. 24) ; Bread, and the principles of bread-making, by Helen W. Atwater (U.S. Department of Agriculture, *Farmer's Bulletin*, No. 112) ; Co-operative experiments with grasses and forage-plants, by Dr. P. Beveridge Kennedy (U.S. Department of Agriculture, Division of Agrostology, *Bulletin* No. 22).

UNDER the title of "Annuaire des Mathématiciens," Messrs. Georges Carré and C. Naud propose to publish a directory containing the names, addresses and academic rank of those interested in the study of mathematics.

MUCH information is given in a clear and concise form in Mr. A. A. C. Swinton's little book on "The Elementary Principles of Electric Lighting" (Crosby Lockwood), the fourth edition of which has just been published. The book only runs into sixty-four pages, but everything in it is to the point ; and electrical artisans, as well as readers unfamiliar with electrical phenomena and effects, will find its pages perfectly intelligible.

THE first part of Dr. Carl Chun's narrative of the cruise of the *Valdivia* and the scientific work accomplished, which has been published by the firm of Gustav Fischer, Jena, shows that the complete work, "Aus den Tiefen des Weltmeeres," will be a most interesting account of a successful expedition. The descriptive matter is untechnical in style, and liberally illustrated with excellent half-tone blocks and plates reproduced from photographs. The complete work will contain six chromolithographs, eight heliogravures, thirty-two full-page plates, and about 180 illustrations in the text. There will be twelve parts in all, two of which will be published every month, and the whole by November next. The work will be a *Challenger* narrative on a small scale, full of interest to all students of natural history and of physical geography in the most comprehensive sense of the term. We propose to review it in detail when all the parts have been received.

PROF. E. B. WILSON's work on "The Cell in Development and Inheritance" (The Macmillan Company) contains a masterly treatment of the facts of cell-structure and division, and is favourably known to many biologists. It originally appeared in 1896, and has already been reviewed in NATURE (vol. lv. p. 530). Since then the aspect of many important questions with which it deals has been greatly changed, more particularly in case of those focused in the centrosome, and involving the phenomena of cell-division and fertilisation. This has necessitated a complete revision of the work, and there is scarcely a page of the second edition, which has just been published, that has not undergone alteration. More than a hundred pages of new matter have also been added. The most important results of modern cell-research, especially on the zoological side, are brought together in the volume, which will continue to be used as a convenient and clear synopsis of a vast amount of knowledge to which additions are constantly being made.

THE additions to the Zoological Society's Gardens during the past week include a Grivet Monkey (*Cercopithecus griseo-viridis*) from North-east Africa, presented by Mr. H. G. F. Stallard; a Campbell's Monkey (*Cercopithecus campbelli*) from West Africa, presented by Miss E. B. Hall; two Palm Squirrels (*Sciurus palmarum*) from India, presented by Mr. W. B. Bingham; two Common Squirrels (*Sciurus vulgaris*), British, presented respectively by Dr. J. L. Williams and Mr. G. S. Johnson; an Egyptian Jerboa (*Dipus aegyptius*) from North Africa, presented by Lady Preston; an Angola Seed-eater (*Serinus angolensis*) from Angola, presented by Miss Long; a Yellow-billed Sheathbill (*Chionis alba*), captured at sea, presented by Captain Bate; ten African Walking Fish (*Periophthalmus koelreuteri*) from West Africa, presented by Dr. H. O. Forbes; a Hocheur Monkey (*Cercopithecus nictitans*), a Moustache Monkey (*Cercopithecus cephus*), a Malbrouck Monkey (*Cercopithecus cynosurus*), an Angolan Vulture (*Gypohierax angolensis*) from West Africa, a Chacma Baboon (*Cynocephalus porcarius*, ♂) from South Africa, a Negro Tamarin (*Midas ursulus*) from Guiana, two Wandering Tree Ducks (*Dendrocygna arcuata*) from the East Indies, four Anderson's Tree Frogs (*Hyla andersoni*), four Changeable Tree Frogs (*Hyla versicolor*) from North America, deposited; an Orinoco Goose (*Chenalopex jubata*), a Blue-fronted Amazon (*Chrysotis aestiva*, var.) from South America, a Little Guan (*Ortalida motmot*) from Guiana, a De Filippi's Meadow Starling (*Sturnella defilippi*) from Argentina, purchased; two Collared Fruit Bats (*Cynonycteris collaris*) from South Africa, received in exchange; three White Ibises (*Eudocimus albus*), six Glossy Ibises (*Plegadis falcinellus*), bred in the Gardens.

OUR ASTRONOMICAL COLUMN.

FRENCH OBSERVATIONS OF THE TOTAL ECLIPSE OF THE SUN.—The *Comptes rendus* of the Paris Academy of Sciences for June 5 (vol. cxxx. pp. 1495-1529) contains the preliminary reports of several of the French astronomers who made observations of the recent total eclipse.

M. le Compte de la Baume-Pluvinel, observing at Elche, near the east coast of Spain, successfully carried out a very extensive programme. Nine photographs of the corona were obtained with objectives of 1·50 metres focal length; on these he says the coronal structure is almost identical with that he observed in 1889 at Salut. The planet Mercury is shown on all these plates, and will be useful for their accurate orientation. Three plates were obtained with a lens of 2·70 metres focal length, in conjunction with a cœlostat. For spectroscopic work three instruments were employed. A single prism spectrograph, with the slit in the line of the solar equator, showed the continuous spectrum of the corona extending to 12′ from the limb. Thirty-five bright lines were recorded, more intense on one side than the other. A second spectrograph had two objective prisms of spar and quartz; plates taken with this showed numerous chromospheric arcs, and a strong one due to the corona, this latter having no definite outer boundary. One interesting plate taken some seconds after totality still shows chromospheric arcs, and will furnish measures of the thickness of the chromospheric layers from the actual limb of the sun. An attempt to observe with a powerful six-prism spectroscope for special examination of the principal coronal radiation was rendered difficult by the feeble intensity of the image.

M. Ch. Trépied, director of the Algiers Observatory, also communicates a number of successful results. In addition to many accurate visual observations, twenty-eight photographs of the partial phases were made; six of the corona during totality, using an objective of 0·16 metre aperture and 1·03 metre focal length; the coronal extensions are recorded to 3·5 lunar diameters from the limb. Spectroscopic photographs were obtained with a Thollon prism spectrograph, an attempt being made to record the spectrum of the corona at diametrically opposite regions.

M. G. Meslin and party at Elche obtained eight photographs of the corona with a Henry lens of 16cm. aperture and 1 metre focus, and wide angle photographs of the region round the eclipsed sun for recording new objects. A photograph of the spectrum of the corona was obtained with a concave Rowland grating of 3 metres radius of curvature, used with a heliostat. The second order was photographed on plates 13 × 18cm., the spectrum extending from F to M; the images of the chromospheric radiations being portions of circles 16mm. in diameter.

THE TOTAL ECLIPSE OBSERVED AT SEA.—In an interesting letter written to the *Gibraltar Chronicle* of May 30, Colonel E. E. Markwick describes the appearance of the recent total eclipse as he and other fortunate passengers observed it from the Orient Steamship Company's R.M.S. *Austral.* The Company had considerately arranged that the vessel should be so navigated as to be near the central line of totality at the time of the eclipse, and, thanks to the skill of those in charge, this was accomplished with perfect success.

The position of the ship during totality was about Long. W., 9° 27′, Lat. N., 41° 3′, this being about 50 miles west of Oporto; the duration of the eclipse was about 1m. 31s. The Orient Company had provided an ample supply of glass plates, which, when smoked, permitted the passengers to view the progress of the partial phases, opera glasses being substituted during totality. During the eclipse the sky near the horizon was a lurid yellow, the clouds visible being reddish; the sea looked dark and sombre against the bright yellow of the sky. Close to the sun, however, the sky was quite blue; the darkness during totality was just sufficient to interfere with distinct vision.

The success of this enterprising project will probably induce many would-be observers in the future to adopt this exceedingly convenient and comfortable style of eclipse expedition; the departure from regular routine, though slight in itself, furnishing opportunity for really important scientific operations without disorganising any of the usual arrangements of the voyage.

NEW VARIABLE STAR IN CEPHEUS.—Prof. W. Ceraski, of Moscow, announces in the *Astronomische Nachrichten* (No. 3644) that Mdme. Ceraski has found a new variable on examination of plates taken by M. Blajko. The star is not in the D.M., and has the following position :—

$$\left. \begin{array}{l} \text{R.A.} \quad \text{oh. 28m.} \\ \text{Decl.} + 79°\ 33' \end{array} \right\} 1855.$$

The brightness varies from between 8–9 to about 12 magnitude. It was increasing in October 1896, and decreasing in October 1897 ; it was almost at minimum during May 1898, April 1899, and at commencement of May 1900.

EPHEMERIS OF EROS.—Herr F. Ristenpart communicates a revised ephemeris of this planet to the *Astronomische Nachrichten* (Bd. 152, No. 3643), as follows :—

Ephemeris for 12h. Berlin Mean Time.

1900.		R.A.		Decl.
		h. m. s.		° ′ ″
June 21	...	0 30 15·44	...	+ 9 57 27·1
23	...	33 42·19	...	10 33 33·6
25	...	37 8·20	...	11 9 49·0
27	...	40 33·44	...	11 46 13·3
29	...	43 57·91	...	12 22 46·6
July 1	...	47 21·57	...	12 59 29·0
3	...	50 44·44	...	13 36 21·0
5	...	0 54 6·51	...	+14 13 22·3

HOWE'S PHOTOGRAPHIC OBSERVATION OF EROS.—Mr. A. C. D. Crommelin writes to point out an error in our riote on the above, in which it was incorrectly stated that Prof. Howe's photographic observation of Eros was obtained during the solar eclipse of May 28. The photograph was taken before sunrise on the morning of the eclipse, some hours before totality. The error was introduced by the report of the observation being included in reports of the eclipse, and if uncorrected might lead to wrong estimates of the comparative brightness of the planet and of the darkness of the sky during totality.

A MODERN UNIVERSITY.
I.

THE granting of a Charter to the University of Birmingham, which has just become an accomplished fact, forms a fitting climax to an educational movement which may turn out to be one of the most momentous of the century. We have seen University Colleges called into existence in the great cities of the land by the perception of leading citizens that culture and scientific education of a high type must be brought to their doors and made accessible to all ; and we have seen the chairs of those colleges occupied by men who have devoted their spare time to the advancement of learning in various ways. All this has been of the greatest interest in the past and is full of hope for the future.

Side by side with these colleges there is now growing up in many cities a Technical School generally under Municipal Government, wherein artisans and hand workers generally may be trained in their craft, and in the main principles underlying it, in a more direct and satisfactory manner than by the old system of apprenticeship.

Such schools can no more turn out a finished artisan than the colleges can turn out a finished scholar. Much remains to be learned in later life and in the actual pursuit of trade or profession, but the early stages are overcome not only more rapidly, but far more thoroughly, by aid of direct instruction ; and in the more favourable cases a substratum of scientific knowledge is laid, and a grasp of principle attained, which must be of the utmost benefit hereafter, and could never have been obtained on the old plan. It is this scientific training in principles which is the really needful thing, when the public is educated enough to perceive it ; it is this which is of interest to the educationist, and not a mere instruction in handicraft : it is the making of men, and not the making of machines, which is of vital importance to the future of a country.

Without a training in principles a man remains ignorant and narrow, limited to the performance of the one thing which he has been trained to do, and incapable of turning his attention profitably to anything else ; inelastic and incapable of devising or of assimilating modifications and developments, which, as they come in, tend to leave him stranded and belated, waiting only for a period of slacker demand to throw him out of employment. And even if the artisan and the foreman are well educated, there remains his employer to be considered. If he is ignorant—too ignorant to turn his enterprise in the right direction when oppor-

tunity offers—his workers must suffer, and the whole nation suffers with them. But though Colleges and Technical Schools impart education on the one hand and instruction on the other ; though they enlarge and make more real the education available to the average citizen, they do not control and modify the educational ideal of the country. That ideal remains in many respects still essentially the same as it was at the beginning of the present century, before all this amazing inrush of new knowledge. The new knowledge has not yet been incorporated into education. The half-hearted effort made by schools to introduce what they term a "modern side" only serves to emphasise the blankness of the prospect. They say, and say truly, no doubt, that the new studies do not answer. They do not pay either for Government appointments or for the university. But a new university, able to set its own standards, select its own faculties, and set its seal on students of its own subjects, has far larger possibilities before it. It can control, and not only impart, education. It may need an effort to rise to its privileges. The easiest plan is to follow the lead of others and establish degrees on the worn old lines, but that is not what we expect and hope from the new university of the Midlands. We hope to see it break away from mediæval traditions and realise the need there is for a new educational ideal.

The aim we have before us is an aim at actualities rather than at artificialities ; at real things rather than at conventions.

There is a stage of thoroughness at which a study of the conventionalities of grammar and orthography is able to convey real information about men and things—the advanced stage when it becomes the science of philology—but as usually learnt by ordinary persons it is little better than a conventional code, and set of rules. If there was little in the world to learn about—as in the middle ages there was but little—it might be well to spend much time in acquiring precisely the gender of nouns and the terminations of irregular verbs in different foreign tongues ; not only for practical purposes but for mental training ; but amid the superfluity of real subjects of the present day, of all of which the ordinary person is densely ignorant, to immerse him for a long period in these barren studies is wasteful of his youth.

On the other hand, History is reality ; and some knowledge of history is necessary for every one. Art, again, and Literature and Music are, or may be, realities ; and the vast majority who have no power of creation should at least learn reverently to appreciate the great work of the greatest masters in all subjects, unless they are deaf and dumb and blind. The things really valuable to the human race should be made in some degree accessible to all, and this part of the work of education the Press and the Stage indirectly in some degree accomplish ; imperfectly, no doubt, but often more really than do the bodies which make the attempt in a more academic way.

Thus we would discriminate between the conventionalities of language and the realities of literature, just as we discriminate between the laws of colour and perspective, the technique of the painter on the one hand—and the great work of art itself, the expression of a thought or of an emotion, or of a beauty or of a fact. To the scholar, as to the painter, the two are inextricably interwoven ; technique is the material in which he works ; but the general human race, who have to do the work of the world, and who constitute the bulk of the nation, are neither scholarly nor artistic, and it is both wasteful and cruel to plunge them into technique, and disgust them with the—to them—dull and meaningless details, instead of educating them in the finished work possible only to masters of the craft.

The same sort of things do we say of science and of mathematics. Here, again, there is too great a tendency to educate youths in subtleties and artifices and minutiæ, as if they were going all to be accomplished mathematicians or men of science. The teacher is himself, perhaps, a mathematician, and so thinks that what was necessary for him is suitable for everybody. More usually, of course, the teacher knows very little about it, and feels only that he was himself taught that way, and that he must pass it on. Only a few stop to think what they are doing, and these are the educationists ; what they have to say is written at large, and there is no need to repeat it. Some of them are faddists, doubtless ; not all are wise ; but it is well at any rate to try and think a matter out ; and the speculative teaching even of a faddist is likely to be more stimulating than the tenth-hand droning of a conventional pedagogue. To indicate our meaning in terms of mathematics and science, as we have tried briefly to indicate it in the domain of more humanistic studies, we would say that a good deal of the teaching of Euclid and algebra and trigonometry is conventional, and unsuitable to the average youth. If a youth is going to be a mathematician, it matters very little how he is taught these things, so long as they are put in his way. He can hardly have too much of them, he can look at them every way and they present no difficulty; but even the young mathematician might be saved the wearisome and long continued grind through the conventional books of Euclid, with the result that at the end he knows about as much geometry as a month of reasonable teaching would have given him. It is not mathematics at all that he is studying when he is doing Euclid in the usual way, it is a piece of old world literature, very admirable in its proper pláce, no doubt, and read through at a fair pace quite interesting. It is at any rate far more interesting than the military despatches with which so large another portion of his time is usually, at the same time, being burdened—a form of literature which is not of the slightest interest, and leaves no residue of real information in his mind, except that Gaul used to be divided into three parts; that Cæsar's army built an ingenious and highly technical bridge; that he had difficult times in conquering a people who are not our ancestors, but who happened to occupy the same plot of ground on the earth's surface as we do now.

The conventional part of algebra we refer to may be illustrated by G.C.M. and other rules which in actual work are never needed or employed even by mathematicians. Factors and Equations and Progressions are well enough,—all those parts that are really used or likely to be used hereafter, and all those parts which give a firmer grasp of principles.

Thus in arithmetic, familiarity with such a subject as scales of notation—with the principles, that is, of Arabian numeration—will be really educative and far more helpful than excessive repetition of a rule called "practice," and much dealing with commercial articles. A variety of problems from mensuration, mechanics and heat might be introduced into arithmetic, and the subject made more living than it is apt to be. Mensuration and practical trigonometry may be made truly educative subjects, and a quantity of arithmetical exercises may be founded upon things actually done in the workshop, the laboratory and the field, in the working out of which boys might readily be got to take a real interest.

But all these are school subjects. What have they to do with a university? They have a great deal to do with it in reality, for it is one of the functions and the privileges of a university indirectly to control, or rather influence, the schools. The influence is quite natural and unavoidable anyhow, but on the side of exhibitions and scholarships it becomes obvious and direct.

The schools must train largely for the universities, and the universities must train largely for life.

Now it is just in this training for life that the universities have proved deficient. The only life they have contemplated has been that of the politician and the lawyer on the one hand, and the scholar and recluse on the other. The kind of training needed or supposed to be needed by the past generation of statesmen has been supplied—with results not wholly and completely satisfactory; the kind of training needed for the highly specialised scholar has likewise and will always be supplied. The ancient universities are the natural homes of this kind of learning, and no modern institutions can hope or should attempt to compete with them. The aroma of centuries is a unique growth, and should be carefully fostered and reverenced by a busy and pushing generation.

It would be a calamity if anything were done to destroy the peace and old world quiet of mediæval institutions, founded on monastic traditions and full of attraction for the few who are called to be learned. That Oxford should specialise in archæology and ancient philosophy is most appropriate; that it should regard with jealous eyes the learning of the present century, and hesitate about letting its old bottles be endangered by the inclusion of new wine, is natural and may be wise. We would urge its custodians jealously to preserve the old learning, and leave experiments in new developments to younger and less fragile growths. We would treat the old universities like old buildings, relics of the past, to be most carefully preserved, and supplying something in the life of the nation which no amount of energy or reforming spirit is competent to supply.

If this old world atmosphere disappears it is an irreparable loss. Let its custodians be jealous and conservative; if they see no

way of engrafting the new learning on the old stock, without ruining it, then it were far better that the new learning should be planted in fresh soil.

Such soil is furnished by natural circumstance at the intersection of great trade routes, at the market places of the world. Here the average man is at his strongest and busiest, here he is most actively in touch with life on this planet, and is serving his day and generation with an energy which is unmistakable. The motives, doubtless, are mixed, and the results are mixed, there is little Utopian about them; yet there is real self-sacrifice for a far-off good instinctively felt. Ugly surroundings are put up with, as a concomitant apparently necessary, and as at any rate temporarily unavoidable; and life is lived hard for an end not often clearly grasped, yet powerfully felt to the uttermost parts of the Empire. It is on such strenuous home industry, of director, of manager, of foreman, of artisan and of salesmen that our empire is established; and if there is one thing that we are more powerfully realising at present than another it is that our empire must be consolidated, that fresh guiding force must be available, not mere energy—of that there is plenty—but more directing force, more intelligent guidance, more discrimination, more breadth of view—in a word, more real education.

The present war will wake the people of these islands out of their comfortable lethargy. They will see that to hold our position in the modern world we require improved training, not only in rifle shooting and artillery practice, but in every department of activity. Other nations are leaping to the front and spending public money lavishly to get their people better educated, better fitted for seizing new ideas and applying them; it will never do for us to lag behind.

A few scholars, a few men of science, a few men of genius in various branches, these will not save a nation. They extend its fame, they adorn it, they stimulate it, and they reward it; but the backbone of a nation is the average man, the average man of affairs, the man who does the business of the world. If he breaks down or is crippled, the ornamental head cannot be supported. He may be a professional man or a merchant, or he may be a manufacturer or a tradesman, but whatever he is, he must not rest on his oars and be content with the tradition of the past. We are entering a new century, many traditions of the past are out of date, and the vital thing for the nation to realise, if it is to maintain its hard won supremacy, is that antique methods of education will no longer serve. They have had their day; they need not yet cease to be, but they must be supplemented by others. The modern university must take care of the average man. Plenty of long-established universities will look after the high honours men, and in every department highly specialised training is already available; our artisans are as skilled, each in his narrow groove, as it is possible for man to be—marvels of mechanical skill they are; but where is the breadth of view, the elasticity, the power to modify, to invent, to reform, to seize new conditions, to adapt one's self to the growing and changing needs of the world? A foreign order comes to an engineering works of the present day with its sizes expressed in decimals. Before the order can be given out it has to be interpreted into clumsy sixteenths and thirty-seconds and sixty-fourths of an inch, which alone the workman understands.

Nor is this portentous ignorance limited to workmen. Directly the domain of science is touched, your ordinary school-trained average man is stranded—he is ignorant even of the scientific alphabet—scientific principles are a sealed book to him: the divorce between science and practice, except in the case of a few leading firms who have already wakened up like their continental confrères, appears to be complete.

The modern university must aim for a long time not at depth so much as at breadth. Depth for the few, breadth for the many. It must seek to turn out all-round men, and not specialists only.

Its graduates should not one of them be illiterate, not one of them ignorant of the fundamental principles of science. Trained scientific men they cannot be, in any numbers—the idea would be absurd—but they should have sufficient education to understand a scientific question and know where to go for the answer. They should have lived for a time—even a short time—in the atmosphere of science, and thereafter it will never be quite strange to them.

The scientific training need not be given solely in an academic manner, aloof from all questions of practical interest.

Some people are best trained in this manner, but other persons with a vivid practical interest or experience in application to life and work are best trained in close touch with practical conditions. Medical training is the best example ; that is thoroughly done. We would have other training arranged on the same practical lines. The modern university will seek, so far as it can, to allow for differences of aptitude, or, it may be, differences of preliminary training. It will not seek to force every undergraduate at first through an arts course, and then through a science course, and then through a technical course. It may be well to do this with professional men, but not with all. Every graduate should pass through these three stages before he can be turned out a useful and educated citizen, fit intellectually to take his share in the work of the world ; but he need not in every case of necessity take them in this logical order.

To force a boy through a course of language or history or literature, at a period when for some reason he is not attracted by it, is doing him but little good. It may, indeed, do him harm, by breeding disgust for subjects which at a later stage he might realise were necessary, and, when properly taught, enthralling. It is love of culture, and not hatred of it, that should be implanted. The so-called " preliminary in arts " course should be taken compulsorily at some stage of a graduate's complete career, but not necessarily at the beginning. A student who has been immersed for a term in purely technical studies will, if he is good for anything, turn to such human subjects with relief ; and it is not fair to turn him out in the world without some worthy human interest and solace. The university has failed in one of its functions if it permits him to depart trained in nothing but unhuman technology.

But then, on the other hand, the arts man, the lawyer, the merchant, the man of business, and still more the teacher—how much better would they not be for a tinge of scientific training. Their ignorance does not come home to all of them, but to many it does ; and probably in middle life they strive, by attending popular lectures and miscellaneous semi-scientific entertainments, to obtain a growth by a top-dressing of superficial information never really assimilated, seldom adequately understood. A manuring of science placed low down when young would have rendered the surface soil fertile, and this later growth easy, just as the youthful smattering of letters renders moderately easy and interesting the subsequent reading of history, or, in some cases, even the learning of a new language ; but to the wholly untrained person these things are, and remain, hopelessly difficult.

A broad training all round can only result in what specialists would call a smattering—what we should prefer to call a leaven ; but so long as it is not confined to a learning of trivial details, but represents a grasp of some of the fundamental principles of a subject, it is all that most men ever have, or can have, in any branch but their own, however highly educated they may be. It takes a very exceptional man to be really learned, or to be able to say anything really worth hearing, off his own subject. There are men who make a large portion of knowledge their province, but the majority of men cannot and should not aim at this. They should know one thing well, and in all else they should not be entirely ignorant.

This absence of entire ignorance is a far more valuable commodity than is usually supposed. It enables the man of affairs to consult specialists with advantage. Special knowledge is always available, if one knows how and where to look for it ; but the man of complete ignorance is at the mercy of every charlatan ; he puts his money into the wildest scheme, on the one hand, and on the other he fails to realise possibilities of sound application lying all about him. His enterprise and power may be great, but the blight of ignorance makes him useless ; and it is just this blight which is endangering our continued industrial and commercial supremacy among the nations of the world.

We look to the new type of university now about to be created to remedy this state of things. If Birmingham succeeds in its high emprise, other great cities will follow suit. The experiment is one that is of interest to the whole British Empire, indeed to the whole Anglo-Saxon race.

In another article we may perhaps enter more into detail concerning some of the features of the scheme ; but it is at present in such extreme infancy that its features are barely recognisable. It does not follow that what is immediately to the front is in reality the most important or the most characteristic.

(*To be continued.*)

THE STEADYING OF SHIPS.[1]

THE evolution of the modern flat-bottomed merchant vessel, with its midship section of approximately rectangular form, from its old pointed-bottomed prototype, with deep central keel, has been a necessary result of commercial competition. The naval architect is called on to increase the carrying capacity of his vessels to the utmost extent, and a limitation is imposed on their draught of water by the limited depth of harbours, docks, rivers and, last but not least, ship canals. The old central keel has had to disappear in order that the extra foot or two of displacement might be utilised for the carriage of cargo, and a substitute has had to be found for it by the attachment of " bilge-keels " or side keels projecting from the ship at the only places where they could be placed without taking up valuable space—namely, at the two rounded-off corners of the rectangular section.

The efficiency of bilge-keels in modifying the rolling oscillation of ships seems to have been for some time a debated point among naval architects, and the experimental fact that the extinction of oscillation produced by these keels may in some instances be many times—possibly as much as ten times—that which would be inferred from determinations of the resistance of a paddle oscillating in water certainly appears at a first glance paradoxical. On reading Mr. Luke's paper in the *Transactions* of the Scotch Shipbuilders, and subsequently Sir William White's account of his experiments on the *Revenge*, it occurred to me that the properties of discontinuous fluid motion, so long a favourite study among mathematicians, might be put to a useful purpose in explaining the high resistances to rolling observed with the use of bilge-keels. So far from these resistances being in contradiction with the principles of hydrodynamics, they appeared to be to a large extent in conformity with our theory of free stream-lines, and this view has been borne out by subsequent calculations, certainly to a far greater degree than I at first anticipated.

According to hydrodynamical theory, if a solid body is set moving through or rotating in an unlimited mass of perfect fluid previously at rest, the motion will continue indefinitely, provided that the body has no sharp edges or corners projecting into the fluid, and that the velocity does not exceed certain limits. The motion involves no continuous expenditure of energy, and if the solid is brought to rest, the fluid will come to rest, and the energy which was expended in starting the motion will be recovered. If, however, the body has any sharp projecting edges, the fluid is unable to flow continuously round these, and discontinuous motion is set up, a mass of dead water being dragged along behind the projecting edges, and this dead water being separated from the moving fluid by a " surface of discontinuity " in crossing which the velocity changes abruptly. In this case the fluid motion is not destroyed when the solid is brought to rest, and energy is absorbed by the fluid. The theory of discontinuous motion is the basis of the well-known calculations of the resistance experienced by a plane lamina moving through a liquid, originally due to Kirchhoff, and subsequently developed by Lord Rayleigh, Love, Michell and others.

The case of a ship floating in water rocking from side to side differs from these ideal cases in the properties that (1) waves are produced on the surface, (2) that water is not a *perfect* fluid ; so that energy is being continuously absorbed by wave-formation, and by the viscosity of the water. If the ship has no sharp keels projecting into the water, these are the only causes which retard the rolling of the ship, but as soon as keels are attached discontinuous motions are set up, which involve a further absorption of the energy of rolling, and the oscillations subside much more rapidly. If we imagine the ideal case of a ship floating in a perfect liquid, the surface of which is coated with a perfectly rigid sheet of ice entirely preventing any waves from forming, but just allowing free play for the ship to roll, the oscillation would continue indefinitely, provided the ship had no sharp projecting keels. If, however, bilge-keels were attached, the oscillations would gradually die down, the energy of rolling being absorbed by the production of discontinuous motions, and being transformed into kinetic energy of the liquid.

The object of this investigation is to show that the efficiency of bilge-keels in modifying the rolling of ships may be greatly increased by the action of the sides of the ship itself, and is so increased in a ship of section approaching to a rectangular form, provided that the bilge-keels are attached at the protruding

[1] Abstract of a paper read before the Institution of Naval Architects.

corners of the section. This increased efficiency is due to two causes :—

(1) The rocking of the ship produces currents in the water, which flow round the corners in the opposite direction to that in which the ship is rolling, thereby increasing the pressure on the bilge-keels.

(2) The discontinuous motion past the bilge-keels alters the distribution of pressure against the sides of the ship, and the differences of pressure thus produced have a moment always tending to retard the rolling motion.

The effect of stream-line motions.—Consider a cylinder of section, such as represented in Fig. 1, rotating in fluid about

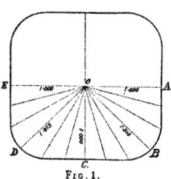

Fig. 1.

an axis through its centre O. It is known that the fluid displaced by the motion of the cylinder will flow past its protruding corners B, D in the opposite directions to that in which the cylinder is moving ; while at the points A, C, E the fluid will be moving in the same direction as the cylinder. Hence if a small lamina representing a bilge-keel be placed at B or D, it will encounter a current of liquid flowing in the opposite direction to that in which it is moving, and the pressure on the lamina will be correspondingly increased.

I made several calculations to form some estimate of the increases produced by these counter-currents on the resistance experienced by a suitably placed bilge-keel, assuming the resistance to vary as the square of the relative velocity. Taking certain sections more or less approximating to the form of a square with rounded corners, in a section where the greatest radius O B exceeded the least radius O A by 13 per cent., the resistance was increased, owing to this cause, by about 36 per cent. ; while in the section actually represented in Fig. 1, O B exceeded O A by 21 per cent., and the resistance on a lamina at B came out 67 per cent. greater than it would be if the lamina had only to encounter the relative velocity of the fluid due to its own motion.

In these cases the fluid was supposed to be indefinitely extended. To estimate the influence of surface-conditions *without*

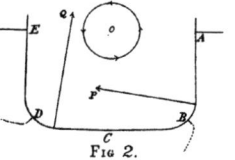

Fig 2.

taking account of waves, I next considered the case of a cylinder partially immersed in a liquid bounded by a rigid horizontal plane, the cylinder itself rocking about an axis in the plane of the surface. The form chosen for the section of the submerged portion was roughly suggested by a diagram of the ship *Revenge*, and the counter-currents past the protruding corners were found to be sufficient to more than double the resistance on a suitably placed bilge-keel.

The effect of pressure-variations against the sides of the ship.—The resistance on a lamina moving in fluid is due to the pressure being greater on the front than on the back of the lamina, this difference of pressure being the result of the discontinuous character of the fluid motion. When the lamina is attached to a

ship as a bilge-keel, this difference of pressure also extends along the sides of the ship, the pressure in front of the bilge-keel being greater than behind it. Whether the pressure in front is increased or the pressure behind is decreased, or both of these effects take place simultaneously, is unimportant, for we may on each of these hypotheses represent the effect by an excess of pressure on the portions in front of the bilge-keels as compared with the portions behind.

Now let Fig. 2 represent a diagrammatic section of a ship approaching to a rectangular section, with bilge-keels at its protruding corners, B, D. Then if the ship be rolling about its centre of gravity, O, in the direction of the circular arrow, the greater pressures in front of the bilge-keels will be distributed over the segments A B and C D, and, as indicated by the arrows P and Q, their moments about O will be in the direction opposed to rotation. When the ship rolls in the opposite direction, the greater pressures will be on the segments B C and D E, and their moments will again tend to retard the rolling. In this way the pressures on the sides of the ship will materially assist the bilge-keels in steadying the ship.

To test the extent to which the pressure on the sides of the ship is likely to be modified by the presence of the bilge-keel, I examined the case of a fluid flowing along a plane, A B, with an edge, B C, projecting at right angles to it (Fig. 3). If from B there be measured off on B A a length = ·927 B C approximately, the thrust on this portion of B A is equal to the thrust on B C, and the average pressure on this portion is therefore a little greater than the average pressure on B C. If, again, we measure off on B A a length = 2·042 B C, the thrust on this portion is equal to twice that on B C. Speaking in general terms, we may say that the

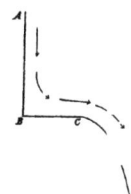

Fig. 3.

pressure is greater along B A than along B C, and that it does not fall off rapidly along the arm B A. If the arm B A, instead of being straight, bends away from B C like the curved side of a ship, we should expect the pressure on it to be even greater than in the case considered. These considerations led me to consider the results of supposing the bilge-keels to produce on the segments in front of them (A B and C D of Fig. 2) uniform increases of pressure equal in intensity to the average pressures on the keels themselves. Applying this hypothesis to a ship, the details of which were kindly furnished me by Dr. Elgar, I found that the total retarding moment came out to be about 3·9 times the retarding moment on the bilge-keels alone. This result, taken in conjunction with the previous result that the currents past the sides of the ship may double the pressure on the bilge-keel, shows that there is nothing paradoxical in the supposition that bilge-keels may have eight times the efficiency in extinguishing oscillations that would be inferred from experiments made with a simple paddle moving freely in fluid.

We learn from Sir William White that in the case of the *Sultan* the agreement between Mr. Froude's estimate of the resistance to rolling, based on the coefficient of resistance of a lamina, and the experimental facts was very close indeed. But the midship section of the *Sultan* was very much more nearly circular than that of the other ships experimented on. For a circular section there would be no counter-currents producing increased resistance, nor would the changes of pressure against the sides of the ship possess any retarding moment.

There are many further points connected with the investigation which want of space prevents us from discussing here. One such point is the fact that the rolling motion is not steady, but oscillatory, and, therefore, account must be taken of the pressures involved in producing changes of motion, in consequence of which the pressures differ from those which would exist at any instant if the motion were steady. These pressures, which I have called the "$d\phi/dt$ pressures," after the term which produces them in the hydrodynamical equation, are completely modified in their action when continuous motion is replaced by discontinuous motion, but their effect can only be discussed from general principles. Other points are the effect of the ship's forward motion in increasing the steadying action, and the influence of bilge-keels in modifying pitching and in improving the steering of ships ; the two last effects are further simple consequences of the properties of discontinuous motion.

We have seen that the action of bilge-keels in steadying a ship is largely modified by the actions of the sides of the ship, and is much augmented when the keels are placed in a suitable position ; and it is interesting to notice how the exigencies of trade, while they have necessitated the removal of keels from the middle line of the ship to the sides, have brought about such a change in the form of the section as to render the new position by far the most effective. In the old pointed-bottomed ship, the central keel was the best, as it not only intercepted the currents flowing round the bottom in each swing from side to side, but also produced a difference of pressure on the two sides whose moment was always opposed to rolling. In the modern flat-bottomed ship of rectangular section a central keel would be unfavourably placed for this purpose, as not only would the water tend to flow in the same direction as that in which the ship is rolling, producing a diminished pressure on the keel, but the differences of pressure on the sides of the ships would have a moment tending to diminish the resistance to rolling. The favourable position now commonly assigned to the bilge-keel is calculated to render its addition to ships of the greatest value in increasing their steadiness.   G. H. BRYAN.

THE "ORTHOSTIGMAT" LENS.

A SPECIMEN of the new series of lenses issued under the joint auspices of Messrs. Beck and Steinheil, has been tested and examined by us. The lens is of the rectilinear symmetrical type with two equal combinations, each consisting of three elements. It is by special construction of the surfaces of these components that the remarkable qualities claimed for, and undoubtedly possessed by, the new lens are attained. The great drawback to the best of the old type lenses was the curvature of field, and it is only in recent years that the discovery of the new varieties of glass has made it possible to correct this, and at the same time perfect the corrections for spherical aberration and astigmatism. The result of the process is that each component of the lens is made up of a positive meniscus, with a double convex lens cemented on one side of it and a double concave one on the other. The centre positive lens will consequently have a lower power than the two outer ones.

When it is understood that each of the twelve surfaces involved in the complete lens is worked with such accuracy that an error of 1/40,000th of an inch is inadmissible, the increased price, compared with the old types of lenses, is amply accounted for in the superior product obtained.

We have tested the lens, which is of about 4¾ inches solar focus, and are quite satisfied that it fulfils to a high degree of perfection the claims made for it by the makers. Although the lens is listed for ¼ plate, the circle of good definition is much larger, and with a stop of $f/16$ excellent definition was obtained over the whole of a ½ plate. The lens at its greatest rapidity works at $f/6\cdot3$, and at this aperture the definition appears very perfect over an area 4¼ inches square. The makers call attention to the special attempts they have made to eliminate astigmatism, and it is not until an oblique pencil falls considerably beyond the listed angle that any distortion shows itself. To make this clearer, let the image of the sun, moon, a star, or any distant object, be focussed at the centre of the plate, and then the camera so tilted that the image falls gradually away to the extreme corner. The slightest distortion can be at once recognised in this manner. With the lens in question no distortion was evident until the image was received at an angle of about 30 degrees from the axis, and for another 10 degrees further the resulting astigmatism, although present, was much less than is present

closer to the axis in a rectilinear of the ordinary type. So that for an angle of view of over 60° the new lens is practically non-astigmatic at the large aperture $f/6\cdot3$. This will recommend it especially for all process and copying work, where critical definition and speed are primary necessities. A word should be said concerning the focussing with these new lenses. This adjustment must be critical, as an almost imperceptible displacement of the plate will suffice to throw it out of the focal plane sufficiently to destroy the definition, and so create false impressions of the capabilities of the optical system. For all such work, therefore, only perfectly rigid apparatus is permissible.

Another important qualification of the lens is its comparative freedom from chromatic aberration, in virtue of which it will be useful for work connected with colour photography, obviating the laborious and uncertain corrections which are necessary in such work with the ordinary lenses, whose chromatic aberrations are only suppressed for the blue and yellow. Critical tests of this correction have not been possible, but sufficient have been made to show that the outstanding error is small.

On account of their covering power, the smaller sizes will be found excellent for low-power microphotography. The new lenses are obtainable with foci varying from 3½ in. to 23½ in., covering plates from 3¼ in. square to 28 × 24 inches.

UNIVERSITY AND EDUCATIONAL INTELLIGENCE.

OXFORD.—The Romanes Lecture will be delivered by Dr. James A. H. Murray on Friday, June 22, upon "The Evolution of English Lexicography."

Sir William Thiselton-Dyer, F.R.S., has been made a permanent elector to the Sherardian Professorship of Botany, in succession to the late Prof. Bartholomew Price.

The Statute instituting Diplomas in Geography has been approved by Convocation, and is to remain in force until October, 1904.

The extensive bequest to the University under Mr. Fortnum's will has made an enlargement of the Ashmolean Museum necessary. The cost is estimated at 1500*l*., and towards this sum Brasenose College has offered a contribution of 500*l*.

At Merton College there will be an election to a fellowship on October 6, after an examination in Animal Physiology and Animal Morphology. Candidates are requested to inform the Warden by September 10 of their choice between these two subjects, and to submit, if they wish, original papers or memoirs.

At a meeting of the Oxford University Junior Scientific Club, held on Wednesday, June 6, Sir John Burdon Sanderson, F.R.S., gave an account of the method he has been lately employing for producing tetanus in muscle, by means of telephone currents produced by musical sounds, showing how the results bear on the vexed question of the inherent rhythmicality of muscle and nerve-cell. The lecture was illustrated by experiments.

At a meeting of the above club held on Friday, June 15, Mr. T. C. Porter, of Eton College, described the growth of the shadow of the Peak of Teneriffe, as witnessed from the summit of the mountain at sunrise and sunset, and its gradual eclipse by the shadow of the earth. He showed photographs taken at the time, and explained how by means of an ordinary watch and pair of opera glasses a rough value of the diameter of the earth might be deduced.

Mr. A. F. Walden, New College, made a preliminary communication to the club on the theory of labile hydrogen atoms.

CAMBRIDGE.—The researches of Mr. J. C. McLennan on electrical conductivity in gases traversed by kathode rays, and of Mr. R. L. Wills, of St. John's College, on the magnetic properties of iron as influenced by temperature and the presence of other elements, have been approved by the Degree Committee as qualifying for the B.A. degree.

In the Natural Sciences Tripos, which is now the largest of the Honour Examinations, forty men and three women are placed in the first class of Part I. In Part II. fourteen men and no women obtain first-class honours.

At St. John's College the following awards in natural science were made on June 18. Foundation Scholars continued or elected : Lewton-Brain, May, Adams, Ticehurst, Fletcher, Browning, Wakely, Gregory, Williams, Harding, Hepworth,

Pascoe, King, Macalister, Mitchell. Exhibitioners : Crocker, Denham, Simpson, Balls. Hockin Prizeman (for electricity) : Browning. Engineering Scholar : Paton.

THE attention of teachers and others engaged in schools is directed to the appeal made by Prof. Karl Pearson in our correspondence columns. Observations of the physical and mental characters of children are required, and measurements of the head, in order to provide material for an investigation of heredity upon which Prof. Pearson is engaged. There should be no difficulty in obtaining the co-operation of masters and mistresses in schools in this work, for the observations and measurements can be made with very little trouble, and they are of as much interest from an educational point of view as they are to biological science.

SCIENTIFIC SERIAL.

Bulletin of the American Mathematical Society, May.— The number opens with four papers read before the Society at the dates annexed : On the geometry of the circle, by Dr. V. Snyder (December 28, 1899) ; isomorphism between certain systems of single linear groups, by Prof. L. E. Dickson (February 24) ; the Hessian of the cubic surface ii., by Dr. J. I. Hutchinson (February 24) ; and note on the group of isomorphisms, by Dr. G. A. Miller (February 24). These papers are short and, in the main, continuations of work previously published by the authors.—Prof. F. S. Woods contributes an interesting sketch of a German translation, by F. Engel, of two articles by Lobachevsky, with the titles "Ueber die Anfangsgründe der Geometrie" and "Neue Anfangsgründe der Geometrie mit einer Vollständigen theorie der Parallellinien." The reviewer's conclusion is that, "while it is remarkable that the solution of a two-thousand-year-old problem should be given almost simultaneously by three men, it should be remembered that these three were not the only mathematicians who had worked upon the problem. More than one had missed the solution by a hair's breadth only ; Lobachevsky, Bolyai and Gauss succeeded in finding it."—Other notices are Vogt's "Algebraic solutions of equations," by J. Pierpont ; the elements of the differential and integral calculus, based on the work by Nernst and Schönflies (translated by W. A. Young and C. E. Linebarger), by L. E. Dickson ; and E. Pascal's "Die Variationsrechnung," by J. K. Whittemore.—University and general mathematical information come into the "Notes" and "New Publications."

SOCIETIES AND ACADEMIES.

LONDON.

Royal Society, April 5.—"The Kinetic Theory of Planetary Atmospheres." By Prof. G. H. Bryan, F.R.S.

The application of the kinetic theory to the atmospheres of planets dates from the paper of Waterston, who gave an investigation based on the then only possible assumption of equal velocities for all molecules, an assumption since known as Clausius' law. Of later papers reference is due in especial to Dr. Johnstone Stoney's memoir "Of Atmospheres on Planets and Satellites" (*Trans.* R. Dublin Soc.), in which the test of permanence of a gas in the atmosphere of a planet is made to depend on the ratio of its velocity of mean square to that relative velocity which would enable a suitably projected body to escape from the planet's attraction. If it be admitted, as Dr. Stoney assumes, that helium cannot exist in our atmosphere, it follows that vapour of water cannot exist on Mars.

The author's object has been to investigate the logical conclusions obtained by applying the Boltzmann-Maxwell distribution to the atmospheres of planets. In 1893 calculations were made, having special reference to the absence of atmosphere from the moon, but these took no account of axial rotation. When this cause is taken into account, the distribution of co-ordinates and *relative* velocities of the molecules is found to be the same as if the planet were at rest, and "centrifugal force" applied to the system. The surfaces of equal density are of the forms originally investigated by Edward Roche, of Montpellier, and they cease to be closed surfaces when passing to the outside of the point on the equatorial plane where centrifugal force just balances the planet's attraction. Calling the surface through this point the "critical surface," the density of molecular distribution over this surface must be very small to ensure permanence.

The ratio of the density at the planet's surface to the density at the critical surface has been called the "critical density ratio," and the author calculates its logarithm for particular gases at different temperatures on the various planets. The use of this logarithm has the advantage that the calculation can at once be extended to any gas at any temperature.

The high value obtained in the case of helium, considered in reference to the earth, appears to afford abundant proof that if helium existed in our atmosphere it would possess a very high degree of permanence at ordinary temperatures. To test this point further, a calculation is made of the total rate at which molecules would flow across the critical surface, this rate being regarded as a superior limit to the rate at which the planet would lose its atmosphere, since it takes no account of molecules which describe free paths beyond the limit and fall back again. To further exhibit the results in a tangible form, the rate of flow is estimated by the number of years in which the total amount of gas escaping across the critical surface would be equal to the amount of the gas in a layer covering the surface of the planet to the depth of 1 cm. This measure is independent of the actual quantity of the gas under consideration existing in the atmosphere, since, if this quantity be increased, the rate of flow across the critical surface and the amount of gas present in the surface layer 1 cm. thick will be increased in the same proportion.

If a gas of molecular weight 2, such as helium, be supposed to exist in the earth's atmosphere, the loss in question would occupy $3 \cdot 5 \times 10^{36}$ years at $-73°$ C., 3×10^{19} years at $27°$, $8 \cdot 4 \times 10^{10}$ years at $127°$ C., 6×10^5 years at $227°$ C., and 222 years at $327°$ C.

If we halve the absolute temperatures, we have the conditions applicable to hydrogen, the losses in question therefore taking place in $8 \cdot 4 \times 10^{10}$ years at $-73°$ C., 6×10^5 years at $-23°$ C., and 222 years at $27°$ C.

For water vapour on Mars, the corresponding results are $1 \cdot 2 \times 10^{33}$ years at $-73°$, $1 \cdot 9 \times 10^{16}$ years at $27°$, $2 \cdot 4 \times 10^{11}$ years at $127°$, $4 \cdot 3 \times 10^5$ years at $227°$, and 106 years at $327°$.

These figures indicate that helium cannot practically escape from our atmosphere at existing temperatures, nor can vapour of water escape from the atmosphere of Mars. A leakage may, and undoubtedly does, take place, which may appear considerable when estimated by the number of actual molecules escaping, but it is wholly inappreciable relative to the mass of gas left behind.

At a future time it is proposed to examine the corresponding results, based on the hypothesis that the atmosphere of a planet is distributed according to the adiabatic instead of the isothermal law.

"On the Weight of Hydrogen desiccated by Liquid Air." By Lord Rayleigh, F.R.S.

In recent experiments by myself and by others upon the density of hydrogen, the gas has always been dried by means of phosphoric anhydride ; and a doubt may remain whether, on one hand, the removal of aqueous vapour is sufficiently complete, and on the other, whether some new impurity may not be introduced. I thought that it would be interesting to weigh hydrogen dried in an entirely different manner, and this I have recently been able to effect with the aid of liquid air, acting as a cooling agent, supplied by the kindness of Prof. Dewar from the Royal Institution. The operations of filling and weighing were carried out in the country as hitherto. I ought, perhaps, to explain that the object was not so much to make a new determination of the highest possible accuracy, as to test whether any serious error could be involved in the use of phosphoric anhydride, such as might explain the departure of the ratio of densities of oxygen and hydrogen from that of 16 : 1. I may say at once that the result was negative.

Each supply consisted of about six litres of the liquid, contained in two large vacuum-jacketed vessels of Prof. Dewar's design, and it sufficed for two fillings with hydrogen at an interval of two days. The intermediate day was devoted to a weighing of the globe *empty*. There were four fillings in all, but one proved to be abortive owing to a discrepancy in the weights when the globe was empty, before and after the filling. The gas was exposed to the action of the liquid air during its passage in a slow stream of about half a litre per hour through a tube of thin glass.

I have said that the result was negative. In point of fact the actual weights found were $\frac{1}{5}$ to $\frac{1}{5}$ milligrams *heavier* than in the case of hydrogen dried by phosphoric anhydride. But I doubt whether the small excess is of any significance. It seems improbable that it could have been due to residual vapour, and it is perhaps not outside the error of experiment, considering that the apparatus was not in the best condition.

May 31.—"Palæolithic Man in Africa." By Sir John Evans, K.C.B., F.R.S.

In April 1896, just four years ago, I ventured to call the attention of the Society (*Roy. Soc. Proc.*, vol. lx. p. 19) to some palæolithic implements found in Somaliland by Mr. H. W. Seton-Karr. In doing so, I pointed out the absolute identity in form of these implements with those from the valley of the Somme and numerous other pleistocene deposits in North-western Europe and elsewhere ; and I cited others from the high land adjoining the valley of the Nile, and from other places in Northern and Southern Africa. I was at the same time careful to point out that though there could be no doubt as to this identity in form, no fossil mammalian or other remains had been found with these African implements. I did not, however, hesitate in claiming them as palæolithic.

Since the publication of my short note, an extensive collection of stone implements formed in Egypt by Mr. H. W. Seton-Karr has been acquired by the Mayer Museum at Liverpool. I have not had an opportunity of examining the specimens, but a detailed account [1] of them, with numerous illustrations, has been published by the Director of the Liverpool Museums, Dr. H. O. Forbes. The majority of the implements are of Neolithic Age or even of more recent date, and with the account of these I need not here concern myself ; but the author is at considerable pains to dispute my view that the instruments of palæolithic forms belong to the Palæolithic Period. As he says, Mr. Seton-Karr's statement that he sometimes found spearheads "on the ground surrounded by a mass of flakes and chips as though the people had dropped their work and fled," is very suggestive and important. He adds, however, that "one such occurrence is almost sufficient in itself, I venture to think, to disprove the high antiquity claimed by Sir John Evans for these implements."

Were it certain that the so-called spear-heads were really of palæolithic form, and had the flakes and chips been fitted on to them so as to reconstitute the original blocks of flint, as has been done in the case of undoubted palæolithic specimens by Mr. Spurrell and Mr. Worthington Smith, the question would still remain to be discussed as to the condition of the localities in relation to subaërial denudation.

It is, however, hardly necessary to discuss these points, as some recent discoveries made in Algeria will, I venture to think, go a long way towards settling the question. I propose, therefore, very briefly to state their nature. About sixty miles to the south-west of the town of Oran, and about ten miles to the north of Tlemcen, on the plateau of Remchi, about a mile to the south of the River Isser, lies a small lake known as Lac Karâr. It occupies a depression in lacustrine limestone of comparatively recent geological date, superimposed on beds of Lower Miocene Age. The level of the water, which is some 15° C. warmer than that of the ordinary springs of the district, and appears to be derived from some deep-seated source, seems to be about 600 feet higher than that of the River Isser. The lake originally filled a much larger part of the depression than it does now, and from its old bed a considerable amount of material has of late years been extracted for the Service des Ponts et Chaussées. This material consists of sand and gravel rich in iron pyrites, in the midst of which lie, pell-mell, bones of animals and stone implements fashioned by the hand of man.

These have for some years been diligently collected by M. Louis Gentil, a geologist, and form the subject of a memoir that has just appeared in *l'Anthropologie* (Tome xl.), by my friend M. Marcellin Boule, of the Galerie de Paléontologie at the Jardin des Plantes, Paris. Some 200 stone implements have been submitted to him, of various sizes, and all or nearly all of well-known palæolithic forms, including several with a broad chisel-like end, of which examples have been found in the laterite of Madras and the gravels of Madrid. They are for the most part formed of an eocene quartzite, though some smaller specimens of the type known as that of "le Moustier" are formed of flint. The *facies* of these latter is not so distinctly palæolithic as that of the former, of which some, through the kindness of M. Marcellin Boule, are exhibited.

The most important part of the discovery is that which relates to the mammalian remains found with the implements. These are of elephant, rhinoceros, horse, hippopotamus, pig, ox, sheep, and certain cervidæ. I will not detain the Society with the details given in M. Boule's memoir, but I may call attention to the fact that the elephant is not the African elephant, but one more nearly related to the quaternary or even pliocene elephants of Europe, to which the designation Atlanticus has been given. Some teeth seem closely allied to those of *E. meridionalis* and even *E. armeniacus*. Having regard to the whole fauna, M. Boule arrives at the conclusion that it is identical with that of the fossiliferous deposits of Algeria, which from their topographical or stratigraphical characteristics have been assigned to the Quaternary or Pleistocene Period. He also cites other instances in Algeria, such as Ternifine and a station near Aboukir, in which palæolithic implements have been found associated with the remains of a similar pleistocene fauna.

Altogether, these recent discoveries in Northern Africa tend immensely to strengthen my position with regard to the truly palæolithic character of the implements found in other parts of that vast continent, and I am tempted to bring for comparison some few specimens from South Africa. One of these, found by Mr. J. C. Rickard at the junction of the Reit and Modder twenty years ago, is almost indistinguishable from those of the Lac Karâr, as is also one from the valley of the Embabaan in Swaziland. But the most remarkable is an implement of typically palæolithic character found in 1873 under 9 feet of stratified beds at Processfontein, Victoria West, by Mr. E. J. Dunn.[1] May the day be not long distant when researches for the implements of palæolithic man may again be carried on, and trenches be dug in South Africa for peaceful instead of warlike purposes.

Anthropological Institute, June 5.—Mr. C. H. Read, President, in the chair.—Dr. J. G. Garson explained in detail the metric system of identification of criminals which is in use in this country. This system, which is a modification of the Bertillon system employed in France, consists in measuring as accurately as possible certain dimensions of the individual, and classifying them, according as they prove severally large, medium or small, in such a way that the search for any single set of measurements at the central office is curtailed to the utmost. Finger prints are used, as an additional proof of identity, on the back of the card which carries the record of the measurements. The paper was illustrated by diagrams and examples of the measurements and of the instruments which are employed ; and was followed by a discussion.

June 12.—Mr. C. H. Read, President, in the chair.—The secretary exhibited, on behalf of Mr. H. Swainson Cowper, a primitive figurine from Adalia in Asia Minor, which presented analogies with the "owl-faced idols" found on the site of Troy by Dr. Schliemann.—Mr. B. H. Pain read a paper on Eskimo craniology, in which he stated that from observations on a number of living Eskimo, lately in London, he had been enabled to extend the comparisons instituted by Virchow between the dimensions of the head and those of the skull in this race. Reference was incidentally made in the paper to the collection of Eskimo crania at Cambridge (of which a descriptive note was published in the *Journal* of the Anthropological Institute, 1895), as well as to the large collection of crania of Greenlanders in the Anatomical Museum at Copenhagen. The paper was fully discussed by M. J. Deniker, Dr. Garson, Mr. Duckworth and Mr. Shrubsall.—Mr. W. L. H. Duckworth read a paper on the skeletal characters of the Mori-ori of the Chatham Islands. The result of his observation and measurement of ten skulls and two complete skeletons of Mori-ori (from the Chatham Islands) is a general corroboration of the earlier results of Turner (*Challenger* Report) and Scott (*Transactions* of the New Zealand Institute) as to the characters of the skeletons of these Pacific Islanders. Special notice was directed to the frequency of occurrence of osteo-arthritis, as evidenced by the condition of the sacrum, innominate bones and femora especially, and to the rare form of occipito-atlantic articulation in one of the specimens. The paper was followed by a discussion.—Mr. J. Gray gave a summary of the anthropometric survey conducted by Mr. James Tocher and himself in East Aberdeenshire, and exhibited diagrams showing the relative frequency and the local

---

[1] *Bull. Liverp. Mus.*, II., Nos. 3 and 4 (Jan. 90); NATURE, vol. lxi. April 19, p. 597.

[1] See also a paper by M. E. T. Hamy in the *Bulletin du Muséum d'Histoire Naturelle*, 1899, No. 6, p. 270.

distribution of various types of complexion, &c.—A paper, by Mr. D. MacIver, on recent anthropometrical work in Egypt, was taken as read.

EDINBURGH.

Royal Society, June 4.—Prof. M'Kendrick, Vice-President, in the chair.—Dr. R. Stewart McDougall read a paper on the biology of certain species of *Pissodes* and *Scolytus*. *Pissodes* is a genus of Coleopterous insects very harmful to pine trees in Great Britain, the eggs being laid under the bark, and the grubs tunnelling in the Cambial layer. In working out the life-history of *P. notatus*, a pest on young pines, and of *P. pini*, which attacks chiefly grown trees, the author found that imagos could be obtained from March to November, and that breeding might take place from April to September inclusive. A remarkable feature was the long life of adult beetles of both sexes, with repeated copulation. Specimens of *P. notatus* had lived 22, 24 and 37 months, hibernating twice in the first two cases and three times in the last case. Marked adults kept in confinement, but otherwise in natural conditions, began hibernation in November, and appeared above ground again in the following March. *Scolytus multistriatus* attacks the elm—almost invariably dead or decaying trunks or branches. Attempts to develop the eggs in living trees had failed. The beetles are late swarmers, appearing chiefly in July. The generation is an annual one.—Sir John Murray read a paper on the physical, chemical, and biological conditions of the Black Sea. Certain peculiar features were pointed out, notably the presence of a cold layer at a depth of 50 fathoms, the deeper waters being warmer ; the lack of vertical circulation, and the consequent stagnation of the deep waters, which can find no outlet through the comparatively shallow straits ; the presence of sulphuretted hydrogen and the absence of oxygen in these depths ; the absence of animal life there ; and the deposit of carbonate of lime on the bottom. This carbonate of lime was not of organic origin, but was formed by chemical action, the sulphuretted hydrogen being one of the products. The special interest of the inquiry arose from the fact that in several of these particulars the Black Sea conditions differed fundamentally from conditions that obtained in oceans and other ocean-connected seas.

Mathematical Society, June 8.—Mr. R. F. Muirhead, President, in the chair.—The following papers were read :—(1) A general proof of the addition theorems in trigonometry ; (2) a slight extension of Euler's theorem on homogeneous functions, by W. Edward Philip ; note on proofs by projection in trigonometry and co-ordinate geometry, by Prof. Gibson.

PARIS.

Academy of Sciences, June 11.—M. Maurice Lévy in the chair.—Reduction of certain problems of heating or cooling by radiation to the more simple case of heating or cooling of the same bodies by contact ; heating of a wall of indefinite thickness, by M. J. Boussinesq.—On the radiation of uranium, by M. Henri Becquerel. The rays from uranium are deviable in a magnetic field, although on account of the comparatively feeble action of the uranium radiations, the time of exposure of the photographic plates has to be very long. Uranium salts treated with barium salts and a sulphate have their radio-activity reduced. The author has not been able to obtain an inactive uranium salt.—Researches on the pressures of saturated mercury vapour, by MM. L. Cailletet, Colardeau and Rivière. An experimental study of the vapour pressure of mercury from its boiling point up to about 880°. At the point where the pressure is about 160 atmospheres, the experiments were stopped by the iron tubes allowing the mercury to pass through them, thus rendering the study of the critical phenomena of mercury impossible. The pressures were read on a metallic manometer which had been directly calibrated against a mercury column, and the temperatures on a thermo-couple.—On the β-phenyl and β-benzyl-a-alkoxy-a-cyanoacrylic acids, by MM. A. Haller and G. Blanc. Starting with phenylacetylcyanacetic and benzylacetylcyanacetic esters, the silver salts are prepared and these treated with alkyl iodide. The esters so obtained are isomeric with the benzoylcyanacetic esters, and are clearly derivatives of crotonic acid.—Note on an earthquake in Mexico on December 19, 1899, by the French Consul in Mexico.—On a photograph obtained at the Observatory of Algiers during the total eclipse of the sun of May 28,

by M. Ch. Trépied.—On the polarisation of the corona of the sun observed at Elche, by M. P. Joubin. The experiments of Prazmowski and Ranyard were confirmed. Further observations with a Bravois bi-plate showed that for all points of the sun's limb between the equator, and about 15° to 20° from the north solar pole, there was no elliptical polarisation. Above this, the colours of the two plates could be clearly distinguished.—The method of Neumann and the problem of Dirichlet.—On the class of primitive continuous finite groups of transformations of Lie, by M. Edmond Maillet.—On the logarithms of the algebraical numbers, by M. Carl Störmer.—On the angular points of solubility curves, by M. H. Le Chatelier.—On the electrical distribution in a Hertz resonator in activity, by M. Albert Turpain.—Permanent modifications in metallic wires and variation of their electrical resistance, by M. H. Chevallier.—On the kathode rays, by M. P. Villard. A study of the heating effect produced upon the kathode. The usual metal electrodes can be replaced by ordinary lamp filaments, the kathode in this case being rapidly raised to a white heat. The fall of potential necessary for the production of light by a filament in this way is much greater than when the carbon is heated by the ordinary Joule effect. — The campylograph, a curve-tracing machine, by M. Marc Dechevrens. A new method of producing Lissajous figures. Instead of being confined to compounding two rectangular motions only, the instrument can combine three simultaneous movements, two rectilinear and oscillatory, the third uniform and circular. Seventeen illustrations of the results obtainable accompany the paper.—Heat of solution of hydrogen peroxide. Thermal value of the hydroxyl group: influence of carbon and hydrogen, by M. de Forcrand.—On the direct production by the wet method of mercuric and mercurous iodides in the crystalline state, by M. F. Bodroux. Mercuric iodide can be crystallised in octahedra from boiling concentrated hydrochloric acid or from a solution of potassium iodide. A better method is to leave mercuric acetate in contact with methyl iodide. The substitution of mercurous nitrate for mercuric acetate yields large crystals of mercurous iodide.—On the impossibility of the primary reduction of potassium chlorate obtained electrolytically, by M. André Brochet. The electrolysis was carried out in presence of large amounts of oxide of cobalt. Since hypochlorites are destroyed by this oxide, whilst chlorates are unaffected, only chlorate which has been formed by primary interaction of the ions will be found. Since no chlorate is produced under these conditions, it follows that in the electrolysis of alkaline chlorides, contrary to the hypotheses of Œttel, Haber and Grinberg, Forster, Torre and Müller, the formation of chlorate is never due to a primary action, but is always due to the intermediate formation of hypochlorites, even in a strongly alkaline medium.—On the decomposition of metallic chlorides, by M. Œchsner de Coninck.—Addition of hydrogen to acetylene in presence of reduced iron or cobalt, by MM. Paul Sabatier and J. B. Senderens. At 180° reduced iron causes the interaction of hydrogen and acetylene, ethane, ethylene, benzene and higher unsaturated hydrocarbons being produced. Cobalt under similar conditions gives a much larger yield of ethane compared with that previously noticed for nickel.—On a product of decomposition of a diiodhydrin of glycerol, by MM. E. Charon and C. Paix-Séailles. The substance formed by the elimination of hydriodic acid from the iodhydrin, $CH_2I.CHI.CH_2OH$, is β-iodopropionaldehyde, most probably the polymer $(CH_2I.CH_2.CHO)_3$.—Action of acetylene upon cuprous chloride dissolved in a solution of potassium chloride, by M. Chavastelon.—On acidimetry and alkalimetry in volumetric analysis, by M. A. Astruc.—Fixation of clay in suspension in water by porous bodies, by M. J. Thoulet. Analysis of marine deposits consisting largely of shells, showed great variations in the amount of clay present, and it appeared possible that this clay might have been abstracted from the water after the death of the animal by mechanical means. Experiments with powdered pumice stone and with wood charcoal confirmed this view.—Preliminary note on the decapod crustacea collected during the Belgian Antarctic expedition, by M. H. Coutière.—The embryos of mummy wheat and barley, by M. Edmond Gain. A microscopical examination showed that in spite of their external appearance of good preservation, the mummy cereals do not possess a cellular organisation compatible with germination.—The ratio of nitrogen to chlorides in the contents of the stomach during digestion, by MM. J. Winter and Falloise.

ST. LOUIS.

Academy of Science, May 23.—A paper by Dr. Adolf Alt, entitled "Original contributions concerning the glandular structures appertaining to the human eye and its appendages," was presented by title.—Dr. M. A. Goldstein read a paper on the physiology of voice production, in which he discussed three essential factors in the production of voice—the motor force, the organ of sound, and the resonators.—Prof. F. E. Nipher read a short communication on the zero photographic plate, to which reference was made at the meeting of May 7 (see pp. 62, 159). The zero plate is one upon which a photographic image has been made, but which will develop no image in a bath placed in light of given candle power, at a distance of one metre from the source. For example, if the developing bath is twenty centimetres from a sixteen candle lamp, a Cramer isochromatic plate, such as is called "instantaneous," held for ninety seconds at a distance of one metre from the lamp, will be a zero plate. With an opaque stencil over the plate when placed in a printing frame during the exposure, there will develop a positive of holes through the stencil, if the exposure is longer, and a negative if the exposure is shorter. If a fresh plate is exposed in our camera with full opening to a brilliantly lighted street scene for one minute, it will develop as a positive in that same bath. This time can be somewhat reduced, but the least time needed has not yet been determined. It is evident that part of this minute is used in producing a zero plate. It is furthermore clear that different parts of the plate will arrive at the zero condition at different times. The exposure may be arrested at a time when the strongly lighted white background of a sign-board will develop white as a positive, and when the black letters will also show white as a negative. It has been found that when a plate is uniformly exposed over its whole surface to the extent that nothing would have developed had it been covered by a stencil, this plate may then be placed in a camera and exposed in the ordinary way, and a perfect positive will develop in the bath to which it has been adapted. This preliminary spoiling of the plate for developing a negative is a very advantageous preparation for taking a positive. It shortens the time of exposure, and ensures that a positive shall be obtained over all parts of the plate. It is not yet known how short the camera exposure may be made, but the present indications are that they will be as short as those now made in the taking of negative pictures. It is currently believed by photographers that in a positive plate the object has " printed its picture " upon the plate. This appears to be a misconception of the process. It is true that in an exposure of long duration an image shows on the plate before it is placed in the bath. But this image is blackest where the light has acted most. It is a negative. This picture disappears in the developing bath when illuminated. The plate becomes perfectly clean. The positive picture then develops exactly as a negative would under ordinary conditions.—Mr. J. B. S. Norton presented some notes on the flora of the south-western United States

---

DIARY OF SOCIETIES.

THURSDAY, JUNE 21.

ROYAL SOCIETY, at 4.30.—On the Effects of Changes of Temperature on the Elasticities and Internal Viscosity of Metal Wires: Prof. A. Gray, F.R.S., Y. J. Blyth, and J. S. Dunlop.—On the Connection between the Electrical Properties and the Chemical Composition of Different Kinds of Glass, Part II.: Prof. A. Gray, F.R.S., and Prof. J. J. Dobbie.—On the Change of Resistance in Iron produced by Magnetisation: Prof. A. Gray, F.R.S., and Prof. E. T. Jones.—Underground Temperature at Oxford in the Year 1899, as determined by Five Platinum Resistance Thermometers: Dr. A. A. Rambaut, F.R.S.—On the Kinetic Accumulation of Stress, illustrated by the Theory of Impulsive Torsion: Prof. K. Pearson, F.R.S.—Lines of Induction in a Magnetic Field: Prof. Hele-Shaw, F.R.S., and A. Hay.—On the Spectroscopic Examination of Colour produced by Simultaneous Contrast: G. J. Burch, F.R.S.—An Experimental Investigation into the Flow of Marble: Dr. F. D. Adams and Dr. J. T. Nicolson.—A Criticism of the Young-Helmholtz Theory of Colour Perception: Dr. F. W. Edridge-Green.—And other Papers.
LINNEAN SOCIETY, at 8.—On some Scandinavian Crustacea: Dr. A. G. Ohlin.—The Subterranean Amphipoda of the British Islands: Chas. Chilton.—On certain Glands of Australian Earthworms: Miss Sweet.—Notes on Najas: Dr. A. B. Rendle.
ZOOLOGICAL SOCIETY, at 4.30.—The Gigantic Sloths of Patagonia: Prof. E. Ray Lankester, F.R.S.
ANATOMICAL SOCIETY (Owens College, Manchester), at 10.30.—Lantern Demonstration on the Comparative Anatomy and Histology of the True Cæcal Apex—the Appendix Vermiformis: Dr. R. J. Berry.—Lantern Demonstration of some Surface Markings of the Calvaria, and their Significance: Prof. Dixon.—Lantern Demonstration of Microphotographs of the Maturation Stages in the Ovum of Echinus: Dr. T. H. Bryce.—

NO. 1599, VOL. 62]

Some Points in the Anatomy of the Digestive System: Prof. Birmingham.—(a) Two Cases of Absent Vermiform Appendix; (b) A Specimen showing Direct Continuity between the Long External Lateral Ligament of the Knee-joint and the Peroneus Longus Muscle; (c) A Supernumerary Bone in the Carpus connected with the Trapezium: Prof. Fawcett.—A Note on the Genital Apparatus of the Jerboa: Dr. Armour.
CHEMICAL SOCIETY, at 8.—Ballot for the Election of Fellows.—Notes on the Chemistry of Chlorophyll: Dr. L. Marchlewski and C. A. Schunck.—Researches on Morphine, I.: Dr. S. B. Schryver and F. H. Lees.—A New Series of Pentamethylene Derivatives, I.: Prof. W. H. Perkin, jun., F.R.S.; Dr. J. F. Thorpe, and C. W. Walker.—Experiments on the Synthesis of Camphoric Acid. III. The Action of Sodium and Methyl Iodide on Ethyl-dimethyl-butanetricarboxylate: Prof. W. H. Perkin, jun., F.R.S.; and Dr. J. F. Thorpe.—On the Oxime of Mesoxamide and some Allied Compounds: Miss M. A. Whiteley.—The Oxyphenoxy- and Phenylenexy-acetic Acids: W. Carter and Dr. W. T. Lawrence.—(1) The Condensation of Ethyl a-Bromo-isobutyrate with Ethyl Malonates and Ethyl Cyanacetates: a-Methyl-a'-isobutylglutaric Acid; (2) Methylisoamylsuccinic Acid, II.: Dr. W. T. Lawrence.

FRIDAY, JUNE 22.

PHYSICAL SOCIETY, at 5.—Notes on Gas Thermometry: Dr. P. Chappuis.—A Comparison of Impure Platinum Thermometers: H. M. Tory.—On the Law of Cailletet and Mathias and the Critical Density: Prof. J. Young, F.R.S.
ANATOMICAL SOCIETY (Owens College, Manchester), at 10.30.—Note on the Configuration of the Heart in a Man and some other Mammalian Groups: Dr. C. J. Patten.—On the Arrangement of the Pelvic Fasciæ and their Relationship to the Levator Ani: Dr. Peter Thompson.—(a) A Preliminary Note on the Development of the Sternum; (b) Specimens of Diaphragmatic Hernia and of a Left Inferior Vena Cava: Prof. Paterson.—Preparations and Lantern Slides illustrating: (a) The Anatomy of the Subclavian and Axillary Arteries; (b) The Position and Relations of the Eustachian Tubes: (c) Stereoscopic Views of Anatomical Preparations: Dr. Arthur Robinson.—A Series of Microscopical Preparations illustrating the Development of the Posterior End of the Aorta: Prof. Young and Dr. Arthur Robinson.—Demonstration of a Series of Preparations of the Posterior End of the Adult Aorta: Prof. Young.

MONDAY, JUNE 25.

ROYAL GEOGRAPHICAL SOCIETY, at 8.30.—Results of the Sir George Newnes Antarctic Expedition: C. E. Borchgrevink.

TUESDAY, JUNE 26.

ROYAL PHOTOGRAPHIC SOCIETY, at 8.—The Selection of Lenses with regard to Photographic Perspective: J. H. Agar Baugh.—How to ascertain the Conjugates of a Lens without Calculation: Rev. F. C. Lambert.

CONTENTS.

THURSDAY, JUNE 28, 1900.

CHRISTMAS ISLAND.

A Monograph of Christmas Island (Indian Ocean); Physical Features ana Geology. By C. W. Andrews, with descriptions of the Fauna and Flora by numerous contributors. Pp. xv + 237. (London: Published by order of the Trustees of the British Museum, 1900.)

TILL 1887 Christmas Island, which is situated in the Indian Ocean nearly a degree to the south-south-west of Batavia, was scarcely known, even by name, to the average Englishman, and only the low-lying shore had been visited by explorers ; the steep cliffs, together with the forest with which the island is clothed, forming a barrier which had hitherto prevented access to the central plateau. In that year the Commander of H.M.S. *Egeria*, with the assistance of a landing party, succeeded in cutting his way into the interior ; and two years later the island was leased by the British Government to a trading company. Since it contains an area of about forty-three square miles, and appears never to have been inhabited by aboriginal tribes, it presented a most favourable opportunity for studying the fauna and flora of an oceanic island of considerable size situated at no very great distance from a considerable land-mass—the Sunda Archipelago. Down to the time mentioned, it appears, indeed, to have been the largest uninhabited tropical island extant ; and as the discovery of valuable deposits of phosphates in the interior indicated that its pristine conditions would soon be rudely disturbed, it was evident that if a biological survey was to be undertaken at all, there was no time to be lost. Fortunately, Sir John Murray interested himself strongly in the matter, and it was eventually arranged that Mr. C. W. Andrews, of the British Museum, who is both a geologist and a zoologist, should undertake the work. He accordingly spent ten months on the island during the years 1897-98 ; and the present volume, in which he has had the assistance of a number of specialists, is the result of his labours.

As is evident by their permitting a member of their staff to undertake the task, the Trustees of the British Museum gave their support to the exploration ; and it is a matter for congratulation that they have seen fit to publish the results in the same form as the Museum "Catalogues." A wise liberality has been exercised in the matter of illustration, the plates (some coloured) being numerous, while a considerable number of reproductions from photographs are given in the text. These latter have, however, received but scant justice at the hands of the printer ; and it is, indeed, with some surprise that we notice the volume bears the name of a local firm of printers.

Situated at a distance of over 190 miles from the nearest land, with the intervening ocean attaining a depth of more than three miles, Christmas Island appears to have derived its limited fauna from the Sunda Archipelago, of which indeed it probably once formed a part. The length of its isolation is, however, indicated by the circumstance that four out of its five indigenous mammals are peculiar species, the fifth—a Shrew—being a local

variety of an Assam and Tenasserim form. The majority of the few land birds are likewise distinct, the most striking being a Goshawk (*Astur natalis*), an Owl (*Ninox natalis*), and a White-eye (*Zosterops natalis*), specimens of all of which were first collected by Mr. J. J. Lister during a flying visit to the island in 1890. As regards the fauna generally, it exhibits no greater evidence of affinity with that of the Mentawei chain of islands, running parallel with Sumatra and Java, than with that of the two islands last named. And the hypothesis that Christmas Island formed the termination of a "Mentawei Peninsula" must accordingly be given up.

One of the main objects of the exploration of the island was to ascertain whether its geological structure would throw any further light on the vexed question of the origin of atolls. As the result of his observations, Mr. Andrews is led to believe that, from the absence of a sufficient thickness of reef-limestone, Christmas Island, although originally an atoll, could not have been formed in the manner required by the Darwinian theory, as the amount of subsidence which has taken place would have been quite insufficient. That a certain amount of subsidence may have occurred in the early history of the island, Mr. Andrews considers to be quite possible.

"It may, of course, be objected," he writes, "that Christmas Island was never a typical atoll, and to this objection no answer is possible ; but since it can be shown that at one time it must have consisted of reefs and islands approximating very nearly to those seen in atolls which are regarded as typical, the determination of the nature of the foundations upon which these reefs and islands rested is at least a step in the right direction. . . . In this case the basis of the island is almost certainly a volcanic peak, the foot of which is now some 2400 fathoms below the level of the sea, and that on its summits and flanks great accumulations of Tertiary limestones have been deposited, and in some cases are interstratified with the products of the eruptions, probably for the most part submarine, which took place from time to time. The oldest of the volcanic rocks are trachytic, the newer basaltic. The last of the eruptions was accompanied by the formation of thick beds of volcanic ash, and it is upon these that the great mass of the Miocene (Orbitoidal) limestones rests."

The occurrence of such a thickness of Tertiary deposits (ranging from the Eocene or Oligocene upwards) is unknown in any other oceanic island. It is important to notice that these rocks, allowing for a difference in the proximity of land at the time of their deposition, are very similar to those of South Java ; but the author considers that there are difficulties in believing that the two series of sediments were deposited in a continuous area, as this would involve great local dislocations. Accordingly the volcanic peak theory is adopted in preference to such a view.

In speaking of elevation and depression, the author is careful to guard himself by stating that such terms are merely used in relation to the sea-level ; and it would appear, from reading between the lines, that he is rather in favour of an actual alteration of the sea-level in these districts. It may further be inferred that he does not intend his conclusions as to the mode of origin of Christmas Island to affect the case of other atolls, his idea apparently being that all atolls are not of precisely similar origin.

K

The rocks which have brought Christmas Island into the most prominent notice are the thick beds of nearly pure lime phosphate capping several of the higher hills. It is inferred that this deposit has been formed by the action of beds of guano on limestone forming the summits of the low islets presumed to have existed previous to the first elevation of the present island. Another phosphatic bed is considered to have been produced by guano acting on volcanic ash. It is for the purpose of working these phosphates that the island has been leased by a commercial company.

Although the greater part of the volume is of a highly technical nature, it must not be inferred that this is the case with the whole of its contents. As an example of its lighter side, the excellent account of the habits of the Frigate-bird may be cited. These birds, which form the main support of the present colony of the island, are of an inquiring and fearless disposition.

"The usual way of obtaining them is," writes the author, "for a man to climb into the topmost branches of a high tree near the coast, armed with a pole eight or ten feet long and a red handkerchief. The latter he waves about, at the same time yelling as loudly as possible. The birds, attracted by the noise and the red colour, swoop round in large numbers, when they are knocked down with the long pole. In this way sufficient birds to supply the small colony with food can usually be obtained in an hour or two; occasionally, however, in unfavourable states of the wind, they are difficult to procure."

From first to last, the exploring, the collecting, and the descriptive and literary portions of the book have been thoroughly well carried out. And, despite the fact that no far-reaching or epoch-making discoveries in either zoology, geology, or distribution have been made, all concerned in the production of the volume before us (save the printer) are to be heartily congratulated on the manner in which they have executed their respective tasks.   R. L.

A NEW WORK ON SILVER.

Metallurgy of Lead and Silver. Part ii. Silver. By Henry F. Collins. Pp. 352. (London: Griffin and Co., Ltd., 1900.)

WE recently had occasion to notice the first volume of the present work, and to speak favourably of its merits. We are pleased to find the second portion equally good. It has been a source of great regret that the distinguished master of metallurgy, the late Dr. Percy, did not live to complete his projected work on Silver, instead of leaving what has been termed a splendid fragment: and as no book claiming to give a full account of the metallurgy of the subject has been published since, we cordially welcome the advent of a further contribution. It is perhaps unnecessary to point out how closely interwoven is the metallurgy of lead with that of silver, or to state that a full treatise on silver cannot be written without considerable reference to lead; and when one author is competent to deal with both branches of the subject, it affords the best means of imparting a sound knowledge of these metals. In the present case we have this additional advantage, that the editor is an

authority on all questions relating to the nature and properties of silver, together with that of assaying. The immense importance of silver in the economic relations of the United States is well known, and many attempts have been made to introduce similar relations into this and other countries; hence it may be considered one of the most important metals known to mankind. The present work is not an exhaustive treatise on silver, and is evidently intended chiefly for those who are connected with the extraction of the metal from its ores. Those ancient methods which are fast becoming obsolete have not escaped notice; for, while they may not possess much practical value at the present time, their chemical and educational value is not to be despised. Numerous references to original sources of information are given throughout the volume, and this will enable the reader to obtain fuller information than is given here. The method of procedure in special works, such as that of matte-smelting at Sunny Corner (p. 268), is described at some length with clearness and precision. The author has followed the same plan as in his first volume, of economising space by giving details of the practice at different localities in the form of tabular statements. This should prove useful for reference and comparison. The book is divided into four main sections, dealing respectively with silver and its ores, amalgamation, lixiviation, and smelting processes. Of these the chapters relating to lixiviation and blast furnace smelting are the best, as they appear to be the branches with which the author is most familiar. The hyposulphite leaching process is described in a more lucid and methodical manner than we have seen elsewhere, and the advantages and disadvantages of calcium sulphide are admirably compared on p. 197. A chapter is specially devoted to hyposulphite leaching practice, in which is given details of plant, mode of working, advantages and disadvantages of lixiviation, cost, and examples of the Russell process in various localities. Data as to cost and results at mills using the Patera and Russell processes respectively are given in the form of tables on pp. 224 to 227. A serviceable chapter on the refining of lixiviation sulphides concludes the section. The fourth section, dealing with the extraction of silver by smelting processes, contains a considerable amount of information in a condensed form. The table of comparison of various systems of smelting is instructive and helpful. The construction of furnaces is made clear by the aid of figures, drawn to scale. The arguments in favour of the hot blast for smelting mattes are pertinent and convincing. Several well-compiled tables are included in this chapter. Pyritic smelting receives only a brief notice in Chapter xv., as this subject has been partly dealt with in the first volume. The subject of matte smelting in reverberatories for silver-copper ores is next considered, and the characteristics of the method, with the points of difference from blast furnace practice, are pointed out. This kind of information is often of great moment to the practical man, who has to decide on the most economic method to adopt in special cases. The final chapters deal with the treatment of argentiferous mattes, which generally require a preliminary concentration to eliminate some of the lead and iron. In some cases a direct method may be adopted, and information is here given for that purpose. The Bessemerising of copper mattes is briefly described. Silver-copper smelting and refining is limited in its application to ores comparatively free from sulphur, arsenic, and lead, and therefore but little used. The plant employed is specified and illustrated by diagrams and tables. The book concludes with a short account of the various wet methods used for argentiferous slimes. The author's attempt to cover the ground embraced by such a wide subject within a moderate compass will, with the aid of tables and summaries, prove most valuable both to practical men and to students.

OUR BOOK SHELF.

The History of Language. By Henry Sweet, M.A. Pp. xi + 148. (London : J. M. Dent and Co., 1900.)

THERE are few living scholars who are so well qualified as Dr. Sweet to write a thoroughly comprehensive introduction to the science of language. He is, as is well known, one of the foremost European authorities on phonetics ; but at the same time he is a profound and original thinker on those psychological aspects of linguistic science in which few phoneticians take any interest. And while possessing a competent knowledge of Indogermanic comparative philology in its latest developments, he is preserved from the narrowness of view of the mere Indogermanist by having made a practical study of Arabic, Finnish and Chinese. Notwithstanding its small size, this "primer" is a very remarkable book. In completeness of outline, it is superior to any elementary manual of the subject known to us ; and it is no mere arid skeleton, but contains a good deal of novel and interesting illustration of the principles expounded. Perhaps it is not quite so easy to master as a "primer" is usually expected to be. Although strictly elementary, in the sense that it assumes no previous philological knowledge on the reader's part, it does undoubtedly demand considerable power of close attention and some training in habits of scientific thought. It will therefore probably be less acceptable to absolute beginners than to those who have already some general knowledge of the subject and desire to render their conceptions of it more systematic and precise. Even by advanced philological scholars it may be studied with interest and profit.

The contents of the book may be said to consist of three portions : an exposition of the general principles affecting the development of language, an outline of the history of the Aryan family of languages, and a statement of the author's views as to the exterior affinities of Aryan and the locality in which it was developed. Perhaps the third part is somewhat out of place in an elementary book, but it is at any rate interesting. Dr. Sweet's hypothesis is that primitive Aryan arose in Scandinavia out of a mixture of the language of Ugrian conquerors with that of the aboriginal population among whom they were absorbed. This is not now such a startling heresy as it would have been a few years ago, though it is not likely at present to find a ready welcome from Indogermanists. The apparent affinities between Aryan and Ugrian certainly seem too striking to be due to mere coincidence, but it is a long step from this admission to the acceptance of the definite theory here propounded. The writers who have hitherto advocated somewhat similar views have always discredited their case by their ignorance of philology and their lack of scientific caution. It is to be hoped that Dr. Sweet will give to the world a full exposition of the grounds on which his conclusions are based. Whether he succeeds

in the establishment of his particular thesis or not, he can hardly fail to make a valuable contribution towards the ultimate solution of the question.

Micro-organisms and Fermentation. By Alfred Jörgensen. Pp. xiii + 318. (London : Macmillan and Co., Ltd., 1900.)

THE study of the biology of fermentation has made considerable progress in recent years. The knowledge that has been gained of the nature and mode of action of the living agents in question is mainly due to the efforts of foreign observers. Through the investigations of Pasteur, and most notably of Hansen, the subject became a recognised branch of methodical and practical inquiry. To be in a position to employ the essential and to exclude the deleterious agents in a fermentative process is to substitute scientific for haphazard methods. This, briefly put, is the aim of technical mycology, and the gain to a given industry is considerable, as *e.g.* in brewing and distilling operations. Of the books dealing with micro-organisms and fermentation, Dr. Jörgensen's has long occupied a leading position, and hardly requires an introduction to the specialist. The new edition just issued has been completely revised, and the English translation has been well done by Dr. A. K. Miller and Mr. A. E. Lennholm. Dr. Jörgensen's reputation as a teacher and investigator, as well as his intimate association with Hansen, place this work above the ordinary run of text-books. The first chapters deal with the methods of microscopical and physiological examination of micro-organisms, and the methods for obtaining and utilising pure cultures of the useful races of saccharomyces are described. The examination of water and air is next dealt with—a subject of importance on account of the injurious organisms that may exist in the air and water of a brewery. The chapter on bacteria is somewhat incomplete. The technical mycologist has commenced to study the bacteria more closely, and a fuller account of this branch of the subject will be found in Lafar's book. An interesting account is given of the alcohol-forming bacteria, and of certain symbiotic ferments, *e.g.* Kephir and the ginger-beer plant. The moulds of importance in technical work are fully dealt with.

Of recent work, Buchner's "Zymase" is shortly alluded to ; but more mention might have been made of Calmette's investigations at Lille and Sèclin upon the symbiotic action of moulds and yeasts in the alcoholic fermentation. The account of the alcoholic ferments in Chapter v. is naturally the main and distinctive feature of this work, and it will be particularly valuable to the English reader on account of the lucid description it contains of Hansen's investigations upon yeasts. The various species of bottom and top fermentation yeasts of interest to the brewing chemist are fully dealt with. The final chapter is devoted to the application of the results of scientific research in practice. The value of the book is added to by a number of illustrations and a very full bibliography. As an introduction to the morphology and biology of the alcoholic ferments, Dr. Jörgensen's work leaves little to be desired, and constitutes a valuable complement to the text-books which deal mainly with the chemical side of the subject.

A. M.

Photography in Colours. By R. C. Bayley. Pp. 74. (London : Iliffe, Sons and Sturmey, Ltd., 1900.)

THIS little book is practically a reprint of a series of articles by the author which have already appeared in a photographic periodical, but the subsequent revisions and convenience of reference occasioned by their collection under one cover should render them more serviceable. The general principle has been to avoid technicalities and purely executive details, aiming rather to give a lucid explanation of the principles governing the various processes, which may be understood by readers not necessarily acquainted with photographic manipulation.

The opening chapters introduce the elementary ideas of the nature of colour and the undulatory theory of light. Following these is a chapter on the Lippmann process, this being the only direct process having a purely physical origin.

The fourth chapter deals with the principles of colour vision, showing how the colour curves of red, green and blue sensitiveness are employed in deciding the screens used in the three-colour photographic process ; two processes of this type, founded by Ives and Joly respectively, being then fully explained.

The work is brought up to date by descriptions of Wood's diffraction grating process, and later improvements on the Joly process. A chapter is also devoted to three-colour photomechanical processes, and another to the method developed by Sanger Shepherd and others of producing lantern slides in three colours.

Leçons nouvelles sur les applications géométriques du calcul différentiel. Par W. de Fannenberg, Professeur à la Faculté des Sciences de l'Université de Bordeaux. Pp. 192. (Paris : A. Hermann, 1899.)

THE geometrical applications of the differential calculus, which are usually given in English treatises on the calculus, are mostly confined to plane curves. In these lessons, on the contrary, the author begins by assuming a knowledge of elementary analytical geometry of three dimensions, and proceeds at once to deal with subjects which occur in the latter part of an English text-book on solid geometry, in chapters on the general theory of curves and surfaces.

Thus we have sections on the descriptive properties of tortuous curves and curved surfaces, followed by sections on the metrical properties of tortuous curves, of ruled surfaces, and of surfaces in general.

The author's treatment of his subject is exceedingly clear and elegant, and there is considerable freshness of method. We may notice, in particular, the early employment of the six co-ordinates of a line ; the use of the system of moving axes formed by the tangent, the principal normal and the binormal at a point on a curve ; the systematic application of Gaussian curvilinear co-ordinates in developing the properties of the several classes of curves that may be traced on a surface.

In fact, a student will find here in small compass a pleasant introduction to some of the most powerful methods of modern analysis as applied to geometry, and if he proceeds afterwards to the "Leçons sur la théorie générale des surfaces," by Darboux, his study of that great classic will have been much facilitated.

Elementary Illustrations of the Differential and Integral Calculus. By Augustus De Morgan. New Edition. Pp. viii + 142. (Chicago : The Open Court Publishing Company. London : Kegan Paul and Co., Ltd., 1899.)

IT is nearly seventy years since De Morgan first published this tractate in the Library of Useful Knowledge. It was afterwards bound up with his large treatise on the differential and integral calculus, but the very inferior typography detracts much from the pleasure of perusing it there. In the present issue we have a very attractive reprint. Although there has been in recent years almost a superabundance of elementary treatises on the calculus, some of them not lacking excellent illustrations of the fundamental principles and processes of the subject, it may still be said that De Morgan's effort to popularisation remains the greatest of its kind, and far above all others in the philosophic spirit which animates it.

LETTER TO THE EDITOR.

[The Editor does not hold himself responsible for opinions expressed by his correspondents. Neither can he undertake to return, or to correspond with the writers of, rejected manuscripts intended for this or any other part of NATURE. *No notice is taken of anonymous communications.]*

A Surface-Tension Experiment.

IF an unbroken vertical jet of falling water is allowed to impinge normally on a smooth circular disc, whose diameter is rather greater than that of the jet, then a phenomenon, illustrated by the accompanying photographs, is observed. These are one-ninth natural size.

A disc about 7 mm. in diameter was supported on the upper end of a knitting-pin, which was held vertically in a clamp.

A jet of water proceeding from a tube of 6 mm. internal diameter was directed downwards, so as to strike the disc centrally.

If the initial velocity of the jet is high, then an umbrella-shaped sheet is formed, which breaks up into a shower of drops at its margin. On diminishing the rate of outflow, the broken

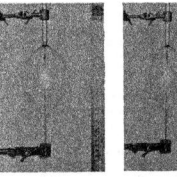

| FIG. 1. | FIG. 2. | FIG. 3. |
| 4000 c.c. per min. | 3000 c.c. per min. | 2100 c.c. per min |

edge of the sheet gathers itself together and closes inwards until it reaches the upright supporting the disc, thus forming a completely closed pear-shaped surface (Fig. 1). The surface-tension of the falling sheet thus drags in the water radially, for if it were in separate drops these would describe parabolic paths.

On further restricting the water supply there is, in general, a tendency for the surface to elongate and at the same time to contract laterally, thus becoming more spindle-shaped (Figs. 2 and 3). In this condition the figure is remarkably steady and well defined.

With a still slower stream of water (Fig. 4), the spindle reaches a certain critical length at which it first begins to

| FIG. 4. | FIG. 5. | FIG. 6. |
| 1600 c.c. per min. | 1000 c.c. per min. | |

oscillate vertically and to pulsate, and then a sudden constriction occurs causing the division of the spindle into two bubbles, one of which rushes down and the other up the vertical support.

The latter bubble persists as a small conical figure immediately beneath the disc (Fig. 5).

Since there is an almost instantaneous transition from Fig. 4 to Fig. 5, it was not found possible to photograph any of the intervening conditions.

The length of the spindle in Fig. 4 (just on the point of undergoing segmentation) is about four times its diameter ; a ratio which is considerably greater than in the three preceding figures, and which approaches that known to hold in the case of an unstable liquid cylinder.

The rate of outflow from the tube is in each case stated under the photograph.

The knitting-pin was then replaced by a glass tube, closed at its upper end, but connected below with a manometer, which was put into connection with the interior of the various bubbles by mean of a small hole previously blown in the side of the tube near its upper extremity. The pressure inside the bubbles was always found to be very nearly atmospheric, an excess of about 1 mm. of water being the greatest noted.

Another form of apparatus gave striking results.

A ⊣ tube was supported with the cross piece vertical, and the upper opening was closed with a cork into which a stiff wire was driven, so that it hung centrally, with its other end projecting a little from the lower opening of the tube.

The circular disc was attached to this projecting wire, and was then exactly in the path of the issuing stream of water which was admitted by the side tube.

By properly regulating the velocity of the water a series of surfaces similar to those produced by the former method can be obtained, but the adjustments are not so easily controlled (Fig. 6).

It is curious to note how the water constituting the walls of the bubble reunites into a single stream which falls from its base.

T. J. BAKER.

King Edward's School, Birmingham, May 28.

THE INTERNATIONAL CATALOGUE OF SCIENTIFIC LITERATURE.

IN view of the proceedings at the recent third International Conference, of which the Acta are printed on another page, there can be little doubt that the ultimate execution of this important enterprise is now assured. Prior to the meeting, some of us, perhaps, vaguely feared that the foreign delegates would come prepared to suggest all sorts of difficulties, if not to announce the unwillingness of the countries they represented to take any part in the work ; but nothing of the kind occurred : all came bent on securing success ; not a word was uttered in depreciation of any of the proposals brought under consideration ; and all present may be said to have taken an enthusiastic interest in carrying the proceedings to a satisfactory issue. Every one was of opinion that if a fair beginning can once be made, the importance of the work is so great ; it will be of such use to scientific workers at large ; that it will rapidly grow in favour and soon secure that wide support which is not yet given to it simply because its character and value are but imperfectly understood. Therefore, all were anxious that a beginning should be made.

It has been estimated that if 300 sets or the equivalent are sold, the expenses of publication will be fully met. As the purchase of more than half this number was guaranteed by France, Germany, Italy, Norway, Switzerland and the United Kingdom, the Conference came to the conclusion that the number likely to be taken by other countries would be such that the subscriptions necessary to cover the cost of the catalogue would be obtained.

The resolution arrived at after this opinion had been formed, " That the catalogue include both an author's and a subject index, according to the schemes of the Provisional International Committee," must, in fact, be read as a resolution to establish the catalogue.

Of the countries represented at the various Conferences, excepting Belgium, not one has expressed any unwillingness eventually to cooperate in the work. Unfortunately, neither the United States nor Russia was officially represented on the present occasion. The attempts that have been made to induce the Government in the United States to directly subsidise the catalogue have not been successful : but that the United States will contribute its fair share, both of material and of pecuniary support, cannot be doubted. There as here private or corporate enterprise must undertake much that is done under Government auspices in Europe. As to Russia, the organisation of scientific workers there has been so little developed that it is very difficult to secure their attention, and probably our Russian colleagues are as yet but very imperfectly aware of what is proposed. The importance of Russian scientific work is so great, however, that it stands to reason that it must be fully considered ; and it may be supposed that Russia will join when she becomes acquainted with what is proposed and what is required of her.

A Provisional International Committee has been appointed, which will take the steps now necessary to secure the adhesion and cooperation of countries not yet pledged to support the scheme.

Originally, it was proposed to issue a card- as well as a book-catalogue, but on account of the great additional expense this would involve, and as the Americans in particular have not expressed themselves in favour of a card issue, it is resolved to publish the catalogue, for the present, only in the form of annual volumes.

From the outset great stress has been laid on the preparation of subject indexes which go behind the titles of papers and give fairly full information as to the nature of their contents. Both at the first and the second International Conference this view met with the fullest approval. Meanwhile, the action of the German Government has made it necessary to somewhat modify the original plan. In Germany, a regional bureau will be established, supported by a Government subvention, and it is intended that the whole of the German scientific literature shall be catalogued in this office ; no assistance will be asked from authors or editors or corporate bodies. In such an office it will for the present be impossible to go behind titles ; consequently, only the titles of German papers will be quoted in the catalogue. In the first instance, some other countries may prefer to adopt this course on the ground of economy. But in this country, at least, the attempt will be made to deal fully with the literature, and the cooperation of authors and editors will be specially invited. An author may not always be best able to judge which are the most important points in his paper to be noted in an index, but the experience gained in the Royal Society during several years past has shown that authors furnish most valuable information, and that their suggestions are easily reduced into shape. A full code of instructions for the use of the regional bureaux is now being prepared under the auspices of the Provisional International Committee.

The catalogue is to be published annually in seventeen distinct volumes. The collection of material is to commence from January 1, 1901. As it will be impossible to print and issue so many volumes at once, it is proposed to publish them in sets of four or five at quarterly intervals. During the first year, parts covering shorter periods will be prepared, so as to make the subsequent regular issue possible of volumes in which the literature published during a previous period of twelve months is catalogued. Valuable opportunity will thus be given from the outset of gaining experience both in the preparation and use of the catalogue.

That many difficulties will be encountered in carrying the work out cannot be doubted ; but if scientific workers generally will but reflect on the inestimable value of accurate classified subject indexes, they cannot but see that it will be to their great advantage to do all in their power to further the enterprise. If the attempt fail, it will only be because those on whose behalf it is undertaken are blind to their own interests.   H. E. A.

NOTES.

THE annual visitation of the Royal Observatory, Greenwich, by the board of visitors, took place on Tuesday last.

A SPECIAL joint meeting of the Royal and Royal Astronomical Societies is being held in the rooms of the Royal Society to-day, to receive preliminary accounts of the observations of the recent eclipse of the sun.

THE Nilson Memorial Lecture will be delivered by Prof. Otto Pettersson, of Stockholm, before the Chemical Society on Thursday next.

THE great kindness and attention shown oy the Alcalde and other authorities at Santa Pola to the astronomical party who went there to observe the recent eclipse, and to the captain and officers of H.M.S. *Theseus*, who conveyed the members of the expedition from Gibraltar, occasioned a very pleasing little episode. On leaving Santa Pola a donation of 10*l*. was collected, and left by Captain Tisdall with the Mayor for the benefit of the poor of the town. This gift was highly appreciated by the local authorities, and the amount has been distributed by a local committee. The children in the schools were not forgotten, and each of them received a packet of sweets and a memorial card relating to the eclipse and the visit of the expedition. We are also able to state that the Mayor of Santa Pola has received from the Spanish Government a decoration of the First Order of the Civil Administration. We heartily congratulate him on his new honour, which all who had any relations with him know was well deserved.

AT a public meeting recently held in Belfast, it was decided to renew the invitation to the British Association to visit Belfast in 1902, and a representative deputation was appointed to present the invitation at the forthcoming Bradford meeting of the Association. The last meeting in Belfast took place in 1874, and was under the presidency of Prof. Tyndall.

SIR WILLIAM MACCORMAC is to receive to-day the honorary degrees of M.D. and M.Ch. from the University of Dublin.

THE death is announced of Prof. Boutan, general inspector of public instruction in France. Prof. Boutan was one of the founders of the Société française de Physique, and was also the author, jointly with M. d'Almeida, of a treatise on physics.

THE death is announced of Dr. Karl Lange, professor of pathological anatomy in the University of Copenhagen ; also of Dr. Wilhelm Kühne, professor of physiology at Heidelberg.

THE new physical laboratory at Owens College, Manchester, will-be opened to-morrow by Lord Rayleigh. The new laboratory will have a larger floor area than that of any other similar institution in the world, with the exception of the Johns Hopkins and the Strasburg laboratories. Great efforts have been made to provide an equipment of the most modern apparatus for use in every branch of physical science, and to maintain conditions which shall ensure their being used to the best advantage. The research laboratories are to be an important feature of the new buildings, and should attract a large number of students. Another feature is the electro-technical wing, which is to constitute a John Hopkinson memorial, and will be formally handed over by the relatives of the late Dr. John Hopkinson, on the occasion of the opening ceremony. It is understood that Dr. C. H. Lees, formerly chief assistant lecturer in the physics department of Owens College, will occupy the post of assistant director of the new laboratories, under Prof. A. Schuster, the director, and that Mr. R. Beattie has been appointed lecturer in electrotechnics.

AT the conversazione to be held at the London Medical Graduates' College and Polyclinic on Wednesday, July 4, Prof. Osler F.R.S., of Baltimore, will deliver an address on "The

Teaching of Practical Medicine," and the museum will be inaugurated.

THE annual general meeting of the Röntgen Society will be held on Thursday, July 5. The Presidential Address will be delivered by Mr. Wilson Noble.

THE second annual meeting of the Astronomical and Astrophysical Society of America is being held in conjunction with the meeting of the American Association at Columbia University. In addition to the papers to be presented, there will be discussions upon the following subjects :—The eclipse of May 28 last ; Observations of Eros to be made at the next opposition ; Spectroscopic determinations of motion in the line of sight.

AN important meeting of the Committee of the Liverpool School of Tropical Medicine was held on the 19th inst., when it was reported that the Government were co-operating with the School in the matter of the despatch of the Yellow Fever Expedition to America and Brazil, and that a letter had been received from the Marquess of Salisbury asking whether the Committee wished him to communicate with the British representatives in the countries to be visited by the expedition. The offer was gratefully accepted, and a further letter was received from Lord Salisbury saying that he had asked the British Ambassador at Washington and H.B.M. Consul at Para to obtain all possible facilities from the United States and Brazilian authorities respectively on behalf of the expedition. Official invitations for the expedition to visit Washington had been received from the heads of the medical departments of the U.S. army, and to visit Baltimore from the authorities of the Johns Hopkins University. As has already been stated in these columns, the expedition consists of Dr. Durham (Grocers Research Scholar) and Dr. Walter Myers (John Lucas Walker Student), both of Cambridge. The expedition, which started on Tuesday last, goes first to Canada, and then proceeds direct to Washington and Baltimore. After conferring with the bacteriological experts there, the expedition will go to New York, and sail from that port to Para. Subsequent movements will be guided by circumstances.

AT a dinner given last Monday in honour of the Yellow Fever Expedition, Mr. A. L. Jones, chairman of the school, announced his intention of giving 1000*l*. towards the erection of a Tropical Diseases Hospital in Liverpool in connection with the Royal Southern Hospital, to be associated with the name of Miss Mary Kingsley. It was also announced that Mr. Blaize, of Lagos, and Mr. John Holt, of Liverpool, had promised 500*l*. each to announced. Two other subscriptions of 100*l*. each were the same object.

THE Summer School of Medicine, which was to have been held at Cambridge from June 25 to June 30, has, unfortunately, had to be abandoned in consequence of the meagre number of acceptances received. The necessity for the taking of this step is the more to be regretted, as very careful preparations had been made to insure a successful session ; demonstrations of the malarial and other blood parasites, and of the most recent work on cancer, had been arranged for, and in addition to these subjects lectures were to have been given by experts in their several lines of work upon various other matters of medical. and surgical interest.

THERE are, it is estimated, about 400 lepers in France. They are scattered about in Brittany, in the Pyrenees, on the shores of the Mediterranean, and in Paris, where they number 150. Among the lepers there are missionaries and nurses who have fallen victims to their devoted care of sufferers in other countries, and officials and soldiers who have contracted the disease in the colonies. An anti-leprosy committee has, says the *British Medical Journal*, recently been formed on the initiative of Dom Santon, a member of the Benedictine Community of Ligugé, who is also a doctor of medicine, for the care of the lepers in France and the prevention of the spread of the disease. Dom Santon has for many years past devoted himself to the study of leprosy, travelling for that purpose in many parts of the world. After conference with the Council of Hygiene he has acquired a property in the Vosges, where he proposes to establish an asylum for lepers to be called the St. Martin Sanatorium. The plans have been approved by the French Government.

THE National Academy of Sciences of the United States has recommended to the trustees of Columbia University that the Barnard medal be given to Prof. Röntgen for his discovery of the X-rays. The medal, of gold, is awarded quinquennially to the person who shall have made such discovery in physical or astronomical science as, in the judgment of the National Academy of Sciences, shall be esteemed most worthy of the honour.

AN entomological expedition is to be sent into Southern Mexico this summer by the University of the State of Missouri. It will be in charge of Prof. J. M. Stedman, head of the Entomological Department, and will have for its object the making of a biological, largely entomological, survey of the region from Vera Cruz on the Gulf, which is in perpetual tropics, to the top of the volcano Popocatepetl, which is far above the perpetual snow line, and down to Acapulco on the Pacific. This will give all the temperature variations from perpetual tropics to perpetual snow, and will allow of the study of life zones under conditions not to be found elsewhere in North America. The collection will become the property of the University, which is to furnish half the expenses, the other half being borne by Prof. Stedman.

AN expedition, consisting of President Jordan and Mr. John O. Snyder, of the Department of Zoology in Stanford University, has sailed for Japan, for the purpose of making a collection of the fishes and insects of that country. Assistance will be given by other graduates of Stanford University at present resident in Japan.

IT is stated in *Science* that Mr. G. B. Gordon has secured the control of the ruins of Copan, and the lands pertaining thereto, for a period of ten years, with the right to make excavations and to remove to Cambridge, Mass., for preservation, a portion of the objects that may be found.

MR. O. A. TITTMANN has been appointed successor to Dr. H. S. Pritchett as superintendent of the United States Coast and Geodetic Survey, Dr. Pritchett having been elected president of the Massachusetts Institute of Technology.

THE American National Geographical Society's prizes for the best essays on Norse discoveries in America have been awarded to Mr. C. B. Dalton, of New York City, and Mr. K. F. Murray, of Norfolk, Va.

Science announces that a donor, who wishes to be anonymous, has presented to the American Museum of Natural History the collection exhibited by Messrs. Tiffany and Co. at the Paris Exposition, consisting of American and foreign cut and uncut precious stones and other objects. The value of the collection is estimated at over 50,000 dollars.

DURING the summer a station will be maintained on Lake Saranac by the New York State Museum, for the study of aquatic insects. The work will be under the direction of Dr. Charles Needham.

THE *Scientific American* states that a new species of petrel has been discovered on the Island of Kauai (Sandwich Islands) by Mr. A. Searle, of the Stanford University. Mr. Searle is

also reported to have found on the same island a new species of seagull. He is about to go to Guam for the purpose of exploring that island, and to make a collection of birds and fishes for the Bishop Museum of Honolulu.

EXCELLENT results have been obtained by the French Government from experiments made with wireless telegraphy. The *Engineer* of June 15 says that the demonstrations showed that communication could be maintained, between ship and shore, to a distance of about sixty miles with comparative ease, only the height of the masts of the Government ship *Utile* preventing longer distances being attained. In consequence of these achievements the French Government have decided to equip their Mediterranean Squadron with the necessary apparatus.

WIRELESS telegraphy stations are, by the instructions of the Chief Signal Service Officer of the United States, to be established in the harbour of San Francisco, in Porto Rico and the Philippines.

AMONG the numerous congresses arranged to take place in connection with the Paris Exposition, in addition to those to which attention has already been called in these columns, may be mentioned the following, dealing respectively with :—Automobiles, on July 9 ; medical electrology and radiology, from July 27 to August 1 ; medicine, from August 2 to 9 ; physics, from August 6 to 11, and on the same dates, technical and industrial education ; chemistry, from August 6 to 11 ; hygiene and demography, from August 10 to 17 ; hypnotism, from August 12 to 15 ; electricity, from August 18 to 25 ; prehistoric anthropology and archæology, from August 20 to 25 ; ethnology, from August 26 to September 1 ; railroads, from September 15 to 23 ; acetylene, from September 23 to 28.

IT is satisfactory to find that the present troubles in South Africa have not interfered with Museum progress in the larger towns of Cape Colony. From the *Report* of the Committee of the Albany Museum for 1899, we learn that it was expected the new buildings would be ready for opening about July 1.

A SEVERE thunderstorm occurred in London on the 25th inst., accompanied by heavy rain and hail. The weather had been very unsettled for some days, with gales on our exposed coasts. On the evening of the 24th a storm area lay off the north-west coast of Ireland ; this subsequently took a somewhat unusual south-easterly direction. At 8 a.m. on the 25th the centre lay over the Midland Counties, and next morning had traversed the south-eastern part of the English Channel. The rainfall on the 24th and 25th amounted to about an inch in several parts of the United Kingdom. The temperature continues low for the season over the whole country.

TWO specimens of the egg of the Great Auk were sold by auction at Stevens's Rooms last week, and realised 315 and 180 guineas respectively. The more important of the two eggs is an unrecorded one from a French collection, and is described as the finest specimen known of a special type of marking. The price just obtained for it establishes a record, 300 guineas having, until this sale, been the highest amount ever received. About seventy-five eggs of the Great Auk are known to be in existence.

ACCORDING to *Science*, the Millinery Merchants' Protective Association of America has proposed to the various Audubon Societies of the country to cease killing or buying any North American birds, except such as are edible and killed in season, if the societies will undertake not to interfere with the use of these birds or with skins imported from countries not in North America.

THE *Pioneer Mail* (Allahabad) of June 1, 1900, has an interesting article on the recent discoveries in the neighbourhood of the previously identified birthplace of Buddha. Mr. W. Peppé, owner of the Birdpur estate on the Nepal frontier, excavated in January 1898 a reliquary (stūpa) of Buddha, and found relics in a casket inscribed in characters not later than the third, and possibly even of the fourth, century B.C. During last winter Prof. Rhys Davids revisited the spot, and gave to the Royal Asiatic Society at its meeting in April last the result of his own local observations and examination of the relics, which is that they have a very fair title to be considered genuine remains of Buddha. These are stated to have been divided after the cremation into eight portions, and distributed amongst sections of the Śakya clan, which inhabited this region. The relics themselves are fully described and illustrated in the Royal Asiatic Society's *Journal* for July 1898, and further notices on the subject are to be looked for in forthcoming numbers of the same journal, and also (by Dr. Hoey) in the *Journal* of the Asiatic Society of Bengal. It is hoped that the Government of India may support Mr. Peppé in further excavations in this evidently promising locality.

THE *Scientific American* for June 9 gives the following interesting particulars of a specially built train used on the Baltimore and Ohio Railway in a series of experiments upon the atmospheric resistance to railroad trains. The trial train was made up of six passenger coaches, such as are used on suburban service. They were provided with four-wheeled trucks, 33-inch cast-iron wheels, and 3¾-inch journals, and the total weight, exclusive of engine and tender, was 325,500 pounds. In preparation for the test all external obstructions were removed from the train. The roofs of the cars were arched; the windows set out flush with the sides of the cars; and the sheathing was laid lengthwise instead of perpendicularly as in other cars. The sheathing extended to within eight inches of the track and covered the trucks. Suitable openings permitted access to the axle boxes, and a sliding door led into the substructure at opposite sides of the car centre. When the cars were coupled, two diaphragms met and enclosed the space between the cars, from edge to edge of the roof line. The platform doors consisted of roller curtains which dropped to the steps and were flush with the sides. Flexible spring curtains completed the vestibule from the roof to the bottom of the car. When the train was coupled it presented the appearance of one long sinuous and flexible car. The tender was of peculiar construction, and continued the unbroken line from the engine cab to the baggage car, to which it was vestibuled. In its entire construction the train complied with the varied demands of practical operation. While the plans called for partial sheathing of the locomotive, it was decided to make the first tests with remodelled cars only, in order to prove how far the existing system of car construction is responsible for the atmospheric resistance of trains. The sheathed train, consisting of six cars and hauled by an engine weighing 57 tons, made the run of 40 miles from Baltimore to Washington in 37 minutes and 30 seconds. One mile was made in 40 seconds, and two miles in 81 seconds. From Beltville to College, a distance of 4½ miles, the time was 3 minutes and 10 seconds, a sustained speed of 85 miles an hour. By far the most remarkable run, however, was from Annapolis Junction to Trinidad, a distance of 20·1 miles in 15 minutes and 20 seconds, at an average speed of 78·6 miles an hour. The first seven miles of this run was up a grade from 25 to 55 feet to the mile, and it was covered in a fraction over 6 minutes; while the last 5 miles on the down grade from Alexander Junction to Trinidad was covered in 2 minutes and 55 seconds, a speed of 102 8 miles an hour. The locomotive used has cylinders 20 × 24, with four coupled 78-inch drivers. The boiler carried 165 pounds of

steam. With ordinary firing the steam never dropped below 160 pounds during the entire run. The best time previously made on the line was a few seconds less than 39 minutes, on which occasion the train consisted of four Pullman cars hauled by the company's fastest and most powerful passenger engine.

THE Report of the Kew Observatory Committee for the year 1899 has been published in the *Proceedings* of the Royal Society in the usual form. From January 1 last the Observatory was incorporated with the National Physical Laboratory, and will no doubt greatly extend its useful work. The Observatory Committee, as hitherto constituted, has ceased to exist, but the work of the Observatory will be carried on by the same staff as heretofore. During the past year the magnetic work is said to have been unusually onerous, as many colonial and foreign institutions have sent their instruments to the Observatory to be verified No very large magnetic disturbances were registered; the mean Westerly Declination was 16° 57′. The electrograph has worked in a satisfactory manner during the year, and, with the sanction of the Meteorological Council, the records for a complete year have been lent to Mr. C. T. R. Wilson, of Cambridge, for investigation. The verification of instruments of all kinds amounted to over 22,000, a falling off of nearly 2400 as compared with the work of the previous year. A seismograph has been in regular operation during the year; a disturbance was particularly noticeable on September 10.

THE dynamical principle of atmospheric circulation is treated by Prof. V. Bjerknes in the *Meteorologische Zeitschrift*, 1900, iii., iv. Starting with the property that the circulation theorems of abstract hydrodynamics (according to which the circulation in any circuit formed by the same particles is constant) only hold good when the pressure is a function of the density alone, Prof. Bjerknes points out that in the atmosphere this condition is not satisfied owing to local differences both in the temperature and in the degree of moisture present in the air. Of these two causes the first seems to be the most important. The conception of "solenoids" is then introduced, a solenoid being an elementary unit tube bounded by pairs of consecutive surfaces of equal volume and equal pressure respectively. The fundamental proposition in connection with circulation asserts that the rate of change of the circulation in any circuit is proportional to the number of solenoids enclosed by that circuit. A number of diagrams are given representing the cases of land and sea breezes, trade-winds, local upward currents, hill and valley winds, cyclones and anticyclones. The omission to take account of the extra complications arising from viscosity and terrestrial rotation probably prevents these investigations from being utilised for calculations in connection with weather prediction; and for this reason Prof. Bjerknes' theory must be rather regarded in the same light as other dynamical theories of physical phenomena, in which certain simplifications not occurring in nature are made in order to bring the calculations within the range of mathematical analysis. But it is only by the aid of such simplifications that order can be evolved out of the chaos of statistics furnished by the experimentalist.

AN account of the seismological observatory of Quarto, near Florence, together with the observations of more than 170 earthquakes made during the meteoric year 1899 (November 1, 1898–October 31, 1899), is published by the director, Mr. D. R. Stiattesi, in the first *Bollettino Sismografico* of the observatory. Through the generosity of Count G. Bastogi, of Florence, this must be one of the most completely equipped observatories in Italy. It contains two Vicentini microseismographs (one with a mass of 500 kg. and a length of 9·28 metres), a pair of horizontal pendulums with mechanical registration, and a pair of geodynamic levels, besides a large number of seismoscopes and tromometers, all of Italian design.

IN the June number of the *Zoologist*, Mr. A. H. Meiklejohn raises the question as to the manner in which the cuckoo carries her egg when about to deposit it in the nest of the bird selected to act as foster-mother. It is commonly supposed that the egg is carried in the beak, and in Prof. Newton's edition of "Yarrell" several instances are quoted where observers state they have actually seen the *modus operandi*. Mr. Meiklejohn, who was fortunate enough to observe a cuckoo in the act of depositing its egg in a robin's nest, is, however, of opinion that the throat of the bird serves as the receptacle for the egg. He states that (1) the cuckoo was constantly opening her mouth during a preliminary encounter with the robins; (2) that the egg was certainly not laid in the ordinary way in the nest; (3) that the egg itself was slightly moist and sticky; (4) that the throat of the bird presented a slightly distended appearance, which might well have been due to the presence of the egg. It will be interesting to note what his fellow-ornithologists think of the auth)r's explanation of the mystery.

IN American laboratories it appears that the place of the common frog is largely taken by the furrowed salamander (*Necturus maculatus*), which forms the general subject for anatomical investigation. Mr. W. S. Miller, assistant professor of anatomy in the Wisconsin laboratory, has accordingly undertaken to describe in detail the anatomy of this amphibian, and papers on its lungs, vascular system and brain appear in the latest issue of the *Bulletin* of the University. The author calls attention to the great amount of individual variation which occurs in the vascular system of *Necturus*.

IN a communication to the latest issue of the *Proceedings* of the Philadelphia Academy, Mr. Witmer Stone shows that the various species of eider-duck, as well as the red-breasted merganser, have a "summer moulting plumage" analogous to that assumed by the mallard after the breeding season. As in the last-named species, this plumage lasts only during the time when the birds are unable to fly, owing to the shedding of their flight feathers, and its dull coloration is doubtless for the purpose of rendering them as inconspicuous as possible during this period. The author calls attention to the circumstance that the feathers of this temporary dress, like those of the first plumage of all birds, are very inferior in their structure. The moulting plumage of the king-eider has hitherto been considered as the ordinary dress of immature birds.

IN a paper on mosquitoes, by Mr. W. R. Colledge, which appears in vol. xv. of the *Proceedings* of the Royal Society of Queensland, the author states that he has succeeded in keeping one of these insects alive for three weeks, and that Dr. J. Bancroft has had some in captivity for eighty or ninety days. Probably their ordinary full term of existence is three months. In cases where the application of kerosene is inconvenient, the introduction of a few minnows into the ponds or pools in which they breed will speedily lead to the destruction of the larvæ and pupæ.

THE origin and formation o. the Red Sea are discussed in a brief article by M. A. Issel (*Bull. Soc. Belge de Géol.*, tome xiii. April). Following Suess, he considers that the lacustrine conditions of the Arabic depression ("Lacus Arabicus") were probably determined in late Miocene times. He maintains that then, or early in Pliocene times, the Nile, a mightier river than it now is, emptied its waters directly into the great lake, the outlet being an immense waterfall. Even in post-Pliocene times the Nile continued to send a portion of its waters into the Red Sea, although it had meanwhile formed new outlets into the Mediterranean area. Traces of this former fluviatile connection are furnished in the actual faunas of the two

seas. The opening of the Straits of Bab-el-Mandeb was caused after a period of volcanic activity, the eruptions being succeeded by subsidence and by erosion of the barrier which separated the Red Sea from the Indian Ocean. It is remarked that the opening of the Suez Canal has sensibly affected the distribution of some forms of life.

IN a paper on the fruiting of the blue flag, or Iris, published in the May number of the *American Naturalist*, Prof. J. G. Needham shows that, in addition to the bees by which they are fertilised, the flowers of this plant are visited by a number of insects of other kinds. The visits of these latter appear for the most part to have been hitherto noticed; and as many of these illicit visitors are of no use for the purpose of fertilisation, the ill-adapted ones are, according to the author, habitually deceived by the flower itself as to its proper entrance. Of the various visitors, two small bees of the genera *Clisodon* and *Osmia* were thoroughly at home in the flower, alighting at the entrance and passing immediately down the narrow passage leading into the nectary, and as quickly emerging and flying off. On the other hand, numerous kinds of Syrphid flies spent a much longer time on the flower, which many of them visited only for pollen. Other visitors were certain small flower-beetles and weevils, which never by any chance succeeded in reaching the nectary.

TWO interesting lectures by Prof. D. T. MacDougal, delivered at the Woods Holl Marine Biological Laboratory, are reprinted from the *Bulletin* of the New York Botanic Garden. In his address on the "Significance of Mycorhizas" a general summary is given of our present knowledge of the occurrence of these organisms, both endotropic and ectotropic. All known species of mycorhizal fungi are stated to belong to the families Oomycetes, Pyrenomycetes, Hymenomycetes, and Gasteromycetes; and it is suggested that their further study and identification may result in a considerable increase in our knowledge of the physiology of vegetable life; and that their culture may not be without importance in the nutrition of a number of perennial flowering plants. The lecture on the "Influence of inversions of temperature, ascending and descending currents of air, upon distribution" is devoted to an explanation of the distribution of the flora in the United States, especially in the region of the great cañons.

APPENDIX III. for 1900 of the *Kew Bulletin of Miscellaneous Information* is entirely occupied with a list of the additions to the Library of the Royal Botanic Gardens, made during the year 1899

THE *West Indian Bulletin*, vol. i. No. 3, published in Barbados, contains the completion of the report on the papers read at the Agricultural Conference held in that island, and of the discussions arising out of them.

THE last two parts which have reached us of Engler's *Botanische Jahrbücher* (vol. xxviii. Hefts 1 and 3) are occupied entirely with the useful description and systematic papers which form so conspicuous a character of the work. Among those relating to flowering plants are the following, or instalments of them:—Compositæ of Ecuador, by Hieronymus; The flora of Central America, by Loesener; Classification of the Calyceraceæ, by Reiche; Revision of the genus *Linnaea*, by Graebner; African Verbenaceæ, Borragineæ, and Labiatæ, by Gürke; *Triplochiton*, a new genus of Malvales from the Cameroons, constituting the type of a new family, Triplochitonaceæ, by Schumann; Report of the botanical results of the Nyassa Lake Expedition, by the Editor. Herr Hennings and Dietel furnish respectively instalments of their papers on the Fungi and on the Uredineæ of Japan.

WE have received a number of papers by different officers of the Observatory of Catania. Most of these we have noticed on their first appearance. Among the others, we may mention a valuable memoir, by Mr. S. Arcidiacono, on the eruptive period of Etna from July 19 to August 5, 1899, in which he points out the approximate coincidence of the great explosion on the former day with the total cessation of the flow of lava in Vesuvius and a strong earthquake in Latium, and also an interesting account of the history of the observatories of Catania and Etna.

THE *Mittheilungen aus dem Roemer-Museum*, Hildesheim (No. 11, April 19), includes a paper, by Mr. A. R. Grote, on the phylogeny of the families of butterflies, with a genealogical tree. It is a continuation and amplification of previous papers on the same subject, published by the author in Germany and America, and is mainly based on neuration. Like many authors, Mr. Grote divides the butterflies into two main super-families, Papilionides and Hesperiades; but it will surprise many entomologists to find that all the butterflies except the true Papilionidæ are referred to the Hesperiades.

A NEW journal has been started in Berlin, the first number of which bears the title "Laboratorium et Museum," while in the second number the words "et Clinicum" are added. The journal is to be of an international character, and includes articles and notes in English, French and German. The title of the journal is sufficiently suggestive in itself of the contents, which comprise descriptions of new apparatus and reagents, methods of preparation, notices of new books, obituary notices, and lists of trade catalogues, of which the publishers will send copies on application.

DR. FRANCESCO FOSSATI has published in the *Memorie del R. Istituto Lombardo* a bibliography of the writings of Volta. Several such lists have already been published: one in 1813 by Prof. Configliachi, containing the titles of forty-four works; one in 1877 by Prof. Pietro Riccardi, containing sixty titles; while the collection procured by Antinori in 1816 contained sixty-seven writings. The present bibliography is partly the outcome of a suggestion made by Prof. Alessandro Volta, junr., at the Como Electrical Congress last year, and it contains the titles of 231 writings.

IN the current number of the *Berichte* (p. 1569) Baeyer and Villiger describe some of the properties of the new hydride of benzoylsuperoxide, $C_6H_5CO\ O.OH$. The substance is obtained by the action of sodium ethylate upon benzoylsuperoxide,

$$C_6H_5CO.O.O.C_6H_5 + NaOC_2H_5 = C_6H_5CO.O.ONa + C_6H_5CO.OC_2H_5.$$

The sodium salt of the new compound is formed together with ethyl benzoate. The ethyl benzoate is removed with ether, and the hydride of benzoylsuperoxide separated by acidifying and extracting with chloroform. On distilling the chloroform, the hydride remains as a colourless crystalline mass, which melts at 41–43°. It is very soluble in the ordinary solvents, with the exception of benzene. The smell is penetrating and pungent, resembling, in the dilute state, hypochlorous acid, but not ozone. In its oxidising action on potassium iodide or aniline, and in its reducing action on permanganate, it stands midway between hydrogen peroxide and Caro's reagent (potassium persulphate dissolved in concentrated sulphuric acid). With benzoyl chloride it forms benzoylsuperoxide; with acetic anhydride, benzoylacetylsuperoxide. The oxidation of benzaldehyde to benzoic acid by exposure to air is shown to be due to he agency of this new compound,

$$C_6H_5COH + O_2 = C_6H_5CO.OOH$$
$$C_6H_5CO.OOH + C_6H_5COH = 2C_6H_5.COOH,$$

which is formed as an intermediate product.

THE additions to the Zoological Society's Gardens during the past week include a Smooth-headed Capuchin (*Cebus monachus*) from South-east Brazil, presented by Mr. F. Wallace; an Indian Desert Fox (*Canis leucopus*) from Persia, presented by Captain D. J. Leiper; a Small Hill Mynah (*Gracula religiosa*) from India, presented by Captain R. York Heriz, R.N.; two Yellow-bellied Liothrix (*Liothrix luteus*) from India, presented by Miss Petrocochino; a Cockateel (*Calopsittacus novae hollandiae*) from Australia, presented by Mrs. Harry Blades; four Ring-necked Parrakeets (*Palaeornis torquatus*) from India, presented by Mr. J. M. G. Baçe; three Chaplain Crows (*Corvus capellanus*) from Southern Persia, presented by Mr. B. T. Ffinch; two Green Lizards (*Lacerta viridis*), four Viperine Snakes (*Tropidonotus viperinus*), a Smooth Snake (*Coronella austriaca*), two Marbled Newts (*Molge marmorata*), European, presented by the Rev. F. W. Haines; an Ourang-outang (*Simia satyrus*, ♂) from Borneo, five —— Mole Rats (*Spalax* sp. inc.) from East Africa, a Grey Parrot (*Psittacus erithacus*) from West Africa, a Yellow-cheeked Amazon (*Chrysotis autumnalis*) from Honduras, nine Mountain Witch Ground Doves (*Geotrygon cristata*) from Jamaica, a Hocheur Monkey (*Cercopithecus nictitans*) from West Africa, seven Brazilian Tortoises (*Testudo tabulata*) from South America, five American Box Tortoises (*Cistudo carolina*) from North America, deposited; two Peba Armadillos (*Tatusia peba*) from South America; three Spotted Owls (*Athene brama*) from Madras; three White-throated Finches (*Spermophila albogularis*) from Brazil, a Thick-billed Seed Finch (*Oryzoborus crassirostris*) from South America, a White-eared Conure (*Pyrrhura leucotis*) from Brazil, a Logger-head Turtle (*Thalassochelys caretta*) from the Tropical Seas, purchased; two Burrhel Wild Sheep (*Ovis burrhel*), a Thar (*Hemitragus enlaicus*), born in the Gardens; two Pied Mynahs (*Sturnopastor contra*), bred in the Gardens.

OUR ASTRONOMICAL COLUMN.

ASTRONOMICAL OCCURRENCES IN JULY.

July 3.	8h. 0m. to 9h. 11m. B.A.C. 4006 (mag. 5·7) occulted by the moon.
4.	1h. Mercury at greatest elongation (26° 2′ east).
5.	Jupiter in conjunction with β Scorpii.
8.	11h. 24m. to 11h. 54m. δ Scorpii (mag. 2·5) occulted by the moon.
8.	13h. Jupiter 1° 35′ north of the moon.
9.	10h. 50m. to 12h. 2m. 24 Ophiuchi (mag. 5·6) occulted by the moon.
9.	Pallas in opposition to the sun.
10.	10h. 18m. to 11h. 10m. 33 Sagittarii (mag. 6·0) occulted by the moon.
10.	12h. 19m. to 13h. 18m. ξ² Sagittarii (mag. 3·5) oc-culted by the moon.
10.	16h. Saturn 0° 48′ south of the moon.
14.	9h. 43m. to 10h. 17m. c¹ Capricorni (mag. 5·2) oc-culted by the moon.
15.	Venus. Illuminated portion of disc, = 0·018. Mars = 0·948.
15.	8h. 29m. to 9h. 21m. κ Aquarii (mag. 5·5) occulted by the moon.
15.	10h. 11m. Minimum of Algol (β Persei).
16.	11h. 3m. 11h. 42m. 16 Piscium (mag. 5·6) occulted by the moon.
21.	13h. 2m. to 13h. 52m. 53 Tauri (mag. 5·5) occulted by the moon.
21.	14h. 53m. to 15h. 12m. D.M. + 20°, 751 (mag. 5·9) occulted by the moon.
25.	Giacobini's comet situated close to a Cygni.
28.	Epoch of the Aquarid meteoric shower (Radiant 340° − 12°).
31.	Ceres in opposition to the sun.

THE NEXT TOTAL ECLIPSE OF THE SUN.—We have recently received *Nautical Almanac Circular* No. 18, issued under the superintendence of Dr. Downing. This pamphlet contains the local particulars of the next total eclipse of the sun, which takes place on May 17, 1901. From inquiries which have been made, it appears that the positions selected in the eastern portion of the shadow track are those which are most easily accessible. These are all situated in the Malay Archipelago, with the exception of Mauritius. The durations of totality at the various stations recommended are as follow :—

Station.	Long.	Lat.	Duration of Totality.
			m. s.
Mauritius	57 33·2 E.	20 6 S.	... 3 35
Padang, Sumatra ...	100 20·5	... 0 58	... 6 14
Pontianak, Borneo	109 20	... 0 1	... 5 40
Fort Victoria, Amboyna	128 11	... 3 41	... 4 15
Port Moresby, New Guinea	147 9	... 9 28	... 3 19

The elements on which the computations are based are those published in the *Nautical Almanac* for 1901. A map of the region is included in the circular, by the aid of which other stations than those specified may be selected if desired.

ANCIENT RECORDS OF METEOR SHOWERS.—In his report for the year 1899, M. D. Eginitis, director of the Athens Observatory, gives a short account of some ancient records of meteor showers which appear to be suggestively consistent with the constants of several conspicuous showers of present times.

A shower was mentioned by the patriarch Nicéphore as lasting all night, but no exact date is given. From the historical statements given, however, M. Eginitis traces the epoch as the autumn of the year 752. This would suggest it being a shower of Andromedes, and, in fact, counting from the conspicuous falls of Bielids in 1852, 1872 and 1892, the twenty years interval corresponding to three periods of the comet, it is seen that the year 752 would be in such a series. He thus considers this apparition of 752 to have been a Bielid shower of Andromedes.

Seven years previously to this, in 745, the appearance of a great comet was recorded by Théophane and Cédrinos.

It may be, however, that the showers of 1852 and later are not from the same swarm as the shower of 752, but that they are the products of slow but continual disintegration of the comet.

Another passage in Cédrinos describes a shower in 558, also occurring in the autumn. Apparently connected with this is the observation of a comet in 518, the interval being almost six times the periodic time of Biela's comet, so that here there would appear evidence of a second series of showers, connected with Biela's comet by similarity of period, but occurring at different epochs from the first series mentioned. The modern showers of 1798 and 1838 would fall in this second group.

Théophane in 763 and Domno Alberico in 1122 record falls of shooting stars in the month of April, and these would correspond to the present showers of Lyrid meteors.

A shower chronicled in April 1094 by Alberico cannot be at present connected with any known radiant.

A MODERN UNIVERSITY.[1]

II.

THE constitution of the new University of Birmingham is on the simplest and broadest lines, and appears to offer scope or great developments in the future, some of which can hardly be foreseen at the present time.

The movement for the foundation of a university arose out of the Mason Science College, founded by Sir Josiah Mason in 1875, just a quarter of a century ago ; though it was not till five years later that the college was open to receive students. In 1892 an amalgamation was effected with the Queen's College Faculty of Medicine, and in 1897 the whole was incorporated as one body under the Mason University College Act. The Senate consists of twenty-seven members, and there are a large number of lecturers and demonstrators ; but that it should have already developed into a university is a very remarkable fact, and a sign of great enterprise and energy on the part of the community among which the college has done its work ; indeed, it is unlikely that this rapid development could have taken place unless it had been fortunate enough to secure the interest and personal influence of a prominent Minister of the Crown.

[1] Continued from p. 136.

The Faculties of the University already provided for are science, arts, medicine and commerce, but provision is made for the addition of other faculties by Statute later on. Each faculty holds its own meetings, and is presided over by its elected dean.

The assemblage of professors constitutes the "Senate," as usual. The "Council," or acting governing body under the Court, consists of the deans of the faculties, five nominees of the Birmingham City Council, twelve members appointed by the Court of Governors, and lastly of the chief officials of the University, *i.e.* the Chancellor, the Vice-chancellor, the Pro-Vice-Chancellor, the Treasurer, the Principal, and the Vice-Principal. The Court of Governors is a very widely representative body, consisting of all the chief officials in the neighbourhood of Birmingham, the head-masters of the principal schools, ten of the Members of Parliament for the boroughs and counties in the Midland district, a nominee of each of the other English Universities (including the University of Wales), a member from each of the Midland County Councils, five nominees by the Birmingham City Council, certain named life governors and donors of certain sums, all the professors of the University, six persons elected by the Guild of Graduates, three by the Guild of Undergraduates, and eleven members appointed by the voluntary elementary schools of the neighbourhood.

It is hardly possible to imagine a wider basis of representation than the one adopted for the Court of Governors of this University.

Among the executive officers there is to be a Secretary, and also a Registrar appointed by the Council ; the Vice-Principal and one of the deans, *i.e.* the Dean of the Faculty of Medicine, are likewise to be appointed by the Council. There is to be a Principal appointed by the Crown ; there is also to be a Vice-Chancellor elected by the Court of Governors, and there is to be a Chancellor ; the first Chancellor being the Right Hon. Joseph Chamberlain, and the first Vice-Principal Prof. R. S. Heath.

Concerning the objects to which all this machinery will be applied, no doubt a good deal will at the beginning be conducted on lines with which we are more or less familiar, though there appears to be no desire to imitate other universities, but rather a hope that it may be possible to strike out on a new line, and develop a broad system of national education suited to modern times, and to the practical requirements of life in an active city of the British Empire.

To this end a committee of inquiry was formed, and a deputation sent to various colleges and universities, chiefly in the United States and Canada, in order to study what was going on there. This body reported to the management committee connected with the establishment of the University of Birmingham, and their report constitutes an important and informing document. In it they say that :—

"Their object has been the teaching of science in its application to industry, and in the first place to the industries of the city and district, coupled with such technical instruction in handicrafts as will enable the students to complete their course in the university itself."

They classify the industries of the district as follows :—mining, metallurgy, engineering, and chemical trades, and non-metallic trades.

They recommend that there shall be chairs of mining, metallurgy, engineering, and applied chemistry. They further recommend that the students should be put through a very thorough course, consisting largely no doubt of a study of mathematics, physics, pure chemistry, and geology, as taught at present, but finishing with a specifically technical course, making it four years in all. A shorter course would likewise be permissible, but it appears would not lead to a degree.

They say the students should be divided into two classes, viz. :—

(1) "Those taking a four-years' course in mechanical (including electrical), civil or mining engineering, metallurgy or applied chemistry, who would study for a master's degree in their respective subjects. At the conclusion of this course facilities would be offered for further study and research to those who could give the time or should wish to proceed to the doctor's degree."

(2) "Those taking a course of from one to three years in any of the above subjects, with a view to the practical application of the teaching to a particular industry. With such students, less time could be devoted to theory, as attention would have to

be concentrated on methods and results. Their work in these courses would be recognised by class certificates."

In addition to the professor of each technical subject there will be an assistant-professor and several instructors, each competent in a particular branch.

They indicate a block plan for the buildings required, their size and their suggested arrangement. These plans suppose a front of two storeys, containing the lecture rooms, library, museum, &c., and at the back a series of blocks, all on the ground floor, and intended for the various laboratories and workrooms which have been described in the report. These could be built to provide accommodation in the first instance for 200 day students, increasing afterwards to 500.

"The space occupied by these buildings, including the necessary yards and roads, a gymnasium, a director's house, and rooms for a caretaker, is about six acres. In view of the future of the university, a total area of not less than twenty-five acres should be provided."

The committee recommend that this land be taken in the outskirts of the city on a main line of, by preference, both rail and road, and they estimate the complete cost as follows :—

Twenty-five acres of land and buildings	...	80,000*l.*
Machinery, apparatus, and instruments	...	66,000
Fittings, utensils, lighting, and heating	...	5,000
Technical library	...	...	...	...	1,500
Museum	...	...	...	...	500
Director's house	...	...	...	...	2,000
	Total	...	155,000*l.*

They estimate the cost of maintenance (including the staff) at 10,450*l.* per annum.

Thus the scheme is a very large one, but it is estimated that the fees from students will ultimately do a good deal towards covering the cost of maintenance.

The committee do not advise night classes, and in this we think they are wise ; they consider that the already existing municipal technical school either does provide or might provide for the need these night classes are intended to meet, and they are sure that the curriculum they propose will absorb all the energies of the teaching staff when employed in the daytime only. They do not clearly indicate the training they propose for mining engineers, but for mechanical, electrical, and civil, they draw up a course the same in the first two years ; in the third year the mechanical and electrical branch off on the one hand, and the civil on the other, while in the fourth year there is more specialisation, but not much distinction drawn even then between mechanical and electrical.

So far, the lines indicated are not very different from what is becoming customary, but they propose to attempt a Faculty of Commerce. Now the establishment of a great commercial school on serious lines is a new experiment, and has not yet been successfully tried anywhere. They propose a capital expenditure of 6000*l.* on class-room accommodation, together with books and apparatus, and an annual expenditure of 2200*l.* on a professor, an assistant professor, an instructor, and some special lecturers. We think that they will find that the addition of certain other chairs will be essential if a commercial faculty is to take its proper position, especially political economy and geography, probably law also. We do not see that any provision has been made for these three subjects. We regard a thorough course in political economy as essential to the well-being of a commercial faculty ; and geography, treated completely, we regard as a much more important subject than the committee at present seem to realise.

The advisory committee enter further into the difficult question of commercial education. They say that modern languages should be learnt when quite young ; which indeed is very true, but it seems to us a counsel of perfection. In practice we feel sure that modern languages would certainly have to form a considerable part of a scheme for commercial education.

Commercial arithmetic, they say, does not go far enough ; and that also is extremely probable ; but a training in elementary mathematics, beyond the immediately practical stage, would be of great advantage to the commercial man in many indirect ways.

As to geography, the committee think that the information is best obtained as wanted from books of reference and consultation with one of the touring agencies ; but in this we entirely

differ from them. To make proper use of books of reference some previous knowledge of the subject is necessary ; and the earth, especially the portion accessible to trade, is not so big but that an adequate knowledge concerning its chief features should be acquired and possessed by a competent man of business, without having to refer constantly to others.

The committee, however, go on to recommend that, in addition to these things, instruction shall be provided in business organisation, the theory and principles of trade unions, associations, trusts, combinations and rings ; that instruction shall be given in commercial law, likewise in accountancy, in shipping and railway practice, and in banking and exchange ; and they say very wisely that " such knowledge as the foregoing is what is required in business, and is usually only learnt bit by bit at a heavy cost, so that the man of business has generally reached the limits of his working life before he has completed his commercial education, and owing to the want of a codified system business men continue from generation to generation to renew the mistakes of their predecessors, and to repeat their experiments, and after much tribulation to re-arrive at their methods, their rules and their conclusions."

They further indicate that this commercial education is not to be taken as a substitute for a more general education, but is to be a supplement to it. They say, " Students in the commercial education course should not be allowed to enter at too early an age. Twenty is quite early enough ; and it would be most desirable that they should have taken a degree in Arts before studying for the commercial degree, and certainly the highest commercial degree should only be given to those already in possession of an Arts degree."

They hope (again in this case) that the fees from students may make it largely self-supporting, but we incline to think that they estimate the fees from students too highly. If they fix the fee for each student at 50*l.* a year, we fear that the expense will exclude a considerable number of those who might otherwise derive special benefit from the course proposed.

They realise that this attempt at a thorough commercial education is a new experiment, and one which, if successful, may have most important consequences on the commercial future of the country, and they conclude as follows :—

" There is no instance elsewhere of any course at once so complete and so valuable ; there is not even, so far as your committee know, any university in the United Kingdom where there is a separate Faculty of Commerce, and as there has not yet been any effort to treat the subject with the thoroughness now proposed, so there is no means of estimating the extent to which advantage would be taken of such teaching. Your committee, however, point to the fact that a Faculty of Commerce so organised and based on the actualities of business experience, would at the present moment stand alone, and would therefore attract to the Birmingham University all who feel the need of such an education, and would also to a much greater extent create a new demand."

There is no doubt, however, that the Arts Faculty in general requires strengthening in many ways, the addition of new chairs being one of them ; and unless this is done as soon as opportunity offers, the scientific and technical training proposed will not acquire its proper university status. The training of the students must not be limited to their immediate fancied needs ; neither students nor their parents are the best judges of what is in the long run really desirable. A much broader training must be given in the university of the future than has been given in the university of the past. Depth without breadth has been the feature of some Honours schools ; shallowness with athletics has been the feature of some Pass schools. The university of the future must mend all this, and secure that all its graduates without exception have had a broad training in many subjects—subjects lying in different departments of human knowledge ; so that they may be really educated and not merely informed. As to the depth possible, that will vary with individual powers, and the standard must not be made impossible for the average man ; but to give the average man a training in some highly specialised practical department, and then turn him out on the world as a university graduate, is not what we expect or hope for from the new university. Such students there will be, doubtless, and they may well receive special diplomas each in his own branch, but they should not be graduates.

Some other students there will be, who, in addition to a broad and liberal culture, have the power of going deeply into some one subject, and these should receive degrees with honour ;

but both these classes will be exceptional. For the average man a broad training in many subjects, well taught and under the most favourable conditions, is what is wanted, in order to leave him adaptable and efficient in the subsequent uncertain calls of actual life ; and such men should constitute the bulk of the pass graduates, and be the backbone of the new scheme.

Annexed to the report is an account of the visit of the committee's deputation to American and Canadian Universities, and the information thus obtained and summarised is of the greatest interest and importance.

A deputation of the Advisory Committee of the University of Birmingham paid a visit to colleges and universities in the United States and Canada at the end of last year, on the suggestion of Mr. Carnegie, who, it is understood, is willing to provide a good round sum for the establishment of an adequate scientific and technical college on this side of the Atlantic. An appendix to the committee's report contains a statement of the condition of affairs which they found in America.

They find that "almost the whole of the students enter on a full four-year course of instruction with a view to graduation. The student on entrance is required either to pass an examination or to present satisfactory evidence that he is qualified to take up the course on which he enters. The entrance examination is not very different, as a rule, from the matriculation examination of the University of London. It is more advanced in mathematics, but probably easier on the literary side."

The working session ranges between thirty-three and thirty-eight weeks ; but outside this there are summer excursion classes, and summer workshop classes usually of about one month.

One remarkable difference they find in the system of lecturing. With us, college lectures form a connected course, almost dispensing with the necessity of a text-book, except for supplementing and extending information. It has often seemed to us that such lectures are perfectly right if the student already partially knows the subject ; it then systematises and organises and more firmly impresses his knowledge ; but if, as too often happens, a student comes to the lecture-room ignorant of the subject, he cannot derive proper benefit from a course of lectures ; he cannot discriminate between the essential and the comparatively unessential ; he cannot without practice watch experiments and take notes at the same time ; he cannot always keep his attention fixed : we have noticed that students who have recently been to a British secondary school, one of the large public schools or indeed any other, cannot as a rule keep their attention long fixed on anything. There are exceptional students, and there are exceptional schools ; but as a rule what they chiefly learn in class work at school is a habit of inattention to what is going on, the average procedure in class being too slow for the quicker boys, too rapid for the slower ones, and too dull for all. This habit of inattention, once firmly acquired, remains with them through the first year and sometimes through the second year of their college life, and they are all the time a perfect curse to any who wish to get on, and who are becoming of an age to realise some of the responsibilities and opportunities of life.

For the college lectures in America it would appear that "a large amount of home preparation and work is required. The student is expected to read up in a text-book the subject matter of the lectures beforehand, the lectures in many cases consisting of exposition and experimental illustration of the text-book. Recitation classes are held in connection with each lecture, in which individual students are questioned on the text-book or lectures, or asked to demonstrate on the blackboard before the rest of the class."

Literary studies are not wholly neglected by the students of science, nor is attention to them confined to the needs of the entrance examination. In addition to the requirements of the entrance examination in languages, grammar, and history, a certain amount of time is given by the science students, especially in the first two years, to what appear to be often called "culture subjects," such as literature, composition and rhetoric, history, political economy, French and German.

But the most important and much-to-be-imitated portion of the system adopted in America, is that whereby the credit given for work does not depend solely on a concluding examination, but is made really to represent the aggregate work of the whole session. There is a Paper examination, and that is quite right, for it is eminently desirable that a student should be able to express what he knows accurately and on demand. Quite half the credit ought to be awarded to this faculty, but not all ; the

remaining half should be awarded for work in class-room and laboratory.

In the States there are no practical examinations as with us. Proficiency in laboratory work is accredited by assigning marks for attendance and for excellence of laboratory and manual work throughout the session. We believe that this system is very successfully in force at such places as the City Guilds Central Technical College at South Kensington, but we have not yet heard of its much to be desired introduction into universities in this country. There is no doubt that it would have the best effect on both student and demonstrator ; and it would have the further advantage that the troublesome practical examinations, especially those in the senior stage, when they become rather farcical, could be dispensed with.

Another desirable innovation is thus expressed :—" The right of dismissal at any stage is maintained and used. Any student who shows that he is unable or unwilling to keep up with the work is excluded by the Faculty from the graduation course. He may be allowed to take on special courses, but usually he is dismissed from the institution. The system has been devised to keep, and succeeds in keeping, the students continuously at work, and the result of the process of exclusion in the earlier stages is that nearly the whole of the final classes are successful in graduating."

One of the most important arrangements in America is the large provision made at some of the institutions for post-graduate work. Only a small proportion of students are able to spare time for it, but it is encouraged by affording every facility for study and research to the post-graduate ; and graduates from one institution frequently work as post-graduates at another. This system of interchange between universities, which already obtains largely in Germany, is surely to be desired in this country, especially in post-graduate stages, where specialisation naturally and properly sets in.

Over-specialisation in undergraduate stages is, we believe, to be deprecated. A certain amount of general knowledge, both literary and scientific, is needed, and should be acquired by all.

The committee found that in America the proportion of staff to students is much greater than with us ; and they further found —what is a matter of great importance—that the subdivision of subjects is, as in Germany, likewise carried much further ; so that, for instance, every important branch of engineering has its own professor, with perhaps an assistant professor, and certainly with instructors ; and no attempt is made, as with us, to place the whole of a gigantic scientific subject in its higher stages under the control of one man.

We observe that in the fundamental subjects of chemistry and physics the general laboratory arrangements and the scope of teaching appear to be much the same in America as in this country. The laboratories are, however, as a rule more spacious, the equipment in apparatus is on a larger scale, greater facilities are given for research, and the size of the laboratories allows most of the physical apparatus to be kept in position—different rooms being used for different subjects. In the more important laboratories many rooms are provided for original research, which is carried on by the staff and post-graduate students. At Cornell there is a special laboratory for physical chemistry. At several colleges there is a department of applied chemistry, through which all students pass who are graduating in chemistry. This is excellent, and tends to make the knowledge much more real and practical. Chemicals are made, instead of being merely purchased ; "the course is short, and generally consists in the production of pure chemicals from commercial articles on a scale in which many kilogrammes are dealt with. The processes are made to resemble, as far as possible, those of manufacturing practice."

In civil engineering we observe that "the work in surveying is very thorough, and includes field work throughout the year, together with a summer course. There is usually an extensive stock of theodolites, levels, and chains, so that each student in the field has his own instrument. During the two last years particular attention is devoted to bridge construction, the student preparing complete drawings and stress sheets in accordance with the practice of the leading railway companies."

Less hostile feeling to academically bred apprentices would be felt in this country if these practical features could be imitated.

In mining engineering, a summer excursion class is sometimes formed to spend some weeks in a mining district, where facilities are given to inspect the actual processes of mining.

Great importance is assigned to engineering-laboratory work, and the whole ground of the engineering lectures is provided for in the equipment of the laboratories. The machines are large in number and in capacity, so that every student performs experiments on an adequate scale. The work is chiefly pursued in the third and fourth years, and occupies from four to six hours weekly. In some laboratories there are full-size locomotives mounted, so that running tests can be made, and special courses are arranged for those who wish to take up the mechanical side of railway practice. Facilities are given by the railroad companies for testing under the conditions of actual running.

The shops gave the impression of being thoroughly practical, and on such a scale that the knowledge acquired there by the student would be of use in his subsequent professional life.

The greater size of the Continent is, perhaps, partially responsible for the following paragraph in the committee's report, although really if education were properly appreciated in this country, our island is large enough for us to follow the example. The paragraph we refer to looks very attractive to those whose work in this country lies in colleges cramped in the middle of great cities. It runs as follows:—

"We were very much struck with the amount of ground occupied by the colleges, each building standing in its own grounds, so that it is well lighted on every side. Usually there is a large entrance hall, a fine staircase, and wide corridors leading to class-rooms and laboratories. The floor space in the laboratories is generally very much greater than with us. The apparatus, instead of being huddled away in dark corners, is set out and classified as if for exhibition, while the machinery occupies a space worthy of its importance."

We observe also that every college possesses departmental libraries and reading-rooms available for the students, in addition to the large central library and reading-room.

The social aspect of university life is not forgotten, and the following glorified edition of a student's union is well worthy of imitation:—

"The University of Pennsylvania at Philadelphia possesses in Houston Hall a fine building given to the University by an old graduate, in memory of his son, who was also a graduate. It is a club-house for the students, any student becoming a member for two dollars per annum. In the building are reading, billiard and smoking rooms, a luncheon-room, a gymnasium, a swimming bath, and rooms for college societies. The hall is entirely and very well managed by the students. It is regarded by the staff as having a most excellent influence on student life."

In concluding this part of the general report of their American visit, the committee make a well-deserved comment, which we will presently quote, for nothing more splendid in the direction of educational endowment has been seen in our times than the magnificent sums which wealthy American citizens are willing to place at the disposal of university authorities. They do, indeed, realise, as we do not, or at least have not yet, the immense, the super-eminent, importance of real education and knowledge, to a country and an empire which has to hold its own against ever-increasing competition, and constantly to make its way in fresh uncivilised regions. The following are the concluding general remarks:—

"We desire to express our admiration alike for the high ideal of scientific education, which is the aim in American universities, and for the enthusiasm in all classes which renders it possible to approach so near that ideal. Everywhere we found evidence that the wealthier citizens realise the importance of university education, and encourage the universities by generous gifts; and everywhere, both by teachers and by students, these gifts are being used for higher learning and research."

THIRD INTERNATIONAL CONFERENCE ON A CATALOGUE OF SCIENTIFIC LITERATURE, LONDON, JUNE 1900.

LIST OF DELEGATES APPOINTED TO ATTEND THE CONFERENCE.

Austria.—Prof. E. Weiss (Kaiserliche Akademie der Wissenschaften, Vienna); Prof. Karl Toldt (Universität, Vienna).

France.—Prof. G. Darboux (Membre de l'Institut de France); Dr. J. Deniker (Bibliothécaire du Museum d'Histoire Naturelle, Paris); Prof. H. Poincaré (Membre de l'Institut de France).

Germany.—Prof. Dr. F. Klein (Geheimer Regierungs-Rath, Universität, Göttingen); Prof. Dr. B. Schwalbe (Direktor, Real-Gymnasium, Berlin); Dr. F. Milkau (Oberbibliothekar, Universität, Berlin).

Greece.—Mons. de Metaxas (Chargé d'Affaires for Greece).

Hungary.—Dr. August Heller (Bibliothekar, Ungarische Akademie, Buda-Pesth); Dr. Theodore Duka (Hon. Member of the Hungarian Academy of Sciences).

Italy.—Prof. Giacomo Ciamician (R. Università, Bologna); Prof. Raffaelo Nasini (R. Università, Padua).

Japan.—Prof. Einosuke Yamaguchi (Imperial University of Kioto).

Mexico.—Señor Don Francisco del Paso y Troncoso.

Norway.—Dr. Jörgen Brunchorst (Secretary, Bergenske Museum).

Switzerland.—Dr. Jean Henri Graf (President, Commission de la Bibliothèque Nationale Suisse, Berne); Dr. Jean Bernoulli (Librarian, Bibliothèque Nationale Suisse, Berne).

United Kingdom.—Representing the Government: The Right Hon. Sir John E. Gorst, Q.C., M.P., F.R.S. (Vice-President of the Committee of Council on Education). Representing the Royal Society of London: Sir Michael Foster, K.C.B., Sec.R.S.; Prof. Arthur W. Rücker, Sec.R.S.; Prof. H. E. Armstrong, F.R.S.; Sir J. Norman Lockyer, K.C.B., F.R.S.; Dr. Ludwig Mond, F.R.S.; Dr. T. E. Thorpe, For.Sec.R.S.

Cape Colony.—Sir David Gill, K.C.B., F.R.S.; Roland Trimen, Esq., F.R.S.

India.—Lieut.-General Sir Richard Strachey, G.C.S.I., F.R.S.; Dr. W. T. Blanford, F.R.S.

Natal.—Sir Walter Peace, K.C.M.G. (Agent-General for Natal).

New Zealand.—The Hon. W. P. Reeves (Agent-General for New Zealand).

Queensland.—The Hon. Sir Horace Tozer, K.C.M.G. (Agent-General for Queensland).

ACTA.

Opening Meeting, Tuesday, June 12, at the rooms of the Society of Antiquaries, at 10 o'clock.

(1) Prof. Darboux moved that Sir John E. Gorst be the President of the Conference. The motion having been carried unanimously—

(2) Sir John Gorst took the chair and welcomed the delegates.

(3) On the motion of Sir M. Foster, seconded by Prof. Darboux, it was resolved that Dr. F. Milkau be the secretary for the German language; that Dr. Jean Bernoulli and Dr. J. Deniker be the secretaries for the French language; that Prof. Giacomo Ciamician be the secretary for the Italian language; that Prof. H. E. Armstrong be the secretary for the English language.

(4) That the secretaries, with the help of shorthand reporters, be responsible for the *procès verbal* of the proceedings of the Conference in their respective languages.

(5) Sir Michael Foster read out the names of delegates appointed to attend the Conference.

(6) On the motion of Sir Michael Foster, it was resolved— (i.) That the meeting adjourn at 1 p.m., and meet again at 2.30 p.m.; (ii.) that on Wednesday, the meeting commence at 11 a.m.

(7) On the motion of Sir Michael Foster, seconded by Prof. Rücker, it was resolved that English, French, German and Italian be the official languages of the Conference, but that it shall be open for any delegate to address the Conference in any other language, provided that he supplies for the *procès verbal* of the Conference, a written translation of his remarks into one or other of the official languages.

(8) Sir Michael Foster presented the Report of the Provisional International Committee, and it was resolved that the report be received.

(9) The following resolutions were then agreed to:—(i.) That the publication of a card catalogue be postponed for the present; (ii.) that the book catalogue be at first issued only in the form of annual volumes.

(10) Sir Michael Foster having moved (iii.) that the catalogue include both an authors' and a subject index, according to the scheme of the Provisional International Committee; Prof. Rücker thereupon explained the financial position, and the delegates of the various countries stated to what extent they were authorised to promise contributions towards the expenses of the catalogue. From these statements it appeared that subscriptions to 163 sets of volumes (or their equivalent) of the

catalogue, to the value of £2,771, would be guaranteed, viz. as follows :—

				£	
Germany	...	...	45 sets	equivalent to 765	
United Kingdom	...	...	45 ,,	,, ,, 765	
France	...	...	35 ,,	,, ,, 595	
Italy	...	...	27 ,,	,, ,, 459	
Switzerland	...	...	6 ,,	,, ,, 102	
Norway	...	...	5 ,,	,, ,, 85	

Other delegates estimated that the probable contributions from their countries would be :—

				£	
Austria	...	...	16 sets	equivalent to 272	
Hungary	...	...	4 ,,	,, 68	
Japan	...	...	5 ,,	,, 85	
Mexico	...	...	5 ,,	,, 85	

It was further estimated that the British Colonies and Dependencies would subscribe for at least 25 sets, equivalent to 425*l*. Taking into account the subscriptions to be expected from the United States, Russia, Holland, Sweden, and a number of other countries, as well as the probability of outside sales, the Conference was of opinion that the necessary subscriptions to cover the cost of 300 sets of the catalogue would be obtained. At the close of the discussion the motion above set forth was unanimously agreed to.

(11) In the course of the discussion, it was stated by delegates from several countries that all the sets subscribed for would be distributed among public institutions, and that they contemplated the private sale of the catalogue in addition.

Second Meeting, Wednesday, June 13.

(12) The following motions, of which notice had been given on the previous day, were considered and adopted :—

(A.) The Conference is of opinion that the financial prospects of the enterprise are sufficiently satisfactory to warrant further steps being taken toward the publication of the catalogue, in view of the fact that the representatives of the various countries have declared that the governments or corporations they represent are willing to subscribe for the number of complete sets of copies at the cost stated in paragraph 10.

La Conférence décide : que le côté financier de l'entreprise est suffisamment élucidé pour justifier les arrangements ultérieurs nécessaires pour la publication du Catalogue, les représentants des différents pays ayant en effet déclaré que les gouvernements ou les corps savants qu'ils représentent sont prêts à souscrire au nombre de séries complètes du Catalogue, et aux prix indiqués dans le § 10.

Angesichts der von den Vertretern der verschiedenen Länder abgegebenen Erklärung, dass die durch sie vertretenen Länder oder Körperschaften entschlossen sind, auf die in § 10 angegebene Zahl vollständiger Exemplare zu dem ebenda festgesetzten Preise zu subskribiren, gibt die Konferenz der Meinung Ausdruck, dass die finanziellen Aussichten des Unternehmens zufriedenstellend genug sind, um weitere Schritte zur Veröffentlichung des Katalogs zu rechtfertigen.

La Conferenza è d'avviso che avendo i rappresentanti dei diversi paesi dichiarato che i governi o i corpi scientifici da loro rappresentati sono pronti a sottoscrivere nella misura indicata al § 10, si può ritenere l'impresa abbastanza soddisfacente dal lato finanziario per autorizzare gli ulteriori passi che sono necessari per la pubblicazione del catalogo.

(B.) That, pending the appointment of the International Council, a Provisional International Committee be appointed which shall be entrusted with the duty of approaching, through the Royal Society, such countries as may be necessary, with the view of obtaining their adhesion to the scheme for the publication of the catalogue, or promises of financial support.

Que jusqu'à la constitution définitive du Conseil International, un comité international provisoire soit nommé. Ce comité sera chargé de se mettre en rapport par l'intermédiaire de la Société Royale, avec les personnes autorisées des différents pays, suivant les nécessités de la situation, afin d'obtenir de ces pays l'adhésion ou l'appui financier à l'œuvre du Catalogue.

Dass bis zur Einsetzung des *International Council* ein *Provisional International Committee* ernannt wird, mit dem Auftrage, sich durch die Vermittelung der Royal Society mit den in Betracht kommenden Ländern in Verbindung zu setzen, um sich ihrer Mitwirkung bei der Veröffentlichung des Katalogs

zu versichern oder ihre Zusagen für finanzielle Unterstützung zu erwirken.

Che sia nominato un comitato internazionale provvisorio sinò a che non sarà costituito definitivamente il Consiglio Internazionale. Questo comitato avrà l'incarico di mettersi in comunicazione mediante la Royal Society con i diversi paesi, secondo che sarà necessario, per ottenere la loro adesione al progetto di pubblicazione del catalogo o la promessa del loro appoggio finanziario.

(C.) The said Provincial Committee is further authorised to make other preparations for the publication of the catalogue, but without incurring financial responsibility.

Inasmuch as it will be necessary for some one corporation to make the necessary contracts and undertake the final financial responsibilities, the Provincial Committee is authorised to include among such preparations, negotiations either with the Royal Society, or with another corporation, or with a government, or with a publisher, but the confirmation of all such preparations, and the carrying out of any final agreement or contract, shall rest with the International Council.

Ce comité sera autorisé en plus à prendre d'autres mesures préliminaires en vue de la publication du Catalogue ; mais il n'aura aucune responsabilité financière.

Comme il sera nécessaire, pour un corps constitué quelconque, de conclure des traités et d'encourir les responsabilités financières définitives, le comité provisoire est autorisé à comprendre parmi les mesures préparatoires de ce genre, les négociations soit avec la Société Royale, soit avec un autre corps constitué, soit avec un gouvernement, soit avec un éditeur. Toutefois la confirmation de toutes ces mesures préparatoires, ainsi que l'arrangement financier définitif ou conclusion d'un traité, doivent incomber au Conseil International.

Das genannte Provisional Committee wird ferner beauftragt, andere Vorbereitungen zur Veröffentlichung des Katalogs zu treffen, ohne jedoch eine finanzielle Verantwortlichkeit einzugehen.

Soweit es notwendig ist, dass eine Korporation die erforderlichen Verträge abschliesst und die endgültige finanzielle Verantwortlichkeit übernimmt, wird das Provisional Committee ermächtigt, derartige Vereinbarungen entweder mit der Royal Society oder mit einer anderen Korporation oder mit einer Regierung oder mit einem Verleger einzuleiten, indem die Bestätigung solcher Verhandlungen wie die Ausführung der endgültigen Vereinbarungen dem International Council vorbehalten bleibt.

Questo comitato sarà inoltre autorizzato a prendere altre misure preliminari per la pubblicazione del catalogo, senza avere tuttavia nessuna responsabilità finanziaria.

E poiché è necessario per un corpo costituito di concludere dei contratti e d'assumere la responsabilità finanziaria definitiva, così il comitato provvisorio è autorizzato ad includere nei lavori preparatori le trattative o colla Royal Society o con altre corporazioni o con un governo o con un editore : ma la conferma di queste trattative e l'approvazione finale del contratto sarà riservata al Consiglio Internazionale.

(13) The "Scheme for the Publication of an International Catalogue of Scientific Literature" was then considered, and it was resolved—

That Article I. be approved—

Omitting the words in paragraph 5 on page 5, lines 9 and 10, "the limits of the several sciences to be determined hereafter," and also the words, page 5, lines 27-29.

That Article II. be approved—

Omitting the words "the same . . . regulations were" in paragraph 10, page 7 ;

Adding Italian to the three languages mentioned in paragraphs 10 (a) and (b);

Altering the word "delegate" to "contracting body (as hereinafter defined)" in paragraph 10 (d) ;

Omitting at the top of p. 8 the words within square brackets ;

Omitting in paragraph 11, p. 8, the words within square brackets : "The . . . Appendix II.," and substituting the following : "Each contracting body shall have one vote in deciding all questions brought before the Council";

And inserting in paragraph 13, before the words : "There shall also be . . ." the words : "If the International Council so decide."

That Article III. be approved without change.

That Article IV. be approved, omitting the opening paragraph in square brackets.

That Article V. be approved, inserting the words : " or as soon after that date as the International Council may decide," in paragraph 29, after "January 1, 1901."

That Article VI. be approved, inserting at the beginning of paragraph 32, the words: "Unless the International Council decide otherwise";

Substituting in paragraph 34, p. 14, line 33, "instructed" for "authorised."

That Article VII., excepting paragraph 37, be approved—

Omitting paragraph 35 and the next paragraph in square brackets, and substituting therefor : "any body which establishes a regional bureau shall be termed a contracting body."

Omitting the words "which takes a complete share" in the first line of paragraph 40, and omitting the whole of the second sentence in this paragraph, and omitting the three appendices.

(14) It was further resolved to substitute for paragraph 37, Section VII., page 15, the following :—

"That it will be an instruction to the Provisional Committee to negotiate with the several contracting bodies with reference to the sale in their respective regions of copies other than those subscribed for by the contracting bodies."

Que les instructions soient données au comité provisoire pour négocier avec les différents corps contractants la question de la vente dans leurs pays respectifs des exemplaires souscrits par ces corps.

Aufgabe der Provisional Committee wird es sein, den verschiedenen contrahierenden Körperschaften (contracting bodies) bezüglich des Verkaufs von Exemplaren in ihren Ländern, ganz abgesehen von der gewahrleisteten Anzahl, bestimmt e Festsetzungen vorzuschlagen.

(15) It was resolved that the Provisional Committee contemplated in Resolution 12 (B) be constituted as follows :—Prof. Armstrong, Dr. Brunchorst, Dr. Graf, Dr. Milkau, Prof. Nasini, Prof. Poincaré, Prof. Weiss ; power being given to the Royal Society, while retaining only a single vote, to nominate further members, and power being given to the Committee to appoint substitutes if any of those named were unable to serve, and also to co-opt two new members.

(16) On the motion of Sir Michael Foster and Prof. Rücker, it was resolved that the Royal Society be requested to appoint the Secretary to the Provisional Committee, and to meet provisionally such expenses as the Committee may incur.

(17) It was resolved that the *procès verbal* of the Conference be signed by the president and secretaries.

(18) The Royal Society was requested to undertake the editing, publication, and distribution of a verbatim report of the proceedings of the Conference.

(19) On the motion of Prof. Schwalbe, a vote of thanks to Sir John Gorst for presiding over the Conference was passed by acclamation.

(20) On the motion of Sir Michael Foster, the thanks of the Conference were accorded to the Society of Antiquaries for the use of their rooms.

(Signed)   John E. Gorst.
Henry E. Armstrong.
Dr. Joh. Bernoulli.
G. Ciamician.
I. Deniker.

THE ROYAL SOCIETY CONVERSAZIONE.

THE exhibitions at the conversazione, which took place on the 20th inst., were numerous and interesting. There was a great wealth of photographs, including a large collection illustrating the results obtained during the last total eclipse. Among the other objects exhibited were the following :—

Prof. W. Ramsay, F.R.S., and Dr. M. W. Travers exhibited the inert atmospheric gases ; their spectra, and some of the apparatus used in determining their physical properties.

The Meteorological Office exhibited North Atlantic weather charts, winter, 1898-99.

Prof. J. W. Judd, C.B., F.R.S., on behalf of the Coral-Reef Committee of the Royal Society, exhibited specimens from the reels of Funafuti.

Mr. J. Mackenzie Davidson exhibited a stereoscopic fluoroscope. The stereoscopic fluoroscope is an instrument to enable an observer to see the shadows cast by the Röntgen rays on the fluorescent screen in *stereoscopic relief.*

Prof. W. E. Dalby exhibited a model to illustrate and experiment upon the balancing of four-cylindered engines.

Prof. Silvanus P. Thompson, F.R.S., showed experiments on the aberration called *Coma.* Coma is an aberration due to the several zones of the lens not having equal focal lengths, and hence, when the lens is transmitting an oblique pencil, the unequal refraction of the different parts gives rise to a singular unilateral distortion of the cones of rays traversing the various zones. In these experiments, the effects are analysed by covering the lens with a series of zones alternately opaque and transparent. Some singular effects can also be produced without the zone plate, by inserting in the oblique pencil, after traversing the lens, objects to cast shadows on a screen. In this way a straight wire can be arranged so that the shadow it casts is a totally-detached circle. Some diagrams of coma, and a string model illustrating their origin, were also shown.

Mr. W. Gowland exhibited Japanese books on botany, intended to show the general character of the work of Japanese botanists from 1759 to 1856.

Mr. W. Gowland, for the Silchester Excavation Committee, exhibited remains of a Roman silver refinery found at Silchester.

Mr. S. Evershed exhibited an electric supply meter (a frictionless motor meter).

Prof. H. S. Hele-Shaw, F.R.S., and Mr. A. Hay showed lines of induction in a magnetic field, represented by stream-line flow.

Prof. E. Ray Lankester, F.R.S., exhibited enlarged models of gnats (mosquitoes) and of human blood-corpuscles infected by the malaria-parasite ; modelled by Miss Delta Emett. (1) Female *Culex pipiens*, Linn., the common brown gnat or mosquito ; enlarged twenty-eight times linear. The insect is shown in the act of alighting. This gnat does not harbour the malaria-parasite. (2) Female *Anopheles maculipennis*, Hoffmansegg, the common spot-winged gnat or mosquito ; enlarged twenty-eight times linear. (3) Human blood-corpuscles infested with the malaria-parasite (æstivo-autumnal or remittent fever) known as *Haemomenas praecox ;* magnified seven thousand five hundred times linear. The blood-corpuscles are transparent, and show the parasites within. The upper row shows the multiplication of the parasite within the corpuscles by fission giving rise to "sporocytes," which creep into other non-infected blood-corpuscles and repeat the process, thus increasing the infection. The lower row shows the formation of a crescent-shaped "gametocyte" within the blood-corpuscle. Instead of breaking up, the parasite enlarges and becomes sausage-shaped. The "gametocytes" thus formed are destined to be swallowed by the gnat Anopheles, when they develop in the gnat's stomach —some into eggs and some into spermatozoa.

The Zoological Society of London showed two living female crowned lemurs (*Lemur coronatus*), each with a young one.

Prof. G. B. Howes, F.R.S., and Mr. H. H. Swinnerton exhibited reconstructional models, built up from microscopic sections, of the developing head skeleton of the New Zealand reptile, *Sphenodon punctatus.*

Sir John Evans, K.C.B., F.R.S., exhibited ancient cameos and gems, and palæolithic implements from Africa.

Prof. Wyndham Dunstan, F.R.S., exhibited the poisonous lotus of Egypt (*Lotus Arabicus*). (a) Living plant grown at Kew. (b) Dried plants from Nubia. (c) Specimens of the new glucoside, *Lotusine*, and its decomposition products.

Mr. Fred Enock exhibited an aquatic walking-stick insect, with eggs (*Ranatra linearis*).

The following demonstrations, with experiments and lantern illustrations, took place :—

Mr. Fred Enock, life-history of the *Cicindela campestris*—the common tiger beetle. This common Coleopteron goes through its metamorphoses in deep vertical burrows made in the sand by the curious larva, which "sits" at the top of the hole, patiently waiting for its prey to come to it, as it does not go in search of it. Three years are passed in its subterranean den, at the lower end of which it remains during the winter months in a semi-torpid state ; activity is resumed at the approach of warm weather.

Prof. J. A. Fleming, D.Sc., F.R.S., demonstrations with an apparatus for the production of short electric waves, and the study of electro-optic phenomena. The apparatus exhibited consists of a radiator for the production of a beam of electric radiation, the wave-length being about eight inches. The radiator is contained in a zinc box, which prevents the diffusion of the radiation in all directions. The receiver consists of a metallic filings tube of the Branly type, associated with a relay and electric bell. The receiver is also contained in a zinc shielding box. The impact of electric waves upon the receiver is indicated by the ringing of the bell. The radiator can be placed at different angular positions. With this apparatus are shown experiments illustrating the opacity of metallic screens, continuous or perforated to electric radiation ; the transparency of insulating screens, and the transparency or opacity of various liquids. Water is found to be particularly opaque even in very thin layers. All damp objects are very impervious to this radiation, such as a wet duster, a moist brick, tobacco having more than the legal amount of water added to it, and the human body or hand. The refraction of electric waves is shown by the use of a paraffin wax prism, the concentration by paraffin lenses, and the polarised quality of the rays by their reflection or stoppage by parallel wire gratings. Also the production of secondary oscillations in linear conductors by holding rods of metal or tubes of liquid in the radiation. The wave-length of the radiation is measured by producing interference as a result of splitting the beam into two portions and transmitting the two portions down two zinc tubes, the relative lengths of which can be adjusted.

Prof. A. C. Haddon, F.R.S., cinematograph photographs of native dances in Torres Straits.

THE RE-ORGANISATION OF THE EDUCATION DEPARTMENT.

IN introducing the Secondary Education Bill to the House of Lords on Tuesday last, the Duke of Devonshire made the following remarks on the re-organisation of the Education Department :—

" Your lordships may remember that on the Bill of last year some discussion took place upon the future organisation of the Education Department. I thought at the time, and I am still more strongly of opinion now, that that discussion was somewhat premature. It proceeded on the assumption that the organisation of the new office would continue on the same lines as those which had existed when the educational departments were separate and distinct, and that there would be in the new office two divisions, one of which would carry on the work of the old Education Office in connection with elementary education, and the other of which would carry on the work of the Science and Art Department. . . . We now propose to revert to a dual organisation of the office, but not entirely upon the lines of the late Education and Science and Art Departments. The principal officers of the department which we propose will be a principal permanent secretary, who will supervise generally the whole work of the department. It must be remembered, when special importance is attached to this or that minor subordinate appointment, that it will be the permanent secretary who will be responsible to the President of the Board in the administration of the whole department, and that it is impossible, and would be undesirable if it were possible, that the office should be divided into what I may call water-tight compartments, the head of each of which would be charged with special duties and no other, and that the idea should be entertained that the work of the office should be carried on in several departments, which should have no connection or relation with each other. We propose that under the principal permanent secretary there shall be two principal assistant secretaries, one mainly charged with duties in connection with elementary and the other with secondary education. We propose to abolish the name ' Science and Art Department.' The Science and Art Department will be merged in the secondary education branch of the office. As soon as it may be possible, we propose to transfer the greater part of the staff of the late Science and Art Department from South Kensington to Whitehall, except such part of it as may be necessary to leave at South Kensington for the administration of the museum and the colleges of science and art. In place of the third division that was contemplated, we now propose to give the principal assistant secretary of secondary education two additional assistant secretaries, one of whom will be chiefly charged with the supervision and control of literary instruction, and the other of technological study. This is not the organisation, I admit, to which I partly committed myself last year ; but I trust that it may, in substance, meet the views, especially the later views, which have been expressed to me by high educational authorities. With the name we hope to get rid of many of the traditions which were supposed to attach to the old Science and Art Department—

traditions which have, I believe, been regarded as opposed to the true interests of education by many of those who have been responsible for the management of the older endowed schools. The original idea of the Science and Art Department was, or at all events was supposed to be, that by means of lectures, classes, and examinations a knowledge of the principles of science and art, which would be valuable to the students themselves and to the nation at large, could be engrafted upon almost any kind of previous elementary or secondary training. It is quite true that this idea has been in recent years very largely modified, but I do not think that it is yet generally known how far the original traditions of the Science and Art Department have been already departed from. We hope and intend that the idea of the future education branch of the office will be to make science and art instruction a part of general education in addition to those classical and literary studies which have hitherto formed its main portion. In the schools and institutions directly assisted by the Board of Education the teaching of science and of art, with the addition, perhaps, of some commercial subjects, will probably remain the principal object. But, on the other hand, in those secondary schools, whether of older or more modern type, which desire to enter into connection with the board, there ought not to be, and there need not be, any interference with the older classical and literary studies so long as there continues to be a demand for them. At the same time, we hope that the scientific resources of the Board will be placed at their disposal if they desire, as many of them do desire, to develop the more modern sides of instruction and education. . . . It may be of interest to the House to know what are the principal appointments which have been made or are proposed to be made in the principal office of the new secondary education branch of the department. Sir George Kekewich, the late secretary of the Board of Education, has become the permanent principal secretary of the new Board, and it is he who will be responsible to the President of the Board and to the Government for the administration of the department as a whole. The principal assistant secretary for secondary education will be Sir William Abney, who has done more than any other man in extending the studies of the schools of science under the Science and Art Department. Under him the assistant secretary to deal with the literary side of instruction will be Mr. Bruce, an assistant commissioner to the Charity Commission under the Endowed Schools Act, who has been chiefly engaged and has obtained much experience in the administration of the Welsh Act. The assistant secretary for technological study has not yet been appointed."

UNIVERSITY AND EDUCATIONAL INTELLIGENCE.

CAMBRIDGE.—Prof. Sir Michael Foster has been nominated by the Council of the Senate as the representative of the University of Cambridge on the Council of the Jenner Institute of Preventive Medicine.

Mr. A. W. Hill, of King's College, and Mr. L. Lewton-Brain, of St. John's College, have been appointed University Demonstrators in Botany.

Mr. E. E. Walker, Trinity College, has been elected to the Harkness Scholarship in Geology and Palæontology.

Prof. Woodhead announces ten courses of lectures and demonstrations in Pathology and Bacteriology to be held during the ensuing Long Vacation.

Mr. Shelford Bidwell, F.R.S., was on June 19 admitted to the degree of Doctor of Science.

Mr. W. N. Shaw, F.R.S., has been elected a Senior Fellow of Emmanuel College. It is a condition attaching to his tenure that he shall give annually in the University not less than three lectures on the Physics of the Atmosphere or some kindred subject. Mr. C. T. R. Wilson, F.R.S., formerly Clerk Maxwell Student in experimental physics, has been elected to a fellowship at Sidney Sussex College.

The following have been awarded scholarships or exhibitions in Natural Science at the several colleges at the end of the academical year :—

Clare College : Bailey, Cartwright, Cassidy, Hughes.
Pembroke College : Lang, Anderson, Hall.
King's College : Kewley, French, Wilde, Mollison, McIntyre.
Christ's College : Fox, Moore, Wilson, Macnab, Muff, Cumberlidge, Sewell.

Emmanuel College: Nixon, Austin, Sutton, Rothera, Banham.

Sidney Sussex College: Bullough, Colt, Drapes, Fearnsides, Harrison, Humphrey, Robinson, Gough.

THE *Appointments Gazette*, which is the journal of the Cambridge Appointments Association, gives in its last issue (June 1900) much valuable information regarding scientific and other posts open to university graduates. It also contains articles on post graduate work in medicine, by Prof. Allbutt ; on training for business, by Mr. G. E. Foster ; and on Long Vacation courses in French, by Mr. H. J. Millar. A list of some two hundred graduates seeking appointments in various departments of industry, with their university qualifications, ages, &c., completes the journal. This list might be consulted with advantage by heads of departments and others in search of suitable candidates for vacant appointments. The Association is doing a useful work in bringing together employers and employed in the various walks of life where university training is of importance, and it already possesses a large and influential membership. The Master of Trinity is chairman, Mr. W. N. Shaw, F.R.S., vice-chairman, and Mr. W. A. J. Archbold, secretary.

PROF. OLIVER J. LODGE, F.R.S., has been appointed Principal of the University of Birmingham.

THE following appointments at the University College of North Wales, Bangor, are announced :—Mr. W. W. Firth to be assistant lecturer in Electrical Engineering, and Mr. Alexander Darroch to be assistant lecturer in the Day Training Department. Mr. W. Cadwaladr Davies was appointed the representative of the Council upon the Central Welsh Board, and Mr. H. Bulkeley Price the representative on the Carnarvonshire County Governing Body.

DR. JOHN WILLIAM WHITE, of Philadelphia, has been elected to the John Rhea Barton Chair of Surgery in the University of Pennsylvania ; Dr. Frank Morley, of Haverford College, has been appointed professor of mathematics in Johns Hopkins University, vice Prof. Thomas Craig, resigned ; Prof. Charles J. Bartlett takes the place of Prof. M. C. White, who for thirty-three years has filled the chair of pathology in the Medical School of Yale University.

THE Report of the Council to the Governors of the City and Guilds of London Institute, dated March 1900, has just been issued, and gives a full account of the work accomplished in connection with the year 1899, and contains verbatim reports of the addresses delivered respectively by Sir Andrew Noble and Sir Douglas Fox at the opening of the session, and at the distribution of certificates and prizes. We notice from the report that during the past twenty years the work of the Examinations Department has developed to an enormous extent : thus in 1888 the number of subjects of examination was 24, the number of centres of examination 89, and the number of candidates 816. In 1899 the number of subjects had increased to 63, the number of centres to 397, and the number of candidates to 14,004 ; the number of registered classes being 1764, and of students in attendance 34,176. These numbers are exclusive of those who receive instruction in manual training. The total number of students last year in the classes registered by the Institute was 36,155, as compared with 34,990 in the previous year.

THE following gifts and bequests for scientific and educational purposes are noticed in *Science*:—By the will of the late Jonas G. Clark, of Worcester, Mass., who founded Clark University in 1889, the entire estate is left to the University, providing the people of Worcester raise a fund of 500,000 dollars. If the sum of 250,000 dollars is raised, he bequeaths 500,000 dollars. If 500,000 dollars is raised, he bequeaths 1,000,000 dollars and makes the University his residuary legatee. He also leaves 100,000 dollars for the University library and a similar sum for a department of art. Messrs. Samuel Cupples and Robert S. Brookings have each given to Washington University one-half of the total capital stock of the St. Louis Terminal Cupples Station and Property Company, which company owns the so-called "Cupples Station." The annual income from this gift to the University will be from 120,000 to 130,000 dollars per year. The gift is to form a permanent endowment fund, the interest of which is to be expended by the Board of Directors in any way which it sees fit. Dr. D. K. Pearsons has offered 50,000 dollars to Carleton College, Northfield, Minn., on condition that the college authorities raise 100,000 dollars before January 1,

1901. By the will of Henry M. Curry the Western University of Pennsylvania receives 10,000 dollars for scholarships ; and the University of Pennsylvania has received 20,000 dollars from Mr. J. D. Lippencott and Mr. J. G. Carruth respectively.

THE second general meeting of the Agricultural Education Committee was held on Friday last at the rooms of the Society of Arts. The report of the executive to the general committee gave a brief account of the constitution and proceedings of the committee from its commencement, and explained that its objects were : (1) to secure systematic and efficient instruction, both theoretical and practical, in agricultural subjects suitable to every class engaged in agriculture ; and (2) to diffuse among the agricultural classes a more thorough appreciation of the advantages of instruction bearing directly or indirectly on their industry. The policy of the committee, the report stated, was largely recognised in the new Day School Code ; and the block grant, the continuous course of rural instruction, lessons in " common things " given through the standards, and the new subject of household management for girls, were all on the lines of the committee's resolutions. Moreover, the executive believed that it was largely due to the representations of the committee that the new Board of Education, shortly after its formation, issued a circular to managers and teachers of rural elementary schools impressing on them the importance of making education in the village school more consonant with the environment of the scholars than was now usually the case, and especially encouraging the children to gain an intelligent knowledge of the common things which surround them in the country. Other provisions of the new Code were referred to with satisfaction, and the report stated that the committee had not failed to co-operate with the Board of Education in bringing them into effect. With regard to the work which remained to be done, attention should be given to organisation. It would seem that the precedent successfully set some years ago in Scotland of handing over the educational work of the Board of Agriculture to an educational authority, while leaving to the Board the inspection of experimental and research work, might well be followed. If that was not done, and the present division of functions continued, the cause of rural education, especially in its higher branches, would undoubtedly suffer. It was also to be hoped that attention would be given to the training of teachers, and that the new Board would introduce some modifications into the curricula of the training colleges to ensure the qualification of a certain number of trained teachers to give instruction on elementary science and common things required by the Code. A good deal also remained to be done in placing evening continuation work on a satisfactory footing.

SCIENTIFIC SERIALS.

American Journal of Science, June.—A method o. studying the diffusion (transpiration) of air through water, and a method of barometry, by C. Barus. The diffusion of air through water is studied by observing the gradual loss of the air contained in a Cartesian diver, and this loss is determined from the change in the temperature coefficient contained in the equation of floatation. The same equation also involves in a simple manner the height of the barometer ; and a Cartesian diver apparatus is, therefore, virtually a water barometer which need only be one foot high instead of thirty feet.—Separation and determination of mercury as mercurous oxalate, by C. A. Peters. The author estimates mercurous salts volumetrically by precipitating with ammonium oxalate, and determining the oxalic acid by potassium permanganate, and gravimetrically by direct weighing of the precipitate.—Electrical resistance of thin films deposited by kathode discharge, by A. C. Longden. The thinnest films have a resistance which is very much higher than is warranted by their thinness. The sign of the temperature coefficient of resistance varies with the thickness, and it is therefore possible to obtain resistances by kathode-ray deposition which do not vary with the temperature. Such resistances form valuable high-resistance standards.—New meteorite from Oakley, Logan county, Kansas, by H. L. Preston. This is a siderite of 61 lbs. found in 1895.—Some observations on certain well-marked stages in the evolution of the Testudinate humerus, by G. R. Wieland. The development of the humerus of the turtle presents a special interest on account of its graduated change of habitat from dry deserts to the ocean.—Geothermal gradient in Michigan, by A. C. Lane. The geothermal gradient at Bay City is 1·5 degrees F. per 100 feet. The Upper Peninsula is a region notorious for its much lower gradient. The author discusses the various hypothesis framed to account for the differences in the gradient. Among these are the cooling action of Lake Superior, a survival of the Ice Age coldness, and differences in the conductivity of rocks. The author favours the last hypothesis.—Production of X-rays by a battery current, by J. Trowbridge. The installation of a plant of 20,000 storage cells at the Jefferson Physical Laboratory has enabled the author to obtain X-rays of exceptional brilliancy, yielding negatives of great contrast. When the X-ray tube is first connected with the battery terminals no current flows. It is necessary to heat the tube, when it suddenly lights up. A distilled-water resistance of about 4,000,000 ohms is inserted in the circuit.

Annalen der Physik, No. 5.—Change of conductivity of gases by a continuous electric current, by J. Stark. The resistance of a gas conveying an electric current is highest near the electrodes, owing to the accumulation of ions of the same sign in this neighbourhood. It has another maximum near the middle, but rather more towards the side of the anode. The resistance is influenced by the heat developed at the electrodes, by the kathode rays, and by the unequal speed at which the two kinds of ions travel through the gas.—Objective presentation of the properties of polarised light, by N. Umow. A beam of parallel plane-polarised light is allowed to fall on various geometrical bodies whose surfaces have peculiar optical properties, such as a cone covered with fuchsine, a quartz plate, or a Babinet compensator. The reflection or transmission of the light gives rise to striking colour phenomena. Peculiar spiral effects are obtained by sending the beam through an opalescent colophonium emulsion.—Magnetic screening, by H. du Bois and A. P. Wills. In this portion of their work, the authors calculate and verify the effect of a triple screen of iron for galvanometers. The external diameters of the three screens are 2·5, 4·3 and 8·0 cm. respectively, and their thicknesses are 0·27, 0·18 and 0·18 cm. The total theoretical " screening ratio," *i.e.* the ratio by which the disturbing magnetic field is reduced, is 60·2, and the observed ratio s 64·6.—Armoured galvanometers, by H. du Bois and H. Rubens. Describes some galvanometers screened in accordance with the results of the previous paper.—Rotating magnetic flag, by G. Jaumann. A small magnet mounted like a flag on a glass rod as an axis may be given a continuous rotation by immersing it in mercury contained in a glass vessel surrounded by a tight-fitting copper vessel, with a current traversing the body of the mercury and returning through the copper vessel. The work spent in overcoming the resistance of the mercury is derived from the current itself. It appears as a counter E.M.F. until the mercury rotates with the magnet.—Thermal deformation of balances, by T. Middel. Delicate balances show a considerable change of sensitiveness with the temperature. The author shows that this is due to the bending of the beam of the balance, owing to the unequal expansion of the upper and the lower portion, and that is due to the unequal working of the metal, the coefficient of expansion for cast brass being less than that of rolled brass.—The additive character of atomic heats, by S. Meyer. The author shows that in the case of twenty-six oxides an excess of the sum of atomic volumes over the molecular volume is accompanied by an excess of the aggregate atomic heats over the molecular heat, and that a defect of atomic volumes is accompanied by a defect of atomic heats in the same manner. Boron and bismuth sesquioxides are the two exceptions.

Bulletin de l'Académie des Sciences de St. Pétersbourg, vol. viii. No. 1.—Yearly report of the Academy.—A newly-discovered Old Turkish inscription, by Dr. W. Radloff, preliminary report. The inscription was discovered by Madame Elizabeth Clements near Urga, and excellent reproductions of it were made. Dr. Radloff found that it was made in honour of the wise Toyukuk, father-in-law of Bilga-khagan, who was born in 646 of our era.—On the elements of earth-magnetism at Kamenets, Khotin and Odessa, by W. Dubinsky. Vol. viii. No. 2.—On the rapid motion of the line of the absides in the system of a' Gimini, by A. Byelopolsky.—On the spectroscopic determination of the movements of γ Virginis, by the same.—Aurora borealis observed at Pavlovsk on December 20, 1897, by V. Kuznetsoff, with two photographs.—Hydrobiological researches at the Sebastopol Biological Station, by A. Ostroumoff.

Vol. viii. No. 3.—On the attempts at reproducing cometary

phenomena by means of experiments, by Th. Bredikhin (in Russian). The recent results obtained by photography permitted us to obtain most exact reproductions of cometary forms. They stimulated the desire of producing theories of comets, and, as far as the author knows, five different theories were proposed lately; they differ essentially in their fundamental principles. No great comets having appeared lately, the earlier drawings, made by previous astronomers, necessarily must be taken into account. Bredikhin found it necessary, therefore, to systematically discuss the facts which relate to the variety of forms of comets, and the passages from the one form to another. These facts can be ignored by no theory, and the author consequently analyses those criteria which must be applied to each theory of the comets.—On the way of building magnetic observatories, by H. Wild.—Description of a very rare case of *Craniopagus parietalis*, by J. Ziematzky (plate).—On the influence of the terms of third order in the perturbations function of the movement of the earth round its centre of gravity on the formulæ of nutation, by A. Ivanof (in French). The author gives a new formula for reducing the length of the second pendulum for any geocentric latitude.

Vol. viii. No. 4.—Ephemerid of the comet of Encke from June 1 to July 31, 1898, by A. Ivanoff.—On the differences of the horizontal intensities of earth magnetism obtained from observations of the unifilar and the bifilar theodolite, by H. Wild.—Positions of 1041 stars of the star-cluster 5 Messier, deduced from photographs, by Madame Shilow. Full list, compiled from careful measurements made on photographic plates.

SOCIETIES AND ACADEMIES.

LONDON.

Royal Society, May 31.—" The Crystalline Structure of Metals." Second Paper. By Prof. J. A. Ewing, F.R.S., and Walter Rosenhain, B.A.

The investigations described in this paper deal principally with the phenomena of annealing. The first section of the paper describes experiments made in the hope of observing under the microscope the process of recrystallisation in strained iron. This attempt to watch the process of recrystallisation failed, although the experimental difficulties of keeping a specimen under microscopic observation while it was being heated were successfully overcome. The specimen was electrically heated in a vessel with a thin glass or mica window, and the microscope-objective was kept cool by directing a strong blast of cold air on it and on the surface of the window.

The next section of the paper deals with the changes of crystalline structure which go on in lead and other metals at comparatively low temperatures. The authors' attention was directed to this by noticing that a piece of plumber's sheet lead, when etched with dilute nitric acid, exhibits a strikingly crystalline structure, with large crystals. The character of this appearance led the authors to the view that a slow process of annealing or recrystallisation was at work in such lead at ordinary atmospheric temperatures, and the authors have satisfied themselves that this is the case. The method of investigation consisted in taking a series of micro-photographs, at low magnifications, of certain marked areas in the surface of a specimen, in order to watch the change which went on through lapse of time, or after application of some thermal treatment.

When a piece of cast lead is severely strained by compression, the originally large crystals, after being considerably flattened, are driven into and through one another, so that the etched surface of a strained specimen presents a fine grain, whose crystalline nature only becomes apparent under considerable magnification (80 to 100 diameters). A piece of lead severely strained in this way, and kept for nearly six months in an ordinary room without any special thermal treatment, was found to be undergoing continuous change during that time. A series of photographs of this specimen, taken at intervals during the six months, show that a great number of the small crystals have grown larger at the expense of their neighbours. In similar specimens which have been kept at 200° C., the growth has been much more rapid and more pronounced. The rate of growth is a function of time and temperature, but some specimens show much more rapid changes than others under similar conditions of temperature; in some cases five minutes' exposure to a temperature of 200° C. is sufficient to alter the crystalline pattern completely. Experiments have also been made at 100° C. and 150° C., leading to the general result that crystalline growth will occur at any temperature from that of an ordinary room, *i.e.* 15° C. or 20° C., up to the melting point of lead, and that in general the higher the temperature the more rapid is the initial rate of change. No numerical data can be given, as the crystals are quite irregular, both in size and shape.

A comparison of micro-photographs of the same specimen at various stages reveals the fact that the growth of an individual crystal occurs, not in uniform layers all round it, but by the formation of arms and branches that invade the neighbouring crystals, the intervening portions sometimes changing at a later stage. This action is analogous to the formation of skeleton crystals in a metal during solidification from the liquid state, the space between the branches filling in as solidification proceeds.

A marked feature observed in several specimens was the large and rapid growth of one or two individual crystals; in several instances such individuals grew until they were some hundreds of times larger than their neighbours. Generally the most aggressive crystals were found near the edges of the specimen. It is noticeable that at times a crystal which has already grown considerably is swallowed up by a more powerful neighbour.

Some light is thrown on the nature of these actions by the fact that this growth only occurs in crystals that have been subjected to severe plastic strain. By casting the metal in a chill mould, specimens of lead can be obtained having a crystalline structure quite as minute as that found in a severely strained specimen, but this structure remains unchanged at temperatures which produce rapid change in a strained specimen.

The investigation of the effects of such comparatively moderate temperatures was extended to other metals, viz. tin, zinc and cadmium. In tin, the various phenomena of crystallisation from the fluid state are strikingly illustrated on a large scale by the thin layer of that metal which constitutes the surface of commercial tin-plate. The effects of rapid and slow solidification in producing small or large crystals respectively are well marked, and an examination of the etched surface of tin-plate under the microscope reveals beautiful geometrical markings or pits, whose oriented facets produce the well-known selective effect of oblique illumination. The study of the crystalline structure affords an explanation of the nature and method of production of patterns in "moirée métallique," a process which has long been in use for the decoration of articles manufactured of tin-plate.

The final section of the paper deals with an hypothesis, which is advanced as an attempt to explain the mechanism of the growth of crystals in apparently solid metal.[1] According to this hypothesis, the metallic impurities which are present in a metal play an important part in the action. When a metal solidifies from the fluid state, the metallic impurities ultimately crystallise as a film of eutectic alloy in the inter-crystalline junctions; when fairly large quantities of such eutectics are present, the microscope reveals their presence as an inter-crystalline cement, such as that formed by "pearlite" in slowly cooled mild steel; very minute quantities of eutectic, however, will be invisible and yet capable of forming a thin film of fusible cement. The authors conceive that the changes of crystalline structure which go on while the piece is in the solid state are accomplished by the agency of eutectic films between the crystals, in dissolving metal from the surfaces of some crystals and depositing it on others. When a metal is severely strained, these films of eutectic will be also strained and in many places broken, thus allowing the actual crystals to come into contact with one another. The difference in the rate of etching of adjacent crystals and the phenomena of the electrolytic transfer, in an acid solution, of lead from one crystal to another in the same mass of metal, support the supposition that there is a difference of electric potential between the crystal faces which are brought into contact by severe strain. If it be assumed that a film of eutectic alloy when fluid, or even when in the pasty condition that precedes fusion, can act as an electrolyte, we may regard any two crystals thus in contact, with a film of eutectic interposed in places, as a very low resistance circuit, and the growth of the positive crystal at the expense of the negative would result. Moreover, such growth would be more rapid at higher temperatures, and its rate at a given temperature would vary in different specimens according to the nature and quantity of the impurities present. That an alloy can act as an electro-

[1] It is proper to say that this hypothesis is due to Mr. Rosenhain.—J. A. E.

lyte has not been established experimentally, but the assumption is supported by the close general analogy between alloys and salt solutions. This analogy extends to the very question of the growth of crystals, as Joly has shown that when crystals of a salt are immersed in their mother-liquor, growth of one at the expense of others will take place.

It should be added that solution of one crystal into the intervening film of eutectic, along with deposit on the neighbouring crystal from the eutectic, may occur as a consequence of differences of orientation, producing differences of "solution pressure" apart from actual electrolysis, but the fact that growth has not been observed to occur except in strained crystals favours the view that the action is electrolytic.

Some further results which have been deduced from the above hypothesis have been verified by experiment. It follows from the hypothesis that an inter-crystalline boundary containing no eutectic would be an impassible barrier to crystalline growth, but if the eutectic could in any way be supplied, growth across the boundary might take place. In an absolutely pure specimen of lead, there would be no eutectic at the inter-crystalline junctions, but as extremely minute traces of impurity would suffice to set up the action, it is almost hopeless to verify the hypothesis in this way. Some experiments on the cold welding of lead have, however, borne out these conclusions. Two clean, freshly-scraped lead surfaces will unite under great pressure in the cold state, and if a piece so welded be annealed, the crystalline growth due to the annealing, with very rare exceptions, never crosses the inter-crystalline boundary formed by the welding surface. To test whether the presence of some eutectic would allow growth to take place, small quantities of a more fusible metal were scattered over the freshly-scraped surfaces of lead before squeezing them together. Then, after a cold weld had been made by pressure, on annealing by exposure to 200° C. it was found that crystal growth frequently crossed the line of the weld, as the above theory led one to expect. This experiment has been repeated many times with the uniform result that whenever a small quantity of eutectic, or of an impurity capable of forming a eutectic with the lead, was scattered over the clean surfaces before welding, a distinct growth of crystals across the boundary took place as a result of annealing. On the other hand, a large number of welds were made without introducing any impurity, and with very rare exceptions they showed no growth across the boundary, even after the annealing process was continued for some weeks. In rare exceptions a minute amount of growth across the boundary was observed, but these may fairly be accounted for by the almost unavoidable presence of traces of impurity. The result as a whole goes far to confirm this solution theory of crystalline growth in annealing.

June 14.—"Static Diffusion of Gases and Liquids in Relation to the Assimilation of Carbon, and Translocation in Plants." By Horace T. Brown, F.R.S., LL.D., and F. Escombe, B.Sc., F.L.S.

This paper is intended to be the first of a series descriptive of the work carried out by the authors in the Jodrell laboratory on the fixation of carbon by green plants, and deals mainly with the purely physical processes by which atmospheric carbon dioxide gains access to the active centres of assimilation.

The new evidence which F. F. Blackman brought forward in 1895 in favour of the gaseous exchanges of leaves taking place exclusively through the stomatic openings, presents at first sight certain difficulties of a physical nature, which have led to an examination of the whole question of the free diffusion of carbon dioxide at very low tension, and under a set of conditions very different from those under which the previous determinations of the coefficient of diffusion of carbon dioxide and air have been made by Loschmidt and others, where the gases were initially of equal tension, and the ratios of mixture departed widely from those of ordinary atmospheric air. The inquiry has led to the discovery of some new facts connected with the static diffusion of gases and liquids, which are of considerable interest, not only from the physical point of view, but from the explanations they suggest of certain natural processes which are primarily dependent on diffusivity.

The method employed in the first instance for the determination of the diffusivity of atmospheric carbon dioxide was one of *static diffusion* down a column of air of a definite length towards an absorptive surface at the bottom of the column When a static condition has been established, there is a steady flux of the carbon dioxide down the air column which may be quantitatively investigated by the same simple mathematical treatment as the "flow" of heat in a bar when the permanent state has been reached, or the "flow" of electricity between any two regions of a conductor maintained at a constant difference of potential.

By a long series of experiments of this nature it was found that the diffusivity constant, k, for very dilute CO_2 does not materially depart from the value assigned to it by Loschmidt and others, when experimenting with much higher ratios of mixture, and that the difference is certainly not of sufficient magnitude to be taken into serious account in the study of the natural processes of gaseous exchange in the assimilating organs of plants.

In the static diffusion of a gas, vapour, or solute, as the case may be, the amount of substance diffusing in a given time, all other conditions being the same, is directly proportional to the sectional area of the column. It is found, however, that if the flow is partially obstructed by interposing at any point in the line of flow a thin septum pierced with a circular *aperture*, the rate of flow across unit area of the aperture is greater than it would be across an equal area of the unobstructed cross-section of the column at this point. If the margin around the aperture has a width of at least three or four times its diameter, the rate of flow is now found to be directly proportional to the *linear dimensions* of the aperture and not to its area, so that the velocity of flow through unit area varies inversely as the diameter.

A large number of experiments on the diffusion of carbon dioxide, water-vapour and sodium chloride in solution are given in support of this proposition. All these show that the rate of diffusion across such a septum, all other conditions being the same, is directly proportional to the diameter of the aperture, and not, as might have been expected, to its area.

Exactly the same result is obtained when small circular discs of an absorbent, such as a solution of caustic alkali, surrounded by a wide rim and exposed to *perfectly still* air, the amount of carbon dioxide absorbed under these conditions being proportional to the *diameters* of the discs.

If, however, there are any sensible air currents the absorption becomes proportional to the areas.

These two sets of phenomena may be explained as follows :—

In the case of the absorbing disc in perfectly still air, the convergent streams of carbon dioxide creep through the air towards the absorbing disc, establishing a steady gradient af density, and this creep will be a flux perpendicular to the lines of equal density, which form curved surfaces or "shells" surrounding the disc and terminating in the rim. The state of things is exactly analogous to the electric field in the neighbourhood of a conductor of the same shape and dimensions as the absorbent disc.[1] In the case of the gas, the curves or "shells" of equal density are the analogues of the similarly curved surfaces of equipotential above the electrified disc, whilst the converging lines of creep or flux of the gas are the analogues of the lines or tubes of force which bend round into the disc as they approach it.

If we consider two such absorbent discs of different diameters, the curved surfaces in each system corresponding to a given density will be found at actual distances from the discs which are in the same proportion to each other as are the diameters of the discs. In other words, the gradient of density on which the rate of flow depends will be proportional to the diameters of the discs, which is exactly what is found experimentally.

This case of an absorbent disc is the exact converse of one which has been theoretically investigated by Stefan, viz. the conditions of evaporation of a liquid from a circular surface. He found that the lines of flux of the vapour proceeding from the surface of the liquid must be hyperbolas, whilst the curved surfaces of equal pressure of the vapour must form an orthogonal system of ellipsoids, having their foci, like the hyperbolas, in the bounding edges of the disc. This was a purely mathematical deduction which has never been verified experimentally, but it will be seen that the exactly converse phenomena of diffusion are in complete agreement with it.

In the other case of a diffusive flow through a circular aperture in a diaphragm, the lines of flow, which are *convergent* as they approach the aperture, bend round their foci situated in the edges of the disc and form a *divergent* system on the other side. If the chamber into which they pass is a perfectly absorbent one,

[1] The authors are indebted to Dr. Larmor for this suggestion of the electrostatic analogy.

and is sufficiently large, there will be formed on the inner side of the diaphragm a system of density shells similar to those outside, but with the gradient of density centrifugally instead of centripetally arranged. This system of shells is termed negative, and is as effective as the outer positive system in regulating the flow according to the "diameter law," so that this law will still hold good even if the outer air currents are sufficient to sweep away the external positive shells altogether.

All the known facts of diffusion through circular apertures in a diaphragm are then discussed from the above point of view, which is fully elaborated in the original paper.

By diffusing colouring matter through apertures in a septum, under such conditions as to prevent convection currents, the "density shells" have been rendered visible, and it has been shown that their ellipsoidal form is exactly that which is demanded by the above hypothesis. Moreover, this method gives an experimental demonstration of the more rapid projection of the diffusing particles from the edges of the aperture than from a point nearer its centre, a fact completely in harmony with the deduction of Stefan regarding the evaporation of liquids under analogous conditions.

The various cases which present themselves in practice with regard to the rate of diffusion through single apertures in a diaphragm are then discussed from the above point of view, and simple formulæ for the determination of this rate for single and double systems of density shells are established : (1) for cases where the thickness of the diaphragm is negligible, and (2) for other cases where the apertures become more or less tubular. In a subsequent section of the paper it is shown how closely the observed facts conform to these deductions, and that in static diffusion through apertures in a septum we have a new and accurate method for the determination of the diffusivity constants of atmospheric CO_2, of the vapours of liquids, and of substances in a state of solution.

Since the velocity of the diffusive flow through unit area of an aperture in a diaphragm varies inversely with the diameter, it might reasonably be expected that a diaphragm could be so perforated with a series of very small holes arranged at suitable distances from each other, as to exercise little or no sensible obstruction when it was interposed in a line of diffusive flow, although the aggregate area of the small holes might represent only a small fraction of the total area of the septum. Multiperforate diaphragms of this kind were found to possess all the remarkable properties which had been anticipated.

The material used for the septa was very thin celluloid, which was perforated at regular intervals with holes of about 0·38 mm. in diameter. Details of a number of experiments with such diaphragms are given, in which it is shown that they may be so arranged as to produce but little obstructive influence on the diffusive flow of a gas when the total area of the apertures amounts only to about 10 per cent. of the area of the septum, and that nearly 40 per cent. of the full diffusive flow may be maintained when the number of the apertures is so far reduced as to represent an area of only 1·25 per cent. of the full area of the septum.

The explanation is to be found in the local intensification of the gradient of density in the immediate neighbourhood of the diaphragm, and which does not extend to the column away from the apertures. This disturbance of gradient is brought about by the rapid convergence of the lines of flux, and their divergence on the other side, with the consequent formation of a system of "density shells" over each aperture. A system of perforations of this kind may be compared with a system of conductors electrified to a common potential, the density of electric potential, and the non-absorbing portions of the diaphragm to a surface formed by lines of electric force. Just as the electric capacity of a plate is not much reduced by cutting most of it away, so also is it possible to block out a large portion of the cross-section of the diffusing column without materially altering the general static conditions on which the flow depends.

The importance of these results in relation to diffusion through porous septa is next considered, diffusion through a thin porous septum being only an extreme case of free diffusion through a multiperforate diaphragm, whose apertures are so far reduced in size as to materially interfere with the mass movement of the diffusing substance.

A section of the paper is devoted to the application of these new observations to the processes of gaseous and liquid diffusion in living plants, and it is pointed out that the structure of a typical herbaceous leaf illustrates in a striking manner all the physical properties of a multiperforate septum. Regarded from this point of view it is shown that the stomatic openings and their adjuncts constitute even a more perfect piece of mechanism than is required for the supply of carbon dioxide for the physiological needs of the plant, and instead of expressing surprise at the comparatively large amount of the gas which an assimilating leaf can take in from the air, we must in future rather wonder that the intake is not greater than it actually is.

From data afforded by actual measurements of the various parts of the stomatal apparatus of the sunflower it is shown that an extremely small difference of tension of the carbon dioxide within the leaf, as compared with that in the outer air, will produce a gradient sufficient to account for the observed intake during the most active assimilation.

It is also shown that the large amounts of water-vapour which pass out of the leaf by transpiration are well within the limits of diffusion, and that it is unnecessary to assume anything like mass movement in the outcoming vapour.

The translocation of solid material from cell to cell in the living plant is next considered, especially with reference to this transference being, at any rate in part, brought about by means of the minute openings in the cell-walls through which the connecting threads of protoplasm pass. Notwithstanding the very small relative sectional area of these perforations they probably exercise an important function in cell-to-cell diffusion, in virtue of their properties as multiperforate septa.

There are two appendices to the paper, one in which a full description is given of a series of experiments on the absorption of carbon dioxide by solutions of caustic alkali from air in movement; the second being devoted to a detailed description of the methods used for accurately determining the carbon dioxide absorbed.

Physical Society, June 22.—Mr. T. H. Blakesley, Vice-President, in the chair.—A paper, entitled "Notes on Gas Thermometry," by Dr. P. Chappuis, was read by Dr. Harker. The author having been led to recognise that hydrogen could not be used as a thermometric substance at high temperatures, on account of its action on the walls of the glass reservoirs, has had recourse to a constant volume nitrogen thermometer with an initial pressure slightly under 800 mm. The value of the coefficient of expansion of nitrogen at constant volume is variable, diminishing up to 80° C. and then increasing slightly. In fact, nitrogen at 100° C. behaves like hydrogen at the ordinary temperatures, its compressibility being less than that required by Boyle's law. A table of corrections was therefore prepared. The readings of the constant volume nitrogen thermometer are too low, but the corrections are small, amounting to about 0·04° C. at the temperature of boiling sulphur. The mean result of the author's experiments for the boiling point of sulphur is 445°·2 under a pressure of 760 mm. Callendar and Griffiths' results obtained with a constant pressure air thermometer is 444°·53. The difference is attributed to the joint action of several causes :—(1) The corrections for a constant pressure thermometer are about double those of a constant volume instrument. This correction applied to Callendar and Griffiths' result would raise it about 0·1°. (2) Callendar and Griffiths have used a value for the gas constant which is larger than that obtained by more recent experiments. Adopting the latter value, the boiling point would be raised to 445°. (3) The divergence may be due to the expansion of the reservoir. The most accurate way of determining this is by the interference method of Fizeau. This method is used with small pieces of the material, and the author has employed it to determine the coefficient of expansion between 0° and 100°. Extrapolation to 450° might cause errors. The linear expansion has recently been determined by Bedford between 0° and 840° by a comparator method. The homogeneity of porcelain is doubtful, especially when glazed, and the great differences occurring between the expansions obtained from the above methods is attributed to the change in form of the tube in Bedford's experiments, brought about by unequal thickness and want of homogeneity and consequent unequal expansion. The author therefore adheres to his value of the boiling point obtained from the expansion by the Fizeau method, whilst recognising the uncertainty attaching to the application of the coefficient of expansion of the reservoir over an interval four times as great as that over which it was determined. —A paper on a comparison of impure platinum thermo-

meters, by Mr. H. M. Tory, was read by Prof. Callendar. The object of this paper is to investigate the probable order of accuracy attainable in the determination of high temperatures by the use of ordinary commercial specimens of platinum wire. Five wires were compared, from 400° to 1000° C. The fundamental coefficients of the wires varied within 40 per cent. of the maximum value, but the temperatures observed by them when calculated on the platinum scale by means of the ordinary simple formula, did not differ by more than 9° at 1000° C. Each wire was directly compared with a pure standard wire, the two being wound side by side in the same tube. Curves have been drawn with the platinum temperatures of the standard wire as abscissæ, and the differences between the temperatures indicated by the two wires compared as ordinates. These curves are all straight lines, within the limits of observation, and hence the determination of two constants is sufficient to enable us to compare an impure platinum thermometer with the standard, and therefore with the scale of the gas thermometer. The two constants can at once be obtained from observations at the boiling point of sulphur and the freezing point of silver, and thus a practical thermometric scale can be established, which between 0° and 1000° never differs by more than two or three degrees from the gas scale.—Prof. Callendar said he was unable to agree with the corrections to his observations suggested by M. P. Chappuis. He considered that the uncertainty in the coefficient of expansion of the gas was due to uncertain changes in the volume of the bulb, and to uncertainty in the coefficient of expansion of mercury. The fundamental coefficient of mercury was ·00018153 according to Regnault, ·00018216 according to the later reduction of Broch, and ·00018256 according to experiments by Chappuis with a hard glass bulb. It made a difference of no less than 4 per cent. in the fundamental coefficient of expansion of the glass, according as the original results of Regnault, or the value found by Chappuis, assuming the linear expansion of the glass, were adopted. The importance of the changes in the volume of the bulb had been fully pointed out, and a method of taking approximate account of these changes had been explained in the paper on the boiling point of sulphur in 1890. Unfortunately the glass employed was rather soft, and the changes of volume which occurred were too great to permit of the most accurate determination of the coefficient. The boiling point, when corrected for the smaller expansion of the bulb, came out lower than 444°·53°. With regard to porcelain, Prof. Callendar did not consider it a good material, on account of the glaze. He did not think that the increase coefficient of a tube or bulb over a large range of temperature could be inferred from a small and possibly asymmetric specimen. The results might be less inconsistent in the case of homogeneous and well-annealed metallic bulbs. The correction for the expansion of the bulb was, he believed, given by the expression $dt = (C + b\theta)t(t - 100)$. He did not agree with M. P. Chappuis that the correction was independent of c, although the value of b was certainly most important at high temperatures. He also wished to take exception to the method adopted by Chappuis of calculating the correction of the nitrogen thermometer. According to Joule and Thomson, the correction should be greater; according to other authorities, it might be less. He hoped to discuss this in a further communication to the Society. Mr. Glazebrook said that, although he placed confidence in Chappuis' formula for a definite piece of porcelain between certain temperatures, he thought further and careful work was necessary before fixing on a formula for ordinary use. Prof. Carhart said he would like to see a comparison made between the results of experiments with gas thermometers and those with platinum and platinum-rhodium couples. Mr. Rose-Innes expressed his interest in the behaviour of nitrogen about 100° C., as mentioned in M. P. Chappuis' paper. Dr. Lehfeldt said the peculiarities of the nitrogen scale between 70° and 80° might be explained by the reversal of the properties of nitrogen between 0° and 100°.— A paper on the law of Cailletet and Mathias and the critical density was read by Prof. S. Young. The law of Cailletet and Mathias is very nearly, though in most cases not absolutely, true. It appears to be only strictly true when the ratio of the actual to the theoretical density at the critical point has the normal value 3·77. The curvature of the "diameter" is generally smaller the nearer this ratio approaches its normal value. The curvature is in nearly every case in opposite directions, according as this ratio is greater or less than 3·77. The curvature is generally so slight that the critical density may be calculated from the mean densities of liquid and saturated vapour at temperatures from about the boiling point to within a few degrees of the critical point with an error generally not exceeding 1 per cent. If, however, the critical density is calculated from the mean densities at low temperatures, the error may be considerable ; in the case of normal decane it is between 5 and 6 per cent. The law does not, as a rule, hold good at all for substances the molecules of which differ in complexity in the gaseous and liquid states. Mr. Rose-Innes said that in his paper the author had used the generalisations of Van der Waals, although the author himself had shown that they were not strictly true. Prof. Young said that the generalisations held in some cases, although they did not in others. In all cases they were approximately true, and it was therefore advisable to use them, and study the results as far as possible.—The Society then adjourned until next October.

Chemical Society, June 7.—Prof. Thorpe, President, in the chair.—The following papers were read.—Condensation of ethyl acetylenedicarboxylate with bases and β-ketonic esters, by S. Ruhemann and H. E. Stapleton. Ethyl acetylenedicarboxylate reacts with benzamidine to form a substance of the probable constitution,

$$\text{CPh:N}\diagdown\text{C:C}\diagup\text{N:CPh}$$
$$\text{NH.CO}\diagup\qquad\diagdown\text{CO.NH}$$

which the authors term glyoxaline red.—Condensation of phenols with ethyl phenylpropiolate, by S. Ruhemann and F. Beddow. Sodium phenoxide reacts with ethyl phenylpropiolate, yielding a substance of the constitution $C_6H_5.C(OC_6H_5):CH.CO_2Et$; this ester is easily hydrolysed, and the acid readily loses carbon dioxide, giving phenoxystyrene, $CH_2:C(OC_6H_5).C_6H_5$.—The constitution of pilocarpine, by H. A. D. Jowett. Isopilocarpine yields on oxidation a lactonic acid of the constitution

$$CH(CH_3)_2.CH.CH.CO_2H$$
$$\qquad\qquad\mid\qquad\mid$$
$$\qquad\qquad O\text{—}CO\qquad;$$

the alkaloid also contains the groups :NH and :NCH$_3$.—The nitrogen chlorides derivable from metachloroacetanilide and their transformations, by F. D. Chattaway, K. J. P. Orton and W. H. Hurtley.—The persulphuric acids, by T. M. Lowry and J. H. West. A quantitative study of the equilibrium established between sulphuric acid, hydrogen peroxide and "persulphuric acid" on mixing the former two substances, affords evidence indicating the existence of the acids $H_2O_2.4SO_3$ and $H_2O_2.2SO_3$ in a mixture of sulphuric acid and hydrogen peroxide.—On diphenyl- and dialphyl ethylenediamines, their nitro-derivates, nitrates and mercurichlorides, by W. S. Mills.—Derivatives of cyanocamphor and homocamphoric acid, by A. Lapworth. The halogen derivatives of cyanocamphor are reduced to cyanocamphor and homocamphoramic acid, $C_8H_{14}'CO_2H).CH_2.CONH_2$, by strong aqueous alkalis. a-Bromo-homocamphoric acid, $C_8H_{14}(CO_2H).CHBr.CO_2H$, made by heating homocamphoric dichloride with bromine, can be converted into homocamphoric acid, $C_8H_{14}\diagup^{\text{CH.CO}_2H}_{\diagdown\text{CO}_2H}$.—The ultra-violet absorption spectra of some closed chain carbon compounds. II. Dimethylpyrazine, hexamethylene and tetrahydrobenzene, by W. N. Hartley and J. J. Dobbie.—A study of the absorption spectra of o-oxycarbanil and its alkyl derivatives in relation to tautomerism, by W. N. Hartley, J. J. Dobbie and P. G. Paliatseas.—Action of formaldehyde on amines of the naphthalene series (II.), by G. T. Morgan. The action of formaldehyde on ethyl-β-naphthylamine in cold acetic acid solution results in the formation of 2 : 2-diethyldiamino-1 : 1-dinaphthylmethane.—The bromination of benzeneazophenol (II.), by J. T. Hewitt and W. G. Aston.—Condensation of ethyl crotonate with ethyl oxalate, by A. Lapworth. Ethyl γ-oxalocrotonate, $CO_2Et.CO.CH_2.CH:CH.CO_2Et$, is formed by the action of sodium ethoxide on a mixture of ethyl crotonate and ethyl oxalate ; it is converted into a-pyrone-a'. carboxylic acid, $\diagup^{\text{CH.CH:C.CO}_2H}_{\diagdown\text{CH.CO.O}}$, by hydrochloric acid.—Researches in silicon compounds. VI. On silicodiphenyldiimide and silicotriphenylguanidine, by J. E. Reynolds. On heating silicophenylamide, $Si(NHC_6H_5)_4$, the diimide, $Si(NC_6H_5)_2$, and silicotriphenylguanidine, $Si(NC_6H_5)_2(NHC_6H_5)_2$, are obtained.—Note on Bach's hydrogen tetroxide, by H. E. Armstrong.

Linnean Society, June 7.—Prof. Sydney H. Vines, F.R.S., President, in the chair.—Mr. R. Morton Middleton exhibited a

letter, dated "London, 13 June 1788," in the handwriting of Sir J. E. Smith, addressed to Charles Louis L'Héritier, at Paris, in which he mentioned a visit to Oxford with Sir Joseph Banks and J. Dryander for the purpose of looking over the plants and drawings of Sibthorp, who was then lecturing there ; and added some critical remarks on several species of *Sida* which L'Héritier had sent him for determination. Mr. Middleton also exhibited an engraved portrait of Sir J. E. Smith from the *Gentleman's Magazine*, 1828, which, with the letter, he presented to the Society.—Mr. F. Enock, with the aid of the lantern, exhibited several photomicrographs and photographs of living insects, and gave an illustrated account of the life-history and metamorphoses of a dragonfly (*Æschna cyanea*).—Mr. E. S. Goodrich read a paper, entitled "Notes on *Syllis vivipara*, Krohn." This worm, which he found in a tank at the Naples Laboratory, appeared to be identical with that described by Krohn in 1869 (*Arch. f. Naturg*. xxxv. p. 197), and in general form resembled Claparède's *Syllis Armandi* (probably *S. prolifera*, Krohn). The peculiar point of interest was its method of reproduction, the embryos growing within the body-cavity of the parent to an advanced stage (when they resemble the adult except in their smaller size and fewer segments), and escaping by the breaking off of the posterior portion of the parent's body.—Dr. Otto Stapf read a paper on the two Melastomaceous genera *Dicellandra*, Hook. f., and *Phæoneuron*, Gilg. He showed that the differences between them are not in the heterandry and homœandry respectively, as was supposed, but in much more important characters which concern all those parts which affect the formation of the fruits and seeds. The diagnoses of the two genera must therefore be revised, with the result that *Phæoneuron* and *Dicellandra* change their character as monotypic genera.—A paper was read by Miss A. L. Embleton giving a full account of the anatomy and histology of *Echiurus unicinctus*, received from Prof. K. Mitsukuri, of Tokyo.

CAMBRIDGE.

Philosophical Society, May 7.—Prof. Clifford Allbutt, F.R.S., in the chair.—Exhibition of anomalous bones from pre-dynastic Egyptian skeletons, by Prof. Macalister, F.R.S.—Ammocoetes a Cephalaspid, by Dr. Gaskell, F.R.S. The paper contained evidence that Ammocoetes was the living representative of the ancient Cephalaspids.—Note on some abnormalities of the limbs and tail of Dipnoan fishes, by H. H. Brindley. *Lepidosiren* and *Protopterus* sometimes exhibit partial bifidity of the limbs and tail. This condition of the limbs of *Protopterus* has received some speculative attention, and it has also been suggested that a branched limb of *Lepidosiren* might have a respiratory function. Boulenger and Howes have since shown that *Protopterus* may regenerate its limb in a branched condition, and sections of branched limbs of *Lepidosiren* show histological features clearly suggesting a reproduced condition. Budgett and Kerr have noticed a considerable tendency to injury and reproduction of limbs and tail in both these fishes—and there can be no doubt that the reproduction is often bifid. A parallel is therefore afforded with the bifidity sometimes seen in lizards' tails, which in all cases examined are reproduced structures. In some of the latter there is evidence that the extra tail is a new growth arising from an injured place, and in others that the new growth is bifid from its commencement. In the cases of *Lepidosiren* examined the latter condition alone seems to hold.—On the standardisation of anti-venomous serum, by W. Myers. It was shown that Calmette's method was based on views which were no longer tenable ; and, further, that a special mixture of snake venoms is required. A more accurate measure of the antitoxin was to test its neutralising power, using ten times the minimal lethal dose of unheated Cobra poison, and mice of 15 grams weight as test animals. With this method it was possible to estimate the serum to within 15 per cent.

Royal Meteorological Society, June 20.—Dr. C. Theodore Williams, President, in the chair.—Mr. W. Marriott read a paper on rainfall in the west and east of England in relation to altitude above sea-level. This was a discussion of the mean monthly and annual rainfall for the ten years 1881-90 at 109 stations which the author had grouped according to the altitude of the stations above sea-level. The western stations were considered to be those which drained to the west, and the eastern stations those which drained to the east of the country. The diagrams exhibited showed that there is a general increase in the annual amount of rain as the altitude increases, and that the rainfall is considerably greater in the west than in the east. The monthly diagrams brought out prominently some interesting features, among which were (1) that the monthly rainfall in the west is subject to a much greater range than in the east ; (2) that in the west the maximum at all altitudes occurs in November, but in the east it is generally in October ; (3) that in the west the spring months, April, May and June, are very dry ; and (4) that both in the west and east there is a very great increase in the rainfall from June to July.—A paper by Mr. J. Baxendell was also read, giving a description of a new self-recording rain-gauge designed by Mr. F. L. Halliwell, of the Fernley Observatory, Southport. This rain-gauge, which the author believes approaches very closely to an ideal standard, has also the merit of being constructed at a moderate price.

PARIS.

Academy of Sciences, June 18.—M. Maurice Lévy in the chair.—On the monument erected to Lavoisier, by M. Berthelot. The monument is now finished, and will be unveiled on July 27.—The problem of the cooling of the earth's crust treated from Fourier's point of view, but by a much simpler method of integration, by M. J. Boussinesq.—Actinometric observations during the eclipse of May 28, by M. J. Violle. A diagram of the results obtained is given which closely approximates to the theoretical curve, the divergence being mainly due to the lag of the instrument, but also apparently in part owing to an absorption of heat by the solar atmosphere. The minimum ratio deducted from the observations was 0·12, distinctly less than the ratio of the radiant surfaces, 0·14. Two sets of observations were carried out, one on the Pic du Midi, at a height of 2860 metres, and the other from a balloon, at a height of about 10,000 metres.—On the formation of nitric acid in the combustion of hydrogen, by M. Berthelot. Hydrogen was burnt from a jet in oxygen containing varying amounts of nitrogen, and also in the calorimetric bomb at pressures of from one to twenty atmospheres, and the amounts of nitric acid formed determined. The proportion of nitric acid formed was greatest in the bomb, and increased with the initial proportion of the gases.—The combustible gases of the atmosphere : the air of towns, by M. Armand Gautier. Air is drawn, after careful purification from dust, moisture and carbon dioxide, over red-hot copper oxide, and the amounts of water and carbonic acid determined. The mean results for twenty-two days was 1·96 mgr of hydrogen and 6·8 mgr. of carbon per 100 litres of air ; but these quantities become 3·96 mgr. and 12·45 mgr. respectively when a correction is applied for the incomplete combustion effected by the particular length of copper oxide used.—The last eclipse of the sun and the zodiacal light, by M. Perrotin.—The occultation of Saturn by the moon of June 13 last, by M. Perrotin.—On the formation of beds of stipite, brown coal and lignite, by M. Grand'Eury. In the formation of brown coals marsh plants were the chief factor, trees only occurring rarely.—M. Dwelshauvers-Dery was elected a Correspondent for the Section of Mechanics, and M. D. P. Ehlert a Correspondent for the Section of Mineralogy.—Observations of the total eclipse of the sun of May 28 last, made at Argamasilla, in Spain, by M. H. Deslandres. The work undertaken included the measurement of the velocity of rotation of the corona, and the examination of its ultra-violet spectrum ; the study of the ultra-violet spectrum of the reversing layer ; the calorific spectrum and the direct photography of the corona.—The partial eclipse of the sun of May 28, at the Observatory of Toulouse, by MM. Monttingerand, Rossard and Besson. The results obtained were confined to direct observation of contacts, measurement of the common chord, photographic observations and the knowledge of meteorological phenomena.—The total eclipse of May 28 studied at Elche, by M. J. C. Sola. Photographs of the spectra of the chromosphere and corona were taken.—Observations of the shadow fringes made during the total eclipse of the sun of May 28, by M. Moye.—On the uniform integrals of the problem of n bodies, by M. Paul Painlevé.—On the general theory of rectilinear congruences, by M. A. Demoulin.—On the expansion of fused silica, by M. H. Le Chatelier. The mean coefficient of expansion of fused silica for a temperature range of 0° to ..... °, is 0·000,0007, the smallest coefficient known for any common substance.—Action of oxidising agents upon alkaline iodides, by M. E. Péchard. A study of the interaction of alkaline iodides with potassium permanganate, sodium periodate, potassium manganate, ozone and hydrogen peroxide.—Study of the viscosity of sulphur at temperatures above the temperature of maximum viscosity, by M. C. Malus.—On the selenides of iron, by M.

Fonzes-Diacon. Several selenides of iron can be prepared of the composition indicated by the formulæ $FeSe_2$, Fe_3Se_4, Fe_7Se_8, Fe_3Se_2, and $FeSe$. They are attacked by hydrochloric acid with difficulty, $FeSe_2$ being unaffected by this reagent.—The true atomic weight of boron, by M. G. Hinrichs. From two analyses of boron carbide made by M. H. Gautier, the author concludes that the true atomic weight of boron is exactly 11.—Action of sulphur dioxide and hydrogen sulphide upon pyridine, by M. G. André. Sulphur dioxide gives a crystalline compound with pyridine, $C_5H_5N.SO_2$, and the action of suphuretted hydrogen upon this gives pyridine trithionate and tetrathionate.—On the αβ-dimethylglutolactonic acids, by M. E. E. Blaise.—On the reserve carbohydrates in the seed of *Trifolium repens*, by M. H. Hérissey. A mannogalactane was isolated from the seeds of *Trifolium repens*, resembling in its properties the carbohydrates obtained from lucerne and fenugreek.—Presence of iodine in the blood, by MM. E. Gley and P. Bourcet. Iodine was found to be present in the blood of dogs in amounts varying from ·013 mgr. to ·06 mgr. per litre of blood. The iodine is in the liquid portion of the blood existing combined with proteid matter, analogous to the iodine in the thyroid gland.—Reality of urinary toxicity and of autointoxication, by M. A. Charrin.—On the anticoagulating power of serum in the pathological state, by MM. Ch. Achard and A. Clerc. Human blood serum, when present in sufficient quantity, prevents the coagulation of milk by rennet, the quantity of rennet solution required to produce coagulation measuring the activity of the serum. The anticoagulating power of the serum is unaffected by many diseases, but in others, such as pneumonia, septicemia with acute nephritis, uterine cancer, and advanced tuberculosis, this power is reduced to one-half or even less.

CAPE TOWN.

South African Philosophical Society, May 2.—L. Péringuey, President, in the chair.—Mr. Sclater exhibited a portion of a bone found at a considerable depth below the surface in Grave Street, and presented to the Museum by Col. Feilden. The bone was obviously the upper portion of the radius and ulna of a large ungulate animal ; it appeared to be too large for an ox, and Mr. Sclater suggested that it might perhaps be that of a hippopotamus.—The Rev. Dr. F. C. Kolbe read his paper, entitled "Ultimate analysis of our concept of matter." The lecturer first briefly stated the four prevailing views on the subject—the mechanical, the dynamic, the vortical, and the scholastic or Aristotelian. The first two theories being for various reasons rejected, the lecturer stated that the purpose of this paper was to reconcile the third and fourth.

CONTENTS.

THURSDAY, JULY 5, 1900.

PROTOPLASM.

Allgemeine Biologie. By Prof. Dr. Max Kassowitz. Vols. i. and ii., Pp. xv + 411, and x + 391. (Vienna: Moritz Perles, 1899).

A BAD hypothesis is better than none at all, is a saying of which many have taken advantage ; but able minds have agreed with the substance of the remark, and few can dissent from it in connection with that everlasting puzzle—Protoplasm. It is in vain to inveigh against the uselessness of speculations as to the structure or constitution, nature—what you will—regarding the physical basis of life, or regarding attempts to picture, however roughly, the movements, rearrangements and evolutions in that veritable witches' dance, the quadrilles of the molecules in which life consists. The inquiring mind is so constituted that it cannot resist the temptation to fashion some rough hypothesis as a tool wherewith to make one more attempt to pick the lock which hides the secret, and the criticism of the serious as to the futility of his efforts is no more powerful than the epigram of the debater to prevent him returning to the ruins of previous speculations, with renewed efforts to rebuild his frail image of something approaching, as nearly as may be, the inconceivably complex, and exposing it once more to the blows of the critic.

To those who are hopeless, let the more hopeful point out the difference in our present ideas of the nature of protoplasm from those of twenty years ago, and say whether no advances have been made.

In this extraordinarily well-written work, it is satisfactory to see that the word Biology is used in the sense of "the science of life," and not in the restricted and often unintelligible way so common in Continental works. Still more satisfactory is it to find here a carefully thought out plan of re-examination of the fundamental phenomena of life, and of "the physical basis of life," on which is erected a hypothesis with bold outlines and stately proportions, yet carefully and minutely fitted details, in conformity with the rapidly advancing knowledge of the last two decades.

The subject of "Allgemeine Biologie," as treated by the Viennese professor, resolves itself under the following headings :—

I. "Aufbau und Zerfall des Protoplasmas," forming the theme of the first volume, which is further subdivided into : (1) "Das Problem des Lebens und die Versuche zur Lösung desselben"; (2) "Aufbau des Protoplasmas"; and (3) "Zerfall des Protoplasmas."

II. The second volume is entitled "Vererbung und Entwicklung."

The third and fourth volumes are not yet to hand, but we are informed that they will deal with "Stoff- und Kraftwechsel der Thiere" and with "Nerven und Seele" respectively. That they will be eagerly looked for by all who have read the two at present under review is a safe prophecy.

Kassowitz takes his stand on the conviction—the reasons for which are given at great length in the first

half of the first volume—that previous hypotheses as to the nature of protoplasm, based on assumptions that any structure visible after treatment can be translated in terms of its structure during life, break down on examination equally with those which would regard protoplasm as a mere emulsion ; and that all attempts to explain what is going on in living protoplasm, which have for their basis the assumption that oxidations, reductions, and metabolic changes generally are carried out in the fluids bathing any such machinery of the protoplasm also fail to withstand criticism. The thermodynamic theory of life fails because the engine itself burns, and the value of a substance as food has no relation to its combustible value. It might have been added that Pfeffer had already shown this in his treatise "Zur Energetik der Pflanze." The osmotic theory of the botanists breaks down, because it attributes to the cell-sap an importance which a mere solution under such conditions does not possess ; the fermentation theory fails to explain more than a few bye-phenomena of life, and has no help for us in questions concerning synthesis ; the electro-dynamic theory breaks down because electric phenomena are least obvious just where we should most expect them. The molecular-physical theory assumes vibrations and the shaking asunder of molecules which are so stable that it is impossible to believe that they could be shattered and others escape under the conditions imposed ; while the vitalistic theory is a mere confession that what is to be explained needs explanation.

Underneath or behind all the assumptions of micellæ, gemmules, biophores, determinants or other formed structural units, as well as all material networks, rods, spherules, fibrils, pellicles and foams, Kassowitz detects the question—Of what are these physiological units and structures composed ? And he regards the fundamental fallacies underlying all previous hypotheses regarding the constitution and working of living protoplasm as chiefly the assumption that protoplasm is composed of proteids built up into some sort of more or less stable machinery, and that the chemical and other changes usually comprised in the term metabolism are carried on outside or merely in contact with this machinery—*e.g.* in a meshwork, or on the surface of the physiological units.

He therefore proposes to examine in detail, and step by step, what comes of deductions made from the hypothesis that protoplasm consists of molecules, in the chemical sense, but of extreme complexity, large volume, and very labile, linked one to another in series, and each requiring for its construction, not only proteids, but also fats, carbohydrates and the mineral salts known to be indispensable for life. And that every vital act consists in destructions and reconstructions of these molecules.

To obtain a coarse picture of this invisible structure we may suppose extremely tenuous fibrils of india-rubber joined up into a complex network and bathed in a fluid which contains the necessary ingredients for putting in new pieces wherever, by stretching the net too far, we break the elastic strands ; such breakages will occur especially between the nodes, and immediately the gaps are bridged over again by new fibrils, or networks of such, further extension, breakage and restitution are possible.

The linked up chains and networks of protoplasm-molecules—which do not correspond to the networks of coagula of other hypotheses—are termed *stereoplasm*, the bath of liquid containing proteids, carbohydrates, fats, minerals, molecular oxygen, etc.—*i.e.* in which are dissolved all the food-materials as well as all the products of shattered molecules—is termed *hygroplasm*. The osmotic attraction for water of the newly formed molecules (imbibition) would set up pressures resulting in such ruptures of the linked up series.

Kassowitz supposes that every molecule of the *stereoplasm* is liable to disruption when stimulated by any mechanical shock, chemical reaction, thermal or electrical radiation, &c., and that the immediate results of the shattering of a given molecule are somewhat as follows. The products of disruption are partly atom-groups which at the moment of disruption display unsaturated affinities of so energetic a character that they split the molecules of atmospheric oxygen brought into the hygroplasm and combine to form saturated compounds such as CO_2 and OH_2; partly atom-groups containing nitrogen, which rearrange themselves into bodies such as proteids, and can be utilised again in building up new protoplasm-molecules, or temporarily stored, or excreted bodies of various kinds.

The combustions involved in the formation of CO_2 and OH_2, and constituting respiration, of course result in the evolution of heat: they are, in fact, explosions, and each such explosion acts as a new stimulus and shatters more protoplasm-molecules, with results and consequences as before, and it is this repeated play which constitutes the propagation of a stimulus—either irregularly in all directions or, if the stereoplasm is linked up more especially along certain tracts (nerves), in definite directions. The accumulation of metabolic products may result in blocking the meshworks, and so impeding the access of oxygen, and the activities slow down accordingly. In the pauses of rest between such destruction changes, the building up of new protoplasm-molecules is accomplished, and this act of restitutive construction is *assimilation*, while the interposition of the new molecules between those already in existence is *growth*.

In illustration of the kind of forces at work in the construction of a new protoplasm-molecule by assimilation, Kassowitz points to such phenomena as selective crystallisation, whereby a minute crystal of Glauber's salt, for instance, in a mixture of the same substance and saltpetre induces the crystallisation of the former only; and to such cases as mixtures of two optically active salts in which the crystallisation of one only is determined by introducing a minute crystal of like optical activity, and to other cases where substitutions of one atomic group by another can be brought about. These illustrations are not intended to serve as examples of what happens, but to show that the forces concerned in chemical attractions may well be those at the bottom of the phenomena of assimilation of like to like in the stereoplasm, or of the building up into the complex molecule of protoplasm of atom-groupings of similar or not very dissimilar nature; and although we have no hope of following the various stages in detail, it is argued that stereochemistry has at least taught that

forms and configurations may result from such molecular phenomena as those indicated.

The arguments to show that the unstable protoplasm-molecule, the lability of which is increased by radiant energy absorbed from without, is itself devoid of oxygen, is capable of reducing highly oxidised food-substances, of absorbing water and setting up osmotic phenomena, of giving rise to metabolites of various kinds, &c., &c., are too long to reproduce here, and I must content myself with one illustration only, of the many given in the remainder of the first volume, to show the application of the hypothesis to special cases.

When the pseudopodium of an *Amoeba* has reached a certain development it suddenly retracts, or rather collapses, for Kassowitz regards the phenomenon as a rapid tumbling to pieces of the molecular structure, owing to stimulation: certain protoplasm-molecules are shattered, atom-groupings of carbon and hydrogen split the molecular oxygen and are at once burnt to CO_2 and OH_2, the heat-vibrations evolved during the combustion shattering more molecules, and so on, throughout that part of the mass. This process exhausted, a period of restitution sets in, and new molecules are built up from the fragments of proteids, carbohydrates, fats and mineral substances at disposal, and become interpolated between those which had escaped destruction, and a new pseudopodium is put out by assimilative growth. Among other arguments for the view that this is really a process of growth, Kassowitz points out that the rate of protrusion of such a pseudopodium, rapid as it appears under a high power, is really not more rapid than the growth of a stem of asparagus, a mushroom or a bamboo.

The most interesting part of the second volume will, for most readers, be those dealing with the questions of variation and evolution.

Having elaborated his theory of the essential structure and mode of working of protoplasm, Kassowitz proceeds to consider the complexities which arise, first, on the differentiation of the nucleus and "germ-plasm," and then on the further divisions of labour involved in multicellular organisation. In these cases the nucleus, internal cells, &c., obtain for their immediate environment, not the outer world, but protoplasm exposed to the action of the latter and modified by it. Whereas undifferentiated protoplasm obtains its supplies of food and energy direct from the environment, the nucleoplasm can never do this, as it never comes in contact with it. Its protoplasm-molecules must select their assimilable materials from the fragments of shattered cytoplasm molecules, and *if any modifications in the modes of disruption and reconstruction of the molecules of the stereoplasm—i.e.* in the "somatoplasm" of authors—*have been brought about by the action of the environment*, the slightly altered atomic groupings and modes of disruption thus put at the disposal of the nucleoplasm—*i.e.* the "germ-plasm" of authors—will affect the building up and modes of disruption of the new molecules of this, and these in their turn react again, and so on.

This short summary of a long argument must serve to indicate the nature of the author's grounds for concluding that Weissmann's contention against the transmission of acquired characters cannot be upheld. It is also the basis on which Kassowitz founds his theory of variation, which latter he regards as always due to the action of the environment, translated in slight differences in the mode of breaking down and reconstruction of protoplasm.

It is impossible to summarise vol. ii. in a review with any hope of doing justice to the criticisms—some of them undoubtedly clever—of contemporary writers on evolution, Weissmann especially coming in for lengthy and severe treatment, particularly with regard to the theory of determinants, and his peculiar views on the meaning of amphimixis, natural selection, and acquired characters. It must suffice to say that Kassowitz offers—assuming the validity of his fundamental hypothesis—what he regards as convincing arguments to prove the essential truth of Darwin's conclusions as to the inheritance of adaptations, of the effects of use and disuse—in short, of the gradually accumulated effects of the environment on the somatoplasm, until the latter is so altered as to affect the germ-plasm, and so fix and hand on the changes. The author regards the theories of pangenesis and their like, equally with such as Haeckel's suggestions as to transmissions of modes of vibrations, as far too complex : his chief objections are that it is to him inconceivable how every structural part could be represented by a physical unit or by a mode of motion transferred to the germ-plasm, and that he cannot see how those who conceive this can get over the difficulty that such vibrations would annul each other as their paths cross, or that the pangenes, biophores, determinants, &c., would get lost on the road. Further, these latter units are, by hypothesis, themselves living, and hence the real difficulty is only shelved.

Kassowitz, however, only demands of his germ-plasm that it be made up of a certain, and probably not a very large number of similarly built and very complex, but not infinitely complex, molecules. He does not suppose that every form-unit of the future organism is represented, but that certain characteristic atom-groupings out of the chemical units of the somatoplasm are utilised in the architecture of the molecule of the germ-plasm, and that in ontogeny these atom-groupings make their effect felt, either directly or indirectly, by the way of correlations of various kinds.

This is true epigenesis. In the developing organism every part is formed anew, from a substance in which none is especially represented. The forms and arrangements which ensue are simply the results of the activities of the atom-groupings already there, working on the materials supplied. When these latter have been assimilated—*i.e.* built up into protoplasm molecules—they exert their cumulative effects on more substance, and also *modify those already present by serving as new environment*—and so the process of evolution proceeds. Every now and again a slight variant gets its play, and the results may be far-reaching ; but, on the whole, the dominant play of the constellations of molecules at work leads to what we term uniformity—a relative term.

While fully appreciating and endorsing Darwin's conclusions as to the importance of artificial selection, Kassowitz appears to undervalue the power of natural selection, curiously enough, because he, like so many others, cannot imagine it to be effective in the early

stages of adaptive changes. He thinks acquired characters must have reached a certain stage of perfection before natural selection can come into play, and argues that when such a stage is reached selection is unnecessary, because so many individuals have already got the adaptation. Kassowitz appears to me to here betray the position of a laboratory philosopher as opposed to a field-naturalist. His own hypothesis points to the laborious accumulation of the effects of repeated stimuli and repeated readjustments : some have survived, others and far more have perished—is this not natural selection ? There seems to be some confusion of thought expressed in implications that natural selection is incompetent to explain the *origin* of variations, which primarily it was never intended to do.

In one or two cases, indeed, the Viennese professor appears to me to have completely misunderstood the position—to an extent so remarkable that the question obtrudes itself, whether the whole argument must not be vitiated into which such misapprehension has crept. To quote one instance only. He admits that the struggle for existence between closely allied varieties, races, or species has resulted in the death of some races, &c., but objects that many plants and animals in the past

" nicht auf diesem Wege ihren Untergang gefunden haben, sondern durch ungünstige äussere Bedingungen und feindliche Einwirkungen, also durch Trockenheit, Ueberschwemmung, Kälte, Nahrungsmangel oder überlegene Feinde vernichtet wurden, dass sie also nicht im Concurrenzkampfe, sondern in einem mit ungenügenden Mitteln geführten Abwehrkampfe unterlegen sind " (vol. ii. p. 131).

But what does all this imply if not selection due to the environment, and the struggle for existence ?

To find a paragraph like this followed by the question —Is it conceivable that such struggle for existence can have led to any adaptive arrangement whatever ? almost takes away one's breath, because it is so totally beside the issues raised by Darwin. The only explanation appears to be that Kassowitz must be combating some foreign misinterpretation of the views of the great master. In spite of these and other faults—I take it, no botanist will accept the explanation of geotropism (vol. i. p. 280)— this remarkable book appears to me to be a valuable contribution to the literature of evolution, well worth reading if only for the numerous criticisms and suggestions scattered throughout its fascinating pages. These, by the bye, are not few—there are nearly 750 pp. of text, and more than that number of notes. It may be that the glamour of the style and the beauty of the theme have led me to pass too lightly over the failings, and to over-estimate the good ; but the good is there.

We have heard much of late about useful knowledge. From the point of view of those who regard all knowledge as "useless" which cannot be directly applied to the material improvement of man, the books before me are indeed of little worth ; but to those who draw distinct lines between knowledge and learning—information and education—no apology will be needed for the conviction that a treatise of this kind is especially welcome at the present time. It is not only instructive, but stimulating to a degree, and of the highest educational value to the biologist of to-day.　　　　H. MARSHALL WARD.

PITMANESE PHONETICS.

Introduction to English, French, and German Phonetics, with Reading Lessons and Exercises. By Laura Soames. New Edition revised and edited by Wilhelm Vietor, Ph.D., M.A. Pp. xxvii + 178 + 89. (London: Swan Sonnenschein and Co., Ltd., 1899.)

THIS new edition of Miss Soames's work, which was designed by the authoress to provide a convenient method of teaching the pronunciation of the English, French and German languages, will no doubt prove useful to those teachers who believe in the advisability of teaching pronunciation by means of Pitmanese. The book is in no sense a scientific treatise on phonetics; the portion which deals with the production of the sounds of the the three languages treated of is simply a very good and useful exposition of the obvious: the main point of the book is the elaboration for teaching purposes of a phonetic alphabet which in many respects falls far short of our ideal of what a phonetic alphabet should be, if such a thing need be constructed for teaching or any other purpose at all, except for the use of scientific students of linguistic phenomena. *E.g.* the authoress uses "*a*" to express the indeterminate vowel-sound: now nobody ever pronounced *the* as "dha"; when it is not fully pronounced "dhǐ," it is pronounced as a German would pronounce "dhö": to write it "dha" is most misleading. Also, the final *-er* in English absolutely = the German ö; *crozier* is pronounced "krözyö," though Miss Soames would tell us to pronounce it "krözhar." She writes *gardener* as "gâdnar": now if we pronounce a true *r* in *gardener* at all, it is most certainly in the first syllable (where it is usually sounded as a faint guttural, a sort of feeble *ayin*), and *not* at the end of the word: *gărdnö* or *gâ'dnö*. Generally speaking, Miss Soames connives at the tendency of modern English to weaken the *r*, and represents it as being far weaker than it really is: in the same way the tendency to lose the distinction between *witch* and *which* is in no way combated by Miss Soames. She spells, most inconsistently, "when," "which," instead of "hwen," "hwich" (hwič), the proper phonetic spelling. Again, to teach a child to pronounce *Sassenach* as "Sasínæk" (Pt. i. p. 109), and *Lochinvar* as "Lokinvar" (Pt. ii. p. 64), is an extremely slipshod proceeding, if it be not a mere solecism on the part of the authoress.

In the German phonetic spelling one or two weak points may also be pointed out. The expression of hard *ch* by *x* is a mistake: this appears to give the ordinary symbol for *ks* in the Latin alphabet a value which it does not possess: every learner cannot be expected to know that the Greek Χ (Russian x), which does possess the value of hard *ch*, has been transported into Miss Soames's phonetic alphabet to express this value. It would have been better to have used the small Greek type and have written *Nackt* "Naχt," not "Naxt." We do not like the adoption of ç to represent final *-g* after front vowels and consonants, as in *Sieg, Berg*, &c., either; a wrong primary impression is again given, and the fact is lost sight of that it is an *h*-sound, not a *k*-sound, which is in question. Why not use the symbol well-known in the transliteration of Egyptian and Assyrian, ḫ, for this sound, keeping χ for hard *ch* and final *-g* in *Tag*, &c.? *Sieg* would

then be phonetically written "Zîḫ." Miss Soames also made ç stand for the *ch*-sound in *manches*; this is incorrect, *ch* here = "hy" ("mănhyez"), a sound quite distinct from the final -ḫ of *Sieg*.

The authoress appended a list of "Loan words used in English," a large portion of which is made up of words and phrases which are not loan words at all; *e.g. ancien régime* (!), *abattoir* (!!), and *Aphrodite* (!!!). On the other hand, such words as *abatis, accolade, aegis*, or *aiguillette* (which is presumably what the authoress means by "aiguille"), *are* loan words. In this list some mistakes of pronunciation occur, *e.g. a fortiori* should be pronounced on Miss Soames's system "ey förtiö'rai," not "fôshiö'rai," a vulgarism which no person with the slightest intelligent knowledge of Latin would ever think of using; *anacoluthon* should be pronounced "ænako'lü'tho'n, sounding the *o*, not "ænako'lyü'than"; *Canaan* "Kanâ'an," not "Keynan"; *Koran* "Karân," not "Kôrân"; and *sheikh* "shêç" (German "Scheech"), not "shîk," which is a terrible mispronunciation. On p. 104, *Eisteddfod*d is given a superfluous final *d*; and on p. 99, the misprint *Bacchas* is noticeable.

On the whole, while this work may be regarded as generally useful for the purpose for which it is intended, it is unluckily marred by a tendency to perpetuate many incorrect and vulgar pronunciations, and even by several mistakes, some of them merely slipshod, others due to ignorance, which the reviser ought to have corrected.

OUR BOOK SHELF.

Psychologie der Naturvölker. By Dr. J. Schultze. Pp. xii + 392. (Leipzig: Veit and Co., 1900.)

IN this study of primitive culture, Dr. Schultze passes under review, from the standpoint of the psychologist, the material which is the common heritage of the anthropologists of to-day. Spite of the suspicions aroused by a sub-title of nineteen words, Dr. Schultze's volume is an unpretentious bit of work by a competent writer, whom no phantasy of construction or love of paradox has led astray from the patient use of authorities and the exercise of a sober judgment. Dr. Schultze's first essays in his subject were printed some thirty years ago. The present contribution is self-contained, though for its author it is but a part of a larger whole, preluded by physiological psychology and a treatise on the psychical life of plant and brute, and to be followed by a study of childhood. It is naturally evolutionist in conception, although the descriptive continuity which the author maintains is accompanied by the refusal to allow that the derivation of apperceptive consciousness from associational, which in the interests of a unitary view of nature he might desiderate, has been adequately made out. A feature of the book is the use made of English authorities. Not only Spencer and Tylor, but McLennan and Lubbock supply the writer with important doctrines, *e.g.* in his account of the evolution of marriage. Mr. Sutherland's "Origin and Growth of the Moral Instinct" is recognised as having anticipated Dr. Schultze in much which he would have been glad to have said, but, far from being dismissed with a *pereat*, is summarised in an appendix. It is on fetichism and animism that Dr. Schultze is most at home. Not that there is not much else of interest on the alleged superiority of vision among savages, on the concreteness of their philology, on the relation of rhythm to melody, on the difference of the sexes in regard to the sense of smell, on the evolution of the sense for landscape, and the like. But to the topics of his earliest studies he returns as to a first love. On the soul-theories of savages and the corresponding eschatology he writes convincingly. The plurality of souls in pulse and blood and breath and shadow, the gradual elimination of some of these and the syncretism of the rest, the place of the dream image in the evolution of the cult of *manes* and in the selection of totems, the literal and unsymbolic character of the latter, the order in which the heavenly bodies enter into primitive worship—these are the points on which Dr. Schultze compresses year-long work into moments of insight and selective description. Believing, as he does, that Germany has a colonial future in direct contact with primitive stocks, Dr. Schultze offers his essay to the understanding of the savage as a help forward to the achievement of the educational mission of his country. A pious gift.　　　　　　　　　　　　　　H. W. B.

The Study of Bird-Life. By W. P. Pycraft. Pp. 240. Illustrated. (London: George Newnes, Ltd. 1900.)

THIS little volume belongs to "The Library of Useful Stories," now in course of issue by the publishers; and although it must have been difficult to compress a general review of the leading facts of bird-life into such a small compass, the author may be congratulated on the success of his attempt. As Mr. Pycraft is a morphologist rather than a systematist, it would naturally be expected that he would incline rather to the morphological and phylogenetic aspects of his subject, and this we find to be the case. We have, for example, an excellent chapter on the morphology of the bird's wing, while two others treat of avian pedigree, and a third is devoted to the distribution of birds in space and time. Perhaps the most specially interesting chapter in the volume is the one dealing with the flightless birds and their fate, since this is a subject on which the author is peculiarly qualified to speak with authority.

Although, of necessity, written from a purely popular standpoint, the volume contains many passages which are well worth the attention of the scientific ornithologist. If there be a fault, it is the introduction of irrelevant matter, the place of which might have been better occupied by details pertaining to the subject in hand. And if a second edition be called for, the author will perhaps be inclined to modify the statement in the tenth chapter, that "the kind of rock" in which bird-remains are found is sufficient to give a notion "of the bird-life of that particular period of the earth's history."　　　R. L.

An Introduction to the Differential and Integral Calculus and Differential Equations. By F. G. Taylor, M.A., B.Sc. Pp. xxiv + 568. (London: Longmans, Green and Co., 1899.)

THE appearance of still another treatise of this kind shows how earnest and how prevalent is the desire to introduce students of physics to a knowledge of the calculus at as early a stage in their career as possible.

The author has studied simplicity of treatment, but has evidently striven to secure accuracy as well as clearness and distinctness in his exposition of the principles of the subject. A special feature, which will be of great advantage to the ordinary student, is the detailed discussion of numerous examples.

Interspersed throughout the several chapters the student will also find an abundance of not too-difficult exercises carefully graduated and with answers appended.

A fair and not excessive amount of space is devoted to the subject of curves, and the illustrative diagrams are distinctly drawn.

The section on the integral calculus concludes with applications to volumes and surfaces of revolution, centroids, and moments of inertia.

The last section of the book forms a good introduction to the methods of dealing with ordinary differential equations of the first and second orders.

ENGLAND'S NEGLECT OF SCIENCE.

JUST before the first movement organised by Lord Roberts there was probably not one thinking person in England who was not ready to vote for an immediate change in all sorts of English methods of doing things. Consequently everybody was willing to listen to the advice of men who had for years been crying in the wilderness and prophesying disaster. Now, however, that we have worried through our military trouble, we shall probably feel so much ashamed of our intense fright as to put aside most of our desire for reform, and even to have less thought of it than before the war began. It is, therefore, the duty of those who have earned the right to a hearing to prevent the nation from sinking down into its sleepy acquiescence with old methods of working; and I am glad to see that Sir Norman Lockyer, in his speech at the Royal Academy dinner, referred to scientific education as a great, necessary line of defence of our country, secondary only to that of our naval and military forces. Again, two articles have appeared in the *Kölnische Zeitung* (March 10 and 11), which criticise our manufacturing and business and military want of method with an unsparing pen. The German writer and many English writers seem to think that we ought to copy Germany. Nobody can feel more than I do the great necessity which exists for reform; but I think that our reform must be far more thorough than anything which can be regarded as a mere copying of Germany; the methods which we adopt must be English methods, invented by Englishmen for Englishmen. If our methods are to help to lead in the future to a history comparable in glory with the history of the past, there must be a great commonsense reform in education in England from top to toe. My friends, Profs. Ayrton and Armstrong, and I have so often pointed out the deficiencies of England in matters which we have carefully studied here and in foreign countries, that I hardly know whether an idea on this subject is my own or one of theirs; I do know, however, that we preach often on this subject, and that we seem to be much attended to.

One thing that seems to be quite exasperating is that almost all the most important, the most brilliant, the most expensively educated people in England; our poets and novelists; our legislators and lawyers; our soldiers and sailors; our great manufacturers and merchants; our clergymen and schoolmasters, are quite ignorant of physical science; and it may almost be said that in spite of these clever ignorant men, and men like them in other countries, through the agency of a few men who are not ignorant, all the conditions of civilisation are being completely transformed. I do not merely mean here ignorance of the principles of science, I mean also ignorance of all those methods of working which come from experimental and observational scientific training. The great men go occasionally to popular scientific lectures (as they go to the Royal Academy), and they think that they comprehend something of the latest scientific discoveries because they have seen some fireworks and lantern slides; they are genial to scientific men when they meet them at dinner parties; but, in truth, scientific men are as much outside their counsels as sculptors or painters, or musicians or ballet-dancers. Among these great men a few visits to Albemarle Street are sufficient to create a reputation for science. I wish to show that this ignorance of our great men tends to create ignorance in our future leaders; is hurtful to the strength of the nation now, and retards our development in all ways.

These great men really direct the building of ships of war, and the creation of munitions of war; that is, they select the men who have to do these things, and they also lay down the unscientific rules which prevent their selected men from doing their work scientifically.

I will give an example. They order that the building

of five line-of-battle ships shall be started immediately. The scientific constructor knows that he ought to throw away—waste—100,000*l.* in making experiments to find out how the older type of ship may be greatly improved. But his superiors have made the rule that for money expended there must be something to look at. Hence no experiments may be made, and the constructor starts at once to expend five millions of pounds on building ships which are nothing like so good as they might be made.

Other examples. For many years huge guns were built of tubes. It was known to the few scientific men who can calculate about such things—the men who are never consulted—that it was not possible to turn and bore those tubes with the accuracy required by the theory. It happens that nature applies a correction to a wrong method of manufacture, and so these guns are not useless. It is quite well known that a little science and expensive experiment would cause the present wire method of manufacture of guns to be discarded for a simpler, quicker, better, cheaper system. The water-tube boilers, so numerous in our Navy, have proved as worthless as the best scientific men thought them from the beginning, and possibly now it is absurdly assumed that all water-tube boilers are useless. The construction of efficient submarine boats was possible thirty years ago. Many electrical and mechanical engineering appliances that might be very useful to an army or navy have never yet been tried under the direction of competent engineers. Above all—and this includes everything—men of scientific training are not chosen for the Government, civil, and naval and military posts where such training is necessary.

If our leaders were merely unscientific—if they were merely like Boers, and had no scientific knowledge—it would not be so bad, for they would probably appoint scientific men to posts in which a knowledge of physical science is needed, and they might accept the opinions of scientific experts. Even if they were like savage chiefs there would possibly be equal chances among all candidates for posts ; but, unfortunately, it is as if our leaders possessed great *negative* knowledge of natural science, and as if a man's chances of being appointed to a scientific post, or of having his advice listened to, were in inverse proportion to his scientific qualifications.

Scientific men look around them and see that everything is wrong in the present arrangements, but they also see that it is useless to give advice which cannot be understood by our rulers. And, indeed, I may say that when by accident a scientific man is appointed on a committee, there is a negative inducement for him to do anything.

Many men enter the services by examination. In some cases the examination is supposed to be in science. In truth, the scientific habit of thought, the real study of science, the very fitness of a boy for entrance to the service, would unfit him for passing these abominable unscientific examinations. For some posts—the Royal Artillery and Engineer services, for example—further scientific food is provided by the Government after a man enters. If one wishes to hear how evil this system of pretended education is, let him ask the opinion of some of the professors who are condemned to help in carrying it out. The whole system is foolishness from bottom to top, and the men prepared by the system cannot see how abominable it is even when they are afterwards trying to improve it.

But however harmful the present state of things may be for the Government services, I think that it is much more pernicious for the country at large. We see that the greatest intellects of our time have been developed through an education other than scientific ; and, as nobody can commend it for the mere knowledge given at school, it is commended for its importance in mental training.

It has been so often asserted by parrots, that many people do really believe that only mere mental training need be given until a boy is sixteen years of age. When one hears such a statement for the hundredth or thousandth time, he sometimes wonders if anybody ever does think for himself. Why, the early period of a boy's life is the time when he is not only getting mental training, but also collecting the largest part of all the knowledge that he ever will possess of the world into which he has come. So great is this stock of facts and theory, that when he looks back upon his life in old age he can hardly find that he has added much to it in the intervening years. Is he a musician in after life ? then he certainly learnt his skill, acquired his touch, trained his ear, and learnt thousands of airs in early youth. Is he a poet ? it is to his earliest efforts that he looks back most fondly, and it was in his early youth that he learnt off by heart all the poetry that he really knows well in after life. He learns to read and write and cypher with ease and readiness ; is this mere training of the mind ? This craze for mind-training is really the worst thing that has happened to the hurt of children. It does not seem to be known to the mind-trainer that a child's mind grows most healthily when let alone—when the child is picking up knowledge in his own way. Give a boy a chance of seeing things for himself, and direct him as little as possible. Is there any kind of knowledge likely to be needed by him in after life ? let him, when quite young, have some chance of picking up something of it for himself. He learns about people ; he cannot help it, as he lives among people.

I take it that whatever kind of knowledge the race has been in the habit of picking up in youth is more easily picked up than any other by a boy himself. A boy takes to thinking for himself so naturally that the greater parts of some vile systems of education seem to be the destruction of this habit. Yes, education often means merely training a boy out of the way he *would* go into the way that we poor creatures think that he *should* go. And hence it is that the boy whose education is neglected, but who has chances of seeing things for himself, has often a much better chance in life than the well-trained prig.

Now there is a kind of knowledge greatly needed in life, that knowledge which is enabling us to fight with and use the powers of nature as they never were fought with or used before our time. The race is not accustomed to picking up this kind of knowledge, and so there is this one case in which artificial help to the child is absolutely necessary. Natural phenomena are complex ; let him have a chance of using apparatus that will simplify these phenomena for him. It seems to me that natural science is almost the only study in which instruction from a father or teacher will not obstruct a boy's own natural method of study. And see how many ways of study are offered by it to a boy. Some of the sciences are greatly observational. If he is fond of abstract reasoning, he attacks things from the mathematical side. If he is fond of fireworks, he can attend popular lectures. If he loves to make and fiddle with apparatus, and use it quantitatively, he has an altogether new method of study. He may choose which method he pleases ; the study is utterly unlike a series of tasks ; he does not get to think of a duty as something disagreeable ; and, above all, he is encouraged to think for himself. Instead of constant correction, criticism, and reproof or punishment because he will think for himself, he is encouraged to consider that opinions which he disagrees with are to be criticised by him. If he feels that it is quite hopeless for him to follow abstract reasoning, say about a whole being greater than its part or the ratio of two incommensurables, or justification by faith, we reply to him—Yes, my boy, you have a good healthy mind like 98 per cent. of all English boys ; it is quite impossible for us to make a seventy-year-old Alexandrian philosopher of you, thank God ! time enough for you to do that for yourself when you have finished your educational course.

I say that this observational and experimental kind of study is almost the only one in which it is possible for a teacher to guide and instruct without doing harm—and it is very important that a boy's studies should be guided. Take the very clever boy who dislikes the study imposed upon him, and who takes earnestly to something else, his own choice, in which be bas no guidance. See how he becomes a "crank." A man who might be of the salt of the earth if he could only co-ordinate his opinions with those of other people, a leader among men of thought ; he loves to differ from all other people, and wastes a valuable life in disputation.

I know that many readers will find it difficult to consider this question ; they will find it impossible to see things from a new point of view. As a rule a man has no point of view of his own, he never thinks for himself except about certain matters that only concern himself. Even a learned man thinks, not on the subject of his learning, but about his special methods of cataloguing his knowledge, and of course it is only from this that he can get any mental enjoyment. The dullest boy thinks a good deal, and even the average man, although thinking for himself has been repressed in him all his life. We ought to call all such people pedants, because they never really think about things of general interest to the world. It is extraordinary how general is the impression of everybody that he really does think for himself and comprehends what he says. At the age of fourteen I wrote an excellent little essay on Chaucer ; I recollect now that my knowledge of Chaucer was confined to a few of the well-known extracts.

The opinions of educated young men change with the moon, or rather with the period of publication of the monthly magazines. A mathematical teacher uses the same fallacious logic in some demonstration year after year, and at length finds out his mistake from somebody else. Learning seems to destroy all power to think. From 500 A.D. to 1453 A.D. the scholars of Constantinople, with all the learning of Greece and Rome, produced not one original work.

I think that for a very clever boy any subject of study is good enough, although not so good as natural science. But Sanscrit, Chinese, or any other language and literature, or astrology or divinity, is just as good a medium as Greek or Latin, if all the best men of his own time happen to use the same medium, and if it enables him to come into mental contact with great men. But what of the other 98 per cent. of all boys—the average boys?

The men who frame schemes of education really frame them for boys such as they themselves were. Anybody who cannot follow such a scheme is said to be stupid, and he is so often called stupid that he actually gets to think himself stupid. In this nineteenth century we do not wish, as in the time of Erasmus, to produce merely a few learned men. At all events, if parents pay largely for education, we do not think it fair to send back 98 per cent. of their sons with the contract unfulfilled on our part. Think of 100 boys being sent to a bootmaker who had only one kind of ready-made boot of one size. He sends ninety-eight of the boys back with feet so hurt by trying on that they can never wear anything but slippers all their life after ; he keeps their money, and compels the boys and their parents to take all the discredit of the transaction. Christ's curse is on the schoolmaster when he calls a boy a blockhead.

It is a very curious thing that when a boy has been called a dunce a number of times he actually gets to think himself a dunce, and in after life never blames his schoolmaster ; he has only praise for the system

of education. Men who have never been able to do more than quote tags from the Latin grammar, or get beyond the Asses' Bridge in Euclid, are usually quite enthusiastic about the value of the orthodox education in the training of the mind, and so we find engineers and other illiterate persons advocating classical education. A donkey might just as well brag of the enormous advantage it was to him in having once been kept about a racing stable. But a much more curious thing is the praise given by clever mathematical physicists to the wretched system of teaching of Euclid which wasted their youth. A well-known and exceedingly able and ingenious scientific man praises the school teaching in physics and chemistry which he had as a boy from a certain master of his, and yet everybody who knew master and pupil knows that the pupil became a scientific man in *spite* of, and not through, the teaching of his master. Even if such clever men were right as to suitability of a system of training for themselves, they have no right to assume that it is right for the other 99 per cent. of boys at school.

Classical education gets all the credit that ought to belong to the other kinds of education that usually accompany it. A boy is at a good public school at which healthy, moral, manly training of all kinds is given to the usual manly type of boy. All the best masters are probably good in classics. The boy's own prizes are for classics, because there are not often scholastic prizes for anything else. Success in classics has been always put before him as the highest kind of success. The boys whom he worships are all good in classics. Of course, classics gets the credit for everything, including those things that are good in *spite* of the classics. Even good manners and tact and amiability, and I might almost say good batting and bowling and fielding, are thought to be due to the classics. The defenders of classics are numerous, and miss no chances. A scientific friend of mine, before a royal commission, commended the study of Greek, because the Greek alphabet is so much used in mathematics. Surely for such a purpose Chinese is ever so much more valuable, as there are many more letters. Again, it is said that the study of classics helps one greatly in the study of modern languages. These defenders forget that Russians and Japanese are the best of linguists, and yet they seldom learn any Latin or Greek. It is strange also to find so many English boys, trained for years in Latin and Greek, who seem to find insuperable difficulties in learning a modern language. In any case, I am inclined to think that there is too much inclination to force boys to learn modern languages. Some boys learn easily ; for them the study may be good. Others learn with extreme difficulty. Had they not better study something else?

Everybody is aware of the enormous difficulty of introducing a new invention, however valuable, if it involves the "scrapping" of much existing machinery. Thus, electric methods of working the District Railway have not yet been introduced. The comfort of railway passengers everywhere is only slowly being attended to. For this reason electric lighting proceeded slowly in England and quickly in America.

Now all the machinery of a school head master is fitted for the teaching of Latin and Greek. Every master is able to teach Latin well to clever boys, and everything good for mental training for clever boys in such teaching is well known to him. These men with capital so invested look with alarm on every new footing gained by science in schools, and with a wisdom gained by experience they introduce what *they* call science teaching, adopting methods which are such as can only disgust boys and their parents with the new study, and then they point to their want of success as a proof that the study of science affords no good education.

The prospect is very dismal ; for the capitalists whom we fight against, whose interests we directly attack, are not only some of the very cleverest men of the country, but they have the ears of nearly all the other clever men.

In the time of Henry VII. the new learning fought and conquered the schoolmen, and England soon became covered with good grammar schools. Then mathematics came gradually in, fighting a hard fight till it has made its way and established itself—not on equality terms, but on terms of sufferance and recognition. To meet modern wants, to equip our men for the fight of to-day, we find that it is absolutely necessary to introduce the study of physical science, and lo! we have opposing us the combined forces of classics and mathematics, each with its own kind of weapon. The weapon of the mathematical pedant is the more dangerous, for he says that he already represents science.

This teaching of pseudo-science in schools has created a manufacture of teachers. At all the universities we are now manufacturing science B.A.'s and B.Sc.'s because there is a new profession where money may be earned by the holder of such a title. This manufacture is called scientific education, and our real scientific men, pleased with the name, pleased at any experiment in scientific education, afraid that if they object there will be no education whatsoever in science, weakly give their countenance to it. To illustrate what I mean:— At the greatest of our universities there is an examination in which experimental physics plays an important part. A friend of mine coaches men for this examination. He tells them: "Listen to my coaching, read the books as I tell you, take care *not* to attend the physics laboratory. For in one day's reading you will get to know all that there is in thirty pages of the book; you may spend a month at the laboratory and you will have gained practically nothing to fit you for any possible kind of examination. The laboratory does not pay." Of course he is right, but if mere learning, if mere knowledge of certain facts, mere power to pass an examination, are what is aimed at, surely there is no scientific education here. My friend asserts that the system by which he earns his living is abominable. The whole thing is so wrong that one wants an earthquake or a fire, one prays for wholesale destruction of the easily working examination machinery.

I remember teaching physics at a school in which the time for science was so limited that only one half-hour's lecture per week could be given to the best men in the school. There were about 100 of them, from the sixth and fifth forms. Some of them are now leaders of English thought. Well, they were actually examined once a fortnight—a paper examination, lasting an hour. Of course, they were not examined on the two lectures; they were really examined on two chapters of the text-book. I am told, and I believe, that in many of the best girls' schools science is supposed to be taught by a teacher reading things from a text-book, the girls taking notes. I should think it an excellent system if girls are required to pass the usual examinations.

Examinations are said to be in mechanics or dynamics, or mathematical physics, or mechanical or civil engineering. They are not; they are fraudulent substitutions of the stupidest kind of mathematics for these sciences.

Assume something or other to be true, that the co-efficient of friction is constant, for example, or that a specific heat is constant, and, after covering the paper with easy mathematical exercise work, arrive at mathematic expressions which are as worthless as the mental training is bad. What a wonderful and useful weapon one possesses in mathematics! In the hands of a man like Rankine, or Kelvin or Maxwell, it removes mountains of difficulty. What a stupefying and useless weapon it is in the hands of a skill-less person who cannot think! And our examination systems and methods of education seem framed to cultivate one Kelvin to 10,000 of the pedantic non-thinking users of mathematics.

My theme has been the necessity for a complete change in our system of early education of everybody. The necessity is specially great in the case of the captains of industry. Many people think that if men are to be taught the scientific principles underlying the proper conduct of business or manufacture, it is only necessary to establish Technical Schools for them. When I was young I remember that there were many agricultural colleges in Ireland; they have all but one been failures. Why? Because the entering pupils were not fit to receive instruction. Instead of their having been prepared for instruction by their earlier education, this had done as much as possible to unfit them. We have just this sort of experience in our Technical Colleges. Great boys enter them, and it is difficult to find out what are the scraps of Euclid and mechanics known to these boys on which one has a chance of building technical instruction. It would almost be better to send such boys direct into practical work; they would probably do as well as the average workman; their fathers' influence and money would get them superior positions, and in a country like England they would do as well as their competitors in business. Yet there can be no doubt that it is of the utmost importance to our country, if we are to retain our supremacy in manufactures, that all managers of works, and many of the superior persons employed in large works, should be scientific men, who are also well experienced in the applications of science to their particular industry. But this is not all. I have heard it said, quite truly, that for a great mechanical engineering works what is needed are well-trained managers and foremen, the best labour-saving tools, and an army of negroes as workmen. I am inclined to think that this statement is true; but there is something to be said for the employment of well-educated, intelligent workmen. First, because they are citizens of the country having votes; second, because I believe that all invention comes up from the common workman These men make thousands of observations, which somehow get to their superiors, and it is through these tha inventions come unconsciously; an inventor makes use of ideas received from hundreds of men; the invention is truly his own, but he receives suggestions, unconsciously, from the men who work with their hands at the bench and in the machine shop. If then, I am right, the manufacturing country that depends upon a few good managers and an army of unintelligent slaves will fall as the Roman Empire fell.

Now a workman's intelligence must come through his trade, else he cannot be happy; and if he is unhappy in his trade he cannot be a good citizen or an efficient workman (from the above point of view). At present we pitchfork many boys into a factory, and depend upon the good nature of the workmen for their learning their trade. It used to be that a master taught such a boy his trade as a member of his own family. This personal teaching is no longer possible; but nothing has taken its place. Attendance of apprentices at evening classes after a hard day's work is quite out of the question for all but a small number of very clever young martyrs who sacrifice, not merely their own health and comfort, but the comfort of their families and their duties as citizens. I have myself publicly suggested several times a remedy for this state of things, which has been praised by competent persons, but it seems to me that it is hopeless to expect any adequate remedy to be applied until the influential people of this country are made to see the gravity of the present position.

The great remedy for all our troubles lies in convincing all influential people in this country that we really must make great radical changes. I have known the subscribers of money to a large technical college in England (the members of its governing board) to laugh, every one of them, in private over the idea that such an institution could do any good to the trade of their town.

How could it possibly do any good when there was such a spirit of unbelief among such people? We must create in England what already exists in Germany and France, and to some extent in America—a belief in the importance of scientific training everywhere. At present there is utter unbelief, and it is due to a bad system of education, which keeps everybody out of sympathy with everything scientific. It is terrible to hear our designers of bridges and steam engines and dynamos and great engineering schemes laugh at science and calculation, especially when one knows that foreign engineers are sneering at our best men ; but it is well to know that, in spite of their laughter, our engineers are doing their very best to make use of all the true science that they have ever learnt ; it is like gold leaf—very thin, but it serves a useful purpose. What they see clearly is the uselessness to God or man of such a so-called scientific training as they themselves had ; they do not dream that there is a real scientific training possible by which useful mathematical and other weapons for solving all sorts of practical problems, handy to use and always ready, may become part of the mental machinery of the average man.

Four hundred years ago, reading, writing and cyphering were taught badly, and practical men looked upon them as things good to forget, things good for priests. If a layman could read or write, he was probably a useless person who, because he could not do well otherwise, took to learning. What a man learnt was clumsily learnt ; if he learnt much, he was fit for nothing but learning ; usually he learnt little with great labour, and made no use of it ; therefore reading, writing and cyphering seemed useless. Do they seem now so very useless ; now that everybody can learn them fairly easily ? It is not so easy now to say that a man is useless merely because he can read, write or cypher. When I was an apprentice, and no doubt it is much the same now, if an apprentice was a poor workman with his hands, he often took to some kind of study, which he called science. In fact, science got to be the sign of a bad workman. But if workmen were so taught at school that they all really knew a little science, science would no longer be laughed at. When a civil engineer or electrical engineer fails because he has no business habits, he takes to calculation and the reading of so-called scientific books, because it is very easy to get up a reputation for science. The man is a bad engineer in spite of his science, but people get to think that he is an unpractical engineer because of his scientific knowledge.

Germany has an enormous advantage just now in this, that all thinking Germans, all influential men, believe that their great success in commerce and manufacture has come through physical science. Every manager and foreman, every captain of any kind of industry in Germany and Switzerland, has passed with honour through the science classes of a great technical school. The money that used to flow towards religious institutions now finds its way towards the greater and greater development of scientific education, so that Germany is getting covered with universities of science.

The open-hearth process has enabled German ores of iron to be used in steel manufacture. The war-earthquakes have stirred up the German people to new life, have produced enthusiasm, and made all kinds of ambition respectable. Any one who saw such a tumble-down, poverty-stricken town as Hanover forty years ago would not recognise it now. There are miles of streets of the brightest shops in Europe ; at any time of the day or night one can read a small print newspaper in these streets ; the streets throng with traffic, and the electric tram-cars have extended the city far into the country ; and so it is in hundreds of towns, and manufactures flourish in thousands of places where the hare and partridge used to have the scenery to themselves. I do not think that the progress of

Germany would have been half so rapid had it not been for the scientific education of the German leaders, but it is absurd to say that all this progress is due to science. The fact is that the whole world is developing its natural resources. England had the start ; every country that has coal and iron, or their equivalents, is competing with England. The countries of greatest natural resources can afford to neglect their scientific education longer than others ; but, sooner or later, knowledge and method and character must tell. If countries are equal in their natural advantages, victory must remain with that one in which there is the best education.

I have hitherto been reviling only the higher education in England. Until quite recently there was no primary education to revile. Let me put before my readers a true contrast. In Scotland, at any time during the last 100 years and more, if in the very poorest parish there was a boy of promise, a boy who showed a fondness for reading, for learning, for taking in what then and now goes under the name of education, a fondness for coming into contact with great minds through books—the success of that boy in life was absolutely sure. However poor his parents might be, however remote his humble home might be from civilisation, he was sent to the university, and got his chance. His nation gloried in his success, even if his own poor country had to be left by him for the richer field of England. Of all the great doctors and ministers and scholars of Scotch blood now to be found in London, only a very few can say that they were not exceedingly poor in their youth. Now contrast such a boy's chances with those of a clever English boy some years ago. Why, until the ever-to-be-praised Science and Art Department gave him a chance, a poor English boy, however promising, was compelled to eat his heart out in unavailing regret, was taught that it was a sin to think of bettering his condition, was taught that a decent education was as remote for him as for the cattle he tended.

I believe that this difference was due to the fact that in Scotland everybody thinks well of a good education, of knowledge, of mental power, because he himself can think, whereas in England education is looked upon with contempt, because there is not one labourer in a thousand who can think.

I do firmly believe that the Prince Consort saved this nation from utter defeat, and that if we are not yet to be defeated we must do as he would have continued to tell us to do. Had he lived till now, this country would not merely have the beginnings of a development of art and science ; it would be covered with educational institutions whose most important object would be scientific study, a secondary leaning to literature not being neglected. As it is, we have the merest dust of his mind expanded into a wonderful Science and Art Department, which is criticised adversely only by the very ignorant or the very prejudiced. Only people like myself, whose whole life has been a pæan of gratitude for what that department has done for me and mine, who have seen in thousands of cases that it has redeemed otherwise wasted lives with enormous benefit to our industries, are really in a position to imagine how much that great man might have done for us if he had lived.

For one thing, just as the Science and Art Department is the envy and admiration of foreigners ; just as it is an English institution, made to fit England and no other country, so he would have developed scientific education in England on lines utterly different from the soul-destroying system of Germany or its imitation in America.

Consider a scientific German as you know him. Say that he is twenty-three or twenty-five years of age, and he is about to enter business. From the age of seven or less he has trudged to school, perhaps at 7 o'clock in the morning, with a bag of books of half his own weight. He had a short interval for dinner, and went on to

6 o'clock at night. And he went on like that till now. There is no fact in all his school books that he has not heard a thousand times. He has had Goethe's maxims so drilled into him that he is "thorough" in every detail. I can imagine one Englishman in a hundred, after such a training, patiently turning over the muck heap of his knowledge; his eye would not gleam with any enthusiasm, but rather would glaze with envy and jealousy at the undeserved success of quite ignorant persons. And yet he would have knowledge, and know in his way how to use it; and it is because Germany has so many thousands of men trained in this way that she is certainly beating us to-day. They may be rather heavily loaded with learning, and I know that decently taught Englishmen who spent less than half the time at studies twenty times more interesting would beat them hollow in manufacture or research, would be the reverse of dull, and would be good citizens; yet the Englishmen I want only exist as yet by ones and twos, and such Germans are numerous. But just think of it! Here we are, a hard-headed, obstinate, cool race of men, who have had no end of chances in our safe little island, whilst our enemies were fighting among themselves, with coal and iron and the influx of good foreigners to set us first in the new field, and we have more than half of all the wealth of the world, and all that is needed for our keeping our good things is that we should believe them to be possibly evanescent; that there really is a chance that some better equipped nation may take them away from us, and therefore that we ought to prepare ourselves to fight for them. We have many chances in our favour and we hardly use them; the competing foreigner is very energetic, and cultivates his smallest chances.   JOHN PERRY.

HUMAN BABIES: WHAT THEY TEACH.

AN investigator anxious to obtain information as to the relationship of a particular species puts the question "What characters do the young stages exhibit?" and in order to answer that question he makes a study of the developmental phases exhibited by those stages. He may argue that if he finds certain characters in the young stages indicative of, and adapted to, habits of life which the adults do not possess, then there must have been a time in the ancestry of the species when such habits of life were of particular value, otherwise they would never have been developed. Or he may simply give, as the reason for his method of research, the concise statement "ontogeny repeats phylogeny," or he may hold to the theory of acceleration of development —which is more than a theory, because it is an actual fact of palæontology—that the characters of youthful descendants tend to become the characters of youthful ancestors, thus producing specific diversity, without the necessity for a theory of natural, or any other form of selection, merely by inequality in the rates of developmental acceleration in different stocks. Wherefore *vice versâ* the characters of youth must at one time have been adult characters; and their differences from those of the adult indicate the degree of different environment under which the adult ancestors lived.

The manner of expressing the reason for a method of research may vary; the method itself remains the same. To know the past history of an organism, study the young. That is a method of universal application. It is the guiding principle of all researches into the past history of organic beings. It becomes then equally applicable to man himself; and in that way the human baby becomes an object of scientific attention. To study the human baby in this manner, the aid of photography is important; it gives a permanent record of what would otherwise be forgotten.

The early attempts of a baby in the matter of progression are particularly instructive. The bipedal gait is not attainable, indicating that the bipedal ability of the human being is of quite recent acquirement. What the child does show is either a truly quadrupedal method of progression, as in Fig. 1, which is also said to be common among children of uncivilised parents, or a kind of falsified quadrupedal progression on the hands and knees, which obtains generally among children of civilised parents, owing no doubt to impediments of clothes, and to over-coddling. Both methods of progression point to the same conclusion, though, of course, the former is the better illustration—that the ancestors of man were animals accustomed to a quadrupedal gait.

The influence of this quadrupedal gait of the ancestors is very strong. The child really has to unlearn it, and to readapt its hind limbs before it can attain the bipedal method of progression. The necessity for such readaptation, and the difficulty of acquiring the balance which progression on hind limbs demands, make the child's early efforts at walking so difficult. Observe a child just able to balance itself momentarily on its hind limbs. The insecurity of the position is shown by the attitude of the arms—outspread to help the balance, and by the feet being planted widely apart. The imperfection of the hind limbs for a bipedal gait is particularly noticeable.

FIG. 1.—Child ten months old, on garden path.

The legs are not straight, but they are considerably bent at the knee. That bending is incorrect for a bipedal gait; but it is a necessity of a quadrupedal progression, and it is just the feature seen when a four-footed animal, such as a cat, is induced to stand on its hind legs. In learning to walk on its hind legs the baby has to make many alterations in the anatomy of its hind limbs to fit them for their new function; and the human ancestors, in order to pass from quadrupeds to bipeds, must have had to do the same.

There is another feature noticeable in regard to such a child in its first attempts at walking—the semi-clasped position of the hands. That is natural, it may be said. Certainly it is, but nevertheless a natural feature requires an explanation, and may be of particular significance. Such is the case here; the semi-clasped position of the hands is naturally and instinctively assumed because the human ancestors had for so many generations been bough-grasping animals, quadrupeds who lived among trees, who particularly used their hands for grasping boughs. Had they used the hands in a manner which always produced extension, then the extended position of the fingers would have become habitual.

Further evidence of the particular character which generations of bough-grasping ancestors have given to the hands of children may be obtained in this way. Get a series of school-children and ask them to hold their hands out straight. The failure of the majority of them to put out the fingers without some indication of the bough-grasping curvature will be very interesting. In some cases, especially among the younger children, the inward curve of the fingers will be very noticeable ; and their inability to fully extend the fingers will be marked. A record of this inward curvature of the fingers may be obtained by photographing the extended hands when held against a dark background.

Even better evidence of the inherited bough-grasping instinct is afforded in Fig. 2. The child, about twelve months old, has picked up a flower-pot, and it has done so by dabbing the hand down upon it in the manner in which a monkey would catch at a branch. It has not made use of the thumb as an adult would do ; but it has caught the rim of the flower-pot between the fingers and the palm of the hand, and in that manner has raised it up to its mouth. The sympathetic grasping attitude of the other hand may not be without significance ; for although an arboreal animal like a monkey can sustain its weight on one hand, yet there would generally be a tendency to grasp with both hands at the same time in order to relieve the one arm of all the weight.

Fig. 2.—Child grasping a flower-pot.

A sympathetic action of this kind is very noticeable among children in regard to the use of the legs, and similarly it may be referred to the habits of arboreal ancestors. If a young child be put to hang on to a rope, which it will do very well long before it can support its weight in the ordinary human manner on its hind legs, or if it be merely lifted up by the hands, it will at once show a disposition to swing up its legs as if to catch at something. And this would be very natural in an arboreal quadruped. As soon as it grasped a bough with its arms, it would swing the legs up in order to grasp with the hind hands (the feet) either another bough, or in many cases the tree-trunk.

The inherited effects of grasping tree-trunks, or limbs with the hind hands are particularly marked in a young child. There is first of all, common to most babies, more or less of the bow-legged character which such trunk grasping would produce in arboreal animals. And then if a quite young child be held up so that its feet touch the ground, it will be seen that the outer portions of the feet rest on the ground, while the soles of the feet are not in position for being put flat, but are more or less opposed to one another in the manner suitable for trunk-grasping. Often, when the baby is lying down, the great flexibility of the ankle joint may be noticed ; and the child will be seen to do, without an effort, what it would be very difficult for an adult to accomplish—it will,

without bending the knees, bring the soles of the feet flat, opposite to one another. It is quite a common thing for a baby to turn the sole of its foot so that across the sole is in a straight line with the inside of the leg.

One habit after another, one action after another which a child performs may be seen to be quite out of keeping with what may be called human instincts, but exactly in accordance with the habits of arboreal animals. And so there is an accumulation of evidence, on the ontogeny repeating phylogeny principle, that the human ancestors were monkey-like animals, arboreal in their habits. One of the first things that the human baby does is to climb, and to climb persistently. It will climb its crib, or a footstool, or the fender, and particularly the stairs. Given a fair chance, and it will develop a perfect mania for stair-climbing and a bump of locality as regards the position of the stairs in the household geography—if such a bull may be permitted. Then it will make for the stairs on all occasions, to climb with crows of pleasure. It may experience tumbles, when it will lie and howl, not so much on account of injury as at the unexpectedness of the catastrophe. But on recovering it will at once make for the stairs again, showing how strongly the climbing instinct is developed.

And the climbing instinct lasts till later in life. Young boys, and girls too, must climb. The stairs themselves have become too small for their efforts then, but the bannisters remain, and they must climb up outside these, and hang on from various points which give any facility for arm exercise. The disposition for arm gymnastics is very marked in children who are not repressed in the unnecessary conventional manner. And it is a pity that it does not receive more systematic encouragement, because it would be beneficial for chest expansion in growing children. As matters stand now such exercise as is permitted favours leg development only, while all school work promotes contracted chests and rounded backs—at any rate with the girls. Boys are rather more fortunate. They are not troubled by an ever-rampant Mrs. Grundy preaching lessons on deportment. They retain the monkey habits of tree-climbing and bird-nesting. If any one reflects how important a prize to a hungry monkey a bird's nest of eggs must be, then he will understand how the inherited instinct can be so strongly developed among boys.

However, I am wandering somewhat from the human baby, and I will return thereto by asking consideration for what should be commonly observed in any family, a child with a pleased expression. There is one point in such expression which has not received due consideration, namely, the raising of lumps of flesh each side of the nose as an indication of pleasure. Accompanying this, though difficult to bring out in a photograph, may be seen small furrows, both in children and adults, running from the eyes somewhat obliquely towards the nose. What these characters indicate may be learnt from the male mandril, whose face, particularly in the breeding season, shows coloured fleshy prominences each side of the nose with conspicuous furrows and ridges. In the male mandril these characters have been developed, because being an unmistakable sign of sexual ardour they gave the female particular evidence of sexual feelings. Thus such characters would come to be recognised as habitually symptomatic of pleasurable feelings. Finding similar features in human beings, and particularly in children, though not developed in the same degree, we may assume that in our monkey-like ancestors facial characters similiar to those of the mandril were developed, though to a less extent, and that they were symptomatic of pleasure, because connected with the period of court-ship. Then they became conventionalised as pleasurable symptoms.

Darwin's idea of Antithesis with regard to the expression of emotions does not commend itself. There is not space now to consider the subject fully ; but it may be broadly stated that methods of expressing pleasure have all arisen from habits and actions employed in self-gratification—the satisfaction of the bodily requirements —either in self-nutriment or in procreation. But they may not be the actions employed by members of the species under its present-day conditions. And in the young they would certainly not be so ; they would be the crystallised epitome, if such a term be allowed, of the habits and actions which proved successful with ancestors when they lived in a very different environment. Striking coloration of the face with ridges and scar-like markings would not now give pleasure to the sexes of the human species in their civilised condition ; but the face of the male mandril is evidence of their having done so and still doing so among monkeys, and the practice of face painting, perhaps also of tattooing among savages, is evidence of the monkey habits having been inherited by the human species, and still finding favour among its members.

In the matter of pain, the idea that the expressions which indicate it go back to ancestors living under very different conditions is excellently brought out. Expressions would be the special muscular actions performed under the stimulus of a feeling of injury—such actions as were necessary to alleviate the pain, those necessary to

FIG. 3.—A child crying.

prevent further pain, or to escape from the danger indicated by the pain, and those which were employed in revenge on the inflicter of pain, on the principle that destruction of the cause of injury would be the surest method of prevention.

Therefore, one of the first things that pain prompts an animal to do is to exhibit and prepare its weapons of offence. In the case of the human baby such weapons of offence would be those which would have been employed by the pre-human ancestors. The picture of the crying child, Fig. 3, illustrates this. The peculiar squareness of the open mouth, caused by retraction of the lips at all four corners, is on purpose to exhibit the fighting weapons, the canine teeth ; although, as a matter of fact, the canine teeth have not yet been developed. But the instinct to open the mouth so as to show canine teeth has been inherited from pre-human ancestors who habitually made use of these teeth in order to fight.

There is another feature in this picture, the tight closing of the eyes. This is to protect the eyes from injury during fighting. I photographed a cat which I pretended to strike. There was the same closing of the eyes ; and, for a similar reason, a throwing back of the ears out of harm's way ; and besides there was the paw ready to strike the assailant. I photographed another cat being teased. There was just the same opening of the mouth as in this picture of the baby, and the canine teeth, which were then disclosed, showed exactly what the cat's inten-

tions were, that they were just the same as that expressed by the throwing open the portholes, and the running out the guns which we so often used to read of in accounts of men-of-war.

The lessons which the human baby can teach as regards the past history of its race are very numerous. I have only been able to glance at some of the more important ; but they are sufficient to show that the subject is one of wide range and considerable interest.

S. S. BUCKMAN.

NOTES.

LAST Thursday a combined meeting of the Royal Society and Royal Astronomical Society took place at Burlington House, when the observers who went away for the recent eclipse communicated the results of their observations. As the reports have not yet been published, we are unable to give an account of them. We have received from Prof. Langley a preliminary account, which we hope to print next week, of the expedition which went under his direction to Wadesboro, U.S.A., to observe the eclipse. The photographs he has obtained surpass any that have ever been taken at an eclipse, and speak volumes for the employment of instruments of great focal-length.

THE new physics laboratory of Owens College, Manchester, was opened on Friday last by Lord Rayleigh. Particulars as to the ceremony and the equipment of the laboratory will appear in our issue for next week.

THE Conference on Malaria, which was to have been held under the auspices of the Liverpool School of Tropical Medicine at the end of the present month, has been postponed in order to avoid clashing with the celebration of the Centenary of the Royal College of Surgeons of England and other gatherings.

WE regret to notice the death, at Manchester, on Monday last, of Dr. Daniel John Leech, a well-known physician, and professor of Materia Medica and Therapeutics in Owens College. As chairman of the Pharmacopœia Committee of the Medical Council he had charge of the publishing of the last edition of the British Pharmacopœia, and his name had been recently mentioned as the probable president of the British Medical Association. Dr. Leech was in his sixty-first year.

THE death is announced of Prof. Corrado Tommasi-Crudeli, secretary of the class of mathematical, physical and natural sciences in the Reale Accademia dei Lincei. Tommasi commenced his career in 1859 as demonstrator of pathological anatomy at Florence, after studying with Claude Bernard, of Paris, and Duchenne In 1862 he went to study pathology under Virchow, at Berlin ; the next year he delivered a course of lectures on pathological histology at Florence, and in 1865 he was appointed professor ordinarius of anatomy at Palermo. During an outbreak of cholera in the following year, Tommasi rendered valuable services by his study of the disease and its mode of propagation, and published a well-known memoir on the subject. In 1870, Tommasi was called to Rome, where he was first appointed head of a newly-formed department of pathological histology. Later, he carried out extensive researches on the propagation of malaria. While his researches, conducted in conjunction with Klebs, have been superseded by recent discoveries, the general conclusions to which he was led have not only been substantially confirmed, but have received their true explanation in the new doctrine of the propagation of malaria by mosquitoes.

DURING the past week the summer meeting of the Institution of Mechanical Engineers has been held in London, and proved a successful gathering, interesting not only because of the various papers read (the titles of which, excluding an additional one, by Mr. E. Goffe, on "The Construction of 'Long Cecil,' a 47-inch Rifled Breechloading Gun in Kimberley during the Siege 1899-1900," have already been given by us), but from the fact of a number of members of the American Society of Mechanical Engineers being present. To these a most hearty welcome was accorded.

THE thirty-second annual convention of the American Society of Civil Engineers was opened on Monday last at the Institution of Civil Engineers. The proceedings were inaugurated by an address of welcome from Sir Douglas Fox, the president of the Institution of Civil Engineers, and the presidential address was delivered and several discussions took place.

THE Audiffret prize, of the value of 15,000 francs, has been awarded by the Academy of Moral and Political Sciences of Paris to Dr. Yersin for 'his discovery of the anti-bubonic serum. The prize is awarded at regular intervals for the "greatest devotion to scientific discovery."

THE Pharmaceutical Society announces that the Salters' Company Research Fellowship is now vacant. The subject of the Fellowship is chemistry considered especially in its relation to pharmacology, that is, the application of the newest methods of scientific chemistry to the elucidation of pharmacological problems. The Fellowship is of the annual value of 100*l.*, and is tenable in the Research Laboratory of the Pharmaceutical Society for one year, but may be renewed under certain conditions, and the holder is expected to devote his whole time to original investigation. Candidates need not necessarily be pharmaceutical chemists or members of the Society. The last day for receiving applications is Saturday next.

THE Balbi-Valier prize, of the value of about 120*l.*, has been awarded by the Venetian Institute of Sciences to Prof. Grassi, of Rome, for his work on the mosquito and its relation to malaria.

A PRIZE of the value of 1000 marks is offered by the Scientific Society of Danzig, in connection with its 150th anniversary, for a paper on the geology of North Germany.

THE German Society of Mechanical Engineers offers a premium of 60*l.* and a gold medal to the designer of the best system of high-speed electric railways for heavy traffic. Designs must be submitted by, at latest, October 6 next.

THE Lavoisier Monument which is being erected on the Place de la Madeleine, Paris, in close proximity to the house in which the famous chemist lived for many years, will, according to the *Chemist and Druggist*, be formally inaugurated on July 27 by the French Minister of Public Instruction. The statue has been erected by international subscription under the auspices of the Paris Academy of Sciences. The sculptor is M. Barrias, and the monument will consist of a bronze statue of Lavoisier, on a pedestal, bearing on two sides bas-reliefs showing Lavoisier working in his laboratory with Mme. Lavoisier writing under his dictation, and Lavoisier expounding the result of his experiments at a meeting of the Academy of Sciences. The scenes have been created from authentic documents.

A MONUMENT has been erected to the memory of Dr. Jean Hameau, the obscure general practitioner of the Gironde, who, in 1836, published a study on viruses, in which he partly anticipated the discoveries of Pasteur. The statue was unveiled recently at La Teste de Buch, where Hameau practised. Addresses were delivered by Dr. Laude, the Mayor of Bordeaux and President of the Medical Syndicates Union of France, Prof. Lannelongue, of Bordeaux, and others. Hameau was born in 1779, and died in 1851. His claim to be considered a precursor of Pasteur has been publicly acknowledged by Prof. Grancher,

and it is probable that had he been possessed of the laboratory accommodation and means of investigation available at the present day, the microbe theory of disease would have been established fifty years sooner than it was.

PROF. H. F. OSBORN, of Columbia University and the American Museum of Natural History, has, according to *Science*, been invited to succeed the late Prof. Cope as vertebrate palæontologist of the Geological Survey of Canada.

MR. W. E. D. SCOTT, curator of the ornithological department in Princeton, announces, says *Science*, that the British Museum has presented to the University two thousand mounted birds, specimens from India, Australia and the Malay Islands. Some time ago the University presented the British Museum with 250 sets of North American birds' eggs.

IT is announced that the repairs to the Arctic steamer *Windward* have now been made, and the vessel was expected to sail by about July 1. The *Windward* will proceed directly, with a call at Disko, to Etah, North Greenland, Lieut. Peary's winter quarters, where instructions from him will doubtless be found, or if not, will be awaited. The vessel will take with her the maximum quantity of coal, additional lumber, oil, sugar, arms, ammunitions, provisions, scientific instruments and everything necessary for Lieut. Peary's work, including two new whale-boats, specially built at New Bedford, and thoroughly equipped in every detail. Upon the arrival of the *Windward* at Etah, Lieut. Peary will assume command, and further movements will be subjected to the conditions of his work and to his instructions. No passengers will be taken on the *Windward*, the Danish Government having qualified their permission to land at the Greenland ports, with conditions that tourists should not be carried. If Lieut. Peary has succeeded in carrying out his plans, that is to say, if he has discovered the North Pole, he will, says the *Scientific American*, return with the ship. If not, the supplies will be landed. It is possible that the *Windward* will bring back the Robert Stein party, which was landed near Cape Sabine by the *Diana* in August last.

A BOTTLE has, it is reported, been found on the shore at Roundstone, co. Galway, containing a printed card directing the finder to forward the contents to Captain Ernest André, Polar Expedition Company, Sweden, and stating that it was thrown from Major André's balloon in the Arctic regions with a view to testing the ocean currents. The bottle has been forwarded to the Board of Trade.

SETS of volumes on naval architecture, and on the history of the British Navy, have been presented by the Institution of Engineers and Shipbuilders of Scotland to Prof. Arnold, of the Sheffield Technical School, as a token of appreciation of the lecture delivered by Prof. Arnold on "The Internal Architecture of Metals."

A MEETING was held in the rooms of the Royal Meteorological Society some time ago to consider the question of a memorial to the late Mr. G. J. Symons, F.R.S., when it was resolved unanimously that the memorial should take the form of a gold medal, to be awarded from time to time by the Council of the Royal Meteorological Society for distinguished work in connection with meteorological science, and an executive committee was appointed to take the necessary steps for the raising of a fund for the purpose. The committee now appeal to the fellows and members of the societies with which Mr. Symons was associated, to the rainfall observers, and to all who have in any way benefited by his advice and assistance, to contribute to this memorial fund, which it is hoped may reach the sum of at least 750*l.* Contributions should be paid to Mr. W. Marriott,

70 Victoria Street, Westminster, S.W. ; or to the "Symons Memorial Fund," Bank of England (Western Branch), Burlington Gardens, W.

AN Anti-rabic Institute for India is, says the special Indian correspondent of the Lancet, at last an accomplished fact. After numerous delays, Government have stepped in and practically settled the difficulties. The Royal Army Medical Corps having an officer in Major D. Semple who had studied in Paris and Lille, determined to utilise his experiences, and the annual expense for sending soldiers to Paris was diverted for the new institute. The central committee of the proposed Pasteur Institute saw their opportunity and took over its control. With a capital of 70,000 rupees and a yearly grant of 19,500 rupees, the expenses of the new institute ought to be fairly well met. Residents in India may be congratulated that at last means are provided whereby European and native patients alike can be offered the best available treatment for the terrible disease of rabies.

DR. L. SAMBON AND DR. LOW, the two medical men entrusted by the British Government with the perilous task of testing the possibility of guarding against malarial infection in the Roman Campagna, have, according to the Lancet, at length found a favourable place for their purpose. After rejecting various other localities as being for one reason or another unsuitable, they have selected a spot about two miles distant from Ostia, between Castel Porziana and Castel Fusano, and within five minutes' walk of the latter place. The site of their hut is on the edge of a "stagno," or swamp, forming part of the royal hunting demesne of Castel Fusano, and left undrained in order to preserve the wild boar, water fowl, &c., which frequent it. The hut will stand close to a canal containing a luxuriant growth of algæ and other aquatic plants, and within a stone's throw of a clump of pine trees, which forms the outskirts of the Castel Fusano pine forest. The few dwellings near are inhabited by peasants who constantly suffer from malaria and are infested by mosquitoes of the anopheles variety. Situated thus in the heart of the swamps surrounding the mouth of a large river, among the haunts of innumerable mosquitoes of the malarial variety, and in a locality notorious as one of the most deadly of the fever-stricken centres of the Roman Campagna, this dread and unhealthy spot appears to offer ideal conditions for the carrying out of the interesting but dangerous experiment now about to be begun. The two daring investigators hope to have everything in readiness early in the present month ; in the meantime, their time is profitably occupied in studying the animal and insect life of the Campagna, collecting and examining frogs, lizards, bats, spiders, mosquitoes, and the like. They have already made some interesting observations, as, for example, that although the larvæ of anopheles are at this season apparently very few, the adult mosquitoes are collected in the houses in great numbers, being especially numerous near byres and stables. The King has graciously given his consent to the erection of the hut in the royal preserves, and the municipality of Rome are doing everything in their power to help the enterprise.

THE annual general meeting of the Jenner Institute of Preventive Medicine was held on Friday last, when Dr. Macfadyen, the director, was able to state that the Institute's work had continued to progress. Among the new features added during the year were a physiological room, a room for incubating purposes, a laundry, a workshop, and a cold storage room. A Hansen apparatus for yeast culture had been presented, and considerable additions had been made to the library. Three papers were communicated to the Royal Society upon the influence of the temperature of liquid air and hydrogen upon bacterial life. Systematic investigations are being carried out in

the bacteriological department upon enteric fever, tuberculosis, and the etiology of cancer. Special investigations were carried out for public authorities during the year, e.g. upon tubercle in milk, glanders, anthrax, &c., and investigations in a number of other directions have been and are being prosecuted with vigour.

THE recent case of the Jenner Institute of Preventive Medicine v. Assessment Committee of St. George's, Hanover Square, was the means of raising once more the important question of the rateability of scientific societies. The Jenner Institute has unfortunately failed to establish its claim to exemption from the payment of rates. The Divisional Court decided that the Institute did not fulfil the conditions of the Act of Parliament exempting "any Society instituted for purposes of science, literature, or the fine arts exclusively." The preparation and sale of preventive and curative medicines was held by Mr. Justice Grantham to be the main object of the Institute, or as Mr. Justice Channell put it, "its main object was to dispense to the public the benefits of science." The Institute was not, therefore, "exclusively" devoted to the advancement of science. The Institute has, as a matter of fact, dispensed for the benefit of the public, and at a considerable loss, certain antitoxins which require the highest scientific skill in their preparation. The preparation of these substances has been at the same time a means of studying and improving the methods for producing immunity to given diseases. The aims in this respect have been of a purely scientific character, and in accordance with the main objects of the Institute, which are not, despite the Court's ruling, of a dispensing nature, except in so far as opportunity is afforded to medical men to test the value of certain antitoxins in the treatment of disease. The eminently useful aims of the Institute, which are carried out at great cost, might have been thought to bring it well within the intention of the Act, but the judicial interpretation of the word "exclusively" has formed the stumbling-block. It is to be feared that the expense of litigation may prevent the Institute from proceeding to an appeal. In any case, it is to be hoped that some steps will be taken to amend an Act, apparently devised for the benefit of scientific societies, but which, as interpreted by the Court, has little or no practical value. It is quite conceivable, as the law stands, that a claim for exemption might be defeated on the ground that a daily newspaper had been admitted to the reading-room of an institute, and that it was not, therefore, "exclusively" devoted to purposes of science.

THE annual general meeting of the Marine Biological Association was held in the rooms of the Royal Society on June 27. The council reported that arrangements had been completed for the supply of sea-water, obtained from the open sea beyond the Plymouth Breakwater, for special experiments on the rearing of sea-fishes and other marine animals. Through the kindness of Mr. J. W. Woodall, the Association has had placed at its disposal a small floating laboratory, which is at present stationed at Salcombe. The periodical surveys of the physical and biological conditions prevailing at the mouth of the English Channel have been continued by Mr. Garstang at quarterly intervals for an entire year. Observations were taken at four fixed stations. They included serial temperature determinations at all depths, filtration of a definite column of water from bottom to surface with a "vertical net," and collections of the floating life at surface, mid-water and bottom, by means of a specially devised closing net. Mr. Garstang has also carried out a series of preliminary experiments on the rearing of sea-fish larvæ under different conditions, with a view to a solution of the difficulties hitherto encountered in regard to the practical work of sea-fish culture. The investigation of the fauna and bottom deposits of the shallow water grounds in the neighbourhood of Plymouth, upon a systematic plan, has been continued during the year.

A SUCCESSFUL trip has at last been made with Count Zeppelin's navigable balloon at Friedrichshafen on Lake Constance. On Saturday evening the ascent was prevented by an explosion of one of the segments of the balloon, and a similar accident is stated to have befallen one of Count Zeppelin's benzine motors. On Sunday evening, Count Zeppelin and four others made their first ascent, and after drifting with the wind, turned and tried to make headway against it ; but the wind appears to have been too strong, and the balloon was quietly lowered to the lake, where the cars floated and the occupants were brought to shore without any damage. From another telegram dated July 2 (Monday) we learn that a successful ascent was made that evening. This must have been a second attempt, and it is stated that the ship travelled safely to Immenstadt, thirty-five miles from Friedrichshafen, and landed all well.

LONDON was again visited by a sharp thunderstorm about midday on Tuesday ; the rainfall and hail during the storm amounted to 0·23 inch in Westminster. The weather has continued in a very unsettled condition, but, nevertheless, the mean temperature in the neighbourhood of London for the month of June exceeded the average by about 0·05. This result has been caused chiefly by the amount of cloud which, while making the days cool, has kept the nights relatively warm. The rainfall for the month was 0·66 inch above the average, and in some parts of England it was double the average.

THE *Lancet*, quoting from a Buenos Ayres review of hygiene, entitled *La Salud*, gives some interesting particulars concerning the plague. Formerly, according to a tradition common amongst certain tribes of South American Indians, wide-spreading fires used to sweep over the land. The inhabitants were in the habit of taking refuge in caves and dens of the earth. From time to time they poked out the branch of a tree, and if this when pulled in again showed no signs of burning they considered it safe to come out. So formerly when plague ravaged and desolated various countries the inhabitants shut themselves up in the cave of isolation and did not come forth until they learned that plague had disappeared. Nowadays, however, just as the Indian tribes possess herds of horses which they did not formerly possess, and are able by these means to stamp out pampas fires, so that there is no need to take refuge in a cave, so also modern cities possess hygienic knowledge and conditions which render isolation unnecessary, and a general dissemination of plague throughout Europe and America is as impossible as a fire which should affect the whole pampas. In India and in China only those persons succumb who live under grossly unhygienic conditions. With reference to the recent outbreak of plague in Argentina, *La Salud* says, "It would be greatly to the honour of the Argentine Republic if she would invite other countries to a conference to consider the question of meeting plague, if not actually by abandoning all international action yet by leaving commerce perfectly free and by treating the disease wherever it appears exactly like any other infectious disease which assumes endemic characters."

THE annual report of Sir George Nares, F.R.S., acting conservator of the Mersey, has just been issued, and shows that considerable work has been done by the sand-pump dredgers *Brancker* and *G. B. Crow* at the Queen's Channel Bar and at certain shoals in the Queen's and Crosby Channels, at the entrance to the river. For several months the surveys show considerable improvement, many of them having no soundings less than 27 ft. below low water spring tides within the dredged cut. Though there has been slight shoaling at the outer end or

the north side of the channel, several good lines of sounding run the full length of the cut with not less than 27 ft. During the year 2,067,000 tons were dredged in Queen's Channel, 1,839,000 tons taken from shoals in that channel, and 2,735,000 tons from shoals in Crosby Channel, making a total of 6,659,000 tons, while since the commencement of the operations 45,148,860 tons of sand have been removed. The number of inward and outward bound vessels passing through Queen's Channel last year was 45,158, against 44,376 in the previous year, and 35,932 in 1893. The daily average of last year was 124 against 98 in 1893. The total using all the channels increased from 41,439 in 1893 to 50,964 in 1898, and 52,216 last year. The sand removed from the river between Liverpool and New Brighton in 1899 was 1,374,670 tons. In the same period 1,375,272 tons of silt and detritus were raised from the Manchester Ship Cana and deposited at sea.

THE idea of substituting electricity for horse-traction on canals has not been so widely developed as one would have expected to see, but some experiments of an instructive nature (*Engineering*, June 22) have been made by a German firm on behalf of the Prussian Government, wherein electric locomotives were employed for this purpose, and the mode of working may be briefly stated as follows :—A section of the Finow Canal, which forms a portion of the waterway between Berlin and Stettin, was chosen, embodying as it does physical difficulties with reverse curves, &c. On the towing-path a meter-gauge track of special design with overhead conductor was laid, on which the electric motor tows the barges ; and owing to the deficiency of adhesive weight a steel rack is bolted to the permanent way, and the rack-rail system is adopted. In spite of many difficulties the experiments proved that the system was capable of meeting all requirements, and worked with apparent ease. This is very satisfactory, because in one place, we are told, the line was raised 9 feet 6 inches above the towing-path with approaches of 1 in 8½ gradients. The electric motor used, we are informed, developed from 14 to 15 horse-power, much more than was necessary ; but this was intentionally provided in view of further experiments to deal with the possibility of electric traction for barges of a heavier type.

IN NATURE for September 1, 1892, Prof. D. Kikuchi announced the foundation by an Imperial ordinance of an earthquake investigation committee in Japan. The objects of the committee were to study the nature of Japanese earthquakes and their distribution in time and space, to discover any means of lessening their disastrous effects, and if possible to ascertain laws by which their occurrence might be predicted. During the eight years of its existence much successful work has been done by the committee in two of these directions. They have accumulated and discussed many series of observations and records, and have conducted numerous experiments on the fracturing and overturning of columns and on the type and material of building best suited to resist a strong shock. The results are printed partly in Japanese papers ; partly, we are glad to see, in their "Publications in Foreign Languages," the third and fourth numbers of which have appeared this year. If abstracts of the former could be given either in English or French, the debt we already owe to Japanese seismologists would be greatly increased.

THE Verein zur Förderung des Unterrichts in der Mathematik und den Naturwissenschaften held its annual meeting at Hamburg in the first week in June. This association, which numbers some 900 members, includes many teachers in the higher schools of Germany among its ranks, and it has taken part in the preparation of reports on physical apparatus suitable for teaching purposes. The programme included, among other subjects of papers, the teaching of geometry of position, wireless telegraphy, experiments with liquid air, flight of birds, the International Catalogue, and the preservation of natural objects of interest in Germany.

WE are indebted to Prof. P. H. Schoute for a paper on the locus of the centre of hyperspherical curvature for the normal curve of n dimensional space. In a previous paper the author pointed out that the characteristic numbers of the locus of the centre of hyperspherical curvature are lowered if some of the points of the given rational curve lying at infinity coincide. At present Prof. Schoute traces for a special case the amount of these lower numbers, namely, for the case where the given curve is the "normal" curve of the n dimensional space in which it is situated. According to the final result, the characteristic numbers of the locus of the centre of hyperspherical curvature for the normal curve are respectively $2n - 1, 3n - 3, 4n - 7, 5n - 13, 6n - 21, \ldots 2n - 1$, from which it follows that they do not change if taken in reverse order.

A REPORT on dietary studies of Harvard and Yale University boat crews, conducted by Prof. W. O. Atwater and his assistant, Mr. A. P. Bryant, forms *Bulletin* No. 75 of the U.S. Department of Agriculture. These studies were undertaken primarily to secure data regarding the food requirements of men performing severe muscular work, and they lead to the conclusions that the actual food consumption of people in general is regulated more or less by the supply at their disposal and their tastes and appetites ; but that it is justifiable to suppose that in a general way the difference between the food of athletes and that of other people represents a difference in actual physical need, even if neither is an accurate measure of that need. The energy of the food consumed per man per day in the dietary studies of university boat crews was found to exceed by 400 calories, or about 10 per cent. the amount found, as the average of fifteen dietary studies among college clubs in different parts of the country, while the protein in the studies of the university boat crews was 48 grams, or 45 per cent. larger in amount.

IN the *Rendiconti del R. Istituto Lombardo,* Dr. Benedetto Corti briefly describes the results of a study of the Diatomaceæ of the lakes of Brianza and Segrino. Of a total of eighteen species of diatoms observed in the lake of Montorfano, two (*Synedra lunaris* and *Stauroneis platystoma*) were Alpine in character, and were supposed by the author to represent the remains of a quaternary diatom flora. This view has been confirmed by a more extended study of the other lakes of Brianza and Segrino, which revealed the presence of fifteen species of diatoms peculiar to the Alpine zone out of a total of seventy-two. Of the Alpines, however, only one, viz. *Navicula firma,* was found in the lake of Sartirana. There is a decided affinity between the diatom flora of these lakes and that of the lake of Palù in the Malenco valley, and that of Poschiavo in the Engadine.

NO. 17 of the "North American Fauna," now in course of publication by the Biological Division of the U.S. Department of Agriculture, is devoted to a revision of the North American voles, or "field-mice," of the genus *Microtus,* by Mr. V. Bailey. As the work is based on the examination of between 5000 and 6000 specimens, including typical representatives of every species, from more than 8.o different localities, it ought to be exhaustive. The genus, which is divided into nine sub-generic groups, is taken to include no less than seventy distinct specific and sub-specific modifications, three of which are described for the first time. It is noticed that the development of oil-glands and musk-glands is most conspicuous in the aquatic members of the group, and least so in those inhabiting the driest regions. Those forms which are most exposed to light and dryness are the palest, while the deepest and

richest tones of colour are developed in those from damp and shaded localities. Attention is directed to the importance of placing every possible check on the increase of these little mammals, and of reducing their numbers when they become unusually abundant.

IN its *Bulletin* No. 12, the department just mentioned publishes a useful report, by Mr. P. S. Palmer, on the legislation for the protection of birds, other than game-birds, now in force in the United States. The author states that many insectivorous birds are still unprotected ; and that the laws relating to such birds in general lack uniformity in different parts of the States ; this diversity in the laws being illustrated by a map. The report closes with a digest of the bird-laws of the different States. The need of further legislation is strongly emphasised.

IN this connection may be noticed a pamphlet on the food of wild birds in this country, issued by the Yorkshire College and the Joint Agricultural Council of Leeds and the East and West Ridings. In this useful publication attention is called to the fact that birds very largely affect both sides of the farmer's balance-sheet ; and that while, unfortunately, the damage they do is readily detected, the great services they render can only be appreciated by those who take pains to investigate the subject. The "pros and cons" in regard to each particular bird seem to be very fairly considered.

THE July number of the *Journal of Conchology* contains an interesting sketch of the life and career of the late Mr. Lovell Reeve, the well-known conchologist, with extracts from his diary and correspondence.

No fewer than three books dealing with the life and work of the late Prof. Huxley are being prepared at the present time. Messrs. Macmillan and Co. will issue the biography of his father by Mr. Leonard Huxley, and a volume on the professor is to be added to Messrs. Putnam's Sons' "Leaders of Science" series from the pen of Mr. P. Chalmers Mitchell ; while a third work is to be contributed to Messrs. W. Blackwood and Sons' "Modern English Writers" series by Mr. Edward Clodd.

MESSRS. CHARLES GRIFFIN AND CO., LTD., have sent us the seventeenth annual issue of the "Year-Book of the Scientific and Learned Societies of Great Britain and Ireland," comprising lists of the papers read during 1894 before societies engaged in no fewer than fourteen departments of research. The serial, which is too well known to need more than a brief reference here, contains much information of service to literary and scientific men. It would be yet more valuable if the officials of certain societies, against whose entry the words "No Return" appears, could be induced to furnish the compiler with the titles of the papers presented to their respective institutions.

A NEW edition (the fifth) of "The Microtomist's Vade-Mecum," by Arthur Bolles Lee, has just been issued by Messrs. J. and A. Churchill. Considerable changes have been made in the present edition, the whole work having been very carefully revised since the last edition appeared nearly four years ago. The text has undergone condensation throughout, making it possible for much new matter to be added without increasing the size of the volume.

MORE and more space, we are glad to see, is being given in the popular magazines to articles dealing in a greater or less degree with scientific subjects, and in the monthlies for July that have reached us we notice the following contributions of this character :—In *Pearson's Magazine* Prof. Simon Newcomb explains to the lay reader in simple language, and by the aid of diagrams, "How the Planets are Weighed" ; while Dr. F. A.

Cook, who was attached to the Belgian Antarctic Expedition, discourses pleasantly on "The Possibilities of Reaching the ·Four Poles." In *Good Words* Mr. E. W. Maunder writes on "The Lords of Cold "(the title, it may be noted, is borrowed from a line in Plumptre's "Dante "), the article being a study in stellar perspective. In the same magazine is also to be found a contribution, by Mr. Aflalo, on "How Wild Creatures Feed." *Chambers's Journal* always contains at least one article of scientific interest ; the present number has in it papers, entitled " Tropical Diseases and Cures " and " Alcohol from Paper and Sawdust."

THE additions to the Zoological Society's Gardens during the past week include a Bonnet Monkey (*Macacus sinicus*) from India, presented by Mr. G. A. S. Bell, R.N. ; a Ring-tailed Lemur (*Lemur catta*) from Madagascar, presented by Miss M. C. Rawcliffe ; a Common Duiker (*Cephalophus grimmi, ♂*) from South Africa, presented by Mr. J. E. Matcham ; five Wild Cats (*Felis catus*) from Inverness-shire, presented by Mr. George J. Bailey ; a Levaillant's Amazon (*Chrysotis levaillanti*) from Mexico, presented by Mr. J. Farmer Hall ; a Royal Python (*Python regius*) from West Africa, presented by Mr. Benjamin Stewart ; an Alpine Newt (*Molge alpestris*), nine Black Salamanders (*Salamandra atra*), two Slowworms (*Anguis fragilis*) from Switzerland, presented by the Rev. J. W. Horsley ; a Common Viper (*Vipera berus*), British, presented by Mr. G. Alan Marriott ; a Common Duiker (*Cephalophus grimmi, ♀*) from South Africa, a Syrian Bear (*Ursus syriacus*) from Western Asia, a Cheetah (*Cynoelurus jubatus*) from India, two Black-faced Kangaroos (*Macropus melanops, ♂, ♀*) from Tasmania, six Wrinkled Terrapins (*Chrysemys scripta rugosa*) from the West Indies, an Amboina Box Tortoise (*Cyclemmys amboinensis*) from the East Indies, five Mississippi Terrapins (*Malacoclemmys geographica*), a Prickly Trionyx (*Trionyx spinifer*) from North America, three Annulated Terrapins (*Nicoria annulata*) from Western South America, deposited ; a Three-toed Sloth (*Bradypus tridactylus*) from British Guiana, purchased.

OUR ASTRONOMICAL COLUMN

EPHEMERIS FOR OBSERVATIONS OF EROS.—The .following computed positions for July are from the ephemeris prepared by Herr F. Ristenpart (*Astronomische Nachrichten*, Bd. 152, No. 3643).

Ephemeris for 12h. Berlin Mean Time.

1900.		R.A.		Decl.
		h. m. s.		° ' "
July 5	...	0 54 6·51	...	+ 14 13 22·3
7	...	57 27·78	...	14 50 33·5
9	...	1 0 48·24	...	15 27 55·3
11	...	4 7·89	...	16 5 27·7
13	...	7 26·71	...	16 43 10·8
15	...	10 44·66	...	17 21 5·0
17	...	14 1·68	...	17 59 10·8
19	...	17 17·70	...	18 37 28·1
21	...	20 32·62	...	19 15 57·1
23	...	23 46·36	...	19 54 38·1
25	...	26 58 84	...	20 33 31·3
27	..	30 9·98	...	21 12 36·9
29	...	33 19·70	...	21 51 55·1
31	...	1 36 27·93	+22	31 26·5

MEASURES OF EROS.—*Harvard College Observatory Circular* (No. 51) contains the results of the measurements of photographs obtained during the years 1893, 1894 and 1896, giving the positions of the planet during those years. The complete discussion of the measures is being prepared for a volume of the Observatory *Annals*, but the numbers here published show that at the Harvard College Observatory there is the means of tracing the path of any object since 1890, during the times in which it was moderately bright, with nearly as great accuracy as if a

series of observations ¡had been taken of it with a meridian circle.

TOTAL ECLIPSE OF THE SUN, MAY 28.—M. Deslandres communicates the report of his work in connection with the recent eclipse to the *Comptes rendus* (vol. cxxx. pp. 1691-1695). His programme comprised four classes of investigation :—(1) velocity of corona ; (2) ultra-violet spectrum of corona and chromosphere ; (3) infra-red spectrum of corona ; (4) photography of corona.

Observing visually with a powerful grating spectroscope, he found by the inclination of the corona line that on the west side of the equator the corona appeared to have a more rapid speed of rotation than the disc. The photographic spectra taken for this purpose are too faint for measurement.

The ultra-violet spectra were obtained with spar quartz prismatic cameras, ten plates being obtained showing good images down to λ 3000.

The investigation of the infra-red radiation from the corona was undertaken with a view of providing a possible means of observing the corona without an eclipse, and the results would indicate that the corona is specially rich in these calorific radiations. M. Deslandres states that at his station, Argamasilla, Spain, totality was five seconds shorter than the calculated time.

THE ROYAL OBSERVATORY, GREENWICH.

IT is customary for the Astronomer Royal to present his annual report to the Board of Visitors of the Royal Observatory on the first Saturday in June, but as it is easier to transfer such a function to another date than to change the time of a total eclipse of the sun, the usual day of meeting was. adjourned until June 26 last. On this day, the weather,. however, did not quite come up to summer standard ; but fortunately the rain held off, and the afternoon proved sufficiently fine to allow the numerous visitors to inspect the buildings and instruments. As is customary, we give below a brief *résumé* of the report.

BUILDINGS.

The building of the new observatory so near to the boundary of the grounds has necessitated an alteration in the position of the old fence, to show the building off more effectively, so that provision has been made in the Navy estimates to put the fence further away, and the plans for this are now under consideration. This building also includes the new library rooms, and we learn that the removal of the books to their new position was completed in March last. The opportunity has also been utilised for their rearrangement and for the preparation of a new catalogue, both of which, we are told, were much needed. Not only is the rearrangement of the books practically complete, but good progress has also been made with the formation of the card catalogue, a system which is to be highly recommended.

TRANSIT CIRCLE.

The sun, moon, planets, and fundamental stars have been. regularly observed on the meridian as in previous years. The number of observations made from 1899 May 11 to 1900 May 10,. is as follows :—

Transits, the separate limbs being counted as one observation	...	...	10,712
Determinations of collimation error	...	...	297
Determinations of level error ...	...	...	684
Circle observations	...	...	10,001
Determinations of nadir point (included in the number of circle observations)	...	...	674
Reflexion observations of stars (similarly included) ...	...	...	637

The number of stars observed in 1899 is about 5000.

An unusually large number of observations was obtained in the three months, August-October, the average number of transits observed being more than 1300 each month. From November to the date of the report, in consequence of the cloudy weather, the average has been only half this number.

The apparent correction for discordance between the nadir observations and stars obtained by reflexion for 1899 was found to be slightly larger—namely, −0″·41—than that of last year, which was −0″·36.

The results of recent years are as follows :—

	Mean	Range in Yearly Means
	"	*"*
1880–1885	− 0″·34	From − 0″·29 to − 0″·45
1886–1891	+ 0·03	„ − 0″·12 to + 0″·09
1892–1898	− 0·30	,, − 0″·25 to − 0″·36

Observations of level and nadir have been made, when practicable, three or more times on the same day, and diurnal changes similar to those referred to in the last report have been found in 1899. The observations of level taken within three hours of noon and midnight give corrections of + 0″·30 and + 0″·18 respectively, to those made within three hours of 6 p.m. Similarly the observations of nadir near noon and midnight give corrections of + 0″·17 to those made within three hours of 6 p.m.

In view of this systematic diurnal movement of the instrument and of the large number of observations of azimuth stars in recent years, it seems probable that the limit of accuracy obtainable by the use of double transits for the determination of the positions of the close polar stars has been reached, as this involves the assumption that the azimuth error remains constant for twelve hours at least. It has therefore been arranged to use these stars for determination of azimuth error, by means of their tabular right ascensions, and to keep the observations for improvement of the tabular place only when the azimuth error has been determined by at least three pairs of close polar stars above and below·pole on the same evening.

The correction for the R–D discordance, found for 1899, is + 0″·080 + 0″·218 sin Z.D. The coefficient of sin Z.D. was about + 0″·6 from 1881 to 1894, diminished to + 0″·41 and + 0″·37 in 1895 and 1896, to + 0″·10 in 1897 and 1898, and has now increased to + 0″·22.

The observations of the zenith distances of pairs of stars directly and by reflexion, alternately on alternate nights, have been discontinued. The observations made in the four years, 1895, 1896, 1897 and 1898, show a satisfactory agreement with the ordinary ·observations, reflexion and direct at the same transit, confirming the striking diminution in the value of the R–D discordance in 1897 and 1898 as compared with 1895 and 1896.

The colatitude of the transit circle, as found from observations of about 600 stars in 1899, is 38° 31′ 21″·76, differing by − 0″·14 from the adopted value. The corresponding values of the correction to the adopted colatitude found in 1897 and 1898 are − 0″·17 and − 0″·15, and it may be noticed that the R–D discordance was very small in these years.

The mean error of the moon's tabular place (computed from Hansen's lunar tables with Newcomb's corrections) is − 0ˢ·099 in R.A. and + 0″·27 in N.P.D., deduced from 116 observations. These are equivalent to an error of − 1″·38 in longitude and 0″·00 in ecliptic north polar distance.

The re-observation of the stars of Groombridge's Catalogue, which was the principal object of the Second Ten-Year Catalogue, furnishes material for determination of the proper motions of more than 4000 stars from observations about eighty years apart, with intermediate positions in the Radcliffe Catalogue of 1845. Provisional proper motions are given in the Introduction and are given, for which the annual proper motion in R.A. or N.P.D. amounts to 0″·1 of a great circle, and had not previously been determined. It is proposed to undertake the determination of the proper motions of all the stars in Groombridge's Catalogue. Before doing this it was considered desirable to re-examine Groombridge's Observations, with special reference to the determination of azimuth error, in view of the large systematic error in Right Ascension. The original MSS. of Groombridge's Observations have been kindly lent by the Council of the Royal Astronomical Society, and the examination is in progress.

THE NEW ALTAZIMUTH.

This instrument is now in good working order. Various repairs have been required and minor improvements have been made. The observations of transits seem quite satisfactory, the accordance in the results for clock error in different positions of the instrument (referred to the transit circle) being very good. For the zenith distance observations further determinations of flexure and division-errors are required, and these are in hand. The investigation of the division errors of both circles has shown that the accordance of two determinations is not very satisfactory, and the cause of the discrepancy is now under investigation.

Among the observations made with this instrument may be mentioned 1729 R.A. observations of the sun, planets and stars, 1418 N.P.D. observations, and 2386 observations for collimation, level, and azimuth errors, and nadir.

THE 28-INCH REFRACTOR.

This instrument has been used throughout the year for micrometric measurements of double stars. 492 stars have been measured ; 268 of these have their components less than 1″·0 apart, and 139 less than 0″·5. The stars whose distance apart is less than 1″·0 have been measured on the average on three nights each, and the wider pairs on two nights. The wider pairs measured consist of bright stars with a faint companion, of third companions to close pairs, and of stars of special interest.

In consequence of Mr. Newall's suggestion that the newly discovered spectroscopic binary Capella might possibly be observed as a double star with large telescopes, it has been examined on fifteen nights (from April 4 to May 10) by a number of observers, who all found the star's image to be sensibly elongated ; while the position angle of the elongation changed during the period of observation (April 4 to May 29) in fair accordance with the period of 104 days given by Mr. Newall.

THOMPSON EQUATORIAL.

THE 26-inch refractor has been in constant use throughout the year. The occulting shutter has been·found of great value in obtaining accurately measurable photographs when one of the objects photographed is considerably brighter than the other objects in the field.

Fifteen photographs of Neptune and its satellite have been obtained, of which seven have been measured. Fifty-four photographs of twenty-six double stars have been obtained, of which forty-seven have been measured. Among these stars are Algol and Aldebaran, with their faint companions of fourteenth magnitude. The measures of distance and position·angle ·of the photographs of double stars are published in the *Monthly Notices of the Royal Astronomical Society* for April 1900. Nineteen photographs have been obtained of Comet Swift, of which fifteen have been measured, and the results published in the *Astronomische Nachrichten*, Nos. 3584–5. In addition, photographs of Polaris and neighbouring stars have been taken for parallax, a few photographs of the moon and some of the major planets with their satellites, and others for testing adjustments and the characters of the images in different parts of the field.

The 30-inch reflector has been used chiefly for the photography of nebulæ and star clusters. The photographs of the nebulosity of the Pleiades and of the Orion nebula are very fine, and show a large amount of detail.

ASTROGRAPHIC EQUATORIAL.

The following statement shows the progress made with the plates for the chart and the catalogue respectively :—

	For the Chart (exposure 40m.).	For the Catalogue (exposures 6m.; 3m., and 20s.).
Number of photographs taken ...	243	236
„ successful plates ...	162	181
„ fields photographed successfully ...	155	175
Total number of successful fields reported 1899 May 10	1027	1030
Number of photographs, previously considered successful, rejected during the year	106	102
Total number of successful fields obtained to 1900 May 10 ...	1076	1103
Number still to be taken... ...	73	46

A comparison of this list with the one published in the last report shows that great progress has been made in this work.

It is satisfactory to note that the plates are now placed in the new observatory, where they are kept dry and not subject to the extremes of temperature as formerly. Those that were previously spoilt through damp are now being gradually replaced.

During the year 88,000 measures of pairs of images (6m. and 3m.), as well as of the diameters of the 6m. images, have been made. The number of quarter plates measured in the twelve months in two positions of the plates is 556.

· At the date of the last report the measurement of the plates was completed from December 64° to 69°; and in Zone 70° from R. A. 0h. to 13h. 48m. During this year Zone 70° has been finished and Zones 71° and 72° have been measured, with the exception of thirty-six quarter plates. Subject to this exception, the measurement is complete from December 64° to 73°.

Good progress has been made with the printing of the measures. Zone 64° is finished and Zone 65° as far as R. A. 21h. 36m. It is estimated that all the measures from December 64° to December 72° will be included in one volume of about 650 pages.

HELIOGRAPHIC OBSERVATIONS.

In the year ending 1900 May 10, photographs of the sun have been taken on 180 days, either with the Thompson or Dallmeyer photo-heliographs. The former, mounted on the Thompson 26-inch refractor, was used as the regular instrument for solar photography up to March 9, when it was temporarily dismounted, the Dallmeyer photo-heliograph being substituted for it. Of the photographs taken with either instrument, 369 have been selected for preservation, besides 11 photographs with double images of the sun, for determination of zero of position angle. Photographs to supplement the Greenwich series have been received from India or Mauritius up to 1900 March 8.

For the year 1899, Greenwich photographs have been selected for measurement on 202 days, and photographs from India and Mauritius (filling up gaps in the series) on 162 days, making a total of 364 days out of 365 on which photographs are at present available.

The chief characteristic of the sun's surface, during the period covered by this report, has been the steady decline in the mean daily number and area of spots observed, August and September 1899 in particular showing a marked sub-minimum.

MAGNETIC OBSERVATIONS.

The variations of magnetic declination, horizontal force, and vertical force, and of earth currents have been registered photographically, and accompanying eye observations of absolute declination, horizontal force and dip have been made as in former years.

The regular observations of magnetic declination have been made since 1899 January 1, in the Magnetic Pavilion, alternating with determinations in the Magnet House (for effect of the iron in the Observatory buildings), the observations in the Magnetic Pavilion being made with a hollow cylindrical magnet mounted in conjunction with the large theodolite.

The determinations of horizontal force and dip have been made with the Gibson deflexion instrument and the Airy dip circle mounted in the new Magnetic Pavilion, since 1898 September.

The principal results for the magnetic elements for 1899 are as follows:—

Mean declination ... ... 16° 34'·2 West.

Mean horizontal force ... { 3·9947 (in British units).
　　　　　　　　　　　　 { 1·8419 (in Metric units).

Mean dip (with 3-inch needles) ... ... 67° 10' 13".

These results depend on observations made in the new Magnetic Pavilion, and are free from any disturbing effect of iron. The correction to the declination, as found in the Magnet House, is − 10'·7, as deduced from the observations made with the new declinometer in the Magnetic Pavilion.

The magnetic disturbances in 1899 have been few in number. There were no days of great magnetic disturbance and sixteen of lesser disturbance. Tracings of the photographic curves for these days, selected in concert with M. Mascart, will be published in the annual volume as usual. The calculation of diurnal inequalities from hourly quiet days in each month has been continued.

The question of the regulations to be enforced for the protection of the Observatory from disturbance of the magnetic registers by electric railways or tramways in the neighbourhood is now under the consideration of the Board of Trade.

METEOROLOGICAL OBSERVATIONS.

Consequent on the changes in connection with the new Observatory buildings, the shed containing the photographic thermometers was moved 15 feet towards the west on May 16 and 17, 1899.

The Kew Committee of the Royal Society has suggested that

steps should be taken to assimilate the methods of registration of atmospheric electricity with the Thomson electrometers at Greenwich and Kew, and the question of the modifications to be introduced into the Greenwich electrometer is now under consideration.

The mean temperature for the year 1899 was 50°·7, being 1°·2 above the average for the fifty years, 1841–90.

During the twelve months ending 1900 April 30, the highest temperature in the shade (recorded on the open stand in the Magnetic Pavilion enclosure) was 90°·0, on August 15. The highest temperature recorded in the Stevenson screen in the Observatory grounds was 88°·8 on the same day.

The month of August was exceptionally warm, the mean temperature for the month being 65°·5, which is 3°·9 above the fifty years' average (1841–1890). This high temperature for the month has only been reached before on one occasion in the previous fifty-eight years, viz. in August 1857. The month of November was also exceptionally warm, the mean temperature for the month being 4°·8 above the average.

The lowest temperature of the air recorded in the year was 18°·0, on February 9. There were fifty days during the winter on which the temperature fell below 32°, a number slightly below the average.

The mean daily horizontal movement of the air in the twelve months ending 1900 April 30 was 268 miles, which is 13 miles below the average for the preceding thirty-two years. The greatest recorded daily movement was 776 miles on April 13, and the least 50 miles on October 22. The greatest recorded pressure of the wind was 27 lbs. on the square foot, on November 3, and the greatest hourly velocity 48 miles, on April 13.

The number of hours of bright sunshine recorded during the twelve months ending 1900 April 30, by the Campbell-Stokes instrument, was 1636 out of the 4454 hours during which the sun was above the horizon, so that the mean proportion of sunshine for the year was 0·367, constant sunshine being represented by 1.

The rainfall for the year ending 1900 April 30 was 21·97 inches, being 2·57 inches less than the average of fifty years. The number of rainy days was 146. The rainfall in the month of August was only 0·354 inch, being the smallest August rainfall on record in the fifty-nine years, 1841–99. The next smallest value was 0·45 inch, in August 1849. The rainfall in February amounted to 3·58 inches, being the largest February rainfall on record in the sixty years, 1841–1900, with the exception of the February rainfalls in 1866 and 1879, which amounted to 4·03 and 3·81 inches respectively.

The remaining portion of the report is devoted to the progress in the printing and distribution of the publications and chronometers, time-signals, longitude operations, &c.

In view of the large additions to and modifications in the instruments and buildings of the Royal Observatory in recent years, it is proposed to prepare a full description of the Observatory, illustrated by photographs.

It may be mentioned that the Observatory equipped and sent out an expedition to observe the total solar eclipse of May 28, having received the sanction of the Admiralty. The Astronomer Royal, with Mr. Dyson and Mr. Davidson, left for Ovar, in Portugal, on May 11, taking with them the Thompson 9-inch photographic telescope, the new 4-inch enlarging lens for large-scale photographs of the corona, a pair of photographic spectroscopes with heliostat, lent by Captain Hills, for photographing the spectrum of the lower chromosphere and of the corona, and a double camera, on one of the photo-heliograph mountings, with lenses of 4 inches and 2½ inches aperture for photographing the coronal streamers.

An examination of the fine photographs that were obtained by the party, which were shown on the day of the visitation, gave one a good idea of the success which had rewarded their efforts.

THE GEOLOGICAL AGE OF THE EARTH.[1]

WHILE, in his efforts to arrive at an estimate of geological time, the geologist himself is seriously hampered by the uncertainty of the data at his disposal, he has followed with expectant interest the successive attempts made by votaries of

1 "An Estimate of the Geological Age of the Earth." By J. Joly, M.A., D.Sc., F.R.S., Hon. Sec. Royal Dublin Society; Professor of Geology and Mineralogy in the University of Dublin. Pp. 44. (*Scientific Transactions* of the Royal Dublin Society, vol. vii. Ser. ii. Dublin, 1899.)

kindred sciences to attain the solution of a problem so fascinating. It is true that past efforts in this direction have taught us to expect at the best a merely approximate result, by whatever method this problem may be attacked ; at the same time, every attempt is welcome which shall tend to more narrowly limit the margin of approximation. In the important treatise before us, Prof. Joly proposes a novel and ingenious method of approaching this difficult question, and if his argument relies for its success on a considerable basis of assumption, he has nevertheless arrived at results of striking interest.

It is, first of all, assumed that the denudation by solution of the land surface, since the first formation of a solid earth-crust, has been on the whole a uniform process ; and further, that the amount of sodium now contained in the ocean has been for the most part transported to it by rivers since the land surface first became exposed to the action of solvent denudation. The reasons which, in the author's view, render these assumptions probable truths are fully discussed in the paper. If, now, we can obtain a correct estimate of the amount of sodium at present contained in the waters of the ocean, and also of the amount annually supplied to the latter by rivers, we have the requisite data whereby the earth's geological age may be determined. Basing his calculations upon the most careful and recent estimates, Prof. Joly finds that the mass of sodium contained in the ocean amounts to $15,627 \times 10^{12}$ tons. In estimating the amount of sodium carried annually by rivers into the sea, Sir John Murray's analyses of nineteen rivers (including many principal ones) are quoted, and a result of 24,196 tons of sodium per cubic mile of river water is obtained. Sir John Murray's estimate of the annual river discharge into the ocean, amounting to 6,524 cubic miles, is also accepted. From these figures the mass of sodium annually carried to the sea is calculated, and this amount divided into the total mass of sodium contained in the ocean, gives a result of about 94,800,000 years, representing the duration of geological denudation. So much is set forth in the first section of the paper. In the succeeding eight sections corrections on the above estimate are discussed, and several possible objections are dealt with.

The author first enters into a speculative discussion regarding the succession of events attendant on the first cooling of the earth's surface, with the object of arriving at conclusions as to the nature of the primitive ocean. Incidentally the view is favoured which supposes that at the first condensation of water upon the surface a greater density was conferred on the sub-oceanic crust than on the sub-aërial tracts ; it is deemed improbable that this distribution of pressure became subsequently seriously modified, and the author gives his support to a belief in the permanency of ocean basins. The early ocean itself is supposed " for want of other known alternative," to have contained " a quantity of hydrochloric acid roughly represented by the chlorine now in the ocean." This being admitted, it is clear that a certain degree of saltness would primarily be acquired by the early hydrosphere, and this must be allowed for in modifying the above estimate of geological time. To accomplish this Prof. Joly first quotes Clarke's average analysis, showing the probable composition of the primitive earth-crust. The action of the heated acid ocean upon such a rock mass, and the apportioning of the acid among the bases, is next considered. It is calculated that of the total amount of chlorine contained in the original ocean, only 14 per cent. could have been taken up to form sodium chloride, and in order to arrive at the actual amount of this first formed sodium chloride, Prof. Joly proceeds to estimate the chlorine of the original ocean. This is done by subtracting from the total amount of chlorine now contained in the ocean the quantity of that element supposed to have been transported to it by rivers during the course of geological time. But of the river-transported chlorine a certain proportion has been derived from the sea itself, and for this a deductive allowance of 10 per cent. is made as probably sufficient. Having estimated that a total of about 76×10^6 tons of chlorine are annually supplied by rivers to the sea, the author *assumes* the duration of geological denudation to have been about 86×10^6 years, and finds that during this period 6536×10^{12} tons of chlorine have been introduced into the ocean. By subtracting this from the total chlorine now contained in the sea (as sodium chloride and magnesium chloride), a total of $21,780 \times 10^{12}$ tons is arrived at, representing the original chlorine of the oceanic waters. If 14 per cent. of this would unite with sodium, then 1972×10^{12} tons of sodium were brought into solution by the action of the primitive acid ocean. This result can now be employed in correcting the original estimate of geological time, which was reckoned on the

supposition that all the sodium now in the ocean had been supplied by rivers. Thus, the total amount of sodium supplied by rivers is reduced to $13,655 \times 10^{12}$ tons, and this deductive correction of 12·6 per cent. reduces the duration of geological time to $86 \cdot 9 \times 10^6$ years. The value of this ingenious correction appears to be lessened, however, by the necessary introduction of an arbitrary assumption for the duration of geological time. Is there any reason, too, to show that at the first condensation of water upon the earth, alkalies may not have been present to neutralise to almost any unknown extent the acid of the primæval ocean? The author believes that the amount of correction necessary in allowing for the action of acids other than hydrochloric, in the primitive ocean, is practically negligible.

By a further slight modification of the figures representing the sodium annually transported into the sea, a final estimate of 89·3,000,000 years·is arrived at, a figure based, we are told, "on the most complete estimate of probabilities." But even in this estimate the author does not claim a degree of accuracy "approximating to so small a time interval as 100,000 years."

Prof. Joly then examines the significance of rock-salt deposits, as possessing a possible bearing on his theory, but having discussed the origin of such deposits he concludes that any error involved by ignoring them must be very slight. But the extent of the saline deposits surely cannot possibly be estimated, and may perhaps have been considerably underrated. Even if it be admitted, as urged by the author, that the salt basins of the present day are in great part not of oceanic origin, this does not necessarily apply in like degree to saline deposits of the past, when earth movements may have played a more vigorous part in aiding their formation as oceanic derivatives.

A point of seemingly great significance in its bearing on Prof. Joly's theory is the retention of salts in the interstices of stratified rocks, the salts derived from the waters in which the rocks were laid down. In 1856 Dr. Sorby drew attention to the soluble salts contained in certain dolomites, and Dr. Sterry Hunt has recently referred to the " fossil sea water " retained in the pores of stratified rocks.

The observations of the Rev. O. Fisher on this point, recorded in a recent review of Prof. Joly's paper (*Geol. Mag.*, March 1900) are of the greatest interest, as showing that some of the sodium of river waters may have been derived not from the rocks, but originally from the ocean itself.[1] In estimating the mass of sodium held in solution by the ocean, should not some allowance be made too for an unknown bulk of highly pervious deep-sea sediments? May not such deposits be in part of great thickness, and by reason of the sea water with which they are impregnated form a store for sodium ?

In the succeeding section the potash and soda percentages of the igneous and sedimentary rocks respectively are considered. Quoting Clarke and Rosenbusch for estimates of these, Prof. Joly attempts to prove that the deficiency of soda in the sedimentary rocks (1·47 per cent. of soda and 2·49 per cent. of potash, as against 3·61 per cent. of soda and 2·83 per cent. of potash in the primitive crust) is accounted for by the amount of sodium calculated to have been supplied to the ocean by rivers. It is claimed "that the estimated amount of sedimentary strata would, in its formation, be adequate to yield to the ocean the sodium that is in it, assuming these sedimentaries to be derived from rocks having the mean composition of the important eruptive masses now known." Allowance is made for a slight deficiency in the sodium of the ocean by the existence of the rock-salt deposits. For the success of this argument it is unfortunately necessary to assume that a correct estimate of the total bulk of the sedimentary rocks is possible. Mr. Mellard Reade's calculation is provisionally taken as a basis. Accepting also Mr. Reade's estimate of the proportion of calcareous to other sediments, the latter are found to be equal to a layer 1·6 miles in depth over all the land area. From this the mass of the detrital sediments is calculated, and the actual amount of their soda is arrived at. To this is now added the amount of sodium (reckoned as soda) contained in the sea. This restoration would bring the soda percentage of the total mass of sedimentary rocks, even allowing for the rock-salt deposits, to little above 3 per cent., and in order that the figures shall be brought into better accordance with Clarke's calculated soda percentage for the primitive crust, the estimate of the amount of detrital sedimentary rocks is ingeniously amended to equal a layer 1·1 miles thick over all the land area, with the result that an amount

[1] Since these lines were written, Prof. Joly has dealt further with this point and with the question of alkalies neutralising the primitive acid ocean (*Geol. Mag.*, May 1900).

is obtained little short of the desired 3·61. This result must appear sufficiently striking, but it may be seriously doubted whether even an approximate estimate of the total bulk of sedimentary strata can possibly be arrived at. Such an estimate must inevitably rest in great part on a basis of pure speculation. Not only are we ignorant, as regards huge areas, of the thickness of these strata, but immense tracts still remain unexplored so far as their geology is concerned. Further, the boldest guess can tell us little of sedimentary strata hidden beneath the surface of the ocean, and it may be looked upon as a lucky coincidence that Prof. Joly is able to attain the above result when restoring to the estimated sedimentary rocks the sodium of the sea. The question also of pre-Cambrian rocks of sedimentary origin appears here to be too lightly passed over, for although so little is known of their actual extent, the trend of recent researches has been to show that they may constitute a not unimportant fraction of the total sediments formed. It is scarcely necessary to recall the fact that the earliest known fossil faunas, including marine forms of comparatively high organisation, clearly indicate that a habitable ocean had already for long ages been in existence.

The unequal ratios of the alkalies in the ocean and in the rivers respectively next receive attention. The fact that the ratio of potash to soda is very much higher in the rivers than in the sea, is believed by the author, not to indicate that the rivers now contain more potash relatively to soda than in former times, but is to be accounted for by the constant abstraction of potash from the ocean, largely in the glauconite now forming on the sea floor, and so extensively distributed in the sedimentary strata. Stress is also laid on the fact that potassium brought from the atmosphere by rain tends to become retained upon the land, while the sodium is more readily returned to the ocean. In arguing for the uniformity of denudation by solution in past times, Prof. Joly brings forward some good reasons to show that the distribution of land and sea can have varied but little. As regards the greater exposure of igneous rocks in early times some interesting points on the nature of weathering and soil formation are noted, and it is concluded that the unequal percentage of sodium in the igneous and sedimentary rocks would, as regards supply to the ocean, be counterbalanced by the different rates of weathering. Sedimentary rocks, poorer in alkalies, allow of more rapid denudation.

In the concluding section of this paper the action of the ocean as an agent in solvent denudation is dealt with. Such action, the author maintains, is carried on chiefly along the coast lines, and is very small as compared with that effected by rain and river waters. Experiments are quoted to show that the power of sea-water to decompose felspar is minute in comparison with that exerted by fresh water. It is further pointed out that the volcanic *débris* of oceanic deposits have the alkali ratio of igneous and not that of sedimentary rocks. A correction of half a million years on the original time estimate is thought to be a sufficient allowance to make for the solvent denudation by the ocean. But even allowing, as held by the author, that chlorides other than sodium chloride may in past times have in some measure retarded solvent denudation by the ocean, it may be suggested that subaqueous volcanic action, at one time more frequent than at present, with its attendant conditions of exceptional temperature and pressure, may by frequent repetition through vast periods of time have played some part in aiding this process.

Prof. Joly does well in finally recognising the uncertainty attending his corrections on the original estimate of geological time, and he certainly allows no too wide a margin for error in the final result when he claims that "a period of between eighty and ninety millions of years" has elapsed since the land first became exposed to denuding agencies. For not only in the data upon which the corrections are founded, but also in the factors employed in the original calculation, there is to be found comparatively little of certainty and much that is purely speculative. In this latter category must be placed the supposed sequence of events at the first cooling of the globe. The relative intensity of geological activities in the past is also unknown to us, and the possibilities as regards the activity of the sun and the influence of the moon in modifying meteorological agencies during the earlier chapters of the earth's history appear to render hopeless the final solution of the time-problem by such a method as that here employed. But in this interesting treatise Prof. Joly has with marked ability and originality attacked a most difficult question, and his novel theory calls for the fullest consideration from all geologists and physicists.

NO. 1601, VOL. 62]

NOTES ON SATURN AND HIS MARKINGS.

THE possessors of telescopes will welcome the reappearance of Saturn as a rather conspicuous object in the evening sky. The planet now rises at 7h. 40m. p.m., and remains visible afterwards throughout the night, but unfortunately his altitude is extremely low. His southern declination being 22⅓°, his position is only 16° above the horizon at Greenwich even at the time of his meridian passage. Notwithstanding these unfavourable conditions, excellent views may, however, occasionally be obtained of his general aspect. From stations in the southern hemisphere the planet may be seen under the best circumstances.

This planet with his rings, belts and moons, forms a picture quite unique of its kind. The globe is greatly compressed at the poles, like that of Jupiter, and the rate of its axial rotation similarly rapid. We recognise also in the dusky bands of Saturn another parallel to the visible lineaments of the "Giant Planet," but there is a marked difference as regards the distinctness with which the details on the two bodies may be viewed. Jupiter's large disc and superior brilliancy enable the markings and their variations of form and motion to be followed with great facility and certainty. Saturn being much smaller and fainter is more difficult, especially as regards the more delicate features. Cassini's division in the rings and the principal belt on the globe may be distinguished with a two inch refractor, but Encke's division in the outer ring is a doubtful, or probably a very variable feature, which at certain times appears to be missing altogether, while on other occasions it is described as faintly outlined as a pencil-like curve at the ansæ.

That there are occasional irregularities on Saturn is proved beyond contention. In 1790 Sir W. Herschel remarked a very dark spot on the limb, and in 1793 noticed some irregularities in a quintuple belt which enabled him to ascertain the planet's rotation period. The large white spot seen by Prof. Hall and others at the close of 1876 affords a good instance of change, and it is well-known that the disposition and number of the belts vary from year to year. We naturally conclude that these belts must occasionally exhibit irregularities like those of Jupiter.

The planet is now presented to us at an angle which permits the ring system to be seen with splendid effect. We now view the northern side of the ball and rings, and this will continue to be the case until 1907.

Perhaps there is no object upon which it is easier to exercise the imagination than upon Saturn. And there is probably no orb in reference to which more errors in detail have been made, though both Mars and Venus have encouraged a large number of observational misconceptions. Many of the abnormal results reported in recent years, and due to small instruments, may be safely dismissed, for they are not only doubtful but, when all the conditions are considered, ridiculous, and palpably the outcome of unconscious suggestions of the imagination. Yet there can be no question as to the good faith of those who are responsible for some of the wonderful seeings lately published. They honestly believe they have seen what they have drawn, and as a matter of fact it is an extremely difficult point to distinguish between real and imaginative features on Saturn. The trembling of the image, its faintness under high power or its smallness under low power encourage much fictitious detail which every observer cannot regard as illusory.

Some of those who claim to have seen many irregular markings on this beautiful planet ascribe their success to special training; but this explanation will scarcely stand, for others of equal experience and using more powerful appliances have quite failed to observe them. The difference is not one of sight, of practice, or of instrumental means. It resolves itself into a question of personal ethics. There are men who will report nothing but what they are absolutely certain is presented to their eyes, and are unbending in their regard for the truth; there are others who, though equally sincere in intention, are not so reliable in their judgment, and accept features which are apparently glimpsed, but which are in reality prompted by the imagination on an unsteady and very delicate object.

It is to be hoped that time will eliminate all the fanciful representations of Saturn which recent observations have so abundantly supplied. The period has now arrived when the planet may be telescopically surveyed with a view to obtain a really sound knowledge of such features as are portrayed in moderately powerful instruments. Those who have employed large and small telescopes in planetary observation aver that the former are more effective than the latter; but it is remarkable that small instruments have been the means by which a large amount of useful work has been done in this field of observation. It is also an unavoidable conclusion that many of the mistakes in planetary work have been due to inadequate power and light in the appliances used. Possibly during the next few months some of the existing discordances may be cleared up, and some new facts learnt concerning this the most beautiful planet of our system.

There was an interesting occultation of Saturn by the moon on June 13, but at Bristol clouds interfered with observation. At Yeovil, Somerset, the Rev. T. E. R. Phillips watched the phenomenon with a 3-inch refractor. There will occur another event of this kind on September 3 next, when the planet will disappear at 7h. 16m. and reappear at 8h. 11m. p.m. Occultations of Saturn are somewhat rare, the last, prior to that of June 13, occurring twelve years ago, viz. on October 1, 1888.

The planet may now be studied with advantage from southern observatories, where his altitude will be considerable and conduce to that excellent definition which is so necessary for the detection of faint and delicate markings. At every opposition it seems necessary that the number and arrangement of the various belts should be noted. A dark polar cap should be looke'l for, and any irregular appearances, such as dark and light spots on the dusky belts or intervening zones, should be carefully recorded. It is unfortunate that the results obtained in previous years are not sufficiently accordant to be of much service. In some cases where one observer has drawn one or two belts, another, equally experienced and with more powerful means, has represented seven or eight. Certain observers see the belts and zones mottled with spots, while others describe the aspect as perfectly smooth and quite devoid of all such irregularities. The evidence is, in fact, so conflicting that new and thoroughly trustworthy observations are greatly needed to set at rest the actual character of the details visible on this exceedingly attractive object.   W. F. DENNING.

UNIVERSITY AND EDUCATIONAL INTELLIGENCE.

OXFORD.—On April 27, 1899, an anonymous gift of 5000*l*. was made towards the building of a pathological laboratory. The donor now allows his name to be made known, and a decree will consequently be proposed on July 7, that the thanks of the University be conveyed to Ewan Richards Frazer, B.M., Balliol College, for his munificent donation to the pathological laboratory.

CAMBRIDGE.—The annual report of the Cambridge Observatory ·appears in the *Reporter* for June 30.· It includes an account of valuable work done with the Newall telescope, and of the steps taken to bring to perfection the Sheepshanks photographic equatorial. The binary character of α Aurigæ was announced as discovered at the Observatory two days before the arrival of Prof. Campbell's independent publication from the Lick Observatory.

Scholarships, or exhibitions, in natural science have been awarded at the following Colleges thus :
· Peterhouse : Lee.
Gonville and Caius : Cleminson, Burne, Garnsey, Lock, Macfie, Rittenberg, Thornton.

The Hopkins Prize for the period 1894-97 has been awarded to Mr. J. Larmor, F.R.S., of St. John's College, for his investigations on the Physics of the Aether and other valuable contributions to mathematical physics.

Mr. William Ritchie Sorley, formerly Fellow of Trinity College, has been chosen Knightsbridge Professor of Moral Philosophy in the place of Dr. Henry Sidgwick, who has resigned in consequence of ill health.

PROF. SIMON NEWCOMB has had the degree of doctor of laws conferred upon him by the University of Toronto.

PROF. J. H. POYNTING, F.R.S., has been elected Dean of the Faculty of Science of the University of Birmingham.

THE honour of knighthood has been bestowed upon Dr. G. Hare Philipson, president of the University of Durham College of Medicine, Newcastle-upon-Tyne.

PROF. J. R. CAMPBELL, head of the Agricultural Department of Yorkshire College, has been appointed under-secretary to the Department of Agriculture and Technical Instruction for Ireland.

THE honorary degree of D.Sc. has been conferred by the University of Oxford upon Prof. J. Mark Baldwin, of Princeton University, New Jersey, U.S.A. The new doctor is professor of psychology at Princeton, and editor of the *Psychological Review*.

HONORARY degrees were on Saturday last conferred by the Victoria University upon Lord Rayleigh, Sir William Huggins, Sir W. C. Roberts-Austen, Sir William Abney, Dr. T. E. Thorpe, Prof. J. Dewar, Prof. A. R. Forsyth, Mr. R. T. Glazebrook, Prof. Pickering, Prof. J. J. Thomson, and Mr. Henry Wilde.

A GRANT of 58*l*. from the Earl of Moray Endowment has been made by the Edinburgh University Court to Dr. J. H. Milroy for purposes of research. At a recent meeting of the Court it was announced that the late Prof. Sir D. Maclagan had bequeathed a marble bust of himself to the University, and that Miss E. A. Ormerod had presented six large volumes of drawings, chiefly by her father, to the library.

THE Drapers' Company offer for competition eight scholarships tenable at the day classes of the East London Technical College in chemistry, physics and engineering. The scholarships are of the value of 25*l*., 10*l*. being paid during the first year and 15*l*. during the second year. They also carry with them free tuition. Particulars may be obtained from the Director of Studies, East London Technical College, People's Palace, E

THE annual meeting terminating the session of the department of engineering in connection with University College, Liverpool, took place on Thursday last, when the William Rathbone Medal and the Rathbone Prizes were distributed by the Lord Mayor of Liverpool. The report, which was of a highly satisfactory character, was read by Prof. Hele-Shaw, after which an address was delivered by Prof. John Perry, F.R.S., upon the value of a thorough scientific education to the engineer.

THE following is a list of the members of the new Board of Education Consultative Committee :—Rt. Hon. Arthur Herbert Dyke Acland, Sir William Reynell Anson, Bart., M.P., Prof. Henry Armstrong, Mrs. Sophie Bryant, Rt. Hon. Sir William Hart Dyke, Bart., M.P., Sir Michael Foster, K.C.B., M.P., Mr. James Gow, Litt.D., Mr. Ernest Gray, M.P., Mr. Henry Hobhouse, M.P., Mr. Arthur Charles Humphreys-Owen, M.P., Sir Richard Claverhouse Jebb, M.P., Hon. and Rev. Edward Lyttelton, Very Rev. Edward Craig Maclure, D.D., Dean of Manchester, Miss Lydia Manley, the Venerable Ernest Grey Sandford, Archdeacon of Exeter, Mrs. Eleanor Mildred Sidgwick, Prof. Bertram Coghill Alan Windle, M.D., Rev. David James Waller, D.D. The draft Order in Council, giving particulars of the duties, &c., of the Committee, has been issued as a Parliamentary paper.

FURTHER munificent gifts for the furtherance of education in the United States are announced in *Science* and are as follows:— The sum of 125,000 dollars has been left to Harvard University by the late Edmund Dwight. The bequest will come into the hands of the University authorities after the death of certain persons who receive the income during their lifetime. The amount (100,000 dollars) promised by Mr. Rockefeller to Denison University, on condition that 150,000 dollars additional be raised before July, has now been claimed, the sum named having been subscribed. The sum of 50,000 dollars has been given to Colorado College by Mr. W. S. Stratton ; Mr. M. K. Jesup has given 25,000 dollars to Princeton University, and Lombard College in Galesburg, Ill., benefits in a like degree by the gift of Mr. W. G. Waterman ; while 10,000 dollars have been subscribed by Messrs. Phelps, Dodge and Co. for the endowment of the department of mining and metallurgy at Columbia University. In addition to the foregoing it is announced that Mr. L. C. Smith will build and equip a civil engineering building in connection with Syracuse University.

SOCIETIES AND ACADEMIES.

LONDON.

Royal Society, June 14.—"The Nature and Origin of the Poison of *Lotus Arabicus*." By Wyndham R. Dunstan, F.R.S., Sec.C.S., Director of the Scientific Department of the Imperial Institute, and T. A. Henry, B.Sc.Lond.

Lotus Arabicus is a small leguminous plant resembling a vetch, indigenous to Egypt and Northern Africa. It grows abundantly in Nubia, and is especially noticeable in the bed of the Nile from

Luxor to Wady Halfa. It is known to the natives as "Khuther," and old plants with ripe seed are used as fodder. The dried plant is unusually green, and possesses the aroma of new-mown hay. At certain stages of its growth it is highly poisonous to horses, sheep and goats, the poisonous property being most marked in the young plant up to the period of seeding. Owing to the trouble which this plant has given to the military and civil authorities in Egypt, the assistance of the Director of Kew was sought in order that the precise nature of the poison might be ascertained, and, if possible, a remedy found. The matter having been referred to the Scientific Department of the Imperial Institute, Mr. E. A. Floyer, Director of Egyptian Telegraphs, collected some of the material for investigation.

It was found that when moistened with water and crushed, the leaves of the plant evolved prussic acid in considerable quantity, the amount being greatest in the plant just before and least just after the flowering period. Further investigation has shown that the prussic acid originates with a yellow crystalline glucoside ($C_{22}H_{19}NO_{10}$), which it is proposed to name *lotusin*. Under the influence of an enzyme, also contained in the plant, lotusin is rapidly hydrolysed, forming *prussic acid, sugar*, and a new yellow colouring matter (*lotoflavin*). The hydrolysis may be effected by dilute acids, but is only very slowly brought about by emulsin and not at all by diastase. The peculiar enzyme, which it is proposed to call *lotase*, appears to be distinct from the enzymes already known. Its activity is rapidly abolished by contact with alcohol, and it has only a feeble action on amygdalin. Old plants are found to contain lotase but no lotusin.

The *sugar* has been proved to be identical with ordinary dextrose.

Lotoflavin, the yellow colouring matter, has the composition expressed by the formula $C_{15}H_{10}O_6$. It belongs to the class of phenylated pheno-γ-pyrones, and is a dihydroxychrysin, isomeric with luteolin, the yellow colouring matter of *Reseda luteola*, and with fisetin the yellow colouring matter of *Rhus cotinus*.

The decomposition which ensues on bringing lotase in contact with lotusin, as happens when the plant is crushed with water, is therefore probably expressed by the following equation :—

$$C_{22}H_{19}NO_{10} + 2H_2O = C_{15}H_{10}O_6 + HCN + C_6H_{12}O_6.$$
$$\text{Lotusin.} \qquad\qquad \text{Lotoflavin. Prussic acid. Dextrose.}$$

Hydrocyanic (prussic) acid occurs in small quantity in many plants, and according to Treub and Greshof is often present in the free state. The only glucoside at present definitely known which furnishes this acid is the well-known amygdalin of bitter almonds, which under the influence of the enzyme emulsin, also contained in the almond, breaks up into dextrose, benzaldehyde and prussic acid.

Owing to the scientific interest which attaches to this new glucoside, its properties and those of its decomposition products have been very fully studied, and the characteristics of the new enzyme have also been investigated.

We are much indebted to Mr. Floyer for the great pains he has taken to collect, in Nubia, the necessary material for this investigation, and also to Sir W. T. Thiselton-Dyer for having grown the plant at Kew from seed obtained from Egypt.

"The Exact Histological Localisation of the Visual Area of the Human Cerebral Cortex." By Joseph Shaw Bolton, B.Sc., M.D., B.S. (Lond.).

Geological Society, June 6.—J. J. H. Teall, F.R.S., President, in the chair.—Mechanically-formed limestones from Junagadh and other localites, by Dr. J. W. Evans. After reviewing the conditions under which granular limestones may be accumulated by current- or wind-action, the author proceeds to describe the limestone of Junagadh, a deposit some 200 feet thick, resembling in hand-specimens the Oolites of this country, though less firmly cemented together. The deposit is situate at a distance of thirty miles from the sea, and contains no large fossils of any kind. Calcareous rocks of similar character are described from other parts of Kathiawad, Kach, the south-eastern coast of Arabia, and the Persian Gulf—some of these contain unbroken marine shells and other fossils. These beds are included by Dr. H. J. Carter under the name of miliolite, on account of the frequent presence in them of the genus *Miliola*. The author discusses the origin of these deposits, and comes to the conclusion that the grains were formed in sea-water saturated with carbonate of lime ; some being deposited by currents in shallow water, and others thrown up as a calcareous beach, from which a portion were sifted out by the wind and blown inland

to form æolian deposits.—Note on the consolidated æolian sands of Kathiawad, by Frederick Chapman. The name miliolite formation was originally given by Dr. H. J. Carter to certain granular calcareous deposits occurring on the coast-line between the peninsula of India and the mouth of the Indus. The foraminifera and other organic remains in the rocks must have inhabited moderately shallow to littoral marine areas. The minute granules are worn and polished ; the prevailing genera of foraminifera are roundish, and would be easily moved by wind ; remains of larger organisms are absent ; and the deposits are false-bedded. All these phenomena are explicable if the deposits represent the accumulation of material derived from littoral calcareous sand of marine origin, mixed with mineral detritus from adjacent hills.—On Ceylon rocks and graphite, by A. K. Coomara Swamy. Ceylon is surrounded by raised beaches, and has been elevated in recent geological times ; fluviatile deposits also occur ; the gems for which Ceylon is famous are obtained from gravels in the Ratnapura district. With the exception of these recent deposits, the island probably consists entirely of ancient crystalline rocks. Graphite occurs chiefly in branching veins in igneous rocks, which at Ragedara are granulites and pyroxene-granulites. The relations to the matrix are described, and are held to favour the idea of the deposition of the mineral as a sublimation-product (Walther), or from the decomposition of liquid hydrocarbons (Diersche). Analysis of several of the minerals, including manganhedenbergite, are given ; and a bibliography of the geology of the island is appended.

Mineralogical Society, June 19.—Prof. N. S. Maskelyne, F.R.S., Past-President, in the chair.—Prof. H. A. Miers presented a communication from Miss Agnes Kelly on conchite, a new form of calcium carbonate. Conchite forms the material of various calcareous secretions in the animal kingdom (more particularly molluscan shells) which have hitherto been referred to aragonite ; it also occurs as the fur in kettles and boilers, and in many concretionary deposits, such as those of Karlsbad. In most of its characters it is intermediate between calcite and aragonite ; like calcite it is uniaxial negative, but shows no cleavage or twinning, and has higher indices of refraction ; and like aragonite it is converted into calcite on heating, but the change takes place at a lower temperature.—Mr. G. F. Herbert Smith described a method for the determination of the three principal indices of refraction from observations made in any arbitrary zone. This method is intended for minerals of low symmetry of which the indices are higher than those of any liquid. Observations are made of the deviations corresponding to different angles of incidence on both faces of a prism, and curves connecting the indices and the angles of orientation are plotted out. As in the method of total reflection, three of the critical values give the principal indices.—Mr. H. L. Bowman described the occurrence of monazite at Tintagel, and gave a detailed account of the crystallographic characters of the associated minerals, albite, quartz, rutile, pyrites and calcite.—Dr. J. W. Evans discussed the alteration of pyrites by underground water, a question which had arisen in connection with the erection of a dam in Mysore. From his experiments the author concludes that, provided the water contain a sufficient amount of carbonate of lime to neutralise the sulphuric acid resulting from the oxidation of the pyrites, exact pseudomorphs of limonite after pyrites are formed ; and as these occupy practically the same volume as the original pyrites, the rock suffers little disintegration by the action of the water.—Petrological notes by Mr. G. T. Prior dealt with the so-called "cancrinite-ægyrine-syenite" of Elfdalen, which he refers to sussexite at the basic end of the grorudite-tinguaite series of Brögger ; with a riebeckite-ægyrine-tinguaite (so-called "proterobase") from the Ruphachthal ; and with melilite-basalts from Madagascar and Siam.—Mr. L. Fletcher discussed the quantitative determination of the action of hydrochloric acid and of soda-solution on the enstatite and felspar of the Mount Zomba meteorite.

CAMBRIDGE.

Philosophical Society, May 21.—Mr. J. Larmor, President, in the chair.—On a certain diophantine inequality, by Major MacMahon, R.A., F.R.S.—On rational square curves of the fourth order, by Mr. Richmond.—On the reduction of quadrics, by Mr. Bromwich.—Experiments upon the rise of temperature of fabrics when moistened, by Dr. L. Cobbett. Dr. Cobbett showed that if expired air is breathed through several layers of dried filter paper wrapped round the bulb of a thermometer, a temperature of 10° C. or more above that of the body may be registered (Dr. Dudgeon's experiment); and that if a roll of flannel, thoroughly dried, be warmed to 96° C. and put into saturated steam at 100° C., the temperature within the roll may rise 30° C. or more above that of the steam (Dr. Parson's experiment). Further, he showed that when a roll of flannel, which has not been artificially dried is put into steam, at atmospheric pressure, heated to 200° C., though the surface of the roll becomes charred, the temperature in its interior rises rapidly to 100° C., but does not exceed this for a long time—indeed, not until all the separable water has been boiled away. He concluded that such substances when quite dry have the property of uniting with water, and of generating heat in the process, and this without becoming damp in the ordinary sense of the word ; and maintained that the source of this heat is not alone the latent heat of the vapour condensed, because a rise of temperature takes place when dried filter paper is wetted with water at the same temperature, but must include also either the latent heat of water converted into the solid state—as Sir W. Roberts has suggested in discussing Dr. Dudgeon's experiment—or else the energy set free in a chemical combination between the material and the water.—Experiments upon striated discharges, by R. S. Willows. The conditions affecting the distance between the striæ were investigated for hydrogen, nitrogen and air. In the first gas, as the current was increased from a very small value, the striæ first separated, attained a maximum distance of separation, and finally approached each other. In nitrogen and air their distance apart at first increased, and finally became constant. The distance apart varies inversely as the pressure until the discharge reaches the walls of the tube. The effect of the nature of the gas, the diameter and length of the tube, and the shape of the electrodes was also investigated. Any variation due to these was found to obey no simple law. The double striæ in hydrogen, noticed by De la Rue and Müller, were found to constitute a normal part of the discharge in this gas, provided a suitable pressure were established.—A method of measuring the retardation produced by a crystal plate, by L. R. Wilberforce. The author described a ready way of approximately determining the retardation produced by a plate of biaxial crystal cut perpendicularly to a mean line. The requisite measurements could be made with an ordinary polariscope.

PARIS.

Academy of Sciences, June 25.—M. Maurice Lévy in the chair.—Problem of the cooling of a wall by radiation, re-duced to the simpler case of cooling by contact, by M. J. Boussinesq.—Note on a series of abnormal contacts in the western region of the lower Pyrenees, by MM. Michel-Lévy and Léon Bertrand.—M. Giard was elected a member of the Section of Anatomy and Zoology in the place of the late M. Milne-Edwards, and M. Bazin was elected correspondent for the Section of Mechanics.—On the large sun-spot observed on June 17 with the great telescope of 1900, by M. Moreux. A sun spot, a drawing of which accompanies the note, had a diameter of 36,000 kilometres, and furnished a good example of the mechanism of segmentation of a sun spot. According to the author's hypothesis, the phenomena are not due to cyclones or volcanoes, but to superheated regions.—Trigonal normal curves, by M. F. Amodeo.—On the motion of a wire in space, by M. G. Floquet.—On two remarkable groups of geometrical loci, by M. E. Mathias. In his experimental re-sults obtained with carbonic acid, M. Amagat has considered the case of the locus of points in the (p, v) plane, such that for a total weight of liquid and vapour equal to unity, the volume of the liquid is constantly equal to that of the vapour. This locus, according to M. Amagat, is a straight line, nearly perpendicular to the axis of abscissæ ; but the author now shows that this locus is a curve constantly convex towards the volume axis.—On the discontinuity of the kathodic emission, by M. P. Villard. The three modes of exciting a Crookes' tube are considered, alternating currents, an induction coil and a static ma-chine, and in each the phenomenon would appear to be dis-continuous.—On the permeability of fused silica to hydrogen, by M. P. Villard. At 1000° fused quartz resembles platinum, in allowing hydrogen to pass through.—On the resistance of fused silica to sudden variations of tempera-ture, by M. Dufour.—On the telegraphone, by M. Valdemar Poulsen. A description of an instrument for automatically re-cording words spoken through a telephone.—On the develop-ment and propagation of the explosive wave, by M. ·H. Le Chatelier. An application of the photographic method to the

study of the explosive wave. Measurements are given for various mixtures of acetylene and oxygen, acetylene and nitric oxide, acetylene and nitrous oxide, and carbon monoxide and oxygen. In the last case the velocities depend upon the mode of ignition, and upon the quantity of the fulminating sub-stance used to start the explosive wave.—On the acidity of the alcohols, by M. de Forcrand. A thermochemical paper.—Addition of hydrogen to ethylene in presence of various reduced metals, by MM. Paul Sabatier and T. B. Senderens. Reduced cobalt effects the combination of ethylene and hydrogen at ordinary temperatures similarly to reduced nickel. A comparison of the results obtained with reduced nickel, cobalt, copper and iron shows that the activity of the metals in causing this reaction is in the order given, nickel being the most energetic.—On the crystalline combinations of acetylene with cuprous chloride and potassium chloride, by M. Chavastilon. It has been previously shown by the author that two kinds of crystals, yellow and colourless, may be obtained from the same copper solution, according to the velocity of the current of acetylene. Further analyses show that the colourless crystals correspond to the formula

$$C_2H_2.(Cu_2Cl_2)KCl,$$

and the yellow crystals,

$$C_2H_2.[(Cu_2Cl_2)_2KCl.]_2.$$

By the action of ether upon the colourless compound, the yellow crystals are obtained.—Oxidation of anethol and analogous substances containing a lateral propenylic chain, by M. J. Bougault. The method of oxidation used is the action of iodine in presence of precipitated mercuric oxide, an aldehyde being obtained. Aldehydes from anethol, isosafrol, isomethyleugenol and isoapiol have been prepared, together with the corresponding acids.—On a new derivative of benzophenone, by MM. Œchsner de Coninck and Derrien.—Composition of the compounds of fuchsine with acid colouring matters, by M. A. Seyewetz.—On the kidney of *Lepadogaster Gouanii*, by M. Frédéric Guitel.—On a fayalite rock, by M. A. Lacroix. The fayalite of Callobrières presents a very remark-able and exceptional mineralogical composition. It is essentially characterised by the association of the fayalite with grünerite, apatite and magnetite.—The function of the cell nucleus in absorption, by M. Henri Stassano. The nucleus, by reason of its chemical composition, plays a predominating part in the absorption of foreign substances.—On the proteolytic diastase of malt, by MM. A. Fernbach and L. Hubert.—Action of high frequency currents upon the elementary respiration, by M. Tripet. In diseases of nutrition, treatment by high frequency currents regulates the activity of reduction of the oxyhæmo-globin.—Influence of extracts of the ovaries upon the modifica-tions of nutrition caused by pregnancy, by MM. Charrin and Guillemonat.—The lake of Ladoga from the thermal point of view, by M. Jules de Schokalsky.—On a balloon ascent made on June 17, by M. Genty.—On an extraordinary halo observed on June 22, by M. Joseph Jaubert.

THURSDAY, JULY 12, 1900.

A MONOGRAPH ON LAND-PLANARIANS.

Monographie der Turbellarien. II. *Tricladida Terricola (Landplanarien).* By Prof. Ludwig von Graff. Pp. i + 574, and an atlas of 58 plates. (Leipzig : W. Engelmann, 1899.)

THE Turbellaria are rapidly becoming one of the most adequately and conveniently described groups of Invertebrates. Practically all that is at present known as to their anatomy, classification and distribution is comprised in three works—the "Monograph," by Prof. von Graff ; the special memoir, by the same author, on the "Acœla" ; and the masterly work, by Prof. Arnold Lang, on the "Polycladida" in the series of monographs on the fauna and flora of Naples Bay. The work before us, a magnificent folio, completes the author's share in the monographic treatment of the group. The first part was published in 1882, and was reviewed in this journal by Prof. Moseley. It is with great pleasure that we notice the dedication of the second part—jointly to Moseley and Fritz Müller—as only one of the many felicitous ways in which Prof. von Graff expresses his admiration for the work of these his fellow-labourers in the anatomy of planarians. We heartily congratulate Prof. von Graff on the appearance of this volume, the conclusion of a work begun twenty-five years ago.

Apart from all other claims upon our notice, this treatise is remarkable as being the first attempt to deal exhaustively with an essentially tropical group of animals, for nothing can be clearer, after reading this account, than that land-planarians, though not restricted to the tropics, have their headquarters in the equatorial forest belt. Von Graff has himself spent some time in Java, Singapore and Ceylon, and the personal acquaintance made in this way with these animals, and the conditions under which they live, gives a vividness and directness to his descriptions. Other naturalists have notably assisted him. Prof. Dendy, whose admirable and continued researches on the land-planarians of Australia and New Zealand receive full acknowledgment in this work, Spencer, Hamilton, Fletcher and others have sent collections of these animals to von Graff from Australia. Strübell, Max Weber, the Sarasins and others have contributed specimens collected by them in the Oriental region. South America is represented by planarians taken by Darwin, Fritz Müller, von Jhering and Plate. Nearly all the chief museums in Europe have contributed their specimens to von Graff, and in this manner he has been able, not only to more than double the number of species which were recognised when he began this work, but also to personally examine all but a very small percentage. To realise the rapidity of the increase in species of land-planarians during the last twenty years, it will be sufficient to state that Moseley's complete list, made in 1877, comprised only 63 forms, while 125 were known when von Graff began his monograph on the group, during the course of which he has added no less than 200 new species, and this, together with increments from other sources, makes a total of 348.

Of this unexpectedly large number (for it is about equal

to all the other Turbellaria put together which have been really adequately described), less than a dozen occur (with the exception of the Manchurian sub-region) in both the Palæarctic and Nearctic regions. The majority come from South America, the Oriental and Australian regions. Even this statement, however, does not represent the richness of the tropics, for Australia is really the only country where land-planarians have been systematically collected and recorded. Our knowledge of the planarian fauna of Africa, India, China, Central and North America is almost a blank. And additions to it will no doubt be made, not only in these countries, but also in places in which naturalists have already sought planarians. The island of Celebes, for example, has been examined by several zoologists, who have searched for land-planarians, but without success. Hickson, and after him Max Weber, searched in vain. More recently, however, the Sarasins have thoroughly explored the island, and have brought to light a most interesting fauna. Von Graff shows that the land-planarians of North Celebes exhibit Oriental characters, those of South Celebes Austro-Malayan features. Altogether twenty-one Celebesian species are now known, and of these all but three are new. We refer to this point particularly as showing that we are only beginning to realise the variety of this element of the tropical fauna, and that years of work are necessary in any one country before the planarian fauna can be fairly estimated. In Ireland a new species has been found near Dublin, and two other additions to the land-planarians of Europe have been made quite recently.

The first part of von Graff's great work is devoted to a full statement of the anatomy and histology of land-planarians. This section must have involved a vast expenditure of labour. Direct observation of the anatomy of living or compressed specimens is impossible, owing to the amount of opaque pigment in the tissues. Even the external apertures are hard to discover. Dissection is precluded by the solidity and tenuity of the body. The only available means in the majority of cases is the laborious one of serial sections, and this method the author has applied to elucidate the structure of no less than eighty-two species.

The chief result obtained in this way is the uniformity of the general anatomical features. Land-planarians form a homogeneous group, and agree closely in structure with the marine Triclad Turbellaria so far as these are at present known. Their distinguishing features appear to be correlated with the terrestrial habit. Among these may be mentioned their greater size and more powerful musculature ; the formation of a "keel" to the foot, and the abundance of glands both for lubricating the foot and for enveloping prey ; their brilliant, often intensely brilliant, colouring ; the presence of sensory thickenings and of sensory pits on the anterior part of the body ; and, perhaps their most significant distinction, the presence of elaborate structures accessory to reproduction. The anatomy of some of the simpler land-planarians is, however, an almost exact repetition of an aquatic Triclad, and the retention of cilia in the epidermis points to the conclusion that in land-planarians we have the first stage in the evolution of a terrestrial group from an aquatic one.

M

The presence of three extreme forms of rod-like secretions and the absence of nematocysts are noteworthy features of the epidermal glands. "Flame-cells" and parts of the canalicular system of excretory vessels have been found. The nervous system exhibits an interesting series of modifications. In the most primitive members of the group—the broad, flattened neotropical Geoplanidæ—the central nervous system consists of a dense plexus forming a horizontal plate lying just above the ventral body-wall. From this plexus nerves are given off, which either at once enter some organ or join with their fellows to form a well-developed cutaneous nerve-plexus right round the body. There is no distinct "brain." In the narrower neotropical and Australian members of the same family, a concentration of this central plexus takes place along two admedian lines, and a marked anterior thickening indicates the "brain." In the other families, as the sense-organs, which are scattered in the Geoplanidæ, become massed in front, so does the individuality of the brain become more and more pronounced. These sense-organs are of four kinds. The tentacles with eyes at their bases, found only in two South American forms. A paired sensory ridge forming a margin to the anterior part of the ventral surface in Geoplanidæ, and to the dorsal and ventral edges of the "cephalic plate" of Bipaliidæ. These ridges are innervated by the cutaneous nerve-plexus. Then the sensory pits which accompany these ridges, but which are supplied direct from the central nervous system. Lastly, the eyes. Of these there are two kinds. One, with the usual Turbellarian type of structure, has the rods directed away from the light, and the nerve entering in front and not, as in most other Invertebrates, from behind. This kind of eye occurs not only down both sides of the whole length of the body in the Geoplanidæ, but also on the dorsal and even the ventral surface. In the Bipaliidæ such eyes are concentrated at the margin and angles of the "cephalic plate." The other type of eye is one common to most Invertebrates, but hitherto unknown in Turbellaria. It consists of a pigment cap with a nerve perforating it behind, and entering the rod-cells, which face outwards towards the light. Such are the large paired eyes of the Rhynchodemidæ, and they are often imbedded in the nervous matter of the "brain."

The most novel and richly illustrated section of the anatomical part of von Graff's volume is, however, that in which the unexpected complexity and variety of the reproductive organs is discussed. This chapter is a most important addition to Turbellarian anatomy, and the results well repay the labour which has been spent on its preparation and illustration.

The next section, a short one, is devoted to the habits of land-planarians. Here, as in the other sections, the author has collected and given *in extenso* all the essentially important information that has been previously obtained. In this section, however, he adds little to the observations of Darwin, Moseley, Dendy, von Kennel and others. Land-planarians, though capable of withstanding considerable variations of temperature, are almost instantly killed by contact with dry objects, and by immersion in water, whether fresh or salt. The majority flourish best in dark, moist places. They are nocturnal, living by day under stones and tree-

trunks, under the sheaths of bananas, and on tree-ferns Some are actually subterranean, and live on earth-worms. These are blind. But most land-planarians are content with a diet of snails, woodlice and insects. *Rhynchodemus vejdowskyi* is one of the few diurnal forms. It not only crawls about by daylight, but moves with such grace and rapidity that when von Graff saw it at Buitenzorg he mistook it at first for a Myriapod. A species of *Geoplana* has been found in some numbers creeping on the pavement of Melbourne in broad daylight.

The coloured plates, which show the appearance and bizarre markings of land-planarians, form quite a feature of von Graff's work. The ground colour is usually diversified by mottling or by brilliant longitudinal stripes. Bold transverse bars of colour are comparatively rare, but they occur in a small group of each of the two families, Bipaliidæ and Geoplanidæ. It is an extremely curious fact that all the barred species of the former family are confined to the islands of the Malay Archipelago ; all those of the latter to the Chilian sub-region, with a single exception found in Brazil.

With reference to any supposed significance of these colours, von Graff suspends his opinion. Dendy, it is true, has shown that *Geoplana* produces an unpleasant taste on the tongue, and that fowls readily pecked at this planarian, but would not swallow it. A casual experiment of this kind is, however, not sufficient to justify the assertion that the colours of the land-planarians are of the "warning" category. The great difficulty is how to explain the prevalence of such brilliant colours and definite patterns in a group which is almost exclusively nocturnal. Yellow is the commonest colour, then orange, red, green, blue and violet. In young specimens, the pattern is more sharply defined, and the pigment (which is present both in large, richly branched connective-tissue cells and in the parenchymatous matrix) relatively more abundant than in the adult. No experiments appear to have been made to test whether land-planarians possess the power of colour-change. As with many other groups of animals upon which elaborate anatomical monographs have been written, the physiology of land-planarians is practically unknown.

The distribution of this group is very interesting, and is clearly illustrated by von Graff both by tables of every species and by a coloured map. To one of the main facts, their rarity north of the Equator and their abundance in the tropics, I have already adverted. Another interesting and suggestive discovery is the large proportion which occur on islands. More than half of the known species (201 out of 348) are purely insular, and each of almost all these (186) is limited to one island. As showing that this is only one of several indications of the local distribution of many species, von Graff points out that only five land-planarians occur in two geographical regions, only twenty in two subdivisions of the six regions, and but eighteen in two parts of the same region separated by an arm of the sea. The land-planarians afford a striking proof of the value of the Sclater-Wallace regions, which accordingly are adopted by the author.

The Oriental region is, perhaps, the richest, certainly the best characterised. Five-sixths of the family Bipaliidæ are confined to this region, and the remainder occur in Madagascar (most of the species being peculiar to this island) and Japan. The land-planarians of the Australian and neotropical regions are alike in one striking feature. The family Geoplanidæ is practically divided between them. The neotropical members of the genus *Geoplana* include most of the flattened primitive ones, and also peculiar forms such as *Leimacopsis* and *Polycladus*. Von Graff goes so far as to share the opinion that this geoplanid fauna has arisen on a lost Antarctic continent, and has spread on the one hand to New Zealand and Australia, on the other to South America. The distribution of earthworms lends strong support to this view, as Mr. Beddard has shown.

The concluding section of the work is composed of full systematic descriptions of the families, genera and species. Von Graff makes five families : the Limacopsidæ, with two tentacles ; the Cotyloplanidæ (an unnatural family), with suckers ; the Geoplanidæ, with scattered eyes ; the Bipaliidæ, with the eyes limited to the flattened " head " ; and the Rhynchodemidæ, with a pair of large eyes. There are now nineteen genera, many of which are new.

This monograph will be of inestimable value to all naturalists interested in land-planarians, and the author is to be congratulated on having completed such a laborious task with unfailing accuracy. The lithographers and publishers deserve a special word of praise for the beautiful plates and printing which adorn this book.

F. W. GAMBLE.

A SCIENTIFIC ENGINEER.

Papers on Mechanical and Physical Subjects. By Osborne Reynolds, F.R.S. Vol. i. Pp. xv + 416. (Cambridge : University Press, 1900.)

THE Cambridge University Press has during some years past contributed very largely to the progress of physical science by the issue of the collected works of great mathematicians and physicists. The volumes which contain the collected writings of Maxwell, Adams and Cayley form a rich storehouse of knowledge ; and the efforts the Press has made to induce living writers, such as Kelvin, Stokes and Rayleigh, to edit their own papers for issue in a collected form deserve the gratitude of all students.

Among the latest of such reprints is the volume before us. Its author, Prof. Osborne Reynolds, has passed a busy life as a teacher in a great commercial and manufacturing city, and his collected papers testify to the breadth of his interests and the wide scope of his work. The papers included in the present volume, some forty in number, were published between 1869 and 1882. They range over a great variety of subjects, from the tails of comets and the solar corona to problems connected with the steering of ships and the bursting of guns. In so varied a collection the relative importance of the different papers differs greatly, and yet all are interesting ; and all have advanced the sum of human knowledge.

Indeed, on reading them, one cannot help regretting that the author's interests have been so widely diffused, and that he has not had the opportunity of concentrating himself on some one or other of the great engineering problems which await solution, applying to it his practical experience and insight and his mathematical skill.

An extract from the author's preface makes the cause of this clear. He writes :

" As affording some explanation of the absence of any connection between many of the subjects in this collection of papers, it may be pointed out that these subjects have not been determined by arbitrary selection, neither have they been the result of following up one line of research. They have for the most part been suggested by the discrepancies between the results obtained in definite mechanical arrangements, such as occur in some parts of the large field of practical mechanics, and the conclusions arrived at as to what those results should be for the same circumstances, by means of geometrical and physical analysis, as far as this analysis was developed at the time."

But to turn to the matter of the papers ; it would take too long to attempt to analyse them all ; and, indeed, the results of the most important are now classic, *e.g.* those on the refraction of sound, the action of a screw propeller, the steering of screw steamers, and the explanation of the radiometer.

The two papers on the refraction of sound are numbered 16 and 22. Stokes had, seventeen years before the date of the first of these papers, suggested the reason why sounds are heard less distinctly against the wind than with it. It is due to the fact that the velocity of the wind rises as we ascend ; hence when a sound-wave is travelling against the wind, the wave-velocity is less in the upper portion of the wave than in the lower ; thus the wave-front is bent upwards, and the sound passes over the head of the observer. The same notion occurred to Reynolds ; he verified it by direct experiment, and pointed out, moreover, that in ordinary conditions of the atmosphere the temperature falls as we ascend ; hence from this cause also the wave-velocity is reduced, and the path of the sound is no longer straight, but curved, with the convexity of the curve turned downwards. If, however, it should happen that the air is warmer above than it is below, the reverse will be the case—the soundwaves will be bent downwards—the sound will thus be audible at a greater distance than previously.

The papers on the action of the screw propeller form an interesting series. The racing of a screw is proved to be due to the admission of air to the screw ; this, it is shown, interferes with the power of the screw to obtain water, and also reduces the resistance which would otherwise be offered by the water the screw would get. For consider a vertical plate, totally immersed in water, which is being pushed forward ; its speed may be such that the water behind cannot remain continuously in contact with it. A vacuum will tend to form behind the plate ; the limiting velocity at which this takes place will depend on the pressure in the water behind the plate ; if no air can reach the plate, this pressure will be the atmospheric pressure, together with that due to the depth of water above the plate ; if air can reach the space behind the plate, the limiting velocity will depend only on the pressure due to the water, and will be much less than in the first case. The blades of the propeller act like the plate ; a stationary screw will be most effective in propelling water when it is turning so fast that a vacuum is just formed behind its floats, and the rate at which the water is driven past depends on the water pressure just close to the floats ; if air can reach the floats, no vacuum can be formed ; the pressure will depend only on the height of the column of water above the screw ; the limiting velocity will be less than when the screw remained free from air.

The steering of screw steamers is dealt with in several papers laid before the British Association ; three of these are reports of a committee appointed in 1875 to investigate the question. Of this committee Prof. Reynolds was secretary. Briefly, their researches confirm the theory he had advanced in a paper published in the *Engineer*, June 4, 1875, explaining the accident to the steamer *Bessemer*, which had failed to enter Calais Harbour on May 8 previously.

Prof. Reynolds pointed out

"when a ship is stopping, the water will be following her stern relatively faster than when she is moving uniformly, and consequently the effect on the rudder will be diminished ; that the longer the ship the greater will be this difference ; also that this effect is greatly increased when a ship is stopping herself with her propellers, as was the *Bessemer*, for since not only is the retardation of the vessel much more rapid, but the water has a forward motion imparted to it by the propellers, which motion, if the propellers are near the rudder, may be greater than that of the ship, in which circumstance the effect of her rudder's action will be reversed."

In the paper on the radiometer, "On the Forces caused by Evaporation from, and Condensation at, a Surface," the true explanation of its action is given in the concluding paragraphs. The paper deals in the main with the effects of evaporation and condensation in causing motion, but near the end the author writes :

"Since writing the above paper, it has occurred to me that, according to the kinetic theory, a somewhat similar effect to that of evaporation must result whenever heat is communicated from a hot surface to a gas. The particles which impinge on the surface will rebound with a greater velocity than that with which they approached, and consequently the effect of the blow must be greater than it would have been had the surface been of the same temperature as the gas."

The longest paper in the collection is that on certain dimensional properties of matter in the gaseous state ; it contains the results of a number of experiments on the thermal transpiration of gases through porous plates, and an extension of the dynamical theory to account for the phenomena.

Enough has been said, perhaps, to show the interest of the volume and the importance of the scientific results it contains. It is got up in the admirable manner which characterises the Pitt Press productions, and in form leaves nothing to be desired.

COUNT SCHEIBLER'S SPORTING TOUR.

Sette Anni di Caccia Grossa é Note di Viaggio in America, Asia, Africa, Europa. By Count Felice Scheibler. Pp. xv + 525. Illustrated. (Milan : U. Hoepli, 1900.)

ENGLISHMEN are, perhaps, somewhat too inclined to believe that great game shooting is a special prerogative of the Anglo-Saxon ; but the publication of the present work, together with the recently issued English translation of Count Potocki's " Sport in Somali-

land," should do something to dissipate this mistaken notion. Count Felice Scheibler may, indeed, be said to be a "mighty hunter," and the frequent mention of his name in Mr. Rowland Ward's " Records of Big Game " will suffice to show that many of the animals that fell to his rifle have yielded trophies of more than usual size. As is indicated in the title of the volume before us, the author's seven years' hunting included experiences of great game of all the four continents of the world although in Asia his travels were limited to India and Ceylon, and in America to the United States and the Dominion of Canada. A well written and well illustrated record of such extensive experiences could not fail to be of interest, not only to his brother sportsmen, but likewise to naturalists ; and the present volume may be truthfully said to fulfil both these conditions. The 250 text-figures with which the work is embellished are for the most part reproductions from photographs taken respectively by the author, Prince di Teano, and Mr. Seton Karr, and are remarkable alike for the manner in which they have been executed and the care with which they have been printed. A large number of these illustrations deal with animals which were shot by the Count, and although most of these were taken after death, yet they frequently portray very clearly some of the more striking characteristics of the particular species. The views of scenery and hunting scenes are, moreover, specially good, and will give to stay-at-home readers an excellent idea of the nature of the districts in which sport was obtained, and of the mode in which various animals are hunted. Of especial interest is the photograph, on p. 176, of recently captured elephants crowded into a *kedda*, while those representing the elephant tamers at work are scarcely less attractive. Some of the titles to the illustrations, such as " Il bufalo record," are perhaps a little comic, but Italian, like French, has not yet evolved a sporting language of its own.

Although the author does not appear to have had the good fortune to discover any new species, his accounts of the habits of many of the less known forms will be found of considerable interest to the naturalist. And a gratifying feature is the attention paid to nomenclature, since this is a point in which sporting works are apt to be very deficient. In the employment of names like *Mazama columbiana* for the Columbian black-tailed deer, and *Taurotragus oryx* for the eland, the Count is, indeed, thoroughly up-to-date and ahead of most works on popular natural history.

Whether, however, the author confined his love for shooting within such limits as would meet with the approval of the recent congress on the preservation of great game is a question which may be left for others to answer. But the plate on p. 457, which represents three individuals of the common African rhinoceros, out of a herd of six, already fallen, while aim is being taken by the author at a fourth, is calculated to give rise to misgivings on this point.

Starting from Liverpool in 1889, Count Scheibler sailed for America, where he soon enjoyed excellent sport in the Rocky Mountains with " grizzly " and wapiti ; afterwards proceeding to British Columbia, where he was successful in obtaining examples of the Rocky Mountain goat. At San Francisco he embarked for India, where his first experiences of sport were obtained in the fever-stricken Sandarbans of Lower Bengal. Proceeding northwards, he had the good fortune to be entertained by the Maharaja of Kuch-Behar, whose territories are now the finest sporting-grounds in India ; and here he obtained, in addition to tiger, the large Indian rhinoceros, the gaur, and the wild buffalo. After a short sojourn in Gya and Ceylon, the party then crossed to Somaliland, which was at that time in its prime as a sporting country. From the Italian province of Erithræa the Count proceeded by sea to Zanzibar, whence he made a journey of considerable length into the interior of Equatorial Africa, obtaining specimens of Coke's hartebeest (*Bubalis cokei*), and the fringe-eared beisa (*Oryx callotis*). The final stage of the tour was Russia, where elk was added to the list of large game.

Although, as the author himself states, the work lays no claim to having advanced either zoological or geographical science, yet it may be commended as a very interesting account of types of animal life which are only too rapidly disappearing from the face of the earth. In fact, it is so interesting that there would seem a considerable probability that an English translation would be well received. R. L.

OUR BOOK SHELF.

Die Moderne Physiologische Psychologie in Deutschland. By W. Heinrich. Pp. iv + 249. (Zürich : Speidel, 1899.)

Zur Prinzipienfragen der Psychologie. By W. Heinrich. Pp. iv + 74. (Zürich : Speidel, 1899.)

An Outline Sketch, Psychology for Beginners. By Hiram M. Stanley. Pp. 44. (Chicago : The Open Court Publishing Company, 1899. London : Kegan Paul and Co., Ltd., 1899.)

MR. HEINRICH'S two little works demand careful study as well thought-out and consistent expositions of a psychological attitude which is in many ways attractive. The author, who may be described as a disciple of Avenarius *minus* his master's metaphysics, holds strongly the necessity of making the principle of psychophysical parallelism, understood in the most rigid sense, the basis of all psychological inquiry, and would consequently recognise no causes or causal laws other than those of the physical and physiological series. He has little difficulty in showing that Wundt and other contemporary writers, who, while professing the doctrine of parallelism, believe in causal sequences between psychical states as such, are inconsistent with their own professions. That the inconsistency can be avoided, or that an intelligible account of human life can be given in terms of purely physiological sequences, is scarcely so clear. As the author himself admits, it is a necessary consequence of his theory that the only difference between rational and purely reflex reaction on stimulus is one of comparative complexity. Whether an account of human life which reduces all activity to the purely reflex type is not like the play of *Hamlet* with the part of Hamlet left out, he does not discuss. The question is, however, directly suggested by his contention that, in treating of the behaviour of our fellow-men, we have no right to introduce the notion of consciousness, but should confine ourselves to establishing physical relations between changes in their environment and their corresponding outward reactions. He seems to forget that language, for instance, loses half its significance if you neglect to observe that it not merely can be understood by a listener, but is meant by the

speaker to be understood. And even if we could agree to take no notice of consciousness in our fellows, it still remains, as the author admits, to examine the relation between the environment, which on his theory all science describes, and ourselves the describers. Thus all the problems about the relation between consciousness and its objects which Mr. Heinrich banishes from our psychological study of our fellows return upon us as soon as we attempt to understand our own relation to our environment. Perhaps the chief value of the author's discussions is that by his insistence on the too often disregarded consequences of the doctrine of parallelism, he compels his readers to ask themselves whether the old belief in the interaction of mind and body is not, with all its difficulties, more satisfactory than the fashionable substitute for it.

It is painful to turn from Mr. Heinrich's able and thoughtful work to such a piece of loose and unsatisfactory popular psychology as Mr. Stanley's essay. If psychology is to be taught in schools at all—in itself a debatable question—it ought, at least, to be taught in a precise and definite form. These scraps of inaccurate chatter are of no more value in psychology than they would be in elementary physics or in any other science. Read, for instance, the light and airy sentences (pp. 8-9) in which Mr. Stanley disposes of the difficult problem of space-perception. What would be thought of a writer on heat or chemistry who should evade all the puzzles of his subject by such loose and flimsy generalisation ? In truth, the only way to treat work of this kind with kindness is to say nothing at all about it. The only words one can find in which to characterise it are that, like a good deal of popular writing on psychological topics, it is quite worthless, because the writer has set no serious standard of scientific accuracy before him.

Rural Wealth and Welfare: Economic Principles illustrated and applied in 'Farm Life. By Geo. T. Fairchild, LL.D. (New York : The Macmillan Company, 1900 ; London : Macmillan and Co., Ltd.)

THE scope of this treatise is perhaps more accurately indicated by its alternative title : it is primarily a text-book of economics, the concrete illustrations being taken preferably from objects and practices familiar to agriculturists. The book is accordingly addressed to this class of the community, though it may be doubted whether the ordinary farmer, at all events in this country, will be competent to make much practical use of the principles expounded. The position of farming, especially in the older civilised States, has perhaps undergone more change during the last thirty years than that of any other great industry, since it is practically within this period that the cultivator has had to learn to face the competition, not merely of his own countrymen, but of the whole world. It is therefore all the more necessary that he should be thoroughly acquainted with the modern conditions under which he has to work ; in this respect, the remarks on the importance, as a factor in prices, of the increased facilities for marketing the enormous quantities of grain and other farm products raised in the United States, are very much to the point.

Lectures on Theoretical and Physical Chemistry. By J. H. van't Hoff. Translated by R. A. Lehfeldt. Part ii. Chemical Statics. Pp. 156. (London : E. Arnold.)

WE welcome the appearance of the English translation of the second part of van't Hoff's lectures. Dr. Lehfeldt has, as before, done his work admirably. It may be regretted, however, that he has adhered so closely to the somewhat uncouth structural formulæ used by the author. We venture to hope that in a future edition a freer use of brackets and points may be made, as the student might have some difficulty in recognising aceto-acetic ether in the formula $H_3CCOCH_2CO_2C_2H_5$.

LETTERS TO THE EDITOR.

[The Editor does not hold himself responsible for opinions expressed by his correspondents. Neither can he undertake to return, or to correspond with the writers of, rejected manuscripts intended for this or any other part of NATURE. *No notice is taken of anonymous communications.]*

Eclipse Photography.

THE writer has obtained results in photography which seem to have an important bearing on the work which should be undertaken in future eclipses.

It is well known that photographic plates exposed in some eclipses have developed no trace of an image of the sun. The astronomer has even been subjected to the mortifying suggestion that he had forgotten to uncap his camera. It is not difficult to reproduce such results at any time by simple over-exposure. In eclipse photography, where it is sought to get the most delicate of details in an object of the most delicate character, the methods now used are hedged in by very peculiar limitations. It requires a very appreciable time to secure delicate details, and, nevertheless, if this time is made too great the plate will fog. The developer must then be given restraining properties, which cause a loss of the very details we are seeking to secure.

In a paper recently published by the Academy of Science of St. Louis, the writer has shown that a plate which, on account of over-exposure will develop as a zero plate in a dark room, will develop as a positive in a light room. The paper contains a half-tone reproduction of a positive obtained by a camera exposure of one minute, and developed within a few inches of a 16-candle incandescent lamp. The plate was an "instantaneous" Cramer plate. Since that time the same results have been reached by first opening up the plate holder and exposing the film to the lamp light until it is all converted into the zero condition. If covered with an opaque punched stencil, no trace of the design will appear on the film when developed in the illuminated bath. The slide is then closed and the plate afterwards exposed in the camera in the usual way. Such a plate cannot be over-exposed in any reasonable time. It may be exposed for a minute or for four hours to a brilliantly-lighted landscape, and the most superb results can be obtained. There is no restraining developer needed. The tendency to fog when the exposure is too short is corrected by taking the developing bath nearer to the light. It seems probable that on very short exposures it might sometimes be advantageous to use a developer which will yield a positive with an under-exposed plate. In the two eclipses of long totality which are now approaching, this method seems to promise very valuable results, and the attention of those who will have the work in charge is earnestly directed to this matter. The results described have been reached but recently, and there is need of preliminary experimenting by any one who wishes to avail himself of these methods.

St. Louis, Missouri, U.S.A.   FRANCIS E. NIPHER.

The Action of Water Upon Glass.

IT is a matter of too frequent observation in India that lenses of optical instruments are liable to serious injury from atmospheric influences. This very often takes the form of injury to the Canada balsam cementing the two lenses of achromatic combinations together; but in other cases, it is due to the solvent action of water on the surface of the glass. As this is a matter of importance necessitating the re-grinding of the lens for its correction, I have thought that the following observations may be of interest and of value to optical instrument-makers, especially as it appears that only particular kinds of glass are attacked in this way. If that is so, it should be possible to avoid using glass of that particular composition; or the edges of the combined lenses may be covered with a coating of cement or varnish so as to prevent moisture getting in between them, and in such a way that it could easily be removed when desired.

My attention was first drawn to some cases of articles of domestic glassware being attacked by water standing in them for some time, and these are recorded to show that a solvent action does take place. The first case that I noticed was that of a cut wine glass which was used—or misused—to hold a few cut flowers. On seeing it dry on one occasion, I noticed it had a dull matt appearance, which I thought was simply a deposit. On examination, however, I found that the surface of the glass had been eaten into up to the level of the water usually put into it.

The next case was that of glass finger-bowls, in which the servants kept water ready for use. These were similarly attacked up to the level of the water. The next was a more remarkable case. A couple of decanters, not required for use, had evidently been washed and drained, more or less, but not dried; possibly during the hot season. The moisture remaining inside had become deposited on the inner surface in droplets, as, indeed, may frequently be seen, and had been standing so for some time. When dried for use the surface was found to be eroded, giving a pattern precisely similar to that formed by condensed moisture: leaving no doubt as to its cause.

Here we have, then, a case of pure water attacking the surface of glass when allowed to stand for some time. Since then, being on the alert, I have met other cases, including some of perfectly new glass articles eroded in like manner, which, without their history, it is impossible to account for.

Now for two instances of physical apparatus being attacked and spoiled by this action. The first noticed was a Newton's Rings apparatus. In this case the two discs of glass were equally attacked, and so much so that the combination gave a dense matt appearance. On opening it out, the discs were found to be firmly adhered, and on inserting a knife edge between the discs and giving a sharp tap on the back to separate them, an irregular piece about 1¼ inch long came from one adhering to the other. The two had thus grown together, and at the junction was actually stronger than in the mass of the glass.

The next was a more serious case, being the object lens of a 3½ inch telescope from a well-known London firm of optical instrument-makers. In this case, the convex lens was badly corroded on its inner surface, though the adjacent face of the concave lens was quite clear. Here we see the difference in action in the case of two different kinds of glass. This, however, would help us little if all kinds of crown glass (of which the convex lens is made) were similarly attacked. But this is not the case, and it is a point of importance to opticians to ascertain what particular kinds of crown glass used in achromatic combinations are liable to this action, and to avoid using them. Of a fairly large number of achromatic combinations I have in the College Laboratory, this is the only one that has been affected, though all are exposed to the same influences, while some belong to old pieces of apparatus. The particular telescope was purchased about six years ago, and the damage took place in one season when it was not much used. Since then I have from time to time opened out the lenses and have frequently found a layer of moisture between them; in one case, of a common piece of apparatus in which the lenses did not fit closely, a complete drop of water was collected, the diameter of the lens being only 1½ inch; and in a Soleil's saccharimeter, clear through vision is obscured by moisture collected and condensed on the surfaces of the lenses in one of the adjusting pieces, which it is very difficult to open out to clean.

All this shows that moisture does collect in the form of water between such layers of glass, and the pattern of the eroded portion of the telescope lens, together with the instances of the action of water on the domestic glass goods mentioned above, leave no doubt that it was moisture alone that caused the damage in this case, although it was not actually seen. I need hardly say that, in both the Newton's Rings apparatus and the telescope lens, the exposed surfaces were perfectly clear and unacted upon.

The causes of moisture collecting in this way would appear to be the excessive moisture in the air for many months in the year, the hygroscopic nature of the glass, and capillary action between the surfaces; while the apparently marked action of water on glass here noticed is probably due to the long-continued higher temperature. It is possible, however, that the above phenomena may not be as new or unusual as they appear to be to me, and that many others could give like experiences.

EDMUND F. MONDY.

Dacca College, Dacca, East Bengal, June 16.

THE TOTAL SOLAR ECLIPSE AS OBSERVED BY THE SMITHSONIAN EXPEDITION.

WADESBORO, in Northern Carolina, was the station selected by the Smithsonian Institution for observing the total solar eclipse of May 28 last. The chances of fine weather at eclipse time were about eight to one, and it is satisfactory that on eclipse day the sky was cloudless and the air clearer than on the average.

The main objects of the investigations undertaken were a photographic and visual study of the structure of the lower corona, and a determination by the bolometer of the heat radiated from it, and lastly an examination of the form of its spectrum energy curve.

Prof. Langley, who was in charge of this expedition, observed the eclipse of 1878 from Pike's Peak (14,000

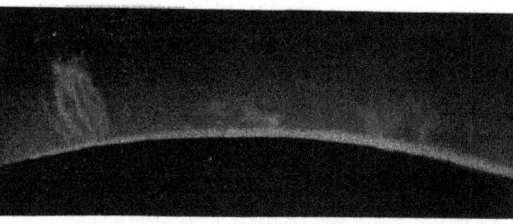

Fɪɢ. 1.—Portion of Smithsonian Astrophysical Observatory Eclipse Camp, showing a part of the large 135-foot telescope (under canvas), the 38-foot coronagraph, and the 5-inch equatorial.

feet), and he was then particularly struck by the remarkable definiteness of filamentary structure close to the sun's limb, a structure which, he remarks, has never been found in any photographs, not even in those beautiful pictures taken by Prof. Campbell at the Indian eclipse of 1898. The eclipse this year afforded him an opportunity of examining this inner corona region with a much more powerful instrument. This instrument was a 12-inch achromatic lens of 135 feet focal length, obtained for the Harvard College Observatory, and lent by Prof. E. C. Pickering. The tube was mounted horizontally in conjunction with a cœlostat of 18-inch aperture, and 30-inch square plates were used, the diameter of the solar image being 15 inches. To supplement this instrument, a 5-inch lens of 38 feet focal length, loaned by Prof. Young, was pointed directly at the sun, and photographs were secured on plates 11 by 14 inches, moved in the focus of the lens by a water clock. For the study of the outer corona and possible intramercurial planets, specially equatorially mounted lenses of 6-, 4- and 3-inch apertures, driven by clock-work, were used.

The accompanying illustration (Fig. 1) shows a small part of the 135-foot telescope. The photographic hut is seen at the end of it, and beyond that the tube contain-

ing the lens of 38 feet focal length pointed at the sun. Prof. Langley is seen observing at the 5-inch equatorial.

For the bolometric work a massive siderostat with a mirror of 7 inches was used in conjunction with a large part of the delicate adjuncts employed at the Smithsonian Institution in recent years.

Further work that was attempted, and for which other apparatus had been taken out, was an automatic method of obtaining photographs of the lower chromosphere at about second contact by means of an objective prism working in connection with the 135-foot lens; visual and photographic observations of times of contact; and sketches of the corona, both from telescopic and naked eye observations.

The observers, under the general charge of Prof. Langley, were distributed as follows:—Prof. Langley used the same 5-inch as he observed with in 1878; Messrs. Abbot and Mendenhall were in charge of the bolometer; Mr. T. W. Smillie made exposures at the 135-foot telescope, and Mr. F. E. Fowle, jun., at the 38-foot telescope. Father Searle, assisted by Mr. P. A. Draper and Mr. C. W. B. Smith, employed four telescopes, mounted on a single polar axis and driven by clock-work, for obtaining photographs of the outer corona and the intramercurial planets. Latitude, longitude, time and contact observations were made by Mr. G. R. Putnam, assisted by Mr. Hoxie. Sketches of the inner corona and contacts were made by Mr. R. C. Child with a 6-inch, and by Father Woodman with a 3½-inch.

Fɪɢ. 2.—Showing prominences at the south-west limb. Taken with a 12-inch lens of 135 feet focal length. Exposure eight seconds. At end of totality. (Natural size of original photograph. Moon, 15 inches diam.).

Among the more general observations made at the time of the eclipse may be mentioned the following:—

Before totality a fall of temperature and a rising breeze were distinctly noticeable. Shadow bands were seen, but their velocity was too rapid and flickering for accurate determination; their size and distance apart (about 5 inches) were also estimated.

During totality the sky to visual observers was notably not dark, and no second magnitude star was seen with the naked eye. Mercury was a conspicuous object

The equatorial streamers were closely observed, and could be followed by the naked eye to 3 or 3½ solar diameters; their structure was likened by Father Woodman to a structure of mother-o'-pearl, and this was generally conceded. Colour estimates, however, varied, and were given as "yellowish green tinge," "straw-coloured" or "golden." (It may be remarked here that the general description of the colour was given by the British observers in Spain and Portugal as "silvery-white.") Prof. Langley's visual telescopic observations gave, as he remarks, "little indication of the finely-divided structure of the inner corona which he had noticed at Pike's Peak. Structure, to be sure, was evident, but not in such minute subdivision as had then been seen; and though one remarkable prominence, as well as several smaller ones, was visible, the coronal streamers did not give to the writer the impression of being connected with these prominences, though the relationship of some of them to the solar poles was abundantly manifest."

The approximate length of totality as observed was 88 seconds, or 4 seconds shorter than the duration as given by the *Nautical Almanac*.

portant result was that the corona gave a positive indication of heat as compared with the moon; this heat though certain, was, we are told, too slight to be subdivided by the dispersion of the prism with the means at hand.

With regard to the negatives depicting the outer corona, these show the extensions reaching to from 3 to 4 solar diameters for the longest streamers.

The plates taken for a search for intramercurial planets have not been carefully examined, but the considerable sky illumination during totality leads Prof. Langley to doubt the possibility of having recorded the images of such faint objects on the plates. Pleione (6·3 magnitude) in the Pleiades, and some fainter stars are, however, recorded on one of the plates.

The expedition seems to have gathered some most valuable data, and to have scored a decided success in every respect; the observations made and the photographs secured promise to be very satisfactory, especially with regard to the primary objects of the expedition.

THE BOARD OF EDUCATION AND ITS CONSULTATIVE COMMITTEE.

IT will be remembered that the Board of Education Act, which received the Royal Assent last year, contained in Section 4 the following provision:

FIG. 3.—North polar coronal region. Taken with a 12-inch lens of 135 feet focal length. Exposure 16 seconds. (Natural size of original photograph. Moon 15 inches diam.)

With regard to the photographs which were found to have been successfully exposed, but of which only a few have as yet been developed, most interesting results will be obtained. During totality six plates were exposed for periods ranging from ¼ to 16 seconds, and three others immediately after third contact; these were all secured by the large 135-foot telescope. We are fortunately able to illustrate two portions (natural size) of the large 15-inch disc. Fig. 2 shows one of the principal prominences with the lower filaments near it (exposure 8 seconds), while Fig. 3 is another portion of the north polar region, with a 16 seconds' exposure. The part near the sun has been intentionally over-exposed, to show more clearly the outer portions of the polar structure, which extended to 6 minutes from the sun. The wealth of detail and imposing magnitude of the scale on which these pictures are taken will no doubt give us much needed information about the structure of the corona just above the chromosphere.

The measurement of the heat of the corona appears to have been successfully performed by Mr. Abbot, with the aid of Mr. Mendenhall, and this is probably the first time that it has really been shown to exist. The im-

"It shall be lawful for Her Majesty in Council by Order to establish a Consultative Committee, consisting, as to not less than two-thirds, of persons qualified to represent the views of universities and other bodies interested in education, for the purpose of:—

(a) framing, with the approval of the Board of Education, regulations for a register of teachers which shall be formed and kept in manner to be provided by the Order in Council; provided that the register so formed shall contain the names of the registered teachers arranged in alphabetical order, with an entry in respect of each teacher showing the date of his registration, and giving a brief record of his qualifications and experience; and

(b) advising the Board of Education on any matter referred to the committee by the Board."

The Order in Council nominating the members of the proposed committee and defining its course of procedure, has just been issued, and is a document of considerable public interest and importance. Advisory Boards are not unknown in other departments of the public service, e.g. in the India Board and at the Admiralty; but a permanent Consultative Committee of unofficial experts, on the scale and with the powers contemplated in the present Order in Council, is a novelty in administration ; and the working of the new experiment will necessarily be watched with much solicitude by all persons who have at heart the improvement and development of our system of public education.

The following are the names of the eighteen persons who are nominated as the first members of the Consultative Committee :—

Right Hon. Arthur Herbert Dyke Acland.
Sir William Reynell Anson, Bart., M.P.
Professor Henry Armstrong.
Mrs. Sophie Bryant.
Right Hon. Sir William Hart-Dyke, Bart., M.P.
Sir Michael Foster, K.C.B., M.P.
Mr. James Gow, Litt.D.
Mr. Ernest Gray, M.P.
Mr. Henry Hobhouse, M.P.
Mr. Arthur Charles Humphreys-Owen, M.P.
Sir Richard Claverhouse Jebb, M.P.
Hon. and Rev. Edward Lyttelton.
Very Rev. Edward Craig Maclure, D.D., Dean of Manchester.
Miss Lydia Manley.
The Ven Ernest Grey Sandford, Archdeacon of Exeter.
Mrs. Eleanor Mildred Sidgwick.
Professor Bertram Coghill Alan Windle, M.D.
Rev. David James Waller, D.D.

It will be noticed that with the exception of the two former Vice-presidents of the Council, and of Mr. Hobhouse, all the persons named in this list may be regarded as representatives of "bodies interested in education." Oxford, Cambridge and London are most appropriately represented by their respective Members of Parliament ; two of the proposed members are head-masters of public schools, one has been a teacher in a public elementary school, one is a High School mistress, another lady is the head of Newnham College, a third is the mistress of a training college for schoolmistresses, and may also be reckoned as a representative of the British and Foreign School Society. Science and technology have their advocates in Prof. Armstrong and Sir Michael Foster ; the Established Church and the National Schools are represented by Archdeacon Sandford the Roman Catholics by Prof. Windle, and the Nonconformists by Dr. Waller, Wales and the Welsh Intermediate Schools by Mr. Humphreys-Owen, and the School Boards of England by Dean Maclure, the chairman of the Manchester School Board. There can be no doubt that an excellent selection of names, typical of various classes, and likely to command the public confidence, has been made by the Lord President and his advisers.

Nevertheless, it was generally hoped and expected that, while two-thirds of the number were very rightly and in fulfilment of the express intentions of the Act to be composed of persons able to express the views of different academic and professional bodies, the remaining third would consist of persons detached from sectional interests, and specially qualified by breadth of view, by large acquaintance with schools and institutions of various classes, both here and in foreign countries, and by a disinterested concern for the interests of national education as a whole, to render service in consultation with the Board of Education. No such proportion has, however, been observed in the composition of this committee. Like some recent Royal Commissions, to which have been entrusted duties especially demanding wide knowledge and judicial impartiality, the chief ingredients in the committee are advocates and partisans specially charged to look after the interests of particular institutions, creeds, or professional bodies. It appears to be assumed that the resultant of all these opposing forces will be a satisfactory conclusion. But when it is considered that one of the first duties of the committee will be to determine the conditions on which teachers shall be ad-

mitted to the official register, and that it will be the task of that committee to determine the kind of qualification which should be recognised, and the relative claims of a great number of different institutions, both public and private, it becomes evident that the list of the proposed committee is seriously incomplete. One of the most important questions which will in due course inevitably demand its attention is the examination and inspection of secondary schools, and it is quite conceivable that on this point professional interests may not prove to be precisely identical with the public interests. It may be hoped that attention will be given to these considerations before October, when the committee is for the first time to be summoned. It is indispensable that a body charged with such novel and weighty responsibilities should from the first command the full confidence of all those who are conscious of the defects in our present system, and who are concerned more with its due expansion and its fulfilment of high national ideals than with the conservation of any traditions and interests, however important and deserving of respect, which belong to particular classes or institutions.

THE INTERNATIONAL ASSOCIATION OF ACADEMIES.[1]

THE Academy will recall the fact that at the conclusion of the mission entrusted to M. Moissan and myself, consent was given to the "Projet de Statuts pour l'Association internationale des Académies," drawn up by the delegates of the nine Academies represented at the Conference held at Wiesbaden early in October last, at the invitation of the Academy of Berlin.

The International Association is now constituted ; and it includes the eighteen following Academies :

1.	Academy of Sciences	...	...	...	Amsterdam.
2.	Prussian Academy of Sciences ...		...	Berlin.
3.	Academy of Sciences, Literature and the Fine Arts			...	Brussels.
4.	Hungarian Academy of Science		...	Budapest.
5.	Academy of Sciences	...	...	...	Christiania.
6.	Society of Sciences	...	...	...	Göttingen.
7.	Academy of Sciences of Denmark	...	Copenhagen.
8.	Academy of Sciences of Saxony	...	Leipzig.
9.	Royal Society	...	...	...	London.
10.	Academy of Sciences of Bavaria	...	Munich.
11.	Academy of Inscriptions and Literature	Paris.
12.	Academy of Sciences	...	...	...	Paris.
13.	Academy of Moral and Political Sciences	Paris.
14.	Academy of Sciences	...	...	...	St. Petersburg.
15.	Academy dei Lincei	...	...	...	Rome.
16.	Swedish Academy of Sciences ...		...	Stockholm.
17.	Academy of Sciences	...	...	...	Washington.
18.	Academy of Sciences	...	...	...	Vienna.

Amongst the Academies invited to join, one only, the Royal Academy of History of Madrid, has as yet not replied to the request of the Wiesbaden Conference.

The provisional rules take into consideration the possibility of the addition of other learned societies, and in § 2 the conditions and formalities are indicated which will be necessary for the admission of a new Academy.

The Association comprises two Sections, the Section of Literature and the Section of Science. The work will be carried out by general meeting and committee. In principle, the general meeting will be held every three years, and each Academy will send as many delegates as it may deem necessary, but each Academy will have only one vote, which should be given by one of the members of the delegation.

In the interval between two general meetings, the Association is represented by the committee, each

[1] Translation of a report made to the Paris Academy of Sciences on July 2, by M. Darboun, permanent secretary of the Academy.

Academy being represented on this by one member only, if it concerns itself with only one of the Sections of Literature or Science; it will send two delegates when it is concerned with both Sections. Amongst the eighteen Academies, twelve belong to both Sections and consequently will send two delegates to the committee. Of the other six, four, namely the Royal Society of London, the Academy of Sciences of Paris, the Academy of Stockholm, and the National Academy of Washington, belong to the Section of Science alone, and two, the Academy of Inscriptions and Literature, and the Academy of Moral and Political Sciences, belong to the Section of Literature. Hence the committee will consist of thirty delegates, of which sixteen will belong to the Section of Science, and fourteen to that of Literature. In full committee the two delegates of one Academy will have only a single vote. After delay, inevitable in such cases, all the Academies, with the exception of two or three, have sent in the names of their delegates. The delegate of the principal Academy will take the chair at the committee of the Association, the principal Academy being that of the place in which it is proposed to hold the next general meeting.

The Conference of Wiesbaden having decided on a resolution to which we can here only draw attention, that the first general meeting of the International Association should be held in Paris this year, a difficulty has arisen not foreseen when the provisional rules were drawn up. Three Parisian Academies having joined the Association, it is necessary to decide to which shall be assigned the Presidency on this occasion. The delegates of the three Academies of the Institute of France have met, and have unanimously decided to confer for this year the presidency of the Association upon the Academy of Sciences, which was the first to join the Association, and, moreover, has taken an active part in the discussions, at the conclusion of which the Association was constituted.

It has been further decided that the first Session of the committee shall be held in Paris towards the end of July, the first meeting being fixed for Tuesday, July 31, at 9.30 a.m., at the Palais de l'Institut.

The agenda for the first meeting will include the preparation of a scheme of government for the committee, the settlement of the exact date and the order of the day for the next general meeting. The Royal Society of London, which has taken so active a part in the formation of the Association, has already announced a scheme which it proposes to submit for approval to this next general meeting; it concerns the measurement of an extended arc of a meridian in the interior of Africa.

The Academy, by the act of joining, has subscribed to the rules of the new Association. There is no occasion to recall here with what prudence and moderation they have been drawn up. The object of the Association is to prepare and promote scientific work of general interest which may be proposed by one of the constituent Academies, and generally to facilitate scientific relations between different countries. In any particular case, each Academy reserves to itself the right to give or refuse its support, or decide the choice of methods and the means to be employed.

If these principles are followed, the Association will become a powerful instrument of study, of concord and of scientific progress; it will rapidly take its place in the front rank of those international scientific associations, the *rôle* of which must necessarily be satisfactory.

Faithful to the principles which they have always followed, the three Academies of the Institute of France, called by the nature of their studies into the Association, will strive to assure it the success and influence which have been desired for it by its promoters.

Finally, attention may be directed to a particular clause in the rules which will interest some of our

colleagues. For taking into consideration the study or preparation of scientific enterprises or researches of international interest, upon the proposition of one or more of the associated Academies, special international commissions may be instituted either by the general meeting or one of its two Sections, or, in the interval between two general meetings, by the committee or one of its two Sections.

THE NEW PHYSICAL LABORATORY AT OWENS COLLEGE.

OWENS COLLEGE recently held high festival on the occasion of the opening, by Lord Rayleigh, of the new physics laboratories. Of these, a preliminary account was given in NATURE of October 27, 1898, on the occasion of the laying of the foundation-stone. As the size of the new building surpasses that of any other physical laboratory in this country, it was fitting that the occasion should be marked by a ceremony of some importance, and dignified by the presence of a number of leading physicists from all parts of the country.

The main features of the new laboratories, as planned by Prof. Schuster, were described in our former article; but it remains to state how they have been carried out. The new building is separated by Coupland Street from the main quadrangle of buildings of which Owens College consists, though it is joined to the older buildings by an underground passage. It is a commodious structure, having three complete storeys above the basement, with simple but effective decorative features both internal and external. The frontage is about 110 feet wide, and the main building extends about 90 feet back. The ground floor is devoted to rooms for electrical measurement, the magnetic testing of iron, electrochemistry, a workshop and a private laboratory. The first floor contains a large laboratory for elementary teaching (36 feet by 44 feet in dimensions), a balance room, a room for chemical physics, two laboratories for electricity and one for optics. On the second floor is a fine lecture theatre with raised auditorium, preparation room, museum and apparatus room, a class-room and two smaller laboratories, and a special room fitted up for physical optics; its special feature being the equipment, designed by Sir Howard Grubb, necessary for working with a 6-inch Rowland grating. From this floor an upper staircase leads to a small astronomical observatory containing an excellent 10-inch equatorial by Cooke, the gift of Sir Thomas Bazley. In the basement are rooms for spectroscopic and photographic work, a cryogenic laboratory and a room for researches at constant temperature. The arrangements for heating, ventilating, and for the supply of gas, electricity, water, steam and compressed air are exceedingly complete. In the ventilation system, the air supplied through a fan and warmed by passage through a flue heated by the gases of the boiler-furnaces, is passed over a surface of oil to deprive it of its dust and prevent blackening effects.

A very important adjunct to this fine building is the John Hopkinson memorial wing for electrotechnics. This consists of two large rooms on the ground floor: one (27 feet by 50 feet) to serve as a dynamo room, the other an electrochemical laboratory (36 feet by 37 feet), together with basement rooms for gas engine, counter-shaft for dynamo-driving, photometers, and heating apparatus. In the dynamo room, where already are placed several of Wilde's dynamos and some more recent types, there is a fine bronze portrait tablet of the late Dr. Hopkinson. The cost of this wing has been defrayed by the parents and relations of the lamented Dr. Hopkinson, who was himself an alumnus of Owens College.

The opening ceremony on the 29th ultimo began with

an academic procession from the Christie library to the lecture theatre of the new laboratories, where the chair was taken by the Treasurer of Owens College, Mr. Alderman Thompson. Amongst those present were Lord Rayleigh, Prof. Schuster, Sir Henry Roscoe, Principal Hopkinson, Prof. Oliver J. Lodge, Prof. Bodington (Vice-Chancellor of Victoria University), Prof. Rücker, Prof. Pickering of Harvard College, Prof. Osborne Reynolds, Prof. Stroud, Prof. J. J. Thomson, Prof. Poynting, Prof. Ramsay, Prof. Core, Archdeacon Wilson, Mr. Wimshurst, Prof. Perry, Mr. W. Mather, M.P., and many others. Lord Rayleigh delivered a short address upon physical laboratory work and research, and formally pronounced the building open. Prof. Schuster gave an account of the aims of the building, and of the various stages in their realisation. Prof. Pickering likened a physical laboratory to a battleship, and enlarged upon the uses of its equipment. The company then adjourned, some to visit the various rooms, others to attend the opening ceremony in the John Hopkinson memorial wing, which was presented in a touching speech by Mr. Alderman Hopkinson on behalf of the family.

A garden-party held in the afternoon in the house of Prof. Schuster was followed in the evening by a reception and conversazione in the new building. In one of the rooms was a very interesting exhibit of some of the apparatus used by Joule, including two "current weighers," a tangent galvanometer, and a mercury pump. These have been presented to the Owens College by Mr. B. A. Joule. In another room Mr. T. Thorp showed his celluloid gratings and celluloid reproductions of Rowland's grating and of his own echelon grating. Mr. Wilde exhibited his magnetarium and a number of lunar photographs. The large electro-magnet presented by him was also shown in operation.

On the morning of the 30th was the annual ceremony of conferring of degrees of the Victoria University. This took place in the Manchester Free Trade Hall, which was crowded with undergraduates and visitors. The Chancellor, Earl Spencer, presided with great dignity. Honorary degrees were conferred on Lord Rayleigh, Sir William Huggins, Sir William Abney, Sir William Roberts-Austen, Dr. T. E. Thorpe, Prof. Dewar, Prof. Forsyth, Mr. R. T. Glazebrook, Mr. Sidney Lee, Prof. E. Pickering, Prof. J. J. Thomson, and last on the father of the profession of electrical engineering, Mr. Henry Wilde. The ordinary degrees were then conferred upon the successful candidates of the year from the three constituent colleges—Owens College, Liverpool University College, and the Yorkshire College. A luncheon in the Town Hall, given by the Lord Mayor, was subsequently partaken of by the Chancellor, the new Honorary Doctors, the University Professors, and a large number of distinguished visitors.

It has been mentioned that the new physics laboratory exceeds in size any other similar building in England. It is, however, smaller than the physics laboratories of Baltimore, Darmstadt and Strassburg. Its cost has been defrayed by the generosity of private individuals.

NOTES.

Two deputations have recently waited upon Mr. Hanbury to put before him the two sides of the question referring to the proposed establishment of the National Physical Laboratory in the Old Deer Park at Richmond. On one side are some naturalists and inhabitants of the neighbourhood, who protest against the proposed buildings as an interference with the amenities of the neighbourhood of Kew Gardens; on the other are the physicists and the members of the Committee, which, after giving great attention to the question of site, decided that

Kew was most suitable. It is a little unfortunate that this difficulty should have arisen, and it could probably have been avoided by the exercise of a little tact and consideration when selecting the site for the laboratory. Much of the misapprehension which at present exists as to the character of a physical laboratory might thus have been removed. Some people seem to think that the fifteen acres required will be covered with buildings in which noisy operations comparable with those of large engineering workshops will be carried on. This, of course, is entirely incorrect. In the first place, the actual area to be covered by buildings is only a quarter of an acre, or the sixtieth part of the whole area proposed to be taken, and secondly, quiet and freedom from all the perturbing characteristics of towns and manufactures are essential for the investigations to be carried on in the laboratories. When this is kept in mind, the alarm of a certain portion of the public, especially those who appreciate the beauties of Kew Gardens, that the buildings would break the present charm, seems a trifle unnecessary. The Observatory being already in the Old Deer Park, it is natural and proper that the laboratory, which is under the same administration, should be there too. As, however, the Park is over 350 acres in extent, it ought not to be difficult to find another suitable site if there is a persistent opposition to the one already selected. In any case, we are convinced that a *modus vivendi* could be arrived at if the representatives of the opposing interests were to meet one another in a conciliatory spirit.

M. ZAMBACO has been elected a correspondent of the Paris Academy of Sciences, in the section of medicine and surgery.

DR. CORFIELD, professor of hygiene and public health at University College, has been elected a Foreign Corresponding Member of the Royal Academy of Medicine of Belgium.

MR. J. H. MAIDEN, director of the Botanic Gardens, Sydney, is expected to arrive in London at the end of the present month, and will be in the United Kingdom and on the Continent for about three months, engaged in special investigations in botany and agriculture.

THE Duke of Northumberland has been elected a trustee of the British Museum.

THE annual meeting of the Victoria Institute will be held on Monday next, July 16, when an address will be delivered by Prof. Hull, F.R.S.

IT is announced in the *Athenæum* that Baron von Richthofen has been nominated Director of the newly founded Museum für Meereskunde of the University of Berlin.

A BOTANIC GARDEN has been established by the Belgian Government at Coquilhautville, Congo Free State. It will be called the Kew Gardens, and is expected to be of great importance to the rubber and other tropical industries.

THE Council of the Royal Geographical Society have decided to award the Murchison Grant for next year to Mr. John Coles, late Map Curator and Instructor to the Society, as an acknowledgment of his services to geography.

THE annual meeting of the Society of Chemical Industry will be held in the lecture theatre of the Royal Institution, Albemarle Street, on Wednesday, July 18, when the presidential address will be delivered, and the officers for the ensuing year appointed. The president-elect is Mr. J. W. Swan, F.R.S.

THE Council of the Sanitary Institute have arranged to hold a meeting in Paris from August 7 to 9, which will immediately precede the meeting of the International Congress of Hygiene and Demography, also to be held in Paris. The Société Française d'Hygiène have offered to the members of the Institute

: a cordial reception, and are providing a reception room, and making arrangements for special visits and excursions for the benefit of members attending.

A PUBLIC HEALTH CONGRESS will be held at Aberdeen from August 2 to 7, under the auspices of the Royal Institute of Public Health. Among the papers promised may be mentioned the following :—" Disinfection," by Prof. Delépine; "Sewage," by Prof. Percy Frankland, F.R.S.; and "The Origin and Treatment of Malarial Fever," by Dr. Patrick Manson. There will also be submitted and discussed a report on the inquiry made into the chemical and bacteriological condition of the air in the London Board Schools.

THE Home Secretary has appointed a committee to inquire into the working of the method of identification of criminals by measurements and finger prints, and the administrative arrangements for carrying on the same, and to report whether any and what changes are desirable. The members of the committee are Lord Belper (Chairman), Mr. F. A. Bosanquet, Q.C., Common Serjeant, Mr. A. De Rutzen, Metropolitan Police Magistrate, and Mr. C. S. Murdoch, C.B., and Mr. C. E. Troup, C.B., of the Home Office, with Mr. C. Lubbock, of the Home Office, as secretary.

AMONG the Civil List pensions granted during the year ended on June 20, we notice the following :—Mr. Benjamin Harrison, in consideration of his researches in the subject of pre-historic flint implements, 26*l.*; Mr. Thomas Whittaker, in consideration of his philosophical writings, 50*l.*; Mr. Charles James Wollaston, in recognition of his services in connection with the introduction of submarine telegraphy, 100*l.*; Mr. Robert Tucker, in consideration of his services in promoting the study of mathematics, 40*l.*; Mrs. Eliza Arlidge, in consideration of the labours of her late husband, Dr. John Thomas Arlidge, in the cause of industrial hygiene, 50*l.*; Miss Emily Victoria Biscoe, in consideration of the services rendered to Antarctic exploration by her late father, Captain John Biscoe, 30*l.*

SOME molluscan remains found in a sandstone from the Malay Peninsula were described by Mr. R. Bullen Newton at the May meeting of the Malacological Society of London. The shells consist of Lamellibranch casts and impressions, many of them being sufficiently well defined to point conclusively to their Triassic origin. The most abundant genus represented is *myophoria*, so characteristic of the Trias period. *Chlamys valoniensis* also occurs, together with other bivalves. These fossils, the first recorded from this area of south-eastern Asia, were collected by Mr. H. F. Bellamy, and subsequently presented by him to the Geological Department of the British Museum. They were obtained from the Pahang Trunk Road, on the Lipis River.

THE annual meeting of the Museums Association was opened at Canterbury on Monday. In an address, Dr. Henry Woodward, F.R.S., the president-elect, referred to his forty-two years' association with the British Museum and to the many changes and improvements which had taken place there during that period. He advocated the publication by the association of a handbook giving an account of every provincial museum throughout the country, with full particulars as to each, not only as to its officers, organisation, and its plan of arrangement, but also what were the chief features of its exhibits and especially any records concerning types and figured specimens preserved in its collections and any other particulars of general public interest. Papers upon museums and related subjects were subsequently read.

A NEW medical institute, having for its object the placing at the disposal of doctors the aids to diagnosis required in many forms of disease, has just been opened in Berlin. The institute

will place at the disposal of the medical profession its laboratories, instruments and apparatus, and its officers will undertake the carrying out of special researches and examinations. It has departments devoted to the study of bacteriology, chemical microscopy, pathological anatomy, and physiology. To the last-named is attached a Röntgen ray room.

GENERAL SIR R. MURDOCH SMITH, K.C.M.G., Director-General of the Museum of Science and Art, Edinburgh, since 1885, died on July 3, after a brief illness. He was born in 1835, and was the executive officer with Sir Charles Newton's archæological expedition in Asia Minor in 1856–59. He explored the Cyrenaica and made successful explanations at Cyrene in 1860–61. Subsequently he became director-in-chief of the Government Indo-European Telegraph Department. He was the author of a "History of the Recent Discoveries at Cyrene," and of a "Handbook of Persian Art."

A PERMANENT committee for the study of tuberculosis as a national scourge has been formed in Russia. The president is Prof. W. D. Scherwinsky, of Moscow. The committee, which has met twice a month since the beginning of April, has, says the *British Medical Journal*, drawn up for itself the following programme of work : (1) Reports on the communications made on tuberculosis to the Pirogoff Congress and other medical societies in Russia ; (2) reports of foreign congresses on tuberculosis; (3) reports on tuberculosis as an infectious disease (diagnosis, etiology—heredity, individual predisposition, external influences, mode of diffusion, economic and social factors) ; (4) statistical data respecting tuberculosis in Russia ; (5) legislative measures and ordinances in regard to tuberculosis of human beings and beasts; (6) sanatoria, koumiss establishments, &c. ; (7) the means actually in use, and which should be used, for the prevention of tuberculosis in the different provinces of Russia ; (8) tuberculosis in animals and its relation to the disease in human beings.

THE new number of the *Geographical Journal* gives further particulars as to the preparations that have been made for the forthcoming National Antarctic Expedition. An executive officer, Lieut. Charles Royds, R.N., of H.M.S. *Crescent*, has been appointed ; and Mr. T. V. Hodgson (of the Marine Biological Station of Plymouth) and Dr. R. Koettlitz (of the Jackson-Harmsworth Expedition) will form part of the scientific staff, which Prof. Pollock (the holder of the chair of physics in the University of Sydney) will, it is stated, be invited to join. The name of the vessel used will be the *Discovery*. As was mentioned in our issue of May 31, the commanding officer of the expedition will be Lieut. R. F. Scott, R.N., and the leader of the scientific staff will be Prof. J. W. Gregory.

FROM information that has reached us from Mr. Rotch's Blue Hill Meteorological Observatory we learn that a kite used in the exploration of the air was on June 19 sent up to the height of 14,000 feet, thus exceeding the greatest height previously obtained there by 1440 feet. The temperature at this height was fifteen degrees below freezing point, the wind velocity was about twenty-five miles an hour from the north-east, and the air was extremely dry, although clouds floated above and below that level. The kites remained near the highest point from 5 to 8 p.m. On the way down the kites passed through a stratum of thin ragged clouds at the height of 1½ miles. These were moving with a velocity of about 30 miles an hour. At this time the wind at the observatory, about 600 feet above the general level of the surrounding country, had fallen to a calm. The highest point was reached with 4½ miles of music wire as a flying line supported by five kites attached to the line at intervals of about ¾ miles. The kites were Hargrave or box kites of the improved form devised at the observatory. They have curved flying surfaces modelled after the wings of a bird. The three kites nearest the top of the line had an area of between 60 and 70 square feet each, and the two others about 25 feet each. The total weight lifted into the air, including wire, instruments and kites, was about 130 lbs.

Mr. E. G. Green, Government entomologist at the Botanic Gardens at Peradeniya, Ceylon, has recently been able to confirm by personal observation the web-spinning habits of the red ant (*Œcophila smaragdina*). He has seen ants actually holding larvæ in their mouths and utilising them as spinning machines. To find what would be done, some leaves which had been newly fastened together by the ants were purposely separated by Mr. Green. The edges of the leaves were quickly drawn together by the ants, and, about an hour later, small white grubs were seen being passed backwards and forwards across the gaps made in the walls of the shelter. Each grub (there were apparently only two of them) was held in the jaws of one of the worker ants, and its movements directed as required. A continuous thread of silk proceeded from the mouth of the larva, and was used to repair the damage. There were no larvæ amongst the occupants of the disturbed inclosures, and the grubs used for spinning were apparently obtained from a nest a short distance away, which probably accounts for the considerable time that elapsed before the rent was repaired.

The temperature of the free air is the title of a paper communicated by Dr. Hergesell to Part V. of *Petermann's Geographische Mitteilungen.* We have frequently referred to the great importance of this subject and to the valuable work performed by Dr. Hergesell in organising ascents of manned and free (or unmanned) balloons, and in discussing the results of the observations obtained. In the present paper he collects and discusses the most recent materials, and deals especially with the daily range and the vertical decrease of temperature in the upper strata of the atmosphere. The observations show that even at a height of a few hundred metres, there is a very small diurnal range; at night-time it amounts, in some ascents, to only a few tenths of a degree, and in the day-time, at about 800 metres, to some 3° or 4° Centigrade, when solar radiation is unobstructed. On cloudy days, and in the mean values, the daily amplitude is much less. With respect to the vertical decrease of temperature, the results of thirty sets of observations show that in all levels up to 10,000 metres an extremely varying temperature obtains, according to the season of the year and the conditions of weather. The decrease at that height reached or exceeded 40° C. in all cases, but no fixed rule could be laid down as to the *regular* decrease with altitude.

A recent number of the *Scientific American* contains a very interesting account of the use of a diver for the collection of zoological specimens that has been made in the Bay of Avalon, California. A large double-ended surf boat, in which the pump was placed, was towed to the scene of operations and anchored securely, bow and stern. Besides this, a number of observation boats, with glass bottoms, were used, and through these every movement of the diver could be observed. As soon as the diver was ready to descend, a scoop-net and a spike were handed to him. Stepping down, round by round, he finally pushed off and slowly sank to the bottom in about twenty-five feet of water. Through the glass bottom of the observation boats every movement could be plainly seen, as the diver walked through the weed, parting it on each side with ease, and collecting such specimens as seemed desirable. In one walk he brought up angel fishes, star fishes, holothurians, echini, a number of large univalve shells, a living shark, and numbers of small shells. The result of two days' work demonstrated the value of this method of collecting specimens, as in using a dredge many of the most delicate forms were injured. Where a diver is used it

is not necessary to take them from the water, the specimens being transferred in the water from the wire collecting-basket to a glass jar. The experiments are stated to have proved beyond question the value of the diver in work of this kind, as the ground covered was a veritable forest of macrosystis, in which groups of rocks were scattered, making work with a dredge impossible.

The Russian steamer *Rurik* has arrived at Tromsö, from Spitsbergen, bringing news from the Russian expedition, which had wintered on the island for the measurement of an arc of the meridian. No news could be sent until now, because the carrier-pigeons which the expedition set free on Spitsbergen refused to fly southwards and obstinately returned to the wintering place. The *Rurik* probably brings in a full report from the chief of the expedition, Prof. Th. Tchernysheff, but from a telegram of the learned geologist, which was sent to the Academy of Sciences from Tromsö, we already learn that all members of the expedition were well. During the winter astronomical and physical observations were made according to the programme. Photographs were taken of aurorae and their spectra, and in the spring observations were made on Mount Keilhaus, at the signal-pillar of the meridian arc. South Spitsbergen was crossed several times. Akhmatoff made pendulum measurements on Mount Keilhaus. The state of ice was still unfavourable in Storfjord, and Prof. Tchernysheff's intention was to make more excursions and, leaving the "ice-breaker" at Storfjord, to try to reach the Swedish party at Seven Islands.

The Transbaikalian Railway will be opened for traffic this month. It begins at Irkutsk, wherefrom a line, forty miles long, goes to Lake Baikal. There the train is placed on an ice-breaker-ferry and is transported to the Mysovskaya Station on the eastern shore of the lake, whence it runs 665 miles pas Verkhneudinsk, Chitá, and Nerchinsk (the town—not the mines) to Sryétensk. Steamers ply regularly during the summer from this little town down the Shilka and the Amur to its mouth. At the station Kaidalóva, near Chitá, begins the railway across Southern Transbaikalia, Mongolia, the Great Khingan Mountains and Manchuria, *via* Tsitsikar (on the Nonni) and Mukden, to Port Arthur. Work is busily carried on along this line, building going on on several sections at once: in Transbaikalia, at Tsitsikar, and at the southern end of the line.

Messrs. Cadett and Neall have sent us a sample of their X-ray paper. It is claimed for this material that a great reduction in exposure is effected as compared with the most rapid dry plates, about one-eighth of the usual exposure being all that is required. The paper has also the advantage over glass plates of freedom from risk of breakage, flexibility and consequent adaptability to the object to be photographed, and portability. The reason for paper of this description requiring so much less length of exposure than ordinary dry plates, is because less density is required for a reflecting surface to show structure than is required for a plate from which prints are required; consequently, with the X-ray paper, and using a 10-inch coil with a good tube, a good print of a hand can be obtained with about two seconds exposure; or, using an electrolytic break with the coil, with less than one second exposure.

An ingenious machine for solving any algebraic equation of the form $px^n + px^{n_1} + p_2x^{n_2} + \&c. = A$, by an application of the principle of Archimedes, is described by M. Georges Meslin in the *Journal de Physique* for June. It consists of a beam balanced on a knife-blade from any point of which may be suspended a solid of revolution, and a series of such solids is provided, constructed in such a manner that in the solid of order *n* the volume cut off by a horizontal plane is proportional to the *n*th power of the distance of the horizontal plane from the lowest point. Thus for orders 1, 2, 3, the forms of the solids are a cylinder ; a paraboloid of revolution, a cone. If the solid of order n is suspended at a distance p from the knife-blade, then when it is immersed to a depth x in liquid, the moment of the resultant upward thrust of the fluid about the knife-edge is proportional to px^n. The operation of solving the equation consists in adjusting the weights at suitable distances, p, p_1, p_2 from the axis, and balancing them, then running water into a trough containing the solids until the fluid thrusts balance a weight A fixed at unit distance from the axis of the beam ; when this is done the equation of moments takes the form of the given algebraic equation and x, the root of the equation is equal to the depth of immersion of the solids.

IN the *Rendiconto del R. Istituto Lombardo*, xxxiii. 11, 12, Prof. Luigi Berzolari considers a generalisation of the problem enunciated by Tanturri, of discovering the number of conics meeting a given algebraic gauche curve in eight points. The generalisation consists in the problem of finding the number of conics meeting one or more given algebraic curves in a points, passing through b given points and touching c given planes, where $a + 2b + c = 8$, and a number of results are given referring to the particular cases when one or more of the algebraic curves are straight lines.

ALTHOUGH it is now about sixty years since Moser published the results of his experiments on the action of light upon various surfaces as revealed by the condensation of vapours upon them, the character of the change produced by light still remains a mystery. Theories have been suggested, guesses have been made, but little or nothing has been proved. Major-General J. Waterhouse, I.S.C., has, during the last year, accumulated some additional interesting facts in connection with this subject. He fully confirms Moser's results as to the production of a change on the surface of metallic silver by exposure to light that can be demonstrated by the condensation of a vapour, such as mercury upon it. But he has gone further, and demonstrated the change by the deposition of silver from solution, after the manner of the development of an exposed wet collodion photographic plate. By some half hour's exposure in bright sunshine "printed out" images were obtained, that is, images visible without any subsequent application of a developer. General Waterhouse shows that these results are not due to pressure against the mask or stencil plate used, nor to the emanation of vapours from it, nor to heat. Usually blue light gives a much stronger effect than red, but in one experiment when the exposure was for three hours to bright sunshine, the effect was reversed, and the patches under red, orange and yellow glasses were developable, while those under the blue and violet glasses were not. But when the silver plate was heated to redness, quenched in dilute sulphuric acid, washed and dried, and the cut out design was also warmed before use, the effect produced by light was so small that it seems doubtful whether there was any effect at all. On the other hand, if the silver plate was exposed to the fumes of certain substances, especially nitric acid, it was rendered very much more sensitive. General Waterhouse, in his communication to the Royal Society, states that he hopes to continue the investigation this summer, and invites others to extend the observations that he has described.

IN the course of the Cavendish Lecture on the "Application of Pathology to Surgery," recently delivered by Mr. H. T. Butlin, of St. Bartholomew's Hospital, to the West London Medico-Chirurgical Society, a good deal was said with reference to research work, especially in relation to pathology. In the course of the lecture the need was pointed out of two species of pathological laboratories for research—one for research in pure pathology, without any reference to its application, which

"need not be attached, so far as its site is concerned, to any hospital. The other for research in applied pathology, the laboratory for inventors, must needs be attached to the hospital ; and those who work in it should have the freest access to the wards, even if they are not in charge of special wards, and should have every opportunity of observing what is done there and in the operating rooms. In order that they may be thoroughly instructed in the science of pathology, they should be taken from among the workers in the laboratory of pure pathology, and should be selected on account of their special aptitude for the work of research and for the originality they have exhibited. They leave the school of discoverers and the science of pure pathology for the school of inventors and the science of applied pathology." After alluding to the advance that has been made during the last few years, the lecturer said : "Money and organisation are necessary if great results are to be secured. The laboratories for research in pure pathology are too small and too scattered, and insufficiently endowed. The laboratories in the hospitals, which ought to be devoted to applied pathology, are used for every kind of microscopical and bacteriological examination and for teaching, so that research is crowded out. And pathological chemistry, from which vast things are to be hoped in future, has taken no proper hold upon the town." An investment of funds for the advancement of medicine and surgery, something like the provisions made in certain industrial establishments in Germany for research, was needed, in the opinion of the lecturer, who had no doubt as to the advantage which would accrue from such a movement.

IN No. 6 of the *Tufts College Studies* appears an important paper, by Mr. J. S. Kingsley, on the ossicles of the ear, which concludes with a suggestive discussion on the origin of mammals. In regard to the latter part of the subject, the author, as might be expected, attaches much importance to the fate of the quadrate bone of the lower vertebrates in mammals. And he arrives at the conclusion that the incus is mainly the representation of that element, although a portion of the latter may be included in the tympanic ring. It is further urged that the articulation of the lower jaw with the skull in mammals does not correspond with the same articulation in the lower vertebrates, but is entirely a new formation.

AS regards the origin of mammals, Mr. Kingsley urges that the ancestral type must certainly have possessed a freely movable quadrate bone ; from which he is led to conclude that the fixed suspensory arrangement of the lower jaw found in the chimæroid fishes, *Ceratodus*, and amphibians, is an acquired, and not a primitive, feature. Hence, the fringe-finned fishes like *Polypterus* indicate the ancestral stock of the higher vertebrates. Reverting to mammals, it is shown that the anomodont reptiles of South Africa are far too specialised to have been the parent stock. From this and other inferences it is concluded that "no reptile has yet been found which will in any way fit the requirements for the ancestor of the mammalia ; but that all known facts point rather to a line of descent from forms allied to the amphibia." There is, however, no amphibian type which conforms to the necessities of the situation, and it is accordingly necessary to go back to the common ancestor of the existing salamanders and cœcilians, or of the extinct labyrinthodonts or stegocephalians. In conclusion, it is stated that the ear-bones negative the view advanced by Mivart as to the egg-laying mammals having developed from a separate stock to that which gave origin to the other members of the class.

MEMOIR 4 of the Australian Museum, Sydney, deals with some of the Crustacea obtained during the trawling expedition of H.M.C.S. *Thetis* off the coast of New South Wales in the early part of 1898. Mr. T. Whitelegge, who has been entrusted with the description of this group, states that the collection of Crustaceans obtained during the cruise is remarkably rich in forms either new to science or to the fauna of New South Wales. Of the forty-five species recorded, twenty come under the latter category and nine under the former. But the present fasciculus applies only to the higher groups of the class, and when the lower forms are worked out a still larger proportion of novelties may be expected. The new types are figured in a well-executed series of plates.

IN its *Bulletin*, No. 180, the Michigan State Agricultural College Experiment Station sets an excellent example by calling attention to the noxious insects which have been most numerous during the past year in that district, and the best means for their destruction.

THE North London Natural History Society's syllabus for the period July to December has just reached us, and gives promise of an interesting session.

THE *Bibliotheca Mathematica*, iii. 1, contains a heliogravure portrait of the late Sophus Lie, together with a descriptive list of his papers by F. Engel, of Leipzig.

MESSRS. WILLIAMS AND NORGATE'S "Book Circular" for June has reached us. In it are to be found notes on new and forthcoming scientific publications, and a list of works on medicine, natural history, chemistry, physics, mathematics, &c.

WE have received Nos. 7 and 9 of *Scientia*. The former, by Dr. Denis Courtade, is entitled "L'Irritabilité dans la Série Animale," while the latter, by Dr. Pierre Bonnier, is called "L'Orientation," and deals with the notion and perception of space by animals, and the localisation of external objects.

THE conclusion of the series of articles on "South African Lepidoptera," by F. Barrett, in the *Entomologist's Monthly Magazine*, appears in the current issue of that periodical, and in it is contained the first instalment of an account of "An Excursion to Egypt, Palestine, Asia Minor, &c., in search of Aculeate Hymenoptera," from the pen of Rev. T. D. Morice.

THE July issue of "Climate, a quarterly journal of Health and Travel" contains several interesting articles, such as "The Art of Travelling" (an interview with Mrs. Bishop, the traveller), and "The Malaria Question," by the editor, in which a good deal of information is given in a compressed form.

IN the new number of "The Journal of the Royal Agricultural Society of England," Mr. W. E. Bear, in an article on "Fumigation for Insect Pests," passes in review the methods of fumigation that are or have been in use in various parts of the world, and the measure of success they have met with.

THE July number of *Knowledge* has as its leading article an account of the recent total solar eclipse, by Mr. E. W. Maunder. It is accompanied by a "process" reproduction of a full-page drawing of the corona, the work of Miss C. O. Stevens. Dr. W. Stanley Smith has commenced in the same periodical an interesting series of articles on Early Theories of Fermentation.

THE current number of *Science Gossip* contains the first of a series of "Geological Notes in Orange River Colony," from the pen of Mayor B. M. Skinner, which probably will appeal to a wider circle of readers just now than would have been the case had war not broken out. The present instalment deals with the country lying between Enslin and Bloemfontein.

THE *Agricultural Journal*, published by the Department of Agriculture, Cape of Good Hope, always contains many items of interest and value to the student of agriculture. The issue for May 10, which has just come to hand, contains, among other things, a good portion of the inaugural address on "The Bearings of Education and Science on Practical Agriculture," which was delivered by Prof. Somerville at Cambridge in November last.

THE Commissioner of Agriculture for the West Indies has issued a handy and useful pamphlet, entitled "Hints and Suggestions on Planting in Tobago." The greater portion deals with the subject of cacao culture, and is written by Mr. E. R. Smart, and revised by Mr. J. H. Hart and others. Short notes on other plants are from the pen of Sir R. B. Llewelyn, formerly Administrator of Tobago.

THE Yorkshire College, Leeds, on behalf of the East and West Ridings Joint Agricultural Council, will provide courses of instruction in the following subjects during the ensuing year:—Results of the Garforth and other experiments in the East and West Ridings; agriculture; veterinary hygiene; horticulture; and poultry keeping. A guide has been issued by the two bodies to experiments at the Manor Farm, Garforth, for the year 1900.

THE *Zambesi Mission Record* is a well-edited quarterly periodical, which contains not only reports of the religious and educational work done by the Catholic Mission under the auspices of which it is brought out, but also from time to time notes and articles on the natural history, botany and meteorology of the area traversed by the society; thus the issue for July contains notes on the weather and climate from observations taken at Bulawayo during 1899, and a lengthy contribution, entitled "By an African Pool," in which there is a good deal of popular science, appealing for the most part to the ornithologist. The latter article is illustrated by well-executed "process" blocks of photographs of specimens from the Albany Museum, Grahamstown.

RECENT successful attempts to prepare tubes and bulbs of fused quartz have led to a more detailed study of the thermal properties of this material. Its low coefficient of expansion and absolute unalterability at high temperatures would point to fused silica as an ideal material for air thermometry, and hence the observation by M. P. Villard in the current number of the *Comptes rendus*, that it resembles platinum in being permeable to hydrogen at high temperatures, is a disappointing one. A manometer connected to a pump and quartz tube, the latter being heated in a Bunsen burner to about 1000° C., shows a slowly increasing pressure, amounting in the course of a day to several centimetres of mercury, and on examination the gas proved to be nearly pure hydrogen. The same number contains a so a note by M. Dufour on the resistance of fused silica to sudden changes of temperature in which it is stated that quartz tubes, even although badly made, may be heated to any temperature and plunged into cold water without showing any signs or breaking.

IN the current number of the *Berichte* is a note by Dr. Vaubel on the phenyl derivative of diimide, NH : NH. This has been isolated in a simple manner from the products of reduction of diazoamidobenzene with zinc dust in alkaline solution. Phenyl-diimide $C_6H_5.N$: NH is an oily liquid of a pale yellowish colour, which can be distilled in steam, and possesses a strong odour of almond oil. Since it cannot be distilled with steam from an alkaline solution, it would appear to possess acid properties; it is very poisonous, and has no reducing action upon Fehling's solution. Contrary to expectation, it explodes neither on heating nor by shock. On account of its great stability towards oxidising agents, the author suggests the formula $C_6H_5.N\vdots NH$ as being the most probable.

THE additions to the Zoological Society's Gardens during the past week include two Tigers (*Felis tigris*, ♂, ♀) from India, presented by H.H. the Maharani Regent of Mysore ; a Black-eared Marmoset (*Hapale jacchus*) from South-east Brazil, presented by Mrs. G. L. Bagnell ; a Pine Marten (*Mustela martes*), British, presented by Mr. C. G. Beale ; a Common Squirrel (*Sciurus vulgaris*), British, presented by Mr. Cecil Slade ; a Yellow-cheeked Amazon (*Chrysotis autumnalis*) from Honduras, presented by Mr. S. Hankings ; two Crimson-crowned Weaver Birds (*Euplectes flammiceps*) from West Africa, presented by Mrs. Charles Green ; a Sharp-nosed Crocodile (*Crocodilus cataphractus*) from West Africa, presented by Mr. J. A. Robb ; a Four-lined Snake (*Coluber quatuorlineatus*), European, presented by Mr. W. R. Temple ; four Natterjack Toads (*Bufo calamita*), European, presented by Mr. Stanley S. Flower ; two Great Wallaroos (*Macropus robustus*, ♂, ♀) from South Australia, three Wrinkled Terrapins (*Chrysemys scripta rugosa*) from the West Indies, deposited ; an Adanson's Sternothere (*Sternothoerus adansoni*), a Common Chameleon (*Chamaeleon vulgaris*) from the Soudan, received in exchange ; a Burrhel Wild Sheep (*Ovis burrhel*), two Black-backed Gulls (*Larus marinus*), a Herring Gull (*Larus argentatus*), bred in the Gardens.

ERRATUM.—We are asked to state that in the report of Prof. S. Young's paper, read before the Physical Society on June 22, on the Law of Cailletet and Mathias, the words " 1 per cent." (p. 215, col. 1' line 3) should be " 0·1 per cent." The 0 was omitted from the report sent to us.

OUR ASTRONOMICAL COLUMN.

COMET GIACOBINI (1900 *a*).—Several observations have been made of this comet since its conjunction with the sun, but it is reported as faint. The following positions are an abridgment from the Ephemeris by Herr Ristenpart in *Astronomische Nachrichten*, No. 3636.

Ephemeris for 12h. Berlin Mean Time.

1900.		R.A.			Decl.
		h. m. s.			° '
July 12	..	22 29 5	...	+46 25·9
14	..	12 29	...	46 50·9
16	..	21 55 5	...	47 5·1
18	..	37 4	...	47 7·1
20	...	18 42	...	46 55 9
22	...	21 0 16	....	46 30·8
24	...	20 42 2	...	45 51·8
26	..	24 19	...	45 0·0
28	...	20 7 20	'....	43 56·4
30	...	19 51 16	...	+42 42·4

The comet attains its maximum north declination on the 18th, to the north-west of α Cygni, afterwards travelling in a south-westerly direction through Cygnus and Lyra.

WALTER PERCY SLADEN.

BY the death of Walter Percy Sladen, the world has lost one of the most lovable of men, and science an earnest devotee—a worker content to spare no effort could he but discover the truth.

Sladen was born on June 30, 1849, at Meerelough House, near Halifax, Yorkshire, and was educated at Hipperholme Grammar School, and afterwards at Marlborough under Dean Bradley. He came of an old Yorkshire family, who have been much respected for many generations ; and ease and refinement of manner were among his marked characteristics, while the charm of his address endeared him to all with whom he came in contact.

He never attended a regular academic course of instruction in the branch of science in which he became eminent ; his elementary training was self-acquired, and his leaning towards zoology innate. The definitive choice of the Echinoderma as the object of his life's work was of his own seeking, after much consideration ; and in this he showed great force of character and a power of self-reliance which there was reason earlier to believe he possessed, for even before he entered Marlborough he evinced an unusual predisposition towards science, in founding for his boy friends a scientific society devoted more especially to the study of astronomy, in connection with which he became known among them as the "Astronomer Royal." Little did he think that he would in later life become for ten years a secretary of a leading scientific society, and that for eighteen he would conduct the affairs of a zoological research committee, as he did in his capacity as Secretary to the British Association Table of the Naples Station.

Sladen's scientific work, so far as his published memoirs and papers are concerned, extended over a period of seventeen years, 1877 to 1893. Of these there are thirty-four in all—twenty-one from his own hand, thirteen in conjunction with his intimate friend and adviser, the late Prof. Martin Duncan. Beyond these there stand to his record certain bibliographical notices and miscellanea. Of the thirty-four published works, fifteen of which he was sole, and four of which he was joint author, deal with the starfishes ; and of the remaining fifteen, nine were conjoint, and devoted, with the exception of two, to fossil forms. Conspicuous among these are reports upon the collections made by the Geological Survey of India ; and among those which he alone produced are Parts i. and ii. of the second volume of the Palæontographical Society's Memoirs on the Fossil Echinodermata, which were his last published works. They deal with the Cretaceous Asteroids, and appeared in the Society's volumes for 1890 and 1893. His first three papers deal with the remarkable creature *Astrophiura*, whose generic name is self-explanatory. The first, a brief description, was published in the *Proceedings of the Royal Society* for 1878 ; the other two, each containing a Latin diagnosis, in the *Zoologischer Anzeiger* and *Annals and Magazine of Natural History*, the year following. His remaining papers appeared in the *Annals* and the *Journal of the Linnean Society*, the publications of the Royal Society of Edinburgh, and elsewhere. They mostly deal with whole collections, and include reports on those made in the Arctic Region in 1875-1876, on those of the *Alert*, *Knight Errant* and *Triton*, as also those made in the Faroe Channel, the Korean Sea, and the Mergui Archipelago. In each Sladen produced good results, as in the discovery of genera such as *Micraster* and *Rhegaster* ; and what more natural, therefore, than that he should have been entrusted with the working out of the Asteroids collected by H.M.S. *Challenger*, the report upon which was the crowning achievement of his life.

This magnificent work of 900 pp., with its accompanying atlas of 118 plates, ranks among the most masterly and exhaustive of the *Challenger* volumes. Before taking it seriously in hand, Sladen visited every museum in Europe (with one exception) which was known to contain starfishes of importance ; and, as pointed out by the editor in its preface, it is a monograph of the whole group. The labour involved in its production was prodigious ; and its interest is enhanced by the fact that the bulk of it was written between the hours of 9 p.m. and those of early morning, often after a day's occupation with other affairs. The extension of the family Pterasteridæ and the great addition to our knowledge of the deep-sea forms are its most salient characters ; but we know not which to admire most, the body of the work, with its laborious descriptions of individual forms, or the supplemental part, in which there is given a list of every known species, with a record of its bathymetric distribution. Elementary student and expert stand alike indebted to him for this monumental work, indispensable to progress in the knowledge of the subject with which it deals. Generic names like *Benthaster* and *Marsipaster* are sufficiently significant in themselves. Proceeding to classification, Sladen made good use of the marginal and ambulacral plates, and his subdivision into the sub-classes *Euasteroidea* and *Palaeasteroidea*, with the ordinal divisions to which he was led, has withstood the test of time and become the adopted classification of the better text-books, as for example those of Lang and Gregory. In this his influence on the progress of science will live, and it is a matter of profound gratification that only a short time before his death he gave expression to the satisfaction this recognition afforded him.

Beyond this magnificent work and those papers more or less immediately associated with it, wholly taxonomic, Sladen produced others of a physiological and developmental order, as for example his Naples Station paper, on the structure and functions of the pedicellariæ, and that announcing his discovery of the "cribriform organ," and his papers on the apical plates of the Astrophiuroids, in which he was obviously in agreement with his friend, the late Dr. Herbert Carpenter, in the belief in a unified ancestry of these. It has been said of his taxonomic work that his descriptions are protracted, and that he deals with specimens as species. There is, however, no reason to believe that he was using the term species in any but a purely conventional sense, without necessarily implying any fixed inter-relationships; and his painstaking accuracy of description was the outcome of an excessive honesty of purpose and desire for thoroughness, in which he was altogether exemplary. There never lived a man with a truer sense of honour.

Some ten years ago Sladen had a bad attack of influenza, followed at intervals by several similar visitations, which unfitted him for serious scientific work, but he always hoped to get better and to take it up again. The last winter was passed in Devonshire with very beneficial results, and he might be said to have been in his usual health when two months ago he started with his wife for Rome. But the wish to return to work was not to be fulfilled; after spending six weeks in Rome he journeyed to Florence, and there after a week of rather active sightseeing, on June 11, he was taken with a fainting fit, and though he quickly recovered consciousness and declared his intention of going to Como that very night, within half an hour he passed away by failure of the heart's action.

He was a Fellow of the Linnean, Zoological, and Geological Societies; for ten years Zoologic Secretary, and later a Vice-President of the former. In his secretarial capacity his genial nature found full sway, and his encouraging attitude to the younger men with whom he came in contact will ever be remembered. As a boy at school he was the captain wrestler. He was a good shot, though never a sportsman or member of a rifle corps, but he belonged to a private Guerilla Club in Yorkshire, of which he was sometime secretary.

In 1890 Sladen married Constance, elder daughter of the late Dr. W. C. Anderson, of York; and about two years ago he inherited from an uncle the estate of Northbrook, near Exeter, and there he has been laid to rest. It will be remembered that he recently gave the sum of 2000*l.* to insure the lives of the Yeomanry and Volunteers of his county going to the front in the Boer campaign; and this is but one among many of his generous acts, the majority of which are known only to the recipients. A loving husband, a trustworthy friend, whose advice was always sound, a keen sympathiser with suffering humanity, he has passed from us; but his memory and tender-loving influence for good will endure.

Among his scientific effects are a large library and some zoological collections of great value. Sladen had always a taste for old books, and one of his last expeditions was to a monastery at Subiaco, to examine some ninth century MSS. there preserved, and he had collected a goodly number of ancient MSS. and examples of early printing. His collection of Echinoderm literature is very complete; while, as to material, he leaves the collections of his friend, the late Herbert Carpenter, rich in Crinoids and other rare animal forms, which include, as a separate historic possession, the materials which formed the basis of the elder Carpenter's book on the microscope. These he purchased. There were also in his possession at the time of death a large series of Cretaceous Echinoderms, upon which he was contemplating a renewal of his Palæontographical Society's work; and the collection of Astrophiuroids of the *Albatross*, entrusted to his hands by Prof. A. Agassiz. There accompanied these a series of superb coloured drawings from the life, like those already published for the Holothuroidea of the expedition; and the very day of his death there reached him a letter from the same distinguished explorer, offering him the materials of his recent Australian cruise. It was Sladen's intention to have returned to these rich possessions; and we could desire no more fitting memorial to his work than that it might be possible to find and train a competent zoologist to continue that which he has left thus unfinished, on the lines on which he would have laboured, and to hand it down to posterity a completed record in his name.   G. B. H.

JEREMIAH HORROCKS AND THE TRANSIT OF VENUS.

WE are indebted to the *Journal* of the Leeds Astronomical Society for 1899, which contains an interesting paper by Mr. A. Dodgson, on the life and work of the illustrious young astronomer, Jeremiah Horrocks. This worthy was born in 1619, 281 years ago, in the reign of James I., at Toxteth, three miles from Liverpool. He received his early education there, but on reaching the age of fourteen, he entered as "sizar" at Emmanuel College, Cambridge. At seventeen he was enabled to become tutor at Toxteth, and two years later, *i.e.* in his nineteenth year, he was appointed curate at Hoole, near Preston. Soon after this he made his memorable astronomical observation of Venus, and only two years later was dead. The life of the young man at Cambridge, as traced by Mr. Dodgson, was one of persistent industry. Imbued at an early age with a love of studying natural phenomena, he was hampered at the outset

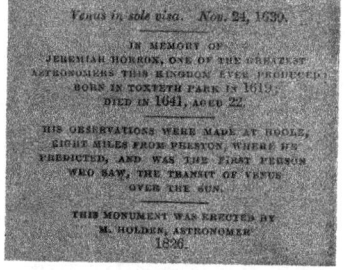

Venus in sole visa. Nov. 24, 1639.

IN MEMORY OF
JEREMIAH HORROX, ONE OF THE GREATEST
ASTRONOMERS THIS KINGDOM EVER PRODUCED
BORN IN TOXTETH PARK IN 1619
DIED IN 1641, AGED 22.

HIS OBSERVATIONS WERE MADE AT HOOLE,
EIGHT MILES FROM PRESTON, WHERE HE
PREDICTED, AND WAS THE FIRST PERSON
WHO SAW, THE TRANSIT OF VENUS
OVER THE SUN.

THIS MONUMENT WAS ERECTED BY
M. HOLDEN, ASTRONOMER
1826.

by the absence of instruction in mathematics and the scarcity of books. This difficulty of getting philosophical and scientific works is clearly shown by the fact that of the thirty-two volumes he possessed later, not one was published in England or written by an Englishman. Lansberg's works he could not make agree with his own observations, and later, having obtained those of Kepler through the advice of his friend Crabtree, of Manchester, he found that even they needed many corrections. His first results in astronomical research were in elucidation of the lunar theory. Sir Isaac Newton confirms that he was the first to state the ellipticity of the moon's orbit; he also stated the causes of "evection" and "annual equation." The experiment of the circular pendulum for illustrating the action of a central force is also due to him. Most interesting, however, is his successful prediction of the transit of Venus in November, 1639. Kepler had stated that the two next transits would occur in 1631 and 1761, but Horrocks found, during his revision of the tables he had in use, that another would take place, the slight errors in Kepler's tables probably accounting for his omitting it. Horrocks made all preparations for observing the phenomenon, arranging the image projected from his telescope (which had cost him half-a-crown) on a sheet of white paper having a circle six inches in diameter traced on it, the circumference being divided into degrees. He watched from sunrise to nine o'clock, and from ten until noon. Resuming his labour again shortly after three, he was overjoyed to find a round black spot just within the limb of the sun, at the internal contact. During some thirty minutes he was enabled to make many observations, which he considered very successful. Besides these definite scientific achievements, he wrote upon many different phenomena connected with the solar system, including the motions of Jupiter, Saturn, and various comets. The illustration on p. 257, reproduced from Mr. Dodgson's paper, shows Carr's house at Hoole, where Horrocks made his transit observation, and also the monumental tablet erected in 1826 to his memory in Hoole Church.

JUBILEE OF THE IMPERIAL GEOLOGICAL INSTITUTE OF VIENNA.

TO celebrate the fiftieth anniversary of the foundation of the Imperial Geological Institute of Vienna, a jubilee meeting was held on June 9 in the Great Hall of the Institute under the presidency of its present director, Hofrath Guido Stache. The meeting was attended by a number of high Government officials, geologists, and representatives of national industries and scientific associations.

The director having welcomed the guests, speeches of congratulation were delivered by his Excellency the Minister for Spiritual and Educational Affairs, Dr. W. Ritter von Hartel, his Excellency the Minister for Railways, Dr. H. Ritter von Wittek, and the Mayor of Vienna, Dr. C. Lueger. These were succeeded by the following representatives of scientific institutions and industries, who presented addresses: Geheimrath von Richthofen, conveying the good wishes of the Prussian Royal Academy of Sciences, the Gesellschaft für Erdkunde, and the German Geological Society; Prof. Dr. Beyschlag, for the Royal Prussian Geological Institution and the Berg Akademie of Berlin; Geheimrath Dr. Lepsius, for the Grand-ducal Institute of Hesse and the Upper Rhine Geological Society at Darmstadt; Prof. Dr. E. Naumann, for the Senckenberg Natural Science Society of Frankfurt a-M.; Sectionsrath Boeck, for the Hungarian Geological Institution and the Hungarian Geological Society; and Chief Geologist Pethö, for the Natural Science Association of Buda-Pesth.

Among Austrian representatives there were: Prof. E. Suess, as President of the Imperial Academy of Sciences; Prof. L. Szajnocha, for the Cracow Academy; Prof. Woldrich, for the Bohemian Francis Joseph Academy; Hofrath Steindachner, for the Court Museum of Natural History; his Excellency Field-Marshal Ritter von Steeb, as Commandant of the Military Geographical Institute; Rector Zeisel, for the Agricultural College; Hofrath Jüraschek, for the Central Statistical Commission; Prof. Doelter, for the Steiermark Scientific Society; and Vice-President Straberger, for the Francisco-Carolineum at Linz.

The good wishes of the Lower Austrian Chamber of Commerce were presented by the President of the Northern Railway, Hofrath Jeitteles, and the congratulations of societies for the advancement of allied sciences were tendered by Custos Marenzeller, Freiherr von Puche, Hofrath Toula, Freiherr von Andrian, and Councillor Karrer. In conclusion, the Chairman read those parts of the Jubilee Report which referred to the advancement of the Institute by the Emperor and the Government.

Among the 264 messages of congratulation received the following are specially mentioned: from the Geological Survey of G₁ea; Britain and Ireland, the Geological Society and the Iron and Steel Institute in London, the Smithsonian Institution and the United States Coast and Geodetic Survey in Washington, the American Philosophical Society in Philadelphia, and the Cincinnati Society of Natural History. Also those of the Imperial Russian Academy of Sciences, the Russian Geological Committee, and the Imperial Russian Mineralogical Society at St. Petersburg; the Naturalists' Society of Moscow, the Royal Swedish Academies at Stockholm and Upsala, the Academia dei Lincei and the Ufficio Geologico in Rome, the Science

Academies of Naples and Turin, the Belgian Geological Society, the Royal Academy of Amsterdam, and scientific associations and institutions at Hallé, Dresden, Leipzig, Breslau and Göttingen.

The Institute, or Geologische Reichsanstalt, was founded in 1849 by the then Minister of Mines and Agriculture, von Thinnfeld, with the object of working out the geology of the whole empire, collecting and arranging the material, and publishing the results in maps and memoirs. Haidinger was its first director, and his chief geologist was Freiherr von Hauer, who was appointed director on Haidinger's death in 1866. In those early days the position of the Institute was not by any means secure. In 1859 an attempt was made to abolish it as a separate institution and to incorporate it with the mathematical and natural science section of the Imperial Academy of Sciences. But the proposed change failed to obtain the approval of the Reichsrath.

Between 1867 and 1871, under von Hauer's direction, a geological map of the Austro-Hungarian Monarchy was published, to a scale of 1 in 576,000. Under the supervision of the present director, Hofrath G. Stache, the publication of a series of detail maps has been commenced. The publications of the Institute comprise the annual *Jahrbuch*, which has now reached its fiftieth volume, the *Verhandlungen*, and the *Abhandlungen*. The latter are in 4to, and up to the present they have an aggregate of 7000 pages and 950 lithographic plates. Besides, explanatory letterpress is issued with each section of the new detailed geological map drawn to a scale of 1 in 75,000.

A chemical laboratory is attached to the Institute, which undertakes geological and industrial analyses. This laboratory was suppressed for several years, owing, it is said, to the overshadowing influence of another laboratory connected with the Vienna Academy of Sciences (see Dr. Tietze's "Life of Franz von Hauer," Vienna, 1900).

The Institute possesses extensive geological and mineralogical collections, chiefly from Austrian and Hungarian districts. These are exhibited in twenty-one rooms, some of which are really halls of great architectural beauty. The library contains over 40,000 volumes.

The Reichsanstalt is under the supervision of the Minister for Spiritual and Educational Affairs. Its annual income is 18,000*l*. Its staff numbers twenty persons, twelve of whom are employed in the Geological Survey.

A PARTIAL EXPLANATION OF SOME OF THE PRINCIPAL OCEAN TIDES.

AT the meeting of the U.S. National Academy of Sciences on April 19, a paper bearing the above title was read by Mr. R. A. Harris, of the United States Coast and Geodetic Survey. An abstract summarising the chief results arrived at has been published by the Academy: the full memoir is to be issued as an Appendix to the Annual Report of the Survey for 1899–1900. The abstract is too short to allow of critical examination of the methods employed in these inquiries, but some of the conclusions stated are very significant and important.

Mr. Harris enunciates the fundamental proposition of his investigation in the following terms: "Considering the actual distribution of land and water, a few computations upon hypothetical cases will suffice to convince one that as a rule the ocean tides, as we know them, are so great that they can be produced only by successive actions of the tidal forces upon oscillating systems, each having, as free period, approximately the period of the forces, and each perfect enough to preserve the general character of its motion during several such periods were the forces to cease their action. . . . having once for all constructed a set of force diagrams for the various latitudes, we have only to discover those regions which have a free period of oscillation about equal to the period of the forces, and to then ascertain at what time the particles should be at elongation in their nearly rectilinear paths."

The main idea underlying this proposition is not altogether new, the novelty in the present paper is rather an attempt to locate and define areas which seem to account for the principal ocean tides, due regard being had to the difficulties arising from irregularities in the natural boundaries of such areas where such exist, or from the absence of natural boundaries. Each *oscillating area* is one of comparatively simple form, of which the free period of oscillation, supposing its boundaries all rigid, would not differ much from twelve lunar hours, and the forces are connected with the dominant ocean tides by applying to such an area, or to a system of such areas, the rule that "if 1 to the particles of water in a given oscillating system, each area of uniform depth, and wherein the resistances are proportional to the velocities of the particles, a series of simple harmonic forces having for period the free period of the body of water be applied and a permanent state established, then must the time of elongation be simultaneous with the time when the virtual work of the external periodic forces upon the system becomes **zero**." Applying this rule, by means of the tidal-force diagrams **the time** can be found when "the aggregate of the elementary **masses, each** multiplied by the intensity of the tidal force in the **direction** of the displacement of the element, and again by a **quantity proportional** to the value of the maximum displacement **(since the oscillation** is harmonic), is zero" : this is the time of **high or low water.** The results of this method appear at once **in a few simple** cases : thus in an east-and-west canal half a **wave-length** long it is high water at the east end at the **component hour 0 or 12,** the Roman numerals being understood to **be the meridian** of the middle point of the canal ; in a **meridional canal** one wave-length long, whose centre lies

UNIVERSITY AND EDUCATIONAL INTELLIGENCE.

OXFORD.—Applications are invited for the new Wykeham professorship of physics, referred to in a note on May 24 (p. 91). The election will take place in November, and applications must reach the Registrar not later than October 24. The following particulars are given in the *University Gazette*:—The subjects on which the professor will chiefly lecture and give instruction will be electricity and magnetism. The professor will have the charge of any laboratory which the University may assign to him. It is expected that rooms, now otherwise occupied, will be assigned to the professor for a laboratory in the course of the year 1901 ; 700*l.* will be appropriated to fitting up the laboratory, and provision has been made for spending 250*l.* a year for the first two years on assistance and maintenance. As soon as the professor is elected, he will be entitled to be admitted to a Fellowship at New College of the annual value of 200*l.* In addition, from January 1, 1901, he will receive from New College (1) an annual payment of 200*l.* ; and (2) a further annual payment of 100*l.* so long as the College has funds available for the purpose. It is anticipated that this further payment will be paid for not less than twelve or thirteen years.

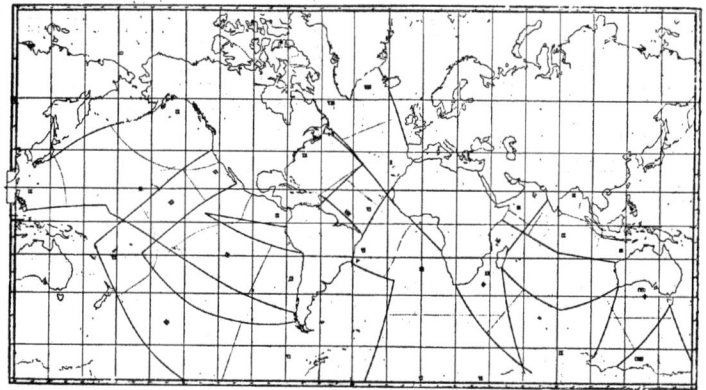

between 45° south and 45° north latitude, it is high water at both ends at the component hour 9 ; if the centre lies beyond these limits, the component hour of high water at the ends is 3.

Before laying down the oscillating areas, Mr. Harris gives a number of lemmas which have to be borne in mind as modifying the motions discussed. To quote one example : "Suppose a stationary oscillation to exist in a canal communicating with a tided sea ; let the length of the canal lie between 0 and ¼ λ, then at the time of high water outside it is high water through-out the canal (*e.g.* many Alaskan canals). If the length lie between ¼ λ and ¾ λ, it is low water for a distance of ½ λ from the head at the time it is high water outside (*e.g.* Irish Sea, node at Courtown ; English Channel, node at Christchurch). If the length be equal, or nearly equal, to ½ λ, then the horizontal motion at the mouth, instead of the vertical motion, determines the time of tide within ; this tide will be three hours later than the tide outside (*e.g.* the Gulf of Maine).

The systems supposed to account for the principal semi-daily movements of the oceans are outlined on the chart which we reproduce in a reduced form ; the Roman numerals indicate the co-tidal hours. The main systems are seven in number : (1) North Atlantic, (2) South Atlantic, (3) North Pacific, (4) South Pacific, (5) North Indian, (6) South Indian, (7) South Australian (solar).

PROF. McCALL ANDERSON, Professor of Clinical Medicine in the University of Glasgow, has been appointed to the chair of Systematic Medicine in the same University, in place of Sir W. Gairdner, resigned.

THE war in South Africa has raised many questions of great national importance which are fortunately receiving the attention of many thoughtful people. Prominent among these subjects of discussion is the urgent problem of how to obtain an improved supply of suitably educated officers, which was recently dealt with in a paper read by the Headmaster of Eton at the Royal United Service Institution. Dr. Warre maintains that a wider diffusion of the knowledge of the *elementa* of military science among the educated youth of the nation would tend, not only to raise the standard of military knowledge in the Army and Auxiliary Forces, but to improve the methods of communicating that knowledge to the rising generation, an indirectly widen the area from which a supply of well-educated officers may constantly be drawn. The great majority of the headmasters of our public schools agree with Dr. Warre, and he has drawn up, at the request of the War Office, a memorandum in which he advocates the need for a new Act of Parliament, the tenor of which should be "that all persons *in statu pupillari* in public secondary schools above the age of fifteen, able and willing to bear arms, should be enrolled for the purpose of instruction in drill, manœuvre, and the use of arms." At the same time the paper makes it quite clear that the Headmaster of Eton thinks mere proficiency in drill is not sufficient —at every step the boy must be taught the reason of everything he is called upon to do, and throughout his training his intelligence must be carefully and steadily developed. Approaching the same question from another point of view, Prof. Armstrong, in a letter to the *Times*, maintains that no amount of mere military training given in schools, or subsequently, will ensure the necessary improvement in our officers, unless the intelligence of boys is more satisfactorily developed in the early years at schools—an end which can best be secured by an adequate training in the scientific method. It may fairly be surmised that the Headmaster of Eton is quite in agreement with Dr. Armstrong as to the paramount importance of early teaching, and that both are equally anxious that intelligent citizens should somehow be produced. Of the value of a familiarity with the methods of science it is here unnecessary to say anything, but it would certainly appear that both contentions are right. What is wanted is Dr. Warre's intelligent military training for public school boys who have all had the advantage of a training in the scientific method for which Dr. Armstrong pleads.

SCIENTIFIC SERIALS.

Transactions of the American Mathematical Society, vol. i. No. 2, April.—On the metric geometry of the plane n-line, by F. Morley. The relations which n-lines of a plane exhibit, when considered in relation to the circular points, have not received, in Prof. Morley's opinion, systematic attention since the important memoirs by Clifford, on Miquel's theorem ("Works," p. 51), and by Kantor (*Wiener Berichte*, vols. lxxvi. and lxxviii.). He applies certain notions which are fundamental in the geometric treatment of the theory of functions, and especially the notion of mapping. The paper is an interesting extension of Clifford's chain, and adds many curious results.—On relative motion, by A. S. Chessin. A memoir extending to 54 pages. The theory developed in it originated in a memoir by Bour in 1863 (*Journal de Liouville*, Ser. 2, vol. viii.). It deals mainly with the so-called "second form" of differential equations of Lagrange, and with the canonical system of differential equations of Hamilton-Jacobi. The first part of the paper deals only with the *theory* of relative motion. The differential equations are derived from one fundamental principle embodied in the so-called "theorem of Coriolis." This enables the author, not only to write down the differential equations of relative motion *immediately* from the corresponding equations of absolute motion, but to obtain equations as *general as those known for absolute motion*. In this first part there are eleven chapters. The second part (promised) is to contain applications of the theory. Among the problems to be discussed is the problem of Foucault's pendulum when the oscillations are not infinitely small, and the problem of Foucault's top, which Gilbert was unable to solve (sur l'application de la méthode de Lagrange à divers problèmes de mouvement relatif). The two problems, our author states, can be easily solved by the theory and formulas given in this first part.—Plane cubics and irrational covariant cubics, by H. S. White.—The paper considers cubics invariant under *partial* transformation by covariants (2, 2), and those invariant under *complete* transformation by covariants (3, 3). There remain for further treatment the two sets of conics invariant under the third transformation (2, 2), and invariant curves of order higher than the third (*cf.* the author's paper in No. 1). The new covariant cubics are eight in number, all of the type called equianharmonics.—A purely geometric representation of all points in the projective plane, by J. L. Coolidge. After some definitions, the writer gives a representation of all points in a real line by lines in a real plane, and then extends the representation so as to include all points in a real plane, noticing in particular those systems of lines which represent points on an imaginary line. He then takes up the subject of chains of points, showing their application to the general theory of projectivity. Finally, he glances briefly at the system of lines which represent points on a real conic, and concludes with remarks as to other possible solutions of the problem and its extension to three dimensions.—The decomposition of the general collineation of space into three

skew reflections, by E. B. Wilson. The paper discusses the question, "Is it possible to decompose the general collineations of space into the product of a number of skew reflections; and if so, what is the least number of skew reflections involved in such a decomposition?"—A new method of determining the differential parameters and invariants of quadratic differential quantics, by H. Maschke, exhibits in a preliminary way a symbolic method in close analogy with the symbolism used in the algebraic theory of invariants, for the construction and investigation of invariants of quadratic differential quantics.— On the extension of Delaunay's method in the lunar theory to the general problem of planetary motion, by G. W. Hill, shows that the tediousness of Delaunay's method disappears when the greatest generality is given to the procedure.—Mr. J. E. Campbell writes on the types of linear partial differential equations of the second order in three independent variables which are unaltered by the transformations of a continuous group.

Bulletin of the American Mathematical Society, June.— Prof. Cole furnishes an account of the *Proceedings* at the New York April meeting of the Society, and abstracts several of the papers read; and Prof. Holgate performs a like office for the April meeting of the Chicago section of the Society.—J. G. Hagen gives a short sketch of the history of the extensions of the calculus. The abstract is confined to those theories that are in close relation to the infinitesimal calculus and the theory of functions, and excludes geometrical methods and methods of demonstration. To name one or two points discussed, they are Cauchy's "Calcul des Résidues," Schell's "Quotial and Instaural," the exponential function of higher order, the logarithmic methods of Bergbohm and Oltramare, and the extension of the calculus of finite differences.—Reviews are given of Burnside's "Theory of Groups," by Dr. G. A. Miller; of D'Ocagne's "Traité de Nomographie," by Prof. Morley; of Barton's "Theory of Equations," by J. Maclay; of Rice's "Theory and Practice of Interpolation," by Prof. E. W. Brown; of Von Braunmühl's "History of Trigonometry," by Prof. Cajori; of M. Boyer's interesting "Histoire du Mathématiques," by the same writer; and of Frischauf's "Vorlesungen über Kreis- und Kugel-Functionen-Reihen," by W. B. Ford.— Varied information is supplied in the "Notes" and "New Publications."

The numbers of the *Journal of Botany* for May, June, and July are almost entirely occupied by articles descriptive of new species, or relating to the geographical distribution of plants, chiefly in the British Islands. Mr. H. N. Dixon records the detection of an addition to British mosses in *Amblystegium compactum*, and Mr. S. M. Macvicar an addition to British Hepaticæ, in *Pellia neesiana*.

SOCIETIES AND ACADEMIES.

LONDON.

Chemical Society, June 21.—Prof. Thorpe, President, in the chair. The following papers were read.—Researches on morphine, I., by S. B. Schryver and F. H. Lees. Morphine readily exchanges an alcoholic hydroxyl group for halogen, yielding the bases chloromorphide, $C_{17}H_{18}O_2NCl$, and bromomorphide; when heated with water these substances give isomorphine, $C_{17}H_{19}O_3N$, and on reduction chloromorphine yields desoxymorphine hydrochloride $(C_{17}H_{19}O_2N, HCl)_2, 3H_2O$. These four new bases are not narcotics.—On the oxime of mesoxamide and some allied compounds, by M. A. Whiteley. Nitrosyl chloride converts malonamide into the isonitroso-derivative, $CONH_2 \cdot C(NOH) \cdot CONH_2$; nitrous acid converts the latter into a pseudonitrole, $CONH_2 \cdot C(NO)(NO_3) \cdot CONH_2$, and hydriodic acid reduces it to aminomalonamide,

$$CONH_2 \cdot CH(NH_2) \cdot CONH_2.$$

—On dimethyldiacetylacetone, tetramethylpyrone and orcinol derivatives from diacetylacetone, by J. N. Collie and B. D. Steele. Disodiodimethylpyrone and methyl iodide react, giving dimethyldiacetylacetone, $C_7H_8O_3(CH_3)_2$, which is converted into tetramethylpyrone, $C_7H_6O_2(CH_3)_2$, by hydriodic acid; the residues from the preparation of dimethyldiacetylpyrone contain trimethylpyrone, $C_8H_{10}O_2$, and an orcinol derivative, $C_9H_{12}O_2$.—Dehydracetic acid, by J. N. Collie. The author has succeeded in preparing dehydracetic acid from triacetic lactone.—The decomposition of hydroxyamidosulphates by copper sulphate, by E. Divers and T.

Haga.—The degradation of glycollic aldehyde, by H. J. H. **Fenton.**—Notes on the chemistry of chlorophyll, by L. Marchlewski and C. A. Schunck.—A new series of pentamethylene derivatives, I., by W. H. Perkin, jun., J. F. Thorpe and C. Walker. Ethoxycaronic acid, $C(CH_3)_2 \begin{matrix} C(OC_2H_5)CO_2H \\ CH.CO_2H \end{matrix}$, is obtained by treating aa'-dibromo-ββ-dimethylglutarate with alcoholic potash, and yields *asym*-dimethylsuccinic anhydride, $\begin{matrix} CMe_2.CO \\ CH_2—CO \end{matrix} O$, with sulphuric acid. Ethyl dibromodimethyl-glutarate and sodiomalonic ether condense, yielding the sodio-derivative, $CMe_2 \begin{matrix} C(CO_2Et).CNa.CO_2Et \\ C(CO_2Et).CO \end{matrix}$; numerous derivatives of this substance are described.—Experiments on the synthesis of camphoric acid. III. The action of sodium and methyl iodide on ethyl dimethylbutanetricarboxylate, by W. H. Perkin, jun., and J. F. Thorpe. Ethyl dimethylbutanetricarboxylate is converted by sodium and methyl iodide into a substance which possibly has the constitution

$$CMe_2 \begin{matrix} CMe(CO_2Et)\ CO \\ CH_2(CO_2Et).CH_2 \end{matrix}$$

which should be easily converted into a substance having the constitution assigned by Bredt to camphoric acid ; the ester is converted on reduction into an isomeride of camphanic acid.—The oxyphenoxy- and phenylenoxy-acetic acids, by W. Carter and W. T. Lawrence.—The condensation of ethyl α-bromoisobutyrate with ethyl malonates and ethyl cyanacetates ; a-methyl a'-isobutylglutaric acid, by W. T. Lawrence. The author shows that the following general equations hold :—(1) CNaRX.CO₂Et + CBrMe₂.CO₂Et = CRX(CO₂Et).CMe₂.CO₂Et + NaBr, when the sodio-derivative is solid, and (2) CNaRX.CO₂Et + CBrMe₂.CO₂Et = CRX(CO₂Et).CH₂.CHMe.CO₂Et + NaBr, when the sodio-compound is dissolved ; R=H or an alkyl group and X=.CN or .CO₂Et.—Methylisoamylsuccinic acid, II., by W. T. Lawrence.—The estimation of furfural, by W. Cormack. Furfural may be estimated by oxidising it to pyromucic acid by standard ammoniacal silver oxide solution, filtering off the reduced silver and titrating the silver left in solution.—The constitution of hydrogen cyanide, by J. Wade.—Inhibiting effect of etherification on substitution in phenols, by H. E. Armstrong and E. W. Lewis. The substitution of benzoyl, phenylsulphonyl, benzylsulphonyl, the radicle C₁₀H₁₈O.SO₃, of Reychler's camphorsulphonic acid, or picryl for phenolic hydrogen in phenolparasulphonic acid renders the substance inert towards bromine.—Bromination of oxyazo-compounds, by H. E Armstrong and P. C. C. Isherwood.—Metasulphonation of aniline, by H. E. Armstrong and W. Berry.—Phenylacetylchloramine and analogous compounds, by H. E. Armstrong.—Benzylanilinesulphonic acids, by I. Smedley.—Benzeneorthodisulphonic acid, by H. E. Armstrong and S. S. Napper.—An isomeride of furfurine, by J. P. Millington and H. Hibbert.—The mono- and di-acetyl and phenacetyl diethyl tartrates, by J. McCrae and T. S. Patterson.

July 5.—Prof. Thorpe, president, in the chair.—The Nilson Memorial Lecture was delivered by Prof. Otto Pettersson, of Stockholm.

Entomological Society, June 6.—Mr. G. H. Verrall, President, in the chair.—Mr. Hedworth Foulkes, B.Sc., and the Rev. H. C. Lang, M.D., were elected Fellows of the Society.—Mr. G. H. Verrall exhibited a species of the genus *Ceratitis*, MacLeay, apparently identical with Bigot's *C. fronicilatus* from the Gold Coast (West Africa), and a very handsome Trypetid reared from the fruit of *Mimusops caffra* by Mr. Claud Fuller at Durban, Natal.—Mr. C. O. Waterhouse exhibited specimens of a Hemipteron, *Aspongopus nepalensis*. They are much sought after by the natives, who use them for food pounded up and mixed with rice.—Mr. Merrifield exhibited a number of pupæ of *Aporia crataegi*, and called attention to the want of correspondence between the markings on the pupal and those on the imaginal wing. As might be expected of an insect whose larva pupates by preference on stems screened by foliage, its colour is not very greatly affected by its surroundings ; on comparing some which had had yellow or orange surroundings with others which had had dark ones, it was shown that the former tended to yellow

ground colour, and the latter to grey, having also an increase of the dark spots with which the thorax and abdomen are thickly strewn. He also exhibited some enlarged coloured photographs of the green and dark pupal forms of *Papilio machaon*, obtained by causing the larvæ to pupate on green, yellow or orange surfaces, and on dark ones respectively.—Sir G. F. Hampson exhibited specimens of *Oligostigma araealis*, from Ceylon, where his correspondent, Mr. J. Pole, had met with a swarm on an island in a river which he estimated at 20,000 ; when disturbed the buzz made by their wings was quite audible, and after three waves of the net 236 specimens were bottled from round its edges, the net still appearing quite full ; as, in the some thirty specimens sent, the sexes were in almost even proportions, this was not a case of male assemblage. He also exhibited cleared wings, showing the neuration of *Diacrisia russula*, *Tyria jacobaeae*, *Callimorpha hera* and *C. dominula*, and contended that the genus *Callimorpha* should therefore be removed from the *Arctiadae* and placed in the *Hypsidae* where it is closely allied to *Nyctemara*, *Callarctia* and other genera as the fully developed proboscis, the non-pectinate antennæ, the smoother sealing, the more diurnal habit, and the larvæ being scantily clothed with hair all bore out the correctness of this association.—Dr. Chapman exhibited a portion of a stem of *Ferula communis* from Ile St. Marguerite, near Cannes, showing burrows and pupa cases of *Loxopera francillonana*. A number of vacant holes were also visible, being the exit of an ichneumon, which affects a large majority of the Tortrix, believed to be *Chelonus inanitus*, Nees.—Mr. F. Enock exhibited living specimens of male and female *Ranatra linearis*, Linn., from Epping, together with the peculiar forked eggs, which he had observed laid by the *Ranatra*, as it rested upon the upper surface of the leaf grasping the edges with its claws. The short anterior legs are held well up close together, in a line with the body, the head raised about an inch from the leaf, while the tip of the abdomen and ovipositor is pressed against the leaf—a downward and forward movement being given. The ovipositor is thus forced through the leaf, then partially withdrawn and the egg extruded and forced into the hole as far as the forked filaments, which prevent it from going right through the leaf.—Mr. H. K. Donisthorpe exhibited cases of *Clythra quadripunctata*, specimens of *Lomechusa strumosa*, with its host *Formica sanguinea*, sent by Father Wassmann from Holland, the insects mounted in the position assumed by the guest and host when the former is being fed by the latter. He also showed *Cossyphodes bewickii*, Woll., a beetle from Cape Colony, with the ants with which it is found—*l'heidola megacephala*, var. *punctulata*, Mayr.—Mr. C. J. Barrett exhibited two females of *Spilosoma mendica* reared by Mr. J. E. Robson, of Hartlepool, tinged with purplish-pink, and ordinary specimens of the same for contrast.—A paper was communicated on life-histories of the Hepialid group of Lepidoptera by Mr. Ambrose Quail, and a note on the habits and structure of *Acanthopsyche opacella*, H. Sch., by Dr. T. A. Chapman.

Zoological Society, June 19. — Prof. G. B. Howes, F.R.S., Vice-President, in the chair.—Dr. Walter Kidd read a paper on the degeneration of the hair-slope in certain mammals, in which reference was made to previous investigations into the hair-slope on the extensor surface of the human forearm, and its bearing upon Weissmann's doctrine of the non-inheritance of acquired characters. Details were given of further observations as to the hair-slope on the nasal and frontal regions of certain mammals. The ordinary type and the exceptional type of slope were described, and lists of animals conforming to the two types were given. These results were held to be opposed to the doctrines of Weissmann, and to be attributable to the habits of the animals in question.—Mr. F. E. Beddard, F.R.S., read a paper on the anatomy of *Bassaricyon alleni*, based on an examination of a specimen of this mammal which had recently died in the Society's gardens. The result arrived at was that this genus was clearly referable to the family Procyonidæ, as had been usually supposed, and allied, especially in external form, to *Cercoleptes*, but distinguished by well-marked characters.—Mr. W. F. Lanchester read the first part of a paper on a collection of crustaceans made at Singapore and Malacca by himself and Mr. F. P. Bedford. It contained a list of the Brachyura comprised in the collection, some notes on the nature of the collecting area, and on the habits of certain of the species, together with descriptions of twelve new forms.—A communication was read from Dr. Einar Lönnberg, of Upsala, on the structure and anatomy of the musk-ox (*Ovibos moschatus*). It contained an account of the development of the horns, descriptions of the hoofs and skull, and a comparison between the skull of the musk-ox and that of the Takin (*Budorcas*).—A communication was read from Mr. A. L. Butler, containing the description of a supposed new species of mountain-antelope from the Malay Peninsula, for which the name *Nemorhoedus swettenhami* was proposed.—The Society then adjourned till November next.

Mathematical Society, June 14.—Lord Kelvin, G.C.V.O., President, in the chair.—Three foreign members being present, the chairman requested them to make communications to the Society. This they did. Prof. Klein spoke on the continuation of the edition of Gauss's collected works ; Prof. Darboux, "Sur différents problèmes relatifs aux transformations de l'espace et aux déformations finis de la matière et sur leurs rapports avec la théorie des systèmes triples orthogonaux" ; and Prof. Poincaré, "Sur quelques théorèmes relatifs à l'analysis situs et sur les propriétés des polyèdres dans l'espace à plus de trois dimensions."—Votes of thanks were passed to them by acclamation.—Prof. Stringham also made a few remarks on a proof by non-Euclidian geometry of the form and directrix property of a plane section of a cone.—Prof. Elliott, F.R.S., communicated some notes on the concomitants of binary quantics.— Lord Kelvin read the titles of the following papers which had been received : Some multiform solutions of the partial differential equations of physical mathematics and their applications, Pt. ii., by H. S. Carslaw ; Some quadrature formulæ, by W. F. Sheppard ; Extensions of the Riemann-Roch theorem in plane geometry, by Dr. Macaulay ; On the invariants of a certain differential expression connected with the theory of geodesics, by J. E. Campbell ; On the transitive groups of degree *n* and class *n* − 1, by Prof. Burnside, F.R.S. ; The invariant syzygies of lowest order for any number of quartics, by A. Young ; Canonical reduction of bilinear forms, Pt. ii., by T. J. Bromwich ; The energy function of a continuous medium, by H. M. Macdonald ; Note on the representation of a circle by a linear equation, by J. Griffiths.

Geological Society, June 20.—J. J. H. Teall, F.R.S., President, in the chair.—On the skeleton of a Theriodont reptile from the Baviaans River (Cape Colony), by Prof. H. G. Seeley, F.R.S. The fossil described in this paper was discovered by Mr. W. Pringle at Ealdon, in the bed of the Baviaans River, a tributary of the Great Fish River. It is now preserved in the Albany Museum. The slab containing it is of hard siliceous sandstone, and is 31 inches long by 10 inches wide. It is split so as to expose a portion of the skull, the vertebral column and ribs as far as the pelvis, the scapula, part of the humerus, the femur, and parts of the tibia and fibula. The tail and left hind-limb, and apparently part of the right fore-limb, are lost, owing to the jointed condition of the rock. The bones have decomposed, and are represented by natural moulds from which a beautiful cast was obtained by means of a jelly mould in the Geological Department of the Natural History Museum, before the specimen was returned to Grahamstown. The remains indicate an animal about 2 feet long, exclusive of the tail, and standing probably about 8 inches high ; it was not more than 6 inches wide in the fore part of the body. The animal was of great mobility, capable of easily bending the body, and, by straightening the limbs, of occasionally raising its height to 10 inches or more. It is a new type of Theriodont reptile, contributing important facts to the osteology of the group, and especially in regard to the natural association of the bones. It is possibly to be included in the Cynodontia, from which it differs in characters of the ilium, scapula, and skull.— Fossils in the Oxford University Museum (IV.) : notes on some undescribed trilobites, by H. H. Thomas. Two new species of *Dalmania* from the Wenlock Shales and one of *Olenus* from the Shineton Shales of Shropshire are described in this paper. The specimens on which the first species of *Dalmania* is founded were collected by the late Dr. Grindrod at Malvern Tunnel. The species has a strong resemblance to certain varieties of *D. caudatus*, especially those more nearly approaching *D. longicaudatus* ; its nearest ally seems to be *D. nexilis*. Among its characters are spines round the head, the height of the head-shield, and the distance between the eyes. The type-specimen of the second species came from the Wenlock Shale of Builth. The Shineton specimen was presented to the Oxford Museum by the Right Rev. Bishop Mitchinson.—On radiolaria from the Upper Chalk at Coulsdon (Surrey), by W. Murton Holmes. The radiolaria described in this paper were contained

in the cavities of two small flints which were thrown out of the new cutting between Coulsdon Station and the new Merstham Tunnel on the L. B. & S. C. Railway. They were probably derived from the zone of *Holaster planus*. After treatment with hydrochloric acid, the material yielded silicified casts of foraminifera as well as radiolaria. The surface of the radiolaria is so much altered by corrosion that specific identification is in most cases impossible. Twenty genera have been recognised, and the organisms appear to belong to forty-one species of these genera. A list of the radiolaria is given, accompanied by a short description of each form, and four new species are described. The Discoidea appear to have the predominance, and the species of *Dictyonistra* come next in numerical order.—The Society then adjourned until Wednesday, November 7.

Linnean Society, June 21.—Dr. A. Günther, F.R.S., Vice-President, in the chair.—Prof. M. Hartog exhibited and made remarks on flowers of new *Abutilon*-seedlings, recently raised by him, and pointed out the extreme variability shown in the form of many of the leaves.—Dr. O. Stapf exhibited fruits of various forms of *Trapa* from Europe, China and India, and discussed the differentiation of the genus into species.—Mr. Clement Reid, F.R.S., exhibited a series of plum-stones recently found in a drain of the Roman baths, and in a rubbish pit, at Silchester. The species identified were Cherry (*Prunus avium*), Damson (*P. domestica*), Bullace (*P. insititia*), Sloe (*P. spinosa*), and Portuguese Laurel (*P. Lauro-cerasus*). Besides these, there was a large variety of Plum, and a very small Sloe, the species of which had not as yet been precisely determined.— On behalf of Dr. O. St. Brody, Mr. B. Daydon Jackson exhibited a small series of British orchids dried by a new process, by which the flexibility of the plant and the natural colours were in a great measure retained.—Mr. R. Morton Middleton exhibited several rush baskets, plaited ropes and dredgers made from *Rostkovia grandiflora*, Hook. f. ; and a crab-catcher and limpet-detacher made from *Berberis ilicifolia*, Forster, all used by the Yahgans south of Beagle Channel, Tierra del Fuego.— Mr. F. Enock exhibited and made remarks upon some living specimens of *Ranatra linearis*, Linn., together with their curious eggs.—A paper by Miss Georgina Sweet, Melbourne, was read, "On the Structure of the Spermiducal Glands and associated parts in Australian Earthworms."—Dr. Charles Chilton read a paper on the subterranean Amphipoda of the British Islands, reviewing the known species of the genus *Gammarus*, and giving some account of the "Well-Shrimp" and its distribution in England so far as he had been able to determine it from specimens collected.—Dr. A. B. Rendle, referring to his recently published "Revision of the genus *Najas*" (*Trans.* Linn. Soc. 2nd Ser., Bot. vol. v. Part 12), read a supplementary paper on the same subject, in which he gave additional information gained from a recent examination of specimens in eleven Continental herbaria, particularly those at Paris, Geneva, Vienna and Berlin.—The Society then adjourned until Thursday, November 1.

DUBLIN.

Royal Dublin Society, May 16.—The Earl of Rosse, F.R.S., in the chair.—Mr. R. J. Moss read a paper on the adhesive and other physical properties of copper preparations used in potato spraying.—Dr. W. E. Adeney read a paper, entitled "Studies in the analysis of fresh and salt waters."

June 20.—Sir Howard Grubb, F.R.S., in the chair.—Mr. J. A. Cunningham read a paper, entitled "A contribution to the theory of the order of crystallisation of the minerals in igneous rocks." The author discussed the theory recently published by Dr. J. Joly, F.R.S. (*Sci. Proc. Roy. Dubl. Soc.* vol. ix. part 3, No. 20, p. 298), and then gave additional facts in support of Bunsen's theory, that the order of melting points of the minerals may be inverted by pressure. Mr. Cunningham showed a rough method of arriving at the relative latent heats of the minerals by means of their fusibilities ; and proceeded to show how the latent heats might be determined by simple chemical measurements. As an example, in the case of quartz, the specific heats of quartz and amorphous silica are already known, and by measuring the difference of the heats of solution of the two substances in HF, the disengagement of heat in passing from the one form of SiO_2 to the other at any temperature is known. Thus, assuming 1425° C. as the melting point of quartz, the number 135·3 was arrived at as a safe minimum for the latent heat of quartz.—Prof. J. Joly, F.R.S., communicated a paper on the order of crystallisation of silicates in igneous rocks. Referring to a previous communication, the author has extended the observations of the viscous yield of quartz fibres to a temperature of 735° C. Dealing with finely powdered rock-crystal, it is found that this when wrapped in strong platinum foil and exposed for twenty-four hours in a Bunsen flame, shows unmistakable evidence of softening. The powder is loosely caked, and although the great mass is apparently optically unaltered, the particles which have been pressed against the platinum have adhered and melted into blebs, which cannot be removed by friction. Their examination is effected by a vertical illuminator and high power. Finely powdered quartz placed in the meldometer and exposed for four hours to a temperature between 1085° C. and 1070° shows similar evidence of fusion. Finely powdered olivine, augite, hornblende and quartz exposed in the meldometer for two and a half hours to a temperature between 1105° C. and 1080° reveal, on subsequent examination, that the evidence of fusion was conspicuously more apparent in the case of quartz. The experiments were repeated in an atmosphere of CO_2, as coloration changes thought to be due to oxidation appear on heating in the case of hornblende and olivine. In CO_2 these changes still appear in less degree. Results otherwise the same as before. The author urges that these results support his view that the softening temperature of the silicates will not be found discordant with the observed normal order of solidification in igneous rocks, but will be found to harmonise with Rosenbusch's law, the depression of the softening point in the scale of temperature being influenced by the amount of silica in combination. On the legitimacy of recent attempts to apply the thermodynamic expression connecting dp/dt with change of volume in reversible systems, the author points out that calculations based on the change of volume of a crystallised silicate to its glass must give erroneous results, and possibly widely erroneous results, seeing that the glass has never existed in the magma at any time, but the crystal was built up from the molecules diffused in the magma. The withdrawal of the molecules from solution may probably have given rise to a volume-change which cannot be ignored.— Mr. S. R. Bennett read a paper containing the results of actinometric observations of the solar eclipse. By exposing an actinometer at intervals of a few minutes throughout the afternoon of the solar eclipse of May 28, it was found that the actinic power of the sun's rays declined rapidly from 2h. 13m. to 3h. 40m. and then increased more rapidly till 4h. 27m. After this there was a regular decrease due to the approach of evening. The exposure at 2h. 13m. was 50s., at 3h. 40m. it was 101s., and 86s. at 4h. 27m. Curves were plotted to represent these results as well as the corresponding ones deducible from theoretical considerations. The curves representing the eclipse effect as found from observation and from theory (on the assumption that the amount of light received from the sun is proportional to the area of his disc exposed) did not agree. That found from observation indicated a greater amount of light received, in the ratio of 2·3 to 1·6 at 3h. 22m., the moment of greatest phase. No satisfactory explanation of this incongruity was given.—Mr. Charles Martin read a paper on heat-radiation observations made at Dunsink Observatory during the eclipse of May 28.

EDINBURGH.

Royal Society, June 18.—Dr. Burgess in the chair.—Prof. Copeland and Mr. Thomas Heath gave descriptions of the preliminary work, the installation of instruments, and the general character of the observations made by them at the recent eclipse. Mr. Heath's communication gave a particular account of the various operations undertaken in connection with the 6-inch Cooke triple object-glass. This object-glass was constructed so as to bring to one focus both the visual and photographic rays. Tested by the most severe tests the instrument was perfectly achromatic. Four photographs of the corona were taken during totality on plates 8½ by 6½ inches. Prof. Copeland manipulated the 40-foot telescope which has already done duty on previous occasions. Mr. Frankland Adams had charge of cameras for taking long exposure photographs, and in the working of these had the valuable assistance of officers of H.M.S. *Theseus*. The navigating officer supplied them with time signals; and by means of beautiful maps, for which they were indebted to the courtesy of the director of the Madrid Observatory, they were able to determine with great

ease and accuracy the latitude and longitude of their station near Santa Pola. The Spanish authorities did everything possible to facilitate their operations; and the members of the expedition experienced nothing but kindness at the hands of the people of the town. Photographs began to be taken 16 seconds before totality, and were continued for 60 seconds after totality. The first contact was observed by Prof. Copeland 10·2 seconds sooner than the time expected, there being a projecting mountain ridge on the limb of the moon which first moved across the sun's edge. The shadow bands which occur just before and just after totality were observed by some of the men of the *Theseus* on a vertical wall which had previously been coated with plaster of Paris.—Dr. Buchan and Mr. Omond reported to the Society the nature of the publication of the Ben Nevis observations. For the satisfactory development of meteorological science it was necessary to publish the continuous daily records, and not merely monthly or weekly means. This was now being done with the Ben Nevis observations, both high-level and low-level; and when the work was completed the meteorologist would be in a position to discuss many problems of the greatest interest and importance. It would require three volumes of the Society's *Transactions* to complete the publication of the observations on the scale that had been determined upon. To meet the expense of publication, the Royal Society of London had voted 500*l.*, and an equal grant had been voted by the Royal Society of Edinburgh.

July 2.—Sir Arthur Mitchell, Vice-President, in the chair.—In a paper on the craniology of the people of India, Part ii, Sir William Turner, F.R.S., described the skulls of the aboriginal hill tribes from the Central Provinces, Chita, Nagpur, and Orissa. Most of the specimens were in the Indian Museum, Calcutta, but others were in the Anatomical Museum of the University of Edinburgh. They belonged to the so-called Kolarian and Dravidian groups of people. From a comparison of skulls the conclusion was drawn which supported the view advocated by Mr. H. H. Risley from the examination of living persons, that these groups did not differ from each other in physical characters, and that they formed a Dravidian type. A comparison was also made of the Dravidian type of skull with the Australians and the Negritos of the Andaman Islands. Skulls of the Uriya speaking people of Orissa were also described, and the presence of dolichocephalic and brachycephalic types, with skulls of intermediate or mesaticephalic proportions, was shown to occur amongst them.—Sir John Murray and Mr. F. P. Pullar communicated the second part of their bathymetrical survey of the fresh-water lochs of Scotland, in which they dealt with Lochs Chon, Ard, Monteith, and Leven of the Forth drainage area, and with Lochs Ericht and Garry in the Tay basin. These lochs differ greatly in elevation, the extremes being : Monteith, 55 feet above the sea; Ericht, 1152 feet above the sea-level. Their areas vary from 277 acres (Chon) to 4690 acres (Ericht). In this most elevated of the larger lochs of Scotland there are two depressions in which the depth exceeds 300 feet, the maximum depth recorded being 512 feet. The deposits in the deeper parts of all the six lochs consist of a dark brown mud containing much organic matter, but in some there is a second layer, three to six inches beneath the upper layer, of a light brown colour and greater consistency, containing less organic matter. Numerous examples were given of the effect of the wind in driving the warmer surface waters towards the leeward end or side of a loch and in drawing up the colder and deeper layers towards the windward end or side. The shallow lochs were warmer in spring and summer than the deeper lochs, and contained more pelagic life. In these discussions it was important to bear in mind the fundamental difference between temperature and amount of heat. It was calculated that the larger lochs with their much smaller change in temperature really stored up more heat than the smaller lochs with their greater change in temperature.—A note by Dr. R. Sydney Marsden was read, drawing attention to a paper he had read in 1880 (see *Proc. R.S.E.*, 1881), which contained an account of his method for the artificial preparation of diamonds. M. Henri Moissan, of Paris, described in 1893 a method for the preparation of adamantine carbon which differed from Dr. Marsden's method in details which did not seem to be essential. The note was a claim for priority in a matter in which the later experimenter was now getting all the credit in the eyes of the scientific world.

PARIS.

Academy of Sciences, July 2.—M. Maurice Lévy in the chair.—Communication from M. Darboux concerning the International Association of Academies (p. 249).—Permanent but unequal heating by radiation of a wall of indefinite thickness reduced to the case of an analogous heating by contact, by M. J. Boussinesq.—The combustible gases of the atmosphere ; the air of woods and of mountains, by M. Armand Gautier. Following up the experiments, previously described, made with the air of Paris, air was examined in the middle of a pine wood, and on the summit of a mountain away from all vegetation. The ratios of carbon to hydrogen found in the three cases were 3·5 for Paris, 2·2 in the air of woods, and 0·33 in the mountain air, the quantities of hydrocarbons per 100 litres of air expressed as methane being 22·6 c.c., 11·3 c.c., and 2·2 c.c. respectively. It was also found that air taken at a high altitude, collected in a place denuded as far as possible of animals, plants and humus, is nearly entirely free from hydrocarbons, but still contains about 2/10,000,000ths of its volume of free hydrogen.— Synthesis of αα-dimethyl-γ-cyanotricarballylic ester and of the corresponding acid, by MM. A. Haller and G. Blanc. Cyanosuccinic ethyl ester is heated with sodium ethylate and α-bromo-isobutyric acid, and the resulting ester separated in the usual way.—M. Zambaco was elected a correspondent for the Section of Medicine and Surgery.—Occultation of Saturn of June 13 observed with the Brunner equatorial at the Observatory of Lyons, by M. J. Guillaume.—On a prerogative of the Gregorian Calendar, by M. Joseph Lais.—On the method of Neumann and the problem of Dirichlet, by M. A. Korn.—On the motion of a wire in space, by M. G. Floquet.—On the propagation of condensed waves in hot gases, by M. H. Le Chatelier.—On the decomposition of harmonics by the ear, by M. F. Larroque.— On the thermo-electricity of some alloys, by M. Emile Steinmann. Nickel steel containing 28 per cent. of nickel gave an electromotive force against lead of 385 microvolts between 20° and 260° C.—On the true atomic weights of ten elements deduced from recent works, by M. G. Hinrichs. By applying the method previously described by the author to some recent determinations of atomic weights, the latter are made to appear as whole numbers exactly.—Attempt at a general theory of acidity, by M. de Forcrand. The theory put forward allows of the prediction of the acidity of a compound containing hydrogen replaceable by a metal when the formula of constitution is known ; and also of the heat of fusion when this cannot be determined directly.—Addition of hydrogen to acetylene and ethylene in presence of finely divided platinum, by MM. Paul Sabatier and J. B. Senderens. A mixture of hydrogen and acetylene, the former being in excess, when passed over platinum black reacts vigorously, ethane together with a little ethylene being produced, the secondary products noticeable with nickel being practically absent. With acetylene in excess, ethylene is the chief product, although ethane is still produced in notable quantities. Working at 180° instead of at ordinary temperatures the reaction becomes more vigorous, but the quantity of secondary condensation products increases.—On the methoxy-hydratropic acid obtained by the oxidation of anethol. Identity of phloretic acid and of hydropara-coumaric acid, by M. J. Bougault.—Method for preparing synthetically higher homologues of acetolacetic ester and acetylacetone, by M. L. Bouveault. By the interaction of acetoacetic ether and the fatty acid chlorides, the β-ketonic ethers and β-diketones are easily obtained.—On the mode of formation of the compounds [C₂H₄.(Cu₂Cl₂).KCl and C₄H₆[(Cu₂Cl₂)₂KCl₂], by M. Chavastelon.—On the metallic compounds of diazoamido-benzene, by M. Louis Meunier.—Action of nitric acid upon trichlor-guaiacol, by M. H. Cousin. The action of nitric acid upon the trichlor-derivative is quite different from that of the tetrachlor- and tetrabromo-derivatives as instead of the orthoquinones produced in the latter case, a complicated condensation product is produced.—On the aloins, by M. E. Léger.—Solubility of cupric chloride in organic solvents, by M. Œschner de Coninck.—On the composition of the albumin of the seed of *Gleditschia triacanthos*, by M. Maurice Goret. The reserve hydrocarbon in this case is a mannogalactase ; hydrolysis yielding only a mixture of mannose and galactose.—Hermaphroditism and parthenogenesis in the Echinoderms, by M. C. Figuer.—Study of the digestive apparatus of *Brachytrupes achatinus*, by M. L. Bordas. —Prehnite considered as a constitutive element of metamorphic limestones, by M. A. Lacroix.—On the combinations of some nucleins with metallic compounds, alkaloids and toxins, by M.

H. Stassano.—The power of selective coloration by methylene blue, possessed by living spore-bearing filaments of *Spirobacillus gigas*, by M. A. Certes.—A preventive remedy against the mannite disease of vines, by M. P. Carles.

CAPE TOWN.

South African Philosophical Society, June 6.—L. Péringuey, President, in the chair.—Mr. W. L. Sclater exhibited a series of photographs of birds and their nests taken by Mr. R. H. Ivy, in the neighbourhood of Grahamstown.—Dr. J. D. F. Gilchrist exhibited :—(1) A Gadoid fish, belonging to the genus *Haloporphyrus* and probably a new species, found by the Government steamer in trawling about 40 miles off Cape Town, in over 100 fathoms. (2) Four fishes showing luminous organs, viz. : a *Monocentris* from shallow water, Mossel Bay ; an *Argyropelecus*, a *Paralisparis* and a *Scopelus* from over 100 fathoms off the Cape Peninsula, probably all new species. (3) A number of new Alcyonarians which have been procured by the Government steamer and described by Prof. Hickson, F.R.S. These included the new genus, *Acrophytum claviger*, and three new species—*Heteroxeina capensis*, *Sarcophytum trochiforme*, *Gorgonia capensis*. (4) Specimens of *Veritillum* illustrating the difference in size of the fauna of the east and west coasts of Africa, the eastern forms being larger than those from the west coast. (5) A specimen of *Agriopus torvus*. (6) A new species of Anchovy from East London, this being the second species of the genus *Engraulis* discovered in South African waters.—Dr. F. Purcell exhibited specimens of all the known South African species of Peripatus, including, in addition to the three previously described forms, four others recently described by himself in the annals of the museum, making seven in all. Dr. Purcell in his remarks on the genus maintained that the supposed great antiquity of Peripatus was very doubtful, depending as it did on the supposition that the tracheæ of the tracheate Arthropods could only have originated once, for it is now known that true tracheæ have originated independently in at least three different ways, for instance, in two ways in spiders and in a third way in insects. It would be reasonable to suppose, therefore, that Peripatus may also have acquired its tracheæ independently of those of the insects.

CONTENTS.

THURSDAY, JULY 19, 1900.

THE RELATIONS BETWEEN ETHER AND MATTER.

Æther and Matter. By Joseph Larmor. Pp. xxviii + 365. (Cambridge University Press, 1900.)

THIS work is essentially the same as an essay to which an Adams Prize was awarded by the University of Cambridge. The subject for which the prize was offered was Aberration, and as this phenomenon, together with the Doppler effect on the frequency of light vibrations are the only ones known due to the motion of matter through the ether (the spelling æther is disagreeably cumbrous),'it naturally led to a discussion of the connection between ether and matter, and the effect of their relative motion on the phenomena of electro-magnetism.

There is a good deal of similarity between the development of this work and that of Maxwell's treatise. If one reads Maxwell's papers, it is pretty evident that he began with a somewhat definite hypothesis as to the nature of the strains in the ether to which he attributed electro-magnetic phenomena ; but in his treatise on electricity and magnetism there is hardly a trace of all this except in the reference to molecular vortices by which he justifies his equations for wave-propagation in magnetised media. In a similar way, Mr. Larmor has published papers in which the relations between ether and matter are developed in connection with a suggestion as to the nature of an electron which is only hinted at in the body of the present work, and is relegated to an appendix with some deprecatory remarks as to its being merely an analogy to show that the properties of an electron are not impossible. The hypothetical structure attributed to an electron requires the medium to possess a very remarkable property which we do not find in matter, namely, an elastic reaction against absolute rotation of its elements ; and although Mr. Larmor shows that gyrostats connected with these elements might confer such a property on them, he does not go so far as to develop any very definite structure for the medium, contenting himself with having shown that such properties as he assumes are not necessarily anti-dynamical. No structure for the medium that depends on gyrostats supported by a rigid framework can possibly be more than a rough working hypothesis, being, in fact, very little better than the brass wheel and india-rubber band, or tubes full of liquid with circulating pumps, that have been suggested as models to show that Maxwell's equations do not necessarily postulate impossible or adynamical properties for the ether.

But just as Maxwell's treatise is really independent of the dynamical analogies from which it grew, so Mr. Larmor's work is really independent of his suggested working analogy as to the structure of an electron. The whole work is based upon the hypothesis that electricity is atomic in its nature, there being only two kinds of atoms, positive and negative electrons. These electrons are, he supposes, essentially centres of strain in the ether, and move from place to place in much the same way as a drop of water might move through ice, melting in

front and freezing up behind. Mr. Larmor leaves it for future investigation to determine whether there is any core, like that of a vortex ring, that accompanies this complex strain wave as it moves through the ether. As to the nature of matter, the only suggestion is that it consists of clusters of electrons in orbital motion round one another ; but as the dynamics of such a system has never been worked out, it is impossible either to assert or deny the possibility of a permanent existence of such clusters. If this be the structure of matter, it certainly makes it probable that the transmutation of the elements is a possible development of chemistry, while a structure such as that of knotted vortices would make it improbable that we would ever be able to untie them and thus transmute one atom into another. There is the alternative possibility that we may find means of transmuting elements within any one of their related groups, but that we may find ourselves unable ever to transmute one group into another. Of course, if we ever found out some means of manufacturing electrons and matter we could probably transmute one kind of matter into another, though this latter might be possible according to Mr. Larmor's hypothesis, while the manufacture of either electrons or matter would be impossible.

All theories that explain electric currents by the motion of electrons are really based upon Rowland's classical experiment, that a moving electric charge produces the same magnetic effects as an electric current, and its converse that the electric force due to changing magnetic induction produces the same effect in moving an electric charge that the electric force due to another electric charge would produce, i.e. that electric force due to these two causes is the same. Our whole treatment of electro-magnetism is practically based upon these same assumptions, but it is remarkable that so few attempts have been made to repeat this fundamental experiment of Rowland's, and no successful attempt seems to have been made to directly verify its converse. In a recent number of the *Comptes rendus* there is an account of a most interesting attempt to measure the electric current that one would expect to be produced in a surrounding coil when a convection current such as Rowland studied is being started and stopped. M. Cremieu has carried out an experiment on this with great care, and in a form in which one would certainly expect that the changing magnetic induction due to the magnetic force Rowland observed should produce an induced current in a coil of wire. He observed no such effect, and concludes that there is no magnetic force such as Rowland observed due to a moving electric charge. These moving electric charges are, however, in some respects, so imperfectly known that there may yet be some difference between driving a current by mechanical and electrical forces, and that it is still possible there may be some other explanation than that drawn by M. Cremieu as the result of his interesting and important experiment. Mr. Larmor's investigations in this treatise of the effects of moving matter hardly touch the question raised by M. Cremieu, for his investigation is concerned with steady states, while M. Cremieu's experiment is essentially concerned with variable ones. If he is right and there is really no magnetic force due to a moving electric charge, and if consequently we must look to some other, possibly accidental, cause for Rowland's observation, it will certainly revolutionise the whole modern treatment of electro-magnetism. The question raised by this experiment is, any way, one of the most fundamental ones in the connection between ether and matter, and it is to be hoped that this question will be settled soon in a conclusive way, either by showing that M. Cremieu's conclusion is not justified by his observation that his experiment really confirms a complete theory, or by overthrowing all our existing views, and leaving a free field for the twentieth century to build a new theory of electro-magnetism on a firmer foundation.

In discussing the result of Michelson and Morley's experiments, from which they concluded that the ether is carried along by the earth in its motion, Mr. Larmor shows that such a hypothesis is quite inconsistent with the fact of aberration and with the untenability of Sir George Stokes's suggestion that ether is like a very soft jelly. How such a soft material could be the means by which tramcars are driven by shearing stresses seems an additional difficulty in the way of this suggestion. Mr. Larmor concludes that the stone support on which the mirrors were borne changed in its dimensions, as it was rotated, by an amount proportional to the square of the ratio of its velocity to the velocity of light, and he justifies this by showing that if matter consists of clusters of electrons, just such a change of dimensions would take place as the experiment shows to take place. There is some difficulty in the hypothesis that the inertia of matter, or any large part of it, is like that of electrons and due to the motion of the neighbouring ether, because this involves the supposition that the inertia would change with the distance between the component electrons. That there may be some very minute effect of this kind is quite possible, though as yet undiscovered, but that any large effect of the kind exists seems extremely improbable. Possibly a careful study of the accuracy of Kepler's laws as applied to the solar system might show some discrepancy depending on a difference between the average distance of the electrons in such different materials as probably constitute Neptune and Mercury.

A previous question to all our explanations of phenomena by analytical dynamics is raised by Mr. Larmor in Appendix B, " On the Scope of Mechanical Explanation : and on the Idea of Force." He has utilised the principle of least action throughout his work, and this appendix is a justification of his doing so, and besides raises questions as to the applicability of dynamical explanation to the growth and decay of vital organisms. Hertz objected to the adequacy of the principle of least action as a complete solution of all possible dynamical systems, because it is not generally applicable when rolling takes place, and we cannot be sure that rolling may not be one of the fundamental facts of the dynamics of the ether. Mr. Larmor dismisses this objection on the doubtful ground that " rolling is foreign to molecular dynamics." Hertz had also objected to the principle of least action for the semi-metaphysical reason that it makes the present state of the system depend on the future as well as on the past. As Mr. Larmor himself uses in his work the vector potential which makes the state at each place

depend on what is simultaneously occurring at all parts of the universe, he naturally finds no objection to the principle of least action, because it makes the present depend on all future time. Neither of these methods has to be most carefully guarded lest it lead us into mistakes. The way in which the vector potential apparently locates the energy in the current instead of in the magnetic field outside it is a most serious objection to its use, although Mr. Larmor seems to have steered clear of the difficulties raised by this curious complication. In a similar way the principle of least action is open to the objection of Hertz of making the present apparently depend on the future to an extent that does not apply to his own principle of the straightest path. It is a question for consideration in connection with Mr. Larmor's discussion on the applicability of dynamics to vital phenomena whether the possibility of determining our actions by considerations as to the future is not connected with the possibility of analytically expressing the dynamics of the present by a formula which involves the future.

It will, from this meagre review, be evident that Mr. Larmor's treatise raises most fundamental and interesting questions, and is one that all who desire to strengthen the foundations of our knowledge of nature should carefully study.　　　　　　　　　　GEO. FRAS. FITZGERALD.

LAND RECLAMATION.

The Reclamation of Land from Tidal Waters. By Alexander Beazeley, M.Instit.C.E. Pp. xii + 314. (London : Crosby Lockwood and Son, 1900)

THE area of this country is gradually diminishing by the continual waste that is going on all round the coast. On the Yorkshire coast it is estimated that two miles have disappeared since the Roman occupation ; and more modern records show that towns and villages have disappeared with their houses and churches, and in some cases the whole parish has been washed away. Along the Norfolk coast the only record of several villages is, " washed away by the sea ᐟ ; and on the Kentish coast, churches and houses have fallen down the cliffs, on which are to be seen the bones formerly deposited in a vanishing churchyard. On the south coast, although the chalk cliffs at the east end of the English Channel are subject to continual falls and slips, more care has been taken to protect them ; but along the clay cliffs of Dorsetshire the waste is continuous ; here twenty acres slipped down seaward in one night from the cliffs at Axminster. On the west coast, the nets of the fishermen are said to become occasionally entangled with the ruins of houses and buildings buried in the sea some distance from the coast off Blackpool.

As some compensation for all this loss due to the ever-continuous operations of nature, the energy of man has succeeded in reclaiming and recovering a large area of rich cultivatable land in estuaries where rivers have discharged great quantities of detritus picked up along their course. At no time in the history of this country were reclamations carried on to a greater extent than in the time of the Romans, and this is the more remarkable as, compared with the population at that time, land must have been plentiful. It was during this era that the great tract of low land lying on the east of England was reclaimed from the sea by the construction of 50 miles of sea-banks, and the 60,000 acres in the district known as Romney Marsh was protected from the sea by a bank 4 miles long and 20 feet high. From the time of the Romans to the Stuart period very little seems to have been attempted in this way, but at that time there are records of innumerable grants made to "undertakers" and "adventurers" who undertook to reclaim the low lands in the Isle of Axholme, Haxey Chase and the Fens of Lincolnshire and Cambridgeshire, and other parts of the country, in return for a certain proportion of the land reclaimed. Another revival took place during the present century at the time when agriculture was prosperous, and land-owners were tempted, by the high rents then paid, to reclaim from the sea numerous intakes of salt marshes by the construction of sea banks in the estuaries of the Humber, the Wash, the Thames, the Severn and other rivers. Since rents have fallen and land-owners have become impoverished by the low rents, and the heavy charges thrown on estates by the payment of the death duties, little or no inclosing has taken place. Land, however, shows signs of recovering something of its former value. The appearance, therefore, of a book dealing with the reclamation of land from tidal waters may be considered as opportune.

The only standard English book on this subject is that of the late Mr. John Wiggins on the "Practice of Embanking Lands from the Sea," which is now out of print. Instead of publishing a new edition with the extensive alterations of the text that would be required to bring this work up to date, the author of the book now under review was invited by the publishers to undertake the preparation of a new treatise, in which all that was applicable to modern practice in Mr. Wiggins' book has been incorporated.

The author has carried out his task efficiently and well, and his book contains a large amount of information that will be of great service to engineers, and also to landed proprietors and others interested in works of reclamation.

The book makes no pretensions to originality; on the contrary, it may be regarded as an epitome of the information and opinions contained in a vast number of papers contained in the *Minutes of Proceedings* of the Institution of Civil Engineers and the papers of allied societies, and in the works of authors on drainage and Fen history.

A careful perusal of a book of this character, and the principles laid down that should be observed in the reclamation of land, might have saved the expenditure of many thousands of pounds on schemes that never came to maturity or have proved financially disastrous. Of these, as examples, may be quoted the great scheme that was at one time entertained, and still has advocates, for the formation of a new county in the Wash, by the enclosure of the sands; an offshoot of which was the abortive scheme of Sir John Rennie for reclaiming 30,000 acres, the greater part of which was bare sands, which experience has since proved would have been utterly unfit for cultivation; and the Norfolk Estuary Scheme, which received parliamentary sanction in 1846 to reclaim

30,000 acres submerged at high water, and of which up to the present time, after an expenditure of nearly 400,000*l.*, there has only been reclaimed 2000 acres of marsh land adjacent to the coast, a great part of which formed the bed of the diverted river. In this case, great benefit has resulted to the drainage of the country by a new direct cut made for the outfall of the river Ouse; but as a land reclamation scheme, it has been a most disastrous failure, owing to the misconception of the promoters as to the action of the sea in forming deposit on the coast, and of the difficulties attending the construction of sea banks.

Mr. Beazeley's book is divided into nine chapters, dealing respectively with: (1) General observations; (2) the site for a bank; (3) the construction of sea banks; (4) the drainage of the land reclaimed; (5) maintenance and repair of sea banks; (6) warping land; (7) cultivation after enclosure; (8) examples of reclamation, value and rents; (9) legal requirements; the text being accompanied by numerous illustrations.

THE MAMMALIAN BRAIN.

Handbuch der Anatomie und vergleichenden Anatomie des Centralnervensystems der Säugetiere. Von Dr. Edw. Flatau und Dr. L. Jacobsohn. I. Makroskopischer Teil, mit 126 Abbildungen im Text, und 22 Abbildungen auf 7 Tafeln. Pp. xvi + 578. (Berlin: Verlag von S. Karger, 1899.)

THE handsome volume before us is a welcome addition to works on the comparative anatomy of the mammalian brain. That the literature of this subject is already vast, may be gathered from the fact that nearly 300 papers are quoted at the end of the volume—this list forming indeed a most useful bibliography. So numerous and scattered are these various works, that only those students who have access to very complete libraries can hope to be able to consult the majority of them, and we have long felt the want of a trustworthy account of the structure of the brains of the various orders of mammalia in a more handy form. This want is to a great extent satisfied by the work of Drs. Flatau and Jacobsohn, which is rather of the nature of an original contribution than of a text-book. For it is no mere compilation; but, on the contrary, almost entirely consists of the description of brains studied by the authors themselves in Prof. Waldeyer's Anatomical Institute in Berlin.

With admirable care the authors describe the structure of the central nervous system of representative examples of all the living orders of mammalia. To give the reader some idea of the thoroughness of their method, one may mention that in the case of the brain of the Chimpanzee, for example, we find paragraphs on the brain weight, the relation of the brain to the skull, the general shape and measurements of the brain, followed by detailed accounts of the convolutions of the cerebral hemispheres, the structure of the corpus callosum, fornix, &c., of the Diencephalon, Mesencephalon, Metencephalon, Myelencephalon, and Medulla spinalis. Naturally the types of all the orders are not treated in quite as much detail as the Chimpanzee. At the end of the chapter on monkeys and apes are elaborate tabular statements of the authors' observations compared with those of previous writers on the subject. Throughout, the text is illustrated by excellent figures, almost all of which are original. The general reader will be especially attracted by the ingenious representations of the brain drawn inside the skull as if seen by transparency, and by the really beautiful series of plates at the end of the volume.

The work is essentially a technical and a practical one. Nevertheless, a final chapter is devoted to a general summary and conclusions. Here Drs. Flatau and Jacobsohn aim, not at bringing forth sensational results, but soberly review such general conclusions as may safely be drawn at present. These, it must be confessed, are somewhat disappointing, not, be it understood, through any fault of the authors, but owing to the inherent difficulties and complications of the subject, and the comparatively few data yet at our disposal.

As to the attempt to homologise the fissures of the cerebral hemispheres with one another in the various orders of mammalia, Drs. Flatau and Jacobsohn freely adopt Gegenbaur's conclusion, that this can only be done to a very limited extent. In most of the orders we generally find some small and lowly organised forms with almost smooth brains ; and it must always be borne in mind that the fissures and convolutions may to a great extent have been independently developed in each group.

Of the usefulness of this volume there can be no doubt, and the appearance of the continuation of the work will be awaited with interest by all workers in the subject of brain anatomy.

OUR BOOK SHELF.

The Origin of the British Flora. By Clement Reid, F.R.S., F.L.S., F.G.S. Pp. vii + 191. (London : Dulau and Co., 1899.)

THIS is a useful contribution to the literature of geographical botany ; but it is unfortunate that the author has given it the ambitious title of " Origin of the British Flora." Any one entering upon the perusal of the book with the expectation engendered by its title will soon meet with disappointment, but must not be blinded thereby to its real merit, which is great, and consists in the historical records, to which two-thirds of it are devoted. The book is essentially a geologist's account of the palæontological evidence of the distribution of plants in Britain during recent geological periods. Every one will agree with the author in thinking that the historical method is the proper one for determining questions of origin, but that the " problem of the origin of our flora is one which can be solved by this method " is surely a sanguine forecast on his part, even allowing for the fact that the flora of our Tertiary deposits has not been worked out yet with much completeness ; his work is emphatic testimony to the fragmentary character of historical evidence in relation to the British flora that has been obtained up to the present time. In his " Table showing the Range in Time of the British Flora," which includes the names of species, remains of which have been found in deposits of pre-Glacial age onwards, there are not three hundred names, and of these not all have as yet been found in deposits within the present area of Britain ; and, moreover, the finds do not touch elements of the flora which have always been a crux in explanations of its origin. The first fifty pages of the book deal, in the slight manner of the magazine article rather than in the detail of a scientific treatise, with some of the problems of the origin of the present British flora. The author is on the side of those who attribute a more important influence to air-transport than to land-connection as a factor in the making of our existing flora. The Watson-Forbes hypothesis is, in a few sentences, put

on one side, and a short chapter is devoted to an account of the transport-mechanism observable in the species of the flora. In Chapter iv. we have an account of the author's idea of the geographical and climatic changes affecting Britain in the late Tertiary times ; the former, the author thinks, " were of no very great importance as bearing on the past history of our flora," although they " must have tended greatly to modify local conditions, and must have sometimes aided, sometimes have hindered, the dispersal of the seeds " ; the latter have left their mark on the flora ; but at the same time " Britain shows signs of a geographical distribution of plants largely independent of that due to climate ; or perhaps we should say not governed by existing climatic conditions." It is not, however, these brief earlier chapters which give value to the book, but the later ones, containing accounts of the deposits in which recent plants have been found and of the positions of these plants.

A Manual of Marine Meteorology for Apprentices and Officers of the World's Merchant Navies. By William Allingham. Pp. viii + 182, and plates. (London : Charles Griffin and Co., Ltd., 1900.)

WE gladly give a word of welcome to this little book, written as it is by a sailor with the view of winning an increase of interest in the subject of meteorology from members of his own profession. The author knows well those for whom he is writing, so that while he has kept his book free from pedantry, he has managed to fill it with practical information and to endow it with the spirit of earnest purpose. The encouragement of a more complete survey of the complicated phenomena manifested, not only in our atmosphere, but in the ocean itself, is highly commendable, and we should imagine the author well qualified by knowledge and experience to interest the class to whom he mainly addresses himself. For he has sailed every ocean in all sorts of weather, and having himself to some extent profited by the systematised experience of others, he seeks now to widen and complete the circle of observation, so that those who come after may have still more trustworthy sources of guidance and readier means for escaping the perilous chances of navigation.

Of course, in many respects marine meteorology goes hand in hand with meteorological inquiries conducted on shore. We may pass over all such details, since the real interest of the book is more closely connected with the practical questions which arise at sea. Among these we may enumerate wave-motion, salinity and temperature of the sea, the direction and velocity of ocean currents, and the construction and use of pilot charts. Such subjects ought to have a profound interest for an intelligent officer, and the method of treatment is likely to call forth the earnest attention of any one who wishes to become really efficient. Some of these subjects may be thought to belong rather to hydrography than to meteorology ; while, again, questions connected with the behaviour of the wind in cyclones, and of the management of the ship in the neighbourhood of cyclonic disturbances, may be said to belong to the domain of seamanship or practical navigation. But there is no fixed line of demarcation between any of these subjects, and trained intelligence is of the greatest service in advancing our knowledge of subjects in which experiment and generalisation play a great part. One can easily conceive that enormous advantages would accrue to science by enlisting the services of a large army of observers, and therefore we welcome any well-considered effort which has for its end so worthy an object. The author knows perfectly well that it is impossible to do justice, within a moderate compass, to the many topics on which he touches ; but his object is served, and well served, if he can arouse an active interest in the many, and induce a few to prosecute inquiries on a more comprehensive basis.

LETTERS TO THE EDITOR.

[*The Editor does not hold himself responsible for opinions expressed by his correspondents. Neither can he undertake to return, or to correspond with the writers of, rejected manuscripts intended for this or any other part of* NATURE. *No notice is taken of anonymous communications.*]

A Surface-tension Experiment.

SANS rien vouloir enlever à l'intérêt de l'expérience d'hydrodynamique signalée par Mr. Baker (P. 196), je crois pouvoir dire qu'elle n'est pas nouvelle, au moins, en tant que phénomène.

Ces sortes de formations de "vasques" liquides etaient très usitées, dans les jardins au XVIIᵉᵐᵉ siècle et nous en rencontrons encore aujourd'hui des exemples dans les parcs où le régime des eaux n'a pas changé depuis cette époque. Pour n'en citer qu'un, que connaissent sans doute beaucoup de vos lecteurs, je rappelerai qui à Burgos le "paseo del Espalon viejo" possède une fontaine où l'on peut voir une belle réalisation de cette expérience. Seulement, là le jet d'eau est dirigé de bas en haut et vient se briser sur un disque placé horizontalement au dessus de lui ; puis retombant, il forme autour du tuyau la surface *fermée* si élégante décrite par M. Baker.

Je crois me souvenir que dans un ouvrage publié en 1663 à Nuremberg par George André Boeckler sous le titre "Architectura curiosa nova," il y a de nombreuses planches représentant des jets d'eau d'effets très variés. Peut-être la forme signalée par Mr. Baker s'y trouve-t-elle ?

Il serait intéressant de le vérifier, comme aussi de chercher la figure mathématique de cette surface fermée.

HENRY BOURGET,
de l'Université de Toulouse.

Duration of Totality of Solar Eclipses at Greenwich.

IN NATURE (vol. lxii. p. 64 and p. 86) will be found an estimate of the maximum duration of totality for a solar eclipse under the most favourable conditions, the result being 7m. 40s for a place in north latitude 4° 52'. For Greenwich I estimate the maximum duration at 5m. 47s. There is good evidence for believing that the "Nautical Almanac" diameter of the moon, used in computing eclipses, is too large. It is almost exactly 2160 miles, and should be reduced probably to 2158 miles. This reduction would alter the above estimates to 7m. 34s. and 5m. 42s. respectively. That all the conditions necessary to produce the maximum totality of 5m. 42s. will ever be simultaneously satisfied for Greenwich is extremely improbable.

Leeds, July 14.　　　　　CHAS. T. WHITMELL.

THE NEW YORK MEETING OF THE AMERICAN ASSOCIATION.

AT the forty-ninth meeting of the American Association for the Advancement of Science, which was held on June 23-30, at Columbia University, New York City, two experiments were tried. The one was a change of date and the other a somewhat radical change in the character of the meeting. Heretofore, it may be remembered, the American Association has met at about the third week in August, approximately at the same time as the meeting of the British Association. The long summer vacation of the American colleges and universities usually lasts from about the end of June until nearly the beginning of October. It therefore resulted that men engaged in educational work were obliged to interrupt their summers at the seaside or the mountains, to attend the Association meetings. This has been found to be very inconvenient to many on account of the long distances in the States and the widely separated places of meeting. The present year was thought to be a particularly favourable one in which to try a change of date, since many members expected to start for Europe after the close of their college terms, and New York, as the principal port of debarkation, was chosen as the place of meeting for much the same reason.

The other experiment was in the doing away, to a large extent, with the social features and entertainments which had characterised previous meetings. It was distinctly

understood that no entertainment fund would be raised in New York, and that the Association would pay its own expenses. It was, therefore, a more distinctively working scientific meeting than has been held before. The attendance was not large, and only 450 members registered. Fifteen affiliated societies held their meetings at the same time, including several which have heretofore not affiliated themselves with the older society. These were the American Mathematical Society, the American Physical Society, and the American Psychological Association. The other societies in attendance were the American Forestry Association, the Geological Society of America, the American Chemical Society, the Society for the Promotion of Agricultural Science, the Association of Economic Entomologists, the Botanical Society of America, the Society for the Promotion of Engineering, Education, the American Folk-Lore Society, and the American Microscopical Society.

The session was opened by the retiring President, Mr. G. K. Gilbert, who was elected at the December meeting of the Council to fill the vacancy caused by the death of Dr. Edward Orton last autumn. Mr. Gilbert introduced the incoming President, Prof. R. S. Woodward, of Columbia University, who thereafter presided over all the general sessions of the Association. A cordial and eloquent address of welcome was made by Mr. Seth Low, President of Columbia University ; and Mr. James Wilson, the Secretary of Agriculture in President McKinley's Cabinet, upon being invited to address the Association, made a strong plea for applied science. On Tuesday afternoon the addresses of five of the Vice-Presidents were given, the other four being postponed until next year.

Vice-President Asaph Hall, junr., addressed the Section of Mathematics and Astronomy on the teaching of astronomy in the United States. Prof. Hall urged that elementary astronomy should be taught in the high schools and preparatory schools as well as in the colleges. Elementary astronomy he defines as meaning such part of the science as can be learned by an intelligent student without mathematical training. He advocated the study of the history of astronomy as a culture study in the colleges, showing that the earliest religious festivals depended upon astronomical observations. An interesting feature of this historical side would be the philosophical study of the different theories of the universe. He advocated the more general teaching of spherical astronomy and the elements of celestial mechanics, and showed that during the past twenty years great advances in astronomical teaching have been made in the States. In his opinion the best equipped observatory for teaching purposes is at Princeton, and the theses in practical astronomy produced in America compare favourably with those presented in Germany and France.

Vice-President Merritt addressed the Section on Physics on the subject of "Kathode Rays and some Related Phenomena," referring to the various views which have been advanced concerning the nature of the kathode rays, and the general adoption of the Crookes' theory of electrified particles. He gave an account of the progress made during the last ten years, and discussed recent experiments concerning the size of the ray particles and the speed at which they travel. Minor difficulties in the present theory were pointed out, and the probable direction of further progress was indicated. Lantern views were shown illustrating various vacuum tube phenomena related to kathode rays.

The address of Vice-President Howe before the Section of Chemistry was on the subject, "The Eighth Group of the Periodic System and some of its Problems." It was pointed out that in the early work of Newlands and of Mendeléeff, which subsequently developed into the periodic law, a serious difficulty was met with in dealing with iron, cobalt, nickel, and the metals of the platinum group.

In his first summing up of the principles of the periodic law in 1869, Mendeléeff concludes that "elements which are similar as regards their chemical properties have atomic weights which are either of nearly the same value (*e.g.* platinum, iridium, osmium) or which increase regularly (*e.g.* potassium, rubidium, cæsium)." So in most schemes for representing the periodic system, each triplet of these elements is considered as a single element, and because even then they do not seem to fall into regular periodic arrangement, they are cast out, Ishmael-like, into an anomalous eighth group. This is doubtless the réason they have been relatively so much neglected by chemists, and possibly it is not incorrect to say that the chemistry of these metals is less known than that of any other group of well characterised elements. Yet there are certainly no nine nearly related elements which present so many interesting chemical problems, the solution of which will so much further our knowledge of chemistry in general. Prof. Howe dealt in detail with this eighth group and some of its many problems.

The ordinary division of these nine metals is into three groups, viz., the common metals, iron, cobalt and nickel, with an atomic weight of from 56 to 59 and a specific gravity of 7·8 to 8·9 ; the lighter platinum metals, ruthenium, rhodium and palladium, with an atomic weight 101·5 to 106·5 and a specific gravity of about 12 ; and the heavy platinum metals, osmium, iridium and platinum, of atomic weight 191 to 195 and specific gravity 21·5 to 22·5. These nine metals are held to fulfil every definition of an element, and are just as much to be looked upon as simple elementary substances as any of those substances which are called elements. Though refined determinations may change, to a slight extent, the atomic weights of some of these elements, especially those of ruthenium and osmium, the weights of these elements relative to each other, and hence their position in the periodic system, will probably remain unchanged. This carries with it the conclusion that in the periodic table an element may have an atomic weight slightly lower than that of the element which precedes it.

Reference was made to the natural grouping of the elements of the eighth group into three triplets, iron, ruthenium, osmium ; cobalt, rhodium, iridium ; and nickel, palladium, platinum. That this is a natural grouping is attested by a comparison of the compounds of these metals. However, in considering now some of these compounds, the evidence of this grouping is only incidentally presented ; Prof. Howe directed attention to some of the more unusual of these compounds, especially with reference to problems which this group presents, and to problems of other groups, suggested by the chemistry of this group.

Vice-President Kemp, before the Section of Geology and Geography, spoke upon the "Pre-Cambrian Sediments in the Adirondacks." He showed that the Adirondack area of ancient crystallines in Northern New York covers about 12,000 square miles. It has long been known that in the gneisses and eruptive rocks which constitute it, crystalline limestones of undoubted sedimentary origin occur in many places. The address presented the results of the work of the last ten years upon these sediments. It has been recently learned that ancient sandstones are also present, and many gneisses, which are doubtless altered shales. The crystalline limestones are in greatest individual areas in the north-west, where the belts have been shown to be from twenty to thirty-five miles long and from two to six miles broad ; but they are most numerous on the east, where the speaker has now located over fifty different localities of relatively thin beds. In their structural relations these narrow beds on the east are interstratified with the gneisses, and are more especially associated with fragmental sediments. From these relations the argument is drawn that the sediments were extensive, that they involved more lime-

stone on the west and more sandstone and shale on the east, and that many gneisses represent former shales. It was further shown that these strata are profoundly metamorphosed, and in such a way that the changes must have been produced while the rocks were under a heavy over-lying load, and were deeply buried. Evidence was brought forward to prove that this burden consisted of pre-Cambrian rocks. The speaker said that, inasmuch as there are abundant eruptive rocks present of a coarsely crystalline type, which were likewise produced under deep-seated conditions, it is assumed that they represent the deeper rocks of an old and very extensive volcanic area, whose tufas and lavas built up the burden of pre-Cambrian rocks, which have now disappeared, and which made possible the metamorphosis of the ancient sediments.

Vice-President Trelease, before the Section of Botany, delivered an address under the title, "Some Twentieth Century Problems," passing in review the great utilitarian development of botanical science during the present century, and indicating its probable greater advancement along utilitarian lines during the next hundred years. He made a general statement of the great problems to be met and solved, and considered in detail the necessity and means of co-operation in the treatment of species and their nomenclature, and in details of publication which are becoming daily of more evident importance for the greatest possible advancement of science. In conclusion he made a strong plea for the establishment of a Government Reservation in the Redwood (*Sequoia sempervirens*) Forests in California, not only as a means of preserving a forest growth which can never be reproduced, but as furnishing the means of solving many problems closely connecting biology and meteorology, which may ultimately be of the greatest economic utility.

The address of the retiring President, Mr. G. K. Gilbert, was delivered on Tuesday evening at the American Museum of Natural History. His subject was "Rhythms and Geologic Time." This address appears in another part of the present number of NATURE.

The programmes of the sectional meetings were very full, and the discussions in the sections of mathematics and astronomy, physics, chemistry and botany were especially animated and prolonged.

Several important matters were decided upon by the Council. Perhaps that of the greatest general interest to members of the Association was the decision to try the experiment during the year beginning January 1, 1901, of publishing all official notices and proceedings of the Association in the journal *Science*, and of sending that journal to all members of the Association at the expense of the Association itself, and without charge to members beyond their annual dues. This will not make *Science* precisely the official organ of the Association, since the management and the editor will remain as before, and the Association will have no strict supervision of the conduct of the journal. The annual volume of proceedings will be reduced during that year, and possibly for future years, should the experiment prove a success, to a business record of the affairs of the Association, including lists of members and fellows, the text of the constitution, and possibly a list of the papers presented at the meetings.

Amendments to the constitution were adopted, establishing a new section of Physiology and Experimental Medicine (Section K), and lengthening the term of office of the Treasurer of the Association from one year to five. A discussion of the new International Association for the Advancement of Science, Art and Education was introduced, and conservative action was taken which simply expressed approval of the idea of international co-operation in the field of science, and promised to designate a delegate to a national conference having that end in view.

Grants were made to the Committee on Anthropometric Measurements ; to the Committee on the Quantitative Study of Biological Variation ; to the Committee on the Study of Blind Vertebrates ; and to the Committee on Study of the Relation of Plants to Climate. The last two committees were established at this meeting. The one on Blind Vertebrates consists of Mr. Theodore N. Gill (chairman), Messrs. A. S. Packard, C. O. Whitman, S. H. Gage, H. C. Bumpus and C. H. Eigenmann. The one on Relation of Plants to Climate consists of Messrs. Wm. Trelease, D. T. MacDougall and J. M. Coulter.

Resolutions were adopted urging upon the Government of the United States (1) the establishment of a bureau of standards in connection with the U.S. Office of Standard Weights and Measures ; (2) the establishment of a Government Reservation in the Primeval Redwood Forest, situated in the Santa Cruz Mountains in California ; and (3) the establishment of a Government Reservation in some portion of the hard wood forests of the Southern Appalachian region.

At the meeting of the General Committee held on the evening of the June 28, the city of Denver, Colorado, was chosen as the place for the next meeting, and the time selected was the week ending August 31. The choice of Pittsburg, Pa., as a meeting place in 1902 was recommended by formal resolution.

On the same evening the following officers for the ensuing year were elected :—For President, Prof. Charles Sedgwick Minot, of the Harvard Medical School ; for Vice-Presidents, as follows :—Section A, Mathematics and Astronomy, Prof. James MacMahon, of Cornell University ; Section B, Physics, Prof. D. T. Brace, of the University of Nebraska ; Section C, Chemistry, Prof. John H. Long, of the North-western University ; Section D, Mechanical Science and Engineering, Prof. H. S. Jacoby, of Cornell University ; Section E, Geology and Geography, Prof. C. R. Van Hise, of the University of Wisconsin ; Section F, Zoology, Prof. D. S. Jordan, of Stanford University ; Section G, Botany, Mr. B. T. Galloway, of the U.S. Department of Agriculture ; Section H, Anthropology, Mr. J. Walter Fewkes, of the Bureau of American Ethnology ; Section I, Social and Economic Science, Mr. John Hyde, Statistician, U.S. Department of Agriculture. General Secretary, Prof. Wm. Hallock, Columbia University ; Secretary to the Council, Dr. D. T. MacDougall, New York Botanical Gardens.

THE WELLCOME RESEARCH LABORATORIES.

IT is a remarkable sign of the times when the head of a firm principally distinguished for the introduction into this country of American methods of dealing with drugs, *i.e.* by putting them up in new and convenient shapes and doses, goes out of his way to fit up extensive research laboratories. This is what Mr. Wellcome has done. In 1896 laboratories were established in the business premises of the firm in Snow Hill. Now, after four years, during which the work continued to grow, it has been found necessary to give a complete house to the department. A well-built modern house has been secured at No. 6 King Street, Snow Hill, and has been converted into a series of three commodious and well-fitted laboratories, a library and office, and a store-room and workshop-laboratory. Each laboratory is self-contained, and each is connected with the other and with the directors' office by means of telephones. The basement contains a good-sized electric motor, and a dark room for polarimetric and photographic work. Use has been made of the electric mains to heat radiators for the distillation of ether, benzene and other inflammable liquids. The whole is under the direction of Dr. T. B. Power, F.I.C., who has a staff

of four assistants, all men who have been carefully selected for their attainments and skill in actual research.

Mr. Wellcome is to be congratulated on his enterprise. His firm, considering the nature of their business, might well have acted on the supposition that research was not strictly within their province. They might have argued, " Research is the business of the drug manufacturer and the manufacturing chemist ; it does not concern the compounder of medicines." Their success in former years is a solid argument in favour of such a view, which can be very easily strengthened by a consideration of the success of many firms who have pursued an exactly similar line of business.

Mr. Wellcome intends to carry on his laboratories in no narrow spirit ; this means, I presume, that he has other views than the conversion of his business into a chemical manufacturing concern. Though much work is done towards the perfection of the firm's preparations, time has been found for several researches which have been published, and other work of this kind is in hand. At present the bulk of the work is carried out on the natural drugs, very little having been undertaken in the direction of investigations leading to the discovery or further knowledge of the properties of artificial medicinal substances. There is undoubtedly a vast field in the direction so far pursued, but every one must hope that the other will not be neglected, and that at length this country may make a contribution to the number of substances of medicinal value derived directly and not through the medium of plant or other life from the carbon compounds of the aromatic series.

The laboratories were informally opened on June 18, when at Mr. Wellcome's invitation a number of gentlemen interested in science, representatives of the Press, were received by Dr. Power and conducted over the building. All interested in the advance of chemistry, whether pure or applied, will wish Mr. Wellcome success, and also that he may find imitators among the numbers of firms who are meditating an advance in the direction of a more scientific method of conducting their manufactures.   R. J. FRISWELL.

NOTES.

IN the House of Commons on Tuesday, Mr. Goschen announced that a committee of experts would be appointed to inquire into the efficiency of water-tube boilers in actual operation in different types of ships of H.M. Navy.

THE Additional Estimate for the Navy for the year 1900–1901 includes 9500*l.* for wireless telegraphy apparatus ; 3600*l.* for telescopic sights for quick-firing guns ; and 16,500*l.* for gyroscopes for Whitehead torpedoes.

THE scientific congresses to be opened in connection with the Paris Exposition during the present month are :—July 19–25, applied mechanics ; July 23–28, applied chemistry ; July 19–21, naval architecture and construction ; July 28–August 3, navigation ; July 28–August 4, chronometry ; July 23–28, photography ; July 18–21, homœopathy ; July 23–28, professional medicine ; July 27–29, medical press ; July 27–August 1, electrology and medical radiology.

WE have been notified that the title of the subject for discussion at the joint meetings of the Institution of Electrical Engineers and the American Institute of Electrical Engineers to be held in the American Pavilion in the Paris Exhibition on the morning of Thursday, August 16, is " The relative advantages of alternate and continuous current for a general supply of electricity, especially with regard to interference with other interests." We understand it is specially desired to discuss how far interference with other undertakings, rather than ordinary commercial and industrial conditions, will come to be a determining factor in the selection between continuous and alternating currents. It is expected that many members of the American Institute will spend a few days in London on their way to the joint meeting in Paris. Arrangements are being made to entertain the visitors, and it is hoped that a large number of the British members will assist in making the visit a memorable one.

A CIRCULAR-LETTER has, this week, been addressed to the students of the Institution of Electrical Engineers informing them that the Council of the Institution propose to grant 5*l.* to each of twenty selected students to assist them to visit the electrical exhibits in the Paris Exhibition. Intending candidates must send in their applications by Saturday, July 28. In the selection, the Council will give preference, other things being equal, to those who, being still students of the Institution, have either, or both, read papers before the students' section, or been members of the committee of that section.

THE Paris Société d'Encouragement has awarded the following medal and prizes :—Gold medal to M. Potier for his work in physics ; 2000 francs to M. Codron for his works on machine tools ; 2000 francs to MM. Charabot Dupont and Pilet for their work on essential oils ; 500 francs to M. Halphen for his work on the analysis of fatty bodies, and to M. Blanc for his work on the constitution of camphor ; 500 francs to M. Granger for his study of the application of tungsten blue to ceramics ; and 1000 francs to MM. Coudon and Boussard for their study of the potato.

THE Paris correspondent of the *Chemist and Druggist* announces that the late M. Milne-Edwards, director of the Paris Museum of Natural History and professor of zoology at the Paris School of Pharmacy, has bequeathed his scientific library, which is exceptionally complete and valuable, to the Paris Museum. The books are to be sold, and the proceeds will be applied towards maintaining the professorship of zoology, which the deceased *savant* occupied with so much distinction. M. Milne-Edwards also bequeathed 20,000 francs to the Paris Geographical Society, of which he was president, and 10,000 francs to the Société des Amis des Sciences.

FOR several days in last week the weather was very warm over a large part of England, and in London the temperature frequently exceeded 80°. This week the temperature has still further increased, and on Monday the thermometer in the screen registered 94° at Greenwich, which is the highest reading in July since 1881, and is higher than in any summer since 1893, while in all there have only been seven days during the last sixty years with so high a temperature there. At Camden Town the shade temperature registered 95°·2, the highest reading there since 1858. Thunderstorms developed at the beginning of the week over a large part of the country, but no appreciable amount of rain has fallen in London for about a fortnight.

WE learn from *Science* that it is proposed to celebrate the 70th birthday of Prof. Wilhelm Wundt, which will occur on August 16, 1902, by the publication of a " Festschrift," to which his former students are invited to contribute. The manuscripts must be forwarded to Prof. Külpe, Würzburg, not later than January 1, 1902.

It was recently stated in the public press that postal packets containing plants for transmission to England were refused at Swiss post-offices on the ground that the plants would not be permitted to enter England. The Board of Customs has, however, just stated that there is no objection to the importation of plants from Switzerland, if they are sent by parcel post or letter post. But plants must not be sent by sample post, and the refusal of packets presented for transmission as samples appears to have produced the impression that the importation of flowers is not allowed.

AN exhibition and conference and other meetings will be held at the Crystal Palace, Sydenham, on July 20 and 21, in celebration of the bicentenary of the introduction of the sweet pea to Britain from Sicily in 1700. Some authorities hold that two forms, having a general relationship one to the other, were introduced, one from Sicily and the other from Ceylon. The history of the sweet pea and its earlier development will be dealt with at the conference meetings which are to be held in connection with the celebration. Many foreign horticulturists are giving the celebration their support in various ways ; and one of the papers at the conference will deal with the culture and development of the sweet pea in the United States, where many fine varieties have been cultivated.

THE *Times* states that the construction of the vessel designed by Mr. W. E. Smith, one of the chief constructors to the Admiralty, for the National Antarctic Expedition, is now in active progress at the yard of the Dundee Shipbuilders' Company. The ship, which is to be named the *Discovery*, is to be barque-rigged and to have three decks. Accommodation for those on board will be provided under the upper deck. The stem will be of the ice-breaker type, with strong fortifications. The length of the vessel between perpendiculars is 172 feet ; beam, 34 feet, and depth, 19 feet. The timbers are of oak dowelled and bolted together, and the keel, deadwoods, the stem, and the stem-posts are also of oak. The planking is of American elm and pitch pine, and the inside beams are of oak. With the object of avoiding the magnetic influence of iron on the scientific instruments on board, it has been decided that for a considerable radius amidships the knees and fastenings shall be of naval brass. In case the *Discovery* should have to winter in the ice, a heavy waggon cloth awning of strong woollen felt is to be provided. The fittings and equipment of the vessel will be of the most modern type. The engines, which are to indicate 450-horse-power, are to be constructed by Messrs. Gourlay Brothers and Co., Dundee.

NEWS has just reached this country of the death of a well-known geologist, Prof. G. H. F. Ulrich, F.G.S., who, since 1878, held the position of director of the School of Mines connected with the Otago University, New Zealand. Prof. Ulrich fell from a cliff while gathering rock specimens at Port Chalmers, and the injuries he received terminated fatally. Prof. Ulrich was born at Clausthal-Zillerfeld, Germany, in 1830, and was educated in his native town at the High School, and subsequently graduated at the Royal School of Mines, Clausthal, Hartz. He went to Forest Creek, Victoria, in 1854, and was appointed in 1857 assistant secretary and draughtsman to the Royal Mining Commission in Victoria. He was afterwards appointed assistant field geologist under Mr. Selwyn in the Geological department of Victoria. He continued an officer of the Geological Survey department until its abolition in 1869, when he became curator of the mining section under Mr. Newbery, superintendent of the industrial and technological museum and lecturer in mining at the University of Melbourne. He was appointed by the South Australian Government to report on their copper mines and goldfields, and in 1875 he paid his first visit to New Zealand and reported on the Otago goldfields. In 1877, the Otago University Council having decided to institute a school of mines, the Chancellor secured the services of Prof. Ulrich for the Otago University. The School of Mines was for some years small, and not very :

fully equipped, but in 1887 additional lecturers were appointed, and as the advantages of the course came to be appreciated, the number of students increased rapidly, and the attendance is now very large. Through the energy of Prof. Ulrich the models and appliances which had been procured from time to time became a valuable collection, especially in the mineral department, to which he was constantly adding from his own private collections of minerals and stones.

THE Committee on Indexing Chemical Literature presented their report of progress at the recent meeting of the American Association. From it we learn that Dr. Alfred Tuckerman has completed and sent to the Smithsonian Institution a supplement to his index to the literature of the spectroscope, which covers the period from 1887 to 1899. Dr. H. Carrington Bolton's second supplement to his select bibliography of chemistry, containing a list of 7500 chemical dissertations, is passing through the press; it will form a volume of the Smithsonian Miscellaneous Collections. Mr. A. G. Smith, of Cornell University, is engaged on an index to the literature of selenium and tellurium, which, it is expected, will be completed this summer. Dr. Frank I. Shepherd proposes to make a bibliography of the alkaloids. Mr. Frank R. Fraprie contemplates preparing an index to the literature of lithium.

IN the *Revue Générale des Sciences*, M. Louis Olivier gives some further particulars of Poulsen's "telegraphone," which is attracting attention at the Paris Exhibition. He describes several devices for increasing the volume of sound, or "intensifying" the record, to use the language of the photographer. The steel band with the consequent poles, which forms the original record, is made to pass between the poles of an electromagnet, which transfers the record to another band. This may be done several times over, and the record taken simultaneously from all the bands. In another arrangement the record is intensified by passing it very rapidly through the second magnetic field, which, as we know, has the effect of increasing the induced currents, and therefore also the intensity of the secondary record.

A NOVEL type of Newton's rings is described by Mr. A. C. Longden in the current number of the *American Journal of Science*. They are prepared by exposing a glass plate to the kathode rays emitted from a small globule of selenium. The film thus deposited is thickest at the point exactly opposite the globule, and tapers off towards the sides. The result is a film in the shape of a very flat lens, the upper and lower surfaces of which reflect light somewhat in the same manner as the upper and lower surfaces of the air film in Newton's device, with the difference, however, that in Mr. Longden's arrangement the film tapers outward instead of inward. Hence the rings increase in breadth and brilliancy away from the centre, and the order of the colours is reversed. The effect is described as very pretty.

THE annual list of the staffs of the Royal Gardens, Kew, and of botanical departments and establishments at home, and in India and the Colonies, in correspondence with Kew, has just been issued as an appendix to the *Bulletin of Miscellaneous Information*. We notice that sixty-six of the officers of the various botanic gardens have been trained at Kew, and seventeen others were appointed on the recommendation of the director of the Royal Gardens. With so many efficient observers distributed over our possessions it is not surprising that Kew is able to be of great service to the Empire as well as to science.

PLAGUE has now been established in Sydney for several months, and in an address delivered before the New South Wales Branch of the British Medical Association, Dr. Frank Tidswell of Sydney recently discussed a variety of interesting questions relating to the disease. Referring to his remarks on rats, the

Lancet points out that there are instances which show that the presence of a plague-rat is often responsible for the illness in man. For example, a number of dead rats found one morning in a cotton factory at Bombay were removed by twenty coolies. Within the three following days about half of them fell sick with plague, whilst those in the store who had not touched the rats were not affected. Again, the coachman of an English family in Bombay found a dead rat in a stable and removed it. Three days later he fell sick with plague and died in a few hours, no other person in the same house being affected. Many persons, however, have caught plague without handling plague rats, and many persons have handled plague rats without catching plague. To explain this difficulty Simond has suggested that the infection is carried by the fleas natural to the rats. Perfectly healthy rats harbour very few fleas, and are very expert in removing them, but fleas are abundant on sick rats. After death, as the body becomes cold, the fleas leave it. In this way Simond accounted for the fact that a plague rat may be handled with impunity some hours after death. If the fleas from the dead rat reach another rat or a human being, they may inoculate the bacilli they acquired by ingesting the blood of their former host. In some of Simond's experiments sick and healthy rats in separate cages were enclosed in a glass jar, and it was found that when no fleas were present the healthy animals did not become infected.

COLOUR photometry is a subject that Sir William Abney has made his own, and in his last communication to the Royal Society he describes a method of estimating the luminosity of coloured surfaces that is especially applicable when the source of light is a large surface, such as the sky. In "Colour Photometry, Part iii." it was shown that only one ray of the spectrum, a greenish-yellow, progressed in luminosity at the same rate as white light. If, for example, red, greenish-yellow, blue and white lights are made of equal luminosity, and the illuminating beams are simultaneously and equally reduced in intensity, the luminosity of the red will diminish the most rapidly, that of the blue the least rapidly, the other two remaining equal. Moreover, the colour disappears more quickly than the luminosity (except in the case of pure red), tending towards greyness, so that colours of feeble luminosity are more easy to match than bright colours. The new method of colour photometry is based upon these facts. By means of concentric rotating discs, which are, when necessary, slit radially and interlaced, the proportion of black and white that matches first a green and then a yellow disc is determined. The comparisons are facilitated by observing the rotating discs through a "black transparent medium," such as an unstained developed photographic film, which may be so dense that the colour practically disappears, giving place to a dull grey. The value of a red disc is ascertained by interlacing it with the green and blue discs to produce a grey, which is then matched with the black and white. Thus, having three standard colours of known values, the luminosity of any other colour can be ascertained by substituting a disc of it for one of the standard colours to produce a grey, and matching the grey as before. The results given by this method agree closely with those obtained by the method previously described by the author. Sir William Abney has in this way determined the luminosities of various coloured surfaces and calculated the amount of black necessary for each, so that they shall be reduced to equal luminosity. He has then prepared a divided disc divided into several annuluses, each partly coloured and partly black, so that when rotated the whole appears of equal luminosity when illuminated by the light for which it is calculated. By the selection of suitable colours such a disc is a very convenient and effective test for any defect in either the colour sensitiveness of a photographic plate, or in the coloured screen used to compensate its inherent deficiencies in this matter. For the rotating disc, which is equally luminous throughout, will give, when the negative is developed, an image of equal density throughout, if the sensitive plate and colour screen are properly adjusted to each other.

THE U.S. Weather Bureau has published a *Bulletin* (No. 29), entitled "Frost fighting," by Mr. A. G. McAdie. A bulletin on the same subject was recently issued by the Bureau, but it is believed that the more recent experiments made in California are sufficiently valuable to extensive fruit interests to justify this second publication, and that the loss due to frosts in that State, hitherto considered unavoidable, can be prevented. The problem is of a two-fold nature : accurate forecasting of the frost period, and efficient methods of raising the temperature at critical times. The various protective methods, based on irrigation, the production of cloud or fog, and devices for screening the fruit trees are photographically illustrated. Of all the methods proposed, with the exception of the use of wire screens, irrigation has the largest amount of evidence in its favour ; hot water from a boiler is forced through a number of furrows, and the temperature of the air is heated by the rising of the water-vapour.

WE have received from the Rev. W. Sidgreaves a copy of the results of meteorological and magnetical observations at Stonyhurst College Observatory, near Blackburn, for the year 1899. This observatory is fully equipped with self-recording instruments, and has for many years published valuable observations both independently and in connection with the Meteorological Office. During the past year a special report of hourly rainfall from 1891 to 1898 was prepared for that office. Much attention is given to solar observations and to the connection of sun-spots with terrestrial phenomena. The movements of the upper clouds, and the determination of the magnetic elements, also occupy the special attention of the small available staff of the observatory. An appendix contains observations taken at St. Ignatius College, Malta.

IN a paper on malformed specimens of the common pond-mussel, published in the last issue of the *Journal of Malacology*, Mr. H. H. Bloomer shows that in certain instances this mollusc is able to repair severe injuries to the mantle-lobes, but cannot make good damage inflicted on the gills.

DR. H. L. BRUNER communicates to vol. xvi. No. 2 of the *Journal of Morphology* the results of observations on the hearts of lungless salamanders, in which it is shown that with the lungs disappears also the septum between the auricles of the heart. Since, however, the normal circulation is not yet fully understood, it would be premature to discuss the reason for this loss.

IN the June issue of the *American Naturalist*, Miss Rathbun continues her invaluable illustrated synopsis of North American invertebrates, dealing in this section with certain groups of crabs. It may be hoped that, when complete, this synopsis will be reissued in book-form.

THE phylogeny of the butterflies of the family Pieridæ (best known by the ordinary British "whites") is discussed by Mr. A. R. Grote in No. 161 of the *Proceedings* of the American Philosophical Society. The author is of opinion that the family is an offshoot from the Hesperlidæ, or skippers, which is itself related to the Nymphalidæ, and that the "blues" may likewise be another offshoot from the same stock. From the scant evidence afforded by fossil forms, it further seems evident that the skippers and the whites are modern types of butterflies, while the skippers and the nymphalids are of greater antiquity.—Anthropologists will find considerable interest in a paper on the divisions of the South Australian Aborigines, by Mr. R. H. Mathews, which appears in the same serial.

IN a paper published in the *Comunicaciones* of the Buenos Aires Museum (vol. i. No. 6) Dr. F. Ameghino describes and figures certain mammalian remains from the areniscan formation of southern Patagonia. These remains are stated to be found in association with those of dinosaurs as well as of fishes of the genera Synechodus, Lepidotus and Ceratodus, and the formation is accordingly correlated with the lower Cretaceous of Europe and the United States. The mammalian remains are, however, of such a highly specialised type that it is almost impossible to believe they can be of such great antiquity ; and it seems probable that some other explanation of their alleged association with Cretaceous types will have to be found.

WE learn from the *American Naturalist* that a school of applied agriculture and horticulture will be established near New York City, to open in September, for study and practical training. Students will have the use of the laboratories and of the extensive collection of plants in the museum and conservatories and in the grounds of the New York Botanic Garden. The work will be under the direction of Mr. George T. Powell.

THE following facilities for the practical study of biology during the summer vacation are offered in the United States, in addition to those already announced. The Biological Laboratory of the Brooklyn Institute of Arts and Sciences at Cold Spring Harbour, Long Island, will be open from July 1 to August 25, under the guidance of Prof. Davenport. The Lake Laboratory of the Ohio State University at Sandusky, Lake Erie, will be open for eight weeks from July 2. Four courses of lectures will be given in zoology, and three in botany. The Rhode Island summer school for nature study is holding its session at Kingston, R.I., from July 5 to 20. Beloit College, Wisconsin, will hold a summer school on Madeline Island, Lake Superior, from July 26 to Aug. 30. The natural science camp for boys will hold its eleventh session at Canandaigua, N.Y., under the management of Mr. Albert L. Aréy. Instruction will be given in biology, entomology, taxidermy, and photography.

THE *Biologisches Centralblatt* for June 15 and July 1 contains a detailed biography of the late eminent diatomist, Comte Abbé F. Castracane, together with a complete bibliography of his very numerous contributions to botanical literature.

Bulletin No. 10 (February 1900) of the Michigan State Agricultural College Experiment Station (Agricultural Department), is devoted entirely to investigations in the cultivation of the sugar-beet, by Mr. J. D. Towar, chiefly in relation to the advantages of different soils and manures.

PROF. L. ERRERA reprints from the *Revue de l' Université de Bruxelles* a paper on spontaneous generation, one of a series of essays on botanical philosophy. After a historical account of the controversy, he sums up thus :—" Si donc la génération spontanée est encore irréalisée dans nos laboratoires, rien ne prouve qu'elle soit à jamais irréalisable."

WE have received the *Transactions* of the British Mycological Society for the season 1898-1899. It contains the address of the President, Dr. C. B. Plowright, on the recent additions to our knowledge of the Uredineæ and Ustilagineæ, with special reference to British species, a report of the New Forest fungus foray, and five papers on new or rare fungi.

THE economic geology of the United States is very amply dealt with in the larger reports of the Geological Survey, while individual States publish reports on particular subjects. One of these on the clays of Alabama, by Dr. E. A. Smith and Dr. H. Ries, has just reached us. The State yields china clay, fire clay, pottery clay, and brick clay, all of which are very fully described with regard to their characters, geological age and distribution, and a number of analyses are given. In addition to the local account, there is also a general discussion of clays, their chemical, physical and mineral characters, such as will be of great use to any one studying the subject from a scientific as well as economic point of view. Mention is made of beds of white pulverulent silica, which when mixed with clay has been used in the manufacture of a paint.

WE have received from the Geological Survey of Canada, Part I of a "Catalogue of Canadian Birds," by Mr. J. Macoren, dealing with water-birds, gallinaceous birds, and pigeons.

THE third volume of Prof. G. O. Sars's "Account of the Crustacea of Norway," dealing with the anomalous group Cumacea, is in course of publication by the Bergen Museum. Parts v. and vi., devoted to the Diastylidæ, have just been issued.

PART 10 of Memoir III. of the Australian Museum, Sydney, on "The Atoll of Funafuti" has now been issued. It is the concluding part of the memoir, and contains lists of the contributors and plates, and an index to the whole work.

MESSRS. ISENTHAL AND Co., have issued a revised edition of their list of apparatus and accessories for work with Röntgen rays. Particular attention is given by this firm to the design and construction of instruments for radiographic work, and any one contemplating an installation for this purpose will find the list just issued well worth examination.

THE additions to the Zoological Society's Gardens during the past week include a Patas Monkey (*Cercopithecus patas*, ♀) from West Africa, presented by Mr. W. B. Davidson Houston; a Rhesus Monkey (*Macacus rhesus*) from India, presented by Mrs. Heigham; a Common Marmoset (*Hapale jacchus*) from Southeast Brazil, presented by Mrs. Alexander Grant; two Grey-headed Love-Birds (*Agapornis cana*) from Madagascar, presented by Mrs. Harry Blades; a Cuckoo (*Cuculus canorus*), European, presented by Mr. L. W. Wiglesworth; an Entellus Monkey (*Semnopithecus entellus*, ♀) from India, a —— Bear (*Ursus*, sp. inc.) from Kuldja, a Himalayan Snow Partridge (*Tetrogallus himalayensis*) from the Himalayas, two Brazilian Tortoises (*Testudo tabulata*) from South America, deposited; a Sharp-nosed Badger (*Meles leptorhynchus*) from China, a Rough Fox (*Canis rudis*) from South America, purchased; a Little Bittern (*Ardetta minuta*), European, received in exchange; a Brindled Gnu (*Connochoetes taurina*, ♀), an Altai Deer (*Cervus eustephanus*), born in the Gardens.

RHYTHMS AND GEOLOGIC TIME.[1]

THE subject to which I shall invite your attention this evening is by no means novel, but might better be called perennial or recurrent; for the problem of our earth's age seems to bear repeated solution without loss of vigour or prestige. It has been a marked favourite, moreover, with presidents and vice-presidents, retiring or otherwise, when called upon to address assemblies whose fields of scientific interest are somewhat diverse—for the reason, I imagine, that while the specialist claims the problem as his peculiar theme of study, he feels that other denizens of the planet in question may not lack interest in the early lore of their estate.

The difficulty of the problem inheres in the fact that it not only transcends direct observation but demands the extrapolation or extension of familiar physical laws and processes far beyond the ordinary range of qualifying conditions. From whatever side it is approached the way must be paved by postulates, and the resulting views are so discrepant that impartial onlookers have come to be suspicious of these convenient and inviting stepping stones.

In giving brief consideration to each of the more important ways by which the problem of the earth's age has been ap-

[1] Abridged from an address to the American Association for the Advancement of Science, at New York, June 26, by the retiring President, Mr. G. K. Gilbert. By the courtesy of the Editor of *Science*, advance proofs of the address were received.

NO. 1603, VOL. 62]

proached, I shall mention first those which follow the action of some continuous process, and afterward those which depend on the recognition of rhythms.

The earliest computations of geologic time, as well as the majority of all such computations, have followed the line of the most familiar and fundamental of geological processes. All through the ages the rains, the rivers and the waves have been eating away the land, and the product of their gnawing has been received by the sea and spread out in layers of sediment. These layers have been hardened into rocky strata, and from time to time portions have been upraised and made part of the land. The record they contain makes the chief part of geologic history, and the groups into which they are divided correspond to the ages and periods of that history. In order to make use of these old sediments as measures of time, it is necessary to know either their thickness or their volume, and also the rate at which they were laid down. As the actual process of sedimentation is concealed from view, advantage is taken of the fact that the whole quantity deposited in a year is exactly equalled by the whole quantity washed from the land in the same time, and measurements and estimates are made of the amounts brought to the sea by rivers and torn from the cliffs of the shore by waves. After an estimate has been obtained of the total annual sedimentation at the present time, it is necessary to assume either that the average rate in past ages has been the same or that it has differed in some definite way.

At this point the course of procedure divides. The computer may consider the aggregate amount of the sedimentary rocks, irrespective of their subdivisions, or he may consider the thicknesses of the various groups as exhibited in different localities. If he views the rocks collectively, as a total to be divided by the annual increment, his estimate of the total is founded primarily on direct measurements made at many places on the continents, but to the result of such measurements he must add a postulated amount for the rocks concealed by the ocean, and another postulated amount for the material which has been eroded from the land and deposited in the sea more than once.

If, on the other hand, he views each group of rocks by itself, and takes account of its thickness at some locality where it is well displayed, he must acquire in some way definite conceptions of the rates at which its component layers of sand, clay and limy mud were accumulated, or else he must postulate that its average rate of accretion bore some definite ratio to the present average rate of sedimentation for the whole ocean. This course is, on the whole, more difficult than the other, but it has yielded certain preliminary factors in which considerable confidence is felt. Whatever may have been the absolute rate of rock building in each locality, it is believed that a group of strata which exhibits great thickness in many places must represent more time than a group of similar strata which is everywhere thin, and that clays and marls, settling in quiet waters are likely to represent, foot for foot, greater amounts of time than the coarser sediments gathered by strong currents; and studying the formations with regard to both thickness and texture, geologists have made out what are called *time ratios*—series of numbers expressing the relative lengths of the different ages, periods and epochs. Such estimates of ratios, when made by different persons, are found to vary much less than do the estimates of absolute time, and they will serve an excellent purpose whenever a satisfactory determination shall have been made of the duration of any one period.

Reade has varied the sedimentary method by restricting attention to the limestones, which have the peculiarity that their material is carried from the land in solution; and it is a point in favour of this procedure that the dissolved burdens of rivers are more easily measured than their burdens of clay and sand.

An independent system of time ratios has been founded on the principle of the evolution of life. Not all formations are equally supplied with fossils, but some of them contain voluminous records of contemporary life; and when account is taken of the amount of change from each full record to the next, the steps of the series are found to be unequal in magnitude. Though there is no method of precisely measuring the steps, even in a comparative way, it has yet been found possible to make approximate estimates, and these in the main lend support to the time ratios founded on sedimentation. They bring aid also to a point where the sedimentary data are weak, for the earliest formations are hard to classify and measure. It is true that these same formations are almost barren of fossils, but biological inference does not therefore stop. The oldest known fauna, the Eocambrian, does not represent the beginnings of life, but a well advanced stage, characterised by development along many divergent lines; and by comparing Eocambrian life with existing life the paleontologist is able to make an estimate of the relative progress in evolution before and after the Eocambrian epoch. The only absolute blank left by the time ratios pertains to an azoic age which may have intervened between the development of a habitable earth crust and the actual beginning of life.

Erosion and deposition have been used also, in a variety of ways, to compute the length of very recent geologic epochs. Thus, from the accumulation of sand in beaches, Andrews estimated the age of Lake Michigan, and Upham the age of the glacial lake Agassiz; and from the erosion of the Niagara gorge the age of the river flowing through it has been estimated. But while these discussions have yielded conceptions of the nature of geological time, and have served to illustrate the extreme complexity of the conditions which affect its measurement, they have accomplished little toward the determination of the length of a geologic period; for they have pertained only to a small fraction of what geologists call a period, and that fraction was of a somewhat abnormal character.

Wholly independent avenues of approach are opened by the study of processes pertaining to the earth as a planet, and with these the name of Kelvin is prominently associated.

As the rotation of the earth causes the tides, and as the tides expend energy, the tides must act as a brake, checking the speed of rotation. Therefore the earth has in the past spun faster than now, and its rate of spinning at any remote point of time may be computed. Assuming that the whole globe is solid and rigid, and that the geologic record could not begin until that condition had been attained, there could not have been great checking of rotation since consolidation. For if there had been, it would have resulted in the gathering of the oceans about the poles and the baring of the land near the equator, a condition very different from what actually obtains. This line of reasoning yields an obscure outer limit to the age of the earth.

On the assumption that the globe lacks something of perfect rigidity, G. H. Darwin has traced back the history of the earth and the moon to an epoch when the two bodies were united, their separation having been followed by the gradual enlargement of the moon's orbit and the gradual retardation of the earth's rotation; and this line of inquiry has also yielded an obscure outer limit to the antiquity of the earth as a habitable globe.

One of the most elaborate of all the computations starts with the assumption that at an initial epoch, when the outer part of the earth was consolidated from a liquid condition, the whole body of the planet had approximately the same temperature; and that as the surface afterward cooled by outward radiation there was a flow of heat to the surface by conduction from below. The rate of this flow has diminished from that epoch to the present time according to a definite law, and the present rate, being known from observation, affords a measure of the age of the crust. The strength of this computation lies in its definiteness and the simplicity of its data; its weakness in the fact that it postulates a knowledge of certain properties of rock—namely, its fusibility, conductivity and viscosity—when subjected to pressures and temperatures far greater than have ever been investigated experimentally.

A parallel line of discussion pertains to the sun. Great as is the quantity of heat which that incandescent globe yields to the earth, it is but a minute fraction of the whole amount with which it continually parts, for its radiation is equal in all directions, and the earth is but a speck in the solar sky. On the assumption that this immense loss of heat is accompanied by a corresponding loss of volume, the sun is shrinking at a definite rate, and a computation based on this rate has told how many millions of years ago the sun's diameter should have been equal to the present diameter of the earth's orbit. Manifestly the earth cannot have been ready for habitation before the passage of that epoch, and so the computation yields a superior limit to the extent of geologic time.

Before passing to the next division of the subject—the computations based on rhythms—a few words may be given to the results which have been obtained from the study of continuous processes. Realising that your patience may have been strained by the kaleidoscopic character of the rapid review which has seemed unavoidable, I shall spare you the recitation of numerical

details, and merely state in general terms that the geologists, or those who have reasoned from the rocks and fossils, have deduced values for the earth's age very much larger than have been obtained by the physicists, or those who have reasoned from earth cooling, sun cooling and tidal friction. In order to express their results in millions of years, the geologist must employ from three to five digits, while the physicists need but one or two. When these enormous discrepancies were first realised, it was seen that serious errors must exist in some of the observational data, or else in some of the theories employed; and geologists undertook with zeal the revision of their computations, making as earnest an effort for reconciliation as had been made a generation earlier to adjust the elements of the Hebrew cosmogony to the facts of geology. But after rediscussing the measurements and readjusting the assumptions so as to reduce the time estimates in every reasonable way—and perhaps in some that were not so reasonable—they were still unable to compress the chapters of geologic history between the narrow covers of physical limitation; and there the matter rests for the present.

The rocks which were formed as sediments show many traces of rhythm Some are composed of layers, thin as paper, which alternate in colour, so that when broken across they exhibit delicate banding. In the time of their making there was a periodic change in the character of the mud that settled from the water. Others are banded on a larger scale; and there are also bandings of texture where the colour is uniform. Many formations are divided into separate strata, as though the process of accretion had been periodically interrupted. Series of hard strata are often separated by films or thin layers of softer material. Strata of two kinds are sometimes seen to alternate through many repetitions. Borings in the delta of the Mississippi show soils and remains of trees at many levels, alternating with river silts. The rock series in which coal occurs are monotonous repetitions of shale and sandstone. Belgian geologists have been so impressed by the recurrence of short sequences of strata that they have based an elaborate system of rock notation upon it.

Passing to still greater units, the large aggregates of strata sometimes called systems show in many cases a regular sequence, which Newberry called a "circle of deposition." When complete it comprises a sandstone or conglomerate, at base, then shale, limestone, shale and sandstone. This sequence is explained as the result of the gradual encroachment, or transgression as it is called, of the sea over the land and its subsequent recession.

In certain bogs of Scandinavia deep accumulations of peat are traversed horizontally by layers including tree stumps in such way as to indicate that the ground has been alternately covered by forest and boggy moss. The broad glaciers of the Ice age grew alternately smaller and larger—or else were repeatedly dissipated and reformed—and their final waning was characterised by a series of halts or partial readvances, recorded in concentric belts of ice-brought drift. Of these belts, called moraines of recession, Taylor enumerates seventeen in a single system.

In explanation of these and other repetitive series incorporated in the structure of the earth's crust, a variety of rhythmic causes have been adduced; and mention will be made of the more important, beginning with those which have the character of original rhythms.

A river flowing through its delta clogs its channel with sediment, and from time to time shifts its course to a new line, reaching the sea by a new mouth. Such changes interrupt and vary sedimentation in neighbouring parts of the sea. Storms of rain make floods, and each flood may cause a separate stratum of sediment. Storms of wind give destructive force to the waves that beat the shore, and each storm may cause the deposit of an individual layer of sediment. Varying winds may drive currents this way and that, causing alternations in sedimentation.

To explain the forest beds buried in the Mississippi silts it has been suggested that the soft deposits of the delta from time to time settled and spread out under their own weight. Various alternations of strata, and especially those of the Coal measures, have been ascribed to successive local subsidences of the earth's crust, caused by the addition of loads of deposit. It has been suggested also that land undergoing erosion may rise up from time to time because relieved of load, and the character of sediment might be changed by such rising. Subterranean forces, of whatever origin, seemingly slumber while strains are accumulating, and then become suddenly manifest in dislocations and eruptions, and such catastrophes affect sedimentation.

A more general rhythm has been ascribed to the tidal retardation of rotation and the resulting change of the earth's form. If the body of the earth has a rather high rigidity, we should expect that it would for a time resist the tendency to become more nearly spherical, while the water of the ocean would accommodate itself to the changing conditions of equilibrium by seeking the higher latitudes. Eventually, however, the solid earth would yield to the strain and its figure become adjusted to the slower rotation, and then the mobile water would return. Thus would be caused periodic transgressions by the sea, occurring alternately in high and low latitudes.

Another general rhythm has been recently suggested by Chamberlin in connection with the hypothesis that secular variations of climate are chiefly due to variations of the quantity of carbon dioxide in the atmosphere.[1] The system of interdependent factors he works out is too complex for presentation at this time, and I must content myself with saying that his explanation of the moraines of recession involves the interaction of a peculiar atmospheric condition with a condition of glaciation, each condition tending to aggravate the other, until the cumulative results brought about a reaction and the climatic pendulum swung in the opposite direction. With each successive oscillation the momentum was less, and an equilibrium was finally reached.

Few of these original rhythms have been used in computations of geologic time, and it is not believed that they have any positive value for that purpose. Nevertheless, account must be taken of them, because they compete with imposed rhythms for the explanation of many phenomena, and the imposed rhythms, wherever established, yield estimates of time.

The tidal period, or the half of the lunar day, is the shortest imposed rhythm appealed to in the explanation of the features of sedimentation. It is quite conceivable that the bottom of a quiet bay may receive at each tide a thin deposit of mud which could be distinguished in the resulting rock as a papery layer or lamina. If one could in some way identify a rock thus formed, he might learn how many half-days its making required by counting its laminæ, just as the years of a tree's age are learned by counting its rings of growth.

The next imposed rhythm of geologic importance is the year. There are rivers, like the Nile, having but one notable flood in each year, and so depositing annual layers of sediment on their alluvial plains and on the sea beds near their mouths. Where oceanic currents are annually reversed by monsoons, sedimentation may be regularly varied, or interrupted, once a year. Streams from a glacier cease to run in winter, and this annual interruption may give a definite structure to resulting deposits. It is therefore probable that some of the laminæ or strata of rocks represent years, but the circumstances are rarely such that the investigator can bar out the possibility that part of the markings or separations were caused by original rhythms of unknown period.

The number of rhythms existing in the solar system is very large, but there are only two, in addition to the two just mentioned, which seem competent to write themselves in a legible way in the geologic record. These are the rhythms of precession and eccentricity.

Because the earth's orbit is not quite circular and the sun's position is a little out of the centre, or is eccentric, the two hemispheres into which the earth is divided by the equator do not receive their heat in the same way. The northern summer, or the period during which the northern hemisphere is inclined toward the sun, occurs when the earth is farthest from the sun, and the northern winter occurs when the earth is nearest to the sun, or in that part of the orbit called perihelion. These relations are exactly reversed for the southern hemisphere. The general effect of this is that the southern summer is hotter than the northern, and the southern winter is colder than the northern. In the southern part of the planet there is more contrast between summer and winter than in the northern. The sun sends to each half the same total quantity of heat in the course of a year, but the difference in distribution makes the climates different. The physics of the atmosphere is so intricate a subject that meteorologists are not fully agreed as to the theoretical consequences of these differences of solar heating, but it is generally believed that they are important, involving differences in the force of the winds, in the velocity and course of ocean currents, in vegetation, and in the extent of glaciers.

[1] An attempt to frame a working hypothesis of the cause of glacial periods on an atmospheric basis. *Journ. Geol.*, vol. vii., 1899.

Now, the point of interest in the present connection is that the astronomical relations which occasion these peculiarities are not constant, but undergo a slow periodic change. The relation of the seasons to the orbit is gradually shifting, so that each season in turn coincides with the perihelion; and the climatic peculiarities of the two hemispheres, so far as they depend on planetary motions, are periodically reversed. The time in which the cycle of change is completed, or the period of the rhythm, is not always the same, but averages 21,000 years. It is commonly called the precessional period.[1]

Assuming that the climates of many parts of the earth are subject to a secular cycle, with contrasted phases every 10,500 years, we should expect to find records of the cycle in the sediments. A moist climate would tend to leach the calcareous matter from the rock, leaving an earthy soil behind, and in a succeeding drier climate the soil would be carried away; and thus the adjacent ocean would receive first calcareous and then earthy sediments. The increase of glaciers in one hemisphere would not only modify adjacent sediments directly, but, by adding matter on that side, would make a small difference in the position of the earth's centre of gravity. The ocean would move somewhat toward the weighted hemisphere, encroaching on some coasts and drawing down on others; and even a small change of that sort would modify the conditions of erosion and deposition to an appreciable extent in many localities.

Blytt ascribed to this astronomical cause the alterations of bog and forest in Scandinavia, as well as other sedimentary rhythms observed in Europe; and it has seemed to me competent to account for certain alternations of strata in the Cretaceous formations of Colorado. Croll used it to explain interglacial epochs, and Taylor has recently applied it to the moraines of recession.

The remaining astronomical rhythm of geological import is the variation of eccentricity. At the present time our greatest distance from the sun exceeds our least distance by its thirtieth part, but the difference is not usually so small as this. It may increase to the seventh part of the whole distance, and it may fall to zero. Between these limits it fluctuates in a somewhat irregular way, in which the property of periodicity is not conspicuous. The effect of its fluctuation is inseparable from the precessional effect, and is related to it as a modifying condition. When the eccentricity is large the precessional rhythm is emphasised; when it is small the precessional rhythm is weak.

The variation of eccentricity is connected with the most celebrated of all attempts to determine a limited portion of geological time. In the elaboration of the theory of the Ice age which bears his name, Croll correlated two important epochs of glaciation with epochs of high eccentricity computed to have occurred about 100,000 and 210,000 years ago. As the analysis of the glacial history progresses, these correlations will eventually be established or disproved, and should they be established it is possible that similar correlations may be made between events far more remote.

The studies of these several rhythms, while they have led to the computation of various epochs and stages of geologic time, have not yet furnished an estimate either of the entire age of the earth or of any large part of it. Nevertheless, I believe that they may profitably be followed with that end in view.

The system of rock layers, great and small, constituting the record of sedimentation, may be compared to the scroll of a chronograph. The geological scroll bears many separate lines, one for each district where rocks are well displayed, but these are not independent, for they are labelled by fossils, and by means of these labels can be arranged in proper relation. In each time line are little jogs—changes in kind of rock or breaks in continuity—and these jogs record contemporary events. A new mountain was uplifted, perhaps, on the neighbouring continent, or an old uplift received a new impulse. Through what Davis calls stream piracy a river gained or lost the drainage of a tract of country. Escaping lava threw a dam across the course of a stream, or some Krakatoa strewed ashes over the land and gave the rivers a new material to work on. The jogs may be faint or strong, many or few, and for long distances the lines may run smooth and straight; but so long as the jogs are irregular they give no clue to time. Here and there, however, the even line will betray a regularly recurring indentation or

[1] Strictly speaking, 21,000 years is the period of the precession of the equinoxes as referred to perihelion; but the perihelion is itself in motion. As referred to a fixed star the precession of the equinoxes has an average period of about 25,700 years.

undulation, reflecting a rhythm and possibly significant of a remote pendulum whose rate of vibration is known. If it can be traced to such a pendulum there will result a determination of the rate at which the chronograph scroll moved when that part of the record was made ; and a moderate number of such determinations, if well distributed, will convert the whole scroll nto a definite time scale.

In other words, if a sufficient number of the rhythms embodied in strata can be identified with particular imposed rhythms, the rates of sedimentation under different circumstances and at different times will become known, and eventually so many parts of geologic time will have become subject to direct calcu-lation that the intervals can be rationally bridged over by the aid of time ratios.

For this purpose there is only one of the imposed rhythms of practical value, namely, the precessional ; but that one is, in my judgment, of high value.   The tidal rhythm cannot be ex-pected to characterise any thick formation.   The annual is liable to confusion with a variety of original rhythms, especially those connected with storms.   The rhythm of eccentricity, being theoretically expressed only as an accentuation of the pre-cessional, cannot ordinarily be distinguished from it.   But none of these qualifications apply to the precessional.   It is not liable to confusion with the tidal and annual because its period is so much longer, being more than 2000 times that of the annual. It has an eminently practical and convenient magnitude, in that its physical manifestation is well above the microscopic plane, and yet not so large as to prevent the frequent bringing of several examples into a single view.   It is also practically regular in period, rarely deviating from the average length by more than the tenth part.

From the greater number of original rhythms it is distinguished, just as from the annual and tidal, by magnitude.   The practical geologist would never confuse the deposit occasioned by a single storm, for example, with the sediments accumulated during an astronomical cycle of 20,000 years.   But there are other original rhythms, known or surmised, which might have magnitudes of the same general order, and to discriminate the precessional from these it is necessary to employ other characters.   Such characters are found in its regularity or evenness of period, and in its practical perpetuity.   The diversion of the mouth of a great river, such as the Hoang Ho or the Mississippi, might recur only after long intervals ; but from what we know of the behaviour of smaller streams we may be sure that such events would be very irregular in time as well as in other ways.   The intervals between volcanic eruptions at a particular vent or in a particular district may at times amount to thousands of years, but their irregularity is a characteristic feature.   The same is true of the recurrent uplifts by which mountains grow, so far as we may judge them by the related phenomena of earthquakes ; and the same category would seem to hold also the theoretically recurrent collapse of the globe under the strains arising from the slowing of rotation.   The carbon dioxide rhythm, known as yet only in the field of hypothesis, is hypothetically a running-down oscillation, like the lessening sway of the cradle when the push is no longer given.

But the precessional motion pulses steadily on through the ages like the swing of a frictionless pendulum.   Its throb may or may not be caught by the geological process which obtains in a particular province and in a particular era, but whenever the conditions are favourable and the connection is made, the record should reflect the persistence and the regularity of the inciting rhythm.

The search of the rocks for records of the ticks of the pre-cessional clock is an out-of-door work.   Pursued as a closet study it could have no satisfactory outcome, because the printed descriptions of rock sequences are not sufficiently complete for the purpose ; and the closet study of geology is peculiarly exposed to the perils of hobby-riding.   A student of the time problem cannot be sure of a persistent, equable sedimentary rhythm without direct observation of the characters of the repeated layers.   He needs to avail himself of every opportunity to study the series in its horizontal extent, and he should view the local problem of original *versus* imposed rhythm with the aid of all the light which the field evidence can cast on the con-ditions of sedimentation.

Neither do I think of rhythm seeking as a pursuit to absorb the whole time and energy of an individual and be followed steadily to a conclusion ; but hope rather that it may receive the incidental and occasional attention of many of my colleagues

of the hammer, as other errands lead them among cliffs of bedded rocks.   If my suggestion should succeed in adding a working hypothesis or point of view to the equipment of field geolo-gists, I should feel that the search had been begun in the most promising and advantageous manner.   For not only would the subject of rhythms and their interpretations be ad-vanced by reactions from multifarious individual experiences, but the stimulus of another hypothesis would lead to the dis-covery of unexpected meanings in stratigraphic detail.

It is one of the fortunate qualities of scientific research that its incidental and unanticipated results are not unfrequently of equal or even greater value than those directly sought.   Indeed, if it were not so there would be no utilitarian harvest from the cultivation of the field of pure science.

In advocating the adoption of a new point of view from which to peer into the mysterious past, I would not be understood to advise the abandonment of old standpoints, but rather to emu-late the surveyor, who makes measurement to inaccessible points by means of bearings from different sides.   Every inde-pendent bearing on the earth's beginning is a check on other bearings, and it is through the study of discrepancies that we are to discover the refractions by which our lines of sight are warped and twisted.   The three principal lines we have now projected into the abyss of time miss one another altogether, so that there is no point of intersection.   If any one of them is straight, both the others are hopelessly crooked.   If we would succeed we should not only take new bearings from each discovered point of vantage, but strive in every way to discover the sources of error in the bearings we have already attempted.

THE RELATION OF STIMULUS TO SENSATION.

NOTHING has done more to place on a scientific footing the discussion of the phenomena which the study of matter and energy presents to the eye of reason, than the establish-ment of a doctrine of quantitative equivalence.   So much oxygen and hydrogen, so much water ; this amount of energy of chemical separation gone, that amount of sensible heat gained. In a similar way, nothing is likely to do more to give support to the hypothesis that sentience or consciousness is a concomi-tant of certain physiological processes than the establishment of a quantitative relation between stimulus and sensation.

It has, indeed, long been obvious that some general relation of this kind holds good.   Increased physical pressure is, within certain limits, increasingly felt ; more light gives a higher degree of visual sensation ; the greater the amplitude of the vibrations of a violin-string the fuller and louder the sound. Such statements are, however, indefinite.   We want to know how much the physical increase must be to give just so much increment in sensation.   If we double the strength of the stimulus, do we double also the strength of the sensation ?   If not, by how much do we increase it ?   Ernst Heinrich Weber sought to express the quantitative relation with some exactness ; Gustav Theodor Fechner and other more recent inquirers have built upon the foundations laid by Weber ; and a provisional law of the relation of physical stimulus to sensation has gradually gained wide acceptance.

Weber's classical experiments dealt with what is termed the "least observable difference."   If, for example, a weight of one pound be laid upon the hand, it gives rise to a sensation of pres-sure.   If, now, an extra ounce be added no difference is felt, nor is the added weight of two or of three ounces perceptible.   The sensation is not increased, and then only just perceivably in-creased, till one-third of a pound is added.   This, then, is said to be the least observable difference.   We now start afresh with a load of two pounds, and add, as before, one-third of a pound. But there is no observable difference ; nor is there any felt in-crease in sensation until two-thirds of a pound are added.   Start-ing once more with an initial load of three pounds, we find that neither the addition of one-third, nor that of two-thirds of a pound affords any observable difference in the sensation experi-enced.   A full pound must be added for the increment to be felt. The least observable differences, therefore, are between

$$1 \text{ lb. and } 1 + \tfrac{1}{3} \text{ lb.}$$
$$2 \text{ lb. ,, } 2 + \tfrac{2}{3} \text{ lb.}$$
$$3 \text{ lb. ,, } 3 + \tfrac{3}{3} \text{ lb.}$$

If, then, we extend and generalise the results of such experi-ments, we find that, within certain limits, to obtain an orderly series of just observable differences in sensation we must always add the same fraction—one-third of the weight—at each constant step of the series.

Now Fechner assumed that these just perceivable increments of sensation are all of the same value, or are constant; in which case they form an arithmetical series—that is to say, one that is produced by successive additions of the same amount. But the corresponding series of stimuli are not in arithmetical progression, since the successive increments are not of the same amount. The increase is, however, always by the same proportional amount. Each successive stimulus has to be multiplied by a constant factor, $\frac{4}{3}$. The series, therefore, forms an orderly sequence in geometrical progression.

We thus reach what is known as the Weber-Fechner formula, by which the relation of stimulus to sensation is expressed in quantitative terms. It may be thus stated:—To obtain an arithmetical series of sensations a geometrical series of stimuli is required. To give the former, equal increments of sensation are added; to obtain the latter we must multiply the successive stimuli by a constant factor.

It must be admitted, however, that the results of a great number of carefully-conducted observations are by no means in satisfactory accordance with this formula. Hering and his pupils have shown that for very small stimuli, lying near the threshold of sensation, both stimulus and sensation increase very nearly *pari passu* in arithmetical progression. The Weber-Fechner formula cannot, therefore, at present be regarded as more than an approximation to the truth.

In extracting the Weber-Fechner formula from the data afforded by observations on the method of least observable difference, it is necessary to piece together the results observed singly and in succession. But from the nature of the field of vision it is possible to obtain a series of increments of stimulus which shall afford a scale of sensation visible as a whole and at a glance. In the current number of the *Psychological Review* (vol. vii. No. 3, p. 217) I have published in detail the results of an investigation "On the relation of stimulus to sensation in visual impressions," by which I have been led to suggest a modification of the Weber-Fechner formula. Stripped as far as possible of technicalities, the method and results may be here briefly described.

It is well known that if a disc with white and black sectors be rapidly rotated, the effect on the eye is a uniform grey. If the white sectors are proportionally small, occupying, for example, only 5 per cent. of the disc, the effect is that of a very dark grey; if they are relatively large, occupying, say, 90 per cent. of the disc, the effect is that of a very light grey. With such sectors the same proportional amount of white is introduced in all parts of the disc, so as to give in each case the same shade of grey throughout its whole extent. But it is possible to introduce varying proportions of white from centre to circumference, and when this is done the rotating disc no longer presents all over its surface the same uniform shade of grey, but shows varying shades. Let us now endeavour to reduce these varying shades to order. Let us arrange the proportions of white stimulus which we introduce, in such a way as to leave a ring of full black (with no white) at the circumference, and to give a ring of full white (with no black) near the centre, and between these extremes to obtain a perfectly smooth and even gradation of shades of grey from one so dark as to be scarcely distinguishable from black, to one so light as to be scarcely distinguishable from white. We may then, when the disc is rapidly rotating, run our eye from white near the centre, through deepening and deepening grey, to black at the circumference, with nowhere any observable jump in sensation—nowhere, so to speak, a steeper slope of change than elsewhere; as if, in fact, we were passing along a perfectly even inclined plane of sensation from the lowest depth of black to the extreme height of white. If we succeed in this—and it is by no means easy of attainment—we shall have secured an arithmetical series of sensation. From one end to the other we have at successively equal distances constant increments of white sensation, just as in ascending a uniform incline we gain equal increments of height for every yard we progress towards our goal. This even slope of sensation is produced by the juxtaposition of all the least observable differences whose sum gives the full scale. Having obtained this result we are able to ascertain, by careful angular measurements of the proportional areas of white at different parts of our disc, the exact amounts of stimulus which are affecting the eye from these several parts. We may, for

example, subdivide the area of the disc lying between the inner white circle and the outer ring of black, by drawing nine concentric circles equidistant from each other, and at these nine distances make angular measurements of the proportional amounts of white to black; and then, by plotting, sweep a curve of stimulus through points representing these measured amounts.

When these amounts are tabulated and dealt with by appropriate mathematical methods, it is found that they are *not* in accordance with the Weber-Fechner formula. Nor does a disc prepared in accordance with this formula give the smooth and evenly-graded incline of an arithmetical series in sensation. For details the reader may be referred to the paper in which the observations and calculations are set forth. The accompanying figure gives the results plotted in a curve on the graphic method. The dotted steps indicate the nine measured increments. The vertical distance of any point on the curve, measured from below, upwards, gives the percentage of sensation. The horizontal distance, measured from left to right, gives the corresponding percentage of white stimulus. The law which results from a discussion of these observations, and of others where red,

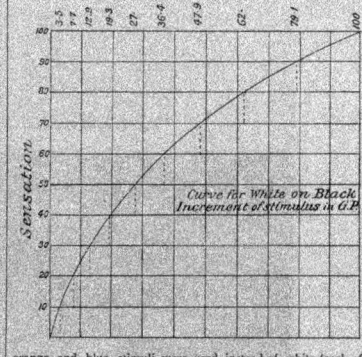

orange and blue stimuli were used instead of white (each of which gives a different curve on the same principle), may be thus formulated:—*For constant increments of sensation the constant increments of stimulus are in geometrical progression.* This differs from the Weber-Fechner formula in assigning the geometrical progression to the successive *increments* of stimulus.

The subjoined table gives the increments and sums of

White on Black

Stage	Sensation		Stimulus		Observed percentage of stimulus
	Increments	Sums	Increments	Sums	
10	10	100	20·90	100	100
9	10	90	17·13	79·10	79
8	10	80	14·03	61·97	62
7	10	70	11·51	47·94	48
6	10	60	9·43	36·43	35·8
5	10	50	7·73	27	27
4	10	40	6·33	19·27	19·5
3	10	30	5·20	12·94	13
2	10	20	4·25	7·74	7·9
1	10	10	3·49	3·49	3·5
0	0	0	0	0	0

stimulus and sensation at ten stages between black and white. A comparison of the last two columns will show the extent of agreement between observation and calculation. The numbers given under the head of stimulus are calculated on the basis of the suggested law, the number 27 per cent. of stimulus, as the concomitant of 50 per cent. of sensation, being taken over from observation as a basis for calculation.

Although I venture to hope that the results of this investigation contribute something towards a solution of the problem, still it will be seen that we have as yet by no means reached the stage at which we can claim that a law expressing the quantitative relation of stimulus to sensation is established beyond question. But from the work of many observers we may at least draw the conclusion that there is some well-defined relation, though its law at present eludes the grasp of our generalisation. And this so far lends support to the doctrine of concomitance.

There has been much discussion as to the true meaning of the relation. Is it primarily a relation between physical stimulus and physiological response, or between physiological response and psychological concomitant? In other words, is the law we seek primarily a physiological or a psychological law? We

were only 1·22 inch in diameter, and the length of the arms, from the centre of cup to the spindle, only 1·96 inch. The author describes at length the whirling apparatus used in making the experiments, and which had been previously used in the year 1888, but in an enclosed space, instead of in the free air. He points out that a whirling apparatus is absolutely necessary for testing anemometers, because we have no other means of accurately measuring the speed of the wind to which the instrument is exposed, unless we employ for that purpose some other anemometer, which must itself be first tested. In the author's view, the effect of using the whirler in the open air is to alternately add to and subtract from the artificial wind resulting from the steady motion of the whirler, so that the actual resultant wind affecting the anemometer acquires a *gusty* character which is analogous to the conditions always existing in the free air, and the artificial gusty wind thus secured affords a highly appropriate test-wind for anemometers that are to be used in the open air. The apparatus employed is shown in a plate, which we reproduce.

The arm, on the extreme end of which the anemometer is placed, is 28 feet long, and is made to rotate either by hand-power or by means of the engine used in the kite experiments.

Fig. 1.—Whirling Machine and Driving Belt for Anemometer Tests.

cannot enter upon the discussion here. Attention may, however, be drawn to two facts :—First, that Prof. Pfeffer claims to have shown that the attractive influence of malic acid on the spermatozoids of ferns is approximately in accordance with the Weber-Fechner formula ; and secondly, that Dr. Augustus Waller's researches on the excitation of muscle and nerve indicate some such relation, though not exactly this relation, between stimulus and physiological response. In view of these facts it seems not improbable, therefore, that the relation may prove to be primarily physiological. In which case we may infer that sensation is directly proportional to the molecular disturbance in the nerve-centres concerned.

C. LLOYD MORGAN.

ANEMOMETER TESTS.

THE U.S. *Monthly Weather Review* for February contains an important contribution by Prof. C. F. Marvin on anemometer tests. The paper gives the results of a series of experiments to determine the factor of an anemometer specially designed for use with kites at considerable altitudes in the free air. For that purpose the anemometer has necessarily to be very small and light, and in the present case the cups

By hand-power any speed up to thirty-five miles an hour could be obtained, and by the engine the velocity could be raised to nearly sixty miles an hour. A good break-circuit seconds pendulum clock was employed, in conjunction with an astronomical chronograph, to record results, and the series of comparisons appears to have been carried out with much care and completeness.

The experiments included a redetermination of the constants for a "standard aluminium cup anemometer," in which the cups were 4·07 inches in diameter, and the arms 6·65 inches in length. This instrument had been used in the investigations of 1888, and the values now obtained gave a slightly lower rate of speed of the cups in a given wind than had been formerly deduced. But as the differences did not exceed 2 per cent., it is fair to conclude that, upon the whole, the agreement was satisfactory.

The author also points out that another result of the experiments is to confirm a conclusion arrived at in 1888, viz. that an anemometer with large cups, as compared with the length of the arms, runs at a speed bearing a more nearly constant ratio to that of the wind than an anemometer with relatively larger arms. In the case of the small kite anemometer now investigated, the factor is practically constant for velocities from ten to fifty miles an hour, the extreme variation being only about 1·5 per cent.

THE GREAT ALPINE TUNNELS.[1]

THE subject for this evening's discourse is that of the three great tunnels through the Alps—viz. the Mont Cenis, the St. Gothard, and that which is now in course of construction, the Simplon.

But before dealing with the details of these particular works, it will be desirable to consider what tunnelling is, and also some of the more remarkable instances of it in bygone days.

One great drawback in connection with the subject—so far as a discourse is concerned—is its unsuitability for the photographic art. Unlike a battleship, or a splendid bridge, or a grand block of buildings which can be made into fine views and pictures, the work of the mole is hardly adapted to the sensitive plate. I therefore propose to make use of the "language of the pencil," and to make a few rough sketches on the blackboard: by these means I trust I may be able to explain some of the difficulties which have to be encountered, and also show how a tunnel is constructed. The child's definition of drawing, "first you think, and then you draw a line round your think," will come to our aid.

The art of tunnelling dates back to very remote ages, and there are records of such works which were constructed 500 to 600 years before the Christian era.

An interesting account is given by one of your most distinguished members, in an article in the "Encyclopædia Britannica," of the tunnel under the river Euphrates at Babylon. This city, similar in some respects to London, lay half on one side and half on the other side of the river. High walls, pene-

Fig. 1.

CROSS SECTION
of the
AQUEDUCT
of EUPALINOS.

In the Island
of
SAMOS.

trated by occasional gates, surrounded the city, and lined each of the banks of the river. These gates (of which a pair of the great hinges can be seen in the British Museum) were closed at night and during war; and a tunnel was constructed below the bed of the river by means of what is technically known as the "cut-and-cover" system. In those days the Greathead shield was unknown, and consequently the river had to be diverted, so that the excavation could be made in the dry bed and cut open to daylight, the arch being built, the ground restored, and the river allowed to resume its former course. The tunnel is said to have been 15 feet in width, and 12 feet in height, built of brick.

Herodotus gives an account of the diversion of the river into a great excavation or artificial lake forty miles square, and states that the besieging enemy, so soon as the water was drawn off, entered into the city by the river bed. It is believed that this same excavation was made use of for the construction of the tunnel. It is, however, desirable to state that doubts have been thrown on the subject, and it is possible that it may have to be relegated to mythology.

The next instance of a tunnel is that referred to by Herodotus in the Island of Samos ("Herodotus," iii. p. 60) (see Fig. 1), and it is satisfactory to know that although very considerable doubts were expressed as to the accuracy of his statements, recent investigations prove that he was exactly correct. The description given by him, when expressed in English words and figures, is as

[1] A discourse delivered at the Royal Institution on Friday, May 25, by Francis Fox, Mem.Inst.C.E.

follows: "They have a mountain which is 910 feet in height; entirely through this they have made a passage, the length of which is 1416 yards. It is, moreover, 8 feet high, and as many wide. By the side of this there is also an artificial canal, which in like manner goes quite through the mountain; and though only 3 feet in breadth, is 30 feet deep. This, by the means of pipes, conveys to the city the waters of a copious spring."

The commentators on this passage say that Herodotus must have made a mistake, but the Rev. H. F. Tozer, in his book "The Islands of the Ægean," p. 167, gives the results of a personal visit.

He says the tunnel is 7 to 8 feet in width; that two-thirds of its width are occupied by a footpath, the other third being a watercourse, 30 feet deep at one end. He and other writers consider that insufficient allowance was made for the fall of the water, and that the water channel had to be deepened. To describe it in more modern language, the resident engineer evidently made a mistake in his levels, necessitating a much deeper excavation than was at first anticipated.

Another, and if possible a more interesting, instance of tunnelling is that described in the *Proceedings* of the Palestine Exploration Society, in connection with the Pool of Siloam, made by Hezekiah, B.C. 710, 2 Kings xx. 20 ("Palestine Exploration," 1882, p. 178). See Fig. 2.

About 710 B.C. a tunnel was driven from the spring to the well—by actual tunnelling—the work being commenced at the two ends, and by shafts, and the workmen met in the middle. The tunnel was only 2 feet in width and 3 feet in height, except at the probable point of meeting, where the height is 4 feet 6 inches. The length is 1708 feet, and there is a fall of 1 foot in this distance. About the middle of its course there are apparently two false cuts, as if a wrong direction had been

Fig. 2.

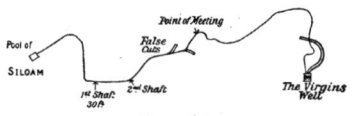

Plan of Tunnel from Spring to Pool of Siloam.

Pool of SILOAM — 1st Shaft 30ft — 2nd Shaft — False Cuts — Point of Meeting — The Virgins Well

(Not to scale.)

taken: but possibly these were intentional, and provided passing places for the workmen and material.

On the soffit of the tunnel is carved an inscription, of which the following is a translation:—

"Behold the excavation. Now this had been the history of the excavation. While the workmen were still lifting up the pick, each towards his neighbour, and while 3 cubits (4 feet 6 inches) still remained to cut through, each heard the voice of the other, who called to his neighbour, since there was an excess of rock on the right hand and on the left. And on the day of the excavation the workmen struck each to meet his neighbour pick against pick, and there flowed the waters from the spring to the pool for a thousand two hundred cubits (1820 feet), and a hundred cubits (151 feet) was the height of the rock over the head of the workmen."

A Roman engineer gives an account of a tunnel which was being driven under his directions for an aqueduct. And as he was only able to visit the work occasionally, he describes how on one of his visits he found the two headings had missed each other, and he says that had his visit been deferred much longer there would have been two tunnels.

The accurate meeting of the headings or driftways of a tunnel can only be attained by the exercise of great care, both as regards direction as well as level.

We need not go very far to find instances of such an error as inaccurate meeting, but there is one well-known case on an important main line in the Midland Counties where the engineers failed to meet, and to this day reverse curves exist in the tunnel to overcome the difficulty.

To attain this accurate meeting fine wires are hung down the shafts of a tunnel, with heavy plumb-bobs suspended from them in buckets of water, or of tar, to bring their oscillations to rest ; the accurate direction being given by means of a theodolite or transit instrument on the surface.

The wires are capable of side movement by means of a delicate instrument, and are gradually brought exactly into the same vertical plane : hence, if they are correct at "bank," or surface, they must also be correct below ground. The engineers below have to drive the galleries or headings so that only *one* wire is visible from their instrument : so long as one wire exactly eclipses the other wire, the gallery is being driven in the right direction.

As regards accuracy in levels, this is done by ordinary levelling ; but it will be seen at once how much depends on care being devoted to both these operations.

Fig. 3.   PLAN.

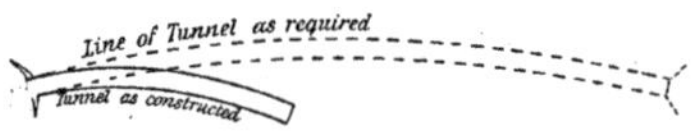

Line of Tunnel as required

Tunnel as constructed

Assume two shafts, 1000 yards apart, between which a gallery has to be driven ; and, allowing a distance of .10 feet between the wires, which are $\frac{1}{10}$th inch in diameter, an error of the diameter of the wire at the shaft will cause a mistake of nearly 4 inches at the point of meeting, or of $7\frac{1}{2}$ inches if a similar error occurs at the other shaft in the opposite direction. The trickling of water down the wires increases their diameter so appreciably, and therefore conduces to further inaccuracy, that it is found necessary to fix a small shield or umbrella on the wire to deflect the water.

Some years ago, a tunnel which had been commenced, but not finished, had to be completed. The first thing to be done by the engineers was to make an accurate survey of the then condition of the work—this rough sketch (see Fig. 3) indicates what was discovered. The explanation given by the former "ganger" was, that he found the rock too hard, and he thought that by bearing round somewhat to the right, he might get into more easily excavated material !

When the wires are hung down the shaft it is sometimes almost impossible to prove that they are not touching, and consequently being deflected from the true vertical line by some rope or pipe, staging or timber in the shaft. To overcome this, an electrical current was passed down the wire—a galvanometer being in circuit. If the wire proved absolutely silent, and no deflection was obtained in the galvanometer, the conclusion could be safely drawn that the wire was hanging freely and truly.

In driving the necessary adit or heading for drainage purposes beneath a sub-aqueous tunnel, a rising gradient from the shaft bottom of 1 in 500 is allowed, to enable the water at the "face" to flow away from the workmen to the pumps in the "sump" or shaft bottom (see Fig. 4).

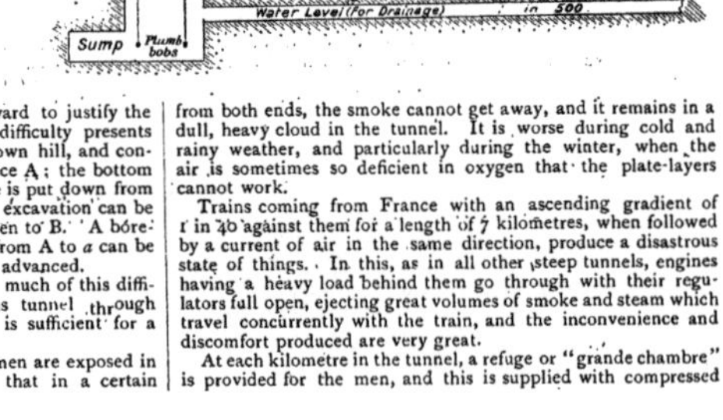

When the heading is driven sufficiently forward to justify the commencement of the main tunnel, a fresh difficulty presents itself. This main tunnel has to be driven down hill, and consequently the water collects at the working face A ; the bottom cannot therefore be removed until a bore-hole is put down from A to *a*. When this is done the remaining excavation can be taken out, and a further length of tunnel driven to B. A bore-hole is now sunk from B to *b*, whilst that from A to *a* can be plugged up : and thus the tunnel is gradually advanced.

By the adoption of the Greathead shield much of this difficulty can be avoided ; but all sub-aqueous tunnel through water-bearing strata, at considerable depth, is sufficient for a lifetime.

As an illustration of the danger to which men are exposed in such work, it is stated, with much regret, that in a certain

tunnel, notwithstanding every precaution being taken, all the men engaged in driving the drainage heading by means of a tunnelling machine have died ; and in the case of the first Vyrnwy tunnel crossing of the River Mersey—driving by Greathead shield under pressure—the mortality was great.

Having explained in very general terms some of the difficulties of tunnel construction, we will proceed to the case of the great tunnels through the Alps, and for the purpose of rendering the subject more easily intelligible, the following particulars may be given :—

	St. Gothard	Mont Cenis	Arlberg	Simplon
Length of tunnel in miles	9·3	7·98	6·36	12·26
North or east portal above sea-level, feet	3639	3766	4296	2254
South or west portal above sea-level, feet	3757	4164	3998	2080
Highest level	3788	4248	4300	2314
Maximum grade in tunnel per 1000 ...	5·82	30	15	7
Maximum height of mountain above tunnel, feet... ...	5598	5428	2362	7005
Possible maximum temperature of rock, deg. Fahr.	85°	85°	65°	104°

MONT CENIS TUNNEL.

The Mont Cenis, or as it is more accurately called, the Frejus Tunnel, is nearly eight miles in length. It is for a double line of way—width 26 feet, and height above rails 20 feet 6 inches.

In consequence of the gradients in the Mont Cenis ascending

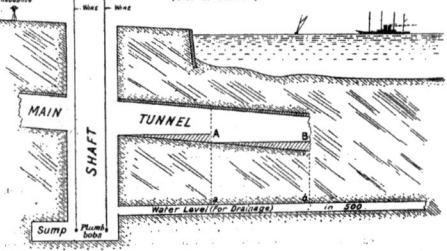

Fig. 4.   Diagrammatic Section to illustrate method of constructing Tunnel below River Bed.
(Not to Scale)

from both ends, the smoke cannot get away, and it remains in a dull, heavy cloud in the tunnel. It is worse during cold and rainy weather, and particularly during the winter, when the air is sometimes so deficient in oxygen that the plate-layers cannot work.

Trains coming from France with an ascending gradient of 1 in 30 against them for a length of 7 kilometres, when followed by a current of air in the same direction, produce a disastrous state of things. In this, as in all other steep tunnels, engines having a heavy load behind them go through with their regulators full open, ejecting great volumes of smoke and steam which travel concurrently with the train, and the inconvenience and discomfort produced are very great.

At each kilometre in the tunnel, a refuge or "grande chambre" is provided for the men, and this is supplied with compressed air, fresh water, a telephone in each direction out, a medicine chest, barometer and thermometer.

The cost of the tunnel was about 3,000,000*l.*, or 220*l.* per yard, and occupied ten years in construction.

The temperature in the middle of the tunnel remains nearly constant, summer and winter, and is about 19° to 20° C. = 66° to 68° Fahr.

' The altitude of the tunnel is 4248 feet above sea-level, and the height of the mountain above the tunnel is 5428 feet ; the temperature of the rock is greatly influenced by this latter fact.

The question of the temperature of the rocks passed through in the construction of a tunnel is one of great interest, as it depends upon several conditions : (1) the character of the rock ;

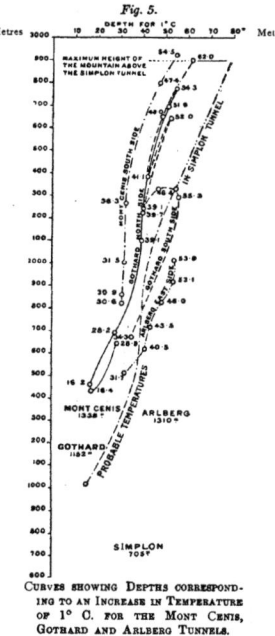

Fig. 5.

CURVES SHOWING DEPTHS CORRESPOND-ING TO AN INCREASE IN TEMPERATURE OF 1° C. FOR THE MONT CENIS, GOTHARD AND ARLBERG TUNNELS.

(With curve of probable temperature for the Simplon Tunnel.)

(2) the inclination of the beds—those which attain a vertical or nearly vertical position being less able to confine the heat than those which are more or less horizontal ; (3) the height of the mountain above the tunnel, or in other words, the thickness of the blanket.

A diagram is shown (Fig. 5), giving the temperature actually encountered in the St. Gothard and Arlberg Tunnels, and from these, aided by the carefully prepared geological section along the centre line of the Simplon Tunnel, an approximate line is given of the temperatures which are expected.

The possibility of cooling the rocks and the air of the tunnel will be dealt with later on, but there is in addition a permanent lowering of the temperature after the tunnel is complete,

particulars of which will be given under the description of the St. Gothard.

For each 144 feet of superincumbent rock or earth the increase is found to be 1° Fahr.

THE ST. GOTHARD TUNNEL.

This, which is at present the longest railway tunnel in the world, is 9·3 miles in length, and constitutes the summit of the "Gothard bahn," that is, the railway which runs from Lucerne to Chiasso near the Italian frontier.

The altitude of the tunnel at its north portal is 3639 feet, and at its south portal 3757 feet above the sea. A gallery of direc-tion was driven throughout, and the gradient of the rails is only such as to provide for efficient drainage, viz. 5·82 per thousand, or about 1 in 172.

The following table may be of interest, giving the result of investigations as to the cooling of the rocks.

TEMPERATURE OF THE ROCK IN THE ST. GOTHARD TUNNEL.

Date.	7·1 kilo. from the North Portal.			7·05 kilo. from the South Portal.		
	Temperature.	Lowering.		Temperature.	Lowering.	
		Successive.	Total.		Successive.	Total.
April and May 1880, the year when the tunnel was pierced	30·46	...	...	30·53	...	...
June 1882	23·73	6·73	...	23·39	7·14	...
July 1883	22·20	1·53	8·26	23·1	0·29	7·43

Above are Centigrade.

Although the works were carried on with energy, and with all the best appliances then known, the time occupied was ten years ; but the most serious feature of the work was the heavy mortality amongst the men : no less than 600 deaths occurred, including those of both the engineer and contractor.

From the experience then gained, great improvements have been introduced into the works of the Simplon, as will be de-scribed later on ; but the heavy loss of life in the St. Gothard was due to insufficient ventilation ; the high temperature ; the ex-posure of the men to the Alpine climate after emerging from the tunnel ; the want of care as to the changing of the men's wet mining clothes ; and the poor character of the food with which the men supplied themselves. All this has been greatly ameliorated, and even in English tunnels certain im-provements have been introduced, which were brought from Switzerland.

The traffic through the tunnel has so largely increased that the question of ventilation became of pressing importance, and the system of Signor Saccardo, the well-known Government in-spector of railways and engineer of Bologna, has been installed, which is an ingenious application of the injector system. One of the first introductions of this method was in the case of the Pracchia Tunnel, on the main line between Florence and Bologna, through the Apennines. This is a railway of single line, and was built many years ago by the late Mr. Brassey. There are 52 tunnels in all, but those on the eastern side are of comparatively little importance. On the western slope the gradient nearly throughout is 25 per thousand (or 1 in 40), and it is here the greatest difficulty exists. There are several tun-nels whose lengths approximate to 1000, 2000 and 3000 yards, and the traffic is both heavy and frequent, the locomotives very powerful, with eight wheels coupled.

Under any conditions of wind the state of the longest tunnel is bad, but when the wind is blowing in at the lower end at the same time that a heavy goods or passenger train is ascending the gradient, a state of affairs is produced which is almost in-supportable, and one might as conveniently travel in a furnace flue.

A heavy train of dining and sleeping carriages, with two engines, conveying one of the crowned heads of Europe and suite, arrived at the exit of Pracchia Tunnel with both enginemen and both firemen insensible ; and in other cases passengers have been seriously affected.

Owing to the height of the mountain no shafts are available ; but Signor Saccardo places a ventilating fan near the mouth of the tunnel, and blows air into it through the annular space which exists between the arch of the tunnel and the gauge of maximum construction (Fig. 6). The results are remarkable ; the volumes of air thrown into the tunnel per minute being as follows :—

	cub. ft.
Direct from the fan	161,860
Induced draught through open tunnel mouth	48,140
Total	210,000

Or 100 cubic metres per second.

The temperature of the tunnel air before the fan was started was 107° F., with 97 per cent. of moisture, whereas, after the fan had been running a few minutes the temperature was 81° F., or a lowering of 26° F., and the tunnel was cool and free from smoke and vapour.

Fig. 6.

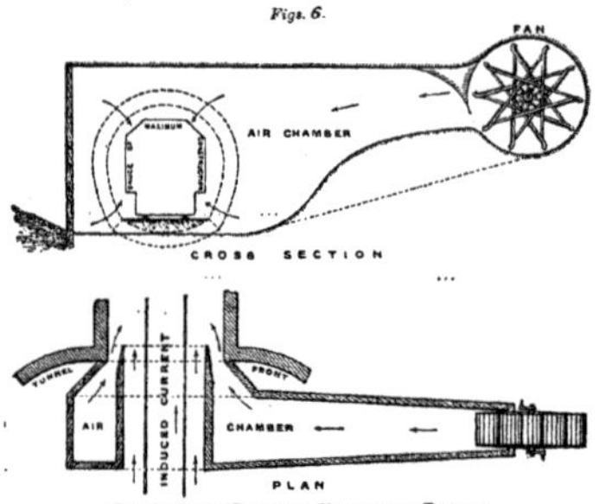

THE SACCARDO SYSTEM OF VENTILATING TUNNELS.

One can travel through with both windows open and feel no inconvenience, the only remark of the brakesman riding on the top of the waggons and carriages being that he finds it almost too cold.

This application is without doubt the solution of the difficult problem of tunnel ventilation under high mountains, and elsewhere where shafts are not available, and where electric traction is not applicable.

This system has within the last twelve months been brought into operation on the St. Gothard, with the most satisfactory results. Careful experiments are being made, but there is no doubt that the problem has been solved.

In addition to these tunnels, the Saccardo system has been applied to the Giovi Tunnel near Genoa—3300 metres in length—and is being installed on the Giovi Tunnel on the Genoa-Ronco Railway, 8303 metres in length, besides on some seven other tunnels in Italy ; and plans are being prepared for the Mont Cenis.

THE SIMPLON TUNNEL.

This tunnel is now in rapid course of construction, the total length of gallery driven up to the end of April being as follows :—

	yards
On the north or Brigue side of the Alps ...	3228
On the south or Iselle	2350

Or over three miles in little more than eighteen months, including the necessarily slow progress at the commencement.

The total distance between the two portals will be 21,564 yards, or 12·26 miles. A gallery of direction has been driven at both ends until the actual tunnels are reached, so as to form

a directly straight line for the accurate alignment of the work from end to end.

This great undertaking will consist of two single-line tunnels running parallel one to the other, at a distance apart from centre to centre of 55 feet 9 inches ; and one of the chief features is the much lower altitude of the rails above sea-level than any of the other Alpine tunnels. This altitude is at its highest point 2314 feet, being 1474 feet lower level than that of the St. Gothard, 1934 feet lower than that of the Mont Cenis, and 1986 feet than that of the Arlberg. This is a matter of great importance in the question of haulage of all the traffic.

The tunnel enters the mountain at the present level of the railway at Brigue, so that no costly approaches are requisite on this side.

Admirable arrangements have been made for the welfare of the men, to avoid the heavy death-rate which occurred on the St. Gothard, and it may be interesting to state what some of these are. For every cubic foot of air sent into the latter tunnel, fifty times as much will be delivered into the Simplon. Special arrangements are made for cooling the air by means of fine jets of water and spray.

The men on emerging from their work, wet through and fatigued, are not allowed to go from the warm headings into the cold Alpine air outside, but pass into a large building which is suitably warmed, and where they change their mining clothes and are provided with hot and cold douche baths. They put on warm dry clothes, and can obtain excellent food at a moderate cost before returning to their homes. Their wet and dirty mining clothes are taken charge of by appointed custodians, who dry and clean them ready for the morrow's work. These and other precautions are expected to reduce the death-rate to a very great extent.

With a view to the rapid advancement of the work, the late M. Brandt, whose death is greatly to be deplored, devised after his long experience on the St. Gothard his now well-known drill. As details of this have been published, and as they would be too technical for this evening's discourse, it will only be necessary to refer to them briefly. This drill is non-percussive, nor is it armed with diamond. It is a rotatory drill 3 inches in diameter with a pressure on the cutting points of 10 tons moving at slow speed, but capable of being acclerated at pleasure, and of being rapidly withdrawn.

The progress of each of the two faces during the month of April last has averaged 17 feet 3½ inches per day, and is a remarkable corroboration of the speed estimated by the engineers four years ago. The estimate was as follows :—

1st year, the daily progress at each face would be ...	8·85 feet
2nd ,, ,, ,, ,, ,, ...	17·22 ,,
3rd ,, ,, ,, ,, ,, ...	19·18 ,,
4th ,, ,, ,, ,, ,, ...	21·32 ,,
5th ,, ,, ,, ,, ,, ...	31·16 ,,

The work is now in its second year, so that the estimated speed is being exceeded. In other words, the tunnel is being driven through granite at a higher speed than is attained in London clay.

It was at one time intended to sink a 20-inch bore-hole from the village of Berisal to the tunnel, a depth of some 2400 feet, for the purpose of delivering water at high-pressure for the works. This may still be done, but the meandering of the tool might result in the awkward dilemma of having to search for it, in solid rock, below ground.

The probable cost of the work now in hand will be about 2,000,000*l.*, and the time occupied in completing the tunnel ready for traffic is estimated to be 5½ years, a penalty or a bonus, as the case may be, for delay or acceleration being fixed at 200*l.* a day.

UNIVERSITY AND EDUCATIONAL INTELLIGENCE.

OXFORD.—The following is the text of the speech with which Prof. J. Mark Baldwin, professor of psychology in Princeton University, was presented for the degree of D.Sc. *honoris causa.* This is the first time the degree has been conferred in Oxford, it having been created only quite recently. The speech was written by Mr. A. C. Clark, of Queen's College, and Prof. Baldwin was presented by Prof. E. B. Elliot, F.R.S., in the absence of Prof. Love, F.R.S., who would in the ordinary course have officiated, being the occupant of the oldest scientific chair in the University :

Adest Jacobus Marcus Baldwin, Academiae de Princeton Graduatus, vir Psychologiae peritissimus. Cujus laudes ut brevissime complectar, primo Psychologiae Professor in Academia de Toronto creatus Psychologiae Experimentalis laboratorium, quod solum adhuc in Academiis Britannicis exstat, instituit, mox ad suam almam matrem reversus in Academia de Princeton Psychologiae Professor factus est. Libri etiam luculentissimi auctor est de ortu et incremento intellegentiae cum in infantibus tum in genere hominum universo, quem summa laude a viris doctis ubique ornatum Academia Havniensis numismate aureo donavit : idem ephemeridis praestantissimae apud Americanos res psychologicas tractantis diu editor, nunc grande Philosophiae et Psychologiae Abecedarium sub prelo nostro excudendum curat. Quo in labore doctissimi cujusque in hoc genere scriptoris opera utitur cum in America tum in Europa, quo in numero Praelector noster Wildianus dux est et fere signifer. Valde, nisi fallor, Academiae nostrae auram redolebit hoc volumen tot Oxoniensium sive sub nostro caelo, sive sub externo degentium, opera diligentia doctrina exquisita ornatum. Quod vero primus Scientiae Doctor in Academia nostra creatus tantam in Psychologia laudem adeptus sit, felicissime profecto accidit, cum adhuc frigere apud nos hujus doctrinae studium externis videatur. Utinam hunc talem virum plurimi ejusdem laureae avidi longo ordine secuturi sint.

Illam vero insignem benevolentiam praetermittere non possum Academiae de Princeton, quae plurimos, qui in hac nostra Academia laude summa floreant ac floruerint, gradibus honoris caus conferendis libentissime auxerit. Cujus liberalitatis non immemor maximo cum gaudio ego hunc virum doctissimum, Academiae suae vivum exemplar, vobis ornandum trado.

The speech delivered by Prof. Love on the occasion of the presentation of Prof. C. F. Chandler, of Columbia College, for the degree of D.Sc. *honoris causa*, was as follows :—

Adest Carolus Fredericus Chandler, chemiae professor apud Americanos, cuius fama extra fines patriae suae iam dudum pervagata est. Hic apud suos litteris humanioribus primo imbutus, dein chemiae deditus doctrinae amplioris appetens ad Germanorum fontes accessit. Ibi doctissimi cuiusque viri, cum in Berolinensi Academia, tum in Gottingensi discipulus, Philosophiae Doctor et Magister Artium apud Gottingenses creatus est. Ex hoc curriculo ad solum suum reversus in Academia de Schenectady primo vicarius erat Professoris optimi, Caroli Joy, dein hoc summo viro ad Academiam de Columbia avocato, ipse Professor factus septem annos de omni chemiae genere, de agris laetificandis, de metallorum ratione, de geologia magnum discipulorum gregem praeclare docebat. Ita laudem insignem adeptus ipse Columbiam accessitus, ubi scholae metallorum novo exemplo instituendae imprimis auctor erat, tres et triginta annos in omni rerum administratione florebat. Per hoc grande mortalis aevi spatium omne genus chemiae felicissime tractavit : idem rude iam donatus a laboribus officiosis nondum recessit sed iuniorum studia adhuc informat. Neque hoc loco silendum arbitror quod huius precibus commoti fratres Havermeyeri, Novi Eboraci cives ornatissimi, aulam pulcherrimam, chemiae sedem, aedificaverunt : Musaei etiam rebus omnibus, quae ad chemiam pertinent, refertissimi ipse auctor est et conditor. Sex et viginti abhinc annos magna chemicorum frequentia ad Doctoris Priestly sepulchrum confluente, ut chemiae inventorem rite salutarent et post centesimum iam annum scientiae suae natalem diem celebrarent, ipse conventus Praeses erat : Societas autem chemicorum Americanorum, quae ex illo coetu orta est, hunc bis Praesidem saepe vice-Praesidem et curatorem habuit. Sodalicii etiam chemicorum, qui Novi Eboraci degunt, Praeses est : hortorum etiam publicorum peritissimus Curator. Idem civitatis suae personam gerens maximo medicorum conventui Havniae interfuit, qui ab omni terrarum orbe missi de valetudine civium conservanda quaererent.

Dies memorare me deficeret si doctissimi viri tot labores enumerare conarer. Illum consulunt populares sui de porcorum fibra unaquaque in quaestum convertenda, de silvis rei navalis causa conservandis, de veneno si quod in vino vel in cervisia delitescat detrahendo, de argentariorum chartulis imperviis aquae et madori reddendis, de oleo e vivis fontibus scaturiente purgando, de plateis igneo vapore nocti illuminandis, de mercibus linteis candore eximio nivem superantibus, de omni re quae ad utilitates, oblectamenta, lucrum denique civium pertinet.

Ut omittam honores quos Academia Gottingensis iuveni dedit,

Doctor Medicinae ab Academia Novi Eboraci, Litterarum etiam Doctor ab Academia sua Columbiensi factus est : neque solum domi clarus est, sed ubicunque terrarum viri docti, chemiae dediti, inveniuntur, hunc sodaliciis et societatibus suis libentissime adsciverunt. Restat ut Academia nostra hoc summo viro in gremium suum accepto suas laudes augeat.

AT a meeting of the Council of the Birmingham University, held on Tuesday, the following letter, received from Lord Calthorpe by Mr. Chamberlain, was read :—"Dear Mr. Chamberlain,—My son and myself beg to offer to the University of Birmingham about twenty-five acres of land on the Bournebrook side of the Edgbaston estate as a site for the new scientific department of teaching and research which it is proposed to establish. There will of necessity be certain conditions, but these will occasion no difficulty.—Yours very truly, CALTHORPE."— It was proposed by the Vice-Chancellor, seconded by Sir James Smith, and unanimously resolved :—"That this Council desires to express to Lord Calthorpe and to the Hon. Walter Calthorpe its high appreciation of their generosity in offering to present twenty-five acres of land to the University of Birmingham as a site for the new scientific department. In gratefully accepting the offer, the Council recognises, not only the value of the gift, but the suitability of the site, which enables it to establish the new department in closer proximity to the centre of the city than would have been possible under any other circumstances."

THE Reports and Prospectuses of Technical Schools, which come under our notice from time to time, show unmistakably that increased provision is being made for practical work in science, and that teachers who have had the advantage of instruction in well-equipped laboratories are in charge of the work. The Municipal Science, Art, and Technical Schools of Plymouth is a case in point. These schools were erected by public subscription as a memorial of the Queen's Jubilee, and on their completion were handed over by the Jubilee Memorial Committee to the town. In the day school department the work is that of the Advanced Section of a "School of Science," that is to say, of a secondary school giving instruction in mathematics, mechanics, physical science, English subjects, French and drawing. There is a laboratory for practice in both chemistry and physics, and for manual instruction in woodwork. Both day and evening classes are held in many science subjects, and pupils whose elementary education is completed may take a two years' course of training in such subjects as will best fit them to become chemists, architects, civil, mechanical, or electrical engineers, or to engage in industrial work of any kind. The increase of institutions of this kind will be the salvation of our national welfare.

FROM the Northampton Institute, one of the youngest of the London Polytechnics, we have received a prospectus of courses in mechanical engineering, electrical engineering, and horological engineering which have just been introduced. The syllabuses of the courses are admirable, and, with the notes upon the objects and character of the work, they indicate that Dr. R. M. Walmsley, the principal, believes in the value of scientific instruction. Students who desire to take up these engineering courses must first show that they are capable of benefiting from it by passing an entrance examination. English and elementary mathematics are obligatory subjects, and we are glad to see that it is not proposed to make the latter a test of ability to perform mathematical gymnastics. The following extract from the prospectus is worth quoting :—"In 'Elementary Mathematics' the examination will aim at ascertaining the candidate's familiarity with simple arithmetical, algebraical and geometrical methods and their application so the solution of ordinary common-sense problems. In arithmetic this will include the use of decimals and abbreviated methods of calculation, with the usual problems of mensuration, including the volumes and surfaces of cylinders, spheres and right cones. In algebra the usual course as far as simple simultaneous equations will be included, but the more academic parts of the subject will not be required. The geometry will include the subjects treated in the first two books of Euclid, with some exercises in the accurate drawing of geometrical figures." The Institute has numerous laboratories and workshops for practical work in mechanics, engineering, metal and woodwork, electrical engineering, physics, electro-chemistry, metallurgy, and instrument making. The attention given to horological theory and mechanism, and horological technology, is a noteworthy characteristic. The courses of work in this as well as the other subjects show that sound instruction in the principles and practice of the chief branches of engineering can be obtained at the Institute.

SCIENTIFIC SERIALS.

· *American Journal of Science*, July.—Energy of kathode rays, by W. G. Cady. This is a translation of a paper already published in the *Annalen der Physik*.—Volcanic rocks from Temisconata Lake, Quebec, by H. E. Gregory. The volcanic rocks consist of fine tuff and coarse amygdaloidal conglomerate or breccia. They are interbedded with Niagara sediments, and this helps to determine the time when widespread volcanic activity gave rise to the numerous small areas of tuffs and lavas in the Maine-Quebec region.—Interpretation of mineral analyses, and a criticism of recent articles on tne constitution of tourmaline, by S. L. Penfield. It is safe to assume that the close approximation of atomic ratio to whole numbers constitutes the strongest argument that can be advanced in support of the excellence of an analysis and to correctness of the derived formula. The author criticises the formulæ proposed by Clarke and Tschermak, and maintains that it is definitely proved that the empirical formula of the tourmaline acid is $H_{26}B_2Si_4O_{21}$.— Carboniferous boulders from India, by B. K. Emerson. The author describes and illustrates some striated carboniferous boulders which remove the doubt as to the former existence of a carboniferous glacial period.—The statement of rock analyses, by H. S. Washington. The author proposes a regular system of stating the results of the chemical analysis of rocks. The oxides are to be enumerated in the following succession : SiO_2, Al_2O_3, Fe_2O_3, FeO, MgO, CaO, Na_2O, K_2O, H_2O, CO_2, and then the rarer oxides, also in definite succession. This will enable the geologist to classify the rocks in a purely chemical system and to pick them out at a glance. They can be advantageously entered upon a card catalogue.—A string alternator, by K. Honda and S. Shimizu. The authors describe a modification of Pupin's string interrupter by means of which a continuous battery current can be converted into an alternating current the frequency of which can be readily varied from 30 to 1000 per second.—Action of light on magnetism, by J. H. Hart. The author failed to obtain the demagnetisation of iron by light acting magnetically like an alternating current, until he adopted the expedient of depositing very fine iron films on glass. He then noticed a small but distinct difference in the magnetic state of the iron according to the plane of polarisation of the incident light.

Bollettino della Società Sismologica Italiana, vol. vi. 1900-1901, N. 1.—Rules and list of fellows (forty-three national and thirteen foreign).—Vesuvian notices (year 1899), by G. Mercalli. A monthly review of the condition of Vesuvius, with sections on the form and state of the crater, the end of the eruptive phase of 1895–1899, the lavic cupola of 1895–1899, the supposed endogenous elevation of the lavic cupola, and the fumaroles of the lavic cupola and fracture.—On the nature of seismic vibrations, by M. P. Rudski (in French). The author contends that superficial, and probably deep-seated, rocks are not isotropic media, and that earthquake waves consist of vibrations which are not entirely longitudinal nor entirely transversal.—Notices of earthquakes recorded in Italy (January 1 to March 14, 1896), by A. Cancani, the most important being the Mexican earthquakes of January 14 and 25, the Greek earthquake of January 22, and distant earthquakes on January 6, 22, and March 7.

SOCIETIES AND ACADEMIES.

LONDON.

· **Royal Society**, May 17.—" The Circulation of the Surface Waters of the North Atlantic Ocean." By H. N. Dickson, B.Sc. Communicated by Sir John Murray, K.C.B., F.R.S.

In this paper an attempt is made to investigate the normal circulation of the surface waters of the Atlantic Ocean north of 40° N. lat., and its changes, by means of a series of synoptic charts showing the distribution of temperature and salinity over the area for each month of the two years 1896 and 1897.

The principal conclusions arrived at with reference to the circulation may be summed up as follows :—

(1) The surface waters along the whole of the eastern seaboard of North America north of (about) lat. 30° N., consisting partly of water brought from the equatorial currents by the Gulf Stream, and partly of water brought down by the Labrador current, are drifted eastward across the Atlantic towards south-western Europe, and banked up against the land outside the continental shelf. This continues all the year round, but it is strongest in summer, when the Atlantic anti-cyclone attains its greatest size and intensity ; and the proportion of Gulf Stream water is greatest at that season.

(2) The drifts in the northern part of the Atlantic area are under the control of the cyclones crossing it. The circulation set up accordingly reaches its maximum intensity in winter, and almost dies out in summer. In winter the drifts tend to the south eastward from the mouth of Davis Strait, eastward in mid-Atlantic, and north-eastward in the eastern region. In spring and autumn the movement is more easterly over the whole distance, and a larger quantity of water from the Labrador stream is therefore carried eastward.

(3) The water banked up in the manner described in (1) escapes partly downwards, partly southwards, and partly northwards. It occupies the whole of the eastern basin of the North Atlantic, and to the north it extends westward to Davis Strait, being confined below 300 fathoms depth by the ridges connecting Europe, the Faeroes, Iceland, and Greenland. Above that level it escapes northward by a strong current through the Faeroe-Shetland Channel and between Faeroe and Iceland, and by the two branches of the Irminger stream, one west of Iceland the other west of Greenland.

(As it seems desirable that this northerly current should have a distinctive name, it might be well to call it the European stream, and its branches the Norwegian, Irminger, and Greenland streams respectively.)

The strength and volume of the European stream is liable to considerable variation, according to the form and position of the Atlantic anti-cyclone, which causes the amount of banked up water, and the proportions escaping northward and southward, to vary. It is also modified by the strength and direction of the surface drifts in its course. It is, however, always strongest in summer.

(4) The Norwegian stream is by far the largest branch of the European, and it traverses the Norwegian Sea and enters the Arctic Ocean. The warm water thus sent northward melts enormous quantities of ice, and the fresh water derived from the ice moves southward in autumn, chiefly in a wide surface current, between Iceland and Jan Mayen, which may entirely cover other parts of the Norwegian stream. Part of the surface water also comes southward through the Denmark Strait, but the amount is much smaller, probably chiefly because the melting of the ice is slower, and the channel is longer blocked.

The Greenland branch of the European current also causes melting of ice in Davis Strait, but the warm winds from the American continent and the water received from the land are probably more effective in increasing the volume of the Labrador current.

(5) The water from the melted ice is spread over the surface of the North Atlantic during late autumn and winter by the increasing drift circulation, and it is gradually absorbed by mixing with the underlying water.

(6) The circulation described is liable to extensive irregular variations, corresponding to variations in the atmospheric circulation.

May 31.—" Influence of the Temperature of Liquid Hydrogen on Bacteria." By Allan Macfadyen, M.D., and Sydney Rowland, M.A. Communicated by Lord Lister, P.R.S.

In a previous communication we have shown that the temperature of liquid air has no appreciable effect upon the vitality of micro-organisms, even when they were exposed to this temperature for one week (about − 190° C.). (*Roy. Soc. Proc.*, February 1 ; *ibid.*, April 5.)

We have now been able to execute preliminary experiments projected in our last paper as to the effect of a temperature as low as that of liquid hydrogen on bacterial life. As the approximate temperature of the air may be taken as 300° absolute, and liquid air as 80° absolute, hydrogen as 21° absolute, the ratio of these temperatures roughly is respectively as 15 : 4 : 1. In other words, then, the temperature of liquid hydrogen is about one-quarter that of liquid air, just as that of liquid air is about one-quarter of that of the average mean temperature. In subjecting bacteria, therefore, to the temperature of liquid hydrogen, we place them under conditions which, in severity of temperature, are as far removed from those of liquid air as are those of liquid air from that of the average summer temperature. By the kindness of Prof. Dewar, the specimens of bacteria were cooled in liquid hydrogen at the Royal Institution. The following organisms were employed : *Bac. acidi lactici*, *B. typhosus*, *B. diphtheriae*, *Proteus vulgaris*, *B. anthracis*, *B. coli communis*, *Staphylococcus pyogenes aureus*, *Spirillum cholerae*, *B. phosphorescens*, *B. pyocyaneus*, a Sarcina and a yeast.

The above organisms in broth culture were sealed in thin glass tubes and introduced directly into liquid hydrogen contained in a vacuum jacketed vessel immersed in liquid air. Under these conditions they were exposed to a temperature of about − 252 C. (21° absolute) for ten hours. At the end of the experiment the tubes were opened, and the contents examined microscopically and by culture. The results were entirely negative as regards any alteration in appearance or in vigour of growth of the micro-organisms. It would appear, therefore, that an exposure of ten hours to a temperature of about − 252° C. has no appreciable effect on the vitality of micro-organisms.

We hope to extend these observations on the influence of the temperature of liquid hydrogen on vital phenomena, and to make them the subject of a future communication, and to discuss their bearing upon problems of vitality.

June 21.—" On the Viscosity of Gases as affected by Temperature." By Lord Rayleigh, F.R.S.

A former paper[1] describes the apparatus by which I examined the influence of temperature upon the viscosity of argon and other gases. I have recently had the opportunity of testing, in the same way, an interesting sample of gas prepared by Prof. Dewar, being the residue, uncondensed by *liquid hydrogen*, from a large quantity collected at the Bath springs. As was to be expected,[2] it consists mainly of helium, as is evidenced by its spectrum when rendered luminous in a vacuum tube. A line, not visible from another helium tube, approximately in the position of D_3 (Neon) is also apparent.[3]

The result of the comparison of viscosities at about 100° C. and at the temperature of the room was to show that the temperature effect was the same as for *hydrogen*.

In the former paper the results were reduced so as to show to what power (n) of the absolute temperature the viscosity was proportional.

	n.	c.
Air	0·754	111·3
Oxygen	0·782	128·2
Hydrogen ⎫	0·681	72·2
Helium ⎭		
Argon	0·815	150·2

Since practically only two points on the temperature curve were examined, the numbers obtained were of course of no avail to determine whether or no any power of the temperature was adequate to represent the complete curve. The question of the dependence of viscosity upon temperature has been studied by Sutherland,[4] on the basis of a theoretical argument which, if not absolutely rigorous, is still entitled to considerable weight. He deduces from a special form of the kinetic theory as the function of temperature to which the viscosity is proportional

$$\frac{\theta^{\frac{3}{2}}}{1 + c/\theta} \quad \dots \dots \dots \dots \dots \dots (1),$$

[1] *Roy. Soc. Proc.*, vol. lxvi. (1900), p. 68.
[2] *Roy. Soc. Proc.*, vol. lix. (1896), p. 207 ; vol. lx. (1896), p. 56.
[3] I speak doubtfully, because to my eye the interval from D_1 to D_3 (helium) appeared about equal to that between D_3 and the line in question, whereas, according to the measurements of Ramsay and Travers (*Roy. Soc. Proc.*, vol. lxiii., 1898, p. 438), the wave-lengths are :
D_1 ... ... 5895·0
D_2 ... ... 5889·0
D_3 ... ... 5875·9
D_5 ... ... 5849·6
so that the above-mentioned intervals would be as 19·1 : 26·3 [*June 23.—* Subsequent observations with the aid of a scale showed that the intervals above spoken of were as 20 : 21. According to this the wave-length of the line seen, and supposed to correspond to D_5, would be about 5855 on Rowland's scale, where $D_1 = 5896·0$, $D_3 = 5890·0$, $D_5 = 5876·0$.] I may record that the refractivity of the gas now under discussion is 0·132 relatively to air.
[4] *Phil. Mag.*, vol. xxxvi. (1893), p. 507.

c being some constant proper to the particular gas. The simple law $\theta^{\frac{1}{2}}$, appropriate to " hard spheres," here appears as the limiting form when θ is very great. In this case, the collisions are sensibly uninfluenced by the molecular forces which may act at distances exceeding that of impact. When, on the other hand, the temperature and the molecular velocities are lower, the mutual attraction of molecules which pass near one another increase the number of collisions, much as if the diameter of the spheres was increased. Sutherland finds a very good agreement between his formula (1) and the observations of Holman and others upon various gases.

If the law be assumed, my observations suffice to determine the values of c. They are shown in the table, and they agree well with the numbers for air and oxygen calculated by Sutherland from observations of Obermayer.

" Underground Temperature at Oxford in the Year 1899, as determined by Five Platinum Resistance Thermometers." By Arthur A. Rambaut. M.A., D.Sc., Radcliffe Observer. Communicated by E. H. Griffiths, F.R.S.

Royal Microscopical Society, June 20.—Mr. Carruthers, F.R.S., President, in the chair.—Mr. G. H. J. Rogers exhibited a modification of the Rousselet compressor, in which two thin indiarubber bands, sunk into grooves, were employed to keep the cover-glass in position, instead of having it cemented, the advantage claimed for this modification being the facility with which a broken cover-glass can be replaced.—Mr. Chas. Baker exhibited an achromatic substage condenser which was a modification of Zeiss's model of the Abbe condenser, the N.A. being 1·0, aplanatic cone 90°, lenses $\frac{7}{8}$-inch diameter, working distance $\frac{1}{18}$-inch. With the front lens removed the condenser is suitable for use with low-power objectives.—A short paper by Mr. E. B. Stringer, on a new projection eye-piece and an improved polarising eye-piece, was taken as read.—A paper by Miss Loraine Smith, on some new microscopic fungi, was also taken as read, the President giving a short *résumé* of it and expressing his opinion that the paper would be an important addition to our knowledge of microscopic fungi. Mr. Bennett said there was one special point with regards to parasitic fungi which might prove to be of considerable practical importance—he referred to the cultivation of fungus parasites on certain insects It had been proposed to do this on the Continent and in Australia and America, with a view of getting rid of insect pests—locusts and others ; and if efforts in this direction were successful they might be the means of producing very beneficial and economic results.—The President then read a paper, and gave a lantern demonstration, on the structure of some palæozoic plants. He said the intelligent study of palæozoic plants was not yet a century old, for although their presence had long been noticed, they appear to have been regarded simply as freaks of nature. The importance of fossils was first recognised by Wm. Smith, who observed that strata could be identified by the organised fossils found in them. He published this important fact in 1816, and thus laid the basis of stratigraphical geology. The majority of fossil plants are found in the shales, and although the tissues had been converted into carbon, the form and venation of the leaves and occasionally the aspect of the fruits had been preserved. The most important information, however, had been obtained from specimens in which the tissues had been replaced by minerals dissolved in the strata enclosing them. He had arranged for the lantern sections of plants from the carboniferous system, but before exhibiting them he wished to point out to what group of plants they belonged. The cellular plants, with few exceptions, had been lost. Sir Wm. Dawson found specimens of a remarkable stem in the lower Devonian rocks of Canada, to which he gave the name of *Protaxites*. From a microscopic study of specimens he, the President, was led to publish a paper in the Society's *Journal* in 1872, in which he demonstrated that the stem was that of a cellular plant belonging to the Algæ, a view which was ultimately accepted by Sir Wm. Dawson. Fungal remains had been detected by Alder, and also by himself. The plants which had been certainly determined were vascular plants belonging to the Equisetaceæ, Filices, and Selaginellaceæ, among Cryptogams, and to the Coniferæ, groups which existed in the present flora of the globe, and were represented in the indigenous flora of Britain. The President proceeded to describe the principal characteristics of the fossil and existing forms of the four orders of plants referred to. In illustration of his remarks a number of preparations were shown on the screen.—Mr. Bennett wished to say a few words to elicit an opinion on a matter of great interest.

He referred to the recent discovery of the mode of impregnation in some of the Cycadeæ by means of active spermatozoids, as in the case of vascular cryptogams. This seemed to suggest the question whether the gymnosperms were not more closely allied to the vascular cryptogams than was usually recognised. Did the evidence of palæontology favour the view that there was a closer affinity to the vascular cryptogams than to the higher section of flowering plants, the angiosperms?—The President said this question deserved careful consideration, but it should be remembered that in these strata they only saw four groups of plants, and that the coniferæ were found alongside the others, and were evidently living at the same period. Brogniart had shown the presence of pollen grains in the apical cavities of fruits which had been preserved in silex. It was not known how these spermatozoids were developed in Salisburia, but if they rendered pollen grains unnecessary, the presence of the pollen in these extinct fruits would be against the idea of including the gymnosperms with the cryptogams.—The President announced that the rooms of the Society would be closed from August 17 to September 17, and that the next ordinary meeting would be held on October 17.

PARIS.

Academy of Sciences, July 9.—M. Maurice Lévy in the chair.—Problem of permanent heating of a sphere by radiation, reduced to the simpler problem of heating the same sphere by contact, by M. J. Boussinesq.—Combustible gases of the atmosphere ; sea air. Existence of free hydrogen in the atmosphere, by M. Armand Gautier. By way of completing his previous researches on the impurities in the air of towns, woods and mountains, the author now gives the results of experiments on sea air. In these experiments no carbon compounds could be detected in 100 litres of air, the amount present, if any, being less than 0·03 mgr. per 100 litres ; hydrogen was still found, however, to the extent of 1·21 mgr. for the same volume. The amount of free hydrogen thus proved to be present in the air, .2 parts by volume in 10,000, is thus about two-thirds of the carbon dioxide normally present.—On two loci relating to the densities of the liquid and saturated vapour of carbonic acid, by M. E. H. Amagat. In remarking on a recent paper by MM. Cailetet and Mathias, the author points out that the conclusions drawn, although opposed to his own, are obtained from the same set of experimental data. The law of rectilineal diameters, although extremely useful when applied within certain limits, is not a mathematical but an empirical law, derived from experiment, and hence its use as rigidly true seems hardly justified.—The chemical constitution of steel ; influence of tempering upon the state of combination of elements other than carbon, by MM. Carnot and Goutal. In manganiferous steels, the state of combination of the sulphur is not altered by tempering, and phosphorus behaves similarly. In steels containing arsenic, the latter is free if the cooling has been slow, tempered steels containing an arsenide of iron, probably Fe₃As.—M. Czerny was elected a Correspondent for the Section of Medicine and Surgery.—On the equations of motion of a wire in any co-ordinates whatever, by M. G. Floquet.—On certain linear partial differential equations of the second order, by M. C. Guichard.—On the instability of certain substitutions, by M. Levi-Civita.—Demonstration of the rotation of the earth by Foucault's experiment with a pendulum 1 metre long, by M. Alphonse Berget. The sensibility of the reading apparatus is increased by viewing the pointer carried by the pendulum in the field of a microscope furnished with cross-wires in the eyepiece. With a pendulum only 1 metre long, the deviation can be clearly observed after four seconds.—On the liquefaction of gaseous mixtures, by M. F. Caubet. The results of experiments upon mixtures of methyl chloride and sulphur dioxide are here given in the form of curves. Two of the curves have the form predicted from theoretical considerations by Gibbs and Konovalow ; these experiments being the first to show these points.—On a new type of mercury pump, allowing of a good vacuum being attained rapidly, by MM. Berlemont and Jouard. A drawing of the pump is given, unaccompanied by any explanation. It is claimed for this pattern that it will work with 12 lbs. of mercury, gives a high vacuum automatically, is not easily broken, and contains neither taps nor rubber connections.—On an ammoniacal chromous sulphate, by M. Ch. Laurent. The salt described, which has the composition Cr(NH₄)₂.(SO₄)₂+6H₂O is analogous to the double sulphate of

iron and ammonia.—On the preparation of gentopicrine and glucoside from fresh gentian root, by MM. Em. Bourquelot and H. Hérissey. The fresh roots are treated with boiling alcohol as soon as possible after picking, in order to prevent the action of the oxydases of the plant upon the glucoside, 22 kilograms of root giving 250 grams of crystallised gentiopicrine.—Experimental parthenogenetic segmentation in Amphibia and Fish, by M. E. Bataillon. The chemical composition of the medium; in which the eggs are placed, has only a secondary effect. The serum of mammals, whether diphtheritic or not, behaves like an isotonic solution of a salt or sugar ; it acts by its osmotic pressure.—Lœb's theory of the chemical fertilisation of the egg, by M. Viguer.—On the cytology of the Hymenomycetes, by M. René Maire.—The experimental origin of a new vegetable species, by M. Hugo de Vries.—Influence of experimental modifications of the organism upon the metabolism of sugar, by MM. A. Charrin and A. Guillemonat.—A new method of measuring stereognostic tactile sensibility, by MM. Ed. Toulouse and N. Vaschide.—Some new facts concerning the subterranean river at Padirac (Lot), by M. E. A. Martel. The work done in 1899 and 1900 has rendered accessible another 400 metres of this underground river.—Combination of the effects of synodic and tropical revolutions of the moon, by M. A. Poincaré.

NEW SOUTH WALES.

Royal Society, May 2.—W. M. Hamlet, President, in the chair.—Annual general meeting. Mr. W. M. Hamlet read an address upon the development of chemistry. Four words of Arabic or Egyptian source were taken as the text or frame work of this address, namely, alchemy, alkali, alkaloid and alcohol. Under the first came a brief review of the most prominent alchemists. The second afforded scope for the derivation of the word denoting the volatile alkali—the alkaline air—ammonia. In the case of the term alkaloid the researches of Fischer were referred to as showing the constitution of such alkaloids as theobromine and caffeine from structural formulæ of uric acid. Under the generic term, alcohol, the fermentation of other substances than those in use for the production of spirits of wine were dealt with. A vote of thanks was passed to the retiring president, and Prof. Liversidge, F.R.S., was installed as President for the ensuing year.

CONTENTS.

THURSDAY, JULY 26, 1900.

TRADE IN ANCIENT ASSYRIA.

Babylonians and Assyrians : Life and Customs. By the Rev. A. H. Sayce. The Semitic Series. Vol. vi. Pp. x + 273. (London : John C. Nimmo, 1900.)

THIS little book belongs to a projected series of volumes which we gather are intended to deal with the Babylonians and Assyrians and other allied Semitic races, "the object of the series" being, according to the prospectus, "to state its results in popularly scientific form." The volume assigned to Prof. Sayce, which is the first of the series to make its appearance, describes the life and customs of the Babylonians and Assyrians, a subject which offers many points of interest to the general reader. Moreover, within recent years much new material has been published which has thrown considerable light on the social condition of the Babylonians and Assyrians during both the earlier and the later periods of their history. Thousands of clay tablets, which were unearthed at Telloh in Southern Babylonia and have found their way into the museums of Europe, contain temple-records, lists and inventories, receipts and tablets of accounts, and furnish a glimpse of the daily life of the early inhabitants of Babylonia at about 2500 B.C. The letters and commercial documents of the period of the First Dynasty of Babylon enable us to form a still more intimate acquaintance with the life of the Babylonians under some of the earliest of their Semitic kings ; while the systematic publication of the legal and epistolary literature in the great collection of tablets from Kouyunjik has increased our knowledge of the social condition of Mesopotamia under the later Assyrian kings. Finally, the large collections of tablets of the Neo-Babylonian and Persian periods, which are now available for study, make it possible to trace the development of laws and customs down to the latest periods of Babylonian history. There is, therefore, no lack of material on which to base a sketch of the manners and customs of the Babylonians and Assyrians.

Prof. Sayce has written many popular books on this subject, and he is well qualified for the task he has undertaken ; but we cannot help wishing that in one important point he had modified the plan on which he has compiled his volume. The book deals largely with small details, containing descriptions of sales of houses and lands and property, deeds of partnership, marriage contracts, receipts, records of loans, and numerous other commercial and legal transactions drawn from the thousands of "contract-tablets" which have been published in various monographs and in the transactions and journals of different societies. Any book describing the life and customs of the Babylonians must necessarily draw upon this large and scattered literature in order to illustrate the general conclusions which the writer formulates. Prof. Sayce has made abundant use of this material, quoting and describing tablets freely ; on p. 16, for instance, in one short paragraph, he refers to no less than nine separate documents of different dates. But

from the first page to the last he has not given a single reference to the works in which the various tablets have been published, or any indication by which they might be identified ; in fact, not including Biblical quotations, we have only found two references in the book (on pp. 1 and 66), and these are to the opinions of modern writers and not to original authorities. This is perhaps not Prof. Sayce's fault, but a defect in the general plan of the series, for it is possible that the editor has ruled that references to authorities are incompatible with "a popularly scientific" treatment. But, whoever may be responsible, this defect detracts largely from the value of the book. No doubt the expert knows already where to look for the original texts quoted, and can control the various statements for himself ; but the series is not meant for the expert. In the editorial preface, we are expressly told that it is intended to "be serviceable to students in colleges, universities, and theological seminaries, to the clergy, and to intelligent lay readers." The object of the work is, therefore, essentially educational ; but without references it can prove but a poor guide or introduction to the study of the subject of which it treats.

Although doubtless hampered by this deficiency in the general plan of the work, Prof. Sayce has produced a very readable, though perhaps a rather rambling, little book. He has written attractively on the general character of Babylonia and its inhabitants, the constitution of the family, the system of education, commercial and social life, laws and government, letter-writing and religion. To treat all these subjects fully in some two hundred and sixty octavo pages is, of course, impossible ; but the author has touched his subjects lightly, and some of his chapters contain valuable summaries, as, for example, that in which he describes the legal condition of women in Babylonia. The plan of writing vaguely without reference to authorities, however, is not conducive to strict accuracy, and we occasionally meet with a rather misleading generalisation. The statement on p. 102, for instance, that

"no deed was valid without the seal or mark of the contracting parties"

is not borne out by the facts, for many deeds of different periods are extant which bear neither seal-impression nor nail-mark. On p. 61 we are told that

"the year was divided into twelve months of thirty days each, an intercalary month being inserted from time to time . . ."

Even for the Assyrian period this statement is probably not accurate, and it takes no account of the changes which the calendar underwent in the long course of Babylonian and Assyrian history. The arrangement of the calendar and the method of harmonising the lunar and solar years are not yet accurately known in many of their details and are still subjects of controversy, but the student would not gather this from Prof. Sayce's statement. The evidence for cremation (pp. 62 ff.) among the Babylonians and Assyrians is far from being conclusive, and many scholars hold that it was not practised in Mesopotamia before the Parthian period. The statement

O

on p. 51 that the writing of the scribes was sometimes so minute that magnifying glasses were used for reading by the Assyrians, and that short sight "must have been common in the Babylonian schools," is, to say the least, rather fanciful, the only evidence for the statement being a circular crystal object found by Layard at Nineveh, and thought to be a lens, but the use of which is unknown. That there was ever "a monotheistic school" at Erech (p. 262) would, we think, be difficult to prove, and the evidence for "human. sacrifice" referred to on p. 103 should surely have been given. It is, no doubt, a consequence of the omission of references that we sometimes come across repetitions in the book, as, for instance, the quotation referring to the Chaldeans and their ships in connection with Eridu on pp. 9 and 183 ; the suggested identification of Sar-ilu with Israel on pp. 17 and 191, the description of the letter referring to a ferry-boat on pp. 186 and 215, &c. Misprints, too, are not uncommon, as, for example, "the eighteen-hundredths part " (p. 114), "I will *lie* up five shekels of silver" (p. 225), "Emu-talum" for "Émutbalum." (p. 211), "weight" for "night" (p. 266), "bears" apparently for "beasts" (p. 52), "cunei-plain " apparently for "plain" (p. 211), and on p. 157 we are told that "Aramaic became the *lingua panca* . . . in the commercial world." Prof. Sayce is probably not to blame for such misprints, for the American editor was doubtless responsible for the correction of the proof-sheets.

ELECTRICAL ORGANS. MUSCLE OR NERVE ?

Beiträge zur Physiologie des elektrischen Organes der Zitterrochen (Torpedo). By Siegfried Garten. Pp. 116, 4 plates. (Leipzig : Teubner, 1899.)

ALTHOUGH electrical fishes have been the object of scientific curiosity and investigation for nearly 300 years, it is only in the last half of this century that physiologists have realised the great importance, for general physiological problems, of the phenomena presented by these remarkable animals. The discovery and investigation of the electrical phenomena accompanying excitation or activity of all the excitable tissues in the animal body have rendered it of supreme importance to attack the problem and the causation of these electrical changes in the organ, where the "electrical function," so to speak, attains its highest degree of development. It seems probable that electrical organs may be developed by the transformation of many different kinds of tissue. In the greater number of these fishes, however, including that which is the subject of the memoir under consideration (Torpedo), the organ is formed by a transformation of embryonic muscle-fibres, accompanied by a disappearance of the cross-striated contractile material, with a great hypertrophy of the nerve-endings. The electrical discharge of the organ, with an E.M.F. probably amounting to 100 to 200 volts (Gotch) and lasting about 6/1000 of a second, may be excited reflexly or by excitation of the nerve to the organ, or, using strong shocks, by stimulation of the organ itself. The direction of the current in the fish is from ventral to dorsal surface. The electrical organ in

the torpedo consists. of. an array of columns, each column being composed of about 400 transverse discs representing the electromotive elements of the organ. On the ventral surface of each of these discs we find the complicated terminal arborisation or network of a nerve-fibre, embedded in granular protoplasm, and separated from the disc by the remains of the primitive muscle-fibre from which the organ was developed.

We must assume that it is in consequence of these structural arrangements that the excitatory electrical change in the whole organ, instead of passing from one end to the other as a wave, and so giving rise to a diphasic variation of small extent, causes merely a change in one direction, which is summated in proportion to the number of discs in the pile, so producing a monophasic variation of considerable E.M.F. It is evident that we could conceive of each disc as consisting of an inferior part, which is excitable and therefore capable of the chemical changes associated with excitation, and of a superior part, structurally and chemically continuous with the inferior, but incapable of excitation. The question at once arises whether these two parts are represented by nerve and muscle, or whether the chief excitatory change takes place in some of the structures derived from the embryonic muscle-fibre. Is the electrical change an action current of nerve-ending or of muscle ?

Du Bois Reymond, for theoretical reasons, supported the latter view, and at the same time laid great stress on a remarkable property of the organ. He found that the electrical conductivity of the organ was greater for homodromous currents, *i.e.* those in the direction of the discharge of the organ, than for heterodromous. It was shown later by Gotch that this irreciprocal conductivity is only apparent, Du Bois Reymond's results being due to the fact that, in measuring the current passing through the organ, he was measuring the algebraic sum of the battery current and the current excited in the organ itself. Naturally, therefore, the homodromous current was greater than the heterodromous.

Gotch has also drawn attention to the fact that in the electrical organ we have an opportunity of deciding the nature of the demarcation-current consequent on injury. Since in this organ the demarcation-current is always in the same direction as the excitatory-current, whatever may be the position of the injury, he concludes that the demarcation-current or current of rest is really in all cases an action-current due to the continued stimulation at the seat of injury.

On these three questions, but especially with regard to the nervous or muscular nature of the excitable tissue, additional evidence is furnished by Garten, whose research is devoted chiefly to the elucidation of three points —the effect of nerve-section and subsequent degeneration on the direct excitability of the electrical organ ; the effect of drugs, such as curare, which are direct poisons for nerve-endings ; and the action of veratrin as a specific muscular poison.

The results of these experiments are a strong confirmation of the views put forward by Gotch. During the first eight days after section of the nerves, the organs can be excited either directly or indirectly ; from the eighth to the eighteenth day after section, the organ can be excited to discharge, but the shocks are weaker than normal. After the nineteenth day, however, no response is obtained to stimulation either of the nerve or of the electrical organ itself. Thus the irritability of the organ disappears with that of the nerve, whereas a muscle is excitable long after the degeneration of its motor nerve with end-plates is complete. It may be mentioned that the organ-current, or current of rest, and the irreciprocal conductivity diminish during the period of lowered irritability of the organ, and are absolutely abolished after the nineteenth day, thus pointing to the excitatory nature of both these sets of phenomena.

The experiments on the action of curare are less satisfactory, owing to the enormous doses (1 grm. for a fish of 1200 grms.) which are necessary to paralyse the indirect excitability of the electrical organs. Since in these large doses the curare excites central discharges, it is necessary to cut all the electrical nerves to prevent paralysis by fatigue. In this case, however, with a sufficient dose of curare, direct and indirect excitability are abolished simultaneously. The same interdependence of direct and indirect excitability is observed in paralysis by fatigue, in marked contrast to the behaviour of voluntary muscle, where a muscle on direct stimulation gives a practically normal contraction after complete fatigue by stimulation of its nerve.

Veratrin, which causes a marked prolongation of the excitatory change in skeletal muscle, was found by Garten to produce a somewhat similar effect on the electrical discharge of the torpedo. This drug, however, diminishes and very rapidly abolishes the direct and indirect excitability of the electrical organ, and no proof is afforded that the prolonged response may not be due to the state of artificial "fatigue" produced by the drug, or that it is in any way specific. Waller's experiments have shown that veratrin has practically no action on the nerve, and although Garten quotes certain of his own experiments which appear to indicate an action of this drug on non-medullated nerve, the strength of the solution employed must be regarded as too great for the demonstration of the specific action of the drug.

We cannot, therefore, regard the experiments with veratrin as detracting in any way from the support afforded to the nerve-ending theory of excitation by the results of nerve-section. It is remarkable that a change which, as the current of action in nerve, needs all the appliances of a well-fitted laboratory for its demonstration, should, by a mere subdivision of the fibrils and their enclosure in compartments separated by non-excitable partitions, be able to produce the strong shocks of high intensity which characterise the discharge of the whole organ. No better demonstration could be afforded of the futility of those hypotheses which would explain the passage of the nerve-impulse as a mere propagated polarisation, and would deny any energy-producing changes in the axis-cylinder itself. The absence of fatigue in medullated nerve does not imply absence of chemical change, but merely equivalence of disintegration and reintegration, an equivalence probably connected, as Waller has suggested, with the presence of a medullary sheath.　　　　　　　E. H. S.

FLUORINE.

Le Fluor et ses Composés. Par M. Henri Moissan. Pp. xii + 396. (Paris: G. Steinheil, 1900).

THERE could scarcely be a greater contrast than that between the gaseous substances most recently added to our list of elements; fluorine on the one hand, argon and its companions on the other. The existence of the hypothetical element fluorine was postulated in many well-investigated compounds as early as the beginning of the present century; yet, on account of its intense chemical activity, fluorine was not prepared as a free element until 1886, despite the numberless attempts which had been made to isolate it in the intervening period. Argon, on the other hand, owing to its absolute inertness, and to the fact of its occurrence along with the very inert nitrogen, led an unsuspected existence until 1894, although it was contained in enormous quantities in the atmosphere—a constant subject of investigation. The compounds of fluorine, then, were known long before the element itself,—compounds of argon are still wanting. Indeed, as has been pointed out (Sedgwick, "Argon and Newton," p. 2), the name element in the ordinary sense cannot properly be applied to argon and its companions at all, since that term implies the existence, or at least the possibility of existence, of compounds concerning which we are still in total ignorance. As yet there is no chemistry of argon.

The time, however, has now arrived when fluorine and its compounds can be brought under review so as to give a picture of the element in itself and in its combinations, which in main outlines, at least, may be looked upon as final. Prof. Moissan was obviously the man to execute this task; he has fortunately undertaken it, and the book before us gives the result of his labours. As evidence of the extent of the author's research, we may adduce the bibliography given as appendix, which occupies eighty-five pages, and contains references to about six hundred books and papers. These references are arranged alphabetically according to authors, and also chronologically, beginning with Agricola, 1558, and ending with the year 1899.

In the book itself the author's investigations have naturally the first place, and one of the chief points of interest is that M. Moissan not only gives us an account of his apparatus, experiments and results, but also of the leading thoughts which guided him from one experiment to another, until the culmination was reached in the liberation of the gaseous element. The student beginning research could not find a more stimulating record of failure and eventual success than that afforded in Chapter i., on the isolation of fluorine. Chapter ii. deals with the most recent apparatus for the production of fluorine by electrolysis. At a temperature of about −80°, attained by the evaporation of a mixture of solid carbonic acid and acetone, it is possible to use an electrolytic vessel of copper instead of platinum, provided that the hydrofluoric acid employed is free from water. This substitution brings elementary fluorine within the scope of any well-equipped chemical laboratory. Chapter iii. deals with the physical properties of fluorine, its liquefaction, and the action of the liquid on various substances. In

Chapter iv. we have a systematic account of the action of fluorine on the non-metallic elements and on some of their compounds, together with a somewhat detailed study of the non-metallic fluorides. The action of fluorine on the metals and their compounds forms the subject-matter of Chapter v., the organic fluorine compounds receiving treatment in Chapter vi. The last chapter in the book deals with the atomic weight of fluorine, the volumetric composition of hydrofluoric acid, the action of fluorine and hydrofluoric acid on glass, and the position of fluorine in the system of the elements. The author definitely places fluorine at the head of the halogen family, sufficient stress, however, being laid on the points in which fluorine resembles the elements of the oxygen family ; such as the ease with which it unites with carbon, and the analogies exhibited by hydrofluoric acid to some dibasic acids. A short summary of the properties of fluorine concludes the volume, and for frontispiece there is an excellent portrait of the author.

The book is as interesting as a monograph can well be, and M. Moissan has earned the gratitude of all chemists by thus placing before them a connected record of one branch of his splendid activity.   J. W.

OUR BOOK SHELF.

A Text-Book of Physical Chemistry. By Dr. R. A. Lehfeldt. Pp. xii+308. (London: Edward Arnold, 1899.)

A FEW years ago the teacher of physical chemistry seeking a suitable elementary text-book, dealing with the more recent developments of the subject, which he could put into the hands of a class of students approaching the study of physical chemistry for the first time, was somewhat embarrassed to find one. This state of things is now changed for the better by the recent appearance of several very excellent works ; among these Dr. Lehfeldt's book will take a high place. The author explains in his preface that the book "is intended to contain what a student—with limited time and many subjects to learn—may usefully read. It is by no means written to suit any examination, but still is written with the practical requirements of students in view."

Dr. Lehfeldt has succeeded in avoiding the unessential and in explaining the fundamental ideas of modern physical chemistry in a thoroughly lucid manner, so that a student who has grasped the contents of this book will experience little difficulty in appreciating the meaning of the larger handbooks or original memoirs.

An introductory chapter on physical units will be useful to chemical students, who are, perhaps, apt to be slipshod in such matters. This is followed by a chapter on molecular weights in gases and solutions, which includes electrolytes and the ionic theory, and by a very well-considered chapter on the connection between physical properties and chemical constitution. The principles of thermodynamics are then explained ; and the two laws (a) of chemical equilibrium in a system of perfect gases at constant temperature, and (b) of the influence of temperature on chemical equilibrium are deduced from them. This chapter presupposes some knowledge of the elements of the calculus, but any student who wishes to understand physical chemistry must make up his mind to acquire the small amount of mathematical knowledge requisite.

The applications of the two thermodynamic theorems

to chemical change and equilibrium in homogeneous and heterogeneous systems are then taken up. This treatment has the great advantage that the whole of the phenomena can be grouped in a very simple way, the close relationship of chemical and physical change is clearly brought out, and the student is not bewildered by the apparent multiplicity of the phenomena. The book concludes with a brief but most interesting chapter on the theory of the galvanic cell, and the connection between electromotive force and chemical affinity. The book may be unhesitatingly recommended as one of the best of its kind.

The only misprint we have noticed occurs on p. 141, line 18, where "increases" is written in place of "decreases."   T. E.

An Introduction to Analytical Chemistry. By G. G. Henderson, D.Sc., M.A., and M. A. Parker, B.Sc. Pp. 228. (London : Blackie and Son, Ltd., 1899.)

THIS is a compact work covering the ground of ordinary qualitative analysis as well as the tests for a number of organic substances, and also containing an account of the most important processes of quantitative analysis.

Without being designed on any new plan or being explanatory to the fullest extent, the book is written in a scientific spirit. The authors state that they have made free use of the works of Dittmar and others, and it is perhaps not uncomplimentary to remark that the influence of that sterling chemist is apparent in the book.

The directions for work are clear and practical, and the analytical methods quite satisfactory. Perhaps the least useful part of the book is that dealing with organic substances and their separation from mixtures. This branch of analytical art is very difficult, and the particular form of it, which has been encouraged by certain examining bodies, has brought disaster to many a good student. It is difficult to understand what useful purpose is served by the efforts of second-year students to prepare for recognising the constituents of, say, a mixture of urea and an inorganic salt. It is of no importance to medical men, it does not help the teaching of organic chemistry, and it crowds out practical work which would be of real value. The examination of such mixtures is a matter for an analyst of mature knowledge and experience.   A. S.

Maryland Weather Service. Vol. i. Pp. 566. (Balti more : The Johns Hopkins Press, 1899.)

THE Maryland State Weather Service was established in 1892, and its reports and climatic charts are favourably known to meteorologists. In 1896 a plan of closer co-operation between the National and State Weather Bureaus was proposed by Prof. W. L. Moore and adopted. This marked the commencement of a new and very important period in the history of the Service, and the present volume is the first published since the two organisations have been in close connection. The energies of the Service are now to be devoted chiefly to the publication of special reports on the climatology of the State, and if the volume before us is to be taken as an earnest of future ones, we may be pardoned a feeling of envy at the sumptuous way in which scientific work of this kind is presented to the public in America. We notice that it is proposed to publish in the near future a full account of the climatic features of Maryland, in which the physiography, meteorology, hydrography, medical climatology, agricultural soils, forestry, crop conditions and the fauna and flora of the State will be considered. The present volume is confined to the physiography and meteorology, and includes an introduction by Prof.

onogeneous and
up. This treat-
it whole of the
simple way, the
rical change is
ot bewildered by
 rena. The book
sting chapter on
connection be-
l affinity. The
led as one of the

cears on p. 141,
en in place of
T. E.

ry. By G. G.
Parker, B.Sc.
Ltd., 1899.)

he ground of
the tests for a
containing an
of quantitative

plan or being
: is written in
at they have
id others, and
sack that the
urrent in the

practical, and
Perhaps the
; with organic
ixtures. This
and the pur-
ed by certain
many a good
at useful pur-
ar students to
say, a mixture
o importance
ng of organic
which would
s mixtures is
idge and ex-
A. S.

566. (Balti

stablished in
e favourably
of closer co-
me Weather
adopted
and very in-
ce, and the
ice the two
ction. The
ed chiefly to
imatology of
he taken as
ed a feeling
tific work of
rvica. We
sur future a
id, in which
ry, medical
conditions
considered.
ysiographic
xt by Prof.

'Bullock Clark, on the establishment and organisation of the Maryland Weather Service; a description of the physiography of Maryland, by Dr. Cleveland Abbe, jun.; a' report on the meteorology of Maryland, by Dr. Abbe, Mr. F. J. Walz and Dr. O. L. Fassig; and a contribution on the aims and methods of meteorology, by Prof. Cleveland Abbe, already noticed in these columns (vol. lxi. p. 448). The illustrations are numerous, instructive, and of a very high class, most of them being full-page collotype plates or lithographs. No State or country has given to the scientific world a volume in which the operations of the "Weather Service" are interpreted more liberally, or the work presented in a more elaborate format.

Volta e la Pila. By Prof. Augusto Righi. Pp. 40. (Milan: Tip. Bernardoni di C. Rebeschini, 1900.)

THIS is an inaugural discourse delivered by Prof. Righi on September 18, 1899, at the National Electrical Congress at Como. It deals with (1) the science of electricity prior to Volta; (2) the scientific work of Volta considered apart from his discovery of the pile; (3) Galvani's discovery of electricity of contact; (4) the pile; (5) the theory of the pile; and (6) conclusions. In an appendix, Prof. Righi gives a note on the theories of the pile, in which he expresses favourable opinions on the "osmotic" theor .

ɩ　　　y

LETTERS TO THE EDITOR.

[The Editor does not hold himself responsible for opinions expressed by his correspondents. Neither can he undertake to return, or to correspond with the writers of, rejected manuscripts intended for this or any other part of NATURE. No notice is taken of anonymous communications.]

An Optical Phenomenon.

IN connection with Prof. Simon Newcomb's letter on "Terrestrial Gegenschein" (NATURE, October 5, 1899) and the subsequent letters of Mr. Mallock (NATURE, October 12 and November 9, 1899), I desire to call attention to an analogous and very beautiful phenomenon of perspective which I should have mentioned at the time but that the winter season of the year is not favourable to its observation in this country.

When the sun is high and shining brightly in a clear sky, let an observer stand so that the shadow of his head falls on the surface of water that is deep, clear, but not quite clear, and slightly agitated by the wind. He will observe that from the place where the shadow of his head falls shafts of light seem to radiate in all directions. When once well observed, the phenomenon is very striking, but it has surprised me to find how few persons have noticed it. I first observed it many years ago, when I used daily, about mid-day in summer, to cross the bridge over the channel leading to the boat store in the Portsmouth Dockyard, near the main entrance. But it was not till a year later, on Ulleswater, that I found the explanation. The lake was then turbid in parts from the washings of mines, but quite clear in others. Standing up in a boat, one could see the phenomenon very clearly where there was very slight turbidity, but not if the water was quite clear, nor if there was much turbidity, and never in a dead calm. This gave the explanation. The convexities of the surface, when there is a slight agitation, acting as lenses, split up the otherwise uniform illumination into separate, parallel shafts of light, each consisting of slightly convergent rays, which, traversing the liquid, are rendered visible by the suspended particles that they illuminate. These shafts seen in perspective have their point of apparent convergence exactly opposite to the sun, *i.e.* in the shadow of his head. If the water is smooth, there are no particles to illuminate and reveal the shafts; if too turbid (or too shallow), the light does not penetrate far enough. If the sun be too low in the sky, too little light enters the water; if it shines through clouds, so that the source is diffuse, a uniform illumination results. Hence the rays are not easily noticed in winter.

After the phenomenon has once been well seen under such circumstances as I have described, one can hardly enter a boat

on a bright day without being haunted by it, and realising that, although the shadow of one's head may not actually fall on the water, yet every streak of light in the water radiates from it.

A. M. WORTHINGTON.

R.N Engineering College, Devonport, July 22.

Temperatures of Recently Killed Chamois.

MR. E. N. BUXTON, in his fascinating "Short Stalks" remarks (p. 38, footnote): "A friend of mine once took the temperature of a freshly killed chamois, and it stood at 130° F.". There is no doubt that many professional chamois hunters believe that the temperature of the animal is considerably higher than that of domestic animals.

During the last three years I have determined the rectal temperature of twenty-nine recently killed chamois.

These may be divided into three classes:

A.—*Those successfully stalked and dropped dead by the first shot.* (12 observations.)

With two exceptions, the temperatures, taken in every case within five minutes of death, lie between 101°·1 and 101°·9, the average being about 101°·5 F., or 38° 6 C.

The two exceptions were (i.) a kid four or five months old, the temperature of which was 103°·2 F., or 39°·6 C., and (ii.) a doe which had received a severe flesh-wound in the back eight days previously, the temperature of which was 102°·4 F., or 39°·1 C.

B.—*Those shot au galop.* (7 observations.) These animals all dropped dead in their tracks, or died almost immediately.

The temperatures on the whole were found to be distinctly higher than in class A, being 101°·5, 102°·3, 102°·4, 102°·9, 102°·9, 103°·5 and 104°·5 respectively.

The first four of these had run from 40 to 50 yards, the fifth about 200 yards, and the last two about 100 yards. The last two were young bucks, which, to judge from the appearance of their incisor teeth, were four and three years old respectively.

C.—*Those wounded at the first shot, but only brought to bag after some interval.* (10 observations.)

Here the temperatures are, on the whole, still higher.

The lowest (101°·7) was that of an animal which ran 50 yards after the first shot, and was then dropped dead by a second.

The next (102°·4) ran about 300 yards. The third (102°·9) was wounded in the stomach, then walked about 250 yards towards me, and was dropped by a second shot at about 30 yards.

The fourth (103°·1) ran about 200 yards. The fifth and sixth (103°·3 and 103°·5) were shot through the kidneys, but were not killed outright by the shot.

The remaining four showed temperatures of 104°·9, 105°·6, 106°·2 and 106°·7.

Of these the first had its fore-leg broken, and was recovered twelve hours later.

The second and fourth were recovered about half an hour after being wounded.

The third was an animal whose hind-leg was broken. It then escaped into another valley, and hid itself in a cave on a rock-wall, where it was spied about four hours later. A second shot failed to hit it, but drove it out of the cave. It then tried to climb the steep rocks above it, and after twice failing to overcome a *mauvais pas*, slipped and fell about 100 feet, and was killed by the fall.

Results similar to these were obtained in 1898 by a Swiss friend of mine. Some of the animals were driven by dogs, and these always showed a higher temperature than those stalked and killed by the first shot, the temperatures of the driven animals varying from 103°·6 to 105°·8, p r t r r

The highest temperature obtained by him was 107°·6 (42° C.). This was an animal which was severely wounded in the back, then lost till twenty-four hours later, when it was found and killed by a dog.

How far the average temperature given under A represents the normal temperature of the living chamois, I am unable to say, because I do not know to what extent sudden death by a bullet would be likely to affect the reading of the thermometer. Perhaps some physiologist would kindly throw some light upon this point.

To save the trouble of calculation to any foreign reader who may see this letter, I may add that 38° C. = 100°·4 F., 39° C. = 102°·2 F., 40° C. = 104° F., 41° C. = 105°·8 F., 42° C. = 107°·6 F.

G. STALLARD.

Rugby, July 12.

The London Mathematical Society.

A FEW months since it was announced in your columns that the Society had directed an index to the first thirty volumes of the *Proceedings*, and a complete list of members, to be drawn up by the secretaries. These have now been issued to members : the general public can have them from the publisher (F. Hodgson, 86 Farringdon Street) at the respective prices, 2s. and 6d. A free distribution of 1000 copies of the first part of the index, which comprises an arrangement of the papers in alphabetical order of authors' names, has been commenced, and upwards of 500 copies have been sent out. In the course of the existence of the Society some 440 persons have been recorded on the roll. This is not a great number, and some younger societies have shown greater vitality. Perhaps this issue may lead to the Society becoming more widely known. R. TUCKER.

London Mathematical Society, July 23.

The Consultative Committee and Technical Education.

THE Council of the Association of Technical Institutions has had under consideration the " Draft Order in Council " constituting the Consultative Committee of the Board of Education.

It welcomes the appointment of the Vice-President of the Association, Mr. Henry Hobhouse, M.P., as a member of the Consultative Committee, and as a representative of agricultural education and of technical education in rural districts. But it views with astonishment and regret the fact that technical education in the great towns of the United Kingdom is wholly unrepresented, although there are upon the Consultative Committee two representatives of elementary education in the persons of the Dean of Manchester and Mr. Ernest Gray, M.P., three heads of secondary schools, viz. Mrs. Bryant, Dr. Gow and the Hon. and Rev. Edward Lyttelton, as well as a large number of persons intimately acquainted with literary education.

It seems to the Council a matter of the greatest national importance that there should be upon the body which is to advise the Board of Education an adequate number of persons who are well acquainted with the applications of scientific knowledge to industries and commerce, and with the best methods of giving such technical training in this country as shall enable us to meet successfully foreign competition.

In view, therefore, of the very serious damage which may be done to technical education, and thereby to the trade and commerce of the country, if the Committee to which the Board of Education will look for advice is composed of persons without adequate knowledge of the matters to which I have referred, I venture to ask you to allow me, through your columns, to draw the attention of Members of Parliament, manufacturers, and merchants to this subject, in the hope that they may take steps to secure that the constitution of the Consultative Committee may be modified in such a way that due provision may be made for the presence of persons possessing special knowledge of trade, manufactures, and technical education.

Merchant Venturers' Technical   J. WERTHEIMER.
   College, Bristol, July 21.   (Hon. Sec.)

THE CENTENARY OF THE ROYAL COLLEGE OF SURGEONS OF ENGLAND.

THE Royal College of Surgeons of England celebrates its centenary on July 25–27. The actual month of course in which George III. founded the College by Royal Charter was March, 1800, but in the spring of 1900 it would have been impossible adequately to marshal the forces of English surgery. Sir William MacCormac, and Mr. Frederick Treves, to name no others, were, if we remember rightly, still in South Africa. The belated birthday of the College is to be fitly commemorated by a grand degree-giving, at which a number of representative European and American surgeons will receive the newly-created distinction of Hon. F.R.C.S. H.R.H. the Prince of Wales has already been presented with the diploma of Honorary Fellowship, a deputation from the College having waited on him on July 24. The form of words used in the Royal diploma is the same as that employed in all cases. " Know all men by these presents, that we, the Royal College of Surgeons of England, do hereby admit his Royal Highness Albert

Edward, Prince of Wales, an Honorary Fellow of the College."

Besides the degree-giving there will be a conversazione, a grand banquet, a Presidential address of welcome, which will deal at length with the history of the surgeon's art, and a reception at the Mansion House. But all such august ceremonial should be regarded neither as an end in itself, nor as specially typical of the progress of surgical education.

The centenary of the Royal College of Surgeons marks, in fact, not so much the hundred and first birthday of a noble institution as the audit-day of English surgery. It is as such that it should be regarded by all thoughtful men. How stands the surgical art of to-day in comparison to that of the opening years of the century? The question requires no long answer : it is not necessary to deal at length with the profound revolution, wrought since the days of Hunter, in surgery, whether intra-cranial, intra-thoracic, or abdominal. It suffices to mention only anæsthesia and antisepsis. In the year 1800 these two great agencies for good were unknown— the surgeon had to arm himself for his task after the manner of a skilled slaughterer, and Death, as often as not, stalked at his elbow through the hospital wards or to the rich man's bedside.

At the beginning of the century, too, science was everywhere in its infancy. The surgeons, though they had ceased to rank with manicurists and barbers, were often little better than bone-setters. They dreaded operations—considered them a confession of weakness, and this through a general ignorance of how safely to operate. Medical etiquette, in those old days, was an affair of various interpretation: quackery preyed unreproved on the general ignorance. To-day surgery has become, as far as may be, scientific. The modern medical man is trained as a man of science ; he is in England also subject to perhaps the severest code of honour known to history. The scientific spirit has so far permeated the public mind that even modern quackery is compelled to pose in the garb of research based on the inductive method. Graham, Buzaglo, and the inventor of the "metallic tractors" appealed in the year 1800 to just such confused instincts as possess the affrighted victims of the savage medicine-men described by Messrs. Spencer and Gillen. To-day the clever impostor takes in vain the sacred name of science, or if he make his appeal to the religious instinct, he is careful to do so almost as philosophically as a Brahmin or a Buddhist.

The progress of social relations is spoken of by jurists as one from status to contract ; the progress of the medical sciences might as fitly be described as from fetich to reason.

In this progress the Royal College of Surgeons has been no unimportant factor. The very conservatism of that great society has been a source of strength. In countries where leading institutions are less tenacious of privilege, less rigidly decorous, the interests they protect tend incessantly to degenerate for lack of ideals, of ethics, and of breeding. The names of countries, especially young ones, will occur to the philosophic, where the medical profession suffers continuously from the unacademic spirit of its academies. Yet there can be no doubt that the conservatism of the College was at one time excessive.

This will be at once apparent to the readers of Sir William MacCormac's centennial address on the " History of Surgery and Surgeons." As a succedaneum to his text, sixty-one carefully prepared biographies of his predecessors in office have been published. Of these presidents certain of the earlier ones constitute an object-lesson in oligarchy and the art and craft of office-holding. Charles Hawkins, first Master of the College in 1800, had for years—since 1790—been Master of the old civic Corporation of Surgeons. Indeed, it died, through inadvertence, under his rule. He belonged to an office-holding race. His father, Sir Caesar Hawkins, the first baronet, and his uncle, Quennell Hawkins, both St. George's men, where office was bought for large sums of money paid to seniors, had been Serjeant Surgeons to George II., and as such were liable to accompany him on his campaigns. Whether they did so is doubtful; it was Ranby who attended the gallant little king at Dettingen. Charles Hawkins enjoyed the honours of the same office, which take us back in thought to Homer's Machaon and Podalirius,[and to the Sanskrit word "Shalya," "an arrow-head," or "surgery." But beyond this we know very little of Charles Hawkins. There are others like unto him whom we need not specify. The "Dictionary of National Biography" knows them not. Their peculiarity was silence, "the fool's best friend." Conjointly they published nothing; an aversion to intellectual exertion seems to have distinguished them. But we can imagine them at least as strict upholders of dignified routine, as courtiers, as men of the world. The delightful eighteenth century died very hard in England—in Latin countries it is not dead yet—and these old gentlemen, with their powdered hair and voluminous cambric cravats, seemed to carry on the tradition of an ample age, where a fine face, a white hand, and a capacity for classical quotation fitted an average great man for any sort of position from the Papal chair to the presidency of the English College of Surgeons. Others there were, however, even in the early days of the College history, who struck a different note. Such were the terrible Abernethy, a man driven into savagery of manner by his innate sense of justice, which abhorred the quacks of his day and generation and their self-indulgent victims, suffering from avoidable ills, chiefly due to the effects of over-feeding and the "*alcoolisme des gens de bon ton.*" Such also the variable Lawrence, an early Darwinian, a passionate reformer and reform journalist, in association with the famous Wakley, an eloquent orator, and, in the end, a conservative College Councillor of the strictest.

In the College Library and Council Room during the centenary celebrations, an exhibition is being held of portraits, busts, relics and manuscripts illustrative of the history of the College, and this in itself bears witness to the changes which a hundred years can bring forth. Among the exhibits never, we believe, shown before, but now sanctioned by the lapse of long years, are papers of importance from the Owen collection. Here, for instance, is the Curator Clift's determined evidence against Sir Everard Home, Bart., who plagiarised from Hunter's papers and then destroyed them. In one exhibited letter Clift quotes Sir Everard's words, "all gone, every Jack of them," in reference to Hunter's descriptions of cases and specimens. This is not the place to discuss the Home–Hunter controversy, which has long ago been given over by the experts, but we may be allowed a postscript. The question between Hunter and Home should be judged from the point of view of 1820. Home was an old-fashioned Scotsman of a proud and ancient stock. Hunter came of a race of "bonnet-lairds." Home looked indulgently down on his brother-in-law, Hunter. These family sentiments are almost incomprehensible to an Englishman, but they rage even in the Scotland of to-day. Home thought he might fairly make use of his humbler connection's notes. He was no academic —had no scholarly regard for literary *meum* and *tuum.* How few have even to-day? Hunter's notes, on the other hand, to judge by the remaining specimens of them, were extraordinarily rough and often illiterate, though at all times they betray the great and ardent mind fretting and hurrying under inadequate powers of expression. William Clift also, John Hunter's amanuensis and subsequent defender, has been described

by one who knew him well as a typical Cornishman, extremely garrulous, prone to repeat himself. There is in the College Library a "solander," *alias* box, full to the brim almost of Clift's repeated indictments against Home. The thing suggests "*idée fixe.*"

Still, though Home acted according to the lights of his day and his order, he committed a crime of magnitude, and owing thereto the history of the great Hunterian Museum since 1800 has been necessarily one of re-construction. Clift began re-writing the Catalogue as it were from memory. He had worked so long with the great John Hunter that he knew how the master would have again spoken of numbers of specimens of which Home had burnt the descriptions. Richard Owen, Clift's son-in-law, was to Clift very much what Clift had been to Hunter. The young man worked ardently under his directions, sometimes aided by Benjamin Brodie, in his youth a zealous comparative anatomist. In one of his Museum Reports Richard Owen yearns for the days when Clift, and he, and a very few others, including Everard Home, worked incessantly in the Museum-room, undisturbed by the visits of students and sight-seers.

The public were indeed discouraged from visiting Hunter's collections, not in any gross spirit of obscurantism, but half-unconsciously, half-hieratically, much as a modern undergraduate reading for a Pass degree is kept at arm's length by the learned Don who is the College librarian. In 1833 Earle, lecturing on the urinary apparatus, gave vent to one of those petulant outbursts which are more illuminating than pages of studied prose. The passage, now often quoted, appears in the *Lancet* of the period, and is a bitter satire on the uncatalogued and dusty condition of the then Museum. Earle, it seems, had searched in vain for hours for pathological specimens with which to illustrate his remarks. The whole amusing tirade, if we remember rightly, was discreetly suppressed by him in his republished lecture.

At the present moment the great Museum—and the Library, too, for that matter—can be read like a book. One of the most notable publications of the centenary is the first portion of the "Physiological Catalogue," which, with its finely executed plates, will remain an enduring monument to the graphic skill and scientific acquirements of the Conservator and his staff.

The Museum, the Library, and the College owe their being, as it were, to John Hunter; but their emergence from the coma of the first three decades of the century is in great measure due to Sir Richard Owen. It is notable that the moment he begins to lecture in 1835, Wardrop's grumbling commentary in the *Lancet* undergoes a change. It seems at first as though the serious young Conservator was not understood. What did he aim at? why should he do so well where others had wrought so indolently? Then gradually the *Lancet* critics change their tone, and bless where before they had cursed.

A lecture by Owen became in time one of the great social and intellectual functions of the London world. Science was not then so specialised as it is to-day, nor perhaps so divorced from the interests of the literary. Bishops presided over the British Association; hereditary peers over the Royal Society; the Prince Consort took an interest in microscopy; the poets had not yet become decadent or æsthetic; the Tractarian movement had not yet replunged the world of women in the ages of faith. The public mind, indeed, would seem to have been more liberal than now. To this mind—alert, interested, deeply curious—Owen addressed himself with zeal. It is singular to note, at this distance of time, that his lecture would end with a debate, in which the Dean of St. Paul's would heckle the professor.

The College Lectures became still more important when Huxley succeeded to the chair Owen had once occupied. Owen retired in 1855, after delivering a course on "The Structure and Habits of Extinct Vertebrate Animals." He had prepared a course for 1856, when, however, lectures were suspended. The Council, it seems, had carped at the long duration of Owen's catalogue-making, and Owen had addressed to them an eloquent *apologia* for his seeming delays. Hence, perhaps, Owen's retirement. In 1863 Thomas Henry Huxley began to lecture, his first course dealing with "The Structure and Development of the Vertebrate Skeleton." His first lecture was devoted to the glyptodon with much-broken carapace, now in the Museum. He continued to deliver a long annual course till succeeded by Flower in 1869. The late Sir William Flower's tenure of the chair, which he shared with the great but somewhat neglected William Kitchen Parker, brings us down to comparatively recent times.

It is as a lecturing body that the College should prove most interesting to the world of Science at large. The names of Owen and the greater Huxley link it with the grand world of Cuvier and Darwin. We might write at length of the beneficent work of the College in pathological anatomy, or serum-therapeutics, a work all the more praiseworthy because it has been sedulously and quietly carried on in despite of the clamours of a stupid section of the public. Of the College examinations it would also be possible to say much. As recently, it should be remembered, as 1860 a doctor could qualify without passing a written examination in medicine. Now, of course, it is scarcely possible, in view of the examinations of the Conjoint Board of the Colleges of Physicians and Surgeons, for any impudent dunderhead to launch himself in practice, and to pocket the fees of a public always a little in love with quackery and mystification. Of the College as a guardian of medical ethics and etiquette, a volume might be written. A hundred years ago the doctor was always satirised by all classes of writers as unscrupulous. Now that charge is only occasionally brought against him by the illiterate, who count for nothing in the long run. That this immense change has been effected is mainly due to the College. And here it is only fair, just reference having been made to the College Museum and Library, to mention the College Office. A long line of secretaries, from Okey Belfour to Mr. Trimmer and Mr. Cowell, have patiently and vigilantly guarded the surgical point of honour. If ever a black sheep has been driven out of the surgical flock it is the College Office that has weighed his demerits and impeached him in the first instance. And this not without deliberation, or, as it was once called, "prayer and fasting." On the other hand, if ever a practitioner has been wrongly accused of malpractice, or unprofessional conduct, it is the College Office that has been at the root of his rehabilitation.

To resume and to conclude—and with the thermometer at 87° it is as well to do so—the surgeon of 1900 is not as his far-off brother of 1800, and the College, in no small degree, has been responsible for the laudable and tremendous transformation. Mere literary men in England have no Academy at their head so drastic and salutary as the College to which surgeons can look up. The doctor in 1800 used occasionally to stipulate, when dealing with workhouse authorities, that he should not be required to treat fever cases. Fever, by the by, in the undrained London of the years prior to Sir John Simon's reforms, was a common cause of death among even the well-to-do. Now, to quote the sestet of an unpublished sonnet,

"To-day skill'd Science runs where bullets hail,
Or cholera's rife, for love of suffering man,—
At the laboratory-table seeks
Plague's grim bacillus, and, if need be, can
Die as did Müller. Nor shall heroes fail:
From Hunter on to Lister their fame speaks !"

VICTOR PLARR.

ELECTRICAL POWER DISTRIBUTION.

IN a lecture on "Electricity as a Motive Power," delivered to the working men of Sheffield, August 23, 1879, the following question was asked: "And why not now?. Why should not the mountain air that has given you workmen of Hallamshire in past times your sinew, your independence of character, blow over your grindstone again? Why should not division of labour be carried to its end, and power be brought to you instead of you to the power? Let us hope then that in the next century electricity may undo whatever harm steam may have done during the present, and that the future workmen of Sheffield, instead of breathing the necessarily impure air of crowded factories, may find himself again on the hill-side, but with electric energy laid on at his command."

The present year sees the dawn of the realisation of this idea of twenty-one years ago. For soon it will no longer be: "If," as I said on that occasion, "a workman could have transmitted to him, just at the time he might require it, a small amount of energy at, say, one halfpenny per hour per horse-power—which would be three or four times the actual cost of production with a very large steam engine—and if he could turn off the power like gas when he did not want it, how many of the smaller workmen of Sheffield would be glad to avail themselves of such a facility?"

To enable such a scheme to be carried out in this country, four Electric Power Distribution Bills have this year been brought before Parliament—one for the county of Durham, one for Tyneside, one for Lancashire, and one for South Wales. And in advocating their second reading on March 1, the President of the Board of Trade expressed the opinion that "the question which the House has to decide is a very important one, perhaps one of the most important ones that have come before the House by means of a private Bill for many years." For he pointed out that "the electrical enterprise of this country is in an exceedingly backward condition," and that :—" It may almost be said that there are villages in North America which are in possession of advantages in connection with electricity which some of our largest towns do not possess."

This opinion was shared by Sir James Kitson and the Committee of the House over which he presided. For from May 3 to well into this month, July, they sat deliberating as to whether, and under what conditions, permission should be given for electric energy to be distributed over nearly 3000 square miles of Great Britain.

A vast amount of evidence was taken regarding the effect on British industry, on the cost of producing manufactured products, and as to the growing up of new factories, and even of new trades, that might come into existence through a general distribution of electric energy. Employer after employer came forward and spoke of his individual need for electrical energy to work scattered tools in his factory, to ventilate and pump his mines, as well as to cut and haul his coal.

"Cheap power is the panacea for the evil effects of foreign competition" was urged again and again by the long stream of manufacturers who occupied the witness box for weeks. The advocates of this cheap power were marshalled in groups like bands of warriors, and, from the various classes of witnesses champions were selected who bombarded the Committee with proofs of the paramount importance of their cause, and overwhelmed the members when they struggled to grasp the arithmetic of "load factors," and begged to know how many Board of Trade units there might be in a horse-power.

At first we recognised many provincial dialects among the crowd in the Committee Room, but when it began to be realised that the inquiry would occupy more weeks than it was at first thought it would need days, the

Northumberland burr was left dominant, and remained so during the whole of May. Then a Lancashire wave rolled in, and it was not until the beginning of the fourth week in June that the "ll" and the "y" formed part of so many of the words that even counsel who had come through the Severn Tunnel to Westminster to plead their cause found some difficulty in pronouncing the names of persons and places.

But the advocates of the universal supply of "electricity in bulk" had not it all their own way, since opposed to them was a band of skirmishers who delivered well-planted criticisms aimed at exposing the grasping character of some of the projects and the desire—not even thinly concealed—of some of the promoters to crush out all small existing systems of electric distribution and to establish huge monopolies for purveying electric energy.

So that one had to moderate the enthusiasm called forth by the near prospect of electric energy being regarded as a public necessity, and therefore being generally distributed like water or gas, with the exercise of a cautious regard for the interests of those undertakers—to use the legal term—who had already been entrusted with spending the money of the ratepayers or of shareholders in establishing electric distribution systems in their own areas. For they contended that this proposal to supply the small and scattered manufacturers with very cheap electric power, which it was alleged would enable them to compete successfully with their more powerful rivals, could not be commercially realised, and that these power distribution schemes had for their object the catching of the popular vote and the passing of Bills which would enable their promoters to pick out those of the customers who were the plums among the consumers already supplied by means of the existing electrical undertakings.

For nine weeks the Committee listened to the arguments *pro* and *con.*, and the tolerance which they showed in patiently hearing questions asked by counsel which had been already asked in their absence and replied to by witnesses who were unaware that the same questions had been answered at length days before, showed how willing the members were to devote their time to a full understanding of the points at issue in order that they might be able to ultimately deliver a sound decision.

The articles of commerce, which we are accustomed to purchase may be divided into those that have weight and volume and those that have not, and it is generally in connection with the former that our system of weights and measures is employed—a pound of mutton, a yard of cloth, four ounces of letter carrying, a thousand cubic feet of gas, a mile of railway journey. But the advance of civilisation has gradually led us to regard as equally suitable for buying and selling other conveniences which it would be far more difficult to meter for the purpose of ascertaining whether we had received our fair supply. A year of police supervision, a length of street improvement, a winter of snow removal, a season of South African campaigning are considered as being furnished at a fair price only when the grumbling in connection with the supply is not too great and the articles in the *Times* are not too severe.

But there remains one commodity which, although it has neither weight, volume, nor linear dimension, can be metered with extreme accuracy, and the public demand for which is daily becoming greater and greater, and that is—energy. Hitherto the working of factories has been associated with water and coal, and either the factories have been built near the stream or in a coal region. When, however, such a site could not be conveniently found, then it has been the custom to carry at a dear rate a black, bulky, dirty substance by rail or water for miles to the factory, and, after strewing a certain portion of its dirt over the neighbourhood in the

form of a descending cloud, to cart the remainder away as dusty ashes.

So accustomed are we to all this—so little does it strike us as incongruous that scuttles full of black lumps should be regularly brought into a drawing-room, no matter how valuable the pictures or rich the curtains and carpets, that we forget that our successors will look with more scorn on our customs than we do on those of our ancestors, seeing that, at least, their floor-coverings of rushes, intermingled with old bones and other refuse, could not be much injured by smoke or by dust.

Electric-lighting, electric-heating, electrically-driven machinery are all undoubtedly clean, but will the two latter pay as well as the former? On this point the evidence before the Committee of the House was somewhat conflicting. What, in fact, is the cost of carrying coal compared with the cost of electrically conveying the energy, or the "essence of the coal" as one of the counsel poetically termed it? Further, what is the saving produced by combining steam-engines, and working one very large engine instead of many small ones at different places?

Briefly, then, apart from all question of dirt, is it cheaper to burn the coal at the pit's mouth, and to convey the energy electrically to each of many machines situated within a radius of, say, ten miles from the electric generating centre, or to load the coal on railway trucks, carry it in different directions to many factories, unload, stoke the furnaces at many places, and distribute the energy from the many steam-engines by shafting, belting, rope-gearing, compressed air, or an electric current generated at the individual factory?

At first sight one would be inclined to answer that without doubt the electric driving of individual machines over an area of, at any rate, fifty square miles from a single centre must be the cheaper. For can we not employ quite thin electric mains and still have only a small percentage loss of energy in transmission, by using a very high electric pressure and sending through the mains a comparatively small current, whereas we have no means of compressing coal, so that not merely its volume, but also its weight and its cost of transport, can be greatly diminished for a given amount of coal-energy conveyed? No doubt; but Great Britain has its Board of Trade, and that body not unnaturally looks with disfavour on the overhead wires in Western America, which are maintained at so high an electric-pressure that it is only the spitting and brush-discharge that occurs which prevent a higher pressure being employed. For that is how the commercial limit of 40,000 volts has been arrived at in the United States for overhead electric transmission of energy.

Eleven thousand volts is the highest potential difference that has hitherto been allowed—even for buried conductors—by our Board of Trade, and even that pressure has been employed in connection with only two systems of transmission, viz. the one from the London Electric Supply Company's generating station at Deptford to their transforming stations at Trafalgar Square, Bond Street, &c., and the other from the Metropolitan Electric Supply Company's new generating station at Willesden to their transforming stations at Amberley Road, Manchester Square, &c.

The promoters of the four Power Bills, therefore, do not contemplate using at the outset a higher pressure than 10,000 volts, or sending more than 1000 kilowatts—that is, 1340 horse-power—through a single underground cable. The evidence as to the cost of such a cable showed that it could be made and laid for something like 1400*l.* a mile; some of the witnesses said 1000*l.*, while others thought that was a "promoter's figure," since they had not succeeded in getting similar cables constructed and laid in trenches in their own districts under 1800*l.* a mile.

It is well known that the relatively high price of electric-lighting arises from the small fraction of the twenty-four hours during which there is any great demand for electric energy, so that it is necessary to fit up in an electric-light central station engines and dynamos which can develop something like ten times as much horse-power as would be necessary to deliver the same total amount of electric energy in the twenty-four hours if the demand were a steady one. This is expressed by saying that the "load factor" is 10 per cent. In one or two English towns the load factor is actually as low as 6, and in very few cases is it higher than 12 per cent.

Now it was urged that if an electric generating station, instead of supplying current simply for electrically lighting a single town, were to supply electric energy for all sorts of purposes throughout a large district, the load factor would be very much higher, and the cost of production would be proportionately diminished. For example, with electric tramway work the load factor is about 40 per cent., that is to say, about 40 per cent. as much horse-power is used by the cars as could be produced by all the engines and dynamos in the tramway station if they were working at full load continuously day and night. With factories again, it was estimated that a load factor of some 30 per cent. might be obtained. Hence it was urged that one of these large electric power stations might rely on a load factor of about 25 per cent.

With this load factor of 25 per cent., it was estimated by Mr. Ferranti, for example, that if a generating station were erected in the coal-fields at a spot with a good supply of water for condensing the steam, and if the plant capacity were about 16,000 horse-power—of which 4000 would be kept as a reserve—the entire cost of generating a Board of Trade unit would be about 0·44*d*., and the cost of transmitting it 0·2*d*. So that it could be sold to the consumer at 1½*d*., and a good business done, as contrasted with the 4*d*. or 6*d*. per unit now charged to private consumers in English towns.

As opposed to this, it was urged by central station engineers and others that this supposed great economy to be obtained by erecting electric generating stations at the pit's mouth and transmitting the energy electrically through considerable lengths of buried conductors was imaginary; that in towns like Manchester, Liverpool, Southport, Bolton, Cardiff, Newport, &c., electric energy for driving machinery was already offered to the public at as low a price as the promoters of these Bills proposed to offer it, that the fraction of the cost of delivering a Board of Trade unit which could be debited to the coal alone was small, and that the proportion arising from the mere cost of carrying the coal was still smaller.

As a matter of fact, although the average prices obtained for private electric lighting are 3·49*d*., 4·04*d*., 4·10*d*., 4·68*d*., 4·69*d*., 5·29*d*. per Board of Trade unit by the Corporations of Manchester, Bolton, Southport, Cardiff, Liverpool and Newport respectively, any person in Manchester who desires to run an electromotor, no matter how small it may be, during the whole of the ordinary factory hours, only pays now 1½*d*. a unit to the Corporation. In Bolton he is charged 1·35*d*. if he takes his full demand for 640 out of the possible 2184 hours in a quarter. Southport charges the Tramway Company 1½*d*. per unit. Cardiff offers the unit at 2*d*. if 4000 units are taken per annum—which means a single motor of only 1 horse-power running for about eight hours a day for 300 days in the year. The Liverpool Tramway Company pays the Corporation only 0·9*d*. per unit; while in Newport, 1½*d*. is the price charged if more than 3000 units per quarter are taken.

The sweeping accusations, therefore, that some of the witnesses levelled against the local authorities of supineness, indifference to the public needs, &c., were

hardly borne out; while such evidence as that of the electrical engineer of Southport, that the carriage of coal to that borough increased the cost of the unit by only one-twelfth of a penny, and of the electrical engineer of Manchester, that the charge for interest on the cost of buried cables added 0·49*d*. to the value of the Board of Trade unit if it were transmitted twenty-five miles, whereas the cost of conveying the equivalent amount of fuel by railway over the same distance only increased the value of a Board of Trade unit of energy by 0·059*d*., that is by only about one-eighth of the former, combined with the evidence that on the Continent and in America cheap *overhead* wire transmission was allowed, and therefore was generally adopted, led the Committee to realise the following, viz.:—that while the assent to the new proposals might confer a great boon on collieries and manufactories in scattered districts, a great wrong might be inflicted if safeguards were not introduced to prevent the introduction of the new electric schemes crippling the natural development of those systems of electric distribution which at present existed in this country, and which had been brought to their present condition by the expenditure of about 40 millions sterling during the past ten years.

In addition to the opposition to all the four Bills, made by private companies who had already obtained provisional orders to supply electrical energy, and by local authorities, some of whom had, and some of whom had not, obtained statutory powers to act as electrical suppliers, two of the Bills, viz. the Durham and the Tyneside, opposed one another. For whereas in the Durham Bill the district up to and including the south side of the Tyne is scheduled, in the other Bill both sides of the Tyne, from the mouth to Ryton, are included.

The promoters of the Tyneside Bill maintained that the *two* sides of the Tyne together naturally formed a *single* supply district, and that a company which had powers to supply the manufacturers on *both* sides with electric energy could do so more economically than if it was confined to the land along the north shore only. But they added that they had no objection to competition on the part of the proposed Durham Company or of any one else.

The Durham promoters urged that they were including in their area a lean portion towards the south of the county of Durham, which they could only undertake to supply if they were given the possession of the south bank of the Tyne undisturbed by competition on the part of the proposed Tyneside Company. It was also alleged that, although these various Bills passed their second reading because the President of the Board of Trade had stated that he had "an assurance from the promoters to the effect that they will undertake to agree to an amendment in Committee which would make it perfectly clear that they do not ask for the power to distribute even in bulk without the consent of the local authority," the Tyneside scheme really aimed at obtaining *private* way-leaves, and by skipping about the district by means of *overhead* wires to supply even private customers in retail without asking for the consent of any local authority. And it was pointed out that the assent to such a proposal would be manifestly unjust in view of the fact that those who were already supplying under "provisional orders," or who might obtain such orders, were compelled under the terms of such an order to supply every one within a certain "compulsory area" within a limited time, as well as being subject to comply with other obligations.

Serious opposition to the Tyneside Bill was also raised by the Newcastle Electric Supply Company. This is the company which for some years has been supplying electric current to the east of the district of Newcastle-upon-Tyne, but which, by arrangement with various local authorities and with the Walker and Wallsend Union Gas Company—which obtained last year a Provisional Order for supplying electric energy to the districts of Wallsend and Willington Quay—has this session been promoting a Bill for distributing electric energy throughout a considerable area along the north of the Tyne to the east of Newcastle.

Between the proposed Tyneside Company and the existing Newcastle-on-Tyne Company a battle royal raged in the Committee Room, not merely because the area proposed by the one included that proposed by the other, but because a certain site on the Tyne bank was scheduled by both companies as the land on which they proposed to erect a generating station.

The advocates of the Tyneside Company, led by Lord Kelvin, asked, in fact, for authority to erect three generating stations—the one just referred to, one immediately opposite on the south bank of the Tyne, and one on the same bank, but much further west.

The advocates of the Newcastle-on-Tyne scheme, on the other hand, pleaded that the first site should be left to them, and urged, not unreasonably, that if the Tyneside Company confined its attention solely to that small bit which lay on the north bank of the Tyne to the west of Newcastle, of the whole of the area it contemplated, it would have ample scope for the spending of its entire capital, viz. a million sterling.

Indeed, one of the grounds of opposition to all the four large schemes before this Committee was that, while enormous areas were scheduled in the Bills, throughout which it was proposed to distribute electric energy, the capital asked for by any one of these companies was only a few hundred thousand pounds, or not more than a million, even when, as in the case of the Tyneside Bill, it was increased to that amount while the Bill was before the Committee. To this the advocates of these four Bills replied that the amounts put down for capital, even as increased during the progress of the Bills through Committee, were only intended to enable a start to be made, and that after a few years the companies would necessarily come to Parliament again for a large increase in capital. Further, that while it was proposed to start with erecting in each of the four districts a single 10,000 horse-power electric generating station, in a few years 50,000 or more horse-power would have to be electrically delivered in each of these districts if the Bills passed.

Excepting the Newcastle-on-Tyne Bill, which came before another Committee, the Lancashire Bill was the most moderate. On the other hand, the South Wales Bill was the most grasping, for it was the only one which professedly aimed at obtaining powers to invade a town and break up its streets without the consent and even against the wish of the local authority if the person whom the company aimed at supplying with electric energy was a "wholesale customer." And such a customer was defined in the Bill as one who was prepared to take 20,000 units a year.

Those who drafted this Bill no doubt were under the impression that there were very few people in the 1050 square miles of the counties of Glamorgan and Monmouth covered by this Bill who at present took more than 20,000 units a year. And no doubt that was the case, for only some six could be cited by those who represented the Corporations of Cardiff and Newport. But hitherto it has been almost exclusively for lighting that people in these boroughs have taken electric energy, and when from the electric supply systems of these Corporations, or from the mains of some outside company—should such a company gain access—manufacturers begin to receive current for working electro-motors, then a 20,000 unit customer would only pay 83l. 6s. 7d. a year for his energy, at 1d. a unit, and therefore in no sense could he come under the category of "wholesale."

In fact, when the promoters of the South Wales Bill realised that even a 3 horse-power motor running continuously day and night—in connection, for example, with a blast furnace—would consume 20,000 units a year, and that a 9 horse-power motor working ten hours a day for 300 days in the year would require the same amount of electric energy, they saw that their "wholesale customer" would have to be thrown overboard.

In despair, however, the promoters still clung to the reed that a local authority could not if it would, and should not if it could, erect plant for supplying factories with electric energy on the large scale contemplated by these Electric Power Distribution Companies, and they urged that the ratepayers' money ought not to be used for speculative purposes, forgetful apparently that, even when a local authority has bought up an electric supply undertaking at twice the sum that it cost the private company to erect it, the rates have been ultimately relieved in consequence of the purchase, and the ratepayers therefore benefited by the local authority becoming a purveyor of electric energy.

Ultimately, on Wednesday, June 27, the chairman, Sir James Kitson, who it is important to remember is not merely an M.P., but what is far more important the head of a great manufacturing firm, a director of a great railway, and the ex-Mayor of a great city, made the following most excellent declaration:—

"A local authority which undertakes and is prepared to give a full and ample supply of electrical energy for all purposes to consumers within its district ought not, without its consent, to be required to give facilities for the supply, within its district, of electrical energy by other undertakers. But if a local authority is unable or unwilling to provide on reasonable terms and within a reasonable time a full and adequate supply of electrical energy for any purpose to any company or person applying for the same within its district, such company or person should be at liberty, after notice to the local authority, to obtain their supply from other authorised undertakers, and the local authority should be required to give all necessary facilities for this purpose. Any difficulty arising out of the above questions should be subject to arbitration as provided by the general Acts."

Doubtless this decision did not please all; but how acceptable was it to those who, like myself, have been hungering for the realisation of our dream of twenty-one years ago—"power brought to the workman, not the workman to the power"—but who have seen with apprehension the growth of obstacles nourished by England's spirit of masterly inactivity and by its not unnatural, nor wholly unwise, veneration of vested interests.

For now local authorities are put on their metal. If you realise, says Sir James Kitson's Committee, what are your duties in providing all your people with "an ample supply of electrical energy for all purposes," we will be no parties to any hindrance through competition being put in your way. But if your district be one in which bumbledom reigns supreme, then our declaration is that no municipal barrier shall be left standing to oppose the free entrance of those who come with offers of cheap electric energy.

Next, on Thursday, June 28, the formal statement was made by the Chairman:—"that the preamble of the South Wales Electrical Supply Bill is proved, also that the preamble of the Durham (County of) Electric Power Supply Bill is proved; and the preamble of the Tyneside Electric Power Bill is not proved to the satisfaction of the Committee, and that the preamble of the Lancashire Electric Power Bill is proved to the satisfaction of the Committee." Then followed the lengthy process of drafting the clauses, and finally, on July 16, these three Bills, of which the preambles had been reported by the Committee as proved, were read a third time in the House.

On the following day, the North Metropolitan Electric

Power Supply Bill, which asks for authority to supply electric energy over a smaller region, consisting of the districts of Hornsey, Hendon, Barnet, St. Albans, Hatfield, Hertford. Ware, &c., was, after being, considered by another Committee of the House of Commons, read for a third time, and finally, on Tuesday, July 24, the Newcastle-on-Tyne Electric Supply Companies scheme already referred to in this article received the sanction of a Committee of the House of Lords.

The era of Electrical Power Distribution on a vast scale in our country has, therefore, begun.

W. E. AYRTON.

THE DAILY WEATHER REPORT OF THE METEOROLOGICAL OFFICE.

THE Meteorological Council has made provisional arrangements for the sale of single copies of the Daily Weather Report at a penny each from the first of August next. The copies will be on sale from about 3 o'clock of the afternoon of the day of issue at the Meteorological Office and at the railway bookstalls of the following terminal railway stations in London: Victoria (S. E. & C. and L. B. & S. C.), Charing Cross, St. Pancras, King's Cross and Euston. Hitherto the issue of the reports has been confined to certain public offices and institutions, and to annual or quarterly subscribers. The distribution has been by hand or by book-post. The area within which delivery can be effected on the day of issue is necessarily very limited, and it is hoped that the facilities afforded by the new arrangement may bring the information which the reports contain within the reach of some of those interested in the subject who live outside the present limits of delivery on the day of issue. If the provisional arrangement should make it apparent that there is any public demand for the accommodation, efforts will be made to continue and extend it.

From the same date some modifications will be introduced into the form of the Report. The morning and evening observations of the telegraphic reporting stations will appear on the first page as usual, but the two charts on the second page, representing the morning distribution of pressure, wind and sea, and of temperature and weather, respectively, will be supplemented by three smaller charts. One will represent the barometric distribution over the whole of Europe at 8 a.m. of the preceding day in order that the general atmospheric changes may be more readily traced. Another will represent mean monthly or bi-monthly morning isotherms for the British Isles, so that the distribution of temperature for the day may be easily compared with the normal distribution for the season as estimated for a period of twenty-five years. The third will represent the distribution of mean maximum temperature estimated in a similar manner.

There will also be several changes on the fourth page of the Report. Instead of "General Remarks on the Weather over Europe" there will be a table giving the latest information in the possession of the Office as to maximum and minimum temperature, rainfall, and weather at selected stations on the Continent and elsewhere which are beyond the area represented by the telegraphic reports. The selection of the stations will be mainly determined by the current interests of travellers, and will be varied from time to time according to the information available.

The information as to the weather in the British Islands will also be supplemented by data as to sunshine for the preceding day from a number of stations which will report by post, and it is intended, in course of time, to replace "yesterday's 2 p.m. reports" by postal reports of maximum and minimum temperature and rainfall for a number of inland stations which are expected to prove

a useful addition to the telegraphic reports of the first page.

For convenience of reference a small supplemental table will give the Greenwich time of sunrise, noon, and sunset for four selected stations in the British Isles, so that the variation in the duration of daylight and the standard times of local noon for any locality may be ascertained.

W. N. SHAW.

THE BRADFORD MEETING OF THE BRITISH ASSOCIATION.

THE local arrangements for the Bradford meeting of the British Association in September next are now rapidly approaching completion. The railway companies have agreed to give the usual special facilities, both to visitors who come to Bradford from long distances, and to members resident in Yorkshire who travel to and fro every day. Those persons who attend the meeting can obtain a return ticket from nearly all the railway companies at a fare and a quarter, provided they present to the booking-clerk a certificate, which can be obtained on application to the local office at 5, Forster Square. Any Members or Associates visiting Bradford day by day, and staying in places within fifty miles of the city, can obtain return tickets at the single fare on presenting their card of membership to the booking-clerk in Bradford. The following railway companies have entered into this arrangement:—the Caledonian, the Great Eastern, the Great Central, the Great Northern, the Great Western, Lancs. and Yorks., London, Brighton and South Coast, London and North Western, London and South Western, Midland, North British, North Eastern, South Eastern, Chatham and Dover, and the other companies belonging to the Associated Railways.

The local programme is now in the press, and will be issued within the next fortnight to the Members and Associates who have notified their intention of being present. The following items, however, will give a brief summary of the information contained in it :—

GENERAL PROGRAMME.

Wednesday, September 5.—4 p.m. : Meeting of General Committee at the Town Hall ; 8.30 p.m. : the President's Address in St. George's Hall.

Thursday, September 6.—3.30 p.m. : Reception at the Technical College (Textile Exhibition) ; 8.30 p.m. : the Mayor's Conversazione in St. George's Hall.

Friday, September 7.—8.30 p.m. : Lecture in St. George's Hall by Prof. Gotch, F.R.S., on "Animal Electricity" ; 9.30 p.m. : Smoking Concert in the Technical College in honour of the President.

Saturday, September 8.—Excursions (half-day) ; 8 p.m.: Artisans' Lecture in St. George's Hall, by Prof. Silvanus Thompson, F.R.S.

Sunday, September 9.—10.30 a.m. : Sermon by the Bishop of Ripon in the Parish Church.

Monday, September 10.—3.30 p.m. : Corporation Garden Party in Lister Park ; 8.30 p.m. : Lecture in St. George's Hall by Prof. W. Stroud, D.Sc., on "Range Finders."

Tuesday, September 11.—8.30 p.m. : Corporation Soirée in St. George's Hall.

Wednesday, September 12.—3.30 p.m.: Private Garden Parties ; 8 p.m. : Full-Dress Concert in St. George's Hall (Festival Choral Society ; Permanent Orchestra ; conductor, Mr. F. H. Cowen ; Miss Ella Russell).

Thursday, September 13.—Excursions (whole day).

The conferences of delegates of corresponding societies will be held on Thursday, September 6, and Tuesday, September 11, at 3 p.m., at the Reception Rooms.

The Reception Room at the Grammar School will be opened on Monday, September 3, at 2 p.m. to 6 p.m., and on the following week-days at 8 a.m. to 6 p.m. On Sunday, September 9, the Reception Rooms will be open from 9 to 10.30 a.m., and from 3 to 5 p.m.

The temporary Museum in connection with the Sections is this year being made a special feature, more particularly in regard to Geology, Botany and Zoology. Joint meetings are being held, on certain days, of the Geological and Botanical Sections, to take up the subject of Carboniferous fossils, and it is proposed to form a collection of fossils found in the neighbourhood to illustrate the papers as much as possible ; and also to display photographs bearing on the subject, taken from the Geological Society's collection in London. These exhibits will form the nucleus of the Museum, but there will also be other collections bearing on the main subjects dealt with by some of the other Sections. At the Municipal Technical College there will be an Exhibition during the week illustrative of the staple trades of the district ; visitors will pass from room to room, and will see the gradual development, through innumerable processes, of the most elaborate fabrics from the unwashed fleeces. On Thursday afternoon, September 6, the Exhibition will be opened by Mr. W. E. B. Priestley, the chairman of the College, and a Reception held, to which all visitors to the meeting will be invited.

The preparations for the Excursions and Garden-parties are nearly complete, and full details will be given in the next article. RAMSDEN BACCHUS.

NOTES.

PROF. R. LIPSCHITZ, professor of mathematics in the University of Bonn, has been elected a correspondant in the section of geometry of the Paris Academy of Sciences.

LORD KELVIN has been elected Master of the Worshipful Company of Clothworkers for the year 1900-1901.

SIR JOHN EVANS, K.C.B., F.R.S., has been elected chairman of the Society of Arts for the ensuing year.

MR. GRANT-OGILVIE, principal of the Heriot-Watt College, has been appointed director of the Museum of Science and Art, Edinburgh.

THE sixty-eighth annual meeting of the British Medical Association will be held at Ipswich during next week, commencing on Tuesday.

DR. M. ARMAND RUFFER, president of the sanitary, maritime, and quarantine board of Egypt, has received from His Majesty the Sultan of Turkey the Order and Insignia of the Medjidjieh of the Second Class.

A CONFERENCE on the housing of the working classes, under the auspices of the Sanitary Institute, will be held on July 30 and 31, in the lecture-room of the Royal Medical and Chirurgical Society. In connection with the conference an exhibition of models and plans will be held in the Parkes Museum.

IT is stated in the Engineer that of the fifty-five ships taking part in the naval manœuvres this year, the Adriĉnne, Camperdown, Jaseur, besides some others, are specially fitted for wireless telegraphy. The Majestic and the Diadem have also been fitted. Torpedo officers have charge of the installation in each case.

THE thirty-seventh annual meeting of the British Pharmaceutical Conference was opened in London on Monday, under the presidency of Mr. E. M. Holmes. An attractive programme containing illustrations of the house of the Pharmaceutical Society where the meetings will be held, the president, and places to be visited, appears as a supplement to the current number of the Pharmaceutical Journal.

WE learn from the British Medical Journal that the Madras Government has passed an order sanctioning the excess expenditure over the original grant of 600 rupees incurred by Capt. R. H. Elliott in connection with the prosecution of his researches into the properties of snake venom, and has made him an additional grant of 200 rupees to cover the cost of further experiments. The Surgeon-General has been requested to report if Capt. Elliott's services will be available for special duty at the end of September when his tour of service terminates.

PROF. HENRY F. OSBORN has been appointed to succeed the late Prof. O. C. Marsh as palæontologist in the United States Geological Survey. Prof. Osborn's special field of work will be to take charge of the vertebrate palæontology of the survey, especially with reference to the completion of the monographs for which the illustrations were prepared under the direction of Prof. Marsh. Prof. Osborn graduated from Princeton in 1877, and was professor of comparative anatomy there until 1890. He was appointed Da Costa Professor of Zoology at Columbia University in 1891, and curator of vertebrate palæontology at the American Museum of Natural History, New York. He is a member of the National Academy of Sciences and other scientific bodies, and is the author of numerous papers and memoirs on fossil mammals and reptiles.

THE seventy-second annual meeting of the German Association of Naturalists and Physicians will be opened at Aachen on Monday, September 17. At the first general meeting, the advances of natural knowledge and medicine during the present century will be surveyed. Prof. van 't Hoff will review the progress of inorganic science ; Prof. O. Hertwig will discourse on the development of biology ; Prof. B. Naunyn will deal with internal medicine, including bacteriology of hygiene ; and Prof. H. Chiari will speak on pathological anatomy in relation to external medicines. At the second general meeting, to be held on September 31, several scientific subjects of current interest will be dealt with. Prof. J. Wolff will speak on the correlation between form and function of individual structures of organisms ; Prof. E. v. Drygalski will describe the plan and purpose of the German Antarctic expedition ; Prof. D. Hansemann will discourse upon cell-problems and their significance in the scientific foundation of the treatment of disease ; and Prof. Holzaphel will take as his subject the development of German coal-measures. On September 19, the naturalists and physicians will meet in separate groups. The questions to be brought before the former group include the circulation of nitrogen in the organic world, by Prof. M. W. Beyerink ; the latest investigations upon steel, by Prof. E. F. Dürre ; and language and technical teaching from a scientific standpoint, by Prof. Pietzker. The chief subject to be discussed in the medical section is the neuron theory in its anatomical, physiological and pathological aspects, by Profs. Verworn and Nissl. The remainder of the meetings will be held in the various sections of the Association, and as more than three hundred communications will be made, there will be no lack of subjects for discussion. In connection with the meeting, an exhibition of physical, chemical, and medical preparations and apparatus will be held.

IN the course of his presidential address, delivered to the Society of Chemical Industry on July 18, Prof. Chandler referred to some of the work of American chemists, and the development of chemical processes of manufacture. Many important investigations in agricultural chemistry have been conducted by the chemical division of the United States Department of Agriculture, among them being the practical determination of the number and activity of the nitrifying organisms in soil, the influence of a soil rich in nitrogen on the nitrogen content of a crop, the manufacture of sugar from the sorghum plant, and the comparative study of typical soils of the United

States. Of agricultural experiment stations there are now 59, and the 148 chemists connected with them have done a large amount of original investigation in subjects more or less closely allied to agricultural and physiological chemistry. Prof. Chandler gave a comprehensive account of the chemical industries. In particular he referred to the progress made in electro-chemistry, and described the methods now adopted for the reduction of aluminium at Niagara, and also for the manufacture of carborundum and artificial graphite. Speaking of water-gas, he described the opposition against its introduction for illuminating purposes. The question came before the Health Department of New York, and, after careful investigation, the department decided that the gas was such an improvement in quality and price, while the increased danger, as compared with that from old-fashioned coal-gas was so slight, that it was not wise to interfere with it. The water-gas industry has now taken almost complete possession of the whole country. There are at least 500 gas companies using water-gas wholly or in part, and it is estimated that in 1899 three-quarters of the entire consumption, or 52,500 million cubic feet, consisted of carburetted water-gas. At the close of the address the Society's medal, which is awarded not oftener than every two years, was presented to Dr. Edward Schunck, F.R.S., in recognition of his classical investigations on natural colouring matters and other researches in connection with technical chemistry.

We learn from the *Times* that the Select Committee to which the Sea Fisheries Bill was referred, presented a special report to the House of Commons on Thursday last expressing the view that it would not be expedient to pass the measure into law without further inquiry and investigation. The committee regards it as proved beyond the possibility of dispute that there is a very great and serious diminution of the fish supply, that the ancient fishing grounds are much depleted, and that in default of a remedy the consequences to the fishing industry and the fish supply will at no very distant future be disastrous. The prohibition of the taking and killing of such fish is described as practically impossible without prohibiting trawling altogether. As regards the prohibition of fishing within certain areas where small fish more particularly abound, the Committee thinks that it is established that there are certain well-known areas in the North Sea where undersized small and young fish congregate, and that to prevent fishing in such areas would be of great value. It is pointed out, however, that such a result could not be obtained without joint international action among the Powers bordering the North Sea, and that the difficulties of such international action and the policing necessarily ancillary thereto are obvious. In conclusion, the Committee considers that no effort ought to be spared—first, to arrange for international treatment of the subject generally, and especially for regulation of the North Sea area ; and, secondly, to provide for the adequate equipment of the Government department in charge of the subject, so that it may effectively pursue scientific investigation, and ascertain with sufficiency and precision what has been done in the way of scientific research or in the matter of practical legislation by other inquirers and by other countries.

MR. E. H. L. SCHWARZ sends us from Cape Town some interesting remarks upon the snake-stone, *apropos* of the facts stated by Mr. Hervey in NATURE of May 24 (p. 79) as to the use of a stone by the Malays as a remedy for poisonous bites. Snake-stones are fairly common in South Africa, and are described as white, porous stones, which, when applied to the place where the snake has bitten a person, adhere till all the poison is drawn out into them, after which they are placed in milk, which in turn draws the poison from the stones, and renders them again fit for use. The farmers firmly believe they are taken from

the head of a snake. It is suggested that snake-stones are made of pumice. To the uneducated, the structure of pumice has a close resemblance to that of bone, and this may possibly explain the popular delusion that snake-stones are made of bone. Mr. Schwarz thinks that the black colour of the stone, described by Mr. Hervey, may have been due to blood, or the stone may have been a black variety of pumice, for there is an instance of originally black pumice having been thrown up near the lighthouse on Cape Agulhas. The fact that the fable of the stone having been taken from the head of a snake is exactly the same in the Malay States as is prevalent in South Africa is interesting, though the Malay slaves which the early Dutch obtained from Batavia in exchange for quaggas, zebras, ivory, &c., may have carried the legend with them. It is not an uncommon custom in Germany for people to carry about with them nuggets of raw gold to draw out of their bodies all the more subtle evils, such as those produced by spirits and devils, while for the grosser evils they carry a potato. Is the snake-stone legend a derivative of these, or are they subsequent to the snake-stone?

MAGNETIC observations were made at several stations upon the day of the recent total solar eclipse (May 28), under the direction of Dr. L. A. Bauer. A brief statement of the results is given in *Terrestrial Magnetism*. Ten observers were engaged in the work, and eight complete series of observations were obtained—seven for declination and one for horizontal intensity. All the stations show a magnetic effect, which cannot be referred to any other cause than that of the eclipse, the principal effect occurring, like the fall in temperature, some minutes after time of totality. The effect is as though part of the night hours were interposed among the day hours, *e.g.* the declination at all of the stations having passed the morning elongation and approaching the mean value of day, is *increased* about 20'-40'' if the declination be east, and *decreased* if the declination be west ; whereas, the horizontal intensity approaching at the time its minimum value for the day, is *increased* for a brief period after time of totality. The observations and results will be published in full in a *Bulletin* of the U.S. Coast and Geodetic Survey.

THE hot and dry spell which set in over a fortnight ago promises still to continue, and although the mid-day temperatures are not generally as high as they were on several days last week, they are far in excess of the average. There have been three days at Greenwich with the shade temperature above 90°, the highest reading as yet being 94° on July 16. The nights are also excessively warm, and on the night of July 22–23 the lowest reading at Greenwich was 67°·6, which is warmer than any night in July or August at Greenwich since August 8, 1846, when the thermometer did not fall below 68°. During the last week there have been five successive nights without the thermometer falling as low as 60°. There has been no rain in London for three weeks, and the same dry weather has been experienced generally in the south-east of England. The conditions have been less settled over the northern and western portions of our islands, where rain has fallen at frequent intervals and no very extreme temperatures have occurred.

FROM Dr. Oliver Lodge, F.R.S., we have received the reprint of a very suggestive lecture on "Modern Views of Matter," delivered before the Literary and Philosophical Society of Liverpool in March last. In it Dr. Lodge discusses the atomic theory, the ether, the conception of "electrons," and the still more recent hypothesis of the existence of "corpuscles."

ARRANGEMENTS have been made for six popular science lectures for young people, under the general title of "The World we Live On," to be delivered in Kensington Town

Hall during October and November next, and we have no doubt they will be, as "instructive and entertaining" as the programme promises, for the lecturers are all experienced exponents of science. The quotation "Pupils trained on books, and books alone, are mere passive recipients of other people's ideas," which we notice on the programme, is not a very happy one; for popular audiences are, after all, only "passive recipients" of the ideas of the lecturer. Popular lectures upon scientific subjects direct attention to natural phenomena, and occasionally induce people to devote serious attention to some branch of science. On this account they are valuable, but there is of course a great difference between listening to an eloquent lecturer, or witnessing striking demonstrations, and actually carrying out the most elementary experiments.

WRITING to Sir Henry Burdett with reference to Mr. Craggs's endowment of a travelling scholarship in connection with the London School of Tropical Medicine, Mr. Chamberlain recently remarked:—"My experience at the Colonial Office daily impresses upon me the extreme importance of doing something to make life in our tropical colonies more healthy for those who are engaged there in the work of civilisation, whether as administrators, missionaries or traders. Science has already given us promise of good results in the near future, and nothing, I believe, can conduce more powerfully to a speedy and satisfactory result than such researches as those which Mr. Craggs has in contemplation. I hope that his munificent action may be followed by other benefactors, so that the work may be simultaneously pursued in different countries." The scholarship is one of 300l. per annum, tenable for three years, and is for research in tropical disease. The first scholar is now attached to an expedition which is engaged in attempting to give practical application to the theory of the inoculation of the human being with the malaria parasite through the medium of the mosquito. The expedition has been equipped by the Colonial Office and is now stationed in one of the most malarious districts of the Roman Campagna. When this experiment is completed, at the expiration of six months, the scholar will proceed to the West Coast, thence to the West Coast and probably to the interior of Africa.

THE fourteenth volume (for the year 1898) of the *Analele* of the Meteorological Institute of Roumania has just been published. Besides the usual tables, it contains several important memoirs. M. St. Murat compares the magnetic instruments of the Institute with those of the Observatory of Parc St. Maur, and describes the observations made during 1898. The director, Dr. Hepites, studies the climatology of Braila and of the Roumanian littoral of the Black Sea, the distribution of rainfall in Roumania during 1898 (this paper being illustrated by a series of monthly maps), and the earthquakes during the same year. Brief accounts are given of eleven shocks, all of them of very slight intensity.

IT appears from Part ii. of the Eighteenth Annual Report of the Fishery Board for Scotland that the salmon fishery for 1899 turned out considerably below the average of recent years. It is true that the weight of salmon forwarded by rail and steamship during the year was slightly in excess of that carried in 1898, but it was still 638 tons below the average; and such slight improvement as took place is attributed to the large run of grilse which occurred during the summer. Adult fish seem to have been comparatively scarce. As the inspector remarks, it is absolutely essential to the continuance of the Scottish salmon fisheries that a stock of breeding fish sufficient to counterbalance the loss caused by fishing, by the salmon's natural enemies and by disease, must be maintained by some means or other. It is satisfactory to learn that some proprietors

have established hatching-stations in order to artificially increase this supply. The inspector is, however, of opinion "that if the present catching power continues to be developed, a very great increase in the number and in the capacity of hatcheries will be necessary to produce noticeable results." In artificially augmenting the stock of salmon we must necessarily be prepared to compete with a vast mortality.

AMONG several papers interesting to entomologists in the June number of the *Agricultural Gazette of New South Wales*, reference may be made to one by Mr. W. W. Froggatt, the Government entomologist, on insects living in figs. The interiors of young wild figs in all countries swarm with minute plant-feeding Hymenoptera of the family Chalcidæ. The males and females of these minute insects differ from one another in colour, size and shape; but the peculiar feature of the group is that, instead of the females being degraded into a wingless condition, it is the males that are devoid of wings, while they are also frequently blind, with abnormally short legs and aborted antennæ. A new Australian species is described and figured.

IN the *Victorian Naturalist* for June, Mr. Robert Hall gives an interesting account of the nesting habits of one of the Australian diamond-birds (*Pardalotus assimilis*). In common with some of their kindred, these birds make their nests at the end of a tunnel drilled by themselves in a bank. "The nest is made to fit in a cavity with a domed ceiling, excavated in the hard subsoil at the end of the tunnel. This tunnel is ten inches long, and is drilled with a slight upward tendency, as is usual in most ground-boring birds. The nest entrance is two feet below the surface of the ground, and in a creek-bank some nine feet above a stream. Both sexes take part in the drilling operations, one excavating while the other removes the rubbish, but it seems that the task of incubation falls to the share of the male.

IN the June issue of the *Johns Hopkins University Circulars* will be found an important communication by Mr. L. E. Griffin on the arterial circulation in the nautilus.

WE learn from the *Bulletin* of the New York Botanic Garden that the herbarium has acquired during the year specimens to the number of 70,000, and that over 4000 species and varieties of plants, belonging to 172 families and 1057 genera, are under cultivation in the Garden.

WE have received Supplement ix. to the *Journal of Reading College*, consisting of the sixth annual report on field experiments (for 1899), viz. :—field experiments in Dorset, Berkshire, Oxfordshire and Hampshire; spraying experiments on charlock; trials with sugar-beet; the manuring of crops; and notes on manures.

CONTRIBUTIONS from the Gray Herbarium of Harvard University, New Series No. 19, consists of a synopsis of the Mexican and Central American species of *Salvia*; a revision of the Mexican and Central American *Solanums* of the sub-section *Torvaria*; and some undescribed Mexican plants (chiefly Labiatæ and Solanaceæ), all by Mr. M. L. Fernald.

HERR PAUL SINTENIS announces that he is undertaking a botanical exploration of the mountain region on the confines of Turkestan and Persia, of the flora of which district very little is at present known. The expedition will probably extend through the present year. Application for sets of the plants collected should be made to Herr Baurath J. Freyn, Smichow-Prague.

THE Agricultural and Mechanical College of Texas, judging from the annual "Catalogue" for the session 1899-1900, is a well-appointed and flourishing institution. Full information as to the various departments of the college, courses of study, &c., is to be found in the "Catalogue," which also contains many full-page illustrations of the college buildings, interiors of the laboratories, &c.

THE *Journal* of the Straits branch of the Royal Asiatic Society for January, 1900, contains, *inter alia*, an important contribution by Mr. H. N. Ridley on the flora of Singapore. The district is a rich one, something like 1900 flowering plants, and 130 ferns being recorded. Mr. Ridley opens with an interesting introduction, in which he gives a sketch of the factors which determine or modify the vegetation. He also describes some interesting phenological facts, and finally gives a sketch of the history of the botanical work in the Island. The chief space is, of course, devoted to an enumeration of the plants, but it contains short notes respecting the more striking individual species.

THE revised edition of "First Records of British Flowering Plants," by Mr. W. A. Clarke, just published by Messrs. West, Newman and Co., is full of extracts of interest to every one who finds pleasure in the study of the British flora. To members of Field Clubs and Natural History Societies the book is particularly valuable. It gives, in the form of extracts from printed botanical works published in Great Britain, the earliest notice of each distinct species of our native and naturalised plants, the last edition of the "London Catalogue" being taken as a basis. The volume thus provides a concise answer to the question which a naturalist often asks, viz. : "How long has this plant been known as British?" An interesting analysis of the "first records" is given at the end of the book. William Turner was the first to record the majority of our native plants. His works, ranging from 1538 to 1568, contain notices of 238 flowering plants, and may be considered the foundation of our British flora. From Lobel (1570, &c.), Mr. Clarke obtains eighty first records and from Gerard's famous Herball (1597), 182 species, so that about 500 species of British plants were known and described three hundred years ago. The book in which these and many other particulars are given is one which every naturalist should keep handy for reference.

THE *Proceedings* of the London Mathematical Society (vol. xxxi.), containing papers read from April to December of last year, have just been published by Mr. Francis Hodgson. The titles and brief abstracts of the papers have already appeared among our reports of societies.

THE Great Eastern Railway Company's "Tourist Guide to the Continent," edited by Mr. Percy Lindley, contains concise notes and numerous illustrations of interesting and easily accessible places in Holland, Belgium, Germany, Switzerland, Norway, Denmark and Sweden. The book is a useful travelling companion for Continental tourists, and is as matter-of-fact as most guide-books.

THE volume of *Proceedings* of the forty-eighth meeting of the American Association for the Advancement of Science, held at Columbus a year ago, has just been received. The presidential addresses, papers and abstracts cover a wide field of scientific work. A noteworthy feature is the series of portraits of former presidents of the Association, accompanying an address by Dr. Marcus Benjamin.

THE value of a well chosen set of inorganic chemical preparations as a part of a course of general chemistry is now generally acknowledged, although the number of elementary text-books dealing with this branch of the subject is comparatively small. The works of Prof. Erdmann, of Halle, in this field are well known, and the English translation of his "Introduction to Chemical Preparations" (Chapman and Hall)

by Dr. F. L. Dunlap, the German edition of which has already been noticed in these columns, will be of great service to students in England and America.

THE seventh part of vol. ii. of the seventh edition of Fresenius's "Quantitative Chemical Analysis," translated from the revised sixth edition by Mr. C. E. Groves, F.R.S., has just been published by Messrs. J. and A. Churchill. This completes the new edition of the work, which has been revised throughout. The special part, dealing largely with applications of chemical analysis to industrial products and other technical matters, has been considerably extended, and many new analytical processes have been introduced. The last section of the work includes sixty exercises especially designed for teaching the theory and practice of quantitative chemical analysis. In addition, there is an appendix containing analytical notes and tables for the calculation of analyses. Practical chemists and teachers are thus now provided with a complete new edition of a standard work on analysis.

THE reaction discovered by Lubawin of the formation of α-amino-acids by the interaction of ammonium cyanide and aldehydes, has been extended by Dr. W. Gulewitsch to ketones, and in the current number of the *Berichte* he describes the details of the preparation of α-amino-isobutyric acid from acetone, the yield under favourable conditions being as high as 74 per cent. of the theoretical. The same number of the *Berichte* also contains a masterly investigation of the action of soda solution upon nitroso-benzene by Prof. Bamberger. No less than twelve substances have been isolated from the products of this extremely complex reaction, including azoxybenzene, nitrobenzene, aniline, *p*-nitroso-phenol, *o*-amidophenol, *p*-amidophenol, hydrocyanic acid, ammonia, and four new acids, and there are still further products awaiting investigation.

THE additions to the Zoological Society's Gardens during the past week include a Bonnet Monkey (*Macacus sinicus*) from India, presented by Mr. P. M. Thornton ; a Rhesus Monkey (*Macacus rhesus*) from India, presented by Miss A. N. Ball ; a Humboldt's Lagothrix (*Lagothrix humboldti*) from the Upper Amazons, presented by Mr. W. S. Churchill ; two Masked Paradoxures (*Paradoxurus larvatus*) from China, presented by Mr. W. T. Lay ; a Senegal Parrot (*Poeocephalus senegalus*) from West Africa, presented by Mr. S. Cordwell ; two Chukar Partridges (*Caccabis chukar*) from North-west India, presented by Mr. Chas. E. Pitman ; a Missel Thrush (*Turdus viscivorus*), European, presented by Mr. J. B. Williamson ; a Common Cuckoo (*Cuculus canorus*), British, presented by Miss Lucy Holland ; two Larger Hill Mynahs (*Gracula intermedia*) from Northern India, a Mauve-necked Cassowary (*Casuarius violicollis*) from the Aru Islands, a Clumsy Tortoise (*Testudo inepta*) from Mauritius, four Elephantine Tortoises (*Testudo elephantina*) from the Aldabra Islands, an Alligator Terrapin (*Chelydra serpentina*), six Blanding's Terrapins (*Emys blandingi*) from North America, deposited ; a Guira Cuckoo (*Guira piririgua*) from Para, six Painted Frogs (*Discoglossus pictus*), South European, purchased.

OUR ASTRONOMICAL COLUMN.

ASTRONOMICAL OCCURRENCES IN AUGUST.

August 4.   11h. 55m.   Minimum of Algol (β Persei).
  "   7.   8h. 43m.   "   "   "
  "   7.   5h.   Conjunction of Mars and Neptune. (Mars 1° 27′ N.)
  "   9.   11h. 34m. to 12h. 40m. Moon occults the star D.M. - 16°, 5609 (mag. 6).
  "   11.   Maximum of August meteoric shower. Perseids. (Radiant 45° + 57°).

Aug. 12. 16h. 10m. to 17h. 10m. Moon occults κ Piscium (mag. 5).

13. 20h. Venus at greatest brilliancy.

15. Venus. Illuminated portion of disc = 0·280.

15. Mars.   ,,   ,,   ,,   = 0·932.

18. Saturn. Outer minor axis of outer ring = 18″·35·

18. 12h. 42m. to 13h. 24m. Moon occults ι Tauri (mag. 4·7).

18. 14h. 47m. to 15h. 39m. Moon occults 105 Tauri (mag. 5·8).

19. Mercury at greatest elongation (18° 32′ W.).

20. 22h. Venus in conjunction with moon. (Venus 1° 49′ S.).

22. Jupiter 26′ S. of β Scorpii.

23. Expected return to perihelion of De Vico-Swift's comet (1844–1894).

27. 10h. 26m. Minimum of Algol (β Persei).

NEW VARIABLE IN HERCULIS.—Prof. W. Ceraski, of the Moscow Observatory, communicates to the *Astronomische Nachrichten* (Bd. 153, No. 3650) the discovery of a new variable by Mdme. Ceraski on photographs taken by M. S. Blajko. The star's position is as follows :—

R.A.	Decl.		Epoch.
h. m. s.	° ′ ″		
18 30 54·8	...	+ 25 55 49	... (1855·0)
18 32 44·1	...	+ 25 57 54	... (1900·0)

The star is not found in the B.D. At maximum the star is slightly brighter than 9th magnitude, decreasing to a minimum of about 12th magnitude. At present its brightness is increasing.

NEW STAR IN AQUILA.—A telegram from Prof. Pickering, Cambridge, Mass., dated 1900 July 9, states that the *Nova* of the 8th magnitude found by Mrs. Fleming in April 1899 is now a nebula of 12 magnitude. Its position is

R.A.		Decl.
h. m. s.		° ′
19 15 16	...	−0 19

A further statement is made in the *Astronomische Nachrichten* (Bd. 153, No. 3651) that the measures are from the photographs.

METEORITIC THEORY OF THE GEGENSCHEIN.—In the *Astronomical Journal*, No. 483 (vol. xxi. pp. 17–21), Mr. F. R. Moulton puts forward a mathematical analysis of the conditions which would appertain if the *Gegenschein* were due to the presence of a more or less condensed region of meteorites. The idea of the problem appears to have been suggested by remarks of Prof. Barnard (who has made consistent observations of the phenomenon during the last sixteen years) to the author.

Discovered by Brorsen about the middle of this century, very few systematic observations are recorded until those of Barnard, who has made careful determinations both of its position and shape. He comes to the conclusion that it is always exactly opposite the sun, or as nearly so as can be determined. Other observers have stated varying positions, but in the case of so difficult an object it is advisable to consider the more systematic records as having greatest truth.

After citing the well-known reasons for considering that interplanetary space is densely occupied by meteoric particles, moving with widely varying velocities in all directions, he supposes that a great multitude will at any time be situated at the opposition point, and that a considerable proportion of these would be under such initial conditions as to remain there for some time. Then the meteors being very small compared with the earth, they are treated as infinitesimal bodies, disturbing neither the earth nor each other. He also neglects the eccentricity of the earth's orbit. Then referring the motion of one meteor to rectangular axes with the origin at the centre of gravity of the sun and the earth, he traces the conditions for stability for a certain time. Then by slightly varying the conditions, he finds the nature of the movement of the infinitesimal body with special reference to the circumstances under which it will make periodic oscillations around certain points. The result of successive integration suggests that meteors passing near one of these selected points with the assumed conditions of motion would be subject to forces directed nearly to this point, and would have a tendency to revolve round it. Although after a few revolutions they might escape, the average result would be a condensation with respect to space, if not with respect to time. The difficult point now to determine is whether a sufficient number would be captured to become visible. If the meteors are revolving round the sun at a distance of about 900,000 miles greater than the earth's mean distance, they will be moving slower than the earth, which will gradually overtake them in longitude. As they approach opposition they will be retarded and drawn in towards the sun, their motion being thereby accelerated. The net result of these actions will be to bring the meteors into the plane of the ecliptic, thus causing the condensation at opposition, and explaining the tendency to an oscillation in latitude which has been observed.

Instead of being exactly opposite the earth, the point of condensation will be nearly opposite the centre of gravity of the earth and moon, and consequently the *Gegenschein* should have a monthly oscillation in longitude of the nature indicated by the observations of Douglass, but much less in extent. The oscillation in latitude would, however, be monthly also, instead of yearly, as the observations tend to indicate.

A phenomenon, observed so far by Barnard alone, is the series of marked changes to which the *Gegenschein* is subjected in short periods of time, being large and round in September and the beginning of October, becoming slightly elongated by the 4th or 5th, very much elongated by the 10th or 11th, and showing merely as a swelling on the zodiacal band by the 18th. Although this is not directly explicable, the shape of the *Gegenschein* will depend on the thickness of the zodiacal disc of meteorites, and if the opposition point should pass through a dense portion of the swarm it is readily conceivable that a change of form would ensue. The distance of the opposition point works out at 930,240 miles from the earth. The period of oscillation would be 183·304 days. It is thus suggested as possible that meteors may move for long periods of time in the vicinity of the opposition point, in sufficient numbers to cause the faint glow of the *Gegenschein* by reflecting the light of the sun. Reference is finally made to a paper by M. Hugo Glydén in the *Bulletin Astronomique*, Tome 1, where similar views are enunciated.

METEOR OF JULY 17.—A bright meteor was seen in many parts of the north of England on the evening of Tuesday, July 17, shortly before nine o'clock. A few particulars concerning the phenomena are given by correspondents in the *Yorkshire Post*. An observer at Menston-in-Wharfedale saw the meteor at a point about N.N.W. from that place, and about forty degrees above the horizon. At Wiseton, Notts, it was seen at 8.47, and at Bramhope at 8.48. At Armley, a hissing noise was heard, and the meteor seen to disappear close by.

THE GREAT EARTHQUAKE OF JUNE 12, 1897.

THE investigation of the great earthquake of June 12, 1897, being the most extensive of which there is historic record, has naturally led to important additions to our knowledge. A detailed report of this earthquake, by Mr. R. D. Oldham, has been published by the Indian Government,[1] and its investigation suggested a line of further research, the results of which have been published in the *Philosophical Transactions* of the Royal Society.[2] The principal results described in these bulky publications are here given in the form of an abstract.

The known extent of the principal seismic area was about 1,200,000 square miles, a figure which will surprise many after the statement that this was the greatest earthquake of which there is historic record. One of the results of this earthquake was, however, a re-examination of the records of the great Lisbon earthquake of 1755, which has shown that the statements regarding it, copied from one text-book to another, are grossly exaggerated. The statement that it was felt in the lead mines of Derbyshire is shown, by reference to the original record, to be an error; the shock that was felt being clearly an independent, local, though possibly sympathetic, shock. Apart from this, there is but one doubtful record of its having been felt so far north as England, though its effects were visible, both in England and in Holland, in disturbances of the water in ponds. The accounts of its having been felt in Iceland and America refer to the sea-waves, which may travel to regions far beyond the utmost limit at which the shock could be felt. Omitting these records, taking only those which refer to the sensible shock, and rounding off the seismic area to an elliptical form, it

[1] "Report on the great earthquake of June 12, 1897." By R. D. Oldham. *Memoirs* of the Geological Survey of India, vol. xxix. 1899, pp. xxx + 379 + xviii; 44 plates, 3 maps, 51 woodcuts in text.
[2] "On the propagation of earthquake motion to great distances." By R. D. Oldham. *Phil. Trans.*, Series A, 1900, pp. 135–174.

is found to cover not more than 1,000,000 square miles; while if the shock of June 12, 1897, is treated in a similar manner, we obtain a total seismic area of over 1,750,000 square miles.

Owing to the paucity of good records, the course of the isoseists could not be traced in detail. The outermost isoseist was, however, determined with approximate accuracy for about half of its circumference. The seismic area presents a peculiarity in that there is a detached area in the alluvium about Ahmedabad over which the shock was felt, though it was unfelt over a tract of about one hundred miles separating this alluvial area from the furthest limit at which the shock was felt on rock. It is also reported to have been felt at Burhanpur, on the border of the Tapti valley alluvium, though it was felt nowhere else in the neighbourhood. Outside the area over which it was felt there are records, in India, of the passage of the earthquake wave as indicated by the swinging of lamps, &c.

Apart from the records in India, there is good evidence that it was felt in Italy; the observers at Catania, Leghorn and Spinea di Mestre all record having felt a slight shock at the exact time when the instruments throughout Italy recorded the advent of the first phase of the disturbance due to this earthquake. Had there been only a single record, it might have been attributed to a distinct local shock; but these three separate records, all agreeing with each other in time, and also with the advent of the first tremors, which, having a period of about '5 second, might have been sensible, leaves little possibility of doubt that the Indian earthquake was actually felt. The observers are, however, to be complimented on their acuteness of observation.

The epifocal area is of a peculiar shape. Situated in Western Assam and North-eastern Bengal, it is bounded on the south by a straight line running about E.S.E. for some 200 miles; on the north it is bounded by a nearly symmetrical double sigmoid curve, the maximum breadth being not less than 50 miles, and possibly as much as twice this amount. Over the whole of this area of not less than 6000 square miles, the intensity of the shock was in excess of 10 degrees of the Rossi-Forel scale, and alterations of level have taken place; while for a year and more afterwards earthquake shocks—some severe, but most feeble and local—were very frequent. The changes of level were not only shown by faults, one of which was traced for a distance of over 12 miles, and had in places a measured throw of over 30 feet, and by differential changes of level, whereby streams were dammed up into lakelets, but also by a remeasurement of some of the triangles of the great Trigonometrical Survey. As the whole of the triangles reobserved lay within the epifocal area, it is not possible to say what amount of actual change has taken place; but changes of position of one hill relative to another were determined, which reach as much as 24 feet in a vertical, and 12 feet in a horizontal, direction.

The results of the triangulation, as published by the Trigonometrical Survey of India, indicate an increase in the horizontal distances between the stations; but in the geological report it is shown that this is probably due to a shortening, by compression, of that side which was assumed as an unaltered base-line. The true nature of the focus is regarded as a thrust plane, from which minor faults branched off, and in places appeared as such, while elsewhere they died out before reaching the surface and merely caused those changes of level which, where other circumstances were favourable, led to the formation of lakes. No less than thirty of these were observed, the largest having a length of 1½ miles and a breadth of ¾ mile, and the smallest a few yards across; the depth varied from 1 to over 20 feet.

Within this epifocal area the violence was everywhere great, though subject to great local increase in the neighbourhood of the fault planes which extended upwards to the surface. Not only were upright stones broken, but sound hardwood trees of a diameter of 6 to 7 inches were snapped across by the violence of the motion they were subjected to; no masonry building was left standing, and the hill sides were scarred by landslips. In many places it was noticed that stones lying on the ground had been projected into and through the air.

The acceleration necessary to cause the fracture of standing monoliths, or sound hardwood trees, must have been great—much greater than the measured accelerations, as determined by West's formula from overturned tombstones, which range up to 32 feet per sec. per sec. It is doubtful, however, whether West's formula is applicable to cases where the height of the overturned column is less than three or four times its diameter; in the earthquake of 1897 all the high accelerations were obtained

in places where there must have been a large vertical component in the wave motion, and the overthrow of squat pillars is regarded as a modified form of projection. It is improbable that accelerations of over 6 feet per sec. per sec. can occur, except in the vicinity of the epicentre, where there is a considerable vertical component in the wave motion, and the excessive accelerations which have been supposed to have been measured in the case of other earthquakes must be regarded with suspicion.

Opportunity was taken to review the various formula for deducing the acceleration and velocity of movement of the wave-particle; these have been all collected in an appendix and discussed. One result of the discussion is in a manner reactionary, for the one quantity which it was believed could be determined with real accuracy, the velocity as deduced from projection, is shown not to be due to wave motion at all. The velocities deduced from observed projections are shown to lead to impossible results if combined either with the deduced accelerations or with any conceivable amplitude or period, and the conclusion is come to that the projection of solid objects was due, not to molecular wave motion, but to a molar displacement of the ground, resulting in permanent changes of level.

Instances of the rotation of objects, both within and without the epicentre, were numerous. As many as possible of these were carefully measured to determine, not only the angular rotation, but also the direction and amount of displacement of the centre of gravity. From a careful examination of the data, it is shown that none of the attempts to explain rotation by simple rectilinear motion are in accordance with the observed facts, and that it is necessary to accept the explanation of vorticose motion. This vorticose motion does not, however, take the form of angular rotation as has been assumed by some investigators; but the whole ground either moves in a more or less circular track, or is subject to a more or less rectilinear to-and-fro motion, whose direction changes continually in azimuth.

Over a large alluvial area the river channels were narrowed, railway lines bent into sharp curves, and bridges compressed and destroyed, much as in the Japanese earthquake of 1891. This compression is shown to have been due, in all cases, to displacement of the superficial alluvium, and not to any general compression. Over this same alluvial area fissures and sand vents were opened in myriads. With regard to the fissures, it is shown that Mallet's explanation of their formation by unsupported masses of clay being thrown off from free surfaces by their own inertia is incomplete, and that they were formed in places where no such action could have taken place. It is suggested that in such cases the fissures were due to the visible surface undulations which were noticed by many observers. The sand vents were formed in such numbers that large areas were temporarily flooded by the volumes of water which issued from them with such force that it rose in solid columns to a height of 3 feet and more from the ground, while splashes and spouts are said to have reached 18 or 20 feet in height. It is noteworthy that in several cases these sand vents are said to have been formed *after* the passage of the shock, and flowed for a period of half an hour or, according to some, several hours. This is attributed to the settling of clay beds on to underlying quicksands, which supported the overlying beds as long as these were continuous, but would not do so after they had been broken up by the earthquake.

Earthquake sounds were very loud and conspicuous, but the data available do not allow of much advance in this difficult branch of seismology. In some cases explosive sounds of short duration were heard after the earthquake had passed, and the connection of these with the "Barisal guns," "mistpœffers," "marina" and other similar phenomena is discussed, all being regarded as probably in the main seismic.

The most important results obtained are probably those connected with the rate of propagation. Numerous time observations in India yield a time curve with double curvature like that of Schmidt's "hodograph," but the curvature is too slight to accord with it, and the true time curve is shown to be most probably a straight line indicating a uniform rate of propagation of 3'0 km. per sec. Turning from the observed rate of propagation of the sensible shock to the distant records, it is shown that the records of the Italian seismographs exhibit three principal phases of motion, after each of which there is a marked diminution of movement. The first of these gives an average rate of propagation of 9'6 km. per sec., the second of 5'6 km. per sec., and the third, the phase of long period undulation, accompanied by marked tilting of the ground, a rate of transmission of 3·0 km. per sec. The agreement of this with the observed rate of transmission of the sensible shock is held to indicate that both are due to a form of wave motion which was propagated at a uniform rate along the surface of the earth. The first two phases, it is suggested, are due to wave motion transmitted through the interior of the earth, and as in the presumably isotropic, or nearly so, material of the interior of the earth a separation of condensational and distortional waves could take place, which Knott and Rudzki have shown to be impossible in the rocks of which the surface of the earth is composed, it is suggested that these two phases are due to the arrival of the condensational and distortional waves respectively, travelling by brachisto-chronic paths through the interior of the earth.

This suggestion is followed up in the second paper. The published records of distant earthquakes were looked up, and those selected of which the time and place of origin were known within a limit of error of 1 minute of time and 1 minute of arc respectively. Further, on account of the known impossibility of separation of the two simple forms of elastic wave motion in the surface crust of the earth, only those records were considered which came from a distance of not less than 20° of arc from the epicentre.

Seven distinct earthquakes were found of which the published records satisfied all these conditions, and as in some of them there was more than one shock, they constituted eleven distinct shocks. From the published records were extracted (1) the time of commencement of the record; (2) the time of any sudden increase of movement, when recorded; and (3) the time of maximum displacement. Tabulating these, it is found that each earthquake exhibits a three-phase character in the record; and, further, that if the times are plotted and a curve drawn through them, the time curves of the first two phases show precisely that curvature which Prof. Rudzki's investigations show to be characteristic of wave motion propagated along brachisto-chronic paths through the earth, where the rate of propagation increases with the depth. Continuing these curves by extrapolation to the origin, they give rates of propagation fairly concordant with the rates of propagation of condensational and distortional waves as experimentally determined for ordinary rock. As a subsidiary part of this investigation, it is shown that the "preliminary tremors" of earthquakes coming from Japan to Europe reach a depth of about ·45 of the radius from the surface, attain there a maximum velocity of 14·5 km. per sec. for the condensational, and 8·8 km. per sec. for the distortional wave, and traverse a medium which has, at that point, a bulk modulus of 17 times, and a rigidity of about 21·5 times that of granite.

The records of the third phase show some irregularity, but the time curve is a straight line, pointing to a uniform rate of transmission along the surface. There is, however, some indication that in the case of the greatest earthquakes it is higher than in the case of lesser ones; in other words, that the rate of transmission is in some way dependent on the magnitude of the earthquake, hence, probably, on the size of the wave. From this it is concluded that the propagation of these surface undulations is, in part at least, gravitational.

EXPLORATIONS OF THE "ALBATROSS" IN THE PACIFIC.[1]

WE left San Francisco in August of last year, and in latitude 31° 10′ N., and longitude 125° W., we made our first sounding in 1955 fathoms, about 320 miles from Point Conception, the nearest land. We occupied 26 stations until we reached the northern edge of the plateau from which rise the Marquesas Islands, having run from station No. 1, a distance of 3800 miles, in a straight line.

At station No. 2 the depth had increased to 2368 fathoms, the nearest land, Guadeloupe Island, being about 450 miles, and Point Conception nearly 500 miles, distant. The depth gradually increased to 2628, 2740, 2810, 2881, 3003 and 3088 fathoms, the last in lat. 16° 38′ N., long. 130 14′ W., the deepest sounding we obtained thus far in the unexplored part of the Pacific through which we were passing. From that point

[1] Abridged from letters written to the Hon. George M. Bowers, U.S. Commissioner of Fish and Fisheries, Washington, D.C., by Mr. Alexander Agassiz, leader of the expedition of the U.S. Fish Commission Steamer *Albatross* to the Pacific.

the depths varied from 2883 to 2690 and 2776, diminishing to 2583, and gradually passing to 2440, 2463 and 2475 fathoms until off the Marquesas, in lat. 7ʰ 58′ S., long. 139° 08′ W., the depth became 2287 fathoms. It then passed to 1929, 1802 and 1040 fathoms in lat. 8° 41′ S., long. 139° 46′ W., Nukuhiva Island being about 20 miles distant. Between Nukuhiva and Houa.Houna (Ua-Huka) Islands we obtained 830 fathoms, and 5 miles south of Nukuhiva 687 fathoms. When leaving Nukuhiva for the Paumotus we sounded in 1284 fathoms about 9 miles south of that island. These soundings seem to show that this part of the Marquesas rises from a plateau having a depth of 2000 fathoms and about 50 miles in width, as at station No. 29 we sounded in 1932 fathoms.

Between the Marquesas and the north-western extremity of the Paumotus we occupied nine stations, the greatest depth on that line being at station No. 31, in lat. 12° 20′ S., and long. 144°·15′ W. The depths varied between 2451 and 2527 fathoms, and diminished to 1208 fathoms off the west end of Ahii, and then to 706 fathoms when about 16 miles N.E. of Avatoru Pass in Rairoa Island.

Between Makatea and Tahiti we made eight soundings, beginning with 1363 fathoms, 2 miles off the southern end of Makatea, passing to 2238, 2363 (the greatest depth on that line), to 2224, 1930, 1585, 775, and finally 867 fathoms off Point Venus.

The deep basin developed by our soundings between lat. 24° 30′ N., and lat. 6° 25′ S., varying in depth from nearly 3100 fathoms to a little less than 2500 fathoms, is probably the western extension of a deep basin indicated by two soundings on the charts, to the eastward of our line, in longitudes 125° and 120° W., and latitudes 9° and 11°.N., one of over 3100 fathoms, the other of more than 2550 fathoms, showing this part of the Pacific to be of considerable depth and to form a uniformly deep basin of great extent, continuing westward probably, judging from the soundings, for a long distance.

I would propose, in accordance with the practice adopted for naming such well-defined basins of the ocean, that this large depression of the Central Pacific, extending for nearly thirty degrees of latitude, be named Moser Basin.

The character of the bottom of this basin is most interesting. The haul of the trawl made at station No. 2, lat. 28° 23′ N., long. 126° 57′ W., brought up the bag full of red clay and manganese nodules with sharks' teeth and cetacean ear-bones; and at nearly all our stations we had indications of manganese nodules. At station No. 13, in 2690 fathoms, lat. 9° 57′ N., long. 137° 47′ W., we again obtained a fine trawl haul of manganese nodules and red clay; there must have been at least enough to fill a 40-gallon barrel.

The nodules of our first haul were either slabs from 6 to 18 inches in length and 4 to 6 inches in thickness, or small nodules ranging in size from that of a walnut to a lentil or less; while those brought up at station No. 13 consisted mainly of nodules looking like mammillated cannon-balls varying from 4½ to 6 inches in diameter, the largest being 6½ inches. We again brought up manganese nodules at the equator in about longitude 138° W., and subsequently—until within sight of Tahiti—we occasionally got manganese nodules.

As had been noticed by Sir John Murray in the *Challenger*, these manganese nodules occur in a part of the Pacific most distant from continental areas. Our experience has been similar to that of the *Challenger*, only I am inclined to think that these nodules range over a far greater area of the Central Pacific than had been supposed, and that this peculiar manganese-nodule bottom characterises a great portion of the deep parts of the Central Pacific where it cannot be affected by the deposits of globigerina, pteropods, or telluric ooze; in the region characterised also by red-clay deposits. For in the track of the great equatorial currents there occur deposits of globigerina ooze in over 2400 fathoms for a distance of over 300 miles in latitude.

We made a few hauls of the trawl on our way, but owing to the great distance we had to steam between San Francisco and the Marquesas (3800 miles) we could not, of course, spend much time either in trawling or in making tows at intermediate depths. Still the hauls we made with the trawl were most interesting, and confirmed what other deep-sea expeditions have realised: that at great depths, at considerable distances from land, and away from any great oceanic current, there is comparatively little animal life to be found.

The bottom temperatures of the deep (Moser) basin varied between 34·6° at 2628 and 2740 fathoms, to 35·2° at 2440 fathoms, and 35° at 2475 fathoms; about 120 miles from the Marquesas.

Our deep-sea nets not having reached San Francisco at the time we sailed, we limited our pelagic work to surface hauls, of which we generally made one in the morning and one in the evening, and whenever practicable some hauls with the open tow nets at depths varying between 100 and 350 fathoms. The results of these hauls were very satisfactory. The collection of surface animals is quite extensive, and many interesting forms were obtained. As regards the deeper hauls with the Tanner net, they only confirm what has been my experience on former expeditions, that beyond 300 to 350 fathoms very little animal life is found, and in the belt above 300 fathoms a great number of many so-called deep-sea crustaceans and deep-sea fishes were obtained. I may mention that we obtained Pelagothuria at about 100 fathoms from the surface.

On our way to Tahiti from the Marquesas we stopped a few days to examine the westernmost atolls of the Paumotus. Striking Ahii we made for Rairoa, the largest atoll of the Paumotu group. Skirting the northern shore from a point a little west of Tiputa Pass, we entered the lagoon through Avatoru Pass, anchoring off the village.

We made an examination of the northern side of the lagoon between Avatoru and Tiputa Passes. The lagoon beach of the northern shore is quite steep, and is composed of moderately coarse broken coral sand at the base, and of larger fragments of corals along the upper face, which is about 5 to 6 feet above high-water mark. These coral fragments are derived in part from the corals living on the lagoon face of the northern shore, and in part of fragments broken by the waves from somewhat below the low-water mark. The ledge which underlies the beach crops out at many places on the lagoon side of the northern shore; we traced it also along the shores of Avatoru Pass, and about half-way across the narrow land running between Avatoru and Tiputa Passes. It crops out also at various points between them in the narrow cuts which divide this part of the northern land of the lagoon into a number of smaller islands. These secondary passes leave exposed the underlying ledge, full of fossil corals.

It became very evident, after we had examined the south shore of the lagoon, that the ledge underlying the north shore is the remnant of the bed, an old Tertiary coralliferous limestone which at one time covered the greater part of the area of the lagoon, portions of which may have been elevated to a considerable height. This limestone was gradually denuded and eroded to the level of the sea. Passages were formed on its outside edge, allowing the sea access to the inner parts of the lagoon. This began to cut away the inner portions of the elevated limestone, forming large sounds, as in the case of Fiji atolls, and leaving finally on the south side only a flat strip of perhaps 2500 to 3000 feet in width which has gradually been further eroded on the lagoon side, and also on the sea face to leave only a narrow strip of land about 1000 feet in width and perhaps 10 to 14 feet in height, the material for this land having come from the disintegration of the ledge of Tertiary limestone, both on the sea face and the lagoon side.

The underlying ledge is not the remnant of a modern reef; its character is identical with that of the elevated limestones of Fiji, which are of Tertiary age, and the rock is in every respect the same as that I observed on many of the elevated islands of Fiji. The atoll of Rairoa is in a stage of denudation and erosion very similar to that of Ngele Levu, in Fiji, only in Ngele Levu the elevated limestone attains a height of about 60 feet. Our visit to the south shore of the lagoon, both on the lagoon side and on the sea face, left us no doubt regarding the character of the underlying ledge of the north shore. As soon as the south shore was sufficiently near, as seen from the lagoon side, for us to distinguish its character, we could see that the entire shore line was formed of a high ledge of limestone, honeycombed, pitted and eroded, both by atmospheric agencies and the action of the waves, in its lower parts both on the lagoon side and on the sea face. The great rollers of the weather side broke through between the columnar masses of the ledge into the lagoon, and as far as the eye could reach there extended a more or less continuous wall.

Crossing over to the weather side of the southern land of Rairoa in one of the passages between two of the islands, we came upon the limestone ledge, from 12 to 14 feet high and about 40 to 50 feet wide, which formed the sea face of the islands and islets, and extended far to the westward as a great

stone wall more or less broken into distinct parts. We found this ledge to consist of elevated limestone as hard as calcite, full of corals, honeycombed and pitted, and worn into countless spires and needles and blocks of all sizes and shapes, separated by deep crevasses or potholes, recalling a similar scene in Ngele Levu on the windward end of the lagoon. In the passages the parts of the ledge which had not been eroded extended as wide buttresses, gradually diminishing in height till they formed a part of the lagoon flat and extended out below the recent beach rock which covered it in short stretches.

The amount of water which is forced into such a lagoon as Rairoa is something colossal, and when we observe that there are but a small number of passages through which it can find its way out again on the leeward side, it is not surprising that we should meet with such powerful currents (7 to 8 knots in several cases) sweeping out of the passages on the lee sides.

The islands and islets of Rairoa are fairly well covered with low trees and shrubs and large groves of palm trees.

It was with great interest that we approached Makatea, as it is the only high elevated island of which Dana speaks as occurring in the western Paumotus. For though he mentions some others as possibly having been elevated 5 to 6 feet, yet he considered them all, as well as Makatea (Metia or Aurora, of Dana) as modern elevated reefs. Yet, from the very description given by him of the character of the cliffs and of the surface of Makatea, I felt satisfied that it was composed of the same elevated coralliferous limestone so characteristic of the elevated reefs of Fiji, and which, from the evidence of the fossils and the character of the rock, both Mr. Dall and myself have been led to regard as of Tertiary age.

The cliffs had the same appearance as those of Vatu Leile, Ongea, Mango, Kambara, and many other elevated islands of Fiji. There were fewer fossils, perhaps, but otherwise the petrographic character of the rock was identical with that of Fiji.

The south-western extremity of the island sloped gradually to the sea, and showed two well-defined terraces. The lines of these two terraces could, as a rule, be traced along the faces of the vertical cliffs by the presence of caverns along the lines of those levels, similar to the lines of caverns indicating the line of present action of the sea at the base of the cliffs.

During our stay in Papeete some time was spent in examining that part of the barrier reef of Tahiti which had been surveyed by the *Challenger*. We found the condition of the outer slope of the reef quite different from its description as given in the *Challenger* narrative. The growing corals were comparatively few in number, and the outer slope showed nothing but a mass of dead corals and dead coral boulders beyond 16 or 17 fathoms, few living corals being observed beyond 10 to 12 fathoms.

We also made an expedition to Point Venus, to determine, if possible, the rate of growth of the corals on Dolphin Bank from the marks which had been placed on Point Venus by Wilkes, in 1839, and by MM. Le Clerk and de Bénazé, of the French navy, in 1869. We found the stones and marks as described, but on examining Dolphin Bank in the steam launch I was greatly surprised to find that there were but few corals growing on it. I could see nothing but sparsely scattered heads, none larger than my fist! the top of the bank being entirely covered by Nullipores, although we sounded across the bank in all possible directions and examined it thoroughly. It is greatly to be regretted that Dolphin Bank was not examined, neither in 1839 nor in 1869, and notes made of what species of corals, if any, were growing on its surface; for an excellent opportunity has been lost to determine the growth of corals during a period of 60 years. The choice of this bank as a standard to determine the growth of corals was unfortunate, as it is in the midst of an area comparatively free from corals.

From Papeete we steamed back to Makatea, and examined the island more in detail. We crossed the island from west to east, the path leading down from the summit of the cliffs bordering the island into a sink at least 40 to 50 feet lower than the rim of either face of the island. The sink occupies a little more than one-third the length of the island. It is deeper at its southern extremity, where it is said to be 75 to 100 feet below the rim of the adjoining cliffs.

It is difficult to determine if this sink is the remnant of the former lagoon of the island, or of a sound formed during its elevation, or if it has been formed by the action of rain and atmospheric agencies. The amount of denudation and erosion to which this island has been subjected is very great, as is clearly indicated by the small cañons, pinnacles, and walls of limestone, as well as by the crevasses which occur in the surface of the basin in all directions. The extent to which this action has penetrated into the mass of the island is also plainly shown by the great number of caverns which crop out at all levels along the sea face of the cliffs, some of which are of great height, and extend as long galleries into the interior of the island.

From Makatea, we visited Niau, Apataki, Tikei, Fakarava, Anaa, Tahanea, Raroia, Takume, Makemo, Tekokota, Hikuero, Marokau, Hao, Aki-Aki, Nukutavake, going as far east as Pinaki, when we turned westward again and made for the Gloucester Islands. These, as well as Hereheretue, proved most interesting ; they formed, as it were, an epitome of what we had seen on a gigantic scale in the larger atolls of the western and central Paumotus. We could see at a glance in such small atolls as Nukutipipi and Anu-Anurunga the connection between structural features which, in an atoll of 40 miles in length and from 10 to 15 miles in width, it was often difficult to determine.

The deepest sounding among the Paumotus was on the line to the northward of Hereheretue in the direction of Mehetia; where we found a depth of 2524 fathoms, and a continuation of the red clay characterising the soundings since we left Pinaki.

We have seen nothing in this more extended examination of the group tending to show that there has anywhere been subsidence. On the contrary, the condition of the islands of the Paumotus cannot, it seems to me, be explained on any other theory except that in their present condition they have been formed in an area of elevation—an area of elevation extending from Matahiva on the west to Pinaki in the east, and from the Gloucester Islands on the south to Tikei on the north.

All the Paumotu Islands we have examined are, without exception, formed of Tertiary coralliferous limestone which has been elevated to a greater or less extent above the level of the sea, and then planed down by atmospheric agencies and submarine erosion, the greatest elevation being at Makatea (about 230 feet), and at Niau, where the Tertiary coralliferous limestone does not rise to a greater height than 20 feet. At Rairoa it was 15 to 16 feet high. At other islands it could be traced only as forming the shore platform.

The appearance of the old ledge and of the modern reef rock is so strikingly different that it is very simple to distinguish the two, even where only comparatively small fragments are found.

In the Paumotus, the islands have been elevated to a very moderate height, and probably to nearly the same height, for the old ledge forming the base of the modern structure is found exposed nearly everywhere at about low-water, when it cannot be traced at a slightly greater elevation. This would readily account for the nearly uniform height of the islands throughout the group.

But there is another element which comes into play in this group, and has an important part in shaping the ultimate condition of these atolls. At the Fijis we have seen the submarine erosion continue until there is little left of many of the atolls beyond the merest small islet or rock to indicate its structure. In the Paumotus, in the great atolls which are evidently only the exposed summits of parts of ridges or spurs of an extensive Tertiary coralliferous limestone bed, the rim of the atoll is, after having been denuded to the level of the sea, again built up from the material of its two faces, which is thrown up on the wide reef flats both from the sea face and from the lagoon side.

Many of the lagoons are filled with shoals or ledges awash or a few feet above the sea-level. These shoals are parts of the old ledge which have not as yet been eroded, and the disintegration of which has gone far to supply the material for the land of the outer rims of the atolls.

The lagoons of these atolls have a general depth of 13 to 20 fathoms. In some cases they are somewhat deeper, as is stated, but there are no measurements, the greater depths, 30 fathoms or more, being due to orogenic conditions. Some of the atolls are quite shallow, as at Matahiva, as well as Pinaki, where the lagoon is not more than 2 to 3 fathoms, and Takume, where it is from 5 to 6 fathoms deep. Some of the smaller islets we visited, among which are Tikei, Aki-Aki, and Nukutavake, have no lagoons.

The only atoll we have seen the lagoon of which is entirely shut off from the sea is Niau. In this case the old ledge forming the rim of the land, which surrounds the nearly circular lagoon, is about a third of a mile in width and sufficiently high, 15 to 20 feet, to prevent any sea from having access to it except in case of a cyclone. It is very difficult in this case to decide whether this lagoon has been gradually filled up after elevation, or whether it is merely a sink on a more or less uneven limestone surface.

Dana and other writers on coral reefs mention a great number of lagoons as being absolutely shut off from the sea. I take it these statements are due to their descriptions being taken from charts, many of which, as in the cause of the Paumotus, are very defective. For nothing is easier than to pass at a short distance by the wide or narrow cuts which give in so many cases the freest access to the sea to the interior of the lagoons, and described as closed because they have no boat passages. I could mention, as instances of such lagoons, those of the atolls of Takume, Hikuero, Anaa, &c., which may be said to be closed, yet into which a huge volume of water is poured at every tide over low parts of the encircling reef flats.

The character of the coral reefs of the Paumotus is very different from that of other coral reef regions I have seen. Nowhere have I seen such a small number of genera, so many small species, and such stunted development of the corals. None of the great heads of the genera so characteristic of the West Indian regions, or of the Great Barrier Reef of Australia, are to be seen, with the exception of a couple of species of alcyonaria they are absent, so far as our experience shows, and there are but few sponges and gorgonians to be found among the corals.

The same paucity of animal life seemed to extend to the deep-water fauna. All the hauls we made off the islands, in from 600 to 1000 fathoms, usually the most productive area of a sea slope, brought nothing, or so little that we came to grudge the time spent in trawling on the bottom, as well as towing on the surface or near it, a great contrast to the conditions in the Atlantic in similar latitudes, and very different from our anticipations.

From Papeete we steamed to Aitutaki, Niue, and for the deep hole of the Tonga-Kermadec Deep, about 75 miles to the eastward of Tonga-Tabu, and in 4173 fathoms made a haul with the Blake beam-trawl, by far the deepest trawl haul yet made. We found in the bag a number of large fragments of a silicious sponge belonging probably to the genus Crateromorpha, which had been obtained by the Challenger in the Western Pacific, but in depths less than 500 fathoms. We also brought up quite a large sample of the bottom ; it consisted of light brown volcanic mud mixed with radiolarians.

On our way back to Papeete from the Paumotus we examined the eastern coast of Tahiti, and from Papeete studied the western coast as far as Port Phaeton, at Tararoa Isthmus. We also examined, in a general way, the Leeward Society Islands ; Murea, Huaheine, Raiatea, Tahaa, Bora-Bora, Motu Iti and Maupiti. There are excellent charts of the Society Islands, so that it was comparatively simple to examine the typical points of the group and to gain an idea of their structure so far as it relates to coral reefs. The Society Islands are all volcanic islands edged with shore platforms, some of great width, upon which the barrier or the fringing reefs of the islands have grown. The structure of the reefs of the Society Islands is very similar to that of the Fiji reefs round volcanic islands. A comparison, for instance, of the charts of Kandavu, Viti Levu, Mbengha, Nairai, and of other volcanic islands in the Fijis, with those of the Society Group, will at once show their identity. Huge platforms of submarine denudation and erosion characterise both, with fringing and barrier reefs determined by local conditions. Perhaps it is easier to follow the changes which have taken place in the Society Islands ; and such islands as Tahaa and Bora-Bora, where we anchored, as well as Maupiti, are admirable examples and epitomes of the structure and mode of formation of the coral reefs of that group.

The only island of the Cook Group which we examined was Aitutaki, as Atiu is composed of elevated limestone, and Rarotonga is volcanic ; I hoped we might find that atoll to be in part volcanic and in part composed of elevated coralliferous limestone ; we found it to be volcanic, an island with the structure of Bora-Bora on a smaller scale.

We anchored at Niue, an island composed of elevated coralliferous limestone showing three well-marked terraces, the lowest of not more than 5 to 16 feet and in many places disappearing completely, the limestone cliffs rising vertically from the sea well into the second or even the third terraces.

The second terrace varies in height from 50 to 60 feet, the third from 90 to 100 feet. The second terrace is deeply undercut; and in the higher vertical cliffs extending into the third terrace from the sea, the former positions of the terraces are usually indicated by lines of caverns.

From Niue we went to the Tongas, which we found a most interesting group. The elevated Tertiary coralliferous limestones take here their greatest development, and are on a scale far beyond that of their development in the Lau Group of the Fijis, or the Paumotus. The first island of the Tongas we visited, Eua, is perhaps the most interesting of the islands composed of Tertiary elevated coralliferous limestone I have visited. From Dana's account of it, evidently given at second hand, I expected to find an island somewhat like Vitī Levu on a very much smaller scale. But as we steamed up to it from the east there could be no mistaking the magnificent face of nearly vertical limestone cliffs forming the whole eastern face of the island, and at points rising to over a thousand feet in height. At all projecting points lines of terraces were plainly marked: at the northern point three could be followed, and at the southern extremity five, with traces of a sixth perhaps.

Upon rounding the southern extremity of the island we could see that the island was composed of two ridges, running north, separated by a deep valley, the western ridge being much lower than the eastern, not rising to a greater height than a little over 500 feet. The western ridge is also composed of limestone, and at the headlands we could trace three terraces.

We anchored at English Roads, opposite the outlet of the drainage of the interior basin, where a small river has cut its way through a depression in the shore terrace. On landing we followed the crest of the western ridge for a few miles and could see the whole valley forming the basin of the island lying between the two ridges, at our feet. Nothing could show more clearly that such an island was not an elevated atoll, but a plateau which had been eroded and denuded for a long period of time by atmospheric and other agencies.

To the westward of the Tonga Islands is a line of volcanic islands extending nearly 200 miles, at a distance of from 15 to 20 miles parallel with the trend of the four irregularly-shaped plateaus upon which rise the Tonga Islands.

The Tonga-Tabu plateau is separated from the Namuka Group plateau by a funnel-shaped channel with a depth passing rapidly into 300 fathoms from the 100-fathom line. The Namuka plateau is rectangular; its principal island is Namuka, where we anchored.

This part of the Tongas is, like the Lau Group in Fiji, made up of islands in part volcanic and in part composed of elevated coralliferous limestone.

The Haapai plateau is triangular, with isolated islands rising on the north-western side from the deep water separating it from the Vavau plateau. On the northernmost plateau of the broad ridge of the Tonga Islands is the Vavau Group, by far the most picturesque of the Tonga Islands. Several parts of the island of Vavau, as at the entrance to the harbour of Neiafu, and at Neiafu, are finely terraced; four terraces are indicated there, and other flat-topped smaller islands show traces of two or three terraces. The northern edge of Vavau Island rises to a height of more than 500 feet, and slopes in a general way southward and inland. The southern shore is deeply indented by bays and sounds, and flanked by innumerable islands and islets, some of considerable height (150 to 250 feet) which gradually become smaller and smaller as they rise towards the southward and eastward, these islands having been formed from the denudation and erosion of the greater Vavau. They form tongues of land and sea and sounds of all shapes and sizes, showing the traces of the former land-connections of the islands and islets, and their disintegration on the eastward and southward by the action of the sea.

It is evident that in the Tonga Group, which is a very extensive area of elevation, the recent corals have played no part in the formation of the masses of land and of the plateau of the Tonga Ridge, and that here again, as in the Society Islands and the Cook Islands, both also in areas of elevation, they are a mere thin living shell or crust growing at their characteristic depths upon platforms which in the one case are volcanic, in the other calcareous, the formation of which has been independent of their growth.

After leaving Suva we steamed to Funafuti, stopping on the way at Nurakita, the southernmost of the Ellice Islands. I was, of course, greatly interested in my visit of Funafuti, where a boring had been made under the direction of a committee of

the Royal Society, in charge of Prof. David, of Sydney, after the first attempt under Prof. Sollas had failed. The second boring reached a depth of more than 1100 feet. This is not the place to discuss the bearing of the work done at Funafuti, as beyond the fact of the depth reached we have as yet no final statement by the committee of the interpretation put upon the detailed examination of the core obtained, and now in the hands of Prof. Judd and his assistants. In addition to the above-named islands, we also examined Nukufetau, another of the Ellice Group.

After leaving Nukufetau we encountered nothing but bad weather, which put a stop to all our work until we arrived under the lee of Arorai, the southernmost of the Gilbert Islands. On our way from Tapateuea we steamed to Apamama and Majana, which we examined, as well as Tarawa. We next examined Maraki. Both Maraki and Taritari, the last island of the Gilberts which we examined, are remarkable for the development of an inner row of islands and sand-bars in certain parts of the lagoon parallel to the outer land-rim, a feature which also exists in many of the Marshall Islands atolls.

We spent about three weeks in exploring the Marshall Islands, taking in turn the atolls of the Ralick Chain to the north of Jaluit: Ailinglab Lab, Namu, Kwajalong and Rongelab; and then the atolls of the Ratack Chain: Likieb, Wojje and Arhno. The atolls of the Marshall Group are noted for their great size and the comparatively small area of the outer land-rims, the land-rims of some of the atolls being reduced to a few insignificant islands and islets. In none of the atolls of the Ellice, Gilbert or Marshall Islands were we able to observe the character of the underlying base which forms the foundations of the land areas of these groups. In this respect these groups are in striking contrast to the Paumotus, the Society Islands, the Cook Group, Niue, the Tongas, and the Fiji Islands, where the character of the underlying foundations of the land-rims is readily ascertained. But, on the other had, these groups give us the means of studying the mode of formation of the land-rims in a most satisfactory manner, and nowhere have we been able to study as clearly the results of the various agencies at work in shaping the endless variations produced in the islands and islets of the different atolls by the incessant handling and rehandling of the material in place, or of the fresh material added from the disintegration of the sea or lagoon faces of the outer land, or of the corals on the outer and inner slopes. It has been very interesting to trace the ever-varying conditions which have resulted in producing so many variations in the appearance and structure of the islands and islets of the land-rims of the different groups.

The boring at Funafuti will show us the character and age of the rocks underlying the mass of recent material of which the land-rim, not only of that atoll, but probably also that of the other atolls of the group and of neighbouring groups, is composed, though, of course, we can only judge by analogy of the probability of the character of the underlying base from that of the nearest islands of which it has been ascertained. When we come to a group like the Marshalls, we have as our guide only the character of the base rock of the islands of the Carolines, which is volcanic; while Naru and Ocean Islands, to the west of the Gilberts and to the south-west of the Marshalls, indicate a base of ancient Tertiary limestone.

The Marshall Islands, as well as the Ellice and Gilbert, seem to be somewhat higher than the Paumotus; but this difference is only apparent, and is due to the difference in the height of the tides, which is very small in the Paumotus, while in these groups it may be five and even six feet.

From Jaluit we visited among the Carolines, the islands and atolls of Kusaie, Pingelap, Ponapi, Andema, Losap, Namu, the Royalist Group, Truk and Namonuito, obtaining thus an excellent idea of the character of the high volcanic islands of the group from our examinations of Kusaie and of Ponapi, having probably a volcanic basis, but this was not observed at any of those we examined.

The reefs of the volcanic islands of the Carolines are similar in character to those of the Society Islands, though there are some features, such as the great width of the platforms of submarine erosion of Ponapi and of Kusaie, and the development of a border of mangrove islands at the base of the volcanic islands, which are not found in the Society Islands.

The Truk Archipelago was perhaps the most interesting of the island groups of the Carolines, and it is the only group of volcanic islands surrounded by an encircling reef that I have thus far seen in the Pacific which at first glance lends any support to the theory of the formation of such island-groups as Truk by subsidence. This group was not visited by either Darwin or Dana ; and I can well imagine that an investigator seeing this group among the first coral reefs would readily describe the islands as the summits, nearly denuded, of a great island which had gradually sunk. But a closer examination will readily show, I think, that this group is not an exception to the general rule thus far obtaining in all the island groups of the Pacific I have visited during this trip ; that we must look to submarine erosion and to a multitude of local mechanical causes for our explanation of the formation of atolls and of barrier and encircling reefs, and that, on the contrary, subsidence has played no part in bringing about existing conditions of the atolls of the South and Central Pacific.

Nowhere have we seen better exemplified than at Truk how important a part is played by the existence of a submarine platform in the growth of coral reefs. The encircling reef protects the many islands of the group against a too rapid erosion, so that they are edged by narrow fringing reefs, and nowhere do we find the wide platforms so essential to the formation of barrier reefs. The effect of the north-east trades blowing so constantly in one direction for the greater part of the year is of course very great ; the disintegration and erosion of islands within its influence is incessant, and their action undoubtedly one of the essential factors in shaping the atolls of the different groups, not only according to the local positions of the individual islands, but also according to the geographical position of the groups. Thus far I do not think any observer has given sufficient weight to the importance of the action of the trades in modifying the islands within the limits of the trades, nor has any one noticed that the coral reefs are all situated practically within the limits of the trades both north and south of the equator.

The soundings made going west from Jaluit to Namonuito indicate that the various groups are, as is the case with the neighbouring groups of the Marshalls and Gilberts, isolated peaks with steep slopes rising from a depth of over 2000 fathoms. The line we ran from the northern end of Namonuito to Guam developed the eastern extension of a deep trough running south of the Ladrones. The existence of this trough had been indicated by a sounding of 4475 fathoms to the southwest of Guam made by the *Challenger*. We obtained, about 100 miles south-east of Guam, a depth of 4813 fathoms, a depth surpassed only, if I am not in error, by three soundings made by the *Penguin* in the deep trough extending from Tonga to the Kermadecs, and by three soundings made by the U.S.S. *Nero* also to the eastward of Guam.

Guam is not wholly volcanic ; the northern half consists of elevated coralliferous limestone. The vertical cliffs bordering the eastern face rise from a height of 100 to 250 or 300 feet at the northern extremity, and resemble in a way similar islands as the Paumotus (Makatea), Niue, Eua, Vavau and others in the Fijis which had made their cliffs a familiar feature in our explorations. In fact, outside of Viti Levu and Vanua Levu, this is the largest island known to me where we find a combination of volcanic rocks and of elevated coralliferous limestone. The *massif* forming the southern half of the island is volcanic, and the highest ridge, rising to about 1000 feet, runs parallel to the west coast, the longest slope being toward the east. This volcanic mass has burst through the limestone near Agana, and the outer western extension of the coralliferous limestone exists only in the shape of a few spurs running out from the volcanic mass, the largest of which are those forming the port of San Luis d'Apra. Near the northern extremity of the island a volcanic mass, Mount Santa Rosa, has burst through the limestone and rises about 150 feet above the general level of that part of the island.

We left Guam in time to reach Rota by day, and found that this island is a mass of elevated coralliferous limestone, the highest cliffs of which reach a height of 800 feet. Perhaps in none of the elevated islands have we been able to observe the terraces of submarine elevation as well as at Rota. It is quite probable that others of the Ladrones, like Saipan, and the islands to the south, are composed in part at least of elevated limestone, judging from the hydrographic charts and the sketches which accompany them. On many of the northern Ladrones there are active volcanoes, so that it is very possible that the volcanic outbursts which have pushed through the limestones, or have elevated parts of the islands of the group, are of comparatively recent date.

During the last part of our cruise, from Suva to Guam, the unfavorable weather greatly interfered with our deep-sea and pelagic work ; in fact with the exception of the soundings made to develop as far as practicable the depths in the regions of the various coral-reef groups we visited, we abandoned all idea of carrying out the deep-sea and pelagic work planned for the district between the Gilbert and Marshall and Caroline groups. To our great disappointment hardly any marine work could be accomplished, and our investigations were limited almost entirely to the study of the coral reefs of the regions passed through.

We were everywhere received with the greatest cordiality and courtesy : by the Governor of the Paumotus, the King of Tonga, Sir George O'Brien (the High Commissioner of the Western Pacific at Suva), Mr. E. Brandeis (the Landes-Hauptmann in charge of the Marshall Islands at Jaluit), and the Governor of the Carolines, and the Japanese authorities.

The work of the expedition was divided between Drs. W. M. Woodworth, A. G. Mayer, and my son Maximilian, who accompanied me as assistants ; and Mr. C. H. Townsend, Dr. Moore, and Mr. Alexander of the Fish Commission, who had also been detailed as members of the expedition.

I must also thank Capt. Moser and the officers of the *Albatross* for the untiring interest shown by them during the whole time of our expedition in the work of the ship, which was so foreign to the usual duties of a naval officer.

UNIVERSITY AND EDUCATIONAL INTELLIGENCE.

MR. R. S. CLAY, late lecturer in physics at the Birkbeck Institution, has been appointed principal of the Wandsworth Technical Institute.

THE Secondary Education Bill was read a second time in the House of Lords on Monday, after a discussion in which objection was raised to the limited character of the measure, and the large powers reserved for the Board of Education. It is not proposed to carry the Bill beyond the second reading this year.

THE first response to Mr. Chamberlain's appeal for further funds for the scientific department of the Birmingham University has been received from Sir James Chance, who has given the sum of 50,000*l.*, subject to conditions to be arranged with the University Council. The endowment fund of the University now amounts to about 400,000*l.*

THE Pass List for the 1900 D.Sc. Examination of the University of London contains the following names :—Experimental Physics : Reginald Stanley Clay, Richard Smith Willows, Harold Albert Wilson. Chemistry : Thomas Slater Price. Botany : Miss Maria Dawson. Zoology : Edgar Johnson Allen, Charles William Andrews.

SOME interesting particulars with regard to chemical and technical education in the United States were given by Prof. Chandler, of New York, in his presidential address to the Society of Chemical Industry last week. The most striking feature of the American system of higher and technical education is the fact that most of the institutions have been founded and maintained by liberal gifts of money from wealthy citizens, in many cases made during the donor's lifetime, and that only a small number have been endowed or supported by the public funds. Thus in 1899 over 33 million dollars were given in this way, the largest sum being the 15 million dollars given by Mr. Leland Stanford, together with large tracts of land, to which as yet no precise value can be attached, to complete the endowment of the Leland Stanford Junior University. There are in all 174 donors, averaging 190,000 dollars each. Schools of chemistry are now so numerous in the United States that it is almost impossible to state their exact number, but Prof. Chandler said it is more than 100. In all there are 480 universities and colleges, and 43 technical schools. In 1899 there were 9784 students pursuing professional courses in the schools of engineering, while 1487 graduated that year, receiving the degree of civil, mechanical, electrical, or mining engineer. The value to the industrial development of the United States of such an army of thoroughly-trained engineers and chemists cannot be too highly estimated.

THE operations of the Technical Instruction Committee of the Cheshire County Council are extensive and satisfactory. All the sums received under the Local Taxation (Customs and Excise) Act of 1890 have been devoted to the promotion of technical instruction in Cheshire from the commencement. The Technical Instruction Committee has framed a scheme of work which has gradually embraced the whole county, and has provided for the various and special requirements of the different districts, as well as of the county at large. The annual report just received records a year's work of steady progress and development, more especially in regard to relatively advanced instruction, and improved methods of carrying on the classes. During the year ending March 31, 1900, the grants made for purposes of technical instruction amounted to nearly 17,000*l.*, and this sum will be considerably increased during the ensuing year. A number of secondary schools receive grants from the Committee, and it is proposed to increase the payments to such schools. As has been before remarked in these notes, assistance thus given is having a very important effect upon the character of the education in secondary schools; for a condition of the grant is that scientific subjects should be taught; and proper laboratory accommodation provided. We read, for instance, in the present report : " All the schools to which grants for building and equipping laboratories and lecture rooms were made have completed these additions, hence they are in a much better position to give sound instruction in science subjects, and especially in the practical stage, than they were formerly." It is well to bear in mind the beneficial influence which Technical Instruction Committees have thus had upon the curricula of Grammar Schools and others of the old-fashioned type. Among other matters dealt with in the report are experiments on tuberculosis in cattle, for which the Committee made a grant of 250*l.*, and experimental work in agriculture.

SOCIETIES AND ACADEMIES.

PARIS.

Academy of Sciences, July 16.—M. Maurice Lévy in the chair.—On the uranium radiation, by M. Henri Becquerel. By mixing uranium chloride with barium chloride and precipitating with sulphuric acid, a precipitate of barium sulphate is obtained which is more or less radio-active according to the quantity of barium salt introduced. The radio-activity of the uranium salt remaining undergoes a corresponding diminution. It cannot be settled from these experiments whether uranium salts possess a radio-active power of their own, or whether this property is due to an admixture of an impurity.—Preparation and properties of two borides of silicon, by MM. Henri Moissan and Alfred Stock. By heating together, with special precautions, in a tube of infusible material a mixture of silicon and boron in the electric furnace, two new borides of silicon are produced, SiB_3 and SiB_6, which can be separated by taking advantage of the facts that SiB_3 is more readily attacked by fused potash, and SiB_6 is more readily destroyed by concentrated nitric acid. Both compounds resist the attack of most reagents and are very hard, scratching ruby with facility.—On the crystallisation of gold, by M. A. Ditte. Gold leaf, heated with a mixture of salt and sodium pyrosulphate or ferrous sulphate, is attacked, and shows traces of crystalline structure, although the temperature has been far below that of the fusion of gold. Platinum gives rise to similar phenomena under the same treatment.—On the solubility of calcium phosphate in the water of soils in presence of carbon dioxide, by M. Th. Schlœsing. Neutral $Ca_3(PO_4)_2$, obtained free from sodium salts, is practically insoluble in water free from dissolved carbon dioxide. The solubility increases with the amount of dissolved carbonic acid, but if this is accompanied in solution with the corresponding amount of calcium bicarbonate, the solvent action is practically destroyed.—New researches on double fertilisation in angiosperms, by M. L. Guignard. In addition to the cases previously given of double fertilisation in monocotyledons, this has now been observed in *Narcissus poeticus* and *Scilla bifolia*. In dicotyledons, *Anemone nemorosa* has been most completely studied.—The movements of the air on encountering surfaces of different forms, by M. Marey.—Observations of the planets (F.G.) and (F.H.) made with the large equatorial of the Observatory of Bordeaux, by MM. G. Rayet and F. Féraud.—On the formation of coal basins, by M. Grand'Eury. Remarks on the mode of formation of the Loire basin.—M. Lipschitz was nominated a correspondent in the section of Geometry.—On the instability of certain periodic solutions, by M. Levi-Civita.—On the ternary bilinear forms of Hermite, by M. Louis Kollros.—On the law of corresponding states, by M. Daniel Berthelot. After discussing various modifications that have been suggested for bringing Van der Waals'

formula into closer agreement with experiment, the author concludes that the three constants f_0, v_0, T_0 are not sufficient to rigorously define the function $f(p, v, T)$ of a substance. It is necessary to add two new constants, T_m and v_m, corresponding to the displacements of the zeros of volume and temperature.—On the temperature of maximum density of aqueous solutions of ammonium chloride and lithium bromide and iodide, by M. L. C. de Coppet. The molecular lowering of the temperature of maximum density varied from 7·16 for ammonium chloride to 8·31 for lithium iodide.—On the electrolytic estimation of bismuth, by M. Dmitry Balachowsky. It is possible to get a coherent metallic deposit of bismuth allowing of washing, provided the following conditions are observed : slight acidity of the solution, absence of large quantities of halogen ions, matt electrodes, and low current density.—On the amalgams of sodium and potassium, by MM. Gunta and Férée. Four amalgams of mercury and sodium were isolated and analysed, Hg_4Na, Hg_6Na, Hg_5Na, Hg_7Na. Similar amalgams, although less clearly defined, were obtained with potassium.—On the reduction of tungstic anhydride by zinc : preparation of pure tungsten, by M. Marcel Delépine. Tungsten of a purity varying from 98·5 to 100 per cent. is obtained by heating zinc with tungstic anhydride or with ammonium tungstate.—Action of reduced nickel upon acetylene, by MM. Paul Sabatier and J. B. Senderens. Acetylene does not react upon reduced nickel in the cold if precautions have been taken to remove all traces of hydrogen from the metal by heating it in a current of nitrogen.—Action of cyanacetic esters with substituted acid radicles upon diazobenzene chloride and tetrazodiphenyl chloride, by M. G. Favrel.—On the limits of grafting in plants, by M. Lucien Daniel.—Action of dry and moist air upon plants, by M. Eberhardt. Compared with dry air, moist air increases the development of the plant, both leaves and stem, the diameter of the latter being reduced. It tends to exaggerate the leaf surface and to diminish the quantity of chlorophyll contained in the leaves.—The volcanic rocks of the Somali Protectorate, by MM. A. de Gennes and A. Bonard.—On a marine formation at the bottom of the Cañon of Regalon, by M. David Martin.—On certain substances specific in pellagra, by MM. V. Babès and E. Manicatide.

CONTENTS.

THURSDAY, AUGUST 2, 1900.

WEAPONS AND WOUNDS.

Les armes blanches ; leur action et leurs effets vulnérants.
By H. Nimier, Professeur au Val-de-Grâce, and Ed.
Laval, médecin aide-major de première classe. Pp. 448.
(Paris : Félix Alcan, 1900.)

*Les projectiles des armes de guerre ; leur action vul-
nérante.* By the same authors. Pp. 212. (Paris :
Félix Alcan, 1899.)

*Les explosifs, les poudres, les projectiles d'exercice ; leur
action et leurs effets vulnérants.* By the same authors.
Pp. 192. (Paris : Félix Alcan, 1899.)

THESE volumes, although their titles are formidable
enough, can scarcely be said to exhaust the subject
of the means invented by man for the special purpose of
destroying his own race. Prof. Nimier, one of the authors,
is well known as a writer on military medical subjects,
and no doubt he has thought it unnecessary to repeat
much of what he has already written on these and cognate subjects. The volumes, however, fill considerable
gaps in our own literature. We have few writers in this
country whose works stand out prominently as works of
importance on the same subjects during the present
century. Guthrie and Ballingall are practically the only
writers whose contributions to the subject cover the
period between the Peninsular and the Crimean Wars.
Since then, Longmore's classical work on gunshot injuries was the sole work of reference until a year or two
ago, when Stevenson, his successor in the Army Medical
School at Netley, brought our knowledge of the injuries
likely to be produced by modern fire-arms up to date.
On the Continent the system of compulsory military
service is responsible for the fact that these subjects excite widespread interest amongst the general and scientific public to a much greater extent than in England ;
and many important additions have been made to the
literature of wounds in war by continental writers within
the last few years.

The bulkiest of these three volumes treats of a class
of weapons which nowadays play a comparatively unimportant part in wars between civilised Powers—namely,
the bayonet, sabre, sword, lance and arrow. This volume
also contains a chapter on defensive armour. We were
much disappointed in finding that its bulkiness, instead
of being due to pages full of historical detail, as we had
anticipated, depends largely upon needless repetitions of
the diagnosis, prognosis and treatment of the wounds
produced in the different regions and tissues of the human
body by the various weapons included in the term *armes
blanches.* These repetitions are wearisome and unnecessary. The wounds produced by side-arms differ in no
way from the contused, incised, punctured or poisoned
wounds described in text-books on general surgery. In
other words, there is no *specialism* in the subject for the
student of physical, military or medical science, except
perhaps that portion of it which deals with arrows and
arrow poisons.

The volume, however, is of much value as a work of
reference for any one desirous of comparing the shape

and construction of the side-arms used by the several
European Powers. The chapters on the bayonet and
mixed types of bayonet are specially interesting in this
respect. The introduction of the magazine rifle has led
to important changes in the length and weight of the
bayonet and to its probable use in future wars. The
short knife-bayonet is now almost universally adopted ;
the shortest being the 21 centimètre long Norwegian
bayonet, used with the Krag-Jörgensön rifle ; as compared with the British Lee-Metford bayonet of 30 centimètres. The Austrian, German, Italian and Spanish
bayonets hold an intermediate position between these
two. Russia and France, on the other hand, still retain
the long, narrow-pointed bayonet. Thus the Russian
bayonet, 1891 pattern, measures 43 centimètres, and the
French Lebel bayonet 52 centimètres with a weight of
466 grammes. Some idea of the slender stiletto-like
proportions of the latter may be formed from the fact
that, although nearly twice as long, it weighs actually
less than the Lee-Metford bayonet. The authors
enter somewhat fully into reasons why the Russians
and French prefer this long weapon of offence
to the shorter bayonet, which they describe as being intended more for lopping branches of trees and digging
trenches than for any other purpose in war. They agree
in thinking that, in modern pitched battles, the last phase,
namely, the charge, *restera à l'état purement platonique*,
one side yielding to the other without waiting for cold
steel. But surprises, night attacks and assaults on con-
voys are circumstances of war which will occur as frequently in the future as in the past ; and it is these that
render the retention of the bayonet as a weapon of offence
of paramount importance. The Russians recognise this
fact so well that their cavalry carry a bayonet for use with
the carbine. The authors also refer to the national temperament of the French as one of the reasons why they have
not followed the example of neighbouring European countries in adopting a bayonet more suitable for camp purposes than as a weapon of offence. The French, they
say, are specially fond of side-arms as weapons, by which
we assume that the national temperament urges them to
get to close quarters as soon as possible. This, however,
seems scarcely sufficient reason for the preference they
have for a long narrow bayonet. Our own soldiers, at
any rate, have amply proved in the present campaign in
South Africa that the short, stout bayonet possesses destructive and moral effects possibly equalled, but certainly
not excelled, by the longer weapon. The authors have
little to say that is of interest with regard to the sabre,
sword and lance. The type of these weapons is practically the same in all civilised countries, and the chapters
on them are mainly descriptive.

Arrows and arrow poisons are fully discussed, the
chapter on them being mainly a *résumé* of the investigations made by the French naval surgeons, Le Dantec,
whose tables of the geographical distribution and
classification of arrow poisons are given in detail. The
subject is occupying much attention at present in this
country in consequence of the rapid extension of
European spheres of influence in the African Hinterland,
where poisoned arrows are so widely used by aboriginal
tribes. Those who are interested in the subject will find
accurate and important details in this chapter, but it

P

must be confessed that English readers have fuller historical and scientific information available on arrow poisons in the inaugural address delivered by Prof. Stockmann, of Aberdeen University, to the North British Branch of the Pharmaceutical Society in 1898 (*Pharmaceutical Journal*, November 26 and December 3, 1898). It is interesting to note how thoroughly Prof. Stockmann's ethnological distribution of arrow poisons—a distribution which is extremely well marked—agrees with that of the French writers. In their description of the methods adopted for the propulsion of arrows and similar projectiles, the authors make no mention of the use of the blowpipe, a somewhat formidable weapon in the hands of Bornean aborigines.

In the chapter on defensive armour there is a guarded reference to what may prove of considerable importance in the future. In the helmet and cuirass we still possess the relics of a period when nations fought with sword and lance ; but the opinion is gradually gaining ground that the use of defensive armour in the form of shields for protection against the projectiles of modern fire-arms, may become a feature in future wars. The Danish Army have already adopted a form of shield for this purpose, and the principle is also recognised in the use of shields with the quick-firing automatic guns of the Maxim type.

The authors' contribution to the subject of projectiles deals with modern fire-arms only ; and, with the exception that the projectiles of the automatic guns are not considered at all, the information on the subject is concise and complete. Modern small-arm projectiles are exceptionally well described. The physical qualities of these projectiles are remarkably similar in the different European countries, the chief variation in form being in the calibre of the bullet, which is between 6·5 and 7 millimètres sectional diameter for Italy, Holland, Norway, Roumania and Spain, and between 7 and 8 millimètres for other countries, the smaller calibre of the former being compensated for by greater length. The dynamic properties, however, have considerable and important variations, which the authors describe with the lucidity and precision characteristic of French writings on subjects of this nature. The chief practical interest in the dynamics of projectiles lies in the relationship between these properties and the surgical results. The principle to which the modern small-arm projectile owes its origin is indicated in the formula $f = mv^2$. In other words, the production of a bullet with a high rate of velocity at the expense of mass has been the object attained in the adoption of magazine rifles. But it is gradually dawning upon the military mind that the equation of work, expressed by the formula $o = \dfrac{mv^2}{2}$, does not express accurately the relative values of velocity and mass in the surgical results. The first occasion on which our own troops used the high velocity small calibre bullet in actual war—namely, in Waziristan in 1895—proved the fallacy of the formula in this respect ; and it is now fairly well recognised that the mass of the projectile is probably as important a factor in producing surgical disaster as its velocity. No doubt the actual power of penetration and the resistance required to bring the projectile to a state of rest is accurately expressed by the

formula ; but it is this very power of penetration, depending so much on increase of velocity combined with reduction of mass, that has earned for the modern bullet the epithet humane. To pursue the subject further would lead to a variety of speculations as to the nature of the weapon of the future. The authors clearly recognise this, and are inclined to regard the action of the United States of America in reducing the diameter of the projectile of the naval small arm to 6 mm. as indicating a tendency to convert modern firearms into *carabines de salon* or *fusils d'enfants*. They are apparently much in sympathy with the use of bullets that deform or expand on impact, or at any rate produce shock, and fear that the agitation against these bullets will only lead to the use of some more deadly projectile in the future. These expanding or deforming bullets and their effects are fully described. The best known example is the soft-nosed Lee-Metford bullet, but the authors refer also to the use of the Lebel bullet with the hard envelope stripped at the apex. They state, however, that the latter does not expand on impact, although it produces shock. Another interesting example of the expanding bullet is the Swiss bullet, which has the lead core naked at the base instead of at the apex. The deformity in this bullet after impact, by the incurving of the soft base, is said to be as great as, if not greater than, the deformity at the apex of the Dum-dum type of bullet. The explosive effects sometimes caused by high velocity bullets are also very clearly discussed, but no new light is thrown upon this very curious phenomenon. The authors adhere to the generally accepted theory that the effects are due to secondary energy transmitted to tissues of a certain nature or in a certain state of tension. The possible part played by ricochets, deformities, and varying angles of impact is not mentioned in this connection. There is also entire absence of any reference to the use of true explosive bullets, which, although abolished by international agreement in 1868, are alleged to have been employed by some Boer commandos in the war that is now being waged in South Africa. The chapter on artillery fire is interesting and valuable, and concludes with a suggestive article on the moral effects of this branch of the service. In other respects the dynamics and ballistics of artillery projectiles and the wounding effects of fragments of shells, projectiles, &c., are worked out on the same lines as in the chapters on small-arm projectiles.

In the volume on explosives there is a variety of details, not readily obtained elsewhere in the same compact form, and on this account it is perhaps the most valuable of the three volumes to the student of military surgery or medical jurisprudence, to whom it is chiefly of interest. The effects of the various explosives in use are amply illustrated by historical incidents, especially incidents connected with anarchist attempts and with explosions in stores, ships and arsenals. The explosives used in the cartridges of the small-arms of different countries are also well described and compared. The authors include in this volume a chapter on the accidents connected with sapping and mining, a subject which we do not remember to have seen noticed in other works of a similar nature. The physical phenomena of a peculiar form of *intoxication* or suffocation to which sappers are liable are fully described and not generally known in this country.

The complete absence of bibliographic references is a notable defect in the volumes, more especially as they are mainly compilations of the works of other writers and investigators, whose names appear frequently in the authors' pages. In fact the reader, who might wish to consult the original works, will have great difficulty in knowing where to look for them. We are always glad, however, to welcome any contribution to a literature which is so meagrely represented in our own country.

W. G. M.

PLANTS OF THE PAST.

Éléments de Paléobotanique. By R. Zeiller. Pp. 421. (Paris : Carré and Naud, 1900.)

SOME of the most striking advances in botanical science during the last two or three decades have been in the domain of Palæobotany. The study of fossil plants is now generally recognised as a science of primary importance, which affords, not merely useful data for the stratigraphical geologist, but furnishes valuable information as to the course of plant evolution, and enables us to connect some of the phyla of the plant-kingdom at points where their common origin is clearly indicated. Prof. Zeiller, of the École des Mines, Paris, has played a prominent part in placing fossil botany on a thoroughly scientific basis ; his work, which embraces a wide field, is characterised by a philosophical handling of facts, a thoroughness of treatment and a breadth of view that are too frequently lacking in scientific writings of the present day. In the book before us Prof. Zeiller has performed a difficult task with considerable success. Within a small compass he has included a systematic though necessarily brief account of the more important types of fossil plants, and concise and clearly-written chapters on various subjects of geological and botanical interest. The illustrations are well executed, and it is a pleasure to note that many of them are new. In the section treating of the preservation of plants as fossils, Zeiller draws attention to a method of examination of "impressions" which he has used with considerable success. It is often possible, after suitable chemical treatment, to examine microscopically the thin carbonaceous film, which may sometimes be detached from the surface of a plant fragment lying on a slab of shale, and in this way to obtain important information as to anatomical details.

Some interesting examples of the Siphoneæ are figured and briefly described ; but one or two of the examples quoted, *e.g.* the supposed *Caulerpa* from the Kimmeridge Clay, are of very doubtful value. The fossil Myxomycetes of Palæozoic age, described by Renault and other authors, should be mentioned with a word of caution as to their acceptance as undoubted Mycetozoa. In describing the vascular cryptogams, Zeiller notes the danger of attaching too much importance to the presence or absence of secondary wood, or to the isosporous or heterosporous character of a genus ; mistakes made in the past, which have persisted for many years, emphasise the need of this caution.

NO. 1605, VOL. 62]

In discussing the systematic position of various extinct generic types of exceptional interest which point to a common origin of cycads and ferns, Zeiller speaks of the collateral form of the vascular bundles as one of those cycadean characters which is met with also in recent ferns. It is, however, important to bear in mind the fact that in the collateral bundles of *Ophioglossum* and other ferns the protoxylem occupies an endarch position, while the cycadean type of bundle is usually mesarch.

The chapter on fossil ferns is particularly well done, and contains much that is new. The genus *Microdictyon*, mentioned by Zeiller as a mesozoic fern closely allied to *Laccopteris*, is hardly sufficiently distinct to be retained as a separate type.

The enlarged photograph of a leaflet of the well-known "fern," *Alethopteris Serlii*, given to illustrate the occurrence of what may possibly be traces of sporangia, does not afford satisfactory evidence that this fern-like frond bore fern-like sporangia. We are still in want of convincing evidence as to the nature of the reproductive organs of both *Alethopteris* and *Neuropteris*, genera in which the characters of ferns and cycads were combined.

A drawing is given of an exceptionally fine example of a rhætic fern—*Clathropteris platyphylla*—from Tonkin ; as Zeiller has shown, this plant may be compared with the recent genus *Dipteris*, which, like *Matonia pectinata*, represents a tropical survival of a widely-spread mesozoic family of ferns. The inclusion of the genus *Sagenopteris* with the Hydropterideæ, rather than with the Filices, is perhaps a little rash, as the evidence so far available as to the reproductive organs is by no means conclusive.

A good description is given of the genus *Sphenophyllum*, but it is to be regretted that exigencies of space prevent full justice being done in this and other cases to the account of anatomical features. Zeiller discusses the possibility of *Sphenophyllum* having lived as a water-plant in the Coal period forests, but it is perhaps more probable that its long and slender stems were supported, like lianas, by the boughs of stouter trees.

In dealing with the Calamarieæ, Zeiller does full justice to the work of English authors, and discusses controversial points with admirable judgment and an open mind. The genus *Sigillaria* is described as a true lycopodiaceous plant, agreeing in certain respects with *Isoëtes*.

In the account of fossil cycads, Zeiller, like other authors, quotes an example of a cretaceous *Cycas* carpophyll, figured by Heer, from Greenland ; the figured specimen, which the writer has seen in the Copenhagen Museum, is not sufficiently well preserved to be determined with certainty, and bears but a distant resemblance to Heer's figure. The genus *Podozamites*, placed by Zeiller among the Cycads, may possibly be more correctly included in the Coniferæ, but it is a type of somewhat doubtful position. The flowers of *Zamites gigas*, usually known as *Williamsonia*, mentioned in the section dealing with the Bennettiteæ, are usually of Inferior Oolite rather than Lower Lias age.

Prof. Zeiller gives a useful summary illustrating our knowledge of fossil angiosperms ; as he points out, the literature on Tertiary plants is in urgent need of revision at the hands of experienced systematists. The concluding chapter, dealing with the bearing of palæo-botanical evidence on plant evolution, is full of interest, and particularly valuable as being written by one who possesses both a wide knowledge of the available data and the power of critically weighing the evidence. Referring to the comparative study of species of fossil plants, Zeiller writes :

"Les Espèces, commes les genres, se succèdent par voie de substitution et non par voie de transformation graduelle, et il en parait être de même à tous les niveaux."

A very useful bibliography of writings referred to in the text is given at the end of the volume.   A. C. S.

PHOTOGRAPHY IN NATURAL COLOURS.

Lehrbuch der Photochromie (Photographie der natürlichen Farben). Von Wilhelm Zenker ; neu herausgegeben von Prof. Dr. B. Schwalbe. Pp. xiii + 157. (Braunschweig : Vieweg und Sohn, 1900)

IN 1868, after long study and repetitions of Edmond Becquerel's experiments on photochromy, Dr. Wilhelm Zenker himself printed and published a " Lehrbuch der Photochromie," which contained a physical explanation of the colour-correctness of these photochromatic images. Zenker's book did not have a wide circulation—it would be difficult, perhaps, to name any one in England who has read it—and it was not until 1890 that Lippmann, by founding a *new* method on the principle suggested by Zenker, drew a slightly increased attention to Zenker's labours. That the attention was only slightly increased was due to two causes : firstly, the rather astonishing *results* of Lippmann and others helped to overshadow the *principle* of Zenker in the eyes of most people ; secondly, among all those whose pursuits have any claim to be considered as scientific, English photographers are especially noticeable for their deliberate ignorance of the creative work of the past in photography.

For the latter reason, chiefly, the present writer has given, during the past year, a rather full analysis of Zenker's work in " Camera Obscura," and we have now a reprint, in good *English* type, of the book. In the words of the preface, " The more modern researches on photography in natural colours have shown that the way and the explanations of modern attempts are connected in many respects with Zenker's ideas." The volume contains besides a portrait of Zenker, a sketch of his life and index of his works by Prof. Gustav Krech ; and, finally, Herr E. Tonn gives (pp. 131-157) an account of the further development of photochromy on the foundations of Zenker's theory. We shall notice these briefly in their order ; but, with regard to the " Lehrbuch " itself, shall abstain from entering at all fully into the subject of its contents, as in the above-cited reference there is already a full account of it in English.

Wilhelm Zenker (1829-1899) cultivated many different branches of knowledge. His first papers (1850-1866) were zoological ; the " Lehrbuch " was his first contribution to photography ; his other papers were on colour-perception (1867), photography and physical optics, astrophysics, and, in later life, meteorology. The " Lehrbuch," however, is probably the most important of

his works, and it is to be hoped that now, with this excellent reprint, his methods will have some influence on English photography.[1]

The book is divided into three parts : (1) Considerations on colours (" Das Wesen der Farben ") ; (2) Account of his predecessors' work in photochromy (" Die Wiedergabe der Farben ") ; and the third part (" Theorie der Photochromie "), after an account of the theories of Seebeck, Becquerel and others, contains Zenker's own ideas (pp. 116-129). There are one or two useful notes to this section.

Herr Tonn's section, with one exception, seems very complete, and full references are given. It is, however, a pity that the very pregnant hint of Lord Rayleigh should be unnoticed (*Phil. Mag.*, 1887). Lord Rayleigh, independently of Zenker, and starting from totally different considerations, indicated in a footnote the Zenker principle, and even went farther ; for not only did he seek to *explain* the results of Becquerel by this principle, but seemed to see the possibility of a *new* method of photochromy based on it. It would be interesting to have some account of Lord Rayleigh's then promised experimental investigations. If Herr Tonn knew this paper, it is difficult to understand how he resisted the temptation of comparing Rayleigh and Zenker—Zenker who was so clearly a non-mathematician.

The chief value of the book, the writer persists in believing, is not historical—for it *has* not had very much influence in the moulding of thought—but is in its spirit ; the influence of its point of view and methods is needed above all at the present time for English photographers ; this does not mean, of course, the small number of English *photo-chemists.*

PHILIP E. B. JOURDAIN.

OUR BOOK SHELF.

Die Harze und die Harzbehälter. By A. Tschirch. Pp. viii + 417. (Leipzig : Gebrüder Borntraeger, 1900.)

THE author has spent eight years in collecting and arranging the scattered facts relating to the obscure group of organic compounds which are classified as resins by virtue of a common physical characteristic.

What Kekulé termed "the chemical lumber room" has occupied at one time a collection of similar obscure groups, such as the alcaloids, colouring matters, tannins, aromatic compounds, &c. ; but since the year when that chemist gave to the world his benzene formula, the lumber room has been industriously ransacked and its contents dragged forth into the light of day. Perhaps the resins have received the scantiest share of attention ; partly, no doubt, owing to the practical difficulties which they offer to the chemist.

We know nothing of the molecular state which finds its physical expression in these amorphous, translucent compounds, nor how to bring them into a condition of ascertained purity. How often does a promising research miscarry by the unwelcome appearance of resinous products ! Nevertheless the mass of research which has accumulated on the subject fills 400 closely printed pages.

A great amount of this research gives very little indication of the nature of the resins themselves. The older chemists distilled them and obtained products such

1 His work for the Paris Academy prize in 1868 stands in close relation to his theory of photochromy (see Fizeau's report, *Compt. rend.*, lxvi., lxvii.). Zenker's memoir was never published, and Otto Wiener (*Wied. Ann.*, 1890, 1895) later and independently followed the same train of thought. (*Cf.* also Cornu, Poincaré, Potier and Berthelot, *Compt. rend.*, cxii. ; and Drude, *Wied. Ann.*, xli., xliii.).

as benzoic acid, toluene, turpentine, &c. At a later date this severe method of treatment was replaced by the milder action of fused potash, with the result that a number of new aromatic acids and phenols were discovered. At the present time the separation of the various constituents of a resin is effected by the use of solvents and the numerous reagents which the resources of modern organic chemistry can offer. The results do **not** carry us very far. As the author says, " our march of conquest has only begun, and the present volume may suggest a successful plan of campaign."	J. B. C.

The Lepidoptera of the British Islands: a Descriptive Account of the Families, Genera and Species Indigenous to Great Britain and Ireland, their Preparatory States, Habits and Localities. By Charles G. Barrett, F.E.S. Vol. vi. Parts 59-70. Heterocera (Noctuina—Geometrina). Pp. 388. Plates 233-280. (London: Lovell, Reeve and Co., Ltd., 1900.)

THE present instalment of Mr. Barrett's great work includes 110 species, from *Hoporina croceago*, Schiff., to *Halix wauaria*, L., and is written in the same exhaustive manner as previous volumes, giving all the information that a collector of British Lepidoptera (as such) is most likely to require. To Continental entomologists who wish to acquire an accurate knowledge of our limited insular fauna. it would also prove very useful ; though it is to be regretted that the bulk of the book, which may be expected to extend to nearly twenty volumes, and the unavoidable costliness of the larger edition issued with plates (which are not included in the cheap edition), must necessarily tend to restrict the sale. Those requiring it may therefore be recommended to obtain it volume by volume, or in monthly parts, as it appears, rather than to wait till the whole work is completed. We need not repeat our comments on earlier volumes, which will equally apply to the one before us ; but the accounts given of the habits of the various moths discussed are always interesting, and sometimes curious ; thus we learn that the rare *Cerastis erythrocephala*, Schiff., after its discovery in 1847, was met with occasionally till 1859, when it seems to have almost disappeared till 1872 and 1873, since when only one specimen, taken in 1894, has been found in England. The periodicity of the appearance of many species in these islands is curious, and has never been fully explained, for the causes which appear applicable to some cases will not explain others ; and, moreover, uncertainty in the appearance of species seems to increase rather than to diminish. English names are not a conspicuous feature in this book, but Mr. Barrett notes that a recent writer has called *Xylina conformis*, Schiff., "The Conformist," and the next species, *X. lambda*, Fab., "The Nonconformist" ! The resemblance of species of *Calocampa* and *Cucullia* to bits of stick is commented on ; in fact, certain moths and larvæ thus fill the gap in our protected fauna caused by the absence of the stick insects proper, or Phasmidæ, which are not found nearer to our shores than the South of France. Several species noted in this volume seem to be now extinct in our islands ; thus, *Chariclea delphinii*, L., does not seem to have been taken in England since about 1815. Their place has been taken by others ; for example, the northern migration of *Plusia moneta*, Fab., reached England in 1890, and is probably still extending. Other moths of interest are those with cannibal larvæ, such as *Scopelosoma satellitia*, L., *Heliothis armigera*, Hübn., &c. There are many other interesting observations, which we have no room to quote, in the present volume, comprising, as it does, the conclusion of the Noctuæ, the Deltoidæ, and the first few species of the Geometræ. We may, however, note that the enigmatical *Sarrothripa revayana*, Schiff., is regarded by Mr. Barrett as a true *Noctua*, and is placed at the end of the *Noctuae Trifidae*.
W. F. K.

LETTER TO THE EDITOR.

[*The Editor does not hold himself responsible for opinions expressed by his correspondents. Neither can he undertake to return, or to correspond with the writers of, rejected manuscripts intended for this or any other part of* NATURE. *No notice is taken of anonymous communications.*]

The Plankton of the Bay of Biscay.

WITH the valuable assistance of Mr. L. A. Borradaile, of Selwyn College, Cambridge, I have just completed a series of observations on the plankton of the Bay of Biscay extending over about three weeks, by means of opening and closing nets, as well as ordinary tow-nets. Our observations point to the fact (unexpected at any rate by myself) that the smaller Mesoplankton practically ceases at a depth of about 1000 fathoms. This conclusion agrees with that reached by Prof. Chun on the basis of the *Valdivia* Expedition (Deutsche Tiefsee Expedition, 1898-99, p. 44). with which, however, we were unacquainted until we had arrived at it independently. Below this limit we almost always captured a few specimens, as to which it was doubtful whether they were alive when captured, or were merely corpses of a shallower fauna sinking to the bottom, but in a few cases we at present incline to assign them to a living Mesoplankton.

We have also taken about 90 hauls under varied conditions at varied depths between 100 fathoms and the surface, which will eventually, we hope, give a fairly accurate basis for the determination of the vertical movements of the Epiplankton.

Our thanks are due to the Lords Commissioners of the Admiralty for placing the ship at our disposal, and to Captain Field and the other officers of the *Research* for their ungrudging assistance.	G. HERBERT FOWLER.
H.M.S. *Research*, Devonport, July 27.

THE TEACHING OF MATHEMATICS.

I THINK it very important to try to get a view of our system of teaching mathematics which is not too much tinted with the pleasant memories of one's youth. Like all the men who arrogate to themselves the right to preach on this subject, I was in my youth a keen geometrician, loving Euclid and abstract reasoning. But I have taught mathematics to the average boy at a public school, and this has enabled me to get a new view. I have seen faces bright outside my room become covered as with a thin film of dulness as they entered ; I have known men, the best of their year in England in classics, lose in half an hour (as men did in the first day of slavery in old times) half their feeling of manhood ; and I have known that, as an orthodox teacher of mathematics, I was really doing my best to destroy young souls. Happily, our English boys instinctively take to athletics as a remedy, and I know of nothing which gives greater proof of the inherent strength, in good instincts and common sense, of our race than this refusal to allow one's soul to be utterly destroyed. I have also mixed much with engineers, who really need some mathematics in their daily work, men who say that they once were taught mathematics, and I know that these men never use anything more advanced than arithmetic, and actually loathe a mathematical expression when it intrudes itself into a paper read before an engineering society. Of all branches of engineering, electrical engineering relies most upon exact calculation. Well, the average electrical engineer in good practice would rather work a week at many separate arithmetical examples than try for an hour to get out the simple algebraic expression, which includes all his week's results and much more. Yet he has passed perhaps certain rather advanced examinations in mathematics. Furthermore, those engineers who can most readily apply mathematics to engineering problems, almost invariably descend to the position of teachers and professors in schools and colleges, and they seem to lose touch completely with the actual life of their profession.

NATURE

I have studied these phenomena very carefully, and I affirm that they are directly traceable to the absurd thing called mathematical teaching in schools and colleges.

The framers of educational methods took in their youth to abstract reasoning as a duck takes to water, and of course they assume that a boy who cannot in one year understand a little Euclid must be stupid. In truth, it is a very exceptional mind, and not, perhaps, a very healthy mind, which can learn things or train itself through abstract reasoning; nor, indeed, is much ever learnt in this way. Do we philosophise about swimming before we know how to swim; or about walking or jumping, or cycling or riding a horse, or planing wood, or chipping or filing metals, or about playing billiards or cricket? Is it through philosophy that we learn a game of cards, or to read or to write? No; we first learn by actual trial; we practice as our mind lets us; we philosophise afterwards—perhaps long afterwards. Then if we are too clever or stupid, we insist on teaching a pupil from the point of view which we have at the end of our studies, and we refuse to look at things from the pupil's point of view.

What a natural but ghastly statement the boy made who said; "Yes, Euclid and Xenophon, the beasts, wrote books for the third and fourth forms"! It is even a ghastlier notion that the jokesomeness of a philosopher, the unessential fringe of a subject, often becomes the soul-destroying, weary, worrying study of a schoolboy.

In a short article I shall not attempt to put forward my views as to how mathematics ought to be taught; I have published some of them in a summary of lectures on "Practical Mathematics," published by the Science and Art Department, and in my "Calculus for Engineers."

We let a Board School boy jump over all the ancient philosophy of arithmetic with its twenty-seven independent Greek characters (for our ten figures), the study of which required a lifetime, so that only old men could do multiplication, and they not only needed many hours to do one easy bit of multiplication, but declared that if the art were not practised every day it could not be remembered. Why not also let a boy jump over all the Euclidian philosophy of geometry, and assume even the forty-seventh proposition of the first book of Euclid to be true? Why not let him replace the second and fifth books of Euclid by a page of simple algebra, and give him much of the sixth book as axiomatic? If you must insist on abstract reasoning, you had better remember that nothing is really axiomatic; but any well-established truths may be looked upon as fundamental or axiomatic, and a system of abstract reasoning may be founded upon them. At present, a man at Cambridge finishes just where the really interesting and useful part of mathematics begins. There would not indeed be much difficulty in framing a course in which he would begin by studies where the studies of good mathematicians now end. This has been tried and proved successful. The present rules of the game are really a little too absurd. A difficult vector subject like geometry must be studied before algebra. Simple exercises on squared paper, well within the capacity of even illiterate persons, must not be approached until one has wasted years on higher algebra and trigonometry and geometrical conics, because they belong to the subject of co-ordinate geometry. It is assumed that it is not until after co-ordinate geometry is thoroughly studied that a man can take in the idea which underlies the calculus, an idea which is possessed by every young boy with absolute accuracy, and by every healthy mind.

Some friends of mine assert that no boy or man ought to be allowed to use logarithms until he knows how to calculate logarithms. They say this, knowing that the calculation is a branch of what is called higher mathematics, and that the average schoolboy, after six

years at mathematics, finds it hopeless to even begin the study of the exponential theorem. It is a hard saying! It is exactly like saying that a boy must not wear a watch or a pair of trousers until he is able to make a watch or a pair of trousers. I am an advocate for the use by all students of all appliances which may be useful to them, whether made by tailors, or watchmakers, or instrument makers, or builders, or pure mathematicians. We need not believe a craftsman when he tells us that we cannot utilise his results without practising his trade. Nevertheless, it is good to be able to do some things for one's self, such as sewing on buttons, or using the lathe or a blowpipe, or the development of a little mathematics. If readers will refer to the above-mentioned *summary* they will see that I consider a good system of mathematics teaching of fundamental importance in the education of all men.

I must not dwell any longer on the imperfections of the existing system, but I hope that even readers who do not quite agree with me that much of the sixth book of Euclid ought to be regarded as axiomatic, will agree that what we usually call arithmetic is useless. For races not troubled with our abominable system of weights and measures, the whole of arithmetic consists merely of multiplication and division. To them a decimal is no more difficult to understand than an ordinary number. It is supposed that an English boy understands at once the meaning of 4,590,000 or 4590 or 459, but that such a number as 45·9 or ·459 or ·00459 is beyond his comprehension. I say that this is a difficulty artificially maintained by our stupid methods of teaching. Like the rest of our stupid methods, it is due to our unscientific ways of thinking. Because the embryo passes through all the stages of development of its ancestors, a boy of the nineteenth century must be taught according to all the systems ever in use and in the same order of time. The decimal system of stating numbers is 700 years old in Europe, but it was not till 280 years ago that Napier invented the use of decimals and the decimal point. Think of compelling all emigrants to pass to America through Cuba, because Cuba was discovered first. Think of making boys learn Latin and Greek before they can write English, because Latin and Greek were the only languages in which there was a literature known to Englishmen 450 years ago.

Again, the ingenious teachers of last century incorporated every kind of arithmetical example in a book and called each kind by a slang name—practice, interest, discount, tare and tret, alligation, position, &c., and we must teach exactly as they did. I do not mind retaining the buttons at the back of my coat; many useless ancient ceremonies may still be practised, and I shall not object. I can even admire them, but the unscientific waste of the valuable youth of millions of our people, now that we are face to face with nations who are determined to destroy England through commerce and war, is so abhorrent to me that I cannot think of it with patience. It is not merely in arithmetic and geometry, but in the higher parts of mathematics that this waste goes on. Newton employed geometrical conics in his astronomical studies, and mechanics was developed; and therefore it is that every young engineer must study mechanics through astronomy, and he dare not think of the differential calculus till he has finished geometrical conics. The young applier of physics, the engineer, needs a teaching of mathematics which will make his mathematical knowledge part of his mental machinery, which he shall use as readily and certainly as a bird uses its wings; and we teach him in such a way that he hates the sight of a mathematical symbol all his life after.

It is just as in classics. Ask the average man if he ever reads anything now in Latin or Greek; ask him about anything to which he devoted ten years of his study at school, and he will answer that the only men he knows of who read the classics are a few famous scholars and the cads who read with delight cribs of the *Odyssey* and the *Iliad* just as if they were novels, because they never had the advantage of a classical education. But, of course, his mind was trained, he can always say that.

The authorities of the Science and Art Department recognise that apprentices and others attending evening classes may possibly benefit by a course of study very different from what is necessary if students are being prepared for university and other examinations. Hence, in addition to their very complete orthodox courses of instruction, they recognise the new method of study, the most elementary part of which is beginning to get crystallised in the following syllabus. There is also an " Advanced " syllabus, which is too long to be published here. I would advise interested persons to write to the Department for copies, and also for the report on the result of last year's examination, as well as for copies of the examination papers and of the above-mentioned summary.

I venture to hope for criticism of this syllabus—first, from men like my Cambridge friends, who are quite sympathetic, but who think the method one fit for evening classes only ; second, from men who think with me that the method is one which may be adopted in every school in the country, and adopted even with the one or two boys in a thousand who are likely to become able mathematicians ; third, from other men. Whatever be the point of view of any critic, he must surely feel that exhaustive criticism is important, for there are many large technical schools in England in which the method has already been adopted, the orthodox system being quite given up. I have been informed that the method is spreading rapidly in Germany also. I can already see from the exceedingly interesting examination results that crystallisation is proceeding rapidly, and if criticism is to be of value, now is the time for it. I hope also that the seemingly bumptious manner in which I criticise orthodox methods of teaching will not induce contemptuous indifference in men of thought. I hold a brief in the interests of average boys and men ; my strong language and possible excess of zeal are due to the fact that nearly all the clever men have briefs on the other side.

　　　　　　　　　　　　　　JOHN PERRY.

PRACTICAL MATHEMATICS.

ELEMENTARY STAGE.

Arithmetic.—The use of decimals ; the fallacy of retaining more figures than are justifiable in calculations involving numbers which represent observed or measured quantities. Contracted and approximate methods of multiplying and dividing numbers whereby all unnecessary figures may be omitted. Using rough checks in arithmetical work, especially with regard to the position of the decimal point.

The use of 5.204×10^5 for 520,400 and of 5.204×10^{-3} for ·005204· The meaning of a common logarithm ; the use of logarithms in making calculations involving multiplication, division, involution and evolution. Calculation of numerical values from all sorts of formulæ.

The principle underlying the construction and method of using a common slide rule ; the use of a slide rule in making calculations. Conversion of common logarithms into Napierian logarithms. The calculation of square roots by the ordinary arithmetical method. Using algebraic formulæ in working questions on ratio and variation.

Algebra.—To understand any formula so as to be able to use it if numerical values are given for the various quantities. Rules of Indices.

Being told in words how to deal arithmetically with a quantity, to be able to state the matter algebraically. Problems leading to easy equations in one or two unknowns. Easy transformations and simplifications of formulæ. The determination of the numerical values of constants in equations of known

form, when particular values of the variables are given. The meaning of the expression " A varies as B."

Factors of such expressions as $x^2 - a^2$, $x^2 + 11x + 30$, $x^2 - 5x - 66$.

Mensuration.—The rule for the length of the circumference of a circle. The rules for the areas of a triangle, rectangle, parallelogram, circle ; areas of the surfaces of a right circular cylinder, right circular cone, sphere, circular anchor ring. The determination of the area of an irregular plane figure (1) by using a planimeter ; (2) by using Simpson's or other well-known rules for the case where a number of equidistant ordinates or widths are given ; (3) by the use of squared paper whether the given ordinates or widths are equidistant or not, the " mid-ordinate rule " being used. Determination of volumes of a prism or cylinder, cone, sphere, circular anchor ring.

The determination of the volume of an irregular solid by each of the *three* methods for an irregular area, the process being first to obtain an irregular plane figure in which the varying *ordinates* or widths represent the varying *cross sections* of the solid.

Some practical methods of finding areas and volumes. Determination of weights from volumes when densities are given.

Stating a mensuration rule as an algebraic formula. In such a formula any one of the quantities may be the unknown one, the others being known.

Use of Squared Paper.—The use of squared paper by merchants and others to show at a glance the rise and fall of prices, of temperature, of the tide, &c. The use of squared paper should be illustrated by the working of many kinds of exercises, but it should be pointed out that there is a general idea underlying them all. The following may be mentioned :—

Plotting of statistics of any kind whatsoever, of general or special interest. What such curves teach. Rates of increase.

Interpolation, or the finding of probable intermediate values. Probable errors of observation. Forming complete price lists by shopkeepers. The calculation of a table of logarithms. Finding an average value. Areas and volumes, as explained above. The method of fixing the position of a point in a plane ; the x and y and also the r and θ, co-ordinates of a point. Plotting of functions, such as $y = ax^n$, $y = ae^{bx}$, where a, b, n, may have all sorts of values. The straight line. Determination of maximum and minimum values. The solution of equations. Very clear notions of what we mean by the roots of equations may be obtained by the use of squared paper. Rates of increase. Speed of a body. Determination of laws which exist between observed quantities, especially of linear laws. Corrections for errors of observation when the plotted quantities are the results of experiment.

In all the work on squared paper a student should be made to understand that an exercise is not completed until the scales and the names of the plotted quantities are clearly indicated on the paper. Also that those scales should be avoided which are obviously inconvenient. Finally, the scales should be chosen so that the plotted figure shall occupy the greater part of the sheet of paper ; at any rate, the figure should not be crowded in one corner of the paper.

Geometry.—Dividing lines into parts in given proportions, and other illustrations of the 6th Book of Euclid. Measurement of angles in degrees and radians. The definitions of the sine, cosine and tangent of an angle ; determination of their values by drawing and measurement ; setting out of angles by means of a protractor when they are given in degrees or radians, also when the value of the sine, cosine or tangent is given. Use of tables of sines, cosines and tangents. The solution of a right angled triangle by calculation and by drawing to scale. The construction of a triangle from given data ; determination of the area of a triangle. The more important propositions of Euclid may be illustrated by actual drawing ; if the proposition is about angles, these may be measured by means of a protractor ; or if it refers to the equality of lines, areas or ratios, lengths may be measured by a scale and the necessary calculations made arithmetically. This combination of drawing and arithmetical calculation may be freely used to *illustrate* the truth of a proposition.

The method of representing the position of a point in space by its distances from three co-ordinate planes. How the angles are measured between (1) a line and plane ; (2) two planes. The angle between two lines has a meaning whether they do or do not meet. What is meant by the projection of a line or a plane figure on a plane. Plan and elevation of a line which is inclined at given angles to the co-ordinate planes. The meaning of the terms " trace of a line," " trace of a plane."

The difference between a *scalar* quantity and a *vector* quantity. Addition and subtraction of vectors.

Slope of a line ; slope of a curve at any point in it. Rate of increase of one quantity y relatively to the increase of another quantity x ; the symbol for this rate of increase, namely, $\frac{dy}{dx}$; how to determine $\frac{dy}{dx}$ when the law connecting x and y is of the form $y = ax^n$. Easy exercises on this rule.

In setting out the above syllabus the items have been arranged under the various branches of the subject.

It will be obvious that it is not intended that these should be studied in the order in which they appear ; the teacher will arrange a mixed course such as seems to him best for the class of students with whom he has to deal.

ANALYTICAL PORTRAITURE.

IT seems well to put on record the principal results of experiments that I have recently made to *isolate the particulars* in which one portrait differs from another. They had a measure of success, but not enough to deserve illustration or lengthy description. The objects I had hoped to attain are important ; namely, to define photographically the direction and degrees in which any individual differs from the race to which he belongs, the race being represented by a composite picture of many individuals belonging to it. Or, again, to define the particulars in which any variety of a plant or animal differs from its parent species. Or to define family features ; or to isolate expressions, recollecting that these consist both of subtractions from, and additions to, the features as seen in repose.

My starting point was that the exact superimposition of a rather faint positive upon its rather faint negative produces an approximately uniform grey, when they are viewed as a single transparency. Thus, I photographed a rotating disc that had been faced with white paper and divided into concentric rings. The innermost disc was left white, the outermost ring was painted black, and the intermediate rings contained successively increasing proportions of black to white. The photographic negative showed rings of graded tints, and from this I took a positive by contact. Subsequently applying the positive to the negative, film to film, and viewing them as a transparency, a nearly uniform grey surface was produced. It was necessary to superimpose them with exactness ; otherwise the edges of the rings were conspicuously dark in one part, and light in the opposite part. Another test experiment was to paste together thicknesses of tracing paper—two-fold, three-fold, &c., up to twelve-fold—to cut distinctively shaped snippets of these and to variously distribute them over the surface of a glass plate, which was then photographed, and a positive taken as well. On treating the positive and negative as above, all the tints between those of the three-fold and the nine-fold inclusive produced a uniform grey.

Let A and B be any two pictures whose respective negatives and positives will be called *neg. a, pos. a, neg. b, pos. b*. My object was to produce photographically a third picture X which should express the difference between A and B ; that is, should be equal to A—B, or else a fourth picture Y which should represent B—A.

It will, however, be simpler to treat the problem at first as an optical one, based on the following equations :—

(I.) *pos. a + neg. a = grey* ; (II.) *pos. a + x = pos. b*

(if treated as a photographic problem, (II.) would be replaced by *pos. a + x = neg. b*). From these we obtain

(III.) *pos. a + {pos. b + neg. a} = pos. b + grey*

and

(IV.) *pos. b + {pos. a + neg. b} = pos. a + grey.*

Calling the terms within brackets by the name of " transformers," the transformer of *b* into *a* is the negative of the transformer of *a* into *b*. The two terms within brackets may be " composited " together on equal terms, then the result may be composited with the first term, allowing two-thirds of the total time of exposure to the transformer, and one-third to the first term. Or, what comes to the same thing in the end, all three terms may be composited in equal shares, allowing one-third of the total time of exposure to each. The transformers in (III.) and (IV.) being respectively $x + grey$ and $y + grey$, are nearly equivalent for the purposes of the inquiry to x and y, because the addition of a uniform shade of grey has little or no effect on pictorial resemblance. A portrait does not cease to resemble the original when it has become somewhat browned by exposure to a London atmosphere, or when it is viewed in shade, or under a tinted glass. Its distinctiveness depends on the *differences* (not the ratios) being preserved between the tints of all adjacent elements of its surface. Of course the grey must not be too dark ; otherwise the deeper tints of the portrait would appear indistinguishably black.

This method of transformation succeeds fairly well. I changed an **F** on a white ground into a good **G** on a grey ground, and I changed with passable success one portrait A on white ground into another portrait B on grey ground, but the transformer itself gave little of that information to the eye which I had expected. It *must* have nearly isolated, but it failed to exhibit in an intelligible form the differences between A and B. Then I photographed two faces, each in two expressions, the one glum and the other smiling broadly. I could turn the glum face into the smiling one, or *vice versa*, by means of the suitable transformer ; but the transformers themselves were ghastly to look at, and did not at all give the impression of a detached smile or of a detached glumness. Part of the ghastliness was due to the different densities of the superimposed positives and negatives, which did not neatly obliterate one another in the unchanged portions of the face, and part was due to their not being superimposed in the best possible way. There can be no doubt of the best fit when engaged in making the transformer of an **I** into an **L** ; but the eye must determine the best fit and proportions of the two components of the transformer of one portrait into another. I cannot yet make up my mind whether or no the process admits of substantial improvement, but feel sure that the only satisfactory experiments now would be those made by two converging lanterns on a screen, one at least of which admits of easy and delicate adjustment in direction and in the intensity of its illumination. The most suitable portraits for the attempt are apparently such as are popularly, and sometimes reproachfully, termed " artistic," that is to say, with blurred outlines and medium tints ; certainly not those which in photographic language are called " plucky." I have no means in my house for experiments of this kind, but perhaps a trial might be made in some laboratory where they exist. The point is to ascertain whether the images of *neg. a* and *pos. b* can be so combined on the screen as to give an intelligible and useful idea of the differences between A and B.

FRANCIS GALTON.

A RECOLLECTION OF KING UMBERTO.

HOW enthusiastically the late King of Italy could devote himself to the welfare of science and art, those of us who were at Como last September had an opportunity of seeing. One very hot day he arrived with the Queen and the Duke of Naples by train from their palace at Monza, near Milan. First they made an official inspection of the galleries and machinery in the Silk and Electricity Exhibition, then they visited the Exhibition of Sacred Art, and, after lunch, they opened the Electrical Congress, held to celebrate the Volta centenary.

This was a no mere regal opening occupying a fraction of an hour, for a solid afternoon's work was done in receiving various addresses and listening to a long lecture on Volta and his pile, in which Volta's work was described at length, and even discussed from the modern standpoint of the ionic theory of voltaic action. Finally, the king had several foreigners presented to him, and he chatted with us about the things in which *we* were interested.

But even this was not enough for one day's work, since, before leaving, the Royal party went to the cathedral to listen to the new oratorio, *The Nativity*, which was exciting so much interest in Como at that time.

Such a keen personal interest in science and art made the king much loved by a people who venerate even the tomb of a worker like Volta. And those of us who saw King Umberto only at Como last year feel that it is not merely a king, but a friend who has now been killed.

W. E. AYRTON.

NOTES.

ON Monday next, August 6, the International Congress of Physics will be opened at Paris with an address by the president, Prof. Cornu. The Congress will then be divided into the seven following sections, which will meet in the rooms of the Société française de Physique : (1) general questions, instruction, measurements ; (2) mechanical and molecular physics ; (3) optics ; (4) electricity and magnetism ; (5) magneto-optics, radio-activity, discharges in gases ; (6) cosmical physics ; (7) biological physics. As many of our readers are aware, much attention has been given to the organisation of the Congress. The secretaries of the committee, Prof. Poincaré and Dr. Guillaume, have been entrusted with the production of three volumes, already in the press, containing more than seventy reports on physical questions of current interest and importance, contributed by physicists of various nationalities. Among the subjects dealt with by British physicists are : the movements produced in an indefinite solid by the displacement of a material body, by Lord Kelvin ; the constant of gravitation, by Mr. C. V. Boys ; the propagation of electricity, by Prof. Poynting ; electric discharges in gases, by Prof. J. J. Thomson ; properties of alloys, by Sir W. C. Roberts-Austen ; and the unit of heat, by Mr. E. H. Griffiths. In addition there are contributions by Profs. Lorentz, van 't Hoff, Warburg, Voigt, van der Waals, H. Poincaré, Cornu, Lippmann, Potier, Becquerel, Arrhénius, Exner, Spring and others. The sectional meetings will partly be held simultaneously and partly at different hours, in order to give members an opportunity of hearing papers of interest to all physicists. In addition to the serious work of the Congress, provision has been made for lighter entertainment. The Municipal Council of Paris will hold a reception on Tuesday, August 7, and the French President will give a reception to the members on August 9. Prince Roland Bonaparte will give a soirée on August 11, and in his splendid library an exhibition of new apparatus and experiments will be held. There is thus every promise that the meeting will be both interesting and pleasant to all who are able to take part in it.

IT is announced that permission has been granted for the Institution of Electrical Engineers to hold a reception in the British Royal Pavilion in the Paris Exhibition from 5 to 7 p.m., on Wednesday, August 22, and that arrangements for the reception are being made accordingly.

WE learn from *Science* that the New York Board of Estimate and Apportionment has authorised the expenditure of 200,000 dollars for the Botanical Garden, and 150,000 dollars for an addition to the American Museum of Natural History.

MR. LEONARD S. LOAT, who is investigating the fishes of Egypt for the British Museum and the Egyptian Government, was last heard of at Korti, where he reports (on May 18) a hot wind and a temperature of 115° in the shade. He had sent home upwards of 2200 specimens of Nile-fishes to the Natural History Museum, and as soon as the river had risen sufficiently would proceed to Senaar and Khartoum.

MR. J. S. BUDGETT, who is engaged in collecting fishes on the River Gambia, dates his last letters (June 22) from McCarthy's Island in the interior. There had been a disturbance in the colony, and one of the Commissioners and a party of police were believed to have lost their lives ; but this had not affected Mr. Budgett's operations, and he had a large number of Polypteri and Protopteri in floating cages in the river. He was in good health, and expected to be home again in September.

THE Rocky Mountain Goat (*Haploceros montanus*) in the Zoological Society's Gardens has now put on its full white summer dress, and is well worthy of inspection. This animal, until lately, was supposed to be the only representative of the Mountain or Goat-like Antelopes in the New World, but a second species of the same genus has recently been discovered in Alaska, and named by Mr. D. G. Elliott, of Chicago, *Oreamnus kennedyi*. The form is no doubt closely allied to *Nemorhaedus* of the mountain ranges of Asia, and probably found its way to the New World in company with the Rocky Mountain Sheep and Wapiti Deer.

THE *Electrician* states that the German Electro-Chemical Society is arranging to hold its seventh annual meeting at Zürich on August 5-7. In addition to the reading of a number of papers, visits are to be paid to the Polytechnic and to the works of the Oerlikon Co.

THE Moxon gold medal of the Royal College of Physicians, founded in 1886 in memory of the late Dr. Walter Moxon, and awarded every third year for distinction in clinical medicine, has been awarded to Sir William T. Gairdner, K.C.B., F.R.S., Emeritus professor of medicine in the University of Glasgow. Prof. Clifford Allbutt will deliver the Harveian Oration on October 18 (St. Luke's Day) ; and Dr. A. E. Garrod, the Bradshaw Lecture in November. Dr. Henry Head has been appointed the Goulstonian, Dr. J. Frank Payne the Lumleian, and Dr. Halliburton the Croonian Lecturer for 1901, and Dr. J. W. Washbourn the Croonian Lecturer for 1902.

WE are indebted to Mr. C. Repington, of Bridge End, Ockham, Surrey, for some eggs of the Wood Leopard Moth (*Zeuzera Æsculi*). They resemble strings of small oval beads, of a yellowish testaceous colour. The moth, although reputed scarce, is commoner round London than is generally supposed, and would be very destructive, if its numbers were not kept down by birds, notably by sparrows and woodpeckers. The eggs might be reared by placing them in chinks of the bark of almost any deciduous tree (apple, elm, &c.). The larvæ feed, like those of the Goat Moth (*Cossus ligniperda*), in the wood of growing trees, but are much less common.

QUESTIONS referring to the Marine Biological Association were asked in the House of Commons on Thursday last, and were replied to by Mr. Ritchie as follows :—"In 1885, the Treasury, when agreeing to a grant to the Plymouth laboratory of the Marine Biological Association, made it a condition 'That the council undertakes to place space in the Plymouth laboratory at the disposal of any competent investigator deputed by a recognised authority to carry out any investigation into fish questions for which the laboratory can give facilities.' The Board of Trade have never employed any naturalist to make investigations on fishes at the laboratory, and they have no staff or funds to devote to such a purpose. I have no information as to what has been done by other Government authorities. The Board of Trade have occasionally consulted the council of the Marine Biological Association on fishery subjects. The latest occasion had reference to the question of a fisheries exhibit at the Paris Exhibition. The inspectors of the Board of Trade have on many occasions consulted the officials of the association in an informal manner. The association were not directly consulted by of the Board Trade as to the Bill dealing with undersized fish, which, however, was founded on the recommendations of the Select Committee of 1893, who took evidence from the association."

THE sixty-eighth annual meeting of the British Medical Association was opened at Ipswich on Tuesday under the Presidency of Dr. W. A. Elliston. An address in medicine was delivered on Wednesday by Dr. Philip Henry Pye-Smith, F.R.S.; an address in surgery will be delivered to-day by Dr. Frederick Treves; and an address in obstetrics will be delivered by Dr. William J. Smyly on Friday. The scientific business of the meeting is being conducted in thirteen sections, as follows, namely: Medicine; surgery; obstetrics and gynæcology; State medicine; psychology; physiology; pathology; ophthalmology; diseases of children; pharmacology and therapeutics; laryngology and otology; tropical diseases; navy, army and ambulance. The exhibits in the annual museum held in connection with the meeting are arranged in the following sections:—Section A: Food and drugs, including prepared foods, chemical and pharmaceutical preparations, &c.; Section B: Instruments, comprising medical and surgical instruments and appliances, electrical instruments, microscopes, &c.; Section C: Books, including diagrams, charts, &c.; Section D: Sanitary appliances and ambulances.

THE address of Mr. E. M. Holmes, the president of the British Pharmaceutical Conference held in London last week, and most of the papers read and discussed at the meetings, are published in full in the current number of the *Pharmaceutical Journal*. Mr. Holmes reviewed the progress of science, so far as it affected pharmacy, during the present century, and indicated some of the changes which have occurred. Referring to the subject of an international Pharmacopœia, he remarked:—"A General Pharmacopœia, that would enable a pharmacist to dispense a prescription with uniformity in any pharmacy on the Continent, may be regarded as a Utopian rather than a practical idea, and could only be attained by alphabetically arranging in a dictionary form all the formulæ in all the known pharmacopœias. But there can be no reason why an approach towards it should not be made by a congress of medical men and pharmacists, limiting their attention, in the first place, to poisonous preparations only, and, in order to avoid international jealousies, adopting as a standard the formulæ that approach nearest to decimal proportions. The comparison of different formulæ is rendered a simple matter by the publication of the different strengths of preparations of the various pharmacopœias in Squire's 'Companion to the British Pharmacopœia.' The next step might be to make uniform the strength of the most generally used preparations that are not poisonous. A really useful International Pharmacopœia cannot be otherwise than a gradual growth!"

AMONG the papers printed in the *Pharmaceutical Journal* are several of interest outside pharmaceutical circles. Mr. E. J. Parry shows that the so-called santalol, which exists to the extent of about 90 per cent. in sandal-wood oil, is a mixture of two or more bodies of an alcoholic nature, one of which is that to which the name santalene has been applied. Mr. T. H Wardleworth deals with some pharmaceutical and economic plants of Jamaica. As the result of a visit to that island he

is of opinion that pharmacists would do well to attempt to obtain from British colonies supplies of many drugs which at present come from other parts of the world. Messrs. T. Tyrer and A. Levy continue their investigation on melting points, the substances more recently examined being salicylic acid, salol, carbolic acid, menthol, and thymol. Messrs. C. T. Tyrer and A. Wertheimer have made a careful physical examination of American, Russian, and French turpentine oils and terebene made therefrom, and propose, at some future date, to investigate similar products from all possible sources. As a general rule they find that the higher the initial rotation of American turpentine the smaller is the product of inactive mixture capable of steam distillation and the higher the specific gravity. French turpentine has a greater tendency to oxidise than American, being intermediate between that and the Russian oil. Dr. F. B. Power summarises the methods which have been advocated for the preparation of mercurous iodide, and gives the results of determinations of the amount of iodine or pure mercurous iodide contained in specimens of the compound made in different ways. These results indicate that precipitated mercurous iodide is quite uniform in composition and also sufficiently stable when properly protected. Mr. E. Dowzard thinks that useful information may be obtained by determining the viscosity of essential oils. A specimen of pure lemon oil had a viscosity of 139·6, whilst that of citrene was found to be 195·8, and that of a mixture of citrene with 7·5 per cent. of citral was 114·9.

H.M.S. *Viper*, which it will be remembered is driven by the "Parsons' steam turbine system" (built by the patentees at their works at Newcastle for the British Government, and described and illustrated in NATURE of March 1), has this month not only broken her own record of 35½ knots, but proved to possess qualifications equally important in marine engineering. On six consecutive runs (says *Engineering*, July 20) the following speeds were attained :—

Time on measured mile						Equivalent speed in knots
m. s.						
1 38¾	...	...	...	...	...	36·585
1 41¼	...	...	...	...	...	35·503
1 37	...	...	...	...	...	37·113
1 38¾	...	...	...	...	...	36·585
1 37	...	...	...	...	...	37·113
1 39½	...	...	...	...	...	36·072

The mean of two runs with and against the tide was 36·845 knots. The Admiralty mean of the six runs over the mile, with and against the tide, was 36·581 knots, which speed was also the mean for the hour's run. The mean revolutions for the hour's run was 1180 per minute. The steam pressure in the turbines ran up to 200 lbs. per square inch, and the mean pressure in the stokeholds was 4½ inches. Another important feature of the trials was that the *Viper* worked up from a speed of 14 knots to 36·585 knots in twenty minutes; almost as much importance is attached to this as to the high speeds attained, both being very valuable considerations in war vessels and cross channel boats. The trials, it is stated, worked without a hitch, and vibration was practically imperceptible in any part of the vessel.

ACCORDING to a writer in the *Times*, several earthquake shocks were felt at Bognor on July 18, between 10 and 11 p.m. Another correspondent suggests that they were merely the reports of the naval salute fired at Cherbourg on the departure of the French President at the times mentioned. The character of the disturbances, as described, certainly bears out this view. Similar movements and rumbling sounds were also observed at Torquay at the same time. Bognor is eighty-nine miles, and Torquay 101 miles, from Cherbourg.

THE Faculty of Sciences of the University of Rome proposes to publish by subscription a complete collection of the works of the late Prof. Eugenio Beltrami. The collection will probably extend to three or four large volumes of 2000 pages in all, and a copy will be sent to subscribers of 2*l*. and upwards. Subscriptions are to be sent to Isaia Sonzogno, secretary of the Scuola d'Applicazione per gli Ingegneri, 5, Piazza San Pietro in Vincoli, Rome.

IN a pamphlet, entitled the "Inidikil System," Mr. A. Lincoln Hyde suggests a decimal system of weights and measures for the English speaking people based on taking the inch as the fundamental unit. One of the author's main arguments for the proposal appears to be the failure of the metric system to obtain public favour in Great Britain and America, and he therefore thinks it desirable to make another attempt at decimalising our weights and measures.

IN the *Proceedings* of the Rochester Academy of Sciences, vol. iii. Brochure 2, Prof. Arthur L. Baker gives a general summary of vector analysis, and a short note on the graphic representation of imaginaries—both suitable for teaching purposes.

PART 13 of the *Rendiconti del R. Istituto Lombardo* contains two mathematical papers—one by Dr. Duilio Gigli, on helicoidal and ruled surfaces in elliptic space ; the other by Signor U. Amaldi, on commutative linear substitutions. In the former, Dr. Gigli, starting with the classical methods of Beltrami, deduces certain theorems relating to ruled surfaces in space of constant curvature, exactly analogous to those known to exist in Euclidean space. The second paper deals with certain generalisations enunciated by Schlesinger in his note, "Ueber vertauschbare lineare Substitutionen" (Crelle, 1899), to which Amaldi applies certain synthetic methods due to Prof. Pincherle.

IN connection with the view that phosphorescence is due to movements of the ether determined by the vibrations of material particles, much interest attaches to the question as to whether the intensity of phosphorescence is modified by a magnetic field. Some experiments described by M. Alexandre de Hemptinne in the *Bulletin de la Classe des Sciences* (Brussels) appear to answer this question in the negative. In one experiment the phosphorescent substance was contained in a tube about 30 cm. long, placed between the poles of an electro-magnet. The middle part of the tube was thus submitted to a field of about 30,000 C.G.S. units, while at the ends the magnetic force was comparatively feeble. The tube contained sulphide of calcium or of zinc, prepared after Becquerel's methods, and it was excited by being exposed to the sun. On observing the tube in a dark room it was seen to be uniformly phosphorescent throughout its length ; it remained phosphorescent for a considerable time, and gradually the intensity diminished, but at no stage of the experiment was any difference of intensity noticeable from one end of the tube to the other. In order to make more exact observations, M. de Hemptinne constructed a phosphoroscope of sufficiently large dimensions to contain an electro-magnet. Although this method was much more sensitive than the preceding one, not the slightest difference could be observed in the behaviour of sulphide of lime, sulphide of zinc, nitrate of uranium, diamond and other more or less phosphorescent substances when submitted to a magnetic field of about 32,000 units.

HIGH summer temperatures have continued to prevail over the southern portion of our islands, but there has been an absence of the excessive heat which was experienced in the preceding week. Heavy thunderstorms occurred over a large part of England on July 27, resulting in a fairly heavy rain over London

and smaller amounts in many parts of the country. Two quite separate storms passed over the metropolis, one in the afternoon and the second late in the evening. The lightning flashes were very frequent and unusually brilliant. At Greenwich, the rainfall accompanying the thunderstorms measured 0·84 inch, while in Westminster it only amounted to 0·42 inch. The weather has been generally cooler since the storms, although the thermometer in the south of England is well above the average. The mean temperature for July was 4° above the average at Greenwich; the mean of the maxima was 78°, and of the minima, or night readings, 57°. The total rainfall for the month at Greenwich was 1·41 inches, which is more than an inch less than the average.

A PAMPHLET on the organisation of the meteorological service in Japan has been published by the Tokio Observatory, for presentation to the Paris Exhibition. This service, which is very complete, consists of eighty stations of the first and second orders, and of about 900 stations at which only rainfall or temperature is recorded. The departmental stations, in accordance with the decree establishing the service, are established in suitable places, chosen by the Ministry of Public Instruction, and any persons wishing to establish additional stations have to obtain the authority of that Ministry. Electrical, earthquake, and other exceptional phenomena are regularly observed, in addition to the usual meteorological observations. All vessels belonging either to the imperial or merchant service, which are over 100 tons burden, are compelled to make observations at regular intervals, six times daily, and the logs are forwarded to the central observatory. There is also a regular service of weather telegraphy and storm warnings. The observations made three times daily are published in *Weather Reports*, together with forecasts for the following day. The average success of these forecasts amounts to 82 per cent., and of the storm warnings to 70 per cent. In addition to the *Daily Weather Report*, monthly and yearly bulletins are issued ; these are naturally written in the Japanese language, but an English translation of the titles and important phrases is added. The present director of the service is Prof. K. Nakamura, graduate of the Tokio University ; the staff and attendants of the central observatory amount to fifty-three in number.

THE ethnology of ancient history, deduced from records, monuments and coins, is a subject in which M. Charles de Ujfalvy has made some important investigations. In *l'Anthropologie*, tome ix., he has published a memoir on the White Huns. The Huns artificially deformed their heads so as to greatly increase their height (*déformation relevée* of Broca). They were nearly related to the Hoa of the Chinese annals (which name is merely the origin of the word Hun), to the Yé-tha of the Chinese (who must not be confounded with the very different Yué-tchi), and to the White Huns or Ephthalites of Byzantine and Armenian authors. The Hûna kings of India practised the same cranial deformation, as is shown by their effigies represented on their coinage. The Ephthalites practised polyandric customs, and their women wore special horned head dresses. Traces of polyandric habits, as well as of these extraordinary coiffures, are still to be met with, after more than twelve hundred years, in certain regions of the old Ephthalitic empire.

THE second part of vol. xxviii. of the *Morphologisches Jahrbuch* is entirely taken up by two profusely illustrated memoirs on the morphological anatomy of Vertebrates. In the first of these Dr. S. Paulli continues his elaborate investigations into the extent and form of the air-chambers in the mammalian skull ; dealing in this section with the morphology of the ethmoid bone and the relations of the aforesaid chambers in Ungulates. Perhaps the most striking feature in this communication is the labour expended in working out the details of the labyrinth formed by the ethmoid and surrounding structures, as is well displayed in the text-figures, which resemble puzzles of an unusually complex type. Very unexpected is the discovery that the structure of the ethmoid divides the more typical Ungulates into two groups, one represented by the Ruminants and the other by the Suina and the Perissodactyla. As this grouping so completely traverses the classification indicated by other parts of the organisation, it may be that the feature in question is purely adaptive. In the general structure of the ethmoid the elephant resembles more typical Ungulates. The second of the above-mentioned memoirs is a continuation of Dr. B. Haller's study of the Vertebrate brain ; the present section dealing with the Pond-Tortoise (*Emys orbicularis*). At the conclusion of his paper the author refers to the structural resemblances between the reptilian brain on the one hand and that of Monotremes and Marsupials on the other. He is led to conclude that a commissure connecting the hemispheres of the brain was developed in an extinct forerunner of the reptiles, which formed the ancestral type of both the Sauropsida and the Mammalia.

TWENTY years ago the late Dr. Dobson described a new species of Australian bat, remarkable for its white head and lower surface of the body, under the name of *Megaderma gigas*. From that time to this the species has been known solely by the type specimen—a male. In No. 7, vol. iii. of the *Records of the Australian Museum*, Mr. E. R. Waite describes a second example, this time a female, obtained in West Australia. To the same journal Mr. Waite likewise contributes a paper on additions to the fish-fauna of Lord Howe Island, in the course of which he describes four new species, one of them being assigned to a new genus. Several of them belong to the coral-eating Chaetodonts. The author draws attention to the circumstance that since the transparent larval form to which the name Leptocephalus was assigned in 1763 is now ascertained to be the young of the Conger-eel, the generic title Conger has to give place to Leptocephalus. As this latter name is now no longer available for other similar larvae of which the adults are unknown, he adopts for them the name Atopoichthys, lately proposed by Garman.

THE July number of the *Biologische Centralblatt* contains an interesting note by Dr. R. Stölzle on the position taken by K. E. von Baer with regard to the origin of the human race. Reference is made to von Baer's opposition to the doctrine of descent from lower animals (1) in pre-Darwinian times ; (2) after the appearance of " The Origin of Species " ; and (3) after the publication of " The Descent of Man."

CAPTAIN R. H. ELLIOTT, who has been for some time conducting researches into the nature and action of snake venom in India, arrives at the following conclusions in the *British Medical Journal* :—(1) The snakemen of South India are certainly ignorant of any method of producing in themselves a highly-developed condition of immunity. (2) Some few of them appear to practise the swallowing of venom, or the injunction of venom into their limbs, but it is doubtful if they do so with any well-defined object. It is possible that they thus obtain some degree of immunisation. (3) They confine themselves almost exclusively to the cobra, and escape harm by their intimate knowledge of the methods of handling this snake.

A COPY of the second edition of a catalogue of the fossils in the students' stratigraphical series of the Woodwardian Museum, Cambridge, by Mr. H. Woods, has been received.

THE plants collected on the Antillean cruise of the yacht *Utowana*, in Bermuda, Porto Rico, the Caymans, Cozumel, Yucatan, and the Alacran shoals, between December 1898 and March 1899, are described, under the title *Plantae Utowanae*, by Dr. Charles Frederick Millspaugh, in vol. ii. No. 1 of the botanical series of the Field Columbian Museum.

PROF. W. H. CORFIELD'S two Harveian lectures on disease and defective house sanitation have been translated into Hungarian by Dr. Frank, of Budapest, for the Royal Society of Public Health of Hungary. Dr. Frank remarks in the preface that the lectures " merit the attention of Hungarian readers because they explain the views of a prominent English hygienist, and also because the sanitary arrangements of dwellings in Hungary are much more unsatisfactory than those in England."

THE additions to the Zoological Society's Gardens during the past week include a White-fronted Capuchin (*Cebus hypoleucus*) from Central America, presented by Mr. W. H. Laws ; a Two-spotted Paradoxure (*Nandinia binotata*) from West Africa, presented by Mr. Robert H. Gush ; a Levaillant's Amazon (*Chrysotis levaillanti*) from Mexico, four Lorikeets (*Trichoglossus rubritorques*) from North-west Australia, six Roofed Terrapins (*Kachuga tectum*) from British India, two Alligator Terrapins (*Chelydra serpentina*), an American Box Tortoise (*Cistudo carolina*), a Sculptured Terrapin (*Clemmys insculpta*) from North America, deposited ; two —— Buntings (*Emberiza sulphurata*) from Japan, purchased ; an Altai Deer (*Cervus eustephanus*), three Crested Pigeons (*Ocyphaps lophotes*), a Spotted Pigeon (*Columba maculosa*), four Vinaceous Turtle Doves (*Turtur vinaceus*), bred in the Gardens.

OUR ASTRONOMICAL COLUMN.

COMET BORRELLY BROOKS, 1900 *b*.—Several telegrams received from the *Centralstelle* at Kiel announce the appearance of a new comet in the constellation Aries. The following are the positions given :—

1900	R.A.	Decl.	Observer.
	h. m. s.		
July 23d. 12h. 50·0m. (Marseilles Mean Time)	... 2 43 33 ⚊ + 11° 51′		...Borrelly.
July 23d. 13·00h. (Geneva Mean Time)	... 2 43 40 ... + 12° 30′		... Brooks.
July 24d. 12h. 57·1m. (Strassburg Mean Time)	... 2 44 26 ... + 14° 32′ 42″		...Kobold.

A later circular from Kiel furnishes an ephemeris for further observations of the comet, prepared by Herr J. Möller from measures of July 24, 25 and 26.

Elements.

$$T = 1900 \text{ Aug. } 3, 298. \text{ Berlin Mean Time.}$$

$$\begin{array}{l} \omega = 12.30\cdot2 \\ \Omega = 328 \quad 1\cdot8 \\ \iota = 62 \ 35\cdot6 \end{array} \Big\} 1900\cdot0.$$

$$\log q = 0\cdot00636$$

Ephemeris for 12h. Berlin Mean Time.

1900.	R.A.	Decl.	Br
	h. m. s.		
Aug. 1	... 2 53 52	+ 38 31·5	... 1·12
3	... 2 57 12	44 37·1	... 1·10
5	... 3 1 8	50 29·1	... 1·08
7	... 3 5 55	56 2·1	... 1·05
9	... 3 11 45	+ 61 11 9	... 0·91

CATALOGUE OF DOUBLE STARS.—The *first* volume (1900) of the *Publications of the Yerkes Observatory* of the University of Chicago has recently been distributed ; it contains a list of 1290 double stars, discovered from 1871–1899 by Prof. S. W. Burnham, now on the staff of the Yerkes Observatory. The majority of the measures have hitherto only been published in sections, comprising portions of nineteen different catalogues, and the work of bringing so large a mass of material together was commenced during the author's connection with the Lick Observatory (1888–1892). While working with the large instrument there, many of the more difficult pairs were re-measured, and their positions carefully re-determined by comparison with the newer star catalogues of the *Astronomischen Gesellschaft*, Córdoba, &c., instead of those of Lalande and Argelander. As, however, in the present work no attempt has been made to supersede other star catalogues with respect to the absolute positions, it has not been thought worth while to bring the co-ordinates past the epoch 1880.

Commencing astronomical observations in 1861 with a very small instrument, Prof. Burnham obtained a six-inch equatorial from Alvan Clark in 1869, with which he commenced systematic work on double stars in 1872. Since that time his observations have been made with instruments of varying aperture, 9·4, 12, 15·5, 16, 18·5, 26, 36 and 40 inches respectively.

Especially interesting is the fact that a great proportion of the pairs discovered have been found to be physically binary, and that these are generally closer and more difficult to measure compared with those in slower motion.

A special list of quadruple stars is given, and various measures have been obtained by the co-operation of other observers with different instrumental equipment. The stars are arranged in order of right ascension ; and besides the present elements, a short description of special particulars with comparative previous measures are added to each where necessary, and several illustrations are given of the instruments used in the course of the work.

SOME RESULTS OBTAINED WITH A STORAGE BATTERY OF TWENTY THOUSAND CELLS.[1]

THE remarkable development of practical employments of electricity have put the professor of physics at a disadvantage, compared with the electrical engineer. The latter has at his service thousands of electrical horse-power, while the college instructor can barely obtain fifty. The engineer can experiment with enormously strong currents and study their effects in chemical industries, and in the production of intense heat. Thus the study of the manifestations of electricity on a great scale seems to be relegated to the electrical engineer.

There is one direction, however, in which the university professor can enter into competition with the engineer and even surpass him in resources. This direction is in the field of high electromotive force ; and I wish to call your attention to some results which I have obtained with a storage battery of twenty thousand cells. For several years I have had at my command ten thousand cells ; and the plant has proved so practical that I resolved last autumn to double the number of cells. The battery is now finished, and you will have an opportunity of seeing its manifestations.

With twenty thousand cells of the Planté type I can obtain forty-two thousand volts, and by the use of Leyden jars I can step up to three million. I cannot go higher, for the very interesting reason that air at atmospheric pressure becomes a fairly good conductor beyond two million volts, and it is impossible to charge Leyden jars to this potential, or to produce sparks in a laboratory of greater length than seven feet. To obtain the greatest manifestations of three million volts, it would be necessary to put the apparatus in an open field at least thirty feet from the ground, and remote from all other objects. Jars and circuits charged to this high voltage emit a luminous discharge to the floor of the room and to the brick walls, and indicate by this inductive discharge the presence of steam pipes twenty feet distant. The air breaks down quickly under this powerful electric stress, and, indeed, acts like a rarified gas.

Nevertheless discharges of electricity six and seven feet long are of interest, especially to many of you who are citizens of Boston, where Benjamin Franklin was born. These discharges closely resemble lightning, and one can reproduce all the photographic effects obtained by students of this astounding natural phenomenon. I have discovered the interesting fact that these long sparks are oscillatory.

The method of proof is this : I connected the condensers which were used in series to produce the high potential of three million volts, in multiple with a known self-induction. The discharge was then photographed. Here is one of the results : The distance between these bead-like figures from centre to centre represents one five-thousandth of a second (Fig 1). When the condensers are connected in series through the same self-induction the discharge still remains oscillatory, but of a much higher period ; we are sure of this fact from Lord Kelvin's discussion of the limits of oscillatory action. You will perceive from Fig. 1 that I have been able, by means of the

[1] Paper read by Prof. John Trowbridge at a meeting of the American Academy of Arts and Sciences, held in the Jefferson Physical Laboratory, Harvard University, Cambridge, U.S.

large battery and the large condenser, to photograph comparatively slow oscillations. I have lately succeeded in obtaining photographs of oscillations eight hundred a second ; and experiments on the permeability of iron wire with powerful discharges with such low periods are now in progress.

That most discharges of lightning are to-and-fro, or oscillatory, I feel sure, and I have outlined my method of proof ; but this was hardly necessary, for the photographs of the long sparks show on mere inspection the to-and-fro motion, for on the line of discharge forks can be observed pointing in opposite direc-

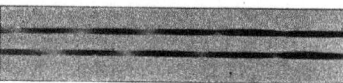

FIG. 1.

tions, showing that the discharge changed from positive to negative. These forks, or branching discharges, have an interesting peculiarity, which was brought out in the following manner. A sheet of plate glass about five feet square was placed between the terminals of the high potential apparatus, and a minute hole was bored in the middle of this plate.

This hole could be made very small by plugging the orifice with paraffin, and making needle-holes in the paraffin. When the spark terminals were opposite the hole, each a foot and a-half

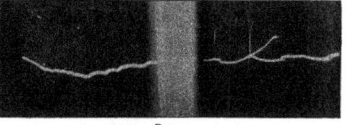

FIG. 2.

from it, the spark sought the hole. A photograph of the spark (Fig. 2) shows an apparent breadth of spark much greater than the diameter of the hole ; indeed, the minute size of the latter cannot be reproduced on the negative ; while the spark seems to the eye to be an eighth of an inch in thickness, and actually measures about a millimetre in diameter on the negative. The reason of this phenomenon, I believe, is that only a portion of the discharge passes through the hole. This can be shown in the following manner. The terminals were not placed oppo-

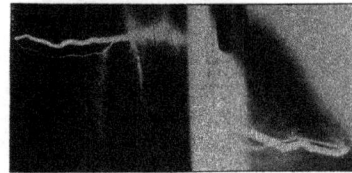

FIG. 3.

site the hole, but to one side of it, about a foot from it; and about half a foot from the glass. The discharge then jumped to the glass (Fig. 3), and pursued a devious way to the hole. When the hole was completely filled with paraffin the spark still jumped to the glass, apparently piercing a hole through it ; but this was impossible, for the thickness prevented this. The discharge was continued evidently by an inductive action. I next restored the orifice, and, keeping the spark terminals in the last position referred to, I hung a large sheet of paraffined paper on the glass, and a photograph of the spark was taken. It was found that an explosion occurred at each change in direction of the spark, or at each fork in it. The two effects are shown in Fig. 4 and Fig. 5 ; and you will see that even the sinuosities are reproduced by a rent in the paper. In the case of a thunderstorm, may not the peculiar rolling of the thunder be due to the successive explosions along the path of a single discharge some hundreds of feet apart?

But I will not dwell longer upon the fascinating study of lightning in a laboratory ; for I wish to call your attention to larger fields of inquiry which the possession of this great battery opens. One of the most promising is that of spectrum analysis. In connection with the battery I have three hundred glass plate condensers, one-eighth of an inch thick, and about ten by eighteen inches coated surface ; this condenser is charged in multiple to a potential of twenty thousand volts. The glass of the thickness of one-eighth of an inch stands this stress ; but I can not use my full voltage of forty thousand, for the glass plates are immediately pierced. To utilise this voltage it will be necessary to employ plates a quarter of an inch in thickness. This is an interesting proof of the large

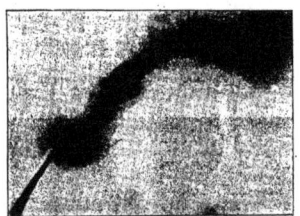

Fig. 4.

Fig 5.

surface density furnished by the battery. The noise of the discharge from this condenser is like the report of a pistol.

Here is an example of its great heating effect. An iron wire was stretched across the spark terminals. This was deflagrated (Fig. 6), while at the same time a spark passed between the terminals. The surrounding air was filled with the scintillating sparks of iron. This shows that it is not impossible that sparks may be formed inside a metallic cage or enclosure ; for we can conceive of such an enclosure as a multiple circuit around a spark gap.

By means of the discharge of these condensers charged to a difference of potential of forty thousand volts, I can produce probably the highest degree of instantaneous temperature which has been reached. I have been obtaining instantaneous photographs of the spectra of gases and of the vapour of metals. One discharge, with a Browning direct vision spectroscope, will give a photograph of the spectra of hydrogen, and ten or twelve discharges are sufficient when a short focus grating is employed with a fairly fine slit. I find it desirable to use a peculiar end-on tube for the study of hydrogen. It is of the nature of a Crookes' tube, one end being blown into a very thin bulb ; this tube can be heated to a very high temperature during the process of exhaustion to drive out the water vapour.

I have thus submitted hydrogen to a higher temperature than it has been possible to reach before, and the study of the spectra promises to be a fruitful one. The advantage of an intense source of light, and consequently of a short time of exposure, is very great ; for a large amount of fog is thus escaped, and faint lines come out which escape observation by the comparatively long exposures hitherto necessary.

There is another direction in which this battery can be used, which promises to be of importance in surgery. It furnishes a new source of the X-rays.

The greatest need in the scientific study, and also in the employment of the X-rays in surgery, is a steady source of them. All the methods now in use give a light which is far from constant ; the electrical impulses which produce the rays are unequal in strength, and even when they are equal they are generally alternating in character. This to-and-fro action tends to produce a blurring of the shadows, for fluctuating electrical impulses sent through an X-ray tube are apt to give a shifting radiant point.

The ideal method of producing the rays is by the employment of a large storage battery, and I have been working toward this much-desired end during the past two years.

Traces of the X-rays can be obtained with a steady current at a voltage of five thousand ; and they are strongly produced at twenty thousand. When forty thousand volts are used with a steady current, the exhibition of rays is surprising ; a fluorescent screen is lighted with extreme brilliancy, and marvellous shadows of the bones of the hand are obtained. A steady current is undoubtedly the ideal current for the production of the

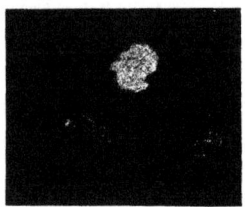

Fig. 6.

X-rays ; for the radiant point of the rays does not fluctuate, and there is no to-and-fro or oscillatory motion which tends to produce what may be called X-rays ghosts. It is well known that these ghosts are often puzzling to the surgeon.

In my experiments I was surprised at the small amount of current necessary with a voltage of forty thousand to produce a strong development of the X-rays. The use of ten milliamperes was dangerous to the tube ; the anode grew white hot, and the Crookes' tube resembled an enclosed arc lamp. It was interesting also to notice that the usual fluorescence ceased to be noticeable ; and although the tube was of a milky white hue, the X-rays were extraordinarily brilliant. In my first experiments the fall of resistance in the tube was so rapid that the anti-kathode was melted. In the case of all the tubes with which I have experimented, the fall in resistance advances very rapidly with the degree of reddening of the anti-kathode. When this becomes red, or when a red spot appears on it, the difference of potential between the terminals of the tube does not in general exceed twenty thousand volts.

It would seem therefore unceonomical to continue the use of a high potential machine when this critical point is reached. At this point, moreover, the rays seem to be given off most vigorously, and at this stage a quantity machine giving a comparatively small voltage could be substituted to advantage for a coil or other apparatus giving six to eight inch sparks. A large storage battery makes it possible to regulate the strength of the current which is at any moment exciting the tube. I accomplish this at present by means of a liquid resistance, which enables me to graduate the strength of the current to any extent. This advantage is a very great one, and is not possessed by any other method. It seems possible, by carefully regulating the strength of the current and the voltage, to obtain photographs of the tendons, and possibly the muscles; for the photographs which I have already obtained show great contrasts, and there are indications of muscular layers and tendons. The contrast between the bones and the flesh is extraordinary, much greater than in the X-ray pictures usually obtained by the Rhumkorf coil.

The investigator, by means of the liquid resistance, can keep the tube at the same point of excitation. For the scientific study of the X-rays nothing seems better adapted than this large battery plant which I have had constructed, and it is not impossible that a smaller plant of the same number of cells, but with less capacity, may be desirable for large hospitals.

The first step in an investigation of the X-rays is to obtain a steady source of these rays: one of the essentials for the accomplishment of this is a steady current which can be regulated. This, I believe, I have secured. The next step will be the proper control of the amount of gas in the tube. At one time I believed that an oscillatory discharge was necessary for the strongest manifestation of the rays. My experiments, however, with a steady current have shown me that an oscillatory discharge is not essential; such a discharge could not take place through the large resistance which I used—4,000,000 ohms. Such are some of the results which can be obtained by the use of this large battery.

THE CRUISE AND DEEP-SEA EXPLOR-ATION OF THE "SIBOGA" IN THE INDIAN ARCHIPEL-AGO.

THE annual summer meeting of the Netherlands Zoological Society, which was held in Amsterdam on July 1, was of more than usual interest ·on account of the fact that it was attended by all the members of the scientific staff of the *Siboga* expedition, who returned only a few weeks before from their one year's cruise in the different basins of the Indian Archipelago, during which they covered a distance of about 12,000 sea miles, *i.e.* about half the circumference of the globe. The track, as indicated on the accompanying Fig. 1, commenced at Soerabaja on March 7, 1899; it ended in the same port on February 27, 1900. The vessel, which is a cruiser belonging to the Dutch Royal Navy, was on its first trip, and before its departure was specially fitted up for the work of the cruise, both with a sounding apparatus of Le Blanc and of Lucas, with some 20 kilometres of wire rope for dredging purposes, and with all modern appliances for pelagic fishing, for plankton collection and for deep-sea work (a "sondeur à clef" of the Prince of Monaco, apparatus for obtaining sea-water from given depths according to Petterson and Sigsbee, Hensen's nets, &c.)

It may here be mentioned that very thorough experiments were made with Mr. G. H. Fowler's net, which is specially intended for plankton from given depths, and which can be opened and shut at will at any moment. About this net, which is of very recent invention, and which has as yet only been used by Mr. Fowler himself, and perhaps on board the *Valdivia*, the members of the *Siboga* expedition are very enthusiastic. It is most trustworthy in its results and fruitful in its catches.

The leader of the *Siboga* expedition, Prof. Max Weber of Amsterdam, well-known by his former expeditions to the East Indies, to the far north and to South Africa, was accompanied by Madame Weber-van Bosse, herself an accomplished naturalist, who made a very complete collection of Algæ during the cruise, and who settled three very important points as a result of the observations made, viz.: (1) the presence in unexpected quantities of calcareous Algæ (Lithothamnion) in the Archipelago, so that they build up reefs of considerable dimensions, in depths of 3 to 40 metres, in one case even at 120 metres. Different circumstances of level, current, &c., must co-operate to render the occurrence of Lithothamnion in such quantities possible: the expedition found them realised in at least thirty different localities, and henceforth the possible contribution of Lithothamnion-remains to the formation of the earth's crust will in many cases have to be reconsidered by the geologists. (2) The presence of a minute vegetal organism about which of late years English and German naturalists have considerably differed in opinion: the Coccosphære. Neither the members of the German Plankton nor those of the *Valdivia* expedition has succeeded in satisfying themselves that these miniature spheres with adherent discs of lime, already known in the Cretaceous

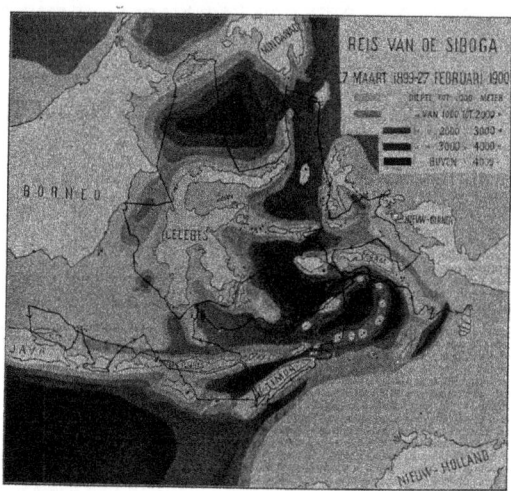

FIG. 1.—Track of the *Siboga*.

period and everywhere present on the bottom of the sea, are organisms and not inorganic concretions or sediments. Mme. Weber has now succeeded in demonstrating the truth of the contrary, and has found this very lowly organised alga in great abundance, and entirely agrees with Mr. George Murray's statements concerning the alga-nature of the coccospheres; she has even found in this alga green chromatophores, and has seen phases of division of the spheres; (3) the presence of shell- and rock-perforating algæ, a group hitherto neglected in the tropics, of which she has brought home a great number.

The zoological collections of the *Siboga* are very extensive, both those collected on the coral-reefs and from the very different depths. Deep-sea animals were met with at depths of about 150 fathoms, where they would hardly have been expected, but where their presence is explained by certain hydrographical circumstances to be mentioned later. Porifera, and among them the most diverse Hexactinellids, were exceedingly numerous. East of the Aru Islands, gigantic specimens of Adeona were captured. This curious Bryozoarian, of leaf-like shape and attached to a segmented stem, has sometimes been considered as one of the Isidinæ.

Of the curious solitary Alcyonarians, the Haimeidæ, which up to now are known as small specimens from the Red Sea and from Algiers, a species of a very considerable size has been met with. Amphianthus, an absolutely flat Actinian, was found on the shell of a Dentalium, and amongst the numerous Echinoderm-finds material abounds to definitely settle the question about the regeneration and the so-called comet forms of Linckia. It could be demonstrated that the regeneration takes place, without any part of the disc being preserved, from a bare arm-fragment. On these Linckias the parasitical molluscs, Thyca and Stylifer, were often present. Various Solenogastres were captured, and many interesting Cephalopods. The fish collection is also very considerable, and a great many deep-sea forms are among them, of which a specimen of Ruvettus attains to a size of several feet.

The most beautifully transparent larval Murænas were at the other end of the scale, and were also exhibited at the meeting. Both they and other pelagic organisms, Medusæ, Heteropods, &c., were most successfully preserved in formalin. On the whole the preservation of all the specimens, for which the most various methods were employed, is first-rate; and Mr. Nierstrasz, to whose supervision this had been more especially entrusted, received due recognition of his merits on this head. Some hundred bottles of plankton have yet to be sorted and worked out. Dr. Versluys communicated to the meeting the results of investigations into the amount of oxygen contained in the sea-water at different points which he had made during the cruise, and Prof. Weber called attention to certain hydrographical results of primary importance obtained by the expedition.

The gist of these is that the communication between the deep water of the Indian and Pacific Oceans and that of the Archipelagian basins is very different from what it was expected to be. The different straits between the lesser Sunda Islands, Bali to Flores, are none of them deep enough to allow of any horizontal passage of the deeper and colder strata (where the temperature is 0·9° C.) into the Banda basin and its continuations between Flores and Timor and between Flores and Celebes. These undoubtedly receive their cold bottom-water from the Pacific Ocean by way of the deep communications indicated on the map to the north of Buru (the so-called Ceram sea), which opens out into the Pacific by a narrow passage (the so-called Moluccan passage). In the deep spurs, to which the name of Bali and Flores sea may be given, the expedition could actually demonstrate the existence of a bottom-current which flows westward and which brings the cold water from the Banda sea into these recesses where the supply from the Indian Ocean through the numerous straits is only superficial and restricted to surface-water of a temperature of more than 12° C. The cold bottom-current of 3° C. just alluded to, which slowly flows westward out of the Banda sea, even rises up along the sloping coasts of the Kangeang-Paternoster-Postillon islands (not indicated on the map) situated north of this deep sea spur, as could be demonstrated both by serial temperatures and even by the net, which, as mentioned above, brought up deep-sea forms from comparatively shallow water, just because of this bottom current, which, being hemmed in, flows towards the surface.

The temperature of 3° C. referred to above is the uniform minimum temperature for the whole of the Banda basin below the depth of 1600 m., and the theoretical conclusion that no deeper communications than this exist with either of the Oceans was practically verified, and also (as indicated above) that the cold water of the greater depths comes from the Pacific and not from the Indian Ocean.

The Banda Sea, *sensu strictiori*, was further found to be different from what was hitherto held. On charts, mention is made of a depth of 7000 metres (4000 fathoms) in the neighbourhood of Banda. This depth has been demonstrated by the *Siboga* to be due to some error, the depth being nowhere below 5500 metres, and the basin itself being most unexpectedly intersected by two shallow ridges, clearly visible on the map, the more westward of which has been named the Siboga Ridge. Geological speculations concerning this part of the earth's crust will undoubtedly be influenced by these results.

For the distribution of deep-sea animals, the difference of a couple of degrees between the bottom-water of these basins and that of the oceans will certainly not have much importance;

and even the ridges will in the long run prevent only very few deep-sea animals from penetrating into the basins in the course of generations, when the difference of pressure can be slowly neutralised. At all events, the catches did not justify expectations that these enclosed deep basins might harbour a deep-sea fauna which, by its isolation from the ocean, had developed into peculiar local deep-sea faunas particular to those basins.

The hydrographical work of the expedition has thus been of very considerable importance, and will soon be also noticeable in improved navigating charts for the regions explored. Even geographical corrections of considerable amount are amongst the results of the cruise. The south coast of the large island of Timor (of which the eastern half is a Portuguese, the western a Dutch possession) will have to undergo a radical alteration, as indicated on the accompanying sketch (Fig. 2). Thus the *Siboga* expedition has not inconsiderably reduced the colonial surface area of Portugal, having anchored in spots which, according to the present maps, lie far inland.

The expedition can thus be complimented on having achieved a most successful piece of work, and it is undoubtedly in the first place due to the undaunted energy of the leader, Prof. Weber, and to the exemplary skill of the officer in command of the vessel, Comm. Tydeman, who for many years has already been one of the leading hydrographers in the Archipelago. The liberality of the Naval Department, and its active co-operation in all that pertained to the expedition, have been especially noticeable.

The results, both hydrographical, botanical, zoological and geological, will, as soon as possible, be worked out by different

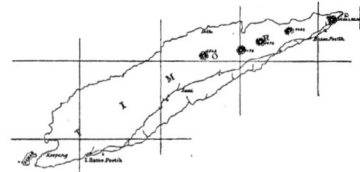

FIG. 2.—The coast-line of Timor. The outside southern coast-line is as indicated upon current maps; the inner line shows the true coast-line as determined by the *Siboga.*

specialists, and be brought together in a series of well-illustrated quarto volumes, the number of which is roughly estimated at about fifteen. Several specialists, both Dutch and foreign, have already promised to co-operate, and what with Alcock's researches in the Bay of Bengal, the *Valdivia's* exploration of the Indian Ocean, the Australian investigations of the Barrier Reef and the Torres Straits, the Belgian Antarctic expedition, and Agassiz's dredging expeditions in the Pacific, we can safely say that, by the time this publication will have appeared, we will have obtained a very thorough knowledge of an important portion of the abyssal regions, towards the exploration of which the *Lightning, Porcupine* and *Challenger* have set the example, and the *Blake, Albatross, Travailleur, Talisman, Gazelle, Vettor Pisani, Willem Barents, Hirondelle* and *Princesse Alice* have so considerably contributed from other parts of the globe.

A. A. W. H.

EXPERIMENTATION ON EMOTION.

OF points where physiology and psychology touch, the place of one lies at the phenomenon ".emotion." Built upon sense-feeling much as cognition is built upon sense-perception, emotion may be regarded almost as a "feeling"—a "feeling" excited, not by a simple unelaborated sensation, but by a group or train of ideas. To such compound ideas it holds relation much as does "feeling" to certain species of simple sense-perceptions. It has a special physiological interest in that certain visceral reactions are peculiarly concomitant with it. Heart, blood-vessels, respiratory muscles and secretory glands play special and characteristic *rôles* in the various emotions. These viscera, though otherwise remote from the general play of psychical process, are affected vividly by the emotional. Hence many a picturesque metaphor of proverb and phrase and name—"the heart is better than the head," anger "swells within the breast," "Richard Cœur de Lion." It was Descartes who first relegated the emotions to the brain. Even this century Bichat wrote, "The brain is the seat of cognition, and is **never** affected by the emotions, whose sole seat lies in the viscera." But brain is now admittedly a factor necessary in all higher animal forms to every mechanism whose working has consciousness adjunct.

What is the meaning of the intimate linkage of visceral actions to psychical states emotional? To the ordinary day's consciousness of the healthy individual, the life of the viscera contributes little at all, except under emotion. The perceptions of the normal consciousness are rather those of outlook upon the circumambient universe than inlook into the microcosm of the "material me." Yet heightened beating of the heart, blanching or flushing of the blood-vessels, the pallor of fear, the blush of shame, the Rabelaisian effect of fright upon the bowel, the action of the lacrymal gland in grief, all these are prominent characters in the pantomime of natural emotion. Visceral disturbance is evidently a part of the corporeal expression of emotion. The explanation is a particular case in that of movements of expression in general. The hypothesis of Evolution afforded a new vantage point for study of that question. Fixed bodily expressions of emotion are hereditary. They are, especially in the "coarser or animal emotions," largely common to man and higher animals. The point of view is exemplified by Darwin's argument concerning the contraction of the muscles round the eyes during screaming. "Children when wanting food or suffering in any way cry out loudly like the young of most animals, partly as a call to their parents for aid, and partly from any great exertion serving as relief. Prolonged screaming inevitably leads to the engorging of the blood-vessels of the eye ; and this will have led at first consciously and at last habitually to the contraction of the muscles round the eyes in order to protect them." Mr. Spencer writes : "Fear, when strong, expresses itself in cries, in efforts to hide or escape, in palpitations and tremblings ; and these are just the manifestations which would accompany an actual experience of the evil feared. The destructive passions are shown in a general tension of the muscular system, in gnashing of the teeth and protrusion of the claws, in dilated eyes and nostrils, in growls : and these are weaker forms of the actions that accompany the killing of prey." In a word, expression of emotion is instinctive action.

Movement of expression, be it facial or vocal, let it involve the skeletal or the visceral musculature, must have an explanation the same in kind as that of other instinctive movement. To enter upon its "why" is to enter upon the "why" of instinct. Suffice it to say here that if we follow the doctrine of evolution we cannot admit any absolute break between man and brute even in the matter of mental endowment. The instinctive bodily expressions of emotion probably arose as attitudes useful in the animal's environment for defence, escape, seizure, embrace, &c. These as survivals have become symbolic for states of mind. Hence the intelligible nexus between the muscular attitude, the pose of feature, &c., and the emotional state of mind. But between action of the viscera and the psychical state the nexus is less obviously explicable. This latter connection adds a difficult corollary to the general problem.

The fact of the connection is on all hands admitted, but as to the manner of it opinion is at issue. Does (1) the psychical part of the emotion arise and its correlate nervous action then excite the viscera ? Or (2) does the same stimulus which excites the mind excite concurrently and *per se* the nervous centres ruling the viscera ? Or (3) does the stimulus which is the exciting cause of the emotion act first on the nervous centres ruling the viscera, and their action then generate visceral sensations ; and do these latter, laden with affective quality as we know they will be, induce the emotion of the mind ? On the first of the three hypotheses the visceral reaction will be secondary to the psychical, on the second the two will be collateral and concurrent, on the third the physical process will be secondary to the visceral.

To examine the last supposition first. It is a view which in recent years has won notable adherents. Prof. William James

writes : "Our natural way of thinking about these coarser emotions (*e.g.* "grief, fear, rage, love") is that the mental perception of some fact excites the mental affection called the emotion, and that this latter state of mind gives rise to the bodily expression. My theory, on the contrary, is that *the bodily changes follow directly the perception of the exciting fact, and that our feeling of the same changes as they occur* is *the emotion.*" "*Every one of the bodily changes, whatsoever it be, is* FELT, *acutely or obscurely, the moment it occurs.*" If the reader has never paid attention to this matter, he will be both interested and astonished to learn how many different local bodily feelings he can detect in himself as characteristic of his various emotional moods." "If we fancy some strong emotion and then try to abstract from our consciousness of it all the feelings of its bodily symptoms we find we have nothing left behind, no "mindstuff" out of which the emotion can be constituted, and that a cold and neutral state of intellectual perception is all that remains." "If I were to become corporeally anæsthetic, I should be excluded from the life of the affections, harsh and tender alike, and drag out an existence of merely cognitive or intellectual form."

Prof. Lange traces the whole psycho-physiology of emotion to certain excitations of the vasomotor centre. For him, as for Prof. James, the emotion is the outcome and not the cause or the concomitant of the organic reaction ; but for him the foundation and corner-stone of the organic reaction is as to physiological quality vascular, namely, vasomotor. Emotion is an outcome of vasomotor reaction to stimuli of a particular kind. This stimulus induces a vasomotor action in viscera, skin, and brain. The change thus induced in the circulatory condition of these organs induces changes in the actions of the organs themselves, and these latter changes evoke sensations which constitute the essential part of emotion. It is by excitation of the vasomotor centre, therefore, that the exciting cause, whatever it chance to be, of emotion produces the organic phenomena which as felt constitute for Lange the whole essence of emotion. The teaching of Prof. Sergi closely approximates to that of Lange.

The views of James, Lange, and Sergi have common to them this, that the psychical process of emotion is secondary to a discharge of nervous impulses into the vascular and visceral organs of the body suddenly excited by certain peculiar stimuli, and that it depends upon the reaction of those organs. Prof. James's position in the matter is, however, not wholly like that of Prof. Lange. In the first place, he does not consider vasomotor reaction to be primary to all the other organic and visceral disturbances that carry in their train the psychological appanage of emotion ; and Prof. Sergi, though more nearly in harmony with Lange, agrees with James in this. In the second place, Prof. James seems to distinctly include other "motor" sensations and centripetal impulses from musculature other than visceral and vascular, among those which causally contribute to emotion. Thirdly, he urges his theory as one completely competent only for the "coarser" emotions, among which he instances "fear, anger, love, grief." For Lange and Sergi the basis of apparition of all feeling and emotion is physiological, visceral, and organic, and has seat for the former authority exclusively, and for the latter eminently, in the vasomotor system.

To obtain some test of this view is not difficult by experiment. Appropriate spinal and vagal transection removes completely and immediately the sensation of all the viscera and of all the skin and muscles below the shoulder (see Fig. 1 on p. 330). The procedure at the same time cuts from connection with the organs of consciousness the whole of the circulatory apparatus of the body. I have had under observation dogs in which this had been carried out. I will cite an animal selected because of markedly emotional temperament. Affectionate toward the laboratory attendants, one of whom had her in charge, toward some persons and toward several inmates of the animal house she frequently showed violent anger. Her ebullitions of rage were sudden. Their expression accorded with a description furnished by Darwin. Besides the utterance of the growl, "the ears are pressed closely backwards, and the upper lip is retracted out of the way of the teeth, especially of the canines." The mouth was slightly opened and lifted ; the eyelids widely parted ; the pupils dilated. The hair along the mid-dorsum, from close behind the head to a point more than half way down the trunk, became rough and bristling.

The reduction of the field of sensation in this animal by the procedure above-mentioned produced no obvious diminution or change of her emotional character. Her anger, her joy, her disgust, and when provocation arose her fear remained as evident as ever. Her joy at the approach or notice of the attendant, her rage at the intrusion of a cat with which she was unfriendly, remained as active and thorough. But among the signs expressive of rage the bristling of the coat along the back no longer occurred. On he other hand, the eyes were well opened, and the pupil distinctly dilated in the paroxysm of anger. Since the brain had been by the transection shut out from discharging impulses *via* the cervical sympathetic the dilatation of pupil must have occurred by inhibition of the action of the oculomotor centre.

The coming of a visitor whose advent months before had elicited violent anger, again provoked an exhibition of wrath significant as ever. The expression was that of aggressive rage. The animal followed each movement of the stranger as though of an opponent, growling viciously. A cat with which she was never friendly, and a monkey new to the laboratory, approaching too near the kennel, excited similar ebullitions. No doubt was left in our minds that sudden attacks of violent anger were still easily excited. But she also gave evidence daily that she had the accession of joyous pleasure and delight she had always shown at the approach of the attendant the first thing of a morning, or at feeding time, or when caressed by him, or encouraged by his voice.

Few dogs even when very hungry can be prevailed on to touch dog's-flesh as food. Almost all turn from it with signs of repugnance and dislike. I had strictly refrained from testing

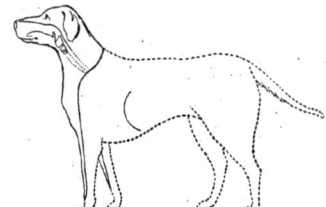

FIG. 1.—Diagram showing the great reduction of the field of sentivity. The head and neck and the diaphragm muscle (indicated by the curved line behind the chest) are practically the only parts left sensitive. The remainder of the body and the limbs, as well as the digestive and respiratory organs behind the throat, and the whole of the circulatory and other organs, are entirely cut off from making any contribution to consciousness.

this animal previously with regard to disgust at dog's-flesh offered in her food. Flesh was given her daily in a bowl of milk, and this she took with relish. The meat was cut into pieces rather larger than the lumps of sugar usual for the breakfast table. It was generally horse-flesh, sometimes, ox-flesh. We proceeded to the observation thus: the bowl was placed by the attendant in the corner of the stall, with milk and meat in every way as usual; but the meat was flesh from a dog killed on the previous day. Our animal eagerly drew itself toward the food; it had seen the other dogs fed, and evidently itself was hungry. Its muzzle had almost dipped into the milk before it suddenly seemed to find something there amiss. It hesitated, moved its muzzle about above the milk, but before actually seizing it stopped short and withdrew again from it. Finally, after further examination of the contents of the bowl (it usually commenced by taking out and eating the pieces of meat), without touching them, the creature turned away from the bowl and withdrew itself to the opposite side of the cage. Some minutes later, in result of encouragement from us to try the food again, it returned to the bowl. The same hesitant display of conflicting desire and disgust was once more gone through. The bowl was then removed by the attendant, emptied, washed, and horse-flesh similarly prepared and placed in a fresh quantity of milk was offered in it to the animal. The animal once more drew

itself toward the bowl, and this time began to eat the meat, soon emptying the dish. To press the flesh upon our animal was of no real avail on any occasion; the coaxing only succeeded in getting her to, as it were, re-examine but not to touch the morsels. The impression made on all of us by the dog's behaviour was that something in the dog's-flesh was repulsive to her, and excited disgust unconquerable by ordinary hunger. Some odour attaching to the flesh seemed the source of its recognition.

Fear appeared clearly elicitable. The attendant, approaching from another room of which the door was open, chid the dog in high scolding tones. The creature's head sank, her gaze turned away from her advancing master, and her face seemed to betray dejection and anxiety. The respiration became unquiet, but the pulse never changed its rate.

In the face of these observations the vasomotor theory of the production of emotion becomes, I think, untenable: also that visceral sensations or presentations are *necessary* to emotion. A mere remnant of all the non-projecting or affective senses was left, and yet emotion persisted. If I understand it aright, Prof. James and Lange's theory lays stress on organic and visceral presentations, but re-presentations of the same species might no doubt be put forward in their place. That would be a somewhat different matter. To exclude the latter hypothesis, the deprivation of vascular and organic sensation might have to date from a very early period of the individual life. Experience early acquires its emotional data. If after that all fresh presentation were precluded, re-presentation might still be possible on the basis of already gained experience. But it is noteworthy that one of the dogs under observation had been deprived of its sensation when only nine weeks old. Disgust for dog's flesh could hardly have genesis in the experience of nine weeks of puppy life in the kennel of the laboratory.

Organic and vascular reaction, though not the actual excitant of emotion, does nevertheless much strengthen it. That is part of the kernel of the old contention about the strength of emotion in the art of the artist. Hamlet's description of the actor, as really moved by his expression, may be accepted as an answer.

But, returning to the main question, we are forced back toward the likelihood that the visceral expression of emotion is *secondary* to the psychical state, or rather to the cerebral nervous action correlate with that. There is plenty of evidence of the strong nexus between emotion and muscular action. As we commonly phrase it, " emotion moves us," hence the word itself. Emotion if developed in intensity, impels toward vigorous movement. Every vigorous movement of the body, though its more obvious instrument be the skeletal musculature of the limbs and trunk, involves also the less noticeable co-operation of the viscera, especially of the circulatory and respiratory. The demand made upon the muscles that move the frame for further expenditure of power involves a heightened action of the nutrient organs which supply to the muscles their material for energy. This increased action of the viscera is therefore colligate with this activity of muscles. We should therefore expect visceral action to occur along with the muscular expression of emotion. The close tie between visceral action and states of emotion need not therefore surprise us.

That emotion is primarily a cerebral process obtains support from observations where the hemispheres of the brain have been removed. Prof. Goltz observed a dog kept many months in that condition. It on no occasion gave any evidence of joy or pleasure in commerce either with man or beast. Anger or displeasure, Goltz says, it repeatedly expressed, both by gesture and by voice. Of sexual emotion it never gave a sign. Save for expression of displeasure when too roughly handled, it was indifferent and supremely neutral to its surroundings. We are, of course, in observations whose basis is such experiment as this, hopelessly cut off from introspective help. It can be urged that the expression of emotion might be provocable, and nevertheless the psychical emotion remain absent. On such an hypothesis the same stimulus which excites the mind must excite concurrently and *per se* excite motor centres producing movement appropriate to an affective process in the mind. This is not improbable. All sensations referred to the body itself rather than interpreted as qualities of objects in the external world, tend to be tinged with "feeling." Sense organs which initiate sensations tinged with feeling tend to excite motor centres directly and imperatively. Hence, in animals reduced to merely spinal condition, stimuli calculated to produce pain normally (of course, unable to do so in a spinal animal incapable of consciousness), evoke movements appropriate for escape from or removal of the stimulus applied. Now "feeling" is implicit in the emotional state; the state is an "affective state." In the evolution of emotion the revival of "feelings" pleasurable and painful must have played a large part. Hence the close relation of emotion with sense organs that can initiate bodily pain or pleasure, and hence its connection with impulsive or instinctive movement. There is no wide interval between the reflex movement of the spinal dog, whose foot attempts to scratch away an irritant applied to its back—both leg and back absolutely detached from consciousness—and the reaction of the decerebrate dog that turns and growls and bites at the fingers holding his hind foot too roughly. In the ormer case the motor reaction occurs, although the mind is not even aware of the stimulus, far less percipient of it as an irritant. The action occurs, and plays the pantomime of feeling; but no feeling comes to pass. In the latter case the motor reaction occurs, and is expressive of emotion; but it is probably the reaction of an organic machine, which can be started working, though the mutilation precludes the psychosis of emotion.

And with the gesture and the attitude will occur the visceral concomitant. It would be consonant with what we know of reflex action if the spur that started the muscular expression should simultaneously and of itself initiate, also, the visceral adjunct reaction. It is almost impossible to believe that with the mere stump of brain that remained to Goltz's dog there could be any elaboration of a percept. All trace of memory was lacking to the creature. Yet though not evincing other emotion, anger it showed as far as expression can yield revelation. Fear, joy, affection seem, therefore, in the observation of this skilled observer of animal mind, to demand higher nervous organisation than does anger. Be that as it may, the retention of its expression by Goltz's dog indicates that by "retrogradation" the complex movement of expression has in certain emotions passed into a simple far the determining motives finally become, even in impulsive acts, weaker and more transient. The external stimulus originally aroused a strongly affective group of ideas, which operated as a motive, but now it causes a discharge of the act before it can be apprehended as an idea. The impulsive movement of a "lower," "coarser," so-called "animal" emotion, has in this case become an automatic reflex process, no longer necessarily combined with the psychical state whence it arose, of which it is normally at once the adjunct and the symbol.   C. S. SHERRINGTON.

THE CENTENARY OF THE ROYAL COLLEGE OF SURGEONS.

MR. VICTOR PLARR'S article, in last week's NATURE, on the celebration of the centenary of the Royal College of Surgeons of England contained a brief statement of the ceremonies which were to commence on the day we went to press. The proceedings were opened on Wednesday morning, July 25, when demonstrations were given in the Hunterian Museum of the College by the conservator, Prof. C. Stewart, F.R.S., who conducted visitors round the galleries, pointing out and describing some of the more important and interesting specimens. At the same time, in the theatre of the Examination Hall, Dr. T. G. Brodie, director of the laboratories of the Conjoint Board, gave an account of some of the work recently carried out in the research laboratories. In the evening a conversazione was held at the College, and was attended by many distinguished guests. Demonstrations were again given by Prof. Stewart and Dr. Brodie on Thursday morning; and in the afternoon, Sir William MacCormack, the president, delivered an address of welcome, and presented the diploma of Honorary Fellow to the Marquis of Salisbury and the Earl of Rosebery. As already stated (p. 294), the Prince of Wales received the diploma on July 24; and the form of the Royal diploma is the same as that employed for all the Honorary Fellowships.

The following is the list of other Honorary Fellows to whom diplomas were presented on Thursday:—E. Albert, professor of clinical surgery, University of Vienna; C. B. Ball, Regius professor of surgery, University of Dublin; E. Bassini, professor of clinical surgery, Royal University of Padua; E. H. Bennett, professor of surgery, Trinity College, Dublin; J. W. Berg, professor of surgery, Royal Caroline Institute of Medicine and Surgery, Stockholm; Prof. von Bergmann, Berlin; O. Bloch, professor

of surgery, University of Copenhagen; E. Bottini, professor of clinical surgery, Royal University of Pavia; I. H. Cameron, professor of clinical surgery, University of Toronto; Dr. Salvador Cardenal Fernandez, vice-president, Royal Academy of Medicine and Surgery, Barcelona; Antonino D'Antona, professor of surgery, Royal University of Naples; Francesco Durante, professor of clinical surgery, Royal University of Rome; Prof. Dr. Friedrich von Esmarch, Kiel; W. S. Halsted, professor of surgery, Johns Hopkins University, Baltimore; Hon. Sir W. H. Ilingston, professor of clinical surgery, University of Laval; Surgeon-General James Jameson, C.B., Director-General, Army Medical Service; W. W. Keen, professor of the principles of surgery and of clinical surgery, Jefferson Medical College, Philadelphia; Theodor Kocher, professor of surgery, University of Bern; Prof. Dr. Franz König, Berlin; Prof. Kosinskij, professor of surgery in the University of Warsaw; Prof. Dr. E. G. F. Küster, Marburg; Elie Lambotte, Brussels; Odilon Marc Lannelongue, professor of surgical pathology, Faculty of Medicine of Paris; Karl Gustaf Lennander, professor of surgery and obstetrics, University of Upsala; W. Macewen, F.R.S., Regius professor of surgery, University of Glasgow; Colonel Kenneth MacLeod, professor of clinical and military medicine, Army Medical School, Netley; Julius Nicolaysen, professor of surgery, Royal University of Christiania; Sir Henry Frederick Norbury, K.C.B., Director-General, Medical Department of the Royal Navy; Leopold Ollier, professor of clinical surgery, University of Lyons; Victor Pachoutine, president, Imperial Military Academy of Medicine, St. Petersburg; Samuel Pozzi, professor in the Faculty of Medicine of Paris; Colonel D. C. O'Connell Raye, Indian Medical Service; T. G. Roddick, professor of surgery, McGill University, Montreal; Federico Rubio y Gali, member of the Royal Academy of Medicine of Madrid; Nicolas Wassilievitch Sklifossovsky, director and Emeritus professor, Imperial Clinical Institute of the Grand Duchess Helena Pavlovna, St. Petersburg; Paul Tillaux, professor of clinical surgery, Faculty of Medicine of Paris; Nicolas Veliaminoff, professor of surgery, Imperial Military Academy of Medicine, St. Petersburg; John Collins Warren, professor of surgery, Harvard University; Robert Fulton Weir, professor of clinical surgery, Columbia University, New York. After the presentation brief addresses of thanks were delivered by Prof. v. Bergmann of Berlin, Prof. Durante of Rome, Dr. W. W. Keen of Philadelphia, Prof. Lannelongue, and Dr. T. G. Roddick of Montreal.

FACTS OF INHERITANCE.[1]

ONE of the distinctive features of the nineteenth century has been a reduction in the number of supposed separate powers or entities—the use of William of Occam's razor, in fact. In view of this progress towards greater precision of phraseology, it cannot be a matter for surprise that a biologist should affirm that to speak of the "Principle of Heredity" in organisms is like speaking of the "Principle of Horology" in clocks. For heredity is certainly no power or force, or principle, but a convenient term for the relation of organic or genetic continuity which binds generation to generation.

Another distinctive feature in scientific progress has been the introduction of precise measurement. In the development of natural knowledge, science begins where measurement begins. This is the case in regard to inheritance. While nothing can take the place of experiment, much has been gained by the application of statistical and mathematical methods to biological results—a new contact between different disciplines—which we may particularly associate with the names of Mr. Francis Galton and Mr. Karl Pearson.

I. THE PHYSICAL BASIS OF INHERITANCE.

What was for so long quite hidden from inquiring minds, or but dimly discerned by a few, is now one of the most marvellous of biological commonplaces—that the individual life of the great majority of plants and animals begins in the union of two minute elements—the sperm-cell and the egg-cell. If inheritance includes all that the living creature is or has to start with in virtue of its genetic relation to its parents and ancestors, then it is

[1] Abridged from a discourse delivered at the Royal Institution on Friday March 30, by Prof. J. Arthur Thomson, F.R.S.

plain that the physical basis of inheritance is in the fertilised ovum. As regards property. there is an obvious distinction between the inheritance and the person who inherits, but there is no such distinction in biology. The fertilised egg-cell *is* the inheritance, and is at the same time the potential inheritor.

An organic inheritance means so much, even when we use the magic word potentiality, that we may consider for a moment the difficulty which rises in the minds of many when they remember that the egg-cell is often microscopic, and the sperm-cell often only 1/100,000th of the ovum's size. Can there be room, so to speak, in these minute elements for the complexity of organisation supposed to be requisite? The difficulty will be increased if the current opinion be accepted that only the nuclei within the germ-cells are the true bearers of the hereditary qualities.

In reference to this difficulty, it may be recalled that the students of physics tell us that the image of a *Great Eastern* filled with framework as intricate as that of the daintiest watches does not exaggerate the possibilities of molecular complexity in a spermatozoon, whose actual size may be less than the smallest dot on the watch's face. Secondly, as we learn from embryology that one step conditions the next and that one structure grows out of another, we are not forced to stock the microscopic germ-cells with more than initiatives. Thirdly, we must remember that the development implies an interaction between the growing organism and a complex environment without which the inheritance would remain unexpressed, and that the full-grown organism includes much that was not inherited at all, but has been acquired as the result of nurture or external influence.

The central problem of heredity is to form some conception of what we have called the relation of genetic continuity between successive generations; the central problem of inheritance is to measure the resemblances and differences in the hereditary characters of successive generations, and to arrive, if possible, at some formula which will sum up the facts. Therefore, while it is interesting to ask how an organisation supposed to be very complex may be imagined to find physical basis in a microscopic germ-cell, the same sort of question may be raised in regard to a ganglion-cell. It is not distinctively a problem of heredity. Similarly, while it is interesting to inquire into the orderly and correlated succession of events by which the fertilised egg-cell gives rise to an embryo, this is the unsolved problem of physiological embryology.

In the preformationist theories, which asserted the pre-existence of the organism and all its parts, in miniature, within the germ—there was a kernel of truth well concealed within a thick husk of error. For we may still say that the future organism is implicit in the germ, and that the germ contains not only the rudiment of the adult organism, but the potentiality of successive generations as well. But what baffled the earlier investigators was the question how the germ-cell comes to have this ready-made organisation, this marvellous potentiality.

An attempt to solve this difficulty of accounting for the complex organisation presumed to exist in the germ-cell is expressed in a theory which occurred at intervals in the long period between Democritus and Darwin, the theory of pangenesis. On this theory, the cells of the body are supposed to give off characteristic and representative gemmules; these are supposed to find their way to the reproductive elements, which thus come to contain concentrated samples of the different components of the body, and are, therefore, able to develop into an offspring like the parent. The theory involves many hypotheses, and is avowedly unverifiable in direct sense-experience, but it is more to the point to notice that there is another theory of heredity which is, on the whole, simpler, which seems, on the whole, to fit the facts better, especially the fact that our experience does not warrant the conclusion that the modifications or acquired characters of the body of the parent affect in any specific and representative way the inheritance of the offspring.

As is well known, the view which most biologists now take of the uniqueness of the germ-cells is expressed in the phrase "germinal continuity." There is a sense, as Mr. Galton says, in which the child is as old as the parent, for when the parent's body is developing from the fertilised ovum, a residue of unaltered germinal material is kept apart to form the future reproductive cells, one of which may become the starting-point of a child. In many cases, from worms to fishes, the beginning of the lineage of germ-cells is demonstrable in very early stages before the differentiation of the body-cells has more than begun. In the development of the threadworm of the horse, according to Boveri, the very first cleavage divides the fertilised ovum into

two cells, one of which is the ancestor of all the body-cells, and the other the ancestor of all the germ-cells. In other cases, particularly among plants, the segregation of germ-cells is not demonstrable until a relatively late stage. Weismann, generalising from cases where it seems to be visibly demonstrable, maintains that in all cases the germinal material which starts an offspring owes its virtue to being materially continuous with the germinal material from which the parent or parents arose. But it is not on a continuous lineage of recognisable germ-*cells* that Weismann insists, for this is often unrecognisable, but on the continuity of the germ-*plasm*—that is, of a specific substance of definite chemical and molecular structure which is the bearer of the hereditary qualities. In development, a part of the germ-plasm, "contained in the parent egg-cell is not used up in the construction of the body of the offspring, but is reserved unchanged for the formation of the germ-cells of the following generation." Thus the parent is rather the trustee of the germ-plasm than the producer of the child. In a new sense, the child is a chip of the old block. The conception of a germ-plasm is hypothetical, just as the conception of a specific living stuff or protoplasm is hypothetical. In the complex microcosm of the cell, we cannot point to any one stuff and say, "this is protoplasm"; it may well be that vital activity depends upon several complex stuffs which, like the members of a carefully constituted firm, are characteristically powerful only in their inter-relations. In the same way, we cannot demonstrate the germ-plasm, even if we may assume that it has its physical basis in the stainable nuclear bodies or chromosomes. The theory has to be judged, like all conceptual formulæ, by its adequacy in fitting facts.

II. DUAL NATURE OF INHERITANCE.

Apart from exceptional cases, the inheritance of a multicellular animal or plant is dual, part of it comes from the mother and part of it from the father.

Prof. E. B. Wilson states the general opinion of experts somewhat as follows:—As the ovum is much the larger, it is believed to furnish the initial capital—including it may be a legacy of food-yolk—for the early development of the embryo. From both parents alike comes the inherited organisation which has its seat (according to many) in the readily stainable (chromatin) rods of the nuclei. From the father comes a little body (the centrosome) which organises the machinery of division by which the egg splits up, and distributes the dual inheritance equally between the daughter-cells.

Recent researches confirm a prophecy which Huxley made in 1878: "It is conceivable, and indeed probable, that every part of the adult contains molecules derived both from the male and from the female parent; and that, regarded as a mass of molecules, the entire organism may be compared to a web of which the warp is derived from the female and the woof from the male." "What has since been gained," Prof. Wilson says, "is the knowledge that this web is to be sought in the chromatic substance of the nuclei, and that the centrosome is the weaver at the loom."

In regard to these conclusions, three notes are necessary. (a) Although inheritance is dual, it is in quite as real a sense multiple, from ancestors through parents. (b) If Loeb is able to induce artificial parthenogenesis in sea-urchins' eggs exposed for a couple of hours to sea-water to which some magnesium chloride has been added; if Delage is able to fertilise and to rear normal larvæ from non-nucleated ovum-fragments of sea-urchin. worm and mollusc, we should be chary of committing ourselves definitely to the conclusion that the nuclei are the exclusive bearers of the hereditary qualities, or that both must be present in all cases. Furthermore, the fact that an ovum without any sperm-nucleus, or an ovum-fragment without any but a sperm-nucleus, can develop into a normal larva points to the conclusion, probable also on other grounds, that each germ-cell, whether ovum or spermatozoon, bears a complete equipment of hereditary qualities. (c) It must be carefully observed that our second fact does not imply that the dual nature of inheritance must be patent in the full-grown offspring, for hereditary resemblance is often strangely unilateral, the characters of one parent being "prepotent," as we say, over those of another.

III. DIFFERENT DEGREES OF HEREDITARY RESEMBLANCE.

One step of progress during the Darwinian era has been the recognition of inheritance as a fact of life which requires no further proof.

Yet this aspect of the study of heredity is by no means worked out. Thus there are some characters, *e.g.* tendency to certain diseased conditions, which are more frequently transmitted than others, and we ought to have, in each case, precise statistics as to the probabilities of transmission.

Again, there are some subtle qualities whose heritability must not be assumed without evidence. Thus it is of very great importance to students of organic evolution that Prof. Karl Pearson has recently supplied, for certain cases, definite proof of the inheritance of fecundity, fertility and longevity.

The familiar saying, "like begets like," should rather read, "like tends to beget like," since variation is quite as important a fact as complete hereditary resemblance. If it seems that in many cases the offspring is practically a facsimile reproduction of the parent, this may be due to absence of variation, or, what comes almost to the same thing, to great completeness of inheritance; but it is more likely to be due to our ignorance, to our inability to detect the idiosyncrasies.

But it will be granted that the completeness with which the characters of race, genus, species and stock are reproduced generation after generation is one of the large facts of inheritance. But this does not sum up our experience, and we must face the task of considering the different degrees of hereditary resemblance. For these a confused classification and a troublesome terminology has been suggested, but it will be enough to restrict attention to three familiar cases—blended, exclusive and particulate inheritance.

A preliminary consideration must be attended to. It is a matter of observation that there are great differences in the degree in which offspring resemble their parents; but it is surely a matter of conjecture that lack of resemblance is necessarily due to incompleteness in the inheritance. Indeed, the fact that the resemblance so often reappears in the third generation makes it probable that the incompleteness is not in the inheritance, but simply in its expression. The characters which seem to be absent, to "skip a generation," as we say, are probably part of the inheritance, as usual. But they remain latent, neutralised, silenced (we can only use metaphors) by other characters, or else unexpressed because of the absence of the appropriate stimulus.

(*a*) In *blended* inheritance, the characters of the two parents, *e.g.* in regard to a particular structure, such as the colour of the hair, may be intimately combined in the offspring. This is particularly well seen in some hybrids, where the offspring often seems like the mean of the two parents; it is probably the most frequent mode of inheritance.

(*b*) In *exclusive* inheritance, the expression of maternal or of paternal characters in relation to a given structure, such as eye-colour, is suppressed. Sometimes the unilateral resemblance is very pronounced, and we say that the boy is "the very image of his father," or the daughter "her mother over again"; though even more frequently the resemblance seems "crossed," the son taking after the mother, and the daughter after the father.

(*c*) It is convenient to have a third category for cases where there is neither blending nor exclusiveness, but where in the expression of a given character, part is wholly paternal and part wholly maternal. This is called *particulate* inheritance. Thus, an English sheep-dog may have a maternal eye on one side, and a maternal eye on the other. Suppose the parents of a foal to be markedly light and dark in colour; if the foal is light brown the inheritance in that respect is blended, if light or dark it is exclusive, if piebald it is particulate. In the last case there is in the same character an exclusive inheritance from both parents.

The facts above referred to may be considered in another aspect, in terms of what is called the quality of prepotency. In the development of a character the paternal or the maternal qualities may predominate, as in unequal blending where there is relative prepotency, or in exclusive inheritance where the prepotency in respect to a given character is absolute. It seems doubtful whether we gain much by using the word, since all these general terms are apt to form the dust particles of intellectual fog; but we have to do with the fact that in respect to certain characters the paternal inheritance seems more potent than the maternal, or *vice versa.*

It seems that one of the ways in which the quality of prepotency may be developed is by inbreeding, as Prof. Ewart and others have maintained.

Therefore, as inbreeding may be frequent in nature, especially in gregarious and isolated groups, and as it tends to develop prepotency, we are able to understand better how new variations may have been fixed in the course of evolution. And we

can appreciate the position maintained by Reibmayr, that the evolution of a human race implies alternating periods of dominant inbreeding, and dominant cross-breeding. The inbreeding gives fixity to character, the cross-breeding averts degeneracy and stimulates new variations which form the raw material of progress.

Until we have more precise statistical data in regard to blended, exclusive and particulate inheritance, we cannot hope to simplify the matter with any security. But perhaps a unified view will be found in the theoretical conception of a germinal struggle in the arcana of the fertilised ovum, a struggle in which the maternal and paternal contributions may blend and harmonise, or may neutralise one another, or in which one may conquer the other, or in which both may persist without combining. We have extended the wide conception of the struggle for existence in many directions; it may be between organisms akin or not akin, between plants and animals, between organisms and their inanimate environment, between the sexes, between the different parts of the body, between the ova, between the spermatozoa, between the ova and the spermatozoa, and Weismann has suggested that it may also be between the constituents of the germ-plasm.

IV. REGRESSION.

We have already referred to the fact that there is a sensible stability of type from generation to generation. "The large," Mr. Galton says, "do not always beget the large, nor the small the small; but yet the observed proportion between the large and the small, in each degree of size and in every quality, hardly varies from one generation to another." In other words, there is a tendency to keep up a specific average. This may be partly due to the action of natural elimination, weeding out abnormalities, often before they are born. But it is to be primarily accounted for by what Mr. Galton calls the fact of "filial regression."

As Mr. Galton puts it, society moves as a vast fraternity. The sustaining of the specific average is certainly not due to each individual leaving his like behind him, for we all know that this is not the case. It is due to a regression which tends to bring the offspring of extraordinary parents nearer the average of the stock. In other words, children tend to differ less from mediocrity than their parents.

This big average fact is to be accounted in terms of that genetic continuity which makes an inheritance not dual, but multiple. "A man," says Mr. Pearson, "is not only the product of his father, but of all his past ancestry, and unless very careful selection has taken place, the mean of that ancestry is probably not far from that of the general population. In the tenth generation a man has [theoretically] 1024 tenth great-grandparents. He is eventually the product of a population of this size, and their mean can hardly differ from that of the general population."

At this point one should discuss reversion or atavism, but it is exceedingly difficult to get a firm basis of fact. The term reversion includes cases where *through inheritance* there re-appears in an individual some character which was not expressed in his parents, but which did occur in an ancestor. The character whose reappearance is called a reversion may be found within the verifiable family, within the breed, within the species, or even in a presumed ancestral species.[1]

The best illustrations of reversion are furnished by hybrids. Thus in one of Prof. Cossar Ewart's experiments a pure white fantail cock pigeon, of old-established breed, which in colour had proved itself prepotent over a blue pouter, was mated with a cross previously made between an owl and an archangel, which was far more of an owl than an archangel. The result was a couple of fantail-owl-archangel crosses, one resembling the Shetland rock-pigeon, and the other the blue rock of India.

But great carefulness is necessary in arguing from the results of hybridisation to those of ordinary mating, and even if some of the phenomena of exclusive inheritance seem to show reversion

[1] Prof. Karl Pearson defines a *reversion* as "the full reappearance in an individual of a character which is recorded to have occurred in a *definite* ancestor of the same race," and *atavism* as "a return of an individual to a character not typical of the race at all, but found in allied races supposed to be related to the evolutionary ancestry of the given race." "In reversion we are considering a variation, normal or abnormal, from the standpoint of *heredity in the individual;* in atavism we are considering an abnormal variation from the standpoint of the *ancestry of the race.*" But the two words seem to be used by some authors in the converse way, or as equivalent, and it is surely difficult to define the field of abnormal variation.

to a near ancestor we need a broader basis of fact than we have at present before we can formulate any law. The recorded cases show that many phenomena are labelled reversions on the flimsiest evidence. Thus the occurrence of a Cyclopean human monster with a median eye has been called a reversion to the sea-squirt, and gout has been called a reversion to the reptilian condition of liver and kidneys. Often there is not the slightest attempt to discriminate between true reversion (*i.e.* the re-expression of latent ancestral characters) and the phenomena of arrested development, or of abnormalities which have been induced from without. Often, too, there has been no scruple in naming or inventing the ancestor to whom the reversion is supposed to occur, although evidence of the pedigree is awanting ; and the vicious circle is not unknown of arguing to the supposed ancestor from the supposed reversion, and then justifying the term reversion from its resemblance to the supposed ancestor. Little allowance has been made for coincidence, and the postulate of characters remaining latent for millions of years is made as glibly as if it were just as conceivable as a throw-back to a great-grandfather.

There seems no way out of the theory that characters may lie latent for a generation or for generations, or in other words that certain potentialities or initiatives which form part of the heritage may remain unexpressed for lack of the appropriate liberating stimulus, or for other reasons, or may have their normal expression disguised. But it does not follow that the reappearance of an ancestral character not seen in the parents is necessarily due to the reassertion of latent elements in the inheritance. It may be a case of ordinary regression ; it may be a case of arrested development ; it may be an extreme variation whose resemblance to an ancestral characteristic is a coincidence ; it may be an individually acquired modification, reproduced apart from inheritance, by a recurrence of suitable external conditions, and so on. What are called reversions are probably in many cases misinterpretations.

V. Galton's Law.

The most important general conclusion which has yet been reached in regard to inheritance is formulated in Galton's Law. Mr. Galton was led to it by his studies on the inheritance of human qualities, and more particularly by a series of studies on Basset hounds. It is one of those general conclusions which have been reached statistically, and I must refer for the evidence and also for its strictest formulation to the revised edition of Mr. Pearson's "Grammar of Science."[1]

As we have seen, it is useful to speak of a heritage as dual, half derived from the father and half from the mother. But the heritable material handed on from each parent was also dual, being derived from the grandparents. And so on, backwards. We thus reach the idea that a heritage is not merely dual, but in a deeper sense multiple.

To appreciate the possible complexity of our mosaic inheritance we must recall the number of our ancestors. We have two parents, four grandparents, eight great-grandparents, about sixteen great-great-grandparents, and so on. But as we go backwards the theoretical number far exceeds the reality ; a reduction in the number of ancestors is brought about by inter-marriage, as this table (from Lorenz) in reference to Kaiser Wilhelm II. clearly shows.

Generations.	I.	II.	III.	IV.	V.	VI.	VII.	VIII.	IX.	X.	XI.	XII.
(1) Theoretical Number.	2	4	8	16	32	64	128	256	512	1024	2048	4096
(2) Actual number known.	2	4	8	14	24	44	74	111	162	206	225	275
(3) Inadequate-ly known.								5	15	50	117	258
(4) Probable total.								116	177	256	342	533

According to Galton's Law, "the two parents between them contribute *on the average* one-half of each inherited faculty, each of them contributing one-quarter of it. The four grandparents contribute between them one-quarter, or each of them one-sixteenth ; and so on, the sum of the series, $\frac{1}{2} + \frac{1}{4} + \frac{1}{8} + \frac{1}{16} + \&c.$, being equal to 1, as it should be. It is a property of this infinite series that each term is equal to the sum of all those that follow ; thus $\frac{1}{2} = \frac{1}{4} + \frac{1}{8} + \frac{1}{16} + \&c.$; $\frac{1}{4} = \frac{1}{8} + \frac{1}{16} + \&c.$, and so on. The pre-potencies or subpotencies of particular ancestors, in any given pedigree, are eliminated by a law that deals only with *average*

[1] Reference should, however, be made to Mr. Pearson's recent paper *Proc. Roy. Soc.*, lxvi. 1900, pp. 140-164) on the law of reversion.

contributions, and the varying prepotencies of sex in respect to different qualities are also presumably eliminated."

The aim of this lecture has been to present in brief compass a statement of the leading facts of inheritance, which should be clear in the minds of all. Nothing has been said in regard to the transmissibility of acquired characters, for this cannot be ranked at present as an established fact, and some other doubtful points have been left unmentioned. The study of inheritance leaves a fatalistic—almost paralysing—impression on many minds, especially perhaps if it be believed that the acquired results of experience and education—of "nurture," in short, cannot be entailed upon the offspring. To some extent this fatalistic impression is justified, but it is well that it should rest upon a sound basis of fact and not on exaggerations. In a sense we can never get away from our inheritance. As Heine said half bitterly, half laughingly, "A man should be very careful in the selection of his parents." On the other hand, although the human organism changes slowly in its heritable organisation, it is very modifiable individually, and "nature" can be bettered by "nurture." If there is little scientific warrant for our being other than sceptical at present as to the inheritance of acquired characters, this scepticism lends greater importance than ever, on the one hand, to a good "nature," to secure which for off-spring is part of the problem of careful mating ; and, on the other hand, to a good "nurture," to secure which for our children and children's children is one of the most obvious of duties, the hopefulness of the task resting upon the fact that, unlike the beasts that perish, man has a lasting external heritage, capable of endless modification for the better.

UNIVERSITY AND EDUCATIONAL INTELLIGENCE.

Mr. A. Rendle Short, of University College, Bristol, has been awarded the gold medal and exhibition in physiology, the gold medal and exhibition in materia medica, and first-class honours in anatomy, upon the results of the recent Intermediate M.B. examination of the University of London. The exhibitions in physiology and materia medica are of the value of 80*l.* and 60*l.*

A discussion on the teaching of geography will be held at Cambridge on Friday, August 24, under the auspices of the Geographical Association, and in connection with the Summer Meeting. Prof. W. M. Davis, of Harvard University, will occupy the chair, and among the subjects to be brought before the Association are class excursions, map drawing, the use of the globe, geography in the grammar school, and possibilities and limitations of geography in a day school. There are several exhibits of interest to teachers of geography in the education exhibition, arranged in connection with the Summer Meeting.

Lord Bute has offered the University of St. Andrews a sum of 20,000*l.*, to be held as a fund for endowing a chair of anatomy, upon the following conditions :—(1) That the said sum of 20,000*l.* shall be paid to the University not later than ten years hence. The exact date cannot be specified, as it will depend upon completion of certain works at Cardiff. Interest at 3 per cent. will be payable to the University from the time of the appointment of the first professor until they receive the principal sum ; (2) that the first presentation to the chair shall be in favour of Mr. Musgrove, the present holder of the lecture-ship in anatomy in St. Andrews ; (3) that the lectures shall be given exclusively in St. Andrews ; (4) that the course shall meet the requirements of the two first *Anni Medici* ; and (5) that before the beginning of the University session, 1901-1902, the approval of the Universities Committee of the Privy Council to the establishment of the chair under the foregoing conditions be obtained, and that the approval of Lord Bute or his representatives be obtained to any further conditions embodied in the ordinance instituting the chair.

The relations between scientific work and industrial progress have been so often described in these columns that there is little new to be said upon the subject. But though readers of Nature may be familiar with many instances of the close connection between science and industry, it will be a long time before the knowledge filters down to the general public and starts a reaction in commercial and manufacturing circles. Every man of science who takes advantage of an opportunity to impress the value of scientific observation and research upon the minds of citizens, is thus doing a service to the nation, as well as extending interest in natural knowledge. Dr. P. Bedson, professor of chemistry at the Durham College of Science, has, we are glad to see, recently shown the Economic Society of Newcastle-on-Tyne some of the lessons taught by the growth of science and industry in Germany during the present century. A reasonable and organised system of education, and schools in which students receive a thorough grounding in the principles of science, and afterwards contribute something to the advancement of knowledge, are the chief factors in Germany's industrial progress. Referring to the system of examinations which still dominates so much of our educational work, and finds its highest development in connection with university teaching, Prof. Bedson points out that it partakes of the character of the training of a stud of racers. He adds :— "Possibly the instinct of sport, so characteristic of the English people, it is which commends the system of competitive examination. Too much is made of what should be regarded as a minor duty of the University, viz. the testing and marking of its students, and too little of the higher function, the training of students under first-rate teaching, with the object that those so trained should help forward the advancement of knowledge." It is satisfactory to know that the movement in favour of rational teaching in elementary schools, and regard for research in institutions of university rank, is gradually affecting scientific education in this country.

SCIENTIFIC SERIALS.

Symons's Monthly Meteorological Magazine, July.—This number contains the completion of two interesting papers, by Mr. E. D. Archibald, on Indian famine-causing droughts, and their prevision. The principal facts are summarised as follows : (1) Extensive droughts occur in the dry area of Southern India at intervals of nine to twelve years, and usually, but not regularly, about a year before the sun-spot minimum. When the conditions are sufficiently acute, famine occurs in the following year. (2) A severe drought in the peninsular of Southern India is followed by a severe drought and ensuing famine in Northern India in about five cases out of seven. (3) Summer droughts tend to occur in Northern India in years of maximum sun-spot, connected in some way with the abnormal high pressure over Western Asia which prevails at such epochs. There is thus a double periodicity of droughts and famine in North India, and a single periodicity in South India in the sun-spot cycle, though the relation between the phenomena is too spasmodic and irregular to be utilised as a trustworthy factor for prevision.

Annalen der Physik, No. 6.—Interruption spark in the alternating circuit with metallic electrodes, by L. Kallir. The author shows that the impossibility of producing an alternate-current arc between metallic electrodes is due to the fact that the spark is confined to one semi-period of the current. Or if it extends over several periods, it is intermittent, and only appears at every alternate semi-period.—Thermoelectric force of some metallic oxides and sulphides, by A. Abt. Pyrolusite, pyrrhotite, pyrites, and chalcopyrite were used in conjunction with various metals or with each other. A pyrites-chalcopyrite couple gave an E.M.F. 10·8 times as high as an antimony-bismuth couple under the same conditions.—Anomalous electromagnetic rotatory dispersion, by A. Schmauss. Measurements of the Faraday effect for various wave-lengths in fuchsine solutions and in didymium glass justify the general conclusion that optical anomaly in dispersion is invariably associated with a corresponding electromagnetic anomaly. In strongly absorbing media the anomaly extends for a considerable distance on both sides of the absorption band, and it increases with the concentration and with the narrowness and sharpness of the absorption band.—Point discharges, by E. Warburg. In carefully purified nitrogen, the current intensity obtained from the discharge of a fine point charged to -3310 volts is 200 times as great as from a point charged to $+5180$ volts. A slight admixture of oxygen reduces the proportion to $4:1$.—Band spectrum of aluminium, by G. A. Hemsalech. The author quotes some experiments which go to show that the band spectrum of aluminium is due, not to the oxide, but to the metal itself.—Behaviour of radium at low temperatures, by O. Behrendsen. Cooling a radium preparation down to the temperature of liquid air reduces its activity by 50 per cent. Restoration to the ordinary temperature produces a considerable but transient increase of activity.—Production of kathode rays by ultra-violet light, by P. Lenard. The discharge of electrified

bodies by ultra-violet light is due to their emitting kathode rays when the ultra-violet light impinges upon them. The author exhausted a vacuum tube until it no longer allowed any discharge to pass. He then exposed the kathode to ultra-violet rays from a zinc spark gap. The discharge set in again immediately, but no discharge could be obtained by similarly illuminating the anode. The rays which produce the discharge across the absolute vacuum can be deflected by a magnet, and their velocity is about one-thirtieth of the velocity of light.

SOCIETIES AND ACADEMIES.

LONDON.

Royal Society, February 8.—"On Electric Touch and the Molecular Changes produced in Matter by Electric Waves." By Jagadis Chunder Bose, M.A., D.Sc., Professor of Physical Science, Presidency College, Calcutta. Communicated by Lord Rayleigh, F.R.S.

It is claimed that the experiments described in the paper show :—

(1) That ether waves produce molecular changes in matter.

(2) That the molecular or allotropic changes are attended with changes of electric conductivity, and this explains the action of the so-called coherers.

(3) That there are two classes of substances, positive and negative, which exhibit opposite variations of conductivity under the action of radiation.

(4) That the production of a particular allotropic modification depends on the intensity and duration of incident electric radiation.

(5) That the continuous action of radiation produces oscillatory changes in the molecular structure.

(6) That these periodic changes are evidenced by the corresponding electric reversals.

(7) That the "fatigue" is due to the presence of the "radiation product," or strained B variety.

(8) That by means of mechanical disturbance or heat, the strained product can be transformed into the normal form, and the sensitiveness may thereby be restored.

June 21.—"An Experimental Investigation into the Flow of Marble." By Frank D. Adams, M.Sc., Ph.D., Professor of Geology in McGill University, Montreal, and John T. Nicolson, D.Sc., M.Inst.C.E., Head of the Engineering Department, Municipal Technical School, Manchester. Communicated by Prof. H. L. Callendar, F.R.S.

The following is a summary of the results arrived at :—

(1) By submitting limestone or marble to differential pressures exceeding the elastic limit of the rock and under the conditions described in this paper, permanent deformation can be produced.

(2) This deformation, when carried out at ordinary temperatures, is due in part to a cataclastic structure and in part to twinning and gliding movements in the individual crystals comprising the rock.

(3) Both of these structures are seen in contorted limestones and marbles in nature.

(4) When the deformation is carried out at 300° C., or better at 400° C., the cataclastic structure is not developed, and the whole movement is due to changes in the shape of the component calcite crystals by twinning and gliding.

(5) This latter movement is identical with that produced in metals by squeezing or hammering, a movement which in metals, as a general rule, as in marble, is facilitated by increase of temperature.

(6) There is therefore a flow of marble just as there is a flow of metals, under suitable conditions of pressure.

(7) The movement is also identical with that seen in glacial ice, although in the latter case the movement may not be entirely of this character.

(8) In these experiments the presence of water was not observed to exert any influence.

(9) It is believed, from the results of other experiments now being carried out but not yet completed, that similar movements can, to a certain extent at least, be induced in granite and other harder crystalline rocks.

"On the Effects of Changes of Temperature on the Elasticities and Internal Viscosity of Metal Wires." By Andrew Gray, LL.D., F.R.S., Professor of Natural Philosophy in the University of Glasgow, and Vincent J. Blyth, M.A., and

James S. Dunlop, M.A., B.Sc., Houldsworth Research Students in the University of Glasgow.
"The Distribution of Molecular Energy." By J. H. Jeans, B.A. Communicated by Prof. J. J. Thomson, F.R.S.

PARIS.

Academy of Sciences, July 23.—M. Maurice Lévy in the chair.—Notice on Charles Friedel, by M. Georges Lemoine.— Visual observations of the corona of May 28, made by Mr. W. H. Wesley at Algiers, with the Coudé equatorial of 0·318 metre aperture, by M. Lœwy.—Phosphoric acid in the presence of saturated solutions of calcium bicarbonate, by M. Th. Schlœsing. Solutions of phosphoric acid were added to saturated solutions of calcium bicarbonate, and carbon dioxide withdrawn by a slow current of air. From the analyses of the precipitated salt, interesting conclusions are drawn as to the use of superphosphates as manure.—Report upon the proposed revision of the arc of the meridian at Quito, by M. Poincaré. The Commission report that it is of opinion that the proposed revision of the arc of meridian at Quito should be carried out. The arc measured should be 6° instead of 4° 5′, the work being done by the staff of the Geographical Service of the Army under the control of the Academy of Sciences.—On the limited problem of three bodies, by M. Lévi-Civita.—On the position and actual appearance of a new star, transformed into a nebula, by M. G. Bigourdan. The nebular constitution ascribed by Prof. Pickering to this new star, discovered by Mrs. Fleming, must have been derived from spectroscopic examination, since at the present time the object appears clearly as a star, without any trace of nebulosity.—Total eclipse of the sun of May 28. Note on the observations made at the Observatory of Algiers, by W. H. Wesley. A description of the study of the lower coronal regions.—Observations of the total eclipse of the sun of May 28, made in Spain, at Hellin, Albacete, and at Las Minas, by M. G. Bigourdan.—Observation of the solar eclipse of May 28, made at Albacete, in Spain, by M. J. Eysséric.—Observation of the total eclipse of the sun of May 28, at Las Minas, in Spain, by M. Salet.—On a system of differential equations equivalent to the problem of n bodies, but admitting of one more integral, by M. W. Ébert.— On the elastic flying machine, by M. L. Lecornu.—On the electrocapillary functions of aqueous solutions, by M. Gouy.— The spectrum of radium, by M. Eug. Demarçay. Mme. Curie has succeeded in obtaining a specimen of radium chloride in which the barium is so far reduced that only a feeble spectrum of three principal rays is obtained. The radium lines, although much stronger than in the specimens previously studied, show no new ray that can be attributed to radium. In its general character the spectrum of radium approximates to those of the metals of the alkaline earths.—Solubility of a mixture of salts having a common ion, by M. Charles Touren.— On a new complex acid and its salts, palladio-oxalic acid and palladio-oxalates, by M. H. Loiseleur. Four new substances are described, palladio-oxalic acid and its silver, sodium, and barium salts. The acid is the first complex acid containing palladium that has been isolated.—On some osmyloxalates, by M. Wintrebert. —Action of some finely divided metals, platinum, cobalt and iron, upon acetylene and ethylene, by MM. Paul Sabatier and J. B. Senderens. Platinum black has no action upon pure acetylene at ordinary temperatures, but at 150° ethylene and hydrogen, together with small quantities of benzene and ethane, are produced. With cobalt, the reaction commences at 200°, ethane and hydrogen being the principal products. Iron behaves similarly to cobalt. With ethylene, platinum and copper produce practically no effect; cobalt gives ethane, hydrogen and methane, and similarly with iron.—Synthesis of paramethoxyhydratropic acid, by M. J. Bougault. The author concludes that in identifying phloretic acid with the synthetical paraoxyhydratropic acid, M. Trinius was in error.—Influence of hydrobromic acid upon the velocity of the reaction of bromine upon trimethylene, by M. G. Gustavson. The organic solutions of ferric chloride, by M. Œchsner de Coninck. The iron salt is removed from solutions in methyl and ethyl alcohols, acetic ether and acetone by repeated filtration through animal charcoal.— On the nature of the reserve carbohydrates of the St. Ignatius bean and nux vomica, by MM. Ém. Bourquelot and J. Laurent. The albumen of these seeds appears to contain several mannane and galactanes of different molecular weights.—On the genera Palythoa and Epizoanthus, by M. Louis Roule.—A teratological process, by M. Étienne Rabaud.—The nepheline rocks of the Puy de Saint-Sandoux, by M. A. Lacroix.—The sub-Pyreneal erosions, by M. L. A. Fabre.—On the existence of carboni. ferous strata in the region of Igli, by M. Ficheur.—On the agglutination of blood corpuscles by chemical agents, and on the conditions of medium which favour or prevent it, by M. E. Hédon.—On the influence of phosphates and of some other mineral substances upon the proteolytic diastase of malt, by MM. A. Fernbach and L. Hubert.—The bacteriolysis of anthrax, by M. G. Malfitano.—On the function of the nucleus in the formation of hæmoglobin and in cellular protection, by M. Henri Stassano.—On the collection of potable water and protection of springs, by M. Léon Janet.

CAPE TOWN.

South African Philosophical Society, June 27.—L. Péringuey, President, in the chair.—Mr. E. H. L Schwarz exhibited copies of some Bushman drawings which he had found near Nieuwoudtville. Along with the usual reproductions of men and animals, there are certain puzzling figures which have not been recorded from other localities. One of these consists of a rude slipper-like form with seven bars across it; another is a circle with seven peripheral radiating bars, and a third shows three concentric circles, from the outer of which there extend twenty-one bars. Mr. Schwarz thought that the first-mentioned figure might be a tally.—Dr. Corstorphine gave a short note on an old beach deposit on the site of the South African Brewery at Woodstock, which had been brought to his notice by Mr. A. W. Ackermann, architect, Cape Town. Some of the sections show a layer of shells and water-worn boulders some three feet thick resting on the slate and covered by about three feet of sand and soil, but within thirty yards the deposit entirely thins out. The shells all belong to species found on the present beach.— A copy of a report on a submarine disturbance, from the magistrate at Walfish Bay, forwarded by Major Stanford, was read by the secretary. The magistrate stated that on May 31 or June 1 last, a new island appeared about 100 yards N.E. of Pelican Point. The island was about 150 feet long by 30 feet wide, and stood 12 feet above high water. It was composed of a tenacious clay; soundings gave 7-10 fathoms around it; steam was observed rising from the clay, and an intense smell of sulphuretted hydrogen was perceptible, even at a distance of five miles.—Mr. Sclater gave an account of some inscribed stones found in Cape Town.

CONTENTS.

THURSDAY, AUGUST 9, 1900.

PRACTICAL NAVIGATION.

Self-instruction in the Practice and Theory of Navigation. By the Earl of Dunraven, Extra Master. Two volumes. Pp. xxv + 354 + 388. (London: Macmillan and Co., Ltd., 1900.)

THE science of navigation, apart from the practical art of seamanship, stands on a very curious footing. Based mainly on mathematical results, it presents probably the only, certainly the most conspicuous, instance of the adaptation of pure science to practical ends. As a consequence nautical astronomy, or those portions of it which are indispensable to navigation, has been systematised to such a pitch of perfection that a mechanical system has been substituted for a reasoning process. Many regard this result with satisfaction as a triumph of scientific simplicity, and pride themselves on the production of navigators capable of producing a definite practical result with the least possible expenditure in training. Perhaps it would be unjust to say that this view is shared by the Earl of Dunraven, the author of the latest book on the theory and practice of navigation. But he is not prepared to throw his known experience as a sailor and his great popularity as a successful yachtsman on the side of those who would make the Board of Trade Regulations more stringent, and would demand from applicants for the various certificates some proof that they have acquired more than a rule-of-thumb acquaintance with the various methods and formulæ that they will have to put into practice. The effect, if not the object, of his book is to show with how little knowledge one may pass the Board of Trade Examinations, and be legally entitled to assume positions of enormous responsibility. But admitting that it is desirable to give the practical seaman every chance in the examination room, and that the accurate solution of a problem is the only point to be regarded, is it easier to teach once for all the ordinary methods for the solution of a spherical triangle, or to burden the memory with a variety of rules which are available only for the solution of the particular family of problems to which these rules have been adapted? Take, for example, the case of the determination of an hour angle from the observation of an altitude in a known latitude. The candidate for a certificate, taught on the lines that Earl Dunraven approves and encourages, has to remember first of all a series of rules about declination and latitude being of the same or different names; then he has to write certain quantities down in a particular order, perform sundry acts of legerdemain, take out four different logarithmic functions of angles, add them up, and is landed in a quantity which his lordship calls "the log. of the hour angle." It is the log. sine squared of half the hour angle, but this is a detail, and if one happens to possess the particular table in which some obliging genius has given this quantity, with argument hour angle, the work is done and it may be, so far as the result is concerned, satisfactorily. To trust to the memory rather than the rigorous process of demonstration is a plan Earl Dunraven thinks admirably adapted to meet the difficulties introduced by "a wet, slippery, and tumbling deck" and the inconveniences

"of a dimly-lit cabin, full of confusion and noise." We fail to perceive the particular advantages of this system, but would express any doubts on this point very modestly, for the author speaks from an actual experience, which we can very inadequately apprehend.

But if our methods of teaching are as far asunder as the poles, it is impossible to escape the influence of the cheerful, breezy style in which the book is written—a model for those who attempt to substitute teaching by written description for oral explanation. The author appears to be sitting at the same table with the student, giving him of his best, and actually pushing him through the examination. If any one has failed to satisfy the examiner that he is competent to do "a day's work," let him take Earl Dunraven for his guide, and he will become fully persuaded of the easiness of the problem, rather than of its difficulty, and will pass the ordeal with success.

The author supposes his pupil to be conversant with the multiplication table, but with practically nothing else, so he gives first a chapter on arithmetic, followed by one on the application of logarithms; the theory is dismissed in a page, and of this short summary the student is told "don't bother to read it unless you have a mind to." This is the keynote of the whole book, only those problems which can have an immediate practical significance, or can be broached in the examination room, are pressed on the student's notice. But to make amends for the lack of theory, on the practical side, the detail is very full and complete. From logarithms we pass to the description of the instruments used at sea, and so arrive at the "sailings" and that troublesome problem of the "day's work," which proves such a stumbling block for so many aspirants for certificates. At this point the author thinks it time to introduce a little algebra and trigonometry, though he advises only extra masters to read it, and we must admit that it contains some hard things, and that we should have some difficulty in solving some of the simple equations proposed by following the rules laid down for our guidance. The author is not seen at his best in these chapters, which are better taught in the schoolroom than on the ship's deck. Tides and charts, so far as their investigation and construction are needed for the examination room, are fully explained. The first volume concludes with the solution of simple problems connected with the determination of latitude, longitude, and azimuth.

We cannot get very far into the second volume without some knowledge of spherical trigonometry, and here, again, we do not find the chapters devoted to this subject altogether satisfactory. Spherical trigonometry covers a very small but well recognised subject of inquiry, and can without much difficulty be made complete. The methods are simple and easily applied, except in one point, and that is the determination of the quadrant in which the various arcs fall. Earl Dunraven has not much assistance to offer on this vexed point. He pins his faith to Haversines, and as a rule keeps free from the employment of auxiliary angles. In this he is no doubt well advised, for the advantages of the method so long insisted upon in elementary treatises are by no means so apparent in actual work. Through the intricacies of the ingenious method known as Sumner Lines, Earl Dunraven conducts us with care, especially dwelling on the use of the

Q

various tables that have been introduced to facilitate the process and hasten the result. After one or two further applications of spherical trigonometry, we are brought face to face with that curious survival, known as a Lunar Distance, and we are quite sure that the author did some violence to his sense of practical utility when he devoted so many wearisome pages to the consideration of this obsolete problem. In the examination room of the Board of Trade, the thorny difficulties of "clearing the distance" may exercise a wholesome effect on the extra master, whose fate it is to attack this problem, and induce him to acquire a greater knowledge of nautical astronomy than he would otherwise do; but we imagine in the great majority of cases the applicant endeavours to forget all about the intricacies of the problem as soon as he is possessed of his qualifying "ticket." The skill of the mechanician has done much to remove the necessity of the ingenious device, but the rapid transit of vessels from port to port, and the numerous time signals in known longitudes, give to the mariner Greenwich Time more accurately than it was ever determined by the method of lunar distances. But for some reason known only to the authorities, an acquaintance with the method is demanded, though the necessary facility in manipulating the sextant cannot so well be required. The whole process affords an interesting case of the resources of analysis outrunning in accuracy the observations to which it is applied.

"Problems," says Earl Dunraven, "will be given you in the examination room on the infernal subject of magnetism and deviation," so he has much to say about the coefficients A to E. To many, we are afraid these coefficients will remain a matter of intricate manipulation, carrying no definite meaning; but if they follow the author's guidance, they ought to issue triumphantly from the examination ordeal. His rules are admirably arranged, and, from a purely mechanical point of view, leave nothing to be desired. We could have wished that the theory had been a little fuller, but we remember, a little regretfully, that the author's object is not to teach magnetism, but to pass the reader or student through an examination of a strictly limited character. We cannot but think that the book is eminently calculated to effect this object. Admirably printed, well and lavishly illustrated, furnished with numerous examples and written in a free and easy, but lucid style, we should imagine that this work is destined to become the most popular book on the subject, and that it will be the one guide and text-book to which the young officer will apply, to help him to meet and defy the terrors of Her Majesty's examiners.   W. E. P.

THE CULTIVATION AND PRODUCTION OF COFFEE.

Le Café, Culture—manipulation—production. Par Henri Lecomte, Agrégé de l'Université, Docteur es Sciences, &c. Pp. vi + 342. (Paris: Georges Carré et C. Naud, 1899.)

COFFEE in its various commercial aspects, whether from the point of view of the planter, the broker, the retail dealer, or the consumer, has from time to time commanded a great deal of attention. Occupying as it does a large and extended area of cultivation within the

tropics, and being an important branch of industrial culture in many of the British possessions, as Jamaica, Ceylon, Southern India, and Borneo, it is but reasonable to expect that treatises on the cultivation, best means of improvement of yield and quality, prevention of disease, &c., would be numerous. In the English language many such works are available, and if this be so, bearing on a culture which though large and important is small in comparison with that of Brazil, Central America, Mexico, Java, and Sumatra, we might also expect to find a large number of books in the languages of the nations to which these extensive coffee growing countries belong.

The work before us is the latest contribution to the French literature of the subject, and extensive as that literature is and for the most part carefully worked out, M. Lecomte's handbook will be a useful and valuable addition not only for its arrangement, but also for the concise character of the information given and the various items of intelligence regarding production in the several countries referred to and exports therefrom.

The first chapter is devoted to the early history of the coffee plant. The botany of the genus *Coffea* is treated of in the second chapter occupying twenty-five pages, and is illustrated by a figure of the so-called Arabian coffee (*Coffea arabica*) in flower and in fruit, and a figure is also given of *C. stenophylla*, the tree which furnishes the wild coffee of Sierra Leone, as well as of the new species from the Congo, *C. canephora*, Pierre. In the enumeration of species given in this chapter thirty-three are referred to, prominence, of course, being given to *C. arabica* and *C. liberica*, the two most important coffee yielding species. The best varieties of *C. arabica* cultivated in various parts of the world are also enumerated. Referring to *Coffea stenophylla* the specific name of which, by the way, is spelt with a capital initial letter, the author gives the following interesting account of it: In 1894 some plants of this new species were received at the Royal Gardens, Kew, from Sierra Leone, and these plants produced flowers in 1895. Seeds were afterwards sent to most of the English colonies where it was thought the plant might flourish. In Ceylon, however, the results have not been satisfactory; but in Dominica, Jamaica, and Trinidad, the case has been different. In the Botanic Garden of Port of Spain, Trinidad, there are some fine fruiting examples of this tree quite free of disease. The author further regrets that this coffee has not yet been introduced into the French colonies. On the climate and elevation suitable for the success of coffee plantations the great coffee-growing country of Brazil has the first consideration. The remaining chapters are devoted to the consideration of soils, the choice of seeds, transplanting, manures, shade trees, &c. The use of simple diagrams showing the different positions in which the coffee plant and its shade trees may be placed will be found useful, as will also the list of trees suitable both for shade and shelter, amongst which we notice such well-known trees as *Albizzia Lebbek, A. stipulata,* and *Exythrina indica.*

On the subject of harvesting or gathering the crop it is pointed out how extremely variable in the period of ripening its seeds the plant is in different countries. Thus in Cuba, Guadaloupe, and other islands of the Antilles, the harvest commences in August and is carried on through

November, while in Brazil it commences in May and ends in September.

Though the broad principles of the preparation of coffee for market are well known, the description here given, especially aided as it is by the practical illustrations, will be of especial value. No book on coffee could possibly be complete without a reference to the diseases to which the plant is subject, whether the disease belongs to the vegetable or animal kingdom. Consequently we find thirty-one pages devoted to this part of the subject. Substitutes for coffee also come under consideration, occupying, however, a comparatively small space, and though no doubt sufficient is said about them, their number might be considerably increased. Perhaps one of the most interesting parts of the book is that treating on production, in which each country is considered separately, the first chapter being devoted to the American Continent, and naturally leading off with Brazil. British, Dutch and French Guiana are also considered, and comparisons made with product and export, as are also those of Paraguay, Venezuela, Columbia, Costa Rica, Mexico and other places. The West Indies, including Jamaica, Porto Rico, Trinidad and other important coffee growing countries, as well as the Eastern countries and Africa, are also referred to. This part of the subject is practically illustrated by a map of the world, showing at a glance the geographical distribution of the coffee plant, together with the production of each country in kilogrammes, and the date to which the figures refer. A comparison of the produce of each country is readily gained by a series of disks of different sizes, with the names of the country beneath each, and the total in figures; from this it will be seen that Brazil is far ahead of any other individual country. An interesting table is also given showing the consumption of coffee in the principal countries of the world, from which it seems that of the European countries Germany consumes by far the largest quantity. The figures in tons for 1897 standing thus—Germany 136,390, France 77,310, England 12,420, while the consumption in the U.S. of America in the same year amounted to 318,170 tons. The book concludes with a table of subjects of the several chapters, but lacks that most necessary adjunct of all books—a good index.

THE BIRDS OF SURREY.

The Birds of Surrey. By J. A. Bucknill. Pp. lvi + 374, illustrated. (London : R. H. Porter, 1900.)

FROM its great extent of open moorland and the presence of several large sheets of water, Surrey occupies an unusually favourable position among the metropolitan counties for the development of a large bird-fauna ; and since a very considerable portion of the county is now undergoing a metamorphosis under the hands of the builder as the area of the metropolis and its suburbs increase, it is most important that a full record should be secured of the species of birds which are fast disappearing from its limits. The compiling of such records, and the careful working out of the past history of locally distributed species within the limits treated of, seem, indeed, to be the chief justification for the publication of county ornithologies. And in this respect, as well as in the careful collection of local bird-

names, the author of the work before us appears to have discharged his task in a thoroughly satisfactory manner. An instance of this is afforded by his account of the occurrence of the black-grouse in Surrey. To many of our readers it will probably come as a surprise to learn that black-cock shooting was a recognised sport on the Surrey moors during the forties, and even to a considerably later date. At the present day there is, however, scarcely a single genuine wild bird of this species to be met with in the county ; and the excellent history of its gradual extermination given by Mr. Bucknill should, therefore, be read with the greatest interest alike by sportsmen and by ornithologists. The raven, the buzzard, the marsh-barrier, and the dotterel are other species which have disappeared from the county, either totally or as nesting birds ; the last record of the occurrence of the dotterel being 1845, when a couple of specimens were purchased from the landlord of an inn at Hindhead.

Of the numerous rare birds that have been noticed from time to time in the county, the great majority have been visitors to the well-known Frensham ponds, the larger of which extends into Hampshire. Here we are practically in Gilbert White's country ; and in these favoured haunts have been seen the osprey, the spoonbill, several of the rarer kinds of duck, the goosander, and the purple heron. Sad to say, the arrival of these wanderers has for the most part been speedily followed by their slaughter ; and, as the author remarks, hundreds of other avian rarities have doubtless been killed and eaten without record. Unhappily, the great increase in game preservation which has taken place of late years in the county appears to have been the cause of the diminution in the numbers of many of the rarer species of birds. But there are many country gentlemen, on the other hand, who are lovers of natural history, and who veto as much as possible the bird-slaying propensities of their game-keepers. It is to such, and to the laws now in force for the protection of wild birds, that we have to look for the commencement of a better state of things in the wilder parts of the county. And the fact that the golden oriole and the hoopoe have been observed of late years on several occasions indicates the probability that these beautiful birds would once more nest in the Surrey groves if only they received adequate protection.

A feature of the book is the beautiful series of illustrations of Surrey scenery ; the views of Frensham Great Pond and of the Surrey Weald being some of the best examples of landscape photogravure that have come under our notice. Although primarily intended for residents in the county (among whom we are glad to see that a long list of subscribers has been enrolled), the book is full of interest to all bird-lovers living in the south of England.　　　R. L.

OUR BOOK SHELF.

Untersuchungen ueber d. Vermehrung d. Laubmoose durch Brutorgane und Stecklinge. Von Dr. Carl Correns, a.ö. Prof. d. Botanik in Tübingen. Pp. xxiv + 472 ; mit 187 abbild. (Jena : Verlag v. Gustav Fischer, 1899.)

FEW people perhaps fully realise how abundantly the mosses are provided with modes of vegetative reproduction, even although they may be fully cognisant of the fact that the protonema—the precursor of the moss-plant—is readily induced to make its appearance from the cut ends of the stems and leaves of these plants. Prof. Correns has done a useful service in bringing together, in a classified manner, the numerous methods employed by mosses to ensure their propagation and dispersal by means less expensive than by the production of spores. The readily friable stems of some species of *Andreaea*, the easily detached branchlets of *Dicranum*, are instances, well known to muscologists, of a large class of propagative bodies. These simpler forms of reproduction are also widely spread amongst plants other than mosses, and in some cases—e.g. *Lycopodium Selago* —the superficial resemblance is rather striking. Less obvious are the subterranean bulbils or buds, such as are met with in *Dicranella*, *Barbula*, or *Funaria*, in which special tuberous bodies are formed. *Dicranella heteromalla* affords a pretty example of a form transitional from the simple to the more complex types, inasmuch as the subterraneous bulbils of this moss are little more than rows of swollen rhizoid-cells arranged somewhat like a string of beads. Many of these bulbils are regarded by Correns rather as of the nature of food reservoirs than as brood bodies ; but it is at least certain that they are in most cases able to function in the latter capacity as well as in that of mere storehouses of food-reserves.

Other and very common cases of brood-bodies are afforded by the so-called "*folia fragilia*"—leaves which readily become detached from the parent plant, and with greater or less intervention of protonematal filaments give birth to new individuals. Oftentimes the leaves destined to this end undergo considerable contraction in size, and, indeed, may assume a totally rudimentary appearance.

Again, as in some species of *Orthotrichum*, cells grow out from the ends of leaves, and the sausage-shaped proliferations, after detachment from the parent plant, grow out to filaments, on which new plants arise.

The above are only a few of the many forms cited by Correns of gametophytic reproductions in the mosses by vegetative means. But as Pringsheim long ago pointed out, it is also possible to reproduce these plants from the sporophyte generation, especially from cut fragments of the seta or stalk of the moss-capsule. These are far more interesting, as they resemble the curious aposporic development met with in a number of ferns. Indeed, these latter offer, perhaps, a means of attacking the details of the phenomena of apospory with a greater chance of success than in the case of the ferns, since they seem more easily induced by simpler experimental devices than is the case with the higher plants.

A general synopsis of the various types and forms of brood-bodies forms a useful adjunct to the main descriptive part of a book on which the author has evidently expended much labour, and which should earn for him the gratitude of all those muscologists who are not merely described as of species, as well as of botanists who seem too often rather to be disposed to ignore an important section of the vegetable kingdom.

Village Notes, and Some Other Papers. By Pamela Tennant. Pp. xii + 204 ; 13 plates. (London : William Heinemann, 1900.)

THESE notes reveal some of the humour and pathos of rural life in South Wilts, and here they lightly touch natural scenes and objects other than human. The plates, which are reproductions from original photographs of Wiltshire views, are excellent, and the book itself is a dainty volume suitable for a drawing-room table. Reference is made to the "pernicious habit of 'underlining' in their letters" which some people adopt, yet we notice an abundance of italicised words in the book, and they are equivalent to the underlined words so severely condemned.

LETTERS TO THE EDITOR.

[*The Editor does not hold himself responsible for opinions expressed by his correspondents. Neither can he undertake to return, or to correspond with the writers of, rejected manuscripts intended for this or any other part of* NATURE. *No notice is taken of anonymous communications.*]

The Conductivity produced in Gases by the Motion of Negatively-charged Ions.

RECENT researches have shown that gases are rendered conductors of electricity when negatively-charged ions move through them with a high velocity. Thus the kathode rays and the Lenard rays possess the property of ionising gases through which they pass (J. J. Thomson, "The Discharge of Electricity through Gases"). Becquerel (*Comptes rendus*, March 26, 1900) also has recently shown that the conductivity produced by radium is due to small negatively-charged particles given off by the radio-active substance. In these cases the charged particles which ionise the gas move with velocities nearly equal to the velocity of light.

Some experiments which I have recently made show that ions which are produced in air by the action of Röntgen rays will produce other ions when they move through the gas with a velocity which is small compared with the velocity of light.

When Röntgen rays are sent through a gas, at atmospheric pressure, the current between two electrodes immersed in the gas increases in proportion to the electric force, when the force is small. For large forces the current attains a value which is practically constant.

When the pressure of the gas is reduced, the connection between conductivity and electromotive force is more complicated. The accompanying tables show the connection between current and electric force for air at 2 and ·8 mm. pressure. At these pressures the current is practically constant for forces of about 10 volts per centimetre, and when forces of this order are acting, all the ions are produced directly by the rays. When the electric force is increased these ions produce others, so that the current again increases.

It appears from the following investigation that the new ions are produced by the collisions between negatively-charged ions and the molecules of the gas.

Let us suppose that n negative ions are moving in a gas between two parallel plates at a distance d apart. Let X be the electric force between the plates $\left(= \dfrac{V_1 - V_2}{d} \right)$, and p the pressure of the gas. In passing a distance dx the n ions produce $a \times n \times dx$ others, where a is a constant depending on X, p, and the temperature, which is constant in these experiments. (The coefficient a is practically zero for small values of X, unless p is also small).

$$\therefore \quad dn = a n \, dx$$

$$\text{and} \quad n = n_0 E^{ax}$$

Hence n_0 ions starting at a distance x from one of the plates will give rise to $n_0(E^{ax} - 1)$ others. When the ions arrive at the plate, the formation of new ions ceases and the current stops, although the electromotive force is kept on. Let n_0 be the number per unit volume produced by the rays. The total number of ions produced will therefore be

$$\int_0^d n_0 E^{ax} dx = \frac{n_0}{a}(E^{ad} - 1)$$

per unit area, $n_0 d$ being the number produced by the rays. Hence

$$\frac{c}{c_0} = \frac{1}{ad}(E^{ad} - 1)$$

where c is the current for a large force X, and c_0 the current composed of ions produced by the rays.

The following experiments were made in order to test the accuracy of this formula for currents produced between two parallel plates whose distance apart could be varied.

The rays fell normally on one of the plates, which was made of thin aluminium, and after passing through the air between the plates, the rays were completely stopped by the second plate, which was of brass. The plates were 10 centimetres in diameter, and the rays were allowed to fall on a circular area at the centre 4 centimetres in diameter. The conductivity was thus confined to a region where the force was constant. A large part of the conductivity (c_0) arises from the secondary radiation from the brass disc. At high pressures the secondary effect is principally confined to a layer of gas near the surface (John S. Townsend, *Camb. Phil. Proc.*, vol. x. Part iv.), but when the pressure is low the secondary rays are not so rapidly absorbed by the gas, and the ionisation (n_0) between the plates is nearly uniform.

The ratios of $\frac{c}{c_0}$ were determined for different forces, the air being at a pressure of two millimetres. When the strength of the rays was reduced to $\frac{1}{3}$ of its original value it was found that the ratios $\frac{c}{c_0}$ were unaltered. This shows that a is independent of n_0 and is some function of X and p.

The plates were then set at one centimetre apart, and the values of c were determined for different forces. The results, corresponding to a pressure 2 and ·8 mm., are given in the second columns of the accompanying tables. The numbers given are the mean between the currents in opposite directions. With this form of apparatus, however, there were only very small differences found in the conductivity when the electromotive forces were reversed. The plates were then set at two centimetres apart, and the currents found in this case for pressures 2·14 and ·8 mm. are given in the third columns of the tables.

The force X is given in volts per centimetre.

TABLE I.—*Air at pressure 2 mm.*

X		$c(d=1)$		$c(d=2)$		Calculated values of $c(d=1)$
20	...	28	...	49·5	...	28
40	...	28·2	...	51	...	28·4
80	...	29·5	...	55	...	29·5
120	...	36	...	81	...	35·5
160	...	51	...	173	...	50
180	...	64·5	...	293	...	63

TABLE II.—*Air at pressure ·8 mm.*

X		$c(d=1)$		$c(d=2)$		Calculated values of $c(d=1)$
10	...	10	...	17·7	...	10
20	...	10·5	...	19	...	10·5
40	...	12	...	24·5	...	12
80	...	17	...	53·5	...	17
120	...	31	...	190	...	29
165	...	61	...	990	...	62·5
186	...	82	..	2180	...	84

The tables show that the current increases more rapidly with X when the plates are two centimetres apart than when they are one centimetre apart. This effect cannot be attributed to a surface action which would be independent of d when X remains constant.

From the formula $\frac{c}{c_0} = \frac{1}{ad}\left(\varepsilon^{ad} - 1\right)$ we can deduce the values of a from the third columns of the tables, by making $d = 2$ and c_0 the smallest value of c. From values of a thus obtained, the ratios $\frac{c}{c_0}$ for the different forces corresponding to plates 1 centimetre ($d = 1$) were calculated. The values of c found in this manner are given in the fourth columns, and they show a good agreement with the experimental determinations.

Other experiments for different pressures have also been made, and they all show an agreement with the present theory.

For the purpose of deciding whether it is the positive or negative ions which produce other ions by their rapid motion through the gas, we may mention the following experimental results. When the lines of force in the gas are not parallel, large differences in current were obtained on reversing the electromotive force. Thus, when the conductivity takes place between two electrodes one inside the other, it was found that for high electromotive forces the current is much greater when the ions go towards the inner electrode.

Thus, with an apparatus consisting of a small spherical electrode surrounded by a large electrode made of thin aluminium, the currents, when the outside electrode was positive, were 14 for a potential difference of 40 volts, and 34 for a potential difference of 300 volts; when the outside electrode was negative the currents were 14 and 174 for the same voltages. In these experiments the pressure was about 2 mm. The positive and negative ions produced by the rays are generated nearly uniformly throughout the area between the electrodes. When the large electrode is positive only a few of the negative ions pass through the region round the small electrode where

the force is big, and the current only increases from 14 to 34. When the electromotive force is reversed all the negative ions produced by the rays come into the region where the force is big, and the current is thereby increased from 14 to 174. It is therefore evident that the increase of conductivity must be attributed to the rapid motion of the negative ions.

I hope in a future paper to give a fuller account of the above experiments, and also to point out some of the applications of this theory to the passage of electricity through gases. I may mention that the high conductivities obtained with ultra-violet light (Stoletow, *Journal de Physique* (2), 9, pp. 463-473, 1890), at pressures of about 1 millimetre, may be explained by this theory.

Approximate values of the energy of translation of the negative ion when producing another ion by a collision can also be obtained from the coefficients a. 　　　　J. S. TOWNSEND.

Trinity College, Cambridge.

A Remarkable Hailstorm.

I HEREWITH enclose you prints, from untouched negatives, of hailstones which fell at Northampton on Friday, July 20.

The drawing board measures 19½" by 17", and the average circumference of the hailstones upwards of five inches. These are by no means the largest that fell, according to the statements of trustworthy persons, but were typical of what fell in my garden.

FIG. 1.—Group of hailstones which fell at Northampton on July 20. Size of board 19½ in. by 17 in.

The majority of the stones were somewhat flattened, as shown in the front of the photograph, but many were nearly spherical like those in my hand (Fig. 1).

The stones were extremely dense and well frozen, and buried themselves in the garden soil. Where they fell on hard surfaces, they usually broke into fragments which rebounded to considerable heights, while glass roofs suffered enormous damage all over

FIG. 2.—Sections of hailstones (Northampton, July 20).

the area, some twelve miles by six, covered by the storm. I have a piece of glass 5/16ths of an inch in thickness many hundred square feet of which were broken at the various factories in the town.

The sections (Fig. 2) were an afterthought and show the structure exceptionally well in two instances. 　　　J. G. ROBERTS.

Northampton and County School, July 30.

THE PHOTOGRAPHY OF SOUND-WAVES AND THE DEMONSTRATION OF THE EVOLUTIONS OF REFLECTED WAVE-FRONTS WITH THE CINEMATOGRAPH.

Introduction.

IN a paper published in the *Philosophical Magazine* for August 1899, I gave an account of some experiments on the photography of sound-waves, and their application in the teaching of optical phenomena. Since writing this paper I have extended the work somewhat and at a meeting of the Royal Society on February 15, 1900, gave an account of this work, and demonstrated certain features of wave motion with the cinematograph.

In the present article I propose to give a somewhat more extended account of the work, paying especial attention to the analogies between the sound-waves and waves of light.

In teaching the subject of optics we are obliged to resort to diagrams when dealing with the wave-front, and in spite of all that we can do, the student is apt to form the opinion that the rays are the actual entities, and that wave-fronts are after all merely conceptions.

The set of photographs illustrating this article will, I think, be of no small use to teachers in ridding the minds of students of the obnoxious rays, and impressing the fact that all of the common phenomena of reflection, refraction and diffraction are due simply to changes wrought on the wave-front.

Sound-waves in air were first observed and studied by Toepler, by means of an exceedingly sensitive optical contrivance for rendering visible minute changes in the optical density of substances. A very full description of the device will be found in Toepler's article (*Wied. Annalen*, cxxxi.), while a brief account of it will be given presently.

The waves in question are the single pulses of condensed air given out by electric sparks. A train of waves would complicate matters too much, and for illustrating the optical phenomena which we are to take up would be useless.

The snap of the spark gives us just what we require, namely, a single wave-front, in which the condensation is considerable.

When seen subjectively, as was the case in Toepler's experiments, the wave-fronts, if at all complicated, as they often are, cannot be studied to advantage, as they are illuminated for an instant only, and appear in rapid succession in different parts of the field. By the aid of photography a permanent record of the forms can be obtained and studied at leisure. The first series of photographs, published in the *Philosophical Magazine*, were made with an apparatus similar to the one to be presently described ; while most of those illustrating this article were made on a much larger scale by employing a large silvered mirror in place of the lens, an improvement due to Prof. Mach, of Prague, who has given much attention to the subject.

As it is a matter of no trouble at all to set up in a few minutes, in any physical laboratory, an apparatus for showing the air-waves subjectively, and as the method does not seem to be as well known as it deserves to be, a brief description of the "Schlieren" apparatus, as Toepler named it, may not be out of place.

The Apparatus.

The general arrangement of the "Schlieren" apparatus is shown in Fig. 1. A good-sized achromatic lens of the finest quality obtainable, and of rather long focus, is the most important part of the device. I have been using the object-glass of a small telescope figured by the late Alvan Clarke. Its diameter is five inches, and the focal length about six feet. I have no doubt but that a smaller lens could be used for viewing the waves, but

one of at least this size is desirable for photographing them.

The lens is mounted in front of a suitable source of light (in the present case an electric spark), which should be at such a distance that its image on the other side of the lens is at a distance of about fifteen feet.

The image of the spark, which we will suppose to be straight, horizontal, and very narrow, is about two-thirds covered with a horizontal diaphragm (*a*), and immediately behind this is placed the viewing-telescope. On looking into the telescope we see the field of the lens uniformly illuminated by the light that passes under the diaphragm, since every part of the image of the spark receives light from the whole lens. If the diaphragm be lowered the field will darken, if it be raised the illumination will be increased. In general it is best to have the diaphragm so adjusted that the lens is quite feebly illuminated, though this is not true for photographic work. Let us now suppose that there is a globular mass of air in front of the lens of slightly greater optical density than the surrounding air (*b*). The rays of light going through the upper portion of this denser mass will be bent down, and will form an image of the spark below the diaphragm, allowing more light to enter the telescope from this particular part of the field ; consequently, on looking into the instrument, we shall see the upper portion of the globular mass of air brighter than the rest of the field. The rays which traverse the under part of "*b*," however, will be bent up on the contrary, forming an image of the spark higher up, and wholly covered by the diaphragm ; consequently this part of the

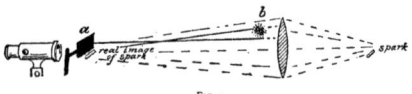

FIG. 1.

field will appear black. It will be readily understood that, with the long path between the lens and the image a very slight change in the optical density of any portion of the medium in front of the lens will be sufficient to raise or depress the image above or below the edge of the diaphragm, and will consequently make itself manifest in the telescope.

The importance of using a lens of first-class quality is quite apparent, since variations in the density of the glass of the lens will act in the same way as variations in the density of the medium before it, and produce unequal illumination of the field. It is impossible to find a lens which will give an absolute even, feeble illumination, but a good achromatic telescope objective is perfect enough for every purpose. A more complete discussion of the operation of the apparatus will be found in Toepler's original paper in the *Annalen*. The sound-waves, which are regions of condensation, and consequent greater optical density, make themselves apparent in the same way as the globular mass of air already referred to. They must be illuminated by a flash of exceedingly short duration, which must occur while the wave is in the field of view.

Toepler showed that this could be done by starting the sound-wave with an electric spark, and illuminating it with the flash of a second spark occurring a moment later, while the wave was still in the field. A diagram of the apparatus used is shown in Fig. 2. In front of the lens are two brass balls (*a*, *a*), between which the spark of an induction coil passes, immediately charging the Leyden-jar *c*, which discharges across the gap at *e* an instant later. The capacity of the jar is so regulated that the interval between the two sparks is about one ten-thousandth of a second. The field of the lens is thus illuminated by the flash of the second spark before the sound-wave started by the first spark has gone beyond the edge of the lens.

To secure the proper time-interval between the two sparks it is necessary that the capacity of the jar be quite small. A good-sized test tube half full of mercury standing in a jar of mercury is the easiest arrangement to fit up. This limits the length and brilliancy of the illuminating-spark, and with the device employed by Toepler I was unable to get enough light to secure photographs of the waves. After some experimenting I found that if the spark of the jar was passed between two thin pieces of magnesium ribbon pressed between two pieces of thick plate-glass, a very marked improvement resulted. With this form of illuminator I found that five or six times as much light could be obtained as by the old method of passing the spark between two brass balls.

The spark is flattened out into a band, and is kept always in the same plane, the light issuing in a thin sheet from between the plates. By this arrangement we secure a light source of considerable length, great intensity, and bounded by straight edges, the three essentials for securing good results. The glass plates, with the ribbon terminals between them, must be clamped in some sort of a holder and directed so that the thin sheet of light strikes the lens : this can be accomplished by darkening the room, fastening a sheet of paper in front of the lens, and then adjusting the plates so that the

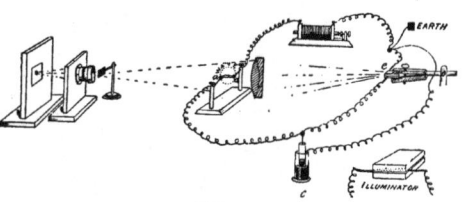

Fig. 2.

paper is illuminated as much as possible. The image formed by the lens will be found to have very sharp straight edges,[1] on one of which the edge of the diaphragm can be set in such a manner as to allow but very little light to pass when the intervening medium is homogeneous ; a very slight change, however, in any portion may be sufficient to cause the entire amount of light passing through that portion to pass below the diaphragm and enter the telescope.

The photographs were made by substituting a photographic objective for the telescope, in the focal plane of which a vertical board was mounted to support the plate. The room was darkened, a plate held in position, and a single spark made to pass between the knobs by pulling a string connected with the hammer of the induction coil. The plate was then moved a trifle and a second impression secured in the same way. This obviated several of the difficulties experienced in the earlier work. The images never overlapped, and the hot air from the spark did not appear in the pictures. About thirty-five images were obtained on each plate in less than a minute, from which it was usually possible to pick a series showing the wave in all stages of its development, owing to the variations in the time-interval between the two sparks.

[1] If more than one image appears it means that the plane of the glass plates of the illuminator does not lie parallel to the optical axis of the system. It is of prime importance to secure a single image.

In the first series the pictures were so small that it was necessary to enlarge them several diameters. Those of the new series, owing to the use of an eight-inch mirror in place of the five-inch lens, and an objective of larger aperture and longer focus, required no enlarging.

The Wave-Front Photographs.

In the study of optics we may treat the subject of regular reflection in two ways, by rays and by wave-

Fig. 3.

fronts. When spherical waves of light are reflected from a plane surface, we know that the reflected waves are also spherical in form, the centre of curvature being a point just as far beneath the reflecting surface as the source of light is above it. In the first of the series of photographs we have the reflection of a spherical wave of sound by a flat plate of glass, the wave appearing as a circle of light and shade surrounding the image of the balls between which the spark passed (Fig. 3). The reflected wave or echo is seen to be spherical, with a curvature similar to the incident wave.

When we have a source of light in the focus of a parabolic mirror, the rays leave the mirror's surface parallel to one another, and move out in an intense narrow beam. Treating this case from the wave-front point of view, we ascertain by the usual geometrical construction that the spherical wave is changed by reflection into a plane or flat wave which moves out of the mirror without further divergence. In the picture (Fig. 4), only a portion of the parabolic reflector is shown near the bottom. The sound-wave starts in the focus, and the reflected portion appears quite flat.[1]

What happens now if we use a spherical mirror in the same way?

Owing to the spherical aberration the reflected rays are not strictly parallel, or the reflected wave is not a true plane. Let us start a sound-wave in the focus of such a

Fig. 4.

mirror, and follow the reflected portion out of the mirror (Fig. 5). We notice that near the axis of the mirror the effect is much the same as in the case of the parabola, that is, the reflected front is plane. Thus we are

[1] In this series and some others left and right have been inadvertently interchanged by the engraver. The series should be followed by the numbers.

accustomed to say that if we confine ourselves to a small area around the axis, a mirror of spherical form acts

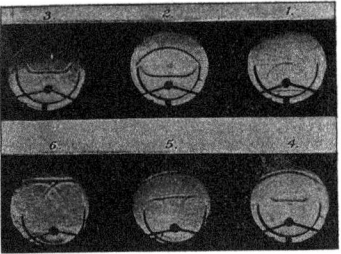

FIG. 5.

almost as well as a parabola. If on the contrary we consider the reflection from the entire hemisphere, we see that the reflected wave curls up at the edges, having a form not unlike a flat-bottomed saucer. The flat bottom moves straight up, travelling everywhere normal to its surface ; but the curled up edges converge inwards, coming to a focus in the form of a ring around the flat bottom. This ring, of course, does not show in the photograph, which is a sectional view, but it will be seen that in one of the views (No. 4) the curved edge has disappeared entirely. In reality it is passing through a ring focus, and presently it will appear again on the other side of the focus, curved the other way, of course, and trailing along after the flat bottom. This curious evolution of the wave can be shown by geometrical construction, and I shall show later how its development can be shown with the cinematograph.

When the spherical waves start in one focus of an elliptical mirror, they are transformed by reflection into converging spheres, which shrink to a point at the other focus, the surface being aplanatic for rays issuing from a point. An elliptical mirror was made by bending a strip (Fig. 6) of metal into the required form, and a sound-wave started at one of the foci. The transformation of the diverging into a converging sphere, and the shrinkage of the latter to a point at the other focus, is well shown (Fig. 6).

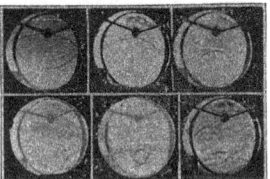

FIG. 6.

We will consider next another case of spherical aberration. When parallel rays of light enter a concave

mirror, those reflected from points of the mirror near its axis converge approximately to a point situated half-way between the surface of the mirror and its centre of curvature. The wave-front in the case of parallel rays is, of course, plane, and is changed by reflection into a converging shell of approximately spherical curvature. If we investigate the case more carefully, we find, how-ever, that the reflected rays do not come accurately to a focus, but envelope a surface known as the caustic—in this case an epicycloid. The connection between the wave-front and the caustic is perhaps not at once apparent. Let us examine the changes wrought on a sound-wave entering a concave hemispherical mirror (Fig. 7).

If we follow the wave during its entrance into the mirror, we see that the reflected portion trails along behind, being united to the unreflected part at the mirror's surface. After the reflection is complete, we find the reflected wave of a form not unlike a volcanic cone with a large bowl-shaped crater (No. 4). This bowl-shaped portion we may regard as a converging shell, which shrinks to point at the focus of the mirror. As it shrinks, the steep sides of the cone run in under the bowl, crossing at about the moment when the con-verging portion is passing through the focus (No. 6). The rim of the crater forms a cusp on the wave-front, and if we follow this cusp we shall see that it traces the

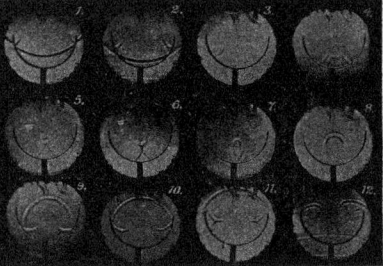

FIG 7.

caustic surface. Hence we may define the caustic as the surface traced by the cusp of the wave-front.

The portion of the wave which comes to a focus at once begins to diverge again, uniting with the sides of the crater, the whole moving out of the mirror in a form somewhat resembling a mushroom or the bell of a Medusa jelly-fish. The turned-under edges of the bell are cusped, and these cusps trace the caustic enveloped by the twice-reflected rays. These forms can also be constructed geometrically.

A much more complicated case is now shown (Fig. 8). Here the wave starts within a complete sphere, or rather cylinder. (Cylindrical surfaces have been used in all these cases for obvious reasons, the sectional views shown in the photographs being the same for both forms of surface.) Starting in the principal focus of the closed mirror, the wave is bounced back and forth, becoming more complicated after each reflection, yet always sym-metrical about the axis. Only a few of the many forms are shown, and, with the exception of the first three or four, are not arranged in order ; for at the time that the series was arranged on the slide this case had not been worked out geometrically, and it was quite impossible to determine the evolution of the different forms. More recently this case has been constructed for five reflections, and all of the forms shown in the photographs found.

We will take up next some cases of refraction, the first

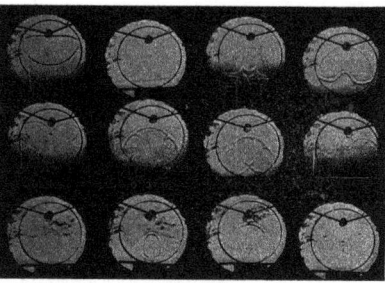

FIG. 8.

being that of a spherical wave at a flat surface of a denser medium. In Fig. 9 we have a rectangular tank with sides made of plane-parallel glass, and covered with a collodion film of soap-bubble thickness made by the method described by Toepler. Ordinary collodion is diluted with about ten parts of ether, poured on a small piece of plate-glass and immediately drained off. As soon as it is quite dry, a rectangle is cut with a sharp knife on the film. Toepler's method of removing the film was to place a drop of water on one of the cuts, and allow it to run in by capillarity ; but I have had better success by proceeding in the following manner :—One end of the plate is lowered into a shallow dish of water, and the plate inclined until the water comes up to one of the cuts. By looking at the reflexion of a window in the water, it is possible to see whether the film commences to detach itself from the glass. If all goes well, it will float off on the surface of the water along the line of the knife-cut, and it should be slowly lowered (one

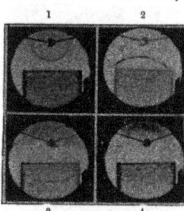

FIG. 9.

end resting on the bottom of the dish) until the rectangular piece detaches itself and floats freely on the surface. The edges of the tank are well greased, and then lowered carefully upon the film, to which they will adhere. The whole must then be lifted from the water in an oblique direction, when the film will be found

covering the tank and exhibiting the most beautiful interference-colours. The tank was filled with carbonic acid and placed under the origin of the sound-wave. On striking the collodion film, the wave is partly reflected and partly transmitted, and it will be seen that the reflected component in air has moved farther than the transmitted component in the carbonic acid. The spherical wave-front is transformed into an hyperboloid on entering the denser medium. This is well shown in No. 3 of the series. In No. 4 the wave is seen in air, having been reflected up from the bottom of the tank.

FIG. 10.

In Fig. 10 we have the refraction of the wave in the same tank under oblique incidence. The bending of the wave within the tank is very marked. The wave-fronts reflected from the side which follows the unreflected portion is also interesting in connection with Lloyd's single mirror interference experiment (No. 2 of series).

After several failures I succeeded in constructing a prism with its two refracting faces of this exceedingly

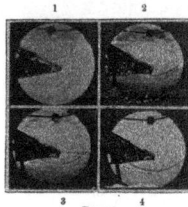

FIG. 11

thin collodion, which, when filled with carbonic acid, showed the bending of the wave-front, exactly as we figure it in diagrams for light. It was necessary to have the collodion thinner than before, since if we are to photograph the wave after twice traversing the film, we must lose as little energy as possible by reflexion.

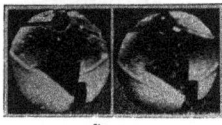

FIG. 12.

Fig. 11 shows the refraction in a carbonic acid prism, the bending being particularly noticeable in No. 4, on which I have, with a pair of dividers, traced out the position which the wave-front would have occupied had it not traversed the prism.

The bending of the wave-front in the opposite direction is shown in Fig. 12, where the same prism is filled with hydrogen gas, in which sound travels faster than in air.

In the next figure we have a very interesting case, though, owing to the experimental difficulties, the photographs are not quite as satisfactory as some of the others. It represents the transformation of a spherical into a plane wave by passage through a double convex lens.

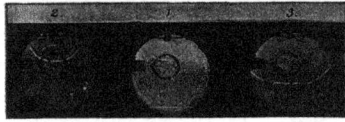

Fig. 13.

The construction of the cylindrical lens of exceedingly thin collodion was a matter of great difficulty. The flat, circular ends were made of thin mica as free from striæ as possible, that the passage of the wave through the lens could be followed. On these discs the collodion film was wound, the whole forming a hollow drum, which

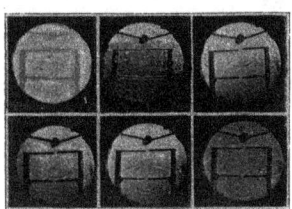

Fig. 14.

was then filled with carbonic acid. The sound-wave, started at the principal focus of this lens, is seen to be quite flat after its emergence (Fig. 13).

We will next take up some cases of diffraction, beginning with the well-known principle of Huygens, that any small portion of a wave-front can be considered as the

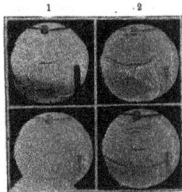

Fig. 15.

centre of a secondary disturbance, and that a small portion of this secondary disturbance can act as a new centre in its turn.

In Fig. 14 we have the wave starting above a plate with a narrow slit in it. This slit is seen to be the centre of a secondary hemicylindrical wave which moves down precisely as if the spark were located at the slit. After

proceeding a short distance this secondary wave encounters a second slit, and the same thing happens as before, the little slice that gets through spreading out into a complete wave, while the intercepted portion bounces back and forth between the plates.

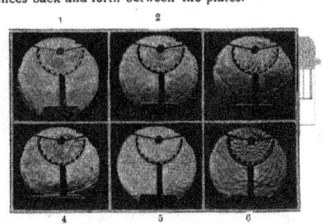

Fig. 16.

Fig. 15 shows the very limited extent to which sound shadows are formed. The wave is intercepted by a small glass plate. Just below the plate in No. 3 of the series a gap in the wave is found, which constitutes a shadow.

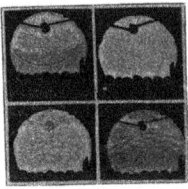

Fig. 17.

But presently, by diffraction, the wave curls in, closing up the gap and obliterating the shadow entirely. In the last one of the series it is interesting to note how the diffracted waves have their centres at the edges of the

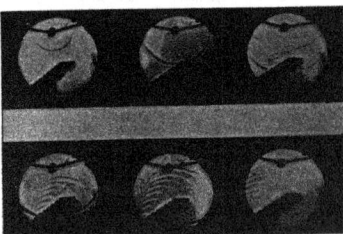

Fig. 18.

obstacle, the edges acting as secondary sources, as in the case of the diffraction of light.

The passage of a wave through a diffraction grating is shown in Fig. 16. The grating is made of strips of glass arranged on a cylindrical surface, the wave starting at the centre of curvature. In No. 2 of the series the union of the secondary disturbances coming from the openings

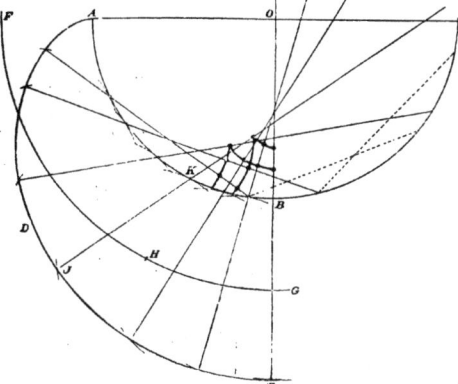

FIG. 19.

into a new wave-front is beautifully shown. In No. 3 the reflected wavelets have converged to the centre, but as each one is a complete hemicylinder, we see them radiating from the centre. This form can be constructed by describing semicircles around points on a circle of such radius that they all pass through the circle's centre. These semicircles represent secondary wavelets starting simultaneously from the various grating elements. In the last three pictures of the series the wave passes down, strikes the table, and is reflected up again, and it is interesting to see how the medium is broken up into meshes by the crossing and recrossing of the secondary waves.

Fig. 17 shows the form of the secondary wavelets formed by the reflection of a wave from a corrugated surface, and is interesting in connection with reflection gratings.

The formation of a musical note by the reflection of a single pulse from a flight of steps is shown photographed in Fig. 18. This phenomenon is often noticed on a still night when walking on a stone pavement alongside a picket-fence, the sound of each footstep being reflected from the pailings as a metallic squeak, which Young has pointed out to be analogous to the power of a diffraction grating to construct light of a definite wave-length.

It occurred to me, while making some geometrical

constructions to aid in unravelling some of the complicated forms reflected from surfaces of circular curvature, that a very vivid idea of how these curious wavefronts are derived one from another could be obtained if a complete series could be prepared on the film of a cinematograph, and projected in motion on a screen.

Having been unable to so control the time-interval between the two sparks that a progressive series could be taken, I adopted the simpler method of making a large number of geometrical constructions, and then photographing them on a cinematograph film.

As a very large number of drawings (100 or so) must be made if the result is to be at all satisfactory, a method is desirable that will reduce the labour to a minimum. I may be permitted to give, as an instance, the method that I devised for building the series illustrating the reflection of a plane wave in a spherical mirror. The construction is shown in Fig. 19.

ABC is the mirror, AOC the plane wave. Around points on ABC as centres describe circles tangent to the wave. These circles will be enveloped by another surface, ADE, below the mirror (the orthogonal surface). If we erect normals on this surface, we have the reflected rays, and if we measure off equal distances on the normals, we have the reflected wave-front. By drawing the orthogonal surface we avoid the complication of having to measure off the distances around a corner. The orthogonal

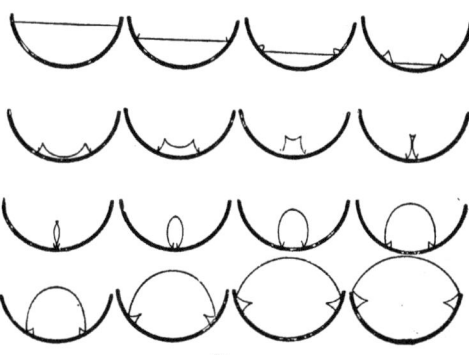

FIG. 20.

surface is an epicycloid formed by the rolling of a circle of a diameter equal to the radius of curvature of the mirror on the mirror's surface, and normals can be erected by drawing the arc FG (the path of the centre of the generating circle), and describing circles of diameter BE around various points on it. A line joining the point of intersection of one of these circles with the epicycloid, and the point of tangency with the mirror, will, when produced, give a reflected ray; for example, JK produced for circle described around H. The construction once

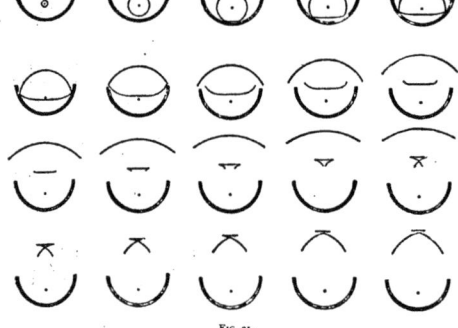

FIG. 21.

prepared, the series of wave-front pictures can be very quickly made. Three or four sheets of paper are laid under the construction, and holes are punched through the pile by means of a pin, at equal distances along each ray (measured from the orthogonal surface).

The centre of the mirror and the point where its axis

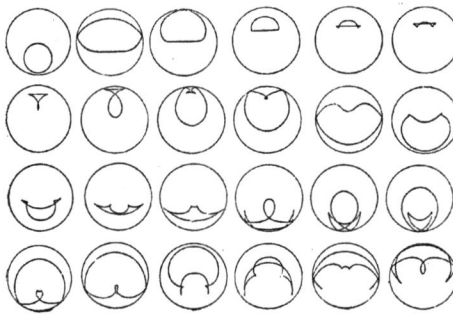

FIG. 22.

meets the surface are also indicated in the same manner. The sheets are now separated, and corresponding pin-holes are united on each sheet by a broad black line, which represents the wave-front. After a time it becomes necessary to consider double reflections, and to do this we are compelled to construct twice-reflected rays (indi-

cated by dotted lines), and measure around a corner each time.

About a hundred pictures are prepared for each series, and the pictures then photographed separately on the film, which, when run through the animatograph, give a very vivid representation of the motion of the wave-front.

Three films have been prepared thus far—reflection of a wave entering a concave hemispherical mirror (Fig. 20); reflection of a spherical wave starting in the principal focus of a concave hemispherical mirror (Fig. 21); and the reflection of a similar wave within a *complete* spherical mirror (Fig. 22). A number of these constructions, taken at

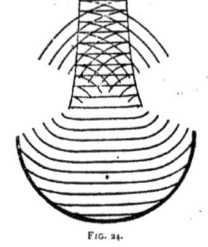

FIG. 23.

intervals along the film, are reproduced, and comparison of them with the actual photographs shows the close agreement between the calculated forms and those actually obtained.

I have already mentioned the fact that the cusps on the wave-fronts trace out the caustic surfaces. This is beautifully shown in Figs. 23 and 24, where the successive fronts are seen superposed. The former is for the reflection of a plane wave in a spherical mirror, the latter for the reflection of a spherical wave starting at the focus of a similar mirror. The caustic curve is shown by a dotted line in Fig. 23, and is seen to be traced by the cusps on the wave-fronts. The construction shows that there is a concentration of energy at the cusp; consequently we may define the cusp as a moving

FIG. 24.

focus, and the caustic as the surface traced by it. Though I hesitate in claiming that this relation, at once so apparent, is at all novel, I may say that, so far as I have been able to find, it is not brought out in any of the text-books, caustic surfaces being invariably treated by *ray* rather than by *wave-front* methods.

The cinematograph series illustrating reflection inside a complete sphere was the most difficult to prepare, as several reflections had to be considered. It has been completed for three reflections, and Mr. Max Mason, of Madison, to whom I am greatly indebted for his patient work in assisting me, is going on with the series. As will be seen, the wave has already become quite complicated, and it will be interesting to see what further changes result after three or four more reflections. I am also under obligations to Prof. A. B. Porter, of Chicago, who prepared the set of drawings illustrating the passage of a wave out from the principal focus of a hemispherical mirror.　　　　　　　　　　　　　　　　R. W. WOOD.

NOTES.

MANY friends and admirers of the late Sir William Flower will be glad to know that a committee has been formed, with Lord Avebury as chairman, to secure the erection of a memorial to him. It is proposed that the memorial shall consist of a bust and a commemorative brass tablet to be placed in the Whale Room of the Natural History Museum—one of the departments in which he was most interested, and to which he devoted special care and attention. There should be a ready response to the invitation for subscriptions to carry out this scheme, for Sir William Flower's services to science are appreciated by every one interested in the extension of natural knowledge. The Natural History Museum ought not, indeed, to be without a memorial of the man who took such an active part in its development. Subscriptions (which must not exceed two guineas) should be paid to Dr. P. L. Sclater, treasurer of the Flower Memorial Fund, 3, Hanover Square, W.

IN the House of Commons on Tuesday, Mr. Goschen gave some particulars with regard to the Committee to inquire into the boilers of her Majesty's ships. The Committee will consist of seven members, and the president will be Vice-Admiral Sir Compton Domvile. The other members of the Committee already chosen are Mr. List, superintending engineer of the Castle Company ; Mr. Bain, superintending engineer of the Cunard Line ; Mr. Milton, chief engineer surveyor of Lloyd's Registry of Shipping ; Prof. Kennedy, formerly professor of engineering at University College ; and, sixthly, an engineer of the Royal Navy holding the rank of an inspector of machinery. The seventh member of the Committee has not yet been selected. The instructions to the Committee are :—To ascertain practically and experimentally the relative advantages and disadvantages of the Belleville boiler for naval purposes as compared with the cylindrical boiler. To investigate the causes of the defects which have occurred in these boilers and in the machinery of ships fitted with them, and to report how far they are preventable either by modifications of details or by difference of treatment, or how far they are inherent in the system. Also to report generally on the suitability of the propelling and auxiliary machinery fitted in recent war vessels, and to offer any suggestions for improvement, stating at the same time the effect as regards weight and space of any alterations proposed. To report on the advantages and disadvantages of the Niclausse and Babcock and Wilcox boilers compared with the Belleville, as far as the means at the disposal of the Committee permit, and also to report whether any other description of boiler has sufficient advantages over the Belleville or the other two types mentioned, as a boiler for large cruisers and battleships, to make it advisable to fit it in any of her Majesty's ships for trial. For the purpose of making direct experiments between ships fitted with Belleville and cylindrical boilers respectively, the *Hyacinth*, fitted with Belleville boilers, will be placed at the disposal of the Committee. A cruiser of similar type fitted with cylindrical boilers will also be placed at the disposal of the Committee when required for the purpose of comparison.

Mr. Goschen added that it is particularly desired that any conclusions the Committee may arrive at should be supported by experimental proof as far as possible, and that they should propose any further experiments which may be considered necessary for this purpose.

WE learn from the *Electrician* that a prize of 1000 francs (40*l*.) is being offered by the Association des Industriels de France contre les Accidents du Travail, 3, Rue de Lutèce, Paris, for the most efficacious insulating gloves for electrical workmen. They should be strong enough to resist not only the electric pressure, but also accidental perforations by copper wires, &c., and must, in addition, be easy to wear by hands of any size and allow the workmen's fingers sufficient freedom to execute their work. The competition is international, and competitors must send two pairs of gloves, accompanied by an explanatory note, to the president of the Association before December 31, 1900. The Association reserves to itself the right to publish descriptions of samples submitted to it, and inventors should therefore take the precaution of protecting their inventions previously.

A GLANCE through the addresses delivered at the meeting of the British Medical Association held at Ipswich last week, and published in the *British Medical Journal*, shows that leading members of the medical profession recognise the close relationship between medicine and other sciences. The president, Dr. W. A. Elliston, in an address in which he traced the developments of the science of British medicine and the evolution of the modern physician, remarked : "I am not unmindful of the up-to-date requirements of general culture—of an accurate knowledge of anatomy, chemistry, physiology, biology, bacteriology, pathology, physics, optics, mechanics, electricity and photography, which are all essential to the well-educated physician ; they are daily called into requisition in order to diagnose and to direct the eye and hand in the treatment of disease." Similar acknowledgment of the dependence of medicine upon other sciences was made by Dr. Pye-Smith in his address abridged in another part of the present number. Mr. Frederick Treves, however, in his address on the progress of surgery during the last hundred years, ended his remarks with a sketch of the surgeon's place in the future, and expressed the hope that surgery might remain a handicraft, and that before all things the surgeon would strive to render his own hands self-sufficing, and not trust too much to diagnoses made for him in the laboratory. Short addresses were delivered by some of the presidents of the thirteen sections of the Association. In the section of pathology, Dr. E. E. Klein spoke upon bacteriology in relation to pathology, giving as illustrations of his theme the bacteriological work bearing upon inflammation, necrosis and cell secretions. Dr. Howard Marsh, in his address to the section of surgery, remarked : "Long a mere matter of routine, the treatment of fractures has lately felt the influence of modern advance in other departments of surgery. The Röntgen process secures an accuracy of diagnosis which formerly was often impossible." Dr. W. G. Smith made some suggestive remarks upon the teaching of pharmacology, pointing out some of the relationships between physiological action and chemical constitution. This fascinating subject has occupied the attention of several physiological chemists, and it offers numerous interesting problems for investigation.

WE learn from the *Athenæum* that the 83rd Annual Meeting of the Swiss Natural Science Society will be held at Thüsis, Canton Grisons, from September 2-4. Three other Swiss scientific societies—the geological, the zoological, and the botanical—will hold their annual meetings at the same time and place. Intending guests are asked to communicate with the president, Dr. Lorenz, at Coire, as soon as possible. Prof. Forel, of Morges, will lecture at the general meeting of the

Society on changes in glaciers ; Prof. Zschokke, of Basle, on the fauna of mountain streams ; and Prof. Schardt on the tectonic conditions on the northern slopes of the Swiss Alps.

A LENGTHY treatise on that mysterious form of atmospheric discharge known as "globe lightning" appears in the current number of the *Annalen der Physik.* It is by Prof. Max Toepler, the inventor of the Toepler machine and the discoverer of the stratified brush discharge. After comparing all the published records of the phenomenon, he comes to the conclusion that the globe is a form of continuous atmospheric discharge analogous to the "brush arc discharge" of the laboratory. A lightning flash leaves behind it a track of heated and possibly ionised air, along which a slow continuous discharge passes for some time after the flash has passed. When this continuous discharge is strong enough, any part of the track which has an exceptionally high resistance may be made to glow, and the glow may continue for several seconds or even half a minute. The track may be blown aside by the wind or driven by electrostatic forces, and then the globe will be seen to wander as is usually described. It often finishes with another lightning flash, and the thunderclap following that is described as an "explosion" of the ball. Considering the size and duration of the globe, Prof. Toepler estimates its current strength at something between 2 and 20 amperes. Considering that a lightning flash often carries 10,000 amperes, the destruction wrought by globular lightning should be inconsiderable.

THE uses of monochromatic light in optical experiments are so numerous that considerable interest attaches to the paper, on the means of producing such light, by MM. Charles Fabry and A. Pérot in the *Journal de Physique* for July. After pointing out the disadvantages of sodium light on account of the proximity of the D lines, the authors divide the methods of producing a beam of monochromatic light into two, viz. : (1) simplification of a beam of white light, and (2) use of light emitted by a gas. Under the latter method are included (a) flames ; (b) gases or vapours rendered luminous by electricity ; (c) induction sparks ; and (d) the electric arc. In connection with (b), it is found that the quality of the rays depends on the nature of the current exciting them, and the authors consider the use of (1) a coil with secondary condenser, (2) alternating currents, (3) continuous currents ; of these methods the last is the best, though the second is better than the first. While the results of these investigations cannot be briefly summarised, we notice that the authors have shown the possibility of improving the action of Michelson's tubes, of using a modification of the mercury arc of Arons as a source of monochromatic light of great intensity, of using the rays of a certain number of metals for interference observations where the difference of path is considerable, and, by measuring the wave-lengths, of adding a number of new fixed points on the spectrum. The paper concludes with a table of wave-lengths determined by MM. Pérot and Fabry, and compared with the determinations of Michelson.

STORMY and boisterous weather has occurred over the greater part of the British Islands during the past week, and exceptionally heavy rains have been experienced in many districts. On August 3 a storm of unusual severity for the time of year swept across England, and a heavy gale blew over the southern portion of the kingdom, occasioning considerable damage on our coasts, as well as to the fruit and corn in the inland districts. A similar disturbance struck our west coasts on August 6, and although it followed very much the same path as the storm of Friday it was more erratic, both in its track and rate. The storm was heavy again in many parts of England, and further damage to the crops has been occasioned. The temperature has fallen considerably during the last few days, and the weather has been cold for the time of year over the whole country, the mid-day readings in many places being below sixty degrees.

THE new Daily Weather Report, the issue of which was announced in NATURE of July 26 (p. 300), is now on sale every day at the Meteorological Office and several railway bookstalls in London. The attempt thus made to create an intelligent interest in meteorological records and forecasts is one to be encouraged, but we are afraid that the method adopted is not very attractive. The weather charts are admirable, and in connection with the statement of the general situation and the forecasts they are most instructive. Too much prominence could not be given to these two pages of the Report, which ought to find a place in the hall of every educational institution in the country. But the tabular matter included in the Report is of too detailed a character to be of public interest, and the Meteorological Council might usefully consider whether it would not be sufficient to publish such statistical information once a week or once a month instead of every day.

THE high kite flight at Blue Hill Meteorological Observatory, of which mention was made a few weeks ago (p. 252), was, Mr. Lawrence Rotch informs us, exceeded on July 19. A line of six kites reached an altitude of 15,900 feet, or three miles and sixty feet, above the sea, which exceeds the highest point ever reached in America by a balloon used for scientific purposes. Prof. Hazen, of the U.S. Weather Bureau, obtained observations in a balloon at a height of 15,400 feet in an ascent from St. Louis in June 1887. This is the highest ascent in America from which observations have been published. Four and three-quarter miles of steel piano-wire were used at Blue Hill as a flying line. The instruments attached to the kites showed freezing temperature at the highest point and a north-west wind with a velocity of twenty-six miles an hour. The air was found to be exceedingly dry.

THE *Proceedings* of the Geologists' Association for May contains an interesting paper, entitled "The Natural History of Phosphatic Deposits," being an address delivered in February last by the retiring President of the Association, Mr. J. J. H. Teall, F.R.S. The phosphates of igneous rocks and mineral veins are first discussed, and it is pointed out that apatite, the most abundant phosphatic mineral, is the principal source from which the phosphorus of the sedimentary rocks and of organic bodies is derived. Attention is drawn to the points of analogy in the vein-occurrence of apatite and tin-stone. Having shortly described the conditions and causes which bring about the formation of modern phosphatic deposits, the author passes in review the principal occurrences of phosphates in the successive geological formations, and concludes by stating that " from the earliest time down to the present day, the physical and chemical conditions under which phosphatic deposits have been formed have remained essentially the same." A useful bibliography of the subject is appended.

IN the *Bulletin de la Classe des Sciences* of the Brussels Academy (1900, No. 4), M. Louis Dollo, curator of the Brussels Museum, describes a new fish, *Racovitza glacialis*, discovered in the Belgian Antarctic Expedition. This is the third new fish discovered by the party, and only one specimen was found, and that a small one, somewhat mutilated at the caudal extremity, the length of the body, exclusive of the tail, being 82 millimetres. It was found on May 28, 1898, in lat. 71° 23' S., long. 87° 32' W., depth 435 metres, and is a member of the family Trachinidæ, distinguished from the genera Bathydraco, Bembrops, Chænichthys, Cryodraco and Gerlachea by well-marked characteristics, which M. Dollo describes. It is, however, most nearly allied to Bathydraco and Gerlachea, and its existence within the Antarctic Circle furnishes a fresh proof of the frequency of the Trachinidæ in the neighbourhood of this circle.

A SUGGESTIVE paper on the driving energy of physico-chemical reaction and its temperature-coefficient is contributed to the *Proceedings* of the American Academy of Arts and Sciences, xxxv. 23, by Prof. Theodore William Richards, in which the author, starting with the close similarity between the equations of Clausius and van 't Hoff, $d\ln P/dT = \lambda/RT^2$ and $d\ln K/dT = U/RT^2$, points out the advantages, previously recognised by Arrhenius, of regarding *pressure* to be the fundamental quantity which determines the progress of chemical reactions, as this aspect affords a more direct method of analysis than the study of volume, concentration or entropy. An expression called the "reaction metatherm" is evolved, which represents in terms of pressure the temperature-coefficient of the equilibrium ratio of ideal physico-chemical reaction. The equation obtained is the mathematical expression of the theorem of Maupertuis or Le Chatelier, and when analysed it shows that the part played by each substance in a reaction may be considered as the logarithm of the product of its "physico-chemical potential" and its actually present pressure. The reaction metatherm may be simplified into a reaction-isobar and a reaction-isochor, according as the pressure or volume is kept constant during the reaction. While, however, the reaction-isobar offers the most convenient basis for calculations to which it is applicable, results under constant volume are more conveniently calculated if the reacting substances are expressed in terms of concentration according to the equation of van 't Hoff.

THE iconography and anthropology of the Irano-Indians is the subject of a recent study by M. Ujfalvy in the current volume of *l'Anthropologie*. He concludes that the ancient Persians had a narrow-faced, dolichocephalic, somewhat flattened head resembling that of the ancient Hindus. They were all fair or reddish, and closely resembled the Macedonians of Alexander. It is impossible to say whether this was the primitive Persian type; at all events, it was their well-characterised type six centuries before our era. Two centuries later the type began to be slightly modified. This took place in the interval between the decline of the dynasty of the Achemanides (328 B.C.) and the rise of that of the Sassanides (240 A.D.). This alteration was probably due to the Semites of Elam and Syria, and to the Turanians of Babylon. The former influence did not affect the dolichocephaly, but it slightly increased the height of the head, and modified the nose considerably. The latter influence shortened the head, and but slightly modified the face. The Sassanide warriors exhibited the new complex type. At the close of the Sassanide period the Arabs reinforced the Semitic characters. M. Ujfalvy accepts and reinforces Houssay's dictum that, in a mixture of Aryans with Mongols or Mongoloids, the latter lose their facial characters, flattening of the nose, prominence of the cheek-bones, absence or sparseness of the beard; but, in exchange, they impose the shape of their skull on the former.

MESSRS. SWIFT AND SON have just patented a very handy little electric lamp for microscopic purposes. The lamp—a 16 candle-power one—is enclosed in a metal cylinder, the inner surface of which is painted white. The light makes its exit through a circular aperture in the side of the cylinder near its free end; the end is closed by a plate set at an angle of 45° and painted white, so as to reflect the light through the circular aperture. The light thus does not pass direct from the incandescent carbon filament to the mirror of the microscope, but is reflected from the white walls of the cylinder. In this way a very even illumination is obtained, which is more uniform than that obtained from the average ground glass lamp. While, however, the light given by this lamp is admirably suited for the ordinary powers such as are attached to the average student's microscope, it is, in our opinion, neither powerful enough nor white enough for high-power work, this being a defect common

to all electric microscope lamps. Where in addition a lamp is required for dissection purposes, as is so often the case, the direct light of the ordinary type of electric lamp will be found more suitable. This lamp is very compact and steady, and its movements, especially those about its horizontal axis, are particularly easy and steady.

THE Botanical Museum of Florence has recently received a donation of considerable interest in connection with the history of botany in Italy, viz. the collections made by Micheli, by Bruno Tozzi, and by G. Targioni-Tozzetti in the 18th century, including the type-specimens of species named by these and other eminent botanists. The donation includes also Micheli's and Targioni-Tozzetti's collections of sea-weeds.

IN No. 4 of vol. xxi. of *Notes from the Leyden Museum*, Dr. J. Büttikofer, the director of the Zoological Garden at Rotterdam, records the birds collected by the Dutch expedition to Central Borneo. Testimony is borne to the thoroughness of the work of the English naturalists, the late Messrs. Everett and Whitehead, and Mr. Charles Hose, by the fact that the author has not been able to add a single new species to the avian mountain fauna of the island. Dr. Büttikofer comes to the conclusion that both the mammalian and the avian faunas of Borneo are remarkably homogeneous, especially so far as the lowlands and the mountain bases up to an elevation of about 1000 metres are concerned. In vol. ii. No. 1 of the same serial, Mr. M. C. Piepers defends his theory of the evolution of colour in Lepidoptera, as explained at the recent Zoological Congress at Cambridge, against the criticisms of Miss Newbiggin and Dr. von Linden.

IN *Nature Notes* for August the Selborne Society refers to the urgent need of a crusade against pigeon-shooting.

THE MS. of the second volume of the late Dr. Stark's "Birds of South Africa" has been found amongst the papers of the deceased naturalist, who was killed at Ladysmith during the siege. It has been revised for the press by Mr. W. L. Sclater, director of the South African Museum, Cape Town, and will be shortly published by Mr. R. H. Porter. It will form part of Mr. Sclater's series of volumes on the fauna of South Africa.

THE report of the Zoological Garden of Calcutta for the year 1898–99, which has recently been received in this country, gives a favourable impression of the present condition and prospects of this establishment, drawn up by Lieut.-Colonel P. A. Buckland, the honorary secretary and treasurer. The superintendent of the Calcutta Garden, Babu R. B. Sanyal, who represented that Institution at the International Congress of Zoology at Cambridge in August 1898, contributes to this report an interesting account of his experiences at the Cambridge meeting, and of his observations on many of the zoological gardens of Europe, which he took the opportunity of visiting on the same occasion.

THE additions to the Zoological Society's Gardens during the past week include a Macaque Monkey (*Macacus cynomolgus*) from India, presented by Mr. T. Forsyth Forrest; a Diana Monkey (*Cercopithecus diana*) from West Africa, presented by Mr. W. Cleaver; two Greater Vasa Parrots (*Coracopsis vasa*) from Madagascar, presented by Mr. G. Barfoot; a Silky Cow-bird (*Molothrus bonariensis*) from South America, presented by Mr. F. Willes; seven Algerian Skinks (*Eumeces algeriensis*), a Spiny-tailed Mastigure (*Uromastix acanthinurus*) from North Africa, presented by Mr. G. H. Fernan; a Common Viper (*Viper berus*), British, presented by Mr. Alfred Cooper; a Green Lizard (*Lacerta viridis*), a Dohl's Snake (*Zamenis dahsi*), European; two Snakes (*Coluber prasinus*) from Upper Burmah, deposited; two Ring-necked Pheasants (*Phasianus torquatus*), two Gold Pheasants (*Thaumalea picta*) from China, a Pheasant (*Phasianus colchicus*), five Barn Owls (*Strix flammea*), British, purchased; a Japanese Deer (*Cervus sika*), born in the Gardens.

OUR ASTRONOMICAL COLUMN.

COMET BORRELLY-BROOKS (1900 *b*).—Several observations of this comet are announced. The comet is at present easily seen with a small telescope, but is becoming fainter.

Ephemeris for 12*h. Berlin Mean Time.*

1900.		R.A.		Decl.		Br.
		h. m. s.		○ ′		
Aug. 9	...	3 11 45	...	+61 11·9	...	0·91
10	...	15 14	...	63 37·6	...	87
11	,,	19 12	...	65 56·8	...	83
12	...	23 41	...	68 9·4	...	79
13	...	28 47	...	70 15·6	...	75
14	...	34 46	...	72 15·3	...	71
15	...	41 48	...	74 8·6	...	67
16	...	3 50 8	...	+75 55·7	...	0·63

During the week the comet passes rapidly northwards from α Persei, across into Camelopardus, and then near the boundary of this constellation and Cassiopeia. Its path is at present so nearly linear that it may be found by sweeping along the direction formed by the stars π, κ and α Persei.

EPHEMERIS OF COMET 1894 IV. (SWIFT).—Mr. F. H. Seares sends the following search ephemeris for the assistance of interested observers :—

Ephemeris for 12*h. Berlin Mean Time.*

1900.		R.A.		Decl.
		h. m. s.		○ ′
Aug. 8	...	15 57 20	...	−24 32·8
12	...	15 59 31	...	36·0
16	...	16 2 10	...	40·2
20	...	16 5 17	...	45·4
24	...	16 8 50	...	51·4
28	...	16 12 50	...	−24 58·1

VARIABLE STARS IN CLUSTERS.—*Harvard College Observatory Circular* (No. 52) contains the results of the measures of a set of photographs of the star cluster Messier 3 (N.G.C., 5272). This object is so low in the sky at Arequipa, and the stars so faint, that satisfactory photographs of it could not be obtained with the 13-inch Boyden refractor with exposures less than 90m. The rate of increase of the light of many of these stars is extremely rapid, and in order to determine such change with the greatest precision, it is necessary to have photographs taken with short exposures. Accordingly, at Prof. E. C. Pickering's request, Prof. J. E. Keeler has taken a series of excellent pictures of the cluster with the 3-foot Crossley reflector of the Lick Observatory. The first of these had an exposure of 60m., while twenty-four others were obtained with exposures of 10m. each. Prof. Bailey has examined these photographs very carefully, devoting attention specially to three of the variable stars. It has previously been stated (*Circular* No. 33) that the proportion of variable stars is greater in this cluster than in any other object of the same class.

The periods of the three variables were found to be : No. 11, 12h. 12m. 25s.; No. 96, 12h. om. 15s.; No. 119, 12h. 24m. 31s. The variations were recorded for intervals of 5m., and are given in a table. From this it appears that the total increase of light takes place in the case of No. 11, within 70m. ; No. 96, within 60m. ; and No. 119, within 80m. The greatest rapidity of increase of light occurs in the star No. 96, which increases during 5m. at the rate of at least 2·5 magnitudes per hour, and during 30m. at the rate of more than 2·0 magnitudes per hour. This rate of change appears to be the most rapid of any known variable. The Algol variable U Cephei, which perhaps undergoes the most rapid change of any variable *not* found in clusters, changes at the rate of about 1·5 magnitudes per hour during about 30m. of its period. In all these stars the rate of change is relatively slow near the beginning and end of the period of increase. In No. 96 the increase is about *ten* times as rapid as the decrease. Generally speaking, the lengths of period and form of light curves of these three stars are similar to those of the variables in the clusters Messier 5 and ω Centauri (*Astrophysical Journal*, vol. x. p. 255).

RECENT INVESTIGATIONS ON RUST OF WHEAT.

RUST, or mildew, is familiar to the agriculturist as a disease destructive to wheat and other cereals, and to the botanist as the subject of important researches relating to fungi. It was known in times of antiquity, as shown by numerous references indicating its destructiveness. Virgil says, " Soon, too, the corn gat sorrow's increase, that an evil blight ate up the stalks " (" Georgics," i. 150-1). In Britain, it is stated that " mildew of wheat-plants has been known for over 300 years, according to the records " (" Report on Mildew on Wheat Plants, 1892," Board of Agriculture, 1893, p. 25). Shakespeare ascribes it to " the foul fiend Flibbergibbet " (*King Lear*, Act iii. Scene 4). The works on husbandry of Hartlib (1655) and Jethro Tull (1731) refer to it. The connection of rust of cereals with a specific fungus is generally ascribed to Fontana (1767), and Persoon, after further investigation, in 1797 named the fungus *Puccinia graminis*. An account of rust, with illustrations of the *Puccinia*, by Sir J. Banks in 1805, is apparently the first important paper on the rust and its fungus in Britain. Since then the epidemic has been the subject of many papers, and of, at least, three organised inquiries. The historical side of the subject is conveniently summarised by Worthington G. Smith (" Diseases of Crops," London, 1884, Chapter xxv.), by C. B. Plowright(" British Uredineæ and Ustilagineæ," London, 1889, p. 46), and in the Board of Agriculture report (" Report on Mildew on Wheat Plants, 1892," Board of Agriculture, 1893, p. 25).

Rust of wheat occurs throughout Britain, especially in the wheat-growing districts, and forms of it are found on oat, barley, rye, and almost all grasses The losses from the form on wheat, reported to the Board of Agriculture in 1892, vary from nine to sixteen bushels per acre of crop. Rust-epidemics have been the subject of special attention in Europe, more particularly in Sweden, France and Austria. A rust conference was formed in 1890 for Australasia, and still continues to meet. In the United States of America, the Department of Agriculture sanctions the statement that " the damage to wheat and oats from rust in this country probably exceeds that caused by any other fungous or insect pest, and in some localities is greater than that caused by all other enemies combined " (Carleton, M. A., " Cereal Rusts of the United States," U.S. Department of Agriculture, *Bulletin* 16, 1899). In India and Japan, substantial losses are ascribed to this disease.

The remedy for this epidemic is a difficult problem, and the aim of recent research has been, in the first place, to obtain a true conception of the fungus causing it. The facts leading up to recent investigations may be briefly reviewed. It is an old and deep-rooted belief amongst growers of wheat that the rust of their crops is influenced by the neighbourhood of barberry bushes. Evidence of this is seen in certain old enactments enforcing destruction of the barberry ; for instance, that passed by a parliament at Rouen in 1660, and others included in the Province Law of Massachusetts (America) between 1738 and 1761. Sir Joseph Banks, in his paper (1805), holds the same opinion.

In 1841 Prof. J. S. Henslow (*Journal* of the Royal Agricultural Society, vol. ii. 1841) suggested that the yellow summer rust of wheat, and the black mildew which comes later, are stages in the life of one and the same fungus. Passing over many papers discussing these relationships, we come to one by De Bary published in 1865 (" Untersuchungen üb. Uredineæ." *Monatsber. d. Berlin Akad.*, 1865). From his experiments De Bary concludes, that the yellow summer rust (*Uredo linearis*, Persoon) on *Gramineæ*, the black autumn rust (*Puccinia graminis*, Persoon) also on *Gramineæ*, and the rust on barberry (*Æcidium berberidis*, Persoon) with its associated " spermogonia " stage, are phases in the life-history of the same fungus, for which the name *Puccinia graminis* is retained. In other words, that three (or four) recognised species of fungi are one and the same. At the same time a new phenomenon in the life of fungi was revealed, namely, that there existed parasitic fungi which required two host-plants in order to develop the forms of reproduction included in their life-cycle ; this De Bary named *metœcism* or (as better known in Britain) heterœcism. The life-history of *Puccinia graminis*, as defined by De Bary, is given in all our text-books. Uredospores (see Fig. 1) are produced on wheat and other *Gramineæ* throughout the summer, and infect the same group of host-plants ; the teleutospores of the *Puccinia* stage hibernate and in the following spring germinate, producing secondary spores (also known as sporidia), which infect barberry foliage and give rise there to the *Aecidium* stage with its aecidiospores; aecidiospores do not infect barberry again, but on *Gramineae* produce the uredospore

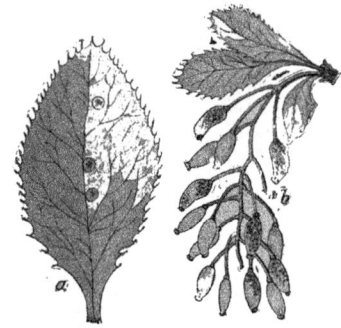

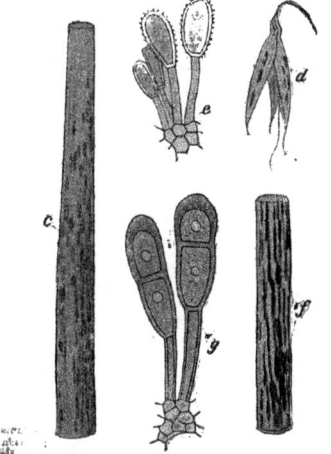

FIG. 1.—Black rust of oats (*Puccinia graminis*, spec. form *Avenae*). *a*, leaf, and *b*, cluster of fruits of barberry with *Aecidium berberidis* (nat. size); *c*, leaf, and *d*, a spikelet of oat with *Uredo* stage (nat. size); *e*, uredospores (× 500); *f*, sheath of oat with *Puccinia* stage (nat. size); *g*, teleutospores (× 500). (J. Eriksson.)

stage, thus completing the cycle. Accompanying the aecidium-cups there occurs constantly a form of reproduction, the spermogónia, which gives off spermatia or spore-bodies whose function neither De Bary, nor any one since, has been able to determine.

These results had an important and direct bearing on rust-epidemics of cereals, and they gave an impetus to further research on the biology of the whole group of rust-fungi or *Uredineae*, and, in fact, of all other parasitic fungi. In 1889, twenty-five years after De Bary's first results, both Plowright (*loc. cit.*, p. 56) and Rostrup published a list of fifty hetercecious rust-fungi. Recently, Dietel (Engler's "Pflanzenfamilien") gave about a hundred cases, including species outside the *Uredineae*. Works like Plowright's "British Uredineæ" are the evidence of this impulse, and a perusal of current botanical periodicals shows that the subject is by no means exhausted.

Ten years ago, three species of rusts occurring on crops of cereals were recognised:

(1) *Puccinia graminis*, with its *Aecidium* stage on barberry and mahonia; its uredospore and teleutospore stages on wheat, barley, oats, rye, and about a hundred species of grasses (see Fig. 1).

(2) *Puccinia rubigo-vera*, with *Aecidium* on many species of *Boragineae*; uredospores and teleutospores on wheat, rye, and a number of grasses (see Fig. 2). A variety, *simplex*, was distinguished on barley.

(3) *Puccinia coronata*, with *Aecidium* on species of buckthorn (*Rhamnus*); uredospores and teleutospores on oats and several grasses.

The four important European cereal-crops were thus known to have each two forms of rust-fungi, distinguished in their external characters, and with distinct *Aecidium* host-plants. Yet it was by no means certain that epidemics of rust were fully traced out. Fortunately the economic importance of rust-epidemics was enough to enforce attention from State departments, notably in Sweden, United States of America, Australasia, and in various parts of Europe and other countries. In Britain, while good work has been and is still done on rust-fungi, there has, in recent times, been no specially organised research relating to cereal rusts, probably because recent developments on the subject have rendered a research too extensive for the resources of any but workers specially retained and remunerated. The investigations on rusts of cereals reviewed here are mainly the outcome of State-aided research.

In Sweden, the Government in 1890 offered ten thousand kröner (about 560*l*.) to the Royal Swedish Academy of Agriculture for an investigation on rust of wheat, &c., intended, at first, to extend over three years, but which has been continued up till now. The grant was placed under the control of Jakob Eriksson, now professor of vegetable physiology at the experimental station of Albano, near Stockholm. The experiments were started in 1890; the first important results of Eriksson and his co-worker, E. Henning, appeared in 1894 (*Zeitsch. f. Pflanzenkrankheiten*, iv. 1894, pp. 66, &c.), and as a bulky volume in 1896 ("Die Getreiderost," Stockholm, 1896; 463 pp. and 14 plates). Other contributions, and re-statements of former work, have been made by Eriksson in almost every existing botanical periodical. The present summary is based chiefly on the latest re-statement (*Revue gén. d. Sciences*, ii. January 15, 1900, pp. 30–39), with aid from other papers.[1]

In Eriksson's experiments test-plants of cereals were grown from seed, or young plants from the open were transferred into pots. The soil used was generally sterilised. After inoculation with rust the plants were watered with distilled water, placed under large glass bell-jars moistened with distilled water, and left undisturbed for twenty-four hours in glass-houses specially constructed for the experiments. After this, observations were made at frequent intervals. The main lines of investigation were: (1) to define the species which cause rusts of cereals and grasses, and to trace their life-history; (2) the propagation of the rusts; (3) germination and vitality of the various forms of spores.

According to Eriksson's results, the three species and one variety of rusts attacking cereals and grasses as recognised in 1890 really represent twelve species and many subdivisions. His list is as follows, but the less important host-plants amongst the grasses are omitted:—

Species 1. *Puccinia graminis*, Pers. (Black Rust), with *Aecidium berberidis*. Specialised form (1) *Secalis*, on *Secale cereale* (Rye), *Hordeum vulgare* (Barley), *H. jubatum; Triticum repens, T. caninum*, &c., *Elymus arenarius*, and *Bromus secalinus*. (2) *Avenae*, on *Avena sativa* (Oat), *A. elatior*, &c., *Dactylis*

[1] J. Eriksson : *Ber. d. deutsch. botan. Gess.*, 1894, p. 292; 1897, p. 183. *Jahrbuch f. wiss. Botanik*, xxix. 1896. *Botan. Centralblatt*, lxxii. 1897, pp. 321–5 and 354–62. *Centralblatt f. Bakter. u. Parasitenkunde*, Abt. ii. 1897, pp. 291–308.

glomerata, Alopecurus pratensis, &c. (see Fig. 1). (3) *Tritici*, on *Triticum vulgare* (Wheat). (4) *Airae*, on *Aira caespitosa.* (5) *Agrostis*, on *Agrostis stolonifera*, &c. (6) *Poae*, on *Poa compressa* and *P. caesia.*

2. *Pucc. Phlei-pratensis*, Er. et Hen., *Aecidium* unknown. On *Phleum pratense* and *Festuca elatior.*

3. *Pucc. glumarum* (Schm.), Er. et Hen. (Yellow Rust), *Aecidium* unknown. Sp. form (1) *Tritici*, on Wheat (see Fig. 2). (2) *Secalis*, on Rye. (3) *Hordei*, on Barley. (4) *Elymi*, on *Elymus arenarius.* (5) *Agropyri*, on *Triticum repens.*

4. *Pucc. dispera*, Er. (Brown Rust of Rye), with *Aecidium Anchusae.* On Rye.

5. *Pucc. triticina*, Er. (Brown Rust of Wheat), *Aecidium* unknown. On Wheats—*Triticum vulgare, compactum, spelta*, and *dicoccum.*

6. *Pucc. bromina*, Er., *Aecidium* unknown. On many species of *Bromus.*

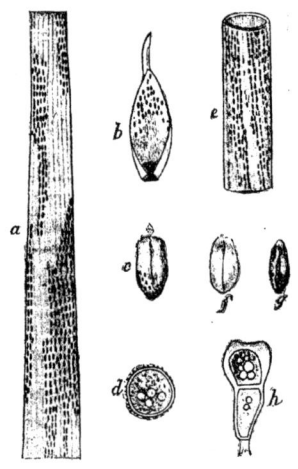

FIG. 2.—Yellow rust of wheat (*Puccinia glumarum*, spec. form *Tritici*) *a*, leaf (nat. size), *b*, outer glume (× 2), and *c*, a grain (× 2). bearing the *Uredo* stage; *d*, uredospore (× 375); *e*, sheath bearing *Puccinia* stage (× 2); *f*, a healthy grain, and *g*, a rusted grain (both × 2); *h*, teleutospore (× 500). (J. Eriksson)

7. *Pucc. agropyrina*, Er., *Aecidium* unknown. On *Triticum repens.*

8. *Pucc. holcina*, Er., *Aecidium* unknown. On *Holcus lanatus*, and *H. mollis.*

9. *Pucc. Trisetii*, Er., *Aecidium* unknown. On *Trisetum flavescens.*

10. *Pucc. simplex* (Kleb.), Er. et Hen., *Aecidium* unknown. On Barley.

11. *Pucc. coronifera*, Kleb., with *Aecidium Catharticae.* Sp. form (1) *Avenae*, on Oat. (2) *Alopecuri*, on *Alopecurus pratensis*, &c. (3) *Festucae*, on *Festuca elatior.* (4) *Lolii*, on *Lolium perenne.* (5) *Glyceriae*, on *Glyceria aquatica.* (6) *Holci*, on *Holcus lanatus* and *H. mollis.*

12. *Pucc. coronata* (Corda), Kleb. (Crown Rust), with *Aecidium Frangulae.* Sp. form (1) *Calamagrostis*, on *Calamagrostis arundinacea*, &c. (2) *Phalaridis*, on *Phalaris arundinacea.* (3) *Agrostis*, on *Agrostis stolonifera*, &c. (4) *Agropyri*, on *Triticum repens.* (5) *Holci*, on *Holcus lanatus* and *H. mollis.*

Comparing these species with those known in 1890, *Puccinia graminis* is now divided into Eriksson's 1 and 2 ; *Puccinia rubigo-vera* into species 3 to 9 : the variety *simplex* is now the species 10 ; and *Puccinia coronata* is divided into 11 and 12.

The species are distinguished by characters of uredospore and teleutospore, and by the host-plants of the *Aecidium*, where known. A species may be subdivided into "specialised forms," which agree in all external characters with the species, but form (1) is tied to one or more species of host, and its uredospores will not infect the hosts of forms (2), (3), &c. Thus *Puccinia graminis* has a specialised form confined to wheat, one to oat, and a third which attacks both rye and barley. The occurrence of the *Aecidium* on the same host, *e.g.* that of *Puccinia graminis* on barberry, might seem to afford a stepping-stone between specialised forms ; yet Eriksson says definitely that, for example, in the form *Avenae* "the form of *Aecidium* on barberry which gives black rust on oat can infect oat only.' It may be mentioned that specialised forms are not peculiar to the rusts of *Gramineae.* Klebahn in a recent paper (H. Klebahn, "Ueber den gegenwärtigen Stand der Biologie der Rostpilze," *Botan. Zeitung*, 1898, pp. 145-58) gives a list of heterœcious fungi, which show subdivision into "biological species" or "species sorores," forms identical with or only slightly different from those indicated by Eriksson's term of specialised forms. Klebahn has also an interesting discussion on the nature of these forms of parasitic fungi, and joins issue with many of Eriksson's results.

Carleton (*loc. cit.*) carried out a long series of experiments for the United States Department of Agriculture. The procedure was almost identical with that of Eriksson, and the results generally support the Swedish observations. Carleton distinguishes the following species and forms on cereals :—

Orange leaf rust of wheat, *Puccinia rubigo-vera tritici* ; identical, or almost so, with Eriksson's *Puccinia triticina.*

Orange leaf rust of rye, *P. rubigo-vera secalis* ; either Eriksson's *P. glumarum secalis* or *P. dispersa.*

Crown rust of oats, *P. coronata*, Corda ; Eriksson's *P. coronifera*, Kleb.

Black stem rust of wheat and barley ; *P. graminis tritici*, Er. et Hen.

   "   "   "   rye, *P. graminis secalis*, Er. et Hen.

   "   "   "   oats, *P. graminis avenae*, Er. et Hen.

The American results differ from the Swedish in two important points :—

(1) Whereas *Puccinia glumarum tritici* is the most destructive rust throughout Europe, it does not seem to occur in America, where the common rust on wheat is *Puccinia rubigo-vera tritici.*

(2) The specialised form *Puccinia graminis tritici* in America appears to infect, not only wheat, but also barley ; this is not the case in Sweden.

These differences indicate that the rust of wheat or other cereal in one country may not be the same as that in another ; and that specialised forms are variable. Eriksson distinguishes degrees of specialisation and classifies the forms thus : (1*a*) Forms restricted to one species of host, or to several species of a genus, *e.g.* species 4, 8, and 1 (5) in the above list. (1*b*) Forms occurring on hosts belonging to different genera, *e.g.* species 1 (1) and 1 (2) in list.

(2) Forms, which, under certain conditions of environment, are less fixed in regard to their hosts ; thus, until recently, Eriksson included species 4, 5, 6, 7 as forms of one species, *Puccinia dispersa*, and it is still doubtful whether they are distinct enough in their external characters to be regarded as species.

Further experiments are necessary to clear up many points regarding the distribution of species of rust. Yet the practical bearing of the results is evident. For example, wheat cannot now be regarded as subject to infection from any cereal or grass showing rust, but only from those which are host-plants of the species or specialised forms peculiar to wheat ; thus a comparison from the above lists shows that no rust from oats has been found to infect wheat. The theoretical bearing of biological species or forms in relation to the classification and phylogeny of fungi is also of the deepest interest.

The propagation of the rusts of cereals has, in the hands of recent investigators, assumed new aspects. The life-cycle of *Puccinia graminis* as defined by De Bary, and already given above, is the one generally described in text-books. Many observers, however, have doubted whether this, the perfect life-history, is always followed entirely. Strong objections of this kind are given by Worthington G. Smith (*loc. cit.*), although at the time (1884) these were raised, the imperfect knowledge regarding the species of rust-fungi somewhat weakens the arguments at the present time. Yet Eriksson, Carleton, and other recent workers show even greater opposition. We are told that rust is a serious epidemic amongst wheat in Australia where there are no native barberries; in India an undoubted black rust (*Puccinia graminis*) occurs on wheat where there are no barberries nearer than 300 miles off in the Himalayas. More conclusive is Eriksson's observation of aecidium-cups on a barberry, yet the rust could only be traced on rye, barley, and couch-grass to a distance from 10 to 25 metres; or, again, tufts of *Festuca elatior* were found with uredospores of *Puccinia coronifera*, while across a road there was a hedge of *Rhamnus* almost free from the aecidial stage, yet easily infected by means of teleutospores from the grass transferred to it by hand. These observations appear to show, firstly, that the aecidial stage is not necessary in the life-history of rust of cereals; secondly, that the range of infection by aecidiospores, or the reverse, is not great. What then remains is one of the following possibilities: (1) The mycelium can hibernate and resume activity in the following spring; (2) the fungus is carried over to a new season by the seed-grain, either through adherent spores, or in some form internally; (3) the uredospores can hibernate; (4) the teleutospores can infect other *Gramineae*. In reviewing these we can only give a few of the leading points relating to cereals.

It is unlikely that the mycelium hibernates in the dead remains of grasses, because it must then be capable of living as a saprophyte for a short time in spring when it awakens to activity; this has not yet been proved. If, however, one examines the undergrowth of an area of grass in winter, green shoots are generally present; the mycelium may winter here. Yet grain-crops are never sown except on ploughed land, and there is no good evidence to show that epidemics of rust, extending over whole acres, are propagated altogether from patches of wild grass. Uredospores adhere to the grain of rusted cereals, and there is a fair amount of evidence to show that these may assist the fungus through the winter. In the United States, Carleton believes that the uredospores of the orange leaf rust of wheat and rye are produced and can germinate late in the autumn, and so infect the sprouted autumn-sown crop for next year. Eriksson, on the other hand, has failed to discover that any one of the rusts of wheat lives all round the year in the *Uredo* stage in Sweden, although Sorauer states that in Germany the *Uredo* mycelium of *Puccinia rubigo-vera* hibernates without injury. There is thus a possibility that the uredospore stage may transmit a rust from year to year. This is more probable if the climate be suited to prolong the growth of grasses late into autumn or early winter, and if the specialised form of rust has more than one host-plant; if it occurs only on one cereal, it seems improbable that enough stray plants are present after harvest to account for a widespread reappearance of rust in the succeeding year. Whether uredospores adhering to straw or grain can survive the winter and germinate has not yet been made quite clear. In laboratory experiments it has been observed that uredospores frequently exhibit deferred germination. After being soaked in water, only a small percentage may produce germ-tubes. Eriksson and others have observed that if the dormant spores are cooled in ice, a further proportion are induced to germinate. Klebahn states that the greater number of all forms of spore germinate if placed on a suitable host-plant; he believes that the proportion which do not germinate at once, do so gradually later on, and sees in this an adaptation for preservation of the race.

The teleutospores of the rusts of cereals have, as a rule, proved incapable of infecting cereals or grasses, the aecidium stage must intervene. There is, however, the objection that observations can only be made under more or less artificial conditions in laboratory or green-house, and may not fulfil all the conditions of infection out-of-doors. Plowright recorded an instance of infection of cereals from the teleutospores of *Puccinia graminis* (*Gardeners' Chronicle*, August 19, 1882), but gives no special prominence to it in his "British Uredineæ" (1889). Indirectly certain facts lead one to suppose that teleutospores may have the power to reproduce other stages in the life-history than the aecidium. A rust-fungus of the group *Leptopuccinia* (e.g. *Puccinia malvacearum* on mallow and hollyhock) produces only teleutospores; these give rise to sporidia, which re-infect the mallow, and form the mycelium from which a new crop of teleuto-

spores arises. Many forms of rust-fungi have only teleutospores and uredospores on the same host: for instance, in Eriksson's list many of the forms have no *Aecidium*; either the hosts of this stage remain to be discovered, or the teleutospore production is fruitless, or the teleutospores are capable of bringing about infection of the cereal or grass host. Species like *Puccinia suaveolens* on thistles have all the forms of spore on one host-plant except aecidiospores, which are unknown; here the teleutospores must, after hibernation, bring about re-infection of the host. The existence of rust-fungi producing all the forms of spore on one host, shows that two hosts are not necessary for the development of the aecidium stage, and suggests that heterœcism may be a later development in the history of the group. It is also noteworthy that the teleutospore is produced by all rust-fungi, with very few exceptions. It is therefore not improbable that the teleutospores of heterœcious rust-fungi may still, through their sporidia, retain the power of infecting the host on which they are produced; in other words, that heterœcism may be facultative.

In investigating the germinative power of teleutospores, Eriksson finds the general opinion, that teleutospores must hibernate, to be true only to a certain degree. As a rule, they do not germinate unless they have passed the winter exposed to all the changes of weather out-of-doors. Spores collected in autumn and kept indoors soon lose the power of germination; hence his conclusion that rusted straw housed in barns will not be in a condition to propagate the disease next spring. In the case of spores left out-of-doors, the germinative power decreases rapidly during the year after their formation, and in October they no longer germinate. There is one exception, teleutospores of *Puccinia graminis tritici* have a feeble power of germination after two winters. On the other hand, Eriksson finds that certain teleutospores (e.g. *Puccinia dispersa* and *P. glumarum tritici*) can germinate in the year of their formation; in the case of the former species, aecidia were produced on *Anchusa* in a short time; in the latter form the host of the aecidia is unknown. Plowright states that a bundle of rusted wheat straw laid near plants of *Anchusa* in August 1885 produced aecidia in September.

Eriksson, after all his experiments, professes to be at a loss to account for epidemics of rust on cereals year after year by external contagion alone, and he adopts the view that *infection is due to an internal germ*. He thus introduces an agent for the propagation of parasitic fungi which has hitherto been received very sceptically by the plant-pathologist. His conclusion is the result both of experiment and examination with the microscope. In the experiments, varieties of cereals were used which are known to be specially liable to rust. Vigorous shoots were taken out-of-doors in spring, and enclosed in long glass tubes with the open ends closed with cotton wool. Seeds were also germinated in sterilised soil and kept in culture boxes, with precautions against entrance of spores by stuffing the ventilators with cotton wool. At first the results were negative, but after (in some way) improving the methods, the test-plants showed rust, especially those shoots taken from out-of-doors. Examination with the microscope failed to reveal any mycelium or other traces of the fungus until a few days before appearance of the rust externally. At this time, however, with the aid of staining reagents, certain protoplasmic bodies were observed in the green cells near the margin of a rust-patch. These plastids occur solitary or in masses in a cell; they are oblong or slightly curved, simple or somewhat branched, and recall the form of bacteroids found in root-tubercles of the *Leguminosae*. In a short time the branch-processes pierce the cell-wall of the host, and develop outside the cell into an intercellular mycelium, part of the original plastid remaining inside the cell as the first haustorium or sucker. Soon after this a rust-pustule appears on the exterior of the host. The plastid is regarded as having passed a period mingled with the cytoplasm of the host "in a kind of symbiosis," till in response to external conditions—nutrition, moisture, heat and light—the "mycoplasm" becomes separated from the cytoplasm, and assumes the form of a plastid. The mycoplasm has its origin from the rust-fungus in the parent host-plant, it becomes located in the embryo of the grain, develops apace with the young plant, and so bridges the period between one crop and the next. The following general facts are said to support this mycoplasm theory: (1) The appearance of rust on plants carefully isolated from contagion; (2) the disease in a field appears regularly four to five weeks after sowing the grain of certain varieties of wheat and barley known to be very liable to yellow rust ; (3) this rust is always more prevalent in sunny parts of the field.

A hypothesis so revolutionary is not likely to be adopted by a cautious fungologist without further evidence. At present, as far as we know, no figures illustrating the development of the mycelium have been published, nor can we obtain details of the staining methods adopted. Klebahn (*loc. cit.*) has entered his protest to the theory, chiefly, however, in general terms. In regard to the prevalence of rust in sunny parts of a field, he points out that Eriksson's own results confirm the fact that dormant spores are induced to germinate by alternate cooling and heating, drought and moisture ; just the conditions to be expected in early summer in sunny parts rather than in shaded parts of a field. Klebahn also supports the view that spores of rusts are capable of wider distribution than Eriksson's results show ; for instance, they have been found in analyses of air. We may recall, in support of this, Robert Hartig's observation in the Tyrol, when, after showers of rain, a yellow dust, coating objects in the neighbourhood, was found to consist almost entirely of the yellow spores of a rust-fungus, *Chrysomyxa* ("Diseases of Plants," Tubeuf and Smith, London, 1897, p. 54). If it be the case, as Eriksson says, that certain rusts of cereals appear regularly in four or five weeks, it seems quite as likely to indicate external infection of young plants at a certain stage in their existence, as to support the theory of an internal germ. The Swedish experiments in isolating test-plants from contagion have been repeated in America by Bolley.[1] Young plants of cereals growing amongst others in a field were enclosed in rust-proof cases ; they grew to maturity without showing any rust, although plants left unenclosed were much attacked. The results are quite negative.

Recent investigations have been directed towards advancing our knowledge regarding the varieties of cereals suited to resist the various forms of rust. Carleton,[2] whose work was aimed in this direction, summarises our general knowledge thus : "as yet there is but little certainty concerning rust resistance, which varies continually under different conditions. Heretofore, in testing varieties for rust resistance, little attention has been paid to the species of rust concerned." For our own part, we feel that our ability in combating the diseases of plants would be greatly strengthened by searching investigations towards attaining disease-proof varieties. A certain amount has been done, much more must yet be done. The results hold good for only small areas of the earth, and there must be thorough and systematic research in many counties before any definite conclusion be arrived at. From a practical point of view the combating of rusts of cereals, and diseases of plants generally, seems likely to be solved sooner in this way than by investigations on the complex conditions of life amongst the rust-fungi. One cannot but feel that the long recent researches have added to what we knew only minor details of practical importance, although they have opened new vistas of the deepest interest to the fungologist ; the outstanding lesson is the close dependence of the fungi on their environment, and the complexity thereby introduced into the study of diseases of plants.

WILLIAM G. SMITH.

MEDICINE AS A SCIENCE AND MEDICINE AS AN ART.[3]

IT has sometimes been disputed whether medicine should be regarded as a science or an art, but there is no doubt that the original meaning of the term medicine, in English and in other languages, is the Art of Healing. Medicine is so defined by Aristotle, and it has all the characters of an art. It depends upon experience and skill ; it deals with individual cases ; and the perfection it aims at is practical, not speculative ; the knowledge how to do, not the knowledge how things happen.

Nevertheless, as practical navigation is founded on astronomy, meteorology and physics ; as the art of agriculture rests on botany, geology and vegetable physiology, so the art of medicine depends on the science of pathology, the practice of physic on the principles of physic.

[1] *Centralblatt f. Bakt. u. Parasitenkunde*, Abt. II., vol. iv,, 1898, pp. 855-9, 889-96, 913-9 (6 figs.). Also *Proc. Amer. Ass. Adv. Science*, 1898, p. 408 (the limits of this paper prevent a longer reference to this research).
[2] *Loc. cit.*, p. 69.
[3] Abstract of the Address in Medicine delivered before the British Medical Association at Ipswich, on August 1, by Dr. P. H. Pye-Smith, F.R.S.

On the one hand, then, we must never forget that we practice an art ; we must never allow theories, or even what appears to be logical deduction, or explanations, however ingenious, or statistics, however apparently conclusive, or authority, however venerable, to take the place of the one touchstone of practical medicine, observation and experience. We must never treat the disease without considering the patient, for the art of healing is the art of healing individually ; nor need we wonder if profound learning and the best scientific training sometimes fail to make a successful practitioner. For beside adequate knowledge to save us from gross blunders, and a strenuous endeavour to do your best for each individual patient, however uninteresting the case or however irksome and unrewarded our toil—beside these first requisites for our art, there is ample room for those personal qualities which ensure success in every department of life ; for power of observation and insight, for the personal influence by which a strong character will secure obedience and inspire hope, for the judgment which divines what kind of remedies are suited to each patient, what kind and of what strength, and for the sympathy which puts one in the patient's place, and not only meets, but anticipates his wants.

On the other hand, however, if medical science without art is inefficient, medical art without science is not only unprogressive, but almost inevitably becomes quackery. As soon as we treat our patients by rule of thumb, by tradition, by dogmas, or by metaphysical axioms, we do injury to ourselves as well as to them. The bone-setter who is ignorant of anatomy ; the wise woman, who cures by charm, are not more irrational or less successful than was the physician of the seventeenth century who, in obedience to the doctrine of signatures, advised an infusion of roses for hæmorrhage, and saffron for jaundice, and lung-wort for consumption ; or the astrologer who prescribed salts of silver, of iron, copper, lead, or mercury in accordance with the horoscope of the patient and the planet under which he was born.[1] Not less mischievous, and in the true sense of the word unscientific, were the systems of medicine known as the Iatromechanical and the Iatrochemical, which in their turn had their vogue. The Brunonian system, explaining all diseases as due to laxity of fibre, was no better ; for indiscriminate use of "corroborants," or as they would now be called "tonics," is irrational. There is no such thing as a tonic or strengthening medicine, the only source of strength is oxidisable food, and bitter medicines only give strength indirectly by improving appetite. The last of the systems of medicine founded on a dogma is homœopathy, of which the theoretical absurdity is somewhat concealed by the more obvious nonsense of infinitesimal doses. It, like the other systems which preceded it, is not a rival to rational medicine ; they are not mistaken answers to a legitimate question, but attempted solutions of a problem which does not exist, attempted answers to a riddle which has none.

Apart from these exploded systems of treatment, our profession has often suffered from lack of the scientific, inquiring, sceptical spirit, and has often been led too easily by authority, by tradition, and by fashion. The reckless abuse of venesection in the last century and the former half of this led to almost complete disuse of a valuable means of treatment ; the misuse of mercury in the treatment of syphilis led to the denial of its unquestionable efficacy ; have we not seen the value of stimulants with fever lead to their indiscriminate use in almost every ailment ? Has not the immense value of careful and thorough nursing led to its absurd exaltation to an independent place, as if good nursing was anything more than an intelligent carrying out of the physician's directions ? Has not the remarkable powers of electrical stimuli led to a blind, unscientific and mischievous employment of this remedy, as if it had some mystic power apart from its demonstrable physiological effects ? May we not say the same of hydropathy, of massage and of hypnotism ? It is significant that the irrational exaltation of any of these particular modes of treatment into a panacea, while it begins in want of scientific intelligence invariably ends in imposture and deceit. Our only safeguard against the spirit of quackery and the deserved loss of public confidence in the

☉[1] Sol	☽ Luna	♂ Mars	☿ Mercurius	♃ Jupiter	♀ Venus	♄ Saturnus
Au	Ag	Fe	Hg	Sn	Cu	Pb
Sunday	Mon-day	Mardi	Mercredi	Thors-day	Vendredi	Saturday

These relations of metals to the planets, and also to the days of the week, are commemorated in the phrases :—*lunar* caustic, *martial* disposition, *mercurial* temperament, ♀ before a prescription, *Cuprum a Cypro* (*divopotens Cypri*) and *saturnine* gout.

profession which it brings with it, is continued recurrence to the scientific basis on which the practice of medicine rests. Our art is most satisfactory and efficient when most closely resting on science. The surgeon is continually guided by anatomy and mechanics in dealing with injuries and deformities. The physician is often able to apply his knowledge of chemistry and natural history to the direct and satisfactory treatment of disease. In general, medical science justifies its claim to the title by the same conclusive argument as astronomy or chemistry—by its predictions coming true. In particular, the detection and treatment of plumbism, the diagnosis and cure of scabies and ringworm, the treatment of poisons by chemical antidotes, and of specific diseases by attenuated inoculations are all instances of strictly scientific medicine. Nor can I refrain from citing the most recent and one of the most remarkable advances of our science in the discovery of the origin of malaria. This heavy tax upon national as well as individual vigour and happiness has been known and treated from the dawn of medicine; but although by a happy accident its efficient treatment was discovered, it is only lately that, by the combined labours of scientific physicians—Frenchmen, Italians, and our own countrymen—the origin of the disease has been discovered, the mode of its transmission traced, the diagnosis of its several forms established, and its prevention brought within reasonable hope.

We know that treatment of symptoms without a diagnosis is always unsatisfactory, and frequently worse; but we know also that diagnosis must rest upon accurate knowledge of morbid anatomy, and of the natural history of the disease. Scientific medicine based on observation and experiment is always practical as well; but empirical medicine, whether based upon fanciful speculation or working by blind rule of thumb, is the most unpractical thing that can be.

Preventive Medicine and Aetiology.—That important and constantly-growing branch of medicine, which deals with the prevention rather than the cure of disease, depends no less upon science, for tracking the dependence of one event upon another is the essence of inductive science. All efficient measures for the preservation of health, whether by individuals or communities, rest upon exact knowledge of the natural course of diseases. In fact, disease may be defined as the reaction of the human organism under conditions which make for its destruction. We must never forget that no irritant will cause inflammation in a lifeless skin; that no bacteria can produce fever without a nervous system to play upon; that no meal, however Gargantuan, and no potations, however deep, can produce their wonted effect without a stomach to react. The infection of small-pox, of diphtheria, or of tubercle exerts a very different influence upon vaccinated or unvaccinated subjects, upon one who has received and one who has not received the prophylactic serum, upon an organism which is predisposed to or refractory against the invasion of the enemy. How closely natural science is related to preventive medicine is shown by the history of Jenner, who was a naturalist, and of Pasteur, who was a chemist. How dependent we are upon science is well illustrated by the history of myxœdema. The cretinoid condition in adults which was discovered by the clinical acumen of Sir William Gull, unintentionally produced by the surgical skill of Prof. Kocher, and reproduced in animals by Mr. Horsley, is now cured by the eminently scientific method due to Dr. Murray, of Newcastle, and to Dr. Hector Mackenzie, of St. Thomas's Hospital. Such examples of accurate tracing of causation by observation and experiment admonish us to give up the perfunctory explanations which so often do duty for investigation. If we ascribe every inflammation to cold, and every vague complaint to gout; if we acquiesce in the popular ascription of disease to over-work, mental strain, and the nervous tension of modern life, we shall make no progress in true aetiology. I see many patients suffering from idleness—few, or none, from hard work. "Nerve-prostration" from "worry" and "brain-tension" often proves a decent synonym for the effects of gambling and drink. Modern life is easier, safer, and smoother than it was a hundred years ago. Our young men and maidens are healthier, stronger, better grown, less hysterical and sounder in mind and body than their great-grandparents. I venture to think that the duty of a physician is not to flatter the self-love of neurotic patients, but to inspire fortitude, and to prescribe regular and steady work as the best cure for a thousand nervous ailments.

As another point in scientific aetiology, allow me to warn against the temptation to assume that because many diseases are now proved to depend upon the presence of bacteria this must be true of all. Science does not anticipate, but waits for proof. We have complete scientific evidence, according to the criteria so well formulated by Koch, of the absolute and constant cause of anthrax, of relapsing fever, of tubercle, and several other diseases in both men and animals; but we must not forget the preliminary difficulty of identifying the specific bacillus—as in the case of enteric fever and diphtheria—nor the difficulty of finding one of the lower animals which is susceptible to the disease, as again in the case of typhoid fever and of cholera; nor the difficulty of the same anatomical and clinical conditions being produced by different organisms, as in the case of pneumonia and ulcerative endocarditis. Moreover, while in some diseases, which are undoubtedly infective and specific, no constant pathogenic microbe has yet been determined—as in typhus, measles, small-pox, and syphilis—we have, on the other hand, in the case of leprosy and of lupus, examples of disease unquestionably specific and bacterial in origin, but very unlike other infective maladies in their clinical course and natural history. At present it is surely undesirable to speak of "the undiscovered microbe of rheumatism." Science has to do with proved facts alone, and our language should never outrun our knowledge.

Experiments in Scientific Medicine.—There is one aspect of scientific medicine so important that it must not be omitted—the necessity of experiments for the progress of pathology, and, through it, for the prevention and cure of disease. It requires no argument to convince any one who is the least acquainted with the principles of inductive science that experiment is no less necessary than observation. In physics and chemistry this is obvious and universally acted on. The same method is indispensable for the progress of animal and vegetable physiology, and to such practical applications of science as engineering, agriculture and medicine. Nor can experiments be carried on in large numbers, by many different experimenters, and under every variety of condition. Any attempt to abolish, to check or to limit this experimental work is, in the degree that it is successful, fatal to progress. Happily it can never be successful, for the impulse to increase knowledge of the works of creation is too deeply implanted in men. Investigation must and will go on by the only path which it can follow. The method which was preached by Bacon and followed out by his great contemporary, William Harvey, which was continued by Lower, Hooke and Mayow in the early days of the Royal Society, by Aselli, Malpighi and Haller, by Hunter, Hewson and Hales, by Edward Jenner, by Sir Charles Bell, by Johannes Müller, by Claude Bernard, by Ludwig, and by the many eminent physiologists and pathologists in Germany, in France and throughout the civilised world, this method of investigation is absolutely necessary for the progress of our science and the improvement of our art. As its objects and methods are better understood, it will secure the enlightened patronage of all who desire the diffusion of human knowledge and the further spread of human happiness. Fortunately this very progress of science has brought with it the removal of the one grave drawback, as every right-thinking man must have felt it, to the benefits of these experiments upon living animals. Inflicting pain upon the humblest of God's creatures is repugnant to our feelings, though no one, unless maintaining a thesis, would contend that it is wrong to exact the most painful efforts, or even the death from exhaustion of a horse in order to carry help to a human being. But the discovery of ether, chloroform, and other anæsthetics, and the improved methods that we owe to the genius of Lister, have not only relieved the surgeon of the most repulsive part of his duties, but have relieved the experimenter also. Except in the investigation of the action of new remedies or in the inoculation of infective diseases, both of which inflict discomfort of a limited degree and duration rather than anything that can be described as pain, the experiments of the laboratory, whether physiological, pathological or therapeutical, are conducted without inflicting pain. The opposition to them has not succeeded, and is sure to diminish. However mistaken our opponents, we are glad to find there is even exaggerated jealousy to avoid anything approaching to cruelty. This legitimate object our more candid critics may be assured is already amply provided for.

MR. BALFOUR ON SCIENTIFIC PROGRESS.[1]

APART altogether from individual likes and dislikes, is there any characteristic note which distinguishes this century from any that have gone before it?

On this point I range myself with those who find the characteristic note in the growth of science. In the last 100 years the world has seen great wars, great national and social upheavals, great religious movements, great economic changes. Literature and art have had their triumphs, and have permanently enriched the intellectual inheritance of our race. Yet, large as is the space which subjects like these legitimately fill n our thoughts, much as they will occupy the future historian, t is not among these that I seek for the most important and the most fundamental differences which separate the present from preceding ages. Rather is this to be found in the cumulative products of scientific research, to which no other period offers a precedent or a parallel. No single discovery, it may be, can be compared in its results to that of Copernicus; no single discoverer can be compared in genius to Newton; but, in their total effects, the advances made by the nineteenth century are not to be matched. Not only is the surprising increase of knowledge new, but the use to which it has been put is new also. The growth of industrial invention is not a fact we are permitted to forget. We do, however, sometimes forget how much of it is due to a close connection between theoretical knowledge and its utilitarian application which, in its degree, is altogether unexampled in the history of mankind. I suppose that, at this moment, if we were allowed a vision of the embryonic forces which are predestined most potently to affect the future of mankind, we should have to look for them, not in the Legislature, nor in the Press, nor on the platform, nor in the schemes of practical statesmen, nor the dreams of political theorists, but in the laboratories of scientific students whose names are but little in the mouths of men, who cannot themselves forecast the results of their own labours, and whose theories could scarce be understood by those whom they will chiefly benefit.

I do not propose to attempt any sketch of our gains from this most fruitful union between science and invention. I may, however, permit myself one parenthetic remark on an aspect of it which is likely more and more to thrust itself unpleasantly upon our attention. Marvellous as is the variety and ingenuity of modern industrial methods, they almost all depend in the last resort upon our supply of useful power; and our supply of useful power is principally provided for us by methods which, so far as I can see, have altered not at all in principle, and strangely little in detail, since the days of Watt. Coal, as we all know, is the chief reservoir of energy from which the world at present draws, and from which we in this country must always draw; but our main contrivance for utilising it is the steam engine, and, by its essential nature, the steam engine is extravagantly wasteful. So that, when we are told, as if it was something to be proud of, that this is the age of steam, we may admit the fact, but can hardly share the satisfaction. Our coalfields, as we know too well, are limited. We certainly cannot increase them. The boldest legislator would hesitate to limit their employment for purposes of domestic industry. So the only possible alternative is to economise our method of consuming them. And for this there would, indeed, seem to be a sufficiency of room. Let a second Watt arise. Let him bring into general use some mode of extracting energy from fuel which shall only waste eighty per cent. of it, and lo! your coalfields, as sources of power, are doubled at once. The hope seems a modest one, but it is not yet fulfilled; and therefore it is that we must qualify the satisfaction with which at the end of the century we contemplate the unbroken course of its industrial triumphs. We have, in truth, been little better than brilliant spendthrifts. Every new invention seems to throw a new strain upon the vast, but not illimitable, resources of nature. Lord Kelvin is disquieted about our supply of oxygen; Sir William Crookes about our supply of nitrates. The problem of our coal supply is always with us. Sooner or later the stored-up resources of the world will be exhausted. Humanity, having used or squandered its capital, will thenceforward have to depend upon such current income as can be derived from that diurnal heat of the sun and the rotation of the earth till, in the sequence of the ages, these also begin to fail. With such

[1] Address delivered by Mr. Balfour, M.P., at the opening of the Cambridge Summer Meeting on August 2. Abridged from the *Times*.

remote speculations we are not now concerned. It is enough for us to take note how rapidly the prodigious progress of recent discovery has increased the drain upon the natural wealth of old manufacturing countries, and especially of Great Britain, and, at the same time, frankly to recognise that it is only by new inventions that the collateral evils of old inventions can be mitigated; that to go back is impossible; that our only hope lies in a further advance.

After all, however, it is not necessarily the material and obvious results of scientific discoveries which are of the deepest interest. They have effected changes more subtle and perhaps less obvious which are at least as worthy of our consideration and are at least as unique in the history of the civilised world. No century has seen so great a change in our intellectual apprehension of the world in which we live. Our whole point of view has changed. The mental framework in which we arrange the separate facts in the world of men and things is quite a new framework. The spectacle of the universe presents itself now in a wholly changed perspective. We not only see more, but we see differently. The discoveries in physics and in chemistry, which have borne their share in thus re-creating for us the evolution of the past, are in process of giving us quite new ideas as to the inner nature of that material whole of which the world's traversing space is but an insignificant part. Differences of quality once thought ultimate are constantly being resolved into differences of motion or configuration. What were once regarded as things are now known to be movement. Phenomena apparently so wide apart as light, radiant heat and electricity, are, as it is unnecessary to remind you, now recognised as substantially identical. From the arrangement of atoms in the molecule, not less than their intrinsic nature, flow the characteristic attributes of the compound. The atom itself has been pulverised, and speculation is forced to admit as a possibility that even the chemical elements themselves may be no more than varieties of a single substance. Plausible attempts have been made to reduce the physical universe, with its infinite variety, its glory of colour and of form, its significance and its sublimity, to one homogeneous medium in which there are no distinctions to be discovered but distinction of movement or of stress. And although no such hypothesis can, I suppose, be yet accepted, the gropings of physicists after this, or some other not less audacious unification, must finally, I think, be crowned with success. The change of view which I have endeavoured to indicate is purely scientific, but its consequences cannot be confined to science. How will they manifest themselves in other regions of human activity, in literature, in art, religion? The subject is one rather for the lecturer on the twentieth century than for the lecturer on the nineteenth. I, at least, cannot endeavour to grapple with it.

SOCIETIES AND ACADEMIES.

LONDON.

Royal Society, June 14.—" The Electrical Effects of Light upon Green Leaves." By Augustus D. Waller, M.D., F.R.S.

In the preliminary communication recently made to the Royal Society, the author shows how, from the study of the electrical effects of light upon the retina, he was led to ask whether the chemical changes aroused by the action of light upon green leaves are also accompanied by electrical effects demonstrable in the same way as the eye currents. The question is tested in the following way :—A young leaf freshly gathered is laid upon a glass plate and connected with a galvanometer by means of two unpolarisable clay electrodes A and B. The half of the leaf connected with A is shaded by a piece of black paper. An inverted glass jar forms a moist chamber to leaf and electrodes, which are then enclosed in a box provided with a shuttered aperture through which light can be directed. A water trough in the path of the light serves to cut out heat more or less. Under favourable conditions there is obtained with such an arrangement a true electrical response to light, consisting in the establishment of a potential difference between illuminated and non-illuminated half of a leaf, amounting to 0·02 volt. The deflection of the galvanometer spot during illumination is such as to indicate current in the leaf from excited to protected part. The deflection begins and ends sharply with the beginning and end of illumination; it is provoked slightly by diffuse daylight, more by an electric arc-light, most by bright sunlight. It is abolished by boiling the leaf, and by the action of an anæsthetic, carbon dioxide.

The first experiments, made at the end of March, were upon iris leaves taken from plants about 6 inches high, and the response to light was then between 0'001 and 0'002 volt in value. Experiments upon similar leaves were resumed early in May, when it appeared that the external condition by which the state of the leaf is most obviously governed is *temperature*. On warm days the response ranged from 0'005 to 0'02 volt; on cold days it did not rise above above 0'005, and was sometimes nil. Some tests upon leaves in a warmed box gave satisfactory results, which may be thus summed up:—The normal response at 15°-20° C. is diminished or abolished at low temperature (10°), augmented at high temperature (30°), diminished at higher temperature (50°), and abolished by boiling.

As the month of May advanced, the iris leaves, even in the warm box, became more and more inert, and by the 23rd inst., when the plants were mostly full grown and in flower, no satisfactory leaf could be found. Leaves of iris appear to give more marked response at or about mid-day, than at or about 6 p.m. Tested by Sach's method the leaves gave no evidence of starch activity during insolation.

On the failure of the iris leaves to react, other leaves were sought for which should give evident differences of reaction in correlation with evident differences of state. Leaves of tropæolum and of mathiola gave a response to light contrary in the main to the ordinary iris response, viz. "positive" during illumination, and subsequently "negative."[1] In these two cases leaves empty of starch acted better than leaves laden with starch. Leaves of begonia gave a variety of responses strongly suggestive of the simultaneous action of two opposed forces effecting a resultant deflection in a + or − direction. Leaves of ordinary garden shrubs and trees, &c., *e.g.* lilac, pear, almond, mulberry, vine, ivy, gave no distinct response ; this is possibly due to a lower average metabolism in such leaves as compared with the activity of leaves of small young plants in which leaf-functions are presumably concentrated within a smaller area. The petals of flowers gave no distinct response, which indicates that chloroplasts are essential to the reaction.

The effect of carbon dioxide upon the iris leaf was abolition of response during and after passage of the gas, with subsequent augmentation. Upon mathiola and tropæolum, augmentation of response followed on applying air containing 1 to 3 per 100 of carbon dioxide, and prompt abolition resulted from a full stream run through the leaf-chamber. On the air supply being kept clear of carbon dioxide there was gradual abolition of response, followed by gradual recovery on the re-admission of a small amount of carbon dioxide.

"Fatigue" effects may be produced if the successive illuminations (of 5 minutes duration) are repeated at short intervals (10 minutes). At intervals of 1 hour, successive illuminations of 5 minutes produce approximately equal effects. With the leaf of mathiola, periods of illumination of 2 minutes at intervals of 15 minutes were used without provoking any obvious sign of fatigue.

June 21.—"Note on Inquiries as to the Escape of Gases from Atmospheres." By G. Johnstone Stoney, M.A., Hon. D.Sc., F.R.S.

Three investigations have been published which profess to supply information about the escape of gases from atmospheres. Two of them, those of Messrs. Cook[2] and Bryan,[3] while differing in other respects, agree in reasoning forwards by the help of the kinetic theory of gas from the supposed causes ; the third[4] pursues a method regarded as trustworthy by the present writer, and reasons backwards by the help of the same theory from the observed effects.

Where, as in the present instance, the *a priori* and *a posteriori* methods have led to inconsistent numerical results, it is incumbent upon us to search for the mistake or mistakes which must somewhere have been made. If these can be found and corrected, an important advantage is gained ; and the present is an attempt to discover some of them by inquiring whether there are conditions or agencies in nature which facilitate the escape of

[1] "Negative" as the term is employed in physiological literature, *i.e.* negative pole of positive element (" zincative").
[2] *Astrophysical Journal* for January 1900.
[3] *Roy. Soc. Proc.*, April 5, 1900, p. 335.
[4] *Scientific Transactions of the Royal Dublin Society*, vol. vi. Part 13 ; or *Astrophysical Journal* for January 1898. And for further evidence that helium is escaping from the earth, see NATURE of May 24, 1900, p. 78.

gaseous molecules from the earth, and which are omitted, or which have not been sufficiently taken into account, in Mr. Cook's and Prof. Bryan's investigations.

Let ΔV be a volume containing at a given epoch a large number *n* of molecules of the atmosphere, and let Δ*t* be a duration commencing at that instant. Also, let *n'* be the number of encounters which each of these molecules on the average meets with in the times Δ*t*. Then will

$$N = nn'$$

be the total number of their free paths in that time ; and the actual number of these free paths, in which the initial speed after an encounter lies at the time *t* between *v* and *v + dv*, must be precisely

$$dN = N(\pi + \delta)dv, \qquad (1)$$

where π is the probability function (that employed by Mr. Cook, or that employed by Prof. Bryan, or some other), and δ (the deviation function) represents whatever is the real divergence of the actual number from that computed by the formula used by them, viz. :

$$dN = N\pi dv ; \qquad (2)$$

in other words, computed on the supposition that δ/π is of negligible amount.

Now π is one fully-determined function in Mr. Cook's investigation, and another fully-determined function in Prof. Bryan's ; but little is known of what δ is in either case, except that it is in both an excessively complex function of N, *v*, *t*, with several other variables, some of which it is difficult even to indicate ; and that by its amount for any given value of *t* and at any given position in the atmosphere it must supply in equation (1) the actual effect, at that time and place, of all natural agencies which had not been taken into account in calculating the expression π.

If due care has been taken in framing the probability law π, it will in many cases be legitimate to assume that δ/π is sufficiently small to warrant our using equation (2) when computing the approximate distribution among the free paths of those speeds which assign *large* values to π, while at the same time it may need proof and may not be a legitimate assumption in reference to those values of *v* which make π *small*. Now it is in this latter case that the assumption has to be made by Mr. Cook and Prof. Bryan.

The conditions under which the assumption is likely not to be true are the following :—

A. Where the events, the law of whose distribution purports to be represented by the π function, are of such a kind that a vast number of the events need to be passed under review in order to secure an approximate conformity to *any* fixed law. Now experiment shows that in ordinary air trillions of the free paths, probably many trillions, must be grouped together in order to make manifest any law in the distribution of the speeds. In all such cases we are not entitled to ignore the δ function, except in estimating the frequency of such speeds as can be shown to assign a sufficient preponderance to the π function. Accordingly it is not legitimate to ignore the δ function when treating of the frequency of speeds which make π excessively small, such as are the speeds which carry molecules away from the earth.

B. But a more important omission occurs where the function π has been arrived at without taking into account agencies in nature which affect the distribution of speeds. Where this has been done the δ function must include the whole effect of these agencies, and this again forbids our relying upon equation (2) in computing the frequency of any speed which makes the value of π small.

B 1. Thus in Mr. Cook's computation no notice is taken of the anisotropic character of the outer strata of the earth's atmosphere, which facilitates the escape of molecules. In Prof. Bryan's this is partly taken into account by treating the molecules as moving in a constant field of force. This may possibly be sufficient, though it ignores the reactions which are also necessarily present. To include them it would be necessary to extend the partition of energy beyond the molecules of the atmosphere to all the other molecules of the earth which attract them.

B 2. Then, again, both computations ignore the incessant turbulence of the atmosphere which, in its lower strata, produces all the phenomena of weather, and in its upper regions phenomena which are swifter and on a larger scale. This turmoil, with all its dynamical, thermal and electrical effects, is due, like most other events upon the earth, to the shiftings about of energy which intervene between the advent of energy and from the sun and its radiation from the earth into space ; and to take it into account in an investigation based on the laws of the partition of energy, it would be necessary to extend that partition beyond the earth to the sun and to the intervening æther.

B 3. So, again, the great absorption of solar radiation which takes place in the outer layers of the earth's atmosphere will have to be taken into account, and as it has not been included under function π, it still further augments the part which the δ function takes in equation (1) and renders equation (2) an insufficient one for the purposes of the investigation.

B 4. The commotion going on in the atmosphere consists in part of electrical phenomena. Some of these—thunderstorms, auroras, the electrical condition of fogs, &c.—can be observed from the stations which men occupy at the bottom of the atmosphere, and are of such a kind that they must be accompanied by a charged condition of that stratum of the atmosphere the density of which renders it a better conductor than the atmosphere above it and below. This stratum, then, and the strata above it receive charges of electricity which, according to the varying condition of the strata further down, will sometimes be disguised electricity and at other times undisguised. This electrified condition of the upper regions, co-operating with ascending currents, which necessarily increase in speed as they advance, will presumably give rise to prominences upon the earth's atmosphere, upon which the density of the electrification will be intensified and from which in consequence gaseous molecules find it easier to escape than from other situations. In this and other ways electricity may help the escape.

Now of these agencies, all of which affect the rate at which gas can escape from the earth, none is included in the investigation which Mr. Cook has made of that phenomenon ; and only the first (B 1) is dealt with by Prof. Bryan. Moreover, it is probable that these are not the only ways in which nature can intervene, and which have been overlooked. The supposition then that either of the probability laws made use of by those investigators can be applied to our actually existing atmosphere, without a large correcting function δ, would appear to be a mistake ; and, if so, the inferences from those laws when so applied are not part of a real interpretation of nature. It need not therefore occasion any surprise that, in the case of helium, the facts of nature seem to negative those inferences. (See NATURE of May 24, 1900 ; the second column of p. 78.)

EDINBURGH.

Royal Society, July 16.—Lord Kelvin, President, in the chair.—Lord Kelvin read a paper on the motion in an infinite elastic solid by the motion, through the space occupied by it, of a body acting on it only by attraction and repulsion. The ideal atom considered in this paper was a region of space in which the ether was changed in density by the action of forces upon it. In the particular case chosen for development the atom was taken as spherical with spherical distributions of density within it, and every element of matter was supposed to act on every element of the ether according to the Newtonian law. The further assumption was made that the average density of the ether within the atom was the same as if the atom were not present. The atom and the ether were then supposed to be in relative motion, and the total kinetic energy of the ether within the atom was calculated, as also the effective inertia of the ether in the space occupied by the matter. On the assumption that the density of the ether at the centre of the atom was 101 times greater than the undisturbed density, it was found that a refractivity was obtained a little smaller than that of oxygen. By assuming that the average density was in excess or defect of the undisturbed density of the ether, we could extend the method so as to include electrical actions.—In a second paper, on the number of molecules in a cubic centimetre of gas, Lord Kelvin pointed out that in the preceding paper he had been obliged to take the number as 4×10^{20} instead of Maxwell's number, 19×10^{18}.—In a paper on the hyperbolic quaternion, Dr. Alex. Macfarlane showed how the introduction of "real" instead of "imaginary" vectors, quaternion theorems of spherical geometry could be generalised so as to be applicable to hyperbolic geometry.— Sir John Murray and Dr. Philippi communicated a preliminary

note on the deep-sea deposits collected during the *Valdivia* expedition of 1898–9. Leaving Hamburg and passing round by the north of Scotland, the *Valdivia* proceeded southwards by the west coast of Africa to the Cape, thence to the Antarctic seas, returning by way of the Indian Ocean and the Suez Canal. Generally speaking, the nature of the deposits agreed with what was already known, but fuller information was gained in many instances. For example, off the mouth of the Congo samples of coprolitic mud had been obtained, largely made up of little oval pellets of mud which had passed through the intestines of echinoderms. These had consolidated and were apparently in the process of being transformed into glauconitic and phosphatic concretions. The study of the formation and distribution of glauconite was geologically of great importance, and a detailed examination of the *Valdivia* collections would probably throw much light on the subject.—Prof. J. C. Beattie communicated a second part of his researches into the leakage of electricity from charged bodies at moderate temperatures. In most of the experiments described, zinc strips resting on insulated iron plates were sprinkled with various salts and then heated to about 350° C., the whole being enclosed in an iron box which was connected to the case of the electrometer. Among the substances used were common salt, alone or with iodine or bromine, and similar combinations with the chlorides of lithium, lead, potassium, &c. Generally a steady negative charge was produced by the heating, but not always. The difference of potential so obtained depended on the nature of the insulated metals, but not on their distance apart. When high voltages were used, the positive charge leaked away, while the negative charge was retained. An explanation was offered founded on Enright's and on Townsend's experiments.—A communication was also presented by Dr. Thomas Muir on the theory of skew determinants and plaffians in the historical order of its development up to 1857.—In a brief review of the session, the President referred to the great losses the Society had sustained through the deaths of the Duke of Argyll and Sir Douglas Maclagan.

CONTENTS.   PAGE

THURSDAY, AUGUST 16, 1900.

A STANDARD TEXT-BOOK OF PHYSICS.

Müller-Pouillet's Lehrbuch der Physik und Meteorologie.
Neunte umgearbeitete und vermehrte Auflage von
Dr. Leopold Pfaundler. In drei Bänden. Erster
Band, Mechanik, Akustik. Pp. xxi + 888 (1886).
Zweiter Band, unter Mitwirkung des Dr. Otto
Lummer, Erste Abtheilung, Optik. Pp. xx + 1192
(1894-1897). Zweite Abtheilung, Wärme. Pp. xiv
+ 768 (1898). Dritter Band, Elektrischen Erschein-
engen. Pp. xvi + 1062 (1888-1890). (Braunschweig :
Friedrich Vieweg und Sohn.)

THE appearance of the second part of the second
volume of this work marks the completion of the
ninth edition of an important treatise on experimental
physics which has for many years been widely used in
Germany. The importance of the work lies in the fact
that it aims at giving a full description of physical apparatus and experimental methods, no attempt being made
to expound mathematical theories, and none but the
most elementary mathematics being employed or
assumed as one of the reader's acquirements.

Herein the work differs from most of our English
text-books of physics, in which the tendency has latterly
been to combine a certain amount of mathematical
theory with short accounts of experiments in illustration
of the theory, both the mathematical and experimental
portions being of necessity very incomplete. This
tendency, probably necessitated by our examination
system, will, as long as it continues, prevent our having
in English such complete works on experimental physics
as that now before us.

Any work on physics, however, in several volumes
produced at different times, must, when completed,
present some lack of uniformity among its parts,
especially if the part dealing with that branch of the
subject which varies most rapidly is not produced last.
This is the case in the present instance. The volume
on magnetism and electricity was published some ten or
twelve years ago, several years before the appearance of
the volumes on light and heat. The reason given is that,
on account of the rapid advance made in electricity, the
volume dealing with this branch in the previous edition
appeared much more out-of-date than the other volumes,
and therefore had more need of revision. For the same
reason, on now reviewing the whole of the present
edition, one cannot help being struck with the fact that
the volume dealing with electricity and magnetism far
less adequately represents the present state of the subject
in this branch than do the other volumes in their own
regions.

In the first volume of the present edition, dealing
with mechanics and sound, after an introductory chapter
on fundamental notions and a short discussion of
uniform and uniformly accelerated motion of a point
in a straight line, the subject of mass and force is immediately taken up, further treatment of kinematics
being postponed to a later stage. It seems to the writer
to be preferable, especially in an elementary book on the
subject, to deal more fully with kinematics before going

on to dynamics proper. The student should first become
well acquainted with the notions of velocity, acceleration,
their composition and resolution, and should give special
attention to cases in which the acceleration is not in the
same direction as the velocity. In this way he is enabled
to acquire a much clearer idea about acceleration as a
quantity with a direction of its own, and is therefore
much better prepared to make the transition from his previous vague notion of force to the more accurate dynamical
meaning of the term.

The subject of mass and its measurement is discussed
at some length, and in a very instructive manner.
The action of a force is finally adopted as the basis of the
dynamical measurement of mass. This system involves the definition of force. A definition of mass
(due to Mach) independent of the definition of force
is referred to in a footnote on p. 85, viz. : bodies
which (by gravitation) produce equal but opposite
accelerations in each other are said to have equal masses.
This includes the definition of the ratio of the masses
of two bodies as the ratio of the accelerations which
they produce in each other, and when a unit of mass
is chosen, the mass of any other body is measured
by the acceleration given to the unit divided by the
acceleration experienced by the body itself.

The phraseology is sometimes not as accurate as one
could wish ; thus on p. 92 we find the expression "an
acceleration of one metre," and in the following sentence,
"a velocity of one metre" ; and again, the kilogramme is
stated to be both the unit of mass and the gravitational
unit of force. Although explanations follow, it must lead
to some confusion in the mind of a beginner to find that
a kilogramme means sometimes a mass and sometimes a
force. It is of the greatest importance in an exposition
of the principles of dynamics that one meaning only
should be attached to every technical term. A similar
confusion arises in connection with the term "weight,"
about which there is a lengthy discussion on pp. 96-99.
The difficulty might have been much diminished by reserving the word kilogramme to mean a mass and
weight to mean a force—viz. the resultant force acting
on a body falling freely near the earth. The common
use of the terms should be explained afterwards.

On pp. 326-333 a short account is given of the behaviour of spinning tops and gyroscopes, with a general
explanation of the couples called into play by a deflection of the axis of rotation. The "drift" of a shell fired
by a cannon is ascribed mainly to gyrostatic action.
The constantly increasing angle between the axis of
rotation and the direction of motion causes the air in
front of the shell to exert a force tending generally to
raise the head of the shell with respect to the centre
of mass ; this produces a deflection of the point of the
shell to the right, and the increased pressure thus introduced on the left side causes a deflection to the right.
It is possible that, with a shell of suitable shape, the
pressure of the air would tend to raise the rear end, and
the gyrostatic deflection would in this case be to the
left. As is remarked in a footnote, however, the greater
friction on the under side of the shell probably plays
an important part, and this always causes a drift to the
right.

R

In Chapter v. a good elementary account of the laws of capillarity is given. On p. 444 Quincke's falling drop method of measuring surface tensions is described, the weight of the drop being stated to be equal to the product of the surface tension and the circumference of the line of contact. Lord Rayleigh has shown that this is not correct even if the liquid motion in the drop at the moment of separation be neglected ; the excess of pressure in the drop corresponding to the curvature of the surface (supposed cylindrical near the plane of contact) has the effect of diminishing the size of the drop to one-half the value stated, and this result agrees more closely with experiment.

The second part of the volume, on sound, resembles in its general mode of treatment most other elementary text-books on the subject. The general nature of wave-motion is made quite clear by numerous diagrams of wave-curves and wave-machines. The deduction given on p. 638 of the expression $\sqrt{E/D}$ for the velocity of propagation of sound-waves is not satisfactory, since it involves the tacit assumption that the whole energy is half potential and half kinetic.

In connection with the experimental measurement of the velocity of sound in water in tubes, referred to on p. 643, the influence of the walls and Kundt's measurements in tubes with walls of different thicknesses should have been mentioned, and in the description of the resonance tube experiment, no method is given for eliminating the end correction.

The last chapter contains an interesting account of the researches of von Helmholtz and others on the vibrations of violin strings, combination tones, analysis of sounds, and the theory of consonance and dissonance.

In the second volume (light and heat) the author is assisted by Dr. Lummer, who, we are told in the preface, is chiefly responsible for the part dealing with optical systems and the theory of optical instruments. This part of the work has been largely re-written for the present edition, and brought well into line with the modern views on image-formation founded by Abbe.

As is the case in doing most things, there are two ways of writing a book on geometrical optics. The first, until recently the usual, method is to begin with very special cases, such as thin lenses, and proceed by degrees to the more general cases of thick lenses and systems of lenses.

The other, and more modern, method is to begin with the general case of a point-point correspondence between two portions of space, of such a kind that to a pencil of rectilinear rays passing through a point in one region corresponds a pencil of rectilinear rays passing through a point (the image) in the other region ; then to introduce the special cases of image-formation by reflecting or refracting surfaces and centred systems, including lenses. The two methods thus proceed on opposite lines.

The latter method has been perfected by Abbe, and is the one adopted by Czapski and, though necessarily in a more elementary and restricted manner, in the present work.

After two chapters dealing with the nature of light, photometry, refraction and reflection at plane surfaces, Chapter iii. treats of the formation of images by refraction

at a single spherical surface ; then the general case of any number of spherical surfaces separating different media, with their centres in a straight line ; and, finally, two co-axial centred systems, with the special case of a " telescopic" system in which the " interval " is zero. The lens is regarded as two centred systems, each consisting of a single spherical surface.

Chapter xii. is devoted to the theory and use of " stops," the calculation of magnifying power and brightness of images in centred systems, and, finally, the laws of formation of images of illuminated objects, as in the ordinary use of the microscope. Here purely geometrical methods break down, and diffraction spectra play an all-important part. A highly interesting account follows o Abbe's theory of microscope images and its remarkable verification by the use of the diffraction plate, in which is shown how the similarity of image to object, as well as the resolving power of the instrument, depends upon the number of diffraction spectra whose rays enter the objective and take part in the final image-formation. How these principles are applied in the construction of microscope-objectives is set forth in the chapter on optical instruments, which also contains details of many of the latest improvements in optical instruments of all kinds.

The second part of vol. ii. (on heat) does not differ from the corresponding part of the previous edition so fundamentally as is the case with the part on optics ; but it is brought more nearly up-to-date by many additions, including the work of Olszewski and Linde on the liquefaction of gases, a chapter on thermochemistry, steam calorimeters, recent determinations of the specific heat of water at various temperatures and of the mechanical equivalent of heat. No reference is made, however, to recent improvements in the choice of a unit of heat.

Thermodynamics does not receive very much attention, few applications being mentioned beyond Kelvin's definition of absolute temperature, and a calculation of the change of melting point produced by pressure. Some details are given, however, of the parts and action of steam, air, and gas engines.

A short chapter on meteorology, dealing with climatic conditions and their changes, brings this volume to a close.

As to the third volume, it suffers, as was remarked before, in comparison with the other volumes from having been written several years ago. Still, it contains a large mass of useful information about electrical and electrotechnical apparatus, much of which is not usually found in text-books on electricity and magnetism.

It is impossible, in the space at our disposal, to give more than a very rough sketch of a work which extends to close upon 4000 pages, and many excellent qualities of the work must for this reason remain unmentioned. One of the chief features is the large number (nearly 3000) of excellent illustrations, and, chiefly in the section on optics, some very beautiful coloured plates. Explanations are, as a rule, given very clearly, and often aided by numerical examples worked out. Further, on account of the very large number of experiments and forms of apparatus described, as well as the numerous references to original papers, the work is certain to prove useful, as it no doubt has already done, to students and teachers of physics.

HUXLEY'S PHYSIOLOGY.

Lessons in Elementary Physiology. By Thomas H. Huxley, LL.D., F.R.S. Enlarged and revised edition. Pp. xxiv + 611. (London : Macmillan and Co., Ltd., 1900.)

HUXLEY'S "Lessons in Elementary Physiology" was probably the best book of its kind which has ever been written. It set forth the elements of human anatomy and physiology in so clear and concise a form, and the little volume formed so complete a compendium of the essential facts which had accumulated in the science with which it dealt, that it was at once welcomed as supplying a want which had long been felt —that of a popular and, at the same time, an authoritative exposition of the subject. Its success was enormous. Edition after edition was sold in rapid succession, and the booklet—for it was nothing more—was not only adopted in schools throughout this country as *the* textbook with which the teaching of physiology was to be begun, but it was soon translated into every civilised language, and even, it is said, into more than one barbaric tongue.

The secret of its success lies on the surface. It was written in the English which was characteristic of the Master : its language trenchant, flowing, and well chosen, its similes apposite, its facts duly marshalled and leading up to their logical conclusion. And the book was what it was intended to be—a popular account, which, while retaining scientific accuracy, should not be burthened by unnecessary details, nor by theories which might or might not ultimately prove correct. Moreover, the ground was clear—where there are now a dozen similar treatises, there was then not one. But it is safe to assert that "Huxley" would in any case have taken the first place.

An entirely new edition of the "Lessons"—the first since the lamented death of the original author—has now made its appearance under the auspices of Sir Michael Foster and Dr. Sheridan Lea. Michael Foster has been associated with the book throughout its whole career. Sheridan Lea's name appears now for the first time in connection with it ; but although the responsibility is joint, the labours of preparation have fallen chiefly upon Dr. Lea's shoulders. We may be sure that the work has been a labour of love to the editors. The intimate friendship which existed between them and Huxley, their veneration for his memory, their desire to maintain the high standard and reputation of the work, must have caused them to put forth their best efforts to ensure its continued success.

In surveying the changes which have been introduced, the point of chief interest appears to be to notice whether the introduction of these changes has tended in any way to modify the original character of the work. We have already seen that this character was that of a popular exposition of the science suitable especially for schools, and the questions naturally arise, is the book still of this nature? Has it been modified to suit it to other purposes than that for which its author originally wrote it ?[1] It must be conceded that the book retains in a measure its character as a popular expositor. This is largely owing to the fact that the editors have preserved "as far as possible the original author's own form of exposition and indeed his own words." But it must also be admitted that its character in this respect has been modified by changes and additions. The purport of these appears to have been to adapt the book for use by students of medicine, a design which may be laudable but cannot fail to affect the general tone of the work. Students of medicine require to learn anatomy and physiology with a minuteness of detail not necessary in a work which is intended to be of a popular nature. Not only is it important that the unquestioned facts of the science should be set before them, but they require also to be made cognizant of statements which, however probable, are not universally accepted as facts, and of theories which may or may not ultimately prove to be correct. And herein it appears to me lies the difference between the new "Huxley" and the old. That the change tends, as the editors claim, to increase the sphere of usefulness of the work, may be perfectly true, but the essential character and original aim of the work has been thereby affected. If there is a gain on the one side there is a loss on the other, and it is impossible that it should not be so ; it is a question of opinion whether the gain counterbalances the loss. For my own part, while recognising the able manner in which the new material is worked up and incorporated with the old and the increased value which is thereby imparted to the work as a text-book preliminary to the study of physiology, I must frankly confess that I regret the change. Students of medicine have already more than one elementary text-book in which the facts and chief theories of physiology are set forth with all the clearness that could be desired, and in one instance at least with a wealth of illustration which cannot be surpassed or even approached in a book of so small a size as "Huxley." On the other hand, the amount of detail which has been introduced into this edition, while valuable for the medical student, is unnecessary or unsuitable for the school boy. Perhaps it was impossible to avoid this change, perhaps it was desirable to make it ; at any rate it has been made, and as years go on the development of the book must proceed along the lines which have been now laid down. That it will be as successful on these lines as it has been upon the old ones may be confidently assumed so long as it remains under the management of the present editors, but I believe that my regret that the change has been introduced will be shared by most of those who remember the appearance of the original book in the late sixties and the enthusiasm with which it was then received.

E. A. Schäfer.

[1] "The following ' Lessons in Elementary Physiology' are primarily intended to serve the purposes of a text-book for teachers and learners in boys' and girls' schools."—*Extract from Preface to the First Edition,* 1866.

THE GLUCOSIDES.

Die Glykoside. By Dr. J. J. I. van Rijn. Pp. xvi + 511. (Berlin : Gebr. Borntraeger, 1900.)

THE student of chemistry or botany, who may have attempted to grope his way through the tangle of chemical facts relating to plant products, will be grateful to the author of this exhaustive monograph on the glucosides, or *glykosides* as he prefers to spell it, where the latest information, with all the necessary references, is easily found.

The study of the glucosides may be said to date from Liebig and Wöhler's remarkable discovery of the decomposition of amygdalin by its own ferment or enzyme. emulsin. The discovery of other glucosides followed in fairly rapid succession. Among these may be mentioned, without reference to chronological sequence : salicin, derived from willow bark ; populin, from aspen leaves ; æsculin, from the bark of the horse chestnut ; daphnin, from *Daphne mezereum* ; phloridzin, from the bark of apple, pear and other fruit trees ; hesperidin, from the fruit of limes, oranges and lemons ; potassium myronate or sinigrin, from black mustard seed ; ruberythric acid, from madder root, &c. They all undergo decomposition by a process of hydrolysis into grape sugar, and at least one other constituent drawn from such very varied groups of compounds as phenols, alcohols, aldehydes, mainly of the aromatic series, and in the case of black mustard seed, from a sulphocyanide. No trustworthy explanation of the constitution of these substances was, or could be, forthcoming until the structure of their proximate constituents had been ascertained.

The most important contribution to our knowledge of the glucosides in recent years has been undoubtedly that of Emil Fischer in his classical researches on the sugars. The formation of the glucosides of the simple alcohols and phenols, and of similar compounds of the mercaptans and ketones, has not only given a valuable clue to the structure of the natural products, but has revealed the close analogy which exists between these compounds and the members of the disaccharoses (cane and milk sugar and maltose). Moreover, the identification of new sugars has led to the successful search for these substances among the glucosides and other plant products. Rhamnose or methyl pentose is found to replace glucose in several glucosides : quercitrin, hesperidin, frangulin, baptisin, datiscin, &c., whilst chinovose, another pentose, is contained in chinovite.

Many other interesting points have arisen. The hydrolysing action of the enzyme accompanying one glucoside has been shown, not to be confined to that glucoside, but to extend to others, although, at the same time, strictly limited to a particular series of compounds. The enzyme of yeast has in a similar way been recognised as not exclusive, although restricted in its hydrolysing power. The action of emulsin and yeast on amygdalin is instructive. Emulsin effects complete hydrolysis of the glucoside into benzaldehyde hydrocyanic and two molecules of glucose, whereas the enzyme of yeast only removes one glucose group. Fischer, who discovered this curious difference between the two enzymes, has allotted the following formula to the second product :

$$C_6H_5CH.CN$$
$$O.CH.CHOH.CHOH.CH.CHOH.CH_2OH.$$

The splitting off of more than one molecule of glucose on hydrolysis occurs with populin, hesperidin, helleborin and others.

All these facts are recognised and carefully recorded in the volume before us.

The work is divided into two parts. The first part deals with the artificial glucosides ; the second, with the

natural products. In the first part, the compounds are arranged in a strictly chemical order ; in the second, according to the natural order of plants. In reference to this arrangement, the author lays stress on the fact that the study of the constituents of plants should not follow a chemical classification, but should include all compounds occurring in the same natural order ; because, he explains, it is only in this way that the chemical, morphological and anatomic properties will appear in their true light. The goal of the chemist should not be determined by purely utilitarian motives, but Rochleder's principle should be borne in mind that "the relationship of plants is determined by compounds of the same chemical nature which they contain."

There are one or two points mentioned in the book which are new to us, and may interest our readers. It appears now that sinigrin, the glucoside of black mustard seed, is not hydrolysed, as usually represented in textbooks, without the addition of the elements of water ; but the glucoside contains one molecule of water $C_{10}H_{16}NS_2KO_9 + H_2O$, and the decomposition then falls into line with the hydrolysis of other glucosides,

$$C_{10}H_{16}NS_2KO_9 + H_2O = C_3H_5NCS + C_6H_{12}O_6 + KHSO_4.$$

According to Beyerinck, indigo does not occur in the oldest of the indigo plants (*Isatis tinctoria*) as the glucoside indican, as usually stated ; but in the form of indoxyl, which rapidly oxidises to indigo in contact with air ; whilst the glucoside indican which is found in *Indigofera leptostachya* and *Polygonum tinctorium* may be extracted and left in contact with air without undergoing any change in the absence of enzymes and bacteria.   J. B. C.

AN OXFORD TEXT-BOOK.

An Introduction to the Study of the Comparative Anatomy of Animals. By G. C. Bourne, M.A., F.L.S. Vol. i. Pp. 258. (London : G. Bell and Sons, 1900.)

THIS admirable little book is designed to meet the requirements of the elementary examinations of the leading universities of Great Britain, and though of necessity largely concerned with creatures upon which laboratory treatises exist in abundance, it has been so framed as to supplement and not supersede certain of these, the author having aimed at "the lessons that may be learned and the conclusions which may be drawn" rather than the detailed description of the facts themselves. The work opens with an "Introduction," in which it is pointed out that in the study of natural science, as in other things, something of the nature of a creed is necessary for action, and there is given a definition of "evolution," on the basis of the principles involved in which the science of comparative anatomy is said to be founded. Passing on to treat of the elementary principles of morphology and physiology, the author proceeds to deal with the anatomy of the frog, the elements of histology of the Vertebrata, the cell and cell theory, and phenomena of development up to the formation of the germinal blastemata. The Protozoa are next dealt with, including the Mycetozoa and Volvocinæ ; and, *apropos* of Volvox and Zoothamnium, there follows a chapter on the Protozoa and Metazoa, with a discussion of their inter-relationships. The Cœlenterata follow based on the study of Hydra and Obelia, with a concluding chapter on classification. The book is novel in conception, accurate, up-to-date, and thoroughly artistic in execution. Bütschli on the "Schaumplasma," Boveri on the Ascaris egg, Maupas on the Ciliata, Keuten on Euglena, Hertwig on Actinosphærium, the mitotic processes in Amœba bi-nucleata, the immortality of the Protozoa, are conspicuous among topics of the times handled in a manner well calculated to arouse the imaginative faculty, which, under our prevailing systems of elementary biological training, is apt to be ignored. Unlike many of its predecessors and contemporaries, the book is written in choice English. It is in places even racy ; and in such paragraphs as those in which the author unfolds the points of dissimilarity between Vertebrate and Invertebrate (dog, fish, and lobster), a perspicuity is noticeable equal to that of a good French writer at his best. It recalls most nearly the irresistible charm of the late M. Paul Bert's " Première Année d'Enseignement Scientifique."

The illustrations, fifty-three in number, are mostly original and altogether admirable, and those of the Hydra, based on the author's unpublished researches, will unquestionably become popular—that of the median longitudinal section of this animal being the best we know. On p. 47 the author gives two new figures of the frog's heart, which, as regards the detailed structure of the pylangium and the ostia of the carotid and pulmocutaneous arteries, are wholly unconventional. It is explained in the preface that these are drawings of reconstructional models from sections, and we dare not doubt their accuracy. The question, however, arises how far the facts they reveal may be true of but one individual ; and the author would have done well to have either intimated this or left the matter aside till further investigated. Again, we regret the too forced introduction of analogy to the inanimate, as, for example, of the nervous system to the telegraphic apparatus. In this, however, the author is but acting in the spirit of the times. His book is simply charming and well worthy his reputation ; and while its literary style should alone ensure for it a wide circulation, it cannot fail to exercise a leavening and humanising influence on the youthful mind. It is to be followed by a second volume, dealing with the Cœlomate Metazoa, and the sooner this appears the better for biological science and culture.

OUR BOOK SHELF.

The Ore Deposits of the United States and Canada. By J. F. Kemp. Pp. xxiv + 462 ; index and 163 illustrations. Third edition. (New York and London : The Scientific Publishing Company, 1900.)

OF Prof. Kemp's industry as a compiler there can be no question. The last edition of his work on ore deposits is teeming with information, and his footnotes alone are a proof of the thoroughness with which he has conducted his search after facts. But it is not a book which appeals to the elementary student, because he is launched into a mass of details without sufficient preparation in the introductory part, which is sadly lacking in woodcuts. And further, there is evidence of haste or want of care in correcting the book for the press. Surely a writer on ore deposits should be able to spell such names as "Pošepný," "Sjögren" and "Příbram" with strict accuracy. Errors in spelling ordinary French and German words are frequent, and when one notes as many

as nine mistakes in seven consecutive lines, there are fair grounds for complaint. It is not only in his spelling that Prof. Kemp evinces carelessness. A mineralogist would not speak of specular iron as " specular hematite " ; the product of a zinc mine should not be called *spelter*, as the word denotes the metal, not the ore. By one of his sentences, one might infer that Prof. Kemp would not admit sulphide of sodium among the metallic sulphides. It is not good English to say : "*Considerable* limonite has also resulted from the weathering of clay-ironstone nodules."

In spite of frequent and unpardonable minor blemishes, which could easily have been avoided by employing a careful proof-reader, the book will be found very useful by those who require a summary of the innumerable memoirs and papers describing American ore deposits.

Prof. Kemp's conclusion that an amendment is needed of the laws regulating the tenure of ore deposits in the Western States will be warmly endorsed by most mining men.　　　　　　　　　　　　　　　　C. L. N. F.

Physiology for the Laboratory. By Bertha Millard Brown, S.B. Pp. viii + 167. (Boston : Ginn and Co., 1900.)

THIS little book sets forth, in twenty-two brief chapters, certain practical directions for the study of the elements of anatomy, histology and physiology of the vertebrate body, and the first principles of bacteriology. Many of the instructions given are in interrogatory form, and for simple experiment and observation of the living in action, in which lies the very essence of the science of physiology, the student is commendably referred to his or her own body. Beyond this, however, there is nothing in the book that is new, or which calls for comment in these pages. The mode of treatment is begotten of a conviction on the part of the authoress, that "there is needed a radical change in the teaching of physiology " ; and we read with astonishment the statement that while the method of teaching botany, chemistry and other sciences "has long been that of going first to the study of the specimen and then to the text-book," this has not been the case for "physiology"—that having apparently been taught from the text-book alone. She is writing, however, of State schools of America, and if the accusation be applicable to them generally, we wish her success in her enterprise.

Michigan Board of Agriculture. Annual Report 1898-99. Pp. 465. (Michigan : State Board of Agriculture, 1899.)

IN this volume are included the thirty-eighth annual report of the Secretary of the Michigan State Board of Agriculture, and the twelfth annual report of the Experiment Station of the State Agricultural College. Many subjects of interest are dealt with in both reports, but only a few can be referred to here. Experiments with Indian corn, to test the influence of thickness of planting upon the character of the crop, show that a gradual increase occurs in the yield of dry matter and protein as the distance between the rows and between individual plants is increased. It appears that, to obtain the greatest yield of valuable nutrients, Indian corn should be planted in rows fully three and a half feet apart, and the seeds six and nine inches apart in the rows. The establishment of several large beet-sugar factories in the State last year has caused increased attention to be given to experiments in beet culture. An interesting detail of some new experimental work, to which reference is made by Prof. C. D. Smith, director of the Experimental Station, is the breeding of bees with longer tongues. It is hoped that, by selection and breeding, a variety of honey bee will be developed capable of extracting nectar from the blossoms of the clover grown in the State.

Among the subjects of *Bulletins* published in the report are :—forestry, strawberry culture, methods of combating disease-producing germs and fruit-growing.

LETTERS TO THE EDITOR.

[*The Editor does not hold himself responsible for opinions expressed by his correspondents. Neither can he undertake to return, or to correspond with the writers of, rejected manuscripts intended for this or any other part of* NATURE. *No notice is taken of anonymous communications.*]

Change of Feeding Habits of Rhinoceros-birds in British East Africa.

THE enclosed extract from a letter just received by me from my friend, Captain Hinde, of the British East Africa Protectorate, will interest all zoologists. It is a curious fact that a bird which is so valuable as Buphaga in clearing parasitic insects from cattle that we lately agreed to give it special protection at the International Conference on the Preservation of African Wild Animals, should now, by a sudden change of conditions induced by man, become a dangerous and noxious creature. This fact shows how difficult is the problem presented by the relations of civilised man to a fauna and flora new to his influence.　　　　　E. RAY LANKESTER.

Natural History Museum, London, August 10.

"The following case of wild birds changing their habits may interest you :—The common rhinoceros-bird (*Buphaga erythropyncha*) here formerly fed on ticks and other parasites which infest game and domestic animals ; occasionally, if an animal had a sore, the birds would probe the sore to such an extent that it sometimes killed the animal. Since the cattle plague destroyed the immense herds in Ukambani, and nearly all the sheep and goats were eaten during the late famine, the birds, deprived of their food, have become carnivorous, and now any domestic animal not constantly watched is killed by them. Perfectly healthy animals have their ears eaten down to the bone, holes torn in their backs and in the femoral regions. Native boys amuse themselves sometimes by shooting the birds on the cattle with arrows, the points of which are passed through a piece of wood or ivory for about half an inch, so if the animal is struck instead of the bird no harm is done. The few thus killed do not seem in any way to affect the numbers of these pests. On my own animals, when a hole has been dug, I put in iodoform powder, and that particular wound is generally avoided by the birds afterwards; but if the birds attack it again, they become almost immediately comatose and can be destroyed. This remedy is expensive and not very effective. Is there any other drug you could suggest that would be less likely to be detected? Perhaps you know that I reported three years ago that these birds rendered isolation under the cattle plague regulations useless in some districts, as I proved beyond doubt they were the only means of communication between clean and infected herds under supervision, a mile or two apart. These birds I have never seen on the great herds of game on the open plains, but I have seen them on antelope and rhinoceros in the immediate neighbourhood of Masai villages, and herds of cattle; on the other hand, I have never seen the small egret on cattle, though often on rhinoceros and gnu."

Atmospheric Electricity.

IN NATURE of June 14 Mr. Wilson replies to the objections raised in my letter of March 29 to his explanation of the origin of atmospheric electricity. Before proceeding to consider Mr. Wilson's reply to my objections it may be well that the point at issue between us should be clearly defined. as Mr. Wilson. in my opinion, somewhat confuses it. Mr. Wilson says, "Mr. Aitken contends there is no such thing as dust-free air in the atmosphere." Now I certainly made no such statement, for the simple reason that I do not know whether such a condition exists to any extent or not, only a few cases being on record. What I did state was, "So far as our knowledge goes, it can hardly be said there is such a thing as dust-free air in our atmosphere, and the cases in which low numbers have been observed are so extremely rare that they can hardly have any bearing on phenomena of such widespread existence as atmospheric electricity, even though we suppose those few particles to be afterwards got rid of." I simply asked for a verdict of "not proven" against Mr. Wilson's theory. I think it will be admitted that it rests with Mr. Wilson, and those who think with him, to prove that the air is generally dust-free at elevations higher than ordinary cumulus and nimbus clouds, as

without this dustless air the supersaturation necessary for condensation on ions is admittedly not possible.

Mr. Wilson discusses the question of the number of dust particles in the atmosphere from Mr. Rankin's Ben Nevis observations and my own at Kingairloch, and points out that practically dust-free air has been observed on Ben Nevis. Such is the case, but so far as I know dust-free air has been observed on only a few occasions, and such isolated instances have evidently no bearing on the case. Mr. Wilson then turns to my observations and says "the mean number of dust particles in a series of 258 observations, extending over nearly five years, amounting to 338 per c.c. ; on one occasion the number was as low as 16 per c.c." The above statement, it must be clearly understood, refers to 258 of the tests made in the purest air, and is not the mean of all the observations. In the tables there are 688 observations for Kingairloch : of these I find there are 41 in which the reading was under 100, 341 were over a 100 but less than 1000 per c.c., whilst the remaining 306 observations were all over 1000 per c.c. The 16 per c.c. referred to by Mr. Wilson only occurred once. In the other years referred to the lowest figures were 38, 43, 67 and 205 per c.c. So that, as already said, the conditions represented by those low figures, such as o on Ben Nevis and 16 at Kingairloch, are so exceptional that they are not likely to play any part in phenomena so universal as atmospheric electricity.

Mr. Wilson, referring to the selected observations taken at Kingairloch on the pure air coming from the Atlantic, says : "Air coming from such a region can hardly be considered as abnormal. Moreover, such observations are necessarily made in air within a few feet of the ground ; at a greater height it is likely to be less contaminated." Taking the last of these points first, an examination of the diagrams given along with the tables, from which Mr. Wilson made his extracts, will show that whenever the air became pure the readings low down and high up were nearly alike. This is shown by the curves in the diagrams for Ben Nevis and Kingairloch being nearly alike during these periods. Further, it may be seen from the curves that there was sometimes less dust at low than at high level when the air came from the Atlantic.

An examination of the tables from which Mr. Wilson took his Kingairloch figures easily refutes his assumption that the air of the Atlantic, as given in these tables, "can hardly be considered as abnormal." In the tables will be found the results of tests made in France, Italy and Switzerland. Observations were made at three places in France on the shores of the Mediterranean, at Hyères, Cannes and Mentone. An analysis of the figures for these places, made during visits extending over five years, shows that the lowest number observed was 725 per c.c., and of eighty-eight tests only ten were under 1000 per c.c., the others being all over 1000. At the Italian Lakes observations were made at Bellagio and Baveno. Many of these observations were made at elevations up to 2000 feet. In all, 188 tests were made : of these the lowest was 300 per c.c. On only thirteen occasions was the number under 1000, and 175 readings gave numbers over 1000 per c.c.

Perhaps it may be objected that all these Continental tests were made in low level polluted air. We shall therefore now examine the result of the observations made on the Rigi Kulm, given in the same tables. The top of the Rigi is 5900 feet above sea-level, but it has only the purifying effect of 4400 feet, as it is only about that height above the surrounding plains. During the tests, made on the visits during the five different years previously referred to, 259 observations were taken on thirty-two days, and the lowest number observed was 210 per c.c. Ninety-seven observations gave readings under 1000 per c.c., whilst the other 162 tests were all over 1000 per c.c. These tests, at both high and low level, give no support to Mr. Wilson's statement that the Atlantic air on the west coast of Scotland "can hardly be considered as abnormal."

Let me further support this point by reference to observations made by others of the air in different parts of the world. Prof. G. Melander, of Helsingfors, in his work, entitled "Sur la condensation de la vapeur d'eau dans l'atmosphère," gives the results of 268 tests made of the air at Salève, Biskra, Torhola, Loimola, Kristianssund and Grip. In all these 268 samples of air tested there were only five with less than 500 per c.c., and no low numbers were observed.

I now turn to the very interesting series of observations made by Mr. E. D. Friedländer and published in the *Quarterly Journal* of the Royal Meteorological Society, vol. xxii. No. 99.

July 1896. In this paper Mr. Fridlander gives the results of his observations made during a voyage round the world from this country to America, across that continent to Santa Cruz Bay, from there across the Pacific Ocean to New Zealand, then to Australia, and homewards by the Indian Ocean, Arabian Sea and Mediterranean, visiting Switzerland on the way. On the western side of the Atlantic the numbers were high, being from 2000 to 4000 per c.c., though the vessel was far from land, its position being 55° 0′ N., 42° 11′ W. Lower numbers were obtained between Labrador and Newfoundland, the readings there being from 420 to 840 per c.c. Readings as low as 280 per c.c. were got in the Gulf of St. Lawrence. On the Pacific coast the lowest was 700 and highest 4500 per c.c. On the Pacific Ocean the lowest reading was 280 per c.c. and highest 2125. Few readings were obtained in New Zealand under 1000 per c.c. In the Indian Ocean the air seems to be rather purer than most places, or at least was so when the observations were made. Readings as low as 200 per c.c. were obtained, and they seldom were over 500. Tests made on only two days in the Arabian Sea gave a minimum of 280 and a maximum of 1375. One day's tests in the Red Sea gave from 383 to 490 per c.c. The result of two days' tests of the Mediterranean air gave a minimum of 875 and a maximum of 2500 per c.c. A result which agrees with that already given for the French coast of the Mediterranean.

Mr. Fridlander's tests of the air in Switzerland give results similar to those already referred to for the Rigi Kulm. At almost all the places the numbers were always over 1000 per c.c., though the observations were made at considerable elevations. But on the Riffelberg (altitude 7400), where Mr. Fridlander spent some days, the numbers varied from 225 to 4000 per c.c. On the summit of the Bieshorn (altitude 13,600) the lowest observation gave 140 per c.c., which, so far as I know, is the lowest number yet observed in Switzerland. When we compare the figures given by Prof. Melander and Mr. Fridlander of the dust particles in the air of different parts of the world with those obtained in the Atlantic air on the west coast of Scotland, we are forced to admit that the latter is abnormally pure.

The rate of fall of cloud particles as given by the calculations of Mr. Wilson seems to be much too rapid. He assumes that the air in which clouds are formed is always rising. This can hardly be said to be the fact. Suppose a large area of the earth's surface to be covered with cloud, forming a vast sea, such as one sometimes sees from the top of a mountain. It is evident that the air over all that area cannot be rising at any considerable rate, and yet the clouds will be seen to keep nearly the same elevation for hours. If the air be still, and if Mr. Wilson's calculations are correct, then the mountains ought to rise out of such a cloudy sea at the rate of nearly 500 feet per hour, a phenomenon which, I venture to say, no one has ever seen.

Mr. Wilson seems to think, though all the dust particles in cloudy air will not become centres of condensation, it is a matter of no importance, as he thinks the cloud will act as a perfect filter, by the descending cloud particles coming in contact with, and absorbing, the inactive dust particles. So that all particles that do not become active centres of condensation will be carried out of the air by the falling drops, and leave the air rising through the cloud particles dustless. He gives no evidence in support of this assumption other than the purification of dusty air in a closed vessel with wet sides. Now dusty air in a closed vessel takes a considerable time to become dust-free, and I think it may be contended that gravitation plays no inconsiderable part in the process, perhaps more than the wet sides referred to by Mr. Wilson. So far as my observations go, there is no evidence of any such powerful purifying effect in clouds. At least when making observations in old clouds, both at top and bottom of them, there were always observed a large number of dust particles, but whether any had been absorbed by the cloud particles or not it would be impossible to say. If any had been absorbed, certainly many were still free.

That clouds have not the purifying effect claimed for them by Mr. Wilson may be best shown by reference to the observations made on the Rigi Kulm on May 21, 1889 (*Proc.* Roy. Soc. Edin., vol. xvii. p. 193). On the morning of that day, when I left Lucerne on my way to the Rigi Kuhn, the sky was covered with cloud, and when ascending the mountain the cloud was entered at an elevation of about 2000 feet below the top. On arriving at the top the clouds were still very dense, and remained so till the evening ; afterwards they settled down to the level of the kulm, when a vast sea of clouds was disclosed stretching in

all directions with the peaks of the higher Alps standing out like islands. Under these conditions the observations made on the top of the Rigi on that day were evidently taken in air just above the upper surface of a uniform stratum of cloud 2000 feet deep, where, according to Mr. Wilson, there ought to have been dustless air, yet the observations showed there were still 210 particles per c.c. Next morning the clouds still extended in most directions and were much thinner, and the number of dust particles had increased to over 800.

I may as a well here call attention to the fact that during the night the upper surface of the clouds had only settled down about 1000 feet. How much of this was due to the cloud particles falling through the air, and how much to evaporation, it would be hard to say. Probably evaporation played the principal part, as the clouds were now much thinner, and the evaporation probably took place from the upper surface, as in the morning the air on the Kulm was dry—the wet-bulb depression being as much as 6°, and a wind of some strength was blowing from the south-east. The rate of descent of the particles in this cloud was therefore much slower than the rate of fall calculated by Mr. Wilson.

Mr. Wilson, in criticising my remarks on the re-evaporation of cloud particles, says : " But all drops that have survived the great tendency to evaporate which accompanies the initial stages of their growth will surely continue to grow so long as the rate of expansion remains the same, or even if it be much reduced." Here again Mr. Wilson assumes that clouds are always rising. Now a great part of the life of a cloud, and the air in which the particles are carried, is spent in moving horizontally, and sometimes even downwards, and occasionally with but little movement in any direction ; and it is during this stable condition that the opportunity is given for the re-evaporation of the smaller drops. Mr. Wilson points out that if a very slight proportion of the water in a drop were to evaporate, it would cool the drop and check the evaporation, a statement with which all will agree. But though the cooling may check evaporation, it will not stop it. The particles in a cloud are close together, and those condensing vapour and growing warmer soon part with their heat by radiation and by contact with the air, so that the heat lost by the evaporating particles is rapidly supplied to them by the condensing ones, and, as we shall see later, this exchange of heat takes place at a much quicker rate than one might imagine.

I do not think that practical chemists will agree with Mr. Wilson's statement that all the ammonia, nitric acid and other impurities, out of which the sun can manufacture nuclei, can be washed out of the air by the rain. The difficulty of removing the last traces of gases by washing is well known.

Are meteorologists prepared to accept that part of Mr. Wilson's theory which necessitates the formation of rain-clouds at an elevation of 7500 feet above the top of the ordinary cumulus and nimbus clouds ? In other words, are meteorologists prepared to affirm that there are two distinct rain zones—one where the ordinary rain-clouds condensed on dust nuclei are formed, then over these clouds clear air for 7500 feet, above which the ion rain-clouds are formed ? This upper ion-cloud must result in rain if the theory is correct, otherwise there will be no separation of the positive and negative ions. I leave it to the meteorologists to say whether rain-clouds have ever been observed at elevations of 20,000 to 30,000 feet—not above sea-level, but above the surface of the ground.

Mr. Wilson does not seem to think that my remarks on the rapid growth of cloud particles in supersaturated air have any bearing on the subject, and objects to my use of the term explosive in reference to the condition of supersaturated air. If I had known a better term I would have used it. Though supersaturated air is in a condition of equilibrium with itself, yet when nuclei are introduced into it there is at once a rapid rush of vapour molecules towards the condensing particles, and a rapid breakdown of conditions all round the nuclei, which seems to me not at all inaptly compared to an explosion—centripetally, of course. Mr. Wilson grounds his objection to the rapid growth of the ion-cloud particles in supersaturated air on the difficulty and slowness with which the condensing drops part with the heat developed by the condensing vapour. I shall not follow Mr. Wilson in his comparison of a condensing with an evaporating drop, as it is not so easy to see the changes taking place in the latter, but will rather refer to an experiment which Mr. Wilson, and others who have experimented on this subject, must often have seen. Take a glass flask in which there is a little water, full of ordinary air, and provided with means of expanding the air in the flask, and either returning the air to the flask, or admitting filtered air. Go on repeating the process of expanding and cloud-making in the flask. After this has been done a number of times, the nuclei become fewer and fewer, and at last only a very few are left in the air. Every one must have noticed when making this experiment that the cloud particles are very small on the first expansion, and that they fall very slowly, almost imperceptibly, but that at the end of the experiment, when the last dust particles become nuclei, the water particles are large and fall rapidly like rain drops. At the beginning of the experiment, with plenty of dust in the air, there is almost no supersaturation, the nuclei being so close the tension is relieved as soon as it is formed. When, however, only a few particles are present, there are large spaces between the nuclei where supersaturation can take place, and it is by falling through this supersaturated air that the drops, when few in number, are able to grow so quickly and become so large. It therefore seems probable that something of the same kind will happen if ions were to become nuclei in supersaturated air. Whenever an ion becomes active it will rapidly grow to the dimensions of a rain-drop in the same manner and for the same reason that the dust-nucleused drops do in supersaturated air. These little drops evidently have a way of parting with the heat of condensation at a very much quicker rate than Mr. Wilson is disposed to admit.

It is this capacity for rapid growth in supersaturated air that makes it so improbable that ions can ever give rise to a cloudy form of condensation. To form a cloud a large number of them would require to become active at the same moment. But this is evidently not possible in a rising column of air. The ions which rise on the top of the ascending column will become active first, and by falling through the lower supersaturated air will grow with great rapidity and give rise to a rainy, but cloudless form of condensation.

There are some points connected with ions about which I think the readers of NATURE would be glad to have some information, and which I think Mr. Wilson, with the aid of the apparatus at his disposal, could give us. For instance, one would like to know (1) how long ions remain in air in an inclosed vessel, when both + and − ions are present ; (2) when only + or − ions are in the air ; (3) whether the presence of dust has any effect on the duration of their life. For practical purposes one would also like to know further (1) how many ions are generally in the air near the ground ; (2) what amount of electricity they carry with them.

Finally, one would like to know how many ions will pass up through a cloud and escape at the top ; as one would almost expect, these ions, with their electric charges, will be more likely to be cleared out of the air by rain than the dust particles, and whether both kinds are equally liable to be washed out by rain. If not, the inequality may help to explain some important electrical phenomena. JOHN AITKEN.
Ardenlea, Falkirk, June 27.

The Melting Points of Rock-forming Minerals.

IN connection with the abstracts of papers read before the Royal Dublin Society by Dr. J. Joly, F.R.S., and myself, given in NATURE for July 12 (p. 262), I might perhaps be permitted to draw attention to a few points. The same subject has been recently dealt with by Mr. C. E. Stromeyer (*Mem.* Manchester Lit. and Phil. Soc., vol. xliv. Part iii. No. 7, 1900) and by Prof. Sollas, F.R.S. (*Geol. Mag.*, July 1900).

In the first place it may be noted that the "melting point" of a substance under a definite pressure has a perfectly definite meaning. The "softening point," on the other hand, obviously depends on the magnitude of the distorting force with which the softness is tested, as well as on the other conditions of experiment.

It is an established fact that the melting points of a very large number of substances vary with the pressure. Bunsen, as far back as 1850, perceived the geological application of this phenomenon. In discussing the crystallisation of plutonic rocks, it is the melting points of the minerals under enormous pressures which really concern us. These pressures are probably sufficient to alter the melting points through several hundred degrees. There are then two ways open for us to ascertain these melting points. Firstly, we might determine them by direct experiment at the necessary large pressures ; or,

secondly, we might measure the melting points at ordinary atmospheric pressure and determine the rate of increase (or decrease) of melting point with increase of pressure ($d\theta/dp$). Considering the gigantic pressures with which we have to deal, it seems decidedly easier to adopt the second method. The agreement between the results obtained from the application of the thermodynamic formula

$$\frac{d\theta}{dp} = \frac{\theta(v_l - v_s)}{L}$$

(where θ = absolute melting temperature ; $(v_l - v_s)$ = the change of volume at the instant of melting ; L = the latent heat in mechanical units) with the results of experiments (*e.g.* M. A. Battelli, *Journal de Phys.*, t. viii. p. 90, 1887), seems to justify the application of that formula to the case of the minerals in question, in the absence of direct experiment. It is true that the formula was deduced for a reversible system, and that no natural process is reversible. But a similar objection would hold against the application of any theoretical formula to the conditions obtainable in experimental work. In the present case it is only claimed for the formula that it will afford an approximate estimate of the melting points of minerals under large pressures ; and after all, even direct measurement of such high temperatures as are involved is always attended with uncertainty. In order to apply this formula we require θ, $(v_l - v_s)$, and L. The melting points of the most important minerals at atmospheric pressure have been determined by Dr. Joly and Mr. R. Cusack (*Proc. Roy. Irish Acad.*, Ser. 3. vol. ii. p. 38 ; vol. iv. p. 399). A large part of the volume change on melting is, I submit, afforded us by the difference in density between the crystalline mineral and its fused glass. Now it is characteristic of amorphous substances to pass gradually and continuously from solid to liquid (*cf.* Preston, "Theory of Heat," pp. 270 and 286) ; and so it is highly probable that such a mineral glass will pass without sudden volume change into the liquid state, and it has, in fact, passed gradually in the inverse direction. It is not contended that any given mineral ever existed as a glass in the molten magma of an igneous rock, but only that it existed as a liquid.

In my paper, above referred to, I have shown how the "fusibility" of a mineral must be connected with its latent heat, and hence by a comparison of relative fusibility and melting temperature we may often deduce the relative latent heats of two minerals. Thus, for example, the "fusibility" of labradorite is 3 on von Kobell's scale, and its melting point is 1229° C., whereas orthoclase has a melting point of only 1175° C., but is much less "fusible," viz. 5 on von Kobell's scale. Hence I infer that the latent heat of orthoclase is decidedly greater than that of labradorite. Similarly, the latent heat of augite is less than that of orthoclase. But the volume-change on melting of augite is greater than that of orthoclase. Therefore $d\theta/dp$ is greater for augite than for orthoclase. It is thus possible to arrive at the order of melting points of minerals under the pressure of rock formation. If, after ascertaining this order, it is still found to be inconsistent with the order of crystallisation, as shown by microscopical examination, it may be necessary to examine the more complicated influences of solution, &c., on the crystallising points of the minerals.

In conclusion, I may point out that it must be a matter of extreme importance in measuring the melting temperature of quartz to make sure that the specimen used is pure, and in particular free from the alkalis. Messrs. Shenstone and Lacell (NATURE, May 3, 1900, p. 20) have found that rock crystal very often contains sodium and lithium, traces of which might be expected to lower the melting point. Further, it has long been known that quartz, with a density of 2·66, passes into the variety of silica with density 2·3 at a temperature below its melting point (*cf.* Fremy, *Enc. Chim.* 6, p. 142). And similar transformations are common among metals. Is it not possible then that the phenomena observed by Dr. Joly may have nothing to do with the fusion point of quartz, but are simply cases of molecular transformation at a temperature below the melting point ? J. A. CUNNINGHAM.
Royal College of Science, Dublin.

Observation of the Circular Components in the "Faraday Effect."

AFTER repeated attempts to determine the nature of the "Faraday effect," I have succeeded in observing that ordinary light, when passing from a surface into a medium in such a way as to be under the influence of a magnetic field, is broken up into two circular components oppositely polarised.

The system used consisted of two rectangular prisms of glass placed with their diagonal faces parallel and separated by a plate of mica of approximately ¼ λ retardation. The lines of force were parallel to this plate. A ray of ordinary light from a sodium flame sent into the system normal to this plate was successively totally reflected parallel to the lines of force and then at right angles to the mica, which served to change the phase and to keep the absolute direction of the circular vibrations the same. The rays passed five times around within this system, giving twenty internal reflections.

The separation of the rays agreed, so far as could be determined, with the calculations based on the assumptions usually made in explaining this phenomenon. When the field was reversed, the direction of vibration of each circular component was reversed. This does not establish the assumption of a relative change in the velocities usually made, as a relative change in the phase of the components, or both, would produce the same effect. It does show, however, that a medium in a magnetic field transmits, in the direction of the lines of force, light vibrations by circular components only.   D. B. BRACE.

Physical Laboratory, University of Nebraska, August 1.

Physical Structure of Asbestos.

CAN any of your readers tell me where I can find a good account of asbestos and its *physical structure?* The ordinary works of reference I am acquainted with give too meagre an account to be of any use.   GEOFFREY MARTIN.

13 Hampton Road, Bristol, August 1.

THE BRADFORD MEETING OF THE BRITISH ASSOCIATION.

IT is now possible to give a forecast of the chief subjects to be brought before the various Sections of the British Association at the forthcoming Bradford meeting. The following outlines of sectional programmes show that many matters of importance and wide scientific interest will be dealt with, so that the Bradford meeting promises to be a memorable one. No particulars as to the probable proceedings of the Physics Section have yet been received.

CHEMISTRY.

Prof. W. H. Perkin, jun., F.R.S., the President of Section B (Chemistry) is this year setting a precedent in the conduct of the sectional meetings. Several members of the Association have been asked to furnish reports on the present state of knowledge in the particular departments of chemistry with which they are especially conversant, and the reading of these reports will be followed by discussion.

In accordance with this programme, Mr. Francis H. Neville, F.R.S., will present a report dealing with the properties and interactions of the metals. The following questions will be brought forward for discussion in connection with the report :—

1. Are the methods usually employed in studying the equilibrium between two or more substances with change of temperature immediately applicable to the study of alloys, and are similar results obtained in the two cases? Thus, with varying conditions of temperature and concentration, a system of ferric chloride and water deposits (1) ice ; (2) $Fe_2Cl_6,12H_2O$; (3) $Fe_2Cl_6,7H_2O$; (4) $Fe_2Cl_6,5H_2O$; (5) $Fe_2Cl_6,4H_2O$; (6) Fe_2Cl_6 or (7) mixtures of the phase numbered n with that numbered $(n + 1)$? Are the solubility curve of ferric chloride and the freezing point curves of metallic mixtures of the same kind?

2. How far does (1) microscopic examination, and (2) change in physical properties, such as electromotive force, &c., enable us to detect the existence of a compound in an alloy?

3. In what definite proportions are metals known to combine? Is any regularity manifest with respect either to their position in the periodic system or to their valency with regard to non-metals?

4. What methods are available for determining the molecular weights of the metals, and can it be asserted in any cases, other than those of mercury, zinc and cadmium, that the molecular weight is satisfactorily determined?

5. Can a definition be given of a metallic element which makes it possible to distinguish between metals and non-metals?

6. Can any explanation be given which will satisfactorily account for (1) the difference between metallic and electrolytic conduction, and (2) the remarkable changes in the electrical conductivity of metals attending admixture?

As some of the questions bearing upon this subject are of as great importance to the physicist as to the chemist, physical members of the Association are to be invited to join in the discussion. Dr. Adolf Liebmann will contribute a report on recent improvements in the treatment of textiles, a subject which acquires peculiar importance from the fact that the 1900 Meeting of the Association is being held in the centre of a district devoted to the textile industry. Dr. Arthur Lapworth will give a report on our knowledge of the chemistry and constitution of camphor. Attention has of late years become so concentrated on the chemistry of the camphor group as to make an authoritative discussion on the constitution of camphor almost a necessity to the organic chemist. Mr. William J. Pope will present a report on our present knowledge of stereochemistry ; it is understood that special attention will be given in this report to the work done during the past twelve months on the optical activity of compounds containing an asymmetric nitrogen, tin or sulphur atom. Among the other papers to be presented at the meeting is one on the specific heat of gases at temperatures above 400°, by Prof. H. B. Dixon, F.R.S. ; and Mr. H. T. Brown, F.R.S., will give an account of his recent work on the diffusion of gases and liquids. The papers of special local interest include one on the treatment of Bradford sewage, by Mr. F. W. Richardson, the City analyst ; and also a paper on the treatment of woolcombers' effluents, by Mr. W. Leach. The title of the sectional address to be delivered by the President is, of course, not yet announced ; it is understood, however, that the address will deal with the teaching of chemistry.

GEOLOGY.

The proceedings of Section C (Geology) will open at 10.30 a.m. on Thursday, September 6, with the delivery of the address of its president, Prof. W. J. Sollas, who has chosen for his subject, "The History of the Earth in relation to a Scale of Time." Prof. Sollas will take a wide scope in discussing this subject, and will introduce such fundamental matters as the constitution of the earth, the relative value of the various geological periods, the origin of ocean basins, the formation of mountain-chains, and the evolution of the organic world. We may be sure that his discourse will be brilliant and suggestive. It is probable that Prof. J. Joly will also treat on the knotty problem of the duration of geological time at the same, or a subsequent, meeting of the Section.

As befits the place of meeting, the geology of the Carboniferous rocks will receive much attention. A joint discussion with the botanists (Section K), on the conditions which existed during the growth of the forests of the Coal Period, will be held on Monday, September 10, when Mr. A. Strahan and Mr. J. E. Marr will open the debate from the geologist's standpoint. The Coal-measures of the West Riding form the subject of a paper by Mr. W. Cash, and those of North Staffordshire of one by Mr. W. Gibson, of H.M. Geological Survey ; while the fossil fishes of the Carboniferous rocks will be discussed by Dr. E. D. Wellburn. Prof. W. B. Scott has promised a paper, with lantern illustrations, on the geology and palæontology of Patagonia, which promises to be of great interest. Prof. A. P. Coleman, of Toronto, brings forward an account of a ferriferous horizon in the Huronian of Lake Superior, and will also present the final report of the Committee for the investigation of Pleistocene deposits in Canada. Mr. J. J. H. Teall will describe a plutonic complex of Sutherland, and its bearing on current hypotheses as to the genesis of igneous rocks. Glacial subjects, as usual, will receive due attention ; papers on the local phenomena will be brought forward by Dr. Monckman and Messrs. Muff, Jowett and others, and on those of Welsh localities by Mr. E. Greenley and Mr. J. R. Dakyns. The concretions of the magnesium limestone of Durham will be discussed by Dr. Abbott. Tidal ripple-marks will be described by Mr. Vaughan Cornish ; and the caves and pot-holes of Ingleborough and district by their explorer, Mr. S. W. Cuttriss, in both cases with lantern illustrations. Mr. A. C. Seward will treat of the Jurassic flora of the Yorkshire coast. Among the reports of committees will be that which deals with the course taken by underground waters in the Ingleborough district, giving the result of recent experiments ; and excursions have been arranged to the sites of the investigation. As usual, short afternoon sectional excursions to places of geological interest in the neighbourhood of Bradford will be included in the arrangements of the Section.

ZOOLOGY (AND PHYSIOLOGY).

Dr. R. H. Traquair, F.R.S., the President of Section D (Zoology and Physiology), will address the Section on "The Bearings of Fossil Ichthyology on the Doctrine of Descent." Major Ronald Ross will (by request) address the Section on "Malaria and Mosquitoes." Messrs. Gamble and Keeble will give an account of their researches on the "Colour-Physiology of Hippolyte," illustrated by lantern projections and practical demonstrations. Prof. L. C. Miall, F.R.S., will read a paper on the "Respiration of Aquatic Insects"; and other papers on the natural history of insects will be given by Messrs. T. H. Taylor, Wilkinson, Walker, and Dr. Munro. Prof. S. J. Hickson, F.R.S., will read a paper on "The Nuclei of Dendrocometes." Among the reports of committees, Mr. Stanley Gardiner's account of his researches on the Coral Islands of the Indian Ocean is awaited with particular interest.

GEOGRAPHY.

In Section E (Geography), the President, Sir George Robertson, will deliver his address on Thursday, September 6, at 11 a.m. The subject of the address is appropriately "Geography and the Empire." Amongst the subjects to which special attention will be directed in the Section may be mentioned that of "Colonial and Foreign Surveys." Papers dealing with these will be read by Mr. E. G. Ravenstein and Mr. B. V. Darbishire ; and Dr. H. R. Mill will contribute a paper on "The Treatment of Regional Geography." Problems of applied commercial geography will be dealt with by Mr. G. G. Chisholm in a paper on "Some Consequences that may be Anticipated from the Development of the Resources of China," and Mr. E. Heawood on the "Commercial Resources of Africa."

An important paper on "Railway connection between Europe and Asia" will be contributed by Sir Thomas Holdich ; and it is hoped that Mr. C. R. Beazeley will return in time to give an account of his journeys on the recently-opened portions of the Siberian Railway.

The excellent work initiated by Mr. T. G. Rooper while H.M. Inspector of Schools at Bradford, is

carried on by Mr. E. R. Wethey, one of the secretaries of the Section, makes papers on "School Geography," and the teaching of elementary geography generally, of special interest at this meeting. Mr. Rooper and Mr. Wethey will each describe parts of their work, and will exhibit some of the maps and models used as illustrations, in the exhibition which forms a novel feature of the Bradford meeting.

In the department of geographical exploration, Mr. C. E. Borchgrevink will give an account of the voyage of the *Southern Cross* in the Antarctic regions. Captain H. P. Deasy will describe his journeys in Central Asia ; Captain E. S. Grogan contributes a paper, "Through Africa from the Cape to Cairo" ; and Mr. Cutliffe Hyne one on "Arctic Lapland."

On special and more technical subjects there will be papers on "Large Earthquakes in 1899," by Prof. John Milne ; on the "Distribution of Relative Humidity," by Mr. E. G. Ravenstein ; on "Snow Ripples," by Mr. Vaughan Cornish ; and on "The Origin of Moels," by Mr. J. E. Marr.

MECHANICAL SCIENCE.

Sir Alexander Binnie, the President of Section G (Mechanical Science), will survey the various stages of scientific progress which have led to the modern conception of natural phenomena. Several interesting papers by local engineers will be read before the Section. One, by Mr. J. Watson, will describe the Bradford waterworks and the very fine reservoirs belonging to that system. In connection with this there will be an excursion on the Saturday to the reservoirs. A paper will be read by Prof. Hele-Shaw on the resistance of road vehicles to traction. A proposal will be made to appoint a committee of the Association to carry out an exhaustive series of experiments on road resistance. Much interest will no doubt be excited by the paper which is to be read by Mr. J. H. Glass on the coal and iron ore fields at Shansi and Honan, and railway construction in China. This paper will be illustrated by a number of lantern slides showing the Chinese methods of working these mineral deposits. In view of the great industry of Bradford, the paper by Prof. Beaumont on the application of photography to textile designing is likely to create great interest. In the department of electrical engineering there is a good programme of papers. The Small Screw Gauge Committee will submit a report descriptive of a series of experiments which have been carried out by Mr. Price in the engineering laboratory at University College, London ; and in connection with this, a paper will be read by Mr. O. P. Clements on screw threads used in cycle construction and for screws subject to vibration. Mr. A. Mallock will give an account of experiments he has made to determine the tractive force, resistance, and acceleration of electric trains. Mr. Aldridge's paper on the automobile for electric street traction will describe a novel process, by means of which, in certain circumstances, it is possible to organise a tram service without tram rails, and this paper will be illustrated by the cinematograph, showing an actual system at work.

ANTHROPOLOGY.

In the Anthropological Section, the President, Prof. J. Rhys, proposes to devote his opening address to "the prehistoric ethnology of the British Isles," a subject full of matter for discussion, on which he is entitled to speak with peculiar authority. Several important papers are expected in the department of anthropometry, especially a note by Dr. Beddoe on the "vagaries of the Cephalic Index," and a paper by Prof. Cunningham on the "Sacral Index." Mr. H. Ling Roth contributes a classification of various modes of ornamenting the skin, such as tattooing, cicatrisation, and the like. Dr. Haddon promises an account of his recent visit to Borneo, with special reference to the industries and daily life of the natives. A discussion is being arranged on the subject of "Animal Worships," with reference to the vexed question of the significance of totemism ; and Mr. David Boyle, of Toronto, will contribute a study of the phenomena of Neo-Paganism among the natives of certain parts of the Dominion of Canada. Among other archæological papers, special interest attaches to Mr. Arthur Evans's account of his recent discovery of tablets inscribed with an Ægean script, in the Mycenæan palace of Gnossus in Crete ; and to Mr. F. Ll. Griffith's discussion of the origin of the Egyptian hieroglyphic system. There will be papers, as usual, on objects of archæological interest in the neighbourhood of Bradford.

BOTANY.

Prof. Vines, the President of Section K (Botany), in his address on Thursday, September 6, will take as his subject the progress of botany in the nineteenth century. It has been arranged to hold a joint discussion with some of the members of the Geological Section on the conditions under which the forests of the Coal Period grew. The origin and manner of formation of Coal, the climatic and physical conditions which prevailed during the deposition of the Coal-measures, the most striking characteristics of the vegetation, and other questions will probably be dealt with. The local committee propose to form a small museum of specimens and photographs to illustrate the botany and geology of the Coal Period.

On Friday afternoon a semi-popular lecture, illustrated by lantern slides, will be delivered by Mr. Percy Groom, on "Plant-form in relation to nutrition."

Among the papers already promised, the following may be mentioned :—On the presence of seed-like organs in certain Palæozoic Lycopods, by Dr. D. H. Scott ; the origin of modern Cycads, by Mr. Worsdell ; the fertilisation of *Caltha palustris*, by Miss Thomas ; on a new type of transition from stem to root in seedlings, by Miss Sargant ; the anatomy of the stem of *Angiopteris evecta*, by Miss Shove ; the structure of the nucleolus, also a demonstration of the structure of the eye-spot and flagellum of *Euglena*, by Mr. Wager ; the biology and cytology of a new species of *Pythium*, by Dr. Trow ; the biology of *Acrospeira mirabilis*, by Mr. Biffen ; the histology and reproduction of the Laminariaceæ, and additional notes on the cytology of the reproductive cells in the Dictyotaceæ and Fucaceæ, by Mr. J. Lloyd Williams ; on the effect of salts on the CO_2 assimilation of *Ulva latissima*, by Mr. Arber ; on fungi found on the scale-insects of Ceylon, by Mr. Parkin ; the structure and affinities of *Dipteris conjugata*, with notes on the geological history of the Dipteridinæ, by Mr. Seward and Miss Dale.

RECORDING TELEPHONES.

NOW that the telephone has become, even in this country, an instrument of such universal commercial and general employment, the advantages of an apparatus that will satisfactorily record the messages transmitted through an ordinary telephone line are so strikingly apparent that it is unnecessary to enlarge upon them. That it should have been possible to construct such an apparatus has been evident since the invention of the phonograph. But the direct combination of the phonograph with the telephone, which seems so simple in theory, has presented difficulties in practice which up to the present have not been successfully overcome, and the phonograph of to-day, over twenty years since its invention, remains little more than a scientific toy, whereas its contemporary, the telephone, has become an almost indispensable adjunct of civilisation. It would

appear, however, that the problem of recording telephone messages is nearing a practical solution, for there have been quite recently put forward, under the names respectively of the "Telephonograph" and the "Telegraphone," two separate inventions of a recording telephone.

The first of these instruments—the "Telephonograph" —is the invention of Mr. E. O. Kumberg, and contains little that is novel in principle, being simply a combination of the phonograph with a loud-speaking telephone receiver, in which the inventor has sought by a suitable design of apparatus to diminish the distortion of voice which is usual with such an arrangement. The invention consists of a phonograph in which a loud-speaking telephone receiver is substituted in place of the ordinary diaphragm to which one speaks. The telephonic currents varying in the receiver set up vibrations in a soft iron diaphragm which is attached by a short stiff wire at its centre to a second diaphragm of mica. The centre of this mica diaphragm is connected by a link with the cutting style, which accordingly traces on the wax cylinder of the phonograph a record of the message transmitted through the telephone. The cylinder can then be subsequently used in connection with the speaking diaphragm of the phonograph to repeat the recorded message. Unfortunately, neither the telephone nor the phonograph is free from distortion, and the "Telephonograph" may be expected to possess in an enhanced degree the imperfection of each of its components ; from what we learn, it seems that Mr. Kumberg's invention is by no means perfect in articulation.

The second instrument which has been brought forward under the name of the "Telegraphone" is, we believe, entirely new in its principle, and if it realises but a part of what is claimed for it by its inventor represents a very great advance in telephony. This instrument is the invention of Herr Valdemar Poulsen, a Danish electrician, and is on view at the Paris Exhibition. It is briefly described in a note contributed by Herr Poulsen to the *Comptes rendus* for June 25, and somewhat more fully in an article which appears in the *Revue Générale des Sciences* for June 30.

It is, of course, perfectly well known that if a piece of steel be placed between the poles of an electromagnet which is excited by a current, a magnetic field is set up in the steel, the strength and direction of which depend upon the strength and direction of the current in the exciting coils of the electromagnet, and the magnetism thus induced in the steel is still retained by it when removed from the inducing magnetic field. This is the principle which Herr Poulsen has utilised in the construction of his new recording telephone. In place of the ordinary telephone receiver he uses a simple electromagnet, the current transmitted through the telephone line passing round the exciting coils of the magnet. When, therefore, one speaks into the transmitting instrument at the far end of the telephone line, the magnetic field due to the electromagnet will vary in strength and direction in accordance with the varying electric currents transmitted through the lines. Between the poles of the magnet is passed a steel wire or band, which is moved forward in the direction of its length at a uniform and rapid velocity. At each point of this wire there will be produced a magnetisation proportional to the current which was flowing through the coils of the electromagnet at the moment when that section of the wire was passing between its poles. There will thus be established in the steel wire a magnetic record of the telephonic message, and just as the varying electric currents have been utilised to produce in the wire a magnetisation varying from point to point along its length, so, by the converse process, may this magnetisation be employed to set up currents in a telephone.

receiver, and thus reproduce the original speech. It is only necessary to connect the coils of the electromagnet in series with a receiving telephone, and to cause the steel wire on which the magnetic record has been made to pass once again at the same speed and in the same direction between the poles of the magnet ; for the variation of the magnetic field which the wire produces as it moves along will generate currents proportional to the rate of variation of this field, and the telephone will respond in the same way, and with the same degree of accuracy, as did the receiving telephone in the early Bell combination of a pair of magneto-telephones prior to the employment of a microphone and a battery.

The diagrams in Figs. 1 and 2 show the arrangement of the apparatus. C is the electromagnet on which are the bobbins of wire B and B', which are connected either to the transmitting or receiving instrument according as it is desired to record a message or to listen to one already recorded. Between the poles P and P' of the electromagnet passes the steel wire F, which is wound in a helix over two drums, T and T', to which the ends of it are attached and which are driven by an electric motor,

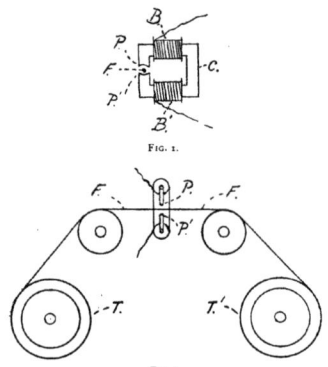

FIG. 1.

FIG. 2.

so that the wire winds off one drum on to the other. The drums rotate at such a speed as to give the wire a linear velocity between the poles of a metre per second. At the same time the magnet C is moved at right angles to the direction of motion of the wire, so that for one revolution of the drums the magnet moves a distance equal to the pitch of the helix.

The message thus recorded can be effaced, according to Herr Poulsen, by the simple process of passing the steel wire between the poles of the magnet when the latter is excited by a steady current from a battery. This operation establishes in the wire a uniform magnetic field at right angles to its length, and this field, so far from interfering with recording a future message, appears to be necessary before any such record can be made. But if it be desired to keep the record, this may be done instead, and the wire used over and over again to repeat the same message. A thousand repetitions can be made, it is said, without any diminution in loudness or distinctness.

If the recorded message is not sufficiently loud, it is possible with a comparatively simple arrangement to

greatly increase its loudness. For this purpose it is necessary to arrange a series of parallel steel wires or bands as shown in Fig. 3, all of which have been previously prepared for receiving a magnetic record by being passed between the poles of a magnet excited by a steady current. The wires are moved at a uniform rate. The first wire, 1, passes first between the poles of an electromagnet M, which is connected to the telephone line, and consequently receives a magnetic record of the transmitted message. The wire next passes between the poles of a second electromagnet, A, which is connected in series with a similar magnet A', between the poles of which the second wire, 2, passes. As the wire 1 passes between the poles of the magnet A currents are induced in the coils of this magnet, which, traversing also the coils of the companion magnet A', produce a magnetic record in the wire 2. A similar action occurs as the first wire passes between the poles of the magnets B, C, ... Z, which are connected in series with the magnets B', C', Z', so that there is established in each of the wires 2, 3, 4, ... n, similar magnetic records. The wires 2, 3, 4, ... n pass finally between the poles of a number of electromagnets, Ω_2, Ω_3, Ω_4, ... Ω_n, which are all joined in series with a telephone receiver, T. If the two magnets which are joined together in series, such as A, A', B, B', ... are arranged in the same perpendicular to the direction of the steel wires, and if all

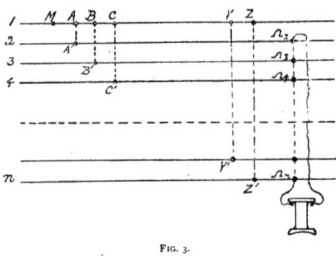

FIG. 3.

the wires are moved forward with the same velocity, then the magnetisations in the various wires—2, 3, 4, ... n— at points lying in the same perpendicular will be similar ; and since all these points will pass at the same instant between the poles of the magnets Ω_2, Ω_3, Ω_4, ... Ω_n, they will there superpose their effects, and the intensity of the sound in the telephone, T, will be increased in proportion to the number of wires. It seems possible, therefore, with this device to indefinitely increase the loudness of the received message, and with quite a simple arrangement to easily double or treble the intensity of the sound.

There is a second ingenious method by which it is suggested that an increase in loudness can be obtained. In this advantage is taken of the fact, already pointed out, that the strength of the currents induced in the coils of the receiving electromagnet depends on the rate of variation of the magnetic field : it is therefore possible to increase the strength of the currents induced by a field altering from one given strength to another by diminishing the time in which such alteration occurs. If, therefore, we have a steel band on which a magnetic record has been made, we may increase the strength of the currents it will induce in the electromagnet connected to the receiving telephone—and consequently the loudness of the repeated message—by simply increasing the speed at which the band passes between the poles of that electromagnet. But in order to obtain the amplification of the sound in this manner, it is apparent that the band must be moved more quickly when used to repeat the message than it was when used to record it, and thus the increased loudness will be accompanied by an increased quickness, and probably at a sacrifice of clearness due to the alteration in pitch. For if the speed of the band when used for recording be increased, the effect will be merely to spread out the message along the wire, as the intensity of the magnetisation set up in the wire depends only on the strength of the currents transmitted through the telephone, and this strength is determined by the loudness and quickness with which the message is spoken at the transmitting end

The invention of Herr Poulsen may be looked upon as the invention of a magnetic phonograph, and must be regarded as an extremely ingenious and beautiful attempt at the solution of the problem of recording telephonic messages. It possesses many advantages over the combination of the telephone and the wax-cylinder phonograph, especially in the fact that the recording is effected by the immaterial agency of magnetism, and not by the mechanical agency of a style writing on wax, so that the imperfections in the articulation due to mechanical causes should be entirely absent. The method of increasing the loudness by the use of a number of parallel bands appears exceedingly simple, and offers a possible means of making a telephonic relay, and thereby increasing the limits of distance to which sound can be transmitted ; it is a method which might be imitated with the ordinary phonograph, by causing the message recorded on the wax cylinder to be repeated to one or more other cylinders, and finally making all repeat their records simultaneously ; but in this case the practical difficulties would be very much greater. That Herr Poulsen's invention is still only in an experimental stage may be gathered from the fact that though the instrument itself is on view at Paris, it has been found too difficult at the Exhibition to make the necessary adjustments to exhibit it in operation publicly ; but we await with interest its further development, for the introduction of a trustworthy recording telephone would be a benefit to the public, for which it is to be hoped they will not have long to wait.

NOTES.

THE French Minister of War has invited the Paris Academy of Sciences to advise as to the precautions to be adopted in selecting and planting trees in the neighbourhood of powder magazines, in order to secure the best protection from lightning.

THE Chancellor of the German Empire has issued an ordinance to the effect that the Réaumur thermometer will not be admitted to official control after January 1, 1901. This will lead to the exclusive use of the centigrade thermometer in Germany.

THE International Congress of Physics, held in Paris last week, appears to have been a complete success, more than a thousand members, including leading physicists of many nationalities, having been obtained. Lord Kelvin received a grand ovation at the opening meeting, and was nominated honorary president of the Congress.

M. OUSTALET and M. Depousarques are the two candidates who have been nominated by the Paris Academy of Sciences for the chair of zoology in the Muséum d'Histoire naturelle, rendered vacant by the death of Prof. Milne-Edwards. The appointment rests with the Minister of Public Instruction.

THE *Athenaeum* announces that Prof. Virchow has been elected an honorary member of the Mathematical and Natural

Science Section of the Vienna Akademie der Wissenschaften, while Prof. Klein, of Göttingen, has been appointed corresponding member of the same section.

THE attention of persons interested in zoological gardens and keeping captive animals should be directed to the passing of the "Act for the Prevention of Cruelty to Wild Animals" (63 and 64 Vict., Ch. 33), which has just become law. This Act extends the provisions of the "Cruelty to Animals Acts 1849 and 1854" (which related to domestic animals only) to all birds, fishes and reptiles not included in that Act. By Sect. 2 of the new Act, "Any person shall be guilty of an offence who, whilst an animal is in captivity or close confinement, or is maimed, pinioned, or subjected to any appliance or contrivance for the purpose of hindering or preventing its escape from such captivity or confinement, shall, by wantonly or unreasonably doing or omitting any act, cause or permit to be caused any unnecessary suffering to such animal ; or cruelly abuse, infuriate, tease, or terrify it, or permit it to be so treated." Any person committing such an offence may be proceeded against under the Summary Jurisdiction Acts, and on conviction is liable to be imprisoned for three months or fined £5.

A REUTER telegram from St. Petersburg states that the Imperial Academy of Science has just received news from the Russian expedition at Spitsbergen stating that in the month of September last the members of the expedition in question had erected, at Horn Sound, observatories for conducting meteorological, magnetic, astronomical and astrophysical researches. On October 20 the sun disappeared for four months, and at the end of October absolute and continuous night set in. The members of the expedition applied themselves constantly to scientific observations after November 17. On February 22 the sun was seen again for the first time. On June 5 and 8 the first boats arrived, ending the complete isolation of the expedition, which had lasted for nine months.

THE annual meeting of the French Association for the Advancement of Science was recently held at Paris. General Sebert, the president, delivered an address on the progress of mechanical industries and the means of developing them. In the course of his remarks he alluded to the value of technical education and research as factors in national advances. "It is noteworthy," he said, "that progress in mechanical industries has always coincided with the development of technical education in the countries in which the industries are carried on. The most rapid progress takes place in the countries where institutions for experiment and research are most numerous. Wherever research laboratories have been established to permit the study of the best conditions of invention, there the most marked advances first take place." The Association's grants for scientific purposes, made at the recent meeting, amounted to 21,241 francs, or about 850l.

MR. A. R. HUNT, writing from Torquay with reference to our note (P. 322) on the rumbling sounds heard at Bognor and Torquay on July 18, says : "I happened to go into my garden a few minutes after ten on the date named, and was at once conscious of a very unusual pulsating rumble. My first idea was earthquake, but the sound came steadily from one point, roughly south-east; and at last died away into distinct taps." The observation is interesting in showing how far the individual reports can be heard (see p. 378).

MR. J. STIRLING, Government geologist of Victoria, New South Wales, is at present in London as the mining representative of the Victorian Government, and during his stay here proposes to address some of the scientific, professional and mining organisations of the United Kingdom on matters of origina research in Australasia. He will give an address at the Convention of Mining Institutes of Cornwall during August, and subsequently at Manchester, Bristol and other centres.

SOME interesting information as to the actual experience of nations who have adopted the metric system is given in a number of reports from Her Majesty's consular and other officers in Europe, which have just been brought together and published by the Foreign Office. H.M. representatives in twenty-two States were asked to give information upon the following points. (1) The ease or difficulty with which the change of systems was made, the manner of introduction of the metric system, and the time occupied in making the change; (2) How far the metric system is satisfactory in its practical operation, and whether there is any desire to return to former systems; (3) As to the effect the adoption of the metric system has had upon the commerce of the nations adopting it. The answers received to these questions go to show that the best way to introduce the metric system is to make it compulsory after a specified period. The change from the old to the new system is slow in country districts, but as new generations come on familiar with the metric measures the old measures gradually drop out of use. In Turkey, the difficulties of enforcing the system upon an ignorant and illiterate people have proved insurmountable; but in the majority of States from which information has been received, the system is becoming more extensively used every day. Once the system has been adopted there is no desire to return to the old measures, and the effect upon commerce is always beneficial. In fact, the reports greatly strengthen the position of those who urge that the metric system should be adopted in England, if only for the sake of British trade.

IT was mentioned in these columns some time ago that the wire fencing of great sheep farms in some parts of Australia was used as telephone wires. A recent report from H.M. Consul at Philadelphia states that this system of communication is being employed by farmers between the towns of Anderson, Pendleton and Ingalls, in Indiana. The top wire of a barb-wire fence is used as the conductor, the continuity of the line being assured by special devices at highways and railway crossings. The line is fourteen miles in length with five stations, two at Anderson, two in Pendleton, and one at Ingalls. Local farmers state that they have used the "fence-line" to converse with friends eight miles distant, and this at a time when the fence posts were still saturated with the morning dew, a condition under which the line is supposed to work with least satisfaction. It is stated that the line has been such a practical success that the farmers of the neighbourhood are organising companies for the purpose of placing themselves in telephonic communication throughout the whole district.

AN excellent article dealing with the photographic side of the suggestions as to analytical portraiture made by Mr. F. Galton in NATURE of August 2 appears in *Photography* of August 9. Illustrations are given of results obtained by combining two portraits of a single person in the same pose, but having different expressions during the two exposures. In one picture the sitter has a normal expression; in the other he is smiling. A transparency was made from the normal negative; and when this positive and its negative were superimposed they neutralised one another. But by placing the positive of the normal expression of face upon the negative of the smiling expression, the two do not, of course, exactly obliterate one another. Certain parts of the features are common to both, and these disappear when the different positive and negative are superimposed, leaving only portions which represent the smile of the sitter's features. In a similar way, by superimposing the positive of a glum portrait upon the negative of a normal

expression, it is possible to obtain differences representing an individual's glumness. Readers of "Alice in Wonderland" will remember that the Cheshire cat gradually disappeared and left only its grin behind. This facetious idea has now been realised, for as our enterprising contemporary points out, Mr. Galton's analytical portraiture shows how the factors of a grin or a scowl can actually be discriminated, so that a grin can be obtained without the face upon which it appeared.

A REUTER telegram from Liverpool, dated August 12, makes the following announcement :—The second malarial expedition of the Liverpool School of Tropical Medicine has just wired home from Bonny, in Nigeria, news of a most important discovery—viz. that the parasite which causes elephantiasis has, like that which causes malaria, been found in the proboscis of the mosquito. Oddly enough, the same discovery has just been simultaneously made by Dr. Low in England in mosquitoes brought from Australia, and by Captain James in India. Elephantiasis is a disease which causes hideous deformity in thousands, or rather millions, of natives in tropical countries, and sometimes in European residents. It is due to a small worm which lives in the lymphatic vessels and occludes them. The fact that this worm can live also in the mosquito has long been known, but the discovery of it in the insect's proboscis shows that it enters the human body by the bites of these pests. Europeans in the tropics are indebted to mosquitoes, not only for much discomfort, but for two dread maladies—malaria and elephantiasis; and it is high time that the authorities should begin seriously to consider Major Ross's advice to destroy these insects or their breeding-places wherever practicable.

THE medical papers contain detailed reports of the Thirteenth International Congress of Medicine, which was held in Paris at the beginning of this month under the presidency of Prof. O. M. Lannelongue. Among the representatives of Great Britain were Sir William MacCormac, Sir T. Lauder Brunton, Sir J. Burdon Sanderson, Sir Dyce Duckworth, Sir Felix Semon, and Prof. Simpson. A banquet in honour of Lord Lister was arranged by Prof. Charles Richet and the "Scientia" social society. Speeches expressing admiration of Lord Lister's work, and describing the influence it has had upon various branches of medical and surgical science, were made by Prof. Richet, Dr. Bouchard, Prof. Guyon. Dr. Lucas Championnière, and Dr. Pinard. In his reply, Lord Lister said he regarded the banquet as being in honour of the noble science of surgery and the Royal Society of London, of which he was the president. It showed that the scientific world knew nothing of the misunderstandings between peoples of different nationalities, and that men of science had mutual respect for one another at all times. Lord Lister added : "I have often said, and it gives me pleasure to repeat it this evening, that I owe much to Pasteur. It is true that I was passionately fond of physiology and surgery. The nature of inflammation was the subject of my first investigations. As a surgeon, I deplored the disastrous results which often followed the most skilful operations, and I saw, what many others had doubtless remarked before me, that the most important troubles of a wound were due to changes in the tissues of the body after the operation, and had an external origin. But all my efforts to avoid these complications were unavailing until Pasteur threw a new and strong light upon the subject, and indicated a possible course of action which I have done my best to follow. That is all. If my efforts have been followed by such beneficial results as have been generously described by speakers this evening, the success must, in a great measure, be ascribed to the fortunate chance of my time."

THE report of the Zoological Gardens of Ghizeh, near Cairo, for the year 1899, gives a good account of the progress of this Institution, which, under the rule of its present director, Captain

Stanley Flower, has become a popular place of resort for the European visitors to Egypt, as well as for the Cairenes. The receipts in 1899 were 3033*l*., of which 968*l*. were for gate-entrances, and the expenditure was 3019*l*. The list of donors includes many well-known names, amongst which we see those of Sir William Garstin, Prince Omar Tousson, Sir F. Wingate and Lord Kitchener. The Government of India presented an elephant. Various new buildings were erected, and others were reconstructed in 1899. The number of animals in the collection on October 1 of that year was 473, against 270 at the corresponding date in 1898. A list of wild birds that inhabit the Ghizeh Gardens, and in many cases breed there, enumerates nineteen species, amongst which is the European song-thrush (*Turdus musicus*). Two proboscis monkeys (*Nasalis larvatus*), presented by the Government of the Netherlands East Indies, unfortunately did not live long. We are informed that since the report was issued Captain Flower has succeeded in bringing to the Ghizeh Gardens from the Sudan a fine young giraffe, presented by the Sirdar.

WRITING from Mashonaland in May, Mr. G. A. K. Marshall raises, in the August number of the *Entomologist*, what appears to be a pertinent question with regard to mosquitoes and malaria. If it be admitted, he observes, that malaria can only be carried by mosquitoes of the genus Anopheles, and that these insects can only acquire the microbes from malarially-infected man, "then we are logically bound to accept the conclusion that if a man, or party of men, free from malarial poison, should penetrate from a healthy area into an unhealthy but uninhabited region, it would be impossible for them to contract fever, however much they might be bitten by mosquitoes. Further, it follows that all uninhabited regions, even of comparatively small size (seeing that the range of individual species of Anopheles is apparently very limited in extent) must be entirely devoid of malaria, even though they may be full of swamps and teem with mosquitoes." Such conclusions are, however, contrary to experience, and if the writer's premises be correct, his difficulty requires an explanation at the hands of specialists.

THE Walcott collection of Hymenoptera, now in the Cambridge University Museum, has yielded to the researches of Mr. R. C. L. Perkins (*Entomologists' Monthly Magazine* for August) a species (*Odynerus tomentosus*) new to the British fauna. Considering that the greater part of the collection was made in the first half of the century, it is not a little remarkable that the species should have escaped notice so long.

THE large scale on which they do things in America has become a proverb. An instance is afforded by Mr. J. B. Smith's description of one hundred new species of moths of the family Noctuidæ in vol. xxii. of the *Proceedings* of the U.S. Museum.

To vol. xxix. No. 13 of the *Proceedings* of the Boston Society of Natural History Dr. H. S. Pratt contributes an important paper on the embryological history of the so-called imaginal discs of the sheep-tick (*Melophagus ovinus*). For the benefit of our non-entomological readers it may be mentioned that these imaginal discs, or folds, are structures in the larva and pupa which do not participate in the general breakdown of tissue at the periodical changes, but undergo continuous development into the corresponding parts of the perfect insect. Hitherto, the author says, these structures have been studied only in the larval and pupal stages; and he for the first time describes their origin and early stages of growth.

FOUR out of the nine papers in Part i. of the *Proceedings* of the Philadelphia Academy for 1900 are devoted to the land and fresh-water molluscs of America. In the first of these Mr.

C. T. Simpson describes a number of new or unfigured river mussels (Unionidæ); the second, by Mr. W. H. Dall, treats of the land-shells of some of the Pacific Islands, more especially those of the Galapagos and Cocos groups; in the third, Mr. H. A. Pilsbry discusses the anatomy of the helicoid genus Ashmunella, and in the fourth the molluscs of the Great Smoky Mountains. This last communication is perhaps the one of most general interest, since the author is of opinion that the cleft in the Appalachian chain formed by the valley of eastern Tennessee indicates the boundary between two zoögeographic provinces. The lists of terrestrial molluscs given by him as respectively characteristic of the eastern and western divisions of this portion of the chain seem to bear out his contention as to the existence of two distinct faunas.

THE sixth of the series of physico-mathematical handbooks published by Messrs. Carré and Naud, of Paris, under the title of "Scientia," is a small treatise by M. Fred. Wallerent on crystalline groups and their optical properties. As an introduction to modern crystallography the little volume should be of much use to those interested in other branches of science who are desirous of acquiring a general knowledge of the history and fundamental principles of the subject, and who do not possess the spare time for mastering a larger treatise.

IN a short note contributed to the *Atti del R. Istituto d'Incoraggiamento* (Naples), Prof. E. Semmola discusses the state of our knowledge of the variations of the electrical potential of the air with the altitude. In reference to Le Cadet's result that the potential decreases with the altitude, Prof. Semmola points out that the late Prof. L. Palmieri, in conjunction with himself, had established a similar property previously. Le Cadet found that the potential decreased from 150 to 44 volts in the first kilometre of altitude, and deduced that the potential decreased much less rapidly at greater altitudes. But Semmola thinks that the high potential found at the surface of the earth was at any rate in part due to the obscurity of the superincumbent air.

A SHORT note on the reflection of light in the neighbourhood of the critical angle is given by Mr. J. G. Coffin in the *Technology Quarterly*, the object being to examine more fully than is done in most text-books the consequences of applying Fresnel's formulæ to refraction from a denser to a rarer medium. Tables are calculated by the author and Prof. Pickering, showing the percentages of light reflected at different incidences in passing from a rare to a dense medium and *vice versâ*, and the results are exhibited graphically by curves. The paper thus contains an amplification of the superficial information contained in the majority of treatises on optics.

IN the *Journal of Proceedings* of the Institution of Electrical Engineers, xxix. 142, 1900, Mr. Alexander Russell discusses the question how condenser and choking-coil currents vary with the shape of the wave of the applied electromotive force. Various forms of wave being considered, the author finds that the sine curve wave produces the least effective current when applied to a condenser, and the largest magnetising current when applied to a choking-coil. Similar results are established for the symmetric wave in the case of a family of waves of equal height. The subject is sufficiently interesting to make us wish for a fuller mathematical investigation, Mr. Russell's note being a mere statement of results.

SOME tests of fire retardent materials are described by Mr. Charles L. Norton in the *Technology Quarterly*, xiii. 2, for June 1900. The tests were made on October 5, 1899, and February 3, 1900, by setting up small buildings previously constructed in the Massachusetts Institute of Technology and building a fire of wood and oil inside. Observations of the progress of the fire and of the subsequent state of the wooden backing led to a number of interesting conclusions as to the value of the protection afforded by various retardent materials. Among these we note the comparative value of a wooden and metallic lath ; the necessity of applying fire retardent material in at least two thicknesses so as to break joints ; the immense superiority of three-ply over two-ply doors; the advantage of the Atkinson composite door as being more gas-tight than a wooden door ; the fire-resisting qualities of three-inch plank as compared with one-inch boards, or lath and plaster ; the excellency of Mississippi wire glass ; and the satisfactory performance of " King's Windsor " cement and " Adamant " plaster.

AN important development of the electron theory has been carried out by Robert Lang in his article on atomic magnetism contributed to the *Annalen der Physik* (No. 7). It may now be said that the phenomena of magnetism have at last been successfully reduced to those of electricity. We know from the work of Thomson and of Drude that an electric current in a wire consists of a stream of very small particles called electrons. These electrons are formed by the splitting up of the metallic atoms into a larger positive and a smaller negative portion. The positive electrons, under the influence of an electromotive force, travel in one direction along the wire, with a velocity of about 1 cm. per second. The negative electrons travel in the opposite direction with the same charge, but with a smaller velocity. The masses are in the ratio of about 9 : 1. Now according to Lang, the negative electrons revolve round the heavier positive electrons in a magnetised metal, like a planet round the sun, and the electric convection currents thus produced are nothing more or less but Ampère's "elementary molecular currents." Lang calculates the speed of the electrons and the diameter of their orbit. The speed is that of light, and the figures obtained lead to conclusions in close agreement with known facts.

AN interesting article, entitled "Cartographie de la Caverne Mammoth," is contributed by Dr. H. C Hovey to the *Bulletin de la Société de Spéléologie*, tome v. 1899. The author gives a short history of the attempts to map the celebrated Mammoth Cave, and points out that, owing to objections on the part of the proprietors, the scientific investigation of these caverns is still incomplete. The paper is accompanied by reproductions of the map by Hovey and Call, and that by C. R. Blackall for purposes of comparison.

A VALUABLE addition to our knowledge of the cretaceous geology of Saxony is furnished by Dr. W. Petrascheck in a paper published in the *Abhandl. der Naturwiss. Gesellsch. Isis* (Dresden, 1900). The author seeks to trace the change of facies developed at various horizons in the cretaceous rocks of this area when followed laterally. He explains, as far as possible, the modifications in the character of the fossil fauna which accompany the changes in petrographical facies. The region discussed comprises the neighbourhood of Dresden and the well known " Saxon Switzerland."

THE August number of the *Journal* of the Chemical Society contains the Friedel Memorial Lecture, delivered before the Society by Prof. J. M. Crafts.

THE second English edition of Prof. Ostwald's "Scientific Foundations of Analytical Chemistry," translated from the second German edition by Dr. George M'Gowan, has been published by Messrs. Macmillan and Co. Since the original work was published in 1894, the principles expounded in it have been steadily gaining acceptance, but, so far as we are aware, no English text-book of chemical analysis has appeared in which he analytical methods and reactions of the laboratory are consistently explained in terms of the theory of ions instead of

being represented by the ideal equation-formulæ. As Prof. Ostwald states, the general standpoint of analytical chemistry has undergone but little change ; nevertheless, the newer ideas are gradually being applied to laboratory work by lecturers and demonstrators who are in touch with modern chemical theory. The new edition just published will be the means of extending the knowledge of the fundamental principles underlying chemical processes, and will be a source of inspiration to teachers who wish to make analytical chemistry a science as well as an art.

A SIMPLE method of preparing free hydroxylamine is given in a recent number of the *Annalen* (311, 117) by Dr. R. Uhlenhuth. When the phosphate of the base is heated gently under reduced pressure, the base distils over in a state of such purity that the distillate solidifies on placing the receiver into melting ice.

THOUGH the need for a universal standard table of atomic weights is recognised by all chemists, the question whether it shall be constructed upon a basis of $O = 16$ or $H = 1$ has yet to be decided. The *Chemical News* publishes a letter from Profs. Bredt, Erdmann, Fischer, Volhard, Winkler and Wislicenus, members of the International Committee on Atomic Weights, upon this point. It is remarked that if cogent reasons necessitate an alteration of the standard of atomic weights, it would be better to start from an element of which the weight can be conveniently ascertained, such an element, for example, as silver or iodine, which also serves as a practical starting-point in consideration of the sharpness of its reactions in numerous analytical operations. But in the opinion of the writers such cogent reasons for an alteration do not present themselves, for the ratio of hydrogen to oxygen has been established with an exactness which fully suffices for all practical purposes. It is felt that the time for an unchangeable table of atomic weights has not yet come ; for each succeeding year brings corrections in the atomic weights of the rarer elements, and at the same time speculations as to their simple or compound nature. Opinions are therefore invited upon the following questions :—(1) Shall the unity of hydrogen be retained as the standard for reckoning atomic weights ? (2) Shall the atomic weights be given approximately with two decimal places in which the uncertain figures can be recognised by the type ? (3) Shall the International Atomic Weight Commission have the current table of atomic weights edited on this basis ? Communications should be sent to Herr Prof. J. Volhard, Mühlpforte 1, Halle-a-S.

THE additions to the Zoological Society's Gardens during the past week include a Lioness (*Felis leo*) from South Africa, presented by the Right Hon. Cecil J. Rhodes ; a Black-backed Jackal (*Canis mesomelas*), a Leopard Tortoise (*Testudo pardalis*), a Puff Adder (*Bitis arietans*) from South Africa, presented by Mr. J. E. Matcham ; a Grey Ichneumon (*Herpestes griseus*) from India, presented by Mr. W. A. Gillett ; a Blue and Yellow Macaw (*Ara ararauna*), a Red and Yellow Macaw (*Ara chloroptera*) from South America, presented by Captain G. H. Arnot ; a Chinese Quail (*Coturnix chinensis*) from China, two Asiatic Quails (*Perdicula asiatica*) from India, two Sparrow Hawks (*Accipiter nisus*), British, presented by Mr. D. Seth-Smith ; a Common Quail (*Coturnix communis*), British, presented by Miss F. E. Burt ; a Lesser White-nosed Monkey (*Cercopithecus petaurista*) from West Africa, a Polar Bear (*Ursus maritimus*, ♀) from the Polar Regions, two Black-headed Caiques (*Caica melanocephala*) from Demerara, a Smooth-headed Capuchin (*Cebus monachus*) from South-east Brazil, a Pleurodele Newt (*Molge walti*), a Leopardine Snake (*Coluber leopardinus*), a Vivacious Snake (*Tarbophis fallax*), six European Pond Tortoises (*Emys orbicularis*), South European ; two Egyptian Mastigures (*Uromastix spinipes*), an Algerian Tortoise (*Testudo ibera*) from North Africa, four Alligator Terrapins

(*Chelydra serpentina*) from North America, a Leopard Tortoise (*Testudo pardalis*) from South America, two Argentine Tortoises (*Testudo argentina*) from the Argentine Republic, deposited ; a Gold Pheasant (*Thaumalea picta*, ♂) from China, two Little Bitterns (*Ardetta minuta*), European, purchased ; a Burrhel Wild Sheep (*Ovis burrhel*), born in the Gardens.

OUR ASTRONOMICAL COLUMN

COMET BORRELLY-BROOKS (1900 *b*).—The following elements and ephemeris are furnished by Herr J. Möller in the *Astronomische Nachrichten* (Bd. 153. No. 3654).

Elements.

T = 1900 Aug. 3·298 Berlin Mean Time.

$$\begin{aligned} \omega &= 12\ 30\cdot2 \\ \Omega &= 328\ \ 1\cdot8 \\ i &= 62\ 35\cdot6 \end{aligned} \Bigg\} 1900\cdot0$$

log *q* = 0·00636

Ephemeris for 12h. *Berlin Mean Time.*

1900.		R.A.			Decl.		Br.
		h. m. s.					
Aug. 16	...	3 50 8	...	+75 55·7	...		0·63
17	...	4 0 12	...	77 36·3	...		60
18	...	4 12 37	...	79 10·3	...		56
19	...	4 28 4	...	80 37·7	...		53
20	...	4 47 52	...	81 57·7	...		50
21	...	5 13 26	...	83 9·5	...		48
22	...	5 47 9	...	84 11·4	...		45
23	...	6 31 8	...	85 0·8	...		42
24	...	7 26 5	...	85 34·6	...		40
25	...	8 28 54	...	+85 49·3	...		0·38

EPHEMERIS FOR OBSERVATIONS OF EROS.—The following is a continuation of co-ordinates computed by Herr F. Ristenpart (*Astronomische Nachrichten*, Bd. 152, No. 3643).

Ephemeris for 12h. *Berlin Mean Time.*

1900.		R.A.			Decl.	
		h. m. s.				
Aug. 16	...	2 0 29·40	...	+27 56 22·5		
18	...	3 19·51	...	28 38 11·6		
20	...	6 6·71	...	29 20 17·4		
22	...	8 50·73	...	30 2 40·0		
24	...	11 31·33	...	30 45 19·5		
26	...	14 8·25	...	31 28 16·1		
28	...	16 41·22	...	32 11 29·5		
30	...	2 19 10·01	...	32 54 59·5		

THE ASTROGRAPHIC CHART CONFERENCE.—The fourth meeting of the International Committee for directing the photographic delineation of the sky has recently been held in Paris, commencing July 19. The first matter taken in hand was the appointment of a sub-committee of nine astronomers to draw up a scheme for the systematic observation of Eros during the coming opposition, for determinations of solar parallax. The reports from the co-operating observatories show that in fifteen of them the work is being vigorously pushed forward ; unfortunately, in the remaining three, Rio de Janeiro, La Plata and Santiago (Chili), the work has entirely fallen through.

Dr. Thome, of the Cordoba Observatory, has been enabled, by the generosity of the Argentine Government, to volunteer for the work assigned to La Plata (- 24° to - 31°), and M. Enrique Legrand stated that he had induced his Government to found an observatory near Monte Video (Uruguay) to carry out the zone (- 17° to - 23°) allotted to Santiago. It was also suggested that the new observatory at Perth, West Australia, might possibly carry out the work on the remaining zone (- 32° to - 40°).

Another important item of the discussion was the advisability of publishing the rectangular co-ordinates of the stars as measured, with, of course, the constants of each plate, or delaying the work until these could be transferred to equatorial co-ordinates. It was considered that in the near future the absolute positions of the comparison stars would be much more accurately known than at present. The only drawback to this scheme is that Dr. Scheiner, of Potsdam, has already started the publication of the catalogue giving R.A. and Decl. of the stars.

In connection with the assignation of photographic magnitudes, it appeared to be generally believed that the estimation of diameters by means of a scale is a surer plan than measurement with a micrometer for this particular branch of work, but as definite ruling was given on this point.

The original plan agreed to in 1896 for taking the chart plates with three exposures of 30m. each has not been followed at all the observatories, and it was resolved at this meeting that in future the method of taking the chart plates shall be decided by the individual directors. In the reproduction of these chart plates, it is unlikely that uniformity will be secured ; the French observatories have made enlarged copies by heliogravure, but as each observatory would have to expend some 10,000*l.* to do this, the actual method of reproduction is left unsettled.

DETERMINATION OF SOLAR PARALLAX.—A circular has been issued by the special committee appointed by the International Astrophotographic Conference held recently at Paris containing the resolutions passed for systematising the work to be done at all the world's observatories during the coming autumn and winter, when it is hoped, by means of observations of the minor planet Eros, to determine the parallax and distance of the sun with a degree of accuracy previously unattainable. The following is a summary of the suggestions adopted :—

(1) That the determination of parallax of Eros be made by micrometric, heliometric and photometric measurements. (*a*) By observations of the planet east and west of the meridian at the same observatory. (*b*) By the co-operation of the observatories of Europe and North America. (*c*) By the co-operation of the observatories of the northern and southern hemispheres.

(2) During the period of parallax observations the diurnal movement of Eros should be determined as accurately as possible by heliometer, micrometer and photography.

(3) (*a*) Observers determining the parallax in right ascension should make measures each night and morning, providing by all favourable circumstances to operate with as large hour angles as possible. (*b*) Observers finding parallax by difference of declination in northern and southern hemispheres, should arrange that the mean instants of observation do not vary much from the meridian passage of the planet at the southern station.

(4) It is necessary that special series of photographs be taken of the region traversed by Eros, in order to furnish accurate determinations of the positions of comparison stars.

As the varying atmospheric conditions will play an exceedingly important part in the observations, particularly those away from the meridian, MM. André and Prosper Henry have been asked to prepare suggestions for eliminating these difficulties.

At the time of writing, the following observatories have signified their intention of helping with the scheme :—Algiers, Athens, Bamberg, Bordeaux, Cambridge (England), Cambridge (U.S.), Cape of Good Hope, Catania, Cordoba, Chicago (Yerkes), Edinburgh, Greenwich, Heidelburg, Leyden, Leipzig, Lyons, Marseilles, Minneapolis (U.S.), Mount Hamilton (Lick), Nice, Potsdam, Rome, San Fernando, Strassburg, Tacuboya, Toulouse, Upsala, Vienna (Ottahring), Vienna (Wahring), Washington.

THE DISTANCE TO WHICH THE FIRING OF HEAVY GUNS IS HEARD.

IN a discussion which took place in NATURE some time ago on the so-called "Barisâl Guns" and other mysterious sounds, Prof. Hughes suggested that it would be desirable to ascertain how far the firing of guns can be heard (vol. liii. p. 31). In connection with another subject, that of spurious earthquakes (see NATURE, vol. lx. pp. 139-141), I have for some time been collecting notes on this point, and I propose here to describe some of the facts obtained, chiefly with regard to the great naval review at Spithead on June 26, 1897, and the operations of the French fleet at Cherbourg on July 18, 1900.

I will mention first a few cases referring to more or less isolated observations of the reports of distant guns. The firing during the battle of Camperdown on October 11, 1797, is said to have been heard in Hull, the distance between the two places being more than 200 miles. A gentleman, formerly resident at Kertch in the Crimea, informs me that he has heard the sound of the guns fired at Sebastopol, distant 158 miles. During the American Civil War, the roar of the guns at the battles of Malvern Hill and Manassas (or Bull Run) was perceptible at

Lexington in Virginia, the distances being about 123 and 125 miles respectively (NATURE, vol. liii. p. 296). When the *Alabama* was sunk nine miles off Cherbourg on the morning of Sunday, June 19, 1864, the sound of the guns was heard in Jersey, at Clyst St. George, near Exeter (108 miles from Cherbourg), and at Brent Tor, near Bridgwater (about 125 miles). The great naval review at Spithead on July 17, 1867, was held during rough, boisterous weather; but the noise of the guns is said to have been heard at Exeter (105 miles), Morebath, near Tiverton (105 miles), Great Malvern (107 miles), and Castle Frome in Herefordshire (110 miles). In all the above cases the sound was, of course, the aggregate of that of many guns of different sizes fired simultaneously. But, in naval reviews, the charge is very much less than in actual warfare; a 6-inch gun, for instance, would fire a blank charge of 7 lbs., whereas the service charge for the same gun would be 48 lbs. fired with shot.

With regard to the distance to which the report of a single gun can be heard, I have very little information. A 110-ton gun fired at Woolwich made a window shake at Chignall St. James (24 miles), and was heard at Witham (32 miles) as a rumbling sound which seemed to deafen the observer slightly (NATURE, vol. xli. p. 369). Time-guns at Bombay have been often heard at the northern Mahim, distant more than 50 miles (vol. lvi. p. 223). The reports of the heavy guns at the battle of Lexington, mentioned above, could be easily distinguished at Malvern Hill from those of the smaller weapons; and a similar observation is recorded below. The subject is evidently one on which useful contributions to our knowledge might be made by residents near the south coast of England.

Naval Review at Spithead on June 26, 1897.

Shortly before the great naval review held in honour of the Queen's Diamond Jubilee, I wrote to the principal London newspapers and to several published in the south of England, and I have to thank the editors of these papers, and the ladies and gentlemen who replied to my inquiries, for the help they have kindly given me. The points to which I directed attention were the times at which the reports were heard, whether the air-vibrations were strong enough to make windows rattle, the direction from which the sound appeared to come, and the direction of the wind.

The fleet collected on this occasion consisted of 165 vessels of war of all classes arranged in five lines about six miles in length. The position of the flag-ship (H.M.S. *Renown*) was about two miles N. 20° E. of Ryde; and the distances given below are all measured from this point. As the Royal yacht entered the lines immediately after 2 p.m., the first shot was fired from the *Renown*, and was taken up by other ships in turn, each firing a Royal salute of twenty-one guns. "The heaviest gun employed," I am informed by the Secretary of the Admiralty, "was probably a 6-inch breech-loading gun, firing a blank charge of 7 lbs."; but others of different sizes were also used. It produced at first a dull crackling noise, according to a correspondent on H.M.S. *Sanspareil*, but, as ship after ship took up the salute, the firing grew more animated and the roll of the guns louder; until, after about five minutes, the report of the last gun died away.

The atmospheric conditions were fairly favourable for the propagation of the sound. Light, but variable breezes, generally between north-east and south-east, prevailed over most of the south of England. The thunderstorms which occurred on that day followed the salute in most places, but nearly all my correspondents (several being retired military officers) agreed that the sound of the guns could be readily distinguished from that of thunder.

In many of the records which I have received, the time is given so roughly that it is difficult to feel confident that they refer to the salute in question, and in several it is omitted altogether. Under the former heading come records from Honiton (90 miles from Spithead) and Shebbear, near Torrington (135 miles); and under the latter from near Rickmansworth (67 miles) and Great Malvern (107 miles). Excluding all such cases, the number of records is reduced to twenty, from nineteen places.

At very few of these places, and at none more distant than about 28 miles, were the vibrations strong enough to shake windows. Distinct reports were heard at the beginning and end of the salute as far as Farnham (34 miles), otherwise the sound was a dull, continuous roar, with occasional booms from the heavier guns. The sound was heard to the east as far as Framfield (57½ miles), to the north-east at Wimbledon (62

miles), to the north at Bloxham Green, near Banbury (88 miles), and to the west at Wellington in Somerset (93 miles). These are more or less isolated places, but there is a fairly continuous series of observations in a north-westerly direction, extending to Melksham (61 miles), Monkton Farleigh, near Bradford-on-Avon (67 miles), Bath (two observations, 69 miles), and Weston, near Bath (71 miles).

In the evening the fleet was illuminated, and a final Royal salute, similar to that at 2 p.m., was fired on the return of the Prince of Wales shortly after 11 p.m. I have only two accounts which may refer to this salute, one from Cosham in Hampshire at 11.30 p.m., the other from Ashburton in Devonshire (116 miles) at 11.59 p.m. The recorded times differ too widely to give much value to these observations.

Naval Review at Cherbourg on July 18, 1900.

About 10 p.m. a sham fight took place between two portions of the French fleet at Cherbourg in honour of the visit of the President, M. Loubet, to that town. The number of vessels engaged was forty-three, including thirteen of the largest and most modern battle-ships in the world. During the next few days accounts appeared in various English newspapers of a series of supposed earthquake-shocks felt shortly after 10 p.m. at different places along the southern coast, from Torquay to Bognor. The long duration of the disturbances and their apparent transmission through the air being opposed to a seismic origin, I wrote letters to a number of London and south-country papers, and the account which follows is chiefly based on the replies which I received to these letters.

As some doubt has been expressed with regard to the connection between the two phenomena, it may be well to mention the evidence in its favour. (1) With two exceptions, not one of the places (forty in number) from which records have come is more than a mile or two from the coast. There are several from the south of the Isle of Wight, but none from that part of Hampshire shielded from Cherbourg by the higher ground of the island. (2) Though a few persons in the open air assert that a tremor was felt, the great majority state that the sound travelled through the air and not through the ground; windows rattled loudly without there being any movement of the floor, and at Lancing (100 miles from Cherbourg) and Seaton in Devon (97 miles) observers placing their hands on the wall felt it distinctly vibrating; the noise caused a drumming in the ears at several places more than a hundred miles from Cherbourg. (3) The sounds were recognised as those of heavy guns by many persons, and with less hesitation the smaller the distance from Cherbourg. (4) The night was very still, hardly a breath of wind could be felt, and the sea perfectly calm; and the sound was heard to the east and west along the English coast at almost equal distances from Cherbourg. (5) Lastly, heavy guns are rarely, if ever, fired from English ships or forts at so late an hour; whereas more than 24,000 charges are said to have been fired in Cherbourg harbour during almost the same interval in which the sounds were heard in England.

Though the times of occurrence are roughly given, they agree for the most part in placing the commencement of the disturbances just after 10 p.m., and the end shortly before 10.30. Clearer evidence as to the identity of the sounds throughout the whole area affected is provided by the similarity in their relative duration and intensity. The fire began about 10.2 or 10.3, and lasted nearly four minutes. Then came a pause of five minutes, when there was another burst of about the same intensity and nearly the same duration. About ten minutes later the third followed, slighter in intensity and of shorter duration, perceived almost as far as the others (at Torquay and Brighton, 101 and 104 miles respectively), though not by all observers.

I have no information as to the size of the guns used on this occasion, but they were probably much heavier than those employed for the salutes at Spithead in 1897. To the west, the sound was heard at Budleigh Salterton, Sidmouth and Torquay (101 miles from Cherbourg), Paignton (102 miles), and Dawlish and Exmouth (104 miles); to the east at Lancing (100 miles), Brighton (104 miles), and Henfield (107 miles, and seven miles from the sea). At all of these places, and at many between, the air-vibrations were strong enough to make windows shake and rattle, and there are accounts of this or a similar effect being observed at a greater distance than the sound—at Plymouth (123 miles), and Menheniot, near Liskeard (136 miles, and five miles from the sea). At the latter place the sudden rattling of a large window was distinctly heard at about 10 p.m., but it was unaccompanied by any sound. Judging from the intensity of the disturbances at Torquay and Brighton, I see no reason to doubt the connection of the latter observation with the firing at Cherbourg.

It is interesting to notice how the character of the sound **changed** with the increasing distance from Cherbourg. At St. **Catherine's** Point (65 miles) and Bonchurch (68 miles), both in **the Isle** of Wight, the sound was described as exactly like that **of heavy guns.** At Bournemouth and Muddiford in Hampshire **(74 miles)** there was a continual rumbling noise, with occasional **heavier** booms. At greater distances, as far as Lancing, Torquay and Paignton, the prominent reports ceased to be audible, and there was merely a deep monotonous throbbing noise, the pulsations recurring with great rapidity and regularity, resembling a very quick beating of a big drum far away, or the beats of the paddles of a distant and unseen steamer. At very great distances the vibrations (or some of them) do not seem to have attained the requisite strength to be audible to certain observers, one at Lancing (100 miles) referring to a most curious throbbing sensation in the air, and a dull sound like that of a distant train ; while another at Brighton (104 miles) remarks that he heard or felt the sound. The rattling of the window and the inaudibility of the vibrations at Menheniot may perhaps be accounted for in this way.   CHARLES DAVISON.

SUBJECTS FOR CONSIDERATION BY ELECTRICAL ENGINEERS.

THE current number (July) of the *Journal* of the Institution of Electrical Engineers contains a list of subjects suggested by the Council as suitable for papers to be read at the meetings or published in the *Journal*. The list is here reprinted, and it should be the means of directing attention to many important problems awaiting solution, as well as eliciting information upon the present position of various branches of electrical engineering.

1. Best methods of generating steam and steam power for variable loads.
2. Comparison of double- and triple-expansion engines for varying load conditions.
3. Automatic handling of fuel in power stations.
4. The present position and applicability of gas or oil engines for electrical power stations.
5. Description of plants for the utilisation of river- or tidal-power in the generation of electrical energy.
6. The present position and prospects of the application of liquid and of powdered fuel in electrical power stations.
7. The utilisation of blast-furnace gases or other waste products of manufactures in the generation of electricity.
8. The application of dust-destructors to the generation of electricity.
9. Electric light and power station chimney shafts ; specialities of their construction and equipment.
10. Experiences with vibrations from electric light and power stations.
11. Bearings of shaft and shafting running at high speed.
12. Improvements in dynamos.
13. Comparison of speed and cost of dynamo.
14. Comparison of single and multiple central stations.
15. The wholesale supply of electricity to towns and factories from centres where very large generating units are employed.
16. The distribution of electrical energy from a distant generating station through districts served from a different source of supply, or under a separate local authority.
17. Electrical distribution by constant current, direct or alternating.
18. Examination of relative advantages and disadvantages of direct-current and alternate-current transmission.
19. Examination of relative advantages and disadvantages of two-phase and three-phase transmission.
20. Methods of controlling speed of alternating current motors.
21. Practical methods of measurement in connection with polyphase distribution.
22. Methods for the conversion of direct current into alternate current.
23. Methods of providing for electrical supply during hours of small demand.
24. Utilisation of lighting plant for other work during the hours of small demand.

25. The electrical equipment of large blocks of offices in a city.
26. Economy of design in the manufacture of small electric fittings.
27. Portable electric lamps of the "safety" type, or otherwise.
28. Enclosed arc lamps.
29. Improvements in incandescence electric lamps.
30. Incandescence electric lamps with filaments other than pure carbon.
31. Application of electrical transmission in factories :—
    (*a*) Detailed description, giving sizes of motors and power provided.
    (*b*) Comparison of separate or combined direct- and alternate-current methods.
    (*c*) Combination of lighting and power for such purposes
32. Electricity meters.
33. Description of electrical methods, or comparison of these with other methods, of propelling vehicles.
34. The supply of electrical energy for tramway purposes.
35. The use of electrical methods of traction on railways served by steam-driven locomotives.
36. The economy and design of electrical elevators.
37. The design and economy of electrically driven pumps.
38. The utilisation of electrical energy in mining.
39. The applications of electrical energy in warfare.
40. The use of electricity in the textile and other industries.
41. The application of electricity in musical instruments.
42. Electro-therapeutics.
43. The establishment of public time-services by electricity.
44. Recent advances in telegraphy.
45. Applications of alternating currents in telegraphy.
46. The transmitting capacity and load factor of telegraph circuits.
47. Hertzian telegraphy.
48. Methods, in aërial telegraphy, of restricting signals to selected stations.
49. Recent improvements in telephony.
50. Descriptions of systems tending to simplify the interchange of telephonic communications.
51. The talking capacity and load factor of telephone circuits.
52. The application of electricity to the generation of heat for domestic purposes (cooking, ventilation, heating, &c.).
53. The construction and use of electric furnaces.
54. The application of electricity to the welding or annealing of metals.
55. The application of electrical heating methods in chemical or metallurgical operations.
56. The applications of electricity in metallurgical processes.
57. The applications of electrolysis in the smelting or refining of metals, or in the chemical industries.
58. The electrical equipment of chemical factories.
59. Improvements in primary batteries.
60. Examination of the present position of secondary batteries in electrical engineering.
61. The direct generation of electrical energy from fuel.
62. The economic employment of thermo-generators.
63. Improvements in the apparatus for producing, and in the applications of, kathode and Röntgen rays.
64. The relative suitability and efficiency of the different materials available for any of the requirements of electrical engineering.
65. The electric strength of di-electrics.
66. Recent advance in the manufacture or use of insulating materials.
67. New insulating materials.
68. Electrical applications of aluminium, sodium, &c.
69. The electrical uses of the rarer metals.
70. The treatment, testing, specifications, or uses of iron or steel, or of iron alloys, for magnetic purposes.
71. The manufacture of permanent magnets.
72. The relation of chemical composition and physical condition to the electrical or magnetic properties of substances, considered in its bearing upon electrical engineering practice.
73. High-resistance metals for instruments or resistance coils.
74. New resistance alloys.
75. The protection of laboratories and observatories against magnetic disturbances due to local causes.
76. Recent legislation in its relation to electrical undertakings.
77. The relations between electric lighting or power corporations and municipal authorities.

PRIZE SUBJECTS OF THE PARIS SOCIÉTÉ D'ENCOURAGEMENT.

THE June number of the *Bulletin de la Société d'Encouragement pour l'Industrie Nationale* contains the programme of prizes and medals proposed by the Society for 1901 and following years. The questions proposed for solution cover a large field ; omitting many which have only a local interest, the chief problems suggested as prize subjects for 1901 are as follows. In Mechanics, prizes of 2000 francs for a motor weighing less than 50 kilograms per horse-power developed ; for an important advance in mechanical methods of transmitting energy ; and for automobiles suitable for use in towns and in the country respectively, the conditions laid down for the motor car suitable for towns requiring the absence of fumes or smell, and in the case of the one for use in the country, only such fuel to be used as can ordinarily be obtained in country towns. In Chemistry, a prize of 1000 francs for the utilisation of any waste product ; of 2000 francs for a publication useful to chemical or metallurgical industry ; two prizes of 500 francs each for scientific researches in chemistry, of which the results can be utilised in industrial work ; a prize of 2000 francs for an improvement in the manufacture of chlorine ; one of 1000 francs for the discovery of a new alloy useful in the arts ; and of 2000 francs for a study of the expansion, elasticity, and tenacity of pottery clays and glazes, for a scientific study of the physical and mechanical properties of glass, for a new method of manufacturing fuming sulphuric acid and sulphur trioxide, and for the manufacture of a steel by the introduction of a foreign element possessing specially useful properties. In the Economical Arts, 2000 francs for an invention of new methods allowing of the utilisation for lighting and heating, either for domestic or industrial purposes, of petroleum, density not less than 0·800 ; 2000 francs for a continuous extractor ; 3000 francs for a method of purifying water for domestic use ; and 2000 francs for a 2-candle power incandescent electric lamp fulfilling certain special conditions.

Other prizes offered include one of 2000 francs for the best study of the diseases of cider and the means of preventing or arresting their development ; of 3000 francs for the invention of a method allowing of the production of an indefinite number of positives in colours either by a direct method or with a Lippmann negative ; of 2000 francs for a memoir on the silk industry in the Lyons region ; of 1500 francs for a memoir on the cycle industry ; and of 3000 francs for a study of commercial syndicates.

According to the general conditions for these prizes, all memoirs must be sent in before December 31, they may be written in the French language, and are open to persons of all nationalities.

UNIVERSITY AND EDUCATIONAL INTELLIGENCE.

TEACHERS in Schools of Science and Technical Schools will find a Diary and Calendar just issued by Messrs. Philip Harris and Co., scientific instrument makers, Birmingham, a convenient little pocket-book. The diary is for the year commencing on September 1, and ending August 31, 1901. The dates are given of examinations in science and technology, and memoranda referring to the days on which official papers must be sent in are brought together in a calendar. The book is thus a real *vade mecum* for science teachers.

THE following Saturday morning courses for teachers have been arranged by the London Technical Education Board. A course of about ten lectures on the teaching of mathematics will be given by Prof. Hudson at King's College. The object of these lectures is to help those who are practically engaged in teaching, and wish to become acquainted with modern methods and improvements in order to render their teaching more effective. A course on physics will be given under the direction of Prof. W. Grylls Adams and Mr. S. A. F. White. The course will consist of practical work in the Wheatstone Laboratory, the object of the instruction being to enable students to obtain an intimate knowledge of the methods employed in physical measurements and familiarity with the use of apparatus. A course of twenty lectures on physiology will be delivered by Prof. Halliburton. The object of the

course is to acquaint teachers with the modern methods of teaching physiology by objective methods. A course of ten lectures on the teaching of physical geography, each lecture followed by a class for practical work, will be given by Miss Catherine A. Raisin, D.Sc., at Bedford College.

THE London Technical Education Board makes provision for advanced students as well as for those of elementary grades. During the coming session evening science courses will be held in connection with the Board at University College, King's College, and Bedford College. At University College, Prof. J. A. Fleming, F.R.S., will give a course of ten lectures, followed by laboratory practice, in advanced, electrical measurements. A course of lectures on the electric motor and its application in electric traction will be given by Prof. C. A. Carus-Wilson, each lecture to be followed by an experimental demonstration or by a class for the practical working of numerical examples in connection with the subject. A course will be given by Prof. E. Wilson, at King's College, on direct and alternating currents. In mechanical engineering, Prof. T. Hudson Beare will give a course of ten lectures, at University College, on the theory of steam engines and boilers, with laboratory work on the testing of steam engines and boilers. Prof. Beare will also give a course of five lectures on the theory of gas and oil engines, combined with laboratory work. A course of five lectures on water-tube boilers will be given by Mr. Leslie Robertson. A course will be delivered by Prof. D. S. Capper and Mr. H. M. Waynworth in the mechanical engineering laboratories of King's College. The course will consist of about twenty demonstrations upon steam and gas engines and general laboratory work. The latter portion of each evening will be devoted to experimental and practical work in the engineering laboratory in illustration of the lectures. A course on civil engineering will be delivered by Prof. Robinson. The methods of producing artificial cold will be the subject of a course of lectures to be delivered at University College by Dr. W. Hampton. At the same college, Mr. E. C. C. Baly will deliver eight lectures dealing with the methods of spectroscopy, especially in connection with the photography of the spectrum.

HER MAJESTY'S Commissioners for the Exhibition of 1851 have made the following appointments to Science Research Scholarships for the year 1900, on the recommendation of the authorities of the respective universities and colleges. The scholarships are of the value of 150*l.* a year, and are ordinarily tenable for two years (subject to a satisfactory report at the end of the first year) in any university at home or abroad, or in some other institution approved of by the Commissioners. The scholars are to devote themselves exclusively to study and research in some branch of science, the extension of which is important to the industries of the country. A limited number of the scholarships are renewed for a third year where it appears that the renewal is likely to result in work of scientific importance. The new scholars and their nominating institutions are as follows :— C. E. Fawsitt, B.Sc. (University of Edinburgh), V. J. Blyth, M.A. (University of Glasgow), J. Moir, M.A., B.Sc. (University of Aberdeen), W. M. Varley, B.Sc. (Yorkshire College, Leeds), J. C. W. Humfrey, B.Sc. (University College, Liverpool), S. Smiles, B.Sc. (University College, London), N. Smith, B.Sc. (Owens College, Manchester), L. L. Lloyd (University College, Nottingham), Alice Laura Embleton, B.Sc. (University College of South Wales and Monmouthshire, Cardiff), J. A. Cunningham, B.A. (Royal College of Science, Dublin), W. S. Mills, B.A. (Queen's College, Galway), J. Patterson, B.A. (University of Toronto), W. C. Baker, M.A. (Queen's University, Kingston, Ontario), J. Barnes, M.A. (Dalhousie University, Halifax, Nova Scotia), J. J. E. Durack, B.A. (University of Sydney). Seventeen scholarships granted in 1898 and 1899 have been continued for a second year on receipt of a satisfactory report of work done during the first year. The names of the scholars and the places where they are studying are as follows :— J. C. Irvine, B.Sc. (University of Leipzig), H. L. Heathcote, B.Sc. (University of Leipzig), Winifred Esther Walker, B.Sc. (University College, London), F. W. Skirrow, B.Sc. (University of Leipzig), C. G. Barkla, B.Sc. (Cavendish Laboratory, Cambridge), Harriette Chick, B.Sc. (Thompson-Yates Laboratories, University College, Liverpool), F. A. Lidbury, B.Sc. (University of Leipzig), W. Campbell, B.Sc. (Royal College of Science, South Kensington), L. Lownds, B.Sc. (University of Berlin), J. T. Jenkins, B.Sc. (University of Kiel and Biological Institution,

Heligoland), R. D. Abell, B.Sc. (University of Leipzig), W. Caldwell, B.A. (University of Würzburg), W. B. McLean, B.Sc. (Owens College, Manchester), B. D. Steele, B.Sc. (University of Breslau), E. J. Butler, M.B. (University of Freiburg), J. W. Mellor, B.Sc. (Owens College, Manchester), L. N. G. Filon, M.A. (King's College, Cambridge). Four scholarships granted in 1898 have been exceptionally renewed for a third year. These scholars and their places of study are :— Dr. A. H. Reginald Buller, B.Sc. (University of Munich), H. T. Calvert, B.Sc. (University of Leipzig), R. L. Wills, B.A. (Cavendish Laboratory, Cambridge), E. H. Archibald, M.Sc. (Harvard University).

SCIENTIFIC SERIALS.

American Journal of Mathematics, vol. xxii. No. 3.— On continuous binary Λ linearoid groups, and the corresponding differential equations and Λ functions, by E. J. Wilczynski. In a previous paper (vol. xxi. 2) the writer has shown that, corresponding to every group of the form

$$\eta_i = \sum_{k=1}^{m} \phi_{ik} (x ; a_1 \dots a_r) y_k (1),$$

where the r parameters a_i are essential, there exists a system of differential equations of order r, whose general solutions are given by (1), if $y_1, \dots y_n$ form a fundamental system. The functions ϕ_{ik} were supposed to be uniform functions of x, and it was found that, if the parameters a_i were properly chosen, ϕ_{ik} were uniform functions of the parameters also. In the present paper he discusses these groups, the corresponding differential equations, and their solutions for the case when $n = 2$. Dr. Lovett, in his note on a property of lines in n-dimensional space, working on the lines of Cesàro's "Lezioni di Geometria Intrinseca," shows that a line of multiple curvature cuts its osculating space of highest dimensions, or lies wholly on one side of that space, according as the number of dimensions of the space necessary to the existence of the curve is odd or even. —Concerning the cyclic sub-groups of the simple group G of all linear fractional substitutions of determinant unity in two non-homogeneous variables with coefficients in an arbitrary Galois field, by Dr. L. E. Dickson (read before the Chicago section of the Mathematical Society, December 1899), leads to a generalisation to the GF[p^n] of results due to Prof. W. Burnside ("On a Class of Groups defined by Congruences," *Proc.* of London Math. Soc. vol. xxvi.). Variations from Burnside's method of treatment are introduced, partly to avoid the separate treatment of the cases $d=1$ and $d=3$, and to take in the exceptional cases $p=2$ and $p=3$, and to reduce the calculations ; and further, on the other hand, to amplify some of the proofs. A few errors are also pointed out and amended.—On some invariant scrolls in collineations which leave a group of five points invariant, by V. Snyder. The writer gives numerous references to memoirs in which the quadric surfaces which are left invariant by cyclical collineations have been exhaustively treated. There is another simple series of scrolls, viz. those contained in a linear congruence, which have not been considered, except one form noticed by Ameseder. The writer confines his attention to such surfaces. There are six collineations which are of essentially different type, which project a set of five points into themselves without leaving every point invariant. In the notation of substitution-groups these may be thus represented :

$$T_2 \equiv (A_1 A_2)(A_3)(A_4)(A_5),$$
$$T_3 \equiv (A_1 A_2)(A_3 A_4)(A_5),$$
$$T_4 \equiv (A_1 A_2 A_3)(A_4)(A_5),\ T_5 \equiv (A_1 A_2 A_3)(A_4 A_5),$$
$$T_6 \equiv (A_1 A_2 A_3 A_4)(A_5)\text{ and }T_7 = (A_1 A_2 A_3 A_4 A_5).$$

—On the reduction of hyperelliptic integrals ($p = 3$) to elliptic integrals by transformations of the second and third degrees, by W. Gilespie. The point of the paper is an application of cubic involution to the problem of the reduction to elliptic integrals, of hyperelliptic integrals of genus $p = 3$ and of the first kind, by a rational transformation of the third degree. It is a continuation of Prof. Bolza's researches on the cubic transformation ("Die Cubische Involution und Dreitheilung, &c.," and "Zur Reduction Hyperelliptischer Integrals, &c.," *Math. Ann.*, Bd. 50, pp. 68 and 314). The closing paper, by Dr. E. H. Moore, was read before the American Mathematical Society at the Buffalo meeting of the summer of 1896, and is entitled "The Cross-ratio Group

of n ! Cremona Transformations of Order $n - 3$ in Flat Space of $n - 3$ Dimensions."

Bulletin of the American Mathematical Society, July.— Some remarks on tetrahedral geometry, by Dr. Timerding, is a paper read at the June meeting. Several properties of a tetrahedral complex are given, viz. the pole curves of such a complex of lines form again another such complex among the cubic space curves circumscribed about the fundamental tetrahedron, the complex curves of such a complex of lines form another tetrahedral complex, &c.—Prof. H. B. Newson's paper on singular transformations in real projective groups was read at the April meeting. It treats of transformations in real projective groups which can not be generated from the real infinitesimal transformations of certain continuous groups. The discussion, which is limited to one and two dimensions, can be readily extended to three and higher dimensions.—Miss Schottenfels, in a paper read at the June meeting, writes on groups of order $8!/2$, and gives a simple proof of a correspondence established by Dr. Dickson (*Proc.* of London Math. Soc., vol xxx.).—Prof. F. S. Woods continues his notes on Lobachevsky's geometry.—Prof. Pierpont reviews H. Burkhardt's "Functionen-theoretische Vorlesungen" (vol. ii. "Elliptische Functionen").—A "correction," notes, new publications, list of papers read before the Society, with references to the places of their publication, and a full index, complete the sixth volume of the second series.

Annalen der Physik, No. 7.—Dispersion of electricity in air, by J. Elster and H. Geitel. Since the sun's rays contain ultra-violet light before they impinge upon the atmosphere, this light must ionise the upper strata, and the ions produced will be gradually distributed through the whole of the atmosphere by diffusion and convection. Hence the atmosphere will contain stray ions of both signs, but chiefly negative ones in the lower strata, owing to their superior mobility. The presence of these ions can be made evident by an electroscope.—Influence of slight impurities upon the spectrum of a gas, by P. Lewis. Very small quantities of hydrogen and nitrogen considerably affect the spectra of helium and argon, but the reverse is not the case.— Fluorescence and phosphorescence in the electric discharge through nitrogen, by P. Lewis. When nitrogen prepared from ammonium nitrate and sulphate, and purified over hot copper is pumped through an H-shaped vacuum tube, the whole wall of the tube shows a brilliant fluorescence lasting a few seconds, which extends for a length of about a yard into the supply and exhaust tubes. The light can be made permanent by keeping the pump at work and thus supplying a continuous stream of fresh nitrogen. Spectroscopic examination shows that the fluorescence is dependent upon the presence of a number of bands in the extreme ultra-violet, due to a combination of nitrogen with a trace of oxygen.—Production of very high notes by Galton's whistle, by M. T. Edelmann. The author gives tables for the pitches of pipes of various dimensions, and instructions how to test the pipes by Kundt's dust figures. He has succeeded in constructing a pipe of only 2 mm. diameter, which gives the enormously high pitch of 170,000 complete vibrations per second, or over two octaves beyond the extreme limit of audibility.—The magnetic force of the atoms, by R. Lang. Magnetism is accounted for by the revolutions of negative about positive electrons.—The air thermometer at high temperatures, by L. Holborn and A. Day. The authors further investigate the properties of the air thermometer consisting of a platinum-rhodium vessel filled with nitrogen, and compare its indications with that of a platinum-iridium thermo-couple, paying particular attention to their regular expansion of the vessel. The corrected value for the melting point of gold is 1064·0° C.—Difference of temperature between the surface and the interior of a radiating body, by F. Kurlbaum. A method is given of determining this difference of temperature by means of two bolometers exposed symmetrically to different surfaces of the same black partition.

SOCIETIES AND ACADEMIES.

LONDON.

Royal Society, June 14.—"Data for the Problem of Evolution in Man. V. On the Correlation between Duration of Life and the Number of Offspring." By Miss M. Beeton, G. U. Yule, and Karl Pearson, F.R.S., University College, London.

According to the Darwinian theory of evolution the members of a community less fitted to their environment are removed by death. But this process of natural selection could not permanently modify a race if the members thus removed were able before death to propagate their species in average numbers. It then becomes an important question to ascertain how far duration of life is related to fertility. In the case of many insects death can interfere only with their single chance of offspring; they live or not for their one breeding season only. A similar statement holds good with regard to annual and biennial plants. In such cases there might still be a correlation between duration of life and fertility, but it would be of the indirect character, which we actually find in a case ot men and women living beyond sixty years of age—a long life means better physique, and better physique increased fertility. On the other hand, there is a direct correlation of fertility and duration of life in the case of those animals which generally survive a number of breeding seasons, and it is this correlation which we had at first in view when investigating the influence of duration of life on fertility in man. The discovery of the indirect factor in the correlation referred to above was therefore a point of much interest. For it seems to show that the physique fittest to survive is really the physique which is in itself (and independently of the duration of life) most fecund.

The data dealt with in this paper consists of four series, the first three collected and reduced by Miss M. Beeton, and the fourth series by Mr. G. U. Yule.

Mothers. Length of Life and Size of Family.

Series I.—Taken from the "Whitney Family of Connecticut," a well-known history of an American Quaker family.

Series II.—Taken from purely English Quaker records. The data for this series were drawn from a great variety of histories and records of the Society of Friends.

Fathers. Length of Life and Size of Family.

Series III.—The great bulk of the data was extracted from the American Whitney Family.

Series IV.—Extracted from Burke's "Landed Gentry."

The following are some of the chief results obtained from the reduction of these series:—

Table of General Results.

Series.	Parent.	Mean age at death.	Mean size of family.	Correlation fertility and duration of life.
I.	Mother	53'292	5'269	0'4943
II.	Mother	61'183	5'811	0'2340
III.	Father	58'086	5'469	0'4764
IV.	Father	63'577	5'336	0'2010

It is shown that the peculiar physique in both men and women which leads to longevity is also associated with greater fecundity. Of two women who both live beyond fifty years, the longer lived is likely to have had before fifty the larger family. The association is, however, much greater for American than English parents, although the American parents dealt with are, in the great majority of cases, of Anglo-Saxon race. Climate, mode of life, in general selection and environment, seem to be differentiating in this respect the English and the Anglo-American. The English Friends, we should suppose, would be as a class very comparable with the American Friends; yet their average life is longer, their fertility greater, and there is less association between longevity and fecundity. In both cases our algebraical formulæ show that American men and women are more alike, and English men and women are more alike than the women to the women or the men to the men of the two races. This is the more remarkable, as the English Friends as a class are by no means identical with the Landed Gentry.

In order to represent the *continuous* change in the regression, which cannot be done by two straight lines, which only enable

us to distinguish the fecund and non-fecund periods of life, the statistics were fitted with cubical parabolas. The regression line at any age in life may then be looked upon as the tangent to the cubical parabola at that age. An inspection of the diagram given below for American mothers shows what an excellent expression such parabolas are for these statistics

For American mothers and fathers we see dy/dx consistently positive throughout life, and we have a most excellent graphical demonstration of the physical characters which tend to longevity being also associated with fecundity.

Weismann has suggested that it may be an advantage to a species that its duration of life should be shortened. This is not, *a priori*, confirmed by the case of a man in the American series: the longer the parents live, the greater the number of their offspring. But if we can lay any stress on a bend-in for the English mothers, and on a similar, but less marked, tendency for the English fathers, we might argue that reproductive selection was possibly in England working against extreme longevity, although favouring parents living till sixty-five or seventy. Indeed, those who rush rapidly to brilliant, but not

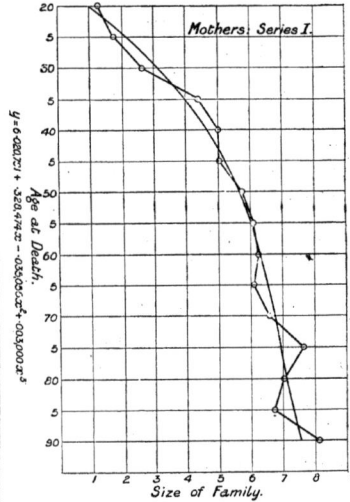

over-stable, conclusions might emphasise Weismann's views by showing how in an old community, with much greater pressure on the material resources, there is a tendency to reduce the fertility of the long-lived parents; while in a new community, with plenty of food and occupation for all, the longest-lived parents are the most fertile! However, all that we can safely say is that there is a marked difference between English and American parents, and that this distinguishing characteristic is almost equally visible if we take opposite sexes of such diverse classes as English Friends and English country gentlemen. We would leave to further investigations its true interpretation.

Admitting a substantial correlation between length of life and fertility, it is of great interest to investigate what effect, other things being equal, reproductive selection would have in modifying the duration of life.

The following table gives the mean length of life of parents taken singly and of parents weighted with their offspring :—

Mean Duration of Life of Parents in Years.

Series.	Unweighted parents.	Weighted parents.	Progression.
I.	53·292	59·920	6·628
II.	61·183	63·839	2·656
III.	58·086	63·082	4·996
IV.	63·577	65·510	1·930

Now these are substantial differences even in the case of the English parents, but they are very large differences in the case of the American parents. If we suppose no assortative mating on the basis of characters tending towards longevity, then it is easy to measure by a rough approximation the effect of reproductive selection in modifying the duration of life.

The increased duration of life would be about two years per generation from the American data, and about 9 to 9·5 months per generation from the English data.

The result for the American series shows us how an especially low expectation of life, due possibly in this case to some family character, will be rapidly raised by reproductive selection, if there be no opposing factor of evolution. The English results on the other hand show us a small, but sensible, tendency in reproductive selection to prolong the duration of life. Allowing three generations to a century, we might expect the duration of life to be raised about two years in a century by this factor of evolution.

A somewhat widespread view of evolution stops at the survival of the fitter without discussing the mode whereby the less fit leave no, or fewer, offspring than the fit. Of course, if the unfit are exterminated before adult life, there is no chance of their reproducing themselves. It has been shown in "Data for the Problem of Evolution in Man (II.)" that a selective death-rate exists for adults, so that the whole work of selection does not take place before the reproductive stage is reached. But Miss Beeton's data for the correlation of duration of life in the case of brethren dying as minors seem to show that the selective death-rate for children is rather less, not greater, than its value for adults.[1] Hence, for the reduction or extermination of stock unsuited to its environment, we should have to look largely to selection in the adult state. In the present paper we have made what we believe to be the first quantitative determination of how a selective mortality reduces the numbers of the offspring of the less fit relatively to the fitter. In the case of life under wild conditions, the correlation between fertility and power of surviving would probably be far greater. But for such life it is almost impossible to get statistics of this nature ; we are thrown back upon measuring the effect in man, and thus obtaining what may well be considered as a minimum value of the influence under discussion.

In the course of our investigations we have found that the relationship between fertility and duration of life does not cease with the fecund period. We thus reach the important result that characters which build up a constitution fittest to survive are also characters which encourage its fertility. This result is of great value from the standpoint of the differentiation of type, where it is absolutely necessary that the fittest to survive should also be the most fertile. On the other hand, we note that duration of life is a character capable of modification by reproductive selection, and we suggest that a considerable part of the increased expectation of life observed in recent years may be due to this cause. In the case of the American statistics, we see at once how reproductive selection can replace a remarkably short-lived stock by a longer-lived stock, for the bulk of the offspring come from the longer-lived members.

PARIS.

Academy of Sciences, July 30.—M. Maurice Lévy in the chair.—On the observatory at Mount Etna, by M. J. Janssen. Remarks on local difficulties due to climatic disturbances and to the peculiar situation of the observatory.—New

[1] The point is still under investigation.

processes of vaccination against symptomatic carbuncle of the ox by means of preventive serum in association with vaccines, by M. S. Arloing. A continuation of former experiments on the subject.—On the age of the sea-shore sands of Dunkirk, by M. J. Gosselet. The formation of these deposits is considered to have commenced since the fourth or fifth centuries.—M. Duhem was elected a corresponding member for the section of mechanics.—Observations of Borrelly's comet (July 23, 1900) at the Paris Observatory, by M. G. Bigourdan.—Provisional elements and ephemerides of the Borrelly-Brooks' comet (July 23, 1900), calculated by M. G. Fayet.—On the spectral images of the chromosphere and protuberances obtained with the prismatic chamber, by M. Georges Meslin. A description of the results obtained with the apparatus previously described.—On two surfaces related to every Weingarten's surface, by M. A. Demoulin.—On artificial radio-active barium, by M. A. Debierne. Many substances become radio-active when brought into intimate contact, by solution or simultaneous precipitation, with radio-active compounds. Artificial radio-active barium chloride, intermediate in character between barium and radium, has thus been obtained.—On the thermo-electricity of steels, by M. G. Belloc. A comparative study of the thermo-electric properties of soft iron, soft steel, and hard steel.—On a means of weakening the influence of industrial electric currents on the terrestrial field in magnetic observatories, by M. Th. Moureaux. An account of the methods whereby the disturbances caused by electric tramways in the neighbourhood of observatories may be removed or corrected for.—On the electrolysis of concentrated solutions of hypochlorites, by M. André Brochet. The electrolysis of hypochlorite resembles, in its later stages, that of alkaline chloride solutions, and tends towards the same limits. There is, therefore, little hope of obtaining concentrated solutions of hypochlorites by the direct electrolysis of chlorides. —On gadolinium, by M. Eug. Demarçay. A study of the spectrum of gadolinium.—On diphenylcarbazide as a sensitive reagent for some metallic compounds, by M. P. Cazeneuve. The conversion of diphenylcarbazide into diphenylcarbazone by the action of salts of copper and mercury and the persalts of iron, as recently described, furnishes a delicate test for these metals. The latter unite with the carbazone to form coloured compounds.—Preliminary study of the chemism of the encephalon, by M. N. Alberto Barbieri. Experiments on the chemical changes occurring in the brain of animals when left for twelve to eighteen hours at a temperature of 45°. —On the dissolution of the nitrogenous constituents of malt, by MM. P. Petit and G. Labourasse. Experiments relating to the existence of a proteolytic enzyme in malt.—Action of the liquid from the external prostate of the hedgehog on the liquid of the seminal vesicles ; nature of this action, by MM. L. Camus and E. Gley.—On some properties and reactions of the liquid from the internal prostate of the hedgehog, by MM. L. Camus and E. Gley. This and the previous paper form a continuation of the authors' researches on the coagulation of the secretion of the seminal vesicles by that of the external prostate, or Cooper's gland. and the coagulation of the latter secretion by that of the internal prostate.—On some Alpheidæ of the American coasts, by M. H. Coutière. An account of some specimens in the collection at the United States National Museum, Washington.

August 6.—M. Maurice Lévy in the chair.—The menstrual function and rut in animals. Function of arsenic in the economy, by M. Armand Gautier. The author has found that the quantities of arsenic and iodine, which in normal blood are hardly estimable, are largely increased during menstruation, the total amount of arsenic eliminated during one period of menstruation representing the whole amount usually present in the thyroid gland. The arsenic and iodine which accumulate in the thyroid gland are eliminated in the male by the hair and nails, and by epithelial desquamation. In the female, this excess is either eliminated by the genital organs or utilised by the growing fœtus.—Observations of the star Capella, considered as a double star, made at the Greenwich Observatory, by Mr. W. H. M. Christie. The independent discovery by Campbell and Newall, by spectroscopic observations, that Capella is a double star, has been confirmed by direct observation with the large Greenwich equatorial. The star appears distinctly elongated in one direction, the distance of the two components being estimated at 0·1 second. Observations of the direction of this elongation, taken between April 4 and July 20, confirm

period of revolution deduced spectroscopically by Newall.—
The comet (1900 *b*) discovered July 23 at the Observatory
of Marseilles, by M. Borrelly.   The comet. is visible to
the naked eye as a star of 6th to 7th magnitude.—
Observations of the comet 1900 *b* (Borrelly-Brooks) made
at the Observatory of Besançon, by MM. A. Sallet and
P. Chofardet.—Observations of the Borrelly-Brooks comet,
made at the Toulouse Observatory with the 25 cm. equatorial,
by M. F. Rossard.—On circuits formed uniquely by
electrolytes, by MM. Camichel and Swyngedauw.   From
the experiments described with circuits consisting wholly
of liquid electrolytes, the authors conclude that an electrolyte
may be traversed by a current without decomposition.—On the
coupling up of alternators from the point of view of the harmonics, and of the effect of synchronised motors, by M. A. Perot.
—On the boiling points of zinc and cadmium, by M. Daniel
Berthelot.   The metals were boiled in an electric furnace specially constructed to avoid the errors due to superheating and
radiation from the walls, the temperature being measured by the
interference refractometer method previously described by the
author.   Zinc boiled at 920°, and cadmium at 778°.—On the
atomic weight of radiferous barium, by Mme. Curie.   First
attempts at determining the atomic weight of the metal in
radiferous barium chloride gave 146 as against 138 for
pure barium chloride.   As the result of prolonged
fractionations, a product has now been obtained in
which the atomic weight is as high as 174.   This, however,
is certainly too low, as the chloride analysed still contains an
unknown amount of barium.—On the electrolytic estimation of
cadmium, by M. Dmitry Balachowsky.   The metal is deposited
upon a dish previously covered with copper.   The solution is
slightly acidified with nitric acid, and the deposition carried out
at 60° under conditions of electromotive force and current
density specified.—On some new spectra of rare earths, by
M. Eug. Demarçay.—On the blue oxide of molybdenum, by
M. Marcel Guichard.   The hydrated blue oxide of molybdenum
has been isolated in a pure state and analysed, and proved to
have the composition $MoO_2.4MoO_3.6H_2O$.—On the normal
proportions of iodine in the organism, and its elimination, by
M. P. Bourcet.   The author, in conjunction with M. Gley, having
previously shown the presence of a trace of iodine in normal
blood, has now determined the amount of this element in various
parts of the body.   The quantities found vary from 0·00 mgr.
in fat, pancreas and bladder, to 0·18 mgr. per 100 grams of
liver and 1·8 mgr. per 100 grams of hair.   The quantities found
are small compared to the amount present in the thyroid gland.
About 0·33 mgr. of iodine is taken into the human system daily
in food ; the thyroid gland contains only about 4 mgr. ; hence
it becomes necessary to discover the means of elimination.   This
is shown to be chiefly effected in man by the skin and epidermal
products, sweat, skin, hair and nails ; in women, by the
menstrual blood, which contains 0·8 to 0·9 mgr. of iodine per
kilogram, as against 0·02 mgr. per kilogram in normal blood.—
On the nitrogenous substances in malt, by MM. P. Petit and G.
Labourasse.—On the origin of the secondary calcareous breccia
of Ariege, and results drawn from the point of view of the age
of the lherzolite, by M. A. Lacroix.—On some temperatures
observed in the park of St. Maur, by M. E. Renou.

NEW SOUTH WALES.

Royal Society, June 6.—The President, Prof. Liversidge,
F.R.S., in the chair.—On the relation, in determining the
volumes of solids, whose parallel transverse sections are
functions of their position on the axis, between the position and coefficients of the section and the (positive) indices of the function, by
G. H. Knibbs.—On the amyl ester of eudesmic acid occurring in
eucalyptus oils, by Henry G. Smith.   In a paper read before
this society, July 1898, on the stringy-bark series of New South
Wales, R. T. Baker and the author show that an ester was
present in the oil of *Eucalyptus macrorhyncha*.   Since then
esters have been found to be present in several eucalyptus oils.
The author shows that esters are present in fair amount in the
oils of *E. botryoides*, *E. Saligna*, and *E. rostrata*, and that an
aromatic alcohol, either linalool or geraniol, is present in the
oil of *E. patentinervis*, over 16 per cent. of free alcohol being
proved.   The saponified oil of *E. patentinervis* has a fine
odour.   Citral also occurs in this oil, proved by its characteristic
reactions.—Note on a new meteorite from New South Wales,

by R. T. Baker.   The meteorite described in this paper was
found early in January of this year, two miles from Bugaldi, a
postal town fifteen miles north-west of Coonabarabran.   It is
pear-shaped and is nearly five inches long and three inches wide
at the broadest part.   It belongs to that class of meteorites
known as siderites, and is probably composed of iron and
nickel.   It has a well-defined, closely adhering "skin" of black
magnetic material, while the metal immediately beneath this
coating is silvery-white in appearance.   On the smooth portion
at the extremity of the larger end can be seen very distinctly
Widmanstatten's figures.   The specimen has an exceedingly
new appearance, as if it had only just arrived upon the earth,
and shows no signs of oxidation.

GÖTTINGEN.

Royal Society of Sciences.—The *Nachrichten* (physico-
mathematical section), Part i. for 1900, contains the following
memoirs communicated to the Society.
February 3.—L. Krüger : Compensation of errors by means
of equations of condition in geodetic determinations of points.
March 3.—E. Marx : Fall of potential and dissociation in
flame-gases.—W. Nernst : On the question of the hydratation
of dissolved substances, Part 1.—H. Lotmar : The same, Part 2.
—C. C. Garrard and E. Oppermann : The same, Part 3.—H.
Minkowski : Theory of the units in algebraic *Zah·körper*.
March 16.—W. F. Osgood : On a theorem of Schönflies re-
lating to the theory of the functions of two real variables.—F.
Bernstein : On the same theorem.—H. E. Timerding : On
linear systems of conics.
Among the official reports of the Society are one (by Prof. F.
Klein) on the publication of Gauss's works ; and one on the
progress of the Encyclopædia of Mathematics.

CONTENTS.

THURSDAY, AUGUST 23, 1900.

A MUSEUM CATALOGUE.

Descriptive and Illustrated Catalogue of the Physiological Series of Comparative Anatomy contained in the Museum of the Royal College of Surgeons of England. Vol. i. Second edition. Pp. xlix + 160. (London : Taylor and Francis, 1900.)

OF the catalogues of the Royal College of Surgeons, rendered famous by the labours of Hunter, Owen and Flower, but one in osteology, by Dr. R. Bowdler Sharpe, and one in teratology, by Mr. B. Thompson Lowne, have appeared during the present conservator's term of office. Prof. Charles Stewart, unlike his predecessor in this office, who concentrated his attention upon one special department, has since his appointment greatly developed all sides of the collection, and with the aid of his competent assistants has added year by year specimens of surpassing value and interest, which have become the admiration of all beholders. The period of his conservatorship has been one of unparalleled activity in all branches of zoology, and in the labour of keeping pace with this he has not been found wanting, as, for example, when, on the discovery of the calcified teeth of the Monotreme, he produced from his rich store of material the famous specimen which has since adorned his shelves, and shows more than all others yet described. With this museum, as with others in our own land, the Englishmen's colonising instinct has come forcibly into play, in the accumulation of objects, not merely of local interest, such as are generally to be found in the museums of other countries, but general and universal, wherefore the present catalogue is of necessity based upon a matchless material.

It is explained in the preface that the specimens registered have been dealt with on the original Hunterian lines, the intention being to bring together examples of such structures in plants and animals as perform the same function ; and comment is further made upon the necessity for a large number of preparations "to supply the places of those that have become worthless, and to serve as illustrations of new discoveries, and phases of thought." At the outset, necessity, begotten of progress and advancement, is met by the propounding of a scheme, clearly explained in the text, under which it is proposed to distinguish, firstly, between "Structures concerned in the preservation of the individual or to its advantage" and those "concerned in the preservation of the race." Under the former of these departments, sixteen headings are included, under the latter eight ; and since the present catalogue deals with but the first three of the former series, those alone need be further remarked upon. They read "Endoskeleton"; "Flexible Bonds of Union and Support"; "Muscular and Allied Systems"; but before passing on to consider them more fully, it may be said that they and the twenty-one headings of sections to follow are, in the preliminary portion of this, the first volume of a series, individually set out in greater detail, each with a concise statement of the order of arrange-

ment to be adopted, and a definition, when necessary, of what is implied in the heading it bears.

The arrangement of each section is so framed as to include both plants and animals, whenever possible, the former being considered first, and each in ascending order. Turning now to the three sections to be specifically considered, we find specimens illustrating, under the first, the chemical composition, structure, and mode of formation of the various endoskeletal systems ; under the second, the various forms of ligaments and joints ; and under the third, the forms, structure and texture of muscles. To particularise in these columns concerning the details of either of these is impossible ; but it may be said that no leading type of tissue or arrangement of parts is unrepresented, and that preparations like that of the cartilages of the cuttle, the elastic honeycomb of the flexor carpi muscle of the elephant, or of the leaves of the sensitive plant fixed in the diurnal position, must be seen to be appreciated.

The most attractive portions of the work are those dealing with the marvellous array of processes occurring in the growth of the coral skeleton, and (as pertaining to the study of joints and jointing) with the question of adaptation in these to the conditions of existence. The study of the general question of origin, detailed constitution and relationship, of the coral skeleton, has for years engrossed the attention of Prof. Stewart ; and to our knowledge of this subject and the allied one of the structural variations of the bony tissues of the osseous fishes, he has in the long run added more than most other investigators since von Kölliker. Beyond laying this magnificent result of his labours before audiences which annually assemble on the occasion of lectures delivered in his official capacity, and occasional "exhibits" before the Linnean and Zoological Societies, he, with a modesty so marked as to be well-nigh depreciative of his talents, has published nothing concerning them ; and the present volume therefore comes rather as a memoir from his hands than as a mere official catalogue, and it is worth all the scattered papers he could have published in the time. It teems with interest and records of beautiful objects, and is illustrated by fourteen magnificent plates, mostly in colour, done from his own drawings by the facile hand of Green, than whom no better English lithographer in zoology exists. Of these plates no praise can be too high, and we expect for them an unprecedented popularity in the future. They must be seen to be appreciated, and, with the exquisite preparations they illustrate, constitute a possession of which even the Royal College of Surgeons may well be proud. Each of the entries in the catalogue bears a registration number, and where desirable a short bibliographic reference, as an aid to the student.

The success of this volume augurs well for the future of the museum and its collections, and knowing the unparalleled excellence of the numerous additions which during the last decade have been made to the series of which it treats, thanks to the curatorial genius of Prof. Stewart and the unrivalled skill of his lieutenant, Mr. R. H. Burne, we can safely predict even a better result for the volumes yet to come. The collection of zootomical preparations arising under their hands is far and away the best in existence, and together with the governors of the college they have ensured a debt of gratitude which it will take generations to repay.

In the pages of this volume the student will find records of structures and relationships undreamt of in the text-books, unrecorded in the best monographs ; and it is a pity that he is not informed of this. The work is a positive storehouse of new facts and intensely interesting details, and will be of inestimable value to zoologists at large.

A TEXT-BOOK OF MAMMALS.

Text-book of Zoology, treated from a Biological Stand- point. Part I., Mammals. By O. Schmeil. Translated by R. Rosenstock, and edited by J. T. Cunningham. 8vo. Pp. vii + 138, illustrated. (London : A. and C. Black, 1900.)

A S stated in the first title-page, this book is intended for the use of schools or colleges, forming, in fact, a portion of the series of School Text-books now in course of issue by the publishers. It is, therefore, essential that it should be written in a popular and attractive style, and also that it should be absolutely accurate and up-to-date, both as regards the facts recorded, and, so far as possible, in nomenclature. So far as this first item is concerned, the present fasciculus appears to fulfil the required conditions fairly well, the anatomical details being treated in a manner which renders them of easy apprehension by the student, while the descriptions of the animals themselves are, if anything, written in a too popular style. The plan adopted is to take a more or less typical member of a group for special treatment, and then to refer to the kindred forms in a more general manner. Illustrations are numerous ; and while many of them are excellent, others, especially the cut of a family of orangs on p. 19, can only be described as hideous caricatures. In a book written primarily for German students, it must be inevitable that the animals of the Fatherland come in for a fuller share of notice than would have been the case had it been the product of an English author, but this is a fault of no special importance.

When, however, we come to the second essential feature of an elementary text-book—accuracy as regards facts, classification and nomenclature—we are bound to confess that the fasciculus before us fails lamentably. Indeed, its appearance is almost a calamity for zoological science in England, since the student who intends to pursue the subject seriously will have much to unlearn ; and even for those who only desire a smattering of the subject, it is most important that they should become acquainted with animals by their proper titles, and that what they are taught as facts should really be such. In his preface the editor tells us that he has practically re- stricted his task to comparing the translation with the original, correcting the proofs, making here and there emendations in detail where a statement seemed open to doubt, or where differences between the faunas of Britain and Germany had to be indicated. For the sake of his own reputation it is a pity that he did not compare the work in detail with a standard English treatise on mam-

mals, when he could scarcely have failed to detect some of the shortcomings of the original text, despite the fact that all the English treatises on the subject are now more or less out of date.

As regards the general classification of the group, although this differs to a certain extent from the one generally adopted in this country, we have no comment to make, except that for some unaccountable reason the order Sirenia is totally omitted, while there appears to be no mention of the animals by which it is represented anywhere in the text !

Turning to some of the ordinal groups, we find the orang taken as a typical representative of the apes, and rightly named *Simia satyrus*. Naturalists will, how- ever, be considerably surprised to see the chimpanzee (p. 22) assigned to the same genus (*Simia*), whereas the gorilla is made the type of a genus by itself ; since if there is one well-established zoological fact, it is the intimate relationship existing between the chimpanzee and the gorilla, and the wide gulf separating both from the orang. Again, under the heading of the Platyrrhine apes, there is no reference to the marmosets, and we quite fail to find a reason for the statement (p. 22) that the howling monkeys are the best known members of that group. In treating of the Lemuroids, the author departs from his rule of selecting one species for special notice, and the space allotted to the group is ludicrously inadequate.

As an instance of careless writing we may refer to the notice of the tiger (p. 33), when, after stating that this animal is found in Amurland and Central Asia, the author proceeds to say that its "favourite haunts are swampy districts of the tropical zone, thickly overgrown with bamboo and similar bushes." Again, on p. 84 we find *Cricetus frumentarius* alluded to as "the marmot or hamster," although the true marmots are noticed in an earlier page. Passing on to p. 105, we meet with the statement that the Indian buffalo is *said* to exist in a wild state in the "East Indies" ; while the European bison is stated to be extinct, although on an earlier page (98) its existence in Lithuania and the Caucasus is alluded to ! Although we do not propose to notice in detail the hope- lessly obsolete generic and specific nomenclature adopted, the statement on p. 106 that "the best-known African antelope is the gazelle (*Antilope dorcas*)" is, however, too ludicrously absurd and incorrect to be passed over. And as a second instance of incorrect nomenclature we may refer to the inclusion of the roe (p. 108) in the same genus as the red deer, from which the fallow deer is excluded. And in this connection it may be mentioned that the editor, who has been recently writing on deer antlers, should have been aware that the brow-tine is not developed in those of the roe.

Before leaving the Placentals, it may be mentioned that the practice of reckoning the carnassial teeth of the land Carnivora as distinct alike from the molar and pre- molar series is not calculated to give the student an idea of the homology of the cheek-teeth throughout the class. And we also venture to think that the statement on page 37, that "in its dentition the wolf very nearly resembles the cat," in spite of the subsequent qualifica- tion that the number of teeth is greater, scarcely accords with the facts.

In the definition of the Marsupials, exception must be taken to the statement that the young are *always* nourished in a pouch ; and when mentioning the occurrence of the group in America no reference is made to *Caenolestes*. Indeed, the account of the whole group is entirely inadequate ; and when the author speaks of the value of American opossum fur, we strongly suspect he had in his mind the product of the so-called opossums of Australia.

Finally, when treating of the Monotremes, the author states that the spiny anteaters are represented solely by *Echidna hystrix* and *E. setosa*. As a matter of fact, these two forms are but local races of a single species whose name is *E. aculeata* ; and the author appears to be totally unacquainted with the very distinct genus commonly known as *Proechidna !*

As already said, we do not intend to criticise in detail the nomenclature employed ; but in the retention of names now discarded by those who have made a special study of the class the author has done his best to put his work out of touch with the present state of science.

In making this statement, we are aware that the author lays stress on the circumstance that he is treating his subject from a biological standpoint. This, however, in our opinion, is no excuse for neglect of the details o classification and nomenclature.

When treating of the adaptation of animals to their environment, the author is always interesting ; and the paragraphs devoted to this part of the subject are, to our mind, the best in the whole fasciculus.　　　R. L.

GOOD AND BAD AIR.

The Carbonic Anhydride of the Atmosphere. By Prof. E. A. Letts, D.Sc., Ph.D., and R. F. Blake, F.I.C., F.C.S., "Scientific Proceedings of the Royal Dublin Society," vol. ix. (N. S.), Part ii. No. 15. Pp. 270. (Dublin : 1900.)

The Air of Rooms. By Francis Jones, F.R.S.E., F.C.S. Pp. 59. (Manchester : Taylor, Garnett, Evans and Co., 1900.)

THE first of these pamphlets would amply justify its publication, if it only served to emphasise the necessity of further investigation into the methods of estimating carbon dioxide in the atmosphere. It is partly experimental, partly bibliographical in character. The authors, finding themselves called upon to make a series of observations on the carbon dioxide of the air, have made a careful study of Pettenkofer's method, and have introduced some necessary corrections, without detracting very much from its simplicity. They take the precaution, suggested by other observers, of preventing the action of the baryta solution on the glass by coating the vessel with a layer of paraffin wax. It may be pointed out that a solution in benzene is more suitable than the melted wax. The thinner film obtained with the solution is less liable to crack. The baryta solution is manipulated very ingeniously out of contact with air. Yet with all these precautions the results show that perfection is far from being attained.

In the set of analyses on page 132 there is a discrepancy between the highest and lowest figures of 10 per

cent., in another set on the same page the difference amounts to 17 per cent., and on the next page to 20 per cent. .

It seems superfluous to introduce the third decimal into the result when the experimental error affects the first decimal place, and equally unnecessary to make a correction for aqueous vapour, which only amounts to about one and a half per cent. on the volume of carbon dioxide, as against 10 per cent. or thereabouts from experimental error.

The authors omit to mention how long the baryta remains in contact with the sample of air. This is an important factor which should not be neglected, for there can be no doubt that the absorption of carbon dioxide by the baryta proceeds at a rapidly decreasing rate and that the final traces of the gas may take many hours to disappear.

The book is full of useful information, drawn from a variety of sources, the collection of which must have cost the authors no little trouble. At the same time, one is inclined to think that the value of the information would have been enhanced if they had gone another step and made a critical selection from the mass of analyses which they reproduce, for the figures cannot all be equally trustworthy, and many of them must be entirely illusive.

The second pamphlet relates to domestic hygiene. It treats of the effects produced on the air of rooms by the use of gas, coal and electric light for heating and lighting purposes. The effect is determined by estimating the amount of carbon dioxide by Pettenkofer's method, and by exposing a layer of permanganate solution to the air and finding the quantity of the salt reduced.

Mr. Jones, unlike the authors of the previous pamphlet, is not troubled by misgivings about Pettenkofer's method, except in the matter of the baryta attacking the glass. He therefore substitutes lime-water as the absorbent, apparently unaware of the fact that its effect on glass is precisely of the same character, which may be easily observed by placing very dilute lime-water coloured with phenolphthalein in any glass vessel ; in a day or two the colour will be completely discharged. As the results here are only required for comparison, great accuracy is not requisite, and the ordinary method may be relied on. The results of the permanganate method will scarcely serve to recommend it. We find, for example, that in two experiments made on July 21 two-thirds the quantity of permanganate was reduced in the one case in double the time. As Mr. Jones points out, the quantity of dust may affect the rapidity of reduction. If this is the case, effective ventilation will produce disturbance of the air and movement of dust as well as local currents from gas-jets, and it will be difficult to differentiate the two. The results which Mr. Jones obtains are precisely what might be anticipated if we take into account the fact that a coal fire produces an enormous air current through a room near the floor level, whereas a gas fire usually only serves to carry away its own products of combustion. Mr. Jones finds that the purest atmosphere is maintained with a coal fire and electric light ; then follow gas fire with electric light, coal fire and gas light, and gas fire and gas light. The worst effect is produced by an open gas cooking stove without flue. The author shows, moreover, that gas light is more deleterious than a gas fire. That the worst atmosphere exists at the top of a room where the heated products of combustion accumulate is only natural. That is the reason, it is to be presumed, why the topsy-turvy method of ventilating at the floor level with a coal fire is the one most generally in use.

J. B. C.

OUR BOOK SHELF.

Lamarckiens et Darwiniens; Discussion de quelques Théories sur la Formation des Espèces. Par Félix Le Dantec, Chargé du Cours d'Embryologie générale à la Sorbonne. Pp. 191. (Paris: Félix Alcan, 1899.)

THIS is a well-intended, but scarcely adequate, endeavour to reconcile the Darwinian with the Lamarckian conception of evolution. While admitting the principle of natural selection as an important factor in organic development, the author seeks to explain the origin of species mainly on a Lamarckian basis. It may be doubted whether his suggested compromise will commend itself to either party. We are of opinion, *pace* M. Le Dantec, that Darwin's estimate of Lamarck was perfectly just; and that if Lamarckian views are to prevail, it must be by dint of facts and arguments other than those adduced by Lamarck himself. The present volume contains nothing approaching a demonstration of the inheritance of acquired characters; and until this is forthcoming, the Lamarckian fabric must be held to lack foundation. It is curious that the author, who has undoubtedly grasped the principle of natural selection, should not see how groundless is his hesitation in applying it. A reason for this failure is doubtless to be found in his tendency to deal with cases of adaptation as if they were ready-made; he has apparently not taken into account the evidence of gradual approximation to the completely adapted condition. How, he asks, can chance have produced the aspect of *Kallima*? A study of allied forms might have shown him that his question was wide of the mark. On the crucial subject of mimicry and protective resemblance, this strange reluctance to carry an admitted principle to its legitimate end produces especially unfortunate results. M. Le Dantec is constrained, not only to suppose that the white of Arctic animals may be a direct result of the colour of their surroundings "as in Poulton's experiments on caterpillars," but to assume the conscious adoption of appropriate habits on the part of protected organisms. It would seem that not much is here gained by the abandonment of the Darwinian standpoint. In the last few chapters of the book the author expounds his "bio-chemical" theory of heredity, but without throwing any new light on the familiar difficulties of the subject. It is open to any one to proclaim his faith in the essentially chemical character of all kinds of protoplasmic activity, but the fact remains that among these phenomena there is a residuum which does not easily relate itself with what is known of the properties of other kinds of matter. This is where the problem was found by M. Le Dantec, and this is where he has left it.   F. A. D.

Helen Keller: Souvenir. Pp. 65. (Washington: Volta Bureau, 1899.)

THE achievements of Miss Helen Keller bear striking testimony to what it is possible to accomplish in the education of the deaf. Though totally blind, as well as deaf, from infancy, she succeeded in passing the examination for admission into Radcliffe College, Harvard University, a year ago. In honour of this remarkable result, the Volta Bureau, which exists for the increase and diffusion of knowledge relating to the deaf, has published this souvenir, containing an account of her career, con-

tributed by Dr. A. Graham Bell, Miss A. M. Sullivan and other instructors, and herself.

Dr. Bell considers that the lesson taught by Miss Keller's case is that books should be used at the earliest stages of a deaf or blind child's education. "I would have a deaf child read books in order to learn the language," he remarks, "instead of learning the language in order to read books." Miss Sullivan describes how she gave Miss Keller books printed in raised letters long before her pupil could read them. Words of particular shapes were associated with particular objects and actions, and in a comparatively short time Miss Keller thus acquired an exceptional knowledge of the English language.

Miss Sullivan employed the manual alphabet exclusively as a means of communication at the commencement of the child's education. She adopted the method of talking to Miss Keller just as she would to a seeing and hearing child, spelling into her hands the words and sentences she would have spoken to her if she could have heard, in spite of the fact that at first much of the language was unintelligible to the child. Three years after beginning to communicate by means of the manual alphabet, Miss Keller began to try to imitate sounds. Some deaf children are taught to speak by imitating the movements of the lips of the teacher. Miss Keller could not see these movements, but she could feel them by touching her teacher's lips, and she was soon able to reproduce the same sounds and articulate words.

How Miss Keller was prepared for admission into Radcliffe College, the entrance examination of which is exactly the same as that of Harvard College, is described by her instructors, and Miss Keller gives a simple chronological account of her studies. The whole statement is a remarkable narrative, and will be of the deepest interest to teachers of the deaf and students of psychological development.

The Psychology of Reasoning. By Alfred Binet. Translated by Adam Gowans Whyte. Pp. 191. (Chicago: Open Court Publishing Company. London: Kegan Paul and Co., Ltd., 1899.)

THIS is a translation of the second edition of M. Binet's well-known book, the main object of which is to show the essential similarity of perception and reasoning, and to illustrate the nature of the latter by our more complete knowledge of the former, especially with reference to the part played in perception by mental images. When the "Psychologie du raisonnement" first appeared, the more or less novel facts about mental imagery discovered by Galton and Charcot were described in a clear and interesting manner, and this feature remains without alteration. In fact no appreciable alterations have been made in the present edition, even when called for; thus Parinaud's evidence in favour of the central seat of after-images is repeated, although generally acknowledged to rest on a misconception of the relations between the two eyes. The book is full of interesting psychological facts; but unfortunately most of these are drawn from hypnotic experiments, and Binet does not yet appear to have recognised that, owing to the influence of unconscious suggestion, it is very dangerous to found psychological theories on such a basis.

Nevertheless M. Binet's work should be very welcome in an English form, and this the more so that the translation has been very well done.

Electric Batteries: How to Make and Use Them. Edited by P. Marshall. Pp. 63. (London: Dawbarn and Ward.)

THE principal forms of primary electric batteries are described in this little book, and some serviceable details are given concerning their working and use. The book will be particularly helpful by amateur electricians; and students of electricity will find in it some information not usually given in the text-books.

LETTERS TO THE EDITOR.

[The Editor does not hold himself responsible for opinions expressed by his correspondents. Neither can he undertake to return, or to correspond with the writers of, rejected manuscripts intended for this or any other part of NATURE. *No notice is taken of anonymous communications.]*

Snow-drifts on Ingleborough in July.

ON July 4 last I was on Ingleborough with a party of geologists examining the swallow-holes which mark where the water, running off the impervious drift and shale above Newby Moss on the southern shoulder of the hill, first reaches the Mountain Limestone. Some of these swallow-holes are what we may call obsolete—that is to say, when new openings have been formed and enlarged as time went on, some of the chasms which obviously at one time carried off the flood-water from a large gathering-ground, now receive only what oozes in from the peat and drift immediately round it, or the rain which falls directly on it : and rain seldom falls vertically up there. Some of them seem to have been developed without any large body of water having ever invaded them. They run to great depths, the open shaft being from 30 to 360 feet deep, below which the cavernous rock carries the water on through caves and crevices and open joints far down to the valley below.

In one of the vertical caves, which lies east of Long Kin and runs down into the limestone rock to a depth of some 70 feet, there were masses of snow to 4 feet in thickness. It was speckled brown on the surface, from the particles of peat which had been blown on to it, but was pure and white within. Obviously the chasm had been filled by drifted snow during the winter, and the summer's sun could not reach that depth to melt it, while the earth temperature was lost on the moist pinnacled rock on which it rested. No flood ever filled this particular chasm with a swirling torrent, such as at times fills Weathercote Cave, Hunt Pot, or even Gaping Ghyll up to the brim, and causes them to overflow.

Here, therefore, we had an accidental combination of conditions favourable for the preservation of snow, long past midsummer, at a height of not more than 1500 feet above the sea, on the flank of a Yorkshire mountain.

This is an interesting fact to bear in mind when speculating upon the causes of glacial conditions having so recently prevailed over the British Isles. We see here that half-way up Ingleborough, in an exceptionally hot summer, the air temperature alone could not remove last winter's snow.

T. McKENNY HUGHES.

The Total Eclipse of the Sun of May 17-18, 1901.

IN the *Nautical Almanac Circular* (No. 18) local particulars of this eclipse are given for four places in the eastern portion of the shadow track, three of which, Padang, Pontianak and Amboyna, are situated in Netherlands India. In the explanations it is mentioned that, from inquiries which have been made, it appears that the positions selected are the most accessible, and that it would probably be impossible for observers to occupy any neighbouring station for which the astronomical conditions might be more favourable.

Surely these inquiries have not been made on the spot, where the information could best have been obtained. Other localities in the Government of Sumatra's west coast are as easily accessible as Padang, from which a railway leads to the interior : and other localities on the banks of the Kapuas as easily as Pontianak. Moreover, many other places may give opportunity to eclipse parties for observation, viz. on the islands of Lingga and Singkep, on the banks of the Barito and the eastern coast of Borneo, in the Gulf of Tomini (Celebes) and in the Moluccas. The conditions, however, will be most favourable in the western part of the Archipelago, both on account of the longer duration of totality and also for local resources. Through the Koninhlijhe Natuurkundige Vereniging at Batavia, data have been gathered referring to the conditions of weather and cloudiness at a number of stations most suitable for the observation of the eclipse, and the data will be published in due time. The general impression is, however, that the chance for fine weather is nowhere very great. The Society will be pleased to procure full information as to the choice of stations, and observers may be sure to receive every available assistance from the local authorities and officials in the Dutch colonies.

J. J. A. MULLER.

(President of the Kon. Natuurk. Ver., Batavia, July 17).

The Reform of Mathematical Teaching.

MANY schoolmasters tell us that boy-nature is so depraved that his time must be fully occupied, and that a "regular hard grind" is the only way to keep him out of mischief. They give him things to grind that do not interest him ; it may be that he does not understand them, or that they have no human interest. And yet every boy has interests, and teaching directed towards those interests would enthral him. The first aim should be to attract the boy's attention, and a subject which no excellence of teaching will make interesting to a particular boy is no fit subject of study for him.

The case of mathematics is bad. The reasoning is too abstract for a boy's mind, but it has worse faults. Long strings of reasoning are employed to deduce fairly obvious conclusions from premises no more obvious, *e.g.* in the theory of parallels. On the other hand, incorrect proofs are given because the boy cannot grasp correct proofs. *e.g.* for the binomial theorem. Geometry is in worse case than algebra. Euclid's interest was logical rather than geometrical ; he wished rather to put together a consistent series of syllogisms than to give the best solution of his problems : witness his bisection of a straight line. In consequence, the natural order of development is lost sight of. A boy ought to be at home with ruler and compasses before he reasons about the constructions possible with them, and yet most schoolboys have never handled compasses. A few weeks ago I asked some hundred boys in a well-taught school (as present teaching goes) to give a certain construction of Euclid's, and also to carry out the construction with ruler and compasses on a given line. Hardly one failed to write out the construction and proof, but only one of the hundred carried out the practical construction. Clearly our present Euclidian teaching has little to do with geometry.

To lay before a boy a proof he does not understand is useless, to prove the obvious is confusing, to give an incorrect proof is immoral. Prof. Perry's plan to abolish proofs in the early stages is a great step in advance of present teaching. For the boy of mathematical ability it would perhaps be well to run theory alongside, at the rate of five or six propositions for Euclid's entire first book. This would, however, interfere with class teaching, and the mathematical boy would lose little by going through a good deal of the practical course before touching theory ; if with a hint here and there he could be got to evolve the theory for himself, he would gain much.

Possibly a theoretical training leads one to look with too favourable an eye on early theory. In any case, that in the hands of a good teacher theory and practice could well go hand in hand for boys even of average ability is shown by two able papers by Mr. Branford, in the *Journal of Education*, on the first teaching of geometry. We may finally reach this stage, but till we have these good teachers practice should precede theory.

DAVID MAIR.

Functions of an Organ of the Larva of the Puss Moth.

THIS season I am breeding, with the object of observing their gradual growth and development, a number of the larvæ of *Cerura* (or *Dicranura*) *vinula*, the Puss Moth ; but I have sought in vain for the function performed by the slender red filaments, ejected, at the insect's will, from its twin tails. They appear to shoot from their sheaths, just on the same principle as do a cat's claws ; but to what purpose?

Surely such a beautiful, delicate organism could not have been appointed to no purpose ! Is this merely an instance of entomic mimicry, simulating, for its own protection, the sting of some venomous insect ; or does this strange organ perform some practical, active function?

I shall be very greatly obliged if you can tell me whether this point has been already decided or not, and, in the latter case, perhaps some of your correspondents will kindly communicate their views upon it. ARTHUR S. THORN.

4 Malcolm Road, Penge, August 10.

Dark Images of Photographed Lightning Discharges.

A VERY clear illustration of the reason why some of the lightning discharges in a photographed thunderstorm appear dark was afforded me at Wednesfield, Staffordshire, about mid-day on Thursday, July 19. There were a number of double flashes, that is, two discharges occurring rapidly in the same apparent region, but following different courses, and separated in time by from one-eighth to one-half of a second. But one flash, quite near to where I stood (one second and a half between flash and sound), gave a repetition following absolutely the same path as the first flash and practically as bright. The only difference was that two faint branches of the first flash were not repeated in the second discharge. The second flash followed so quickly (about an eighth of a second, I estimate), that the impression on the retina of the first discharge had not died out when the second exactly covered it, so that I could appreciate the absolute coincidence. A few kinematographic records of thunderstorms would show whether or not such repetitions are common, and whether they are the cause of dark flashes on the photographic plate.

Cave Castle, Dumbartonshire, N.B.   J. B. HANNAY.

THE LAVOISIER MONUMENT.

THE monument erected by international subscription in honour of Lavoisier was unveiled on July 27, in the presence of M. Leygues, French Minister of Public

FIG. 1.—The statue of Lavoisier.

Instruction, and many eminent men of science, including most of the members of the International Congress of Chemistry. The committee entrusted with the raising of the fund for the statue succeeded in obtaining a sum of 100,000 francs, which was subscribed by admirers of Lavoisier in most parts of the civilised world. M. Moissan was the secretary of the committee, and he acknowledged at the unveiling ceremony that there had

not been the slightest difficulty in obtaining the means to erect the monument—many subscribers, indeed, were astonished to learn that Paris, where monuments abound, did not possess a statue of the eminent chemist whose investigations helped to lay the foundations of modern chemistry. It is true that appreciation of the great chemist has been shown by the publication of his complete works, but these are only known to a limited number of students, and the people who live in the present are likely to forget how much they owe to the past unless they are reminded of their indebtedness by some striking monument in bronze or stone. For this reason, it is well that a permanent memorial of Lavoisier's greatness has now been erected.

The statue, which is represented in the accompanying illustration from La Nature, is erected in the open space behind the Madeleine Church, close to the house where Lavoisier lived for some years. It is of bronze, and stands upon a granite pedestal ornamented with bas-reliefs. The statue is by M. Barrais, and the pedestal is due to M. Gerhardt. Upon the front of the statue the following inscription appears, in French, "Antoine Laurent Lavoisier, 1743-1794, founder of modern chemistry. Erected by public subscription, under the patronage of the Academy of Sciences. M. Berthelot, Permanent Secretary of Physical Sciences, 1900." One of the bas-reliefs represents Lavoisier explaining his discovery of the composition of air to his colleagues of the Academy of Sciences, of which he was president, the characters introduced into the scene being Monge, Lagrange, Condorcet, Berthollet, Vicq d'Azyr, Laplace, Lamarc and Guyton de Morveau. On the other bas-relief Lavoisier is shown in his laboratory dictating notes to his wife. The statue appears to be a real work of art, worthy of the sculptor and of the subject.

M. Berthelot was to have presided at the ceremony of the unveiling, but illness prevented him from being present, and his address was read by M. Darboux.

Reference was made to the fact that the inauguration took place under the auspices of the Institute of France, the City of Paris and the French Government, and stress was laid upon the truly international character of the homage to the genius of Lavoisier, as testified by the subscriptions. The following is a free translation of parts of the address :—

The names of Galileo, Newton, Leibnitz and Lavoisier show that science has no nationality, a monopoly of pure or applied science being the property of no one nation. The erection of a statue in a public place is an honour usually reserved for statesmen and warriors, men who have spattered the earth with blood, too often without lasting profit to the nation devoted to them. To-day the famous savant, thinker, artist, is put in the first rank by enlightened people, and posterity will doubtless continue to show an increasing respect for the memory of those men who have served the human race, and to relegate to obscurity the men of blood and intrigue who have enslaved it.

The work of Lavoisier is epoch-making from two points of view, from that of philosophy, because he established the fundamental law which governs the chemical transformations of matter, and from the practical point of view, because this law has become the base of innumerable industries founded on these transformations, and the origin of the rules of hygiene and therapeutics which follow from it. The fundamental discovery of Lavoisier was the distinction between matter and the imponderable agents, such as heat, light, electricity, which influence it. The discovery of this distinction overthrew all the old ideas dating from antiquity, and which continued up to the end of the last century. According to the ideas which were current when Lavoisier started his work, there were four elements —earth, air, fire and water—from which all substances existing in nature were said to be built up. By associating these elements in different proportions and by different methods, it ought to be possible to produce all bodies and transform any one into any other. As a matter of fact, the prolonged researches of serious workers had never succeeded in establishing this transformation, nor has this been accomplished since. But preformed ideas are tenacious, especially when supported by mysticism.

An equally grave mistake was committed in supposing that bodies submitted to the influence of heat alone could vary in weight, a variation apparently proved by innumerable observations with the balance in chemistry. It is, in fact, a most singular error, although one frequently held, that the use of the balance in chemistry dates only from the end of the last century. In reality its use is sixteen centuries old. The balance was used both in chemistry and in trade; it may be seen represented on the monuments of ancient Egypt. Bodies such as coal, oils and organic substances under the action of heat were known to lose their weight, hence was drawn the conclusion that matter may be transformed into heat and disappear; whilst heat, on the other hand, under inverse conditions, could be fixed, becoming visible ponderable matter. These opinions gave way to the views of Stahl, according to whom combustible bodies were rich in phlogiston, or fixed heat. Such was the state of science about 1772, when Lavoisier appeared on the scene. Ten years were sufficient for him to effect a complete transformation. He established, by the most precise experiments, a fundamental distinction, previously unknown, between the nature of bodies which we know, and heat and other agents capable of modifying these; it is the distinction between ponderable bodies and the imponderable heat, light, electricity, the intervention of which causes no change of weight in ponderable matter.

It could hardly have been expected that one man alone should make all the researches establishing the properties of gases, the composition of air and of hot water, and in this respect there can be no doubt that Lavoisier profited by the partial work of his predecessors and of contemporary workers; but to him alone belongs the merit of demonstrating the connecting links, and of giving the facts their true interpretation.

Two fundamental problems were first attacked by Lavoisier, the gain in weight of metals on calcination, and the apparent loss of weight of carbon, sulphur and oils on combustion. His first discovery was to put these phenomena upon a proper experimental basis. He demonstrated that in all such cases a weighable substance contained in the air takes part in the change, the addition of which explains the increase of weight of the calcined metals, an increase equal to the loss of weight sustained by the air. The same ponderable element in the air was shown to take part in the burning of carbon, sulphur and oils, forming gaseous compounds, the weight of which was also determined by Lavoisier. It was thus established, what had never been done before, that the materials of bodies possessing weight kept this weight constant throughout a series of chemical changes, heat and other agents of the same order having no effect either in increasing or decreasing the weight of the original bodies. This fundamental distinction between ponderable matter and imponderable agents is one of the greatest discoveries that has ever been made; it lies at the base of physical, chemical and mechanical science. Lavoisier, however, went farther than this, and attempted to penetrate the constitution of ponderable matter itself. He recognised that in all known experiments it presents itself as constituted by a certain number of undecomposable elements or simple bodies, which, combining amongst themselves, form all known compounds.

The two fundamental laws of nature once established—the distinction between matter and imponderable agents, and the invariability of the nature and weight of the simple bodies— Lavoisier went on to draw important conclusions on the composition of the acids and metallic oxides, the composition of air, water and organic substances, on the *rôle* of heat in chemistry, on animal heat and on the nature of respiration in physiology.

What share ought to be now attributed to Lavoisier in the classical discovery of the compound nature of air and water, a

discovery in which he competed with Priestley and Cavendish? The matter would take too long to give here in detail. Suffice it to say that he alone swept from the composition of air and water the erroneous notion of phlogiston maintained by his contemporaries.

All these discoveries, accumulated in the course of only a dozen years, and carried out with wonderful ardour and energy, were not simple proofs of isolated facts; on the contrary, they were the consequences logically deduced and experimentally demonstrated of the two fundamental laws due to the genius of Lavoisier. The physiological questions relating to respiration were also answered completely and successfully; given a correct knowledge of the elementary composition of carbonic acid, of food materials and of air, respiration was then obviously a slow combustion of food by the oxygen of the air, a combustion producing carbonic acid and developing at the same time sufficient heat to maintain the human body at a nearly constant temperature.

A complete account of the after effect of Lavoisier's work would require almost a history of physical science during the nineteenth century; but an attempt will be made to recapitulate the more immediate consequences upon existing knowledge. The notion of the invariability of the weights of the simple bodies dominates the whole of chemistry at the present time; it is the scientific basis of all our chemical equations of composition and constitution, the origin of the new and singular algebra which, from its origin in the works of Lavoisier, so struck the mathematicians of his time. It is also the solid foundation of all our analyses, and is a certain starting-point in industries the most diverse, the manufacture of acids, alkalis, colouring matters, scents, drugs, in metallurgy and in agriculture.

And here a necessary reflection occurs. It cannot be pretended that Lavoisier was the direct and personal author of the vast array of discoveries here enumerated; but it is certain that it is he who has established the solid base upon which the modern chemical edifice is constructed, and without which these discoveries could not have been made; it is he who has raised the flaming torch of truth which we daily invoke, and for that reason it is just and equitable to give to him a part of the glory of the inventions of science and modern industry.

NILE FLOODS AND MONSOON RAINS.

THE practice or science of weather forecasting will evidently proceed on two very different lines— according to the relative importance of local or seasonal changes in the general meteorological conditions, and whether the prediction has reference to a long or a short period. The machinery employed in cases where the forecast aims at great minuteness over a small area consists mainly of the synoptical chart, based on information supplied by rapid telegraphic communication, and in the hands of experts this means probably proves sufficient, and furnishes a fair percentage of accurate predictions. But in the more difficult, as certainly the more important, problem of predicting the weather some time in advance, and over a considerable area, a problem which regularly recurs in the monsoon forecast for India, one must evidently depend on the more general physical conditions that are produced by the motions of the earth and the distribution of land and water on its surface. These causes, it is true, are always operative, and to a certain extent meteorological phenomena, broadly considered, must be periodic in their main features. The causes of deviation from periodicity, and the extent of the area affected by such abnormal conditions, are problems which the professional meteorologist has to encounter, and it is to be feared with insufficient means. But it seems not unlikely that, in proportion as the problem becomes more general, by bringing wider areas within the scope of the discussion, the prospects of greater success will become more assured; and it cannot but be considered a most significant feature that indications are not wanting that in the two considerable areas, India and Egypt, the respective climates betray peculiarities which may either react upon each other, or the origin of which must be sought in a common source.

From two independent investigations come attempts to trace a connection between the amount of the Nile floods and the abundance or deficiency of the south-west monsoon rainfall in India. Mr. Willcocks broached this subject in a paper read before the meteorological congress at the World's Exposition in Chicago, and there suggested that famine years in India are generally years of low flood in Egypt, and that when the summer supply of the Nile had been deficient and late, a high flood might well follow, since the drought in the valley of the White Nile must create a powerful draught from the Indian Ocean or the Arabian Sea, a district in which is to be sought the origin of the massive current of the south-west monsoon. Unfortunately, any exact data to establish this interesting connection are not forthcoming, and can hardly be expected, since the Nile is supplied from two distinct sources, and it is impossible to separate and trace the effect of either contribution. Of the great lakes of Central East Africa which constitute a reservoir for the Nile waters, little is known as to the variation in their relative height due to the rainfall in their vicinity, which lasts from March to December. At Port Alice, on the Victoria Nyanza, and at some other stations, observations, more or less regular, are made of the variation in the heights of the water, but in the absence of any common datum level these heights are referred to that of the mean lake. Much surveying work and long-continued observations will have to be made before these scanty statistics can be turned to full account. Of the second source of supply to the Nile, viz. the flood waters in the Atbara, the Blue Nile and other rivers, fed during the rainy season from June to November, we know practically nothing as to their amount. But it is this seasonal supply which is probably the greatest factor in causing variations in the Nile floods, and where a connection with the causes of the Indian rains is closest. Whatever influences the flow of the monsoon current from the equator northwards over the Indian seas towards the heated regions of India and the Malay Peninsula must have a proportional effect on East Africa and South Arabia. With heavy monsoon rains, therefore, it is not unlikely that the contributing rivers add materially to the volume of the Nile waters, but it is not altogether a trustworthy guide to gauge the amount of water that enters the Nile by measuring the quantity that passes a particular station. Much water enters the Nile that never contributes to the irrigation of Egyptian lands. Of the amount lost by evaporation no account can be taken, but a source of greater error arises from the peculiar flatness of the ground about Shambé, which forms the apex of the swamp delta. Here the Nile can spread its waters over a large area, and practically lose itself as a river among the beds of reeds and rushes, which form a veritable swamp. Engineering works, already projected or actually begun, aim at clearing some or other of the feeding streams, such as the Bahr el Gebel or the Bahr el Zarab, and the effect must be, when completed, to break the continuity of such observations as have been made. Other sources of error are to be found in the varying quantity and character of the "sudd" which may interrupt the flow or diminish the amount of evaporation ; but without insisting on too much accuracy, there exists a certain amount of evidence that the two great agricultural countries of Egypt and India are likely to be prosperous together or to suffer in common.

Mr. Eliot, the meteorological reporter to the Government of India, in his recent forecast of the probable character of the south-west monsoon rains of 1900, not only fully endorses Mr. Willcocks' statement, but adds some statistics which render a connection highly probable. Omitting a few local particulars from Mr.

Eliot's statistical summary, the broad features are shown below.

Year.	Variation of mean rainfall of year from normal.	Character of Nile flood.
	Inches	
1876	− 4·49	Good high flood.
1877	− 4·28	Poor flood.
1891	− 3·54	Late flood.
1896	− 4·83	Low Nile.
1899	− 11·14	Very low flood : lowest of century.

The years of excess of Indian rainfall tell a similar tale, even more distinctly.

Year.	Rainfall variation in inches.	Character of Nile flood.
1878	+ 6·34	Very severe flood : banks of river carried away in October.
1886	+ 3·02	High flood.
1892	+ 5·09	Very high and late flood.
1893	+ 9·07	High flood.
1894	+ 6·47	High flood.

Having mentioned some of the causes which prevent a rigorous comparison between the Nile floods and the Indian rainfall, one is not unprepared to find some discrepancies ; but Mr. Eliot certainly does not overstate his case when he contends that these tables indicate that in at least four out of five seasons in which there was a partial failure of the rains in India there was a low Nile, and that generally the two countries are similarly affected by the meteorological conditions and the variations of those conditions. The causes of these variations are obscure, and at present very imperfectly recognised, for a complete solution, as Mr. Eliot points out, demands a much more intimate knowledge of the atmospheric conditions that prevail over a large area. The meteorology of Australia and the Indian Ocean, and perhaps also of the Antarctic Ocean, must be linked on to that of the Indian monsoon area " before it will be possible to ascertain the missing factors necessary to complete the explanations of the relations between the chief features of the monsoon currents and rainfall of India and the antecedent and concurrent conditions in the Indian area and the regions to the south." To trace and anticipate the effect of weather conditions over the area that embraces both India and Egypt, in which our interests are so largely involved, should stimulate further inquiry, with the result of placing at the command of science additional means for dealing with so grave a problem.

THE FORTHCOMING MEETING OF THE BRITISH ASSOCIATION AT BRADFORD.

IN the last article on the subject of the forthcoming meeting of the British Association an account was given of the handbook that is to be published in connection with the visit, and some information was furnished in regard to hotel and lodging accommodation. In the present article it is proposed to give a description of the excursions arranged by the local committee.

Following the custom of former years, it has been arranged that half-day excursions only shall take place on the Saturday, and that the whole-day excursions shall be reserved until the Thursday, when the serious work of the Association will be completed. The only exceptions to this are that the Mayor and Corporation are inviting a small party of engineers to visit their waterworks at Gowthwaite, in the Nidd Valley, and that a party exclusively for geologists will travel to Pateley Bridge by the same train in order to visit the Brimham Rocks. These two excursions will occupy the whole of Saturday. The excursions, then, arranged for Saturday, September 8, are as follows :—

BOLTON PRIORY.—The party will leave the Bradford (Midland) Station at 1.32. Drive from Bolton to the Priory, where they will be received by the vicar, the Rev. A. P. Howes, who will give a brief description of its history and architecture. They will then drive forward along the banks of the Wharfe to the Wooden Bridge, where tea will be provided : an opportunity will be given for a visit to be made to the Strid (celebrated by Wordsworth), and then the party will be driven back to the station in order to reach Bradford in time for dinner. Mr. Geoffry Fison will be the leader of the excursion.

FARNLEY HALL.—The residence of the Fawkes family, which contains a wonderful collection of Turner's pictures. The Hall is of great historical interest, as it was the residence of Lord Fairfax in the time of the Civil War, and many relics of the period are shown. The party will leave the Midland Station at 1.15 for Otley, where they will be met by Major Mitchell, of Cayley Hall, the leader ; they will then be driven to Farnley, and as much time as possible will be spent in inspecting the Turner pictures and the beautiful old Hall. Major Mitchell will afterwards entertain the party to tea in his grounds, and they will then be driven back to the station.

ILKLEY.—The excursion will start from the Midland Station at 1.32, under the leadership of Mr. Mortimer Wheeler ; at Ilkley the party will be divided into several smaller bodies, who will in turn visit the Roman camp and fortifications, some curious Saxon crosses that are to be seen in the churchyard, and some remarkable instances of cup and ring marks, which are to be seen on Rombald's Moor above the village. At 4.30 the different parties will reassemble in the beautiful grounds of the Wells House Hydro, where they will be entertained to tea by the invitation of the directors. They will arrive back in Bradford about 7 o'clock.

HAWORTH.—The train will leave the Midland Station at 1.20, and the party will be met at Haworth by Mr. F. Greenwood, the president of the Brontë Society, who will escort them to the church and the Brontë Museum, and show them many places which will be familiar, from description, to the readers of "Shirley." The leader of the party will be Mr. J. A. Clapham.

KNARESBOROUGH.—Major H. D. Sichel will conduct a party to Knaresborough, the train leaving at 1.15. On arrival, the visitors will be divided into two parties, and, under the leadership of Major Sichel and Mr. Arthur Harris respectively, they will be taken by opposite routes to visit the Castle, the petrifying Dropping Well, and Eugene Aram's Cave. Afterwards they will be driven to Plumpton Rocks, where tea will be provided, and they will return by a train reaching Bradford about 7.30.

KIRKSTALL ABBEY AND ADEL.—The train will leave the Midland Station at 1.25, and Kirkstall Abbey will be described by Mr. E. Kitson Clarke, the leader. The visitors will then be driven to Adel Church, which is almost a unique instance of Saxon architecture, and which will be described by the vicar. They will then drive back to the Yorkshire College, Leeds, where they will be entertained to tea by the principal, Dr. Bodington, one of the vice-presidents of the Association.

PATELEY BRIDGE.—As indicated above, this is the only excursion extending over the whole day. Two parties will leave by a special train at 1.15, the one conducted by the Mayor (Mr. Wm. C. Lupton, J.P.), for a small party of engineers, who will be driven from Pateley to the Nidd Valley Waterworks ; the other, exclusively for geologists, who, under the leadership of Mr. J. Lower Carter, will walk to the Brimham Rocks, and visit other places of geological interest.

For the week-end (September 8-10), the Yorkshire Naturalists' Union are organising a specially interesting excursion. The district which has been selected is the neighbourhood of Grassington, in Upper Wharfedale, which is not merely a romantically picturesque region, but a remarkably good district for nearly all branches of natural history and geology. The excursion is intended, as far as possible, to be one strictly for practical working naturalists, and as accommodation is very limited, it will be needful to give preference to such members of the British Association as are likely to investigate in their own particular department. The arrangements will be under the direction of leading Yorkshire naturalists, who hope to introduce their comrades from other parts of the country to a remarkably interesting district. There will be the usual fully descriptive circular prepared, which will be sent to any one who may apply for it to the hon. secs. of the Yorkshire Naturalists' Union, Leeds.

THURSDAY, SEPTEMBER 15.—The whole-day excursions arranged for the concluding day of the meeting are as follows :—

THE ACKTON COLLIERY.—This excursion, which will be under the leadership of Mr. C. J. Cutcliffe-Hyne, is intended for a limited number of botanists, geologists and engineers, in order that some opportunity may be given them of examining the Yorkshire coal-measures. The party will be divided into two on arriving at Featherstone : the one will be taken down the pit, while the other will examine the machinery and various interesting material on the bank. They will then meet at one o'clock and will be entertained to lunch by Lord Masham, the owner of the mine, after which the proceedings will be reversed, and the respective parties will be taken round the bank and down the pit ; they will then reunite, and after partaking of afternoon tea will return to Bradford.

BOLTON PRIORY.—This is an amplification of the previous half-day visit, again under the leadership of Mr. Geoffry Fison. Fuller opportunities will be furnished of seeing the Priory and the Strid, and lunch will be provided at the Wooden Bridge. In the afternoon a visit will be made to Barden Tower, the ancient keep of the Lord Clifford, of the Wars of the Roses fame, and of his son, the Shepherd Lord.

RIPON AND FOUNTAINS ABBEY.—Mr. Mortimer Wheeler will conduct a party to Ripon ; after a special musical service at the Cathedral, they will be driven to Fountains Abbey, and lunch will be provided in the Refectory. They will then be taken over the ruins by the Dean of Ripon, after which the Marquis of Ripon will entertain the party to tea. On returning to Ripon, if time permits, they will be conducted to the crypt and the more interesting parts of the Cathedral by the Dean before leaving for Bradford.

SWINTON PARK.—By the invitation of Lord Masham and under the leadership of the Mayor, Mr. William Lupton, a party will visit Masham. On arrival, they will be driven to Jervaulx Abbey, the ruins of which, of course, possess great historical interest, and will then return to Lord Masham's residence, Swinton Hall, the drive each way being of extraordinary beauty. At the Hall they will be entertained to lunch by Lord Masham, after which the afternoon will be spent in inspecting the very fine collection of Old Masters and modern pictures, and the party will drive to the station to join the special train, which will convey also the party from Ripon.

MALHAM.—A party, under the leadership of Mr. Cecil Slingsby, will leave at an early hour for Bell Busk ; thence they will drive across country to Malham, and after lunch they will visit Gordale Scar, and, if time permits, at the invitation of Mr. Walter Morrison, M.P., they will go on to Malham Tarn and Malham Cove. They will leave Malham about 5 p.m. and drive to Skipton, visiting Skipton Castle on the way, and thence by train back to Bradford.

SETTLE AND CLAPHAM.—By the same train which conducts the party to Malham, another party will leave for Settle under the guidance of Mr. J. J. Brigg. After visiting the Victoria Caves, they will drive to Ingleton and lunch. From there they will walk through the beautiful grounds of Mr. J. A. Farrer and explore the Clapham Caves, in which most extraordinary specimens of stalactites and stalagmites are to be seen.

The two last excursions are specially intended for geologists.

YORK.—It is, of course, essential that York, where the first meeting of the British Association was held seventy years ago, should be visited. The party will arrive in York about 11 o'clock, under the leadership of Mr. J. A. Clapham. The visitors will immediately proceed to see the walls, the museum, and St. Mary's Abbey. Then, after lunch at the Station Hotel, they will visit the Minster, where most of the afternoon will be spent. By the invitation of the Lord Mayor, they will afterwards be entertained to tea at the Guildhall before leaving for the station.

For all the half-day excursions a uniform charge will be made, and similarly for Thursday's excursions there will also be a uniform charge. Visitors applying for excursions will be required to hand in this fee, together with the application form ; and tickets, as nearly as possible in accordance with their preferences, will be allotted to them. By making all the excursions of equal cost, it is expected that the work of allotment will be simplified.

The next article will deal with the mayoral and civic functions that have been arranged, and some account will be given of the large garden-party which the municipality will hold on Monday, September 10, and of the various private garden-parties to be given on September 12.

RAMSDEN BACCHUS.

NOTES.

WE regret to announce the death of Dr. John Anderson, F.R.S., the distinguished zoologist.

DR. D. MORRIS, C.M.G., the Imperial Commissioner of Agriculture for the West Indies, has just arrived in this country.

PROF. G. CAREY FOSTER, F.R.S., has been appointed Principal of University College. Prof. Foster is a Fellow of the College, and was formerly professor of Experimental Physics and Quain Professor of Physics ; he is also a Fellow of the University of London, in which University he acted as examiner previous to his election to the Senate.

THE International Geological Congress is now in session at Paris. Among the items included in the programme are discussions on international co-operation in geology, fundamental researches for the establishment of a definitive classification, scheme for an international lexicon of petrology, and the photography of types of fossil species.

REUTER reports that Major Gibbons, the African traveller, reached Omdurman on August 20. The line of route traversed by the expedition represents a distance of 13,000 miles. Among the objects attained were the mapping of Barotseland ; the accomplishment of the first steam navigation of the Middle Zambesi ; and the tracing of the whole course of the river, the discovery of its source, and the determination of its watershed. Thence the route of the expedition was eastward, and by way of the Great Lakes and the Nile.

THE annual meeting of the English Arboricultural Society was held at Manchester last week. Prof. Somerville was appointed president for the ensuing year. Reports were read from the judges upon essays on "Foreign *versus* Native Timber," "Agricultural and Woodland Drainage," and "Thinning." The silver medal for the first essay was awarded to Mr. George Cadell, late of the Indian Forest Department, and bronze medals for the other essays were given to Mr. D. A. Glen, of Kirby, near Liverpool, and Mr. A. Dean, of Egham.

THE third annual report of the Council of the Röntgen Society shows that the society is making satisfactory progress. The demonstrations at the meetings are very valuable to all workers with Röntgen rays, and the papers and abstracts published in the *Archives* enable members who are unable to attend the meetings to keep well in touch with the latest developments of radiographic work. Dr. J. B. Macintyre, one of the earliest and most prominent investigators with Röntgen rays, has consented to be nominated as the next president of the society.

SIR WILLIAM STOKES, the eminent surgeon, died suddenly at Pietermaritzburg on Saturday. He filled the post of President of the Royal College of Surgeons of Ireland in 1896 ; and among his other appointments was the professorship of surgery in the Royal College of Surgeons in 1872, senior surgeon of the Government Hospital of Ireland in 1868, president of the Pathological Society of Ireland, and Surgeon in Ordinary to the Queen in Ireland from 1892. He was the author of a number of addresses, and contributions to the medical press, on clinical and operative surgery.

A REUTER telegram from St. Petersburg states that news has been received there from Dr. Sven Hedin, showing that his expedition this spring to Lob Nor to settle the various questions in dispute regarding that lake and its surroundings has resulted in discoveries exceeding his expectations. He found, in fact, that the lake known to previous explorers no longer exists, having dried up, leaving its bottom strewn with shells and marine growths. Around this old basin, however, a regular system of new lakes has been formed, which Dr. Sven Hedin has explored and mapped. In connection with this announcement, it is worth remark that at the time of the visit of Prince Henry of Orleans to Lob Nor, towards the end of 1889, the lake consisted of a number of interlacing lakes and river-arms, the contraction of the former large water-area being probably due to the using up of the waters of the Tarim for irrigation by the increasing population of the river basin.

THE *Scientific American* announces that the U.S. Congress has granted funds for the inauguration of agricultural experiment stations in the islands of Hawaii and Porto Rico. Prof. S. A. Knapp has been selected to investigate the agricultural conditions and possibilities of Porto Rico. He went to the island in June, and will study the lines of experimental investigation which should be undertaken there, places suitable for stations, and the approximate expense of inaugurating and maintaining the work. Dr. W. C. Stubbs will make a preliminary survey of the conditions in the Hawaiian Islands. He sailed for Hawaii about the middle of July, and will spend the month of August in the islands. The conditions there are somewhat different from those of Porto Rico, as a station for experiments in sugar production has been maintained by private beneficence for a number of years.

THE Berlin Academy of Science has (says *Science*) made the following grants for scientific work : Prof. Adolf Schmidt, of Gotha, for the collating and publication of material on terrestrial magnetism, 750 marks ; Dr. Leonhard Schultze, of Jena, for investigations on the heart of invertebrates, 2000 marks ; Prof. Emil Ballowitz, of Greifswald, for investigations on the structure of the organs of smell of vertebrates, 800 marks ; Dr. Theodore Boveri, of Würzburg, for experiments in cytology, 500 marks ;

Prof. Maxime Braun, of Königsberg, for studies on the Trematoden, 970 marks; Dr. Paul Kuckuck, of Heligoland, for investigations on the development of Phæosporeæ, 400 marks; Dr. Wilhelm Solomon, of Heidelberg, for his geological and mineralogical investigations in the Adamello mountains, 1000 marks; Prof. Alexander Tornquist, of Strassburg, for the publication of his work on the mountains of Vicenza, 1100 marks; Prof. Alfred Voltzkow, of Strassburg, for the drawings of his work on the development of the crocodile, 1000 marks: Prof. Johannes Walther, of Jena, for the publication of his work on deserts, 1000 marks.

WE learn from the *Daily Graphic* that the Norwegian Government has built and fitted out a steam vessel for the express purpose of marine scientific research, and has placed her, as well as a trained staff of assistants, in charge of Dr. J. Hjort as leader of the Norwegian Fishery and Marine Investigations. The vessel herself, the *Michael Sars*, has been constructed in Norway on the lines of an English steam trawler—that type of boat being regarded as the most seaworthy and suitable for such an expedition—but considerably larger, being 132 feet in length, 23 feet beam, and fitted with triple expansion engines of 300 horse-power. The fishing gear includes, *inter alia*, trawls, nets, and lines of all kinds, with massive steel hawsers and powerful steam winches to work the heavy apparatus, while the numerous scientific instruments are of the very best and latest description. The expedition left Christiania in the middle of July, on what may be termed its trial trip along the Norwegian coast (accompanied for part of the time by Dr. Nansen, who was desirous of testing various instruments in which he had made improvements), and has just sailed from Tromsö on a lengthy cruise to the North Atlantic and Arctic Oceans. Dr. Hjort has already added so much to the knowledge of pelagic fishes, their life, habits, and the causes affecting their migrations, that, with the means now at his disposal, a considerable amount of valuable information will probably be gained which will prove of service to the fishing industry of all nations.

MESSRS. JOCHELSON AND BOGORAS, of the Jesup North Pacific Expedition of the American Museum, recently started for the north-eastern part of Asia, by way of San Francisco and Vladivostok, to continue the work of the expedition in Siberia. A few particulars of the investigations undertaken are given in the *American Museum Journal*. The region to be visited is situated north-east of the Amur River. The explorers will study the relations of the native tribes of that area to the inhabitants of the extreme north-western part of America, and also to the Asiatic races visited by Dr. Laufer, under the auspices of the Museum, and to those living farther west. It is expected that in this manner they will succeed in clearing up much of the racial history of these peoples, and it is hoped that the question as to the relations between the aborigines of America and Asia will be definitely settled. Thus the work proposed is part of the general plan of the Jesup North Pacific Expedition, which was organised for the investigation of the relations between the tribes of Asia and America. It is fortunate that this inquiry has been taken up at the present time, since the gold discoveries along the coast of Bering Sea are rapidly changing the conditions of native life; so that within a few years their primitive customs, and perhaps the tribes themselves, will be extinct. It is expected that the journey, which will extend over a period of two years, will result in a series of most interesting additions to the collections of the Museum, and in an important advancement of our knowledge of the peoples of the world.

IT has already been noted (vol. 61, p. 451) that Prof. A. Heilprin has brought forward evidence which throws doubt

upon the permanence of the waters of Lake Nicaragua, the fountain head of the San Juan River. His conclusions have been criticised, but he gives further reasons for them in the *Bulletin* of the Geographical Society of Philadelphia (July), and shows that this new factor will have to be taken into consideration in connection with the proposed Nicaragua Canal. The full conclusions now drawn by Prof. Heilprin from data furnished by the Nicaragua Canal Commission of 1897-99, and the special reports of the chief engineer and hydrographer appended thereto, are:—(1) Lake Nicaragua has undergone marked shrinkage during the period of the last twenty-five to fifty years. (2) The shrinkage is a progressive one, and there are no known conditions by which the loss incurred can be made good. (3) The assumption is well founded that the earlier measurements of the altitude of the lake surface, made by Galisteo and Baily, indicating an abasement of the waters by 20 to 30 feet, were accurate. The relations of these conditions to canal construction become immediately apparent, and it may well be agreed that a region subject to the changes which have been indicated "would offer serious obstacles to the construction of a canal of the magnitude of the one proposed or to its permanency after construction."

MR. W. N. SHAW, F.R.S., informs us that Mr. W. Kennedy, the observer for the Meteorological Office at Roche's Point, co. Cork, notes that at 9.15 p.m. (G.M.T.) on August 13 a very large meteor shot into view eastward, going E.S.E. At about an altitude of 70° it exploded with a brilliant flash, and a noise was heard like that of a rocket fired off at some distance. The meteor left a long luminous track visible for some seconds after the explosion. The trail would have been very brilliant but that the eastern sky was lit up by the moon at the same time.

IN the afternoon of Friday, August 17, some parts of the south of London were visited by one of the sharpest thunderstorms that have occurred for some time. The weather was very close, the thermometer reaching 82°, and the distribution of barometric pressure was of a complex character. During the storm, which lasted about an hour, and was accompanied by a heavy hail-squall, the amount of rainfall at the central part, near Herne Hill, was 1·2 inch. In some parts of the suburbs the roads were completely flooded, while in others comparatively little rain fell. At Westminster there was none, at Brixton 0·4 inch, and at Greenwich only five-hundredths of an inch. During the same afternoon a severe thunderstorm occurred at Ilford, Essex.

A DISCUSSION of the thunderstorm observations recorded in 1897 at ten selected stations in India, by Mr. W. L. Dallas, is contained in Part ix. vol. vi. of the *Indian Meteorological Memoirs*. The results for the year have been divided into five-day periods. The storm-frequency varies considerably in different parts, but, generally speaking, the number of storms is unimportant during February and the early part of March; but after the middle of March the thunderstorm season commences, and continues until the middle of October, the maxima occurring towards the end of May and September. After October 23 no storms are reported. Storms are much more frequent in the afternoon than in the morning, and when a storm occurs in the forenoon it is followed, almost without exception, by another in the afternoon. There is a belief that the damage done by lightning in the tropics is slight compared with that done in temperate zones, and the fact that at ten observatories in the year in question only four instances of damage being recorded gives support to this belief.

PROF. CANCANI remarks in a recent paper (*Ital. Soc. Sismol. Boll.*, vi. pp. 37-42) that seismology stands almost alone among the sciences of observation and experiment in that so far no pattern instrument and no comparable apparatus have been introduced. He admits that the Seismological Committee of the British Association have taken a step in the right direction, but considers that the instrument used by them possesses several defects which prevent its general adoption. The conditions which should be satisfied by the type apparatus, he describes as follows: It must be astatic or possess a stationary mass, and must be equally capable of recording the very small and rapid preliminary vibrations and the subsequent undulations of long period; it must have the sanction of experience, the cost of erection and maintenance must be small, and the construction so simple that it does not easily get out of order; it must allow the continuous inspection of the traces, and its sensibility must lie within convenient limits.

ABOUT two years ago Dr. Sambon brought forward evidence that sunstroke was an infectious disease, and consequently due to microbic influences. This view has not met with general acceptance, and Mr. E. H. Freeland, who has had exceptional opportunity of observing cases of sunstroke, both ashore and afloat, shows in the *Middlesex Hospital Journal* (July) that all the phenomena of this affection can be explained on general physiological principles without reference to germs at all. He concludes his paper as follows:—"Whether sunstroke be due to external physical causes, or whether it be an infectious disease and due primarily to a micro-organism which has yet to be isolated, must be decided in the future. For the present it seems to me that there is ample evidence for believing that sunstroke is due primarily to thermic influences—the exposure of the body to a hot moisture-laden atmosphere—and secondarily to the circulation in the blood of certain toxic poisons, the result of perverted tissue metabolism; and that, until more tangible evidence is brought forward to prove that the affection is due to microbic influence, one may safely accept the older doctrine with regard to its causation as a sound working hypothesis, if nothing else."

PROF. F. E. NIPHER, of Washington University, St. Louis, Missouri, has sent some further particulars with reference to the methods he uses to obtain a "zero" plate. His observations upon photographic reversal have already been noticed in these columns (pp. 62, 159), and he has pointed out the bearing of his work upon eclipse photography (p. 246). The following details of the operations he follows may enable other photographers to repeat his experiments. "The plate is placed under a punched stencil in a printing frame. It is exposed at *l* cm. from a 16 c.p. lamp. By a few trials one can find the time-interval of exposure, so conditioned that nothing will develop on the plate in a developer of fixed composition, strength and temperature, and at a fixed distance from the 16 c.p. lamp. This is a standard developer. With a shorter time of exposure than that giving the zero plate, a negative will result, and with a longer time, a positive. A plate to be used in taking any picture to be developed in the standard developer (as a positive) is first exposed to the 16 c.p. light at a distance *l* cm. for a time which experiment has shown will put the film into the zero condition when developed in the standard bath. It is then put into the plate-holder, and given a camera exposure in the usual way, after which it is developed. It is not important that the developing bath should be at any particular distance from the lamp. The plate is to be pre-exposed so that a zero plate will result in that particular bath, at any fixed distance from the lamp. I usually make this distance about eight inches."

ACCORDING to Maxwell's electromagnetic theories, a moving body charged electrically produces a magnetic field. In the *Bulletin* of the French Physical Society, M. V. Crémieu

gives a brief note on certain experiments destined to test the actual existence of such a field, as well as the converse result that a moving charge placed in a variable magnetic field experiences a certain ponderomotive force. Having, at the suggestion of M. Lippmann, conducted some experiments for the purpose of investigating the latter effect, with negative results, M. Crémieu now gives an account of certain investigations made with a disc of 37 cm. in diameter, rotating at the rate of 100 to 130 revolutions per second in the centre of an annular coil connected with a highly sensitive galvanometer. If the disc is suddenly charged, the convection current thus produced should give rise to an induced current through the galvanometer, and the magnitude of the convection current being determined by the number of revolutions and the density of the charge, the amount of the expected deflection of the galvanometer could be calculated. No deviation of the predicted magnitude was obtained, and the author concluded that a moving charge does not produce a magnetic field. Such a conclusion leads logically to the rejection of existing theories of the electric current, and M. Crémieu proposes to conduct further experiments with the object of throwing more light on this difficult question. The author does not, in this note, say anything about the effects of the self-induction of the rotating disc, and further information on this point appears desirable in criticising the results.

A FEW interesting details referring to the use of wireless telegraphy in the French navy are given by a naval correspondent of the *Daily Graphic*. It is stated that half-a-dozen ships in the combined French squadron recently at Cherbourg were fitted for wireless telegraphy, and the clicking, crackling, and sparking of the big coils was heard on board all day. Messages have been taken in and sent out at distances quite twice or three times as great as anything achieved with the instruments in use in the British ships. The French do not fit the wire to a gaff as in our ships; it is suspended between the funnels to the triatic stay, and is much less conspicuous. The manœuvring of the submarine boats, *Morse* and *Narval*, is described as marvellous; they are, it is stated, much ahead of the American *Holland* boat, which is considered to be a formidable weapon.

AN interesting and detailed account of Count von Zeppelin's successful trial trip of his navigable balloon on July 2 is given in *Die Umschau* by an anonymous author, who has endeavoured to dispel the somewhat exaggerated reports which have been circulated as to the success or failure of the experiment. It is pointed out that the delay in the ascent, which some persons attributed to an accident, was really caused by the wind being too strong at the time originally proposed for the trip. The wind-velocity at the time of starting was 5·5 metres per second, and the balloon was actually driven forwards for a short distance in the face of this wind. But after a short time the path deviated till it made an angle of 30° with the wind-direction. This deviation, the writer explains, was due to several causes. In the first place the rope broke which supported the movable mass necessary for the maintenance of longitudinal balance, and to restore equilibrium it was necessary to stop or even reverse one of the machines, so that the balloon could no longer be driven full ahead. Moreover, the framework was found to have undergone a little deformation, which gave the machine a slight bias to one side, interfering with the steering. The wind causing the balloon to drift towards the shore, a descent was made in order that Count Zeppelin might land on the water (to use an Irishism), and thus have his machine towed back by steamer. The descent was very gradual, the cars gently sinking down to the water without the sudden jerk which is commonly experienced in an ordinary balloon. This result is attributed to the favourable form of the balloon, a cylinder experiencing greater resistance than a sphere. The performance of the motors and screws is described as brilliant.

We have already referred to the great loss anthropology has sustained in the death of Mr. Frank Hamilton Cushing on April 10. In the current number of the *American Anthropologist* are memorial notices by various leading American anthropologists, from which it is evident that a peculiarly gifted and winning personality has passed away. Mr. Cushing had great manual dexterity and an acute appreciation of how things were made, and he had practised himself to do anything an American Indian could accomplish, and with the same limited resources. For five years he lived with the Zuñi Indians, living their life and familiarising himself with their ideas and modes of thought, and he rose high in the social Pueblo life, taking part in their councils and in their sacred ceremonies. An intense eagerness to learn more and more of aboriginal thought and deed was the mainspring of his life, and his kindly sympathetic nature and keen intelligence and dexterity placed him in the front rank of field investigators. We understand that Mr. Cushing left an immense amount of MS. material, which it is to be hoped will be fully published, for his published works by no means do justice to the extent and value of his researches.

Our contemporary *Science*, for July 27, contains a summary of the "Lacey Act," recently passed by Congress for the protection of game and other birds in the United States, and for the regulation of the importation of foreign birds and mammals. The carrying out of this important Act has been confided by the Secretary of the Department of Agriculture to the Division of Biology, Dr. T. S. Palmer being the officer selected to supervise its actual administration. Dr. Palmer has lost no time in making known the principles of the new law, having already published a *Bulletin* of the Department, entitled "Protection and Importation of Birds, under Act of Congress, approved May 25, 1900." As regards the importation of wild animals and birds, an absolute veto is placed on certain injurious species ; and importers must in all cases obtain special permits from the Secretary of Agriculture before a single individual can be landed. These permits should be applied for in advance. No permits are issued for shipping birds from one State to another, although in certain States the Commissioners of Fish and Game have authority to allow the shipment of a limited number for breeding purposes. No permits are necessary for domesticated birds, and the same applies to natural history specimens for museums. In the case of the larger ruminants special permits will be issued, as heretofore, in the form prescribed for domesticated mammals. The prohibited species include the European house-sparrow, the starling, fruit-bats or flying foxes, and the mongoose, or ichneumon. Special inspectors are appointed to carry out the law, and to give advice in cases of difficulty. The attention of all concerned is drawn to those sections which make it unlawful to ship from one State to another animals or birds taken in contravention of local laws, and which require all packages containing live birds and animals to be clearly marked with the name and address of the shipper, and with the nature of their contents.

Whether or no the inferior animals have souls, forms the subject of an article by Herr S. von Uexküll in the *Biol. Centralblatt* of August 1.

In Part 3 of vol. xxviii. of the *Morphologisches Jahrbuch*, Dr. B. Haller publishes his third memoir on the vertebrate brain, treating specially of that of the mouse, but adding some observations in regard to Echidna. The second article in the same

number is by Dr. Fürbringer, and treats of the systematic position of the Myxinoids. The author is of opinion that vertebrates should be subdivided as follows :—

  I. Acrania (Amphioxina).
 II. Craniota :
    (1) Distoma (Myxinoides).
    (2) Cyclostoma (Petromyzontes).
    (3) Gnathostoma :
      (*a*) Anamia (Pisces, Dipneusta, Amphibia).
      (*b*) Amniota (Reptilia et Aves, Mammalia).

We have received the *Report* of the Manchester Museum for 1899-1900. From this we learn that the Museum has been enriched during the period in question with two collections of first-class importance, one of these being Mr. C. H. Schill's cabinet of Lepidoptera, and the other the Layard collection of weapons and implements.

In the concluding part of his "Ornithological Notes," published in the July issue of the *Victorian Naturalist*, Mr. Robert Hall, of Melbourne, discusses the question whether a tree-building diamond-bird (Pardalotes) is the foster-parent of a cuckoo. In the case referred to the young cuckoo was actually seen to be fed by the diamond-bird, one of whose own young was brought up with it. The incident is at present quite unique.

The Library of the Patent Office is an institution known and appreciated by many students of science, both pure and applied. A series of classified catalogues of the contents of the Library has just been started by the publication of a "Subject List of Works on Photography and the Allied Arts and Sciences." Each volume of the series will contain (1) a general alphabet of subject headings, with descriptive entries, in chronological order, of the works arranged under these headings ; (2) a key or a summary of these headings shown in class order. The present list comprises 557 works (73 serials, 484 text-books, &c.) wholly or in part photographic—representing 1300 volumes. The catalogue is really a valuable little bibliography of photography as well as a guide to the contents of the Library.

The additions to the Zoological Society's Gardens during the past week include a Green Monkey (*Cercopithecus callitrichus*), a —— Monkey (*Cercopithecus*, sp. inc.) from West Africa, presented by Mr. L. J. Sparrow ; a Mozambique Monkey (*Cercopithecus pygerythrus*) from East Africa, presented by Mr. C. Mackay ; three Pheasants (*Phasianus colchicus*), British ; a Common Peafowl (*Pavo cristatus, ♂*) from India, presented by Captain G. H. Arnot ; a Long-legged Buzzard (*Buteo ferox*), a Black Kite (*Milvus migrans*), two Lesser Kestrels (*Tinnunculus cenchris*), European, two American Kestrels (*Tinnunculus sparverius*) from America, presented by Mr. J. Simonds ; a Bengal Weaver Bird (*Ploceus bengalensis*), a Manyar Weaver Bird (*Ploceus manyar*), four Black-throated Weaver Bird (*Ploceus atrigula*), an Indian Roller (*Coracias indica*) from India, presented by Mr. E. W. Harper ; a Spiny-tailed Iguana (*Ctenosaura acanthura*) from Central America, presented by Mr. C. Hagenbeck ; a Common Lizard (*Lacerta vivipara*), British, presented by Mr. Stanley S. Flower ; a Military Macaw (*Ara militaris*) from South America, a Roseate Cockatoo (*Cacatua roseicapilla*), six Blue Lizards (*Gerrhonotus coeruleus*) from Australia, three Blue-tongued Lizards (*Tiliqua scincoides*) from Western North America, a White-collared Kingfisher (*Halcyon chloris*) from India, a Saddle-backed Tortoise (*Testudo ephippium*), three Albemarle Tortoises (*Testudo vicina*), two Thin-shelled Tortoises (*Testudo microphyes*) from the Galapagos Islands, deposited ; an Argali Sheep (*Ovis ammon, ♀*) from the Altai Mountains, two Black Storks (*Ciconia nigra*), European ; a Ring-necked Pheasant (*Phasianus torquatus*) from China, purchased ; four Indian Crows (*Corvus splendens*), a

Little Cormorant (*Phalacrocorax javanicus*), a Green-winged Dove (*Chalcophaps indica*) from India, received in exchange ; a Japanese Deer (*Cervus sika*), five Rosy-billed Ducks (*Metopiana peposaca*), bred in the Gardens.

OUR ASTRONOMICAL COLUMN

VELOCITIES OF METEORS.—At the second annual meeting of the Astronomical and Astrophysical Society of America, recently held at Columbia University, New York, Dr. W. L. Elkin described the apparatus and results of photographs obtained at the Yale Observatory for the determination of the velocity of meteors (*Science*, vol. xii. pp. 125-6). The idea of using photography for this purpose appears to have been first suggested by J. H. Lane in 1860, but it was not until 1885 that Zenker made the next practical attempt in Berlin, and attention has again been recently called to the matter by Prof. Fitzgerald. The Yale apparatus consists of a bicycle wheel fitted with twelve radial opaque screens, fixed so that, while rotating, the screens are brought intermittently in front of the cameras. The wheel as at present worked makes about 50-60 revolutions per minute, but it would be better to increase this speed in future apparatus. A check on the velocity is afforded by records made each revolution on a chronograph. The length of interruption of the meteor trail and the consequent velocity are then determinable if a second observation of the meteor from a distant station has been obtained. In November and December 1899, five such duplicate trails were secured. The apparent velocities of these are given as 50'4, 12'2, 50'3, 20'2, 36'5 kilometres per sec.; their altitudes varying from 45 to 100 kilometres. Correcting the apparent velocities for the attraction of the earth and the diurnal rotation by Schiaparelli's formulae, the true velocities with respect to the sun are 34'4, 32'0, 32'4, 39'8, 34'0 kilometres per sec.

Comparing these velocities with those calculated on assumption of parabolic or elliptic orbits, the real velocities are in all cases smaller, indicating that the atmospheric retardation has amounted to 8-15 kilometres per sec. The elements deduced for one meteor, an Andromedid, agree remarkably closely with those of Biela's comet, showing the method to be capable of considerable accuracy.

STANDARDS FOR FAINT STELLAR MAGNITUDES.—Prof. E. C. Pickering announced at the above-mentioned conference that a grant of 500 dollars had been made from the Romford Fund for the purpose of carrying out an investigation on the brightness of faint stars by the co-operation of several observatories possessing large telescopes. The point immediately desirable is the accurate measurement of a few stars which shall serve as standards for future work on a larger scale. Five photometers have been constructed, each having a photographic wedge which may be interposed between the eye and the star as seen by the telescope. Thirty-six regions have been carefully selected in different parts of the sky, and twenty stars (five of each of magnitudes 12, 15, 16, 17) are to be chosen in each region, the faintest to be selected and measured with the Lick and Yerkes telescopes. The stars of the 16th magnitude will be measured with the 26-inch of the University of Virginia, and perhaps also with the 23-inch Princeton refractor ; those of the 15th magnitude will be measured by the 15-inch Harvard telescope. All of these are to be then compared with stars of the 12th magnitude, whose *absolute* magnitudes will finally be determined with the 12-inch Harvard meridian photometer. After the work is properly got in hand, it is hoped that it may be reduced to a simple routine without sacrificing the quality of the results.

THE TOTAL SOLAR ECLIPSE, MAY 28, 1900.—As more detailed reports of the results obtained by the American observers during the recent total eclipse come to hand, it is interesting to note the increased use which has been made of large diffraction gratings, both concave and plane. In *Science* (vol. xii. pp. 174-184), Mr. L. E. Jewell describes the work at Pinehurst, N.C., and Griffin, Georgia, of the two parties organised by the physical department of the Johns Hopkins University. At each station there were installed two spectroscopes, one having a plane diffraction grating, surface 3 × 5 inches, 15,000 lines to the inch, used in conjunction with a quartz lens to photograph the spectrum of the first order ; the other having a concave grating of 10 feet radius and 15,000 lines to the inch, mounted

in the usual Rowland form, with a large quartz lens to throw an image of the sun on the slit-plate from a heliostat. The photographs were very successful, and show the spectrum from wave-lengths 3000 to 6000, even the exposures of only one second giving good negatives.

In the same number of *Science* Profs. E. B. Frost and E. E. Barnard describe the apparatus they successfully used during the same eclipse at Wadesboro, N.C.

REPORT OF THE CAPE OBSERVATORY.—In his report for the year 1899 Sir David Gill, Her Majesty's Astronomer at the Cape of Good Hope Observatory, makes special mention of the completion of the new record room, providing storage for manuscripts, the safe preservation and orderly arrangement of the precious astrographic plates, and also serving as the place where the measurements of these plates are undertaken.

The pier and foundations for the new transit circle are completed, but the delay in obtaining the sheet steel dome has kept the work at a standstill. The observations with the transit instrument have been mainly those of the standard stars for the reduction of the Catalogue Astrographic plates. When the new transit circle arrives it will be entirely devoted to the systematic meridian observations of the sun, Mercury, Venus and fundamental stars. With the heliometer, observations of all the oppositions of major planets have been continued.

The 24-inch object glass of the McClean equatorial was returned to Sir Howard Grubb for refiguring, and this instrument has hitherto only been used with a slit spectroscope for stellar spectra. Since the photographic objective was dismounted the 18-inch visual lens has been used for measurements of twenty-one close double stars. The 7-inch equatorial has been used in the revision of the Cape Photographic Durchmusterung, in the observation of suspected variable stars, and in the detection of double stars.

The 6-inch instrument with a Zöllner photometer has been used for determining the visual magnitudes of stars in selected areas of different galactic latitudes, the photographic magnitudes of which are already determined. A comparison between the visual and photographic magnitudes will subsequently be made. With the astrographic equatorial 152 *chart* plates and 184 *revision catalogue* plates have been passed. 103 plates, containing 38,785 stars, have been measured during the year, all observations showing an error of 0″·6 being repeated.

Seventy-eight photographs of *Iris* were taken during the period July 11-December 31, with *six* exposures on each plate. In conjunction with meridian observations of comparison stars, it is intended to use the results of the measurements of these plates for determining the mass of the moon.

The geodetic survey of South Africa and Rhodesia has been considerably advanced, but was interrupted by the outbreak of the Transvaal war. The Anglo-German boundary survey has been hindered by the waterless character of the Kalihari Desert, but the work is now completed as far as Arahoab, from which an offset chain will be carried to the 20th meridian.

ROUSDON OBSERVATORY (DEVON).—Sir C. E. Peek sends a pamphlet of sixteen pages containing the sixth contribution of systematic observations of variable stars made at his observatory at Rousdon, Lyme Regis, Devonshire. The present report furnishes the details of the variability of T Cassiopeiæ for the ten years 1889-1898, and of R Cassiopeiæ for the twelve years 1887-1898. The light curves of both stars are also plotted at the end of the pamphlet.

INDEPENDENT DAY NUMBERS FOR 1902.—A small pamphlet has been issued from the Cape Observatory giving the independent day numbers for correcting the places of stars given in the *Nautical Almanac* for 1902. The values of the constants of precession, aberration and nutation employed in these tables are those recommended by the Paris International Conference of 1896.

THE AUGUST PERSEIDS OF 1900.

OBSERVATIONS of this well-known annual display were much hindered by moonlight, though the weather was generally clear at about the time of the maximum. Our satellite was full on the evening of August 10, and obscured all the smaller meteors. Apart, however, from this interference, the shower of 1900 seems to have been a somewhat scanty one. It furnished a considerable number of large meteors it is true, and of these it is hoped the real paths may be computed ; but on the nights of August 10 and 11 observers were somewhat disappointed with the character of their results. The effect of the full moon's influence in practically obliterating a meteoric shower may not, however, have been given sufficient weight. The best night appears to have been August 12, when shooting stars were tolerably frequent considering the circumstances.

But if moonlight presented an obstacle to success in the second week of August, there was no such impediment early in the month and during the last fortnight of July. The earlier stages of the shower were therefore well observed. In fact, it is questionable whether the Perseids have ever been more successfully observed in the month of July. Among those who participated in the observations were Prof. A. S. Herschel, Messrs. J. R. Bridger, W. E. Besley, A. King, and many others. At Cambridge a large number of meteors were recorded. The results show that the first Perseids were noticed on about July 16, and gradually increased in numbers until the date of maximum. The radiant showed the usual E.N.E. motion in a most decided manner.

At Bristol, between July 15-30, in 17¼ hours of observation, 177 meteors were seen, including about 24 Aquarids (radiant 338° – 10°) and 20 Perseids. But the only night on which a sufficient number of Perseids were registered to indicate a good radiant was July 30, when the position was at 31° + 54° from 10 paths. Several of the most prominent of the minor showers of the epoch were observed, and their radiants accurately determined as follows :—

24° + 43°, 7 meteors	305° – 12°, 8 meteors
291° + 59°, 6 ,,	315° + 47°, 9 ,,
292° + 52°, 7 ,,	335° + 73°, 5 ,,

Other showers were indicated at 53° + 63°, 245° + 72°, 333° + 28°.

Mr. W. E. Besley, at Clapham Common, London, registered the paths of 110 meteors between July 14 and 24, and the great majority of these were seen on the 23rd (30 meteors) and 24th (51 meteors). His results are important, for on the former date he found the radiant point of the Perseids at 23° + 51°, and on the latter date at 25° + 52½°, from 5 and 7 meteors respectively.

Prof. A. S. Herschel, at Slough, during a series of short watches between July 17 and August 1, recorded 53 meteors, including some very interesting early Perseids and several Aquarids. The position of the latter radiant was placed at 339° – 12°, from about 7 paths.

Some of the meteors seen by Prof. Herschel were also noted by the writer at Bristol. The earliest Perseid of which duple observations were secured appeared on July 19, at 11h. 49m., and it was a fine object, estimated to equal Jupiter by the Bristol observer. The radiant from the combined paths was at 17° + 50°, and the height of the meteor varied during its descent from 81 to 54 miles. Another Perseid was seen at Slough and Bristol on July 23, 12h. 12m., of 1st magnitude. Its radiant was at 24° + 52°, and height 84 to 55 miles. These radiants, together with those determined by Mr. Besley on July 23 and 24, and that by the writer at Bristol on July 30, agree very satisfactorily with the ephemeris place of the radiant given by the writer in *Ast. Nach.*, 3546, and *Memoirs* R.A.S., vol. liii. p. 210.

Fairly bright Aquarids were recorded at Slough and Bristol on July 28 and 30, with heights from 65 to 44 miles and 56 to 40 miles respectively. These meteors are usually lower in the atmosphere than the Perseids, and move slower. If we take the radiant of the former shower in 1900 as 339° – 11°, we shall probably have a position which is certainly within 1° of probable error.

On July 15, at 10h. 13m., a Capricornid fireball was seen at Bristol and four other places. It was a splendid object, about three times brighter than Venus, in the northern part of England. It fell from heights of 51 to 21 miles, along a path of 78 miles ; velocity, 16 miles per second.

On July 17, at 8h. 47m., a magnificent fireball appeared over the northern part of England and Scotland. Though the sun had not long set, the brilliancy of the meteor was described as very dazzling, and the nucleus left a streak which remained visible for three-quarters of an hour. The meteor was directed from a radiant at 249° – 20° in Scorpio, and fell from a height of 58 to 15 miles, along a path of about 175 miles.

On July 24 another fireball appeared, and was rated at about three times the brightness of Venus. It was seen at Bristol and

at several stations in the eastern counties. It fell from 68 to 27 miles, along a path of 103 miles ; velocity, 19 miles per second, and was directed from a well-known July radiant at 280° – 15°.

But the number of brilliant meteors which have recently appeared is so large that the objects cannot be alluded to in detail. Many ordinary shooting stars have also been doubly observed, and these will be tabulated and published at a later period. Among these there was an interesting θ Perseid on July 23, 11h. 13m., with heights of 83 to 59 miles, and a radiant at 30° + 47°, quite distinct from the true Perseids.

On about August 10–12 the radiant of the Perseids was found far east of its place in July. On August 12, Mr. King, at Leicester, determined the position as 48½° + 58° from 16 Perseids, and Mr. Besley derived it at 47° + 56½° on the same night from 4 meteors. On August 16 the writer at Bristol saw 5 Perseids from a radiant at 54° + 58°.

Though the shower was partially obliterated by moonlight just at the important time, it has this year furnished some interesting materials for discussion as regards its earlier and later stages.   W. F. DENNING.

WHAT PRESSURE IS DANGEROUS ON ELECTRIC RAILWAYS WITH OVERHEAD TROLLEY WIRES.[1]

THE following investigations were set on foot on account of a dissension between the firm of Messrs. Brown, Boveri and Co., Switzerland (Baden), and the authorities regarding the proper pressure for two different electric railways to be worked by three-phase alternating current, namely, the lines Stansstad-Engelberg and Fernatt-Garnergratt, which lines it was proposed to work at a pressure of 750 volts. But this pressure being regarded as dangerous, the authorities refused to allow one exceeding 500 volts to be actually employed.

In these circumstances the firm communicated with Prof. H. F. Weber, of the Zürich Polytechnic, asking him to express his opinion on this matter. In view, however, of his own want of experience on this particular point, Prof. Weber commenced a long series of investigations of the physiological effects of the electric current on the human body, and he used himself as the measuring instrument, thus exposing himself to great danger.

The experiments were made with reference to the special circumstances of the above railways, where the current was supposed to be supplied through two overhead leads, the rails being used as the third conductor of the three-phase system.

Two series of experiments were made corresponding with the cases—

(*a*) A person seizes the two bare leads with both hands simultaneously, or both of the leads fall on a bare part of the human body.

(*b*) A bare part of a person standing on the railway or on a car comes into contact with one of the leads.

The apparatus used in the case of experiments (*a*) consisted of an iron ring wound with 630 turns of wire, through which was sent an alternating current, the frequency of which was 50 per second. The voltage between the first and the last turn was kept at 210 volts. To every thirtieth turn was soldered a copper wire of 10 cm. length, and 6 mm. diameter, and consequently the pressure between the first and the second wire was 10 volts, that between the first and third 20 volts, and so on, up to 210 volts.

Prof. Weber tried these pressures successively on himself, constantly holding with one hand the first wire and seizing with the other hand each of the wires in succession. The experiments were made three times, his hands being each time wetted to begin with, and afterwards being used dry. The results of each of the three series so obtained were consistent with one another.

When experimenting with *wet hands* he obtained the following results :—

P.D.	Effect.
10 volts.	Very feeble trembling of the muscles of the fingers ; the current from hand to hand was measured and found to be 0·001 ampere.
20 volts.	Very considerable trembling of the hands, wrists and forearms ; the hands and the arms were able to be moved freely, and the wires could be

1 By William Rung, C.E., of the firm of Brown, Boveri and Co., Switzerland. Translated from the Danish *Civilengeneer* by F. Lehmann, M. F. Danish C.E.

released easily. The current was from 0·0020 to 0·0027 ampere.

30 volts. The fingers, hands, wrists, the forearms and upper arms nearly paralysed ; the fingers or hand could scarcely be moved ; serious pains in the fingers, hands and arms, and the experiment not endurable for more than 10 seconds. The wires could, however, be released, but only by using the greatest determination. Current, 0·015 ampere.

40 volts. The fingers, hands and arms were instantaneously paralysed, and the pain was almost unbearable. The wires could hardly in any case be released. The pain could not be endured longer than 5 seconds.

50 volts. Again instantaneous paralysis of all the muscles of the fingers, hands and arms ; the wires could not be released ; the state endurable for 2 seconds at most, whence it was impossible to measure the current.

Having obtained the above results, the experimenter did not find it advisable to let the pressure exceed the 50 volts ; the fact that when the hands were wet, it was impossible on 50 volts pressure to release the wires, seemed to prove to him that serious danger was just beginning at this point.

With *dry hands* he formed the following results :—

P.D.   Effect.

40 volts. The fingers only tingle slightly ; the current too feeble for measurement.

The effects gradually increasing and extending to the arms up to the shoulders, until at

80 volts. The fingers, hands and arms were almost cramped and aching in every part ; great effort was required to release the wires ; current from 0·009 to 0·011 ampere.

90 volts. At the same moment in which the wire is seized, the hands are absolutely paralysed and the wire cannot be released again. The pains in the hands and arms were so violent that they caused the experimenter to call out involuntarily ; the effects could not be endured for more than 1–2 seconds.

The experimenter now went back to 80 volts, and the difference was so great that the effects of this pressure seemed to be extremely feeble relatively to the effects at 90 volts pressure ; this fact prevented him from trying the effects of pressure higher still

From these experiments Prof. Weber draws the following conclusions :—

"A simultaneous touching of both of the poles of an alternating current circuit is dangerous as soon as the pressure exceeds 100 volts ; and since it is impossible to set one's self free, the case must be regarded as fatal whenever immediate help is not at hand."

These results are consistent with several disasters which have happened in practical life.

In 1896, in Horgen (Switzerland), a man, to prevent his falling down from a ladder, seized with both his hands two non-insulated leads with a P.D. of 240 volts between them, and was immediately killed.

In a mine in Silesia a workman seized in the same manner some non-insulated leads and was killed on account of his being unable to release them, the P.D. being 300 volts.

In the Electric Central Station in Olten a workman, desirous of proving to his companions that a pressure of 500 volts was quite safe, seized both of the leads and was killed instantly. From this it is obvious that the general opinion of a pressure of 500 volts not being dangerous does not hold good, the limit being much lower. In spite of the great number of disasters which have already happened, the danger does not seem to have been generally appreciated, and workmen and erectors are often seen to deal with leads and apparatus of relatively high pressures in the most careless manner. That disasters have not taken place far more often may be due to the fact that in most cases help has been at hand instantly.

Entirely differing from these are the results of the other series of experiments (*b*). In this case the person is supposed to stand at one of the poles itself, namely, the earth, from which

he is, however, rather well insulated by means of his shoes ; and, as it will be evident from the results, the danger is in this case very small even at high pressures.

The arrangement used for this series of experiments consisted of twenty glow-lamps, each for 100 volts pressure, connected in series and all well insulated, the total alternating pressure between the first and the last lamp being 2000 volts. The free terminal of the first lamp was earthed, and between every two consecutive lamps a 6 mm. copper wire was soldered to the main connecting the lamps. Between the earth and the first, second, third to the twentieth of the 6 mm. wires, the pressure was consequently 100, 200, 300, . . . 2000 volts.

Standing on the ground, Prof. Weber touched the different wires—firstly, merely by a slight touch ; secondly, by firmly gripping them in his hand. The experiments were made under two conditions, the experimenter standing firstly on moist gravel soil, and afterwards on clay covered by a thin layer of coal-dust.

Standing *on moist gravel soil*, he obtained the following results :—

P. D.	Effect when the wire was	
	Slightly touched.	Firmly gripped.
800 volts.	Feeble stinging of the skin.	No effect.
	Gradually increasing until at	
2000 volts.	Violent stinging of the skin.	Intense trembling of the fingers.

Standing on clay covered with coal-dust, he obtained the following results :—

P. D.	Effect when the wire was	
	Slightly touched.	Firmly gripped.
200 volts.	Scarcely sensible stinging of the skin.	No effect whatever.
500 „ 700 „	} Gradually increasing.	The fingers begin to tingle feebly. Intense trembling of the fingers, hands, arms and ankle-joints.
1000 „	Stinging like burning by a flame.	The effects in the fingers, hands, arms and feet not endurable longer than 1 to 2 seconds ; difficulty in releasing the wire.
1300 „	Same effect.	The fingers, hands, and the arms are entirely paralysed, and the wire cannot be released.

From the last series of experiments it will be obvious that to touch one of the poles is not dangerous as long as the pressure does not exceed about 1000 volts ; the intense stinging which appears at the first slight touching serves as a protection against the danger, for the hand is instinctively drawn back rapidly.

The main result of these experiments is, then, that all pressures between 100 and 1000 volts must be regarded as equally dangerous, and consequently there is no reason for not using the higher pressures between 500 and 1000 volts, especially as they lead to greater economy in the working of the electric railway. Further, there is only a very little chance of the passengers or other persons coming into contact with both of the leads. To this danger the employés only are exposed, and being generally people with some electric training, they are acquainted with the danger and may be supposed to be sufficiently careful.

Finally, it is to be remarked that the authorities after these investigations allowed the use of a working pressure of 750 volts as originally proposed.

SEA COAST DESTRUCTION AND LITTORAL DRIFT.

THE increasing number of seaside resorts that are constantly being established all round the coast of this country, and the necessity of protecting the sea front from the devastation of the waves, has led recently to greater interest being shown in the protection of the shores and cliffs.

The means taken to preserve our coasts are as diverse as many of them are ineffectual ; and in many cases are designed without any proper consideration being given to the way in which the waves act, or to the physical conditions which have to be dealt with in the management of the littoral drift ; while frequently the amenities of the beaches of seaside resorts, and their use and enjoyment by visitors, are impaired by structures as ugly as they are useless to attain the object in view, and in other cases the construction of costly works is rapidly followed by their destruction by the sea.

It is proposed as shortly as possible in this article to state the conditions that have led to the present state of the cliffs and coast, and the conditions under which the material is drifted along the shore. For the purpose of illustration, the coast-line of the south-west of England between Start Point and the Solent has been selected, as this presents features of unusual interest for the study of coast destruction and the drift and accumulation of beach material. The cliffs between these two points consist of a series of rocks of varying degrees of hardness, showing in many places almost vertical faces to the sea, and ranging in height up to 500 or 600 feet. The destruction of these cliffs leads to the deposit, on the beach, of fragments of rock, or inland gravels derived from their summits, which are converted by wave action into shingle, consisting of pebbles of varying character and size, but generally shaped into the form of flattened ovoids, readily distinguishing them from the angular gravels due to glacial drift, or the rounded pebbles rolled down inland rivers.

Originally, no doubt, these cliffs descended to the bed of the English Channel with the same slopes as characterise their land faces, and were washed by the deep water of the sea without the intervention of the sand beaches which now stretch from them, and which, where they exist, have an almost uniform inclination along all parts of the coast.

The present form of these cliffs is due to the destructive action of the waves, or to landslips and weathering from rain and frost. The wearing away has not been regular. Headlands composed of hard rocks project out boldly to low-water mark and beyond, while the softer rocks which formerly adjoined them have been gradually worn away, leaving indents of various shapes and depths.

Some indication of the original position of the coast, and the distance to which it extended beyond the present line, is afforded by the remains of a raised beach, portions of which, consisting of pebbles which have been subjected to marine action, are to be found at Portland Bill on the east of Lyme Bay, at Hopes Nose, near Torquay, at Brixham and in Start Bay on the west. These beaches are at a much higher level than the present water-line. The direction of this old beach was located by the late Sir J. Prestwich as running in an unbroken line at ten miles outside the present shore between the thirteen- and twenty-fathom contour. There are also patches of gravel near the nine-fathom contour extending all round the bay at about a mile from the shore.

The coast-line has been broken up into three principal islands between the two headlands, namely, Lyme Bay, Weymouth Bay and Bournemouth Bay ; the contour of which again is split up into numerous minor bays and coves, the greater number of which have their own peculiar characteristics and contain their own peculiar accumulations of shingle. Two of these shingle banks, namely, the Chesil Bank in Lyme Bay, and that at Hurstcastle at the entrance to the Solent, are, perhaps, the most remarkable accumulations to be found anywhere round the coast of Great Britain ; while at Axmouth is one of the most extensive landslips of which there is any record. The fight between sea and land is continuous and unceasing, with the result that the area of this country is gradually being reduced.

The rivers which discharge into the sea along the south coast are few and insignificant in character, and are utterly incapable of transporting from the land the large amount of stones or sand now found on the coast. In some cases they have been blocked up by the littoral drift. A careful consideration of all the circumstances can therefore only lead to the conclusion that some of the results which have been attained must be due to other and mightier forces than those now in existence. These forces may probably be ascribed to the same agencies that gave to this country the shape which it now assumes, and by which the valleys and rivers were scooped out.

At the close of the last great Ice Age the melting of the vast bed of snow which then covered this country must have led to large torrents of water escaping seaward, which would carry with them the debris from the rocks broken up by frost and ice, in the shape of boulders, gravel and sand ; and besides leaving deposits in the valleys and those which are to be found on the summit of the cliffs, would carry the degraded material to the sea and form a talus at the level of the water. . This deposit,

after the wear and tear caused by the waves during long ages, resulted in the present sand beaches.

It is certain that the enormous mass of sand which now covers the littoral of the sea cannot have been deposited by existing agencies. The degradation of the cliffs that takes place is quite inadequate to account for its existence ; more especially as only the harder rocks have afforded the material of which the sand is composed, the softer detritus having been carried away in suspension to the depths of the ocean. The sand of the seashore consists almost entirely of grains of quartz of a nearly uniform size, and even where flints abound in the sea cliffs and in the shingle on the beach, this material is conspicuous by its absence in the sand of the shore.

As pointed out in my letter in Nature of November 30 last, only about one-third of the flints lying on the beaches along the south coast, not only in the part dealt with but also on those bordering on the chalk cliffs extending from Brighton to the North Foreland, and in the large accumulations at the Chesil Bank and Dungeness, are derived from the chalk, the colour of the majority of the flints being different shades of brown, grey, white and red, the former being most prevalent, whereas flints from the chalk are invariably black with a white exterior coating. If the above assumption as to the deposit of land detritus on the coast be correct, it affords a reasonable explanation of this phenomenon.

Another proof that the large accumulations of shingle along the coast are not due to agencies at present operating, is afforded by the fact that the Chesil Bank and Hurst Bank, where the supply of new material drifting along the coast is limited, have not varied materially in shape or increased in size during the time to which the most ancient records relating to them extend, the fresh supply coming from the cliffs being only sufficient in these cases to make up the wear and tear caused by the waves.

The supply afforded by the degradation of the cliffs is after all limited in quantity, and only about equal to making good the waste due to the constant wave action on the shingle. If from any cause an abnormal accumulation takes place on any particular part of the coast, denudation immediately sets in on the coast beyond. Instances of this are afforded by the extension of the Point at Dungeness and the banking up of the shingle on the west side, which has led to a diminished supply all along the Dymchurch and Hythe coasts. The construction of the Admiralty Pier at Dover has led to the denudation of the coast to the northward along St. Margaret's Bay. The pier at Shoreham Harbour for a time denuded the supply at Hove and Brighton ; and it is found universally to be the case that where the drift has been stopped by the debris from the fall of the cliffs acting as a groyne extending out to low water, or where artificial works have arrested the progress of the drift, the coast beyond has suffered from denudation.

There is no continuous drift of shingle throughout the whole length of this coast, but the material is confined to the various bays and the banks where it has been accumulated. It has been stated that the land gravel found along the foot of the chalk cliffs at the east end of the English Channel may have been derived from the waste of the gravel beds of the cliffs of Dorset and Hampshire, and that pebbles found on the Chesil Bank have been derived from the coast of Devon ; but an examination of the coast shows that under present conditions, at any rate, this is neither the case nor physically possible. For this to have occurred the shingle must not only have passed the numerous headlands which project into the water, but also the approaches to Southampton and Portsmouth, in which the depth of the channels is from five to ten fathoms.

Nearly every bay and cove along the coast here dealt with has its own peculiar accumulation of shingle, which does not travel beyond the projecting headland, and in many cases there are long intervals along the shore where the rock is bare. Thus the shingle in Slapton Bay is of a different character from that found in any other part of the same coast. It consists almost entirely of round white quartz pebbles, resembling peas in shape, and averaging from an eighth to a quarter of an inch in diameter. This shingle not only covers the beach, but has been thrown up into a bank, the top of which is above the level of high tides, and has drifted across a deep indent in the bay, into which two fresh-water streams discharge, entirely closing this from the sea and forming it into a fresh-water lake about two miles long. The quartz pebbles of the Slapton beach do not drift beyond the eastern horn of the bay, and are not to be found in the next recess. All along the Devonshire coast the numerous bays and coves have beaches on which are accumulated shingle derived from the limestone, slate, greenstone, and other rocks which surmount them, while in others there is an absence of shingle and only sand is found. In the bay lying between the headland of the Exe and that at Otterton-Point, the beach at Budleigh Salterton is strewn with quartzite boulders and pebbles derived from a large bed contained in the cliffs bordering this part of the bay. These pebbles are of a pink colour, some having marks on them like blood spots. No stones of a similar character are found in the next bay, the drift being stopped by some rock ledges which project out from Otterton Point and form a natural groyne. The shingle in front of Seaton consists almost entirely of the chert and flint derived from the rocks at Beer Head. Beyond this, for several miles there is no continuous bank of shingle, but accumulations are to be found in the bight of the bays, the pebbles being derived almost entirely from the gravel beds in the cliffs. The shingle on the east side of Bridport Harbour is of a different character from that on the west shore, and resembles in size and shape that at Slapton, but the colour of the pebbles is different, these consisting of flint instead of quartz. This small shingle continues all along the coast, and up to the commencement of the Chesil Bank.

The drift of shingle along the shore only takes place above the line of low water, and within the zone covered by the horizontal range of the tide, and it does not accumulate below the line of mean tide level, except where its progress is stopped by encountering an obstruction, and when the quantity has become so great as to extend out into deep water. When the shingle encounters a river of any magnitude, it extends out in a spit across the entrance to the estuary, causing the tides to be diverted from their direct course, and to flow round the end; or else the channel becomes diverted from its course to the leeward, and made to flow in a course parallel with the coast for some considerable distance. Examples of this are found along the coast here dealt with, in the Spits across the estuaries of the Teign and the Exe, and that across Christchurch Harbour, and in the diversion of the streams at Seaton and Charmouth.

The Chesil Bank, which commences near Abbotsbury and extends in a south-easterly direction to the island of Portland, a distance of 10¾ miles, has in its course diverted several small streams, which now flow in a channel running parallel with the bank. The width of this great mound of shingle is about 500 feet, and its height varies from 32 to 53 feet, its top being from 23 to 43 feet above high tides.

Hurst Castle shingle bank extends out from the mainland at the entrance to the Solent for 1½ miles, terminating in a hook-like formation on which stands Hurst Castle, erected in the reign of Henry VIII. The bank slopes down across the Solent for a distance of three miles, leaving only a deep narrow passage between its foot and the Needles, in which is from 4 to 9 fathoms of water. This shingle bank forms on its southern side a steep submarine cliff from 20 to 70 feet in height, the face being very steep and dropping almost suddenly from a dry bank to several fathoms of water. So far as any record exists, this bank has not increased or diminished in size or undergone any material alteration since the castle was built. Eastward of the bank there is no drift of shingle, the foreshore for several miles consisting of a wide belt of alluvial deposit.

Another lesson this stretch of coast appears to teach is, that the theory which has generally received acceptance, that the prevailing direction in which the shingle is drifted along any given coast is always in the same direction as that of the prevailing wind, is not founded on fact. This theory may be said to have been settled on the facts brought forward in a paper read at the Institution of Civil Engineers in 1853 on the Chesil Bank, and the discussion which followed.

A careful examination of the facts mentioned in that article do not appear to warrant any such conclusion, but, on the contrary, tends to disprove it. The local movement of shingle along the south-west coast, and also along the other parts of the seashore, are certainly not uniformly in accord with the direction of the prevailing winds of this country. Approximately, the wind in England blows for two-thirds of the year from the south-west. On the east coast the general direction of the drift is from north to south. On the south-west coast the general direction is from west to east. From Beachy Head to Dover it is north-east; from Dover to the North Foreland, northerly; from the North Foreland to the mouth of the Thames, westerly; and north of the Thames south-westerly.

On the west coast, the drift is from south to north, up to the middle of the Irish Channel; and north of this, from south to north; and up the Bristol Channel from west to east. In all these instances the movement is in the same direction as the set of the flood-tide.

Although this is the general direction of the drift, there are numerous instances where, owing to the varying set of the tides, the drift moves in three or four different directions within very short ranges. Taking the example of Lyme Bay and the Chesil Bank, the locality where the prevailing wind theory was established, the facts as given by the author of the paper were : That the prevailing direction of the wind on this part of the coast varied between S.S.W. and S.W.¼W., which is practically at right angles to the Chesil Bank. If then the drift is in the direction of the prevailing wind, this should lead to a north-easterly movement. The bulk of the materials of which the bank is composed are stated in the paper to have come from the cliffs to the west of the bank, and therefore must have travelled in a south-easterly direction. At the east end of the bank the shingle derived from the debris of Portland moves in a northerly direction. On the other side of the island of Portland, in Weymouth Bay, the shingle is moved in a south-westerly direction; therefore, within the space of five or six miles, the drift is in three opposite directions, not one of which is in the direction of the prevailing winds, but all of which are in the direction at which the flood-tide strikes the shore. Further, it is correctly stated that the effect of winds from the south-west tends to pull down the bank, which is restored again to its normal condition during calms and north-east winds.

Along all tidal coasts it will be found that the general direction of the drift is the same as that of the flood-tide, and that the beach material in bays is moved in the same direction as that in which the wavelets due to the flowing tide break on the beach. It is not contended that shingle is never drifted by waves due to wind. On the contrary, it is a well-known fact that shingle is frequently drifted, first in one direction and then in the opposite, during the occurrence of gales blowing from different quarters obliquely on the shore, and that the beaches are alternately heaped up with material at one place and denuded at another. This process, however, is only occasional and intermittent, and beyond it there is a regular and continuous drift in one given direction, the main course along the coast being in the same direction as the flood-tide, the building up of shingle banks being most active during calms and off-shore winds.

After a long and careful investigation the writer has satisfied himself that the building up of shingle banks and the regular and continuous drift that takes place along the coast are due to wave action caused by the flow and ebb of the tides. As the great tidal wave moves along the deep water of the channels surrounding the coast, its crest is in advance of the sides, which encounter the friction of the shallower water. The swelling tide therefore meets the water at an angle oblique to its central course. As the lateral flow of the swelling water comes in contact with the shore it is checked and reflected back, causing a series of small undulations or wavelets, which break at the margin of the water.

Although these waves are small, varying in height, according to the condition of the tide and the slope of the beach, from 6 to 24 inches, they are constant, and never cease during the time that the beach is covered by the tide, the number of them varying from ten to twenty a minute. Allowing a mean of fifteen, this gives a total of 3600 impulses during the period that each tide is acting on the shore. These wavelets are never absent from the shore, except when absorbed by larger waves due to wind. As the wavelets break, the water attains a horizontal movement, and aided by the flood current lifts up and carries forward coarse sand and pebbles in a movement oblique to the coast line, and so gives them a slow but continuous forward movement.

The constant murmur that is heard on a shingle beach on days when there is a total absence of wind, and when the sea is perfectly smooth and calm in the offing, attests the fact that the pebbles on the face of a shingle beach are in constant movement. These tidal wavelets are capable of moving and pushing up the face of a shingle bank pebbles weighing from 1 or 2 ounces up to 5 or 6 lbs. A calculation as to the mechanical power of the water contained in an average sized wavelet shows that the kinetic energy developed amounts to 165 foot lbs., which is capable of lifting 9900 pebbles, each weighing 4 ounces, to a height of 1 foot.   W. H. WHEELER.

RECENT STUDIES IN GRAVITATION.[1]

THE studies in gravitation which I am to describe to you this evening will perhaps fall into better order if I rapidly run over the well beaten track which leads to those studies, the track first laid down by Newton, based on astronomical observations, and only made firmer and broader by every later observation.

I may remind you, then, that the motion of the planets round the sun in ellipses, each marking out the area of its orbit at a constant rate, and each having a year proportional to the square root of the cube of its mean distance from the sun, implies that there is a force on each planet exactly proportional to its mass, directed towards, and inversely as the square of its distance from the sun. The lines of force radiate out from the sun on all sides equally, and always grasp any matter with a force proportional to its mass, whatever planet that matter belongs to.

If we assume that action and reaction are equal and opposite, then each planet acts on the sun with a force proportional to its own mass; and if, further, we suppose that these forces are merely the sum totals of the forces due to every particle of matter in the bodies acting, we are led straight to the law of gravitation, that the force between two masses M_1 M_2 is always proportional to the product of the masses divided by the square of the distance r between them, or is equal to

$$\frac{G \times M_1 \times M_2}{r^2}$$

and the constant multiplier G is the constant of gravitation.

Since the force is always proportional to the mass acted on, and produces the same change of velocity whatever that mass may be, the change of velocity tells us nothing about the mass in which it takes place, but only about the mass which is pulling. If, however, we compare the accelerations due to different pulling bodies, as for instance that of the sun pulling the earth with that of the earth pulling the moon, or if we compare changes in motion due to the different planets pulling each other, then we can compare their masses and weigh them, one against another and each against the sun. But in this weighing our standard weight is not the pound or kilogramme of terrestrial weighings, but the mass of the sun.

For instance, from the fact that a body at the earth's surface, 4000 miles, on the average, from the mass of the earth, falls with a velocity increasing by 32 ft. / sec.², while the earth falls towards the sun, 92 million miles away, with a velocity increasing by about ¼ inch / sec. ², we can at once show that the mass of the sun is 300,000 times that of the earth. In other words, astronomical observation gives us only the acceleration, the product of G × mass acting, but does not tell us the value of G nor of the mass acting, in terms of our terrestrial standards.

To weigh the sun, the planets, or the earth, in pounds or kilogrammes, or to find G, we must descend from the heavenly bodies to earthly matter and either compare the pull of a weighable mass on some body with the pull of the earth on it, or else choose two weighable masses and find the pull between them.

All this was clearly seen by Newton, and was set forth in his "System of the World" (third edition, p. 41).

He saw that a mountain mass might be used, and weighed against the earth by finding how much it deflected the plumb line at its base. The density of the mountain could be found from specimens of the rocks composing it, and the distance of its parts from the plumb line by a survey. The deflection of the vertical would then give the mass of the earth.

Newton also considered the possibility of measuring the attraction between two weighable masses, and calculated how long it would take a sphere a foot in diameter, of the earth's mean density, to draw another equal sphere, with their surfaces separated by ¼-inch, through that ¼-inch. But he made a very great mistake in his arithmetic, for while his result gave about one month, the actual time would only be about 5¼ minutes. Had his value been right, gravitational experiments would have been beyond the power of even Prof. Boys. Some doubt has been thrown on Newton's authorship of this mistake, but I confess that there is something not altogether unpleasing even in the mistake of a Newton. His faulty

[1] A discourse delivered at the Royal Institution of Great Britain on Friday, February 23, by Prof. John H. Poynting, F.R.S.

arithmetic showed that there was one quality which he shared with the rest of mankind.

Not long after Newton's death the mountain experiment was actually tried, and in two ways. The honour of making these first experiments on gravitation belongs to Bouguer, whose splendid work in thus breaking new ground does not appear to me to have received the credit due to it.

One of his plans consisted in measuring the deflection of the plumb line due to Chimborazo, one of the Andes peaks, by finding the distance of a star on the meridian from the zenith, first at a station on the south side of the mountain, where the vertical was deflected, and then at a station to the west, where the mountain attraction was nearly inconsiderable, so that the actual nearly coincided with the geographical vertical. The difference in zenith distances gave the mountain deflection. It is not surprising that, working in snowstorms at one station, and in sandstorms at the other, Bouguer obtained a very incorrect result. But at least he showed the possibility of such work, and since his time many experiments have been carried out on his lines under more favourable conditions. Now, however, I think it is generally recognised that the difficulty of estimating the mass of a mountain from mere surface chips is insurmountable, and it is admitted that the experiment should be turned the other way about and regarded as an attempt to measure the mass of the mountains from the density of the earth known by other experiments.

These other experiments are on the line indicated by Newton in his calculations of the attraction of two spheres. The first was carried out by Cavendish.

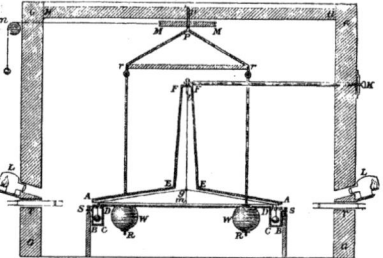

FIG. 1.—Cavendish's apparatus.

In the apparatus (Fig. 1) he used two lead balls, B B, each 2″ in diameter. These were hung at the end of a horizontal rod 6′ long, the torsion rod, and this was hung up by a long wire from its middle point. Two large attracting spheres of lead, W W, each 12″ in diameter, were brought close to the balls on opposite sides so that their attractions on the balls conspired to twist the torsion rod round the same way, and the angle of twist was measured. The force could be reckoned in terms of this angle by setting the rod vibrating to and fro and finding the time of vibration, and the force came out to less than 1/3000 of a grain. Knowing M_1 M_2 and r the distance between them and the force G M_1 M_2/r^2, of course Cavendish's result gives G, or knowing the attraction of a big sphere on a ball, and knowing the attraction of the earth on the same ball, that is, its weight, the experiment gives the mass of the earth in terms of that of the big sphere, and so its mean density. This experiment has often been repeated, but I do not think it is too much to say that no advance was made in exactness till we come to quite recent work.

By far the most remarkable recent study in gravitation is Prof. Boys' beautiful form of the Cavendish experiment, a research which stands out as a model in beauty of design and in exactness of execution (Fig. 2). But as Prof. Boys has described his experiment already in this theatre (*Proc. R.I.*, xiv. Part ii. 1894, p. 353), it is not necessary for me to more than refer to it. It is enough to say that he made the great discovery, obvious perhaps when made, that the sensitiveness of the apparatus is increased by reducing its dimensions. He therefore decreased the scale as far as was consistent with exact measurement of the parts of the apparatus, using a torsion rod, itself a mirror, only 2″ long, gold balls, m m, only ⅛″ in diameter. and attracting lead masses, M M, only 4¼″ in diameter. The force to be measured was less than $1/5 \times 10^8$ grain.

The exactness of his work was increased by using as suspending wire one of his quartz threads. It would be difficult to over-estimate the service he has rendered in the measurement of small forces by the discovery of the remarkable properties of these threads.

One of the chief difficulties in the measurement of these small gravitational pulls is the disturbances which are brought about by the air currents, which blow to and fro and up and down inside the apparatus, producing irregular motions in the torsion rod. These, though much reduced, are not reduced in proportion to the diminution of the apparatus.

A very interesting repetition of the Cavendish experiment

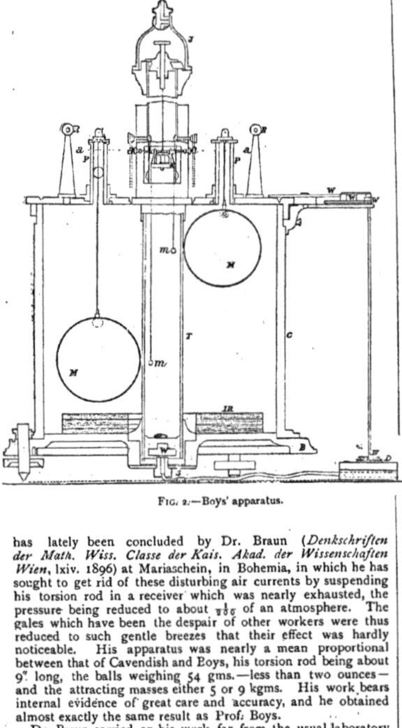

in place of the somewhat untrustworthy metal wire which he used in the work already published.

Prof. Boys has almost indignantly disclaimed that he was engaged on any such purely local experiment as the determination of the mean density of the earth. He was working for the Universe, seeking the value of G, information which would be as useful on Mars or Jupiter or out in the stellar system as here on the earth. But perhaps we may this evening consent to be more parochial in our ideas, and express the results in terms of the mean density of the earth. In such terms, then, both Boys and Braun find that density 5·527 times the density of water, agreeing therefore to 1 in 5000.

There is another mode of proceeding which may be regarded as the Cavendish experiment turned from a horizontal into a vertical plane, and in which the torsion balance is replaced by the common balance. This method occurred about the same time to the late Prof. V. Jolly and myself. The principle of my own experiment (*Phil. Trans.*, 182, 1891, A, p. 565) will be sufficiently indicated by Fig. 3. A big bullion balance with a 4-foot beam had two lead spheres, A B, each about 50lbs. in weight, hanging from the two ends in place of the usual scale

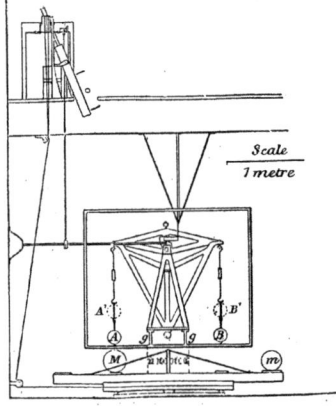

FIG. 2.—Boys' apparatus.

FIG. 3.—Common balance experiment (Poynting).

has lately been concluded by Dr. Braun (*Denkschriften der Math. Wiss. Classe der Kais. Akad. der Wissenschaften Wien*, lxiv. 1896) at Mariaschein, in Bohemia, in which he has sought to get rid of these disturbing air currents by suspending his torsion rod in a receiver which was nearly exhausted, the pressure being reduced to about 1/90 of an atmosphere. The gales which have been the despair of other workers were thus reduced to such gentle breezes that their effect was hardly noticeable. His apparatus was nearly a mean proportional between that of Cavendish and Boys, his torsion rod being about 9″ long, the balls weighing 54 gms.—less than two ounces—and the attracting masses either 5 or 9 kgms. His work bears internal evidence of great care and accuracy, and he obtained almost exactly the same result as Prof. Boys.

Dr. Braun carried on his work far from the usual laboratory facilities, far from workshops, and he had to make much of his apparatus himself. His patience and persistence command our highest admiration.

I am glad to say that he is now repeating the experiment, using as suspension a quartz fibre supplied to him by Prof. Boys

pans. A large lead sphere, M, 1′ in diameter and weighing about 350 lbs., was brought first under one hanging weight, then under the other. The pull of the lead sphere acted first on one side alone and then on the other, so that the tilt of the balance beam when the sphere was moved round was due to twice the pull. By means of riders the tilt, and therefore the pull, was measured directly as so much increase in weight. This increase, when the sphere was brought directly under the hanging weight with 1′ between the centres, was about ¼ mgm. in a total weight of 20 kgm., or about 1 in 100,000,000. If, then, a sphere 1 foot away pulls with $1/10^8$ of the earth's pull, the earth being on the average 20,000,000 feet away, it is easy to see that the earth's mass is calculable in terms of the mass of the sphere, and its density is at once deduced. The direct aim of this experiment, then, is not G, but the mass of the earth.

It is not a little surprising that the balance could be made to indicate such a small increase in weight as 1 in 100 million. But not only did it indicate, it measured the increase, with variations usually well within 1 per cent. of the double attraction, or to 1 in 5000 million of the whole weight, a change in weight which would occur merely if one of the spheres were moved 1/18 inch nearer the earth's centre. This accuracy is only attained by never lifting the knife edges and planes during an experiment, thus keeping the beam in the same state of strain throughout, and, further, by taking care that none of the mechanism for moving the weights or riders shall be attached in any way to the balance or its case, two conditions which are absolutely essential if we are to get the best results of which the balance is capable.

Quite recently another common balance experiment has been brought to a conclusion by Prof. Richarz and Dr. Krigar-Menzel ("Anhang zu den Abhandlungen der Königl," *Preuss. Akad. der Wissenschaften zu Berlin*, 1898) at Spandau, near Berlin. Their method may be gathered from Fig. 4. A balance of 23 cm., say 9-inch beam, was mounted above a huge lead pile about 2 metres cube, and weighing 100,000 kgm. Two pans were supported from each end of the beam, one pan above, the other pan below the lead cube, the suspending wires of the lower pans going through narrow vertical tubular holes in the lead. Instead of moving the attracting mass, the attracted mass was moved. Masses of 1 kgm. each were put first, say, one in the upper right-hand pan, the other in the lower left-hand pan, when the pull of the lead block made the right hand heavier and the left hand lighter. Then the weights were changed to the lower right hand, and the upper left hand when the pulls of the lead pile were reversed. When we remember that in my experiment a lowering of the hanging

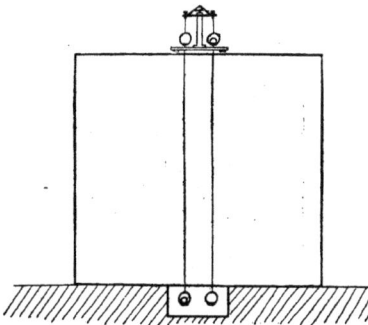

FIG. 4.—Common balance experiment (Richarz and Krigar-Menzel).

sphere by 1½ inches would give an effect as great as the pull I was measuring, it is evident that here the approach to and removal from the earth by over 2 metres would produce very considerable changes in weight, and, indeed, these changes masked the effect of the attraction of the lead. Preliminary experiments had, therefore, to be made before the lead pile was built up, to find the change in weight due to removal from upper to lower pan, and this change had to be allowed for. The quadruple attraction of the lead pile came out at 1·3664 mgm., and the mean density of the earth at 5·505.

This agrees nearly with my own result of 5·49, and it is a curious coincidence that the two most recent balance experiments agree very nearly at, say, 5·5 ; and the two most recent Cavendish experiments agree at, say, 5·53. But I confess I think it is merely a coincidence. I have no doubt that the torsion experiment is the more exact, though probably an experiment on different lines was worth making. And I am quite content to accept the value of 5·527 as the standard value for the present.

And so the latest research has amply verified Newton's celebrated guess that "the quantity of the whole matter of the earth may be five or six times greater than if it consisted all of water."

I now turn to another line of gravitational research. When we compare gravitation with other known forces (and those

which have been most closely studied are electric and magnetic forces) we are at once led to inquire whether the lines of gravitative force are always straight lines radiating from or to the mass round which they centre, or whether, like electric and magnetic lines of force, they have a preference for some media and a distaste for others. We know, for example, that if a magnetic sphere of iron or cobalt or manganese is placed in a previously straight field, its permeability is greater than the air it replaces, and the lines of force crowd into it, as in Fig. 5.

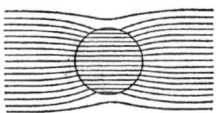

FIG. 5.—Paramagnetic sphere placed in a previously straight field.

The magnetic action is then stronger in the presence of the sphere near the ends of a diameter parallel to the original course of the lines of force, and the lines are deflected. If the sphere be diamagnetic, of water, or copper, or bismuth, the permeability being less than that of air, there is an opposite effect, as in Fig. 6, and the field is weakened at the end of a diameter parallel to the lines of force, and again the lines are deflected. Similarly, a dielectric body placed in an electric field gathers in the lines of force, and makes the field where the lines enter and leave stronger than it was before.

If we enclose a magnet in a hollow box of soft iron placed in a magnetic field, the lines of force are gathered into the iron and largely cleared away from the inside cavity, so that the magnet is screened from external action.

Now common experience might lead us at once to say that there is no very considerable effect of this kind with gravitation. The evidence of ordinary weighings may, perhaps, be rejected, inasmuch as both sides will be equally affected as the balance is commonly used. But a spring balance should show if there is any large effect when used in different positions above different media, or in different enclosures. And the ordinary balance is used in certain experiments in which one weight is suspended beneath the balance case, and surrounded, perhaps, by a metal case, or perhaps by a water bath. Yet no appreciable variation of weight on that account has yet been noted. Nor does the direction of the vertical change rapidly from place to place, as it would with varying permeability of the ground below. But perhaps the agreement of pendulum results, whatever the block on which the pendulum is placed, and whatever the case in which it is contained, gives the best evidence that there is no great gathering in, or opening out of the lines of the earth's force by different media.

Still, a direct experiment on the attraction between two masses with different media interposed was well worthy of trial, and such an experiment has lately been carried out in America

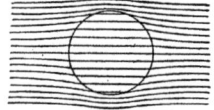

FIG. 6.—Diamagnetic sphere placed in a previously straight field.

by Messrs. Austin and Thwing (*Physical Review*, v. 1897, p. 294). The effect to be looked for will be understood from Fig. 7. If a medium more permeable to gravitation is interposed between two bodies, the lines of force will move into it from each side, and the gravitative pull on a body, near the interposed medium on the side away from the attracting body, will be increased.

The apparatus they used was a modified kind of Boys'

apparatus (Fig 8). Two small gold masses in the form of short vertical wires, each ·4 gm. in weight, were arranged at different levels at the ends virtually of a torsion rod 8 mm. long. The attracting masses M_1 M_2 were lead, each about 1 kgm. These were first in the positions shown by black lines in the figure, and were then moved into the positions shown by dotted lines. The attraction was measured first when merely the air and the case of the instrument intervened, and then when various slabs, each 3 cm. thick, 10 cm. wide and 29 cm. high, were interposed. With screens of lead, zinc, mercury, water, alcohol or glycerine, the change in attraction was at the most about 1 in 500, and this did not exceed the errors of experiment. That is, they found no evidence of a change in pull with change of medium. If such a change exists, it is not of the order of the change of electric pull with change of medium, but something far smaller. Perhaps it still remains just possible ·that there are variations of gravitation permeability comparable with the variations of magnetic permeability in media such as water and alcohol.

Yet another kind of effect might be suspected. In most crystalline substances the physical properties are different along different directions in a crystal. They expand differently, they conduct heat differently, and they transmit light at different speeds in different directions. We might, then, imagine that the lines of gravitative force spread out from, say, a crystal sphere unequally in different directions. Some years ago, Dr. Mackenzie (*Physical Review*, ii. 1895, p. 321) made an experiment in America in which he sought for direct evidence of such unequal distribution of the lines of force. He used a form of apparatus like that of Prof. Boys (Fig. 2), the attracting masses being calc spar spheres about 2 inches in

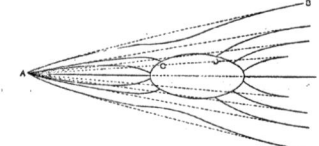

FIG. 7.—Effect of interposition of more permeable medium in radiating field of force.

diameter. The attracted masses in one experiment were small lead spheres about ½ gm. each, and he measured the attraction between the crystals and the lead when the axis of the crystals were set in various positions. But the variation in the attraction was merely of the order of error of experiment. In another experiment the attracted masses were small calc spar crystal cylinders weighing a little more than ½ gm. each. But again there was no evidence of variation in the attraction with variation of axial direction.

Practically the same problem was attacked in a different way by Mr. Gray and myself (*Phil. Trans.*, 192, 1899, A, p. 245). We tried to find whether a quartz crystal sphere had any directive action on another quartz crystal sphere close to it, whether they tended to set with their axes parallel or crossed.

It may easily be seen that this is the same problem by considering what must happen if there is any difference in the attraction between two such spheres when their axes are parallel and when they are crossed. Suppose, for example, that the attraction is always greater when their axes are parallel, and this seems a reasonable supposition, inasmuch as in straightforward crystallisation successive parts of the crystal are added to the existing crystal, all with their axes parallel. Begin, then, with two quartz crystal spheres near each other with their axes in the same plane, but perpendicular to each other. Remove one to a very great distance, doing work against their mutual attractions. Then, when it is quite out of range of appreciable action, turn it round till its axis is parallel to that of the fixed crystal. This absorbs no work if done slowly. Then let it return. The force on the return journey at every point is greater than the force on the outgoing journey, and more work will be got out than was put in. When the sphere is in its first position, turn it round till the axes are again at right angles.

Then work must be done on turning it through this right angle to supply the difference between the outgoing and incoming works. For if no work were done in the turning, we could go through cycle after cycle, always getting a balance of energy over, and this would, I think, imply either a cooling of the crystals or a diminution in their weight, neither supposition being admissible. We are led, then, to say that if the attraction with parallel axes exceeds that with crossed axes, there must be a directive action resisting the turn from the crossed to the parallel positions. And conversely, a directive action implies axial variation in gravitation.

The straightforward mode of testing the existence of this directive action would consist in hanging up one sphere by a wire or thread, and turning the other round into various positions, and observing whether the hanging sphere tended to twist out of position. But the action, if it exists, is so minute, and the disturbances due to air currents are so great, that it would be extremely difficult to observe its effect directly. It occurred to us that we might call in the aid of the principle of forced oscillations, by turning one sphere round and round at a constant rate, so that the couple would act first in one direction and then in the other, alternately, and so set the hanging sphere vibrating to and fro. The nearer the complete time of vibration of the applied couple to the natural time of vibration of the

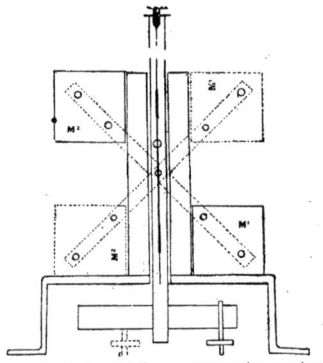

FIG. 8.—Experiment on gravitative permeability (Austin and Thwing).

hanging sphere, the greater would be the vibration set up. This is well illustrated by moving the point of suspension of a pendulum to and fro in gradually decreasing periods, when the swing gets longer and longer, till the period is that of the pendulum, and then decreases again. Or by the experiment of varying the length of a jar resounding to a given fork, when the sound suddenly swells out as the length becomes that which would naturally give the same note as the fork. Now, in looking for the couple between the crystals, there are two possible cases. The most likely is that in which the couple acts in one way while the turning sphere is moving from parallel to crossed, and in the opposite way during the next quarter turn from crossed to parallel. That is, the couple vanishes four times during the revolution, and this we may term a quadrantal couple. But it is just possible that a quartz crystal has two ends like a magnet, and that like poles tend to like directions. Then the couple will vanish only twice in a revolution, and may be termed a semicircular couple. We looked for both, but it is enough now to consider the possibility of the quadrantal couple only.

Our mode of working will be seen from Fig. 9. The hanging sphere, ·9 cm. in diameter and 1 gm. in weight, was placed in a light aluminium wire cage with a mirror on it, and suspended by a long quartz fibre in a brass case with a window in it opposite the mirror, and surrounded by a double-walled tinfoiled wood case. The position of the sphere was read in the usual way by scale and telescope. The time of swing of this little sphere was 120 seconds.

A larger quartz sphere, 6·6 cm. diameter and weighing 400 gms., was fixed at the lower end of an axis which could be

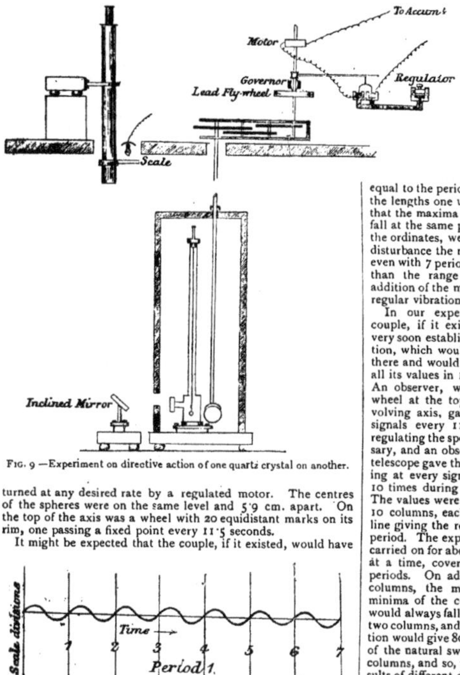

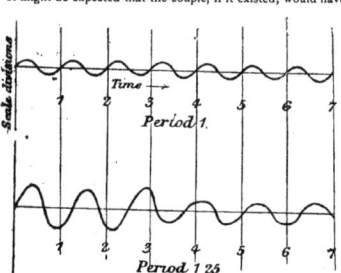

FIG. 9—Experiment on directive action of one quartz crystal on another.

turned at any desired rate by a regulated motor. The centres of the spheres were on the same level and 5·9 cm. apart. On the top of the axis was a wheel with 20 equidistant marks on its rim, one passing a fixed point every 11·5 seconds.

It might be expected that the couple, if it existed, would have

FIG. 10.—Upper curve a regular vibration. Lower curve a disturbance dying away.

the greatest effect if its period exactly coincided with the 120 second period of the hanging sphere—*i.e.* if the larger sphere revolved in 240 seconds. But in the conditions of the experiment the vibrations of the small sphere were very much damped, and the forced oscillations did not mount up as they would in a

freer swing. The disturbances, which were mostly of an impulsive kind, continually set the hanging sphere into large vibration, and these might easily be taken as due to the revolving sphere. In fact, looking for the couple with exactly coincident periods would be something like trying to find if a fork set the air in a resonating jar vibrating when a brass band was playing all round it. It was necessary to make the couple period, then, a little different from the natural 120 second period, and, accordingly, we revolved the large sphere once in 230 seconds, when the supposed quadrantal couple would have a period of 115 seconds.

Figs. 10 and 11 may help to show how this enabled us to eliminate the disturbances. Let the ordinates of the curves in Fig. 10 represent vibrations set out to a horizontal time scale. The upper curve is a regular vibration of range ± 3, the lower a disturbance beginning with range ± 10. The first has period 1, the second period 1·25. Now cutting the curves into lengths equal to the period of the shorter time of vibration, and arranging the lengths one under the other as in Fig. 11, it will be seen that the maxima and the minima of the regular vibration always fall at the same points, so that, taking 7 periods and adding up the ordinates, we get 7 times the range, viz. ± 21. But in the disturbance the maxima and minima fall at different points, and even with 7 periods only, the range is from + 16 to − 13, or less than the range due to the addition of the much smaller regular vibration.

In our experiment, the couple, if it existed, would very soon establish its vibration, which would always be there and would go through all its values in 115 seconds. An observer, watching the wheel at the top of the revolving axis, gave the time signals every 11·5 seconds, regulating the speed, if necessary, and an observer at the telescope gave the scale reading at every signal, that is, 10 times during the period. The values were arranged in 10 columns, each horizontal line giving the readings of a period. The experiment was carried on for about 2½ hours at a time, covering, say, 80 periods. On adding up the columns, the maxima and minima of the couple effect would always fall in the same two columns, and so the addition would give 80 times the swing, while the maxima and minima of the natural swings due to disturbances would fall in different columns, and so, in the long run, neutralise each other. The results of different days' work might, of course, be added together.

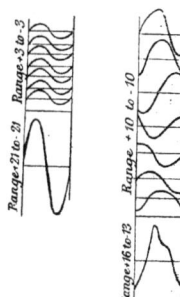

FIG. 11.—Results of superposition of lengths of curves in Fig. 10 equal to the period of the regular one.

There always was a small outstanding effect such as would be produced by a quadrantal couple, but its effect was not always in the same columns, and the net result of about 350 period observations was that there was no 115 second vibration of more than 1 second of arc, while the disturbances were sometimes 50 times as great.

The semicircular couple required the turning sphere to revolve in 115 seconds. Here, want of symmetry in the apparatus would come in with the same effect as the couple sought, and the outstanding result was, accordingly, a little larger.

But in neither case could the experiments be taken as showing a real couple. They only showed, that, if it existed, it was incapable of producing an effect greater than that observed.

Perhaps the best way to put the result of our work is this: Imagine the small sphere set with its axis at 45° to that of the other. Then the couple is not greater than one which would take 5¼ hours to turn it through that 45° to the parallel position, and it would oscillate about that position in not less than 12 hours.

The semicircular couple is not greater than one which would turn from crossed to parallel position in 4½ hours, and it would oscillate about that position is not less than 17 hours.

Or, if the gravitation is less in the crossed than in the parallel position, and in constant ratio, the difference is less than 1 in 16,000 in the one case and less than 1 in 2800 in the other.

We may compare with these numbers the difference of rate of travel of yellow light through a quartz crystal along the axis and perpendicular to it. That difference is of quite another order, being about 1 in 170.

As to other possible qualities of gravitation, I shall only mention that quite indecisive experiments have been made to seek for an alteration of mass on chemical combination,[1] and that at present there is no reason to suppose that temperature affects gravitation. Indeed, as to temperature effect, the agreement of weight methods and volume methods of measuring expansion with rise of temperature is good, as far as it goes, in showing that weight is independent of temperature.

So while the experiments to determine G are converging on the same value, the attempts to show that, under certain conditions, it may not be constant, have resulted so far in failure all along the line. No attack on gravitation has succeeded in showing that it is related to anything but the masses of the attracting and the attracted bodies. It appears to have no relation to physical or chemical condition of the acting masses or to the intervening medium.

Perhaps we have been led astray by false analogies in some of our questions. Some of the qualities we have sought and failed to find, qualities which characterise electric and magnetic forces, may be due to the polarity, the + and −, which we ascribe to poles and charges, and which have no counterpart in mass.

But this unlikeness, this independence of gravitation of any quality but mass, bars the way to any explanation of its nature.

The dependence of electric forces on the medium, one of Faraday's grand discoveries for ever associated with the Royal Institution, was the first step which led on to the electromagnetic theory of light now so splendidly illustrated by Hertz's electromagnetic waves. The quantitative laws of electrolysis, again due to Faraday, are leading, I believe, to the identification of electrification and chemical separation, to the identification of electric with chemical energy.

But gravitation still stands alone. The isolation which Faraday sought to break down is still complete. Yet the work I have been describing is not a failure. We at least know something in knowing what qualities gravitation does not possess, and when the time shall come for explanation all these laborious and, at first sight, useless experiments will take their place in the foundation on which that explanation will be built.

SOCIETIES AND ACADEMIES.

PARIS.

Academy of Sciences, August 13.—M. Maurice Lévy in the chair.—On the neogenic regions of Lower Egypt and the Isthmus of Suez, by MM. C. Depéret and R. Fourtau. Of the Miocene formation the following were recognised :—The Burdigalian, or first Mediterranean stratum, by the presence of *Echinolampas amplus, Scutella Innsi, Lovenia, Cidaris avenionensis, Amphiope truncata* and other fossils; the Vindobonian, or second Mediterranean stratum, by a blue lime containing *Pecten cristato-costatus* and numerous echinoderms. Of the Pliocene, in the neighbourhood of Cairo, are layers of yellowish sand containing *Clypeaster aegyptiacus* and other characteristic fossils.—The area of the basins of Russia in Asia, by M. J. de Schokalsky. The map is made upon the scale of 1 in 4,200,000, and the area evaluated by means of a sheet of celluloid divided in square millimetres. The area found is 16,085,000 sq. kilometres.—On a hypsometric map of European Russia, by M. J. de Schokalsky. The previous map of M. de Tillo was limited by the latitude 60° N. ; the present map includes the whole of European Russia upon a scale of 1 in 15,300,000.—Observations of the Borelly comet, made at the Observatory of Algiers with the 31'8 cm. equatorial, by M. F. Sy.—A new arrangement of apparatus serving to measure geodesic bases, by M. Alphonse Berget. Ruled plates of iron floating in a bath of mercury are used instead of the ordinary scales. The method has the advantage of securing without trouble the absolute horizontality of the rules ; two consecutive rules are necessarily in the same horizontal plane, since their mercury baths are connected ; there is no correction necessary for the flexure of the rules, and the temperature correction is much more certain.—Stereoscopic vision of curves traced by a phase apparatus, by M. Marc Dechevrens.—Properties of magnetic

[1] Landolt, *Zeit. für Phys. Chem.*, xii. 1, 1894. Sanford and Ray, *Physical Review*, v. 1897, p. 247.

deposits obtained in a magnetic field, by M. Ch. Maurain. Iron was deposited in a magnetic field either from a solution of ferrous and ammonium chlorides, or from a solution of ferrous sulphate in sodium pyrophosphate. It was found that the intensity of magnetisation of different layers of the deposit growing in a uniform field has the same value, and that the uniform magnetisation acquired by a deposited strip increases with the strength of field in which the deposit is obtained.—The *rôle* of discontinuities in the propagation of explosive phenomena, by M. Paul Vieille. On the assumption of an adiabatic elasticity, together with continuity, the velocities of wave propagation found in certain cases are too great. It is necessary to assume that the phenomenon is discontinuous.—Action of hydrogen upon the sulphides of arsenic, by M. H. Pélabon. Details of experiments of the interaction of realgar and hydrogen in sealed tubes at 610° C. The reaction is a reversible one, and the limit is affected by the introduction of an excess of arsenic.—The properties of the blue oxide of molybdenum, by M. Marcel Guichard. The blue oxide is a molybdate, and could not be obtained free from water, two oxides of molybdenum only existing in the anhydrous state, MoO_2 and MoO_3.—On the colouring matter of *Echinus esculentus*, by M. Griffiths.—On the composition of the ashes of some medicinal plants, by M. Griffiths.—On a cause of error in the examination of wines for salicylic acid, by M. J. Ferreira da Silva. The method of Petlet and Grobert will indicate the presence of salicylic acid in a pure wine that is really free from it. The official German method gives exact results.—On a variety of the anthrax bacillus ; a short asporogenic form, *Bacillus anthracis brevigemmans*, by M. C. Phisalix. In the organism of the dog the *B, anthracis* undergoes important modifications, becoming shorter with a rapid and complete segmentation. It is still uncertain whether this should be regarded as a variety or a new species.—Anti-hepatic serum, by M. C. Delezenne.—Application to man of the regeneration of confined air by means of sodium peroxide, by MM. A. Desgrez and V. Balthazard. The apparatus described weighs twelve kilograms, and by its means a man can penetrate easily into an irrespirable atmosphere.

CONTENTS.

THURSDAY, AUGUST 30, 1900.

RIGHT- AND LEFT-HANDEDNESS.

Rechts- und Linkshändigkeit. Von Dr. Fritz Lueddeckens. Pp. vi + 82. (Leipzig : Engelmann, 1900.)

THE chief interest of the treatise under consideration consists in the exposition of a variety of conditions which, in the author's opinion, are associated with that commonly known as right-handedness, a condition which is generally assumed to refer to a greater physical development and functional employment of that side of the body. Moreover, the author seeks to refer all the phenomena upon which his extended conception of right- (or left-) handedness is based to a common cause, which he finds in the existence of a higher degree of blood-pressure in the vessels of one side of the head (and in the common instance, viz. the right-handed one, in the vessels of the left side) than in those of the other. Dr. Lueddeckens is thus firstly concerned to prove the existence of such an inequality in blood-pressure as has just been mentioned, and the earlier pages of his book are devoted to this part of the subject. From an account of the embryological history of the arterial system, and the departures from original bilateral symmetry which that system presents, we are led (p. 8) to a study of the hydrodynamic conditions obtaining in the aortic arch in the living subject, and in this connection it is argued that the circumstances are such as will produce a higher blood-pressure in the left than in the right common carotid artery. The greater frequencies of cerebral hæmorrhage, and of embolism of the arteria centralis retinæ on the left side of the brain, are urged in further support of this view.

If it be conceded that the cerebral hemispheres may differ in respect of the pressure in their arterial systems, it becomes possible to divide individuals into three classes, viz. those in whom the blood-pressure is higher on the left side ; those in whom the pressure is higher on the right side ; and, finally, there must be a category in which will be ranged such cases as present a degree of blood-pressure which is the same in each hemisphere. The majority of cases will, it is believed, fall within the first of these divisions, and to such cases (ordinary right-handed persons) our attention is first directed : evidence is drawn from a comparison of the structures supplied by the chief branches of the common carotid arteries of each side, and firstly of the structures in the area supplied by the external carotid artery, including various superficial structures and also the ear (internal maxillary artery), and the conclusion is arrived at that there is a predominance in growth and a more easily excitable nervous sensibility (auditory sensations being especially observed) on the *left* side, such preponderance being directly associated with the higher blood-pressure on that side.

In the second place, a number of observations on the shape of the eyeball, and on the comparative dimensions of the two eyes in the same individual, are discussed.

For it is argued that a difference in blood-pressure will find expression in a difference in the shape of the eyeball on the same side, and that this difference in shape will, in turn, be manifested by differences between the two eyes, in respect of sight. And it is finally submitted that observations on the respective refractive powers in the two eyes of a number of persons examined with regard to this point bear out the conclusion which was thus arrived at on *a priori* grounds.

Thirdly, differences in the size of the pupil on the two sides are taken as criteria of differences in blood-pressure, the smaller pupil corresponding to the higher blood-pressure, and *vice versâ*. Thus we should, according to the author's argument, expect normally to observe differences in this respect. But inasmuch as such difference between the pupils is, by many authorities, considered to accompany pathological conditions only, the author is at considerable pains to show that a purely physiological difference in size may exist. And so again, his arguments that the difference in blood-pressure will be manifested by a difference in the pupils, and further that the difference is normally (in right-handed persons, that is) indicative of a higher blood-pressure on the left side of the head, are considered to be justified by the number of cases in which the smaller of two "physiologically" unequal pupils was observed in the left eye.

Turning from the special domain of ophthalmic anatomy and physiology, the relative weights of the cerebral hemispheres next claim attention, and Hamarberg's results are quoted as indicative of a slight excess of weight in favour of the left hemisphere. The conformation of the several cortical areas is then described, in allusion to their well-known connection with the voluntary production of movements (and of speech in particular). Sensory phenomena are next considered, and evidence of a right-sided predominance in nervous excitability is adduced from the results of work by Biervliet (on the muscular sense, taste, hearing, sight) ; and lastly, psychical events are dealt with, though with much brevity owing to the difficulty of obtaining relevant evidence.

Passing briefly over the category of subjects in whom an equal blood-pressure is presumed to obtain on both sides of the head, the remaining class in which the blood-pressure on the right side of the head exceeds that on the left is dealt with much in the same way, and in nearly as much detail, as the first class ; and with very similar results, *mutatis mutandis* : in other words, Dr. Lueddeckens finds in the majority of left-handed persons the various sources of evidence which have previously been detailed, and which indicate in the left-handed persons a higher blood-pressure on the right side of the head, just as they indicated this condition on the left side of the head in right-handed persons. Interesting observations on the psychical phenomena of young left-handed individuals are recorded, and in particular their difficulties in learning to write, their tendency to adopt mirror-writing, and the greater frequency of impediments to speech among the left-handed may here be noted. Finally, the tendency of the left-handed to lie on the left side during sleep is commented upon.

The foregoing sketch will, it is thought, render the

T

following comments intelligible. Firstly, the demonstration of an essential and fundamental point, viz. the higher degree of blood-pressure in the area supplied by the left common carotid artery, leaves a good deal in the way of direct evidence to be desired : the most important point urged in support being perhaps the comparatively greater frequency of cerebral hæmorrhage on the left side. The author admits that, as regards the brain, the confluence of the two vertebral arteries (to form the basilar) equalises the conditions on the two sides so far as the parts (medulla, pons, and posterior parts of hemispheres) supplied by these are concerned, whereas the equalising effect is not supposed to be felt in other parts of the circle of Willis. We regret that we can find no direct guidance on this point in Hill's important work on the cerebral circulation.

The arrangement of the great vessels springing from the aortic arch is also a subject that admits of a good deal of discussion in the present connection.

With reference to the auditory nerve (p. 16), and the greater sensibility of the auditory centre in the left hemisphere, it may be mentioned that some support is afforded to this view by the earlier date at which the auditory fibres running up to the first temporal gyrus in the left hemisphere acquire their medulla, and presumably attain a fully functional state (Flechsig). In his observations on the eye, the author is to be congratulated on having devised new applications of routine clinical methods to the elucidation of the questions with which he deals. As regards actual differences in the dimensions of the eyes, it is remarkable that no evidence on this subject is forthcoming from the otherwise exhaustive work by L. Weiss on the anatomy of the eye (*Anatomische Hefte*, Bd. viii. 1897). The recognition of non-pathological differences in the size of the pupils is a point on which it is worth while to insist ; moreover, the phenomenon will lose little, if any, of its importance as a physical sign in the early diagnosis of certain nervous diseases (*e.g.* general paralysis of the insane). As regards the weights of the hemispheres of the brain, it may be well to remark that there appears to be a mis-quotation on p. 49, where the weight of a left cerebral hemisphere is stated to be 218 gm., and that of the corresponding right hemisphere 133 gm. only ; at any rate, if there is not a mistake in quoting Hamarberg's figures, the brain could hardly be regarded as other than pathological, and consequently valueless in this connection. But more important than this is the fact that Braune's extensive weighings show that the difference between the two hemispheres is quite negligible. At the same time we may mention that, according to Bastian, the specific gravity of the left hemisphere exceeds that of the right. Finally, we do not feel inclined to agree with the author in explaining instances of the existence of double personalities on the supposed presence of equal blood-pressure in right and left common carotid arteries.

On the whole, we think that while the amount of evidence in support of the author's main assumption might well be increased, at the same time the clear record of observations, and the deliberate discussion of their significance, will render Dr. Lueddeckens' volume of much interest to biologists.

W. L. H. DUCKWORTH.

MODERN VIEWS ON THE CHARACTERS OF THE CELLULAR ELEMENTS IN THE BLOOD.

Histology of the Blood : Normal and Pathological. By P. Ehrlich and A. Lazarus. Edited and translated by W. Myers, M.A., M.B., B.Sc. Pp. xiii + 216. (Cambridge : At the University Press, 1900.)

NOT much more than a year has elapsed since the first part of "die Anæmie," by Ehrlich and Lazarus, appeared in Nothnagel's "System of Pathology and Therapeutics" ; but during that short time the work has taken a foremost place among those dealing with the histology of the blood. Perhaps the most striking feature of the book is its originality, broad lines being laid down along which future investigators may work, and no subject is taken up without being enriched by some suggestive hypothesis based on interesting observations made by Ehrlich himself or some of his pupils. Although comparatively a small book, it may be said, without disparagement to the many other works on hæmatology, to be the one to which the term "epoch-making" may, without exaggeration, be applied. It is only possible to refer shortly to some of the most important subjects discussed in its pages. Although it is undoubtedly with reference to the leucocytes that the most important observations are made, there are also points of great interest treated of in the chapter dealing with the morphology of the erythrocytes. This is especially the case with regard to the authors' views on polychromatophilia as a sign of degeneration, and on the method of transformation of megaloblasts to megalocytes and normoblasts to normocytes. Not less important are the paragraphs dealing with the megaloblastic type of the blood and marrow in pernicious anæmia. But it is when the authors come to discuss the normal and pathological histology of the white blood corpuscles that we find on every page observations that shed light on points that have been long in obscurity.

Although the authors belong to a comparatively small school that believes in the absolutely distinct characters of two types of white blood corpuscles, lymphocytes and granular leucocytes, yet no one, whatever his own opinions may be, can rise from a perusal of these pages without granting that no stronger case could have been presented in support of this view than the one placed before us in this book. Perhaps it is mainly to Ehrlich and Ribbert in Germany, and Muir in this country, that we are indebted for the most powerful arguments against the view that all leucocytes are developed from the lymphocyte. The arguments presented in this book in favour of the view that there are two great types of white cells, are obtained from morphological, experimental, pathological and clinical data. The morphological characters of the different forms of white cells are first described in a very lucid manner. There is an exceedingly valuable contribution to our knowledge of the functions of the spleen in Kurloff's work on the effects of removal of that organ from guinea-pigs. The functions of the lymph glands and bone marrow are described, and additional evidence is given in favour of the two-fold type of the white blood corpuscle.

The chapter dealing with the demonstration of the cell granules and their significance is, of course, one in which Ehrlich, as a pioneer in this subject, naturally finds himself at home. He repudiates Altmann's claims to priority with regard to the importance of cell granules. The description of the different types of leucocytosis and leucocythæmia is exceedingly good, and perhaps constitutes the most valuable section of the work. It is unfortunate that the translator has not seized the opportunity, as he has done in the case of Kanthack and Hardy's investigations, of referring to the very important work done by Muir on experimental leucocytoses and leucocythæmia. It is to him that we are indebted in the first place for the recognition of the "leucoblastic" type of marrow in experimental leucocytosis.

Ehrlich's chemiotactic theories with regard to the emigration of different cells from their seat of formation, the marrow, into the blood, and from the blood into the tissues, &c., are presented in a most interesting fashion, although, unfortunately, it is still impossible to speak about the etiology of medullary leucocythæmia in anything but the most indefinite way. To Dr. Myers' translation one can only refer in terms of praise. Perhaps it errs at places by being rather too literal. References to Jenner's eosin-methylene blue mixture, and to Kanthack and Hardy's work, are welcome additions made by the translator and editor. Confirmation and amplification of the very important investigations of Kanthack and Hardy, and Hardy alone, on the solution of oxyphil granules when cells containing the latter come in contact with chains of *B. anthracis*, &c., would be heartily welcomed by all who are interested in the subject of leucocyte secretions.   T. H. Milroy.

BIOLOGY AT WOODS' HOLL, U.S.A.

Biological Lectures from the Marine Laboratory at Woods' Holl, U.S.A., for 1899. Pp. 282. (Boston : Ginn and Co., 1900.)

THE present volume, like all its predecessors, is replete with interest and full of testimony to the activity and good work of the Whitman School. It contains the reports of sixteen lectures, of which as many as four are for the first time botanical ; and although among the zoological writers we miss the names of Whitman and one or two of the most tried among his earlier collaborateurs, the effects of their teaching and example are all evident. More especially is this the case with the lectures by C. M. Child on "The Significance of the Spiral Type of Cleavage," and by E. Thorndike on "Instinct," in which certain of Whitman's most famous conclusions receive support.

Conspicuous lectures are those by C. B. Davenport on "The Aims of the Quantitative Study of Variation," and by Jacques Loeb on "The Nature of the Process of Fertilisation," each in extension of work for which these investigators are now well known. The latter writer, dealing with facts which show that the process of fertilisation and development may be produced in the egg cell by the action of certain salts, to an advanced stage, would have us believe he has transferred the problem of fertilisation from the realm of morphology into the realm of physical chemistry. There is an important address by Alphæus Hyatt on "Some Governing Factors usually neglected in Biological Investigations," in which the uniformitarian hypothesis receives a check and a defence

is set up of a law of "Tachygenesis" or "abbreviated development " ; and there is incorporated in it a discussion on heredity, in its bearings on Ribot's argument that it is a "specific memory," and that a form of automatism is the link between memory and habit.

T. H. Morgan continues to write on "Regeneration," and among the lectures there are two which are noteworthy as embodying full bibliographies, of service for reference—viz. those by A. G. Mayer on "The Development of Colour in Moths and Butterflies," and by G. N. Calkins on "Nuclear Division in Protozoa." Interest amounting to novelty is greatest as concerns the work of C. H. Eigenmann on the breeding habits of the blindfishes, the Amblyopsidæ, of the Mississippi Valley, in which the discovery that the bleached condition is assumed by the young even when reared in the light, is brought forward as evidence of hereditary establishment of an effect of the environment ; and as concerning a lecture by H. S. Jennings on "The Behaviour of Unicellular Organisms," in which, from the fact that a multiplicity of causes may bring about similar reactions, it is argued that organisation and not the nature of the stimulus determines the result of experiment. Of the botanical lectures, that by D. H. Campbell on "The Evolution of the Sporophyte" furnishes an argument in favour of the abandonment of aquatic life having had a potent influence in its higher development, while another by D. P. Penhallow will be useful, as giving a succinct account of the alteration and carbonisation processes undergone by vegetable organisms during fossilisation. The remaining lectures are upon the effects of temperature and currents of air upon distribution, the significance of mycorrhizas, the associative processes in animals, and the "Physiology of Secretion "; and the *tout ensemble* gives promise of increased attention in the future to questions of cytology, in both their experimental and physiological aspects, with a leaning to those which involve philosophic principles and abstract ideas. No doubt much of the biological work of the next generation will be of this type, but in view of the probability that that may stand in danger of being overdone, and of the idea that nothing remains possible on the old lines, it may be said that in the very book under review there is reached the conclusion that "it is the individual which is the unit and not the cell." In the future, when everything will need to be gone over again under an advance in methods and a better understanding, the facts of mere anatomy—the value of which there is a growing tendency to depreciate—will assuredly prove as important and instructive as in the past. Our American brethren may do well to bear this in mind.

OUR BOOK SHELF.

Brief Guide to the Commoner Butterflies of the Northern United States and Canada. Being an Introduction to a Knowledge of their Life-histories. With Illustrations of all the Species. By Samuel Hubbard Scudder. Pp. xi + 210 ; 22 plates. (New York : Henry Holt and Co., 1899.)

OUR notice of the first edition of this work appeared in NATURE for August 10, 1893. This is not before us while writing ; but as far as we can tell without actual comparison, the present edition, as regards the letterpress, is little more than a reprint of the first. But the plates are a welcome addition. They represent seventy-three species, without colour, carefully drawn and easily recognisable, though sometimes badly printed. The small size of the book renders it very convenient for handy reference. A European entomologist will recognise one or two old friends, such as the Camberwell Beauty, the Painted Lady, Red Admiral, and a Small Copper, hardly distinguishable from our own; but the proportions of the various families and genera are very different from what obtains in Europe. A single plate, representing five species, and another representing only six species, are enough to illustrate the Satyridæ, and the Blues and Coppers together; while a much more crowded plate is required for the Hair-Streaks, and two for the Skippers. There are also several very large and conspicuous species, including six large Swallow-Tails, and the northern representatives of several tropical genera. But although the average size of the North American butterflies is much larger than ours, and much of the settled part of the country lies much further south, the number of species in the Northern States is much smaller than in Europe, owing to the comparative absence of Satyridæ and Lycænidæ; and it is not till we reach the frontiers of Mexico that the vast wealth of the tropical American butterfly fauna (almost equalling that of all the other continents put together) begins to dawn upon us.

W. F. K.

Elements of Qualitative Analysis. By G. H. Bailey, D.Sc., Ph.D., and G. J. Fowler, M.Sc. Pp. 115. (Manchester: J. E. Cornish, 1900.)

AMONG the distinctive characteristics of this addition to the already numerous volumes on practical chemistry are: the prominence given to the recognition of common elementary substances by an examination of their simple physical and chemical properties, the attention given to dry methods of analysis, and the series of flame-reactions. These sections provide students of practical chemistry with excellent exercises in manipulation, and will counteract the belief that the best way to analyse a substance is always to dissolve it and go through the usual routine treatment of solutions and precipitates. There is little sympathy with ordinary qualitative analysis at the present time, but where the subject is taught it should be taught intelligently; and as this little book provides a reasonable course of laboratory work, it merits a trial.

LETTERS TO THE EDITOR.

[*The Editor does not hold himself responsible for opinions expressed by his correspondents. Neither can he undertake to return, or to correspond with the writers of, rejected manuscripts intended for this or any other part of* NATURE. *No notice is taken of anonymous communications.*]

Railways and Moving Platforms.

ABOUT twenty years ago I was in the habit of speaking with Prof. Ayrton and other friends about a scheme which might increase ten-fold the carrying capacity of the Underground Railway. I prepared a letter for the *Times* newspaper about two years ago, but at the earnest entreaty of a friend I applied for patent protection for the scheme, and did not publish the letter. I have not proceeded with the patent, and wish now that I had published the letter. Indeed, I wish that, instead of merely talking the matter over with friends twenty years ago, I had published what I had to say.

Travelling now on the new Central London Railway, one feels that there is enormous waste of energy and of time in starting and stopping the trains. Again, a train must not be longer than the platform. On my scheme the train does not stop, and the longer it is the better. Indeed, I can imagine an endless train keeping a perfectly constant speed all the time.

My scheme is easy enough to understand now that moving platforms are common. After passengers enter a station I get

them gradually into a state of motion, so that moving alongside the train and at the same speed, they may enter and other passengers may leave. There are many ways in which the scheme may be carried out. From a wayside station passengers may enter an express train which does not stop, in the following way. They enter a small train at the station; this train gradually gets up a speed equal to that of the express; it runs alongside the express at a particularly well-laid part of the line; there is an exchange of passengers, and the local train gradually comes to rest again at the station.

For the Underground Railway, the method which most commended itself to Prof. Ayrton and me long ago was this. At a station, say St. James's Park, the platform was a carefully constructed turntable, 500 feet in diameter, the rim of which travelled at 8 miles per hour. The whole was not really a floor; it was only a skeleton of a turntable, being an outer rim 8 feet broad and many radial passages. The very long train to Mansion House, travelling at 8 miles per hour, was close to the rim of the turntable; indeed geared with it in a rough, simple manner for less than half its circumference; the train from Mansion House did the same on the other side. I need not speak of the automatic opening and closing of the doors of the train.

A passenger, let us say second class for Mansion House, takes his ticket and descends a spiral staircase, which revolves so slowly that even the frailest and most timid of old ladies is not frightened; in fact, it revolves on its own axis once in 134 seconds. At the bottom the passenger sees a few notices; one of them saying second class, Mansion House, has a hand pointing along a radial passage, and this is followed. As the passenger moves radially, he does not notice that he is gradually getting up speed circumferentially. He does not notice that the floor gets slightly inclined as he moves out, to counteract the small effect of centrifugal force. When he reaches the outside of the platform he probably finds a train there, seemingly at rest, with the doors open, and he enters it, moving perhaps along the platform, choosing one compartment rather than another. If he is lucky he has about one minute in which to make his choice. But he will notice near him on the platform an altering time signal which tells him how much more time he has to waste: 50 seconds, 40, or 30, or 20, or 10; if he delays after the signal says 0, an iron railing will come between him and the train; he will see the train moving laterally away from the platform, and he must wait seventy-four seconds before he sees a train coming laterally towards him; the railing goes away, and he has again sixty seconds in which to enter.

If he had a third class ticket to South Kensington, he would have proceeded in exactly the same way. Also every passenger wanting to leave the train at St. James's Park had sixty seconds in which to do it. Trains at 16 miles an hour give only half these times. A platform of only 250 feet diameter would give only half the time if the train speed was 8 miles an hour. I need not dwell upon the details of this and other methods which suggest themselves. It may be soon or syne, but I feel sure—I have felt sure for many years—that my method will have to be adopted.   JOHN PERRY.

August 11.

Snow-drifts on Ingleborough.

IN his interesting letter on "Snow-drifts on Ingleborough in July," Prof. Hughes describes what may be called the first stage in the formation of a glacière. These ice-caves, not very rare in parts of the Alps and Jura, were made by the present Bishop of Bristol the subject of an attractive book (published thirty-five years ago), and have been occasionally noticed in the earlier volumes of NATURE and elsewhere. I have always believed that snow, drifted into caves during the winter, was the initial cause of these natural ice-houses (about half a dozen of which I have visited), and can quote a case from the Alps which is a slight variation of that described by Prof. Hughes. On July 24, 1873, I went up the Pic d'Arzinol (9845 feet) from Evolena in the Val d'Hérens, and on the way down—so far as I remember between five and six thousand feet above sea-level—my guide diverged from the track to show me what he called the Pertuis Freiss. These were two fissures, apparently joints, opened by a slight subsidence. A description of one will serve for both, except that there was hardly any descent to its floor. The fissure extended some four yards into the hill, and was at widest about as many feet. Ice was patched about the floor, and in places formed a plaster on the walls, its thickness being at most three inches. It showed prismatic structure, though rather small. The air within was cold (I had no thermometer) ; but as the surface of the ice was wet, it was above 32° F., though I think not much. The guide told me that the fissures in winter-time were filled with snow. This accumulation, probably owing to the shape of the fissure, no longer remained as snow, but was represented by the ice on the floor and walls, which the guide said seldom, if ever, disappeared. The absence of ice from the walls of the Ingleborough "swallow hole" was probably due to some exceptional dryness of the rock ; but Prof. Hughes has undoubtedly found a "baby" ice-cave, like that I have described, and it will be worth examining some more of these dry shafts to see whether a slightly better developed specimen may not be lurking in the neighbourhood.   T. G. BONNEY.

Permeability of Iron under the influence of the Oscillatory Discharge from a Condenser.

IN your issue of August 2 there is an abstract of a very interesting paper, read by Prof. Trowbridge, on his experiments with a battery of 20,000 secondary cells. In it he mentions that the permeability of Iron when under the influence of a very powerful discharge from a large condenser is now under observation.

I should like to draw attention to some experiments I was making over a year ago in Lord Blythswood's laboratory (an account of which has not yet been published), in which I have gone into the subject in some detail.

In my experiments the lowest frequency used was about 5000 a second. I enclose two photographs of sparks taken in

FIG. I.

the usual way with a revolving mirror. The discharge in photograph (1) took place through a coil of about 5 millihenrys self-induction from a battery of Leyden jars of a total capacity of ·06 microfarads, the potential difference between the coatings, before discharge, being 13,500 volts. In photograph (2) a fine wire core, consisting of 550 No. 28 soft Iron wires, was inserted

FIG. 2.

into the coil (which was wound on a hollow paper spindle of about 1·3 cm. internal diameter). The other conditions of discharge were identical in the two cases ; the speed of the mirror, however, was 19 revolutions per second for photograph (1), and ·16 per second for photograph (2), thus tending to draw out the spark more in the first photograph.

The essential differences are, however, well marked. At the beginning of the discharge we have the "pilot" spark, first noticed by Prof. Boys ; and then in the photograph (2), taken with the iron wire cores, a series of oscillations gradually increasing in length. The first half-oscillation, however, is nearly twice as long as the half-oscillations in photograph (1), when there were no iron wire cores in the coil. The increase in the time for a half-oscillation is due, of course, to the increased self-induction of the coil on account of the iron ; and the gradually-increasing length is due to the increase in permeability of the iron as the intensity of the discharge dies away. In photograph (1) the frequency of oscillation of the spark taken with the coil having air cores is about 9000 per second, and in (2) the approximate magnitude of the current during the first discharge with the iron cores, 15 amperes.

It would be impossible in the course of a short note to describe in detail the work that has been done, but in numerous experiments (over three hundred spark photographs have been

taken) that have been made, the iron has been found to behave in the same way under these oscillating magnetisations as it does when steady currents are used to produce magnetising forces of the same intensity. In most experiments single layer coils have been used in which the magnetising forces due to a given current can be calculated, and it has been possible to determine approximately the forces acting on the iron. From the results, curves showing the variation in permeability with magnetising force have been plotted. In some experiments, the magnetising current due to the discharge has been as large as 1000 amperes. In order to obtain discharges as powerful as this, a very large glass-condenser has been used with a total capacity of 1·5 microfarads, made up of plates of glass (coated with shellac) 1·6 mm. (1/16″) thick. The conducting surfaces are of tinfoil. The glass appears a great deal stronger than that used by Prof. Trowbridge, as it has been tested repeatedly at 20,000 volts. It is possible, however, that the suddenness with which his condenser is charged from his cells may account for the readiness with which the glass breaks. In my experiments the condenser was charged by a large Wimshurst machine of 160 plates, which took almost half a minute to get up the full potential of 20,000 volts. The glass used is known technically as 15 oz. 3rds selected flat sheet, and was obtained from Messrs. Malloch, of Glasgow.   E. W. MARCHANT.

Blythswood Laboratory, Renfrew, August 7.

Function of the Whips of the Larva of the Puss Moth.

YOUR correspondent (p. 389) will find a detailed account of the various defensive appliances of the larva of *Cerura vinula* in Prof. Poulton's work on the "Colours of Animals" (International Science Series), and in papers published by him in the *Transactions* of the Entomological Society of London for 1886 and 1887, the latter papers being illustrated by beautiful coloured plates.

It is usually believed by entomologists that the function of the "whips" in the caudal appendages of the larva is to drive away, or frighten away, Ichneumon Flies or other enemies, but there is still room for further inquiry ; and although the larva is highly protected, it is liable to the attacks of some species of Ichneumon Flies, though it may be able to defend itself against others, for the protection of no animal is absolutely complete.

The appendages are doubtless homologous with the retractile fleshy fork in the neck of the larvæ of the Swallow-tailed Butterflies (Papilionidæ), which probably fulfils a similar function.   W. F. KIRBY.

Hilden, Sutton Court Road, Chiswick.

The Migration of Swifts.

ON the morning of Friday, August 10, I witnessed a large flight of Swifts travelling westward along the Sussex coast. The birds were passing this place in a continuous though thin stream for several hours ; I saw them myself from 10 a.m. when I first visited the shore, and watched them till 12 noon. A few birds were also noted travelling in the same direction between 5 p.m. and 6 p.m. The day was bright but showery, and a fresh W.N.W. breeze was blowing at the time, so that the birds were flying almost against the wind ; they flew low, seldom rising fifteen feet in the air, and often passing within two feet of my head as I lay on the shingle ; they kept to the coast-line and for the most part over the top of the fringe of tamarisks that here stretch for miles just above the shingle. Since that day I have not seen a single Swift in the neighbourhood, in spite of having travelled on my bicycle as far west as the mouth of Chichester Harbour along the coast, and to various points north of this line as far as Chichester and Arundel inland. It would be interesting to know if other observers witnessed any similar flights on August 10, and also if Swifts are still to be seen in any places in our islands at the present time. I have on two previous occasions seen Swifts arrive on the east coast of Norfolk as late as the first week in September (after a complete dearth of the birds for some three weeks), and depart again after a few days' sojourn—these perhaps are migrants from the European continent. As many of your readers are now doubtless at the seaside, it seems a favourable opportunity to ask them to keep their eyes open and record any facts that they may observe bearing on the movements of these birds.   OSWALD H. LATTER.

East Preston, near Worthing, August 19.

UNITS AT THE INTERNATIONAL ELEC-TRICAL CONGRESS.

AT the suggestion of Prof. Hospitalier, Section I. of the Congress agreed that the following should be the members of the Commission on Units :—Messrs. Ayrton (Great Britain), De Chatelain (Russia), Dorn (Germany), De Fodor (Hungary), Eric Gérard (Belgium), Hospitalier (France), Lombardi (Italy), Kennelly (United States) ; and at the first meeting of the Commission, on August 21, which was attended also by Prof. F. Kohlrausch and Sir W. Preece—whose names had been added to the list of the Government delegates for Germany and England—a report presented to the Congress by the American Institute of Electrical Engineers was taken into consideration. This report had been drawn up for that Institute by a committee appointed for this purpose, and it contained the following resolutions :—

(1) We consider that it is necessary to give names to the absolute units in the electromagnetic and electro-static systems, as well as convenient prefixes to designate the decimal multiples and submultiples of these units in addition to those already in use.

(2) The International Congress of Electricians, which will take place this year in Paris, should be invited to choose the names and the prefixes.

(3) A great advantage would be gained by a rational-isation of the electric and magnetic units, and the Congress should be invited to find ways and means to obtain such a rationalisation.

The proposition to rationalise the units—that is, to change them so that the coefficient 4π should not appear —was withdrawn by Dr. Kennelly on behalf of the United States ; as well as the suggestion regarding the employment of prefixes, and it was resolved that :—

The Commission will only deal with propositions that will introduce no change in the decisions arrived at at previous congresses.

A long discussion then took place as to whether it was really necessary to give names to the C.G.S. units either in the electrostatic or the electromagnetic systems, and finally it was agreed to withdraw the proposition so far as it dealt with the electrostatic system.

The desirability of giving a name to the unit of magnetic field and to the unit magnetic flux was strongly urged, and as the names of *Gauss* and *Weber* had been employed for some years in America for these units respectively, the advantage of adopting these names for the C.G.S. units of field and flux was advocated. On the other hand, the resolution arrived at by the Electrical Standards Committee of the British Association in 1895 to employ these names respectively for other units was pointed out. Finally, the Commission, at the end of their second sitting, on August 22, recommended the following :—

"The Commission is not of opinion that it is necessary to give names to all the electromagnetic units.

"However, in view of the use already of practical instruments which give the strength of a magnetic field directly in C.G.S. units, the Commission recommends that the name *Gauss* be assigned to this unit in the C.G.S. system.

"The Commission proposes to assign to the unit of magnetic flux, of which the magnitude will be subse-quently defined, the name of *Maxwell*."

These resolutions were brought before Section I. of the Congress on August 24, and led to a long discussion. M. Mascart opposed the giving a name to the C.G.S. unit of magnetic field. The employment of practical instruments for the direct measurement of the strength of magnetic fields in C.G.S. units was not, in his opinion, a sufficient reason for assigning a name to that unit. Besides, this decision of the Commission appeared to be

contrary to the spirit of the Congresses of 1881 and 1889, which did not give the names of men to the C.G.S. units. He admitted that the name of a man might be given to the practical unit. In any case, the name of "Gauss" seemed to him liable to give rise to confusion, for Gauss was the originator of the first absolute system employed, viz. that of the "millimetre-milligramme-second" system, and that system, as distinguished from the "centimetre, gramme, second" system, was still in actual use in certain cases—for the measurement of the earth's field, for example.

Prof. Kohlrausch said that the "absolute units" were enough for the physicists, but that, if the engineers felt the need of practical units, Dr. Dorn and he did not see that any inconvenience would arise from names being given to them, such as those of Gauss and of Maxwell, for example. The German delegates could not, how-ever, commit their Government in the matter, and they considered that the Congress should limit its recom-mendations to the use of these new names without seeking that legal sanction should be given to them.

Prof. Ayrton agreed with M. Mascart, and mentioned that during the past five years many "Ayrton-Mather Field Testers" had been constructed to read off the strength of a magnetic field directly in C.G.S. units, but that no need for any special name for that unit had been felt in connection therewith. He added, however, that, while holding the opinion expressed by M. Mascart that it was not desirable to give the names of persons to the C.G.S. units, the units of field and flux had this peculiarity, that without any multipliers they were the practical units adopted.

To this M. Mascart replied that the word "practical" in this connection was ambiguous, since, although it was true that the C.G.S. units of magnetic field and flux were employed in practice, they did not belong to the so-called "practical system."

M. Hospitalier appealed to the Section to give names to the unit of field and the unit of flux. He did not ask for any legal decision in the matter, for the names were put forward as a simple recommendation to the Section.

After a discussion in which Messrs. Ayrton, Carpentier, Dorn, Hospitalier, Kohlrausch, Mailloux, Mascart, A. Siemens, Silvanus Thompson and others took part, Prof. Eric Gérard stated that in his opinion it was desirable to come first to a decision that names should be given to the C.G.S. units of magnetic field and to flux of magnetic induction.

M. Mascart, expressing his approbation of this idea, the president of the Section, M. Violle, put the following proposition formally to the meeting :—

"The Section recommends the adoption of specific names for the C.G.S. units of magnetic field and of magnetic flux." This proposition being adopted, with only two dissentients, the meeting was adjourned for a short time to enable the members to exchange their views regarding the exact names that should be employed. On the meeting reassembling, the president put the two following propositions successively :—

(1) *The Section recommends the adoption of the name of* GAUSS *for the C.G.S. unit of magnetic field.*

(2) *The Section recommends the adoption of the name of* MAXWELL *for the C.G.S. unit of magnetic flux,*

both of which were adopted with only two dissentients.

On the same afternoon these resolutions of Section I. were submitted to the Chamber of Government Delegates to the Congress and adopted, and finally, at the closing meeting of the Congress on Saturday, August 25, the action which had been taken in the matter was formally reported by M. Paul Janet, one of the two secretaries of the Congress.

THE AMERICAN INSTITUTE AND THE ENGLISH INSTITUTION OF ELECTRICAL ENGINEERS IN PARIS.

STARTING with a trip in electric launches up the Thames on Sunday, August 19, a lunch at Henley, visits to electric works in London and its neighbourhood on Monday, a dinner in the evening with many Anglo-American patriotic speeches, a trip to Chatham on Tuesday, inspection of the dockyard, a second lunch, more speeches, and a reception by General and Mrs. Fraser in the afternoon, the members of the two electrical societies prepared themselves to encounter a somewhat blowy passage in journeying together to Paris.

On Thursday, August 16, the formal joint meeting was held in the large hall of the American Pavilion at the Exhibition, with Mr. Carl Hering, the president of the American Institute, and Prof. Perry, the president of the English Institution, as joint chairmen. The American, unlike the British Royal Pavilion, is a large circular building stretching uninterruptedly from floor to dome with a series of galleries running round it, and it is fitted up as a kind of huge commercial club, whereas the British Pavilion has been designed to represent an old English manor house, and contains a loan collection of the finest examples of the British school of painting, chiefly of the eighteenth and early nineteenth centuries.

When one remembers the invasion of England with American machinery—especially electric machinery—one envies the commercial instincts that have produced the American Pavilion, with all its facilities for aiding commerce, its lifts, the doors of which magically glide open and shut again on touching a button, and in which you are rapidly and noiselessly wafted to any of the many galleries.

In our Pavilion, on the contrary, commerce has been relegated to a top room, reached by a back staircase, entered literally through a back door, and the lift connected with this commercial room is not advanced—and never will advance—beyond the construction of the well for it. But walk in at the front door, and you can feast your eyes on the work of Gainsborough, Reynolds, Romney, Constable, Turner, Lawrence, Hoppner, Opie, Hogarth and of others ; and, after the roar of the Exhibition, the grinding of the moving platform running all round it, and the rumbling of the electric railways, you feel as if you had passed out of the whirl and money making of a factory into the peace and grandeur of Westminster Abbey. Why, however, has the British Royal Commission made so little use of this treasure on the Quai d'Orsay?

Mr. Hering welcomed the members of the two electrical societies present, and expressed the hope that this meeting might be the forerunner of many joint meetings, the next of which he hoped to see held in the United States, and an invitation to attend that meeting he daintily expressed in English, French and German.

Prof. Perry followed, and stated that, although no minutes could be read of any previous joint meeting, minutes of the present meeting were being taken, as he felt sure that there would be another joint meeting at which they would have to be read.

Prof. Mascart rose to express the thanks for the honour which the English Institution had done him in electing him one of their four vice-presidents some months ago. He hoped that not only might there be a joint meeting of the two societies in the United States—at Philadelphia, for example—but that it would be one at which all the Institutions of Electrical Engineering in the world would be represented. And although he feared that advancing age might prevent his being present, he would none the less co-operate in spirit.

The special subject dealt with at the present joint meeting was :—"The Relative Advantages of Alternate and Continuous Current for a General Supply of Electricity, especially with regard to Interference with other Interests," and the discussion was opened by Mr. Ferranti. He stated that this was not a continuation of the old contention between the relative advantages of direct and alternating current, for the rivalry which formerly existed between the two systems, and which led the advocates of the one to regard everything as absolutely wrong which was done by the advocates of the other, was luckily dying out. Engineers had begun to realise that the direct and the alternate current systems of electric distribution had each their separate functions, and the object of the present discussion was to elicit an expression of opinion as to whether the "interference with existing interests" did not furnish an important consideration in the choice of the system to be adopted in a particular case. It was not merely, he urged, the damage to water and gas pipes that was *now* being caused by the employment of the direct current that had to be taken into account, but they had to bear in mind the value of the underground property that might be injured ten years hence if the great development of the distribution of electric energy, which must necessarily take place in that period, were carried out on a wrong plan. He concluded by expressing the opinion that the difference in the magnitude of the disturbance caused by the two kinds of current was very great.

Mr. Arnold next spoke as a member of the American Institute—it being arranged that representatives of the two bodies should speak alternately. He drew attention to the difficulty of using the alternate current for general distribution arising from the inability to satisfactorily balance the load, and he considered, therefore, that the direct current system was the better. And, in view of the difficulties which attended the employment of the alternating current for driving electric tramcars, he considered that in this case also the direct current was the one to be adhered to.

Sir William Preece reminded the meeting that he has not given his adhesion publicly to either the direct or the alternating current system, and, therefore, that he was in a position to speak quite impartially. He considered that the interference of alternating current circuits with telephone lines could be entirely overcome by the employment of a metallic return for the telephone, but it had to be admitted that the surgings which occasionally took place in alternate current circuits disturbed the block signalling on railways. He referred to a case in France where the triphase alternate system of working had supplanted the direct current one, and suggested that this was an indication of the increasing appreciation of the former method, and that the capacity of long cables introduced a serious difficulty with alternate current transmission.

The variety of frequencies employed by the various companies—the London Electric Supply Company, for example, using a frequency of 67, while the City of London Company employed 97—he regarded as objectionable, and he hoped that this joint meeting would deal with the importance of arriving at a uniform standard of frequency. He also suggested that the relative advantages of underground and overhead conductors might well occupy the attention of the meeting.

Dr. Kennelly spoke of the relative fields for direct and alternating currents, and gave as an example that with an isolated plant of moderate size a direct current at a pressure of 100 volts might be employed, while if the area to be dealt with was larger, the pressure might still be direct, but a pressure of 200 or 240 volts would have to be resorted to, whereas when the area became large, transformation became necessary, and for that the alternating current was, of course, especially well adapted.

He referred to the growing use of high pressure alternating currents for transmitting power to tramways, and performing a double transformation for supplying the low pressure direct current for driving the electromotors on the cars ; and he considered that this unnecessary complication arose from the tramway motor having been developed as a direct current motor, and from the difficulty that would now be experienced in replacing the many tens of thousands of direct current tramway motors with an alternating current type.. In the case of new tramways and railways, at any rate, he looked forward to the time when the alternating current would alone be employed, but he admitted that the electric simplification would be accompanied with greater risk of shock and danger to life.

As to the interference that might be caused by electric tramways to magnetic observatories, he thought that, in view of the far greater commercial importance of the tramway, the magnetic observatory would have to give way, and remove its apparatus to a place where electric tramways were not required by the public.

Prof. Ayrton expressed the view that, since no doubt existed as to the considerable damage that electrical undertakings had caused to underground pipes, telephones, submarine cables and magnetic observatories, the question arose whether an endeavour was to be made to prevent the attack or to strengthen the defence. In the case of telephone circuits the Joint Committee of the two Houses of Parliament in England had decided that since—wholly apart from the advent of electric tramways—the Telephone Companies had realised that, in order to prevent interference between the telephone lines themselves, as well as to prevent the disturbance caused by neighbouring telegraph lines, it was necessary to abandon the earth-return and employ a metallic return, and since such a metallic return would shield the telephone circuit from disturbance that might otherwise be caused by electric tramways, there was no necessity to debar the tramway from employing the earth.

But as regards the electrolytic destruction of gas and water pipes the matter was quite different, and, therefore, the Board of Trade had imposed a regulation forbidding the difference of potential between any part of the rail and the terminal of the dynamo being allowed to exceed 7 volts. Prof. Ayrton pointed out, however, that this limit was too high even to prevent electrolysis, and certainly would not prevent the mutilation of messages received through a submarine cable which was landed in the neighbourhood of an electric tramway, as instanced at the Cape of Good Hope.

He questioned whether the security anticipated by Mr. Ferranti and others that would follow from a general substitution of alternating for direct current would be nearly as great as was imagined, and he referred to the experiments which he had published some years ago on the comparatively rapid production of separated hydrogen and oxygen that could be obtained in an ordinary sulphuric acid voltameter, through which an ordinary *alternating* current was passing. The specimen of a pipe corroded with an *alternating* current of one ampere passing for six weeks lying on the table, and which had been sent to the meeting by Mr. Trotter, of the Board of Trade, was an important illustration of the electrolytic action that could be produced with the commercial alternating current supplied by the London Electric Supply Company.

A magnetic observatory was in a more serious position still, since, as the undisturbed magnetism of the earth had to be measured, no system of defence could be utilised, and nothing short of the absence of attack could be satisfactory. He was glad, therefore, to say that the Electric Tramway Companies in London, thanks to the action of the Board of Trade in appointing a joint committee to represent the commercial and the tramway interests, and thanks to the experiments and the negotiations carried out by this committee during the past eight months, had not regarded the preservation of magnetic records from the drastic point of view advocated by Dr. Kennelly. In fact, the president of their Institution, Prof. Perry, in co-operation with Prof. Rücker, had succeeded in inducing the London Tramway Companies to propose a scheme in which, first, all the lines within a radius of two miles round the Kew Observatory should be divided up into absolutely *distinct one mile sections* ; secondly, that the current should be led to the trolley wire and withdrawn from the rails at the *middle* of each of these sections ; and, thirdly, the difference of potentials between the rails and the earth within this two miles radius should never be allowed to exceed one-fifth of a volt. And with these conditions, calculation showed that, although the protection afforded would not, of course, be as good as that obtained with a wholly insulated system, it would be probably sufficiently great to prevent any appreciable interference being caused with the magnetic observations regularly taken at the Kew Observatory.

M. Corda thought the adoption of the alternate or the direct current was mainly a matter of cost, and since the Fire Insurance Companies allowed the maximum pressure to be used with the alternating current to be only half as great as with the direct current, he considered that as long as that regulation lasted the direct current must gain the day.

Prof. Crocker said that the interference produced by an electric circuit on another undertaking might be divided into that produced by induction and that produced by leakage. The disturbance of the apparatus in a magnet observatory was due to both causes, but, as there were so few magnetic observatories in the world, that particular disturbance might be dismissed from consideration. With alternating currents the disturbance produced by induction was the more serious because this induction set up currents in other wires, and it was, therefore, very difficult to avoid. With direct currents the leakage disturbance was the more serious, but it was possible to prevent this. Some time ago he had had occasion to test the insulation of the whole of the New York electric lighting system, which was split up into sections for this purpose, and he found that the current which leaked to earth did not exceed one per cent. of the current that was supplied to the houses, whereas with the gas system in New York from 10 to 20 per cent. of the gas was lost by leakage. Consequently, since very high insulation could be obtained with the type of underground cables that were employed with high pressure work, it followed that the leakage on the low pressure electric light system employed in New York could be reduced to a still lower value than one per cent. Further, that if it were possible in London to reduce the potential difference between the rails and the earth to only half a volt, he would imagine that electrolysis might be avoided even with the ordinary trolley wire tramway. He was, therefore, in favour of employing the direct current for the purpose of avoiding interference with other interests.

But he considered that the considerations of economy and efficiency were more important than those regarding interference, and, while the three-phase and the direct current motors of the same power had the same efficiency from half up to full load, the direct current motor was the more efficient for small loads. Further, while for constant speed the regulation with both types of motor was about the same, the direct current motor had a distinct advantage in regulation when the speed was variable. On the whole, therefore, he was in favour of the use of a direct current system of electric supply.

Mr. Mordey, on account of lack of time, dealt shortly with the drop of pressure along the rails of an electric tramway, and stated that he had found that when the length was even 28 miles, the difference of potential between any parts of the rails and the generating station could be kept down to 7 volts; and he referred to the much greater attention that was given in England than in America to reducing the maximum drop of pressure along tramway rails. The employment of rotary transformers, as on the new Central London Railway, he deprecated as a makeshift, and suggested that, if the cost of all the transformers employed along the 6 miles of the route had been capitalised, it would have paid the company to have employed far thicker conductors. As regards the difficulty arising from the capacity of long underground cables traversed with alternate current, he pointed out that no difficulty in overcoming the effects of capacity had ever been met with in dealing with the 250 miles of underground cable in St. Petersburg. The Board of Trade had succeeded in using such instruments in their laboratory at Westminster that no interference could be caused by the construction of any electric tramway in the neighbourhood; therefore, he deplored the resistance that had been successfully offered a few years ago by a London college to the passing of a Bill for the construction of an underground electric railway near that college.

Mr. Mailloux pointed out that the small power-factor obtainable with alternate current motors, and the greater change in speed with a change in the E.M.F. that was experienced with alternate current than with direct current motors, was a serious objection to the employment of the former, and he instanced a case where the large current that was necessary for starting an alternate current motor had led him to adopt a direct current system in a sugar factory where 2000 horse-power was employed. The Fire Insurance Rules in the United States, which compelled the use of iron conduits, but which did not require that both the going and return conductors should be enclosed in the same iron tube—a condition, however, rendered necessary if alternate currents were employed —led to an important economy being obtained by using two separate conductors in separate iron tubes, which was, of course, quite possible with a direct current.

Prof. S. P. Thompson expressed his surprise that in wiring ships for electric lighting, where the possible disturbance of the compasses was a vital consideration, the direct current and two pole machines, the worst type to use, had been frequently employed even by the best firms, like that of Messrs. Siemens. He looked forward to seeing the use of multipolar machines on board ship, and of the alternating current; for not only would the compasses be then secure from disturbance, but there would be much greater freedom from electrolysis in damp places, and therefore of fire. He pointed out that the alternate current lent itself so readily to the use of efficient *low* voltage glow lamps combined with economic *high* voltage transmission; and finally that, since it was impossible to employ any device to screen a magnetic observatory from magnetic disturbance, since such a device would cut off the effects produced by variations of the earth's magnetism which the observatory existed to measure, there was a strong reason for running electric tramways with alternating current in any city where a magnetic observatory existed.

At the close of the preceding discussion, M. Hospitalier, Mr. Gavey, Mr. Hering and General Webber referred to points of special and novel interest in the several electrical sections of the Exhibition, in connection with which they had served as jurors; and in the afternoon these gentlemen acted as guides in taking parties of members of the two electrical societies to view the exhibits which had been specially mentioned.

THREE BOOKS OF POPULAR NATURAL HISTORY.[1]

MR. HUDSON has never written any book that is not extremely pleasant to read, though since he settled in England he has never had so much to tell us as was told in his "Naturalist in La Plata." That book, though it may not be his own favourite, will always, if we are not mistaken, be reckoned as his best; and the reason is simply that it treated of animal life among which *he* was entirely at home, and of which *we* knew little or nothing. His English books have not this quality, though they have many other excellences. The one before us, for example, is charmingly written, full of grace and feeling, touched with a tender and sympathetic imagination, made piquant by a certain quite inoffensive egoism; but, as we read in his pages of the South Downs, we are forced to recognise the fact that he is not of them. He is a stranger there—a most appreciative one, it is true—but still a stranger. It is perhaps given to few who have not been bred among the Downs to enter fully into their spirit, and we will not deny that Mr. Hudson, rambling alone through their sweet air and lying on their delicious turf, has caught it as none could do without rare gifts of sympathy and observation; yet there is something missing.

It is not pleasant to have to find fault with a book so readable; but a naturalist cannot but regret that Mr. Hudson should have given himself up so entirely to *impressions* throughout a volume of just three hundred pages, that no real contribution to natural history is to be found in them. He notices an interesting point, writes a charming paragraph about it, and leaves it, sometimes without making it clear what plant or creature he is talking about. To take an example: he has observed that the banded variety of *Helix nemoralis* is almost the only one to be found on the high downs, and that its bright coloration does not save it from the thrushes; but he does not pursue this fact, which has attracted the attention of conchologists and suggested at least one interesting explanation. Snail life on the downs is, indeed, so extraordinarily abundant, that a book which contains so much pleasant reading about the down turf is hardly complete without a chapter specially devoted to it. The same may be said of his remarks on insect life; he tells us of the common blue butterfly, and its habit of clinging to the bents, but of other blues he says nothing; a skipper is mentioned, but we are left in the dark as to the species. In writing of a certain fly, he declares that neither books nor entomologists have been able to tell him its name, and leaves it with a few words of good-natured contempt for the specialism of the present age. A little more exactness in a book by a naturalist, which naturalists may be expected to read, would have greatly added to its permanent value. Even men of letters may complain when they find an allusion to Arthur Young's famous "Tour through Great Britain in 1727." What book can this be?

The best chapters are those which deal with the birds and the human beings of the downs. Shepherds and shepherd boys are delightfully pictured; and Mr. Hudson has discovered for himself the pleasing habit of the ruddy-faced shepherd lads in adorning themselves with wild flowers. About the birds he has plenty to tell us—it is his own subject; and the chapter on "Shepherds and Wheat-ears" will be read by all ornithologists with mingled pleasure and pain. All that he writes of the singers of the downland is beautiful and true; perhaps the songs of the stonechat and whinchat have never

---

[1] "Nature in Downland." By W. H. Hudson. Pp. xii + 307. (London: Longmans, Green and Co., 1900.)
"The Birds of Cheshire." By T. A. Coward and Charles Oldham. Pp. 278. (Manchester: Sherratt and Hughes, 1900.)
"In Birdland, with Field-glass and Camera." By Oliver G. Pike. Pp. xvi + 280. (London: T. Fisher Unwin, 1900.)

been so well described. Of the linnet, too, he says most truly that it has one note, and only one, of almost unapproachable musical beauty. The singing of the skylarks, that invariable accompaniment of down life, is described with all Mr. Hudson's wonderful sympathy and delicacy of language; but what are we to say of his belief that the highest notes of this bird may be heard on the downs at a distance of three miles? It is a belief which it would hardly be possible to test.

"The Birds of Cheshire" is an excellent book of its kind. The first essential of such a compilation is that it should be unimpeachable as a record; and, so far as we can discover, the compilers have here used both pains and judgment in testing the records of others, while their own experiences are recorded simply and faithfully. Thus a real step is gained in the collection of valuable material for that comprehensive work on the distribution of birds in these islands which we may hope to see in due time. There is no superfluous matter in this volume, and no

Fig. 1.—Bearded Tit feeding young.

attempt at fine writing; and excellent paper, print and binding combine to make it a very pleasant book to handle. The half-dozen plates of Cheshire scenery are very effective, and nothing is wanting, unless it be a rather better map of the county.

The avifauna of Cheshire, as the authors remark, is surprisingly poor; the county does not lie upon any regular line of migration. It is too far north for the nightingale, which has seldom occurred, though we note that it has been recorded by that excellent observer, Rev. C. Wolley-Dod. The lesser whitethroat, as might be expected, is not common, nor is the grasshopper warbler. We should have expected the pied flycatcher to be more common than seems to be the case; the tree sparrow, a bird of peculiar distribution, has probably been often overlooked. The goldfinch and linnet are decreasing in numbers, but the opposite is the case with the turtle-dove. The list of waders, gulls, and birds of the coast, is not very large, and we regret to find that

the ubiquitous golfer is contributing to its further diminution. The characteristic bird of the county is a noble one—the great crested grebe, which is widely distributed; and in dealing with it the authors have allowed themselves some half-a-dozen pages, which will be welcome to all ornithological readers.

Mr. Pike's little book bears the same relation to his photographs as a popular lecture does to its lantern illustrations: *i.e.* it is of secondary interest. Photography, applied to birds and their nests and eggs, seems to be a most attractive pursuit, leading its votary often to spend hours in the endeavour to catch a bird at some opportune and interesting moment, or to find the nest on which he has set his heart. It should certainly be useful in training the faculty of observation, and in assisting the memory; and it may become a most welcome substitute for the predatory habits of private egg-collectors, who are perhaps the most dangerous enemies of our rarer birds. The actual contribution to zoology, however, does not seem as yet to have been great, and it is quite possible that before long we may have too many books on the subject. Mr. Pike's is, however, so unpretending and so pleasantly written, that it will no doubt be welcome to many beginners in ornithology who wish to learn where and how to look for nests, and a few of his experiences and his photographs will be interesting even to the more experienced. Part iv., on Norfolk birds, is perhaps the most valuable section of the book, and of the three photographs which Mr. Pike succeeded in taking of the nest of the bearded tit we select one for reproduction, as a favourable specimen of his work.

THE INTERNATIONAL CONGRESS OF MATHEMATICIANS.

A CONGRESS of mathematicians was held at Chicago during the World's Fair; but this was an isolated one. The series of international congresses was inaugurated at Zürich in 1897, and the second congress of this series met in Paris from the sixth to the eleventh of the present month. About 225 mathematicians of various nationalities, with 25 members of their families, were present. It had been expected that the numbers would be very much greater, as many as one thousand provisional acceptances having been received before last December; the diminished attendance was doubtless due partly to the great heat of the preceding month, but probably in greater measure to the fear of exhibition crowds and exhibition extortions. It had been supposed that the Exhibition would attract people to the Congress; on the contrary, it seems to have kept them away. The composition of the Congress was certainly international; the numbers of members from the different countries were approximately as follows:—France, 90; Germany, 25; United States, 17; Italy, 15; Belgium, 12; Russia, 9; Austria and Hungary, 8; Switzerland, 8; England, 7; Sweden, 7; Denmark, 4; the remainder being from South America (4), Holland, Spain, Roumania, Servia, Portugal, Turkey, Armenia, Greece, Canada, Mexico, Japan.

The actual business was preceded by a *réunion* at the Café Voltaire, on the evening of August 5, when about half the members were present. The proceedings proper consisted of two general meetings on Monday and Saturday, with sectional meetings on the four intervening days. The opening general meeting had been announced for 2.30 p.m., August 6, in the Palais des Congrès in the Exhibition grounds; but unfortunately some action on the part of the Exhibition authorities necessitated changing the hour to the morning, and this change was decided upon too late to be communicated to all the members, many of whom had not even arrived in Paris at that hour. Thus a considerable number of the members were unable to be present at the first general meeting, which was held on August 6 at 9.30 a.m.   M. Hermite was acclaimed président d'honneur ; M. Poincaré, president ; the vice-presidents (some *in absentia*) were announced as MM. Czuber, Gordan, Greenhill, Lindelöf, Lindemann, Mittag-Leffler, Moore, Tikhomandritzky, Volterra, Zeuthen, Geiser. The secretaries were MM. Bendixson, Capelli, Minkowski, Ptaszycki, Whitehead ; the general secretary, M. Duporcq.

M. Poincaré, on taking the chair, spoke a very few words of greeting, and then called upon the speakers of the day.   M. Cantor, in his address, "Sur l'historiographie des mathématiques," sketched the development of this subject through Montucla (toujours un modèle que tout historiographe des sciences doit suivre), Kaestner, Cossali, Bossut, Chasles, Libri, Nesselmann, Gerhardt, Arneth, Hankel, Boncompagni, up to authors of the present day.   He expressed the firm conviction that the history of mathematics, from the beginning of Lagrange's work, can only be written as a series of special histories, with a final volume (Histoire des Idées) co-ordinating the whole.   M. Volterra, "Trois analystes italiens, Betti, Brioschi, Casorati, et trois manières d'envisager les questions d'analyse," compared and contrasted the work of these three mathematicians, and considered the influence their differing lines of thought and expression have had on the development of Italian analysis.

Six sections had been arranged, with meetings extending over four days. While in general two sections were sitting at the same hours, yet matters were so arranged as to avoid, as far as possible, the conflict of interests that had been felt at Zürich, where only one day was devoted to the sectional meetings. These six sections, with their presidents and secretaries, were as follows :—
(1) Arithmetic and Algebra : Hilbert, Cartan ; (2) Analysis : Painlevé, Hadamard ; (3) Geometry : Darboux, Niewenglowski ; (4) Mechanics and Mathematical Physics : Larmor, Levi-Civita ; (5) Bibliography and History : Prince Roland Bonaparte, d'Ocagne ; (6) Teaching and Methods : Cantor, Laisant.

Owing, however, to the unavoidable absence on some days of the president of Section 5, and the small number of papers in that section, Sections 5 and 6 sat together, under the presidency, first of M. Cantor (Wednesday), and then of M. Geiser (Friday) ; and at the Wednesday morning sitting the two papers of most general interest in the Congress were read. These were Hilbert's address on the future problems of mathematics, valuable as assisting the mathematician to orientate himself, and Fujisawa's account of the mathematics of the old Japanese school, of special interest as giving information, not readily accessible otherwise, about a system of mathematics that is now entirely obsolete. It appears that the Japanese invented zero for themselves, and employed the circle as a symbol for zero ; that they used imaginaries and complex numbers, and calculated the value of π correctly to forty-nine places of decimals. In connection with this, M. Cantor remarked that the use of zero is probably Babylonian, and dates from about 1700 B.C.

M. Hilbert considered the origin and nature of the problems of mathematics the study of which is most likely to prove profitable ; the characteristics of a proper solution ; and the methods of attacking any problem that offers special difficulties. If the problem is really insoluble, then for the advance of mathematics it is essential that the impossibility be rigorously demonstrated. He illustrated his argument by means of selected problems that invite attack—problems regarding the axioms of arithmetic and of physics, prime and transcendental numbers, questions in the theory of functions, and the determination of the arrangement of the circuits that an algebraic curve can possess ;

referring to a paper about to appear in the *Nachrichten der Kgl. Gesellschaft der Wissenschaften zu Göttingen*, 1900, for a more complete list of definite problems that demand investigation.

Much interest was displayed in the papers read by M. Mittag-Leffler at the Tuesday morning sitting of Section 2, "Sur fonction analytique et expression analytique," "Sur une extension de la série de Taylor." The domain of an ordinary power-series is a circle that reaches to the nearest singular point ; at all points inside this there is convergence, at all points outside there is divergence ; this the author generalised so as to obtain a certain expression convergent within a particular region (an étoile), and divergent without. He raised the question whether an analytic expression can be found which shall represent, throughout its domain of definition, an assigned analytic function. A discussion followed between MM. Borel, Hadamard and Painlevé, as to the nature of the connection between "analytic expression in a complex variable x" and "analytic function in x." At the Thursday sitting of Section 1, M. Padé read a paper, "Aperçu sur les développements récents de la théorie des fractions continues " ; in this he showed the dependence of the expression of a function of x as a continued fraction on a certain diagram, in which each convergent is represented by a point whose co-ordinates are the degrees of the numerator and denominator of the convergent ; and, referring to the discussion that followed Mittag-Leffler's paper, suggested that a continued fraction may be found to be a suitable analytic form for any assigned analytic function.

The Friday morning combined sitting of Sections 5 and 6 was to a great extent occupied by the discussion of a resolution offered by M. Leau, urging the Academy to consider favourably the adoption of a universal language, not with a view to displacing any of the existing languages, but as a scientific medium auxiliary to these. Some such resolution has been brought forward lately on several similar occasions by the advocates of the latest artificial language, Esperanto. The discussion showed, on the part of mathematicians, very little sympathy with the suggestion, and very little recognition of a need for any such medium. As one speaker remarked, mathematics already has a universal language, the language of formulæ ; and the general sense of the sections was evidently that the existing diversity of languages need cause no real difficulty, so long as writers are willing to confine themselves to English, French, German and possibly Italian, this view of the case being formulated by a Russian, M. Vassilief. The only result of the discussion was the rejection of M. Leau's motion, and the recording of a wish that the Academy would discountenance any unnecessary diversity in the languages employed for scientific purposes. The four languages enumerated by M. Vassilief are those officially recognised in the meetings of the Congress, though it was noticeable that a great many of the speakers chose to speak in French, possibly out of compliment to their hosts.

Other communications of value, though of less general interest, were the following :—In Section 1, M. Stephanos, Sur la séparation des racines des équations algébriques ; in Section 2, M. Tikhomandritzky, Sur l'évanouissement des fonctions Θ de plusieurs variables ; M. Bendixson, Sur les courbes définies par les équations différentielles ; M. Jahnke, Zur Theorie der Thetafunctionen von Zwei Argumenten ; in Section 3, M. Lovett, On contact-transformations between the elements of space ; M. d'Ocagne, Sur les divers modes d'application de la méthode graphique à l'art du calcul ; M. Stringham, Orthogonal transformations in elliptic or in hyperbolic space ; M. Jamet, Sur le théorème de Salmon concernant les cubiques planes ; in Section 4, M. Hadamard, Relations entre les caractéristiques réels et les caractéristiques imaginaires pour les équations différentielles à plusieurs variables indépendantes ; M. Volterra, Comment on passe de l'équation de Poisson à caractéristique imaginaire à une équation semblable à caractéristique réel ; in Sections 5 and 6, M. Padoa, Un nouveau système irréductible de postulats pour l'algèbre ; M. Capelli, Sur les opérations fondamentales de l'arithmétique. The attendance at these sectional meetings, all of which were held at the Sorbonne, varied from 50 to 120.

The concluding general meeting was held at the Sorbonne at 9 a.m. on Saturday. The proceedings opened with the sending of a message of greeting to M. Hermite, the président-d'honneur of the Congress. It was then unanimously voted that the next Congress be held in Germany, in 1904, at the beginning or end of the summer vacation, the place mentioned as probable being Baden-Baden. M. Mittag-Leffler then delivered his address, " Une· page de la vie de Weierstrass," and M. Poincaré spoke briefly on the " Rôle de l'intuition et de la logique en mathématiques," closing the proceedings immediately afterwards with the few words, " La séance est levée ; le congrès est clos."

On the conclusion of the Tuesday afternoon sectional meetings, members were received at the École Normale Supérieure, where a pleasant opportunity for social intercourse was enjoyed ; and at noon on the day after the closing of the Congress a banquet was held at the Salle de l'Athénée-Saint-Germain, when about 160 members sat down. In the absence of M. Poincaré, the proceedings were conducted by M. Darboux ; speeches were made also by MM. Geiser, J. Tannery, Stephanos and Vassilief. A considerable number of members of this and other scientific congresses accepted the invitation of Prince Roland Bonaparte to a scientific *soirée* on Saturday. A *fête* had been arranged by President Loubet for Thursday evening, but could not be held on account of the funeral of the King of Italy ; the invitations were consequently transferred to the *fête* in honour of the Shah on August 10.

It will be seen that very little business was transacted, apart from the reading of papers. At the joint sitting of Sections 5 and 6, it was asked what steps had been taken to put into effect the resolutions of the Zürich Congress as to the formation of a committee to consider certain questions of bibliography, &c., these having been adopted with the hope of ultimately consolidating mathematical enterprise, and directing it into profitable channels. No very satisfactory answer was forthcoming ; M. Laisant, on behalf of the French Mathematical Society, replying that they had done nothing in this line, having been entirely occupied with making material provision for the Congress. He drew the attention of members, however, to the announcement of the *Annuaire des Mathématiciens*, undertaken by Carré et Naud, 3, Rue Racine, which is designed to be a complete register of all mathematicians, with their addresses. It is much to be hoped that these questions, raised at Zürich, will be dealt with in a business-like manner at the Congress of 1904.

NOTES.

THE Scientia Club gave a banquet to Lord Kelvin during the International Physical Congress at Paris. M. Louis Olivier presided over a distinguished company, and speeches in appreciation of Lord Kelvin's scientific work were made by him and by Profs. Mascart and Cornu.

FROM the official report of the International Congress of Electricians at Paris, we see that two communications, by Mrs. Ayrton and M. Blondel, were received with great appreciation. Mrs. Ayrton's paper was on the luminous intensity of the electric arc with continuous current, and she showed that the best result, both from the point of view of luminosity and expenditure of

energy, was obtained from an arc only a millimetre in length. Demonstrations in illustration of this conclusion, and showing the absorbing and cooling effects of carbon vapour produced in the arc, as well as the production and absorption of green and yellow radiations, were given by Mrs. Ayrton at a special meeting in the École supérieure d'Électricité. M. Blondel reviewed the progress of electric lighting during the past ten years, and made some very valuable remarks on arc lamps with alternating currents, and on the carbons commonly used in arc lights.

IN opening the business of Section A (Mathematical and Physical Science) of the British Association at the forthcoming Bradford meeting, we understand that Dr. Larmor will review the change of ideas which has recently become current regarding the nature of electric actions and their dependence on the æther ; but there has been a strong tendency to eliminate from the exposition of the theory those dynamical explanations which formed a main feature of its development in the hands of Clerk Maxwell. It is of fundamental importance to consider how far purely descriptive methods can thus avail towards an effective formulation of general physical theory, without appealing to a dynamical foundation of some kind. In all branches of the subject the discrete atomic constitution of matter is reached when we probe deep enough ; thus the method of representation of the physical activities of the material atoms, so far as they can be known to us, is of the essence of a dynamical treatment. This leads on to the cognate question, whether denial of direct action at a distance necessarily implies the passing on of all electric effects from element to element of the medium entirely by simple stress ; if that be too narrow a scheme, the efforts that have been made towards formulation on this basis were foredoomed to failure. The scope and limitations of the method of statistical enumeration of the activities of the atoms, which is the only one now available in ultimate thermodynamic discussions, depend on considerations of a different order. The modern extension of the range of the principle of Carnot also requires us to face the question how far the processes of chemical interaction between atoms, as distinct from the properties of the molecules when formed, are amenable to dynamical representation.—The general scheme for the business of the section is to take physical papers of a mathematical nature on Friday, September 7. On Monday, September 10, the section will divide into two, dealing with mathematics and meteorology respectively. On Tuesday a discussion on ions will be opened by Prof. Fitzgerald. It is also hoped to arrange discussions on the partition of molecular energy, and on the relation of radiation to temperature, under the thermodynamic aspect.

A PASTEUR Institute has just been opened at Kasauli, a hill station in the Punjab district, about thirty miles from Simla. It is thus no longer necessary for a person bitten by a rabid animal in India to journey to Paris for treatment by inoculation. The treatment at the Kasauli Institute is to be given free of charge.

PROF. J. C. BOSE, who has been attending the recent International Congress of Physics at Paris as the delegate of the Government of Bengal, will also attend the British Association meeting at Bradford in the same capacity, and will there describe some new electrical investigations with which he has lately been engaged.

THE following international congresses upon scientific subjects will be held in connection with the Exposition at Paris during September : 3-8, History of Religion ; 3-5, Basque Studies ; 3-4, Pharmacy Specialities ; 10-16, Meteorology ; 10-12, Agriculture; 10-12, Folklore; 13-14, Fish Culture and Arboriculture; 14-19, Agriculture and Fisheries; 15-20, Aeronautics; 22-28, Acetylene.

PROF. GIARD, director of the biological station at Wimereux, has been made Chevalier of the Order of Leopold by the Belgian Government, as a recognition of the hospitality which he has extended to Belgian students and naturalists at his laboratory for many years.

DR. J. W. B. GUNNING, the director of the State Museum at Pretoria, speaks in high terms of the way in which he has been treated by the new British authorities. He has not only been confirmed in his post at the museum, but also materially assisted in his efforts to add to it a zoological garden, which he had planned before the outbreak of hostilities. It was to this (incipient) garden that the celebrated lioness (now in the Regent's Park Garden) was presented by Mr. Rhodes, but subsequently returned to the donor by Mr. Kruger's order.

AN international association for the promotion of psychical research has been established under the title : Société internationale de l'Institut Psychique. A *Bulletin* has been issued containing a report of the inaugural meeting held at Paris on June 30, and explaining the objects of the organisation. The Comité de Patronage includes the names of Prof. Mark Baldwin, Prof. W. F. Barrett, Sir William Crookes, Prof. O. J. Lodge, and Mr. W. H. Myers. The general secretary is M. Yourievitch, Russian Embassy, Paris.

A MONUMENT to Bertrand Pelletier and J. B. Caventou, renowned as pharmaceutical chemists, was unveiled at Paris, by M. Moissan, during the recent International Congress of Pharmacy. Caventou was born in 1795, and studied at the Paris School of Pharmacy. While pharmacist at the St. Antoine Hospital he met Pelletier, and their fruitful collaboration began. Two years after discovering brucine and strychnine they were able to announce the discovery of quinine, and with rare disinterestedness they made their work public by presenting an account of their methods and results to the Paris Academy of Sciences on September 11, 1820. In their memoir they stated that they had succeeded in isolating cinchonine and quinine from both yellow and red cinchona bark, and described the therapeutic properties of these substances. In 1827 the Montyon prize of the Academy of Sciences was awarded to them in recognition of their valuable discovery, and now the monument, representing the two investigators together, stands to remind observers of their joint services to science and humanity.

THE International Meteorological Congress will be opened at Paris on Monday, September 10, under the presidency of M. E. Mascart, and will continue during the week. The International Meteorological Committee, which met last year at St. Petersburg, decided that it would convene, at the same time as the Congress, the various committees appointed by the Paris Conference in 1896. These committees are the following :—(1) Terrestrial Magnetism and Atmospheric Electricity. (2) Aeronautics. (3) Clouds. (4) Radiation and Insolation. The first of these committees held an important meeting at Bristol in 1898, the proceedings and resolutions of which have been published in the reports of the British Association. A great number of ascents, both with free as well as captive balloons, have been made in different countries, for the systematic investigation of the upper regions of the atmosphere. Finally, the publication and discussion of the international observations of clouds, made in 1896-97, will probably be completed in 1900 in the majority of the countries taking part in the same. From this it will be seen that communications of very high interest will be brought before the Congress. The

questions which will be dealt with are not restricted exclusively to meteorology properly so-called : they will include, generally, everything which affects the physics of the globe. The meetings of the Congress and of the committees will be held at the House of the Société d'Encouragement, 44, Rue de Rennes, where the International Conference met in 1896. Communications relating to the organisation or to the programme of the Congress should be addressed to Mons. Angot, general secretary, 12, Avenue de l'Alma, Paris.

A REUTER telegram, dated Madrid, August 24, states that twelve fragments of a meteorite have fallen on the boundary of the provinces of Jaen, Cordova and Granada. The fall was preceded by a series of loud detonations. One fragment, weighing about a pound, which was picked up at Val, in the province of Jaen, is said to be of hexagonal shape, grey on the surface, and of a greenish colour inside.

As attempts are being made to found a domestic science, and to introduce exactitude into the operations of the kitchen, a note in the *Monthly Weather Review* recording the actual experience of a housekeeper at Albuquerque, New Mexico, is of interest. It appears that cooking recipes and practices which are trustworthy not far from sea-level are worthless at Albuquerque, the altitude of which is 4933 feet. Water boils there at 202° F., instead of 212° F.; hence articles of food, the cooking of which depends upon heat applied through the medium of water, require a longer time for cooking than is given in the cookery books. On account of the extreme dryness of the atmosphere, farinaceous foods, such as beans, corn, &c., lose so much of their moisture that they have to be left for a long time in water before cooking, in order to be softened. But the worst difficulty is with cake-making. Ordinary recipes as to number of eggs and amount of baking powder break down altogether, and housekeepers have to modify them if they wish their operations to be successful. As the barometric pressure determines to what extent the disengaged carbon dioxide shall expand and aerate the dough, this may explain the different action of baking soda and egg batter. In any case, the observation is interesting, and chemists may find it worthy of their attention.

La Nature of August 18 contains an article, by M. E. Roger, director of the meteorological station at Châteaudun, near Paris, entitled "The Greatest Heat of the Century." A temperature of 103°·6 in the screen was observed there on July 27 of this year. The nearest temperature to this hitherto recorded in the vicinity of Paris during the last hundred years was 101°·5 at Montsouris Observatory on the 20th of the same month. At Poitiers in July 1870, a temperature of 106°·2 was recorded. Among the highest temperatures recorded in or near London are 95°·2 at the late Mr. Symons' station, Camden Town, on July 16 last, and 97°·1 at Greenwich in July 1881 ; in that month a reading of 101° was obtained at Alton, Hants.

THE Pilot Chart of the North Atlantic Ocean for August, issued by the U.S. Hydrographic Office, contains a diagram showing the path of the noteworthy cyclonic tropical storms of the years 1898 and 1899, together with the time of their duration, which varies from 2 to 39 days ; two of the storms were traced entirely across the Atlantic. Taken collectively, the several tracks exhibited show the doubtful accuracy of generalised statements concerning certain characteristics of these storms, such as their velocity, the latitude of their recurvature, &c. Thus the statement is often made that in the higher latitudes, after recurvature, the velocity along the track will average 25 or 30 miles an hour. However true this may be as the statement of an average, its untrustworthiness with regard to a particular storm is well shown by one of the tracks laid down, in which the velocity of the centre after recurvature off the coast of Florida fell to about three miles an hour through three degrees of latitude.

WE have received a copy of the meteorological observations made at Sir Cuthbert Peek's observatory at Rousdon, Devon, for the year 1899, the sixteenth year of the series. This observatory is a second order station of the Royal Meteorological Society, and possesses a very complete equipment of instruments, both astronomical and meteorological, including various patterns of standard anemometers, the observation and comparison of which form a special and valuable feature of the regular work of the station. The mean temperature of the year exceeded the average by more than 2°; but the year was free from extremes in either direction, although at Greenwich on August 12 a temperature of 90° was recorded. The rainfall was about two and a half inches below the average, and amounted to only 29·31 inches; falls of an inch or more occurred on five days. A daily comparison is made between the actual weather and the forecasts of the Meteorological Office; as regards wind, the percentage of success has increased from 69 in 1884 to 93 in 1899, and in the case of weather, from 73 to 92 in the same period.

THE following notes from a report by Mr. H. A. Byatt, assistant collector, Fort Alston, are published with others in the *British Central Africa Gazette*:—"After passing over the ridge of hills which culminates about two miles to the east of Ndonda, some forty miles from the lake shore, the appearance of the country and the nature of its soil changes very considerably. In place of the low-lying marshy expanses along the coast, one finds a monotonous series of undulating grassy plains, covered almost exclusively with a growth of tall rank grass. The soil generally, though occasional small deposits of clay are found, consists of a layer of coarse porous sand, apparently of no great depth, lying upon a substratum of hard rock, and may well have been washed down by centuries of rain from the low hills above mentioned. The country is remarkably waterless. Judging by the appearance of the vegetation which it supports, the soil is of poor quality, and offers but little hope of successful cultivation. Large timber is conspicuous by its absence, and it is only at rare intervals that the raphia-palm and other trees requiring a copious supply of moisture are found; but possibly such woods as the Mlanje cedar might be introduced with success. Owing to the rank growth of grass, it is an ideal cattle country; but the true reason of the excellent condition of cattle in this country is to be found, I believe, in the presence of a certain salt in the earth—possibly a nitrate or phosphate of soda. In many places it is so abundant that upon the evaporation of the water it is left as a thick white deposit on the surface of the soil, whence it is gathered up by the natives and used as a condiment. Of other minerals, beyond the existence of graphite, I have so far found no trace."

THE progress of work on the new wheel-pit of the Niagara Falls Power Company, at Niagara Falls, N.Y., which is intended to supplement the present hydraulic installation of the same company, is described and illustrated in the New York *Electrical Review* (August 15). It is only a few years since the company began operations with a plant capable of being extended to 50,000 horse-power. Both the rapid growth of electrochemical industries at Niagara, and of electric power applications in Buffalo, twenty-six miles away, have rapidly carried the plant up to the limit of its former hydraulic equipment. Now the new one, which is slightly larger than the old, is under construction, and it is expected that within a year 105,000 horse-power will be generated and distributed from this one plant. The growth of such industries in the United States has been extraordinary. In New York State there is another plant under construction which will be finished within a year,

and will develop the enormous total of 150,000 horse-power. Practically all of the latter installation will be used in electrochemical work in the manufacture of carbides and caustic and bleaching powder.

THE manufacture of artificial dye-stuffs in Germany was referred to in a recent report from H. M. late Consul-General at Frankfort-on-Main. The endeavours of manufacturers and industrial chemists are directed, generally speaking, to producing the organic natural products, such as those of colour plants, dye woods, insects, molluscs, &c., by artificial and even cheaper and purer means, and in a more serviceable form for dyeing; also to producing new colours, which not only approach in brilliancy and effectiveness the natural kinds, but even surpass them. Since the discovery that the important dye-stuff madder —alizarine—could be produced in an easy and cheaper manner from the carburetted hydrogen of coal-tar, the use of dye-stuffs obtained by coal-tar distilling has gradually grown to such an extent that in Germany about five times as many artificial colours are made as in all other countries combined. According to the census in 1895, there existed twenty-five factories for the manufacture of aniline and aniline colours, and forty-eight factories (with seven branches) for the production of other coal-tar products (i.e. not only for colours, but also for other commodities, such as picric acid, &c.). The aniline works employ 7266 hands, the latter factories 4194; in all, 11,460 men.

IN connection with the foregoing note, the *Board of Trade Journal* gives some particulars as to the manufacture of artificial indigo in Germany, from a report to the Foreign Office. The importance of indigo is evidenced by the fact that the production of natural vegetable indigo equals in value the entire world's production of artificial dye-stuffs. The present artificial indigo of commerce represents almost pure indigotin. It is sold in the form of a 97 per cent. powder, whereas the indigotin contained in vegetable indigo fluctuates between 70 and 80 per cent. It contains no indigo-red, no indigo-brown and no indigo-blue. The lack of indigo-red and indigo-blue, which both seem to be of some importance in the relation of the dye-stuff to the fibre, are its special disadvantages. The indigo-red seems to be of importance in the production of darker shades of colour. There is no doubt that at some time not too far off it will be possible to produce this ingredient also. Artificial indigo is used by dyers in the same way as vegetable indigo. If it is possible to render the process of manufacture materially cheaper, and thereby to considerably reduce the price of artificial indigo, the danger to natural indigo will be greatly increased; it is, indeed, to be feared that with the increase of chemical knowledge the same fate awaits this dyeing plant, which is extensively cultivated in British territories, as overtook the Krapp plant, the cultivation of which nowadays no longer pays. Artificial indigo affords a new example of the manner in which applied science revolutionises the most varied spheres and destroys as well as creates great wealth.

THE *Atlantic Monthly* for August contains an account, by Mr. Sylvester Baxter, of a method devised by Mr. Arthur J. Mundy, whereby a ship may be guided into port in stormy weather which prevents ordinary signals from being of service. The method is called "Acoustical Triangulation." It is based on the property that sound travels under water with a velocity that is unaffected by the disturbances such as winds, which have so large an influence on the propagation of sound-waves in air; and the putting of this principle into practice depends on the invention of a successful apparatus for ringing a bell under water by electrical connections. Three bells placed at the corners of a triangle, preferably equilateral, are sounded at known intervals of time. By noting the intervals of time between the instants when the first and second bells are heard, the locus of the ship's position is known to be one of the branches of a certain hyperbola the foci of which are the two bells. By noting the apparent interval between the second and third bell, the ship is similarly located on another hyperbola, and the intersection of the two curves gives the required position of the ship. The only objection to the method appears to us to be that a pair of hyperbolic branches may intersect in *two* points, so that for given intervals between the bell sounds the position of the ship may be ambiguous. This could be avoided by having four bells instead of three.

AT the annual meeting of the Physical and Natural History Society of Geneva, Prof. A. Pictet surveyed the work which had been brought before the society during 1899. Eighteen meetings were held, and no less than seventy-three communications or reports were read. M. F. L. Perrot and Prof. Guye have made a series of measures of surface tension of various liquids by the method of falling drops. The conclusion arrived at as a result of their observations was that these tensions are not proportional to the weight of the drops. A new recording telephone was described by M. F. Dussaud. M. T. Tommasina has studied the variations of conductivity of coherers, and M. E. Steinmann has contributed a note on the thermo-electricity of various alloys. To the section of chemistry and mineralogy, M. Louginine has contributed an important memoir on the latent heat of evaporisation of some organic liquids ; MM. Dutoit and Friderich have determined the molecular weights of some organic liquids by the method of capillary ascensions ; Prof. A. Pictet and M. Athanasescu have presented a note on the constitutional relation between two alkaloids of opium-papaverine and laudanine ; M. Duparc has described his researches on the Liparite rocks of Algeria ; and M. H. Auriol has made a detailed study of the agricultural soils of the Canton of Geneva. In the section of botany, M. de Candolle stated that grains of wheat which he had kept for four years under mercury had germinated and produced normal plants ; and Prof. Chodat has described several micro-organisms of plants. Among the subjects of papers contributed to the section of zoology and anthropology are the development of the wings of Lepidoptera, by M. A. Pictet ; and a comparative study of a series of skulls from old burying-places in the Valais district, by M. Pitard. In the section of physiology and medicine, Prof. J. L. Prevost and Dr. Battelli have described their detailed researches on the action of electric currents upon animals ; and M. Babel has given an account of his work on the comparative toxicology of aromatic amines.

THE present trend of legislation in the interests of fish preservation in the United States is, Dr. Whitten remarks (*State Library Bulletin*, No. 12, New York), to place more reliance on methods of fish propagation than on a multiplicity of vexatious restrictions, and to obtain through scientific research the knowledge essential to enlightened regulation. In 1871 the United States commission of fish and fisheries was created to undertake scientific investigations, collect information, and to further the introduction and multiplication of food fishes, particularly in waters under national jurisdiction. In 1898 the commission maintained 34 fish-cultural stations and distributed $57,509,546 eggs, fry and adult fish. Fish commissions have been created in every State except Kentucky. Many of the commissions exist primarily for protective purposes, but others carry on valuable scientific work and maintain hatcheries and stock local waters with the most valuable food fishes. Illinois has a zoological station, and Oregon has created the office of State biologist for the investigation of the animal resources of the State and the development of such as have economic value.

THE relation of the cell to the enzymes, or soluble ferments which originate from cells, was touched upon by Sir J. Burdon-Sanderson, Bart., in the address he delivered before the recent

International Medical Congress at Paris. Formerly, he pointed out, each kind of cell was regarded as having a single special function proper to itself, but the progress of investigation has shown that each species of cell possesses a great variety of chemical functions and that it may act on the medium which it inhabits, and be acted upon by it, in a variety of ways. Thus, for example, the colourless corpuscles of the blood (or, as they are now called, leucocytes) are considered not merely as agents in the process of suppuration or as typical examples of contractile protoplasm, but rather as living structures possessing chemical functions indispensable to the life of the organism. Similarly, the blood disc, which formerly was thought of merely as a carrier of hæmoglobin, is now regarded as a living cell possessed of chemical susceptibilities which render it the most delicate reagent which can be employed for the detection of abnormal conditions in the blood. The tendency of recent research is to show that the reactions referred to as chemical functions of the cell (action of the cell on its environment—action of the environment on the cell) are the work of ferments—intrinsic or extrinsic—which are products of the evolution of the living cell, and therefore to which the term enzymes may be applied.

RECENT researches have plainly indicated that in the case of the disease-producing micro-organisms, the specific functions which for years were regarded as proper to, and inseparable from, the cell belong essentially to the enzymes which they contain. It has been further shown that similar statements can be made as regards ferment-processes which differ widely from each other and no less widely from those induced by bacteria. So that in the domain of microbiology the enzyme may in a certain sense be said to have "dethroned the cell." For if, as M. Duclaux has said, it is possible to extract from the cell a substance which breathes for it, another which digests for it, another which elaborates the simple from the complex, and finally another which reconstitutes the complex from the simple, the cell can no longer be considered as *one*, but rather as a complicated machine, the working of which is for the most part dependent on enzymes, which, however numerous and varied may be the processes in which they are engaged, all follow and obey the universal law of adaptation, and all contribute to the welfare and protection of the organism.

IN our last week's issue reference was made to Dr. Haller's views as to the relationships of the different groups of the Vertebrata, based on his study of the hag-fishes and lampreys. And in the July number of the *Journal of Anatomy and Physiology* the subject of the origin of Vertebrates, as deduced from the study of the larval lamprey (Ammocœtes), is resumed by Dr. Gaskell. In this important communication the author arrives at the conclusion that Ammocœtes is a representative of the Devonian Cephalaspids, and also that a larval form of the latter group must have existed which was of the nature of the Eurypterid Crustaceans. Again, judging from the development of Limulus, it would appear that the larval Eurypterid resembled a Trilobite, and there is evidence that Trilobites are Phyllopods, which are almost certainly derived from Chœtopod Crustaceans. Admitting the derivation of the lampreys from Cephalaspids, we find that the latter, in their adult condition, approximate to larval Amphibians ; and we hence pass from the latter to the lower Mammals, and so on to Man. Thus, according to Dr. Gaskell, the study of Ammocœtes, owing to the importance of larval forms, enables us to bridge the gulf between the Annelid and Man.

THE greater portion of the July number of the *Quart. Journ. Microscopical Science* is occupied by a communication from Messrs. F. W. Gamble and J. H. Ashworth on the anatomy and classification of the Sandworms (Arenicolidæ). This is followed by a most interesting series of diagrams illustrating the life-history of the parasites of malaria, by Messrs. Ross and

Fielding-Ould, accompanied by a short explanatory text. The diagrams were originally intended to illustrate a lecture delivered at the Royal Institution, to which reference is made in NATURE of March 29. The authors adopt the name Hæmamœbidæ for the intracorpuscular amœba-like bodies which occur in the blood of certain animals. Of these, three species occur in human beings (producing the various types of malarial fever), one in monkeys, three in bats, and two in birds. Only the human and avian forms are illustrated. The development of four of these has been followed in gnats; the three human forms living in Anopheles, while the bird-infesting species dwells in the common *Culex pipiens.* To this paper Prof. E. Ray Lankester appends a separate communication describing the generative process in the aforesaid " Hæmamœbids" and in the allied Coccidiidæ, which are parasitic in cuttle-fish. Sexual conjunction, or " zygosis," has recently been demonstrated to occur in the former group, and shortly before certain peculiar bodies known as microgametes and macrogametes were found to occur in the latter. These Prof. Lankester now shows respectively correspond to the spermatozoa and ova of higher organisms, specimens of the Coccidiidæ being figured in which the process of fertilisation by the microgametes is actually taking place.

To vol. xiii. Part 1 of the *Annals* of the New York Academy of Sciences, Prof. H. F. Osborn contributes an important paper on the correlation between the Tertiary mammalian horizons of Europe and America. The author is of opinion that the Puerco Eocene of the United States has no parallel in the European series, and that the Egerkingen beds of Switzerland are newer than the Wasatch. The three main divisions of the European Miocene are correlated with the Loup Fork and a portion of the John Day groups of America. But the most generally interesting portion of the paper is that in which Prof. Osborn enunciates his views with regard to former land connections. The theory of an extensive Antarctic continent synchronously connecting South America and Australia, and also communicating at some epoch with Africa, is deemed to be demonstrated. And it is considered that South America has experienced four distinct streams of faunal migration. In the first it received its peculiar Ungulates and Edentates; in the second it yielded the ancestors of Aard-varks, Pangolins, and perhaps Hyraces, to Africa; during a third land connection Marsupials immigrated from Australasia; and in the fourth the modern North American types effected an entrance. In contradistinction to the general view that Africa received its fauna from the north, the author is of opinion that the Dark Continent was itself the great dispersing centre and theatre of evolution; but whether South America received its original fauna from Africa or from North America is left an open question.

AMERICAN zoologists continue to devote their attention to the mammals of the Old World; and in the *Proceedings* of the Washington Academy of Sciences for July, Mr. G. S. Miller publishes two papers on the squirrels of Siam and Malacca, as well as a third on the European red-backed field-mice. Mr. Bonhote has just been writing on the former subject in the *Annals of Natural History,* and it seems a pity that naturalists cannot agree to divide their work so as to avoid overlapping and consequent unnecessary multiplication of names.

THE *Zoologist* for August contains an interesting account of a visit to Lundy Island during the nesting season, by Mr. F. L. Blathwayt. In the course of the paper allusion is made to the tradition that the Great Auk, or some equally large unknown bird, formerly inhabited the island. Only one or two pairs of this bird were known to the islanders, but an egg (subsequently broken) was secured in 1839. This subject seems worthy of further investigation.

WE are glad to see that in its September issue the *Girl's Realm* is endeavouring to awake an interest in the animal life of the sea-shore among its numerous juvenile readers, by publishing an illustrated article, entitled " An Hour in a Drang," by Mr. E. Step. " Drang," we learn, is Cornish for a deep cleft; and in his admirable description of such a cleft among the rocks at low tide, the author introduces his young friends to its living inhabitants in such a delightful manner that he can scarcely fail to gain many converts to the study of natural history. The photographs of crabs and lobsters with which the article is illustrated are admirably presented.

A *Bulletin* (Technical Series, No. 8) just issued by the U.S. Department of Agriculture (Division of Entomology) contains contributions towards a monograph of the American Aleurodidæ, by Mr. A. L. Quaintance, and a paper on the Red Spiders of the United States (Tetranychus and Stigmæus), by Mr. Nathan Banks. We have only one or two species of the interesting homopterous family Aleurodidæ in England. They are garden insects, which have a superficial resemblance to small white moths. In the present monograph forty-two American species of Aleurodes (Latreille) are described, most of them for the first time, and ten others belonging to the genus Aleurodicus (Douglas). To these the plates refer. The second paper, which is illustrated by wood-cuts, relates to the mites improperly called Red Spiders, which are equally troublesome in gardens and greenhouses in Europe and America. Of the two genera here discussed, ten species of Tetranychus (Dufour) and one of Stigmæus (Koch) are described; several as new. It is rather a pity that the term entomology is used in England so narrowly as practically to exclude mites, spiders, centipedes, &c., from entomological publications, and thus to hinder the popularisation of knowledge respecting them. In America, entomology is given the wider extension which it possessed at the beginning of the century, as may he seen by the inclusion of mites in the present publication.

THE members of the Manchester Microscopical Society deserve a word of encouragement for the efforts they make to extend a knowledge of natural history. One section of the society is entirely concerned with this work, and the members of it propagate the gospel of natural history by lecturing and demonstrating wherever their services are required. A programme containing a list of nearly fifty subjects has been issued, and the honorary secretary, Mr. George Wilks, 56, Brookland Street, Eccles New Road, Manchester, will arrange for lectures or demonstrations upon any of them if a communication is made to him.

THE two last numbers of the *Bulletin* of the Free Museum of Science and Art of Philadelphia show that this institution is growing rapidly under the care of Mr. Culin. The more important recent additions are figured. In vol. ii. No. 3 is an account of the historical Dickeson collection from the Mississippi mounds, and in the following number is a descriptive catalogue of the Berendt collection of books and manuscripts on the languages of Central America in the Museum Library, carefully compiled by the late Dr. Brinton.

THOSE interested in the decorative art of primitive folk should consult two fully illustrated papers in the *American Anthropologist* (N.S. vol. ii. No. 2, 1890). One, by Mr. R. B. Dixon, deals with basketry designs of the Maidu Indians of California, in which animal and plant forms, feathers, arrow heads, mountains and clouds are plaited in a very conventional manner. The author makes the significant remark, " The knowledge of the designs is almost exclusively confined to the older women, the younger generation knowing only very few." The second paper is one by Mr. B. Laufer, on the Amoor tribes, and is a preliminary account of the work done by this observer on the Jesup North Pacific

Expedition. Mr. Laufer gives a careful analysis of zoomorphic patterns, mainly of the Gold tribe; their decorative art shows distinct traces of Chinese influence, but the designs have been evolved in an original and interesting manner.

THE catalogue of bacteriological and pathological apparatus, just published by Messrs. J. J. Griffin and Sons, contains several new instruments and accessories, and will well repay inspection. Among the apparatus we notice several spirit Bunsen burners, which can be used instead of ordinary Bunsen burners where gas is not available. These are, of course, suitable for any laboratory, and not merely for bacteriological work. Of special interest are a number of new centrifuges for use in the examination of blood, sputum, milk. In water, urine and milk analysis a comparatively low rate of revolution is required, and a hand centrifuge giving up to 2000-3000 revolutions a minute is sufficient. When examining blood or sputum it may be necessary to make upwards of 10,000 revolutions a minute, which rate can be obtained by a water-power centrifuge manufactured by Messrs. Griffin. Another noteworthy addition is a special test-tube possessing characteristics always required for bacteriological work, but rarely found.

THE additions to the Zoological Society's Gardens during the past week include a Javan Mynah (*Gracula javanensis*) from Malacca, presented by Mr. George Smith; an Indian Crow (*Corvus splendens*) from India, presented by Mr. E. A. Williams; a Rose-coloured Pastor (*Pastor roseus*) from India, an Indigo Finch (*Cyanospiza cyanea*), a Nonpareil Finch (*Cyanospiza ciris*) from North America, presented by Mr. L. Ingram Baker; a Raven (*Corvus corax*), European, presented by Mr. G. St. Leger Hopkinson; three Blackish Sternotheres (*Sternothoerus nigricans*) from Madagascar, two Prasine Snakes (*Coluber prasina*) from Upper Burmah, eleven American Box Tortoises (*Cistudo carolina*) from North America, deposited; an Occipital Blue Pie (*Urocissa occipitalis*) from the Western Himalayas, ten Common Chameleons (*Chamaeleon vulgaris*) from North Africa, purchased; a Brush-tailed Kangaroo (*Petrogale penicillata*), born in the Gardens.

OUR ASTRONOMICAL COLUMN

ASTRONOMICAL OCCURRENCES IN SEPTEMBER.
Sept. 1. 8h. Jupiter in conjunction with the moon. Jupiter, 0° 51' North.
   3. 7h. 16m. to 8h. 11m. Moon occults the planet Saturn.
   4. 7h. 35m. to 8h. 50m. Moon occults the star ξ¹ Sagittari (mag. 5·0).
   5. 7h. 24m. Transit (ingress) of Jupiter's Sat. III.
  12. 12h. 35m. to 13h. .43m. .Moon occults π Arietis (mag. 5·6).
  12. 16h. 27m. to 17h. 40m. Moon occults ρ³ Arietis (mag. 5·5).
  13. 9h. 43m. to 10h. 34m. Moon occults 13 Tauri (mag. 5·4).
  14. 8h. 39m. to 9h. 18m. Moon occults D.M. + 20°, 785 (mag. 5·8).
  15. Venus. Illuminated portion·of disc = 0·493.
  15. Mars.   „   „   „   = 0·915.
  16. 12h. 8m. Minimum of Algol (*β* Persei).
  17. 6h. Venus at greatest elongation. 46° 1' West.
  18. 14h. 48m. to 15h. 40m. Moon occults 29 Cancri (mag. 5·9).
  19. 8h. 57m. Minimum of Algol (*β* Persei).
  23. 0h. Sun enters Libra, autumn commences.
  27. Saturn. Outer minor axis of outer ring = 17"·25.
  28. 21h. Jupiter in conjunction with the moon. Jupiter, 0° 13' North.

RING NEBULA IN LYRA.—It is interesting to find in the *Bulletin* de la Société Astronomique de France, August 1900, an account of the first published work done with the great 50-inch refractor of the Paris Exposition while that exhibition is still in progress. M. Eugène Antoniadi, of the Juvisy Observ-

NO. 1609, VOL. 62]

atory, has been for some time making systematic observations of nebulæ with the instrument, and a drawing showing a considerable amount of detail accompanies his paper on the Ring Nebula, the first of the series he has undertaken to study. He mentions that the lens used is the photographic one, the other, specially corrected for the visual rays, not yet being in position. The focal length of this glass is about 186 feet (57 metres).

OCCULTATION OF SATURN.—On Monday evening next, September 3, there will be an occultation of Saturn by the moon, for which the following particulars for Greenwich may be useful:—

	Sidereal Time.	Mean Time.	Angle from	
			North point.	Vertex.
	h. m.	h. m.		
Disappearance ...	18 6 ...	7 16 ...	128 ...	126
Reappearance.....	19 1 ...	8 11 ...	217 ...	206

Providing the weather be favourable, this should be an excellent opportunity for observing the occultation of the planet, as the altitude will be almost at its maximum, meridian passage at Greenwich occurring at 7h. 7m. G.M.T. Moreover, from its being such a bright object, observations may be made with instruments of the lowest optical power.

In the *Bulletin* de la Société Astronomique de France for August 1900, M. M. Honorat gives an illustrated description of his observation of the last occultation of Saturn on June 13. He mentions the conspicuous contrast between the slightly yellowish colour of the moon and the greenish tint of the planet. During the occultation the planet appeared separated from the lunar limb by a narrow shadow about 5" of arc in width, probably a contrast effect.

At the reappearance of Saturn at the terminator, he could not perceive any trace of penumbral shadow cast on the planet's disc.

OPPOSITION OF EROS.—Two additional circulars have been issued by the special committee appointed by the Astrographic Conference to direct the observations of Eros during the coming opposition. Special attention is drawn to the work which may be commenced at once, such as micrometric observations with all equatorials of large aperture, for furnishing definite positions for the theory of the planet's movement, and that these should be published as soon as possible, to perfect the ephemerides for the actual parallax work later. An ephemeris is included from the computations of M. Millosevich, and tables showing the limiting times between which the planet will have an altitude greater than 20° at various latitudes, and also a table indicating the proper regions to be included on the photographs on dates extending from September 19 to January 7.

In the *Astronomische Nachrichten* (Bd. 153, No. 3656), Prof. S. J. Brown, of the U.S. Observatory at Washington, calls attention to the many opportunities for simultaneous micrometer observations at widely separated stations, and as many observatories are not equipped with the photographic instruments necessary for the more general programme contemplated, gives data for assisting micrometer observers to co-operate for this type of work alone. The high declination of the planet makes it possible to secure simultaneous observations at all the Eastern stations west of Pulkowa, and at all the American observatories east of Denver. He also gives a table showing the Greenwich Mean Time at which the planet will be simultaneously visible at the observatories of Pulkowa, Königsberg, Vienna, Evanston, Madison, Yerkes and Denver for intervals of ten days from 1900 October 1–1901 January 19. Careful sketches of the comparison stars in the field should be made to facilitate subsequent identification. Owing to the rapid orbital motion of Eros rendering observations for position angle and distance very troublesome, measures should be made in rectangular co-ordinates referred to the true equatorial position of the fixed micrometer wire.

THE INTERNATIONAL PHYSICAL CONGRESS.

THE first International Congress of Physics, which has just finished its sittings, has been a brilliant success. The number of participators exceeded a thousand, and, in spite of the attractions which Paris always offers, in spite of the simultaneous rivalry of the Universal Exhibition itself, sectional and general meetings were closely followed up to the last day by a great number of visitors.

The cause of this unexpected success must no doubt be sought in the idea underlying the plan of the Congress, worked out as it was with the greatest care by a committee of the Société Française de Physique. That committee deliberately rejected the method of simply presenting personal memoirs, or notes on limited subjects, and concentrated all its efforts upon the preparation of a well-arranged summary of the actual state of physical science, in the branches in which, within the last few years, the greatest progress has been made, and the actual stage of progress of which at the end of the nineteenth century it was considered most important to investigate. Once the list of subjects was completed, the work was divided among the physicists who seemed best qualified to give a complete representation of their special subject. This plan gave rise to a series of reports,[1] many of which are works of a very high value, and which, in their entirety, constitute the most complete representation of any science at a given epoch yet made. These reports number about 80. To summarise these here would be, so to speak, repeating the work of the Congress on a small scale, and that could not be thought of. I shall, therefore, confine myself to referring to them by groups which are obviously related and mutually supplement each other.

For this considerable task, a preface was necessary. M. H. Poincaré provided such a preface, and brought it before the Congress amid great applause, showing how mathematical generalisation could render experimental work infinitely more fertile. Experimental physics is a library. Mathematical physics arranges it and prepares the catalogue. It does not enrich it, but if it is well prepared it enables one to draw a greater profit from the former. The celebrated mathematician then showed how hypotheses have succeeded each other, in the form of physical images or simply mathematical images, where the symbol often remains true even when the mechanism is no longer accepted. Mathematical analysis also alone gives the true sense of the simplicity hidden under complexity, as in the case of Newton's law—which is always rediscovered in the most complicated movements of the heavenly bodies—or the kinetic theory of gases, where the law of large numbers hides the isolated individuals, only permitting the appearance of an aggregate for which the laws of Mariotte-Boyle and Gay-Lussac, long considered simple, are only the destruction of the action of individual molecules. Starting from these now well-known facts, M. Poincaré showed how the same methods and ideas apply to theories now being evolved concerning the interaction of matter and ether. His speech will no doubt be read and studied for a long time to come, and will remain one of the most perfect expressions of the state of mind of the masters of modern science.

To the organisers of the Congress, Lord Kelvin's promise of a personal contribution of work had been a powerful and valued encouragement. But what they hardly dared hope for was to see him, after the fatigues of a voyage, take a very active part in the Congress, and to see him hold spell-bound by the charm of his discourse a respectfully attentive audience bent upon seizing every thought of the great physicist. M. Poincaré's speech gave him an occasion for a brilliant improvisation on the constitution of the ether; and he also dealt with the subject in a paper on the waves produced in an elastic solid by the motion of a body acting upon it by attraction and repulsion. But it was not only in that speech that the illustrious honorary president of the Congress showed the interest he felt in the assembly. Presiding every day at sectional meetings, he clothed both reports and debates with a very special authority.

To facilitate work, the Congress had been divided beforehand into seven sections, the work of which I propose briefly to review.

In the measurement section, presided over by M. Benoît, the chief work was that of determining the actual state of metrology properly so called. After a very complete recapitulation, by the president, of the history of standards and methods employed in the measurement of length and the progress so far made, detailed attention was devoted to the complete metrological definition of standards and their legal definition; the legal status of the electrical units; and some improvements which might conveniently be made in a number of insufficient definitions, or definitions referring to conceptions recently introduced into science, such as the different abscissæ of the spectrum, &c. Some resolutions were passed, such as that recommending the adoption

[1] These reports presented to the Congress have been translated into French. They were printed for purposes of discussion, and will be shortly published in three volumes.

of the mechanical C.G.S. units (erg and joule) for the expression of calorimetric quantities, comprising, naturally, the solar constant, to be reduced by the meteorologists to the calorie per minute per sq. cm. Also that in the expression of elastic constants, the C.G.S. unit of pressure, the *barie*, be adopted, of which the multiple by 10^6, the *megabarie*, is sufficiently represented by the pressure exercised by a column of mercury 75 cm. long at o° and under normal gravity. The Congress further supported the sectional resolution that national laboratories be created in countries which do not as yet possess any.

The interferential methods of measurement brought out an excellent paper by M. Macé de Lépinay; there were also four contributions relating to thermometry of precision (Chappuis), pyrometry (Barus), the mechanical equivalent of heat (Ames), and a special study of the variation of the specific heat of water (Griffiths). All these showed that great progress has been made in these various departments. Thus at present the divergencies among the various gas thermometers are known over a long interval, and it is also known that though the hydrogen and nitrogen thermometers, for instance, may still differ between o° and 100°, their divergence at the higher temperatures is insignificant if care is taken to slightly correct the mean coefficient of expansion between o° and 100°. The difficulty of employing hydrogen at high temperatures lends a great importance to this provision. Pyrometry also is rapidly advancing, and as regards the mechanical equivalent, the great divergences which existed a few years ago have disappeared owing to a more complete correction of thermometric values and a better knowledge of electric standards.

Some very fine work has also been done in connection with gravitation. The measurement of the Newtonian constant, admirably expounded by Mr. Boys, whose special work in this department is now classical, and the announcement of anomalies of gravitation by Messrs. Bourgeois and Eötvös, gave rise to very interesting discussions. A few years ago these anomalies were placed beyond a doubt, and it is already possible to study the details with the aid of apparatus which, like that of M. Eötvös or that of Messrs. Threlfall and Pollock, indicates the most minute details, whereas the pendulum formerly employed only gave the more considerable anomalies. The Congress expressed a hope that the study of these anomalies will be pursued by the new methods, not only for the sake of knowing the gravitational acceleration in every place, but also for the better knowledge of the constitution of the globe.

Finally, M. Leduc presented to the section a report on the electro-chemical equivalent of silver, and M. Gouy another on the standard of E.M.F. It appears from the latter that the cadmium standard is preferable to every other.

The measurement of the velocity of sound, dealt with by M. Violle, forms in a manner the transition between the section of measurement and that of mechanical and molecular physics. In the latter, presided over by M. Violle, after a very complete treatise by M. Amagat on the whole of his work, and an admirable paper by M. van der Waals on the statics of mixed fluids, M. Mathias showed, in a paper well provided with references, how the critical point may be determined by various methods. Specialising further, Prince Galitzine dealt with the refractive index, and, finally, M. Battelli exhibited the relations between the statics of fluids and their calorimetry. Except as regards mixtures, the ideas on these various subjects are well fixed nowadays, and new light can only come from the experimental side. Mixtures are less known, and the paper in which the celebrated Amsterdam physicist condensed our actual knowledge of this question will no doubt powerfully contribute to make them known. Having created the idea of continuity between the liquid and the gaseous states, he has had the satisfaction of seeing it become classical. But it is in another direction that this evolution advances nowadays. Does this continuity also exist between the liquid and the solid state? The diffusion of solids, their flow under pressure, the constitution of alloys, so well studied, notably by M. Spring and Sir W. Roberts-Austen, might lead to that belief, especially since M. Schwedoff has proved the rigidity of liquids. M. Tammann raises some doubts concerning this idea, and recommends a careful distinction between the amorphous and the crystalline states. In any case, the presence at the Congress of the eminent physicists mentioned, with the exception of M. Tammann, who was represented by M. Weinberg, contributed greatly to the interest of the subject and to its future progress.

The study of the permanent or temporary deformations of

solids naturally furnishes interesting data. A work by M. Mesnager, and another by the author of the present article, were devoted to these two questions. It is interesting to note that the last experimental researches are all in favour of the chemical theory of temporary deformations.

M. Voigt has devoted a great amount of indefatigable activity to the study of the elasticity of crystals. His summary of this question was a great boon to the second section. It was a considerable piece of work, in which, naturally, the mathematical formula was predominant. This work will serve as a base for all those interested in the elasticity and the piezo-electricity of crystals, as well as in questions of symmetry.

M. van't Hoff was not present at the Congress, but he showed his interest by sending a work on the formation of crystals in a mother-liquor containing a mixture of salts. In this case questions of equilibrium play an important part, the form of the crystals depending, not only upon the solubility of each salt in the mixture, but also upon the quantity of each.

It was very interesting to learn the ideas arrived at by M. van der Mensbrugghe, after a long career devoted to the study of capillarity. The report presented by Joseph Plateau's son-in-law constitutes a precious document on capillarity, a subject which has been somewhat eclipsed by other subjects, but which has formed the object of investigation of the greatest spirits, and continues to do so.

We must be short, and can only mention the report by M. Brillouin on gaseous diffusion, by M. Perrin on osmosis, and by M. Bjerknes on hydrodynamical actions at a distance. The latter derives its interest more especially from the fact that a hydrodynamical model may be constructed which possesses all the characteristics of a world subjected to actions at a distance.

The third section, presided over by M. Lippmann, dealt with optics. The recent researches on the laws of radiation naturally formed part of its programme, opened by those inseparable reports, on the theoretical laws of radiation, by M. W. Wien; on the radiation of solids, by M. Lummer; and on gaseous radiation, by M. Pringsheim. The practical realisation of the black body, the verification of Stefan's law for a large range of temperature, and certain simple relations between the temperature and the position of the maximum in the spectrum, are the salient facts which the experimental work of recent years has brought out. For gases, a doubtful point is the validity of Kirchhoff's law, but according to M. Pringsheim that does no seem to be in any danger if only the purely thermal radiati is considered.

Recently, the spectrum has been greatly extended in the infra-red. M. Rubens, to whom the greatest progress in this direction is due, had undertaken to give a summary of this question, showing how the dispersion formulæ agreed with experiment, and demonstrating experimentally the connection between long light waves and electrical waves. This work again called forth a discussion on the formulæ and theories of dispersion opened by M. Carvallo.

The kinematics of the spectrum has also made great progress since Balmer showed for the first time that the hydrogen rays are represented by a very simple formula. The researches of Kayser and Runge and other physicists, Rydberg among them, have shown that the distribution of the spectrum lines is governed by laws, some of which are clearly established, while others are as yet unknown. Of all this work, M. Rydberg gave an excellent summary.

The velocity of light has, as we know, given rise to metrological work of the first rank, and of extreme difficulty. It fell to the distinguished president of the Congress, M. Cornu, to give a review of this subject, and during the remarkable speech which he delivered at the École Polytechnique, the physicists from all parts had the privilege of seeing the original apparatus of Fizeau and of Foucault, who were the first to give an approximate value of that velocity by measurements confined to the earth.

It is this characteristic velocity which for Maxwell was the touchstone of the theory involving the identity of luminous and electrical oscillations. As the instruments become more perfect, and the sources of error disappear, this identity is more and more emphasised. It was very interesting to co-ordinate the numbers furnished by light proper with those furnished by the comparison of units and the direct measurement of the velocity of electric waves. M. Abraham undertook the first part of this work, and MM. Blondlot and Gutton the second. This brings us to the electrical section, presided over by M.

Potier, and in his absence by M. Bouty. The line of demarcation, however, is becoming more and more difficult to draw. The extremely interesting work of M. Lebedef on the pressure produced by radiations, has its origin in the great work of Maxwell; but it might also arise from pure thermodynamics, as shown by Bartoli and Boltzmann. As regards Hertzian waves, treated in a masterly manner by M. Righi, they approach so closely to the work of M. Rubens, that the small interval which still separates them is probably the only reason—and a very artificial one—for keeping them separate at all. In a supplementary note, M. Branly gave an account of some of his own researches on coherers. The reports just mentioned furnished the experimental side of an idea, the theoretical aspect of which was treated of in a paper by Prof. Poynting on the propagation of electrical energy.

We encounter another group of questions in the gaseous dielectrics, studied by M. Bouty, as well as electrolysis and ionisation, which have made such vast progress during the last decade, and which were dealt with by M. Arrhenius, one of the promoters of the new ideas, in a paper which will remain a model of clearness. Finally, we have M. Christiansen's theories of contact electricity, M. L. Poincaré's theories of the electric cell, and the exposition of Nernst's ideas, which had not been contemplated in the programme of the Congress, but which enabled their founder to give to the meeting a review admirably completing this group of questions.

The presentation of present ideas on magnetism had been excellently prepared by two fundamental reports, one, by M. du Bois, on the general magnetic properties of bodies, and another, by M. Warburg, on hysteresis, which he was the first to observe, and which, in the hands of Ewing, Hopkinson and others, attained such great importance. Two particular aspects of magnetism, viz. magneto-striction and the E.M.F. of magnetisation, which could not form part of the general reports, were treated separately by M. Nagaoka and M. Hurmuzescu.

Although the applications of electricity are almost entirely beyond the subject-matter of the Congress, there are some which are connected so closely with general physics that it seemed very desirable to have them dealt with. This was done by M. von Lang, whose work on the electric arc is well known, while M. Potier gave an exhaustive paper on the theory of polyphase currents, and M. Blondel the description of apparatus for tracing the curves of rapidly varying currents.

In a few years' time the work of the fifth section—ionisation and magneto-optics, presided over by M. Becquerel—will no doubt fall naturally into one of the preceding sections. But at the present moment they are still so undefined, they open up such new horizons, that it appeared well to collect them in a special section. The idea proved very fruitful, for the section was largely attended, and the discussions at it proved very fascinating.

M. Lorentz had prepared an admirable report on magneto-optics, with special reference to the Zeeman phenomenon. He expounded both his own ideas and those of M. Voigt. The presence of the latter gave the section the privilege of an exposition at first hand of his latest ideas.

The absence of Prof. J. J. Thomson could not but be severely felt. But the work which he had sent in, concerning the ratio of the electric charges to the masses carrying them, was read amid great interest after the general exposition made by M. Villard of the state of our knowledge of kathode rays.

The phenomena of actino-electricity, somewhat forgotten now, though much studied ten years ago, gave rise to a report by MM. Bichat and Swyngedauw. Perhaps increased attention will be devoted to them now that the researches of M. Becquerel and those of M. and Mme. Curie have proved so fertile in the examination of new bodies.

The speeches in which first M. Becquerel and then M. Curie expounded the disconcerting properties of uranium, polonium and radium rays, were for many a revelation. These extraordinary bodies, discovered by their radio-active properties, which were first announced by M. Becquerel, and then followed up with such startling success by M. and Mme. Curie, were known to the majority of those present, but only a few had seen those few decigrammes of material extracted from several tons of the mineral richest in it, pitchblende, and certainly the effects produced surprised by their intensity those who saw them for the first time.

Several hundred persons at a time could see this light, which appears everlasting, radiated perpetually by radium, the clear patch which it produces even across a sheet of metal on a screen of barium platino-cyanide, the instantaneous discharge of an electrified body brought near to the substance, and the sparks passing when radium is brought within a few centimetres of the spark gap. The magnetic deflection of the rays could, of course, not be made evident to such a large audience. But the original negatives could be projected, and they showed the curvilinear propagation of the rays in a magnetic field.

The new bodies constantly project matter endowed with a great velocity. Neighbouring bodies are impregnated with it, and become radio-active in turn. These particles attach themselves, not only to objects, but to persons as well, so that M. Curie will be condemned for some time to abandon every kind of electrostatic research. No electrometer remains charged in his neighbourhood, and it is certain that if radium had only been as plentiful as gold, static electricity would never have been discovered.

In the same domain, important generalisations have been made, such as the theory of dispersion in metals, founded by M. Drude upon the electron theory, of which the author gave an account to the section.

The sixth section, under the presidency of M. Mascart, occupied itself with cosmical physics. Terrestrial magnetism should undoubtedly have formed part of the work of this section, but the Meteorological Congress which will shortly meet intends to make that the principal object of its studies, and it was evidently necessary to leave it aside.

Yet the work of this section was very fruitful. Here, naturally, observation still holds a predominant place, as in the work of the Swiss physicists, with M. Hagenbach at their head, on glaciers; and the detailed study of oscillations of lakes by MM. Sarasin and Forel, who brought their results before the Congress.

In the department of atmospheric electricity, a very good account was given by M. F. Exner, and Mr. Paulsen gave an account of the Danish expedition to Iceland for the study of the aurora. The evaluation of the solar constant by M. Crova, according to recent researches, and a very ingenious theory of suns-pots established by M. Birkeland after troublesome calculations, were heard with much interest. Finally, M. C. Dufour showed how, without the help of any laboratory apparatus, the approximate brightness of the stars could be determined.

It had seemed useful to collect in a seventh section some works relating to biology. In the absence of M. d'Arsonval, this section, presided over by M. Charpentier, did a great deal of good work, and justified the idea of the organisers of the Congress. The application of physical and mathematical methods to the transmission of energy in organisms, to which M. Broca has devoted attention for a considerable time, and the curious retina phenomena studied by M. Charpentier, gave this section a vast field of discussion. Finally, the new theory of accommodation established by M. Tscherning received the sanction of a very largely attended meeting, while M. Hénocque spoke of the spectroscopic methods used in biology.

The proceedings of the Congress were not confined to sectional work and general meetings. A visit to the laboratories of the Sorbonne and the École Polytechnique showed many experiments in progress, installed by the professors of these establishments or their provincial and foreign colleagues. These could only be properly appreciated by observing them closely and in small groups.

Shall I speak of the reception in the Jardin de l'Elysée, whither the President of the Republic invited several Congresses to witness a theatrical performance? Or of the charming *soirée* for which Prince Roland Bonaparte had placed at the disposal of the organisers his vast and magnificent library for a number of interesting experiments? This *soirée*, which will leave in the minds of all who were present the most agreeable memories, would itself deserve a lengthy description. But I cannot conclude this already lengthy article without saying how much the French physicists have been touched by the sympathetic action of the foreign secretaries of the Congress, who deposited a magnificent crown on the modest tomb of the great Fresnel, of which the Société Française de Physique has constituted itself the guardian. A moving speech by M. Warburg, and a warm expression of thanks by M. Cornu, president both of the Congress and of the Society, referring in a few words to the life of that great physicist, ended this first Congress, where so many new thoughts have been born, and so many friendships made or consolidated. CH. ED. GUILLAUME.

ORIENTATION OF THE FIELD OF VIEW OF THE SIDEROSTAT AND COELOSTAT.

OBSERVERS who have practical acquaintance with the siderostat and heliostat are familiar with the fact that while the reflected image of a star may be kept stationary, the images of surrounding stars have a rotation around it; while if the sun is the object viewed in the mirror, the image will rotate about the axial ray. It is on account of this rotation of the field that neither the siderostat nor the heliostat can be used with a fixed telescope for celestial photography, except for objects which can be photographed with short exposures.

Certain unexpected peculiarities of this motion have recently led Prof. Cornu to investigate the general laws governing the rotation of the field in both instruments (*Comptes rendus*, vol. cxxx. No. 9, 1900; *Bulletin Astronomique*, February 1900). Some of the results at which he has arrived are of great interest, and we believe attention has not been previously drawn to them, although they could have doubtless been derived from Orbinsky's formula for the orientation of the field ("Die totale Sonnenfinsternisse am 9 Aug. 1896"), or from other formulæ which have been employed by observers as occasion required.

Prof. Cornu first discusses the general question of the orientation of the field, irrespective of the mechanical means of retaining the reflected image in a fixed position. In Fig. 1, NESW represents the horizon, z the zenith, P the pole, PD the hour circle of the star D, and D' the point of the horizon towards which the rays are reflected. PN is equal to the latitude of the

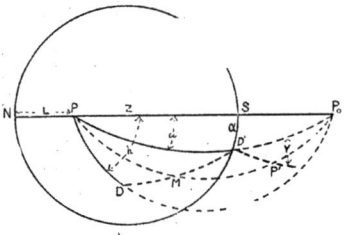

FIG. 1.—Orientation of field of siderostat.

place, = L; PD is the polar distance of the star, = δ; SPD is the hour angle of the star, = *h*. For the purposes of calculation the point D' is defined by its polar distance PD' = ρ, and by the angle SPD' = ω which the plane PD' makes with the meridian; ρ and ω can be determined in terms of the azimuth of D' (= SD' = α, reckoned positive towards the west) and the latitude, by solving the right-angled triangle PSD', in which PS = 180° − L ; thus

$$\cos \rho = \cos \alpha \cos L; \ \tan \omega = \frac{\tan \alpha}{\sin L}.$$

The normal to the mirror must always bisect the arc DD' of a great circle, at M, so that the position of the reflected ray from any part of the sphere can be easily determined. Thus the image of P is at P' in the continuation of the hour circle PM, MP' being equal to PM. To determine the orientation of the field, it is most convenient to ascertain the direction; after reflection, of the point P, since it is a fixed point on the sphere. Taking the plane of PD'P₆ as the reference plane, and its trace on the sphere as a fixed direction, the orientation of the reflected pole is conveniently defined by the angle P₆D'P' = V, which can be readily calculated, as also D'P', the distance of the reflected pole from the centre of the field.

Applying this in the first place to the siderostat, where the reflected rays are south or nearly so, and the angle α consequently small, Prof. Cornu obtains the following results :—

(1) The reflected image of the pole describes a circle round the centre of the field, with a radius equal to the polar distance of the star observed.

(2) Since the angle Y is equal to the supplement of the angles at the base of the triangle PDD', P being the apex, the orientation of the reflected pole (that is, the direction of the north point of the field) is given by the equation

$$\tan \tfrac{1}{2} Y = \frac{\cos \tfrac{1}{2}\,(\rho+\delta)}{\cos \tfrac{1}{2}\,(\rho-\delta)} \tan \tfrac{1}{2}\,(h-\omega)^{1}$$

The law of rotation readily follows. The interval from the passage of the star over the hour circle PD' being expressed by t, with a day as the unit of time, $h-\omega = 2\pi t$, and the equation becomes

$$\tan \tfrac{1}{2} Y = K \tan \tfrac{1}{2}\, 2\pi t.$$
$$\text{where } K = \frac{\cos \tfrac{1}{2}\,(\rho+\delta)}{\cos \tfrac{1}{2}\,(\rho-\delta)}$$

Hence :—(a) The rotation of the field has the same period as the diurnal motion.

(b) The motion is continuous and in the same direction, direct or inverse according to the sign of K.

(c) The plane of reference is a plane of symmetry, since the angle Y has equal values of contrary sign at equidistant intervals of time from passage across the reference plane.

Prof. Cornu illustrates the rotation by a diagram similar to those in Fig. 2.

(3) The angular velocity of rotation at the epoch t is given by

$$\frac{d Y}{d t} \propto 2\pi\,\frac{K}{\cos^{2}\pi t + K^{2}\sin^{2}\pi t}.$$

The denominator is always positive, so that the velocity has always the same sign as K ; its value varies from 2πK (when $t = 0$) to $\frac{2\pi}{K}$ (when $t = \tfrac{1}{2}$), and is equal to the diurnal motion when the conditions make the denominator equal to K. The velocity varies so slowly for small values of t, that it may be sufficient to regard it as constant and equal to 2πK. Since the northern meridian passage cannot be observed with the siderostat, the value $\frac{2\pi}{K}$ is not observable.

(4) The apparent motion of the field, as seen in the mirror, will evidently be in the same direction as the apparent motion of D'P' seen from outside the sphere, and it will not be reversed by an astronomical telescope. When the polar distance of the star observed is less than the supplement of the polar distance of the reflected ray, the apparent direction of rotation of the field of a siderostat is clockwise ; it is in the contrary direction if the polar distance of the star is greater than this supplement.

(5) When $\cos\tfrac{1}{2}(\rho+\delta)=0$, we have $K=0$, and $Y=0$ for all values of t. Hence there is no rotation of the field when the polar distance of the star observed is equal to the supplement of the polar distance of the direction of the reflected ray.

In this case, $\rho+\delta=180°$, and $PM=90°$, so that the mirror is parallel to the earth's axis, and the instrument thus behaves like a cœlostat.

[1] This gives the angle reckoned from the direction D'P. To obtain the inclination to a vertical line passing through the mirror, it would be necessary to calculate the angle PD'z.

(6) When the reflected ray is in a horizontal and southerly direction, as is usually the case, $\omega = 0$, and $\rho = \pi - L$, so that the formula for orientation becomes

$$\tan \tfrac{1}{2} Y = K \tan \tfrac{1}{2}\, h$$

where

$$K = \frac{\sin \tfrac{1}{2}\,(L-\delta)}{\sin \tfrac{1}{2}\,(L+\delta)}.$$

It readily follows that there is no rotation of the field in this case when the polar distance of the star observed is equal to the latitude of the place of observation ; the rotation is clockwise if the polar distance be less than the latitude, and contrary if greater. Fig. 2 illustrates the varying conditions of rotation in the latitude of London (a) for the position of the sun at the winter solstice, (b) for the position of the sun at the summer solstice, and (c) for a star which passes through the zenith. In each case the numbers are placed to represent the position angles of the north point of the field at corresponding hour angles.

In the case of the heliostat, where the rays are reflected in a northerly direction, a similar method of computation is adopted by Prof. Cornu ; but as the instrument is so little used in work of precision, it is unnecessary to give the details. The important result is that the field of view under ordinary conditions has an angular velocity of rotation greater than that of the diurnal motion.

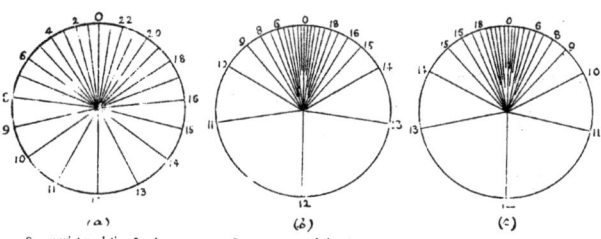

(a) Sun at winter solstice, London.   (b) Sun at summer solstice, London.   (c) Star through zenith, London.

FIG. 2.—Illustrating rotation of field of siderostat.

A knowledge of the orientation of the field as reflected by a mirror is so frequently required that it may be useful to refer briefly to other ways of treating the problem.

Orbinsky proceeds in the same manner as Prof. Cornu, but considers the more general case in which the reflected rays are neither in the meridian nor horizontal. The direction of the normal is midway between the direction of the star and that of the reflected ray, on a great circle, so that the direction of the reflected ray from any other point of the celestial sphere can at once be determined.

In this way the position of the zenith point of the field (vertex) is derived with respect to the vertical circle in the plane of the reflected ray. A calculation of the angle between the vertex and the north point is then all that is required to give the direction of the north point of the field with respect to a vertical line through it.

Another method of representing the orientation was adopted by Mr. Shackleton in connection with the eclipse of 1896. This can be applied to a reflection in any direction, but it will suffice to indicate its application to a siderostat with the reflected ray in the meridian. Using Prof. Cornu's notation so far as possible, in Fig. 3 NESW is the horizon, NPS the meridian, P the pole, D the star, M the mirror, MS the direction of the reflected ray from D, and SDN the trace of the plane of reflection. Representing the direct field by $an\delta$, n is the north point. The field of the mirror appears behind the mirror as $a'n'b'$, $a'b'$ remaining in the plane of reflection, and $b'Nn'$ being equal to δbn. Since Nv is a vertical line through the field of the mirror, and $vNa' = DSP$, it is evident that $vNn' = 180° - (PSD + PDS)$.

vNn' thus corresponds with the angle Y in Prof. Cornu's formula, and its value is derived by precisely the same formula.

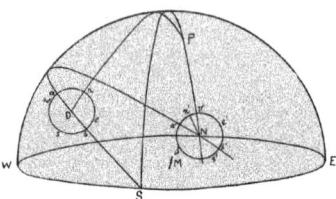

FIG. 3.—Orientation of field of siderostat.

The orientation of the field of a cœlostat is very readily derived. In this instrument the mirror turns on a polar axis in its own plane, so that the normal is always on the equator, and the polar distance of the reflected ray is always equal to the supplement of the polar distance of the star. Thus, in Fig. 4, PD′ is the supplement of PD. The reflection of the hour circle through the star, PD, will coincide in direction with that through the reflected ray, PD′, so that n will become n', and it only remains to determine the angle PD′z to ascertain the position of the north point with regard to a vertical line through the field of the mirror. If we suppose the rays to be reflected in a horizontal direction, in the triangle PZD, PZ = the co-latitude, zD′ = 90° and PD′ = 180° − PD, so that the required angle can be at once derived. In this case it is convenient to know the azimuth of the reflected ray, that is, PzD′; and the simplest solution is to calculate this angle first by the formula

$$\cos PzD' = \cos (180° - PD) \sec L.$$

The required angle is then derived from the formula

$$\sin PD'z = \sin PzD' \cos L \operatorname{cosec} (180° - PD).$$

The position of the north point having been determined, the remaining points can at once be placed, noting that the east and west points are reversed as compared with the direct view in the sky.

It is important to note that although there is no rotation of the field so long as the telescope remains in one position, the

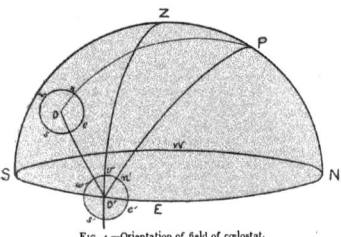

FIG. 4.—Orientation of field of cœlostat.

whole field is turned when the telescope is set in a different direction. Thus, if the telescope is directed west for observa-

tions of the morning sun, the orientation of the field will be different when the telescope is pointed in an easterly direction for observations of the sun in the afternoon; in the former case the north point lies to the right of the vertical, and in the latter case to the left.

Other investigations relating to the cœlostat, including the determination of the best position for the telescope under given conditions, have been made by Prof. H. H. Turner (*Monthly Notices R.A.S.*, vol. lvi. p. 408).

As the cœlostat has not yet come into very general use, it may be of interest to add a few remarks as to the arrangements which have been made by Sir Norman Lockyer at the Solar Physics Observatory for utilising this instrument in a permanent observatory (Fig. 5). On account of the varying declinations of the heavenly bodies, the position of the observing telescope must admit of corresponding changes, either in inclination or azimuth, or both. When special instruments, such as the spectro-heliograph, are to be used with the cœlostat, as at the Solar Physics Observatory, motion in azimuth is the only motion permissible, and this is provided for by fixing the receiving instrument on a platform which runs on circular rails, with the cœlostat at the centre. The platform carrying the telescope or spectroscope is covered with a travelling hut, the roof of which

FIG. 5.—The cœlostat of the Solar Physics Observatory.

is inclined so as not to obstruct the mirror. The cœlostat itself is provided with a hut, which is removed to the north when the instrument is in use; this is shown to the left in the illustration.

A. FOWLER.

THE ANNIVERSARY MEETING OF THE REALE ACCADEMIA DEI LINCEI.

A MELANCHOLY interest attaches to the anniversary meeting of that ancient scientific society, the Reale Accademia dei Lincei, held in June, from the fact that the society was then mourning the loss of its distinguished president, Prof. Beltrami, and has since been plunged into deeper mourning by the untimely and unexpected loss of its patron, King Humbert, who with Queen Margherita had for many years taken part in these yearly meetings. It is, moreover, largely due to the munificence of the late King of Italy that the society is enabled to further the advancement of science by the award of prizes for theses dealing with some subject of scientific research.

From the presidential report of Prof. Mesadaglia, we learn that the society's losses have included, besides Beltrami, the names of Capasso, De Simoni, Ferrara, Nestore and Tommasi-Crudeli among the ordinary members, and, of foreign members, Bertrand, Bunsen, Janet and Liais. The *Atti*, or "Proceedings" contain for the year 147 papers, in addition to which several longer papers are being printed in the *Memorie* or "Transactions," and the corresponding societies with which an exchange of publications is made now number not less than 500. Under the title of *Notizie degli Scavi*, the society brings out accounts of archæological discoveries in Italy, the material for which is furnished monthly by the Minister of Public Instruction. Of recent publications, we note the issue of three volumes of the "Codex Atlanticus" of Leonardo da Vinci, a magnificent work, in the cost of publishing which the late King gave material assistance ; also the "Forma Urbis Romæ" of Signor Lanciani, consisting of a large scale archæological map of Rome.

For the Royal prize of 1000 francs for normal and pathological physiology six candidates entered, and a large number of essays of considerable merit were submitted by them. The prize has been adjudged to Prof. Giulio Fano, of Florence, for sixteen papers, dealing, amongst other subjects, with the physiology of the embryonic heart, the doctrine of experimental psychology, the organ of hearing, the graphic registration of respiratory chimism and reflex movements, the latter being a continuation of previous researches on the organs of *Emys Europea*. Of the six candidates for the Royal prize for geology and mineralogy, two were considered worthy of the award, which was therefore divided equally between them. One of the successful candidates, Prof. De Lorenzo, chose geological subjects, and sent in about twenty essays, the most important of which dealt with the trias of the environs of Lagonegro, the mesozoic mountains of Lagonegro, geological observations on the Apennines of the southern Basilicate, and geological studies of the southern Apennines. Prof. Giorgio Spezia's work, on the other hand, was entirely mineralogical, dealing with the influences of temperature and pressure, respectively, on the chemical metamorphism of rocks and minerals. From a long and laborious series of experiments, many of them occupying five or six months, the author concluded that pressure has little or no effect, while the influence of temperature is considerable. The results have a special bearing on the theory of quartz formation. The Royal prize for advances in archæological science was adjudged to Dr. Paolo Orsi, of Rovereedo, for his investigations of the antiquities of Eastern Sicily, Dr. Orsi has thrown quite a new light on the prehistoric development of the people known as the Siculi, from the neolithic epoch down to the period of expansion of the Greek colonies. A special prize for philosophy and moral science had been offered for an essay dealing with either the theory of consciousness or the foundations of practical philosophy. This prize has been divided equally between Prof. Bernardino Varisco and Prof. Francesco de Sarlo. The Minister of Public Instruction offered a sum of 3400 lire for two prizes in physical and chemical sciences, and a like sum for two prizes in philological sciences, the prizes being confined to teachers in secondary schools. The committee for the prizes in physical and chemical sciences have awarded two equal prizes—one to Prof. O. Marco Corbino, more especially for his work on light traversing metallic vapours in a magnetic field, and the other to be divided between Profs. Carlo Bonacini and Riccardo Malagoli, more especially for their joint papers on Röntgen rays. In philology, the prizes have been divided up into a number of minor awards, distributed between Signori Giuseppe Vandelli (whose work stood first), Antonio Belloni, Astorre Pellegrini, Giuseppe Rua, Giuseppe Lisio, Augusto Balsano, Giovanni Negri and Guglielmo Volpi.

At the conclusion of the awards a biographical commemoration of the late Prof. Beltrami was delivered by Prof. Luigi Cremona. In the number of the *Atti*, this is followed by a chronological list of Beltrami's scientific works, in compiling which use has been made of the previously published lists by Prof. Dini, and by Signori Pinti and Brambilla in the *Annali di Matematica* and the *Rendiconto* of the Naples Academy respectively.

The proceedings terminated with an address by Signor Giuseppe Colombo on the progress of electrotechnics in Italy. Signor Colombo briefly traced the gradual development of the theory of the electrical transmission of energy, from the discovery of Volta, through the various stages indicated by Pacinotti's invention of the first dynamo, Galileo Ferraris' principle of the rotating magnetic field, and a number of intermediate inventions, down to the principle of wireless telegraphy, to the development of which two Italians, Righi and Marconi, have so largely contributed. The absence of coal has long been a serious bar to the progress of Italy in commercial competition, but Signor Colombo proves by statistics that Nature has provided a source

of energy more than sufficient to fill the deficiency, in the water-power with which the country has been well endowed, and it only needs the development of plant for the electrical transmission of power, aided, moreover, by the best means for minimising waste of energy, to raise Italy to a condition of commercial prosperity.

UNIVERSITY AND EDUCATIONAL INTELLIGENCE.

THE *Pall Mall Gazette* states that Miss Cruickshank has given to Aberdeen University, in memory of her brother, Dr. Alexander Cruickshank, the botanic garden at Chanoury, Old Aberdeen, extending to six acres, and capable of accommodating nearly six thousand specimens. Miss Cruickshank has devoted to its endowment the sum of 15,000*l*.

MR. GILBERT R. REDGRAVE, Senior Chief Inspector in the South Kensington branch of the Board of Education, has been appointed an Assistant Secretary for Technology. Announcement is made that in the ensuing autumn the Duke of Devonshire will appoint a departmental committee, on which the county councils and the City and Guilds of London Institute will be represented, to consider, *inter alia*, the co-ordination of the technological administration of the Board of Education with the technological work at present carried on by educational bodies other than that Board.

A GOOD idea of the scope and value of the work of the examinations department of the City and Guilds of London Institute can be obtained from the "Programme of Technological Examinations (1900–1901)," published by Messrs. Whittaker and Co. Examinations are held in seventy technological subjects, and also in manual training (wood-work and metal-work). For each examination a syllabus is given, and a useful list of works of references ; and the questions and practical exercises set at the recent examinations are all reprinted. Several of the syllabuses have been revised, notably those of photography, pottery and porcelain, silk throwing and spinning, and silk weaving, electric lighting, watch and clock making, typography, lithography, carpentry and joinery.

THE Redruth School of Mines, of which the syllabus for 1900–1901 is before us, offers exceptional facilities for studying the principles of mining in the Cornish mining district. One wing of the school building is occupied by a large mineral gallery, erected to the memory of the late Dr. Robert Hunt, F.R.S. The museum, which contains a valuable collection of mineral specimens, and is the property of the Mining Association and Institute of Cornwall, is at all times accessible to students of the school. The mining course consists of practical underground work, including the timbering of shafts and levels, and of lectures on geology, the principles of mining, the raising and mechanical preparation of ores, and of practical work in gold panning and vanning. Students, in addition, are taught the methods of prospecting for minerals in all possible positions, and are trained to detect favourable indications on the surface. There is thus a reasonable combination of science with practice in subjects essential to the training of mining engineers.

WHAT school gardens are to children, allotments are to adults in agricultural districts, and both provide valuable means of experiment. The Report of the Technical Instruction Committee of the Oxfordshire County Council shows that this is well recognised is several parts of the county. For instance, at the Chipping Norton Agricultural Class there were fifty-four students of an average age of thirty-eight. They were factory hands, labourers, mechanics and small tradesmen, who all cultivated allotments, and were thus able to put principles to a practical test, and determine the causes thus affecting growth. At Reading College, which is connected with the Oxfordshire Committee, various insects and plants were received from different parts of the county for identification, and advice was given in many localities. Field experiments were made on sainfoin and lucerne, rotation, "finger and toe," mangel, and different manures for barley. Charlock spraying was investigated at three farms, and other experimental work had been done under the auspices of the College and the Technical Education Committee.

SCIENTIFIC SERIAL.

American Journal of Science, August.—Rowland's new method for measuring electric absorption and losses of energy due to hysteresis and Foucault currents, and on the detection of short curcuits in coils, by L. M. Potts. Rowland's method, in which the condenser is placed in one arm of a Wheatstone bridge, together with the fixed coils of an electrodynamometer, while the movable coil is mounted in the cross connection, is practically useful. The electric absorption always acts as a resistance in series with a capacity. The resistance is independent of the current, but the temperature has a decided effect on both.—Some new Jurassic vertebrates, by W. C. Knight.· The author describes two new species, called *Plesiosaurus shirleyensis* and *Cimoliosaurus laramiensis* respectively. They are in the collection of the University of Wyoming.—Carnotite and associated vanadiferous minerals in Western Colorado, by W. F. Hillebrand and F. Leslie Ransome. Carnotite is probably a mixture of minerals of which analysis fails to reveal the exact nature. Instead of being the pure uranyl-potassium vanadate, it is to a large extent made up of calcium and barium compounds. Near Placerville, Colorado, certain sandstones show a green colouring and cementing material which contains nearly 13 per cent. of V_2O_5. It is intended to work this sandstone for vanadium.—Restoration of Stylonurus Lacoanus, a giant arthropod from the Upper Devonian of the United States, by C. E. Beecher. The arthropod described takes equal rank with the Giant Spider Crab of Japan and the great "Seraphim" (*Pterygotus anglicus*). The animal has a length of nearly 5 feet, and with the legs extended it would measure about 8 feet.— Iodometric estimation of arsenic acid, by F. A. Gooch and Julia C. Morris.—Further notes on pre-glacial drainage in Michigan, by E. H. Mudge. The author discusses the present and former levels in the vicinity of the village of Saranac.

metres the air would consist of oxygen 0·3, nitrogen 4·6 and hydrogen 95·1 per cent.—On the dielectric cohesion of gases, by M. Bouty. When a gas contained in an insulating vessel is placed in a constant electric field, there is a certain critical pressure above which the gas acts as a dielectric, and below which the discharge passes. The relation between this critical pressure (p) and the field (y volts per centimetre) has been studied for three gases—hydrogen, air and carbon dioxide. For low pressures the relation found is $y = a + \dfrac{c^2}{p^2}$. For higher pressures the curve is practically coincident with the asymptote, $y = a + b(p + \pi)$.—On the extraction of oxygen from the air by solution at a low temperature, by M. Georges Claude. Various solvents for air have been tried at low temperatures in the hope of discovering a liquid in which the difference of solubility of the two gases would be very marked. The experiments, however, were unsuccessful, as it was found that at low temperatures the solubility of the nitrogen increased, so that starting with a mixture containing 65 per cent. of oxygen, after solution and boiling out, the amount of oxygen was practically unchanged, amounting in no case to more than 70 per cent.—On the pyrogallol-sulphonic acids, by M. Marcel Delage.—On the dextrins of saccharification, by M. P. Petit. The results obtained by the action of diastase upon starch were very divergent, depending upon the age of the diastase and the conditions under which it had been preserved. —On the use of sodium peroxide for making wholesome wells containing carbonic acid, by M. E. Derennes. The use of milk of lime for the absorption of dangerous amounts of carbonic acid contained at the bottom of a well has the disadvantage that the residual gas may consist almost entirely of nitrogen. The substitution of sodium peroxide for lime would ensure as much oxygen being given off as carbon dioxide absorbed.

SOCIETIES AND ACADEMIES.

PARIS.

Academy of Sciences, August 20.—M. Maurice Lévy in the chair.—New observations on the high valley of Dordogne, by M. A. Michel·Lévy. Owing to the cuttings recently made for the railway between Queuille and Mont-Dore, some new facts on the geology of this valley have been discovered. · On the left flank of the valley the deposit of labradorite can be traced up to the Capucin. More to the south an outcrop of trachyte, rich in black mica and amphibole, can be followed up to near the ravines of Riveaugrand. The right flank of the Mont-Dore valley shows clearly the prolongation of the lower andesite of the Grand Cascade. · A trachytic dyke has also been recently discovered by M. Paul Gautier in the first ravine west of Compissade, which is rich in granitic inclusions.—On the existence of *Ceratitis capitata*, var. *hispanica*, in the neighbourhood of Paris, by M. Alfred Giard. During the present spring a large proportion of the apricots at Courbevoie, near Paris, fell off the trees in a green state, and the remainder, although apparently exceptionally fine when ripe, where found to be honeycombed with larvæ. This larvæ were found, on development, to give rise to *Ceratitis capitata*, a species that has already been found to be very destructive to many kinds of fruit in the Azores, at Madeira, the Cape of Good Hope, Algeria and Malta. This is its first appearance near. Paris, possibly owing to an exceptionally favourable spring. Means for combating this scourge are suggested, as it is of the first importance that it should not become acclimatised in Paris.—Observations on shooting stars made from August 11 to August 14 at the Observatory of Paris, by Mlle. D. Klumpke. About thirty meteors were observed during four nights, of which some came from Perseus and others from the polar region. The former were white, short and very rapid, the latter luminous and coloured.—Observations of the sun made at the Observatory of Lyons with the Brunner equatorial during the first quarter of 1900, by M. J. Guillaume. The results are given in three tables showing the number of spots, their distribution in latitude, and the distribution of faculæ in latitude.—On the composition of the air in a vertical section, and on the composition of the upper layers of the terrestrial atmosphere, by M. G. Hinrichs. By applying a formula of Laplace, the composition of the air is deduced at different levels. From these calculations, carbon dioxide would disappear at 30,000 metres, argon at 60,000 metres. At 100,000

THURSDAY, SEPTEMBER 6, 1900.

A NEW DEPARTURE IN THE TEACHING OF ZOOLOGY.

Introduction to Zoology: a Guide to the Study of Animals, for the use of Secondary Schools. By C. B. Davenport, Ph.D., and Gertrude C. Davenport, B.S. Pp. xii + 412. (New York : The Macmillan Co. London : Macmillan and Co., Ltd., 1900.)

THE senior author of this book is well known in zoological circles for his two-volume work on "Experimental Morphology"—one of the most novel and ambitious of modern text-books ; and his wife, whose aid he acknowledged in its preface, now appears as joint author of the present equally ambitious production, which has for its object nothing short of a revolution of the methods of zoological teaching in vogue in the secondary schools of the United States. The key to the plan of the work and nature of its contents lies in the prefatory pronouncement that the "vast majority of secondary students are not to be zoologists, but rather men of affairs," and that "what the ordinary citizen needs" zoologically is (not a course in comparative anatomy but) "an acquaintance with the commonest animals"—a knowledge of " where else over the world the common animals of his State are to be found, and of how animals affect man," and that to know these matters is for him " more important than to know the location of the pedal ganglion of the snail." There can be little doubt that in this resolve the authors are in agreement with a large section of active teachers, but it must not be forgotten that the didactic system of laboratory instruction in vogue, against which they are in the long run entering a protest, has in its development become modified beyond the conceptions and intentions of its founders, and that as originally planned it did not ignore non-anatomical considerations to the extent their attitude implies. In their forcible recognition of the later tendency towards so doing, however, and their bold attempt to overcome it, they have performed a useful task, but experience can alone decide upon the wisdom of the remedy they propose.

Their book is of 336 pages, excluding appendices, and is divided into twenty-one chapters. The first four chapters deal with the Insecta, and then follow one each devoted respectively to the Myriapoda and Spiders, two each to Crustacea, Worms and Molluscs, one each to the Echinodermata, Cœlenterata and the Protozoa, and then a series on the Vertebrates taken in ascending order, the whole closing with a novel chapter on the frog's egg as a study in development. The plan adopted in each chapter is much as follows :—Opening with a concise statement of the systematic position and relationships of an order or other great group of animals conveniently selected (with a definition of its name usually in a footnote), there follows a very brief description of the habitus, and if so be the food and other special topics of interest, of one or more of its familiar species. There is then given a short descriptive account of its more familiar allies, and it may be of its development ; and the whole

chapter is brought to a close by an appendix, in the form of a key to the families of the order to which the type chosen belongs, or to the orders of the class or other great divisions of the group under consideration, while in places an accompanying key to the identification of members of allied subfamilies may be added or incorporated. The plainest and most concise terms are adopted, and there is a tolerably free use of illustration, preference being given to photographs of entire animals, often with their natural surroundings, in many cases with marked success ; and it cannot be denied that the authors have been desperately earnest in the task of selection and compilation. The body of the work is followed by three main appendices, of which the second is a bibliographic list embodying a none too fortunate selection of books of reference, the third a synopsis of the "animal kingdom," and the first an outline of a course of laboratory work upon the type-organisms selected as titular for the main chapters. Novelty here is as great as with the rest of the book, for in the "Exercises" prescribed, after each type-organism has been referred to its habitat, with brief directions for its capture and preservation in the living state where necessary or desirable, there follow instructions for drawing, and series of questions, framed with a view of compelling the observer to determine details for himself, and not of pointing out the precise nature and limits of the observation he is expected to make, as is customary with most laboratory treatises current. "Hints for observations on the living animal" usually follow, as do " Topics for further study." This very novel scheme is the outcome of experience gained while aiding in the conduct of the zoological affairs of the Harvard University, and as here delimited it is prescribed "for use in schools that can give to the subject five periods per week for half a year," at discretion and with modification determinable by local needs. The book thus embraces a very ambitious programme, and we question if the most hopeful aspect of the undertaking is not simply the better encouragement of field-work and of observation of nature in the open, in respect to which our existing methods do perhaps stand in need of reform.

It appears to us, however, that too much has been attempted within the limits of the book, that there is danger in its too frequent brevity of statement, and that it stands in need of a greater uniformity of treatment. What, for example, is to be gained by merely referring to the Tunicata as "Chordata which are either attached or form colonies, or both " (which is an erroneous statement), and as a group of Invertebrates which " lie nearest to the stem from which the Vertebrates arose," when whole paragraphs are given to far less generally important assemblages of forms? What also the use of defining the Stomatopoda as including " only Squilla," and then Cumacea, Isopoda and Amphipoda as embracing a number of forms? The inclusion of the Sponges in the Cœlenterata ; the old-fashioned classification of birds, with the Ratitæ referred to the order " Cursores"; the inclusion of the Bryozoa on one page among the Gephyrea and Leeches ("Annelida"), and on another among the " Scolecida"! can only be cited as examples of classificatory treatment sorely in need of revision ; while among definitions given we note a frequent lack of accuracy and precision, as, for example, with that of the Ophidia as having their "eyelids absent." The senior author's previous work explains the introduction of experimental observations of the antenniform-ophthalmite order, and brief note is taken of "variation" and abnormality. The social life and "language" of ants, protective resemblance and mimicry among the Lepidoptera, the habits of the spiders, and many other similarly fascinating topics receive in due course passing consideration. The reader will put down the book feeling the better for its perusal and with a desire to know more, while its "keys." to the identification of the common forms of life, oft overlooked because always present, but withal foremost in their claims on our attention, will prove useful and encouraging. We are doubtful, however, whether the authors would not have done better to have attempted less and that more uniformly, and whether they are justified in their refrain that in matters of elementary scientific education the mere "needs" of the ordinary citizen are to be alone gratified. We are by no means convinced that this argument is sound. Their method would seem likely to discount the teacher's important function of deciding what is to be left untaught—a matter of the utmost urgency in elementary scientific work. We shall watch with interest the development of their scheme.

COLOUR PHOTOGRAPHY.

A Handbook of Photography in Colours. By Thomas Bolas, Alexander A. K. Tallent and Edgar Senior. Pp. viii + 343. (London : Marion and Co., 1900.)

THE preface or introduction is written by the publishers, and is immediately followed by an index. Then follow three "sections." (1) 85 pages, by Mr. Bolas, on the "Historical Development of Heliochromy. General Survey of Processes. Direct Heliochromes on Silver Chloride." (2) 205 pages, by Mr. Tallent, on "Three-colour Photography." (3) 27 pages, by Mr. Senior, on "Lippmann's Process of Interference Heliochromy." Each section is quite distinct from the others, except that they are bound into one volume and indexed together ; there is therefore much repetition. For example, Maxwell's colour-sensation curves and Abney's revised curves are each given twice (the two renderings, by the way, are not identical), and Lippmann's formula for his emulsion is given at p. 55 and also at p. 332. Careful editing would have avoided such waste of space. Some of the diagrams are drawn with exceedingly thick lines, and are provided with very large heavy lettering, while others incline rather in the opposite direction. Some of the spectra as drawn for showing absorption, sensitiveness and so on, have the red to the left-, and others the red to the right-hand side ; some are normal, and others are as produced by prisms. It may be said that these are quoted from various sources ; but in a volume in which it is thought necessary to explain with a large diagram the refraction of light on passing from air into water, surely a little explanation of these differences is desirable. At p. 180, a spectrum which is normal is described as "prismatic." The volume appears not to have been edited at all, therefore the only way to

do justice to the authors is to regard it as three distinct books.

Mr. Bolas gives an excellent summary of the whole subject, both historical and practical. As a careful compiler should do, he has erred, if it be an error, in including too much rather than too little. Carey Lea's highly coloured partial reduction products of the halogen salts of silver might have been passed by, by some writers, as well as other references to conjectures. Zenker's work and Wiener's investigations are described intelligibly, although concisely ; indeed, the author has evidently spared no pains to give every one his due, and to use to the best advantage the small space at his disposal.

Mr. Tallent begins his section with several pages on the properties of light and the construction of ordinary spectroscopes—matter which, we think, might well have been omitted in order to make room for the treatment of subjects for which the reader is often referred to other books or articles. The peculiar firework-like diagram at p. 112, given to illustrate dispersion, is more likely to mislead than assist the student ; and some of the other diagrams might have been made more clear, in spite of the extraordinary boldness of the drawing and lettering. Mr. Tallent has gathered together a great deal of information about three-colour work, which he presents in the form of notes rather than as a treatise. It is doubtless advantageous in some cases to supply the raw material only, but the possession of bricks and mortar does not enable every one to build himself a house. If the very popular style of description sometimes adopted were given up in favour of more technical details, and if the practical applications of the various principles were more closely associated with the enunciation of the principles themselves, we think that the book would be more useful to the large majority of those who will read it. But we must be grateful to Mr. Tallent for having made a beginning in the getting together of the hitherto widely scattered items of the subject. His work must be of considerable assistance to any one following him, and we hope that later on he himself may be able to give us a treatise founded on these notes.

Mr. Senior treats only of Lippmann's interference process, and he writes on this with authority, for he has given the matter much practical attention, and has produced some of the best examples that have been seen. He gives his formulæ and methods apparently without any reserve, as well as the published formulæ of other notable workers. He precedes the practical details with a few pages on the optical principles involved, setting forth clearly the character of "stationary waves." We think that most people reading p. 323 would consider it as showing that the colours reflected from a Lippmann photograph are complementary to those of the objects photographed, but it is quite obvious that Mr. Senior does not intend to convey this impression.

The publishers, in their preface, state that thirty-one years ago they published the pioneer work on photography in colours (by Ducos du Hauron), and they feel satisfaction now in following up the line they "opened up over a quarter of a century ago." All who are interested in the subject will feel thankful to Messrs. Marion and Co. for having done so.   C. J.

OUR BOOK SHELF.

Probleme Kritische Studien über den Monismus. Von Dr. H. v. Schoeler. Pp. viii + 107. (Leipzig : Engelmann, 1900.)

DR. VON SCHOELER'S critique of monism is the work of a mind which, with all reverence for scientific fact, has nothing but disdain for post-scientific theory. The monism of which Haeckel, for instance, or Romanes is the exponent, has captured many scientific intellects, but it is in Dr. von Schoeler's view an arbitrary conceptual construction. It leaves the problems still with us. Of the ultimates of physics, of biology, of psychology, we know but little, and that little makes against such monism. Life is more than knowledge, and will find emancipation neither in science nor in religion, but in art.

Dr. von Schoeler's discursive criticisms of a variety of attempts at construction, or steps towards construction, are of somewhat unequal value, but not uninteresting. They enter into some detail, and might well give pause to any one who is inclined to build a system without straw to his bricks. But, if it be true that a man's *nay* has no meaning apart from the underlying and implied *yea*, the unsatisfactory character of our author's positive teaching must reflect upon his polemics. The way in which he couples Plato and W. K. Clifford on p. 95 suggests doubts as to his insight. Those who allow that current mechanical theories are tending more and more to immaterialism, that organic evolution reduced to its facts cannot assert even descriptive continuity unhypothetically, and that for a deduction of consciousness from the non-conscious we have not so much as the point of departure, will be left cold by Dr. von Schoeler's appeal for Diploïsmus, and his enthusiasm for Goethe's *Weltanschauung* made complete in the light of Kant's. The duality for which he goes to Bruno's *gemini efficientes* falls even for Dr. von Schoeler within a monism (v. p. 97, ll. 1–4). And Kant's provisional dualism does not exclude a hypothetical monism, idealist or agnostic. Only the resolution of it from the standpoint of understanding is not achievable, and from any other standpoint it cannot be more than hypothetical.

It is doubtful, finally, whether Dr. von Schoeler quite understands in what sense a system admits certain unexplained points as truly problems, and in what sense it claims to solve them. Is the origin of motion a problem recognised by monism as one which it must attempt to solve? And is not metaphysic always, so to speak, a *post mortem* examination? The "gray in gray" of philosophy is a commonplace : "the owl of Athene wings for flight only when twilight falls."   H. W. B.

Diamond Drilling for Gold and other Minerals. By G. A. Denny. Pp. x+158. (London : Crosby Lockwood and Son, 1900.)

AT the Paris Exhibition of 1867 much interest was aroused by Leschot's method for cutting through hard rock by diamonds in rapid rotation. Originally intended for use on a small scale, this method was soon applied by Major Beaumont and others to deep boring ; and the great improvements made of recent years in the construction of the instruments used, and the large amount of experience that has been gained by their general use in mining districts, have added so much to the importance of the subject of boring, that it is no longer possible to deal with it adequately in a chapter of a general treatise on mining. An independent work is needed. In German, this exists in Tecklenburg's monumental work. In English, however, Mr. Denny's handbook is the first to give a detailed account of the use of modern diamond core-drills in searching for mineral deposits. The work, which covers 158 pages, contains much information of a practical character, including particulars of the cost of

apparatus and of working. It is, unfortunately, limited in its scope. South African conditions are alone considered, and the descriptions of the drills are confined to machines made by two American firms. The numerous well-designed drills of English and Continental make are not mentioned. The work cannot, however, fail to be of value to any one contemplating using diamond-drilling machines for the examination of mineral lands in South Africa.

The author gives some interesting results, deduced from his own experience, of the rate of progress of machine diamond drilling in various rocks. For holes up to 1000 feet he finds that, including all normal delays, the rate averages per week : in limestone, 150 to 200 feet ; in Carboniferous sandstone, 150 feet ; in slate, 100 to 150 feet ; in greenstone, 110 feet ; in basalt, conglomerate, diabase, diorite and dolomite, 100 feet ; in porphyry, 90 feet ; in quartz, 85 feet ; in granite, 73 feet ; and in chert, 60 feet. As regards the cost of drilling, the author points out that diamond drilling on the Witwatersrand is almost always done by contract, the reasons being that men with the requisite experience are not easily secured, and that a mining company rarely has sufficient work to justify the outlay upon the plant. The average tender for a hole 3000 feet deep would work out at about 37*s*. 6*d*. per foot, the price being fixed on a sliding scale, say 25*s*. per foot for the first 500 feet, and rising by 5*s*. per foot every 100 feet. If the company undertook the work on its own account, the cost per foot would be about 26*s*. 8*d*., or 3950*l*. for the 3000 feet. This with 2340*l*., the cost of drilling outfit, brings the total cost of the hole to 6290*l*., as compared with 5625*l*., the contractors' price. It is to be regretted that the author has not compared these prices with those obtaining elsewhere. The borehole at Paruschowitz in Upper Silesia, for example, the deepest in the world, completed to a depth of 6566 feet in 1893, was bored by the diamond drill. The average rate of progress was 16½ feet a day, and the cost was 3761*l*., or 35*s*. a yard. Details of the average working cost of diamond drills in Victoria and New South Wales, which are published annually in the Government reports, might advantageously have been cited for purposes of comparison.   B. H. B.

Symons's British Rainfall, 1899. Compiled by H. Sowerby Wallis. Pp. 251. (London : Edward Stanford, 1900.)

A PORTRAIT of the late Mr. G. J. Symons, the founder of the British Rainfall Organisation, and an appreciative tribute to his memory, are naturally found in this volume —the first to appear without his name upon the title-page. The thirty-nine volumes for which Mr. Symons was responsible form a real monument to his industry and scientific work. Mr. Sowerby Wallis, who was associated with him for thirty years, will continue the work along the lines which have hitherto proved so successful.

The usual particulars are given concerning the rainfall and meteorology of various parts of the British Isles during the year 1899, as observed at about 3500 stations. The average rainfall of the ten years 1890–1899 is discussed in an article, the values being given for a hundred stations well distributed over the three kingdoms. The discussion is only provisional, but so far as it has gone it indicates that over a large part of the kingdom the rainfall in the period considered was deficient by from 5 to 10 per cent. and upwards. Over an area of about 300 miles long by 100 miles wide, stretching right across the country from south-west to north-east, the fall for the period shows a deficiency of 10 per cent. or more. In other words, accepting the values discussed, it appears that little more than eight and a half years' rain fell over a large part of Central England in the ten years 1890 1899.

LETTERS TO THE EDITOR.

[*The Editor does not hold himself responsible for opinions expressed by his correspondents. Neither can he undertake to return, or to correspond with the writers of, rejected manuscripts intended for this or any other part of* NATURE. *No notice is taken of anonymous communications.*]

Railways and Moving Platforms.

IN reference to Prof. John Perry's letter in NATURE of Aug. 30 on the subject of "Railways and Moving Platforms," will you kindly allow me to state that I worked out a scheme for moving platforms for railways in India in the year 1877, and that a paper of mine on the subject was published in the professional papers of the Thomason Civil Engineering College at Rurki in India in 1877 or 1878?

My plan was to have alongside of the main line a length of about a mile of level line, with a steep incline at either end, for the moving platform.

The moving platform would acquire sufficient speed on these inclines to run alongside of, and be made fast to, a train on the main line running in the same direction at reduced speed.

After the platform had been made fast, and passengers and crates of luggage transferred to a similar platform on the train, the train was to increase its speed, and finally release the platform with sufficient way on to carry it up the incline at the other end of its run, so as to be in readiness for the next train from the opposite direction.

This represented the case for moving platforms in its simplest form to suit Indian traffic. Details, of course, can readily be supplied.   W. SEDGWICK

September 3.   (Lt.-Col. late R.E.).

The Migration of Swifts.

REFERRING to the letter of Mr. O. H. Latter in NATURE of August 30, there are swifts' nests in the roof of my house, and the birds went to their nests as usual on the evening of August 11, but on the following evening, Sunday, August 12, they had gone. In other houses close at hand swifts remained on the 13th and 14th, but by the evening of the 15th all had disappeared. On one occasion some years ago a solitary swift remained after his companions had departed, and consorted with the neighbouring house martins until the last week in September.

A curious occurrence was observed by me on May 26 last, when the swifts had already been for some time with me. But on the afternoon of that day I was on board the Transatlantique Company's steamer crossing the Mediterranean from Marseilles to Algiers in order to observe the total eclipse of the sun. The boat left Marseilles at one o'clock, and at 4h. 40m., by which time we were seventy miles from the land, a flock of birds was seen following the ship in the distance. In a few minutes they were flying round the ship, and turned out to be common swifts. They were estimated to be about 200 in number. They gradually forged ahead, leaving the ship behind, and in a few minutes were lost in the distance. Several of the passengers made a note of the occurrence. The course followed by the swifts was one or two points to the right of the ship's course, or about S.S.W., in the direction of the Balearic Islands.

       WILLIAM ANDREWS.

Steeple Croft, Coventry, September 1.

The Reform of Mathematical Teaching.

THE deeply interesting letter of Mr. David Mair on the above subject in a recent number of NATURE will no doubt be attentively perused. Many, I imagine, could echo his experience when he says, speaking of 100 boys at a "well-taught school," "hardly one failed to write out the construction and proof, but only one of the hundred carried out the practical construction. Clearly our present Euclidian teaching has little to do with geometry."

Grave dissatisfaction in regard to the study of Euclidian geometry has long been voiced in Great Britain and in France, and the reason for the continued lamentable state of things in that special scientific direction is not, I think, far to seek. When Euclid, as a practical teacher, made use of his Propo-

sitions, theory and practice had doubtless that close association which Mr. Mair appears rightly to desire as a remedy for the present extraordinary condition of Euclidian study.

Now, if we assume that Euclid's own pupils were first of all drilled in a thorough knowledge of the sub-geometric properties of the cube, we are led largely to modify the supposed indictment against Euclid's method.

Again, Euclid seems to me to have used constructions, not only of a strict, but of a particular or edificial nature ; so that, properly viewed, each succeeding proposition carried, not only its specific teaching or lesson, but beautiful groups also of surrounding truths of an implicit character.

If the foregoing remarks be valid, it is easy to see that only two serious faults can be charged against Euclid's method—namely, the use of false diagrams and indirect demonstrations. The former, of course, being misleading, are not in harmony with the modern Herbartian cumulative principles of "educative instruction."

As to the second fault, the splendid argumentation of the famous Simson in his "Notes" will not, I fear, greatly avail in Euclid's favour. Mr. Mair rightly vindicates for the pupil the power of "educational interest," another Herbartian doctrine, by the by ; and I agree with him that "geometry is in worse case than algebra," because with the latter, I opine, the notion of "zero" inevitably leads the mind towards the necessary cubical standard.

No one could reasonably wish the true Euclidian or edificial geometry to be suppressed, but might without presumption press for a reform in the way of presenting all geometrical truths by a direct reference to those sub-geometric conditions which have equal rank with the oldest of nature's immutable laws.

If for any school or pupil, instead of enlarging the mental vision, and creating intellectual joys, the Euclidian tasks savour of prolonged drudgery, mystery and confusion, as they assuredly often do, such toil should perforce be abandoned to prevent, at least, waste of time and energy, not to speak of the gain to human happiness.

Possibly in the broad expanse of the educational economy the universities of the world may eventually suggest practical reforms of a vital kind in the teaching methods that affect mathematical science.

I am not competent to pursue the subject to any length, but I might remind students that, although by a sort of repartee to a kingly personage Euclid said there was "no royal road to geometry," he did not say there was no royal gate to it, and that I trust I have, however faintly, herein indicated.

August 25.   HENRY WOOLLEN.

The Trembling of the Aspen Leaf.

IT is well known that the vibratory motion of the leaf of the aspen and other poplar trees is caused by a flattening of the petiole at its junction with the lamina. The lower part of the leaf-stalk is elongated and rigid, thus forming a basis upon which the flattened portion of the stalk can, in virtue of its elasticity, move to and fro as the wind acts upon the leaves of the tree. It is stated by Kerner that this adaptation prevents the leaves striking against each other and the branches of the tree, whereby they might get bruised. In this connection it is a noticeable fact that the poplars which exhibit this property of leaf-vibration most strongly have sparsely distributed leaves, and the foliage is scanty in comparison with other trees of the genus, especially the abele, or white poplar.

With reference to the abele, it may be noticed that the leaves which possess the trembling motion in a slight degree are covered with a white felting, which Kerner asserts is an adaptation to protect the stomata from the excess of atmospheric moisture which prevails in the damp situations and river-sides where this tree commonly grows. Yet, it may be noted, the aspen is found in precisely similar conditions, and must be affected injuriously by any influence that would act unfavourably upon the abele. Now, it has occurred to me that the real use of the vibratory motion of the aspen leaf may be an adaptation to throw off rapidly the excess of aqueous condensation liable to take place upon the foliage of trees growing in marshy situations. I should very much like to hear the opinions of some of your readers on this point, which, if a true solution of the matter, is interesting from the fact that the same end is secured by very different means in two plants of the same genus, and so very much alike in all other particulars.

Hobart, July 25. HENRY J. COLHOURN.

Electricity direct from Coal.

WITH reference to the announcement made in the *Daily Mail* of September 1, that Thomas A. Edison had completed a machine for the generation of electric power direct from coal without the use of engines or dynamos, may I ask you to reprint a few lines from an article on electric traction which I wrote for you, and which you published on April 12, 1894?

"Before electric traction can be employed on a very large scale, we must possess a means of producing the electricity on the spot and at the time it has to be used, or, in other words, we must possess a battery in which the energy of coal can be transformed directly into electric current, so that we may do without storage batteries in which to carry electric energy about, or heavy copper conductors through which to convey it at moderately low tension from the spot where it is produced to where it is used, or light aerial conductors through which to convey it at high tension.

"How long we shall be without this, or how many minds are engaged in the solution of this or some such problem, we know not, but the moment it is solved, and solved doubtless it will be, there will be such a transformation scene in the industrial applications of electricity as one can hardly conceive. It would mean that for almost all purposes except those in which heating is required electricity would or could be used. An electric light-producing battery in every house, quite independently of any mains in the streets; an electric power-producing battery to carry us whither we would on rails or on the streets; and in every house to put an end to all the evils attendant on crowded factories and workshops, in crowded streets and towns; such and other advantages would result from turning electricity from a servant into a master, from a mere transformer of energy into a source of energy."

E. F. BAMBER.

48 St. James's Square, W., September 3.

Artificial Deformations of Heads, and some Customs connected with Polyandry.

WITH reference to your note on M. Charles de Ujfalvy's recent article in *l'Anthropologie* (p. 323, *ante*), I may be allowed to call your attention to the ancient Korean practice of artificially deforming their heads, which was apparently similar to the method adopted by the Huns as well as the Hûna kings of India. Thus, the Chinese "History of the Later Han Dynasty," written in the fifth century, *sub.* "Eastern Barbarians," says: "The people of Ma-Kan (in the south-western part of the Korean peninsula) wish their heads flat; so the head of every child just born they compress with stone to deform it."

The special horned head-dresses worn by the polyandric women of the White Huns put in mind the old Japanese usages, described by Fujioka and Hirade in their "History of the Japanese Customs and Manners," Tokio, 1897, vol. i. p. 169: "In the festival of the god of Tsukuma, every woman had to go in procession after the holy sedan-chair, with a number of pans on her head proportionate to her immoralities. In the temple of Usaka, while the priest was praying in a feast-day, every woman was scourged on similar principles."

KUMAGUSU MINAKATA.

1 Crescent Place, South Kensington, August 11.

Huxley and his Work.

ON p. 13 of "One Hundred and One Great Writers," issued by the *Standard*, and presumably edited by Dr. Garnett, occurs the following remarkable account of Huxley and his work. "Huxley's work is that of the populariser, *the man who makes few original contributions to science or thought*, but states the discoveries of others better than they could have stated them themselves." Comment is needless. F. W. HENKEL.

Markree Observatory, Collooney, Ireland, August 23.

THE CAUSES OF FRACTURE OF STEEL RAILS.

WHEN the down Scotch express was running through St. Neots station, on the Great Northern Railway, on December 10, 1895, a rail broke into seventeen pieces, part of the train left the metals, and a serious accident resulted. Several features of the report on the mishap, drawn up in the ordinary course by the late Sir Francis Marindin, might well have occasioned deep thought, notably the conclusion that the first fracture of the rail took place over a chair at a minute induced flaw, which did not exist when the rail was manufactured.

The whole report, however, is suggestive rather than explanatory, and the result was the appointment by the Board of Trade of a committee to investigate the question of the loss of strength of steel rails caused by prolonged use. The committee was a very strong one, and contained some distinguished steel manufacturers, engineers, metallurgists and chemists. They collected a vast amount of information, much of it apparently considered unsuitable for publication, and made long series of experiments, many of them, judging from the report, more easily made than their results explained.

Finally, after four years' work, they have issued a report with the satisfactory feature that practically no change is recommended to be made in the mode of management of the permanent way by the railway companies.

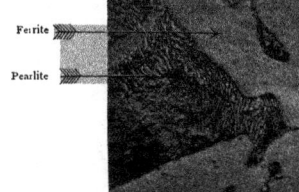

FIG. 1.—Steel rail. × 850 D. Showing pearlite and ferrite.

Nevertheless, although no legislation seems likely to result from the labours of the committee, the evidence that has been collected and published in the appendices to the report is of great scientific interest. The experimental work that was undertaken was divided among the members of the committee. A number of rails found broken on the road, or discarded as worn out, were selected for examination. Prof. Unwin took charge of the tests on their hardness, tensile strength and bending strain, and Mr. Windsor Richards of those on their resistance to the shock of falling weights; Sir William Roberts-Austen made micrographical examination of the rails, and Dr. Thorpe analysed them. Sir Lothian Bell includes in his comprehensive memorandum details of a number of mechanical tests on rails, and Prof. Dunstan gives an interesting account of the effects of atmospheric corrosion.

Interest naturally centred around the St. Neots rail. It was found to be of ordinary composition, and the mechanical tests applied to it showed that the steel was of variable, but, on the whole, of good quality. It was only on microscopic examination that the extraordinary character of the rail became evident. Good rail steel, according to Sir W. Roberts-Austen, consists of "ferrite," or iron free from carbon, and "pearlite," which is a mixture of alternate bands of ferrite and "cementite" (the carbide corresponding to the formula Fe_3C). The structure is shown in Fig. 1, a reproduction of a micrograph of rail steel magnified 850 diameters, in which the light constituent is ferrite, and the alternate bands of ferrite and cementite together make up the constituent pearlite. Well-developed pearlite with a conspicuous banded structure is characteristic of good rail steel. It is the form of carbide produced by slow cooling. When, however, steel is hardened by "quenching," pearlite is no longer to be found, and the whole mass consists of interlacing crystalline fibres devoid of banded structure, and is called "martensite."

With regard to this, Sir William Roberts-Austen says : "The detection of martensite in a rail should at once cause it to be viewed with extreme suspicion, as showing that the rail is too hard locally to be safe in use." An examination of the running edge of the St. Neots rail (Fig. 2) showed that a surface layer of about 1/100th of an inch thick existed, in which the carbide was mainly in the form of martensite.

This surface layer forms the lighter upper part of the figure, while the darker portion below it shows the usual assemblage of pearlite and ferrite granules. The actual running edge is shown against the dark space at the top

shock to spread through the mass. The induced flaws of Sir Francis Marindin might thus be explained.

On the other hand, a warning note is struck by Prof. Unwin, who points out that minute transverse fissures are common on the rolling surfaces of old rails ; and as only one rail in 2000 or 3000 breaks on the road, the others being discarded as worn out, it must be rare for such fissures to spread into the substance of the rail. It is suggested that if much silicon is present, the spreading of fissures becomes more rapid ; but the St. Neots rail contained only 0·09 per cent. of silicon, an amount well within the limits of composition put forward by Messrs. Windsor Richards and Martin as suitable for rail steel.

On considering the problem of how martensite can be formed on the rolling surfaces, it is again evident that the St. Neots rail is a remarkable exception. All old rails become "hammer-hardened" on the surface by long use, and their strength and percentage of elongation may be increased by annealing, but no clear case of any production of martensite in this way could be obtained. The effect produced by the "cold-rolling" action of

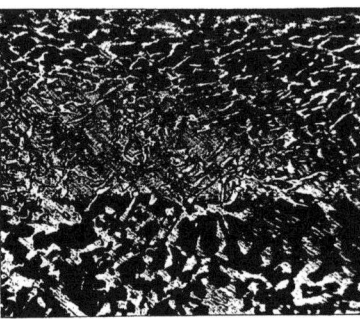

FIG. 2.—St. Neots rail, running edge. Pearlite passing into martensite.
X 140 D.

FIG. 3.—St. Neots rail, rolling surface.   X 5 D.

of the figure. Martensite was also found in small patches in some other worn and broken rails, but to a far less extent, the St. Neots rail being unique in this respect.

The questions that naturally presented themselves on this discovery were, first, How far would this structure account for the brittleness of the rail ? and secondly, How had the martensite been produced ? With regard to the first question, the rolling surface of the St. Neots rail was found to be traversed by a number of transverse cracks (Fig. 3), some just passing through the hardened skin, others running into the substance of the rail. The upper surfaces of rails are subject to tension over chairs by the weight of passing trains applied between the chairs, and cracks are formed in this way. To realise the importance of these cracks, it is only necessary to turn to Mr. Martin's memorandum. He found that a heavy steel rail nicked with a chisel to a depth of 1/64th of an inch broke under the weight of six hundredweight let fall from a height of twelve feet, while the same rail, if not previously nicked, resisted successfully the fall of a ton weight from a height of twenty feet. The loss of strength due to these minute cracks is therefore amazing, and can only be accounted for on the hypothesis that shallow nicks are readily induced by

passing trains on steel is shown in Fig. 4, which is a photograph enlarged 140 diameters of the running edge of a rail taken up after ten years' wear. The direction of the granules is changed in the surface layer, but otherwise the structure is unaltered. Roberts-Austen succeeded in producing a structure like that of the St. Neots rail, but only by local heating with an electric arc. With regard to this, he observes that this experiment "points to the probability that local heating of a rail by skidding, followed as it would be by the rapid abstraction of the heat by the mass of the cold rail, can produce patches of martensite, though it may be very difficult in the laboratory to imitate the actual conditions by mechanical means." He seems to think that "the structure of the St. Neots rail would point to the changes having been effected while the rail was actually in use."

Although the St. Neots rail has thus baffled the committee, inasmuch as they cannot say definitely whether other rails are likely to be altered in the same way, it is evident that one of the most important results of their labours will prove to be the full realisation of the fact that steel possesses a complex structure which can be studied with the microscope, that this structure varies greatly with the mechanical and thermal treatment to which it has been subjected, and that the durability of the rail depends on its structure.

Apart from the micrographical appendices, much interesting information may be obtained from a study of the mechanical tests, and some of the conclusions drawn by Prof. Unwin from these are among the most definite in the report, though they do not go far to explain the St. Neots mishap. It is found, for instance, that rails

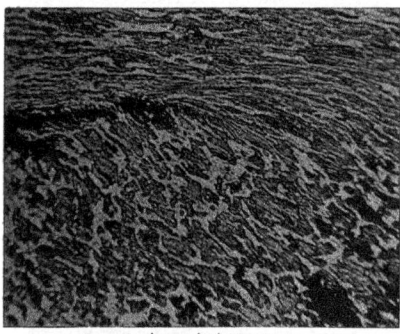

Fig. 4 —Running edge showing "flow." × 140 D.

generally break near their ends "owing to greater straining action due to discontinuity at the joint," and that the fish-joints are an unavoidable source of danger. It is also found that a rail is more liable to break when its worn head is turned down, as usually happens after a few years' use in the case of double-headed rails. Considerations of space alone prevent these points from being dealt with at greater length.

THE BRADFORD MEETING OF THE BRITISH ASSOCIATION.

THE final arrangements for the Bradford meeting of the British Association are now complete, and there is every indication that the gathering will be one of the largest that has been held in recent years. Representatives of scientific institutions are coming from nearly every country in the world; and there are delegates from the United States, Canada, the Cape, New Zealand, the West Indies, India, France, Germany, Russia, Denmark and Sweden, Spain, Italy and Greece. The Bradford people have come forward in a most willing manner to offer hospitality to the visitors, and a large proportion of the strangers will at any rate have had the opportunity of accepting private hospitality.

In our last article we dealt with the excursion programme. We propose now to say something about the various social functions which have been organised for the week.

The first social gathering will be a reception at the Municipal Technical College this afternoon (September 6). Mr. W. E. B. Priestley, the chairman of the Technical Instruction Committee, will welcome the visitors; and, after partaking of afternoon tea, they will be escorted in small bodies round the building, to see the textile exhibition, and the various processes of the textile industries. On the evening of the same day (Thursday)

the Mayor and Mayoress have invited the Association to a conversazione in St. George's Hall. The building will be elaborately decorated, and music will be provided by a large string band, under the conductorship of Mr. I. Shepherd. There will be exhibits of various scientific novelties in different parts of the building; and the galleries will be specially levelled up for refreshment and supper buffets. The 2nd West York Artillery Volunteers are providing a Guard of Honour to line the staircase.

At the conclusion of Prof. Gotch's lecture on Friday night there will be a smoking concert at the Technical College in honour of the President, for which various well-known elocutionists have been engaged.

On Monday, September 10, the Mayor and Corporation are inviting all persons attending the meeting to a large garden-party in Lister Park. The portion of the park where the guests will chiefly collect will be that above the lake. Around or in this space there will be several refreshment tents, and in front of each will be little tables in the open, surrounded by chairs, after the style of the foreign *cafés*. The lake will be decorated by means of Venetian masts and flags; while some new boats will be out, with boatmen in suitable garb in charge of them. The Black Dyke Band plays near the lake, and the band of the Bradford Rifles at the high end of the park. Archery and other amusements will be provided; and in one corner there will be some ballooning experiments under the direction of the Rev. J. M. Bacon, who is well known just now in connection with the trials which, in conjunction with Admiral Fremantle, he has been making in improved military signalling from balloons. Probably Mr. Bacon will be accompanied by the Admiral. It is proposed to erect a 70-foot pole about 30 yards from the balloon, and another, of equal height, in the furthest corner of the park; and, somewhere between the two, it will be attempted to explode a mine by means of wireless telegraphy, an experiment which was recently successfully performed at Newbury. Mr. Nevil Maskelyne will take part in the wireless telegraphic experiments, in order to exhibit the new receiver which he has patented and sold to Lloyd's.

On Tuesday the Mayor and Corporation are inviting the Association to a soirée in St. George's Hall. The arrangements will be somewhat similar to those on the occasion of the Mayor's function on the preceding Thursday, excepting that the music will be provided by the band of the Artillery Volunteers.

On Wednesday various private garden-parties will take place. Mrs. Henry Illingworth has invited 150 members of the Association to visit her grounds, for tea, tennis and croquet; and there will be music for those who prefer to rest after their labours. There will also be a garden-party at Ferniehurst, Baildon, by the kind invitation of Mr. and Mrs. G. C. Waud; a procession of prize-winning hackneys will take place in the course of the afternoon; and also sheep-dog trials, for which most of the celebrated dogs in the North Country have been brought together.

Messrs. Wm. Fison and Co. have also invited a hundred members to a garden-party at Greenholme, where, after tea, they will have an opportunity of visiting the turbine machinery and their textile works. Another function on the same day will be a garden-party at Royds Hall, by the kind invitation of the Low Moor Iron Company. The visitors will first be taken round the foundries to see some smelting work, and to examine some of the most striking parts of the machinery.

On the evening of Wednesday (the 12th) there will be

a grand concert in St. George's Hall, at which Madame Ella Russell will be the *prima donna*. The Bradford Permanent Orchestra, under the conductorship of Mr. Frederick Cowen, will render a very interesting programme ; and the Festival Choral Society are to perform certain celebrated items from the works of Handel, Wagner and Sullivan.

A temporary museum, illustrative of papers read before the Sections, will be provided in the Girls' Grammar School. Its primary object is to afford a space where those who read papers can deposit any specimens or apparatus which they may show, and which thus can be examined more at leisure than in the Section Room. In addition, the Bradford officials have endeavoured to get together some choice geological specimens of local interest, which will illustrate the discussions to take place in the Geological and Botanical Sections on the origin of coal ; and also specimens illustrating the reef-knolls of the Craven district, which have been the subject of controversy in recent years. Further, Mr. Butler Wood, the chief librarian under the Bradford Corporation, has taken under his care a collection of local pre-historic specimens.

Amongst the readers of papers, Mr. J. J. Stead, of Middlesbrough, will exhibit specimens of metals, treated by a peculiar method which he has discovered. There will be numerous maps, principally geological—several dealing with the investigation of the underground watercourses of Malham and Ingleborough. There will also be a contribution, by Mr. A. D. Ellis, of a number of rare and valuable atlases, of the sixteenth to the eighteenth centuries ; while various lantern slides used in the Sections will be shown on transparent screens. In addition, the Science and Art Department have promised the loan of the best work that was done in the recent National Competition ; and the collection from South Kensington is expected to be of an exceptionally attractive character, representing, as it does, the cream of the best work of the Art Schools of the country. It is intended that the exhibits shall be largely illustrative of the products of Bradford, and that they shall also illustrate, to a certain extent, the work that is being done at the Municipal Technical College.　　　　RAMSDEN BACCHUS.

INAUGURAL ADDRESS BY PROF. SIR WILLIAM TURNER, M.B., D.C.L., D.Sc., F.R.S., PRESIDENT OF THE ASSOCIATION.

TWENTY-SEVEN years ago the British Association met in Bradford, not at that time raised to the dignity of a City. The meeting was very successful, and was attended by about 2000 persons—a forecast, let us hope, of what we may expect at the present assembly. A distinguished chemist, Prof. A. W. Williamson, presided. On this occasion the Association has selected for the presidential chair one whose attention has been given to the study of an important department of biological science. His claim to occupy, however unworthily, the distinguished position in which he has been placed rests, doubtless, on the fact that, in the midst of engrossing duties devolving on a teacher in a great University and School of Medicine, he has endeavoured to contribute to the sum of knowledge of the science which he professes. It is a matter of satisfaction to feel that the success of a meeting of this kind does not rest upon the shoulders of the occupant of the presidential chair, but is due to the eminence and active co-operation of the men of science who either preside over or engage in the work of the nine or ten sections into which the Association is divided,· and to the energy and ability for organisation displayed by the local Secretaries and Committees. The programme prepared by the general and local officers of the Association shows that no efforts have been spared to provide an ample bill of fare, both in its scientific and social aspects. Members and Associates will, I feel sure, take away from the Bradford Meeting as pleasant memories as did our colleagues of the corresponding Association Française, when, in friendly collaboration at Dover last year, they testified to the common citizenship of the

Universal Republic of Science. As befits a leading centre of industry in the great county of York, the applications of science to the industrial arts and to agriculture will form subjects of discussion in the papers to be read at the meeting.

Since the Association was at Dover a year ago, two of its former Presidents have joined the majority. The Duke of Argyll presided at the meeting in Glasgow so far back as 1855. Throughout his long and energetic life, he proved himself to be an eloquent and earnest speaker, one who gave to the consideration of public affairs a mind of singular independence, and a thinker and writer in a wide range of human knowledge. Sir J. Wm. Dawson was President at the meeting in Birmingham in 1886. Born in Nova Scotia in 1820, he devoted himself to the study of the Geology of Canada, and became the leading authority on the subject. He took also an active and influential part in promoting the spread of scientific education in the Dominion, and for a number of years he was Principal and Vice-Chancellor of the M'Gill University, Montreal.

Scientific Method.

Edward Gibbon has told us that diligence and accuracy are the only merits which an historical writer can ascribe to himself. Without doubt they are fundamental qualities necessary for historical research, but in order to bear fruit they require to be exercised by one whose mental qualities are such as to enable him to analyse the data brought together by his diligence, to discriminate between the false and the true, to possess an insight into the complex motives that determine human action, to be able to recognise those facts and incidents which had exercised either a primary or only a secondary influence on the affairs of nations, or on the thoughts and doings of the person whose character he is depicting.

In scientific research, also, diligence and accuracy are fundamental qualities. By their application new facts are discovered and tabulated, their order of succession is ascertained, and a wider and more intimate knowledge of the processes of nature is acquired. But to decide on their true significance a well-balanced mind and the exercise of prolonged thought and reflection are needed. William Harvey, the father of exact research in physiology, in his memorable work, "De Motu Cordis et Sanguinis," published more than two centuries ago, tells us of the great and daily diligence which he exercised in the course of his investigations, and the numerous observations and experiments which he collated. At the same time he refers repeatedly to his cogitations and reflections on the meaning of what he had observed, without which the complicated movements of the heart could not have been analysed, their significance determined, and the circulation of the blood in a continuous stream definitely established. Early in the present century, Carl Ernst von Baer, the father of embryological research, showed the importance which he attached to the combination of observation with meditation by placing side by side on the title page of his famous treatise, "Ueber Entwickelungsgeschichte der Thiere" (1828), the words *Beobachtung und Reflexion*.

Though I have drawn from biological science my illustrations of the need of this combination, it must not be inferred that it applies exclusively to one branch of scientific inquiry ; the conjunction influences and determines progress in all the sciences, and when associated with a sufficient touch of imagination, when the power of seeing is conjoined with the faculty of foreseeing, of projecting the mind into the future, we may expect something more than the discovery of isolated facts ; their co-ordination and the enunciation of new principles and laws will necessarily follow.

Scientific method consists, therefore, in close observation, frequently repeated so as to eliminate the possibility of erroneous seeing ; in experiments checked and controlled in every direction in which fallacies might arise ; in continuous reflection on the appearances and phenomena observed, and in logically reasoning out their meaning and the conclusions to be drawn from them. Were the method followed out in its integrity by all who are engaged in scientific investigations, the time and labour expended in correcting errors committed by ourselves or by other observers and experimentalists would be saved, and the volumes devoted annually to scientific literature would be materially diminished in size. Were it applied, as far as the conditions of life admit, to the conduct and management of human affairs, not require to be told, when critical periods in our welfare as a nation arise, that we shall muddle through somehow. Recent experience has taught us that wise discretion and careful prevision are as necessary in the direction of public affairs as in the pursuit of science, and in both instances, when properly exercised, they enable us to reach with comparative certainty the goal which we strive to attain.

Improvements in Means of Observation.

Whilst certain principles of research are common to all the sciences, each great division requires for its investigation specialised arrangements to insure its progress. Nothing contributes so much to the advancement or knowledge as improvements in the means of observation, either by the discovery of new adjuncts to research, or by a fresh adaptation of old methods. In the industrial arts, the introduction of a new kind of raw material, the recognition that a mixture or blending is often more serviceable than when the substances employed are uncombined, the discovery of new processes of treating the articles used in manufactures, the invention of improved machinery, all lead to the expansion of trade, to the occupation of the people, and to the development of great industrial centres. In science, also, the invention and employment of new and more precise instruments and appliances enable us to appreciate more clearly the signification of facts and phenomena which were previously obscure, and to penetrate more deeply into the mysteries of nature. They mark fresh departures in the history of science, and provide a firm base of support from which a continuous advance may be made and fresh conceptions of nature can be evolved.

It is not my intention, even had I possessed the requisite knowledge, to undertake so arduous a task as to review the progress which has recently been made in the great body of sciences which lie within the domain of the British Association. As my occupation in life has required me to give attention to the science which deals with the structure and organisation of the bodies of man and animals—a science which either includes within its scope or has intimate and widespread relations to comparative anatomy, embryology, morphology, zoology, physiology and anthropology—I shall limit myself to the attempt to bring before you some of the more important observations and conclusions which have a bearing on the present position of the subject. As this is the closing year of the century it will not, I think, be out of place to refer to the changes which a hundred years have brought about in our fundamental conceptions of the structure of animals. In science, as in business, it is well from time to time to take stock of what we have been doing, so that we may realise where we stand and ascertain the balance to our credit in the scientific ledger.

So far back as the time of the ancient Greeks it was known that the human body and those of the more highly organised animals were not homogeneous, but were built up of parts, the *partes dissimilares* (τὰ ἀνόμοια μέρη) of Aristotle, which differed from each other in form, colour, texture, consistency and properties. These parts were familiarly known as the bones, muscles, sinews, blood-vessels, glands, brain, nerves and so on. As the centuries rolled on, and as observers and observations multiplied, a more and more precise knowledge of these parts throughout the Animal Kingdom was obtained, and various attempts were made to classify animals in accordance with their forms and structure. During the concluding years of the last century and the earlier part of the present, the Hunters, William and John, in our country, the Meckels in Germany, Cuvier and St. Hilaire in France, gave an enormous impetus to anatomical studies, and contributed largely to our knowledge of the construction of the bodies of animals. But whilst by these and other observers the most salient and, if I may use the expression, the grosser characters of animal organisation had been recognised, little was known of the more intimate structure or texture of the parts. So far as could be determined by the unassisted vision, and so much as could be recognised by the use of a simple lens, had indeed been ascertained, and it was known that muscles, nerves and tendons were composed of threads or fibres, that the blood- and lymph-vessels were tubes, that the parts which we call fasciæ and aponeuroses were thin membranes, and so on.

Early in the present century Xavier Bichat, one of the most brilliant men of science during the Napoleonic era in France, published his "Anatomie Générale," in which he formulated important general principles. Every animal is an assemblage of different organs, each of which discharges a function, and acting together, each in its own way, assists in the preservation of the whole. The organs are, as it were, special machines situated

in the general building which constitutes the factory or body of the individual. But, further, each organ or special machine is itself formed of tissues which possess different properties. Some, as the blood-vessels, nerves, fibrous tissues, &c., are generally distributed throughout the animal body, whilst others, as bones, muscles, cartilage, &c., are found only in certain definite localities. Whilst Bichat had acquired a definite philosophical conception of the general principles of construction and of the distribution of the tissues, neither he nor his pupil Béclard was in a position to determine the essential nature of the structural elements. The means and appliances at their disposal and at that of other observers in their generation were not sufficiently potent to complete the analysis.

Attempts were made in the third decennium of this century to improve the methods of examining minute objects by the manufacture of compound lenses, and by doing away with chromatic and spherical aberration, to obtain, in addition to magnification of the object, a relatively large flat field of vision with clearness and sharpness of definition. When in January 1830 Joseph Jackson Lister read to the Royal Society his memoir "On some properties in achromatic object-glasses applicable to the improvement of microscopes," he announced the principles on which combinations of lenses could be arranged, which would possess these qualities. By the skill of our opticians, microscopes have now for more than half a century been constructed which, in the hands of competent observers, have influenced and extended biological science with results comparable to those obtained by the astronomer through improvements in the telescope.

In the study of the minute structure of plants and animals the observer has frequently to deal with tissues and organs, most of which possess such softness and delicacy of substance and outline that, even when microscopes of the best construction are employed, the determination of the intimate nature of the tissue, and the precise relation which one element of an organ bears to the other constituent elements, is in many instances a matter of difficulty. Hence additional methods have had to be devised in order to facilitate study and to give precision and accuracy to our observations. It is difficult for one of the younger generation of biologists, with all the appliances of a well-equipped laboratory at his command, with experienced teachers to direct him in his work, and with excellent text-books, in which the modern methods are described, to realise the conditions under which his predecessors worked half a century ago. Laboratories for minute biological research had not been constructed, the practical teaching of histology and embryology had not been organised, experience in methods of work had not accumulated; each man was left to his individual efforts, and had to puzzle his way through the complications of structure to the best of his power. Staining and hardening reagents were unknown. The double-bladed knife invented by Valentin, held in the hand, was the only improvement on the scalpel or razor for cutting thin, more or less translucent slices suitable for microscopic examination; mechanical section-cutters and freezing arrangements had not been devised. The tools at the disposal of the microscopist were little more than knife, forceps, scissors, needles, with acetic acid, glycerine and Canada balsam as reagents. But in the employment of the newer methods of research, care has to be taken, more especially when hardening and staining reagents are used, to discriminate between appearances which are natural characters, and those which are only artificial productions.

Notwithstanding the difficulties attendant on the study of the more delicate tissues, the compound achromatic microscope provided anatomists with an instrument of great penetrative power. Between the years 1830 and 1850 a number of acute observers applied themselves with much energy and enthusiasm to the examination of the minute structure of the tissues and organs in plants and animals.

Cell Theory.

It had, indeed, long been recognised that the tissues of plants were to a large extent composed of minute vesicular bodies, technically called cells (Hooke, Malpighi, Grew). In 1831 the discovery was made by the great botanist, Robert Brown, that in many families of plants a circular spot, which he named areola or nucleus, was present in each cell; and in 1838 M. J. Schleiden published the fact that a similar spot or nucleus was a universal elementary organ in vegetables. In the tissues of animals also structures had begun to be recognised comparable with the cells and nuclei of the vegetable tissues, and in

1839 Theodore Schwann announced the important generalisation that there is one universal principle of development for the elementary part of organisms, however different they may be in appearance, and that this principle is the formation of cells. The enunciation of the fundamental principle that the elementary tissues consisted of cells constituted a step in the progress of biological science which will for ever stamp the century now drawing to a close with a character and renown equalling those which it has derived from the most brilliant discoveries in the physical ' sciences. It provided biologists with the visible anatomical units through which the external forces operating on, and the energy generated in, living matter come into play. It dispelled for ever the old mystical idea of the influence exercised by vapours or spirits in living organisms. It supplied the physiologist and pathologist with the specific structures through the agency of which the functions of organisms are discharged in health and disease. It exerted an enormous influence on the progress of practical medicine. A review of the progress of knowledge of the cell may appropriately enter into an address on this occasion.

Structure of Cells.

A cell is a living particle, so minute that it needs a microscope for its examination ; it grows in size, maintains itself in a state of activity, responds to the action of stimuli, reproduces its kind, and in the course of time it degenerates and dies.

Let us glance at the structure of a cell to determine its constituent parts and the *rôle* which each plays in the function to be discharged. The original conception of a cell, based upon the study of the vegetable tissues, was a minute vesicle enclosed by a definite wall, which exercised chemical or metabolic changes on the surrounding material and secreted into the vesicle its characteristic contents. A similar conception was at first also entertained regarding the cells of animal tissues ; but as observations multiplied, it was seen that numerous elementary particles, which were obviously in their nature cells, did not possess an enclosing envelope. A wall ceased to have a primary value as a constituent part of a cell, the necessary vesicular character of which therefore could no longer be entertained.

The other constituent parts of a cell are the cell plasm, which forms the body of the cell, and the nucleus embedded in its substance. Notwithstanding the very minute size of the nucleus, which even in the largest cells is not more than 1/500th inch in diameter, and usually is considerably smaller, its almost constant form, its well-defined sharp outline, and its power of resisting the action of strong reagents when applied to the cell have from the period of its discovery by Robert Brown caused histologists to bestow on it much attention. Its structure and chemical composition ; its mode of origin ; the part which it plays in the formation of new cells and its function in nutrition and secretion have been investigated.

When examined under favourable conditions in its passive or resting state, the nucleus is seen to be bounded by a membrane which separates it from the cell plasm and gives it the characteristic sharp contour. It contains an apparently structureless nuclear substance, nucleoplasm or enchylema, in which are embedded one or more extremely minute particles called nucleoli, along with a network of exceedingly fine threads or fibres, which in the active living cell play an essential part in the production of new nuclei within the cell. In its chemical composition the nuclear substance consists of albuminous plastin and globulin ; and of a special material named nuclein, rich in phosphorus and with an acid reaction. The delicate network within the nucleus consists apparently of the nuclein, a substance which stains with carmine and other dyes, a property which enables the changes, which take place in the network in the production of young cells, to be more readily seen and followed out by the observer.

The mode of origin of the nucleus and the part which it plays in the production of new cells have been the subject of much discussion. Schleiden, whose observations, published in 1838, were made on the cells of plants, believed that within the cell a nucleolus first appeared, and that around it molecules aggregated to form the nucleus. Schwann again, whose observations were mostly made on the cells of animals, considered that an amorphous material existed in organised bodies, which he called cytoblastema. It formed the contents of cells, or it might be situated free or external to them. He figuratively compared it to a mother liquor in which crystals are formed. Either in the cytoblastema within the cells or in that situated external to them, the aggregation of molecules around a nucleolus to form

a nucleus might occur, and, when once the nucleus had been formed, in its turn it would serve as a centre of aggregation of additional molecules from which a new cell would be produced. He regarded therefore the formation of nuclei and cells as possible in two ways : one within pre-existing cells (endogenous cell-formation), the other in a free blastema lying external to cells (free cell-formation). In animals, he says, the endogenous method is rare, and the customary origin is in an external blastema. Both Schleiden and Schwann considered that after the cell was formed the nucleus had no permanent influence on the life of the cell, and usually disappeared.

Under the teaching principally of Henle, the famous Professor of Anatomy in Göttingen, the conception of the free formation of nuclei and cells in a more or less fluid blastema, by an aggregation of elementary granules and molecules, obtained so much credence, especially amongst those who were engaged in the study of pathological processes, that the origin of cells within pre-existing cells was to a large extent lost sight of. That a parent cell was requisite for the production of new cells seemed to many investigators to be no longer needed. Without doubt this conception of free cell-formation contributed in no small degree to the belief, entertained by various observers, that the simplest plants and animals might arise, without pre-existing parents, in organic fluids destitute of life, by a process of spontaneous generation ; a belief which prevailed in many minds almost to the present day. If, as has been stated, the doctrine of abiogenesis cannot be experimentally refuted, on the other hand it has not been experimentally proved. The burden of proof lies with those who hold the doctrine, and the evidence that we possess is all the other way.

Multiplication of Cells.

Although von Mohl, the botanist, seems to have been the first to recognise (1835) in plants a multiplication of cells by division, it was not until attention was given to the study of the egg in various animals, and to the changes which take place in it, attendant on fertilisation, that in the course of time a much more correct conception of the origin of the nucleus and of the part which it plays in the formation of new cells was obtained. Before Schwann had published his classical memoir in 1839, von Baer and other observers had recognised within the animal ovum the germinal vesicle, which obviously bore to the ovum the relation of a nucleus to a cell. As the methods of observation improved, it was recognised that, within the developing egg, two vesicles appeared where one only had previously existed, to be followed by four vesicles, then eight, and so on in multiple progression until the ovum contained a multitude of vesicles, each of which possessed a nucleus. The vesicles were obviously cells which had arisen within the original germ-cell or ovum. These changes were systematically described by Martin Barry so long ago as 1839 and 1840 in two memoirs communicated to the Royal Society of London, and the appearance produced, on account of the irregularities of the surface occasioned by the production of new vesicles, was named by him the mulberry-like structure. He further pointed out that the vesicles arranged themselves as a layer within the envelope of the egg or zona pellucida, and that the whole embryo was composed of cells filled with the foundations of other cells. He recognised that the new cells were derived from the germinal vesicle or nucleus of the ovum, the contents of which entered into the formation of the first two cells, each of which had its nucleus, which in its turn resolved itself into other cells, and by a repetition of the process into a greater number. The endogenous origin of new cells within a pre-existing cell and the process which we now term the segmentation of the yolk were successfully demonstrated. In a third memoir, published in 1841, Barry definitely stated that young cells originated through division of the nucleus of the parent cell ; instead of arising, as a product of crystallisation, in the fluid cytoblastema of the parent cell or in a blastema situated external to the cell.

In a memoir published in 1842, John Goodsir advocated the view that the nucleus is the reproductive organ of the cell, and that from it, as from a germinal spot, new cells were formed. In a paper, published three years later, on nutritive centres, he described cells the nuclei of which were the permanent source of successive broods of young cells, which from time to time occupied the cavity of the parent cell. He extended also his observations on the endogenous formation of cells to the cartilage cells in the process of inflammation and to other tissues undergoing pathological changes. Corroborative observations on endogenous formation were also given by his brother, Harry Goodsir, in 1845. These observations on the part which the nucleus plays by cleavage in the formation of young cells by endogenous development from a parent centre—that an organic continuity existed between a mother cell and its descendants through the nucleus—constituted a great step in advance of the views entertained by Schleiden and Schwann, and showed that Barry and the Goodsirs had a deeper insight into the nature and functions of cells than was possessed by most of their contemporaries, and are of the highest importance when viewed in the light of recent observations.

In 1841 Robert Remak published an account of the presence of two nuclei in the blood corpuscles of the chick and the pig, which he regarded as evidence of the production of new corpuscles by division of the nucleus within a parent cell ; but it was not until some years afterwards (1850 to 1855) that he recorded additional observations and recognised that division of the nucleus was the starting-point for the multiplication of cells in the ovum and in the tissues generally. Remak's view was that the process of cell-division began with the cleavage of the nucleolus, followed by that of the nucleus, and that again by cleavage of the body of the cell and of its membrane. Kölliker had previously, in 1843, described the multiplication of nuclei in the ova of parasitic worms, and drew the inference that in the formation of young cells within the egg the nucleus underwent cleavage, and that each of its divisions entered into the formation of a new cell. By these observations, and by others subsequently made, it became obvious that the multiplication of animal cells, either by division of the nucleus within the cell, or by the budding off of a part of the protoplasm of the cell, was to be regarded as a widely spread and probably a universal process, and that each new cell arose from a parent cell.

Pathological observers were, however, for the most part inclined to consider free cell-formation in a blastema or exudation by an aggregation of molecules, in accordance with the views of Henle, as a common phenomenon. This proposition was attacked with great energy by Virchow in a series of memoirs published in his "Archiv," commencing in vol. i. 1847, and finally received its death-blow in his published lectures on Cellular Pathology, 1858. He maintained that in pathological structures there was no instance of cell development *de novo* ; where a cell existed, there one must have been before. Cell-formation was a continuous development by descent, which he formulated in the expression *omnis cellula e cellulâ*.

Karyokinesis.

Whilst the descent of cells from pre-existing cells by division of the nucleus during the development of the egg, in the embryos of plants and animals, and in adult vegetable and animal tissues, both in healthy and diseased conditions, had now become generally recognised, the mechanism of the process by which the cleavage of the nucleus took place was for a long time unknown. The discovery had to be deferred until the optician had been able to construct lenses of a higher penetrative power, and the microscopist had learned the use of colouring agents capable of dyeing the finest elements of the tissues. There was reason to believe that in some cases a direct cleavage of the nucleus, to be followed by a corresponding division of the cell into two parts, did occur. In the period between 1870 and 1880 observations were made by Schneider, Strasburger, Bütschli, Fol, van Beneden and Flemming, which showed that the division of the nucleus and the cell was due to a series of very remarkable changes, now known as indirect nuclear and cell division, or karyokinesis. The changes within the nucleus are of so complex a character that it is impossible to follow them in detail without the use of appropriate illustrations. I shall have to content myself, therefore, with an elementary sketch of the process.

I have previously stated that the nucleus in its passive or resting stage contains a very delicate network of threads or fibres. The first stage in the process of nuclear division consists in the threads arranging themselves in loops and forming a compact coil within the nucleus. The coil then becomes looser, the loops of threads shorten and thicken, and somewhat later each looped thread splits longitudinally into two portions. As the threads stain when colouring agents are applied to them, they are called chromatin fibres, and the loose coil is the chromosome (Waldeyer). As the process continues, the investing membrane of the nucleus disappears, and the loops of threads arrange themselves within the nucleus so that the closed ends of the loops are directed to a common centre, from which the loops radiate outwards and produce a starlike figure (aster). At the same time clusters of extremely delicate lines appear both in the nucleoplasm and in the body of the cell, named the achromatic figure, which has a spindle-like form with two opposite poles, and stains much more feebly than the chromatic fibres. The loops of the chromatic star then arrange themselves in the equatorial plane of the spindle, and bending round turn their closed ends towards the periphery of the nucleus and the cell.

The next stage marks an important step in the process of division of the nucleus. The two longitudinal portions, into which each looped thread had previously split, now separate from each other, and whilst one part migrates to one pole of the spindle, the other moves to the opposite pole, and the free ends of each loop are directed towards its equator (metakinesis). By this division of the chromatin fibres, and their separation from each other to opposite poles of the spindle, two starlike chromatin figures are produced (dyaster).

Each group of fibres thickens, shortens, becomes surrounded by a membrane, and forms a new or daughter nucleus (dispirem). Two nuclei therefore have arisen within the cell by the division of that which had previously existed, and the expression formulated by Flemming—*omnis nucleus e nucleo*—is justified. Whilst this stage is in course of being completed, the body of the cell becomes constricted in the equatorial plane of the spindle, and, as the constriction deepens, it separates into two parts, each containing a daughter nucleus, so that two nucleated cells have arisen out of a pre-existing cell.

A repetition of the process in each of these cells leads to the formation of other cells, and, although modifications in details are found in different species of plants and animals, the multiplication of cells in the egg and in the tissues generally on similar lines is now a thoroughly established fact in biological science.

In the study of karyokinesis, importance has been attached to the number of chromosomes in the nucleus of the cell. Flemming had seen in the Salamander twenty-four chromosome fibres, which seems to be a constant number in the cells of epithelium and connective tissues. In other cells again, especially in the ova of certain animals, the number is smaller, and fourteen, twelve, four, and even two only have been described. The theory formulated by Boveri that the number of chromosomes is constant for each species, and that in the karyokinetic figures corresponding numbers are found in homologous cells, seems to be not improbable.

In the preceding description I have incidentally referred to the appearance in the proliferating cell of an achromatic spindle-like figure. Although this was recognised by Fol in 1873, it is only during the last ten or twelve years that attention has been paid to its more minute arrangements and possible signification in cell-division.

The pole at each end of the spindle lies in the cell plasm which surrounds the nucleus. In the centre of each pole is a somewhat opaque spot (central body) surrounded by a clear space, which, along with the spot, constitutes the centrosome or the sphere of attraction. From each centrosome extremely delicate lines may be seen to radiate in two directions. One set extends towards the pole at the opposite end of the spindle, and, meeting or coming into close proximity with radiations from it, constitutes the body of the spindle, which, like a perforated mantle, forms an imperfect envelope around the nucleus during the process of division. The other set of radiations is called the polar, and extends in the region of the pole towards the periphery of the cell.

The question has been much discussed whether any constituent part of the achromatic figure, or the entire figure, exists in the cell as a permanent structure in its resting phase ; or if it is only present during the process of karyokinesis. During the development of the egg the formation of young cells, by division of the segmentation nucleus, is so rapid and continuous that the achromatic figure, with the centrosome in the pole of the spindle, is a readily recognisable object in each cell. The polar and spindle-like radiations are in evidence during karyokinesis, and have apparently a temporary endurance and function. On the other hand, van Beneden and Boveri were of opinion that the central body of the centrosome did not disappear when the division of the nucleus came to an end, but that it remained as a constituent part of a cell lying in the cell plasm near to the nucleus. Flemming has seen the central body with its sphere in leucocytes, as well as in epithelial cells and those of othe tissues. Subsequently Heidenhain and other histologists have recorded similar observations. It would seem. therefore, as if there were reason to regard the centrosome, like the nucleus, as a permanent constituent of a cell. This view, however, is not universally entertained. If not always capable of demonstration in the resting stage of a cell, it is doubtless to be regarded as potentially present, and ready to assume, along with the radiations, a characteristic appearance when the process of nuclear division is about to begin.

One can scarcely regard the presence of so remarkable an appearance as the achromatic figure without associating with it an important function in the economy of the cell. As from the centrosome at the pole of the spindle both sets of radiations diverge, it is not unlikely that it acts as a centre or sphere of energy and attraction. By some observers the radiations are regarded as substantive fibrillar structures, elastic or even contractile in their properties. Others, again, look upon them as morphological expressions of chemical and dynamical energy in the protoplasm of the cell body. On either theory we may assume that they indicate an influence, emanating, it may be, from the centrosome, and capable of being exercised both on the cell plasm and on the nucleus contained in it. On the contractile theory, the radiations which form the body of the spindle, either by actual traction of the supposed fibrillæ or by their pressure on the nucleus which they surround, might impel during karyokinesis the dividing chromosome elements towards the poles of the spindle, to form there the daughter nuclei. On the dynamical theory, the chemical and physical energy in the centrosome might influence the cell plasm and the nucleus, and attract the chromosome elements of the nucleus to the poles of the spindle. The radiated appearance would therefore be consequent and attendant on the physico-chemical activity of the centrosome. One or other of these theories may also be applied to the interpretation of the significance of the polar radiations.

Cell Plasm.

In the cells of plants, in addition to the cell wall, the cell body and the cell juice require to be examined. The material of the cell body, or the cell contents, was named by von Mohl (1846) protoplasm, and consisted of a colourless tenacious substance which partly lined the cell wall (primordial utricle), and partly traversed the interior of the cell as delicate threads enclosing spaces (vacuoles) in which the cell juice was contained. In the protoplasm the nucleus was embedded. Nägeli, about the same time, had also recognised the difference between the protoplasm and the other contents of vegetable cells, and had noticed its nitrogenous composition.

Though the analogy with a closed bladder or vesicle could no longer be sustained in the animal tissues, the name "cell" continued to be retained for descriptive purposes, and the body of the cell was spoken of as a more or less soft substance enclosing a nucleus (Leydig). In 1861 Max Schultze adopted for the substance forming the body of the animal cell the term "protoplasm." He defined it to be a particle of protoplasm in the substance of which a nucleus was situated. He regarded the protoplasm, as indeed had previously been pointed out by the botanist Unger, as essentially the same as the contractile sarcode which constitutes the body and pseudopodia of the Amœba and other Rhizopoda. As the term "protoplasm," as well as that of "bioplasm" employed by Lionel Beale in a somewhat similar though not precisely identical sense, involves certain theoretical views of the origin and function of the body of the cell, it would be better to apply to it the more purely descriptive term "cytoplasm" or "cell plasm."

Schultze defined protoplasm as a homogeneous, glassy, tenacious material, of a jelly-like or somewhat firmer consistency, in which numerous minute granules were embedded. He regarded it as the part of the cell especially endowed with vital energy, whilst the exact function of the nucleus could not be defined. Based upon this conception of the jelly-like character of protoplasm, the idea for a time prevailed that a structureless, dimly granular, jelly or slime destitute of organisation, possessed great physiological activity, and was the medium through which the phenomena of life were displayed.

More accurate conceptions of the nature of the cell plasm soon began to be entertained. Brücke recognised that the body of the cell was not simple, but had a complex organisation. Flemming observed that the cell plasm contained extremely delicate threads, which frequently formed a network, the interspaces of which were occupied by a more homogeneous substance. Where

the threads crossed each other, granular particles (mikrosomen) were situated. Bütschli considered that he could recognise in the cell plasm a honeycomb-like appearance, as if it consisted of excessively minute chambers in which a homogeneous, more or less fluid, material was contained. The polar and spindle-like radiations visible during the process of karyokinesis, which have already been referred to, and the presence of the centrosome, possibly even during the resting stage of the cell, furnished additional illustrations of differentiation within the cell plasm. In many cells there appears also to be a difference in the character of the cell plasm which immediately surrounds the nucleus and that which lies at and near the periphery of the cell. The peripheral part (ektoplasm) is more compact and gives a definite outline to the cell, although not necessarily differentiating into a cell membrane. The inner part (endoplasm) is softer, and is distinguished by a more distinct granular appearance, and by containing the products specially formed in each particular kind of cell during the nutritive process.

By the researches of numerous investigators on the internal organisation of cells in plants and animals, a large body of evidence has now been accumulated, which shows that both the nucleus and the cell plasm consist of something more than a homogeneous, more or less viscid, slimy material. Recognisable objects in the form of granules, threads or fibres can be distinguished in each. The cell plasm and the nucleus respectively are therefore not of the same constitution throughout, but possess polymorphic characters, the study of which in health and the changes produced by disease will for many years to come form important matters for investigation.

Function of Cells.

It has already been stated that, when new cells arise within pre-existing cells, division of the nucleus is associated with cleavage of the cell plasm, so that it participates in the process of new cell-formation. Undoubtedly, however, its *rôle* is not limited to this function. It also plays an important part in secretion, nutrition, and the special functions discharged by the cells in the tissues and organs of which they form morphological elements.

Between 1838 and 1842 observations were made which showed that cells were constituent parts of secreting glands and mucous membranes (Schwann, Henle). In 1842 John Goodsir communicated to the Royal Society of Edinburgh a memoir on secreting structures, in which he established the principle that cells are the ultimate secreting agents; he recognised in the cells of the liver, kidney, and other organs the characteristic secretion of each gland. The secretion was, he said, situated between the nucleus and the cell wall. At first he thought that, as the nucleus was the reproductive organ of the cell, the secretion was formed in the interior of the cell by the agency of the cell wall; but three years later he regarded it as a product of the nucleus. The study of the process of spermatogenesis by his brother, Harry Goodsir, in which the head of the spermatozoon was found to correspond with the nucleus of the cell in which the spermatozoon arose, gave support to the view that the nucleus played an important part in the genesis of the characteristic product of the gland cell.

The physiological activity of the cell plasm and its complex chemical constitution soon after began to be recognised. Some years before Max Schultze had published his memoirs on the characters of protoplasm, Brücke had shown that the well-known changes in tint in the skin of the Chamæleon were due to pigment granules situated in the cells in the skin which were sometimes diffused throughout the cells, at others concentrated in the centre. Similar observations on the skin of the frog were made in 1854 by von Wittich and Harless. The movements were regarded as due to contraction of the cell wall on its contents. In a most interesting paper on the pigmentary system in the frog, published in 1858, Lord Lister demonstrated that the pigment granules moved in the cell plasma, by forces resident within the cell itself, acting under the influence of an external stimulant, and not by contractility of the wall. Under some conditions the pigment was attracted to the centre of the cell, when the skin became pale; under other conditions the pigment was diffused throughout the body and the branches of the cell, and gave to the skin a dark colour. It was also experimentally shown that a potent influence over these movements was exercised by the nervous system.

The study of the cells of glands engaged in secretion, even when the secretion is colourless, and the comparison of their appearance when secretion is going on with that seen when the cells are at rest, have shown that the cell plasm is much more granular and opaque, and contains larger particles during activity than when the cell is passive ; the body of the cell swells out from an increase in the contents of its plasm, and chemical changes accompany the act of secretion. Ample evidence, therefore, is at hand to support the position taken by John Goodsir, nearly sixty years ago, that secretions are formed within cells, and lie in that part of the cell which we now say consists of the cell plasm ; that each secreting cell is endowed with its own peculiar property, according to the organ in which it is situated, so that bile is formed by the cells in the liver, milk by those in the mamma, and so on.

Intimately associated with the process of secretion is that of nutrition. As the cell plasm lies at the periphery of a cell, and as it is, alike both in secretion and nutrition, brought into closest relation with the surrounding medium, from which the pabulum is derived, it is necessarily associated with the nutritive activity. Its position enables it to absorb nutritive material directly from without, and in the process of growth it increases in amount by interstitial changes and additions throughout its substance, and not by mere accretions on its surface.

Hitherto I have spoken of a cell as a unit, independent of its neighbours as regards its nutrition and the other functions which it has to discharge. The question has, however, been discussed, whether in a tissue composed of cells closely packed together cell plasm may not give origin to processes or threads which are in contact or continuous with corresponding processes of adjoining cells, and that cells may therefore, to some extent, lose their individuality in the colony of which they are members. Appearances were recognised between 1863 and 1870 by Schrön and others in the deeper cells of the epidermis and of some mucous membranes which gave sanction to this view, and it seems possible through contact or continuity of threads connecting a cell with its neighbours, that cells may exercise a direct influence on each other.

Nägeli, the botanist, as the foundation of a mechanico-physiological theory of descent, considered that in plants a network of cell plasm, named by him idio-plasm, extended throughout the whole of the plant, forming its specific molecular constitution, and that growth and activity were regulated by its conditions of tension and movements (1884).

The study of the structure of plants with special reference to the presence of an intercellular network has for some years been pursued by Walter Gardiner (1882-97), who has demonstrated threads of cell plasm protruding through the walls of vegetable cells and continuous with similar threads from adjoining cells. Structurally, therefore, a plant may be conceived to be built up of a nucleated cytoplasmic network, each nucleus with the branching cell plasm surrounding it being a centre of activity. On this view a plant would retain to some extent its individuality, though, as Gardiner contends, the connecting threads would be the medium for the conduction of impulses and of food from a cell to those which lie around it. For the plant cell therefore, as has long been accepted in the animal cell, the wall is reduced to a secondary position, and the active constituent is the nucleated cell plasm. It is not unlikely that the absence of a controlling nervous system in plants requires the plasm of adjoining cells to be brought into more immediate contact and continuity than is the case with the generality of animal cells, so as to provide a mechanism for harmonising the nutritive and other functional processes in the different areas in the body of the plant. In this particular, it is of interest to note that the epithelial tissues in animals, where somewhat similar connecting arrangements occur, are only indirectly associated with the nervous and vascular systems, so that, as in plants, the cells may require, for nutritive and other purposes, to act and react directly on each other.

Nerve Cells.

Of recent years great attention has been paid to the intimate structure of nerve cells, and to the appearance which they present when in the exercise of their functional activity. A nerve cell is not a secreting cell ; that is, it does not derive from the blood or surrounding fluid a pabulum which it elaborates into a visible, palpable secretion characteristic of the organ of which the cell is a constituent element, to be in due course discharged into a duct which conveys the secretion out of the gland. Nerve cells, through the metabolic changes which take place in them in connection with their nutrition, are associated with the production of the form of energy specially exhibited by

animals which possess a nervous system, termed nerve energy. It has long been known that every nerve cell has a body in which a relatively large nucleus is situated. A most important discovery was the recognition that the body of every nerve cell had one or more processes growing out from it. More recently it has been proved, chiefly through the researches of Schultze, His, Golgi and Ramon y Cajal, that at least one of the processes, the axon of the nerve cell, is continued into the axial cylinder of a nerve fibre, and that in the multipolar nerve cell the other processes, or dendrites, branch and ramify for some distance away from the body. A nerve fibre is therefore an essential part of the cell with which it is continuous, and the cell, its processes, the nerve fibre and the collaterals which arise from the nerve fibre collectively form a neuron or structural nerve unit (Waldeyer). The nucleated body of the nerve cell is the physiological centre of the unit.

The cell plasm occupies both the body of the nerve cell and its processes. The intimate structure of the plasm has, by improved methods of observation introduced during the last eight years by Nissl, and conducted on similar lines by other investigators, become more definitely understood. It has been ascertained that it possesses two distinct characters which imply different structures. One of these stains deeply on the addition of certain dyes, and is named chromophile or chromatic substance ; the other, which does not possess a similar property, is the achromatic network. The chromophile is found in the cell body and the dendritic processes, but not in the axon. It occurs in the form of granular particles, which may be scattered throughout the plasm, or aggregated into little heaps which are elongated or fusiform in shape and appear as distinct coloured particles or masses. The achromatic network is found in the cell body and the dendrites, and is continued also into the axon, where it forms the axial cylinder of the nerve fibre. It consists apparently of delicate threads or fibrillæ, in the meshes of which a homogeneous material, such as is found in the cell plasm generally, is contained. In the nerve cells, as in other cells, the plasm is without doubt concerned in the process of cell nutrition. The achromatic fibrillæ exercise an important influence on the axon or nerve fibre with which they are continuous, and probably they conduct the nerve impulses which manifest themselves in the form of nerve energy. The dendritic processes of a multipolar nerve cell ramify in close relation with similar processes branching from other cells in the same group. The collaterals and the free end of the axon fibre process branch and ramify in association with the body of a nerve cell or of its dendrites. We cannot say that these parts are directly continuous with each other to form an intercellular network, but they are apparently in apposition, and through contact exercise influence one on the other in the transmission of nerve impulses.

There is evidence to show that in the nerve cell the nucleus, as well as the cell plasm, is an effective agent in nutrition. When the cell is functionally active, both the cell body and the nucleus increase in size (Vas, G. Mann, Lugaro) ; on the other hand, when nerve cells are fatigued through excessive use, the nucleus decreases in size and shrivels ; the cell plasm also shrinks, and its coloured or chromophile constituent becomes diminished in quantity, as if it had been consumed during the prolonged use of the cell (Hodge, Mann, Lugaro). It is interesting also to note that in hibernating animals in the winter season, when their functional activity is reduced to a minimum, the chromophile in the plasm of the nerve cells is much smaller in amount than when the animal is leading an active life in the spring and summer (G. Levi).

When a nerve cell has attained its normal size it does not seem to be capable of reproducing new cells in its substance by a process of karyokinesis, such as takes place when young cells arise in the egg and in the tissues generally. It would appear that nerve cells are so highly specialised in their association with the evolution of nerve energy, that they have ceased to have the power of reproducing their kind, and the metabolic changes both in cell plasm and nucleus are needed to enable them to discharge their very peculiar function. Hence it follows that when a portion of the brain or other nerve-centre is destroyed, the injury is not repaired by the production of fresh specimens of their characteristic cells, as would be the case in injuries to bones and tendons.

In our endeavours to differentiate the function of the nucleus from that of the cell plasm, we should not regard the former as concerned only in the production of young cells, and the latter as the exclusive agent in growth, nutrition, and, where gland cells are concerned, in the formation of their characteristic products. As regards cell reproduction also, though the process of division begins in the nucleus in its chromosome constituents, the achromatic figure in the cell plasm undoubtedly plays a part, and the cell plasm itself ultimately undergoes cleavage.

A few years ago the tendency amongst biologists was to ignore or attach but little importance to the physiological use of the nucleus in the nucleated cell, and to regard the protoplasm as the essential and active constituent of living matter ; so much so, indeed, was this the case that independent organisms regarded as distinct species were described as consisting of protoplasm destitute of a nucleus ; also that scraps of protoplasm separated from larger nucleated masses could, when isolated, exhibit vital phenomena. There is reason to believe that a fragment of protoplasm, when isolated from the nucleus of a cell, though retaining its contractility and capable of nourishing itself for a short time, cannot increase in amount, act as a secreting structure, or reproduce its kind : it soon loses its activity, withers, and dies. In order that these qualities of living matter should be retained, a nucleus is by most observers regarded as necessary (Nussbaum, Gruber; Haberlandt, Korschelt), and for the complete manifestation of vital activity both nucleus and cell plasm are required.

Bacteria.

The observations of Cohn, made about thirty years ago, and those of De Bary shortly afterwards, brought into notice a group of organisms to which the name "bacterium" or "microbe" is given. They were seen to vary in shape : some were rounded specks called cocci, others were straight rods called bacilli, others were curved or spiral rods, vibrios or spirillæ. All were characterised by their extreme minuteness, and required for their examination the highest powers of the best microscopes. Many bacteria measure in their least diameter not more than 1/25000th of an inch, 1/10th the diameter of a human white blood corpuscle. Through the researches of Pasteur, Lord Lister, Koch, and other observers, bacteria have been shown to play an important part in nature. They exercise a very remarkable power over organic substances, especially those which are complex in chemical constitution, and can resolve them into simpler combinations. Owing to this property, some bacteria are of great economic value, and without their agency many of our industries could not be pursued ; others again, and these are the most talked of, exercise a malign influence in the production of the most deadly diseases which afflict man and the domestic animals.

Great attention has been given to the structure of bacteria and to their mode of propagation. When examined in the living state and magnified about 2000 times, a bacterium appears as a homogeneous particle, with a sharp definite outline, though a membranous envelope or wall, distinct from the body of the bacterium, cannot at first be recognised ; but when treated with reagents a membranous envelope appears, the presence of which, without doubt, gives precision of form to the bacterium. The substance within the membrane contains granules which can be dyed with colouring agents. Owing to their extreme minuteness it is difficult to pronounce an opinion on the nature of the chromatine granules and the substance in which they lie. Some observers regard them as nuclear material, invested by only a thin layer of protoplasm, on which view a bacterium would be a nucleated cell. Others consider the bacterium as formed of protoplasm containing granules capable of being coloured, which are a part of the protoplasm itself, and not a nuclear substance. On the latter view, bacteria would consist of cell plasm inclosed in a membrane and destitute of a nucleus. Whatever be the nature of the granule-containing material, each bacterium is regarded as a cell, the minutest and simplest living particle capable of an independent existence that has not yet been discovered.

Bacteria cells, like cells generally, can reproduce their kind. They multiply by simple fission, probably with an ingrowth of the cell wall, but without the karyokinetic phenomena observed in nucleated cells. Each cell gives rise to two daughter cells, which may for a time remain attached to each other and form a cluster or a chain, or they may separate and become independent isolated cells. The multiplication, under favourable conditions of light, air, temperature, moisture and food, goes on with extraordinary rapidity, so that in a few hours many thousand new individuals may arise from a parent bacterium.

Connected with the life-history of a bacterium cell is the

formation in its substance, in many species and under certain conditions, of a highly refractile shiny particle called a spore. At first sight a spore seems as if it were the nucleus of the bacterium cell, but it is not always present when multiplication by cleavage is taking place, and when present it does not appear to take part in the fission. On the other hand, a spore, from the character of its envelope, possesses great power of resistance, so that dried bacteria, when placed in conditions favourable to germination, can through their spores germinate and resume an active existence. Spore formation seems, therefore, to be a provision for continuing the life of the bacterium under conditions which, if spores had not formed, would have been the cause of its death.

The time has gone by to search for the origin of living organisms by a spontaneous aggregation of molecules in vegetable or other infusions, or from a layer of formless primordial slime diffused over the bed of the ocean. Living matter during our epoch has been, and continues to be, derived from pre-existing living matter, even when it possesses the simplicity of structure of a bacterium, and the morphological unit is the cell.

Development of the Egg.

As the future of the entire organism lies in the fertilised egg cell, we may now briefly review the arrangements, consequent on the process of segmentation, which lead to the formation, let us say in the egg of a bird, of the embryo of the young chick.

In the latter part of the last century, C. F. Wolff observed that the beginning of the embryo was associated with the formation of layers, and in 1817 Pander demonstrated that in the hen's egg at first one layer, called mucous, appeared, then a second or serous layer, to be followed by a third, intermediate or vascular layer. In 1828 von Baer amplified our knowledge in his famous treatise, which from its grasp of the subject created a new epoch in the science of embryology. It was not, however, until the discovery by Schwann of cells as constant factors in the structure of animals and in their relation to development that the true nature of these layers was determined. We now know that each layer consists of cells, and that all the tissues and organs of the body are derived from them. Numerous observers have devoted themselves for many years to the study of each layer, with the view of determining the part which it takes in the formation of the constituent parts of the body, more especially in the higher animals, and the important conclusion has been arrived at that each kind of tissue invariably arises from one of these layers and from no other.

The layer of cells which contributes, both as regards the number and variety of the tissues derived from it, most largely to the formation of the body is the middle layer or mesoblast. From it the skeleton, the muscles, and other locomotor organs, the true skin, the vascular system, including the blood and other structures which I need not detail, take their rise. From the inner layer of cells or hypoblast, the principal derivatives are the epithelial lining of the alimentary canal and of the glands which open into it, and the epithelial lining of the air-passages. The outer or epiblast layer of cells gives origin to the epidermis or scarf skin and to the nervous system. It is interesting to note that from the same layer of the embryo arise parts so different in importance as the cuticle—a mere protecting structure, which is constantly being shed when the skin is subjected to the friction of a towel or the clothes—and the nervous system, including the brain, the most highly differentiated system in the animal body. How completely the cells from which they are derived had diverged from each other in the course of their differentiation in structure and properties is shown by the fact that the cells of the epidermis are continually engaged in reproducing new cells to replace those which are shed, whilst the cells of the nervous system have apparently lost the power of reproducing their kind.

In the early stage of the development of the egg, the cells in a given layer resemble each other in form, and, as far as can be judged from their appearance, are alike in structure and properties. As the development proceeds, the cells begin to show differences in character, and in the course of time the tissues which arise in each layer differentiate from each other and can be readily recognised by the observer. To use the language of von Baer, a generalised structure has become specialised, and each of the special tissues produced exhibits its own structure and properties. These changes are coincident with a rapid multiplication of the cells by cleavage, and thus increase in size of the embryo accompanies specialisation of structure. As the process continues, the embryo gradually assumes the shape characteristic of the species to which its parents belonged, until at length it is fit to be born and to assume a separate existence.

The conversion of cells, at first uniform in character, into tissues of a diverse kind is due to forces inherent in the cells in each layer. The cell plasm plays an active though not an exclusive part in the specialisation ; for as the nucleus influences nutrition and secretion, it acts as a factor in the differentiation of the tissues. When tissues so diverse in character as muscular fibre, cartilage, fibrous tissues, and bone arise from the cells of the middle or mesoblast layer, it is obvious that, in addition to the morphological differentiation affecting form and structure, a chemical differentiation affecting composition also occurs, as the result of which a physiological differentiation takes place. The tissues and organs become fitted to transform the energy derived from the food into muscular energy, nerve energy, and other forms of vital activity. Corresponding differentiations also modify the cells of the outer and inner layers. Hence the study of the development of the generalised cell layers in the young embryo enables us to realise how all the complex constituent parts of the body in the higher animals and in man are evolved by the process of differentiation from a simple nucleated cell—the fertilised ovum. A knowledge of the cell and of its life-history is therefore the foundation-stone on which biological science in all its departments is based.

If we are to understand an organ in the biological sense a complex body capable of carrying on a natural process, a nucleated cell is an organ in its simplest form. In a unicellular animal or plant such an organ exists in its most primitive stage. The higher plants and animals again are built up of multitudes of these organs, each of which, whilst having its independent life, is associated with the others, so that the whole may act in unison for a common purpose. As in one of your great factories each spindle is engaged in twisting and winding its own thread, it is at the same time intimately associated with the hundreds of other spindles in its immediate proximity, in the manufacture of the yarn from which the web of cloth is ultimately to be woven.

It has taken more than fifty years of hard and continuous work to bring our knowledge of the structure and development of the tissues and organs of plants and animals up to the level of the present day. Amidst the host of names of investigators, both at home and abroad, who have contributed to its progress, it may seem invidious to particularise individuals. There are, however, a few that I cannot forbear to mention, whose claim to be named on such an occasion as this will be generally conceded.

Botanists will, I think, acknowledge Wilhelm Hofmeister as a master in morphology and embryology, Julius von Sachs as the most important investigator in vegetable physiology during the last quarter of a century, and Strasburger as a leader in the study of the phenomena of nuclear division.

The researches of the veteran Professor of Anatomy in Würzburg, Albert von Kölliker, have covered the entire field of animal histology. His first paper, published fifty-nine years ago, was followed by a succession of memoirs and books on human and comparative histology and embryology, and culminated in his great treatise on the structure of the brain, published in 1896. Notwithstanding the weight of more than eighty years, he continues to prosecute histological research, and has published the results of his latest, though let us hope not his last, work during the present year.

Amongst our own countrymen, and belonging to the generation which has almost passed away, was William Bowman. His investigations between 1840 and 1850 on the mucous membranes, muscular fibre, and the structure of the kidney, together with his researches on the organs of sense, were characterised by a power of observation and of interpreting difficult and complicated appearances which has made his memoirs on these subjects landmarks in the history of histological inquiry.

Of the younger generation of biologists, Francis Maitland Balfour, whose early death is deeply deplored as a loss to British science, was one of the most distinguished. His powers of observation and philosophic perception gave him a high place as an original inquirer, and the charm of his personality—for charm is not the exclusive possession of the fairer sex—endeared him to his friends.

General Morphology.

Along with the study of the origin and structure of the tissues of organised bodies, much attention has been given during the century to the parts or organs in plants and animals, with the view of determining where and how they take their rise, the order of their formation, the changes which they pass through in the early stages of development, and their relative positions in the organism to which they belong. Investigations on these lines are spoken of as morphological, and are to be distinguished from the study of their physiological or functional relations, though both are necessary for the full comprehension of the living organism.

The first to recognise that morphological relations might exist between the organs of a plant, dissimilar as regards their function, was the poet Goethe, whose observations, guided by his imaginative faculty, led him to declare that the calyx, corolla, and other parts of a flower, the scales of a bulb, &c., were metamorphosed leaves, a principle generally accepted by botanists, and indeed extended to other parts of a plant, which are referred to certain common morphological forms although they exercise different functions. Goethe also applied the same principle in the study of the skeletons of vertebrate animals, and he formed the opinion that the spinal column and the skull were essentially alike in construction, and consisted of vertebræ, an idea which was also independently conceived and advocated by Oken.

The anatomist who in our country most strenuously applied himself to the morphological study of the skeleton was Richard Owen, whose knowledge of animal structure, based upon his own dissections, was unrivalled in range and variety. He elaborated the conception of an ideal, archetype vertebrate form which had no existence in nature, and to which, subject to modifications in various directions, he considered all vertebrate skeletons might be referred. Owen's observations were conducted to a large extent on the skeletons of adult animals, of the knowledge of which he was a master. As in the course of development modifications in shape and in the relative position of parts not unfrequently occur and their original character and place of origin become obscured, it is difficult, from the study only of adults, to arrive at a correct interpretation of their morphological significance. When the changes which take place in the skull during its development, as worked out by Reichert and Rathke, became known and their value had become appreciated, many of the conclusions arrived at by Owen were challenged and ceased to be accepted. It is, however, due to that eminent anatomist to state from my personal knowledge of the condition of anatomical science in this country fifty years ago, that an enormous impulse was given to the study of comparative morphology by his writings, and by the criticisms to which they were subjected.

There can be no doubt that generalised arrangements do exist in the early embryo which, up to a certain stage, are common to animals that in their adult condition present diverse characters, and out of which the forms special to different groups are evolved. As an illustration of this principle, I may refer to the stages of development of the great arteries in the bodies of vertebrate animals. Originally, as the observations of Rathke have taught us, the main arteries are represented by pairs of symmetrically arranged vascular arches, some of which enlarge and constitute the permanent arteries in the adult, whilst others disappear. The increase in size of some of these arches, and the atrophy of others, are so constant for different groups that they constitute anatomical features as distinctive as the modifications in the skeleton itself. Thus in mammals the fourth vascular arch on the left side persists, and forms the arch of the aorta ; in birds the corresponding part of the aorta is an enlargement of the fourth right arch, and in reptiles both arches persist to form the great artery. That this original symmetry exists also in man we know from the fact that now and again his body, instead of corresponding with the mammalian type, has an aortic arch like that which is natural to the bird, and in rarer cases even to the reptile. A type form common to the vertebrata does therefore in such cases exist, capable of evolution in more than one direction.

The reputation of Thomas Henry Huxley as a philosophic comparative anatomist rests largely on his early perception of, and insistence on, the necessity of testing morphological conclusions by a reference to the development of parts and organs, and by applying this principle in his own investigations. The principle is now so generally accepted by both botanists and anatomists that morphological definitions are regarded as depending essentially on the successive phases of the development of the parts under consideration.

The morphological characters exhibited by a plant or animal tend to be hereditarily transmitted from parents to offspring, and the species is perpetuated. In each species the evolution of an individual, through the developmental changes in the egg, follows the same lines in all the individuals of the same species, which possess therefore in common the features called specific characters. The transmission of these characters is due, according to the theory of Weismann, to certain properties possessed by the chromosome constituents of the segmentation nucleus in the fertilised ovum, named by him the germ plasm, which is continued from one generation to another, and impresses its specific character on the egg and on the plant or animal developed from it.

As has already been stated, the special tissues which build up the bodies of the more complex organisms are evolved out of cells which are at first simple in form and appearance. During the evolution of the individual, cells become modified or differentiated in structure and function, and so long as the differentiation follows certain prescribed lines the morphological characters of the species are preserved. We can readily conceive that, as the process of specialisation is going on, modifications or variations in groups of cells and the tissues derived from them, notwithstanding the influence of heredity, may in an individual diverge so far from that which is characteristic of the species as to assume the arrangements found in another species, or even in another order. Anatomists had indeed long recognised that variations from the customary arrangement of parts occasionally appeared, and they described such deviations from the current descriptions as irregularities.

Darwinian Theory.

The signification of the variations which arise in plants and animals had not been apprehended until a flood of light was thrown on the entire subject by the genius of Charles Darwin, who formulated the wide-reaching theory that variations could be transmitted by heredity to younger generations. In this manner he conceived new characters would arise, accumulate, and be perpetuated, which would in the course of time assume specific importance. New species might thus be evolved out of organisms originally distinct from them, and their specific characters would in turn be transmitted to their descendants. By a continuance of this process new species would multiply in many directions, until at length from one or more originally simple forms the earth would become peopled by the infinite varieties of plant and animal organisms which have in past ages inhabited, or do at present inhabit, our globe. The Darwinian theory may therefore be defined as Heredity modified and influenced by Variability. It assumes that there is an heredity quality in the egg, which, if we take the common fowl for an example, shall continue to produce similar fowls. Under conditions, of which we are ignorant, which occasion molecular changes in the cells and tissues of the developing egg, variations might arise in the first instance probably slight, but becoming intensified in successive generations, until at length the descendants would have lost the characters of the fowl and have become another species. No precise estimate has been arrived at, and indeed one does not see how it is possible to obtain it, of the length of years which might be required to convert a variation, capable of being transmitted, into a new and definite specific character.

The circumstances which, according to the Darwinian theory, determined the perpetuation by hereditary transmission of a variety and its assumption of a specific character depended, it was argued, on whether it possessed such properties as enabled the plant or animal in which it appeared to adapt itself more readily to its environment, *i.e.* to the surrounding conditions. If it were to be of use the organism in so far became better adapted to hold its own in the struggle for existence with its fellows and with the forces of nature operating on it. Through the accumulation of useful characters the specific variety was perpetuated by natural selection, so long as the conditions were favourable for its existence, and it survived as being the best fitted to live. In the study of the transmission of variations which may arise in the course of development it should not be too exclusively thought that only those variations are likely to be preserved which can be of service during the life of the individual, or in the perpetuation of the species, and possibly available for the evolution of new

species. It should also be kept in mind that morphological characters can be transmitted by hereditary descent, which, though doubtless of service in some bygone ancestor, are in the new conditions of life of the species of no physiological value. Our knowledge of the structural and functional modifications to be found in the human body, in connection with abnormalities and with tendencies or predisposition to diseases of various kinds, teaches us that characters which are of no use, and indeed detrimental to the individual, may be hereditarily transmitted from parents to offspring through a succession of generations.

Since the conception of the possibility of the evolution of new species from pre-existing forms took possession of the minds of naturalists, attempts have been made to trace out the lines on which it has proceeded. The first to give a systematic account of what he conceived to be the order of succession in the evolution of animals was Ernst Haeckel, of Jena, in a well-known treatise. Memoirs on special departments of the subject, too numerous to particularise, have subsequently appeared. The problem has been attacked along two different lines: the one by embryologists, of whom may be named Kowalewsky, Gegenbaur, Dohrn, Ray Lankester, Balfour and Gaskell, who with many others have conducted careful and methodical inquiries into the stages of development of numerous forms belonging to the two great divisions of the animal kingdom. Invertebrates, as well as vertebrates, have been carefully compared with each other in the bearing of their development and structure on their affinities and descent, and the possible sequence in the evolution of the Vertebrata from the Invertebrata has been discussed. The other method pursued by palæontologists, of whom Huxley, Marsh, Cope, Osborne and Traquair are prominent authorities, has been the study of the extinct forms preserved in the rocks and the comparison of their structure with each other and with that of existing organisms. In the attempts to trace the line of descent the imagination has not unfrequently been called into play in constructing various conflicting hypotheses. Though from the nature of things the order of descent is, and without doubt will continue to be, ever a matter of speculation and not of demonstration, the study of the subject has been a valuable intellectual exercise and a powerful stimulant to research.

We know not as regards time when the fiat went forth, "Let there be Life, and there was Life." All we can say is that it must have been in the far-distant past, at a period so remote from the present that the mind fails to grasp the duration of the interval. Prior to its genesis our earth consisted of barren rock and desolate ocean. When matter became endowed with Life, with the capacity of self-maintenance and of resisting external disintegrating forces, the face of nature began to undergo a momentous change. Living organisms multiplied, the land became covered with vegetation, and multitudinous varieties of plants, from the humble fungus and moss to the stately palm and oak, beautified its surface and fitted it to sustain higher kinds of living beings. Animal forms appeared, in the first instance simple in structure, to be followed by others more complex, until the mammalian type was produced. The ocean also became peopled with plant and animal organisms, from the microscopic diatom to the huge leviathan. Plants and animals acted and reacted on each other, on the atmosphere which surrounded them and on the earth on which they dwelt, the surface of which became modified in character and aspect. At last Man came into existence. His nerve-energy, in addition to regulating the processes in his economy which he possesses in common with animals, was endowed with higher powers. When translated into practical activity it has enabled him throughout the ages to progress from the condition of a rude savage to an advanced stage of civilisation; to produce works in literature, art, and the moral sciences which have exerted, and must continue to exert, a lasting influence on the development of his higher Being; to make discoveries in physical science; to acquire a knowledge of the structure of the earth, of the ocean in its changing aspects, of the atmosphere and the stellar universe, of the chemical composition and physical properties of matter in its various forms, and to analyse, comprehend, and subdue the forces of nature.

By the application of these discoveries to his own purposes Man has, to a large extent, overcome time and space; he has studded the ocean with steamships, girdled the earth with electric wire, tunnelled the lofty Alps, spanned the Forth with a bridge of steel, invented machines and founded industries of all kinds for the promotion of his material welfare, elaborated systems of government fitted for the management of great communities, formulated economic principles, obtained an insight into the laws of health, the causes of infective diseases, and the means of controlling and preventing them.

When we reflect that many of the most important discoveries in abstract science and in its applications have been made during the present century, and indeed since the British Association held its first meeting in the ancient capital of your county sixty-nine years ago, we may look forward with confidence to the future. Every advance in science provides a fresh platform from which a new start can be made. The human intellect is still in process of evolution. The power of application and of concentration of thought for the elucidation of scientific problems is by no means exhausted. In science is no hereditary aristocracy. The army of workers is recruited from all classes. The natural ambition of even the private in the ranks to maintain and increase the reputation of the branch of knowledge which he cultivates affords an ample guarantee that the march of science is ever onwards, and justifies us in proclaiming for the next century, as in the one fast ebbing to a close, that Great is Science, and it will prevail.

SECTION A.

MATHEMATICS AND PHYSICS.

Opening Address by Joseph Larmor, M.A., D.Sc., F.R.S., Pres. C.P.S., President of the Section.

It is fitting that before entering upon the business of the Section we should pause to take note of the losses which our department of science has recently sustained. The fame of Bertrand, apart from his official position as Secretary of the French Academy of Sciences, was long ago universally established by his classical treatise on the Infinitesimal Calculus : it has been of late years sustained by the luminous exposition and searching criticism of his books on the Theory of Probability and Thermodynamics and Electricity. The debt which we owe to that other veteran, G. Wiedemann, both on account of his own researches, which take us back to the modern revival of experimental physics, and for his great and indispensable thesaurus of the science of electricity, cannot easily be overstated. By the death of Sophus Lie, following soon after his return to a chair in his native country Norway, we have lost one of the great constructive mathematicians of the century, who has in various directions fundamentally expanded the methods and conceptions of analysis by reverting to the fountain of direct geometrical intuition. In Italy the death of Beltrami has removed an investigator whose influence has been equally marked on the theories of transcendental geometry and on the progress of mathematical physics. In our own country we have lost in D. E. Hughes one of the great scientific inventors of the age ; while we specially deplore the removal, in his early prime, of one who has recently been well known at these meetings, Thomas Preston, whose experimental investigations on the relations between magnetism and light, combined with his great powers of lucid exposition, marked out for him a brilliant future.

Perhaps the most important event of general scientific interest during the past year has been the definite undertaking of the great task of the international co-ordination of scientific literature ; and it may be in some measure in the prolonged conferences that were necessitated by that object that the recently announced international federation of scientific academies has had its origin. In the important task of rendering accessible the stores of scientific knowledge, the British Association, and in particular this Section of it, has played the part of pioneer. Our annual volumes have long been classical, through the splendid reports of the progress of the different branches of knowledge that have been from time to time contributed to them by the foremost British men of science ; and our work in this direction has received the compliment of successful imitation by the sister Associations on the Continent.

The usual conferences connected with our department of scientific activity have been this year notably augmented by the very successful international federation of mathematicians and of physicists which met a few weeks ago in Paris. The three volumes of reports on the progress of physical-science during the last ten years, for which we are indebted to the initiative of the French Physical Society, will provide an admirable conspectus

of the present trend of activity, and form a permanent record for the history of our subject.

Another very powerful auxiliary to progress is now being rapidly provided by the republication, in suitable form and within reasonable time, of the collected works of the masters of our science. We have quite recently received, in a large quarto volume, the mass of most important unpublished work that was left behind him by the late Prof. J. C. Adams ; the zealous care of Prof. Sampson has worked up into order the more purely astronomical part of the volume ; while the great undertaking, spread over many years, of the complete determination of the secular change of the magnetic condition of the earth, for which the practical preparations had been set on foot by Gauss himself, has been prepared for the press by Prof. W. G. Adams. By the publication of the first volume of Lord Rayleigh's papers a series of memoirs which have formed a main stimulus to the progress of mathematical physics in this country during the past twenty years has become generally accessible. The completed series will form a landmark for the end of the century that may be compared with Young's "Lectures on Natural Philosophy ". for its beginning.

The recent reconstruction of the University of London, and the foundation of the University of Birmingham, will, it is to be hoped, give greater freedom to the work of our University Colleges. The system of examinations has formed an admirable stimulus to the effective acquisition of that general knowledge which is a necessary part of all education. So long as the examiner recognises that his function is a responsible and influential one, which is to be taken seriously from the point of view of moulding the teaching in places where external guidance is helpful, test by examination will remain a most valuable means of extending the area of higher education. Except for workers in rapidly progressive branches of technical science, a broad education seems better adapted to the purposes of life than special training over a narrow range ; and it is difficult to see how a reasonably elastic examination test can be considered as a hardship. But the case is changed when preparation for a specialised scientific profession, or mastery of the lines of attack in an unsolved problem, is the object. The general education has then been presumably finished ; in expanding departments of knowledge, variety rather than uniformity of training should be the aim, and, the genius of a great teacher should be allowed free play without external trammels. It would appear that in this country we have recently been liable to unduly mix up two methods. We have, been starting students on the special and lengthy, though very, instructive, processes which are known as original research at an age when their time would be more profitably employed in, rapidly acquiring a broad basis of knowledge. As a result, we have been extending the examination test from the general knowledge to which it is admirably suited into the specialised activity which is best left to the stimulus of personal interest. Informal contact with competent advisers, themselves imbued with the scientific spirit, who can point the way towards direct, appreciation of the works of the masters of the science, is far more effective than detailed instruction at second hand, as regards growing subjects that have not yet taken on an authoritative form of exposition. Fortunately there seems to be now no lack of such teachers to meet the requirements of the technical colleges that are being established throughout the country.

The famous treatise which opened the modern era by treating magnetism and electricity on a scientific basis appeared just 300 years ago. The author, William Gilbert, M.D., of Colchester, passed from the Grammar School of his native town to St. John's College, Cambridge : soon after taking his first degree, in 1560, he became a Fellow of the College, and seems to have remained in residence, and taken part in its affairs, for about ten years. All through his subsequent career, both at Colchester and afterwards at London, where he attained the highest position in his profession, he was an exact and diligent explorer, first of chemical and then of magnetic and electric phenomena. In the words of the historian Hallam, writing in 1839, " in his Latin treatise on the Magnet he not only collected all the knowledge which others had possessed, but he became at once the father of experimental philosophy in this island " ; and no demur would be raised if Hallam's restriction to this country were removed. Working nearly a century before the time when the astronomical discoveries of Newton had originated the idea of attraction at a distance, he established a complete formulation of the interaction of magnets by what we now call the exploration of their fields of force. His analysis of the facts of magnetic influence, and incidentally of the points in which it differs from electric influence, is virtually the one which Faraday re-introduced. A cardinal advance was achieved, at a time when the Copernican Astronomy had still largely to make its way, by assigning the behaviour of the compass and the dip needle to the fact that the earth itself is a great magnet, by whose field of influence they are controlled. His book passed through many editions on the Continent within forty years; it won the high praise of Galileo. Gilbert has been called "the father of modern electricity" by Priestley, and "the Galileo of magnetism" by Poggendorff.

When the British Association last met at Bradford in 1873 the modern theory which largely reverts to Gilbert's way of formulation, and refers electric and magnetic phenomena to the activity of the æther instead of attractions at a distance, was of recent growth; it had received its classical exposition only two years before by the publication of Clerk Maxwell's treatise. The new doctrine was already widely received in England on its own independent merits. On the Continent it was engaging the strenuous attention of Helmholtz, whose series of memoirs, deeply probing the new ideas in their relation to the prevalent and fairly successful theories of direct action across space, had begun to appear in 1870. During many years the search for crucial experiments that would go beyond the results equally explained by both views met with small success; it was not until 1887 that Hertz, by the discovery of the æthereal radiation of long wave-length emitted from electric oscillators, verified the hypothesis of Faraday and Maxwell and initiated a new era in the practical development of physical science. The experimental field thus opened up was soon fully occupied both in this country and abroad; and the borderland between the sciences of optics and electricity is now being rapidly explored. The extension of experimental knowledge was simultaneous with increased attention to directness of explanation; the expositions of Heaviside and Hertz and other writers fixed attention, in a manner already briefly exemplified by Maxwell himself, on the inherent simplicity of the completed æthereal scheme, when once the theoretical scaffolding employed in its construction and dynamical consolidation is removed; while Poynting's beautiful corollary specifying the path of the transmission of energy through the æther has brought the theory into simple relations with the applications of electrodynamics.

Equally striking has been the great mastery obtained during the last twenty years over the practical manipulation of electric power. The installation of electric wires as the nerves connecting different regions of the earth had attained the rank of accomplished fact so long ago as 1857, when the first Atlantic cable was laid. It was largely the theoretical and practical difficulties, many of them unforeseen, encountered in carrying that great undertaking to a successful issue, that necessitated the elaboration by Lord Kelvin and his coadjutors of convenient methods and instruments for the exact measurement of electric quantities, and thus prepared the foundation for the more recent practical developments in other directions. On the other hand, the methods of theoretical explanation have been in turn improved and simplified through the new ways of considering the phenomena which have been evolved in the course of practical advances on a large scale, such as the improvement of dynamo armatures, the conception and utilisation of magnetic circuits, and the transmission of power by alternating currents. In our time the relations of civilised life have been already perhaps more profoundly altered than ever before, owing to the establishment of practically instantaneous electric communication between all parts of the world. The employment of the same subtle agency is now rapidly superseding the artificial reciprocating engines and other contrivances for the manipulation of mechanical power that were introduced with the employment of steam. The possibilities of transmitting power to great distances at enormous tension, and therefore with very slight waste, along lines merely suspended in the air, are being practically realised; and the advantages thence derived are increased manifold by the almost automatic manner in which the electric power can be transformed into mechanical rotation at the very point where it is desired to apply it. The energy is transmitted at such lightning speed that at a given instant only an exceedingly minute portion of it is in actual transit. When the tension of the alternations is high, the amount of electricity that has to oscillate backwards and forwards on the guiding wires is proportionately diminished, and the frictional waste reduced. At the terminals the direct transmission from one armature of the motor to the other, across the

intervening empty space, at once takes us beyond the province of the pushing and rubbing contacts that are unavoidable in mechanical transmission; while the perfect symmetry and reversibility of the arrangement by which power is delivered from a rotatory alternator at one end, guided by the wires to another place many miles away, where it is absorbed by another alternator with precise reversal of the initial stages, makes this process of distribution of energy resemble the automatic operations of nature rather than the imperfect material connections previously in use. We are here dealing primarily with the flawless continuous medium which is the transmitter of radiant energy across the celestial spaces; the part played by the coarsely constituted material conductor is only that of a more or less imperfect guide which directs the current of æthereal energy. The wonderful nature of this theoretically perfect, though of course practically only approximate, method of abolishing limitations of locality with regard to mechanical power is not diminished by the circumstance that its principle must have been in some manner present to the mind of the first person who fully realised the character of the reversibility of a gramme armature.

In theoretical knowledge a new domain, to which the theory as expounded twenty years ago had little to say, has recently been acquired through the experimental scrutiny of the electric discharge in rarefied gaseous media. The very varied electric phenomena of vacuum tubes, whose electrolytic character was first practically established by Schuster, have been largely reduced to order through the employment of the high exhaustions introduced and first utilised by Crookes. Their study under these circumstances, in which the material molecules are so sparsely distributed as but rarely to interfere with each other, has conduced to enlarged knowledge and verification of the fundamental relations in which the individual molecules stand to all electric phenomena, culminating recently in the actual determination, by J. J. Thomson and others following in his track, of the masses and velocities of the particles that carry the electric discharge across the exhausted space. The recent investigations of the circumstances of the electric dissociation produced in the atmosphere and in other gases by ultra-violet light, the Röntgen radiation, and other agencies, constitute one of the most striking developments in experimental molecular physics since Graham determined the molecular relations of gaseous diffusion and transpiration more than half a century ago. This advance in experimental knowledge of molecular phenomena, assisted by the discovery of the precise and rational effect of magnetism on the spectrum, has brought into prominence a modification or rather development of Maxwell's exposition of electric theory, which was dictated primarily by the requirements of the abstract theory itself; the atoms or ions are now definitely introduced as the carriers of those electric charges which interact across the æther, and so produce the electric fields whose transformations were the main subject of the original theory.

We are thus inevitably led, in electric and æthereal theory, as in the chemistry and dynamics of the gaseous state which is the department of abstract physics next in order of simplicity, to the consideration of the individual molecules of matter. The theoretical problems which had come clearly into view a quarter of a century ago, under Maxwell's lead, whether in the exact dynamical relations of æthereal transmission or in the more fortuitous domain of the statistics of interacting molecules, are those around which investigation is still mainly concentrated; but as the result of the progress in each, they are now tending towards consolidation into one subject. I propose—leaving further review of the scientific aspect of the recent enormous development of the applications of physical science for hands more competent to deal with the practical side of that subject—to offer some remarks on the scope and validity of this molecular order of ideas, to which the trend of physical explanation and development is now setting in so pronounced a manner.

If it is necessary to offer an apology for detaining the attention of the Section on so abstract a topic, I can plead its intrinsic philosophical importance. The hesitation so long felt on the Continent in regard to discarding the highly-developed theories which analysed all physical actions into direct attractions between the separate elements of the bodies concerned, in favour of a new method in which our ideas are carried into regions deeper than the phenomena, has now given place to eager discussion of the potentialities of the new standpoint. There has even appeared a disposition to consider that the Newtonian dynamical principles, which have formed the basis of physical explanation for nearly two centuries, must be replaced in these deeper subjects by a method of direct description of the mere course of phenomena, apart from any attempt to establish causal relations ; the initiation of this method being traced, like that of the Newtonian dynamics itself, to this country. The question has arisen as to how far the new methods of æthereal physics are to be considered as an independent departure, how far they form the natural development of existing dynamical science. In England, whence the innovation came, it is the more conservative position that has all along been occupied. Maxwell was himself trained in the school of physics established in this country by Sir George Stokes and Lord Kelvin, in which the dominating idea has been that of the strictly dynamical foundation of all physical action. Although the pupil's imagination bridged over dynamical chasms, across which the master was not always able to follow, yet the most striking feature of Maxwell's scheme was still the dynamical framework into which it was built. The more advanced reformers have now thrown overboard the apparatus of potential functions which Maxwell found necessary for the dynamical consolidation of his theory, retaining only the final result as a verified descriptive basis for the phenomena. In this way all difficulties relating to dynamical development and indeed consistency are avoided, but the question remains as to how much is thereby lost. In practical electro-magnetics the transmission of power is now the most prominent phenomenon ; if formal dynamics is put aside in the general theory, its guidance must here be replaced by some more empirical and tentative method of describing the course of the transmission and transformation of mechanical energy in the system.

The direct recognition in some form, either explicitly or tacitly, of the part played by the æther, has become indispensable to the development and exposition of general physics ever since the discoveries of Hertz left no further room for doubt that this physical scheme of Maxwell was not merely a brilliant speculation, but constituted, in spite of outstanding gaps and difficulties, a real formulation of the underlying unity in physical dynamics. The domain of abstract physics is in fact roughly divisible into two regions. In one of them we are mainly concerned with interactions between one portion of matter and another portion occupying a different position in space ; such interactions have very uniform and comparatively simple relations ; and the reason is traceable to the simple and uniform constitution of the intervening medium in which they have their seat. The other province is that in which the distribution of the material molecules comes into account. Setting aside the ordinary dynamics of matter in bulk, which is founded on the uniformity of the properties of the bodies concerned and their experimental determination, we must assign to this region all phenomena which are concerned with the unco-ordinated motions of the molecules, including the range of thermal and in part of radiant actions ; the only possible basis for detailed theory is the statistical dynamics of the distribution of the molecules. The far more deep-seated and mysterious processes which are involved in changes in the constitution of the individual molecules themselves are mainly outside the province of physics, which is competent to reason only about permanent material systems ; they must be left to the sciences of chemistry and physiology. Yet the chemist proclaims that he can determine only the results of his reactions and the physical conditions under which they occur ; the character of the bonds which hold atoms in their chemical combinations is at present unknown, although a large domain of very precise knowledge relating, in some diagrammatic manner, to the topography of the more complex molecules has been attained. The vast structure which chemical science has in this way raised on the narrow foundation of the atomic theory is perhaps the most wonderful existing illustration both of the rationality of natural processes and of the analytical powers of the human mind. In a word, the complication of the material world is referable to the vast range of structure and of states of aggregation in the material atoms ; while the possibility of a science of physics is largely due to the simplicity of constitution of the universal medium through which the individual atoms interact on each other.

The reference of the uniformity in the interactions at a distance between material bodies to the part played by the æther is a step towards the elimination of extraneous and random hypotheses about laws of attraction between atoms. It also places that medium on a different basis from matter, in that its mode of activity is simple and regular, whereas intimate material interactions must be of illimitable complexity. This gives strong ground for the view that we should not be tempted towards ex-

plaining the simple group of relations which have been found to define the activity of the æther, by treating them as mechanical consequences of concealed structure in that medium ; we should rather rest satisfied with having attained to their exact dynamical correlation, just as geometry explores or correlates, without explaining, the descriptive and metric properties of space. On the other hand, a view is upheld which considers the pressures and thrusts of the engineer, and the strains and stresses in the material structures by which he transmits them from one place to another, to be the archetype of the processes by which all mechanical effect is transmitted in nature. This doctrine implies an expectation that we may ultimately discover something analogous to structure in the celestial spaces, by means of which the transmission of physical effect will be brought into line with the transmission of mechanical effect by material framework.

At a time when the only definitely ascertained function of the æther was the undulatory propagation of radiant energy across space, Lord Kelvin pointed out that, by reason of the very great velocity of propagation, the density of the radiant energy in the medium at any place must be extremely small in comparison with the amount of energy that is transmitted in a second of time ; this easily led him to the very striking conclusion that, on the hypothesis that the æther is like material elastic media, it is not necessary to assume its density to be more than 10^{-18} of that of water, or its optical rigidity to be more than ten 10^{-8} of that of steel or glass. Thus far the æther would be merely an impalpable material atmosphere for the transference of energy by radiation, at extremely small densities but with very great speed, while ordinary matter would be the seat of practically all this energy. But this way of explaining the absence of sensible influence of the æther on the phenomena of material dynamics lost much of its basis as soon as it was recognised that the same medium must be the receptacle of very high densities of energy in the electric fields around currents and magnets.[1] The other mode of explanation is to consider the æther to be of the very essence of all physical actions, and to correlate the absence of obvious mechanical evidence of its intervention with its regularity and universality.

On this plan of making the æther the essential factor in the transformation of energy as well as its transmission across space, the material atom must be some kind of permanent nucleus that retains around itself an æthereal field of physical influence, such as, for example, a field of strain. We can recognise the atom only through its interactions with other atoms that are so far away from it as to be practically independent systems ; thus our direct knowledge of the atom will be confined to this field of force which belongs to it. Just as the exploration of the distant field of magnetic influence of a steel magnet, itself concealed from view, cannot tell us anything about the magnet except the amount and direction of its magnetisation, so a practically complete knowledge of the field of physical influence of an atom might be expressible in terms of the numerical values of a limited number of physical moments associated with it, without any revelation as to its essential structure or constitution being involved. This will at any rate be the case for ultimate atoms if, as is most likely, the distances at which they are kept apart are large compared with the diameters of the atomic nuclei ; it in fact forms our only chance for penetrating to definite dynamical views of molecular structure. So long as we cannot isolate a single molecule, but must deal observationally with an innumerable distribution of them, even this kind of knowledge will be largely confined to average values. But the last half-century has witnessed the successful application of a new instrument of research, which has removed in various directions the limitations that had previously been placed on the knowledge to which it was possible for human effort to look forward. The spectroscope has created a new astronomy by revealing the constitutions and the unseen internal motions of the stars. Its power lies in the fact that it does take hold of the internal relations of the individual molecule of matter, and provide a very definite and

[1] We can here only allude to Lord Kelvin's recent most interesting mechanical illustrations of a solid æther interacting with material molecules and with itself by attraction at a distance ; unlike the generalised dynamical methods expounded in the text, which can leave the intimate structure of the material molecule outside the problem, a definite working constitution is there assigned to the molecular nucleus. It is pointed out in a continuation that is to appear in the *Phil. Mag.* for September, that a density of æther of the order of only 10^{-9}, which would not appreciably affect the inertia of matter, would involve rigidity comparable with that of steel, and thus permit transmission of magnetic forces by stress ; this solid æther is, however, as usual, taken to be freely permeable to the molecules of matter.

detailed, though far from complete, analysis of the vibratory motions that are going on in it ; these vibrations being in their normal state characteristic of its dynamical constitution, and in their deviations from the normal giving indications of the velocity of its movement and the physical state of its environment. Maxwell long ago laid emphasis on the fact that a physical atomic theory is not competent even to contemplate the vast mass of potentialities and correlations of the past and the future, that biological theory has to consider as latent in a single organic germ containing at most only a few million molecules. On our present view we can accept his position that the properties of such a body cannot be those of a " purely material system," provided, however, we restrict this phrase to apply to physical properties as here defined. But an exhaustive discovery of the intimate nature of the atom is beyond the scope of physics ; questions as to whether it must not necessarily involve in itself some image of the complexity of the organic structures of which it can form a correlated part must remain a subject of speculation outside the domain of that science. It might be held that this conception of discrete atoms and continuous æther must stands, like those of space and time, in intimate relation with our modes of mental apprehension. into which any consistent picture of the external world must of necessity be fitted. In any case it would involve abandonment of all the successful traditions of our subject if we ceased to hold that our analysis can be formulated in a consistent and complete manner, so far as it goes, without being necessarily an exhaustive account of phenomena that are beyond our range of experiment. Such phenomena may be more closely defined as those connected with the processes of intimate combination of the molecules : they include the activities of organic beings which all seem to depend on change of molecular structure.

If, then, we have so small a hold on the intimate nature of matter, it will appear all the more striking that physicists have been able precisely to divine the mode of operation of the intangible æther, and to some extent explore in it the fields of physical influence of the molecules. On consideration we recognise that this knowledge of fundamental physical interaction has been reached by a comparative process. The mechanism of the propagation of light could never have been studied in the free æther of space alone. It was possible, however, to determine the way in which the characteristics of optical propagation are modified, but not wholly transformed, when it takes place in a transparent material body instead of empty space. The change in fact arises on account of the æther being entangled with the network of material molecules ; but inasmuch as the length of a single wave of radiation covers thousands of these molecules the wave-motion still remains uniform and does not lose its general type. A wider variation of the experimental conditions has been provided for our examination in the case of those substances in which the phenomenon of double refraction pointed to a change of the æthereal properties which varied in different directions ; and minute study of this modification has proved sufficient to guide to a consistent appreciation of the nature of this change, and therefore of the mode of æthereal propagation that is thus altered. In the same way, it was the study and development of the manner in which the laws of electric phenomena. in material bodies had been unravelled by Ampère and Faraday that guided Faraday himself and Maxwell—who were impressed with the view that the æther was at the bottom of it all—in their progress towards an application of similar laws to æther devoid of matter, such as would complete a scheme of continuous action by consistently interconnecting the material bodies and banishing all untraced interaction across empty space. Maxwell in fact chose to finally expound the theory by ascribing to the æther of free space a dielectric constant and a magnetic constant of the same types as had been found to express the properties of material media, thus extending the seat of the phenomena to all space on the plan of describing the activity of the æther in terms of the ordinary electric ideas. The converse mode of development, starting with the free æther under the directly dynamical form which has been usual in physical optics, and introducing the influence of the material atoms through the electric charges which are involved in their constitution,[1] was hardly employed by him ;

[1] In 1870 Maxwell, while admiring the breadth of the theory of Weber, which is virtually based on atomic charges combined with action at a distance, still regarded it as irreconcilable with his own theory, and left to the future the question as to why "Theories apparently so fundamentally opposed should have so large a field of truth common to both."—"Scientific Papers," ii. p. 228.

in part, perhaps, because, owing to the necessity of correlating his theory with existing electric knowledge and the mode of its expression, he seems never to have reached the stage of moulding it into a completely deductive form.

The dynamics of the æther, in fact the recognition of the existence of an æther, has thus, as a matter of history, been reached through study of the dynamical phenomena of matter. When the dynamics of a material system is worked up to its purest and most general form, it becomes a formulation of the relations between the succession of the configurations and states of motion of the system, the assistance of an independent idea of force not being usually required. We can, however, only attain to such a compact statement when the system is self-contained, when its motion is not being dissipated by agencies of frictional type, and when its connections can be directly specified by purely geometrical relations between the co-ordinates, thus excluding such mechanisms as rolling contacts. The course of the system is then in all cases determined by some form or other of a single fundamental property, that any alteration in any small portion of its actual course must produce an increase in the total "Action" of the motion. It is to be observed that in employing this law of minimum as regards the Action expressed as an integral over the whole time of the motion, we no more introduce the future course as a determining influence on the present state of motion than we do in drawing a straight line from any point in any direction, although the length of the line is the minimum distance between its ends. In drawing the line piece by piece we have to make tentative excursions into the immediate future in order to adjust each element into straightness with the previous element ; so in tracing the next stage of the motion of a material system we have similarly to secure that it is not given any such directions as would unduly increase the Action. But whatever views may be held as to the ultimate significance of this principle of Action, its importance, not only for mathematical analysis, but as a guide to physical exploration, remains fundamental. When the principles of the dynamics of material systems are refined down to their ultimate common basis, this principle of minimum is what remains. Hertz preferred to express its contents in the form of a principle of straightness of course or path. It will be recognised, on the lines already indicated, that this is another mode of statement of the same fundamental idea ; and the general equivalence is worked out by Hertz on the basis of Hamilton's development of the principles of dynamics. The latter mode of statement may be adaptable so as to avoid the limitations which restrict the connections of the system, at the expense, however, of introducing new variables ; if, indeed, it does not introduce gratuitous complexity for purposes of physics to attempt to do this. However these questions may stand, this principle of straightness or directness of path forms, wherever it applies, the most general and comprehensive formulation of purely dynamical action : it involves in itself the complete course of events. In so far as we are given the algebraic formula for the time-integral which constitutes the Action, expressed in terms of any suitable co-ordinates, we know implicitly the whole dynamical constitution and history of the system to which it applies. Two systems in which the Action is expressed by the same formula are mathematically identical, are physically precisely correlated, so that they have all dynamical properties in common. When the structure of a dynamical system is largely concealed from view, the safest and most direct way towards an exploration of its essential relations and connections, and in fact towards answering the prior question as to whether it is a purely dynamical system at all, is through this order of ideas. The ultimate test that a system is a dynamical one is not that we shall be able to trace mechanical stresses throughout it, but that its relations can be in some way or other consolidated into accordance with this principle of minimum Action. This definition of a dynamical system in terms of the simple principle of directness of path may conceivably be subject to objection as too wide ; it is certainly not too narrow ; and it is the conception which has naturally been evolved from two centuries of study of the dynamics of material bodies. Its very great generality may lead to the objection that we might completely formulate the future course of a system in its terms, without having obtained a working familiarity with its details, of the kind to which we have become accustomed in the analysis of simple material systems ; but our choice is at present between this kind of formulation, which is a real and essential one, and an empirical description of the course of phenomena combined with explanations relating to more or less isolated groups. The list of great names, including Kelvin, Maxwell, Helmholtz, that have been associated with the employment of the principle for the elucidation of the relations of deep-seated dynamical phenomena is a strong guarantee that we shall do well by making the most of this clue.

Are we then justified in treating the material molecule, so far as revealed by the spectroscope, as a dynamical system coming under this specification? Its intrinsic energy is certainly permanent and not subject to dissipation; otherwise the molecule would gradually fade out of existence. The extreme precision and regularity of detail in the spectrum shows that the vibrations which produce it are exactly synchronous, whatever be their amplitude, and in so far resemble the vibrations of small amplitude in material systems. As all indications point to the molecule being a system in a state of intrinsic motion, like a vortex ring, or a stellar system in astronomy, we must consider these radiating vibrations to take place around a steady state of motion which does not itself radiate, not around a state of rest. Now not the least of the advantages possessed by the Action principle, as a foundation for theoretical physics, is the fact that its statement can be adapted to systems involving in their constitution permanent steady motions of this kind, in such a way that only the variable motions superposed on them come into consideration. The possibilities as regards physical correlation of thus introducing permanent motional states as well as permanent structure into the constitution of our dynamical systems have long been emphasised by Lord Kelvin;[1] the effective adaptation of abstract dynamics to such systems was made independently by Kelvin and Routh about 1877; the more recent exposition of the theory by Helmholtz has directed general attention to what is undoubtedly the most significant extension of dynamical analysis which has taken place since the time of Lagrange.

Returning to the molecules, it is now verified that the Action principle forms a valid foundation throughout electrodynamics and optics; the introduction of the æther into the system has not affected its application. It is therefore a reasonable hypothesis that the principle forms an allowable foundation for the dynamical analysis of the radiant vibrations in the system formed by a single molecule and surrounding æther; and the knowledge which is now accumulating, both of the orderly grouping of the lines of the spectrum and of the modifications impressed on these lines by a magnetic field or by the density of the matter immediately surrounding the vibrating molecule, can hardly fail to be fruitful for the dynamical analysis of its constitution. But let it be repeated that this analysis would be complete when a formula for the dynamical energy of the molecule is obtained, and would go no deeper. Starting from our definitely limited definition of the nature of a dynamical system, the problem is merely to correlate the observed relations of the periods of vibration in a molecule, when it has come into a steady state as regards constitution and is not under the influence of intimate encounter with other molecules.

It may be recalled incidentally that the generalised Maxwell-Boltzmann principle of the equable distribution of the acquired store of kinetic energy of the molecule, among its various possible independent types of motion, is based directly on the validity of the Action principle for its dynamics. In the demonstrations usually offered the molecule is considered to have no permanent or constitutive energy of internal motion. It can, however, be shown, by use of the generalisation aforesaid of the Action principle, that no discrepancy will arise on that account. Such intrinsic kinetic energy virtually adds on to the potential energy of the system; and the remaining or acquired part of the kinetic energy of the molecule may be made the subject of the same train of reasoning as before.

Let us now return to the general question whether our definition of a dynamical system may not be too wide. As a case in point, the single principle of Action has been shown to provide a definite and sufficient basis for electrodynamics; yet when, for example, one armature of an electric motor pulls the other after it without material contact, and so transmits mechanical power, no connection between them is indicated by the principle such as could by virtue of internal stress transmit the pull. The essential feature of the transmission of a pull by stress across a medium is that each element of volume of the medium

[1] For a classical exposition see his Brit. Assoc. Address of 1884 on "Steps towards a Kinetic Theory of Matter," reprinted in "Popular Lectures and Addresses," vol. i.

acts by itself, independently of the other elements. The stress excited in any element depends on the strain or other displacement occurring in that element alone; and the mechanical effect that is transmitted is considered as an extraneous force applied at one place in the medium, and passed on from element to element through these internal pressures and tractions until it reaches another place. We have, however, to consider two atomic electric charges as being themselves some kind of strain configurations in the æther; each of them already involves an atmosphere of strain in the surrounding æther which is part of its essence, and cannot be considered apart from it; each of them essentially pervades the entire space, though on account of its invariable character we consider it as a unit. Thus we appear to be debarred from imagining the æther to act as an elastic connection which is merely the agent of transmission of a pull from the one nucleus to the other, because there are already stresses belonging to and constituting an intrinsic part of the terminal electrons, which are distributed all along the medium. Our Action criterion of a dynamical system, in fact, allows us to reason about an electron as a single thing, notwithstanding that its field of energy is spread over the whole medium; it is only in material solid bodies, and in problems in which the actual sphere of physical action of the molecule is small compared with the smallest element of volume that our analysis considers, that the familiar idea of transmission of force by simple stress can apply. Whatever view may ultimately commend itself, this question is one that urgently demands decision. A very large amount of effort has been expended by Maxwell, Helmholtz, Heaviside, Hertz, and other authorities in the attempt to express the mechanical phenomena of electrical action in terms of a transmitting stress. The analytical results up to a certain point have been promising, most strikingly so at the beginning, when Maxwell established the mathematical validity of the way in which Faraday was accustomed to represent to himself the mechanical interactions across space, in terms of a tension along the lines of force equilibrated by an equal pressure preventing their expansion sideways. According to the views here developed, that ideal is an impossible one; if this could be established to general satisfaction the field of theoretical discussion would be much simplified.

This view that the atom of matter is, so far as regards physical actions, of the nature of a structure in the æther involving an atmosphere of æthereal strain all around it, not a small body which exerts direct actions at a distance on other atoms according to extraneous laws of force, was practically foreign to the eighteenth century, when mathematical physics was modelled on the Newtonian astronomy and dominated by its splendid success. The scheme of material dynamics, as finally compactly systematised by Lagrange, had therefore no direct relation to such a view, although it has proved wide enough to include it. The remark has often been made that it is probably owing to Faraday's mathematical instinct, combined with his want of acquaintance with the existing analysis, that the modern theory of the æther obtained a start from the electric side. Through his teaching and the weight of his authority, the notion of two electric currents exerting their mutual forces by means of an intervening medium, instead of by direct attraction across space, was at an early period firmly grasped in this country. In 1845 Lord Kelvin was already mathematically formulating, with most suggestive success, continuous elastic connections, by whose strain the fields of activity of electric currents or of electric distributions could be illustrated; while the exposition of Maxwell's interconnected scheme, in the earlier form in which it relied on concrete models of the electric action, goes back almost to 1860. Corresponding to the two physical ideals of isolated atoms exerting attraction at a distance, and atoms operating by atmospheres of æthereal strain, there are, as already indicated, two different developments of dynamical theory. The original Newtonian equations of motion determined the course of a system by expressing the rates at which the velocity of each of its small parts or elements is changing. This method is still fully applicable to those problems of gravitational astronomy in which dynamical explanation was first successful on a grand scale, the planets being treated as point-masses, each subject to the gravitational attraction of the other bodies. But the more recent development of the dynamics of complex systems depends on the fact that analysis has been able to reduce within manageable limits the number of varying quantities whose course is to be explicitly traced, through taking advantage of those internal relations of the parts of the system that are invariable, either geometrically or dynamically. Thus, to take the simplest case, the dynamics of a solid body can be confined to a discussion of its three components of translation and its three components of rotation, instead of the motion of each element of its mass. With the number of independent co-ordinates thus diminished, when the initial state of the motion is specified the subsequent course of the complete system can be traced ; but the course of the changes in any part of it can only be treated in relation to the motion of the system as a whole. It is just this mode of treatment of a system as a whole that is the main characteristic of modern physical analysis. The way in which Maxwell analysed the interactions of a system of linear electric currents, previously treated as if each were made up of small independent pieces or elements, and accumulated the evidence that they formed a single dynamical system, is a trenchant example. The interactions of vortices in fluid form a very similar problem, which is of special note in that the constitution of the system is there completely known in advance, so that the two modes of dynamical exposition can be compared. In this case the older method forms independent equations for the motion of each material element of the fluid, and so requires the introduction of the stress—here the fluid pressure—by which dynamical effect is passed on to it from the surrounding elements : it corresponds to a method of contact action. But Helmholtz opened up new ground in the abstract dynamics of continuous media when he recognised (after Stokes) that, if the distribution of the velocity of spin at those places in the fluid where the motion is vortical be assigned, the motion in every part of the fluid is therein kinematically involved. This, combined with the theorem of Lagrange and Cauchy, that the spin is always confined to the same portions of the fluid, formed a starting-point for his theory of vortices, which showed how the subsequent course of the motion can be ascertained without consideration of pressure or other stress.

The recognition of the permanent state of motion constituting a vortex ring as a determining agent as regards the future course of the system was in fact justly considered by Helmholtz as one of his greatest achievements. The principle had entirely eluded the attention of Lagrange and Cauchy and Stokes, who were the pioneers in this fundamental branch of dynamics, and had virtually prepared all the necessary analytical material for Helmholtz's use. The main import of this advance lay, not in the assistance which it afforded to the development of the complete solution of special problems in fluid motion, but in the fact that it constituted the discovery of the types of permanent motion of the system, which could combine and interact with each other without losing their individuality,[1] though each of them pervaded the whole field. This rendered possible an entirely new mode of treatment ; and mathematicians who were accustomed, as in astronomy, to aim directly at the determination of all the details of the special case of motion, were occasionally slow to apprehend the advantages of a procedure which stopped at formulating a description of the nature of the interaction between various typical groups of motions into which the whole disturbance could be resolved.

The new train of ideas introduced into physics by Helmholtz was thus consolidated and emphasised by Helmholtz's investigations of 1858 in the special domain of hydrodynamics. In illustration let us consider the fluid medium to be pervaded by permanent vortices circulating round solid rings as cores : the older method of analysis would form equations of motion for each element of the fluid, involving the fluid pressure, and by their integration would determine the distribution of pressure on each solid ring, and thence the way it moves. This method is hardly feasible even in the simplest cases. The natural plan is to make use of existing simplifications by regarding each vortex as a permanent reality, and directly attacking the problem of its interactions with the other vortices. The energy of the fluid arising from the vortex motion can be expressed in terms of the positions and strengths of the vortices alone ; and then the principle of Action, in the generalised form which includes steady motional configurations as well as constant material configurations, affords a method of deducing the motions of the cores and the interactions between them. If the cores are thin they in fact interact mechanically, as Lord Kelvin and Kirchhoff proved, in the same manner as linear electric currents would do ; though the impulse thence derived towards a direct hydrokinetic explanation of electro-magnetics was damped by the fact

[1] We may compare G. W. Hill's more recent introduction of the idea of permanent orbits into physical astronomy.

that repulsion and attraction have to be interchanged in the analogy. The conception of vortices, once it has been arrived at, forms the natural physical basis of investigation, although the older method of determining a distribution of pressure-stress throughout the fluid and examining how it affects the cores is still possible ; that stress, however, is not simply transmitted, as it has to maintain the changes of velocity of the various portions of the fluid. But if the vortices have no solid cores we are at a loss to know where even this pressure can be considered as applied to them ; if we follow up the stress, we lose the vortex ; yet a fluid vortex can nevertheless illustrate an atom of matter, and we can consider such atoms as exerting mutual forces, only these forces cannot be considered as transmitted through the agency of fluid pressure. The reason is that the vortex cannot now be identified with a mere core bounded by a definite surface, but is essentially a configuration of motion extending throughout the medium.

Thus we are again in face of the fundamental question whether all attempts to represent the mechanical interactions of electro-dynamic systems, as transmitted from point to point by means of simple stress, are not doomed to failure ; whether they do not, in fact, introduce unnecessary and insurmountable difficulty into the theory. The idea of identifying an atom with a state of strain or motion, pervading the region of the æther around its nucleus, appears to demand wider views as to what constitutes dynamical transmission. The idea that any small portion of the primordial medium can be isolated, by merely introducing tractions acting over its surface and transmitted from the surrounding parts, is no longer appropriate or consistent : a part of the dynamical disturbance in that element of the medium is on this hypothesis already classified as belonging to, and carried along with, atoms that are outside it but in its neighbourhood—and this part must not be counted twice over. The law of Poynting relating to the paths of the transmission of energy is known to hold in its simple form only when the electric charges or currents are in a steady state ; when they are changing their positions or configurations their own fields of intrinsic energy are carried along with them.

It is not surprising, considering the previous British familiarity with this order of ideas, that the significance for general physics of Helmholtz's doctrine of vortices was eagerly developed in this country, in the form in which it became embodied through Lord Kelvin's famous illustration of the constitution of the matter, as consisting of atoms with separate existence and mutual interactions. This vortex-atom theory has been a main source of physical suggestion because it presents, on a simple basis, a dynamical picture of an ideal material system, atomically constituted, which could go on automatically without extraneous support. The value of such a picture may be held to lie, not in any supposition that this is the mechanism of the actual world laid bare, but in the vivid illustration it affords of the fundamental postulate of physical science, that mechanical phenomena are not parts of a scheme too involved for us to explore, but rather present themselves in definite and consistent correlations, which we are able to disentangle and apprehend with continually increasing precision.

It would be an interesting question to trace the origin of our preference for a theory of transmission of physical action over one of direct action at a distance. It may be held that it rests on the same order of ideas as supplies our conception of force : that the notion of effort which we associate with change of the motion of a body involves the idea of a mechanical connection through which that effort is applied. The mere idea of a transmitting medium would then be no more an ultimate foundation for physical explanation than that of force itself. Our choice between direct distance action and mediate transmission would thus be dictated by the relative simplicity and coherence of the accounts they give of the phenomena : this is, in fact, the basis on which Maxwell's theory had to be judged until Hertz detected the actual working of the medium. Instantaneous transmission is to all intents action at a distance, except in so far as the law of action may be more easily formulated in terms of the medium than in a direct geometrical statement.

In connection with these questions it may be permitted to refer to the eloquent and weighty address recently delivered by M. Poincaré to the International Congress of Physics. M. Poincaré accepts the principle of Least Action as a trustworthy basis for the formulation of physical theory, but he imposes the condition that the results must satisfy the Newtonian law of equality of action and reaction between each pair of bodies concerned, considered by themselves; this, however, he would allow to be satisfied indirectly, if the effects could be traced across the intervening æther by stress, so that the tractions on the two sides of each ideal interface are equal an l opposite.[1] As above argued, this view appears to exclude *ab initio* all atomic theories of the general type of vortex atoms, in which the energy of the atom is distributed throughout the medium instead of being concentrated in a nucleus; and this remark seems to go to the root of the question. On the other hand, the position here asserted is that recent dynamical developments have permitted the extension of the principle of Action to systems involving permanent motions, whether obvious or latent, as part of their constitution; that on this wider basis the atom may itself involve a state of steady disturbance extending through the medium, instead of being only a local structure acting by push and pull. The possibilities of dynamical explanation are thus enlarged. The most definite type of model yet imagined of the physical interaction of atoms through the æther is, perhaps, that which takes the æther to be a rotationally elastic medium after the manner of MacCullagh and Rankine, and makes the ultimate atom include the nucleus of a permanent rotational strain-configuration, which as a whole may be called an electron. The question how far this is a legitimate and effective model stands by itself, apart from the dynamics which it illustrates; like all representations it can only cover a limited ground. For instance, it cannot claim to include the internal structure of the nucleus of an atom or even of an electron; for purposes of physical theory that problem can be put aside, it may even be treated as inscrutable. All that is needed is a postulate of free mobility of this nucleus through the æther. This is definitely hypothetical, but it is not an unreasonable postulate because a rotational æther has the properties of a perfect fluid medium except where differentially rotational motions are concerned, and so would not react on the motion of any structure moving through it except after the manner of an apparent change of inertia. It thus seems possible to hold that such a model forms an allowable representation of the dynamical activity of the æther, as distinguished from the complete constitution of the material nuclei between which that medium establishes connection.

At any rate, models of this nature have certainly been most helpful in Maxwell's hands towards the effective intuitive grasp of a scheme of relations as a whole, which might have proved too complex for abstract unravelment in detail. When a physical model of concealed dynamical processes has served this kind of purpose, when its content has been explored and estimated, and has become familiar through the introduction of new terms and ideas, then the ladder by which we have ascended may be kicked away, and the scheme of relations which the model embodied can stand forth in severely abstract form. Indeed, many of the most fruitful branches of abstract mathematical analysis itself have owed their start in this way to concrete physical conceptions. This gradual transition into abstract statement of physical relations in fact amounts to retaining the essentials of our working models while eliminating the accidental elements involved in them; elements of the latter kind must always be present because otherwise the model would be identical with the thing which it represents, whereas we cannot expect to mentally grasp all aspects of the content of even the simplest phenomena. Yet the abstract standpoint is always attained through the concrete; and for purposes of instruction such models, properly guarded, do not perhaps ever lose their value: they are just as legitimate aids as geometrical diagrams, and they have the same kind of limitations. In Maxwell's words, "for the sake of persons of these different types scientific truth should be presented in different forms, and should be regarded as equally scientific whether it appear in the robust form and the vivid colouring of a physical illustration, or in the tenuity and paleness of a symbolical expression." The other side of the picture, the necessary incompleteness of even our legitimate images and modes of representation, comes out in the despairing opinion of Young ("Chromatics," 1817), at a time when his faith in the undulatory theory of light had been eclipsed by Malus's discovery of the phenomena of polarisation by reflection, that this difficulty "will probably long remain, to mortify the vanity of an

ambitious philosophy, completely unresolved by any theory": not many years afterwards the mystery was solved by Fresnel.

This process of removing the intellectual scaffolding by which our knowledge is reached, and preserving only the final formulæ which express the correlations of the directly observable things, may moreover readily be pushed too far. It asserts the conception that the universe is like an enclosed clock that is wound up to go, and that accordingly we can observe that it is going, and can see some of its more superficial movements, but not much of them; that thus, by patient observation and use of analogy, we can compile, in merely tabular form, information as to the manner in which it works and is likely to go on working, at any rate for some time to come; but that any attempt to probe the underlying connection is illusory or illegitimate. As a theoretical precept this is admirable. It minimises the danger of our ignoring or forgetting the limitations of human faculty, which can only utilise the imperfect representations that the external world impresses on our senses. On the other hand such a remainder has rarely been required by the master minds of modern science, from Descartes and Newton onwards, whatever their theories may have been. Its danger as a dogma lies in its application. Who is to decide, without risk of error, what is essential fact and what is intellectual scaffolding? To which class does the atomic theory of matter belong? That is, indeed, one of the intangible things which it is suggested may be thrown overboard, in sorting out and classifying our scientific possessions. Is the mental idea or image, which suggests, and alone can suggest, the experiment that adds to our concrete knowledge, less real than the bare phenomenal uniformity which it has revealed? Is it not, perhaps, more real in that the uniformities might not have been there in the absence of the mind to perceive them?

No time is now left for review of the methods of molecular dynamics. Here our knowledge is entirely confined to steady states of the molecular system: it is purely statical. In ordinary statics and the dynamics of undisturbed steady notions, the form of the energy function is the sufficient basis of the whole subject. This method is extended to thermodynamics by making use of the mechanically available energy of Rankine and Kelvin, which is a function of the bodily configuration and chemical constitution and temperature of the system, whose value cannot under any circumstances spontaneously increase, while it will diminish in any operation which is not reversible. In the statics of systems in equilibrium or in steady motion, this method of energy is a particular case of the method of Action; but in its extension to thermal statics it is made to include chemical as well as configurational changes, and a new point appears to arise. Whether we do or do not take it to be possible to trace the application of the principle of Action throughout the process of chemical combination of two molecules, we certainly here postulate that the static case of that principle, which applies to steady systems, can be extended across chemical combinations. The question is suggested whether extension would also be valid to transformations which involve vital processes. This seems to be still considered an open question by the best authorities. If it be decided in the negative a distinction is involved between vital and merely chemical processes.

It is now taken as established that vital activity cannot create energy, at any rate in the long run, which is all that can from the nature of the case be tested. It seems not unreasonable to follow the analogy of chemical actions, and assert that it cannot in the long run increase the mechanical availability of energy—that is, considering the organism as an apparatus for transforming energy without being itself in the long run changed. But we cannot establish a Carnot cycle for a portion of an organism, nor can we do so for a limited period of time; there might be creation of availability accompanied by changes in the organism itself, but compensated by destruction and the inverse changes a long time afterwards. This amounts to asserting that where, as in a vital system or even in a simple molecular combination, we are unable to trace or even assert complete dynamical sequence, exact thermodynamic statements should be mainly confined to the activity of the existing organism as a whole; it may transform inorganic material without change of energy and without gain of availability, although any such statements would be inappropriate and unmeaning as regards the details of the processes that take place inside the organism itself.

In any case it would appear that there is small chance of reducing these questions to direct dynamics; we should rather

---

[1] *Cf.* also Hertz on the electro-magnetic equations, § 12, *Wied. Ann.* 1890. The problem of merely replacing a system of forces by a statical stress is widely indeterminate, and therefore by itself unreal; the actual question is whether any such representation can be co-ordinated with existing dynamics.

regard Carnot's principle, which includes the law of uniformity of temperature and is the basis of the whole theory, as a property of statistical type confined to stable or permanent aggregations of matter. Thus no dynamical proof from molecular considerations could be regarded as valid unless it explicitly restricted the argument to permanent systems; yet the conditions of permanency are unknown except in the simpler cases. The only mode of discussion that is yet possible is the method of dynamical statistics of molecules introduced by Maxwell. Now statistics is a method of arrangement rather than of demonstration. Every statistical argument requires to be verified by comparison with the facts, because it is of the essence of this method to take things as fortuitously distributed except in so far as we know the contrary; and we simply may not know essential facts to the contrary. For example, if the interaction of the æther or other cause produces no influence to the contrary, the presumption would be that the kinetic energy acquired by a molecule is, on the average, equally distributed among its various independent modes of motion, whether vibrational or translational. Assuming this type of distribution to be once established in a gaseous system, the dynamics of Boltzmann and Maxwell show that it must be permanent. But its assumption in the first instance is a result rather of the absence than of the presence of knowledge of the circumstances, and can be accepted only so far as it agrees with the facts; our knowledge of the facts of specific heat shows that it must be restricted to modes of motion that are homologous. In the words of Maxwell, when he first discovered in 1860, to his great surprise, that in a system of colliding rigid atoms the energy would always be equally divided between translatory and rotatory motions, it is only necessary to assume, in order to evade this unwelcome conclusion, that "something essential to the complete statement of the physical theory of molecular encounters must have hitherto escaped us."

Our survey thus tends to the result, that as regards the simple and uniform phenomena which involve activity of finite regions of the universal æther, theoretical physics can lay claim to constructive functions, and build up a definite scheme; but in the domain of matter the most that it can do is to accept the existence of such permanent molecular systems as present themselves to our notice, and fit together an outline plan of the more general and universal features in their activity. Our well-founded belief in the rationality of natural processes asserts the possibility of this, while admitting that the intimate details of atomic constitution are beyond our scrutiny and provide plenty of room for processes that transcend finite dynamical correlation.

NOTES.

M. FAYE has been elected a Foreign Member of the Reale Accademia dei Lincei of Rome.

DR. OUSTALET has been appointed professor of zoology in the Paris Natural History Museum, in succession to the late Prof. Milne-Edwards.

WE regret to see the announcement of the death of Mr. Henry Sidgwick, late professor of moral philosophy at Cambridge.

SIR JOHN B. LAWES, BART., F.R.S., whose agricultural experiments at Rothamsted are of world-wide renown, died on Friday last, at eighty-six years of age.

THE announcement in *Science* that Prof. J. E. Keeler, director of the Lick Observatory, and the author of many important papers on astrophysics, died in San Francisco on August 12, from the effects of heart disease, will be received by astronomers with much regret. Prof. Keeler was only forty-three years of age.

IT has been officially notified that a death which occurred in hospital at Glasgow on Monday in last week was due to true bubonic plague. The presence of the disease is suspected in several cases of illness under treatment.

THE Committee on Water-tube Boilers in the Navy has now been completed by the selection of Dr. John Inglis, lately

president of the Institution of Engineers and Shipbuilders in Scotland, and vice-president of the Institution of Naval Architects.

THE Melbourne correspondent of the *Times* states that, in compliance with a request of the Royal Geographical Society and other British scientific bodies, Prof. Baldwin Spencer has received leave of absence from the Melbourne University for one year, to enable him to study the customs and beliefs of the natives of the northern portion of South Australia.

THE Berlin Academy of Sciences has made the following grants, in addition to those already announced (p. 394): Dr. Holtermann, Berlin, for a botanical expedition to Ceylon, 4000 marks; Prof. Ludolf Krehl, Greifswald, for experiments on respiration, 1500 marks; Prof. Julius Tafel, Würzburg, for the continuation of his work on electrolysis, 100 marks; Dr. Benno Wandolleck, Dresden, for the investigation of the morphology of diptera, 800 marks.

THE names of one hundred eminent Americans no longer living are to be engraved in the Hall of Fame of the New York University. *Science* states that the following names of men of science have been proposed: John Adams Audubon, Spencer F. Baird, Alexander D. Bache, Nathaniel Bowditch, William Chauvenet, Henry Draper, James P. Espy, Asa Gray, Robert Hare, Joseph Henry, Edward Hitchcock, Isaac Lea, Matthew Fontaine Maury, Maria Mitchell, Benjamin Peirce, David Rittenhouse, Benjamin Silliman, Benjamin Thompson, John Torrey.

THE Marconi Wireless Telegraph Company have contracted to supply the Admiralty with Marconi apparatus for thirty-two ships and stations. The test of efficiency which has to be satisfied is that the instruments shall enable communication to be carried on between a fitted ship in Portsmouth Harbour and a fitted ship at Portland, a distance of about sixty-five miles, with a good deal of land between, including the Dorsetshire Hills, making it about ninety miles by sea. A trial set of the apparatus successfully fulfilled the conditions a few days ago.

THE death is announced of Dr. W. H. Lowe, formerly president of the Royal College of Physicians of Edinburgh. Dr. Lowe held several important positions in Edinburgh, among others those of president of the Royal Medical Society, and vice-president, subsequently president, of the Royal Botanic Society. He was elected a Fellow of the Royal College of Physicians of Edinburgh in 1846, and president of that college in 1873. At the meeting of the British Medical Association in Edinburgh in 1875 he presided over the section of psychology, and delivered the address before that section.

THE eleventh annual general meeting of the members of the Institution of Mining Engineers will be held at Bristol on Tuesday, September 18. Among the papers to be read, or taken as read, are the following:—The geological features of the Somerset and Bristol coal-field, with special reference to the physical geology of the Somersetshire Basin, by Mr. James McMurtrie; methods of working the thin coal-seams of the Bristol and Somerset coal-field, by Mr. George E. J. McMurtrie; the analogy between the gold "cintas" of Columbia and the auriferous gravels of California, by Mr. Edward Gledhill; the theory of the equivalent orifice treated graphically, by Mr. H. W. Halbaum; development and working of minerals in the Leon district, Spain, by Mr. J. A. Jones; and the geological age of the gold-deposits of Victoria, Australia, by Mr. James Stirling.

THE programme of the meeting of the Iron and Steel Institute, to be held in Paris on September 18-21, under the presidency of Sir William Roberts-Austen, has just been issued.

The following are subjects of papers to be brought before the meeting :—The development of the iron and steel industries in France since 1889, by H. Pinget ; iron and steel from the point of view of the " phase-doctrine," by Prof. Bakhuis-Roozeboom ; iron and steel at the Paris Exhibition, by Prof. H. Bauerman ; American methods of testing iron and steel, by Mr. Albert Ladd Colby ; rolling-mills, by Mr. Louis Katona ; the constitution of slags, by Baron H. von Jüptner ; a new method of producing high temperatures, by Mr. Ernest F. Lange ; the action of aluminium on the carbon of cast iron, by Messrs. Godfrey Melland and H. W. Waldron ; the present position of the solution theory of carburised iron, by Dr. A. Stansfield ; iron and phosphorus, with appendixes on (1) eutectics, (2) solid solutions, (3) method of determining free phosphide of iron in iron and steel, and (4) heat-tinting metal sections for microscopic examination, by Mr. J. E. Stead.

. WE learn from the *Forres, Elgin and Nairn Gazette* of August 29 that a serious flood, due to the bursting of the Sanquhar reservoir on the morning of August 23, wrought great havoc over the western part of Forres. Since the great Moray floods of 1829, described by Sir Thomas Dick Lauder, the district has not suffered such a disaster. In that year the Findhorn was the main cause of flooding ; in the present case injury was done by the breaking down of the embankment which dammed up the waters in a valley on the Sanquhar estate to the east of the Findhorn and a little south of Forres. The dam formed a reservoir of from eight to twelve acres. On Wednesday morning, August 22, the area was only partially covered with water, from eight to ten feet below the level of the overflow. Within twelve hours an inch and a half of rain fell. The reservoir filled rapidly, and by 3 a.m. on Thursday the water was rushing down the overflow, which was only thirty feet wide. Shortly before 5 a.m. the immense breastwork burst outwards from top to bottom in one mass, about twenty feet wide, close to the overflow, and the waters rushed wildly out. Near by an iron bridge with a concrete pier, 30 feet broad and 4 feet thick, were carried away, an ash tree was uprooted, and the waters spread rapidly over the low grounds in a wave that gathered to a height of three or four feet. Sheaves of barley and oats were carried off, wooden outhouses were torn away, stone walls, iron railings, gates and glasshouses were broken down, doors were driven in and a number of villas and cottages were submerged for some time to a depth of from three to five feet. Fortunately no lives were lost.

WITH reference to the inquiry as to the functions of the protruding filaments of the caterpillar of the Puss Moth (p. 385), we have received communications from several correspondents, who all agree with Mr. W. F. Kirby (p. 413) in regarding the appendages as chiefly intended for driving away Ichneumon Flies.

A NOTE in the *Electrician* refers to a curious effect produced by severe thunderstorms upon the glow lamps on the circuits of the Calcutta Electric Supply Co. It appears that immediately following each lightning flash the brightness of the glowing lamps has been observed to increase suddenly, gradually returning to the normal incandescence. This phenomenon has so frequently been observed that the engineers of the company have sought every possible explanation of the curious phenomenon, but have been unable to find any defect in their circuits—which are on the overhead wire system—that might offer an explanation. Indeed, the only conceivable explanation is one which appears so extraordinary that many may find considerable difficulty in accepting it. It is well known that carbon, acting as a coherer in a wireless telegraph apparatus, undergoes the usual sudden decrease in resistance when subjected to electric radiation. It is suggested that the carbon filaments

of a glowing lamp may undergo a similar change when exposed to the influence of a tropical thunderstorm in its immediate vicinity. This sudden decrease in the resistance of the filament would, of course, produce a correspondingly rapid increase in its candle-power, after which the gradual self-decoherence of the carbon would account for the return of the lamp to its normal incandescence.

WE have received an interesting account of the climate of Norway, by Mr. A. S. Steen, being a reprint from the Official Publication for the Paris Exhibition, 1900. As that country stretches through more than 13 degrees of latitude and extends nearly 300 miles beyond the Arctic Circle, the most varied shades of continental and maritime climates are represented within its confines. Mr. Steen has divided the country into south-east, west and north sections, this being in fact in accordance with nature's own division. In the inland districts of south-east Norway and Finmark we have examples of the most typical inland climate, viz. severe winter and relatively high temperature maxima in summer, and small rainfall ; while along the whole length of coast-line the winter is unusually mild, the summer cool, and rain falls in abundance. The influence of the Gulf Stream can be traced all over the country, and it is one of the chief agencies to which Norway owes its condition as a civilised inhabited State to its farthest bounds on the shores of the Polar Sea. The following are quoted as some of the highest summer temperatures : in the south-east 86° and upwards, and 93° at Christiania (once only) ; on the south coast no higher temperature than 80°·5 has ever been recorded. In the west, temperatures of 88°·5, and once 93° at Vossevangen, have been recorded. In the northern section temperatures of 85° to 88° have been recorded, but in the most southern of the Lofoten Isles (in the middle of the ocean) the thermometer has never risen above 68°.

A PAPER, by Mr. A. E. Sunderland, on applications of electrochemistry in dye and print works, is published in the Society of Arts *Journal* (August 24). The requirements which should be fulfilled by a machine for electrical dyeing are considered to be as follows :—(1) The poles must not be of metal, but of carbon or biscuit porcelain, which conduct by becoming saturated with the electrolyte. (2) They must be as near to one another as possible. (3) The cloth must pass between the poles in the open width. (4) The poles may be perfectly smooth, and preferably cylindrical, revolving freely. These particulars are necessary, because in the ordinary passage of the electric current across any dye solution, the tendency of the dye is to concentrate itself around the negative pole, and not to circulate freely in the whole dye vessel ; there is thus always a great danger of unevenness. In the finishing of goods the peculiar effect which is produced by calendering a piece in two different directions, one impression upon another, is well known. This is technically termed water-marking or moire, and is due to the irregular reflection from the surface of the material, one part of the light being totally reflected, and the other part dispersed. The effect can be introduced in several ways, one of which depends upon electricity. This process, Mr. Sunderland remarks, resolves itself practically into the local application of electrolysis. A platinum plate of suitable size is connected with the positive pole of the source of current. On this conducting surface is placed some absorbent material saturated with a solution of common salt. On this pad is placed the fabric to be water-marked, and the plate engraved with the water-mark connected with the negative terminal is pressed down upon it. The salt solution is decomposed, and a facsimile of the water-mark is printed on the cloth. To produce opaque designs, the absorbent material is saturated with a solution of barium chloride, which is decomposed on passing the current.

In the *Atti dei Lincei*, ix. 2, Dr. A. Campetti describes experiments made with common salt and copper sulphate tending to prove that there exists a difference of potential between a solid salt and its unsaturated solution, this difference of potential being of the same order of magnitude and sign as the difference of potential between a more concentrated and a less concentrated solution of the same salt.

THE question as to whether evaporation from the surface of an electrified liquid produces a loss of electricity is one of considerable interest in connection with theories of atmospheric electricity. An investigation of this point is given by Signori A. Pochettino and A Sella in the *Atti dei Lincei*, ix. 1. The method employed was to examine the rate at which an electrified plate lost its charge under the varying conditions when its surface was dry or was covered with a layer of water, and when it was exposed to a current of dry air or air saturated with watery vapour. The results tabulated show that the loss is more rapid in dry air than in saturated air, that with saturated air the presence or absence of the layer of water makes no practical difference, but that with dry air the discharge is actually slightly less rapid when the plate is wet than when it is dry. It is inferred that evaporation does not produce loss of charge; that the difference between dry air and saturated air is due to the fact that the dry air was ionised, while all trace of ionisation had been removed from the saturated air, and that the greater insulation obtained with dry air by wetting the plate was due to the ionisation being partially removed by the evaporation from the plate.

IN a paper published in the *Proceedings* of the Cambridge Philosophical Society, Mr. Barrett-Hamilton suggests that the (pathological) changes of colour and form which occur in certain Salmonoids during the breeding season may afford a clue to the origin of secondary sexual characters in animals in general. "Once," writes the author, "the existence of such a primitive state of things characterised by growth or discoloration of the whole or part of the body is admitted, we have therein the starting-point whence natural selection by alteration, suppression or accentuation of the details might easily produce many or all the nuptial changes of animals as we now see them, evolving in each a structure suitable to its own particular need, whether in eye, as in the Eel, in snout, as in the Salmon, or in hind-limb, as in Lepidosiren.

Indian Museum Notes (vol. v. No. 1) contains an interesting paper, by Mr. E. E. Green, on Indian Scale-insects (Coccidæ), showing the great increase which has recently taken place in our knowledge of their various groups. So late as 1886 only seven Indian representatives of the family were recorded, the well-known Wax-insect (*Ceroplastes ceriferus*) being one. Now thirty-seven species, distributed among fourteen genera, are known from Continental India, although this represents only a very small proportion of the real number. Not only are these insects interesting from their structure and their beauty of form and colour, but some are of commercial importance. The remainder of the number treats of various insect pests, notably those infesting tea and coffee plants, and those destructive to cereals and crops.

THE distribution of the Ruff in Ireland forms the subject of an interesting paper, by Mr. C. L. Patten, in the *Irish Naturalist* for August.

THE July number of the *Agricultural Gazette* of New South Wales maintains the usual high and useful character of this journal, an article on the important part played by bacteria in soil being of especial interest to the scientific agriculturist.

FROM the Indian Museum we have received "Illustrations of the Shallow-Water Ophiuroidea collected by the *Investigator*," by Dr. R. Koehler, published by the trustees. The specimens described are figured in eight plates.

To the last issue of the *Journal* of the Asiatic Society of Bengal for 1899, Mr. F. Finn contributes a paper on Indian Weaver Finches (Ploceidæ), in the course of which he shows how a supposed new species has been named on a specimen of a well-known bird in its summer plumage.

MALACOLOGISTS will find much to interest them in the sheets of the *Proceedings* of the Philadelphia Academy last to hand, Mr. H. A. Pilsbry communicating four papers dealing respectively with the land snails of Japan, South America, Australia and Polynesia, and India.

Science Gossip for September contains an interesting article, by Mr. R. J. Hughes, on the colouring of shells, in which he demonstrates that the most common pigment among those of northern Europe is the sesquioxide of iron. Another paper in the same number forms the continuation of "Geological Notes in the Orange River Colony," by Major Skinner, R.A.M.G.

IN the July number of the *American Naturalist*, Prof. H. L. Osborn describes a remarkable Axolotl from Dakota, which appears different from any named form, and may indicate a new type. In the course of his paper the author raises the question whether we yet know the adult of the true Mexican Axolotl, the specimens that have developed into Salamanders being from the United States and perhaps specifically distinct.

WE have received three fascicules of the "Results of the Branner-Agassiz Expedition to Brazil," in course of publication in the *Proceedings* of the Washington Academy. Two of these, written by ladies, are devoted to Crustacea, while the third, by Mr. C. H. Gilbert, deals with the fishes. When ladies appear as authors of papers, it is much to be desired that they should prefix "Mrs." or "Miss" to their names, as it is otherwise often difficult to ascertain their proper titles.

MENTION in these columns has already been made of Mr. G. S. Miller's work on Old World mammals, and we have now received a paper, communicated by that naturalist to the *Proceedings* of the Washington Academy of Sciences (vol. ii. pp. 203-246), in which he describes a very large number of new species, mainly Rodents, collected by Dr. W. L. Abbott on islands in the North China Sea. Many of these are rats and mice.

THE publication of Prof. E. Morselli's free course of lectures on man from an evolutionary point of view is still proceeding, and the fascicules when bound together will form an interesting volume on physical anthropology, or, as the author terms it, *Antropologia générale*. The last number to hand (No. 45) concludes the section on the brain, and commences an account of the progenitors of man.

THERE is always something of interest in our well illustrated contemporary, *The Reliquary and Illustrated Archaeology*. The July number contains some architectural notes from Monmouthshire, by Mr. J. Russell Larkby, illustrated by numerous sketches, and a short paper, by Mr. R. E. Head, on lace bobbins; these are often decorated in various ways, and different parts of the country furnish local types.

THOSE who are interested in criminal anthropology will find in a recent number of the *Bulletin de la Société d'Anthropologie de Paris* (Tome x. 4e série, p. 453) a psycho-physiological, medico-legal, and anatomical study of an atrocious criminal named Vacher, by MM. J.-V. Laborde, Manouvrier, Papillaut and Gellé. It is strange that studies of this kind are never made in this country. It is quite time that physical anthropology and psychology were more directly recognised by persons interested in criminology.

SIR ARCHIBALD GEIKIE's "Outlines of Field Geology" (Macmillan) has been the counsellor and friend of many young geologists and intelligent observers of the earth's features. A new edition—containing numerous alterations and additions, while retaining the original form—has just been published, and it should be possessed by every lover of country rambles or teacher of earth-knowledge.

A BULKY volume, containing "Agricultural Statistics of British India for the years 1894-95 to 1898-99," has just been distributed by the Department of Revenue and Agriculture of the Government of India. The tables show (1) total average, classification of areas, irrigation, fallow land, area under crops, and stock ; (2) prices of produce ; (3) incidence of the land revenue on area and population ; (4) varieties of tenure held direct from Government ; (5) register of transfers of landed property ; and (6) yields of principal crops.

THE additions to the Zoological Society's Gardens during the past week include a Green Monkey (*Cercopithecus callitrichus*, ♂) from West Africa, presented by Mr. C. A. Gilbert ; two Boschboks (*Tragelaphus sylvaticus*) from South Africa, presented by Dr. A. MacCarthy Morrough ; a Rufous-necked Wallaby (*Macropus ruficollis*) from New South Wales, presented by Miss Seymour ; a Germain's Peacock Pheasant (*Polyplectron germaini*) from Cochin China, presented by Mr. Arthur Yates ; a Common Boa (*Boa constrictor*) from South America, presented by Mr. G. R. Fairbanks ; two Red-bellied Squirrels (*Sciurus variegatus*) from South America, a Yellow-fronted Amazon (*Chrysotis ochrocephala*) from Guiana, ten Roofed Terrapins (*Kachuga tectum*) from India, deposited ; a Wapiti Deer (*Cervus canadensis*), two Collared Fruit Bats (*Cynonycteris collaris*), born in the Gardens ; two White Ibises (*Endocimus albus*), bred in the Gardens.

OUR ASTRONOMICAL COLUMN.

EPHEMERIS FOR OBSERVATIONS OF EROS.—In the last issue of the *Astronomische Nachrichten* (Bd. 153, No. 3660), Signor E. Millosevich gives a revised ephemeris of this asteroid for the next few weeks :—

Ephemeris for 12h. Berlin Mean Time.

1900.		R.A.		Decl.
		h. m. s.		° ′ ″
Sept. 6	...	2 27 24·16	...	+ 35 29 53·7
7	...	28 28·18	...	35 52 13·0
8	...	29 30·73	...	36 14 36·1
9	...	30 31·77	...	36 37 2·8
10	...	31 31·22	...	36 59 33·0
11	...	32 29·03	...	37 22 6 8
12	...	33 25·13	...	37 44 43·9
13	...	2 34 19·43	...	+ 38 7 24·2

COMET SWIFT (1894 IV).—Mr. F. H. Seares has calculated the osculating elements, and from them computed a finding ephemeris for this comet, which may possibly have some connection with the lost comet of De Vico. As it is important, however, that the comet should again be observed before any further attempt is made to establish such connection, he hopes that all possessing the necessary optical power will prosecute the search for it (*Astronomische Nachrichten*, Bd. 153, No. 3656).

Osculating Elements.
Epoch and Osculation 1900 July 23·0 Berlin Mean Time.

$$M = 317 \ 16 \ 15·0$$
$$\pi = 348 \ 56 \ 56·0$$
$$\Omega = 24 \ 50 \ 38·8$$ 1900·0
$$i = 3 \ 35 \ 17·0$$
$$\phi = 31 \ 2 \ 30·2$$
$$\mu = 554''·3823$$

Ephemeris for Berlin Mean Midnight.

1900.		R.A.		Decl.
		h. m. s.		° ′
Sept. 9	...	16 27 21	...	− 25 21·9
13	...	32 57	...	25 30·5
17	...	38 55	...	25 39·1
21	...	45 15	...	25 47·5
25	...	51 56	...	25 55·6
29	...	16 58 57	...	− 26 3·1

THE NEW SPECTROGRAPHS FOR THE POTSDAM GREAT REFRACTOR.—In the *Astrophysical Journal*, vol. xi. pp. 393-399, Prof. H. C. Vogel describes the two new spectrographs which have recently been completed for the great Potsdam refractor of 80 cm. aperture.

(a) *Three-prism spectrograph.*—This is designed so that the combined deviations of the three prisms shall be nearly 180°, thus bringing the collimator and camera almost parallel. These are then mounted on a massive steel plate 78 cm. long, 41 cm. broad, and 7 mm. thick, which in its turn is firmly attached to the tail-piece of the telescope by an elliptical base plate 10 mm. thick, lateral flexure being guarded against by several intermediate metal ribs. The slit has only one movable jaw, and the whole can be rotated round the telescope axis, and the position angle recorded to 1′ of arc. For comparison spectra the arc light has been found most convenient, the difficulty of spectral displacement of comparison lines due to imperfect adjustment of source having been overcome by interposing a translucent diffusing screen between the light and slit. The collimator lens (Steinheil) has an aperture of 3·2 cm. ; focal length, 48 cm. One of the camera objectives is a Zeiss anastigmat of 56 cm. focus ; the other a triple cemented lens by Steinheil of 4·1 cm. aperture ; focus, 41 cm. The prisms are of very white Jena glass, and with the Zeiss camera lens a spectrum of uniform focus from *b* to K is obtained. Delicate arrangements have been made for securing constant temperature conditions, &c. The weight of the complete spectrograph is 31 kilog.

(b) *Single-prism spectrograph.*—In this instrument the collimator lens is 3·5 cm. aperture, focus 53 cm. ; the camera lens, 4 cm. aperture, focus 72 cm., both being triple cemented objectives by Steinheil. The prism is by Zeiss, and has a refracting angle of 60°, with faces 61 mm. long and 45 mm. high. The spectrum is uniformly sharp from D to N. The whole instrument weighs 20 kilograms.

Prof. Vogel gives the results of the application of tests instituted by Prof. W. W. Campbell for the Mills' spectrograph of the Lick Observatory, showing that the performance of both instruments is very trustworthy. Three plates are given showing the instruments in position as attached to the telescope.

STRUCTURE AND CONSTITUTION OF TWO NEW METEORITES.—Messrs. G. P. Merrill and H. N. Stokes recently communicated a paper to the Washington Academy of Sciences (*Proceedings*, vol. ii. pp. 41-68, July 1900), describing the results of their examination of two fragments of newly-fallen meteorites. One, a stony meteorite, fell on July 10, 1899, in Allegan, Michigan, U.S.A., the largest fragment weighing 62½ lbs. To the unaided eye this stone shows on the broken surface a quite even granular structure of grey colour, and, on closer examination, numerous beautifully spherulitic chondrules, averaging not more than one or two millimetres in diameter. In some cases these chondrules have pitted surfaces. More critical inspection indicates that they are composed of enstatite and olivine. Numerous brilliant metallic points of a silver-white colour indicate the presence of disseminated iron, so that the stone may be said to be made up of the chondrules, iron and dark grey silicate minerals, imbedded in a light ashy grey matrix. This Allegan stone is exceedingly friable. Microscopically, the ground mass of the stone is seen to be made up of a confused agglomerate of olivine and enstatite particles with interspersed metallic iron, iron sulphide and chromic iron. An important feature is that in no cases do the silicates occur with perfect crystallographic outlines, both olivine and enstatite being fragmental. The presence of alumina and alkalis suggested a search for felspar, but it was decided that this mineral was not present. A considerable proportion of the ground mass was found microscopically to be composed of a dark glassy material. Careful chemical analysis showed that 77 per cent. of the meteorite was of non-metallic origin, the remainder being chiefly iron and nickel. The second meteorite examined is known as the Mart Iron, having been found early in 1898 near Mart, in Texas. This originally weighed 15¾ lbs., from which a slice weighing 456 grams was presented to the National Museum at Washington. The etched surface shows the iron to belong to the octahedral variety, and is of moderately coarse crystallisation. Chemical analysis showed that 98·3 per cent. of the meteorite was composed of the metals iron, nickel, copper, cobalt, the remainder being made up of schreibersite and a small quantity of troilite.

Photographs of both meteorites in their present condition are given, and numerous drawings indicating the microscopical structure.

LATITUDE-VARIATION, EARTH-MAGNETISM AND SOLAR ACTIVITY.

IN the *Astronomische Nachrichten* (No. 3619) I have published the results of an investigation dealing with the effects of periodic changes in solar activity on the motion of our planet. It is there shown that these changes, as indicated by the frequency of sun-spots, exert a subtle but pregnant influence on the secular variations of the earth's elements ; and, moreover, that disturbances precisely similar to those which appear in the observations of the obliquity and of the sun's longitude are distinctly exhibited in the variation of terrestrial latitude.

In the further pursuit of these researches I have been led to

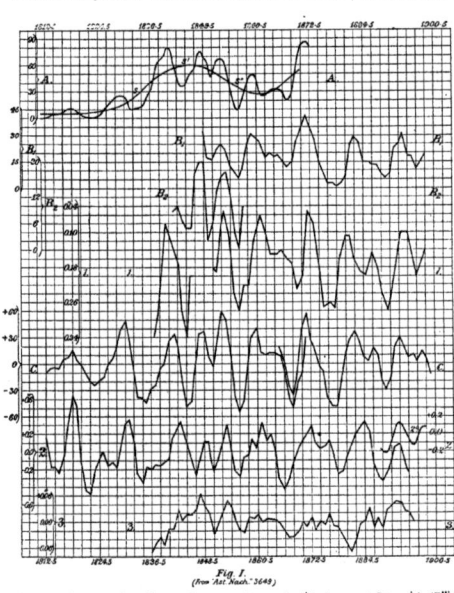

Fig. I.
(from Ast. Nach. 3649.)

A, frequency of auroræ (Loomis); B₁, frequency of magnetic disturbances at Greenwich (Ellis); B₂, frequency of magnetic disturbances at Greenwich (Airy); c, frequency of auroræ (Europe south of Arctic Circle) from 1812–1871 (Fritz), and in Scotland from 1865–1899. The "great" period has been eliminated in both curves in order to exhibit more clearly the eleven-years' fluctuations. 1, semi-amplitudes of latitude-variation (Chandler, Nyrén and Albrecht); 2, residuals of obliquity (Fig. c in *A.N.* 3619); 3, Greenwich corrections to the R.A. of stars derived from observations of the sun (Thackeray).

conclude that the anomalies existing in the observations of the sun's right-ascensions and declinations are to be attributed exclusively to changes in the position of the earth's axis of rotation with regard to the axis of maximum moment of inertia, and that these changes in their turn are intimately connected with the varying display of forces on the solar surface. In a subsequent article which appeared in NATURE (No. 1584, March 8) I made a suggestion as to the nature of this connection, and advanced the hypothesis that the magnetism of the earth is probably the medium through which the changes of solar energy react upon the motion of the earth's pole.

In view of the results of Joule's experiments regarding

magnetic strain in iron rods, it seems reasonable to believe that something similar to the molecular displacement in the rod may take place in a magnetic body like our earth with respect to its magnetic axis. As has long been known, this axis is by no means coincident with the earth's axis of figure, but is inclined to it, according to Gauss, at an angle of about 12°. It is therefore not unnatural to conclude that a molecular strain in the direction of the magnetic axis will occasion an asymmetric change of the earth's figure, and will thereby produce a displacement of the axis of figure relative to the instantaneous axis of rotation. Such a displacement could remain constant only so long as the total magnetic potential of the earth was not subject to alteration, in which case the pole of rotation would describe a circle of constant radius round the pole of figure as centre. But, as already stated, various facts compel us to believe that the magnetic potential of the earth does alter, and, indeed, changes synchronously with the state of solar activity—the most striking instance being the regular increase of auroræ and of magnetic disturbances with the increasing frequency of sun-spots. Now, if we consider auroræ as discharges of the earth's electric force, it follows that the strain in the direction of the magnetic axis should abate after an auroral display, and the pole of figure should therefore approach the pole of instantaneous rotation.

The outcome of this hypothesis would be that changes in the state of solar activity, since they produce a measurable effect on the terrestrial magnetic forces, should also be accompanied by corresponding changes in the motion of the earth's axis. In *A.N.* 3649 I have endeavoured to test this theory by such facts as have up to the present been discovered regarding the complex phenomenon of latitude-variation, and I shall here briefly describe the results there obtained.

In the first place, from the material afforded by the investigations of Loomis and Fritz, and from observations made in Scotland during the past thirty years, I constructed curves exhibiting the frequency of auroral displays from 1812 to 1899. (Fig. 1, curves A and C.) In these the influence of the eleven-years solar period is most conspicuous, but there is also exhibited an intimate connection with the "great" cycle of sun-spots. In contrast with the spot-curve, minima of auroræ are found to be of unequal heights. Accordingly, if a second curve *aa'a''* be drawn through the aurora-curve A in such a way as to bisect as nearly as possible every wave appearing in it, this new curve will be seen to attain its maxima and minima simultaneously with the "great" spot period. It is therefore beyond dispute, as already proved by Wolf and Fritz, that the existence of a "great" period of solar activity is an established fact, and that it is clearly brought out by the curve of auroræ.

In the next place, curves were formed to show the variations in the frequency of terrestrial magnetic disturbances. For this purpose I took the annual number of days on which such disturbances were observed at Greenwich, as recorded by Mr. Ellis in *Monthly Notices* 60, December 1899. (Fig. 1, curve B₁.)

Mr. Ellis subdivides his records according to the intensity of the disturbance, but only three of his subdivisions were taken into account ; those, namely, showing the frequency of moderate, active, and great disturbances respectively. For the satisfactory combination of these different sets, the principle was adopted that the weight assigned to each set should be inversely proportional to the mean frequency—unit weight being given to the "active" disturbances—on the ground that the small frequency of "great" disturbances is probably compensated for by their much more energetic character. In order to extend the record as far back as possible, a similar curve (Fig. 1. II.) was constructed, based on Airy's statistics from 1841–1857 (*Trans. Roy. Soc.*, vol. cliii.).

These magnetic curves are in perfect harmony with those of the aurore, and both sets bring out a remarkable and, for our purpose, eminently important fact, namely, that although for the most part they correspond closely with the eleven-years spot period, there nevertheless appear waves which at first sight seem to have nothing to do with the display of solar activity. Most noticeable in this respect are the maxima of 1852 and 1864, of which the former especially seems almost to contradict the existence of a connection between sun-spots and aurore. A closer examination of the spot-curves, however, reveals the fact that in these two instances there occurred, at exactly the same times, peculiar disturbances or irregularities in the exhibition of spots on the solar surface, so that, after all, the waves exhibited at these points in the magnetic and auroral curves may be supposed to have been caused by solar influence, although they appear for some reason or other on a greatly exaggerated scale. This is decidedly the opinion of Prof. Fritz and of Mr. Ellis, both of whom expressly mention instances of this character, the latest having occurred in 1898 following the spot-maximum of 1894.

The succeeding part of this research may be conveniently divided into two parts. In the first, the earth-magnetic variations are compared and contrasted, in relation to the eleven-years period of sun-spots, with certain phenomena associated more or less directly with latitude-variation : and in the second with similar phenomena relative to the "great" sun-spot and aurora period.

These phenomena are :—

I. With regard to the eleven-years period—

i. The semi-amplitudes of latitude-variation.

ii. The periods of latitude-variation.

iii. The changes in the values of the obliquity as observed at Greenwich from 1812–1896.

iv. Mr. Thackeray's corrections to the right-ascensions of stars derived from observations of the sun during the years 1836–1895.

II. With regard to the "great" period—

v. Dr. Chandler's long period inequality in the latitude-variation.

vi. The observed residuals of the obliquity from 1753–1896 as compared with Leverrier's tabular values.

vii. The Greenwich corrections to the sun's right-ascensions relative to a fundamental system of fixed stars, according to Prof. Newcomb.

The data requisite for the formation of the curve of semi-amplitudes (Fig. 1, curve 1) was taken from Dr. Chandler's paper in *A.J.* No. 277, and from the publications of Dr. Nyrén (*A.N.* 3166), Mr. Wunsch (*A.N.* 3112) and Prof. Albrecht ("Berichte über den Stand der Erforschung der Breitenvariation"). It has to be borne in mind, as pointed out in my previous papers, that the latitude phenomenon lags behind the comparison magnetic curves by about 1·5 years. Such a lag appears indeed to be a characteristic feature common to most terrestrial phenomena which have hitherto been found to be influenced by solar activity. In this connection I must refer to the highly important and interesting discovery made by Sir Norman Lockyer some years ago, that the curve representing the frequency of the "unknown" lines widened in sun-spot spectra during a cycle of solar activity follows the spot-curve by exactly the same interval, viz. 1·5 years. I have received a

communication from Sir Norman, stating that this curve of "unknown" lines goes excellently with my curve 1 of Fig. 1, the maxima and minima of the spectroscopic curves showing a perfect synchronism with those of the curve of latitude-variation. I take this opportunity of expressing my indebtedness to Sir Norman Lockyer for drawing my attention to this most significant and singular fact.

The lag in the case of curve 1 has been allowed for by shifting this curve one and a half years in the backward direction. When it has thus been made to coincide with the comparison curves, their correspondence becomes most striking—a remarkable feature being the exactness with which certain secondary maxima in 1865, 1887 and 1895 are represented in each case.

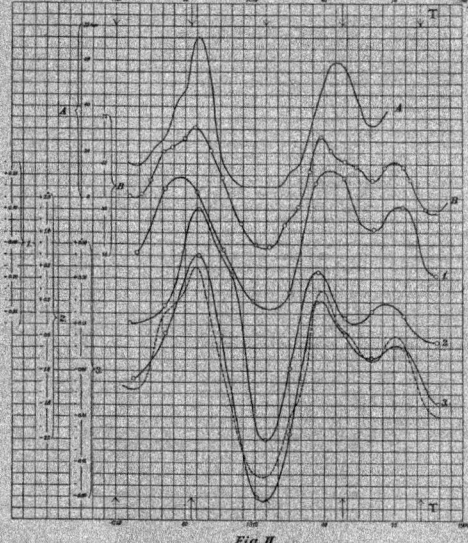

Fig. II.
(*From Nat. Mach.*)

τ, turning points of Dr. Chandler's long period inequality of latitude-variation ; a, great period of aurore (Loomis). (The ordinates represent eleven-years means of Loomis' annual aurora numbers.) B, great period of sun-spots (Wolf); t, twelve-years' means of residuals of obliquity (Greenwich observations) after elimination of purely secular change ; e, twelve-years' means of corrections to the sun's R. A. relative to a system of fixed stars (Greenwich observations) after elimination of purely secular change ; 3, combined curve of obliquity and sun's R.A., the dotted curve representing Wolf's great period of sun-spots on a somewhat different scale.

Indeed, so accurately are the motions of the amplitude-curve shadowed in those of the magnetic variations, that to any unprejudiced astronomer this fact in itself must indicate with sufficient force the existence of a *vera causa*.

The periods of latitude-variation afford a new and almost as certain proof of the existence of an intimate connection between the polar motion and earth-magnetic phenomena.

Dr. Chandler has already shown that a small amplitude corresponds to a great period and *vice versâ* (*A.J.* 277, p. 98). How well this influence is borne out may be seen from the following analysis of the average values of τ according to the lengths of the period :—

Period in days.	Observed r.	Computed r.
Under 390	0·20	0·20
390-420	0·18	0·19
420-450	0·15	0·13
Over 450	0·10	0·08

This statement in itself constitutes a proof of my assertion, and renders it unnecessary for me to add anything further on this point. Any one who cares to plot down the values for the periods given by Dr. Chandler must arrive at the conclusion that the comparison of the curve so obtained, with the magnetic and auroral curves, gives indeed a convincing argument in favour of the earth-magnetic hypothesis.

I have next to consider the changes in the values of the obliquity as observed at Greenwich due to the eleven-years period of solar activity.

In my paper, *A.N.* 3619, I have discussed fully the reduction of these values to a uniform and homogeneous system, as well as the elimination from them of the secular variation and the influence of the "great" sun-spot and aurora period. The resulting curve (Fig. 1, curve 2) exhibits the utmost conformity with those of the earth-magnetic and latitude phenomena.

This fact is of the highest significance, inasmuch as it affords added testimony to the accuracy of the data on which my research is founded. The curves communicated show that at times, when the amplitude of latitude-variation reaches maximum values (1½ years after minimum displays of magnetic disturbance), the Greenwich obliquity attains small values; while at times, when the amplitude is at a minimum (1½ years after maximum displays of magnetic disturbance), the obliquity appears to be excessively great. This leads at once to the conclusion that whenever the amplitude is great, the minimum latitude for the Greenwich meridian must occur near the time of the winter solstice, and that when the amplitude is small just the reverse ought to take place. Now Dr. Chandler's statistics in *A.J.* 277 afford ample means of testing this conclusion. In point of fact they show that at times of maximum amplitude the epochs of minimum latitude for the Greenwich meridian have always occurred on some date between the beginning of November and the end of February, while at times of amplitude-minima these epochs, with the exception of the first in 1840, are comprised within the interval from May to August. The mean date in the former case is January 10, and in the latter July 16; and the mean deviation of a single epoch from these two dates is not more than about ± 40 days.

I consider that, in spite of the great uncertainty which naturally attaches to researches of so delicate a character, the evidence afforded by these results is to be taken as a proof that the residuals in the obliquity, far from being accidental, are really caused by latitude-variation. Thus, owing to the great extension of the series of Greenwich solar observations, these residuals form an excellent test of my assertion that the motion of the pole depends on the intensity of the earth-magnetic forces.

As regards the corrections to the right-ascensions of the stars derived from Greenwich observations of the sun, I need only state that after subtracting the secular variation found by Mr. Thackeray (*M.N.* June 1896), the resulting values give curve 3 of Fig. 1, which, in spite of somewhat large accidental discrepancies, is in general agreement with all the others, especially with that of the obliquity.

Having thus shown that my contention with regard to a connection between the eleven-years period of auroral displays and magnetic disturbances and the motion of the earth's pole of rotation appears to be borne out by all the facts which constitute the sum of our present knowledge of the peculiar phenomena relating to latitude-variation, I next consider the "great" period of aurorae, which, as already stated, is synchronous with the great period of solar activity.

For this purpose I exhibit in Fig. 2 curves showing the great aurora-period according to Loomis' annual numbers, and the great spot-period in Wolf's relative numbers.

The interval comprised by this great period is according to Wolf equal to six small cycles, *i.e.* sixty-six years. Now this is exactly the period of Dr. Chandler's long inequality of latitude-variation. The smallest amplitudes and greatest periods of latitude-variation, according to Chandler's formula, fell in 1782 and 1848, almost exactly at the times of greatest auroral displays; whereas the greatest amplitudes and smallest periods

occurred in 1815 and 1881, *i.e.* just at the times when the display of aurorae reached a minimum.

But in addition to this there are other facts which point to an influence on the earth's motion exercised by some force varying with the great period of solar activity. In my previous papers I have discussed at some length the evidence afforded by the curves representing the observed residuals of the obliquity, and Prof. Newcomb's corrections to the right-ascension of the sun relative to a fundamental system of fixed stars. I therefore need not here dwell upon their importance as strongly supporting my hypothesis.

A reference to curves 1, 2 and 3 of Fig. 2 will show how exquisitely parallel are their courses, and how complete is their agreement, not only with the changes in the displays of aurorae and solar activity, but also with Dr. Chandler's long period inequality. It seems utterly inconceivable that a correspondence so consistent can be attributed merely to accident.

It remains to state briefly one or two very important and interesting deductions made from the results of the last ten years' researches into the phenomena of latitude-variation.

The frequency of aurorae and magnetic disturbances, as is well known, shows, in addition to the variations associated with changes of solar activity, other fluctuations depending on the season of the year—a fact which has been closely investigated and corroborated by Mr. Ellis. It appears that the magnetic disturbances recorded at Greenwich reveal decided maxima at the equinoxes and minima at the solstices, thus betraying, like the aurorae, a half-yearly period.

Now the foregoing results point to the conclusion that the distance of the pole of instantaneous rotation from the pole of figure depends on the display of earth-magnetic forces. Hence in the course of a year this distance must become twice comparatively short and twice comparatively long; *i.e.* instead of being circular, the path described by the pole of rotation round the pole of figure must be elliptical—the mean pole being situated at the centre of the ellipse. If the period of polar motion were exactly one year, the position of the axes of this ellipse referred to a fixed meridian would remain unaltered. But from Dr. Chandler's investigations we know the period of latitude-variation to be on the average 428 days. Hence the effect of seasonal change in the earth-magnetic forces must consist in continuously rotating the axes of the polar ellipse in a direction opposite to that of the motion of the pole. These conclusions are well corroborated by the observed facts, and are clearly revealed in the plate appended to Prof. Albrecht's latest "Bericht." The comparatively great eccentricity of the ellipses admits of a tolerably accurate determination of the angles between their major axes and the Greenwich meridian. If the magnitudes of these angles be computed (Table vi. of my paper, *A.N.* 3649), it will be found that they exhibit quite unmistakably the progressive change of position of the ellipse with regard to the meridian, the average angular distance between two successive positions of the major axes being about 33°.

But a closer examination of these figures shows that they indicate fluctuations in this average amount which stand in a remarkable connection with the varying display of magnetic disturbances.

The motion of the ellipse appears to have been largely progressive in 1892, 1894 and 1895; while it has been very slight, and at times even retrograde, in 1890, 1893 and 1896. In the first-mentioned years the ellipses are also more irregular and distorted than in the others, indicating a more vehement and spontaneous character of the forces acting on the motion of the pole. Now, according to Mr. Ellis, these years were the *only ones* in which magnetic disturbances of the character "great" occurred at Greenwich, while in the last-named years his statistics show that there prevailed a decided magnetic calm. While leaving the final confirmation of this interesting fact to future observations, it does not seem too much to say that in the face of existing evidence it is difficult to retain the idea that a coincidence so peculiar can possibly be ascribed to mere accident.

The results of my researches may be thus shortly summarised:—

i. The changes in the motion of the pole of rotation round the pole of figure are in an intimate connection with the variations of the earth-magnetic forces.

ii. Inasmuch as the latter phenomena are in a close relation with the state of solar activity, the motion of the pole is also indirectly dependent on the dynamical changes taking place at the sun's surface.

iii. The distance between the instantaneous and mean poles decreases with increasing intensity of earth-magnetic disturbance.

iv. The length of the period of latitude-variation increases with increasing intensity of earth-magnetic disturbance.

v. In strict analogy with the phenomena of aurorae and of magnetic disturbance, the influence of the eleven-years period of sun-spots, as well as of the "great" period, is clearly exhibited in the phenomenon of latitude-variation ; and the same deviations from the solar curve as are manifested by the aurorae are also evident in the motion of the pole.

vi. The half-yearly period of the earth-magnetic phenomena influences the motion of the pole of rotation in such a way that its path, instead of being circular, assumes the form of an ellipse, having the mean pole at its centre.

vii. The half-yearly period also explains the conspicuous fact of a rotation of the axes of the ellipse in a direction opposite to that of the motion of the pole.　　J. HALM.

UNIVERSITY AND EDUCATIONAL INTELLIGENCE.

PROF. J. G. MacGREGOR, of Dalhousie University, Halifax, Nova Scotia, has been appointed professor of physics in University College, Liverpool, in succession to Prof. Lodge.

THE Calendar of the Glasgow and West of Scotland Technical College, for the session 1900-1901, has been received. Among the contents of the volume we are glad to notice schemes of courses of study, extending over three years, for students who intend to take up some branch of applied science or engineering as a profession. Students are permitted to attend single classes, but they are encouraged to follow one of the regular courses in the several departments of study. This is the only way to derive any real advantage from a Technical College, desultory attendance at classes without any definite object being of little value.

AT a special meeting of the University Court of St. Andrews, held on Saturday last, the proposal of the Marquis of Bute, who offered a sum of 20,000*l.* to be held as a fund for endowing a chair of anatomy in the University, was considered. After deliberation, the Court resolved cordially to accept the proposed gift on the conditions as stipulated by his lordship, and to request the Lord Rector to inform Lord Bute of the Court's decision. The Court further resolved to proceed at once with the creation of a professorship of anatomy at St. Andrews, to be endowed by Lord Bute's gift, the first presentation to the chair being Dr. Musgrove, the present lecturer in anatomy, such presentation to be made as soon as the ordinance creating the chair is approved by her Majesty in Council.

THE mission of science in education was recently considered in some detail by Prof. J. M. Coulter in an address delivered at the University of Michigan, and published in *Science.* The claims set forth in the paper are formulated as follows :—The introduction of science among the subjects used in education has revolutionised the methods of teaching, and all subjects have felt the impulse of a new life ; it has developed the scientific spirit, which prompts to investigation, which demands that belief shall rest upon a foundation of adequate demonstration, which recognises that the sphere of influence surrounding facts may be speedily traversed and that everything beyond is as uncertain as if there were no facts ; it has introduced a training peculiar to itself, in that it teaches the attitude of self-elimination, an attitude necessary in order to reach ultimate truth, and thus supplements and steadies the other half of life, which is to appreciate. To obtain these results, there must be teachers who can teach, whose background and source of supply is the investigator. Moreover, the results are immensely desirable, inasmuch as they do not interfere with anything that is fine and uplifting in the old education, but simply mean that the possibilities of high attainment and high usefulness are open to a far greater number.

MESSRS. S. Z. DE FERRANTI, the electrical engineers at Hollinwood, near Oldham, have just adopted an educational scheme for their apprentices. Success at evening classes, combined with steady work, are to be the chief recommendations for promotion from one department to another. And the apprentice who obtains the highest position in the South Kensington Examinations in subjects of importance to the theoretical training of an engineer will be awarded a scholarship tenable in the day engineering department of the Manchester Municipal Technical School. His fees will be paid by Messrs. Ferranti, and also the wages he would receive if working in their shops. Mr. F. Brocklehurst takes this generous scheme as the text of a pamphlet upon "Technical Education," issued by the Technical Instruction Committee of Manchester, and he hangs upon it some instructive remarks as to the responsibilities of manufacturers and the nation at large, if England is to maintain her position in the industrial world. Referring to education in Switzerland, he points out that at Winterthur, a small engineering town near Zurich, the technical school is attended by 400 day students who have voluntarily left their employment (sacrificing their wages in so doing) for one or two sessions in order to devote themselves to technical study. The town, the canton and the State combine to assist the realisation of their ambitions by bearing the burden of cost, and in keeping the fees of the technical school low. In the same way the great Polytechnic of Zurich is crowded in its day department with hundreds of young men preparing themselves for the engineering, electrical and chemical industries. Germany provides many similar examples. In the Technical High School of Darmstadt there are to be found 1100 day students, all of them over eighteen years of age, many of them graduates of universities, and the remainder having received a splendid high-class education in secondary schools. These are engaged in the study of electrical, chemical or mechanical science directly bearing upon industrial pursuits. This is only one of many technical high schools in Germany, the culmination of which is seen in the Charlottenberg Technical High School, near Berlin—the finest institution of its kind in the world—with its more than 2000 day students. These young men are being prepared for the highest positions, as technical chemists, mechanical, naval, civil and railway engineers, shipbuilders and architects. There are now in the German Technical High Schools no fewer than 11,000 day students. In connection with the figures given it must be noted that (1) they are exclusive of science students taking university courses ; (2) the pupils are without exception youths of over eighteen years of age ; and (3) each technical high school insists upon an entrance examination of an exacting character.

The great advance of the United States in engineering is, as Mr. Brocklehurst remarks in his pamphlet referred to above, largely due to the fact that during the last forty years very important engineering schools have been founded. The chief of these is the Massachusetts Institute of Technology at Boston. This is attended by 1171 day students, whose average age at entrance is eighteen years and nine months, and who are either graduates from other colleges or have attended the public high schools for at least four years. The Worcester Engineering Polytechnic has 823 day students. Nearly 1000 are in the Lehigh Engineering College. The Stevens Institute of Technology, New Jersey, has 214 ; and the Case School of Applied Science in Cleveland, Ohio, 218. Five hundred and ninety-seven day students attend the classes of the Sheffield Scientific School in Connecticut, while the Sibley College of Engineering —part of Cornell University, New York—has 492 day students. There are 242 day students in the Engineering Department of the University of Michigan. A recent report shows that in the Engineering Colleges of the United States the number of day students enrolled is 9659, and that their growth since 1878 is 516 per cent. ! Fifty-one per cent. of these students have had a three-year high school course, which would bring them to seventeen years of age. The number of engineering students graduated in 1899 was 1413, and the number of institutions providing an education in this branch of technical instruction (engineering) is 89. This is exclusive of evening work altogether. It is also exclusive of what America is doing in the fields of chemistry and textiles. Little wonder is it that this wealth of educational opportunity is producing its crop of skilled craftsmen trained to compete on more than equal terms with the Briton. The Manchester Technical Institution Committee is doing a service to the nation by placing these facts prominently before the manufacturers of the district.

SOCIETIES AND ACADEMIES.

PARIS.

Academy of Sciences, August 27.—M. Faye in the chair.
—On Egyptian gold, by M. Berthelot. Analyses of specimens
of gold of different epochs show that at the time of the sixth
and twelfth Egyptian dynasties the art of separating the silver
from native gold was not known. Some gold leaf of the
Persian epoch was pure, the silver having been separated. As,
however, there is a period of about twelve centuries between the
dates of the last two specimens analysed, specimens of intermediate dates must be examined if the date of this metallurgical
discovery is to be fixed.—Observations of the comet 1900 *b*
(Borelly-Brooks) made with the large equatorial of the Observatory of Bordeaux, by MM. G. Rayet and A. Férand. The
nucleus of the comet on July 31 was of about the 9th or 10th
magnitude, the head having a diameter of 3' to 4' of
arc.—The apparent semi-diameter of the sun and its position relative to the moon, deduced from the eclipse
of May 28, by MM. Ch. André and Ph. Lagrula. The final
result for the apparent semi-diameter of the sun is given as
15' 59"·24 ± 0"·30.—On an anomaly of the dichotomous phase
of the planet Venus, by M. E. Antoniadi. The edge of the
planet is always more brilliant than the central regions; thence
irradiation ought to produce the prolongations actually observed.
The phenomenon would thus appear to be of purely physiological origin.—Dielectric cohesion and explosive fields, by M.
E. Bouty. The term explosive field is applied to the minimum strength of field between two nearly plane electrodes
required to produce sparking. The curves relating to critical
fields, as described in a preceding note, show many analogies
with those of explosive fields. Thus both the critical and
explosive fields are linear functions of the pressure of the gas,
and the constants for the gases hydrogen, air and carbonic
acid are arranged in the same order of magnitude.—On
the composition of the combinations obtained with fuchsine
and the sulphonated azo-colouring matters, by M. Seyewetz.—
—On lighting by the cold physiological light called living light,
by M. Raphael Dubois. By the growth of certain micro-
organisms in suitable media, details of which are given, a room
may be illuminated with an intensity about equal to moonlight.
—Action of the total pressure upon the assimilation by chlorophyll, by M. Jean Friedel. Although the influence of the
partial pressure of the carbon dioxide in the atmosphere upon
chlorophyll assimilation has been well investigated, the effect of
changing the total pressure of the air has not yet been examined.
It was found that the lowering of the total pressure, even to
¼ atmosphere, does not modify the nature of the chlorophyll
assimilation, but that its intensity diminishes in a regular manner
with the pressure. Four species of plants were used, and the
numbers obtained for the variations were of the same order in
all of them.—On the ancient extent of the glaciers in the region
discovered by the Belgian Antarctic Expedition, by M. Henryk
Arctowski.

CAPE TOWN.

South African Philosophical Society, August 1.—L.
Péringuey, President, in the chair.—The secretary read a
second report on the mud island which appeared off Pelican
Point at the beginning of June, from Mr. Cleverly, R.M.,
Walfish Bay, and showed the photographs taken by Mr.
Waldron, Public Works Department. Mr. Cleverly reported
that the island no longer existed on June 7, it having then entirely
subsided, as, on steaming over the site, soundings of six and
seven fathoms were obtained. The sea was much discoloured,
and a distinct odour of sulphur was still to be distinguished.
Small quantities of dead fish were found on Pelican Point, but
this is a not unusual occurrence. About the time of the island's
appearance heavy rollers set in along the coast, and though these
did not affect Walfish Bay, thirty yards of the new breakwater
at Swakop Mouth were totally destroyed, a derrick carried
away, and two men drowned. Though these rollers are usually
experienced on this coast in the winter months, Mr. Cleverly
understands that the engineer in charge at Swakop Mouth had
set up a theory that the damage to his works resulted from an
earthquake wave, and that he pointed to the appearance of the
mud island at Walfish Bay in support of his theory, but in Mr.
Cleverly's opinion the cause of the upheaval must have been
extremely local as no disturbance whatsoever was felt at the
settlement or in the confined waters of Walfish Bay. Mr.
Waldron, on the invitation of the president, gave an account of

his visits to the island. It was visited on June 1, 2 and 4. At
the next visit, on June 7, there was no island. On June 1 one
member of the party landed and noticed a small basin-shaped
hollow containing water and emitting gas bubbles. The odour
was distinctly that of sulphuretted hydrogen. Dr. Corstophine
agreed with Dr. Marloth as to there being no need for volcanic
activity to explain the phenomenon; nor was there any evidence
of such. He compared the appearance of the island at Walfish
Bay with the "mud lumps" known to arise in the Gulf of
Mexico, and quoted Sir Charles Lyell's account of these. The
Walfish Bay island was evidently a quite similar phenomenon.
As to the gas, the Gulf of Mexico "mud lumps" usually gave
off marsh-gas, and the sulphuretted hydrogen perceived as being
emitted at Walfish Bay, was probably due to the decomposition
of animal as against plant material. The fine mud from
Walfish Bay, under the microscope, was found to contain
diatoms, fish scales, bones, and other remnants of animal
matter.—Notes on stone implements of palæolithic type
found at Stellenbosch and the vicinity, by L. Péringuey and
G. S. Corstophine. The discovery of stone implements of
a particularly ancient type at Bosman's Crossing, Paarl and
Malmesbury, was described. From the rude character of the
chipped stones, Mr. Péringuey was disposed to regard them as
being equal in age to the palæolithic implements of Europe, but
Dr. Corstophine had shown him the difficulty of accepting this
theory owing to the geological deposits in or on which the
stones are found. So far no implements have been found in any
deposit that can be regarded as of great antiquity. In the
Stellenbosch district the implements are found imbedded either
in the rain-wash of weathered granite or in the laterite, or
simply on the surface, so that no geological evidence has yet
been discovered as to the presumable antiquity of the
implements. One feature of this occurrence, which Dr.
Corstophine pointed out, is that as yet no implements have
been found on the recent alluvial terraces of the Eerste River,
but only on the hill slopes round about. The implements are
formed from water-worn boulders of Table Mountain Sandstone,
and often retain a considerable portion of the water-worn
surface.

CONTENTS.

THURSDAY, SEPTEMBER 13, 1900.

BACTERIOLOGY.

The Structure and Functions of Bacteria. By Alfred Fischer, Professor of Botany at the University of Leipzig. Translated into English by A. Coppen Jones. Pp. viii + 198. (London : Clarendon Press, 1900.)

THE first two chapters are concerned with the morphology of bacteria, and the subject is most ably dealt with, as might be expected from so talented a botanist as Prof. Fischer. Nevertheless, these are not the chapters which strike one as of exceptional interest or importance, because they treat of matters discussed in every text-book of bacteriology, and afford but little new information.

Chapters iii. and iv., under the title of " Taxonomy," deal with the question of species and variability among bacteria ; the involution and attenuation of microbes ; the systematic position of bacteria and their classification. Prof. Fischer points out that the heated controversy on the " species question " rests on our different conception of *pleogony* and *pleomorphism* among bacteria. Pleomorphism in its true sense does not exist among bacteria according to Fischer, and mutability of function (other than of a temporary character) is also denied by him. Thus he asserts positively that " It has not been hitherto possible to entirely suppress a single biological character in any species." We are inclined to take a less dogmatic position as regards the *suppression* of a biological character, while fully agreeing with the author that the attainment in our cultures of *acquired* characters that are permanent and hereditary in bacteria is not to be expected. The classification of bacteria proposed by Fischer has much to commend it, but it is natural to shrink from so sweeping a change in our existing nomenclature as would be necessary if his views were universally adopted. The tetanus bacillus, according to Fischer, would fall under the genus Plectridium, the subfamily Plectridieæ, the family Bacillaceæ, and the order Haplobacterinæ. The typhoid bacillus would come under the same order, but the genus would be Bactridium, the sub-family Bacilleæ, and the family Bacillaceæ.

Fischer suggests that

"the names of the genera might be formed conveniently in such a way that the root of the word indicated the shape of the cell, and the termination the arrangement of the cilia. The root-words might be *baktron* (rod), *kloster* (spindle), and *plectron* (drum-stick), and the terminations *-inium* for monotrichous, *-illum* for loptotrichous, and *-idium* for peritrichous types."

Chapter v. deals with the distribution of bacteria ; their modes of life, and spontaneous generation. Using the modes of nutrition as a basis for classification, Fischer divides bacteria into the following groups :—

I. *Prototrophic Bacteria.*

Nitrifying bacteria, bacteria of root-nodules, sulphur and iron bacteria, occur only in the open in nature—never parasitic, always monotrophic.

II. *Metatrophic Bacteria.*

Zymogenic, saprogenic and saprophile bacteria occur in the open and upon the external and internal surfaces of the body—sometimes parasitic (facultative parasites), monotrophic or polytrophic.

III. *Paratrophic Bacteria.*

Occur only in the tissues and vessels of living organisms—true (obligatory) parasites.

The author remarks that it is worthy of note that not only the bacteria but all other organisms can be arranged in these three biological divisions.

Chapters vi. to ix. deal with the physiology of nutrition; general principles of culture, respiration of bacteria, influence of physical agents, and the action of chemicals. It is impossible in the limits of this review to do justice to these chapters, which although treating of matters described in every text-book of bacteriology, yet succeed in presenting the subject to the reader in a new and attractive light.

Chapters x. and xi. treat of the circulation of nitrogen in nature. It may be said without fear of contradiction that the author has dealt with this complex problem in a masterly manner. No other writer that we are aware of (unless, perhaps, Lafar) has placed the subject before the reader in so comprehensive and intelligible a form. To indicate the scope of these chapters we cannot forbear quoting from the introductory remarks as follows :—

"Apart from the activity of organisms like the pigment and phosphorescent bacteria, and other remarkable metabolism of the sulphur—and iron—bacteria, the work of bacteria in nature embraces three great processes :
(1) The circulation of nitrogen : effected by putrefaction, the formation of nitrates, and the assimilation of atmospheric nitrogen.
(2) The circulation of carbon by the fermentation of carbohydrates and other non-nitrogenous products of animals and plants.
(3) The causation of disease in other organisms, particularly in man and the higher animals.
There are in nature five sources of nitrogen open to plants and animals :
(1) The atmosphere (79 per cent. by volume of free nitrogen).
(2) The nitrates of the soil and the traces of nitrous acid formed in the air during thunderstorms.
(3) Ammonia, which occurs in minute quantities in the air, and is set free abundantly by the putrefaction and decay of dead organisms.
(4) Animal excreta, which contains nitrogen compounds of many kinds, even down to ammonia ; and
(5) The tissue of plants and animals."

Chapters xii. to xiv. deal with the circulation of carbon dioxide in nature, and are full of interest to the biologist, and will doubtless appeal very strongly to students of agricultural chemistry.

Chapters xv. to xvii. treat of bacteria in relation to disease. In no captious spirit we venture to offer the criticism that here the author treads on less familiar ground, and although the subject is discussed in a scholarly and instructive manner, there is some evidence that a pure botanist is apt to fall into error when invading the domain of the pathologist. We do not agree with the author when he says that *B. coli* is almost indistinguishable from the parasite of typhoid fever. It is easily distinguished—the difficulty lies in differentiating between certain phase-forms or allies of *B. coli* and the typhoid germ. Again, Prof. Fischer would seem to be in error when he says that the dimensions of the two are about the same, that both are actively motile and peritrichously ciliated, and that the cilia are too delicate for their number to be of determinative value. As a matter of fact, *B. coli* is a short rod, hardly longer than broad, frequently showing only very feeble motility, and usually having only 1 to 3 flagella, which stain with difficulty ; whereas the typhoid bacillus occurs as long, thin, slender rods and filaments, which (the rods) are actively motile and move about in a fashion quite different from the colon bacillus. Moreover, the flagella average ten in number, and stain readily. The statement that *B. coli* is frequently present in dirty water must be accepted with reserve, unless it be assumed that the word "dirty" is meant by the author to convey the idea of fouling with matter of an excremental sort. Again, the author, speaking of the staphylococci (s.p. aureus, citreus and albus), says that in nature these germs are found everywhere. We venture to dispute the truth of this remark, which is stated as if it were a fact ; yet in our judgment it is merely a supposition, and an erroneous one.

These few criticisms are made in no carping spirit ; indeed, the book as a whole strikes us as being one of the best that has been written on the subject, and in many respects it is quite unique. The chapters dealing with the circulation of nitrogen and carbon in nature are altogether admirable. We can find no words sufficiently strong to recommend this book to the perusal of all students of bacteriology, and particularly to those interested in biology from the technical point of view.

Unstinted praise must be given to the translator ; in offering to English readers a translation of Prof. Alfred Fischer's "Vorlesungen über Bakterien" he has placed us under a deep debt of gratitude.　A. C. HOUSTON.

OUR BOOK SHELF.

A Walk Through the Zoological Gardens.　By F. G. Aflalo, F.R.G.S., F.Z.S.　Pp. 232. (London : Sands and Co., 1900.)

IT is not by any means abundantly clear that a guide to the Zoological Society's Gardens is needed, inasmuch as there already exists the well-known and accurate guide to the Society's collection by Mr. Sclater. Although it is true that the author does not call his book a "guide" in the title, he nevertheless observes in the preface that it is his object "to conduct the reader from house to house and from paddock to paddock, pointing out the chief features of interest" on the way. We must, therefore, consider the book as intended to be a guide. As such it does not appear to us to be at all informing ; it would have been well, too, to avoid positive error. The author calls a sea-lion a seal, which—seeing that true seals are often exhibited—is confusing. The African Mudfish, *Protopterus*, often on view in the Reptile house, is dubbed *Lepidosiren*, which, we need scarcely explain, is a South American Dipnoan. There are other errors of fact, and certain statements which are so loose and confused that they are practically erroneous. It is naturally impossible in a small book like the present to give an exhaustive account of all the animals to be seen in the course of a year or two in the Gardens. But the author leaves out so many important beasts that he fails to convey a real notion of the extent and variety of the collection. By cutting out the tale of how he rescued a blue pencil from a cormorant, which afterwards swallowed a lady's parasol, and by forbearing to mention that porcupines "pare their teeth on elephants' tusks" (!), and generally by avoiding gossip of a totally uninteresting and equally uninstructive kind, Mr. Aflalo might have grappled more successfully with the immense amount of material at his disposal.

LETTER TO THE EDITOR.

[*The Editor does not hold himself responsible for opinions expressed by his correspondents. Neither can he undertake to return, or to correspond with the writers of, rejected manuscripts intended for this or any other part of* NATURE. *No notice is taken of anonymous communications.*]

The Reform of Mathematical Teaching.

To your issue of August 2 Prof. Perry contributes an indictment of the present system of mathematical teaching in this country. As he invites criticism, one need not apologise either for defending existing methods or for criticising his suggested improvements. His main attack seems to be directed against mathematics as an educational subject, and in particular the teaching of Euclid falls under his ban. The elements of abstract reasoning are, he says, quite unnecessary to a boy's mental development. Why does he not add that common-sense is of no value also?

Do we always first learn by actual trial, as is stated in his article? Do we throw ourselves into deep water and learn to swim forthwith? Do we set about jumping, cycling, billiards or cards without any previous explanation? Surely, as a rule, in these matters we are taught, not only what to do at the start, but also, if we can grasp them, the guiding principles. In a game of whist, who does not dread the unreasoning partner who has learnt the rule "Third player plays highest," and blindly acts upon it?"

Euclid, though it might advantageously be shortened by the assumption of a few more axioms and postulates, is not, I venture to say, at all a "soul-destroying, weary, worrying study for the schoolboy." Of course it may be made so, but to every boy, with care, it may become interesting, and, in the experience of many teachers, it proves a more engrossing subject to their classes than either arithmetic or algebra.

Prof. Perry very properly points out some of the weak spots in present-day arithmetic. He instances "our abominable system of weights and measures." One may suggest that that system is hardly the fault of our system of mathematics ; it is entirely its misfortune. Will he not, instead of girding at the unfortunate teachers of mathematics, agitate for a conference of delegates from all bodies interested in this most important question ?

Later on in his article it is stated that practice, interest, discount, tare and tret, alligation, position, &c., are at this day taught exactly as during the last century. This statement is absurd.

It is true that discount, percentages, stocks, areas, &c., are all dependent on the rule of proportion ; but for purposes of explanation and of interest it is certainly as well not to lump these together in one heterogeneous muddle under the head of "Proportion." If such a method were in vogue, or if the whole of arithmetic were, by means of formulæ, reduced to multiplication and division, one would certainly see "the film of dulness covering a boy's face as he entered the class-room."

As regards the syllabus quoted by Prof. Perry, it is easy to agree with him thus far—that it is admirably adapted for a *technical* training. In practical mathematics, where mental training is of minor importance, exigencies of time will compel the teacher to omit explanations, or only to give them roughly, for his chief object is to enable his pupils to apply mathematical results, as distinct from reasoning, to problems in engineering, science, or kindred subjects.

On the other hand, the average boy's mathematical education up to the age of fifteen or sixteen is an absolutely different matter ; to put it crudely—the teacher's main effort is to enable his pupil to ask and to answer reasonably the question "Why?"

At present there is really no orthodox system, but, in my opinion, the methods enunciated in the principal text-books of the day do, with slight exceptions, tend to develop a boy's mental powers.

When the boy has decided on his profession, then by all means continue his education on the lines suggested by Prof. Perry.

Finally (if one may misquote his opening words), "it is very important to try to get a view of our system of teaching mathematics, which is not too much tinted with pleasant (or possibly unpleasant) memories of science and engineering."
　　　　　　　　　　　　　　　　　　　W. F. BEARD.

¹ *SIR JOHN BENNET LAWES, BART., F.R.S.*

ONE of our truly great men has passed away. Born in 1814, his life extended over the greater part of a century full of great men and great deeds ; yet among this goodly company he will surely be placed in the front rank if we have regard to his personal qualities, and to the far-reaching and beneficial character of his achievements. An only son, he was left at the age of eight without a father, and owed much of his bringing up to the care of his mother, to whom he was extremely attached. Educated at Eton and Oxford, he entered in 1834, at the age of twenty, on the management of the paternal estate at Rothamsted, Herts. How many youths placed thus early in the possession of a beautiful home, and a sufficient income, would have fallen into habits of easy enjoyment, and left nothing behind them worth recording ? But these circumstances, so full of danger to the average lad, were exactly suited for the development of the work which the youthful squire was to accomplish. His active mind obtained at once perfect freedom of action. His love of work and high sense of duty led him to devote himself to the management of the home farm, and under his keen observation its fields became to him pages of nature, which year by year told him new facts, answered new questions, and revealed fresh wonders.

In boyhood Sir John Lawes had somehow imbibed a taste for chemistry. He has told in his own graphic way how he delighted to write with a stick of phosphorus on the door of a dark room, and to give electric shocks to the old housekeeper. On taking up his residence at Rothamsted he proceeded to fit up one of the bedrooms as a laboratory and commenced a variety of experiments. Chemistry was at that time to a large extent a department of medical science, and some of his earliest work was directed to the isolation of the alkaloids of medicinal plants. The manufacture of calomel by a new process—the combustion of mercury in chlorine gas—also engaged his attention ; an old barn was converted into a laboratory, and the process carried out for some time on a commercial scale. We have here his first decided step as a chemical manufacturer.

Three or four years elapsed before any experiments were made on agricultural subjects. In 1837 Sir John Lawes commenced growing plants in pots with various substances applied as manure, and more numerous experiments of this kind were carried out in the two following years. His attention had been drawn to the uselessness of bones when applied as a manure to turnips on heavy soils, although they proved extremely effective on light soils. He tried treating fresh bone and burnt bone with sulphuric acid, and speedily remarked the great increase in the manurial effect of the bone so treated. Mineral phosphates, as apatite, were treated with sulphuric acid, and were also found to yield a most effective manure. The value of these acidified manures having been confirmed by trials in the field, both at Rothamsted and elsewhere, it became evident that a fact of the greatest importance to agriculture had been discovered. The farmer was no longer dependent on bone or guano for a supply of phosphate, the vast deposits of phosphatic rocks and minerals were equally available for his use, and could be converted into a potent manure by treatment with sulphuric acid ; moreover, the new manure was as effective upon heavy as upon light land.

In 1842 Sir John Lawes took out his patent for treating apatite and other phosphatic minerals with sulphuric acid, thus producing superphosphates. This event laid the foundation of the vast manufacture of artificial manures which now supplies the varied wants of the farmer, and enables him to restore or increase the fertility of his land in the most effective and economical

manner. At the present time the annual production of superphosphates in the United Kingdom reaches the enormous total of 900,000 tons ! In other countries the quantities manufactured are equally prodigious. The influence of these manures on the productiveness of the soil in civilised countries is incalculable.

It was a serious step for a young country squire to start a large manure factory, and to enter himself as a wrestler in the keen competitions of commercial life. The difficulties which beset a successful, rapidly growing business are especially great, as the continually increasing demand for capital more than swallows up the accruing profits. But the owner of Rothamsted had already developed his mental powers. His cool head, his quick grasp of a situation, his methodical habits, his enterprising spirit, and his scientific training made him a most successful man of business. In 1843 a manure factory was started by Sir John Lawes at Deptford Creek. In 1857 the business had so increased that 100 acres of land were purchased at Barking Creek, and a much larger and more complete plant erected. The whole of the manure business was sold in 1872 for 300,000*l.*

Having entered upon London commercial life, Sir John Lawes did not confine his enterprise to the production of manures ; his activity was displayed in many directions. In 1867 a large factory at Millwall for the production of tartaric and citric acid was acquired. Here the same tale has to be told as that relating to the manure business. The scientific enterprise of the new owner, his carefulness for economy of production, and the advantages he obtained from command of capital, soon placed him also at the head of this branch of English chemical manufacture.

We need not in this place refer at greater length to Sir John Lawes' commercial career, which continued with activity up to his death ; it has been due to the man to depict the results of this large portion of his life's work, and also to note the remarkable character which he obtained in the commercial world. His enterprise and command of capital made him at all times a formidable competitor ; but his name was still better known as associated with scrupulous fairness of dealing, and in cases of difficulty it became recognised that it was better to trust to Sir John Lawes' generosity than to fight for individual rights.

One might readily suppose that the commercial life of such a successful man of business would have engrossed nearly the whole of his time and energy ; this was far from being the case. The wonder of the man was that he carried on so long and so successfully a dual life. He seemed to have the power of putting aside at once all one class of thoughts and interests, and of embarking with perfect freshness and enthusiasm on another subject ; and of all subjects, agriculture, chiefly regarded as presenting a series of problems in agricultural chemistry, was the one in which his greatest interests were centred. Business might occupy his mind, but agriculture was always the mistress of his heart. This would appear in a striking manner in the midst of his London work. On the occasion of his weekly visit to the chemical laboratory at Millwall, a very few minutes would generally suffice for the despatch of immediate business, and Sir John Lawes would then frequently spend an hour in eloquent talk with the writer of the present notice upon the agricultural problems which then most interested him. Visitors to the Rothamsted experiments might well be ignorant of the existence of the vast London business. The spirit of the counting-house never invaded Rothamsted ; here the experimental fields, and the reports preparing on the results they yielded, seemed to engage his whole attention. No one knew the fields as well as he did, and an afternoon spent at home generally included a walk, spud in hand, through the whole of them.

For the sake of clearness in our narration, we have brought together in one view the principal facts relating to Sir John Lawes' career as a chemical manufacturer ; we have next to regard him as the founder of the Agricultural Experiment Station at Rothamsted, and as its presiding genius during a period of nearly sixty years.

We must go back again to 1842. When Sir John Lawes had taken out his patent, and had determined to start a factory in London, it would have seemed natural if his entire energies had been transferred to this new and promising sphere of labour ; but now the truly scientific character of the man was made manifest. The agricultural investigations in the fields at Rothamsted, which had become so fascinating, were not to be given up, but extended. This could only be done by engaging scientific assistance. A young chemist, Dr. J. H. Gilbert, was engaged to superintend the Rothamsted experiments ; and thus, in 1843, began that partnership in labour which has yielded such a rich harvest of results.

The station thus founded at Rothamsted began its work long before any of the agricultural stations now existing in other countries, of which there are at present several hundreds. The investigations carried out have proceeded to a considerable extent on lines peculiar to the place, and have generally been of a very laborious character. The most striking characteristic of Rothamsted is its experimental fields, covering nearly forty acres. Here the various crops of a four-course rotation are grown, both separately and in their usual order of succession ; the influence of different manures upon the quantity and composition of the crops is studied ; the alterations in the composition of the soil brought about by different treatment are determined, and in some cases the composition of the drainage water from the different plots is ascertained. The Rothamsted investigations have also included many important and laborious experiments on farm animals.

It is quite impossible to attempt to enumerate the various investigations made at Rothamsted. The first formal report appeared in 1847. The collected reports now occupy nine volumes. We may, however, note some stages in the development of this great enterprise. By 1848 most of the experimental fields had commenced their work ; by 1856 the whole of the present series was in operation. The chemical work required soon exceeded the capacity of the old barn first used, and in 1854-5 a handsome new laboratory was built and presented to Sir John Lawes as a testimonial from the agriculturists of England for his services to agriculture. Large additional buildings for preparing and storing samples have since been added. In 1889 Sir John Lawes transferred the whole of the laboratories and experimental fields to trustees, with an endowment of 100,000*l.*, so that the agricultural investigations might be permanently continued. The management is now vested in a committee, nominated by the Royal, the Royal Agricultural, the Chemical and the Linnean Societies. The jubilee of the Rothamsted Station was celebrated in 1893, and on this occasion Dr. J. H. Gilbert received the honour of knighthood. A full account of the celebration will be found in NATURE of July 27 and August 3, 1893.

But we must turn once more to the man himself. Sir John Lawes received many honours. The Queen created him a baronet in 1882. Universities gave him their degrees ; societies bestowed upon him their medals. Prosperity could not spoil him. Quite free from personal ambition, he was always ready to give the credit of success to his fellow-workers. Visitors to the Rothamsted experiments—and they were many—were delighted when Sir John himself was the pilot of the party ; the two hours' talk to which they listened was a treat to be remembered. In terse, vigorous sentences the practical results of each trial were brought before them, while the whole was illuminated by many a flash of humour. In

his middle life Sir John Lawes wrote a great number of short articles for the agricultural press. In these he excelled. His thorough knowledge of the details of farming, and his practical mind, prevented him from ever writing as a mere doctrinaire ; the facts ascertained by investigation were presented by him in their concrete aspect as things to be reckoned with by the farmer in his daily life on the farm.

Our notice would be incomplete without some reference to his local beneficence. As Lord of the Manor he did much to maintain and increase the charms of a pretty village now rapidly transforming itself into a town. He was the agricultural labourers' best friend. He provided them with an ample supply of allotment gardens, and in 1857 built a club room for their benefit ; this was visited and described by Charles Dickens in 1859. Sir John Lawes also tried to introduce several co-operative schemes for the labourers' benefit. He was always seen to great advantage on the occasion of the annual allotment club dinner at which he presided, when he carved a huge piece of beef provided by himself, and afterwards made a humorous speech to the labourers. As a generous donor to public and private charities he will be long remembered.

As he passed into old age his powers seemed to suffer little diminution. A few days before his last illness he went as usual to London, and thence down to the factory at Millwall. He died on August 31, in his eighty-sixth year, full of days and full of honours, and venerated by all who knew him. R. WARINGTON.

THE BRADFORD MEETING OF THE BRITISH ASSOCIATION.

IN the midst of the turmoil of the Association week it is difficult to give any careful compression of the results of the meeting ; but, from the point of view of the Local Committee, the visit has been an unqualified success. Apprehensions were felt, in making the preliminary arrangements, lest there might be some incongruity in certain cases between the guests and the hosts : but, judging from the absence of rumours to that effect, the fears were groundless.

One point upon which at first some individual soreness arose was due to a slight misunderstanding in the matter of the excursions. These had long presented an exceedingly difficult problem to the local organisers of Association meetings ; so much so, that the entire abandonment of all excursions has at times been contemplated. It is probably inevitable that, in any scheme put forward to ensure the success of the excursionists as a whole, some individual hardship must be occasioned, leading to a certain amount of unpleasantness. In the circumstances, it is scarcely surprising that, at first, a few sufferers should have complained that the excursion arrangements had been planned in a new and inferior manner ; but, eventually, they fell in readily with the innovations. At former meetings it has been found impossible to organise excursions upon any exact system, owing to the fact that persons applied for tickets in a vague and undecided spirit, and often failed to take them up when they were allotted to them, to the consequent deprivation of other persons. Accordingly, in the Bradford scheme, the following regulations were issued :—

(1) That out of the seven excursions for the day three should be selected in the order of preference. (2) That the fee (made practically uniform) should be handed in with the application form ; and (3) that persons complying with these requirements should in due course receive a ticket for one of the trips selected, or have their money returned. Those applicants who stated that they would only name a single excursion were

advised not to hand in any application or fee, as no guarantee could be given that they would receive one particular ticket. Despite the individual irritation which has been occasioned by an imperfect comprehension of the scheme, and by occasional unwillingness to comply with the conditions, the new arrangement has received a widespread approval, and has formed another instance of the willing and courteous acceptance of the Bradford arrangements, which has so conspicuously characterised the present meeting of the Association.

RAMSDEN BACCHUS.

Meetings of the General Committee.

At the first meeting of the General Committee, Prof. Schäfer reported that the following resolutions had been considered and acted upon by the Council of the Association.

I. That in view of the opportunities of ethnographical inquiry which will be presented by the Indian census, the Council of the Association be requested to urge the Government of India to make use of the census officers for the purposes enumerated below, and to place photographers at the service of the census officers :—(1) To establish a survey of the jungle races, Bhils, Gonds, and other tribes of the central mountain districts ; (2) to establish a further survey of the Naga, Kuki, and other cognate races of the Assam and Burmese frontiers ; (3) to collect further information about the vagrant and criminal tribes, Haburas, Beriyas, Sansiyas, &c., in North and Central India ; (4) to collect physical measurements, particularly of the various Dravidian tribes, in order to determine their origin ; and of the Rajputs and Jats of Rajputana and the Eastern Punjab, to determine their relation with the Yu-echi and other Indo-Scythian races ; (5) to furnish a series of photographs of typical specimens of the various races, of views of archaic industries, and of other facts interesting to ethnologists.

II. That the Council be requested to represent to Her Majesty's Government the importance of giving more prominence to botany in the training of Indian forest officers.

A resolution was adopted, that in future women should be eligible as members of the Sectional Committees.

At the second meeting of the General Committee it was decided that the meeting of the Association in 1902 should be held in Belfast. The meeting next year will be held in Glasgow, with Prof. A. W. Rücker as president, and will commence on Wednesday, September 11, 1901. The vice-presidents for the meeting will be : The Earl of Glasgow, Lord Blythswood, Lord Kelvin, the Lord Provost of Glasgow, the Principal of the University of Glasgow, Sir John Stirling-Maxwell, M.P., Sir Andrew Noble, Sir Archibald Geikie, Sir W. T. Thiselton-Dyer, Mr. James Parker Smith, M.P., Mr. John Inglis, and Mr. Andrew Stewart. Alluding to the arrangements for the Glasgow meeting, the Lord Provost of Glasgow said the University had been placed at the disposal of the Association, and all the Sections will probably be accommodated under one roof. There will also be an exhibition in Glasgow next year, and the 450th anniversary of the University will be celebrated.

Prof. Schäfer has retired from his position as one of the general secretaries, and Dr. D. H. Scott has been elected his successor.

The following is a synopsis of the grants of money appropriated to scientific purposes by the General Committee :—

Mathematics and Physics.

		£
*Rayleigh, Lord—Electrical Standards		45
*Judd, Prof. J. W.—Seismological Observations		75
*Rücker, Prof. A. W.—Magnetic Force on board Ship (renewed) ... ... ,...		10

* Re-appointed.

Chemistry.

	£
Hartley, Prof. W. N.—Relation between Absorption Spectra and Constitution of Organic Substances (balance, £6 8s. 9d. in hand)	—
*Roscoe, Sir H. E.—Wave-length Tables	5
*Miers, Prof. H. A.—Isomorphous Sulphonic Derivatives of Benzene	35

Geology.

	£
*Hull, Prof. E.—Erratic Blocks (£6 in hand)	—
*Geikie, Prof. J.—Photographs of Geological Interest (balance, £10 in hand)	—
*Lloyd-Morgan, Prof. C.—Ossiferous Caves at Uphill (renewed)	5
*Watts, Prof. W. W.—Underground Water of North-West Yorkshire	50
*Scharff, Dr.—Exploration of Irish Caves (renewed) ...	15
*Marr, Mr. J. E.—Life-zones in British Carboniferous Rocks	2

Zoology.

	£
*Herdman, Prof. W. A.—Table at the Zoological Station, Naples	100
*Bourne, Mr. G. C.—Table at the Biological Laboratory, Plymouth	20
*Woodward, Dr. H.—Index Generum et Specierum Animalium...	75
*Newton, Prof. A.—Migration of Birds	10
Poulton, Prof. E. B.—Life-history of the Marble Gall-fly	—

Geography.

	£
Keltie, Dr. J. Scott—Terrestrial Surface Waves ...	5
Mill, Dr. H. R.—Changes of Land-level in the Phlegræan Fields	50

Economic Science and Statistics.

	£
*Giffen, Sir R.—State Monopolies in other Countries (£13 13s. 6d. in hand)	—
Brabrook, E. W.—Legislation regulating Women's Labour	15

Mechanical Science.

	£
*Preece, Mr. W. H.—Small Screw Gauge (balance in hand) and...	45
Binnie, Sir A.—Resistance of Road Vehicles to Traction	75

Anthropology.

	£
*Evans, Mr. A. J.—Silchester Excavation	10
*Penhallow, Prof. D. P.—Ethnological Survey of Canada	30
Garson, Dr. J. G.—Age of Stone Circles (balance in hand)	—
*Read, Mr. C. H.—Photographs of Anthropological Interest (balance of £10 in hand)	—
*Tylor, Prof. G. B.—Anthropological Teaching	5
Evans, Sir John.—Exploration in Crete	145

Physiology.

	£
*Schäfer, Prof. E. A.—Physiological Effects of Peptone...	30
Schäfer, Prof. E. A.—Chemistry of Bone Marrow ...	15
Starling, Prof. E. H.—Suprarenal Capsules in the Rabbit	5

Botany.

	£
*Farmer, Prof. J. B.—Fertilisation in Phæophyceæ ...	15
Marshall Ward, Prof.—Morphology, Ecology, and Taxonomy of Podostemaceæ	20

Corresponding Societies.

	£
*Whitaker, Mr. W.—Preparation of Report	15
	£945

* Re-appointed.

SECTION A.

DEPARTMENT OF ASTRONOMY.

OPENING ADDRESS BY DR. A. A. COMMON, F.R.S., F.R.A.S., CHAIRMAN OF THE DEPARTMENT.

IT has been decided to form a Department of Astronomy under Section A, and I have been requested to give an address on the occasion. In looking up the records of the British Association to see what position Astronomy has occupied, I was delighted to find, in the very first volume, "A Report on the Progress of Astronomy during the Present Century," made by the late Sir George Airy, so many years our Astronomer Royal, and at that time Plumian Professor of Astronomy at Cambridge. This report, made at the second meeting of the Association, describes, in a most interesting manner, the progress that was made during the first third of the century, and we can gather from it the state of astronomical matters at that time. The thought naturally occurred to me to give a report, on the same lines, to the end of this century, but a little consideration showed that it was impossible in the limited time at my disposal to give more than a bare outline of the progress made.

At the time this report was written we may say, in a general way, that the astronomy of that day concerned itself with the position of the heavenly bodies only, and, except for the greater precision of observation resulting from better instruments and the larger number of observatories at work, this, the gravitational side of astronomy, remains much as it was in Airy's time.

What has been aptly called the New or Physical Astronomy did not then exist. I propose to briefly compare the state of things then existing with the present state of the science, without dealing very particularly with the various causes operating to produce the change; to allude briefly to the new astronomy; and to speak rather fully about astronomical instruments generally, and of the lines on which it is most probable future developments will be made.

In this report (*Brit. Assoc. Report*, 1831–32, p. 125) we find that at the beginning of the century the Greenwich Observatory was the only one in which observations were made on a regular system. The thirty-six stars selected by Dr. Maskelyne, and the sun and moon, were observed on the meridian with great regularity, the planets very rarely and only at particular parts of their orbits; small stars, or stars not included in the thirty-six, were seldom observed.

This state of affairs was no doubt greatly improved at the epoch of the report, but it contrasts strongly with the present work at Greenwich, where 5000 stars were observed in 1899, in addition to the astrographic, spectroscopic, magnetic, meteorological, and other work.

Many observatories, of great importance since, were about that time founded, those at Cambridge, Cape of Good Hope and Paramatta having just been started. A list is given of the public observatories then existing, with the remark that the author is "unaware that there is any public observatory in America, though there are," he says, "some able observers."

The progress made since then is truly remarkable. The first public observatory in America was founded about the middle of the century, and now public and private observatories number about 150, while the instrumental equipment is in many cases superior to that of any other country. The prophetic opinion of Airy about American observers has been fully borne out. The discovery of two satellites to Mars by Hall in 1877, of a fifth satellite to Jupiter by Barnard in 1892, and the discovery of Hyperion by Bond, simultaneously with Lassell, in 1848, are notable achievements.

The enormous amount of work turned out by the Harvard Observatory and its branches in South America, all the photographic and spectroscopic work carried out by many different astronomers, and the new lines of research initiated show an amount of enthusiasm not excelled by any other country. A greater portion of the astronomical work in America has been on the lines of the new astronomy, but the old astronomy has not been at all neglected. In this branch pace has been kept with other countries.

From this report we gather that the mural quadrant at most of the observatories was about to be replaced by the divided circle. Troughton had perfected a method of dividing circles, which, as the author says, "may be considered as the greatest improvement ever made in the art of instrument making."

Two refractors of 11 and 12 inches aperture had just been

imported into this country; clockwork for driving had been applied to the Dorpat and Paris equatorials, but the author had not seen either in a state of action.

The method of mounting instruments adopted by the Germans was rather severely criticised by the author, the general principle of their mounting being "telescopes are always supported at the middle, not at the ends."

"Every part is, if possible, supported by counterpoises."

"To these principles everything is sacrificed. For instance, in an equatorial the polar axis is to be supported in the middle by a counterpoise. This not only makes the instrument weak (as the axis must be single), but also introduces some inconvenience into the use of it. The telescope is on one side of the axis; on the other side is a counterpoise. Each end of the telescope has a counterpoise. A telescope thus mounted must, I should think, be very liable to tremor. If a person who is no mechanic and who has not used one of these instruments may presume to give an opinion, I should say that the Germans have made no improvement in instruments except in the excellence of the workmanship."

I have no doubt that this question had often occupied Airy's mind, for in the Northumberland Equatorial Telescope which he designed shortly after for Cambridge he adopted what has been called the English form of mounting, where the telescope is supported by a pivot at each side, and a long polar axis is supported at each end. This telescope is in working order at the present time at Cambridge.

When he became Astronomer Royal he used the same design for what was for many years the great equatorial at Greenwich, though the wooden uprights forming the polar axis were in the Greenwich telescope replaced by iron. It says much for the excellence of the design and workmanship of this mounting, designed as it was for an object-glass of about 13 inches diameter, when we find the present Astronomer Royal, Mr. Christie, has used it to carry a telescope of 28 inches aperture, and that it does this perfectly.

Notwithstanding the greater steadiness of the English form of mounting, the German form has been adopted generally for the mounting of the large refractors recently made.

There is much interesting matter in this report of an historical character.

As I have already said, the new astronomy, as we know it, did not exist; but in a report (*Brit. Assoc. Report*, 1831–32, p. 308) on optics, in the same volume, by Sir David Brewster, we find that spectrum analysis was then occupying attention, and the last paragraph of this report is well worth quoting: "But whatever hypothesis be destined to embrace and explain this class of phenomena, the fact which I have mentioned opens an extensive field of inquiry. By the aid of the gaseous absorbent we may study with the minutest accuracy the action of the elements of material bodies in all their variety of combinations, upon definite and easily recognised rays of light, and we may discover curious analogies between their affinities and those which produce the fixed lines in the spectra of the stars. The apparatus, however, which is requisite to carry on such inquiries with success cannot be procured by individuals, and cannot even be used in ordinary apartments. Lenses of large diameter, accurate heliostats, and telescopes of large aperture are absolutely necessary for this purpose; but with such auxiliaries it would be easy to construct optical combinations, by which the defective rays in the spectra of all the fixed stars down to the *tenth* magnitude might be observed, and by which we might study the effects of the very combustion which lights up the suns of other systems."

Brewster's words are almost prophetic, and it would almost appear as if he unknowingly held the key to the elucidation of the spectrum lines, for it was not until 1859 that Kirchhoff's discovery of the true origin of the dark lines was made.

Fraunhofer was the first to observe the spectra of the planets and the stars, and to notice the different types of stellar spectra. In 1817 he recorded the spectrum of Venus and Sirius, and later, in 1823, he described the spectrum of Mars; also Castor, Pollux, Capella, Betelgeux and Procyon.

Fraunhofer, Lamont, Donati, Brewster, Stokes, Gladstone and others carried on their researches at a time when the principles of spectrum analysis were unknown, but immediately upon Kirchhoff's discovery great interest was awakened.

With spectrum analysis thus established, aided as it was later by the greater development of photography, the new astronomy was firmly established.

The memorable results arrived at by Kirchhoff were no sooner published than they were accepted without dissent. The works of Stokes, Foucault and Ångström at that period were all suggestive of the truth, but do not mark an epoch of discovery.

Astronomical spectroscopy divided itself naturally into two main branches, the one of the sun, the other of the stars, each having its many offshoots. I shall just mention a few points relating to each. The dark lines in the solar spectrum had already been mapped by Fraunhofer, and now it only needed better instruments and the application of laboratory spectra with Kirchhoff's principle to advance this work still further.

Fraunhofer had already pointed out the way in using gratings, and these were further improved by Nobert and Rutherfurd.

Kirchhoff's Map of the Solar Spectrum, published in 1861–62, was the most complete up to that time; but the scale of reference adopted by him was an arbitrary one, so that it was not long before this was improved upon. Ångström published in 1868 his map of the "Normal Solar Spectrum," adopting the natural scale of wave-lengths for reference, and this remained in use until quite recent times.

The increased accuracy in the ruling of gratings by Rutherfurd materially improved the efficiency of the solar spectroscope, but it was not until Prof. Rowland's invention of the concave grating that this work gained any decisive impetus. The maps (first published in 1885) and tables (published in the years 1896–98) of the lines of the solar spectrum are now almost universally accepted and adopted as a standard of reference. These tables alone record about 10,000 lines in the spectrum of the sun, which is in marked contrast to the number 7 recorded by Wollaston at the beginning of the century (1802). Good work in the production of maps has also been done in this country by Higgs.

Michelson has also recently invented a new form of spectroscope called the "Echelon" (*Ast. Phys. Journ.*, vol. viii. 1898, p. 37), in which a grating with a relatively small number of lines is employed, the dispersion necessary for modern work being obtained by using a high order (say the hundredth) into which most of the light has been concentrated.

Besides lines recorded in the visual and ultra-violet portions of the solar spectrum, maps have been made of the lines in the infra-red, the most important being that of Langley's, published in 1894, prepared by the use of his "bolometer." Good work had, however, been done in this direction previously by Becquerel, Lamansky and Abney; the last, indeed, succeeded even in photographing a part of it.

The recording of the Fraunhofer lines in the solar spectrum is not all, however. The application of the spectroscope to the sun has several epoch-marking events attached to it, notably those of proving the solar character of the prominences and corona, the rendering visible of the prominences without the aid of an eclipse by the discovery of Lockyer and Janssen in 1868, the photography of the prominences both round the limb and those projected on the solar disc by the invention of the spectra-heliograph by Hale and Deslandres in 1890.

Success has not yet favoured the many attempts to photograph the corona without an eclipse by spectroscopic means; but even now this problem is being attacked by Deslandres with the employment of the calorific rays.

Spectroscopic work on the sun has led to the discovery of many hundreds of dark lines, the counterparts of which it has not yet been possible to produce on the earth.

But besides those unknown substances which reveal their presence by dark lines, there were two others discovered, which showed themselves only by bright lines, the one in the chromosphere, to which the name of Helium was given, and the other in the corona, to which the name of Coronium was applied.

The former was, however, identified terrestrially by Ramsay in 1895, though the latter is still undetermined. The revision of its wave-length, brought about by the observations of the eclipse of 1898, may, however, result in this element being transferred from the unknown to the known in the near future.

The study of stellar spectra was taken up by Huggins, Rutherfurd and Secchi. Rutherfurd (*Am. Journ.*, vol. xxxv. 1862, p. 77) published in 1862 his results upon a number of stars, and suggested a rough classification of the white and yellow stars; but Secchi deserves the high credit of introducing the first systematic differentiation of the stars according to their spectra, he having begun a spectroscopic survey of the heavens for the purposes of classification (*Comptes rendus,*

t. lvii. 1853), whilst Huggins devoted himself to the thorough analysis of the spectra of a few stars.

The introduction of photography marks another epoch in the study of stellar spectra. Sir William Huggins applied photography as early as 1863 (*Phil. Trans.*, 1864, p. 428), and secured an impression of the spectrum of Sirius, but nearly another decade elapsed before Prof. H. Draper (*Am. Journ. of Soc. and Arts*, vol. xviii. 1879, p. 421) took a photograph of the spectrum of Vega in 1872, which was the first to record any lines. With the introduction of dry plates this branch of the new astronomy received another impetus, and the catalogues of stellar spectra have now become numerous. Among them may be mentioned those of Harvard College, Potsdam, Lockyer, McClean and Huggins. The Draper Catalogue (*Annals Harvard Coll.*, vol. xxvii. 1890) of the Harvard College, which is a spectroscopic Durchmusterung, alone contains the spectra of 10,351 stars down to the 7–8 magnitudes, and this has further been extended by work at Arequipa, whilst Vogel and Müller of Potsdam (*Astro-Phys. Obs. zu Potsdam*, vol. iii. 1882–83) made a spectroscopic survey of the stars down to the 7·5 magnitude between −1° and +20° declination. This has again been supplemented by Scheiner (*ibid.*, vol. vii. 1895: "Untersuchungen über die Spectra der helleren Sterne"), and by Vogel and Wilsing (*ibid.*, vol. xii. 1899: "Untersuchungen über die Spectra von 528 Sternen"). Lockyer (*Phil. Trans.*, vol. clxxxiv. A, 1893) in 1892 published a series of large-scale photographs of the brighter stars, and more recently McClean (*Phil. Trans.*, vol. cxci. A, 1898) has completed a spectroscopic survey of the stars of both hemispheres down to the 3½ magnitude. For the study and investigation of special types of stars, the researches of Dunér on the red stars, made at Upsala, and those of Keeler and Campbell on the bright-line stars, made at the Lick Observatory, deserve mention. For the study of stellar spectra the use of prisms in slit or objective-prism spectroscopes has predominated, though more recently the use of specially ruled gratings has been attended by some degree of success at the Yerkes Observatory.

Several new stars have also been discovered by their spectra by Pickering in his routine work of charting the spectra of the stars in different portions of the sky. The photographic plate containing their peculiar spectra was, however, not examined in many cases until the star had died down again.

Spectrum analysis also opened up another field of inquiry, viz. that of the motion of the stars in the line of sight, based on the process of reasoning due to Doppler, and accordingly named Doppler's Principle ("Ueber das farbige Licht der Doppelsterne," . . . *Abhandl. der K. Böhmischen Ges. d. Wiss.* V. Folge, 2 Bd. 1843.)

The observatories of Greenwich and Potsdam were among the first to apply this to the stars, and more recently Campbell at Lick, Newall at Cambridge, and Belopolsky at Pulkowa have made use of the same principle with enormous success.

It was also discovered that there are certain classes of stars having a large component velocity in the line of sight, which changes its direction from time to time, and in many such cases orbital motion has been proven, as in the case of Algol.

Another class of binary stars has also been discovered spectroscopically and explained by Doppler's principle. I refer to the stars known as spectroscopic binaries, in which the spectrum lines of one luminous source reciprocate over those from the other source of light, according as one is moving towards or away from the earth. This displacement of the spectrum lines led to the discovery of the duplicity of β Aurigæ and ζ Ursæ Majoris by Pickering (*Am. Jour.* [3], 39, p. 46, 1890).

Several other such stars have now been detected, notably β Lyræ, and lastly Capella, discovered independently by Campbell (*Astro-Phys. Jour.*, vol. x. p. 177) at Lick, and Newall (*Monthly Notices*, vol. lx. p. 2, 1899) at Cambridge.

The progress of the new astronomy is so closely bound up with that of photography that I shall briefly call to mind some of the many achievements in which photography has aided the astronomer.

Daguerre's invention in 1839 was almost immediately tried with the sun and moon, J. W. Draper and the two Bonds in America, Warren de la Rue in this country, and Foucault and Fizeau in France, being among the pioneers of celestial photography; but no real progress seems to have been made until after the introduction of the collodion process. Sir John Herschel in 1847 suggested the daily self-registration of the sun-spots to supersede drawings; and in 1857 the De la Rue photo-heliograph was installed at Kew. From 1858–72 a daily record was maintained by the Kew photo-heliograph, when the work was discontinued. Since 1873 the Kew series has been continued at Greenwich, and is supplemented by pictures from Dehra Dûn in India and from Mauritius. The standard size of the sun's disc on these photographs has now been for many years 8 inches, though for some time a 12-inch series was kept up.

The first recorded endeavour to employ photography for eclipse work dates back to 1851, when Berowsky obtained a daguerreotype of the solar prominences during the total eclipse. From that date nearly every total eclipse of the sun has been studied by the aid of photography.

In 1860 the first regularly planned attack on the problem by means of photography was made, when De la Rue and Secchi successfully photographed the prominences and traces of the corona, but it was not until 1869 that Prof. Stephen Alexander obtained the first good photograph of the corona.

In recent years, from 1893 until the total eclipse which occurred last May, photography has been employed to secure large-scale pictures of the corona. These were inaugurated in 1893 by Prof. Schaeberle, who secured a 4-inch picture of the eclipsed sun in Chili: these have been exceeded by Prof. Langley, who obtained a 15-inch picture of the corona in North Carolina during the eclipse of May 1900.

Photography also supplied the key to the question of the prominences and corona being solar appendages, for pictures of the eclipse work taken in Spain in 1860 terminated this dispute with regard to the prominences, and finally to the corona in 1871.

In 1875, in addition to photographing the corona, attempts were made to photograph its spectrum, and at every eclipse since then the sensitised plate has been used to record both the spectrum of the chromosphere and the corona. The spectrum of the lower layers of the chromosphere were first successfully photographed during the total eclipse of 1896 in Nova Zembla by Mr. Shackleton, though seen by Young as early as 1870, and a new value was given to the wave-length of the coronal line (wrongly mapped by Young in 1869) from photographs taken by Mr. Fowler during the eclipse of 1898 (India).

Lunar photography has occupied the attention of various physicists from time to time, and when Daguerre's process was first enunciated, Arago proposed that the lunar surface should be studied by means of the photographically produced images. In 1840 Dr. Draper succeeded in impressing a daguerreotype plate with a lunar image by the aid of a 5-inch refractor. The earliest lunar photographs, however, shown in England were due to Prof. Bond, of the United States. These he exhibited at the Great Exhibition in 1851. Dancer, the optician, of Manchester, was, perhaps, the first Englishman who secured lunar images, but they were of small size (Abney, "Photography").

Another skilful observer was Crookes, who obtained images of 2 inches diameter, with an 8-inch refractor of the Liverpool Observatory. In 1852 De la Rue began experimenting in lunar photography. He employed a reflector of some 10 feet focal length and about 13 inches diameter. A very complete account of his methods is given in a paper read before the British Association. Mr. Rutherfurd at a later date having tried an 11½-inch refractor, and also a 13-inch reflector, finally constructed a photographic refracting telescope, and produced some of the finest pictures of the moon that were ever taken until recent years. Also Henry Draper's picture of the moon taken Sept. 3, 1863, remained unsurpassed for a quarter of a century. Admirable photographs of the lunar surface have been published in recent years by the Lick Observatory and others. I myself devoted considerable attention to this subject at one time; but only those surpassing anything before attempted have been published in 1896–99 by MM. Loewy and Puiseux, taken with the Equatorial Coudé of the Paris Observatory.

Star prints were first secured at Harvard College, under the direction of W. C. Bond, in 1850; and his son, G. P. Bond, made in 1857 a most promising start with double-star measurements on sensitive plates, his subject being the well-known pair in the tail of the Great Bear. The competence of the photographic method to meet the stringent requirements of exact astronomy was still more decisively shown in 1866 by Dr. Gould's determination from his plates of nearly fifty stars in the Pleiades. Their comparison with Bessel's places for the same objects proved that the lapse of a score of years had made no difference in the configuration of that immemorial cluster; and

Prof. Jacoby's recent measures of Rutherfurd's photographs taken in 1872 and 1874 enforce the same conclusion.

The above facts are so forcible that no wonder that at the Astrophotographic Congress held in Paris in 1887 it was decided to make a photographic survey of the heavens, and now eighteen photographic telescopes of 13 inches aperture are in operation in various parts of the world, for the purpose of preparing the international astrographic chart, and it was hoped that the catalogue plates would be completed by 1900.

Photography has been applied to the discovery of minor planets, and that something like 450 are now known, the most noteworthy, perhaps, as regards utility being the discovery of Eros (433) in 1898 by Herr Witt at the Observatory Urania, near Berlin.

With regard to the application of photography to recording the form of various nebulæ, it is interesting to quote a passage from Dick's "Practical Astronomer," published in 1845, as opposed to Herschel's opinion that the photography of a nebula would never be possible.

"It might, perhaps, be considered as beyond the bounds of probability to expect that even the distant nebulæ might thus be fixed, and a delineation of their objects produced, which shall be capable of being magnified by microscopes. But we ought to consider that the art is only in its infancy, and that plates of a more delicate nature than those hitherto used may yet be prepared, and that other properties of light may yet be discovered, which shall facilitate such designs. For we ought now to set no boundaries to the discoveries of science, and to the practical applications of scientific discovery, which genius and art may accomplish."

It was not, however, until 1880 that Draper first photographed the Orion Nebula, and later by three years I succeeded in doing the same thing with an exposure of only thirty-seven minutes. In December 1885 the brothers Henry by the aid of photography found that the Pleiades were involved in a nebula, part of which, however, had been seen by myself (*Monthly Notices*, vol. xl. p. 376) with my 3-foot reflector in February 1880, and later, February 1886; it was also partly discerned at Pulkowa with the 30-inch refractor then newly erected.

Still more nebulosity was shown by Dr. Roberts's photographs (*ibid.*, vol. xlvii. p. 24), taken with his 20-inch reflector in October and December 1886, when the whole western side of the group was shown to be involved in a vast nebula, whilst a later photograph taken by MM. Henry early in 1888 showed that practically the whole of the group was a shoal of nebulous matter.

In 1881 Draper and Janssen recorded the comet of that year by photography.

Huggins (*Proc. Roy. Soc.*, vol. xxxii. No. 213) succeeded in photographing a part of the spectrum of the same object (Tebbutt's Comet 1881, II.) on June 24, and the Fraunhofer lines were amongst the photographic impressions, thus demonstrating that at least a part of the continuous spectrum is due to reflected sunlight. He also secured a similar result from Comet Wells (*Brit. Assoc. Report*, 1882, p. 442).

I propose to consider the question of the telescope on the following lines: (1) The refractor and reflector from their inception to their present state. (2) The various modifications and improvements that have been made in mounting these instruments, and (3) the instrument that has lately been introduced by a combination of the two, refractor and reflector, a striking example of which exists now at the Paris Exhibition.

At a meeting of the British Association held nearly half a century ago (1852) (Belfast) Sir David Brewster showed a plate of rock crystal worked in the form of a lens which had been recently found in Nineveh. Sir David Brewster asserted that this lens had been destined for optical purposes, and that it never was a dress ornament.

That the ancients were acquainted with the powers of a magnifying lens may be inferred from the delicacy and minuteness of the incised work on their seals and intaglios, which could only have been done by an eye aided by a lens of some sort.

There is, however, no direct evidence that the ancients were really acquainted with the refracting telescope, though Aristotle speaks of the tubes through which the ancients observed distant objects, and compares their effect to that of a well from the bottom of which the stars may be seen in daylight ("De Gen. Animalium," lib. v.) As an historical fact without any equivocations, however, there is no serious doubt that the telescope was invented in Holland.

The honour of being the originator has been claimed for three men, each of whom has had his partisans. Their names are Metius, Lippershey and Janssen.

Galileo himself says that it was through hearing that some one in France or Holland had made an instrument which magnified distant objects that he was led to inquire how such a result could be obtained.

The first publisher of a result or discovery, supposing such discovery to be honestly his own, ranks as the first inventor, and there is little doubt that Galileo was the first to show the world how to make a telescope (Newcomb's "Astronomy," p. 108). His first telescope was made whilst on a visit to Venice, and he there exhibited a telescope *magnifying three times*: this was in May 1609. Later telescopes which emanated from the hands of Galileo magnified successively four, seven and thirty times. This latter number he never exceeded.

Greater magnifying power was not attained until Kepler explained the theory and some of the advantages of a telescope made of two convex lenses in his "Catoptrics" (1611). The first person to actually apply this to the telescope was Father Scheiner, who describes it in his "Rosa Ursina" (1630), and Wm. Gascoigne was the first to appreciate practically the chief advantages by his invention of the micrometer and application of telescopic sights to instruments of precision.

It was, however, not until about the middle of the seventeenth century that Kepler's telescope came to be nearly universal, and then chiefly because its field of view exceeded that of the Galilean.

The first powerful telescopes were made by Huyghens, and with one of these he discovered Titan (Saturn's brightest satellite): his telescopes magnified from forty-eight to ninety-two times, were about 2½ inches aperture, with focal lengths ranging from 12 to 23 feet. By the aid of these he gave the first explanation of Saturn's ring, which he published in 1659.

Huyghens also states that he made object-glasses of 170 feet and 210 feet focal length; also one 300 feet long, but which magnified only 600 times; he also presented one of 123 feet to the Royal Society of London.

Auzout states that the best telescopes of Campani at Rome magnified 150 times, and were of 17 feet focal length. He himself is said to have made telescopes of from 300 to 600 feet focus, but it is improbable that they were ever put to practical use. Cassini discovered Saturn's fifth satellite (Rhea) in 1672, with a telescope made by Campani, magnifying about 150 times, whilst later, in 1684, he added the third and fourth satellites of the same planet to the list of his discoveries.

Although these telescopes were unwieldy, Bradley, with his usual persistency, actually determined the diameter of Venus in 1722 with a telescope of 212 feet focal length.

With such cumbersome instruments many devices were invented of pointing these *aërial telescopes*, as they were termed, to various parts of the sky. Huyghens contrived some ingenious arrangements for this purpose, and also for adjusting and centreing the eye-piece, the object-glass and eye-piece being connected by a long braced rod.

It was not, however, until Dolland's invention of the achromatic object-glass in 1757–58 that the refracting telescope was materially improved, and even then the difficulty of obtaining large blocks of glass free from striæ limited the telescope as regards aperture, for even at the date of Airy's report we have seen that 12 inches was about the maximum aperture for an object-glass.

The work of improving glass dates back to 1784, when Guinand began experimenting with the manufacture of optical flint glass.

He conveyed his secrets to the firm of Fraunhofer and Utzschneider, which he joined in 1805, and during the period he was there they made the 9.6 inches object-glass for the Dorpat telescope.

Merz and Mädler, the successors of Fraunhofer, carried out successfully the methods handed down to them by Guinand and Fraunhofer.

Guinand communicated his secrets to his family before his death in 1823, and they entered into partnership with Bontemps. The latter afterwards joined the firm of Chance Bros., of Birmingham, and so some of Guinand's work came to England.

At the present day MM. Feil, of Paris, who are direct descendants of Guinand and Messrs. Chance Bros., of Birmingham, are the best known manufacturers of large discs of optical glass.

It is related in history that Ptolemy Euergetes had caused to be erected on a lighthouse at Alexandria a piece of apparatus for discovering vessels a long way off; it has also been maintained that the instrument cited was a concave reflecting mirror, and it is possible to observe with the naked eye images formed by a concave mirror, and that such images are very bright.

Also the Romans were well acquainted with the concentrating power of concave mirrors, using them as burning mirrors, as they were called. The first application of an eye lens to the image formed by reflection from a concave mirror appears to have been made by Father Zucchi, an Italian Jesuit. His work was published in 1652, though it appears he employed such an instrument as early as 1616. The priority, however, of describing, if not making, a practical reflecting telescope belongs to Gregory, who, in his "Optica Promota," 1663, discusses the forms of images of objects produced by mirrors. He was well aware of the failure of all attempts to perfect telescopes by using lenses of various curvature, and proposed the form of reflecting telescope which bears his name.

Newton, however, was the first to construct a reflecting telescope, and with it he could see Jupiter's satellites, &c. Encouraged by this, he made another of 6½ inches focal length, which magnified thirty-eight times, and this he presented to the Royal Society on the day of his election to the Society in 1671.

To Newton we owe also the idea of employing pitch, used in the working of the surfaces.

A third form of telescope was invented by Cassegrain in 1672. He substituted a small convex mirror for the concave mirror in Gregory's form, and thus rendered the telescope a little shorter.

Short also, from 1730–68, displayed uncommon ability in the manufacture of reflecting telescopes, and succeeded in giving true parabolic and elliptic figures to his specula, besides obtaining a high degree of polish upon them. In Short's first telescopes the specula were of glass, as suggested by Gregory; but it was not until after Liebig's discovery of the process of depositing a film of metallic silver upon a glass surface from a salt in solution that glass specula became almost universal, and thus replaced the metallic ones of earlier times.

Shortly after the announcement of Liebig's discovery Steinheil (*Gaz. Univ. d'Augsburg*, March 24, 1856)—and later, independently, Foucault (*Comptes rend.*, vol. xliv. February 1857)—proposed to employ glass for the specula of telescopes. and, as is well known, this is done in all the large reflectors of to-day.

I now propose to deal with the various steps in the development of the telescope, which have resulted in the three forms that I take as examples of the highest development at the present time. These are the Yerkes telescope at Chicago, my own 5-foot reflector, and the telescope recently erected at the Paris Exhibition, dealing not only with the mountings, but with the principles of construction of each. When the telescope was first used all could be seen by holding it in the hand. As the magnifying power increased, some kind of support would become absolutely necessary, and this would take the form of the altitude and azimuth stand, and the motion of the heavenly bodies would doubtless suggest the parallactic or equatorial movement, by which the telescope followed the object by one movement of an axis placed parallel to the pole. This did not come, however, immediately. The long focus telescopes of which I have spoken were sometimes used with a tube, but more often the object-glass was mounted in a long cell and suspended from the top of a pole, at the right height to be in a line between the observer and the object to be looked at; and it was so arranged that by means of a cord it could be brought into a fairly correct position. Notwithstanding the extreme awkwardness of this arrangement, most excellent observations were made in the seventeenth century by the users of these telescopes. Then the achromatic telescope was invented and mechanical mountings were used, with circles for finding positions, much as we have them now. I have already mentioned the rivalry between the English and German forms of mountings, and Airy's preference for the English form. The general feeling amongst astronomers has, however, been largely in favour of the German mounting for refractors, due, no doubt, to a great extent, to the enormous advance in engineering skill. We have many examples of this form of mounting. A list of the principal large refracting and reflecting telescopes now existing is given at the end of this paper. All the refractors in this list, with the exception of the Paris telescope of 50 inches, and the Greenwich telescope of 28 inches, are mounted on the German form. Some of these carry a reflector as well, as, for instance, the telescope lately presented to the Greenwich Observatory by Sir Henry Thompson, which, in addition to a 26-inch refractor, carries a 30-inch reflector at the other end of the declination axis, such as had been previously used by Sir William Huggins and Dr. Roberts; the last, and perhaps the finest, example of the German form being the Yerkes telescope at Chicago.

The small reflector made by Sir Isaac Newton, probably the first ever made, and now at the Royal Society, is mounted on a ball, gripped by two curved pieces, attached to the body of the tion. We have not much information as to the mounting of early reflectors. Sir William Herschel mounted his 4-foot telescope on a rough but admirably-planned open-work mounting, capable of being turned round, and with means to tilt the telescope to any required angle. This form was not very suitable for picking up objects or determining their position, except indirectly ; but for the way it was used by Sir William Herschel it was most admirably adapted : the telescope being elevated to the required angle, it was left in that position, and became practically a transit instrument. All the objects passing through the field of view (which was of considerable extent, as the eyepiece could be moved in declination) were observed, and their places in time and declination noted, so that the positions of all these objects in the zone observed were obtained with a considerable degree of accuracy. It was on this plan that Sir John Herschel made his general catalogue of nebulæ, embracing all the nebulæ he could see in both hemispheres ; a complete work in the history of astronomy.

Sir William Herschel's mounting of his 4-foot reflector differs in almost every particular from the mountings of the long focus telescopes we have just spoken of. The object-glass was at a height, the reflector was close to the ground. There was a tube to one telescope, but not to the other. The observer in one case stood on the ground, in the other he was on a stage at a considerable elevation. One pole sufficed with a cord for one ; a whole mass of poles, wheels, pulleys and ropes surrounded the other. In one respect only were they alike—they both did fine work.

Lassell seems to have been the first to mount a reflector equatorially. He, like Herschel, made a 4-foot telescope, and this he mounted in this way. Lord Rosse mounted his telescopes somewhat after the manner of Sir William Herschel. The present Earl has mounted a 3-foot equatorially.

A 4-foot telescope was made by Thomas Grubb for Melbourne, and this he mounted on the German form. The telescope being a Cassegrain, the observer is practically on the ground level. A somewhat similar instrument exists at the Paris Observatory. Lassell's 4-foot was mounted in what is called a fork mounting, as is also my own 5-foot reflector, for this in some ways seems well adapted for reflectors of the Newtonian kind.

We now come to the Paris telescope. This is really the result of the combination of a reflector and a refractor. I cannot say when a plane mirror was first used to turn the light into a telescope for astronomical purposes. It seems first to have been suggested by Hooke, who, at a meeting of the Royal Society, when the difficulty of mounting the long focus lenses of Huyghens was under discussion, pointed out that all difficulties would be done away with if, instead of giving movement to the huge telescope itself, a plane mirror were made to move in front of it (Lockyer, "Star-gazing," p. 453).

The Earl of Crawford, then Lord Lindsay, used a heliostat to direct the rays from the sun, on the occasion of the transit of Venus, through a lens of 40 feet focal length, in order to obtain photographs, and it was also largely used by the American observers on the same occasion.

Monsieur Loëwy at Paris proposed in 1871 a most ingenious telescope made by a combination of two plane mirrors and an achromatic object-glass, which he calls a Coudé telescope, which has some most important advantages. Chief amongst these are that the observer sits in perfect comfort at the upper end of the polar axis, whence he need not move, and by suitable arrangements he can direct the telescope to any part of the visible heavens. Several have been made in France, including a large one of 24 inches aperture, erected at the Paris Observatory. and which has already made its mark by the production of perhaps the best photographs of the moon yet obtained. I have already spoken of Lord Lindsay and his 40-foot lens, fed, as it were,

with light from a heliostat. This is exactly the plan that has been followed in the design of the large telescope in the Paris Exhibition. But in place of a lens of 4 inches aperture and a heliostat a few inches larger, the Paris telescope has a plane mirror of 6 feet and a lens exceeding 4 feet in diameter, with a focal length of 186 feet. The cost of a mounting on the German plan and of a dome to shelter such an instrument would have been enormous. The form chosen is at once the best and cheapest. One of the great disadvantages is that from the nature of things it cannot take in the whole of the heavens. The heliostat form of mounting of the plane mirror causes a rotation of the image in the field of view which in many lines of research is a strong objection. There is much to be said on the other side. The dome is dispensed with, the tube, the equatorial mounting and the rising floor are not wanted. The mechanical arrangements of importance are confined to the mounting of the necessary machinery to carry the large plane mirror and move it round at the proper rate. The telescope need not have any tube (that to the Paris telescope is, of course, only placed there for effect), as the flimsiest covering is enough if it excludes false light falling on the eye-end ; and, more important than all, the observer sits at his ease in the dark chamber. This question of the observer, and the conditions under which he observes, is a most important one as regards both the quality and quantity of the work done.

We have watched the astronomer, first observing from the floor level, then mounted on a high scaffold like Sir William Herschel, Lassell and Lord Rosse ; then, starting again from the floor level and using the early achromatic telescope ; then, as these grew in size, climbing up on observing chairs to suit the various positions of the eye-end of the telescope, as we see in Mr. Newall's great telescope ; then brought to the floor again by that excellent device of Sir Howard Grubb, the rising floor. This is in use with the Lick and the Yerkes telescopes, where the observer is practically always on the floor level, though constant attention is needed, and the circular motion has to be provided for by constant movement, to say nothing of the danger of the floor going wrong. Then we have the ideal condition, as in the Equatorial Coudé at the Paris Observatory, where the observer sits comfortably sheltered and looks down the telescope, and from this position can survey the whole of the visible heavens. The comfort of the observer is a most important matter, especially for the long exposures that are given to photographic plates, as well as for continued visual work. In such a form of telescope as that at Paris the heliostat form of mounting the plane mirror is most suitable, notwithstanding the rotation of the image. But there is another way in which a plane mirror can be mounted, and that is on the plan first proposed by Auguste many years ago, and lately brought forward again by Mons. Lippman, of Paris, and that is by simply mounting the plane mirror on a polar axis and parallel therewith, and causing this mirror to rotate at half the speed of the earth's rotation. Any part of the heavens seen by any person reflected from this mirror will appear to be fixed in space, and not partake of the apparent movement of the earth, so long as the mirror is kept moving at this rate. A telescope, therefore, directed to such a mirror can observe any heavenly body as if it were in an absolutely fixed position so long as the angle of the mirror shall not be such as to make the reflected beam less than will fill the object-glass. There is one disadvantage in the cœlostat, as this instrument is called, and that is its suitability only for regions near the equator. The range above and below, however, is large enough to include the greater portion of the heavens, and that portion in which the solar system is included. Here the telescope must be moved in azimuth for different portions of the sky, as is fully explained by Prof. Turner in vol. lvi. of the *Monthly Notices*, and it therefore becomes necessary to provide for moving the telescope in azimuth from time to time as different zones above or below the equator are observed. No instrument yet devised is suitable for all kinds of work, but this form, notwithstanding its defects, has so many and such important advantages that I think it will obviate the necessity of building any larger refractors on the usual models. The cost of producing a telescope much larger than the Yerkes on that model, in comparison with what could be done on the plan I now advocate, renders it most improbable that further money will be spent in that way. It may be asked, What are the lines of research which could be taken up by a telescope of this construction, and on what lines should the telescope be built ? I will endeavour to answer this. All the work that is usually done by an astronomical telescope, excepting very long-continued observations, can be equally well done by the fixed telescope. But there are some special lines for which this form of research is admirably suited, such as photographs of the moon, which would be possible with a reflecting mirror of, say, 200 feet focal length, giving an image of some 2 feet diameter in a primary focus, or a larger image might be obtained either by a longer focus mirror or by a combination. It might even be worth while to build a special cœlostat for lunar photography, provided with an adjustment to the polar axis and a method of regulating the rate of clock to correct the irregular motion of the moon, and thus obtain absolutely fixed images on the photographic plate.

The advantage of large primary images in photography is now fully recognised. For all other kinds of astronomical photography a fixed telescope is admirably adapted ; and so with all spectroscopic investigations, a little consideration will show that the conditions under which these investigations can be pursued are almost ideal. As to the actual form such a construction would take, we can easily imagine it. The large mirror mounted as a cœlostat in the centre ; circular tracts round this centre, on which a fan-shaped house can be travelled round to any azimuth, containing all the necessary apparatus for utilising the light from the large plane mirror, so as to be easily moved round to the required position in azimuth for observation. In place of a fan-shaped house movable round the plane mirror, a permanent house might encircle the greater portion round the mirror, and in this house the telescope or whatever optical combination is used might be arranged on an open framework, supported on similar rails, so as to run round to any azimuth required. The simplicity of the arrangement and the enormous saving in cost would allow in any well-equipped observatory the use of a special instrument for special work. The French telescope has a mirror about 6 feet in diameter and a lens of about 4 feet. This is a great step in advance over the Yerkes telescope, and it may be some time before the glass for a lens greater than 50 inches diameter will be made, as the difficulty in making optical glass is undoubtedly very great. But with the plane mirror there will be no such difficulty, as 6 feet has already been made ; and so with a concave mirror there would be little difficulty in beginning with 6 feet or 7 feet. The way in which the mirror would be used, always hanging in a band, is the most favourable condition for good work, and the absence of motion during an observation, except of course that of the plane mirror (which could be given by floating the polar axis and suitable mechanical arrangements, a motion of almost perfect regularity).

One extremely important thing in using silver or glass mirrors is the matter of resilvering from time to time. Up to quite recently the silvering of my 5-foot mirror was a long, uncertain, and expensive process. Now we have a method of silvering mirrors that is certain, quick, and cheap. This takes away the one great disability from the silver or glass reflecting telescope, as the surface of silver can now be renewed with greater ease and in less time than the lenses of a large refracting telescope could be taken out and cleaned. It may be that we shall revert to speculum metal for our mirrors, or use some other deposited metal on glass ; but even as it is we have the silvered glass reflector, which at once allows an enormous advance in power. To do justice to any large telescope it should be erected in a position, as regards climate, where the conditions are as favourable as possible.

The invention of the telescope is to me the most beautiful ever made. Familiarity both in making and in using has only increased my admiration. With the exception of the microphone of the late Prof. Hughes, which enabled one to hear otherwise inaudible sounds, sight is the only sense that we have been able to enormously increase in range. The telescope enables one to see distant objects as if they were at, say, one-five-thousandth part of their distance, while the microscope renders visible objects so small as to be almost incredible. In order to appreciate better what optical aid does for the sense of sight, we can imagine the size of an eye, and therefore of a man, capable of seeing in a natural way what the ordinary eye sees by the aid of a large telescope, and, on the other hand, the size of a man and his eye that could see plainly small objects as we see them under a powerful microscope. The man in the first case would be several miles in height, and in the latter he would not exceed a very small fraction of an inch in height.

Photography also comes in as a further aid to the telescope,

as it may possibly be to the microscope. For a certain amount of light is necessary to produce sensation in the eye. If this light is insufficient nothing is seen ; but owing to the accumulative effect of light on the photographic plate, photographs can be taken of objects otherwise invisible, as I pointed out years ago, for in photographs I took in 1883 stars were shown on photographic plates that I could not see in the telescope. All photographs, when closely examined, are made up of a certain number of little dots, as it were, in the nature of stippling, and it is a very interesting point to consider the relation of the size and separation of these dots that form the image, and the rods and cones of the reckoner which determines the power of the eye.

Many years ago I tried to determine this question. I first took a photograph of the moon with a telescope of very short focus (as near as I could get it to the focus of the eye itself, which is about half an inch). The resulting photograph measured one two-hundredth of an inch in diameter, and when examined again with a microscope showed a fair amount of detail, in fact, very much as we see the moon with the naked eye ; making a picture of the moon by hand on such a scale that each separate dot of which it was made corresponded with each separate sensitive point of the retina employed when viewing the moon without optical aid, I found, on looking at this picture at the proper distance, that it looked exactly like a real moon. In this case the distance of the dots was constant, making them larger or smaller forming the light or shade of the picture.

I did not complete these experiments, but as far as I went I thought that there was good reason to believe that we could in this way increase the defining power of the eye. It is a subject well worthy of further consideration.

I know that in this imperfect and necessarily brief address I have been obliged to omit the names of many workers, but I cannot conclude without alluding to the part that this Association has played in fostering and aiding Astronomy. A glance through the list of money grants shows that the help has been most liberal. In my youth I recollect the great value that was put on the British Association Catalogue of Stars ; we know the help that was given in its early days to the Kew Observatory ; and the Reports of the Association show the great interest that has always been taken in our work. The formation of a separate Department of Astronomy is, I hope, a pledge that this interest will be continued, to the advantage of our science.

List of Large Telescopes in existence in 1900.

Refractors 15 inches and upwards		Refractors 15 inches and upwards	
	Inches		Inches
Paris (Exhibition) .	50	Mount Etna . .	21·8
Yerkes . . .	40	Strassburg . .	19·1
Lick . . .	36	Milan . . .	19·1
Pulkowa . . .	30	(Dearborn) Chicago .	18·5
Nice . . .	29·9	Warner Observatory,	
Paris . . .	28·9	Rochester, U.S. .	16·0
Greenwich . .	28·0	Washburn Observa-	
Vienna . . .	27·0	tory, Madison,	
Washington, U.S. .	26·0	Wisconsin . .	15·5
Leander, McCormick		Edinburgh . .	15·1
Observatory, Vir-		Brussels . .	15·1
ginia . .	26·0	Madrid . . .	15·0
Greenwich . .	26·0	Rio Janeiro . .	15·0
Newall's, Cambridge.	25·0	Paris . . .	15·0
Cape of Good Hope .	24·0	Sir William Huggins.	15·0
Harvard . . .	24·0	Paris . . .	15·0
Princeton, N.J., U.S.	23·0		

Reflectors 2 feet 6 inches and upwards			Reflectors 2 feet 6 inches and upwards		
	Ft.	In.		Ft.	In.
Lord Rosse . .	6	0	South Kensington .	3	0
Dr. Common . .	5	0	Crossley (Lick). .	3	0
Melbourne . .	4	0	Greenwich . .	2	6
Paris . . .	4	0	South Kensington .	2	6
Meudon . . .	3	3			

SECTION B.

CHEMISTRY.

OPENING ADDRESS BY PROF. W. H. PERRIN, JUN., PH.D., F.R.S, PRESIDENT OF THE SECTION.

The Modern System of Teaching Practical Inorganic Chemistry and its Development.

IN choosing for the subject of my Address to-day the development of the teaching of practical inorganic chemistry I do so, not only on account of the great importance of the subject, but also because it does not appear that this matter has been brought before this Section, in the President's Address at all events, during the last few years.

In dealing generally with the subject of the teaching of chemistry as a branch of science it may be well in the first place to consider the value of such teaching as a means of general education, and to turn our attention for a few minutes to the development of the teaching of science in schools.

There can be no doubt that there has been great progress in the teaching of science in schools during the last forty years, and this is very evident from the perusal of the essay, entitled " Education : Intellectual, Moral, and Physical," which Herbert Spencer wrote in 1859. After giving his reasons for considering the study of science of primary importance in education, Herbert Spencer continues : " While what we call civilisation could never have arisen had it not been for science, science forms scarcely an appreciable element in our so-called civilised training."

From this it is apparent that science was not taught to any appreciable extent in schools at that date, though doubtless in some few schools occasional lectures were given on such scientific subjects as physiology, anatomy, astronomy, and mechanics. Herbert Spencer's pamphlet appears to have had only a very gradual effect towards the introduction of science into schemes of education. For many years chemical instruction was only given in schools at the schoolroom desk, or at the best from the lecture table, and many of the most modern of schools had no laboratories.

The first school to give any practical instruction in chemistry was apparently the City of London School, at which, in the year 1847, Mr. Hall was appointed teacher of chemistry, and there he continued to teach until 1869.[1] Besides the lecture theatre and a room for storing apparatus, Mr. Hall's department contained a long room, or rather passage, leading into the lecture theatre, and closed at each end with glass doors. In this room, which was fitted up as a laboratory, and used principally as a preparation room for the lectures, Mr. Hall performed experiments with the few boys who assisted him with his lectures. As accommodation was at that time strictly limited, he used to suggest simple experiments and encourage the boys to carry them out at home, and afterwards he himself would examine the substances which they had made.

From this small beginning the teaching of chemistry in the City of London School rapidly developed, and this school now possesses laboratories which compare favourably with those of any school in the country.

The Manchester Grammar School appears to have been one of the first to teach practical chemistry. In connection with this school a small laboratory was built in 1868 ; this was replaced by a larger one in 1872, and the present large laboratories, under the charge of Mr. Francis Jones, were opened in 1880.

Dr. Marshall Watts, who was the first science master in this school, taught practical chemistry along with the theoretical work from the commencement in 1868.

As laboratories were gradually multiplied it might be supposed that boys were given the opportunity to carry out experiments which had a close connection with their lecture-room courses. But the programme of laboratory work which became all but universal was the preparation of a few gases, followed by the practice of qualitative analysis. The course adopted seems to have been largely built up on the best books of practical chemistry in use in the colleges at that time ; but it was also, no doubt, largely influenced by the requirements of the syllabus of the Science and Art Department, which con-

[1] Mr. A. T. Pollard, M.A., Head Master of the City of London School, has kindly instituted a search among the bound copies of the boys' terminal reports, and informs me that in the School form of Terminal Report a heading for Chemistry was introduced in the year 1847, the year of Mr. Hall's appointment.

tained a scheme for teaching practical chemistry.[1] Even down to quite recent times it was in many schools still not considered essential that boys should have practical instruction in connection with lectures in chemistry.

A Report issued in 1897 by a special Committee appointed by the Technical Education Board of the London County Council adduces evidence of this from twenty-five secondary schools in London, in which there were 3960 boys learning chemistry. Of these 1698 boys, or 43 per cent., did no practical work whatever ; 955 boys, or 24 per cent., did practical work, consisting of a certain amount of preparation of gases, together with qualitative analysis ; but of these latter 743, or 77 per cent., had not reached the study of the metals in their theoretical work, so that their testing work can have been of little educational value. It was also found that in the case of 655, or 68 per cent. of the total number of boys taking practical work, the first introduction to practical chemistry was through qualitative analysis.

But some years before this Report was issued a movement had begun which was destined to have a far-reaching effect. A Report " on the best means for promoting Scientific Education in Schools " having been presented to the Dundee Meeting of this Association in 1867, and published in 1868, a Committee of the British Association was appointed in 1887 " for the purpose of inquiring and reporting upon the present methods of teaching chemistry." The well-known Report which this Committee presented to the Newcastle Meeting in 1889 insisted that it was worth while to teach chemistry in schools, not so much for the usefulness of the information imparted as for the special mental discipline it afforded if the scientific method of investigating nature were employed. It was argued that " learners should be put in the attitude of discoverers, and led to make observations, experiments, and inferences for themselves." And since there can be little progress without measurement, it was pointed out that the experimental work would necessarily be largely of a quantitative character.

Prof. H. E. Armstrong, in a paper read at a conference at the the Health Exhibition five years before this, had foreshadowed much that was in this Report. He also drew up a detailed scheme for " a course of elementary instruction in physical science," which was included in the Report of the Committee, and it cannot be doubted that this scheme and the labours of the Committee have had a very marked influence on the development of the teaching of practical chemistry in schools. That this influence has been great will be admitted when it is understood that schemes based on the recommendation of the Committee are now included in the codes for both Elementary Day Schools and Evening Continuation Schools. The recent syllabuses for elementary and advanced courses issued by the Incorporated Association of Headmasters and by the Oxford and Cambridge local boards and others are evidently directly inspired by the ideas set forth by the Committee.

The Department of Science and Art has also adopted some of the suggestions of the Committee, and a revised syllabus was issued by the Department in 1895, in which qualitative analysis is replaced by quantitative experiments of a simple form, and by other exercises so framed " as to prevent answers being given by students who have obtained their information from books or oral instruction." This was a very considerable advance, but it must be admitted that there is nothing in the syllabus which encourages, or even suggests, placing the learners in the attitude of discoverers, and this, in the opinion of the Committee of this Association, is vital if the teaching is to have educational value.

Many criticisms have been passed upon the 1889 Report. It has been said that life is much too short to allow of each individual advancing from the known to the unknown, according to scientific methods, and that even were this not so too severe a tax is made upon the powers of boys and girls. In answer to the second point it will be conceded that while it is doubtless futile to try to teach chemistry to young children, on the other hand experience has abundantly shown that the average schoolboy of fourteen or fifteen can, with much success, investigate such problems as were studied in the researches of Black and Scheele, of Priestley and Cavendish and Lavoisier, and it is quite remarkable with what interest such young students carry out this class of work.

It may be well to quote the words which Sir Michael Foster

[1] I find, on inquiry, that examinations in the Advanced Stage and Honours of Practical Chemistry were first held by the Science and Art Department in 1878, the practical examination being extended to the Elementary Stage in 1882.

used in this connection in his admirable Presidential Address to this Association in 1899. He said: "The learner may be led to old truths, even the oldest, in more ways than one. He may be brought abruptly to a truth in its finished form, coming straight to it like a thief climbing over a wall; and the hurry and press of modern life tempt many to adopt this quicker way. Or he may be more slowly guided along the path by which the truth was reached by him who first laid hold of it. It is by this latter way of learning the truth, and by this alone, that the learner may hope to catch something at least of the spirit of the scientific inquirer."

I believe that in the determination of a suitable school course in experimental science this principle of historical development is a very valuable guide, although it is not laid down in the 1889 Report of the British Association.

The application of this principle will lead to the study of the solvent action of water, of crystallisation, and of the separation of mixtures of solids before the investigation of the composition of water, and also before the investigation of the phenomena of combustion. It will lead to the investigation of hydrochloric acid before chlorine, and especially to the postponement of atomic and molecular theories, chemical equations, and the laws of chemical combination, until the student has really sufficient knowledge to understand how these theories came to be necessary.

There can be no doubt that this new system of teaching chemistry in schools has been most successful. Teachers are delighted with the results which have already been obtained, and those whom I have had the opportunity of consulting, directly and indirectly, cannot speak too highly of their satisfaction at the disappearance of the old system of qualitative analysis, and the institution of the new order of things. Especially I may mention in this connection the excellent work which is being carried on under the supervision of Dr. Bevan Lean at the Friends' School in Ackworth, where the boys have attained results which are far in advance of anything which would have been thought possible a few years since.

It is, of course, obvious that if a schoolboy is made to take the attitude of a discoverer, his progress may appear to be slow. But does this matter? Most boys will not become professional chemists; but if while at school a boy learns how to learn, and how to "make knowledge"[1] by working out for himself a few problems, a habit of mind will be formed which will enable him in future years to look in a scientific spirit at any new problems which may face him. When school-days are past the details of the preparation of hydrogen may have been forgotten; but if it was really understood at the time that it could not be decided at once whether the gas was derived from the acid or from the metal, or from the water, or in part from the one and in part from the other, an attitude of scepticism and of suspended judgment will have been formed, which will continue to guard from error.

In the new system of teaching chemistry in schools much attention must necessarily be given to weights and measurements; indeed, the work must be largely of a quantitative kind, and it is in this connection that an important note of warning has been sounded by several teachers.[2] They consider, very rightly, that it is important to point out clearly to the scholar that science does not consist of measurement, but that measurement is only a tool in the hand of the inquirer, and that when once sufficient skill has been developed in its use it should be employed only with a distinct object. Measurements should, in fact, be made only in reference to some actual problem which appears to be really worth solving, not in the accumulation of aimless details.

And, of course, all research carried out must be genuine and not sham, and all assumption of the "obvious" must be most carefully guarded against. But the young scholar must, at the same time, remember that although the scientific method is necessary to enable him to arrive at a result, in real life it is the answer to the problem which is of the most importance.[3]

Although, then, there has been so much discussion, during the last ten years, on the subject of teaching chemistry in schools, and such steady progress has been made towards a really satisfactory system of teaching the subject to

young boys and girls, it is certainly very remarkable that practically nothing has been said or written bearing on the training which a student who wishes to become a chemist is to undertake at the close of his school-days at the college or university in which his education is continued.

One of the most remarkable points, to my mind, in connection with the teaching of chemistry, is the fact that although the science has been advancing year by year with such unexampled rapidity, the course of training which the student goes through during his first two years at most colleges is still practically the same as it was thirty or forty years ago. Then, as now, after preparing a few of the principal gases, the student devotes the bulk of his first year to qualitative analysis in the dry and wet way, and his second year to quantitative analysis, and, although the methods employed in teaching the latter may possibly have undergone some slight modification, there is certainly no great difference between the routine of simple salt and mixture followed by quantitative analysis practised at the present day and that which was in vogue in the days of our fathers and grandfathers.

Since, then, the present system has held the field for so long, not only in this country but also on the Continent, it is worth while considering whether it affords the best training which a student who wishes to become a chemist can undergo in the short time during which he can attend at a college or university. In considering this matter I was led in the first place to carefully examine old books and other records, with the object of finding out how the present system originated, and I think that valuable and interesting information bearing on the subject may be obtained from a very brief sketch of the rise and development of the present system of teaching chemistry, and especially in so far as it bears on the inclusion of qualitative analysis. Unfortunately, it is not so easy to gain a good historical acquaintance with the matter as I at first imagined would be the case, and this is due in a large measure to the fact that so few of the laboratories which took an active part in the development of the present system of chemical training have left any record of the methods which they employed. In this connection I may, perhaps, be allowed to suggest that it would be a valuable help to the future historian if all prominent teachers of chemistry would leave behind them a brief record of the system of teaching adopted in their laboratories, showing the changes which they had instituted, the object of these changes, and the results which followed their adoption.

There is no doubt that the progress of practical chemistry went largely hand in hand with the progress of theoretical chemistry, for as the latter gradually developed, so the necessity for the determination of the composition, first of the best known, and then of the rarer minerals and other substances, became more and more marked.

The analytical examination of substances in the dry way was employed in very early times in connection with metallurgical operations, and especially in the determination of the presence of valuable constituents in samples of minerals. Cupellation was used by the Greeks in the separation of gold and silver from their ores and in the purification of these metals. Geber knew that the addition of nitre to the ore facilitated the separation of gold and silver, and subsequently Glauber (1604–1668) called attention to the fact that many commoner metals could easily be separated from their ores with the aid of nitre.

But it was not till the eighteenth century that any marked progress was made in analysis in the dry way, and the progress which then became rapid was undoubtedly due to the discovery of the blowpipe, and to the introduction of its use into analytical operations. The blowpipe is mentioned for the first time in 1660, in the transactions of the Accademia del Cimento of Florence, but the first to recommend its use in chemical operations was Johann Andreas Cramer in 1739. The progress of blowpipe analysis was largely due to Gahn (1745–1818), who spent much time in perfecting its use in the examination of minerals, and it was he who first used platinum wire and cobalt solution in connection with blowpipe analysis. The methods employed by Gahn were further developed by his friend Berzelius (1779–1848), who gave much attention to the subject, and who with great skill and patience gradually worked out a complete scheme of blowpipe analysis, and published it in a pamphlet, entitled "Ueber die Anwendung des Löthrohrs," which appeared in 1820. After the publication of this work blowpipe analysis rapidly came into general use in England France and Germany, and the scheme devised by Berzelius is essentially that employed at the present day.

[1] *Cf.* Prof. J. G. Macgregor in NATURE, September 1899.
[2] *Cf.* H. Picton in *The School World*, November 1899; Bevan Lean, *ibid.*, February 1900.
[3] *Cf.* Mrs. Bryant, "Special Reports on Educational Subjects," vol. ii. p. 113.

Indeed, the only notable additions to the methods o. analysis in the dry way since the time of Berzelius are the development of flame reactions, which Bunsen worked out with such characteristic skill and ingenuity, and the introduction of the spectroscope.

The necessity for some process other than that of analysis in the dry way seems, in the first instance, to have arisen in quite early times in connection with the examination of drugs, not only on account of the necessity for discovering their constituents, but also as a means of determining whether they were adulterated. In such cases analysis in the dry way was obviously unsuitable, and experience soon showed that the only way to arrive at the desired result was to treat the substance under examination with aqueous solutions of definite substances, the first reagent apparently being a decoction of gallnuts, which is described by Pliny as being employed in detecting adulteration with green vitriol.

The progress made in connection with wet analysis was, however, exceedingly slow, largely owing to the lack of reagents ; but as these were gradually discovered wet analysis rapidly developed, especially in the hands of Tachenius, Scheele, Boyle, Hoffman, Margraf, and Bergmann. Boyle (1626-1691) especially had an extensive knowledge of reagents and their application ; and, indeed, it was Boyle who first introduced the word "analysis" for those operations by which substances may be recognised in the presence of one another. Boyle knew how to test for silver with hydrochloric acid, for calcium salts with sulphuric acid, and for copper by the blue solution produced by ammonia.

Margraf (1709-1782) introduced prussiate of potash for the detection of iron, and Bergmann (1735-1784) not only introduced new reagents and new methods for decomposing minerals and refractory substances, such as fusion with potash, digestion with nitric acid or hydrochloric acid, but he also was the first to suggest the application of tests in a systematic way, and, indeed, the method of analysis which he developed is on much the same lines as that in use at the present day. He paid special attention to the qualitative analysis of minerals, and gave careful instructions for the analysis of gold, platinum, silver, lead, copper, zinc, and other ores. The work of Scheele (1742-1786) had indirectly a great influence on qualitative analysis, as, although he did not give a general systematic method of procedure in the analysis of substances of unknown composition, yet the methods which he employed in the examination of new substances were so original and exact as to remain models of how qualitative analysis should be conducted.

Great strides in analytical chemistry in the wet way were made through the work of Berzelius, who, by the discovery of new methods, such as the decomposition of silicates by hydrofluoric acid and the introduction of new tests, greatly advanced the art. He paid special attention to perfecting the methods of analysis of mineral waters, and these researches, as well as his work on ores, and particularly his investigation of platinum ores, stamp Berzelius as one of the great pioneers in qualitative and quantitative analytical chemistry.

By the labours of the great experimenters whom I have mentioned qualitative analysis gradually acquired the familiar appearance of to-day, and many books were written with the object of arranging the mass of information which had accumulated, and of thus rendering it available for the student in his efforts to investigate the composition of new minerals and other substances. Among these books may be mentioned the "Handbuch der analytischen Chemie," by H. Rose, and especially the well-known analytical text-books of Fresenius, which have had an extraordinarily wide circulation and passed through many editions.

The work of the great pioneers in analytical chemistry was work done often under circumstances of great difficulty, as before the end of the seventeenth century there were no public institutions of any sort in which a practical knowledge of chemistry could be acquired. Lectures were, of course, given from very early times, but it was not until the time of Guillaume François Rouelle (1703-1770), at the beginning of the eighteenth century, that lectures began to be illustrated by experiments. Rouelle, who was very active as a teacher, numbered among his pupils many men of eminence, such as Lavoisier and Proust, and it was largely owing to his influence that France took such a lead in practical teaching. In Germany progress was much slower, and in our country the introduction of lectures illustrated by experiments seems to have been mainly due to Davy.

When it is considered how slowly experimental work came to be recognised as a means of illustration and education, even in connection with lectures, it is not surprising that in early times practical teaching in laboratories should have been thought quite unnecessary.

The few laboratories which existed in the sixteenth century were built mainly for the practice of alchemy by the reigning princes of the time, and, indeed, up to the beginning of the nineteenth entury, the private laboratories of the great masters were the only schools in which a favoured few might study, but which were not open to the public. Thus we find that Berzelius received in his laboratory a limited number of students who worked mostly at research ; these were not usually young men, and his school cannot thus be considered as a teaching institution in the ordinary sense of the word.

The earliest laboratory open for general instruction in Great Britain was that of Thomas Thomson, who, after graduating in Edinburgh in 1799, began lecturing in that city in 1800, and opened a laboratory for the practical instruction of his pupils. Thomson was appointed lecturer in Chemistry in Glasgow University in 1807, and Regius Professor in 1818, and in Glasgow he also opened a general laboratory.

The first really great advance in laboratory teaching is due to Liebig, who, after working for some years in Paris under Gay-Lussac, was appointed in 1824 to be Professor of Chemistry in Giessen. Liebig was strongly impressed with the necessity for public institutions where any student could study chemistry, and to him fell the honour of founding the world-famed Giessen Laboratory, the first public institution in Germany which brought practical chemistry within the reach of all students.

Giessen rapidly became the centre of chemical interest in Germany, and students flocked to the laboratory in such numbers as to necessitate the development of a systematic course of practical chemistry, and in this way a scheme of teaching was devised which, as we shall see later, has served as the foundation for the system of practical chemistry in use at the present day.

When the success of this laboratory had been clearly established, many other towns discovered the necessity for similar institutions, and in a comparatively short time every university in Germany possessed a chemical laboratory. The teaching of practical chemistry in other countries was, however, of very slow growth ; in France, for example, Wurtz in 1869 drew attention to the fact that there was at that time only one laboratory which could compare with the German laboratories, namely, that of the École normale supérieure.

In this country the provision of suitable laboratories for the study of chemistry seems to date from the year 1845, when the College of Chemistry was founded in London, an institution which under A. W. Hofmann's guidance rapidly rose to such a prominent position.

In 1851 Frankland was appointed to the chair of chemistry in the new college founded in Manchester by the trustees of John Owens, and here he equipped a laboratory for the teaching of practical chemistry. Under Sir Henry Roscoe this laboratory soon became too small for the growing number of chemical students, a defect which was removed when the new buildings of the college were opened in 1873. In 1849 Alexander Williamson was appointed Professor of Practical Chemistry at University College, London, where he introduced the practical methods of Liebig.

Following these examples, the older universities gradually came to see the necessity for providing accommodation for the practical teaching of chemistry, with the result that well-equipped laboratories have been erected in all the centres of learning in this country.

Since Liebig, by the establishment of the Giessen Laboratory, must be looked upon as the pioneer in the development of practical laboratory teaching, it will be interesting to endeavour to obtain some idea of the methods which he used in the training of the students who attended his laboratory in Giessen. From small beginnings he gradually introduced a systematic course of practical chemistry, and a careful comparison shows that this was similar in many ways to that in use at the present day. The student at Giessen, after preparing the more important gases, was carefully trained in qualitative and quantitative analysis ; he was then required to make a large number of preparations, after which he engaged in original research.

Although there is, as far as I have been able to ascertain, no printed record of the nature of the quantitative work and the preparations which Liebig required from his students, the course of qualitative analysis is easily followed, owing to the existence of a most interesting book published for the use of the Giessen students.

In 1846, at Liebig's request, Henry Will, Ph.D., Extraordinary Professor of Chemistry in the University of Giessen, wrote a small book, for use at Giessen, called "Giessen Outlines of Analysis," which shows clearly the kind of instruction given in that laboratory at the time in so far as qualitative analysis is concerned. This book, which contains a preface by Liebig, is particularly interesting on account of the fact that it is evidently the first Introduction to Analysis intended for the training of elementary students which was ever published. In the preface Liebig writes: "The want of an introduction to chemical analysis adapted for the use of a laboratory has given rise to the present work, which contains an accurate description of the course I have followed in my laboratory with great advantage for twenty-five years. It has been prepared at my request by Prof. Will, who has been my assistant during a great part of this period."

This book undoubtedly had a considerable circulation, and was used in most of the laboratories which were in existence at that time, and thus we find, for example, that the English translation which Liebig " hopes and believes will be acceptable to the English public " was the book used by Hofmann for his students at the College of Chemistry. In this book the metals are first divided into groups much in the same way as is done now; each group is then separately dealt with, the principal characteristics of the metals of the group are noted, and their reactions studied. Those tests which are useful in the detection of each metal are particularly emphasised, and the reasons given for selecting certain of them as of special value for the purposes of separating one metal from another.

Throughout this section of the book there are frequent discussions as to the possible methods of the separation, not only of the metals of one group, but of those belonging to different groups; and the whole subject is treated in a manner which shows clearly that Liebig's great object was to make the student think for himself. After studying in a similar manner the behaviour of the principal acids with reagents, the student is introduced to a course of qualitative analysis comprising (1), preliminary examination of solids (2), qualitative analysis of the substance in solution.

Both sections are evidently written with the object, not only of constructing a system of qualitative analysis, but more particularly of clearly leading the student to argue out for himself the methods of separation which he will ultimately adopt. The book concludes with a few tables which differ considerably in design from those in use at the present day, and which are so meagre that the student could not possibly have used them mechanically.

The system introduced in this book, no doubt owing to the excellent results obtained by its use, was rapidly recognised as the standard method of teaching analysis in most of the institutions existing at that time. Soon the course began to be further developed, book after book was published on the subject, and gradually the teaching of qualitative analysis assumed the shape and form with which we are all so well acquainted. But the present-day book on qualitative analysis differs widely from "Giessen Outlines" in this respect, that whereas in the latter the tables introduced are mere indications of the methods of separation to be employed, and are of such a nature that the student who did not think for himself must have been constantly in difficulties, in the book of the present day these tables have been worked out to the minutest detail. Every contingency is provided for; nothing is left to the originality of the student; and that which, no doubt, was once an excellent course has now become so hopelessly mechanical as to make it doubtful whether it retains anything of its former educational value.

The question which I now wish to consider more particularly is whether the system of training chemists which is at present adopted, with little variation, in our colleges and universities is a really satisfactory one, and whether it supplies the student with the kind of knowledge which will be of the most value to him in his future career.

Those who study chemistry may be roughly divided as to their future careers into two groups—those who become teachers and those who become technical chemists. Now, whether the student takes up either one or the other career, I think that it is clear that the objects to be aimed at in training him are to give

him a sound knowledge of his subject, and especially to so arrange his studies as to bring out in every possible way his capacity for original thought.

A teacher who has no originality will hardly be successful, even though he may possess a very wide knowledge of what has already been done in the past. He will have little enthusiasm for his subject, and will continue to teach on the lines laid down by the text-books of the day, without himself materially improving the existing methods, and, above all, he will be unable, and will have no desire, to add to our store of knowledge by original investigation.

It is in the power of almost every teacher to do some research work, and it seems probable that the reason why more is not done by teachers is because the importance of research work was not sufficiently insisted on, and their original faculty was not sufficiently trained, at the schools and colleges where they received their education.

And these remarks apply with equal force to the student who subsequently becomes a technical chemist.

In the chemical works of to-day sound knowledge is essential, but originality is an even more important matter. A technical chemist without originality can scarcely rise to a responsible position in a large works; whereas a chemist who is capable of constantly improving the processes in operation, and of adding new methods to those in use, becomes so valuable that he can command his own terms.

Now, this being so, I think it is extraordinary that so many of the students who go through the prescribed course of training —say for the Bachelor of Science degree—not only show no originality themselves, but seem also to have no desire at the conclusion of their studies to engage in original investigation under the supervision of the teacher. That this is so is certainly my experience as a teacher examiner, and I feel sure that many other teachers will endorse this view of the case.

If we inquire into the reason for this deficiency in originality, we shall, I think, be forced to conclude that it is in a large measure due to the conditions of study and the nature of the courses through which the student is obliged to pass.

A well-devised system of quantitative analysis is undoubtedly valuable in teaching the student accurate manipulation, but it has always seemed to me that the long course of qualitative analysis which is usually considered necessary, and which generally precedes the quantitative work, is not the most satisfactory training for a student.

There can be no doubt that to many students qualitative analysis is little more than a mechanical exercise : the tables of separation are learnt by heart, and every substance is treated in precisely the same manner : such a course is surely not calculated to develop any original faculty which the student may possess. Then, again, when the student passes on to quantitative analysis, he receives elaborate instructions as to the little details he must observe in order to get an accurate result ; and even after he has become familiar with the simpler determinations he rarely attempts, and indeed has no time to attempt, anything of the nature of an original investigation in qualitative or quantitative analysis. It indeed sometimes happens that a student at the end of his second year has never prepared a pure substance, and is often utterly ignorant of the methods employed in the separation of substances by crystallisation ; he has never conducted a distillation, and has no idea how to investigate the nature and amounts of substances formed in chemical reactions ; practically all his time has been taken up with analysis. That this is not the way to teach chemistry was certainly the opinion of Liebig, and in support of this I quote a paragraph bearing on the subject which occurs in a very interesting book on "Justus von Liebig : his Life and Work," written by W. A. Shenstone (pp. 175, 176).

"In his practical teaching Liebig laid great stress on the producing of chemical preparations; on the students preparing, that is to say, pure substances in good quantity from crude materials. The importance of this was, even in Liebig's time, often overlooked ; and it was, he tells us, more common to find a man who could make a good analysis than to find one who could produce a pure preparation in the most judicious way."

"There is no better way of making one's self acquainted with the properties of a substance than by first producing it from the raw material, then converting it into its compounds, and so becoming acquainted with them. By the study of ordinary analysis one does not learn how to use the important methods of crystallisation, fractional distillation, nor acquire any considerable experience in the proper use of solvents. In short, one does not, as Liebig said, become a chemist."

One reason why the present system of training chemists has persisted so long is no doubt because it is a very convenient system : it is easily taught, does not require expensive apparatus, and, above all, it lends itself admirably for the purpose of competitive examination.

The system of examination which has been developed during the last twenty years has done much harm, and is a source of great difficulty to any conscientious teacher who is possessed of originality, and is desirous, particularly in special cases, of leaving the beaten track.

In our colleges and universities most of the students work for some definite examination—frequently for the Bachelor of Science degree—either at their own University or at the University of London.

For such degrees a perfectly definite course is prescribed and must be followed, because the questions which the candidate will have to answer at his examination are based on a syllabus which is either published or is known by precedent to be required. The course which the teacher is obliged to teach is thus placed beyond his individual power of alteration, except in minor details, and originality in the teacher is thereby discouraged : he knows that all students must face the same examination, and he must urge the backward man through exactly the same course as his more talented neighbour.

In almost all examinations salts or mixtures of salts are given for qualitative analysis. " Determine the constituents of the simple salt A and of the mixture B " is a favourite examination formula ; and as some practical work of this sort is sure to be set, the teacher knows that he must contrive to get one and all of his students into a condition to enable them to answer such questions.

If, then, one considers the great amount of work which is required from the present-day student, it is not surprising that every aid to rapid preparation for examination should be accepted with delight by the teacher ; and thus it comes about that tables are elaborated in every detail, not only for qualitative analysis in inorganic chemistry, but, what is far worse, for the detection of some arbitrary selection of organic substances which may be set in the syllabus for the examination. I question whether any really competent teacher will be found to recommend this system as one of educational value or calculated to bring out and train the faculty of original thought in students.

If, then, the present system is so unsatisfactory, it will naturally be asked, How are students to be trained, and how are they to be examined so as to find out the extent of the knowledge of their subject which they have acquired ?

In dealing with the first part of the question—that is, the training best suited to chemists— I can, of course, only give my own views on the subject—views which, no doubt, may differ much from those of many of the teachers present at this meeting. The objects to be attained are, in my opinion, to give the student a sufficient knowledge of the broad facts of chemistry, and at the same time so to arrange his practical work in particular as to always have in view the training of his faculty of original thought.

I think it will be conceded that any student, if he is to make his mark in chemistry by original work, must ultimately specialise in some branch of the subject. It may be possible for some great minds to do valuable original work in more than one branch of chemistry, but these are the exceptions ; and as time goes on, and the mass of facts accumulates, this will become more and more impossible. Now a student at the commencement of his career rarely knows which branch of the subject will fascinate him most, and I think, therefore, that it is necessary, in the first place, to do all that is possible to give him a thorough grounding in all branches of the subject. In my opinion the student is taken over too much ground in the lecture courses of the present day : in inorganic chemistry, for example, the study of the rare metals and their reactions might be dispensed with, as well as many of the more difficult chapters of physical chemistry, and in organic chemistry such complicated problems as the constitutions of uric acid and the members of the camphor and terpene series, &c., might well be left out. As matters stand now, instruction must be given on these subjects simply because questions bearing on them will probably be asked at the examination.

And here, perhaps, I might make a confession, in which I do not ask my fellow-teachers to join me. My name is often attached to chemistry papers which I should be sorry to have to

answer ; and it seems to me the standard of examination papers, and especially of Honours examination papers, is far too high. Should we demand a pitch of knowledge which our own experience tells us cannot be maintained for long ?

In dealing with the question of teaching practical chemistry, it may be hoped, in the first place, that in the near future a sound training will be given in elementary science in most schools, very much on the lines which I mentioned in the first part of this address. The student will then be in a fit state to undergo a thoroughly satisfactory course of training in inorganic chemistry during his first two years at college. Without wishing in any way to map out a definite course, I may be allowed to suggest that instead of much of the usual qualitative and quantitative analysis, practical exercises similar to the following will be found to be of much greater educational value.

(1) The careful experimental demonstration of the fundamental laws of chemistry and physical chemistry.

(2) The preparation of a series of compounds of the more important metals, either from their more common ores or from the metals themselves. With the aid of the compounds thus prepared the reactions of the metals might be studied and the similarities and differences between the different metals then carefully noted.

(3) A course in which the student should investigate in certain selected cases : (a) the conditions under which action takes place ; (b) the nature of the products formed ; (c) the yield obtained. If he were then to proceed to prepare each product in a state of purity, he would be doing a series of exercises of the highest educational value.

(4) The determination of the combining weights of some of the more important metals. This is in most cases comparatively simple, as the determination of the combining weights of selected metals can be very accurately carried out by measuring the hydrogen evolved when an acid acts upon them.

Many other exercises of a similar nature will readily suggest themselves, and in arranging the course every effort should be made to induce the student to consult original papers, and to avoid as far as possible any tendency to mere mechanical work.

The exact nature of such a course must, however, necessarily be left very much in the hands of the teacher, and the details will no doubt require much consideration ; but I feel sure that a course of practical inorganic chemistry could be constructed which, while teaching all the important facts which it is necessary for the student to know, will, at the same time, constantly tend to develop his faculty of original thought.

Supposing such a course were adopted (and the experiment is well worth trying), there still remains the problem of how the student who has had this kind of training is to be examined.

With regard to his theoretical work there would be no difficulty, as the examination could be conducted on much the same lines as at the present time. In the case of the practical examination I have long felt that the only satisfactory method of arriving at the value of a student's practical knowledge is by the inspection of the work which he has done during the whole of his course of study, and not by depending on the results of one or two days' set examination. I think that most examiners will agree with me that the present system of examination in practical chemistry is highly unsatisfactory. This is perhaps not so apparent in the case of the qualitative analysis of the usual simple salt or mixture ; but when the student has to do a quantitative exercise, or when a problem is set, the results sent in are frequently no indication of the value of the student's practical work. Leaving out of the question the possibility of the student being in indifferent health during the short period of the practical examination, it not infrequently happens that he, in his excitement, has the misfortune to upset a beaker when his quantitative determination is nearly finished, and as a result he loses far more marks than he should do for so simple an accident.

Again, in attacking a problem he has usually only time to try one method of solution, and if this does not yield satisfactory results he again loses marks ; whereas in the ordinary course of his practical work, if he were to find that the first method was faulty, he would try other methods until he ultimately arrived at the desired result.

It is difficult to see why such an unsatisfactory system as this might not be replaced by one of inspection which I think could easily be so arranged as to work well.

A student taking, say, a three years' course for the degree of Bachelor of Science might be required to keep very careful notes of all the practical work which he does during this course, and in order to avoid fraud his notebook could from time to time be initialled by the professor or demonstrator in charge of the laboratory. An inspection of these notebooks could then be made at suitable times by the examiners for the degree, by which means a very good idea would be obtained of the scope of the work which the student had been engaged in, and if thought necessary a few questions could easily be asked in regard to the work so presented. Should the examiners wish to further test the candidate by giving him an examination, I submit that it would be much better to set him some exercise of the nature of a simple original investigation, and to allow him two or three weeks to carry this out, than to depend on the hurried work of two or three days.

The object which I had in view in writing this Address was to call attention to the fact that our present system of training in chemistry does not appear to develop in the student the power of conducting original research, and at the same time to endeavour to suggest some means by which a more satisfactory state of things might be brought about. I have not been able, within the limits of this Address, to consider the conditions of study during the third year of the student's career at college, or to discuss the increasing necessity for extending that course and insisting on the student carrying out an adequate original investigation before granting him a degree, but I hope on some future occasion to have the opportunity of returning to this very important part of the subject. If any of the suggestions I have made should prove to be of practical value, and should lead to the production of more original research by our students, I shall feel that a useful purpose has been served by bringing this matter before this Section. In concluding I wish to thank Prof. H. B. Dixon, Prof. F. S. Kipping and others, for many valuable suggestions, and my thanks are especially due to Dr. Bevan Lean for much information which he gave me in connection with that part of this Address which deals with the teaching of chemistry in schools.

SECTION C.

GEOLOGY.

OPENING ADDRESS BY PROF. W. J. SOLLAS, D.SC., LL.D., F.R.S., PRESIDENT OF THE SECTION.

Evolutional Geology.

THE close of one century, the dawn of another, may naturally suggest some brief retrospective glance over the path along which our science has advanced, and some general survey of its present position from which we may gather hope of its future progress ; but other connection with geology the beginnings and endings of centuries have none. The great periods of movement have hitherto begun, as it were, in the early twilight hours, long before the dawn. Thus the first step forward, since which there has been no retreat, was taken by Steno in the year 1669 ; more than a century elapsed before James Hutton (1785) gave fresh energy and better direction to the faltering steps of the young science ; while it was less than a century later (1863) when Lord Kelvin brought to its aid the powers of the higher mathematics and instructed in it the teachings of modern physics. From Steno onward the spirit of geology was catastrophic ; from Hutton onward it grew increasingly uniformitarian ; from the time of Darwin and Kelvin it has become evolutional. The ambiguity of the word "uniformitarian" has led to a good deal of fruitless logomachy, against which it may be as well at once to guard by indicating the sense in which it is used here. In one way we are all uniformitarians, *i.e.* we accept the doctrine of the "uniform action of natural causes," but, as applied to geology, uniformity means more than this. Defined in the briefest fashion it is the geology of Lyell. Hutton had given us a "Theory of the Earth," in its main outlines still faithful and true ; and this Lyell spent his life in illustrating and advocating ; but as so commonly happens the zeal of the disciple outran the wisdom of the master, and mere opinions were insisted on as necessary dogma. What did it matter if Hutton as a result of his inquiries into terrestrial history had declared that he found no vestige of a beginning, no prospect of an end? It would have been marvellous if he had ! Consider that when Hutton's

"Theory" was published William Smith's famous discovery had not been made, and that nothing was then known of the orderly succession of forms of life, which it is one of the triumphs of geology to have revealed ; consider, too, the existing state of physics at the time, and that the modern theories of energy had still to be formulated ; consider also that spectroscopy had not yet lent its aid to astronomy and the consequent ignorance of the nature of nebulæ ; and then, if you will, cast a stone at Hutton. With Lyell, however, the case was different : in pressing his uniformitarian creed upon geology he omitted to take into account the great advances made by its sister sciences, although he had knowledge of them, and thus sinned against the light. In the last edition of the famous "Principles" we read : "It is a favourite dogma of some physicists that not only the earth, but the sun itself, is continually losing a portion of its heat, and that as there is no known source by which it can be restored we can foresee the time when all life will cease to exist on this planet, and on the other hand we can look back to a period when the heat was so intense as to be incompatible with the existence of any organic beings such as are known to us in the living or fossil world. . . . A geologist in search of some renovating power by which the amount of heat may be made to continue unimpaired for millions of years, past and future, in the solid parts of the earth . . . has been compared by an eminent physicist to one who dreams he can discover a source of perpetual motion and invent a clock with a self-winding apparatus. *But why should we despair of detecting proofs of such regenerating and self-sustaining power in the works of a Divine Artificer?*" Here we catch the true spirit of uniformity ; it admittedly regards the universe as a self-winding clock, and barely conceals a conviction that the clock was warranted to keep true Greenwich time. The law of the dissipation of energy is not a dogma, but a doctrine drawn from observation, while the uniformity of Lyell is in no sense an induction : it is a dogma in the narrowest sense of the word, unproved, incapable of proof ; hence perhaps its power upon the human mind ; hence also the transitoriness of that power. Again, it is only by restricting its inquiries to the stratified rocks of our planet that the dogma of uniformity can be maintained with any pretence of argument. Directly we begin to search the heavens the possibility, nay even the likelihood, of the nebular origin of our system, with all that it involves, is borne in upon us. Lyell therefore consistently refused to extend his gaze beyond the rocks beneath his feet, and was thus led to do a serious injury to our science : he severed it from cosmogony, for which he entertained and expressed the most profound contempt, and from the mutilation thus inflicted geology is only at length making a slow and painful recovery. Why do I dwell on these facts? To depreciate Lyell? By no means. No one is more conscious than I of the noble service which Lyell rendered to our cause : his reputation is of too robust a kind to suffer from my unskilful handling, and the fame of his solid contributions to science will endure long after these controversies are forgotten. The echoes of the combat are already dying away, and uniformitarians, in the sense already defined, are now no more ; indeed, were I to attempt to exhibit any distinguished living geologist as a still surviving supporter of the narrow Lyellian creed, he would probably feel, if such a one there be, that I was unfairly singling him out for unmerited obloquy.

Our science has become evolutional, and in the transformation has grown more comprehensive : her petty parochial days are done, she is drawing her provinces closer around her, and is fusing them together into a united and single commonwealth— the science of the earth.

Not merely the earth's crust, but the whole of earth-knowledge is the subject of our research. To know all that can be known about our planet, this, and nothing less than this, is its aim and scope. From the morphological side geology inquires, not only into the existing form and structure of the earth, but also into the series of successive morphological states through which it has passed in a long and changeful development. Our science inquires also into the distribution of the earth in time and space ; on the physiological side it studies the movements and activities of our planet ; and not content with all this it extends its researches into ætiology and endeavours to arrive at a science of causation. In these pursuits geology calls all the other sciences to her aid. In our commonwealth there are no outlanders ; if an eminent physicist over our territory we do not begin at once to prepare for war, because the very fact of his undertaking a geological inquiry of itself confers upon him all the duties and privileges of citizenship. A physicist studying geology is by definition a geologist. Our only regret is, not that physicists occasionally invade our borders, but that they do not visit us oftener and make closer acquaintance with us.

Early History of the Earth : First Critical Period.

If I am bold enough to assert that cosmogony is no longer alien to geology, I may proceed further, and taking advantage of my temerity pass on to speak of things once not permitted to us. I propose, therefore, to offer some short account of the early stages in the history of the earth. Into its nebular origin we need not inquire—that is a subject for astronomers. We are content to accept the infant earth from their hands as a molten globe ready made, its birth from a gaseous nebula duly certified. If we ask, as a matter of curiosity, what was the origin of the nebula, I fear even astronomers cannot tell us. There is an hypothesis which refers it to the clashing of meteorites, but in the form in which this is usually presented it does not help us much. Such meteorites as have been observed to penetrate our atmosphere and to fall on to the surface of the earth prove on examination to have had an eventful history of their own of which not the least important chapter was a passage through a molten state; they would thus appear to be the products rather than the progenitors of a nebula.

We commence our history, then, with a rapidly-rotating molten planet, not impossibly already solidified about the centre and surrounded by an atmosphere of great depth, the larger part of which was contributed by the water of our present oceans, then existing in a state of gas. This atmosphere, which exerted a pressure of something like 5000 lb. to the square inch, must have played a very important part in the evolution of our planet. The molten exterior absorbed it to an extent which depended on the pressure, and which may some day be learnt from experiment. Under the influence of the rapid rotation of the earth the atmosphere would be much deeper in equatorial than polar regions, so that in the latter the loss of heat by radiation would be in excess. This might of itself lead to convectional currents in the molten ocean. The effect on the atmosphere is very difficult to trace, but it is obvious that if a high-pressure area originated over some cooler region of the ocean, the winds blowing out of it would drive before them the cooler superficial layers of molten material, and as these were replaced by hotter lava streaming from below, the tendency would be to convert the high into a low-pressure area, and to reverse the direction of the winds. Conversely under a low-pressure area the in-blowing winds would drive in the cooler superficial layers of molten matter that had been swept away from the anticyclones. If the difference in pressure under the cyclonic and anticyclonic areas were considerable, some of the gas absorbed under the anticyclones might escape beneath the cyclones, and in a later stage of cooling might give rise to vast floating islands of scoria. Such islands might be the first foreshadowings of the future continents. Whatever the ultimate effect of the reaction of the winds on the currents of the molten ocean, it is probable that some kind of circulation was set up in the latter. The universal molten ocean was by no means homogeneous: it was constantly undergoing changes in composition as it reacted chemically with the internal metallic nucleus; its currents would streak the different portions out in directions which in the northern hemisphere would run from north-east to south-west, and thus the differences which distinguish particular petrological regions of our planet may have commenced their existence at a very early stage. Is it possible that as our knowledge extends we shall be able by a study of the distribution of igneous rocks and minerals to draw some conclusions as to the direction of these hypothetical lava currents? Our planet was profoundly disturbed by tides, produced by the sun; for as yet there was no moon; and it has been suggested that one of its tidal waves rose to a height so great as to sever its connection with the earth and to fly off as the infant moon. This event may be regarded as marking the first critical period, or catastrophe if we please, in the history of our planet. The career of our satellite, after its escape from the earth, is not known till it attained a distance of nine terrestrial radii; after this its progress can be clearly followed. At the eventful time of parturition the earth was rotating, with a period of from two to four hours, about an axis inclined at some 11° or 12° to the ecliptic. The time which has elapsed since the moon occupied a position nine terrestrial radii distant from the earth is at least fifty-six to fifty-seven millions of years, but may have been much more. Prof. Darwin's story of the moon is certainly one of the most beautiful contributions ever

made by astronomy to geology, and we shall all concur with him when he says, "A theory reposing on *vera causa*, which brings into quantitative correlation the length of the present day and month, the obliquity of the ecliptic, and the inclination and eccentricity of the lunar orbit, must, I think, have strong claims to acceptance."

The majority of geologists have long hankered after a metallic nucleus for the earth, composed chiefly, by analogy with meteorites, of iron. Lord Kelvin has admitted the probable existence of some such nucleus, and lately Prof. Wiechert has furnished us with arguments—"powerful" arguments Prof. Darwin terms them—in support of its existence. The interior of the earth for four-fifths of the radius is composed, according to Prof. Wiechert, chiefly of metallic iron, with a density of 8·2 ; the outer envelope, one fifth of the radius, or. about 400 miles in thickness, consists of silicates, such as we are familiar with in igneous rocks and meteorites, and possesses a density of 3·2. It was from this outer envelope when molten that the moon was trundled off, twenty-seven miles in depth going to its formation. The density of this material, as we have just seen, is supposed to be 3·2 ; the density of the moon is 3·39, a close approximation, such difference as exists being completely explicable by the comparatively low temperature of the moon.

The outer envelope of the earth which was drawn off to form the moon was, as we have seen, charged with steam and other gases under a pressure of 5000 lb. to the square inch ; but as the satellite wandered away from the parent planet this pressure continuously diminished. Under these circumstances the moon would become as explosive as a charged bomb, steam would burst forth from numberless volcanoes, and while the face of the moon might thus have acquired its existing features, the ejected material might possibly have been shot so far away from its origin as to have acquired an independent orbit. If so we may ask whether it may not be possible that the meteorites, which sometimes descend upon our planet, are but portions of its own envelope returning to it. The facts that the average specific gravity of these meteorites which have been seen to fall is not much above 3·2, and that they have passed through a stage of fusion, are consistent with this suggestion.

Second Critical Period. "Consistentior Status."

The solidification of the earth probably became completed soon after the birth of the moon. The temperature of its surface at the time of consolidation was about 1170° C., and it was therefore still surrounded by its primitive deep atmosphere of steam and other gases. This was the second critical period in the history of the earth, the stage of the "consistentior status," the date of which Lord Kelvin would rather know than that of the Norman Conquest, though he thinks it lies between twenty and forty millions of years ago, probably nearer twenty than forty.

Now that the crust was solid there was less reason why movements of the atmosphere should be unsteady, and definite regions of high and low pressure might have been established. Under the high-pressure areas the surface of the crust would be depressed ; correspondingly under the low-pressure areas it would be raised ; and thus from the first the surface of the solid earth might be dimpled and embossed."[1]

Third Critical Period. Origin of the Oceans.

The cooling of the earth would continuously progress, till the temperature of the surface fell to 370° C., when that part of the atmosphere which consisted of steam would begin to liquefy ; then the dimples on the surface would soon become filled with superheated water, and the pools so formed would expand and deepen, till they formed the oceans. This is the third critical stage in the history of the earth, dating, according to Prof. Joly, from between eighty and ninety millions of years ago. With the growth of the oceans the distinction between land and sea arose—in what precise manner we may proceed to inquire. If we revert to the period of the "consistentior status," when the earth had just solidified, we shall find, according to Lord Kelvin, that the temperature continuously increased from the surface, where it was 1170° C., down to a depth of twenty-five miles, where it was about 1430° C., or 260° C. above the fusion point of the matter, forming the crust.

[1] It would be difficult to discuss with sufficient brevity the probable distribution of these inequalities, but it may be pointed out that the moon is possibly responsible, and that in more ways than one, for much of the existing geographical asymmetry.

That the crust at this depth was not molten but solid is to be explained by the very great pressure to which it was subjected—just so much pressure, indeed, as was required to counteract the influence of the additional 260° C. Thus if we could have reduced the pressure on the crust we should have caused it to liquefy; by restoring the pressure it would resolidify. By the time the earth's surface had cooled down to 370° C. the depth beneath the surface at which the pressure just kept the crust solid would have sunk some slight distance inwards, but not sufficiently to affect our argument.

The average pressure of the primitive atmosphere upon the crust can readily be calculated by supposing the water of the existing oceans to be uniformly distributed over the earth's surface, and then by a simple piece of arithmetic determining its depth; this is found to be 1·718 miles, the average depth of the oceans being taken at 2·393 miles. Thus the average pressure over the earth's surface, immediately before the formation of the oceans, was equivalent to that of a column of water 1·718 miles high on each square inch. Supposing that at its origin the ocean were all "gathered together into one place," and "the dry land appeared," then the pressure over the ocean floor would be increased from 1·718 miles to 2·393 miles, while that over those portions of the crust that now formed the land would be diminished by 1·718 miles. This difference in pressure would tend to exaggerate those faint depressions which had arisen under the primitive anti-cyclonic areas, and if the just solidified material of the earth's crust were set into a state of flow, it might move from under the ocean into the bulgings which were rising to form the land, until static equilibrium were established. Under these circumstances the pressure of the ocean would be just able to maintain a column of rock 0·886 miles in height, or ten twenty-sevenths of its own depth. It could do no more; but in order that the dry land may appear some cause must be found competent either to lower the ocean bed the remaining seventeen twenty-sevenths of its full depth, or to raise the continental bulgings to the same extent. Such a cause may, I think, be discovered in a further effect of the reduction in pressure over the continental areas. Previous to the condensation of the ocean, these, as we have seen, were subjected to an atmospheric pressure equal to that of a column of water 1·718 miles in height. This pressure was contributory to that which caused the outer twenty-five miles of the earth's crust to become solid; it furnished, indeed, just about one fortieth of that pressure, or enough to raise the fusion point 6° C. What, then, might be expected to happen when the continental area was relieved of this load? Plainly a liquefaction and corresponding expansion of the underlying rock.

But we will not go so far as to assert that actual liquefaction would result; all we require for our explanation is a great expansion; and this would probably follow whether the crust were liquefied or not. For there is good reason to suppose that when matter at a temperature above its ordinary fusion point is compelled into the solid state by pressure, its volume is very responsive to changes either of pressure or temperature. The remarkable expansion of liquid carbon dioxide is a case in point: 120 volumes of this fluid at −20° C. become 150 volumes at 33° C.; a temperature just below the critical point. A great change of volume also occurs when the material of igneous rocks passes from the crystalline state to that of glass; in the case of diabase[1] the difference in volume of the rock in the two states at ordinary temperatures is 13 per cent. If the relief of pressure over the site of continents were accompanied by volume changes at all approaching this, the additional elevation of seventeen twenty-sevenths required to raise the land to the sea-level would be accounted for.[2] How far down beneath the sur-

[1] C. Barus so names the material on which he experimented; apparently the rock is a fresh dolerite without olivine.

[2] Prof. Fitzgerald has been kind enough to express part of the preceding explanation in a more precise manner for me. He writes: "It would require a very nice adjustment of temperatures and pressures to work out in the simple way you state it; but what is really involved is that in a certain state diabase (and everything that changes state with a considerable change of volume) has an enormous isothermal compressibility. Although this is very enormous in the case of bodies which melt suddenly, like ice, it would also involve very great compressibilities in the case of bodies even which melted gradually, if they did so at all quickly, *i.e.* within a small range of temperature. What you postulate, then, is that at a certain depth diabase is soft enough to be squeezed from under the oceans, and that, being near its melting point, the small relief of pressure is accompanied by an enormous increase in volume which helped to raise the continents. Now that I have written the thing out in my own way it seems very likely. It is, anyway, a suggestion quite worthy of serious consideration, and a process that in some places must almost certainly have been in operation, and maybe is still operative. Looking at it again, I hardly think it is quite

face the unloading of the continents would be felt it is difficult to say, though the problem is probably not beyond the reach of mathematical analysis; if it affected an outer envelope twenty-five miles in thickness, a linear expansion of 4 per cent. would suffice to explain the origin of ocean basins. If now we refer to the dilatation determined by Carl Barus for rise in temperature in the case of diabase, we find that between 1093° and 1112° C. the increase in volume is 3·3 per cent. As a further factor in deepening the ocean basins may be included the compressive effect of the increase in load over the ocean floor: this increase is equal to the pressure of a column of water 0·675 mile in height, and its effect in raising the fusion point would be 2° C., from which we may gain some kind of idea of the amount of compression it might produce on the yielding interior of the crust. To admit that these views are speculative will be to confess nothing; but they certainly account for a good deal. They not only give us ocean basins, but basins of the kind we want, that is, to use a crude comparison once made by the late Dr. Carpenter, basins of a tea-tray form, having a somewhat flat floor and steeply sloping sides; they also help to explain how it is that the value of gravity is greater over the ocean than over the land.

The ocean when first formed would consist of highly heated water, and this, as is well known, is an energetic chemical reagent when brought into contact with silicates like those which formed the primitive crust. As a result of its action saline solutions and chemical deposits would be formed; the latter, however, would probably be of no great thickness, for the time occupied by the ocean in cooling to a temperature not far removed from the present would probably be included within a few hundreds of years.

The Stratified Series.

The course of events now becomes somewhat obscure, but sooner or later the familiar processes of denudation and the deposition started into activity, and have continued acting uninterruptedly ever since. The total maximum thickness of the sedimentary deposits, so far as I can discover, appears to amount to no less than 50 miles, made up as follows:—

		Feet		
Recent and Pleistocene	...	4,000	...	Man.
Pliocene	...	5,000	...	Pithecanthropus.
Miocene	...	9,000	...	
Oligocene	...	12,000	...	
Eocene	...	12,000	...	Eutheria.
Cretaceous	...	14,000	...	
Jurassic	...	8,000	...	
Trias	...	13,000	...	Mammals.
Permian	...	12,000	...	Reptiles.
Carboniferous	...	24,000	...	Amphibia.
Devonian	...	22,000	...	Fish.
Silurian	...	15,000	...	
Ordovician	...	17,000	...	
Cambrian	...	16,000	...	Invertebrata.
Keeweenawan	...	50,000	...	
Penokee	...	14,000	...	
Huronian	...	18,000	...	

Geologists, impressed with the tardy pace at which sediments appear to be accumulating at the present day, could not contemplate this colossal pile of strata without feeling that it spoke of an almost inconceivably long lapse of time. They were led to compare its duration with the distances which intervene between the heavenly bodies; but while some chose the distance of the nearest fixed star as their unit, others were content to measure the years in terms of miles from the sun.

Evolution of Organisms.

The stratified rocks were eloquent of time, and not to the geologist alone; they appealed with equal force to the biologist. Accepting Darwin's explanation of the origin of species, the likely that there is or could be much squeezing sideways of liquid or other viscous material from under one place to another, because the elastic yielding of the inside of the earth would be much quicker than any flow of this kind. This would only modify your theory, because the diabase that expands so much on the relief of pressure might be that already under the land, and raising up this latter, partly by being pushed up itself by the elastic relief of the inside of the earth and partly by its own enormous expansibility near its melting point. The action would be quite slow, because it would cool itself so much by its expansion that it would have to be warmed up from below, or by tidal earth-squeezing, or by chemical action, before it could expand isothermally."

present rate at which form flows to form seemed so slow as almost to amount to immutability. How vast then must have been the period during which by slow degrees and innumerable stages the protozoon was transformed into the man! And if we turn to the stratified column, what do we find? Man, it is true, at the summit, the oldest fossiliferous rocks 34 miles lower down, and the fossils they contain already representing most of the great classes of the Invertebrata, including Crustacea and Worms. Thus the evolution of the Vertebrata alone is known to have occupied a period represented by a thickness of 34 miles of sediment. How much greater, then, must have been the interval required for the elaboration of the whole organic world! The human mind, dwelling on such considerations as these, seems at times to have been affected by a sur-excitation of the imagination, and a consequent paralysis of the understanding, which led to a refusal to measure geological time by years at all, or to reckon by anything less than "eternities."

Geologic Periods of Time.

After the admirable Address of your President last year it might be thought needless for me to again enter into a consideration of this subject; it has been said, however, that the question of geological time is like the Djin in Arabian tales, and will irrepressibly come up again for discussion, however often it is disposed of. For my part I do not regard the question so despondingly, but rather hope that by persevering effort we may succeed in discovering the talisman by which we may compel the unwilling Djin into our service. How immeasurable would be the advance of our science could we but bring the chief events which it records into some relation with a standard of time!

Before proceeding to the discussion of estimates of time drawn from a study of stratified rocks let us first consider those which have been already suggested by other data. These are as follows:—(1) Time which has elapsed since the separation of the earth and moon, fifty-six millions of years, minimum estimate by Prof. G. H. Darwin. (2) Since the "consistentior status," twenty to forty millions (Lord Kelvin). (3) Since the condensation of the oceans, eighty to ninety millions, maximum estimate by Prof. J. Joly.

It may be at once observed that these estimates, although independent, are all of the same order of magnitude, and so far confirmatory of each other. Nor are they opposed to conclusions drawn from a study of stratified rocks; thus Sir Archibald Geikie, in his Address to this Section last year, affirmed that, so far as these were concerned, one hundred millions of years might suffice for their formation. There is then very little to quarrel about, and our task is reduced to an attempt, by a little stretching and a little paring, to bring these various estimates into closer harmony.

Prof. Darwin's estimate is admittedly a minimum; the actual time, as he himself expressly states, "may have been much longer." Lord Kelvin's estimate, which he would make nearer twenty than forty millions, is founded on the assumption that since the period of the "consistentior status" the earth has cooled simply as a solid body, the transference of heat from within outwards having been accomplished solely by conduction.[1]

It may be at once admitted that there is a large amount of truth in this assumption; there can be no possible doubt that the earth reacts towards forces applied for a short time as a solid body. Under the influence of the tides it behaves as though it possessed a rigidity approaching that of steel, and under sudden blows, such as those which give rise to earthquakes, with twice this rigidity, as Prof. Milne informs me. Astronomical considerations lead to the conclusion that its effective rigidity has not varied greatly for a long period of past time.

Still, while fully recognising these facts, the geologist knows —we all know—that the crust of the earth is not altogether solid. The existence of volcanoes by itself suggests the contrary, and although the total amount of fluid material which is brought from the interior to the exterior of the earth by volcanic action may be, and certainly is, small—from data given by Prof. Penck, I estimate it as equivalent to a layer of rock uniformly distributed 2 mm. thick per century; yet we have every reason to believe that volcanoes are but the superficial manifestation of far greater bodies of molten material

[1] The heat thus brought to the surface would amount to one-seventeenth of that conveyed by conduction.

which lie concealed beneath the ground. Even the wide areas of plutonic rock, which are sometimes exposed to view over a country that has suffered long-continued denudation, are merely the upper portion of more extensive masses which lie remote from view. The existence of molten material within the earth's crust naturally awakens a suspicion that the process of cooling has not been wholly by conduction, but also to some slight extent by convection, and to a still greater extent by the bodily migration of liquid lava from the deeper layers of the crust towards the surface.

The existence of local reservoirs of molten rock within the crust is even still more important in another connection, that is, in relation with the supposed "average rate of increase of temperature with descent below the ground." It is doubtful whether we have yet discovered a rate that in any useful sense can be spoken of as "average." The widely divergent views of different authorities as to the presumed value of this rate may well lead to reflection. The late Prof. Prestwich thought a rise of 1° F. for every 45 feet of descent below the

Fig. 1.—Map of the British Isles, showing the distribution of rates of increase of temperature with descent. The rates are taken from the "British Association Report," except in the case of those in the south of Ireland.

zone of constant temperature best represented the average; Lord Kelvin in his earlier estimates has adopted a value of 1° F. for every 51 feet; the Committee of this Association appointed to investigate this question arrived at a rate of 1° F. for every 60 feet of descent; Mr. Clarence King has made calculations in which a value of 1° F. for 72 feet is adopted; a re-investigation of recorded measurements would, I believe, lead to a rate of 1° F. in 80 or 90 feet as more closely approaching the mean. This would raise Lord Kelvin's estimate to nearly fifty millions of years.

When from these various averages we turn to the observations on which they are based, we encounter a surprising divergence of extremes from the mean; thus in the British Isles alone the rate varies from 1° F. in 34 feet to 1° F. in 92 feet, or in one case to 1° F. in 130 feet. It has been suggested, and to some extent shown, that these irregularities may be connected with differences in conductivity of the rocks in which the observations were made, or to the circulation of underground water; but many cases exist which cannot be explained away in such a manner, but are suggestive of some deep-seated cause, such as the distribution of molten matter below the ground. Inspection of the accompanying map of the British Isles, on which the rates of increase in different localities have been plotted, will afford some evidence of the truth of this view. Comparatively low rates of increase are found over Wales and in the province of Leinster, districts of relatively great stability, the remnants of an island that have in all probability stood above the sea ever since the close of the Silurian period. To the north of this, as we enter a region which was subject to volcanic disturbances during the Tertiary period, the rate increases.

It is obvious that in any attempt to estimate the rate at which the earth is cooling as a solid body the disturbing influence of subterranean lakes of molten rock must as far as possible be eliminated; but this will not be effected by taking the accepted mean of observed rates of increase of temperature: such an average is merely a compromise, and a nearer approach to a correct result will possibly be attained by selecting some low rate of increase, provided it be based on accurate observations.

It is extremely doubtful whether an area such as the British Isles, which has so frequently been the theatre of volcanic activity and other subterranean disturbance, is the best fitted to afford trustworthy results; the Archæan nucleus of a continent might be expected to afford surer indications. Unfortunately the hidden treasures of the earth are seldom buried in these regions, and bore-holes in consequence have rarely been made in them. One exception is afforded by the copper-bearing district of Lake Superior, and in one case, that of the Calumet and Hecla mine, which is 4580 feet in depth, the rate of increase, as determined by Prof. A. Agassiz, was 1° F. for every 223·7 feet. The Bohemian "horst" is a somewhat ancient part of Europe, and in the Przibram mines, which are sunk in it, the rate was 1° F. for every 126 feet of descent. In the light of these facts it would seem that geologists are by no means compelled to accept the supposed mean rate of increase of temperature with descent into the crust as affording a safe guide to the rate of cooling of a solid globe; and if the much slower rate of increase observed in the more ancient and more stable regions of the earth has the importance which is suggested for it, then Lord Kelvin's estimate of the date of the "consistentior status" may be pushed backwards into a remoter past.

If, as we have reason to hope, Lord Kelvin's somewhat contracted period will yield to a little stretching, Prof. Joly's, on the other hand, may take some paring. His argument, broadly stated, is as follows. The ocean consisted at first of fresh water; it is now salt, and its saltness is due to the dissolved matter that is constantly being carried into it by rivers. If, then, we know the quantity of salt which the rivers bring down each year into the sea, it is easy to calculate how many years they have taken to supply the sea with all the salt it at present contains. For several reasons it is found necessary to restrict attention to one only of the elements contained in sea salt: this is sodium. The quantity of sodium delivered to the sea every year by rivers is about 160,000,000 tons; but the quantity of sodium which th᷉ sea contains is at least ninety millions of times greater than this. The period during which rivers have been carrying sodium into the sea must therefore be about ninety millions of years. Nothing could be simpler; there is no serious flaw in the method, and Prof. Joly's treatment of the subject is admirable in every way; but of course in calculations such as this everything depends on the accuracy of the data, which we may therefore proceed to discuss. Prof. Joly's estimate of the amount of sodium in the ocean may be accepted as sufficiently near the truth for all practical purposes. We may therefore pass on to the other factor, the annual contribution of sodium by river water. Here there is more room for error. Two quantities must be ascertained: one the quantity of water which the rivers of the world carry into the sea, the other the quantity or proportion of sodium present in this water. The total volume of water discharged by rivers into the ocean is estimated by Sir John Murray as 6524 cubic miles. The estimate being based on observations of thirty-three great rivers, although only approximate, it is no doubt sufficiently exact; at all events such alterations as it is likely to undergo will not greatly affect the final result. When, however, we pass to the last quantity to be determined, the chemical composition of average

river water, we find that only a very rough estimate is possible, and this is the more unfortunate because changes in this may very materially affect our conclusions. The total quantity of river water discharged into the sea is, as we have stated, 6524 cubic miles. The average composition of this water is deduced from analyses of nineteen great rivers, which altogether discharge only 488 cubic miles, or 7·25 per cent. of the whole. The danger in using this estimate is twofold: in the first place 7·25 is too small a fraction from which to argue to the remaining 92·75 per cent., and, next, the rivers which furnish it are selected rivers, *i.e.*, they are all of large size. The effect of this is that the drainage of the volcanic regions of the earth is not sufficiently represented, and it is precisely this drainage which is richest in sodium salts. The lavas and ashes of active volcanoes rapidly disintegrate under the energetic action of various acid gases, and among volcanic exhalations sodium chloride has been especially noticed as abundant. Consequently we find that while the proportion of sodium in Prof. Joly's average river water is only 5·73 per million, in the rivers of the volcanic island of Hawaii it rises to 24·5 per million (Walter Maxwell, "Lavas and Soils of the Hawaiian Islands," p. 170). No doubt the area occupied by volcanoes is trifling compared with the remaining land surface. On the other hand the majority of volcanoes are situated in regions of copious rainfall, of which they receive a full share owing to their mountainous form. Much of the fallen rain percolates through the porous material of the cone, and, richly charged with alkalies, finds its way by underground passages towards the sea, into which it sometimes discharges by submarine springs.

Again, several considerations lead to the belief that the supply of sodium to the ocean has proceeded, not at a uniform, but at a gradually diminishing rate. The rate of increase of temperature with descent into the crust has continuously diminished with the flow of time, and this must have had its influence on the temperature of springs, which furnish an important contribution to river water. The significance of this consideration may be judged from the composition of the water of geysers. Thus Geyser, in Iceland, contains 884 parts of sodium per million, or nearly 160 times as much as Sir John Murray estimates is present in average river water. A mean of the analyses of six geysers in different parts of the world gives 400 parts of sodium per million, existing partly as chloride, but also as sulphate and carbonate.

It should not be overlooked that the present is a calm and quiet epoch in the earth's history, following after a time of fiery activity. More than once, indeed, has the past been distinguished by unusual manifestations of volcanic energy, and these must have had some effect upon the supply of sodium to the ocean. Finally, although the existing ocean water has apparently but slight effect in corroding the rocks which form its bed, yet it certainly was not inert when its temperature was not far removed from the critical point. Water begins to exert a powerful destructive action on silicates at a temperature of 180° C., and during the interval occupied in cooling from 370 to 180° C. a considerable quantity of sodium may have entered into solution.

A review of the facts before us seems to render some reduction in Dr. Joly's estimate imperative. A precise assessment is impossible, but I should be inclined myself to take off some ten or thirty millions of years.

We may next take the evidence of the stratified rocks. Their total maximum thickness is, as we have seen, 265,000 feet, and consequently if they accumulated at the rate of one foot in a century, as evidence seems to suggest, more than twenty-six millions of years must have elapsed during their formation.

Obscure Chapter in the Earth's History.

Before discussing the validity of the argument on which this last result depends, let us consider how far it harmonises with previous ones. It is consistent with Lord Kelvin's and Professor Darwin's, but how does it accord with Professor Joly's? Supposing we reduce his estimate to fifty-five millions; what was the earth doing during the interval between the period of fifty-five millions of years ago and that of only 26½ millions of years ago, when, it is presumed, sedimentary rocks commenced to be formed? Hitherto we have been able to reason on probabilities; now we enter the dreary region of possibilities, and open that obscure chapter in the history of the earth previously hinted at. For there are many possible answers to this question. In the first place the evidence of the stratified rocks may have been wrongly interpreted, and two or three times the amount of time we have demanded may have been consumed in their formation. This is a very obvious possibility, yet again our estimate concerning these rocks may be correct, but we may have erroneously omitted to take into account certain portions of the Archæan complex, which may represent primitive sedimentary rocks, formed under exceptional conditions, and subsequently transformed under the influence of the internal heat of the earth. This, I think, would be Prof. Bonney's view. Finally Lord Kelvin has argued that the life of the sun as a luminous star is even more briefly limited than that of our oceans. In such a case if our oceans were formed fifty-five millions of years ago, it is possible that after a short existence as almost boiling water they grew colder and colder, till they became covered with thick ice, and moved only in obedience to the tides. The earth, frozen and dark, except for the red glow of her volcanoes, waited the coming of the sun, and it was not till his growing splendour had banished the long night that the cheerful sound of running waters was heard again in our midst. Then the work of denudation and deposition seriously recommenced, not to cease till the life of the sun is spent. Thus the thickness of the stratified series may be a measure rather of the duration of sunlight than of the period which has elapsed since the first formation of the ocean. It may have been so—we cannot tell—but it may be fairly urged that we know less of the origin, history, and constitution of the sun than of the earth itself, and that, for aught we can say to the contrary, the sun may have been shining on the just-formed ocean as cheerfully as he shines to-day.

Time required for the Evolution of the Living World.

But, it will be asked, how far does a period of twenty-six millions satisfy the demands of biology? Speaking only for myself, although I am aware that eminent biologists are not wanting who share this opinion, I answer, Amply. But it will be exclaimed, Surely there are "comparisons in things." Look at Egypt, where more than 4,000 years since the same species of man and animals lived and flourished as to-day. Examine the frescoes and-study the living procession of familiar forms they so faithfully portray, and then tell us, how comes it about that from changes so slow as to be inappreciable in the lapse of forty centuries you propose to build up the whole organic world in the course of a mere twenty-six millions of years? To all which we might reply that even changeless Egypt presents us with at least one change—the features of the ruling race are to-day not quite the same as those of the Pharaohs. But putting this on one side, the admitted constancy in some few common forms proves very little, for so long as the environment remains the same natural selection will conserve the type, and, so far as we are able to judge, conditions in Egypt have remained remarkably constant for a long period.

Change the conditions, and the resulting modification of the species becomes manifest enough; and in this connection it is only necessary to recall the remarkable mutations observed and recorded by Prof. Weldon in the case of the crabs in Plymouth Harbour. In response to increasing turbidity of the sea water these crabs have undergone or are undergoing a change in the relative dimensions of the carapace, which is persistent, in one direction, and rapid enough to be determined by measurements made at intervals of a few years.

Again, animals do not all change their characters at the same rate: some are stable, in spite of changing conditions, and these have been cited to prove that none of the periods we look upon as probable, not twenty-five, not a hundred millions of years, scarce any period short of eternity, is sufficient to account for the evolution of the living world. If the little tongue-shell, *Lingula*, has endured with next to no perceptible change from the Cambrian down to the present day, how long, it is sometimes inquired, would it require for the evolution of the rest of the animal kingdom? The reply is simple: the cases are dissimilar, and the same record which assures us of the persistency of the *Lingula* tells us in language equally emphatic of the course of evolution which has led from the lower organisms upwards to man. In recent and Pleistocene deposits the relics of man are plentiful: in the latest Pliocene they have disappeared, and we encounter the remarkable form *Pithecanthropus;* as we descend into the Tertiary systems the higher mammals are met with, always sinking lower and lower in the scale of organisation as they occur deeper in the series, till in the Mesozoic deposits they have entirely disappeared, and their place is taken by the lower mammals, a feeble folk, offering

little promise of the future they were to inherit. Still lower, and even these are gone; and in the Permian we encounter reptiles and the ancestors of reptiles, probably ancestors of mammals too; then into the Carboniferous, where we find amphibians, but no true reptiles; and next into the Devonian, where fish predominate, after making their earliest appearance at the'close of the Silurian times; thence downwards, and the vertebrata are no more found—we trace the evolution of the invertebrata alone. Thus the orderly procession. of organic forms follows in precisely the true phylogenetic sequence; invertebrata first, then vertebrates, at first fish, then amphibia, next reptiles, soon after mammals, of the lowlier kinds first, of the higher later, and these in increasing complexity of structure till we finally arrive at man himself. While the living world was thus unfolding into new and nobler forms, the immutable *Lingula* simply perpetuated its kind. To select it, or other species equally sluggish, as the sole measure of the rate of biologic change would seem as strange a proceeding as to confound the swiftness of a river with the stagnation of the pools that lie beside its banks. It is occasionally objected that the story we have drawn from the palæontological record is mere myth or is founded only on negative evidence. Cavils of this kind prove a double misapprehension, partly as to the facts, partly as to the value of negative evidence, which may be as good in its way as any other kind of evidence.

Geologists are not unaware of the pitfalls which beset negative evidence, and they do not conclude from the absence of fossils in the rocks which underlie the Cambrian that pre-Cambrian periods were devoid of life; on the contrary, they are fully persuaded that the seas of those times were teeming with a rich variety of invertebrate forms. How is it that, with the exception of some few species found in beds immediately underlying the Cambrian, these have left behind no vestige of their existence? The explanation does not lie in the nature of the sediments, which are not unfitted for the preservation of fossils, nor in the composition of the then existing sea water, which may have contained quite as much calcium carbonate as occurs in our present oceans; and the only plausible supposition would appear to be that the organisms of that time had not passed beyond the stage now represented by the larvæ of existing invertebrata, and consequently were either unprovided with skeletons, or at all events with skeletons durable enough for preservation. If so, the history of the earlier stages of the evolution of the invertebrata will receive no light from palæontology; and no direct answer can be expected to the question whether, eighteen or nineteen millions of years being taken as sufficient for the evolution of the vertebrata, the remaining available eight millions would provide for that of the invertebrate classes which are represented in the lowest Cambrian deposits. On *a priori* grounds there would appear to be no reason why it should not. If two millions of years afforded time enough for the conversion of fish into amphibians, a similar period should suffice for the evolution of trilobites from annelids, or of annelids from trochospheres. The step from gastrulas to trochospheres might. be accomplished in another two millions, and two millions more would take us from gastrulas through morulas to protozoa.

As things stand, biologists can have nothing to say either for or against such a conclusion: they are not at present in a position to offer independent evidence; nor can they hope to be so until they have vastly extended those promising investigations which they are only now beginning to make into the rate of the variation of species.

Unexpected Absence of Thermal Metamorphosis in Ancient Rocks.

Two difficulties now remain for discussion: one based on theories of mountain chains, the other on the unaltered state of some ancient sediments. The latter may be taken first. Prof. van Hise writes as follows regarding the pre-Cambrian rocks of the Lake Superior district: "The Penokee series furnishes an instructive lesson as to the depth to which rocks may be buried and yet remain but slightly affected by metamorphosis. The series itself is 14,000 feet thick. It was covered before being upturned with a great thickness of Keweenaw rock., This series at the Montreal River is estimated to be 50,000 feet thick. Adding to this the known thickness of the Penokee series, we have a thickness of 64,000 feet. . . . The Penokee series were then buried to a great depth, the exact amount depending upon their horizon and upon the stage in Keweenaw time, when the tilting and erosion, which brought them to the surface, commenced.

" That the synclinal trough of Lake Superior began to form before the end of the Keweenaw period, and consequently that the Penokee rocks were not buried under the full succession, is more than probable. However, they must have been buried to a great depth—at least several miles—and thus subjected to high pressure and temperature, notwithstanding which they are comparatively unaltered " (*Tenth Annual Report U.S. Geol. Survey*, 1888-89, p. 457).

I select this example because it is one of the best instances of a difficulty that occurs more than once in considering the history of sedimentary rocks. On the supposition that the rate of increment of temperature with descent is 1° F. for every 84 feet, or 1° C. for every 150 feet, and that it was no greater during these early Penokee times, then at a depth of 50,000 feet the Penokee rocks would attain a temperature of nearly 333° C. ; and since water begins to exert powerful chemical action at 180° C. they should, on the theory of a solid cooling globe, have suffered a metamorphosis sufficient to obscure their resemblance to sedimentary rocks. Either then the accepted rate of downward increase of temperature is erroneous, or the Penokee rocks were never depressed, in the place where they are exposed to observation, to a depth of 50,000 feet. Let us consider each alternative, and in the first place let us apply the rate of temperature increment determined by Prof. Agassiz in this very Lake Superior district : it is 1° C. for every 402 feet, and twenty-five millions of years ago, or about the time when we may suppose the Penokee rocks were being formed, it would be 1° C. for every 305'5 feet, with a resulting temperature at a depth of 50,000 feet of 163° C. only. Thus the admission of a very low rate of temperature increment would meet the difficulty ; but on the other hand it would involve a period of several hundreds of millions of years for the age of the " consistentior status," and thus greatly exceed Prof. Joly's maximum estimate of the age of the oceans. We may therefore turn to the second alternative. As regards this it is by no means certain that the exposed portion of the Penokee series ever was depressed 50,000 feet ; the beds lie in a synclinal the base of which indeed may have sunk to this extent, and entered a region of metamorphosis ; but the only part of the system that lies exposed to view is the upturned margin of the synclinal, and as to this it would seem impossible to make any positive assertion as to the depth to which it may or may not have been depressed. To keep an open mind on the question seems our only course for the present, but difficulties like this offer a promising field for investigation.

The Formation of Mountain Ranges.

It is frequently alleged that mountain chains cannot be explained on the hypothesis of a solid earth cooling under the conditions and for the period we have supposed. This is a question well worthy of consideration, and we may first endeavour to picture to ourselves the conditions under which mountain chains arise. The floor of the ocean lies at an average depth of 2000 fathoms below the land, and is maintained at a constant temperature, closely approaching 0° C., by the passage over it of cold water creeping from the polar regions. The average temperature of the surface of the land is above zero, but we can afford to disregard the difference in temperature between it and the ocean floor, and may take them both at zero. Consider next the increase of temperature with descent, which occurs beneath the continents : at a depth of 13,000 feet, or at same depth as the ocean floor, a temperature of 87° C. will be reached on the supposition that the rate of increase is 1° C. for 150 feet, while with the usually accepted rate of 1° C. for 108 feet it would be 120° C. But at this depth the ocean floor, which is on the same spherical surface, is at 0° C. Thus surfaces of equal temperature within the earth's crust will not be spherical, but will rise or fall beneath an imaginary spherical or spheroidal surface according as they occur beneath the continents or the oceans. No doubt at some depth within the earth the departure of isothermal surfaces from a spheroidal form will disappear ; but considering the great breadth both of continents and oceans this depth must be considerable, possibly even forty or fifty miles. Thus the sub-continental excess of temperature may make itself felt in regions where the rocks still retain a high temperature, and are probably not far removed from the critical fusion point. The effect will be to render the continents mobile as regards the ocean floor ; or, *vice-versâ*, the ocean floor will

be stable compared with the continental masses. Next it may be observed that the continents pass into the bed of the ocean by a somewhat rapid flexure, and that it is over this area of flexure that the sediments denuded from the land are deposited. Under its load of sediment the sea-floor sinks down, subsiding slowly, at about the same rate as the thickness of sediment increases ; and, whether as a consequence or a cause, or both, the flexure marking the boundary of land and sea becomes more pronounced. A compensating movement occurs within the earth's crust, and solid material may flow from under the subsiding area in the direction of least resistance, possibly towards the land. At length, when some thirty or forty thousand feet of sediment have accumulated in a basin-like form, or, according to our reckoning, after the lapse of three or four millions of years, the downward movement ceases, and the mass of sediment is subjected to powerful lateral compression, which, bringing its borders into closer proximity by some ten or thirty miles, causes it to rise in great folds high into the air as a mountain chain.

It is this last phase in the history of mountain making which has given geologists more cause for painful thought than probably any other branch of their subject, not excluding even the age of the earth. It was at first imagined that during the flow of time the interior of the earth lost so much heat, and suffered so much contraction in consequence, that the exterior, in adapting itself to the shrunken body, was compelled to fit it like a wrinkled garment. This theory, indeed, enjoyed a happy existence till it fell into the hands of mathematicians, when it fared very badly, and now lies in a pitiable condition neglected of its friends.[1]

For it seemed proved to demonstration that the contraction consequent on cooling was wholly, even ridiculously, inadequate to explain the wrinkling. But when we summon up courage to inquire into the data on which the mathematical arguments are based, we find that they include several assumptions, the truth of which is by no means self-evident. Thus it has been assumed that the rate at which the fusion point rises with increased pressure is constant, and follows the same law as is deduced from experiments made under such pressures as we can command in our laboratories down to the very centre of the earth, where the pressures are of an altogether different order of magnitude ; so with a still more important coefficient, that of expansion, our knowledge of this quantity is founded on the behaviour of rocks heated under ordinary atmospheric pressure, and it is assumed that the same coefficient as is thus obtained may be safely applied to material which is kept solid, possibly near the critical point, under the tremendous pressure of the depths of the crust. To this last assumption we owe the terrible bogies that have been conjured out of "the level of no strain." The depth of this as calculated by the Rev. O. Fisher is so trifling that it would be passed through by all very deep mines. Mr. C. Davison, however, has shown that it will lie considerably deeper, if the known increase of the coefficient of expansion with rise of temperature be taken into account. It is possible, it is even likely, that the coefficient of expansion becomes vastly greater when regions are entered where the rocks are compelled into the solid state by pressure. So little do we actually know of the behaviour of rock under these conditions that the geologist would seem to be left very much to his own devices ; but it would seem there is one temptation he must resist—he may not take refuge in the hypothesis of a liquid interior.

We shall boldly assume that the contraction at some unknown depth in the interior of the earth is sufficient to afford the explanation we seek. The course of events may then proceed as follows. The contraction of the interior of the earth, consequent on its loss of heat, causes the crust to fall upon it in folds, which rise over the continents and sink under the oceans, and the flexure of the area of sedimentation is partly a consequence of this folding, partly of overloading. By the time a depression of some 30,000 or 40,000 feet has occurred along the ocean border the relation between continents and oceans has become unstable, and readjustment takes place, probably by a giving way of the continents, and chiefly along the zone of greatest weakness, *i.e.* the area of sedimentation, which thus becomes the zone of mountain building. It may be observed that at great depths readjustment will be produced by a slow flowing of solid rock, and it is only comparatively near the surface, five or ten miles at the most below, that failure of

[1] With some exceptions, notably Mr. C. Davison, a consistent supporter of the theory of contraction.

support can lead to sudden fracture and collapse ; hence the comparatively superficial origin of earthquakes.

Given a sufficiently large coefficient of expansion—and there is much to suggest its existence (*vide*, p. 483)—and all the phenomena of mountain ranges become explicable : they begin to present an appearance that invites mathematical treatment ; they inspire us with the hope that from a knowledge of the height and dimensions of a continent and its relations to the bordering ocean we may be able to predict when and where a mountain chain should arise, and the theory which explains them promises to guide us to an interpretation of those world-wide unconformities which Suess can only account for by a transgression of the sea. Finally it relieves us of the difficulty presented by mountain formation in regard to the estimated duration of geological time.

Influence of Variations in the Eccentricity of the Earth's Orbit.

This may perhaps be the place to notice a highly interesting speculation which we owe to Prof. Blytt, who has attempted to establish a connection between periods of readjustment of the earth's crust and variations in the eccentricity of the earth's orbit. Without entering into any discussion of Prof. Blytt's methods, we may offer a comparison of his results with those that follow from our rough estimate of one foot of sediment accumulated in a century.

Table showing the Time that has elapsed since the Beginning of the Systems in the first column, as reckoned from Thickness of Sediment in the second column, and by Prof. Blytt in the third :—

		Years		Years
Eocene	...	4,200,000	...	3,250,000
Oligocene	...	3,000,000	...	1,810,000
Miocene	...	1,800,000	...	1,160,000
Pliocene	...	900,000	...	700,000
Pleistocene	...	400,000	...	350,000

It is now time to return to the task, too long postponed, of discussing the data from which we have been led to conclude that a probable rate at which sediments have accumulated in places where they attain their maximum thickness is one foot per century.

Rate of Deposition of Sediment.

We owe to Sir Archibald Geikie a most instructive method of estimating the existing rate at which our continents and islands are being washed into the sea by the action of rain and rivers ; by this we find that the present land surface is being reduced in height to the extent on an average of 1/2400 foot yearly (according to Prof. Penck 1/3600 foot). If the material removed from the land were uniformly distributed over an area equal to that from which it had been derived, it would form a layer of rock 1/2400 foot thick yearly ; *i.e.* the rates of denudation and deposition would be identical. But the two areas, that of denudation and that of deposition, are seldom or never equal, the latter as a rule being much the smaller. Thus the area of that part of North America which drains into the Gulf of Mexico measures 1,800,000 square miles ; the area over which its sediments are deposited is, so far as I can gather from Prof. Agassiz' statements, less than 180,000 square miles ; while Mr. McGee estimates it at only 100,000 square miles. Using the largest number, the area of deposition is found to measure one-tenth the area of denudation ; the average rate of deposition will therefore be ten times as great as the rate of denudation, or 1/240 foot may be supposed to be uniformly distributed over the area of sedimentation in the course of a year. But the thickness by which we have measured the strata of our geological systems is not an average but a maximum thickness ; we have therefore to obtain an estimate of the maximum rate of deposition. If we assume the deposited sediments to be arranged somewhat after the fashion of a wedge with the thin end seawards, then twice the average would give us the maximum rate of deposition ; this would be one foot in 120 years. But the sheets of deposited sediment are not merely thicker towards the land, thinner towards the sea, they also increase in thickness towards the rivers in which they have their source, so that a very obtuse-angled cone, or, better, the down-turned bowl of a spoon, would more nearly represent their form. This form tends to disappear under the action of waves and currents, but a limit is set to this disturbing influence by the subsidence which marks the region opposite the mouth of a large river. By

this the strata are gradually let downwards, so that they come to assume the form of the bowl of a spoon turned upwards. Thus a further correction is necessary if we are to arrive at a fair estimate of the maximum rate of deposition. Considering the very rapid rate at which our ancient systems diminish in thickness when traced in all directions from the localities where they attain their maximum, it would appear that this correction must be a large one. If we reduce our already corrected estimate by one-fifth, we arrive at a rate of one foot of sediment deposited in a century.

No doubt this value is often exceeded ; thus in the case of the Mississippi River the bar of the south-west pass advanced between the years 1838 and 1874 a distance of over 2 miles, covering an area 2·2 miles in width with a deposit of sediment 80 feet in thickness ; outside the bar, where the sea is 250 feet in depth, sediment accumulates, according to Messrs. Humphreys and Abbot, at a rate of 2 feet yearly. It is quite possible, indeed it is very likely, that some of our ancient strata have been formed with corresponding rapidity. No gravel or coarse sand is deposited over the Mississippi delta ; such material is not carried further seawards than New Orleans. Thus the vast sheets of conglomerate and sandstone which contribute so largely to some of our ancient systems, such as the Cambrian, Old Red Sandstone, Millstone Grit, and Coal Measures, must have accumulated under very different conditions, conditions for which it is not easy to find a parallel ; but in any case these deposits afford evidence of very rapid accumulation.

These considerations will not tempt us, however, to modify our estimate of one foot in a century ; for though in some cases this rate may have been exceeded, in others it may not have been nearly attained.

Closely connected with the rate of deposition is that of the changing level of land and sea ; in some cases, as in the Wealden delta, subsidence and deposition appear to have proceeded with equal steps, so that we might regard them as transposable terms. It would therefore prove of great assistance if we could determine the average rate at which movements of the ground are proceeding ; it might naturally be expected that the accurate records kept by tidal gauges in various parts of the world would afford us some information on this subject ; and no doubt they would, were it not for the singular misbehaviour of the sea, which does not maintain a constant level, its fluctuations being due, according to Prof. Darwin, to the irregular melting of ice in the polar regions. Of more immediate application are the results of Herr L. Holmström's observations in Scandinavia, which prove an average rise of the peninsula at the rate of 3 feet in a century to be still in progress ; and Mr. G. K. Gilbert's measurements in the great Lake district of North America, which indicate a tilting of the continent at the rate of 3 inches per hundred miles per century. But while measurements like these may furnish us with some notion of the sort of speed of these changes, they are not sufficient even to suggest an average ; for this we must be content to wait till sufficient tidal observations have accumulated, and the disturbing effect of the inconstancy of the sea-level is eliminated.

It may be objected that in framing our estimate we have taken into account mechanical sediments only, and ignored others of equal importance, such as limestone and coal. With regard to limestone, its thickness in regions where systems attain their maximum may be taken as negligible ; nor is the formation of limestone necessarily a slow process. The successful experiments of Dr. Allan, cited by Darwin, prove that reef-building corals may grow at the astonishing rate of six feet in height per annum.

In respect of coal there is much to suggest that its growth was rapid. The carboniferous period well deserves its name, for never before, never since, have Carbonaceous deposits accumulated to such a remarkable thickness or over such wide areas of the earth's surface. The explanation is doubtless partly to be found in favourable climatal conditions, but also, I think, in the youthful energy of a new and overmastering type of vegetation, which then for the first time acquired the dominion of the land. If we turn to our modern peat-bogs, the only Carbonaceous growths available for comparison, we find from data given by Sir A. Geikie that a fairly average rate of increase is six feet in a century, which might perhaps correspond to one foot of coal in the same period.

The rate of deposition has been taken as uniform through the whole period of time recorded by stratified rocks ; but lest it should be supposed that this involves a tacit admission of uniformity, I hasten to explain that in this matter we have no choice; we may feel convinced that the rate has varied from time to time, but in what direction, or to what extent, it is impossible to conjecture. That the sun was once much hotter is probable, but equally so that at an earlier period it was much colder; and even if in its youth all the activities of our planet were enhanced, this fact might not affect the maximum thickness of deposits. An increase in the radiation of the sun, while it would stimulate all the powers of subaërial denudation, would also produce stronger winds and marine currents; stronger currents would also result from the greater magnitude and frequency of the tides, and thus while larger quantities of sediment might be delivered into the sea they would be distributed over wider areas, and the difference between the maximum and average thickness of deposits would consequently be diminished. Indications of such a wider distribution may perhaps be recognised in the Palæozoic systems. Thus we are compelled to treat our rate of deposition as uniform, notwithstanding the serious error this may involve.

The reasonableness of our estimate will perhaps best appear from a few applications. Fig. 2 is a chart, based on a map by De Lapparent, representing the distribution of land and sea over the European area during the Cambrian period. The strata of this system attain their maximum thickness of 12,000 feet in Merionethshire, Wales; they rapidly thin out northwards, and are absent in Anglesey; scarcely less rapidly towards Shropshire, where they are 3000 feet thick; still a little less rapidly towards the Malverns, where they are only 800 feet thick; and most slowly towards St. David's Head, where they are 7400 feet thick. The Cambrian rocks of Wales were in all probability the deposits of a river system which drained some vanished land once situated to the west. How great was the extent of this land none can say; some geologists imagine it to have obliterated the whole or greater part of the North Atlantic Ocean. For my part I am content with a somewhat large island. What area of this island, we may ask, would suffice to supply the Cambrian sediments of Wales and Shropshire? Admitting that the area of denudation was ten times as large as the area of deposition, its dimensions are indicated by the figure *a b c d* on the chart. This evidently leaves room enough on the island to furnish all the other deposits which are distributed along the western shores of the Cambrian Sea, while those on the east are amply provided for by that portion of the European continent which then stood above water.

If one foot in a century be a quantity so small as to disappoint the imagination of its accustomed exercise, let us turn to the Cambrian succession of Scandinavia, where all the zones recognised in the British series are represented by a column of sediment 290 feet in thickness. If 1,600,000 years be a correct estimate of the duration of Cambrian time, then each foot of the Scandinavian strata must have occupied 5513 years in its formation. Are these figures sufficiently inconceivable?

In the succeeding system, that of the Ordovician, the maximum thickness is 17,000 feet. Its deposits are distributed over a wider area than the Cambrian, but they also occupied longer time in their formation; hence the area from which they were derived need not necessarily have been larger than that of the preceding period.

Great changes in the geography of our area ushered in the Silurian system: its maximum thickness is found over the Lake district, and amounts to 15,000 feet; but in the little island of Gothland, where all the subdivisions of the system, from the Landovery to the Upper Ludlow, occur in complete sequence, the thickness is only 208 feet. In Gothland, therefore, accord-

ing to our computation, the rate of accumulation was one foot in 7211 years.

With this example we must conclude, merely adding that the same story is told by other systems and other countries, and that, so far as my investigations have extended, I can find no evidence which would suggest an extension of the estimate I have proposed. It is but an estimate, and those who have made acquaintance with "estimates" in the practical affairs of life know how far this kind of computation may guide us to or from the truth.

This Address is already unduly long, and yet not long enough to glance backwards over the past we see catastrophism yield to uniformitarianism, and this to evolution, but each as it disappears leaves behind some precious residue of truth. For the future of our science our ambition is that which inspired the closing words of your last President's Address, that it may become more experimental and exact. Our present watchword is Evolution. May our next be Measurement and Experiment, Experiment and Measurement.

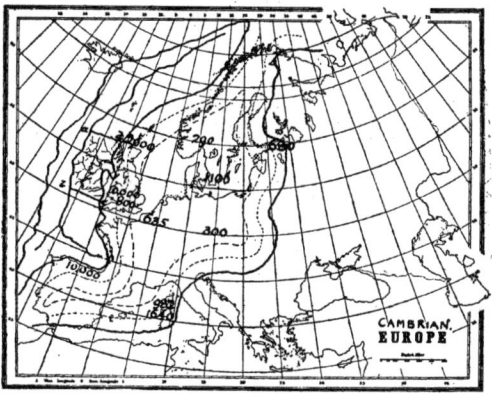

Fig. 2.—Chart of the distribution of land and sea, and of the thickness of deposits of the Cambrian system. The dotted lines indicate distances of 100 and 200 miles from the shore.

NOTES.

ANOTHER of those disastrous hurricanes which occasionally visit the West Indies and United States at this season of the year has to be recorded. On the 8th inst. a storm of great violence struck the coasts of Louisiana and Texas, and, owing to the thickly populated districts over which it swept and to the high water wave which accompanied it, immense destruction to property and lamentable loss of life ensued. The fury of the storm is said to have been felt for at least a hundred miles inland, but up to the present time scarcely any details have arrived as to its character and the exact path that it followed. This part of America is one of the three regions referred to in the works of Prof. W. M. Davis from which tropical storms move into temperate latitudes in the northern hemisphere; but we must wait for further details before it can be stated whether the one in question was of the nature of a tornado, which differs from an ordinary hurricane chiefly in its excessive violence over a small, instead of a large, area. From the description so far, given, and from its duration, the storm would appear to have been of the nature of the worst West India hurricanes.

THE *Stella Polare*, with the Duke of the Abruzzi and members of his Arctic expedition on board, arrived at Christiania on Tuesday. A Reuter telegram from Tromsö states that the ship was pushed by the ice in Table Bay right on to the land, and four separate parties were sent out from it. The first was despatched northwards to establish depôts, and had to travel two days overland. The second party, consisting of a Norwegian and two Italians, was to have been out twelve days, but the men never returned. The third party were away twenty-four days, and the fourth 105 days. This last reached lat. 86° 33′ N., thus penetrating a little further north than Dr. Nansen and Lieut. Johansen, who reached lat. 86° 14′ N. The scientific results of the expedition are said to be satisfactory, but no information concerning them, or of the latitude observations, is yet available.

THE recent Congress of British Chambers of Commerce at Paris adopted resolutions urging the adoption of the metric system of weights and measures in our Government departments, and the teaching of decimals in public elementary schools at an early stage, as an essential part of arithmetic.

THE *Entomologist* for September contains an interesting account of the artificial ant-hills in the Paris Exhibition. These are shown by M. C. Janet, who has devoted many years to studying the habits of the social Hymenoptera. They are made of porous pink plaster covered with glass, through which visitors may observe the movements of their busy inhabitants, and are constructed after the plan of a natural hill in a garden near Beauvois. Several species of ants are exhibited, one of which has "slaves." M. Janet is of opinion that ants have a language of sounds, and that at any rate they produce grating noises, probably by rubbing together their bodies ; while he is fully assured that they possess an acute sense of hearing.

To the current issue of the *Entomologist's Monthly Magazine*, Sir George Hampson communicates a notice of certain malformed specimens of moths recently acquired by the British Museum. The object of the communication is to draw attention, not only to peculiarities of the insects themselves, but likewise to the fact that the authorities of the Museum have recently started a collection to display the abnormalities and "sports" which occur among insects, and, it may be added, among all other groups of animals as well. Sir George Hampson appeals to all entomologists to assist in the formation of this series.

THE chapters on "Nature and Science for Young Folks" in the August and September numbers of *St. Nicholas* maintain, under the able editorship of Mr. E. F. Bigelow, their high standard of excellence and interest, many of the illustrations being unusually attractive and instructive. In the August number we have first a delightful dissertation on "Flowers of the Sea," with an exquisite photograph of a dried sea-weed ; this being followed by an account of the manner in which gnat-larvæ maintain their breathing apparatus at the surface of the water during the process of respiration. In the September issue the attention of the young reader is at first attracted to the intrinsic beauty of leaves—such as can be met with in any copse or wood —by an account of how to make leaf-garlands and chains, from which the transition is easy to the numerous points of interest displayed by microscopic sections of these beautiful structures, and to the part they play in the economy of nature. A prize competition in which figures of birds and insects are given for identification by the reader strikes us as an especially good feature, and in every way superior to the useless and hackneyed "missing word competition."

THE meteorological reporter to the Government of India has issued his usual "Brief memorandum on the weather in India during the months of June and July, and forecast of the general distribution of the rainfall in India during the months of August and September, 1900." The character of the monsoon rains in June and July has been in fair accordance with the forecast issued in the beginning of June. From a consideration of the weather and of the snowfall in the mountains during those months, it is assumed that the general distribution of rainfall for the months of August and September, will be favourable. A comparison with several previous years also suggests that it is probable that the rainfall of the retreating south-west monsoon (October to December) will be favourably distributed in the Peninsula.

THE French Meteorological Office has recently issued its *Annales* for the year 1897 : the first volume, under the title of "Memoirs," contains, in addition to the usual articles relating to thunderstorms and magnetic observations, a discussion by M. Angot of the temperature at a number of stations since 1851. This valuable paper forms the first part of an investigation of the climate of France during forty years ending with 1890. A second memoir by M. Angot contains the principal results of the simultaneous observations made at the Meteorological Office in Paris and on the Eiffel Tower during six years. M. Teisserenc de Bort, who established and maintains at Trappes an observatory specially devoted to the study of clouds and the movements of the upper air, publishes the results hitherto obtained from numerous ascents of unmanned balloons. The second volume contains the results of observations at stations in France and its Colonies. The publication of the valuable series of rainfall observations, which has hitherto formed the third volume, is postponed for the present, from motives of economy. The Central Office in Paris receives 167 telegraphic reports daily, and in the year in question 89 per cent. of the weather forecasts were successful. The observers receive encouragement in the way of medals, sixty-eight of these were awarded for observations on land, and eighteen for observations at sea. Much attention is now given to the development of observations in the French Colonies, and in distant parts generally, a matter which, until a few years ago, had been somewhat overlooked.

MANY attempts have been made to introduce Euclid's Fifth Book, or its equivalent, in the teaching of geometry, and recently a work on the Fifth and Sixth Books appeared from the pen of Prof. M. J. M. Hill, F.R.S. A later contribution to this subject is a paper, by Prof. George A. Gibson, on "Proportion," a substitute for the Fifth Book of Euclid's Elements, published as an appendix to the *Proceedings* of the Edinburgh Mathematical Society. A deductive treatment of proportion such as this, which takes account of incommensurability, is doubtless a good mental training for the advanced student ; but seeing that in an elementary course on arithmetic and algebra the rules for multiplying and dividing negative and fractional quantities are practically never logically proved by students, but rather assumed and applied to working examples, there is surely sufficient precedent for taking it as granted that the laws of proportion apply to incommensurable magnitudes. At any rate the matter is one for discussion between the two schools, one of which seeks to put the teaching of mathematics on a rigorously deductive basis, while the other would abolish Euclid's lengthy deductions and rather teach the results of mathematical reasoning and how to work with them in practical applications.

WE have received a copy of a rectorial address, delivered by Prof. Brückner before the University of Berne, entitled "Die Schweizerische Landschaft einst und jetzt." The topographical changes which have taken place in Switzerland since the glacial period and within historic times are summarised, and apart from the intrinsic interest of the subject, the address is an admirable illustration of the methods of descriptive geography.

DR. A. LORENZEN contributes two articles to *Die Natur* on the Danish Expedition to East Greenland in 1898–1899. The expedition was successful in closely following the plan of work with which it set out. Its chief results consist of mapping the coast of Greenland from lat. 65° ¾′ N. to 67° 22′ N., and sketching it as far as 68° N. ; making botanical, zoological and geological collections ; ethnographic observations ; observations of the ice north of the Angmagsalik district ; meteorological and other observations in winter quarters.

OUR German contemporary, *Naturwissenschaftliche Wochenschrift* for August 19 and September 2 contains a full account of the discovery of the remains of *Grypotherium listai* in Ultima Esperanza cavern, Patagonia.

Perhaps the most interesting article in the September number of the *Irish Naturalist* is an account, by Mr. R. Warren, of a visit to Loch Erne in search of the Sandwich tern, which has hitherto been known to breed in Ireland only in a single locality in county Mayo.

THE report of the expeditions organised by the British Astronomical Association to observe the total solar eclipse of May 28, 1900, will be contained in a volume shortly to be issued from the office of *Knowledge*. The work will be edited by Mr. E. Walter Maunder, and will contain many photographs of the various stages of the eclipse.

AMONG the scientific instrument makers who exhibited in the British Section at the Paris Exhibition, the Grand Prix was awarded to four firms, namely :—Class 15 (*Instruments de Précision*), the Cambridge Scientific Instrument Co., Ltd., Cambridge, and Messrs. Ross, Ltd., London. Class 16 (*Médecine et Chirurgie*), Messrs. Down Bros., London. . Class 27 (*Applications diverses de l'Électricité*), Mr. James White, Glasgow. Mr. W. Duddell received a gold medal (Class 27) for the oscillograph exhibited by the Cambridge Scientific Instrument Company, and Mr. Wayne, the inventor of both the Wayne and Simplex Steam-engine Indicators, and now engaged at the Cambridge works, received a silver medal. A silver medal was also awarded in Class 16 to the company itself. Two gold and two silver medals were awarded to Mr. J. J. Hicks, and one to Messrs. Crompton and Co., Messrs. Negretti and Zambra, and Messrs. Watson and Sons ; and silver medals were also awarded to Mr. A. Higgins, Mr. E. Wellings, Mr. W. Sims and Mr. W. Barton of Mr. Hicks' firm.

THE additions to the Zoological Society's Gardens during the past week include a Lion (*Felis leo*) from Uganda, presented by Captain Delme Radcliff ; a Macaque Monkey (*Macacus cynomolgus*) from India, presented by Miss K. Bishop ; a Ring-tailed Coati (*Nasua rufa*) from South America, presented by Mr. G. Percy Ashmore ; two Cunning Bassaris (*Bassaris astuta*) from Mexico, presented by Miss Franklin ; two Chilian Sea Eagles (*Geranoaëtus melanoleucus*) from South America, presented by Mr. Tom Simonds ; a Puma (*Felis concolor*) from the Argentine Republic, presented by Mr. Maurice F. Dennis ; a Nilotic Crocodile (*Crocodilus niloticus*) from Omdurman, presented by Major H. B. Weatherall ; two Tenrecs (*Centetes ecaudatus*) from Madagascar, a Cunning Bassaris (*Bassaris astuta*) from Mexico, three Cardinal Eclectus (*Eclectus cardinalis*) from Moluccas, deposited ; two Purple Herons (*Ardea purpurea*), two Common Cormorants (*Phalaerocorax carbo*), three Common Spoonbills (*Platalea leucorodia*), European, purchased.

OUR ASTRONOMICAL COLUMN

EPHEMERIS FOR OBSERVATIONS OF EROS.—The following is continued from the new data given by E. Millosevich in the *Astronomische Nachrichten* (Band 153, No. 3660) :—

Ephemeris for 12h. Berlin Mean Time.

1900.		R.A.		Decl.
		h.　m.　s.		° ′ ″
Sept. 13	...	2 34 19·43	...	+38 7 24·2
14	...	35 11·87	...	38 30 7·3
15	...	36 2·36	...	38 52 52·9
16	...	36 50·83	...	39 15 40·7
17	...	37 37·20	...	39 38 30·3
18	...	38 21·38	...	40 1 21·4
19	...	39 3·30	...	40 24 13·9
20	...	2 39 42·89	...	+40 47 7·2

The following elements for two epochs some two years apart are also given in the same periodical :—

I. Epoch 1898 August 2·5 Berlin.

$$M = 205\ 21\ 41\cdot83$$
$$\pi = 121\ 10\ 51\cdot40$$
$$\Omega = 303\ 31\ 56\cdot17 \Big\} 1900\cdot0$$
$$i = 10\ 49\ 35\cdot35$$
$$\phi = 12\ 52\ 14\cdot44$$
$$\mu = 2015''\cdot26908$$
$$\log a = 0\cdot1637824$$

II. Epoch 1900 October 31·5 Berlin.

$$M = 304\ 24\ 40\cdot34$$
$$\pi = 121\ 9\ 47\cdot82$$
$$\Omega = 303\ 30\ 50\cdot02 \Big\} 1900\cdot0$$
$$i = 10\ 49\ 38\cdot97$$
$$\phi = 12\ 52\ 40\cdot61$$
$$\mu = 2015''\cdot23324$$
$$\log a = 0\cdot1637875$$

THE DAYLIGHT METEOR OF SUNDAY, SEPTEMBER 2.

JUST before sunset on September 2 a magnificent meteor was observed in the north of England and Scotland. A large number of descriptions of the object have appeared in the newspapers, and it appears that notwithstanding broad daylight the spectacle was a very brilliant one.

At St. Anne's, Lancashire, the meteor fell in a northerly direction, and left a column of white smoke, which remained visible ten minutes. At Hunt's Cross the time was noted as 6h. 52m., and the object is said to have fallen near Halewood, leaving a long trail of white dust for several minutes. As seen from Birkenhead the meteor appeared at 6h. 54m. in the N.E., and looked like a descending rocket. Its path was nearly vertical, and it left a "dust trail" for nearly six minutes. At Wetherby, Yorks, the smoke-like cloud left by the nucleus remained visible until 7h. 30m. At Overton, Ellesmere, the object is said to have apparently fallen on a field on the left bank of the Dee, about a mile from Bangor Isycoed. At Ulverstone it passed over Morecambe Bay, in a southerly direction towards Blackpool. At Penton, Cumberland, the time was noted as 6h. 54m., and the direction was due south. It remained visible two seconds, and was falling towards the earth.

At Keswick, Mr. Lawson Dykes saw the fireball at 6h. 55m., and says it fell through an arc of about ten degrees, the altitude of appearance being 35° and disappearance 25°. It was pear-shaped and of immense size, with a distinct tail. The line of flight was almost due N. to S. At Warkton, Northamptonshire, Dr. Herbert Spencer noted the time as 6h. 55m., and says the track of the meteorite was afterwards marked by a narrow white streak, which persisted for more than five minutes.

At and near Edinburgh the fireball was witnessed by many persons. One observer says that at 6h. 55m. there was a sudden flash, and what appeared to be a streak of molten silver followed by a train of sparks whizzed past, apparently falling into a large field of turnips on his right hand. Its direction was due S.E. At Inveresk the meteor appeared to be in the direction of Dalkeith. It resembled a large ball of fire with a tail, and seemed to fall to the earth. At Earlswood, nine miles S.S.E. of Birmingham, the time was noted as 6h. 55m., and the end point of the flight occurred in altitude 20° N. and was directed from N.N.E. At Blackwall, Alfreton, an observer noted the time as 6h. 53m., and says the meteor left a trace in the sky of a sinuous form and in colour a silver-grey. The trace remained distinctly visible in the sky for thirty minutes. Its direction was N.W. At West Kirby, Birkenhead, the meteor was seen to fall into a wood on the east side of the hill there, and apparently so close that the observer thought it would possibly set fire to the trees.

In Wiltshire a party of Bathonians saw it while driving, and describe it as a ball of fire with a comet-like tail falling direct from heaven to earth and alighting apparently in a field about a mile distant.

A number of other accounts have come to hand, but for the most part they do not supply any details which would be useful for computing the real path of the fireball. It probably disappeared over Lancashire at a height of twenty-five miles, and was directed from a radiant point high in the northern sky. The long-enduring streak or cosmic cloud was no doubt illuminated by the sun's rays. It is not likely that the exact heights of appearance and disappearance of the meteor can be ascertained, though there are plenty of descriptions. The object having appeared in daylight, there were no stars or other celestial objects visible by which its path could be determined. The observers content themselves with giving rough estimations as to the general direction, but these are of little use in any endeavour to compute the real path of the object. It is hoped, however, that some further observations of a more satisfactory character will come to hand and enable a fairly trustworthy result to be obtained. The fireball was a very exceptional one to have created so brilliant an effect just before sunset.

It will be remembered that a large fireball was seen on January 9 at 2h. 55m. in the afternoon, and that on July 17 last another of these striking objects appeared soon after sunset and was observed by many persons in Scotland and the north of England. The present year is likely to be a notable one as regards the number and brilliancy of the fireballs which have appeared. W. F. DENNING.

UNIVERSITY AND EDUCATIONAL INTELLIGENCE.

THE new Technical School at Bootle will be opened by the Earl of Derby on Thursday, September 27.

AFTER a discussion upon technical and commercial education at the recent Congress of British Chambers of Commerce, in Paris, the meeting endorsed the following resolution, which was adopted at the Congress of Chambers of Commerce of the Empire in June 1900:—"That it is most desirable to take steps to urge the extension of technical and commercial education throughout the Empire, and that wherever possible this education should be placed under efficient public control; and that this congress is of opinion that the utmost effort should be made throughout the Empire to encourage and furnish facilities for commercial education as a branch of technical and scientific study, and that the Home and Colonial Governments be moved to give aid thereto and ample powers of contribution out of local resources; and, further, it is very desirable that Chambers of Commerce should be represented on Boards of Education in order to advance the interests of commercial education."

THE following is a list of candidates successful in the recent competition for the Whitworth Scholarships and Exhibitions. Scholarships of 125*l.* a year each (tenable for three years) : George W. O. Howe, Harold B. Philpot, Harry Noble, William M. Wallace. Exhibitions of 50*l.* (tenable for one year) : Alfred W. Steed, Charles E. Stanier, Benjamin Moss, Isaac V. Robinson, Herbert G. Tisdall, Leonard Southerns, Charles A. King, John McCulloch, William P. Chandler, Charles W. Price, Harry B. Matthews, Leonard G. Crawford, James Wilson, George Stow, Joseph H. Dobson, Alec P. Simpson, Fredk. G. Rappoport, Arthur J. Butler, Alfred L. Oke, James M. Macintosh, William H. Cumner, Thomas A. Goskar, Leopold D. Coueslant, John C. Gardner, James C. Metcalfe, Harold Shatwell, Walter A. Turnbull, John E. Grant, Albert S. Raworth, Henry H. Thorne.

THE following Royal Exhibitions, National Scholarships and Free Studentships in Science have been awarded by the Board of Education, South Kensington. Royal Exhibitions : James C. Macfarlane, William T. S. Butlin, Louis D. Stansfeld, Leonard A. V. Webb, Isaac V. Robinson, Arthur Baker, Benjamin Moss. National Scholarships for Mechanics : Albert E. Dodridge, Albert Wilson, Charles E. Stanier, Frederick Bowen, Robert R. Cormack. Free Studentships for Mechanics : Fred. G. Rappoport, Harry B. Matthews, John Alexander. National Scholarships for Physics : Ernest Nightingale, Royden C. Wale, Frederick P. Rolfe, William Tannock, Frank E. Glover. Free Studentship for Physics : Leonard R. Broome.

National Scholarships for Chemistry : George H. Green, Philip S. Pomeroy, William H. Stephens, Harold Leadbetter, Frederick P. Leach. Free Studentship for Chemistry : Hamilton McCombie. National Scholarships for Biology : Charles Martin, Archibald D. Hogg, Cosby T. Nesbitt, Hamilton E. Quick, Horace A. Wager. National Scholarships for Geology : Hubert C. Jones, William Rawson. Free Studentship for Geology : Stanley R. Jones.

SOCIETIES AND ACADEMIES.

PARIS.

Academy of Sciences, September 3.—M. Maurice Lévy in the chair.—Physiological action and therapeutical applications of compressed oxygen, by M. A. Mosso. The author has verified and extended the observations of Haldane upon the simultaneous action of compressed oxygen and carbonic oxide upon various animals. Where at the ordinary pressure of the atmosphere 0·5 per cent. or less of the carbonic oxide is fatal, animals are not poisoned in an atmosphere of oxygen at two atmospheres containing 6 per cent. of the gas. This result is of interest from the physiological point of view as showing that animals may live, without red corpuscles, on the oxygen dissolved in the blood plasma, provided that the amount in solution is sufficiently increased by pressure.—The last sign of life, by M. Augustus D. Waller. Living matter responds to an electrical stimulus by a current in the same direction. The same substance, killed by heat, either gives no response or gives a polarisation current in the opposite sense. This method is applied to determine the last sign of life.—On the Laplace equations with quadratic solutions, by M. Tzitzeica.—On singularities of analytical functions, and in particular of functions defined by differential equations, by M. Paul Painlevé.—The effects of work of certain muscular groups on other groups doing no work, by MM. Kronecker and Cutter. The muscles of the lower limbs exercised in climbing were found to exert an influence upon the biceps of the arm. A moderate amount of work done by one group of muscles appears to have a strengthening effect upon another group not taking part in the action, the effect being probably due to an increase in the circulation of the blood and lymph.—On a perpetual calendar, by M. l'abbé Salvatore Franco.

CONTENTS.

THURSDAY, SEPTEMBER 20, 1900.

A MAGNETIC THEORY OF THE UNIVERSE.

From Matter to Man; a New Theory of the Universe.
By A. Redcote Dewar. Pp. 289+viii. (London:
Chapman and Hall, Ltd., 1898.)

WE do not recommend this work to the serious attention of our readers, but as a study in word-stringing it is not devoid of interest. It appears to be the production of a writer who has acquired a knowledge of scientific terms by extensive reading, without having any real grasp of the vast range of subjects over which he travels. The result is such as might have been arrived at by a student who had been through a hurried course of cram, and who at his final examination had been set some such question as this :—Given, a vocabulary of scientific terms, construct a theory of the universe.

The author is strong on magnetism. We have never met with such a liberal use of this term as is indulged in by Mr. Dewar. Like the "vapours" which afflicted our ancestors in the last century, and which accounted for all their ailments, the word magnetism accounts for nearly everything in Mr. Dewar's universe. It is in the fifth chapter that the stupendous importance of this form of energy is first sprung upon the reader :—

"The inference from this basis [that every atom is a magnet] is astounding, for not only does it involve the magnetism of the earth as well as the magnetism of all the constituents of the earth, but it establishes beyond a doubt the *magnetism of all the products of the earth—mineral, vegetal and animal.* Every crystal, plant, animal and man is thus a magnet, whose every energy—muscular, nervous, vital, or mental—resolves itself into the familiar operations of magnetism."

"Still further, as all the planets and heavenly bodies are alike in nature, so far as we can judge from analogy, so must they be governed by similar energy to the earth. Hence we reach the final conclusion, that *the bottom energy of the universe is also magnetism*" (pp. 72-73).

Having once become impregnated with this universal magnetic cult, the reader, who may otherwise have been unprepared for the series of mental shocks which is in store for him, will learn with comparative calmness that the difference of gravity at the equator and the poles is because the earth is a magnet (p. 69); he will feel sure that sooner or later "animal magnetism" mesmerism, &c., are bound to appear on the scene, and in this he will not be disappointed (p. 87); and he may even learn with equanimity that he possesses "a virtual magnetic battery" in his "vacuole or stomach" (p. 85). In the chapter on the causes of vegetal evolution we read :—

"These huge internal fires [of the earth] are virtually *the earth's magnetic battery*, through which it is kept in life as a living planet tenanted with vegetal and animal life; hence, when all conditions are suitable, and the soil is properly saturated with water, thereby inducing suitable chemical action, the resultant magnetic forces throw up a clotted vegetation on every available spot of ground on the globe. *This vegetation is but a bristling beard of earthly material ejected by the earth's magnetism;* filaments of matter, having the same relation to the earth as a man's beard to his chin, or as the bristling iron filings on a horse-shoe magnet" (p. 162).

It must be admitted that the force of magnetism can no further go, and the attributing to this agency of the vegetable-like accumulation of snow-drift on a projecting obstacle (p. 124), or the turning of a sun-flower towards the sun (p. 137), may be accepted as a mere bagatelle. Truly, as the author says, after having evolved "vegetal molecules" by a "chance flux of suitable atoms," the "magnetic laws are equal to all emergencies" (p. 159). The subtle distinction between a horse-shoe magnet as a *dry magnet*, and a plant as a *wet magnet* (p. 161) is too fine for any but ultra-metaphysical minds to grasp, but since it leads to the practical conclusion that a flower-pot full of moist earth is an artificial plant battery, it may be allowed to pass by virtue of its horticultural merits.

We confess that, in turning over the pages of this astonishing production with the object of endeavouring to learn the author's views, we have been so fascinated by his glib manipulation of the affairs of the universe, that we have allowed ourselves but little time for a critical examination of his scientific data. In fact, the work may be said to consist mainly of generalities, so that there are but few detailed statements to grasp. Here are a few specimens :—

The law of combination in definite proportions is illustrated (p. 58) by the statement that "if 20 parts of oxygen be mixed with 6 of hydrogen, only 16 parts of oxygen and 2 of hydrogen unite, 8 parts remain uncombined." This error is driven home in the following page :—

"Innumerable elements, for instance, may often meet in suitable conditions for combination, but if unsuitable in *proportion*, no selection ensues ; consequently there is no production," &c.

Hydrogen is said to be a constituent of saltpetre (p. 79).

"If a lot of chips be thrown into the water they all attract each other and form a mass" (p. 87). This is explained by the statement that sticks and stones require "stick and stone magnets to magnetise them."

"Contrasted with chemical combination, chemical decomposition has been almost ignored by chemists" (p. 101).

"Magnetic induction" is used to explain the crystallisation of a solid from a solution (p. 118).

After describing the movements of a "geometer" caterpillar, the author says : "Other caterpillars and centipedes develop feet on each ring" (p. 198).

"The fire-flies of America, which, Diogenes-like, carry a lamp with them," are classed with the bombardier beetle, both the light of the former and the discharge of the latter being described as "undoubtedly the result of electric action" (p. 201).

It need hardly be said that the propounder of a new philosophy—such as Mr. Dewar claims to be—must clear away the rubbish of previous thinkers before he can lay the foundations of his own system. Many of the current doctrines are accordingly denounced in no measured terms, and the author's emendations launched at the reader. For example, the classification of matter into simple and compound substances by chemists is scoffed at (p. 55), and here is Mr. Dewar's amendment :—

"From unlimited corroborative evidence we believe this [duality or sexuality in elementary substances] to be the case ; hence we enunciate as one of the fundamentals of the new Materialism that the normal chemical

Y

division of the elements into metallics and non-metallics is the bottom classification of matter, the only one with confidence to be designated great."

Again (p. 56):

"With this alteration (hydrogen considered as metallic) are we warranted in ascribing to this dual classification of substance that importance, both scientifically and philosophically, which we assign it? The evidence is overwhelmingly affirmative, for no known natural product exists which does not contain both classes of these elements in combination. *Matter itself must thus be sexual.*"

There is a very widely spread mineral substance composed of silicon and oxygen which forms no inconsiderable a proportion of the earth's rocky constituents, and of which the author has no doubt heard. Silica in its various forms is certainly a natural product ; and so Mr. Dewar will no doubt insist upon classifying silicon with the metals. There is also a gas composed of carbon and oxygen which is present in the atmosphere, and which is of vital importance for plant life. We should like to know how Mr. Dewar brings carbon dioxide under his "fundamental principle, which embodies one of the most salient truths in the science of the century" (p. 57).

The reader who is anxious to know how the "New Materialism" deals with the problem of life will find it disposed of in a light and airy way that might even be provocative of mirth were it not evident on every page that the author intends us to take him seriously. There is absolutely no mystery about it at all—there is no unknown force, there is no impulse different from the ordinary laws of matter. The animal is "a mere mass of conjoined magnets," containing "a virtual magnetic battery in its stomach" (p. 222). Elsewhere we are told that the plant differs from the animal in having its magnetic battery outside instead of within, and the author seems quite proud of having discovered a distinction between animals and plants that has hitherto eluded the men of science (p. 164). As for the appearance of life on the earth, it is a mere trifle to the "New Materialism":

"Under suitable conditions of heat, light and moisture, a chance flux of suitable atoms combines sexually into vegetal molecules" (p. 159) [magnetism as before].

"Even, as on a frosty night, the surface of the ground is whitened with crystals of rime, so in many a river and ocean bed the water must often coagulate with millions of vegetal and animal cells" (p. 209).

"But as igneous activity subsided to solid quiescence, and water, soil, light and heat interacted, the protoplasmic elements—oxygen, hydrogen, carbon, nitrogen, &c.—would meet in suitable proportion, and [magnetism as before] spontaneous production of simple organisms—protophyta, protozoa and the lowest kinds of fungi and algæ—would ensue as a matter of course, &c." (p. 246).

The origin of man is described (p. 247) in a manner that can only make the reader exclaim that the New Materialism, like a certain historical character, is *capable de tout* :—

"Man's first progenitors thus probably appeared on the earth as spontaneously produced protoplasmic cells or ovules, hundreds or thousands in number, developed by sexual and magnetic affinities from a flux of the chemical elements in some ambrosial inlet of water."

No further extracts need be given, and no further criticism is necessary to justify the opinions expressed at the beginning of this notice.   R. MELDOLA.

OUR BOOK SHELF.

Untersuchungen über Philons und Platons Lehre von der Weltschöpfung. Von Jakob Horovitz. Pp. xiii + 124. (Marburg : N. G. Elwart, 1900.)

DR. HOROVITZ' essay is the outcome of his thesis approved for the degree of philosophy in the University of Marburg. Its purpose is to focus the rays of light which close exegesis of the *Timaeus* throws upon the cosmogonic scheme wherein Philo effected the synthesis of Plato and Moses. While recognising the Stoic and Neopythagorean elements in Philo's teaching, Dr. Horovitz has little difficulty in showing that in both style and matter the dominant influence was Plato's. It is to the analysis, then, of Plato's creation-myth that we must turn if we would understand Philo with his enormous influence on the development of the doctrine of the Logos in Christian literature.

The ζῷον νοητόν of the *Timaeus* is no mythical duplicate of the demiurge, but distinguished as *das ewige Urbild* from the latter, whose real causal activity leads to an identification with the creative reason and ideal good of earlier dialogues. The subordinate artificers of physical creation are not the ideas as distinct from the idea of good, but in part a concession to popular theology, in part perhaps due to the place of evil in Plato's system, and the fact that dualism, though overruled, is not extinguished. In his valuable and textually supported discussion of the problems, Dr. Horovitz perhaps tends to overestimate the consistency and continuity of Plato's writings, and to underestimate the *mythus* element in the *Timaeus*.

Now Philo's intelligible world or order, the work of the one day of creation before time was or the serial "days" of the production of the world of sense began, is to be assimilated to the intelligible ζῷον of Plato as modified in conception by a use of the Stoics' metaphor of architect and supra-sensual city. It is not the Logos, save in the sense in which his plan is the mind of the architect. Dr. Horovitz moves familiarly among the conceptions of Logos, intelligible world, ideal man and the like, and by adjustment of the emphasis on the various clauses of Philo's commentary produces a construction which might carry conviction. The *mutata* of Philo, and the reasons why they were *mutanda* from the Platonic theory, are well brought out. The ideal man is the work of God, the physical man is the work of God in conjunction with subordinate agents, and these powers find their natural analogue in the angels of the Jewish scheme. Platonic scholars, or those of them who have not despaired of the ζῷον as unintelligible, will find food for reflection in the one side of Dr. Horovitz' study. Theologians, students of Neoplatonism, persons who take an interest in the Hegelian *Religionsphilosophie*, may well take their starting-point from the other.   H. W. B.

Fungus Diseases of Citrus Trees in Australia, and their Treatment. By D. McAlpine. Pp. 132 ; 19 plates. (Melbourne : Brain, Government Printer, 1899.)

THIS is one of the many useful publications dealing with plant diseases issued by the Victoria Department of Agriculture. According to statistics given, the cultivation of orange and lemon trees is extending rapidly, and one successful lemon grower considers that instead of paying 62,498l. annually for oranges and lemons, the colony could not only produce sufficient for home consumption, but could also supply the half of Europe. Under these circumstances the appearance of a work of the kind under consideration is most opportune, more especially as it is stated to be written for the benefit of growers. It is therefore somewhat disappointing to find that a considerable portion of the text is devoted to technical descriptions of new species of fungi, a subject of no interest whatever to cultivators, more especially as many of the species enumerated are simply saprophytic forms, whose presence can do no injury. The enumeration of such species is, from a scientific standpoint, of great value; but they are altogether out of place in a work which should place before practical men the outcome of scientific research in language divested of scientific technicalities. The author considers it essential that each fungus should possess a popular name in addition to its scientific one, and there is some justification for this idea, especially when such names are of local origin, and express a definite idea, as "collar-rot," "wither-top," &c., but it is more than doubtful whether the English rendering of the scientific name, as "West Australian Septoria," or "Glœosporium-like Colletotrichum," will be adopted by the fruit grower. Fifty-one species of fungi found on citrus trees are described as new to science; this is a somewhat daring piece of work in the comparative absence of literature and herbaria. It must be borne in mind that the fact of a fungus not agreeing with any species recorded in Saccardo's "Sylloge Fungorum" by no means justifies an author in describing it as new to science.

In defining parasites and saprophytes respectively, the author states that it is not always easy to decide between the two, and the crucial test, by means of pure cultures, is not alluded to. This, however, may not be due to lack of knowledge or desire on the part of the author, who, as vegetable pathologist, is probably expected to cover too much ground; hence fundamentals, which consume time, are apt to be neglected in favour of less exact methods, which may meet with approval for the time being.

The twelve coloured plates illustrating the most pronounced and destructive forms of disease attacking lemons and oranges are excellent in every respect, and should prove of great service in enabling planters to recognise at an early stage the appearance of a disease which, if neglected, might prove disastrous. The most approved methods of treating the various diseases are given in tabular form.

Missouri Botanical Garden. Eleventh Annual Report. Pp. 144; 58 plates. (St. Louis, Missouri, 1900.)

THIS volume is almost entirely made up of four scientific papers representing work carried out in connection with the Missouri Botanical Garden. The papers are: a disease of *Tascodium distichum* known as "peckiness," also a similar disease of *Libocedrus decurrens* known as pin-rot, by Dr. H. von Schrenk; Agaves flowering in the Washington Botanic Garden in 1898, by Mr. J. N. Rose; A Revision of the American species of Euphorbia of the section Tithymalus occurring north of Mexico, by Mr. J. B. S. Norton; and a Revision of the species of Lophotocarpus of the United States, and a description of a new species of Sagittaria, by Mr. J. G. Smith. Dr. von Schrenk's paper has already been noticed (vol. lxi. p. 452).

Mr. Trelease, the director of the Garden, shows by these papers and his report that valuable work is being done. A small synoptical collection representative of the principal natural orders of flowering plants has been installed in the central part of the Garden, where it is proposed to continue it as a convenient means of enabling teachers in elementary schools to demonstrate to their pupils the characters of the larger plant groups. The total number of species and varieties now cultivated in the Garden is nearly ten thousand.

Mr. Trelease devoted a couple of months last summer to the study of the botany of the Alaskan coast region and the islands of Bering Sea, as a member of the Harriman Alaska Expedition. The scientific results of his work will, no doubt, be published after the large amount of material collected has been subjected to critical study.

LETTERS TO THE EDITOR.

[The Editor does not hold himself responsible for opinions expressed by his correspondents. Neither can he undertake to return, or to correspond with the writers of, rejected manuscripts intended for this or any other part of NATURE. No notice is taken of anonymous communications.]

Atmospheric Electricity and Dew-ponds.

IT is not my intention to enter into controversy with such authorities as Mr. Aitken or Mr. Wilson. I wish only to describe certain phenomena which have come under my notice, in the hope that I may help to throw some light on a subject of great importance, theoretical as well as practical.

On the chalk hills in the south-east of England there are a number of ponds known as dew-ponds. One of these was described by Gilbert White ("Natural History of Selborne," Letter 29). There is a pond of considerable size at one of the highest points of the main ridge of the South Downs close to Chanctonbury Ring. From its position it is obvious that this pond can only be fed by water precipitated into it directly from the air. Yet it always contains a considerable volume of water. At the end of the dry weather last year, when most of the ponds in the valleys were empty, this dew-pond still contained several thousand gallons. How does this pond obtain the enormous quantities of water necessary to compensate for the rapid evaporation in such an exposed position, and also to supply large flocks of sheep?

It appeared to me that there was but one possible explanation: a difference of electrical potential must cause an attraction between the particles of moisture and the summit of the hill upon which the pond is situated. It is, of course, well known that drops of rain, &c., usually have an electrical charge, but it was necessary to ascertain whether this was capable of producing such a great effect. In order to test this point I took two porcelain basins of equal size and suspended them by means of silk threads from stakes driven into the ground at a high part of the ridge of the South Downs. In each of these basins was fixed an upright piece of sheet-copper. The two pieces of apparatus were placed in exactly similar positions and were in every respect identical, except that in the one case the copper screen was connected to earth by means of a wire, whereas in the other case it was insulated by the silk threads.

The apparatus was left thus during the night of April 1, 1899. There was a thick mist on the hills, so much so that I was unable to select the most favourable position for the apparatus. In the morning the amount of water in the two basins was measured. In the basin with the insulated screen there were 15·5 c.c. of water; in that with the screen connected to earth there were 18·0 c.c. This clearly confirmed my theory, for the insulated apparatus would tend to acquire an electrical charge from the particles of moisture. Consequently the attraction would be less than in the case of the apparatus which was not insulated. The insulation must have been very imperfect, for the silk became saturated with moisture as soon as the apparatus was erected. The position chosen, also, was not so favourable as it might have been. Nevertheless there was a difference of 16 per cent. in the quantities of moisture collected.

I intended to repeat and extend the experiment, but I have been unable to find an opportunity. I hope that this letter will call attention to a matter of considerable interest.

ARTHUR MARSHALL.

Chemical Department, Woolwich Arsenal, September 10.

Huxley and his Work.

MR. HENKEL's quotation from "One Hundred and One Great Writers" (NATURE, September 6) reminds one that the taunt of being a populariser was familiar enough to Huxley himself. It recalls also the little side-thrusts with which in return the detractors were sometimes honoured. In the preface to vol. viii. of the "Collected Essays," for instance, Huxley remarks:—

"The popularisation of science has its drawbacks. Success in this department has its perils for those who succeed. The people who fail take their revenge, as we have recently had occasion to observe, by ignoring all the rest of a man's work and glibly labelling him a mere populariser."

The belittling process, though unjustifiable, was understandable enough in those old days of controversy. To-day it seems ra'her uncharitable.   C. SIMMONDS.

Thurlow Hill, West Dulwich, September 17.

A Large Tasmanian Crab.

I AM sending you a photograph of a large crab (*Psilocranium gigas*), caught in the Tasmanian waters during the present month. The crab weighed 30 lbs. It is one of the largest that has been caught in these waters. We have several specimens in the Tasmanian Museum weighing from 16 to 22 lbs. They

are generally caught by the fishermen in very deep water, from fifteen to thirty fathoms, while fishing for the fish known as the Tasmanian Trumpeter (*Latris hecatcia*).

I shall be glad to know whether readers of NATURE have ever known of a larger *Psilocranium gigas* having been caught.
     ALEX. MORTON.

Tasmanian Museum and Art Gallery, July 30.

Large Puff Balls.

HAVING seen in some papers lately notices of large puff balls, it may probably be of interest to record the measurements of one far exceeding in size any I ever heard of.

It was found by my daughter, Mrs. Pole-Carew, in a small park belonging to me near this place, where she is residing. I took careful measurements of it at the time it was found, of which I send you a copy.

It differed in no respect except size, either inside or out, from the ordinary smooth puff ball.

Measurements of a large Puff Ball found in Chipley Park, near Wellington, Somerset, June 12, 1900.

Horizontal circumference	57	inches
Vertical ditto, greatest	51	,,
,, ,, smallest	46	,,
Height	14	,,
Greatest width	18½	,,
Smallest ,, ...	17	,,
Weight	14 lb. 10 oz.	

     W. A. SANFORD.

Nynehead Court, Wellington, Somerset, September 11.

"A Tour through Great Britain in 1727."

Is not the "Tour" queried by your reviewer (p. 417, column 2) that of Defoe, which has frequently been reprinted? and yet the first edition (1724–27) is still the best, in spite of re-editors and its extension from three vols. to four.
Ulverston.   S. L. PETTY.

PROF. HENRY SIDGWICK.

THE death of Henry Sidgwick entails the removal of one of the most potent influences that have been felt in Cambridge for the last forty years. Only a brief allusion can be made here to the time and energy which he devoted to University affairs, and to the constant and generous aid given by him to departments handicapped by poverty. As one of the strongest leaders in University policy, his power depended on a practical insight and decision of view for which those who know him only through his writings would be unlikely to credit him.

He was born in 1838. His father, the Rev. William Sidgwick, was headmaster of the Skipton Grammar School. Having been at Rugby under Dr. Goulburn, he entered Trinity College in October 1855. In 1859 he took his degree as Senior Classic and 33rd Wrangler, was elected to a Trinity Fellowship, and soon afterwards appointed Lecturer in Classics and Assistant Tutor. His interest in literary criticism and in problems of practical ethics was indicated, at this early stage, by various minor writings, of which we may specially mention an article on "The Prophet of Culture" (*Macmillan's Magazine*, 1867), in which he made a very characteristic examination of Matthew Arnold's closing lecture at Oxford. In 1868 was founded a Society, called "The Free Christian Union," of which Sidgwick was vice-president. His famous essay on "The Ethics of Conformity and Subscription" (1870) was written at the Society's request. This period of his life culminated in 1869 in the determination to give up his Trinity Fellowship on the ground that he no longer believed in the explicit creed to which the holders of Fellowships were required to subscribe under the old system of "tests." This action of Sidgwick's and the similar acts of some of his friends and contemporaries were undoubtedly important forces in the promotion of the abolition of the tests. Not long after, Sidgwick was made Lecturer and Examiner in the Moral Sciences, and later, Prælector in Moral and Political Philosophy at Trinity. In 1881 he was elected to an Honorary Fellowship there, and in 1883 he succeeded Birks in the Knightbridge Professorship of Moral Philosophy, which he resigned at the beginning of his illness last June.

As a teacher, Sidgwick exerted a profound and enduring influence, largely due to the extraordinary patience and quick perception with which he recognised and criticised the efforts of his pupils at independent thought. He presented to them an ideal of conscientious thoroughness in the pursuit of speculative truth, which has impressed and inspired even those who have developed their thought in directions far removed from his own.

Sidgwick's most important work, "The Methods of Ethics," was published in 1874 (2nd ed., 1877; 3rd, 1884; 4th, 1890; 5th, 1893). Its purpose is unlike that of most other modern works in philosophy. Not aiming directly at the construction of an ethical system, it adopts rather the Socratic method of stimulating the plain man to examine his own principles, and by self-criticism to free them from vagueness, obscurity and inconsistency. By many readers the unimpassioned, elaborately reasoned, judiciously balanced criticism is found unprofitable. But its penetrating subtlety and breadth of view are characteristics which have been recognised by all who have come under its influence, and have won for it a place amongst the philosophical classics. In general spirit it carries on the tradition of English common-sense empiricism; and, while to Sidgwick all forms of transcendentalism were repellent, yet unlike many of his predecessors in English philosophy, his criticism of opposed schools of thought was always keen and thoroughly scholarly. A different side of Sidgwick's intellectual character is shown in his work on "Practical Ethics," a collection of essays and addresses (1898), in which his speculations are applied to the very definite solution of actual problems of conduct in modern life.

In the "Outlines of the History of Ethics for English Readers," Sidgwick has supplied a most useful guide to the study of the subject. In the "Principles of Political Economy" (1st ed., 1883; 2nd, 1887) and in "The Scope and Method of Economic Science" (1885), there is a return to the older English thought, but the subject is treated with an acuteness and originality specially characteristic of Sidgwick's intellect, which have given to these works real value as contributions to economic science. The third book of the principles contains the "Art of Political Economy,' which, together with "The Elements of Politics" (1st ed., 1891; 2nd, 1897), shows the keen interest always felt by Sidgwick in political and social questions, and the practical sagacity with which he handles these problems. In politics, Sidgwick combined the freedom from prejudice of the Radical with the caution of the Conservative.

Perhaps the most important practical work with which the name of Sidgwick has been associated is in connection with the higher education of women. He was the virtual founder of Newnham College, through the scheme of lectures for women which he initiated in 1869, and the house of residence which he started and persuaded Miss Clough to take charge of in 1871. In 1880, Mrs. Sidgwick having consented to become vice-principal of the second Hall of the College just opened, they both came to live there for two years; and when, after Miss Clough's death in 1892, Mrs. Sidgwick became principal of the College, they made it their permanent home.

In 1882 Sidgwick accepted the presidency of the newly formed Society for Psychical Research, in the subject-matter of which he had been interested for many years. The spirit which has characterised the proceedings of the Society, and the success which it has achieved, have been largely due to the sobriety and wisdom of Sidgwick's constant counsel and control.

PROF. JAMES EDWARD KEELER.[1]

THE sudden death of Prof. James E. Keeler, director of the Lick Observatory, which occurred at San Francisco on August 12, removes one who stood at the very forefront of astrophysical research.

James Edward Keeler was born at La Salle, Illinois, on September 8, 1857. His qualifications for scientific work clearly showed themselves at the Johns Hopkins University, where he took an undergraduate course, and served as assistant to Prof. Hastings, with whom he observed the total solar eclipse of 1878 in Colorado.

Shortly after this he was appointed assistant at the Allegheny Observatory, where he had an important part in the long series of bolometric investigations carried on by Prof. Langley, then director of the Observatory. In July 1881 he was a member of Prof. Langley's well-known expedition to Mount Whitney, in Southern California, where an extensive region in the extreme infra-red of the solar spectrum was discovered with the bolometer. Later he studied for two years in Berlin and Heidelberg under Helmholtz and Quincke, and returned to the Allegheny Observatory, where he remained until appointed a member of the staff of the Lick Observatory. His work on Mount Hamilton commenced in 1885, and for some time he was the only astronomer at the Observatory, which was still in process of construction. In May 1891 he was elected professor of astrophysics in the Western University of Pennsylvania and director of the Allegheny Observatory.

Keeler's work at the Lick Observatory was continued in a most effective manner with the modest instrumental resources at Allegheny. With a full understanding of the art of making the most of his means, he took up photo-

graphy for the first time, made himself thoroughly familiar with photographic processes, and then, with the aid of a spectrograph whose general design has been followed in the construction of the great modern spectrographs at Mt. Hamilton, Potsdam, Pulkowa and Williams Bay, he obtained the photographs of the spectra of red stars which excited so much interest at the dedication of the Yerkes Observatory. He also made an admirable series of drawings of Mars, which was published in the *Memoirs* of the Royal Astronomical Society.

In the spring of 1898 Keeler had practically decided to accept a position on the staff of the Yerkes Observatory, and would have done so had he not just then been appointed director of the Lick Observatory. His recent work on Mt. Hamilton has not been confined to the direction of the affairs of a great observatory. The remarkable success of his experiments with the Crossley reflector, of which a full account is fortunately preserved in the June number of the *Astrophysical Journal*, has impressed every one who has seen the wonderful photographs of nebulæ and star clusters made with this instrument.

Of Keeler's other contributions to science two in particular deserve present mention : his determination with the Lick telescope of the motion in the line of sight of the planetary nebulæ, and his demonstration of the meteoric constitution of Saturn's rings. The memoir which describes the first of these investigations already ranks as a classic of astrophysical literature ; while the spectroscopic demonstration of the meteoric constitution of Saturn's rings is perhaps the most striking of the many effective applications which have been made of Doppler's fruitful principle.

Much more might be said of Keeler's work, but this should suffice to indicate its lasting value. It is a satisfaction to add that its merit has been widely appreciated, as has recently been evidenced by the award of the Draper and Rumford medals. He was president of the Astronomical Society of the Pacific and a councillor of the Astronomical and Astrophysical Society of America. He was elected an Associate of the Royal Astronomical Society in 1898 and a member of the National Academy of Sciences at its last meeting. His kindly and genial manner, combined with unusual tact and rare judgment, drew to him many friends, who will long mourn his loss.

NOTES.

THE annual meeting of the Iron and Steel Institute was opened at Paris on Tuesday with an address by the president, Sir W. Roberts-Austen, K.C.B., F.R.S. It was announced that Mr. Andrew Carnegie has offered to the Institute the sum of 6500l. for the purpose of founding a medal and scholarship to be awarded for any piece of work that may be done in any works or University, and to be open to either sex.

IT is stated by the Paris correspondent of the *Times* that M. Yersin, to whom the Academy of Moral Sciences recently awarded a prize of 15,000 francs for philanthropic acts, has devoted the sum to his anti-plague serum establishment at Nha-trang.

THE *British Medical Journal* announces that the prize of 4000 marks voted by the Berlin Congress of Tuberculosis for the best popular work on tuberculosis as a social scourge, and the means of preventing its ravages, has been awarded to Dr. S. A. Knopf, of New York. The work will be published by the German Central Committee.

A TABLE of standard sizes of conductors for electric supply mains has been drawn up by the Cable Makers' Association and sent to electrical engineers. The table shows the nominal

[1] Abridged from an obituary notice contributed to *Science* of September 7 by Prof. George E. Hale.

NO. 1612, VOL. 62]

size of conductors, in square inches, number and diameter of strands, resistance in standard ohms per 1000 yards, and weight in lbs. per 1000 yards. It is proposed to adopt the sizes and combinations of conductors shown in the table as the basis for tenders, beginning on October 1.

THE list is now completed of the subjects to be discussed at the International Botanical Congress to be held in Paris from the first to the tenth of October, in connection with the Exposition, and promises a time of varied interest from both a structural and an economical point of view. Those who wish to become members, and thus to obtain the results of the Congress, should send their subscriptions (20 fr.) to M. H. Hua, Treasurer to the Congress, 2 Rue de Villersexel, Paris.

FIVE additional cases of plague, of a mild type, were reported in Glasgow on Monday. This raises the total to twenty-two plague cases, one suspected case, and 115 persons under observation. The attacked persons had been in contact with plague cases. Prof. Muir states in his report on the new cases that his experiments show without doubt that the bacillus was that of bubonic plague. He examined nine cases, both microscopically and by means of cultures, and found the same results.

AT the Geographical Congress at Berlin in October 1899 it was decided to form an International Seismological Society. The first meeting of the delegates from different countries will be held at Strassburg on April 11, 1901. The principal subjects chosen for discussion are : The organisation and extension of macroseismic investigations in all countries, the organisation of international microseismic observations, the selection of apparatus for international and local seismic observations, the annual publication of international seismic reports, and the statutes of the new society.

THE Liverpool Marine Biology Committee's station at Port Erin has been very fully occupied during the greater part of the summer, and there are at present half-a-dozen workers doing original research in the laboratory. On Saturday last a party of the Isle of Man Natural History and Antiquarian Society proceeded to Port Erin on a visit to the laboratory, and were heartily welcomed by Prof. Herdman, F.R.S. Mr. Isaac C. Thompson gave a lecture, "On the Place of the Copepoda in Nature." It was pointed out that the copepoda are of the utmost value as scavengers, as they live on the products of decomposition, putrefaction, drainage matter, &c., and by their internal laboratories convert refuse matter into most valuable food material, some copepoda constituting one of the chief sources of food for fishes, and therefore of man. Mr. Thompson said that no less than 200 species have been found in Liverpool Bay. Their beautiful organisation illustrates the truth that the wonderful structure of some animals, which can only be studied with the microscope, shows them to be as full of interest as those familiar to our ordinary vision. Besides the many free swimming copepods, there are many species found as fish parasites, living on the gills and on other external parts of our common fishes ; some of these are nourished by the fish and do harm, while others do not, their presence being rather an advantage than otherwise.

THE importance of an organised and continuous system of rainfall observations is obvious to every one possessing sufficient knowledge of physical geography to know the relation of rainfall to agriculture, water supply, and all questions in which the development of natural resources is concerned. It is as essential that such observations should be systematically carried on in a thickly populated country like our own as it is that they should be made in all parts of the British Empire in which observers can be found. And when stations have been established, it is again essential that there should be a central bureau in which the observations can be collected, and their relation to one another, and to

the natural features of the district determined. A report, by Mr. F. R. Johnson, upon irrigation and water supply, rainfall, and water rights in Cape Colony, published in the *Cape Agricultural Journal* (August 2), issued by the Department of Agriculture, reminds us of the value of continuous rainfall records in connection with works for irrigation and water supply. It is unusual for stream flow measurements, owing to their cost, to be available, even in the most favourable circumstances, to anything like the same extent, and it is therefore very necessary that the rainfall records should be carefully discussed, and full advantage taken of all they are capable of teaching, so that when considered and compared with shorter periods of stream flow measurement (possibly only available for an adjoining catchment) the significance of the whole may be appreciated and understood. So far this has not been done in Cape Colony, from an engineering point of view, and the need of the information is felt now that the hydrographic conditions of the Colony are being investigated. From the rainfall observations so far examined, it appears that sufficient water should be available to irrigate about five million acres of arable land ; and when it is considered that this means an enhanced land value in the Colony of upwards of a hundred million pounds, the advantage of setting to work at once to digest and apply the data available to specific proposals for irrigation is apparent.

IN an article on "The Amount of the Circulation of the Carbonate of Lime and the Age of the Earth," by Prof. Eug. Dubois (*K. Akad. van Wetenschappen te Amsterdam*, 1900), it is conceded that the ocean, which derives all of its carbonate of lime from rivers or the waste of sea-cliffs, holds as much of it as it can, and that rivers are incessantly bringing a surplus. A considerable amount of carbonate of lime is often to be found in the matter carried in suspension by large rivers to the ocean, and it is obvious that in these river-waters the solution must be saturated. The quantity of carbonate of lime in river-waters is naturally determined by the rocks in the drainage areas. The author gives reasons which lead him to conclude that not more than one-thirtieth part of the carbonate of lime which rivers now discharge into the ocean is newly formed from silicates, although originally all was so derived. His calculations, based on the amount of carbonate of lime annually carried to sea by rivers, show that the formation of the whole estimated minimum amount of carbonate of lime on the earth would require about 45,000,000 of years, and that of the real amount a much larger lapse of time. He estimates that 1/2,770,000th of the total quantity of carbonate of lime of the earth participates annually in the present circulation. The final result of his investigation, though admittedly only suggestive, is that the real lapse of time since the formation of a solid crust and the appearance of life upon the globe may be more than a thousand million of years.

MR. A. GIBB MAITLAND, Government geologist of Western Australia, has issued, as *Bulletin* No. 4 of the survey reports, a general account of "The Mineral Wealth of Western Australia." This work is intended to replace the useful mining handbook which was prepared by Mr. Harry P. Woodward, and has long been out of print. The author gives a sketch of the geological features of the Colony, and then deals specially with gold, lead and copper, tin, iron, miscellaneous minerals (such as antimony, zinc, &c.), coal and graphite, guano and artesian water (with records of borings). A full list of minerals is appended, and there is a map showing the distribution of useful minerals in Western Australia, and five geological maps of particular mineral districts. The guano, which is obtained from the Abrolhos Islands and elsewhere in the north, is an important product. In 1899 upwards of two thousand tons were obtained, the total value being over 5000l. The amount raised last year was, however, small compared with some previous records.

A REPORT on the mineral statistics and mines of Canada for 1898 (1900), drawn up by Mr. E. D. Ingall, contains records of a great variety of mineral products. It is interesting to note that the output of coal, supplied mainly from Nova Scotia (nearly two-thirds), and from British Columbia (nearly one-third), shows an increase : the total production being nearly four million tons. Natural gas is obtained from wells in southern Ontaria. Gold shows a large increase, due to the output from the Yukon ; but silver, which is almost wholly derived from British Columbia, shows a decrease. On the whole, however, the growth of Canada's mineral industries is stated to be very encouraging.

IN the memoirs of the Society "Antonio Alzate" (vol. xiv.), Señor M. Moreno publishes a discussion of the sunshine values at the Observatory of Leon (Mexico), situated in latitude 21° ·07′ N., showing the daily amounts recorded by a Campbell-Stokes instrument from June 1892 to December 1898. These values are all the more acceptable from the fact that out of some thirty observatories in the Mexican Republic, Leon appears to be the only station which furnishes a complete record of sunshine. The following figures, giving the average percentages of the possible amounts, show that the locality enjoys a large amount of bright sunshine :—winter, 71 ; spring, 72 ; summer, 59 ; autumn, 69 ; and for the year, 68. In Dr. Scott's discussion of ten years' sunshine in the British Isles, the annual average for London (City) is 24 ; for Greenwich, 25 ; and for Jersey, the sunniest part of the British Isles, 40.

THE U.S. *Monthly Weather Review* for May last contains an interesting article, by Prof. C. Abbe, on the history of modern weather prediction. He considers that the first effort towards this end was the publication of the "Mannheim Ephemerides," a series of thirteen volumes, for the years 1780–92, containing detailed meteorological observations for thirty-six stations in Europe and for three stations in America. After many years, Prof. H. W. Brandes first compiled from those observations daily weather maps for 1783 ("Beiträge zur Witterungskunde," Leipzig, 1820). In 1826 Prof. J. P. Espy organised a joint committee for the purpose of studying storms ; numerous maps were constructed and many published in four successive reports (1845–60). The labours of Espy and Redfield established the fact that individual features of the weather, as well as storms, move in such a manner that their approach can be predicted by means of maps. Prof. J. Henry constructed daily weather maps from telegraphic reports, for personal study, for several years from 1848, and from 1856 onwards they were exhibited at the Smithsonian Institution. These maps were made the basis of frequent special predictions of the weather for the benefit of members of Congress and others. This date brings us down to the time of the establishment of the meteorological offices in this country and abroad.

MANY interesting points referring to the work of the Government Laboratory are mentioned by the director, Dr. T. E. Thorpe, F.R.S., in the report recently issued. A number of tinned meats were examined for the Admiralty for food preservatives, but no antiseptic other than common salt was detected. Numerous butters contained boric preservative and were artificially coloured. As usual, the use of boric acid is most prevalent in butters from France, Belgium and Australia, and is very common also in Holland. The most frequent colouring-matter is annatto, but the use of coal-tar yellows appears to be on the increase, and is especially prevalent in Holland, the United States and Australia. In the course of the year it was decided by the Board of Trade that all passenger ships should be required to carry a filter capable of delivering water free from micro-organisms. With Dr. Thorpe's assistance, a filter which satisfactorily fulfils this condition has been decided upon. As a

supplement to the work of the Steel Rails Committee (see p. 437), an investigation was undertaken with the object of elucidating the mode in which the phosphorus in steel is chemically combined. The inquiry clearly showed, as has already been surmised, that the phosphorus present is, like the carbon, not infrequently in more than one form of combination. The greater part of the work done in connection with the Home Office arose out of the inquiries instituted by the Home Secretary, relative to the prevalence of lead-poisoning arising from the use of lead compounds in pottery manufacture. A considerable number of "fritts" and "glazes" have been examined, and the conditions determining the ease with which lead compounds may be extracted from them by dilute acids, comparable as regards their action with that of the gastric juice and other animal solvents, have been ascertained. As the result of the inquiry made last year, the Home Secretary has required the manufacturers of pottery to abandon the use of raw lead, and in view of the facts brought to light by the examination of the fritts and glazes, he has expressed his intention of prescribing that in future such glazes shall conform to a standard of insolubility as regards lead.

THE evidence for presence of totemism in various parts of the world is now being carefully examined, as it is beginning to be realised that a cult of animals is not necessarily the same as totemism. This is the attitude taken by Dr. C. Hose and Mr. W. MacDougall in their paper read before the recent meeting of the British Association, and to which reference will be made elsewhere in our columns. Dr. E. Westermarck, the author of the well-known work on "Human Marriage," has published some of the results of his investigations in Morocco in a paper in the *Journal* of the Anthropological Institute (vol. xxix. p. 252), entitled "The nature of the Arab *Ginn*, illustrated by the present beliefs of the people of Morocco," in which he adopts the same conclusion. The *Gnun*, as they are called in Morocco, form a special race of beings, created before Adam. They have no fixed forms, but may assume almost any shape they like. Usually they are hurtful to man. The bad *Gnun* being always ready to attack human beings, various means are used for keeping them at a distance. The *Gnun* are afraid of salt and steel, which are consequently employed as prophylactics ; the best and, from a religious point of view, the correct preventive against their attacks is the recital of passages of the Koran. Dr. Westermarck adversely criticises Robertson Smith's explanation that the *Ginn* are modernised representatives of totem animals, and states that they are beings invented to explain the wonderful and mysterious in nature. They are, in fact, survivals of the early indigenous animistic beliefs of a saltless and ironless antiquity, which, at a later date, were absorbed and developed under the influence of Islâm.

THE Society for the Protection of Birds has just issued, in pamphlet form, a communication from Sir C. Lawson, which appeared in the *Madras Mail* of March 27 and April 11, pleading for the adequate protection of insectivorous birds in India. It appears that for more than a decade a law has been in force in Madras for the protection of birds, but that when, some time ago, steps were taken to extend this enactment to the other presidencies, the responsible advisers to the Indian Government did not consider that the time was ripe for such legislation. Sir C. Lawson now urges that, in great part owing to the famine, the need of bird protection by law demands immediate recognition. At present he pleads only for those insectivorous species whose wholesale slaughter for the sake of their plumage leaves grain and cotton-fields at the mercy of insect pests, thus causing "a deplorable sacrifice of human food and the materials of human raiment, besides inflicting penury on individuals and great loss to the State." The society is endeavouring to form an Indian branch, and applies for support to all interested in India. Of the excellency of the object in view we are fully assured, but it must be borne in mind that legislation of the nature proposed entails many difficulties in India, and should not be introduced without very mature consideration.

WE have received a *Bulletin* (vol. iii. No. 2) of the Madras Government Museum containing an illustrated account by the superintendent, Dr. E. Thurston, of the sea fisheries of Malabar and South Canara. A considerable portion of the pamphlet is taken up with an account of the instructions which have been recently drawn up by the author for the guidance of the officials at the various fish-curing establishments in attaining statistics of the life-history and migrations of the more abundant species of fish. As an instance of the difficulties encountered in India in obtaining statistics of this nature, it may be mentioned that "at the fish-census, 1889, the officer who was told off to make the record of fishes brought ashore, was at first driven away by the fishermen, who refused to give him the requisite information, from fear that the census was being taken with a view to increased taxation." The remainder of this valuable report is taken up by a diary made by the author during a tour of inspection of the fish-curing yards of the districts in question during the autumn of last year. Some of the most striking modes of fishing are illustrated in the plates, and details given of the amount and value of the catches at the different stations. Dr. Thurston is of opinion that a much greater variety of fish might be introduced with advantage at the dinner tables of Europeans residing in Madras than is at present the case.

As nest-building fishes are comparatively few, naturalists will read with interest an account given in the August issue of the *American Naturalist*, by Messrs. Young and Cole, of the manner in which the brook-lamprey (*Lampetra wilderi*) makes a structure of this nature. It is believed that the males precede the females at spawning time and commence nest-building before the arrival of the latter. The nest is made among pebbles, but it does not seem that the lampreys follow any definite plan in its construction. They affix themselves to such pebbles as require removing from the nest, and then endeavour to swim straight away with them. In the case of a heavy stone two lampreys may join forces. The number of fish in a nest may vary from one to thirty or forty; but there are generally between three and twenty-five.

A PAIR of fenestræ covered with membrane have for some time been known to occur in the head of the common cockroach, and represent functional ocelli in other species of the same group. In the *American Naturalist* for August, Mr. C. Kochi records the existence of a pair of spots in the former insect, just below the aforesaid fenestræ. These spots he believes also represent the sites of another pair of ocelli which in other insects have shifted their position and coalesced to form the unpaired median ocellus.

THE *Sunday Magazine*, like many other popular journals, publishes occasional articles on scientific subjects, the one in its September issue being devoted to swimming crabs. Excellent illustrations are given of several of the species to be met with on the British coasts, while the letterpress describes their distinctive features and the leading peculiarities in their habits.

A PAPER on the life-histories of the mosquitoes of the United States, by Dr. L. O. Howard, recently published in one of its *Bulletins* by the U.S. Department of Agriculture, appears opportunely. The writer gives concise but clear descriptions, accompanied by enlarged illustrations, of all the members of the group met with in the States, devoting special attention to those of the malaria-producing genus, Anopheles. Dr. Howard calls

attention to the circumstance that he advocated the employment of kerosene for the destruction of the larvæ as far back as 1894, and claims that this mode has proved more effectual, when used on a sufficiently large scale, than any other yet suggested. In certain cases, however, as in the instance of tanks containing water intended for drinking purposes, the employment of kerosene may be undesirable, and the introduction of fish, where none previously existed, is then advocated. The value of most small fishes as destroyers of mosquito larvæ is well illustrated by a natural experiment which recently took place in Connecticut. "In this case a very high tide broke away a dyke and flooded the salt meadows of Stratford, a small town a few miles away from Bridgeport. The receding tide left two small lakes, nearly side by side and of the same size. In one lake the tide left a dozen or more small fishes, while the other was fishless. An examination in the summer of 1891 showed that while the fishless lake contained tens of thousands of mosquito larvæ, that containing the fish had no larvæ."

WE have received from the publishers two numbers of the *Zeitschrift für wissenschaftliche Zoologie*—the last of vol. lxvii. and the first of lxviii. The former contains an article, by Herr E. Wasmann, describing a new genus (Termitoxenia) of wingless Dipterous insects parasitic in the nest of white ants or termites. Four species of these remarkable insects are recognised, of which one is Indian, while the other three are from Africa. Another article in the same issue, which should be of considerable interest to stock-beeders, treats of the histology of certain infusians found in the stomachs of cattle and in the cœcum of the horse. The three articles forming the first part of vol. lxviii. are all devoted to invertebrate anatomy and morphology.

THE August issue of the *Journal* of the Royal Horticultural Society contains, in addition to numerous articles dealing with fruits, vegetables and flowers, a paper treating of the scale and mealy-bug, and a second discussing the black currant-mite and its ravages. In the latter, which should be especially valuable to horticulturists, it is stated that the origin of the pest in question is unknown, but that its first recorded occurrence in the British Isles is about fifty years ago, when it was found in Scotland.

THE Natural History and Ethnographical Museum of Pará, Brazil, has commenced to issue a series of memoirs; the first of these is an account by Dr. Emilio A. Goeldi, the director of the museum, of some archæological excavations which he made in 1895 of some artificial burial caves of an extinct tribe of Indians on the Rio Cunany (Goanany). Each cave consists of a circular shaft, 8 feet 2½ inches deep and 3 feet 4½ inches in diameter, the opening of which was closed by a large disc of granite. A crescentic chamber had been cut in the soil at the base of the shaft, in which were deposited a number of remarkable earthen vessels of very varied form, most of which were painted in red with peculiar designs and patterns. Some of the funeral vases were conventionally modelled to represent the human form, and others had on them various animals moulded in high relief. The memoir is illustrated by capital plates.

MESSRS. G. W. WILSON, of Aberdeen, have just issued a catalogue of more than seven hundred new lantern slides representing scenes and objects in Spain. The list should be of particular interest to teachers of geography.

WE have received the Annual Reports of the Royal Botanic Garden, Calcutta, from the superintendent, Major Prain, for the years 1898-1899, 1899-1900. Close attention has been given by the garden staff throughout the past year to the propagation and distribution of plants of economic importance,

AFTER an interval of about two years, Sir George King, late superintendent of the Royal Botanic Garden, Calcutta, continues, in the *Journal* of the Asiatic Society, Bengal, his materials for a flora of the Malayan Peninsula. The present part consists of a monograph of the Malayan species of Melastomaceæ, prepared with the assistance of Dr. O. Stapf, of the Kew Herbarium.

IN an article on the fertilisation of *Peronospora parasitica* in the *Annals of Botany* for June, Mr. Harold Wager points out that there are at present known three distinct types of fertilisation in the Peronosporeæ. In the first, represented by *Peronospora parasitica*, the oosphere and oospore are uninucleate, and fusion takes place between two nuclei only. In the second type the oosphere is uninucleate and the oospore multinucleate, and fusion is effected between two nuclei only. In the third type the oosphere and oospore are both multinucleate, and fusion takes place between a number of nuclei in pairs.

IN a paper recently read before the Linnean Society, on the origin of the Basidiomycetes, Mr. G. Massee points out the connection between the Hyphomycetes and the Protobasidiomycetes. The conidial forms of many Hyphomycetes are true Protobasidiomycetes. There is no evidence that the Autobasidiomycetes are in any way descended from the Protobasidiomycetes ; while, on the other hand, there are indications that the Autobasidiomycetes may probably have been derived by gradual modifications of the spore-bearing organs or oasids of conidial forms of certain ascigerous fungi.

PROF. J. J. THOMSON'S inspiring work on "The Discharge of Electricity through Gases" (Constable and Co.) has been translated into French by Dr. L. Barbillion, and published by MM. Gauthier-Villars under the title "Les Décharges électriques dans les Gaz." Dr. Barbillion adds a few notes, and Dr. C. E. Guillaume contributes a preface. The original volume was reviewed in NATURE of January 12, 1899 (vol. lix. p. 241), and the translation will doubtless be received by French physicists with the same appreciation as the work has commanded in Great Britain.

SEVERAL articles of real interest to students of science and philosophy have appeared in recent numbers of *The Open Court*. The August number contained an account of Galileo in which his work is presented in new aspects, and the opposition to his conclusions as to the movements of the earth and the character of the visible universe is in some part explained. The current number contains an instructive contribution on Greek religion and mythology, by the editor, Dr. Paul Carus, and one on animism in popular thought and in science, by Prof. E. Mach. Many of the articles in the magazine are excellently illustrated.

AN atlas for druggists and students of pharmacy, by Prof. Ludwig Koch, is in course of publication by the firm of Gebrüder Borntraeger, Leipzig, under the title of "Die mikroskopische Analyse der Drogenpulver." The first volume is to be devoted to barks and woods, and the second part of it, containing six plates, has just appeared.

THE third and fourth parts of Dr. Chun's elaborate account of the German *Valdivia* expedition have just been published by the firm of Gustav Fischer, Jena. The parts are illustrated with numerous half-tone figures and some very fine plates, and the work promises to be a very attractive narrative of an extensive voyage.

THE additions to the Zoological Society's Gardens during the past week include a Sooty Mangabey (*Cercocebus fuliginosus*) from West Africa, presented by Mr. B. Stewart ; a Squacco Heron (*Ardea ralloides*), South European, presented by Mr. A. F. Putz ; a Black-headed Terrapin (*Damonia reevesi unicolor*) from China, an Algerian Skink (*Eumeces algeriensis*) from North-west Africa, a Common Chameleon (*Chamaeleon vul-*

garis) from North Africa, presented by Mr. F. J. Bridgman ; an European Pond Tortoise (*Emys orbicularis*), European, presented by Miss F. M. Weippert ; a Wall Lizard (*Lacerta muralis*), a Tessellated Snake (*Tropidonotus tessellatus*), European, presented by Mr. Walter Hunter ; two Badgers (*Meles taxus*), British ; two Indian Fruit Bats (*Pteropus medius*) from India, three Black-spotted Teguexins (*Tupinambis nigropunctatus*) from South America, two Antillean Boas (*Boa diviniloqua*) from the West Indies, five Undulated Lizards (*Sceloporus undulatus*) from South-east United States, deposited.

OUR ASTRONOMICAL COLUMN

EPHEMERIS FOR OBSERVATIONS OF EROS :—

1900.		R.A.		Decl.
		h. m. s.		° '
Sept. 20	...	2 39 43	...	+40 47·1
21	...	40 20	...	41 9·9
22	...	40 55	...	41 32·8
23	...	41 27	...	41 55·7
24	...	41 56	...	42 18·6
25	...	42 23	...	42 41·4
26	...	42 46	...	43 4·2
27	...	43 7	...	43 27·0
28	...	43 25	...	43 49·7
29	...	43 40	...	44 12·3
30	...	2 43 51	...	+44 34·8

SWIFT'S COMET (1892 I.).—In the *Annals of Harvard College Observatory* (vol. xxxii. Part ii. pp. 267–295), Prof. W. Pickering describes the photographs obtained of this comet at Arequipa during March and April 1892, illustrating his remarks by reproductions from nine of the negatives.

The plates were taken with four instruments—the 13-inch Boyden telescope, 8-inch and 2·5-inch doublets, and a 20-inch reflector ; two photographs were also obtained with the 8-inch refractor provided with an objective prism of 13° refracting angle. The exposures varied from 5 to 133 minutes.

While the comet was easily visible to the naked eye, it was carefully examined with a double image prism, but no traces of polarisation could be detected in either the head or tail. The nucleus was yellowish-green in colour, giving out a triangular jet towards the sun.

The head was distinctly divisible into three parts—nucleus, bright primary envelope and an outer fainter one. The tail was composed of two sets of rays having distinctly different origins. The brighter of these sets, forming what may be called the "inner" tail, took its origin from the rear side of the inner envelope, and, in some of the photographs this attains the great length of over 20° of arc. The rays were absolutely straight so far as could be determined from the photographs, and were inclined to each other about 10°. The outer tail sprung from the external faint envelope, and, in contrast to the other, was marked by conspicuous deep and wide rifts between the rays composing it.

Prof. Pickering thinks that certain periodic differences in appearance are caused by a rotation of the comet about an axis passing longitudinally through the tail. Comparisons of the Arequipa photographs with others obtained by Dr. Wolf and Prof. Barnard show that it is quite possible to detect changes from one hour to another, and from a detailed examination of the angular deviation of the rays it is probable that the rotation period is about 94–97 hours.

The photograph taken April 14 shows a strong deflection of the inner tail, but the absence of other photographs near that date render it impossible to trace the cause, and the phenomenon was not subsequently repeated.

In general, it was impossible to identify any particular feature on two successive days, but on April 6, 7, 8, a bright condensation was noticed each day, and its distance from the nucleus of the comet was found to increase day by day. These displacements were carefully measured, converted into kilometres by reference to the comet's elements, and an estimate made of the amount of the repulsive force exerted upon the comet's tail by the sun. This indicated the total repulsive force to be about 39·5 times the gravitational force. The spectrum photographs have been difficult to reduce, but the brightest region of the spectrum appears as an intense and very narrow line about λ 3890. No indications of the hydrogen lines were seen.

THE BRADFORD MEETING OF THE BRITISH ASSOCIATION.

SECTION D.

ZOOLOGY.

Opening Address by Ramsay H. Traquair, M.D., LL.D., F.R.S.

IN opening to-day the sittings of the Zoological Section, I must first express my sense of the honour which has been conferred on me, in having been chosen as your President on this occasion, and I may add that I feel it not only as an honour to myself personally but also as a compliment to the field of investigation in which the greater part of my own original work has been done. It is a welcome recognition of the doctrine, which I, and much more important men indeed than I, have always maintained, namely, that Palæontology, however valuable, nay, indispensable, its bearings on Geology may be, is in its own essence a part of Biology, and that its facts and its teachings must not be overlooked by those who would pursue the study of Organic Morphology on a truly comprehensive and scientific basis. As I have asked on a previous occasion, "Does an animal cease to be an animal because it is preserved in stone instead of spirits? Is a skeleton any the less a skeleton because it has been excavated from the rock, instead of prepared in a macerating trough?" And I may now add—Do animals, because they have been extinct for it may be millions of years, thereby give up their place in the great chain of organic being, or do they cease to be of any importance to the evolutionist because their soft tissues, now no longer existing, cannot be imbedded in paraffin and cut with a Cambridge microtome?

These are theses which I think no one denies theoretically; but what of the practical application of the rule? For though cordially thanking my biological brethren for the honour they have done me in placing me in this chair to-day, I must ask them not to be offended if I say that in times past I have a few things against some of them at least. I refer first to the apathy concerning palæontological work, more especially where fishes are concerned, which one frequently meets with in the writings of biologists, as seen in the setting up of classifications and theories and the erection of genealogical trees without any, or with at least inadequate, inquiry as to whether such theories or trees are corroborated by the record of the rocks. But more vexatious still are the offhand proceedings of some biologists who, when they wish to complete their generalisations on the structure of a living organism, or group of organisms, by allusion to those which in geological time have gone before, do not take the trouble to consult the original palæontological memoirs or papers, or to make themselves in any way practically acquainted with the subject, but derive their knowledge at second or third hand from some text-book or similar work, which may not in every case be exactly up to date on the matters in question. Nay, more than this, I think I have seen the authors of such text-books or treatises credited with facts and illustrations which were due to the labours of hard-working palæontologists years before.

But a better time, I am convinced, is not far off, when the unity of all biological science will be recognised, not merely theoretically, but also practically by workers in every one of its branches.

Of one thing I must, however, warn those who have hitherto devoted their time exclusively to the investigation of things recent, namely, that a special training is necessary for the correct interpretation of fossil remains, especially those of the lower Vertebrata and many groups of Invertebrata. So it comes that what looks to the uninitiated eye a mere confused mass of broken bones or plates may to the trained observer afford a flood of valuable light on questions of structure previously undetermined. We must take into account the condition of the fossil as regards mineralisation and crushing; we must learn to recognise how the various bones may be dislocated, scattered, or shoved over each other, and to distinguish true sutures from mere fractures. We must carefully correlate the positive results obtained from one specimen with those afforded by others, and in this way it happens that to make a successful restoration of the exo- or endo-skeleton of a fossil fish or reptile may require years of patient research. But the thought sometimes does come up in my mind, that some people imagine that fossils, such as fishes, occur in the rocks all restored and ready, so that the author of

such a restoration has no more scientific credit in his work than if he were an ordinary draughtsman drawing a perch or a trout for an illustrated book! But the student of fossil remains must not only learn to see what does exist in the specimen he examines, but also to refrain from seeing things which are not there —to know what he does not see as well as what he does see. For many grave errors have arisen from want of this necessary training, as, for instance, where the under surface of a fish's head has been described as the upper, or where markings of a purely petrological character have been supposed to indicate actual structures of the greatest morphological importance. Or we may find the most wonderful details described, which *may* indeed have existed, but for which the actual evidence is only the fertile imagination of the writer.

From this it will be apparent that though Palæontology is Biology and Biology includes Palæontology, yet as regards original research a division of labour is in most cases necessary. For though palæontological investigations are absolutely impossible without an adequate knowledge of recent zoology, yet the nature of the remains with which the palæontologist has to deal renders their interpretation a task of so different a character from that allotted to the investigation of the structure and development of recent forms that he will scarcely have time for the successful carrying out of a second line of research. Conversely, the same holds regarding the sphere of work of the recent biologist.

Now these last remarks of mine may perhaps tend to confirm an idea which I have at least been told is prevalent in the minds of recent biologists, namely, that the results of Palæontology are so uncertain, so doubtful, and so imperfect, that they are scarcely worthy of serious attention being paid to them. And the best answer I can make to such an opinion, if it really does exist, is to try to place before you some evidence that Palæontology is not mere fossil shell hunting, or the making up of long lists of names to help the geologists to settle their stratigraphical horizons, but may present us with abundance of matter of genuine biological interest.

Since the days of Darwin, there is one subject which more than all others engrosses the attention of scientific biologists. I mean the question of Evolution, or the Doctrine of Descent. Time was when controversies raged round the very idea of Evolution, and when men of science were divided among themselves as to whether the doctrine to which Darwin's theory of Natural Selection gave so mighty an impetus was or was not to be accepted. Times have, however, changed, and I hardly think that we should now find a single true scientific worker who continues to hold on by the old special creation idea. Philosophic zoologists now busy themselves either with amassing morphological evidences of Descent or with the discussion of various theories as to the factors by which organic evolution has been brought about—whether Natural Selection has been the all-sufficient cause or not, whether acquired peculiarities are transmissible, and so on.

From the nature of things it is clear that the voice of the palæontologist can only be heard on the morphological aspect of the question, but to many of us, including myself, the morphological argument is so convincing that we believe that even if the Darwinian theory were proved to-morrow to be utterly baseless, the Doctrine of Descent would not be in the slightest degree affected, but would continue to have as firm a hold on our minds as before.

Now as Palæontology takes us back, far back, into the life of the past, it might be reasonably expected that it would throw great light on the descent of animals, but the amount of its evidence is necessarily much diminished by two unfortunate circumstances. First, the terrible imperfection of the geological record, a fact so obvious to any one having any acquaintance with Geology that it need not be discussed here; and secondly, the circumstance that save in very exceptional cases only the hard parts of animals are preserved, and those too often in an extremely fragmentary and disjointed condition. But though we cannot expect that the palæontological record will ever be anything more than fragmentary, yet the constant occurrence of new and important discoveries leads us to entertain the hope that, in course of time, more and more of its pages will become disclosed to us.

Incomplete, however, as our knowledge of Evolution as derived from Palæontology must be, that is no reason why we should not appraise it at its proper value, and now and again stop for a moment to take stock of the material which has accumulated.

You are all already acquainted with the telling evidence in favour of Evolution furnished by the well-known Mammalian limbs, as well as of teeth, in which the progress, in the course of time, from the more general to the more special is so obvious that I cannot conceive of any unprejudiced person shutting his eyes to the inference that Descent with modification is the reason of these things being so. Suppose, then, that on this occasion we take up the palæontological evidence of Descent in the case of fishes. This I do the more readily because what original work I have been able to do has lain principally in the direction of fossil ichthyology ; and again, because it does seem to me that it is in this department that one has most reason to complain of want of interest on the part of recent biologists, even, I may say, of some professed palæontologists themselves.

But the subject is really of so great an extent that to exhaust it in the course of an address like the present would be simply impossible, so I shall in the main limit myself to the consideration of Palæozoic forms, and this more especially seeing that we may hope for a large addition to our light on the fishes of the more recent geological formations from the fourth volume of the "Catalogue of Fossil Fishes" in the British Museum, which will soon appear from the pen of my friend, Dr. A. Smith Woodward. I need scarcely say how much his previous volume has conduced to a better knowledge of the Mesozoic forms.

Here I may begin by boldly affirming that I include the Marsipobranchii as fishes, in spite of the dictum of Cope that no animal can be a fish which does not possess a lower jaw and a shoulder-girdle. Why not ? The position seems to me to be a merely arbitrary one ; and it is, to say the least, not impossible that the modern Lampreys and Haggs may be, as many believe, the degenerate descendants of originally gnathostomatous forms.

To the origin of the Vertebrata, Palæontology gives us no clue, as the forerunners of the fishes must have been creatures which, like the lowest Chordata of the present day (Urochorda, Hemichorda, Cephalochorda), had no hard parts capable of preservation. And though I shall presently refer again to the subject, I may here affirm that, so far as I can read the record at least, it is impossible to derive from Palæontology any support to the view, recently revived, that the ancient fishes are in any way related to Crustacean or merostomatous ancestors.

What have we then to say concerning the most ancient fishes with which we are acquainted ?

The idea that the minute bodies, known as Conodonts, which occur from the Cambrian to the Carboniferous, are the teeth of fishes and possibly even of ancient Marsipobranchs may now be said to be given up. They are now accepted by the most trustworthy authorities as appertaining to Invertebrata such as Annelides and Gephyrea.

More recently, however, Rohon[1] has described from the Lower Silurian of the neighbourhood of St. Petersburg small teeth (*Palaeodus* and *Archodus*) associated with Conodonts, and which seem to be real fish teeth, but not of Selachians, as is shown by the presence of a pulp cavity surrounded by non-vascular dentine. It is impossible to say anything more of their affinities.

Obscure and fragmentary fish remains have been obtained by Walcot, and described by Jaekel, from rocks in Colorado supposed to be of Lower Silurian or Ordovician age.[2] But doubts have been thrown on their age, and the fossils themselves, which have, it must be owned, a very Devonian look about them, are so extremely fragmentary that they do not help us much in our present purpose.

It is not till we come to the Upper Silurian rocks that we begin to feel on the ground securely under our feet, though we may be certain, from the degree of specialisation of the forms which we there find, that fishes lived in the waters of the globe for long ages previously.

Characteristic of the "Ludlow bone-bed" are certain minute scales on which Pander founded the family Cœlolepidæ, having a flat or sculptured crown, below which is a constricted "neck," and then a base usually perforated by an aperture leading into a central pulp cavity. As these little bodies, looked upon by Agassiz as teeth, were shown by McCoy to be scales, and as they occurred at Ludlow in England and Oesel in Russia along with small Selachian spines (*Onchus*), they were usually considered as appertaining, with the latter, to small Cestraciont sharks. The genera *Thelodus*, *Coelolepis* and others were

founded on these dermal bodies, but it is doubtful if any but the first of these names will stand.

But the aspect of affairs was altogether changed by the discovery three years ago, by the officers of the Geological Survey, of entire specimens of *Thelodus* in the Upper Silurian rocks of the South of Scotland, from which it was evident that the fish, though somewhat shark-like, could hardly be reckoned as a true Selachian.[1] *Thelodus Scoticus*, Traq., has a broad flattened anterior part corresponding to the head and forepart of the body, very bluntly rounded in front, and passing behind into right and left angular flap-like projections, which are sharply marked off from the narrow tail, which is furnished with a deeply cleft heterocercal caudal fin. Unless the flap-like lateral projections are representatives of pectorals, no other fins are present, neither do we find any teeth or jaws, nor any trace of internal skeleton ; and it is only a few days since Mr. Tait, collector to the Geological Survey of Scotland, pointed out to me in a recently acquired specimen a right and left dark spot at the outer margins of the head near the front, which spots may indicate the position of the eyes.[2] A previously unknown genus, *Lanarkia*, Traq., also occurred, in which the creature had the very same form, but instead of having the skin clothed with small shagreen-like scales, possessed, in their place, minute sharp conical hollow spines, without base and open below. What we are to think of those two ancient forms, apparently so primitive, and yet undoubtedly also to a great extent specialised, we shall presently discuss.

Let us now for a moment look at the genus *Drepanaspis*, Schlüter, from the Lower Devonian of Gmunden in Western Germany.[3] We have here a strange creature whose shape entirely reminds us of that of *Thelodus*, having the same flat broad anterior part, bluntly rounded in front, and angulated behind, to which is appended a narrow tail ending in a heterocercal caudal fin, which is, however, scarcely bilobate. But here the dermal covering, instead of consisting of separate scales or spinelets, shows a close carapace of hard bony plates, of which two are especially large and prominent—the median dorsal and the median ventral—their large ones being placed around the margins, while the intervening space is occupied by a mosaic of small polygonal pieces. The position of the mouth, a transverse slit, is seen just at the anterior margin ; it is bounded behind by a median mental or chin-plate, but no jaws properly so called are visible, nor are there any teeth. Then on each margin near the front of the head is a small round pit, exactly in the position of the dark spot seen in some examples of *Thelodus*, which, if not an orbit, must indicate the position of some organ of sense. Again, the tail is covered with scales after the manner of a "ganoid" fish, being rhombic on the sides, but assuming the form of long deeply imbricating fulcra on the dorsal and ventral margins. The position of the branchial opening, or openings, has not yet been definitely ascertained.

All these plates are closely covered with stellate tubercles, and we cannot escape from the conclusion that they are formed by the fusion of small shagreen bodies like those of *Thelodus*, and united to bony matter developed in a deeper layer of the skin.

If the angular lateral flaps of *Thelodus* represent pectoral fins, then we would have the exceedingly strange phenomena of such structures becoming functionally useless by enclosure in hard unyielding plates, though still influencing the general outline of the fish. Be that as it may, can we doubt that in *Drepanaspis* we have a form derived by specialisation from a Cœlolepid ancestor ?

This *Drepanaspis* throws likewise a much desired light on the fragmentary Devonian remains known since Agassiz's time as *Psammosteus*. These consist of large plates and fragments of plates, composed of vaso-dentine, and sculptured externally by minute closely-set stellate tubercles, exactly resembling the scales of some species of *Thelodus*. These tubercles are also frequently arranged in small polygonal areas, reminding us exactly of the small polygonal plates of *Drepanaspis*, and, like them, often having a specially large tubercle in the centre.

1 "Ueber untersilurische Fische," *Mélanges Géol. et Paléont.* vol. i. (St. Petersburg, 1890), pp. 9-14.
2 *Bulletin Geol. Soc. America.* vol. iii. 1892, pp. 153-171.

1 R. H. Traquair, " Report on Fossil Fishes collected by the Geological Survey in the Silurian Rocks of the South of Scotland," *Trans. Roy. Soc. Edin.*, vol. xxxix. 1899, pp. 827-864.
2 I am indebted to Sir A. Geikie, F.R.S., Director-General of the Geological Survey, for permission to make use of this and other facts disclosed by Mr. Tait's work in the Lesmahagow Silurians during the present summer.
3 R. H. Traquair, *Geol. Mag.*, April 1900.

That *Psammosteus* had an ancestry similar to that of *Drepanaspis* can also hardly be doubted.

Finally, in the well-known *Pteraspis* of the Upper Silurian and Lower Devonian formations we have a creature which also has the head and anterior part of the body enveloped in a carapace, to which a tail covered with rhombic scales is appended behind, and, though the caudal fin has never been properly seen, such remains of it as have occurred distinctly indicate that it was heterocercal in its contour. The plates of the carapace have a striking resemblance in general arrangement to those of *Drepanaspis*, though the small polygonal pieces have disappeared, and there is a prominent pointed rostrum in front of the mouth ; and it is to be noted that the small round apertures usually supposed to be orbits are in a position quite analogous to that of the sensory pits in *Drepanaspis*. The plates of the carapace of *Pteraspis* are not, however, tuberculated, but ornamented by fine close parallel ridges, the microscopic structure of which, along with their frequent lateral crenulation, leaves no doubt in our minds that they have been formed by the running together in lines of *Thelodus*-like shagreen grains. An aperture supposed to be branchial is seen on the plate forming the posterior angle of the carapace on each side.

Until these recent discoveries concerning the Cœlolepidæ and Drepanaspidæ, *Pteraspis* and its allies, *Cyathaspis* and *Palæaspis*, constituted the only family included in the order Heterostraci of the sub-class Ostracodermi, distinguished, as shown by Lankester, by the absence of bone lacunæ in the microscopic structure of their plates. It is now, however, clear that we can trace them back to an ancestral family in which the external dermal armature was still in the generalised form of separate shagreen grains or spinelets.

But the Ostracodermi are usually made to include two other groups or orders, namely the Osteostraci and the Asterolepida.[1] The Osteostraci are distinguished from the Heterostraci by the possession of lacunæ in their bone structure, and by having the eyes in the middle of the head-shield instead of at the sides. *Cephalaspis*, which occurs from the Upper Silurian to the top of the Devonian, is the best known representative of this division. Instead of a carapace, we find a large head-shield of one piece, though its structure shows evidence of its having been originally composed of a mosaic of small polygonal plates, and it is also to be noted that the surface is ornamented by small tubercles, these frequently being one larger in size in the centre of each polygonal area. The posterior-external angles of the shield project backward in a right and left pointed process or *cornu*, scarcely developed in *C. Murchisoni*, internal to which, and also organically connected with the head-shield, is a rounded flap-like structure, which strongly reminds us of the lateral flaps of the Cœlolepidæ. The body is covered with scales, which on the sides are high and narrow ; there is a small dorsal fin, and the caudal, though heterocercal, is not bilobate. It is scarcely necessary for me to add that we find just a little evidence of jaws or of teeth as in the case of the Heterostraci.

The association of the Heterostraci and Osteostraci in one sub-class of Ostracodermi has been strongly protested against by Prof. Lankester and Dr. O. M. Reis, but here the Scottish Silurian strata come to the rescue with a form which I described last year under the name of *Ateleaspis tessellata*, and of which some more perfect examples than those at my disposal at that time have recently come to light through the labours of Mr. Tait, of the Geological Survey of Scotland.

Here we have a creature whose general form reminds us strongly of *Thelodus*, but whose close affinity to *Cephalaspis* is absolutely plain, were it only on account of the indications of orbits on the top of the head.

The expanded anterior part which here represents the head-shield of *Cephalaspis* shows not the slightest trace of cornua, but forms posteriorly a gently rounded lobe on each side, clearly suggesting that the cornual flaps of *Cephalaspis* are homologous with and derivable from the lateral expanses in the Cœlolepidæ. This cephalic covering is composed of numerous small polygonal plates like those of which the head-shield in *Cephalaspis* no doubt originally consisted, and the minute tubercles which cover their outer surfaces also suggest that the superficial layer was formed by the fusion of Cœlolepid scales.

[1] To these I myself recently added a fourth, the Anaspida, for the remarkable Upper Silurian family of Birkeniidæ, but as these throw no light as yet on the problem of Descent they may at present be only mentioned.

The body is covered with rhombic scales, sculptured externally with tubercles and wavy transverse ridges, and arranged in lines having the same general direction as the scutes of *Cephalaspis*, from which we may infer that the latter originated from the fusion of scales of similar form. The fins are as in *Cephalaspis*, there being one small dorsal situated far back, and a heterocercal caudal, which is triangular in shape, and not deeply cleft into upper and lower lobes as in the Cœlolepidæ. Finally, the scales, on microscopic examination, show well-developed bone lacunæ in their internal structure.

That *Ateleaspis* belongs to the Osteostraci there is thus not the smallest doubt, but its general resemblance to the Cœlolepidæ in its contour anteriorly led me to regard it as an annectent form, and consequently to believe that there is after all a genuine genetic connection between the Heterostraci and the Osteostraci. And I have not seen reason to depart from that opinion even th ugh *Ateleaspis* turns out to be still closer to *Cephalaspis* than was apparent in the original specimens.

If this be so, then *Cephalaspis*, as well as *Pteraspis* and its allies, is traceable to the Cœlolepidæ, shark-like creatures in which, as we have already seen, the dermal covering consists of small shagreen-like scales, or of minute hollow spines, and consequently all theories as to the arthropod origin of the Ostracodermi, so far as they are founded on the external configuration of the carapace in the more specialised forms, must fall to the ground. And from the close resemblance of these scales of *Thelodus* to Elasmobranch shagreen bodies—for forty-five years they had been, by most authors, actually referred to the Selachii—I concluded that the Cœlolepidæ owed their origin to some form of primitive Elasmobranchs. That is, however, not in accordance with the view of the late Prof. Cope, that the Ostracodermi are more related to the Marsipobranchii, and that, from the apparent absence of lower jaw, they should be placed along with the last-named group in a class of Agnatha, altogether apart from the fishes proper. And Dr. Smith Woodward, who is inclined to favour Cope's theory, has expressed his view that the similarity of the Cœlolepid scales to Elasmobranch shagreen is no proof of an Elasmobranch derivation, but that such structures, representing the simplest form of dermal hard parts, may have originated independently in far distant groups.[1] Knowing what we do of the occurrence of strange parallelisms in evolution, it would not be safe to deny such a possibility. But as to a Marsipobranch affinity, I would point out that the apparent want of lower jaw among the hard parts which nature has preserved for us is no proof of the absence of a Meckelian cartilage among the soft parts which are lost to us for ever ; and also, as Prof. Lankester has remarked, that there is no evidence whatever that any of the creatures classed together as Ostracodermi were monorhinal like the Lampreys. The only fossil vertebrate having a single median opening, presumably nasal, in the front of the head is *Palæospondylus*, but, whatever be the true affinities of this little creature, at present the subject of so much dispute, I think we may be very sure that it is not an Ostracoderm.

The Devonian "Antiarcha" or Asterolepida, of which *Pterichthys* is the best known genus, are also usually placed in the Ostracodermi, with which they agree in the possession of a carapace of bony plates, in the absence of distinct lower jaw or teeth, in the non-preservation of internal skeleton, and in having a scaly tail furnished with a heterocercal caudal fin, and, as in the Cephalaspidæ, also with a small dorsal. But they have in addition a pair of singular jointed thoracic limbs, evidently organs of progression, which are totally unlike anything in the Osteostraci or in the Heterostraci, or indeed in any other group of fishes. These limbs are covered with bony plates and hollow inside, but though I once fancifully compared them in that respect with the limbs of insects, I must protest strongly against this expression of mine being quoted in favour of the arthropod theory of the derivation of the Vertebrata !

Nor do I think that there is any probability in the view published by Simroth nine years ago,[2] namely, that *Pterichthys* may have been a land animal which used its limbs for progression on dry ground, and that the origin of the heterocercal tail was the bending up of the extremity of the vertebral axis caused by its being dragged behind the creature in the act of walking. That view was promulgated before the discovery of the membranous expanse of the caudal fin in this genus.

But though the Asterolepida are apparently related to and inclusible in the Ostracodermi, the geological record is silent as

[1] *Geol. Mag.*, March 1900.
[2] "Die Entstehung der Landthiere," Leipzig, 1891.

to their immediate origin, no intermediate forms having been found connecting them more closely with either the Heterostraci or the Osteostraci. In the possession of bone lacunæ and of a dorsal fin, a greater resemblance to the latter, but it may be looked upon as certain that they could have had no direct origin from that group.

As regards the Ostracodermi as a sub-class, they become extinct at the end of the Devonian epoch, and cannot be credited with any share in the evolution of the fishes of more recent periods, not even if we restore the Coccosteans or Arthrodira to their fellowship. To the latter most enigmatical group, which I shall still continue to look upon as fishes, I shall make some reference further on.

Coming now to say a word regarding the Elasmobranchii, it is plain from the fin-spines found in Upper Silurian rocks that they are of very ancient origin, and that if we only knew them properly they would have a wonderful tale of evolution to tell. But their internal skeleton is from its nature not calculated for preservation, and for the most part we only know those creatures from scattered teeth, fin-spines and shagreen, specimens showing either external configuration or internal structure being rare, especially in Palæozoic strata. But from what we do know, there is no doubt that the ancient sharks were less specialised than those of the present day, and that the recent Notidanids still preserve peculiarities which were common in the Selachii of past ages.

If we ask whether the fossil sharks throw any light on the disputed origin of the paired limbs, whether from the specialisation of right and left lateral folds, or whether that type of limb called "archipterygium" by Gegenbaur, consisting of a central jointed axis with pre- and post-axial radial cartilage attached, was the original form, I fear we get no very definite answer from Elasmobranch palæontology. The paired fins of the Upper Devonian shark, *Cladoselache*, as described by Bashford Dean, Smith Woodward and others, seem to favour the lateral fold theory, and Cope pointed to the right and left series of small intermediate spines which in some Lower Devonian Acanthodei (*Parexus* and *Climatius*) extend between the pectorals and ventrals as evidence of a former continuous lateral fin. So also, if I am right in looking on the lateral flaps of the Cœlolepidæ as fins, the evidence of these ancient Ostracodermi would be in the same direction.

But, on the other hand, we have the remarkable group of Pleuracanthidæ, extending from the Lower Permian back to the Upper Devonian, in which the paired fins are represented by an "archipterygium" which in the pectoral at least is biserial.

From this biserial "archipterygium" in the Pleuracanthidæ, Prof. A. Fritsch, ten years ago,[1] derived the tribasal arrangement of modern sharks, much according to the Gegenbaurian method, effecting, however, a compromise with the lateral fold theory by assuming that the Pleuracanth form originated from one, consisting of simple parallel rods, like that described in *Cladoselache*.

In my description of the pectoral fin of the Carboniferous *Cladodus Neilsoni*[2] I have shown that the cartilaginous structures apparently present an uniserial archipterygium intermediate between the arrangement in *Pleuracanthus* and that in the modern sharks, but I felt compelled to acknowledge that the specimen might also be interpreted in exactly the opposite way, namely, as an example of a transition from the "ptychopterygium" of *Cladoselache* to the Pleuracanth and Dipnoan limb. And so in fact this fin of *Cladodus* is claimed in support of their views by both parties in the dispute.

When we add that Semon emphatically denies that there is any proof for considering that the pectoral fin of *Cladoselache* is primitive in its type,[3] and that Campbell Brown, in his recent paper on the Mesozoic genus *Hybodus*,[4] supports Gegenbaur's theory, it will be seen that Elasmobranch palæontology has not as yet uttered any very clear or decided voice on the question as to whether the so-called archipterygium is the primary form of paired fin in the fish, or only a secondary modification. We shall now inquire if we can obtain any more light on the subject from the Crossopterygii and Dipnoi.

---

[1] "Fauna der Gaskohle und der Kalksteine der Permformation Böhmens," vol. iii. Pt. i. (Prague, 1890), pp. 44-45.
[2] *Trans. Geol. Soc.* (Glasgow), vol. xi. Pt. i. 1897, pp. 41-50.
[3] "Die Entwickelung der paarigen Flossen des Ceratodus Forsteri." (Jena, 1898.)
[4] "Ueber das Genus *Hybodus* und seine systematische Stellung,' *Palaeontographica*, vol. xlvi. 1900.

The Crossopterygii are a group of Teleostomous fishes, characterised externally by their jugular plates and lobate paired fins, and represented in the present day only by the African genera *Polypterus* and *Calamoichthys*, which together form the peculiar family Polypteridæ. The Crossopterygii appear suddenly in the middle of the Devonian period, their previous ancestry being unknown to us.

Four families[1] are known to us in Palæozoic times—the Osteolepidæ, Rhizodontidæ, Holoptychiidæ and Cœlacanthidæ, but it is only with the first three that we have at present to deal. The Osteolepidæ and Rhizodontidæ, which appear together in Middle, and die out together in Upper Palæozoic times, resemble each other very closely. In both we have the paired fins, more especially the pectoral, obtusely or subacutely lobate ; there are two separate dorsal fins, one anal, and the caudal, which is usually heterocercal, though in some genera it is more or less diphycercal. In both the teeth are conical and have the same complex structure, the dentine being towards the base thrown into vertical labyrinthic folds, exactly as in the Stegocephalian Labyrinthodonts, and this along with the lung-like development of the double air-bladder in the recent Polypteridæ has given rise to the view from these forms the Stegocephalia have originated. The nasal openings must have been on the under surface of the snout, as in the Dipnoi.

Of these two so closely allied families we must conclude that the Osteolepidæ are the more primitive, as in them the scales are acutely rhombic and usually covered with a thick layer of ganoine, while in the Rhizodontidæ they are rounded, deeply imbricating, and normally devoid of the ganoine layer, which, however, occasionally recurs on the scales of *Rhizodopsis* and the fin-rays of *Gyroptychius*.

What then of the structure of the paired fins? Fortunately in the Rhizodont genera *Tristichopterus* and *Eusthenopteron* the internal skeleton of the lobe was ossified, and what we see clearly exhibited in the pectoral of some specimens is striking enough. We have a basal piece attached to the shoulder-girdle and followed by a median axis of four ossicles placed end to end. The first of these shows on its postaxial margin a strong projecting process, while to its preaxial side, close to its distal extremity, a small radial piece is obliquely articulated, and a similar one is joined also to the second and third segments of the axis. The arrangement in the ventral fin is essentially similar.

In fact, we have in the Rhizodontidæ a short uniserial "archipterygium," and the question is, Has this been formed by the shortening up and degeneration of an originally elongated and biserial one, or on the other hand do we find here a condition in which the stage last referred to has not yet been attained? This question is inseparable from the next, whether the Rhizodonts or the Holoptychians form the most advanced type?

The Holoptychidæ resemble the Rhizodontidæ extremely closely in their external head-bones, in their rounded, deeply imbricating scales, and in the form and arrangement of their median fins. But the teeth show a more complex and specialised structure than those of the Rhizodontidæ ; the simple vertical vascular tubes formed by the repeated folding of the dentine in that family being connected by lateral branches around which the dentine tubules are grouped in such a way as to give rise in transverse sections to a radiating arboreseent appearance ; hence the term "dendrodont." In this respect, then, the Holoptychidæ show an advance on the Rhizodontidæ—what then of the paired fins? While the ventral remains subacutely lobate, as in the previous family, the pectoral has now assumed an elongated *acutely lobate* shape, with the fin-rays arranged along the two sides of a central scaly axis exactly as in the Dipnoi ; and though the internal skeleton has not yet been seen, yet, judging by analogy, we cannot escape the belief that it was in the form of a complete biserial "archipterygium."

What, then, is the condition of affairs in the oldest known Dipnoan ?

The oldest member of this group with whose configuration we are acquainted is *Dipterus*, which likewise appears in the middle of the Devonian period simultaneously with the Osteolepidæ, Rhizodontidæ and Holoptychiidæ. In external form it closely resembles a Holoptychian, having a heterocercal caudal fin, two similarly placed dorsals, one anal, and circular imbricating scales, which, however, have the exposed part covered

---

[1] *Five*, if we include the singular and still imperfectly known Tarrasiidæ of the Lower Carboniferous.

with smooth ganoine. But now we have the ventrals as well as the pectorals acutely lobate in shape, and presumably archipterygial in structure ; the top of the head is covered with many small plates, there is no longer a dentigerous maxilla, the skull is autostylic, and the palatopterygoids and the mandibular splenial are like those of *Ceratodus* and bear each a tooth-plate with radiating ridges.

Now, comparing *Dipterus* with the recent *Ceratodus* and *Protopterus*, the first conclusion we are likely to draw is, that the older Dipnoan is a very specialised form, that its heterocercal tail and separate dorsals and anal are due to specialisation from the continuous diphycercal dorso-ano-caudal arrangement in the recent forms, that the Holoptychiidæ were developed from it by shortening-up of the ventral archipterygium, as well as by the changes in cranial structure, and that the Rhizodontidæ and Osteolepidæ are a still more specialised series in which the pectoral archipterygium has also shared the fate of the ventral in becoming shortened up and uniserial.

Five years ago, however, M. Dollo, of the Natural History Museum at Brussels, the well-known describer of the fossil reptiles of Bernissart, proposed a new view to the effect that the process of evolution had gone exactly in the opposite direction ;[1] and after long consideration of the subject I find it difficult to escape from the conclusion that this view is more in accordance with the facts of the case, though, as we shall see, it also has its own difficulties.

I have already indicated above that we are, on account of the more specialised structure of the teeth, justified in considering the Holoptychians, with their acutely lobate pectorals, a newer type than the Rhizodonts, even though they did not survive so long in geological time. What, then, of the question of autostyly?

We do not know the suspensorium of *Holoptychius*, but that of the Rhizodontidæ was certainly hyostylic, as in the recent *Polypterus*. Now as there can be no doubt that the autostylic condition of skull is a specialisation on the hyostylic form, as seen also in the Chimæroids and in the Amphibia, to suppose that the hyostylic Crossopterygii were evolved from the autostylic Dipnoi is, to say the least, highly improbable ; in my own opinion, as well as in that of M. Dollo, it will not stand. And if we assume a genetic connection between the two groups it is in accordance with all analogy to look on the Dipnoi as the children and not as the parents of the Crossopterygii.

M. Dollo adopts the opinion of Messrs. Balfour and Parker that the apparently primitive diphycercal form of tail of the recent Dipnoi is secondary, and caused by the abortion of the termination of the vertebral axis as in various "Teleostei," so that no argument can be based on the supposition that it represents the original "protocercal" or preheterocercal stage. Very likely that is so, but it is not of so much importance for the present inquiry, as both in the Osteolepidæ and Rhizodontidæ we find among otherwise closely allied genera some which are heterocercal, others more or less diphycercal. *Diplopterus*, for example, differs from *Thursius* only by its diphycercal tail, and in like manner among the Rhizodontidæ *Tristichopterus* is heterocercal, *Eusthenopteron* is nearly diphycercal, and there can be no doubt that, in spite of this, their caudal fins are perfectly homologous structures.

But of special interest is the question of the primitive or non-primitive nature of the continuity of the median fins in the recent Dipnoi. Like others I was inclined to believe it primitive, and that the broken-up condition of these fins in *Dipterus* was a subsequent specialisation, and in fact gave the series *Phaneropleuron, Scaumenacia, Dipterus macropterus* and *D. Valenciennesii* as illustrating this process of differentiation. This view of course draws on the imperfection of the geological record in assuming the existence of ancient pre-Dipterian Dipnoi with continuous median fins, which have never yet been discovered. But Dollo, using the very same series of forms, showed good reason for reading it in exactly the opposite direction.

The series is as follows :—

(1) *Dipterus Valenciennesii*, Sedgw. and Murch., from the Orcadian Old Red, where the oldest Dipnoan with whose shape we are acquainted, has two dorsal fins with *short* bases, a heterocercal caudal, and one short-based anal.

(2) *Dipterus macropterus*, Traq., from a somewhat higher horizon in the Orcadian series, has the base of the second dorsal much *extended*, the other fins remaining as before.

[1] "Sur la Phylogénie des Dipneustes," *Bulletin Soc. belge géol. paléont.* Aydr., vol. ix. 1895.

(3) In *Scaumenacia curta* (Whiteaves), from the Upper Devonian of Canada, the first dorsal has advanced considerably towards the head, and its base has now become elongated, while the second has become still larger and more extended, though still distinct from the caudal posteriorly.

(4) In *Phaneropleuron Andersoni*, Huxley, from the Upper Old Red of Fifeshire, the two dorsal fins are now fused with each other and with the caudal, forming a long continuous fin along the dorsal margin, while the tail has become nearly diphycercal, with elongation of the base of the lower division of the fin. But the anal still remains separate, narrow, and short-based.

(5) In the Carboniferous *Uronemus lobatus*, Ag., the anal is now also absorbed in the lower division of the caudal, forming now, likewise on the hæmal aspect, a continuous median fin behind the ventrals. There is also a last and feeble remnant of a tendency to an upward direction of the extremity of the vertebral axis.

(6) In the recent *Ceratodus Forsteri*, Krefft, the tail is diphycercal (secondary diphycercy), the median fins are continuous, the pectorals and ventrals retain the biserial archipterygium, but the cranial roof-bones have become few.

(7) In *Protopterus annectens*, Owen, the paired fins are more eel-like, and the paired fins have lost the lanceolate leaf-like appearance which they show in *Ceratodus* and the older Dipnoi They are like slender filaments in shape, with a fringe on one side of minute dermal rays ; internally they retain the central jointed axis of the "archipterygium," but according to Wiedersheim the radials are gone, except it may be one pair at the very base of the filament.

(8) Finally, in *Lepidosiren paradoxa*, Fitz., the paired fins are still more reduced, having become very small and short, with only the axis remaining.

From this point of view, then, *Dipterus*, instead of being the most specialised Dipnoan, is the most archaic, and the modern *Ceratodus, Protopterus* and *Lepidosiren* are degenerate forms, and instead of the Crossopterygii being the offspring of *Dipterus*-like forms, it is exactly the other way, the Dipnoi owing their origin to Holoptychiidæ, which again are a specialisation on the Rhizodontidæ, though they did not survive so long as these in geological time. Consequently the *Ceratodus* limb, with its long median segmented axis and biserial arrangement of radials, is not an archipterygium in the literal sense of the word, but a derivative form traceable to the short uniserial type in the Rhizodonts. But from what form of fin *that* was derived is a question to which palæontology gives us no answer, for the progenitors of the Crossopterygii are as yet unknown to us.

Plausible and attractive as this theory undoubtedly is, and though it relieves the palæontologist from many difficulties which force themselves upon his mind if he tries to abide by the belief that the Dipnoan form of limb had a selachian origin, and was in turn handed on by them to the Crossopterygii, yet it is not without its own stumbling-blocks.

First, as to the dentition, on which, however, M. Dollo does not seem to put much stress, it is impossible to derive *Dipterus* directly from the Holoptychiidæ, unless it suddenly acquired, as so many of us have to do as we grow older, a new set of teeth. The dentodont dentition of *Holoptychius* could not in any way be transformed into the ctenodont or ceratodont one of *Dipterus* : both are highly specialised conditions, but in different directions. Semon has recently shown that the tooth-plates of the recent *Ceratodus* arise from the concrescence of numerous small simple conical teeth, at first separate from each other.[1] Now this stage in the embryo of the recent form represents to some extent the condition in the Uronemidæ of the Carboniferous and Lower Permian, which stand quite in the middle of Dollo's series.

Again, the idea of the origin of the Dipnoi from the Crossopterygii in the manner sketched above cuts off every thought of a genetic connection between the biserial archipterygium in them and in the Pleuracanthidæ, so that we should have to believe that this very peculiar type of limb arose independently in the Selachii as a parallel development. It may be asked, Why not? We may feel perfectly assured that the autostylic condition of the skull in the Holocephali arose independently of that in the Dipnoi, as did likewise a certain amount of resemblance in their dentition. But those who from embryological grounds oppose any notion of the origin of the Dipnoi from "Ganoids" might here say, if they chose, If so, why should not also the same form of limb have been independently evolved in Crossopterygii ?

Accordingly, while philosophic palæontology is much indebted

[1] "Die Zahnentwickelung des *Ceratodus Forsteri*." (Jena, 1899.)

to M. Dollo for his brilliant essay, and though we must agree with him in many things, such as that the Crossopterygii were not derived *from* the Dipnoi, and that the modern representatives of the latter group are degenerate forms, yet as to the *immediate* ancestry of the Dipnoi themselves, and the diphyletic origin of the so-called archipterygium, we had best for the present keep an open mind.

In his "Catalogue of the Fossil Fishes" in the British Museum (vol. ii. 1891), Dr. Smith Woodward, following the suggestion of Newberry in 1875, classified the Coccosteans or "Arthrodira" as an extremely specialised group of Dipnoi. At first I was much taken with that idea, but after looking more closely into the subject I began to doubt it extremely. My own opinion at present is that the Coccosteans are Teleostomi belonging to the next order, Actinopterygii ; but Prof. Bashford Dean, of New York, will not have them to be even "fishes," but places them in a distinct class of "Arthrognatha," which he places next to the Ostracophori (= Ostracodermi), even hinting at a possible union with them, whereby the old "Placodermata" of McCoy would be restored. It will, therefore, be better to leave them out of consideration for the present, pending a thorough re-examination of their structure and affinities.

We come then to the great order of Actinopterygii, to which a large number of the fishes of later Palæozoic age belong, as well as the great mass of those of Mesozoic, Tertiary and Modern times. Of these we first take into consideration the oldest sub-order, namely, the Acipenseroidei or Sturgeon tribe, in which the dermal rays of the median fins are more numerous than their supporting ossicles, while the tail is, in most, completely heterocercal. And the oldest family of Acipenseroids with which we are acquainted is that of the Palæoniscidæ, which, in addition to well-developed cranial and facial bones, has the body normally covered with rhombic ganoid scales furnished with peg-and-socket articulations. Of this family one genus, *Cheirolepis*, appears in the same Devonian strata (Orcadian series) with the earliest known Crossopterygii, and of its immediate ancestry we know no more than we do of theirs. *Cheirolepis* is a fully evolved palæoniscid, as shown by its oblique suspensorium, wide gape, and other points of its structure. In the Lower Carboniferous rocks of Scotland, where the family attains an enormous development, we find one or two genera, *e.g.* *Canobius*, which appear less specialised, as the suspensorium is nearly vertical, and the mouth consequently smaller.

This endures up to the Purbeck division of the Jurassic formation, and in the Carboniferous *Cryphiolepis*, the Lower Permian *Trissolepis* and the Jurassic *Coccolepis* we find the same degeneration of the rhombic scales into those of a circular form and imbricating arrangement, which we find repeated in other groups of "Ganoids." In fact, in one Carboniferous genus, *Phanerosteon*, the scales disappear altogether with the exception of those on the body prolongation in the upper lobe of the caudal fin, and a few just behind the shoulder-girdle.

And in these Palæozoic times we notice also a side branch of the Palæoniscidæ, constituting the family Platysomidæ, in which, while the median fins acquire elongated bases, the body becomes shortened up and deep in contour. The scales become high and narrow, their internal rib and articular spine coincident with the anterior margin ; the suspensorium, too, instead of swinging back as in the typical Palæoniscidæ, tends to be directed obliquely forward, while the snout becomes simultaneously elongated in front of the nares.

A most interesting series of forms can be set up, beginning with *Eurynotus*, which, though it has the platysomid head contour and a long-based dorsal, has only a slight deepening of the body, and still retains the palæoniscid squamation and a short-based anal fin. In *Mesolepis*, which resembles *Eurynotus* in shape, being only slightly deeper, we have now the characteristic platysomid squamation, and the base of the anal fin is considerably elongated. *Platysomus* has a still more elongated anal fin, and the body is rhombic ; while in *Cheirodus* the body is still deeper in contour, with peculiar dorsal and ventral beaks, long fringing dorsal and anal fins, while the ventrals seem to have disappeared altogether. Here also, as in the allied genus *Cheirodopsis*, the separate cylindro-conical teeth characteristic of the family are, on the palatal and splenial bones, replaced by dental plates, reminding us of those of the Dipnoi. Certainly the Platysomidæ seem to me to form a morphological series telling us as strongly in favour of Descent as any other in the domain of palæontology.[1]

[1] R. H. Traquair, " Structure and Affinities of the Platysomidæ," *Trans. Roy. Soc. Edin.,* xxix. 1879, pp. 343-391.

If we now return to the Palæoniscidæ we find that they dwindled away in numbers in the Jurassic rocks, and finally became extinct at the close of that epoch. But already in the Lias (leaving the Triassic Catopteridæ out of consideration for the present) we find that they have sent off another offshoot sufficiently distinct to be reckoned as a new and separate family, namely, the Chondrosteidæ, in which the path of degeneration, in all but the matter of size, seems to have been entered on.

In the genus *Chondrosteus*, though the palæoniscid type is clearly traceable in the cranial structure, there is marked degeneration as regards the amount of ossification, and though the suspensorium is still obliquely directed backward the toothless jaws are comparatively short, and the mouth seems now to have become tucked in under the snout as in the recent sturgeon. Then the scales have entirely disappeared from the skin except on the upper lobe of the heterocercal caudal fin, where they are still found arranged exactly as in the Palæoniscidæ.

Chondrosteus in fact conducts us to the recent Acipenseroids—the Polyodontidæ (Paddle-fishes) and Acipenseridæ (Sturgeons).

The first of these resembles *Chondrosteus* in the nakedness of the skin, except on the upper lobe of the caudal fin,[1] the more palæoniscid aspect of the external cranial plates, such of them as remain, for they are now still further reduced. But in front of the mouth and eyes there is an addition in the form of an enormous vertically flattened paddle-shaped snout covered above and below with a large number of small ossifications.

The sturgeons have, however, nearly altogether lost the palæoniscid arrangement of the cranial roof-bones, which, strange to say, now exhibit an arrangement reminding us of that in *Dipterus*, and the external facial plates are still more reduced than even in *Polyodon* ; but we may note a very strong resemblance to *Chondrosteus* in the position of the mouth, the edentulous jaws, and the jugal bone, indeed also in the palatal apparatus.

So the sturgeons and paddle-fishes of the present day would seem to be the degenerate, though bulky, descendants of the once extensively-developed group of Palæoniscidæ, even as the modern Dipnoi are degenerated from those of Palæozoic times.

We now notice another *apparent* offshoot of the Palæoniscidæ, namely, the family of Catopteridæ (*Catopterus* and *Dictyopyge*), which is limited to rocks of Triassic age. Unfortunately the osteology of the head is not well known, but Dr. Smith Woodward's observations are to the effect that both the head and shoulder-girdle are of palæoniscoid type. The relationship of these small fishes to the Palæoniscidæ is shown by the general shape, the number and position of the fins, the rhombic ganoid scales, and the close arrangement of the rays of the median fins. But the rays of the dorsal and anal fins are now almost equal in number to their supporting ossicles, and the tail has become only abbreviate heterocercal. That is to say, the caudal body prolongation no longer proceeds to the termination of the upper lobe, which is reduced in size and in the number of its rays. The Catopteridæ are obviously an annectent group, as although from their abbreviate heterocercal tail they have usually been placed in the next sub-order, Dr. Smith Woodward prefers to look upon them as Chrondrostei (*i.e.* Acipenseroidei).[2] Wherever we place them they express the beginning of a set of changes towards a more modern type of fish, which are emphasised in the great series of Lepidosteoid fishes (Protospondyli + Æthospondyli of Smith Woodward), being the fishes more or less allied to the recent Bony Pike of North America.

But these changes must have been well advanced before the Triassic era, for already in the Upper Permian occurs the genus *Acentrophorus*, whose fellowship with *Semionotus*, *Lepidotus*, and all the rest of the series of Mesozoic semi-heterocercal "Ganoids" is at once obvious.

If we look at the configuration of a typical Jurassic member of this series, such as *Lepidotus* or *Eugnathus*, we shall at once see that we are a stage nearer the modern osseous fish. Though the scales are bony, rhombic, and ganoid we are struck by the "Teleostean"-like aspect of the external bones and plates of

[1] Collinge has, however, found rudimentary scales in the skin of the recent *Polyodon folium* (*Journ. Anat. and Phys.,* ix. pp. 458-487), and Cope has described an allied Eocene genus, *Crossopholis,* in which minute scales are seen (*Mem. Nat. Acad. Sciences,* iii. 1886, pp. 161-163).
[2] Dr. Smith Woodward also refers the singular Belonorhynchidæ of the Trias to the same sub-order on account of the excess of the number of the dermal rays of the dorsal and anal fins, over that of their supporting ossicles, even although the tail is here abbreviate diphycercal.

the head, the rays of the dorsal and anal fins are fewer and correspond in their number to that of the internal supports or "interspinous" bones, while in the caudal we see again the semi-heterocercal or abbreviate-heterocercal condition we noticed above in *Catopterus*.

Then if we refer to the tail of *Lepidosteus* itself we shall observe how few are its rays and how evident it is that we have here to do only with the lower lobe of the original palæoniscoid caudal fin. For a convincing corroboration of this we have only to look at the tail of the embryo *Lepidosteus* as described and figured by Prof. A. Agassiz to see that it in reality passed through an Acipenseroid stage, and the last we see of the upper lobe of this tail is in the form of a filament which projects from the top of the original lower lobe and then disappears.

Again, in these Lepidosteid forms we have a repetition of the same tendency for the thick rhombic, peg-and-socket articulating scales to become rounded and imbricating as we saw in the Crossopterygii and again in the Palæoniscidæ. So, for instance, in *Caturus*, which has been shown by Dr. Smith Woodward to resemble *Eugnathus* so closely in structure, the scales are deeply overlapping, and most of them "cycloidal" in shape. To such an extent does this go that in the recent *Amia*, whose skeletal structure so clearly shows it to belong to this group, the rounded scales are so thin and flexible that after it was removed from the Clupeoid family, or Herrings, and placed among the "Ganoids" it was considered to be the type of a distinct sub-order of "Amioidei." Ten years ago, however, Dr. Beard came to the conclusion, from anatomical and embryological data, that this division could no longer be maintained, and that the Amioids must in fact be united with the Lepidosteids.[1] Dr. Smith Woodward has, therefore, in the third volume of his catalogue, done well to reduce the "Amioidei" to the rank of a family, including also the Jurassic genera *Liodesmus* and *Megalurus*, and to place this family close to the Eugnathidæ.

As the Asipenseroids dwindled away after the close of the great Palæozoic era, and are now scantily represented only by the degenerate paddle-fishes and sturgeons, so the Lepidosteid series, flourishing greatly in the Trias and Jura, in their turn declined in the Cretaceous, and in the Tertiary period became about as much a thing of the past as they are now, the North American *Lepidosteus* and *Amia*, of which remains of extinct species have also been found in Eocene and Miocene rocks, only remaining. These two genera can, however, hardly be called "degenerate."

But that the fishes which succeeded the Lepidosteids in populating the seas and rivers of the globe were evolved from them there can be no reasonable doubt, while it is equally clear that they branched off at an early period, as already in the Trias we find the first representatives of the modern type of Isospondyli, which contains our familiar Herrings, Salmonids, Elopids, Scopelids, &c. For Dr. Smith Woodward has not only definitely placed the Jurassic Leptolepidæ and Oligopleuridæ in the Isospondyli, but also the Pholidophoridæ, which appear in the Trias and extend to the Purbeck. And it is of special interest that in the Pholidophori the scales are still brilliantly ganoid and mostly retain the peg-and-socket articulation, while in the allied Leptolepidæ, although they have become thin and circular, a layer of ganoine mostly remains.

With the Isospondyli we now get fairly among the bony fishes of modern type—Teleostei as we used to call them—to which other sub-orders are added in Cretaceous and Tertiary times, and which in the present day have assumed an overwhelming numerical preponderance over all other fishes. The prevalent form of scale among these is thin, rounded, deeply imbricating, and with the posterior margin either plain (cycloid) or serrated (ctenoid). But that these "cycloid" and "ctenoid" scales are modifications from the rhombic osseous "ganoid" type we cannot doubt after what we have seen. It is indeed strange that the same tendency to the change of rhombic into circular overlapping scales should have occurred independently in more than one group.

For reasons given at the beginning, and also because I fear I have already exceeded the limit of time usually allotted to such an Address, I must now stop.

But in conclusion I may allude to a well-known fact regarding the tail of these modern fishes, the bearing of which on the doctrine of Descent is sufficiently clear and has long been recognised.

[1] "The Inter-relationships of the Ichthyopsida," *Anatomischer Anzeiger*, 1890. Smith Woodward arrived at the same result in 1893 from the study of the Jurassic genera *Lepidotus* and *Dapedius*. See *Proc. Zool. Soc. Lond.*, June 20, 1893, pp. 559–565.

We have seen that the completely heterocercal tail of the typical Acipenseroid becomes, by abortion of the upper lobe and shortening of the axis, the semi-heterocercal one of the Lepidosteids, in most of which, however, the want of symmetry is still perceptible externally by a short projection or "sinus" of scales which is directed obliquely upward at the beginning of the top of the fin. In the ordinary bony fishes and in some Lepidosteids also the caudal fin becomes likewise symmetrical, as seen from the outside ; generally also bilobate, though the upper lobe is not that of a Palæoniscid or Sturgeon. This condition of tail has been long known as "homocercal." But in many such homocercal tails, when we dissect away the skin and soft parts, the upward bend of the vertebral axis is revealed, and in some, as in the Salmon, the extremity of the vertebral axis is continued as a cartilaginous style among the rays near the upper margin of the fin. But there are many others, such, for instance, as the peculiarly specialised group of Pleuronectidæ or flat fishes, in which the skeleton of the caudal extremity *looks* quite symmetrical, but yet in the embryo the extremity of the notochord is seen to have an upward bend, showing that the homocercal tail is indeed a specialisation on the old heterocercal one. It is strange that though this embryological fact was long ago pointed out by Agassiz, and though he noted its great interest in connection with the prevalence of heterocercy among the Palæozoic fishes, yet he remained to the end an opponent of evolution. But this is just one of these instances in which Phylogeny and Ontogeny mutually illustrate each other. Why, otherwise, should the tail of the embryo stickleback or flounder be heterocercal?

Incompletely as I have treated the subject, it cannot but be acknowledged that the palæontology of fishes is not less emphatic in the support of Descent than that of any other division of the animal kingdom. But in former days the evidence of fossil ichthyology was by some read otherwise.

It is now a little over forty years since Hugh Miller died : he who was one of the first collectors of the fossil fishes of the Scottish Old Red Sandstone, and who knew these in some respects better than any man of his time, not excepting Agassiz himself. Yet his life was spent in a fierce denunciation of the doctrine of evolution, then only in its Lamarckian form, as Darwin had not yet electrified the world with his "Origin of Species." Many a time I wonder greatly what Hugh Miller would have thought had he lived a few years longer, so as to have been able to see the remarkable revolution which was wrought by the publication of that book.

The main argument on which Miller rested was the "high" state of organisation of the ancient fishes of the Palæozoic formations, and this was apparently combined with a confident assumption of the completeness of the geological record. As to the first idea, we know of course that evolution means the passage from the more general to the more special, and that although as the general result an onward advance has taken place, yet specialisation does not always or necessarily mean "highness" of organisation in the sense in which the term is usually employed. As to the idea of the perfection of the geological record, that of course is absurd.

We do not and cannot know the oldest fishes, as they would not have had hard parts for preservation, but we may hope to come to know many more old ones, and older ones still, than we do at present. My experience of the subject of fosil ichthyology is that it is not likely to become exhausted in our day.

We are introduced at a period far back in geological history to certain groups of fishes some of which certainly are high in organisation as animals, but yet of generalised type, being fishes and yet having the potentiality of higher forms. But, because their ancestors are unknown to us, that is no evidence that they did not exist, and cannot overthrow the morphological testimony in favour of evolution with which the record actually does furnish us. We may therefore feel very sure that fishes, or "fish-like vertebrates," lived long ages before the oldest forms with which we are acquainted came into existence.

The modern type of bony fish, though not so "high" in many anatomical points as that of the Selachii, Crossopterygii, Dipnoi, Acipenseroidei and Lepidosteoidei of the Palæozoic and Mesozoic eras, is more specialised in the direction of the fish proper, and, as already indicated, specialisation and "highness" in the ordinary sense of the word are not necessarily coincident.

But ideas about these things have undergone a wonderful change since those pre-Darwinian days, and though we shall never be able fully to unravel the problems concerning the descent of animals, we see many things a great deal more clearly now than we did then.

SECTION F.

ECONOMIC SCIENCE AND STATISTICS.

ABRIDGED FROM THE OPENING ADDRESS BY MAJOR P. G. CRAIGIE, V.P.S.S., PRESIDENT OF THE SECTION.

OF all statistical work the enumeration of the units of population must ever take the foremost place, and on the eve of the census to be taken before many more months have passed a reference to that great impending task could hardly be omitted on this occasion. In common with all students of the machinery of census-taking, I am sure I echo the feelings of the Section—as I do those of the Royal Statistical Society, who have long laboured in this direction—in deeply regretting that the first census of the twentieth century is not to possess the distinction many had hoped to see conferred upon it of being by preliminary announcement—as I hope it may prove to be in ultimate fact—the first of a series not of decennial but of quinquennial countings of the people.

The growing complexity of social conditions and speed of life in all its functions at the present date, contrasted with the leisurely movements of a hundred years ago, would alone and amply justify a more frequent stock-taking of the inhabitants of Great Britain than has been the practice in the past. The practical wants of our much multiplied system of local government cannot fail, I believe, ere long to bring about the granting of an intermediate numbering, even if for the moment other considerations overrule the more academic pleas of statisticians for this reform, or the arguments, sound as I believe them to be, for a permanent Census Office, a permanent Census Act, and a trained and continuous Census Staff, to whom preparation of the machinery beforehand and detailed elaboration of the results after the actual census year might with real economy be entrusted.

Although the proposal which has been before the International Statistical Institute in one form or another for a sychronous "world's census," at the moment of passing from one century to another, is hardly likely, for administrative reasons, and in view of the previous fixtures of the great census-taking Governments of the earth, to be literally realised, the dates of the great countings of the nations will nevertheless come sufficiently close for all practical comparisons. The great Russian enumeration, on the success of which M. Troinitsky is so heartily to be congratulated, is not yet long accomplished. The twelfth census of the United States is now being taken. The Scandinavian inquiry coincides with the century's end, the Italian and the Spanish censuses are already overdue, and both France and England take their count within a few months after the twentieth century has begun.

Attempts to utilise statistical data, to determine the relative development of agriculture in different parts of the world and at different periods of time, are sometimes made with regard solely to what is described as the world's aggregate of one or two leading individual products as typical as the rest ; or, again, one or two typical countries, or at least countries where the available information is more complete than elsewhere, are chosen, and the course of development or decline of their crop areas or the several descriptions of their animal produce is traced and compared.

Certain obvious objections, which it is well to recognise, impede the student of figures who resolves to proceed on the first of these methods. At the outset he is arrested by embarrassment attending the choice of what single products are to be held as representative of agricultural outturn. The most usual of all selections is that which restricts inquiries to the case of wheat. This course appears to be rendered, comparatively speaking, easy, as more has probably been written and more statistics, official or unofficial, theoretical or commercial, actual or imaginary, have been compiled with regard to this bread grain than for any other crop. But it is time we recognised that wheat has too much and too exclusive attention directed to it as a type of agricultural production. Very widely as it is undoubtedly used in the form of bread, even as food its place is occupied at one time or another, and in one country or another, by other substitutes, and its cultivation, is, after all, not the employment which demands the most attention and most skill at the hands of the agriculturist. Not only do rye and even maize serve as substitutes or supplements in feeding man, but other crops, such as oats, barley, millet, rice, and so on, have claims to greater notice than they receive, and play a direct as well as indirect part in providing food. Cotton, flax and wool are other typical products, the use of which for clothing is all-important to an

enormous population, and the extension or retrogression of such crops deserves some of the attention of the agricultural statistician. Tea, coffee, wine, spirits and beer are, it is not to be forgotten, agricultural products in one clime or another, either directly or indirectly ; and crops so important as sugar or tobacco are almost to be classed as necessaries of existence. Of yearly growing importance is it also, in these days, when the animal portion of our food supply bulks so much more fully than before in the daily rations of populations as they grow in wealth and increase in consumptive power, that we should closely follow the fluctuations in the live stock maintained for food and learn the teaching of the agricultural returns on the manufacture of beef, of mutton, of pig meat, or of milk.

Although the attempt to grasp the relative magnitude of the agricultural production of one State as compared with another, or to note the growth or decline of its prominence in the cultivation of particular staples, or the manufacture of particular kinds of human food, is always an enterprise of difficulty in existing statistical conditions, it is one which has fascination for many classes of economists and politicians. If attempted at all, it is well to recognise that there are inevitable dangers in the task, and that if any figures are relied on as conclusive their meaning must be interpreted by some knowledge of the demographic conditions of each State and its geographical, climatic and agricultural circumstances.

Taking a few of the most conspicuous products of the soil, it will generally be found that a very few leading States are so particularly identified with one or other type of production that the examination of their records are therefore available as guides to the course of a single crop.

Probably quite two-thirds of the cotton of the world is grown in the United States alone, where the surface so employed reaches 25,000,000 acres as compared with under 9,000,000 acres in British India, the next largest cotton-growing region of which statistical record exists. In wool the produce of the Australasian Colonies of Great Britain—with flocks which still exceed 100,000,000 head—makes much the largest contribution to the total. In rice, so far as statistics carry us, our Indian possessions head the list of producers. In hops the English crop still probably exceeds the German in production, although the latter with larger area closely contests the place. In tobacco, while the acreage apparently employed in British India is nearly double the 595,000 acres in the United States, no other country in our statistical records comes within one-seventh of the American area. The vineyards of Italy are returned as covering 8,500,000 acres, and those of France 4,300,000 acres, while those of Austria and Hungary, next in magnitude, cover but a seventh part of the last-mentioned figure. Russia bulks largely as a grower of flax, and alone shows a whole third of the area of barley recorded in all the countries which supply returns, and if in the case of potatoes the Russian acreage is not very different from that of Germany the total production of the latter happens to reach the largest aggregate of any single country.

If the subject of inquiry be the place of wheat-growing in the world at one date or another, it would not be to the older European countries, other than Russia at all events, we should turn to see where the surface so utilised was extending. Reckoned by the percentage of her cereal area which she still devotes to wheat, France, with 47 per cent. under the crop, or Italy, with 55 per cent., would naturally be selected as typical wheat-growers ; but both are practically in a stationary or, collectively, even in a slightly retrograding position. It is on the other side of the Atlantic where the most noteworthy movements have occurred. In comparatively new exporting countries, such as Argentina and Canada, though the statistics from neither are complete, wheat areas still extend, and that of the United States, though fluctuating with great sensitiveness under varying price conditions, and moving from one centre to another westward or north-westward across the American continent, is now reported as covering 44,600,000 acres. This total, it must be allowed, whatever views may be held as to future progress, makes the United States a typical grower of this particular cereal, to which it gives an importance second only to the still more extensive product of American soil, to which we give the name of maize, but to which alone in American parlance is allowed the title of corn.

The leading changes in the production of typical crops as measured by the acreage, and the stock of cattle, sheep and swine recorded at or near the commencement, the middle, and the close of the past thirty years, may be contrasted for exporting countries with expanding populations and growing agriculture, and in countries where these conditions are absent, or in a typical consuming centre like our own country. Relying on the agricultural returns of the United States, a table could be constructed, as under, for three dates within the past thirty years which furnish the following indication of agricultural changes :—

United States	1870	1885	1899
Population, in million persons ...	38·6	56·1	76·0
Area under maize, in million acres	38 6	73·1	82·1
Area under wheat ,,	19·0	34·2	44·6
Area under oats ,,	8·8	22·8	26·3
Area under cotton ,,	9·9	18·3	25 0
Cattle (million head)	25·5	43·8	43 9
Sheep ,, ...	40·9	50·4	41·9
Swine ,, ...	26·8	45·1	38·7

In 1870 the United States held, it would thus appear, a population of 38,600,000, and grew an acre of maize for each unit of the population, and an acre of wheat for every two persons, and somewhat more than an acre of cotton for every four. At this period the surplus exported to other nations, it may be added, represented two-thirds of the cotton, rather more than one-fifth of the wheat, but less than one per cent. of the maize.

In 1885 the population had augmented to an estimated total of 56,000,000, or by 45 per cent. The area under the crops above quoted had meantime been extended in nearly twice this ratio. The United States exported still about two-thirds of the cotton grown ; the wheat export was slightly greater in proportion to the product than before, or 26 per cent. ; while nearly 3 per cent. of the maize crop found a market abroad.

The population of the States is now estimated to have risen to 76,000,000, or twice what it was thirty years ago, although the census has yet to say if this calculation has been realised. The cultivation of maize had meantime reached 82,000,000 acres, wheat was reported to cover 44,000,000 acres, and cotton 25,000,000 acres, while the foreign market received 65 per cent. of the cotton, 33 per cent. of the wheat, and now as much as 9 per cent. of the maize grown on these areas.

In none of these cases, it will be noted, has the area under crop failed to increase, but in all the rate of increase was distinctly slower in the second than in the first half of the period. If time sufficed to trace the annual course of movement between the contrasted dates, it might be well remembered that from 1871 onward to 1889, with only a single slight check in 1887, the growth of the maize acreage has been continuous. From 1889 to 1894 fluctuations were reported yearly, ending in the latter year at a total acreage no higher than that of 1880, but returning again in a single year, if the record can be trusted, to the highest point reached. The wheat acreage movement has been more irregular, and the latest figures are complicated by the admitted corrections which were made to an' amount of 5,000,000 acres for too low previous estimates in 1897. Allowing for this,' the regular upward movement of the wheat acreage was apparently checked in 1880, and has only begun again since 1898 under the stimulus of higher prices in that year.

In live stock the development would seem to have been arrested altogether between 1885 and the end of the century in the case of cattle, and turned into an absolute decline in the number of sheep and swine, although in the fifteen years before 1885 cattle had increased more than 71 per cent., swine 74 per cent., and sheep 25 per cent. As a matter of fact the maximum number of cattle was reached in 1892, when the numbers were 54,000,000, or ten millions more than at present, the stock of swine declining in a still greater ratio from the same year, and sheep declining and rising again in the separate periods between 1883 and 1889, and between 1893 and 1897. If the ratio under each head to population is considered, it would appear that the United States possessed 661 cattle for every 1000 of her citizens in 1870. This was raised to 829 per 1000 persons in 1885, while the ratio now has fallen again below the starting-point, or to 604 per 1000 persons. Sheep have fallen in the thirty years from 1060 in 1870 to 880, and now to

537 head only per 1000 inhabitants. These remarkable changes are worthy of note in connection with the exports of living animals and animal products, which last have been maintained at a still higher level than before.

Turning to a country of nearly stationary population, provided for in the main from its own agricultural produce with only slight assistance from abroad, a like contrast for the beginning, the middle, and the end of the period under review will give roughly the results shown below. Here, although we are provided with an annual figure, the start has to be made after the Franco-German war with the data two years later, or in 1872. (For table, see below.)

Thus in France, where wheat-growing has always had such a predominance among the cereals, the area is neither increasing nor diminishing. The total of 17,000,000 acres falls, however, somewhat short of the provision of an acre to two persons, which held good in the United States ; but this is more than corrected by the higher average yield, which is nearly 5 bushels per acre greater in France than in America. Taking wheat and rye together, there are a million acres less of bread corn grown in France than there was when her slow-moving population was two millions smaller, or less than 58 acres to 100 persons now as against 60 acres to the 100 twenty-eight years ago.

France	1872	1885	1899
Population, in million persons ...	36·1	38·2	38 5
Area under wheat, in million acres ...	17·1	17·2	17·1
Area under oats ,, ...	7·9	9·1	9·7
Area under rye ,, ...	4·7	4·1	3·6
Area under vineyards ,, ...	6·5	4·9	4·3 [1]
Cattle (million head)	11·3	13·1	13·4 [1]
Sheep ,, ...	24·6	22·6	21·3 [1]
Swine ,, ...	5·4	5·8	6·2 [1]

[1] 1898.

The changes which the last quarter of the nineteenth century has seen in the leading features of French agriculture may be easily summarised. The population of 1872 but little exceeded 36,000,000, that of 1885 reached 38,000,000, and the latest data only bring it up to little over 38,500,000. The wheat-growing area remains, it would appear, under all conditions practically at 17,000,000 acres, the only break to the general uniformity of the cultivation of this cereal (with which the returns include spelt) occurring in the season of 1891, when, under exceptional climatic conditions, only 14,000,000 acres were harvested.

There is one typical French agricultural product—wine—which has materially declined under circumstances which are well known. The vineyards of 1872, which were reported as covering 6,500,000 acres, are now returned as less by a third of that area, and covering 4,300,000 acres only.

In cattle a material growth up to 1885, but a very small increase since that year, is reported ; while if sheep, as in all European countries, are fewer, the fall is less than in Germany, and it is most marked in the first half of the period. Swine in France have steadily increased. As regards the cattle, it may be noted that France had 313 cattle to each 1000 of her people in 1872, 345 in 1885, and 352 per 1000 now. Of sheep the number per 1000 is 560, against 681 at the earlier date.

Treating a few of the distinctive points of our own agriculture in the same way as the beginning, middle, and end of the past thirty years, the statistics of the United Kingdom would give these results :—

United Kingdom	1870	1885	1899
Population, in million persons ...	31·2	36·0	40·7
Area under wheat, in million acres	3·8	2·6	2·1
Area under oats ,, ,,	4·4	4·3	4·1
Area under other corn crops ,,	3·6	3·1	2·6
Cattle (million head)	9·2	10·9	11·3
Sheep ,, ...	32·8	30·1	31·7
Swine ,, ...	3·7	3·7	4·0

Here the most striking contrast with France is in the growth of population. From being a country with 5,000,000 fewer inhabitants the United Kingdom is now one actually greater by 2,000,000 persons than is France. This is an increase of more than 30 per cent., while the surface under wheat has heavily fallen, the main loss occurring under circumstances which have been amply discussed between 1879 and 1895. With some revival, as in America, consequent on an improvement of price in recent years, the slight apparent decline I have shown in the cultivation of oats is in fact confined to Ireland, the area in Great Britain being greater than at the beginning of the period. The cattle stock of the United Kingdom is increased by some 23 per cent., and the swine by about 8 per cent., while our flocks of sheep have been maintained at a level far exceeding that of other European States, and distinctive in a peculiar manner of the agriculture of Great Britain, for they still represent, as it appears, on the average 400 sheep to every 1000 acres of land, against 164 in France, 81 in Germany, 32 in Belgium, and 17 in the United States.

Passing to a comparison with another great country, which, like the United States, is a typical exporter of more than one form of agricultural produce, it may be asked how far the available statistics of Russia allow such information to be furnished. For the earliest of the three years contrasted the dates from the Russian empire are meagre and unsatisfactory. Poland must be excluded as blank in our statistics at that time, while as regards animals no figures at all would appear to have been made public for any of the last twelve years. With such qualifications as these, the available data for the nearest year in the larger crops stood as under :—

Russia in Europe (*ex* Poland)	1870	1885	1899
Population, in million persons ...	65.7	81.7	94.2 [3]
Area of rye, in million acres ...	66.4 [1]	64.6	63.4
Area of wheat ,, ...	28.7 [1]	28.9	38.0
Area of oats ,, ...	32.8 [1]	34.9	36.1
Area of other cereals, in million acres	?	31.4	34.2
Cattle (million head)	22.8	23.6 [2]	(24.6) [4]
Sheep ,, ...	48.1	46.7 [2]	(44.5) [4]
Swine ,, ...	9.1	9.4 [2]	(9.2) [4]

[1] In 1872. [2] In 1883. [3] Census of 1897. [4] In 1888.

Thirty years ago. the population of European Russia, *ex* Poland, would appear from such data as we possess to have been estimated in round numbers at under sixty-six million persons. It is given as somewhere about eighty-two millions in 1885, and according to the recent census it is ninety-four millions now. The bread corn of the country continues to be much more largely rye than wheat, and the area in the year 1872, for which statistics are available, occupied the former crop was practically an acre to the person, or in all 66,400,000 acres, less than half an acre per inhabitant, or 29,000,000 acres, being under wheat. The combined surface devoted to these two bread grains together was thus 95,000,000 acres in the aggregate, or 145 acres to every 100 persons.

Fifteen years later, when the population was apparently greater by 16,000,000 persons, or 24 per cent., the statistics of rye acreage indicate 2,000,000 acres less than before, or 64,600,000 acres. The wheat acreage, if the official data be accepted, was little if at all in excess of the 1872 figure, the rye and wheat together roughly giving 115 acres to 110 persons. The suggestion of this decline, while the exports of both grains were maintained or extended, affords an opportunity for closer inquiry into the basis of the published returns which are received from that country.

But carrying the review of the official figures further, the very latest data for this section of the Russian territory would appear to indicate a yet further shrinkage in the acreage of rye, but accompanied now, as was apparently not the case until lately, by a considerable increase in land under wheat. The total of this cereal is now put as high as 38,000,000 acres, but the net available area of bread-stuffs, although brought up to 101,000,000 acres, represents a still diminishing ratio to population, or 107 acres to every 100 persons. Moreover, as Russia must be regarded as growing both wheat and rye for export as well as consumption, the larger proportions of her acreage which is employed in feeding a non-Russian population deserve to be

specially marked in this connection, when the low yields of both cereals are remembered.

Whether the foregoing figures do indeed represent the facts of each period is, I think, a worthy object of inquiry for some of our younger statisticians, and it is a problem one would like to see solved as regards this particular country before venturing on any too confident conclusion as to what is the real meaning of the changes of the past, and what may be the future position in regard to the growth of bread-stuffs and the growth of population in the world as a whole.

Calculations, however, such as those just quoted cannot fail to remind the student how very different in productive power the "acre" of wheat may be, and is, in different countries. Assuming that we take the existence of 38,000,000 acres as reported of wheat land in Russia in Europe (*ex* Poland) to be proved, a comparison of the estimated yields shows that such an area represents less than 12,000,000 acres of the productive power we are accustomed to in Great Britain. So, too, for the vast wheat area of the United States, it takes two and a third acres to produce what is now our average yield in this country. Three Indian or three Italian acres of wheat of the calibre now in use would in the same way be required to supply the number of bushels that a single acre of our soil in the climate we enjoy, and worked under the system of farming that we practise here, would in ordinary seasons produce. In other extensive areas of wheat-growing the yields, though greater than the above, are very considerably below our own, the Austrian, Hungarian and French yields standing at 16, 17 and 18 bushels respectively, against the 30 bushels which is apparently the average yield of the last five years in the United Kingdom. Only when we come to very small total areas do we find instances where the average wheat yields approach or over any considerable periods exceed our own. When Denmark, for example, is referred to as reaching 42 bushels per acre in the season of 1896, it is not to be forgotten that only a minute area of selected land, in this case only 84,000 acres, is devoted to this cereal. Results realised on this small scale can hardly be spoken of as an average in contrast with those of countries where millions of acres are grown, and can usually be paralleled in some sections of the bigger country.

Nor should it be forgotten, if the agricultural position of one State be compared with another, how widely the conditions of different parts vary from the picture presented by the average figures credited to the State as a unit, and how often sections of one country differ more from each other agriculturally than from the country with which they are contrasted. Within the United Kingdom alone we are, or ought to be, familiar with essential local differences of this type, which have to be kept in mind. Even in respect of the relative density of population and the number of mouths to be sustained in a given area, it may be quite correct to describe every 1000 acres in the United Kingdom as carrying on their surface on the average 519 persons, but it may be remembered with advantage that, considered geographically apart, Scotland, for example, is a country of but 220 persons, and Ireland of but 219, to the 1000 acres of area.

Such a position suggests that it might be fair to draw our agricultural comparisons between Scotland or Ireland as units of area, and such a country as Denmark, where the population is 248 to the 1000 acres. Thus one-third of the cereal area of England is still devoted to the growth of wheat, while Denmark has but 3 per cent. so occupied, thereby resembling Scotland or Ireland, where some 4 per cent. only of the corn is wheat. Similarly, on this population basis, Austria with 320 persons, or Switzerland with 312, to the 1000 acres may be not inappropriately classed with Wales, where the density is 345. In particular, an examination of the live stock maintained by each 1000 acres of the surface in all these cases affords parallels and contrasts which are both interesting and instructive. (For table, see p. 512.)

Thus Wales bears easily the palm as regards the total stock of sheep carried, while Ireland, with a population practically bearing a similar ratio to that of Scotland to her surface, has more than three times as dense a stock of cattle and more than eight times as many pigs, although not much more than half as many sheep to the 1000 acres. Although beaten as regards the number of pigs maintained on a given area by Denmark and by Hungary, Ireland's cattle are more than twice as numerous relatively as those of France, where the population is not so very different in proportion to the soil.

Country	Per 1000 Acres of Total Area			
	Persons	Cattle	Sheep	Swine
Ireland	219	217	207	61
Scotland	220	64	390	7
Hungary	232	85	102	92
Denmark	248	186	115	88
France	293	103	164	48
Switzerland	311	132	27	57
Austria	320	117	43	48
Wales	345	147	685	50

Among countries where the areas are still greater in proportion to the resident population it may not be without interest to group together—as regards their present density—persons, cattle, sheep and swine.

Countries	Per 1000 Acres of Total Area			
	Persons	Cattle	Sheep	Swine
New South Wales	7	10	221	1
New Zealand ...	11	18	294	3
Victoria ...	21	32	234 [1]	6
Norway ...	26	13 [1]	18 [1]	2 [1]
United States ...	32	19	17	17
Sweden ...	49	25	13	8
Russia (ex Poland)	66	20 [2]	36 [2]	7 [2]

[1] In 1890. [2] In 1888.

Such figures serve to emphasise the vast difference between the flocks maintained in our Australasian Colonies and the other countries in this group.

The animal wealth of England by herself, omitting the Celtic fringes above quoted, may be compared with a nearer competitor. Belgium has 893 persons to 1000 acres, England 925; and Belgium has 195 head of cattle and 160 head of swine, but only 32 sheep, on an average area of this size in her little kingdom, against 144 cattle, 64 pigs, and as many as 488 sheep in England. Were the comparison to be made more closely yet, the cattle stock of Belgium agrees closely in point of density with, say, the particular division of our area comprising the north-western counties of England, which have 194 cattle to 1000 acres, or considerably more than the great butter-exporting country of Denmark, and at least a very close approach to the 197 head per 1000 acres which are to be found in the fat pastures of the Netherlands.

These limited comparisons on single points of agricultural production do not, I know, satisfy the demands which are often made for world-wide surveys and comparisons on a larger scale. I confess I somewhat distrust the strength and due coherence of the statistical bricks on which these heroic conclusions are built up. It is most usual in corn trade journals, and the practice is sometimes followed in serious debate and reproduced in the year-books of the United States Government, to give a yearly picture of at least the world's wheat crop. For the close comparison of one season with another much must depend on the sufficiency of the weakest item in the account, and weakness is sure to creep in somewhere when crops are estimated on varying systems, at different dates, and on authorities of unequal value. The definitions adopted by one calculator as to the limits of the "world" vary from those of another, and commercial estimates, as they are called, may be, at the discretion of the computer, substituted for or adopted in the absence of official data, so that the guesses at a single country's harvest may differ more widely from each other than would account for the total margin between one year's aggregate supply and another, to the confounding of satisfactory conclusions as to what is really happening. Last but not least of the obstacles to uniform grouping of harvests in complete years—ending as these years do at different periods—is the fact, not to be overlooked, that wheat harvests are being gathered somewhere in every month in the twelve.

One is driven back then to the attempt to rest opinions on the growth of one form of culture or another on recorded acre-

age, rather than assumed production. Yet even here a good illustration of the difficulty of any extensive compilation may be found in the tentative memorandum Sir Robert Giffen put before the last Royal Commission on Agriculture as indicating, with many necessary reservations and qualifications, the relative movements of grain area, live stock, and population in the twenty years before 1893. Briefly, the earlier totals brought into conjunction for this purpose were made up, as regards the population figures taken to represent the starting-point of 1873, from the statistics of groups of countries and colonies at dates for the most part about 1871-3, but in some instances ranging back to 1866 and on to 1881, and aggregating 365,800,000 persons. Against these were set a total of 461,800,000 persons, enumerated, for the most part, about 1890-93, but in a few instances, where later data were wanting, going back to 1880-88, the growth of population between the totals being 26 per cent.

The acreage about 1873 and about 1893, contrasted with these figures, included wheat, rye, barley and oats, but not maize—a larger crop than any of the last three. The countries contrasted were limited necessarily by the extent of information, and the list did not include all of which the population was accounted for, the increases per cent. being 28 per cent. in the case of oats, 19 per cent. in the case of wheat, 5 per cent. in the case of barley, with a decrease of 5 per cent. in rye. It should be observed, however, that the calculation as to the increase of wheat would have been much closer to that of population had not a very large area, nearly stationary in amount, been credited to India and Japan at both dates; the local population of these Asiatic countries being disregarded as, generally speaking, non-wheat-eating.

It was only as an outline pointing the direction in which inquiry might be useful that Sir Robert Giffen called attention to these figures, which, as he acknowledged, were of the roughest possible description, and rather suggestive of a closer inquiry, which should take account of the difference between the consumptive power of the countries aggregated, the varying productive power of nominally equal areas of surface, and the varying type of live stock maintained.

If the wheat acreage table, in the memorandum referred to, is examined in detail, a very effective picture of the difficulty of exact comparison as between any two given dates is incidentally presented. Out of twenty-four countries enumerated (including Canada and Australasia as units) a twenty or twenty-one years' comparison is only really effected in five cases—Russia, the United States, France, United Kingdom and Australasia. In five other instances the period dealt with is only from seventeen to eighteen years; in three other cases only fourteen or fifteen years. In Canada, Egypt and Denmark, the comparison will be found to be more limited still, and only to cover eleven or twelve years; while in the Argentine Republic, where the recent expansion of wheat-growing has been prominent, the available statistics allowed only of a comparison of two periods, no more than nine years apart. For seven other countries the wheat acreage was necessarily either omitted or inserted as presumably the same as both the earlier and the later date. Had the retrospect been confined to the cases where a twenty or twenty-one years' comparison was possible—and these, after all, included the most important and typical wheat-growing communities—the increase would have stood, not at 19, but at 24 per cent., or scarcely below that of the growth of population generally. This result is reached without taking account of any South American figures, where the increase of area is relatively much greater, or of those of India, where the comparison is difficult and the acreage-growing but slightly. But, further, it is to be remembered that if the comparison of the memorandum were to be continued up to 1899, instead of stopping at 1893, the figures would have shown that wheat-growing had apparently made a new start in the five important countries for which the long comparison was possible, as many million acres having been added in the past six years as in the whole preceding twenty—a result which may afford much occasion for suspending our final judgment and no little warning of the danger of single-year contrasts.

Since the above calculations were before the Commission there has been an extension of 10,000,000 acres in the official estimates of wheat areas in the United States, and 5,400,000 acres in Russia, while, although official details are still wanting beyond 1895 for Argentina, nearly 3,000,000 acres more were in that year accounted for in that republic; and there is an impression, apparently well founded, that by the present time the total may have reached 8,000,000 acres, or nearly five million acres more than the final figure in Sir Robert Giffen's calculation. If anything like 20,000,000 acres have thus been added to the wheat-growing surface of the globe in the last five or six years, which these further figures suggest, even if no correction be made for the Indian quota, there may be much less difference than was suggested in the memorandum between the growth of population and wheat-growing.

Without attempting in any way to controvert what was one of the lessons of the memorandum I have been examining, as to the tendency to increase the numbers of cattle at a ratio above that of population, it has also to be remembered that the apparent 37 per cent. increase there shown between 1873 and 1893 may have to be discounted by subsequent deductions in the United States, in Australasia. and at the Cape in recent years; while it is one of the problems I have never yet seen satisfactorily answered, why in almost all old countries except our own the diminution of the stock of sheep seems contizuous and remarkable. I mention these matters only, however, to suggest the amount of uncertainty which must attend the efforts to arrive at conclusions, made even by the highest authorities, on the only data which exist. If there is, as I have shown, such uncertainty still in the facts on which a conclusion could be built as to the past history of the relative growth of live stock, or of cereal culture and the supply of bread-stuffs, how much greater must the difficulty be of those who attempt, on the basis of such data, to forecast the course of events for a generation yet to come! I confess I am not intrepid enough to follow some of the conjectures which have been hazarded on this point, and can only, in concluding this address. recur once more to the prime qualifications for safe statistical deductions with which I opened my remarks—redoubled caution in handling calculations, a very guarded use of data giving records of single and isolated years, and a wise reservation in any prophetic pictures of the future of agricultural production, whether of wheat or cotton, in meat or in wool, of the contingency, always present, of altered conditions which ever and anon in the past have altered and falsified the predictions of earlier observers.

SECTION H.
ANTHROPOLOGY.
OPENING ADDRESS BY PROF. JOHN RHYS, M.A., LL.D., PRESIDENT OF THE SECTION.

PERHAPS I ought to begin by apologising for my conspicuous lack of qualification to fill this chair, but I prefer, with your permission, to dismiss that as a subject far too large for me to dispose of this morning. So I would beg to call your attention back for a moment to the excellent address given to this Section last year. It was full of practical suggestions which are well worth recalling: one was as to the project of a Bureau of Ethnology for Greater Britain, and the other turned on the de-sirability of founding an Imperial Institution to represent our vast Colonial Empire. I mention these in the hope that we shall not leave the Government and others concerned any peace till we have realised those modest dreams of enlightenment. People's minds are just now so full of other things that the interests of knowledge and science are in no little danger of being overlooked. So it is all the more desirable that the British Association, as our great parliament of science, should take the necessary steps to prevent that happening, and to keep steadily before the public the duties which a great and composite nation like ours owes to the world and to humanity, whether civilised or savage.

The difficulties of the position of the president of this Section arise in a great measure from the vastness of the field of research which the Science of Man covers. He is, therefore, constrained to limit his attention as a rule to some small corner of it; and, with the audacity of ignorance, I have selected that which might be labelled the early ethnology of the British Isles, but I propose to approach it only along the precarious paths of folk-lore and philology, because I know no other. Here, however, comes a personal difficulty: at any rate I suppose I ought to pretend that I feel it a difficulty, namely, that I have committed myself to publicity on that subject already. But, as a matter of fact, I can hardly bring myself to confess to any such feeling; and this leads me to mention in passing the change of attitude which I have lived to notice in the case of students in my own position. Most of us here present have known men who, when they had once printed their views on their favourite subjects of

study, stuck to those views through thick and thin, or at most limited themselves to changing the place of a comma here and there, or replacing an occasional *and* by a *but*. The work had then been made perfect, and not a few great questions affecting no inconsiderable portions of the universe had been for ever set at rest. That was briefly the process of getting ready for posterity, but one of its disadvantages was that those who adopted it had to waste a good deal of time in the daily practice of the art of fencing and winning verbal victories; for, metaphorically speaking,

"With many a whack and many a bang
Rough crab-tree and old iron rang."

Now all that, however amusing it may have been, has been changed, and what now happens is somewhat as follows: AB makes an experiment or propounds what he calls a working hypothesis; but no sooner has AB done so than CD, who is engaged in the same sort of research, proceeds to improve on AB. This, instead of impelling AB to rush after CD with all kinds of epithets and insinuating that his character is deficient in all the ordinary virtues of a man and a brother, only makes him go to work again and see whether he cannot improve on CD's results; and most likely he succeeds, for one discovery leads to another. So we have the spectacle not infrequently of a man illustrating the truth of the poet's belief,

"That men may rise on stepping-stones
Of their dead selves to higher things."

It is a severe discipline in which all display of feeling is considered bad form. Of course every now and then a spirit of the ruder kind discards the rules of the game and attracts attention by having public fits of bad temper; but generally speaking the rivalry goes on quietly enough to the verge of monotony, with the net result that the stock of knowledge is increased. I may be told, however, that while this kind of exercise may be agreeable to the ass who writes, it is not conducive to the safety of the publisher's chickens. To that it might suffice to answer that the publisher is usually one who is well able to take good care of his chickens; but, seriously, what it would probably mean is, that in the matter of the more progressive branches of study, smaller editions of the books dealing with them would be required, but a more frequent issue of improved editions of them or else new books altogether, a state of things to which the publisher would probably find ways of adapting himself without any loss of profits. And after all, the interests of knowledge must be reckoned uppermost. It is needless to say that I have in view only a class of books which literary men proper do not admit to be literature at all; and the book trade has one of its mainstays, no doubt, in books of pure literature, which are like the angels that neither marry nor give in marriage: they go on for ever in their serene singleness of purpose to charm and chasten the reader's mind.

My predecessor last year alluded to an Oxford don said to have given it as his conviction that anthropology rests on a foundation of romance. I have no notion who that Oxford don may have been, but I am well aware that Oxford dons have sometimes a knack of using very striking language. In this case, however, I should be inclined to share to a certain extent that Oxford don's regard for romance, holding as I do that the facts of history are not the only facts deserving of careful study by the anthropologist. There are also the facts of fiction, and to some of those I would now call your attention. Recently, in putting together a volume on Welsh folklore, I had to try to classify and analyse in my mind the stories which have been current in Wales about the fairies. Now the mass of folklore about the fairies is of various origins. Thus with them have been more or less inseparably confounded certain divinities or demons, especially various kinds of beings associated with the rivers and lakes of the country. They are creations introduced from the workshop of the imagination; then there is the dead ancestor, who also seems to have contributed his share to the sum total of our notions about the Little People. In far the greater number of cases, however, we seem to have something historical, or, at any rate, something which may be contemplated as historical. The key to the fairy idea is that there once was a real race of people to whom all kinds of attributes, possible and impossible, have been given in the course of uncounted centuries of story-telling by races endowed with a lively imagination.

When the mortal midwife has been fetched to attend on a fairy mother in a fairy palace, she is handed an ointment which she is to apply to the fairy baby's eyes, at the same time that she is gravely warned not to touch her own eyes with it. Of course any one could foresee that when she is engaged in applying the ointment to the young fairy's eyes one of her own eyes is certain to itch and have the benefit of the forbidden salve. When this happens the midwife has two very different views of her surroundings : with the untouched eye she sees that she is in the finest and grandest that she has ever beheld in her life, and there she can see the lady on whom she is attending reposing on a bed, while with the anointed eye she perceives how she is lying on a bundle of rushes and withered ferns in a large cave, with big stones all round her and a little fire in one corner, and she also discovers that the woman is a girl who has once been her servant. Like the midwife we have also to exercise a sort of double vision, if we are to understand the fairies and see through the stories about them. An instance will explain what I mean : Fairy women are pretty generally represented as fascinating to the last degree and gorgeously dressed : that is how they appear through the glamour in which they move and have their being. On the other hand, not only are some tribes of some fairies described as ugly, but fairy children when left as changelings are invariably pictured as repulsive urchins of a sallow complexion and mostly deformed about the feet and legs : there we have the real fairy with the glamour taken off and a certain amount of depreciatory exaggeration put on.

Now when one approaches the fairy question in this kind of way, one is forced, it strikes me, to conclude that the fairies, as a real people, consisted of a short, stumpy, swarthy race, which made its habitations underground or otherwise cunningly concealed. They were hunters, probably, and the fishermen ; at any rate, they were not tillers of the ground or eaters of bread. Most likely they had some of the domestic animals and lived mainly on milk and the produce of the chase, together with what they got by stealing. They seem to have practised the art of spinning, though they do not appear to have thought much of clothing. They had no tools or implements made of metals. They appear to have had a language of their own, which would imply a time when they understood no other, and explain why, when they came to a town to do their marketing, they laid down the exact money without uttering a syllable to anybody by way of bargaining for their purchases. They counted by fives and only dealt in the simplest of numbers. They were inordinately fond of music and dancing. They had a marvellously quick sense of hearing, and they were consummate thieves ; but their thievery was not systematically resented, as their visits were held to bring luck and prosperity. More powerful races generally feared them as formidable magicians who knew the future and could cause or cure disease as they pleased. The fairies took pains to conceal their names no less than their abodes, and when the name happened to be discovered by strangers the bearer of it usually lost heart and considered himself beaten. Their family relations were of the lowest order : they not only reckoned no fathers, but it may be that, like certain Australian savages recently described by Spencer and Gillen, they had no notion of paternity at all. The stage of civilisation in which fatherhood is of little or no account has left evidence of itself in Celtic literature, as I shall show presently ; but the other and lower stage anterior to the idea of fatherhood at all comes into sight only in certain bits of folklore, both Welsh and Irish, to the effect that the fairies were all women and girls. Where could such an idea have originated ? Only, it seems to me, among a race once on a level with the native Australians to whom I have alluded, and of whom Fraser of "The Golden Bough" wrote as follows in last year's *Fortnightly Review* : "Thus, in the opinion of these savages, every conception is what we are wont to call an immaculate conception, being brought about by the entrance into the mother of a spirit, apart from any contact with the other sex. Students of folklore have long been familiar with the notions of this sort occurring in the stories of the birth of miraculous personages, but this is the first case on record of a tribe who believe in immaculate conception as the sole cause of the birth of every human being who comes into the world. A people so ignorant of the most elementary of natural processes may well rank at the very bottom of the savage scale." Those are Dr. Fraser's words, and for a people in that stage of ignorance to have imagined a race all women seems logical and natural enough— but for no other. The direct conclusion, however, to be drawn from this argument is that some race—possibly more than one—which has contributed to the folklore about our fairies, has passed through the stage of ignorance just indicated ; but as an indirect conclusion one would probably be right in

supposing this race to have been no other than the very primitive one which has been exaggerated into fairies. At the same time it must be admitted that they could not have been singular always in this respect among the nations of antiquity, as is amply proved by the prevalence of legends about virgin mothers, to whom Frazer alludes, not to mention certain wild stories recorded by the naturalist Pliny concerning certain kinds of animals.

Some help to make out the real history of the Little People may be derived from the names given them, of which the most common in Welsh is that of *y Tylwyth Teg* or the Fair Family. But the word *cor*, "a dwarf," feminine *corres*, is also applied to them ; and in Breton we have the same word with such derivatives as *korrik*, "a fairy, a wee little wizard or sorcerer," with a feminine *korrigan* or *korrigen*, analogously meaning a she-fairy or a diminutive witch. From *cor* we have in Welsh the name of a people called the Coranians figuring in a story in the fourteenth-century manuscript of the Red Book of Hergest. There one learns that the Coranians were such consummate magicians that they could hear every word that reached the wind, as it is put ; so they could not be harmed. The name Coranians of those fairies has suggested to Welsh writers a similar explanation of the name of a real people of ancient Britain. I refer to the *Coritani*, whom Ptolemy located, roughly speaking, between the river Trent and Norfolk, assigning to them the two towns of *Lindum*, Lincoln, and *Ratae*, supposed to have been approximately where Leicester now stands. It looks as if all invaders from the Continent had avoided the coast from Norfolk up to the neighbourhood of the Humber, for the good reason, probably, that it afforded very few inviting landing-places. So here presumably the ancient inhabitants may have survived in sufficient numbers to have been called by their neighbours of a different race " the dwarfs " or *Coritani*, as late as Ptolemy's time in the second century. This harmonises with the fact that the Coritani are not mentioned as doing anything, all political initiative having long before probably passed out of their hands into those of a more powerful race. How far inland the Coritanian territory extended it is impossible to say, but it may have embraced the northern half of Northamptonshire, where we have a place-name *Pytchley*, from an earlier *Pihtes lea*, meaning "The Pict's Meadow," or else the meadow of a man called Pict. At all events, their country took in the fen district containing Croyland, where towards the end of the seventh century St. Guthlac set up his cell on the side of an ancient tumulus and was disturbed by demons that talked Welsh. Certain portions of the Coritanian country offered, as one may infer, special advantages as a home for retreating nationalities : witness as late as the eleventh century the resistance offered by Hereward in the Isle of Ely to the Norman Conqueror and his mail-clad warriors.

In reasoning backwards from the stories about the Little People to a race in some respects on a level with Australian savages, we come probably in contact with one of the very earliest populations of these islands. It is needless to say that we have no data to ascertain how long that occupation may have been uncontested, if at all, or what progress was made in the course of it : perhaps archæology will be able some day to help us to form a guess on that subject. But the question more immediately pressing for answer is, with what race outside Wales may one compare or identify the ancient stock caricatured in Welsh fairy tales ? Now, in the lowlands of Scotland, together with the Orkneys and Shetlands, the place of our fairies is to some extent taken by the Picts, or, as they are there colloquially called, "the Pechts." My information about the Pechts comes mostly from recent writings on the subject by Mr. David MacRitchie, of Edinburgh, from whom one learns, among other things, that certain underground—or partially underground—habitations in Scotland are ascribed to the Pechts. Now one kind of these Pechts' dwellings appear from the outside like hillocks covered with grass, so as presumably not to attract attention, an object which was further helped by making the entrance very low and as inconspicuous as possible. But one of the most remarkable things about them is the fact that the cells or apartments into which they are divided are frequently so small that their inmates must have been of very short stature, like our Welsh fairies. Thus, though there appears to be no reason for regarding the northern Picts themselves as an undersized race, there must have been a people of that description in their country. Perhaps archæologists may succeed in classifying the ancient habitations in the North accordingly: that is, to tell us what class of them were built by the Picts and what by the Little People whom they may be supposed to have found in possession of that part of our island.

In Ireland and the Highlands of Scotland the fairies derive their more usual appellations from a word *síd* or *síth* (genitive *síde*), which may perhaps be akin to the Latin *sēdes* and have meant a seat, settlement, or station; but whatever its exact meaning may have originally been, it came to be applied to the hillocks or mounds within which the Little People made their abodes. Thus, *Aes Síde* as a name for the fairies may be rendered by mound people or hill folk; *fer síde*, "a fairy man," by a mound man; and *ben síde* by a mound woman or banshee. They were also called simply *síde*, which would seem to be an adjective closely allied with the simpler word *síd*.

But to leave this question of their names, let me direct your attention for a moment to one of the most famous kings of the fairies of ancient Erin: he was called Mider of Brí Leith, said to be a hill to the west of Ardagh, in the present county of Longford. There he had his mound, to which he once carried the queen of Eochaid Airem, monarch of Ireland. It was some time before Eochaid could discover what had become of her, and he ordered Dálan, his druid, to find it out. So the druid, when he had been unsuccessful for a whole year, prepared four twigs of yew and wrote on them in Ogam. Then it was revealed to him through his keys of seership and through the Ogam writing that the queen was in the *síd* of Brí Leith, having been taken thither by Mider. By this we are probably to understand that the druid sent forth the Ogam twigs as letters of inquiry to other druids in different parts of the country; but in any case he was at last successful, and his king hurried at the head of an army to Brí Leith, where they began in earnest to demolish Mider's mound. At this Mider was so frightened that he sent the queen forth to her husband, who then departed, leaving the fairies to digest their wrath; for it is characteristic of them that they did not fight, but bided their time for revenge, which in this case did not come till long after Eochaid's day. Now, with regard to the fairy king, one is not told, so far as I can call to mind, that he was a dwarf, but the dwarfs were not far off; for we read of an Irish satirist who is represented as notorious for his stinginess; and to emphasise the description of his inhospitable habits he is said to have taken from Mider three of his dwarfs and stationed them around his own house, in order that their truculent looks and rude words might repel any of the men of Erin who might come seeking hospitality or bring any inconvenient request. The word used for dwarf in this story is *corr*, which is usually the Irish for a crane or heron, but here, and in some other instances, which I cannot now discuss, it seems to have been identical with the Brythonic *cor*, "a dwarf." It is remarkable, moreover, that the *rôle* assigned to the three Irish *corrs* is much the same as that of the dwarf of Edeyrn son of Nudd in the Welsh story of *Geraint and Enid* and Chrétien de Troies' *Erec*, which characterises him as *fel et de put'eire*, "treacherous and of an evil kind."

By way of summarising these notes on the Mound Folk I may say that I should regard them as isolated and wretched remnants of a widely spread race possessing no political significance whatever. But, with the inconsistency characteristic of everything connected with the fairies, one has on the other hand to admit that this strange people seems to have exercised on the Celts—probably on other races as well—a sort of permanent spell of mysteriousness and awe stretching to the verge of adoration. In fact, Irish literature states that the pagan tribes of Erin before the advent of St. Patrick used to worship the *síde* or the fairies. Lastly, the Celt's faculty of exaggeration, combined with his incapacity to comprehend the weird and uncanny population of the mounds and caves of his country, has enabled him, in one way or another, to bequeath to the great literatures of Western Europe a motley train of dwarfs and little people, a whole world of wizardry, and a vast wealth of utopianism. If you subtracted from English literature, for example, all that has been contributed to its vast stores from this native source, you would find that you left a wide and unwelcome void.

But the question must present itself sooner or later, with what race outside these islands we are to compare or identify our mound-dwellers. I am not prepared to answer, and I am disposed to ask our archæologists what they think. In the meantime, however, I may say that there are several considerations which impel me to think of the Lapps of the North of Europe. But even supposing an identity of origin were to be made out as between our ancient mound-inhabiting race and the Lapps, it would remain still doubtful whether we could expect any linguistic help from Lapland. The Lapps now speak a language belonging to the Ugro-Finnic family, but the Lapps are not of the same race as the Finns; so it is possible that the Lapps have adopted a Finnish language, and that they did so too late for their present language to help us with regard to any of our linguistic difficulties. One of these lies in our topography: take for instance only the names of our rivers and brooks—there is probably no county in the kingdom that would be too small to supply a dozen or two which would baffle the cleverest Aryan etymologist you could invite to explain them; and why? Because they belong in all probability to a non-Celtic, non-Aryan language of some race that had early possession of our islands. Nevertheless it is very desirable that we should have full lists of such names, so as to see which of them recur and where. It is a subject deserving the attention of this Section of the British Association.

We have now loitered long enough in the gloom of the Pecht's house: let us leave the glamour of the fairies and see whether any other race has had a footing in these islands before the coming of the Celts. In August 1891 Prof. Sayce and I spent some fine days together in Kerry and other parts of the southwest of Ireland. He was then full of his visits to North Africa, and he used to assure me that, if a number of Berbers from the mountains had been transferred to a village in Kerry and clad as Irishmen, he would not have been able to tell them by their looks from native Irishmen such as we saw in the course of our excursions. This seemed to me at the time all the more remarkable, as his reference was to fairly tall blue-eyed persons whose hair was rather brown than black. Evidence to the same effect might now be cited in detail from Prof. Haddon and his friends' researches among the population of the Arran Islands in Galway Bay. Such is one side of the question which I have in my mind: the other side consists in the fact that the Celtic languages of to-day have been subjected to some disturbing influence which has made their syntax unlike that of the other Aryan languages. I have long been of opinion that the racial interpretation of that fact must be, that the Celts of our islands have assimilated another race using a language of its own in which the syntactical peculiarities of Neo-Celtic had their origin; in fact that some such race clothed its idioms in the vocabulary which it acquired from the Celts. The problem then was to correlate those two facts. I am happy to say this has now been undertaken from the language point of view by Prof. J. Morris Jones, of the University College of North Wales. The results have been made public in a book on The Welsh People recently published by Mr. T. Fisher Unwin. The paper is entitled "Pre-Celtic Syntax in Insular Celtic," and the languages which have therein been compared with Celtic are old Egyptian and certain dialects of Berber. It is all so recent that we have as yet had no criticism, but the reasoning is so sound and the arguments are of so cumulative a nature, that I see no reason to anticipate that the professor's conclusions are in any danger of being overthrown.

At the close of his linguistic argument, Prof. Morris Jones quotes a French authority to the effect that, when a Berber king dies or is deposed, which seems to happen often enough, it is not his son that is called to succeed him, but the son of his sister, as appears to have been usual among certain ancient peoples of this country; and this more anon. In the next place my attention has been called by Prof. Sayce to the fact that ancient Egyptian monuments represent the Libyans of North Africa with their bodies tattooed, and that even now some of the Touaregs and Kabyles do the same. These indications help one to group the ancient peoples of the British Isles to whose influence we are to ascribe the non-Aryan features of Neo-Celtic. In the first place one cannot avoid fixing on the Picts, who were so called because of their habit of tattooing themselves. For as to that fact there seems to be no room for doubt, and Mr. Nicholson justly lays stress on the testimony of the Greek historian Herodian, who lived in the time of Severus, and wrote about the latter's expedition against the natives of North Britain a long time before the term *Picti* appears in literature. For Herodian, after saying that they went naked, writes about them to the following effect: "They puncture their bodies with coloured designs and the figures of animals of all kinds, and it is for this reason that they do not wear clothes, lest one should not behold the designs on their bodies." This is borne out by the names by which the Picts have been known to the Celts. That of *Pict* is itself in point, and I shall have something to say of it presently; but one of the other names was in Irish *Cruithni*, and in Welsh we have its etymological equivalent in *Prydyn* or *Prydain*. These vocables are derived respectively from Irish *cruth* and Welsh *pryd*, both meaning shape, form, or figure, and it is an old surmise that the Picts were called by those names in allusion to the animal forms pricked on their bodies, as described by Herodian and others. The earlier attested of these two names may be said to be Prydyn or Pryûain, which the Welsh used to give in the Middle Ages to the Picts and the Pictland of the North, while the term *Ynys Prydain* was retained for Great Britain as a whole, the literal meaning being the island of the Picts: that is the only name which we have in Welsh to this day for this island in which we live— *Ynys Prydain*, "The Picts' Island." Now one detects this word Prydain in effect in the Greek Πρεταυικαί Νῆσοι given collectively to all the British Isles by ancient authors, such as Strabo and Diodorus. It may be rendered the Pictish Islands, but a confusion seems to have set in pretty early with the name of the Brittanni or Brittones of South Britain: that is to say, *Pretanic*, "Pictish," became *Brittannic* or British; and this is, historically speaking, the only known justification we have for including Ireland in the comprehensive term "The British Isles," to which Irishmen are sometimes found jocularly to object.

In the next place may be mentioned the Tuatha Dé Danann of Irish legend, which cannot always be distinguished from the Picts, as pointed out by Mr. MacRitchie. The tradition about them is, that, when they were overcome in war by Mil and his Milesians, they gave up their life above ground and retired into the hills like the fairies, a story of little more value than that of the extermination of the Picts of Scotland. In both countries doubtless the more ancient race survived to amalgamate with its conquerors. There was probably some amount of amalgamation between the Tuatha Dé Danann or the Picts and the Little Moundsmen; but it is necessary not to confound them. The Tuatha shared with the Little People a great reputation for magic; but they differed from them in not being dwarfs or of a swarthy complexion: they are usually represented as fair. In the case of Mider, the fairy king, who comes in some respects near the description of the heroes of the Tuatha Dé Danann, it is to be noticed that he was a wizard, not a warrior.

Guided by the kinship of the name of the Tuatha Dé Danann on the Irish side of the sea and that of the Sons of *Dôn* on this side, I may mention that the Mabinogion place the Sons of Dôn on the seaboard of North Wales, in what is now Carnarvonshire: more precisely their country was the region extending from the mountains to the sea, especially opposite Anglesey. In that district we have at least three great prehistoric sites all on the coast. First comes the great stronghold on the top of Penmaen Mawr; then we have the huge mound of Dinas Dinlle, eaten into at present by the sea south-west of the western mouth of the Menai Straits; and lastly there is the extensive fortification of Tre'r Ceiri, overlooking Dinlle from the heights of the Eifl. By its position Tre'r Ceiri belonged to the Sons of Dôn, and by its name it seems to me to belong to the Picts, which comes, I believe, to the same thing. Now the name Tre'r Ceiri means the town of the Keiri, and the Welsh word *ceiri* is used in the district in the sense of persons who are boastful and ostentatious, especially in the matter of personal appearance and fine clothing. It is sometimes also confounded with *cewri*, "giants," but in the name of Tre'r Ceiri it doubtless wafts down to us an echo of the personal conceit of the ancient Picts with their skins tattooed with decorative pictures; and Welsh literature supplies a parallel to the name *Ynys Prydain* in one which is found written *Ynys y Ceuri*, both of which may be rendered equally the Island of the Picts, but more literally perhaps some such rendering as "the Island of the Fine Men" would more nearly hit the mark. Lastly, with the Sons of Dôn must probably be classed the other peoples of the Mabinogion, such as the families of Llyr, and of Pwyll and Rhiannon.

All these peoples of Britain and of Ireland were warlike, and such, so far as one can see, that the Celts, who arrived later, might with them form one mixed people with a mixed language, such as Prof. Morris Jones has been helping to account for.

Let us now see for a moment how what we read of the state of society implied in the stories of the Mabinogion will fit into the hypothesis which I have roughly sketched. In the first place I ought to explain that the four stories of the Mabinogion were

probably put together originally in the Goidelic of Wales before they assumed a Brythonic dress. Further, in the form in which we know them, they have passed through the hands of a scribe or editor living in Norman times, who does not always appear to have understood the text on which he was operating. To make out, therefore, what the original Mabinogion meant, one has every now and then to read, so to say, between the lines. Let us take, for example, the Mabinogi called after Branwen, daughter of Llyr. She was sister to Brân, king of Prydain, and to Manawyddan, his brother: she was given to wife to an Irish king named Matholwch, by whom she had a son called Gwern. In Ireland, however, she was, after a time, disgraced, and served in somewhat the same way as the heroine of the Gudrun Lay; but in the course of the time which she spent in a menial position, doing the baking for the Court and having a box on the ear administered to her daily by the cook, she succeeded in rearing a starling, which one day carried a letter from her to her brother Brân at Harlech. When the latter realised his sister's position of disgrace, he headed an expedition to Ireland, whereupon Matholwch tried to appease him by making a concession, which was, that he should deliver his kingdom to the boy Gwern. Now the question is, wherein did the concession consist? The redactor of the Mabinogi could, seemingly, not have answered, and he has not made it the easier for any one else to answer. In the first place, instead of calling Gwern son of Matholwch, he should have called him Gwern son of Branwen, after his mother, for that is to be the sense is, that, in a society which reckoned birth alone, Gwern was not recognised as any relation to Matholwch at all, whereas, being Brân's sister's son, he was Brân's rightful heir. No such idea, however, was present to the mind of a twelfth-century scribe, nor could it be expected.

Let us now turn to Irish literature, to wit, to one of the many stories associated with the hero Cúchulainn. He belonged to Ulster, and whatever other race may have been in that part of Ireland, there were Picts there: as a matter of fact Pictish communities survived there in historical times. Now Cúchulainn was not wholly of the same race as the Ultonians around him, for he and his father are sharply marked off from all the other Ultonians as being free from the periodical illness connected with what has been called the couvade, to which the other adult braves of Ulster succumbed for a time every year. Then I may mention that Cúchulainn's baby name was *Setanta Beg*, or the Little Setantian, which points to the country whence Cúchulainn's father had probably come, namely, the district of Ireland, there were Picts there: as a matter of fact Pictish near the mouth of the Ribble, in what is now Lancashire. At the time alluded to in the story I have in view, Cúchulainn was young and single, but he was even then a great warrior, and the ladies of Ulster readily fell in love with him; so one day the nobles of that country met to consider what was to be done, and they agreed that Cúchulainn would cause them less anxiety if they could find him a woman who should be his fitting and special consort. At the same time also that they feared he might die young, they were desirous that he should leave an heir, "for," as it is put in the story, "they knew that it was from himself his rebirth would be." The Ulster men had a belief, you see, in the return of the heroes of previous generations to be born again; but we have here two social systems face to face. According to the one to which Cúchulainn as a Celt belonged, it was requisite that he should be the father of recognised offspring, for it was only in the person of one of them or of their descendants that he was to be expected back. The story reads as if the distinction was exceptional, and as if the prevailing state of things was wives more or less in common, with descent reckoned according to birth alone. Such is my impression of the picture of the society forming the background to the state of things implied by the conversation attributed to the noblemen of Ulster. Here again one experiences difficulties arising from the fact that the stories have been built up in the form in which we know them by men who worked from the Christian point of view, and it is only by scrutinising, as it were, the chinks and cracks that you can faintly realise what the original structure was like.

Among other aids to that end one must reckon the instances of men being designated with the help of the mother's name, not the father's: witness that of the King of Ulster in Cúchulainn's time, namely Conchobar mac Nessa, that is to say, Conor, son of a mother named Nessa; similarly in Wales with Gwydion son of Dôn. Further we have the help of a considerable number of ancient inscriptions, roughly guessed to date from the fifth or sixth century of our era, and commemorating persons traced back to a family group of the kind, perhaps, which Cæsar mentions in the fourteenth chapter of his fifth book. Within these groups the wives were, according to him, in common (*inter se communes*). Take for instance an inscription from the barony of Corcaguiny in Kerry, which commemorates a man described as " Mac Erce, son of *Muco Dovvinias*," where *Muco Dovvinias* means the clan or family group of *Dovvinis* or *Dubin* (genitive *Duibne*), the ancestress after whom Corcaguiny is called *Corco-Duibne* in Medieval Irish. We have the same formula in the rest of Ireland including Ulster, where as yet very few Ogams have been found at all. It occurs in South Wales and in Devonshire, and also on the Ogam stone found at Silchester, in Hampshire. The same kind of family group is evidenced also by an inscription at St. Ninian's, in Galloway; and, to go further back—perhaps a good deal further back—we come to the bronze discovered not long ago at Colchester, and dating from the time of the Emperor Alexander Severus, who reigned from 222 to 235. This is a votive tablet to a god Mars Medocius, by a Caledonian Pict, who gives his name as Lossio Veda, and describes himself further as *Nepos Vepogeni Caledo*. He alludes to no father, and *Nepos Vepogeni* is probably to be rendered Vepogen's sister's son. At any rate, the Irish word corresponding etymologically to the Latin *nepos* has that sense in Irish; but so far as I know it has never been found meaning a nephew in the sense of brother's son. That may serve as an instance how the ideas of another race penetrated the fabric of Goidelic society; for here we must suppose a time to have come when there was no longer any occasion for a word meaning a brother's son, which, of course, there never was in the non-Celtic society which ranked men and women according to their birth alone.

Now this Caledonian Pict was not exceptional among his kinsmen, for they succeeded in observing a good deal of silence concerning their fathers down, one may say, to the twelfth century. It is historical that the king of the northern Picts was not wont to be the son of the previous king. In short, when the Celtic elements there proved strong enough to ensure that the son of a previous king should succeed, a split usually took place, the purer Picts being led by the rule of succession by birth to set up a king of their own. The fact is not so well known that the same succession prevailed also some time or other at Tara in Ireland: it is proved by a singular piece of indirect evidence, the existence of a tragic story to explain why " no son should ever take the lordship of Tara after his father, unless some one came between them." The last clause is due, I should ray, to somebody who could not understand such a prohibition based on the ancient rule that a man's heir was his sister's son. This would be, according to Irish legend, in the lifetime of Conor mac Nessa.

It is curious to notice how the stories about the Pictish *ménage* seemed to have puzzled ancient authors. I will only cite one instance, to wit, from Golding's sixteenth century translation of what then passed as the production of Solinus, and what may now pass, even according to Mommsen, as quite old enough for my present purpose. It runs thus: " From the Promontorie of *Calydon* to the Iland *Thule* is two dayes sayling. Next come the Iles called *Hebudes*, five in number, the inhabiters whereof know not what corne meaneth, but liue onely by fishe and milke. They are all vnder the gouernment of one King. For as manie of them as bee, they are seuered but with a narrowe groope one from another. The King hath nothing of hys own, but taketh of euery mans. He is bounde to equitie by certaine lawes: and least he may start from right through couetousnesse, he learneth Justice by pouertie, as who may have nothing proper or peculiar to himselfe, but is found at the charges of the Realme. Hee is not suffered to haue anie woman to himselfe, but whomsoeuer he hath minde vnto, he borroweth her for a tyme, and so others by turnes. Wherby it commeth to passe that he hath neither desire nor hope of issue."

The man who wrote in that way presumably failed to see that the king was not subject to any special hardship as compared with the other men in his kingdom, where none of them had any offspring that he could individually call his own. This, be it noticed, refers to the Hebrides, not, as sometimes happens, to the more distant island of Thule, where there was also a king, as any reader of "Faust" will tell us.

We now come to the Celts, and begin with Pliny's version of Cæsar's words about the division of Gaul into three parts, as follows: *Gallia omnis Comata uno nomine appellata in tria*

populorum genera dividitur, amnibus maxime distincta. A Scalde ad Sequanam Belgica, ab eo ad Garunnam Celtica eademque Lugdunensis, inde ad Pyrenaei montis excursum Aquitanica, Aremorica antea dicta. We may for the present dismiss the third or Aquitanic Gaul from our minds; but Belgic and Celtican Gaul may be taken as representing the two sets of Celts of our own islands. The Belgic Gauls began last to come to this country, and their advent seems to fall between the visits of Pytheas and Julius Cæsar: that is, roughly speaking, between the middle of the fourth century and that of the first century B.C. In this country they came to be known collectively as Brittanni or Brittones, the linguistic ancestors of the people who have spoken Brythonic or the *Lingua Brittannica*, such as the Welsh, the Cornish, and the Strathclyde Britons. As to the others Celts, it is much harder to say when or whence exactly they came—I mean the linguistic ancestors of the Gaels of Ireland, Man, and Scotland, that is to say, the peoples whose language has been Goidelic. Some scholars are of opinion that there were no Goidelic-speaking peoples in Britain till some such came here from Ireland on sundry occasions, beginning with the second century, in the time of the Roman occupation, but how the Goidels would be supposed by them to have reached Ireland I do not exactly know. My own notion is that the bulk of them reached that country by way of Britain, and that they arrived in Britain, like the Belgic Gauls later, from the nearest parts of the Continent; for this would be previous to the appearance of the Belgic Gauls on the western sea-board of Europe: that is to say, at a time when Celtica extended not merely to the Seine, but to the Scheldt or to the Rhine, if not further. Then as to the time of the coming of the ancestors of the Goidels, it has been supposed coincident with a period of great movements among the Celts of the Continent, in particular the movements which resulted, among other things, in some of them reaching the shores of the Mediterranean and penetrating to the heart of the Iberic peninsula. Perhaps one would not be far wrong in fixing on the seventh and the sixth centuries B.C. as covering the time of the coming of the earlier Celts to our shores.

In Britain I should suppose these earlier hordes of Celts to have conquered most of the southern half of the island; and the Brythonic Celts, when they arrived, may have overrun much the same area, pushing the Goidelic Celts more and more towards the west. Under that pressure it is natural to suppose that some of the latter made their way to Ireland, but it is quite possible that their emigration thither had begun before. Some time or other previous to the Roman occupation the Brythonic people of the Ordovices seem to have penetrated to the sea between the rivers Dovey and Mawddach, displacing probably some Goidels who may have gone to the opposite coasts of Ireland; but more traces in Irish story appear of invasions on the part of the Dumnonii, who possessed the coast between Galloway and Argyle. These were so situated as to be able to assail Ireland both in front and from behind, and this is countenanced to some extent by Irish topography, not to mention the long legends extant as to great wars in the west of Ireland between the Tuatha Dé Danann and invaders including the Fir Domnann. I suspect also that it was the country of these northern Dumnonians which was originally meant by Lochlinn, a name interpreted later to mean Norway.

Such are some of the faint traces of the Goidelic invasions of Ireland from Britain, but it is possible—perhaps probable—that Ireland received settlers on its southern coast from the north-west of Gaul at a comparatively late period, at the time, let us say, when Cæsar was engaged in crushing the Veneti and the Aremoric League. This has been suggested to me by the name of the Usdiæ, which probably survives in the first syllable of *Ossory*, denoting a tract of country now, roughly speaking, covered by the county of Kilkenny, but which may have been considerably larger before the Déisi took possession of the baronies of the two Decies and other districts now constituting the county of Waterford, not to mention possible encroachments on the part of Munster on a boundary which seems to have been sometimes contested. Now the Continental name which invites comparison with that of the Usdiæ is that of the Ostiæi, who in the time of Pytheas appear to have occupied the north-western end of what afterwards came to be called Brittany; they were also called Ostiones, and more commonly Osismi. I see no reason to suppose that the ships of the Aremoric League could not make the voyage from Brittany to the principal landing-places on the south of Ireland from the Harbour of Cork to that of Waterford, and I gather from Ptolemy's Geography that Ireland was relatively better known on the Continent than Britain, although the latter had been in a manner, connected with the Roman world. This I should explain somewhat as follows :—Cæsar, who knew very little about the west of Britain and probably less about Ireland, says that in his time the great druidic centre of Gaul was in the country of the Carnutes, somewhere, let us say, near the site of the present town of Chartres, that druidism had been introduced from Britain to Gaul, and that those who wished to understand it had to go to Britain to study. The authors of antiquity tell us otherwise nothing about druids in Britain except that Tacitus speaks of such in the Annals, in his well-known passage as to Suetonius Paulinus landing with his troops in Anglesey and the scene of slaughter which ensued. Indeed, one may go further and say that there is no proof that any Belgic or Brythonic people ever had druids : they belonged to the Celtican Gauls and the Goidelicising Celts of Britain and Ireland, who had probably accepted the institution from the Pictish race. At any rate it is significant that the Life of St. Columba introduces the reader to a genuine druid at the court of the Pictish king, near Inverness, where, as well as on Loch Ness, the saint had to contend with him. In any case, it is highly probable that druidism was no less a living institution in Ireland than in the Goidelic and Pictish parts of Britain. Presumably it was more so, and it may be conjectured that Gaulish students of druidism visited Ireland no less than Britain ; also, vice-versâ, that Irish druids paid visits to the Celtican part of Gaul where druidism flourished on the Continent, and in a word that there was regular intercourse between Gaul and the south of Ireland. If the druids of Ireland, who, among other rôles, played that of schoolmasters and teachers in that country, travelled to Celtica, they must have spread on the Continent some information about their native country, while generations of them cannot have returned to Ireland, with their druidic pupils, without bringing with them some of the arts of civilised life as understood in Gaul : among these one must rank very decidedly the art of writing, which the druids practised. Now you know the usual account given of the ordinary Latin for Ireland, namely *Hibernia* —to wit, that it was suggested by such native names as that of one of the greatest tribes of that country, namely the Ἰούερνοι or *Iverni*, and that it had its *v* ousted when Latin began about the fourth century to write *b* for *v*, and that an *h* was then prefixed to make the word *Hibernia* properly connote the wintry climate which our sister island had always been supposed to enjoy. But now comes the question, where did Pomponius Mela, who flourished about the middle of the first century, get his *Iuverna*, which Juvenal also used? Doubtless from a druid like Dâlan, or some other educated native of Ireland, for what the editors print as *Iuverna*, *Iuuerna*, or *Juverna* would appear in ancient manuscripts as *IVVERNA* or *iuuerna*, in which the first two syllables are spelt correctly with *v v* according to a system of spelling well known in Ogmic writing centuries later. But a particular system of spelling seems to me to imply writing, and thus one is encouraged to think that the Ogam alphabet may have been invented no later than the first century in the intercourse I have conjectured to have been going on between the north-west of Gaul and the south of Ireland, where the majority of Ogam inscriptions are now found. But what has archæology to say on the question of such intercourse?

After this digression I come back to the two main streams of Celtic immigration from the same parts of the Continent in two different periods of time. The later of these introduced the *Lingua Brittannica*, which was practically a dialect of old Gaulish ; but the affinities of the other Celtic language of these islands, the Goidelic, are not so easy to determine. I have long thought that I can identify traces of it on the Continent, and that its principal home was in the region which Pliny called Celtica, between the Garonne and the Seine. I ventured accordingly to call it *Celtican*, as the simpler word *Celtic* had already been wedded to a wider signification. Since then the existence of that language has been placed beyond doubt by the discovery of fragments of a calendar engraved on bronze tablets. This find was made about the end of 1897 at a place called Coligny, in the department of the Ain, and the pieces are now in the museum at Lyons. It is difficult to say for certain whether Coligny is within the territory once occupied by the Sequani, or else by the

Ambarri, a people subject to the Ædui, who were rivals of the Sequani and Arverni. The name of the Sequani would seem to have belonged to the Celtican language, and Mr. Nicholson, in his interpretation of the calendar, has ventured in this instance to call it Sequanian. But two inscriptions in what appears to be the same language have come to light also at a place called Rom, in the Deux Sèvres and on the Roman road from Poitiers to Saintes. This Celtican language is to be carefully distinguished from Gaulish, but it is not exactly what I expected it to be : it is better. For several of the phonetic changes characteristic of Goidelic had not taken place in Celtican. Among other things it preserves intact the Aryan consonant *p*, which has since mostly disappeared in Goidelic, as it had even then in Gaulish. This greater conservatism of Celtican enables one to refer to it the national appellation of the people of the region in question, namely, that of the *Pictones*, from which it is impossible to sever the name of the Picts of Britain and Ireland, who are found also called *Pictones* and *Pictanei*. Here I may mention that Mr. Nicholson calls attention to instances of tattooing on some of the faces on ancient coins belonging to Poitou and other parts of western France. In the light of the names here in question one sees that *pictos* was a Celtican word of the same etymology, and approximately, doubtless, of the same meaning, as the Latin word *pictus*, that the Celticans had applied it at an early date to the Picts on account of their habit of tattooing themselves, and that the Picts had accepted it (with its derivative *Pictones*) so generally that by the time when the Norsemen arrived in the North of Scotland, it was the name which the natives gave them as that by which they called themselves. That is practically proved by the Norsemen calling Caithness and Sutherland *Petta-land* or the Land of the Picts, and the sea washing its northern shore *Pettalands fiorth*, which survives modified into Pentland Firth.

Another Celtican word of great interest here has by a mere chance come down in a high German manuscript written before the year 814 ; it is *Chortonicum*, and occurs among a number of geographical names, several of which refer to Gaul, so that Chortonicum may very well have meant the country of the Pictones. At all events, the great German philologist, Pott, at once saw that it was to be explained by reference to the word Cruithne, " a Pict," with which it decidedly goes as distinguished from its Brythonic equivalent *Prydyn* (or the older *Priten*), with an initial *p*. The Celtican form originally meant was some such vocable as Qurtonico-n, with the *qu* which was usual in Celtican and early Goidelic, where it formed, in fact, one of the most conspicuous distinctions between those languages and Brythonic or Gaulish, in which *qu* had been changed into *p*.

My remarks have again run into tiresome details, but it is only by attending to such small points that one can hope to force language to yield us any information in the matter of ethnology. It may perhaps help in some measure if I sum up what I have been trying to say, thus :

The first race we have found in possession of the British Isles consisted of a small, swarthy population of mound-dwellers, of an unwarlike disposition, much given to magic and wizardry, and perhaps of Lappish affinities : its attributes have been exaggerated or otherwise distorted in the evolution of the Little People of our fairy tales.

The next race consisted of a taller, blonder people, with blue eyes, who tattooed themselves and fought battles. These tattooed or Pictish people made the Mound Folk their slaves, and in the long run their language may be supposed to have been modified by habits of speech introduced by those slaves of theirs from their own idiom. The affinities of these Picts may be called Libyan, and possibly Iberian.

Next came the Celts in two great waves of immigration, the first of which may have arrived as early as the seventh century before our era, and consisted of the real ancestors of some or our Goidels of the Milesian stock, and the linguistic ancestors of all the peoples who have spoken Goidelic. That language may be defined as Celtican so modified by the idioms of the population which the earlier Celts found in possession that its syntax is no longer Aryan.

Then, about the third century B.C., came from Belgica the linguistic ancestors of the peoples who have spoken Brythonic ; but, in the majority of cases connected with modern Brythonic, they are to be regarded as Goidels who adopted Brythonic speech, and in so doing brought into that language their Goidelic idioms, with the result that the syntax of insular Brythonic is no less non-Aryan than that of Goidelic, as may be readily seen by comparing the thoroughly Aryan structure of the few sentences of old Gaulish extant.

Those are the races which have been inferred in the course of these remarks, in which I have proceeded on the principle that each successive band of conquerors has its race, language and institutions eventually more or less modified by contact with the race, language and institutions of those whom it has conquered. That looks simple enough when stated so, but the result which we get proves complicate. In any case I have endeavoured in this address to substitute for the rabble of divinities and demons, of fairies and phantoms that disport themselves at large in Celtic legend, a possible series of peoples, to each of which should be ascribed its own proper attributes. But that will only be possible if we can enlist the kindly aid of the Muse of Archæology.

THE INTERNATIONAL CONGRESS OF APPLIED CHEMISTRY.

THIS congress was held in Paris during the last week of July, M. Moissan being president and M. Berthelot honorary president. The work was divided into ten sections : analytical chemistry, chemical industry of inorganic products, metallurgy, mines and explosives, chemical industry of organic products, the sugar industry, chemical industry of fermentation, agricultural chemistry, hygiene, food analysis, medical and pharmaceutical chemistry, photography, and electrochemistry. More than two hundred papers were read and discussed, and numerous resolutions were passed, of which the following were the most important. In view of the great inconvenience caused commercially by uncertainty in the atomic weights used by analytical chemists, the congress, hoping that the adoption of the atomic weight of oxygen as a base ($O = 16$) would lead to a greater certainty and to a simplification in the calculation of atomic weights, agreed to work in unison with the International Commission on atomic weights. It further suggested the necessity for an International Commission for fixing methods and coefficients of analysis in commercial work. Committees were also appointed to deal with questions of indicators in volumetric work, analysis of manures, potash estimation, and the use of sulphurous acid in wine. In the second section the chief questions dealt with were the determination of high temperatures, construction of glass and porcelain furnaces, the manufacture of sulphuric acid, and of barium and hydrogen peroxides. In the section of metallurgy, mines and explosives, papers were read dealing with the sampling of minerals, the constitution of iron and steel, the use of the microscope in the study of metals, utilisation of waste heat, and the estimation of sulphur, manganese and phosphorus in metals. In the section dealing with the industry of organic substances the most important discussion was on the use of alcohol for other than drinking purposes, and a series of resolutions was passed stating that in the opinion of the congress no duty should be charged upon alcohol used in the preparation of pharmaceutical and chemical products. In the case of alcohol intended for use as fuel, the substances added should be of a character appropriate to its use, not too costly, and not containing any non-volatile substance. Any attempt to recover pure alcohol from methylated spirit should be liable to severe penalties, and all makers of stills should be compelled to give particulars to the excise authorities of stills sold or repaired. In the other sections discussions were held on the relation of the sugar industry to the State, the methods of analysis of wines and spirits, the carbide industry, manufacture of percarbonates, and numerous other papers of interest.

UNIVERSITY AND EDUCATIONAL INTELLIGENCE.

IT is officially announced that Mr. L. R. Wilberforce, demonstrator in physics at the Cavendish Laboratory, Cambridge, and University lecturer in physics, has been appointed to the Lyon Jones chair of experimental physics at University College, Liverpool, vacated by the removal of Dr. Oliver Lodge to the University of Birmingham.

THE Admiralty has created an important new post in the Dockyard staff, namely, that of electrical engineer, to rank next to the four chief assistant engineers. To fill this post the Admiralty has chosen Mr. Louis J. Steele, M.I.E.E., late chief engineer of Messrs. Verity, and formerly assistant engineer with Messrs. Johnson and Phillips. Mr. Steele received his

training at the Technical College, Finsbury, under Prof. Silvanus Thompson and Prof. Perry, and carried off the certificate of the College in 1890. He will be attached to the Dockyard staff at Portsmouth.

INSTRUCTION in chemistry is well provided for at the Goldsmiths' Institute, New Cross. During the session about to commence, Mr. W. J. Pope will give courses of lectures on oils, fats and waxes, organic chemistry with special reference to recent work and current views, inorganic chemistry and stereochemistry. In this last course of lectures, the principles which form the foundation of stereochemistry will be discussed, together with the methods which have led to the discovery of stereoisomerism amongst compounds of carbon, nitrogen, tin and sulphur. Particular attention will be paid to the bearing of stereochemistry upon current chemical problems. Lectures in chemistry will also be given by Mr. Stanley J. Peachey.

DURING the past year the Degree of Doctor of Philosophy was conferred by twenty-two Universities in the United States upon 233 candidates. The distribution of these degrees among the various Universities, and the subjects taken, are dealt with in an article in *Science*. It appears that 120 of the degrees were granted to students of the humanities, and 113 for scientific subjects. The tables show that the humanities are favoured at Havard and Yale Universities, and the sciences at Johns Hopkins, Columbia and Cornell Universities. Last year Johns Hopkins gave more than its proportionate share of degrees in chemistry, physics, zoology and physiology, Chicago in mathematics, geology, sociology and education, Harvard in physics, zoology and anthropology, Columbia in astronomy, botany, zoology and education, Yale in palæontology and psychology, Cornell in botany and psychology, and Clark in mathematics, psychology and education. The six science subjects in which most students presented theses are as follows :—Chemistry 26, physics 15, botany 12, mathematics 11, zoology 11, psychology 9.

EVER since the funds were provided for technical education in this country, it has been insisted upon in these columns, and by men of science generally, that such education could only be profitably carried on by giving rational instruction in scientific principles instead of attempting to teach actual processes and trade methods, which are constantly in a state of flux on account of new developments. The most gratifying characteristic of educated opinion at the present time is the acceptance of this view ; and it is especially noteworthy in connection with the substitution of nature study for agriculture in rural schools. In an address recently delivered before the Cheshire College of Agriculture, Prof. Robert Wallace dwelt upon the relation between the work of an agricultural school and actual farm work, and showed himself in complete sympathy with the view which has been expressed over and over again in these columns. Here is the case in a few words :—"What a young farmer should learn is not ordinary farm work, viz. to plough and harrow a given area in the day. He can become an expert at that kind of thing at home to greatest advantage, without cost for instruction, and at the same time prove a valuable aid to his father. He requires to be taught just those things which are not to be learned on an ordinary farm, to have explained to him the meaning of processes which are founded on scientific principles, and to become familiar with the common facts of those sciences which bear upon agricultural practice." If this had been borne in mind by Technical Instruction Committees in rural districts from the time they came into existence, their efforts would have received more encouragement from practical men, and have been attended with better results, than have been attained in many cases.

SCIENTIFIC SERIALS.

Transactions of the American Mathematical Society, vol. i. No. 3.—Wave propagation over non-uniform electrical conductors, by M. I. Pupin, is a paper read before the society in December last. The main object of it is the solution of a problem which, looked at from a purely mathematical point of view, can be stated as follows :—Find the integral of the partial differential equation $L \frac{d^2 y}{dt^2} + R \frac{dy}{dt} = \frac{1}{C} \frac{\delta^2 y}{\delta s^2}$, and determine it so as to satisfy $k + 2$ boundary conditions, where $k + 1$ is the number of coils. The principal difficulty is to determine the proper mathematical formulation of these sundry conditions so as to obtain a system of equations which can be readily solved. The paper is illustrated by diagrams which put the problems discussed in a clear light.—" Ueber systeme von differentialgleichungen dessen vierfach periodische functionen genüge leisten," by M. Krause, was presented at the Chicago (April) meeting of the present year. References are given to Hermite (" Sur quelques applications de la théorie des fonctions elliptiques," 1885), and to a paper by Picard (*Comptes rendus*, Band 89), and to previous work by the author.—E. B. van Vleck follows with a paper on linear criteria for the determination of the radius of convergence of a power series. Its object is to establish criteria for the convergence of a power series when the $(n + 1)$th coefficient A_n is connected with the preceding coefficients by a linear relation which tends to take a limiting form as n increases indefinitely. The criteria include Cauchy's ratiotest as a special case, and may be looked upon as an extension of the test, and are applicable in cases in which the simple ratiotest fails. The paper closes with two theorems which are an extension for the case of two variables, criteria for the convergence of power series in such a case are stated to be very rare.—On the existence of the Green's function for the most general simply connected plane region, by W. F. Osgood.—A short but suggestive note—" D " lines on quadrics, by A. Pell. These lines, so named by Cosserat, were originally considered by Darboux. They are the lines drawn upon a surface in such a way that the osculating sphere at every point is tangent to the surface at that point. In addition to the above, the lines have been studied by Enneper and Ribaucour (for surfaces in general). In the present paper the author applies the theory of elliptic functions to the integration of Darboux's differential equation, and obtains an idea of the appearance of the lines and also some of their properties.—Starting from an article, by Prof. F. Morley, in the previous number of the *Transactions*, F. H. Loud gives sundry metric theorems concerning n lines in a plane. By giving a different interpretation to formulæ got by Prof. Morley, Mr. Loud obtains a new series of theorems and other results of some interest.—An application of group theory to hydrodynamics, by E. J. Wilczynski. It was observed by Sophus Lie that the stationary motion of a fluid can serve as a perfect picture of a one-parameter group in three variables. Apparently this fact has not been utilised for the purposes of hydrodynamics. This paper does this. Amongst other advantages, the treatment, from the new standpoint, leads to special cases of exceptional interest and importance, which otherwise appear to be difficult and unpromising.—Dr. L. E. Dickson, following up work recently published in the *Proceedings* of the London Mathematical Society (vol. xxxi. pp. 30, 351), contributes an article on the determination of an abstract simple group of order $2^7 \cdot 3^6 \cdot 5 \cdot 7$, holohedrically isomorphic with a certain orthogonal group and with a certain hyperabelian group (contributed to the Chicago [April] meeting of the society).

In the *Journal* of the Royal Microscopical Society for August, Mr. E. M. Nelson has one of his useful technical articles on the "lag" in microscopic vision, as well as a historical account of the improvements in the structure of the microscope introduced by the firm of Ross. Mr. E. B. Stringer describes a new form of fine adjustment. Miss A. Lorrain Smith gives a description of some new microscopic fungi, including a new species of Entomophthora, not parasitic, but saprophytic on dead animal tissues. There is, in addition, the usual summary of current researches relating to zoology, botany and microscopy.

In the *Journal of Botany* for August, Messrs. W. and G. S. West have a second instalment of their notes on freshwater algæ, in which some new species and varieties are described. The remaining papers are descriptive or geographical.

SOCIETIES AND ACADEMIES.

PARIS.

Academy of Sciences, September 10.—M. Maurice Lévy in the chair.—Occultation of Saturn by the moon on September 3 observed at the Observatory of Lyons, by MM. J. Guillaume, G. Le Cadet and M. Luizet.—On differential systems with a general uniform integral, by M. Paul Painlevé. Four types of systems are examined, problems in mechanics such as the movement of a heavy body fixed by a point, the inversion of total differentials, the case where the general integral of a differential system does not admit of transcendental singularities, and the

study of the integrals of a differential system in a real field.—On the liquefaction of air by expansion with production of external work, by M. Georges Claude.—On the dielectric cohesion of gases and vapours, by M. E. Bouty. The experiments previously described upon the relation existing between the distance at which insulation breaks down and the pressure of the gas have been extended to vapours of liquids. Results of measurements for water, and eleven organic liquids, are given in the present paper. —On the modification of the electrical and organic properties of cables under the prolonged action of currents, by M. Georges Rheins. When a cable is submitted to the action of an alternating current it preserves its electrical and organic properties intact. With a continuous current in one direction the cable gradually loses its electrical properties, this effect being produced by the slow penetration of the copper from the wire into the sheath. The effect is similar with both gutta and paper coatings.—New researches on the absorptive power of hæmoglobin for oxygen and carbonic acid, by M. L. G. de Saint-Martin. As the result of numerous experiments quoted, the author is of opinion that, contrary to the views generally held, it is impossible, especially in pathological cases, to estimate hæmoglobin by means of the absorbing power of the blood.—On the nitrocelluloses, by M. Léo Vignon. Both the nitrocelluloses and the nitro-oxycelluloses energetically reduce Fehling's solution, their reducing power being apparently independent of the degree of nitration. The reducing powers of the nitration products of cellulose and oxycellulose are of the same order, about one-fifth that of inverted sugar.—On the wood of the Conifers of peat bogs, by M. L. Géneau de Lamarlière. In the wood of Conifers taken from a peat bog, the intercellular layer formed of lignin and pectic compounds is intact, whilst the internal portion has been strongly attacked by microbial action. The lignin and cellulose have disappeared, an amorphous substance remaining behind which is soluble in potash after the action of chlorine. The material resembles callose.—Influence of a dry or moist medium upon the structure of plants, by M. Eberhardt. Compared with normal air, the effect of dry air is to increase the thickness of the epidermal cuticle and the number of stomata, to make the cork layer form earlier, to increase the production of ligneous tissue, and to cause an increase in the amount of pallisade tissue in the leaf.

CONTENTS.

THURSDAY, SEPTEMBER 27, 1900.

THE MAMMALS OF SOUTH AFRICA.

The Fauna of South Africa; Mammals. Vol. i. *Primates, Carnivora and Ungulata.* By W. L. Sclater. Pp. xxx + 324; illustrated. (London : Porter, 1900.)

IN our review of the "Birds of South Africa" (vol. i.), published earlier in the year, reference was made to the scope of the present series of volumes and the peculiarities of the South African fauna; and it will therefore be unnecessary to recapitulate what has been there written. In the introduction to the volume before us Mr. Sclater remarks that since 1832 no one has attempted to give a complete account of the mammals of South Africa, attention having been concentrated by writers on this subject on the larger forms which constitute the chief attraction to sportsmen and travellers. Accordingly, in the case of the smaller representatives of the class the author has practically a clear field before him, much labour being necessary to collect and collate the numerous papers which have been written of late years on the Rodents and other small mammals of Africa. This portion of his subject is, however reserved for the second volume; and at present we have only to consider how Mr. Sclater has treated the section dealing with the larger types of mammalian life.

As he himself admits, his task in this respect has been a comparatively easy one; the "Book of Antelopes" clearing the way in regard to that very important group of the Ungulata, while the Zebras have been carefully worked out by Mr. Pocock and other naturalists, and the Carnivora have attracted the attention of numerous writers. On all these valuable sources of information Mr. Sclater has drawn largely; and it is no discredit to him if the work partakes to a very considerable degree of the nature of a compilation, and contains comparatively little that is new and original. Indeed, this is fully acknowledged in the Introduction, where the author takes care to state that in his account of the habits of the different animals he has relied on the observations of others, and endeavoured to compile from published writings and manuscript letters an adequate and readable account of each.

By these observations we by no means intend to imply that Mr. Sclater's work is in any sense a superfluous or unnecessary one; the "Book of Antelopes" and other works of that description are expensive and accessible only to the few; and, as already said, there is no modern and up-to-date work treating of South African mammals as a whole.

Both in respect to his treatment of the aforesaid life-histories and in his description of the species themselves Mr. Sclater may, indeed, be fairly congratulated on the result of his labours; the volume before us being sufficiently popular and interesting to attract the attention of the sportsman, while at the same time it contains a sufficient amount of technical detail to satisfy the needs of the working naturalist. Whether, however, in these days of cheap natural histories and zoological text-books it is necessary that every work on local faunas should contain a hackneyed recapitulation of the characteristics of the

orders and other large groups of animals may be a question which many would, we think, be inclined to answer in the negative. As regards the numerous illustrations in this volume, it is much to be regretted that the majority, which were executed in the Colony, are of a very inferior description, and in no wise worthy to stand alongside those borrowed from the "Book of Antelopes" and other well-known works. Probably there is not time to alter the arrangements made for illustrating the second volume; but if there be, it is most desirable that the drawings should be made and photographed in this country.

As regards the local variations presented by species, the author is perhaps a little too conservative; and although he gives full details in regard to the numerous races of Burchell's zebra, we venture to think that more might have been said, for instance, with regard to the local phases of the Kaffir cat and some of the jackals. Among the Carnivora, it is interesting to find that the author recognises the black-footed cat (*Felis nigripes*), which has been so unaccountably overlooked by recent writers, as entitled to rank as a species. And, in another group, we think he is decidedly well advised in adopting the Colonial term "Dassie" (an abbreviation of the Boer *klip-dass* = rock-badger) as the popular name for those animals which used to be scientifically known as Hyrax, until that name was displaced by the earlier Procavia.

Generally speaking, Mr. Sclater is, indeed, well up to date as regards nomenclature, both popular and scientific. He does not, however, in all cases give credit for recent emendations in nomenclature to those to whom it is due. For instance, the reader would be led to imagine that the author was the first in modern times to replace the ordinary scientific title for the eland by *Taurotragus oryx*, whereas the change was initiated last year by Mr. Rowland Ward in his "Records of Big Game." And it cannot be urged in the author's defence that the omission is due to the fact of his quoting only references from works bearing directly on South Africa, since he departs from this rule in the case of *Cephalophus grimmi* (p. 157), as well as in other instances. A change of name for which the author appears to be really responsible occurs in the substitution of *Strepsiceros capensis* for *S. kudu*; but this directly raises the question of the advisability of adopting the alliterative *S. strepsiceros*, which some would now regard as the proper name of the kudu.

On the whole, the volume appears to be remarkably free from misprints and slips. On p. 317 the author has, however, given *Elephas planifrons* as the type of Falconer's subgenus *Euelephas*, and *E. hysudricus* as that of *Loxodon*; whereas the two specific names should be transposed. But this unfortunate slip is not all, for in making these two species the respective types of the subgenera the author has totally misrepresented Falconer. Mr. Sclater has, of course, taken them as the types because they occur first in Falconer's table. But *Euelephas* of Falconer is merely the typical subgenus of *Elephas* (*Elephas* proper it would now be called), and therefore the type of the one is the type of the other; this being, of course, the Indian elephant. Again, in the paper to which Mr. Sclater refers, Dr. Falconer, in writing of the Loxodons, says that "the existing type of this group

Z

is the African elephant, which Fred. Cuvier, in 1835, proposed to erect into a distinct genus under the name of Loxodonta." A more unfortunate error, complicated by a more unfortunate slip, could scarcely be conceived.

In one other passage where the author ventures into the domain of palæontology he has scarcely been more successful, since (p. 308) he unhesitatingly accepts the alleged Cretaceous age of presumed Hyracoid remains discovered in the Argentine. Possibly, however, his omission to mention that fossil "dassies" occur in the European Pliocene may be due to the time. that the volume has taken in passing through the press, although the fact was announced at the Zoological Congress held at Cambridge in 1898.

Much general interest will attach to Mr. Sclater's account of the two large mammals which have undoubtedly become extinct in South Africa in modern times. With regard to the first of these, the author remarks that the last blaauw-bok (*Hippotragus leucophoeus*) was probably killed in 1799 ; and that, in addition to several pairs of horns, five complete mounted specimens are known to be preserved. The quagga (*Equus quagga*) he believes to have survived in the Orange Colony till at least 1878, although it is difficult to obtain exact information owing to the Boers confounding this species with Burchell's zebra. Of the white rhinoceros it is considered not improbable that a few may still survive in Zululand, although it is sad to learn that no less than six are reported to have been killed so lately as 1894, one of these being exhibited in the museum at Pretoria. The latest information with regard to the white-tailed gnu is that a few herds were, till recently, preserved on some farms in the Orange Colony and the Transvaal ; while it is suggested that a few stragglers may survive in the Kalahari, Gordonia and German South-west Africa. Much anxiety will now be felt by naturalists as to what has happened to the gnus, and also to the blesboks, till lately preserved in the Boer Republics ; and it is to be hoped that those responsible for the settlement of these districts will do all in their power to protect such remnants as the war may have left.

We hope ere long to have the pleasure of congratulating Mr. Sclater on the completion of his task.   R. L.

OUR BOOK SHELF.

Acetylene, a Handbook for the Student and Manufacturer. By Vivian B. Lewes, F.I.C., &c. Pp. xxvi + 978. (Westminster : Archibald Constable and Co., Ltd., 1900.)

IN this handsome volume of nearly 1000 pages, Prof. Lewes has presented the English reader with a handbook on the manufacture and use of acetylene which in completeness of scope and wealth of illustration will compare with its French and German rivals.

In the first part (consisting of four chapters) the scientific history of acetylene and its properties is set forth with considerable detail ; useful summaries of many researches are given, and references to the original memoirs are added. The question of the discovery of "commercial calcium carbide" is discussed with discrimination, the chief credit being assigned to the Canadian engineer, Mr. T. L. Willson. The reactions of

acetylene, especially with metallic salts, are fully considered.

Part ii., the most important in the book, describes the development of the electric furnace, and its special adaptation to the manufacture of calcium carbide. The generation of acetylene by the action of water on the carbide is next considered, and then the question of impurities and their removal is discussed. Most of the figures illustrating this portion of the book are clear and satisfactory, but a few are indistinct and on too small a scale. The chapter on the combustion of acetylene is illustrated by a number of useful drawings of burners and flames, and full data are given for a comparison between acetylene and other methods of illumination, both as regards prime cost and working expenses. We think Prof. Lewes has shown himself eminently fair in the discussion of this subject.

The method of treatment adopted by the author naturally leads to some repetition, but in a book of reference this will not be felt an inconvenience. It was perhaps hardly necessary to give the author's "acetylene theory of luminosity" twice over. In a new edition we hope that the number of small inaccuracies will be reduced. We did not expect to find a chief gas-examiner saying that "sulphur dioxide, in ill-ventilated apartments, will absorb oxygen and moisture from the air, and will in this way become converted into minute traces of sulphuric acid, which, concentrating themselves upon any cold surface in the room, give rise to corrosion," &c. The Harcourt pentane standard is not approved of, apparently, by Prof. Lewes, who states that it was first described in 1887. It was described ten years earlier. The specific heats of gases given on page 609 are incorrect, and several names are wrongly spelled, *e.g.* Vieille should be Vieille (p. 68). Smithell should be Smithells (563). In spite of small errors, the book is a mine of information, and will be useful, both to chemical students and to others interested in the making and use of acetylene.

Wireless Telegraphy and Hertzian Waves. By S. R. Bottone. Pp. 113. (London : Whittaker and Co., 1900.)

THERE are many whose interest in wireless telegraphy will take the form of a desire to experiment for themselves, and who, whether from inclination or necessity, will prefer to do so with home-made apparatus. To these the little book before us will especially commend itself.

The first half of Mr. Bottone's work is devoted to "preliminary notions," "historical considerations," and to a chapter on electric waves. This earlier half seems to us to leave much to be desired. Thus a clear elementary description of the fundamental experiments of electrical science is followed (p. 12) by a very obscure summary of the properties of electric charges and currents. Again, the confusing of the words "stress" and "strain" will not please the reader accustomed to the modern strict usage of these terms.

The description of apparatus in these earlier chapters is often involved, and many sentences will be found which through faulty punctuation or other small errors are not at once intelligible. A considerable amount of repetition also seems to occur, apart from deliberate recapitulation.

The later part of the book includes a number of really good descriptions in detail of how to make such apparatus as a small induction coil, a Wimshurst machine, a relay, or a coherer ; and the author is evidently familiar with the little practical difficulties which arise. Possibly the importance of making a dimensioned drawing before starting work might have been emphasised ; but in all other respects these "workshop recipes" seem very complete and well suited to the wants of those about to make such apparatus.   D. K. M.

LETTERS TO THE EDITOR.

[*The Editor does not hold himself responsible for opinions expressed by his correspondents. Neither can he undertake to return, or to correspond with the writers of, rejected manuscripts intended for this or any other part of* NATURE. *No notice is taken of anonymous communications.*]

Vibrissæ on the Forepaws of Mammals.

IT is well enough known that carnivorous and other—especially nocturnal—animals are provided with numerous long hairs, generally called vibrissæ, upon various regions of the face. The "whiskers" of the cat are a familiar example. But it is not so widely known that there exists very commonly in those same creatures a tuft of long hairs upon the wrist, which are connected with a large nerve. There have been incidental references to these structures; thus Mr. Bland Sutton described and figured them in several Lemurs. But it is not, I believe, a matter of common knowledge that they are present in a great variety of mammals. I have examined members of the groups, Lemuroidea, Carnivora, Rodents, and Marsupials, and invariably found these structures in those members of the groups in question which use their forepaws as climbing or grasping organs, or in both ways. They are generally not very conspicuous, as the individual hairs are often not markedly thicker than those of the surrounding fur. But often they contrast by their colour. In a pale, almost albino, example of the squirrel *Sciurus maximus*, the hairs were especially obvious, owing to their being black, and thus contrasting with the pale brown of the surrounding part of the pelage. In a black cat the same vibrissæ were white. It is always, however, easy to assure oneself of their presence by the sense of touch. The bundle of these rather stiff hairs and the thick nerve termination cannot be missed, if the skin be gently squeezed. In a newly born phalanger this structure was particularly obvious; but in a kangaroo of corresponding age there were no signs of an elevation of the skin bearing thick hairs. It will be remembered that the mode of life of these two marsupials is very different. Although I have examined up to the present but few genera of mammals, it appears to me that this structure will be found to be pretty universal. I have of course not detected these arm vibrissæ in Ungulates.

Zoological Society's Gardens.   FRANK E. BEDDARD.

The Distance to which the Firing of Heavy Guns is Heard.

IN the number of NATURE for August 16, there is an article by Mr. Charles Davison on the distance to which the firing of heavy guns is heard. The writer of the article seems to wish to collect facts bearing upon this question: I can supply one bit of information of the kind desired.

In the summer of 1863, during the siege of Charleston, S.C., by the Federal forces, being at the time an officer in the Confederate Army, I went, under orders, from Charleston by way of Millen, Augusta and Branchville. It was just at the time of the first heavy naval bombardment of Port Sumtra. The train stopping at a water tank a few miles (I do not now remember just how many) on the Macon side of Millen, and therefore somewhat farther than this place from Charleston, I heard distinctly, not only the general, more or less varying, roar of the bombardment, but also the low boom of individual guns. The sound was faint, but unmistakable in the stillness while the engine was taking water, but was lost as soon as the train got into motion again and its noise began. At Augusta, during the stop made there, I could catch the sound of the guns again, though it was interfered with a good deal by the confused noise of a large town. At Branchville, a hamlet of a few houses, the sound was easily recognised by any one, and was accompanied by a general *feeling* of tremor.

Millen is nearly due west from Charleston, and distant about 117 miles in a direct line. Augusta is approximately 25° north of west from Charleston, and about 122 miles distant. Branchville is about 35° north of west from Charleston, and at a distance of about sixty miles.

Mr. Davison says that he has but little information as to the distance at which the discharge of *single guns* has been heard. I may therefore add that the heaviest guns in use in the bombardment I refer to were the 15-inch smooth bore muzzle-loading guns carried by the Federal turreted "Monitors." I do not remember now what was reported to be the charge of powder used, but they were, of course, firing shotted cartridges—some solid shot, but more frequently shell.   J. W. MALLET.

The Solidification of Alloys.

IN a recent discussion on alloys, which took place at the Bradford Meeting of the British Association in the Section of Chemistry, a curious uncertainty was alluded to, which occurs in the cooling of certain alloys from the liquid state, as to the relative proportions of different varieties of crystals which form, depending on the rate that the cooling is proceeding with.

I would wish to draw the attention of those more particularly interested in the matter to a direction in which to look for what may be one of the causes of this peculiarity, namely, to the effect that different conductivities for heat in the different kinds of crystals may exercise in determining the relative proportions in which they form, where, as in this case, two or more varieties are possible. Where there is a difference in the conductivity of two possible varieties, the more of the better conducting material that is formed the faster in general the cooling can proceed.

The matter might be looked upon as a kind of inorganic evolution. Suppose that in the first instance round the boundaries, through which heat is passing out, of the cooling material, the two varieties form with equal facility, where the better conducting material forms heat escapes fastest and solidification of the molten material proceeds fastest, we may suppose this to follow in composition the lines of the crystals in proximity, namely, of the better conducting kind. Thus, by a kind of survival of the fittest, one of the varieties prevails.

When the cooling is very slow, where in the limit the temperature is at any moment the same throughout, this controlling influence is a vanishing quantity.

A similar principle is probably the cause of the radiating structure seen often in a cooled mass of certain materials, such as bismuth and possibly ice, which have different conductivities in different directions in the crystal.   FRED. T. TROUTON.

Physical Laboratory, Trinity College, Dublin.

The Reform of Mathematical Teaching.

AS I am in full sympathy with Prof. Perry's views, my own training, somewhat on the lines suggested by him, may be of interest. I was once taught Euclid and thoroughly hated the subject. At thirteen I was sent to school in Germany, where I was taught geometry; it had so little resemblance to Euclid that I looked on it as a new subject and was delighted with it. After eighteen months I returned to England to serve my apprenticeship, but not before I had advanced as far as solid geometry, quadratic equations and trigonometry, and I believe that this early and rapid mathematical training was of inestimable advantage to me in the works. It seems unconsciously to have led me to look on practical subjects with so much of a mathematical feeling that even now my fellow engineers consider me very mathematical, yet all the subsequent mathematical training at college (Germany) only extended over another eighteen months, and I admit that I would have liked to have had more.

I now come in contact with many engineers, both old and young, and almost invariably find that they are unmathematical, *i.e.*, they cannot look at an engineering problem with an analytical eye; and no wonder, if they have been brought up on Euclid. To me these volumes seem to be a collection of mathematical puzzles, which the ancient Greeks sent each other for solution, and which are most excellently edited by Euclid. A similar collection might nowadays be made of the trying problems in the chess columns of our daily literature, and these might be so pieced together as to afford most excellent mental training, but such a work would never teach good chess playing. It would be an excellent reference book for past masters, and that is what Euclid would still be if higher mathematics had not been invented.   C. E. STROMEYER.

Lancefield, West Didsbury.

Leaf Decay and Autumn Tints.

"OBSERVATION shows," says Emile Mer, "that in most cases where wood dies in contact with living wood there is produced from the second towards the first a migration of starch and tannin, and (in a conifer) of resin; there is thus produced from the portions remaining living towards the dying or dead portions a drainage of substances, &c." These remarks refer to the formation of secondary periderm and of the duramen, but their scope and tenour may perhaps, I think, be extended to the case of forest leaves approaching the end of their existence as living organs. How does it come to pass that in autumn the leaves of some of our forest trees exhibit a brilliant livery of crimson, while others exhibit only a yellow or golden glory? Take, for instance, the case of the ash constituents in the dry substance of the leaf. It is known by analysis that the percentage of ash increases through nearly the whole life of the leaf in beech, sycamore, elm, but not in oak, larch, cherry, &c.; it depends a good deal on whether some one ash constituent (generally lime or silica) is being steadily stored up. For example, the dry leaf of *Acer campestre* on May 1 has 6 per cent. ash, and in October 16·2 per cent. ash; the dry leaf of *Prunus avium* has on April 28, 7·8 per cent., and on October 2, 7·2 per cent. ash. Now the leaf of the former tree is only yellow in autumn and never red, while that of the latter is very often beautifully crimson. In the former case there is a kind of gradual decay or death of some of the cells (mostly of the upper epidermis) which occasions a drainage of mineral and organic substances to these parts from the still living tissues. This drainage and accumulation attest, in fact, such a decay; and what is more, they seem to have a distinct influence over the ultimate autumnal coloration of the leaf itself. It is easy to understand, in fact, that the leaves which exhibit such a decay and approach to dissolution are just these wherein the chromogen precursive of the brilliant red coloration would likewise suffer an analogous kind of change, *i.e.* it would tend to become brown, to produce phlobaphene, just as it does in the outer bark which is the practically dead portion of the rind. Where this accumulation of mineral matter and all which it implies does not take place, as in cherries, currants, American oaks, pears, wild vine, barberry, &c., then the chromogen does not deteriorate; it evolves its proper pigment, and assumes the flush and glow of active living colour. On the other hand, in elms, chestnut, linden, birch, poplars, &c., which are never red but only yellow, it is only the vivid carotin attached to the last faded and now exhausted chlorophyll which gleams forth, but only for a time, and if not too much obstructed by the dull browns of decomposed carbohydrates and superoxidised tannic chromogens.   P. Q. KEEGAN.

Patterdale, Westmorland.

Homochronous Heredity and Changes of Pronunciation.

SEEING that in ancient German, or rather Gothic, Swedish, Danish, probably in French, and possibly in Sardinian, the *th* sound surviving in English (though much less frequent than it used to be) was once largely used, but nowadays Frenchmen and Germans find a difficulty with it, I should like to know whether systematic experiments have been made as to whether children of various ages of these two nationalities can pronounce it more exactly and spontaneously than their compatriots of a maturer age? I should like to make the same inquiry concerning English children and their pronunciation of the gutturals discarded or altered in such words as *night, bough* or *laugh?*

CHARLES G. STUART-MENTEATH.

23 Upper Bedford Place, W.C.

Authorities:—Helfenstein, "Comparative Grammar of the Teutonic Languages," 1870, pp. 156–9; G. Koerting, "Neugriechisch und Romanisch," 1896, p. 23; W. Meyer-Luebke, "Grammatik der Romanischen Sprachen," 1890, p. 428.

The Daylight Meteor of Sunday, September 2.

As Mr. Denning expresses a wish in your issue of September 13 for further information concerning this meteor, I write to inform you of what I saw myself.

I observed the time at which the meteor fell, and made it 6.50, but my watch is no chronometer. I saw the meteor from the road, between Deganwy and Llandudno, and it appeared to fall over the Little Orme's head. If you joint this point to Leyburn in Yorkshire, you have the line as near as I can give it, and I do not think it is very far out. I did not note any column of smoke or cloud after the meteor fell. Its path was vertical. Some one says its angle of appearance was 35°, and disappearance 25°, which I should say is about correct. The sun was shining brightly, though low down in the west. The brilliance was greatest just before disappearance. I have never before seen any meteor to compare with it in brightness.

38 Hillfield Road, Hampstead.   T. ROOKE.

THE meteor of September 2, described in your issue o. September 13, was seen in Ireland also, in even brighter daylight.

I noted the time, 6.27 p.m. (Irish), and the direction, E.N.E., from a point near Enniskerry, co. Wicklow. There was a possible error of a couple of minutes in my watch, and a considerable error possible in the estimated direction, which was a rough approximation made without a compass.

B. ST. G. LEFROY.

THE THEORY OF IONS.

EVER since Faraday enunciated the law of electrolysis, that the same quantity of electricity passed when chemically equivalent masses of different substances were produced, it has been a matter of speculation whether this may not be due to atomic charges of electricity. Every one, in describing electrolysis and explaining how the substances evolved appeared at the electrodes without any apparent action in between them, based his description and explanation upon the supposition of electric charges on the atoms. Some substances, such as hydrogen, were given positive, and some, such as chlorine, were given negative charges, and the electric current through the liquid was explained as due to the convection of these charges by the moving atoms or groups of atoms, and the movements of these were ascribed to the electric force acting on these charges. The amount of the charge on each atom or group of atoms was proportional to its valency, and as this has with good reason always been taken as a whole number, the charges ascribed to the moving elements were all simple multiples of the charge ascribed to a monovalent atom, such as hydrogen or chlorine. All this has naturally led to the hypothesis that electricity itself is atomic. In electrolysis, at least, there is a certain minimum quantity that corresponds to a single atomic bond, and quantities of electricity transferred by electrolysis are always multiples of this unit. It was surely natural, then, to give a name to this important physical unit quantity of electricity, and it has consequently been called an "electron."

Further, in electrolysis, the electrons always appear connected with, and travelling with, certain atoms or groups of atoms. For example, in copper sulphate solutions, the positive electrons travel in pairs with the divalent copper atoms, and the negative electrons with the divalent atomic group SO₄. These charged atoms, or groups of atoms, playing such an important part in electrolysis, have been called "ions."

Now there is a very important difference between different liquids in their behaviour when we try to pass an electric current through them. Some are quite easily decomposed, others offer a very great resistance, and it has been a matter of most interesting speculation as to the cause of this. In the first place, most of the easily decomposed liquids are solutions in water, of acids, alkalis or salts, and this has naturally attracted attention. In the second place, these solutions are all ones in which double decompositions, and such-like chemical actions take place with facility. Can a common explanation be given of this remarkable coincidence of electric conductivity and chemical activity? Electric conductivity is due to two causes—first, the electric charges on the ions; and second, the independent mobility of these oppositely charged ions under electric force. Without entering upon the very interesting questions involved in innumerable speculations as to the causes of these charges and of the mobility of the ions, all modern theories acknowledge that, in some way or another, water, and some other liquids in a less degree, have the very remarkable property of conferring upon certain substances dissolved in them the wonderful independent mobility of the ions which we see in electrolysis.

In consequence of this mobility of these differently electrified ions, it is easy to understand their chemical activity in these conducting solutions, and thus these two important properties of these solutions receive a common explanation. No really satisfactory explanation of how the solvent water confers on substances dissolved in it this wonderful independent mobility of the ions has yet been proposed. Some writers have described the phenomena as if all that was needed was to assert that the ions move about independently, that the material, $CuSO_4$ for example, is simply dissociated into Cu and SO_4, and that these ions gad about freely and independently in the liquid. Such writers consequently speak of the substance as being dissociated in solution. In what is recognised as ordinary chemical dissociation there is no different electrification of the constituents into which the substance is dissociated, and there is thus an essential distinction between the independent mobility of electrified ions in solution and what is recognised as chemical dissociation. Whatever be the true account of the matter, it is almost certainly very much more complicated than ordinary chemical dissociation, and the action of the water is evidently of the first importance. There is great difficulty in explaining this independent mobility, on account of two things which have not been satisfactorily explained as yet. In the first place, it is very hard to understand why these oppositely electrified ions do not combine together in pairs as they would do if they were merely under the action of the electric forces due to their opposite electrification. In the second place, it is very hard to understand where the energy comes from that is required to separate these independent ions and keep them free from one another. When copper sulphate is dissolved in water there is very little change of temperature; if it be the anhydrous salt, there is a rise of temperature; while we would naturally expect an enormous absorption of heat to account for all the energy that would be required to separate the Cu and SO_4 from one another. This all shows how very different this phenomenon is from ordinary dissociation, and in consequence this peculiar action of water, and in a less degree of some other solvents, has been called "ionisation." The two most important properties of an ionised fluid are conductivity for electricity, and a remarkably chemical activity which has been shown to be directly proportional to the conductivity.

Besides liquids there are gaseous conductors of electricity. Gases do not usually conduct electricity at all. Even under circumstances when one would naturally expect them to carry electric charges on their molecules, they seem quite incapable of doing so. When the surface of a liquid is electrified and it evaporates, one would naturally expect the escaping molecules of gas to carry away with them some of the electric charge from the surface of the liquid. It is not known whether an electrified metal volatilising would or would not carry away with it some of the superficial electric charge, but ordinary liquids volatilising certainly do so to a very small extent, if at all. This may be, of course, because the charge is carried by superficial *ions*, and it is not the ions that escape, but the *molecules* of the liquid itself. But why these extremely movable ions cannot escape as a gas from the surface of the liquid is a matter still requiring explanation.

Under a great variety of circumstances, however, gases are able to conduct electricity. Leaving aside the spark, glow and arc discharges in gases at high pressures, and the well-known discharges in gases at low pressures, in all which cases there is evidently something like a breaking up of the gas itself under intense electric force, there are a number of cases in which a gas can conduct electricity under quite feeble electric forces.

Within any space at all close to a spark discharge of any kind, in flames and in the gases escaping from them, in the neighbourhood of surfaces of solids illuminated by ultra-violet light, in the neighbourhood of surfaces of solids being acted on chemically by the gas, in a gas traversed by kathode rays, and in a gas traversed by X-radiations and by those various, most curious and remarkable radiations that have been classed under the name of Becquerel rays—in all these cases gases conduct electricity, apparently quite freely.

Under these circumstances it has been usual to describe a gas in this condition as "ionised," and to seek for a separate and independent mobility of its ions. Great success has attended these investigations. The difficulties that surround ionisation in liquids are mostly absent here. The ions, if left for a time to themselves, do combine together in pairs, and it requires a continuous and considerable expenditure of energy to keep them apart. The diffusion of the independent electricities has been studied, and many quantitative results that were to be expected from the theory have been proved to exist.

There is, however, an important difference between the conductivity of a gas and that of a liquid. In the case of a liquid, the electricity always travels along with matter in the form of an atom or a group of atoms; in a gas there is every reason to believe that we are often dealing with electric charges which, if connected with matter at all, are connected with masses which are about 500 times smaller than a hydrogen atom. So far no good reason has been given for believing that the electric charges that move about among the molecules of a gas carry any matter along with them. There does not seem any difficulty in supposing that the electric charge of an atom can exist independently of the atom. All theories of electrolysis have supposed that electric charges are transferred within the liquid along with material atoms, but from the liquid to the electrodes all theories have supposed that the electric charges jump from the liquid atom to the electrode; and if it can jump from one atom to another, there seems no reasonable objection to believing in its independent existence. On account of the difference between the nature of the conductivity of a gas and of a liquid, it would be well to confine the term ionisation to the case of conductivity due to the mobility of charged atoms or groups of atoms, and to call the conductivity due to the existence of these mobile electric charges which are not connected with atoms by another name, such as "electronisation."

One of the most remarkable results of the study of these mobile electric charges which are unconnected with atoms is that only negative electric charges have as yet been discovered to be free from atoms; the corresponding positive charges seem to be always attached to atoms or groups of atoms. This has naturally led to a rehabilitation of the old single fluid theory of electricity in which matter plays the part of the positive fluid in the old double fluid theories, and the phenomena of electronisation, so far known, certainly lend support to this hypothesis. But we really know so little about the subject, that it is rather too soon to form anything beyond a rough working hypothesis. So long as we know that there exist outstanding, second order effects like gravitation, we are premature in concluding that the connection between positive electrification and matter proves any first order difference between positive and negative electricity, as it may be a second order effect.

The conductivity of gases produced by the pressure of these movable electric charges is in some respects more analogous to that of metals than to that of liquids. In liquids an electric current is accompanied by streams of matter, while in the electronised gas, so far as it has been observed, there may be no stream of matter attached to the electric charges which carry the current. It has consequently been suggested, and with good reason, that in solids and melted metals, which conduct metallically, the electrons are freely movable, and that this is the cause of their conductivity. There is some reason to believe that in this case also it is the negative electron which is most freely movable. Some most interesting calculations have been made upon this hypothesis, in which it has been supposed that there was something like a gaseous pressure of these mobile electrons in the metal. Thermo-electric effects have been attributed to the dependence of this pressure upon temperature, and the convection of heat accompanying electric currents has been attributed to the convection of energy of irregular motion by these electrons. The Hall effect has also been shown to be a possible consequence of a different mobility of the positive and negative electrons.

Upon these principles it is natural to attribute the magnetic properties of iron and other substances to electrons describing orbits round the atoms. These revolving electrons, in this case, represent the amperean atomic currents to which magnetisation has long been attributed. A remarkable confirmation of this has been derived from the Zeeman effect, which can be explained by the supposition that negative electrons are describing orbits round the atoms. Further, the mass that moves with these electrons has been shown to be of the same order of magnitude as the 500th part of the mass of an atom of hydrogen, which, from experiments on gaseous electronisation, seems to accompany the free electrons in a gas when it conducts.

There seems to be some reason to think that in a highly magnetisable material, such as iron, either there are more than the four electrons corresponding to its atomicity in rotation, or else that these are rotating very much more rapidly than corresponds to the vibrations of ordinary light. Some objection may be taken to the latter hypothesis from the difficulty of explaining why enormously rapid ether waves are not thereby generated in the surrounding medium, and the energy of the motion thereby lost by radiation. There are suggested explanations of this difficulty, but the other hypothesis, that matter has in it many more electrons than correspond to its atomicity, and that these latter are merely peculiar in being removable, agrees with a very interesting suggestion that all matter is built up of electrons. That an atom of hydrogen, for example, consists of some 500 electrons, one of oxygen of some 8000, and so forth. This is a natural deduction from these speculations, and receives some confirmation from its being consistent with the change in dimensions of a body as it moves in different directions through the ether which has been assumed in order to explain the experiment on the motion of the earth through the ether, which Michelson and Morley conducted. A supposition such as this naturally suggests that atoms could be built up of electrons as well as the electrons separated from matter; and if that be so, there seems no impossibility in the dreams of the alchemist, and an element of one kind may some day be transmitted into that of another. What is as yet known is, however, a very slender foundation for these speculations, and it is quite likely that matter and electricity are distinct in kind, and cannot be transmuted into one another in the way suggested.

Enough has been said in this very sketchy description of ionic theory to show how far-reaching it is ; how it touches upon the confines of our knowledge and upon the borderland between physics and chemistry. Advances in our knowledge of ionic theory are likely to dispel many of the clouds surrounding the connection of matter and ether, and may lay the foundations for an intelligible structure of the physical universe.　　　　G. F. F. G.

THE RECENT CRETAN DISCOVERIES AND THEIR BEARING ON THE EARLY CULTURE AND ETHNOGRAPHY OF THE EAST MEDITERRANEAN BASIN.

WHILE recently excavating the prehistoric Palace of Knossos, which lies in the great central gap between the higher ranges of Crete, mid-way between the peaks of Ida and Dicta, I was much struck by the almost continuous dualistic style of the elements. But in this case the "eternal struggle" was not between East and West. It was North and South that here fought it out. The boreal blasts which have collected from the steppes of Eastern Europe sweep almost unopposed across the Ægean, and find their first obstacle in the long mountain wall of Crete. They pour through the central gap. Not unopposed, however ; they are beaten back, and their place triumphantly taken for weeks at a time, by the parching South wind—the *Notios* of the Cretan natives—which is really the *K'hamsīn* of the Libyan Desert. Owing to the fact that the shoot and dumping-ground of the excavations was, perforce, at the southern end, the works were interrupted for days at a time by an overwhelming dust-cloud due to this cause, for the *K'hamsīn* seems to have an affinity for dust out of proportion to its actual strength. Disagreeable, however, as were these hindrances to the work of the spade, one had at least leisure to reflect on the historic lessons supplied by these natural phenomena. Crete certainly stands geographically in closer relation to Asia Minor than it does to Africa. Carpathos and Rhodes, not to speak of minor islands, afford natural stepping-stones of intercourse. The actual relations between Crete and Anatolia, ethnic and other, must not be underrated. Yet in a broad historic point of view Crete stands apart from it. It was not like Cyprus, which, although at different times it has become an outpost of Egypt and of Europe, has always remained essentially a part of Western Asia. But the main currents of Cretan history, like those of its two prevalent winds, have been Northern and Southern—European and African. Of its two direct geographical connections, that with Greece and that with Anatolia, it has consistently held to the former. On the other hand, its' intercourse with the opposite Libyan coast—the Cyrenaica—and with Egypt has been singularly continuous from a very remote period. And in this lies the high importance of the part played by the island in the early history of European culture. Germs received here from the Nile Valley and its borderlands, at a time when the greater part of Europe was still in its Stone Age, were propagated northwards and westwards, and seedlings hence derived spread in prehistoric times, and by more than one channel, as far as the British islands.

During five successive campaigns of preliminary exploration in Crete, I was able to collect a variety of evidence establishing the very early derivation of certain indigenous forms of stone vases and decorative motives, from those of Egypt. A series of archaic Cretan seals exhibited designs copied almost directly from those of: Twelfth Dynasty scarabs, and approximately dating, therefore, from the middle of the third millennium before our era, while steatite vases were found almost indistinguishable in form from Old Empire types of considerably earlier date. The primitive three-sided seal-stones, on which appear the first rudiments of Cretan script, reproduce the type of a three-sided seal, apparently of Libyan origin, which, from its analogy with a special class of Egyptian cylinders, approximately date from the middle of the fourth millennium B.C. So long, however, as the early archæological strata of the Cyrenaica are left as at present wholly unexplored, a great blank is still left in the materials for comparison on the Libyan side. It remains to be seen whether the Danish expedition now organising will be able to overcome the hitherto insuperable obstacles to the thorough scientific exploration of that region, but the fanatical spirit of the Senoussi is of ill omen.

What the results of these Cretan observations have certainly ascertained is that whether directly from the Nile Valley, or indirectly through Libyan intermediaries, Egyptian elements were making their way into Crete at a period which must carry back by over a thousand years the materials for approximate chronology in the Ægean world. The derivation, on steatite seals and vases of Egyptian forms, of the Twelfth Dynasty spiral ornament (only at the beginning of the Mycenæan period taken over upon metal work) is of extraordinary importance as supplying the "missing link" in the origin and diffusion of the spiral system in the early European Metal Ages. By the Danube Valley and the course of the Elbe, the old route of the amber traffic brought this spiraliform system to the Bronze Age population of North Germany and Scandinavia, and was by them in turn diffused, as has been shown by Mr. Coffey, to Ireland, whose wealth in gold made it the Rand of prehistoric Europe. On the other side, survivals of the Mycenæan adaptations of the primitive spiral ornament, which had lingered on among the Illyrian tribes of the North-West corner of the Balkan peninsula, gained a new vitality in contact with the artistic genius of the invading Celtic tribes. Assimilated by these, and transported on the wave of Belgic conquest to the North-West, the spiraliform system of design re-entered the British Isles in another form ; and in Ireland, where the elder spiral branch of the Bronze Age had long since expired—lived on to supply designs to St. Columba and his missionary fellow-workers. The chains are long ones that connect the carvings of New Grange on the one side and the illuminations of the Book of Durrow on the other with the art of Twelfth Dynasty Egypt ; but they run through prehistoric Crete.

Of the intercourse between Crete and the Egypt of the Middle Kingdom the Palace of Knossos has supplied a new and striking piece of evidence in a diorite figure with hieroglyphic inscriptions, which give the character of the names it bears ; its good style and material have been recognised by Egyptologists as a Twelfth, or at most, early Thirteenth Dynasty work. In other words, the latest date to which it can safely be referred hardly comes down to 2000 B.C. We have here therefore a valuable indication for the approximate chronology of the earlier elements of the Palace of Knossos itself, which in any case go back beyond the period to which the remains of Mycenæ have given a name. The high level of civilisation, however, already attained in the City and House of Minos at this remote date is shown, not only by such an artistic importation from the land of the Pharaohs as the diorite figure, but by fragments of wall-painting in an already fully developed style—one represents a boy placing crocus-like flowers in an ornamental vase—and by ceramic fabrics of great beauty. In order not to confuse the evidence, I endeavoured in this year's excavations within the Palace walls, as far as possible, to confine myself to the upper and purely Mycenæan layer, and the relics found of this earlier period have therefore been comparatively limited in number. But beneath the floors of houses immediately below the Palace and on the opposite hill, Mr. D. G. Hogarth, the Director of the British School at Athens, found a whole series of vases of this early painted class, many of them showing naturalistic designs of lilies, tulips, and other flowers, presenting shapes in some cases so graceful as never to have been surpassed in any later age of Greece. This style of Cretan pottery, which has received the name of Kamáræs from the grotto where its first occurrence was described by Mr. J. L. Myres, has been found by Mr. Petrie at Kahun in Egypt, again in a Twelfth Dynasty

connection. The intercourse between Crete and the Nile Valley in the third millennium before our era has thus left its traces on both shores of the Libyan sea. The approximate date thus ascertained for the earlier part of the Palace at Knossos gives additional interest to the fact that this in turn overlays a vast Neolithic settlement, for which it supplies a chronological *terminus à quo.* In the Central Court a trial shaft was excavated, which went down 24 feet through continuous Stone Age deposits containing incised, chalk-inlaid pottery, axes and mace-heads of serpentine and other materials, obsidian knives and cores, and primitive images of clay and marble akin to those from the earliest settlement of Troy.

But the great bulk of the remains of the Palace of Knossos as yet brought to light belong to the most flourishing days of the better-known Mycenæan civilisation, and are contemporary with the Eighteenth and Nineteenth Dynasties of Egypt. The building itself is of vast extent—about two acres have already been uncovered, and beside it the Palaces of Mycenæ itself, of Tiryns, and of all other such buildings on the mainland of Greece shrink into comparative insignificance. We have not here the same mighty bastions, though the megalithic gypsum blocks of the lower part of the walls are sufficiently imposing. What we see here is the island capital of a great maritime power, the memory of which survives in that of the traditional "thalassocracy" of Minos, and which seems to have rather relied on its "wooden walls." Here are vast paved courts, propylæa, spacious corridors, and successions of magazines, and, amidst a maze of lesser passages and rooms, the actual council chamber of the prehistoric kings, with its curiously carved gypsum throne in the centre. There can be little doubt that this building was the prehistoric original of the fabled "Labyrinth," the etymological meaning of which is the house of the *labrys* or double-axe, the emblem of the Cretan Zeus. This symbol is carved on the principal blocks and corner-stones, and repeated on every side of every slab of what appear to be the sacred columns of two inner shrines. The legendary fame of Dædalus, to whom both the building itself and the works of art it contained were traditionally ascribed, is fully borne out by the actual remains. Both in painting and sculpture we see here a higher level than was reached either at Mycenæ or Tiryns. For monuments of Mycenæan painting, indeed, the Palace of Knossos stands almost alone. On many of the walls the frescoes were still found adhering, almost as brilliant as when they were executed, and we have here a new revelation of ancient painting. Quite new in ancient art were certain miniature groups of ladies in fashionably dressed though somewhat *décolleté* attire, seated in animated conversation apparently in the courts and balconies of the Palace itself. In the decorative designs and the fabulous animals, such as the griffins and sphinxes, the influence of Eighteenth Dynasty Egyptian models is evident ; but these foreign elements are adapted in an independent manner. Of more special interest are life-size processions of youths bearing various vases, who display a singular general resemblance to the procession of the tribute-bearing Keft chieftains on the tomb of Rekhmara at Thebes, which dates from the first half of the fifteenth century B.C. It is known that the Kefts of the Egyptian monuments represent the Mycenæan race of the Ægean isles and coast-lands. On the Knossian wall-painting, we see them in their home.

The upper part of one of these Knossian figures, which is well preserved, is of the highest ethnographic interest, as presenting for the first time a careful naturalistic pourtrayal of a Mycenæan man. The profile is of a pure European character, almost classically Greek in its regularity. The lips are somewhat full ; the eyes and hair are dark—the latter somewhat curly. The head is of the high brachycephalic type. The skin shows the reddish-brown hue of Egyptian convention, just as the women are in the Egyptian manner painted white. The type of head delineated is essentially that of the race which, through all the changes of Cretan history, still remains predominant in the island. The finely-cut profile, the dark hair, the high brachycephalic skull, are as characteristic now as they were over three thousand years ago when the painting was executed. It is interesting to note that the physiognomy is distinct from the more hawk-like Armenoid type which, as Von Luschan has shown, represents the underlying ethnic element of a large part of Anatolia. It is equally non-Semitic. That this Cretan type represents that of the pre-Hellenic occupants of mainland Greece is highly probable. It still survives intact in the Illyric part of the peninsula, and I have myself been again and again struck in Cretan mountain villages with resemblances to the highland population of Albania. The Slav-speaking Montenegrins, like the Herzegovinians, so far as race and physique go, largely represent the same Illyrian element ; and it was curious to notice among the Montenegrin gensdarmes recently established by the Powers in Crete the striking points of similarity to the natives of the island. Here and there, in Crete and elsewhere, varieties of the predominant "Mycenæan" type take a more aquiline cast, and show points of transition to the Armenoid race of Anatolia. The cranial type is essentially the same, and on the whole the finely-cut European physiognomy, of which this Mycenæan fresco supplies the first authentic record, may be regarded as a Western- differentiation of the more Eastern form.

This ethnographic result curiously corresponds with the earliest philological evidence at our disposal. A whole series of local names in Crete, Greece proper and the Macedonian and Thoracian lands to the North, represent allied but differentiated elements common to Caria and a large tract of Asia Minor. Thus, to take a conspicuous instance, the Western area supplies a variety of names in -*nth*, like Korinthos, Erymanthos, Perinthos, Labyrinthos, answering to others on the Anatolian side having the -*nd*- sound, such as Kalandos, Oromandos, Pyrindos, and Labrandos. In this, as in its physical type and in other respects, Crete, it will be seen, cleaves to the Greek and Thraco-Illyrian world.

My own previous researches had been a good deal occupied with a class of early Cretan seal-stones containing signs both linear and pictographic in which I ventured to detect the rudiments of a pre-Phœnician form of writing. In regard to this matter the excavation of the Palace at Knossos produced a real revelation. In chamber after chamber whole deposits came to light of inscribed clay tablets undoubtedly representing the Royal archives. The character of the writing was of two altogether distinct classes—one hieroglyphic and one linear. The hieroglyphic script answered to that of the groups of characters that I had already noticed on a series of seal-stones of Mycenæan fabric chiefly found in Eastern Crete. The ruder prototypes of these with simple pictographic designs go back on Cretan soil to a much more remote period, and find, as already noticed, very early analogies on the other side of the Libyan sea. There can be no doubt that this script in its conventionalised form is the property of the old indigenous race of the island, the Eteocretes of the Odyssey. The linear writing, on the other hand, which forms the bulk of these Knossian archives, is of a very much more developed form. It is upright, of great elegance and curiously European in aspect, a certain proportion of the signs—some seventy of which were in common use—showing correspondences with the syllabic characters of Cyprus and also with the later Greek.

The pictorial illustrations which not infrequently accompany the linear inscriptions enable us in many

cases to learn the purport of these clay documents. They thus are often seen to refer to the Royal stores and arsenals, and show a decimal system of numbers akin to the Egyptian. Others, no doubt, are deeds and correspondence like the contemporary cuneiform tablets of Babylonia. Those relating to the Royal treasure show ingots, vases and ox-heads of precious metal identical with those borne by the Keft tributaries on the wall-paintings of Rekhmara's tomb belonging to the early part of the fifteenth century B.C., a valuable indication as to date. The Palace of Knossos contains no element as late as the latest prehistoric period represented at Mycenæ itself, and the date of its destruction can hardly be brought down later than, at most, the twelfth century B.C. The most recent of the clay documents contained within it lie at least behind that date.

The result of these discoveries is therefore to carry back the existence of written documents on Greek soil some eight centuries beyond the earliest known monuments of Greek writing, and five even beyond the earliest dated Phœnician record, as seen on the Moabite stone. The whole question of the origin of writing is thus placed on a new basis. The hieroglyphic Cretan forms supply, in fact, exact correspondence with what in virtue of their names we must suppose to have been pictorial originals of the Phœnician letters. *Aleph*, the ox's head ; *Beth*, the house ; *Daleth*, the door ; *He*, the window ; *Vaa*, the peg—and indeed over two-thirds of the Phœnician series find obvious prototypes among the Cretan forms. The ingenious theory of De Rougé, which has so long held the field, and by which the Phœnician letters were derived by a selected process from early hieratic Egyptian forms signifying quite different objects, becomes henceforth untenable. The analogy supplied by the Cretan hieroglyphs in favour of a simple and natural derivation is at all events overwhelming.

It does not necessarily follow that the Phœnician letters were directly derived from the Cretan ; some signs, like that of the camel's head, certainly point to the accretion of Syrian elements. But the correspondences are still so great as to point at any rate to some kind of collateral relationship. Elsewhere I have ventured to suggest that these points of community may be due to the great Ægean settlement on the coast of Canaan, of which the Philistines stand forth as the representatives, and which has left its abiding record in the name of Palestine. The Biblical traditions, as is known, give the name of "Cherethim," or Cretans to a branch of the Philistine race ; and Caphtor, the isles or coastlands from which the Philistines traditionally came, has been plausibly identified with Keftô, the Ægean maritime realm of the Kefts, who on Egyptian monuments appear as the representatives of the Mycenæan civilisation. Of this special connection with Crete, the finds at Knossos already referred to afford convincing evidence. Other recent discoveries afford a singular support to these conclusions. It has been pointed out by Dr. Wilhelm Max Müller that in an Egyptian list of Keft names, going back to the Eighteenth Dynasty, appears the most characteristic of all Philistine name-forms, Achish ; and it thus appears that the name was known in prehistoric Knossos earlier than in Gath. Not less significant in its way is the discovery made during the recent excavations of the Palestine Exploration Fund on, or near, the site of Gath, of imported Mycenæan pottery in the pre-Israelite stratum. More and more it appears that the high early Ægean civilisation, of which Crete is now seen to be the centre, was exercising a far-reaching influence on the coasts of Canaan before the rise of the Phœnician commercial power. Cadmus had sat at the feet of Minos, and the priceless gift which in darker days of her history he bore to Hellas, was in some respects at least a restitution of what Greece herself had given long before.

Gaza, the chief Philistine emporium—the crossing

point of the caravan routes between Egypt, Syria and Southern Arabia—owed its traditional foundation to Minos, and continued down to Roman times to worship the Cretan Zeus. The great cave on Mount Dicta, which was the legendary scene of the infancy of this indigenous divinity, to whom, as we have seen, the Palace of Knossos was also consecrated, has now been thoroughly explored by Mr. Hogarth, and has produced a vast mass of votive relics illustrating the prehistoric culture of Crete from the earliest Metal Age onwards. The crevices of the stalactite columns of the lower part of the cave were found to have been utilised for the insertion of bronze offerings, especially miniature figures of the double axe, which was the particular symbol of this God. Many stone libation tables were also found representing the adaptation of early Egyptian forms, and among the votive bronzes an Egyptian figure of the god Amon Ra, whose personality presents some points of affinity to the chief Cretan God. Another bronze from this site, a miniature chariot, drawn by an ox and a ram, has a special interest as an early example of a series of votive bronzes on wheels, in the shape of cars and tripods, supporting bowls, birds and other objects, which form a feature in the remains of a wide European zone during the Late Bronze and Early Iron Age. That their ultimate source was Egypt appears probable from the four-wheeled car with the silver boat of Queen Aah-hotep ; but here again we see among Cretan remains what is probably the earliest European example of the class. Once more the archæological phenomena bring home to us the fact that we stand here at the meeting-place of the North and South wind.   ARTHUR J. EVANS.

THE ASCENT OF MOUNT ST. ELIAS (ALASKA).[1]

THE Italian original of this work was reviewed in our columns a short time ago (see NATURE, May 3), and we now welcome the English translation. In the preface we are informed that "the whole profit on the sale of the Italian edition, together with all royalties and rights on foreign editions, will be dedicated to an Insurance Fund for Italian Guides."

In its present garb the story of the expedition is told in simple and straightforward language, with only here and there an unaccustomed term to show its foreign origin ; *e.g.* "In September snow-storms continue almost

[1] "The Ascent of Mount St. Elias (Alaska)." By H.R.H. Prince Luigi Amedeo di Savoia, Duke of the Abruzzi ; narrated by Filippo de Filippi ; illustrated by Vittorio Sella : and translated by Signora Linda Villari with the author's supervision. Pp. xii + 241. 34 photogravure plates, 4 panoramic views, and 117 illustrations in text. (Westminster : Archibald Constable and Co., 1900.)

without *cease*" (p. ix.), and (in reference to rock-systems) "the different components of the *soil* of South Alaska are all stratified" (p. 232). The picturesque passages in the descriptions of the scenery have, however, lost their glow and read somewhat flat, as indeed can scarcely be avoided in a close translation. The distinctiveness of Prof. Israel C. Russell's name seems lost under the unfamiliar initials J. C., which are used throughout the book (except in the appendix, p. 232), although the full name is given correctly on p. 3. Considering the high estimation in which the citizens of San Francisco hold their business energy, it is rather amusing to read Dr. Filippi's impression that their city "being an agricultural centre, is very quiet and exempt from the feverish turmoil of the industrial Eastern States" (pp. 9-10).

FIG. 1.—Mount St. Elias from the third Newt on Cascade.

The profuse illustrations of the original are all reproduced ; and in other respects this English edition is almost, but not quite, as sumptuous as its Italian forerunner. In fact so handsome is it, that in spite of the great mountaineering achievement which it chronicles, one cannot help harbouring, like a well-known essayist under similar circumstances, a lurking desire to strip it of its fine coat to re-clothe some ragged veteran of greater intrinsic consequence.   G. W. L.

JOHN ANDERSON, M.D., LL.D., F.R.S., &c.

BY the death, on August 15, of Dr. John Anderson, in his sixty-seventh year, a serious loss has been inflicted on zoological science. Amongst the zoologists of this and other countries, Dr. Anderson was widely known and warmly esteemed. The particular branch of inquiry to which for many years before his death he had devoted himself, the investigation of the Vertebrata of Egypt, could only be successfully carried on by a naturalist who, in addition to experience in collecting, had both time and funds at his command, and who also possessed sufficient energy and tact to ensure the assistance of highly-placed Government officials. All these advantages Dr. Anderson combined in an unusual degree, and although it is to be hoped that the work he left unfinished will not be brought to an end by his death, there can be no question that the want of his guiding hand in the enterprise will be severely felt.

Dr. Anderson's scientific work consisted of two distinct parts. From 1865 to 1886 he was at the head of the Indian Museum, Calcutta, and chiefly engaged in the collection, arrangement and study of Indian and Burmese Vertebrata. After his retirement from India, in 1886, the subject which occupied him principally, and of late years exclusively, was, as already mentioned, the study of the fauna inhabiting Egypt and the Nile valley.

He was the son of Thomas Anderson, a banker of Edinburgh, and was born on October 4, 1833. His elder brother, Dr. T. Anderson, was in the medical service of the East India Company, became well known as a botanist, and was for some years superintendent of the Botanical Gardens, near Calcutta. After passing through the medical course in the University of Edinburgh, John Anderson received a gold medal and the degree of Doctor of Medicine in 1861. For a couple of years he held the Professorship of Natural Science at the Free Church College, Edinburgh, and he went to Calcutta in 1864.

His arrival in Calcutta was at a fortunate time. The Asiatic Society of Bengal had gradually come into the possession of a large collection, not only of the archæological remains, manuscripts, coins and similar objects, for the study of which the Society was originally established, but also of zoological and geological specimens in large numbers. In the course of the preceding quarter of a century the collections had increased, chiefly through the work of Edward Blyth, the curator, until the Society's premises were crowded, and the Society's funds no longer sufficed for the proper preservation and exhibition of the specimens collected. After long negotiations, interrupted by the disturbances of 1857, arrangements were completed in 1864 by which the archæological and zoological collections of the Society (the geological specimens had been previously transferred) were taken over by the Government of India, who undertook to build a new museum in Calcutta, of which the Society's collections would form the nucleus. The trustees appointed by the Government to manage the new museum asked the Secretary of State for India to select a curator, and Dr. J. Anderson was nominated for the post early in 1865. His status was changed, a few years later, to that of superintendent of the museum, and in addition to his museum work he became Professor of Comparative Anatomy at the Medical College, Calcutta. He held both offices until his retirement from India in 1886.

The time at which Dr. Anderson arrived in India was fortunate in another respect. It coincided with a great impulse given to Indian zoology by the publication of Jerdan's "Birds of India," the last volume of which appeared in 1864, and with the presence in Calcutta of a larger number of men interested in the study of the fauna than were assembled there at any time before or since. Amongst these men were Jerdan himself, Ferdinand Stoliczka, Francis Day, and Valentine Ball, all of whom have now passed away. Probably at no time has so much progress been made in the study of Indian Vertebrata as in the years 1864-74, and in this work Dr. Anderson took an important part.

The new Indian Museum, which now towers over the other buildings of Chowringhee, was not ready for occupation till 1875, but meantime Dr. Anderson had been busily engaged in adding to the zoological collections and in getting them into order. One of his first tasks was the bringing together of an ethnological series, for which the conditions of Calcutta are favourable. Amongst other important additions made by him was that of a fine series of human skulls representing various Indian

races. Another very valuable museum series brought together by him consisted of a good collection of Indian Chelonia; skeletons, carapaces and stuffed specimens.

The work in Calcutta was interrupted by two important expeditions to Upper Burma and Yunnan, to both of which Dr. Anderson was attached as naturalist and medical officer. Both expeditions were designed to pass through China to Canton or Shanghai, but in neither case was it found practicable to carry out the original plan. The first expedition, commanded by Colonel E. B. Sladen, left Calcutta at the end of 1867, proceeded as far as Momein in Yunnan, and returned to India in November 1868; the second, under the command of Colonel Horace Browne, left in January 1875, but was treacherously attacked by the Chinese before it had proceeded more than three marches beyond the Burmese frontier, and compelled to return, Mr. Margery, of the Chinese Consular Service, who had been despatched to accompany the mission, and who had preceded it by a march, being murdered with several of his followers. The difficulties experienced by both missions from the time they crossed the frontier between Burma and China, and the opposition of the inhabitants of the country, seriously interfered with zoological observations, and the collection of specimens was generally impossible; but still some important additions were made to the previous knowledge of the fauna. A full account of the journey was given in Dr. Anderson's reports and in a work by him, entitled "Mandalay to Momein," published in 1876. The detailed observations on zoology, supplemented by important notes on some Indian and Burmese mammals and chelonians, were published in 1878-9, under the title of "Anatomical and Zoological Researches, comprising an Account of the Zoological Results of the two Expeditions to Western Yunnan in 1868 and 1875, and a Monograph of the two Cetacean Genera, Platanista and Orcella." The work appeared in two quarto volumes, one consisting of plates. Dr. Anderson was the first who succeeded in obtaining specimens of the porpoise (*Orcella*) inhabiting the Irrawaddi, and the examination of this previously undescribed form led him to make a thorough anatomical investigation of an allied species occurring in the Bay of Bengal and in the estuaries of rivers flowing into the bay, and also of the remarkable cetacean, *Platanista*, inhabiting the Ganges, Brahmaputra and Indus.

The only other important collecting expedition undertaken by Dr. Anderson during his tenure of the superintendentship of the Indian Museum was to Tenasserim and the Mergui Archipelago in 1881-2. This journey was chiefly, though by no means exclusively, undertaken for the collection of marine animals, and the descriptions of the results, to which several naturalists contributed, were published first in the *Journal* of the Linnean Society, and subsequently as a separate reprint in two volumes, under the title of "Contributions to the Fauna of Mergui and its Archipelago." This appeared in 1889. Dr. Anderson's share was the description of the Vertebrata and an account of the Selungs—a curious tribe inhabiting some of the islands; but in connection with his visit to Mergui, and as part of a general description of the fauna which he had at first proposed to publish, he prepared an account of the history of Tenasserim, formerly belonging to Siam. This historical *résumé*, which deals especially with British commercial and political intercourse with Siamese and Burmese ports, was compiled mainly from the manuscript records of the East India Company, preserved in the library of the India Office, and was published in 1889 in a separate volume, entitled "English Intercourse with Siam." The book forms a well-written and interesting chapter of the history of British progress in Southern Asia.

Besides the works already mentioned and many papers, descriptive of mammalia and reptiles, which were published in the *Journal* of the Asiatic Society of Bengal and in the *Proceedings* of the Zoological Society of London, Dr. Anderson wrote two catalogues on very different subjects for the museum under his charge in Calcutta. Of these, one was the first part of the "Catalogue of Mammals," published in 1881, the other the 'Catalogue and Handbook of the Archæological Collection" which appeared in 1883.

Dr. Anderson was elected a Fellow of the Royal Society in 1879, and retired from the Indian Service in 1886. He had married a few years previously, and after retiring he travelled with his wife to Japan. Finally he settled in London, but for the remainder of his life his health was somewhat precarious, and he passed several winters in Egypt. Here he took up the study of the mammals and reptiles, which had received but scant attention since the early part of the century, when the great and superbly illustrated French work on Egypt appeared—a work which, brilliantly begun by Savigny and others, was never adequately completed.

To the work of collecting, examining, figuring and describing the Mammalia, Reptilia and Batrachia of Egypt, the later part of Dr. Anderson's life, when he was well enough for work, was mainly devoted. He also paid some attention to the fauna of the neighbouring countries, and in 1898 published "A Contribution to the Herpetology of Arabia," founded on the collections of the late Mr. J. T. Bent and others. The first part of the important work he had intended to produce on the zoology of Egypt, containing an account of the physical features of the country and descriptions of the Reptilia and Batrachia, appeared in 1898. It is a fine quarto volume with excellent figures, many of them coloured. He had made large collections and notes for the volume on Mammalia, and these it is hoped will be published in due course.

One of the last undertakings in which Dr. Anderson engaged, as soon as the Upper Nile valley was once more thrown open to civilisation, was the systematic collection and description of the fish inhabiting the river and its tributaries. That this important work (of which a notice appeared in NATURE of February 23, 1899) is now being carried out with warm interest and assistance from the Egyptian Government, must be attributed to Dr. Anderson's foresight, zeal and skilful advocacy. Both in our Indian Empire and in North-eastern Africa, Dr. Anderson contributed much to the solution of one of the chief biological questions of the present day, an accurate knowledge of the distribution of animal life.

W. T. B.

NOTES.

A NEW instance of the want of encouragement, and often opposition, which scientific work receives in this country is given by Major Ronald Ross in a letter in Monday's *Times*. It appears from a correspondence just published, that in 1898 the Secretary of State for India refused to permit officers and soldiers to undergo voluntary inoculation against typhoid. It is known to our readers that Dr. Wright, professor of pathology at Netley, elaborated the system of inoculation against typhoid so long ago as 1896. The treatment is based on the soundest scientific principles, and substantial evidence of its value as a preventive measure had been obtained by laboratory experiments. It is entirely free from danger, and there would have been no difficulty in obtaining numerous soldiers to undergo inoculation with Dr. Wright's typhoid vaccine. From the results of the inoculations which might thus have been made three years ago, results would have been obtained which could have been utilised in the recent war in South Africa, and might have been the means of saving hundreds of lives. But unfortunately for the army as well as for science, officers and soldiers appear to have been forbidden

to submit themselves for inoculation. In other words, a real success against disease might have been scored, and in any case the information gained would have been of value in making further efforts to diminish mortality from typhoid, but the officials who should have done everything in their power to assist the work, deliberately stopped it by hampering the freedom of the persons who would most benefit by the treatment. It is difficult to understand this singular action, and Major Ross has done a public service by directing attention to it.

IT was announced in NATURE several months ago (P. 230) that Dr. L. Sambon and Dr. G. C. Low, of the London School of Tropical Medicine, had arranged to live from May to the end of October—that is, during the malarial season—in a part of the Roman Campagna, near Ostia, where scarcely a person spends a night without contracting malarial fever of a virulent type. No quinine or other drug was to be taken as a precautionary measure, but the investigators were to live in a mosquito-proof hut from an hour before sunset to an hour after sunrise, so as to avoid being bitten by mosquitoes, which only feed during the night. The experiment was planned to test the reality of the connection between malaria and mosquitoes, and we learn from the *British Medical Journal* that it has been most successful. On September 13, Prof. Grassi visited the residence of the investigators with several other men of science, and gave his testimony as to the value of the experiment in the following telegram to Dr. Manson : " Assembled in British mosquito-proof hut, having verified perfect health experimenters amongst malarial stricken inhabitants, I salute Manson who first formulated mosquito malarial theory." So far as the experiment has gone, therefore, the result is entirely satisfactory, and affords the strongest support to the mosquito theory of malaria. Additional evidence is given by Dr. Elliott, a member of the Liverpool expedition sent to Nigeria some time ago to investigate the subject of malarial fever, who has recently returned to this country. He reports that the members of the expedition have been perfectly well, although they have spent four months in some of the most malarious spots. They lived practically amongst marshes and other places hitherto supposed to be the most deadly, and they attribute their immunity to the careful use of mosquito nets at night.

ANOTHER experiment arranged in connection with their malarial investigation in the Campagna is described in the *British Medical Journal*. Drs. Sambon and Low have shown that by avoiding mosquitoes they avoid malaria ; but this is, after all, only negative evidence, and its full value can only be appreciated in connection with the actual production of malaria in a healthy person in this country by the bites of mosquitoes containing the germ of the disease. This evidence is now forthcoming. We learn from our contemporary that a consignment of mosquitoes which had been fed on the blood of a sufferer from malaria in Rome, under the direction of Prof. Bastianelli, was received in London early in July. A son of Dr. Manson, who offered himself as a subject for experiment, allowed himself to be bitten by these insects, and, though he has never been in a malarious country since he was a child, he is now suffering from well-marked malarial infection of double tertian type, and microscopical examination shows the presence of numerous parasites in his blood. Full details of the experiments will be published in due course ; meanwhile, they must be regarded as affording the most striking confirmation of the transmission of malaria by mosquito bites that has yet been obtained.

DR. L. A. BAUER, in charge of magnetic work of the U.S. Coast and Geodetic Survey, has gone to Alaska and to the Hawaiian Islands, in order to select the sites for the magnetic observatories in those regions. The principal or standard magnetic observatory is now being erected sixteen miles to the south-east of Washington City, and a fourth observatory is, temporarily, in operation at Baldwin, Kansas. The last named observatory is central to the area being surveyed by four magnetic parties, and it will be shifted about in the western States according to the requirements of the magnetic survey. It is the intention to have the four observatories ready in time to co-operate with the Antarctic expeditions.

IN connection with the usurpation of swallows' nests by house sparrows, Mr. J. H. Allchin sends a description of a swallow-cum-sparrow's nest seen by him at Dymchurch, in the Romney Marsh. The original nest was built on a beam immediately under the corrugated iron roof of a shed, but the usurpers had so completely covered it with straw, grass, feathers, fibres and other materials, that it was almost impossible to see any portion of it. Mr. Allchin remarks : " I have seen other nests of swallows which had been taken possession of by sparrows, but in those instances the only evidences of occupation were bits of straw or grass sticking out of the entrance ; this is the first one I have seen covered over so thoroughly as to completely hide the work of the original builders.

ANOTHER successful experiment with electric traction on railways is reported from Germany, the line being from Berlin to Zehlendorf on the new Wannsee railway. The train in question (says *Feilden's Magazine* for September) was equipped as if actually running to scheduled time. It was furnished with a motor car at each end, the work of propulsion being divided equally between them, the advantage claimed for this being that the reversing of the train becomes unnecessary at the end of each journey. Eight ordinary cars were employed in addition, seating in all 400 passengers. These experiments are to be continued over a period of one year, at the termination of which it is expected that the question will be decided whether or not electric propulsion is to be wholly substituted for steam power, while at the same time much useful data will be gathered. An advantage already claimed is that electric motive power is about 15 per cent. cheaper than steam, and also at higher velocities the chance of accidents is supposed to be less. A train of this description is at present on trial in this country, and it will be useful to compare notes from each when the material is available.

FROM all quarters we learn that the present season has been remarkable for the appearance of numerous specimens of the clouded yellow butterflies (*Colias edusa* and *C. hyale*), as well as the holly-blue (*Lycaena argiolus*). During one country walk of three miles in Cambridgeshire, on August 13, the present writer saw three *hyale* and one *edusa* ; in a garden near Brighton a holly-blue was seen on September 4, and many collectors report having obtained fair series of one or both of the two yellows in a day's hunting. From *Science Gossip* we learn that the variety *helice* of *C. edusa* has occurred in some numbers in clover fields in east Essex. The year 1892 will be remembered as the last occasion on which *C. edusa* occurred in abundance, but the present season is characterised by the comparative frequency of the pale species *hyale*, which was far less plentiful in 1892. The humming-bird moth (*Macroglossa stellatarum*) appears to have been gaining rather more than the usual notoriety in the daily papers which it has received ever since, some thirty years ago, the late Rev. J. G. Wood, in his " Common British Moths," wrote : " This moth, which is tolerably common, has been very familiar to the public of late years on account of the many letters which have appeared in the daily journals, much to the amusement of practical entomologists, who have been too familiar with the insect in question to think it worth a special notice."

DR. ANTONIO PORTA communicates to the *Rendiconti del R. Istituto Lombardo* certain studies on the anatomy of the common frog-hopper (*Aphrophora spumaria*, L.) having especial reference to the secretion of froth, so well known to all gardeners. The author finds that the apparatus which secretes the frothy liquid in *A. spumaria*, and possibly in other species, consists of hypodermal glands scattered over the back and especially near the stigma, that the *corpus ovalis* is perhaps in relation with the secretion of froth, that the mass of cells found in the latero-ventral position collect and perhaps produce material of which the animal makes use in the elaboration of the secretions, and that the glandular epithelium of the seventh and eighth segments serve as supports for minute appendages of a branchial character, which have disappeared in *Cicada* and *Nepa*, thus confirming the hypothesis of Wheeler.

IN view of our knowledge of the influence of radiant energy on electrically charged bodies, much interest attaches to the question whether a solar eclipse has any marked effect on atmospheric electricity. Dr. Julius Elster made observations during the last total eclipse at Algiers, and remarked an important fall of the potential of atmospheric electricity at and slightly after the totality. The observations are given in the last number of the *Memoirs* of the Società degli Spettroscopisti Italiani. On the other hand, Dr. Emilio Oddone describes, in the *Rendiconti del R. Istituto Lombardo*, observations made with an electrometer at Pavia, where during the last eclipse eight-tenths of the solar diameter were obscured. The results were of a negative character. Before the eclipse, high negative potentials were observed, which were attributable to clouds accompanying a distant thunderstorm ; but during the eclipse the variations in the electrostatic potential seem to have been similar to the ordinary diurnal variations. It thus appears that the eclipse exercised no very marked influence on the electric state of the air ; but whether any portion of the observed variations was attributable to this cause is a question which it would be difficult to answer.

MR. DAVID ROBERTSON has communicated to the *Proceedings* of the Philosophical Society of Glasgow a short note on the equilibrium of a column of air and the atmospheric temperature gradient, in which the adiabatic formula for the maximum gradient consistent with stability is established in a simple manner.

PARTS 10 to 12 of the *Meddelanden från Lunds Astronomiska Observatorium* contain several papers on mathematical astronomy. One, by T. Brodén, deals with some probability considerations relating to the convergence of certain continued fractions, a problem treated by Gyeldén in 1898. Certain librations in the planetary system are the subject of a paper by C. V. L. Charlier, while G. Norén and J. A. Wallberg contribute lengthy formulæ for the development of the disturbing function in its canonical elements.

No. 110 of Ostwald's " Klassiker der exacten Wissenschaften " (Leipzig, Wilhelm Engelmann, 1900) is a reprint of J. H. van 't Hoff's papers on the laws of chemical equilibrium. The three papers in question are those communicated in French to the Swedish Academy of Sciences about the year 1885, and deal with the laws of chemical equilibrium in attenuated systems, a general property of attenuated media, and the electric conditions of chemical equilibrium. The present book is a translation of these papers by Georg Bredig, and an appendix of twenty pages contains a brief biographical notice of van 't Hoff and numerous notes, both historic and explanatory.

IN the course of a paper on the various forms of phosphorescence, in the *Revue Scientifique* for September 8, M. Gustave Le Bon describes a dark lamp (" lampe noire ") for the production of invisible radiations of great wave-length in connection with the study of phosphorescence. Among other experiments performed with this lamp, the following is very striking :—In an absolutely dark room, a dark lamp is placed on a table, this lamp not transmitting any trace of visible light. In front of it, M. Le Bon places a statuette covered with sulphide of lime that has been left in darkness for several days, and consequently retains no trace of phosphorescence. After about a couple of minutes the statuette becomes luminous, and appears to emerge from the darkness.

THE director of the Meteorological Observatory at Ponta Delgada, St. Michael, has published an interesting report on the proposed establishment of an international meteorological service in the Azores, including a history of the observations in those islands, and a chart showing the tracks of a number of storms which have visited that part of the North Atlantic during the last five years. The first regular observations were made at Angra (Terceira) in 1864, at Ponta Delgada in 1865, and at Santa Cruz (Flores) in 1897. The observations at Ponta Delgada are now regularly published in the Daily Weather Report issued by the Meteorological Council. Since the year 1893 six of the islands have been in telegraphic communication with Lisbon, and eventually cables will be laid to England, Germany, and the United States, and Flores will be connected with the other islands. The direct communication of observations between America, the Azores and this country cannot fail to be most useful both to science and to shipping ; and, although the chart above referred to shows that most of the depressions passing the archipelago strike the coasts of Europe considerably south of the British Islands, a knowledge of the positions and movements of the larger areas of high and low barometric pressures in the North Atlantic must be of prime importance for the purpose of storm prediction.

IT is well known that while country-folk adhere to the old idea that adders when frightened are in the habit of protecting their young by swallowing them, a large number of naturalists regard the feat as an impossibility. In the September number of *The Zoologist* Mr. G Leighton, a well qualified anatomist, has set himself the task of ascertaining whether there is any foundation for the objection. And he arrives at the conclusion that there is no anatomical reason why the oft-repeated statement of country observers should not be founded on fact. The author concludes by stating that the objection raised on the ground that the swallowing is unnecessary is a mere matter of opinion, adding that all that is now necessary is for a competent authority to dissect an adder which has been observed to swallow its young. "Until this is done scientific naturalists will continue to regard the question as one capable of proof, if true, but hitherto unproved."

THE eminent physiologist Dr. Gustave Loisel has communicated to the *Revue générale des Sciences* of September 15 a long and able letter urging the importance of establishing a course of instruction in practical embryology in the new French Universities. For a considerable time it appears that this subject has been taught to a certain extent in some of these institutions ; but, for various reasons, it has not hitherto been made a part of the regular curriculum in all. After pointing out its extreme importance to students of medicine, anatomy, and gynæcology, Dr. Loisel formulates his appeal as follows : (1) That a single course of elementary embryology, embracing both that of man and of other vertebrates, should be established in each University, and that the necessary apparatus should be provided ; (2) that this course should be instituted in a manner which would serve the needs of all students to whom a knowledge of this subject is of importance in their future career. These resolutions, we are glad to see, have been unanimously adopted by

the Section of Medicine at the recent Congress, and we may therefore hope that this important addition to the teaching of the Universities may shortly be in working order.

WE have received vol. vii. pt. 1 of the *Transactions* of the Norfolk and Norwich Naturalists' Society, which contains a number of papers on local topics.

IN the *Victorian Naturalist* for August, Mr. A. J. North describes a new genus and species of Australian Passerine bird as *Eremiornis carteri*, while Mr. R. Hall continues his valuable notes on the distribution of the birds of Australia.

WE have received the autumn number of *Bibby's Quarterly*, a journal issued at Liverpool ostensibly for the advertisement of certain agricultural and other commodities, but which contains a number of very interesting and well illustrated articles dealing with stock-raising and kindred subjects. Among these, one treating of ostrich-farming should attract general attention.

THE September issue of the *Annals* of the South African Museum is devoted to the commencement of a synopsis of the moths of South Africa, by Sir G. F. Hampson. South Africa is the oldest British possession of any considerable size which has hitherto never had a catalogue of its indigenous moths, and as there are now many collectors in the country, Sir George Hampson has been well advised in endeavouring to supply an acknowledged want.

THE "British Anti-Dubbing Association" has forwarded to us an influentially signed letter respecting the cruel practice of cutting the combs and wattles of game-fowls. In spite of the fact that the practice is already illegal, and that birds which have been "dubbed" are ineligible for prizes at the British Dairy Farmers' Association show, it is still largely prevalent. It is now hoped that by bringing the matter into prominent notice, the pressure of public opinion may be brought to bear upon the promoters of poultry-shows, so as to disqualify all mutilated birds from being classed.

THE present boundaries in North-west Bohemia between the districts in which pure German, pure Tschech (Chekh), and the various mixtures of these languages are spoken, are clearly indicated by Dr. J. Zemmrich on a map in *Globus* (Bd. lxxviii. p. 101) which illustrates his paper on that subject.

THE disposal of the dead is an important subject of ethnographical inquiry ; therefore thanks from students are due to Mr. W. Crooke for his paper on "Primitive rites of disposal of the dead, with special reference to India," in the *Journal* of the Anthropological Institute (vol. xxix. p. 271). Nearly every form of burial is practised in India, and Mr. Crooke has given full references for every statement he has made.

FROM Dr. Thurston's report on the administration of the Madras Museum for the year 1899-1900, we learn that the general progress of that institution is satisfactory. Anthropologists will be pleased to hear that the superintendent has found time to continue his valuable investigations concerning the various races met with in the Presidency, those which have recently engaged his attention being the Pathan, Sheik and Saiyad Muhamadans of Madras city.

IN the *Abhandlungen der Naturwiss. Gesellsch., Isis*, 1900, Prof. J. Deichmüller describes a find of three broken urns and a stone axe of Neolithic age from near Dresden ; these urns and two others described in the paper are decorated with incised lines. The same author also describes a late Slavic cemetery at Niedersedlitz of a date about 1100 of our era. The single measurable skull was meso-orthocephalic, with a cephalic index of about 78·7.

A RECENT number of the *Abhandlungen* of the Vienna Geographical Society consists of an important paper, by Prof. Dr. J. Cvijik, of Belgrade, forming the first part of a study of the glaciation and morphology of parts of Bosnia, Herzegovina and Montenegro. The memoir, which it is impossible to summarise in a note, is illustrated by nine maps.

CHIEF CONSTRUCTOR KRETSCHMER publishes in the *Marine-Rundschau* a paper on the German Antarctic Expedition. The paper deals first with the chief difficulties of Antarctic exploration, the achievements of former expeditions, and the general scheme of work to be undertaken by the expeditions now being fitted out. The second part is of special interest from the minute details and numerous drawings given of the design and construction of the vessel now being built for the German Expedition. We have also received a reprint of Mr. W. S. Bruce's paper in the June number of the *Scottish Geographical Magazine*, giving an account of the proposed Scottish National Antarctic Expedition.

A DESCRIPTIVE catalogue of a collection of the economic minerals of Canada, exhibited at the Paris Exhibition, has been prepared under the direction of Dr. G. M. Dawson. It is interesting to note that the collection includes samples of lithographic stone.

THE *Proceedings* of the Geologists' Association for August 1900 contains some highly interesting notes on the geology of the English Lake District, by Mr. J. E. Marr. The notes, which were prepared for the summer excursion of the Association, embody the results of work carried out for many years by Mr. Marr, partly in conjunction with Mr. A. Harker. While supporting the generally accepted views of the succession of the older Palæozoic rocks, the facts now brought forward indicate that the disturbances to which these rocks have been subjected are due to the pushing forward of the strata in a northerly direction *at unequal rates.* Under these conditions the Skiddaw Slates moved furthest forward, causing the Green Slates and Porphyries to "lag behind," and the Upper Slates (Silurian, with Coniston Limestone at base) to lag behind the Green Slates and Porphyries. The peculiar faulting attending these disturbances is specially described. The intrusive igneous rocks and their metamorphic effects and other subjects are also dealt with.

IN the same number of the *Proceedings* there is a paper, by Mr. G. E. Dibley, on zonal features of the chalk pits in the Rochester, Gravesend and Croydon areas. The author has laboured long and enthusiastically in collecting from the various zones, and the results which he now publishes in notes, and in a carefully arranged list of fossils, form an important addition to our knowledge of the life-history of the chalk. An interesting bone, which he obtained from the Middle Chalk of Cuxton, is described by Mr. E. T. Newton as probably belonging to the Rhynchocephalia, a group of lizard-like animals, which includes the living New Zealand *Hatteria* and the Triassic *Hyperodapedon.*

THE *Transactions* of the American Microscopical Society for 1899 (vol. xxi.), contains a number of interesting articles on microscopic objects, zoological and botanical, together with a smaller number on microscope construction and laboratory apparatus.

IN the *Agricultural Gazette* of New South Wales for August, we notice a number of papers of interest and value for farmers and horticulturists in the Colony. Much information is contained in this and in previous numbers on the diseases to which domestic animals and cultivated crops are liable, and on the best methods for their treatment.

THE parts most recently received of Engler's *Botanische Jahrbücher* are Heft 4 of vol. xxviii. and Heft 2 of vol. xxix. Besides a few shorter articles, these parts are almost entirely occupied by two important descriptive papers—a continuation of the editor's report on the results of the German Nyassa expedition, and one by D. Diels on the flora of Central China.

THE *Bulletin* of the Imperial Society of Naturalists of Moscow, No. 4 for 1900, contains several interesting botanical papers in German. Of these the most important is the second of a series by W. Arnoldi on the morphology and history of development of the Gymnosperms. The present paper is devoted to the process of fertilisation in *Sequoia* (*Wellingtonia*), and is a link in the chain of the numerous and most important observations of recent years which connect the process of impregnation in Gymnosperms with that in Vascular Cryptogams on the one hand, and that in Angiosperms on the other hand.

THE third annual dinner of the association of old students of the Central Technical College will be held on Tuesday, October 2nd, at the Restaurant Frascati, Oxford-street. Old students can obtain further particulars from the honorary secretary, Mr. M. Solomon, 12, Edith-road, West Kensington, W., to whom all applications for tickets should be made.

THE three parts of vol. xxxix. of the *Transactions* of the Royal Society of Edinburgh, which have just been issued, contain several very valuable papers read before the Society during the sessions 1897-98 and 1898-99. All the papers have been published separately, and most of them have been reviewed in NATURE, or briefly described in the reports of the meetings of the Society.

MESSRS. BAILLIÈRE, TINDALL AND COX have published the fifth edition of "A Synopsis of the British Pharmacopœia," compiled by Mr. H. Wippell Gadd, with analytical notes and suggested standards by Mr. C. G. Moor. This little pocket-book is widely appreciated: it contains a complete table of chemicals, drugs and preparations in the official "Pharmacopœia," with their character, doses, &c., as well as other information arranged in a convenient form.

THE additions to the Zoological Society's Gardens during the past week include a Mona Monkey (*Cercopithecus mona*, ♀) from West Africa, presented by Mrs. C. Campbell; a Red-footed Ground Squirrel (*Xerus erythropus*) from West Africa, presented by Dr. Oswald Horrocks; a Grey Ichneumon (*Herpestes griseus*) from India, presented by Captain W. H. Rotheram, R.E.; a Plantain Squirrel (*Sciurus plantani*) from Java, presented by Mr. H. H. Goodwin; two Dusky Ducks (*Anas obscura*) from North America, presented by Mr. W. H. St. Quintin; a Peregrine Falcon (*Falco peregrinus*), European, presented by Mr. A. L. Jessopp; three Jays (*Garrulus glandarius*), British, presented by Dr. R. B. Sharpe; four Pheasants (*Phasianus colchicus*), British, presented by Mr. F. Larratt; two Western Yellow-winged Laughing Thrushes (*Trochalopterum nigrimentum*), a Rufous-chinned Laughing Thrush (*Ianthocincla rufigularis*), two Tickell's Ouzels (*Merula unicolor*), a Spotted wing (*Psaroglossus spiloptera*) from British India, presented by Mr. E. W. Harper; a Blue and Yellow Macaw (*Ara ararauna*) from South America, presented by Mr. Randolph Berens; a Red Tiger Cat (*Felis chrysothrix*), a Leopard (*Felis pardus*), two Rose-ringed Parrakeets (*Palæornis docilis*) from West Africa, a Yellow-crowned Troupial (*Icterus chrysocephalus*), a Yellow-backed Troupial (*Icterus croconotus*) from South America, an Alpine Marmot (*Arctomys marmotta*), two Cross-bills (*Loxia curvirostra*), European; ten Elephantine Tortoises (*Testudo elephantina*) from the Aldabra Islands, deposited.

OUR ASTRONOMICAL COLUMN

ASTRONOMICAL OCCURRENCES IN OCTOBER.

Oct. 6. 13h. 35m. to 14h. 29m. Moon occults κ Piscium (mag. 5).
9. 10h. 42m. Minimum of Algol (β Persei).
11. 6h. 51m. Transit (egress) of Jupiter's Sat. III.
11. 8h. 47m. to 9h. 25m. Moon occults ω³ Tauri (mag. 4·6).
12. 7h. 30m. Minimum of Algol (β Persei).
12. 18h. 36m. to 19h. 23m. Moon occults ζ Tauri (mag. 3).
13. 15h. to 15h. 43m. Moon occults ν Geminorum (mag. 4).
15. Venus. Illuminated portion of disc = 0·637.
15. Mars. ,, ,, = 0·902.
16. 17h. 26m. to 18h. 30m. Moon occults κ Cancri (mag. 5).
17. Saturn. Outer minor axis of outer ring = 16″·68.
19. 10h. Conjunction of Jupiter and Uranus. Jupiter, 0° 25′ N.
19-21. Epoch of Orionid meteoric shower. (Radiant 91° + 15°.)
26. 12h. Conjunction of Jupiter and moon. Jupiter, 0° 27′ S.
28. 6h. 21m. Jupiter's Sat. IV. in conjunction S. of planet.
28. Probable date of perihelion of Barnard's comet (1884 IL).
29. 8h. 27m. to 8h. 46m. Moon occults d Sagittarii (mag. 4·9).
29. 16h. Mercury at greatest elongation (23° 46′ E.).

THE FIREBALL OF SUNDAY, SEPTEMBER 2, 6h. 54m.— A very large number of observations of this brilliant object were made, but they were not very exact, as the meteor appeared in daylight. The radiant point was probably in Cepheus at about 334° + 57°. The object, during its visible flight, appears to have descended from a height of eighty-five miles over Richmond, Yorks., to twenty miles over Fleetwood, Lancs.; and to have traversed a path of eighty-four miles. Another fine meteor was observed on Sunday evening, September 16, at 8h. 44m., and descriptions have come from London, Birmingham, Oxford and Llanelly. The radiant was in the southern sky between Capricornus and Piscis Australis at 324° − 25°. The meteor fell from about fifty miles over Bewdley to thirty-two miles over Wigan, and had a visible course of eighty-six miles. The velocity is somewhat doubtful.

EPHEMERIS FOR OBSERVATIONS OF EROS :—

1900.		R.A.		Decl.	
		h. m. s.		° ′ ″	
Sept. 27	...	2 43 7·41	...	+43 27 2·0	
28	...	43 25·24	...	43 49 43·2	
29	...	43 40·01	...	44 12 20·4	
30	...	43 51·63	...	44 34 52·7	
Oct. 1	...	43 59·96	...	44 57 18·9	
2	...	44 4·92	...	45 19 38·2	
3	...	44 6·44	...	45 41 50·2	
4	...	2 44 4·44	...	+46 3 53·8	

The co-operative observations for determinations of parallax will commence about the beginning of October. The planet is at present in the constellation Perseus, and passes the meridian of London about 2·40 a.m.

EPHEMERIS OF COMET BORRELLY-BROOKS (1900*b*).—This comet is now rapidly becoming fainter, and the following abridgment from a complete Ephemeris furnished by Herr A. Scheller (*Astronomische Nachrichten*, Bd. 153, Nos. 3660, 3663) will doubtless suffice for observers possessed of the necessary optical power :—

Ephemeris for 12h. Berlin Mean Time.

1900.		R.A.		Decl.		Br.	
		h. m. s.		° ′ ″			
Sept. 29	...	14 26 34	...	+69 7·7	...	0·07	
Oct. 3	...	32 53	...	68 11·0	...	·06	
7	...	38 55	...	67 24·5	...	·06	
11	...	44 50	...	66 47·3	...	·05	
15	...	50 38	...	66 18·7	...	·04	
19	...	14 56 27	...	65 58·3	...	·04	
23	...	15 2 16	...	65 45·6	...	·03	
27	...	8 8	...	65 40·4	...	·03	
31	...	15 · 14 2	...	+65 42·4	...	0·03	

AUTOMATIC PHOTOGRAPHY OF THE CORONA.—Mention has often been previously made of Prof. C. Burckhalter's in-genious apparatus for obtaining photographs of the solar corona during an eclipse, and it now appears that he was extremely suc-cessful at the eclipse in May last. *Popular Astronomy*, vol. viii., contains reproductions from two negatives of the corona secured by him, one uncontrolled as has hitherto been usual, the other the result of intercepting part of the coronal light for varying periods of time during the total exposure. The total exposure in each case was 8·0 seconds, but by means of a system of re-volving diaphragms arranged in one of the cameras, the image was shielded in various regions for different times, thus permitting the details of the inner corona to be photographed on the same plate as the outermost faint streamers. The following are the calculated effective exposures at the several stated distances from the moon's centre (moon's semi-diameter = 15′ 58″).

Distance from

moon's centre	16′	...	20′	...	32′	...	50′	...	110
Exposure	... 0·04s.	...	0·23s.	...	1·76s.	...	3·20s.	...	8·00s.

The photograph shows the inner coronal detail close to the limb of the moon, the outer streamers extending for more than a lunar diameter. Several of the inner coronal tufts appear to be projected on the long broader streamers as background.

THE IRON AND STEEL INSTITUTE.

THE Iron and Steel Institute held its autumn meeting in Paris on September 18 and 19, under the presidency of Sir William Roberts-Austen, K.C.B., F.R.S. Besides a long pro-gramme of ten papers, visits to the Exhibition, to the works at St. Chamond, at Hayange in Lorraine, and at St. Denis near Paris, were arranged by an influential reception committee, of which Mr. Robert de Wendel was president and Mr. Henri Vastin honorary secretary. The attendance was unusually large, and the meeting was in every respect a successful one. The proceedings began on September 18 at the house of the Société d'Encouragement, with an address of welcome by Mr. Robert de Wendel, president of the French Association of Iron-masters. Sir William Roberts-Austen, having acknowledged the welcome, delivered a presidential address dealing in fault-less literary style with the history of metallurgy in France.

The first paper read by the secretary, Mr. Bennett Brough, was by Mr. H. Pinget, secretary of the Comité des Forges, and dealt with the development of the iron industry in France since the Institute's last visit to Paris in 1889. The increase in out-put of iron and steel has been much greater than it was in the interval between the two previous exhibitions in Paris. No striking technical invention has been made, but great progress has been effected in increasing the power of the appliances used and in improving the quality of the products. There is a marked tendency to replace cast iron by cast steel, and success has attended endeavours to cast complicated forms in metal which is both tough and of high tensile strength. Moreover, special steels are now available for the requirements of particular appli-cations, such as the growing exigencies of armour plate. The discussion on this paper was confined to complimentary remarks from Sir Lowthian Bell, F.R.S., Mr. Greiner and others.

The second paper, the most important submitted to the meeting, was that by Mr. J. E. Stead on iron and phosphorus. It is typical of modern metallurgical research, and contains a mass of original observations showing how phosphorus occurs in iron and steel. The subject is dealt with in four sections : (1) the constitution, properties and microstructure of iron contain-ing phosphorus ; (2) the effect of carbon when introduced by the fusion or cementation process into iron containing phosphorus ; (3) the microstructure of pig iron containing phosphorus ; and (4) the diffusion of solid phosphide of iron into iron. There are appended to the paper useful notes on eutectics, on solid solutions, on the method of determining free phosphide of iron in iron and steel, and on heat-tinting metal sections for microscopic examination. The observations recorded show that iron will retain as much as 1·75 per cent. of phosphorus as phosphide in solid solution, and that when more than that is present, the excess separates and is found as free phosphide of iron mixed up with the mass of iron. It is also shown that carbon added to solid solutions of phosphorus in iron throws out of solution the dissolved phos-phide, which appears in a separate state. The most remarkable result given indicates that when carbon is added by the cementation process, the phosphide, when in large quantity, is thrown, not only out of solution, but escapes entirely out of the metal as a liquid eutectic leaving a constant residuum behind. A method is described by which phosphorus compounds in pig iron can be identified by means of the microscope. This consists in simply heating the polished surfaces to about 300° C. for a few minutes, when each constituent takes a different oxidation tint. The iron acquires a sky-blue colour, the carbide a red-brown and the phosphide compound a pale yellow. The coloured sections are of great beauty. Many results are given showing how the solid phosphide diffuses in solid iron, and showing that under suitable conditions well-formed crystals will grow in solid metal.

Mr. H. Bauerman's paper on iron and steel at the Universal Exhibition, Paris, 1900, was prepared mainly for the use of the members of the Institute visiting the Exhibition during the meeting. It contained a critical description of the more prominent metallurgical exhibits, and forms a valuable record of the condition of the metallurgical industry at the close of the century.

On September 19, the remaining papers on the programme were dealt with. Chief among these was that by Mr. E. F. Lange, on a new method of producing high temperatures. The principle underlying the process, which is the outcome of researches made by Dr. H. Goldschmidt of Essen, is not new, as It is based upon the heat energy developed by the chemical action of aluminium upon oxygen, or rather that between aluminium and certain metallic oxides. The practicability of the process was clearly shown by the welding together during the meeting of two short lengths of heavy girder rails. The method not only opens up a new field for aluminium but also promises to be of considerable importance in engineering work. In the discussion Sir William Roberts-Austen pointed out the extreme precision with which the reduction took place, and Sir Lowthian Bell dwelt on the value of the process if it should prove that carbonless iron could be obtained by it for electrical purposes.

The paper by Mr. A. L. Colby, of Bethlehem, United States, on American standard specifications and methods of testing iron and steel, embodied the results of over a year's work by a committee of American experts, conducted with a view to the adoption of international standards. Some of the specifications were criticised by Mr. R. A. Hadfield. The engineer, he thought, was encroaching on the field of the metallurgist. Interesting contributions to the discussion were made by Mr. C. P. Sandberg and by Dr. Dudley, of Pennsylvania.

In a paper on the influence of aluminium on the carbon in cast-iron, Mr. G. Melland and Mr. H. W. Waldron gave the results of an elaborate research in which they endeavoured to determine the amount of aluminium which is necessary to produce the maximum separation of graphite in a white pig-iron as free as possible from silicon and other impurities, and to ascertain, by casting every melting both in sand and in chill moulds, the effect produced by slow and rapid cooling upon the mode of existence of the carbon in the metal with amounts of aluminium varying from 0·02 to 12 per cent.

In the paper by Mr. Louis Katona, of Resicza, Hungary, the various disadvantages of the rolling-mills now in use were discussed, and suggestions were made for obviating them with a view to increasing the output and lessening the fuel consumption.

In a lengthy paper on the constitution of slags, which was taken as read, Baron H. von Jüptner discussed iron slags from a modern point of view, and described the varying reactions which take place between them and iron. The slags considered are divided into three groups—silicate slags, phosphate slags and oxide slags. The results of the investigation tend to show that slags should be regarded as solutions, and not as complicated chemical compounds.

The "phase-rule" of Gibb has served as a guide to the authors of two well-reasoned papers of great scientific interest—one on iron and steel from the point of view of the phase doctrine, by Prof. Bakhuis-Roozeboom, of Amsterdam, and the other on the present position of the solution theory of carburised iron, by Dr. A. Stansfield. The phase rule says in effect that in a system such as that of the carburised irons, in which two distinct substances (carbon and iron) are involved, but in which certain forms or phases of carbon or iron, or carbon-iron solution, or carbon-iron compound, are present; no more than two of these phases can exist in equilibrium with each other at a particular temperature. In the case of a solution of salt in water, this

would mean that there could only be salt and ice and solution together at a particular temperature (the eutectic temperature), and that at any other temperature there could only be ice and solution or salt and solution (at temperatures above the eutectic), or ice and salt (at temperatures below the eutectic). In the case of a salt solution this is quite evident, but the value of the phase rule is that we can apply it with equal confidence in cases where we do not, to begin with, know the answer to our question. Applying the rule to the case of solid carburised iron at temperatures above that of all the known allotropic charges—we have the four possible substances of iron, graphite, cementite and solid solution of carbon (either graphite or cementite) in pig-iron. The rule states that only two of these can in general exist permanently together. The general conclusions to be drawn from Dr. Stansfield's researches are :—

(1) That carbon is less soluble in iron when presented in the form of graphite than when presented in the form of cementite.

(2) That the apparent reversal of this in steel is due partly to the absence of nuclei of graphite on which further deposits might take place ; partly to the length of time required for the separation of the graphite, involving, as it does, the gradual passage of carbon through the iron to reach the nuclei, and partly to the mechanical pressure which must oppose the formation of graphite in solid steel.

The meeting was brought to a close by a vote of thanks to the French authorities and societies, whose hospitality had been enjoyed, proposed by the president and seconded by Mr. W. Whitwell, president-elect. A vote of thanks to the president was proposed by Mr. Greiner, of Seraing, Belgium, and seconded by Mr. Nordenfelt. The social functions in connection with the meeting were of a very attractive character. They included an operatic entertainment organised by the Comité des Forges, a reception by the Commissioner-General and Mrs. Jekyll at the British Royal Pavilion, a banquet at the Hôtel Continental, a reception by Mr. E. Schneider in the Le Creusot pavilion, a reception at the Hôtel de Ville by the president of the Municipal Council, and a reception on September 24 by the Minister of Public Works.

THE BRADFORD MEETING OF THE BRITISH ASSOCIATION.

SECTION K.

BOTANY.

Opening Address by Prof. S. H. Vines, M.A., D.Sc., F.R.S., President of the Section.

There has been considerable difference of opinion as to whether the present year marks the close of the nineteenth or the beginning of the twentieth century. But whatever may be the right or the wrong of this vexed question, the fact that the year-date now begins with 19, instead of with 18, suggests the appropriateness of devoting an occasion such as the present to a review of the century which has closed, as some will have it, or, in the opinion of others, is about to close. I therefore propose to address you upon the progress of Botany during the nineteenth century.

I am fully conscious of the magnitude of the task which I am undertaking, more especially in its relation to the limits of time and space at my disposal. So eventful has the period been that to give in any detail an account of what has been accomplished during the last hundred years would mean to write the larger half of the entire history of Botany. This being so, it might appear almost hopeless to attempt to deal with so large a subject in a Presidential Address. But I trust that the very restrictions under which I labour may prove to be rather advantageous than otherwise, inasmuch as they compel me to confine attention to what is of primary importance, and thus to give special prominence to the main lines along which the development of the science has proceeded.

Statistics.

We may well begin with what is, after all, the most fundamental matter, viz. the relative numbers of known species of plants at the beginning and at the end of the century. It might appear that the statistics of plants was a subject susceptible of very simple treatment, but unfortunately this is not the case. It must be remembered that a "species" is not an invariable standard unit, like a pound or a pint, but that it is an idea dependent upon the subjectivity of individual botanists. For instance, one botanist may regard a certain number of similar plants as all belonging to a single species, whilst another may find the differences among them such as to warrant the distinction of as many species as there are plants. It is this inevitable variation in the estimation of specific characters which renders it difficult to deal satisfactorily with plants from the statistical point of view. However, the following figures may be regarded as giving a fair idea of the increase in the number of "good" species of living plants.

It is generally stated that about 10,000 species of plants were known to Linnæus in the latter half of the eighteenth century, of which one-tenth were Cryptogams ; but so rapid was the progress in the study of new plants at that time that the first enumeration of plants published in the nineteenth century, the "Synopsis" of Persoon (1807), included as many as 20,000 species of Phanerogams alone. Turning now to the end of the century, we arrive at the following census, for which I am indebted mainly to Prof. Saccardo (1892) and to Prof. de Toni who has kindly given me special information as to the Algæ :—

Species of Phanerogams indicated in Bentham and Hooker's " Genera Plantarum " (Durand, " Index," 1888).

Dicotyledons	...	...	...	...	78,200
Monocotyledons	...	...	...	...	19,600
Gymnosperms	...	...	...	...	2,420
					100,220
Estimated subsequent additions (Saccardo)	...				5,011
		Total Phanerogams	...		105,231

Species of Pteridophyta (indicated in Hooker and Baker's " Synopsis " ; Baker's " New Ferns." and " Fern Allies").

Filicinæ (including Isoëtes), about	...			3,000
Lycopodinæ, about	...	...	...	432
Equisetinæ, about	...	...	...	20
	Total Pteridophyta	...	...	3,452

Species of Bryophyta (Saccardo's Estimate).

Musci	...	...	...	...	4,609
Hepaticæ	...	...	..	...	3,041
		Total Bryophyta	...	...	7,650

Species of Thallophyta.

Fungi (including Bacteria) (Saccardo)	...	39,663
Lichens (Saccardo)	...	...	5,600
Algæ (incl. 6000 Diatoms) (de Toni)	...	14,000
	Total Thallophyta	...	...	59,263

Adding these totals together—

Phanerogams		...	...	...	105,231
Pteridophyta	...	...	...	...	3,452
Bryophyta	...	...	...	...	7,650
Thallophyta	...	...	...	...	59,263

we have a grand total of 175,596

as the approximate number of recognised species of living plants.

These figures are sufficiently accurate to show how vast have been the additions to the knowledge of plants in the period under consideration, and they afford much food for thought. In the first place, they indicate how closely connected has been the growth of this branch of Botany with the exploration and opening-up of new countries which has been so characteristic a feature of the century. Again, no one can consider these figures without being struck by the disparity in the numbers of species included in the different groups ; a most interesting topic, which cannot, however, be entered upon here. It must suffice to point out in a general way that the smaller groups represent families of plants which attain their numerical zenith in long past geological periods, and are now decadent, whilst the existing flora of the world is characterised by the preponderating Angiosperms and Fungi.

We may venture to cast a forward glance upon the possible future development of the knowledge of species. Various partial estimates have been made as to the probable number of existing species of this or that group, but the only comprehensive estimate with which I am acquainted is that of Prof. Saccardo (1892). He begins with a somewhat startling calculation to the effect that there are at least 250,000 existing species of Fungi alone, and he goes on to suggest that probably the number of species belonging to the various other groups would amount to 150,000 ; hence the total number of species now living is to be estimated at over 400,000. On the basis of this estimate it appears that we have not yet made the acquaintance of half the contemporary species ; so that there remains plenty of occupation for systematic and descriptive botanists, especially in the department of Fungology. It is also rather alarming, in view of the predatory instincts of so many of the Fungi, to learn that they constitute so decided a majority of the whole vegetable kingdom.

In spite of the great increase in the number of known species, it cannot be said that any essentially new type of plant has been discovered during the century. So far as the bounds of the vegetable kingdom have been extended at all, it has been by the annexation of groups hitherto regarded as within the sphere of influence of the zoologists. The most notable instance of this has occurred in the case of the Bacteria, or Schizomycetes, as Naegeli termed them. These organisms, discovered by Leeuwenhoek 200 years ago, had always been regarded as infusorian animals until in 1853, Cohn recognised their affinity with the Fungi. These plants have acquired special importance, partly on account of the controversy which arose as to their supposed spontaneous generation, but more especially on account of their remarkable zymogenic and pathogenic properties, so that Bacteriology has become one of the new sciences of the century.

Classification.

Having gained some idea of the number of species which have been recognised and described during the century, the next point for consideration is the progress made in the attempt to reduce this mass of material to such order that it can be intelligently apprehended ; in a word, to convert a mass of facts into a science ; " Filum ariadneum Botanices est systema, sine quo chaos est Res Herbaria " (Linnæus).

The classification of plants is a problem which has engaged attention from the very earliest times. Without attempting to enter into the history of the matter, I may just point out that, speaking generally, all the earlier systems of classification were more or less artificial, the subdivisions being based upon the distinctive features of one set of members of the plant. When I say that of all these systems that proposed by Linnæus (1735) was the most purely artificial, I do not imply any reproach : if it was the most artificial, it was at the same time the most serviceable, and its author was fully aware of its artificiality. This system is generally regarded as his most remarkable achievement ; but the really great service which Linnæus rendered to science was the clear distinction which he for the first time drew between systems which are artificial and those which are natural. Recognising, as he did, his inability to frame at that period a satisfactory natural system, he also realised that with the increased number of known plants some more ready means of determining them was an absolute necessity, and it was for this purpose that he devised his artificial system, not as an end, but as a means. The end to be kept in view was the natural classification : " Methodus naturalis est ultimus finis Botanices " is his clearly expressed position in the " Philosophia Botanica."

There is a certain irony in the fact that the enthusiastic acceptance accorded to his artificial system throughout the greater part of Europe contributed to postpone the realisation of Linnæus's cherished hopes with regard to the attainment of a natural classification. It was just in those countries, such as Germany and England, where the Linnean system was most readily adopted that the development of the natural system proceeded most slowly. It was in France, where the Linnean system never secured a firm hold, that the quest of the natural system was pursued ; and it is to French botanists more particularly that our present classification is due. It may be traced from its first beginnings with Magnol in 1689, through the bolder attempts of Adanson and of Bernard de Jussieu (1759), to the relatively complete method propounded by Antoine Laurent de Jussieu in his "Genera Plantarum," just 100 years later.

The nineteenth century opened with the struggle for pre-dominance between the Jussiean and the Linnean systems. In England the former soon obtained considerable support, notably that of Robert Brown, whose "Prodromus Floræ Novæ Hollandiæ," published in 1810, seems to have been the first English botanical work in which the natural system was adopted; but it did not come into general use until it had been popularised by Lindley in the 'thirties.

Meantime the Jussiean system had been extended and improved by Auguste Pyrame de Candolle (1813-24). It is essentially the Candollean classification which is now most generally in use, and it has been immortalised by its adoption in Bentham and Hooker's "Genera Plantarum," one of the great botanical monuments of the century. In Germany, however, it has been widely departed from, the system there in vogue being based upon Brongniart's modification (1828, 1850) of de Candolle's method as elaborated successively by Alex. Braun (1864), Eichler (1876-83) and Prof. Engler (1886, 1898). It must be admitted that for the last fifty years the further evolution of the natural system, at any rate so far as Phanerogams are concerned, has been confined to Germany.

One of the most important advances in the classification of Phanerogams was based upon Robert Brown's discovery in 1827 of the gymnospermous nature of the ovule in Conifers and Cycads, which led Brongniart (1828) to distinguish these plants as "Phanérogames gymnospermes"; and although the system-atic position of these plants has since then been the subject of much discussion, the recognition of the Gymnospermæ as a distinct group of archaic Phanerogams is now definitely accepted.

Moreover, the greatly increased knowledge of the Cryptogams has involved a considerable reconstruction in the classification of that great sub-kingdom. One of the most striking discoveries is that first definitely announced by Schwendener (1869) concerning Lichens, to the effect that the body of a Lichen consists of two distinct organisms, an Alga and a Fungus, living in symbiosis; a discovery which was so nearly made by other con-temporary botanists, such as de Bary, Berkeley and Sachs, and which can be traced back to Haller and Gleditsch in the eighteenth century.

But the discoveries which most affected the classification of the Cryptogams are those relating to their reproduction. Whilst it had been recognised, almost from time immemorial, that Phanerogams reproduce sexually, sexuality was denied to Cryptogams until the observations on Liverworts and Mosses by Schmidel and by Hedwig (of whom it was said that he was born to banish Cryptogamy) in the eighteenth century; and even as late as 1828 we find Brongniart classifying the Fungi and Algæ together as "Agames." But in the middle third of the nine-teenth century, by the labours of such men as Thuret, Pringsheim, Cohn, Hofmeister, Naegeli and de Bary, the sexuality of all classes of Cryptogams was clearly established. It is worthy of note that, although the sexuality of the Phanerogams had been accepted for centuries, yet the details of sexual reproduction were first investigated in Cryptogams. For it was not until 1823 that Amici discovered the pollen-tube, and it was more than twenty years later (1846) before he completed his discovery by ascertaining the true significance of the pollen-tube in relation to the development of the embryo; whilst it remained for Strasburger to observe, thirty years later, the actual process of fertilisation.

The discovery of the reproductive processes in Cryptogams not only facilitated a natural classification of them, but had the further very important effect of throwing light upon their relation to Phanerogams. Perhaps the most striking botanical achieve-ment of the nineteenth century has been the demonstration by Hofmeister's unrivalled researches (1851) that Phanerogams and Cryptogams are not separated, as was formerly held, by an impassable gulf, but that the higher Cryptogams and the lower Phanerogams are connected by many common features.

The development of the natural classification, of which an account has now been given, proceeded for the most part on the assumption of the immutability of species. As Linnæus ex-pressed it in his "Fundamenta Botanica," "species tot numer-amus, quot diversæ formæ in principio sunt creatæ." It is difficult to understand how, with this point of view, the idea of affinity between species could have arisen at all; and yet the establishment of genera and the attempts at a natural system

prove that the idea was operative. The nature of the prevalent conception of affinity is well conveyed by Linnæus's aphorism, "Affines conveniunt habitu, nascendi modo, proprietatibus, viribus, usu."

But a conviction had been gradually growing that the assumed fixity of species was not well founded, and that, on the contrary, species are descended from pre-existent species. This view found clear expression in Lamarck's "Philosophie Zoo-logique," published early in the century (1809), but it did not strongly affect public opinion until after the publication of Darwin's "Origin of Species" in 1859. Regarded from this point of view, the problems of classification have assumed an altogether different aspect. Affinity no longer means mere similarity, but blood-relationship depending upon common descent. We no longer seek a "system" of classification; we endeavour to determine the mutual relations of plants. The effect of this change has been to stimulate the investigation of plants in all their parts and in all stages of their life, so as to attain that complete knowledge of them without which their affinities cannot be accurately estimated. If the classification of Cryptogams is, at the present moment, in a more satisfactory position than that of Phanerogams, it is just because the study of the former group has been, for various reasons, more thorough and more minute than that of the latter.

Palaeophytology.

The stimulating influence of the new doctrine was not, how-ever, confined to the investigation of existing plants; it also gave a remarkable impulse to the study of fossil plants, inasmuch as the theory of descent involves the quest of the ancestors of the forms that we now have around us. Marvellous progress has been made in this direction during the nineteenth century, by the labours more especially of Brongniart, Goeppert, Unger, Schimper, Schenck, Saporta, Solms-Laubach, Renault, on the Continent, and in our own country of Lindley and Hutton, Hooker, Carruthers, and more especially of Williamson. So far-reaching are the results obtained that I can only attempt the barest summary of them. I may perhaps best begin by saying that only a small proportion of existing species have been found in the fossil state. In illustration I may adduce the statement made by Mr. Clement Reid in his recent work, "The Origin of the British Flora," that only 270 species, that is, about one-sixth of the total number of British vascular plants, are known as fossils. Making all due allowances for the im-perfection of the geological record, for the limited area investi-gated, and for the difficulty of determination of fragmentary specimens, it may be stated generally that the number of exist-ing species has been found to rapidly diminish in the floras of successively older strata; none, in fact, have been certainly found to persist beyond the Tertiary period. Certain existing genera, belonging to the Gymnosperms and to the Pteridophyta, have, however, been traced far down into the Mesozoic period. Similarly, the distribution in time of existing natural orders does not coincide with that of existing genera; thus the Ferns of the Carboniferous epoch apparently belong, for the most part, if not altogether, to the order Marattiaceæ, but they are not referable to any of the existing genera.

Moreover, altogether new families of fossil plants have been discovered: such are, among Gymnosperms, the Cordaitaceæ and the Bennettitaceæ; among Pteridophyta, the Calamariaceæ, the Lepidodendraceæ, the Sphenophyllaceæ and the Cycado-filices. It is of interest to note that all these newly discovered families can be included within the main subdivisions of the existing flora; in fact, no fossil plants have been found which suggest the existence in the past of groups outside the limits of our Phanerogamia, Pteridophyta, Bryophyta and Thallophyta.

It cannot be said that the study of Palæobotany has as yet made clear the ancestry and the descent of our existing flora. To begin with the angiospermous flowering plants, it has been ascertained that they make their first appearance in the Cretaceous epoch, but we have no clue as to their origin. The relatively late appearance of Angiosperms in geological time suggests that they must have sprung from an older group, such as the Gymnosperms or the Pteridophyta; but there is no evidence to definitely establish either of these possible origins. Then as to the origin of the Gymnosperms, whilst it cannot be doubted that they were derived from the Pteridophyta, the existing data are insufficient to enable us to trace their pedigree. The most ancient family of Gymnosperms, the Cordaitaceæ, can be traced as far back as any known Pteridophyta, and cannot, therefore, have been derived from them; but the fact that the Cordaitaceæ exhibit certain cycadean affinities, and the discovery of the Cycadofilices, suggest that what may be termed the cycadean phylum of Gymnosperms (including the Cordaitaceæ, Bennettitaceæ, Cycadaceæ, and perhaps the Ginkgoaceæ) had its origin in a filicineous ancestry, of which, it must be admitted, no forms have as yet been recognised.

Turning to the Pteridophyta, the origin of the Ferns is still quite unknown: the one fact which seems to be clear is that the eusporangiate forms (Marattiaceæ) are more primitive than the leptosporangiate. With regard to the Equisetinæ, the Calamariaceæ were no doubt the ancestors of the existing and of the fossil Equisetums. Similarly, in the Lycopodinæ, the palæozoic Lepidodendraceæ were the forerunners of the existing Lycopodiums and Selaginellas. The discovery of the Sphenophyllaceæ seems to throw some further light upon the phylogeny of these two groups, inasmuch as these plants possess characters which indicate affinity with both the Equisetinæ and the Lycopodinæ, thus suggesting the possibility that they may have sprung from the same ancestral stock.

To complete the geological survey of the vegetable kingdom I will briefly allude to the Bryophyta and the Thallophyta. Owing no doubt to their delicate texture, the records of these plants have been found to be very incomplete. So much is this the case with the Bryophyta that I forbear to make any statement concerning them. The chief point of interest with regard to the Fungi is that most of those which have been discovered in the fossil state were found in the tissues of woody plants on which they were parasitic. In this way it has been possible to ascertain, with some probability, the existence of Bacteria and of mycelial Fungi in the Palæozoic period. The records of the Algæ are more satisfactory; they have been traced far back into the Palæozoic age, where they are represented by siphonaceous forms and by the somewhat obscure plants known as *Nematophycus* and *Pachytheca*.

In a general way the study of Palæobotany has proved the development of higher forms from lower forms in the successive geological periods. Thus the Tertiary and Quaternary periods are characterised by the predominance of Angiosperms, just as the Mesozoic period is characterised by the predominance of Gymnosperms, and the Palæozoic by the predominance of Pteridophyta. And yet, as I have been pointing out, we are not able to trace the ancestry of any one of the larger groups of plants. The chief reason for this is that the geological record, so far as it is known, has been found to break off with such surprising abruptness that the earliest, and therefore the most interesting, chapters in the evolution of plants are closed to us. After the wealth of plant-forms in the Carboniferous epoch there is a striking falling-off in the Devonian, in which, however, plants of high organisation, such as the Cordaitaceæ, the Calamariaceæ and the Lepidodendraceæ, still occur. In the Silurian epoch vascular plants are but sparingly present—but it is remarkable that any such highly organised plants should be found there—together with probable Algæ, such as *Nematophycus* and *Pachytheca*. The Cambrian rocks present nothing but so-called "Fucoids," such as *Eophyton*, &c., some of which may be Algæ. The only known fossil in the oldest strata of all, the Archæan, is the much-discussed *Eozoon canadense*, probably of animal origin; but the occurrence here of large deposits of graphite seems to indicate the existence of a considerable flora which has, unfortunately, become quite undeterminable. Thus, whilst there is some evidence that the primitive plants were Algæ, there is at present no available record of the various stages through which the Silurian and Devonian vascular plants were evolved from them.

Morphology.

If inquiry be made as to the cause of the great advance in the recognition of the true affinities of plants, and consequently in their classification, which distinguishes the nineteenth century, I would refer it to the progress made in the study of morphology. The earlier botanists regarded all the various parts of plants as "organs" in relation to their supposed function; hence their description of plants was simply "organography." The idea of regarding the parts of the plant-body, not in connection with their functions, but with reference to their development and their mutual relations, seems to have originated with Jung in the seventeenth century (1687): it was revived by C. F. Wolff about seventy years later (1759), but it did not materially affect the study of plants until well on in the nineteenth century, after

Goethe had repeatedly written on the subject and had devised the term "morphology" to designate it. For a time this somewhat abstract mode of treatment led to mere theorising and speculation, so much so that the years 1820–1840 will always be stigmatised as the period of the "Naturphilosophie." But fortunately this time of barrenness was succeeded by a veritable renascence. Robert Brown and Henfrey in England; Brongniart, St. Hilaire and Tulasne in France; Mohl, Schleiden, Naegeli, A. Braun, and, above all, Hofmeister in Germany, led the way back from the pursuit of fantastic will-o'-the-wisps to the observation of actual fact. Instead of evolving schemes out of their own internal consciousness as to how plants ought to be constructed, they endeavoured to discover by the study of development, and more particularly of embryogeny, how they actually are constructed, with the result that within a decade Hofmeister discovered the alternation of generations in the higher plants; a discovery which must ever rank as one of the most brilliant triumphs of morphological research.

With the knowledge thus acquired it became possible to determine the true relations of the various parts of the plant-body: to distinguish these parts as "members" rather than as "organs"; in a word, to establish homologies where hitherto only analogies had been traced—which is the essential difference between morphology and organography.

The publication of the "Origin of Species" profoundly affected the progress or morphology, as of all branches of biological research: but it did not alter its trend; it confirmed and extended it. We are not satisfied now with establishing homologies, but we go on to inquire into the origin and phylogeny of the members of the body. In illustration I may briefly refer to two problems of this kind which at the present time are agitating the botanical world. The first is as to the origin of the alternation of generations. Did it come about by the modification of the sexual generation (gametophyte) into an asexual (sporophyte); or is the sporophyte a new formation intercalated into the life-history? In a word, is the alternation of generations to be regarded as homologous or as antithetic? I am not rash enough to express any opinion on this controversy; nor is it necessary that I should do so, since the subject has twice been threshed out at recent meetings of this Section. The second problem is as to the origin of the sphorophylls, and, indeed, of all the various kinds of leaves of the sporophyte in the higher plants. It is suggested, on the one hand, that the sporophylls of the Pteridophyta have arisen by gradual sterilisation and segmentation from an unsegmented and almost wholly reproductive body, represented in our day by the sporogonium of the Bryophyta; and that the vegetative leaves have been derived by further sterilisation from the sporophylls. On the other hand, it is urged that the vegetative leaves are the more primitive, and that the sporophylls have been derived from them. It will be at once observed that this second problem is intimately connected with the first. The sterilisation theory of the origin of leaves is a necessary consequence of the antithetic view of the alternation of generations; whilst the derivation of sporophylls from foliage-leaves is similarly associated with the homologous view. Here, again, exercising a wise discretion, I will only venture to express my appreciation of the important work which has been done in connection with this controversy—work that will be equally valuable, whatever the issue may eventually be.

I will conclude my remarks on morphology with a few illustrations of the aid which the advance in this department has given to the progress of classification. For instance, Linnæus divided plants into Phanerogams and Cryptogams, on the ground that in the former the reproductive organs and processes are conspicuous, whereas in the latter they are obscure. In view of our increased knowledge of Cryptogams this ground of distinction is no longer tenable; whilst still recognising the validity of the division, our reasons for doing so are altogether different. For us, Phanerogams are plants which produce a seed; Cryptogams are plants which do not produce a seed. Again, we distinguish the Pteridophyta and the Bryophyta from the Thallophyta, not on account of their more complex structure, but mainly on the ground that the alternation of generations is regular in the two former groups, whilst it is irregular or altogether wanting in the latter. Similarly, the essential distinction between the Pteridophyta and the Bryophyta is that in the former the sporophyte, in the latter the gametophyte, is the preponderating form. It has enabled us further to correct in many respects the classifications of our predecessors by altering the systematic position of various genera, and sometimes of larger groups. Thus the Cycadaceæ have been removed from among the Monocotyledons, and the Coniferæ from among the Dicotyledons, where de Candolle placed them, and have been united with the Gnetaceæ into the sub-class Gymnospermæ. The investigation of the development of the flower, in which Payer led the way, and the elaboration of the floral diagram which we owe to Eichler, have done much, though by no means all, to determine the affinities of doubtful Angiosperms, especially among those previously relegated to the lumber-room of the Apetalæ.

Anatomy and Histology.

Passing now to the consideration of the progress of knowledge concerning the structure of plants, the most important result to be chronicled is the discovery that the plant-body consists of living substance indistinguishable from that of which the body of animals is composed. The earlier anatomists, whilst recognising the cellular structure of plants, had confined their attention to the examination of the cell-walls, and described the contents as a watery or mucilaginous sap, without determining where or what was the seat of life. In 1831 Robert Brown discovered the nucleus of the cell, but there is no evidence that he regarded it as living. It was not until the renascence of research in the 'forties, to which I have already alluded, that any real progress in this direction was made. The cell-contents were especially studied by Naegeli and by Mohl, both of whom recognised the existence of a viscous substance lining the wall of all living cells as a "mucous layer" or "primordial utricle," but differing chemically from the substance of the wall by being nitrogenous ; this they regarded as the living part of the cell, and to it Mohl (1846) gave the name "protoplasm," which it still bears. The full significance of this discovery became apparent in a somewhat roundabout way. Dujardin, in 1835, had described a number of lowly organisms, which he termed Infusoria, as consisting of a living substance, which he called "sarcode." Fifteen years later, in a remarkable paper on *Protococcus pluvialis*, Cohn drew attention to the similarity in properties between the "sarcode" of the Infusoria and the living substance of this plant, and arrived at the brilliant generalisation that the "protoplasm" of the botanists and the "sarcode" of the zoologists are identical. Thus arose the great conception of the essential unity of life in all living things, which, thanks to the subsequent labours of such men as de Bary, Brücke, and Max Schultze, in the first instance, has become a fundamental canon of Biology.

A conspicuous monument of this period of activity is the cell-theory propounded by Schwann in 1839. Briefly stated, Schwann's theory was that all living bodies are built up of structural units which are the cells : each cell possesses an independent vitality, so that nutrition and growth are referable, not to the organism as a whole, but to the individual cells. This conception of the structure of plants was accepted for many years, but it has had to give way before the advance of anatomical knowledge. The recognition of cell-division as the process by which the cells are multiplied—in opposition to the Schleidenian theory of free cell-formation—early suggested doubts as to the propriety of regarding the body as being built up of cells as a wall is built of bricks. Later the minute study of the Thallophyta revealed the existence of a number of plants, such as the Myxomycetes, the phycomycetous Fungi, and the siphonaceous Algæ, some of them highly organised, the vegetative body of which does not consist of cells. It became clear that cellular structure is not essential to life ; that it may be altogether absent or present in various degree. Thus in the higher plants the protoplasm is segmented or septated by walls into uninucleate units or "energids" (Sachs), and such plants are well described as "completely septate." But in others, such as the higher Fungi and certain Algæ (e.g. *Cladophora*, *Hydrodictyon*), the protoplasm is septated, not into energids, but into groups of energids, so that the body is "incompletely septate." Finally there are the Thallophyta already enumerated, in which there is complete continuity of the protoplasm : these are "unseptate." Moreover, even when the body presents the complete cellular structure, the energids are not isolated, but are connected by delicate protoplasmic fibrils traversing the intervening walls ; a fact which is one of the most striking discoveries in the department of histology. This was first recognised in the sieve-tubes by Hartig (1837) ; then by Naegeli (1846) in the tissues of the Florideæ. After a long period of neglect the matter was taken up once more by

Tangl (1880), when it attracted the attention of many investigators, as the result of whose labours, especially those of Mr. Gardiner, the general and perhaps universal continuity of the protoplasm in cellular plants has been established. Hence the body is no longer regarded as an aggregate of cells, but as a more or less septated mass of protoplasm : the synthetic standpoint of Schwann has been replaced by one as distinctively analytic.

Time does not permit me to do more than mention the important discoveries made of late years, mainly on the initiative of Strasburger, with regard to the details of cytology, and especially to the structure of the nucleus and the intricate dance of the chromosomes in karyokinesis. Indeed, I can do but scant justice to those anatomical discoveries which are of more exclusively botanical interest. One important generalisation which may be drawn is that the histological differentiation of the plant proceeds, not in the protoplasm, as in the animal, but in the cell-wall. It is remarkable, on the one hand, how similar the protoplasm is, not only in different parts of the same body, but in plants of widely different affinities ; and, on the other, what diversity the cell-wall offers in thickness, chemical composition, and physical properties. In studying the differentiation of the cell-wall the botanist has received valuable aid from the chemist. Research in this direction may, in fact, be said to have begun with Payen's fundamental discovery (1844) that the characteristic and primary chemical constituents of the cell-wall is the carbohydrate which he termed cellulose.

The amount of detailed knowledge as to the anatomy of plants which has been accumulated during the century by countless workers, among whom Mohl, Naegeli, Unger and Sanio deserve special mention as pioneers, is very great—so great, indeed, that it seemed as if it must remain a mere mass of facts in the absence of any recognisable general principles which might serve to marshal the facts into a science. The first step towards a morphology of the tissues was Hanstein's investigation of the growing point of the Phanerogams (1868), and his recognition therein of the three embryonic tissue-systems. This has lately been further developed by the promulgation of van Tieghem's theory of the stele, which is merely the logical outcome of Hanstein's distinction of the plerome. It has thus become possible to determine the homologies of the tissue-systems in different plants and to organise the facts of structure into a scientific comparative anatomy. It has become apparent that, in many cases, differences of structure are immediately traceable to the influence of the environment ; in fact, the study of physiological or adaptive anatomy is now a large and important branch of the subject.

The study of Anatomy has contributed in some degree to the progress of systematic Botany. It is true that some of the more ambitious attempts to base classification on Anatomy have not been successful ; such, for instance, as de Candolle's subdivision of Phanerogams into Exogens and Endogens, or the subdivision of Cormophyta into Acrobrya, Amphibrya, and Acramphibrya, proposed by Unger and Endlicher. Still it cannot be denied that anatomical characters have been found useful, if not absolutely conclusive, in suggesting affinities, especially in the determination of fossil remains. A large proportion of our knowledge of extinct plants, to which I have already alluded, is based solely upon the anatomical structure of the vegetative organs ; and although affinities inferred from such evidence cannot be regarded as final, they suffice for a provisional classification until they are confirmed or disproved by the discovery and investigation of the reproductive organs.

Physiology.

The last branch of botanical science which I propose to pass in review is that of physiology. We may well begin with the nutritive processes. At the close of the eighteenth century there was practically no coherent theory of nutrition ; such as it was it amounted to little more than the conclusion arrived at by van Helmont a century and a half earlier, that plants require only water for their food, and are able to form from it all the different constituents of their bodies. It is true that the important discovery had been made and pursued by Priestley (1772), Ingen-Housz (1780) and Sénébier (1782) that green plants exposed to light absorb carbon dioxide and evolve free oxygen ; but this gaseous interchange had not been shown to be the expression of a nutritive process. At the opening of the nineteenth century (1804) this connection was established by de Saussure, in his classical "Recherches Chimiques," who demonstrated that, whilst absorbing carbon dioxide and evolving oxygen, green plants gain in dry weight; and he further contributed to the elucidation of the problem of nutrition by showing that, whilst assimilating carbon dioxide, green plants also assimilate the hydrogen and oxygen of water.

Three questions naturally arose in connection with de Saussure's statement of the case: What is the nature of the organic substance formed? What is the function of the chlorophyll? What is the part played by light? It was far on in the century before answers were forthcoming.

With regard to the first of these questions the researches of Boussingault (1864) and others established the fact that the volume of carbon dioxide absorbed and that of the oxygen evolved in connection with the process are approximately equal. Further, the frequent presence of starch in the chloroplastids, to which Mohl first drew attention (1837), was subsequently found by Sachs (1862) to be closely connected with the assimilation of carbon dioxide. The conclusion drawn from these facts is that the gain in dry weight accompanying the assimilation of carbon dioxide is due to the formation, in the first instance, of organic substance having the composition of a carbohydrate; a conclusion which may be expressed by the equation

$$CO_2 + H_2O = CH_2O + O_2.$$

The questions with regard to chlorophyll and to light are so intimately connected that they must be considered together. The first step towards their solution was the investigation of the relative activity of light of different colours, originally undertaken by Sénébier (1782) and subsequently repeated by Daubeny (1836), with the result that red and orange light was found to promote assimilation in a higher degree than blue or violet light. Shortly afterwards Draper (1843), experimenting with an actual solar spectrum, concluded that the most active rays are the orange and yellow; a conclusion which was generally accepted for many years. But in the meantime the properties of the green colouring matter of plants (to which Pelletier and Caventou gave the name "chlorophyll" in 1817) were being investigated. Brewster discovered in 1834 that an alcoholic extract of green leaves presents a characteristic absorption spectrum; but many years elapsed before any attempt was made to connect this property with the physiological activity of chlorophyll. It was not until 1871–72 that Lommel and N. J. C. Müller pointed out that the rays of the spectrum which are most completely absorbed by chlorophyll are just those which are most efficient in the assimilation of carbon dioxide. Subsequent researches, particularly those of Timiriazeff (1877), and those of Engelmann (1882–84) based on his ingenious Bacterium-method, have confirmed the views of Lommel and of Müller, and have placed it beyond doubt that the importance of light in the assimilatory process is that it is the form of kinetic energy necessary to effect the chemical changes, and that the function of chlorophyll is to serve as the means of absorbing this energy and of making it available for the plant.

These are perhaps the most striking discoveries in relation to the nutrition of plants, but there are others of not less importance to which brief allusion must be made. We owe to de Saussure (1804) the first clear demonstration of the fact that plants derive an important part of their food from the soil; but the relative nutritive value of the inorganic salts absorbed in solution was not ascertained until Sachs (1858) reintroduced the method of water-culture which had originated centuries before with Woodward (1699) and de Saussure. Special interest centres around the question of the nitrogenous nutrition of plants. It was long held chiefly on the authority of Priestley and of Ingen-Housz, and in spite of the contrary opinion expressed by Sénébier, Woodhouse (1803) and de Saussure, that plants absorb the free nitrogen of the atmosphere by their leaves. This view was not finally abandoned until 1860, when the researches of Boussingault and of Lawes and Gilbert deprived it of all foundation. Since then we have learned that the free nitrogen of the air can be made available for nutrition—not indeed directly by green plants themselves, but, as Berthelot and Winogradsky more especially have shown, by Bacteria in the soil, or, as apparently in the case of Leguminosæ, by Bacteria actually enclosed in the roots of the plants with which they live symbiotically.

We now turn from the nutritive or anabolic processes to those which are catabolic. The discovery of the latter, just as of the former, was arrived at by the investigation of the gaseous interchange between the plant and the atmosphere. In the eighteenth century Scheele and Priestley had found that, under certain circumstances, plants deteriorate the quality of air; but it is to Ingen-Housz that we owe the discovery that plants, like animals, respire, taking in oxygen and giving off carbon dioxide. And when Sénébier (1800) had ascertained for the inflorescence of *Arum maculatum*, and later de Saussure (1822) for other flowers, that active respiration is associated with an evolution of heat, the connection between respiration and catabolism was established for plants as it had been long before by Lavoisier (1777) in the case of animals.

Among the catabolic processes which have been investigated none are of greater importance than those which are designated by the general term *fermentations*. The first of these to be discovered was the alcoholic fermentation of sugar. Towards the end of the seventeenth century Leeuwenhoek had detected minute globules in fermenting wort; and a century later Lavoisier had ascertained that the chemical process consists in the decomposition of sugar into alcohol and carbon dioxide; but it was not until 1837–38 that, almost simultaneously, Cagniard de Latour, Schwann and Kützing discovered that Leeuwenhoek's globules were living organisms, and were the cause of the fermentation. Shortly before, in 1833, Payen and Persoz extracted from malt a substance named *diastase*, which they found could convert the starch of the grain into sugar. These two classes of bodies, causing fermentative changes, were distinguished respectively as *organised* and *unorganised* ferments. The number of the former was rapidly added to by the investigation more especially of the Bacteria, in which Pasteur led the way. The extension of our knowledge of the unorganised ferments, or enzymes, has been even more remarkable; we now know that very many of the metabolic processes are effected by various enzymes, such as those which convert the more complex carbohydrates into others of simpler constitution (diastase, cytase, glucase, inulase, invertase); those which decompose glucosides (emulsin, myrosin, &c.); those which act on proteids (trypsins) and on fats (lipases); the oxidases, which cause the oxidation of various organic substances; and the zymase, recently extracted from yeast, which causes alcoholic fermentation.

The old distinction of the micro-organisms as "organised ferments" is no longer tenable; for, on the one hand, certain of the chemical changes which they effect can be traced to extractable enzymes which they produce; and, on the other, as Pasteur has asserted, every living cell may become an "organised ferment" under appropriate conditions. The distinction now to be drawn is between those processes which are due to enzymes and those directly effected by living protoplasm. Many now definitely included in the former class were, until lately, regarded as belonging to the latter; and no doubt future investigation will still further increase the number of the former at the expense of the latter.

The consideration of the metabolic processes leads naturally to that of the function of transpiration and of the means by which water and substances in solution are distributed in the plant. This is perhaps the department of physiology in which progress during the nineteenth century has been least marked. We have got rid, it is true, of the old idea of an ascending crude sap, and of a descending elaborated sap, but there have been no fundamental discoveries. With regard to transpiration itself, we know more of the details of the process, but that is all that can be said. As for root-pressure, Hofmeister (1858–82) discovered that "bleeding"—as the phenomena of root-pressure were termed by the earlier writers—is not confined, as had hitherto been thought, to trees and shrubs; but the current theory of the process, allowing for the discovery of protoplasm and of osmosis, has advanced but little upon that given by Grew in the third book of his "Anatomy of Plants" (1675). Again, the mechanism of the transpiration-current in lofty trees remains an unsolved problem. To begin with, there is still some doubt as to the exact channel in which the current travels. Knight (1801–8) first proved that the current travels in the alburnum of the trunk, but not, he thought, in the vessels, for he found them to be dry in the summer, when transpiration is most active; a view in which Dutrochet (1837) subsequently concurred. Meyen (1838) then suggested that the water must travel, not in the lumina, but in the substance of the cells of the vessels, and was supported by such eminent physiologists as Hofmeister (1858), Unger (1864, 1868) and Sachs (1878); but it has since been strongly asserted by Boehm, Elfving, Vesque, Hartig and Strasburger that the young vessels always contain water, and that the current travels in the lumina and not in the walls of the vessels.

Now as to the force by which the water of the transpiration-current is raised from the roots to the topmost leaf of a lofty tree. From the point of view that the water travels in the substance of the walls the necessary force need not be great, and would be amply provided by the transpiration of the leaves, inasmuch as the weight of the water raised would be supported by the force of imbibition of the walls. From the point of view that the water travels in the lumina, the force required to raise and support such long columns of water must be considerable. Dismissing at once as quite inadequate such purely physical theories as those of capillarity and gas-pressure, there remain two theories as to the nature of this force which resemble each other in being essentially vitalistic, but differ in that the one involves pressure from below, the other suction from above. In the one, suggested by Godlewski and by Westermaier (1884), the cells of the medullary rays and of the wood-parenchyma are supposed to absorb liquid from the vascular tissue at one level and force it back again by a vital act at a higher level : this theory was disposed of by the fact that the transpiration-current can be maintained through a considerable length of a stem killed by heat or by poison. In the other, suggested by Dixon and Joly (1895–99), and also by Askenasy (1895–96), it is assumed that there are, in the trunk of a transpiring tree, continuous columns of water which are in a state of tensile stress, the tension being set up by the vital transpiratory activity of the leaves. Some idea of the enormous tension thus assumed is given by the following simple calculation relating to a tree 120 feet high. Not only has the liquid to be raised to this height, but in its passage upwards a resistance calculated to be equal to about five times the height of the tree has to be overcome. Hence the transpiration-force in such a tree must at least equal the weight of a column of water 720 feet in height ; that is, a pressure of about twenty-four atmospheres, or 360 lb. to the square inch. But there is no evidence to prove that a tension of anything like twenty atmospheres exists, as a matter of fact, in a transpiring tree ; on the contrary, such observations as exist (*e.g.* those of Hales and of Boehm) indicate much lower tensions. Under these circumstances we must regretfully confess that yet one more century has closed without bringing the solution of the secular problem of the ascent of the sap.

The nineteenth century has been, fortunately, rather more fertile in discovery concerning the movements and irritability of plants. But it is surprising how much knowledge on these points had been accumulated by the beginning of the century : the facts of plant-movement, such as the curvatures due to the action of light, the sleep-movements of leaves and flowers, the contact-movements of the leaves of the sensitives, were all familiar. The nineteenth century opened, then, with a considerable store of facts ; but what was lacking was an interpretation of them ; and whilst it has largely added to the store, its most important work has been done in the direction of explanation.

The first event of importance was the discovery by Knight, in 1806, of the fact that the stems and roots of plants are irritable to the action of gravity and respond to it by assuming definite directions of growth. Many years later the term "geotropism" was introduced by Frank (1868) to designate the phenomena of growth as affected by gravity, and at the same time Frank announced the important discovery that dorsiventral members, such as leaves, behave quite differently from radial members, such as stems and roots, in that they are diageotropic.

It was a long time before the irritability of plants to the action of light was recognised. Chiefly on the authority of de Candolle (to whom we owe the term "heliotropism"), heliotropic curvature was accounted for by assuming that the one side received less light than the other, and therefore grew the more rapidly. But the researches of Sachs (1873) and Müller-Thurgau (1876) have made it clear that the direction of the incident rays is the important point, and that a radial stem, obliquely illuminated, is stimulated to curve until its long axis coincides with the incident rays. Moreover, the discovery by Knight (1812) of negative heliotropism in the tendrils of *Vitis* and *Ampelopsis* really put the Candollean theory quite out of court ; and further evidence that heliotropic movements are a response to the stimulus of the incident rays of light is afforded by Frank's discovery of the diaheliotropism of dorsiventral members.

The question of the localisation of irritability has received a good deal of attention. The fact that the under surface of the

pulvinus of *Mimosa pudica* is alone sensitive to contact was ascertained by Burnett and Mayo in 1827 ; and shortly after (1834) Curtis discovered the sensitiveness of the hairs on the upper surface of the leaf of *Dionaea*. After a long period of neglect the subject was taken up by Darwin. The irritability of tendrils to contact had been discovered by Mohl in 1827 ; but it was Darwin who ascertained, in 1865, that it is confined to the concavity near the tip. In 1875 Darwin found that the irritability of the tentacles of *Drosera* is localised in the terminal gland ; and followed this up, in 1880, by asserting that the sensitiveness of the root is localised in the tip, which acts like a brain. This assertion led to a great deal of controversy, but the researches of Pfeffer and Czapek (1894) have finally established the correctness of Darwin's conclusion. It is interesting to recall that Erasmus Darwin had suggested the possible existence of a brain in plants in his "Phytologia" (1800). But the word "brain" is misleading, inasmuch as it might imply sensation and consciousness : it would be more accurate to speak of centres of ganglionic activity. However, the fact remains that there exist in plants irritable centres which not only receive stimuli but transmit impulses to those parts by which the consequent movement is effected. The transmission of stimuli has been found in the case of *Mimosa pudica* to be due to the propagation of a disturbance of hydrostatic equilibrium along a special tissue ; in other cases, where the distance to be traversed is small, it is probably effected by means of that continuity of the protoplasm to which I have already alluded.

Finally, as regards the mechanism of these movements, we find Sénébier and Rudolphi, the earliest writers on the subject in the nineteenth century, asserting, as if against some accepted view, that there is no structure in a plant comparable with the muscle of an animal. Rudolphi (1807) suggested, as an alternative, that the position of a mobile leaf is determined by the "turgor vitalis" of the pulvinus, and thus anticipated the modern theory of the mechanism. But he gives no explanation of what he means by "turgor" ; and the term is frequently used by writers in the first half of the century in the same vague way. Some progress was made in consequence of the discovery of osmosis by Dutrochet (1828), and more especially by his observation (1837) that the movements of *Mimosa* are dependent on the presence of oxygen, and are therefore vital. But it was not, and could not be, until the existence of living protoplasm in the cells of plants was realised, and the movements of free-swimming organisms and naked reproductive cells had become more familiar, that the true nature of the mechanism began to be understood ; and then we find Cohn saying, as long ago as 1860, that "the living protoplasmic substance is the essentially contractile portion of the cell." This statement may, perhaps, seem to put the case too bluntly, and to savour too much of animal analogy ; but the study of the conditions of turgidity has shown more and more clearly that the protoplasm is the predominant factor. The protoplasm of plant-cells is undoubtedly capable of rapid molecular changes, which alter its physical properties, more particularly its permeability to the cell-sap. It may be that these changes cannot be directly compared with those going on in animal muscle ; but if we use the term "contractility" in its wider sense, as indicating a general property of living matter, that the position of a mobile leaf is determined by Cohn's statement is fully justified. This is borne out by the observations of Sir J. Burdon-Sanderson (1882–88) on the electrical changes taking place in the stimulated leaf of *Dionaea*, and by Kunkel's (1878) corresponding observations on *Mimosa pudica* : in both cases the electrical changes were found to be essentially the same as those observable on the stimulation of muscle. We find, then, that the advances in Physiology, like those in Anatomy, teach the essential unity of life in all living things, whether we call them animals or plants.

With this in our minds we may go on to consider in conclusion, and very briefly, that department of physiological study which is known as the Bionomics of Œcology of plants. In the earlier part of the century this subject was studied more especially with regard to the distribution of plants, and their relation to soil and climate ; but since the publication of the "Origin of Species" the purview has been greatly extended. It then became necessary to study the relation of plants, not only to inorganic conditions, but to each other and to animals ; in a word, to study all the adaptations of the plant with reference to the struggle for existence. The result has been the accumulation of a vast amount of most interesting information. For instance, we are now fairly well acquainted with the adaptations of water-plants (hydrophytes) on the one hand, and of desert-plants (xerophytes) on the other ; with the adaptations of shade-plants and of those growing in full sun, especially as regards the protection of the chlorophyll. We have learned a great deal as to the relations of plants to each other, such as the peculiarities of parasites, epiphytes and climbing plants, and as to those singular symbioses (Mycorhiza) of the higher plants with Fungi which have been found to be characteristic of saprophytes. Then, again, as to the relations between plants and animals : the adaptation of flowers to attract the visits of insects, first discovered by Sprengel (1793), has been widely studied ; the protection of the plant against the attacks of animals, by means of thorns and spines on the surface, as also by the formation in its tissues of poisonous or distasteful substances, and even by the hiring of an army of mercenaries in the form of ants, has been elucidated ; and finally those cases in which the plant turns the tables upon the animal, and captures and digests him, are now fully understood.

Conclusion.

Imperfect as is the sketch which I have now completed, it will, I think, suffice to show how remarkable has been the progress of the science during the nineteenth century, more particularly the latter part of it, and how multifarious are the directions in which it has developed. In fact Botany can no longer be regarded as a single science ; it has grown and branched into a congeries of sciences. And as we botanists regard with complacency the flourishing condition of the science whose servants we are, let us not forget, on the one hand, to do honour to those whose life work it was to make the way straight for us, and whose conquests have become our peaceful possession ; nor, on the other, that it lies with us so to carry on the good work that when this Section meets a hundred years hence it may be found that the achievements of the twentieth century do not lag behind those of the nineteenth.

UNIVERSITY AND EDUCATIONAL INTELLIGENCE.

As was explained at length in our issue of March 22, in accordance with the new statutes of the University of London, a reconstituted Senate is to be elected shortly. The new Senate will be composed of the Chancellor, the Chairman of Convocation and fifty-four Senators, of whom sixteen are to be elected by Convocation. These sixteen members of the Senate will have, it would appear from the statutes, two distinct functions. They will, in addition to their general duties as senators, be required to form a special council for external students. This council, which is to consist of twenty-eight members of the Senate, will include the chancellor, the vice-chancellor, the chairman of Convocation, the sixteen senators elected by Convocation, and nine other members of the Senate elected by the Senate. Members of Convocation will, in a few days, proceed to choose their sixteen representatives ; and, not unnaturally, there is considerable diversity of opinion as to the suitability of the nominated candidates. Two rival associations have sprung up. One body of graduates insists that the duties to be performed upon the ouncil for external students should be considered of paramount importance in electing senators ; the other, that their responsibilities as members of the Senate should be kept continually in view, because the work of the new University as a whole, but more especially the development of its teaching facilities, is of the most pressing nature. While admitting the necessity of safe-guarding the interests of the external student, and of ensuring the high value of the degrees of the University, it is desirable that every possible means of improving the higher education of London should receive primary consideration. It would be nothing less than a calamity were Convocation to elect sixteen irreconcilables with no ideas outside that of introducing the peculiar, though somewhat circumscribed, needs of the external student into all deliberations of the Senate. It is therefore to be hoped that the common-sense which attended the election of their representative in Parliament will characterise the selection of the sixteen senators chosen by Convocation. It is easily possible to find members of the University who, while fully aware of the needs, and in sympathy with the aims of the external student, have also broad views as to the work of a great teaching University.

Dr. A. P. Laurie, lecturer in physics and chemistry at St. Mary's Hospital Medical School, has been appointed principal of the Heriot Watt College, Edinburgh.

Dr. Spencer W. Richardson, lecturer on physics at the University College, Nottingham, has been appointed principal and professor of physics at the Hartley College, Southampton.

The Birkbeck Institution, London, which has now completed seventy-seven years of educational work in the metropolis, commences its new session on Monday, October 1. The Institution has had many additions to its appliances in recent years, and the physical, chemical and metallurgical laboratories are now very thoroughly equipped. The day classes provide courses in chemistry, biology, physics and mathematics for the science degrees of London University. During the recess considerable additions and improvements have been made by the aid of a gift of 2000 guineas from Mr. F. Ravenscroft, to commemorate his completion of a membership of fifty years.

Addresses will be given at the opening of many of the metropolitan and provincial medical schools at the beginning of October. At Middlesex Hospital on October 1, Dr. T. Clifford Allbutt, F.R.S., will distribute the prizes gained during the previous year and deliver an address. At St. George's Hospital the introductory address will be delivered by Dr. Francis G. Penrose. At University College the session of the faculty of medicine will be opened by Prof. G. Vivian Poore ; the session of the faculty of arts and laws, and of science, will be opened with an address by Prof. F. W. Oliver on October 2. At St. Mary's Hospital the introductory address will be given by Mr. H. S. Collier. At St. Thomas's Hospital the session will open on Tuesday, October 2, when the prizes will be distributed by Sir William MacCormac. At the opening of the session at Charing Cross Hospital on October 2, Lord Lister will deliver the third biennial Huxley Lecture. The London School of Tropical Medicine will open on October 1, and the introductory address will be delivered by Sir William MacGregor, K.C.M.G., C.B., on Wednesday, October 3. At the London School of Medicine for Women the introductory address will be given on October 1 by Miss Aldrich Blake, M.S., M.D., after which the prizes for the past year will be distributed. At the Royal Veterinary College the introductory address will be delivered by Prof. McFadyean. The winter session at the University of Birmingham will begin on October 1 with an address by Prof. B. C. A. Windle. At University College of South Wales and Monmouthshire, Cardiff, the address will be delivered on October 1 by Sir John Williams. At University College, Liverpool, the Bishop of Liverpool will deliver an address on October 13 and distribute the prizes.

A summary of the scheme of work carried on by the Essex Technical Instruction Committee for the promotion of interest in the science of agriculture and other branches of knowledge bearing upon rural industries, has been prepared by Messrs. T. S. Dymond and J. H. Nicholas. The work is in every respect satisfactory, and should do much to broaden the views of the practical farmers of the county as to the value of agricultural education and experiment. Every year an educational excursion extending over several days is organised, the one this year being to Denmark to study dairy farms and dairying, high school and agricultural education, co-operation and organisation of agricultural industry there. Field experiments are carried out by arrangement with farmers distributed in all parts of the county, the advantage being that as demonstrations of the effect of manures, &c., they receive wider attention, and also that the experiments can be made on each of the different classes of land occurring in the county. Meetings of farmers are held in the experimental fields in each district at the season most suitable for studying the results of the experiments. The County Technical Laboratories at Chelmsford are now recognised as a centre from which information upon agricultural matters can be obtained. The advice of the staff is frequently sought on insect and fungoid pests, on difficulties met with in the dairy, &c., and their opinion asked on the value of foods and of fertilisers, and the best manurial treatment of land. As occasion arises, inquiries are undertaken on matters of agricultural importance, such as the chemical and physical effect of the salt water inundation upon agricultural land on the coast of Essex, and the best method for its amelioration. The agricultural work of the Essex Technical Instruction Committee is thus of the same character as that carried on by the Government

Agricultural Experiment Stations in the United States and elsewhere.

THIS is the time of year when prospectuses and calendars of Technical Colleges, Schools, and Institutes are received from various parts of the country in such numbers that it is impossible to do justice to them in a short note. Several publications of this character recently received must, however, be mentioned. The Northampton Institute, Clerkenwell, the principal of which is Dr. R. M. Walmsley, has greatly developed, and has commenced a set of day courses in mechanical engineering, electrical engineering and horological engineering. These courses have already been announced in NATURE, and their scope described. Other changes tending to the greater efficiency of the Institute have been introduced. A noteworthy point is that in many parts of the prospectus notes are given which should be of real value in making students understand what true education means, and in directing their energies in proper channels. The notes are in complete accord with rational methods of instruction.

THE Merchant Venturers' Technical College at Bristol has for many years been prominent among the the technical schools of the country. It aims at providing a sound, continuous, and complete preparation for an industrial career, and has developed with the times. Among recent improvements mentioned in the calendar we notice that a much larger physical laboratory has been equipped and will be opened this session, and also an additional special laboratory for heat and mechanical physics.

THE prospectus of the Municipal Science, Art and Technical Schools of Devonport has been received. Remembering the tendency of students to skim over many subjects, instead of concentrating their attention on a few, we are glad to see among the regulations of the school the following note :—"Students are strongly advised not to attempt more than three subjects, one of which should be practical geometry or mathematics, and they should consult the teacher as to the course of study most suitable to their profession."

THE Municipal Technical School of Manchester is one of the finest in the country, and its syllabus for the session 1900-1901 is proportionally attractive. The following extract from the syllabus shows the relation of the work of the school to that of a University College. "The chief object of the school is to provide instruction in the principles of those sciences which bear directly or indirectly upon our trades and industries, and to show by experiment how these principles may be applied to their advancement. The aim of the school is distinct from that of the University Colleges, inasmuch as it is designed to teach science solely with a view to its industrial and commercial applications, and not for the purpose of educating professional scientific men. It, however, offers to students of the University Colleges the opportunity of technical instruction in the industrial applications of certain branches of science."

THE Calendar of the Royal Technical Institute, Salford, contains much good advice to students, and many sound remarks upon objects and methods of study. In the day classes of the Institute, the number of hours per week allotted to each subject in the first year is as follows :—mathematics 6 ; general physics (including mechanics) 4 ; practical physics 3 ; electricity (theoretical) 2 ; electricity (practical) 2 ; theoretical chemistry 2 ; practical chemistry 3 ; practical, plane and solid geometry 2½ ; drawing (freehand, model, &c.) 3 ; workshop practice 2 ; English and French 4 ; total 33½. The second and third years' courses become more specialised according to the department which the student proposes to enter. There is no compulsory course of instruction for evening students. Students are free to select those classes which will help them to make progress in their particular trade or business. They are warned, however, against strictly confining themselves to such classes ; it is pointed out that if they desire to gain a thoroughly sound knowledge in technical subjects, the study of them should be preceded by several of the pure and applied sciences. Thus, for example, little real progress can be made in applied mechanics without a knowledge of theoretical mechanics ; or in machine or building construction without geometry ; and unless the student undergoes systematic instruction in mensuration, arithmetic and mathematics, he will derive very little benefit from such subjects as steam, machine design, physics, &c. Mathematics has been aptly termed the alphabet of science, and students should not fail to acquire mathematical knowledge if they wish to make satisfactory progress in science and technology. The work of an Institute inspired with this spirit cannot fail to be of value.

SOCIETIES AND ACADEMIES.

PARIS.

Academy of Sciences, September 17.—M. Maurice Lévy in the chair.—Remarks relating to the decomposition of nitric esters and of nitroglycerine by alkalis, and on the relative stability of explosive materials, by M. Berthelot. In certain cases, instead of the production of the alcohol and nitrate as in the normal reaction, an aldehyde is formed, together with some nitrite. The results of M. Leo Vignon upon the nitrocelluloses confirm these views.—On the nomographic resolution of the equation of the seventh degree, by M. Maurice d'Ocagne.—On the deformations of contact of elastic bodies, by M. A. Lafay. Spheres of bronze and steel were studied and the amount of compression under varying loads measured by optical arrangements analogous to the Fizeau apparatus for the measurement of the expansion of crystals. The application of the theory developed by Hertz showed differences between the calculated and observed values which increased with the radius of the sphere. Since this divergence might possibly be due to the mutual friction of the surfaces in contact, experiments were made with oiled spheres, but the results were not affected by the lubrication.—Action of iodine and yellow oxide of mercury upon styrolene and safrol, by M. J. Bougault. Styrolene with iodine and mercuric oxide yielded an addition product, not obtainable pure, but apparently $C_6H_5 \cdot CHI \cdot CH_2 \cdot OH$, from which phenylacetic aldehyde was obtained by the action of silver nitrate. Safrol gives a similar addition product, but no aldehyde could be obtained from this by the action of silver nitrate.—On the reduction of the nitrocelluloses, by M. Leo Vignon. It has been shown in a previous paper that the nitration of cellulose yields, not nitrocelluloses, but nitro-oxycellulose containing an aldehyde group. With ferrous chloride, these bodies are reduced, the nitro-group being eliminated but the aldehyde group left intact. With ammonium sulphide, the reduction takes place in a different manner, cellulose or hydrocellulose being produced, substances without reducing action.

CONTENTS.

THURSDAY, OCTÓBER 4, 1900.

A MANUAL OF THE ECHINODERMS.

A Treatise on Zoology. Edited by E. Ray Lankester, F.R.S. Part iii. The Echinoderma. By F. A. Bather, J. W. Gregory and E. S. Goodrich. Pp. ix + 344. (London : A. and C. Black, 1900.)

THE first instalment of the long-expected "Oxford Zoology"—first, that is to say, in order of publication—will be heartily welcomed as filling a distinct gap in zoological literature, and not of this country alone. During the latter half of the nineteenth century scientific literature has accumulated with such rapidity as to render it practically impossible for a zoologist at the present day to master thoroughly more than a limited part of his subject. To acquire a knowledge of the results gained in fields other than that which he has made his speciality, he must be dependent to a large extent upon the manuals and guide-books compiled by those who are sufficiently familiar with the latest discoveries in particular branches of zoology to be able to give a clear and critical account of the present state of knowledge in these departments. Nowhere is this necessity more strongly felt than in dealing with the Echinoderms, a group in which the student is confronted, on the one hand, with intricate morphological problems and with phylogenetic questions of a most puzzling kind ; and, on the other hand, with such a vast array of extinct types that the non-expert feels at once out of his depth when attempting to obtain an adequate knowledge of them. In the Pelmatozoa, practically half the phylum, we find a group of the greatest historical and phylogenetic importance, but one in which the existing forms teach us no more about the race in the past, and regarded as a whole, than do the modern Egyptians about the former dynasties whose remains are entombed in their land. The abundance of forms unearthed by the palæontologist has called forth a literature which exemplifies fully the danger of something like a deadlock in zoological science, as the result simply of its fertility. The student soon loses his way and finds himself struggling with a mass of hard facts and contradictory hypotheses, due on the one hand to the great diversity of form and structure in the objects themselves, and on the other hand to difficulties inseparable from the study of animals known almost entirely as fossils. Any one who has endeavoured, for instance, to gain an acquaintance with the structure and evolution of fossil Crinoids from the voluminous works of Messrs. Wachsmuth and Springer and other writers must have felt the urgent need for a guide and interpreter, failing whom it was necessary either to study deeply or to pass lightly by, to become an expert or to be content with ignorance. Yet no one with even a superficial acquaintance with the problems of Echinoderm morphology and phylogeny would willingly pass over the extinct forms, and least of all the more ancient Pelmatozoa, such as the Cystids and their allies, since it is obvious that here, if anywhere, is to be found in a concrete form the solution of many puzzles in the evolution of the phylum. Nowhere is palæontology, as a source of material evidence for theories of phylogeny, given so fair a trial as in the case of Echinoderms with

their complete skeleton and consequent abundance of well-preserved fossil types, and it must be conceded that palæontology, if it condescends to speak clearly, can give the only final judgment in questions of evolution and ancestral history.

For many reasons, therefore, a plain and intelligible account of the Echinoderms, and especially of the Pelmatozoa, by those who have an expert knowledge of them, both as fossils and as recent forms, was greatly to be desired, and in the present volume we have the first complete treatise that has been published under these conditions in any language. The intention of the authors is to give a systematic account of the Echinoderms, including every known genus, living or extinct, and at the same time to trace as far as possible the evolution and relationships of the forms comprised under each class or order, as inferred both from their structural affinities and from their succession in time. The aim in view is therefore to effect a happy combination of the older styles of systematic treatise with the modern methods of comparative morphology, developmental history and phylogenetic speculation.

An introductory chapter, giving a general description of the organisation and development of Echinoderms, from the pen of Mr. Bather, attempts to trace the origin of the characteristic radiate symmetry from the bilateral ancestor represented by the Dipleurula larva. Like most other recent authorities on the group, Mr. Bather supports the opinion that the radiate symmetry was acquired in all Echinoderms during an ancestral fixed stage, in which the animal fed by means of currents produced by cilia and directed along special food-grooves towards the mouth. In all animals with this mode of nutrition, which was probably the primitive method in each of the principal phyla, except perhaps the Cnidaria and the Arthropods, the general tendency of evolution is towards a reduction or loss of active locomotion, and frequently towards fixation, which certainly occurred in the Echinoderms. The common ancestor of the phylum was, in fact, to all intents and purposes, a Pelmatozoon, fixed by the aboral pole, the original right side of the bilateral ancestor, and with ciliated grooves converging to the mouth on the upper side. Amongst the Cystids ancestral stages are to be found showing the gradual acquisition of a radiate pentamerous symmetry, first by the food-grooves and then by the skeleton and other organs of the body, last of all by the gonads. The Pelmatozoa retained permanently this mode of life, continually adapting and perfecting their organisation to the necessities entailed by it. The other Echinoderm classes, on the other hand, grouped together as Eleutherozoa, and including the modern starfishes, sea-urchins and holothurians must have become free again at a very early period after the acquisition of radiate symmetry, giving up their method of nutrition by means of ciliary currents, and losing in consequence their food-grooves, which atrophy as such, the condensation of the nerve-plexus at the base of the grooves persisting, and being further specialised as the "superficial" nervous system. The holothurians were the first stock to become Eleutherozoic, radiate symmetry in their case not having extended to the gonads, as it has in the case of the starfish and urchins.

A A

The direct and positive evidence which is available may seem at first sight an insufficient foundation for the hypothesis of a Pelmatozoic ancestor of all Echinoderms, that is to say, a pre-Cambrian form in which the food-grooves initiated a radiate symmetry with which all other systems of organs gradually fell into line. But the necessity of some such assumption becomes irresistible when we realise by careful reflection the inadequacy of any other theory to account for the evolution of the characteristic radiate symmetry and the complete hold it has taken upon all organs of the echinoderm body. In the ontogeny of existing types it always seems as if it were the hydrocœle or water vascular system which actually set the tune to which all the other systems of organs dance, but it is difficult, if not impossible, to imagine clearly a course of ancestral evolution, limited and guided, as it must have been, by the necessities of the struggle for existence, in which the hydrocœle took the initiative in this respect, and did not itself follow the lead of some other system. The hydrocœle of the Pelmatozoic ancestor was probably at its first origin simply a compartment of the cœlom which had the function of furnishing tactile tentacles, formed as hollow outgrowths of the body-wall, in connection with the food-grooves. On this hypothesis it is easy to understand why the hydrocœle was the first system of organs to be affected by the radiate symmetry initiated by the primitive nutritive system, and consequently why, in the Eleutherozoa, after atrophy of the food-grooves, the symmetry should apparently start from the hydrocœle itself.

In the present volume the Pelmatozoa are also undertaken by Mr. Bather, who recognises four classes—Cystids, Blastoids, Crinoids and Edrioasterids. Another and perhaps more natural (*i.e.* phylogenetic) classification is hinted at (P. 39), but the arrangement quoted above is adopted as involving the least disturbance of established names and ideas. The Pelmatozoa occupy about two-thirds of the volume, and the treatment of this most difficult group cannot be too highly praised. An expert in this branch of zoology might perhaps find details to criticise or ideas with which to disagree ; the worker in other fields can only express his appreciation of the erudition displayed and the labour expended in setting forth the structure and evolution of this vast series of forms. In a group which is to a large extent represented by fossils, and in which so little material is available at the present day for the scalpel and the microtome, it is natural that less space and attention should be given to the anatomy and morphology of the soft tissues than to that of the skeleton and its never-ending complications of plates and ossicles. A simple Crinoid is taken as a type of Pelmatozoic organisation, and its anatomy is briefly described. One small point, at least, in this description is open to criticism. The author identifies Ludwig's blood-vessel and ring in the Crinoid as the "pseudhæmal," *i.e.* perihæmal, system (pp. 100 and 102). This seems to be an oversight, as elsewhere (p. 26) he states that this system "is so much reduced in Crinoidea that its existence is denied by some authors." Since the perihæmal system of canals, where well developed, as in the starfish, has been shown very clearly to be of cœlomic origin, it is impossible to identify

with it the Crinoid "blood-vessel," which has all the characters of the canals termed in this work the "lacunar" or "hæmal" system. If anything in the Crinoid arm is to be identified as perihæmal (a term we much prefer to pseudhæmal), then probably the sub-tentacular canals have the most right to this title, as being cœlomic canals which occupy approximately the same position as the perihæmal canals in the starfish, and which have also the same relation to the latéral nerve cords that the perihæmal canals have to "Lange's nerves." On this view we should have to regard the perihæmal system as a portion of the cœlom, which in the Pelmatozoa has reached only an incipient degree of specialisation, being in the region of the disc completely merged in the general body-cavity.

The account of the Holothurians has been written by Mr. E. S. Goodrich, who gives a useful summary of our present knowledge of the group. The remaining Eleutherozoa—Stelleroidea and Echinoidea—have been undertaken by Prof. J. W. Gregory, whose researches on these groups are well known to zoologists, and who gives us a most valuable and complete account of them. It is necessary, however, to point out a few errors or over-sights which have crept in, some of which are important, though they do not detract from the value of the work as a whole. On p. 261 it is pointed out that we are indebted to Sladen for a memoir on the aberrant form *Astrophiura*, and the work is quoted in due course amongst the literature of Stelleroidea, but nowhere else is any reference made to *Astrophiura* and its peculiarities ; it is omitted from the classification, does not appear in the index, and is, in fact, ignored altogether. The genus *Ophioteresis* is used as an argument for uniting the Asteroids and Ophiuroids on the ground that "the radial ambulacral vessels and nerve trunks lie in shallow grooves on the ventral surface of the anus" (p. 262 ; also pp. 270 and 274). The author gives no definite authority for this statement, but leaves us to infer that he obtains the fact from Bell's description of the genus. Bell, however, did not describe any such condition as that which Gregory dwells upon so often and makes the basis for such important deductions, and it is highly improbable that it occurs at all. It is much more probable the ambulacral vessels and nerve trunks pass in *Ophioteresis* through the aperture in the centre of the vertebral ossicle which Gregory figures plainly enough (Fig. xiv.), while maintaining a discreet silence about it. Finally, it must be mentioned that the peristomial plates in the Ophiuroid mouth skeleton are *not* "between the mouth frames and the buccal shields" (p. 264), but are above, *i.e.* to the aboral side of, the former, according, at least, to the careful descriptions and figures of Ludwig ; the "mouth frames" are between the buccal shields and the peristomial plates.

A conscientious reviewer does his best to find mistakes in the works submitted to his scrutiny and judgment. In the present instance it cannot be said that we have been very successful in our search, having regard to the size and scope of the work. In conclusion, we can but congratulate heartily the editor, authors and publishers on the very valuable treatise they have produced, a work which reflects credit on all concerned, and is a triumph for English zoology.   E. A. M.

THE BOTANY OF CAPTAIN COOK'S FIRST VOYAGE.

Illustrations of the Botany of Captain Cook's Voyage Round the World in H.M.S. "Endeavour" in 1768–71. By the Right Hon. Sir Joseph Banks and Dr. Daniel Solander, with Determinations by James Britten. Part I.: Australian Plants. 101 Plates, with descriptive letterpress. (London: Printed by order of the Trustees of the British Museum. All Booksellers. 1900.)

"BETTER late than never" may be said of the book the title of which is given above. It is a curious fact that the scientific results of several of the most important and most costly voyages of discovery, both English and foreign, have either not been published at all, or only in part, and in a fragmentary manner. Cook's first voyage is, perhaps, the most notable example of unfinished works of this kind in the history of British exploration. This is the more to be deplored, because collecting and methodical investigation were carried out on a scale previously unknown, and an immense sum was subsequently expended by Sir Joseph Banks in preparing the botanical results for publication. This is not the place to enter into the causes of the cessation of this part of the work; but it was not the only part that was long belated. It was not till 1893 that Captain Cook's own "Journal" was published, edited by Sir William Wharton; and three years later appeared Banks's "Journal" of that memorable voyage, edited by Sir Joseph Hooker. Although I have said "better late than never," it is obvious that the illustrations now in course of being issued have been, to some extent, forestalled, and the letterpress is historically interesting, rather than a contribution to science. According to the prospectus the complete work will comprise 800 plates; these will include a series illustrating the botanical collections of Cook's second voyage, when the Forsters, father and son, were the naturalists. Sydney Parkinson was the botanical artist on the first voyage, but he and the two other artists all died on the voyage, and their work was left in an unfinished condition. So much has been written about the plates now being issued and the desirability of their publication, that something superior to what they really are was probably expected by most people. Indeed it is difficult to suppress a feeling of disappointment. Compared with the botanical illustrations of other expeditions of discovery of a little later date, they are hard and unattractive, and floral dissections are almost entirely wanting. They lose, too, in effect, as they are transfers and not direct impressions of the original engravings on copper. The majority of the plates were engraved from drawings by F. P. Nodder, prepared from Parkinson's sketches and the dried specimens, and only the former name appears on the plates. Our remarks on this point, however, should be regarded in the light of explanation rather than criticism, because after all we must not forget that their publication has been delayed more than a century. Of course, it is highly regrettable that they were not published at the time, so that they might have been more fully utilised in the many publications which have appeared during the last century and a quarter on Australasian and Pacific Islands botany. A fact of great importance is that a comparatively small number of the plants here depicted had previously been figured. Mr. Britten has most con-

scientiously reproduced Solander's descriptions and remarks, even to the extent of palpable errors. Thus the locality Endeavour River is given throughout as Endeavour's River, and "petioli ½-uncialia," instead of unciales. But perhaps this course is more satisfactory than any attempt at improving the original ; and errors of the latter kind may be due to slips of the transcriber. The keenest reader may overlook false terminations in Latin descriptions, and the most ready writer is apt to make them.

On the other hand, our thanks are due to Mr. Britten for much valuable information, and the correction of many current errors. Doubtless when the time comes for the "Introduction," some account will be given of the countries or districts explored, and the botanical results summarised.

With regard to nomenclature, it is fortunate that, although the rule of priority has been strictly followed, there are few suppressions of familiar names ; but that is because there were few opportunities. Of course, the familiar names appear, but only as synonyms. Mr. Britten is an uncompromising disciple of the school of reformers, and he has been permitted to exercise his will in this national publication. Thus *Ionidium* becomes *Calceolaria;* and the calceolarias that everybody is familiar with have *Fagelia* for their generic name. *Cosmia* takes the place of *Calandrinia; Damapana* that of *Smithia;* and *Caulinia* that of *Kennedya*. The complications that such changes cause are almost interminable, as the revival of one name may affect half-a-dozen other well-established generic appellations. But this is not the place to discuss the question. Botanists will be thankful to the Trustees of the British Museum for this valuable addition to their pictorial books, which is at the same time a monument to some of the scientific pioneers in British exploration.

W. BOTTING HEMSLEY.

OUR BOOK SHELF.

Fancy Water-Fowl. By F. Finn. Pp. 45. Illustrated. (London : *Feathered World* Office, 1900.)

MR. FINN, especially to Indian readers, is such a well-known writer on popular ornithology in more than one journal that the reproduction of a series of his articles in book-form can scarcely fail to be welcomed by a wide circle. And in selecting ornamental, or "fancy," water-fowl as a subject, he has hit upon one which appeals to a large number of bird lovers, if for no other reason than the facility with which these handsome birds can be reared and kept in confinement, even when the available space is limited.

The author has confined himself, on the advice of a lady friend, to well-known species, and in the selection he has made he is, on the whole, to be congratulated. We should, however, have liked to see mention made of the so-called Coscoroba Swan of South America, on account of its very peculiar organisation, although we are well aware that, chiefly owing to its delicate constitution, it is seldom seen in European collections.

Both the illustrations and the text have been reproduced in their original guise from the *Feathered World*. With regard to the page plates there is considerable individual variation in their degrees of excellence, the figure of the Spotted-bill Duck, forming the frontispiece, being decidedly superior to that of Rosy-billed Pochards which comes later, the last-mentioned being somewhat coarse and blurred in outline. Indeed, we venture to think that if a second edition be called for it would be a decided improvement if the plates were photographed down to octavo size, while at the same time the text might be printed in larger type.

As it is, however, the book is decidedly attractive, and ought to prove indispensable to all breeders of ornamental water-fowl.

Catalogue of Eastern and Australian Lepidoptera Heterocera in the Collection of the Oxford University Museum. Part ii. Noctuina, Geometrina and Pyralidina. By Col. C. Swinhoe. Pterophoridæ and Tineina. By the Right Hon. Lord Walsingham and John Hartley Durrant. Pp. vi + 630; with 8 plates. (Oxford : Clarendon Press, 1900.)

THE first volume of this important work was published as long ago as 1892 ; it included the Sphinges and Bombyces ; and the second and concluding volume, which is nearly twice as thick as the first, has at length been issued.

A great number of *Lepidoptera Heterocera* (moths) were described by the late Francis Walker, not only from the British Museum, but from various private collections, chiefly from that of W. Wilson Saunders. After the death of the latter, large portions of his collection found their way into the Oxford Museum, and the types have now been carefully identified, and a considerable number figured. This is extremely important, as it will enable lepidopterists at a distance to identify species with more certainty than by descriptions alone ; and a figure also helps to fix the identity of a species in case the type should be lost or destroyed.

About 2340 species of moths are enumerated in the present volume, and we note that in addition to Walker's types many described by Mr. F. Moore and other entomologists are likewise contained in the Oxford Museum ; nor must we omit to mention that several new genera and species are described and figured by the authors of the Catalogue for the first time. However, the work is one which, notwithstanding its importance, appeals so exclusively to specialists that a more lengthy notice is hardly required in the columns of NATURE.

W. F. K.

Sir Stamford Raffles : England in the Far East. By H. E. Egerton, M.A. Pp. xx + 290. (London : Unwin, 1900.)

THIS volume, which is one of a series, entitled "Builders of Greater Britain," and edited by Mr. H. F. Wilson, does not call for much comment in a journal devoted to science. The author of the biography naturally deals mainly with Sir Stamford Raffles as an administrator in the Straits Settlements and the Malay Archipelago, and only incidentally, and that very briefly, refers to him as a zoologist. Raffles was, as everybody knows, one of the founders, and the first president, of the Zoological Society of London ; and his bust adorns the lion house of that society. Mr. Egerton, in narrating this fact, is chiefly impressed by "how much innocent pleasure this distinguished child-lover has given to countless thousands of children " by his successful efforts in this direction. He mentions, however, the collections which he took care to make, and which were largely reported upon by Dr. Horsfield. In those days much that was brought back from the East in the way of zoological specimens was quite new to science, and the animals had to have names given to them ; it is not such a great compliment as Mr. Egerton seems to think to name a species *Gymnura rafflesii,* after Sir Stamford. This compliment is usually paid to the capturer of a new form, and it is ridiculous to say that "Raffles' reputation in the scientific world is attested by the fact that the great French naturalist, M. Geoffroy St. Hilaire, described a new variety of animal under the specific name ' Rafflesii.'"

LETTERS TO THE EDITOR.

[The Editor does not hold himself responsible for opinions expressed by his correspondents. Neither can he undertake to return, or to correspond with the writers of, rejected manuscripts intended for this or any other part of NATURE. No notice is taken of anonymous communications.]

The Teaching of Mathematics.

PROF. JOHN PERRY has asked me to write something in criticism of the views he has lately expressed about the teaching of mathematics. I am inclined to ask, What is the use? He knows my views pretty well, and others too ; and those who don't can learn them if they want to by buying my books. That is the best way, as it brings in one-and-threepences, and so does some good. I think there is a great deal to be said on both sides, and that if you are a born logic-chopper you will think differently from Faraday. The subject is too large, and I will only offer a few remarks about the teaching of geometry, based upon my own experience and observations. Euclid is the worst. It is shocking that young people should be addling their brains over mere logical subtleties, trying to understand the proof of one obvious fact in terms of something equally, or, it may be, not quite so obvious, and conceiving a profound dislike for mathematics, when they might be learning geometry, a most important fundamental subject, which can be made very interesting and instructive. I hold the view that it is essentially an experimental science, like any other, and should be taught observationally, descriptively and experimentally in the first place. The teaching should be a natural continuation of that education in geometry which every child undergoes by contact with his surroundings, only, of course, made definite and purposeful. It should be a teaching of the broad facts of geometry as they really exist, so as to impart an all-round knowledge of the subject. It should be Solid as well as Plane ; the sphere and cube, &c., as well as the usual circle and square ; models, sections, diagrams, compasses, rulers, &c., every aid that is useful and practical should be given. And it should be quantitative as well. The value of π should be *measured* ; it may be done to a high degree of accuracy. So with the area of the circle, ellipse and all sorts of other things. The famous 47th. The boy who really measures and finds it true will have grasped the fact far better than by a logical demonstration without adequate experimental knowledge ; for it happens that boys, who are generally very stupid in abstract ideas, learn a demonstration without knowing what it is all about in an intelligent manner. It may be said by logicians that you do not *prove* anything in this way. I differ. It might equally well be said that you prove nothing by *any* physical measurements. You have really proved the most important part. What a so-called rigorous proof amounts to is only this, that by limitation and substitution, arguing about abstract perfect circles, &c., replacing the practical ones, you can be as precise as you please. Now when a boy has learnt geometry, and has become competent to reason about its connections, he may pass on to the theory of the subject. Even then it should not be in Euclidean style ; let the invaluable assistance of arithmetic and algebra be invoked, and the most useful idea of the vector be made prominent. I feel quite certain that I am right in this question of the teaching of geometry, having gone through it at school, where I made the closest observations on the effect of Euclid upon the rest of them. It was a sad farce, though conducted by a conscientious, hard-working teacher. Two or three followed, and were made temporarily into conceited logic-choppers, contradicting their parents ; the effect upon most of the rest was disheartening and demoralising. I also feel quite certain about the experiential and experimental basis of space geometry, though that opinion has been of slow growth. If I understand them rightly, it is generally believed by mathematicians that geometry is pre-existent in the human mind, and that all we do is to look at nature and observe an approximate resemblance to the properties of the ideal space. You might assert the same pre-existence of dynamics or chemistry. I think it is a complete reversal of the natural order of ideas. It seems to me that geometry is only pre-existent in this limited sense ; that since we are the children of many fathers and mothers, all of whom grew up and developed their minds (so far as they went) in contact with nature, of which they were a part, so our brains have grown to suit. So the child takes in the facts of space geometry naturally and easily. The experience of past generations makes the acquisition of present experience easier, and so it comes about that we cannot help seeing it. But it is all experience, after all; although learned philosophers, by long, long thinking over the theory of groups and other abstruse high developments, may perhaps come to what I think is a sort of self-deception, and think that their geometry is pre-existent in themselves, whilst nature's is only a bad copy. Like the old Indian pundit, whose name was something like Bhatravistra, who, after fifty years inward contemplation, discovered God;—where—it would not be polite to mention.　　　　OLIVER HEAVISIDE.

September 22.

The New Senate of the University of London.

IN your paragraph (NATURE, September 27, p. 543) on the new Senate about to be elected in the University of London, you have put the issue as it has occurred to me. I have not been able to give my support to either of the two bodies which have set their electoral machinery in motion, for the simple reason that neither of them has produced a list of names of candidates in which higher educational work is adequately represented. I thoroughly endorse your remark that "It would be nothing less than a calamity were Convocation to elect sixteen irreconcilables with no idea outside that of introducing the peculiar needs of the external student into all deliberations of the Senate."

The University may boast of the value of the degree; but this is only to say that as an organism its *cell*-life is strong. As an organism, however, its *somatic* life is weak; and the *summation and co-ordination of function* is the main idea for the new Senate of the University to keep before it, if the University is to be a factor of real power in our national and imperial life in the centuries to come. An experience as a teacher of over a quarter of a century (Wellington College and Nottingham) entitles me, I think, to speak on this matter.

Bishop's Stortford, September 28.　　　　A. IRVING.

The Peopling of Australia.

IN the issue of NATURE dated December 28, 1899, there appeared a notice of my book, "Eaglehawk and Crow," from the pen of Prof. A. C. Haddon. A copy did not reach me till the end of February, and for that and other reasons which need not be mentioned I delayed replying to the criticisms passed. With your kind permission I shall now endeavour to meet the principal objections raised to my work, with a desire of advancing, if even in a very small measure, our knowledge of Australian ethnology. All ethnologists are agreed upon the difficulty of the Australian problem, and no one who attempts to solve it will be surprised at their agreement.

I regret that, owing to my omitting to define my use of the term Melanesian, Prof. Haddon misapprehended one of my fundamental positions. In a note on page 5 I say, "Papuan is applied, not in its narrowest application (dark New Guinean), but as the equivalent of Melanesian, and is meant to include the Tasmanian aborigines, &c." From this Prof. Haddon inferred that I excluded the Papuans proper from my Papuan race. Nothing was further from my intention. I included them as a sub-race under the wider term Melanesian, as many writers have done, as even the latest writer on the subject, Deniker, has done in his "Races of Man," page 285, and elsewhere. The basis of my ethnological position may be thus represented:—

Papuan or Melanesian Race. { Papuan Proper.
　　　　　　　　　　　　　{ Malanesian Proper.
　　　　　　　　　　　　　{ Tasmanian Papuan. { Primitive Australian.
　　　　　　　　　　　　　　　　　　　　　　{ Tasmanian.

This classification underlies my whole book. I confess that I would now prefer to restrict the name Melanesian to the Melanesians proper as less liable to ambiguity, but in making Melanesian the general name I followed the lead of others much more competent than I am. That I recognised the narrower application of Papuan is evident from the above quotation from page 5, and such a passage as the following shows that I recognise Melanesians proper. "There are indications of groups of Melanesians having reached Australia on the eastern Queensland coast," page 73. Further, I invariably refer to

the Tasmanians as Papuans, with occasionally some such qualifying word as *primitive*.

My solution of the Australian racial problem having received the approval of Prof. Keane ("Ethnology," pp. 291-2), I may state it briefly here. The now extinct Tasmanians represent the primeval Australian aborigines. They were probably not a pure race, but embraced Negrito and Papuan elements. At the time of their arrival in Australia they probably occupied the islands to the north, and their congeners were the first to occupy Melanesia. Upon the primitive Papuans there was a strong graft of what, for want of a better name, and following the example of others, I have called "Dravidians," using this as a term of convenience to indicate likeness to the people of southern and central India. Then followed a further migration, in a desultory manner, of people of Malay stock; the precise locality whence these came is indeterminable, but I give evidence of distinctly Sumatran influence in the north-west. Concurrently, or subsequently, companies of Melanesians proper and Papuans proper have mingled with the Australians on the north and east of Queensland.

The two earliest immigrations entered Australia from New Guinea or neighbourhood. The population became distributed by streams diverging from the base of Cape York Peninsula.

When allowance has been made for Prof. Haddon's misconception of my use of the term Papuan, there is little more in his notice that needs to be referred to, as he concedes my main positions.

Mr. S. H. Ray, having been invited by Prof. Haddon to offer observations upon the linguistic part of the work, criticised it in a manner which seems to be unnecessarily caustic, fastening attention upon petty points which he objected to, and ignoring the main issues. He begins by asserting that I belong to a school of Australian pseudo-philologists who believe that a likeness of words in sound and meaning is a proof of common origin, and this in spite of my explicit disavowal of such a position, and my exposure of the unsoundness of it on page 44, where I show that on such a principle the Australian languages might be derived from the English. Having made so fair a start with a *petitio principii*, by gross misrepresentation of my statements, he proceeds to buttress his assertion. "We are asked to believe," he continues, "that Malay immigrants, presumably from various parts of the Archipelago, entered Australia from the north, and wandering about the interior, scattered 'astonishing relics' of the speech of one of these sections all over the island continent." He is not asked to believe any such ridiculous nonsense, and it is singularly disingenuous to say so in the face of my sober statements on page 57, "Either the Malay inroad, if made at the north, took place in long past ages, or now and again parties of Malays, either from choice or necessity, landed and became naturalised at various spots on the east, north and west, and modified the speech of the people, first immediately round them, and then landwards"; and on page 61, "This last influx (the Malay) may have come by several little rills, entering at places widely apart and gradually losing themselves in the life-lake." The "wandering about the interior" is a pure invention of Mr. Ray's. When the universal practice of exogamy is taken into account, along with the general pressure and movement of people, language, customs, &c., from north to south, my theory of Malay influence on the Australian people and language will be accepted as reasonable by unprejudiced minds. In the *Journal* of the Anthropological Institute for 1894-5, in a paper on "The Languages of British New Guinea," this very Mr. Ray uses language, and language alone, as a basis of classification for proving racial distinctions and affinities and movements. I do not say that this was an improper use of the linguistic argument, but it differs from mine in this, that I rarely rely upon language alone. I back up the linguistic evidence by that of other ethnological characters.

To come to particulars: my identifying an explicit type of Australian words for "Head" with the Malay "Kapala" is objected to because "Kapala" is a word of Indian origin. But the word has been current in Malay for five or six centuries, and is in use in that very part of Sumatra from which, according to my hypothesis, came the authors of the best Australian rock-paintings. It is quite possible that I may be mistaken in relating certain Australian words to "Kapala," but Mr. Ray's ground of objection has little or no cogency.

"Mama" and "bapa" are terms for mother and father of

wide currency in Australia. The former I connect with early Papuan influence, the latter more especially with Malay. He objects on the ground that connectives of "mama" are more common in the Malay districts of the Eastern Archipelago than "bapa." But in Australia the word "mama" occurs only in the extreme S.W. and S.E., among the purest modern representatives of the earliest occupants of Australia, thus affording ground for the conclusion that the term "mama" preceded the term "bapa." The wide prevalence of "bapa" forms in other countries I myself refer to on page 44 ; but the question is, What race was specially influential in giving such forms currency in Australia? As against my position it is not sufficient for Mr. Ray to say that "mama" variants are of more frequent occurrence in Malay centres than "bapa" variants, he will have to prove that the words of "mama" type are not adopted words in Malay, were not earlier in use in the East Indian Archipelago than the other type of words, and are not more markedly Papuan than these.

Mr. Ray complains that individual words in the languages quoted "are not always accurately given or properly understood." This may be ; but like himself I am dependent upon my authorities. When further on he suggests that I might have attempted uniformity of spelling in the foreign words, he is like the "children sitting in the market-place." A desire to be free from suspicion of tampering with my borrowed materials kept me from applying to them a uniform system of spelling, and evidently my caution was not unnecessary.

Mr. Ray's harshness is all the more indefensible since he himself falls demonstrably into error on the very point upon which he proposes to correct me. As proof of my mistaking the form and meaning of words, he cites the New Guinea numerals (pp. 165, 169). He says they are explainable compounds. He does not, however, attempt to explain them. But even if they are, this fact alone does not prove that they could not be transmitted to Australia. One feature about Australian numerals is clearly shown in my tables, viz. that they occur geographically in lines that converge on Cape York Peninsula. Some of them are most certainly identical with forms in use on Saibai Island on the New Guinea coast, *e.g.* "woorba," with variants traceable along the Queensland coast from a point about 1000 miles S.E. of Cape York, and represented in the form "warapune" Prince of Wales Island, "woorapoo" at Warrior Island, and in "urapon" at Saibai. One numeral, "luadi" (two), used by the Kalkadoon tribe, whose territory is about 150 miles south of the Gulf of Carpentaria and some 600 miles S.W. of Cape York, is a Melanesian numeral. It did not fly that distance through the air. And there is just as little doubt about the identity of at least several of the other Australian numerals with the New Guinea forms to which I have related them. My table of numerals was not formed rashly. It will be worth Mr. Ray's while to examine and test it carefully. The convergence of numerals upon Cape York Peninsula is only one striking illustration of what occurs in the case of other words, and words thus traced to the very coast must have come from New Guinea and adjacent islands.

As another example of my misunderstanding words, Mr. Ray refers to my "ori kaiza," pp. 66–7. He says: "Ori kaiza" is mongrel, "ori" (bird) is Toaripi, Papuan Gulf, and "kaiza" (big thing) is Saibai, West Torres Straits. This is, for himself, a most unfortunate example. Although he speaks so authoritatively, he is utterly at fault. Sir W. MacGregor's reports give "uroi" (bird) as a Saibai word ; and even Mr. Ray himself, in his paper already quoted from, gives "urui" as Saibai for 'bird,' a fact he appears to have forgotten. Besides, in the "Voyage of the *Rattlesnake*," containing vocabularies obtained in 1849 from a white woman who had been among the natives for four and a half years, McGillivray gives "wuroi" as a Cape York word, and "ure" as a Kowrarega word, both meaning bird. Mr. Ray's assertion, therefore, that "ori kaiza" is mongrel, is contrary to fact, and my tracing of this compound word across Australia from S.W. to N.E., and to the New Guinea coast, is not in the least invalidated by Mr. Ray's groundless and inconsistent statement that the word is mongrel.

Mr. Ray characterises my comparison of Australian words with Malay and New Hebridean as "absurd and misleading." This may be so to one with his pre-conceptions, but certainly not from the point of view which I have taken of the relation subsisting between the races whose words are compared. If the Tasmanians were the original occupants, both of Australia and the greater part of Melanesia, which is my hypothesis, it

is not unreasonable to suppose that certain radicals would be common to Tasmanians, Australians and Melanesians proper. And further, one of the most competent authorities on the Oceanic languages, the Rev. Dr. MacDonald, of Efate, is of opinion that Malay, Melanesian and Polynesian are sister languages derived originally from one mother tongue. If he be right, there would be no absurdity in affirming analogies between Malay and New Hebridean words. But I have included the Malay with a note almost like an apology. I only cite eight Malay words, and the only conclusions I draw concerning the Malay in this connection is "The terms for father, skin, are the same in Malay, Australian and New Hebridean" (page 156).

I would have liked to have shown that the Melanesians proper have had much more influence upon the Australians than Mr. Ray seems to have any conception of, but I have already taken up so much space that I must content myself with saying that this proposition can be successfully maintained, and with your indulgence I hope in a future letter to make good my words. In conclusion, I would just say that I welcome fair and sound criticism based on accurate knowledge for its influence in promoting truth, but mere fault-finding and ridicule can benefit neither authors nor readers. One sentence from my reviewer in the *Saturday Review* may not be out of place here :—"If Mr. Mathew has not proved his theories to the satisfaction of all his readers, it is not from lack of knowledge or scientific methods, but from the imperfection of his materials."

Coburg, Victoria, August 16.   JOHN MATHEW.

THE PRESERVATION OF BIG GAME IN AFRICA.

PAST experience in America and South Africa shows how rapidly the teeming millions born of the soil may be shot out. Writers of half a century ago describe on the veldt in South Africa a paradise of varied life, which is now irretrievably lost, through the carelessness and wastefulness of white men. Some species have absolutely disappeared, never to be seen again on the face of the earth. Others are so scarce that it is doubtful whether their power of reproduction can save the race. The fact that an International Conference, attended by delegates from Germany, France, Italy, Portugal and the Congo Free State, on the subject of the preservation of the game from destruction in Africa, met recently in London, under the auspices of our Foreign Office, shows that a widespread interest is now taken in this subject. Let us see how the matter stood previous to the meeting of the Conference—at least as regards British territory.

Excluding the settled parts of South Africa which were outside the purview of the Conference, we may observe, in the first place, that our Foreign Office appears to be thoroughly alive to the urgency of the question in those territories under their jurisdiction. They had enacted game regulations which ought to have been effective for their purpose. A 25*l.* license was imposed upon strangers, and one of 3*l.* upon residents and officials, as a necessary condition of shooting, while the licensees were limited to two specimens in the case of elephants, rhinoceros, hippopotamus, buffalo and giraffe. Fines up to 500 rupees, and imprisonment for two months, were the maximum penalties. Above all, Reserves for the game were defined. Similar regulations to the above were in force in British East Africa ; but let us confine our attention to British East Africa as an example with which I am familiar. Here, on the best feeding grounds, there are vast herds of wildebeest, hartebeest, impala, zebras, gazelles of several species, and in lesser numbers waterbuck, giraffes and rhinoceros. All these, and others, may be seen from the windows of the train as it traverses the new Uganda railway, which has now been constructed to a point about two-thirds of the way to Lake Victoria. The Kenia province, which is about 100 miles by 40, has been constituted a game Reserve. Other Reserves have been established in Uganda and British Central Africa. Each of the Foreign Powers engaged in the Conference have bound themselves to provide similar Reserves where they have not already done so, and to maintain them as such with strictness, and much depends upon the interpretation of that word. Now this is just what we had not until recently done in the case of the Kenia Reserve.

One of the regulations provides "That public officers may be specially authorised to kill, &c." in that Reserve. Unfortunately the words "may be authorised" in this regulation were interpreted by many of the Protectorate and railway officers stationed at Nyrobe, Kikuyu and elsewhere as "are authorised," and thus as making them free of the Reserve. This laxity of interpretation had a tendency to spread, and large quantities of game were at first killed there after the arrival of the railway. A Reserve is no true Reserve which is subject to personal exceptions, and in the circumstances which I have detailed was a delusion and of little value. We may rest assured that, now that this defect has been pointed out, the Foreign Office will not be backward to remedy it; and even if it were not so, they are under an international obligation to make the Reserve a reality. We may, therefore, confidently expect that the words I have quoted, which admit of a serious leakage, will disappear. It must not be thought that the officers to whom I have referred are indifferent to the preservation of game. It is in their interest, above all others, that these regulations should be maintained, and I am confident that the good sportsmen, of whom there are many among them, are anxious to be protected against those who cannot be so described. Nothing can be stronger than their reprobation of the worst transgressors, as, for instance, of a gentleman wearing her Majesty's uniform, who, I was told, killed approximately a score of wildebeest in a day, and left them rotting on the ground. The author of this disgusting butchery was brought to book, but he passed into Uganda, and thus, sheltered by a technicality, escaped the payment of the fine. It is to be hoped that the long arm of the autocratic committee which governs both territories will ultimately reach this glaring offender.

It remains to be considered in what respects the recommendations of the Convention will strengthen the game laws in their present form. The principal recommendations of the Conference may be summarised as follows. A special and select list of animals are to be absolutely protected at all times. Another schedule comprises the species which are to receive protection for immature animals and breeding females. The sale of tusks of elephants weighing less than eleven pounds is forbidden, and finally each Power undertakes to establish adequate Reserves and to protect them from encroachment. It will be seen that these recommendations impose upon them certain obligations, and we may thus expect that the new regulations for the British territories will include a schedule of animals as sacred from molestation as the bulls of Apis. The giraffe, eland and buffalo are, at any rate, among those which are sure to enjoy this royal distinction. It is a little difficult to see why vultures, owls and rhinoceros birds, which are exceedingly useful, but are not sought for food, should have been added by the Conference to such a distinguished list. The second list, of which the breeders and young are to be protected, will doubtless include such animals as rhinoceros, hippopotamus, waterbuck, sable, greater and lesser koodoo. The importance of this will be seen when it is remembered how slowly these larger animals breed. Apart from these restrictions a limit will doubtless be placed on the numbers of all the game animals allowed to be killed under each license, a high limit being given for the common species, and a much lower one, probably not exceeding two specimens, for those in most danger of disappearance.

Infractions of this rule may be somewhat difficult to

detect, but every licensee, at the expiry of his license, should be required to furnish a return of what he has killed. This would impose a certain restraint on thoughtless sportsmen, and when the returns are collated would form a basis for a valuable tabulation of the numbers of each species killed from year to year, and serve as an indication of the increase or diminution of any species in a given area. A small export duty on skins and horns would be a useful assistance to such a return.

The maintenance of Reserves is of the highest importance for the preservation of the various species. In my opinion, the position and boundaries of the Kenia Reserve, which is perhaps the most important of all, should be reconsidered. These boundaries were selected because they happened to be the defined limits of a Province, and not because they represented the real needs of the game. A large portion of the area is densely populated and cultivated. Another considerable area is at a high elevation and covered with forest which harbours some elephants; but is of no use to the great families of grass feeders, such as the zebras, the numerous kinds of antelope and gazelle, the rhinoceros and ostrich. The great bulk of these are confined to the grass plains along the Athi River, and unfortunately its left bank only is within the Reserve. This feeding area is thus but a small fraction of the whole Reserve, and is quite inadequate to feed the vast herds; nor does it, as a matter of fact, cover their frequent migrations in search of fresh grazing, which extends to both sides of the Athi, and southwards to the plains of Kilimanjaro. The limited belt of grazing ground within the Reserve has been still further curtailed by the location of the important railway centre of Nyrobe in the midst of it, since the Reserve was constituted. This will necessarily drive the game from that part of the protected plains. It is therefore desirable that the boundaries of the Reserve should be reconsidered by competent officers on the spot, not forgetting the important assistance which would be rendered by the railway in safeguarding and watching it, provided it traverses it or skirts its boundary on one side.

Then as to the difficult question of elephants, difficult because of the high money value of their tusks. I am personally opposed to the destruction of elephants at all, on the ground that, valuable as they are for their ivory, that will soon come to an end at the present rate of destruction, and that they might be still more valuable as weight-carriers. That is, perhaps, a counsel of perfection, but that they require some far more effective protection is obvious to every one who has studied the subject. Recently an Englishman sold in Mombasa the produce of his trip in ivory for 8000*l.* The hundreds of elephants necessary to produce this amount were, of course, not in the main killed by his own rifle. Some of the ivory may have been bought, but numbers of native hunters were said to have been hired for this purpose and attached to his staff, and were sent far and wide over the country. Thus this caravan must have left a broad trail of destruction for hundreds of miles. When the wealthy and powerful set such an example, how can the law be enforced against those who have the excuse of poverty. It is to be hoped that the Foreign Office will be able to devise means for the arrest of wholesale destruction like this. Although the Convention has not recommended it, is it too much to hope for the imposition of an adequate export duty, uniform at all the ports of exit, to whatever Power they may belong, and the total prohibition of the export of cow ivory?

The question of bringing resident natives under the prohibition which extend to Europeans requires to be carefully weighed. In my opinion, it is neither possible or just to stop their hunting so long as they are confined to their primitive weapon, the poisoned arrow. From time immemorial the destruction caused by the indigenous inhabitants has not appreciably diminished the stock. The land and the animals upon it are their birthright, and to interfere with it would surely cause trouble. We are not bound, however, to furnish them with civilised weapons, and every precaution should be taken to prevent their obtaining them.

Finally, the best of rules are useless without two things—a sound public opinion among the resident whites whom they chiefly affect, and a firm and knowledgable man to carry them out. The first exists, and I am convinced is on the increase. How should it be otherwise, unless one presupposes the most shortsighted selfishness? As to enforcing the rules, that which is the business of several officials, all of whom are engaged in office work, is practically no one's business. Let there be one man on the spot—that is to say one in each great game district, and especially in each Reserve—whose duty it is to know and to act.   E. N. BUXTON.

NOTES.

As was announced in our last issue, many of the medical schools in London and elsewhere were re-opened this week, and addresses were delivered by well-known medical men and men of science. At the Charing Cross Hospital Medical School the third Huxley Lecture was delivered by Lord Lister on Tuesday.

A COURSE of twelve "Swiney" lectures on "Extinct and Persistent Types" will be delivered in the lecture theatre of the Victoria and Albert Museum, South Kensington, by Dr. R. H. Traquair, F.R.S., on Tuesdays, Wednesdays, and Fridays from October 9 to November 2. No charge is made for admission to the lectures.

At the meeting of the Royal Photographic Society to be held on Tuesday next, October 9, the President will deliver his annual address, and present the medals awarded at the Society's Exhibition.

The Lettsomian lectures will be delivered before the Medical Society of London in March and April next, and the oration will be given in May by Mr. F. Richardson Cross.

THE seventeenth annual meeting of the Association of Official Agricultural Chemists is to be held at the Columbian University of Washington, commencing on Friday, November 16 next.

THE fourteenth International Medical Congress will be held at Madrid early in 1903, under the presidency of Prof. Julien Calleja.

THE annual "Fungus Foray" of the Essex Field Club will take place on Saturday next from High Beach, Epping Forest. The prospective arrangements of the club include the opening of the Essex Museum of Natural History by the Countess of Warwick, on the 18th inst. The scientific winter evening meetings will be resumed in the Physical Lecture Theatre of the West Ham Technical Institute on October 27.

Science announces that Prof. H. T. Todd, having reached the age limit, has retired from the Directorship of the U.S. Nautical Almanac. Prof. S. J. Brown, astronomical director of the U S. Naval Observatory, has undertaken the duties of the office.

ACCORDING to the *Lancet*, a scheme has been sanctioned by the Charity Commissioners by which 644*l.* left to the Royal College of Surgeons of England in 1884 will be devoted to providing every four years a "Cartwright Medal" for an essay on dental surgery. The medal will be accompanied by an honorarium.

THE gold medal of the American Philosophical Society, known as the Magellanic, will be awarded in December next for the best discovery or most useful invention in the physical sciences brought before the Society before November 1.

PROF. F. KLEIN has been awarded 800 marks by the Göttingen Society of Sciences for his Mathematical Encyclopædia, and the same society has awarded 500 marks to Prof. Wiecherts for that worker's seismological recording instruments.

IN addition to the medals and prizes given for communications discussed at the meetings of the Institution of Civil Engineers in the past session, the Institution has made a number of other awards in respect of other papers dealt with during the same period, *e.g.* a George Stephenson medal and a Telford premium to Mr. L. F. Vernon Harcourt, and Telford premiums to seven other gentlemen. For students' papers, the James Forrest medal and a Miller prize have been awarded to Mr. C. B. Fox, the James Prescott Joule medal and a Miller prize to Mr. J. W. Smith, and Miller prizes to four other students. The council have nominated Mr. R. F. Whitehead to the Palmer scholarship at the University of Cambridge in succession to Mr. A. H. Kirby.

A NEW technical school is to be erected in Belfast on a portion of the grounds purchased from the Royal Academical Institution, and a principal is to be appointed shortly at a salary of 600*l.* per annum, whose experienced practical advice will, it is hoped, be of much value in making the interior arrangements of the building. and in organising the work of the institution while the building operations are in progress.

PROPERTY valued at upwards of 200,000 dollars has been left by Mr. Charles H. Smith for the maintenance of the botanical specimens in the city park of Providence, Rhode Island.

WE have had pleasure on more than one occasion to refer to the good work that is being done to the cause of scientific education by the Essex Technical Instruction Committee, and are glad now to call attention to two new courses of lectures that are about to be inaugurated by the committee. A first-year's course of instruction in botany for teachers will commence at Chelmsford on Saturday, October 6, and will be continued on successive Saturdays until about the middle of May, 1901, and an elementary course of practical instruction in dairy bacteriology will commence on Thursday, October 11, and will be continued on ten consecutive Thursdays.

THE *Windward*, according to *Science*, was expected to reach St. John's by about the middle of September, but a short delay would not be surprising as the vessel started late, owing to some difficulty with the machinery, and was subsequently delayed by ice along the coast of Labrador. The arrival of the steamship is awaited with interest and some anxiety, as news will be brought, not only of the return of Peary, but also of Captain Sverdrup and Dr. Stein. The former has the *Fram* provisioned for five years, with a crew of twelve men. He planned to round the northern boundary of Greenland and to make his way down its unknown east coast to Cape Bismarck. It is said that the expedition under Dr. Robert Stein of the U.S. Geological Survey, who is accompanied by Mr. Leopold Kann, of Cornell University and Mr. Samuel Warmbath of Harvard University, was poorly equipped and left in a dangerous position. Lieut. Peary himself expected to establish his last depôt at Cape Hecla, the most northerly point of Grinnell Land just beyond the 82nd parallel, whence he intended to advance with Eskimo and sleds as far north as possible.

AT a meeting held on June 12 last at the University o. Melbourne, it was unanimously resolved to form a society to be called the Society of Chemical Industry of Victoria, the objects for the establishment of the society being : (a) to afford its members opportunities of meeting and discussing matters connected with applied and industrial chemistry ; (b) generally to advance the cause of chemical industry in Victoria. It was further resolved that meetings shall be held at monthly intervals during the greater part of the year, at which papers on special branches of industrial chemistry are to be read and discussed. At a subsequent meeting, Prof. Orme Masson was elected president. The first paper to be read before the Society was one by Mr. D. Avery on September 4, on the cyanide process for gold extraction. The Society has made an encouraging start, as up to the end of August 118 members had been enrolled. We trust a lengthy and prosperous career lies before the new arrival.

THE sixteenth session of the Queensland branch of the Royal Geographical Society of Australasia was inaugurated at Brisbane on August 17, when a paper was read by the secretary—Mr. J. R. Thomson—on the geographical evolution of the Australian Continent. In connection with the society it has been decided to award a medal annually for the best and most scientific paper on some subject dealing with Queensland, and a fund for this purpose has been opened.

THE seventy-second annual meeting of the Association of German Naturalists and Physicians opened on September 17 at Aix-la-Chapelle with an attendance of about two thousand members. Of the thirty-eight sections, seventeen are devoted to such subjects as natural history, geology, geography, education, &c., the remaining twenty-one dealing with all the special subjects of medicine, including balneology, accidents, history of medicine and medical geography, and finally veterinary matters. A special correspondent of the *British Medical Journal* states that at the opening meeting the usual speeches of welcome were delivered by the Mayor and others, and the introductory addresses this year were by arrangement devoted, not only to giving a retrospect of the subject, but also to a sketch of its development during the nineteenth century. Dr. J. H. van 't Hoff spoke on the development of the exact natural sciences (natural history, chemistry and allied subjects). Dr. G. Hertwig delivered an address on the evolution of biology, in which, after relating anatomical discoveries, he came to the large question of the natural origin of the organic world. He considered that theories as to inheritance and natural selection still rested on the uncertain basis of hypothesis. He pointed out, however, that the difficulty arose from the absence of sufficient prehistoric records, and expressed his agreement with the opinion of Huxley that Darwin's teaching as to evolution will survive, apart from his principles of selection. Prof. Naunyn gave an address on the evolution of medicine, connecting the progress of the science with the names of Schwann, Pasteur, and Lister. The fourth and last address was given by Prof. Chiari, whose subject was the evolution of pathological anatomy.

THE British Mycological Society and the Cryptogamic Society of Scotland held a most successful meeting at the Boat of Garten from September 17 to 22. Various portions of the old forests of Rothiemurchus and Abernethy were worked from day to day, and a rich collection of scarce Hydnei and Cortinarii was secured. Prof. Marshall Ward (President of the British Society) gave an address, entitled "Nutrition of Fungi," and also contributed a paper on "Naematelia." Exeter was selected as the centre for next year's foray in the last week in September, and Prof. Marshall Ward was re-elected President.

AT the annual meeting of the Hull Scientific and Field Naturalists' Club, on September 26, an active and successful year of work was reported. A committee has been formed to work in connection with the National Trust for Places of Historic Interest or Natural Beauty; several important "finds" have been made during the weekly excursions; an exhibition of local natural history, geological and archæological specimens

has been held in the Technical Schools; and the Club has become a Corresponding Society of the British Association.

EVERY one interested in astronomy will welcome the new publication *Astronomischer Jahresbericht*, the first volume of which, for the year 1899, has recently made its appearance. This important yearly volume is published by Herr Walter F. Wislicenus with the aid of the *Astronomischen Gesellschaft*, and printed in Berlin (Druck und Verlag von Georg Reimer, 1900, pp. 536). The object of this volume is to present to astronomical readers a brief summary of the contents of every publication, whether it be in book, article or pamphlet form, which treats of any matter connected with theoretical or practical astronomy, or with researches in astrophysics. The project is a great one, and with careful attention could be carried out successfully. This, the initial volume, reflects great credit on Herr Wislicenus, who, although associated with five other workers, seems to have laboured nobly and undertaken the greater part of the volume. The subject-matter is divided into four main sections, namely :—General and historical ; astronomy, which includes spherical, orbit determinations, celestial mechanics, instruments and methods of observation, and, lastly, observations ; astrophysics ; and geodesy and nautical astronomy. The work is made complete by an excellent table of contents, and an index of names and the full titles of works referred to are given in each case. The author hopes that for future volumes he will have the help of all well-wishers of this work, and that such help will take the form of either references to published works or the works themselves, especially when they appear in transactions of societies which are published too late for insertion in the yearly volume, or other publications which are not specially devoted to astronomical matters. A glance at the present volume is sufficient to show the utility and value of this work, and it should be found in every astronomical observatory and laboratory.

QUOTING from the *Botanical Gazette*, *Science* says that the private herbarium of Mr. Harry N. Patterson, of Oquawka, Illinois, containing about 30,000 sheets, has been secured by the Field Columbian Museum, and will be installed as promptly as careful cataloguing will admit. The botanical department of the museum is, says our contemporary, to be congratulated upon this accession of one of the notable private herbaria of the country ; one that will add a complete collection of Pringle's Mexican plants to its already excellent representation of the flora of that region and the Antillean islands. Mr. Patterson's herbarium is more or less contemporaneous with that of the late Mr. Bebb, which the museum secured some three years ago, and as Mr. Patterson made it his aim to secure a complete series of the species of North America, its addition to the collections of the museum will be of great value to botanical students and specialists in the west.

THE Royal Italian Institute of Military Geography has thoroughly revised the old map of the region round about Vesuvius, issued by the institute in 1876, on the scale of 1/10,000. It has also completed a new plan in relief of the cone of Vesuvius which has been subject in recent times to considerable changes in its configuration owing to the repeated eruptions. Both map and plan have been prepared under the direction of Prof. Matteucci, who for years has made a study of Vesuvius. The correction of the map has been rendered necessary, not only by the eruptions, but also by the number of new roads and buildings.

THE discovery of a new gutta-percha is reported from Zanzibar. This substance is derived from a tree which grows principally at Dunga. When tapped with a knife, a white fluid exudes, which, when placed in boiling water, coagulates into a substance which in character bears a very striking resemblance to gutta-percha. As the material cools it becomes exceedingly hard, but while soft it can be moulded into any required shape. The fruit of the tree resembles a peach in shape, but grows to the size of a small melon. Experts have experimented with this new product to see if it in any way possesses the qualities of gutta-percha, and although it is not expected to prove equal to the genuine article, it is considered that it will be quite suitable for some purposes for which gutta-percha is at present utilised, and it will thus become a marketable article. It is said to abound in Zanzibar, and will be a very cheap product.

WRITING to the *British Medical Journal* on the subject of "Mosquitoes and Malaria," Mr. H. J. Elwes, F.R.S., says:— "The connection between mosquitoes and malaria seems to be now so clearly proved that some experiments should be undertaken by the Indian Medical Department to find out under what conditions mosquitoes do not produce malaria. Some years ago when on a hunting expedition in a very malarious district in the Bhotan Terai, I succeeded in escaping malaria by keeping within mosquito curtains till after sunrise, and getting into them again as soon as possible after dark, smoking freely at the same time within the curtains of my camp bed. Two out of the four Europeans of my party, and nearly all the natives who did not take these precautions, suffered so severely from malaria that our camp was unable to march after three weeks in the district. I may mention that it was then observed by experienced officers that from fourteen to eighteen days was the time which elapsed between exposure to infection and the appearance of severe fever. But there are places in Eastern Bengal and no doubt elsewhere where mosquitoes are very numerous and annoying, which do not seem to be subject to severe malaria, and I remember that Dacca, the only place where I was kept from sleeping a whole night by mosquitoes, was looked on as a station free from severe malaria, and I certainly, though I had previously been suffering from fever in Assam, never had a touch of it there. The great importance of finding out as soon as possible what precautions should be enforced by those responsible for the health of soldiers and others who are obliged to live in malarious districts cannot be overrated."

IN his report on the work of the Government Laboratory, Dr. T. E. Thorpe refers to the examination of some ordinary writing ink which was submitted to him by the Stationery Office, on a complaint that it thickened excessively and clogged the pen, and, in illustration, a sample of the contents of the ink-wells in use in the particular public office were forwarded, together with a sample of the ink as supplied. It was found that after the deposition of the separated solid matter of the ink, collected from the ink-wells in use, the fluid portion had a specific gravity twice that of the ink supplied. In other words, the ink had been allowed to become concentrated by evaporation to practically double its original strength through the use of excessively large ink-wells and inattention to the supply. It is, of course, necessary that the ink supplied shall be capable of furnishing a record which may be relied upon as permanent. Ink made with tannin and iron salts has had the advantage of very extended and prolonged use, with the result that complete confidence is felt as to the permanence of writing, for which it is used. But ink of this character possesses the undoubted disadvantage that it rapidly thickens on exposure, and Dr. Thorpe points out that it is specially advisable that such ink should be used in ink-wells of small size which receive regular attention at short intervals.

IT may safely be said that as petrol stands to-day as the paramount means of propulsion for automobiles accommodating passengers and of a light character, so steam has forced its way

(at least in this country) as the means adopted for heavy motor vehicles for road service, carrying a load varying from three to ten tons. In support of this argument, an interesting article is given in the *Engineering Magazine* for September, describing these heavier types of vehicles, and although all typical designs are mentioned in every case, not petrol, but steam, represents the power used. As can be well imagined, these heavier class of waggon have had many difficulties to overcome, and with the exception of one type, a ten ton steam motor waggon by Messrs. C. and A. Musker, of Liverpool, the general designs are practically the same. The Thorneycroft waggon is fully described and illustrated in its different applications, ranging from the tipping dust van to a steam delivery waggon, and provided with a "trailer," by which is meant a vehicle towed behind. The ratios of gearing between the engine and the driving axle are 10.1 and 17.7 to 1. On all ordinary gradients five or six tons can be taken, two of which are conveyed on the trailer. Several other waggons by different makers are illustrated, with their dimensions graphically stated. The chief differences lie in the position and type of the engine, the power transmitted to the driving wheels in different ways, various kinds of boilers and different working pressures employed, and slight external appearance. The "Musker," already referred to, is only in its experimental stage, its chief features being an efficient liquid fuel burner combined with a flash-type boiler built up of three cylindrical coils of strong steel tubes, and the flame circulating in the annular space between them. All machinery is placed beneath the "body," thus affording a larger loading area than any other vehicle. It remains to be seen, however, whether the advantage claimed will be realised ; if so, and considering its great load capacity (ten tons), it is indeed an important step in this branch of engineering.

A REPORT on the geology of the West Moreton or Ipswich coal-field in Queensland, by Mr. W. E. Cameron, has been published at the Geological Survey Office, Brisbane. It is accompanied by an appendix on the economic value of Queensland coal by Mr. Robert Wilson. The Ipswich coal-field is estimated to cover an area of about 12,000 square miles, and the coal has been most extensively worked in the neighbourhood of the town of Ipswich, which is about twenty-five miles south-west of Brisbane. The strata are of Jura-Trias age, and they are a good deal folded and faulted. They yield workable coals from two to four feet and more in thickness. Experiments made on the Government steamer *Otter* by Mr. Wilson show that some varieties are very good and useful steaming coals ; and that generally the coals of Queensland "are well able to hold their own with any others at present found in Australia." The report is illustrated by a detailed geological map on a scale of an inch to twenty chains, and also by a geological map of a large area on the scale of an inch to a mile.

THE third volume of the *Annales* of the French Meteorological Office, containing rainfall values and completing the observations for the year 1897 (see p. 490), has been published in a greatly reduced form. The daily rainfall values are given for three hundred stations only, instead of nine hundred; and the scale of the rainfall charts has also been reduced. The valuable series of monthly and annual summaries are given for all stations, as before.

IN the *Botanical Gazette* for August, Prof. D. G. Fairchild is enthusiastic as to the advantages presented by the Botanic Garden at Rio de Janeiro for the study of tropical botany, although at present no facilities are afforded for teaching or study. He regards Rio, with its fashionable suburb Petropolis, as the most picturesque city in South America. To any botanist who wishes to study tropical vegetation, Petropolis and the other suburbs of Rio will prove the most attractive place in the world. As compared with the mountains of Java or Sumatra, they are civilised, and have a much more salubrious climate and all the conveniences of modern civilised life. The south island of Hawaii or the South Pacific Islands have no such stretches of virgin forest, or such a flora or fauna ; to explore Ceylon is hot and uncomfortable in comparison ; and the mountains of Jamaica and Trinidad are uninhabited except by scattered planters. Prof. Fairchild reckons the hotel expenses at Petropolis as about two dollars *per diem.*

FREQUENT as earthquakes are in the Philippine Islands, those of the year 1897, being unusual both in number and in violence, form the subject of an important memoir by P. José Coronas, which we have just received from the Observatory of Manila. He estimates the total number of shocks at 307, occurring in 108 groups. No part of the archipelago was entirely free from earthquakes, though less than five were felt in Mindoro, Paragua and the central part of Luzon. In the northeast of Samar, where more than a hundred were felt, they were most frequent and most destructive. Full descriptions are given of the three most important earthquakes—those of Luzon, on August 15 ; Zamboanga, on September 21, with the accompanying sea-waves and long series of after-shocks ; and Samar, on October 19-20. Four of these earthquakes were recorded at distant stations, both Shide and Edinburgh being more than 11,000 kms. from the origins. The mean velocities of the waves of the two principal Zamboanga earthquakes are estimated at 8·7 and 8·1 kms. per second along the surface, or 7·6 and 7·1 kms. per second along the chords.

DETERMINATIONS of the rate of increase of underground temperature, apart from their scientific interest, have an important practical application in fixing the limit of depth at which mining operations can be carried on successfully. In this connection a report has been lately issued by the Department of Mines of the Government of Victoria, dealing with observations of underground temperature at Bendigo, the author being Mr. James Stirling, Government Geologist. The rise of temperature of the rocks with the depth varies in different parts of the earth's surface, thus making it difficult in any mining district to determine what the rate of increase is without actual experiment. Thus, if we accepted the hitherto recognised formula for the Bendigo field of 1° Fahr. for every 60 feet in depth, we should have a temperature of 125° at the 3,500 feet level. The observations already made prove that this temperature is not reached. It has been asserted in some quarters that mining might extend to as great a depth as 10,000 feet, if the difficulties of haulage could be overcome ; but when we consider the effect of compressing the air at such a depth (*i.e.* the compression caused by its own weight), it will be seen that ventilation under ordinary conditions would be practically unattainable. At a depth of 10,000 feet the ventilating current entering the shaft at, say, a temperature of 60° Fahr., would attain a temperature of 90° by its own weight, altogether apart from the additional heat acquired by contact of the air with the heated rock surfaces. It is possible, however, to imagine a limit of 5000 feet as a workable depth, although the present observations as to the normal rate of increase of temperature of the rocks at Bendigo—1° Fahr. for every 135 feet—suggest 4000 feet as a convenient practical limit to healthy working. Mr. Stirling's report is accompanied by charts illustrating the temperature and pressure gradients in No. 180 mine. In connection with the composition of the air, Mr. Stirling calls attention to the very defective ventilation of many mines, and to the necessity of owners and directors of mines taking steps to remedy the existing evils.

IN NATURE, vol. lix. p. 133, we briefly referred to the very interesting investigations of MM. Hildebrandsson and Teisserenc de Bort into the history and present conditions of dynamical meteorology. Part ii. of this important work has now been issued, dealing generally with revolving storms, and the organisation of the international meteorological services, and particularly with the parts taken by Le Verrier, FitzRoy, and Buys Ballot, and reproducing specimens of the earliest reports and charts issued by each. Le Verrier seems to have been the first in Europe to conceive the idea of telegraphic weather forecasts, although, owing to inadequate support, he was the last of the three to introduce a regular working service. It is interesting to read, thirty-five years after the death of Admiral FitzRoy, the judgment of the eminent authors upon his work in this country, viz. that the criticism of his weather service was both severe and unjust, and Le Verrier's opinion is quoted that if he did not arrive at sufficiently practical results, probably on account of the limited area dealt with, no one else in his place could have done better. In another chapter, dealing with the fundamental works in the different countries between 1865 and 1872, the laborious investigations of Dr. Buchan occupy a prominent place. The publication of his remarkable memoirs and charts at this early epoch were of the highest importance in the development of dynamical meteorology, and the early researches made subsequently in other countries have been, to a great extent, simply verifications of his ideas. The Storm Atlas of Prof. Mohn, the present chief of the Norwegian Meteorological Service, the publications of the Meteorological Office, and the Synoptic Charts of the late Captain Hoffmeyer and of the Copenhagen and Hamburg institutes, are also specially referred to as having contributed greatly to the development of meteorological science.

THE remarkable colour-changes exhibited by a familiar prawn (*Hippolyte varians*) form the subject of an extremely interesting and most beautifully illustrated paper by Dr. Gamble and Mr. Keeble, which appears in the *Quarterly Journal of Microscopical Science* for September. The species in question may be met with commonly in the lower tidal pools along the shore, or may be obtained by trawling in deeper water. It has long been known that different individuals exhibit variations in colour ranging from one end of the spectrum to the other, and also that many specimens display a protective resemblance to the particular seaweeds on which they may be resting. It is now demonstrated that all the different colour-variations are capable of passing into one another, and the protective resemblances of individuals to their environment are most admirably displayed in the coloured plates with which the paper is illustrated. But this is not all. Twice during the twenty-four hours every specimen is living in deeper water than ordinary, and this includes a certain change in coloration to harmonise with the stronger or weaker light. But a much more important colour-change is induced by the daily alternation of light and darkness, and as the shades of evening approach every single individual of the species gradually loses its distinctive diurnal hue and becomes of a full transparent azure blue. The change is heralded by a reddish glow followed by a green tinge, which finally melts into the azure. And it is not a little remarkable that the day-and-night change has been so long established that it has become periodic and occurs whether the specimens are kept in perpetual darkness or *vice versâ.*

To the same journal Monsieur E. L. Bouvier communicates a supplemental paper on the results of his examination of the series of examples of Peripatus in the British Museum. He deals especially with the specimens described as *P. jamaicensis*, which are shown to include two perfectly distinct species.

OUR German contemporary, *Naturwissenschaftliche Wochenschrift*, of September 23, contains a long digest of Prof. G. Siebert's translation of Lydekker's " Geographical History of Mammals," which was published so long ago as 1897.

THE additions to the Zoological Society's Gardens during the past week include two Macaque Monkeys (*Macacus cynomolgus*) from India, presented respectively by Mrs. Woods and Mrs. Sassoon ; a Plantain Squirrel (*Sciurus plantani*) from Java, a Vulpine Phalanger (*Trichosurus vulpecula*) from Australia, presented by Mrs. A. Jeffrey ; a Ground Hornbill (*Bucorvus abyssinicus*), a Bell's Cinixys (*Cinixys belliana*) from West Africa, presented by Mr. Henry Strachan ; a Peregrine Falcon (*Falco peregrinus*), European, presented by Mr. W. R. Bryden ; a Brazilian Tapir (*Tapirus americanus*), two Snowy Egrets (*Ardea candidissima*), six Ring-necked Lizards (*Tropidurus torquatus*), three Surinam Lizards (*Ameiva surinamensis*), a —— Lizard (*Crocodilurus lacertinus*), two Tuberculated Iguanas (*Iguana tuberculata*), six Giant Toads (*Bufo marinus*) from Para, presented by Captain A. Pam ; a Vivacious Snake (*Tarbophis fallax*), European, presented by Mr. W. H. St. Quintin ; a Spix's Macaw (*Cyanopsittacus spixi*) from Brazil, a Large Grieved Tortoise (*Podocnemis expansa*) from the Amazons, six Florida Tortoises (*Testudo polyphemus*) from North America, four Elegant Snakes (*Tropidonotus ordinatus infernalis*), four Couch's Snakes (*Tropidonotus ordinatus couchi*) from California, deposited ; a Bristly Ground Squirrel (*Xerus setosus*) from South Africa, a Pink-headed Duck (*Rhodonessa caryophyllacea*) from India, purchased.

OUR ASTRONOMICAL COLUMN

EPHEMERIS FOR OBSERVATIONS OF EROS :—

1900.		R.A.		Decl.
		h. m. s.		° ' "
Oct. 4	...	2 44 4·44	...	+46 3 53·8
5	...	43 58·85	...	46 25 48·2
6	...	43 49·57	...	46 47 32·2
7	...	43 36·50	...	47 9 5·1
8	...	43 19·53	...	47 30 25·4
9	...	42 58·56	...	47 51 32·1
10	...	42 33·51	...	48 12 23·6
11	...	2 42 4·23	...	+48 32 58·9

THE ROYAL PHOTOGRAPHIC SOCIETY'S EXHIBITION.

THE Royal Photographic Society hold their annual exhibition this year in the New Gallery, Regent Street, instead of, as heretofore, at the Water Colour Society's Gallery in Pall Mall. The result of the change to the larger galleries is certainly a matter for congratulation, because the very restricted accommodation of previous years crowded out professional and trade work, and gave very little space indeed for the exhibition of scientific and technical photography. This year, if any branch of photography is not represented, it is because of other difficulties than want of space. The only notable omission that occurs to us is that of cinematography, and this is accounted for by the very stringent regulations now enforced making a practical demonstration impossible.

The pictorial section occupies about as much of the walls as usual, and the greater part of the remaining space is taken up by professional and trade work, and apparatus exhibits, many of which, however, are not entirely devoid of scientific interest. But upstairs, in the gallery that runs round the central hall, there will be found a very excellent collection of "scientific, technical and photomechanical exhibits."

The Royal Observatory, Greenwich, contribute some of their most recent work with the 30-inch reflector, the 26-inch Thompson photographic refractor and other instruments. The photograph of the great nebula in Orion, taken last December, appears to be especially noteworthy. Two plates of the planet Eros are shown. A photograph of ξ Ursæ Majoris, taken with the 28-inch equatorial, the object-glass being corrected for photography by the separation of the lenses and reversal of the crown lens, as proposed by Sir G. G. Stokes, testifies to the value of this method of correction. Examples of work with the occulting shutter and several recent eclipse photographs will be

examined with interest. Among several other astronomical photographs may be mentioned a paper enlargement of very considerable dimensions of the transit of Venus in 1882, by Prof. David P. Todd, and a series of photographs by the Rev. John M. Bacon illustrating his balloon ascent to search for the Leonids last November.

There are several contributions of photomicrographs. As examples of skill in this direction, the series by Mr. E. M. Nelson, of diatoms, exhibited by the Royal Microscopical Society, will probably attract the most attention. The natural history and biological photographs of all kinds are too numerous to refer to in detail. As notable illustrations of the value of a series of photographs illustrating biological changes, the sixteen lantern slides, by Mr. Martin F. Woodward, from his photomicrographs, showing the fertilisation and segmentation of the egg of *Ascaris megalocephala*, and a frame of photographs, by Mr. Edgar Scamell, showing the different stages in the growth of a nasturtium, will well repay careful study. The photographs in the latter series are so numerous that they would almost serve to illustrate the growth of the plant as a "living picture" by means of a cinematograph. It is very usual to slow down a rapid movement that its details may be recognised, and there is doubtless much to be learnt from the representation in a few seconds of changes that naturally require days or even weeks for their completion.

The applications of photography in many other directions are well illustrated. The automatic recording of the variations of scientific instruments, spectroscopic work, surveying, mining, engineering, the production of metal reliefs, are a few of the subjects that occur to us. Dr. W. J. Russell shows prints to illustrate the photographic activity of the radiations from "the metals radium and polonium," and also from uranium salts, which he finds do not lose any of their activity by keeping them for three years in the dark.

Photography itself, as distinguished from its applications, has received considerable attention, and we would point out that if exhibits of this character could be kept together in future exhibitions, it would much facilitate their study. A print from the enlargement (four thousand diameters), by Dr. Neuhauss, of a section of a film of a Lippmann interference photograph, copies of which have already been seen here, is exhibited by the doctor himself, and shows very clearly that the silver is deposited in layers, as the theory of the process indicates. Several examples of the Lippmann process may be seen in another part of the exhibition. An interesting demonstration of the possible range of exposure is given by the Kodak Company. They show seven negatives exposed under the same conditions, but for periods of from one to fifty, all of which were developed for the same length of time in the same developer. The longer the exposure the denser the negative, but the prints from them are scarcely distinguishable from one another. They show clearly that a small variation in exposure, or even none at all, will serve for very different subjects if negatives of various densities are not objected to.

Mr. Thomas Manly, the inventor of the "Ozotype" process of pigment printing, shows some examples of his method, one of which was exposed and washed thirteen months before the pigment was applied to it, proving that the power of the exposed bichromated paper to render gelatine insoluble and so fix the pigment does not sensibly change by keeping it. The process which in this country has hitherto been associated with Prof. Joly's name is illustrated by the Colour-photo Company of Chicago, and called the "McDonough-Joly process," referring to Mr. McDonough, who worked out the method in America simultaneously with Dr. Joly. They show that there is still room for improvement in the ruling of the triple coloured lines, and also in the nearer approximation of the photographic plate and the coloured screen. By looking at various angles across the ruling, the colour of the different parts of many of the pictures alternate between green, red and blue. This, we take it, is due to the distance between the colour screen and the photographic plate. Mr. Sanger Shepherd shows some striking examples of his triple film three-colour photographs.

The most notable novelty in apparatus is the "panoram kodak," for which the Kodak Company have been awarded a medal. All forms of projection have their advantages and their disadvantages. By adopting the cylindric or panoramic perspective, many subjects are possible for photography that could not be rendered by the plane perspective given by the ordinary fixed lens and plate. The arrangements necessary for a rotating lens have invariably been costly and heavy, and the Kodak Company have made quite a new departure in cameras in designing one that is light and cheap, and rapid enough in action to serve as a hand camera. The sample shown is called the "No 1 Panoram Kodak," from which we suppose that larger cameras of the same pattern will be issued in due course. It gives a picture seven inches long with a lens of about three and a half inches focal length. Film is used, and the drawing of it over the curved guides, to bring a new piece into position, is no more difficult than changing the film in any of the other 'kodaks. It has no shutter as usually understood, but the lens with its cone behind it swings beyond the sensitive surface and past a little flap, so that in its position of rest light cannot pass through the lens to the film. The apparatus is very ingeniously constructed, simple and effective.

There are many other exhibits of technical interest that 'might be noticed in detail, particularly, perhaps, photographs of living creatures of all kinds ; but to enumerate them would be to reproduce a considerable portion of the catalogue. Many are evidence of the great skill and perseverance of the exhibitors. Some fine examples of photogravure show this process at its best. Some photogravures in colour, by Messrs. Ignatz Herbst and Theodore Reichs, show what can be done by a single printing after the various colours have been applied to the plate by the hand of one or more artists.

The exhibition is open daily until November 3.

THE INTERNATIONAL GEOLOGICAL CONGRESS.

THE eighth International Geological Congress was held this year in France. The work of the Congress consisted of papers read at the meetings at Paris, which were followed by discussions, and by excursions into different parts of the country, conducted by French geologists.

The meetings of the Congress took place from August 10-27 at the Palais des Congrès, within the enclosure of the International Exhibition.

The inauguration was held on Friday, August 16, under the presidency of M. Leygues, Minister of Public Instruction and the Fine Arts. M. Karpinsky, president of the last session of the Congress at St. Petersburg, gave an address ; he then read the following list of the members of the Committee, proposed by the Council :—Ex-presidents, MM. Capellini and Karpinsky. President, M. Albert Gaudry. General secretary, M. Charles Barrois. Vice-presidents—Germany : MM. H. Credner, Lepsius, Schmeisser, Zirkel, von Zittel. Austria and Hungary : MM. Böckh, Mojsisovics of Mojsvar, Tietze. Belgium : MM. Mourlon, Renard. Bulgaria : M. Zlatarski. Canada : Dr. Frank Adams. United States : Messrs. Hague, Osborn, Stevenson. France : MM. Michel-Lévy, Marcel Bertrand. Great Britain : Sir Archibald Geikie, Sir John Evans. India : Dr. Blanford. Italy : MM. Cocchi, Mattirolo. Japan : M. Kochibe. Mexico : M. Aguiléra. Norway : Dr. Brögger. The Netherlands : M. Martin. Portugal : MM. Choffat, Mendés-Guerreiro. Roumania : M. G. Stéfanescu. Russia : MM. Loewinson-Lessing, A. P. Pavlow, Sederholm, Tschernyschew. Sweden : M. Högbom. Switzerland : MM. Baltzer, C. Schmidt. Secretaries : MM. Zimmermann, W. Pavlow, von Arthaber, Gäbert, Croma, Cayeux, Thévenin, Thomas. Treasurer : M. Léon Carez. This list was voted with applause.

M. Albert Gaudry, the new president, then read the inaugural address. In the warmest terms the eminent geologist welcomed the assembly of scientific men who had come from all parts of the world, and then proposed that they should rise to show honour to the memory of the learned geologists who had passed away since the last Congress. The president referred to the principal propositions submitted during the preceding sessions, and enumerated the four sections of the present Congress :—

I. Section of general and tectonic geology.
II. Section of stratigraphy and palæontology.
III. Section of mineralogy and petrography.
IV. Section of applied geology and hydrology.

M. Charles Barrois, general secretary, read his report on the work of the Committee of Organisation. M. Leygues, Minister of Public Instruction and the Fine Arts, welcomed the foreign members of the Congress in the name of the Government.

Section I. (General and Tectonic Geology). President : Sir Archibald Geikie.

Papers :—Presidential Address on international co-operation in geological investigation ; *Chamberlin*, the assistance of the Congress in the fundamental investigations of geology ; *J. Joly*, the geological age of the earth fixed by the amount of sodium in the sea ; on the experiments relative to erosion in fresh water and salt water ; order of the formation of silicates in igneous rocks ; mechanical structure of marine sedimentation ; *A. de Lapparent*, definition for each of the periods of the history of the globe, of the regions where by preference arguments should be sought on which the precise delimitation of the geological strata and substrata could be founded ; *Munier-Chalmas*, Parisian Tertiary strata, delimitation of the Secondary and Tertiary formations ; *Stanislas-Meunier*, phenomena of subterranean sedimentation ; *Bleicher*, denudation of the Lorraine plateau and its results ; *Richter*, reading of the report of the Commission on Glaciers ; *H. F. Reid*, on the movements of glaciers ; *Arctowski*, remarks relating to the former extent of glaciers in the land regions discovered by the Belgian Antarctic expedition ; *Popovici-Hatzeg*, presentation of the new geological map of Roumania on the scale of 1/300,000 ; *Vorweg*, proposition tending to simplify the observation of the inclination and strike of the strata ; *l'Abbé Parat*, geological observations in the caves of La Cure (Morvan).

Section II. (Stratigraphy and Palæontology). President : Dr. von Zittel. Discussion on the report of the International Commission on Stratigraphic Classification.

Papers :—*Scott*, fauna of Patagonia ; *Raulin*, Tertiary districts of Aquitania ; *C. Eg. Bertrand*, charbons gélosiques et charbons humiques ; *Grand'Eury*, formation of coal-seams in the coal basins of Central France ; *Lemière*, transformation of vegetables into fossil fuel ; *Osborn*, progress of the methods of palæontology ; relations between the mammal fauna and the Tertiary horizons of Europe and America ; *E. Ficheur*, presentation of the third edition of the geological map of Algeria on the scale of 1/800,000 ; *Flamand*, on the geology of the south of Algeria and the regions of the Sahara ; *Douvillé*, on the Jurassic formation of Madagascar ; on the results of the exploration of M. de Morgan in Persia ; *Zeiller*, fossil plants of Tonquin ; *Malaise*, the Cambrian and Silurian of Belgium ; *Dr. P. Œhlert*, on the reproduction of fossil types ; *W. F. Hume*, the rift valleys of Sinaï ; *T. Barrow* and *W. F. Hume*, on the geology of the eastern desert of Egypt.

Section III. (Mineralogy and Petrography). President : Dr. Zirkel. Honorary Presidents : MM. Rosenbusch and Fouqué.

M. Lacroix announced the views adopted by the International Commission of Petrography in its meetings of October 25 and 26, 1899.

The following proposals were adopted by the Assembly :—

(1) The names of the authors should always be given after the names of the rocks, as is the custom in zoology and botany.

(2) It is proposed to the Congress of 1900 to appoint an International Commission charged to publish the names of all new rocks with their descriptions as concisely as possible, with also their chemical analysis and, if necessary, a drawing representing their structure. This publication is to appear in the volume of the reports of the International Congresses.

(3) It is, above all, desirable to regulate the nomenclature of the eruptive rocks, where the want of unity is particularly felt. Different authors attribute a different sense and signification to one and the same name, while different terms are employed to designate the same rock, the same group of rocks, or the same structure. All the inconveniences of the present nomenclature can, and should be, avoided, at least for the large groups.

(4) The characteristics of the large groups, for example, of the families should be founded on the mineralogical composition, supported by the chemical composition and the structure.

(5) The large groups ought to be fixed from the present without disturbing the subsequent development of the classification, and the separation of these groups into subdivisions.

(6) It is desirable to designate the principal types of structure by special names.

(7) It is necessary to avoid the employment of the same term in different senses.

(8) One should avoid as much as possible the employment and introduction of different terms to designate the same notion, the same rock, or the same group of rocks.

(9) It is necessary to avoid as much as possible for new types of rocks the employment of pre-existing names, and assigning to them a new sense, or restricting or enlarging their meaning.

Dr. Zirkel was elected president of the Committee of Petrography.

Papers:—*Sacco*, attempt at a general classification of rocks; *Salomon*, attempts at a nomenclature of the metamorphic rocks; *Weinschenk*, on dynamo-metamorphism and piézo-crystallisation; on the formation of graphite; *Hague*, on the Tertiary volcanoes of the Absaroka Range; *Sabatini*, the present state of our knowledge of the volcanoes of Central Italy.

Section IV. (Applied Geology and Hydrology). President: M. Schmeisser.

Papers:—*Mourlon*, the new methods of Belgian geology; *Gosselet*, mineralisation of deep-seated waters; *Van der Veur*, on the enlargement of the kingdom of the Netherlands by the draining of the Zuyder Zee; *L. Fabre*, the plateaux of the Hautes-Pyrenees and the dunes of Gascony; *Van den Broeck*, the applications of geology; *Kunz*, progress of the production of precious stones in the United States; *Léon Janet*, utilisation and protection of sources of drinking water; *De Launay*, the teaching of practical geology; *A. de Richard*, origin of petroleum.

General meetings. Presentation of works:—*E. de Margerie* and *L. Raveneau*, cartography at the Universal Exhibition of 1900; *Louis Raveneau*, ninth annual geographical bibliography of the annals of geography, 1899. Presentation of the reports and proposals of general interest adopted by the Council; the Assembly adopted successively:—

(1) Report of the Committee of Geological Nomenclature, presented by M. Tschernyschew, with the benefit of the remarks made at the meeting of the Section.

(2) Report of the Committee of the Geological Map of Europe, by M. Capellini.

(3) Report of the Committee of Petrography, by Dr. Zirkel.

(4) Report of the Glacier Committee, by M. Richter.

(5) Proposal by Sir A. Geikie on international co-operation in geological investigations.

(6) Proposition by M. Œhlert on the reproduction of types.

M. Tietze proposed to the meeting, on the part of the Austro-Hungarian Government, to organise in three years a new Session of the International Geological Congress at Vienna. He informed them of the advanced state of the preparatory work for such a congress, and enumerated the many excursions which would be arranged for the members of the Congress.

The invitation of the Austro-Hungarian Government was unanimously accepted, and M. Tietze thanked the Congress for the warm reception given to his proposal.

Papers:—*Matthew*, on the most ancient Palæozoic fauna; *Walcott*, the pre-Cambrian fossiliferous formations; *Cayeux*, on the radiolaria and sponges of the pre-Cambrian rocks of Brittany; *Pavlow*, the Portlandian rocks of Russia compared with those of the Boulonnais; on some means which would contribute to the determination of the genetic classification of fossils; *Van den Broeck*, on the age of the deposits of the Iguanodons of Bernissart; *Guébhard*, disturbances and fractures of the folds in the Alps of France; *Stanislas-Meunier*, structure of the diluvium of the Seine; *Hull*, sub-oceanic terraces and valleys of the rivers of the western coast of Europe; *Hudleston*, the eastern shores of the Atlantic; *E. Martel*, on the recent discovery of large caverns and fissures.

During the Congress receptions were offered to its members, first by the Geological Society of France, at their new rooms in the Hôtel des Sociétés Savantes. The president of this society, M. A. de Lapparent, of the Institute, inaugurated this reception by an address, which was warmly applauded. M. and Mdme. Albert Gaudry invited the members of the Congress to their house to a most brilliant soirée. Prince Roland Bonaparte received at his hotel the united members of the Geological and Anthropological Congresses, who were also received together by the Municipal Council at the Hôtel de Ville of Paris.

The Committee of Organisation offered a most brilliant banquet at the Hôtel du Palais d'Orsay; the addresses of M. Albert Gaudry, Sir Archibald Geikie, and MM. Tietze, Credner and de Lapparent were warmly applauded. Finally, cards for a reception at the Elysée, and tickets for the National Theatre, were placed at the disposal of the president by the Minister of Public Instruction and the Fine Arts, for distribution among the foreign members. Visits were arranged by the aid of the Committee, to the International Exhibition, the National collections of geology and mineralogy, to the Museum of Natural History, to the Sorbonne, and to the School of Mines.

The excursions of the Congress were well attended. The programme submitted to the geologists of the whole world was of the most tempting description. A pocket-guide, prepared by the united efforts of the French geologists, gave in several numbers a complete account of the geology of France.

In order to allow every one to take part in the greatest number of excursions, they were divided into three periods: before, during, and after the Congress.

(1) Excursions before the Congress: Ardennes, conducted by M. Gosselet; Gironde, by M. Fallot; Touraine, by M. G. Dollfus; Pyrenees (crystalline rocks), by M. Lacroix; Aquitania (Charente et Dordogne), by M. Glangeaud; Turonian of Touraine and Cenomanian of Le Mans, by M. de Grossouvre; Mayenne, by M. D. P. Œhlert; Brittany, by M. Barrois.

(2) Excursions during the Congress: Tertiary basin of Paris, MM. Munier-Chalmas, Léon Janet, Stanislas-Meunier and G. Dollfus.

(3) Excursions after the Congress: Boulonnais and Normandy, MM. Gosselet, Munier-Chalmas, Pellat, Rigaux, Bigot, Cayeux; Central Rocks, MM. Michel-Lévy, Marcellin Boule, Fabre; Coal-basin of Central France, MM. Fayol, Grand'Eury; Tertiary basins of the Rhone; Secondary and Tertiary rocks of the Lower Alps, MM. Deperet, Haug; Alps of Dauphiny, MM. Marcel Bertrand, Kilian, Lory, Paquier, Sayn, Léenhardt, Termier; Picardy, MM. Gosselet, Cayeux, Ladrière; Range of the Black Mountains, M. Bergeron; Pyrenees (sedimentary deposits), M. L. Carez; Lower Provence, MM. Marcel Bertrand, Vasseur, Zürcher.

These excursions, beginning on August 3, ended on October 2, and have had therefore a duration of three months.

The next meeting of the International Geological Congress will be held at Vienna in 1903.　　　　　　　L. Gentil.

FORTHCOMING BOOKS OF SCIENCE.

MR. F. ALCAN (Paris) announces:—"De l'Infection en chirurgie d'armée. Evolution des Blessures de Guerre," by Dr. Nimier; and a new edition of volume i. of "Manuel d'Histologie pathologique," by Profs. Cornil and Ranvier.

The Australian Book Company (of West Smithfield) announce:—"The Geology of Sydney and the Blue Mountains; A Popular Introduction to the Study of Australian Geology," by Rev. J. Milne Curran.

The announcements of Messrs. Baillière, Tindall and Cox include:—"The Hair in Health and Disease," by Dr. David Walsh; "Infantile Syphilis," by Dr. G. Carpenter; "Microscopy of the Starches," by Prof. Hugh Galt; "Standards of Foods and Drugs," by C. G. Moor; and new editions of Rose and Carless's "Manual of Surgery," Walsh's "Manual of Physiology," Walsh's "Röntgen Rays in Medical Work," Ilimes's "Guide to Public Health Acts," Hutchinson's "Aids to Ophthalmic Surgery and Medicine," Sparke's "Artiolic Anatomy of Man," Dennis's "Second-Grade Perspective."

Mr. Batsford promises:—"Waterworks Distribution," by J. A. McPherson, and "Sanitary Engineering," by Colonel Moore.

Messrs. Bemrose and Sons, Ltd., call attention to:—"Decimal Calculator and Multiplier," by C. Barker; and a new edition of "The Scientific Angler," by D. Foster.

Messrs. A. and C. Black will publish:—"The Human Ear: its Identification and Physiognomy," by Miriam A. Ellis; "Introduction to the Study of Physics," by A. F. Walden and J. J. Manley; vol. i. "General Physical Measurements—a Text-book of Zoology," by Dr. Otto Schmeil, translated R. Rosenstock; part iii. "Invertebrates."

Messrs. W. Blackwood and Son's list includes:—"Khurasan and Sistan," by Lieut.-Colonel C. E. Yate, illustrated; "The Sovereignty of the Sea," by Dr. T. Wemyss Fulton, illustrated; "A Manual of Classical Geography," by John L. Myres; "Physical Maps for the Use of History Students, (Greece, British Isles)" by Bernhard V. Darbishire; "Exercises in Geometry," by J. A. Third.

In the Cambridge University Press's list we notice:—"Scientific Papers," by Lord Rayleigh, F.R.S., vol. ii.; "Scientific Papers," by the late Dr. John Hopkinson, F.R.S.; "Scientific Papers," by Prof. Osborne Reynolds, F.R.S., vol. ii.; "The Scientific Papers of John Couch Adams," vol. ii., edited by Prof. W. G. Adams and R. A.

Sampson; "Lectures on the Lunar Theory," by John Couch Adams, from his collected Papers, edited by R. A. Sampson; "A Treatise on Spherical Astronomy," by Prof. Sir Robert S. Ball, F.R.S. ; "A Treatise on Geometrical Optics," by R. A. Herman ; "Advanced Exercises in Practical Physics," by Prof. Arthur Schuster, F.R.S., and Dr. Charles Lees ; "The Prevention of Valvular Disease of the Heart," by Dr. Richard Caton ; "Zoological Results based on material from New Britain, New Guinea, Loyalty Islands, and Elsewhere," collected during the years 1895, 1896 and 1897, by Dr. Arthur Willey. Part v., an account of the Entozoa, by A. E. Shipley ; of the Nemertina, by R. C. Punnett ; the development of the Robber Crab (*Birgus*), by L. A. Borradaile ; new genera and species of Entomostraca, by the Rev. T. R. R. Stebbing, F.R.S. ; anatomy of *Neohelia porcellana* (Moseley), by Edith M. Pratt, illustrated. "Fauna Hawaiiensis," or the Zoology of the Sandwich Islands : being results of the explorations instituted by the Joint Committee appointed by the Royal Society of London for promoting Natural Knowledge and the British Association for the Advancement of Science, and carried on with the assistance of those bodies and of the trustees of the Bernice Pauahi Bishop Museum, edited by Dr. David Sharp, F.R.S., vol. ii., part v. Arachnida, by Mons. Eugène Simon ; Crustacea, Isopoda, by M. Adrien Dollfus ; Amphipoda, by Rev. T. R. R. Stebbing, F.R.S. ; (Cambridge Natural Science Manuals—Biological Series).— "Zoology," by Prof. E. W. MacBride and A. E. Shipley ; "Fossil Plants ; a Manual for Students of Botany and Geology," by A. C. Seward, F.R.S. In 2 vols. Vol. ii. (Physical Series)— "Electricity and Magnetism," by R. T. Glazebrook, F.R.S. (The Cambridge Series for Schools and Training Colleges)— "The Teacher's Manual of School Hygiene," by E. W. Hope and Edgar Browne ; "An Introduction to Logic," by W. E. Johnson ; "Euclid : Books I.–III., with Simple Exercises," by R. T. Wright ; "An Introduction to Physiography," by W. N. Shaw, F.R.S. ; "A New Primer of Astronomy," by Prof. Sir Robert S. Ball, F.R.S. ; "A New Primer of Mechanics," by Prof. L. R. Wilberforce ; "A New Primer of Physics," by the same author ; "A New Primer of Physiology," by Dr. Alex. Hill ; "A Brief History of Geographical Discovery since 1400," by Dr. F. H. H. Guillemard. (Pitt Press Mathematical Series)—"The Elements of Hydrostatics," by Prof. S. L. Loney.

Messrs. Carré and Naud (Paris) announce :—"Les Terres rares," by A. Job ; "Les Nouveaux gaz," by Raveau ; "Les sucres et leurs principaux dérivés," by Prof. L. Maquenne ; "Essais du Commerce et de l'Industrie," by Cuniasse et Zwilling ; "Chimie des matières colorantes," by Rudolf Nietzki, translated by C. Favre and Vaucher ; "La Chimie photographique," by Namias Rodolf, translated by Jaquez ; "La Vinification dans les pays chauds," by Dugast ; "La Pratique industrielle des courants alternatifs," by Chevrier ; "Microbiologie de la distillerie (Ferments, Microbes)," by Levy.

Messrs. Cassell and Co., Ltd., give notice of :—"Our Bird Friends," by R. Kearton, illustrated ; "Cyclopædia of Mechanics," edited by P. N. Hasluck ; "Practical Gas-Fitting and Practical Draughtsmen's Work," edited by P. N. Hasluck ; "A Practical Method of Teaching Geography," by J. H. Overton. part ii.

Messrs. Chapman and Hall, Ltd., announce a new edition of "What is Heat? a Peep into Nature's most Hidden Secrets," by Frederick Hovenden, illustrated.

The list of Messrs. J. and A. Churchill includes :—"A Treatise on Physics," by Prof. A. Gray, F.R.S., in three parts, illustrated ; and new editions of Notter, Firth and Horrocks's "Hygiene," and "Carpenter's Microscope and its Revelations," edited by Rev. Dr. W. H. Dallinger, F.R.S., illustrated.

Messrs. T. and T. Clark (Edinburgh) will publish :—"The Herschells," by James Sime.

In the list of Messrs. Archibald Constable and Co., Ltd., we notice "Through Siberia," by J. Stadling, edited by Dr. F. H. H. Guillemard, illustrated ; "Across and About the Black Republic of Hayti," by Hesketh Prichard ; "Travels in the East of Nicholas II., 1890–1," written by Prince E. Ookhtomsky, and translated by Robert Goodlet, edited by Sir George Birdwood, vol. ii. ; "Motor Vehicles and Motors," by W. W. Beaumont, illustrated ; "Modern Astronomy," by Prof. H. H. Turner, F.R.S., illustrated : "Practical Electro-Chemistry," by B. Blount, illustrated.

Messrs. Dent and Co. announce :—"Birds that come to our

Houses and Gardens," by the Rev. H. D. Astley, illustrated ; White's "Natural History of Selborne "; "Modern Chemistry," 2 vols., by Prof. Ramsay, F.R.S. ; "Plants, their Structure and Life," by Dr. Dennert ; "Primitive Man," by Dr. Homes ; "First Aid to the Injured," by Dr. Drinkwater.

Messrs. Duckworth and Co. call attention to :—"Problems of Evolution," by F. W. Headley.

Mr. Wilhelm Engelmann (Leipzig) announces :—"Pompeji in Leben und Kunst," von A. Mau ; "Die Rohstoffe des Pflanzenreichs. Versuch einer technischen Rohstofflehre. 2. Gänzlich umgearbeitete und erweiterte Auflage," I. Band, von J. Wiesner (Wien) ; "Monographien afrikanischer Pflanzenfamilien und -Gattungen," herausgegeben von A. Engler. V. R. Schumann, *Sterculiaceae africanae* ; A. de Bary's "Vorlesungen über Bakterien." Dritte Auflage, durchgesehen und teilweise neu bearbeitet von W. Migula ; "Hoffmann v. Fallersleben, Unsere volkstümlichen Lieder." Vierte Auflage, herausgegeben und neu bearbeitet von Karl Hermann Prahl ; "Die Assanierung von Paris" (Assanierung der Städte in Einzeldarstellungen, I. Band, Heft 1), von Dr. Th. Weyl ; "Physikalisch-chemische Propaedeutik, unter besonderer Berücksichtigung der medizinischen Wissenschaften und mit historischen und biographischen Angaben," I. Band. von Prof. H. Griesbach.

Messrs. R. A. Everett and Co. give notice of :—"The Veterinary Manual for Horse Owners," by Frank T. Barton, illustrated ; "The Stable Key, or Stud and Stable Studies," by Captain W. A. Kerr, V.C., illustrated.

The announcements of Messrs. C. Griffin and Co., Ltd., include :—"Central Electrical Stations," by C. H. Wordingham, illustrated ; "The Metallurgy of Steel," by F. W. Harbord, illustrated ; "A Dictionary of Dyestuffs," by C. Rawson, W. M. Gardner and W. F. Laycock ; "A Dictionary of Textile Fibres," by W. J. Hannan, illustrated ; "Pernicious Anæmia," by Dr. William Hunter ; "The Construction and Maintenance of Vessels built of Steel," by T. Walton, illustrated ; and new editions of "A Short Manual of Inorganic Chemistry," by Dr. A. Dupré, F.R.S., and Dr. Wilson Hake ; "Tables and Data for the use of Analysts, Chemical Manufacturers and Scientific Chemists," by Prof. J. Castell Evans ; and "Ore and Stone Mining," by Prof. C. Le Neve Foster, F.R.S., illustrated.

Messrs. Gurney and Jackson give notice of :—"The Birds of Ireland," by Richard J. Ussher and Robert Warren, illustrated ; and a new edition of "Lunge's Coal-tar and Ammonia."

Mr. W. Heinemann's list includes :—"The Regions of the World, 1900," a series of twelve volumes descriptive of the physical environment of the nations, with maps by J. G. Bartholomew, edited by H. J. Mackinder, illustrated ; vol. i. "Britain and the British Seas," by the editor ; vol. ii. "Western Europe and the Mediterranean," by Elisée Reclus ; vol. iii. "Central Europe," by Dr. Joseph Partsch ; vol. iv. "Scandinavia and the Arctic Region," by Sir Clements R. Markham ; vol. v. "the Russian Empire," by Prince Kropotkin ; vol. vi. "The Near East," by D. G. Hogarth ; vol. vii. "Africa," by Dr. J. Scott Keltie ; vol. viii. "India," by Col. Sir T. Holdich ; vol. ix. "The Far East," by Archibald Little ; vol. x. "North America," by Prof. Israel C. Russell ; vol. xi. "South America," by Dr. John C. Branner ; vol. xii. "Australasia and Antarctica," by Dr. H. O. Forbes ; "The Life of William Cotton Oswell," by his son, W. E. Oswell, illustrated ; "The First Ascent of Mount Kenya," by H. J. Mackinder, illustrated ; "Mount Orin and Beyond," by Archibald Little : "Nature's Garden : An Aid to Knowledge of Wild Flowers and their Insect Visitors," by Neltje Blanchan, illustrated ; "Pompei : The City, its Life and Art," by Pierre Gusman. Translated by Florence Simmonds and M. Jourdain, illustrated.

Messrs. Hodder and Stoughton will issue in their "Self Educator" series :—"Botany," by R. S. Wishart ; "Chemistry," by J. Knight.

Messrs. Hutchinson and Co. announce :—"The Living Races of Mankind," by the Rev. H. N. Hutchinson, Prof. J. W. Gregory, and R. Lydekker, F.R.S., illustrated ; "Disciples of Aesculapius, Biographies of Leaders of Medicine," by the late Sir Benjamin Ward Richardson, F.R.S., with a biography by his daughter, Mrs. George Martin, in 2 vols., with portraits and illustrations.

In the list of Messrs. Isbister and Co. Ltd., we notice :—"By Land and Sky," by Rev. J. M. Bacon.

The announcements of Messrs. Longmans and Co. include :— "Armenia : Travels and Studies," by H. F. B. Lynch. 2 vols. illustrated ; "Diseases of the Anus and Rectum," by D. H.

Goodsall and W. Ernest Miles, two parts, illustrated ; "Living Anatomy," by Cecil L. Burns and Dr. Robert J. Colenso, 40 plates, with descriptive letterpress ; "Essays in Illustration of the action of Astral Gravitation in Natural Phenomena," by William Leighton Jordan ; "A Practical Guide to Garden Plants," by John Weathers, illustrated ; "Human Personality, and its Survival of Bodily Death," by Frederic W. H. Myers, 2 vols.

In the list of Messrs. Sampson Low and Co., Ltd., we observe :—"Golden Tips, a Description of Ceylon and its Great Tea Industry," by H. W. Cave ; "The Inhabitants of the Philippines," by Frederic H. Sawyer, illustrated ; "Lepcha Land, or Six Weeks in the Sikhim Himalayas," by Florence Donaldson, illustrated ; "Textile Machinery, Recent Improvements," by E. A. Posselt, part ii. ; and a new edition of "On the Manufacture of Vinegar, Cider and Fruit Wines, &c.," by W. T. Brannt.

The science list of Messrs. Macmillan and Co., Ltd., is :— "Life and Letters of Thomas Henry Huxley, F.R.S.," by Leonard Huxley, with portraits and illustrations, in 2 vols. ; "Studies Scientific and Social," by Dr. Alfred R. Wallace, F.R.S., in 2 vols., vol. i. Scientific, vol. ii. Social, illustrated ; "The Cambridge Natural History," vol. viii. Amphibia and Reptiles, by Dr. H. Gadow, F.R.S., illustrated ; "The Scientific Memoirs of Thomas Henry Huxley," edited by Sir M. Foster, K.C.B., F.R.S., and Prof. E. Ray Lankester, F.R.S., in 4 vols., vol. iii. ; "Dictionary of Philosophy and Psychology," edited by Prof. J. Mark Baldwin, in 3 vols. ; "Cyclopedia of Horticulture," vols. iii. and iv., edited by Prof. L. H. Bailey, illustrated ; "Botany, a Text-Book for Schools," by Prof. L. H. Bailey, illustrated ; "The Principles of Vegetable-Gardening," by Prof. L. H. Bailey, illustrated ; "Principles of Stock-breeding, the Application of Biological Laws to the Breeding of Domestic Animals (including Poultry), whether for 'Fancy' or Profit," by Prof. W.H. Brewer ; "First Experiments in Psychology," by Prof. Edward B. Titchener, in 2 vols., vol. i. Qualitative Experiments, vol. ii. Quantitative Experiments ; "Foundations of Knowledge," by Prof. Alexander T. O. McCosh ; "School Geography," by Prof. R. S. Tarr, vol. iii. Europe, &c. ; "The Principles of Mechanics," by Prof. Frederick Slate ; "Design and Construction of Electric Power Plants," by Bion J. Arnold ; "Elementary Electricity and Magnetism," by Prof. D. C. Jackson and Prof. J. P. Jackson, illustrated ; "An Introduction to Celestial Mechanics," by Prof. Forest Ray Moulton ; Surgical Technique, being a Hand-book of and Operating Guide to all Operations on the Head, Neck, and Trunk," by Prof. Fr. von Esmarch and Dr. E. Kowalzig, translated by Prof. Ludwig H. Grau, edited by Prof. Nicholas Senn, illustrated ; and new editions of "West African Studies," by Mary H. Kingsley ; "The Golden Bough, a Study in Magic and Religion," by Dr. J. G. Frazer, 3 vols. ; and "Modern Perspective, a Treatise upon the Principles and Practice of Plane and Cylindrical Perspective," by Prof. William R. Ware, with a portfolio of plates.

Messrs. Methuen and Co. give notice of :—"The Science of Hygiene," by W. C. C. Pakes, illustrated ; "The Principles of Magnetism and Electricity : an Elementary Text-book," by P. L. Gray, illustrated.

Mr. Murray's announcements include :—"A Treatise on Medical Jurisprudence," by Dr. G. Vivian Poore ; "The Life of Gilbert White," based on letters, journals, and other documents in the possession of the family and not hitherto published, by his great grand-nephew, Rashleigh Holt-White, 2 vols., illustrated ; "The Birds of Siberia," by the late Henry See-bohm, with the author's latest corrections, edited by Dr. F. H. H. Guillemard, illustrated ; "The Life of Sir John Fowler, Bart., K.C.M.G.," a record of engineering work, 1834-1898, by Thomas Mackay, illustrated ; "The Natural History of Re-ligion," based on the Gifford lectures delivered in Aberdeen in 1889-90 and 1890-91, by Prof. E. B. Tylor, F.R.S., illustrated ; 'A Handy Book of Horticulture," by F. C. Hayes, illustrated ; "Heredity," by Prof. J. Arthur Thomson, illustrated ; a popular edition of "The Origin of Species by means of Natural Selection," by Charles Darwin, F.R.S., with a photogravure portrait of the author ; and a new edition of "Scrambles Amongst the Alps in the Years 1860-69," including the history of the first ascent of the Matterhorn, by Edward Whymper, illustrated.

Messrs. George Newnes, Ltd., promise :—"The Story of Thought and Feeling " (an elementary book on Psychology), by F. Ryland ; "The Story of Animal Life," by B. Lindsay ; "In Nature's Workshop," by Grant Allen, illustrated.

Mr. J. C. Nimmo's list comprises :—"Babylonians and Assyrians. Excavations and Account of Decipherment of Inscriptions," by Prof. A. V. Hilprecht ; "Syria and Palestine, Important Discoveries in Recent Years" ; "Reminiscences of a Falconer," by Major C. H. Fisher ; and new editions of "British Game Birds and Wild Fowl," by Dr. Beverley R. Morris ; "Fern Growing, Fifty Years' Experience in Crossing and Cultivation, with a list of the most important varieties and a History of the Discovery of Multiple Parentage," by E. J. Lowe, F.R.S. ; Rev. F. O. Morris's "A History of British Birds," 6 vols. ; "A Natural History of the Nests and Eggs of British Birds" ; "A History of British Butterflies" ; "A Natural History of British Moths" ; and "A Handbook of British Birds," by J. E. Harting.

Mr. D. Nutt's list contains :—"Mythology and Folktales, their Relation and Interpretation," by E. Sidney Hartland.

The announcements of the Oxford University Press include :— "The Structure and Life-History of the Harlequin Fly," by Prof. L. C. Miall, F.R.S., and A. R. Hammond ; "A Text-book of Arithmetic," by Richard Hargreaves ; "The 'Junior' Euclid," by S. W. Finn, Books III. and IV.

Mr. Y. J. Pentland announces :—"Text-book of Physiology," edited by Prof. Schäfer, F.R.S. ; vol. ii. "Text-book of Medicine," edited by Dr. G. A. Gibson ; "Text-book of Pharmacology and Therapeutics," edited by Dr. W. Hale White ; and "Diseases of the Throat, Nose and Ear," by Dr. P. McBride.

Messrs. G. P. Putnam's Sons promise :—"Care of the Consumptive," by Dr. Charles Fox Gardiner ; "Thomas Henry Huxley, a Sketch of his Life and Work," by P. Chalmers Mitchell, with portraits ; "Medical and Surgical Nursing," edited by Dr. H. J. O'Brien ; and a new edition of "Materia Medica for Nurses," by L. L. Dock.

The list of the Religious Tract Society includes :—"The Royal Observatory, Greenwich, a Glance at its History and Work," by E. Walter Maunder.

Mr. Grant Richards promises :—"Flame, Electricity, and the Camera," by George Iles, illustrated.

Messrs. Sands and Co. will add to their Library for Young Naturalists, edited by F. G. Aflalo, "The Animals of Africa," by H. A. Bryden, illustrated ; "Types of British Plants," by C. S. Colman, illustrated.

Messrs. Walter Scott, Ltd., will add to their "Contemporary Science" Series :—"The Mediterranean Race," by Prof. Sergi, illustrated.

Messrs. Seeley and Co., Ltd., promise a new edition of "The Chemistry of Paints and Painting," by Prof. A. H. Church, F.R.S.

The list of the Society for Promoting Christian Knowledge contains :—"Among the Birds," by Florence Anna Fulcher ; "Sounding the Ocean of Air," being six lectures delivered before the Lowell Institute of Boston in December, 1898, by A. Lawrence Rotch, illustrated.

Messrs. Swan Sonnenschein and Co., Ltd., call attention to :— Aristotle's "Psychology, including the Parva Naturalia," translated and edited with commentary and introduction by Prof. William A. Hammond ; "A History of Contemporary Philosophy," by Dr. Max Heinze, translated by Prof. William Hammond ; "Ethics," by Prof. W. Wundt, vol. iii. ; The Principles of Morality and the Sphere of their Validity, translated by Prof. E. B. Titchener ; "Physiological Psychology," by Prof. W. Wundt, translated by Prof. E. B. Titchener, 2 vols., illustrated ; "Text-book of Palæontology for Zoological Students," by Theodore T. Groom, illustrated ; "Text-book of Embryology : Invertebrates," by Dr. E. Korschelt and Dr. K. Heider, translated by Mrs. H. M. Bernard, and edited (with additions) by Martin J. Woodward, vol. iv., illustrated ; "The Romance of the Earth," by Prof. A. W. Bickerton, illustrated ; "Biological Types in the Vegetable Kingdom," by Wilfred Mark Webb ; "Mammalia," by the Rev. H. A. Macpherson ; "Birds' Eggs and Nests," by W. C. J. Ruskin Butterfield ; "Inductive Geometry," by H. A. Nesbitt ; and a new edition of "Handbook of Practical Botany, for the Botanical Laboratory and Private Student," by Prof. E. Strasburger, edited by Prof. W. Hillhouse, illustrated.

The following is the science list of the University Corre-

spondence College Press :—" Algebra, The Tutorial, Part I.,
Elementary Course," by Rupert Deakin ; " Arithmetic, The
Tutorial," by W. P. Workman ; " Building Construction (Science
and Art)," by Brysson Cunningham ; " Machine Construction,
First Stage (Science and Art)," by J. Handsley Dales ; " Mathematics, First Stage (Science and Art)" ; " Physiography,
Section One (Science and Art)," by Fabian Rosenberg ; " Practical Plane and Solid Geometry, First Stage (Science and Art),"
by G. F. Burn.

Mr. T. Fisher Unwin will add to his " Masters of Medicine"
Series, " Thomas Sydenham," by J. F. Payne, and " Andreas
Vesalius," by C. L. Taylor.

Messrs. Frederick Warne and Co. will issue new editions of :—
" The Cattle D-ctor," by Geo. Armatage ; " Wayside and
Woodland Blossoms, First and Second Series," by Edward Step.

Messrs. Wells Gardner, Darton and Co.'s list includes a new
edition of " Playing at Botany," by Phœbe Allen.

Messrs. Whittaker and Co.'s announcements are :—" Periodic
Classification and the Problem of Chemical Evolution," by
George Rudorf ; " Inspection of Railway Material," by G. R.
Bodmer ; " Electric Wiring Tables," by W. Perren Maycock ;
"Telephone System of the British Post Office," by T. E.
Herbert ; and " Horseless Road Locomotion," by A. R.
Sennett.

MATHEMATICS AT THE BRITISH ASSOCIATION.

THE mathematical communications to this year's meeting of
the British Association were made on Monday, September
10, in one of the halls assigned to the Mathematical-Physical-
Astronomical Section. Major P. A. MacMahon, F.R.S., took
the chair.

The committee appointed in 1888 to calculate tables of certain
mathematical functions opened the proceedings by presenting a
report on their year's progress. The work on which they have
for some time been engaged, namely, the preparation of a new
" Canon Arithmeticus," is now almost completed. The calculations have been made by Lieut.-Colonel Allan Cunningham,
who, in presenting the report, announced that the liberality of
the British Association and of the Royal Society had enabled
the committee to undertake the publication of the tables as a
separate volume. Before the Association meets next year this
will probably have been given to the world, and the committee,
after an existence of thirteen years, will (unless some other work
is found for it) cease to exist.

Another report was taken next—this time not of a committee,
but of a single worker, Miss F. Hardcastle, of Cambridge, who
was commissioned two years ago to prepare an account of " The
present state of the theory of point-groups " for the Association.
In the absence of Miss Hardcastle, one of the secretaries stated
that a first instalment of the work is to be published in this
year's annual report ; this, however, will give only the general
classification of the subject, and an account of those memoirs on
the theory of elimination which are of importance in it. The
greater part of Miss Hardcastle's report will not be ready until
next year.

The chair was then taken by Prof. Forsyth, while Major
MacMahon read a paper on " A property of the characteristic
symbolic determinant of any n quantics in n variables." Let

$$\xi_1 \quad \xi_2 \qquad \xi_n$$
$$a_{x1}, a_{x2}, \ldots a_{nx} = \ldots + C_{\xi_1 \ \xi_2} \ldots \xi_n x_1 x_2 \ldots x_n + \ldots$$

be (in symbolic notation) any n quantics in m variables, and let

Major MacMahon arrives at the remarkable result that

$$\Sigma\Sigma \ldots \Sigma C_{\xi_1 \ \xi_2} \ldots \xi_n$$

(where the summation is extended over all positive integral values

of $\xi_1, \xi_2, \ldots \xi_n$) has the value $\dfrac{(-1)^n}{f(1)}$, where $f(\theta)$ is the

characteristic determinant of the umbræ $a_{11}, a_{12} \ldots$

The next communication was made in French by Prof.
Cyparissos Stephanos, of the University of Athens, " Sur les
relations entre la géometrie projective et la mécanique." The
fundamental thought of this paper may be explained as follows.
Consider a system of forces in equilibrium. What geometrical
transformations of space will transform this system into another
system of forces also in equilibrium ? Prof. Stephanos solves
this problem, and finds that the only transformations which
thus conserve equilibrium are those which, considered as performed on the Pluckerian co-ordinates of the forces, are linear and
homogeneous. When the system of forces is coplanar, these transformations are homographies in the plane. This train of thought
is of some importance in Graphical Statics.

Mr. H. S. Carslaw (Fellow of Emmanuel College, Cambridge)
followed with a paper on " The use of multiple space in applied
mathematics." The method of images, so powerful in electro-
statical problems, can in its original form be applied only when
the fundamental angles of the problem are submultiples of π.
Prof. Sommerfeld pointed out a year or two ago that by introducing the idea of a branched space, analogous to the branched
planes used in Riemann's theory of Functions, the method of
images can be freed from this limitation. Mr. Carslaw's work
is an extension and development of this suggestion, which is
applied by him to the solution of several of the standard problems
of the potential theory.

Lieut.-Colonel Cunningham then gave some results obtained
by himself and Mr. H. J. Woodall in the " Determination of
successive high primes." As an example of a new process due
to the authors, the factors of all numbers between 16 776 196,
and 16 778 236 have been determined. 117 of the numbers
in this series are found to be primes, a fact which led to some
discussion on Riemann's work in the theory of prime numbers.

This was followed by a paper on " The construction of magic
squares," by Dr. J. Willis, of Bradford, in which some new
modes of formation were described and exemplified in diagrams.
Major MacMahon then communicated two papers in succession.
The first was entitled " The asyzygetic and perpetuant
covariants of systems of binary quantics " ; it was concerned
with the extension, to a system containing any number of binary
quantics, of the work which has already been done in connection
with the semivariant forms of a single binary quantic.

In the second paper, " On the symbolism appropriate to the
study of orthogonal and Boolian invariant systems which appertain to binary and other quantics," Major MacMahon explained a
new and most remarkable method which he has discovered in
the invariant theory, which promises to revolutionise the treatment of that subject. Previous writers have considered those
forms associated with a quantic, which are invariant when the
variables are subjected to the general linear
transformation. When the variables are subjected only to
linear transformations of special types, such as the orthogonal
and Boolian transformations, the family of invariant forms
associated with a given quantic is, of course, much larger ; but
these special classes of transformations have hitherto been,
comparatively speaking, ignored, as forming a tedious and outlying branch of the subject. Major MacMahon's discovery is
a new symbolic method for obtaining the forms which are
invariant for orthogonal and Boolian transformations, in the
same way as Aronhold's symbolic method enables the investigator
to obtain the forms which are invariant for the general linear
transformation. Major MacMahon obtains six symbolic factors
analogous to Aronhold's symbolic factors a_x and (ab), and the
ordinary invariant-theory can be derived as a particular case of
the new theory, by simply rejecting those forms which contain
any one of a certain four of these factors.

A paper by Mr. A. B. Basset, F.R.S., in which the result
that " a quintic curve cannot have more than 15 real points
of inflexion "—an extension of a theorem of Zeuthen's on
quartic curves—is obtained, was briefly communicated by the
chairman ; and a remarkably interesting session closed with two
communications by Prof. J. D. Everett, F.R.S., " On Newton's
contributions to central-difference interpolation," and " On a
central-difference interpolation formula." In the former of
these papers the author observed that certain formulæ in the
calculus of finite differences, usually attributed to Stirling,
were really known to Newton ; in the second, a formula of interpolation was obtained which is less unsymmetrical than those
generally given.   E. T. WHITTAKER.

PHYSICS AT THE BRITISH ASSOCIATION.

THE interesting way in which Dr. Larmor, in his Presidential Address to Section A, touched on some of the problems of theoretical physics appears to have had a considerable influence on the subsequent proceedings of that section during the meeting at Bradford. At few recent meetings has the number of impromptu discussions of theoretical questions been so great, and even although these discussions may not always have been ended in the settlement of some question previously in dispute, they have provided in a way that only the Association meetings can, opportunities for exchanges of opinions so necessary in these days of specialisation, and so valued by those who have the advance of their subject at heart.

The large section room was well filled for the President's address. After a vote of thanks moved by Prof. FitzGerald and seconded by Principal Oliver Lodge had been carried, a large proportion of the audience left to hear the address of Prof. Perkin, the President of Section B, and the reading of papers commenced. In what follows they are given in order of subject and not of reading.

Dr. Trouton gave a short account of his experiments on the creeping of liquids, and on the surface tensions of mixtures. He has found that the tendency of certain liquids to creep up the sides of their containing vessels is due to such liquids being mixtures. The more volatile constituent creeps in advance of the other, and the action is stopped if evaporation is prevented. Zinc surfaces seem more favourable to the process than surfaces of other metals or of glass. The surface tensions of mixtures of liquids are as a rule less than the values calculated from the surface tensions and proportions of their constituents, while those of salt solutions increase with the number of gram equivalents of the salt present at a rate nearly independent of the nature of the salt, a fact to which Quincke was the first to draw attention.

Prof. G. H. Bryan, in a note on the partition of molecular energy, explained how, in his endeavour to build up irreversible thermal phenomena from reversible dynamics, he had been led to a novel method of investigating the mean distribution of energy amongst a number of particles moving in an external field having a potential. He found that two such particles do not follow Maxwell's law of partition of energy, and concluded that the law would not be followed in a general assemblage of particles. Prof. FitzGerald considered that two particles in an external field did not sufficiently represent the molecules of a gas, and suggested that if the case of three particles had been worked out, they would have been found to follow Maxwell's law. He hoped that physicists would accept that law as valid for gases till a system had been constructed for which it could be proved conclusively not to hold. Prof. Bryan, on the other hand, challenged physicists to construct a simple system of particles which *would* tend towards Maxwell's distribution of energy.

Dr. Larmor gave some results of his application of the principle of least action to the statistical dynamics of gas theory, as illustrated by meteor swarms and optical ray systems. He finds that if a swarm of meteors is moving under its own mutual attractions and conservative outside forces, and if from some point vectors be drawn equal and parallel to the velocities of the meteors, the product of the volume marked out by the ends of these vectors into the volume occupied by the meteors themselves will remain constant throughout the motion. If the mutual attractions are insensible, the product of the solid angle bounded by the velocity vectors into the square of the mean velocity of the swarm will remain constant. In optics this corresponds to the concentration in cross section of a beam being proportional to the solid angular divergence of the beam, into the square of the refractive index of the medium in which it is travelling. In the case of a gas where encounters between the particles may take place, the above distribution of particles and velocities is found to be a possible steady state.

The report of the Seismological Committee was presented by the secretary, Mr. Milne. During the past year he has analysed the records of the earthquakes which occurred during 1899, and has found that the earthquake wave takes about 110 minutes to travel from its origin to the opposite end of the earth's diameter, but whether it is propagated through the centre of the earth or as a surface wave cannot at present be decided. He suggests that earthquakes may be connected with the small changes of latitude known to occur, and that earth-

quake waves may have a disturbing effect on the timepieces of observatories. Messrs. Clement Reid and Horace Darwin are engaged in an attempt to detect movements at a geological fault owing to earthquakes.

The Committee on the sizes of pages of periodicals reported that it had succeeded in some cases in inducing societies publishing proceedings of exceptional sizes to conform to the rules the committee laid down in its 1895 report. It did not seek re-appointment.

A paper on the relation of radiation to temperature was contributed by Dr. Larmor. The late Prof. Balfour Stewart pointed out at the meeting of the Association in 1871 that if an enclosure at constant temperature contained a moving body at the same temperature, the radiation received from the body at a point in advance would, by Döppler's principle, differ from that received by a point behind the body. Dr. Larmor applies this principle to the case of a spherical enclosure shrinking in size, and, therefore, the wave-lengths of all radiations will decrease as the radius. Further, there will be a pressure of the radiation on the inside surface of the sphere, which will require work to be performed during the shrinkage. This work is converted into radiation, and changes the temperature of the radiation inversely as the radius, and the energy of the radiation inversely as the fourth power of the radius. From this Stefan's law that radiation is proportional to the fourth power of the temperature follows ; and further, the energy of radiation between λ and $\lambda + d\lambda$ is of the form $\lambda^{-5} f(\lambda T) . d\lambda$. Prof. FitzGerald pointed out the great simplification which Dr. Larmor had introduced into the treatment of the problem by the consideration of the radiation in the ether only, a method of which the legitimacy could not be doubted.

Dr. S. P. Langley sent over from America a chart of the infra red spectrum from ·7 to 5·3 μ, obtained by the bolometric method described in his communication to the Association at Oxford in 1894. His bolometer is now arranged so that a difference of temperature of one-millionth of a degree centigrade is detected ; and the whole operation of producing the charts is automatic. They show distinctly the variation of atmospheric absorption with the seasons, and may possibly, he thinks, lead to a new method of weather forecasting.

The Committee on Meteorological Photography reported that as the result of about 400 photographic observations of clouds made from two stations near Exeter, the following mean heights have been found :—Cirrus, . 10,200 ; cirro-cumulus, 8600 ; cumulus top, 3000 ; base, 1300 ; strata-cumulus, 2200 metres. During the early part of the day the clouds rise, attain their maximum altitudes about 2 or 3 p.m., and fall during the afternoon and evening. The greatest altitudes are associated with thunderstorms and the lowest with cyclones.

The Committee on Solar Radiation reported that experiments had been made under the direction of Prof. Callendar, with a view to testing the modified copper-cube actinometer and reducing its records to an absolute scale. During the course of these experiments it has been found necessary to introduce further changes, and the instrument now used consists of a blackened copper disc provided with thermojunctions, suspended within a tubular water jacket around which a stream of water at constant temperature is maintained. The radiation to be measured passes down the tube and falls normally on the copper disc. This instrument has been tested by exposing it to the radiation from an electric lamp at a known distance, and has been found capable of giving consistent results for weak radiation, but the intensity of solar radiation is too great to permit the elementary theory of the instrument to be applied. It is, therefore, proposed at present to record only the *vertical* component of the radiation from sun and sky by means of the bolometric method described in the 1898 report. Two flat platinum thermometers, one bright and the other blackened, are placed horizontally side by side and exposed to the radiation from the sky. Their difference of temperature is automatically recorded, and is taken to be proportional to the radiation to which they are exposed. By means of an observation with an electric lamp at a fixed distance, the indications of the instrument can be converted into absolute measure.

Mr. A. S. Davies described a novel form of mercurial barometer, in which a fixed volume of the gas the pressure of which is to be determined, is compressed isothermally by a column of mercury of known length and the compression measured. The instrument consists of a glass bulb, from which a tube of small bore projects downwards, and ends in another bulb open to the air and containing a little mercury. When a reading is required, the instrument is inverted so that a column of mercury runs down the tube towards the first bulb and compresses the air in it. From the position taken up by the end of this column, when the compressed gas has cooled to its original temperature, the original pressure of the gas can, if the gas is dry, be found. In the instrument this pressure is read off on a scale alongside the tube. The air enclosed is dried by passage through a plug containing calcium chloride. The instrument is very compact and portable.

Mr. A. L. Rotch contributed a note on the use of kites for meteorological observations at Blue Hill Observatory, Mass. Observations with kites have been made up to 16,000 feet above sea level, and have been reduced and published in abstract in NATURE, July 12 and August 9.

Captain Campbell Hepworth exhibited and discussed some charts illustrating the weather of the North Atlantic Ocean during the winter of 1898-9. At sea this period was one of violent storms, while the weather in America was exceptionally cold, and in Europe very mild. Some of the cyclones crossed quickly from the American coast to the British Isles, while others—in particular the worst one in February—made slow progress. Much damage was done to shipping, and even powerful vessels, like the *Lucania* and the *Fürst Bismarck*, were unable to make headway, and arrived at their destinations several days late.

Mr. J. W. Thomas, in a communication on the physical effects of wind in towns and their influence upon ventilation, pointed out that the well-known effect of currents of air in diminishing the air pressure in vessels across the openings of which they passed, was generally neglected by writers on ventilation. A gusty wind, during its period of maximum velocity, reduces the air pressure in a room by the withdrawal of air through the chimney. When the wind lulls, the air passes down the chimney into the room, and the chimney "smokes."

Mr. J. Hopkinson gave an account of the rainfall of the northern counties of England. The means for the ten years ending 1890 are :—.

Cumberland, 57·9 ; Westmoreland, 55·9 ; Derbyshire, 40·2 ; Lancashire, 38·3 ; Yorkshire, 33·4 ; Cheshire, 31·3 ; Northumberland, 31·0 ; Durham, 28·1 ; Nottinghamshire, 24·4 ; Lincolnshire, 24·3 inches per annum. These numbers show distinctly the effect of highlands in increasing the average rainfall.

Mr. G. E. Petavel described the apparatus he is using in his experiments on the explosive pressures of gases. He measures the maximum pressure attained in his explosion vessel by means of a piston which is forced out and makes a telephone contact if the pressure exceeds a certain value. By means of the compression of a cylinder he measures also the rate of change of the pressure. He finds that in the case of hydrogen and oxygen the maximum is about ten times the initial pressure, and that inert gases delay but do not greatly diminish the maximum pressure.

Mr. J. W. Gifford gave an account of a quartz-calcite lens he had designed, having the same focal length for wave-lengths 5607 and 2761, which he considers may be taken as the centres of the visual and photographic portions of the spectrum respectively.

Messrs. A. Dufton and W. M. Gardner exhibited at the Technical College an arrangement they had devised for the production of an artificial light of the same character as daylight. Such an artificial light has been much wanted by those engaged in dealing with coloured stuffs, and the practical demonstration given by the authors showed that they have successfully supplied this want. They use an enclosed arc lamp, and surround the translucent bulb of the lamp by a tank containing a solution of copper sulphate of the proper strength, or by a box with sides of glass of the same colour and the requisite thickness.

Mr. H. Ramage described his method of investigating correspondences between spectra. He takes wave frequency as abscissæ, and atomic weights of the elements whose spectra are to be compared as ordinates, and joins by lines the "corresponding" points of the various spectra. These lines are in general curved, and in the case of the components of a doublet their distance apart increases with the atomic weight. If the squares of the atomic weights are taken as ordinates, they become straight lines intersecting on the axis of wave frequency. He proposes, therefore, to introduce a term aW^2, where a is a constant, and W the atomic weight, into Rydberg's formula, which will thus become $n = n_0 - aW^2 - \dfrac{N_0}{(m-\mu)^2}$.

Mr. G. J. Burch exhibited an experiment on simultaneous contrast. One half of the slide of a stereoscope consists of blue and the other of red glass. By means of diffraction gratings in the eyepiece of the stereoscope, two spectra are produced which appear to cover two patches of black paper on the two glasses. Under these circumstances, that seen by the eye which looks at the red glass appears to lack red, the eye being partially blinded for red, the other for a similar reason lacks blue. In Mr. Burch's opinion these facts confirm the views of Thomas Young on colour contrast.

The Committee for improving the method of determining Magnetic Force on Board Ship reported that an instrument had been constructed according to the designs of Captain Creak, which gave promise of overcoming many of the difficulties met with in using Fox's circle.

The work of the Committee on Radiation in a Magnetic Field had been interrupted by the death of Mr. Preston, but the committee now proposed to issue copies of Preston's photographs showing how the various types of lines are affected by the magnetic field.

The Electrical Standards Committee reported that the standards had been removed to Kew, where an outbuilding had been fitted up for the temporary use of the committee of the National Physical Laboratory. The sub-committee on platinum thermometry has decided that platinum thermometers shall be constructed of a selected sample of platinum wire, and be used as standards for high temperature measurements. The selection of wire is still under the consideration of the committee. Arrangements have been made for the construction of a mercury resistance standard and an ampere balance. The Committee approves of the adoption of the names *Gauss* and *Maxwell* for the C.G.S. units of magnetic field and flux respectively.

Mr. R. S. Whipple gave an account of his improved standard resistance coils. Alongside the platinum-silver wire of the standard coil a second coil of platinum is wound. The difference of resistance of the two coils depends on their temperature, which may therefore be regulated to have any required value. Dr. R. T. Glazebrook pointed out that the method had been used by Messrs. Crompton in constructing their standard resistances.

Mr. E. H. Griffiths described the form of Wheatstone bridge he has devised for determining the freezing-points of dilute solutions by platinum thermometry. A platinum thermometer of about 18 ohms resistance placed in the solution, and another similar one in ice, form two of the arms of a Wheatstone bridge. The rest of the bridge is of platinum. The galvanometer is connected to the bridge by means of two sliders, each of which moves along a pair of platinum wires, one of the pair forming part of the bridge, and the other connected permanently to the galvanometer. The readings of these sliders for a balance determine the difference of temperature of the two thermometers. Using a Paschen galvanometer giving a deflection of 1 mm. on a scale 1 metre distant for a current of 10^{-12} ampere, Mr. Griffiths can determine temperature to one-millionth of a degree centigrade. Mr. R. Threlfall pointed out that although the temperature of the platinum wire of the thermometer could be determined to this degree of accuracy, the temperature of the solution could not. In reply to Dr. Glazebrook, Mr. Griffiths stated that the mercury contacts in Carey Foster's method introduced changes which prevented this high degree of accuracy being attained with it.

Prof. F. G. Baily described a lecture-room form of volt and ammeter which he had devised. By means of a series of resistance coils all contained in a small box, the deflections of a galvanometer of the d'Arsonval type are made to correspond to simple multiples or submultiples of a volt or an ampere.

Prof. W. B. Morton communicated some results he had obtained by applying J. J. Thomson's and Sommerfeld's solution of the propagation of an electric wave along a single wire, to the approximate solution of cases of several parallel wires, some of which may be returns, when the square of the ratio of the radii of the wires to their distances apart may be neglected. His results agree with the more complete investigations given by Mie in the June number of the *Annalen der Physik*.

A communication from Mr. S. H. Burbury on the vector potential of electric currents in a field where disturbances are propagated with finite velocity, was, in the absence of the author, taken as read. There are difficulties in the way of the usual definition of the vector potential due to electric currents when these currents are changing. These difficulties Mr. Burbury proposes to obviate by substituting in the definition, for the current at a given point at the given instant, the current which existed at that point r/V seconds before, where r is the distance of the point from that at which the vector potential is to be measured, and V is the velocity of propagation of an electric disturbance.

The communication which most attracted the attention of the members of the Association, and produced a great addition to the attendance at Section A, was that of Sir William H. Preece on wireless telephony. By a series of experiments carried out at Lock Ness, the Menai Straits, the Skerries and at Rathlin Island, he has shown conclusively that wireless telephony is a practical and commercial system. At the Skerries a line half a mile long terminated by earth plates placed in the sea, at a mean distance of nearly three miles from a similar wire three and a half miles long on the mainland, was quite sufficient to enable telephonic messages to be transmitted with the ordinary instruments. At Rathlin Island the wire is eight miles from the mainland and communication is readily maintained. Endeavours are to be made to extend the system to ships and there seems every probability of success.

Prof. G. F. FitzGerald, in a note on Crémieu's experiment, described the arrangement adopted by Mr. Crémieu and the negative result he had obtained, and contrasted them with the arrangement and result obtained by Rowland in his experiments on electric convection made in 1876. He considered that the discrepancy of the results of experiments, which appeared to have been carried out with great care, did not necessarily disprove our theory of electromagnetism, but rather signified that there was some action of a moving ion not hitherto included in our equations which was well worth investigating. Dr. J. Larmor pointed out that any want of symmetry of the revolving disc and fixed case in Crémieu's apparatus would tend to cause some part of the charge on the disc to remain stationary. Prof. A. Gray announced that he had already commenced work with a view to repeating both Rowland's and Crémieu's experiments with the same apparatus.

Prof. J. Chunder Bose gave an account of his work on the effect of electrical stimulus on inorganic and living substances. By measurements of conductivity he determines the magnitudes of the changes produced in the molecular structure of substances due to an electric stimulus. Taking time of exposure to stimulus, or time of recovery from effect of stimulus, as abscissæ, and change of conductivity as ordinates, he draws curves for numerous substances under varying conditions. He finds that the curves for organic and inorganic substances are similar. On this as a basis he has constructed an artificial retina, which has enabled him to explain many obscure phenomena of vision.

The Committee on Electrolysis and Electro-chemistry reported that the experiments on the freezing points of the solutions whose electrical conductivities had been found by Mr. Whetham were still in progress. Some experiments on the consumption of carbon anodes in electrolysis have been made by Mr. Skinner, who has found that the anion produced by electrolysis of any highly oxidised material consists partly of carbonic acid. The committee now lapses.

Prof. G. F. FitzGerald opened a joint discussion with the chemical section, on ions. While acknowledging that the dissociation theory of electrolysis had proved a useful hypothesis, he wished to draw attention to the fact that there were phenomena which it was incapable of explaining, and that dissociation itself had not been dynamically explained. Why should water dissociate a dissolved salt into its ions? where does the necessary energy come from? how can the dissociated ions wander about in the solvent without recombining? and why do some ions travel faster than others? seemed to him questions which the supporters of the theory had never satisfactorily answered. The recent work on conduction in gases seemed to render it necessary to restrict the term "ionisation" in future to the process of producing atoms differently electrified, and to introduce a new term "electronisation" for the production of conductivity by the motion of particles of apparent mass about 1/500 of that of a hydrogen atom. In gases conductivity was probably due to both causes, and in liquids to the former only. In the case of metals, he should like to ask, how thick was the layer of electricity on the surface? did the thickness of a thin metal plate alter its capacity? and would the electrons fly to the surface of a metal when it revolved? He thought the questions still open to discussion might be summarised as follows:—

(1) The cause and nature of ionisation.

(2) The source of the energy in dissociation in a liquid.

(3) The cause of the failure of the law of dissociation as the concentration increased.

(4) The reason for the different rates of migration of the ions.

(5) The nature of the double layers, or why different metals should attract electricity differently.

(6) Are the processes of ionisation the same in liquids and gases? and, if so, why?

(7) Do electrons gravitate, *i.e.* have they a material nucleus or not?

(8) Is magnetism due to rotation of the electrons?

Dr. Larmor, before calling on the chemists present for their remarks, pointed out that the large dielectric constant of water meant a large electric moment for the water molecule, and therefore a considerable separation of the positive and negative charges on the molecule. A molecule of dissolved salt might therefore readily come under the influence of one of these charges alone.

Prof. H. E. Armstrong stated that in the opinion of chemists the atoms were permanent and stable, and that the removal of 1/500 of the mass of a hydrogen atom along with its negative charge seemed to them impossible. He thought that the same process produced conductivity in gases which produced it in liquids; that in gases the vapour of water played the part of the water in electrolysis of a dissolved salt, and that in all cases it was necessary to form a "triplet" by the presence of a third substance, before any chemical or other action could result. This third substance was generally one having one of its constituents in an "unsatisfied" condition, like the oxygen in water or the nitrogen in ammonia, and in which there was in consequence a tendency towards "association" of molecules.

Mr. Whetham stated by letter that he did not think the difficulties of the dissociation theory were as great as they were represented. The ions might be free from each other but be connected with the molecules of the solvent.

Principal Oliver Lodge thought that although in a liquid the charges apparently travelled with the atoms, while in a gas the electrons appeared to be free, in neither case was conduction by means of molecular aggregates excluded. He considered metallic conduction the handing on of the electrons from one atom to another. He looked forward to an electrical theory of matter, in which the hydrogen atom would consist simply of 500 electrons without nuclei.

Mr. W. J. Pope pointed out that the dissociation theory only held up to concentrations of 5 per cent., and that there was a difficulty in the case of salts which on account of their asymmetry rotated the plane of polarisation of light.

Dr. H. C. Pocklington gave an account of his work on the radiation of a black body on the electro-magnetic theory. Assuming that the energy of the total radiation emitted by a black body at any temperature is the product of powers of the temperature θ, the velocity of propagation of the radiation v, and the atomic charge Q, Dr. Pocklington finds by the theory of dimensions that the power of the temperature is 4, *i.e.* Stefan's law, and that the radiation between λ and $\lambda + \delta\lambda$ is proportional to

$$\theta^4 \cdot v \cdot Q^{-6} \frac{d\lambda}{\lambda} \cdot f\left(\frac{\theta\lambda}{Q^2}\right),$$

in agreement with Wien's law.

Mr. C. E. S. Phillips gave an account of his experiments on the apparent emission of kathode rays from an electrode at zero potential. He has found that the green flecks which make their appearance on the inner surface of a partially exhausted vacuum bulb when a discharge passes, are due to the emission from the kathode of jets of occluded gas, which continue even when the two electrodes of the bulb are both earth connected. These jets produce shadows of opaque bodies held in their path, and although their velocity is probably not greater than that of sound, they can cause the green fluorescence in the glass on which they impinge.

Mr. J. B. B. Burke communicated a paper on the phosphorescent glow in gases. He uses electrodeless tubes, and finds that the glow begins to appear at a pressure of ·7 mm., is a maximum at ·1, and disappears at ·02 mm. of mercury. It seems to be composed of two parts, one carrying the charge, the other uncharged, but capable of producing conductivity in those parts of the tube to which it penetrates. The conductivity effect is propagated quickly, but the glow appears to be propagated by diffusion along the tube. Prof. A. Smithells mentioned that his experiments on flame showed that the emission of light from the flame was in the same way independent of the conductivity of the flame.

At the close of the meeting of the section on Wednesday morning, September 12, votes of thanks to the president and secretaries were passed, and the section adjourned to Glasgow.

C. H. LEES.

ASTRONOMY AT THE BRITISH ASSOCIATION.

ASTRONOMY this year constituted a distinct department of Section A, with its own chairman and secretaries, and a separate room was provided for the reading of papers on this subject. The new departure was sufficiently justified by the attendance at the meetings, and in future years, when the formation of the Department of Astronomy becomes more widely known, increased success may be confidently expected. The department met on Friday, September 7, and Tuesday, September 11, and altogether sixteen papers were read.

At Friday's meeting, after the chairman's address, three papers by Prof. Todd, relating to eclipse work, were read by one of the secretaries in the absence of the author. In one of these attention was called to the "application of the electric telegraph to the furtherance of eclipse research." In 1878, the idea first occurred to Prof. Todd to telegraph eastward in advance of the lunar shadow in order to enable the immediate verification of any possible discovery, as of an intramercurian planet, without waiting for another eclipse. The feasibility of the method was demonstrated in January 1889, and again more completely during the eclipse of May 28, 1900. At the station occupied by Mr. Douglas in Georgia, totality preceded the same phenomena in Tripoli, where Prof. Todd himself was observing, by 2h. 45m., and the outcome of the experiment was that, through the generous help afforded by the various telegraph companies, an account of the American observations was received by Prof. Todd more than two hours before totality occurred at Tripoli. Abundant time for special preparations to verify any important discovery was thus available.

In his second paper Prof. Todd dealt with a variety of methods of operating eclipse instruments automatically. The "mechanical system," as distinct from the pneumatic and electric arrangements which he had previously devised, was first tried during the recent eclipse. The instruments being set up on the roof of the British Consulate, gravity was utilised for the mechanical operation of slides and shutters. One hundred photographs were secured at Tripoli by seven instruments operated in this manner. Experience indicates that the gravity method is the best where the number of instruments is not great.

Another paper by Prof. Todd described the use of a wedge of yellow optical glass in giving correctly graduated exposures of the partial phases and corona on a single biograph film.

An important paper on the classification of sun-spots was read by the Rev. A. L. Corti, S.J., and illustrated by a fine series of lantern slides selected from the thousands of drawings made at Stonyhurst during the last twenty years. Five types, with a certain number of sub-divisions, were found sufficient to denote the characters of all the spots which had so far been examined.

The chief type, of which the others are probably but phases, is the two-spot formation. The faculæ associated with the different types are also of different characters, and it may be possible to foretell the outburst of a spot by the observation of a certain kind of facula. As an illustration of the use of the type numbers, the life-history of a composite disturbance which crossed the solar disc five times between May 14 and September 4, 1887, was thus described :—

I., II.*b* | IV.*d*, IV.*a* | IV.*a*, IV.*d*, IV.*a* | IV.*a*, I., II.*a* | I.

In the course of the discussion on this paper it became evident that the need for some short system of notation had long been felt by observers of sun-spots, and that, providing the scheme suggested would cover all cases, it would be very valuable. The chairman remarked that possibly a still better system, which would tax the memory less, might be devised on the plan of Herschel's notation for nebulæ.

Prof. Turner exhibited and explained "a cheap form of micrometer for determining star positions on photographic

plates." The essential features are a wooden frame to support the photograph, with an attachment carrying a simple microscope containing a scale in the eye-piece. For less than thirty shillings an efficient instrument can be constructed, capable of yielding measures of practical utility. It thus becomes possible for any one to undertake important researches at a much smaller outlay than would be involved in the purchase of a telescope, since there is no lack of material to work upon. Among the investigations mentioned by Prof. Turner as possible with such a machine, were the determination of the positions of nebulæ and comets, and measurements to ascertain the forms of the trails of meteors. Considerable interest in the proposal was displayed, and the hope was expressed that many who now spend their time in fruitless star gazing with small instruments may be induced to undertake these micrometric measurements.

Thursday's meeting was opened with an interesting paper by Dr. Lockyer, in which a comparison was made of the details of the prominences and corona as shown in photographs taken during the recent eclipse at intervals of 2½ hours, by Prof. Langley and Sir Norman Lockyer, in America and Spain respectively. While enormous changes in the prominences were revealed, no change was detected in the structure of the corona in the region of the North Pole, which had been specially investigated. An interesting feature of one of the photographs taken in Spain with an exposure of 40 seconds was the extreme hardness of the moon's limb, notwithstanding the relative motion of the moon during the exposure ; the explanation of this unexpected appearance was based on the rapid diminution in intensity of the corona as the outer layers are reached, so that the momentary exposures of the lower corona on the advancing limb of the moon at the beginning of the exposure, and on the following limb at the end, were sufficient to give a strong impression.

The new form of refracting telescope recently erected at Cambridge, chiefly for photographic work, was described by Mr. A. R. Hinks, and illustrated by lantern slides. The object-glass is a Taylor triple lens of 12½ inches aperture, and the chief peculiarity of the mounting is that the portion which is usually the lower half of the tube forms the polar axis, with the eye-end, at the top, while the object-glass end is hinged to the other piece, and a plane mirror is placed at the junction. In another paper Mr. Hinks referred to the preparations which are being made for determining the solar parallax by observations of Eros during the coming winter, and exhibited a series of diagrams showing the path of the Cambridge Observatory as seen from that planet at various times. With the aid of such diagrams the observer can see at a glance the most favourable times for making micrometric measurements or taking photographs for the purpose in hand. The importance of the observations of Eros was emphasised by Prof. Turner, who remarked that at the present time the probable error of the adopted value of the solar parallax was equivalent to the thickness of a wicket in the length of a cricket pitch. Unlike the transit of Venus, the observations of Eros would be easily reduced, and the results of the observations would soon be accessible.

A paper on "some points in connection with the photography of a moving object," by Mr. W. E. Plummer, had an important bearing on the photographic method of ascertaining the position of such a rapidly moving object as the planet Eros. A comparison of measurements of the positions of a comet made with a micrometer and those determined from photographs indicated that considerable errors might be introduced in the photographic results on account of the difficulty of determining the epoch of observation. Since the first few moments of exposure on the moving object leave no impression, the middle of the trail does not correspond to the middle of the exposure. In exposures of ten minutes on Eros the danger of error was very considerable. Mr. Hinks remarked that it was hoped to obtain sufficiently strong impressions of the field containing Eros with exposures of one or two minutes, under favourable circumstances, and, moreover, special precautions to eliminate this difficulty had been arranged at the Paris Conference.

Mr. John Herschel described in detail his method of observing and recording the paths of meteors. Special maps are constructed in which the brighter stars are represented by perforations made with needles of various sizes, the side of the paper away from the observer being blue, while that towards him is white. The map being laid on a sloping desk of ground glass illuminated by a night light, the paths of the meteors are ruled in by means of a transparent celluloid ruler having a black edge,

The duration of flight is estimated by repeating the letters of the alphabet, minus w, at the rate of five per second, after experience gained from previous practice.

Among other papers presented to the astronomical department was one by Mr. C. T. Whitwell, on "The Duration of Totality of the Solar Eclipse of May 28, 1900." A table which was given illustrated very forcibly the discrepancies between the calculated and observed durations at various observing stations. It was pointed out that to reconcile the observations and calculations by supposing that there were errors in the adopted value of the moon's diameter, or in the position of the observing station, involved the assumption of greater errors than were probable, though each may account in part for the discordance. Another suggestion, due to Mr. Crommelin, was put forward—namely, that, on account of the irregularities of the moon's limb, the beginning of totality is retarded by an amount corresponding to the movement of the moon required to bring the lowest depressions to the edge of the sun's disc after the assumed geometrical boundary has made contact, while for a similar reason the end of totality would be hastened.   A. FOWLER.

CHEMISTRY AT THE BRITISH ASSOCIATION.

ALTHOUGH the president of section B, Prof. W. H. Perkin, junr., is mainly known as a specialist on polymethylene compounds, his address upon the teaching of inorganic chemistry proved to be of very general interest and was enthusiastically received by a large audience. His contention that the present system of examinations could be advantageously superseded by an inspection of the students' laboratory notebooks was favourably commented upon by Sir H. E. Roscoe and Dr. H. E. Armstrong, although it was admitted that the practical difficulties in the way of such a method are very considerable. The presidential address was followed by the report of the committee on the teaching of science in elementary schools, of which Dr. J. H. Gladstone is chairman; the report consisted principally of a discussion of the returns of the Education Department in so far as they concern the teaching of elementary science. The debate which ensued materially assisted the strong case which was subsequently made out in favour of establishing a separate section of the Association for dealing with educational matters. A paper was next read by Dr. Letts and Mr. R. F. Blake on some problems connected with atmospheric carbonic anhydride and on a new and accurate method for determining its amount, suitable for scientific expeditions; attention was drawn to the variations in the amount of atmospheric carbonic anhydride, and possible explanations of the variations were considered. The authors determine carbonic anhydride in air by absorbing it from about six litres with caustic potash solution, subsequently liberating it by boiling the potash solution with acid in a vacuum and measuring the volume of the carbonic anhydride in a suitable eudiometer. Mr. W. Ackroyd contributed papers on the distribution of chlorine in West Yorkshire and on a limiting standard of acidity for moorland waters. Water from the upper reaches of the West Yorkshire rivers contain from 0·7 to 1·3 parts of chlorine per 100,000, but as the sea or a more populous district is approached, the chlorine number becomes much greater. No cases of plumbism have yet been traced to the solvent action upon lead pipes of water of which the acidity is less than the equivalent of 0·5 part of sulphuric acid per 100,000; this acidity value is therefore tentatively proposed as a limiting standard for potable waters of moorland origin. Dr. T. W. Hime read a paper on the effects of copper on the human body, in which he sought to show that the agitation against the use of articles of food containing small quantities of copper salts is unjustifiable, because a large number of well-known food stuffs contain copper as a normal constituent and because such articles of food exert no poisonous action at all. Reports were received from the committees on the bibliography of spectroscopy and on the preparation of a new series of wave-length tables of the spectra of the elements. Prof. H. B. Dixon and Mr. F. W. Rixon, in a paper on the specific heat of gases at temperatures up to 400°, showed an apparatus for making such determinations at constant volume in which a steel cylinder containing the gas is heated and dropped into a calorimeter; the preliminary results obtained with carbonic anhydride were stated. Mr. F. H. Neville communicated a report on the chemical com-

pounds contained in alloys of which the following is a brief abstract. Intermetallic compounds may be compared with the unstable compounds of the halogens with each other and with sulphur; they often bear a great superficial resemblance to their constituent elements and appear to show marked dissociation, or to form systems in true equilibrium with the liquid mixture of their components.

The intermetallic compounds may be isolated from an alloy (1) by filtration, (2) by volatilisation of excess of a volatile metal, or (3) by removing the excess of metal by means of a suitable solvent. Method (1) has been used by Heycock and Neville, who, on filtering a partially solidified solution of gold and cadmium in tin, obtained a crystalline residue having the composition AuCd; method (2) was applied to the preparation of the same compound by distilling the excess of cadmium from an alloy of gold and cadmium. Lebeau prepared the compounds $SbNa_3$, $BiNa_3$ and $SnNa_4$ by distilling the excess of sodium from alloys in ammonia and nitrogen gas. Debray isolated the compounds $PtSn_4$, $RhSn_3$ and $RuSn_3$, and Le Chatelier obtained Cu_3Sn by the action of dilute hydrochloric acid on alloys containing excess of tin; by methods of a similar nature Heycock prepared $PtAl_3$ and Stead isolated the crystalline substances Au_4Pb, Au_2Pb_3, SnSb and Sn_3As_2. There is, in the application of this method, considerable risk of the solvent attacking the crystals, and Stead has found that the formation of crystals having a core differing in composition from the outside constitutes a serious drawback to the method of partial solution regarded as an independent method of investigation. In the systematic study of intermetallic compounds may be placed first that of the chemical equilibrium of the binary system; this is generally expressed by the freezing-point curve, and has been mainly investigated by Roozeboom and Le Chatelier. Next, and perhaps of equal importance, is placed the microscopic examination of the solid alloys; whilst thirdly, and more limited in applicability, comes the determination of the difference of electrical potential existing between a metal and its alloys. On determining the freezing-points of a series of mixtures of two metals and plotting the freezing-points as ordinates against the compositions as abscissæ, a freezing-point curve is obtained which in its simplest form consists of two branches meeting at an angle—the eutectic angle—lying at a minimum point on the curve. In other cases the freezing-point curve shows a maximum point, but this is not cusped and lies on a gradual change of curvature; the freezing-point curve may thus consist of a series of branches connected by summits and eutectic depressions. It is pretty generally recognised that the eutectic alloy is merely a conglomerate and not a compound, but it is a remarkable fact that the position of the eutectic point on the curve often corresponds closely with some simple molecular composition; cases of this have been observed, not only with alloys, but also amongst organic compounds, by Paternò and Ampolla. The branches of the curve upon which summits lie are caused by the separation of compounds of definite chemical composition from the solidifying magma; the maximum points lie at positions corresponding to the composition of the compound, but Le Chatelier considers that the summit does not necessarily lie exactly at the point indicated by the molecular composition, owing to dissociation occurring in the liquid state. This point, however, needs further investigation. The points upon the freezing-point curve merely denote the temperatures at which solid begins to separate from the magma, but Roozeboom has shown that valuable results may be obtained by plotting, not only the temperatures at which solid begins to separate, but also the temperatures at which complete solidification occurs; in general, the one curve lies below the other, but they intersect or become one whenever the alloy solidifies as a whole. The microscopic examination of the pattern shown by the polished surface of an alloy which has, if necessary, been etched or heated to produce oxidation colours has been worked at principally by Osmond, Charpy and Stead. The existence of coated crystals is made evident by this method, as in the case of the bronzes rich in tin, in which Stead has shown that the Cu_3Sn crystals are coated with CuSn. Le Chatelier has pointed out that in these cases the solid alloy is not in equilibrium, and that annealing will, in general, cause considerable change. Charpy and Stead also consider that evidence of the existence of series of mixed crystals is obtained by microscopic examination. Röntgen ray photographs of thin sections of alloys which contain one transparent metal and one more opaque often give good views of the crystals in the alloy; they were introduced by Heycock and Neville. Laurie and Herschkowitz have studied the potential difference set up between an alloy and the more electro-positive metal contained in it, using a salt of the electro-positive metal as the electrolyte. It is shown that if the alloy consists of a conglomerate of the two metals, the potential of the alloy is that of the more electro-positive constituent; if, however, the two metals are mutually soluble in the solid state, the potential of the alloys will change very gradually with change of composition. Lastly, the existence of intermetallic compounds is indicated by a sudden and large change in potential when the composition of the alloy attains that of the compound. The author notes that although the molecular depressions of the freezing-point of one metal by solution in it of a second point in general to the molecular and atomic weights of the second metal being identical, the evidence is not complete because the second metal may exist combined with the first in the solution. The reading of this report was followed by a lively discussion, in which Sir W. Roberts-Austen, Mr. J. E. Stead, Mr. W. J. Pope, Mr. Stansfield and Dr. H. E. Armstrong took part. Mr. J. E. Stead then read a paper on the mutual relations of iron, phosphorus and carbon when together in cast iron and steel, which was illustrated by a very excellent series of drawings and photomicrographs. Prof. J. A. Ewing and Mr. W. Rosenhain gave a paper on the crystalline structure of metals, in which it was shown that the crystalline character of metals like lead, zinc, tin and cadmium is altered by subjecting them to a severe plastic strain at moderate temperatures; evidence was also adduced in favour of the solution theory of annealing. Prof. Barrett read a paper on the electric conductivity of the alloys of iron, and Mr. C. S. Bradley spoke on some new chemical compounds discovered by the use of the electric furnace. The sixth report of the Committee on electrolytic methods of quantitative analysis was presented; it consisted of papers on the determination of bismuth by Prof. J. E. Reynolds, and on the electro-deposition of iron, by Dr. C. A. Kohn and others. A paper on a simple method for comparing the "affinities" of certain acids was contributed by Messrs. H. J. H. Fenton and H. O. Jones. Oxalacetic acid is decomposed by dilute sulphuric acid into phenylpyrazolonecarboxylic acid in accordance with the equation :—

$$C_{10}H_{10}N_2O_4 = C_{10}H_8N_2O_3 + H_2O,$$

whilst water converts it into the hydrazone of pyruvic acid with evolution of carbonic anhydride. Using decinormal solutions of various acids, it is found that the amounts of carbonic anhydride evolved are inversely proportional to the concentration of the hydrogen ions and hence afford a measure of the affinity constants of the various acids. Mr. H. J. H. Fenton and Miss M. Gosling gave a paper on derivatives of methylfurfural, and Mr. H. M. Dawson spoke on the influence of pressure on the formation of oceanic salt deposits. A paper on recent developments in stereochemistry was read by Mr. W. J. Pope, in which it was pointed out that until a year ago the only known substances exhibiting optical activity in the amorphous state contained an asymmetric carbon atom. Last year, however, Pope and Peachey described a compound which owes its optical activity to the presence in the molecule of an asymmetric nitrogen atom, that is to say, a nitrogen atom which is directly attached to five different groups of atoms. On treating optically inactive methylallylphenylbenzylammonium iodide with silver dextrocamphorsulphonate in a nearly water-free solvent and evaporating the solution, a crystalline mixture of the dextrocamphorsulphonates of dextro- and levo-methylallylphenylbenzylammonium is obtained which is easily resolved by fractional crystallisation; on treating the aqueous solutions of these salts with potassium iodide solution, crystalline precipitates of the iodides of the two optically active substituted ammonium iodides are obtained. This result proves that ammonium salts are not mere molecular compounds of ammonia with an acid, but are true atomic compounds, in which five atoms or groups of atoms are directly attached to the nitrogen atom. The use of strong optically active acids has also been applied during the present year to the preparation of compounds owing their optical activity to the presence of an asymmetric tin atom. On treating methylethylpropylstannomethyl iodide with silver dextrocamphorsulphonate and evaporating the solution, dextromethylethylpropylstannomethyl dextrocamphorsulphonate is obtained in the crystalline state. On treating the aqueous solution of this salt with potassium iodide solution, dextromethylethylpropylstannomethyl iodide separates as a

yellow oil under certain conditions, although under others the iodide becomes inactive owing to the occurrence of racemisation. Similarly, dextromethylethylthetine platinochloride was prepared from optically inactive methylethylthetine bromide, proving that the asymmetric sulphur atom gives rise to optical activity in the same way that the asymmetric carbon, nitrogen or tin atom does. Further, these results prove that the sulphonium compounds contain quadrivalent sulphur, and are true atomic compounds. Since the four elements which we now recognise as able to give rise to optical activity in appropriate compounds are representatives of three groups of the periodic classification, it may be concluded that all the quadri- and quinque-valent elements of the carbon, oxygen and nitrogen families can act as centres of optical activity. Dr. J. B. Cohen, Dr. Divers, Mr. W. Barlow, Dr. H. E. Armstrong, and Dr. F. S. Kipping took part in the discussion which followed the reading of this paper. Dr. A. Lapworth presented a report on our knowledge of the constitution of camphor, in which he showed that the constitutional formula proposed for camphor by Bredt,

$$\begin{array}{c} CH_2\text{---}CH\text{-----}CH_2 \\ | \qquad | \\ | \quad CMe_2 \quad | \\ | \qquad | \\ CH_2\text{---}CMe\text{---}CO \end{array}$$

is the only one which is in accordance with the facts, and that the Perkin-Bouveault formula must be considered as erroneous. The President, Dr. H. E. Armstrong, Dr. F. S. Kipping and Mr. W. J. Pope joined in the ensuing discussion. A paper was read in which Prof. J. Bredt quoted further evidence in support of the constitution which he has proposed for camphor. Prof. Ossian Aschan, of Helsingfors, gave a paper in which it was shown that on replacing the ketonic oxygen atom in the camphor molecule by two hydrogen atoms the material becomes optically inactive, as it should do if Bredt's formula is correct. The Committee on isomeric naphthalene derivatives, of which Dr. H. E. Armstrong is secretary, reported that Mr. W. A. Davies has continued the study of the action of bromine on betanaphthol, and has obtained two isomeric tribromonaphthols, melting at 155° and 159° respectively. The report of the Committee on isomorphous derivatives of benzene, drawn up by Dr. H. E. Armstrong, was presented. A number of series of homologues of formanilide of the composition $C_6H_5 \cdot NX \cdot COY$, where X and Y are alkyl groups, have been crystallographically examined and numerous crystallographic relationships established by Mr. L. P. Wilson. Dr. J. Jee has further investigated the isotrimorphous series of 1 : 3 : 4-dihalogenbenzenesulphonic chlorides and bromides, and has proved a relation between the stability of the crystalline modifications of the various compounds and their position in the series. Dr. S. Ruhemann and Mr. H. E. Stapleton gave papers on the synthesis of benzo-γ-pyrone and on the combination of thiophenol and guaiacol with the esters of the acids of the acetylene series. Dr. J. B. Cohen and Mr. H. D. Dakin read a paper on the chlorination of the aromatic hydrocarbons and the constitution of the dichlorotoluenes, in which it is shown that the chief products of the chlorination of toluene are the 1 : 2 : 3- and 1 : 2 : 4- and possibly a little of the 1 : 2 : 5-dichlorotoluene. Mr. C. F. Cross gave a paper showing that Caro's reagent acts on furfural with formation of a hydroxypyromucic acid. Mr. H. T. Brown gave an account of his recent work on the diffusion of gases and liquids. Dr. J. B. Cohen read a paper on smoke prevention, contending that the production of smoke should be regulated by some system of Government inspection. Mr. T. Fairley read a paper on the heating and lighting power of coal gas, and stated that in populous districts from 20 to 50 per cent. of the gas produced is consumed for heating purposes or by gas engines. Dr. A. Liebmann contributed a report on recent improvements in the textile industries, in which he observed that the inflammability of artificial silk, which constituted so serious an objection to the use of the material, has now been entirely prevented; the use of artificial silks is, however, limited by their brittleness and susceptibility to damage by damp. Major-General Waterhouse gave a paper on the sensitiveness of silver to light, whilst Dr. J. H. Gladstone and Mr. G. Gladstone contributed some thoughts on atomic weights and the periodic law. Mr. F. W. Richardson gave a paper on Bradford sewage and its treatment, in which it was noted that the presence of large quantities of wool-grease and nitrogenous impurities make the Bradford sewage peculiarly difficult to deal with; the grease soon chokes up the filters and, if it were absent, the sewage could readily be treated biologically. Mr. W. Leach, in a paper on wool-combers effluents, also referred to the unsatisfactory character of the purification methods at present applied to the sewage. Mr. W. B. Bottomley discussed the utilisation of the sewage sludge, and contended that the sludge should be pressed and dried, when it forms a valuable manure. Dr. Letts and Mr. R. F. Blake gave a simple and accurate method for estimating the dissolved oxygen in fresh water, sea water, sewage effluents, &c.

SCIENTIFIC SERIALS.

American Journal of Science, September.—The gas thermometer at high temperatures, by L. Holborn and A. L. Day. This is a further study of the nitrogen thermometer with platinum-iridium bulb, which is superior to the porcelain bulb. The correction for expansion is 10° at 500°, 30° at 1000°, and 40° at 1150°. The authors make an elaborate comparison of the gas thermometer with the thermocouples, and determine anew the melting points of a number of metals. Those of silver and gold are 955° and 1064° respectively —Monazite, by O. A. Derby. A single granule of the mineral, no matter how minute, can be securely identified by moistening it with sulphuric acid on a slip of glass and burning off the sulphuric acid over a spirit lamp, when the residue shows the characteristic crystallisation of cerium in radiating needles or isolated crystals of the shape of cucumber seeds.—The spectra of hydrogen and the spectrum of aqueous vapour, by J. Trowbridge. When a condenser discharge is sent through a rarefied gas confined in a glass vessel, the gas cannot be considered dry, for aqueous vapour is liberated from the glass. The four-line spectrum of hydrogen in the solar atmosphere is an evidence of aqueous vapour, and therefore of oxygen in the sun. Conclusions in regard to the temperature of the stars exhibiting hydrogen spectra are misleading if purely based upon conditions of pressure and temperature, for electric dissociation plays a determining part. X-Ray phenomena produced by a steady battery current strongly suggest an electrical theory of the origin of the sun's corona.—A new effect produced by stationary sound-waves, by B. Davis. When a small cylinder, closed at one end, is placed in the stationary sound-wave of an organ pipe, it will not only arrange itself perpendicularly to the motion of the wave, but will move across the wave in a direction perpendicular to the stream-lines. When four such cylinders are mounted in the shape of an anemometer on a needle point, they rotate while the pipe is sounded.—Some interesting developments of calcite crystals, by S. L. Penfield and W. E. Ford. The crystals described show a great diversity of habit, often on a single hand specimen, due to different methods of twinning, together with peculiarities in the development of certain crystal faces. Some peculiar cases of rhombohedral twinning are described.—Method of measuring surface tension, by J. S. Stevens. The surface tension is measured by floating an iron wire on the surface of the liquid, and suspending a piece of soft iron by it. The iron is pulled into a magnetising coil immersed in the liquid by currents which increase until the surface is broken through.

Annalen der Physik, No. 8.—Structure, system and magnetic behaviour of liquid crystals, and their mixture with solid ones, by O. Lehmann. The author has succeeded in proving that all the characteristics of crystallisation which the "liquid crystals" described by him do not possess, cannot logically be made part of the definition of a crystal. The only general characteristics of crystals are that they are not isotropic, and that they possess a molecular directive force which governs their shape, and the manner in which their constituent particles are deposited. The directive force is preserved by means of the surface tension, and crystals may therefore be liquid or solid, but they cannot be gaseous. Liquid crystals may be produced by depositing solid crystals on the cover glass of a microscope and gently heating them above the fusing point.—Generation of electricity in liquid air, by H. Ebert and B. A. Hoffmann. A body suspended above liquid air acquires a strong negative charge. This electrification is due to the friction of minute particles of very cold ice suspended in the air vapour. The authors constructed a kind of electrifying machine by means of a tube containing a piece of wire opaque through which the vapour of liquid air was driven.—Spectrum of radium, by C. Runge. The author has located three of Demarçay's lines with the precision necessary to distinguish them from neighbouring solar lines. The lines located have wave-lengths of 4826·14, 4682·346 and 3814·591

respectively.—Influence of a spark-gap upon the generation of Röntgen rays, by A. Winkelmann. The maximum gaseous pressure at which X-rays can be produced may be raised by introducing a spark-gap into the circuit, the best position for it being next the kathode. Hydrogen yields X-rays at greater pressures than air or carbonic acid.—Fall of potential and dissociation in flame gases, by E. Marx. The author proves that an apparent failure of Ohm's law in flame gases is due to the fact that owing to the scarcity of ions the saturation current is soon attained.—Hall effect in flame gases, by the same author. Owing to the great speed of the ions in flame gases, and the difference in the velocities of the positive and negative ions, a Hall effect is much more appreciable in flames than in electrolytes. The author demonstrates the existence of such a Hall effect in the case of a flat Bunsen flame into which a fine spray of a solution of some alkaline salt is blown.

SOCIETIES AND ACADEMIES.

PARIS.

Academy of Sciences, September 24.—M. Maurice Lévy in the chair.—Nature of the combustible gases found in the air of Paris, by M. Armand Gautier. The author has shown in previous papers that the ratio of carbon to hydrogen found by his method of combustion in dilute mixtures of methane and air is 2·4, instead of the theoretical 3. The much higher value of this ratio found in the air of Paris proves that there must be present gaseous substances richer in carbon than methane, such as benzene vapour or its analogues. The experimental results obtained are in accord with the assumption that in 100 litres of Paris air there are 19·5 c.c. of hydrogen, 12·1 c.c. of methane, 1·7 c.c. of benzene vapour and 0·2 c.c. of carbon monoxide.— Experiment in wireless telegraphy with the human body and metallic screens, by MM. E. Guarini and F. Poncelet. The electric waves were generated by a Wimshurst influence machine and were allowed to act directly upon a coherer. It was found that the human body acted perfectly as a screen.—On crystallised calcium aluminate, M. Em. Dufau. The crystallised aluminate is obtained by heating a mixture of calcined alumina and lime in an electric furnace. Its formula is $CaAl_2O_4$; it forms transparent needles which do not scratch glass.—On Russian flour, by M. Balland. Proximate analyses of three samples of Russian flour are given.

CONTENTS.

THURSDAY, OCTOBER 11, 1900.

A NEW GEOGRAPHY OF PLANTS.

Pflanzen- und Tierverbreitung. Von Alfred Kirchhoff. Hann, Hochstetter, Pokorny, Allgemeine Erdkunde. Fünfte Auflage, von J. Hann, Ed. Brückner und A. Kirchhoff. iii. Abteil. Mit 157 Abbild. im Texte, u. 3 Karten in Farbendruck. Pp. 327. (Prag u. Wien : F. Tempsky. Leipzig : G. Freytag, 1899.)

TO see a portion of my special domain surveyed by an authority in another branch of science appeals to me in a particularly interesting and instructive light, especially if it comes from a man of judgment and broad views. Such a survey may expose any bias, and is likely to open up new vistas. Dr. Kirchhoff, professor in the University of Halle, is a geographer of repute, and naturally approaches the facts of distribution of plants and animals from a point of view different to that of a botanist or zoologist. It is true his book is intended as an introduction to phytogeography and zoogeography for the student of physical geography, and not as a critical essay on these branches ; yet its merits seem to demand that we should rank it higher and judge it accordingly. It is not a mere abstract of one of the few well-known treatises on the distribution of plants and animals, made to serve as a text-book for the beginner, but the product of a mind evincing a considerable power of assimilation of matter, necessarily foreign to it in many of its details, and with an admirable grasp of that which is essential. With these accomplishments are combined the gift of a lucid exposition and of a language which is, apart from certain idiosyncrasies of expression, clear and pleasant.

The author reminds the reader in the preface that in the earlier editions of Hann, Hochstetter and Pokorny's "Allgemeine Erdkunde," the corresponding part was written by Dr. A. Pokorny ; that the progress of phyto- and zoo-geography has rendered it necessary to recast the whole ; and that it is written more especially for the students of geography, and excursions into the domain of the naturalist had therefore to be avoided. Yet geography on the one side, and botany and zoology on the other, overlap to such an extent within the compass of phyto- and zoo-geography, that the author was naturally obliged to fall back over and over again on the botanist or zoologist, who supplied him, if I may say so, with the woof to his warp. It is here where, as one might expect, the weak points of the book become evident ; and it is a pity that the author did not avail himself of the aid of a competent botanist for final revision. I suppose the same applies to the zoological details. Reckless statements, as, for instance, that a single individual of *Sisymbrium sophia* produces 720,000 seeds, or that most of the 854 species of the Egyptian flora are introduced, or that Southern Italy (exclusive of Sicily) possesses 3132 species against 9000 for the whole of British India, &c., can hardly be excused by the desire of putting the case as emphatically as possible. Instances of a decided looseness of expression in describing certain facts would have been discovered at once by a botanical critic, *e.g.* when the author says, on p. 18, that the plants derive

their "principal food" from the soil, while he applies the same term, on p. 24, to the carbon dioxide of the atmosphere ; when he quotes the orchids and aroids of the equatorial zone as examples of parasites (p. 19), although he describes them quite correctly as epiphytes in another place ; and again, when he attributes "genuine grass leaves" to the Australian *Xanthorrhoea* (p. 112). Slips, also, like that on p. 162, where the "Rose of Jericho" is figured as *Asteriscus pygmaeus*, a member of the order of Compositæ, but described in the text as a "little Cruciferous plant (*Anastatica hierochuntica*)," or on p. 198, where the author speaks of the "Kompositengeschlecht der . . . Cæsalpinien," call for the helping hand of the botanist. Willows are not wind-fertilised (p. 51) ; *Rhododendron ponticum* is by no means killed by – 2° C. (p. 86) ; the peach has not originated in Southern Asia (p. 135) ; there is no transformation of our ordinary grape-vine into an evergreen plant in the tropics ; these are statements which should be repeated no longer in text-books. It would be easy to quote a good many more mistakes of this order, but I do not wish to dwell on them more than is necessary to show where, in a future edition, careful revision must be undertaken.

The book is divided into three parts. The first part occupies 139 pages, and deals with general telluric conditions in relation to the organised world, the second (88 pages) with the phytogeographical divisions (Florareiche), the third (100 pages) with the zoogeographical divisions (Faunareiche). The first part is subdivided into five sections, dealing with (1) the reproductive and migratory capacities of the organisms, (2) the natural conditions of vegetable and animal life, (3) the variability of organisms, (4) the Theory of Descent and its geographical proofs, and (5) the general principles of the distribution of plants and animals. To condense this abundance of matter into 139 pages is a very difficult task, and that it has been done on the whole so satisfactorily says much for the judgment of the author.

In the following two parts, Prof. Kirchhoff has set himself the task of characterising the principal geographical divisions of the organic world, relying, of course, on the researches of the recognised authorities in phyto- and zoo-geography, but viewing them from the more comprehensive standpoint of the geographer, as it appeared to him "desirable for a more vivid comprehension of the nature of the countries of the globe." The separation of the vegetable and the animal kingdom into floras and faunas, their distribution in the present, their harmony with the physical character of their respective areas, and their mutual adaptation within each area, are the result of a process of evolution in which the shaping, selecting, separating and shifting forces have been, for both kingdoms, the same to such an extent that a far-reaching parallelism in their geographical differentiation is to be expected *a priori*. This has been, perhaps, too often lost sight of by specialists.

On the other hand, the modern geographer, from the very fact that he starts with conditions bringing about that parallelism, would be naturally led to a more uniform conception of the differentiation of the organic world into floras and faunas. The result of this has been, in Prof. Kirchhoff's case, the almost complete congruence of the two maps representing the "Florareiche" and the "Faunareiche" respectively. Much might be said with respect to the divisions adopted by the author, but space forbids.

The book is abundantly illustrated, and most of the illustrations are well selected and quite to the point ; but exception might perhaps be taken to some of the pictures in which the author attempts to represent as many types as possible in one plate, with the usual consequence that the *ensemble* looks unnatural and untrue. The map showing the distribution of the European species of *Asplenium* (p. 89) is not very illustrative and scarcely in place ; whilst Kerner's maps dealing with *Tubocytisus* (pp. 90, 91) require thorough revision, although they are excellent so far as method is concerned.

OTTO STAPF.

FOUNDATIONS OF AGRICULTURE.

Agricultural Botany—Theoretical and Practical. By John Percival, M.A.(Cantab.), F.L.S. Pp. xii + 798. (London : Duckworth and Co., 1900.)

THE professor of botany at the South-Eastern Agricultural College at Wye has done well to depart from the utterly inefficient standard of text-books in this subject hitherto set and followed in this country ; and, although we do not think the best possible has yet been produced, the present work is so distinctly an improvement, and so clearly sounds the right note, that we have no hesitation in recommending it as *the* elementary handbook for the agricultural student. What it lacks most conspicuously is a clear enunciation of the principle underlying the teaching of botany to students of agriculture, and it will be just as necessary for the teacher using this book, as it is for him who uses others, to emphasise the point of view (lost sight of in nearly all our text-books) that the plant is a focussing centre in which are concentrated the materials gathered by roots and leaves and the solar energy fixed by the chlorophyll-action, so that plant substance—be it in a cabbage, a potato, a crop of wheat or an oak forest—represents a real gain of energy from the surrounding universe, stored up with an equally real recovery of material which would otherwise have been lost to us because dissipated into the atmosphere in an unavailable form. It is this which makes farming, planting, forestry and other branches of agriculture so fundamentally different from the mining industries, where the coal, iron, &c., brought from their storehouse in mother earth are merely temporary sources of wealth representing expenditure of capital.

As regards features of technical detail, there are several interesting departures from the repetitions of previous text-books, and our chief regret is that these are not more original in conception and treatment. For instance, the section on recognition of trees and shrubs by means of twigs in winter is a very welcome one, but it might have been made far better. Again, the part dealing with our common grasses could have been improved by bolder departures from, and less reliance on, Continental and other authorities in common use, though it should be pointed out that the author has, at any rate, provided new drawings of the "seeds" of most of

the grasses. This, however, not always with advantage —*e.g.* the very bad figure (210) of Yorkshire fog. Nor do we regard the summary of characters leading to the recognition of grasses by their leaves as either adequate or worthy of the scope of the book ; it might have been made much better with a little attention to points not included in ordinary pamphlets on the subject.

These are faults to be remedied in later editions, and must not be allowed to outweigh the really excellent portions of the book dealing with the various large groups of cultivated farm-plants—*e.g.* Chapter xxv., dealing with the hop, is well done, as are Chapters xxxiv.–xxxviii., dealing with those very difficult subjects, the varieties of our cereals. Indeed, we may commend the whole of this part of the work which treats of the classification and special botany of farm crops, with few reservations, such as those hinted at above, as an admirable summary of what the student should direct his attention to in this department of his studies. The general botany is also fairly well done ; and although we do not consider the section on "Internal Morphology" quite happy either as regards selection of subjects or treatment of details, we have little but praise for the part dealing with physiology, which is so markedly in advance of the stuff we are too apt to meet with in existing agricultural text-books in this country, that we prefer to dwell only on its merits. The chapters on weeds and on diseases of farm-plants are also distinctly better than those in any previous English works dealing with agricultural botany, and we heartily congratulate the author on his exhibition of capacity in the *rôle* of a teacher of elementary students of agriculture. At the same time, we would point out that much may be done in future editions to improve this subject, and still more in improving and extending the account of the doings of bacteria in the soil. The agricultural student ought to be made to realise that the soil is a matrix, in which the rocks and salts, water and other lifeless constituents, play little more than the subordinate parts of a skeleton or scaffolding, on and between which the real work of conversions, transferences, destructions and constructions of materials necessary for the life of higher plants are being carried out by lower organisms of many different kinds. A vivid picture of the struggles of root-hairs for salts and oxygen, of the relations between anaërobic and aërobic organisms, of the dangers of attack from parasites here, and of the missing of advantageous connections with symbiotic helpmeets there, and of the mutual interactions of the living and non-living factors in keeping up the "fertility," moisture, heat, &c., of the complex soil, would be a fitting subject for a chapter designed to knit together the enormous number of facts here thrown down before the unwary student, and among which he is sure to stumble and flounder.

The ideal here sketched is not an easy one to attain, and we are aware that facts are coming in every day, and that our knowledge of the factors concerned is still in its infancy. Nevertheless, it is no empty compliment to the author to point out that there are indications in the present book that he would be quite capable of putting the crown to his really excellent attempt at an elementary text-book for agricultural students, by adding a chapter which should drive home to this somewhat apathetic class of learners that botany is no longer to be looked upon by them as a luxury or hobby, or as an interesting adjunct to the study of agriculture, but it must be regarded as *the* fundamental science on which all agricultural operations must be based ; and such a chapter should make it perfectly clear that the neglect of the principles and facts it embraces is going to spell ruin in the future, just as the intelligent appreciation of its teachings is going to render the properly trained and equipped planter, forester or farmer, master of the situation his forefathers misunderstood.

OUR BOOK SHELF.

Surveying and Exploring in Siam. By James McCarthy, F.R.G.S., Director-General of the Siamese Government Surveys. Pp. xii + 215. (London : John Murray, 1900.)

ABOUT the year 1880 the Siamese Government became convinced of the necessity of accurate surveys for frontier delimitation, and then it was that Mr. McCarthy commenced the long series of explorations which are recorded in the present work, and which have won for him the gold medal of the Royal Geographical Society. To the student of Indo-China, Mr. McCarthy's book is full of extremely valuable information regarding the aboriginal and mountain races of the highlands of the interior, with whom the nature of the author's work brought him into constant contact. Mr. McCarthy has a sympathetic eye for his fellow travellers, and a kindly word for all but the most obstructive of the native officials. From obstruction by this class, the officers of the Siamese Survey have indeed suffered probably more than any other European officials of the Government ; inasmuch as the Survey was practically the pioneer department of the modern régime, and it had to contend against the whole of the forces of conservatism, superstition and suspicion which were at the outset arrayed against all innovation of the kind. Against these, for many years, Mr. McCarthy battled almost single-handed, carrying out meantime slowly and laboriously the triangulation of the frontier districts, and himself training his own assistants. The physical difficulties of the country, which can only be thoroughly appreciated by those who have experienced them, and the inevitable sickness which attacks all who spend the wet season in the jungle, further delayed and hampered the work. The author makes light of the difficulties which had to be overcome, but those who read between the lines will see how formidable they were.

As may be supposed, the book is in no way popular or sensational, and the author's dry, matter-of-fact style does not lend itself to picturesque narrative. Yet politics on the north-eastern frontier of Siam during the incursions of the Haw bandits, in the 'eighties, were exciting enough. If one desired to be critical, one might say that the book is composed of short sentences and scrappy and incomplete descriptions. Yet these faults will be condoned by all who take an interest in scientific geography for the sake of the admirable scientific results of Mr. McCarthy's work. And those who seek to know more of the magnificent plateau of Teng, the highest peaks of Indo-China, or the very interesting hill tribes, such as the Ka, Lamet, Meo and Yao, and the Southern Shān races generally, will find more accurate information in the present work than in any other we are acquainted with.

An excellent index, triangulation charts, and a map of Siam in two sheets, with a number of illustrations, complete a work which forms an important addition to the bibliography of Eastern Asia.

Church Stretton. Vol. i. *Geology*, by E. S. Cobbold ; *Macro-Lepidoptera*, by F. B. Newnham ; *Molluscs*, by Robert A. Buddicom. Edited by C. W. Campbell-Hyslop. Pp. 196. (Shrewsbury : L. Wilding.)

THIS is an excellent piece of work, and reflects much credit upon those who originated the idea of preparing an account of the scientific features of the Church Stretton district, and also upon the contributors, editor and publisher of the present volume. Church Stretton is a market-town about twelve miles south by west of Shrewsbury, Shropshire, and has a population of about 2000. The district is interesting from a geological point of view, and Mr. Cobbold's notes (which occupy the greater part of the book) will be valuable to geologists visiting it for the first time, and will also give residents a new interest in their rambles. Most of the fossiliferous localities and the main rock exposures are mentioned or described, so that any one interested in the geological and topographical characteristics can readily find them.

Mr. Newnham gives a descriptive catalogue of the macro-lepidoptera found in the neighbourhood of Church Stretton. The district is a fair field and good hunting-ground for the entomologist, many insects being found in it which do not occur in the lower-lying parts of Shropshire. Future collectors will find the catalogue exceptionally valuable, and will doubtless be able to supplement it.

A list of the land and fresh-water molluscs, with notes on the habits of each species and its comparative local scarcity and abundance, is given by Mr. R. A. Buddicom. The total number of species of British land and fresh-water molluscs is reckoned at 138 (not counting slugs) of which 42 have been found in or near Stretton. A plate containing illustrations of 37 species, natural size, photographed from actual specimens, accompanies Mr. Buddicom's paper.

Other monographs, on the botany, archæology, climatology and ornithology of the district, are in preparation, and if they are of the character of this one they will afford pleasure to every resident or visitor in Church Stretton who has an interest in the study of outdoor nature. The district is fortunate in possessing such a useful guide to its natural characteristics.

Surveying with the Tacheometer. By N. Kennedy. Pp. vi + 104. (London : Crosby Lockwood and Son, 1900.)

THIS handy little volume is put forward in the hope of bringing the tacheometer into more general use among land surveyors, its present position in the background being due chiefly, the author thinks, to the fact of the Continental instruments having hitherto been provided with circles divided with 100° to a right angle, instead of 90°, thus necessitating special reduction tables. The publication of a universal method of reduction, no matter what the division value, by Mr. G. Gilman removes the greater part of these objections.

The tacheometer is first minutely described, excellent illustrations being provided for reference, the only essential difference from a good transit theodolite being the insertion of a subsidiary lens between the objective and eye-piece, which, by special adjustment, enables the angular distance between two wires in the eye-piece to be made equal to any desired quantity, decided by calibration on a previously measured base. Subsequent sections deal with the variations introduced by working on inclined ground, details of actual field and office work, concluding with some suggestions on possible methods of utilising existing transit theodolites for tacheometric work. Examples of entries in field-book, plans of surveys, &c., are given at the end of the book. The work is very clearly written, and should remove all difficulties in the way of any surveyor desirous of making use of this useful and rapid instrument.

LETTERS TO THE EDITOR.

[The Editor does not hold himself responsible for opinions, expressed by his correspondents. Neither can he undertake to return, or to correspond with the writers of, rejected manuscripts intended for this or any other part of NATURE. *No notice is taken of anonymous communications.]*

Ascent of Sap.

PROF. VINES, in his interesting address to the Botanical Section of the British Association, has referred to the problem of the ascent of sap.

We believe Prof. Vines is under a mistake when he states : "Now as to the force by which the transpiration-current is raised from the roots to the topmost leaf of a lofty tree. From the point of view that the water travels in the substance of the walls the necessary force need not be great, and would be amply provided by the transpiration of the leaves, inasmuch as the weight of the water raised would be supported by the force of imbibition of the walls. From the point of view that the water travels in the lumina, the force required to raise and support such long columns of water must be considerable." If we gather the sense aright, this statement involves perpetual motion, as may be seen by imagining both cell-walls and lumina filled with water. According to Prof. Vines, water may be obtained from the cell-walls in the higher parts of the plant with the exertion of a less force than from the lumina. If now we establish a connection, the lumina will draw from the cell-walls, and with a second connection below, an endless circulation will arise. The error arises from supposing that water can be withdrawn from the cell-walls and maintained in upward motion without opposition from the entire gravitational pull. Or, stating the matter in another way, the force which is assumed to uphold the water will also act to resist its withdrawal from the walls. Indeed the withdrawal of water from the cell-walls must be necessarily attended by much higher frictional resistance than would obtain if the supply were received from the lumina. The same objection may, in our opinion, be urged against any exposition of the "imbibition" theory. The underlying fallacy is, in fact, essentially the same as that on which the theory of capillarity and gas-pressure is founded, and which Prof. Vines rejects as "quite inadequate."

Prof. Vines, in further discussing the question, speaks of a tensile force of 360 lbs. to the square inch as being required to bring the sap to the summit of a tree 120 feet high, and states that, not only is there no evidence for the existence of such a force, but that it is even negatived by the indications of the experiments of Hales and Boehm. Without discussing the validity of the supposition that such a force is anywhere required (beyond stating, as our opinion, that the grounds upon which this estimate has been obtained are very doubtful), we would certainly like to know to which of the experiments of Hales and Boehm Prof. Vines refers. So far as we are aware, the indications of Hales' and Boehm's experiments were of necessity limited by the difficulty of putting the water (external to the branches experimented upon) into a condition capable of bearing tension. These investigators, however, did not clearly understand this point. The experiments made by us in the same direction certainly all failed from this cause. The most, then, that can be admitted is that direct observation has never revealed the full state of tension of the sap of a transpiring tree, although, as in the case of some of Boehm's experiments, indications of the existence of tension were conclusively obtained. This is a very different thing from assuming that negative indications have been experimentally obtained. It is hard to see why Prof. Vines should consider the existence of a transpiration force of 20 atmospheres as improbable. It has, indeed, been shown by experiment that the turgescence of the cells of the leaves of many trees is capable of exerting a tractional force of over 20 atmospheres on the water in the conduits.

Prof. Vines dismisses the tension theory as offering no solution to the problem. But how does the matter stand ? The more important points may be stated in a very few words.

In the theory of the tensile sap we find full reason for the subdivided structure of the water-conduits and for the structure of their lignified cell-walls (especially as seen in the ingenious mechanism of the bordered pit); stability is conferred on a liquid in tension and liable to the evolution of gas-bubbles by the first, and a minimum of resistance with safety against rupture is secured to the wall by the second. To raise the water through this system

the turgescence of the leaf-cells is fully adequate, even were the tension greater than what Prof. Vines demands. Again, in the light of this theory, the advantage of the periodic recurrence of root pressure becomes apparent, as a safeguard against the multiplication of functionless lumina destroying the continuity of the system. On the other hand, those who have discussed this theory have as yet brought to light no fact in vegetable physiology or anatomy opposed to its validity ; while many points, *e.g.* the collapse of protoxylem elements, and the occurrence of year-rings, have received in it an explanation.

From the physical point of view, the theory is not only adequate to meet all the requirements of the plant, but the existence of tension in a system of minute chambers having walls at once permeable to water and impermeable to free gas, whether altogether or partially filled with dust-free water, is inevitable. The *onus* of proof does not here lie with the upholders of the tension theory merely because it has come late upon the scene, but its opponents must show how the tensile state is evaded before they can dismiss the existence of the tensile stress in the sap at such times as root-pressure is not the uplifting force.

If, then, the sap is in tension from the nature of the conditions and the leaves active in withdrawing water from above, why deny the adequacy of the explanation ?

With regard to the date of publication of our theory, Prof. Vines is slightly in error. Our paper was communicated to the Royal Society in Nov. 1894, and an abstract of it appeared in NATURE in the same month. HENRY H. DIXON.
Trinity College, Dublin. J. JOLY.

Homochronous Heredity and the Acquisition of Language.

THE question raised by Mr. Stuart-Menteath in NATURE of September 27 (page 524) is one of such general interest to all students of heredity that it is to be hoped that some authoritative expression of opinion will be forthcoming. Even in its present form the query involves the subject of the heredity of acquired characters, and it would be of the greatest importance to have experimental systematic observations carried out if such observations have not already been recorded. So far as my very limited acquaintance with the subject extends, I know of no such experiments. It would be desirable perhaps to widen the scope of the question, and to put it in this form : Take children of different nationalities, say German, French and English ; allow them from infancy to hear all three languages indiscriminately. Is there any reason for believing that each child would show a predilection or greater facility for acquiring the language of its country ? R. MELDOLA.
October 8.

Autotomic Curves.

BRITISH mathematicians have usually employed the phrase "non-singular curve" to designate a curve which has no double points. This phrase is an exceedingly infelicitous and misleading one, since a *point of inflexion* is just as much a singularity as a *double point*.

The word *autotomic* (self-cutting) has occurred to me as a suitable one to designate a curve which has double points ; but the objection to this word is that the phrase "an anautotomic curve" is somewhat offensive to the ear. In the case of media which are not isotropic, mathematicians have evaded a similar difficulty, which would be caused by the use of the word *anisotropic*, by employing the term *aeolotropic*.

Perhaps some of your readers, who have kept up their classics, may be able to suggest a suitable word to convey the idea of a "not-self-intersecting-curve." A. B. BASSET.
Fledborough Hall, Holyport, October 5.

THE OPENING OF THE MEDICAL SCHOOLS.

THE subject-matter of the studies comprising the medical curriculum lends itself exceptionally well to the delivering of inaugural addresses. Every October produces its crop of young men and women beginning the study of medicine, and to these are addressed with never-failing regularity an almost constant number of introductory lectures. To those who watch from a distance the perhaps somewhat monotonous rhythm of medical academics, it would appear at least probable that these opening addresses, actuated, as they certainly are, by a perpetually similar motive, would be in imminent danger of suffering from a monotony almost approaching boredom. Medicine, however, and its allied sciences are never at rest, and the constant change in them from year to year forms an almost inexhaustible subject-matter at once interesting to the initiated, and stimulating to the novice. The latter learns, as a rule even in his first lecture, that in embracing medicine as his profession he has not to tread a rigid scholastic entity, but to pick a somewhat circuitous way over a plastic science, which is capable and willing to receive the intellectual footprints of all who are strong enough to impress them.

The third Huxley lecture was delivered this year by Lord Lister, and the authorities of Charing Cross Hospital are to congratulated upon having heard one of the most interesting discourses which it will probably ever be their lot to listen to. Although none of us can for any length of time forget the scientific work for which Lord Lister is celebrated, perhaps few of us are cognisant of his early researches, which, indeed, although perhaps to the superficial observer remote from his later work, according to him—and what better authority can we want—led up to it. All those who have the opportunity of reading this address *in extenso* should do so ; it forms another of the many instances of how work of a more or less erudite character led up to results most emphatically utilitarian. This should be remembered and taken to heart by those in authority who are apt to look askance at work, whatever it may cost in the way of perseverance or intellectual effort, which is not immediately productive of utilitarian result.

At the relatively new School of Tropical Medicine, Sir W. MacGregor delivered an inaugural address, taking for his subject some problems of tropical medicine. An interesting point in his discourse was the importance he attached to the study of dysentery. According to him, epidemic dysentery is a scourge of tremendous magnitude, carrying off in some cases 50 to 75 per cent. of the labourers upon certain plantations in Polynesia. His concluding remarks were directed to malaria, and he emphasised in this connection the importance of investigating certain equine maladies on the West Coast of Africa, apparently of a malarial type.

At University College an inaugural lecture was delivered by Dr. Vivian Poore upon science and practice. The lecturer addressed himself mainly to those students who were actually beginning the study of medicine as distinct from the so-called elementary medical sciences. He warned his hearers against adopting the view that all that could not be submitted to laboratory methods was *ipso facto* not scientific, and in this connection drew attention to certain discoveries made by physicians by mere observation, which were from the highest standpoint scientific. Many facts, although explained by bacteriological research, had been discovered by physicians by methods which were purely clinical. Dr. Poore finally pointed out the advantages that should accrue to the students at University College Hospital from the reconstruction it was undergoing, and also how the reconstitution of the University of London might be expected to affect the medical lectureships in the metropolis.

King's College Hospital began the medical session by an old students' dinner, at which Sir John Cockburn presided. The speeches which ensued, although limited, were mainly directed to two subjects, the share which has been taken by the staff in the surgical work in South Africa, and the manner in which the school had been improved during the past year by the establishment of largely increased laboratory accommodation and new lectureships, special reference being made to the subject of pharmacology.

At the inaugural dinner at Guy's considerable interest was attached to Mr. Fripp's speech, which dealt with the work done by the Imperial Yeomanry Hospital in South Africa. This hospital had broken a deal of new ground, and many in a position to judge of the organisation and management of hospitals had expressed the hope that from the demonstration which it had been enabled to give of the way that modern scientific methods could be introduced into the service of the sick and wounded in the time of war, reorganisation of military hospitals upon the lines of the large civil institutions might be effected.

The medical session at St. Thomas's was opened by Sir William MacCormac. In the course of his remarks Sir William pointed out that the present time was a favourable one for entering the medical profession, as London was about to have a great university in reality, and not one in name only, and also that the supply of medical men was not keeping pace with the increase in the population, as shown by the list of medical students entered on the rolls of the General Medical Council.

At St. Mary's Hospital the introductory lecture was delivered by Mr. Stansfield Collier upon the future of the medical student. At the London School of Medicine for Women Mr. Aldrich Blake addressed the students. The subject-matter of both these lectures consisted mainly of advice to the student with regard to the mental habits he should cultivate in approaching his work. An interesting lecture was delivered at the Royal Veterinary College by Prof. MacFadyean. According to the lecturer, the century approaching its conclusion embraces practically the whole history of the veterinary profession in Great Britain. One hundred years ago the Royal Veterinary College was only in the tenth year of its existence, and since then great changes had come over veterinary opinions and practice. With regard to the progress that had been made concerning the causes of disease, glanders and tuberculosis were taken as types. As recently as twenty years ago the opinion that tuberculosis was an extremely hereditary disease was universal among veterinary surgeons, and it was only within the last decade that the erroneousness of this view had been generally recognised. Discoveries which threw light on the cause and nature of maladies necessarily influenced methods of treatment and prevention, and the past century had witnessed great changes in the means adopted to counteract disease. As an instance of this, the almost complete extent to which bleeding has become obsolete may be given. Firing and blistering, although still extensively used, were employed now with more discrimination than formerly, and it seemed upon insufficient evidence.

The inaugural address in the medical faculty of the University of Birmingham was delivered by Prof. Windle, who chose for his theme the very appropriate one of the needs, aspirations and ideals of the Birmingham Medical School. Those who have carefully followed the stages in the development of the great Midland University will not have failed to observe that the main spirit which actuated it when it was, so to speak, on paper, and will continue to guide and distinguish it when architects' plans have been replaced by well-equipped laboratories, is the spirit of scientific research. It will be a great centre for teaching students how to prosecute research in all branches of technical industry. The generous manner in which land has been given by Lord Calthorpe and funds have been supplied by the wealthy citizens of Birmingham has secured opportunities for this purpose, which, when the whole is complete, will probably be unrivalled. Such opportunities are certain to attract students, and especially medical students. It is to be hoped that what is being done at Birmingham will stimulate the formation and endowment of laboratories at the London schools, which, if they wish to attract the mass of students they have done in the past, will be compelled to offer them advantages at any rate not inferior to those that they can find much nearer home.

The students of the Middlesex Hospital were addressed by Prof. Clifford Allbutt upon abstractions and facts in medicine. His concluding remarks with regard to the value of research laboratories are significant, the lecturer confessing that mere observation of disease and morbid anatomy have taken us almost as far as these means can do. Morbid processes should be tracked in their earliest dynamic initiation in order that they can be arrested in these stages. The clinical laboratory of a county hospital should be the centre of enlightenment to all the private practitioners of the district.

The space at our command has only permitted us to reproduce a relatively small fraction of the many interesting and instructive addresses that were delivered during the course of last week. It is sincerely to be hoped that teachers and students alike have profited by them, and that their united efforts will result in the addition to the profession of a body of workers who will be in the truest sense medical imperialists, and who, while working to the fullest advantage the store of learning they have inherited, will not rest content with it, but extend in all directions the empire of medical knowledge, even up to the threshold of the temple of truth.

F. W. TUNNICLIFFE.

A NIGHT WITH THE GREAT PARIS TELESCOPE.

SINCE the final decision was made some years ago to commemorate the Paris Exhibition of 1900 by the installation of a giant telescope which should surpass in size and power any other then in existence, so many varied and contradictory statements have been quoted in the Press, and even in many scientific journals, that a considerable amount of scepticism has been inherent in the minds of most persons interested in the matter. Much of the inaccuracy is traceable to a rather loose estimate being given of the magnification which it was hoped to employ, it being stated that the moon would be apparently brought so close that any object of 1 square metre area could be distinguished. By the extreme kindness and courtesy of M. François Deloncle, to whose initiative the entire instrument is due, the writer was enabled, not only to thoroughly examine all parts of it during the day, but also to take part in the practical astronomical use to which it is already being put during every clear night. A general view of the siderostat is shown in Fig. 1, and the inclusion of the attendant's figure in the upper balcony will give some idea of its relative size. The masonry foundation is about 5 feet 6 inches high, the extreme height of the curved casting carrying the mechanism at the back being about 34 feet. The circular glass mirror seen between the upright fork in front is 6·5 feet (2 metres) in diameter, and about 11·8 inches (30 centimetres) thick, being silvered on the upper exposed side. When not in use a large glass plate is lowered over the silvered surface by a windlass worked from the gallery. As the glass mirror weighs some 3600 kilogrammes, and the iron cell and forked support about 3100 kilogrammes, the friction on the pivot allowing rotation would have been too great for accurate driving if some provision had not been made for eliminating it. This has been successfully done by immersing the base of the fork casting in a bath of mercury, contained in the circular part of the front half of the main base plate, thereby relieving the pivots of about 9/10ths of the total weight. The rotation of the mirror in a vertical plane is also facilitated by the counterpoise weights shown at the ends of the levers acting on each extremity of the horizontal axis passing through the centre of the mirror.

At the western side of the siderostat, above the

handles moving the instrument in right ascension and declination, are two telescopes, which by a system of lenses and mirrors enable the observer to read the divisions on two graduated circles without leaving his position. By his side there is also a standard sidereal clock and a telephone.

FIG. 1.—The great siderostat, Paris, 1900.

Leaving now the siderostat, and mounting the staircase seen behind it, access is gained to the upper balcony which runs round both sides of the whole length of the building. Fig. 2 is a view taken from the eye-end of the telescope, 200 feet away to the south, and the siderostat

FIG. 2.—Eye-end of the refractor, Paris, 1900.

can just be seen under the arch at the north end. Above, on the gallery, the circular object-glass is clearly shown in its case. Two of these lenses are to be provided, one specially corrected for visual work, the other for photographic purposes. The lens completed and in position is the latter. The carriages for holding these lenses in their massive cells are enclosed in the large rectangular glass case shown in the figure, and being mounted on rails, it is a matter of but a few moments to interchange the position of the lenses with respect to the long tube seen extending from end to end of the gallery. This enormous (and, from an astronomical point of view, probably superfluous) tube is composed of twenty-four sections of sheet steel, each about 8 feet 3 inches (2·50 metres) long, and 59 inches (1·50 metres) in diameter, which are supported on six braced columns rising from the floor below.

At the near end of the tube is shown the massive tail-piece, with the various arrangements for focussing, clamping and rotation of the photographic plate. The focussing is done by traversing the whole of the tail-piece on the two short rails, a motion of about 5 feet (1·50 metres) being allowed, as it is unlikely that the focal lengths of the two lenses will turn out identical.

The lens in position, corrected for the photographic rays, is 49·2 inches (1·25 metres) aperture, and 187 feet (57 metres) focal length. The diameter of the image of the sun or moon at the principal focus will therefore be about 21 inches in diameter.

Rotation of the photographic plate, about 30 inches square, during exposure is necessary on account of the fact that when a siderostat is used, only the central point of the field of view remains stationary, all the surrounding parts having a motion round this as centre. To eliminate this M. Gautier has provided a subsidiary clock, placed to the rear of the eye-end, which, by means of connecting gear to the milled wheel seen on the circumference of the end of the tube, turns the whole tail-piece of the telescope at any desired rate. For visual observations the plate-holder is removed, and an adapter carrying an eye-piece is inserted. The whole of this eye-end section is now covered in by a temporary dark-room.

To admit the light from celestial objects to the mirror of the siderostat, the roof and walls of the building for some 70 or 80 feet are made in two sections, one of which slides to the north, the other to the south, allowing of a clear view from the zenith southwards to within some few degrees of the horizon. This done (requiring some six or eight attendants to work the pulley blocks), all the subsequent movements and adjustments are easily made by two observers. One is stationed at the base of the siderostat, with the handles for working the instrument, telescopes for reading the circles, sidereal clock and telephone all within his reach. Whatever object is selected from the previously prepared working list for the evening, the declination handle is turned until the scale reading seen through the telescope gives the correct declination of the object. Then the hour-angle, or difference between the sidereal time indicated by the clock and the right ascension of the star, is set by means of another handle and the second small telescope. This done, the required object will be near the centre of the field of the large telescope, and the clock being set running to keep it there, the astronomer takes his place at the eye-end to bring it exactly to the centre of the eye-piece. He also has a telephone by his side, and for a space of some two or three minutes there is a continual cross-fire of such terms as "Déclinaison," "Ascension droit," "Doucement," "Au sens contraire," with various endearing terms of admonition in cases of overdoing the movements. Considering, however, the comparatively high power (500) which is the lowest used, it is astonishing how quickly an object is obtained after the setting of the circles. This in itself furnishes an incontestable proof of the extremely accurate adjustment of the siderostat and telescope, both as regards the angle of its polar axis and its position in the meridian. The object being found, all lights are extinguished and a drawing or photograph made as carefully as possible. On the evening it was my privilege to be present, our

first object was the Ring Nebula in Lyra (Messier 57). The astronomer in charge was M. Eugène Antoniadi, of the Juvisy Observatory, near Paris, who has started a systematic study of nebulæ with the telescope. This view of the nebula surpassed anything seen before, although frequent observations of it have been made by the writer with a telescope of 36 inches aperture. The great increase of light given by the great glass made it possible to use a highly magnified image, which was at the same time bright enough for the eye to detect detail without any strain.

We next turned to the coloured double star β Cygni, after that α Lyræ (Vega), the companion of which was a very conspicuous object. During all this time the instrument was being used in a building containing several hundred people, electric lights all over heating the air, and the huge searchlights and illuminations from the surrounding buildings causing considerable atmospheric glare—conditions under which none but first-class apparatus would be workable. However, owing probably in a great measure to the large proportion of the roof which is opened, the star images were not inconveniently unsteady.

M. Deloncle appears indefatigable in doing his utmost to entertain any one having special interest in his *protégé*. Parties of guests are often there to listen to a short address from him on its construction and installation, after which they go to the eye-piece, and in turn see whatever may be on view. This goes on till after midnight, and then, the last visitor away, all lights are turned out, the lens case opened to permit of air circulation, and about half an hour or more allowed to elapse for the general temperature to be equalised throughout the various parts. We turned next to an object which M. Antoniadi had not previously observed (a small planetary nebula in Sagitta), found it quite easily by the circles and slight subsequent sweeping, and then occupied about an hour and a half in careful drawing. This nebula is G.C. 4572, and M. Antoniadi's drawing appears in the *Bulletin de la Société Astronomique de France*, September 1900.

In sweeping from one object to another, thousands of stars cross the field of view, and it was specially noticeable that no distortion of the star images was to be detected as the mirror was moved to different angles of incidence. With such a high power this is in itself a severe test of the planeness of the silvered surface, and it is worth drawing attention to the fact that the figuring of this mirror by M. Gautier has been done entirely by mechanical means, controlled at every step by the most delicate optical tests.

In connection with the actual work of observing, nothing is so important as accurate clock driving, in order that the astronomer may not be troubled by constantly having to bring the object into the centre of the field. How efficient this instrument is in this respect will be understood when it is stated that during the period for which the clock runs at one winding, about forty-five minutes, the star images do not move sufficiently in the field of a power of 500 to necessitate any adjustment. The angular diameter of the field of view is about 3′ of arc.

By the time M. Antoniadi had finished his drawing of this planetary nebula it was about 3.0 a.m., and approaching daylight. The mirror cover was lowered into its place, the two sliding sections of the roof pulled together over the mirror, and the clock stopped. In a few minutes M. Antoniadi and myself were the only occupants of the gallery, and the institution being already in such regular use that a couple of beds are provided for observers staying all night, we decided that, as it was so terribly hot indoors, we would take up our beds and camp out in the courtyard. We did this, and sleep being somewhat out of the question, spent some time gazing upward in the hope of seeing some forerunners of the August

Perseids, with the usual result of latter-day meteor watching—we saw none. However, it was a novel and exceedingly pleasant experience to be there lying under the stars, the greatest telescope on earth immediately to one's side, the highest building in the world towering over our heads.

It is to be hoped that after the Exhibition is over the telescope will find a resting-place under the Home Government at some station out of the city, where the purity of the atmosphere will allow of its power being efficiently used.   C. P. BUTLER.

TOBACCO.

WHEN Columbus landed in 1492 in the West Indies he found the natives smoking a herb wrapped in a maize leaf, and the name of the herb was Tobago. In 1560 Jean Nicot distributed plants raised from seed to various parts of Europe. These two events give us the clue to the popular and scientific names of a drug the cultivation and preparation of which have now attained such enormous importance that Governments are supported by the revenue derived from its taxation, and colossal fortunes are made by its sale. Some idea of the scale on which the industry is carried on may be gathered from the statistics recently published in the "Year-book of the United States Department of Agriculture for 1899," where we read that during that year 266,661,752 pounds of tobacco, 4,542,016,570 cigars and 4,590,388,430 cigarettes were prepared in the United States alone, yielding a revenue to the Government of 52,043,859·05 dollars.

Small wonder then that the cultivators of so valuable a plant have shown great interest in all the processes of raising, planting, manuring and gathering the crop, and of drying, curing and preparing it for market; or that consternation has arisen in their midst at the origin and spread of a disease which attacks the golden leaf, and bids fair to ruin the crop in some districts. It happens, moreover, that biological problems of wide significance are arising in connection with the complex art of fermenting the leaf so as to obtain the best flavour and strength, as well as in regard to the "Mosaic disease" above referred to, and the experience of Dutch growers, of which an excellent account is now to hand in Koning's "Der Tabak, Studien ueber seine Kultur und Biologie" (Amsterdam and Leipzig : W. Engelmann, 1900), shows that the employment of scientifically trained botanists in the technical laboratories of tobacco plantations is likely to be as usual an event in the future as in breweries and bacteriological laboratories.

The tobacco plant is exceedingly small in the seedling stage—eighteen thimblefuls of seed suffice for a hectare, *i.e.* two and a half acres of land—and is very carefully raised in pots and manured with pigeon's dung, planted out and weeded with extraordinary precautions against numerous enemies, and the leaves eventually picked by hand, sorted, tied into bundles and hung to dry. It is a very exhausting crop, and requires much potash ; and an astonishing amount of information has accumulated concerning the effects of different soils, manures, climate and other factors of the environment on the properties of the leaves. Moreover, there are numerous cultivated races in existence in the various tobacco-growing countries, as always occurs with planted crops.

During the process of slow drying the leaf may remain alive for two to three weeks, and the contained starch is converted into sugar, and further alterations result in an increase of acids. Proteids diminish and amines increase, but the nitrates and alkaloids (nicotin) should undergo no change. The slow alterations referred to are essential, and due to enzyme and other actions in the still living leaf ; in artificially or rapidly dried leaves

the arrest of such changes materially affect the flavour and burning of the tobacco, and naturally much turns on the age and quality of the leaf itself, the soil and season and other conditions of growth, &c.

The dried or "cured" leaves are next submitted to fermentation, a process of vital importance in the opinion of the tobacco expert, since it is this which determines the finer flavours and odours of the manufactured product. Fermentation is started by damping heaps of 15,000 to 30,000 lbs. of the dried leaves, packed in a special manner, and carefully watched by experienced workmen as the temperature rises. The process occupies three to four months, and the leaves are turned about once a month. The temperature rises to about 50–56° C., and a loss of vapour, accompanied by a sweet and sharp odour, is noticed. The reaction may be neutral, though in some cases ammonia is given off, due to the action of undesired bacteria.

As would be expected, the fermentation is always accompanied by bacteria ; but it has long been in dispute whether the essentials of the process are due to bacteria or to the action of special enzymes in the cells of the leaves.

Suchsland's researches had convinced him, not only that the fermentation is due to bacteria, but that a peculiar species of bacteria was specially concerned in the production of the approved flavour, and that the desirable properties of Cuban tobaccos could be imparted to inferior growths by introducing this species into the fermentation. Loew, on the other hand, maintained that the aroma and flavour depend simply on the action of enzymes or other cell-contents in the leaf itself.

Koning has investigated the various bacteria found in the fermenting heaps, and followed the changes induced in the tobacco.

Put generally, the fermented tobacco undergoes little or no change as regards the total nitrogen or the nicotin, but organic acids diminish, and the sugars and nitrates are destroyed, and various aromatic substances are formed which affect the quality of the product.

Among the bacteria isolated Koning claims to have found the species concerned in this remarkable neutral fermentation, and which imparts the aroma and flavour desired, and thus confirms Suchsland's results. He states that tobacco infected with the specific bacteria, fermented and made up, and then handed to experts, was selected by the latter as the superior from specimens containing a pair of cigars each, in packets of two, and labelled *a* and *b, c* and *d*, &c., only ; but the evidence appears conclusive.

During the last ten years increased attention has been drawn to a disease of tobacco leaves, which causes irregularly alternating light and dark patches, and is known as the "Mosaic disease." Koning has established that this is infectious, and is carried through the fields by the fingers of the workmen who "top" the growing plants by pinching off the buds. He has examined the various fungi known to cause leaf-diseases in tobacco, and cannot refer it to these, and the presumption that it is a bacterial disease was strengthened by finding that certain manured soils were almost sure to have badly diseased plants on them ; and that experiments showed that if a bit of diseased leaf, or a little of the sap from such is rubbed into a wound, the young leaves formed above the wound contract the disease. The same result follows if such sap is placed at the roots of healthy plants. But infection fails in all these cases if the sap is previously boiled.

Here may be mentioned that Adolf Mayer had proved the infectious nature of the filtered sap in 1885; and Beijerinck, working at this disease a short time ago (1898), had come to the conclusion that since no organisms could be isolated from the sap—the infectious nature of which he also proved—which will reproduce the disease, and since the same sap *filtered through porcelain* still infects the plants, unless it was previously sterilised by heating, the causal agent must be a *contagium vivum fluidum*—a something of the nature of a poisonous enzyme, which not only diffuses through the plant-membranes—*e.g.* the cell-walls of root-hairs—but increases as it passes from cell to cell.

Koning confirms Beijerinck's principal results, but concludes that since the infecting fluid may be heated to 100° C. for a few minutes without losing its powers, whereas alcohol and glycerine destroy the virulence, as also does *repeated* filtration through porcelain, the active agent is *an extremely minute organism*, which can traverse the pores of a filter. He compares the results with those obtained with the virus of various animal diseases from which no organism has as yet been isolated.

It should be borne in mind that the existence of organisms small enough to pass through a porcelain filter has been accepted by several authorities.

When we reflect that well-studied micrococci are only 0·5 – 0·8 μ in diameter, and that the wave-length of those light rays corresponding to the sodium-line D is about 0·6 μ, some of these matters become less astounding : organisms 1/5th to 1/10th this size would probably be well beyond the powers of our best lenses, and would roll through the pores of a filter as shot through the meshes of a sieve.

It thus appears that—without regarding the work as quite conclusive, which it is not—we have here important contributions to several most weighty biological questions centred about the culture of an economic plant.

NOTES.

The International Congress of Botany was opened in Paris on the 1st inst., and was in session until Tuesday last. M. Prillieux was the president.

The new science laboratories at King's College, London, are to be opened on the 30th inst. by Lord Lister.

Another death from plague has occurred in Glasgow, bringing the number of fatal cases in hospital since the outbreak up to six. A fatal case of plague is also reported from Llandaff.

A Reuter telegram announces the arrival at Copenhagen, on October 4, of Lieut. Amdrup and all the members of his expedition. From July 18 to September 2 the expedition, while engaged on the coast of Greenland, explored and mapped out a stretch of land hitherto entirely unknown and extending from Cape Dalton, 69°28′, to Aggas Island, 67°22′. Lieut. Amdrup is reported to have brought with him important collections, the results of his researches. The *Antarctic* reached Tasiusak on September 11, and sailed thence on her return journey on September 18.

The *Athenaeum* states that the Kolthoff Arctic Expedition has succeeded in bringing to Sweden a male and a female calf of the musk ox (*Ovibos moschatus*, Gmelin). As soon as the animals appear to be acclimatised they are to be set free in the northern mountain regions, where it is thought they will speedily increase in number, as they are very prolific. Herr Kolthoff has great faith in the future importance of the musk ox, not so much as an article of food as on account of its thick brown wool, which is said to be remarkably strong.

The petrified remains of the extinct rhamphorhynchus have been discovered in the stone quarries of Eichstätt, Bavaria. It is stated that the teeth and fingers are very distinct, and that the membrane between the fingers is visible in places.

According to the *Exchange Gazette* of St. Petersburg, the question of the official introduction of the metric system of weights and measures into Russia has been decided in principle in an affirmative sense. The Ministry of Finance is now considering in what manner, and when, the projected reform shall be carried out.

The trustees of the American Medical Association have established a fund of 500 dollars, to be expended annually for the encouragement of scientific research ; but no individual is to receive more than 100 dollars at one time.

The lecture arrangements of the London Institution for the session terminating on February 28 next have now been completed. The science lectures are as follows :—" The Rise of Egyptian Civilisation," by Prof. Flinders Petrie ; " The Earth's Beginning," by Sir Robert Ball ; " The Earth's Earliest Inhabitants," by Prof. Grenville Cole ; " The Caves of Jenolan," by Mr. F. Lambert ; " The Tercentenary of the Science of Electricity," by Prof. Silvanus Thompson ; " The Evolution of the Brain," by Dr. Alex. Hill ; " Modern Aeronautics," by Mr. Eric S. Bruce ; " The First Ascent of Mount Kenya," by Mr. H. J. MacKinder ; The Effect of Alcohol on the Nervous System," by Prof. Victor Horsley ; " The Decorative Art of Primitive Peoples," by Prof. A. C. Haddon ; and " Aquatic Autocrats and Fairies," by Mr. F. Enock. The Christmas course, intended for juveniles, is to be delivered by Prof. W. B. Bottomley, and will be devoted to " Structure and Colour," " Insect Visitors," " Unbidden Guests," and " Place in Nature."

The next meeting of the Royal Microscopical Society will be held on Wednesday, the 17th inst., when Part ix. of a report on the recent Foraminifera of the Malay Archipelago will be presented. Preceding the meeting there will be an exhibition of slides and models of skin structure, by Mr. F. W. Watson Baker.

The first monthly general meeting of the new session of the Institution of Mechanical Engineers will take place on Friday, October 19, when a paper, entitled " Observations on an Improved Glass Revealer for Studying Condensation in Steam Engine Cylinders and rendering the Effects Visible," will be read by Mr. Bryan Donkin, and discussed.

A new monthly meteorological journal has recently made its appearance in Holland, and bears the name of *Nederlandsch Tijdschrift voor Meteorologie*. The style of the journal is popular in character.

The current *Geographical Journal* publishes further details as to the programme of Dr. Sven Hedin's journeys in Northern Tibet and neighbouring regions for the present year. At the time of sending his last letter, on June 27 last, the explorer was about to start for the Chamen Tagh, whither his caravan had already preceded him, his intention being to cross the Astyn Tagh and Koto-Shili ranges, so as to obtain a geological section of the country, and correct his route with that of his former Tibetan journey. After returning to his headquarters in the Chamen Tagh, he hoped to make his way across Northern Tsaidam to Sachu, and thence west to the old bed of Lob Nor, continuing his investigations of the latter end of the ruins in its vicinity. Thence he proposed to carry a chain of altimetric observations to Kara-koshun and Chaklik. He hoped to arrive at the last named point by about January 1, 1901.

A SLIGHT earthquake disturbance in Bombay, on Monday, September 17, is noted in the *Pioneer Mail*, Allahabad. Only one of the instruments at the Colaba Observatory recorded it. The disturbance began at about 3h. 48m. a.m., Bombay time, and reached its maximum at about 3h. 54m. The larger movements ceased at 4h. 2m., and the after-tremors at 4h. 16m. Thus the whole disturbance lasted fully 28 minutes. It was not a distant earthquake, nor was the movement large. The apparent distance of the origin from Bombay may have been about 500 miles. The same journal also records the occurrence, on September 10, of a slight earthquake shock in Madras.

THE New York correspondent of the *Lancet* states that the Chicago Board of Education has established a department called "Child-study and Pedagogic Investigation." The examination is undertaken for the purpose of determining the mental and physical status of the school-children. Examinations were at first limited to the determination in each pupil of the following points: Height, height sitting, weight, ergograph work, strength of grip right and left, hearing right and left, and acuity of vision. In addition to this, obvious developmental defects have been noted. The number of children examined down to the present time is 5636. The conclusions thus far reached are that there is a physical basis of precocity, that dull children are lighter and precocious children heavier than the average child, and that mediocrity of mind is associated with mediocrity of physique. A similar result was obtained in the examination of 33,500 school-children in St. Louis in 1892 by Dr. W. T. Porter. This is the first instance of a municipal board in America appropriating money for research work, and its effect may be far-reaching.

THE German Consul in Payta-Piara (Peru) reports the discovery of large rubber forests on the Niera River, a branch of the Amazon. An expedition has been organised to start for the interior to secure the right to collect the rubber. The increasing demand for rubber has drawn attention to the advantages of cultivating gutta—a leading product of Java and several of the neighbouring islands. A recent number of the *Straits Budget* points out that gutta trees growing wild cannot meet the growing demand which must soon outrun the supply unless gutta plantations extensive enough to meet future needs are laid out. Gutta leaves have been freely resorted to in order to eke out the supply. A company has recently been formed at Batavia to develop this branch of industry.

A LENGTHY account (based on the preliminary report presented to the R. Accademia dei Lincei) of Prof. Grassi's malaria experiment appears in the last number of the *British Medical Journal*, from which we glean the following particulars. In making the experiment, which took place in the plain of Capaccio, near Salerno, two objects were kept in view, viz., (1) To afford an absolute proof of the fact that malaria is transmitted exclusively by the bite of *Anopheles* mosquitoes ; (2) to found on the results of recent research a code of rules to be adopted for freeing Italy from malaria in a few years. The experiment consisted in protecting from malaria railway employés and their families living in ten railway cottages and at the stations of S. Nicolò, Varco and Albanella, situated along the Battipaglia-Reggio Railway. They numbered 104 persons, including thirty-three children under ten years of age. Of these 104 individuals, at least eleven, including four children, had never suffered from the disease, not having previously lived in a malarious district ; a certain number, it appeared, had not suffered from it for two or three years, and all the others, that is to say, the large majority, had suffered from it during the last malarial season, some of them even in the winter. During the malarial season, the health of the protected individuals was

good, with the exception of a few cases of bronchitis and a case of acute gastro-enteritis. None of these cases were treated with quinine. The 104 persons, with three exceptions, had remained free from malaria up to September 16th, the date of the report. From the report it is evident that the twofold object of the experiment has met with every success, and it certainly looks as if it will be possible to free Italy in a short time from malaria, and that the much dreaded plain of Capaccio (excepting for the *Anopheles* infected by malaria germs, from which one can easily protect himself) is one of the healthiest places in Italy.

THE current issue of the *British Medical Journal* gives particulars of the new bacteriological laboratory of the Melbourne University, at present in course of erection. The large hall of the laboratory will afford accommodation for eighty students. Situated on the ground floor there are four research laboratories and one preparation room specially fitted for the professor of pathology, who has, in addition, one room set apart for pathological-histological work and one for chemical pathology. Steam is laid on all over the building for both heating and sterilising purposes. One room has been set apart as a plague laboratory, having germ-proof windows and doors. The walls are covered with tiles made of opaque glass and the floor with lead, so that it can be flooded with antiseptics in case of accidents. In the building there are rooms devoted to micro-photography, the preparation of media of all kinds for the sterilising and disinfecting of all utensils, and departments for numerous other purposes. The department is subsidised by the Board of Public Health, by the Ministry of Agriculture, by the Metropolitan Board of Works, by the City Council, and by other municipalities, who contribute between them about 800*l.* per annum. The duties of this bacteriological department consist in the examination of water and milk supply, of investigating diseases which are discovered at the abattoirs and amongst stock, and analysing the effluence from the sewage farm. It is also prepared to assist the Board of Public Health, on behalf of the medical profession, in diagnosing diphtheria, typhoid, tubercle and cases of plague. It is expected that the building will be opened in March next.

A NEW use for kites is brought under notice in the latest issue of the United States *Monthly Weather Review*. An exhibition has recently been given in Chicago showing how those within a besieged town or other inaccessible place can use the kite line to carry a telephone, with its separate telephone wire, through the air, and let it drop from the kite upon a distant place while the kite still remains in the air. By using a very large box kite and attaching to the kite line a little way below the kite a pulley through which runs the telephone wire, the telephone may be dropped from the pulley while the insulated wire keeps up the connection with the man at the kite reel. Of course, at the present time (as is pointed out in the *Review*), when kites have rarely been sent out with more than two miles of wire, which corresponds to a horizontal distance of much less than two miles, this method does not promise to put into communication persons separated by a great distance, but it may be very useful for short distances.

IN order to enable Essex dairy farmers and ladies engaged in dairying to gain an insight into agricultural education and the organisation and practice of the agricultural industries of Denmark, the Essex Technical Instruction Committee arranged a visit to that country in May last, when thirty-one persons spent a week there in visiting the agricultural schools, dairy farms, butter factories, &c. The excursion (judging from the careful report just issued from the County Technical Laboratories, Chelmsford) seems to have been in every way a great success. The members of the party were impressed with the thriving condition of agriculture in Denmark, the happy position of which is attributed in the report to legislation, education and co-operation. Visits such as that under consideration cannot, if properly organised, fail to bring about useful results.

THOSE interested in the phylogeny of the Vertebrata will be pleased to hear of the discovery in the seas of Alaska of a new representative of the Enteropneusta, in the form of a species allied to the typical group of Balanoglossus. This new form, which it is proposed to call *Harrimania maculosa*, is described by Mr. W. E. Ritter in Part 2 of "Papers from the Harriman Alaska Expedition," now in course of publication in the serial last quoted. The new form leads its describer to conclude that the notochord of the Enteropneusta is undoubtedly homologous with that of the Vertebrata. It is specially noticeable in that it consists of two parts ; the anterior corresponding with the same structure in the other members of the group, while the additional posterior moiety is connected with the æsophagus. This latter portion is peculiar in that it persists throughout life. Harrimania, which is considered to be the most primitive member of the group, instead of burrowing like Balanoglossus, lives under stones, where it often makes its way through the mud at the plane of contact between the latter and the stone beneath which it is concealed. While some examples were only found at extreme low tide, others occurred much nearer high-water mark than is the case with any other Enteropneusta.

THE last published number of the U.S. *Monthly Weather Review* (June) contains a note by Prof. E. B. Garriott on the extension of the Weather Bureau work, which constitutes one of the most substantial advances in the history of that institution. The West Indian branch was established in 1898, and at the present time practically all the cable islands and ports of the West Indies and the Carribean coast of South America receive advices regarding tropical storms. The West Indian observing stations number thirteen, and hurricane warnings are displayed at more than a hundred points. Telegraphic reports are now also received from well-distributed Mexican stations. It is believed that the reports received from the northern and western parts of Mexico will lead to a better understanding of the important storms which sweep north-eastward from the tropical Pacific and cross the United States to the Atlantic. Reports from the extreme north-west of Canada have been added within the last two years and have furnished valuable data regarding the movements of North Pacific storms. The Weather Bureau has consequently reports from an area extending over more than 42° of latitude and 65° of longitude. The advantage afforded by these widespread telegraphic stations can hardly be over estimated.

THE *Philosophical Magazine* for October contains an interesting paper by Mr. R. J. A. Barnard, of Melbourne, on the annual march of temperature. The author has examined the observations for forty years at Melbourne and has divided them into two groups, 1859–78 and 1879–98. In each group the average for every day is obtained and smoothed twice by replacing the temperature of each day by the mean of the five consecutive days of which the particular day is the middle. The results of the second smoothing show that in the second week of March the temperature begins to drop rapidly, reaching a secondary minimum for each curve on the 19th. It then rises again to the extent of 2°·5 during the next week, reaching a maximum on the 25th and 26th, the data being the same again for both groups. The spells are not so marked as those found by Dr. Rijkevorsel for Europe, owing probably to the more uniform conditions of the southern hemisphere. The results show that a period of less than forty years is not likely to give any trustworthy information about spells, and that the division of

the observations for any particular place into groups of twenty years and the comparison of them in the way specified appear to be satisfactory methods of finding out whether such spells really exist.

THE rise and fall in the level of a lake, produced by the mechanical action of wind, is strikingly shown in a short paper contributed by Prof. A. J. Henry to the U.S. *Monthly Weather Review* for May. Continuous records of lake level at four points, viz. Amherstburg, Ontario, mouth of the Detroit River, and Buffalo Harbour, Lake Erie, were considered in conjunction with continuous records of wind direction and velocity and atmospheric pressure made at the Weather Bureau offices in Detroit and Buffalo ; and a relation between the wind and water is clearly shown when the two sets of records are plotted under one another. It has been known for some years that general winds, as distinguished from local winds, blowing parallel to the longer axis of the main body of the lake, have a tendency to heap up the water at the end of the lake toward which they blow, and to depress it at the opposite end. Owing to the convergence of the shore lines at Buffalo, the heaping up of the waters in that harbour, under the influence of a south-westerly wind, becomes a serious menace to the safety of wharf and dock property. Likewise, owing to the shoal water at either end of the lake, a decrease in the available depth in the harbours and channels produces vexatious delays and frequent groundings.

PROF. HENRY'S examination of the facts referred to in the foregoing note shows that when high water exists at Amherstburg it is always low water at Buffalo. The synchronism of the times of high water and low water at the two places is almost perfect. The period of oscillation is likewise fairly constant, ranging from six to eight hours for a half oscillation, and from twelve to sixteen hours for a whole oscillation. The computed time of a whole oscillation, assuming the lake to have a mean depth of 50 feet, is, roughly speaking, seventeen hours. While the information on the subject is as yet too fragmentary to admit of drawing trustworthy conclusions, this much seems to be apparent : the oscillations are stationary rather than progressive. A wave of water is not propagated, in the ordinary sense of that word, from one end of the lake to the other, but the whole lake oscillates about a pivotal or nodal line, which, in the case of longitudinal oscillations, may be said to cross the lake about the longitude of Fairport, Ohio. Although there is no instrumental evidence of the fact, it may be assumed that, as in the case of similar oscillations in other land-locked bodies of water, the oscillations at the nodal line are zero, increasing to a maximum at the respective ends of the lake. Prof. Henry concludes by saying that it is within the range of probability that the occurrence of the more pronounced oscillations can be forecast by the Weather Bureau at no distant period.

To the *Entomologists' Monthly Magazine* for October, Mr. R. McLachlan contributes an abbreviated translation of an article, by M. A. Lancaster, on the swarms of a species of dragon-fly (*Libellula quadrimaculata*) which were observed in Belgium on June 5 and 10 last. On both occasions the temperature was very high, and the insects flew against the wind. The translator is of opinion that nothing certain in regard to the causes of these remarkable migrations has hitherto been ascertained. "As a rule," he writes, "the multitudes are so vast as to make it difficult to believe that all can have been bred within a very limited area. On the contrary, it rather looks as if the individuals in a certain initial locality, being seized with an uncontrollable migratory impulse, were progressively joined by others till the accumulations formed the ultimate swarm. A part of the second swarm seems to have reached England.

IN an exhaustive paper read before the Manchester Literary and Philosophical Society (*Memoirs*, vol. xliv. Pt. 4, pp. 1–8), Mr. Thomas Thorp describes his modification of the diffraction process of colour photography, first put forward by Prof. Wood. The difficulty of determining the correct degrees of rulings necessary for giving certain colours by Prof. Wood's arrangement is obviated by adjusting several copies of a grating having the same spacing at angles to each other, the angles giving the best colour combinations being found by experiment. Full details are given of the method of obtaining the celluloid copies from an original metal reflection grating. A diagram is included to illustrate the procedure for obtaining stereoscopic views with the apparatus.

IN the *Astrophysical Journal* (vol. xii. pp. 30–48), Herr J. Hartmann, of the Astrophysical Observatory at Potsdam, gives a useful and interesting series of suggestions on the design and critical adjustments of photographic spectrographs intended for observational work of a high degree of accuracy.

IN *Science* for September 28 is to be found an article on the "International Catalogue of Scientific Literature" from the pen of Sir Michael Foster, in which it is stated that more than forty-five complete sets of the work have already been subscribed for in the United States, and that, therefore, the catalogue will be begun at once.

MANY of our readers will be pleased to know that the recently delivered Huxley lecture of Lord Lister is to be had *in extenso* in both the *Lancet* and the *British Medical Journal* for Saturday last.

THE last part published (vol. xii. part 4) of the *Journal* of the College of Science, Tokyo, gives evidence of the continued activity of the Japanese University in zoological research, in two valuable papers:—Further observations on the nuclear division of Noctiluca, by C. Ishikawa ; and Notes on some exotic species of Ectoparasitic Trematodes, by Prof. S. Goto, both well illustrated. This number also contains the commencement of an enumeration, by T. Ito and Prof. J. Matsumura, of the flowering plants of the Lûchû Islands, the rich and interesting flora of which is at present but imperfectly known.

THE second volume of the elaborate "Cyclopedia of American Horticulture," by Prof. L. H. Bailey, Dr. W. Miller and others, has been published by Messrs. Macmillan and Co., Ltd. The volume extends from E to M, and is to be followed by two others. There will be more than two thousand original illustrations in the complete work, and the text will be on a proportionally large scale. The scope of this remarkable undertaking may be judged from the sub-title, which certifies that the complete work will comprise "suggestions for cultivation of horticultural plants, descriptions of the species of fruits, vegetables, flowers and ornamental plants sold in the United States and Canada, together with geographical and biographical sketches." Our review of the work will be deferred until the appearance of the final volume.

NEW editions of several established works have been recently received. The fifth edition of "Quantitative Chemical Analysis," by Dr. Frank Clowes and Prof. J. B. Coleman, has been issued by Messrs. J. and A. Churchill. The book provides a sound course of work both in manipulation and analysis, and is in touch with modern methods. Among the additional matter we notice a description of a new and ingenious method of determining melting-points, a special apparatus for the rapid filtration and ignition of precipitates, and an improved form of condenser for use in the distillation of liquids.—The principles and practice of paper manufacture are presented in a form suitable for students in "A Text-book of Paper-Making," by C. F. Cross and E. J. Bevan (E. and F. N.

Spon). The second edition of the book has just appeared.—Another second edition is "Agricultural Zoology," by Prof. J. R. Bos, translated by Prof. J. R. Ainsworth Davis. An appendix has been added, containing an instructive statement of conditions which determine the appearance of harmful animals ; also the general principles as to the means to be employed against them, and lists of pests classified according to their habitat. A full index has also been added.—Now that attention is being given to nature study in rural schools, Mrs. Brightwen's writings upon animal and plant life should receive additional admirers. Her books are of a kind that cannot be too widely known, so we are glad to see that Mr. Fisher Unwin has just issued new editions of "Wild Nature Won by Kindness " and "Glimpses into Plant Life."

THE "Memoirs and Correspondence of Lyon Playfair " was fully noticed in our issue for December 7 last, so that it is now unnecessary for us to do more than to state that a "popular " edition of the book has just been issued by Messrs. Cassell and Co., Ltd.

THE new issue of the *Journal* of the Royal Agricultural Society of England, contains, as usual, a number of interesting and valuable articles, among which may be mentioned an account, by Dr. Fream, of the York meeting of the society, "The Trials of Steam Diggers at York," and an obituary notice (with a page portrait after Herkomer) of Sir John Bennet Lawes.

Nature Notes for October mentions that an avocet was shot near Penzance in the spring.

IN the October number of the *Journal of Conchology*, Mr. A. G. Stubbs concludes his synopsis of the freshwater and land molluscs of the Tenby district.

IN the last annual "General Report of the Geological Survey of India," a voluminous book of 258 pages, the results of much industrious labour by the officers of the Survey are summarised. Considerable activity is being displayed in the prosecution of chemical and palæontological research, and the field-work has been directed towards elucidating economic questions as well as others purely geological. Part iii. of this work contains individual "progress reports" by various officers of the Survey, which consist of papers dealing, not only with economic matters such as auriferous reefs and coal-fields, but also with various questions of considerable geological interest.

UNDER the title of "A List of Works on North American Entomology," the U.S. Department of Agriculture, Division of Entomology, have issued, as No. 24 of the new series of their *Bulletin*, an extremely useful classified index to the most important publications relating to the various orders and families of North-American insects. It has been compiled under the direction of the entomologist, Dr. L. O. Howard, by his assistant, Mr. Nathan Banks. The idea is excellent and might well be adopted for other publications of a similar character respecting the insects of other countries. Such an index cannot be complete, but so many of the most important works on the subject are included that it will be easy for a student taking up the study of any special branch of North American insects to feel his way by this bibliography at the commencement and to enlarge his knowledge on a good foundation as he progresses.

THE latest part of the *Journal* of the Asiatic Society of Bengal (that issued on July 9) is wholly taken up with Sir George King's "Materials for a Flora of the Malayan Peninsula," a portion of which has been written by Dr. O. Stapf, of Kew.

IN addition to the usual "notes" and proceedings of Irish Societies, the October issue of the *Irish Naturalist* contains an interesting communication by Mr. C. B. Moffat, entitled "The Habits of the Hairy-armed Bat."

WE have received a copy of the *Papers* and *Proceedings* of the Royal Society of Tasmania for 1898-1899 (issued June 1900). It contains a useful list of Tasmanian mollusca, by Miss Lodder ; several petrological papers, by Mr. W. H. Twelvetrees and Mr. W. F. Petterd, in which limurite, haüyne-trachyte, felsites, nephelinite and other rocks are described ; also a note by the same authors on bones of Tasmanian labyrinthodonts. Numerous other natural history subjects are dealt with.

AMONG recent American and Colonial reprints we have to note the following :—Contributions to the U.S. National Herbarium, the plant covering of Ocracoke Island, by Thomas H. Kearney, jun. ; and Stigmonose, a disease of carnations and other pinks, by Albert F. Woods ; both issued by different divisions of the U.S. Department of Agriculture ; Progress of plant breeding in the United States, by Herbert J. Webber and E. A. Bessey, reprinted from the year book of the U.S. Department of Agriculture for 1899 ; and Observations on the eucalypts of New South Wales, parts 5 and 6, by H. Deane and J. H. Maiden, from the *Proceedings* of the Linnean Society of New South Wales.

A SMALL pamphlet, issued by the Joint Agricultural Council of the East and West Ridings and of the Yorkshire College, containing a list of one hundred Yorkshire weeds, should be useful to farmers. It will be supplemented by a herbarium containing entire plants of the hundred weeds, and a cabinet containing their seeds or fruits, and will be followed by others.

IN the *Bulletin*, No. 3 (Petroleum Series), issued by the School of Mines, University of Wyoming, Mr. W. C. Knight gives an account of the oil-fields of Crook and Uinta counties. The geological features of the oil-yielding districts are briefly sketched and illustrated by maps and sections in the text. Tables showing the results of testing in various samples of oil are contributed by Mr. E. E. Slosson.

"NOTES on some Jurassic Plants in the Manchester Museum," by Mr. A. C. Seward, are published in the *Memoirs* of the Manchester Literary and Philosophical Society, vol. xliv. 1900, and reprinted in "Notes from the Manchester Museum," No. 6. The collection dealt with includes the plants figured by Lindley and Hutton, in addition to other specimens in the museum. The synonymy of numerous species is given, and a list of Inferior Oolite plants contained in the museum collections is also added. These notes are accompanied by four well-executed plates and a useful bibliography.

IN the *Mémoires du Musée Roy. d'Hist. Nat. de Belgique*, t. i. 1900, Mr. Seward describes the Wealden flora of Bernissart (Belgium). The plants described are obtained from a fresh-water deposit and are of a fragmentary nature ; the flora is remarkable for the relatively large number of ferns included, while the cycads are absent and the conifers rare. Twenty species of plants are recognised, more than half of which are known also from England and Germany. The memoir is well illustrated by four plates and several text-figures.

SINCE the publication in our last number of a list of forthcoming science books, we have received the list of announcements of Mr. Gustav Fischer, of Jena. It is as follows :—"Atlas der topographischen Anatomie des Menschen," by Profs. Bardeleben and Haeckel, Zweite völlig umgearbeitete und vermehrte Auflage ; "Organographie der Pflanzen, insbesondere der Archegoniaten und Samenpflanzen," by Prof. Goebel, Zweiter Teil, 2 Heft, Erster Teil ; "Klinisches Jahrbuch," Siebenter Band, Fünftes Heft ; "Lehrbuch der vergleichenden Anatomie der wirbellosen Tiere," by Prof. A. Lang, Zweite umgearbeitete Auflage, Erste Lieferung, Mollusca ; "Die Grundlagen und die Methoden für die mikroskopische Untersuchung von Pflanzenpulvern," by Prof. A. Meyer ; "Aetiologie und Prophylaxe der Lungentuberkulose," by Dr. J. Ruhemann ; "Anatomisch-klinische Vorträge aus dem Gebiete der Nervenpathologie," by Prof. K. Schaffer ; "Praktikum der physiologischen Chemie," by Prof. Fr. N. Schulz ; "Lehrbuch der Histologie und der mikroskopischen Anatomie des Menschen mit Einschluss der mikroskopischen Technik," by Prof. P. Stoehr, Neunte verbesserte Auflage ; "Das Neuron in Anatomie und Physiologie," by Prof. Max Verworn.

Mr. H. K. Lewis promises the following new books :—"Blood and Blood Pressure," by Dr. G. Oliver ; and new editions of "Hygiene and Public Health," by Dr. Louis Parkes : "Medical Electricity," by Dr. Lewis Jones ; "The Student's Medical Dictionary" and "A Pocket Medical Dictionary," by Dr. G. M. Gould.

Messrs. Lovell Reeve and Co., Ltd., have in preparation a "Monograph of the Membracidæ," by G. B. Buckton, F.R.S., who will be glad to hear, through the publishers, from entomologists and others as to specimens which they have reason to believe are as yet unknown to science.

Messrs. Williams and Norgate announce :—"The Opus Majus of Roger Bacon," edited with introduction and analytical table by J. H. Bridges.

MR. MARTINUS NIJHOFF, bookseller, of the Hague, has just issued the first part of his classified natural science catalogue containing nearly 2500 entries.

THE additions to the Zoological Society's Gardens during the past week include a Vervet Monkey (*Cercopithecus lalandii*) from South Africa, presented by Lieut. Sullivan ; a Black-backed Jackal (*Canis mesomelas*) from South Africa, presented by Mr. J. E. Matcham ; a Two-spotted Paradoxure (*Nandinia binotata*) from West Africa, presented by Mr. R. G. Pointer ; a Vulpine Phalanger (*Trichosurus vulpecula*) from Australia, presented by Miss Bartlett ; a Blue-faced Amazon (*Chrysotis versicolor*) from St. Lucia, West Indies, presented by Miss M. Moon ; two Greek Tortoises (*Testudo graeca*), European, presented by Sister Heather Grey ; a Chameleon (*Chamaeleon vulgaris*) from North Africa, presented by Mrs. E. Putz ; a Bonnet Monkey (*Macacus sinicus*) from India, two Pucheran's Guinea Fowls (*Guttera pucherani*) from Somaliland ; a Large Grieved Tortoise (*Podocnemis expansa*) from the Amazons ; a Greek Tortoise (*Testudo graeca*), European, deposited ; two Common Rattlesnakes (*Crotalus durissus*), two Water Vipers *Ancistrodon piscivorus*), two Copper-head Vipers (*Ancistrodon contortrix*), two Mocassin Snakes (*Tropidonotus fasciatus*), a Hog-nosed Snake (*Heterodon platyrhinos*) from North America, received in exchange.

OUR ASTRONOMICAL COLUMN.

Ephemeris for observations of Eros.

1900.		R.A.		Decl.
		h. m. s.		
Oct. 11	...	2 42 4·23	...	+48 32 58·9
12	...	41 30·67	...	48 53 16·3
13	...	40 52·76	...	49 13 14·2
14	...	40 10·47	...	49 32 50·9
15	...	39 23·66	...	49 52 5·1
16	...	38 32·33	...	50 10 54·8
17	...	37 36·44	...	50 29 18·2
18	...	2 36 36·00	...	+50 47 13·4

THE STABILITY OF A SWARM OF METEORITES AND OF A PLANET AND SATELLITE.

THE problem of the stability of a swarm of meteorites which is under the action of its own gravity and the attraction of the sun, that is to say the determination of the condition under which the swarm will remain unbroken up by the tidal action of the sun, has been dealt with by Schiaparelli, M. Luc Picard and M. Charlier (*Bulletin de l'Académie de St. Petersbourg*, t. xxxii. No. 3). The result obtained by the first assigns a much wider limit of stability to such a system than that arrived at by the other two investigators mentioned ; but there cannot, I think, be any doubt of the greater correctness in actual cases of the narrower limit. ·The problem is intimately related to the still more interesting question of the stability of the earth-moon system, which was treated by Mr. G. W. Hill in a remarkable paper in the *American Journal of Mathematics* for 1878. This again is, in another form, the problem treated still earlier with a special object in view by Edouard Roche (*Mém. Acad. de Montpellier*, vol. i. 1847-50 ; see also *Annales de l'Observatoire*, t. v.), when he arrived at his result concerning the limiting relation between the distance of a satellite from a primary and the diameter of the primary, which must hold in order that the satellite, held together by its own gravitation only, may just not break up under the tidal forces due to the primary, and his corresponding result for a planet's or satellite's atmosphere.

These investigations, though of great general interest, are not so well known as might be expected, and one object of this paper is to give some slight account of them. An abstract of the work of Charlier and Hill is given also in Dr. Routh's recently published work on "Dynamics of a Particle." But I wish also to point out how the main conclusions of Charlier, that of Roche with respect to a planet's atmosphere, and more indirectly the result of Hill, can be obtained by means of elementary considerations.

The problems just referred to have been treated by very different methods. Schiaparelli's discussion is a direct attack of a somewhat involved nature ; those of MM. Picard and Charlier (*Bulletin de l'Académie de St. Petersbourg*, t. xxxii. No. 3 ; see also Tisserand, *Mécanique Céleste*, t. iv.) make use of the method of revolving axes. The radius vector from the centre of the sun to the centre of the meteoric swarm is supposed to revolve with angular velocity, n say, about the centre of the sun as a fixed point ; then the motion of a particle of the swarm is referred to three directions at right angles to one another having their origin at the centre of the swarm, and turning with the radius vector just specified. These axes may be taken as an axis of ξ towards the sun, an axis of η at right angles to this in the plane of motion of the centre, and an axis of ζ at right angles to this plane. Then equations of motion relative to these moving axes are written down for a particle the component distances of which from the centre are ξ, η, ζ, it being supposed in the first place that the distance r of the centre of the swarm from the sun, and the angular velocity n of the radius vector are both variable. Approximate values of the forces are obtained by supposing that ξ, η, ζ are small in comparison with r, and that r, and therefore also n, is constant. When account is taken of the condition that must hold for the central particle, the equations assume the very simple form

$$\ddot{\xi} - 2n\dot{\eta} - (3n^2 - \mu)\xi = 0,\ \ddot{\eta} + 2n\dot{\xi} + \mu\eta = 0,\ \ddot{\zeta} + (n^2 + \mu)\zeta = 0.$$

The value of μ is $\tfrac{4}{3}\pi ks$, where k is the gravitation constant and s is the average density of the portion of the swarm within the spherical surface on which the particle lies, supposed symmetrical about the centre. Considering only particles in the plane of ξ, η, the values of these co-ordinates are supposed to oscillate about certain constant values, so that $\xi = a \cos(\omega t + \epsilon)$, $\eta = b \sin(\omega t + \epsilon)$. That is, each particle is supposed to revolve in an ellipse, the centre of which is the centre of the swarm, and of which one axis is along the line of centres and the other perpendicular to that line. The ellipse is a circle if $a = b$, and ω is then the angular velocity of the *relative* motion of the particle about the centre. These values substituted in the first two equations of motion lead to the condition

$$(\omega^2 - \mu)(\omega^2 + 3n^2 - \mu) - 4\omega^2 n^2 = 0.$$

Now ω is $2\pi f$, if f be the frequency of oscillation ; and if the oscillation be stable, f will have a positive real value. The roots of the quadratic in ω^2 just written must therefore be real and positive ; and it is not hard to see that the required condition

for this is $\mu > 3n^2$. This gives, when μ is replaced by $\tfrac{4}{3}\pi ks$, and n^2 by kM/r^3, where M is the mass of the sun (for we have for the central particle $n^2 r = kM/r^2$), the inequality

$$\tfrac{4}{3}\pi s r^3 > 3M.$$

In order, therefore, that the swarm of small particles may keep together, it is necessary that its average density be greater than that of a spherical distribution of matter of radius equal to the sun's distance and of three times the sun's mass. The problem for an elliptic orbit of eccentricity e has been considered by M. Callandreau (*Bulletin Astronomique*, 1896). The condition $\mu > 3n^2$ is in this case replaced by $\mu > 3n^2 + 5e^2 n^2$. The swarm is therefore rendered less stable by the eccentricity.

It is to be remembered that the effect of the distortion of the swarm by the tidal force of the sun is neglected, and it does not seem of much importance to consider eccentricity of orbit so long as the assumption of sphericity of figure is maintained.

Since the equation of condition stated above may be satisfied, only values of ω consistent therewith, and with the inequality $\mu > 3n^2$, are admissible. Thus if $\omega = 0$, that is if there be no revolution of any particle about the centre of the swarm, the equation gives $\mu = 3n^2$, and the inequality is not fulfilled. This is a limiting case between stability and instability.

Now let the differential equations of motion referred to above (from which the more roughly approximate equations quoted are derived) be modified for the case in which the swarm is replaced by a planet of given mass m, and the particle considered by a satellite of mass m' at the external point ξ, η, ζ, then let them be multiplied by $\dot{\xi}$, $\dot{\eta}$, $\dot{\zeta}$ respectively, integrated, and added. Thereby will be obtained the equation of kinetic energy for the relative motion, commonly called Jacobi's equation. This has, if $\dot{\xi}^2 + \dot{\eta}^2 + \dot{\zeta}^2$, the square of the resultant relative velocity, be denoted by v^2, the form

$$v^2 - \left[2\frac{\mu}{\rho} + 2\frac{r^2 n^2}{r_1} + n^2\left\{(r - \xi)^2 + \eta^2\right\}\right] + C = 0$$

where

$$\mu = k(m + m'),\ \rho = \sqrt{\xi^2 + \eta^2 + \zeta^2},\ r = \sqrt{(r - \xi)^2 + \eta^2 + \zeta^2},$$

C is a constant, and r is, as before, the distance of the centre of the sun from that of the planet.

Now when v^2 has a given value, the satellite must have its centre on the surface of which the equation is obtained by placing that value in the equation just written. Hence, since v^2 is positive, the satellite cannot pass across the surface for which $v^2 = 0$, that is the surface for which

$$2\frac{\mu}{\rho} + 2\frac{r^3 n^2}{r_1} + n^2\left\{(r - \xi)^2 + \eta^2\right\} - C = 0 ;$$

by putting $\zeta = 0$ in this we obtain the equation of the curve in which the surface intersects the plane of ξ, η. An investigation of the surface shows that if C be positive the surface consists of three sheets, of which two are closed and surround the sun and the planet respectively ; and the third is asymptotic to a surface of revolution about an axis passing through the sun's centre perpendicular to the ecliptic, and surrounds the two closed surfaces. Within the closed surfaces, or outside the third surface, v^2 is positive ; between the closed surfaces and the outer asymptotically cylindrical surface v^2 is negative, and therefore v is imaginary. The satellite must therefore be within one of the closed surfaces, or beyond the outer surface ; in either case it cannot cross the surface of zero velocity.

When the proper values of the quantities for the earth-moon system are inserted, it is found that the moon is within the closed sheet surrounding the earth, from which, therefore, it cannot escape. The distance of the moon's centre from the earth, Mr. Hill has calculated, cannot exceed 109'694 equatorial radii of the earth. The result is based, of course, on the assumption that the eccentricity of the earth's orbit may be neglected.

If, besides neglecting the eccentricity, we suppose the moon to move in the plane of the ecliptic, and to be so distant that we may neglect terms in η, the equation of the curve of no velocity in the plane of the ecliptic is

$$\frac{\mu}{\rho} + \frac{3}{2}n^2 \xi^2 = c,$$

or if $\xi = \rho \cos \theta$

$$3n^2 \cos^2\theta \cdot \rho^3 - c\rho + 2\mu = 0,$$

where c is another constant.

The roots of this cubic in ρ are all real if $\cos^2\theta > c^3/81 n^2\mu^2$, and the rule of signs shows that there is only one negative root. The curve of no velocity consists then of a closed branch round the origin of co-ordinates, the centre, E, of the earth in the present case. Besides this there are two infinite branches which are asymptotic to the parallel lines AB, A'B' represented by

$$3n^2\xi = c.$$

Thus the curve is as roughly represented in Fig. 1. The line CD shows the direction of the radius vector from the sun.

Between the closed curve and the infinite branches v is imaginary, and the satellite must be either within the closed branch or beyond the boundary represented by the infinite branches. The calculation gives very approximately 110 equatorial radii of the earth for the greatest distance of any point of the closed branch from the centre. The form of this branch is that of an oval, being slightly longer in the direction towards the sun than in the transverse direction.

The theorem of Roche which we discuss here is contained in the statement that the atmosphere of a satellite cannot be held together merely by the gravitational attraction of the satellite unless the inequality

$$\frac{m}{M} > (2 + c)\frac{a^3}{r^3}$$

is fulfilled, in which m is now the mass of the satellite, M that of the planet, c the ratio of the square of the angular velocity

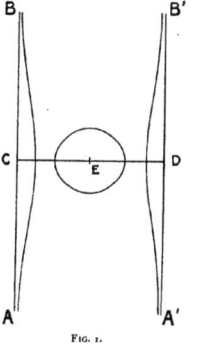

FIG. 1.

of the satellite's axial rotation to the square of the angular velocity of its orbital revolution, a denotes the satellite's radius, and r the distance of the centre of the satellite from the planet. If the densities of the planet and satellite be s_1,s, and their radii a_1,a, and c be unity, that is if the satellite turns always the same face to the primary, we have for the inequality

$$\frac{r^3}{a_1{}^3}\frac{s}{s_1} > 3, \text{ or } r > 1\cdot44 a_1 \sqrt[3]{s_1/s}.$$

It should be mentioned that Roche's investigations embraced much more than this; they included the determination of the figure of a fluid satellite, and entered into other matters which cannot be discussed here.

To deal with these questions in an elementary way, consider the important particular case of a spherical swarm of radius a moving round the sun, and turning as a whole about an axis perpendicular to the orbit in the period of revolution, so that it turns the same face always towards the sun. This is, of course, a less general problem than that considered above ; it is indeed the case of that problem in which ω is zero. It is interesting to see from the more general investigation that the condition obtained by the consideration of this case is sufficient to

give stability for any value of ω provided it fulfils the equation of condition stated above. We shall obtain also by the elementary process a wider condition for the case in which ω is not zero. This will give the inferior limit assigned by Roche to the distance of a satellite from its primary.

A particle, of unit mass, say, at the centre, C (Fig. 2), at distance SC $(=r)$ from the sun, is in relative equilibrium under the sun's attraction and the so-called centrifugal force. That is, we have for that particle

$$\frac{kM}{r^2} - n^2 r = 0.$$

Again, a particle on the outside of the swarm at the point nearest the sun is at a distance $r - a$, and under attraction $kM/(r - a)^2$. Hence there is a preponderance of attraction over the acceleration $n^2(r - a)$ towards S. This excess is

$$\frac{kM}{(r - a)^2} - n^2(r - a) = kM\left\{\frac{1}{(r - a)^2} - \frac{1}{r^2} + \frac{a}{r^3}\right\}$$
$$= 3kM\frac{a}{r^3}$$

nearly. This must at least be balanced by the attraction towards the centre, C, exerted by the swarm, if the particle is not to leave the swarm. Hence we must have $\frac{4}{3}\pi k a^3/a^2 > 3kMa/r^3$, or

$$\frac{4}{3}\pi s r^3 > 3M,$$

as before. The same result would be obtained for a particle at B. In that case the attraction of the sun $kM/(r + a)^2$ would be insufficient to supply the acceleration $n^2(r + a)$ towards the sun. The condition that this should be supplied by the attraction of the swarm is that $\frac{4}{3}\pi r^3$ should be at least equal to 3M.

This result holds, of course, for all particles within the swarm on the line SC, for no particle experiences any force on the whole from the spherical layer outside it.

It is to be observed that a particle at A or B (or on the line SC) is in greater danger of leaving the swarm from the causes

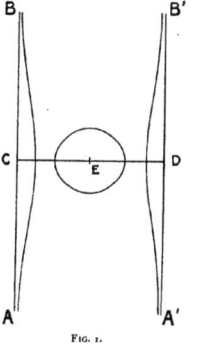

FIG. 2.

just explained, than a particle elsewhere on the spherical surface.

If the particles of the swarm have other angular velocities than that supposed above about an axis through its centre perpendicular to the plane of the orbit, the investigation will run as follows. Suppose applied to each of the particles a force per unit mass equal and opposite to that, $n^2 r$, exerted by the sun on the central particle. This will have no effect on the relative motions of the particles or on the figure of the swarm. Upon the particle nearest the sun the force per unit mass toward the sun is now

$$\frac{kM}{(r - a)^2} + \omega_1{}^2 a - n^2 r = 2kM\frac{a}{r^3} + \omega_1{}^2 a$$

if ω_1 be the angular velocity *in space* of the radius vector drawn from the centre to the particle (that is, not the relative angular velocity ω above, but $\omega + n$). This must at least be balanced by the attraction towards the centre exerted by the swarm if the particle is not to leave it. Thus we have $\frac{4}{3}\pi k s a > 2kMa/r^3 + \omega_1{}^2 a$, or since $kM = n^2 r^3$,

$$\frac{4}{3}\pi s r^3 > \left(2 + \frac{\omega_1{}^2}{n^2}\right)M.$$

Thus if the swarm as a whole make one rotation in the period of revolution round the sun, $\omega_1{}^2/n^2 = 1$, and we obtain the same result as before.

Let now the swarm of particles be replaced by a spherical planet with an atmosphere composed of discrete small particles, the whole being held together by gravitational attraction alone. Then if the mass of the planet be denoted by m, the inequality $\frac{4}{3}\pi s r^3 > 3M$ becomes $\frac{4}{3}\pi s a^3 > 3Ma^3/r^3$, that is $m/M > 3a^3/r^3$. This is to be fulfilled if the atmosphere is not to be dissipated by tidal action. The same thing will, of course, hold for a primary and the atmosphere of a satellite.

In the more general case, that in which the satellite has rotational velocity ω about its axis (supposed perpendicular to the plane of the orbit), we have, assuming that the satellite is spherical and denoting $\omega_1{}^2/n^2$ by ι, $m/M > (2+\iota)a^3/r^3$. This agrees with the former result when $\iota = 1$. These results were first obtained by M. Roche.[1]

The figure of a fluid satellite is determined by finding a surface to which the resultant of the gravitational pull of the primary on unit mass, a force n^2r equal and opposite to the gravitational pull on unit mass at the centre, the gravitational force per unit mass exerted by the matter of the satellite itself, and the centrifugal force of unit mass, is everywhere perpendicular. A first approximation to the force due to the satellite itself is obtained by neglecting the deviation from sphericity, as is done above. But into this discussion we cannot here enter. It can only be stated that the final result, taking into account the distortion of the satellite, is that the satellite will be broken up if it approaches closer to the primary than the limit given by the inequality

$$r > 2\cdot 44 \sqrt[3]{s_1/s_1}.$$

Now imagine a planet and a satellite moving round the sun, the satellite being destitute of relative velocity. The satellite may, for example, be regarded as a particle (of unit mass say) of a ring of small mass composed of particles surrounding the primary at a distance a, the whole turning, if that were possible, with angular velocity equal to that of the primary round the sun. By what we have seen above, the excess of solar attraction over the sunward acceleration is, for a particle on the side nearest the sun, $3kMa/r^3$. This must be balanced by the attraction km/a^2, so that we have the equality $km/a^2 = 3kMa/r^3$. From these expressions for the forces we see that the potential energy, with a term for centrifugal force included, may be taken, for unit mass of the infinitesimal satellite at the point nearest the sun, as $-km/a - \tfrac{3}{2}kMa^2/r^3$. This is an example of the almost self-evident principle, known as the theorem of Coriolis, that if there be included a term in the potential energy which will give the components of centrifugal force, we may write down the equation of relative kinetic energy, just as if the rotating axes were fixed. The potential energy thus required for the centrifugal force on a particle of unit mass at a point at distances ξ, η, ζ from the planet's centre, measured respectively along the line of centres, perpendicular to this line in the plane of motion, and perpendicular to the plane of motion, is $-\tfrac{1}{2}n^2\{(r-\xi)^2+\eta^2\}$, or, since $n^2r = kM/r^3$,

$$-\tfrac{1}{2}kM\{(r-\xi)^2+\eta^2\}/r^3.$$

The total potential energy being taken as

$$-\left[\frac{km}{\rho}+\frac{kM}{r_1}+\frac{1}{2}\frac{kM}{r^3}\left\{\left(r-\xi\right)^2+\eta^2\right\}\right]$$

(with, as before,

$$v^2 = \dot\xi^2+\dot\eta^2+\dot\zeta^2,\quad \rho^2 = \xi^2+\eta^2+\zeta^2,\quad r_1{}^2 = (r-\xi)^2+\eta^2+\zeta^2,$$

and $n^2 = kM/r^3$), the equation of relative kinetic energy is

$$\tfrac{1}{2}v^2 - \left[\frac{km}{\rho}+\frac{kM}{r_1}+\frac{1}{2}\frac{kM}{r^3}\left\{\left(r-\xi\right)^2+\eta^2+\zeta^2\right\}\right]+C=0,$$

which, for an infinitesimal satellite, is Hill's equation as given above.

For the moon, which has mass m' sensible in comparison with the mass, m, of the earth, the first term in the square brackets should be $k(m+m')/\rho$.

It may be noticed by the dynamical student that if the above expression for the potential energy be denoted by V, we have not $\ddot\xi = -\partial V/\partial\xi$, &c., for the equations of motion, but

$$\ddot\xi - n\dot\eta = -\partial V/\partial\xi,\quad \ddot\eta + n\dot\xi = -\partial V/\partial\eta,\quad \ddot\zeta = -\partial V/\partial\zeta.$$

A. GRAY.

---

[1] I have found since the above was written that the same elementary view of this matter is given by Roche himself in his paper "Recherches sur les Atmosphères des Comètes," *Annales de l'Observatoire*, t. V. 1859. Perhaps I may here direct attention to a valuable paper by Roche (which may, however, though I have not seen it referred to, be well known to astronomers), entitled "Essai sur la Constitution du Système Solaire," *Mém. de l'Acad. de Montpellier*, t. viii. This gives a general account of the author's cosmogonic researches.

ANTELOPES AND THEIR RECOGNITION MARKS.

THE Tragelaphine Antelopes hold a unique position amongst the hollow-horned ruminants. No other group can show species so sharply contrasted in size and build as the massive eland rising over sixty inches at the withers. and the dainty little bush-buck which falls short of half that height. Only the Indian black buck amongst the gazelles can match the nylghaie and nyala for diversity of sexual colouring ; and for elegance of form, coupled with beauty of marking and grandeur of carriage, the kudu is surpassed by no species of mammal.

Apart from certain features presented by the skull and horns, the affinity between the species here mentioned is attested by the markings of the skin. On a ground-colour shading from slate to chestnut are distributed certain white spots, patches or stripes, which crop up so persistently in the different genera as to leave no doubt they are a heritage from a common ancestor. A comparison between the skins of the existing species suggests that this ancestor was coloured somewhat as follows :—Body and head yellowish red ; flanks and hind-quarters striped with white ; on the throat two white patches, one at each end ; one or two spots on the cheeks, a V-shaped stripe between the eyes, a white chin, a white upper lip ; legs paler on the inner side, quite white at base close to chest and groin, and with two white spots on the pasterns in front.

Some or all of these markings have been inherited with scarcely an exception by every known species of Tragelaphine. Sometimes the spots on the head, sometimes the stripes on the body, sometimes the patches on the throat are suppressed ; but even in extreme cases of suppression, a spot here, a stripe there, persists as a tell-tale sign of descent. The usefulness of characters so constant may be taken for granted. The nature of their usefulness has been discussed by both Wallace and Darwin ; but so great is the discord between the opinions of these authorities that one cannot think both are right.

Referring to the importance of special marks for recognition where many species of nearly the same size and general form inhabit the same region, Mr. Wallace says : "It is interesting to note that these markings for recognition are very slightly developed in the antelopes of the woods and marshes. . . . The wood-haunting bosch-bok (*T. sylvaticus*) goes in pairs, and has hardly any distinctive markings on its dusky chestnut coat, but the male alone is horned. The large and handsome kudu frequents brushwood, and its vertical white stripes are no doubt protective, while its magnificent spiral horns afford easy recognition. The eland, which is an inhabitant of the open plain, is uniformly coloured, being sufficiently recognisable by its large size and distinctive form ; but the Derbyan eland is a forest animal, and has a protectively striped coat. In like manner, the fine Speke's antelope, which lives entirely in the swamps and among reeds, has pale vertical stripes on the sides (protective), with white markings on face and breast for recognition" ("Darwinism," p.220).

It may be inferred from this passage that the interest attached to the slight development of recognition marks in the antelopes of the woods and marshes lies in the needlessness of such marks for species living apart and not herding with others of the same general size and form. If, however, there is no likelihood of confusion, it is not quite clear from what species the horns of the kudu serve to distinguish their owner, nor what significance in this connection is to be attributed to the occurrence of horns only in the male of the bosch-bok. Similarly, it is not clear what use Speke's marsh-buck can have for recognition marks. If, however, the spots on the face and throat subserve recognition in this species, we must also conclude they are retained for that purpose in the bongo (*T. euryceros*), the lesser kudu, the nyala (*T. angasi*), in which they are very conspicuous, as well as in the various smaller kinds of bush-buck, which in other parts of Africa live the same life as the bosch-bok of the Colony. Surely, too, Derby's eland is at least as recognisable by its large size and distinctive form as the Cape species ; yet it is adorned with a conspicuous V-shaped stripe between the eyes, and the lower throat-patch forms a white collar, standing boldly out against the black hue of the neck.

In short, if the marks in question have been preserved for recognition, it is singular that they are exceptionally well developed in the species that live in pairs or small parties by themselves in thick bush—species which, according to the hypothesis, have little, if any, need of them. It is conceded, of course, that the spots on the head and throat, like the stripes on the body, patches on the rump or any other visible external feature, including large ears, may, if required, serve as marks of identity; but in the case of the Tragelaphines, at least, it is hard to believe that that is any more their primary function than it is the primary function of large ears.

A mass of evidence can be brought forward in favour of Wallace's view that the body-stripes of these antelopes are protective; but there appear to me to be equally strong reasons for classifying the face and foot markings in the same category, and for regarding them as representing spots or streaks of sunlight passing through foliage or reflected from leaves.

It is possible, perhaps probable, that the other white patches on a typical Tragelaphine serve the same end; but their situation forcibly suggests that they have a still deeper signification. I believe they come within the scope of Thayer's hypothesis of concealment by the counteraction of light and shade. A convex body stands out amid surroundings of its own colour on account of the contrast between the light that is reflected from its upper surface and the shadow that pervades it below. Take away the light by darkening the upper side and the shadow by lightening the lower, and the body will vanish from sight with the destruction of its visible shape. By applying this principle to a typically marked Tragelaphine, the lesser kudu for example, it will be seen that the white is laid on where shadows are thrown; that the white rim on the upper lip and the white chin must counteract the shadows caused by the fold of the

FIG. 1.—The Lesser Kudu, ♂.

mouth and by the muzzle; the two white blotches on the neck must counteract the shadows thrown by the head and by the curvature of the throat, and the shadows cast by the breast and groin must be similarly obliterated by the white patches on the inner side of the base of the limbs.

That the white patches must have the effect here assigned to them will be obvious, I think, to any one who, with Thayer's hypothesis in mind, looks at a lesser kudu when it is standing full-face. The reason for the presence of marks concealing the animal from this point of view will be referred to later on.

If the markings of the Tragelaphines have the significance here attached to them, they should be better developed in the species that live in the bush than in those that frequent the open. Let us see in a few cases to what extent they are correlated with habit. Two well-marked species of eland live in Africa, to wit, Derby's eland (*T. derbianus*) from Senegambia and the commoner form (*T. oryx*) which, with its subspecies, ranges throughout the whole of East and South Africa. The former, according to Winwood Reade, "lives in the forest, and never of its own accord enters the plain." It is reddish in colour, with a black neck, a white collar, eye-stripes and many white stripes on the flanks and hind-quarters. Of the common eland the typical "dun coloured" desert form is, according to Selous, "particularly plentiful in the dry desert country through which the Chobe runs," and examples "from the Kalahari desert have no sign of a

stripe." Farther to the north, both in Angola, South-east and East Africa, this unmarked type is replaced by its ancestral form, Livingstone's eland, which in colour and habits is intermediate between the Cape and Senegambian species. The skin is always marked with narrow white stripes, and the V-shaped mark between the eyes is often present. In British East Africa this eland, according to Jackson, was found in "sparsely timbered country and open bush bordering the plain rather than the plains themselves," and in Angola (Penrice) "it seems most partial to a thinly timbered country." There is thus a complete gradation from the strongly marked forest species through the weakly marked species frequenting the open bush to the unmarked desert species.

Take again the kudus. Both the species are well marked with white stripes on body and head, but the smaller (*S. imberbis*) is much more strikingly marked than the larger (*S. strepsiceros*), having more stripes on the body and two patches on the throat. In Somaliland, where both species occur, the larger lives, according to Swayne, in the mountains, on very broken ground where there is plenty of bush; and sometimes indeed ventures into the open plain (Inverarity). The lesser kudu, on the contrary, "is found in thick jungles . . . especially where there is an undergrowth of the slender pointed aloe which grows from four to six feet high" (Swayne). Both Swayne and Inverarity, moreover, bear witness that this species will allow a hunter to get within a few yards before dashing away—a notorious habit with protectively coloured animals. Evidence of a like kind is furnished by other species of Tragelaphines. The beautifully marked nyala (*T. angasi*) and bongo (*T. euryceros*) live in dense thickets; and the lovely little bush-bucks (*T. sylvaticus, scriptus*, &c.) seldom venture out of cover except at night-time to feed. On the other hand, the nylghaie, an aberrant member of the same tribe, is without body-stripes, and lives for the most part in more or less open country in India, and is not a typical denizen of the thick jungle at all.[1]

Further evidence on this head is supplied by another set of facts. Ungulates which live in thickets or rough ground affording cover to enemies have larger ears and a keener sense of hearing than those of the plains or high mountains where intruders have little chance of concealment. Note the small ears of the camel, a typically desert form; or of goats and sheep which from the mountain peaks can sweep the surrounding country for miles with their eyes and seek safety in flight long before the foe gets within ear shot. Compare also the small equine ears of Burchell's zebra, which herds in the open plain free from obstacles to interfere with vision, with the longer asinine ears of the mountain zebra which frequents rocky broken ground well fitted for the hiding of carnivores. In all the brilliantly marked Tragelaphines the ears are long and expanded, but in the nylghaie, and especially in the common eland, they are short and narrower. Indeed, one of the chief structural differences between Derby's eland and the Cape species is found in the size of the ears.[2]

The co-existence of white marks with long ears and a bush-life bears out the supposition that the marks, like the ears, are primarily for protection, and that if subservient to purposes of recognition they are merely of secondary importance in that capacity.

I strongly suspect, too, that the markings of the sable, roan, gemsbok and bontebok are for concealment, and not for recognition as Mr. Wallace supposes. The theory of recognition marks as applied to these antelopes assumes the need of some patch or spot to enable the members of a species to identify their own kind amongst the herds of other sorts living in the same place. The theory would rest upon a securer basis, if it could be shown that closely allied species feed together. But nothing, I suppose, is more certain than that, as a very general, perhaps invariable, rule, closely allied and similar species are not found together. If, for example, the gemsbok and the

[1] An apparent exception to the rule that the development of white stripes and spots is correlated with a jungle life is found in the Situtunga and Speke's marsh-buck, which "live in vast reed beds and papyrus swamps, and only come into the open at night " (Selous). Yet the stripes fade away in the adults of both sexes. Why is this? Possibly because these animals depend for concealment, not so much upon coloration that harmonises with that of the vegetation, as upon a newly acquired and efficient habit of hiding under the water itself, with only the end of the nose jutting above the surface (Selous and Gedge).

[2] Compare in this connection the small ears of the orang with the large ears of the chimpanzee. The former lives a more arboreal and therefore safer life than the latter, which requires quick hearing to enable it to escape to the trees when feeding on the ground.

beisa lived side by side in the Kalahari, or Peters's palla and the common species in Rhodesia, there would perhaps be strong reasons for thinking that the differences in the facial bands, which enable us to recognise these species apart, serve the same end where the antelopes are themselves concerned. But the gemsbok and the beisa, the common and Angolan palla, never cross each other's path. Again, in cases where the geographical areas of two forms, closely allied, but distinguishable by bands or patches, meet, the two forms frequently interbreed, and so falsify the contention that marks keep like to like.

That ungulates of different sorts herd together is well known ; we read, for example, of zebras, gnus, pallas, spring-boks and buffaloes feeding in each other's company on the veldt. But so distinct from each other in form are these animals and others that might occur with them, that it is rating their visual powers very low—much lower, indeed, than our own—to hold that they require special patches to keep them from committing the errors of identification which the hypothesis assumes they are liable to fall into. I believe, then, that the need for recognition marks in the case of antelopes has been much over-rated,[1] and is too slight to warrant the belief that the facial and other stripes of, say, the gemsbok or sable have been perfected by their usefulness as such. On the other hand, when we see that the pattern of the zebra is for concealment, that the network of white stripes on the giraffe blends with the lights passing

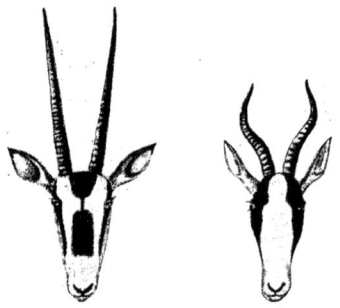

Fig. 2.—A, Face of the Beisa ; B, face of the Bontebok.

through the intercrossing branches of trees, that the colour and shape of the feeding hartebeest are like those of the ant-hills, all these and other facts attesting the importance of concealment, we are justified in suspecting that the white blaze of the bontebok and the facial bands of the gazelles and oryxes are developed for concealment and not for recognition.

The markings take the form of strongly contrasted bands of white and black, or brown. Objects banded in this way are, as a rule, more, and not less, difficult to see in their natural surroundings than those that are uniformly coloured. There is little of the gloss on the coat of a grey or white horse that is seen on a bay or black, because white hair reflects the light less vividly than dark. Hence alternating bands of these hues impart a blurred irregular aspect to a body, destroy the apparent evenness of its surface and break up the continuity of its outline. In an uncertain light a zebra's stripes[2] "merge into a grey tint," and mutually counteract each other, so that the animal is nearly invisible.

The stripes on the head of a gemsbok or sable are in a general way so like those on a zebra's coat that they must, one would

[1] If the American prong-buck were an inhabitant of Africa, I presume that its conspicuous patterns, possibly indeed the unique shape of its horns, would be cited as evidence supporting the theory of recognition marks. But in the prairies of the United States there are no species that resemble it in size and form, so as to create confusion as to identity. The species furnishes a good instance of Thayer's principle.

[2] Mr. Wallace is surely "putting the cart before the horse" in the passage where he speaks of the coloration of the zebra as an instance of a style of marking for recognition becoming also protective.

think, have the effect of making the head inconspicuous. To explain the prevalence of such marks upon the head and fore-part of the body, of which the quagga furnishes an illustration, the following suggestions may be made.

Once aware of an enemy's presence, an antelope has three chances of escape—concealment, flight and self-defence. Concealment is often the wisest course to pursue, especially where females and young are concerned. For concealment perfect stillness is of all things most important. Movement means detection, and detection may mean death. But it is necessary at the same time for every movement of the enemy to be scrutinised, so that the right moment for flying may be seized, when the necessity for flight becomes apparent. For this purpose the face must be turned towards the enemy and both eyes be kept upon him. In this watchful attitude little of the foreshortened body is visible from the enemy's point of view, and if the head of the antelope be carried low with the nape in a line with the spine, practically nothing of the animal is exposed but the head and the fore-legs. Hence the special importance of protective markings on these parts.[1] Again, when lying on the ground the body will often be hidden by low scrub or, if cover be absent, may simulate a mound of earth or a termite's nest[2] ; but the head, if protectively coloured, may with safety be raised to keep an eye on the surrounding country and guard against surprise.

The sexual colours of the Tragelaphines still remain to be touched upon. Darwin believed that the markings of the kudu, harnessed bush-buck, &c., were in the first instance acquired by the male, then intensified by sexual selection, and partially transferred to the female. Against this hypothesis may be urged the evidence already adduced in favour of their protective[3] value, and the distinctness they exhibit in the newly-born or even foetal young. Nevertheless a difference of colour, small or great, but as yet unexplained, does exist between the males and females of all the species of the group. It is noticeable, too, that the deviation affects the male ; that it takes the direction of nigrescence, but by no means always of beauty, and that the female adheres to the typical coloration of the group.

There is no evidence, so far as I am aware, that the assumption of a dark coat by the male is connected with any peculiarity in mode of life of this sex, which would attest its use for concealment. On the contrary, since the colour—at least, in the case of the nylghaie ("Descent of Man," p. 535)—becomes intensified at the breeding season, without the growth of new hair, and has its appearance arrested by emasculation, its significance appears to be purely sexual and the outcome or accompaniment of " male katabolism." If so, it may have been intensified and fixed by the exercise of choice on the part of the females, or by the destruction or expulsion of their paler, less vigorous rivals by the stronger and darker males, which thus secured the females for themselves and left the greatest number of offspring.

It is significant that the three species of antelopes, namely the nyala, the nylghaie and the Indian black buck, in which the sexes differ strikingly from each other—differ indeed to an extent that is equalled by few species of mammals and surpassed by none—the female is without horns or other weapons of defence. This defencelessness, coupled with the exigencies of maternity, has compelled an adherence on her part to the normal protective coloration of the group ; whereas the males, powerfully built and strongly horned, have been able to dispense to a great extent with colours that harmonise with those of the environment.

Warning characters are rare in mammals ; but the difference in colour between the bull and cow of the species just mentioned may conceivably benefit the former at the expense sometimes of the latter by serving to distinguish him, the horned powerful fighter and dangerous antagonist when brought to bay, from her, the weak and defenceless one who may be attacked and pulled down with impunity. Why not ?   R. I. POCOCK.

[1] Presumably it is for an analogous reason that the tiger has a pair of sunlight patches on its face, so that its chances of concealment may be increased when watching for prey or creeping towards it.
[2] In the case of the gemsbok, beisa and some gazelles the black longitudinal stripe passing along the side may enhance this resemblance by representing the shadow that is often seen where a boulder or mound meets the soil. The darkening of the coat on this area of the body, such as is seen, for instance, in the Asiatic wild ass (Kiang), should have the effect of deadening the light reflected from the bulging flank.
[3] Here and elsewhere in this article I have purposely used the word "protective" as the equivalent of procryptic or celative, which are less familiar to general readers.

GEOLOGY AT THE BRITISH ASSOCIATION.

THE geologists had a busy and profitable week at Bradford, and an air of business-like application to work pervaded their meetings from first to last. The programme ought to have been long enough to satisfy even the most devoted adherent of Section C; but apparently there was no sense of satiety, since on two or three occasions when it was proposed from the platform that communications should be taken as read, there were protests raised by the audience, who seemed determined to carry matters through with true North-country thoroughness, and wished to hear everything. And indeed it may be said that there was scarcely a paper in the long list which did not contain scientific matter well worthy of discussion, though it must be acknowledged that in several instances the matter was not particularly novel. It was only by the strict enforcement of a time-limit upon readers of papers and debaters, and by sessions on the mornings of Saturday and Wednesday as well as lengthy sittings on the other days of the meeting, that the business was got through. Under these circumstances it was inevitable that some excellent papers scarcely received full justice; but the discussions were nevertheless unusually full of vigour, and what was still better, entirely lacking in acrimony.

The fine weather of the week, which was so favourable to the short afternoon excursions, now a recognised and highly valued feature in the affairs of the section, had probably much effect in fostering the prevailing good-humour, while the personality of the president was a strong influence in the same direction, especially in the discussions.

To particularise all the papers within the space-limits of this article is impossible, and we can only attempt to convey a general impression of the proceedings, with brief reference to the points of main interest.

On Thursday, as a fitting appendix to the wide-reaching generalisations of the president's address, already printed in these pages, we had a series of papers from Prof. J. Joly dealing with geological problems from the standpoint of the physicist. In one of these, "On the geological age of the earth, as indicated by the sodium contents of the sea," Prof. Joly reiterated the calculations and conclusions which have recently attracted so much attention in geological circles; in another, "On the inner mechanism of marine sedimentation," he showed the chemical and physical reasons for the rapid precipitation of fine matter brought down in suspension by rivers into the sea; a third of kindred character gave the result of "experiments on denudation by solution in fresh and salt water"; and a fourth, which was especially attractive to the petrologists and mineralogists, dealt with "the viscous softening of rock-forming minerals at temperatures below their normal melting point," showing how certain minerals could be observed to attain a plastic state some time before actually melting.

At the same meeting Prof. W. B. Scott, of Princeton, gave a highly interesting account, with lantern illustrations, of recent explorations in Patagonia conducted under the direction of Mr. J. B. Hatcher. Besides correcting previous errors as to the age of the deposits, the records and the rich collections of fossils obtained by this expedition have sufficed to prove a close connection between Australia, New Zealand and South America in Miocene times, and in several other respects to modify profoundly our previous ideas of South American geology, and incidentally to show how much geologists have still to learn in every way from the unexplored tracts of the earth's surface.

On Friday, Prof. J. Milne led off in his usual happy vein with an account of the year's work of the Seismological Committee, and was followed by Mr. Clement Reid, who showed how well-chosen, from geological reasons, was the site for instrumental observation, by the same committee, of the Upway disturbance. There then followed a series of papers and reports on the Mountain-limestone district of north-west Yorkshire and its underground waters, Mr. S. W. Cuttriss giving an account of the adventurous exploration of the deep pot-holes and caves of this district by himself and other members of the Yorkshire Ramblers' Club, and the Rev. W. Lower Carter and Mr. A. R. Dwerryhouse presenting the results obtained by a local committee and by a committee of the Association in the investigation of the subterranean drainage of the limestone. Being well-illustrated by lantern slides, these papers besides attaining their more direct purpose served to give the strangers an idea of the general characteristics of the district which was afterwards to be visited by geological excursion parties. By the use of suitable

chemical reagents the course of the water from its disappearance in "sink holes" of the limestone to its reappearance in springs at lower levels has in several instances been traced; it has been shown that the main direction of underground flow is along the master-joints of the limestone; and a subterranean watershed of which there is no indication at the surface has been traced for some distance. These experiments are to be continued, and a grant of 50*l*. was made by the Association towards this end.

Among other papers taken on Tuesday were two by Mr. E. Greenly, giving further results of his painstaking researches in Anglesey. In one of these he dealt with the ancient surfaces or peneplains which he thinks can be recognised in North Wales; the older plain he is inclined to regard as of sub-Carboniferous age, and the later as Mesozoic, possibly Cretaceous. There was an interesting discussion on this paper, in the course of which Mr. Greenly acknowledged that his views were only tentatively held, and might require modification. Dr. G. Abbott then gave an account of his investigation of the concretionary structures of the Magnesian Limestone of Durham, illustrating his subject by lantern-slides and the exhibition of a large series of specimens.

Saturday's business began with a paper by the President, "On a concealed coalfield beneath the London basin," in which it was urged, on data not altogether convincing, that if a boring were made in the vicinity of Enfield Lock on the Lea, it might be expected to reach Coal-measures. As a speaker in the discussion remarked, such a boring would no doubt reveal something interesting, but whether Coal-measures was another matter. Then followed a paper by Mr. R. H. Tiddeman "On the formation of reef-knolls," which was practically a criticism of Mr. J. E. Marr's views as to the development of these structures in the Mountain-limestone of West Yorkshire and Lancashire by earth-movement, and a reiteration of the author's earlier contention that they were originally formed as mounds on a slowly sinking sea-bottom. As Mr. Marr was present to champion his own cause, the paper was followed by a brisk but friendly discussion, which was prolonged on a later day in the open air, when some of the mounds at Cracoe near Skipton were visited by a few members interested in the subject. No definite conclusion was reached, but the necessity for further investigation was made evident, and it was suggested that the truth might lie in a combination of the two hypotheses.

Another paper taken on Saturday was that of Mr. W. Gibson, "On rapid changes in the thickness and character of the Coal-measures of North Staffordshire," in which it was shown that the areas of maximum and minimum deposit in these rocks correspond respectively with a syncline and anticline, thereby suggesting that local areas of deposit were being marked out by contemporaneous movements of elevation and depression, thus fulfilling in North Staffordshire the conditions characteristic of the Carboniferous rocks of the Midlands generally. These results have an important practical application, inasmuch as the unexplored coal-field to the westward, which occupies a syncline, may thereby be expected to exhibit an increase in the thickness of the strata. At the same meeting, Rev. J. F. Blake brought forward some revolutionary suggestions in regard to the registration of type-specimens, among other proposals urging that a new class of "adopted" types should be recognised and registered where the original types were missing or inadequate, and that the type should consist of a single specimen. As Prof. Blake has now been elected a member of the committee of the Association at present in existence for furthering the registration of type-specimens, we may hope that his interest in the matter may bear fruit.

On Monday there was a crowded audience to hear the joint discussion with the botanists on the conditions during the growth of the forests of the Coal-measures. The discussion was opened by Mr. R. Kidston, who gave a succinct account of the plant-life of the period, illustrated by fine lantern slides. Mr. A. Strahan then dealt with the physical conditions, and gave his adherence to the "drift" as opposed to the "growth-in-place" theory of the origin of coal-seams, summing up the normal sequence of events in the formation of a seam as follows:—First, the out-spreading of sand and gravel with drifted plant remains; followed by shale as the currents lost velocity; and then a growth of presumably aquatic vegetation in extremely shallow water into which wind-borne vegetable dust and floating vegetable matter was carried; after which renewed subsidence brought in the sand and mud-laden currents again and the whole process was re-commenced. Mr. A. C. Seward followed with a clear statement of the botanical evidence bearing on the climatic and other physical conditions under which coal was formed; and Mr. J. E. Marr continued with a general outline of the geological evidence, laying stress on the peculiar coincidence during the Carboniferous period of a dominant vegetation of giant cryptogams with extensive plains of sedimentation and suitable climatic conditions. The debate thus initiated was then thrown open to the meeting and was carried on briskly by numerous speakers, among whom were Dr. Horace Brown, who gave the result of his experiments on the growth of plants in an atmosphere containing a slight excess of carbonic acid gas, and showed a series of lantern slides illustrating these experiments; Prof. P. F. Kendall, who supported the growth-in-place theory for most coals except cannel-coal; Mr. R. D. Oldham, who referred to the absence of seat-earths or under-clays to the seams in the Indian coal-fields; Dr. D. H. Scott, Dr. H. Woodward, Dr. II. O. Forbes, Dr. Wheelton Hind, Dr. Le Neve Foster, Mr. W. Cash, and others. In winding up this somewhat discursive debate, which had occupied the whole of the morning, the president leant strongly towards the growth-in-place theory, and this view was evidently also in favour with the greater portion of the audience.

Dr. E. D. Wellburn next gave two papers on the fossil fish of the Yorkshire Coal-field and of the Millstone Grits. Mr. J. J. H. Teall, President of the Geological Society, then described the plutonic complex of Cnoc-na-Sroine (Sutherlandshire), and discussed the three possible ways in which it may have originated, viz. by (1) successive intrusions; (2) differentiation *in situ*; or (3) modification of the original magma by the absorption of adjacent basic rocks, the conclusion being that the first method has not in this case played an important part, and that the second, coupled perhaps to some extent with the third, has been the main agent in forming the complex. Prof. K. Busz, of Münster, followed with a paper on a granophyre dyke intrusive in gabbro at Ardnamurchan (Scotland), in which it was shown that the granophyre in question has absorbed a considerable quantity of basic material from the previously consolidated gabbro, and has thereby added hornblende and mica to its proper constituents. Both papers provoked lively discussion.

Tuesday was essentially the glacialists' day and they made vigorous use of it, occupying nearly the whole session. Time was found, however, at the opening for a paper by Miss Igerna B. J. Sollas, " On *Naiadites* from the Upper Rhætic of Redland, Bristol"; and there was another break at the close, when Prof. A. P. Coleman, of Toronto, gave an account of the recent discovery of a ferriferous horizon in the Huronian north of Lake Superior, where a band of iron-bearing sandstone and jasper has already been traced for sixty miles in the Michipicoton district, and promises to be of great value from both the economic and the scientific standpoints, as it furnishes an easily-recognised horizon, probably equivalent to that containing the most famous iron mines of the United States, and affords an excellent clue to the stratigraphy.

Of the glacial papers, the first, by Mr. F. W. Harmer, was a theoretical discussion of the influence of winds upon climate during past epochs, in which it was sought to restore hypothetically the distribution of cyclonic and anti-cyclonic areas during the Pleistocene period, and to explain in this manner the phenomena of interglacial periods, which the author believes to have occurred alternately in the eastern and western continents, the conditions of comparative warmth and cold during this period having been local and due directly to meteorological causes. Then followed a series of excellent papers on the glacial phenomena of the West Riding, by Dr. Monckman, Mr. E. Wilson, and Messrs. A. Jowett and II. B. Muff, in which particular attention was drawn to the former existence of glacially-dammed lakes in the side valleys draining to the Aire, and to the overflow channels cut by the streams which had their source in these lakes. The glaciation of the East Riding was afterwards dealt with in two papers by Mr. J. W. Stather; and Mr. R. H. Tiddeman brought forward evidence proving that the raised beach of Gower in South Wales, with the bone-beds which rest upon it in the caves, must be either of pre- or inter-Glacial age, since they are overlain by glacial drift; this matter is of much consequence in the correlation of Pleistocene deposits of the unglaciated parts of our island with those of the glaciated tracts.

At the final meeting on Wednesday morning, Mr. R. D. Oldham discussed the mode of formation of the Basal Carboniferous Conglomerate of Ullswater in the light of his Indian

experience; and suggested that it was a torrential deposit formed on dry land near the foot of a range of hills, in a generally dry and hot climate varied by seasonal or periodical bursts of rain. In a second paper Mr. Oldham called attention to good examples of new beach-formation on the shores of Thirlmere Reservoir, and recommended that a photographic survey should be made from time to time to record the progress and growth of this beach. Mr. W. H. Crofts followed with a careful and well-illustrated account of sections in Glacial and post-Glacial deposits in a new dock at Hull; and Mr. A. C. Seward gave a summarised description of the Jurassic flora of the Yorkshire coast, with many fine lantern illustrations. Mr. G. W. Lamplugh afterwards reviewed the evidence as to the age of the English Wealden series, and supported the long-accepted but recently questioned view that the whole of the time-interval between the closing stages of the Jurassic and the commencement of the Aptian is represented.

The reports of committees of research read during the meeting included, among others, Prof. W. W. Watts', on the collection and preservation of geological photographs; Prof. P. F. Kendall's, on erratic blocks of the British Isles; Dr. Wheelton Hind's, on life-zones in British Carboniferous rocks; and Prof. A. P. Coleman's, on the Pleistocene beds of Canada.

The short afternoon excursions were under the leadership of Mr. J. E. Wilson, Mr. H. B. Muff and Dr. Monckman, who were thus able to show in the field some of the phenomena which they had described in their papers. These excursions were well attended and much appreciated by the visitors from a distance, who in this way were enabled rapidly and pleasantly to gain a grasp of the leading features of the local geology.

A well-arranged temporary museum, under the supervision of Mr. J. E. Wilson, for the exhibition of specimens illustrating the papers and the coal-discussion, was located in a large room adjoining the section room, and was especially serviceable in enabling those interested in the particular subjects illustrated to examine the material at their leisure and to compare notes upon it with the exhibitors. The lantern, so often a source of annoyance at the sectional meetings, was ably managed throughout; and indeed the whole of the local arrangements for the accommodation of the section were admirably planned and carried out, the only drawback being that the noise of heavy traffic on the stone pavement outside was at times troublesome.

To sum up the week's work, it may be remarked that there was an unusual number of papers dealing with subjects of broad general interest and therefore well suited for public discussion, and a scarcity of those detailed studies in stratigraphy or classification which, though probably of more permanent scientific value, are ill-adapted for presentation at these meetings; the local papers also were numerous and well above the average in character; petrology and palæontology were both adequately represented; but systematic geology received little attention. The morning meetings were well attended throughout, but, as usual, in the afternoons only the devoted nucleus of the section remained.

ZOOLOGY (AND PHYSIOLOGY) AT THE BRITISH ASSOCIATION.

THE opening day (Thursday) was devoted to the president's address in the morning and the reports of various committees in the afternoon. The reports were as follows:—

(1) Bird migration in Great Britain and Ireland.—Mr. Eagle Clarke has completed the extraction of the voluminous records of occurrences of birds in Great Britain and Ireland from the periodical literature of 1880-1887. The information thus provided supplements in a most useful manner the original Lighthouse data, and renders it possible for the first time to write an authoritative history of the migrations of each British bird. Mr. Clarke begins the series with a summary of details of the various migratory movements of (i) the Song-Thrush (*Turdus musicus*) and (ii) the White Wagtail (*Motacilla alba*).

(2) Investigations at the Naples Zoological Station.—The utility of the British Association's table was again demonstrated by the number of naturalists who had occupied it during the year. Reports on work done there were submitted by Mr. H. M. Kyle (anatomy of flat-fishes), Mr. E. S. Goodrich (structure of certain polychæte worms), Prof. W. A. Herdman (Compound Ascidians), Mr. R. T. Günther (anatomy of *Phyllirhoë* and certain Cœlenterates), Dr. A. H. R. Buller (fertilisation process

in Echinoidea) and Prof. Ramsay Wright (methods of preservation of specimens).

(3) Investigations at the Plymouth Marine Laboratory.—The British Association table was occupied by Mr. A. D. Darbishire, who investigated the natural history of *Pinnotheres* and the myology of *Calanus*; and by Mr. W. M. Aders, who studied the spermatogenesis of Cœlenterates.

(4) Index Animalium.—Mr. C. Davies Sherborne has made great progress in this important work, and the first part of his catalogue of post-Linnæan names up to the year 1800 is now ready for printing.

(5) Plankton of the English Channel.—Mr. Garstang has completed the five quarterly surveys provided for, and a final report will be presented at the Glasgow meeting.

(6) Zoology of the Sandwich Islands.—Mr. R. C. L. Perkins is again at work in the islands, and reports that already the forests are being extensively destroyed and replaced by sugar cane. It is fortunate for science that the committee foresaw this event, and were enabled to begin their investigations before it was too late. Four parts of the second volume of the "Fauna Hawaiiensis" have been published during the year.

On Friday, Prof. W. B. Scott, of Princeton University, U.S.A., gave an account of the Miocene fauna of Patagonia, based on an elaborate investigation of the Santa Cruz beds. The fauna was characterised by the abundance and variety of Marsupials of Australian type, of Edentates (ground-sloths, glyptodons, and armadillos), of porcupine-like Rodents and primitive Ungulates. There was no trace of tree-sloths and anteaters, of rats, mice, squirrels, hares and rabbits, or of carnivorous Eutheria. The place of the latter was taken by flesh-eating Marsupials, as in Australia to-day. South America was usually regarded as having no Insectivora, but some of the small mammals examined by him appeared to belong to this class. Among the hoofed animals one series was of particular interest in showing the complete evolution of a one-toed type from three-toed ancestors. This monodactyle positively "out-horsed the horse," for even the splints had gone. Yet morphologically it was no horse. It furnished the most conclusive instance he knew of convergent evolution in widely separated groups of animals. In conclusion, Dr. Scott showed that, after the removal of the Miocene barrier, the true carnivora of the modern fauna, together with the llama, deer, tapir, peccary, and the hares and rats, immigrated from North America, while the giant sloths and glyptodons extended their range to the northward.

Dr. Gregg Wilson exhibited a number of eggs and embryos of *Ornithorhynchus*, and described the water-side burrows and nests made by this lowly mammal. The duckmole protects its eggs and nest by blocking the passages between the nest and the entrances with solid walls of earth.

Prof. W. C. McIntosh communicated (through the secretary) a paper on some points in the life-history of the littoral fishes, in which he discussed the mortality of certain shore fishes at different stages of growth.

Major Ronald Ross then delivered a formal lecture on malaria and mosquitoes, dealing more particularly with the life-history of the sporozoan whose reproduction in the blood is the cause of malarial fever, and with the part played by the mosquito *Anopheles* in transferring the parasite by means of its so-called "salivary" secretion to the blood of fresh human hosts. Native children were the chief source of infection, since their blood swarmed with the parasites. The prevalence of malaria, however, might be reduced by efficient surface-drainage, which would check the multiplication of mosquitoes by destroying the pools and ditches in which their larvæ were developed.

In the afternoon Prof. S. J. Hickson exhibited microscopic preparations of *Dendrocometes*, demonstrating the existence of micronuclei in this suctorian, and the remarkable fusion of the macronuclei during conjugation. He advocated the employment of brazilin with iron-alum as a convenient substitute for the iron hæmatoxylin method of staining.

Dr. J. F. Gemmill described the anatomy of the head in cyclopean trout embryos. The cerebral lobes are more or less united, and the *trabeculæ cranii* are fused together anteriorly, and bent down below the median eye or eyes. The infundibulum and pituitary body are entirely absent; the optic nerves are rudimentary or absent, and the eyes, though provided with retina and choroid, have no choroidal fissure. In some specimens the mouth opening is absent, and the lower jaw arch greatly shortened.

Prof. R. Burckhardt of Basel communicated two papers on some causes of brain-configuration in selachians, and on the systematic value of the brain in selachians. He showed the profound influence of the position of the eyes and other superficial organs upon the shape of the brain, and advocated the employment of cerebral characters in the classification of cartilaginous fishes.

On Saturday, the papers were of a more or less physiological character, as follows :—

Prof. Marcus Hartog : "On a peptic zymase in young embryos," in which the author announced the confirmation of his discovery in 1896 of a peptic zymase in young embryos of the frog, in the entire embryo of the chick after twenty-four hours, and in the extra-vascular blastoderm of the three days' chick. He concluded that the law holds good for animals, as well as plants, that the cell cannot directly utilise the reserves it contains, but only the products of their hydrolysis, and this hydrolysis is not a function of the living protoplasm, but of the zymases it forms. These facts also explain apparent exceptions to Herbert Spencer's law of division at the doubling of the volume. A cell that is only accumulating reserve material has no need to constantly readjust its surface to its volume. When, however, the formation of a zymase enables it to utilise its reserves, and its protoplasm grows at the expense of the products of their digestion, the need for augmented surface declares itself, and we get the repeated cell divisions so marked in the "segmentation" of the embryo.

Dr. R. Irvine : "On the mechanical and chemical changes which take place during the incubation of eggs." Hen's eggs during incubation lose weight daily, principally through the oxidation of their carbon and hydrogen, parts of which pass off as CO_2 and H_2O through the shell. The percentage of ash is increased by absorption of lime from the shell.

Prof. Gotch described some recent experiments on the physiological effect of local injury in nerve which led to the important conclusion that an electrical disturbance was not always a concomitant of the passage of a nervous impulse.

In addition to the above, Prof. Johnson Symington read papers on the articulations between occipital bone and atlas and axis in the mammalia, and observations on the development of the cetacean flipper, and exhibited a convenient hand-magnifier for demonstrating slide-preparations to lecture-classes (Erbe, Tübingen).

The reports of committees on the following subjects were also communicated :—(1) The physiological effects of peptone when introduced into the circulation (Prof. W. H. Thompson). (2) Comparative histology of the suprarenal capsules (Mr. Swale Vincent). (3) The vascular supply of secreting glands (Dr. J. L. Bunch). (4) Electrical changes in mammalian nerve (Dr. J. S. MacDonald), and (5) The comparative histology of cerebral cortex (Dr. G. Mann).

On Monday, Mr. R. T. Günther read a note on *Mnestra parasites*, Krohn, in which he submitted reasons for referring this parasitic medusa to the family Cladonemidæ (Anthomedusæ), owing to its possession of compound tentacles with clavate appendages and other cladonemid characters.

Prof. L. C. Miall reviewed the respiratory organs of aquatic insects. He contrasted the slight nature of the adaptations to aquatic life which are exhibited by adult insects with the remarkable modifications for the same end which occur in insect larvæ. He explained the difference as probably due to the fact that profound structural changes in adult insects would interfere with their powers of flight, which were of importance for mating and other purposes. Among larvæ there were two principal lines of modification, (1) specialisation of the spiracular apparatus by which air could be inspired directly from the atmosphere through the surface film of water, and (2) development of a closed tracheal system, by which air was extracted from its solution in the surrounding water. This latter series culminated in a purely vesicular system, destitute of tracheæ, and finding its nearest parallel in the air-bladder apparatus of physoclist fishes.

Mr. T. H. Taylor described the tracheal gills of *Simulium*, whose mode of respiration presented peculiar difficulties still unsolved.

Mr. J. J. Wilkinson described the pharynx of *Eristalis*, and Mr. N. Walker the structure and life-history of the gooseberry sawfly.

In the afternoon Mr. Stanley Gardiner opened with the interim report of the committee appointed to investigate the structure, formation and growth of the coral reefs of the Indian Ocean. Special attention was given to the island of Minikoi, in the Laccadive group. It was clear from their observations that in this atoll there had been an elevation of the original reefs to a height of at least 25 feet above low tide level. All their evidence showed that the lagoons of atolls were generally formed by the solution of the central rock of originally more or less flat reefs.

Prof. R. Burckhardt followed with a paper on the anatomy and systematic position of the Læmargidæ. He recorded the discovery of luminous organs in nine species of Læmargidæ and Spinacidæ. The affinity of these families of sharks was further evidenced by his discovery of a cartilage hook in the dorsal fins of *Laemargus*.

Prof. Burckhardt also showed photographs and other illustrations of the nestling kagu (*Rhinochetus*), a rare flightless bird of New Caledonia.

Prof. R. J. Anderson described the dentition of the seal ; and Mr. Graham Kerr, on behalf of Mr. G. E. H. Barrett-Hamilton, exhibited some skulls of Antarctic seals (chiefly Phocidæ) brought home by the Belgian expedition.

On Tuesday, Mr. N. Annandale exhibited a number of photographic slides illustrating the appearance and habits of some Malay insects under natural conditions. One striking series represented the pupa of a Mantis (*Hymenopus bicornus*) seated on an inflorescence of the so-called "Straits Rhododendron" (*Melastoma polyanthum*), a detailed resemblance to which is brought about by the colour and shape of the insect, and by the extraordinary attitude which it adopts upon the flower.

Prof. E. B. Poulton also showed a large number of slides, representing the collections of insects made by Mr. G. A. K. Marshall in Mashonaland and Mr. R. Shelford in Borneo, as arranged in the Oxford Museum to illustrate the general principles of Müllerian mimicry. An interesting series of mutilated butterflies caught at large showed the comparative rarity of indiscriminate injuries by birds, and the frequency with which enemies aimed at the conspicuously marked tips of the forewings and at the back of the hind-wings, where tail-like processes were so commonly developed, these being just the places where the bites would be least dangerous to the insects.

Other slides, illustrating mimicry and protective resemblance, were exhibited by Mr. Mark L. Sykes ; and Prof. Lloyd Morgan described some recent experiments upon newly-hatched chicks, which showed that the avoidance of distasteful forms by birds is not instinctive, but the fruit of experience. Chicks fed for a time on palatable food placed on black-and-orange banded slips of glass did not hesitate to attack the distasteful caterpillars of the cinnabar moth when these were eventually offered them ; whereas chicks which had been accustomed to associate the same coloured slips with bad food refused to attack the similarly striped caterpillars. These observations provided a sound experimental basis for the Müllerian theory of mimicry.

Mr. F. W. Gamble described the results of investigations made by Mr. F. W. Keeble and himself on the colour changes of various prawns, especially *Hippolyte varians*, his paper being illustrated by a series of living specimens as well as by lantern slides. The prawns adapted their colours to those of surrounding weeds ; but, whatever their colour during the day, they always assumed a characteristic blue colour at night. This change, in newly-caught specimens, came on at the proper time quite independently of the darkness, and the morning phase would be resumed at daybreak, even when the animal was kept in the dark. After a few days under such unnatural conditions, however, the periodicity became altered.

A paper, by Dr. Æneas Munro, on the locust plague and its suppression concluded the business of the section.

GEOGRAPHY AT THE BRITISH ASSOCIATION.

THE work in Section E at the Bradford meeting was somewhat limited in amount, but its quality was in no way below the average. In fact, the number of "popular" papers was smaller than usual, while those of a more serious character predominated. The section was excellently housed in the Church Institute, and the meeting began with a presidential address of quite a novel character. Sir George Robertson took the British Empire as his text, and laid great stress on the relative shrinkage of distances by the improvement of means of communication by land and sea, a fact which in great measure

neutralised such ill effects as might arise from continuous expansion of territory.

The keynote struck by the president was geography as the science of distances, and in unison with it a series of papers dealt with problems of which distances and means of transport were the essential features. Mr. E. G. Ravenstein discussed the question of foreign and colonial surveys in a comprehensive paper, in which he pointed out the manner and extent of the official surveys of the chief countries of the world. While recognising that the British Ordnance Survey fell short of perfection, he considered that its accuracy was not equalled by the maps of any other country. He strongly urged the adoption of a more systematic method of surveying in Africa, in many parts of which the only existing maps were produced by travellers with inadequate assistance and many other things to do. In commenting on the paper, Colonel Johnston, the Director-General of the Ordnance Survey, explained the position of South Africa with regard to its surveys. He said that a nearly perfect system of triangulation had been carried out, but this has not yet been utilised by being made the foundation of a detailed survey.

Mr. B. V. Darbishire read a paper on military maps, with special reference to the use of the Ordnance Survey Maps in field manœuvres.

Colonel Sir Thomas H. Holdich discussed the question of a railway connection between Europe and India. He considered the northern approaches to India across Kashmir or the Hindu Kush from the Oxus valley to be impracticable. On the other hand there appeared to be no insurmountable difficulty in the way of a connection by the Hari-rud valley, through which approach a distance of only 500 miles intervened between the farthest outposts of the existing railways, Kushk on the Russian side and New Chaman on the Indian. The new line would pass by Kandahar. This line could, in the opinion of the author, be made to pay by local traffic, and he believed it would strengthen rather than weaken the defences of India.

Mr. C. Raymond Beazley read a paper, which was largely historical and statistical, on the Trans-Siberian railway.

Mr. G. G. Chisholm gave a very timely forecast of the probable economic changes which may be expected to result from the imminent development of the resources of China by modern methods. These would include, in his opinion :—A rise in prices in China, especially in the industrial regions ; a demand for food-stuffs not likely to be supplied by China itself ; a great stimulus to the food-producing regions most favourably situated for meeting this demand, more particularly Manchuria, Siberia, and western North America ; and the creation of a tendency to a gradual but prolonged rise in wheat and other grain prices all the world over, reversing the process that has been going on since about 1870.

Mr. Edward Heawood treated of the commercial resources of tropical Africa, and his paper also partook of the spirit of forecast, his expectations being that Africa will greatly increase in importance by the cultivation of tropical plantations.

The travel papers which excited the most interest were those contributed by Mr. Borchgrevink on his expedition to the Antarctic regions and by Captain H. H. P. Deasy on his journeys in Central Asia. Both were illustrated by remarkably fine lantern slides. As the facts which they recounted have already been published, it is unnecessary to summarise them here.

Physical geography occupied a large part of the time of the section, and, with regard to this part of the work, it is impossible to refrain from expressing the desire that some arrangement might be come to with regard to the section in which papers lying on the borderland between different subjects should be treated. With regard to meteorology, for instance, might it not be arranged to read all climatological papers—the essential principle of which is geographical distribution—at Section E, and only the theoretical papers or those dealing with instruments and atmospheric physics at Section A ?

On this occasion the report of the committee on the climate of tropical Africa, of which Mr. H. N. Dickson is secretary, was read to Section E, and a remarkable discussion of the geographical distributions of relative humidity was presented by Mr. E. G. Ravenstein to the same section. In this he said that, notwithstanding the paucity of available material, he had ventured, in 1894, to publish in Philip's "Systematic Atlas," a small chart of the world showing the distribution of humidity, and he now placed the results before this meeting with some diffidence. His charts brought out the broad features of the subject, and to reduce the sources of error be had limited himself to indicating four grades of mean annual humidity, the upper limits of which were respectively 50 per cent. (very dry), 65 per cent., 80 per cent., and 100 per cent., (very damp). The relative humidity over the oceans might exceed 80 per cent., but in certain regions ("horse latitudes") it was certainly much less, and in a portion of the Southern Pacific it seemed not to exceed 65 per cent. One chart exhibited the Annual Range of Humidity, viz. the difference between the driest and the dampest months of the year. In Britain, as in many other parts of the world, where the moderating influence of the ocean was allowed free scope, this difference did not exceed 16 per cent., but in the interior of the continents it occasionally exceeded 45 per cent., spring or summer being exceedingly dry, whilst the winter was excessively damp, as at Yarkand, where a humidity of 30 per cent. in May contrasted strikingly with a humidity of 84 per cent. in December. This great range directed attention to the influence of temperature (and of altitude) upon the amount of relative humidity, for during temperate weather we were able to bear a great humidity with equanimity, whilst the same degree of humidity, accompanied by great heat, may prove disastrous to men and beasts. Hence, combining humidity and temperature, the author suggested mapping out the Earth according to sixteen *hygrothermal types*, as follows :—(1) Hot (temperature 73° and over) and very damp (humidity 81 per cent. or more) : Batavia, Camaroons, Mombasa. (2) Hot and moderately damp (66–80 per cent.) : Havana, Calcutta. (3) Hot and dry (51–65 per cent.): Bagdad, Lahore, Khartum. (4) Hot and very dry (50 per cent. or less) : Disa, Wadi Halfa, Kuka. (5) Warm (temperature 58° to 72°) and very damp: Walvisch Bay, Arica. (6) Warm and moderately damp : Lisbon, Rome, Damascus, Tokyo, New Orleans. (7) Warm and dry : Cairo, Algiers, Kimberley. (8) Warm and very dry : Mexico, Teheran. (9) Cool (temperature 33° to 57°) and very damp : Greenwich, Cochabambo. (10) Cool and moderately damp : Vienna, Melbourne, Toronto, Chicago. (11) Cool and dry : Tashkent, Simla, Cheyenne. (12) Cool and very dry : Yarkand, Denver. (13) Cold (temperature 32° or less) and very damp : Ben Nevis, Sagastyr, Godthaab. (14) Cold and moderately damp : Tomsk, Pike's Peak, Polaris House. (15) Cold and dry. (16) Cold and very dry : Pamirs.

The actual mean temperature of the Earth amounted, according to his computation, to 57° F., and this isotherm, which separated types 8 and 9, also divided De Candolle's "Mikrothermes" from the plants requiring a greater amount of warmth.

Mr. Vaughan Cornish described his recent observations on snow ripples with beautiful photographic illustrations, and Prof. J. Milne gave an account of the large earthquakes recorded in 1899. Mr. R. T. Günther described the peculiar character of the coast of the Phlegræan Fields near Naples, and showed that by observations of the numerous submerged buildings of that district it might be possible to determine the date and duration of the fluctuations of the land and sea level during the last twenty centuries. The Association subsequently voted a money grant to assist him in carrying out the researches which he had suggested.

Dr. H. R. Mill exhibited and described the new insulating water-bottle designed by Profs. Pettersson and Nansen, and made by Messrs. Ericsson, for obtaining water-samples from any desired depth and bringing them up without change of temperature. The new apparatus was tested by Prof. Nansen last August on board the *Michael Sars* in the North Atlantic, and found to be completely satisfactory.

Dr. Mill also read a paper on the treatment of regional geography, in which he laid down the general principle that the fixed conditions of the land surface had first to be described, and then the mobile distributions, which were modified by the fixed forms. As an example, he dwelt at some length on the configuration of a section of the South Downs and the effect of this configuration in determining the distribution of rainfall in the district, a problem which he hoped to treat in greater detail at a future date.

Mr. J. E. Marr described the typical land form known as a moel, with special reference to the forms it assumed when dissected by sub-aerial erosion.

Two educational papers of much interest were read. One by Mr. T. G. Rooper dealt with the progress made in teaching of geography in the elementary schools of the West Riding since 1883. He illustrated it by the exhibition of a series of remark-

able relief models on different scales produced by school teachers and used by them in their regular work. Some of these were of typical features, such as the Red Tarn, to typify a mountain lake, others of the actual school district taken from the Ordnance map, and others, on a small scale, of large parts of the country. The second paper was by Mr. E. R. Wethey, who gave a demonstration of his method of teaching commercial geography by the use of lantern maps, diagrams and pictures, a large number of which, in novel and striking forms, he showed upon the screen. Educational questions have always occupied a considerable share of the time of Section E, and the committee very cordially supported the proposal to recommend the Council of the Association to form a new Section for the discussion of education in a more complete and technical manner than could be secured in a gathering of votaries of one isolated branch of science.

UNIVERSITY AND EDUCATIONAL INTELLIGENCE.

OXFORD.—Mr. E. S. Goodrich has been elected to a fellowship in natural science at Merton College.

CAMBRIDGE.—In his annual address to the Senate at the opening of the term, the Vice-Chancellor announced that the Benefactor Fund amounted to 55,000*l.*, and that the Squire Trustees had agreed to contribute 15,000*l.* towards the erection of the Law School. The plans for the Botanical and Medical Departments have been approved, and building will shortly commence ; but fresh benefactions are still needed to meet the urgent demands for further accommodation.

The new Department of Agriculture, under the able guidance of Prof. Somerville, is now well started. The funds at its disposal have enabled it to secure an efficient staff, and it is provided with an excellent experimental farm. The University has sought to encourage the study by establishing a special amination in agricultural science for the B.A. degree.

Dr. L. Humphry has been appointed assessor to the Regius Professor of Physic ; Sir G. G. Stokes and Prof. Darwin electors to the Isaac Newton Studentship in Physical Astronomy ; and Dr. Tatham an examiner for the diploma in Public Health. Mr. Leathem (St. John's) and Mr. Grace (Peterhouse) have been appointed moderators, and Mr. Whitehead (Trinity) an examiner, for the Mathematical Tripos.

Rooms for work in clinical pathology, bacteriology, &c., have just been erected by the staff and presented as a gift to Addenbrooke's Hospital. They will be open for work, under the direction of Prof. Sims Woodhead, during the present term.

At Emmanuel College a research studentship of 100*l.* has been awarded to Mr. J. Mellanby. Grants have been made from the studentship fund of 60*l.* to Mr. G. F. Abbott, and of 40*l.* to Mr. D. G. Hall. At Queen's College the Rev. C. H. W. Johns has been elected to the office of lecturer in Assyriology.

MR. C. R. P. ANDREWS, of St. John's Training College, Battersea, has been appointed first principal of the new Government training college to be opened at Perth, Western Australia.

DR. SAMSON GEMMELL, of Anderson's College, Glasgow, has been appointed professor of clinical medicine in the University of Glasgow, in succession to Prof. McCall Anderson.

DR. CULLIS, professor of mathematics at the Hartley College, Southampton, has been appointed professor of mathematics at the Presidency College, Calcutta.

MR. J. F. HUDSON, late lecturer in mathematics at Jesus College, Oxford, has been appointed professor of mathematics at the Hartley College, Southampton.

MR. J. STUART THOMSON, formerly demonstrator of zoology at the School of Medicine of the Royal Colleges, Edinburgh, has been appointed lecturer in botany and zoology at the Municipal Science, Art and Technical Schools, Plymouth.

THE School of Engineering of Columbia University, New York, announces a new course of study dealing with the construction of automobiles, self-propelling road engines and railway cars.

PROF. GOSS has been made dean of the Engineering Schools of Purdue University, Lafayette, Ind., and Prof. L. C. Glen, of South Carolina College, has been appointed to the chair of geology in Vanderbilt University.

MR. PERCY H. FOULKES has been elected first principal of the Harper Adams College, Newport, Salop. He will enter upon his duties on January 1, 1901, soon after which date the college will, it is expected, be ready to receive pupils.

AT a general meeting of Convocation of the University of London, held on Tuesday last, the following were elected to serve as members of the Senate under Section 12 of the statute of the reconstructed University:—Mr. John Fletcher Moulton, Dr. J. D. McClure, Sir A. Kaye Rollit, Dr. T. B. Napier, Dr. J. B. Benson, Dr, T. L. Mears, Sir H. H. Cozens-Hardy, Dr. T. Barlow, Mr. J. F. Payne, Sir Philip Magnus, Dr. S. Bryant, Dr. C. W. Kimmins, Dr. F. Clowes, Prof. Silvanus Thompson, Dr. F. S. Macaulay, and Mr. J. W. Sidebotham.

DURING the past week very many addresses have been delivered to students at the opening of the winter sessions of the various science, technical and medical schools in London and the provinces, in the course of which much excellent advice has been given. An article dealing with some of the utterances made to medical students is to be found in another part of the present issue. In this column we refer, and only very briefly, to two addresses given to students of other branches of knowledge, viz. those by Sir Alexander Binnie at the opening of the Central Technical College, on October 2, and by Prof. Le Neve Foster at the distribution of medals, prizes, &c., to the students of the Royal College of Science, on October 4. The subjects chosen for their addresses by both speakers were well suited to the occasion, and should prove of much service to the audiences who listened to them. Prof. Foster took as his topic "Common Sense," in the course of which he referred to the remark of Prof. Huxley that science was organised common sense, and the two or three years' training in science which students received at the college was, therefore, simply training in ordinary common sense. If they wished to succeed in any calling they must exercise the faculty of thought. It was difficult to realise that times were changing, but change was everywhere taking place, and they must throw aside the idea that in the production of British manufactures the methods that had come down to them from their forefathers were necessarily the best. In Lancashire it was said that what Lancashire did to-day Great Britain would do to-morrow. They might say that what the scientific man did to-day the manufacturing man would do to-morrow. The laboratory experiment of to-day was, in fact, the manufacturing process of to-morrow. But if the student desired to take an active part in the improvement of the industrial life of the country and of manufacturing processes, he must work hard and not place too much reliance on his teacher. All that the professor could do was to give the student a general ground-work upon which afterwards by his own experiments he could build up his frame-work of knowledge. Sir A. Binnie in his address contrasted the advantages which students of to-day have over those educated in the middle of the present century, and urged upon his hearers not to confine themselves merely to the curriculum of study laid before them, or to take too narrow a view or devote themselves exclusively to one particular branch of learning. The aim of the speaker was to impress upon his audience that to be a true student of science the mind must be opened out and widely cultivated by observation to grasp every detail, as it often occurs that it is among the almost unnoticed minutiæ of a particular science that those wonderful correlations that lead in the future to wide results are to be found. He spoke of the necessity of acquiring a wide and broad view of the subjects which should engage the student's attention for the reason that he felt that education could only be complete when studied as a whole, and the beauty of all the different sciences brought clearly before the mind. Further, one can never tell, when entering upon active work, into what avenues or by-paths of practice he may be led, and to illustrate this Sir A. Binnie referred to his own experience. He also urged upon his hearers to study the history of their profession, and of the various discoveries which have been made in the different branches of science to which they would apply themselves. Altogether the students are to be congratulated upon the helpful advice tendered to them.

SOCIETIES AND ACADEMIES.

PARIS.

Academy of Sciences, October 1.—M. Maurice Lévy in the chair.—On the absorption of free oxygen by normal urine, by M. Berthelot. Normal urine absorbs free oxygen in amounts larger than those corresponding to the solubility of oxygen in

water. The acidity is not altered by the absorption.—Remarks on the acidity of urine, by M. Berthelot.—On the distribution of the horizontal component of the earth's magnetism in France, by M. E. Mathias. As the result of work spread over a period of six years in the neighbourhood of Toulouse, it was found that a very simple formula would combine the results of all the observations, namely : $\Delta H = -1\cdot26$ (Δ long.) $-7\cdot42$ (Δ lat.), in which ΔH was the difference between the measurement for an element at a place X and that of the corresponding element at Toulouse. It was further found that the above formula applies to the whole of France.—On the selenides of nickel, by M. Fonzes-Diacon. Nickel leaflets heated in a current of nitrogen carrying small quantities of selenium vapour give cubical crystals of a selenide of the composition NiSe. Another selenide approximating in composition to Ni_3S_4 is obtained by heating anhydrous nickel chloride in a current of hydrogen selenide at a dull red heat. At 300° C. the diselenide NiS_2 is obtained as a greyish-black, friable mass. All these products heated to a white heat in a current of hydrogen give a sub-selenide, Ni_2Se.—Oxycelluloses from cotton, flax and hemp, by M. Leo Vignon. Purified fibres of various textile material were submitted to the oxidising action of hydrochloric acid and potassium chlorate; the yield in all cases was the same, about 70 per cent. ; phenylhydrazine furnished the same osazone. Small differences were observed in the reducing powers of the oxycelluloses from different sources.—On the mutability of *Œnothera Lamarckiana*, by M. Hugo de Vries. This furnishes an example of the rare phenomenon of a state of mutability in a pure species. The new species appears suddenly without preliminary or intermediate stages; the transformed individual shows all the characters of a new type, although the parents and grandparents are absolutely normal. The seeds of the transformed individuals give rise to the new type only, no tendency being observed to revert to the characters of *Œ. Lamarckiana*.—On the Eocene of Tunis and Algiers, by M. L. Pervinquière.—The ravine of Chevalleyres and the retrogression of torrents, by M. Stanislas Meunier. Attention is drawn to the mode of formation of this *col*, the size of which would appear out of all proportion to the small stream to which the ravine is undoubtedly due. The transfer of rock masses, and other effects usually ascribed to glacier action, may be traced to this torrent.—Observations of a meteor which fell on the evening of September 24, by M. Jean Mascart. The meteor, the nucleus of which was star-like and very bright, was seen at 10.16 p.m. on September 24 between Meudon and Bellevue.

CONTENTS.

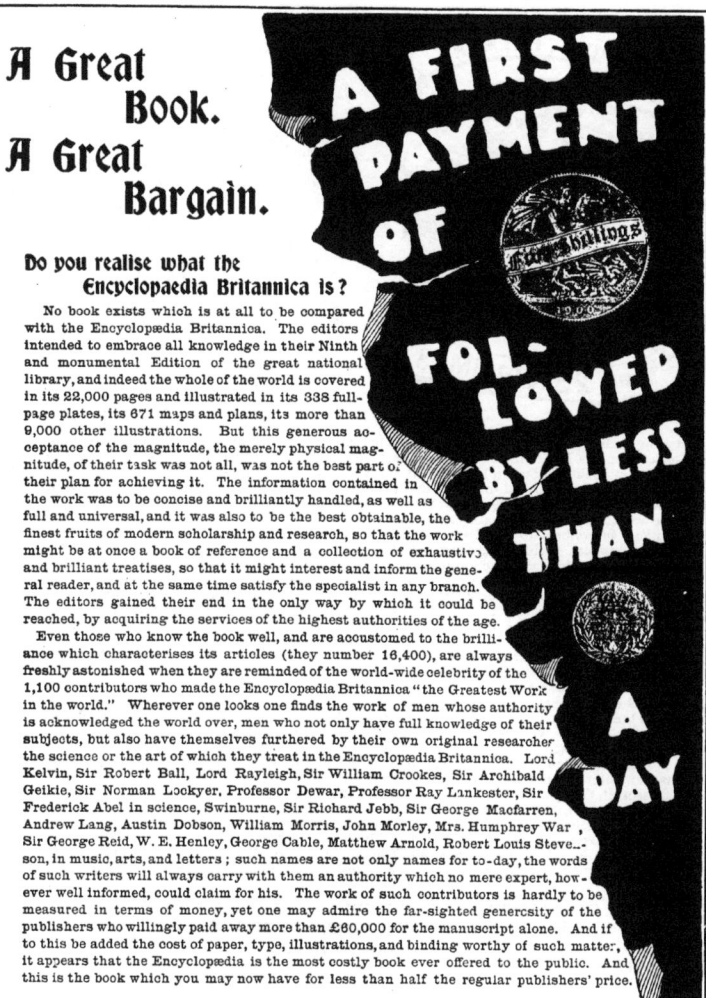

A Great Book.
A Great Bargain.

A FIRST PAYMENT OF Five Shillings FOLLOWED BY LESS THAN A PENNY A DAY

Do you realise what the Encyclopaedia Britannica is?

No book exists which is at all to be compared with the Encyclopædia Britannica. The editors intended to embrace all knowledge in their Ninth and monumental Edition of the great national library, and indeed the whole of the world is covered in its 22,000 pages and illustrated in its 338 full-page plates, its 671 maps and plans, its more than 9,000 other illustrations. But this generous acceptance of the magnitude, the merely physical magnitude, of their task was not all, was not the best part of their plan for achieving it. The information contained in the work was to be concise and brilliantly handled, as well as full and universal, and it was also to be the best obtainable, the finest fruits of modern scholarship and research, so that the work might be at once a book of reference and a collection of exhaustive and brilliant treatises, so that it might interest and inform the general reader, and at the same time satisfy the specialist in any branch. The editors gained their end in the only way by which it could be reached, by acquiring the services of the highest authorities of the age.

Even those who know the book well, and are accustomed to the brilliance which characterises its articles (they number 16,400), are always freshly astonished when they are reminded of the world-wide celebrity of the 1,100 contributors who made the Encyclopædia Britannica "the Greatest Work in the world." Wherever one looks one finds the work of men whose authority is acknowledged the world over, men who not only have full knowledge of their subjects, but also have themselves furthered by their own original research the science or the art of which they treat in the Encyclopædia Britannica. Lord Kelvin, Sir Robert Ball, Lord Rayleigh, Sir William Crookes, Sir Archibald Geikie, Sir Norman Lockyer, Professor Dewar, Professor Ray Lankester, Sir Frederick Abel in science, Swinburne, Sir Richard Jebb, Sir George Macfarren, Andrew Lang, Austin Dobson, William Morris, John Morley, Mrs. Humphrey Ward, Sir George Reid, W. E. Henley, George Cable, Matthew Arnold, Robert Louis Stevenson, in music, arts, and letters ; such names are not only names for to-day, the words of such writers will always carry with them an authority which no mere expert, however well informed, could claim for his. The work of such contributors is hardly to be measured in terms of money, yet one may admire the far-sighted generosity of the publishers who willingly paid away more than £60,000 for the manuscript alone. And if to this be added the cost of paper, type, illustrations, and binding worthy of such matter, it appears that the Encyclopædia is the most costly book ever offered to the public. And this is the book which you may now have for less than half the regular publishers' price.

[*See Next Page.*]

A FIRST PAYMENT OF FIVE SHILLINGS BRINGS THE 25 VOLUMES

1900

Do you Realise that the Book is a Book for You?

Whatever may be your profession, whatever may be the nature of your interests and the direction of your ambitions, you will find in the Encyclopædia Britannica much to help you at your work, to forward you in your hobby, to encourage you in your ambitions. No man, however wide may be his activities, however quick his understanding, can satisfactorily live off his own personal experience. The Encyclopædia Britannica takes the reader over the whole world, giving him the experience of others who have travelled further afield or whose path in life has fallen in other places. It is a book calculated to appeal to everyone, and since it has been brought within the reach of everyone, it is interesting to remark what a comprehensive list is to be made from those who have purchased ".THE TIMES" Reprint. The fifty-two peers include Lord Salisbury; there are 20 bishops, and 80 officers in the Army and Navy, starting with Lord Roberts. Mr. Kipling is among the long list of authors, and the late Lord Russell appears in the roll of judges. Finance and manufacture are represented by some of the biggest names. Such a list shows in what respect the Encyclopædia Britannica is held by the most distinguished men in the country. But there is another list which may be made, a list no less significant, for it shows how the work is rated by those who still have their way to make. There are clergymen who are not yet bishops (sixty out of the first thousand subscribers to the DAILY MAIL offer were clergymen), officers who are not yet generals, lawyers, doctors, engineers, architects, who still have names and a fortune to make. Perhaps you think that these professions account for most of the names upon the registers, but this is not so. Manufacturers, men engaged in business of every sort, accountants, land agents, auctioneers, Government inspectors, make up a total as great as that of the learned professions. The man of business is no less keen to possess a book which may be of practical use to him. The third highest figure in the list belongs to merchants. They are followed by manufacturers and managers in every sort of works, iron foundries, chemical works, cycle manufactories, breweries, lace factories, collieries, boot and shoe factories, cotton mills, laundries. The appeal of the book to practical men, moreover, could hardly be better exemplified than by the number of farmers and market gardeners who are down among the subscribers.

In the last announcement in which we put before readers of *Nature* the unique offer of the great library, we attempted to suggest the grasp, the fulness, the brilliance which characterises the scientific side of the many-sided book. The contributors whose names we mentioned (such names as Lord Kelvin, Sir William Crookes, Professor Newton, Professor Ray Lankester, Sir Robert Ball, Sir Archibald Geikie, Sir Norman Lockyer, Lord Rayleigh, Professor St. George Mivart, Professor Dewar) are much more than experts fully equipped with information, they are men who have not only written the history of their various branches, but also helped to make that history by their own original researches. They are men whose names are not only for to-day. It was on such a noble scheme that the editors prepared the whole of the famous 9th Edition. In every department of knowledge they claimed the services of men who were making history, of original thinkers. There is a very

A Book that can never again be sold on such extraordinary terms.

A Book that, if you are to get it while it is to be had so easily, you must secure at once.

real sense in which the Encyclopædia Britannica may be called a "practical book." Very rarely in the history of learning could such a collection of talent be matched. The contributors were men who could speak with the authority of Discoverers, setting forth in every province of knowledge the new spirit of the age. For if in the Encyclopædia Britannica we have the best fruits of the new learning in science, so also we have the new movement in decoration set forth by Marris, the contemporary admiration for Elizabethan drama in the work of Mr. Swinburne, the new founder for the period of the Renaissance presented by no less an authority than John Addington Symonds. All the collected learning of the past is in the pages of the Encyclopædia Britannica, and there also in its fulness is the  new knowledge which most men only half know, the new point of view from which we are ready to see things if only we may have authoritative direction. Such direction is to be found on every subject in the Encyclopædia Britannica. For the editors went the way to make not only a work of reference, fit to answer a question, but also a collection of treatises covering the whole world. That no book of reference has ever been put together by such authorities as made the Encyclopædia Britannica "the greatest work in the world" goes without saying. But it is equally true that no man, even if he could gather from the ends of the earth the best text books written on every subject, could make a library at all to be compared with the Encyclopædia Britannica. The charm and usefulness of the Encyclopædia Britannica is that its possessor has not only his own subject brilliantly treated to refer to, but everybody else's subject at his service, and the naturalist may read Sir Richard Jebb upon some classical point with as much confidence as he would turn to the contributions of Professor Cayley if he were a mathematician. It is, in fact, a book calculated to interest everyone, and especially those to whom the scientific method, as it is used by the greatest men, appeals strongly.

Do you Realise the Bargain Offered You?

The arrangement effected between the DAILY MAIL and "THE TIMES" in the matter of "THE TIMES" Reprint of the Ninth and latest Edition of the Encyclopædia Britannica, makes it possible to put before the public a bargain unprecedented in the history of bookselling. An invaluable work, a complete library, which everyone desires to possess, is offered at less than half the published price. That is the first point in the remarkable bargain. The second point is this—that the price, the "less than half-price," need not be paid at once; it may be paid in small monthly instalments of 12s. each. But there is a third, and most important, point in the bargain. It is this. Although you need not pay the price all at once, you get all the 25 volumes at once, for they are sent you entire upon receipt of a preliminary payment of 5s. For the sum of 5s., therefore, the 25 volumes of "THE TIMES" Reprint of the Encyclopædia Britannica is put into your hands to use from A to Z. While you and your household are enjoying the possession of the work, you complete your purchase by sending in monthly payments of 12s., and when the purchase is completed you will have paid something less than half the price which the publishers put upon exactly the same book. Is this not a unique thing in book-selling? The novel arrangement places the great national library well within reach; it is no longer an unattainable treasure, it is no longer a work which necessitates a journey to a public library for the man who would consult it. The Encyclopædia Britannica may now be, where such an invaluable treasury of all knowledge should be, at everyone's elbow. And you may have it at your elbow, for constant reference and study, by spending the small sum of 12s. a month, which means saving from your other expenses a little less than 6d. a day.

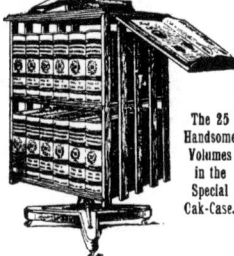

The 25 Handsome Volumes in the Special Oak-Case.

An Enquiry Form is given on the next page.

Specimen Volumes, Bindings and Bookcase can be seen, and full details obtained, at the following places:

LONDON.

NEAR BOUVERIE ST., FLEET ST.
"DAILY MAIL" OFFICE
("Encyclopædia" Dept.), 4, Harms-
worth Buildings, Tallis Street.

NEAR THE BANK.
Messrs. Cramer & Co., Ltd., 46-40,
Moorgate Street, E.C.

NEAR MARK LANE.
Messrs. Eyre & Spottiswoode,
101, Leadenhall Street.

NEAR CANNON STREET STATION.
Messrs. William Dawson & Sons,
Ltd., 121, Cannon Street, E.C.

WEST END.
Messrs. Cramer & Co., Ltd., 207 &
209, Regent Street, W.

NEAR CHARING CROSS.
Messrs. William Dawson & Sons,
Ltd., 23, Northumberland Avenue.

PROVINCES.

MANCHESTER.
Messrs. Forsyth Brothers, 126 &
128, Deansgate.

CARDIFF.
Messrs. Dawson, Hayes Buildings,
Working Street.

EXETER.
Messrs. Dawson, 22, Gandy Street.

LEICESTER.
Messrs. Dawson, 7B, Halford St.

Do You Realise the Need for Haste?

We want you to realise that what we are telling you is absolutely true, and what we are telling you and have told you before is just this, that if you wish to make sure of not missing this great opportunity that we are giving to you you must make haste, you must send in your order promptly. You will realise at once the need for haste when we tell you that in the first place the present offer can remain open only for a limited time, and only a limited number of sets can be sold on these present terms. Judging from the very large demand that has already been made for sets of the Encyclopædia Britannica *upon the* DAILY MAIL *terms, the limited space of time will be still further limited by the fact that before the date upon which this offer must be withdrawn has been reached, the limited number of sets will have been taken up. It is not only that after the date upon which this offer must be withdrawn we shall not be able to accept any orders, but that also (and this promises to be long before that time), after we have accepted an order for the last set of which we can dispose, all other orders will be too late. These* 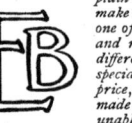 *plain facts will surely make you realise that if you wish to be in time you must make haste. Of course, in every case we should like you to be able to come up to one of the many offices that we have opened where you can see for yourself, handle, and read, the volumes of the Encyclo-pædia Britannica, where you can see the different bindings, where you can see the revolving bookcase, which has been specially manufactured for subscribers and is sold to them at an extremely low price, and where also you can obtain fuller details of the bargain offer now made to you than we can possibly give you in any advertisement. But if you are unable to make it convenient to call you can obtain equally full information by using our Enquiry Form, or by sending us a post card, in both cases the cost being to you only a halfpenny, and you may fully rely upon the information which will be conveyed to you by the illustrated prospectus of the* Encyclopædia Britannica, *this prospectus containing not only an account of the book, but facsimile pages, extracts from numerous articles, and specimens of the coloured plates, maps, and other illustrations. In fact, this illustrated prospectus may very well be termed a "sample book." Every statement in it you may rely upon as being an under, rather than an over statement of the facts. Of this we feel so sure that we know that when you receive your books you will be astonished at three things: you will be astonished that we have been so moderate in what we have said about them, you will be astonished at finding them an even more valuable and handsome possession than even you expected them to be, and, above all, you will be amazed that you have been able to acquire this book, not only at such an extraordinary bargain price, but upon an equally extraordinary and convenient system of payment.*

ENQUIRY FORM.
To be sent to the "DAILY MAIL" OFFICE.

Please send me the Illustrated Prospectus of the "Encyclopædia Britannica," order form, and full details of the "Daily Mail" offer, and provisionally *reserve me a set of the Volumes.*

[Name]..

NA 50   [Address]..

> ☛ Please address to the "**Daily Mail**" (*"ENCYCLOPÆDIA" DEPARTMENT*),
> 4, HARMSWORTH BUILDINGS, TALLIS STREET, LONDON, E.C.
> **A full description of the Book is given in the preceding pages.**

This Enquiry Form can be sent in an open Envelope with only a Halfpenny stamp

THURSDAY, OCTOBER 18, 1900.

THE SUBORDINATION OF THE INDIVIDUAL TO THE WELFARE OF THE SPECIES.

The Foundations of Zoology. By William Keith Brooks, Ph.D., LL.D., Professor of Zoology in the Johns Hopkins University. Pp. viii + 339. (Columbia University Press. New York : The Macmillan Co. London : Macmillan and Co., Ltd., 1899.)

THIS volume forms the fifth of the Columbia University Biological Series edited by Prof. H. F. Osborn and Prof. E. B. Wilson, and it is appropriately placed beside the well-known earlier memoirs which deal with historic, phylogenetic and ontogenetic evolution. The traditions of the series are sufficient warrant for the admirable editing, printing and general appearance of the volume.

The author arranges his work in fourteen chapters corresponding to thirteen lectures, the sixth being divided into two parts. The subjects which follow the introductory lecture are " Huxley, and the Problem of the Naturalist," " Nature and Nurture," " Lamarck," " Migration in its Bearing on Lamarckism," forming the titles of the second, third, fourth and fifth lectures. The sixth deals with " Zoology and the Philosophy of Evolution," and its second part with the .vįe.ws of Galton and Weismann. " Galton, and the Statistical Study of Inheritance " is the subject of the seventh, and " Darwin and the Origin of Species " that of the eighth lecture, the subjects of the remaining lectures being " Natural Selection and the Antiquity of Life," " Natural Selection and Natural Theology," " Paley, and the Argument from Contrivance," " The Mechanism of Nature " and " Louis Agassiz and George Berkely." The titles are quoted in full, inasmuch as it will be recognised. that the author's arrangement is unusual, both as regards treatment and the choice of some of the subjects. The same observation is true' of the separate lectures : we everywhere meet with interesting views and modes of statement which are individual and original, and evidently represent the deep personal convictions of the author upon subjects to which he has devoted much time and thought. It may be questioned, however, whether the printed lecture is not an inconvenient form in which to address a wider audience than can be gathered in any hall or theatre. The spoken lecture is the best of all forms of communicating ideas, because we have the speaker's personality associated with his thoughts. But the form of a lecture is in large part determined because its substance is conveyed so easily and rapidly by speech and hearing. The same idea must often be repeated in different words, in order that it may be grasped and remembered before passing to others ; and an argument may, and often should, be drawn out and enforced at a length which would be unnecessary and even tedious in a printed memoir. The lecturer has the great advantage that he can omit or expand according as he realises the extent to which his audience is in touch with him. When ideas are conveyed in print, the conditions are, of course, entirely different. When the reader does not fully understand, he can pause and reflect, and can read

again without losing the sequent ideas. Hence the form can, and should, be far more terse and condensed, and the argument does not need the same enforcement, while the repetition so necessary in a lecture is apt to become irritating.

Allowing for these qualities, which are essential to a lecture, the chapters are most interesting and stimulating.

In estimating the life-work of Huxley, the author rightly places in the foreground the great and successful struggle for intellectual liberty.

" To what nobler end could life be devoted than the attempt to show us how we may ' learn to distinguish truth from falsehood, in order to be clear about our actions, and to walk sure-footedly in this life.' If he has succeeded, and every zoologist who is free to follow nature wherever she may lead is a witness that he has succeeded—if, as the end of his lifelong labour, intellectual freedom is established on a firmer basis—this is his best monument, even if the man should quickly be forgotten in the accomplishment of his end. No memorial could be more appropriate than the speedy establishment of that intellectual liberty which is not intellectual licence on a basis so firm that the history of the struggle to obtain it shall become a forgotten antiquity " (p. 35).

Space prevents further allusion to the interesting criticism of Huxley's philosophy, and the statement of the particular parts of it which have proved to be of the highest value to the author.

" The interminable question whether ' acquired characters ' are inherited " is not directly attacked by the author ; but it is indirectly attacked in an extremely interesting and effective way. Granted that such inheritance is possible, the author inquires how far it is of value in accounting for the facts of natural history, and concludes that it is of no importance. The third lecture especially deals with this subject, although it recurs in various places throughout the volume. The discussion opens with a most appropriate reference to the teachings of Aristotle.

" Herbert Spencer tells us that the segmentation of the backbone is the inherited effect of fractures, caused by bending ; but Aristotle has shown (' Parts of Animals,' I., i.) that Empedocles and the ancient writers err in teaching that the bendings to which the backbone has been subjected are the cause of its joints, since the thing to be accounted for is not the presence of the joints, but the fitness of the joints for the needs of their possessor. It is an odd freak of history that we of the end of the nineteenth century are called upon to reconsider a dogma which was not only repudiated two thousand years ago, but was even then antiquated."

The writer warns us that the tendency of exclusive laboratory teaching may be to lead us to forget Aristotle's principle ; and he devotes the whole of this most important chapter to the demonstration, from the discussion of numerous examples, that the problem of fitness is the real problem which confronts the naturalist, and that it is entirely untouched by the explanation of nature as inherited nurture. The chapter concludes with a most convincing reply to the opposing arguments of an English writer. The author unfortunately omits a reference to the publication from which he quotes. The same omission is to be noted in other cases, as in the quotation from Agassiz on p. 16.

C C

In the chapter on Lamarck a powerful argument is derived from "adjustments to the life of other beings than the ones which exhibit the adjustment," such as the poison-fang or sting, which are valuable to the possessor because of their effect on other species. The author finds "the production of adaptations of this sort by the inheritance of the beneficial effects of use, or in any way except by selection, quite unthinkable." Henslow's volume, "The Origin of Floral Structures through Insect and other Agencies" (Internat. Scientific Series) does not appear to be known to the author, although by a few well-chosen examples he shows the futility of the supposed origin which is therein suggested.

"For all I know, the Lamarckian may claim that the visits of insects have, in some way, modified the flower, to its own good, by their mechanical action, by pulling down this part, and by pushing up that, generation after generation, until they have caused adaptive modification in the flower. I do not know how much his ingenuity may be able to make out of this hypothesis ; but no one can believe that the hooks and spines, which are so obviously adapted for distributing burrs and seeds, by fastening them to the fur of passing mammals, have been produced by the inheritance of the effects of this sort of mechanical contact ; for these structures do not come into use until they are dead ; and, most assuredly, dead things cannot transmit 'acquired characters' to their descendants. When a drop of rain or dew falls on the dead, dry, twisted glume of the animated oat (*Avena sterilis*), it untwists in such a way as to push like the leg of a grasshoper, and, raising the seed, to send it off with a jump. After the seed has fallen, this process is repeated again and again, until the heavy end, where the seed is placed, falls at last into some roughness in the ground, when the glumes begin to kick and to struggle, and, catching in the grass and roots, or on the rough ground, to push the seed down and to plant it. The seed is alive, but the glumes are dead and dry, and as completely out of the line of descent to future generations as the dead leaves which drop from the tree."

This quotation illustrates the very effective manner in which the Lamarckian principle is dealt with. In certain striking cases it is shown to be obvious that the hypothesis of Lamarck *cannot* supply an explanation, while selection offers a probable solution. At first sight these examples may appear to be exceptional and rare, but the author shows us that

"all the adaptations of nature are of this sort. *In all cases*, the structure, habits, instincts and faculties of living things, from the upward growth of the plumule of the sprouting seed to the moral sense of man, are primarily for the good of other beings than the ones which manifest them."

In support of this conclusion, the evidence of "the insignificance of the individual, as compared with the welfare of the species" is marshalled and illustrated in a peculiarly convincing and striking manner. Of all the examples, the most wonderful is certainly that of the queen-bee in her relation to the other members of the royal family and to the hive. A hive requires a queen, but would be disorganised by the presence of more than one queen at the same time. Until the queen-mother has led out a swarm, the workers will not permit a young queen, although mature, to leave her cell. In order to preserve her from the reigning queen, she is walled up with layers of wax and fed through a small opening.

When swarming has occurred, a young queen is allowed to escape : she in her turn is impelled to kill the rest of the royal brood, but is prevented by the workers. Later on in the season, however, when it is no longer possible to swarm, the attitude of the workers entirely changes, and they now "incite her to destroy her rivals." And here we meet with a most wonderful adaptation. It is obvious that any royal larva may, under certain circumstances, benefit the hive by producing a reigning queen, or, on the other hand, under different circumstances, may be killed in order to prevent a danger to the community. The instincts of the royal larva are such that it prepares beforehand for the latter alternative, and facilitates its own murder without inconvenience or danger to the queen, by spinning an incomplete cocoon which exposes the soft abdomen to the sting. Darwin pointed out in the "Origin of Species" that the social Hymenoptera afford the most complete evidence of instincts which cannot be due to use-inheritance inasmuch as they are exhibited by the sterile workers, the offspring of drones and queens with quite different instincts. Brooks has used the same example with great effect to emphasise "the supreme importance of the species, and the relative insignificance of the individual." Darwin's conclusion is also put with remarkable force on p. 95. This most interesting and convincing chapter concludes as follows :

"Some may ask whether it may not be possible that while natural selection is the chief factor in the origin of species, there may still be a residuum to be accounted for by the 'inheritance of acquired characters.' For all I know this may be not only possible, but actually the case. I have never felt the slightest interest in *a priori* demonstrations of the impossibility of this sort of inheritance ; and for all I know to the contrary, proof of its occurrence may be found at any time, although I know no good evidence of its occurrence. I had satisfied myself, long before the recent revival of interest in the matter, that whether it be a real factor or not, the so-called Lamarckian factor has little value as a contribution to the solution of the problem of the origin of species ; and renewed study has strengthened this conviction."

It must be remembered, on the other hand, that such inheritance would require an inconceivably elaborate mechanism, which can hardly have arisen and been sustained in order to account for a factor which is of little value in evolution.

"Migration in its bearing on Lamarckism" is the title of the succeeding lecture. The same subject was treated of in one of the most fascinating of Wallace's classical essays upon natural selection. It is interesting to compare the two, and to recognise how very greatly the interpretation of this difficult problem has been elucidated by the younger zoologist. Wallace dwells upon the lines of bird migration in their relation to past geographical change, and to the special need for insect food during the breeding season. Brooks treats the problem as a part of the wide principle of the subordination of the individual to the welfare of the species ; he doubts the dependence on geological change and the great importance of food, and makes the illuminating suggestion that security from the enemies of eggs and young is the controlling factor alike of bird and fish migration, and he dwells on the risk to parents involved in the process.

"Long journeys are hazardous. Every Californian salmon which enters upon the long journey to the breeding ground is destroyed, and the whole race is wiped out of existence for the good of generations yet unborn. Very few shad ever return to the ocean, and storm and accident and ruthless enemies work their will on the migrating birds and decimate them without mercy; yet the dangerous return to safe breeding grounds still goes on, in order that children which are yet unborn may survive to produce children in their turn."

Want of space prevents any further criticism of this most interesting volume. Enough has been said to prove that all the lectures demand the serious consideration of every student of evolution.

It is a peculiar pleasure to the British naturalist to find the Darwinian principle illustrated and defended with such remarkable force and success by a distinguished American zoologist.   E. B. P.

A MODERN TEXT-BOOK OF OPTICS.

Lehrbuch der Optik. Von Dr. Paul Drude, Professor des Physik au des Universitat Giessen. Pp. xiv + 498. (Leipzig: Verlag von S. Hirzel, 1900.)

PROF. DRUDE'S name is well known to English physicists. As a careful and exact worker, the author of a book on the Physics of the Ether, and the successor of Gustav Wiedemann in the editorship of the *Annalen des Physik*, he has already made a high reputation for himself, and the book now under consideration will serve to add to it. Text-books of optics, it is true, are numerous, and the reviewer is apt to think that of the making of many books there is no end. Prof. Drude's book, however, contains much that is novel—at any rate, to English text-books—and the student will find up-to-date information on many points of interest.

In some respects the book has much in common with the late Prof. Preston's well-known text-book; it gains, however, in the end as a treatise on the subject by the definite adoption of the electromagnetic theory, although it is, of course, in consequence, less complete in that it gives no account of elastic solid theories.

The first hundred pages deal with geometrical optics. After a clear statement of the fundamental laws, including the law of the minimum path, and Malus' law of orthotomic systems, we have a chapter on the geometrical theory of optical images. A definition of an optical image is given; it is then shown that the image of a plane is a plane, and hence the analytical relation between the position of a point and its image is found. From this, following Abbé and Czapski, the geometrical theory of a perfect image is developed clearly and concisely. Throughout this part the book runs on similar lines to Dr. Moritz von Rohr's "Geschichte des Photographischen Objectivs," recently reviewed in these pages (NATURE, vol. lxi. p. 511), though, of course, the more technical part is dealt with much more briefly than in Dr. von Rohr's book.

Further chapters deal with the formation of images by real rays and the effects produced by the limitations in the size of the pencils in the case of actual instruments.

The chapter on optical instruments is perhaps rather brief, but it is not the main object of the author to describe these. Throughout this part the book is very different from anything yet published in English, and will well repay study; it is interesting to read and clearly written; at the same time, it is commendably brief, and contains little long or cumbersome analysis.

The remaining four hundred pages are devoted to physical optics. In the first section of this, which deals with the general properties of light, there is, with one exception, nothing particularly novel. The treatment of interference, diffraction, the geometrical theory of double refraction and the colours of polarised light follow the usual lines; it could hardly be otherwise. The whole is brought up to date, however; there is, for example, an excellent account of Michelson's echelon spectroscope, while the theory of the resolving power of an optical instrument is given in some fulness; it is all well done, though the English reader will not find much to make him prefer the book, as a text-book, to Preston. The one exception is the chapter on Huyghens' principle. In his elementary discussion on the rectilinear propagation of light, Dr. Drude makes a distinct step by adopting the methods given by Dr. Schuster (*Phil. Mag.*, vol. xxxi. 1891), while he completes the discussion by giving Kirchhoff and Voigt's solution of the problem of finding the disturbance at a given point due to disturbances existing at some previous time over a surface surrounding the point. To do this, he has, of course, to make use of the differential equation satisfied by the disturbances, and this is not found till a later stage in the book; but the student who has read sufficient mathematics to follow the proof will probably be acquainted with the fact that the differential equation quoted does represent wave motion, and will not find any logical difficulty in the order adopted, while the proof will put the whole theory of diffraction before him on a sounder basis. An English reader, however, who realises what he owes to Stokes in this matter, may be allowed to express surprise that there is no reference in Prof. Drude's work to the great paper on the dynamical theory of diffraction, published in 1849 in the ninth volume of the *Transactions* of the Cambridge Philosophical Society.

The second section of this part deals with the optical properties of bodies, and here the distinctive points of Prof. Drude's method show themselves. After a brief reference to the elastic solid theory of the ether and the difficulties to which it leads, he adopts formally the electromagnetic theory.

The optical disturbance at any point through which light-waves are passing can be represented by the periodic variations of a vector quantity, the light-vector, as Drude calls it, and in a transparent isotropic medium this vector follows the same laws as do the electric or magnetic force in an insulating body. The electromagnetic theory of light identifies the light vector either with the electric or the magnetic force. Drude adopts the first of the two alternatives.

In an æolotropic medium, a third vector, the rate of change of the electric displacement, or the electric current, needs to be considered—in an isotropic body this coincides in direction with, and is proportional to the electric force. For reasons which are stated, however, in a crystal, this third vector, the electric current, is taken to represent the light vector. The consequences of this theory are then worked out fully. The general equations of the electromagnetic field are obtained from the two laws (1) that the work done in carrying a unit magnetic pole once round an electric current i is $4\pi i$; and (2) that the work done in carrying a unit quantity of electricity once round a magnetic current j is $4\pi j$.

The phrase magnetic current is perhaps not a very common one, though some English writers have used it. The magnetic current multiplied by 4π is equal to the rate of change of magnetic induction; thus the second law is merely Faraday's law of induction of electric currents.

In forming the equations care must be taken to measure throughout in the same units, electrostatic or electromagnetic, as the case may be. Prof. Drude assumes that electric inductive capacity and permeability have no dimensions and introduces a quantity, which he tells us is of the dimensions of a velocity and equal to the velocity of light, as representing the ratio of the units. The same result would have been reached more simply by introducing two symbols, κ_0, μ_0, of unknown dimensions to represent the inductive capacity and permeability of a vacuum, and then showing that $1/(\kappa_0\mu_0)^{\frac{1}{2}}$ was of the dimensions of a velocity.

From the equations thus found, together with the known electromagnetic laws expressing the action which takes place at the common surface of two media, the laws of transmission, reflexion and refraction in isotropic and crystalline transparent bodies can, as is well known, be deduced so long at least as we avoid phenomena of dispersion. They lead to Fresnel's sine and tangent laws for reflexion, and these in reality are not accurately satisfied; but it is shown that the small amount of elliptic polarisation observed can be accounted for by the supposition that the transition across the interface is not sudden. On this point a reference to a paper in the *Phil. Trans.* for 1894, Part ii., by G. A. Schott, would not have been misplaced. In fact, we may say that so long as the difference between the properties of a refracting body and those of the ether can be completely expressed by a change in the inductive capacity, the simple equations of the electromagnetic field suffice for the co-ordination of optical effects; but when this is no longer the case, when the supposition of a mere change in refractive index is not sufficient to express the action of the matter upon ether, modifications in the equations which can not be entirely justified by reference to known electromagnetic laws become necessary. Absorption and dispersion, aberration and the action of magnetism on light require further hypotheses for their explanation, and the part of the book in which Prof. Drude deals with these and cognate phenomena is of great interest.

The phenomena of absorption and of metallic reflexion are explained by the hypothesis that absorbing media are conductors like the metals.

The total current in such media is composed of two parts, that of displacement or polarisation depending on the rate of change of the electric force, and that of con-

duction proportional to the force. From this it follows that in the equation for a component of the electric force, X, for example, a term in aX/dt appears; we have a viscous as well as an elastic resistance to the motion.

Prof. Drude points out, as Lord Rayleigh had done nearly thirty years before (*Phil. Mag.*, 1872), that the numerical results derived from experiments on the metals cannot be reconciled with such a simple theory; it needs modification, and the direction of the requisite change is indicated by the theory of dispersion which is discussed next.

Up to this point the theory has not been mechanical. We know from purely electrical observations the laws of electromagnetic force without needing to know the mechanism, ætherial or material, to which that force is due. Changes in the electric force give rise in a dielectric, to an electric current, Maxwell's displacement current, and the laws obeyed by this current in transparent bodies are exactly those of light.

The light vector may be electric displacement, or it may be some periodic change in the ether, *e.g.* a twist or a displacement of the ether particles, which obeys exactly the same laws as electric displacement; we do not know, and, so far as the theory is concerned, we do not need to know, which of these hypotheses is true.

When we are dealing with the action of matter, however, it becomes necessary to introduce some mechanical conceptions. Thus Prof. Drude supposes that the molecules of a dielectric are composed of charged ions which are set in motion by the electric force when a train of light waves traverse the medium. The current in this case across any section is made up of the displacement current, together with the convection current due to the displacement of the ions; thus a new variable, expressing the displacement of the matter, is brought into Maxwell's simple equations. In consequence a new set of equations, determining the motion of the ions, become necessary.

Now, the external force on an ion will be proportional to its charge and to the electric force. Drude supposes that, in addition, its motion is retarded by a force proportional to its displacement, and by a frictional force. Of course, since we are dealing only with harmonic motion, this is merely equivalent to saying that the force of restitution can be expressed by a series of harmonic functions. In this way equations are obtained similar to those given by Sellmeyer's mechanical theory—a theory, as Lord Rayleigh has recently shown, originally due to Maxwell—from which the phenomena of dispersion can be deduced. The same hypotheses serve to overcome the difficulties of a theory of metallic reflexion based on conductivity.

Fairly obvious modifications of the equations of motion of the ions lead to explanations of the rotatory polarisation of sugar and quartz. The action of magnetism on light is more complex; it is deduced from an hypothesis of molecular vortices. The ionic charges are supposed to be in a state of rotation about the lines of magnetic force, and the consequences of this on the equations of motion are examined. This leads to a rational explanation of the magnetic rotation of the plane of polarisation, and of the Hall effect, while in another section the Zeeman effect is touched on. The last chapter of the section deals with aberration; it is supposed that the ether in a moving body remains, so far as the motion of the body is concerned, at rest. Thus another term has to be added to the expression for the current; the ions are carried with the body, and give rise to a convection current.

This assumption appears, however, open to criticism. Since the total charge in any element of volume is zero, the total convection current due to the motion of that element, as a whole, must also be zero. The case differs from that in which the oppositely charged ions are set in motion in opposite directions by electric force. The fact that the axes to which we refer the relative motions of the ions are themselves in motion, introduces new terms into the equations which are sufficient to account for aberration without assuming the existence of this convection current.

The consequences of this relative motion are examined, following H. A. Lorentz, to whose labour on this subject so much of our knowledge is due, and an explanation given of aberration and of Fizeau's celebrated experiment on the effect of moving water on the velocity of light.

In all this work Prof. Drude has been most successful; the electromagnetic theory, supplemented by the one additional hypothesis of the moving ions, serves to co-ordinate in a satisfactory way very many of the phenomena of light.

Further knowledge may modify our views, but up to the present Prof. Drude's book contains the most rational account of the phenomena of optics which we possess; it is a book which should be read by all students, and he is to be congratulated on having written it.

And now having said this, in conclusion a grumble and a suggestion may be permitted. There is no index, and though the table of contents is a full one, this can never replace an index. Again, the book would be more interesting and more valuable, and would give a fairer account of the subject, if the references to original papers, especially papers published some time back and in other countries besides Germany, were more complete. A second edition will be called for before long. Will Prof. Drude increase the gratitude due to him for his work by remedying these two defects?

AGRICULTURAL EDUCATION IN THE UNITED STATES.

Year-book of the United States Department of Agriculture, 1899. Pp. 880; 63 plates. (Washington: Government Printing Office, 1900.)

THE present volume is a special one, the Secretary of Agriculture desiring "that the Year-book for 1899, the distribution of which will occur during the last year of this century, shall present to the reader a picture of the development of agriculture in the United States during the nineteenth century, and of its condition at the present time." The volume contains twenty-six reports, from the various bureaux and divisions under the Department of Agriculture. These reports are followed by an appendix giving particulars respecting the various agricultural organisations now at work in the country. The whole is copiously illustrated.

The various reports on the development of knowledge

and of work during the past century are of course written in a popular style, being primarily intended for the information of the general community in the United States; we must not, therefore, expect to find in them much exact science. They are, nevertheless, of great permanent value, and should be carefully studied by all those who desire that the agriculture and the agriculturist of Great Britain should exhibit the rapid progress in improvement which this volume shows to be taking place on the other side of the Atlantic.

As the subject of agricultural education is now occupying the public mind in England, it will perhaps be of service if we briefly mention what is at present being done in America, as set forth in the volume now before us.

The Report dealing with education informs us that the attempts to introduce instruction in agriculture into elementary rural schools have failed. Now, however, a hopeful movement has been started by the College of Agriculture at Cornell University, and taken up by some other State colleges, for the introduction of "nature studies" into elementary schools. To accomplish this object leaflets containing suitable matter for lessons have been issued, and model lessons are given in the schools by travelling inspectors. The first difficulty to be surmounted is, in fact, the teaching of the teachers. Up to the present time little has been done toward the establishment of second grade agricultural schools, and agricultural subjects are not as yet taught in the High Schools.

In America, the State College or University, with the Experiment Station attached to it, have been the prime movers in agricultural education. The colleges have by no means confined their work to their own students, but have actively carried on a large amount of external teaching of various kinds. Thus, besides the full course of instruction, lasting two or four years, provided for the members of the college, short winter courses of twelve weeks' instruction are in many cases provided for the special requirements of young farmers, and in some States these short courses have been very successful. The staff of the college and experiment station also do much good by lecturing at farmers' institutes. These institutes will meet for a session of three days in various places, the time being occupied by a series of papers and discussions. It is estimated that about 2000 of these meetings were held in the United States during 1898, attended by half a million farmers. In Wisconsin the best papers are issued as an annual volume, 60,000 copies of which are distributed, one being placed in the library of every elementary school. The practical influence of these institutes has been very great. Several State colleges have also commenced correspondence classes in agriculture, and have enrolled a large number of readers who receive assistance and advice from the college. The influence of the experiment stations has also been very great; their investigations have produced a local interest in the study of agricultural problems, and afforded examples of the aid which science can render to the farmer. Without the work of the station the teaching of the college would have appeared academic and theoretical, and would have failed to commend itself to the practical man. The farmer's bulletins issued by each experiment station, and distributed post-free throughout the State to every farmer asking to receive them, are of considerable educational value. The work of all the State colleges and experiment stations is unified by the Association of American Agricultural Colleges and Experiment Stations. This Association consists of delegates appointed by the colleges and stations, and by the United States Department of Agriculture, and meets for several days once a year to hear reports and discuss methods of work. The Association has permanent executive committees, which carry out the work initiated by the Association.

Although both colleges and experiment stations are State institutions, they are more or less under the influence of the National Department of Agriculture, as every institution receives annual grants from Government funds, for the proper use of which the Department of Agriculture is made responsible. The United States Department of Agriculture is on a very large scale ; the sum appropriated to its use by Congress in 1899 was 2,829,702 dollars. It includes many sub-departments, provided with a numerous staff of scientific workers. It has excellent laboratories, a botanic garden, museum and library containing 68,000 volumes, three-quarters of which are on agricultural subjects. It undertakes investigations of all kinds. It publishes in the *Experiment Station Record* summaries of all the work done by the experiment stations. The publications it issues for gratuitous circulation are most voluminous, and embrace all subjects with which it is thought the farmer or student should be acquainted. In 1899, 26,420 pages were published, and 7,075,975 copies printed. Of the present year-book the edition is 500,000 copies, with 20,000 extra copies for the Paris Exhibition.

We have already mentioned the sum annually spent by Congress on the Department, we may conclude by saying that the annual income of the State agricultural and mechanical colleges is stated in the year-book to be 6,008,379 dollars, while the income of the experiment stations amounts to 1,143,334 dollars. Such is in brief the provision made in the United States for the improvement of the science and practice of agriculture in the country. R. WARINGTON.

OUR BOOK SHELF.

Lehrbuch der Anorganischen Chemie. Von Dr. A. F. Holleman. In gemeinschaft mit dem Verfasser bearbeitet und herausgegeben von Dr. Wilhelm Manchot. Pp. xii + 440. (Liepzig : Veit und Co., 1900.)

THIS is an advanced text-book of inorganic chemistry, distinguished from others chiefly by the embodiment in it of chapters of modern physical chemistry. The book, indeed, gives the impression of having been produced by shuffling the detached chapters of two others—one, an ordinary treatise on inorganic chemistry, the second on physical chemistry.

It is almost impossible to discern the system which has guided the compilers. The book begins with some generalities about the scope of science, and the differences between physics and chemistry. It then proceeds to describe some chemical operations, such as dissolving, filtering and distilling. This is done in language suitable for children, and illustrated by two diagrams, in one of which a filter paper is seen to project considerably above the rim of the funnel. The elements having been named,

oxygen is next described—such terms as critical temperature being taken as understood by the reader, who has just been told how to separate salt from sand. After a description of hydrogen, the indestructibility of matter is discussed, and then comes water. The laws of chemical combination and the atomic theory occupy the next few pages, then chlorine and its compounds. We now come upon the laws of Gay Lussac and Avogadro, ozone and hydrogen peroxide, then modern methods of determining molecular weights, with a discussion of semi-permeable membranes. And so the book proceeds. Dissociation is discussed between iodine and fluorine, electrolytic dissociation between the halogens and sulphur, the phase rule under sulphur, thermochemistry, including thermodynamics between sulphur and nitrogen.

It is impossible to say anything in praise of this arrangement or want of arrangement. It can hardly be defended on logical or didactic grounds, and one is tempted to think that there is nothing more than a striving for novelty at the bottom of it.

The book does not aim at teaching how chemists do their work, discover facts, and establish theories ; and surely if it were desired to present descriptive inorganic chemistry on the basis of the general theories of modern physical chemistry, it would have been better to have begun with an account of these theories and to have woven them into the descriptive part throughout.

Whilst speaking thus of the general scope of Prof. Holleman's book, it is right to add that in detail there are features that call for commendation. The descriptive part is well abreast of the times, and many of the intercalated chapters on physical chemistry are clearly and concisely written. A concluding chapter summarising Werner's voluminous papers on the metal-ammonium compounds is a valuable addition.

On the whole, it may be said that as a work of moderate dimensions conveying the chief facts of inorganic chemistry and an account of those physicochemical theories which bear especially on inorganic chemistry, Prof. Holleman's book will probably find considerable acceptance in Germany, but it is neither to be expected nor desired that it will set a fashion in its plan of construction. A. S.

Flora of Bournemouth, including the Isle of Purbeck. By E. F. Linton, M.A. With map. New edition. Pp. vii + 290. (Bournemouth : Sold by H. S. Commin, Bright's Stores, and W. Mate and Sons.)

THE local flora embodied in the pages of the book before us appears to be usefully compiled, though perhaps the volume as a whole would have been improved had it been printed on thinner paper, so as to form a more convenient pocket companion. Opening with a short introduction on the physical and geological characters of the district, the author gives a list of some 1137 plants (flowering plants and ferns) as occurring within the area treated of, and adds localities, as is usual in works of this nature. The book should prove useful to those lovers of wild flowers who are visiting the Bournemouth district, to many of whom it may perhaps be a matter of surprise that so large a percentage of the British flora occurs within a twelve-mile radius from the town.

Carnations and Picotees for Garden and Exhibition. By H. W. Weguelin, F.R.H.S. Pp. viii + 125. (London : George Newnes, Ltd., 1900.)

THIS is a book which will be useful to those who are fond of carnations. The cultural hints are clear, and lists are given of many of the best sorts. The text is a little diffuse in places, but in a work of this character that is a pardonable characteristic. The author is enthusiastic on his subject, and his book is worth reading, if only to show what can be done with the flowers as materials for open borders.

LETTERS TO THE EDITOR.

[The Editor does not hold himself responsible for opinions expressed by his correspondents. Neither can he undertake to return, or to correspond with the writers of, rejected manuscripts intended for this or any other part of NATURE. *No notice is taken of anonymous communications.]*

Collateral Heredity Measurements in Schools.

As a result of the appeal, made in NATURE last June, for aid in the measurement of pairs of brothers and sisters, I have received friendly help from a number of masters and mistresses up and down the country. I think I have received between 400 and 500 data forms properly filled in. Considerable as this assistance has been, I would still beg for further aid, as I want the collection to reach, if possible, 1000 pairs for each fraternal relationship. I have at the present time several head-spanners free, and shall only be too glad to send one to any teacher who will undertake the necessary observations on six to ten pairs of brothers or sisters. As I said in my former letter, the determination of the intensity of hereditary resemblance is a very important matter, and it can, at any rate in the case of man, only be achieved by co-operative effort on the part of those interested in science.   KARL PEARSON.
University College, London, October 9.

The White Rhinoceros on the Upper Nile.

IT may interest your readers to learn that during his recent notable traverse of Africa from South to North, Major A. St. Hill Gibbons shot on the Upper Nile, near Lado, a rhinoceros which he considered to be the white or square-mouthed rhinoceros (*R. simus*), hitherto only known from south of the Zambesi, and now, unhappily, nearly extinct there. His determination is fully borne out by the skull, which I have had the pleasure of examining, and which shows all the many characters that distinguish *R. simus* from the common species, *R. bicornis*.

That a rhinoceros of this group existed in Central Africa has been suspected before. Dr. Gregory in "The Great Rift Valley," mentions having seen in Leikipia, but failed to shoot, three specimens which he believed to be *R. simus*. Some years earlier Count Teleki shot a "White Rhinoceros" in the same district, but his account has more reference to the colour than to the specific determination of the animal, and his specimen may only have been a pale-coloured *R. bicornis*.

Now, however, Major Gibbons has fortunately set the matter at rest, as there can be no question that his animal is not *R. bicornis*, but belongs to the rarer southern form, hitherto supposed to be practically extinct.

The discovery of this animal in the Nile watershed brings it geographically nearer to its European and Siberian ally, the Pleistocene *R. antiquitatis*, both species being in turn, no doubt, offshoots of the Pliocene *R. platyrhinus* of the Siwaliks.
        Natural History Museum,   OLDFIELD THOMAS.
            October 12.

P.S.—This find has an interesting parallel in Mr. W. Penrice's discovery in Angola of a zebra allied to the true Cape Zebra (*Equus zebra*), now also nearly extinct there. But in that case the species proves different by its shorter hair, and much broader white striping, and has been named *Equus penricei*.

Disease of Birch Trees in Epping Forest and Elsewhere.

IN Epping Forest, and in other districts around London, birch trees have been attacked during the late summer by a disease which causes them to die very rapidly. In a portion of the Forest known as Lord's Bushes, thirteen diseased and twenty-four completely dead trees were noted on June 10 within an area of about one and half acres.

A few were attacked in the Forest in the summer of 1899, but it was not till this year that the disease appeared in such a destructive form. On Chiselhurst Common, Hayes Common and Keston Common no signs of the disease were evident in the early summer, but now dead or diseased trees may be found in great numbers. Trees attacked in a similar manner occur at Walton-on-Thames, by the canal between Weybridge and Woking, at Lewisham and at Westerham.

The disease is probably due to a micro fungus, *Melanconis stilbostoma*, Tul., for it appears on the branches of both living and dead trees. The diagnosis of the disease is almost precisely that of *Valsa oxystoma*, described as the destroyer of *Alnus viridis* in some parts of the Tyrol.

It would be interesting to know if any of your readers have observed the disease in the Midlands or in the north of England.
                                            ROBT. PAULSON.
10, Denholme Road, Maida Hill, October 8.

Sunspots and Frost.

IN the study of winter cold, we find, I think, some striking contrasts associated with different parts of the sunspot-curve. These contrasts, whether they are really due to sunspot variations or no, seem worthy of attention as a practical matter, and an occasion for observing whether such relations are maintained in future.

Taking the Greenwich records since 1841, let us see how many frost days there were in each three-year group following the sunspot maxima 1848, 1860, 1870, 1883 and 1893 ; and how those sums are related to the average (which is 164 in three years). The following table shows this :—

Three-year groups		*a* Frost days		Relation to average
1849–51	...	147	...	− 17
1861–63	...	118	...	− 46
1871–73	...	131	...	− 33
1884–86	...	160	...	− 4
1894–96	...	133	...	− 31
		689		− 131

Thus, each of those three-year groups was mild, in respect of frost days, and there was a total deficiency of 131 days.

Now, let us do the same with the three-year groups following the minima, 1843, 1856, 1867, 1878, 1889 :—

Three-year groups		*b* Frost days		Relation to average		*c* *b − a*
1844–46	...	166	...	+ 2	...	+ 19
1857–59	...	180	...	+ 16	...	+ 62
1868–70	...	170	...	+ 6	...	+ 39
1879–81	...	210	...	+ 46	...	+ 50
1890–92	...	201	...	+ 37	..	+ 68
		927		+ 107		+ 238

In this case, each three-year group is over average, and the total excess is 107 days. The added column (*b − a*) shows that the three-year groups after minima had altogether 238 frost days more than the groups after maxima, giving an average of 47·6 for each pair of groups compared.

If we group together the fourth, fifth, sixth and seventh years after maxima (*i.e.* '52–55, '64–67, '74–77 and '87–90), and count the frost days in those four-year groups, we find that the latter share the character of the three-year groups after minima, each having an excess of frost days over the (four-year) average. We are now in the last year of another of these groups (viz. 1897–1900).

Analysing those mild three-year groups after maxima, we find out of a total of fifteen years only four with more than the average of frost days, and only one group (1884–86) in which two of the three years had an excess.

It occurred to me to examine what kind of summers we had in those mild groups, and the following curious table was arrived at :—

		M.T. Summers.		Relation to average.
1849–51	...	61·2	...	av.
1861–63	...	60·4	...	− ·8
1871–73	...	61·7	...	+ ·5
1884–86	...	61·2	...	av.
1894–96	...	61·6	...	+ ·4

Thus, the divergence from the average never gets beyond a decimal value. Analysing, one finds only three of those fifteen summers in which the divergence gets beyond a decimal value (viz. − 2·4, − 1·1 and + 1·4). The summers of three-year groups after minima might be shown to have a distinctly opposite character. But I do not lay stress on this.
                                        ALEX. B. MACDOWALL.

Simple Experiments on Phosphorescence.

IN consequence of reading your note in NATURE of September 27, on M. Gustave le Bon's paper on various forms of phosphorescence, the following experiments were tried. A surface, previously dark, of the sulphide of calcium, was exposed to the radiation of a blackened vessel of boiling water ; this gave no decisive result.

On repeating the experiment with a smoothing iron at the temperature ordinarily used, the surface in about a minute glowed brightly. There is this difference from the excitement by bright daylight, or gaslight, that the glow is comparatively transient.

This renders it probable that a cylinder of iron heated by a spirit flame duly concealed would act as M. le Bon's dark lamp does. A. M. M.

MEXICAN SYMBOLISM.[1]

A RESIDENCE for some years among the Huichol Indians of Mexico has enabled Dr. Carl Lumholtz to enrich ethnology with a wonderfully detailed and exhaustive memoir on their symbolism, and our thanks are due, not only to the author, but to the authorities of the American Museum of Natural History for the appearance of this most valuable study, which is lavishly illustrated by more than three hundred figures in the text and four plates, three of which are coloured heliotypes.

It is extremely fortunate for students of American archæology and comparative religion that the symbolism of pagan Mexican Indians should be minutely studied, as this will throw light on the meaning of the inscriptions on ancient Mexican monuments, and will afford illustrations for the comparative studies of cults.

All sacred things are symbols to primitive man, writes Dr. Lumholtz, and the Huichols seem literally to have no end of them. Religion is to them a personal matter, not an institution, and therefore their life is religious, and from the cradle to the grave wrapped up in symbolism. From their symbolism it may be inferred that the main thought of their prayers is food—corn, beans and squashes. Even in the hunting of the deer, the primary consideration is that the success of the chase means good crops of corn. Agriculture depends upon rain, therefore most of the symbolic objects express, first of all, prayers for rain, and, by implication, for food, and then prayers for health, good fortune and long life. In many cases the supplicant himself is represented by symbolic objects in the shape of a human figure or a heart ; but in others the god is thus depicted.

The act of sending a prayer to a god is symbolised by attaching a representation of the prayer to an arrow, the painting of the rearshaft of the arrow is symbolic of the special deity to whom the prayer is offered. In other cases, the prayer is directed to the god by placing the symbolic object representing the prayer to the temple of the deity, or by tying it to his chair, or placing it in his votive bowl.

Speaking in a general way, individual or personal prayers are conveyed by arrows or back-shields ; these latter are symbols of the rectangular shield that the Huichol warrior wore to protect his back. The main idea of the back-shield is that it protects against the heat of the sun, and prayers expressed by it are largely for health, but also for protection against evil, sickness, accident, &c. Back-shields represent prayers of all kinds, such as prayers for rain, good crops, and even that the supplicant may have children ; it should be remembered that the same mat served the warrior as back-shield and bed. Tribal prayers were mostly conveyed by the usually circular front-shields. Personal and tribal prayers may also be conveyed by "eyes." These are crosses of bamboo splints, or straw interwoven with coloured threads in the form of a diamond, The eye is the symbol of the power of seeing and understanding unknown things ; the prayer expressed by this symbolic object is that the eye of the god may rest on the supplicant.

The diminutive sandals of an ancient pattern that are

attached to a prayer-arrow may be taken as an example of symbolism. Such sandals are now only worn by shamans at the greatest feast of the Huichols—that which is held for the underworld. They therefore become the symbol of a prayer that this feast may come off ; also that nothing untoward may happen to the shaman at this feast ; but as the feast cannot be celebrated unless a deer has been killed, a pair of such sandals also expresses a prayer for luck in killing deer. In olden time only men wore sandals, which at that time were of the ancient pattern referred to ; thus these sandals are also used to express a woman's prayer for a husband.

Practically the same design may be the symbol of various objects, for example, curved lines in general indicate serpents, but when there are dots between curved lines they mean ears of corn in the fields. Bands of curved lines with dots between them are the tracks of wind, rain and water in the fields. Zigzag lines stand not only for rain-serpents but also for lightning, the sea surrounding the world, hills and valleys projected on the horizon, bean plants and squash vines. A cross refers to the four cardinal points, but also signifies money, sparks, &c.

There is a further complication in the strong tendency to see analogies, even the most heterogeneous phenomena are considered as identical. For instance, the following are some of the objects that are believed to be serpents : most of the gods and all the goddesses, the pools of water and springs in which the deities live, the wind sweeping through the grass, the moving sea and ripples of water, flowing rivers, darting lightning, rain, fire, smoke, clouds, their own flowing hair, their girdle ribbons, pouches, wristlets, anklets, maize, bow, arrow, tobacco gourd, trails of men on the land—all are considered as serpents.

On reading this suggestive memoir, one is struck with the fact that the religion of the Huichols contains elements appropriate to two distinct stages of culture. In former ages their ancestors were evidently nomad hunters, who subsisted mainly on the meat of deer, which they killed with bows and arrows. Probably at this period they shot their arrows in the air in magical rites, so as to ensure the killing of deer ; possibly also they attached pictographs or symbols to the arrows as messages or prayers to the gods, but this was almost certainly a later phase. On acquiring the art of agriculture, they continued the old practices for ensuring a sufficient food supply. According to the Huichol myths, corn was once deer, and at the feast preparatory to the clearing of the cornfields the Huichols drink the broth of deer-meat, which they call "making corn," and the blood of deer is sprinkled on the grains of corn before they are sown, that they may become equally sustaining, for the deer is the symbol of sustenance and fertility.

Departmental gods generally originate when a people become settled and take to agriculture. The prayer arrows would then be deposited in the houses of the gods. At this time, as at present, the moving principle in the religion of the Huichols was the desire of producing rain, and thus successfully raising corn, which now is their principal food ; therefore is it that most of the symbolic objects express first of all prayers for rain and then for other blessings. Since the deer represents sustenance, it may easily be perceived why in their myths water sprang from the forehead of a deer.

There is no space to enter into the cult of that remarkable plant the "Hikuli" (*Anhalonium lewinii*), which is to them the plant of life—the life of the deer and the corn—and adds a further mystical element to this instructive transitional religion. The philosophy of life of these people may be best summed up in a statement by one of themselves. "To pray for luck to the god of fire and to put up snares for the deer—that is, to lead a perfect life." ALFRED C. HADDON.

[1] "Symbolism of the Huichol Indians," by Carl Lumholtz. Memoirs of the American Museum of Natural History. Vol. iii. Anthropology II. 4to. Pp. 228. (1900.)

FURTHER INVESTIGATIONS ON XENIA IN MAIZE.

PROBABLY few botanical discoveries of recent years have aroused more interest than the remarkable observations of Nawaschin upon the fusion of one of the generative nuclei of the pollen tube with the definitive nucleus of the embryo sac. Since further investigations have rendered it not improbable that the process is of general occurrence, its bearing upon some curious phenomena met with in hybrids is of great interest as affording an explanation, not only as to how hybrid embryos, but also how truly hybrid endosperms can be produced by crossing different races of plants. De Vries' beautiful observations upon maize, which were made almost simultaneously with those of Correns, have already formed the subject of an article in this journal, and they have recently been considerably extended by some experiments conducted by Webber[1] in America. As a result of his investigations, Webber concludes, in all cases in which the hybrid corn shows a change of colour, that this is due to the endosperm alone, the translucent pericarp retaining, as might have been theoretically anticipated, the character properly appertaining to the corn of the female parent. But in a large number of instances it was found that, although the embryo on germination showed that hybridisation had occurred, there was no evidence of the transference of the qualities of the male parent to the accompanying endosperm. On the other hand, in some hybrid corns the endosperm exhibited a spotted appearance, which might even (*e.g.* when Gilman Flint was crossed with Stowell's Evergreen) be restricted to only a portion of its substance. The author suggests that the former case might be explained as being the result of failure on the part of the generative nucleus to unite with any nucleus within the embryo sac. The spotted endosperms, on the other hand, might be due to an independent segmentation of the second pollen nucleus, which had failed to unite with the polar nuclei, in which case the portion of endosperm so arising might be expected to retain the characters of the male parent.

There is no inherent improbability in such a suggestion, nor need it necessarily affect any views which may be entertained as to the sexuality of the fusion which we are (perhaps rather hastily) beginning to regard as general; the investigations of Boveri and of Hertwig respectively have shown that the nuclei of both male and female reproductive cells of sea-urchins can be made to segment by appropriate means, and even produce larvæ, and this without any preceding fusion of the sexual cells themselves.

It is clear, however, that much more investigation is required before the points raised by Dr. Webber can be cleared up, but it is to be hoped that so promising a field of research will not be left to lie fallow, although the work itself will necessarily prove arduous.

PORTABLE GAS PRODUCERS.

AIR-GAS, as it is popularly called, consists of an admixture of ordinary atmospheric air with the vapour of one of the volatile hydrocarbons, such as pentane, gasolene or petroleum spirit. Travellers and others having called attention to the production of a natural gas in the petroleum-bearing districts, as at Baku and elsewhere, it was not long before attempts were made to imitate the workings of nature by producing from the petroleum of commerce a combustible gas. The carburetting of ordinary air by forcing a current over liquid petroleum first seems to have been proposed by Lowe in 1831, as in that year he took out a patent for his apparatus.

[1] "Xenia, or the Immediate Effect of Pollen in Maize," by Herbert J. Webber, U.S. Department of Agriculture, *Bulletin* 22, Washington, 1900.

Air-gas producers may be roughly classified as follows :—

(1) Apparatus in which air is *forced under pressure* either through or over liquid petroleum, in which class mention may be made of the apparatus of Jackson, Müller, Weston and Maxim.

(2) Apparatus in which, on account of the danger of using large quantities of liquid petroleum, an absorbent is used to take up the hydrocarbon, either in part or entirely, the air being, as in the first case, *forced under pressure* over or through the absorbent.

Considerable commercial importance attaches to certain of the apparatus mentioned under the two classes given above, both the "Alpha" apparatus of Müller and the "Sun" apparatus of Hearson having had a considerable success, both here and in America, for the lighting of country houses and the like. It will be easily seen, however, that the necessity of having some *motive power* to actuate the current of air introduces complex mechanism which militates against the general adoption of such apparatus. This disadvantage has, however, been met by the apparatus comprising the third and last class of

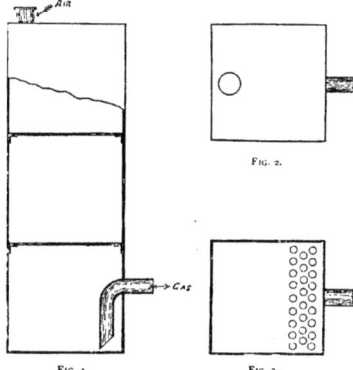

Fig. 2.

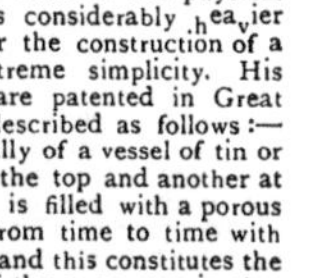

Fig. 1.   Fig. 3.

aero-gas generators, which possess a peculiar interest on account of their simplicity and efficiency.

In 1895 Mr. Naum Notkin, of Moscow, was struck with the idea that use might be made of the physical property that carburetted air is considerably heavier than ordinary atmospheric air for the construction of a gas-producing apparatus of extreme simplicity. His method and apparatus, which are patented in Great Britain (No. 29667/94), may be described as follows :—

The apparatus consists essentially of a vessel of tin or other material, with an orifice at the top and another at the bottom. This simple vessel is filled with a porous material which is impregnated from time to time with one of the lighter hydrocarbons, and this constitutes the whole apparatus. The action of the apparatus is that ordinary atmospheric air enters at the upper orifice, and taking up a certain proportion of hydrocarbon vapour becomes heavier and *gravitates* through the mass of absorbents, taking up more and more of the hydrocarbon vapour, until it finally issues from the lower orifice in the form of a gas capable of lighting, heating, and all other uses to which ordinary gas is put. Not only does the gaseous mixture become heavier, but it becomes cooler by the rapid volatilisation of the petroleum, and this cooling action is greater the more rapid the passage of air through the receptacle. As the absorbents may be regarded as solids, there is no danger either from the presence of loose petroleum or of explosion.

These carburetters (or, as they are termed, aero-gas fountains), in the form best adapted for lighting and heating purposes, consist of a reservoir of ordinary tin, with an air admission regulator at the top and a bent draw-off pipe at the bottom, the pipe being so designed as to syphon out the gas and prevent the possible over-flow of any loose petroleum that might be left on the bottom from an overcharge. Fig. 1 is a vertical section of such a carburetter. The carburetter is divided

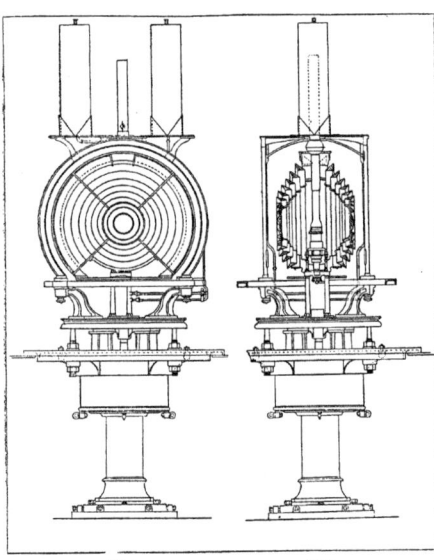

FIG. 4.   FIG 5.

horizontally by two perforated shelves, the object of which is to produce a longer travel of the gas, and to distribute it through the perforations. Fig. 2 is a plan of the top of this fountain, while Fig. 3 is a drawing of one of the perforated shelves. The absorbent is a species of wood pulp which is entirely unaffected by the petroleum, and acts merely as a means of holding it in suspension.

With the carburetter as applied to table and other lamps, the burners used are argands, with steatite centre and very wide gas ways. The light is of high illuminating power and of remarkable purity. For street lighting the carburetter forms part of the lamp, which has a hinged top, so that when the carburetter is exhausted it can be

lifted out and a fresh one put in its place. For ships' lights, as well as in railway and other signal lights, the system offers peculiar advantages. With respect to heating, all classes of stoves can be adapted for this system.

There is one other most important branch of lighting for which the carburetter is designed, namely, light-houses, beacons and buoys. The advantages of gas as an illuminant were early apparent to lighthouse authorities, and in the Government inquiry into the relative advantages of paraffin, gas and electricity as sources of light for lighthouse illumination, the superiority of gas was clearly pointed out, but owing to the necessity of elaborate plant needing to be installed in the vicinity of each lighthouse to be lit by gas, it was pointed out that, despite its intrinsic advantages, it could not be recommended on account of the expense and difficulty entailed in the production. Since those days, how-ever, the Pintsch system of vaporising oil for gas, despite its costliness both as regards the gas produced and the plant required, has been largely made use of by the lighthouse services both at home and abroad. The simple automatic car-buretters that have just been described will, it is clear, place within reach of the lighthouse authorities the possibility of making use of gas-light in place of the paraffin lamps now in common use. Figs. 4 and 5 show section and elevation of a third order dioptric apparatus in which the carburetters are placed above. In place of the oil tanks required for the storing of the paraffin, the tin carbu-retters can be served out to the various stations ready charged, and these can be returned when exhausted and fresh ones supplied. As the absorbent takes up about three-quarters of its own volume of liquid, it is seen that the room required for storing the fountains or carburetters is little more than that needed for the present paraffin supply. As the flame given from this aero-gas is steady and constant, the trouble of maintaining the old paraffin burners of many wicks, so as to give a constant light, is obviated. By doing away with the constant level and pressure arrangements now in vogue, a considerable economy will be effected in light-house apparatus, while at the same time the risk will be lessened of a failure of some part of the mechanism.

In their application to engines for motor-cars, launches, &c., these fountains have a very wide field of usefulness, in which they offer advantages that cannot be secured without them.

J. A. PURVES.

NOTES.

WE regret to see the announcement of the death of Sir Henry Wentworth Dyke Acland, K.C.B., F.R.S., late Regius professor of medicine in the University of Oxford. The funeral will take place to-morrow (Friday) at Holywell Cemetery, Oxford.

PROF. T. G. BONNEY, F.R.S., has resigned the chair of geology which he has worthily occupied at University College, London, for a period of twenty-three years. The chair will become vacant at Christmas.

, THE works of the late Prof. E. Beltrami (consisting of three or four large volumes) are to be issued on subscription by the Faculty of Science of the University of Rome.

A REUTER telegram from Kingston, Jamaica, states that the scientific expedition sent by Harvard University, to observe the minor planet Eros during the approaching opposition, has arrived there under the leadership of Prof. Pickering.

SIR LOWTHIAN BELL, F.R.S., has been elected president of the Institution of Junior Engineers in succession to the Hon. C. A. Parsons, F.R.S.

THE anniversary meeting of the Mineralogical Society will be held on Tuesday, November 13, at 8 p.m., when a new set of bye-laws will be recommended for adoption by the committee and council of the Society.

A REUTER telegram reports that Mr. William Zeigler, a wealthy citizen of New York, will supply the funds for an expedition to start early in 1901 under Mr. Evelyn Baldwin, in the hope of reaching the North Pole. The expedition will sail in two steamers.

AT the recent meeting in Paris the International Geodetic Association discussed the difference of longitude between Paris and Greenwich, with special reference to the discordant results obtained by the French and English astronomers in 1888 and 1892. General Bassot attributed the want of agreement to an imperfect knowledge of the constant of electrical transmission of the signals. The difference of longitude will be measured again next year.

A NEW departure, which should be of much assistance to lecturers, has been made by the Sanitary Institute. Frequent applications having been made to the Institute for the loan of lantern slides and diagrams for lecture purposes, the council have collected a large number of such slides relating to sanitary arrangements and appliances, diseases, &c., which can be borrowed by members and associates for lecture purposes at a small charge. A list of 611 slides at present available can be obtained from the secretary of the institute.

THE *Board of Trade Journal* says that information has just been received, by the Imperial Academy of Science, of the discovery of diamondiferous deposits on the Kamenka, a tributary of the Sanarka. This, it is reported, is the first time that diamonds have been discovered in this region, although the existence of such deposits in the neighbourhood of the Sanarka had already been indicated. It is stated that in structure and colour the diamonds found resemble those of Brazil.

THE annual general meeting of the London Mathematical Society will be held on November 8, at 5.30 p.m. The following nominations for the new council have been made :—Dr. Hobson, F.R.S., president. Lord Kelvin, Prof. Burnside, F.R.S., and Major MacMahon, F.R.S., vice-presidents. Other members :—J. E. Campbell, Lieut.-Colonel Cunningham, R.E., Prof. Elliott, F.R.S., Dr. Glaisher, F.R.S., Prof. M. J. M. Hill, F.R.S , A. B. Kempe, F.R.S., H. M. Macdonald, A. E. Western and E. T. Whittaker. The treasurer (Dr. Larmor, F.R.S.) and hon. secretaries (R. Tucker and Prof. Love, F.R.S.) are renominated. Lord Kelvin will probably not be able to give a valedictory address.

THE foundation stone of the Imperial " Limes " Museum, which is to be erected in the restored Roman fort of the Saalburg in the vicinity of Homburg, was laid by the German Emperor on Thursday last. The museum is to contain the Roman relics which have been discovered in the excavations in the neighbourhood of the " Limes Transrhenanus," the great Roman wall which extended from the Danube to the Rhine. The

excavations were begun in 1873, and have brought to light many objects of great interest, which have hitherto been placed in the Saalburg Museum at Homburg. They will be removed to the new museum as soon as it is completed. The Emperor sent a congratulatory telegram in Latin to Prof. Mommsen, who was unable to be present at the ceremony.

THE Philosophical Faculty of the University of Göttingen has (says *Science*) proposed the following subject for prizes on the Benecke Foundation : A critical investigation, based upon experimental research, of those complex chemical compounds which cannot be explained upon the ordinarily received theory of valence, or can be so explained only by a forced interpretation of the theory. This investigation should specially consider how far the phenomena of molecular addition play a part in the formation of these compounds and as to whether it is possible to formulate a comprehensive theory of these complex compounds. The first prize is 3400 marks and the second prize 680 marks. Papers in competition must be written in a modern language, and be accompanied by a sealed envelope containing the name, a motto on the outside of the envelope corresponding to the same motto on the paper. They should be sent to the Faculty of the University of Göttingen not later than August 30, 1902.

IN the year 1895 the Academy of Sciences of Berlin announced the following problem for the Steiner prize :—" To completely solve any important hitherto unsolved problem relating to the theory of curved surfaces, taking into account, so far as possible, the methods and principles evolved by Steiner. It is required that sufficient analytical explanations shall accompany the geometrical investigations to verify the correctness and completeness of the solution. Without wishing to limit the choice of subject, the Academy takes the opportunity to call attention to the special problems to which Steiner has referred in his general remark at the end of his second paper on maximum and minimum in figures in a plane, on a sphere, and in space." The foregoing problem having remained unsolved up to the present, the Academy again announces it for the year 1905. For its solution a prize of 4000 marks is offered, with an additional sum of 2000 marks. Papers sent in competition may be written in German, French, English, Italian, or Latin, and must be submitted before December 31, 1904, to the Bureau of the Academy, Universitäts-Strasse 8, Berlin N.W. The result will be announced at the Leibnitz meeting of 1905. Each manuscript submitted must bear a mark or *nom de plume*, and be accompanied by a sealed envelope containing the name and address of the author, and bearing outside the corresponding mark or assumed name.

THE Society for the Protection of Birds is offering two prizes, of 10*l.* and 5*l.* respectively, for the best papers on the protection of British birds. The mode of dealing with the subject is left entirely to competitors, but among the points suggested for treatment are the utilisation and enforcement of the present Acts and County Council Orders ; the modification or improvement of the law ; educational methods ; and the best means of influencing landowners and gamekeepers, agriculturists and gardeners, collectors, bird-catchers and bird-nesters. Essays are to be sent in by November 30. Particulars may be obtained from the hon. sec., at the Society's offices, 3, Hanover Square, London, W.

DURING the last few weeks even the scientific recluse, occupied as he mostly is in the *recherche de l'absolu*, has had forced upon him, by serious and comic papers alike, the question of electioneering cries. Perhaps, however, it will be news to him that the subject of vivisection, so-called, has been pushed into the forefront of electioneering polemics. Warning was indeed given months ago that this might be the case as is evidenced by the contents of certain letters published in Mr. Paget's book. These letters, addressed to several of Her Majesty's Ministers, threatened, in no unmistakable terms, should these officials not become anti-vivisectionists, to use against them at the next election organised opposition; which would probably prevent them being returned to Parliament by their respective constituencies. To what extent this has actually been done we have no means of knowing, but our attention has been drawn to a letter in the *North Down Herald and County Down Independent* in comparison with which the diatribes of Mr. Coleridge sink into insignificance. The interest, however, of Miss Margaret Alder's letter does not centre in its actual abuse, but in the fact that she places vivisection first among various causes which have rendered Englishmen "fit to kill, murder and rob the peaceable and pious people of South Africa." This conclusion possibly explains why during the past few months many physiologists—even those whose problems lie, for the most part, outside the field of actual animal experiments—have received daily papers and magazines in which attention has been directed by means of blue pencil to letters and articles in which the ingenuity of the pamphleteer has been used to distort the aims and results of the physiologist. Henceforth the anti-vivisectionist societies, one and all, had better be known under their true colours. They are not honest organisations sustained by conscientious thinkers, or even artistic sentimentalists, but pigmy political cliques for turning the trend of political opinion one way or another.

THE lecture session at the Imperial Institute will be opened on October 29 with a lecture entitled "The Federal Family," by Sir John A. Cockburn, K.C.M.G. This lecture is the first of a special series of eight illustrated public lectures, relating to the Australasian Colonies, to be given on Monday evenings before Christmas. The remaining lectures will be as follows :—"Golden Victoria, its scenery, geological features, and mines," by Mr. James Stirling ; "Western Australia in 1900," by Mr. George Berry ; "The coal resources of Victoria," by Mr. James Stirling ; "The work of the Queensland weather bureau, in its relation to the natural resources and commerce of Australasia," by Mr. Clement L. Wragge ; "The Australian Alps, scenery, native vegetation, and mineral wealth," by Mr. James Stirling ; "New Zealand," by Mr. J. Carthcart Wason ; "Sunny Tasmania for English Invalids," by the Hon. Sir Philip Oakley Fysh, K.C.M.G.

AN important addition to the British Museum (Natural History) has just been made in the form of mounted specimens of two beautiful antelopes from the swamps of the White Nile, belonging to species hitherto very imperfectly represented in the collection. They are, in fact, the first complete specimens of their kind which have ever been exhibited in England. The species are *Cobus maria* and *C. leucotis*, both remarkable for their sable hue (at least in the males), relieved by white on the ears, and also the elegant and peculiar curvature of their heavily ringed horns. Of the former species the Museum possessed the heads of a male and female presented by the late Consul J. Petherick in 1859, which are in such a bad condition that they have not been exhibited to the public for several years, while the latter was best represented by a stuffed head (also the gift of the same gentleman), which is, however, so faded that its true colours are completely lost. For the new specimens (which have been set up by Rowland Ward) the Museum is indebted to Captain Dunn, now stationed, we believe, at Omdurman, by whom they were presented. Acknowledgments are, however, also due to Captain Stanley Flower, by whom the skins were brought to this country. The specimen of *C. leucotis* is exhibited in the case devoted to new acquisitions, but, on account of its larger size, the example of *C. maria* is placed

in the case in the West Corridor which will eventually form the home of both.

THE meteorological subcommittee of the Croydon Microscopical and Natural History Club has just published its report for the year 1899. It contains valuable information relating to the daily and monthly rainfall statistics for eighty stations in Kent and Surrey, together with notes relating to the temperature and weather for each month, by Dr. F. C. Bayard, the hon. sec. of the subcommittee. The observations show that the deficiency of rainfall throughout the district was about two inches. The deficit does not appear very large, but some tables showing the total departures from the average during the last ten years reveals a serious state of things. For Greenwich, for instance, the departures from the average of eighty years show a deficit of twenty-eight inches, which is practically three inches above a year's average rainfall. And again, for Surbiton, on the western side of the club's district, the departures during the same period show a deficiency of 19·5 inches, compared with an average of forty years, or a deficiency of practically only five inches below the average rainfall for a year.

THE report on the administration of the Meteorological Department of the Government of India in 1899-1900 has appeared in the same form as in previous years ; the first part gives a general account of the results of the more important sections of the work of the department, and the second part gives the usual details of administration, chiefly in the form of tables. Seismological observatories have been established at three stations. The international and special cloud observations referred to in previous reports will be shortly published, with a brief discussion. The arrangements for registration of snowfall in the mountain districts and the measurement of rainfall continue to form an important part of the work of the Indian Meteorological office ; monthly returns from 2300 rainfall stations were published during the year. The storm-warning work was carried out satisfactorily ; ample and timely warning appears to have been given of all the more important storms. The special warnings of floods also appear to have given general satisfaction.

THE current issue of the *National Geographic Magazine* contains an article, illustrated by diagrams, on "The West Indian Hurricane of September 1-12, 1900," by Prof. E. B. Garriott, of the U.S. Weather Bureau, and another by Mr. W. J. McGee, entitled "The Lessons of Galveston." The Lessons in question are four in number, three of which are physical and one is human. The former are as follows :—1. The danger of building on sand, Galveston being founded on a sand-bank ; 2. "The bank on which Galveston was built is something more than a simple heap of silicious grains and dust ; it is a record of past wave work ;" and "it is the duty of the nature student to interpret natural records and guard against the building on the storm records." Lesson 3 is that of coast subsidence, and, in the opinion of the author, "it is the business of the geologist to detect and weigh the evidences of subsidence or elevation of coasts, and to estimate the rates of movement for the guidance of local residents and investors, and it behoves such citizens to avail themselves of the scientific researches."

To the *Journal* of the Franklin Institute, Mr. Lewis M. Haupt, an American engineer of reputation in matters relating to river training and harbour work, contributes a paper dealing with the present condition of the navigation in the lower reach of the Mississippi, in which he advocates that the principle of reaction-jetties should be applied for dealing with the contemplated improvement of the South-west Pass. The Mississippi affords an outlet into the Gulf of Mexico for 15,000 miles of navigable waterways. In the delta the main stream divides into three principal branches, and although the water is of great depth in the channel before it enters the delta, in their natural condition these channels are so shoal as only to afford sufficient depth of water for the navigation of the smaller class of steamers. About a quarter of a century ago Captain Eads entered into a contract with the United States Government to deepen one of these passes so as to give 27 feet at low water, and to maintain this depth for a fixed period. Contrary to his strong remonstrances the South Pass, the smallest of the three outlets, was selected. Through this pass, by means of two parallel training walls, the water was confined to a width of 700 feet, and by the scour thus created, aided to a large extent by dredging, a channel having 26 feet at low water was made, and has been maintained up to the present time. This contract has now expired, and the dimensions of vessels in the meantime having outgrown the channel, the Government have had to consider the question of providing a deeper waterway. The Board of Engineers to whom the matter has been referred have advised that the South-west Pass should now be improved and deepened so as to give a depth of 35 feet at low water.

THE Board of Agriculture has published its annual report for the year 1899–1900 on the distribution of grants for agricultural education and research, with statements respecting the several colleges and institutions allied and the experiments conducted. The larger portion of the funds distributed in grants by the Board consists of subventions of a general character awarded to eight collegiate centres of agricultural education in England and Wales. Subsidiary grants have also been made to three dairy institutes, and in aid of the cost of certain specific experiments undertaken under arrangement with the Board. Examinations have also been conducted under the joint auspices of the Royal Agricultural Society and the Highland and Agricultural Society of Scotland for the recently established national diploma.

MR. HAROLD WAGER reprints from the *Journal* of the Linnean Society an interesting paper on the eye-spot in *Euglena*. He finds it to consist of a mass of pigment-granules apparently imbedded in a protoplasmic matrix. The light absorbed by the eye-spot seems to act upon a swelling near the base of the flagellum, and thus to modify its movements. *Euglena* appears, therefore, to possess a very simple form of light-organ, consisting of a sensitive region—the swelling on the flagellum—and a light-absorbing pigment-spot.

IN a note contributed to the *Rendiconto* of the Naples Academy, vi. 5–7, Dr. Giuseppe de Lorenzo discusses the probable causes of the increased activity of Vesuvius at the beginning of May last. This activity assumed the form of "Strombolian" explosions audible as far away as Posilipo, by which masses of incandescent lava were hurled into the air to an altitude of about 500 metres. These explosions Dr. de Lorenzo attributes to the exceptional rainfall, which, filtering through the volcanic cone, has penetrated to the column of lava. This hypothesis appears in conformity with the observations of Spallanzani, von Rath, Dana and others, and with the experimental researches of Daubrée.

TWO papers on the figure of the earth have recently appeared, one, by M. Marcel Brillouin, in the *Revue générale des Sciences*, xiv., and the other, by Ingeguere Ottavio Zanotti Bianco, in the *Atti* of the Royal Academy of Turin, xxxi. M. Brillouin discusses the different geoïdic surfaces adopted in the problem of reduction to sea level, and points out the relative advantages of the geoïds of Pratt and Helmert. Bianco's paper contains extracts from the writings of Pratt and Helmert, showing the relative part played by these investigators and by Bruns in developing the general theory of geoïdic surfaces.

THE Selborne Society's magazine, *Nature Notes*, contains in its September issue an interesting account of a mirage, seen last June over the Needles (Isle of Wight) from the opposite shore,

by Captain Giles A. Daubeny. It is not uncommon when looking at a distant headland to see the appearance of a pointed nose jutting out over the water—an effect caused by the formation of an inverted image near the water-line ; but in the present instance four different horizons appear to have been observed when viewing the rocks through a telescope.

THE admirable series of memoirs published by the U.S. Department of Agriculture on the harmfulness or otherwise to the agriculturist and horticulturist of the commoner birds of North America has recently been augmented by one from the pen of Mr. F. E. L. Beal, dealing with the food of the bobolink, blackbirds and grackles. This memoir forms *Bulletin* No. 13 of the Department. The bobolink is an exception to most birds in that, both at seed-time and harvest, it inflicts immense damage on the rice-crops of the Southern States. At present, therefore, the harm it does far outweighs such benefits as it may confer ; but as the bird could exist perfectly well without touching a grain of rice, hopes are entertained that means may be found of checking its depredations on that crop. On the other hand, most or all of the so-called blackbirds (which are not to be confounded with the species of the same name in Europe) feed largely upon noxious insects and weed-seeds, and are therefore highly beneficial to the cultivator. Much the same may be said of the grackles. As usual, the *Bulletin* is illustrated with good figures of the species described, and the whole publication does the greatest credit to the Government by whom it is issued.

Bulletin No 67 of the West Virginia University Experiment Station is devoted to a communication by Dr. A. D. Hopkins on the Hessian fly in West Virginia, and how to prevent losses by its ravages. As the result of his investigations, the author finds that the date of the appearance of the swarms of this insect depends upon the latitude and altitude of the place, and he gives a formula by means of which the former may be approximately determined for any particular locality. From this the dates may be calculated at which it is reasonably safe to sow wheat in order to escape loss from the ravages of the pest. The approximate limits of the best wheat-sowing period, and also the approximate normals for the disappearance of the fly in different districts, are graphically illustrated by means of a map.

THE September issue of the *American Naturalist* commences with an interesting paper, by Miss (or ? Mrs.) Sampson, on unusual modes of development among frogs and toads. Commencing with a *résumé* of the normal mode of breeding as exemplified in the common frog, the author goes on to show how different members of the group depart from this mode of procedure. Two species, for example, the one from West Africa and the other from Brazil, deposit their spawn in nests formed of leaves stuck together, the tadpoles moving in a mass of froth, recalling that of the cuckoo-spit insect. In both these instances the spawn is deposited in the neighbourhood of water, into which the tadpoles ultimately fall ; but in a tree-frog from Rio, in which the eggs are likewise hatched in a frothy mass among leaves, the larvæ actually die if they are put into water. In another Brazilian tree-frog the tadpoles frequent cracks in rocks, and adhere to the surfaces of the latter by means of an abdominal sucker. Full reference is made to the mode of development in the Surinam toad, and also to that of the marsupial frogs, in which the young are hatched in a dorsal pouch. But perhaps the most extraordinary "nursery" arrangements in the entire group are those of the Chilian Rhinoderma, in which the tadpoles undergo their development in an enormous pouch on the throat of the male. In the same journal Mr. F. Russell has a paper on cranial abnormalities in the American races, among some of whom the persistence of the frontal suture may occur as frequently as 2·9 per cent.

THE October issue of the *Entomologist* contains a summary of the capture of rare British insects during the past summer and autumn. In addition to the swarms of the pale clouded yellow butterfly (*Colias hyale*), to which allusion has been already made, no less than ten examples of the Camberwell beauty (*Vanessa antiopa*) are recorded as having been taken in the south-east and east of England. Caterpillars of the death's-head moth (*Acherontia atropos*) have been extraordinarily abundant in potato crops during the season, and many other rarities are recorded.

THE *Bulletin* of the American Mathematical Society for October contains a useful list of courses in mathematics announced by seventeen German universities for the 1900–1901 session.

THE Department of Mines, Victoria, has issued No. 7 of the Reports of the Victorian Coal-fields, by Mr. James Stirling, the Government geologist. It consists of descriptions, with illustrations, of the fossil flora of the Jurassic beds of South Gippsland.

A SIMPLE description of the movements and obvious characteristics of the members of the solar system, and other celestial bodies, is given in Mr. W. T. Lynn's "Astronomy for the Young" (pp. 51), the second edition of which has just been published by Mr. G. Stoneman, London, E.C.

THE lecture syllabus of the Hull Scientific and Field Naturalists' Club for the winter session ranging from October to March has just reached us, and gives promise of a full and interesting winter's work. Judging by the contents of the *Transactions* of the Club for 1900, a copy of which has also been sent to us, the institution is in a healthy and vigorous condition.

COUNTY floras have at present been pretty much confined to flowering plants and fern allies. We welcome the precedent set by the Yorkshire Naturalists' Union in issuing an Alga-Flora of Yorkshire, being a complete account of the known freshwater algæ of the county, by Mr. W. West and Mr. G. S. West. The present publication, which is only a first instalment, includes 208 species.

THE October number of the *Contemporary Review* contains two articles of scientific interest, one, by Prof. Marcus Hartog, on "Interpolation in Memory," and one by Mr. A. Shadwell, entitled "The true aim of Preventive Medicine." The current issue of the *Humanitarian* has in it a very readable contribution on "Heredity as a Factor in the Interpretation of Disease," from the pen of Prof. D. J. Hamilton of Aberdeen.

A PROOF of old Semitic influence in South Africa is afforded, according to K. Meinhof (*Globus*, Band lxxviii. p. 203), by the occurrence of the word "darama" or "ndalama" in various Bantu dialects for "gold." An ancient Arabic word for gold was "dirhem," pl. "darāhim." According to the phonetic system of the Bantu languages this would necessarily be transformed into "ndalama."

MR. C. FOX-STRANGWAYS contributes some interesting notes on Spitsbergen and Iceland in the *Transactions* of the Leicester Philosophical Society for April, 1900. Having spent only a short time on these islands, the author does not claim to record much that is new regarding them, and his article is written chiefly in explanation of a series of photographs which are reproduced to accompany the notes.

THE additions to the Zoological Society's Gardens during the past week include a Brown Capuchin (*Cebus fatuellus*) from Guiana, presented by Mrs. W. L. Gower; two Yellow-whiskered Lemurs (*Lemur xanthomystax*) from Madagascar, presented by Mr. J. B. Joel; a Common Genet (*Genetta vulgaris*), European, presented by Baron de Soutellinlio; an Alligator (*Alligator*

mississippiensis) from North America, presented by Mrs. Bazalgette; a Porose Crocodile (*Crocodilus porosus*) from the East Indies, presented by Miss Gwendoline Waite; a Broad-nosed Lemur (*Hapalemur simus*) from Madagascar, four Indian Fruit Bats (*Pteropus medius*), six Starred Tortoises (*Testudo elegans*) from India, a Dusky Sloth (*Bradypus infuscatus*) from Colombia, an Illiger's Macaw (*Ara maracana*) from Brazil, two Salvin's Amazons (*Chrysotis salvini*), an Annulated Terrapin (*Nicoria annulata*), a Brazilian Tortoise (*Testudo tabulata*), an Electric Eel (*Gymnotis electricus*) from South America, four Wrinkled Terrapins (*Cyclemmys scripta rugosus*) from the West Indies, a Common Water-Buck (*Cobus ellipsiprymnus*) from South Africa, deposited; a Violaceous Night Heron (*Nycticorax violaceus*) from South America, purchased.

OUR ASTRONOMICAL COLUMN.

EPHEMERIS FOR OBSERVATIONS OF EROS:—

1900.		R.A.		Decl.
		h. m. s.		" "
Oct. 18	...	2 36 36 00	...	+50 47 13.4
19	...	35 30.95	...	51 4 38.6
20	...	34 21.37	...	51 21 32.0
21	...	33 7.26	...	51 37 51.5
22	...	31 48.74	...	51 53 35.1
23	...	30 25.82	...	52 8 41.0
24	...	28 58.65	...	52 23 7.4
25	...	2 27 27.35	...	+52 36 52.0

NEW PLANETARY NEBULA.—Mr. R. G. Aitken writes to the *Astronomische Nachrichten* (Bd. 153, No 3667) announcing that the object catalogued as star BD + 83°.357 is a small planetary nebula. With the 36 inch Lick telescope the object appears to have a stellar nucleus of about 10.5–11 magnitude centrally placed in a circular nebulous envelope about 5″–6″ of arc in diameter. The complete object is about 9.5 magnitude, and its position is:—

$$\left. \begin{array}{l} \text{R.A.} \quad \text{12} \ \text{29} \ \overset{\text{s.}}{\text{10}} \\ \text{Decl.} + 83° \ 21'\!\cdot\!8 \end{array} \right\} (1855\cdot0).$$

PARIS OBSERVATORY, ANNUAL REPORT.—In his report of the work accomplished at the Paris Observatory during the year 1899, M. Loewy, the director, states that a considerable part of the time was spent in preparing for the Exposition. Special photographs on a large scale were taken of the moon about the time of first and last quarter, giving an image about 1·38m. in diameter; considerable difficulty was encountered in the preparation of the plates for these, and special mention is made of the services rendered by MM. Lumière in this matter. Among the new instruments adopted are (*a*) the mercury bath designed by M. Hamy for registering earth tremors, (*b*) a new micrometer by Gautier for measuring the chart plates, (*c*) a new form of chronograph designed by M. l'Abbé Verschaffel, director of the Abbadia Observatory. This latter has been introduced for use in the proposed new determination of the difference in longitude between Paris and Greenwich.

For the chart photographs ninety-six sheets have been issued, and the first part of the catalogue, giving the exact positions of stars down to the eleventh magnitude, will be issued during the present year. An investigation is in progress for determining more accurately the photographic magnitudes of stars. Valuable help has been given by M. l'Abbé Verschaffel, of Abbadia, who has determined the coordinates of 3700 fundamental stars of reference for the chart photographs. A fourth volume of the "Atlas de la Lune," containing seven plates, has been published, accompanied by a descriptive memoir.

M. Bigourdan has continued his study of the nebulæ, having now measured the positions and made detailed study of 6000 of them; to complete his programme of work 400 more still remain to be examined.

The small equatorial Coudé, which was the first instrument of this type, has been entirely remodelled. A new objective has been made by MM. Henry, of longer focus than the old one (5·25 metres instead of 4·22), the silvered mirrors protected more from the action of atmospheric gases, and the whole instrument encased in a thick layer of felt. These modifications have removed all the defects existing in the old telescope.

RANGE-FINDERS.[1]

NAVAL and military authorities are agreed that for accurate shooting almost everything depends upon the range being known with sufficient exactitude. It is not surprising, therefore, that an immense amount of attention should have been devoted to methods of rapidly determining the ranges of distant objects, whether stationary or in motion. These methods may be classified as follows:—

(1) *Mechanical.*—In this method a trial shot is fired from a gun, and so far as is practicable observation is made as to whether the shot strikes the ground (or the sea) on the near or the far side of the target. The results of the observation are used to correct the next trial shot, and so on. The method is clumsy and (at all events on board ship) costly; it is inapplicable in a naval engagement where the ranges frequently alter with great rapidity, and is, of course, totally inapplicable for purposes of navigation. On the other hand, it involves the use of no instrument beyond those ordinarily used in warfare.

(2) *Flash and Sound Method.*—Here the time-interval between seeing the flash and hearing the report from one of the enemy's guns is measured by a suitable chronograph or, by starting to count at the rate of eleven in three seconds immediately on seeing the flash; the number arrived at on hearing the report will roughly give the number of hundreds of yards in the range. This method is, of course, inapplicable amid the din of a general engagement, and besides it permits the enemy to have the first shot.

(3) *Optical Range-finders.*—In the determination of a distance by an optical range-finder, *length of base* and *time* available are very important factors. Given plenty of base-length and plenty of time, a theodolite, such as is used in surveying, satisfies all the conditions except that of portability. However, *time* is usually so important a factor for military operations that the theodolite is completely excluded even for field service. If the time is restricted, but the extent of base unrestricted, recourse may be had to long-base, two-observer instruments, such as those now in use by infantry and field artillery. When both base and time are restricted, as they are on board ship—and moreover the observer's station is itself in motion—the problem becomes at once much more difficult and much more interesting. We need not then be surprised that while military range-finders are usually simple and cheap, naval range-finders are comparatively costly and complicated.

The operation of all optical range-finders depends upon the measurement, by some device or other, of the angle subtended by a known base-length at the distance to be determined, the base being almost invariably "broadside-on" to the point. Where the known base is at the target and the point is at the observer's station, the operation of determining the distance consists simply in measuring the angle subtended by the base at the point. Instruments of this class measure, say, the angle subtended by the height of a man (assumed to be of mean stature) at the observer's station, and require for their successful employment the co-operation, willing or otherwise, of the enemy, and are therefore seldom applicable under the conditions of modern warfare. Men who have returned from active service in South Africa have stated that sometimes they never saw a Boer in an engagement. Where the known base is at the observer's station, which is an essential characteristic of all range-finders of general application (the instruments previously described being in reality mere angle-measurers), the operation of determining the range consists almost invariably in swinging the base round until it is "broadside-on" to the target, and then measuring the angle subtended by the base at the target. The measurement of the angle at the target is very commonly effected by making one of the base-angles a right angle, and observing how much the other is "off" the right angle (mekometer, &c.), or else by observing how much the sum of the base-angles is less than two right angles (naval range-finder, &c.).

Since a mere enumeration of the names of inventors of range-finders would occupy the whole of the available time for a lecture, it will be necessary to restrict the selection of examples to those instruments which are, or promise to be, in actual operation for warlike purposes. Attention then will be confined to the following instruments:—

(1) The Watkin mekometer, used by our troops in South Africa.

[1] An evening lecture delivered at the British Association meeting at Bradford by Prof. W. Stroud.

(2) The Watkin depression range-finder, used at certain stations for coast defence.

(3) The Zeiss tele-stereoscope.

(4) The Barr and Stroud range-finder, used in the British and other navies, in coast fortifications, and to some extent in the field.

Restrictions of time unfortunately render it impossible to describe the Watkin artillery range-finder, as well as the Fiske instrument which is used to some extent by the U.S. Navy.

(1) *Mekometer.*—This instrument consists of two parts connected by a cord 25 or 50 yards long, which is kept tightly stretched by two observers, each of whom supports one of the parts. These two observers are designated respectively the right-angle man and the range-taker. The former carries a small optical square or instrument in which two mirrors are fixed at 45° so as to set out a right-angle.

The part carried by the range-taker consists practically of a box-sextant in which one of the mirrors is adjustable by means of a graduated micrometer-screw, on which the ranges are marked for the base specified, the infinity mark corresponding to the case when the mirrors are exactly at 45°. Each of the parts of the instrument carries a prominent mark in a suitable position, and the operation of taking a range on a fixed object is as follows:—The two observers supporting the instruments connected by the taut cord set themselves in such positions that the cord is nearly "broadside-on" to the target; the range-taker now stands still, while the right-angle man—by moving forwards or backwards, with the cord always taut—adjusts his position accurately, so that the image of the mark on the range-taker's instrument, seen through his optical square, coincides with the target viewed directly. He now shouts "On," when the range-taker adjusts his micrometer-screw so as to bring the mark on the other instrument into coincidence with the target, and then the reading on the graduated scale gives the range.

For objects in motion the range-taker remains steady, while the right-angle man, as far as possible, continuously shifts his position so as to keep the mark on the range-taker's instrument and the target in coincidence. In cases of rapid motion at right angles to the line of sight, difficulties are experienced in keeping continuously "on," and in such cases the right-angle man shouts "On" whenever the mark passes the target, and the range-taker then seizes the opportunity of taking the range.

There are several serious objections to two-observer instruments. In the first place, they cannot be used for the measurement of the distances of sky-lines, trenches, hedges, &c., of a more or less horizontal character, with no prominent vertical features about them. Secondly, it not infrequently happens that the two observers are working on different objects altogether or on different portions of the same object, in which case an entirely fictitious range is obtained. Most important of all, they necessarily expose the range-takers. It has been stated that several casualties have occurred in South Africa among men engaged in range-finding. Under such conditions as those pertaining in South Africa, it is obviously most risky to expose men in the open. Single-observer instruments are free from all these defects.

(2) *Depression Range-finder.*—If we have a vertical base, and if the target is constrained to move in a horizontal plane, we may dispense with the right-angle man altogether. The only case where this plan can be adopted in practice is when we wish to determine the range of a ship from a hill near the coast. The base-line is now the vertical height of the observing station above sea-level, and the range is obtained very simply by observing the angular depression of the water-line of the ship. Allowance has to be made for the effect which the tides produce in varying the equivalent height of the base. Unfortunately, the water-line of a ship—especially if the waves are at all high—forms an exceeding bad object upon which to observe, and this uncertainty can only be compensated by having a very long base—*i.e.* by having the range-finder at a very considerable altitude, say 150 feet or so. We therefore cannot use such an instrument on a flat coast, nor can we use it for determining distances at night, unless we can sufficiently well illuminate the water-line of a ship by means of a search-light.

(3) *Zeiss Range-finder.*—This is the first of the two single-observer instruments which it is proposed to describe in the prescribed limits of time. The credit of the idea underlying the construction of the instrument is due to the late Herr Grousillier of the German Army, but as the firm of Zeiss, of Jena, have spent many

years in working out the details of the instrument, it will probably be known in the future by their name. In the telestereoscope of Helmholtz two parallel reflectors are placed with reference to each eye, in such a position as to produce the optical equivalent of an increase in distance between the two eyes.

Then when we look into the instrument at a distant landscape we shall get the appearance shown in Fig. 1. If this picture be viewed stereoscopically we shall obtain a mental impression as to the distance of any part of the landscape by comparison with the marks, and in this way get its approximate range.

FIG. 1.—Field of view in the Grousillier-Zeiss stereoscopic range-finder.

This increases the stereoscopic effect. In the next place, by placing a telescope before each eye (and the two telescopes may conveniently be incorporated in the frame-piece supporting the reflectors), we multiply the stereoscopic effect still further, e.g. if the distance between the eyes has been artificially increased tenfold by the reflectors, and if the telescopes magnify tenfold, the stereoscopic effect will be increased altogether one-hundred-fold. In this way immense stereoscopic solidity is imparted to the picture in the field of view.

FIG. 2.—Bridge of H.M.S *Royal Sovereign*, showing Barr and Stroud range-finder on the top of the chart-house.

This instrument may be adapted to range-finding in the following way:—Suppose we imagine for the moment the instrument fixed, and that we see in the field of view of each eye the image of a pole 1000 yards away. Let permanent marks be made in the focal plane of each telescope exactly coincident with these images. Then, whenever we look into the instrument we shall apparently see a pole at a distance of 1000 yards. Let a similar pair of marks be fixed corresponding to 1100 yards, and so on.

The instrument is highly ingenious and very pretty, and it will no doubt offer a solution of the problem of military range-finding should it prove sufficiently accurate in practice.

(4) *B. and S. Naval Range-finder.*—This instrument, with a base of 4½ feet, has been adopted in her Majesty's Navy, and

FIG. 3.—Barr and Stroud range-finder on fortress mounting.

nearly all the larger ships are now equipped with one or more of them. Fig. 2 shows a view of the charthouse of H.M.S. *Royal Sovereign*, on the top of which the range-finder will be seen near the Kelvin compass.

This instrument, in favourable circumstances of weather, will measure ranges in a few seconds of time with an accuracy of something like 3 yards at 1000 yards, 30 at 3000, 120 at 6000, and so on. Prof. Barr, of the University of Glasgow, and the lecturer devised the instrument in its main features in 1888 ; it took, however, a period of five years to make a satisfactory naval instrument, and for the past seven years the improvement of the optical and mechanical details has been going on.[1]

Fig. 3 shows the same instrument mounted for fortress observation.

It is claimed for these instruments that they offer a solution —of course not necessarily the only or the best—of the problem of range-finding in all cases where want of portability is not a drawback.

B. and S. Field Range-finder.—Fig. 4 shows a smaller instrument of the same type with a base of 3 feet, which weighs 12 lbs. One of these instruments has been used by Major Guinness in South Africa since the beginning of February, and that officer reports that after carrying the instrument on his ammunition wagon over all sorts of ground for six months it was in no way damaged or deranged. The figure shows how

FIG. 4.—Barr and Stroud field range-finder.

the range-finder can be used for taking a distance with practically no exposure of the man.

The lecturer concluded by describing and exhibiting the electrical telegraph for naval use which Prof. Barr and he had devised at the request of the Admiralty, so as to enable the captain of the ship in the conning tower to receive from the range-taker continuous records of the enemy's distance and to transmit the same and also orders to the guns. The importance of trustworthy means for transmitting orders and other communications from one station to another on a warship is now fully realised. Thus, for example, it is reported that in the opinion of Admirals Fournier and de Beaumont and the officer in command, the loss of the French torpedo-boat destroyer *Framee*, which occurred about a month ago, was due to the fact that the apparatus on board the vessel for the transmission of orders was inadequate. The need for continuous and almost instantaneous transmission of ranges to the gunners will be obvious when it is remembered that in naval engagements the range is continually and rapidly altering.

The lecturer concluded by expressing his thanks to Mr. J. J. Hicks, of London, Mr. Steward, of London, and Messrs. Zeiss, of Jena, for the loan of range-finders to illustrate the lecture.

[1] Readers unacquainted with the instrument are referred for details of its construction to *Transactions* of Inst. of Mech. Eng. (1896), or *Engineering*, 1896 Part I. p. 233, &c.

MECHANICS AT THE BRITISH ASSOCIATION.

ALTHOUGH no very striking paper was presented to the section at this meeting, still several papers of value and of considerable interest were dealt with.

In the committee of the section two very important pieces of work were carried out. The committee on small screw gauges, which has now been at work for some years, presented an interim report in which the difficulties the committee had met with in obtaining standard gauges were very fully discussed, and an account of some experiments on different forms of threads, made by Mr. T. M. Gorham and Mr. W. A. Price in the laboratory of Prof. Hudson Beare at University College, London, were described. The committee stated in their report that they have now every hope of bringing their inquiry to a successful conclusion, and the committee was therefore re-appointed and a grant was secured for the necessary expense of completing the work.

A committee was also appointed at Bradford to deal with the question of the resistance of road vehicles to traction. Prof. Hele Shaw read a short paper before the section, drawing attention to the need of modern experiments on the nature of the resistances encountered by vehicles on the common road. He pointed out how the growth of the cycle and motor car industry made information upon this point a matter of the greatest importance.

There is no doubt that we are on the eve of a very considerable increase in mechanical propulsion on common roads, and at present designers of such vehicles have to rely largely upon old experiments with solid steel tyres, and carried out on roads very different indeed from the modern ones. The powerful auto cars which can now be obtained make it comparatively an easy matter to determine the tractive power necessary to move a vehicle with any load upon any type of road, and no doubt the work of the committee will largely consist—after a suitable dynamometer and speed indicator have been arranged for—in carrying out exhaustive experiments with different types of vehicles and different types of tyres on all the various classes of roads now in use.

A grant of money was secured from the Association for the purposes of this committee, and we have every hope that when the committee submits its report it will justify its appointment.

In the work of the section papers by local engineers bulked largely. Perhaps the two most interesting and valuable were a paper by Mr. J. Watson, the Waterworks engineer at Bradford, in which the new Nidd Valley Waterworks were described, and the paper by Mr. J. MacTaggart, the Superintendent of the Cleansing Department of the City of Bradford, entitled "The Disposal of House Refuse in Bradford."

In Mr. Watson's paper a short historical summary of the various schemes for supplying Bradford with water was given, followed by a very exhaustive and complete account of the Nidd Valley scheme. This scheme, now rapidly approaching completion, is one which will cost the City of Bradford nearly 1,500,000*l.* and will afford a supply of about twenty million gallons of water per day, and in addition will provide a large compensation reservoir (Gouthwaite) for the land-owners along the Nidd Valley.

It is essential for such a city as Bradford, where the chief industry is that of the woollen trade, that the water shall be very soft, and, of course, this led to some difficulty in finding a suitable collecting ground for the city supply.

The section had the opportunity of visiting, under the guidance of Mr. Watson, portions of the works on the Saturday of the meeting.

In Mr. MacTaggart's paper very valuable information was given as to the most modern methods in a big manufacturing city for the disposal of the daily city refuse. So successful have the various arrangements been, mostly due to the author, that it is hoped in time the destruction of the whole of the refuse of the city will be carried out, not only without creating any nuisance, but without any cost to the ratepayers. The refuse is chiefly dealt with by destructors, and the great merit of the Bradford system is in the utilisation of the clinker produced in the destructor furnaces for various useful purposes, the power to work the machinery required for these purposes, and the lighting of the works all being obtained from the steam generated by the surplus heat of the destructors. A large number of specimens were shown to the section of concrete paving-slabs, bricks and encaustic tiles manufactured from the clinker, and some interesting figures as to the strength of these materials were also given.

Another paper by a Yorkshire engineer, Mr. E. K. Clark, dealt with the subject of the shop buildings in large engineering works. The author had collected a large amount of statistics and figures, both as to the method of construction now generally adopted—viz., the shops all on one level—and as to the materials commonly used in their construction, and the paper will form a very valuable reference for any engineer engaged either in laying out new engineering workshops, or in reconstructing old buildings.

Mr. Glass contributed a lengthy paper dealing with the coal and iron ore fields of Shansi and Honan, and railway construction in China. The author was engaged in 1898 by a syndicate to proceed to China, and to examine and make a complete report on the coal and iron-ore fields of these two important provinces of China, and also to make surveys for the railways which it would be necessary to construct in order to utilise these deposits.

An interesting description was given of the general features of the country, illustrated by some beautiful lantern slides from photographs taken by the author, and also a very complete account of the Chinese method of working these mineral deposits. The author stated that it was somewhat difficult to arrive at very exact estimates of the quantity of coal available in these great fields, but it is believed there are more than 33,000 square miles of coal-fields in Shansi alone, and that the present output of Great Britain, which is more than 200 million tons a year, could be maintained from the anthracite coal-fields of Eastern Shansi alone for a period of 3000 years.

Samples of the coal have been analysed and show that it is a good steam coal.

Similar favourable accounts were given in regard to the iron ore deposits, and the author computed that it would be possible to produce a ton of pig-iron from these ores at the very low cost of 12s. 1½d. per ton, assuming labour to be at the same rate as it is now at Middlesbrough in England. The lowest price at which the pig-iron was being sold at the foundries visited by the author was a little over 20s. per ton.

The extraordinary richness of these mineral deposits and the enormous area awaiting development show how pressing the problem of reorganising peacefully the internal government of China is for the civilised world.

On the day devoted to electrical engineering, two short communications were read by Sir Wm. Preece and Mr. F. J. Behr, dealing with the proposed Monorail High Speed Electric Line between Manchester and Liverpool.

It will be remembered this scheme came before Parliament last session and was rejected, largely owing to the strenuous opposition of the existing railways.

The interesting feature in these papers was the account given of the brakes and signals which it is proposed to adopt, which must be an important matter on a line where it is proposed to run the trains at frequent intervals, and at such an excessively high speed as 110 miles per hour. Perhaps if the promoters had been content to reduce the proposed speed somewhat at the beginning, their scheme might have been more favourably considered.

A valuable paper dealt with on this day was one on the measurement of the tractive force, resistance and acceleration of trains, by Mr. A. Mallock. The author described the apparatus which he used for the purpose, and gave an account of some experiments which he has recently carried out on these important questions. He concluded from his results that pendulum observations combined with a record of speed and power offer a simple and effective means of determining the resistance to and efficiency of electric or other kinds of motor vehicles.

On this day also, a communication from Mr. W. T. E. Binnie, a son of the president, was read, describing a new form of self-registering rain-gauge which he has invented. The accuracy of the gauge depends on the fact that all drops falling from a tube are of constant size, provided that the tube is either very small so that the water passing down the interior chokes the bore, or that some special device is provided to spread out the water so as to wet the entire circumference of the tube. If, therefore, the weight of each drop were ascertained, it is clear that a measure of the amount of water passing down the tube would be obtained by counting the number of drops, and the electrical appliances are concerned with this part of the work.

An instrument made on this principle has been in operation for some time, and the records obtained from it in a period of five months give a total excess of 1·6 per cent. over the register of an ordinary rain-gauge.

Mr. W. Dawson contributed a paper descriptive of the Demerbe system of tramway construction, a system which has been tried on the Bradford Corporation tramways and found very successful in the reduction of the cost of permanent way repairs. In the Demerbe system the rail consists of a hollow trough and the fish plate is placed inside under the ends of the rails and exactly fits its contour, the fish-plate being forced into close contact with the under side of the rail by driving in two cotters. The rail, when laid in position, is completely filled, by means of specially-designed tools, with concrete. The tie bars are flat and very simply arranged, and the gauging of the rails can, therefore, be done rapidly with almost mathematical exactitude.

The system certainly seems to have considerable merits compared to the girder system of tram rails.

Two other instruments described in short communications were a combination integrating wattmeter and maximum demand indicator, the invention of Mr. J. H. Barker and Prof. Ewing, and a new form of calorimeter for measuring the wetness of steam, designed by Prof. Goodman.

The former instrument is designed to measure two quantities, the total amount of electricity used by a consumer and the maximum number of lamps or their equivalent ever lighted at one time. By this means it is possible to grade the charges for electric energy, and it enables the consumer to be charged at a lower rate for current he may use over and above the units which would have been used had the largest number of lamps in his installation burnt for one hour every day during the whole period the current was in use.

Prof. Goodman's instrument is intended to get over some of the more serious difficulties in measuring the wetness of steam supplied by any boiler. He discards entirely the wire-drawing system, and determines the wetness by condensing a known weight of the mixed steam and water in a known weight of cold water, and measuring the rise of temperature.

Prof. Beaumont, of Leeds, in a most interesting paper described the photographic method of preparing textile designs due to Szczepanik, a number of designs being shown which had been produced by this process.

In continuation of a paper which was read before the Association at Liverpool in 1896, Mr. A. T. Walmisley gave further information as to the use of expanded metal in concrete work, and gave particulars of a number of important tests which have been carried out to determine the increase of strength and the adhesion between the concrete and the metal.

The attendance at the sectional meetings was very good, and the president, Sir Alexander Binnie, may be congratulated on a successful and useful meeting.

BOTANY AT THE BRITISH ASSOCIATION.

IN the absence of Prof. Vines the presidential address was read by Dr. D. H. Scott. On the motion of Prof. Bayley Balfour, supported by Prof. Marshall Ward, Prof. Bower and other speakers, it was unanimously agreed to ask the Recorder of the Section to convey to Prof. Vines the sincere regret of the Botanists that he was prevented by illness from presiding over their meeting.

The customary semi-popular lecture was this year delivered by Prof. Percy Groom, who chose for his subject "Plant-form in relation to nutrition."

On Monday, September 10, the Section of Geology joined Section K in a discussion on the conditions under which the plants of the Coal Period grew. The discussion was opened by Mr. Kidston (Stirling), who gave a general account of the flora of the Coal-measures, illustrated by a series of excellent photographs of the various types of Upper Carboniferous plants. Mr. Seward dealt with the botanical evidence bearing on the climatic and other physical conditions under which coal was formed. On the geological side the discussion was opened by Mr. Strahan and Mr. Marr. Dr. Horace Brown discussed the question of the possible richness in CO_2 of the coal period atmosphere, and gave an account of some of his recent experiments with plants grown in an atmosphere containing twice or thrice the present amount of carbonic acid gas. Dr. Scott, Dr. Blackman, and Prof. Hartog also took part in the discussion from the botanical standpoint.

GENERAL.

Prof. Bower, F.R.S., gave an account, illustrated by several excellent photographs, of sand-binding plant as seen in the dunes on the Scotch coast in the neighbourhood of North Berwick.

British sylviculture, by Samuel Margerison. In this communication attention was called to the large importation of foreign timber, and the urgent need of Government aid in the production of British timber. The author spoke of the existence of much land in this country at present unproductive, or only slightly productive, which is suitable for giving a native supply of timber. He compared the results of Continental and British sylviculture, and pointed out that in Britain the natural conditions are not less favourable, but the management is generally inferior. The author urged the importance of encouraging forestry schools which should afford opportunities for detailed research and teaching, with equipment, scientific and practical, worthy of the subject.

The great smoke-cloud of the North of England and its influence on plants, by Albert Wilson. The author spoke of the extent of the great smoke-producing district of the North of England and the miserable condition of the vegetation in some parts of the area. Among the various points dealt with in the paper, the following may be mentioned; the long distance reached by the smoke of large towns; the discoloration of rain-water ("black rain"); the effect of smoke on mosses and hepatics as compared with that on higher plants; the threatened extinction of *Ulota* and *Orthotricha*; the influence of smoke on sunshine and air-temperature in calm summer weather and in anti-cyclonic weather during autumn or winter.

Embryonic tissues, by Prof. Marshall Ward, F.R.S. The author urged the advisability of improving the current terminology with regard to the nature and growth of the tissues termed embryonic. Sachs termed all the tissues of the growing-points, cambium, pericycle, &c., embryonic tissues. Prof. Ward would restrict the term *embryonic tissue* to that of the embryo alone before the desmogen strands are developed, the other tissues being designated *derived* or *secondary tissues*. The tissues of the growing-points are derived from embryonic tissue, and differ from it in that, instead of being capable of developing all or any part of the plant, they are more or less restricted to the power of developing shoots, leaves, &c., or only roots. The proposed classification would apply equally to the lower organisms; some of the Algæ and Schizomycetes appear to be always in the embryonic stage. Prof. Ward also urged the desirability of distinguishing between the *assimilatory* growth of true embryonic tissues and the *vacuolar growth* of the derived tissues.

PHYSIOLOGY.

Dr. F. F. Blackman and Miss Matthaei communicated the results of their recent work on the effect of the closure of stomata on assimilation; Dr. Blackman also gave an account of his investigations on the so-called optimum strength of CO_2 for assimilation.

Formation of starch from glycollic aldehyde by green plants, by Henry Jackson. Glycollic aldehyde has lately been isolated in a crystalline state, and more recently it has been shown by the author that this substance, under the influence of dilute alkalis, very quickly condenses to two synthetic hexoses. Leaves of *Tropaeolum* and clover, which had been depleted of their starch by growing in the dark, were floated in a three per cent. aqueous solution of diose, control experiments being made with cane-sugar, glycerine and distilled water, the whole series being kept in the dark for six days. They were then tested by Sach's method; those floating in pure water were quite starchless, those in glycerine almost so, but those growing in diose had accumulated starch in the tissues, though not to the same extent as those placed in cane-sugar.

On the effect of salts on the CO_2 assimilation of *Ulva latissima*, L., by E. A. Newell Arber. It was found that an inhibition of the power of CO_2 assimilation could be caused by the presence or absence of certain salts in the medium. *Ulva* was obtained free from starch and exposed to light in various media. In distilled water only a very small amount of starch was formed, while in tap-water containing traces of nutrient salts the inhibition was only partial. The presence of NaCl in the medium was found to be essential in order to obtain the maximum of CO_2 assimilation. A total or almost total absence of NaCl caused a very marked inhibition, and no other salt could be found to replace NaCl in regard to CO_2 assimilation. The absence from sea-water of any one of the following salts,

$MgCl_2$, $MgSO_4$, $CaSO_4$, or KCl, did not inhibit the assimilation. The presence of a nitrate in appreciable quantity in the medium caused an inhibition.

The sea-weed *Ulva latissima* and its relation to the pollution of sea-water by sewage, by Prof. Letts and J. Hawthorn. For a number of years past a very serious nuisance has arisen from the sloblands of the upper reaches of Belfast Lough during the summer months, the stench at low-tide being quite overpowering, and the air heavily charged with sulphuretted hydrogen.

The nuisance is caused by deposits of the green sea-weed, *Ulva latissima*, which in the two localities mentioned grows in abundance, and during high winds or gales is washed ashore. In Belfast Lough the quantity thus deposited is enormous. Once deposited, these layers of sea-weed often remain more or less stationary for months in the shallow bays or pools of the neighbourhood, and in warm weather rapid putrefaction occurs, and a perfectly intolerable stench arises, which is perceptible over a wide area and seriously affects, not only the comfort of the inhabitants of the district, but also the value of their property.

The evidence which the authors have collected tends to the conclusion that the occurrence of *Ulva latissima* in a given locality is an indication of sewage contamination, and there can be no doubt as to the power which the weed possesses of absorbing nitrogen compounds from polluted sea-water. While thus acting as scavenger it may itself give rise to a very extensive nuisance.

Further investigations on the intumescences of *Hibiscus vitifolius*, L., by Elizabeth Dale. In a previous paper (*Proc. Phil. Soc. Camb.* vol. x. 1900, p. 192) the author gave the results of some experiments, which pointed to the conclusion that the conditions determining the formation of the outgrowths were moisture, warmth and light. More recent work has given the following results: (1) In a moist atmosphere, bright sunlight and a high temperature, large numbers of intumescences were formed in two or three days; (2) outgrowths were produced under red, yellow and white-washed glass, but not under blue or green glass; (3) the distribution of outgrowths is dependent upon that of the stomata; (4) the checking of transpiration in a damp atmosphere is one cause of the development of the outgrowths, but this in itself is insufficient. There is further evidence that an altered course of metabolism is also involved.

ANATOMY, PALÆOBOTANY, &c.

On a fourth type of transition from stem to root-structure occurring in certain monocotyledonous seedlings, by Ethel Sargant. Van Tieghem dercribed three types of transition from a stem to a root-structure (*Traité de Botanique*, 1891, p. 782). Miss Sargant found a fourth type in certain monocotyledonous seedlings. The best example is *Anemarrhena asphodeloides*, but there are very clear traces of the same structure in some allied genera. In *Anemarrhena asphodeloides* there are two bundles in the cotyledon which pass downwards through the hypocotyl into the primary root. During the transition each phloem group divides into two. Each xylem group branches in three directions. It sends a group of protoxylem elements to divide its own two phloëm groups from each other. Two lateral protoxylem groups are also formed from the xylem at each bundle in the space dividing the bundles from each other. The four lateral protoxylem groups thus formed are reduced to two by the fusion of adjacent groups in pairs. In the end, there are four phloem groups and four protoxylem groups in the root-stele.

On the presence of seed-like organs in certain Palæozoic lycopods, by Dr. D. H. Scott, F.R.S. Specimens discovered by Messrs. Wild and Lomax in the Lower Coal-measures of Lancashire prove that the seed-like bodies described by Williamson under the name of *Cardiocarpon anomalum* were borne on strobili, agreeing with *Lepidostrobus*. Each megasporangium, which was seated on the upper surface of the sporophyll, became enclosed, when mature, in an integument springing from the tissue of the sporophyll-pedicel. The integument closed in over the top of the sporangium, leaving only a narrow crevice or micropyle, which differed in its elongated, slit-like form from the more or less tubular micropyle of an ordinary seed. Within the megasporangium four megaspores were produced, one of which occupied almost the whole of the sporangial cavity, while the other three remained small, and were evidently abortive. The integumented megasporangium, containing the single functional megaspore or embryo-sac, became detached, together with the remains of its sporophyll, from the cone. It appears to have been indehiscent, and presents close analogies with a true seed. In a male strobilus, probably of the same species as the specimens above described, the microsporangia were found to be provided with integuments, resembling those of the megasporangia, but more widely open.

It is proposed to give the generic name *Lepidocarpon* to this Lepidostroboid fructification.

The primary structure of certain Palæozoic stems referred to *Araucarioxylon*, by D. H. Scott, F.R.S. The Palæozoic forms of *Araucarioxylon* have been shown to belong in most cases to the stems of the extinct Gymnospermous order Cordaiteæ, which was in some respects intermediate between Cycadales and Coniferæ. The Cordaitean stems hitherto investigated resemble Coniferæ in the development of their wood, for the spiral first-formed tracheides were in contact with the pith, the whole of the wood, primary as well as secondary, having thus been developed in centrifugal order. The specimens of Lower Carboniferous age now illustrated are peculiar in possessing distinct strands of primary wood in the pith. In one, *Araucarioxylon fasciculare*, sp. nov., the pith is small, but the primary strands of xylem are of large size, attaining their maximum diameter when about to pass out towards a leaf. Their structure is mesarch, and they closely resemble the corresponding strands in *Lyginodendron Oldhamium*. The secondary wood has narrow medullary rays, and resembles that of an araucarian Conifer. The other species is identical with *Araucarioxylon antiquum* of Witham. The interest of the two species (described from specimens in Mr. Kidston's collection) consists in their affording a link between certain of the Cycadofilices and the Cordaiteæ.

On the structure and affinities of *Dipteris conjugata*, with notes on the geological history of the Dipteridineæ, by A. C. Seward, F.R.S., and Elizabeth Dale. The genus *Dipteris* is represented by four recent species : *D. conjugata*, Reinw. [= *Polypodium (Dipteris) Horsfieldii*, R. Br.], *D. Wallichii*, R. Br., *D. Lobbiana*, Hk., and *D. quinquefurcata*, Baker. Among Mesozoic ferns the genera *Protorhipis, Dictyophyllum* and *Camptopteris* afford examples of extinct types closely allied to *Dipteris*, and widely spread geographically during the Jurassic epoch.

The sporangial characters of *Dipteris* do not conform precisely to those typical of the Polypodiaceæ, and the anatomical features afford additional evidence in favour of placing *Dipteris* in a special subdivision of the leptosporangiate ferns.

The paper dealt with the structure of the stem, which possesses a single annular stele, the roots, leaves and sporangia of *Dipteris conjugata*, the comparison of the anatomical features with those of the Cyatheaceæ and other ferns, and concluded with an account of the geological and geographical range of such fossil ferns as may reasonably be placed in the family Dipteridineæ.

On the structure of the stem of *Angiopteris evecta*, Hoffm., by R. F. Shove, Girton College, Cambridge. This paper dealt with the anatomy of the stem and roots of a plant of *Angiopteris evecta* from Ceylon.

The steles of the stem are both mesarch and endarch in structure, but the protoxylem groups occupy for the most part a peripheral position. The earliest protoxylem appears along the inner edge of the steles, while the protophloem arises on the outer edge of each stele as a discontinuous arc of small and rather thick-walled elements. This arc of protophloem is never completed round the stele, but the next stage in the development of the tissues after the appearance of the protoxylem is the differentiation of large sieve-tubes external to the protophloem.

The conducting tissues of Bryophytes, by A. G. Tansley. The most important part of our present knowledge of these tissues is due to Haberlandt, who, in the Polytrichaceæ, distinguished a *hadrom* (*hydrom*) or water-conducting system from a *leptom* system, conducting plastic, especially nitrogenous substances.

In the present investigation the lignified strand of prosenchyma in the thallus of certain Liverworts was shown to be a hydrom strand, and its development was considered to be correlated to some extent with the localisation of the absorptive region of the thallus.

The rhizome of four species of *Polytrichum* was investigated, and was found to possess the distribution of tissues characteristic of the root of a vascular plant. The transition to the structure of the aërial stem was followed, and some new points in the

structure and course of the leaf-traces were observed ; new light was thrown also on the constitution of the Polytrichaceous stele, which is thought to consist of two regions distinct in function and by descent. An attempt was made to trace out the course of evolution of these conducting tissues in the Bryophytic series.

The origin of modern Cycads, by W. C. Worsdell. The author's conclusion is that the Cycads are descended directly from some cycado-filicinean type possessing the structure exhibited especially by such forms as the Medulloseæ and Lyginodendreæ, the chief point being that the *collaterally*-constructed one or more vascular cylinders of modern cycads have been derived from one or more *concentrically*-constructed cylinders of some cycado-filicinean form. Those characters in the modern plants which approximate most nearly to the primitive ancestral type are found in those parts of the plant where they would most naturally be expected, viz. :—The *axial* organs : the *primary node* or transitional region between stem and root, and the *flowering axis* ; the *foliar* organs : the *cotyledon*, the *sporophyll*, and the *integument* of the sporangium. The author discussed the evidence derived from an anatomical study of recent cycads, and dealt with certain fossil types which he regarded as supporting his conclusions.

CYTOLOGY, &c.

On the osmotic properties and their causes in the living plant and animal cell, by Prof. Overton. A very great number of experiments on the permeability of the living protoplasm of plant and animal cells has led to the conclusion that the general osmotic properties of the cell depend on a phenomenon of *elective solubility*, certain layers of the protoplasm being impregnated with a mixture of lecithin and cholesterin. All substances that are soluble in this mixture, and they include by far the greater number of organic compounds, being able to penetrate into the living cell. The rapidity of the passage of different compounds into the cell depends on their relative solubility in water and in a mixture of cholesterin and lecithin. A knowledge of the osmotic properties of the living protoplasm throws much light on the action of many poisons and other drugs.

Demonstration of the structure and attachment of the flagellum in *Euglena viridis*, by Harold Wager. The flagellum of *Euglena viridis* possesses a bifurcate base, which is attached to the wall of the excretory reservoir at the anterior end of the body (*Journ.* Linn. Soc. Zool. vol. xxvii. p. 463). As it passes to the exterior through the gullet, an enlargement occurs in the region of the eye-spot. This structure can be seen in very favourable cases in the living condition, but usually only after the action of reagents. The best reagents for this purpose are either a 1 per cent. solution of osmic acid or a 2 per cent. solution of bichromate of potash with a 1 per cent. solution of osmic acid. The structure may be obscured by small grains of paramylon, which sometimes accumulate at the anterior end of the body.

The behaviour of the nucleolus during karyokinesis in the root-apex of Phaseolus, by Harold Wager. From a study of the changes undergone by the nucleolus during karyokinesis in cells of the root-apex of *Phaseolus multiflorus*, the following chief results have been obtained.

(1) The nucleolus is the most conspicuous object in the nucleus of the young meristematic cells. The nuclear network forms a delicate peripheral layer only in the resting nucleus. (2) The nucleolus stains deeply in hæmatoxylin, the nuclear network slightly ; in safranin and gentian violet the nucleolus stains red, the nuclear network light blue. (3) In the resting condition of the nucleus the nucleolus is suspended to the nuclear network by delicate filaments. (4) The nucleolus often shows a vacuolar structure. (5) In the process of nuclear division the nucleolus first of all becomes irregular in shape, and the nucleolar substance appears to pass, by means of the connecting strands, into the nuclear network, which thereby becomes more prominent. (6) As the chromosomes are formed the nucleolus disappears, but a portion of the nucleolus is often visible in the equatorial plate. (7) The chromatic substance of the chromosomes appears to be derived almost entirely from the nucleolus. (8) As the daughter-nuclei are being formed the chromatic substance of the chromosomes runs together into small spheres, which ultimately fuse to form the single large nucleolus.

On double fertilisation in a dicotyledon, *Caltha palustris*, by Ethel N. Thomas. The polar nuclei of this plant unite before fertilisation, but that there is no absolutely fixed period is shown by the very different appearance of sacs in which polar fusion is taking place. The male generative nuclei, when first set free in the embryo-sac, are extremely small and heavily stained. Their chromatic substance is so densely aggregated as to render the spermatozoid to all appearance homogeneous. Of the two spermatozoids one passes to the middle of the sac and there *fertilises the definitive nucleus*; the other fertilises the nucleus of the oosphere. By the time the male generative nucleus or spermatozoid has reached the definitive nucleus, it has enlarged immensely, and shows a light spongy structure with scattered chromatin granules. The other spermatozoid increases very little in size, and always remains dark and dense.

When the spermatozoids leave the pollen-tube they are somewhat short and thick, and only slightly curved, but when the one has approached the definitive nucleus, it has the typical vermiform shape, with one or several coils.

THALLOPHYTA.

Germination of the zoospore in Laminariaceæ, by J. Lloyd Williams. The zoospore comes to rest and becomes spherical. The single chloroplast divides in two. A tube is produced, the spore-contents pass into it. At the end of the tube a swelling is formed, into which the contents migrate and are shut off from the empty spore-case and tube by a wall. This has been wrongly described by Areschoug in the case of *Dictyosiphon* as an instance of sexual fusion. In the enlargement, the chloroplasts multiply, and additional eyespots appear on several, which, however, disappear after a few days. The newly-separated cell now divides, and forms a branched protonema-like structure.

Notes on *Dictyota*, by J. Lloyd Williams. The factors concerned in the production of the fortnightly crops of sexual cells were discussed. Experiments on the liberation of antherozoids show the importance of bright light and cool temperature. *Dictyota* is particularly responsive to changes in the environment.

The nuclear changes in the unfertilised eggs are peculiar. The chromosomes are differentiated, a very irregular multipolar spindle is formed; this separates into a number of nuclei of various sizes, in which at first the chromosomes are scattered. These soon disappear, the nucleoli are formed, and the nuclei appear in the resting condition.

The *Azygospores* of *Entomophthora gloeospora*, by Prof. Vuillemin (Nancy) (communicated by Prof. Hartog). The genus *Entomophthora*, as seen in the two species *E. Delpiniana* and *E. gloeospora*, with its uninucleated segments, and *Empusa*, with its continuous hyphæ with scattered nuclei. The resting-spores of *Entomophthora* may be terminal, lateral or intercalary. The youngest spores contain a single nucleus, which undergoes a series of four successive binary divisions until there are sixteen; there may, however, be irregularities as regards the number of spores. In the next stage the nuclei approach so as to form eight pairs, and the two nuclei of each pair then fuse; this fusion is repeated until there are only two left. These last two may then fuse at once, so as to leave the now maturing azygospore with a single nucleus, or they may remain apart. This manner of development is interpreted as a case of true apogamy, and regarded as corresponding to the sexual process in *Basidiobolus*.

Fungi found in Ceylon growing upon scale-insects (*Coccidae* and *Aleurodidae*), by J. Parkin. Fungi associated with scale-insects have till recently been little studied. A few species have been mentioned from time to time as growing upon scales of dead coccids, but, till within the last few years, hardly any attention has been called to their probable parasitic character, or to the possibility of their being employed to check the ravages of scale-pests. Webber in 1897 pointed out for the first time the parasitic habit of certain species—five in all—of *Aschersonia* on scale-insects infesting the orange and other plants in Florida. Zimmermann (Java) in the following year gave a preliminary account of a fungus (*Cephalosporium*) attacking the green bug (*Lecanium viride*), so harmful to the coffee, and described how it may be artificially cultivated for infecting experiments.

The various kinds dealt with were referred to the following genera :—*Nectria, Torrubiella, Aschersonia, Cephalosporium, Verticillium, Microcera, Camptotrichum* (?).

Mr. Parkin drew attention to the wide distribution, especially in and near the tropics, of fungi infecting scale-insects, and referred to them as the true cause of death of the insects. The

paper was illustrated by a series of carefully-prepared specimens and drawings.

On the life-history of *Acrospeira mirabilis* (Berk. and Br.), by R. H. Biffen. Loose brown masses of the spores of this fungus are occasionally found in Spanish chestnuts. These spores are developed from the apices of hyphæ coiled into a spiral of, at the most, two turns, which becomes septate into three cells; the cell next below the apical one swells and becomes thick-walled, thus forming a chlamydospore. The coiled hypha may also develop into a spiral resembling the ascogonium of *Eurotium*, which, after investment by branches arising from its apex, breaks down into chlamydospores. In this way bodies very suggestive of the spore-masses of some of the Ustilagineæ are formed. Endoconidia are found in old cultures. Some evidence has also been obtained for the existence of an ascigerous stage.

On the structure of the root-nodules of *Alnus glutinosa*, by T. W. Woodhead. The nodules are traversed by a central strand of short, thick-walled fibres, with transverse pits in the walls. Surrounding this are 4–5 layers of cubical cells, rich in protoplasm, followed by a small-celled bulky cortex. On the outside of this is a phellogen, which produces a layer of cork several cells deep. The cortical cells are largely occupied by the organism which produces the nodule.

The organism is usually present as a globular sporangium at the end of a short hypha. Towards the base of the nodule are strands of cells occupied by disorganised contents indicating a previous tract of growth of the organism : this is succeeded by groups of cells filled with the organism in various stages. Towards the apex, and immediately behind the growing-point, the cells containing the sporangia are immediately followed by cells filled with fine hyphal filaments, which may be seen to penetrate the walls of the young adjacent cells.

A Gymnosporangium from China, by Prof. F. E. Weiss. This fungus was first observed by Dr. A. G. Parrott in the spring of 1899 in Lao-ho-kou, in North Central China. Its spore masses made their appearance in April after a few days' continuous rain on the branches of *Juniperus chinensis* in the form of bright yellow, gelatinous masses. The teleutospores are of the usual type, two-celled, tapering towards both ends and somewhat rounded at the apex. They possess eight germ-pores. What is in all probability the *Roestelia* stage of this fungus was observed during the summer on the leaves of the pear, *Pyrus sinensis*, Ldl. A tree of this species growing in proximity to the infected junipers was attacked by a fungus of the *Roestelia* type, producing typical æcidiospores.

In the appearance of its teleutospore masses this fungus appears most nearly related to *Gymnosporangium Sabinae* (Dicks), a widely distributed form occurring in Europe and in America, and to *Gymnosporangium Cunninghamianum* (Barclay), a Himalayan form, both of which have their *Roestelia* stage on a pear.

The biology and cytology of Pythium, by Prof. Trow. The species described by the author was cultivated from conidia and oospores found in rotten cress seedlings. The study of pure cultures led to the following among other conclusions :—(1) No zoospores are produced under any circumstances. (2) The species is new and ranks as the most highly developed of the genus. (3) The fertilisation-tube penetrates the wall of the oogonium at a spot prepared for it, passes through the periplasm and penetrates deeply into the egg. One male nucleus passes down the tube and enters the egg. The oosphere clothes itself with a delicate wall and increases in size. (4) The fusion of the male and female nuclei is delayed until a thick oospore wall has been developed. (5) The nuclei multiply by indirect division in the mycelium and sexual organs. The only nuclear fusion is that of the male and female nuclei in fertilisation.

Observations on Pythium, by M. Poirault and E. J. Butler. The authors examined seven species, two of which were underscribed forms. In two species, *Pythium gracile* and *P. intermedium*, sexual organs were observed for the first time. Klebs' results on the dependence of spore-formation in *Saprolegnia* on external conditions were carried a step further, it being shown that a given spore could be induced to develop zoospores or vegetative hyphæ on appropriate treatment. The authors conclude that *Pythium* represents a stage in the colonisation of the land, by saprolegniaceous ancestors resembling *Aphanomyces*. It is closely linked to the *Peronosporaceæ* through *Pythium intermedium*, which possesses chains of gonidia, suckers, and a thick-walled mycelium.

Observations on some Chytridineæ, by M. Poirault and E. J. Butler. Four undescribed forms occur parasitic on *Pythium*. Their life-history has been worked out by the authors. *Chytridium gregarium* was found on the eggs of the rotifer *Metopidia Lepadella*; the unknown resting-spores were discovered. Observations were made on *Olpidiopsis Saprolegniæ*. The infection takes place in the zoospore-stage of *Saprolegnia*, and is often multiple. Penetration takes place by a fine tube, through which the protoplasm of the parasitic zoospore enters the host, leaving behind an empty capsule.   A. C. S.

UNIVERSITY AND EDUCATIONAL INTELLIGENCE.

OXFORD.—The following lectures are announced for the present term : Prof. R. B. Clifton, Acoustics ; J. Walker, Double refraction ; R. E. Baynes, Elementary mechanics of solids and fluids, and Mathematical theory of heat ; Prof. W. Odling, Organic chemistry, metallic bodies ; W. W. Fisher, Inorganic chemistry (preliminary course) ; J. Watts, Organic chemistry (honours course) ; V. H. Veley, Physical chemistry ; J. E. Marsh, Stereochemistry ; A. G. Vernon-Harcourt, The subjects of the preliminary examination (chemistry) ; P. Elford, Mendeléeff's periodic system. Introduction and group I ; P. Elford, Chemists and their work ; A. F. Walden, Origin, meaning and use of chemical symbols, formulæ and equations (elementary course) ; Prof. W. J. Sollas, Physical geography ; Prof. W. J. Sollas, Jurassic fossils ; Prof. H A. Miers, Elementary crystallography ; H. L. Bowman, The metallic minerals ; Prof. W. F. R. Weldon, General course of morphology (Coelentera) ; J. W. Jenkinson, Elementary morphology ; E. S. Goodrich, Morphology of fishes ; R. W. T. Günther, Polyzoa and brachiopoda ; J. B. Thompson, Morphology of the ichthyopsida ; J. B. Thompson, Ichthyopsidan palæontology ; Prof. F. Gotch, General course of physiology, Part I. Animal physiology ; Prof. F. Gotch, Advanced course on muscle ; J. S. Haldane, Subjects of the final honour school (physiology) ; G. Mann, Histology : G. J. Burch, Physiological physics ; W. Ramsden, Introduction to physiological chemistry ; G. Mann, Practical histology ; W. Ramsden, Elementary physiological chemistry ; Prof. S. H. Vines, Short elementary course (revision) with practical work ; Prof. E. B. Tylor, Development of language, writing, arithmetic ; H. J. Mackinder, The historical geography of the British Islands ; H. J. Mackinder, The development of geographical ideas ; H. N. Dickson, The atmospheric circulation ; A. J. Herbertson, The geographic cycle ; Sir J. Burdon-Sanderson, General pathology ; Prof. A. Thomson, Anatomy of the nervous system ; Prof. H. H. Turner, Elementary mathematical astronomy ; Prof. A. E. H. Love, Gravitational attraction, and theory of the potential ; and the theory of sound ; Prof. E. B. Elliott, Theory of numbers ; and substitutions and resolvents ; Rev. F. J. Jervis-Smith, Dynamo and motor machinery and electrical testing ; G. F. Stout; Child psychology ; and the psychological development of the categories of subject, cause and end.

The electors to the newly instituted Wykeham Professorship of Physics will proceed to an election in November. Candidates are requested to send in their applications by October 24. The electoral body consists of the following :—The President of the Royal Society, Sir George Stokes, Prof. Esson, Prof. Odling, Mr. Hayes.

The Rev. E. C. Spicer, of New College, has been elected to the University Scholarship recently instituted in connection with the new School of Geography.

CAMBRIDGE.—The following is the speech delivered on October 11, by the Public Orator, Dr. Sandys, in presenting for the degree of Doctor in Science, *honoris causa*, Mr. Samuel Pierpont Langley, Keeper of the National Museum, Secretary of the Smithsonian Institution, and Director of the Astrophysical Observatory in Washington ; the inventor of the "bolometer" and the "aërodrome."

Trans aequor Atlanticum ad nos nuper advectus est vir scientiarum in provincia insignis, qui etiam de astronomia recentiore librum pulcherrimum conscripsit. In urbe quod reipublicae maximae transmarinae caput est, viri huiusce curae multa mandata sunt : primum museum maximum rerum naturae spoliis quam plurimis ornatum ; deinde institutum celeberrimum scientiae et augendae et divulgandae destinatum ; denique ars et specula quaedam stellarum lumini in partes suas distribuendo

dedicata. Luminis in spectro, ut aiunt, infra radios rubros radii alii qui oculorum aciem prorsus effugiunt, viri huiusce ingenio, instrumenti novi auxilio quod βολόμετρον nominavit, paulatim proditi et patefacti sunt. Nemo mirabitur virum stellarum observandarum amore tanto affectum, etiam e terra volandi desiderio ingenti esse commotum,—adeo ut, quasi alis novis adhibitis, plus quam trium milium pedum per spatium, etiam avium volatum aemulari potuerit. Fortasse aliquando, Icari sortem non veritus, etiam Horati praesagia illa sibi ipsi vindicabit :—

> "non usitata nec tenui ferar
> penna biformis per liquidum aethera."

Fortasse rerum terrestrium impatiens, rerum caelestium avidus, ausus erit e terris "volare
> sideris in numerum, atque alto succedere caelo."

THE Senate of the reorganised University of London is now complete, and is constituted as follows :—Chancellor—The Right Hon. the Earl of Kimberley. Vice-Chancellor—Sir H. E. Roscoe, F.R.S. Chairman of Convocation—Edward Henry Busk. *Crown Members*—The Hon. W. Pember Reeves, Sir H. E. Roscoe, F.R.S., Mrs. E. M. Sidgwick, Sir John Wolfe Wolfe-Barry, F.R.S. *Faculty Members*—Theology—The Rev. Principal Alfred Cave. Arts—Prof. M. J. M. Hill, F.R.S., Prof. W. Paton Ker, Miss Emily Penrose, Prof. G. C. Warr. Laws—Lord Davey, appointed by the Crown. Music—Sir Charles Hubert Hastings Parry. Medicine—Dr. J. R. Bradford, F.R.S., Dr. J. Kingston Fowler, Dr. E. C. Perry. Science—Sir Michael Foster, Sec.R.S., Dr. William D. Halliburton, F.R.S., Prof. William Ramsay, F.R.S., Prof. A. W. Rücker, F.R.S. Engineering—Prof. W. C. Unwin, F.R.S. Economics, &c.—Prof. W. A. S. Hewins. Royal College of Physicians—Dr. W. H. Allchin, Dr. P. H. Pye-Smith, F.R.S. Royal College of Surgeons—Dr. A. P. Gould, Dr. H. G. Howse. University College—Prof. G. C. Foster, F.R.S., Lord Reay. King's College—Lord Lister, P.R.S., the Rev. Principal A. Robertson, D.D. Lincoln's Inn—Lord Macnaghten. Inner Temple—Judge Sir Alfred Marten, Q.C. Middle Temple—Mr. C. M. Warmington, Q.C. Gray's Inn—Mr. C. A. Russell, Q.C. Incorporated Law Society—Mr. W. Godden, Mr. R. Pennington. Corporation of London—Dr. T. B. Crosby. London County Council—Dr. W. J. Collins, Mr. Sidney Webb. City and Guilds of London Institute—Sir Frederic Abel. *Convocation Members*—Arts—Dr. J. Bourne Benson, Dr. J. D. McClure, Dr. T. Lambert Mears, Mr. J. Fletcher Moulton, M.P., F.R.S., Dr. T. B. Napier, Sir A. K. Rollit, M.P. Laws—Mr. Justice Cozens-Hardy. Music—Mr. J. W. Sidebotham. Medicine—Dr. Thomas Barlow, Dr. J. F. Payne. Science—Mrs. Sophia Bryant, Prof. Frank Clowes, Dr. C. W. Kimmins, Dr. F. S. McAulay, Sir Philip Magnus, Prof. Silvanus Thompson, F.R.S.

THE installation of the Earl of Rosebery as Lord Rector of Glasgow University has been fixed to take place on November 16.

PROF. BRUNHES, professor of physics in the University of Dijon, has been appointed director of the observatory on the Puy-de-Dôme.

MR. HOLBROOK GASKELL has given 1000*l.* towards the building and equipment of a new physics laboratory for University College, Liverpool.

MAJOR R. H. FIRTH has been selected to succeed Colonel J. Lane Notter, R.A.M.C., as professor of Military Hygiene at the Army Military School, Netley.

A FELLOWSHIP of the annual value of 100*l.*, for three years, will be awarded at Newnham College, Cambridge, in June 1901. Applications from former students of the college should be sent by May 1 to the Principal, from whom further information may be obtained.

THE Essex Museum of Natural History will be opened at West Ham this evening (October 18) by the Countess of Warwick ; and the Municipal Technical Institute, which was destroyed by fire a few months ago, will be reopened by Mr. J. Passmore Edwards.

MR. T. GRAHAM YOUNG, son of the late Dr. James Young, F.R.S., has offered the sum of 10,000*l.* to the West of Scotland Technical College Building Fund, provided that certain conditions are fulfilled as to the site of the new College, the construction of a chemical department, and the completion of the building within five years of next January.

THE trustees of the College of the City of New York are, according to *Science*, considering the lengthening of the course to seven years. They have asked that the appropriation of 200,000 dollars made to the institution by the Board of Estimate and Apportionment be increased by 25,000 dollars.

THE Berlin University has, it is stated, decided to alter the conditions permitting foreigners to take the title of doctor of philosophy. Foreigners are only to be allowed to graduate if they hold certificates equivalent to that of the Humanistischer Gymnasium, Realgymnasium, or the Oberrealschule of the German Empire.

THE *Pioneer Mail*, Allahabad, states that a petition is about to be presented to the Lieutenant-Governor of Bengal asking that the Behar School of Engineering may be affiliated to the Calcutta University. "The one thing," the petitioners observe, "which has hampered the progress of the school in the past has been the uncertainty regarding its future. If Government will now settle the matter finally by raising the school to the status of a College, all obstacles in the way of the development of technical education in Behar will be for ever removed."

THE new Municipal Technical School which was opened by the Earl of Derby a few days ago, at Bootle, an enterprising

be developed by the teaching of the grammar of iron-work in the way indicated. In time it is hoped that this branch of the school will need extension and that it may then be possible to equip a mechanical laboratory provided with testing machines and demonstration appliances on a larger scale than can be used in an ordinary class room. As to the cost of carrying out this work under the new conditions, a sum of something like 700*l.* per annum will be required from the rates. Fees, grants and Imperial funds will in all probability contribute about 3,700*l.* per annum. Hitherto no rate has been levied for purposes of technical instruction, and the educational work has been entirely paid for from students' fees and Imperial grants.

SOCIETIES AND ACADEMIES.
LONDON.

Royal Society, June 21.—"On the Spectroscopic Examination of Colour produced by Simultaneous Contrast." By George J. Burch, M.A., Reading College, Reading. Communicated by Francis Gotch, F.R.S., Professor of Physiology, University of Oxford.

It is well known that a neutral grey looks blue-green against

The Municipal Technical School, Bootle.

borough on the borders of Liverpool, is shown in the accompanying illustration. Classes in science and technology have been held in Bootle since 1891, but the work has been very restricted on account of lack of adequate accommodation. The question of providing a special building for the work to be carried on and developed soon became a pressing one, and in 1897 a suitable site was obtained. In due course tenders were invited, and the handsome and convenient building here shown was erected at a cost of 22,065*l.* Four schools are to be accommodated in the building, viz. (1) the evening school, consisting of many science and commercial classes and some trade or technological classes; (2) a school of art, located in the upper rooms on the Balliol Road frontage; (3) a school of domestic economy placed in a special portion of the west wing; (4) a day-school for boys of twelve years of age and upwards, giving a course of instruction as nearly as possible on the lines of a good secondary or modern high school. The engineering department contains modern machines of workshop size, fitted up conveniently for demonstrating and teaching the essential workshop processes in engineering on a small but useful scale. Of course, a trade will not be taught, for commercial and industrial conditions cannot hold in a school; but care, accuracy, thought, possibly invention, may

a red ground and orange against a blue ground. This phenomenon may be spectroscopically investigated as follows:—

A square of red glass is inserted on one side of the central partition of an ordinary stereoscope, and a square of blue glass on the other. Over each eye-lens is fixed one of Thorp's replicas of Rowland's gratings having 15,000 lines to the inch. Two slits are held in a frame in front of the aperture by which light is usually admitted when using the stereoscope for opaque photographs. The spectra of the first order of these slits appear in the middle of the two glasses. In order to prevent direct admixture of the colours of each spectrum with those of the opposite background, two opaque squares of black material are cemented to each of the coloured glasses, so shaped as to appear of the exact size and position of the spectra. On looking through the stereoscope, two spectra are seen side by side in a field, the colour of which continually oscillates from red through purplish-grey to blue. That connected with the red glass shows little or no red, but a splendid green and an equally splendid violet; while that belonging to the blue glass has the red well developed, the green pale and dingy, and the blue almost absent. The effect of varying the nature of the blue screen is very instructive. With cobalt glass the red is not very bright, owing probably to the transmission of some red rays by the cobalt glass, but the addition of a film stained with Prussian blue, by which these rays are absorbed, greatly improves the red. On the other hand, a pale yellow film which cuts off the violet causes the violet of the spectrum on the blue ground to stand out brightly, while a purple film brings out the green, which, owing to the green light transmitted by ordinary cobalt glass, is generally a good deal enfeebled. In each case the contrast of the two spectra seen by different eyes is so well marked that the experiment seems likely to be of service in teaching.

The complementary colour to red is shown to consist, not of one simple colour-sensation, but of two at least, namely, green and violet, and in the author's case of blue also. Against a magenta background the complementary colour is seen to be spectral green. But in this case the physical stimulus is complex. On adding to the magenta a yellow glass, to cut out the violet, or using candle light, the violet reappears in the complementary spectrum, while if a blue glass is added instead, the violet vanishes, and red stands out brightly in the spectrum. It may be thus shown that the colour which has green for its complementary is not spectroscopically simple, and since the spectral elements of it have each a different and independent effect upon the spectrum of the complementary colour, the author concludes that the green sensation has no special connection with the red, or indeed with any single colour sensation.

MANCHESTER.

Literary and Philosophical Society, October 2.—Prof. Horace Lamb, F.R.S. (President), in the chair.—Mr. Thomas Thorp described a method of producing a spectrum-like band from a volumetric curve by the use of a photographic camera with a cylindrical lens, and also gave a brief account of the solar eclipse of May last as seen in Algiers.—A paper on plumbism in pottery workers was given by Mr. William Burton, who dealt with the subject from the technical point of view, showing how the plumbism was caused by the inhalation or swallowing of dust containing soluble lead compounds. It was pointed out that the abolition of lead from the glaze was not a practicable remedy, as leadless glazes suited to the conditions of the general run of English pottery are not yet within the reach of the manufacturers. Further, the abolition of lead from the glaze would not affect those cases which arise from the use of enamel or on glaze colour used in the form of dust or of spray, and no one has yet ventured to suggest that the use of lead fluxes for this purpose could be abolished. The existing remedies in the shape of mechanical means for dealing with the dust in such a way that it should neither be swallowed nor inhaled, the periodic medical examination of the workers, and the proposed further safeguards of converting all the lead used into compounds of considerably lower solubility than those now employed, were also treated of. The final opinion of the author was that the combination of all these safeguards would, within a reasonable time, render the operations of glazing and decorating pottery with substances containing lead compounds as free from risk to the operative as it was possible for any industrial occupation to be.

PARIS.

Academy of Sciences, October 8.—M. Maurice Lévy in the chair.—Note on the thirteenth conference of the International Geodetic Association, by M. Bouquet de la Grye. From the reports to the conference it appears that the whole of Europe and North America will be shortly completely covered by a network of triangles. Special observations have been carried out under the direction of the central office in several observatories to determine exactly the line of displacement of the poles. An arc of meridian is being measured in Spitsbergen by Swedish and Russian observers, and Great Britain has commenced the determination of an arc from the Cape to Alexandria, which will join on to the European network through Asia Minor.— Remarks by M. Guyon on the *Connaissance des Temps* for 1903.— Observations of the sun made at the Observatory of Lyons with the Brunner equatorial during the second quarter of 1900, by M. J. Guillaume. The results are summarised in three tables showing the number of sunspots, their distribution in latitude and the distribution of the faculae in latitude.—The total eclipse of the sun of May 28, 1900, observed at Elche (Spain), by M. Lebeuf. The results of the observations on the times of contact are given.—Researches on the inverse effect of the magnetic field which ought to produce movement in an electrified body, by M. V. Crémieu. The application by Lippmann of the principle of the conservation of energy to the experiments of

Rowland on electric convection shows that, reciprocally, magnetic variations ought to produce a movement of electrified bodies placed in the field. The experiments carried out by the author, a detailed description of which, with two diagrams, is given in the paper, show that the expected effect is not produced. The deflection upon the scale should according to the theory have been of the order of 100 to 140 mm. No measurable deflection occurred. Telegraphy without wires with relays. Inconvenience of the Guarini relays, by MM. Guarini and Poncelet.—On iron silicide $SiFe_2$, and on its presence in commercial ferrosilicons, by M. P. Lebeau. A mixture of iron and copper silicide is heated for several hours to the highest temperature of a wind furnace. The mass is extracted with 10 per cent. nitric acid, when a crystalline mass of iron silicide, Fe_2Si, is left behind, which is further purified by successive treatment with soda solution, nitric acid and water.—On a new pyrogenous product from tartaric acid, by M. L. J. Simon. An acid, $C_7H_8O_3$, was obtained, differing both in melting point and solubility from the acid of the same composition previously obtained from the same source by MM. Wislicenus and Stadnicki. Its potassium and silver salts were prepared.—Acetyl derivatives of cellulose and oxycellulose, by MM. Léo Vignon and F. Gerin. Cellulose acetylated by means of acetic anhydride and zinc chloride gives a tetra-acetyl derivative ; oxycellulose behaves similarly, and the derivative thus obtained clearly retains its aldehydic functions.—The Albien and Cenomanian of Hainaut, by M. Jules Cornet.

DIARY OF SOCIETIES.

TUESDAY, OCTOBER 23.
ROYAL PHOTOGRAPHIC SOCIETY, at 8.—A Demonstration of the Ozotype Printing Process : Thomas Manly.
FRIDAY, OCTOBER 26.
PHYSICAL SOCIETY, at 5.—Exhibition of Experiments illustrating certain Phenomena of Vision : Dr. Shelford Bidwell, F.R.S.—On the Concentration at the Electrode in a Solution, with special reference to the Liberation of Hydrogen by the Electrolysis of a Mixture of Copper Sulphate and Sulphuric Acid : Dr. J. S. Sand.—Electromotive Force and Osmotic Pressure ; Dr. R. A. Lehfeldt.
SATURDAY, OCTOBER 27.
ESSEX FIELD CLUB, at 6.30.—Contributions to the Pleistocene Geology of the Thames Valley. The Grays Thurrock Area, Part I. : Martin A. C. Hinton and A. S. Kennard.

CONTENTS.

THURSDAY, OCTOBER 25, 1900.

THE ENGLISH GAULT AND UPPER GREENSAND.

Memoirs of the Geological Survey of the United Kingdom. The Cretaceous Rocks of Britain. Vol. I. The Gault and Upper Greensand of England. By A. J. Jukes-Browne, B.A., F.G.S. With contributions by William Hill, F.G.S. Pp. xiv + 499. (Published by order of the Lords Commissioners of H.M. Treasury, 1900.)

TO review this book is no easy task. To select is difficult when details are so full ; to criticise requires one to have studied, at any rate, certain districts of Cretaceous deposits as thoroughly as Mr. Jukes-Browne, by whom it has been written. So, in regard to the latter, we reserve the right of private judgment only on one or two small matters. A slip of the pen on p. 411, line 47, substituting eastern for western, and the reference to the map at the bottom of the same page (to a less degree) will cause a passing perplexity to readers. But who can help occasionally nodding in passing five hundred pages through the press ? To have included the Cretaceous deposits of Scotland and Ireland, would, we think, have made the subject more complete, and not materially increased the number of pages. We think also that Neocomian is big enough and distinct enough, physically and palæontologically, to be released from its subject position of Lower Cretaceous. Surely its claims to systematic independence are as good as those of Oligocene. Yet that has been officially welcomed while Neocomian is slighted. Mr. Jukes-Browne has, however, coined a new, though more subordinate name—that of Selbornian—to include the Gault and Upper Greensand, from the place hallowed by the memory of Gilbert White. We confess to retaining something of primitive man's prejudices against strangers, especially in nomenclature, but must admit that its author makes out a very strong case for the novel term. Briefly stated it is this : During the last twenty years it has been gradually ascertained that the Gault and Upper Greensand in this country are not two distinct stages of the Cretaceous system, but that " the greater part of the Folkestone Gault and the greater part of the so-called Upper Greensand are correlative deposits formed at the same time in different parts of the same sea." As the author shows, uncertainties in regard to their usage prevent us from borrowing either Albien or Cenomanien from the French, and, as a comprehensive term is so much needed, we have no choice but to coin a new one.

In a series of chapters Mr. Jukes-Browne, aided by Mr. William Hill, whose services in the field as well as in microscopic investigations have been of the highest value, describes the Selbornian in different districts. First, however, by way of clearing the ground, comes an introductory chapter on the Upper Cretaceous system as a whole ; a second contains an historical account of the Chalk, Upper Greensand and Gault. For purposes of reference, this epitome of various opinions will be very useful, and has probably cost Mr. Jukes-Browne more

labour than any other chapter in the book, for tracking down misconceptions and errors is a longer business and less exciting than hunting a fox, while the prey is equally malodorous and worthless. A chapter follows on the value of zones in the Cretaceous system, and another with a general account of the Selbornian. The Lower Gault, the most persistently argillaceous member, is a little over 34 feet thick at Folkestone ; it increases to 90 feet at Devizes, is proved by borings to be 150 feet in Bucks, and in Bedfordshire may perhaps be still thicker ; thence it thins gradually towards the north-east, till at Roydon, in Norfolk, it is only 7 feet. Afterwards it disappears as a clay, being represented by part of the Red Chalk from Hunstanton to Speeton. It is separable into three zones—that of *Ammonites mammillatus* at the base, thin, sandy, perhaps sometimes absent ; then that of *A. interruptus ;* the third being that of *A. lautus.* The Upper Gault with part of the Upper Greensand, the zone of *A. rostratus,* is lithologically variable, consisting of marly clays in the south-east, in Bucks and in South Norfolk, it becomes more sandy elsewhere (as in the Isle of Wight, and especially in Dorset and Devon) while in most other parts it is largely composed of the rock known as malmstone. The Warminster Beds, or the zone of *Pecten asper,* arenaceous and containing chert, form the uppermost division of the Selbornian. All have their equivalents in the Red Chalk of Norfolk and the north-east. The third and second are missing in the neighbourhood of Cambridge ; while in the west the Haldon Beds begin and end a little later than those of Blackdown.

Details are given of the sections in different parts of England, including those pierced by borings under London and in the Eastern counties, with lists of fossils and information of economic value, while Mr. W. Hill supplies two very interesting chapters on the lithology of the Gault and Upper Greensand. From these and the palæontological data Mr. Jukes-Browne concludes that the Gault clays were probably deposited in a sea, increasing in depth from about 150 to 200 fathoms, the sands of the upper part indicating stronger currents, possibly without any shallowing. He also discusses the physical geography of the period and the direction from which the sediment came. Here perhaps he touches on questions too speculative for an official publication, which should be restricted as far as possible to facts and to such inferences as follow directly from them. In this matter, while accepting Mr. Jukes-Browne's general conclusions as to physical geography, we doubt, for that very reason, whether the mud can have been brought from the south-east. If the region in which it was deposited was then an elongated gulf, open in that direction to the sea, we are unable to understand how an inflowing current of any strength could have been produced. If the gulf resembled the Red Sea, the inset from without would be sufficient to balance evaporation but not strong enough to carry the mud very far ; while if it received important rivers, the flow and consequent supply of material would be from the opposite direction. Mr. Jukes-Browne, it is true, refers to the possibility of a return coast current from the north, but we fail to see how, in a gulf of this form, the "blind-cord" action could be set up enough to be a primary agent in the transportation of sediment.

Quitting this very debatable question, we heartily congratulate Mr. Jukes-Browne and the Survey on this instalment of a complete memoir on the Cretaceous rocks of England. The possession of a synoptic view of any one formation is a great boon to geologists, as it saves them from the labour of hunting through a number of separate Survey memoirs. In future it might perhaps be well to shorten those explanatory of the maps by reserving all broader question for volumes like the present one. In this we note with especial pleasure the inclusion of chapters written by a geologist without any official position as an indication that the Survey now welcomes external help. The "get-up" of the volume shows improvement, but there is room for more. The illustrations, for instance, suffer from the thinness of the paper, through which the type can be seen. This defect spoils an excellent outline sketch on p. 152. A few plates, however, are printed on separate paper, and yet the book is issued at a moderate price. Difficult as it notoriously is to overcome the love of saving a ha'porth of tar so characteristic of Treasury officials, we wish the Director-General still greater success in persuading them to come nearer to the level of the volumes issued by the Geological Survey of the United States. T. G. BONNEY.

THE PRINCIPLES OF PATENT LAW.

The Law and Practice relating to Letters Patent for Inventions. By R. W. Wallace and J. B. Williamson. Pp. lxv + 922. (London: W. Clowes and Sons, Ltd., 1900.)

THE subject of inventions is always an interesting one whether to the manufacturer or to the man of science, nearly all new improvements in commercial chemistry or mechanics being, in these days of vigorous competition, sought to be protected by Letters Patent. Whether the English system of patent law is entirely satisfactory in all its details is a matter on which there are many opinions, and the Committee now inquiring into this may possibly suggest some alterations being made.

As a guide to the existing state of the law up to the most recent decisions in the Courts, Messrs. Wallace and Williamson have issued this volume, which may be called a treatise rather than a text-book. The arrangement and division of the subject is clear; starting from the granting of the Letters Patent, the reader is carried on to what is required of an inventor to render the grant valid and up to the petition for extension. Very little attention is given to the past history of the law, only a few pages being devoted to the well-known cases of the early part of the seventeenth century and a few remarks made on the history of claims. On the difficult subject, in which every discoverer of some new process must be interested, of what is necessary to constitute a patentable invention, the authors have not attempted to lay down any definition of their own, but have devoted two chapters to a careful collection of the important decisions on the point. In fact, throughout the book

there is given the material for forming an opinion rather than definitions.

For those who have not a well-stocked library of law-books there is the very great advantage that this volume gives verbatim extracts of nearly every important case, and even for those who have the books at command it will often save them the trouble of hunting up the passages they most often need. For those professionally interested in the subject there is a full and accurate account of the various steps in an action for infringement, which is very clearly set out; and after 600 pages of text there are some 250 of appendices containing the various statutes, together with forms and precedents for almost every conceivable case.

The principles upon which the specifications and claims should be drafted are adequately dealt with, and the question of amendment is gone into. There is a short chapter on the procedure on petitions for compulsory licenses, which procedure, curiously enough, does not appear to have attracted the attention of those who would be likely to benefit by it until within the last two or three years.

The printing of the work has been well done in large type and on good paper, the headings to the various paragraphs being sufficiently clear. A conspicuous feature of the work is the full index and the table of cases, which gives with each case the date of the decision and the subject-matter of the patent decided on. The book may confidently be recommended to any one desiring a complete account of the principles upon which our system of patent law is founded, as well as to those who constantly require a trustworthy book of reference.

HISTORICAL CHEMISTRY.

Lectures on the History of the Development of Chemistry since the Time of Lavoisier. By Dr. A. Ladenburg. Translated from the second German edition by Leonard Dobbin, Ph.D., with additions and corrections by the author. Pp. xvi + 373. (Edinburgh: Published for the Alembic Club by W. F. Clay. London: Simpkin, Marshall, Hamilton, Kent and Co., Ltd., 1900.)

THE small knot of chemists in Edinburgh who constitute the Alembic Club have already earned the thanks of chemists by placing at the disposal of English readers their valuable reprints of important chemical memoirs, a series which it is hoped may run on long and, if possible, at an increasing rate. The present volume is a more ambitious undertaking, being the translation of a substantial work which has long enjoyed much favour in Germany as a lucid and not too bulky account of the development of modern chemistry.

English chemical literature is not rich in original historical writing, though there are at least one or two British chemists who may be looked to with confidence for the occasional production of a scholarly and readable contribution. Going back a long way, it may be said that Thomas Thomson's "History" is inferior to no book of the kind published since—in respect to literary style and readableness. But in those days the science was narrower, and it was well within the ability of one man to do justice to the whole subject. The exhaustive historical writings of Hermann Kopp, and the more recent contributions of Berthelot, leave little to be desired in completeness, and provide a repository of information invaluable for purpose of reference. This, however, is literature for the fully fledged chemist or chemical author.

The chemical student requires something different. The importance to him of attending to the historical aspect of chemistry is recognised by most teachers. It is indeed maintained by some that there is no other satisfactory way of approaching even the elements of chemistry, than by performing experiments in historical order. A Board School might be cited where the older boys are given the Alembic Club reprints, and asked to do the experiments as there described. Whatever may be thought of this, it cannot be denied that a study of chemical history is most important, not only for a clear grasp of the origin and growth of our present theories, but because of that more subtle influence on the mind and imagination which perhaps may be included in the much-abused word culture.

The full advantage of historical study is not to be obtained by the reading of such a work as the one under notice, but rather by the careful study of those original memoirs or books which will ever remain landmarks in the track of scientific progress. At the same time, a connected history is a useful and perhaps a necessary adjunct to these partial studies, and this want is met extremely well by the book under notice.

Prof. Ladenburg has cast his story in the form of lectures, and for the purpose in view this is a satisfactory arrangement. In tracing the history of chemistry from the time of Lavoisier to the present day a vast amount of material has, of course, to be dealt with ; and of the prodigious amount of reading and critical examination entailed upon the author there is abundant evidence both in the text and in the numerous references which are appended. As to the general balance of the book it may be said that the earlier part is fuller and more explanatory than the later. The account, for example, of the controversy between Berthollet and Proust is very clear and interesting, whilst the accounts of the controversies that raged later in the century in regard to fundamental questions of organic chemistry are much more compressed and difficult to follow. The last chapter of the book is little more than an enumeration of the chief chemical topics that have engaged attention during the past fifteen years.

However, looking at the book as a whole, it must be said that Prof. Ladenburg has produced a most useful history, extremely readable considering the inevitable compression, remarkably free from the bias of personal opinions, and giving a connected view of the progress of chemical science which will be of great benefit to students.

Dr. Dobbin has succeeded admirably in the arduous work of translating narrative German into narrative English. Here and there sentences are to be found which declare their origin ; but on the whole the English (or should one say British ?) flows smoothly, and there is a remarkable absence of typographical errors or mistakes of a more serious kind. Dr. Dobbin and the Alembic Club may certainly be congratulated on their latest contribution to chemical literature.　　　A. S.

OUR BOOK SHELF.

Untersuchungen über Mikrostrukturen des erstarrten Schwefels nebst Bemerkungen über Sublimation, Überschmelzung und Übersättigung des Schwefels und einiger anderer Körper. By O. Bütschli. Pp. iv +96 ; 4 plates. (Leipzig : W. Engelmann, 1900.)
Untersuchungen über die Mikrostruktur künstlicher und natürlicher Kieselsäuregallerten (Tabaschir, Hydrophan, Opal). By O. Bütschli. Pp. 287-348 ; 3 plates. (Reprinted from *Verhandl. d. Naturhist.-Med. Vereins zu Heidelberg*, N.F. Band vi. 1900.)

A PREVIOUS work by the professor of zoology at Heidelberg (" Untersuchungen über Strukturen," 1898), reviewed in NATURE (vol. lx. p. 124), dealt more especially with the microstructure of organic substances, comparing them with the supposed alveolar structure of protoplasm. In the first of the present pamphlets the author describes in minute detail his observations in the same direction made on inorganic substances, more particularly sulphur. Amongst the various globular and crystalline forms produced by the sublimation and subsequent transformations of sulphur, he describes some which have a radial or concentric arrangement of vacuities or air-spaces suggesting an alveolar structure. The subject is, however, treated throughout from a crystallographic rather than from a biological point of view, and much the same ground has been covered in a more concise and earlier paper by Dr. R. Brauns, the professor of mineralogy at Giessen (" Beobachtungen über die Krystallisation des Schwefels aus seinem Schmelzfluss," *Neues Jahrb. f. Mineralogie, &c.*, 1899, Beil.-Bd. xiii. pp. 39-89 ; 7 plates).

The second pamphlet describes with equal minuteness the appearances shown under the microscope by chips and thin sections of dried gelatinous silica, as well as of the natural forms of colloidal silica, tabasheer and opal (including hydrophane and precious opal), which are all very similar in their minute structure.

Both pamphlets are admirably illustrated with numerous well-prepared microphotographs.

The School Journey. A Means of Teaching Geography, Physiography and Elementary Science. By Joseph H. Cowham. With additional "Journeys" by G. G. Lewis and Thomas Crawshaw. Pp. 79. (London : Westminster School Book Depôt, 1900.)

FOR many years the study of geography at the Westminster Training College has been supplemented by an excursion from Croydon to Godstone, under the guidance of Mr. Cowham, the lecturer on education at the college, and the author of several excellent educational works. In this volume a description is given of the chief characteristics observable during the ramble ; and horizontal and vertical sections, as well as photographic illustrations, elucidate the physical geography of the district traversed. In addition, the book contains accounts of excursions to Greenwich and Woolwich, and along a river bank in Lancashire, contributed by two of Mr. Cowham's former pupils.

The book appears at the right psychological moment ; for the feeling that geography should, whenever possible, be made an outdoor study, is spreading, and every statement of experience is of value to teachers who want to improve methods of instruction in geography but are unable to see clearly how to carry out schemes which have been put on paper by persons who may not have given full consideration to ways and means. Here, however, we have notes upon actual excursions and how they were planned and performed, and with these before them, teachers should have no difficulty in arranging others if they have some knowledge of physical geography. The Geologists' Association and Prof. Seeley's Geological

Field Class provide teachers in London or the neighbourhood with exceptional opportunities for acquiring a knowledge of the significance of the geological structures and formations in the home counties, and Mr. Cowham's book will show them how the facts can usefully be applied to school excursions.

Air, Water and Food. By Ellen H. Richards and Alpheus G. Woodman. Pp. 226. (London: Chapman and Hall, Ltd. New York: John Wiley and Sons, 1900.)

OF the many volumes which have been written on these subjects, there are few which, in the opinion of the writer, can be more safely recommended to the student of sanitary science. Each of the three subjects is introduced and fairly discussed in language which is clear, trenchant and concise.

The authors are, moreover, no mere theorists, but describe the operations of the laboratory in a business-like fashion which leaves no doubt about their practical knowledge. The diagrams are more successful as illustrations than the photographs, in which, as frequently happens, the glass apparatus has such an ill-defined and ghost-like appearance as to be unrecognisable by the unprofessional eye. In other respects the book is well got up. J. B C.

Elementary Physics and Chemistry, ii., iii. By R. A. Gregory and A. T. Simmons. Second stage, pp. vi + 140; third stage, pp. vi + 114. (London: Macmillan and Co., Ltd., 1900.)

THESE two volumes complete a work of three parts, consisting of a course of experimental illustration of the elementary principles of chemistry and physics. The syllabus of subjects considered is based on the new Code issued by the Education Department, but the descriptions are by no means confined to it. The subject-matter is arranged in the form of a succession of separate graduated lessons, each consisting of description of apparatus, method of conducting experiment, results obtained and the reasons for them, short summary, and a set of exercises. The books thus arranged seem especially valuable to teachers having to give a comprehensive course in a definite number of lessons—in Evening Continuation Schools, for instance—as the whole work to be gone through may be at once divided into sections. Numerous excellent illustrations add considerably to the utility of the volumes. C. P. B.

Principes D'Hygiene Coloniale. Par Le Dr. Georges Treille. Pp. iv + 272. (Paris: Georges Carré et C. Naud, 1899.)

THIS useful little volume is addressed particularly to those who wish to inform themselves of the physical conditions of life in the tropics with a view to living there, and to those who have an indirect interest in tropical regions. The earlier portion of the volume deals with tropical climatology in general, and in particular with the climatology of the French colonies. A chapter is devoted to considering the action of the climate on bodily functions. The latter portion deals with public and domestic hygiene. In the discussion on European habitations in the tropics, one would have wished to see more stress laid on the importance of Europeans living apart from the natives—a custom which has been so universally adopted in India, and which no doubt accounts to a large extent for the comparative freedom of Europeans from malaria in that country.

We fully endorse the indictment of the use of alcohol specially in the form of absinthe, but we should have liked to see more information on measures to be taken to ensure a supply of good water for domestic purposes. C. B. S.

LETTERS TO THE EDITOR.

[The Editor does not hold himself responsible for opinions expressed by his correspondents. Neither can he undertake to return, or to correspond with the writers of, rejected manuscripts intended for this or any other part of NATURE. *No notice is taken of anonymous communications.]*

Genesis of the Vertebrate Column.

IN the review of "The Foundations of Zoology," by Prof. W. Keith Brooks, which was contained in the last number of NATURE, the reviewer quotes from him the following sentence :— "Herbert Spencer tells us that segmentation of the backbone is the inherited effect of fractures caused by bending."

Before the reader accepts this version of my view, he would do well to read §257 of "The Principles of Biology." The simplest expression of that view is contained in the criticism of Prof. Owen's "Theory of the Vertebrate Skeleton," originally published by me in the *British and Foreign Chirurgical Review* for 1858, and now appended to "The Principles of Biology." The sentence setting it forth runs thus :—

"The production of a higher, more powerful, more active creature of the same type, by whatever method it is conceived to have taken place, involved a change in the notochordal structure. Greater muscular endowments presupposed a firmer internal fulcrum—a less yielding central axis. On the other hand, for the central axis to have become firmer while remaining continuous, would have entailed a stiffness incompatible with the creature's movements. Hence, increasing density of the central axis necessarily went hand in hand with its segmentation; for strength, ossification was required; for flexibility, division into parts."

There is here no mention or thought of "fracture"—no implication of a dense part formed and then broken, but the implication of dense matter being deposited in successive separate portions, in such way as to fulfil the two requirements of strength and flexibility. HERBERT SPENCER.

Brighton, October 21.

Albinism and Natural Selection.

A CASE of partial albinism in fishes which has recently come under my notice is likely to be of general interest from the evidence it apparently affords of the value of the normal specific coloration of precaceous fishes, and of the serious disadvantage of conspicuous abnormalities.

A white-skinned specimen of the common hake (*Merluccius merluccius*, L.) was trawled in the Bristol Channel last week amongst a catch of normal hake, and was sent to me from Milford immediately on landing, owing to the fishermen's impression that it belonged to some rare species unknown to them.

It was, however, perfectly normal in all respects except its remarkable leanness and the absence of all pigmentation from the external skin and the inner lining of the buccal cavity and gill-covers. The pigmentation of the retina and peritoneum was normal.

In a normal hake there is a profuse black pigmentation over the upper part of the body, as well as over the inside of the mouth and gill-covers. The general appearance of a normal hake is consequently dusky; that of the abnormal specimen was white.

The lean and emaciated condition of the white hake was very striking, especially in the head region, where not only the bony ridges of the skull and cheeks projected sharply beneath the thin layer of skin, but even the lines of sculpture of the superficial bones were plainly recognisable. In a normal hake, of approximately equal length, with which I compared the specimen, these details were quite invisible, and the bony ridges were rounded off or hidden by the plumpness of the integument. In girth and weight the albino was far inferior to the normally pigmented fish. The albino measured 26¼ ins. in length to the base of the caudal fin, and 6¾ ins. in length of head (from snout to opercular spine). Its girth round the back of the head was 9 ins., and just behind the 10th anal finray 9½ ins. The normal hake measured 27½ ins. in length, and had the same length of head as the albino. Its girth in the same two regions was 10½ ins. and 10½ ins. respectively. The albino weighed 4 lb. 5 oz., the normal hake 5 lb. 9½ oz., both fish being gutted in the same way.

That is to say, although the length of the albino was only 4½ per cent. less than that of the normal hake, the deficiency in girth amounted to 11 per cent. and the deficiency in weight to 23 per cent.

The question arises whether the emaciation of body, and lack of pigmentation, should be regarded as results of some disease (which was not otherwise apparent) ; or whether the lean condition should be attributed to the insufficient nutrition of a predaceous fish whose stalking powers had been reduced by its conspicuous appearance.

The hake is a predaceous and nocturnal fish, which preys on mackerel, herring and other active fish, especially at night.

The bulk of evidence appears to favour the view that albinism in fishes is a congenital, and not an acquired character (*cf.* colour variation in flat fishes) ; and I am not aware that leanness of body is specially correlated with the albino condition.

Perhaps some of your readers could refer me to other records which would throw light on this case?

Plymouth, October 10.   WALTER GARSTANG.

Tenacity of Life of the Albatross.

SIR WILLIAM CORRY told me some time ago that on one of his steamships coming from New Zealand, an albatross, supposed to have been choked dead, kept in an ice box at a temperature which was always much below freezing point, was found to be alive at the end of fourteen days.   He has been kind enough to obtain for me the following statement in writing from Captain Reed.   Of course the birds mentioned in this statement could not really have been choked dead, but I think the facts are very interesting.   JOHN PERRY.

October 11.

THE bird referred to was supposed to be killed by being strangled with twine tied as tightly as possible round the neck. This twine was not removed.   The beak was closed and tied and the legs crossed behind the tail and.   It was then wrapped in an old meat cloth and put with three other birds in the return box at the end of the port snow trunk.   It remained there for certainly not less than ten days at a temperature of from zero when machine blowing on that side to 18° F. when blowing on the starboard side.   The snowboy complained that the bird "grunted" when he went near it with his lamp, and Mr. Coombes, the 1st Ref. Eng. brought it out.   When put down on the engine-room floor, it could move its neck about and open its beak, and the eyes were open and lifelike.   The lower half of the body and the legs were frozen hard.   The fastening on the beak had come off.   It was alive for two hours after being taken out, and was then strangled and put in the snowbox.

There was another bird treated in the same way, and hung up by the beak in the meat chamber for over four days, and then found to be alive and able to make a "grunting" noise. The temperature of the chamber was never higher than 4° F., and often 8° to 10° below zero.   Mr. Coombes, then 1st Ref. Eng., now in *Star of Australia*, and Mr. Boyes, then 2nd, now 1st Ref. Eng., both declare this to be quite true.

If opportunity offers on the passage home I will try how long it is possible for these birds to live in these low temperatures,

S.S. Star of New Zealand, Wellington,   WM. J. REED.
August 22.

The Peopling of Australia.

IN the issue of NATURE for October 4, Mr. J. Mathew has questioned the accuracy of certain observations upon the linguistic part of his book, "Eaglehawk and Crow," which were made by me at the request of Prof. A. C. Haddon, and included by him in the review of Mr. Mathew's book in NATURE for December 28, 1899.

I shall be glad if you will permit me to reply as briefly as possible to the complaints in Mr. Mathew's letter.

Mr. Mathew charges me with being "unnecessarily caustic" in my remarks on his theories, and with attending to "petty points " instead of the main issues.   To the former charge I must plead zeal for accuracy, and fear of the formation of hasty conclusions.   To the second I may be allowed to say, that as the whole of Mr. Mathew's theory (linguistically at least) is based upon the "petty points," their accuracy is vital to the whole structure.   Although on p. 44 of his book Mr. Mathew

disavows the fallacy that "likeness of words in sound and meaning is a proof of a common origin," he nevertheless adopts it in very many of his comparisons.   Take, for example, the Malay and Central Australian words on p. 59; the south-west Australian and New Guinea words on p. 72, the examples in his chapter on the Malay element in Australian, and the satisfaction expressed in his letter to NATURE at a comparison between Australian, Malay and New Hebridean, because the "terms for father and skin are the same."

My summary of his chapter on the Malay element in Australian is quoted by Mr. Mathew in his letter as "ridiculous nonsense."   I maintain that it is a perfectly fair summary of his actual words.   He states on p. 5 that "Malay refers generally to the people of that race to the north of Australia without distinguishing nationality," and on page 61 that the Malay invasion came from the north.   Speaking of the invaders, he says on page 61, "The straggling stream winds about here and there, touches the shore at various places and recoils back inwards," but when I state that the meaning of this is "wandering about the interior," he says the latter phrase is a "pure invention of Mr. Ray's."

Although Mr. Mathew declares in his letter that the Malays came from an indeterminable (though probably Sumatran) locality, all the Malay words in his proofs are those of the current literary or colloquial Malay, and several of them (*e.g.* tangan, gigi, kapala, bapa, rambut), are by no means the common words used by the Malayan peoples of the Archipelago. In two instances his words are incorrect : *kaka* is wrongly given *kakis* (p. 59), so as to agree with Australian words like *koko, kahbooja*, and '*duwan*' (p. 60), said to mean 'ear' is probably meant for *daun*, "leaf," which only means the "external ear," *i.e.* the 'leaf of the ear,' when conjoined with *telinga*.

That Mr. Mathew believes the Malay words were "scattered all over the island continent" plainly appears from his examples. He shows so-called Malay words on the coast of New South Wales, East Queensland, and the extreme east (p. 58); others across the centre of Australia from the Gulf of Carpentaria southward, and on Cloncurry River (p. 59), and others in West Australia (p. 60).

Mr. Mathew states that in the *Journal* of the Anthropological Institute for 1894–5, I have used languages as the basis of a classification of the New Guinea Islanders.   That is so, but my method is not comparable with Mr. Mathew's.   I showed that certain New Guinea languages (Motu, Keapara, &c.) should be called Melanesian because they agreed with the languages of the Melanesian islands, *almost entirely in grammar, and very largely in vocabulary*, and that others should be called non-Melanesian because they had *no agreements whatever* with the Melanesian. Can Mr. Mathew show by a similar grammatical and lexical comparison, that the Australian is related to any other group of languages?   With regard to terms like 'bapa' and 'mama' for 'father' and 'mother,' my argument was that no dependence can be placed on these words to show a connection of languages.   They are among the earliest vocables uttered by a human being, and in very many languages of the world have become appropriated to the earliest recognised human relationships.

This is not the time or place to reply to the somewhat contradictory propositions in Mr. Mathew's letter.   He wishes me to prove : (1) That words of 'mama' type are not adopted words in Malay ; (2) that they were not earlier in use in the East Indian Archipelago ; (3) that they are not more markedly Papuan than the 'bapa' type.   I may, however, be permitted to remark : (1) That words for father containing the syllable *ma* (of which *mama* is a reduplication) are the commonest in the vocabularies of the tribes of Borneo, Celebes, Philippines, &c., least subject to Malay influence, whilst words containing the syllable *ba* or *pa* are confined to the nearest connections of the Malay.   Hence the words of *ma* (or *mama*) type are the original, not adopted words, and (2) are necessarily the earlier in use.   Mr. Mathew's second proposition thus contradicts his first.   Also (3), the languages of the Papuans in West New Guinea have forms of *bapa* for 'father,' those in Central New Guinea have *babe* or *apai*.   One Papuan and all the Melanesian have forms of *ma* (*ama* or *tama*).

Mr. Mathew complains that I have not explained the New Guinea numerals.   Could I do this within the limits of a review? The convergence of Australian forms towards Cape York, stated by Mr. Mathew, does not necessarily imply that the words came from New Guinea, and his examples only show that the Saibai numeral *may be* connected with the Australian. On the opposite New Guinea coast the numerals for 'one' are very different to the Saibai *urapon.* They are : *naubi* (Morehead River), *ambior* (Wasi Kasa), *tarangesa* (Bugilai Tribe), *icpa* (Kunini Tribe), *atanok* (Dabu Tribe), *netat* (Murray Island), *nao* (Kiwai Island), *monon* (Purari Delta), *farakeka* (Papuan Gulf), *aia* (Yule Island). Mr. Mathew asks us to believe that the Kalkadoon numeral " *luadi* " (two) is a Melanesian numeral used in Australia 150 miles south of the Gulf of Carpentaria, when the only comparable form in Melanesia is the Duke of York Island *ruadi* (second). This is a flagrant example of the method adopted, though disavowed, by Mr. Mathew.

The point missed by Mr. Mathew in discussing the phrase " *ori kaisa*," which I called mongrel, is that he has no proof that ' *ori* ' and ' *uroi* ' are the same words. Considering that the Gulf tribes have a word (*uiva*) for 'cassowary' (the emu is not found), and that nowhere else in New Guinea is there a word similar to *ori*, meaning ' bird,' and also that it requires a Torres Straits word to give ' *ori* ' an Australian meaning, it is highly improbable that it can explain words like *waitch, wadgie, warritch* for " emu." The Saibai for ' big bird ' is ' *koi urui,*' for ' cassowary,' ' *tamu.*'

In his letter, Mr. Mathew objects to my calling comparisons of Australian with Malay and New Hebridean " absurd and misleading." He suggests, without any warrant, that I find the absurdity in analogies between Malay and New Hebridean, whereas I have directly affirmed the connection of Malay with Melanesian (including New Hebridean) in the *Journal* of the Polynesian Society for 1896. The real absurdity is that of supposing that there is a relationship between Australian, Malay and New Hebridean, because " the terms for father, skin are the same."

In conclusion, I must again express my regret that Mr. Mathew should regard my criticism of his work as ' mere fault-finding and ridicule.' I have studied these languages for many years (without postulating theories), and have much material yet unpublished. To point out the weakness of Mr. Mathew's argument, with regard to method and deductions is fair criticism, and should not lead to a charge of unsoundness and inaccurate knowledge. If, as Mr. Mathew states, his materials are imperfect, why found a theory upon them ?

218, Balfour Road, Ilford, Essex,   SIDNEY H. RÁY.
    October 14.

RECENT AND PROPOSED GEODETIC MEASUREMENTS.

IN the history of geodesy and the discussion of the problem of the figure of the earth, the measurement of three arcs of meridian stands out conspicuously. These are the Peruvian and Lapland arcs in the Northern Hemisphere, and that of Lacaille in the neighbourhood of the Cape of Good Hope. Those who took part in the original measures worthily distinguished themselves, but it is inevitable that with the demand for greater accuracy in geodetic inquiries the necessity for repeating the ancient measures should be acknowledged. Without experience and with imperfect instruments, it is remarkable that the old astronomers accomplished what they did. To determine the approximate figure of the earth, and to derive a fairly accurate value of the compression was no mean achievement. But Lacaille's arc of meridian soon fell into disrepute, and other measures have had to be substituted by Maclear and others. Svanberg repeated the original work of Clairant and his colleagues, in Lapland, soon after its completion, and within these last few years a new determination of the Peruvian arc has been imperatively called for. There is no doubt but that the French nation, who have so honourably distinguished themselves in the difficult task of geodetic measurement, will be able to undertake this work, and increase their scientific reputation. From 1734, when the enlightened Government of the day undertook the measurement of two of the aforementioned arcs, down to our own time, when we have seen the Mediterranean successfully bridged by a geodetic survey, and the interior of Africa connected in one unbroken chain with our own

Shetland Islands, French men of science have played a conspicuous part in all questions connected with the true figure of the earth. At a moment therefore when the re-measurement of the famous arc of Peru has been forced upon us, it would have been with a feeling akin to disappointment if we did not find the French nation eager to repeat the historic work with all the skill that long experience has suggested, and all the accuracy that modern science demands. The report of a committee of the French Academy of Sciences, however, assures us that the ardour displayed by the French in the past is no whit abated, and that though the sacrifice of time and money is considerable, these drawbacks will be cheerfully borne. For the arc in Lapland another has been substituted on the Island of Spitsbergen, and the necessary work of triangulation has been for some time quietly carried on under the auspices of the Russian and Swedish Governments ; while at the same time, the reports of Her Majesty's Astronomer at the Cape of Good Hope tell us what is being done in the way of supplementing Lacaille's arc in the Southern Hemisphere. With this activity before us, it seems a fitting opportunity to compare the aims and the motives that inspired those who inaugurated the earliest geodetic expeditions with those that will influence and guide the latest surveys in the same or similar regions.

In the middle of the last century a distinct issue was presented to the scientific mind. Then the general figure of the earth was an undetermined problem, and whether the polar or the equatorial axis was the longer was a vexed subject of controversy. It is amusing enough now to read of the disputes between the partisans of defective observations on the one hand, and the upholders of an incomplete theory on the other. To-day no such broad issue is before us, the differences are of a more subtle character, demanding great nicety of observation and more effective analysis. Then the true difficulties of the problem were scarcely apprehended. Only a few years earlier, Fernel had attempted to determine the length of a degree by counting the number of revolutions of his carriage-wheel between Paris and Amiens, a method which recalls the earliest attempts of Eratosthenes. Picard, it is true, had recognised the necessity of employing means of greater accuracy, and had taught the true principles of geodetic measurement ; but the methods pursued by his descendants, and the precautions taken to ensure accuracy and recognise the surface to which the measurements are referred, constitute almost a new science. For with accumulating materials and greater experience, it has become necessary to distinguish between three surfaces. First, the ellipsoid of revolution, which corresponds most nearly to the form of the earth ; secondly, the true geoid, that is to say, the surface of equilibrium of water at rest under the influence of centrifugal force, and the attraction of the earth's mass ; and, thirdly, a corrected geoid, differing but slightly from the true, in which it is attempted to eliminate the effect of unequal masses on the earth's surface and in its interior. Theory shows that the true length of the arc of meridian, measured on the corrected geoid, will be given in terms of the measured base, if the effect of local attractions has been correctly determined. It is precisely in overcoming the difficulty of correctly eliminating the effect of local attractions, and of reducing the length of the measured base to the level of the water surface of the geoid, that the measurements of this century will mainly differ from those of the last.

A preliminary survey of the district undertaken by Captains Maurain and Lacombe has shown that the country of Peru possesses peculiar difficulties for an accurate geodetic survey. The levels vary very considerably with the distance from the coast, while here and there mountains of a volcanic character rise to a height of 6000 metres. But for the interest attaching to a historic site, and the desirability of continuing measurements outside middle latitudes to which they have hitherto been almost entirely confined, a project involving so many hardships might well be abandoned. Not the least interesting part of the report of Captain Maurain is his description of the country in which the pioneers of the last century carried out their observations, often on the slopes of mountains rising abruptly to the height of some 3000 metres. Of this monumental work scarcely a vestige remains, and the original line of route can be traced only from the written records. The care with which the fundamental points in a triangulation are now marked was not appreciated in those early days, and even the pyramids constructed to commemorate the ends of the arc, and the successful

Fig. 1.—The shaded portion of the map shows the district in which the measurements will be made.

termination of the work, have been demolished by the Indians, who hoped to find buried treasure under the monuments, though the jealousy of the Spaniards has led to almost equally deplorable results.

But the enterprise of the French will not be contented with the simple re-measurement of the arc of Bouguer, which extended from the environs of Quito on the north to Cuenca, somewhat south of Guayaquil (Fig. 1). Captain Maurain's instructions led him to examine the country north and south of these stations, with the view of extending the arc to six degrees of latitude. In

neither direction does the task become more simple. Towards the north the two chains of the Cordilleras unite in a confused mass, bristling with numerous summits, rising to an altitude. only less than that of Chimborazo or Cotopaxi. On the southern side the country becomes more open, but covered with forests, and with a wet climate, suggestive of fever. Nevertheless, it seems possible to push northwards as far as Pasto in Columbia, and southwards to the Peruvian town of Sullana, an arc of six degrees, or about double that originally measured. The Finance Minister asks, as Finance Ministers will, whether it is absolutely necessary to carry the arc beyond the limits of the ancient survey, and the answer of the French Academicians, as might have been anticipated, is to insist on the maintenance of the whole scheme as contemplated in the preliminary examination.

The entire programme is vast, and worthy the best traditions of French science. Three bases of about eight or nine kilometres in length will be measured, one approximately central, and two of verification at the north and south ends respectively. The difference of level between the central base and the sea at Guayaquil, where tide gauges will be erected, will be determined with the greatest nicety. Throughout the arc fifty-two stations will be selected for observation, of which three will be fundamental, and the longitudes be determined by telegraph. Magnetic observations will be carried on as a matter of course, and in a country marked by so many mountain masses special care will be taken to determine their extent and density, with the view of eliminating the effect of local attraction. But, after all precautions, it is not impossible that the measures be made on a lengthened protuberance on the surface of the geoid, and that the curvature of the line of route should differ sensibly from that which would be found along a line nearer to the Pacific or further inland. To decide this point, two methods are proposed—one by means of pendulum observations, which will give the variation of gravity throughout the whole region ; the other, by determining the difference of geodetic and astronomical longitudes between a point on the coast and the observatory at Quito. The army of experts who have examined the plan and arranged the details assures us that no difficulty has been overlooked, and as a result an admirably equipped expedition will leave France next spring, to take up quarters on the equator, where four years' hard and anxious work awaits the members.

As already mentioned, towards the polar end of the quadrant the expedition under the Russians and Swedes has already made good progress. The arc measured by Maupertuis and Clariaut extended through only fifty-seven minutes in the latitude $66°$, and Svanberg attempted no greater distance than $1\frac{1}{2}°$. But the modern scheme includes an arc of more than $4°$ in length, in the much higher latitude of $77°-81°$. The general control is in the hands of H. Sergieffsky, and trained as he has been in the accurate school of Pulkowa, admirable results may be expected. The difficulties to be encountered are probably not less, though of a different kind to those that await the French in the tropics. Fifty stations will be occupied in the course of the triangulation, which compares satisfactorily with the fifty-two of the French scheme. Two base lines only will be measured, in which it is proposed to use Jäderin's steel tape line twenty metres long. Very little is known of the success that has attended this method, though its accuracy is vouched for by Dr. Döllen's careful examination, and the French prefer to rely upon the same apparatus that was used in the determination of the French meridian. It was expected that the field-work in Spitsbergen would be finished this summer, but we are still waiting information concerning the amount of progress that has been made.

Lacaille's arc of meridian, measured in 1752, and

practically the only measurement effected in the Southern Hemisphere, was long a subject of perplexity in all theoretical investigations of the figure of the earth, since the result indicated that the earth's surface was less curved in the Southern than in the Northern Hemisphere ; but though the verification of the arc was an urgent necessity, it was not till 1840 that Sir Thomas Maclear commenced the work that solved the difficulty. The apparently anomalous result offers a good instance of the effect of local attraction in disturbing astronomical latitudes. The astronomical amplitude of Lacaille's arc (1° 12') was proved to be very nearly correct, but a large local disturbance of the direction of gravity at the northern end caused a zenithal error of some 8″. Maclear enlarged the arc to nearly 5°, and here geodetic operations practically ceased, till the present astronomer, Sir David Gill, developed a scheme which dwarfs into insignificance all previous measures, and which, if it can be carried out, will prove of the utmost scientific value. He regards all that has been accomplished in South Africa as the first step in a chain of triangulation which, approximately traversing the 30th meridian of east longitude, shall extend continuously through Africa to the mouth of the Nile. He would make his chain follow, or rather precede, the line of that taken by Mr. Rhodes's transcontinental telegraph, proceeding northwards along Lake Tanganyika, through the region of the Lakes Albert and Victoria Nyanza, and along the Nile Valley. The definite survey of Egypt has not yet been undertaken, but commercial and political motives will doubtless soon bring this within the domain of practical science, and assist the onward progress of the scheme. Sir David Gill does not stop at the shores of the Mediterranean. Onwards, by an additional chain of triangles from Egypt along the coast of the Levant and through the isles of Greece, he would connect the African arc with the existing European systems, till it reached an amplitude of 105°. Sir David Gill puts before us a number of considerations by which such a scheme might be carried to a successful issue, without by any means minimising the difficulties which his experience teaches him stand in the way. It is not necessary to dwell on these obstacles, some of which are sufficiently obvious. It should, however, be remembered that no other meridian offers greater, if equal, facilities, or furnishes a better prospect for the realisation of this magnificent scheme. Sir David Gill has not confined his attention merely to the elaboration of schemes ; he has accomplished much good work himself, often with straitened means, and by his personal influence and indefatigable energy aided and encouraged others. Under his auspices a chain of triangles has been carried eastward from Cape Town to Port Elizabeth, whence two branches have been carried to the north, one ending at Kimberley, while the other, crossing Natal, stops for the present at Newcastle in the extreme north of the colony. Much exploring work, hardly inferior in point of accuracy, has been carried through Bechuanaland and northwards along the 20th meridian, marking the boundary between German and British South Africa. His latest report tells us that on the eastern side of the continent stations were occupied from Bongwe (Lat. 19° 51' S., Long. 30° 19' E.) to a point in 16° 30' S., and within sixty miles of the Zambesi. There the smoke from extensive grass fires compelled stoppage of field-work for the season, and his party retired to the observatory to occupy themselves in the work of reduction. The outbreak of the war and interruption by the Boers of telegraphic communication with Cape Town have for the moment delayed the determination of longitude of distant stations, but we may be sure that once the country has settled down to normal conditions, Sir David Gill will be ready to prosecute his scheme with the ardour that has ever characterised his undertakings.

RECENT ANTARCTIC BOOKS.[1]

THE long-continued neglect of Antarctic exploration has given place to a period of great activity. Several expeditions have, during the last five years, been hovering on the margin of the unknown, and penetrating within it a few steps farther than their predecessors. Great preparations are being made, for what ought to prove the most determined effort to explore and study those regions by means of simultaneous national expeditions from this country and from Germany, and the public will soon begin to ask where the Antarctic regions may be and why people wish to go there.

The forerunners of the inquiring public have hitherto been obliged to cull such information as may be obtainable by the tedious process of consulting the records of original voyages which have been "out of print" for a generation at least ; but in 1898 Dr. Karl Fricker came to their aid by producing his admirable compilation, "Antarktis," the introductory volume of Kirchhoff and Fitzner's "Bibliothek der Länderkunde." Of this book it is impossible to speak too highly, and we must congratulate the editors and publishers of the series on their choice of a beginning. We must congratulate the English publishers also on having the courage to show the British public how much better the results of British pluck and enterprise are appreciated in Germany than at home. We know nothing less pleasant to contemplate with regard to the long Antarctic record than the apathy of the British public and publishers alike to the pioneer work of Cook, the public-spirited enterprise of the Enderbys, and the great achievements of Ross. The fact that the account of Ross's Antarctic voyage has never appeared in a popular edition is a mystery in the face of the frequent inquiries for that delightful book.

Dr. Fricker passes lightly over the early history of the Antarctic regions, not mentioning, we may note, Rainaud's excellent historical summary, "Le Continent Austral," in which the myth of the Great South Land is traced to its source. He takes up the work of Cook in the eighteenth century, of the Russian expedition under Bellingshausen, of Weddell, Biscoe, Balleny and the other sealers sent out by Messrs. Enderby Brothers, and deals in a thoroughly satisfactory manner with the simultaneous expeditions at the dawn of the Victorian era, when the French under Dumont d'Urville, the Americans under Wilkes and the British under Ross competed in the investigation of the south polar seas. Probably it was wise to pass lightly over the acrid controversy between Ross and Wilkes, although the English reader might have liked to see how such Homeric heroes assailed each other in the pages of the *Athenaeum* half a century ago.

The history of recent voyages stops with the trip of the *Antarctic* to Victoria Land in 1895. This was inevitable in the case of Dr. Fricker's German edition ; but the translator might have endeavoured to convey, in a supplementary chapter, some idea of the great results which have been achieved since the first publication. It would only have been fair to the author to have given him the opportunity of revising his section on Bouvet Island, with which the detailed description of the various known lands of the Antarctic regions commences. The translator might at least have added a note to let the reader know that this interesting group, which was sought in vain by Cook and by Ross, was re-discovered by the *Valdivia* on November 25, 1898, during the first scientific voyage conducted by Germans to the edge of the Antarctic (see NATURE, vol. lx. p. 114).

[1] "The Antarctic Regions." by Dr. Karl Fricker. Pp. xii + 292. Maps and Plates. (London : Swan Sonnenschein and Co., Ltd., 1900.)
"Through the First Antarctic Night, 1898-99. A Narrative of the Voyage of the *Belgica* among newly-discovered lands and over an unknown sea about the South Pole." By Frederick A. Cook, M.D., Surgeon and Anthropologist of the Belgian Antarctic Expedition. Illustrated. Pp. xxiv + 478. (London : William Heinemann, 1900.)

The description of the various portions of land seen by Antarctic explorers is well done, and the critical remarks of the author appear to be judicious and likely to be of service to subsequent expeditions. Then follow accounts of the ice-conditions, on which Dr. Fricker has

Fig. 1.—Curious weather-worn iceberg (300 feet high). (From Dr. Cook's "First Antarctic Night.")

made himself an authority, climate and life. The book ends with a chapter on the future of Antarctic exploration, excellent when it was written, but now happily out of date.

Mr. A. Sonnenschein has translated the book very well indeed from the literary point of view. We could scarcely have believed it possible that a translation could be made so literal, and yet so free from constraint, as this one is. Still a scientific man cannot help noticing some slips in the rendering of technical expressions, and it may prove useful to the general reader to correct some of these. On p. 104, line 32, the translator interpolates "Mount" before "Erebus," not noticing that the author refers to the temporary position of the ship which was the mountain's godmother; similarly, on p. 117, the objectionable word "insects" is introduced after "coral" without Dr. Fricker's authority. On p. 120 "the lower parallel" scarcely conveys the idea "a great-circle course" which the author expressed. In several places the geological *dip* of rocks is rendered by *slope*, a totally different thing. On p. 175 "layers of secondary formation" suggests Mesozoic rocks, but drift, without regard to the geological character of the stones, is the true meaning. On p. 176 "precipitate rock" should be "sedimentary rock." In several places the word *schären* is translated "dunes," but it really refers to skerries or rocky islets like those of the Skärgård of Sweden. The phrase "relative moisture" is used throughout instead of the familiar "relative humidity." On p. 248 the translator suggests the use of the German word *firn* in English; but it seems to us that the French equivalent *névé* has received too general currency to be

displaced from English writings. On p. 261 the word "Translator" has been inadvertently added to one of the author's footnotes.

The editorial note prefixed to the English edition is not very satisfactory. It is gratifying news, which we have not seen before, that Belgium is actively fitting out an expedition for Antarctic exploration ; but the statement that a Belgian expedition was sent out in 1897 should have been supplemented by the fact that it returned in 1899 with rich results. The *Valdivia* expedition is not noticed, although Mr. Borchgrevink's return properly finds a place. It would have been useful if the numerous recent papers on Antarctic exploration in English had been added to the bibliography, and if the efforts of Sir Clements Markham and the councils of the Royal Society and the Royal Geographical Society in promoting the British Antarctic Expedition had been specifically referred to.

Dr. Cook is the first of the staff of the *Belgica* to place his experiences on record in book form, and his description is intended for the general rather than the scientific reader. Its great value lies in the frankness with which the subjective side of exploration in the polar regions is dealt with, and in the professional observations on the health of the explorers. It will be remembered that the *Belgica*, after surveying part of the coasts of the channel which continues Bransfield Strait to the south between 64° and 65° S., sailed west and south, and wintered in the Antarctic pack, where for thirteen months the ship was fast in the ice. The claims as to geographical discovery, and the results of the scientific observations, may be left for

Fig. 2.—Making soundings. (From Dr. Cook's "First Antarctic Night.")

discussion when the official report of the expedition is published. Dr. Cook says very little about the leader, M. de Gerlache, whose resolution to push as far as possible to the south does not seem to have met with the approval of his subordinates, and it is noticeable that de Gerlache's portrait is not given in the admirable series showing the officers and scientific staff before and after their experiences.

The preliminaries of the expedition when one might almost think time was wasted in Tierra del Fuego, are described in considerable detail ; but the interest of the reader will be mainly attracted by the description of the first winter night (a night of seventy days) ever lived through by human beings in the Antarctic regions. It is described with a restrained realism that suggests many thoughts. We do not admire the author's style in such a passage as—"Even the sailors cannot resist the temptation to stand still and drink, with awe-inspiring amazement, the strange wine of action which hangs over the mysterious whiteness of the new world of ice" ; but when he comes to deal with the details of every-day monotony in the narrow limits of the lonely ship, the narrative acquires an intensity of interest which the simplest and most correct expression could hardly increase. The efforts of the scientific staff to carry on observations in most unfavourable conditions deserve the greatest praise.

Dr. Cook attributes the terrible depression of spirits and the circulatory troubles which affected every one on board the *Belgica* to the absence of sunlight and the monotony of the food. He never mentions scurvy ; but the symptoms described read not unlike the incipient stages of that disease. With regard to food, he raises a strong protest against essences and "artificial" foods of every kind. However nourishing these may be, their softness and want of flavour excite repulsion. Something with a taste, and tough enough for the teeth to have some work, was what the officers of the *Belgica* sighed for. Of all the foods on board, the Norwegian *Fiskeballar*, or "Fiskabolla," as it is written, were the objects of the heartiest detestation. Either the supply must have been of inferior quality or the abundance produced disgust, for only a few weeks ago we heard a person of intelligence declare spontaneously, on first tasting this delicacy, that with a supply of fiskeballar he could face a polar winter with equanimity. Sugar and milk ran short, and their loss was very severely felt. The experience of the *Belgica* should be very carefully considered by those responsible for victualling the new Antarctic expeditions, and compared with that of the *Fram*. Dr. Cook, by the way, throws doubt on the perfect health and general serenity of Dr. Nansen's expedition ; but it appears possible that with a small company of one nationality, personally selected by the leader, and living together, the chance of harmony is greater than with a larger number divided into cabin and forecastle, composed of five nationalities, and speaking as many languages.

Both the books which we have brought together in this review are good, splendidly illustrated, and full of interest ; but each would have been better of careful revision. Dr. Cook is unhappy with his proper names ; we note *Grand* (for Gand), *Recluse* (for Reclus), Bismark, Monacho, Bellany (for Balleny), Jessup (for Jesup), and there is also carbon diolide, all of which are wrong. In both works the comparison of temperatures on the centigrade and Fahrenheit scales is sometimes at fault, and in one between the hours of 4 a.m. on Sunday and 8 a.m. on Monday several gentlemen succeeded in obtaining thirty-six hours of continuous sleep. HUGH ROBERT MILL.

NOTES.

LORD KELVIN proposes to give a valedictory address to the London Mathematical Society on November 8. The subject will be "The Transmission of Force through a Solid."

THINGS scientific are beginning to move in Egypt a little. A notice has been published in the official journal that on and after September 1 universal time will be adopted in Egypt,

and the noonday signal given at mean noon of the 30th Meridian East of Greenwich, *i.e.* East European time. The Ports and Lights Administration have also notified that the time balls at Alexandria and Port Said will on and after October 1 be dropped according to the same time, and not local time as heretofore. At present these time balls are dropped by local arrangements, but before the end of the year the midday signal ball at each place will be dropped automatically by electric signal from Abbassia Observatory. Regarding meteorology, there are now eight stations between Alexandria and Khartum forwarding daily telegraphic weather reports, and these will be increased shortly. Abbassia has now a complete self-registering equipment, and hourly observations for 1900 will be published.

MR. J. S. BUDGETT, of Trinity College, Cambridge, who, it will be remembered, accompanied Mr. Graham Kerr on his journey in search of *Lepidosiren*, and who last year spent several months investigating the zoology of the Gambia region, has just returned to England from a second expedition to that river. Mr. Budgett's main object was to obtain material for studying the development of the Crossopterygian fish *Polypterus*. In his first expedition he obtained eggs and larvæ which were said to be those of this fish, but which, as it turned out, belonged apparently to a Teleost. Mr. Budgett has in his recent expedition failed to obtain the Polypterus material, but he is to a certain extent compensated for this by having obtained a mass of other embryological material which appears to be of great interest. Amongst these is a practically complete series of eggs and larvæ of the Dipnoan *Protopterus* whose developmental history had hitherto remained quite unknown. It is an interesting fact that the developmental stages of all three surviving members of the important group Dipnoi—*Ceratodus, Lepidosiren* and *Protopterus*, belonging to Queensland, South America and Africa respectively—owe their discovery and first observation to workers of the Cambridge school of Zoology.

At the meeting of the Röntgen Society on November 1, Dr. J. B. Macintyre will deliver his presidential address.

LIEUT. C. LECOINTE, who was second in command of the Belgian Antarctic Expedition, has been appointed director of the astronomical work at Brussels Observatory, in succession to M. Lagrange, retired.

A REUTER correspondent at Friedrichshafen describes another ascent made with Count Zeppelin's air-ship on October 17. The balloon remained for three-quarters of an hour at an elevation of 600 metres, and, after carrying out a number of successful steering manœuvres, alighted safely on the lake shortly before 6 o'clock, half a mile from Manzell. Herr Eugen Wolf, who took part in the ascent, has given the following account of his experience :—"The trial lasted an hour and twenty minutes. The start upwards was first-rate. The air-ship moved at an almost unvaried height of 300 metres and went against the wind. All the steering tests proved the efficacy of the new gear, and the air-ship satisfactorily answered the movements of the steering apparatus. The horizontal stability of the vessel was wonderful. Any list was easily counteracted by shifting the sliding weight. The speed of the air-ship was such that, when going against the wind, it outstripped the motor boats on the lake. In still air its own speed was at least eight metres per second. We descended at full speed in the direction of the air-ship's shed rather faster than we expected, owing to an as yet unexplained escape of the whole of the gas in one of the balloons in the forward part of the ship. No damage of any importance was sustained in the descent. The King and Queen of Württemberg and Princess Maria Theresa of Bavaria watched the trial on private steamers."

SIR HENRY DYKE ACLAND, whose death we recorded last week will probably be remembered more on account of the influence he exerted on behalf of scientific education at Oxford than for his contributions to natural knowledge. He was born in 1815, and graduated in medicine in the University of Oxford in 1846, having previously been appointed Lee's Reader in Anatomy. He became Radcliffe Librarian in 1851, and Regius Professor of Medicine in 1857, which post he held until the year 1894. He was a member of the General Medical Council from 1854 to 1874, and president in the years 1874-1887, when Sir Richard Quain succeeded him. Referring to his work on the Council, the *Lancet* says it was invaluable, and as he was likewise a member of the Medical Education Committee of the Hebdomadal Council of the University of Oxford, his influence on the scope and direction of the course of studies of a medical student was very great indeed, and was invariably directed towards the enlargement of the scope of scientific training. Not only did he use his influence for the good of the medical profession in his own country, but he extended his interests to foreign countries, and in 1879 sent an eloquent and encouraging letter to the authorities of the Johns Hopkins University, urging his readers to higher things and to the raising of the standard of medical education. He always placed the greatest stress upon general culture as a necessary qualification for the successful medical man, and being himself of very wide interests and a man of science, displayed an excellent example of a scientific and scholarly physician. In 1869 he was appointed a member of the Commission to inquire into the sanitary laws of England and Wales, and did valuable work in connection with it. He was the author of several works on medical and scientific subjects, including an important memoir on the visitation of cholera in Oxford in 1854, and another on village health and village life written in 1884 for the International Health Exhibition.

A LITTLE more than a year ago the attention of the Council of the Manchester Literary and Philosophical Society was directed to the fact ¡that Dalton's tomb in Ardwick cemetery, Manchester, was in a very bad condition, owing to neglect. A committee was appointed to take steps to put the monument in a thorough state of repair, and there was no difficulty in obtaining subscriptions for this purpose. A full-page illustration of the tomb in its restored condition appears in the latest number of the *Memoirs and Proceedings* of the Society.

REFERRING to the age of the big trees of California, Prof. C. E. Bessey records in *Science* that he once counted with much care the rings of growth of the tree of which the stump constitutes the floor of the so-called dancing pavilion. This count was made from circumference to centre, and every ring in all that distance was counted, no estimates or guesses being made. The result was that 1147 rings were counted, and accordingly it is safe to say that this tree, which was fully 24 or 25 feet in diameter, and considerably more than 300 feet in height, acquired these dimensions in eleven hundred and forty-seven years. Prof. Bessey doubts whether any of the existing trees approach the age of two thousand years.

A DESCRIPTION of the condition of gases, materials and food in a mine which had been tightly closed for fifteen months was given by Mr. F. G. Meacham at the recent meeting of the Institution of Mining Engineers. When the mine was reopened, the air was analysed and was found to consist of 84 per cent. of nitrogen, 12 per cent. of fire-damp, and 4 per cent. of carbon dioxide. The gases were greatly compressed, and it is estimated that about 1,500,000 cubic feet escaped from the first bore hole in twenty-four hours. When the mine was entered, it was found that the gases had had no deleterious effect upon the food, or the materials left in the mine ; in fact, everything left in the mine was found practically undamaged. Bread had become as dry as

biscuit, cooked bacon was as fresh as when left, and water in the horses' tubs had not evaporated, although surrounded by perfectly dry coal-dust. Previous to the fire, oatmeal was supplied to men working in hot places to mix with their drinking-water, and this was found to be as sweet as when sent down the pit. The rails and ropes were not rusted. Men's clothing was dry, and in practically the same condition as when left. In the stables, the chaff was unimpaired, and the horses readily ate it. The timber in the pit did not seem to have undergone any change whatever. In the three months that had elapsed since the reopening of the mine, greater decay had taken place than during the fifteen months that the pit was closed.

MALARIA is not the only disease which is propagated by mosquitoes. In the *Atti dei Lincei*, ix. 5, Prof. Grassi and G. Noè describe observations on the transmission of the filariæ of the blood by mosquito bites. The same species, *Anopheles claviger*, which is mainly responsible for the dissemination of malaria, also plays the part of host to *Filaria immitis*. The present investigation deals with the mode of exit adopted by the filariæ in passing from the mosquito to the punctured animal, and it would appear that the parasites make their escape by means of a rupture in the integument of the labium. In the succeeding part of the same journal, Prof. Grassi describes experiments carried out by a committee with the assistance both of the Italian Government and of the Mediterranean Railway Company, with a view to the prevention and cure of malaria in infected districts. The experiments were carried out in the plains about Pæstum, which have long been known as a hot-bed of malaria ("malaricissima" is the epithet Grassi applies to the region), and fell into two categories, namely, cure of the disease by the use of quinine, and protection from the bites of *Anopheles claviger* by the use of wire gauze as a covering for windows, doors and even chimneys of houses, the inhabitants of which were required to remain indoors from before sunset till after sunrise, or to go about covered with veils at night. By thus preventing mosquito bites, it was found that the malarial regions could be safely inhabited even at the season when the fever was at its height, and under such conditions the district might be made as healthy as any part of Italy.

IN opening the recent International Aeronautical Congress at the Meudon Observatory, the president, M. Janssen, rapidly and very eloquently reviewed the most important points of the progress made since the meeting of the last congress held at Paris in 1889. During the interval, progress has been considerable in all directions, and new and important questions have been dealt with. The military authorities of several of the European countries have rendered much assistance in allowing their balloons and requisite appliances to be used in scientific investigations. In Germany alone no less than seventy-five ascents have been made during the last five years, the results of which have recently been discussed in a valuable work by MM. Assmann and Berson. Since the last congress in 1889, M. Le Monnier's idea of employing unmanned balloons has been realised ; the success of these ascents and the results obtained by their means, notably in the investigations of MM. Violle and Cailletet, have given rise to the creation of the International Aeronautical Committee, which recently met in Paris under the presidency of Dr. Hergesell. M. Janssen also referred to the important results obtained from kite observations, especially by Mr. Rotch and M. L. Teisserenc de Bort. At the Berlin Meteorological Office a new service has been established for experiments, both with kites and balloons. The use of balloons for astronomical observations was also discussed, and recommended for observing the Leonid showers in November next. This method was successfully used under M. Janssen's directions by M. Hausky, in 1898, and was adopted by other countries in the following year.

THE U.S. pilot chart of the North Atlantic Ocean for October shows the track of the Galveston hurricane. The storm was first noted to the east of Martinique on August 30. Next morning its centre passed slightly to the northward of Antigua, where the barometer fell to 29·84 inches ; it traversed the southern portion of Haiti on September 1, and reached the southern part of Cuba on the 3rd. The barometric depression, which had been quite shallow, began to deepen over western Cuba, where the barometer read 29·79 inches on September 5. To the west of southern Florida the storm increased rapidly in area and strength, a reading of 28·10 inches and gales of hurricane force being noted on September 7 in lat. 26° 40′ N., long. 90° W. The storm-centre passed slightly south ward of Galveston on September 8. The destruction of life and property at and near this city was unprecedented in the history of West India hurricanes. The strength of the storm decreased rapidly to the northward of Galveston, again increasing in the region of the Great Lakes, Newfoundland and the Grand Banks, where it attained great violence, force 12 being frequently reported. The storm moved to the north of the 60th parallel in about 20° W. on September 16. The recurving so far westward, long. 98°, is quite unusual. Before recurvature, the storm moved in a W.N.W. direction, and after recurving it took an E.N.E. course, its progressive movement increasing greatly in velocity.

NATURALISTS will read with much interest a paper by Mr. R. Hall in the October number of the *Zoologist*, describing his experiences among the elephant seals of Kerguelen Island. The visit took place during the winter of 1897–98, when Mr. Hall found these huge animals in great numbers. He believes that they arrive in August on the island, whence, after breeding, they depart in February or March. A large male may measure as much as 20½ feet in length, and the weight of many of the animals is estimated at between two and six tons. The finest herd seen included a couple of dozen males averaging about 19 feet in length. In disposition these seals are sluggish and peaceful, although when attacked many of them will show fight. On several days from sixty to seventy were killed, but forty *per diem* was considered a good average. It is to be hoped that steps will be taken by Government to prevent the extermination of these remarkable seals. Mr. Hall gives a characteristic photograph of a group on shore. In the same journal, Mr. E. Selous brings to a close his diary of the habits of the thicknee, or great plover (*Œdicnemus crepitans*), in the course of which he notices that these birds indulge in dances comparable to those so graphically described by Mr. W. H. Hudson in the case of an Argentine plover.

THE latest issue (vol. xiii. pt. 2), of the *Journal* of the College of Science at Tokyo, contains a coloured plate and description of a gigantic and gorgeously coloured medusiform hydroid recently captured in deep water off Misaki. It is identified by its describer, Mr. Miyajima, with a form obtained in Japanese waters during the *Challenger* expedition, and named by Prof. Allman *Monocaulus imperator*, the generic title being now changed to Branchiocerianthus. There are, however, certain differences from the type-specimen of the latter, and other examples are much needed in order to determine the value of these variations.

THE Yorkshire College, Leeds, together with the conjoint Agricultural Council of the East and West Ridings, have just published a pamphlet on sheep-breeding experiments in the county, forming No. 13 of their series. It is a common custom in Yorkshire to cross ordinary ewes with pedigree rams of other breeds ; and the object of the experiments has been to ascertain whether such crosses are profitable, and which are the best. The results are tabulated in the pamphlet.

THE October number of the *Biologisches Centralblatt* includes a paper by Herr Stempell on the formation and growth of the shells of molluscs ; and another, by Herr Wesenberg-Lund, on the relation between the fresh-water plankton and the specific gravity of the water in which it occurs.

WE have received from the Trustees of the Indian Museum, Calcutta, a fasciculus of the "Illustrations of the Zoology of the R.I.M.S. *Investigator*," containing the plates to Fishes (Part vii.) and Crustacea (Part viii.), and also the index to Part i. (1892–1900).

THE U.S. Department of Agriculture, in *Bulletin* No. 24, has just issued a list of works on North American entomology, compiled by Mr. N. Banks. With the exception of a few dealing with the general subjects, the various memoirs are catalogued under the headings of the different groups to which they refer.

WE have received from Prof. Jamshedje Edalji a paper on "Reciprocally related figures and the principle of continuity," which is remarkable as a collection of exercises in polar reciprocation. It contains reciprocal theorems corresponding to the properties of the circle contained in Euclid's Third Book, as also to many of the exercises in Todhunter's Euclid.

IN the *Berichte der naturforschenden Gesellschaft* (Freiburg i. Br), Dr. Otto Berg discusses the significance of kathodic rays in connection with the mechanism of discharge. In connection with the heating effects produced when kathodic rays fall on a solid body, experiments with a thermo-element show that (1) for given potential the heat produced is proportional to the quantity of electricity carried over ; (2) the ratio of the quantities of heat and electricity decreases as the potential increases. The same journal also contains a paper by Dr. F. Himstedt on observations with Becquerel and Röntgen rays. Dr. Himstedt has observed no action of radium on a coherer, but has found a noticeable reduction of resistance of a selenium cell due to these rays. A similar diminution of resistance amounting to as much as 50 per cent. is produced when Röntgen rays fall on a selenium cell, and this effect might be conveniently used to measure the intensity of Röntgen rays. The same action is also produced by ultra-violet, but not, according to Dr. Himstedt's experience, by ultra-red light.

THE *American Naturalist* states that the discontinuance of the Italian scientific journal *Erythea* has been immediately followed by the reappearance of *Zoe*, a journal of very much the same scope, after a suspension of several years.

AN interesting report is printed by the U.S. Department of Agriculture (Division of Vegetable Physiology and Pathology) by Mr. Hermann von Schrenk, on two diseases of the red cedar (*Juniperus virginiana*), caused by the attacks of two parasitic fungi, *Polyporus juniperinus* and *P. corneus*, the former new to science. The paper is copiously illustrated by seven plates.

PROF. F. PÉCHOUTRE, of the Lycée Buffon, Paris, contributes to the *Revue générale des Sciences* (for September 30) a very interesting epitome of recent researches in vegetable cytology and the process of impregnation in flowering plants. A very useful summary is given of all the most important papers—and they have been very numerous—published on the Continent, in England, in Japan, and in America, during the last three years, under the following heads :—Centrosomes and blepharoplasts ; Chromatic Reduction ; Centrosomes and kinetic centres ; the influence of organic substances on the action of nitrifying microbes ; the Antherozoids of Angiosperms and double impregnation ; the phenomenon of *Xenia* and the hybrid impregnation of the endosperm. Under the last heading the observations of De Vries and Correns are referred to, but not the most recent ones by Webber ; and several of the sections are illustrated by excellent wood-cuts.

MESSRS. C. BAKER, of High Holborn, send us their illustrated catalogue of microscopes and accessories, and stains, reagents, &c., for use in pathological and bacteriological research, including necessaries for the study of tropical diseases and examination of blood. The slide-lending department existing in connection with this firm appears to meet a distinct want.

THE Hampstead Astronomical and Scientific Society encourages interest in natural knowledge by popular lectures and instructive papers on scientific subjects. A course of six lectures upon the astronomy of the spectroscope and photographic camera will be delivered on Monday evenings at the Hampstead Library by Mr. P. E. Vizard, commencing on November 12. The programme of papers to be read at the meetings is an attractive one, and should be the means of increasing the membership of the Society.

FOUR public lectures will be given in the library of the Sanitary Institute under the auspices of the Childhood Society, which exists for the scientific study of the mental and physical conditions of children. The lectures will be as follows : " Treatment of Feeble-Minded Children in Asylums," by Rev. T. W. Sharpe, C.B. ; " The Training of Teachers," by Prof. W. H. Woodward ; " Physiology for Teachers," by Prof. C. S. Sherrington ; " Causes of Failure in the Health of School Girls," by Mrs. D. Colman.

THE list of announcements of the firm of Gebrüder Borntraeger, Berlin, has just reached us, and is as follows :—" Sammlung geologischer Führer " : vol. v. Elsass (Vogesen), by Drs. Benecke, Bücking, van Werveke and Schumacher ; vol. vi. Riesengebirge, by Dr. Gürich ; vol. vii. Schonen (Schweden), by Dr. Hennig ; " Lehre von den Erzlagerstätten," by Prof. R. Beck (Part i.) ; " Flora der Deutschen Schutzgebiete in der Südsee," by Drs. C. Lauterbach and C. Schumann ; " Werden und Vergehen," by Carus Sterne, fourth edition, vol. ii.

THE British South Africa Company has issued a pamphlet on the rubber industry of its territories. The rubber-producing plants of the territory are described as being mostly gigantic creepers belonging to the natural order Apocyneæ. The pamphlet is chiefly occupied with hints on the administrative policy desirable for the protection and encouragement of the industry. Apparently no serious effort has yet been made either to ascertain the rubber-producing value of the native trees and shrubs, or to encourage the cultivation of those species which are found to be most valuable.

THE edition of Darwin's " Origin of Species," just published by Mr. John Murray for half-a-crown, is the cheapest scientific book we have had before us for many a day. The volume is clearly printed, has more than seven hundred pages, and a collotype portrait of Darwin appears as a frontispiece. The first edition of the work was published on November 24, 1859, so the copyright will shortly expire, and probably other editions will be issued by various publishers, but the book which Mr. Murray has brought out will be able to hold its own against all that follow it. If there is a person who claims to be a naturalist, or even to have an interest in natural history, and does not possess a copy of Darwin's immortal work, he should make haste to add the new book to his library.

THE Cambridge Scientific Instrument Company have issued a list of apparatus designed and used by Prof. Ewing, F.R.S., for the teaching of mechanics in engineering laboratories, and now manufactured in their works. Several of these pieces of apparatus relate to experiments on the elasticity of materials, and to measurements of the moduluses of elasticity, by various methods. Amongst these are the latest forms of Prof. Ewing's

microscope extensometer. The remaining instruments and devices are designed to enable students to make quantitative experiments in mechanics. The importance of such mechanical laboratory work, carried out by the students themselves, as a supplement to their study of theoretical mechanics by books or lectures, is now generally recognised. Well-made instruments such as are supplied by the Cambridge Company are essential to ensure accurate work by advanced students.

IN the last number of the *Berichte*, E. Fischer gives an account of further investigations on the division of racemic amido-acids into their optical components. He found previously that by replacing hydrogen in the amido-group in these acids by a benzoyl group, compounds of much more strongly marked acid characters are produced, which are capable of forming well crystallised salts with bases. By crystallising such salts of the active alkaloids, strychnine, brucine, cinchonine, the amido-acids in the form of their benzoyl derivatives have been divided. Alanine, aspartic acid and glutaminic acid were the first to be resolved by this method, and these have now been followed by leucine, amidocaproic acid, phenylalanine and a-amidobutyric acid.

IN the same journal, v. Baeyer and Villiger discuss the action of permanganate on hydrogen peroxide and assail the views of Berthelot and Bach on the existence of oxides of hydrogen higher than the dioxide. Berthelot found that at a low temperature permanganate is decolourised without evolution of oxygen, which he ascribes to the formation of hydrogen trioxide (H_2O_3). The authors, on the other hand, find that at $-16°$, though more slowly, the same volume of oxygen is evolved as at the ordinary temperature. Bach concluded that, as an excess of oxygen above the calculated quantity was evolved with permanganate, " Caro's acid " (hydrogen peroxide in sulphuric acid) contained hydrogen tetroxide. The authors find this observation correct, but the interpretation at fault. They ascribe the decomposition to the catalytic decomposition of Caro's acid, due to the presence of manganous sulphate. Of the nature of the process which occurs when hydrogen peroxide and permanganate react, the authors bring facts in support of the view of Weltzien and M. Traube, who consider that the permanganate oxidises the hydrogen of the peroxide, thereby liberating the oxygen of the latter, and not that the free oxygen is made up of oxygen atoms derived from both peroxide and permanganate.

THE additions to the Zoological Society's Gardens during the past week include a Patas Monkey (*Cercopithecus patas*) from West Africa, presented by Mrs. Creighton Hall ; a Green Monkey (*Cercopithecus callitrichus*) from West Africa, presented by Mr. Cecil T. Reaney ; a Bonnet Monkey (*Macacus sinicus*) from India, presented by Mr. Anthony J. Smith ; a Macaque Monkey (*Macacus cynomolgus*) from India, presented by Mr. G. H. Jalland ; two Muscat Gazelles (*Gazella muscatensis*) from Arabia, an Indian Desert Fox (*Canis leucopus*) from India, presented by Mr. P. Z. Cox ; a Bonnet Monkey (*Macacus sinicus*) from India, a Sooty Mangabey (*Cercocebus fuliginosus*) from West Africa, a Ruffled Lemur (*Lemur varius*), a Black-headed Lemur (*Lemur brunneus*) from Madagascar, a Short-tailed Wallaby (*Macropus brachyurus*), a Great Kangaroo (*Macropus giganteus*), four Brown's Parrakeets (*Platycercus browni*) from Australia, a Blue-necked Cassowary (*Casuarius intensus*), two One-Wattled Cassowaries (*Casuarius uniappendiculatus*) from New Guinea, seventeen Speckled Terrapins (*Clemmys guttata*), three Painted Terrapins (*Chrysemys picta*), ten Alligator Terrapins (*Chelydra serpentina*) from North America, an Elephantine Tortoise (*Testudo elephantina*) from the Aldabra Island, and Oldham's Terrapin (*Cyclemys dhor*) from the Malay Peninsula, a Missel Thrush (*Turdus viscivorus*), European, deposited.

OUR ASTRONOMICAL COLUMN.

EPHEMERIS FOR OBSERVATIONS OF EROS :—

1900.	R.A.	Decl.
	h. m. s.	° ' "
Oct. 25 ...	2 27 27·35 ...	+52 36 52·0
26 ...	25 52·13 ...	52 49 53·2
27 ...	24 13·05 ...	53 2 9·2
28 ...	22 30·39 ...	53 13 38·2
29 ...	20 44·30 ...	53 24 18·5
30 ...	18 55·10 ...	53 34 8·5
31 ...	17 2·99 ...	53 43 6·5
Nov. 1 ...	2 15 8·32 ...	+53 51 11·3

OPPOSITION OF EROS.—M. Loewy has distributed a fifth circular containing additional information intended to secure uniformity of observation among the many institutions which have now commenced their work of determining the parallax of Eros. It is advised that the positions relative to neighbouring stars be measured in rectangular co-ordinates, and also that the eleventh magnitude be adopted as the inferior limit of brightness for the comparison stars. For those undertaking photographic determinations, it is recommended—

(1) That on each plate there be made two exposures of different length, so that each star may be recorded by two images some twenty seconds of arc apart in declination. This procedure will fulfil the two functions of eliminating spurious stars, and of enabling two series of measures to be made on star-discs of very different diameters, so that the photographic spreading effect may be allowed for to some extent. For instruments of the type employed for the international chart (0·33 metre aperture), exposures of six and three minutes are recommended.

(2) On the plates especially for the planet's position, two exposures should also be made, one while the guiding star is accurately followed, the other keeping the planet on the cross wires, or if this be too difficult on account of its faintness, the guide star should be given a motion equal, and in the contrary direction to that of the planet. The result of these operations will be two images of Eros, one showing as a short faint line, the other as a circular patch. This is now known to be quite possible, as the planet has recently been photographed, October 4 and 6, both at Paris and Algiers, in three minutes, the magnitude being estimated at 10·5. Prof. Joly also states that with the 15-inch reflector at Dublin, images were obtained with exposures of six and two minutes.

Another list of comparison stars is provided, and the ephemeris extended to March 1901. The brightness of Eros is now slowly increasing, being 9·81 magnitude on October 29, reaching its maximum of 9·02 on December 18, and then decreasing to 10·48 in March at the close of the time of these special observations. M. Loewy asks all collaborating in the enterprise to forward regularly the successive progress made in the various sections.

NEW DOUBLE STARS.—In the *Astronomische Nachrichten* (Bd. 153, No. 3668), Mr. R. G. Aitken gives the particulars relating to 62 new double stars discovered by him with the 12-inch equatorial of the Lick Observatory. This list is supplementary to that previously published in *Astronomische Nachrichten*, No. 3635. Each star after discovery has been measured with the 36-inch telescope on at least one night ; 39 of the pairs are under 2" distance, 24 under 1", and twelve under 0"·5. The list has been checked by comparison with Prof. Burnham's General Catalogue of Double Stars.

ASTRONOMICAL WORK AT DARAMONA OBSERVATORY.—We are in receipt of an interesting volume from Mr. W. E. Wilson, containing reprints of the astronomical and physical researches made at his observatory at Daramona, Westmeath, since 1892. Mr. Wilson started in 1871 with a 12-inch equatorial, by Grubb, but did little more with this instrument than lunar photography with wet plates and determinations of solar radiation. In 1881, however, he built the present large observatory attached to the house, and installed therein a new 24-inch silver-on-glass mirror of 10 feet 6 inches focus. The old 12-inch mounting proving too light for the extra load, it was replaced in 1892 by one of the best pattern with Grubb driving clock and electrical control. In 1889, an additional laboratory was added for the physical investigations which have formed so large a portion of the observer's programme.

The purely astronomical work with the 24-inch has practically been confined to the photography of star clusters and

nebulæ. Very beautiful reproductions in collotype of some of these are included at the end of the volume.

The astrophysical portions deal with experimental investigations on the heat of the sun, absorption of heat in the solar atmospheres, thermal radiation of sun (both photosphere and spots) and electrical measurement of starlight.

HISTORICAL ASPECTS OF THE DISCOVERY OF THE CIRCULATION OF THE BLOOD.[1]

THE discovery of the circulation of the blood by William Harvey is commonly regarded among scientific discoveries as eminent if not unique ; and this in the judgment not of Englishmen only. My purpose to-day is to show that at any rate it was made against enormous difficulties.

To put this discovery in right perspective we must have some vision of the history of philosophy, science and medicine. Medicine, herein in contrast with Theology and Law, had its sources almost wholly in the Greeks ; from them for good or evil it took its first scheme of thought ; and in the schools of Hippocrates and of Alexandria it was based, more soundly, on natural history and anatomy. The noble figure of Galen, the first physiologist and the last of the great Greek physicians, portrayed for us by Dr. Payne in the Harveian Oration of 1896, stood eminent upon the brow of the abyss when, as if by some convulsion of nature, Medicine was overwhelmed for fifteen centuries. Galen practised the method of verification by experiment, first introduced perhaps by Archimedes, but after him it was lost till the time of Gilbert, Galileo and Harvey.

In the growth of societies small civilisations have been sacrificed to the formation of larger aggregates, whereby stable equilibrium may be attained for the highest ends. Perhaps because of her very freedom of thought Greece never became a nation. Even the Roman peace, bought as it was at the cost of learning and the arts, was but a mechanical peace. In the wilder regions of the empire the bodies but not the wills, of men were in subjugation, and even in Rome itself the sanction of patriotism was failing. Under the Frankish invasion the very traditions of learning and obedience seemed to be broken up. Then Europe was saved by the inspiration of the Christian religion which, entering as a new element into the ancient fabric of Roman empire, was now to hold men's service in heart and soul as well as in body ; but to this end no mere mystic or personal religion could suffice ; clothing itself with the political and ritual pride, and even with the mythology of the pagan empire, it inspired a new adoration ; but it imposed upon men also a universal and elaborated creed. In the third century philosophy was born again in neo-platonism, the offspring of the coition of East and West in Alexandria, where all religions and all philosophies met. The world and the flesh were crucified that by the spirit man might enter into God. Pure in its ethical mood, neo-platonism, says Harnack, led surely to intellectual bankruptcy ; the irruption of the barbarians was not altogether the cause of the eclipse of natural knowledge. Yet even then, as again and again, the genius of Aristotle came to save the human mind. Proclus, ascetic as he was, was also versed in Aristotle ; he compelled the Eastern mysteries into peripatetic categories, and bequeathed a formal philosophy to the faith. Thus the first Scholastic period was fashioned, and the objects and methods of inquiry were determined for thirty generations. Rationalised dogma lived upon dialectic, and conflicted with mysticism ; but logic, dogma and mysticism alike disdained experience. The Faith, then, was the first adversary of Copernicus, Galileo and Harvey.

It was the fortune of the Faith that, of all the treatises of Plato, the Timæus, the most fantastic and least scientific, should have survived to instruct the mediæval world ; while those works of Aristotle which might have made for natural knowledge fell out of men's hands : moreover, the Categories and the " Interpretation " made for more than Aristotelian nominalism, and turned men's minds rather to rhetoric and dialectic than to natural science. Thus Plato's chimæra of the human microcosm, a reflection of his theory of the macrocosm, stood beside the Faith as the second great adversary of physiology.

The influence of authority, whereby Europe was to be welded together, penetrated into all human ideas. As was

[1] Abstract of the Harveian Oration of 1900, delivered by Prof. Clifford Allbutt, F.R.S.

the authority of the Faith, so was that of Plato; and, in the second period of the Middle Ages, of the Arabian versions of Aristotle and Galen. It is not easy for us to realise a time when intellectual progress, which involves the successive abandonment of provisional syntheses, was unconceived; when truths were regarded as absolute; when reasons were not tested but counted; when even Averroists found final answers either in Aristotle or in Galen. Thus in the irony of things was Harvey withstood by the dogma of that Galen who, in his own day, had earnestly appealed from dogma to nature.

In the Isagoge of Porphyry is set forth distinctly a problem, which during the Middle Ages rent Western Europe asunder; a problem, says John of Salisbury, which engaged more of the time and passions of men than for the house of Cæsar to conquer and govern the world; a problem, indeed, which in our own day is not wholly resolved. This was the controversy of the Realists and Nominalists, first brought to a clear issue by William of Champeaux and Roscellinus respectively. Now Plato held ideas not as mere abstractions but as creative forces; and we shall see how potent was this function in medieval thought. Every particular, every thing, was regarded by the realist as the product of universal matter and individual form. Now form might be regarded, and variously was regarded, as a shaping, determinative force or principle, pattern or mould, having a real existence apart from stuff; or, on the other hand, as an abstract principle or pattern having no existence but as a conception of the mind of the observer. And for the Realist, not individuals only, but genera and species also have their forms; either preexistent (universalia ante rem) or continuously evolved in the several acts of creation (universalia in re). For instance, the Church for the Realist is a thing apart from the wills of successive generations of individual men; Man has fallen, not only in many or all individual cases, but also as a kind—a kind having an independent existence; in the Sacraments again there may be a change of hypostasis without change in sensible matter. Now, if forms pre-exist (ante rem), the will of God must be predetermined; or if form be an immanent function acting in re, we are landed in pantheism. Thus Erigena, the brilliant prophet and protestant of the first period of the scholastic philosophy, was virtually a pantheist, as Spinoza was the last great realist. Aquinas, who determined the philosophy now ruling in Rome, brought about a compromise, which covers up rather than solves the difficulty. The problem, it is evident, was no hair-splitting; it dealt with the very nature and origin of being, and it agitated the minds of earnest men at a time of fervid and widespread enthusiasm for knowledge.

Now closely allied to the argument concerning universals was that concerning "matter and form." Whether the terms used were "form and matter," "force, energy or pneuma and matter," "soul, archæus or life and body," "determinative essence and determinate subsistence," "male principle and female element," "the potter and the clay of the potter"; or whether again they were "effect and cause," "nature and law," "being and becoming," the riddle lay in the contrast of the static and dynamic aspects of things; in the incessant formation of variable individuals in the eternal ocean of existence. Even Francis Bacon never got out of the tangle of Form, Cause and Law. It has been the temptation of philosophers of all times, and even of Harvey himself, than whom none had put better the conditions of scientific method, to suppose that by means of abstraction kinds may be apprehended; that thus they may get nearer to the inmost core of things; that by purging away the characters of individuals they may detect the essence and the cause of individuation; not perceiving that the content of notions is indeed in inverse proportion to their universality. We see this error continually to-day. For instance, we may discuss the causes of typhoid fever, and bewilder ourselves by forgetting that there is no such thing as typhoid fever, and that the only causes of a general notion are the psychological causes of its generation in the mind of the thinker at the time; that which is due to objective causes is of course not the notion, but the particular case—a very different affair.

Before motionless stuff—before the problem of the "primum mobile," even Harvey himself, when he had come to the end of his admirable experiments and began to indulge in contemplation, stood helpless. In his need for a motor for his machine he was not able to divest himself of the language nor even of the philosophy of his day. In his day he could not help regarding rest and motion as different things, and motion as a

superadded quality. The motion he attributes, not to a property of the heart, but of the blood—to its "innate heat," which is as far as he could possibly have got. But, by way of explanation, he adds that the innate heat of the blood "is not fire, nor derived from fire"; nor is the blood occupied by a spirit, but is a spirit; it is also "celestial in nature, the soul, that which answers to the essence of the stars . . . is something analogous to heaven, the instrument of heaven." In denying that a spirit descends and stows itself in the blood or elsewhere, as an "extraneous inmate," he bravely says: "I cannot discover this spirit with my senses, nor any seat of it"; and yet, in the treatise "De Generatione," he propounds a theory of the impregnation of the female, not by any material from the male, but by the influence upon her of a "general immaterial idea"; which, even for his own time, was very substantial realism. The riddle which oppressed the great thinkers, from the Greeks to Lavoisier, was, then, the nature of the "Bildungstrieb"—of the "impetum faciens." What makes the ball to roll? Does heart move blood, or blood move heart? and, in either case, what bestows and perpetuates the motion? Telesius, the first of the brilliant band of natural philosophers in Italy of the sixteenth and seventeenth centuries, still sought this principle of nature in the "form" of the peripuretics. Gilbert regarded his magnetic force as "of the nature of soul, surpassing the soul of man." Galileo, although willing to conceive circular motion as perpetual, and even self-existent, was unable thus to conceive rectilineal motion. All these naturalists, including Harvey, and even Descartes, followed the mediæval world and Aristotle in deriving the source of motion directly from that of the spheres—from the quintessence (vid. Arist. De Coelo; and Met. xii.). Till Copernicus transfigured the cosmos, and Galileo and Newton carried terrestrial physics into the celestial worlds, the heavenly bodies were regarded as animated beings, themselves active, and, by propagation from sphere to sphere, animating all "sublunary" matter, wheels within wheels, even to its innermost particles.

Of the origin of energy we have not solved the riddle; we have given it up; but instead of finding its sources without we find them within. Harvey's contemporary, Francis Bacon, sagaciously guessed that heat is an expansive motion of particles; but he regarded heat and cold as two contrary principles. Almost in the same generation the brilliant John Mayow perceived a substance in the air "allied to saltpetre," which passed in and out of the blood by the way of the lungs or placenta. So innate heat gave way to phlogiston, and soon afterwards oxygen and the conservation of energy turned out to be the "form," "spirit," "essence," "primum mobile," "causa efficiens," "potentiality," and the rest of them; so by Lavoisier, a vast pile of metaphysics was blown into the air. But to kill a strong theory outright takes many a generation; realism, shaken by Abelard, and scotched by Hales and Ockham, not only survived to mislead Harvey, but it stretches its withered hand over us still—in the nursery, in the schoolroom, in the university, and in the great arguments of life.[1]

As strong as realism was a third adversity—the pride of the human mind. The asceticism derived from the East, disdainful of carnal things, brought the dualism of matter and spirit into monstrous eminence; and in respect of medicine, in a few generations it turned the cleanest people in the world into the most filthy.[2] Almost to this day the mechanical arts, presumably concerned with lower categories, have been regarded as base; and the crafts, even of the laboratory, as unworthy of great souls.

Anatomy had to labour also against both ecclesiastical and popular antipathy; chemistry and mechanics were gross pursuits unless endowed with the perilous distinction of alchemy and sorcery. Unfortunately, this charge upon the dignity of man was made heavier rather than lighter by Petrarch and the humanists of the Renaissance; and in Oxford of the seventeenth century we find that Boyle was bantered by his friends as one "given up to base and mechanical pursuits." In a certain important respect medicine suffered greatly from this prejudice. It is obvious that, speaking generally, medicine would find its most positive and direct control in those diseases and in those

[1] The readers of NATURE know how effectively this mischievous survival has been attacked recently by Prof. Perry, Mr. Heaviside and other contributors. But even greater men, whose blows still resound through the centuries, have attacked it in vain.
[2] Those curious in such things will notice that the mediævalising clergy of our own day have discarded in their persons that fair linen which in their fathers was the emblem and example of cleanliness.

therapeutical experiments which are manifest on the outside of the body. Yet surgery fell under the proscription of a handicraft, and as such was eliminated from the colleges of physicians both in London and Paris. Thus the genuine work of such men as Paré and Gale were without influence upon medicine, and thus it came about that Francis Bacon said of the physicians of Harvey's day that they saw things from afar off, as if from a high tower. From Erasistratus to Celsus, physicians practised medicine as one art. Galen taught, not the simplicity, but the unity of medicine ; and Littré points out that this unity is consistent in the Hippocratic writings. Surgery, by virtue of its imperative methods, was kept clear of philosophy on the one hand and of humanism on the other. Fortunately for Harvey, his master, Fabricius, was as great a surgeon as anatomist ; and such was Fallopius. Thus it was that medicine, at the end of the Middle Ages, had not recovered the standard of Alexandria. And against this adversity, also, had the founder of physiology to contend.

Happily the Arabian scholastic philosophy took its root in Alexandria when neo-platonism had veered towards Aristotle, and it was therefore more uniformly peripatetic than the Christian scholasticism. It is one of the signs of the greatness of Aristotle that, thus garbled and glossed, his power made itself felt in the thirteenth century, chiefly by the great Franciscans Alexander Hales, Roger Bacon, and William Ockham—Roger Bacon, whom we may call the first of the natural philosophers of the West. This former renascence determined the second period of the Middle Ages : the period distinguished by the Arabian version of Aristotle, by a check to the chimæras of realism, by some liberty of secular knowledge—for even bishops came out of the school of Toledo—and again by the coming of the Friars, whose influence upon the thought of the Middle Ages was a curious proof that, as all ways are said to lead to Rome, so all systems of thought, in spite of the thinkers, led to natural science. The logic and rhetoric of the Dominicans, by their rationalism, defined, and in defining restricted, the dominion of the Faith. Men got used to reason, and made a language for thought. And in the history of the unlearned Friars Minors we find, as elsewhere in history, that mysticism is more favourable to natural knowledge than the passionate dogmatism of Clairvaux or the dogmatic rationalism of St. Thomas. The Victorians, as Gerson after them, despised reason rather than feared it ; mysticism makes for individual religion, as in Glisson and Newton, rather than for the Church. Hence it may have been that independent thinkers, like Hales and Bacon and Ockham, entered the Franciscan Order. The former renascence bred also a more tolerant spirit. Albert of Cologne owed as much to Avicenna as St. Thomas to Averroes : sages technically damnable, yet "mighty spirits," worthy of reverence. Dante put in hell, but on green meadows, in an open place, lofty and luminous, not only Aristotle, Plato and Socrates, but also Euclid, Ptolemy, Hippocrates, Avicenna, Galen and "Averroes who made the great commentary." Universities were founded in France, England and Italy. But the natural science which made the second renascence irresistible was absent from the former ; and at the end of the century a reaction set in. During the two following centuries in Spain freedom of thought was crushed out by the Church ; but in the conflagration of books of philosophy, medical works, such as the "Colliget" and the Commentary on Galen of Averroes, were largely spared ; yet in the fourteenth and fifteenth centuries the very name of Averroes, "the mad dog that barked against the Christ," not only became ecclesiastically accursed, but also began to signify loose life as well as free thought ; a resentment of which there was no trace in Albert or Aquinas.

Averroism, however, held its ground at Padua, which had become celebrated for medicine as Bologna for law ; and although Averroism, like any other philosophy taught as a separate study, decayed, yet, effeteias it was, it kept the ground open at a time when the tide was turning against free thought ; when the commercial supremacy of Venice was declining, when the Spanish Inquisition was established in Rome, and when even the influence of the Florentine humanists was rather against natural knowledge than for it. No doubt the coarse and disingenuous scepticism of the physicians of North Italy and their pretentious manners alienated the humanists, not only from themselves, but also from natural philosophers such as Telesius and Galileo ; and Averroists and humanists alike stood by at the burning of Bruno. Harvey entered Padua at a fortunate time ;

he found Galileo engaged in teaching, and also in methodical research ; and Galileo was not only a great discoverer, but was the first to formulate fully and clearly that method which we know under the name of the inductive method. The discovery of Greek texts had destroyed the conventional Aristotle and the conventional Galen ; Gregory, by the reform of the calendar, had put the axe to the root of astrology ; Newton was soon to carry terrestrial physics into celestial spheres ; and Boyle was soon to create chemistry ; while anatomy was fully awake already. In England, moreover, with the accession of Elizabeth more spacious times were assured, and Charles protected Harvey. Clinical teaching had been established at Padua by Fracastorius and Montanus, to be pursued in Heidelberg, Leyden and Vienna. Physiology, however, awaited Harvey. Servetus had buried his conception of the lesser circulation under a pile of theology ; Columbo and Fabricius had prepared the way, not so much by the value of their discoveries as by their practice of the experimental method in this science ; for the anatomists, Galenists to a man, had done next to nothing for physiology.

The genius and courage required to make discoveries like that of the circulation of the blood cannot be measured directly ; there is no method of determining the specific gravity of such adventures ; I have tried, however, to shadow forth the weight of the social systems, opinions, prejudices and habits against which Harvey's gigantic effort was made. Almost in the year of the publication of the "De Motu Cordis" (A.D. 1628), the Parliament of Paris issued an edict that no teacher shall promulgate anything contrary to the accepted doctrines of the ancients. Under such conventions Harvey's discovery burst like an earthquake ; under corrupt Galenism, venerable sophistries, current abstractions bequeathed by realism, and long-winded dialectics on critical days, coctions, derivatives or revulsives, and dogmas based on uncritical subservience to texts. His work stood out even more ascendant against a lurid background of folk superstitions—of vampyres, witch-burning, magic, cabbalism, astrology, alchemy, chiromancy and water-casting. In terrestrial and celestial physics, Galileo, persecuted as he was, had some strong current with him ; Copernicus was before him, Kepler was beside him : but in physiology upon the path of Galen the waters had closed as upon the track of a great ship, and among Harvey's contemporaries and immediate forerunners there was none to claim a share with him in the discovery of the central fact of physiology, or in his application of the method which opened the way to Pecquet, Glisson, Steno, Wharton, Willis, Haller and Bernard.

THE ANNUAL CONGRESS OF THE GERMAN ANTHROPOLOGICAL SOCIETY.

THE thirty-first Congress of the German Anthropological Society was held in the University town of Halle from September 24-27. In addition to its rich University collections, a special interest is attached to Halle as being the seat of the oldest German society for encouraging the study of natural science, viz. the Leopoldina-Carolina Academy, which is thus comparable to the Royal Society in this country. To students of prehistoric archæology, the Prussian province of Saxony is chiefly interesting from the fact of the existence of copper-mines at Eisleben, some little distance from Halle. The meetings were held under the presidency of Prof. Virchow, assisted by Prof. Ranke. At the opening session on Monday, September 24, the presidential address (dealing with the general progress of anthropological study and teaching) was followed by a series of addresses from representatives of the University and town of Halle, of which that of the local secretary, Dr. Förtsch, is particularly noteworthy as containing a sketch of local prehistoric archæology, a field of research in which Dr. Förtsch has been particularly active, and which he has popularised with evident success. Of the subsequent communications to the Congress, the majority of which dealt with archæology, there appear to us most worthy of mention the discussion opened by Prof. Virchow on the "Earliest appearance of the Slavs in Germany," and the account (illustrated with excellent lantern slides) given by Dr. Birkner (Munich) of the investigation of the graves of the German Emperors in Speyer. Prof. v. Fritzsch (Halle) and Dr. Lehmann-Nitsche (La Plata) rendered interesting accounts of discoveries of prehistoric man in Thüringia and in the Argentine respectively, the latter record being still the subject of investigation as regards the exact antiquity (Tertiary period) claimed for the find.

It is a matter of some surprise that the department of Physical Anthropology should not have been the subject of more papers than were actually presented at Halle, which University claims the two Meckels and Welcker among its former professors of anatomy. The chief contributions to this subject were those of Dr. Schmidt-Monnard (Halle) on the relation between the growth and the weight of children of both sexes ; of Dr. Eisler (Halle) on the *Musculus sternalis* ; and of Prof. Klaatsch (Heidelberg) on the method of research adopted by anatomists, illustrated specifically by observations on the "short head" of the Biceps femoris muscle in the mammalian series.

The chief excursion of the Congress was made on Wednesday, September 26, to Eisleben, where the copper-mines already referred to were visited, and demonstrations of copper-smelting were given by representatives of the Mansfeld Co. Subsequently the local collection of prehistoric pottery, &c., was inspected.

The concluding session was held on September 27, when the presidency (for the ensuing year) was assumed by Prof. Waldeyer (Berlin). It is a matter of interest to note that the Congress was made the occasion of circulating "special inquiry" sheets regarding the structure and building of boats in all parts of Germany. General proposals regarding cartography and systematic records for provincial localities were brought forward by Dr. Voss (Berlin).

In addition to the anthropologists already mentioned in the foregoing notes, there were present Freiherr v. Andrian-Werburg (Vienna), Prof. Hein (Vienna), Prof. Montelius (Stockholm), Prof. Koganei (Tokio), and others to the number of about one hundred and twenty.

ANTHROPOLOGY AT THE BRITISH ASSOCIATION.

THE Section of Anthropology had a very successful session under the presidency of Prof. Rhys ; indeed, it was one of the very best meetings of Section H in the history of the Association. Nearly every department of the subject was represented and that too by new and original contributions. It is interesting to note the different lines which members of the Universities of Oxford and Cambridge are at present taking up. The field of Classical Archaeology is offering rich prizes to Arthur Evans, Hogarth, and Myres ; while the expeditions of Hose, Stanley Gardiner, Haddon, Skeat, and others are providing material for a more complete knowledge of primitive peoples.

I.—PHYSICAL AND EXPERIMENTAL ANTHROPOLOGY.

(1) *Somatology.*

Dr. John Beddoe, in a short paper on the vagaries of the cephalic index, described two long-headed skulls which had the general characters of dolichocephaly, but the one that appeared the more typical had a latitudinal index (living) of 82·3, owing to retarded ossification of the posterior part of the temporo-parietal suture ; but for this the author thought the index would not have exceeded 77. Prof. Macalister, as at the last meeting of the Association, deprecated the importance usually ascribed to the cephalic index.

Prof. A. Francis Dixon read a paper, entitled "On certain markings on the frontal part of the human cranium and their significance." An examination of the frontal region of the cranium shows that, in many cases, grooves or channels are present on the bone, corresponding to the branches of the supra-orbital nerves. Their presence indicates a want in proportion between the growth in length of the nerves and the amount of expansion of the underlying part of the cranium. The nerves might be looked upon as constricting cords which become depressed in the developing bone as the cranium expands. In races in whom the grooves are common, and strongly marked, we would expect the presence of a tendency towards increased development and capacity of the frontal part of the cranium ; while, on the other hand, in races in whom the grooves do not occur, or are rare, and but feebly marked, we would expect to find much uniformity in the shape and size of the cranium, indicating that none of its various parts are tending towards an increased development. In this connection it is interesting to note that the frontal grooves are almost never found in Australian and Tasmanian skulls, that they are rare among Melanesians,

slightly more common among Polynesians, while among Bushmen and negroes, especially in Zulus and Kaffirs, they are very common, and often extraordinarily well marked.

Mr. W. L. H. Duckworth described nine crania collected by Mr. J. Stanley Gardiner in his expedition to Rotuma. As might have been expected from the position of the island, the skulls could be resolved into two types, one of the form of cranium usually found among Polynesian peoples, though possessing something of a Mongolian aspect ; the other was of distinctly Melanesian affinities. The paper was well illustrated with lantern slides.

A second paper by Mr. Duckworth was on some anthropological observations of the Pangan tribe of aborigines in the Malay Peninsula made by Mr. J. Laidlaw in the Skeat Expedition. Mr. Duckworth measured one adult male skeleton the stature of which was about five feet, which is the average height of the men measured by Mr. Laidlaw. The latter describes the people as having a skin colour of varying shades of dark brown, and the black hair is in some cases frizzly and in others more or less curly or wavy. Mr. Duckworth regarded the skull as of a negroid type with infantile characteristics. It is probable that the skeleton in question belonged to a half-breed Negritto and Malayan individual.

The developmental changes in the human skeleton from the point of view of anthropology were described by Dr. David Waterson. A comparison was instituted between the bones of the embryo and those of the lower races of mankind and of the higher apes, both as regards their relative length and their characters. As it has been shown that the curvature of the spine in the lumbar region is a post-natal development, and one adapted to the assumption of the erect attitude by the infant, it can also be shown that in a similar way the configuration of the bones of the lower extremity alters after birth, before the infant can stand erect.

Prof. A. Macalister discoursed on perforate humeri from ancient Egyptian skeletons. In examining those from Libyan graves, he was struck with the large number of humeri which had a supra-articular perforation, the proportion of such among these old Egyptian remains being much greater than among other races. He found that these perforations reached 57 per cent. among the ancient Egyptians, whereas among average English people they ranged about 3 per cent., and 10 per cent. among Neolithic skeletons, while the percentage rose to 53 in the skeletons of Indians from Arizona.

A paper on the sacral index was read for Prof. D. J. Cunningham. Inasmuch as the true length of the sacral portion of the vertebral column is not indicated by the shortest distance between the apex and base of the sacrum, but rather by the length of the curve formed by the sacral vertebræ, it is proposed that, in making measurements for the determination of a sacral index, "length" should be measured by using a tape along the concavity of the sacral curve, and not by calipers, one limb of which is placed upon the base and the other on the apex of the sacrum. Breadth (measured by calipers in the ordinary manner) multiplied by 100 and divided by length, measured in the manner indicated, gives the true sacral index. The curvature of the sacrum may be conveniently plotted by taking a tracing from a strip of soft metal which has been previously adapted by pressure to the front of the sacrum along its middle line. The index of curvature may be expressed by the number derived by multiplying the height of this plotted curve by 100 and dividing by the number corresponding to the true length of the sacrum.

Dr. Cunningham communicated a second paper on the microcephalic brain, which also was read by Dr. Dixon, and illustrated with a large number of lantern slides. The brain of the microcephalic idiot may exhibit features which do not merely represent a "fixed" embryonic condition. In one specimen the arrangement of the fissures and sulci is found to approach more closely the ape than the human type, and in almost every furrow some simian character can be detected. These simian characters must not be considered mere fœtal conditions rendered permanent. The ape-like condition existing in this brain does not as a whole correspond to that of any one ape, or group of apes, but there is a complicated mixture of features some of which are characteristic of high apes, while others find a parallel in the brain of low apes. The microcephalic brain may be regarded as a partial "atavism." So far as its surface markings are concerned, the specimen noted has reverted in part, or wholly, to an arrangement which, in all probability, existed in some early stem-form of man

Dr. J. G. Garson read a paper, with lantern illustrations, on a system of classification of finger-prints.

(2) *Psychology.*

Prof. G. J. Stokes read a paper on the sense of effort and the perception of force, and Prof. Marcus Hartog contributed another on interpolation in memory. In the discussion on the latter paper, Prof. Lloyd Morgan contended that psychologists had not so completely overlooked the practical perceptual judgment as Prof. Hartog seemed to suppose.

Mr. E. W. Brabrook presented the final report of a committee which has been acting in conjunction with the Childhood Society for the scientific study of the mental and physical conditions of children.

II.—ETHNOLOGY.

(1) *Sociology.*

Mr. E. S. Hartland read a paper that attracted a good deal of attention, on the imperfection of our knowledge of the black races of the Transvaal and Orange River Colonies.

(2) *Technology.*

Dr. A. C. Haddon read a paper, illustrated by lantern slides and specimens, on the textile patterns of the Sea-Dayaks. The Sea-Dayak women weave short cotton rep petticoats and cotton sleeping wraps which are covered with beautiful and often intricate patterns. The patterns are made in the following manner : the warp is stretched on a frame, the woman takes the first fifteen to thirty strands and ties them tightly with strips of leaves at irregular intervals, according to the design, which she carries in her memory. The next fifteen to thirty strands are similarly tied, and this process is repeated until all the threads have been utilised. The warp is then removed from the frame and dipped in a reddish dye, which colours the free portions of the warp, but the tied-up portions remain undyed ; thus a light pattern is left on a coloured background, when the lashing is untied. If a three-colour design is required, as is usually the case, the first lashing is retained, and various portions of the previously dyed warp are tied up ; the whole is immersed in a black dye, and then both sets of lashing are untied. The pattern is thus entirely produced in the warp. By far the greater number of these designs are based upon animals, whereas most of the patterns carved by the men on wooden and bamboo objects are derived from plant motives. The decorative art of the Sea-Dayaks of Sarawak differs in character from that of the Kayans, Kenyahs, and other inland tribes.

A paper was read by Mr. W. Rosenhain on the making of a Malay "Kris." This, it was explained, was a species of native sword, and the paper dealt with some specimens of Malay metal work which had been submitted to the writer for microscopic and other examination by Mr. W. W. Skeat. By means of many lantern views Mr. Rosenhain exhibited samples of this kind of weapon, showing especially the "damask" pattern of the "Kris." Swords of this description were composed of many strands of two kinds of metal. The body of the blade was made of steel, a layer of laminated "damask" iron was welded upon each side of the central layer. A thin layer of steel was then welded on, outside the "damask" iron. The author was of opinion that the striated "damask" effect was due to the opening of the loose welds in the "damask" iron during the forging of the blade, steel being driven between the laminæ. The outside layer of steel was entirely ground away during the process, and when the compound surface so produced was "etched" by the pickling process employed, the more readily corroded steel was attacked, leaving the edges of the layers of iron as a series of narrow projecting ledges. The tools of the Malay goldsmith were then carefully described, and subsequently a description was given of the micro-structure and composition of Malay bronzes. The concluding part of the paper dealt with the Malay method of producing chains by casting, a practical illustration of the method being given by Mr. Rosenhain on the platform.

Prof. Macalister said that the older Universities of this country were frequently reproached for not doing anything practical. He felt, however, that Mr. Rosenhain had that afternoon removed a great deal of that reproach by practically demonstrating for them the process of casting metal chains practised by the Malays.

Prof. H. Louis read a "Note" on the "Kingfisher" type of a Malay "Kris," which was illustrated by specimens. This was a type of "Kris" used only in a very limited

area in the north-east of the Malay Peninsula ; it was, however, rare even in its own home. The Malay legend of its origin, he remarked, was that a party of Malays from the Bugis Islands invaded that portion of the Peninsula many centuries ago. One of their leaders was known as "the Kingfisher "—presumably on account of his rapid movements. The invasion was successful, but the leader fell in one of the last engagements, and after his death his followers carved their Kris handles into shapes resembling the Kingfisher's head and beak. Under Chinese influence the pattern became more and more ornate, being modified by the Chinese "Dragon," until it reached the present fixed type.

Dr. Haddon gave an afternoon lantern demonstration on the houses and family life in Sarawak. He exhibited a series of about fifty lantern slides of photographs taken during his recent expedition to Sarawak, selected to illustrate the type of houses common among the settled inland tribes of Borneo and the everyday life of the people.

Mr. W. Law Bros delivered an afternoon lantern lecture on some Indian monuments, illustrated by numerous beautiful slides taken by himself.

A paper, with lantern illustrations, on permanent artificial skin marks was read by Mr. H. Ling Roth. Whatever might have been the original idea, ultimately its objective became manifold. Mr. Ling Roth described four systems or methods of tattooing. The Tahitian method of "tattaow" was performed by pricking or tapping with pigment, and produced the smooth result such as was seen in our soldiers and sailors. The Maoris of New Zealand adopted this pricking method, and also a cutting method performed with a sharp, adze-like edge like a narrow chisel. This produced slight but permanent grooves in the skin. The third method was that adopted in West Africa, and was similar to the second, but the grooves produced were deeper and wider, and generally no pigment was used. Lastly, there were the curious raised marks of the Tasmanians, Australians and Melanesians generally, and the Central African tribes. In this case the cuts were made with sharp edged stones, and were continually reopened or irritated by the insertion of vegetable juices or sand. Hence was produced an abnormal amount of reparative action, and the wounds did not heal naturally as healthy concave scars, but developed into nodulous growths of sometimes considerable size. For these four methods Mr. Ling Roth suggested a nomenclature—namely, tatu, moko, moko and keloid.

Mr. Davis read a paper on the system of writing in ancient Egypt for Mr. F. Ll. Griffith, and whilst doing so illustrated his remarks with very clever sketches on the blackboard. Egyptology has now reached a position among the sciences from which it may contribute trustworthy information for the benefit of kindred researches. Egyptian writing consists of Ideographic and Phonetic Elements, the signs serving as :— (1) Word-signs ; (2) Phonograms ; (3) Determinatives. The highest development shown is an alphabet, which, however, is never used independently of other signs : it is apparently not acrophonic in origin ; it represents consonants and semiconsonants only, vocalisation not being recorded by Egyptian writing. No advance can be detected in the system from the beginning of the historic period to the end, notwithstanding some improvements in practical working which facilitated the use of cursive writing. Phonograms derived from word-signs. The end of the native system was brought about by the gradual adoption of the Greek character—beginning, perhaps, in the second century A.D. If any radical improvement was ever made in the Egyptian form of writing, that improvement must have taken place at or after adoption by another people : e.g. some have supposed that our alphabet was derived by the Phœnicians from Egypt ; but any such derivations are at present entirely hypothetical.

A paper was read by Mr. Arthur J. Evans on the new scripts he has recently discovered in Crete ; an account of his researches has already appeared in NATURE (p. 526).

(3) *Religion.*

Some peculiar features of the animal-cults of the natives of Sarawak, and their bearing on the problems of totemism, were discussed in a paper by Dr. C. Hose and W. McDougall.

Customs had previously been observed that seemed to indicate the existence of a well-developed totemism, either at the present time or in recent times, among the natives of Sarawak. Information bearing on this subject was therefore collected as diligently as possible from various tribes.

A great number and variety of peculiar rites and customs were found to be observed by the people of the different tribes in their dealings with animals and plants. This paper was confined to giving (1) a general account of the customs of one of the inland tribes, the Kenyahs; (2) to describing the "Nyarong," or spirit-helper of the Sea-Dayaks, and some similar institutions among the other tribes; and (3) to pointing out the bearing of our observations on the totem problem.

The Kenyahs are a warlike, agricultural people living as isolated communities of twenty to fifty or more families, each community inhabiting a single long house built on the river-bank. Their religion is peculiar, in that they believe in a beneficent Supreme Being and a group of departmental deities, while they attribute to every agent that affects their lives a spirit that must he properly respected and, if necessary, propitiated. Most important to them of all the animals is the common white-headed hawk. He brings messages of warning and advice from the Supreme Being to those who know how to read the signs he gives, and he is consulted before every undertaking of importance, and sacrifices of fowls and pigs are made to him. A wooden image of the hawk stands before every house. Several other birds give them omens of lesser importance, and none of these may be killed or eaten. The domestic fowl is killed as a sacrifice to the hawk or other powers, and its blood is sprinkled on the altar-posts of the gods and on the persons taking part in various ceremonies, especially peace-making ceremonies. The domestic pig is sacrificed in much the same way. The spirit of a pig is always charged with some prayer to be carried to the Supreme Being, and the answer is read from the markings of its liver. The crocodiles are regarded as a friendly and allied tribe, and may be killed in retaliation only. No Kenyah will kill a dog, and the dead body of a dog is regarded with fear. Kenyahs will not eat the flesh of deer or horned cattle, and there are many restrictions on touching or using any parts of them Only old or renowned warriors will wear or touch the skin of a tiger. One house is decorated with carvings of the gibbon on every large beam, and all Kenyahs have a dread of the Maias and the long-nosed monkey. There thus seems to be every degree of regard paid to the different beasts, from the mere uneasy feeling in the presence of the uncanny, long-nosed monkey to the elaborate cult of the hawk, and the nature of the respect paid to any species seems in nearly every case to be the direct expression of the impression made on the barbarian's mind by the behaviour of the beasts.

The Spirit-Helper.—Every Sea-Dayak hopes to be guided and helped all through his life by a spirit which announces itself to him in dreams and takes up its abode in some peculiar natural object or in some animal. In the latter case the Dayak will never kill or eat one of the same species of animal, and will lay the same prohibition on all his descendants, so that a whole family may come to pay especial regard to one species of animal for many generations. A similar institution occurs, though less commonly, among the other tribes. In such cases we seem to be able to trace sometimes the actual origin and growth of a totem ; but neither among the Sea-Dayaks, nor the inland tribes of Sarawak, could the people be said to be in a totemistic stage of culture, nor was there sufficient evidence of an earlier totemistic cult. Mr. Hartland complimented the author on his caution and carefulness. He agreed with Dr. McDougall in regarding these animal cults as affording little proof of totemism as a stage in social or religious custom.

Dr. E. B. Tylor read a paper, by Mr. W. G. Aston, on the Japanese *gohei* and the Ainu *inao*." The leading idea of the paper was the illustration of the principle in religious development by which an object which was in the first instance simply an offering to a god has in the lapse of time been conceived as the embodiment of the god, or even as a distinct and independent deity. In ancient Japan the offerings to the gods were hemp and bark fibre, with cloth made from these materials. In later times there was substituted a small quantity of paper made of the same bark material and attached to a wand in the form known to us as the *gohei*. With the change of form the original character of the *gohei* as offerings was forgotten. They were looked upon as receptacles of the god, or embodiments of him, and honour was paid to them accordingly. At festivals it was supposed that the god descended into the *gohei* on a certain formula being pronounced by the priest. Hypnotic practitioners also used these objects, through which the deity who inspired them was supposed to enter their bodies. In other cases the devotees went further, and constituted the object which was

originally an offering into a distinct and independent deity. The Ainus of Yezo use in their worship whittled sticks called *indo*, which have a general resemblance to an old form of the *gohei*, and are doubtless a cheaper substitute. The *inao*, like the *gohei*, are primarily offerings, but in some cases they ultimately gain direct worship as gods, having become, in short, genuine fetishes. Another link between the *inao* and the *gohei* is found in certain whittled sticks which a century ago were in use in Northern Japan for striking women with, as at the Roman *Lupercalia*, in order to secure fertility. Similar sticks, after consecration by the Shinto priests, were formerly used at Kiota to kindle the household fire at the new year to avert possible pestilence.

Mr. David Boyle, curator of the Toronto Museum, read a paper on the paganism of the civilised Iroquois of Ontario. Notwithstanding the contact of the Iroquois, or Six Nation Indians, with white people for more than three hundred years, a very considerable number of the former have retained many of their old-time beliefs with the appropriate ceremonies. Of four thousand Caniengas (Mohawks), Senecas, Cayugas, Onondagas, Oneidas and Tuscaroras now residing in the Grand River Reserve, within sixty miles of Toronto, Ontario, fully one-fourth continue to observe the ancient feasts or dances connected with the growth and ingathering of corn and fruits, and for desired changes in the weather, as well as for the curing of disease. Some modification in the ceremonies was made about a century ago by an Onondaga named Ska-ne-o-dy-o, who announced himself as a prophet who had paid a visit to the abode of the Great Spirit. The changes introduced by him, however, have not by any means removed the pagan character of the native beliefs, although he certainly did attempt to imitate some Christian observances. Still the addresses of the medicine-men retain most of the old-time forms, although their significance in many cases is lost and even the meaning of numerous words is no longer known. The leading idea of the present form of worship is that of a great spirit ; but this has been acquired from missionary sources, and although the Indians have adopted the idea of a heaven, they do not believe in any hell. The quoted examples of petitions addressed to Rawen Niyoh, the Creator, illustrated the lack of assimilation of the old and new forms. One of the most characteristic ceremonies connected with the Iroquois paganism is that of the sacrifice or burning of the White Dog at the new year feast during the February moon, when the spirit of the dog, accompanied by offerings of tobacco, conveys to *Niyoh* information respecting the condition of his "own people" on the Grand River Reserve.

III.—ETHNOGRAPHY.

The Anthropology of West Yorkshire was the subject of a communication by the venerable Dr. John Beddoe. He remarked that the most striking qualities of typical Englishmen had been thought to be strongly developed in Yorkshire. Among these, he feared, was the defect of imagination so often found in those who called themselves, with some pride, practical men. Such men entertained a positive dislike, and even contempt, for knowledge of which they did not see the immediate use. This character was not British, Celtic, or Welsh.

Dr. Beddoe's impression, acquired by simple inspection, is that in the central parts of the West Riding, and notably at Leeds, a prevailing type is characterised by an oblong, or rather trapezoidal, head, inclining to be broad rather than narrow, with a vertical forehead, smooth and not prominent brows, and a straight profile, with a straight or sometimes concave nose. The smooth brows dissociated this type from that of the Bronze race and the squareness from the smoothly elliptic or oval one of the Southern Saxon. He was inclined to call it Anglian. Light hair was prevalent hereabouts, and also in the mountainous regions to the north and south.

On the whole, he thought the eastern and central regions of Yorkshire, judging by physique, less purely Teutonic than Teesdale or the Wapentake of Morley, though more so than Craven. The author discussed the question whether any considerable British or pre-Anglian element remained in the country around Bradford. Without coming to any positive conclusion, he was disposed to consider the inhabitants of these parts as mainly Anglian in type. More British blood remained further north, in Craven. A prevalent type about Leeds seemed to him to resemble the Burgundian Belair type of His and Rütimeyer.

Mr. J. Gray read a paper on physical characteristics of the

population of West Aberdeenshire from observations made and statistics compiled by himself and Mr. J. F. Tocher. These observations, he said, were made at a Louach gathering in Strathdon, an isolated district lying right at the head of the Don Valley, with the principal object of ascertaining if any difference existed between the people in the upper ends of the river valleys and those on the eastern seaboard. It was reasonable to conclude, from the anthropological statistics compiled, that in Aberdeenshire at some distant date an early, tall, broad-headed, dark-haired and blue-eyed people, descendants of men of the Bronze age, who had perhaps come from South-east Europe, had been driven inland to the upper ends of the valleys and hills by later immigrants from North Germany who were shorter, had narrower heads, and were of a blonde type.

Mr. J. L. Myres then read a communication from Mr. D. Randall-MacIver on the present state of our knowledge of the modern population of Egypt, in which it was pointed out that whereas we had from statues and paintings, and especially from the skeletons found in recent excavations, a very fair idea of the composition and changes in the ancient population of Egypt, the interpretation of that evidence and of the problems in Egyptian history which it might be expected to solve was seriously hampered by the absence of corresponding evidence in regard to the modern population.

The Committee for the Ethnological Survey of Canada continue its most useful and necessary work, the report presented at this meeting being of exceptional value. The work of the past year has furnished conspicuous evidence of the great importance of securing ethnological data with as little delay as possible. While this is eminently true with respect to the white population, which is experiencing new and marked changes almost every year, in consequence of the introduction of foreign elements, often in large numbers, it is more particularly true with respect to the native Indian population. In many localities the original blood has become so diluted by intermarriage with whites that it is often a matter of great difficulty to find an Indian of pure blood. Proximity to settlements of white people has resulted in a more or less profound impress upon the social life and tribal customs, which are fast becoming obsolete and forgotten. The old chiefs who have served as the repertories of traditional knowledge are rapidly passing away, and with their death there disappears the last possibility of securing reliable data of the greatest value. Conspicuous instances of this kind have been brought to notice during the past year, especially in the case of the British Columbia Indians, whose ethnology is of the greatest interest and importance in consequence of their possible connection with the people of Eastern Asia. At present the great difficulty of securing competent and willing investigators is one of the most serious obstacles to be contended with, and it is believed that the often considerable expense involved in the prosecution of such work is largely accountable for this condition of affairs.

It is gratifying to note that the Department of Education for Ontario has taken a very practical and active interest in ethnological studies in that province, and that it provides for the publication of the results of research in its annual reports. Evidence has latterly been accumulating to indicate the presence at one time of numerous aboriginal settlements in localities which were very sparsely inhabited when first visited by the white explorers.

The committee appointed to carry out investigations on the natural history and ethnography of the Malay Peninsula presented their report, which had been drawn up by Mr. W. W. Skeat, the leader of the expedition.

The expedition itself was composed of members of the Universities of Oxford and Cambridge. The objects of the expedition were to carry out a scientific survey, in which ethnology, zoology, botany and geology should be included, of the little known Malay provinces of Lower Siam. The report stated that some very curious racial problems were found in the district arising out of the fusion of two antagonistic race elements. The most interesting problems were found in connection with the very primitive jungle tribes of the interior, concerning which much valuable information was obtained. The inhabitants were found to be for the most part Malay, who had become subject to Siamese influence. One little known tribe—the sacred tribe of the Prams—claimed to have come from India, and to have established themselves in the country previous to the coming of the Malays. The expedition was

successful in obtaining a copy of their sacred book, from which it is believed an account of the origin of the tribe may be obtained. One small jungle tribe of Pangans was heard of, but though a forced march was made to reach them, the wild men had heard of the approach of the expedition, and had taken flight. Their late dwelling-place, a cave under a projecting rock, was photographed, as also were the very curious "tree-graves" used by the Siamese. This "tree-grave" burial, however, was now condemned by the Bankok authorities, and would become extinct before long. The most interesting of Malay industries observed was the manufacture of damasked "krisses." Near the head-waters of the Muda a tribe of from twenty to thirty individuals was found living in a long barrack-like shelter of palm leaves. From them and from a neighbouring tribe valuable information as to their manners, customs and language, as well as full measurements of a few individuals, and some probably unique phonographic records of their songs, had been obtained, which were of an extremely simple and primitive character. It was found also that many of the leading Malay industries were being rapidly modified by the introduction of European methods and appliances, and it was now the rarest and most difficult thing to obtain cloth actually made of home-spun thread, the use of Singapore silk and aniline dyes being already the fashion everywhere.

IV.—ARCHÆOLOGY.

Mr. A. M. Bell then contributed a paper on the occurrence of flint implements of palæolithic type on an old land-surface in Oxfordshire, near Wolvercote and Pear-tree Hill, together with a few implements of various plateau types. He stated that a large section of the quaternary river-gravel there produced the usual fauna and many fine implements of human workmanship. This gravel cut into and was therefore newer than a previous land surface, a portion of which was found at Wolvercote and another within half a mile at Pear-tree Hill. In both places flint instruments of palæolithic type, together with bulbed flakes and a few implements of plateau-type had been found. In every case these flints were vitreous, a point which distinguished them from those belonging to the river-gravels at Wolvercote. The older surface has been previously described as Northern Drift. Mr. Bell said he supposed it to be a *remaniement*—*i.e.* a re-handling or working over of the true Northern Drift, but deposited under semi-frozen conditions. It must be anterior to the river-valley, and consequently its relics of man were the oldest as yet obtained from the Thames valley. The drift in question most resembles the drifts of Caddington described by Mr. G. Worthington Smith and some sections on the Lower Greensand near Limpsfield, both of which are implementiferous, and the author would correlate the Wolvercote and Pear-tree Hill surface with these drifts.

A paper by Mr. J. Paxton Moir on stone implements of the natives of Tasmania, was read by Prof. E. B. Tylor. Examples of tools prepared with the help of grinding, and furnished with handles, were very rare, and were evidently of Australian origin. The Tasmanian implements were variously used—as, for example, by the women to cut notches in the bark of trees as an assistance in climbing; in the making of the Tasmanian spear, by scraping it straight and smooth; in grooving the handle of the Tasmanian club; and in sawing bones.

Prof. Tylor then proceeded further to add his own views on the Stone age in Tasmania as related to the history of civilisation. Inasmuch as the stage of civilisation attained by any people was of necessity closely associated with the nature of the implements in their possession, we might fairly assume that the development and habits of Europeans of the earliest authentic Stone age were essentially similar to those of the Tasmanians. What then, could we infer as to the earliest human races—the earliest, that is, after we had crossed the great gulf of the unknown which, on the evolutionary hypothesis, existed between the animal and the man? We found, taking the Tasmanians as representative of the earliest grade of society, that these people had no bows and arrows or throwing stones, but they had spears and clubs. They had houses and boats, but of the rudest imaginable type. They knew of fire, and could make it by the fire stick. With their stone implements they prepared and utilised the skins of animals. They made basket-work—in fact, the basket-weaving art had not substantially advanced from the earliest ages of which we had any knowledge until the present. They had a mythology, with a Greek-like legend, as to the origin of fire ; and lastly, they had religious ideas, lower and ruder than those of the North American Indians, but on substantially the same lines. This circumstance—that the Tasmanian natives represent the most ancient known beginnings of human civilisation, while at the same time they were so recently a real, living and known people—rendered all knowledge which could be acquired concerning them a contribution of priceless value to the history of mankind.

Mr. H. Ling Roth said that some eighteen months ago he had received from a Yorkshire gentleman, Mr. J. Backhouse Walker, an account given by an old Australian settler, who in his youth had come across a group of black fellows whilst they were actually engaged in making these stones. The first process was simply to split them by hurling them violently on the rocky ground, and some stones were at once used in this rough shape for cutting up kangaroo meat, whilst other stones were prepared by chipping. At one period, doubtless the Tasmanians covered the whole of Australia ; and they were subsequently almost swept away—only scattered representatives being left in small areas—by another race.

Dr. A. C. Haddon made a communication on relics of the Stone age of Borneo (illustrated by lantern slides and specimens). Dr. C. Hose, the Resident of the Baram District of Sarawak, obtained numerous examples of stone implements from various interior tribes in his district ; these he has generously presented to the University of Cambridge. The implements are made of various rocks, including fibrolite, impure sandstone, arkose, silicified limestone, shale, andesite and chalcedony. The form, too, varies greatly ; some are obviously axe heads, others adze blades, while certain cylindrical forms, with a more or less cup-shaped cutting end, were probably used to extract the pith from the sago palm. In the collection are several stones of irregular form ; the former use of some of them is problematical, but they have recently been used as touchstones. The natives have a high regard for these stone implements, which have in their eyes a sacred character, and it is very difficult to persuade their owners to part with them. In all cases fowls had to be sacrificed to appease the spirits. The implements are stored with other sacred objects, and most of them are believed to be teeth, or toe-nails, of Baling Go, the Thunder God.

Mr. Butler Wood read a paper, with a map illustration, on the Prehistoric Antiquities of Rumbald's Moor, between Bradford and Ilkley. The stone implements found there, he said, consisted of axe-heads, arrow-heads, awls, spear-heads, knives and scrapers. Several fine specimens were exhibited. Baskets full of flint chippings scattered about in a very limited area seemed to prove that it was a place of fabrication of flint instruments, for flint *in situ* was not found within fifty miles. About half a dozen bronze axe-heads had been found at lower elevations. So far as evidences of interment went, the barrows were unsatisfactory. There was a double stone circle within four miles of Bradford Town-hall containing in the centre a boulder with a cup and ring mark on it. He hoped the Corporation of Bradford would not permit it to continue neglected, but would preserve it as a valuable specimen of antiquity. Mr. Wood also briefly mentioned the earthworks and the so-called pit dwellings which seemed to have been attempted excavations for iron. Springs of water were always found near these excavations.

Dr. Haddon emphasised Mr. Wood's observation as to the necessity for the preservation of local antiquities. It was the duty of the section to bring to the notice of local authorities the fact of the existence of valuable archæological remains in the district, and he hoped the Press would help him to insist upon the fact that it was the duty of all the local authorities to preserve their antiquities as well and as long as possible. It had been observed by Mr. Wood that if steps were not taken the double stone circle would soon be worn away and disappear owing to pedestrians walking recklessly over and about it ; and he hoped that as a result of this paper and of the feeling expressed by the section, the proper authorities would take such steps as would effectually preserve this ancient stone structure. He also thought it a pity that the three Saxon crosses at Ilkley were not placed inside the parish church, and that the cup and ring markings on boulders preserved at Ilkley would be better protected if a light shed was erected over them.

An excellent paper by Mrs. Armitage on some Yorkshire earthworks describes a particular kind of earthwork, very common in Yorkshire and in other parts of England, consisting of a moated hillock with a banked and moated court attached. This type of fort has been attributed in turn to the Britons, Romans, Saxons and Danes, with equal improbability. The theory most general at present is that it is Saxon. But Saxon strongholds were built to shelter all the people of the neighbourhood, and were therefore of large area, while these earthworks are evidently intended to protect some individual chieftain and his personal following, as is shown by their small area. There is positive evidence that the Normans built earthworks of this kind in the eleventh century as the bases of wooden castles, and these moated hillocks are still very numerous in Normandy. They are called *mottes* in Norman-French, and this word is found in various parts of England in the form *mote*. An inquiry into the castles known to have been built by the Normans when they first came to England shows that almost all these castles had mottes, while the *burhs* or *boroughs* built by the Saxons never have these appendages, unless a Norman castle-builder has been at work there. The recognition of the Norman origin of these castles would help to solve an historical puzzle—how the Normans were able to hold England down. It was by a system, of small fortified posts scattered all over the country that the action of the central machinery was carried into the remotest parts of the kingdom.

Mr. D. G. Hogarth read a paper on the cave of Psychró in Crete, which was copiously illustrated by excellent lantern slides. It has been known for some years that a large cave above the village of Psychró, in the Lasithi district of Crete, was a repository of primitive votive objects in bronze, terra-cotta, &c. As this cave is situated in the eastern flank of the mountain which dominates the site of ancient Lyttos, and is the only important cave known in the neighbourhood, it was conjectured that it was the Lyttian grotto connected with the story of the infancy of Zeus in the legend, whose earliest version is preserved by Hesiod. A thorough exploration of it has served fully to confirm this view. The cave is double. A rude altar was discovered in the middle of the upper grotto, surrounded by many strata of ashes, pottery and other refuse, among which many votive objects in bronze, terra-cotta, iron and bone were found, together with fragments of some thirty libation tables in stone, and an immense number of earthenware cups used for depositing offerings. The lowest part of the Upper Grotto was found to be enclosed by a wall partly of rude Cyclopean character, and partly rock-cut ; and within this Temenos the untouched strata of deposit ranged from the early Mycenæan age up to the Geometric period of the ninth century B.C. or thereabout. Only very slight traces were found of later offerings. The earliest votive stratum belongs to the latest period of the pre-Mycenæan age, that marked by the transition between the "Kamáraes" fabric of pottery and the earliest Mycenæan lustre-painted ware. But below all is a thick bed of yellow clay, containing scraps of primitive hand-burnished black and brown pottery, mixed with bones of animals. This bed seems to be water-laid, and to be prior to the use of the cave as a sanctuary.

The southern or Lower Grotto falls steeply for some 200 feet to a subterranean pool, out of which rises a forest of stalactite pillars. Traces of a rock-cut stairway remain. In the chinks in the lowest stalactite pillars, a great many of which were found still to contain toy double axes, knife-blades, needles, and other objects in bronze, placed there by dedicators, as in niches. The knife-blades and *simulacra* of weapons are probably the offerings of men ; the needles and depilatory tweezers of women. The frequent occurrence of the double axe, not only in bronze, but moulded or painted on pottery, found in the cave, leaves no doubt that its patron god was the "Carian" Zeus of *Labranda*, or the *Labyrinth*, with whom perhaps his mother, the Nature goddess, was associated, and the statuettes probably represent the two deities. Here was the primitive scene of their legend, afterwards transferred in classical times to a cave on Mount Ida.

Mr. Arthur J. Evans remarked that the cave was one of the most ancient shrines of the classical world bound up with the earliest cult of Zeus. Mr. J. L. Myres suggested that the eras of Cretan civilisation covered by the objects found in the cave might go back, at furthest, to the twelfth Egyptian dynasty, and extend down to the eighteenth dynasty.

A report of the committee appointed to co-operate with the Silchester Excavation Fund Committee was read by Mr. E. W. Brabrook.

UNIVERSITY AND EDUCATIONAL INTELLIGENCE.

CAMBRIDGE.—Dr. J. Larmor, F.R.S., has been appointed a member of the General Board of studies. Mr. F. C. Kempson and Dr. G. F. Rogers have been appointed demonstrators of anatomy; Mr. C. T. R. Wilson and Mr. J. S. E. Townsend have been appointed demonstrators of physics.

The Gedge Prize for physiology has been awarded to Mr. J. Barcroft, King's, and Mr. H. H. Dale, Trinity.

The Biennial election to the Council of the Senate will take place on November 7. Dr. Hill, Mr. Austen Leigh, Sir Richard Jebb. Dr. Kirkpatrick, Dr. Langley, F.R.S., Mr. Mollison, Mr. Shipley, and Dr. Keynes are the retiring members. In the absence of any acute issue it is not unlikely that most of these may be re-elected.

A syndicate is about to be appointed to consider what steps, if any, should be taken towards the better organisation of instruction in military subjects within the University. The question has been brought up by a memorial signed by a large number of influential residents in Cambridge, who desire the establishment of a closer connection between the University and the Army.

The High Court has issued an order varying the conditions of the Gedge Bequest for the encouragement of physiological research, whereby advanced students are admitted to the competition, provided they are of not less than three or more than five years' standing.

The examiners in sanitary science announce that sixteen candidates have, at the recent examination, become qualified for the University Diploma in Public Health.

On October 22, the number of freshmen who matriculated was 841, including fifteen advanced students. Last year the number was 883.

Prof. Darwin, F.R.S., has been elected a member of the Financial Board, Dr. Larmor a member of the Observatory Syndicate, Mr. Berry an examiner for the Mathematical Tripos Part I., and Prof. Lamb, Mr. Richmond, Mr. Baker and Mr. Macdonald examiners for Part II.

A VALUABLE address on "Famine in India" was given by Prof. Robert Wallace at Edinburgh University on October 18, to inaugurate the course of lectures on Colonial and Indian agriculture, specially endowed by Mr. Robert and Mr. John Garton, of Newton-le-Willows, Lancashire, and permanently attached to the chair of agriculture and rural economy in the University. The course is to be a part of the regular work to be done during the winter session of five months, and is to be delivered free to present and past students of the department of agriculture. As Prof. Wallace will be largely concerned with agricultural conditions in India and the Colonies, and their possible development, the subject of his inaugural lecture at what may be the closing epoch of the most prolonged, if not the most disastrous, of famines to which the Indian peoples have been periodically subjected was an appropriate one. "One of [the greatest problems of the future," he remarks, "will be the supply of food for the rapidly-increasing, teeming millions of population. The haphazard method of production by which the accumulated resources of temporary fertility have been drawn upon as successive new unpopulated areas of virgin soil have been placed under requisition must sooner or later cease, and more scientific methods of cultivation and better systems of management must extract more bountiful results from new and improved breeds of plants and of domesticated animals. The expected era implies a more accurate knowledge of agricultural details, and a wider and more Imperial conception of the greater kindred questions than the present time affords."

THE museum which was opened by the Countess of Warwick at Stratford on October 18 will, we trust, lead to the establishment of many similar local museums. Prof. R. Meldola is largely responsible for the erection of the museum, for he advocated the formation of a collection of objects, and a permanent home for it, in his inaugural address to the Essex Field Club twenty years ago. The suggestion was taken up by the more energetic members of the Club, and specimens of scientific interest gradually accumulated, but it was felt that a museum building was essential to make the collections of wide value. By the munificence of Mr. Passmore Edwards, and the enlightened policy of the Town Council of West Ham, a building, which has cost about £4000, has been erected adjoining, and communicating with, the Municipal Technical Institute. This is the building which was opened hy the Countess of Warwick last week. The Corporation has agreed to warm, light and provide for the care-taking of the building, and to make a grant of not less than £100 per annum towards the curatorial expenses. The county collections, cases and cabinets of the Essex Field Club (excepting the Epping Forest collections, which are to be retained in the Forest Museum at Chingford) are placed in the museum. The Club undertakes the selection and scientific control of the collections, and will devote a sum of £50 per annum towards the curatorial expenses. From a pamphlet by Mr. William Cole, the honorary secretary and curator to the Club, we learn that the plan and scope of the museum has been clearly defined, and will be rigidly adhered to in order to avoid the error of gathering together a miscellaneous collection of incongruous specimens. The museum will be a local (Essex) one, supplemented by short series having an educational value, and designed to show the place of the local forms in the general scheme of classification of animal and vegetable organisms. The promoters aim at its eventually fulfilling three main purposes: (1) The instructive recreation of the ordinary visitor by means of carefully arranged sets of the chief forms of life inhabiting the district, and examples showing the nature and meaning of fossils and geological formations. (2) Collecting and preserving authentic series of all forms of life, recent and extinct, occurring in Essex, as well as geological and anthropological specimens. This is a matter of really great scientific importance in view of the changes in our fauna and flora now so rapidly being brought about by the increase of population and the consequent effacement of natural conditions in many parts of the county. (3) Assisting students and field-naturalists in identifying and studying the groups in which they are interested. It would facilitate the advancement of education and natural knowledge if a museum of this kind were connected with every Municipal Technical Institute.

THE parliamentary vote for the University of London for the year 1900-1901 was £18,840 gross, but it only amounted to a net payment of £10, the difference being received from fees paid by candidates presenting themselves for examination by the University. Viewed as a strictly business concern, therefore, the University is practically self-supporting. Many people both at home and abroad will be astonished at the trivial amount actually paid by the Government in aid of its greatest University.

SCIENTIFIC SERIALS.

American Journal of Science, October.—Notes on the Colorado Canyon District, by W. M. Davis. The Kaibab section of the canyon discloses the nearly even floor on which the horizontal Palæozoic strata rest. The floor is of complex structure. The fundamental schists with granitic dikes are overlaid in the eastern section by the heavy Unkar and Chuar series dipping eastward. The wedge in which the tilted formations terminate westward is a most remarkable geological structure, alike for its distinctness and for its significance.— Determination of minerals in thin rock sections by their maximum birefringence, by L. V. Pirsson and H. H. Robinson. The method described is an adaptation of Michel-Levy's colour diagram. The thickness of the rock section having been determined, the highest colour given in any of the numerous sections of the unknown mineral is observed, and by means of the diagonals the numerical value is noted which corresponds to the given colour in a section of the determined thickness. The maximum birefringence having thus been determined, the table of birefringences is referred to and the mineral usually found to be one of a group of several, among which it is easily distinguished by cleavage, colour and other optical properties. The authors add a table of birefractive powers, ranging from 0·287 (rutile) to 0·001 (chlorite).—Experiments on high electrical resistance, by O. N. Rood. An electric current travelling along a bad conductor has many analogies to a stream of pitch. It attains the end of the channel after a considerable interval, and if the resistance is very high, the potential at the end remains at zero. The author describes a number of experiments carried out with glass, silk, mica, jade, guttapercha, ebonite, amber and rosin. When glass, silk and mica were connected with one coating of a charged Leyden jar, it was found that within fifteen minutes the part farthest from the jar had assumed its potential. This was not the case with the other substances.

Ebonite showed a slight change of the opposite sign at the furthest end, probably due to prolonged inductive action.—New occurrences of corundum in North Carolina, by J. H. Pratt. The new occurrences are in an amphibole schist and a quartz schist respectively.—Products of the explosion of acetylene, and of mixtures of acetylene and nitrogen, by W. G. Mixter. Acetylene and ammonia yield hydrocyanic acid at a much lower temperature than is required to cause nitrogen to combine. It may be that ammonia is the first compound of nitrogen formed in the bomb, but the fact that a little ammonia is found among the products is not conclusive, as that may have resulted from the decomposition of hydrocyanic acid.

Annalen der Physik, No. 9.—Electric conductivity of pressed powders, by F. Streintz. Fine powders of platinum black, lamp-black and graphite were prepared and subjected to great pressure. The resistance of 1 cubic mm. of platinum black was found to be 0·92 ohms at zero, as against 0·14 for ordinary platinum. The increase of resistance on heating was only half that of platinum. Lampblack showed a corresponding resistance of 40,000 ohms, and a very high negative temperature coefficient, therein resembling the electrolytes ; while graphite, with its positive temperature coefficient, ranges itself among the metals.—Resistivity of bismuth in a variable magnetic field, by H. Eichhorn. Bismuth does not instantaneously lose the high resistance it acquires in a strong magnetic field. This is proved by mounting a bismuth coil on a rotating disc so that it traverses a magnetic field once during each revolution, and measuring, by means of contact pieces, the instantaneous resistance at various points of the orbit when at rest and in motion respectively.—Ratio of the thermal and electric conductivities, by E. Grüneisen. A very small proportion of arsenic added to copper suffices to reduce the thermal conductivity to one-tenth, and the electric conductivity to one-twelfth of its original value. In iron, the electric conductivity is much more sensitive to impurities than the thermal conductivity.—Reflection of kathode rays, by H. Starke. Kathode rays were made to impinge upon a metallic plate enclosed in a cylinder of the same metal, through a small opening in the end of which the rays entered. Another inner cylinder was used to measure the rays reflected by the plate and diffused by the gas, on the principle of Faraday's ice-pail. By putting the plate into two different positions, the diffusion and reflection could be separately estimated. The reflective power of aluminium was found to be 28·2, and that of copper 45·5.—Mechanical effect of kathode rays, by H. Starke. The author employs a kathode shaped like a propeller, but fixed. The rays impinge at an angle of 45° upon a thin plate of aluminium suspended above the kathode by means of a thin platinum wire. The results are negative so far.—Hardness of metals, by F. Auerbach. The author determines the hardness of metals by his method of finding the greatest pressure between a plate and a lens which the substance will stand without permanent deformation. Mild steel was found to have a hardness of 361, or that of quartz, hard copper 143 (like apatite), brass 107 (like fluorspar), gold 97, silver 91, aluminium 52, and lead 10.—Thermal conductivity of gases, by P. A. Eckerlein. The conductivities of air, hydrogen and carbonic acid are as 1 : 6 8 : 0·73, and the temperature coefficients are 0·00362, 0·00422 and 0·00352 respectively.

SOCIETIES AND ACADEMIES.
LONDON.

Entomological Society, October 3.—Mr. G. H. Verrall, president, in the chair.—Mr. G. C. Champion exhibited specimens of *Trogophloeus anglicanus*, Sharp, from Plymouth ; *Pachyta sexmaculata*, L., from Nethy Bridge, and *Anchomenus quadripunctatus*, De Geer, from Woking. Mr. M. Jacoby exhibited an ichneumon from Blandford parasitic on *Sirex*—*Rhyssa persuasoria*, and Col. Yerbury said that he had met with the same species in some numbers in Scotland. One female observed in the act of oviposition had thrust her ovipositor, which is about the consistency of a human hair, through an inch of fir trunk. Col. Yerbury exhibited :—(1) a rare sawfly, *Xyphidria camelus*, taken in Scotland this year at Nethy Bridge. The species is mentioned in the old books as extinct in the United Kingdom, and there are no modern specimens in the South Kensington Museum collection. (2) Rare diptera from Scotland including (a) *Laphria flava*, from Nethy Bridge ; (b) *Chamaesyrphus scaevoides*, new to the fauna of Great Britain,

from the Mound Sutherland, where it was common on Umbelliferæ under fir trees, one female also being taken on the path up Cairngorm near Glenmore Lodge ; (c) *Microdon devius ;* and (d) *Chilosia chrysocoma* at mountain-ash blossom, Nethy Bridge ; (e) *Stomphastica flava*, two males from Golspie, September, 1900.—Mr. H. K. Donisthorpe exhibited (1) a specimen of *Drusilla canaliculata*, with the dead body of a *Myrmica* in its mouth, captured at Chiddingfold on July 17 ; (2) Specimens of *Myrmedonia collaris* and its larva taken in Wicken Fen with *M. loevinodis* in August, 1900.—The Rev. F. D. Morice exhibited a remarkable hermaphrodite of the bee *Podalirius* (= *Anthophora*) *retusus*, in which the male characters were confined to the left side of the head and genitalia, the right side of the thorax and the abdominal segments. The antennæ and hind (polliniferous) legs were those of a female, and the genitalia half of each sex.—Dr. Chapman exhibited beetles of the genus Orina, some of them alive, and remarked on the fact that while some were Viviparous others were oviparous, in the case of the former the larvæ being developed in the ovaries.—Mr. H. J. Elwes exhibited a collection of lepidoptera from Greece, taken this season in conjunction with Miss Fountaine in the Morea, and in the Parnassus region, including *Coias baldreichi, G. rhamni*, var. *farinosa*, and *Lycaena ottomanus*, with a var. of *L. semiargus*, probably a distinct species.—Mr. H. H. May exhibited a variety of *Strenia clathrata* not unlike *Syricthus alveolus* on the wing.—Mr. F. Enock exhibited a male bee *Stelis aterrima*, one of the bees parasitic in the nests of *Osmia fulviventris*, usually considered a rare insect.—Papers were communicated on "Descriptions of new species and a new genus of South American Eumolpidæ with remarks on some of the genera," by Mr. M. Jacoby, and on "Lepidoptera Heterocera from Northern China, Japan and Corea" (Part iv.), by Mr. J. H. Leech, &c.

PARIS.

Academy of Sciences, October 15.—M. Maurice Lévy in the chair.—Preparation and properties of the carbides of neodidymium and praseodidymium, by M. Henri Moissan. The oxides of neo- and praseodidymium heated with carbon in the electric furnace give crystallised carbides of the formula RC_2, like the carbides of cerium and lanthanum. These carbides are decomposed by cold water, giving a mixture of acetylene, ethylene and paraffins, the first-named predominating. At 1200° the carbides are superficially attacked by ammonia, some nitride being formed.—Observations of the planet Eros made with the large equatorial of the Observatory of Bordeaux, by MM. G. Rayet and A. Féraud. The planet is of about the ninth magnitude, and leaves a clear trace upon the photographic plate.—On the general equation which gives the integral of Jacobi as a particular case, by M. Gruey.—Observations of the Borrelly-Brooks Comet, made with the Brunner equatorial at the Observatory of Lyons, by M. J. Guillaume.—The problem of stationary temperatures, by M. W. Stekloff.—On the explosive mixtures formed by air and by hydrocarbon vapours of the principal organic series, by M. J. Meunier.—On the elimination of the harmonics from alternating currents by the use of condensers, and on the interest of this elimination from the point of view of security of human life, by M. Georges Claude.—On the accessory reactions of electrolysis, by M. A. Brochet. In the electrolysis of sodium hypochlorite, the loss of hypochlorite in four hours is much greater than that calculated from the current used ; and in the preparation of chlorate yields greater than those calculated are obtained. The author traces these anomalous results to the fact that the immediate neighbourhood of the anode is always acid, and hence the hypochlorous acid in that region is transformed spontaneously into chlorate without using electrical energy, even when the bulk of the liquid is alkaline.—On isopyrotritaric acid, a new pyrogenous product from tartaric acid, by M. L. J. Simon. The ferric salt of this acid, the isolation of which was described in an earlier note, is highly characteristic, possessing a deep violet colour in solution. Pyrotritaric acid and similarly constituted furfurane acids do not give this reaction, which is so sensitive that it may be used for the detection of ferric salts and also as an indicator in acidimetry.—On the morphology of the respiratory apparatus of the larva of *Bruchus ornatus*, by M. L. G. Seurat. The larva of *Bruchus ornatus* presents some peculiarities in the morphology of its respiratory apparatus which clearly distinguishes it from the Curculionidæ, the most important being the rounded form of the stigmata, the existence of a prothoracic ring completely uniting the lateral trunks, and of ten transversal latero-ventral anastomoses, of which three are thoracic.—The proteolytic ferment of seeds during germination, by M. V. Harlay. The proteolytic ferment in lentils during germination is analogous in its behaviour to the animal ferment trypsine.—On early tuberculisation in plants, by M. Noel Bernard.—On the Cretacian of the *massif* of Abou-Roach (Egypt), by M. R. Fourtau.—Fixation by porous bodies of clay in suspension in water, by M. J. Thoulet.

NEW SOUTH WALES.

Linnean Society, June 27.—The President, the Hon. James Norton, in the chair.—Notes on some Australian and New Zealand parasitic Hymenoptera, with descriptions of new genera and species, by William H. Ashmead. Sixty-four species were represented in two collections brought together by Mr. W. W. Froggatt and Mr. A. Koebele, formerly of the U.S. Department of Agriculture. Of these forty-nine are described as new.—On the *Carenides* (Fam. *Carabidæ*), Part iii., by Thomas G. Sloane. Nine species referable to the genera *Laccopterum, Carenum, Eutoma* and *Carenidium* are described as new, from Queensland, North-west, West and Central Australia. A synoptic table of the groups of species into which the genus *Carenum* may be subdivided is given, with notes thereon.—Descriptions of two new species of Diptera from Western Australia, by D. W. Coquillett. A species of *Phytomyxa*, the larvæ of which mine the leaves of the beet, and one of *Myiophasia*, parasitic upon the Scarabæid *Anoplostethus opalinus*, Burm., are described. The second of these, founded upon male-specimens, may indeed be congeneric with *Neophasia picta*, Brauer and Bergenst., founded on a female specimen without antennæ from West Australia.—Descriptions of two new blind weevils from Western Australia and Tasmania, by Arthur M. Lea. Only two species of blind Coleoptera have hitherto been recorded from Australia, namely, *Halorhynchus cæcus*, Woll., from West Australia, and *Illaphanus stephensi*, Macl., from New South Wales, both dwelling close to sea-beaches. An additional species of *Halorhynchus* from the "outer beach" at Geraldton, Western Australia, is described in the present paper, together with an insect for which a new genus is proposed, and of which the type-specimen was found in the nest of a small red ant near Hobart.—The double staining of spores and bacilli, by R. Greig Smith. An improvement upon Klein's method of double staining spores and bacilli is described. The spore-bearing material is distributed in normal saline in a small test-tube, an equal volume of carbol-fuchsin is added and the mixture placed in boiling water for fifteen minutes. A loopful is then withdrawn, spread over a coverglass, dried and fixed in a flame. The bacilli are decolorised in 1½ per cent. (by volume) of alcoholic hydrochloric acid, washed in water and counterstained in methylene blue. Even the most refractory spores are stained deep red, the bacilli blue.

[uly 25.—The President, the Hon. James Norton, in the chair. — Descriptions of new Australian Lepidoptera, by Oswald B. Lower. Forty species, referable to the *Bombycina, Geometrina, Pyralidina, Tortricina, Tineina* (*Œcophoridæ, Gelechiadæ, Elachistidæ, Tineidæ*), are treated of, thirty-seven being described as new.—On *Didymorchis*, a Rhabdocoele Turbellarian inhabiting the branchial cavities of New Zealand crayfishes, by Prof. William A. Haswell, F.R.S. *Didymorchis* attracted notice during a search for allies of the *Temnocephaleæ*, and is probably the nearest known relative of the group in question. The animal is about 1 mm. long and less than ⅔ mm. in greatest breadth; and as far as observed is practically an invariable companion of the crayfish *Paranephrops cetosus*, though not occurring in large numbers. A remarkable feature is that cilia are developed only on a portion of the ventral surface of the body, and are entirely absent round the margin and on the dorsal surface. On the whole the animal seems to make a nearer approach to the *Vorticida* than to any of the other known groups.—Supplement to a monograph of the *Temnocephaleæ*, by Prof. William A. Haswell, F.R.S. Three additional species of *Temnocephala* are described—*T. tasmanica*, allied to the much larger *T. quadricornis*, occurring in the branchial cavities, and occasionally on the external surface of *Astacopsis tasmanicus*; *T. aurantiaca*, found upon the lower surface of the abdomen of a Tasmanian Astacopsis, at present undetermined; and *T. cœca*, found upon the surface of the remarkable burrowing Isopod, *Phreatoicopsis terricola*, Spencer and Hall. The paper concludes with some remarks on certain points in the structure of the members of the family, mainly suggested by Monticelli's recent paper (*Bolletino della Soc. di Nat. in Napoli*, xii. 1898). —Observations on the Tertiary flora of Australia, with special

reference to Ettingshausen's theory of the Tertiary cosmopolitan flora, Part i., by Henry Deane.—On the bacterial flora of the Sydney water supply, Part i., by R. Greig Smith. Thirty-two species of micro-organisms commonly occurring in Sydney water are described. These include six new species and four new subspecies.

DIARY OF SOCIETIES.

FRIDAY, OCTOBER 26.

PHYSICAL SOCIETY, at 5.—Exhibition of Experiments illustrating certain Phenomena of Vision : Dr. Shelford Bidwell, F.R.S.—On the Concentration at the Electrode in a Solution, with special reference to the Liberation of Hydrogen by the Electrolysis of a Mixture of Copper Sulphate and Sulphuric Acid : Dr. J. S. Sand.—Electromotive Force and Osmotic Pressure; Dr. R. A. Lehfeldt.

SATURDAY, OCTOBER 27.

ESSEX FIELD CLUB, at 6.30.—Contributions to the Pleistocene Geology of the Thames Valley. The Grays Thurrock Area, Part I. : Martin A. C. Hinton and A. S. Kennard.

THURSDAY, NOVEMBER 1.

CHEMICAL SOCIETY, at 8.—Dehydrohomocamphoric Acid and its Oxidation Products : Arthur Lapworth.—Derivatives of Ethyl a-methyl-a-phenylcyanglutarate : W. Carter and W. Trevor Lawrence.—The Nitration of Acetamino-o-phenylacetate (diacetyl-o-aminophenol)—a Correction : R. Meldola, F.R.S., and Elkan Wechsler.—Rhamnazin and Rhamnetin : A. G. Perkin and J. R. Allison.—(1) Luteolin, Part III. ; (2) Genistein, Part II. : A. G. Perkin and L. H. Horsfall.—Colouring Matter of the Flowers of *Delphinium consolida* : A. G. Perkin and E. J. Wildon.—The Action of Alkalis on the Nitro-compounds of the Paraffin Series. Part II. : Wyndham R. Dunstan, F.R.S., and Ernest Goulding.—Hexachlorides of Benzonitrile, Benzamide and Benzoic Acid : F. E. Matthews.—The Influence of Solvents on the Rotation of Optically active Compounds, Part I. : T. S. Patterson.—Note on Galinek's Amidomethylnaphthimidazole : R. Meldola, F.R.S., and F. H. Streatfeild.— The Action of Heat on Ethyl-Sulphuric Acid : W. Ramsay and G. Rudorf.—The Amount of Chlorine in Rain-water collected at Cirencester : Edward Kinch.

RÖNTGEN SOCIETY, at 8.—Presidential Address : Dr. J. B. Macintyre.

CONTENTS.

Q   Nature   ʔ♦
1   v. 62
N2
v. 62
cop. 2

Sci.
Serials

PLEASE DO NOT REMOVE
CARDS OR SLIPS FROM THIS POCKET

UNIVERSITY OF TORONTO LIBRARY

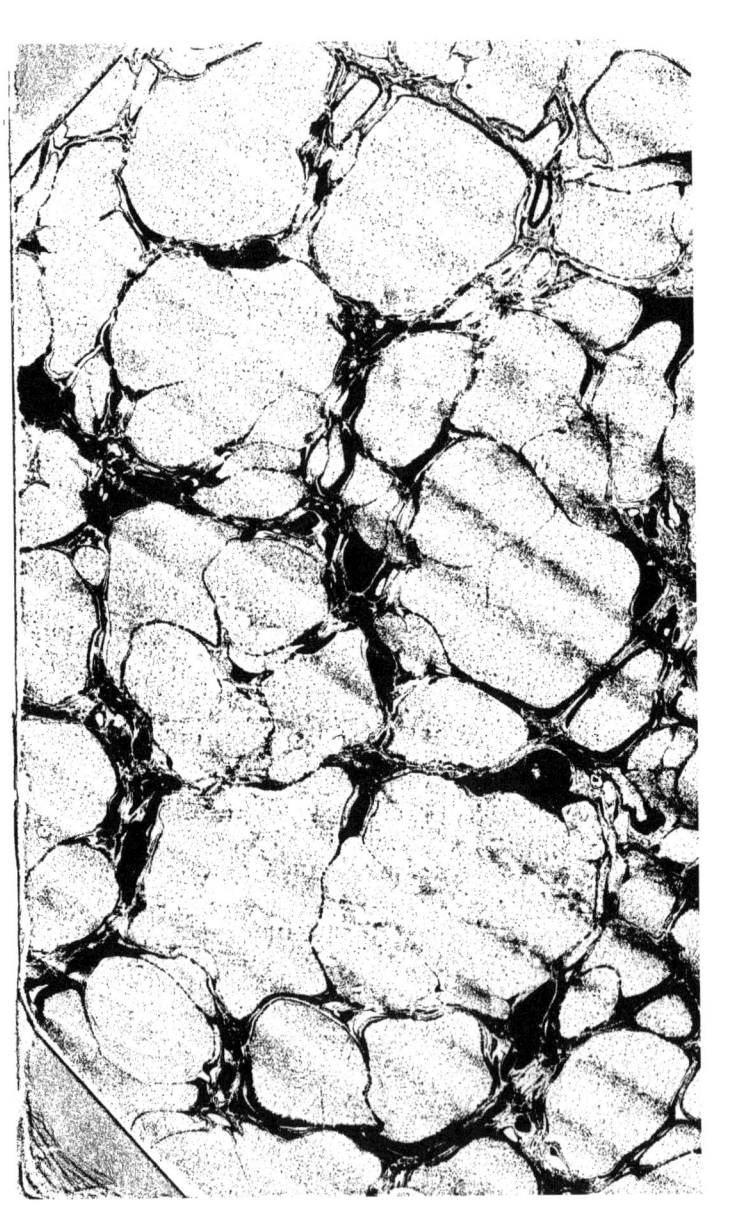

Lightning Source UK Ltd.
Milton Keynes UK
UKHW02n1238120218
317657UK00007B/1177/P

9 780483 528994